VIe CONGRÈS INTERNATIONAL DE RHÉOLOGIE LYON 1972
VIth INTERNATIONAL CONGRESS OF RHEOLOGY LYON 1972
VI. INTERNATIONALER KONGRESS FÜR RHEOLOGIE LYON 1972

Editors:
G. VALLET (Lyon) · W. MESKAT (Leverkusen)

Rheological Theories · Measuring Techniques in Rheology
Test Methods in Rheology · Fractures
Rheological Properties of Materials · Rheo-Optics · Biorheology

Organizing Committee: G. Vallet (President), B. Persoz and M. Joly (Vice-Presidents), C. Smadja (Secretary), A. Larcan, Ph. Comte, J. F. May, and Ph. Berticat (Members)

(Special Issue from Rheologica Acta, Vols. 12 and 13)

With 1203 figures, 6 diagrams and 162 tables

SPRINGER-VERLAG BERLIN HEIDELBERG GMBH 1975

ISBN 978-3-7985-0424-0
DOI 10.1007/978-3-662-41458-3

ISBN 978-3-662-41458-3 (eBook)

Originally published by Dr. Dietrich Steinkopff Verlag · Darmstadt in 1975

Manufactured by Universitätsdruckerei Mainz GmbH, Mainz, and
Dr. Alexander Krebs, Hemsbach/Bergstr. und Bad Homburg v. d. H.

CONTENTS · INHALT

---

*) Die Seitenzahlen beziehen sich auf die fetten Zahlen am Fuße.
The page numbers quoted in these contents are that to be found **bold typed** *at the bottom of each page!*

Presentation of the Poiseuille gold medal of the international society of biorheology to Prof. A. L. Copley, Lyon, France, 1972

By G. W. Scott Blair

Mesdames et Messieurs; Ladies and Gentlemen; Meine Damen und Herren:

It is for me a great honour to be invited to present the *Poiseuille Gold Medal* of our Society to my old friend, *Al Copley*, Professor *Copley* was rightly chosen by an overwhelming majority of our members to be the third recipient of this medal.

Alfred Lewin Copley was born in 1910 near Dresden in Germany and studied in a number of German universities, including Heidelberg, where he took his M.D. in 1935. It is good to learn that this famous university has acknowledged the world-wide distinction of its alumnus by bestowing on him the degree of Doctor of Medicine, *Honoris Causa*, this year. This is a great honour, well deserved.

As a result of his active opposition to the Nazi party in Germany, he was obliged to leave his native country and he first went to Switzerland, where he took a doctorate in Medicine at the University of Basel, as well as studying Physical Chemistry, Philosophy, Neurology and other subjects under various distinguished teachers. Later he moved to the United States and became a citizen of that country.

As a young man he wrote a play and was much interested in the stage; in fact, he might well have made this his career. In travelling around with a Repertory Company, he found that many of the casts came to him for personal counsel. This led him to feel, rightly, that he would make a good psychiatrist. However, his interest moved from psychiatry to neurology and I am told by a mutual friend that he had started on a most successful career as a clinical neurologist. (He was much influenced by *V. von Weizsäcker*.) However, his later inclination lay in a different direction and he made his career in scientific research.

Some time about 1942 or 1943, I first came across his papers on rheology of blood, which much impressed me, and I invited him to come to a Symposium to be held in Oxford; but, because of the war situation, he was unable to accept. He had been in touch with Professor *Eugene C. Bingham* (popularly named the "Founder of Modern Rheology") who worked, among many other systems, on blood and there was a possibility that they might have worked together, but this did not materialize.

Copley introduced the name "*Biorheology*" at the First International Congress on Rheology in Holland in 1948 and it was there that we first met. I had been interested by a paper he had written on the forces needed to pull hair out of the scalp, in relation to baldness. It was at this Congress that we immediately became friends. He worked for 5 years in Paris and I visited him there on several occasions.

In 1957, he came to London and worked for nearly 3 years at the Charing Cross Hospital. Although I had worked before that in other fields of biorheology, it was *Al* who introduced me to the problems of blood flow and coagulation.

After 3 years of pleasant and fruitful co-operation, I was indeed sad when he felt that it was time for him to return to his own country. I had spent many happy evenings with him and his wife and their daughter in their flat in London.

We all know that "*A. L. Copley*" has an *alter ego*; he is also "*Alcopley*", the painter. (He chose this "*nom de plume*" to avoid confusion with the other painter called "*Copley*".) We also know that he married *Nina Tryggvadottir*, the very distinguished Icelandic artist, a lady whose charm and kindliness made her beloved by all who knew her. Her premature death was a great blow, not only to the world of art, but to her many friends in our Society. This is not the place to pay tribute to *Als* work as an artist; his paintings are to be found all over the world and he also finds time to be co-Editor of the Journal *Leonardo*. Indeed, although *Al* has been one of my closest friends for many years, there is one puzzle about him that I have never solved; how does he find time to sleep and eat? I know that, although he is appreciative of good cuisine, he eats very fast and I suspect that he must sleep very little!

I need not say very much here about his distinguished researches on blood and blood vessels: these are already known to you all. Perhaps his most useful contribution to the promotion of further research has been his explanation for the fact that blood starts to clot on contact with all surfaces except (other than in pathological conditions) the vessel wall. His theory of the existence of a fibrin-like "endoendothelial layer" on the vessel wall, by the mere fact of its being controversial (like all original ideas) has led to much further research. He also introduced the term "hemorheology" (in Britain "haemo-").

On the strength of his researches alone, he would be a strong candidate for our Medal but he also has another claim. He has played, I think it would be fair to say, not *a* leading role, but *the* leading role in the organization our Science and our Society. In his modest way he has tried to insist that he shared this role with me, but I feel that he has done far more than I have. He is co-Editor in chief of our official journal, founder and co-Editor in chief of the new journal *Thrombosis Research*, and is also on the Editorial Board of *Rheologica Acta*. Each year he finds time to come to Oxford, both to discuss the affairs of *Biorheology* with me and also to visit the headquarters of our publishers. He has also undertaken the editing of the Proceedings of many congresses.

I know also that when help is needed in the organization of conferences and congresses concerned with Biorheology, he is ready to fly to any part of the world to lend a hand, at very short notice if need be. His help is always appreciated, partly because of his remarkable powers of organization and his memory for detail and partly because of that trait in his character which would have made him such a good psychiatrist: kindliness, sympathy and understanding.

In a few moments, you will see him wearing the *Poiseuille Gold Medal*, bearing the portrait of that great Frenchman, which was designed by his dear wife.

Mais, avant de faire la présentation officielle, je ne peux pas oublier que nous sommes en France, et je voudrais terminer avec quelques mots en français, par longue tradition, la langue internationale européenne. *Al Copley*, qui a travaillé en France, en Allemagne, en Suisse et en Angleterre, ainsi qu'aux Etats Unis, n'aura aucune difficulté de suivre mes quelques mots en français.

Alors, avec le plus grand plaisir du monde, je vous offre, mon très cher ami, *Al*, la médaille *Poiseuille* si richement méritée.

In memoriam Arthur Tobolsky

Arthur Tobolsky, aged 53, died unexpectedly on September 7, 1972 while attending a conference in Utica, N.Y. His entire professional life was devoted to teaching and research in polymer science and rheology. He graduated from Columbia in 1940, and received his PhD in Princeton in 1944. Since then, with the exception of one year at the Brooklyn Polytechnic Institute, he has been on the staff of the Chemistry Department at Princeton, where he has directed the work of scores of graduate students and postdoctoral coworkers. He has made fundamental contributions in the fields of rheology, rubber elasticity, polymerization kinetics and thermodynamics, degradation, and in many theoretical aspects. He has served on the Board of Editors of *American Scientist*, the *Journal of Polymer Science*, and the *Journal of Applied Physics*. He has received the *Witco Award* in Polymer Chemistry (Amer. Chem. Soc.), the *Ford Prize* in Polymer Physics (Amer. Phys. Soc.), and the *Bingham Award* of the Society of Rheology (U.S.). He was a brilliant scientist, inspiring teacher, and friend, and he will be missed by all of his colleagues.

Joint inaugural address: Biorheology as an organized science

By *Alfred L. Copley*

It is my privilege and great pleasure as President of the *International Society of Biorheology* and President of the *First International Congress of Biorheology* to give you the greetings of the membership of our Society. We are happy and proud to be here in Lyon and have our first Congress convened jointly with the VI. International Congress on Rheology. We are grateful to the members of the International Committee on Rheology for having sponsored this joint venture. We are grateful as well to the members of the Organizing Committee of the VI. International Congress on Rheology and in particular we thank its President Professor *Vallet* and its Secretary Dr. *Smadja* for everything they have done to make us welcome here in Lyon and for taking over the chores in the administration of our Congress in association with that of the Congress on Rheology.

With deep sorrow, we are missing here in Lyon two eminent scientists. We are mourning the deaths of Professor *A. Policard* and of Professor *Aharon Katchalsky*. Professor *Policard*, Membre de l'Institut et de l'Académie de Médecine, Paris, who was a cytologist and experimental pathologist, kindly consented to serve as Honorary President of the First International Congress of Biorheology, but, to our great regret, he passed away last Spring at the great age of 91. Professor *Aharon Katchalsky*, Director of the Polymer Department, The Weizmann Institute of Science, Rehovot, Israel, who was a biologist, biophysicist and chemical physicist, was senselessly killed on May 30 on his return to Israel from scientific meetings in Boston and Goettingen.

Tomorrow our Congress will honor the memory of *Aharon Katchalsky*. Two weeks prior to this great tragedy, which shocked the world of science, *Aharon* phoned me from Boston to tell me how much he was looking forward to participating in our Congress here at Lyon. *Aharon* hoped to present to us his new theory on membranes as applied to biorheology.

Biorheology, as an organized science, had its beginning at the First International Congress on Rheology at Scheveningen, Holland, in 1948, when the term biorheology was introduced (1). *Biorheology comprises the study of the deformation and flow of all living matter and of materials of biological significance derived directly from living organisms.*

There have been scientific conferences devoted to biorheology or to its, at present, most vigorous branch, *hemorheology* (2). The latter denotes the *flow properties of blood, its components and of the vessels in which blood is contained*, i.e., the structures with which the flowing blood comes into contact.

In 1958, a meeting took place at Charing Cross Hospital Medical School of the University of London on "*The Flow of Blood in Relation to the Vessel Wall*" (3). As a result of the great interest of the participants in this meeting, a conference, entitled "*Flow Properties of Blood and Other Biological Systems*", was held the following year at the Physiology Laboratory of Oxford University. This conference was convened jointly by the Faraday Society and the British Society of Rheology (4). Its Proceedings were published by Pergamon Press, the English publishing house which, three years later, in 1962, began publication of "*Biorheology. An International Journal*". Needless to say that with the publication of this journal, the science of biorheology became more known. The journal "Biorheology" continues to serve as a stimulus to research in different branches of biorheology, including hemorheology.

In 1963 the "*Symposium on Biorheology*" was held as part of the IV. International Congress on Rheology at Brown University, Providence, Rhode Island, U.S.A. (5). At that time, it was proposed, at a round table discussion on hemorheology, by Professor *Hartert* to organize an *International Society of Biorheology* (6). An *International Society of Hemorheology* came into being three years later, when the First International Conference of Hemorheology was held under the auspices of the University of Iceland at Reykjavik (7).

During the conference at Reykjavik, the *Poiseuille Gold Medal* was awarded to its first recipient *Robin Fåhraeus*, Emeritus Professor of Pathology of the University of Uppsala, Sweden. This award honors both the recipient and the French physiologist *Jean-Léonard-Marie Poiseuille* (8), who may be considered as the first biorheologist. *Poiseuille*, who lived and worked in Paris, died there in 1869. The second person so honored was the physicist *George W. Scott Blair* of Oxford, England, who received this award during the Second International Conference of Hemorheology, held at Heidelberg in 1969 under the auspices of the University of Heidelberg (9).

At this Heidelberg Conference, it was decided by the membership of the International Society of Hemorheology to widen the scope of the activities of our Society to include all branches of biorheology. Accordingly, the name of the Society was changed to the *International Society of Biorheology*.

The decision to widen the scope of our Society has already proved advantageous. Many investigators, both from the non-biological and biological sciences, continue to be attracted to work on biorheological problems. The broader aims of our Society promise to make others aware of the growing importance of biorheology both in the biological and physical sciences. Since the beginning of this year, the journal "Biorheology", which enters its eleventh year, has become the official journal of our Society.

Three years ago, Professor *Aharon Katchalsky*, as President of the International Union of Pure and Applied Biophysics, at the III. International Biophysics Congress in Boston, initiated measures for the International Society of Biorheology to become an Affiliated Commission of the Union. As a result of this action the Society became affiliated with the *International Union of Pure and Applied Biophysics* and we were fully accepted by its General Assembly last month during the IV. International Biophysics Congress at Moscow, held from August 7–14. On August 8th, a Symposium on Biorheology was held as part of this Congress (10) with many persons in attendance. I am glad to say that this Symposium was highly successful.

The main reason for holding jointly both our Congresses here at Lyon is to acquaint the non-biologists, participating in the VI. International Congress on Rheology, with biological problems. Biorheology offers a framework to connect the sciences of biorheology with rheology. This frame, which proved to be secure, permits the application of a number of rheological treatments to biological systems. It is my hope that the biorheological findings, which will be presented here at Lyon, will provide stimuli to rheologists.

I am convinced that biorheology, in spite of its brief history as an organized science, is of growing importance in the biological and medical sciences. I believe that new knowledge gained in biorheology will be of significance to rheology as well as to the practice of medicine and surgery, and thus serve the well-being of our species.

We participants are particularly fortunate that our two Congresses have two great institutions as our hosts: l'Université Claude-Bernard et l'Institut National des Sciences Appliquées.

The University is proudly named after a son of France, the physiologist *Claude Bernard*. He was born in 1813 near Villefranche, not far from Lyon. *Claude Bernard* (11) introduced in 1878 the concept that animals have two milieus, "an external milieu in which the organism is located and an internal milieu in which the elements of its tissues live. The real existence of a living thing does not take place in the external milieu, the atmosphere for airbreathing creatures fresh or salt water for aquatic animals, but in the liquid internal milieu, formed by the circulating organic fluid which surrounds and bathes all the anatomical elements of the tissue". He considered a complex organism as "an assembly of simple forms which are the anatomical elements living in the internal milieu".

In 1865 his "*Introduction to the Study of Experimental Medicine*" was published (12). Anyone who reads this book will be struck with admiration and astonishment about *Claude Bernards* insight. Let me cite the following: "I believe, in a word, that the true scientific method confines the mind without suffocating it, leaves it as far as possible face to face with itself, and guides it, while respecting the creative originality and the spontaneity which are its most precious qualities. Science goes forward only through new ideas and through creative or original power of thought." *Claude Bernard* concludes this remarkable book with the following words: "I have had to limit myself by forewarning biological science and experimental medicine against exaggerating the importance of erudition and against invasion and domination by systems; because sciences submitting to these would lose their fertility and would abandon the independence and freedom of mind essential to the progress of humanity."

These thoughts of *Claude Bernard* are bound to guide workers in all sciences, including rheology and biology. With this guidance of *Claude Bernard*, it gives me great pleasure to open, as President, the First International Congress of Biorheology, held here in this great city of Lyon jointly with the VI. International Congress on Rheology. My best wishes go to the participants for the scientific success of both Congresses. Thank you.

References

1) *Copley, A. L.*, Proc. Internat. Congress on Rheology, Scheveningen, Holland, 1948, Vol. 1, p. 47 (Amsterdam and New York 1949).

2) *Copley, A. L.*, J. Colloid Sci. 7, 323 (1952).

3) *Scott Blair, G. W.*, Nature **182**, 90 (1958).

4) *Copley, A. L.* and *G. Stainsby* (Eds.), Flow Properties of Blood and Other Biological Systems (Oxford and New York 1960).

5) *Copley, A. L.* (Ed.), Symposium on Biorheology, Proc. 4. Internat. Congress on Rheology, Providence, R. I., U.S.A., 1963, part 4 (New York 1965).

6) *Frasher jr., W. G.* and *H. Wayland*, Biorheology **2**, 55 (1964).

7) *Copley, A. L.* (Ed.), Hemorheology. Proc. 1. International Conference, University of Iceland, Reykjavik, 1966 (Oxford and New York 1968).

8) *Joly, M.*, in: Hemorheology. Proc. 1. Internat. Conference, University of Iceland, Reykjavik, 1966, p. 29, Ed. *A. L. Copley* (Oxford and New York 1968).

9) *Hartert, H. H.* and *A. L. Copley*, Theoretical and Clinical Hemorheology. Proc. 2. Internat. Conf., Internat. Soc. Hemorheology, University of Heidelberg, 1969, Heidelberg (Berlin–Heidelberg–New York 1971).

10) *Copley, A. L.* and *V. I. Vorob'ev*, Symposium XI. Biorheology. Program IV. Internat. Biophysics Congress, Moscow, U.S.S.R., 1972, p. 90; Proceedings, Academy of Sciences of the U.S.S.R. (Moscow, in press).

11) *Bernard, C.*, Leçons sur les phenomènes de la vie communs aux animaux et aux vegetaux. p. 113 (Paris 1878).

12) *Bernard, C.*, An Introduction to the Study of Experimental Medicine. Translated by *H. Copley Green*, p. 226 (New York 1957).

Rheol. Acta **12**, 92–99 (1973)

From the Department of Engineering Mathematics, The University of Newcastle upon Tyne (England)

Memory effects in a non-uniform flow: A study of the behaviour of a tubular film of viscoelastic fluid

By C. J. S. Petrie

With 1 figure

(Received October 27, 1972)

1. Introduction

1.1. The film blowing process

The process is used for the manufacture of a thin sheet or film of a thermoplastic material (e.g. polyethylene) from molten material supplied under pressure by a screw extruder. The polymer melt is forced through an annular die and the tubular film so formed is thinned both by blowing and by axial drawing. The tube is formed into a closed bubble by flattening it when it is cool enough to avoid blocking (the tendency of the film to stick to itself) and then the flattened film is wound onto take-up rolls. The axial tension is provided by the driven nip rolls which close the bubble at the top (the process is usually run vertically with the die at the bottom). The blowing is caused by maintaining an air pressure slightly above atmospheric inside the bubble, and this causes an increase in the radius of the tube and stretching of the film in a circumferential direction. Fig. 1 illustrates the process schematically.

1.2. Mathematical modelling of the process

The mathematical models discussed below are concerned with the region where the polymer is molten, between the die and the freeze-line. The basic assumptions are that the flow is steady and axisymmetric and that the film is thin. The thin film approximation has two significant aspects; the flow variables are taken to be independent of distance across the film so that we have a problem with one independent variable (distance in the axial direction or the direction of flow), and the radii of curvature of the film are assumed to be large compared with the film thickness so that we may treat the film as locally plane to obtain rates of strain and other kinematic variables at a point.

The forces which may act on the film are the applied pressure and axial tension, surface tension, gravity, inertia forces (required to accelerate the fluid), air drag and internal (viscous or viscoelastic) forces. The equilibrium of these forces gives us two equations relating the forces and the geometrical variables. The constitutive equation for the fluid will give more such equations (one for each component of the stress tensor) and with the continuity equation we have enough to obtain the components of the stress tensor, the film thickness, the bubble radius, and the fluid velocity. The order of the system of differential equations we have to solve for this depends on the constitutive equation used.

The boundary conditions required in the simplest case (i.e. for a *Newton*ian fluid) are the film thickness and bubble radius at the die and the magnitude and direction (axial) of the force applied to take up the film. As we are

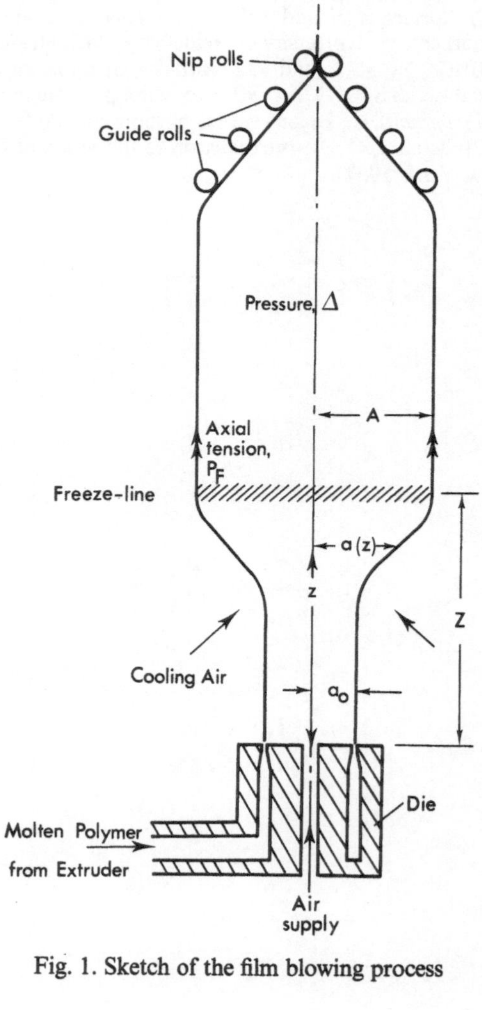

Fig. 1. Sketch of the film blowing process

considering the region where the polymer is molten, we assume this force to be given at the freeze-line. For a viscoelastic fluid additional boundary conditions are needed, which in some sense correspond to what the fluid "remembers" of its deformation history upstream of the die exit.

Details of the flow upstream and downstream of the region of interest are ignored, it being assumed that their effect on the flow may be represented by the values chosen for the boundary conditions. (For example the initial value of the film thickness will be the die gap multiplied by a correction factor to allow for the die swell effect.) Further simplification is effected by neglecting inertia, gravity and air drag and by assuming that the flow is isothermal. Enough of the physically significant quantities are retained to give the essential features of the flow and to illustrate the behaviour of a viscoelastic fluid in this flow.

Details of the work on this problem for a *Newton*ian fluid have been published elsewhere (*Pearson* and *Petrie*, 1970a, b, c).

1.3. Definition of symbols and dimensionless variables

The principal variables are the film thickness, H, the bubble radius, a, and the fluid velocity, U, which are functions of the axial distance from the die, z. Characteristic lengths are the film thickness at the die, H_0, and the bubble radius at the die, a_0. The ratio of these, $\varepsilon = H_0/a_0$, is small. If Q is the volumetric flow rate, we may define a characteristic velocity, $U_0 = Q/2\pi a_0 H_0$, which is the fluid velocity at the die, and have a characteristic shear rate, U_0/a_0. Hence, taking a characteristic viscosity μ^*, we have a typical viscous stress, $\mu^* U_0/a_0$.

We use these quantities to make the variables dimensionless and define

r a/a_0, dimensionless bubble radius,
h H/H_0, dimensionless film thickness,
u U/U_0, dimensionless fluid velocity;
all functions of
x z/a_0, dimensionless axial distance.

We shall use $'$ to denote differentiation with respect to x, and define θ by
$\tan\theta = r'$.

Important dimensionless parameters are

X Z/a_0, the freeze-line height,
R A/a_0, the blow ratio,
B $\varepsilon^{-1}\pi a_0^2 H_0 \Delta/\mu^* Q$, the pressure inside the bubble,
T $a_0 P_F/\mu^* Q$, the axial tension at the freeze-line,
S $\varepsilon^{-1} 2\pi a_0 H_0 \Gamma/\mu^* Q$, the surface tension,
T $T_F - R^2 B$, the total axial force at any cross-section.

(The lengths Z and A are the actual freeze-line height and bubble radius at the freeze-line respectively, and Δ, P_F and Γ are the physical variables corresponding to the dimensionless quantities B, T_F, S.) Apart from the surface tension, material parameters will be defined when the constitutive equation is discussed in the next section.

2. Viscoelastic fluid models

2.1. Choice of constitutive equation

We may note three important features of the behaviour of non-*Newton*ian and viscoelastic fluids, any of which may be important in determining the flow of such a fluid. These are a shear viscosity which varies with shear rate, normal stress differences which are apparent in simple viscometric flows, and elasticity or fluid "memory" as shown by elastic recoil and stress relaxation phenomena. From the enormous variety of constitutive equations which have been suggested, some generalisation of the *Maxwell* model seems most likely to exhibit all these phenomena while retaining a reasonable degree of simplicity.

The formulation chosen here is in terms of a differential equation rather than an integral equation, and the most general equation we consider (briefly) is the eight-parameter *Oldroyd* equation (*Oldroyd*, 1961)

$$p + \lambda_1 \frac{\mathscr{D}p}{\mathscr{D}t} - \mu_1 (p \cdot e + e \cdot p) + \mu_0 \operatorname{tr}(p)\, e + \nu_1 \operatorname{tr}(p \cdot e)\, \mathrm{I} = 2\eta_0 \left\{ e + \lambda_2 \frac{\mathscr{D}e}{\mathscr{D}t} - 2\mu_2 e \cdot e + \nu_2 \operatorname{tr}(e \cdot e)\, \mathrm{I} \right\} \quad [1]$$

where $\mathscr{D}/\mathscr{D}t$ is the appropriate convected derivative defined by

$$\frac{\mathscr{D}p}{\mathscr{D}t} = \frac{\partial p}{\partial t} + v \cdot \nabla p + \omega \cdot p - p \cdot \omega$$

p is the viscous (or extra-)stress tensor, $e = \frac{1}{2}(\nabla v + v\nabla)$ is the rate of the strain tensor, $\omega = \frac{1}{2}(\nabla v - v\nabla)$ is the vorticity tensor. We allow the material parameters $\eta_0, \lambda_1, \ldots$, to be functions of the second invariant of e,

$$\mathrm{II} = \operatorname{tr}(e \cdot e)$$

a generalisation of the original *Oldroyd* equation. This allows us to include the models of *White* and *Metzner* (1963) and *Tanner* (1965) as well as the power-law fluid ($\eta_0 = K(\mathrm{II})^n$) in our general formulation. The second-order fluid and *Oldroyd* fluids A and B are of course also special cases of this. The dimensionless forms of the material parameters are obtained by dividing η_0 by μ^*, and the other parameters by a_0/U_0 since they each have the dimensions of a time. The same symbols will be used for the dimensionless parameters in the next section.

2.2. Kinematics of the flow

The kinematics have been discussed from a simple physical viewpoint in *Pearson* and *Petrie* (1970b) using the local approximation that the film is plane. This has been formally justified

[*Pearson* and *Petrie* (1970a)] by an asymptotic series approach, which may be extended to deal with viscoelastic fluids.

If we define a local *Cartes*ian coordinate system (ξ_1, ξ_2,, ξ_3) at a point in the film then, with the approximation of a plane film and of the variables being independent of distance across the film, ξ_2, and around the bubble, ξ_3, we have the rate of strain tensor

$$e = \cos\theta \begin{pmatrix} u' & 0 & 0 \\ 0 & uh'/h & 0 \\ 0 & 0 & ua'/a \end{pmatrix} \qquad [2]$$

(using $dx/d\xi_1 = \cos\theta$). Continuity, in the dimensionless form

$$urh = 1 \qquad [3]$$

gives

$$u' = -u(h'/h + r'/r)$$

and will be used to eliminate u and u'.

For a plane film being stretched biaxially (in the ξ_1 and ξ_3 directions) the vorticity tensor, ω, is zero (we have a purely extensional flow) and it turns out that this naive approximation gives the correct result (for the *Oldroyd* fluid, at least). This is discussed a little more fully in Appendix 1.

Following from this approximation we have

$$\frac{\mathscr{D}e}{\mathscr{D}t} = u\cos\theta\, e'$$

and

$$\frac{\mathscr{D}p}{\mathscr{D}t} = u\cos\theta\, p'.$$

The other terms in [1] are obtained readily and it can be seen that a diagonal stress tensor will satisfy [1], which then gives three equations

$$\begin{aligned} p_k &+ \lambda_1 u\cos\theta p_k' - 2\mu_1 e_k p_k \\ &+ \mu_0(p_1 + p_2 + p_3)\, e_k \\ &+ \nu_1(p_1 e_1 + p_2 e_2 + p_3 e_3) \\ &= 2\eta_0(e_k + \lambda_2 u\cos\theta e_k' - 2\mu e_k^2 \\ &+ \nu_2(e_1^2 + e_2^2 + e_3^2); \quad k = 1, 2, 3 \end{aligned} \qquad [4]$$

where p_k and e_k are the diagonal components of the tensors p and e respectively (in dimensionless form).

2.3. Dynamics of the flow

It has been shown [*Pearson* and *Petrie* (1970a)] that the pressure difference across the film, Δ, and the surface tension, Γ, must both be of order ε (since otherwise we get the static surface tension governed problem). Then, to order 1 the normal stress in the ξ_2 direction, $p_2 - p$, must be constant across the film, and if we measure pressures relative to atmospheric pressure we have

$$p_2 - p = 0$$

where p is the hydrostatic pressure in the fluid. We replace the viscous stress components p_1 and p_3 by the total stresses in the longitudinal (or flow) direction,

$$L = p_1 - p$$

and in the circumferential direction,

$$C = p_3 - p.$$

Then equilibrium of forces in the axial direction gives

$$r(Lh + 2S)\cos\theta - r^2 B = T \qquad [5]$$

and for equilibrium in the radial direction we have

$$-(Lh + 2S)\,\theta'\cos\theta + (Ch + 2S)/r\sec\theta = 2B. \qquad [6]$$

[The terms $-\theta'\cos\theta$, which is $-d\theta/d\xi_1$, and $1/r\sec\theta$ are the principle curvatures of the film. Details of this are given in *Pearson* and *Petrie* (1970b).]

From these two equations, the three eqs. [4], eqs. [2] and [3] and the equations defining θ, L and C we obtain a sixth order system of differential equations. This can be written with the first derivatives of the six variables r, θ, h, h', C and π by eliminating L and u from the system

$$r' = \tan\theta \qquad [7]$$

$$(T + r^2 B)\,\theta' = (hC + 2S)\cos\theta - 2rB \qquad [8]$$

$$\begin{aligned} &\lambda_1 u\cos\theta\pi' + \pi + (3\nu_1 - 2\mu_1)\, u\cos\theta \\ &\quad\times\left\{\frac{r'}{r}C - \left(\frac{r'}{r} + \frac{h'}{h}\right)L\right\} \\ &\quad= 2\eta_0(3\nu_2 - 2\mu_2)\,\mathrm{II} \end{aligned} \qquad [9]$$

$$\begin{aligned} &\lambda_1 u\cos\theta C' + C - \frac{2}{3}\mu_1 u\cos\theta \\ &\quad\times\left\{\left(\frac{2r'}{r} + \frac{h'}{h}\right)C - \left(\frac{r'}{r} - \frac{h'}{h}\right)L\right\} \\ &\quad-\left(\frac{2}{3}\mu_1 - \mu_0\right)u\cos\theta\left(\frac{r'}{r} - \frac{h'}{h}\right)\pi \end{aligned}$$

$$= 2\eta_0 u \cos\theta\left[\frac{r'}{r} - \frac{h'}{h}\right.$$
$$+ \lambda_2\left\{u\cos\theta\left(\frac{r'}{r} - \frac{h'}{h}\right)\right\}'$$
$$\left. - 2\mu_2 u\cos\theta\left(\frac{r'^2}{r^2} - \frac{h'^2}{h^2}\right)\right] \qquad [10]$$

$$\lambda_1 \cos\theta L' + L + \frac{2}{3}\mu_1 u\cos\theta$$
$$\times\left\{\left(\frac{2r'}{r} - \frac{h'}{h}\right)L - \left(\frac{r'}{r} + \frac{2h'}{h}\right)C\right\}$$
$$+ \left(\frac{2}{3}\mu_1 - \mu_0\right)u\cos\theta\left(\frac{r'}{r} + \frac{2h'}{h}\right)\pi$$
$$= 2\eta_0 u\cos\theta\left[-\left(\frac{r'}{r} + \frac{2h'}{h}\right)\right.$$
$$- \lambda_2\left\{u\cos\theta\left(\frac{r'}{r} + \frac{2h'}{h}\right)\right\}'$$
$$\left. - 2\mu_2 u\cos\theta\left(\frac{r'^2}{r^2} + \frac{2r'h'}{rh}\right)\right] \qquad [11]$$

$$u = 1/rh \qquad [12]$$

$$L = (T + r^2 B)\sec\theta/rh - 2S/h \qquad [13]$$

$$\mathrm{II} = 2\frac{\cos^2\theta}{r^2h^3}\left(\frac{r'^2}{r^2} + \frac{r'h'}{rh} + \frac{h'^2}{h^2}\right) \qquad [14]$$

$$\pi = 3p + L + C. \qquad [15]$$

If λ_2 is zero there is no term in h'' and the system is of fifth order, while if λ_1 is zero the order is reduced to four (if $\lambda_2 \neq 0$). If λ_1 and λ_2 are both zero we have a third order system as for the *Newton*ian fluid. In principle the parameters in these equations may all be functions of the second invariant of the rate of strain tensor, II, but this does complicate the actual solution.

We shall consider two special cases of the constitutive equation, the simple *Maxwell* model with $\lambda_1 = \mu_1 = \lambda$, $\eta_0 = \mu$ and all other material parameters zero, and the model obtained from the condition that the trace of the viscous stress tensor, π, be identically zero. The *Maxwell* model (which is the *Oldroyd* model B with retardation time $\lambda_2 = 0$) includes the *Tanner* and *White* and *Metzner* models as long as μ and λ are allowed to be functions of II. Eqs. [9], [10] and [11] give us the differential equations

$$\lambda p' = -rh\sec\theta p + 2(\lambda p + \mu)\frac{h'}{h} \qquad [16]$$

$$\lambda C' = -rh\sec\theta C + 2\lambda C\frac{r'}{r}$$
$$+ 2(\lambda p + \mu)\left\{\frac{r'}{r} - \frac{h'}{h}\right\} \qquad [17]$$

$$\{\lambda L + 4(\lambda p + \mu)\}\frac{h'}{h} = -rh\sec\theta L$$
$$- \{2(\lambda p + \mu) + \lambda(L + C)\}\frac{r'}{r} \qquad [18]$$

for p, C and h (with L given by [13]). Thus, with eqs. [7] and [8] we have a fifth order system.

If we impose the condition $\pi = 0$, we require $\nu_1 = \frac{2}{3}\mu_1$, $\nu_2 = \frac{2}{3}\mu_2$, and eq. [9] is identically satisfied. Then eq. [11] is essentially an equation for h'', provided that $\lambda_2 \neq 0$ (using [13] for L and L'). We can express the other equation best in terms of a variable $(2C - L)$ (in order to eliminate h'' from [10]):

$$\lambda_1(2C - L)' + rh\sec\theta(2C - L)$$
$$- \mu_1\left\{\frac{r'}{r}(2C - L) + \left(\frac{2h'}{h} + \frac{r'}{r}\right)L\right\}$$
$$= 2\eta_0\left[3\frac{r'}{r} + 3\lambda_2\left\{\frac{\cos\theta r'}{r^2h}\right\}'\right.$$
$$\left. - 2\mu_2\frac{\cos\theta}{rh}\left\{\frac{r'^2}{r^2} - \frac{2r'h'}{rh} - \frac{2h'^2}{h^2}\right\}\right]. \qquad [19]$$

3. Solution of the equations for a viscoelastic film

3.1. Numerical solution

Preliminary computations have been carried out with the two *Maxwell* type of models, that described by eqs. [7], [8], [16], [17] and [18] which we shall call B (by analogy with the fluid B of *Oldroyd*) and that described by eqs. [7], [8], [11] and [19], with μ_2 and λ_2 set to zero, which we shall call C. For simplicity surface tension has been ignored and constant material parameters considered so that we may take μ (model B) and η_0 (model C) to be 1. As far as numerical results are concerned, we can report briefly that results for model C are generally similar to those for model B. There are difficulties in the numerical solution with both models arising from the predictable inherent instability of the systems of equations when we try to integrate in the direction of x decreasing. (This is the most convenient direction of integration in the *Newton*ian case, but here the stresses increase exponentially as x decreases). Also for model C we show later that, for a particular case, there only is a solution for a limited range of values of x, and this appears to cause additional computational difficulty, at least for $\mu_1 = 0$.

The general pattern of the numerical results for model B, integrating in the direction of x increasing shows two types of bubble shape. The first is that of r decreasing with increasing rapidity until the numerical solution becomes unreliable and r is so small that the model is not useful anyway. This occurs for small values of T, large values of B and also is affected by the initial values of θ and the stresses (C, π or p). The effect of the initial values of the stresses is small except where it produces a change to the second type of bubble shape, with r increasing to a maximum (possibly after a slight initial decrease) before decreasing towards zero. Larger values of T and smaller values of B give this type of behaviour which can give shapes similar to actual bubble shapes, unlike the first type of solution.

The desirable second type of solution is favoured by the initial stress conditions $2C - L = 0$, $\pi = 0$ (which are the conditions for a *Newton*ian fluid if $\theta = 0$) as opposed to $C = 0$, $p = 0$ which approximate the conditions inside the die. (The latter might be appropriate for a viscoelastic fluid on the argument that the fluid "remembers" the stresses in it upstream of $x = 0$.) The effect of increasing λ from zero is generally to decrease r at any x and to reduce the value of x at which r reaches its maximum value.

If we consider for a moment the two-point boundary-value problem with T_F prescribed rather than T the following results illustrate this effect (of λ) in a different way.

B	T_F	X	λ	R	$h(0)/h(X)$
1	1.6	6	0	4	8
			1	2.8	17
.2	2.9	5	0	3.5	20
			.1	3.2	45

Thus the prediction (based on a limited amount of computation) is that the effect of fluid elasticity is to reduce the blow ratio and to increase the thickness reduction, both favouring an increase in machine direction drawing as opposed to transverse drawing.

The magnitude of the parameters are reasonable for some materials and processes, e.g., $\lambda = .1$, $U_0 = .015$ m/sec, $a_0 = .04$ m corresponds to a relaxation time of .3 sec and a small film-blowing apparatus with a throughput of $Q = 4 \times 10^{-6}$ m^3/sec and an initial film thickness of 10^{-3} m. For a viscosity of 5×10^4 Nsm^{-2} (and $a_0 = .04$ m) the value $B = 0.1$ corresponds to a pressure of 100 Nm^{-2} inside the bubble, and $T_F = 1.6$ to an axial force of 8 N. Clearly more computation is needed before either comparison with real processes or a detailed theoretical analysis of the various interacting effects is undertaken.

3.2. *Qualitative results*

For the *Newton*ian fluid one important solution, which approximates the initial shape of a long bubble, is that of a bubble of constant radius. The *Newton*ian solution is

$$2C - L = 0 \qquad [20]$$

for which we require r to satisfy

$$3r^2 B - T - 2rS = 0. \qquad [21]$$

Then we have

$$\frac{u'}{u} = \frac{-h'}{h} = \frac{T + r^2 B - 2rS}{4\mu}. \qquad [22]$$

We wish to see whether such a solution is possible for the viscoelastic fluid models we have been considering. Details of this are set out in Appendix 2, and we present some conclusions here.

In general we do not expect such a solution to be possible, and can prove this for three cases. These cases are model B with λ constant, model C with $\mu_1 \neq 0$, and the second-order type of fluid ($\lambda_1 = \mu_0 = \mu_1 = \nu_1 = 0$, η_0, λ_2, μ_2, ν_2 constant) with $\mu_2 = 0$. On the other hand we can find special cases where the *Newton*ian solution is possible. Such cases include model C if $\mu_1 = 0$, the second-order fluid if $\mu_2 = 0$ as well as obvious cases such as the purely viscous fluid with variable viscosity (μ in [22] a function of u'^2) and an *Oldroyd* fluid with the same operator on both sides of the constitutive equation (i.e., $\lambda_1 = \lambda_2$, $\mu_1 = \mu_2$, $\nu_1 = \nu_2$, $\mu_0 = 0$).

These special cases raise the question as to whether there is any particular reason to single out models for which μ_1 and μ_2 are zero. The main role of these for the experimental or phenomenological rheologist is as adjustable parameters which allow any combination of primary and secondary normal stress differences in simple shear. We shall not pursue this point further, but shall look instead at the solutions for model C with $\mu_1 = 0$, λ_1 and η_0 constant. We set $\eta_0 = 1$, $r = 1$, $h(0) = 1$ and have

$$2C - L = 0$$

and

$$(4 - \lambda_1 L)\, h' = -Lh^2.$$

Here

$$L = (T + B)/h$$

$$2C - L = (3B - T)/h$$

so that $T = 3B$ and

$$\left(1 - \frac{\lambda_1 B}{h}\right) h' = -Bh.$$

The solution is not the decaying exponential of the *Newton*ian case, but

$$\log h + \lambda_1 B(1/h - 1) = -Bx.$$

For $0 < \lambda_1 < B < 1/h$ decreases from 1 to $\lambda_1 B$ as x increases from 0 to $(1/B)\log(1/\lambda_1 B) - (1 - \lambda_1 B)/B$. As x approaches this value $h' \to -\infty$, and there are no solutions for larger x. We remark in passing that this solution is not found numerically as it is inherently unstable both for x decreasing and for x increasing.

4. Discussion

This report of the first stage of a project raises a number of questions which have been pointed out in the text, and reaches few conclusions. It is a valuable indication of the validity of the approach that the upstream stress boundary conditions do not affect the solution greatly, and indeed their effect does seem to decay in many cases as we go away from the die. Thus we are led to believe that even for a fluid with memory the precise details of what happens at the die exit will have a significant effect only on the flow near the die exit.

It is perhaps disappointing that the choice of constitutive equations within the general class of *Maxwell*-type equation does not seem crucial. The hope that analysis of a complex flow will help to distinguish between different constitutive equations is an attractive one, and it may be that some useful conclusions will be obtained from further work along these lines.

Among the many lines for future work, one of mathematical interest which might aid a general understanding of the numerical results is the study of solutions for small λ (i.e. for a nearly *Newton*ian fluid). This gives us a singular perturbation problem, and when the unperturbed problem needs to be solved numerically this may not turn out to be feasible. An idea of the behaviour of solutions for large x would also be useful, if indeed solutions exist for large x. One could go on to ask about the existence and uniqueness of solutions of the two-point boundary-value problem, but this is an unresolved question even for the *Newton*ian case.

The principal value of this study is probably that it shows that a complex flow can be analysed for a complicated fluid. The results are on the whole reassuring and the difficulties predictable though not avoidable.

Appendix 1

The kinematic approximations

Care must be taken in making the approximations to the kinematics discussed in Section 2.2, as the following more detailed discussion shows. The simple plane approximation using *Cartes*ian coordinates presents no problem, but requires formal justification. One reason for doubting the validity of this approach is that if we are dealing with viscoelastic fluid it may be expected to "remember" that the local *Cartes*ian coordinates we are using are in fact rotating from the point of view of a material element, and the curvature of the film might be expected to have an effect for a viscoelastic fluid even though it does not for a *Newton*ian fluid.

These doubts are reinforced when we go back to the asymptotic series approach and calculate the vorticity tensor ω. It turns out that it has two components of order 1, namely

$$\omega_{12} = -\omega_{21} = u\theta' \cos\theta.$$

These are of the same order as the diagonal terms e_1, e_2, e_3 in the rate of strain tensor, and so cannot be neglected. We thus have terms

$$\begin{pmatrix} 2\omega\tau & \omega(p_2 - p_1) & 0 \\ \omega(p_2 - p_1) & -2\omega\tau & 0 \\ 0 & 0 & 0 \end{pmatrix}$$

in $\mathscr{D}p/\mathscr{D}t$, where we write ω for $u\theta' \cos\theta$ and τ for the shear stress component, p_{12}. This could lead to serious difficulties in the asymptotic expansions, but fortunately there is a compensating term.

This compensating term arises because we are now using curvilinear coordinates, and so all derivatives must be covariant derivatives to maintain the correct tensorial character of the expressions we write down. In particular, the $v \cdot \nabla p$ term is, in component form, $v^k p^{ij}{}_{,k}$ (using the summation convection on a suffix repeated above and below) and instead of $u \cos\theta (p^{ij})'$, we have

$$v^k p^{ij}{}_{,k} = u \cos\theta (p^{ij})' + u \left\{ {i \atop 1\ k} \right\} p^{kj} + u \left\{ {j \atop 1\ k} \right\} p^{ik}.$$

The terms involving the *Christoffel* symbols contribute an amount which exactly cancels the term $\omega \cdot p - p \cdot \omega$ displayed above, and the result is that the operator $\mathscr{D}./\mathscr{D}t$ is equal to $u \cos\theta\, d./dx$ as for a plane film.

The asymptotic analysis follows much the same course as in the case of the *Newton*ian fluid and we reach the same conclusion as to the orders of magnitude of the various terms. Thus the equations presented above are justified formally to the same extent as for the *Newton*ian fluid; i.e., in the limit of small ε provided that the bubble pressure and the surface tension are both of order ε.

Appendix 2

Cylindrical bubble solutions

If we set $r' = \theta' = 0$ in the equations for the general *Oldroyd* fluid, seeking solutions with r constant, we obtain

$$C = (2r^2 B - 2rS)\, u \qquad [23]$$

$$L = (T + r^2 B - 2rS)\, u \qquad [24]$$

$$\lambda_1 u \pi' + \pi + (3\nu_1 - 2\mu_1)\, u' L = 4\eta_0 (3\nu_2 - 2\mu_2)\, u'^2 \qquad [25]$$

$$\lambda_1 u (2C - L)' + (2C - L) + 2\mu_1 u' L = 8\eta_0 \mu_2 u'^2 \qquad [26]$$

$$\lambda_1 u L' + L + \tfrac{2}{3}\mu_1 u' (2C - L) + 2(\mu_0 - \tfrac{2}{3}\mu_1)\, u' = 4\eta_0 (u' + \lambda_2 u u''). \qquad [27]$$

From these we may obtain three differential equations for the two functions u and π and one adjustable constant, r.

In general these three equations will not have a solution, though in special cases a particular choice of r makes them consistent. A proof in general terms would be extremely complicated and, in view of the arbitrary nature of the choice of constitutive equation, has not been sought. We consider three special cases, models B and C discussed above, and the second-order fluid.

For model B eqs. [23], [24] and [26] give

$$u' = -(2C - L)/\lambda(2C + L)$$

so that, for constant λ, u' is constant. Using this, eq. [27] gives us

$$\pi = (2C^2 - 2LC - L^2)/(2C - L) - 3\eta_0/\lambda$$

and if η_0 is constant also, eq. [25] requires

$$\eta_0/\lambda = 4LC(C - L)/(2C + L)$$

a contradiction (since the right-hand side is proportional to u^2, and the trivial solution $u' = 0$ requires $B = T = 0$, contrary to our implicit assumption that B, at least, is non-zero).

If we allow η_0 to vary we reach the same conclusion, since we obtain

$$u\eta_0' + \eta_0/\lambda = 4LC(C - L)/(2C + L)$$

(where $\eta_0' = d\eta_0/dx$) and since η_0 is a given (material) function of u'^2 this again is not compatible with the result that u' is constant. The same conclusion almost certainly holds for λ a function of u'^2 (i.e. of II).

For model C we have $\pi = 0$ and eqs. [26] and [27] give two expressions for u' which are not, in general, consistent, namely

$$u'(\lambda_1(2C - L) + 2\mu_1 L) = -(2C - L)$$

and

$$u'(\lambda_1 L + \tfrac{2}{3}\mu_1(2C - L) - 4\eta_0) = -L.$$

If $\mu_1 = 0$ and r is chosen to make $2C - L$ zero then the first of these is identically satisfied and the second gives

$$u' = -L/(\lambda_1 L - 4\eta_0).$$

Finally the second-order fluid ($\lambda_1 = \mu_1 = \mu_0 = \nu_1 = 0$) gives us

$$\pi = 4\eta_0(3\nu_2 - 2\mu_2)\, u'^2$$

$$2C - L = 8\eta_0 \mu_2 u'^2$$

$$L = 4\eta_0(\lambda_2 u u'' + u').$$

If $\mu_2 \neq 0$ and η_0, λ_2 and μ_2 are constant the second of these gives, on differentiation,

$$u'' = (2C - L)/16\eta_0 \mu_2 u$$

which is constant, and using this in the third equation gives u'/u equal to a constant, and again there is no non-trivial consistent solution. If $\mu_2 = 0$ then the second equation is satisfied identically for r such that $2C - L = 0$, and then we have two differential equations from which we can determine the functions u and π.

Summary

Formal use of constitutive equations such as that of *Oldroyd* in the mathematical model of a flow leads, in general, to a higher order differential equation than is obtained for a purely viscous fluid, and so we expect to need more boundary conditions in order to specify the problem completely. (These extra boundary conditions may be thought of as arising from the need to specify what the fluid "remembers" of the flow outside the region of interest.) In flows which are uniform spatially, or uniform with time for a material element, the uniformity will provide the extra information and so no extra conditions are needed. Similarly for confined flows, where no new fluid enters the region of interest, no information about flow outside this region is needed.

Here the steady flow of a tubular film of a viscoelastic fluid is studied with the particular aim of examining the effect of these extra boundary conditions in a situation where they may be expected to have some significant influence on the flow as a whole. The flow, while being geometrically complex, is essentially an elongational free-surface flow involving the biaxial stretching of a thin axisymmetric tubular film. Features of the constitutive equations studied are the presence of a non-zero relaxation time and the possibility of a variable viscosity. One effect of the non-zero relaxation time is that a tube of constant radius (possible but unstable for a *Newtonian* fluid) is not dynamically possible. Preliminary computational results suggest that the effect of the extra upstream boundary conditions is not large, and also have failed to show any major difference between the two generalisations of the *Maxwell* model which have been used.

References

Oldroyd, J. G., Rheol. Acta **1**, 337–344 (1961).

Pearson, J. R. A. and *C. J. S. Petrie*, J. Fluid Mech. **40**, 1–19 (1970a); J. Fluid Mech. **42**, 609–625 (1970b); Plastics and Polymers **38**, 85–94 (1970c).

Tanner, R. I., A. S. L. E. Trans. **8**, 179–183 (1965).

White, J. L. and *A. B. Metzner*, J. App. Poly. Sci. **7**, 1867–1889 (1963).

Author's address:

Dr. *C. J. S. Petrie*
Dept. of Engineering Mathematics
Claremont Road
The University
Newcastle upon Tyne NE 1 7 RU (England)

Rheol. Acta **12**, 100–105 (1973)

From the Department of Applied Mathematics the University, and the Department of Mathematics, Liverpool Polytechnic, Liverpool (England)

Unsteady flows in elastico-viscous liquids

By N. D. Waters and M. J. King

With 2 figures

(Received October 27, 1972)

1. Introduction

During recent years the unsteady flow of elastico-viscous liquids in various geometries has attracted much attention. *Ting* (1963) considered a number of unsteady flow problems for the "second order" fluid of *Coleman* and *Noll* (1960) but found that bounded solutions could not be obtained for the cases of physical interest. *Waters* and *King* (1970, 1971) found that bounded solutions could be obtained to these problems for liquids with equations of state of the type proposed by *Oldroyd* (1950). The problems considered by these authors all involved the generation or decay of steady flow in fairly simple geometries. *Waters* and *King* found that the flows were strongly affected by the presence of elasticity and that, for quite realistic values of the elastic parameters, the velocity profiles oscillated about their final steady form before tending to it. Of course it has been known for some time that the presence of elasticity has a considerable effect in unsteady flow situations. Often experimentalists try to eliminate such effects from their experiments. Sometimes unsteady behaviour is actually used to determine rheological parameters, for example, in oscillatory experiments and Balance Rheometers (*Walters*, 1970). It would be useful if unsteady problems of the generation or decay type could also be used to such an end. Unfortunately the solutions obtained for such problems are rather complicated although they do give information about the general nature of the flows. For example, the solution (*Waters* and *King*, 1971) for the generation of flow in a pipe of circular cross-section has given an explanation of some anomalous behaviour in jet-thrust experiments carried out by *Oliver* and *MacSporran* (1970). They found unexpected peaks in jet-thrust measurements for certain ranges of the tube lengths. Although the unsteady nature of this "entry length" problem is with respect to distance, not time as for the problem considered by *Waters* and *King*, such behaviour could be explained qualitatively by the analysis of *Waters* and *King*.

2. Unsteady translation of a sphere starting from rest under the action of a constant force

In 1966 *Thomas* and *Walters* considered the motion of a sphere starting from rest under the action of a constant force (which could be gravity) in an elastico-viscous liquid. They used a linearized form of equation of state and found solutions for small and large values of the time which did not demonstrate any oscillatory behaviour. *Fielder* and *Thomas* (1967) considered the similar problem for an infinite lamina and found the solution for all values of the time, showing that the lamina oscillated about its terminal velocity before tending to it. *King* and *Waters* (1972) showed that this oscillatory behaviour could also be predicted for the sphere. They obtained the solution for all values of the time and found that restrictions on the material parameters in the analysis of *Thomas* and *Walters* were unnecessary.

The unsteady motion of a sphere moving in an elastico-viscous liquid has been used experimentally by *Hwang* et al. (1969) to determine rheological parameters. These authors devised a magnetic rheometer in which the sphere was moved horizontally through a very thick mucus by a magnetic force and then allowed to come to rest. By neglecting the liquid inertia and using a *Fourier* transform technique they were able to determine the complex viscosity directly.

Broadbent and *Walters* (Department of Applied Mathematics, University College of Wales, Aberystwyth, private communication) have set up a falling sphere apparatus. Use is made of an optical technique to obtain a trace of the vertical displacement of the sphere, which falls from rest under the action of gravity in an elastico-viscous liquid, against time. The predicted oscillatory behaviour can be clearly demonstrated using this apparatus and it is desirable to utilize these results to characterize elastico-viscous liquids. *Bullivant* (1971) has considered the possibility of using the method of *Hwang* et al. to this end but found that because the falling sphere is not brought to rest, as in their experiment, the *Fourier* transforms are unbounded. However, if inertia is neglected the theory of *King* and *Waters* (1972) simplifies greatly and can be used to interpret the experimental data to give a relaxation and a retardation time for the liquid [*King* and *Waters* (1972)].

It is not always possible to neglect the inertia of the liquid and some other, more satisfactory, method must be found for most liquids. *King* and *Waters* (1972) noticed that for the *Maxwell* liquid the maximum velocity attained by the sphere is linearly related to the square root of the non-dimensional relaxation time S_1, for values of $S_1 \geq 50$, i.e.

$$U_{max} = k S_1^{\frac{1}{2}}. \qquad [1]$$

They found that the constant k depended on the relative densities of the liquid and the sphere and could be determined from their solution by computation. Since this result is only true for large values of S_1 it suggests that an analytical expression might be obtained for k. This possibility is explored in the next section.

3. Approximate solution for the unsteady translation of a sphere

The *Maxwell* liquid is characterized by

$$p_{ik} = -p g_{ik} + p'_{ik} \qquad [2]$$

$$p'_{ik} + \lambda_1 \frac{\partial}{\partial t} p'_{ik} = 2\eta_0 e^{(1)}_{ik} \qquad [3]$$

where p_{ik} is the stress tensor, p an arbitrary isotropic pressure, g_{ik} the metric tensor, η_0 the viscosity, $e^{(1)}_{ik}$ the rate-of-strain tensor and t the time. S_1 is related to the relaxation time λ_1 by

$$S_1 = \frac{\eta_0 \lambda_1}{\varrho a^2} \qquad [4]$$

where ϱ is the density of the liquid and a is the radius of the sphere.

By considering a non-dimensional form of [3] for large values of S_1 and large changes in the stress with respect to time, we shall neglect the first term compared to the second in the left hand side of eq. [3]. If the sphere is being driven by a constant force F through the elastico-viscous liquid then the velocity of the sphere $U(t_1)$ can very easily be shown to be given by

$$\frac{U(t_1)}{U_0} = \frac{2S_1^{\frac{1}{2}}}{c} \exp\left(\frac{-t_1}{2\alpha S_1^{\frac{1}{2}}}\right) \sin\left(\frac{c t_1}{2\alpha S_1^{\frac{1}{2}}}\right) \qquad [5]$$

where $U_0 = \dfrac{\varrho F}{6\varrho_1 \eta_0 a}$ is the terminal velocity,

$$\alpha = \frac{2\varrho + \varrho_1}{9\varrho_1}, \quad c = \sqrt{4\alpha - 1}, \quad t_1 = \frac{\eta_0 t}{\varrho a^2} \qquad [6]$$

and ϱ_1 is the density of the sphere.

$U(t_1)$ has stationary values when

$$t_1 = \frac{2\alpha S_1^{\frac{1}{2}}}{c}\left(\tan^{-1} c + n\frac{\pi}{2}\right), \qquad n = 0, 1, 2 \ldots \qquad [7]$$

giving

$$\frac{U_{max}}{U_0} = \sqrt{\frac{S_1}{\alpha}} \exp\left(-\frac{1}{c}\tan^{-1} c\right). \qquad [8]$$

We require that $\alpha > \frac{1}{4}$. If $\alpha \leq \frac{1}{4}$ no oscillatory behaviour takes place.

We observe that by comparing [1] and [8]

$$k = \left(\alpha^{\frac{1}{2}} \exp\left(\frac{1}{c}\tan^{-1} c\right)\right)^{-1}. \qquad [9]$$

By integrating [5] similar expressions can be obtained for the displacement of the sphere.

Table 1 shows the values of k calculated from [9] compared to those calculated from the full solution for the *Maxwell* liquid obtained by *King* and *Waters* (1972). We see that the agreement is within 3%.

α	k *King* and *Waters* (1972)	k eq. [9]
1.00	0.555 ± 0.007	0.546
1.82	0.473 ± 0.009	0.461
2.50	0.430 ± 0.009	0.417

It is also interesting to compare the velocity of the sphere predicted by the two methods of solution. Fig. 1 (a) and (b) show the velocity of the sphere against t_1 predicted by eq. [5] compared to that predicted by the theory of *King* and *Waters* (1972), for $\alpha = 2.00$ and $S_1 = 500$ and 1500 respectively. We see that up to and beyond the first velocity peak the two curves are in very good agreement. It should be noted that $U(t_1)/U_0$ computed from [5] tends to zero for large t_1, not unity as for the full solution. The approximate solution is a special case of the analysis of *Hunter* (1968) who considered the translation of a sphere through a viscoelastic medium. The velocity of the sphere will obviously decay to zero since we are essentially regarding the medium as a viscoelastic solid. The

*Newton*ian solution is not easily shown on the scale of fig. 1 but, in fact, tends quickly and uniformly to unity from below.

In practice it may be possible to use any one of the formulae [7–9] to determine S_1. Expressions involving the displacement of the sphere will give less accurate results since the first displacement peak occurs when the velocity is zero and from fig. 1 we see that at this point the approximation is not quite as good.

Thus we have observed that the approximation is a good one for large values of S_1 while rapid changes are taking place in the liquid. The approximation simplifies the analysis and the final form of the solution enormously. It has the advantage over the usual "small time" *Laplace* transform technique in that it produces a solution not unlike the full solution instead of the first few terms of a series which do not demonstrate any oscillatory effects. This approximation can perhaps be best thought of as assuming that a highly elastic liquid responds like a solid to very rapid changes.

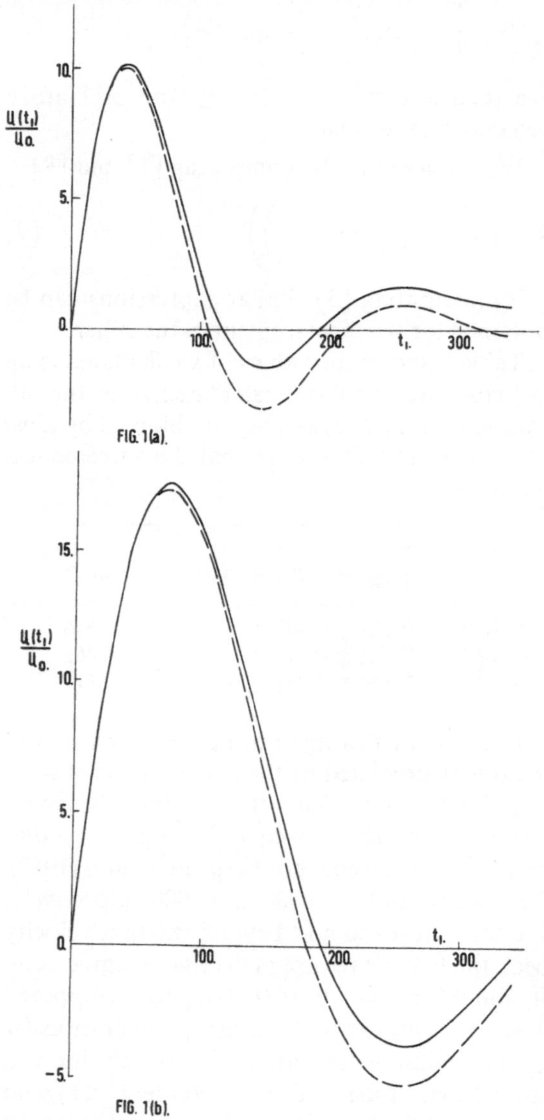

Fig. 1. Approximate sphere velocity (broken line) compared with that for the *Maxwell* liquid (full line) against t_1 for $\alpha = 2.00$ and
(a) $S_1 = 500.0$ (b) $S_1 = 1500.0$

4. Unsteady decay of the motion of a sphere rotating about vertical diameter

In some of the problems considered by *Ting* (1963) and *Waters* and *King* no advantage is gained by using the approximation described in section 3; either no solution is obtained or its final form is still too complicated for easy use. In this section we consider a problem which can very easily be set up experimentally i.e. a sphere rotating slowly about its vertical diameter in an elastico-viscous liquid, being allowed to come freely to rest when the driving torque is suddenly removed. *Jones* and *Walters* (1966) have considered a similar problem of an elastico-viscous liquid contained between two concentric spheres; the inner one, suspended by a torsion wire, oscillating due to a small angular displacement while the outer sphere is at rest.

We shall first obtain the solution for the *Maxwell* liquid and in the next section obtain the approximate solution using the method of section 3.

The *Maxwell* liquid is characterized by eqs. [2] and [3]. We define a set of spherical polar co-ordinates r, θ, φ whose origin is at the centre of the sphere, $\theta = 0$ is vertically upwards and the sphere has radius $r = a$. We assume that the flow is slow and thus the unsteady velocity distribution has the form,

$$(0, 0, r\omega^*(t)\sin\theta).$$

The only non-trivial equation of state is

$$p'_{(r\varphi)} + \lambda_1 \frac{\partial}{\partial t} p'_{(r\varphi)} = 2\eta_0 e^{(1)}_{(r\varphi)} \quad ^{1)} \qquad [10]$$

and the only equation of motion not satisfied automatically is

$$\varrho r \sin\theta \frac{\partial\omega^*}{\partial t} = \frac{1}{r^3}\frac{\partial}{\partial r}(r^3 p'_{(r\varphi)}) \qquad [11]$$

where ϱ is the density of the liquid, subject to the initial and boundary conditions

[1]) Brackets placed around suffixes denote *physical* components of tensors.

(i) $\omega^*(r,t) = \Omega_1$ for $t \leq 0$

(ii) $\omega^*(r,t) = \Omega^*(t)$ at the sphere $r = a, t > 0$

(iii) $\omega^*(r,t) \to 0$ as $r \to \infty, t > 0$. [12]

Lastly, we use the fact that the rate of change of angular momentum of the sphere must equal the couple on the sphere due to the liquid.

$$I \frac{d\Omega^*}{dt}(t) = 2\pi a^3 \int_0^\pi [p'_{(r\varphi)}]_{r=a} \sin^2\theta \, d\theta \qquad [13]$$

where I is the moment of inertia of the sphere about the axis of rotation.

We now use the method of *Laplace* transforms and write

$$\bar{f}(\gamma) = \int_0^\infty e^{-\gamma t} f(t)\, dt$$

as the *Laplace* transform of $f(t)$. If we also non-dimensionalize using

$$a r_1 = r, \quad t_1 = \frac{\eta_0 t}{\varrho a^2}, \quad S_1 = \frac{\eta_0 \lambda_1}{\varrho a^2}, \quad s = \frac{\varrho a^2 \gamma}{\eta_0}$$

$$\omega(r,t) = \Omega_1 \omega^*(r_1, t_1), \quad \Omega(t) = \Omega_1 \Omega^*(t_1) \qquad [14]$$

then [10] and [11] give

$$\frac{\partial^2 \bar{\omega}}{\partial r_1^2} + \frac{4}{r_1}\frac{\partial \bar{\omega}}{\partial r_1} - \beta^2 \bar{\omega} = -\frac{\beta^2}{s r_1^3} \qquad [15]$$

where $\beta^2 = s(1 + S_1 s)$. [16]

Conditions [12] become

$$\left.\begin{array}{ll} \text{(i)} & \bar{\omega}(r_1, s) = 1 \quad \text{for} \quad t_1 \leq 0 \\ \text{(ii)} & \bar{\omega}(r_1, s) = \bar{\Omega}(s) \quad \text{at} \quad r_1 = 1 \\ \text{(iii)} & \bar{\omega}(r_1, s) \to 0 \quad \text{as} \quad r_1 \to \infty. \end{array}\right\} \qquad [17]$$

If we assume that the sphere is uniform and has density $\varkappa\varrho$ the transform of [13] becomes

$$\varkappa s \bar{\Omega} - \varkappa = \frac{5s}{\beta^2}\left\{\left(\frac{\partial \bar{\omega}}{\partial r_1}\right)_{r_1=1} - 3S_1\right\}. \qquad [18]$$

A solution satisfying [15], [17] and [18] is

$$\bar{\omega}(r_1, s) = \frac{1}{s r_1^3} - \frac{\varkappa_1(1 + sS_1)(1 + \beta r_1)\exp\{-\beta(r_1 - 1)\}}{s r_1^3(\beta^3 + \varkappa_2\beta^2 + \varkappa_1\beta + \varkappa_1)} \qquad [19]$$

where $\varkappa_1 = 15/\varkappa$ and $\varkappa_2 = 1 + 5/\varkappa$.

Thus the transformed angular velocity of the sphere becomes

$$\bar{\Omega}(s) = \frac{1}{s}\left\{1 - \frac{\varkappa_1(1 + sS_1)(1 + \beta)}{\beta^3 + \varkappa_2\beta^2 + \varkappa_1\beta + \varkappa_1}\right\}. \qquad [20]$$

Before evaluating the *Laplace* inversion integral for $\bar{\Omega}(s)$ we notice that the denominator has only one real root for β with $\varkappa_1$ positive. We shall use a method similar to that used by *King* (1970) and *King* and *Waters* (1972). The integrand will have simple poles at $s=0$ and the values of s obtained solving [16] for each root of the cubic for β occurring in [20], two of these roots being complex conjugates. Following *King* (1970) we choose that branch of β such that the square root of a positive real number is positive and obtain the solution

$$\Omega(t_1) = W(t_1) + \frac{2\varkappa_1^2}{3\pi}\int_0^{1/\sqrt{S_1}} \frac{(1 - \mu^2 S_1)^{5/2} \quad \mu^2 e^{-\mu^2 t_1}\, d\mu}{\{\varkappa_1 - \mu^2\varkappa_2(1 - S_1\mu^2)\}^2 + \mu^2(1 - \mu^2 S_1) \quad \{\varkappa_1 - \mu^2(1 - \mu^2 S_1)\}^2} \qquad [21]$$

where

$$W(t_1) = -2\varkappa_1 \sum \frac{\beta^2(1 + \beta)e^{st}}{s^2(1 + 2sS_1)(3\beta^2 + 2\beta\varkappa_2 + \varkappa_1)}$$

the summation is evaluated over the values of β given by the roots of the cubic

$$\beta^3 + \varkappa_2\beta^2 + \varkappa_1\beta + \varkappa_1 = 0. \qquad [22]$$

It is possible to combine the residues from the two complex conjugate roots in [21] to obtain a real function. However, since the cubic must be solved numerically the complex numbers can be handled as an ordered pair by the computer.

The corresponding *Newton*ian solution involves no poles for this choice of branch and has solution

$$\Omega(t_1) = \frac{2\varkappa_1^2}{3\pi}\int_0^\infty \frac{\mu^2 e^{-\mu^2 t_1}\, d\mu}{(\varkappa_1 - \mu^2\varkappa_2)^2 + \mu^2(\varkappa_1 - \mu^2)^2}. \qquad [23]$$

5. Approximate solution for the decay of the motion of the rotating sphere

If we apply the approximation used in section 3 for large S_1 and large time rates of change we essentially use $\beta = s^2 S_1$ in [19] and [20]. We can invert [20] by using tables [*Bateman* (1954)] to obtain

$$\Omega(t_1) = e^{-\xi t_1}(A \cos\eta t_1 + B \sin\eta t_1) + (1 - A)\, e^{-\zeta t_1} \qquad [24]$$

where

$$A = 1 - \frac{\{\zeta^2 - \zeta(\varkappa_2 \varepsilon - \varkappa_1) - \varkappa_1 \varepsilon\}}{(\xi - \zeta)^2 + \eta^2}$$

$$B = -\frac{1}{2\eta}\{2\varkappa_1 - 2\zeta - (\varkappa_2 \varepsilon - 3\zeta)\, A\}$$

$$\varepsilon = 1/\sqrt{S_1}, \quad \xi = \tfrac{1}{2}(\varkappa_2 \varepsilon - \zeta),$$

$$\eta^2 = \tfrac{1}{4}(4\varepsilon^2 \varkappa_1 - \varepsilon^2 \varkappa_2^2 + 3\zeta^2 - 2\zeta\varepsilon\varkappa_2)$$

and $-\zeta$ is the real root of eq. [22].

6. Numerical calculations and conclusions

A *Newton-Raphson* technique was used to obtain the real root of eq. [22]. The complex conjugate roots of the equation were then determined analytically. Both the solution for the *Maxwell* liquid [21] and the approximate solution [24] were evaluated numerically and compared for $\varkappa = 8.5$ (equivalent to $\alpha = 2.00$ in section 3), $S_1 = 500$ and $S_1 = 1500$ in fig. 2. The *Newton*ian solution is not shown but tends quickly and uniformly to zero. All numerical work was carried out on the Liverpool Polytechnic ICL 1902 A computer.

It can be seen from fig. 2 that the approximate velocity is very close to that for the full solution.

The problem in section 4 has been solved for the *Maxwell* liquid in order to compare the solution with that obtained by using the approximation outlined in section 3. The problem can also be solved for a more complicated liquid like *Oldroyds* (1950) liquid *B*.

We have demonstrated a very useful approximation for solving unsteady flow problems for elastico-viscous liquids of the *Maxwell* type for large values of S_1 where there are large changes in the stress with time. The complexity of both the analysis and the final solution is reduced

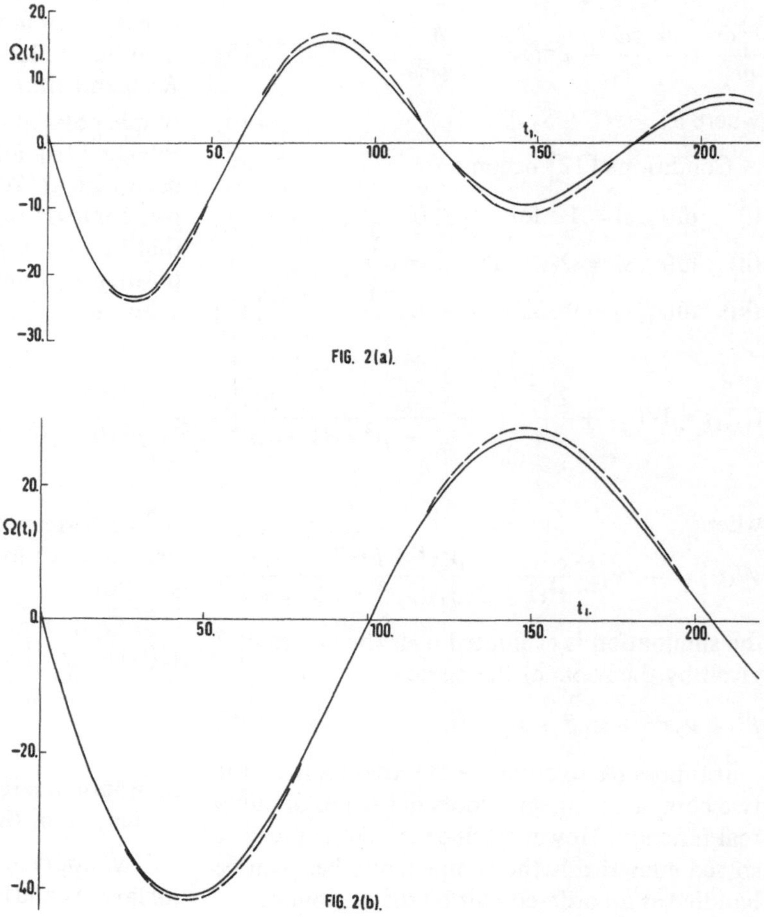

Fig. 2. Approximate sphere angular velocity (broken line) compared with that for the *Maxwell* liquid (full line) against t_1 for $\varkappa = 8.50$ and
(a) $S_1 = 500.0$
(b) $S_1 = 1500.0$

dramatically, giving formulae that are easily used by experimentalists. In section 3 formulae are given which may be used directly to determine S_1 from experimental observations. By comparing figs. 1 and 2 we see that there is better agreement in fig. 2. This is because our approximation essentially regards the liquid as responding to sudden changes like a solid and thus problems in which the velocity dies out with large time are solved more accurately.

This technique may allow the approximate solution of unsteady flow problems which hitherto have been too complicated to solve analytically. The approximate solution will probably reveal the essential features of the full solution.

Summary

Consideration is given to two problems involving the unsteady motion of a sphere in a *Maxwell* liquid. In both problems the full solutions are obtained and compared in each case to the solution obtained using an approximation procedure. Good agreement is obtained for large values of the elastic parameter. The approximation reduces the mathematics considerably and thus allows experimentalists to determine the elastic parameter involved.

References

Bateman, H., Tables of Integral Transforms, Vol. 1 (New York 1954).

Bullivant, S., M. Sci. Dissertation University of Liverpool (1971).

Coleman, B. D. and *W. Noll*, Arch. Rat. Mech. Anal. **6**, 355 (1960).

Fielder, R. and *R. H. Thomas*, Rheol. Acta **6**, 306 (1967).

Hunter, S. C., Proc. Edin. Math. Soc. **16**, 56 (1968).

Hwang, S. H., M. Litt, and *W. C. Forsman*, Rheol. Acta **8**, 438 (1969).

Jones, J. R., and *T. S. Walters*, Mathematika **13**, 83 (1966).

King, M. J., Ph. D. Thesis University of Liverpool (1970). – *King, M. J.* and *N. D. Waters*, J. of Phys. D. **5**, 141 (1972).

Oldroyd, J. G., Proc. Roy. Soc. **A 200**, 235 (1950).

Oliver, D. R. and *W. C. MacSporran*, Can. J. Chem. Eng. **48**, 243 (1970).

Tsian Wu Ting, Arch. Rat. Mech. Anal. **14**, 1 (1963).

Walters, K., Proc. I.U.M.T.A.M. Conference III, 507 (Israel 1962). – *Walters, K.* J. Fluid Mech. **40**, 191 (1970).

Waters, N. D. and *M. J. King*, Rheol. Acta **9**, 345 (1970). – *Waters, N. D.* and *M. J. King*, J. Phys. D. **4**, 204 (1971).

Authors' addresses:

Dr. *N. D. Waters*
Dept. of Applied Mathematics, The University,
P.O. Box 147, Liverpool, L69 3BX (England)

Dr. *M. J. King*
Dept. of Mathematics, Liverpool Polytechnic,
Byron Street, Liverpool, L3 3AE (England)

Rheol. Acta 12, 106–113 (1973)

From the Department of Theoretical and Applied Mechanics, Cornell University, Ithaca (U.S.A.)

A continuum theory for viscoelastic composite beams*)

By M. Cengiz Dökmeci and Mg. AlpD.

With 1 figure

(Received October 27, 1972)

1. Introduction

Recently, a new line of attack (1) has been pursued in order to study stresses and deformations theoretically in laminated and fibrous composite structures [see, e.g., (2–6) and references therein]. This theory takes into account both elastic and geometric properties of each constituent. Thus, it differs conceptually from the classical elastic modulus theory of composites. The classical theory has been shown to be satisfactory for static problems only, whereas the new theory yields good results for static as well as dynamic problems of composites. The reader is referred to *Herrmann* (7) for a comprehensive discussion of these theories.

The well-known *Bernoulli-Euler* and *Timoshenko* beam models ignore the deformation and skin effects and like in composites. Also, they automatically abrogate the cross-sectional distortion, and thereof the coupling of adherent layers and the dispersion of free harmonic waves. These models are evidently unable to describe the mechanical behavior of laminated composite beams. With this paper, a layered beam model is introduced to develop the theory of viscoelastic composite beams, retaining all essential features, within the frame of the effective stiffness theory of composites.

In this paper, following a review of the fundamental equations of linear thermoviscoelastodynamics in Art. 2, we first study the geometry and kinematics of laminated composite beams in Art. 3. The laminated beam is assumed to be consist of N-anisotropic viscoelastic layers of uniform thickness. A displacement field linearized with respect to the lateral coordinates of each layer is chosen apriori as a starting point. This field eliminates the *Bernoulli-Euler* hypothesis of beams, and it leads to include the effects of transverse shear, normal strains and rotatory inertia in a systematic and consistent manner. Further, the dynamic interactions between the adjacent layers are taken into account. In Art. 4, the resultants of various field quantities are introduced. The appropriate set of the constitutive equations for the stress resultants is then established. The macroscopic beam equations of motion and heat conduction, the initial and boundary conditions are extensively studied on the basis of the linearized temperature and displacement fields in Art. 5. The governing equations for some special cases of interest are presented in Art. 6. The last article is devoted to a brief conclusion.

Notation

Standard indicial notation in a Euclidean 3-space is used throughout the text. *Einstein*'s summation convention is implied for repeated indices, unless indices are put within parantheses. Except for the indices m and n, Latin indices take the values 1, 2, 3, while Greek indices the values 2, 3. The indices m and n are respectively used to denote the m-th and n-th layers from the bottom layer of the beam, and they take the values $1, 2, \ldots, N$. Partial differentiations with respect to the indicated coordinates x_i are indicated by a comma and, in particular, to the coordinate of the beam axis x_1 by a prime. Overbars are used to designate the *Laplace* transformations, stars to denote the *Stieltjes* convolution and superposed dots to represent partial differentiation with respect to time, t. Also, the prescribed field quantities are indicated by a superscript or subscript zero.

2. Résumé of basic equations of viscoelastodynamics

To begin with, we state the relevant equations which govern the linear nonisothermal theory of nonpolar thermoviscoelastodynamics. A more elaborate account of these equations might be found in (8–10).

Consider a nonpolar thermoviscoelastic body B with the material volume v and its surface s. The body B is referred to a fixed rectangular *Cartes*ian coordinates x_i in a *Euclid*ian 3-space. Let σ_{kl} denote the symmetric stress tensor, u_k the displacement vector, ϱ the mass density, f_i the body force vector per unit volume, e_{ij} the infinitesimal strain tensor, Θ the infinitesimal temperature deviation from the reference temperature T_0, q_i the heat flux vector measured per unit area per unit time and t_i the surface traction vector, all defined for $(x_i, t) \in v \times (0, \infty)$. Then the governing equations might be expressed as follows.

---

*) Supported by the Office of Naval Research.

Equations of motion

$$\sigma_{ij,i} + f_j - \varrho \ddot{u}_j = 0, \quad \sigma_{ij} = \sigma_{ji} \quad \text{in} \quad v \times (0, \infty). \quad [2.1]$$

Strain-displacement relations

$$e_{ij} = \tfrac{1}{2}(u_{i,j} + u_{j,i}). \quad [2.2]$$

Boundary conditions

$$u_i - u_i^0 = 0 \quad \text{on} \quad s_u \times (0, \infty) \quad [2.3]$$

$$t_j^0 - n_i \sigma_{ij} = 0 \quad \text{on} \quad s_\sigma \times (0, \infty) \quad [2.4]$$

$$\Theta - \Theta_0 = 0 \quad \text{on} \quad s_\theta \times (0, \infty) \quad [2.5]$$

$$q_i n_i = 0 \quad \text{on} \quad s_K \times (0, \infty) \quad [2.6]$$

with

$$s_u \cap s_\sigma = 0$$

$$s_\theta \cap s_K = 0$$

$$s_u \cup s_\sigma = s_\theta \cup s_K = s. \quad [2.7]$$

Here, s_u and s_σ stand for the disjoint surface parts of s where the displacement and traction vectors are prescribed, respectively. s_θ is the surface part upon which the temperature field is prescribed and s_K is the complementary part of s over which the surface is considered to be perfectly insulated against heat flow. n_k is the outward unit normal vector to s. All the surface parts s_u, s_σ, s_θ and s_K are required to remain constant with time.

Initial conditions

$$\Theta(x_k\, t) = u_i(x_k, t) = \sigma_{ij}(x_k, t) = 0 \quad \text{in} \quad v \times (-\infty, 0^-). \quad [2.8]$$

Constitutive equations

$$\sigma_{ij} = \tau_{ij} + G_{ijkl} * de_{kl} - \varphi_{ij} * d\Theta$$

$$q_i = -K_{ij}\Theta_{,j} \quad \text{in} \quad v \times (0, \infty) \quad [2.9a]$$

with

$$\varphi_{ij} = \varphi_{ji}, \quad G_{ijkl} = G_{jikl} = G_{ijlk} \quad \text{in} \quad v \times (0, \infty)$$

$$\varphi_{ij} = G_{ijkl} = 0 \quad \text{in} \quad v \times (-\infty, 0^-) \quad [2.9b]$$

for anisotropic materials. In these equations, a *Stieltjes* convolution denoted by a star is given as

$$\varphi * d\psi = \psi * d\varphi = \int_{0-}^{t} \varphi(t - \tau)\, d\psi(\tau). \quad [2.10]$$

G_{ijkl}, φ_{ij} and K_{ij} indicate respectively the relaxation functions and thermal conductivity of the thermoviscoelastic material, and τ_{ij} the initial stress tensor which is henceforth omitted.

For isotropic materials, the relaxation functions and the thermal conductivity are given by

$$K_{ij} = K\delta_{ij}, \quad \varphi_{ij} = \varphi\delta_{ij}$$

$$G_{ijkl} = \tfrac{1}{3}(G_2 - G_1)\,\delta_{ij}\delta_{kl} + \tfrac{1}{2}G_1(\delta_{ik}\delta_{jl} + \delta_{il}\delta_{jk}) \quad [2.11]$$

where $G_1(t)$, $G_2(t)$ and $\varphi(t)$ are the independent functions, respectively, appropriate to the states of shear, dilatation and temperature, and δ_{ij} is the *Kronecker* delta. The relaxation functions can be expressed in terms of the *Lamé* constants $\lambda(t)$ and $\mu(t)$ by

$$G_1 = 2\mu, \quad G_2 = 3\lambda + 2\mu. \quad [2.12]$$

The corresponding constitutive relations are written in the form

$$\sigma_{ij} = (\lambda * de_{kk} - \varphi * d\Theta)\,\delta_{ij} + 2\mu * e_{ij}. \quad [2.13]$$

Heat conduction equations

$$q_{i,i} + T_0 \frac{\partial}{\partial t}(M * d\Theta + \varkappa_{ij} * de_{ij}) = 0 \quad [2.14]$$

and

$$q_{i,i} + T_0 \frac{\partial}{\partial t}(M * d\Theta + \varkappa * de_{ii}) = 0 \quad [2.15]$$

for anisotropic and isotropic materials, respectively. In these equations, M, $\varkappa_{ij}$, $\varkappa$ stand for the relaxation functions of material.

It has been shown that the governing equations collected above possess a unique solution provided that the initial values of the relaxation functions are positive definite and the thermal conductivity is positive-semidefinite [see, e.g. (11–13)].

3. Displacement and temperature fields

Consider a laminated composite beam which consists of N reinforcing and matrix layers. The layers are made of viscoelastic materials, and each of them may possess different uniform thickness, orientation, and/or anisotropic properties. The composite beam is embedded in a *Euclidian* 3-space. It is referred to a x_i-system of *Cartesian* coordinates in this space. The axis x_1 is chosen in a parallel direction to the centroidal axis of the composite beam. The axes x_2 and x_3 are located at the initial cross-section of the first

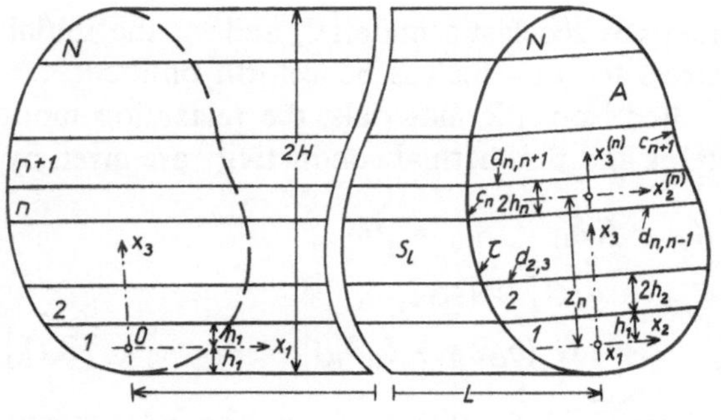

Fig. 1

layer; they are respectively chosen in a parallel direction, and a perpendicular direction to the layering (see fig. 1). Further, a local *Cartesian* coordinate system $x_i^{(n)}$ is defined by

$$x_1^{(n)} = x_1^{(1)} = x_1$$
$$x_2^{(n)} = x_2^{(1)} = x_2$$
$$x_3^{(n)} = x_3^{(1)} - z_n = x_3 - z_n \quad [3.1]$$

at the initial cross-section of the n-th layer. The distance between the midplanes of the first and n-th layers is denoted by z_n and it is expressed by

$$z_n = \sum_{m=1}^{n} z_{nm} h_{(m)} \quad [3.2a]$$

with

$$z_{nm} = 2 - \delta_{1m} - \delta_{nm}. \quad [3.2b]$$

We denote the length of the laminated composite beam by L, its entire volume by v and its surface by s. The area of cross-sections of the beam, A is assumed to be bounded by a smooth and nonintersecting *Jordan* curve τ. The bonding between the n-th and $(n + 1)$-th layers is taken as a plane denoted by $s_{n,n+1}$. The intersection of A and $s_{n,n+1}$ is a straight line and it is indicated by $d_{n,n+1}$. Let the area of the cross-section of the n-th layer be A_n. The contour of A_n is designated by τ_n which consists of $d_{n,n-1}$, $d_{n,n+1}$ and c_n as shown in the figure. Thus, we write

$$\tau_n = c_n \cup d_{n,n-1} \cup d_{n,n+1}. \quad [3.3]$$

Under the well-known ad hoc assumptions for beams, all field quantities are continuous in the beam space v. Also, these assumptions allow to treat the beam as a one-dimensional continuum model of a three-dimensional body. Based on these considerations, the displacement and temperature of each layer can be expanded into a product of series in terms of the aerial coordinates x_2 and x_3. The truncated form of these series, namely

$$u_i^{(n)}(x_k, t) = v_i^{(n)}(x_1, t) + x_2 \alpha_i^{(n)}(x_1, t) + x_3 \beta_i^{(n)}(x_1, t) \qquad n = 1, 2, \ldots, N \quad [3.4]$$

$$\Theta^{(n)}(x_i, t) = \theta^{(n)}(x_1, t) + x_2 \varphi^{(n)}(x_1, t) + x_3 \psi^{(n)}(x_1, t) \qquad n = 1, 2, \ldots, N \quad [3.5]$$

will be used in the present analysis. Eq. [3.4] is sufficiently general to abrogate the usual *Bernoulli-Euler* hypotheses for each layer and to take into account all the effects of shear and normal strains as well as the rotatory inertia.

In this paper, no relative deformation and temperature are considered at the bonding planes $s_{n,n+1}$. Hence, the continuity of the displacement and temperature fields on these planes is implied, that is,

$$u_i^{(n)} - u_i^{(n+1)} = 0, \quad \Theta^{(n)} - \Theta^{(n+1)} = 0 \quad \text{on} \quad s_{n,n+1}. \quad [3.6]$$

With the aid of eqs. [3.4] and [3.5], these continuity requirements can be expressed by

$$v_i^{(n)} + (z + h)_{(n)} \beta_i^{(n)} - v_i^{(n+1)} - (z - h)_{(n+1)} \beta_i^{(n+1)} + x_2 (\alpha_i^{(n)} - \alpha_i^{(n+1)}) = 0$$

$$\theta^{(n)} + (z + h)_{(n)} \psi^{(n)} - \theta^{(n+1)} - (z - h)_{(n+1)} \psi^{(n+1)} + x_2 (\varphi^{(n)} - \varphi^{(n+1)}) = 0 \quad [3.7]$$

which lead to the $8(N - 1)$ continuity constraints as follows.

$$\alpha_i^{(n)} = \alpha_i$$
$$v_i^{(n)} + (z + h)_{(n)} \beta_i^{(n)} = v_i^{(n+1)} + (z - h)_{(n+1)} \beta_i^{(n+1)}$$

and [3.8]

$$\varphi^{(n)} = \Phi$$
$$\theta^{(n)} + (z + h)_{(n)} \psi^{(n)} = \theta^{(n+1)} + (z - h)_{(n+1)} \psi^{(n+1)} \quad [3.9]$$

By virtue of the continuity constraints [3.7], the number of the independent displacement and temperature functions, $v_i^{(n)}$, $\alpha_i^{(n)}$, $\beta_i^{(n)}$ and $\theta^{(n)}$, $\varphi^{(n)}$, $\psi^{(n)}$ is now reduced from $12N$ to $4(N + 2)$. Accordingly, the independent unknown functions are selected as

$$\alpha_i(x_1, t), v_i(x_1, t) = v_i^{(1)}, \beta_i^{(n)}, \theta = \theta^{(1)}, \Phi, \psi^{(n)} \quad [3.10]$$

and the remaining functions are expressed in terms of these functions by

$$\alpha_i^{(n)} = \alpha_i, \quad v_i^{(n)} = v_i + \sum_{m=1}^{n} b_{nm}\beta_i^{(m)}$$

$$\varphi^{(n)} = \Phi, \quad \theta^{(n)} = \theta + \sum_{m=1}^{n} b_{nm}\psi^{(m)} \qquad [3.11a]$$

where

$$b_{nm} = z_{nm}h_{(m)} - \delta_{nm}z_{(n)}. \qquad [3.11\ ]$$

It should be noted that in eq. [3.4], $v_1^{(n)}$ represents the extensional motions, $\alpha_3^{(n)}$ and $\beta_2^{(n)}$ the torsional motions, and $v_\alpha^{(n)}$, $\alpha_1^{(n)}$ and $\beta_1^{(n)}$ the flexural and shear motions of layers. If the terms $\beta_i^{(n)}$ are dropped out in eq. [3.4], we arrive at the equation

$$u_i^{(n)} = v_i^{(n)} + x_2\alpha_i^{(n)}. \qquad [3.12]$$

In view of the continuity conditions [3.6], this equation yields

$$v_i^{(n)} = v_i, \quad \alpha_i^{(n)} = \alpha_i; \quad u_i^{(n)} = u_i \qquad [3.13]$$

for all layers; and it displays no coupling among the displacement functions. Thus, eq. [3.13] leads to the usual effective modulus theories of composites.

4. Loads, resultants of stress and heat fluxes, and constitutive relations

In this section, as a first phase of the governing equations we define the resultants for various field quantities and give the strain distribution and the constitutive relations.

Body force resultants

$$\{E_i, F_i, G_i\}^{(n)} = \int_{A_n} \{1, x_2, x_3\}\, f_i dA. \qquad [4.1]$$

Stress resultants

$$\{\underset{\sim}{N}_{ij}\}^{(n)} = \{N_{ij}, M_{ij}, K_{ij}\}^{(n)} = \int_{A_n} \{1, x_2, x_3\}\, \sigma_{ij} dA. \qquad [4.2]$$

Acceleration resultants

$$\{\underset{\sim}{\ddot{A}}_i\}^{(n)} = \{\ddot{A}_i, \ddot{B}_i, \ddot{C}_i\}^{(n)} = \int_{A_n} \{1, x_2, x_3\}\, \ddot{u}_i dA \qquad [4.3]$$

which may be written as

$$\{\underset{\sim}{A}_i\}^{(n)} = \{\underset{\sim}{v}_i\}^{(n)}\{\underset{\sim}{I}\}^{(n)} \qquad [4.4a]$$

with

$$\{\underset{\sim}{v}_i\}^{(n)} = \{\underset{\sim}{v}_i, \underset{\sim}{\alpha}_i, \beta_i\}^{(n)}$$

$$\{\underset{\sim}{I}\}^{(n)} = \begin{Bmatrix} A & I_{10} & I_{01} \\ I_{10} & I_{20} & I_{11} \\ I_{01} & I_{11} & I_{02} \end{Bmatrix}^{(n)}. \qquad [4.4b]$$

Here, eq. [3.4] is used and the aerial quantities are defined by

$$I_{ij}^{(n)} = \int_{A_n} x_2^i x_3^j dA, \quad A = I_{\infty}. \qquad [4.5]$$

Load resultants

$$\{L_j, P_j, R_j\}^{(n)} = \oint_{c_n} \nu_\alpha \sigma_{\alpha j}\{1, x_2, x_3\}\, ds \qquad [4.6]$$

where ν_α is the outward unit normal vector to c_n. In view of eq. [3.3], the load resultants may be expressed by

$$\begin{Bmatrix} L_j \\ P_j \\ R_j \end{Bmatrix}^{(n)} = \begin{Bmatrix} l_j + a_j - e_j \\ p_j + b_j - d_j \\ r_j + (z+h)\,a_j - (z-h)\,e_j \end{Bmatrix}^{(n)} \qquad [4.7]$$

where

$$\{a_j, b_j\}^{(n)} = \int_{d_{n,n+1}} \{1, x_2\}\, \sigma_{3j} dx_2$$

$$\{e_j, d_j\}^{(n)} = \int_{d_{n,n-1}} \{1, x_2\}\, \sigma_{3j} dx_2$$

$$\{l_j, p_j, r_j\}^{(n)} = \int_{c_n} \nu_\alpha \sigma_{\alpha j}\{1, x_2, x_3\}\, ds. \qquad [4.8]$$

Effective surface loads

$$\{N_j, K_j\} = \oint_{\tau} \nu_\alpha\{1, x_2\}\, \sigma_{\alpha j} ds$$

$$\mathcal{M}_j^{(n)} = r_j^{(n)} + \sum_{m=n}^{N} b_{mn} l_j^{(m)}$$

$$\{N_j^0, M_j^0, K_j^0\}^{(n)} = \int_{A_n} n_1\{1, x_2, x_3\}\, \sigma_{1j}^0 dA \qquad [4.9]$$

and

$$\{N_j^0, M_j^0, K_j^0\} = \int_{A} n_1\{1, x_2, x_3\}\, \sigma_{1j}^0 dA \qquad [4.10]$$

with

$$\{N_j, M_j, K_j\}^{(n)} = \int_{A_n} n_1\{1, x_2, x_3\}\, \sigma_{1j} dA$$

$$\{N_j, M_j, K_j\}^{(n)} = n_1\{N_{1j}, M_{1j}, K_{1j}\}^{(n)} \qquad [4.11]$$

in which eqs. [3.3] and [4.2] are considered.

Viscoelastic stiffnesses

$$\{\underset{\sim}{G}_{ijkl}\} = G_{ijkl}\{\underset{\sim}{I}\}, \quad \{\underset{\sim}{\varphi}_{ij}\} = \varphi_{ij}\{\underset{\sim}{I}\}$$
$$\{\underset{\sim}{\varkappa}_{ij}\} = \varkappa_{ij}\{\underset{\sim}{I}\}, \quad \{\underset{\sim}{K}_{ij}\} = K_{ij}\{\underset{\sim}{I}\}. \qquad [4.12]$$

For isotropic materials, these stiffnesses are written as

$$\{\underset{\sim}{\lambda}\} = \lambda\{\underset{\sim}{I}\}, \quad \{\underset{\sim}{\varphi}\} = \varphi\{\underset{\sim}{I}\}$$
$$\{\underset{\sim}{\mu}\} = \mu\{\underset{\sim}{I}\}, \quad \{\underset{\sim}{\varkappa}\} = \varkappa\{\underset{\sim}{I}\}, \quad \{\underset{\sim}{K}\} = K\{\underset{\sim}{I}\}. \qquad [4.13]$$

Temperature resultants

$$\{\Theta_1^0, \Theta_2^0, \Theta_3^0\} = \int_A \Theta^0\{1, x_2, x_3\}\, dA \qquad [4.14]$$

and

$$\{\underset{\sim}{\Theta}\}^{(n)} = \{\Theta_1, \Theta_2, \Theta_3\}^{(n)} = \int_A \Theta\{1, x_2, x_3\}\, dA \qquad [4.15]$$

which may readily be written as

$$\{\underset{\sim}{\Theta}\}^{(n)} = \{\underset{\sim}{\theta}\}^{(n)}\{\underset{\sim}{I}\}^{(n)} \qquad [4.16]$$

where

$$\{\theta\}^{(n)} = \{\theta, \varphi, \psi\}^{(n)}. \qquad [4.17]$$

Here, eq. [3.5] is used.

Heat flux resultants

$$\{\underset{\sim}{Q}_j\} = \{Q_j, H_j, S_j\}^{(n)} = \int_{A_n} \{1, x_2, x_3\}\, q_j dA$$
$$\{Q, H, S\}^{(n)} = \oint_{C_n} \nu_\alpha q_\alpha \{1, x_2, x_3\}\, ds \qquad [4.18]$$

and

$$\begin{Bmatrix} Q \\ H \\ S \end{Bmatrix}^{(n)} = \begin{Bmatrix} 0_1 + 0_2 - 0_3 \\ h_1 + h_2 - h_3 \\ s_1 + (z+h)\ \ 0_2 - (z-h)\ \ 0_3 \end{Bmatrix}^{(n)}$$
$$\{Q_0, H_0, S_0\} = \oint_\tau \nu_\alpha q_\alpha^0 \{1, x_2, x_3\}\, ds \qquad [4.19\text{a}]$$

with

$$\{0_1, h_1, s_1\}^{(n)} = \int_{C_n} \nu_\alpha q_\alpha \{1, x_2, x_3\}\, ds$$
$$\{0_2, h_2\}^{(n)} = \int_{d_{n,n+1}} \{1, x_2\}\, q_3 dx_2$$
$$\{0_3, h_3\} = \int_{d_{n,n-1}} \{1, x_2\}\, q_3 dx_2. \qquad [4.19\text{b}]$$

Effective surface fluxes

$$\{H, L\} = \oint_\tau \nu_\alpha \{1, x_2\}\, q_\alpha ds$$

$$M^{(n)} = s_1^{(n)} + \sum_{m=n}^{N} b_{mn} 0_1^{(m)}. \qquad [4.20]$$

Strain distribution

By the use of eqs. [2.2] and [3.4] the strain distribution is found to be

$$e_{ij}(x_k, t) = \nu_{ij}(x_1, t) + x_2 \lambda_{ij}(x_1, t) + x_3 \gamma_{ij}(x_1, t) \qquad [4.21]$$

where

$$\nu_{\alpha\beta} = \tfrac{1}{2}(\alpha_\alpha \delta_{2\beta} + \alpha_\beta \delta_{2\alpha} + \beta_\alpha \delta_{3\beta} + \beta_\beta \delta_{3\alpha})$$
$$\lambda_{\alpha\beta} = 0, \quad \gamma_{\alpha\beta} = 0$$
$$\nu_{\alpha 1} = \tfrac{1}{2}(v'_\alpha + \alpha_1 \delta_{2\alpha} + \beta_1 \delta_{3\alpha})$$
$$\lambda_{\alpha 1} = \tfrac{1}{2}\alpha'_\alpha, \quad \gamma_{\alpha 1} = \tfrac{1}{2}\beta'_\alpha$$
$$\nu_{11} = v'_1, \quad \lambda_{11} = \alpha'_1, \quad \gamma_{11} = \beta'_1. \qquad [4.22]$$

Constitutive equations

In conjunction with eqs. [2.9] and [4.2], the constitutive relations for an anisotropic thermo-viscoelastic beam material are readily obtained and they are recorded as follows.

$$\{\underset{\sim}{N}_{ij}\}^T = \{\underset{\sim}{G}_{ijkl}\} * \{d\underset{\sim}{e}_{ij}\}^T - \{\underset{\sim}{\varphi}_{ij}\} * \{d\underset{\sim}{\theta}\}^T \qquad [4.23]$$

and

$$\{\underset{\sim}{Q}_j\}^T = -\{\underset{\sim}{K}_{j1}\}\{\underset{\sim}{\theta}'\}^T - \{\underset{\sim}{K}_{j\alpha}\}\{\underset{\sim}{\delta}_\alpha\}^T \qquad [4.24]$$

where

$$\{\underset{\sim}{e}_{ij}\} = \{\nu_{ij}, \lambda_{ij}, \gamma_{ij}\}$$
$$\{\underset{\sim}{\delta}_i\} = \{\delta_{2i}\varphi + \delta_{3i}\psi, 0, 0\}. \qquad [4.25]$$

In the case of isotropy, these equations become as

$$\{\underset{\sim}{N}_{ij}\}^T = \langle\{\underset{\sim}{\lambda}\} * \{d\underset{\sim}{e}_{rr}\}^T - \{\underset{\sim}{\varphi}\} * \{d\underset{\sim}{\theta}\}^T\rangle\, \delta_{ij} + 2\{\underset{\sim}{\mu}\} * \{d\underset{\sim}{e}_{ij}\}$$
$$\{\underset{\sim}{Q}_j\}^T = -\{\underset{\sim}{K}\}\{\underset{\sim}{\theta}'\delta_{j1}\}^T - \{\underset{\sim}{K}\}\{\underset{\sim}{\delta}_j\}^T \qquad [4.26]$$

where T stands for transpose of matrices.

5. Beam equations

In order to complete the governing equations of viscoelastic laminated composite beams, the macroscopic equations of motion and heat conduction, and the initial and boundary conditions are now presented. For the convenience of analysis, the one-sided *Laplace* transform of all time variables, namely

$$L[\varphi(x_i, t)] = \int_0^\infty \varphi(x_i, t)\, e^{-st} dt = \bar{L}(x_i, s) \qquad [5.1]$$

is assumed to be exist and then all the equations are given below in terms of the *Laplace* transformed field quantities.

Macroscopic equations of motion

The local equations of motion [2.1] are *Laplace* transformed, and they are given as

$$\bar{\sigma}_{ij,i} + \bar{f}_j - s^2 \varrho \bar{u}_j = 0 \quad [5.2]$$

where the initial conditions [2.8] are taken into account. This equation is first integrated over the cross-sections of layers, and it is found

$$\{\bar{N}'_{1j} + \bar{L}_j + \bar{E}_j - \varrho s^2 \bar{A}_j\}^{(n)} = 0. \quad [5.3]$$

By multiplying eq. [5.2] by x_2 and x_3 and again performing integrations over the cross-sections, we respectively arrive at the equations

$$\{\bar{M}'_{1j} - \bar{N}_{2j} + \bar{P}_j + \bar{F}_j - \varrho s^2 \bar{B}_j\}^{(n)} = 0$$
$$\{\bar{K}'_{1j} - \bar{N}_{3j} + \bar{R}_j + \bar{G}_j - \varrho s^2 \bar{C}_j\}^{(n)} = 0 \quad [5.4]$$

for the n-th layer. In eqs. [5.2–5.4], the resultants defined in the previous section are employed, and the *Gauß* transformation theorems converting some surface integrals over A_n into the line integrals around the contour c_n are applied.

Eqs. [5.2–5.4] represent the macroscopic balance equations of motion for the system of discrete layers. In view of eqs. [3.10], these equations are now combined to obtain the macroscopic composite beam equations. In so doing, we add all eq. [5.2] to get

$$\sum_{n=1}^{N} \{\bar{N}'_{1j} + \bar{L}_j + \bar{E}_j - \varrho s^2 \bar{A}_j\}^{(n)} = 0. \quad [5.5]$$

Taking into account the continuity of tractions at the interfaces, this equation becomes

$$\sum_{n=1}^{N} \bar{V}_j^{(n)} + \bar{N}_j = 0 \quad [5.6a]$$

where

$$\bar{V}_j^{(n)} = \{\bar{N}'_{1j} + \bar{E}_j - \varrho s^2 \bar{A}_j\}^{(n)}. \quad [5.6b]$$

Like manner, eqs. [5.3] and [5.4] are combined to arrive at

$$\sum_{n=1}^{N} \bar{\Phi}_j^{(n)} + \bar{K}_j = 0 \quad [5.7]$$

$$\bar{\Psi}_j^{(n)} + \sum_{m=n}^{N} b_{mn} \bar{V}_j^{(m)} + \bar{\mathcal{M}}_j^{(n)} = 0 \quad [5.8]$$

where

$$\bar{\Phi}_j^{(n)} = \{\bar{M}'_{1j} - \bar{N}_{2j} + \bar{F}_j - \varrho s^2 \bar{B}_j\}^{(n)}$$
$$\bar{\Psi}_j^{(n)} = \{\bar{K}'_{1j} - \bar{N}_{3j} + \bar{G}_j - \varrho s^2 \bar{C}_j\}^{(n)}. \quad [5.9]$$

Macroscopic equations of heat conduction

The *Laplace* transformed equation of heat conduction is written in the form

$$\bar{q}_{i,i} + s^2 T_0(\bar{M}\bar{\Theta} + \bar{\varkappa}_{ij}\bar{e}_{ij}). \quad [5.10]$$

As in the local equations of motion, we multiply this equation by 1, x_2 and x_3, and integrate over the cross-sections of layers A_n. Thus, we obtain

$$\{\bar{Q}'_1 + \bar{Q} + s^2 T_0(\bar{M}\bar{\Theta}_1 + \bar{\varkappa}_{ij}\bar{a}_{ij})\}^{(n)} = 0$$
$$\{\bar{H}'_1 + \bar{H} - \bar{Q}_2 + s^2 T_0(\bar{M}\bar{\Theta}_2 + \bar{\varkappa}_{ij}\bar{b}_{ij}\}^{(n)} = 0$$
$$\{\bar{S}'_1 + \bar{S} - \bar{Q}_3 + s^2 T_0(\bar{M}\bar{\Theta}_3 + \bar{\varkappa}_{ij}\bar{c}_{ij}\}^{(n)} = 0 \quad [5.11a]$$

where

$$\{\bar{a}_{ij}, \bar{b}_{ij}, \bar{c}_{ij}\}^{(n)} = \{\bar{e}_{ij}\}^{(n)} \{\underline{I}\}^{(n)} \quad [5.11b]$$

for the system of discrete layers, in terms of the heat flux resultants [4.18] and [4.19].

Using the continuity of heat fluxes at the interfaces, the above macroscopic equations of heat conduction are combined to yield

$$\sum_{n=1}^{N} \bar{\xi}^{(n)} + \bar{H} = 0$$

$$\sum_{n=1}^{N} \bar{\Gamma}^{(n)} + \bar{L} = 0$$

$$\bar{\Lambda}^{(n)} + \sum_{m=n}^{N} b_{mn} \bar{\xi}^{(m)} + \bar{M}^{(n)} = 0 \quad [5.12]$$

where

$$\bar{\xi}^{(n)} = \bar{Q}'_1 + s^2 T_0(\bar{M}\bar{\Theta}_1 + \bar{\varkappa}_{ij}\bar{a}_{ij})$$
$$\bar{\Gamma}^{(n)} = \bar{H}'_1 + s^2 T_0(\bar{M}\bar{\Theta}_2 + \bar{\varkappa}_{ij}\bar{b}_{ij}) - \bar{Q}_2$$
$$\bar{\Lambda}^{(n)} = \bar{S}'_1 + s^2 T_0(\bar{M}\bar{\Theta}_3 + \bar{\varkappa}_{ij}\bar{c}_{ij}) - \bar{Q}_3. \quad [5.13]$$

Boundary and initial conditions

The *Laplace* transformed equations for the boundary conditions [2.3–2.6] are of the form

$$\bar{u}_k - \bar{u}_k^0 = 0 \quad \text{on} \quad s_u = s_l \quad [5.14]$$
$$\bar{t}_k^0 - n_i \bar{\sigma}_{ij} = 0 \quad \text{on} \quad s_\sigma = s_e \quad [5.15]$$
$$n_i \bar{q}_i = 0 \quad \text{on} \quad s_K = s_e \cup s_q \quad [5.16]$$
$$\bar{\Theta} - \bar{\Theta}_0 = 0 \quad \text{on} \quad s_\theta = s_l \cap s_q. \quad [5.17]$$

That is to say, the displacements are taken to be prescribed on the lateral surface s_l of the beam, the tractions on the edge faces s_e, the heat fluxes on the edge faces and some portion of the lateral surface, and the temperature on some portion s_q of the lateral surface.

In view of eqs. [3.10], [5.14], and [5.17], the displacement and temperature boundary conditions are given by

$$\bar{v}_i^{(1)} - \bar{v}_i^0 = 0, \quad \bar{\alpha}_i - \bar{\alpha}_i^0 = 0, \quad \bar{\beta}_i^{(n)} - \bar{\beta}_i^{0\,(n)} = 0$$
$$\text{on} \quad s_l$$
$$\bar{\theta}^{(1)} - \bar{\theta}^0 = 0, \quad \bar{\varphi} - \bar{\varphi}^0 = 0, \quad \bar{\psi}^{(n)} - \bar{\psi}^{0\,(n)} = 0$$
$$\text{on} \quad s_\theta . \qquad [5.18]$$

As before, multiplying eqs. [5.15] by 1, x_2 and x_3, and integrating over A_n, we obtain

$$\{\bar{N}_k^0 - \bar{N}_k\}^{(n)} = 0, \quad \{\bar{M}_k^0 - \bar{M}_k\}^{(n)} = 0$$
$$\{\bar{K}_k^0 - \bar{K}_k\}^{(n)} = 0. \qquad [5.19]$$

Combining these equations in accordance with [3.10] the tractions boundary conditions become

$$\bar{N}_k^0 - \sum_{n=1}^{N} \bar{N}_i^{(n)} = 0, \quad \bar{M}_i^0 - \sum_{n=1}^{N} \bar{M}_i^{(n)} = 0$$

$$\bar{K}_i^{0\,(n)} - \bar{K}_i^{(n)} + \sum_{m=n}^{N} b_{mn}(\bar{N}_i^0 - \bar{N}_i)^{(m)} = 0. \qquad [5.20]$$

Similarly, we write the boundary conditions for heat flux on s_e as

$$\bar{Q}_1^{(n)} = 0, \quad \bar{H}_1^{(n)} = 0, \quad \bar{S}_1^{(n)} = 0 \qquad [5.21]$$

and combining these equations, we get

$$\sum_{n=1}^{N} \bar{Q}_1^{(n)} = 0, \quad \sum_{n=1}^{N} \bar{H}_1^{(n)} = 0$$

$$\bar{S}_1^{(n)} + \sum_{n=1}^{N} b_{mn} \bar{Q}_1^{(m)} = 0. \qquad [5.22]$$

The boundary conditions on s_q are found to be

$$\bar{Q}_0 = 0, \quad \bar{H}_0 = 0, \quad \bar{S}_0 = 0 \quad \text{on} \quad s_q . \qquad [5.23]$$

Further, we record a set of initial conditions based on eq. [2.8] in the form

$$v_i^{(1)} = \alpha_i = \beta_i^{(n)} = 0, \quad \theta^{(1)} = \varphi = \psi^{(n)} = 0$$
$$\text{for} \quad -\infty < t < 0 \qquad [5.24]$$

which are already considered in the derivation of macroscopic equations of motion and heat conduction.

6. Special cases

Besides the isotropic thermoviscoelastic laminated beams whose constitutive equations are given in Art. 4, we now mention two more particular cases of interest. The first one is the thermoelastic laminated beams. For this case, the relaxation functions in the *Stieltjes* convolutions become constant elastic moduli, namely,

$$G_{ijkl}(t) = h(t) \cdot G_{ijkl}, \quad \varphi_{ij}(t) = h(t)\,\varphi_{ij}$$
$$\varkappa_{ij}(t) = h(t)\,\varkappa_{ij}, \quad M(t) = h(t)\,M \qquad [6.1]$$

where $h(t)$ stands for the *Heaviside* unit step function. Substitution of these relations into eqs. [4.23] and [4.24] yields the actual constitutive equations. Eqs. [6.1] should be replaced by

$$K_{ij} = K\delta_{ij}$$
$$G_{ijkl} = [\lambda\delta_{ij}\delta_{kl} + 2\mu(\delta_{ik}\delta_{jl} + \delta_{il}\delta_{jk})]\,h(t)$$
$$\varphi_{ij} = \varphi\delta_{ij}h(t)$$
$$\varkappa_{ij} = \varkappa\delta_{ij}h(t) \qquad [6.2]$$

for an isotropic thermoelastic case.

Another case of interest is the isothermal viscoelastic beams. In this case, the coupling term involving with e_{ij} and the heat conduction equation can be neglected, i.e., $\varkappa_{ij} = \varphi_{ij} = 0$. Thus, the mechanical and thermal response can be treated separately.

Lastly, we note that all inertia terms can be neglected for quasi-static problems.

7. Conclusion

In closing, a laminated beam model is introduced in lieu of the *Timoshenko* and *Bernoulli-Euler* beam models in the analysis of composites. With this model, the delamination and skin effects (14) being an important feature of anisotropic solids are all taken into account. Then a linear nonisothermal theory of viscoelastic laminated composite beams is consistently developed on the basis of the effective stiffness theory of composites. The theory consists of the macroscopic beam equations of motion [5.6–5.8] and heat conduction [5.12], the kinematic relations [4.21], the constitutive eqs. [4.23], [4.24], the boundary conditions [5.18–5.23] and the initial conditions [5.24]. Further, two special cases for isotropic and isothermal elastic [cf. (15)] and viscoelastic beams are presented.

To introduce the *Laplace* transforms into the formulation is quite expedient as it is evident from the foregoing analysis. Moreover, we note that the governing equations can be obtained by means of *Hamilton*'s principle [cf. (15)]

together with the *Lagrange* multipliers through which the continuity constraints are included. It is also noteworthy that the interface stresses are abrogated in the governing equations; they can be computed with the aid of the stress resultants.

Lastly, to prove the uniqueness of solutions of the boundary value problem defined by the governing equations of the composite beam is straight forward. This can be achieved by the use of the standard methods (15–17).

Summary

A dynamical continuum theory is developed for laminated composite beams. Starting with an assumed displacement- and temperature field, the one-dimensional approximate theory is consistently constructed within the frame of the three-dimensional theory of linear, non-isothermal, anisotropic, coupled viscoelasticity. Each constituent of the beam may possess different constant thickness and mechanical properties. All dynamic interactions between the adjacent constituents are included. Further, the effects of transverse shear and normal strains and rotatory inertia as well as those of cross-sectional distortion are all taken into account. The resulting equations consist of the macroscopic beam equations of motion and heat conduction, the kinematical relations, the initial and boundary conditions and the constitutive equations, and they govern the extensional, flexural and torsional motions of laminated composite beams. The special cases of constituents which made of either isotropic thermoviscoelastic or anisotropic thermoelastic materials are discussed briefly.

References

1) *Herrmann, G.* and *J. D. Achenbach*, Proc. AIAA/ASME Eight Structures, Structural Dynamics, and Material Conference, p. 112 (New York 1967).

2) *Achenbach, J. D.* and *G. Herrmann*, Dynamics of Structured Solids, p. 23 (New York 1968).

3) *Grot, R. A.* and *J. D. Achenbach*, Acta Mechanica **9**, 245 (1970).

4) *Sun, C. T., J. D. Achenbach*, and *G. Herrmann*, J. Appl. Mech. **35**, 467 (1968).

5) *Herrmann, G.* and *J. D. Achenbach*, Mechanics of Composite Materials, p. 337 (New York 1970).

6) *Dökmeci, M. C.*, Development in Theoretical and Applied Mechanics, p. 109 (Univ. of North Carolina Press 1971).

7) *Herrmann, G.*, Proc. Soc. Experimental Stress Analysis **29**, 235 (1972).

8) *Christensen, R. M.* and *P. M. Naghdi*, Acta Mechanica **3**, 1 (1967).

9) *Truesdell, C.* and *R. A. Toupin*, Handbuch der Physik**III/1** (Berlin-Heidelberg-New York 1960).

10) *Eringen, A. C.*, Mechanics of Continua (New York 1967).

11) *Onat, E. T.* and *S. Breuer*, Progress in Applied Mechanics, p. 349 (London 1963).

12) *Sternberg, E.* and *M. E. Gurtin*, ibid., p. 373.

13) *Christensen, R. M.*, Theory of Viscoelasticity (New York 1971).

14) *Biot, M. A.*, Int. J. Solids Struct. **2**, 645 (1966).

15) *Dökmeci, M. C.*, J. Elasticity **3**, 27 (1973).

16) *Love, A. E. H.*, A Treatise on the Mathematical Theory of Elasticity (Dover 1944).

17) *Knops, R. J.* and *L. E. Payne*, Uniqueness Theorems in Linear Elasticity (Berlin-Heidelberg-New York 1971).

Authors' address:
M. Cengiz Dökmeci and *Mg. AlpD.*
Theoretical and Applied Mechanics
Dept. of Cornell University, Thurston Hall, Ithaca, N.Y. 14850 (U.S.A.)

Rheol. Acta **12**, 114–115 (1973)

L'influence du temps dans les solutions statiques

Par Eva Doležalová (Praha, ČSSR)

Avec 4 figures

(Reçu p. p. le 27 octobre 1972)

Supposons d'abord les relations linéares entre les composantes de la contrainte et celles de la déformation instantanée et différée. Dans ce cas nous pouvons écrire

$$\varepsilon(t,\tau) = \frac{\sigma(\tau)}{E(\tau)} + \sigma(\tau)\,C(t,\tau) \qquad [1]$$

si la sollicitation est constante, et

$$\varepsilon(t) = \frac{\sigma(\tau_0)}{E(\tau_0)} + \sigma(\tau_0)\,C(t,\tau_0) + \int_{\tau_0}^{t} \frac{\partial\sigma(\tau)}{\partial\tau}\left(\frac{1}{E(\tau)} + C(t,\tau)\right) d\tau \qquad [2]$$

si la sollicitation varie avec le temps.

La formule [2], intégré, donne l'équation *Volterrienne*:

$$\varepsilon(t) = \frac{\sigma(t)}{E(t)} - \int_{\tau_0}^{t} \sigma(\tau)\frac{\partial}{\partial\tau}\left(\frac{1}{E(\tau)} + C(t,\tau)\right) d\tau \qquad [3]$$

$$\sigma(t) = E(t)\,\varepsilon(t) + E(t)\int_{\tau_0}^{t} \varepsilon(\tau)\,\Gamma(t,\tau)\,d\tau. \qquad [4]$$

Dans le cas général, quand la relation entre la contrainte et la déformation n'est pas linéare, on obtient

$$\varepsilon(t) = f(\sigma(t)) - \int_{\tau_0}^{t} f(\sigma(\tau))\,K(t,\tau)\,d\tau. \qquad [5]$$

Pour déterminer les fonctions $f(\sigma)$ et $\varkappa(t,\tau)$ il faut étudier les résultats des essais de longue durée. La fonction $f(\sigma)$ coïncide en forme avec l'équation du diagramme classique; la fonction $K(t,\tau)$ est proportionnelle (pour les matériaux avec stabilité mécanique suffisante), à la vitesse de déformation. On la peut exprimer par exemple par les formules [6] ou [7]:

$$C_1(t,\tau) = a \ln\frac{t-\tau+\tau_j}{\tau_j}$$

$$\Rightarrow K_1(t,\tau) = \frac{\partial C_1(t,\tau)}{\partial\tau}$$

$$= -\frac{a}{t-\tau+\tau_j} \qquad [6]$$

$$C_2(t,\tau) = b(t-\tau)^{\alpha}, \quad 0<\alpha<1$$

$$\Rightarrow K_2(t,\tau) = \frac{\partial C_2(t,\tau)}{\partial\tau}$$

$$= -\frac{b\alpha}{(t-\tau)^{1-\alpha}}. \qquad [7]$$

Le type [6] correspond aux déformations et vitesses montrées sur la fig. 1, le type [7] est illustré par la fig. 2. La différence est surtout au moment de l'établissement de la sollicitation, $t=\tau_0$. Dans le cas deuxième on obtient l'équation intégrale singulaire.

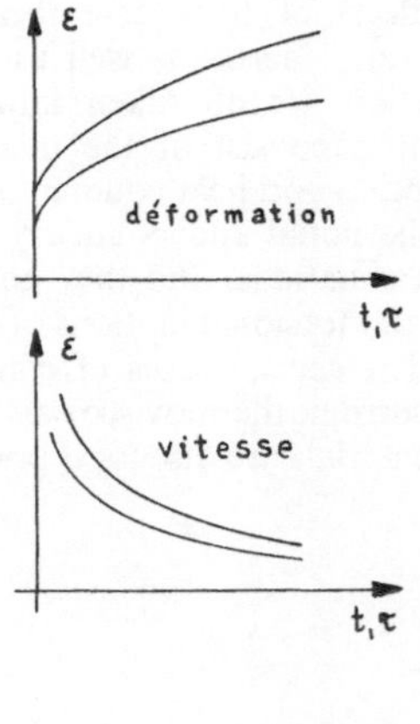

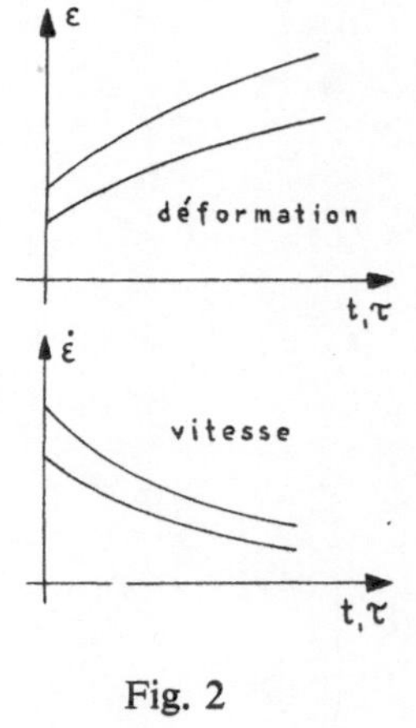

Fig. 2

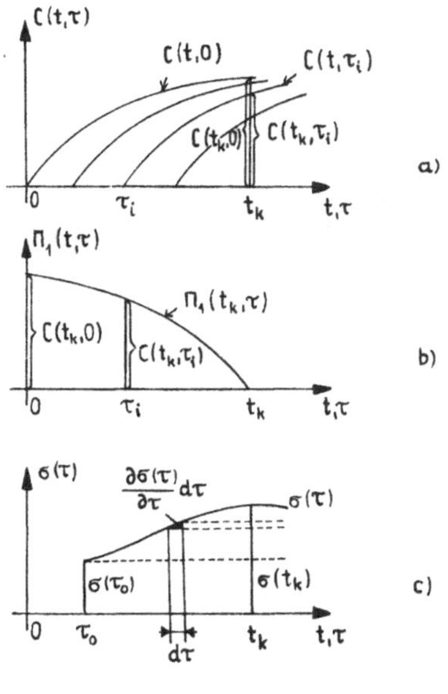

Fig. 3

Les relations précédentes nous permettent de trouver la déformation aussi par la méthode graphique. Si les courbes de déformation $C(t, \tau)$ sont connues avec une exactitude suffisante (fig. 3a), nous pouvons construire les «lignes d'influence» $\Pi_1(t, \tau)$ (fig. 3b). L'ordonnée de la courbe $\Pi_1(t_k, \tau)$ signifie la déformation au moment fixe t_k, correspondante à la sollicitation constante, égale à l'unité, maintenue du moment variant τ_i.

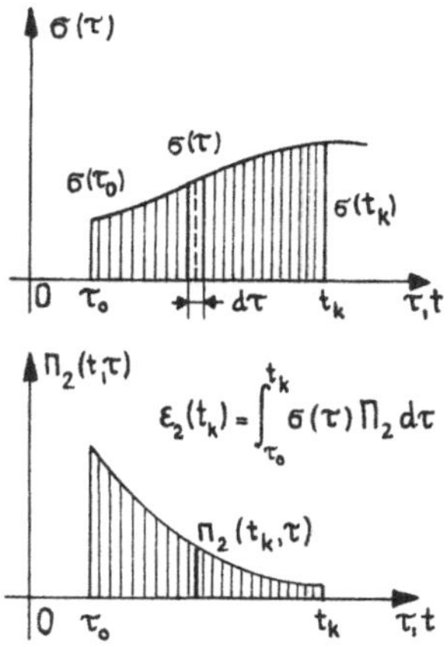

Fig. 4

Considérons une sollicitation variante généralement avec le temps (fig. 3c). D'après le principe de superposition, la déformation globale est égale à la somme de la composante instantanée et des effets de sollicitations partielles appliquées séparément. La fig. 4 montre un autre type des «lignes d'influence» $\Pi_2(t, \tau)$. Ici on trouve la deuxième composante de la déformation directement en combinant les deux surfaces correspondantes.

Adresse de l'auteur:

Ing. *Eva Doležalová* CSc
Belohorska 139, Praha 6 (ČSSR)

Rheol. Acta **12**, 116–126 (1973)

Département de Mathématique, Faculté des Sciences, Université Libre de Bruxelles (Belgique)

Une théorie de l'effet Faraday dans un diélectrique viscoélastique isotrope polarisable et magnétisable

Par Philippe Boulanger

(Reçu p. p. le 27 octobre 1972)

1. Introduction

Dans un travail remarquable (1), *R. A. Toupin* a étudié, dans le cadre de la mécanique rationnelle des milieux continus, l'interaction d'un diélectrique polarisable élastique avec un champ électromagnétique et différents effets optiques qui en résultent. Il envisage notamment l'effet *Faraday*. Celui-ci consiste en une rotation du plan de polarisation des ondes lumineuses qui traversent le diélectrique dans la direction d'un champ magnétique. Toutefois, cet effet est mis en évidence grâce à l'introduction d'un terme supplémentaire dans une des équations constitutives du modèle, ce qui est en contradiction avec le principe d'équiprésence de *C. Truesdell*. Il y a donc lieu d'étudier un modèle plus général de diélectrique si on veut rendre compte de l'effet *Faraday* par une théorie conforme aux méthodes de la mécanique des milieux continus.

Une étude des diélectriques viscoélastiques polarisables s'est également révélée insuffisante à rendre compte de l'effet *Faraday*. Dès lors, il nous a semblé intéressant de prendre en considération le phénomène de magnétisation. Pour généraliser la théorie de *R. A. Toupin* à des continus magnétisables, il est indispensable de disposer d'une expression de la force et du supplément d'énergie électromagnétiques en présence de magnétisation. De telles expressions ont été proposées par *G. Mayné* et *Ph. Boulanger* (2), et appliquées à l'étude des effets optiques dans un diélectrique élastique isotrope polarisable et magnétisable (3). Ces résultats sont exposés dans une conférence spécialisée du présent congrès (4). On constate cependant que le diélectrique étudié ne présente pas l'effet *Faraday*.

Dans cette communication, j'envisage un modèle plus général de diélectrique isotrope polarisable et magnétisable: j'introduis dans les équations constitutives, en plus des variables thermodynamiques qui caractérisent le modèle élastique, les dérivées de *Lie* du tenseur de déformation *d'Almansi*, de la polarisation, et de la magnétisation. Ce modèle sera appelé viscoélastique. Il constitue une généralisation aux diélectriques polarisables et magnétisables du modèle envisagé dans (5). J'étudie la propagation des ondes planes électromagnétiques et mécaniques dans ce diélectrique en présence d'un champ magnétique longitudinal.

2. Equations generales

Pour étudier l'interaction d'un diélectrique polarisable et magnétisable avec un champ électromagnétique, on se basera sur les bilans de quantité de mouvement, de moment de quantité de mouvement, d'énergie, sur les équations de *Maxwell*, et sur l'inégalité de *Clausius-Duhem*.

Bilan de quantité de mouvement

$$\begin{aligned}\varrho \ddot{x}_\lambda = \sigma^{\mu}_{\lambda,\mu} - E'_\lambda \operatorname{div} \bar{P}' + \frac{1}{c}(\underset{v}{\mathscr{L}} \bar{P}' \times \bar{B}')_\lambda \\ - \frac{1}{c}(\underset{v}{\mathscr{L}} \bar{P}' \times \bar{M}')_\lambda + \frac{1}{c}(\bar{E}' \times \underset{v}{\mathscr{L}} \bar{M}')_\lambda \\ - (\operatorname{rot} \bar{M}' \times \bar{M}')_\lambda - (\bar{B}' \cdot M')_{,\lambda} - B'_\lambda \operatorname{div} \bar{M}' \\ + \frac{1}{c} \operatorname{div} \bar{v}(\bar{E}' \times \bar{M}')_\lambda + f_\lambda \qquad [2.1]\end{aligned}$$

où $x^\lambda = x^\lambda(X^A, t)$ sont, dans un système cartésien, les coordonnées de la position d'une particule du milieu à l'instant t, en fonction de ses coordonnées X^A dans une configuration de référence; ϱ est la masse volumique à l'instant t, reliée à celle de la configuration de référence ϱ_0 par $\varrho = \varrho_0 \,|(x/X)|^{-1}$, $|(x/X)|$ étant le jacobien des x^λ par rapport aux X^A;

$$\bar{E}' = \bar{E} + \frac{1}{c}\bar{v} \times \bar{B}, \quad \bar{B}' = \bar{B} - \frac{1}{c}\bar{v} \times \bar{E}$$

$$\bar{P}' = \bar{P} - \frac{1}{c}\bar{v} \times \bar{M}, \quad \bar{M}' = \bar{M} + \frac{1}{c}\bar{v} \times \bar{P},$$

$\bar{E}$, $\bar{B}$ étant les champs magnétique et électrique, $\bar{P}$ la polarisation, $\bar{M}$ la magnétisation, et $\bar{v}$ la vitesse du milieu;

$\underset{v}{\mathscr{L}} \bar{P}'$, $\underset{v}{\mathscr{L}} \bar{M}'$ sont les dérivées de *Lie* de $\bar{P}'$, $\bar{M}'$, $\bar{P}'$ étant considéré comme densité vectorielle (polaire), $\bar{M}'$ comme vecteur (axial) c'est-à-dire

$$\underset{v}{\mathscr{L}} \bar{P}' = \frac{\partial \bar{P}'}{\partial t} + (\operatorname{div} \bar{P}')\,\bar{v} + \operatorname{rot}(\bar{P}' \times \bar{v})$$

$$\begin{aligned}\underset{v}{\mathscr{L}} \bar{M}' = \frac{\partial \bar{M}'}{\partial t} + (\operatorname{div} \bar{M}')\,\bar{v} - \bar{M}'(\operatorname{div} \bar{v}) \\ + \operatorname{rot}(\bar{M}' \times \bar{v})\end{aligned}$$

f_λ sont les composantes de la force volumique extérieure autre que la force électromagnétique; σ^λ_μ sont les composantes du tenseur des contraintes de *Cauchy* $\boldsymbol{\sigma}$.

Cette forme du bilan de quantité de mouvement résulte de l'expression de la force électromagnétique en présence de magnétisation proposée et justifiée dans (2), (4) par des considérations sur le tenseur impulsion-énergie électromagnétique.

Bilan de moment de quantité de mouvement

$$\boldsymbol{\sigma} = {}^\tau\boldsymbol{\sigma}. \qquad [2.2]$$

Le tenseur des contraintes de *Cauchy* est symétrique. Ceci est dû au fait que l'expression de la force électromagnétique intervenant dans [2.1] provient d'un tenseur impulsion-énergie électromagnétique symétrique [(2), (4)].

Bilan d'énergie interne

$$\varrho\dot{\varepsilon} = \sigma^\mu_\lambda v^\lambda_{;\mu} - \varphi^\lambda_{;\lambda} + \bar{E}' \cdot \underset{v}{\mathscr{L}}\bar{P}' + \bar{B}' \cdot \underset{v}{\mathscr{L}}\bar{M}' + \varrho r \qquad [2.3]$$

où ε est l'énergie interne spécifique;
φ^λ sont les composantes du flux de chaleur $\bar{\varphi}$;
r le supplément d'énergie non mécanique et non électromagnétique.

Cette forme du bilan d'énergie interne résulte de l'expression du supplément d'énergie électromagnétique en présence de magnétisation proposée et justifiée dans (2), (4), qui va de pair avec celle de la force électromagnétique. Signalons que ces expressions peuvent être obtenues à partir de tenseurs impulsion-énergie proposés par différents auteurs *(Minkowski, Abraham, Eringen, De Groot* et *Suttorp)* moyennant un changement de définition du tenseur des contraintes et de l'énergie interne.

Equations de Maxwell

Pour un diélectrique non conducteur, polarisable et magnétisable, les équations de *Maxwell* sont

$$\mathrm{rot}\,\bar{E} + \frac{1}{c}\frac{\partial\bar{B}}{\partial t} = 0 \quad \mathrm{div}\,\bar{B} = 0 \qquad [2.4]\ [2.5]$$

$$\mathrm{rot}\,(\bar{B} - \bar{M}) - \frac{1}{c}\frac{\partial(\bar{E} + \bar{P})}{\partial t} = 0 \quad \mathrm{div}\,(\bar{E} + \bar{P}) = 0 \qquad [2.6]\ [2.7]$$

Inégalité de Clausius-Duhem

$$\varrho\dot{\eta} + \mathrm{div}\,\frac{\bar{\varphi}}{\theta} - \varrho r \geq 0 \qquad [2.8]$$

où η est l'entropie spécifique du milieu;
θ est la température absolue (>0).

Un *processus thermodynamique* pour un diélectrique est un ensemble de 12 fonctions de X^A et de t : $x^\lambda = x^\lambda(X^A, t)$, $\theta > 0$, $\boldsymbol{\sigma} = {}^\tau\boldsymbol{\sigma}$, ε, $\bar{f}$, r, $\bar{E}$, $\bar{B}$, $\bar{P}$, $\bar{M}$, η, $\bar{\varphi}$, satisfaisant à [2.1], [2.3], [2.4] à [2.7].

3. Equations constitutives

Les équations générales du paragraphe 2 doivent être complétées par des équations constitutives qui caractérisent le modèle de diélectrique étudié. Nous supposerons ici que ε, $\boldsymbol{\sigma} = {}^\tau\boldsymbol{\sigma}$, $\bar{E}'$, $\bar{B}'$, η $\bar{\varphi}$, sont des fonctions de tenseurs $x^\lambda_{;A}$, $\dot{x}^\lambda_{;A}$, des densités vectorielles (polaires) $\bar{P}'$, $\dot{\bar{P}}'$, des vecteurs axiaux $\bar{M}'$, $\dot{\bar{M}}'$, du scalaire θ, et du vecteur (polaire) gradient de température $\bar{g} = \mathrm{grad}\,\theta$. De telles équations constitutives satisfont au principe d'équiprésence de *C. Truesdell* (6). Nous appellerons viscoélastique le modèle ainsi défini. Il généralise le modèle élastique étudié dans (3), (4). En l'absence de magnétisation, il se réduit au modèle envisagé dans (5).

Nous exigerons que les équations constitutives adoptées soient conformes au principe d'objectivité (7), c'est à dire qu'elles soient indépendantes de l'observateur. On peut alors démontrer que les équations constitutives du modèle viscoélastique envisagé, doivent être de la forme

$$\begin{aligned}
\varepsilon\ &= \hat{\varepsilon}(\boldsymbol{E}, \dot{\boldsymbol{E}}, \bar{\Pi}', \dot{\bar{\Pi}}', \bar{\mathscr{M}}', \dot{\bar{\mathscr{M}}}', \theta, \bar{G})\\
\sigma^{\lambda\mu} &= \varrho\, x^\lambda_{;A} x^\mu_{;B} \hat{\Sigma}^{AB}(\boldsymbol{E}, \dot{\boldsymbol{E}}, \bar{\Pi}', \dot{\bar{\Pi}}', \bar{\mathscr{M}}', \dot{\bar{\mathscr{M}}}', \theta, \bar{G})\\
&\quad \times\ \hat{\Sigma}^{AB} = \hat{\Sigma}^{BA}\\
E'_\lambda &= \varrho_0 X^A_{;\lambda} \hat{\mathscr{E}}_A(\boldsymbol{E}, \dot{\boldsymbol{E}}, \bar{\Pi}', \dot{\bar{\Pi}}', \bar{\mathscr{M}}', \bar{\mathscr{M}}', \theta, \bar{G})\\
B'_\lambda &= \varrho X^A_{;\lambda} \hat{\mathscr{B}}_A(\boldsymbol{E}, \dot{\boldsymbol{E}}, \bar{\Pi}', \dot{\bar{\Pi}}', \bar{\mathscr{M}}', \dot{\bar{\mathscr{M}}}', \theta, \bar{G})\\
\eta\ &= \hat{\eta}(\boldsymbol{E}, \dot{\boldsymbol{E}}, \bar{\Pi}', \dot{\bar{\Pi}}', \bar{\mathscr{M}}', \dot{\bar{\mathscr{M}}}', \theta, \bar{G})\\
\varphi^\lambda &= x^\lambda_{;A} \hat{\Phi}^A(\boldsymbol{E}, \dot{\boldsymbol{E}}, \bar{\Pi}', \dot{\bar{\Pi}}', \bar{\mathscr{M}}', \dot{\bar{\mathscr{M}}}', \theta, \bar{G}) \qquad [3.1]
\end{aligned}$$

où $\boldsymbol{E}$ est le tenseur de *Green* défini par $E_{AB} = \frac{1}{2} \times (\delta_{\lambda\mu} x^\lambda_{;A} x^\mu_{;B} - \delta_{AB})$;
$\bar{\Pi}'$ est la densité vectorielle (polaire) définie par

$$\Pi'^A = |(x/X)|\, X^A_{;\lambda} P'^\lambda$$

$\bar{\mathscr{M}}'$ est le vecteur axial défini par $\mathscr{M}'^A = X^A_{;\lambda} M'^\lambda$;
$\bar{G}$ est le vecteur (polaire) définie par $G_A = x^\lambda_{;A} g_\lambda = \theta_{,A}$.

On se limitera à l'étude de diélectriques isotropes: les fonctions tensorielles $\hat{\varepsilon}$, $\hat{\Sigma}^{AB} = \hat{\Sigma}^{BA}$, $\hat{\mathscr{E}}_A$, $\hat{\mathscr{B}}_A$, $\hat{\eta}$, $\hat{\Phi}^A$ seront supposés isotropes. On pourra donc se servir du théorème de *C.-C. Wang* (8) pour obtenir une représentation de ces fonctions tensorielles.

Un *processus thermodynamique* sera dit *admissible* pour le diélectrique étudié, s'il satisfait aux équations constitutives [3.1].

Nous exigerons que les équations constitutives [3.1] satisfassent au *principe de dissipation:* elles devront être telles que pour tout processus thermodynamique admissible, l'inégalité de *Clausius-Duhem* [2.8] soit vérifiée à tout instant et en tout point. En portant l'expression de ϱr que l'on tire de [2.3] dans cette inégalité, et en introduisant l'énergie libre de *Helmholtz* $\Psi = \varepsilon - \theta\eta$, on obtient

$$-\varrho\dot{\Psi} - \sigma^{\mu}_{\lambda}v^{\lambda}_{,\mu} - \frac{\bar{\varphi}\cdot\bar{g}}{\theta} + \bar{E}'\cdot\underset{v}{\mathscr{L}}\bar{P}' + \bar{B}'\cdot\underset{v}{\mathscr{L}}\bar{M}' - \varrho\dot{\theta}\eta \geq 0. \qquad [3.2]$$

On exigera donc que les équations constitutives soient telles que pour tout processus thermodynamique admissible, l'inégalité [3.2] soit vérifiée à tout instant et en tout point.

En vertu de $[3.1]_1$ $[3.1]_5$, $\Psi = \hat{\Psi}(E, \dot{E}, \bar{\Pi}', \dot{\bar{\Pi}}', \bar{\mathscr{M}}', \dot{\bar{\mathscr{M}}}', \theta, \bar{G})$ et en portant les équations constitutives dans [3.2], on obtient donc

$$\left(-\varrho\frac{\partial\hat{\Psi}}{\partial E_{AB}} + \varrho\hat{\Sigma}^{AB}\right)\dot{E}_{AB} - \varrho\frac{\partial\hat{\Psi}}{\partial\dot{E}_{AB}}\ddot{E}_{AB} + \left(-\varrho\frac{\partial\hat{\Psi}}{\partial\Pi'^A} + \varrho\hat{\mathscr{E}}_A\right)\dot{\Pi}'^A - \varrho\frac{\partial\hat{\Psi}}{\partial\dot{\Pi}'^A}\ddot{\Pi}'^A + \left(-\varrho\frac{\partial\hat{\Psi}}{\partial\mathscr{M}'^A} + \varrho\hat{\mathscr{B}}_A\right)\dot{\mathscr{M}}'^A - \varrho\frac{\partial\hat{\Psi}}{\partial\dot{\mathscr{M}}'^A}\ddot{\mathscr{M}}'^A - \varrho\left(\frac{\partial\hat{\Psi}}{\partial\theta} + \hat{\eta}\right)\dot{\theta} - \varrho\frac{\partial\hat{\Psi}}{\partial G_A}\dot{G}_A - \frac{1}{\theta}\hat{\Phi}^A G_A \geq 0 \qquad [3.3]$$

Pour tirer des conséquences du principe de dissipation, on admettra qu'il existe toujours au moins un processus thermodynamique admissible tel que les variables intervenant dans [3.3] aient des valeurs arbitrairement données. Du principe de dissipation et de la forme [3.3] de l'inégalité [3.2], on déduit alors

$$\frac{\partial\hat{\Psi}}{\partial\dot{E}_{AB}} = \frac{\partial\hat{\Psi}}{\partial\dot{\Pi}'^{AB}} = \frac{\partial\hat{\Psi}}{\partial\dot{\mathscr{M}}'^A} = \frac{\partial\hat{\Psi}}{\partial G_A} = \hat{\eta} + \frac{\partial\hat{\Psi}}{\partial\theta} = 0 \qquad [3.4]$$

ce qui signifie que $\hat{\Psi}$, $\hat{\eta}$, et donc $\hat{\varepsilon}$ ne dépendent pas de $\dot{E}$, $\dot{\bar{\Pi}}'$, $\dot{\bar{\mathscr{M}}}'$, $\bar{G}$. En décomposant les fonctions $\hat{\Sigma}^{AB}$, $\hat{\mathscr{E}}_A$, $\hat{\mathscr{B}}_A$, $\hat{\Phi}^A$ en partie d'équilibre et partie complémentaire, c'est à dire

$$\hat{\Sigma}^{AB} = \hat{\Sigma}^{AB}_{(e)} + \hat{\Sigma}^{AB}_{(c)}$$

avec

$$\hat{\Sigma}^{AB}_{(e)} \overset{\text{def}}{=} \hat{\Sigma}^{AB}(E, 0, \bar{\Pi}', 0, \bar{\mathscr{M}}', 0, \theta, 0)$$

et de même pour $\hat{\mathscr{E}}_A$, $\hat{\mathscr{B}}_A$, $\hat{\Phi}^A$, on obtient encore les restrictions suivantes

$$\hat{\Sigma}^{AB}_{(e)} = \frac{\partial\hat{\Psi}}{\partial E_{AB}}, \quad \hat{\mathscr{E}}^{(e)}_A = \frac{\partial\hat{\Psi}}{\partial\Pi'^A}, \quad \hat{\mathscr{B}}^{(e)}_A = \frac{\partial\hat{\Psi}}{\partial\mathscr{M}'^A}, \quad \hat{\Phi}_{(e)} = 0 \qquad [3.5]$$

$\hat{\Psi}$ étant en vertu de [3.4] une fonction de E, $\bar{\Pi}'$, $\bar{\mathscr{M}}'$, θ seulement. Compte tenu de ces restrictions, l'inégalité [3.3] se réduit à

$$\hat{\Sigma}^{AB}_{(c)}\dot{E}_{AB} + \hat{\mathscr{E}}^{(c)}_A\dot{\Pi}'^A + \hat{\mathscr{B}}^{(c)}_A\dot{\mathscr{M}}'^A - \frac{1}{\varrho\theta}\hat{\Phi}^A_{(c)}G_A \geq 0 \qquad [3.6]$$

qui devra être vérifiée pour tout processus thermodynamique admissible.

4. Linéarisation des équations autour d'un état d'équilibre homogène

Dorénavant, nous supposerons que la seule force extérieure agissant sur le milieu est celle d'origine électromagnétique, c'est-à-dire que $f_\lambda = 0$, et que le milieu est en évolution isotherme (θ uniforme et constante), le bilan d'énergie [2.3] étant vérifié grâce à un choix approprié de r.

Soit

$$_0x^\lambda(X^A),\ _0\boldsymbol{\sigma},\ _0\bar{E},\ _0\bar{B},\ _0\bar{P},\ _0\bar{M} \qquad [4.1]$$

une solution indépendante du temps du bilan de quantité de mouvement [2.1], des équations de *Maxwell* [2.4] à [2.7] et des équations constitutives mécaniques et électromagnétiques $[3.1]_2$, $[3.1]_3$, $[3.1]_4$ dans lesquelles on a tenu compte des restrictions apportées par le principe de dissipation. Nous supposerons que cette solution représente un état homogène du diélectrique, c'est-à-dire que $_0\boldsymbol{\sigma}$, $_0\bar{E}$, $_0\bar{B}$, $_0\bar{P}$, $_0\bar{M}$ sont constantes et que $_0x^\lambda(X^A)$ sont des fonctions linéaires des X^A.

Soit une autre solution caractérisée par

$$x^\lambda = {}_0x^\lambda(X) + \delta_X x^\lambda(X, t),$$

$$\begin{aligned} T^A_\lambda &= {}_0T^A_\lambda + \delta_X T^A_\lambda(X, T) \\ B_\lambda &= {}_0B_\lambda + \delta_x B_\lambda(x, t) \\ E_\lambda &= {}_0E_\lambda + \delta_x E_\lambda(x, t) \\ \Pi^A &= {}_0\Pi^A + \delta_X \Pi^A(X, t) \\ \mathscr{M}^A &= {}_0\mathscr{M}^A + \delta_X \mathscr{M}^A(X, t) \end{aligned} \qquad [4.2]$$

où T^A_λ est le tenseur des contraintes de *Piola-Kirchoff* défini par $T^A_\lambda = \sigma^\mu_\lambda X^A_{,\mu} |(x/X)| \cdot \delta X$ désigne une variation à X fixé, δ_x une variation à x fixé.

Nous allons linéariser les équations qui régissent les variations introduites en [4.2]. En ce qui concerne le bilan de quantité de mouvement et les équations de *Maxwell*, on obtiendra le même résultat que dans le cas élastique (4) car l'état non perturbé est également un état d'équilibre homogène:

$$\begin{aligned} {}_0\varrho \ddot{u}_\lambda = {} & z^\mu_{\lambda,\mu} - {}_0E_\lambda \operatorname{div} \bar{p}' + \frac{1}{c}(\dot{\bar{p}}' \times {}_0\bar{B})_\lambda \\ & - \frac{1}{c}(\dot{\bar{p}}' \times {}_0\bar{M})_\lambda - {}_0M^\mu(m'_{\lambda,\mu} - m'_{\mu,\lambda}) \\ & - {}_0M^\mu(u_{\lambda,\alpha\mu}\, {}_0M^\alpha - u_{\mu,\alpha\lambda}\, {}_0M^\alpha) \\ & - {}_0M^\mu b'_{\mu,\lambda} - {}_0B^\mu\, {}_0M^\alpha u_{\mu,\alpha\lambda} - {}_0B^\mu m'_{\mu,\lambda} \\ & + \frac{1}{c}({}_0\bar{E} \times \dot{\bar{m}}')_\lambda + \frac{1}{c} \operatorname{div} \dot{\bar{u}}({}_0\bar{E} \times {}_0\bar{M})_\lambda \\ & - {}_0B_\lambda(u^\mu_{,\alpha\mu}\, {}_0M^\alpha + \operatorname{div} \bar{m}') \end{aligned} \qquad [4.3]$$

$$\operatorname{rot} \bar{e} + \frac{1}{c}\dot{\bar{b}} = 0 \qquad \operatorname{div} \bar{b} = 0 \qquad [4.4]\ [4.5]$$

$$\begin{aligned} \operatorname{rot}(\bar{b} - \bar{m}) - \frac{1}{c}(\dot{\bar{e}} + \dot{\bar{p}}) = {} & \operatorname{rot}({}_0M^\lambda \dot{\bar{u}}_{,\lambda}) \\ & + \frac{1}{c}\, {}_0P^\lambda \dot{\bar{u}}_{,\lambda} - \frac{1}{c}\, {}_0\bar{P} \operatorname{div} \dot{\bar{u}} \end{aligned} \qquad [4.6]$$

$$\operatorname{div}(\bar{e} + \bar{p}) = 0 \qquad [4.7]$$

où

$$u_\lambda = \delta_X x_\lambda, \qquad z^\mu_\lambda = |({}_0x/X)|^{-1}\, {}_0x^\mu_{,A} \delta_X T^A_\lambda$$

$$\bar{e}' = \bar{e} + \frac{1}{c}\dot{\bar{u}} \times {}_0\bar{B}, \qquad \bar{b}' = \bar{b} - \frac{1}{c}\dot{\bar{u}} \times {}_0\bar{E}$$

$$\bar{p}' = \bar{p} - \frac{1}{c}\bar{u} \times {}_0\bar{M}, \qquad \bar{m}' = \bar{m} + \frac{1}{c}\bar{u} \times {}_0\bar{P}$$

$$e_\lambda = \delta_x E_\lambda, \qquad b_\lambda = \delta_x B_\lambda,$$

$$p^\lambda = |({}_0x/X)|^{-1}\, {}_0x^\lambda_{,A} \delta_X \Pi^A, \qquad m^\lambda = {}_0x^\lambda_{,A} \delta_X \mathscr{M}^A.$$

Il reste à linéariser les équations constitutives $[3.1]_2$, $[3.1]_3$, $[3.1]_4$ avec les restrictions $[3.5]_1$, $[3.5]_2$, $[3.5]_3$. Le calcul se fait de manière analogue à celle du cas élastique, mais conduit à un resultat plus général. On obtient

$$\begin{aligned} z^{\lambda\mu} = {} & {}_0\sigma^{\mu\nu} u^\lambda_{,\nu} + {}_0C^{\lambda\mu\nu\gamma} u_{\nu,\gamma} + {}_0D^{\lambda\mu\nu\gamma} \dot{u}_{\nu,\gamma} \\ & + {}_0S^{\lambda\mu}_\nu \dot{p}'^\nu + {}_0R^{\lambda\mu}_\nu \dot{P}'^\nu + {}_0X^{\lambda\mu}_\nu m'^\nu \\ & + {}_0U^{\lambda\mu}_\nu \dot{m}'^\nu \end{aligned} \qquad [4.8]$$

$$\begin{aligned} e'^\lambda = {} & -{}_0E^\nu u_{\nu,\lambda} + {}_0S^{\nu\gamma}_\lambda u_{\nu,\gamma} + {}_0Q^{\nu\gamma}_\lambda \dot{u}_{\nu,\gamma} \\ & + {}_0T_{\lambda\mu} p'^\mu + {}_0Z_{\lambda\mu} \dot{p}'^\mu + {}_0W_{\lambda\mu} m'^\mu \\ & + {}_0M_{\lambda\mu} \dot{m}'^\mu \end{aligned} \qquad [4.9]$$

$$\begin{aligned} b'_\lambda = {} & -{}_0B_\lambda \operatorname{div} \bar{u} - {}_0B_\mu u^\mu_{,\lambda} + {}_0X^{\nu\gamma}_\lambda u_{\nu,\gamma} \\ & + {}_0K^{\nu\gamma}_\lambda \dot{u}_{\nu,\gamma} + {}_0W_{\mu\lambda} p'^\mu + {}_0L_{\lambda\mu} \dot{p}'^\mu \\ & + {}_0Y_{\lambda\mu} m'^\mu + {}_0N_{\lambda\mu} \dot{m}'^\mu \end{aligned} \qquad [4.10]$$

où

$$\begin{aligned} {}_0C^{\lambda\mu\nu\gamma} &= {}_0C^{\mu\lambda\nu\gamma} = {}_0C^{\lambda\mu\gamma\nu} = {}_0C^{\nu\gamma\lambda\mu} \\ &= {}_0\varrho \left(\frac{\partial^2 \hat{\Psi}}{\partial E_{AB} \partial E_{CD}}\right)_0 {}_0x^\mu_{,A}\, {}_0x^\lambda_{,B}\, {}_0x^\nu_{,C}\, {}_0x^\gamma_{,D} \end{aligned}$$

$$\begin{aligned} {}_0D^{\lambda\mu\nu\gamma} &= {}_0D^{\mu\lambda\nu\gamma} = {}_0D^{\lambda\mu\gamma\nu} \\ &= {}_0\varrho \left(\frac{\partial \hat{\Sigma}^{AB}}{\partial \dot{E}_{CD}}\right)_0 {}_0x^\mu_{,A}\, {}_0x^\lambda_{,B}\, {}_0x^\nu_{,C}\, {}_0x^\gamma_{,D} \end{aligned}$$

$${}_0S^{\lambda\mu}_\nu = {}_0S^{\mu\lambda}_\nu = \varrho_0 \left(\frac{\partial^2 \hat{\Psi}}{\partial E_{AB} \partial \Pi'^C}\right)_0 {}_0x^\mu_{,A}\, {}_0x^\lambda_{,B}\, {}_0X^C_{,\nu}$$

$${}_0R^{\lambda\mu}_\nu = {}_0R^{\mu\lambda}_\nu = \varrho_0 \left(\frac{\partial \hat{\Sigma}^{AB}}{\partial \dot{\Pi}'^C}\right)_0 {}_0x^\mu_{,A}\, {}_0x^\lambda_{,B}\, {}_0X^C_{,\nu}$$

$${}_0X^{\lambda\mu}_\nu = {}_0X^{\mu\lambda}_\nu = {}_0\varrho \left(\frac{\partial^2 \hat{\Psi}}{\partial E_{AB} \partial \mathscr{M}'^C}\right)_0 {}_0x^\mu_{,A}\, {}_0x^\lambda_{,B}\, {}_0X^C_{,\nu}$$

$${}_0U^{\lambda\mu}_\nu = {}_0U^{\mu\lambda}_\nu = {}_0\varrho \left(\frac{\partial \hat{\Sigma}^{AB}}{\partial \dot{\mathscr{M}}'^C}\right)_0 {}_0x^\mu_{,A}\, {}_0x^\lambda_{,B}\, {}_0X^C_{,\nu}$$

$${}_0Q^{\nu\gamma}_\lambda = {}_0Q^{\gamma\nu}_\lambda = \varrho_0 \left(\frac{\partial \hat{\mathscr{E}} A}{\partial \dot{E}_{CD}}\right)_0 {}_0x^\nu_{,C}\, {}_0x^\gamma_{,D}\, {}_0X^A_{,\lambda}$$

$${}_0T_{\lambda\mu} = {}_0T_{\mu\lambda} = \frac{\varrho_0^2}{{}_0\varrho} \left(\frac{\partial^2 \hat{\Psi}}{\partial \Pi'^A \partial \Pi'^B}\right)_0 {}_0X^A_{,\lambda}\, {}_0X^B_{,\mu}$$

$${}_0Z_{\lambda\mu} = \frac{\varrho_0^2}{{}_0\varrho} \left(\frac{\partial \hat{\mathscr{E}}_A}{\partial \dot{\Pi}'^B}\right)_0 {}_0X^A_{,\lambda}\, {}_0X^B_{,\mu}$$

$$W_{\lambda\mu} = \varrho_0 \left(\frac{\partial^2 \hat{\Psi}}{\partial \mathscr{M}'^A \partial \mathscr{M}'^B}\right)_0 {}_0X^A_{,\lambda}\, {}_0X^B_{,\mu}$$

$${}_0M_{\lambda\mu} = \varrho_0 \left(\frac{\partial \hat{\mathscr{E}}_A}{\partial \dot{\mathscr{M}}'^B}\right)_0 {}_0X^A_{,\lambda}\, {}_0X^B_{,\mu}$$

$$
\begin{aligned}
{}_0K^{\nu\gamma}_{\lambda} &= {}_0K^{\gamma\nu}_{\lambda} = {}_0\varrho \left(\frac{\partial \hat{\mathscr{B}}_A}{\partial \dot{E}_{CD}}\right)_0 {}_0x^{\nu}_{,C}\, {}_0x^{\gamma}_{,D}\, {}_0X^{A}_{,\lambda} \\
{}_0L_{\lambda\mu} &= \varrho_0 \left(\frac{\partial \hat{\mathscr{B}}_A}{\partial \dot{\Pi}'^B}\right)_0 {}_0X^{A}_{,\lambda}\, {}_0X^{B}_{,\mu} \\
{}_0Y_{\lambda\mu} &= {}_0Y_{\mu\lambda} = {}_0\varrho \left(\frac{\partial^2 \hat{\Psi}}{\partial \mathscr{M}'^A \partial \mathscr{M}'^B}\right)_0 {}_0X^{A}_{,\lambda}\, {}_0X^{B}_{,\mu} \\
{}_0N_{\lambda\mu} &= {}_0\varrho \left(\frac{\partial \hat{\mathscr{B}}_A}{\partial \dot{\mathscr{M}}'^B}\right)_0 {}_0X^{A}_{,\lambda}\, {}_0X^{B}_{,\mu}
\end{aligned}
$$

Comme l'état initial [4.1] est homogène, tous ces tenseurs ne dépendent pas des X^A.

5. Effet Faraday

Nous allons envisager le cas particulier où dans [4.1]

$$ {}_0x^{\lambda} = X^{\lambda}, \quad {}_0\bar{P} = 0, \quad {}_0\bar{M} \neq 0. \qquad [5.1] $$

Comme [4.1] doit vérifier les équations constitutives $[3.1]_2$, $[3.1]_3$, $[3.1]_4$ avec les restrictions [3.5], et comme ${}_0\bar{M}$ ne dépend ni de X^A ni de t,

$$ {}_0\sigma_{\lambda\mu} = \varrho_0 \left(\frac{\partial \hat{\Psi}}{\partial E_{\lambda\mu}}\right)_0, \quad {}_0E_{\lambda} = \varrho_0 \left(\frac{\partial \hat{\Psi}}{\partial \Pi'^{\lambda}}\right)_0 $$

$$ {}_0B_{\lambda} = \varrho_0 \left(\frac{\partial \hat{\Psi}}{\partial \mathscr{M}'^{\lambda}}\right)_0. $$

Comme le diélectrique est supposé isotrope, $\hat{\Psi}(\boldsymbol{E}, \boldsymbol{\Pi}', \boldsymbol{\mathscr{M}}', \theta)$ est (pour une valeur fixée de θ) une fonction de 14 invariants que l'on obtient à partir du théorème de *C.-C. Wang* (8) en considérant le tenseur antisymétrique associé au vecteur axial $\boldsymbol{\mathscr{M}}'$. Ces invariants sont

$$
\begin{aligned}
I_1 &= \operatorname{tr} \boldsymbol{E}, \quad I_2 = \operatorname{tr} \boldsymbol{E}^2, \quad I_3 = \operatorname{tr} \boldsymbol{E}^3 \\
I_4 &= \boldsymbol{\Pi}'^2, \quad I_5 = \boldsymbol{\mathscr{M}}'^2, \quad I_6 = \boldsymbol{\Pi}' \cdot \boldsymbol{E}\boldsymbol{\Pi}' \\
I_7 &= \boldsymbol{\Pi}' \cdot \boldsymbol{E}^2 \boldsymbol{\Pi}', \quad I_8 = \boldsymbol{\mathscr{M}}' \cdot \boldsymbol{E}\boldsymbol{\mathscr{M}}' \\
I_9 &= \boldsymbol{\mathscr{M}}' \cdot \boldsymbol{E}^2 \boldsymbol{\mathscr{M}}', \quad I_{10} = \boldsymbol{E}\boldsymbol{\mathscr{M}}' \cdot (\boldsymbol{E}^2 \boldsymbol{\mathscr{M}}' \times \boldsymbol{\mathscr{M}}') \\
I_{11} &= (\boldsymbol{\Pi}' \cdot \boldsymbol{\mathscr{M}}')^2, \quad I_{12} = \boldsymbol{E}\boldsymbol{\Pi}' \cdot (\boldsymbol{\mathscr{M}}' \times \boldsymbol{\Pi}') \\
I_{13} &= \boldsymbol{E}^2 \boldsymbol{\Pi}' \cdot (\boldsymbol{\mathscr{M}}' \times \boldsymbol{\Pi}') \\
I_{14} &= (\boldsymbol{\Pi}' \cdot \boldsymbol{\mathscr{M}}')(\boldsymbol{E}\boldsymbol{\mathscr{M}}' \cdot (\boldsymbol{\mathscr{M}}' \times \boldsymbol{\Pi}')).
\end{aligned}
$$

De cette représentation de $\hat{\Psi}$ on déduit

$$ {}_0\sigma_{\lambda\mu} = \varrho_0 \left(\frac{\partial \hat{\Psi}}{\partial I_1}\right)_0 \delta_{\lambda\mu} + \varrho_0 \left(\frac{\partial \hat{\Psi}}{\partial I_8}\right)_0 {}_0M_{\lambda}\, {}_0M_{\mu} $$

$$ {}_0\bar{E} = 0, \quad {}_0\bar{B} = 2\varrho_0 \left(\frac{\partial \hat{\Psi}}{\partial I_5}\right)_0 {}_0\bar{M}. \qquad [5.2] $$

L'état non perturbé envisagé ici est donc un état sans champ électrique. Les équations [2.1] et [2.4] à [2.7] sont trivialement satisfaites, car ${}_0\bar{E}$, ${}_0\bar{B}$, ${}_0\bar{P}$, ${}_0\bar{M}$, ${}_0\boldsymbol{\sigma}$ ne dépendent ni de X^A ni de t et ${}_0x^{\lambda}$ ne dépend pas de t.

La représentation de Ψ permet d'obtenir

$$
\begin{aligned}
{}_0C^{\lambda\mu\nu\gamma} &= \varrho_0 \left(\frac{\partial^2 \hat{\Psi}}{\partial I_1^2}\right)_0 \delta^{\lambda\mu}\delta^{\nu\gamma} \\
&\quad + \varrho_0 \left(\frac{\partial \hat{\Psi}}{\partial I_2}\right)_0 (\delta^{\lambda\gamma}\delta^{\mu\nu} + \delta^{\lambda\nu}\delta^{\mu\gamma}) \\
&\quad + \varrho_0 \left(\frac{\partial^2 \hat{\Psi}}{\partial I_1 \partial I_8}\right)_0 \\
&\quad \times (\delta^{\lambda\mu}\, {}_0M^{\gamma}\, {}_0M^{\nu} + \delta^{\gamma\nu}\, {}_0M^{\lambda}\, {}_0M^{\mu}) \\
&\quad + \frac{1}{2} \varrho_0 \left(\frac{\partial \hat{\Psi}}{\partial I_9}\right)_0 (\delta^{\lambda\nu}\, {}_0M^{\mu}\, {}_0M^{\gamma} \\
&\quad + \delta^{\mu\gamma}\, {}_0M^{\lambda}\, {}_0M^{\nu} + \delta^{\lambda\gamma}\, {}_0M^{\mu}\, {}_0M^{\nu} \\
&\quad + \delta^{\mu\nu}\, {}_0M^{\lambda}\, {}_0M^{\gamma}) \\
&\quad + \varrho_0 \left(\frac{\partial^2 \hat{\Psi}}{\partial I_8^2}\right)_0 {}_0M^{\lambda}\, {}_0M^{\mu}\, {}_0M^{\nu}\, {}_0M^{\gamma} \\
{}_0S^{\lambda\mu}_{\nu} &= 0 \\
{}_0X^{\lambda\mu}_{\nu} &= 2\varrho_0 \left(\frac{\partial^2 \hat{\Psi}}{\partial I_1 \partial I_5}\right)_0 \delta^{\lambda\mu}\, {}_0M_{\nu} \\
&\quad + 2\varrho_0 \left(\frac{\partial^2 \hat{\Psi}}{\partial I_8 \partial I_5}\right)_0 {}_0M^{\lambda}\, {}_0M^{\mu}\, {}_0M_{\nu} \\
&\quad + \varrho_0 \left(\frac{\partial \Psi}{\partial I_8}\right)_0 (\delta^{\lambda}_{\nu}\, {}_0M^{\mu} + {}_0M^{\lambda}\delta^{\mu}_{\nu}) \qquad [5.3] \\
{}_0T_{\lambda\mu} &= 2\varrho_0 \left(\frac{\partial \hat{\Psi}}{\partial I_4}\right)_0 \delta_{\lambda\mu} \\
&\quad + 2\varrho_0 \left(\frac{\partial \hat{\Psi}}{\partial I_{11}}\right)_0 {}_0M_{\lambda}\, {}_0M_{\mu} \\
{}_0W_{\lambda\mu} &= 0 \\
{}_0Y_{\lambda\mu} &= 2\varrho_0 \left(\frac{\partial \hat{\Psi}}{\partial I_5}\right)_0 \delta_{\lambda\mu} \\
&\quad + 4\varrho_0 \left(\frac{\partial^2 \hat{\Psi}}{\partial I_5^2}\right)_0 {}_0M_{\lambda}\, {}_0M_{\mu}.
\end{aligned}
$$

D'autre part, à partir du théorème de *C.-C. Wang*, en associant des tenseurs antisymétriques aux vecteurs axiaux $\boldsymbol{\mathscr{M}}'$, $\dot{\boldsymbol{\mathscr{M}}}'$, on peut obtenir des représentations pour les fonctions tensorielles isotropes $\hat{\Sigma}^{AB}$, $\hat{\mathscr{E}}_A$, $\hat{\mathscr{B}}_A$ dépendant de $\boldsymbol{E}$, $\dot{\boldsymbol{E}}$, $\boldsymbol{\Pi}'$, $\dot{\boldsymbol{\Pi}}'$, $\boldsymbol{\mathscr{M}}'$, $\dot{\boldsymbol{\mathscr{M}}}'$, θ ($\bar{G} = 0$ car on s'est limité à des processus isothermes). De ces représentations, on déduit que

$$
\begin{aligned}
{}_0D^{\lambda\mu\nu\gamma} = {} & (A_1 - A_2)\,\delta^{\lambda\mu}\delta^{\nu\gamma} + A_2(\delta^{\lambda\nu}\delta^{\mu\gamma} + \delta^{\lambda\gamma}\delta^{\mu\nu}) \\
& + A_3(\delta^{\lambda\mu}\,{}_0M^{\nu}\,{}_0M^{\gamma} + \delta^{\nu\gamma}\,{}_0M^{\lambda}\,{}_0M^{\mu}) \\
& + G_3(\delta^{\lambda\mu}\,{}_0M^{\nu}\,{}_0M^{\gamma} - \delta^{\nu\gamma}\,{}_0M^{\lambda}\,{}_0M^{\mu}) \\
& + \tfrac{1}{2}A_5(\varepsilon^{\lambda\nu\alpha}\varepsilon^{\gamma\mu\beta}\,{}_0M_{\alpha}\,{}_0M_{\beta} \\
& + \varepsilon^{\lambda\gamma\alpha}\varepsilon^{\nu\mu\beta}\,{}_0M_{\alpha}\,{}_0M_{\beta}) \\
& + G_1(\varepsilon^{\lambda\nu\alpha}\,{}_0M_{\alpha}\,{}_0M^{\gamma}\,{}_0M^{\mu} \\
& + \varepsilon^{\lambda\gamma\alpha}\,{}_0M_{\alpha}\,{}_0M^{\nu}\,{}_0M^{\mu} \\
& + \varepsilon^{\mu\nu\alpha}\,{}_0M_{\alpha}\,{}_0M^{\gamma}\,{}_0M^{\lambda} \\
& + \varepsilon^{\mu\gamma\alpha}\,{}_0M_{\alpha}\,{}_0M^{\nu}\,{}_0M^{\lambda}) \\
& + A_{12}\,{}_0M^{\lambda}\,{}_0M^{\mu}\,{}_0M^{\nu}\,{}_0M^{\gamma} \\
& + \tfrac{1}{2}G_2(\delta^{\lambda\nu}\varepsilon^{\gamma\mu\alpha}\,{}_0M_{\alpha}\ N\ \delta^{\mu\gamma}\varepsilon^{\lambda\nu\alpha}\,{}_0M_{\alpha} \\
& + \delta^{\lambda\gamma}\varepsilon^{\nu\mu\alpha}\,{}_0M_{\alpha} - \delta^{\mu\nu}\varepsilon^{\lambda\gamma\alpha}\,{}_0M_{\alpha})
\end{aligned}
$$

$$
{}_0R^{\lambda\mu}_{\nu} = {}_0Q^{\lambda\mu}_{\nu} = 0, \qquad {}_0M_{\lambda\mu} = {}_0L_{\lambda\mu} = 0
$$

$$
\begin{aligned}
\tfrac{1}{2}({}_0U^{\lambda\mu}_{\nu} + {}_0K^{\lambda\mu}_{\nu}) = {} & A_4(\delta^{\lambda}_{\nu}\,{}_0M^{\mu} + \delta^{\mu}_{\nu}\,{}_0M^{\lambda}) \\
& + A_6\delta^{\lambda\mu}\,{}_0M_{\nu} + A_7\,{}_0M^{\lambda}\,{}_0M^{\mu}\,{}_0M_{\nu} \\
& + A_{13}(\varepsilon^{\lambda}_{\gamma\nu}\,{}_0M^{\gamma}\,{}_0M^{\mu} + \varepsilon^{\mu}_{\gamma\nu}\,{}_0M^{\gamma}\,{}_0M^{\lambda})
\end{aligned}
$$

$$
\begin{aligned}
\tfrac{1}{2}({}_0U^{\lambda\mu}_{\nu} - {}_0K^{\lambda\mu}_{\nu}) = {} & G_4(\delta^{\lambda}_{\nu}\,{}_0M^{\mu} + \delta^{\mu}_{\nu}\,{}_0M^{\lambda}) \\
& + G_6\delta^{\lambda\mu}\,{}_0M_{\nu} + G_7\,{}_0M^{\lambda}\,{}_0M^{\mu}\,{}_0M_{\nu} \\
& + G_{13}(\varepsilon^{\lambda}_{\gamma\nu}\,{}_0M^{\gamma}\,{}_0M^{\mu} + \varepsilon^{\mu}_{\gamma\nu}\,{}_0M^{\gamma}\,{}_0M^{\lambda})
\end{aligned}
$$

$$
\begin{aligned}
{}_0Z_{\lambda\mu} &= A_9\,\delta_{\lambda\mu} + G_9\,\varepsilon_{\lambda\mu\nu}\,{}_0M^{\nu} + A_{10}\,{}_0M_{\lambda}\,{}_0M_{\mu} \\
{}_0N_{\lambda\mu} &= A_8\,\delta_{\lambda\mu} + G_8\,\varepsilon_{\lambda\mu\nu}\,{}_0M^{\nu} + A_{11}\,{}_0M_{\lambda}\,{}_0M_{\mu}
\end{aligned}
\qquad [5.4]
$$

où les coefficients $A_1, \ldots, A_{13}$, $G_1, \ldots, G_4$, $G_6, \ldots, G_9$, G_{13} sont des fonctions de ${}_0\bar{M}^2$ (θ étant fixé).

En tenant compte de [5.1], [5.2], [5.3], [5.4] dans les éqs. [4.3] à [4.10] et en portant l'expression de $z^{\lambda\mu}$ dans [4.3], on obtient le système d'équation suivant en $\bar{u}$, $\bar{e}$, $\bar{b}$, $\bar{p}$, $\bar{m}$:

$$
\begin{aligned}
\varrho_0\,\ddot{u}_{\lambda} = {} & C_1\,u^{\mu}_{,\mu\lambda} + A_1\,\dot{u}^{\mu}_{,\mu\lambda} + C_2\,u^{\mu}_{\lambda,\mu} + A_2\,\dot{u}^{\mu}_{\lambda,\mu} \\
& + \frac{1}{2}G_2(\varepsilon^{\nu\mu\alpha}\,{}_0M_{\alpha}\,\dot{u}_{\nu,\lambda\mu} + \varepsilon^{\nu\lambda\alpha}\,{}_0M_{\alpha}\,\dot{u}^{\mu}_{\nu,\mu} \\
& + \varepsilon^{\alpha\nu\lambda}\,{}_0M_{\alpha}\,\dot{u}^{\mu}_{,\nu\mu}) \\
& + G_1(\varepsilon^{\lambda\nu\alpha}\,{}_0M_{\alpha}\,{}_0M^{\gamma}\,{}_0M^{\mu}\,\dot{u}_{\nu,\gamma\mu} \\
& + \varepsilon^{\lambda\gamma\alpha}\,{}_0M_{\alpha}\,{}_0M^{\nu}\,{}_0M^{\mu}\,\dot{u}_{\nu,\gamma\mu} \\
& + \varepsilon^{\mu\nu\alpha}\,{}_0M_{\alpha}\,{}_0M^{\gamma}\,{}_0M^{\lambda}\,\dot{u}_{\nu,\gamma\mu}) \\
& + C_3\,u_{\mu,\nu\lambda}\,{}_0M^{\mu}\,{}_0M^{\nu} \\
& + (C_3 - 1)\,{}_0M^{\mu}\,u_{\nu,\nu\mu}\,{}_0M_{\lambda} \\
& + (A_3 + G_3)\,{}_0M^{\nu}\,{}_0M^{\mu}\,\dot{u}_{\mu,\nu\lambda} \\
& + (A_3 - G_3)\,{}_0M^{\lambda}\,{}_0M^{\mu}\,\dot{u}_{\nu,\nu\mu} \\
& + C_4\,u_{\lambda,\mu\nu}\,{}_0M^{\mu}\,{}_0M^{\nu} + C_5\,u^{\nu}_{\mu,\nu}\,{}_0M^{\mu}\,{}_0M_{\lambda} \\
& + \frac{A_5}{2}\,\varepsilon_{\lambda\gamma\alpha}\varepsilon_{\nu\mu\beta}\,{}_0M^{\alpha}\,{}_0M^{\beta}\,\dot{u}^{\nu}_{,\gamma\mu} + C_6\,{}_0M_{\mu}\,m^{\mu}_{,\lambda} \\
& + (A_6 + G_6)\,{}_0M_{\mu}\,\dot{m}^{\mu}_{,\lambda}\,C_7\,m^{\mu}_{,\nu}\,{}_0M_{\mu}\,{}_0M^{\nu}\,{}_0M_{\lambda} \\
& + (A_7 + G_7)\,{}_0M_{\lambda}\,{}_0M_{\mu}\,{}_0M^{\nu}\,\dot{m}^{\mu}_{,\nu} \\
& + (C_4 - C_5)\,m_{\lambda,\mu}\,{}_0M^{\mu} \\
& + (C_4 - C_5 - C_8 + 1)\,m^{\mu}_{,\mu}\,{}_0M_{\lambda} \\
& + (A_4 + G_4)\,(\dot{m}^{\mu}_{,\mu}\,{}_0M_{\lambda} + \dot{m}_{\lambda,\mu}\,{}_0M^{\mu}) \\
& - b_{\mu,\lambda}\,{}_0M^{\mu} + C_{12}\,{}_0M^{\lambda}\,{}_0M^{\mu}\,{}_0M^{\nu}\,{}_0M^{\gamma}\,u_{\nu,\mu\gamma} \\
& + (A_{13} + G_{13})\,(\varepsilon^{\mu}_{\gamma\nu}\,{}_0M^{\gamma}\,{}_0M^{\lambda}\,\dot{m}^{\nu}_{,\mu} \\
& + \varepsilon^{\lambda}_{\gamma\nu}\,{}_0M^{\gamma}\,{}_0M^{\mu}\,\dot{m}^{\nu}_{,\mu}) \\
& + A_{12}\,{}_0M^{\lambda}\,{}_0M^{\mu}\,{}_0M^{\nu}\,{}_0M^{\gamma}\,\dot{u}_{\nu,\gamma\mu} \\
& + \frac{1}{c}(C_8 - 1)\,(\dot{\bar{p}} \times {}_0\bar{M})_{\lambda} \\
& - \frac{1}{c^2}(C_8 - 1)\,(\ddot{\bar{u}} \times {}_0\bar{M}) \times {}_0\bar{M})_{\lambda}
\end{aligned}
$$

$$
\operatorname{rot}\bar{e} + \frac{1}{c}\dot{\bar{b}} = 0 \qquad \operatorname{div}\bar{b} = 0
$$

$$
\operatorname{rot}(\bar{b} - \bar{m}) - \frac{1}{c}(\dot{\bar{e}} + \dot{\bar{p}}) = \operatorname{rot}({}_0M^{\lambda}\bar{u}_{,\lambda})
$$

$$
\operatorname{div}(\bar{e} + \bar{p}) = 0
$$

$$
\begin{aligned}
\bar{e} = {} & -\frac{1}{c}(C_8 + C_9)\,\ddot{\bar{u}} \times {}_0\bar{M} + C_9\,\bar{p} \\
& + C_{10}(\bar{p}\cdot{}_0\bar{M})\,{}_0\bar{M} + A_9\,\dot{\bar{p}} + A_{10}(\dot{\bar{p}}\cdot{}_0\bar{M})\,{}_0\bar{M} \\
& + G_9\,\dot{\bar{p}} \times {}_0\bar{M} - \frac{1}{c}G_9(\dot{\bar{u}} \times {}_0\bar{M}) \times {}_0\bar{M}
\end{aligned}
$$

$$
\begin{aligned}
b_{\lambda} = {} & (C_6 - 1)\,u^{\mu}_{,\mu}\,{}_0M_{\lambda} + (A_6 - G_6)\,{}_0M_{\lambda}\,\dot{u}^{\mu}_{,\mu} \\
& + C_7({}_0M^{\mu}\,{}_0M^{\nu}\,u_{\mu,\nu})\,{}_0M_{\lambda} \\
& + (A_7 - G_7)\,({}_0M^{\mu}\,{}_0M^{\nu}\,\dot{u}_{\nu,\mu})\,{}_0M_{\lambda} \\
& + (C_4 - C_5 + 1)\,u_{\lambda,\mu}\,{}_0M^{\mu} \\
& + (A_4 - G_4)\,(\dot{u}_{\lambda,\mu} + \dot{u}_{\mu,\lambda})\,{}_0M^{\mu} \\
& + C_8\,m_{\lambda} + A_8\,\dot{m}_{\lambda} + G_8(\dot{\bar{m}} \times {}_0\bar{M})_{\lambda} \\
& + C_{11}({}_0\bar{M}\cdot\bar{m})\,{}_0M_{\lambda} + A_{11}({}_0\bar{M}\cdot\bar{m})\,{}_0M_{\lambda} \\
& + (A_{13} - G_{13})\,(\varepsilon^{\nu\alpha}_{\lambda}\,{}_0M_{\alpha}\,{}_0M^{\gamma}\,\dot{u}_{\nu,\gamma} \\
& + \varepsilon^{\gamma\alpha}_{\lambda}\,{}_0M^{\nu}\,\dot{u}_{\nu,\gamma}) + (C_4 - C_5 + 1 - C_8) \\
& \times u_{\mu,\lambda}\,{}_0M^{\mu}
\end{aligned}
\qquad [5.5]
$$

où les coefficients C_i, fonctions de ${}_0\bar{M}^2$, sont définis par

$$
C_1 = \varrho_0 \left(\frac{\partial^2 \hat{\Psi}}{\partial I_1^2}\right)_0 + \varrho_0 \left(\frac{\partial \hat{\Psi}}{\partial I_2}\right)_0
$$

$$C_2 = \varrho_0 \,{}_0\!\left(\frac{\partial\hat{\Psi}}{\partial I_2}\right) + \varrho_0 \,{}_0\!\left(\frac{\partial\hat{\Psi}}{\partial I_1}\right)$$

$$C_3 = \varrho_0 \,{}_0\!\left(\frac{\partial^2\hat{\Psi}}{\partial I_1 \partial I_8}\right) + \frac{\varrho_0}{2} \,{}_0\!\left(\frac{\partial\hat{\Psi}}{\partial I_9}\right) - 2\varrho_0 \,{}_0\!\left(\frac{\partial\hat{\Psi}}{\partial I_5}\right) + 1$$

$$C_4 = \frac{\varrho_0}{2} \,{}_0\!\left(\frac{\partial\hat{\Psi}}{\partial I_9}\right) + \varrho_0 \,{}_0\!\left(\frac{\partial\hat{\Psi}}{\partial I_8}\right) - 1$$

$$C_5 = \frac{\varrho_0}{2} \,{}_0\!\left(\frac{\partial\hat{\Psi}}{I_9}\right)$$

$$C_6 = 2\varrho_0 \,{}_0\!\left(\frac{\partial^2\hat{\Psi}}{\partial I_5 \partial I_1}\right) - 2\varrho_0 \,{}_0\!\left(\frac{\partial\hat{\Psi}}{\partial I_5}\right) + 1$$

$$C_7 = 2\varrho_0 \,{}_0\!\left(\frac{\partial^2\hat{\Psi}}{\partial I_8 \partial I_5}\right), \qquad C_8 = 2\varrho_0 \,{}_0\!\left(\frac{\partial\hat{\Psi}}{\partial I_5}\right)$$

$$C_9 = 2\varrho_0 \,{}_0\!\left(\frac{\partial\hat{\Psi}}{\partial I_4}\right), \qquad C_{10} = 2\varrho_0 \,{}_0\!\left(\frac{\partial\hat{\Psi}}{\partial I_{11}}\right)$$

$$C_{11} = 2\varrho_0 \,{}_0\!\left(\frac{\partial^2\hat{\Psi}}{\partial I_5^2}\right) \qquad C_{12} = \varrho_0 \,{}_0\!\left(\frac{\partial^2\hat{\Psi}}{\partial I_8^2}\right).$$

Dans la première équation de ce système, on doit négliger le dernier terme car le bilan [2.1] n'est valable qu'à l'approximation non relativiste où l'on néglige notamment les produits de champs ${}_0\bar{E}$, ${}_0\bar{B}$, ${}_0\bar{P}$, ${}_0\bar{M}$ divisés par $\varrho_0 c^2$ devant 1.

Afin de voir si le diélectrique présente l'effet *Faraday*, recherchons pour le système [5.5] des solutions $\bar{u}, \bar{e}, \bar{b}, \bar{p}, \bar{m}$ du type onde plane

$$\bar{a} = \mathcal{R}_e \left\{ \tilde{a}\, e^{i\omega\left(t - \frac{n}{c}\,\bar{s}\cdot\bar{x}\right)} \right\} \qquad [5.6]$$

dont le vecteur unité normal à l'onde $\bar{s}$ est parallèle à ${}_0\bar{B}$ et ${}_0\bar{M}$: ${}_0\bar{M} = {}_0M\,\bar{s}$. $\tilde{a}$ désigne le vecteur complexe amplitude du vecteur $\bar{a}(\bar{u}, \bar{e}, \bar{b}, \bar{p}, \bar{m})$, $\omega > 0$ la pulsation, n l'indice de réfraction complexe (caractérisant la vitesse et le coefficient d'absorption de l'onde).

En portant les expressions [5.6] de $\bar{u}, \bar{e}, \bar{b}, \bar{p}, \bar{m}$ dans le système [5.5], on obtient le système algébrique linéaire homogène suivant:

$$\varrho_0 \omega^2 \tilde{u} = (D_1 + i\omega B_1)\,\omega^2 \frac{n^2}{c^2} (\bar{s}\cdot\tilde{u})\,\bar{s} + (D_2 + i\omega A_2)\,\omega^2 \frac{n^2}{c^2}\tilde{u} + i\omega\frac{n}{c}(D_4 + i\omega(A_4 + G_4))\,{}_0M\tilde{m} + i\omega\frac{n}{c}(D_3 + i\omega(B_3 + H_3))\,(\bar{s}\cdot\tilde{m})\,\bar{s} + \frac{i\omega}{c}(C_8 - 1)\,{}_0M\bar{s}\times\tilde{p} + i\omega^3\frac{n^2}{c^2}\,{}_0MH_2\bar{s}\times\tilde{u} - \frac{\omega^2 n}{c}\,{}_0M^2(A_{13} + G_{13})\,\bar{s}\times\tilde{m}$$

$$\tilde{b} - n\bar{s}\times\tilde{e} = 0$$

$$n\bar{s}\times(\tilde{b} - \tilde{m}) + \tilde{e} + \tilde{p} = -\frac{i\omega}{c}\,{}_0Mn^2(\bar{s}\times\tilde{u})$$

$$\tilde{e} = (C_9 + i\omega A_9)\,\tilde{p} + \frac{i\omega}{c}\,{}_0M \times (C_8 + C_9 + i\omega A_9)\,\bar{s}\times\tilde{u} - \omega^2 G_9 \frac{{}_0M^2}{c}(\tilde{u} - (\bar{s}\cdot\tilde{u})\,\bar{s}) - i\omega\,{}_0MG_9\bar{s}\times\tilde{p} + (C_{10} + i\omega A_{10})\,{}_0M^2\bar{s}(\bar{s}\cdot\tilde{p})$$

$$\tilde{b} = i\omega\frac{n}{c}\,{}_0M(1 - D_3 + i\omega(B_3 - H_3))\,\bar{s} \times (\bar{s}\cdot\tilde{u}) - i\omega\frac{n}{c}\,{}_0M(D_4 + 1 + i\omega \times (A_4 - G_4))\,\tilde{u} + (C_8 + i\omega A_8)\,\tilde{m} + {}_0M^2(C_{11} + i\omega A_{11})\,\bar{s}(\bar{s}\cdot\tilde{m}) - (A_{13} - G_{13})\,\omega^2\frac{n}{c}\,{}_0M^2\bar{s}\times\tilde{u} - i\omega\,{}_0MG_8\bar{s}\times\tilde{m} \qquad [5.7]$$

où

$$D_1 = C_1 + {}_0M^2 2C_3 - {}_0M^2 + {}_0M^2 C_5 + {}_0M^4 C_{12}$$

$$D_2 = C_2 + {}_0M^2 C_4$$

$$D_3 = C_6 + {}_0M^2 C_7 + C_4 - C_5 - C_8 + 1$$

$$D_4 = C_4 - C_5$$

$$B_1 = A_1 + {}_0M^2 2A_3 + {}_0M^4 A_{12}$$

$$B_3 = A_4 + A_6 + {}_0M^2 A_7$$

$$H_2 = \tfrac{1}{2}\, G_2 - {}_0M^2 G_1$$

$$H_3 = G_4 + G_6 + {}_0M^2 G_7.$$

Nous n'avons pas écrit les équations aux amplitudes correspondant à $[5.5]_3$, $[5.5]_5$ car elles résultent de la multiplication scalaire de $[5.7]_2$, $[5.7]_3$ par $\bar{s}$.

Pour étudier le système [5.7], on élimine $\tilde{e}$ et $\tilde{b}$ grâce à $[5.7]_3$, $[5.7]_4$, et on obtient ainsi trois équations vectorielles en les inconnues $\tilde{u}$, $\tilde{p}$, $\tilde{m}$.

En multipliant scalairement ces trois équations par $\bar{s}$, on obtient un système homogène en $\bar{s}\cdot\tilde{u}$, $\bar{s}\cdot\tilde{p}$, $\bar{s}\cdot\tilde{m}$ qui n'admet de solutions non triviales que si son déterminant est nul, c'est-à-dire si

$$\frac{1}{n^2} = \frac{D_1 + D_2 + i\omega(B_1 + A_2)}{\varrho_0 c^2} - \frac{(D_4 + D_3)^2 - \omega^2((B_3 + A_4)^2 - (H_3 + G_4{}^2)}{C_8 + {}_0M^2 C_{11}} \frac{+ 2i\omega(D_4 + D_3)(B_3 + A_4)}{+ i\omega(A_8 + {}_0M^2 A_{11})} \frac{{}_0M^2}{\varrho_0 c^2} \qquad [5.8]$$

ce qui donne la vitesse et le coefficient d'absorption des ondes élastiques longitudinales ($\tilde{u}$ est suivant $\bar{s}$ si on ne retrouve pas la même valeur de $\frac{1}{n^2}$ lors de la recherche de solutions non triviales avec $\tilde{u}$, $\tilde{p}$, $\tilde{m}$ perpendiculaires à $\bar{s}$). Si ${}_0M = 0$, [5.8] se réduit à

$$\frac{1}{n^2} = \frac{C_1(0) + C_2(0) + i\omega(A_1(0) + A_2(0))}{\varrho_0 c^2}$$

qui est la valeur classique de $\frac{1}{n^2}$ pour les ondes élastiques longitudinales augmentée d'un terme imaginaire pur dû à la viscosité du milieu ($C_1(0)$ et $C_2(0)$ sont les paramètres de *Lamé*).

Supposons maintenant que $\bar{s}\cdot\tilde{u} = \bar{s}\cdot\tilde{p} = \bar{s}\cdot\tilde{m} = 0$. Multiplions vectoriellement par $\bar{s}$ les deux premières équations du système en $\tilde{u}$, $\tilde{p}$, $\tilde{m}$ (l'éq. $[5.7]_1$ et celle déduite de $[5.7]_2$ par substitution des expressions de $\tilde{e}$ et $\tilde{b}$), et gardons la troisième (celle déduite de $[5.7]_3$). De la première équation du système ainsi obtenu, on tire, si $(C_8 - 1)\,{}_0M \neq 0$,

$$\tilde{p} = -\frac{ic}{\omega\,{}_0M(C_{8-1})} \times \Bigg[\left(-\varrho_0 + (D_2 + i\omega A_2)\frac{n^2}{c^2}\right)\omega^2 \bar{s}\times\tilde{u} - i\omega^3 \frac{n^2}{c^2}\,{}_0MH_2\tilde{u} + \frac{i\omega n}{c}(D_4 + i\omega(A_4 + G_4)) \times {}_0M\bar{s}\times\tilde{m} + \frac{\omega^2 n}{c}\,{}_0M^2(A_{13} + G_{13})\tilde{m}\Bigg] \qquad [5.9]$$

et en portant cette expression de $\tilde{p}$ dans les deux autres, on obtient un système du type suivant en $\tilde{u}$, $\tilde{m}$:

$$A\bar{s}\times\tilde{u} + B\tilde{u} + C\bar{s}\times\tilde{m} + D\tilde{m} = 0 \qquad (\bar{s}\cdot\tilde{u} = \bar{s}\cdot\tilde{m} = 0)$$

$$E\bar{s}\times\tilde{u} + F\tilde{u} + G\bar{s}\times\tilde{m} + H\tilde{m} = 0 \qquad (\bar{s}\cdot\tilde{u} = \bar{s}\cdot\tilde{m} = 0) \qquad [5.10]$$

où A, B, C, D, E, F, G, H sont des coefficients complexes fonctions de n. Ce système n'admet que la solution triviale $\tilde{u} = \tilde{m} = 0$, sauf si

$$\begin{vmatrix} B & -A & D & -C \\ A & B & C & D \\ F & -E & H & -G \\ E & F & G & H \end{vmatrix} = \begin{vmatrix} A - iB & C - iD & 0 & 0 \\ E - iF & G - iH & 0 & 0 \\ 0 & 0 & A + iB & C + iD \\ 0 & 0 & E + iF & G + iH \end{vmatrix} = 0$$

c'est-à-dire si une des deux relations

$$(A + i\varepsilon B)(G + i\varepsilon H) - (C + i\varepsilon D)(E + i\varepsilon F) = 0 \qquad (\varepsilon = \pm 1) \qquad [5.11]$$

est vérifiée.

Dans le cas présent, $A + i\varepsilon B$, $C + i\varepsilon D$, $E + i\varepsilon F$, $G + i\varepsilon H$ ont pour expressions

$$A + i\varepsilon B = i\omega\frac{n}{c}\varrho_0 c^2 \frac{n^2}{C_8 - 1} \times \Bigg[(C_9 + i\omega A_9 - \varepsilon\omega\,{}_0MG_9) \times \left(\frac{1}{n^2} - \frac{D_2 + i\omega A_2 + \varepsilon\omega\,{}_0MH_2}{\varrho_0 c^2}\right) + \frac{{}_0M^2}{\varrho_0 c^2}\frac{1}{n^2}(C_8 - 1)(C_9 + C_8 - D_4 - 1 + i\omega(A_9 - (A_4 - G_4)) + \varepsilon\omega\,{}_0M(A_{13} - G_{13} - G_9))\Bigg]$$

$$C + i\varepsilon D = \frac{{}_0Mn^2}{C_8 - 1}\Bigg[(C_8 - 1) \times (C_8 + i\omega A_8 - \varepsilon\omega\,{}_0MG_8)\frac{1}{n^2} + (C_9 + i\omega A_9 - \varepsilon\omega\,{}_0MG_9) \times (D_4 + i\omega(A_4 + G_4) + \varepsilon\omega\,{}_0M(A_{13} + G_{13}))\Bigg]$$

$$E + i\varepsilon F = \frac{i\omega\varrho_0 c^2}{c(C_8 - 1)}$$
$$\times \Big[-(C_9 + 1 + i\omega A_9 - \varepsilon\omega_0 M G_9)$$
$$\times \left(-\frac{1}{n^2} + \frac{D_2 + i\omega A_2 + \varepsilon\omega_0 M H_2}{\varrho_0 c^2} \right)$$
$$- \frac{{}_0M^2}{\varrho_0 c^2}(D_4 + i\omega(A_4 - G_4)$$
$$- \varepsilon\omega_0 M(A_{13} - G_{13}))(C_8 - 1)$$
$$+ \frac{1}{n^2}\frac{{}_0M^2}{\varrho_0 c^2}(C_8 - 1)$$
$$\times (C_9 + C_8 + i\omega A_9 - \varepsilon\omega_0 M G_9)\Big]$$

$$G + i\varepsilon H = n_0 M$$
$$\times \Big[(C_8 - 1 + i\omega A_8 - \varepsilon\omega_0 M G_8)$$
$$+ \frac{1}{C_8 - 1}(D_4 + i\omega(A_4 + G_4)$$
$$+ \varepsilon\omega_0 M(A_{13} + G_{13}))$$
$$\times (C_9 + 1 + i\omega A_9 - \varepsilon\omega_0 M G_9)\Big]$$

et les relations [5.11] sont donc deux équations du second degré en $\frac{1}{n^2}$. Nous ne chercherons pas les solutions exactes de ces équations: nous chercherons les solutions à l'ordre le plus bas en les paramètres $\frac{{}_0M^2}{\varrho_0 c^2}$, $\frac{C_2 + i\omega A_2}{\varrho_0 c^2}$, $\frac{{}_0MH_2\omega}{\varrho_0 c^2}$.

Ceci se justifie car $\frac{{}_0M^2}{\varrho_0 c^2}$ est un terme relativiste, $\frac{C_2 + i\omega A_2}{\varrho_0 c^2}$ diffère par des termes en $\frac{{}_0M^2}{\varrho_0 c^2}$ de la valeur $\frac{C_2(0) + i\omega A_2(0)}{\varrho_0 c^2}$ de $\frac{1}{n^2}$ pour les ondes élastiques transversales en l'absence de champ électromagnétique, et car on suppose que le coefficient du terme en $\bar{s} \times \tilde{u}$ de $[5.7]_1$ est du même ordre de grandeur que ceux des termes en $\tilde{u}$ et $(\bar{s} \cdot \tilde{u})\,\bar{s}$.

On obtient ainsi pour chacune des éq. [5.11] une solution d'ordre 0 (onde rapide) et une solution d'ordre 1 (onde lente) en les paramètres considérés:

$$\frac{1}{n^2} = \frac{(C_9 + i\omega A_9 - \varepsilon\omega_0 M G_9)}{(C_9 + 1 + i\omega A_9 - \varepsilon\omega_0 M G_9)} \frac{\times (C_8 - 1 + i\omega A_8 - \varepsilon\omega_0 M G_8)}{\times (C_8 + i\omega A_8 - \varepsilon\omega_0 M G_8)} \qquad [5.12]$$

$$\frac{1}{n^2} = \frac{C_2 + i\omega A_2 + \varepsilon\omega_0 M H_2}{\varrho_0 c^2} + \frac{{}_0M^2}{\varrho_0 c^2} \times \left[C_4 - \frac{(D_4 + i\omega A_4 + \varepsilon\omega_0 M G_{13})^2 + \omega^2(\varepsilon G_4 - i_0 M A_{13})^2}{C_8 - 1 + i\omega A_8 - \varepsilon\omega_0 M G_8} \right]. \qquad [5.13]$$

Ces valeurs de $\frac{1}{n^2}$ correspondent respectivement à des ondes électromagnétiques et élastiques transversales.

Montrons que pour ces ondes, $\tilde{u}$, $\tilde{p}$, $\tilde{m}$, sont en général polarisés circulairement. Le système [5.10] ne peut admettre, dans le cas présent, au maximum qu'une simple infinité de solutions. En effet il ne peut admettre plus de solutions que si tous les mineurs d'ordre 3 de son déterminant sont nuls ce qui implique soit que les deux éq. [5.11] soient vérifiées simultanément, soit que $A + i\varepsilon B = C + i\varepsilon D = E + i\varepsilon F = G + i\varepsilon H = 0$. Or la première éventualité n'est pas réalisée en général puisque les valeurs [5.12], [5.13] de $\frac{1}{n^2}$ dépendent de la valeur de ε, et la seconde est irréalisable car $\frac{1}{n^2}$ devrait être à la fois d'ordre 1 et 0. Le système [5.10] admet donc une simple infinité de solutions lorsque $\frac{1}{n^2}$ a une des valeurs [5.12], [5.13]. Or si $\tilde{u}$, $\tilde{m}$ est solution de ce système, $\bar{s} \times \tilde{u}$, $\bar{s} \times \tilde{m}$ l'est également, et on doit donc avoir

$$\bar{s} \times \tilde{u} = \lambda\tilde{u}, \quad \bar{s} \times \tilde{m} = \lambda\tilde{m} \quad \text{avec} \quad \lambda \neq 0,$$

d'où il résulte que $\tilde{u} \cdot \tilde{u} = \tilde{m} \cdot \tilde{m} = 0$. $\tilde{u}$, $\tilde{m}$ sont donc polarisés circulairement, et on déduit de [5.9] qu'il en est de même de $\tilde{p}$. L'une des valeurs de ε correspond à une polarisation circulaire droite, et l'autre à une polarisation circulaire droite. Les expressions [5.12] et [5.13] montrent que les valeurs de $\frac{1}{n}$ correspondent aux ondes polarisées circulaires gauche et droite sont en général distinctes, à la fois pour les ondes rapides (ondes électromagnétiques) et les ondes lentes (ondes élastiques).

En résumé, dans le modèle envisagé, 5 types d'ondes peuvent en général se propager dans la direction d'un champ magnétique uniforme et constant:

- une onde élastique longitudinale absorbée;
- deux ondes électromagnétiques polarisées circulaires gauche et droite, ayant des vitesses et des coefficients d'absorption distincts (effet *Faraday* et dichroisme);
- deux ondes élastiques transversales polarisées circulaires gauche et droite ayant des vitesses et des coefficients d'absorption distincts.

Remarquons que dans le cas particulier où tous les coefficients A_i sont nuls, les ondes ne sont pas absorbées. Dans celui où tous les G_i sont nuls, l'état de polarisation des ondes électromagnétiques et des ondes élastiques transversales peut être quelconque, et il n'y a qu'une valeur de $\frac{1}{n^2}$ pour chacune de ces ondes.

Lorsqu'il n'y a pas de dépendance à $\dot{\boldsymbol{E}}$ dans les équations constitutives [3.1], on peut voir que [5.12] et [5.13] fournissent toujours chacune deux valeurs distinctes de $\frac{1}{n^2}$; s'il n'y a pas de dépendance à $\dot{\bar{\Pi}}'$, $\dot{\bar{\mathcal{M}}}'$, [5.12] ne fournit plus qu'une seule valeur de $\frac{1}{n^2}$ et l'effet *Faraday* disparait pour les ondes électromagnétiques, mais il subsiste un effet *Faraday* pour les ondes élastiques transversales. Enfin, les deux effets disparaissent si le diélectrique est non magnétisable. (${}_0M \to 0$, $\bar{m} \to 0$, $C_8 \to \infty$, de telle manière que ${}_0MC_8 = {}_0B$ et $C_8\bar{m} = \bar{b}$.)

6. Conclusion

Sur la base des principes généraux de la mécanique des milieux continus, l'effet *Faraday* a été mis en évidence dans un diélectrique isotrope, grâce à la prise en considération à la fois de la magnétisation et de dérivées par rapport au temps de la polarisation et de la magnétisation dans les équations constitutives.

Remerciements

Je tiens à remercier Monsieur *G. Mayné* dont les remarques et les conseils me furent très utiles.

Résumé

Dans ce travail, on étudie l'effet *Faraday* dans le cadre de la mécanique des milieux continus. On se base sur les équations de *Maxwell*, ainsi que sur les bilans de quantité de mouvement et d'énergie interne pour un continu polarisable et magnétisable sous la forme obtenue par *G. Mayne* et *Ph. Boulanger* (4). A ces équations on adjoint des équations constitutives où l'on introduit, en plus des variables qui caractérisent un diélectrique élastique, les dérivées par rapport aux temps de la déformation, de la polarisation et de la magnétisation. Grâce à une linéarisation des équations autour d'un état d'équilibre, on étudie la propagation des ondes planes électromagnétiques et mécaniques dans la direction d'un champ magnétique constant. Cette étude met en évidence l'effet *Faraday* ainsi qu'un effet analogue pour les ondes mécaniques transversales.

Summary

In this work, the *Faraday* effect is investigated within the frame of continuum mechanics. We use here *Maxwell* equations, and the momentum balance and internal energy balance equations in the form obtained by *G. Mayne* and *Ph. Boulanger* (4). To these equations are added constitutive equations which contain the time derivatives of the deformation, the polarization and the magnetization, in addition to the variables which caracterize the elastic dielectric. By linearizing the equations about an equilibrium state, we study the propagation of electromagnetic and mechanical plane waves in the direction of a constant magnetic field. By this means, we are able to derive the *Faraday* effect and an analogous one for transverse mechanical waves.

Zusammenfassung

Die Arbeit befaßt sich mit der Erforschung des *Faraday*-Effekts im Rahmen der Kontinuumsmechanik. Als Basis dienen hier die *Maxwell*schen Gleichungen sowie die Impulsbilanz und Energiebilanz in einem polarisierbaren und magnetisierbaren Kontinuum in der von *G. Mayne* und *Ph. Boulanger* festgelegten Form (4). Benützt werden ferner Materialgleichungen, in denen die Zeitableitungen der Deformation, Polarisation und Magnetisierung, sowie die Veränderlichen, die ein elastisches Dielektrikum bestimmen, eingeführt werden. Dank einer linearen Anordnung der Gleichungen um einen Gleichgewichtszustand wird die Ausbreitung elektromagnetischer und mechanischer ebener Wellen, in der Richtung eines magnetischen Feldes, erforscht. Auf diese Weise ist es möglich, den *Faraday*-Effekt zu überprüfen sowie einen analogen Effekt in mechanischen Transversalwellen.

Littérature

1) *Toupin, R. A.*, J. Eng. Sci. **1**, 101 (1963).

2) *Boulanger, Ph.* et *G. Mayné*, C. R. Acad. Sci. **274**, 591 (Paris 1972).

3) *Boulanger, Ph., G. Mayné* et *R. van Geen*, Int. J. Solids Struct. (soumis pour publication).

4) *Mayné, G.* et *Ph. Boulanger*, Conférence spécialisée présentée au VI Congrès International de Rhéologie (Lyon, France 1972).

5) *Boulanger, Ph., G. Mayné, A. Hermanne, J. Kestens* et *R. van Geen*, L'effet photoélastique dans le cadre de la mécanique rationnelle des milieux continus. Cahier du Groupe Français de Rhéologie. Tome **2**, n⁰ 5 (1971).

6) *Truesdell, C.* et *R. A. Toupin*, Handbuch der Physik III/1 (Berlin-Heidelberg-New York 1960).

7) *Truesdell, C.* et *W. Noll*, Handbuch der Physik III/3 (Berlin-Heidelberg-New York 1965).

8) *Wang, C.-C.*, Arch. Rat. Mech. Anal. **36**. 166 (1970).

Adresse de l'auteur:

Philippe Boulanger
Département de Mathématiques
Université Libre de Bruxelles
Av. F. D. Roosevelt, 50
B 1050 Bruxelles (Belgique)

Rheol. Acta 12, 127–132 (1973)

From the Chemical Engineering California Institute of Technology, Pasadena (U.S.A.), and the DAMTP, Cambridge University, Cambridge (U.K.)

Theoretical studies of a suspension of rigid particles affected by Brownian couples

By *L. G. Leal* and *E. J. Hinch*

With 3 figures

(Received October 27, 1972)

Introduction

In this paper we briefly review the present theoretical understanding of a suspension of particles whose orientation is affected by rotational *Brown*ian motion. The particles are rigid, axially symmetric, sufficiently small so that they and their disturbance flow are inertialess and not acted upon by external body forces or couples. The suspension is dilute so that there are no important hydrodynamic interactions between the particles and is examined for a flow with a sufficiently large length scale that the concept of an equivalent homogeneous material – the bulk suspension – is meaningful. The orientation state of the particles in the suspension is determined by a competition between the random *Brown*ian rotations and the tendency to preferred alignment induced by the gradient in the bulk deformation rate for the suspension. We are interested in the resulting rheological behaviour.

The constitutive relation for this suspension may be expressed by three mathematical statements. The first two describe the microstructural dynamics; the rotational motion of single particles in the absence of *Brown*ian effects, and the evolution of the orientation statistics when random rotations are superimposed onto this undisturbed motion. The remaining equation relates the macroscopically observable bulk stress tensor to the instantaneous microstructural state and to the bulk strain rate.

We begin by considering the bulk suspension to be undergoing a homogeneous time-dependent general linear flow

$$\underline{u}(\underline{x}, t) = \underline{\underline{\Gamma}}(t) \cdot \underline{x}.$$

The deformation rate gradient $\underline{\underline{\Gamma}}$ is split into its symmetric and antisymmetric parts, the strain rate $\underline{\underline{E}}$ and vorticity $\underline{\underline{\Omega}}$. The rotation of an isolated particle in such a flow, with no *Brown*ian couples acting, is obtained from the solution of the low *Reynolds* number equations of motion using the conditions of zero net force and couple. This basic problem was solved by *Jeffery* (1922) for rigid spheroidal particles. If a particle has its symmetry axis in the direction of the unit vector $\underline{p}$, then *Jefferys* solution yields the rate of change of the particle's orientation,

$$\dot{\underline{p}} = \underline{\underline{\Omega}} \cdot \underline{p} + \frac{r^2 - 1}{r^2 + 1} [\underline{\underline{E}} \cdot \underline{p} - \underline{p}(\underline{p} \cdot \underline{\underline{E}} \cdot \underline{p})] \qquad [1]$$

where r is the aspect of the particles ($r > 1$, prolate).

In the presence of rotary *Brown*ian motion, the orientation of any particular particle is determinate only in a statistical sense. The statistics of the particle orientation are quantified by the differential probability density function $N(\underline{p})$. This positive function is normalized so that the total probability of the particle having some orientation is unity. The *Fokker-Planck* equation for the temporal development of the orientation distribution, first stated for shear flow by *Boeder* (1932), is

$$\frac{\partial N}{\partial t} + \operatorname{div}(N\dot{\underline{p}} - D\nabla N) = 0. \qquad [2]$$

The flux of probability in the orientation space contains two contributions; the undisturbed advection governed by (1), and the *Brown*ian rotations which are modelled as a diffusion process with a *Stokes–Einstein* diffusion coefficient D.

The macroscopically observable bulk stress is defined as the ensemble average of the instantaneous stress at a point (cf. *Batchelor*, 1970). This definition reduces to a simple volume average in our suspension. Separating off the *Newtonian* stress for the pure solvent, the contribution to the bulk stress due to the presence of the suspended particles, first evaluated by *Giesekus* (1962) from *Jefferys* hydrodynamic solution, is

$$\underline{\underline{\sigma}} = 2\mu\Phi\{A\langle\underline{p}\underline{p}\underline{p}\underline{p}\rangle : \underline{\underline{E}} + B\langle\underline{p}\underline{p}\rangle \cdot \underline{\underline{E}} + \underline{\underline{E}} \cdot \langle\underline{p}\underline{p}\rangle] + C\underline{\underline{E}} + F\langle\underline{p}\underline{p}\rangle D\}. \qquad [3]$$

Here μ is the viscosity of the solvent, Φ the volume fraction of the suspended particles, and A, B, C, and F material coefficients which only depend on the particle aspect ratio. The angle brackets denote statistical averages of the included quantitiy with respect to the distribution function. Eq. [3] is of the general form for a tensorial relation between $\underline{\underline{\sigma}}$ and $\underline{\underline{E}}$ which is linear in $\underline{\underline{E}}$ and D and

*) Presented by *L. G. Leal* at the VIe Congrès International de Rhéologie, September 1972.

which is also linear in N, i.e. each particle contributes independently (a dilute suspension).

Eq. [1]–[3] are the constitutive relation for the suspension. This structured material is nonlinear: reflecting the degree of preferred alignment permitted by the *Brown*ian disorientations, $pppp$ and pp in eq. [3] depend on the directional character of $\underline{\underline{\Gamma}}$ and its magnitude relative to D. The material has a memory: the time derivative in eq. [2] represents the finite rate at which the particles can reorientate in a viscous solvent. Further, the memory possesses the property of fading which is expressed by the *Brown*ian diffusion-like term in eq. [2]. We proceed to investigate the constitutive relation, first considering the calculation of the orientation distribution and then discussing the corresponding rheological behavior.

Solution of the equations

A knowledge of the distribution N, which is governed by eqs. [1] and [2] and the normalization constraint, clearly allows an evaluation of the bulk stress through eq. [3]. Unfortunately in the time-dependent general linear flow neither a general explicit solution nor a simpler implicit representation than [1] and [2] is available either for the full distribution or the required second and fourth moments with respect to $\underline{p}$. It is therefore necessary to build up a picture of the suspension by studying a variety of limiting conditions. The appropriate non-dimensional groups which define the possible regimes are: the flow type $\underline{\underline{\Gamma}}/\|\underline{\underline{\Gamma}}\|$, the strength of the flow $\|\Gamma\|/D$, the frequency of imposed oscillations ω/D, and the particle aspect ratio as in the coefficient $(r^2-1)/(r^2+1)$.

The first approximate solutions were obtained by restricting attention to situations in which the orientation distribution is almost uniform. The corresponding limiting conditions can be easily perceived from eqs. [1] and [2]. The advection term of [2] from [1] consists of two parts; a rotation with the vorticity which can only relocate any anisotropy, and a variable rotation associated with the straining motion. The diffusion term in [2] acts to reduce peaks in N. Hence the nearly isotropic limit is due to a small effective strain rate, a condition achieved by restricting either $\|\underline{\underline{\Gamma}}\|/D$, $\|\underline{\underline{E}}\|/\omega$ or $(r^2-1)/(r^2+1)$ to a low value. In each of these three cases the distribution function is found in a perturbation expansion in the particular small parameter by iterating about the uniform state. The limit of dominant *Brown*ian motions, $\|\underline{\underline{\Gamma}}\|/D \to 0$, was first studied by *Burgers* (1938) who (incorrectly) calculated the intrinsic viscosity for steady shear flow. Later *Giesekus* (1962) evaluated the full bulk stress tensor in steady shear flow, and then re-expressed his results in the appropriate terms of the second order fluid approximation. The limit of rapidly oscillating flows, $\|E\|/\omega \to 0$, leads to the linear viscoelastic approximation in which the fading memory function has a single time constant. The near-sphere limit, $(r^2-1)/(r^2+1) \to 0$, was commenced by *Peterlin* (1938) and *Peterlin* and *Stuart* (1939) for steady shear flow. Their unnecessary additional restriction to dominant *Brown*ian motions was removed by *Sadron* (1953) for the birefringence functions, and by *Hinch* and *Leal* (1972) for the shear viscosity and normal stress differences. A large number of terms in this asymptotic expansion have been numerically summed to obtain, under a wide range of the parameters $\|\underline{\underline{\Gamma}}\|/D$ and $(r^2-1)/(r^2+1)$, tables of the birefringence functions (*Scheraga, Edsall* and *Gadd*, 1951) and the intrinsic viscosity (*Scheraga*, 1955) in steady shear flow. *Leal* and *Hinch* (1972) have recently given the asymptotic form of the full constitutive equation in the limit $(r^1-1)/(r^2+1) - 0$.

The remaining approximate solutions are strictly anisotropic and correspondingly more complex. The special circumstance of a steady irrotational flow does admit a useful solution which is valid for all flow strengths and particle aspect ratios. The key to this interesting case is the observation that when there is no vorticity the right hand side of eq. [1] may be written as the gradient of a potential function. Thus the distribution function takes a *Maxwell*ian form and the bulk stress is given in terms of error functions. *Brenner* (1972) has studied this case, correcting the prior investigation of extensional motion by *Takserman-Krozer* and *Ziabicki* (1963) which also failed to exploit the potential function.

The majority of our own contribution has been directed to the limit of weak *Brown*ian motions for steady flows. Because of the complexity of this regime, a general solution cannot be given. Instead we have developed the necessary techniques so that now any particular flow can be tackled. A prerequisite to solving the weak *Brown*ian motion limit is an understanding of the particle behavior when no random couples are present, i.e. the solution of the initial value problem represented by eq. [1]. *Bretherton* (1962) has discussed how this non-linear problem may be replaced by a linear one. According to the relative magnitudes of the real parts of the eigenvalues of the fundamental matrix $\underline{\underline{\Omega}} - (r^2-1)/(r^2+1)\,\underline{\underline{E}}$

for the prescribed flow, the particle motion falls into one of four possible classes.

In class I (one eigenvalue exceeding the others, e.g. extensional motion for $r > 1$) the particles all tend to a particular orientation which is independent of their initial positions. Weak *Brown*ian motions cannot resist this tendency, but just locally smooth out the developing delta function in the distribution into a *Gauss*ian.

In class II (two eigenvalues with equal positive real parts, e.g. reversed extensional [compressional] motion for $r > 1$) the particles all tend to some plane which is independent of their initial positions. Weak *Brown*ian motions cannot resist the tendency to the plane, but in the plane where the motion is less purposeful they are influential. To the first approximation the equilibrium distribution is restricted to the plane with a variation there which is inversely proportional to the velocity around the plane – there is more chance of finding the particle where it is rotating slower. At the next approximation, the restriction to the plane is smoothed into a *Gauss*ian distribution, and the peaks around the plane are decreased.

In class III (two purely imaginary eigenvalues, e.g. shear flow) the particles execute closed periodic orbits. The selection of the orbit out of a one parameter family and the relative phase about that orbit depend for each particle on its initial orientation. Removing this dependence upon the initial state, weak *Brown*ian motions have the most drastic effect for these flows, despite the fact that alone for this class no large gradients in the distribution develop under the limiting process. At the first approximation the *Brown*ian motions have two effects. First the temporal oscillations are damped out by a mixing of the phases about each orbit. Thus the equilibrium distribution shows variations around each orbit which are inversely proportional to the local rotation speed and to the degree of crowding with nearby orbits. Second, the diffusion process provides the mechanism by which the relative population of different orbits is determined. The equilibrium orbit distribution is found from an integral condition that there is not net diffusion across each orbit. Detailed calculations for this first approximation in the example of shear flow were given by *Leal* and *Hinch* (1971). Higher order approximations can be obtained through a perturbation expansion as shown by *Hinch* and *Leal* (1972).

In class IV [all the eigenvalues vanish, e.g. shear flow for dumbbells $((r^2 - 1)/(r^2 + 1) = 1)$] the particles all tend to some orientation which is independent of their initial positions. This class differs from class I through the approach to the final orientation from only one side. On the other side the particles first move away from the final orientation: they execute a half turn before returning on the correct side. Weak *Brown*ian motions cannot resist the inevitable accumulation of the particles near the final orientation. Gradients in the distribution build up there until an equilibrium is achieved between the diffusional leakage onto the outgoing side and the rate at which the leakage can be returned on the incoming side. Following earlier attempts by *Burgers* (1938) and *Brenner* (1971), the distribution function for the whole class of flows was given in terms of a matched asymptotic expansion by *Vinch* and *Leal* (1972). *Stewart* and *Søprenson* (1972) have recently reported numerical calculations using a *Galerkin* technique for the distribution at large but finite flow strength.

The analysis of time-dependent anisotropic situations is scantily developed. In many limiting anisotropic situations, the *Brown*ian motions are negligible at the lowest order. In these cases, a first approximation to the evolution of the distribution function could be obtained by a straightforward integration of eq. [1] for several representative orientations (though such solutions may not be uniformly valid at large times). This would be particularly simple when the distribution is essentially concentrated in a single direction. Another simplifying circumstance, which enables small *Brown*ian effects not to be entirely neglected, is that of a strong flow with constant directional properties but a time-dependent magnitude. We have recently examined an important example of the more mathematically trying class III flows – a time-dependent shear flow. The assumption that the *Brown*ian rotations are weak implies that they can only have an effect on a time scale much longer than the inverse shear time. A two-timing asymptotic expansion is obviously applicable for the distribution function. On the short time scale the distribution oscillates as each particle effectively orbits according to eq. [1]. These oscillations decay as the phases around the orbits become mixed on the longer diffusional time scale, when also the relative distribution of the different orbits drifts towards an equilibrium. The diffusion across the orbits is enhanced by local crowding of nearby orbits, while the rate of phase mixing is additionally increased by the crowding around the orbits caused whenever the local orbiting speed falls.

Rheological effects

The numerous approximate solutions of the constitutive equations will now be brought together to produce a fairly complete picture of the suspension's rheological behavior in two simple common flows.

We begin with steady axially symmetric extensional motion, strain rate E. The case $E > 0$ is sometimes referred to as uniaxial extension, and $E < 0$ as biaxial extension. The axial symmetry means that the bulk stress tensor can be totally represented by a strain-dependent effective viscosity, which we have schematically plotted in fig. 1 (for details see *Brenner*, 1972). The viscosity strain-thickens[1]). The detailed dependence upon particle aspect ratio and flow strength is explicitly given by the steady irrotational solution. As indicated in fig. 1, the constant limiting values of the

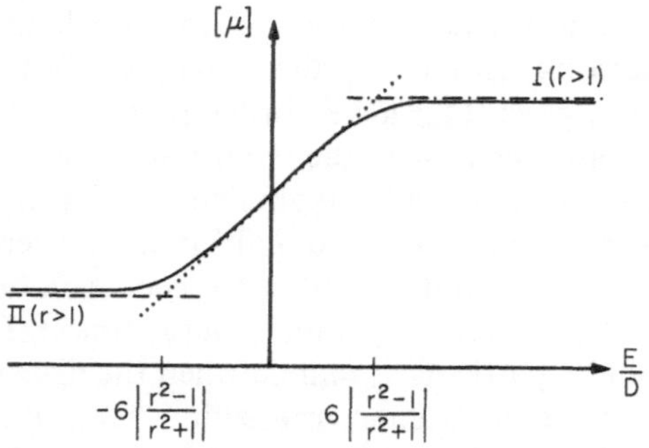

Fig. 1. The strain-thickening viscosity. . . . . . Dominant *Brownian* motion, ——— Weak *Brownian* motion (class II $[r > 1]$, class I $[r < 1]$), —·—· Weak *Brownian* motion (class I $[r > 1]$, class II $[r < 1]$)

viscosity can also be obtained (more simply) from the steady weak *Brownian* motion theory, the positive/negative strong flows being class I/II if $r > 1$ (or class II/I if $r < 1$). In addition, the linear variation for weak flows, $|(r^2 - 1)\, E/(r^2 + 1)\, D| \ll |$, is given by both the dominant *Brownian* motion (second order fluid) approximation and the near-sphere theory. The extent of strain-thickening of the intrinsic viscosity is small for near-spheres $0(r - 1)$, and increases to a factor of 4 for rods and 7/4 for disks. The physical explanation for strain-thickening lies in the change in particle orientation as E varies. A rod tends to align in the direction of the principle strain axis when E is large and positive, and to align in an orthogonal direction when E is large and negative. The alignment of a disk is the reverse order. In either case, as E increases, the particle reaches out further into the largest flow thus causing an increasing disturbance. The largest velocity of the undisturbed flow evaluated on the particle varies by a factor of two between the two limiting orientations. Thus the dissipation (and hence the effective viscosity) can vary at most by a factor of 4, a figure only achieved by the rod-like particles.

In a steady shear flow, shear rate γ, the effective viscosity must be supplemented by two normal stress differences for a full rheological description. The qualitative variation of these quantities with shear rate is shown in fig. 2; the details may be found in *Hinch* and *Leal* (1972). The viscosity shear-thins. The primary normal stress difference, $\sigma_{11} - \sigma_{33}$ for $u_1 = \gamma x_2$, is positive while the secondary difference, $\sigma_{22} - \sigma_{33}$, is negative

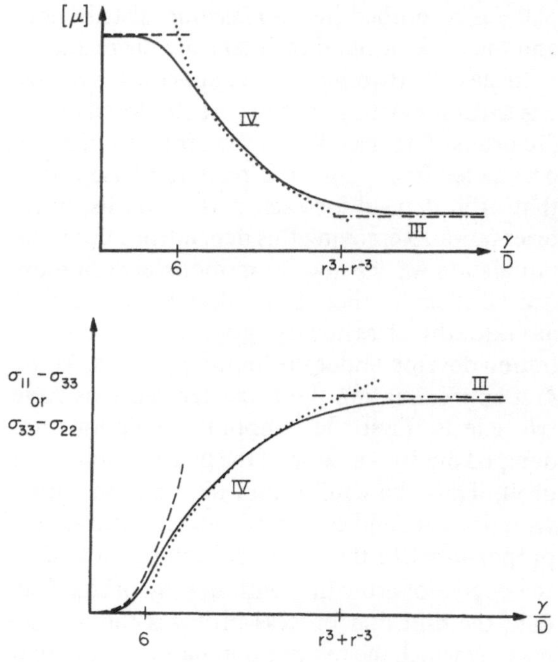

Fig. 2. The shear-thinning viscosity and normal stress functions. ——— Dominant *Brownian* motion, ... Weak *Brownian* motion (class IV) – $D/\gamma(r^3 + r^{-3}) \ll 1$, —·—·— Weak *Brownian* motion (class III) – $D/\gamma(r^3 + r^{-3}) \gg 1$

and numerically smaller. An explicit form for the shear dependence is only available from the near-sphere theory in which the deviation from *Einsteins* particle contribution to the stress tensor is

[1]) Some authors have referred to the decrease in (μ) for E decreasing towards $-\infty$ as biaxial strain thinning.

small, $0(r-1)^2$. For particles which are not near-spheres the qualitative variations are similar but there are only analytic results in various limiting situations. The dominant *Brown*ian motion (second order fluid) approximation gives a constant viscosity and the quadratic departures from zero of the normal stresses. There is some indication from the near-sphere solution that this expansion does not converge beyond $\gamma/D > 6$. The weak *Brown*ian motion theory (class III) gives the constant large shear-rate limits of the normal stress differences and of the shear-thinned viscosity. This analysis is only valid for $\gamma/D \gg r^3 + r^{-3}$. Thus for extreme aspect ratios there is a significant gap between the strong and weak shear approximations. The hole is filled by an intermediate regime in which the flow is effectively class IV, i.e. $|(r^2-1)/(r^2+1)| = 1$. This approximation yields a $(\gamma/D)^{-1/3}$ power law shear-thinning intrinsic viscosity and normal stresses increasing according to $\gamma^{2/3} D^{1/3}$. The extent of the shear-thinning increases from nothing for spheres to an order of magnitude in r for extreme aspect ratios. The physical explanation of the shear-thinning is that, when not greatly disorientated by *Brown*ian motions, the shear flow causes the particles to spend most of their orbit, $1 - 0(r + r^{-1})^{-1}$, within $0(r + r^{-1})^{-1}$ of full alignment with the flow. In such an orientation, especially for extreme aspect ratios, the particles are least disruptive and this leads to a relatively low effective viscosity.

A study by *Leal* and *Hinch* (1972) of several time-dependent situations using the near-sphere asymptotic form of the constitutive equation identified two temporal rheological characteristics. One is a natural oscillatory response with a frequency either equal to, or half, the magnitude of the vorticity Ω. This behavior results from the oscillating flow seen by the effectively spherical particles as they spin with the vorticity. The other is an exponential decay with a rate $6D$ which is due to the *Brown*ian diffusion. In a flow with an imposed frequency ω, these two effects typically produce a linear impedance containing factors like $(\omega \pm \Omega + i6D)^{-1}$.

These two temporal characteristics are quantitively modified when the particles are not near-spheres. The frequency of the oscillations, equal to the imaginary part of the eigenvalues of the fundamental matrix $\underline{\underline{\Omega}} + (r^2-1)/(r^2+1)\,\underline{\underline{E}}$, is slowed down from the vorticity rate by components of any straining motion, the oscillations disappearing when all the eigenvalues are real. The effective straining also creates a new drift feature associated with the real part of the eigenvalues. In contrast to the slowed oscillations, the diffusional decay rate $6D$ is enhanced in anisotropic situations by an amount proportional to the square of population gradients in the orientation space.

A simple illustration of the modification of these two temporal rheological characteristics is given by the transients associated with the start up of a steady shear flow from rest. The near-sphere analysis (*Leal* and *Hinch*, 1972) yields a time-dependent intrinsic viscosity

$$\frac{5}{2} + (r-1)^2 \left\{ \frac{26}{147} + \frac{3}{5}\frac{36D^2}{\gamma^2 + 36D^2} \times \left[(1 - e^{-6Dt}\cos\gamma t) + \frac{\gamma}{6D} e^{-6Dt} \sin\gamma t \right] \right\}$$

which displays the decaying natural oscillation, decay rate $6D$ and frequency γ. From recent progress, we sketch the form of the time varying intrinsic viscosity for rod-like particles $r \gg 1$ in fig. 3. The first two stages show a non-linear

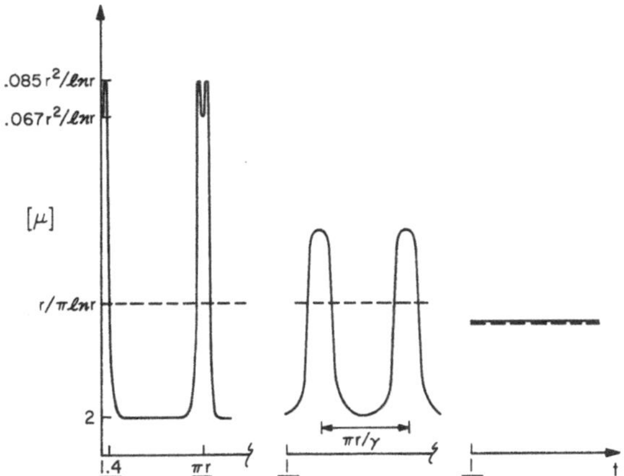

Fig. 3. The transient intrinsic viscosity for rods. ——— Average value, ——— Actual value

oscillation with the reduced frequency of $2\gamma/r$. In the initial stage the intrinsic viscosity peaks strongly at $0(r^2/\ln r)$ during the short time $0(\gamma^{-1})$ at which many particles are in the principle strain directions. The averaged viscosity over the oscillation is the *Eisenschitz* value $r/\pi \ln r$. By the time of the second stage $0(D^{-1}r^{-4})$, the peaks have been reduced to $0(r/\ln r)$ by phase mixing but the average value is unchanged. The enhanced decay rate is due to a crowding around the orbits associated with their points of slowest rotation. The remainder of the oscillation decays and the average value drifts to its steady value $0.315 r/\ln r$ on the longer diffusion time scale of $D^{-1} r^{-2}$. The enhancement above the $6D$ rate in the latter case is due to the crowding together of nearby orbits.

In conclusion we would remark that this suspension exhibits many rheological behaviors that have been experimentally observed in more complex materials. This is not so surprising if the orientation of their microstructure is an important physical process.

Acknowledgement is made to the donors of the Petroleum Research fund, administered by the American Chemical Society, for partial support of *L. G. Leal*.

References

Batchelor, G. K., J. Fluid Mech. **41**, 545 (1970).
Boeder, P., Z. Physics. **75**, 258 (1932).
Brenner, H., J. Coll. Sci. **34**, 258 (1970). — *Brenner, H.*, J. Chem. Eng. Sci. **27**, 1069 (1972).
Bretherton, F. P., J. Fluid Mech. **14**, 284 (1962).
Burgers, J. M., Kon. Ned. Akad. Wet. Verhand. (Eerste Sedic) **16**, 113 (1938).
Giesekus, H., Rheol. Acta **2**, 50 (1962).
Hinch, E. J. and *L. G. Leal*, J. Fluid Mech. **52**, 683 (1972).
Jeffery, G. B., Proc. Roy. Soc. ≫ **102**, 161 (1922).
Leal, L. G. and *E. J. Hinch*, J. Fluid Mech. **46**, 685 (1971). — *Leal, L. G.* and *E. J. Hinch*, J. Fluid Mech. (to appear 1972).
Peterlin, A., Z. Phys. **112**, 1 (1938). — *Peterlin, A.* and *M. A. Stuart*, Z. Phys. **112**, 129 (1939).
Sadron, C., Flow Properties of Disperse Systems, Chap. **4**. Ed.: *J. J. Hermans* (Amsterdam 1953).
Scheraga, H. A., J. Chem. Phys. **23**, 1526 (1955). — *Scheraga, H. A.*, *J. T. Edsal*, and *J. O. Gadd*, J. Chem. Phys. **19**, 110 (1951).
Stewart, W. E. and *J. P. Sorensen*, Trans. Soc. Rheol. **16/1**, 1 (1972).
Takserman-Krozer, R. and *A. Ziabicki*, J. Polymer Sci. **1**, 491 (1963).

Authors' addresses:

Dr. *L. G. Leal*
Chemical Engineering California Institute of Technology
Pasadena, California 91109 (U.S.A.)

Dr. *J. E. Hinch*
DAMPT, Cambridge University
Cambridge (UK)

Rheol. Acta 12, 133–140 (1973)

Institut de la Mécanique Théorique et Appliquée de l'Académie Tchécoslovaque des Sciences, Prague (CSSR)

Structure, anisotropie et modèles à deux dimensions en rhéologie

Par Z. Sobotka

Avec 6 figures

(Reçu p. p. le 27 Octobre 1972)

1. Introduction

L'auteur a introduit les modèles rhéologiques à deux dimensions pour l'état plan de contraintes ou de déformations des corps viscoélastiques, élastoplastiques, viscoplastiques et élastoviscoplastiques. Ces modèles consistent en bidimensionnelles régions élastiques, visqueuses et plastiques qui peuvent être isotropes ou anisotropes.

Les modèles rhéologiques à deux dimensions représentent directement la bidimensionalité du comportement rhéologique, la distribution et la configuration des phases rhéologiques, l'anisotropie et la cohésion interne.

Ces modèles peuvent représenter deux espèces de l'anisotropie totale des corps rhéologiques causée soit par l'anisotropie des phases soit par la configuration des phases. Ils ont été la base de la découverte du cisaillement asymétrique en rhéologie, qui est analogue aux effets *Cossérat* en milieux élastiques.

Les bornes entre les régions rhéologiques sont formées soit par les droites perpendiculaires ou inclinées soit par les courbes. La forme et la configuration des régions rhéologiques correspond aux types différents de l'anisotropie. Nous avons ainsi les modèles orthogonaux et non-orthogonaux.

Les modèles rhéologiques à deux dimensions représentent l'union des points de vue phénoménologiques et structuraux. Ils correspondent soit à la surface unité soit ils ont les dimensions différentielles de dx et dy. Dans le dernier cas ils peuvent représenter directement l'arrangement réel de la microstructure.

Le modèle aux dimensions finies peut être considéré aussi comme un résultat de l'intégration d'éléments différentiels et correspond à la non-homogénéité uniforme ou résultante dans la zone en considération. Les parties d'un corps à la constitution rhéologique non-uniforme peuvent être représentées par les modèles rhéologiques différents.

Les modèles rhéologiques à deux dimensions sont beaucoup plus rapprochés à la configuration structurale réelle que les modèles usuels et c'est pourquoi ils donnent les équations rhéologiques correspondant à la réalité d'une manière plus précise et représentant de nouveaux phénomènes.

L'auteur va considérer l'état plan de contraintes des corps viscoélastiques. Dans le cas de matières orthotropes, les déformations de régions élastiques sont données par la loi de *Hooke*:

$$\varepsilon_x = a_{11}\sigma_x + a_{12}\sigma_y, \qquad [1.1]$$

$$\varepsilon_y = a_{21}\sigma_x + a_{22}\sigma_y, \qquad [1.2]$$

$$\varepsilon_{xy} = a_{33}\sigma_{xy} \qquad [1.3]$$

et les vitesses de déformations de régions visqueuses sont définies par la loi généralisée de *Newton*:

$$\frac{d\varepsilon_x}{dt} = b_{11}\sigma_x + b_{21}\sigma_y, \qquad [1.4]$$

$$\frac{d\varepsilon_y}{dt} = b_{21}\sigma_x + b_{22}\sigma_y, \qquad [1.5]$$

$$\frac{d\varepsilon_{xy}}{dt} = b_{33}\sigma_{xy}. \qquad [1.6]$$

2. Structure, anisotropie et la distribution des régions rhéologiques

Le modèle rhéologique orthogonal composé de deux phases, représenté fig. 4 peut être considéré comme le plus simple des modèles à deux dimensions. La distribution des deux regions caractérisée par les longueurs α_1, α_2, β_1 et β_2 peut être déterminée d'une manière purement phénoménologique, c'est-à-dire par une appréciation correspondant approximativement aux résultats des essais mécaniques macroscopiques du corps viscoélastique.

Mais on peut sortir aussi de différents points de vue considérant la configuration structurale. Si les particules orientées de la phase visqueuse sont dispersées dans la phase élastique (fig. 1), on

Fig. 1

peut remplacer leurs sections par les rectangles dont les côtés sont α_k et β_k et dont la surface S_k est égale à la surface de la section réele. Désignons par S_i la surface du rectangle inscrit et par S_c la surface du rectangle circonscrit à la section d'une particule, puis par $\chi_i = \frac{\beta_i}{\alpha_i}$ et $\chi_c = \frac{\beta_c}{\alpha_c}$ les relations correspondantes de côtés. La relation des côtés du rectangle équivalent $\chi_k = \frac{\beta_k}{\alpha_k}$ peut être déterminée au moyen de l'interpolation linéaire comme il suit

$$\chi_k = \chi_i + \frac{S_k - S_i}{S_c - S_i}(\chi_c - \chi_i). \qquad [2.1]$$

Il y a différentes manières possibles de déterminer de la distribution résultante équivalente des phases élastiques et visqueuses. La plus simple distribution est indiquée fig. 4 où il y a seulement une section rectangulaire de la phase visqueuse. La relation des côtés du rectangle visqueux $\chi = \frac{\beta_2}{\alpha_2}$ peut être déterminée par la formule suivante

$$\chi = \frac{\sum_{k=1}^{n} S_k \chi_k}{\sum_{k=1}^{n} S_k}, \qquad [2.2]$$

où $\chi_k = \frac{\beta_k}{\alpha_k}$ sont les relations de longueurs des côtés de rectangles remplacant les sections de particules orientées dispersées de la phase visqueuse et $\sum_{k=1} S_k = S$ la somme d'aires des sections de ces particules. Les longueurs caractéristiques du rectangle résultant sont données par

$$\alpha_2 = \sqrt{\frac{S}{\chi}} \qquad [2.3]$$

$$\beta_2 = \sqrt{S\chi} \qquad [2.4]$$

L'auteur a introduit aussi la notion des modèles rhéologiques quadrillés qui sont composés de nombreux champs carrés élastiques et visqueux. Un tel modèle est représenté sur la fig. 2. Ces modèles peuvent exprimer les phénomènes rhéologiques assez complexes de la déformation viscoélastique et les effets de la configuration structurelle. Ils sont très convenables pour les ordinateurs.

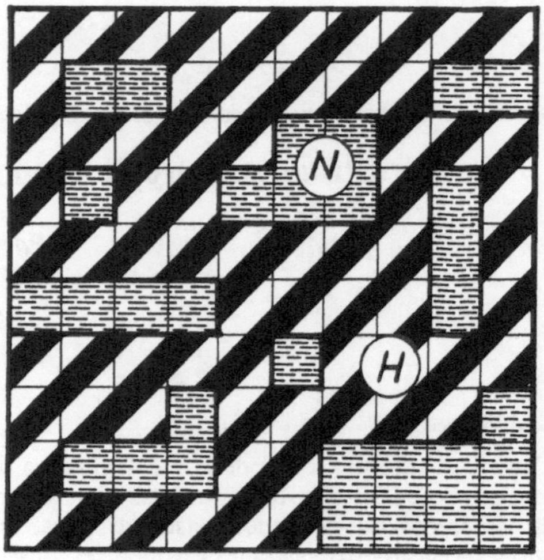

Fig. 2

Les sections planes des corps viscoélastiques avec la non-homogénéité non-uniforme peuvent être représentées par un ensemble de différents modèles bidimensionels qui ont les dimensions finies ou différentielles. Si la configuration de leur structure est orientée dans deux directions perpendiculaires, l'ensemble des modèles rhéologiques partiels est orthotrope (fig. 3). Dans le cas d'une variabilité continue de modèles partiels différentiels, on peut faire l'intégration analytique sur toute la surface représentative.

Fig. 3

3. Modèle orthogonal pour les corps viscoélastiques orthotropes

Les modèles rhéologiques orthogonaux à deux dimensions finies correspondent aux surfaces unités des sections planes des corps rhéologiques. Ils sont divisés par les lignes droites perpendiculaires en régions rhéologiques.

La fig. 4 représente un simple modèle rhéologique orthogonal à deux dimensions pour les corps viscoélastiques orthotropes. Ce modèle consiste en une région élastique *Hookée*nne H et en une région visqueuse *Newton*ienne N. La distribution de la surface unité représentative en régions rhéologiques est indiquée par les longueurs caractéristiques α_1, α_2, β_1 et β_2.

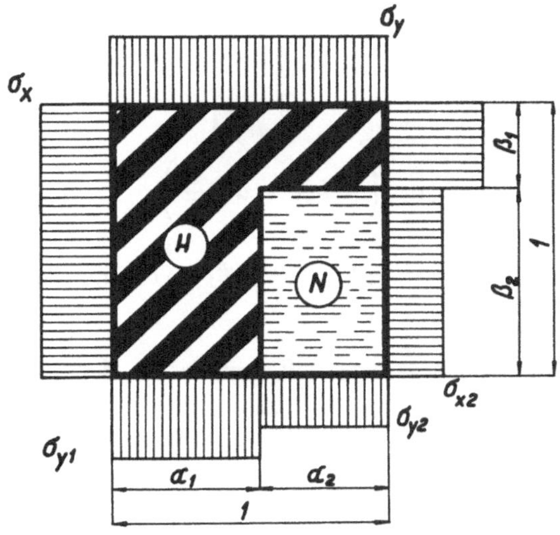

Fig. 4

La condition de la cohésion entre les deux régions exige l'égalité de déformations et de vitesses de déformation des parties voisines de ces régions. La déformation ε_x de la surface unité représentative consiste en composante ε_{x1} de la partie gauche purement élastique et en composante ε_{x2} de la partie droite viscoélastique d'après la relation suivante

$$\varepsilon_x = \alpha_1 \varepsilon_{x1} + \alpha_2 \varepsilon_{x2}. \qquad [3.1]$$

La déformation ε_y dans la direction de y s'exprime d'une manière analogue

$$\varepsilon_y = \beta_1 \varepsilon_{y1} + \beta_2 \varepsilon_{y2}. \qquad [3.2]$$

En tenant compte de contraintes internes causées par la cohésion entre les régions rhéologiques produisant les deformations égales des parties voisines de ces régions, on a

$$\varepsilon_{x1} = a_{11}\sigma_x + a_{12}(\beta_1\sigma_y + \beta_2\sigma_{y1}), \qquad [3.3]$$

$$\varepsilon_{y1} = a_{21}(\alpha_1\sigma_x + \alpha_2\sigma_{x1}) + a_{22}\sigma_y. \qquad [3.4]$$

L'égalité de vitesses de déformations de la région visqueuse et des parties voisines de la région élastique est donnée par

$$\frac{d\varepsilon_{x2}}{dt} = a_{11}\frac{d\sigma_{x1}}{dt} + a_{12}\frac{d\sigma_y}{dt} = b_{11}\sigma_{x2} + b_{12}\sigma_{y2}, \qquad [3.5]$$

$$\frac{d\varepsilon_{y2}}{dt} = a_{21}\frac{d\sigma_x}{dt} + a_{22}\frac{d\sigma_{y1}}{dt} = b_{21}\sigma_{x2} + b_{22}\sigma_{y2}. \qquad [3.6]$$

Comme on peut le voir à partir de la distribution simplifiée des contraintes fig. 4, les contraintes normales résultantes consistent en composantes σ_{x1}, σ_{x2}, σ_{y1} et σ_{y2} d'après les relations

$$\sigma_x = \beta_1\sigma_{x1} + \beta_2\sigma_{x2}, \qquad [3.7]$$

$$\sigma_y = \alpha_1\sigma_{y1} + \alpha_2\sigma_{y2}, \qquad [3.8]$$

d'où

$$\sigma_{x2} = \frac{\sigma_x - \beta_1\sigma_{x1}}{\beta_2}, \quad \sigma_{y2} = \frac{\sigma_y - \alpha_1\sigma_{y1}}{\alpha_2}. \qquad [3.9]$$

Introduisant les relations précédentes dans les éqs. [3.5] et [3.6], nous obtenons le système de deux équations différentielles

$$\left(\frac{\beta_1}{\beta_2}b_{11} + a_{11}\frac{d}{dt}\right)\sigma_{x1} + \frac{\alpha_1}{\alpha_2}b_{12}\sigma_{y1} = \frac{b_{11}}{\beta_2}\sigma_x + \left(\frac{b_{12}}{\alpha_2} - a_{12}\frac{d}{dt}\right)\sigma_y. \qquad [3.10]$$

$$\frac{\beta_1}{\beta_2}b_{21}\sigma_{x1} + \left(\frac{\alpha_1}{\alpha_2}b_{22} + a_{22}\frac{d}{dt}\right)\sigma_{y1} = \left(\frac{b_{21}}{\beta_2} - a_{21}\frac{d}{dt}\right)\sigma_x + \frac{b_{22}}{\alpha_2}\sigma_y. \qquad [3.11]$$

Ce système peut être résolu par rapport aux composantes σ_{x1} et σ_{y1} de la manière suivante

$$\sigma_{x1} = \tilde{H}^{-1}\left\{\left(\frac{\alpha_1}{\alpha_2}b_{22} + a_{22}\frac{d}{dt}\right) \times \left[\frac{b_{11}}{\beta_2}\sigma_x + \left(\frac{b_{12}}{\alpha_2} - a_{12}\frac{d}{dt}\right)\sigma_y\right] - \frac{\alpha_1}{\alpha_2}b_{12}\left[\left(\frac{b_{21}}{\beta_2} - a_{21}\frac{d}{dt}\right)\sigma_x + \frac{b_{22}}{\alpha_2}\sigma_y\right]\right\}, \qquad [3.12]$$

$$\sigma_{y1} = \tilde{H}^{-1}\left\{\left(\frac{\beta_1}{\beta_2}b_{11} + a_{11}\frac{d}{dt}\right) \times \left[\left(\frac{b_{21}}{\beta_2} - a_{21}\frac{d}{dt}\right)\sigma_x + \frac{b_{22}}{\alpha_2}\sigma_y\right] - \frac{\beta_1}{\beta_2}b_{21}\left[\frac{b_{11}}{\beta_2}\sigma_x + \left(\frac{b_{12}}{\alpha_2} - a_{12}\frac{d}{dt}\right)\sigma_y\right]\right\}, \qquad [3.13]$$

où $\tilde{H}^{-1}$ est un opérateur inverse à l'opérateur différentiel du deuxième ordre

$$\tilde{H} = \left(\frac{\beta_1}{\beta_2} b_{11} + a_{11} \frac{d}{dt}\right)\left(\frac{\alpha_1}{\alpha_2} b_{22} + a_{22} \frac{d}{dt}\right) - \frac{\alpha_1 \beta_1}{\alpha_2 \beta_2} b_{12}^2. \qquad [3.14]$$

Portant la relation [3.3] et la relation [3.5] intégrée dans l'éq. [3.1], nous obtenons

$$\varepsilon_x = \alpha_1 [a_{11} \sigma_x + a_{12}(\beta_1 \sigma_y + \beta_2 \sigma_{y1})] + \alpha_2 (a_{11} \sigma_{x1} + a_{12} \sigma_y) \qquad [3.15]$$

et introduisant l'expression [3.4] et la relation [3.6] intégrée dans l'éq. [3.2], nous avons

$$\varepsilon_y = \beta_1 [a_{22} \sigma_y + a_{21}(\alpha_1 \sigma_x + \alpha_2 \sigma_{x1})] + \beta_2 (a_{22} \sigma_{y1} + a_{21} \sigma_x). \qquad [3.16]$$

Introduisant les expressions [3.12] et [3.13] dans les éqs. [3.15] et [3.16], nous obtenons deux équations déformations-contraintes

$$\varepsilon_x = \tilde{H}^{-1}(\tilde{K}_{11} \sigma_x + \tilde{K}_{12} \sigma_y), \qquad [3.17]$$

$$\varepsilon_y = \tilde{H}^{-1}(\tilde{K}_{21} \sigma_x + \tilde{K}_{22} \sigma_y), \qquad [3.18]$$

où

$$\tilde{K}_{11} = a_{11}\left[\alpha_1 \tilde{H} + \frac{\alpha_2}{\beta_2} b_{11}\left(\frac{\alpha_1}{\alpha_2} b_{22} + a_{22} \frac{d}{dt}\right) - \alpha_1 b_{12}\left(\frac{b_{21}}{\beta_2} - a_{21} \frac{d}{dt}\right)\right] + \alpha_1 \beta_1 a_{12}\left(\frac{\beta_1}{\beta_2} b_{11} + a_{11} \frac{d}{dt}\right) \times \left(\frac{b_{21}}{\beta_2} - a_{21} \frac{d}{dt}\right) - \alpha_1 \beta_1 a_{12} b_{11} b_{21}, \qquad [3.19]$$

$$\tilde{K}_{12} = (\alpha_1 \beta_1 + \alpha_2) a_{12} \tilde{H} + a_{11}\left(\alpha_1 b_{22} + \alpha_2 a_{22} \frac{d}{dt}\right)\left(\frac{b_{12}}{\alpha_2} - a_{12} \frac{d}{dt}\right) - \frac{\alpha_1}{\alpha_1} a_{11} b_{12} b_{22} + \frac{\alpha_1}{\alpha_2} a_{12} b_{22}\left(\beta_1 b_{11} + \beta_2 a_{11} \frac{d}{dt}\right) - \alpha_1 \beta_1 a_{12} b_{21}\left(\frac{b_{12}}{\alpha_2} - a_{12} \frac{d}{dt}\right), \qquad [3.20]$$

$$\tilde{K}_{21} = (\alpha_1 \beta_1 + \beta_2) a_{21} \tilde{H} + a_{22}\left(\beta_1 b_{11} + \beta_2 a_{11} \frac{d}{dt}\right)\left(\frac{b_{21}}{\beta_2} - a_{21} \frac{d}{dt}\right) - \frac{\beta_1}{\beta_2} a_{22} b_{11} b_{21} + \frac{\beta_1}{\beta_2} a_{21} b_{11}\left(\alpha_1 b_{22} + \alpha_2 a_{22} \frac{d}{dt}\right) - \alpha_1 \beta_1 a_{21} b_{12}\left(\frac{b_{21}}{\beta_2} - a_{21} \frac{d}{dt}\right), \qquad [3.21]$$

$$\tilde{K}_{22} = a_{22}\left[\beta_1 \tilde{H} + \frac{\beta_2}{\alpha_2} b_{22}\left(\frac{\beta_1}{\beta_2} b_{11} + a_{11} \frac{d}{dt}\right) - \beta_1 b_{21}\left(\frac{b_{12}}{\alpha_2} - a_{12} \frac{d}{dt}\right)\right] + \alpha_2 \beta_1 a_{21}\left(\frac{\alpha_1}{\alpha_2} b_{22} + a_{22} \frac{d}{dt}\right) \times \left(\frac{b_{12}}{\alpha_2} - a_{12} \frac{d}{\mathrm{dt}}\right) - \alpha_1 \beta_1 a_{21} b_{12} b_{22} \qquad [3.22]$$

sont les opérateurs différentiels du deuxième ordre.

Les opérateur conjugués $\tilde{K}_{12}$ et $\tilde{K}_{21}$ ne sont pas égaux en général. Leur égalité exige les conditions spéciales de la configuration de phases exprimée par les longueurs caractéristiques α_1, α_2, β_1 et β_2 comme on le peut le voir à partir des expressions [3.20] et [3.21].

Après avoir multipliè les éqs. [3.17] et [3.18] par l'opérateur différentiel $\tilde{H}$ donné par l'expression [3.14], nous obtenons pour les déformations ε_x et ε_y deux équations différentielles suivantes

$$\frac{d^2 \varepsilon_x}{dt^2} + \left(\frac{\beta_1 b_{11}}{\beta_2 a_{11}} + \frac{\alpha_1 b_{22}}{\alpha_2 a_{22}}\right) \frac{d\varepsilon_x}{dt} + \frac{\alpha_1 \beta_1}{\alpha_2 \beta_2} \cdot \frac{b_{11} b_{22} - b_{12}^2}{a_{11} a_{22}} \varepsilon_x = \frac{1}{a_{11} a_{11}} (\tilde{K}_{11} \sigma_x + \tilde{K}_{12} \sigma_y), \qquad [3.23]$$

$$\frac{d^2 \varepsilon_y}{dt^2} + \left(\frac{\beta_1 b_{11}}{\beta_2 a_{11}} + \frac{\alpha_1 b_{22}}{\alpha_2 b_{11}}\right) \frac{d\varepsilon_y}{dt} + \frac{\alpha_1 \beta_1}{\alpha_2 \beta_2} \cdot \frac{b_{11} b_{22} - b_{12}^2}{a_{11} a_{22}} \varepsilon_y = \frac{1}{a_{11} a_{22}} (\tilde{K}_{21} \sigma_x + \tilde{K}_{22} \sigma_y). \qquad [3.24]$$

Parce que ces équations ont les mêmes coefficients dans leur parties gauches, elles ont aussi la même équation caractéristique

$$L^2 + \left(\frac{\beta_1 b_{11}}{\beta_2 a_{11}} + \frac{\alpha_1 b_{22}}{\alpha_2 a_{22}}\right) L + \frac{\alpha_1 \beta_1}{\alpha_2 \beta_2} \cdot \frac{b_{11} b_{22} - b_{12}^2}{a_{11} a_{22}} = 0, \quad [3.25]$$

d'où

$$L_{1,2} = -\frac{1}{2}\left(\frac{\beta_1 b_{11}}{\beta_2 a_{11}} + \frac{\alpha_1 b_{22}}{\alpha_2 \alpha_{22}}\right) \pm \frac{1}{2}\sqrt{\left(\frac{\beta_1 b_{11}}{\beta_2 a_{11}} - \frac{\alpha_1 b_{22}}{\alpha_2 a_{22}}\right)^2 + \frac{4\alpha_1 \beta_1 b_{12}^2}{\alpha_2 \beta_2 a_{11} a_{22}}}. \quad [3.26]$$

Après la solution des deux équations différentielles [3.23] et [3.24], les relations déformations-contraintes peuvent être écrites sous la forme

$$\varepsilon_x = \frac{e^{L_1 t}}{L_1 - L_2} [\int_{t_0}^{t} (\tilde{K}_{11} \sigma_x + \tilde{K}_{12} \sigma_y) e^{-L_1 \tau} d\tau + (L_2 \varepsilon_{x0} - \mathring{\varepsilon}_{x0}) e^{-L_1 t_0}] + \frac{e^{L_2 t}}{L_2 - L_1} [\int_{t_0}^{t} (\tilde{K}_{11} \sigma_x + \tilde{K}_{12} \sigma_y) e^{-L_2 \tau} d\tau + (L_1 \varepsilon_{x0} - \mathring{\varepsilon}_{x0}) e^{-L_2 t_0}], \quad [3.27]$$

$$\varepsilon_y = \frac{e^{L_1 t}}{L_1 - L_2} [\int_{t_0}^{t} (\tilde{K}_{21} \sigma_x + \tilde{K}_{22} \sigma_y) e^{-L_1 \tau} d\tau + (L_2 \varepsilon_{y0} - \mathring{\varepsilon}_{y0}) e^{-L_1 t_0}] + \frac{e^{L_2 t}}{L_2 - L_1} [\int_{t_0}^{t} (\tilde{K}_{21} \sigma_x + \tilde{K}_{22} \sigma_y) e^{-L_2 \tau} d\tau + (L_1 \varepsilon_{y0} - \mathring{\varepsilon}_{y0}) e^{-L_2 t_0}], \quad [3.28]$$

où ε_{x0} et ε_{y0} sont les déformations initiales et $\mathring{\varepsilon}_{x0}$ et $\mathring{\varepsilon}_{y0}$ les vitesses de déformations initiales.

Les expressions pour les contraintes peuvent être obtenues par l'inversion des éqs. [3.17] et [3.18] sous la forme opérationnelle

$$\sigma_x = (\tilde{K}_{11} \tilde{K}_{22} - \tilde{K}_{12} \tilde{K}_{21})^{-1} \tilde{H} (\tilde{K}_{22} \varepsilon_x - \tilde{K}_{12} \varepsilon_y), \quad [3.29]$$

$$\sigma_y = (\tilde{K}_{11} \tilde{K}_{22} - \tilde{K}_{12} \tilde{K}_{21})^{-1} \tilde{H} (\tilde{K}_{11} \varepsilon_y - \tilde{K}_{21} \varepsilon_x), \quad [3.30]$$

où $(\tilde{K}_{11} \tilde{K}_{22} - \tilde{K}_{12} \tilde{K}_{21})^{-1}$ est l'opérateur inverse à l'opérateur différentielle du quatrième ordre

$$\tilde{N} = \tilde{K}_{11} \tilde{K}_{22} - \tilde{K}_{12} \tilde{K}_{21}. \quad [3.31]$$

C'est pourquoi nous avons pour les contraintes après la multiplication des éqs. [3.29] et [3.30] par l'opérateur $\tilde{N}$ deux équations différentielles du quatrième ordre.

4. Effets de cisaillement dans les corps viscoélastiques orthotropes

Les modèles rhéologiques à deux dimensions peuvent représenter de types différents de la déformation causée par le cisaillement.

Si l'on suppose les extrémités du modèle rigides en flexion, la surface représentative subit les déformations du cisaillement symétrique. Les déformations de cisaillement des deux régions du modèle sur la fig. 4 sont égales. Les contraintes de cisaillement agissant sur la région élastique sont $\sigma_{xy1} = \sigma_{yx1}$ et sur la région visqueuse $\sigma_{xy2} = \sigma_{yx2}$. Les contraintes résultantes sont données par

$$\sigma_{xy} = \sigma_{yx} = (1 - \alpha_2 \beta_2) \sigma_{xy1} + \alpha_2 \beta_2 \sigma_{xy2}. \quad [4.1]$$

Portant dans l'équation précédente les relations [1.3] et [1.6], nous avons

$$\sigma_{xy} = (1 - \alpha_2 \beta_2) a_{33} \varepsilon_{xy} + \alpha_2 \beta_2 b_{33} \frac{d\varepsilon_{xy}}{dt}, \quad [4.2]$$

d'où

$$\varepsilon_{xy} = e^{-L_{xy} t} \left(\int_{t_0}^{t} \frac{\sigma_{xy}}{\alpha_2 \beta_2 b_{33}} e^{L_{xy} \tau} d\tau + \varepsilon_{xy0} e^{L_{xy} t_0} \right), \quad [4.3]$$

où

$$L_{xy} = \frac{(1 - \alpha_2 \beta_2) a_{33}}{\alpha_2 \beta_2 b_{33}} \quad [4.4]$$

est le temps réciproque de retardation en cisaillement et ε_{xy0} la déformation initiale de cisaillement.

Si les extrémités de la surface unitaire représentative sont fléchissables, les déformations de cisaillement

$$\varepsilon_{yx} = \beta_1 \varepsilon_{yx1} + \beta_2 \varepsilon_{yx2}, \quad [4.5]$$

$$\varepsilon_{xy} = \alpha_1 \varepsilon_{xy1} + \alpha_2 \varepsilon_{xy2} \quad [4.6]$$

et les contraintes de cisaillement

$$\sigma_{yx} = \alpha_1 \sigma_{yx1} + \alpha_2 \sigma_{yx2}, \quad [4.7]$$

$$\sigma_{xy} = \beta_1 \sigma_{xy1} + \beta_2 \sigma_{xy2} \quad [4.8]$$

sont différentes dans les directions de x et y comme on le peut voir sur la fig. 5.

Les relations entre les composantes des déformations et des contraintes du cisaillement non-symétrique peuvent être exprimées sous la forme

$$\varepsilon_{yx1} = a_{33} \sigma_{yx}, \quad \varepsilon_{yx2} = a_{33} \sigma_{yx1}, \quad [4.9]$$

$$\frac{d\varepsilon_{yx2}}{dt} = b_{33} \sigma_{yx2}, \quad [4.10]$$

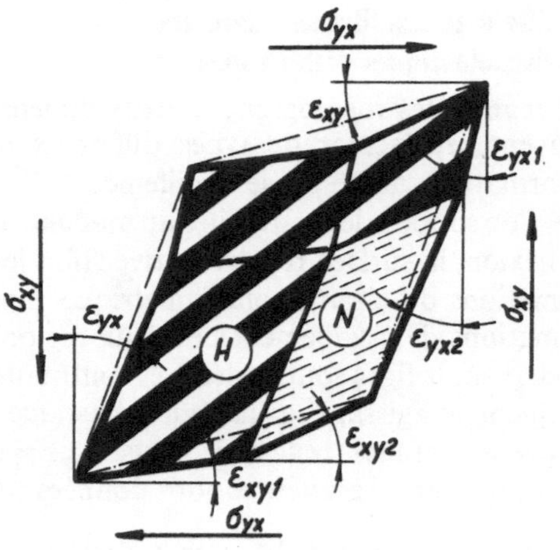

Fig. 5

$$\varepsilon_{xy1} = a_{33}\sigma_{xy}, \quad \varepsilon_{xy2} = a_{33}\sigma_{xy1}, \qquad [4.11]$$

$$\frac{d\varepsilon_{xy2}}{dt} = b_{33}\sigma_{xy2}. \qquad [4.12]$$

En comparant les éqs. [4.9] et [4.10], nous avons

$$a_{33}\frac{d\sigma_{yx1}}{dt} = b_{33}\sigma_{yx2}. \qquad [4.13]$$

Introduisant σ_{yx2} de l'éq. [4.7] dans la relation [4.13] nous obtenons l'équation différentielle du premier ordre

$$a_{33}\frac{d\sigma_{yx1}}{dt} + \frac{\alpha_1 b_{33}}{d_2}\sigma_{yx1} = \frac{b_{33}}{\alpha_2}\sigma_{yx} \qquad [4.14]$$

pour la composante

$$\sigma_{yx1} = e^{-L_{yx}t}\left(\int_{t_0}^{t}\frac{b_{33}}{\alpha_2 a_{33}}\sigma_{yx}e^{L_{yx}\tau}d\tau + \sigma_{yx10}e^{L_{yx}t_0}\right), \qquad [4.15]$$

où

$$L_{yx} = \frac{\alpha_1 b_{33}}{\alpha_2 a_{33}} \qquad [4.16]$$

est le temps réciproque de rétardation et σ_{yx10} est la composante initiale de la contrainte de cisaillement dans la direction de x.

Introduisant les expressions [4.9] et [4.15] dans l'éq. [4.5], nous obtenons la relation déformation-contrainte pour le cisaillement horizontal

$$\varepsilon_{yx} = a_{33}\left[\beta_1\sigma_{yx} + \beta_2 e^{-L_{yx}t}\left(\int_{t_0}^{t}\frac{b_{33}}{\alpha_2 a_{33}}\sigma_{yx}e^{L_{yx}\tau}d\tau + \sigma_{yx10}e^{L_{yx}t_0}\right)\right]. \qquad [4.17]$$

Les relation pour le cisaillement vertical peuvent être dérivées d'une manière tout à fait analogue.

La relation entre les composantes σ_{xy1} et σ_{xy2} s'ensuit des éqs. [4.11] et [4.12]:

$$a_{33}\frac{d\sigma_{xy1}}{dt} = b_{33}\sigma_{xy2}. \qquad [4.18]$$

Introduisant σ_{xy2} de l'éq. [4.8], nous obtenons

$$a_{33}\frac{d\sigma_{xy1}}{dt} + \frac{\beta_1 b_{33}}{\beta_2}\sigma_{xy1} = \frac{b_{33}}{\beta_2}\sigma_{xy}, \qquad [4.19]$$

d'où

$$\sigma_{xy1} = e^{-L_{xy}t}\left(\int_{t_0}^{t}\frac{b_{33}}{\beta_2 a_{33}}\sigma_{xy}e^{L_{xy}\tau}d\tau + \sigma_{xy10}e^{L_{xy}t_0}\right), \qquad [4.20]$$

où

$$L_{xy} = \frac{\beta_1 b_{33}}{\beta_2 a_{33}}. \qquad [4.21]$$

Portant les éqs. [4.11] et [4.20] dans l'expression [4.6], nous avons la relation déformation-contrainte pour le cisaillement vertical

$$\varepsilon_{xy} = a_{33}\left[\alpha_1\sigma_{xy} + \alpha_2 \mathrm{e}^{-L_{xy}t}\left(\int_{t_0}^{t}\frac{\mathrm{b}_{33}}{\beta_2 a_{33}}\sigma_{xy}e^{L_{xy}\tau}d\tau + \sigma_{xy10}e^{L_{xy}t}\right)\right]. \qquad [4.22]$$

5. Modéle à deux dimensions pour les corps viscoélastiques anisotropes

Un simple modèle rhéologique à deux dimensions pour les corps viscoélastiques anisotropes est représenté sur la fig. 6. Il consiste en une région élastique et en une region visqueuse qui sont séparées par deux lignes droites inclinées. La distribution des phases rhéologiques est définie par les longueurs caractéristiques α_1, α_2, β_1, β_2 et par les angles φ_x et φ_y.

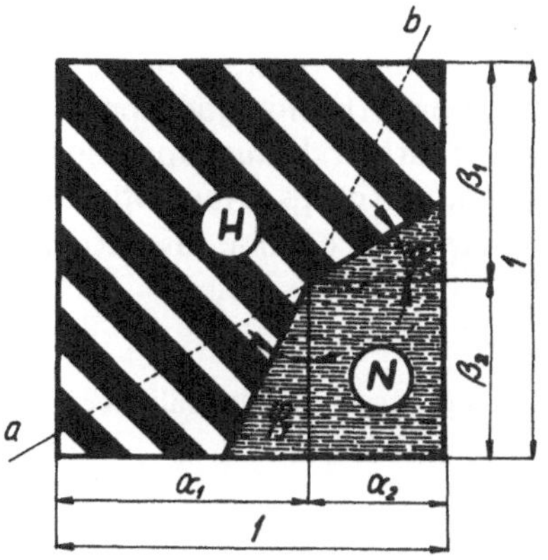

Fig. 6

Les déformations des sections inclinées *a* et *b* sont

$$\varepsilon_a = \alpha_1 \varepsilon_{a1} + \alpha_2 \varepsilon_{a2}, \qquad [5.1]$$

$$\varepsilon_b = \beta_1 \varepsilon_{b1} + \beta_2 \varepsilon_{b2}, \qquad [5.2]$$

ε_{a1} et ε_{b1} étant les déformations des parties élastiques et ε_{a2} et ε_{b2} celles des bornes entre les régions élastiques et visqueuses. Les relations entre les déformations le long des droites *a* et *b* et dans les directions des coordonnées s'expriment de la manière suivante

$$\varepsilon_a = \varepsilon_x \cos^2\varphi_x + \varepsilon_y \sin^2\varphi_x + (\varepsilon_{yx} + \varepsilon_{xy}) \sin\varphi_x \cos\varphi_x, \qquad [5.3]$$

$$\varepsilon_b = \varepsilon_x \sin^2\varphi_y + \varepsilon_y \cos^2\varphi_y + (\varepsilon_{yx} + \varepsilon_{xy}) \sin\varphi_y \cos\varphi_y. \qquad [5.4]$$

La cohésion exige l'égalité de ces déformations dans les parties voisines des régions élastiques et visqueuses.

Pour les déformations de cisaillement, on peut écrire

$$\varepsilon_{yx} = \beta_1 \varepsilon_{yx1} + \beta_2 \varepsilon_{yx2}, \qquad [5.5]$$

$$\varepsilon_{xy} = \alpha_1 \varepsilon_{xy1} + \alpha_2 \varepsilon_{xy2}. \qquad [5.6]$$

ou l'indexe 1 correspond a la déformation de la partie purement élastique et l'index 2 correspond à la partie viscoélastique.

Les conditions d'équilibre sur les droites *a* et *b* s'expriment par les quatre équations suivantes:

$$\sigma_x \sin^2\varphi_x + \sigma_y \cos^2\varphi_x + (\sigma_{yx} + \sigma_{xy}) \sin\varphi_x \cos\varphi_x = \alpha_1 [\sigma_{x1} \sin^2\varphi_x + \sigma_{y1} \cos^2\varphi_x + (\sigma_{yx1} + \sigma_{xy1}) \sin\varphi_x \cos\varphi_x] + \alpha_2 [\sigma_{x2} \sin^2\varphi_x + \sigma_{y2} \cos^2\varphi_x + (\sigma_{yx2} + \sigma_{xy2}) \sin\varphi_x \cos\varphi_x], \qquad [5.7]$$

$$(\sigma_x - \sigma_y) \sin\varphi_x \cos\varphi_x + \sigma_{yx} \cos^2\varphi_x - \sigma_{xy} \sin^2\varphi_x = \alpha_1 [(\sigma_{x1} - \sigma_{y1}) \sin\varphi_x \cos\varphi_x + \sigma_{yx1} \cos^2\varphi_x - \sigma_{xy1} \sin^2\varphi_x] + \alpha_2 [(\sigma_{x2} - \sigma_{y2}) \sin\varphi_x \cos\varphi_x + \sigma_{yx2} \cos^2\varphi_x - \sigma_{xy2} \sin^2\varphi_x], \qquad [5.8]$$

$$\sigma_x \cos^2\varphi_y + \sigma_y \sin^2\varphi_y + (\sigma_{yx} + \sigma_{xy}) \sin\varphi_y \cos\varphi_y = \beta_1 [\sigma_{x1} \cos^2\varphi_y + \sigma_{y1} \sin^2\varphi_y + (\sigma_{yx1} + \sigma_{xy1}) \sin\varphi_y \cos\varphi_y] + \beta_2 [\sigma_{x2} \cos^2\varphi_y + \sigma_{y2} \sin^2\varphi_y + (\sigma_{yx2} + \sigma_{xy2}) \sin\varphi_y \cos\varphi_y], \qquad [5.9]$$

$$(\sigma_x - \sigma_y) \sin\varphi_y \cos\varphi_y + \sigma_{yx} \sin^2\varphi_y - \sigma_{xy} \cos^2\varphi_y = \beta_1 [(\sigma_{x1} - \sigma_{y1}) \sin\varphi_y \cos\varphi_y + \sigma_{yx1} \sin^2\varphi_y - \sigma_{xy1} \cos^2\varphi_y] + \beta_2 [(\sigma_{x2} - \sigma_{y2}) \sin\varphi_y \cos\varphi_y + \sigma_{yx2} \sin^2\varphi_y - \sigma_{xy2} \cos^2\varphi_y]. \qquad [5.10]$$

Ce système donne les composantes σ_{x2}, σ_{y2}, σ_{yx2}, σ_{xy2} exprimées par les contraintes σ_x, σ_y, σ_{yx}, σ_{xy} et σ_{x1}, σ_{y1}, σ_{yx1}, σ_{xy1}.

Portant les relations déformations-contraintes pour la phase élastique

$$\varepsilon_x = a_{11}\sigma_x + a_{12}\sigma_y + a_{13}\sigma_{yx} + a_{14}\sigma_{xy},$$

$$\varepsilon_y = a_{21}\sigma_x + a_{22}\sigma_y + a_{23}\sigma_{yx} + a_{24}\sigma_{xy}, \qquad [5.11]$$

et pour la phase visqueuse

$$\frac{d\varepsilon_x}{dt} = b_{11}\sigma_x + b_{12}\sigma_y + b_{13}\sigma_{yx} + b_{14}\sigma_{xy},$$

$$\frac{d\varepsilon_y}{dt} = b_{21}\sigma_x + b_{22}\sigma_y + b_{23}\sigma_{yx} + b_{24}\sigma_{xy}, \qquad [5.12]$$

dans les éqs. [5.3], [5.4], [5.5] et [5.6] et tenant compte de l'égalité des déformations normales des parties voisines des régions élastique et visqueuse, nous obtenons le système des quatre équations différentielless pour les composantes σ_{x1}, σ_{y1},

D'une manière analogue au cas du modèle orthogonale, nous obtenons enfin les relations déformations-contraintes sous la forme

$$\varepsilon_x = \tilde{H}_a^{-1}(\tilde{K}_{11}\sigma_x + \tilde{K}_{12}\sigma_y + \tilde{K}_{13}\sigma_{yx} + \tilde{K}_{14}\sigma_{xy}), \qquad [5.13]$$

$$\varepsilon_y = \tilde{H}_a^{-1}(\tilde{K}_{21}\sigma_x + \tilde{K}_{22}\sigma_y + \tilde{K}_{23}\sigma_{yx} + \tilde{K}_{24}\sigma_{xy}), \qquad [5.14]$$

$$\varepsilon_{yx} = \tilde{H}_a^{-1}(\tilde{K}_{31}\sigma_x + \tilde{K}_{32}\sigma_y + \tilde{K}_{33}\sigma_{yx} + \tilde{K}_{34}\sigma_{xy}), \qquad [5.15]$$

$$\varepsilon_{xy} = \tilde{H}_a^{-1}(\tilde{K}_{41}\sigma_x + \tilde{K}_{42}\sigma_y + \tilde{K}_{43}\sigma_{yx} + \tilde{K}_{44}\sigma_{xy}), \qquad [5.16]$$

où $\tilde{H}_a$ et $\tilde{K}_{ij}$ sont les opérateurs différentiels du quatrième ordre.

Après avoir multiplié les équations précédentes par l'opérateur $\tilde{H}_a$, nous obtenons les équations différentielles du quatrième ordre dont la solution s'exprime sous la forme

$$\begin{aligned}
\varepsilon_x = \int_{t_0}^{t} (\tilde{K}_{11}\sigma_x + \tilde{K}_{12}\sigma_y + \tilde{K}_{13}\sigma_{yx} + \tilde{K}_{14}\sigma_{xy}) \\
\times \Bigg[\frac{e^{L_1(t-\tau)}}{(L_1 - L_2)(L_1 - L_3)(L_1 - L_4)} \\
+ \frac{e^{L_2(t-\tau)}}{(L_2 - L_1)(L_2 - L_3)(L_2 - L_4)} \\
+ \frac{e^{L_3(t-\tau)}}{(L_3 - L_1)(L_3 - L_2)(L_3 - L_4)} \\
+ \frac{e^{L_4(t-\tau)}}{(L_4 - L_1)(L_4 - L_2)(L_4 - L_3)} \Bigg] d\tau \\
+ C_{11} e^{L_1 t} + C_{12} e^{L_2 t} + C_{13} e^{L_3 t} \\
+ C_{14} e^{L_4 t}, \text{ etc.,} \qquad [5.17]
\end{aligned}$$

où L_k sont les racines de l'équation caractéristique et C_{ij} les constantes d'intégration.

Les relations rhéologiques pour les corps anisotropes sont alors beaucoup plus compliquées que celles pour les corps orthotropes qui peuvent être obtenues aussi comme cas particulier.

Littérature

1) *Mandel, J.*, Mécanique des milieux continus (Paris 1966).

2) *Reiner, M.*, Deformation, Strain and Flow (London 1960).

3) *Sobotka, Z.*, Contributions to Mechanics, pp. 391 à 435 (Oxford, London, Edinburgh, New York, Toronto, Sydney, Paris, Braunschweig 1969).

4) *Sobotka, Z.*, Proc. 5th Congress Rheology, Kyoto 1968, Vol. 1, pp. 175–200 (Tokyo 1969).

Adresse de l'auteur:

Institut de la Mécanique Théorique
et Appliquée de l'Académie Tchécoslovaque des Sciences
Prague (ČSSR)

Rheol. Acta **12**, 141–149 (1973)

From the Bulgarian Academy of Sciences, Sofia (Bulgaria)

Electromechanical interaction in viscoelastic micropolar medium

By G. Brankov and N. Petrov

(Received October 27, 1972)

The endeavour at a more precise description of the mechanical properties of the real deformable bodies imposed the invention of new rheological models in which except the mechanical factor other physical factors are also taken into consideration.

The first works in which the temperature deformations were taken into account are related to *Duhamel* (1). Later on *Voigt* (2) and *Jeffreys* constitute a generalized equation of the heat transmission in which the strain tensor is taken into consideration.

The influence of the electric fields on the mechanical properties of crystals is examined by *Voigt* (4), *Cady* (5), *Mason* (6) and others.

The investigations show that a great part of the organic substances exhibit electromechanical interactions (7–14). In many of them and especially in the low-molecular ones exist inner degrees of rotation. The so called "liquid crystals" also belong to that group. In the recent years they are subject of a very intensive investigation (15–20).

The present investigations are a continuation of the work (21) in which the authors examine the non-polar case at some limitations of the constitutive operators.

I. Mathematical apparatus (22, 23)

We introduce the space H:

$$\langle a| \in H, \quad |a\rangle \in H$$

$|a\rangle$ ket *Dirac* vector,
$\langle a|$ brack *Dirac* vector.

1^0 $|\lambda a\rangle + |\mu b\rangle = \mu|b\rangle + \lambda|a\rangle \in H$

$$\langle \lambda a| + \langle \mu b| = \bar{\mu}\langle b| + \bar{\lambda}\langle a| \in H$$

$\mu, \lambda \in C$ complex numbers field
$\bar{\mu}, \bar{\lambda}$ are complex conjugate of μ and λ.

2° $\exists\,|0\rangle, \quad \langle 0| \in H = \rangle 0|a\rangle = |0\rangle,$

$$0\langle a| = \langle 0|$$

$$|a\rangle + |0\rangle = |a\rangle, \quad \langle a| + \langle 0| = \langle a|$$

3^0 $(|a\rangle + |b\rangle) + |c\rangle = |a\rangle + (|b\rangle + |c\rangle)$

$$(\langle a| + \langle b|) + \langle c| = \langle a| + (\langle b| + \langle c|)$$

4^0 In $H = U_\alpha H_\alpha$ the bilinear form is introduced:

$$\mathscr{S}(\langle a| \in H_\alpha;\ |b\rangle \in H_\beta) = \langle a|b\rangle \in C$$

satisfying the condition

$$\langle a|b\rangle = 0 \quad \text{at} \quad \alpha \neq \beta \Rightarrow H = \otimes_\alpha H_\alpha$$

5^0 In $H_\alpha = H^0 \otimes L_\alpha$

$$= H^0 \otimes L_\alpha^1 \otimes L_\alpha^2 \ldots \otimes L_\alpha^n$$

there is a basis

$$|t\, i_1 i_2 \ldots i_{n_\alpha} \alpha\rangle = |t\rangle\, |i_1 \alpha\rangle\, |i_2 \alpha\rangle \ldots |i_n \alpha\rangle$$

$$|t\rangle \in H^0, \quad |i_k \alpha\rangle \in L_\alpha^k$$

under the condition of normalization

$$\langle \alpha\, i_{n_\alpha} \ldots i_1 t|t' j_1 \ldots j_{n_\alpha} \alpha\rangle$$
$$= \langle t|t'\rangle\, \langle \alpha\, i_1|j_1 \alpha\rangle \ldots \langle \alpha\, i_{n_\alpha}|j_{n_\alpha} \alpha\rangle$$
$$= \delta(t - t')\, \delta_{i_1 j_1}\, \delta_{i_2 j_2} \ldots \delta_{i_{n_\alpha} j_{n_\alpha}}$$

and under the condition of fullness

$$\exists\, \psi_{i_1 i_2 \ldots i_{n_\alpha}}(t) \in 0 \Rightarrow |\psi\rangle$$
$$= \psi_{i_1 i_2 \ldots i_{n_\alpha}}(t)\, |t\, i_1\, i_2 \ldots i_{n_\alpha} \alpha\rangle, \quad \forall\, |\psi\rangle \in H_\alpha$$

0 is the plurality of the generalized functions.

We shall use *Einsteins* rule for summarizing of repeatable indexes which we shall generalize also for the case when the indexes are continuous. The line under the indexes shows that the underlined indexes are not summarized.

More often we shall use the following projection operators:

$$\mathscr{P}_\alpha = |\underline{\alpha}\rangle\langle\underline{\alpha}| = |i_1 i_2 \ldots i_{n_\alpha}\underline{\alpha}\rangle \times \langle\underline{\alpha} i_{n_\alpha} \ldots i_2 i_1|: \quad H \to H_\alpha$$

$$\mathscr{P}_t = |\underline{t}\rangle\langle\underline{t}|: \quad H \to L = \oplus_\alpha L_\alpha$$

$$\mathscr{P}_{\alpha t} = |\underline{\alpha}\underline{t}\rangle\langle\underline{t}\underline{\alpha}|: \quad H \to L_\alpha .$$

According to the condition of fullness on the basis

$$\sum_\alpha \mathscr{P}_\alpha = I, \quad \sum_t \mathscr{P}_t = I, \quad \sum_{\alpha t} \mathscr{P}_{\alpha t} = I \qquad [1]$$

where I is the identified operator.

In H^0 is defined by the differentiation operator D:

$$\langle t| D |t'\rangle = -\frac{\partial}{\partial t'}\delta(t-t'), \quad -\infty < t < \infty . \qquad [2]$$

Let us examine the problem of the proper numbers of the operator D:

$$D|\omega\rangle = i\underline{\omega}|\underline{\omega}\rangle, \qquad \omega \in C.$$

We pass into $|t\rangle$ representation by $[1_2]$:

$$D|t'\rangle\langle t'|\omega\rangle = i\underline{\omega}|\underline{\omega}\rangle$$

$$\langle t| D |t'\rangle\langle t'|\omega\rangle = i\underline{\omega}\langle t|\omega\rangle$$

$$2 \Rightarrow \frac{\partial}{\partial t}\langle t|\omega\rangle = i\underline{\omega}\langle t|\underline{\omega}\rangle$$

$$\Rightarrow \langle t|\omega\rangle = \lambda e^{i\omega t}, \qquad \lambda \in C. \qquad [3]$$

We multiply the last equation on the left side with $|t\rangle$

$$|t\rangle\langle t|\omega\rangle = \lambda e^{i\omega t}|t\rangle \Rightarrow |\omega\rangle = \lambda e^{i\omega t}|t\rangle$$

$$\Rightarrow |\omega\rangle\langle\omega| = \lambda\bar{\lambda} e^{i|\omega t - \bar{\omega} t'|}|t\rangle\langle t'| .$$

In the case when $\lambda = \frac{1}{\sqrt{2\Pi}}$ and $\omega \in R$ – the plurality of the real numbers we obtain

$$\sum_\omega \mathscr{P}_\omega = I, \quad \text{where} \quad \mathscr{P}_\omega = |\underline{\omega}\rangle\langle\underline{\omega}| .$$

II. Basic conceptions, principles and assumptions

The system which will be subject of the present investigation represents the material volume ΔVbelonging to the K body. It is evident that the behaviour of $\Delta V$is determined by the interaction of the remaining part of the body, the internal volume interactions in ΔV and of the external fields, representing the interaction of the body K with the remaining part of the universe. In the case when the examined system $\Delta V$is with very small volume the outer effect may be expressed by local variables representing selfcongruent fields defined by the body and the external fields. In the concrete case subject of our investigations these variables are:

stress tensor σ_{ij}
couple-stress tensor m_{ij}
local electric field $\mathscr{E}_i$
and the temperature T

(the temperature may be examined also as a generalized force because it is a gradient of the internal energy of the entropy).

The reaction of the system is represented by the corresponding to the thermodynamical forces generalized coordinates:

strain tensor ε_{ij}
micro-rotation tensor x_{ij}
electric polarization P_i
entropy density S.

In the case of small deformations where the linear theory is valid we assume analogically to (24) that the strain tensor and the micro-rotation tensor are related with the displacement vector and the angle of micro-rotation by the equations

$$\varepsilon_{\varkappa l} = U_{\varkappa, l} - \varepsilon_{\varkappa lm}\mathscr{S}_m$$

$$x_{\varkappa l} = \mathscr{S}_{\varkappa, l}$$

where
$U_\varkappa$ displacement vector
$\mathscr{S}_\varkappa$ angle of micro-rotation
$\varepsilon_{\varkappa lm}$ *Levi-Chevitta* symbol.

In the case of finite deformations we suppose that ε_{ij} and x_{ij} are combinations of the displacement and micro-rotation gradients, satisfying the invariance principle with respect to the group of the orthogonal rotation.

Under time state of the system we shall understand temporal section of the plurality of the variables forming a basis with respect to all variables (25, 26). Under the evolution of the system we shall understand time-order plurality of the time states.

In the case when we examine the concept "system state" it will be identified with the concept-thermodynamical state. If the studied system has no internal degrees of freedom then the kinematic values describing the system reaction appeared as basical and coincide with the variables describing the state of the system. But as it is known the viscoelastic bodies have internal degrees of freedom. In the case of finite number internal degrees of freedom the system may be considered as composed of finite number separate subsystems each one of them having no internal degrees of freedom. According to *Nagels* definition (25, 26) the state of the system in this case is determined by the states of the separate subsystems. That idea, for the case of viscoelastic bodies is treated in the concluding remarks of the paper (27).

We shall use the following basic principles and assumptions:

Principles

I Principles of causality.

II First Principle of Thermodynamics.

III Second Principle of Thermodynamics.

IV In time representation all physically observable variables are real.

V The physical variables which independently one of the other describe the state of the input and of the output belong to the orthogonal subspaces of H.

Assumptions for the system

1′. The system is linear.

2′. The system is smooth.

3′. The system is invariant with respect to time translation.

4′. The system is composed by finite number subsystems each one of them having no internal degrees of freedom.

We suppose that each subsystem is a pattern of micro-ranges with equivalent physical properties space-spread in ΔV.

The input of the system is represented by the form

$$|\zeta\rangle = \sigma_{ij}(t)\,|tij\varepsilon\rangle + m_{ij}(t)\,|tijx\rangle + \mathscr{E}_i(t)\,|tip\rangle + \Delta T(t)\,|ts\rangle.$$

The output of the system is represented by the form

$$|\xi\rangle = \varepsilon_{ij}(t)\,|tij\varepsilon\rangle + x_{ij}(t)\,|tijx\rangle + P_i(t)\,|tip\rangle + S(t)\,|ts\rangle$$

where

$|tij\varepsilon\rangle \in H_\varepsilon \equiv H^0 \otimes L_\varepsilon$

$|tijx\rangle \in H_x \equiv H^0 \otimes L_x$

$|tip\rangle \in H_p \equiv H^0 \otimes L_p$

$|ts\rangle \in H_s \equiv H^0 \otimes L_s$

$|\zeta\rangle, \quad |\xi\rangle \in H \equiv H_\varepsilon \oplus H_x \oplus H_p \oplus H_s$

$\Delta T = T - T_0$ (T_0 is the equilibrium temperature)

The output of the kth subsystem is represented by the

$$|\xi^k\rangle = \varepsilon^k_{ij}(t)\,|t_{ij}\varepsilon\rangle + x^k_{ij}(t)\,|tijx\rangle + P^k_i(t)\,|tip\rangle + S^k(t)\,|ts\rangle$$

where $\varepsilon^k_{ij}(t)$; $x^k_{ij}(t)$; P^k_i; S^k are the thermodynamical coordinates representing its state.

The state of the system is given in the form

$$|X\rangle = (|\xi^1\rangle, |\xi^2\rangle, \ldots, |\xi^N\rangle) \in X \equiv \oplus_k X^k; \qquad (0, 0, \ldots, |\xi^k\rangle, 0, \ldots, 0) \in X^k$$

where

$|X\rangle$ is the system state

N the number of the subsystems

X states space.

III. Constitutive equations

According to the general system theory each linear, smooth, finite, continuous in time and invariant to time translation dynamic system may be described by the eq. (28):

$$\frac{d}{dt}|X(t)\rangle = F|X(t)\rangle + G|\zeta(t)\rangle$$
$$|\xi(t)\rangle = \Gamma|X(t)\rangle \qquad [4]$$

where F, G, Γ are independent of the time operators.

In operator form [4] is represented by

$$D|X\rangle = F|X\rangle + G|\zeta\rangle$$
$$|\xi\rangle = \Gamma|X\rangle. \qquad [5]$$

Analogically for the kth subsystem we have:

$$D|\xi^k\rangle = F^k|\xi^k\rangle + G^k|\zeta\rangle \qquad [6]$$

where it is taken into account that:

a) k^{th} subsystem has no internal degrees of freedom

$$|X^k\rangle = (0, 0, \ldots, |\xi^k\rangle, 0, \ldots, 0).$$

b) Because of the small size of the range and the continuity of the examined local variables σ_{ij}; m_{ij}; $\mathscr{E}_i$; ΔT, all subsystems are characterized with one and the same input ($|\zeta^k\rangle = |\zeta\rangle$).

Consequently the complex system may be written in the form

$$D(|\xi^1\rangle, |\xi^2\rangle, \ldots, |\xi^N\rangle) = \begin{pmatrix} F^1 & & & 0 \\ & F^2 & & \\ & & \ddots & \\ 0 & & & F^N \end{pmatrix} \times \begin{pmatrix} |\xi^1\rangle \\ |\xi^2\rangle \\ \vdots \\ |\xi^N\rangle \end{pmatrix} + (G^1, G^2, \ldots, G^N)\,|\zeta\rangle$$

$$|\xi\rangle = \Gamma \begin{pmatrix} |\xi^1\rangle \\ |\xi^2\rangle \\ \vdots \\ |\xi^N\rangle \end{pmatrix}. \qquad [7]$$

We shall prove that from the requirement the work of the system to be equal to the sum of the work done by the separate subsystems and from the entropy additivity it follows that:

$$\Gamma = (l_1, l_2, \ldots, l_N).$$

l_k is the relation of the volume ΔV^k of the kth subsystem and the volume of the total system ΔV. At small deformations l_k does not depend on the body state.

Let us project [7_2] in L_s;

$$S(t) = \langle st|\,\Gamma\,|X\rangle = \langle st|\,\Gamma^k\,|\xi^k\rangle$$
$$= \langle s|\,\Gamma^k\,|ij\varepsilon\rangle\,\varepsilon_{ij}^k + \langle s|\,\Gamma^k\,|ijx\rangle\,x_{ij}^k + \langle s|\,\Gamma^k\,|ip\rangle\,P_i^k + \langle s|\,\Gamma^k\,|s\rangle\,S^k. \qquad [8]$$

On the other hand according to the principles of thermodynamics ΔS is extensive function and is presented by the sum

$$\Delta S = \sum_{k=1}^{N} \Delta S^k \Rightarrow S = \sum_{k=1}^{N} l_k S^k \qquad [9]$$

ΔS total entropy of ΔV
S^k entropy density of kth subsystem.

Comparing [9] and [8] we obtain

$$\langle s|\,\Gamma^k\,|s\rangle = l_k, \quad \langle s|\,\Gamma^k\,|\alpha\rangle = 0 \quad \text{at} \quad s \neq \alpha.$$

The work done by the system for unit time in unit volume is

$$\frac{dA}{dt} = \sigma_{ij}\frac{d\varepsilon_{ij}}{dt} + m_{ij}\frac{dx_{ij}}{dt} + \mathscr{E}_i\frac{dP_i}{dt}$$
$$= \left\langle \zeta(t) \middle| \frac{d\xi(t)}{dt} \right\rangle - T\frac{dS}{dt}. \qquad [10]$$

Substituting $|\xi(t)\rangle$ with its equal from [7_2] we obtain:

$$\frac{dA}{dt} = \left\langle \zeta(t)|\,\Gamma^k \middle| \frac{d\xi^k(t)}{dt} \right\rangle - T\frac{dS}{dt}$$
$$= \left\langle \zeta^k(t)|\,\Gamma^k \middle| \frac{d\xi^k(t)}{dt} \right\rangle - T\frac{dS}{dt}.$$

On the other hand the work done by the system in unit volume and expressed by the work of the separate subsystems is

$$\frac{dA}{dt} = \left\langle \zeta^k(t) \middle| \frac{d}{dt}\,\xi^k(t) \right\rangle l_k - \sum_k T\frac{dS^k}{dt}\,l_k.$$

The comparison of the last two equations shows

$$|\xi(t)\rangle = \sum_k l_k|\xi^k(t)\rangle \Rightarrow \Gamma = (l_1, l_2, \ldots, l_N). \qquad [11]$$

After using that result in [7] we obtain the system of equations

$$D(|\xi^1\rangle, |\xi^2\rangle, \ldots, |\xi^N\rangle) = \begin{pmatrix} F^1 & & & \\ & F^2 & & 0 \\ & & \ddots & \\ 0 & & & F^N \end{pmatrix} \begin{pmatrix} |\xi^1\rangle \\ |\xi^2\rangle \\ \vdots \\ |\xi^N\rangle \end{pmatrix} + (G^1, G^2, \ldots, G^N)\,|\zeta\rangle$$

$$|\xi\rangle = (l_1, l_2, \ldots, l_N) \begin{pmatrix} |\xi^1\rangle \\ |\xi^2\rangle \\ \vdots \\ |\xi^N\rangle \end{pmatrix}. \qquad [12]$$

Differential form of the constitutive equations

In $|\omega\rangle$ representation we obtain from [6]

$$|\xi^k(\omega)\rangle = R^k(\omega)\,G^k|\zeta(\omega)\rangle \qquad [13]$$

where we assume that the resolvent R^k of the operator F^k is defined.

From [12_2] and [13] we obtain

$$|\xi(\omega)\rangle = l_k R^k G^k|\zeta(\omega)\rangle. \qquad [14]$$

Let us multiply the left and the right sides of [14] with the product

$$\prod_{k=1}^{N} (i\omega\eta^k + E^k)$$

where we have

$$\eta^k = (l_k G^k)^{-1}, \quad E^k = -(l_k G^k)^{-1} F^k$$

(it is supposed that G^k has inversion operator). In the case when the commutation relations are fulfilled

$$[(i\omega\eta^k + E^k), \quad (i\omega\eta^l + E^l)] = 0$$
$$\varkappa, l = 1, 2, \ldots, N$$

we obtain

$$\prod_{k=1}^{N} (i\omega\eta^k + E^k) |\xi(\omega)\rangle = \sum_{k=1}^{N} \prod_{l \neq k}^{N} (i\omega\eta^l + E^l) \times |\zeta(\omega)\rangle.$$

Developing the products we obtain:

$$\sum_{k=0}^{N} A^k (i\omega)^k |\xi(\omega)\rangle = \sum_{k=0}^{N-1} B^k (i\omega)^k |\zeta(\omega)\rangle \qquad [15]$$

where the operators A^k and B^k are products and sums of the operators η^k and E^k.

$$[15] \Rightarrow \sum_{k=0}^{N} A^k (i\omega)^k |\xi(\omega)\rangle |\omega\rangle = \sum_{k=0}^{N} B^k (i\omega)^k |\zeta(\omega)^k |\zeta(\omega)\rangle |\omega\rangle.$$

Using that

$$(i\omega)^k |\xi(\omega)\rangle |\omega\rangle = (i\omega)^k |\omega\rangle |\xi(\omega|\rangle = D^k |\omega\rangle |\xi(\omega)\rangle = D^k |\xi\rangle$$

we obtain the operator equation:

$$\sum_{k=0}^{N} A^k D^k |\xi\rangle = \sum_{k=0}^{N-1} B^k D^k |\zeta\rangle.$$

Multiplying both sides of that equation on the left side with $\langle t|$ and using $[1_2]$ we obtain:

$$\sum_{k=0}^{N} A^k \langle t| D^k |t'\rangle \langle t'|\xi\rangle = \sum_{k=0}^{N-1} B^k \langle t| D^k |t'\rangle \langle t'|\zeta\rangle. \qquad [16]$$

Then using in [16] that

$$\langle t| D^k |t'\rangle = \langle t| D |t^1\rangle \langle t^1| D |t^2\rangle \times \langle t^2| D |t^3\rangle \ldots \langle t^{k-1}| D |t'\rangle$$
$$= (-1)^k \frac{\partial}{\partial t^1} \delta(t - t^1) \frac{\partial}{\partial t^2} \delta(t^1 - t^2) \ldots \ldots \frac{\partial}{\partial t'} \delta(t^{k-1} - t')$$
$$= (-1)^k \frac{\partial^k}{\partial t'^k} \delta(t - t')$$

we obtain

$$\sum_{k=0}^{N} A^k \frac{d^k}{dt^k} |\xi(t)\rangle = \sum_{k=0}^{N-1} B^k \frac{d^k}{dt^k} |\zeta(t)\rangle.$$

This equation is projected in L_ε, L_x, L_p. (We examine the case $H = H_\varepsilon \oplus H_x \oplus H_p$.)

As a result of that after using $[1_1]$ we obtain:

$$\sum_{k=0}^{N} \frac{d^k}{dt^k} [A_{ijrs}^{\varepsilon\varepsilon k} \varepsilon_{rs} + A_{ijs}^{\varepsilon x k} x_{rs} + A_{ijr}^{\varepsilon p k} P_r]$$
$$= \sum_{k=0}^{N-1} \frac{d^k}{dt^k} [B_{ijrs}^{\varepsilon\varepsilon k} \sigma_{rs} + B_{ijrs}^{\varepsilon x k} m_{rs} + B_{ijr}^{\varepsilon p k} \mathscr{E}_r]$$
$$\sum_{k=0}^{N} \frac{d^k}{dt^k} [A_{ijrs}^{x\varepsilon k} \varepsilon_{rs} + A_{ijrs}^{\varepsilon x k} + A_{ijr}^{x p k} P_r]$$
$$= \sum_{k=0}^{N-1} \frac{d^k}{dt^k} [B_{ijrs}^{x\varepsilon k} \sigma_{rs} + B_{ijrs}^{x x k} m_{rs} + B_{ijr}^{x p k} P_r]$$
$$\sum_{k=0}^{N} \frac{d^k}{dt^k} [A_{ijr}^{p\varepsilon k} \varepsilon_{jr} + A_{ijr}^{p x k} x_{jr} + A_{ij}^{p p k} P_j]$$
$$= \sum_{k=0}^{N-1} \frac{d^k}{dt^k} [B_{ijr}^{p\varepsilon k} \sigma_{jr} + B_{ijr}^{p x k} x_{jr} + B_{ij}^{p p k} P_j] \qquad [17]$$

where

$$A_{i_1 i_2 \ldots i_n \, j_1 j_2 \ldots j_n}^{\alpha\beta k} = \langle \alpha i_1 \ldots i_{n_\alpha}| A^k |j_{n_\beta} \ldots j_{1_\beta}\rangle.$$

From [17] system and in the particular case $H \equiv H_\varepsilon$, $N = 3$ and $\|\eta^1\| \to 0^+$ *Brankovs* model is obtained describing the rheological properties of a large group polymers with net structure:

$$\sum_{k=0}^{2} A_{ijrs}^k \frac{d^k}{dt^k} \varepsilon_{rs} = \sum_{k=0}^{2} B_{ijrs}^k \frac{d^k}{dt^k} \sigma_{rs} \qquad [18]$$

where

$$A_{ijrs}^2 = \langle ij| E^1 \eta^2 \eta^3 |\varkappa l\rangle$$
$$A_{ijrs}^1 = \langle ij| E^1 \eta^2 E^3 + E^1 E^2 \eta^3 |\varkappa l\rangle$$
$$A_{ijkl}^0 = \langle ij| E^1 E^2 E^3 |kl\rangle$$
$$B_{ijkl}^2 = \langle ij| \eta^2 \eta^3 |\varkappa l\rangle$$
$$B_{ijkl}^1 = \langle ij| E^1 \eta^2 + E^2 \eta^3 + \eta^2 E^3 + E^2 \eta^3 |kl\rangle$$
$$B_{ijkl}^0 = \langle ij| E^1 E^2 + E^1 E^3 + E^2 E^3 |kl\rangle.$$

Analogically from [17] *Zeners*, *Bürgers*, *Maxwells*, *Voigts* models are obtained.

In the case $H \equiv H_\varepsilon \oplus H_x$, $N = 1$ we obtain the micropolar generalization of *Voigts* model which is done by *Eringen* (24).

Integrated form of the constitutive equations

According to [14]

$$|\xi(\omega)\rangle = l_k R^k(\omega) G^k |\zeta(\omega)\rangle.$$

Let us multiply that equation with $|\omega\rangle$.
We obtain:

$$|\xi\rangle = l_k R^k(\omega)\, G^k |\omega\rangle \langle\omega|\zeta\rangle$$

passing into $|t\rangle$ representation

$$|\xi(t)\rangle = \langle t|\xi\rangle = l_k R^k(\omega)\, G^k \langle\omega|t'\rangle \langle t'|\zeta\rangle \langle t|\omega\rangle$$

$$\Rightarrow |\xi(t)\rangle = \frac{1}{2\Pi} \int_{-\infty}^{\infty} \int_{-\infty}^{\infty} l_k R^k(\omega)\, G^k l^{i\omega(t-t')} \times |\zeta(t')\rangle\, dt'\, d\omega. \quad [19]$$

That equation may be written in the compact form

$$|\xi(t)\rangle = \int_{-\infty}^{\infty} U(t-t')\, |\zeta(t')\rangle\, dt' \quad [20]$$

where

$$U(t-t') = \frac{1}{2\Pi} \int_{-\infty}^{\infty} l_k R^k(\omega)\, G^k l^{i\omega(t-t')} d\omega.$$

From [20] is became clear that the time state of the output $|\xi(t)\rangle$ is determined by the time states of the input in the temporal interval $t' \in (-\infty, \infty)$.

But according to the principle of causality $|\xi(t)\rangle$ is determined simply by the state of the input $\{|\zeta(t')\rangle, t' \in (-\infty, t)]\}$.

As it is shown in (21) from that follows that

$$U(t-t') = 0 \quad \text{at} \quad t' > t. \quad [21]$$

Last condition leads to limitations on the F^k operator spectrum.

At $t - t' < 0$

$$U(t-t') = -\lim_{R\to\infty} \frac{1}{2\Pi i} \oint_{L^+} l_k R^k(z)\, G^k l^{z(t-t')} dz$$

where the integration is accomplished on the closed curve L^+ consisting of imaginary axis and the right infinite semicircumference.

According to *Cauchy* generalized theorem

$$\lim_{R\to\infty} \frac{1}{2\Pi i} \oint_{L^+} l_k R^k(z)\, G^k l^{z(t-t')} dz = [\mathrm{lxp}(t-t')\, F^k_{\sigma+}]\, G^k l_k \quad [22]$$

where the narrowing of the operator F^k on the spectrum subplurality $\sigma^+ \equiv \{\lambda^k_m \in \sigma^+, R_l \lambda^k_m > 0\}$ is indicated by $F^k_{\sigma+}$.

It is visible from [22] that the principle of causality can be fulfilled then and only then, when $\sigma^+ \equiv \varnothing$.

At $t - t' > 0$

$$U(t-t') = \lim_{R\to\infty} \frac{1}{2\Pi i} \oint_{L^-} l_k R^k(z)\, G^k l^{z(t-t')} dz = [\exp(t-t')\, F^k_{\sigma-}]\, G^k l_k \quad [23]$$

where with σ^- the spectrum subplurality $\{\lambda^k_m \in \sigma^-, R_l \lambda^k_m < 0\}$ is denoted. The integration is accomplished on the imaginary axis and on the left infinite semicircumference.

From $\sigma^+ \equiv \varnothing \Rightarrow F^k_{\sigma-} = F^k$.

Following "*Dunford*" (29)

$$l^{(t-t')F^k} = \sum_{i=1}^{P} \sum_{m=0}^{v_i-1} \frac{(F^k - \lambda^k_i I)^m}{m!} (t-t')^m \times l^{\lambda^k_i(t-t')} E_{\lambda^k_i} = \Psi^{ki}(t-t')\, l^{\lambda^k_i(t-t')} \quad [24]$$

where v^k_i are the indices of the spectral points λ^k_i.

From the requirement $U(t-t')$ to be a real operator and from $\sigma^+ \equiv \varnothing$ it follows that the spectral points λ^k_i lay on the negative semiaxis in the complex plane. It is in accordance with the principle of the fading memory used by many authors.

At $t = t'$

$$U(0) = \frac{1}{2\Pi} \int_{-\infty}^{\infty} l_k R^k(\omega)\, G^k d\omega \neq \lim_{R\to\infty} \frac{1}{2\Pi i} \oint_{L^-} l_k R^k(z)\, dz.$$

Immediately it is checked up that

$$\|U(0)\| \leq \|l_k G^k\|.$$

In the case when $l_k G^k$ is finite operator the point $t = t'$ may be separated from the integral [20] and consequently we obtain:

$$|\xi(t)\rangle = \int_{-\infty}^{t} \Psi^{ki}(t-t')\, l^{\lambda^k_i(t-t')} |\zeta(t')\rangle\, dt'. \quad [25]$$

At $\|l_k G^k\| \to \infty$, $U(t-t')$ has physical meaning only in the case $\lambda^k_i \to -\infty$. Furthermore at that case it is possible for $U(t-t')$ to have also a singular part with particularity in $t' = t$ point. That fact may be symbolically expressed by putting middle bracket in the upper end of the integral [25]

$$|\xi(t)\rangle = \int_{-\infty}^{t]} \Psi^{ki}(t-t')\, l^{\lambda^k_i(t-t')} |\zeta(t')\rangle\, dt'. \quad [26]$$

After projecting [26] in the subspaces L_ε, L_x, L_p and L_s we obtain;

$$\varepsilon_{ij}(t) = \int_{-\infty}^{t]} l^{-\frac{t-t'}{\tau_{kl}}} [\Psi^{\varepsilon\varepsilon kl}_{ijrs}(t-t')\, \sigma_{rs}(t')$$

$$+ \Psi^{\varepsilon xkl}_{ijrs}(t-t')\,m_{rs}(t') + \Psi^{\varepsilon pkl}_{ijr}(t-t')\,\mathscr{E}_r(t') + \Psi^{\varepsilon skl}_{ij}(t-t')\,\Delta T(t')]\,dt'$$

$$x_{ij}(t) = \int_{-\infty}^{t]} l^{-\frac{t-t'}{\tau_{kl}}}\,[\Psi^{x\varepsilon kl}_{ijrs}(t-t')\,\sigma_{rs}t') + \Psi^{xxkl}_{ijrs}(t-t')\,m_{rs}(t') + \Psi^{xpkl}_{ijr}(t-t')\,\mathscr{E}_r(t') + \Psi^{xskl}_{ij}(t-t')\,\Delta T(t')]\,dt'$$

$$P_i(t) = \int_{-\infty}^{t]} l^{-\frac{t-t'}{\tau_{kl}}}\,[\Psi^{p\varepsilon kl}_{irs}(t-t')\,\sigma_{rs}(t') + \Psi^{pxkl}_{irs}(t-t')\,m_{rs}(t') + \Psi^{ppkl}_{ir}(t-t')\,\varepsilon_r(t') + \Psi^{pskl}_{i}(t-t')\,\Delta T(t')]\,dt'$$

$$S(t) = \int_{-\infty}^{t]} l^{-\frac{t-t'}{\tau_{kl}}}\,[\Psi^{s\varepsilon kl}_{rs}(t-t')\,\sigma_{rs}(t') + \Psi^{sxkl}_{rs}(t-t')\,m_{rs}(t') + \Psi^{spkl}_{r}(t-t')\,\mathscr{E}_r(t') + \Psi^{sskl}(t-t')\,\Delta T(t')]\,dt' \qquad [27]$$

where

$$\Psi^{\alpha\beta kl}_{i_1 i_2 \dots i_n\, j_1 \dots j_n} = \langle \alpha\, i_{n_\alpha} \dots i_1 |\, \Psi^{kl} | j_1 \dots j_{n_\beta} \rangle, \qquad \tau_{kl} = -\frac{1}{\lambda^k_m}.$$

IV. Basic thermodynamic relations

According to the basic thermodynamic concepts the internal energy ΔH and the dissipative function $\Delta\mathscr{D}$ are additive magnitudes. Consequently, for their density we have

$$H = \sum_{\kappa=1}^{N} l_k H^k(\varepsilon^k_{ij};\ x^k_{ij};\ P^k_i;\ S^k)$$

$$\mathscr{D} = \sum_{k=1}^{N} l_k \mathscr{D}^k\left(\frac{d\varepsilon^k_{ij}}{dt};\ \frac{dx^k_{ij}}{dt};\ \frac{dP^k}{dt}\right) \qquad [28]$$

where $\varepsilon^k_{ij}, x^k_{ij}, P^k_i, S^k$ are the internal system parameters. H^k and $\mathscr{D}^k$ are the density of the internal energy and the dissipative function of the kth subsystem respectively.

In quadratic approximation we have:

$$H^k = \tfrac{1}{2} H^{k\varepsilon\varepsilon}_{ijrs}\varepsilon^k_{ij}\varepsilon^k_{rs} + \tfrac{1}{2} H^{kxx}_{ijrs}x^k_{ij}x^k_{rs} + \tfrac{1}{2} H^k_{ij}P^k_iP^k_j + \tfrac{1}{2} H^{kss}S^kS^k + H^{k\varepsilon x}_{ijrs}\varepsilon^k_{ij}x^k_{rs} + H^{k\varepsilon p}_{ijr}\varepsilon^k_{ij}P^k_r + H^{k\varepsilon s}_{ij}\varepsilon^k_{ij}S^k + H^{kxp}_{ijr}x^k_{ij}P^k_r + H^{kxs}_{ij}x^k_{ij}S^k + H^{kps}_{i}P^k_iS^k + \Pi^{k\varepsilon}_{ij}\varepsilon^k_{ij} + \Pi^{kx}_{ij}x^k_{ij} + \Pi^k_iP_i + \Pi^kS^k$$

$$H^{k\alpha\alpha}_{ijrs} = H^{k\alpha\alpha}_{rsij}, \qquad H^{k\alpha\alpha}_{ij} = H^{k\alpha\alpha}_{ji} \qquad [29_1]$$

$$\mathscr{D}^k = A^{k\varepsilon\varepsilon}_{ijrs}\frac{d\varepsilon^k_{ij}}{dt}\frac{d\varepsilon^k_{rs}}{dt} + A^{kxx}_{ijrs}\frac{dx^k_{ij}}{dt}\frac{dx^k_{rs}}{dt} + A^{kpp}_{ij}\frac{dP^k_i}{dt}\frac{dP^k_j}{dt} + 2A^{k\varepsilon x}_{ijrs}\frac{d\varepsilon^k_{ij}}{dt}\frac{dx^k_{rs}}{dt} + 2A^{k\varepsilon p}_{ijr}\frac{d\varepsilon^k_{ij}}{dt}\frac{dP^k_r}{dt} + 2A^{kxp}_{ijr}\frac{dx^k_{ij}}{dt}\frac{dP^k_r}{dt} + A^{k\varepsilon}_{ij}\frac{d\varepsilon^k_i}{dt} + A^{kx}_{ij}\frac{dx^k}{dt} + A^k_i\frac{dP^k_i}{dt}$$

$$A^{k\alpha\alpha}_{ijrs} = A^{k\alpha\alpha}_{rsij}, \qquad A^{k\alpha\alpha}_{ij} = A^{k\alpha\alpha}_{ji}. \qquad [29_2]$$

Due to the fact that each subsystem has not internal degrees of freedom it is possible the thermodynamical forces to be divided into equilibrium and non-equilibrium part:

$$\zeta^{k(r)} = \frac{\partial H^k}{\partial \xi^k}, \qquad \zeta^{k(ir)} = \frac{1}{2}\frac{\partial \mathscr{D}^k}{\partial\left(\frac{d\xi^k}{dt}\right)}$$

$$\zeta^k = \zeta^{k(r)} + \zeta^{k(ir)}.$$

The thermodynamical forces are symbolically indicated with ζ^k and the thermodynamical co-ordinates with ξ^k. The equilibrium part of the thermodynamical forces is denoted with (r) and the non-equilibrium part $-(ir)$.

Correspondingly for each subsystem of [29] we obtain

$$\sigma^{k(r)}_{ij} = H^{k\varepsilon\varepsilon}_{ijrs}\varepsilon^k_{rs} + H^{k\varepsilon x}_{ijrs}x^k_{rs} + H^{k\varepsilon p}_{ijr}P^k_r + H^{k\varepsilon s}_{ij}S^k$$

$$m^{k(r)}_{ij} = H^{kxx}_{ijrs}x^k_{rs} + H^{k\varepsilon x}_{ijrs}\varepsilon^k_{rs} + H^{kxp}_{ijr}P^k_r + H^{kxs}S^k$$

$$\mathscr{E}^{k(r)}_i = H^{kpp}_{ij}P^k_j + H^{kp\varepsilon}_{rsi}\varepsilon^k_{rs} + H^{kpx}_{rsi}x^k_{rs} + H^{kps}_iS^k$$

$$T = H^{kss}S^k + H^{k\varepsilon s}_{ij}\varepsilon^k_{ij} + H^{kxs}_{ij}x^k_{ij} + H^{ksp}_iP_i + T_0$$

$$\sigma^{k(ir)}_{ij} = A^{k\varepsilon\varepsilon}_{ijrs}\frac{d\varepsilon^k_{rs}}{dt} + A^{k\varepsilon x}_{ijrs}\frac{dx^k_{rs}}{dt} + A^{k\varepsilon p}_{ijr}\frac{dP^k_r}{dt}$$

$$m^{k(ir)}_{ij} = A^{kxx}_{ijrs}\frac{dx^k_{rs}}{dt} + A^{kx\varepsilon}_{ijrs}\frac{d\varepsilon^k_{rs}}{dt} + A^{kxp}_{ijr}\frac{dP^k_r}{dt}$$

$$\varepsilon^{k(ir)}_i = A^{kpp}_{ij}\frac{dP^k_j}{dt} + A^{kp\varepsilon}_{rsi}\frac{d\varepsilon^k_{rs}}{dt} + A^{kxp}_{rsi}\frac{dx^k_{rs}}{dt}. \qquad [30]$$

When writing these formulae it is accepted that

$$\Pi^{k\varepsilon}_{ij} = \Pi^{kx}_{ij} = \Pi_i = A^{k\varepsilon}_{ij} = A^{kp}_{ij} = A^k_i = 0,$$

$$\Pi^k = T_0.$$

Multiplying [6] on the left side with $(G^k)^{-1}$ we obtain:

$$|\zeta\rangle = l_k E^k |\xi^k\rangle + D\, l_k \eta^k |\xi^k\rangle .$$

$|\zeta\rangle$ may be represented as a sum of the two parts

$$|\zeta\rangle = |\zeta^{k(r)}\rangle + |\zeta^{k(ir)}\rangle$$

the first of which does not depend on the differentiation operator. After projecting $|\zeta^{k(r)}\rangle$ and $|\zeta^{k(ir)}\rangle$ in L_ε, L_x, L_p, L_s correspondingly and comparing the obtained results with [30] we find out the following relations:

$$
\begin{aligned}
\langle \varepsilon j i|\, l_k E^k |r l \varepsilon\rangle &= \langle \varepsilon l r|\, l_k E^k |i j \varepsilon\rangle = H^{k\varepsilon\varepsilon}_{ijrl} \\
\langle x j i|\, l_k E^k |r l x\rangle &= \langle x l r|\, l_k E^k |i j x\rangle = H^{kxx}_{ijrl} \\
\langle p i|\, l_k E^k |r p\rangle &= \langle p r|\, l_k E^k |i p\rangle = H^{kpp}_{ir} \\
\langle \varepsilon j i|\, l_k E^k |r s x\rangle &= \langle x s r|\, l_k E^k |i j \varepsilon\rangle = H^{k\varepsilon x}_{ijrs} \\
\langle \varepsilon j i|\, l_k E^k |r p\rangle &= \langle p r|\, l_k E^k |i j \varepsilon\rangle = H^{k\varepsilon p}_{ijr} \\
\langle \varepsilon j i|\, l_k E^k |s\rangle &= \langle s|\, l_k E^k |i j \varepsilon\rangle = H^{k\varepsilon s}_{ij} \\
\langle x j i|\, l_k E^k |r p\rangle &= \langle p r|\, l_k E^k |i j x\rangle = H^{kxp}_{ijr} \\
\langle x j i|\, l_k E^k |s\rangle &= \langle s|\, l_k E^k |i j x\rangle = H^{kxs}_{ij} \\
\langle p i|\, l_k E^k |s\rangle &= \langle s|\, l_k E^k |i p\rangle = H^{kps}_{i} \\
\langle \varepsilon j i|\, l_k \eta^k |r l \varepsilon\rangle &= \langle \varepsilon l r|\, l_k \eta^k (i j \varepsilon\rangle = A^{k\varepsilon\varepsilon}_{ijre} \\
\langle \varepsilon j i|\, l_k \eta^k |r l x\rangle &= \langle x l r|\, l_k \eta^k |i j \varepsilon\rangle = A^{k\varepsilon x}_{ijre} \\
\langle \varepsilon j i|\, l_k \eta^k |r p\rangle &= \langle p r|\, l_k \eta^k |i j \varepsilon\rangle = A^{k\varepsilon p}_{ijr} \\
\langle x j i|\, l_k \eta^k |r p\rangle &= \langle p r|\, l_k \eta^k |i j x\rangle = A^{kxp}_{ijr} \\
\langle s|\, l_k \eta^k |\alpha\rangle &= \langle \alpha|\, l_k \eta^k |s\rangle = 0.
\end{aligned}
\qquad [31]
$$

These equations represent limitations for the operators E^k and η^k deriving from the requirement the examined system to be thermodynamical. After using the properties of the operators E^k and η^k [31] in the expression of the internal energy H^k we obtain the compact form:

$$H^k = \tfrac{1}{2}\langle \xi^k(t)|\, l_k E^k |\xi^k(t)\rangle + T_0 S^k(t). \qquad [32]$$

For the dissipative function $\mathscr{D}^k$ we obtain correspondingly:

$$\mathscr{D}^k = \left\langle \frac{d\xi^k}{dt}(t) \middle|\, l_k \eta^k \middle| \frac{d\xi^k}{dt}(t) \right\rangle . \qquad [33]$$

After using in [33] that

$$|\xi^k(t)\rangle = \int_{-\infty}^{t]} \Psi^{kl}(t-t')\, l^{-\frac{t-t'}{\tau_{kl}}} |\zeta(t')\rangle\, dt' \qquad [34]$$

and summarizing index "k" we obtain the following expressions for the density of the total internal energy and the dissipative function:

$$
\begin{aligned}
H = l_k H^k = \int_{-\infty}^{t]} \int_{-\infty}^{t]} &\langle \zeta(t')|\, \Psi^{+kl}(t-t') \\
&\times l_k E^k l_k \Psi^{km}(t-t'')\, |\zeta(t'')\rangle \\
&\times l^{-\frac{t-t'}{\tau_{kl}}}\, l^{-\frac{t-t''}{\tau_{km}}}\, dt'\, dt'' + T_0 S
\end{aligned}
$$

$$
\begin{aligned}
\mathscr{D} = l_k \mathscr{D}^k = \int_{-\infty}^{t]} \int_{-\infty}^{t]} &\left\langle \frac{d\zeta}{dt'}(t') \middle|\, \Psi^{+kl}(t-t') \right. \\
&\times \left. l_k \eta^k l_k \Psi^{km}(t-t'') \middle| \frac{d\zeta}{dt''}(t'') \right\rangle \\
&\times l^{-\frac{t-t'}{\tau_{kl}}}\, l^{-\frac{t-t''}{\tau_{km}}}\, dt'\, dt'' .
\end{aligned}
\qquad [35]
$$

With Ψ^{+kl} is denoted the adjoint operator of Ψ^{kl}.

According to the second principle of thermodynamics $\mathscr{D} \geq 0$. Consequently the operator η^k in [35] is positive.

V. Conclusion

The purpose of the authors of the present paper was the creation of the basic rheological equations referred to continuous media and admitting electro and thermomechanical effects.

Today it is known that these effects play fundamental role in the biological functions of the biological tissues (9).

In the process of the investigations were used different approaches.

The best results were achieved with the application of the methods of the general system theory.

The authors hope that by a system approach the rheological properties of the nearer to reality non-linear and non-invariant with respect of time continuous media may be examined.

References

1) *Duhamel, J.*, Second mémoire sur les phénomènes thermomécaniques. J. de l'Ecole Polytechnique **15** (1837).

2) *Voigt, W.*, Lehrbuch der Kristallphysik (Leipzig 1910).

3) *Jeffreys, H.*, The thermodynamic of an elastic solid. Proc. Cambr. Phil. Soc. **26** (1930).

4) *Voigt, W.*, Die fundamentalen physikalischen Eigenschaften der Kristalle (Leipzig 1898).

5) *Cady, W.*, Piezo-electricity. Inter. Ser. in Pure and Applied Physics (New York-London 1946).

6) *Mason, W.*, Bell System Techn. **26**, 80–138.

7) *Palonsky, J., P. Douzon,* and *C. Sadron*, Compt. Rend **250**, 3414 (1960).

8) *Duchesne, J., J. Depireux, A. Bertinchamps, N. Cornet,* and *J. van der Kaa*, Nature **188**, 405 (1960).

9) *Shamos, M.* and *L. Lavin*, Nature **213**, 267 (1967).

10) *Pohl, H., A. Rembaum*, and *A. Henry*, J. Amer. Chem. Soc. **84**, 2699 (1962).
11)
12) *Fukada, E., M. Date*, and *M. Hirac*, Nature **211**, 1079 (1966).
13) *Furukawa, T., Y. Uematsu, K. Asakowa*, and *Y. Wada*, J. Appl. Polymer Sci. **12**, 2675 (1968).
14) *Kawai, H.*, Japan J. Appl. Phys. **8**, 975 (1969).
15) *Leslie, F.*, Arch. Ration Mech. Anal. **28**, 265 (1968).
16) *Davison, L.*, Phys. Fluids **10**, 2333 (1967).
17) *Davison, L.* and *A. Donald*, Phys. Rev. **183**, 288 (1969).
18)
19) *Parodi, O.*, Le journal de physique **31**, 581 (1970).
20) *Lee, J.* and *A. Eringer*, J. Chem. Phys. **54**, 5027 (1971).
21)
22) *Dirac, P.*, The Principles of Quantum Mechanics (Oxford 1947).
23) *Porter, W.*, Modern Foundations of System Engineering, Russian translation: (1971).
24) *Eringen, A.*, Int. J. Eng. Sci. **5**, 191–204 (1967).
25) *Nagel, E.*, The Causal Character of Modern Physical Theory, pp. 421–422 (New York 1953).
26) *Nagel, E.*, The Structure of Science, pp. 279–281 (London 1961).
27) *Zigler, H.*, Some Extremum Principles in Irreversible Thermodynamics with Application to Continuum Mechanics. In: Progress in Solid Mechanics Vol. **IV**, (Amsterdam 1963).
28) *Kalman, R., P. Falb*, and *M. Arbib*, Topics in Mathematical System Theory (New York 1969).
29) *Dunford, N.* and *J. Schwartz*, Linear Operators part II – spectral theory (New York 1963).

Authors' address:
G. Brankov and *N. Petrov*
Institute of Technical Mechanics
Bulgarian Academy of Sciences
Geo Milev IV klm IV bl, Sofia (Bulgaria)

Rheol. Acta 12, 150–154 (1973)

From the Department of Materials Engineering, University of Illinois at Chicago Circle, Chicago, Illinois (U.S.A.)

A moving punch on an infinite viscoelastic layer *)

By T. C. T. Ting

With 1 figure

(Received October 27, 1972)

1. Introduction

When a smooth rigid punch of arbitrary shape is pressed onto an elastic body, an area of contact D is formed within which the contact pressure p is non-negative. Outside of the contact area D, $p = 0$. The problem is to find the contact pressure p and the shape of the contact region D for a given depth of indentation, or a given total force applied to the punch. In general, the governing equation can be reduced to the following integral equation:

$$w(x_1, x_2) = \iint_D K(x_1, x_2, \xi_1, \xi_2)\, p(\xi_1, \xi_2)\, d\xi_1 d\xi_2 \qquad [1a]$$

in which $K(x_1, x_2, \xi_1, \xi_2)$ is known and w is given by

$$w(x_1, x_2) = \alpha - f(x_1, x_2) \qquad [1b]$$

where α is the depth of indentation and $f(x_1, x_2)$ is the function which describes the shape of the punch. Unless K and f are simple in form, the solutions of eq. [1] for p and D in general require a numerical approximation. Since the physical observation demands that $p > 0$ inside D and $p = 0$ outside D, the method of linear programming is probably the most effective one in solving this problem numerically. The purpose of this paper is to show that not only the contact problem of a rigid punch on an elastic body yields eq. [1], many contact problems of a rigid punch on a viscoelastic body and the contact problems of a moving punch on a viscoelastic body also reduce to eq. [1]. It should be noticed that depending on the shape of the punch, D may not be a simply-connected region. It should also be mentioned that for a viscoelastic body, D is in general not a constant but varies with time even in the case of a punch which is held stationary on the surface of a viscoelastic body.

2. Rigid punch on an elastic body

The equations for the static equilibrium of a homogeneous, isotropic, linearly elastic material can be written as (1)

$$\sigma_{ij,j} = 0 \qquad [2]$$

$$\varepsilon_{ij} = \tfrac{1}{2}(u_{i,j} + u_{j,i}) \qquad [3]$$

$$\sigma_{ij} = 2\mu\varepsilon_{ij} + \lambda\delta_{ij}\varepsilon_{kk} \qquad [4]$$

where u_i is the displacement, σ_{ij} and ε_{ij} are the stress and strain tensor respectively, μ and λ are the *Lame* constants, δ_{ij} is the *Kronecker* delta, and a comma stands for the partial differentiation. Consider a body with a flat surface but otherwise arbitrary configuration subjected to a concentrated vertical load of magnitude P on the flat surface, fig. 1. Aside from the load P, the rest of the boundary surface of the body is either traction free or has zero displacement. Let the x_1 and x_2 axes lie on the flat surface and the x_3 axis point toward inside of the body. The position of the load P will be denoted by $\underset{\sim}{\xi} = (\xi_1, \xi_2)$. If w is the vertical displacement of a point $\underset{\sim}{x} = (x_1, x_2)$ on the flat surface, i.e., $w = u_3(x_1, x_2, 0)$, it is not difficult to show by the dimensional analysis that

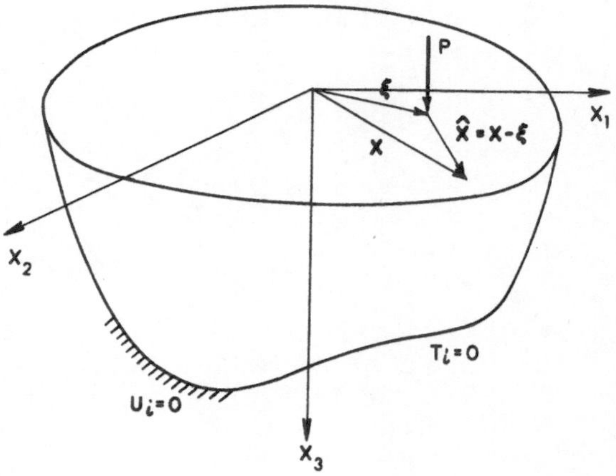

Fig. 1

*) The work presented here was supported by the National Science Foundation under Grant GK 35163 with the University of Illinois.

$$w(\underset{\sim}{x}) = \frac{1-\nu}{\mu l} PF(\underset{\sim}{x}/l, \underset{\sim}{\xi}/l, \nu) \qquad [5]$$

where l is a typical length, ν is the *Poissons* ratio which is related to λ and μ by

$$\frac{\lambda}{\mu} = \frac{2\nu}{1-2\nu} \qquad [6]$$

and F is a function of $\underset{\sim}{x}/l$, $\underset{\sim}{\xi}/l$ and ν. In view of the reciprocal theorem (1), we have

$$F(\underset{\sim}{x}/l, \underset{\sim}{\xi}/l, \nu) = F(\underset{\sim}{\xi}/l, \underset{\sim}{x}/l, \nu). \qquad [7]$$

The factor $(1-\nu)$ in eq. [5] can actually be absorbed in the function F. However, with $(1-\nu)$ separated from F, F becomes independent of ν for certain geometries of the body. For instance, for an infinite layer of thickness h rested on a frictionless rigid surface (2, 3),

$$F = \frac{1}{2\pi}\int_0^\infty \frac{2\sinh^2\beta h}{\sinh(2\beta h) + 2\beta h} J_0(\beta|\underset{\sim}{x} - \underset{\sim}{\xi}|)\, d\beta \qquad [8]$$

where J_0 is the zero order *Bessel* function of the first kind. If $h \to \infty$, we have a half-space and

$$F = \frac{1}{2\pi|\underset{\sim}{x} - \underset{\sim}{\xi}|},$$

$$|\underset{\sim}{x} - \underset{\sim}{\xi}| = \sqrt{(x_1 - \xi_1)^2 + (x_2 - \xi_2)^2}. \qquad [9]$$

Notice that in eqs. [8] and [9] F is a function of $|\underset{\sim}{x} - \underset{\sim}{\xi}|$ only.

If the load applied on the flat surface is a distributed load with intensity p per unit surface area, we obtain, in view of the linearity of the problem,

$$w(\underset{\sim}{x}) = \frac{1-\nu}{\mu l} \int_D F(\underset{\sim}{x}/l, \underset{\sim}{\xi}/l, \nu)\, p(\underset{\sim}{\xi})\, d\underset{\sim}{\xi} \qquad [10a]$$

where the integration is over the region D where the loads are applied.

If, instead of applying a distributed load, a rigid punch with a smooth surface is pressed onto the flat surface of the body, eq. [10a] furnishes an integral equation for the contact pressure $p(\underset{\sim}{x})$ in which $w(\underset{\sim}{x})$ is replaced by

$$w(\underset{\sim}{x}) = \alpha - f(\underset{\sim}{x}) \qquad [10b]$$

where α is the depth of indentation and $f(\underset{\sim}{x})$ is the function which describes the shape of the punch (3). It should be noted that the domain of contact D is an unknown. D is determined by the condition that p vanishes along the perimeter of D [see (4)].

Eqs. [10] belong to the type of equations given in eqs. [1] and hence in general require a numerical solution. In the particular case of a half-space pressed by a smooth axisymmetric punch, eqs. [10] have the following explicit solution [see (4)].:

$$p(r) = \frac{2\mu}{(1-\nu)\pi}\int_r^a \frac{1}{\sqrt{\beta^2 - r^2}} \frac{\partial}{\partial\beta}\int_0^\beta \frac{\beta}{\sqrt{\beta^2 - m^2}} \times \frac{\partial}{\partial m} f(m)\, dm\, d\beta \qquad [11]$$

$$\alpha = \int_0^a \frac{a}{\sqrt{a^2 - r^2}} \frac{d}{dr} f(r)\, dr \qquad [12]$$

where $r = |\underset{\sim}{x}|$ is the radial distance from the origin, and a is the radius of the contact area.

In the rest of this paper, we will consider the class of problems in which F is independent of ν and is a function of $|\underset{\sim}{x} - \underset{\sim}{\xi}|$ only. Thus eq. [10a] can be written as

$$w(\underset{\sim}{x}) = \frac{1-\nu}{\mu} \int_D F(|\underset{\sim}{x} - \underset{\sim}{\xi}|)\, p(\underset{\sim}{\xi})\, d\underset{\sim}{\xi}. \qquad [13]$$

3. Rigid punch on a viscoelastic layer

If the body considered in fig. 1 is a linear viscoelastic solid, eqs. [2] and [3] remain the same while eq. [4] is replaced by, (16, 17),

$$\sigma_{ij}(t) = 2\int_{0-}^t \mu(t-\tau)\, d\varepsilon_{ij}(\tau) + \delta_{ij}\int_{0-}^t \lambda(t-\tau)\, d\varepsilon_{kk}(\tau)$$

or [14]

$$\sigma^*_{ij} = 2s\mu^*\varepsilon^*_{ij} + \delta_{ij} s\lambda^*\varepsilon^*_{kk} \qquad [15]$$

where a superscript * stands for the *Laplace* transform of the function:

$$g^*(s) = \int_{0-}^\infty g(t)\, e^{-st}\, dt. \qquad [16]$$

By applying the correspondence principle [see (5, 16, 17)] to eq. [13], we obtain the displacement w, now a function of $\underset{\sim}{x}$ and time t, as [see (3)],

$$w(\underset{\sim}{x}, t) = \int_{0-}^t \varphi(t-\tau) \frac{\partial}{\partial\tau} \int_{D(\tau)} F(|\underset{\sim}{x} - \underset{\sim}{\xi}|)\, p(\underset{\sim}{\xi}, \tau)\, d\underset{\sim}{\xi}\, d\tau$$

where [17]

$$s\varphi^* = \frac{1 - s\nu^*}{s\mu^*}. \qquad [18]$$

Notice that D is now dependent on time. Thus for a given pressure distribution $p(\underset{\sim}{x}, t)$ in the region $D(t)$, eq. [17] furnishes the vertical displacement $w(\underset{\sim}{x}, t)$. If, instead of the distributed load, a smooth rigid punch is pressed onto the surface, w in eq. [17] is replaced by, (3),

$$w(\underset{\sim}{x}, t) = \alpha(t) - f(\underset{\sim}{x})\, H(t) \qquad [19]$$

and eq. [17] becomes an integral equation for $p(\underset{\sim}{x}, t)$. In eq. [19], $H(t)$ is the *Heaviside* step function. We have assumed that the punch is pressed onto the body at $t = 0$.

The case in which the contact region $D(t)$ is an increasing function of time t is relatively simple (6, 15). In this case, eq. [17] can be reduced to, (3),

$$w(\underset{\sim}{x}, t) = \int_{D(t)} F(|\underset{\sim}{x} - \underset{\sim}{\xi}|) \times \left[\int_{0-}^{t} \varphi(t - \tau) \frac{\partial}{\partial \tau} p(\underset{\sim}{\xi}, \tau)\, d\tau \right] d\underset{\sim}{\xi}. \qquad [20]$$

This is a particular case of eq. [1 a]. The problem becomes complicated when $D(t)$ has a single maximum (7). Nevertheless, eq. [17] can be reduced to eq. [1 a], (3, 4, 8). For an arbitrary $D(t)$, the problem has been studied in (3, 4, 9, 10, 11).

The case of a viscoelastic layer pressed by a rigid punch was considered by *Tsai* (12).

4. Moving punch on a viscoelastic layer

We will now consider the problem of a rigid punch pressed against and moved on the surface of a viscoelastic layer. We will assume that the punch has been moving on the surface since time $\tau = -\infty$. Moreover, the motion is quasi-static and hence the inertia can be ignored.

To this end, first consider a single concentrated load of magnitude P which has been moving on the surface of the layer since $\tau = -\infty$. If the layer were elastic, we would have,

$$w(\underset{\sim}{x}, t) = \frac{1 - \nu}{\mu} P F(|\underset{\sim}{x} - \underset{\sim}{\xi}(t)|) \qquad [21]$$

where $\underset{\sim}{\xi}(t)$ is the position of the load P at time t. For a viscoelastic layer, we obtain by using the correspondence principle the following result:

$$w(\underset{\sim}{x}, t) = P \int_{-\infty}^{t} \varphi(t - \tau) \frac{\partial}{\partial \tau} F(|\underset{\sim}{x} - \underset{\sim}{\xi}(\tau)|)\, d\tau. \qquad [22]$$

If we integrate the right hand side by parts and change the variable τ to $(t - \tau)$, we obtain,

$$w(\underset{\sim}{x}, t) = P \int_{0-}^{\infty} F(|\underset{\sim}{x} - \underset{\sim}{\xi}(t - \tau)|)\, \dot{\varphi}(\tau)\, d\tau \qquad [23]$$

where

$$\dot{\varphi}(\tau) = \frac{d}{d\tau} \varphi(\tau). \qquad [24]$$

Now, consider the special case in which the load P has been moving on a straight line with a constant velocity V, i.e.,

$$\underset{\sim}{\xi}(t) = V t \underset{\sim}{e} \qquad [25]$$

where $\underset{\sim}{e}$ is a constant unit vector. Eq. [23] then reduces to

$$w(\underset{\sim}{x}, t) = P K(\hat{\underset{\sim}{x}}) \qquad [26]$$

where

$$\hat{\underset{\sim}{x}} = \underset{\sim}{x} - V t \underset{\sim}{e} \qquad [27]$$

and

$$K(\hat{\underset{\sim}{x}}) = \int_{0-}^{\infty} F(|\hat{\underset{\sim}{x}} + V \tau \underset{\sim}{e}|)\, \dot{\varphi}(\tau)\, d\tau. \qquad [28]$$

If we use a moving coordinate system with the load P at the origin of this moving coordinate system, $\hat{\underset{\sim}{x}}$ is the position vector of the point $\underset{\sim}{x}$ referred to this moving coordinate system, fig. 1.

If instead of a single load, a distributed load p is moved along the direction $\underset{\sim}{e}$ at the constant speed V, p will be a function of $\hat{\underset{\sim}{\xi}} = \underset{\sim}{\xi} - V t \underset{\sim}{e}$, and we obtain from eq. [26],

$$w(\underset{\sim}{x}, t) = w(\hat{\underset{\sim}{x}}) = \int_{D(\hat{\underset{\sim}{\xi}})} K(\hat{\underset{\sim}{x}} - \hat{\underset{\sim}{\xi}})\, p(\hat{\underset{\sim}{\xi}})\, d\hat{\underset{\sim}{\xi}}. \qquad [29\,a]$$

Now, consider a rigid punch pressed against and moved on the surface of the layer at the constant velocity $V\underset{\sim}{e}$. Eq. [29 a] now becomes an integral equation for $p(\hat{\underset{\sim}{x}})$ in which $w(\hat{\underset{\sim}{x}})$ is given by

$$w(\hat{\underset{\sim}{x}}) = \alpha - f(\hat{\underset{\sim}{x}}). \qquad [29\,b]$$

Equation [29 a] is a particular case of eq. [1 a].

Consider next the case in which the punch is moved along a circular path indefinitely at a constant speed. As before, we first consider a single load P moving at a constant angular velocity ω along a circular path of radius r'. If we use a polar coordinate system with origin at the center of the circular path, we have,

$$|\underset{\sim}{x} - \underset{\sim}{\xi}(t)| = \sqrt{r^2 + r'^2 - 2 r r' \cos(\theta - \omega t)} \qquad [30]$$

where $r = |\underset{\sim}{x}|$, $r' = |\underset{\sim}{\xi}(t)|$, and θ is the angle $\underset{\sim}{x}$ makes with the x_1 axis. By substituting eq. [30] into [23], we obtain

$$w(\underset{\sim}{x}, t) = PK(r, \hat{\theta}, r') \qquad [31]$$

where

$$\hat{\theta} = \theta - \omega t \qquad [32]$$

$$K(r,\hat{\theta},r') = \int_{0-}^{\infty} F(\sqrt{r^2 + r'^2 - 2rr'\cos(\hat{\theta} + \omega\tau)}) \times \dot{\varphi}(\tau)\, d\tau. \qquad [33]$$

If, instead of a single load P, a distributed load $p(r, \hat{\theta})$ is moved at a constant angular speed ω around a circular path, we then have,

$$w(\underset{\sim}{x}, t) = w(r, \hat{\theta}) = \iint_D K(r, \hat{\theta} - \hat{\theta}', r')\, p(r', \hat{\theta}')\, r' \times dr'\, d\hat{\theta}'. \qquad [34a]$$

Again, for a rigid punch moving around a circular path, eq. [34a] becomes an integral equation for $p(r, \hat{\theta})$ in which

$$w(r, \hat{\theta}) = \alpha - f(r, \hat{\theta}). \qquad [34b]$$

Eqs. [34] are similar to eqs. [1].

5. Moving punch on a standard linear viscoelastic layer

In this section, we will consider the particular case of a rigid punch moving along a straight path on a layer of standard linear viscoelastic material. For a standard linear viscoelastic solid, $\varphi(t)$ can be expressed as

$$\varphi(t) = \gamma[1 + b(1 - e^{-t/\eta})] \qquad [35]$$

where γ, b and η are constants. In other words, $\varphi(t)$ satisfies the differential equation

$$\dot{\varphi} + \eta\ddot{\varphi} = 0. \qquad [36]$$

Now, let us rewrite eq. [29a] as, with the use of eq. [28],

$$w(\hat{\underset{\sim}{x}}) = \int_{D(\hat{\underset{\sim}{\xi}})} p(\hat{\underset{\sim}{\xi}}) \int_{0^-}^{\infty} F(|\hat{\underset{\sim}{x}} - \hat{\underset{\sim}{\xi}} + V\tau\underset{\sim}{e}|)\, \dot{\varphi}(\tau)\, d\tau\, d\hat{\underset{\sim}{\xi}}$$

or

$$w(\hat{\underset{\sim}{x}}) = \varphi(0) \int_{D(\hat{\underset{\sim}{\xi}})} F(|\hat{\underset{\sim}{x}} - \hat{\underset{\sim}{\xi}}|)\, p(\hat{\underset{\sim}{\xi}})\, d\hat{\underset{\sim}{\xi}} + \int_{D(\hat{\underset{\sim}{\xi}})} p(\hat{\underset{\sim}{\xi}}) \int_0^{\infty} F(|\hat{\underset{\sim}{x}} - \hat{\underset{\sim}{\xi}} + V\tau\underset{\sim}{e}|)\, \dot{\varphi}(\tau)\, d\tau\, d\hat{\underset{\sim}{\xi}}. \qquad [37]$$

Without the loss of generality, we assume that $\underset{\sim}{e}$ is a unit vector pointing in the direction of the x_1 axis. Then, since

$$p(\hat{\underset{\sim}{\xi}}) \frac{\partial}{\partial \hat{x}_1} F(|\hat{\underset{\sim}{x}} - \hat{\underset{\sim}{\xi}}|) = -p(\hat{\underset{\sim}{\xi}}) \frac{\partial}{\partial \hat{\xi}_1} F(|\hat{\underset{\sim}{x}} - \hat{\underset{\sim}{\xi}}|) = -\frac{\partial}{\partial \hat{\xi}_1}[p(\hat{\underset{\sim}{\xi}}) F(|\hat{\underset{\sim}{x}} - \hat{\underset{\sim}{\xi}}|)] + F(|\hat{\underset{\sim}{x}} - \hat{\underset{\sim}{\xi}}|) \frac{\partial}{\partial \hat{\xi}_1}(p\hat{\underset{\sim}{\xi}})$$

and

$$\dot{\varphi}(\tau) \frac{\partial}{\partial \hat{x}_1} F(|\hat{\underset{\sim}{x}} - \hat{\underset{\sim}{\xi}} + V\tau\underset{\sim}{e}|) = \frac{1}{V} \dot{\varphi}(\tau) \frac{\partial}{\partial \tau} F(|\hat{\underset{\sim}{x}} - \hat{\underset{\sim}{\xi}} + V\tau\underset{\sim}{e}|) = \frac{1}{V} \frac{\partial}{\partial \tau}[\dot{\varphi}(\tau) F(|\hat{\underset{\sim}{x}} - \hat{\underset{\sim}{\xi}} + V\tau\underset{\sim}{e}|)] - \frac{1}{V} \ddot{\varphi}(\tau) F(|\hat{\underset{\sim}{x}} - \hat{\underset{\sim}{\xi}} + V\tau\underset{\sim}{e}|),$$

differentiation of eq. [37] with respect to $\hat{x}_1$ yields

$$\frac{\partial}{\partial \hat{x}_1} w(\hat{\underset{\sim}{x}}) = \int_{D(\hat{\underset{\sim}{\xi}})} F(|\hat{\underset{\sim}{x}} - \hat{\underset{\sim}{\xi}}|) \times \left[\varphi(0) \frac{\partial}{\partial \hat{\xi}_1} p(\hat{\underset{\sim}{\xi}}) - \frac{1}{V} \dot{\varphi}(0)\, p(\hat{\underset{\sim}{\xi}})\right] d\hat{\underset{\sim}{\xi}} - \frac{1}{V} \int_{D(\hat{\underset{\sim}{\xi}})} p(\hat{\underset{\sim}{\xi}}) \int_0^{\infty} F(|\hat{\underset{\sim}{x}} - \hat{\underset{\sim}{\xi}} + V\tau\underset{\sim}{e}|) \times \ddot{\varphi}(\tau)\, d\tau\, d\hat{\underset{\sim}{\xi}}. \qquad [38]$$

The rest of the terms vanish because $p(\hat{\underset{\sim}{\xi}}) = 0$ on the perimeter of $D(\hat{\underset{\sim}{\xi}})$ and $\dot{\varphi}(\tau) = 0$ for $\tau = \infty$. From eqs. [37] and [38], we obtain, in view of eqs. [35] and [36],

$$w(\hat{\underset{\sim}{x}}) - V\eta \frac{\partial}{\partial \hat{x}_1} w(\hat{\underset{\sim}{x}}) = \gamma \int_{D(\hat{\underset{\sim}{\xi}})} F(|\hat{\underset{\sim}{x}} - \hat{\underset{\sim}{\xi}}|)\, \Pi(\hat{\underset{\sim}{\xi}})\, d\hat{\underset{\sim}{\xi}} \qquad [39]$$

where

$$\Pi(\hat{\underset{\sim}{x}}) = (1 + b)\, p(\hat{\underset{\sim}{x}}) - V\eta \frac{\partial}{\partial \hat{x}_1} p(\hat{\underset{\sim}{x}}). \qquad [40]$$

Eq. [39] is an integral equation for $\Pi(\hat{\underset{\sim}{x}})$, and is a particular case of eq. [1a]. Once $\Pi(\hat{\underset{\sim}{x}})$ is found, eq. [40] is an ordinary differential equation for $p(\hat{\underset{\sim}{x}})$.

Notice that eq. [39] is identical to the equation for a punch of modified shape, i.e. of shape $f(\hat{\underset{\sim}{x}}) - V\eta \frac{\partial}{\partial \hat{x}_1} f(\hat{\underset{\sim}{x}})$ pressed on an *elastic* layer.

In the special case of two-dimensional problem in which the layer is of infinite thickness (i.e. a half-plane), and pressed by a circular cylinder, eqs. [39] and [40] reduce to the results obtained by *Hunter* [13].

6. Concluding remarks

If the viscoelastic *Poissons* ratio $v(t)$ does not depend on t, and hence is a constant, the results obtained in Sections 3, 4 and 5 apply also to the case when F of eq. [5] depends on v. Hence when $v(t)$ is a constant, problems such as a punch on a viscoelastic layer which is rigidly glued to a plane surface are reduced to the integral equation, eq. [1a].

The following contact problems considered in (13, 14) are particular cases of the problem studied here. The problem studied here can be modified to include other motions such as the motion of a punch which is moving up and down periodically and indefinitely on the surface of the viscoelastic layer.

Summary

The problem of a rigid punch pressed against and moved on the surface of an elastic or viscoelastic layer is studied. It is shown that the governing equations reduce to the same integral equation for the elastic contact problem. Two particular motions of the punch are considered. In the first case the punch moves at a constant speed along a straight line on the surface of a viscoelastic layer. In the second case the punch moves at a constant speed along a circular path. Finally, the special case of a punch moving on a layer of a standard linear viscoelastic solid is studied. The equation is identical to a punch of modified shape pressed on an elastic layer.

References

1) *Sokolnikoff, I. S.*, Mathematical Theory of Elasticity (New York 1956).
2) *Sneddon, I. N.*, Fourier Transforms (New York 1951).
3) *Ting, T. C. T.*, J. Appl. Mech. **35**, 248–254 (1968).
4) *Ting, T. C. T.*, J. Appl. Mech. **33**, 845–854 (1966).
5) *Lee, E. H.*, Viscoelastic Stress Analysis, Structural Mechanics, Proceedings of the First Symposium on Naval Structural Mechanics (New York 1960).
6) *Lee, E. H.* and *J. R. M. Radok*, J. Appl. Mech. **27**, 438–444 (1960).
7) *Hunter, S. C.*, J. Mech. Physics Solids **8**, 219–234 (1960).
8) *Graham, G. A. C.*, Int. J. Eng. Sci., **3**, 27–45 (1965).
9) *Graham, G. A. C.*, Int. J. Eng. Sci. **5**, 495–514 (1967).
10) *Ting, T. C. T.* and *C. H. Wu*, J. Appl. Mech. **39**, 461–468 (1972).
11) *Efimov, A. B.*, Vestnik Moskovskogo Universiteta, Seriya 1, Matematika-Mekhanika, No. **2**, 120–127 (1966).
12) *Tsai, Y. M.*, Q. Appl. Math. **27**, 371–380 (1969).
13) *Hunter, S. C.*, J. Appl. Mech. **28**, 611–617 (1961).
14) *Morland, L. W.*, J. Appl. Mech. **29**, 345–352 (1962).
15) *Yang, W. H.*, J. Appl. Mech. **33**, 395–401 (1966).
16) *Christensen, R. M.*, Theory of Viscoelasticity (New York 1971).
17) *Gurtin, M. E.* and *E. Sternberg*, Arch. Rat. Mech. Analysis **11**, 291–356 (1962).

Author's address:

Dr. *T. C. T. Ting*
Dept. of Materials Engineering
University of Illinois, P.O. Box 4348
Chicago, Illinois 60680 (U.S.A.)

Rheol. Acta 12, 155–159 (1973)

De Centre de Recherches Physiques, Marseille (France)

Ecoulement à travers un milieu poreux isotrope *)

Par M. Chezeaux

Avec 4 figures

(Reçu p. p. le 27 octobre, 1972)

Introduction

On étudie l'écoulement d'un fluide visqueux à travers un milieu poreux[1]): le champ de vitesse qui caractérise le mouvement du fluide se déduit, à chaque instant, d'une fonction ψ biharmonique appelée « fonction de courant ». On montre que ψ dépend de la nature du fluide et de la structure du milieu poreux traversé. Pour définir cette structure on compare un tel milieu à un réseau géométriquement bien défini: les nœuds du réseau sont les particules solides et la forme géométrique de la maille est liée à la structure du milieu. On détermine le champ de vitesse du fluide, s'écoulant à travers les pores, en considérant, en premier lieu, l'écoulement autour d'un obstacle et en généralisant ensuite, par la théorie du potentiel à un ensemble de n obstacles régulièrement répartis. Pour simplifier le problème, on suppose le milieu isotrope (un axe de symétrie) ou stationnaire (un plan de symétrie).

A. Ecoulement autour d'un obstacle

Dans la plupart des milieux poreux étudiés, les particules solides restent immobiles lorsque s'écoule le fluide (laine de verre, éponge, roches) ou bien se déplacent, au cours du mouvement du fluide, avec une vitesse très faible et donc négligeable devant celle du fluide (sable par exemple): on considère donc l'obstacle fixe et on admet qu'il existe, sur la surface de séparation solide-fluide, une vitesse relative V_g du fluide par rapport à la paroi solide; V_g est fonction d'un coefficient de frottement de glissement μ (dans le cas où le fluide adhère aux parois $\mu \equiv 0$).

*) Communication présentée au VIe Congrès International de Rhéologie, Lyon, 4–8 Septembre 1972.

[1]) On appelle milieu poreux (1) un ensemble de particules solides (ou grains) entourées de particules fluides (ou pores) présentant une certaine cohésion; c'est en cela qu'il se distingue d'une suspension de particules solides dans un fluide. Pour exprimer cette cohésion, on suppose que l'ensemble des grains forme un domaine connexe indéformable.

I. Hypothèses

1. Le fluide est supposé incompressible visqueux et vérifiant l'équation de *Navier* qui, dans le cas d'un écoulement stationnaire et lent, s'écrit sous la forme vectorielle suivante:

$$\varrho(V \cdot \nabla)\, V = -\nabla p + \eta \nabla^2 V\,. \qquad [1]$$

ϱ et η sont respectivement la densité et la viscosité du fluide, p la pression, V la vitesse d'écoulement.

De plus, le mouvement vérifie l'équation de continuité:

$$\varrho(\nabla, V) = 0. \qquad [2]$$

De l'équation [2] on déduit qu'il existe une fonction ψ telle que, dans le cas d'un mouvement plan par exemple, on ait:

$$\vec{V}(x, y) = -\frac{\vec{K}}{y} \Lambda \psi(x, y). \qquad [3]$$

En prenant le rotationnel de l'équation [1] et en considérant la formule suivante:

$$\nabla \cdot (\nabla, V) = \nabla \Lambda (\nabla \Lambda V) + (\nabla, \nabla)\, V.$$

On obtient:

$$\nabla_3 \Lambda V = 0 \Rightarrow \nabla^4 \psi = 0 \qquad [4]$$

la *fonction de courant* ψ est *biharmonique;* la détermination de la fonction ψ permet à partir de l'éq. [3] de calculer le champ de vitesse V du fluide en tout point des pores.

2. Le mouvement du fluide admet un axe de symétrie (l'axe $z'z$ par exemple). Dans ce cas particulier la fonction de courant est symétrique par rapport à $z'z$: le mouvement est donc le même dans chaque plan perpendiculaire à cet axe; on définit ainsi la fonction $\psi(M)$ considérée au point M au moyen des deux coordonnées: z et $\varpi = (x^2 + y^2)^{1/2}$, ϖ représentant la distance du

point M considéré à l'axe $z'z$ (2). Si u et v désignent les composantes de la vitesse V suivant la direction $z'z$ et le plan $\bar{\omega}$ d'abscisse z, on déduit de l'éq. [3]

$$u = -\frac{1}{\bar{\omega}} \frac{\partial \psi}{\partial \bar{\omega}}$$

$$v = \frac{1}{\bar{\omega}} \frac{\partial \psi}{\partial z}. \qquad [5]$$

Si on appelle $V(n)$ et $V(T)$ les composantes de la vitesse considérées respectivement suivant le rayon vecteur et la direction perpendiculaire à ce rayon vecteur (fig. 1), on a:

$$u = V(n) \cos\theta - V(T) \sin\theta$$

$$v = V(n) \sin\theta + V(T) \cos\theta$$

une combinaison des relations [5] et [6] donne:

$$V(n) = -\frac{1}{r^2 \sin\theta} \frac{\partial \psi}{\partial \theta}$$

$$V(T) = -\frac{1}{r \sin\theta} \frac{\partial \psi}{\partial r}. \qquad [6]$$

Cas particulier: écoulement irrotationnel

Si

$$(\nabla \Lambda V) = 0 \Rightarrow V = (\nabla \cdot \varphi)$$

et avec la condition:

$$(\nabla, V) = 0 \Rightarrow \nabla^2 \varphi = 0. \qquad [7]$$

Si l'écoulement est irrotationnel, le champ de vitesse V se déduit d'une fonction potentiel de vitesse φ harmonique et la fonction de courant ψ est orthogonale à φ.

II. Conditions aux limites

On cherche donc à déterminer la fonction biharmonique $\psi(M)$ définie et régulière à l'intérieur du domaine $\mathscr{D}_1$ (pore) sachant que l'on a en tout point P de la frontière Σ_0 (surface de séparation solide-fluide) (fig. 2) les conditions de vitesse:

$$V(n) = 0$$

$$V(T) = V_g. \qquad [8]$$

La condition $V(n) = 0$ se déduit de l'éq. [1]: le fluide ne peut pénétrer à l'intérieur de l'obstacle, la condition $V(T) = V_g$ suppose qu'il existe sur Σ_0 une vitesse de glissement du fluide par rapport à la paroi ⇒ le fluide n'adhère pas à la paroi. On a donc le problème suivant: trouver une fonction biharmonique ψ connaissant sur Σ_0 les dérivées partielles de ψ: c'est le *problème de Lauricella* et *Mathieu* (analogue au problème de *Neumann* des fonctions harmoniques); ψ est ainsi déterminée à une constante près. On peut calculer cette constante si on considère les valeurs des dérivées

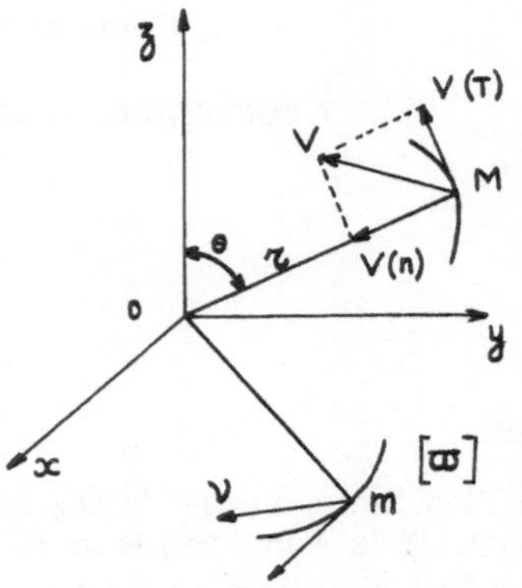

Fig. 1

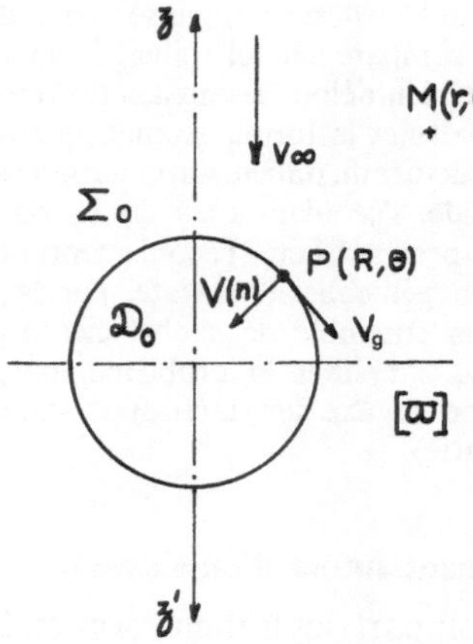

Fig. 2

partielles de ψ à l'infini (ou loin de l'obstacle). Par raison de symétrie, on pose que, loin de l'obstacle, le fluide s'écoule avec une vitesse V_∞ constante et dirigée parallèlement à l'axe $z'z$ d'écoulement, on a donc [7]:

$$r \to \infty \qquad V = (\nabla \cdot \varphi) \Rightarrow \nabla^2 \varphi = 0. \qquad [9]$$

Loin de l'obstacle, le champ de vitesse dérive d'une fonction φ potentiel de vitesse et la fonction de courant ψ lui est orthogonale: φ et ψ sont reliées entre elles par les conditions de *Cauchy*.

III. Détermination de la fonction de courant

Pour exprimer la fonction $\psi(M)$ biharmonique dans le domaine $\mathscr{D}_1$ de frontière Σ_0, on applique la *formule d'Almansi:*

Soit $\psi(M)$ une fonction biharmonique, régulière dans un certain domaine $\mathcal{D}_1$, M_0 étant un point fixe intérieur à $\mathcal{D}_1$, posons $M_0M = r$. *E. Almansi* a montré qu'il existe deux fonctions harmoniques $\psi_1(M)$ et $\psi_2(M)$ régulières dans $\mathcal{D}_1$ telles que l'on ait identiquement, à l'intérieur de $\mathcal{D}_1$ de frontière Σ_0 (fig. 2):

$$\psi(M) = (r^2 - R^2)\psi_1(M) + \psi_2(M) \qquad [10]$$

pour M compris dans le plus grand domaine étoilé inscrit dans $\mathcal{D}_1$ et ce développement est unique (3); R est une constante qui caractérise la frontière Σ_0 (par exemple dans le cas de l'obstacle sphérique R est le rayon de la sphère) sur Σ_0, on écrit:

$$V(n) = -\frac{1}{R^2 \sin\theta}\left[\frac{\partial\psi(\theta, R)}{\partial\theta}\right] = 0$$

$$V(T) = -\frac{1}{R \sin\theta}\left[\frac{\partial\psi(\theta, r)}{\partial r}\right]_{r=R} = V_g \qquad [11]$$

$\psi_1(M)$ et $\psi_2(M)$ sont des fonctions harmoniques solutions de l'équation de *Laplace*: pour un écoulement de révolution autour de l'axe $z'z$ et dans l'espace R^3, la solution partielle de première espèce de l'équation de *Laplace*, en coordonnées cylindriques, s'écrit:

$$\psi_1(M) = (\alpha_1 r^2 + \beta_1 r^{-3})\sin^2\theta$$

$$\psi_2(M) = (\alpha_2 r^2 + \beta_2 r^{-3})\sin^2\theta.$$

α_i et β_i sont des constantes déterminées à partir des conditions aux limites [9] et [11] et de la forme de l'obstacle, les calculs donnent:

$$V(n) = -2\cos\theta\left[(r^2 - R^2)\left(\alpha_1 + \frac{\beta_1}{r^3}\right) + \alpha_2 + \frac{\beta_2}{r^5}\right]$$

$$V(T) = -\sin\theta\left[2\alpha_1(2r^2 - R^2) - \frac{\beta_1}{r^5}(r^2 - 3R^2) + 2\alpha_2 - \frac{3\beta_2}{r^5}\right]. \qquad [13]$$

B. Application de la théorie des réseaux à un milieu poreux

On compare la structure d'un milieu poreux isotrope ou stationnaire à un réseau d'objets diffringents à maille régulière dont les noeuds sont les particules solides et la forme de la maille dépend de la répartition des grains (ou obstacles) et des pores. On considère que le ε ième pore est extérieur au domaine $\mathcal{D}_\varepsilon$ de frontière Σ_ε (définie par $r = R_\varepsilon$) et l'ensemble des pores forme le domaine $\mathcal{D}$ de frontière Σ (définie par $r = R$) et extérieur à l'ensemble des domaines $\mathcal{D}_\varepsilon$ (fig. 3).

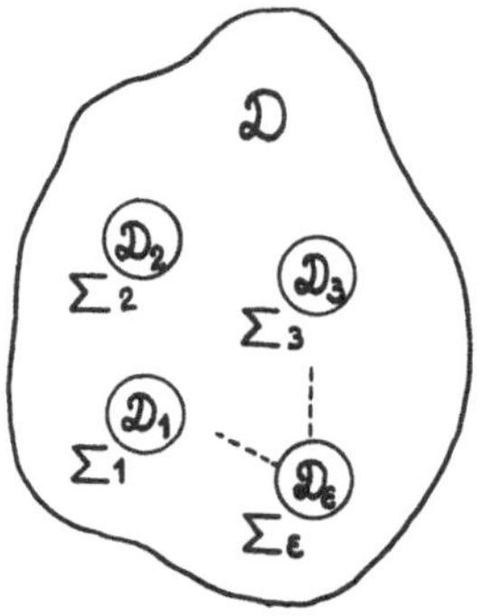

Fig. 3

I. Hypothèses

1. Dans chaque pore le mouvement du fluide vérifie l'éq. [1] et admet l'axe $z'z$ pour axe de symétrie: pour chaque obstacle considéré isolément, on définit donc une fonction de courant ψ biharmonique qui peut être caractérisée au moyen de la formule d'*Almansi* [10]. Ainsi pour le ε ième obstacle pris isolément, on a:

$$\psi_\varepsilon(M) = (r_\varepsilon^2 - R_\varepsilon^2)\psi_{1\varepsilon}(M) + \psi_{2\varepsilon}(M) \qquad [14]$$

avec:

$$\psi_{1\varepsilon}(M) = (\alpha_{1\varepsilon} r_\varepsilon^2 + \beta_{1\varepsilon} r_\varepsilon^{-3})\sin^2\theta_\varepsilon$$

$$\psi_{2\varepsilon}(M) = (\alpha_{2\varepsilon} r_\varepsilon^2 + \beta_{2\varepsilon} r_\varepsilon^{-3})\sin^2\theta_\varepsilon.$$

II. Conditions aux limites

On cherche la fonction de courant $\psi(M)$ solution de l'équation $\nabla^4\psi = 0$ définie et régulière dans un domaine $\mathcal{D}$ extérieur à des domaines $\mathcal{D}_\varepsilon$ et vérifiant sur les frontières Σ_ε de ces domaines les conditions de *Lauricella*:

$$\varepsilon \text{ ième obstacle} \quad \frac{\partial\psi(R_\varepsilon, \theta_\varepsilon)}{\partial\theta_\varepsilon} = 0$$

$$\frac{1}{R_\varepsilon}\left[\frac{\partial\psi(r_\varepsilon, \theta_\varepsilon)}{\partial r}\right]_{r_\varepsilon = R_\varepsilon} = V_g. \qquad [15]$$

III. Expression de la fonction de courant d'un réseau

1. Réseau rectiligne d'obstacles parallèles

Soit un ensemble d'obstacles identiques $\mathcal{D}_\varepsilon$ disposés suivant l'axe $z'z$ de telle façon qu'ils puissent se déduire les uns des autres par des translations de module constant $2d$ (fig. 4), on déduit (4):

$$\psi_\varepsilon = (r_\varepsilon^2 - R^2)\,\psi_{1\varepsilon}(M) + \psi_{2\varepsilon}(M)$$
$$\varepsilon = 0, \pm 1, \pm 2, \ldots$$

$$\frac{\partial\psi(R,\theta_\varepsilon)}{\partial\theta_\varepsilon} = 0$$

$$\frac{1}{R}\left[\frac{\partial\psi(r_\varepsilon,\theta_\varepsilon)}{\partial r}\right]_{r_\varepsilon = R_\varepsilon} = V_g$$
$$\varepsilon = 0, \pm 1, \pm 2, \ldots \qquad [16]$$

ce qui entraîne:

$$\frac{\partial\psi(R,\theta_0)}{\partial\theta_0} \equiv \frac{\partial\psi(R,\theta_1)}{\partial\theta_1} \cdots \equiv \frac{\partial\psi(R,\theta_\varepsilon)}{\partial\theta_\varepsilon} = 0$$

$$\left[\frac{\partial\psi(r_0,\theta_0)}{\partial r}\right]_{r_0 = R} \equiv \left[\frac{\partial\psi(r_1,\theta_1)}{\partial r}\right]_{r_1 = R} \cdots$$
$$\equiv \left[\frac{\partial\psi(r_\varepsilon,\theta_\varepsilon)}{\partial r}\right]_{r_\varepsilon = R} = V_g \qquad [17]$$

avec

$$r_0^2 = y_0^2 + z_0^2$$
$$r_1^2 = y_0^2 + z_1^2 = y^2 + (zc - 2d)^2$$
$$\vdots$$
$$r_\varepsilon^2 = y_0^2 + (z_0 - 2\varepsilon d)^2.$$

Les relations [17] montrent qu'il suffit d'écrire les conditions aux limites de *Lauricella* sur un seul obstacle. Par suite de la propriété de linéarité de la formule d'*Almansi* et des fonctions harmoniques, la fonction de courant totale du réseau est la somme des fonctions de courant calculées pour chaque obstacle supposé centré sur l'axe perpendiculaire à $z'z$ (en O)

$$\psi = \sum_{n=-\varepsilon}^{n=+\varepsilon} \psi_n \qquad [18]$$

soit:

$$\psi = \sum_{n=-\varepsilon}^{n=+\varepsilon} \left[(r_n^2 - R^2)(\alpha_{1n} r_n^2 + \beta_{1n} r_n^{-3}) + \alpha_{2n} r_n^2 + \beta_{2n} r_n^{-3}\right]$$

avec

$$r_n^2 = y_0^2 + (z_0 - 2nd)^2$$

de cette valeur de ψ on déduit les composantes de vitesse $V(n)$ et $V(T)$ calculées dans les éq. [13].

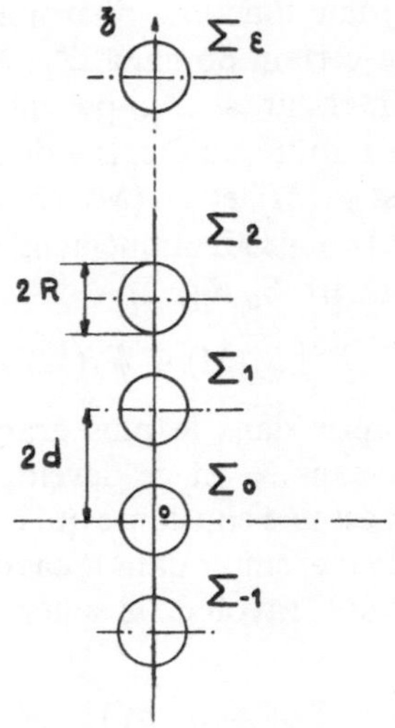

Fig. 4

2. Réseau à maille trirectangle (milieu stationnaire)

Soit dans le cas d'un réseau régulier à trois dimensions un ensemble d'obstacles qui se déduisent les uns des autres par des translations de module $2b$ (suivant l'axe Oz), $2c$ (sur Oy) et $2d$ (sur Oz). Par analogie avec les équations précédentes, on a de même (4)

$$\psi = \sum_{n_1=-\zeta}^{n_1=+\zeta} \sum_{n_2=-\tau}^{n_2=+\tau} \sum_{n_3=-\varepsilon}^{n_3=+\varepsilon} \psi_{n_1 n_2 n_3} \qquad [19]$$

Posons pour simplifier l'écriture:

$$\psi_{n_1 n_2 n_3} = \psi_{\varepsilon'}.$$

on a:

$$\psi_{\varepsilon'} = (r_{\varepsilon'}^2 - R^2)\,\psi_{1\varepsilon'} + \psi_{2\varepsilon'}$$

et

$$r_{\varepsilon'}^2 = (x_0 - 2n_1 b)^2 + (y_0 - 2n_2 c)^2 + (z_0 - 2n_3 d)^2.$$

C. Applications:

On applique les résultats précédents à un échantillon de milieu poreux à grains sphériques de rayon R. On calcule d'abord les constantes α_i et β_i de l'éq. [12].

I. Calcul des constantes α_i et β_i

A l'infini, on a

$$V(n) = V_\infty = -\frac{\partial\varphi}{\partial r}$$

$$\Rightarrow -2[\alpha_1(r^2 - R^2) + \alpha_2] = V_\infty$$

$$\alpha_1 \equiv 0, \quad \alpha_2 = -\frac{V_\infty}{2}. \qquad [20]$$

Sur

$$\Sigma_0(r = R): V(n) = 0, \quad V(T) = V_g$$

$$\left.\begin{array}{l} \alpha_2 + \beta_2 R^{-5} = 0 \\ 2\beta_1 R^{-3} + 2\alpha_2 - 3\beta_2 R^{-5} = V_g \end{array}\right\} \Rightarrow$$

$$\Rightarrow \begin{cases} \beta_1 = \frac{1}{2}(V_g - 5V_\infty) R^3 \\ \beta_2 = +\frac{1}{2} V_\infty R^5. \end{cases} \qquad [21]$$

II. Applications à un réseau isotrope ou stationnaire

Exemple 1

Le réseau est de type cubique centré (isotrope): la maille élémentaire est un rhomboèdre de côté $a\sqrt{3}$ (si les longueurs des côtés du cube sont $2a$), les vecteurs de translation fondamentaux sont tous égaux à $a\sqrt{3}$. De l'éq. [19], on déduit.

$$\psi = \sum_{\zeta=-l_1}^{\zeta=+l_1} \sum_{\tau=-l_2}^{\tau=+l_2} \sum_{\varepsilon=-l_3}^{\varepsilon=+l_3} \tfrac{1}{2}(r^2 - R^2)(V_g - 5V_\infty) - \tfrac{1}{2} V_\infty(1 - R^2)$$

avec

$$r^2 = (x - a\sqrt{3}\zeta)^2 + (y - a\sqrt{3}\tau)^2 + (z - a\sqrt{3}\varepsilon)^2.$$

Exemple 2

Le réseau est de type hexagonal: dans ce cas on a deux vecteurs fondamentaux dans le plan xOy égaux à a (si $2a$ est la longueur du côté du losange de base) et un vecteur fondamental suivant Oz de longueur c. La fonction de courant est identique à la formule [22] avec, cette fois-ci,

$$r^2 = (x - a\zeta)^2 + (y - a\tau)^2 + (z - c\varepsilon)^2.$$

D. Conclusion

La vitesse de glissement V_g est reliée à la viscosité du fluide par l'équation de *Navier* (V_g est inversement proportionnel à η). Dans le cas d'adhésion parfaite ($\mu \equiv 0$) on obtient un écoulement indépendant de cette viscosité. Par contre si μ existe et si on calcule le débit Q en volume du fluide traversant le milieu poreux: $Q = \int_v |V|\,dv$ on peut vérifier que dans le cas de vitesses lentes le débit est inversement proportionnel à la viscosité du fluide: on retrouve ainsi la loi de *Darcy* et la notion d'analogie milieu poreux-réseau semble au départ valable si on admet qu'il existe dans le milieu poreux des liaisons entre le particules solides et fluides.

Littérature

Biot, M. A., J. acoust. Soc. Amer. **28**, 168–191 (1956).
Lamb, H., Hydrodynamics. 6e edit. (Cambridge 1932).
Nicolesco, M., Les fonctions polyharmoniques (Paris 1936).
Dumery, G., Thèse de Docteur, Faculté des Sciences (Marseille 1967).

Adresse de l'auteur:

Mme *Michèle Chezeaux*
Centre de Recherches Physiques
31, chemin Joseph-Aiguier (9 arr.)
F-13274 Marseille Cedex 2 (France)

Rheol. Acta **12**, 160–164 (1973)

Lehrstuhl und Institut für Mechanische Verfahrenstechnik der Universität Erlangen-Nürnberg

Une solution similaire pour l'écoulement dans la couche limite d'un fluide polaire

Par F. Ebert

Avec 3 figures

(Reçu p. p. le 27 octobre, 1972)

Au génie chimique on trouve souvent des fluides compliqués comme les suspensions, les solutions des haut-polymères, les émulsions etc. En général c'est impossible de décrire l'écoulement de ces fluides par la hydrodynamique classique. C'est pourquoi on a essayé d'élargir la conception classique d'un fluide pour être capable à traiter aussi des fluides avec des structures compliquées.

La théorie des «fluides polaires» représente une généralisation du concept classique d'un fluide. Il semble convenable de décrire les fluides dont on vient de parler par cette théorie. On peut résumer l'essentiel de la théorie, dont les principes sont donnés par *E.* et *F. Cossérat* (1), comme suit: On considère le mouvement des éléments fluides dans un petit volume situé dans la région de l'écoulement. La dimension caractéristique de l'élément du fluide est supposée suffisamment petite relative aux dimensions du volume, pour que la matière puisse être considérée comme continue. Dans la théorie des fluides classiques – par exemple pour le fluide *Newton*ien – on attribue au mouvement d'un particule fluide les degrés de liberté pour la transition translatoire. En plus, dans la théorie des «fluides polaires» des degrés de liberté relatif à la rotation et la déformation de l'élément sont rajoutés. La description cinématique est alors reliée à deux champs des vecteurs, l'un indépendant de l'autre:

a) On a le champ des vitesses qui donne la vitesse moyenne d'un élément en un point quelquonque dans la région de l'écoulement. Il est relié à la rotation d'un petit élément fluide entourant le point P dans la manière bien connue:

$$\omega = \operatorname{rot} v. \quad [1]$$

La déformation d'un petit volume de fluide peut être décrite par le tenseur de la déformation $\tilde{d}$, que l'on obtient du champ des vitesses par

$$d = \tfrac{1}{2}(\nabla v + (\nabla v)^T). \quad [2]$$

b) Il est possible d'attacher un trièdre à chaque élément du fluide. Ce trièdre permet de fixer les dimensions de l'élément fluide et son orientation.

La méthode donnée de décrire un continu, on peut attribuer aux particules de fluide une orientation moyenne, représentative pour le point P situé dans le volume considéré. La variation de la géométrie du trièdre au courant du mouvement est décrite par deux quantités cinématiques:

La première exprime la rotation du trièdre. Elle peut être définie par un tenseur antisymétrique ou par un vecteur axial.

Ce vecteur constitue le champ de spin ν.

La deuxième donne la déformation du trièdre et par conséquent celle du particule correspondant. La quantité convenable est le tenseur de deformation du particule $\tilde{W}$.

Dans ce qui suit on considère des fluides, dont les éléments sont rigides. En ce cas les éléments possèdent un spin, tandis que l'on n'observe pas de déformation.

La présence d'un spin cause des moments de déformation à la surface d'un élément. Ces moments sont représentés par le tenseur, $\tilde{m}$. En plus, il faut tenir compte de la quantité de mouvement angulaire dans une manière analogue au traitement de l'impulsion qui caractérise la transition translatoire d'un particule de fluide.

Sur ce sujet, sur la théorie des fluides polaires, un ouvrage a été publié par *Cowin* (2) et *Eringen* (3, 4).

Les premières calculations des très simples champs des vitesses sur la base de cette théorie ont démontré, que l'influence des moments de déformation est particulièrement effective dans des zones discontinues comme dans la zone de la couche limite ou dans la région d'une plaine séparant deux domaines des vitesses.

Les équations constitutives du fluide considéré sont données par:

$$t_{ij} = (-p + \lambda_v d_{nn}\delta_{ij} + (2\mu_v + \varkappa_v)\, d_{ij} + \varkappa_v \varepsilon_{ijk}(\omega_k - \nu_k) \quad [3]$$

$$m_{ij} = \alpha_v \nu_{n,n}\delta_{ij} + \beta_v \nu_{i,j} + \gamma_v \nu_{j,i}. \quad [4]$$

δ_{ij} composantes du tenseur unitaire
t_{ij} composantes du tenseur des contraintes
m_{ij} composantes du tenseur des moments de deformation
ω_k composantes du tourbillon donné par éq. [1]
ν_k composantes du spin
$\alpha_v, \beta_v, \gamma_v, \varkappa_v, \lambda_v, \mu_v$ coéfficients de viscosité

Supposons que le courant soit incompressible l'équation de continuité s'écrit:

$$v_{n,n} = 0. \quad [5]$$

Le bilan des forces donne

$$\varrho \frac{d}{dt} v_i = t_{ni,n} + \varrho f_i. \quad [6]$$

f_i composantes d'une force extérieure

Le bilan des quantités de mouvement angulaire s'écrit

$$2\varrho j \varepsilon_{rki} \frac{d}{dt} \nu_r = t_{[ik]} - 2\varepsilon_{ikr} m_{nr,n} + \varrho l_{[ik]}. \quad [7]$$

l_{ik} composantes des extérieures quantités de mouvement angulaire
j moment d'inertie des particules du fluide

On peut attribuer le comportement donné par ces équations constitutives à une suspension des fines particules dispersées dans un fluide visqueux.

Aux parois fixes il faut que les conditions d'adhérence soient satisfaisantes. Cela veut dire que les éléments du fluide se meuvent comme la paroi.

Modèle de la couche limite

Considérons le permanent écoulement le long d'une paroi fixe. On utilise les coordonnées cartésiennes x, y, z; les composantes de vitesse soient u, v, w, celles du spin o, p, q.

Si les approximations pour la couche limite sont introduites dans la manière connue dans l'équation de continuité et le bilan des forces, le rapport de l'épaisseur de la couche limite par une longueur L, caractérisant la configuration de l'écoulement, est donné par:

$$\frac{\delta}{L} \sim \frac{1}{\sqrt{Re^*}} \sim \frac{1}{\mathrm{Re}_M}. \quad [8]$$

Le nombre de *Reynolds* Re* est différent du nombre de *Reynolds* utilisé pour des fluides classiques dans le terme de la viscosité.

$$\mathrm{Re}^* = \frac{\varrho u_\infty L}{\mu_v + \varkappa_v}. \quad [9]$$

Le nombre de *Reynolds* Re_M est défini à l'aide de la quantité Q caractérisant le spin. Le décisif coéfficient de viscosité est $\varkappa_v$:

$$\mathrm{Re}_M = \frac{\varrho u_\infty^2}{\varkappa_v Q}. \quad [10]$$

L'éq. [8] implique que les forces visqueuses sont du même ordre de grandeur que les forces d'inertie, ce qui s'applique aussi aux moments de déformation.

Les bilans des forces s'écrivent alors

$$u\frac{\partial u}{\partial x} + v\frac{\partial u}{\partial y} = -\frac{1}{\varrho}\frac{\partial p}{\partial x} + \frac{1}{\varrho}(\mu_v + \varkappa_v)\frac{\partial^2 u}{\partial y^2} + \frac{\varkappa_v}{\varrho}\frac{\partial q}{\partial y}; \quad \frac{\partial p}{\partial y} \approx 0. \quad [11]$$

L'équation de continuité donne

$$\frac{\partial u}{\partial x} + \frac{\partial v}{\partial y} = 0. \quad [12]$$

La signification du paramètre $\varkappa_v$ est évident quand on considère l'éq. [11]. Elle donne l'accouplement dynamique du champ de spin et du champ des vitesses.

Tenant compte des hypothèses sur la couche limite, la balance de la quantité de mouvement angulaire est donnée par la relation

$$u\frac{\partial q}{\partial x} + v\frac{\partial q}{\partial y} = \frac{\gamma_v}{\varrho j}\frac{\partial^2 q}{\partial y^2} - \frac{\varkappa_v}{\varrho j}\frac{\partial u}{\partial y} - 2\frac{\varkappa_v}{\varrho j} q. \quad [13]$$

Il faut remarquer que les hypothèses de la couche limite nécessitent la rélation suivante:

$$j \quad \frac{\gamma_v}{(\mu_v + \varkappa_v)} = l^2. \quad [14]$$

Cela signifie, que le moment d'inertie d'un élément du fluide est de l'ordre de grandeur du rapport du coéfficient γ_v sur $\mu_v + \varkappa_v$. γ_v est donné seulement dans le bilan de la quantité de mouvement angulaire. $\mu_v + \varkappa_v$ détermine les termes des contraintes dans le bilan d'impulsion. La racine carrée de ce quotient est homogène à une longueur. On peut l'interpréter comme longueur caractéristique l d'une élément du fluide. On voit, que le moment d'inertie des éléments par raport à la densité est de l'ordre de la longueur caractéristique au carré.

Disposition d'une solution similaire

Nous cherchons une solution similaire du système des éqs. [11, 12, 13]. Une solution similaire est caractérisée par le fait que les distributions du spin et de la vitesse à différentes positions x le long de la paroi deviennent congruentes si elles sont normalisées avec des facteurs d'échelle appropriés. Ces facteurs d'échelle ne dépendent que de la coordonnée x. En ce cas les équations différentielles partielles deviennent des équations différentielles ordinaires.

L'équation de continuité est satisfaite par la fonction de courant $\psi(x, y)$.

Nous proposons la forme suivante non dimensionelle pour la fonction de courant,

$$f(\xi, \eta) = \frac{\sqrt{\mathrm{Re}^*}}{L u_\delta(x) g(x)} \psi(x, y),$$

$$\xi = \frac{x}{L}, \quad \eta = \frac{y\sqrt{\mathrm{Re}^*}}{L g(x)}. \qquad [15]$$

La fonction de spin non-dimensionelle est définie par

$$k(\xi, \eta) = \frac{\varkappa_v L}{(\mu_v + \varkappa_v)\, h(x)\, u_\delta(x) \sqrt{\mathrm{Re}^*}} q(x, y). \qquad [16]$$

$u_\delta(x)$ est la vitesse extérieure de la couche limite. $g(x)$ et $h(x)$ sont des facteurs d'échelle non dimensionels.

Si l'on porte ces définitions dans le système d'équations on obtient

$$f''' + f''f + 1 - (f')^2 + k' = o, \qquad [17]$$

$$k'' + k'f - (E + f')k - Gf'' = o, \qquad [18]$$

$$(') = \frac{d}{d\eta}$$

où les coéfficients E et G sont donnés par

$$G = \left(\frac{\delta}{l}\right)^2 \left(\frac{\varkappa_v}{\varkappa_v + \mu_v}\right)^2 \left(\frac{d(u_\delta/u_\infty)}{d(x/L)}\right)^{-1} = \frac{1}{2} E \frac{\varkappa_v}{\varkappa_v + \mu_v}. \qquad [19]$$

En calculant les éqs. [17, 18] on a tenu compte de l'hypothèse de la similarité. Les fonctions f et k ne dépendent que de la coordonnée non dimensionelle qui donne la distance de la paroi. Les coéfficients dans les équations sont constants. Cela nécessite que la vitesse extérieure u_δ soit proportionelle à la coordonnée x (« mouvement dans le voisinage du point d'arrêt»). Il faut que f et k satisfassent les conditions suivantes:

$$\eta = o: \quad f = f' = k = o; \quad f'' = W_f, \; k' = W_q;$$

$$\eta \to \infty: \quad f' = 1, \quad f'' = f''' = k' = k'' = o.$$

W_f resp. W_q est la pente de la distribution de la vitesse resp. du spin sur la paroi fixe.

Le résultat de l'intégration numérique des équations est représenté dans la fig. 1. La distribution de la vitesse non dimensionelle est portée en ordonnée et la coordonnée en abscisse. Le quotient de l'épaisseur de la couche limite sur la longueur caractéristique 1 des particules du fluide est constant. Le paramètre $\mu_v/\varkappa_v$ détermine l'accouplement du champ des vitesses au champ de spin. Pour $\mu_v/\varkappa_v \to \infty$ nous obtenons le cas d'un fluide *Newton*ien qui a été calculé par

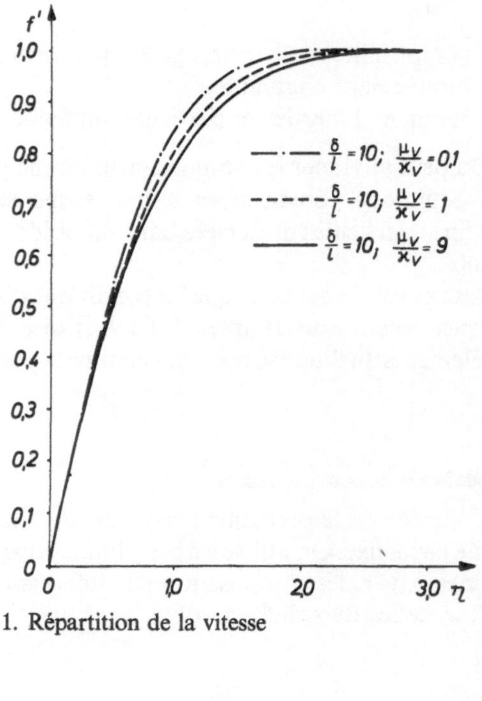

Fig. 1. Répartition de la vitesse

Hartree (5, pour $\beta = 1$). La fig. 2 représente la distribution du spin non dimensionel. Les courbes démontrent un maximum prononcé pour des petites distances de la paroi. La distance η croissante, le spin tend vers zéro. Le spin est particulièrement effectif dans le voisinage immédiat de la paroi ce qui a été mentionné au commencement.

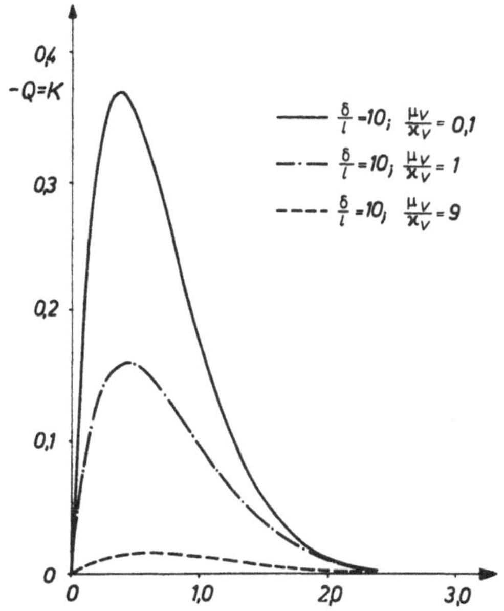

Fig. 2. Répartition du spin

Les fonctions f' et k connues, on peut calculer les contraintes effectuées par le fluide polaire dans la couche limite. Il est alors intéressant de comparer le comportement d'un fluide polaire à celui d'un fluide *Newton*ien. Nous considérons tout d'abord le frottement à la paroi que l'on obtient par l'éq. [3].

$$(\tau_{xy})_{y=0} = (\mu_v + \varkappa_v)\left(\frac{\partial u}{\partial y}\right)_{y=0} + \varkappa_v(q)_{y=0}. \quad [20]$$

$(\tau_{xy})_{y=0}$ donne la contrainte tangentielle effectuée sur une plaine $y = \text{const}$ en direction de la coordonnée x. Si l'on suppose des conditions équivalentes du courant on reconnait que le frottement effectué par un fluide polaire est toujours plus grand que celui du fluide *Newton*ien.

Si l'on compare les composantes de la contrainte tangentielle en direction de la paroi avec les contraintes normales on voit que les composantes ne sont pas égales. Par conséquent, le tenseur des contraintes est asymétrique dans le cas d'un fluire polaire. On constate que l'expression

$$\frac{\tau_{yx}}{\tau_{xy}} < 1$$

est toujours valable pour l'écoulement considéré.

Il est intéressant de voir l'influence du paramètre δ/l illustrée pour le frottement à la paroi caractérisée par $(\tau_{xy})_{y=0}$ (fig. 3). Supposons que les paramètres de l'écoulement soient constants. Tant que la longueur caractéristique l du particule du fluide est petite comparée avec l'épaisseur de la couche limite δ, la variation du frottement reste sans importance. Quand la longueur se rapproche de l'épaisseur le frottement à la paroi augmente fortement. (Supposons pour le cas

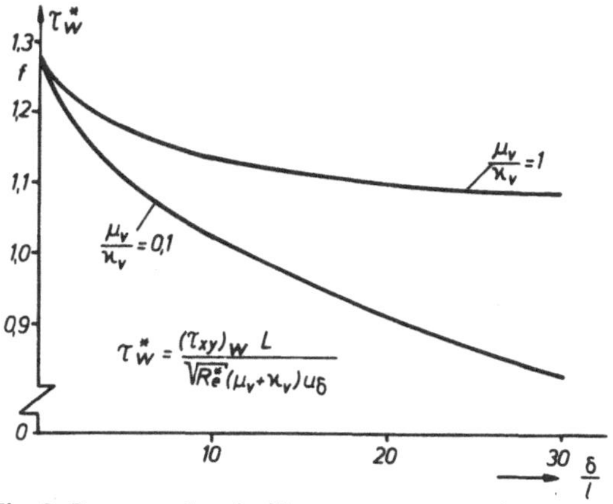

Fig. 3. Rapport des cisaillements correspondants

technique une couche limite de l'ordre de grandeur de 1 cm, le diamètre des particules effectuant les moments de déformation de 1 μm – 1 mm. Cela signifie que δ/l varie de 10 à 10^3.) Il faut encore dire que l'écoulement des fluides polaires cause des contraintes normales. On obtient pour la contrainte normale en direction parallèle à la paroi la formule suivante

$$\tau_{xx} = (2\mu_v + \varkappa_v)\frac{\partial u}{\partial x} = -\tau_{yy}. \quad [21]$$

Cette quantité représente une contrainte à traction tandis que la contrainte normale sur la paroi constitue une contrainte à compression. Dans cet ordre d'idées le fluide polaire démontre le comportement d'un fluide viscoélastique.

Summary

The micropolar fluid defined by the constitutive equations of *Eringen* (3, 4) is to be considered. This fluid is characterized by describing the motion of the fluid elements by two kinematic fields: The velocity field and the spin field. The corresponding equations of motion are specialized for boundary layer flow. A similar solution can be given for "flow near stagnation" as outer flow.

Concerning the flow properties of the polar fluid two parameters are of main interest: First, the ratio of boundary layer thickness to characteristic length scale of the fluid elements; secondly, a parameter describing the coupling between velocity and spinfield. The solution is discussed comparing the behaviour of the micropolar fluid to that of a *Newton*ian liquid.

Zusammenfassung

Für die mikropolare Flüssigkeit nach *Eringen* (3, 4) werden die Grenzschichtgleichungen abgeleitet. Diese Flüssigkeit ist dadurch gekennzeichnet, daß die Bewegung ihrer Elemente durch zwei kinematische Felder gegeben ist, das Geschwindigkeits- und das Spinfeld. Für die Staupunktsströmung als Außenströmung kann man eine Ähnlichkeitslösung der Grenzschichtgleichungen angeben. Für das Strömungsverhalten der polaren Flüssigkeit erweisen sich zwei Parameter als wesentlich: Das Verhältnis von Grenzschichtdicke zu charakteristischer Abmessung des Fluidelements sowie ein Parameter, der die Kopplung zwischen Geschwindigkeitsfeld und Spinfeld wiedergibt. Die Lösung der Grenzschichtgleichungen wird diskutiert, wobei die Strömungseigenschaften von polarer und *Newton*scher Flüssigkeit einander gegenübergestellt werden.

Littérature

1) *Cossérat, E.* et *F. Cossérat,* Théorie des corps déformables (Paris 1909).

2) *Cowin, S. C.,* Mechanics of Cossérat Continua. Ph.-D.-Thesis, The Pennsylvania State University 1962.

3) *Eringen, C. A.,* Int. J. Engng. Sci. **2**, 205–217 (1964).

4) *Eringen, C. A.,* J. Math. Mech. **16**, 1–18 (1966).

5) *Hartree, D. R.,* Proc. Cambr. Phil. Soc. **33**, Part II.

Adresse de l'auteur:

Dipl.-Ing. Dr. *F. Ebert*

Lehrstuhl f. Mech. Verfahrenstechnik
D-8520 Erlangen
Egerlandstraße 5

Rheol. Acta **12**, 165–169 (1973)

From the School of Physical Sciences, The New University of Ulster, Coleraine (Northern Ireland)

The orientation of liquid crystals by temperature gradients

By P. K. Currie

(Received October 27, 1972)

1. Introduction

It was reported by *Stewart* (1) that when the liquid crystal *p*-azoxyanisole was placed in a vertical temperature gradient the molecular orientation adopted was vertical when the higher temperature was at the bottom and horizontal when the higher temperature was at the top. These observations were confirmed by *Holland* and *Stewart* (2), *Stewart, Holland,* and *Reynolds* (3), and *Stewart* (4), the latter stressing that the horizontal orientation was not produced by convection currents; in fact the horizontal orientation was observed only when the vertical convection currents were reduced as much as possible. More recently, *Picot* and *Fredrickson* (5) und *Fisher* and *Fredrickson* (6) have doubted that a temperature gradient can exert an orienting influence, while *Patharkar, Rajan,* and *Picot* (7) give experimental evidence suggesting that such an influence may exist.

It is clear from the experiments of *Stewart, Holland,* and *Reynolds* that the orienting effect, if it exists, must be due to a coupling between the vertical gravitational body force and the temperature effects; otherwise there would be no change of orientation when the direction of the temperature gradient is reserved. This may explain why *Fisher* and *Fredrickson*, who studied the effect of a horizontal temperature gradient, could find no evidence of an orienting influence, while *Patharkar, Rajan,* and *Picot*, with a vertical temperature gradient, found some such evidence.

A theoretical explanation of this phenomenon is suggested here, using the continuum theory for liquid crystals proposed by *Leslie* (8–10). An examination of the static configuration adopted by an unbounded liquid crystal in the presence of a vertical temperature gradient suggests that the stable orientations are either vertical or almost horizontal, as found experimentally. A further discussion leads to criteria determining the onset of instability for a liquid crystal confined between two parallel plates for the cases of vertical and horizontal boundary orientation.

2. Basic equations

The equations governing the behaviour of an incompressible nematic liquid crystal are taken to be those proposed by *Leslie* (8–10). In *Cartes*ian tensor notation the equations are:

$$v_{i,i} = 0, \quad d_i d_i = 1 \qquad [2.1]$$

$$\varrho \dot{v}_i = F_i - p_{,i} - \left(\frac{\partial W}{\partial d_{k,j}} d_{k,i}\right)_{,j} + \tilde{\sigma}_{ij,j} \qquad [2.2]$$

$$\varrho_1 \ddot{d}_i = G_i + \gamma d_i + \left(\frac{\partial W}{\partial d_{i,j}}\right)_{,j} - \frac{\partial W}{\partial d_i} + d_j(\tilde{\sigma}_{ji} - \tilde{\sigma}_{ij}) \qquad [2.3]$$

$$T\left\{\frac{\partial^2 W}{\partial T \partial d_i}\dot{d}_i + \frac{\partial^2 W}{\partial T \partial d_{i,j}}\dot{d}_{i,j} + \frac{\partial^2 W}{\partial T^2}\dot{T}\right\} + (\varkappa_1 T_{,i} + \varkappa_2 d_j T_{,j} d_i)_{,i} = v_{i,j}\left(\frac{\partial W}{\partial d_{k,j}} d_{k,i} - \tilde{\sigma}_{ij}\right) + \dot{d}_i d_j(\tilde{\sigma}_{ji} - \tilde{\sigma}_{ij}). \qquad [2.4]$$

Here $\underset{\sim}{v}$ is the velocity field and $\underset{\sim}{d}$ the director field specifying the local molecular orientation. The superposed dot denotes material derivative. ϱ is the density, ϱ_1 a positive inertial constant, p the pressure and γ the director tension. $\underset{\sim}{F}$ and $\underset{\sim}{G}$ are the external body forces. T is the temperature and $\varkappa_1$ and $\varkappa_2$ are thermal conductivities. The free energy per unit volume W and the non-equilibrium stress tensor $\underset{\sim}{\tilde{\sigma}}$ have the following forms:

$$2W = 2W_0 + \alpha_2 d_{i,j} d_{i,j} + (\alpha_1 - \alpha_2 - \alpha_4) d_{i,i} d_{j,j} + \alpha_4 d_{i,j} d_{j,i} + (\alpha_3 - \alpha_2) d_p d_q d_{i,p} d_{i,q} \qquad [2.5]$$

$$2\tilde{\sigma}_{ij} = 2\mu_1 d_k d_p v_{k,p} d_i d_j + \mu_2 d_j d_k (v_{k,i} - v_{i,k}) + \mu_3 d_i d_k (v_{k,j} - v_{j,k}) + \mu_4 (v_{i,j} + v_{j,i}) + \mu_5 d_j d_k (v_{k,i} + v_{i,k}) + \mu_6 d_i d_k (v_{k,j} + v_{j,k}) + 2\mu_2 d_j \dot{d}_i + 2\mu_3 d_i \dot{d}_j. \qquad [2.6]$$

All the material functions $\varrho, \varrho_1, \varkappa_1, \varkappa_2, W_0, \alpha_1 \ldots \alpha_4$ and the viscosities $\mu_1 \ldots \mu_6$ are functions of temperature alone.

3. Static solution and linearized equations

We consider first the case of an unbounded liquid crystal in equilibrium in the presence of a vertical temperature gradient. We take *Cartes*ian axes, with z vertical and x and y horizontal and let $\underset{\sim}{k}$ be the unit vector in the z-direction. The only external body force present is that of gravity, so that

$$F_i = -\varrho g k_i, \quad G_i = 0, \quad T_{,i} = \varphi k_i \qquad [3.1]$$

12

where g is the acceleration due to gravity and the vertical temperature gradient $\varphi = \varphi(z)$. An obvious equilibrium solution is that with $\underset{\sim}{d}$ constant. Eqs. [2.1–2.4] give

$$p = p_0 - g \int \varrho dz, \quad \gamma = 0,$$
$$\{\varkappa_1 + \varkappa_2 (d_j k_j)^2\} \varphi = \text{constant}. \qquad [3.2]$$

This equilibrium state is now disturbed by a small amplitude velocity field $\underset{\sim}{v}$, associated with which is the director field $\underset{\sim}{d} + \underset{\sim}{n}$ and the temperature field $T + s$. Here and in the subsequent analysis, $\underset{\sim}{d}$ and T denote the equilibrium fields given by [3.1] and [3.2]. We linearize eqs. [2.1–2.4] about the equilibrium state. In the process of linearization we ignore all variations of material parameters with temperature except where associated with gravity. This is the approximation usually adopted in problems of hydrodynamic stability (11) and will be valid provided φ is not too large. The linearized equations are

$$v_{i,i} = 0, \quad d_i n_i = 0 \qquad [3.3]$$

$$\varrho \frac{\partial v_i}{\partial t} = -\varrho' g k_i s - \bar{p}_{,i} + A_{ijkm} v_{j,km} + B_{ijk} \frac{\partial n_{j,k}}{\partial t} \qquad [3.4]$$

$$\varrho_1 \frac{\partial^2 n_i}{\partial t^2} = \bar{\gamma} d_i + C_{ijkm} n_{j,km} + D_{ijk} v_{j,k} + \lambda_1 \frac{\partial n_i}{\partial t} \qquad [3.5]$$

$$C \frac{\partial s}{dt} = \varkappa_1 s_{,ii} + \varkappa_2 d_j d_i s_{,ij} + \varphi \varkappa_2 (k_j d_i n_{j,i} + k_j d_j n_{i,i}). \qquad [3.6]$$

Here, φ is now a constant, representing some average temperature gradient, $p' = \partial \varrho / \partial T$, $\bar{p}$ and $\bar{\gamma}$ are the perturbations to the pressure and director tension, and

$$\begin{aligned}
2 A_{ijkm} &= 2\mu_1 d_i d_j d_k d_m + (\mu_2 + \mu_5) \delta_{ik} d_j d_m \\
&\quad + (\mu_3 + \mu_6) \delta_{km} d_i d_j + \mu_4 \delta_{ij} \delta_{km} \\
&\quad + (\mu_5 - \mu_2) \delta_{ij} d_k d_m \\
B_{ijk} &= \mu_2 \delta_{ij} d_k + \mu_3 d_i \delta_{jk} \\
C_{ijkm} &= \alpha_2 \delta_{ij} \delta_{km} + (\alpha_1 - \alpha_2) \delta_{jk} \delta_{im} \\
&\quad + (\alpha_3 - \alpha_2) \delta_{ij} d_k d_m \\
2 D_{ijk} &= (\lambda_2 - \lambda_1) \delta_{ij} d_k + (\lambda_2 + \lambda_1) \delta_{ik} d_j \\
\lambda_1 &= \mu_2 - \mu_3, \quad \lambda_2 = \mu_5 - \mu_6 \\
C &= -T \partial^2 W_0 / \partial T^2. \qquad [3.7]
\end{aligned}$$

C is the specific heat at constant volume.

4. Plane waves

It is consistent with the assumptions already made to assume that the ratios of the material functions ϱ, ϱ', ϱ_1, α_i, $\varkappa_i$, μ_i, C to one another are constants, independent of temperature. We can then seek plane wave solutions of [3.3–3.6] of the form

$$(\underset{\sim}{n}, \underset{\sim}{v}, s) = (\underset{\sim}{a}, \underset{\sim}{b}, c) \exp \{ i m \underset{\sim}{\nu} \cdot \underset{\sim}{x} + \omega t \}$$
$$\underset{\sim}{\nu} \cdot \underset{\sim}{\nu} = 1 \qquad [4.1]$$

with similar forms for $\bar{p}$ and $\bar{\gamma}$. We examine waves travelling with their direction of propagation $\underset{\sim}{\nu}$ lying in the plane of $\underset{\sim}{d}$ and $\underset{\sim}{k}$. Substitution from [4.1] into [3.3–3.7] and the subsequent elimination of $\underset{\sim}{a}$, $\underset{\sim}{b}$, c, $\bar{p}$ and $\bar{\gamma}$ yields the secular equation

$$\begin{aligned}
&2(C\omega + m^2 \varkappa) ([2\varrho\omega + m^2 \mu] \\
&\quad \times [\varrho_1 \omega^2 - \lambda_1 \omega + m^2 \alpha] \\
&\quad + \omega m^2 [\lambda_1 - \lambda_2 \cos 2\psi] \\
&\quad \times [\mu_3 \sin^2 \psi - \mu_2 \cos^2 \psi]) \\
&\quad + m^2 \varrho' g \varphi \varkappa_2 (\cos 2\theta + \cos 2\psi) \\
&\quad \times (\lambda_1 - \lambda_2 \cos 2\psi) = 0 \qquad [4.2]
\end{aligned}$$

where $\cos \psi = \underset{\sim}{\nu} \cdot \underset{\sim}{d}$ and $\sin \theta = \underset{\sim}{d} \cdot \underset{\sim}{k}$ and $\alpha = \alpha_1 \sin^2 \psi + \alpha_3 \cos^2 \psi$, $\varkappa = \varkappa_1 + \varkappa_2 \cos^2 \psi$ and $\mu = \mu_4 + (\mu_5 - \mu_2) \cos^2 \psi + (\mu_3 + \mu_6 + 2\mu_1 \cos^2 \psi) \sin^2 \psi$. The inertial coefficient ϱ_1 is very small, and for all values of ω for which the continuum theory is valid $\varrho_1 \omega^2$ can be neglected. The secular equation [4.2] is then a cubic for ω. If the configuration is to be stable this cubic must have no roots with positive real part. A necessary condition for this to be so is that the coefficient of ω^3 and the constant coefficient should be of the same sign, i.e.

$$-\varrho \lambda_1 C (\varkappa \mu \alpha - \varrho' g \varkappa_2 \lambda_2 \varphi [\cos 2\theta + \cos 2\psi] \times [\cos 2\theta_0 + \cos 2\psi] / m^4) \geq 0 \qquad [4.3]$$

where $\cos 2\theta_0 = -\lambda_1 / \lambda_2$. From basic thermodynamic principles it is known that (12, 13)

$$\varrho > 0, \quad C > 0, \quad \varkappa > 0, \quad \mu > 0,$$
$$\alpha > 0, \quad \lambda_1 < 0, \quad \lambda_2 > 0, \qquad [4.4]$$

while from experimental observations on p-azoxyanisole (14, 15) (see however section 6 below)

$$\varkappa_2 > 0, \quad \varrho' < 0. \qquad [4.5]$$

Thus [4.3] is satisfied for all wave numbers m provided that for all ψ

$$\varphi (\cos 2\theta + \cos 2\psi)(\cos 2\theta_0 + \cos 2\psi) \geq 0. \qquad [4.6]$$

If φ is positive then [4.6] requires that $\theta = \theta_0$. The experimental evidence given by *Porter* and *Johnson* (16) suggests that in simple shearing flow *p*-azoxyanisole aligns with the direction of the flow. The analysis of *Leslie* (8) for shearing flow then implies that $\lambda_1 = -\lambda_2$ and $\theta_0 = 0$. We thus conclude that if $\varphi > 0$ then $\theta = 0$, i.e. *if the temperature increases with height the molecular orientation must be horizontal.*

With $\theta_0 = 0$ and φ negative, we conclude from [4.6] that $\theta = \pi/2$. Thus *if the temperature decreases with height the molecular orientation must be vertical.*

It is now necessary to show that all waves of the type [4.1], not just those propagating in the plane of $\underset{\sim}{d}$ and $\underset{\sim}{k}$, are damped provided that the orientation is either vertical or horizontal as appropriate. This can be shown to be the case. The analysis is not given here, being similar to that given by *Currie* (17) and *Martinoty* and *Candau* (18).

The above conclusions depend on the assumption that $\theta_0 = 0$. If $\theta_0 \neq 0$ we still have the conclusion that *if the temperature increases with height the molecular orientation must be inclined at an angle θ_0 to the horizontal.* This may be compared with the experimental results of *Stewart, Holland* and *Reynolds* (3). If we assume that the ratio of vertical and horizontal scattered X-ray intensities is equal to $\tan\theta_0$ then we find θ_0 has a value of about 11°. Light scattering experiments by the *Orsay Liquid Crystal Group* (19) give a larger value for θ_0 of about 17°, but this would seem to be rather too large to be consistent with the experiments of *Porter* and *Johnson* (16). Recently, *Gähwiller* (20) has reported that for the liquid crystals MBBA and HBAB the angle θ_0 seems to be temperature dependent, varying from 5° to 17.5°, the largest angle occuring near the nematic-isotropic transition.

If φ is negative and $\theta_0 \neq 0$, then [4.6] cannot be satisfied for all ψ. However, if $\theta = \pi/2$, then [4.6] is satisfied for all ψ except $\pi/2 - \theta_0 < \psi \leq \pi/2$, that is for all disturbances other than those propagating in an almost vertical direction. In a real experiment these disturbances may well be inhibited by the presence of boundaries. Moreover, vertical convection currents, which are ignored here, will tend to have a stabilizing effect on a vertical orientation. Thus we expect the orientation to be approximately vertical when the temperature decreases with height, as is found experimentally.

5. Stability between parallel plates

The discussion in the previous section has been concerned with an unbounded material. Of more interest is the case when the liquid crystal is confined between parallel plates at $z = \pm h$, with the upper plate held at a temperature T_2 and the lower at temperature T_1. The plates will exert some orienting influence on the liquid crystal, and we shall concern ourselves with just the two cases of horizontal and vertical boundary orientation. The equilibrium solution given by [3.1] and [3.2] is a possible configuration of the material, with $\underset{\sim}{d}$ taking the boundary orientation throughout the material.

We linearize the governing equations about the equilibrium configuration to give equations [3.3] to [3.6], φ being the average temperature gradient given by

$$\varphi = (T_2 - T_1)/2h. \qquad [5.1]$$

For the purposes of the stability analysis we simplify the equations by making the following assumptions about the functions $\underset{\sim}{A}$, $\underset{\sim}{B}$, $\underset{\sim}{C}$ and $\underset{\sim}{D}$ defined in [3.7]:

$$2A_{ijkm} = \mu\delta_{ij}\delta_{km}, \quad C_{ijkm} = \alpha\delta_{ij}\delta_{km},$$
$$B_{ijk} = -D_{ijk} = \lambda_1\delta_{ij}d_k. \qquad [5.2]$$

These approximations seem not unreasonable in the light of the measurements of the constants α_i by *Saupe* (21) and the μ_i by the *Orsay Liquid Crystal Group* (19). The assumed form for $\underset{\sim}{D}$ implies that $\theta_0 = 0$. We also neglect terms involving ϱ_1 and put $\varkappa_2 = 0$ except in the term multiplying the temperature gradient φ. Equations [3.3–3.6] become, after eliminating $\bar{\gamma}$,

$$v_{i,i} = 0, \quad d_i n_i = 0 \qquad [5.3]$$

$$\varrho\frac{\partial v_i}{\partial t} = -\varrho' g k_i s - \bar{p}_{,i} + \frac{1}{2}\mu v_{i,jj} + \lambda_1\frac{\partial n_{i,k}}{\partial t}d_k \qquad [5.4]$$

$$\alpha n_{i,jj} + \lambda_1\frac{\partial n_i}{\partial t} = \lambda_1(v_{i,k} - d_i d_j v_{j,k})\,d_k \qquad [5.5]$$

$$C\frac{\partial s}{\partial t} = \varkappa_1 s_{,ii} + \varphi\varkappa_2(k_j d_i n_{j,i} + k_j d_j n_{i,i}). \qquad [5.6]$$

As far as the predictions of section 4 are concerned, these equations seem to retain all the important characteristics of the full linearized equations, with the assumption that $\theta_0 = 0$.

We examine the stability of a basic vertical orientation with respect to disturbances of the form

$$(\underline{n}, \underline{v}, s) = (n\underline{i}, v\underline{k} + u\underline{i}, s) \exp\{imx + \omega t\} \quad [5.7]$$

where n, v, u and s are functions of z, and $\underline{i}$ is the unit vector along the horizontal x-axis. The boundary conditions applied are those of constant orientation and temperature and zero velocity, so that

$$n = v = u = s = 0 \quad \text{at} \quad z = \pm h. \quad [5.8]$$

If we put $\zeta = z/h$, substitution from [5.7] into [5.3–5.6] and the elimination of n, s, u and $\bar{p}$ yields

$$(D^2 - a^2)(D^2 - a^2 - P\omega)(D^2 - a^2 - Q\omega) \times (D^2 - a^2 - R\omega)\, v + a^2 M \varphi D^2 v = 0 \quad [5.9]$$

where $D = d/d\zeta$, the non-dimensional wave number $a = mh$ and $P = Ch^2/\varkappa_1$, $Q = -\lambda_1 h^2/\alpha$, $R = 2\varrho h^2/\mu$ and $M = 2h^4 \varrho' g \varkappa_2 \lambda_1/\mu \varkappa_1 \alpha$ are positive constants. In deriving this equation, terms involving λ_1 have been neglected in comparison with similar terms involving μ, as is consistent with previous approximations.

The boundary conditions on v derived from [5.8] are

$$v = Dv = (D^2 - a^2)(D^2 - a^2 - R\omega)\, v = (D^2 - a^2)(D^2 - a^2 - R\omega) \times (D^2 - a^2 - P\omega)\, v = 0 \quad \text{on} \quad \zeta = \pm 1. \quad [5.10]$$

Similarly, for the case of horizontal orientation $\underline{d} = \underline{i}$ we examine the stability with respect to disturbances of the form

$$(\underline{n}, \underline{v}, s) = (n\underline{k}, v\underline{k} + u\underline{i}, s) \exp\{imx + \omega t\} \quad [5.11]$$

where n, u, v, s are functions of z. With the same assumptions as before we find the equation

$$(D^2 - a^2)(D^2 - a^2 - P\omega)(D^2 - a^2 - Q\omega) \times (D^2 - a^2 - R\omega)\, v + a^4 M \varphi v = 0 \quad [5.12]$$

together with the boundary conditions [5.10].

We seek the critical values of φ for which [5.9] and [5.12] have solutions for ω with positive real part. As is customary, we adopt the principle of exchange of stabilities and assume that these critical values are given by $\omega = 0$. We can then use the variational methods described by *Chandrasekhar* (22). Let the function F be defined by $F = (D^2 - a^2)^2 v$. It then follows that if v satisfies [5.9] with $\omega = 0$ then φ is given by

$$M\varphi = \{\int_{-1}^{1} ([D^2 - a^2] F)^2 \, d\zeta\} \times \{a^2 \int_{-1}^{1} ([D^2 - a^2] Dv)^2 \, d\zeta\}^{-1} \quad [5.13]$$

while if v satisfies [5.12] with $\omega = 0$

$$M\varphi = -\{\int_{-1}^{1} ([D^2 - a^2] F)^2 \, d\zeta\} \times \{a^4 \int_{-1}^{1} ([D^2 - a^2] v)^2 \, d\zeta\}^{-1}. \quad [5.14]$$

It is clear from these expressions that the vertical orientation becomes unstable at a positive value of φ and the horizontal orientation at a negative value. To find the critical values of φ [5.13] and [5.14] must be minimized for all choices of trial function F and all values of a. Similar problems have been studied by *Chandrasekhar* (22) and he shows that it is generally sufficient for the accuracy desired here to minimize the expressions with respect to the wave number a using the trial function $F = \cos(\pi\zeta/2)$. The following critical values are then found:

$M\varphi > 45$, vertical orientation unstable,
$M\varphi < -158$, horizontal orientation unstable.

For p-azoxyanisole at 125 °C, $\varkappa_2/\varkappa_1 \approx 0.2$ (14), $\lambda_1/\mu \approx -1$ (19), $\varrho^1 \approx -10^{-3}$ gm/cm³ °C (13), $\alpha \approx 10^{-6}$ dynes (21). Thus, substituting for M and using [5.1] we find the following conditions for instability:

$(T_2 - T_1)\, h^3 > 3.10^{-4}$ cm³ °C,
vertical orientation unstable,
$(T_2 - T_1)\, h^3 < -10^{-3}$ cm³ °C,
horizontal orientation unstable.

6. Discussion

The analysis presented here would seem to give a theoretical explanation of the experimental observations of *Stewart*, *Holland* and *Reynolds*. The stability analysis of section 5 suggests that the uniform orientation of a liquid crystal confined between parallel plates will become unstable for relatively small temperature differences provided the gap width is not too small.

These conclusions depend on the assumption made in [4.5] that $\varkappa_2$ is positive. There would seem to be some doubt about this; *Patharkar*, *Rajan* and *Picot* (7) suggest that $\varkappa_2$ is negative, while *Longley-Cook* and *Kessler* (14) cite experimental evidence leading to the conclusion that $\varkappa_2$ is positive. Clearly, if $\varkappa_2$ were negative our conclusions would be reversed and we would expect a horizontal orientation with the temperature decreasing vertically. The experimental observations seem therefore to support the conclusion that $\varkappa_2$ is positive.

Summary

It has been known for some time that the molecular orientation of liquid crystals can be affected by the presence of a temperature gradient. *Stewart* reported in 1936 that, when *p*-azoxyanisol was placed in a vertical temperature gradient, the orientation adopted was vertical when the higher temperature was at the bottom and horizontal when the higher temperature was at the top. The continuum theory for nematic liquid crystals, proposed by *Leslie*, is used to explain this phenomenon. An examination of the stability of the static orientation adopted by the director in the presence of a vertical temperature gradient shows that the stable orientations are either vertical or horizontal, as found experimentally.

Zusammenfassung

Es ist längst bekannt, daß die Molekularorientierung kristalliner Flüssigkeiten durch die Anwesenheit einer Temperatursteigung beeinflußt werden kann. *Stewart* zeigte 1936, daß die angenommene Orientierung von *p*-Azoxyanisol in einer vertikalen Temperatursteigung bei höherer Temperatur unten vertikal war, sowie horizontal bei höherer Temperatur oben. Die Kontinuumstheorie für nematische Kristallinflüssigkeiten – wie von *Leslie* vorgeschlagen – wird benutzt, um dieses Phänomen zu erklären. Eine Untersuchung der Stabilität der statischen Orientierung, die der Direktor in Anwesenheit einer vertikalen Temperatursteigung annimmt, zeigt, wie auch das Experiment feststellt, daß die stabilen Orientierungen entweder vertikal oder horizontal sind.

References

1) *Stewart, G. W.*, J. Chem. Phys. **4**, 231 (1936).

2) *Holland, D. O.* and *G. W. Stewart*, Phys. Rev. **51**, 62 (1937).

3) *Stewart, G. W., D. O. Holland*, and *L. M. Reynolds*, Phys. Rev. **58**, 174 (1940).

4) *Stewart, G. W.*, Phys. Rev. **69**, 51 (1946).

5) *Picot, J. J. C.* and *A. G. Fredrickson*, I and EC Fundamentals **1**, 84 (1968).

6) *Fisher, J.* and *A. G. Fredrickson*, Mol. Cryst. and Liq. Cryst. **6**, 255 (1969).

7) *Patharkar, M. N., V. S. V. Rajan*, and *J. J. C. Picot*, Mol. Cryst. and Liq. Cryst. **15**, 225 (1971).

8) *Leslie, F. M.*, Arch. Rat. Mech. Anal. **28**, 265 (1968).

9) *Leslie, F. M.*, Proc. Roy. Soc. **A 307**, 359 (1968).

10) *Leslie, F. M.*, Mol. Cryst. Liq. Cryst. **7**, 407 (1969).

11) *Stuart, J. T.*, Hydrodynamic stability. Chap IX of Laminar Boundary Layers. Ed.: *L. Rosenhead* (Oxford 1963).

12) *Leslie, F. M.*, Quart. J. Mech. Appl. Math. **19**, 357 (1966).

13) *Ericksen, J. L.*, Phys. Fluids **9**, 1205 (1966).

14) *Longley-Cook, M.* and *J. O. Kessler*, Mol. Cryst. and Liq. Cryst. **12**, 315 (1971).

15) *Hoyer, W. A.* and *A. W. Nolle*, J. Chem. Phys. **24**, 803 (1956).

16) *Porter, R. S.* and *J. F. Johnson*, J. Phys. Chem. **66**, 1826 (1962).

17) *Currie, P. K.*, The propagation and adsorption of small-amplitude waves in incompressible nematic liquid crystals (unpublished).

18) *Martinoty, P.* and *S. Candau*, Mol. Cryst. Liq. Cryst. **14**, 243 (1971).

19) *Orsay Liquid Crystal Group*, Mol. Cryst. Liq. Cryst. **13**, 187 (1971).

20) *Gähwiller, Ch.*, Phys. Rev. Letters **28**, 1554 (1972).

21) *Saupe, A.*, Z. Naturforschung **15A**, 815 (1960).

22) *Chandrasekhar, S.*, Hydrodynamic and hydromagnetic stability **4** (Oxford 1961).

Author's address:

Dr. *P. K. Currie*
School of Physical Sciences
The New University of Ulster
Coleraine (Northern Ireland)

Rheol. Acta 12, 170–176 (1973)

From the Micromechanics Laboratory, McGill University, Montreal (Canada), and the Applied Physics Division, Pulp & Paper Research Institute of Canada, Quebec (Canada)

Microrheology of fibrous systems

By D. R. Axelrad, D. Atack, and J. W. Provan

With 4 figures

(Received October 27, 1972)

1. Introduction

Previous work (1–4) dealt with a random theory of deformation based on the concepts of statistical mechanics and the theory of probability. This paper is concerned with the microrheology of a 2-dimensional fibrous system as indicated in fig. 1 a, b. A mathematical model is proposed that contains the aspects of bonding between fibres.

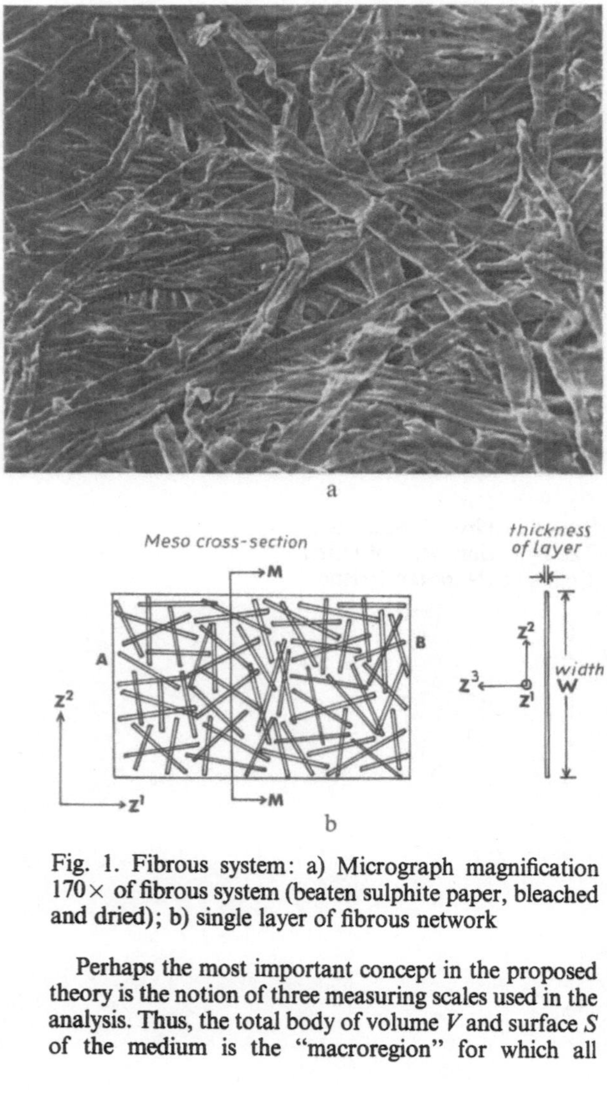

Fig. 1. Fibrous system: a) Micrograph magnification 170× of fibrous system (beaten sulphite paper, bleached and dried); b) single layer of fibrous network

Perhaps the most important concept in the proposed theory is the notion of three measuring scales used in the analysis. Thus, the total body of volume V and surface S of the medium is the "macroregion" for which all boundary conditions are specified. An intermediary scale is then used termed "mesoregion" which is much smaller than the macroregion and which has a volume ${}^MV$ and surface ${}^MS$. It is postulated that the mesoregions within the macrodomain are denumerable and non-intersecting such that ${}^MV \cap {}^LV = \Phi$; $L \neq M$, and that they are large enough to contain a statistical ensemble of microelements of volume ${}^\alpha v$ and surface ${}^\alpha s$. In the present case, each fibre in the 2-dimensional network is considered as a microelement. The analysis is further limited to a fibrous system that has a strong bonding in the z^1 and z^2 directions (fig. 1 b), but a much weaker one in the z^3 direction. This is the case of a fibrous system such as paper, for instance. It is further assumed that a "material functional" that reflects the geometrical and rheological properties of the network exist. It is used in lieu of the conventional constitutive relations in the present formulation. In a reduced and simplified form, this functional will be as follows:

$$ {}^M\mathfrak{M} = {}^M\mathfrak{M}\{p_1, p_2, \underline{\underline{E}}, \underline{D}, \psi_0, b, \pi(\underline{Q}), \ldots\} \quad [1.1] $$

where the meaning of the various parameters and functions will be fully discussed in subsequent paragraphs.

For the classification of single fibres and bonding areas, it should be noted that each layer forming a 2-dimensional array of randomly arranged fibres or microelements (fig. 1 b) will be cut into parts A, B by an arbitrary cross-section MM. The latter represents in the 2-dimensional case a "mesodomain". Each mesodomain is denumerable with varying z^1 and contains intersections with N^M single fibres ($\alpha = 1, \ldots, N^M$) and n^M junctions or bonds ($i = 1, \ldots, n^M$) where α denotes an individual fibre in the mesodomain and i designates each bond area.

Whenever a meso-section MM intersects a fibre α, it does so across a certain length ${}^\alpha b(\alpha = 1, \ldots, N^M)$ and a bond area across a distance ${}^ic(i = 1, \ldots, nM)$. Denoting by ${}^MW$ the sum of all ${}^\alpha b$ and ${}^ic$ in the meso cross-section in the z^2 direction and the total width of the layer in that direction by W, then evidently:

$$ {}^MW = \sum_{\alpha=1}^{N^M} {}^\alpha b + \sum_{i=1}^{n^M} {}^ic \quad [1.2] $$

and ${}^MW/W =$ an approximate void ratio. The probability that the meso-section contains $\alpha = 1, \ldots, N^M$ sections of single fibres can be stated as:

$$ p_1 = p_1\{MM \cap {}^\alpha b \,|\, {}^MW\} \quad [1.3] $$

and that the cross-section contains $i = 1, \ldots, n^M$ sections through bond areas by:

$$p_2 = p_2\{MM \cap {}^i c \,|\, {}^M W\} \qquad [1.4]$$

in which also

$$p_1 + p_2 = 1. \qquad [1.5]$$

Hence, two cases arise, in the formulation:

1. where the meso-section MM intersects fibres only;
2. where the meso-section MM intersects bond areas.

These two cases will be treated separately and then joined together to give the response characteristics of the fibrous system.

2. Single fibre behaviour

(i) Deformation kinematics of the single fibre

The basic kinematic relations pertaining to a single fibre will be briefly discussed in this section. Fig. 2 indicates the undeformed configuration of the αth microelement or fibre in the network.

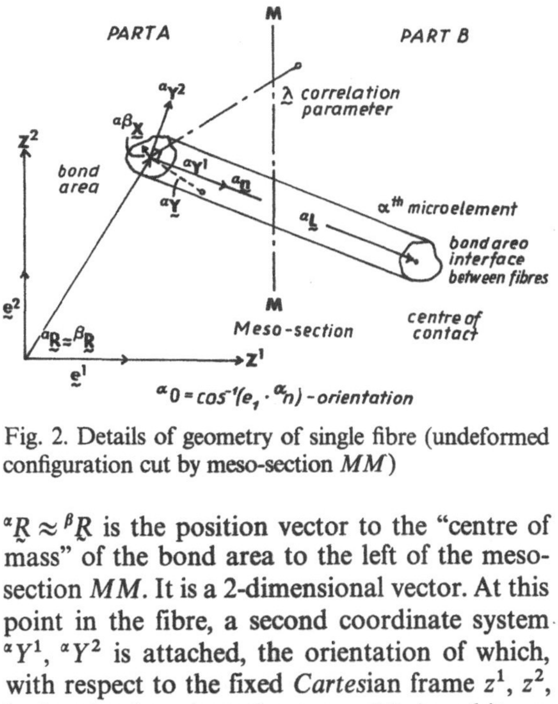

Fig. 2. Details of geometry of single fibre (undeformed configuration cut by meso-section MM)

${}^\alpha \underset{\sim}{R} \approx {}^\beta \underset{\sim}{R}$ is the position vector to the "centre of mass" of the bond area to the left of the meso-section MM. It is a 2-dimensional vector. At this point in the fibre, a second coordinate system ${}^\alpha Y^1$, ${}^\alpha Y^2$ is attached, the orientation of which, with respect to the fixed *Cartesian* frame z^1, z^2, is given by the orientation tensor ${}^\alpha \underset{\approx}{O}$. An arbitrary point within the bond area relative to the frame ${}^\alpha Y$ is denoted by ${}^{\alpha\beta}\underset{\sim}{X}$ and a point in the fibre is described by ${}^\alpha \underset{\sim}{Y}$. The distance between centres of the bond areas is denoted by ${}^\alpha \underset{\sim}{L}$. It is seen that the orientation ${}^\alpha O$ is described by ${}^\alpha O = \cos^{-1}(\underset{\sim}{e}_1 \cdot {}^\alpha \underset{\sim}{n})$ where $\underset{\sim}{e}_1$, $\underset{\sim}{e}_2$ are the base vectors of the fixed frame z^1, z^2, ${}^\alpha \underset{\approx}{O}$ being the associated orientation tensor. All kinematic parameters related to the undeformed configuration of the fibre are denoted by majuscules, whilst those in the deformed configuration will be designated by minuscules.

Thus, for instance, it is assumed that:

$$ {}^\alpha \underset{\sim}{r} = {}^\alpha \underset{\sim}{r}({}^\alpha \underset{\sim}{R}, t), \quad {}^\alpha \underset{\sim}{y} = {}^\alpha \underset{\sim}{y}({}^\alpha \underset{\sim}{Y}, t), $$
$$ {}^{\alpha\beta} \underset{\sim}{x} = {}^{\alpha\beta} \underset{\sim}{x}({}^{\alpha\beta} X, t) \qquad [2.1]$$

are stochastic processes whose time-dependent probability distributions are of the *Gaussian* form so that correlation theory may be applied [see *Yaglom* (5)]. The displacements of these points can be written respectively as:

$$ {}^\alpha \underset{\sim}{u} = {}^\alpha \underset{\sim}{r} - {}^\alpha \underset{\sim}{R}; \quad {}^\alpha \underset{\sim}{v} = {}^\alpha \underset{\sim}{y} - {}^\alpha \underset{\sim}{Y}; $$
$$ {}^{\alpha\beta} \underset{\sim}{w} = {}^{\alpha\beta} \underset{\sim}{x} - {}^{\alpha\beta} \underset{\sim}{X}. \qquad [2.2]$$

In analogy to continuum mechanics, the primitive strain measure in the fibre can be expressed in terms of the "microdeformation gradient", i.e.:

$$ {}^\alpha \underset{\approx}{F} = \frac{\partial^\alpha \underset{\sim}{y}}{\partial^\alpha \underset{\sim}{Y}}; \quad {}^\alpha \underset{\approx}{f} = \frac{\partial^\alpha \underset{\sim}{Y}}{\partial^\alpha \underset{\sim}{y}} \qquad [2.3]$$

where these quantities are assumed to be affine and from which, in a well-known manner, the "micro" *Lagrang*ian and *Euler*ian strains may be derived so that:

$$ 2\,{}^\alpha \underset{\approx}{E} = {}^\alpha \underset{\approx}{F}^c \cdot {}^\alpha \underset{\approx}{F} - \underset{\approx}{\delta}; \quad 2\,{}^\alpha \underset{\approx}{e} = \underset{\approx}{\delta} - {}^\alpha \underset{\approx}{f}^c \cdot {}^\alpha \underset{\approx}{f} \qquad [2.4]$$

in which the superscript c on $\underset{\approx}{F}$ or $\underset{\approx}{f}$ indicates the dyadic conjugate and $\underset{\approx}{\delta}$ is the *Kronecker* Delta.

This strain measure is evidently an oversimplification to the actual case since other structural effects may contribute to the deformation. However, in the present analysis, the strain will be assumed to be linear such that:

$$ {}^\alpha \underset{\approx}{E} \doteq {}^\alpha \underset{\approx}{\epsilon} \doteq {}^\alpha \underset{\approx}{e}. \qquad [2.5]$$

Again analogously to continuum mechanics, deformation rates can be derived for the single fibre where:

$$ {}^\alpha \dot{\underset{\approx}{e}} = \frac{1}{2}\left\{\frac{\partial^\alpha \dot{\underset{\sim}{v}}}{\partial^\alpha \underset{\sim}{Y}} + \left(\frac{\partial^\alpha \dot{\underset{\sim}{v}}}{\partial^\alpha \underset{\sim}{Y}}\right)^T\right\} \qquad [2.6]$$

in which T means the transpose of the relevant quantity and a superimposed dot indicates the time derivative. Each of these kinematic variables is assumed to behave as a stochastic process during deformation. These processes, for the simplification of the analysis, are further restricted to statistically homogeneous, non-isotropic and stationary *Gaussian* processes. These restrictions permit the application of correlation theory. The statistical mean over all single fibres in a meso-section will be indicated by $\langle\cdot\rangle_{N^M}$, whilst fluctuation from this mean will be shown by a star on the quantity. Hence, the strain and strain rates can be expressed by:

$$^{\alpha}\varepsilon = \langle \varepsilon \rangle_{N^M} + {}^{\alpha}\varepsilon^*$$

$$^{\alpha}\dot{\varepsilon} = \langle \dot{\varepsilon} \rangle_{N^M} + {}^{\alpha}\dot{\varepsilon}^* \quad [2.7]$$

For the formulation of a correlation theory, the centred correlation or covariance functions for the strain and deformation rates, respectively, are also required. These can be expressed by using a correlation parameter λ, i.e., the distance vector from the centre of the bond area to any point in the 2-dimensional layer (see fig. 2) as follows:

$$B_{\varepsilon}(\lambda, s) = \langle \varepsilon^* [{}^{\alpha}Q\,{}^{\alpha}Y + {}^{\alpha}R, t]\, \varepsilon^* \times [{}^{\alpha}Q\,{}^{\alpha}Y + {}^{\alpha}R + \lambda, t + s] \rangle_{N^M}$$

$$B_{\dot{\varepsilon}}(\lambda, s) = \langle \dot{\varepsilon}^* [{}^{\alpha}Q\,{}^{\alpha}Y + {}^{\alpha}R, t]\, \dot{\varepsilon}^* \times [{}^{\alpha}Q\,{}^{\alpha}Y + {}^{\alpha}R + \lambda, t + s] \rangle_{N^M}. \quad [2.8]$$

(ii) Energy potentials and microstress

It is assumed that the viscoelastic response of the fibrous network is due to the behaviour of the fibres only and that the bond between fibres is an elastic one. From thermodynamic principles, the elastic completely reversible response of each fibre will be described by a free-energy potential $^{\alpha}U^e$, if the temperature is also an independent variable. Hence, the elastic components of the "microstress" tensor $^{\alpha}\xi^e$ can be stated as:

$$^{\alpha}\xi^e = \frac{\partial^{\alpha} U^e}{\partial^{\alpha}\varepsilon} \quad [2.9]$$

where the intrinsic energy

$$^{\alpha}U^e = \tfrac{1}{2} E\,{}^{\alpha}\varepsilon\,{}^{\alpha}\varepsilon \quad [2.10]$$

and E is a fourth order elastic coefficient tensor. Combining [2.9] and [2.10] gives:

$$^{\alpha}\xi^e = E\,{}^{\alpha}\varepsilon. \quad [2.11]$$

The non-elastic response of a fibre can be described in terms of a dissipation function $^{\alpha}U^v$ which, in general, will depend not only on the deformation rates, but also on the state of the fibre and its history. Using extremum principles (6), it can be shown that the irreversible part of the microstress, i.e., $^{\alpha}\xi^v$ can be obtained from a dissipation potential $^{\alpha}\Phi^v$ such that:

$$^{\alpha}\xi^v = \frac{\partial^{\alpha}\Phi^v}{\partial^{\alpha}\dot{\varepsilon}}. \quad [2.12]$$

If the dissipation function is further restricted to the case of a homogeneous polynomial of degree 2, i.e., to a linear relationship between $^{\alpha}\xi^v$ and $^{\alpha}\dot{\varepsilon}$, [2.12] becomes simply:

$$^{\alpha}\xi^v = D\,{}^{\alpha}\dot{\varepsilon} \quad [2.13]$$

which is analogous to [2.11] and where D is again a fourth order material coefficient tensor (3). Employing these potentials, the total stress acting at point $^{\alpha}Y$ in the fibre for the simple linear viscoelastic case, becomes:

$$^{\alpha}\xi = {}^{\alpha}\xi^e + {}^{\alpha}\xi^v = \frac{\partial^{\alpha}U^e}{\partial^{\alpha}\varepsilon} + \frac{\partial^{\alpha}\Phi^v}{\partial^{\alpha}\dot{\varepsilon}}. \quad [2.14]$$

Combining [2.11] and [2.13] gives:

$$^{\alpha}\xi = E\,{}^{\alpha}\varepsilon + D\,{}^{\alpha}\dot{\varepsilon}. \quad [2.15]$$

Alternatively, using an after-effect theory approach as shown in reference (3), the conventional creep-type relation can be obtained as follows:

$$^{\alpha}\varepsilon(t) = C\,{}^{\alpha}\xi(0) + \int_{0^+}^{t} F(t-\tau)\,{}^{\alpha}\xi(\tau)\,d\tau \quad [2.16]$$

in which C is the compliance tensor of the αth fibre, F a fluidity tensor and $F(t-\tau)$ represents the stress history of the fibre.

The correlation statistical analysis follows immediately upon assuming that the intrinsic energy function and dissipation function are the same for all fibres. Thus, substituting [2.7] into [2.15] yields:

$$^{\alpha}\xi = \langle \xi \rangle_{N^M} + {}^{\alpha}\xi^* \quad [2.17]$$

and

$$^{\alpha}\xi = E\langle \varepsilon \rangle_{N^M} + D\langle \dot{\varepsilon} \rangle_{N^M} + E\,{}^{\alpha}\varepsilon^* + D\,{}^{\alpha}\dot{\varepsilon}^*. \quad [2.18]$$

The mean value of the stress in each viscoelastic fibre is obtained by taking the average of [2.18], i.e.:

$$\langle \xi \rangle_{N^M} = E\langle \varepsilon \rangle_{N^M} + D\langle \dot{\varepsilon} \rangle_{N^M}$$

and

$$^{\alpha}\xi^* = E\,{}^{\alpha}\varepsilon^* + D\,{}^{\alpha}\dot{\varepsilon}^*. \quad [2.19]$$

Taking the statistical mean of the product of two microstresses [2.18] and upon the assumption that there is no interdependence between $^{\alpha}\varepsilon$ and $^{\alpha}\dot{\varepsilon}$, yields the stress correlation function as follows:

$$B_{\xi}(\lambda, s) = \langle \xi^* [{}^{\alpha}Q\,{}^{\alpha}Y + {}^{\alpha}R, t]\, \xi^* \times [{}^{\alpha}Q\,{}^{\alpha}Y + {}^{\alpha}R + \lambda, t + s] \rangle_{N^M} = EEB_{\varepsilon}(\lambda, s) + DDB_{\dot{\varepsilon}}(\lambda, s). \quad [2.20]$$

Alternatively, substituting [2.17] into [2.16] yields the following:

$$^{\alpha}\underset{\approx}{e}(t) = \langle \underset{\approx}{e} \rangle_{N^M} + {}^{\alpha}\underset{\approx}{e}^* = \underline{\underline{C}} \langle \underset{\approx}{\xi}(0) \rangle_{N^M} + \int_{0^+}^{t} \underline{\underline{F}}(t-\tau) \langle \underset{\approx}{\xi}(\tau) \rangle_{N^M} \, d\tau + \underline{\underline{C}}\, \underset{\approx}{\xi}^*(0) + \int_{0^+}^{t} \underline{\underline{F}}(t-\tau)\, {}^{\alpha}\underset{\approx}{\xi}^*(\tau)\, d\tau .$$

Thus, the mean value of the microstrain and the fluctuating component of the microstrain tensor for the αth fibre will be as follows:

$$\langle \underset{\approx}{e} \rangle_{N^M} = \underline{\underline{C}} \langle \underset{\approx}{\xi}(0) \rangle_{N^M} + \int_{0^+}^{t} \underline{\underline{F}}(t-\tau) \langle \underset{\approx}{\xi}(\tau) \rangle \quad d\tau \qquad [2.22]$$

$$^{\alpha}\underset{\approx}{e}^* = \underline{\underline{C}}\, {}^{\alpha}\underset{\approx}{\xi}^*(0) + \int_{0^+}^{t} \underline{\underline{F}}(t-\tau)\, {}^{\alpha}\underset{\approx}{\xi}^*(\tau)\, d\tau . \qquad [2.23]$$

The correlation function for the microstrains following the definition given in [2.8] will be:

$$B_e(\underset{\sim}{\lambda}, s) = \underline{\underline{C}}\,\underline{\underline{C}} B_\xi(\underset{\sim}{\lambda}, 0) + \underline{\underline{C}} \int_{0^+}^{t} \underline{\underline{F}}(t-\tau) B_\xi(\underset{\sim}{\lambda}, \tau)\, d\tau + \underline{\underline{C}} \int_{0^+}^{t+s} \underline{\underline{F}}(t+s-\tau) B_\xi(\underset{\sim}{\lambda}, \tau)\, d\tau + B_{\int F\xi d\tau}(\underset{\sim}{\lambda}, s) \qquad [2.24]$$

where the stress correlation function is defined in [2.20]. The above exposition indicates, on the assumption of the existence of a *Gauss*ian distribution for stresses and strains, the possibility of establishing a "failure criterion" for fibrous systems.

3. Fibre-fibre interaction

It is readily noticed from fig. 3 showing a micrograph of unbeaten, low-yield sulphite SF paper paper (magnification: 1120X) that the bond areas between fibres show a distinct discontinuous structure of "fibrils" within these areas. Evidently, no mathematical model is as yet available for dealing with such a system. However, it is proposed here, in analogy to the coincidence lattice theory of metals (7), to substitute for the actual network a continuous lattice such that the entire bonding zone between the αth and β^{th} fibre consists of "coincidence lattice cells" whose equilibrium distance between "matching points" on the αth and β^{th} side is ${}^{\alpha\beta}\Delta^q$, $q = 1, \dots, {}^{\alpha}p$ (see fig. 4a, b). The actual number of lattice points and the size of the cell ${}^{\alpha}\Delta^q S = {}^{\beta}\Delta^q S$ will depend on the relative orientation of the two fibres $0 \leq {}^{\alpha\beta}\Theta < \pi$ as well as on the size of the bond area ${}^{\alpha\beta}S$ (fig. 4a, b).

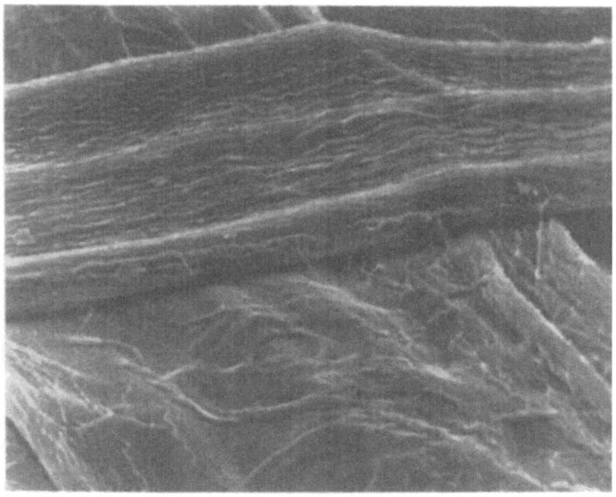

Fig. 3. Micrograph of actual fibrous network (unbeaten, low-yield sulphite SF paper, magnification 1120×)

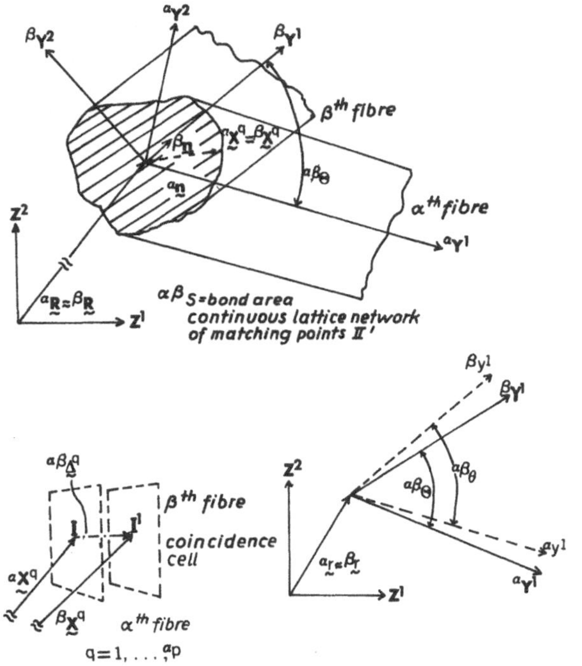

Fig. 4. Fibre-fibre interface: a) Fibre-fibre interface; b) coincidence cell model and matching points II; c) relative rotation of fibre-fibre interface

In the present analysis, the matching points whose "undeformed" position relative to the *Y*-axes are ${}^{\alpha}X^q \approx {}^{\beta}X^q$ (fig. 4a) serve as the two points required for the pair potential formulation of the hydrogen bonding in a fibrous material. Using the previous notation for undeformed and deformed configurations, then the deformed distance vector between matching points is $^{\alpha\beta}\delta^q$ and its displacement relative to the undeformed or equilibrium distance is:

$$^{\alpha\beta}\underset{\sim}{d}^q = {}^{\alpha\beta}\underset{\sim}{\delta}^q - {}^{\alpha\beta}\underset{\sim}{\Delta}^q = \left|^{\alpha\beta}\underset{\sim}{d}^q\right| \cdot \underset{\sim}{e} \qquad [3.1]$$

where $\underset{\sim}{e}$ is the base vector and $\underset{\sim}{d}$ a two-dimensional vector in the z^1, z^2 plane, since only shear and torsion of an interface are assumed to occur. The hydrogen bonding between fibres is considered in this analysis to be adequately described by a potential of the *Morse* type (8), i.e.:

$$^{\alpha\beta}\Psi^q = {}^{\alpha\beta}\Psi^q_0 \{1 - \exp(-b\left|^{\alpha\beta}\underset{\sim}{d}^q\right|)\}^2 \qquad [3.2]$$

in which Ψ^q_0 and b are material parameters to be determined from spectroscopic data and the size of the coincidence cell being considered. Differentiating this potential with respect to $\left|^{\alpha\beta}\underset{\sim}{d}^q\right|$, a "discrete" interaction force between matching points is obtained as follows:

$$\begin{aligned}^{\alpha\beta}\underset{\sim}{F}^q &= \underset{\sim}{e} \cdot \frac{d\,^{\alpha\beta}\Psi^q}{d\left|^{\alpha\beta}\underset{\sim}{d}^q\right|} \\ &= \underset{\sim}{e} \cdot 2\,^{\alpha\beta}\Psi^q_0 b\{1 - \exp(-b\left|^{\alpha\beta}\underset{\sim}{d}^q\right|)\} \\ &\quad \times \exp(-b\left|^{\alpha\beta}\underset{\sim}{d}^q\right|).\end{aligned} \qquad [3.3]$$

In order to join this interfacial force with those discussed previously and since any force system must be balanced at all times, a "continuous interaction force" can be introduced symbolically by following *Gel'fand* and *Vilenkin* (9) such that:

$$^{\alpha\beta}\underset{\sim}{\tau} = \delta(^{\alpha\beta}\underset{\sim}{d} - {}^{\alpha\beta}\underset{\sim}{d}^q) \cdot {}^{\alpha\beta}\underset{\sim}{F}^q \qquad [3.4]$$

where δ is the *Dirac* Delta function. This generalized force $\underset{\sim}{\tau}$ can be used to define a "pseudo-stress" in the fibre in the neighbourhood of the bonding area in such a manner that:

$$^{\alpha}\underset{\sim}{\zeta}^i = {}^{\alpha\beta}\underset{\sim}{\tau}\,^{\alpha}\underset{\sim}{n}; \qquad {}^{\beta}\underset{\sim}{\zeta}^i = {}^{\alpha\beta}\underset{\sim}{\tau}\,^{\beta}\underset{\sim}{n} \qquad [3.5]$$

in which $^{\alpha}\underset{\sim}{n}$ is the unit vector in direction of $^{\alpha}y^1$ (fig. 4c) and similarly, $^{\beta}\underset{\sim}{n}$ that in the direction $^{\beta}y^1$. As a first approximation, this stress may be considered as the boundary stress acting at the extremities of a span of the fibre. Thus, the balance of forces is maintained at all times. Further, to associate this generalized force with the bonding potential and since the bonding has been assumed to be an entirely elastic one, a binding elastic energy can be defined as a function of the potential so that:

$$^{\alpha\beta}\mathfrak{F}(\Psi) = {}^{\alpha\beta}\underset{\sim}{\tau} \cdot {}^{\alpha\beta}\underset{\sim}{d} \qquad [3.6]$$

from which, after some calculations, a "bond strain" may be defined by:

$$\underset{\sim}{e}^i(^{\alpha\beta}d) = \frac{\partial^{\alpha\beta}\mathfrak{F}}{\partial^{\alpha\beta}\underset{\sim}{\zeta}^i} \qquad [3.7]$$

which can be used to express to a first approximation, the binding elastic energy as a quadratic function of the bond strain, namely:

$$^{\alpha\beta}\mathfrak{F} = \tfrac{1}{2}\,^{\alpha\beta}\underset{\sim}{\Delta}\,\underset{\sim}{e}^i\,\underset{\sim}{e}^i. \qquad [3.8]$$

4. The material functional $^M\mathfrak{M}$ and the mesoscopic response

The single fibre and the fibre-fibre interaction behaviour have been briefly discussed in the previous sections. The main objective of the present theory is, however, to outline a procedure for establishing a characteristic energy or material functional $^M\mathfrak{M}$, which characterizes the response behaviour of the two-dimensional fibrous network from the analysis of the single fibre response and the nature of the fibre-fibre bond. From the foregoing exposition, it is seen that such a functional will have the form:

$$^M\mathfrak{M} = {}^M\mathfrak{M}\{p_1, p_2, \underset{\sim}{E}, \underset{\sim}{D}, \Psi_0, b, \pi(\underset{\sim}{Q}), \ldots\} \qquad [4.1]$$

in which $\pi(\underset{\sim}{Q})$ is the probability distribution of fibre orientation with respect to the z^1, z^2 axes and the other parameters have been dealt with in sections 2 and 3.

As mentioned earlier, the consideration of the single fibre behaviour and that of the bonded fibres present two distinct problems, which were analysed separately. Combining their individual effects on the overall response of the network results in the formulation of the functional $^M\mathfrak{M}$ as follows. First, the functional $^M\mathfrak{M}^f$ describes the response characteristic of the single fibre and then the functional $^M\mathfrak{M}^b$ will represent the bond characteristics, viz:

$$\begin{aligned}^M\mathfrak{M} &= p_1\,^M\mathfrak{M}^f(\underset{\sim}{E}, \underset{\sim}{D}, \pi(\underset{\sim}{Q})) \\ &\quad + p_2\,^M\mathfrak{M}^b(\Psi_0, b, \pi(\underset{\sim}{Q}))\end{aligned} \qquad [4.2]$$

where the probabilities p_1 and p_2 have been defined in [1.3] and [1.4] with the condition [1.5]. They must be considered as material parameters. Heuristically, $^M\mathfrak{M}^f$ can be expressed by:

$$^M\mathfrak{M}^f = \sum_{\alpha=1}^{N^M} \left\{^{\alpha}O \int_{\alpha_b} (^{\alpha}U^e + {}^{\alpha}\Phi^v)\,dz^2\right\}. \qquad [4.3]$$

However, since it is assumed that $\pi(\underset{\sim}{Q})$, $\underset{\sim}{E}$ and $\underset{\sim}{D}$ can be assessed from experimental observations, the relation [4.3] may be written as:

$$ {}^{M}\mathfrak{M}^{f} = \frac{1}{2W} \int_{W} \int_{-\pi/2}^{\pi/2} f(\underset{\approx}{O}) \, [\underset{=}{E}{}^{\alpha}\underset{\approx}{e}{}^{\alpha}\underset{\approx}{e} + \underset{=}{D}{}^{\alpha}\dot{\underset{\approx}{e}}{}^{\alpha}\dot{\underset{\approx}{e}}] \, d\underset{\approx}{O} \, dz^{2} \quad [4.4] $$

in which use has been made of [2.9] and a combination of [2.12] and [2.13]. Further, in relation [4.4] f (O) is the density of the probability distribution of fibre orientations with the range as indicated earlier between $-\pi/2 \leq {}^{\alpha}O \leq \pi/2$, on account of the designation of ${}^{\alpha}\underset{\approx}{R} \cdot {}^{\alpha}O$ is chosen to indicate a direction parallel to z^1. The integration over z^2 is a result of the choice of the mesodomain only. The second characteristic functional of relation [4.2] can be considered in either form, i.e.:

$$ {}^{M}\mathfrak{M}^{b} = \sum_{i=1}^{n^M} \{ \sum_{q=1}^{{}^{\alpha}p} {}^{\alpha}\Delta^{q} S^{\alpha\beta}\Psi^{q} \} $$

or

$$ {}^{M}\mathfrak{M}^{b} = \frac{1}{2W} \int_{W} \int_{0}^{\pi/2} \sum_{q=1}^{{}^{\alpha}p} g({}^{\alpha\beta}\Theta) \, {}^{\alpha\beta}\Psi^{q\alpha}\Delta^{q} S \, d\Theta \, dz^{2} \quad [4.5] $$

where, in the second form of [4.5], $g({}^{\alpha\beta}\Theta)$ is the probability density of angles between fibres. It can be uniquely determined from the fibre orientation probability density $f(\underset{\approx}{O})$. Furthermore, each coincidence cell size and the potential ${}^{\alpha\beta}\Psi_{0}^{q}$ will depend on the relative angle ${}^{\alpha\beta}\theta$ (see also fig. 4c). For counting purposes, ${}^{\alpha\beta}\theta$ is limited to the range $0 \leq \theta \leq \pi/_{2}$. The characteristic functional significant for the response of the network can now be formulated by combining [4.3] or [4.4] with [4.5]. Hence, from a knowledge of the material functional as defined above, the mesoscopic response behaviour can be computed. It is most important to note that this representation is not a continuum interpretation of stress or strain, but follows from the micromechanical theory (1–4). To clarify this statement further, it should be noted that the mesoscopic stress is defined by:

$$ {}^{M}\underset{\approx}{\xi} = \frac{p_1}{N^M} \sum_{\alpha=1}^{N^M} {}^{\alpha}\underset{\approx}{O}{}^{\alpha}\underset{\approx}{O}{}^{\alpha}\underset{\approx}{\xi} + \frac{p_2}{n^M} \sum_{i=1}^{n^M} {}^{\alpha}\underset{\approx}{O}{}^{\alpha}\underset{\approx}{O}{}^{\alpha}\underset{\approx}{\xi}^{i} \quad [4.6] $$

where ${}^{\alpha}\underset{\approx}{\xi}$ is defined in [2.15] and ${}^{\alpha}\xi^{i}$ in [3.5]. The orientation angle is ${}^{\alpha}O = \cos^{-1}(e_1 \cdot \underset{\approx}{n})$. In a similar manner, a mesoscopic strain and strain rate will be of the form:

$$ {}^{M}\underset{\approx}{e} = \frac{p_1}{N^M} \sum_{\alpha=1}^{N^M} {}^{\alpha}\underset{\approx}{O}{}^{\alpha}\underset{\approx}{O}{}^{\alpha}\underset{\approx}{e} + \frac{p_2}{n^M} \sum_{i=1}^{n^M} {}^{\alpha}\underset{\approx}{O}{}^{\alpha}\underset{\approx}{O}\underset{\approx}{e}^{i} \quad [4.7] $$

and

$$ {}^{M}\dot{\underset{\approx}{e}} = \frac{p_1}{N^M} \sum_{\alpha=1}^{N^M} {}^{\alpha}\underset{\approx}{O}{}^{\alpha}\underset{\approx}{O}{}^{\alpha}\dot{\underset{\approx}{e}}. \quad [4.8] $$

From these strict interpretations of the mesoscopic quantities, the following response characteristics of the fibrous network can be given. It should be noted that, in general, the mesostress will be determined from:

$$ {}^{M}\xi = \frac{\partial^{M}\mathfrak{M}}{\partial^{M}\underset{\approx}{e}} + \frac{\partial^{M}\mathfrak{M}}{\partial^{M}\dot{\underset{\approx}{e}}} = \frac{p_1}{N^M} \sum_{\alpha=1}^{N^M} {}^{\alpha}\underset{\approx}{O}{}^{\alpha}\underset{\approx}{O}\underset{=}{E}{}^{\alpha}\underset{\approx}{e} + \frac{p_2}{n^M} \sum_{i=1}^{n^M} {}^{\alpha}\underset{\approx}{O}{}^{\alpha}\underset{\approx}{O}{}^{\alpha\beta}\underset{=}{\Delta}\underset{\approx}{e}^{i} + \frac{p_1}{N^M} \sum_{\alpha=1}^{N^M} {}^{\alpha}\underset{\approx}{O}{}^{\alpha}\underset{\approx}{O}\underset{=}{D}{}^{\alpha}\dot{\underset{\approx}{e}}. \quad [4.9] $$

Hence, considering a constant stress input, the contribution of the binding energy available in each bond area to the elastic stress and stress history of the single fibre will be obtained from the combined probabilities as follows:

$$ {}^{M}e = p_1 \left\{ \frac{1}{N^M} \sum_{\alpha=1}^{N^M} [{}^{\alpha}\underset{\approx}{O}{}^{\alpha}\underset{\approx}{O}\underset{=}{C}{}^{\alpha}\underset{\approx}{\xi}(0) + \int_{0^+}^{t} {}^{\alpha}\underset{\approx}{O}{}^{\alpha}\underset{\approx}{O}\underset{=}{F}(t-\tau)\,{}^{\alpha}\underset{\approx}{\xi}(\tau)\, d\tau] \right\} + \frac{p_2}{n^M} \sum_{i=1}^{n^M} \frac{\partial^{\alpha\beta}\mathfrak{F}}{\partial^{\alpha}\underset{\approx}{\xi}^{i}} {}^{\alpha}\underset{\approx}{O}{}^{\alpha}\underset{\approx}{O}. \quad [4.10] $$

Finally, separating the conservative, i.e., instantaneous contribution from the dissipative or relaxation part, gives:

$$ {}^{M}\underset{\approx}{e} = \frac{p_1}{N^M} \sum_{\alpha=1}^{N^M} {}^{\alpha}\underset{\approx}{O}{}^{\alpha}\underset{\approx}{O}\underset{=}{C}{}^{\alpha}\underset{\approx}{\xi}(0) + \frac{p_2}{n^M} \sum_{i=1}^{n^M} {}^{\alpha}\underset{\approx}{O}{}^{\alpha}\underset{\approx}{O} \frac{\partial^{\alpha\beta}\mathfrak{F}}{\partial^{\alpha}\xi^{i}} + \frac{p_1}{N^M} \int^{t} \sum_{\alpha=1}^{N^M} {}^{\alpha}\underset{\approx}{O}{}^{\alpha}\underset{\approx}{O}\underset{=}{F}(t-\tau)\,{}^{\alpha}\xi(\tau)\, d\tau. \quad [4.11] $$

5. Conclusions

It is seen from the above presentation that the overall response of the fibrous network may be formulated. The response behaviour of the network is composed of a purely elastic contribution to an instantaneous stress application and the relaxation part as stated in [4.11]. If the single fibres are considered as elastic media, the elastic microstress can be defined in the *Cauchy* sense. However, the microstresses in the bond area must be assessed from a careful study of the hydrogen bonding between fibres. The contribution to the rheological response due to the single fibre behaviour only is expressed in terms of the stochastic microstress and its stress history. An assessment of the latter can only be made on the assumption that the deformation process is *Gauss*ian and statistically homogeneous so that correlation theory may be applied. In general, the quantities involved in the proposed theory can be assessed in part from their *Gauss*ian distributions which in turn may be obtained from corresponding experiments.

Acknowledgements

The authours would like to acknowledge the Scanning Electron Microscope Group at the Pulp and Paper Research Institute of Canada, Pointe Claire, Quebec, for their kind permission to use fig. 1a and 3.

Summary

The present work is concerned with the microrheology of a 2-dimensional fibrous system. A mathematical model is proposed that contains the aspects of bonding between fibres. The notion of three measuring scales, as well as the concept of a "Material Functional" that reflects the geometrical and rheological properties of the network is introduced in the analysis. The deformation kinematics is treated first for single fibres and then the fibre-fibre interaction by using a generalised force concept is considered. A formal presentation of the material functional required for the characterization of the fibrous network response is given.

Zusammenfassung

In der vorliegenden Arbeit wird die Mikrorheologie einer 2-dimensionalen Faser-Struktur behandelt. Es wird ein mathematisches Modell vorgeschlagen, welches den Aspekt einer atomaren Bindung zwischen den Fasern einbezieht. Es werden in dieser Theorie drei Maßskalen eingeführt sowie eine „Materialfunktion", welche die geometrischen und rheologischen Eigenschaften der Struktur widerspiegelt. Erst wird die Deformationskinematik der einzelnen Fasern behandelt, sodann wird die Faser-Faser-Einwirkung mit Hilfe einer generalisierten Kraft formuliert.

Schließlich wird die entsprechende Materialfunktion formuliert, die für die Beschreibung der Reaktion einer Faserstruktur zu einer mechanischen Beanspruchung benötigt wird.

References

1) *Axelrad, D. R.* and *L. G. Jaeger*, Random Theory of Deformation in Heterogeneous Media. In: Structure Solid Mechs. and Engineering Design, Part **1**, p. 571–578 (J. Wiley, London-New York 1970).

2) *Axelrad, D. R.*, Random Theory of Deformation of Structured Media. Int. Centre of Mech. Sciences, Udine, Italy (1972).

3) *Axelrad, D. R.*, Proc. 5th Int. Cong. on Rheology **2**, 221–231 (Tokyo 1970).

4) *Axelrad, D. R.*, Arch. Mech. Stosowanej **23**, 131–140 (1971).

5) *Yaglom, A. M.*, Theory of Stationary Random Functions (1962).

6) *Ziegler, H.*, Some Extremum Principles in Irreversible Thermodynamics with Application to Continuum Mechanics. In: Progress in Solid Mechs. **4** (Amsterdam 1963).

7) *Goux, C.* et al., Theoretical and Experimental Determinations of a Grain Boundary. Int. Conf. on the Structure and Properties of Grain Boundaries, IBM Watson Research Center, N.Y. State, Aug. 23–25, 1971.

8) *Sternstein, S. S.* and *A. M. Nissan*, A Molecular Theory of the Visco-Elasticity of a Three-Dimensional Hydrogen-Bonded Network. In: Formation and Structure of Paper **1**. Ed.: *F. Bolam*, Technical Sec. (London 1962).

9) *Gel'fand, I. M.* and *N. Ya. Vilenkin*, Generalized Functions Vol. **4** (London-New York 1964).

Authors' addresses:

D. R. Axelrad
Professor of Mechanical Eng., Micromechanics Lab.
McGill University, P.O. Box 6070, Montreal (Canada)

D. Atack
Director of Applied Physics Division, Pulp & Paper Research Institute of Canada, Pointe Claire, Quebec (Canada)

and *J. W. Provan*
Assistant Professor of Mechanical Eng., Micromechanics Lab., McGill University, Montreal (Canada)

Rheol. Acta 12, 177–182 (1973)

From the Micromechanics Laboratory, McGill-University Montreal (Canada)

Microrheology of crystalline media

By D. R. Axelrad and J. W. Provan

(Received October 27, 1972)

1. Introduction

A random theory of deformation and flow of crystalline media, which is based upon statistical mechanics and the theory of probability, has been proposed in previous publications (1–3). Furthermore, for the visco-elastic relaxation of structured systems, a linear stochastic model approach, based upon this deformation theory, has been developed by introducing the *Green*'s impulse transfer function of system theory (4). All these dynamic models, however, only consider interaction effects in a very general form. It is the purpose of this paper to formulate interaction effects between constituents of a crystalline media in a more detailed manner and to present an analysis incorporating time-dependent surface effects.

Hence, a model is proposed in which a rate-dependent internal surface potential is used to account for the interaction effects. This potential can be formulated in terms of the stochastic kinematic parameters and time-dependent variables at the interface. The concept of three measuring scales, i.e., the micro-, meso- and macrodomain is again used here [see ref. (1–5)]. Thus, the model describes a statistical ensemble of crystals of volume α_v $(\alpha = 1, \ldots, N)$ contained in a mesodomain $M(M = 1, \ldots, P)$. The crystals are close packed and separated by grain boundaries of finite width. It may be noted that this model is visualized as one in which the rheological properties of the medium manifest themselves in the crystals as well as in the grain boundaries. However, since the latter occupies a fraction of the total volume, this fraction σ enters as a parameter in the formulation of a "Material functional" ${}^{M}\mathfrak{M}$. The mesodomains of initial volume ${}^{M}\mathring{V}$ are denumerable and non-intersecting. They represent the smallest region of the macroscopic domain, with the initial volume $\mathring{\sigma}$, upon which the boundary conditions may be considered as phenomenologically specified. Further, $\mathring{V} = \cup\, {}^{M}\mathring{V}$; ${}^{M}\mathring{\sigma}\,\mathrm{D}\,{}^{L}\mathring{\sigma} = \Phi$; $L \neq M$. Apart from these restrictions, the choice of mesodomains is completely arbitrary and serves essentially as a transition mechanism between the stochastic microparameters and those applying to the total macrodomain.

In the subsequent kinematic and stress formulation, majuscules will denote undeformed position vectors and orientation tensors while their corresponding interpretations in the deformed configuration will be denoted by minuscules, stochastic dependence on time being implicitly implied. Greek letters to the left of a parameter indicate reference to microsystems while capital Latin superscripts on the left refer to mesoscopic parameters. For brevity, vectors will be indicated by "$\sim$", second order tensor quantities by "$\approx$" and fourth order material coefficient tensors by "$=$".

2. Microkinematics and microstresses

(i) Basic relations

${}^{\alpha}\underset{\sim}{R}$ is the position vector relative to a fixed *Cartes*ian frame to the centre of mass of the αth crystal in a mesodomain containing N such crystals or microdomains. The position vector to any undeformed point in the crystal is denoted by:

$$ {}^{\alpha}\underset{\sim}{X} = {}^{\alpha}\underset{\approx}{Q}\,{}^{\alpha}\underset{\sim}{Y} + {}^{\alpha}\underset{\sim}{R} \qquad [2.1] $$

where ${}^{\alpha}\underset{\approx}{Q}$ represents the orientation relative to the *Cartes*ian frame of crystallographic axes of the particular crystal and ${}^{\alpha}\underset{\sim}{Y}$ is the position ot the point in question relative to the centre of mass and to the crystal axes. Note that ${}^{\alpha}\underset{\approx}{Q}\,{}^{\alpha}\underset{\sim}{Y}$ is equivalent to the quantity $\underset{\sim}{D}^{\alpha 1}$ in reference 5. In the deformed configuration:

$$ {}^{\alpha}\underset{\sim}{x} = {}^{\alpha}\underset{\approx}{q}\,{}^{\alpha}\underset{\sim}{y} + {}^{\alpha}\underset{\sim}{r}. \qquad [2.2] $$

In order to distinguish between the coordinates of a point interior to the crystal and those referring to a point at the surface of the crystal, the following notation is used:

$$ {}^{\alpha}\underset{\sim}{G} = {}^{\alpha}\underset{\approx}{Q}\,{}^{\alpha}\underset{\sim}{H} + {}^{\alpha}\underset{\sim}{R}; \quad {}^{\alpha}\underset{\sim}{g} = {}^{\alpha}\underset{\approx}{q}\,{}^{\alpha}\underset{\sim}{h} + {}^{\alpha}\underset{\sim}{r}. \qquad [2.3] $$

Hence, the displacement of each of the above points is simply:

$$ {}^{\alpha}\underset{\sim}{u} = {}^{\alpha}\underset{\sim}{r} - {}^{\alpha}\underset{\sim}{R} $$

$$ {}^{\alpha}\underset{\sim}{w} = {}^{\alpha}\underset{\sim}{x} - {}^{\alpha}\underset{\sim}{X} = {}^{\alpha}\underset{\approx}{q}\,{}^{\alpha}\underset{\sim}{y} - {}^{\alpha}\underset{\approx}{Q}\,{}^{\alpha}\underset{\sim}{Y} + {}^{\alpha}\underset{\sim}{u} $$

$$ {}^{\alpha}\underset{\sim}{\mu} = {}^{\alpha}\underset{\sim}{g} - {}^{\alpha}\underset{\sim}{G} = {}^{\alpha}\underset{\approx}{q}\,{}^{\alpha}\underset{\sim}{h} - {}^{\alpha}\underset{\approx}{Q}\,{}^{\alpha}\underset{\sim}{H} + {}^{\alpha}\underset{\sim}{u}. \qquad [2.4] $$

The microdeformation gradients are defined as:

$$ {}^{\alpha}\underset{\approx}{F} = \frac{\partial^{\alpha}\underset{\sim}{y}}{\partial^{\alpha}\underset{\sim}{Y}}; \quad {}^{\alpha}\underset{\approx}{f} = \frac{\partial^{\alpha}\underset{\sim}{Y}}{\partial^{\alpha}\underset{\sim}{y}} \qquad [2.5] $$

and the micro-*Lagrangian* and *Euler*ian strains as:

$$2\,{}^{\alpha}\underset{\approx}{E} = {}^{\alpha}\underset{\approx}{F}^{c} \cdot {}^{\alpha}\underset{\approx}{F} - \underset{\approx}{\delta}; \quad 2\,{}^{\alpha}\underset{\approx}{\varepsilon} = \underset{\approx}{\delta} - {}^{\alpha}\underset{\approx}{f}^{c} \cdot {}^{\alpha}\underset{\approx}{f}. \qquad [2.6]$$

In this analysis, the linear strain tensor ${}^{\alpha}\underset{\approx}{e}$ is such that:

$${}^{\alpha}\underset{\approx}{e} = {}^{\alpha}\underset{\approx}{E} = {}^{\alpha}\underset{\approx}{\varepsilon}. \qquad [2.7]$$

In a familiar manner, these strains can also be written in terms of the displacements defined by [2.4] and the strain rate of a microelement can be expressed by:

$${}^{\alpha}\dot{\underset{\approx}{e}} = \frac{1}{2}\left[\frac{\partial^{\alpha}\dot{\underset{\sim}{w}}}{\partial^{\alpha}\underset{\sim}{Y}} + \left(\frac{\partial^{\alpha}\dot{\underset{\sim}{w}}}{\partial^{\alpha}\underset{\sim}{Y}}\right)^{T}\right]. \qquad [2.8]$$

(ii) Coincidence cell model

The coincidence cell model of a crystalline interface is developed from that studied by *Goux* (6) and *Bollman* (7). The undeformed distance between two crystal surfaces or the thickness of the grain boundary can be stated as:

$${}^{\alpha\beta}\underset{\sim}{\Delta} = {}^{\beta}\underset{\sim}{G} - {}^{\alpha}\underset{\sim}{G}$$

$$= {}^{\beta}\underset{\approx}{Q}\,{}^{\beta}\underset{\sim}{H} - {}^{\alpha}\underset{\approx}{Q}\,{}^{\alpha}\underset{\sim}{H} + {}^{\beta}\underset{\sim}{R} - {}^{\alpha}\underset{\sim}{R} \qquad [2.9]$$

and for the deformed state by:

$${}^{\alpha\beta}\underset{\sim}{\delta} = {}^{\beta}\underset{\sim}{g} - {}^{\alpha}\underset{\sim}{g}$$

$$= {}^{\beta}\underset{\approx}{\varrho}\,{}^{\beta}\underset{\sim}{h} - {}^{\alpha}\underset{\approx}{\varrho}\,{}^{\alpha}\underset{\sim}{h} + {}^{\beta}\underset{\sim}{r} - {}^{\alpha}\underset{\sim}{r}. \qquad [2.10]$$

The relative displacement of the grain boundary, when under the action of externally applied loads, becomes:

$${}^{\alpha\beta}\underset{\sim}{d} = {}^{\alpha\beta}\underset{\sim}{\delta} - {}^{\alpha\beta}\underset{\sim}{\Delta} = {}^{\beta}\underset{\sim}{\mu} - {}^{\alpha}\underset{\sim}{\mu} \qquad [2.11]$$

where β represents any grain contiguous to the αth crystal and ${}^{\alpha}\underset{\sim}{\mu}$ is defined in terms of the surface coordinates by eq. [2.4]. The relative velocity or the time rate of change of ${}^{\alpha\beta}\underset{\sim}{d}$ is given by:

$${}^{\alpha\beta}\dot{\underset{\sim}{d}} = {}^{\alpha\beta}\dot{\underset{\sim}{\delta}}. \qquad [2.12]$$

Interaction effects between individual microelements are taken into consideration by adopting a probabilistic surface coincidence cell model, a schematic of which is shown in the accompanying presentation (8). Since the orientations ${}^{\alpha}\underset{\approx}{Q}$, ${}^{\beta}\underset{\approx}{Q}$ are random parameters, the coincidence areas $\Delta^{\varrho}S = {}^{\alpha}\Delta^{\varrho}S = {}^{\beta}\Delta^{\varrho}S$ and their number $\varrho = 1, \ldots, {}^{\alpha}p$ are random dependent functions of the orientations, the lattice parameter $\underset{\sim}{a}$, the initial volume of the crystal ${}^{\alpha}\hat{v}$ as well as of surface irregularities of each crystal. If the centre of each coincidence cell is considered as the coincidence point in the two grain surfaces forming the grain boundary (fig. 3, ref. 8) then the undeformed and deformed distances, as well as the relative velocity of this coincident point, form the three basic parameters, i.e., ${}^{\alpha\beta}\underset{\sim}{\Delta}^{\varrho}$, ${}^{\alpha\beta}\underset{\sim}{\delta}^{\varrho}$ and ${}^{\alpha\beta}\dot{\underset{\sim}{d}}^{\varrho} = {}^{\alpha\beta}\dot{\underset{\sim}{\delta}}^{\varrho}$ which are significant for the subsequent analysis.

(iii) Microstresses and interfacial potentials

In this section, first the definition of a microstress will be given. Following conventional thermodynamic principles, the elastic and reversible response characteristics of each individual crystal will be described by either the intrinsic energy or by a free energy potential, if the temperature is also an independent parameter. If this elastic energy potential is denoted by ${}^{\alpha}U^{e}$, the components of the microstress tensor ${}^{\alpha}\underset{\approx}{\xi}^{e}$ relative to the ${}^{\alpha}\underset{\sim}{y}$-axes can be expressed by:

$${}^{\alpha}\underset{\approx}{\xi}^{e} = \frac{\partial^{\alpha}U^{e}}{\partial^{\alpha}\underset{\approx}{e}} \qquad [2.13]$$

where, for the linear range of response, it is usually assumed that:

$${}^{\alpha}U^{e} = \tfrac{1}{2}\,\underset{\equiv}{E}\,{}^{\alpha}\underset{\approx}{e}\,{}^{\alpha}\underset{\approx}{e}. \qquad [2.14]$$

$\underset{\equiv}{E}$ being the fourth order elastic coefficient tensor for the αth crystal. For simplicity of the present analysis, it is assumed that $\underset{\equiv}{E}$ does not vary from crystal to crystal and furthermore, that the elastic influence of dislocations, etc., in the microelement is absorbed in $\underset{\equiv}{E}$. Combining [2.13] and [2.14] gives:

$${}^{\alpha}\underset{\approx}{\xi}^{e} = \underset{\equiv}{E}\,{}^{\alpha}\underset{\approx}{e}. \qquad [2.15]$$

For the inelastic and irreversible response of each crystal, it is evident that a dissipation potential ${}^{\alpha}\Phi^{v}$ will be required which, following *Ziegler* (9), will depend in general not only on the strain rates but also upon the state of the crystal and its history. Applying, however, the extremum principle of least irreversible force, the actual quasistatic strain rate ${}^{\alpha}\dot{\underset{\approx}{e}}$ minimizes the magnitude of the irreversible stress ${}^{\alpha}\underset{\approx}{\xi}^{v}$ so that:

$${}^{\alpha}\underset{\approx}{\xi}^{v} = \frac{\partial^{\alpha}\Phi^{v}}{\partial^{\alpha}\dot{\underset{\approx}{e}}}. \qquad [2.16]$$

Further restricting the dissipation function to a homogeneous polynomial of degree 2, i.e., permitting only a linear relationship between ${}^{\alpha}\underset{\approx}{\xi}^{v}$ and ${}^{\alpha}\dot{\underset{\approx}{e}}$, then:

$${}^{\alpha}\underset{\approx}{\xi}^{v} = \underset{\equiv}{D}\,{}^{\alpha}\dot{\underset{\approx}{e}} \qquad [2.17]$$

in which $\underset{\equiv}{D}$ is a fourth order dissipative material coefficient tensor with, in general, 21 independent terms. Hence, the total microstress acting at a point in the crystal, when the latter is regarded as a continuum, will be obtained from:

$$^{\alpha}\underset{\approx}{\xi} = {}^{\alpha}\underset{\approx}{\xi}^{e} + {}^{\alpha}\underset{\approx}{\xi}^{v} = \underline{\underline{E}}^{\alpha}\underset{\approx}{\varepsilon} + \underline{\underline{D}}^{\alpha}\dot{\underset{\approx}{\varepsilon}}. \qquad [2.18]$$

In a completely analogous linear formulation, by using an aftereffect theory (1), it can be shown that the inverse of [2.18] may be written in the following manner:

$$^{\alpha}\underset{\approx}{\varepsilon} = \underline{\underline{C}}\,{}^{\alpha}\underset{\approx}{\xi}(0) + \int_{0^+}^{t} \underline{\underline{F}}(t-\tau)\,{}^{\alpha}\underset{\approx}{\xi}(\tau)\,d\tau \qquad [2.19]$$

in which $\underline{\underline{C}}$ is the compliance and $\underline{\underline{F}}$ the fluidity tensor of the αth crystal.

The interfacial response behaviour can be described in a similar manner as the microstresses by using a conservative and dissipative potential. Both take the form of "pair potentials" between coincidence points of a coincidence lattice cell. Considering the elastic response first, one can specify Ψ^e as a *Lennard-Jones*, *Morse*, *Coulomb* or hydrogen bond type of potential. For example, a suitable potential for the present analysis is a slightly modified *Morse* potential based on the pair potential Ψ^e as follows:

$$^{\alpha\beta}\Psi_e^{\varrho} = {}^{\alpha\beta}\Psi_0^{\varrho}\{1 - \exp[-b|^{\alpha\beta}\underset{\sim}{d}^{\varrho}|]\}^2; \quad \varrho = 1,\ldots {}^{\alpha}\psi \qquad [2.20]$$

which is based on the coincidence areas $\Delta^{\varrho}S$ being dependent random functions and where without loss of generality, one may specify $^{\alpha\beta}\Psi_0^{\varrho}$ to account for these random parameters. The constant b in [2.20] is a metallographic parameter to be assessed from micrographic studies. Hence, $^{\alpha\beta}\Psi_e^{\varrho}$ can be formulated from a conventional pair potential Ψ^e associated with a specific material. $^{\alpha}p$ is the number of coincidence cells surrounding α.

Following *Yvon* (10), the local reversible surface force between centres of the coincidence areas is obtained from:

$$^{\alpha\beta}\underset{\sim}{F}_e^{\varrho} = \underset{\sim}{e}\,\frac{d\,{}^{\alpha\beta}\Psi_e^{\varrho}}{d|^{\alpha\beta}\underset{\sim}{d}^{\varrho}|}; \qquad \varrho = 1,\ldots,{}^{\alpha}p \qquad [2.21]$$

where $\underset{\sim}{e}$ is the unit vector in the direction of $^{\alpha\beta}\underset{\sim}{d}^{\varrho}$. This discrete surface force will act on each coincidence cell contained in the boundary between the αth crystal and its βth contiguous neighbour.

Similarly, for the inelastic response of the interface, a dissipative potential $^{\alpha\beta}\Psi_v^{\varrho}$ may be postulated that is based on a known dissipative pair potential Ψ^v, so that analogously to the elastic case:

$$^{\alpha\beta}\underset{\sim}{F}_v^{\varrho} = \underset{\sim}{\varepsilon}\,\frac{d\,{}^{\alpha\beta}\Psi_v^{\varrho}}{d|^{\alpha\beta}\dot{\underset{\sim}{d}}^{\varrho}|} \qquad [2.22]$$

represents the dissipative component of the surface interaction force $^{\alpha\beta}\underset{\sim}{F}^{\varrho}$ acting between coincidence points and where $\underset{\sim}{\varepsilon}$ is the unit vector in the direction of $^{\alpha\beta}\dot{\underset{\sim}{d}}^{\varrho}$. The total interaction force may, therefore, be written as:

$$^{\alpha\beta}\underset{\sim}{F}^{\varrho} = {}^{\alpha\beta}\underset{\sim}{F}_e^{\varrho} + {}^{\alpha\beta}\underset{\sim}{F}_v^{\varrho} = \underset{\sim}{e}\,\frac{d\,{}^{\alpha\beta}\Psi_e^{\varrho}}{d|^{\alpha\beta}\underset{\sim}{d}^{\varrho}|} + \underset{\sim}{\varepsilon}\,\frac{d\,{}^{\alpha\beta}\Psi_v^{\varrho}}{d|^{\alpha\beta}\dot{\underset{\sim}{d}}^{\varrho}|}. \qquad [2.23]$$

However, this force $^{\alpha\beta}\underset{\sim}{F}^{\varrho}$ is still a discrete quantity. In order to find the totality of interaction effects, a summation can only be carried out in a generalized form. Using the concept of generalized functions [see *Gel'fand* and *Vilenkin* (11)], a generalized surface force may be written symbolically as follows:

$$^{\alpha\beta}\underset{\sim}{\tau} = \delta[^{\alpha\beta}\underset{\sim}{d} - (^{\alpha\beta}\underset{\sim}{d}^{\varrho} - {}^{\alpha\beta}\dot{\underset{\sim}{d}}^{\varrho})]\,{}^{\alpha\beta}\underset{\sim}{F}^{\varrho} \qquad [2.24]$$

where δ is the 3-dimensional delta function whose integral is equal to 1 when $^{\alpha\beta}\underset{\sim}{d} = (^{\alpha\beta}\underset{\sim}{d}^{\varrho} + {}^{\alpha\beta}\dot{\underset{\sim}{d}}^{\varrho})$ and equal to zero when $^{\alpha\beta}\underset{\sim}{d} \pm (^{\alpha\beta}\underset{\sim}{d}^{\varrho} + {}^{\alpha\beta}\dot{\underset{\sim}{d}}^{\varrho})$. This relation has meaning only when under an integral sign. On the basis of this generalized surface force, the "surface microstress tensor" $^{\alpha}\underset{\sim}{\xi}^{i}$ having meaning, say one atomic layer within the microelement and the generalized surface traction $^{\alpha\beta}\underset{\sim}{\tau}$, can be related as follows:

$$\int_{\Delta^{\varrho}S} {}^{\alpha\beta}\underset{\sim}{\tau}\,ds = \int_{^{\alpha}\Delta^{\varrho}S} {}^{\alpha}\underset{\approx}{\xi}^{i}\,{}^{\alpha}\underset{\sim}{n}\,ds = \int_{^{\beta}\Delta^{\varrho}S} {}^{2}\underset{\approx}{\xi}^{i}\,{}^{\beta}\underset{\sim}{n}\,ds \qquad [2.25]$$

giving a balance of forces at all times, n being the normal to the surface of the crystal. Eq. [2.25] may be considered as the boundary conditions of each microelement or crystal.

It is of interest to note that a change of surface energy may be written in the following form:

$$d\mathfrak{F}^{\varrho} = \int_{\Delta^{\varrho}S} {}^{\alpha\beta}\underset{\sim}{\tau}\cdot{}^{\alpha\beta}\underset{\sim}{d}\,ds + \int_{\Delta^{\varrho}S} {}^{\alpha\beta}\underset{\sim}{\tau}\cdot{}^{\alpha\beta}\dot{\underset{\sim}{d}}\,dt\,ds \qquad [2.26]$$

which may be regarded in a first approximative form as the quadratic:

$$d\mathfrak{F}^{\varrho} = \tfrac{1}{2}\,{}^{\alpha\beta}\underline{\underline{\Lambda}}^{\varrho}\,{}^{i}\underset{\approx}{\varepsilon}\,{}^{i}\underset{\approx}{\varepsilon} + \tfrac{1}{2}\,{}^{\alpha\beta}\underline{\underline{\Gamma}}^{\varrho\varrho}\,{}^{i}\dot{\underset{\approx}{\varepsilon}}\,{}^{i}\dot{\underset{\approx}{\varepsilon}}. \qquad [2.27]$$

Alternatively, an inverse may be written in the form:

$$^{i}\underset{\approx}{\varepsilon} = {}^{\alpha\beta}\underline{\underline{G}}^{\varrho}\,{}^{\alpha}\underset{\approx}{\xi}^{i}(0) + \int_{0^+}^{t} {}^{\alpha\beta}\underline{\underline{H}}^{\varrho}(t-\tau)\,{}^{\alpha}\underset{\approx}{\xi}^{i}(\tau)\,d\tau \qquad [2.28]$$

where $^{\alpha}\underline{\underline{\xi}}^{i}$ is determined from [2.25].

3. Application of probability distribution and correlation theory

For the determination of the basic quantities described in section 2, it is necessary to find their distributions. It is assumed in the present analysis that these distributions are of the *Gauss*ian type and that they are statistically homogeneous and non-isotropic so that the related correlation is a function, not of the actual position of the microelement in the macrodomain, but of the relative distance between points of interest (12). This parameter is designated as the correlation parameter $\underset{\sim}{\lambda}$. All probability distributions in this analysis can be categorized into three groups. The first group concerns the distributions which are determined prior to any test and are of a morphological nature. Hence, from the analysis of an actual crystalline material, the following *Gauss*ian distributions are assumed to be known:

(i) $\pi(\underset{\approx}{Q})$: the distribution of crystallographic axes orientations.

(ii) $\pi(\underset{\approx}{v})$: the size distribution of crystals or microelements.

Combining these distributions with basic parameters of the crystal, i.e., a, Ψ^e and Ψ^v, distributions of the second type, or those of a specific dependent nature, can be calculated. These distributions are:

(i) $\pi\{{}^{\alpha\beta}\Psi^{\varrho}_{e}\}$: the distribution of elastic coincidence potentials

(ii) $\pi({}^{\alpha\beta}\Psi^{\varrho}_{v}\}$: the distribution of dissipative coincidence potentials

(iii) $\pi\{p\}$: the distribution of the number of coincidence cells surrounding the crystal.

Finally, distributions of the third group are those determined from observations made on actual material specimens under test conditions [see *Axelrad* and *Kalousek* (13)]. Thus, upon specifying the external boundary conditions on the macroscopic body, the following distributions are considered determinable either by direct observation or by calculation from such observations.

(i) $\pi(\underset{\approx}{\varrho})$: distribution of orientations of crystals at time t.

(ii) $\pi(\underset{\approx}{e})$: distribution of microstrains (eq. [2.6]).

(iii) $\pi(\dot{\underset{\approx}{e}})$: distribution of microstrain rates (eq. [2.8]).

(iv) $\pi(\underset{\sim}{d})$: distribution of crystal surface displacements (eqs. [2.4] and [2.11]).

(v) $\pi(\dot{\underset{\sim}{d}})$: distribution of crystal surface relative velocities from eq. [2.12].

For simplification of the present analysis, it is assumed that $\pi(\underset{\approx}{Q}) = \pi(\underset{\approx}{\varrho})$ so that the introduction of "micromoments" into the analysis is alleviated and, further, that the balance of forces (eq. [2.25]) is valid. The latter assumption implies that the distributions $\pi({}^{\alpha\beta}\Psi^{\varrho}_{e})$, $\pi({}^{\alpha\beta}\Psi^{\varrho}_{v})$, $\pi(p)$, $\pi(\underset{\sim}{d})$ and $\pi(\dot{\underset{\sim}{d}})$ need not be specified, at present, provided that a distribution of microstresses ${}^{\alpha}\underset{\approx}{\xi}$ can be established from the remaining distributions. This can be done by application of correlation theory as shown below.

It is evident from the above considerations that the basic kinematic quantities will be determined from the distributions $\pi(\underset{\approx}{e})$ and $\pi(\dot{\underset{\approx}{e}})$. These distributions describe the stochastic processes:

$$ {}^{\alpha}\underset{\approx}{e} = \langle \underset{\approx}{e} \rangle_N + {}^{\alpha}\overset{*}{\underset{\approx}{e}}; \quad {}^{\alpha}\dot{\underset{\approx}{e}} = \langle \dot{\underset{\approx}{e}} \rangle_N + {}^{\alpha}\dot{\underset{\approx}{e}} \qquad [3.1] $$

where the symbol $\langle \cdot \rangle_N$ indicates the statistical mean over the N crystals in the mesodomain and * their fluctuating component. Since *Gauss*ian distributions have been assumed to exist throughout the analysis, application of correlation theory requires only the statement of second moments or correlation functions in addition to the above mean values, viz:

$$ B_e(\underset{\sim}{\lambda}, s) = \langle \overset{*}{e}[{}^{\alpha}\underset{\sim}{X}, t]\, \overset{*}{e}[{}^{\alpha}\underset{\sim}{X} + \underset{\sim}{\lambda}, t + s] \rangle_N \qquad [3.2] $$

$$ B_e(\underset{\sim}{\lambda}, s) = \langle \overset{*}{\dot{\underset{\approx}{e}}}[{}^{\alpha}\underset{\sim}{X}, t]\, \overset{*}{\dot{\underset{\approx}{e}}}[{}^{\alpha}\underset{\sim}{X} + \underset{\sim}{\lambda}, t + s] \rangle_N. \qquad [3.3] $$

Upon substituting [3.1] into [2.18] and writing the microstress in an analogous manner to the strain, i.e.:

$$ {}^{\alpha}\underset{\approx}{\xi} = \langle \underset{\approx}{\xi} \rangle_N + {}^{\alpha}\underset{\approx}{\xi} \qquad [3.4] $$

it is readily seen that:

$$ \langle \underset{\approx}{\xi} \rangle_N = \underset{\equiv}{E} \langle \underset{\approx}{e} \rangle_N + \underset{\equiv}{D} \langle \dot{\underset{\approx}{e}} \rangle_N $$

$$ {}^{\alpha}\underset{\approx}{\xi} \quad = \underset{\equiv}{E}\, {}^{\alpha}\overset{*}{\underset{\approx}{e}} + \underset{\equiv}{D}\, {}^{\alpha}\dot{\underset{\approx}{e}}. \qquad [3.5] $$

Furthermore, the stress correlation function is obtained by taking the statistical mean of the products of two microstress fluctuations as follows:

$$ B_{\xi}(\underset{\sim}{\lambda}, s) = \langle \underset{\approx}{\xi}[{}^{\alpha}\underset{\sim}{X}, t]\, \underset{\approx}{\xi}[{}^{\alpha}\underset{\sim}{X} + \underset{\sim}{\lambda}, t + s] \rangle_N \qquad [3.6] $$

$$ = \underset{\equiv}{E}\underset{\equiv}{E} B_e(\underset{\sim}{\lambda}, s) + \underset{\equiv}{D}\underset{\equiv}{D} B_e(\underset{\sim}{\lambda}, s). \qquad [3.7] $$

With the estimate of $\langle \underset{\approx}{\xi} \rangle_N$ from [3.5] and $B\xi(\lambda, s)$ of [3.6], a unique distribution $\pi(\underset{\approx}{\xi})$ for microstresses can be established. Using this distribution, a distribution $\pi(\underset{\sim}{\tau})$ for the internal surface tractions can be calculated using the balance of forces relation [2.25]. It should be noted that a knowledge of these two internal stress distributions is of utmost significance in attempting to analyse fatigue, microcrack formations and relaxation phenomena in polycrystalline media from a probabilistic mechanics point of view.

4. Response characteristics of crystalline media

It has been mentioned in the introduction to this lecture that a material functional ${}^M\mathfrak{M}$ will be used to describe the response characteristics of a polycrystalline solid. Such a functional will be formed from two parts, one of which will relate to the viscoelastic response of the crystals themselves and another part which is associated with the effect of the grain boundaries. The functional will be of the form:

$${}^M\mathfrak{M} = {}^M\mathfrak{M}\{\pi(\underset{\approx}{Q}), \pi(v), E, \underset{\approx}{D}, \Psi^e, \Psi^v, \underset{\approx}{g}, \sigma;\ \theta, t\} \quad [4.1]$$

$$= (1-\sigma)\,{}^M\mathfrak{M}^c + \sigma\,{}^M\mathfrak{M}^i \quad [4.2]$$

in which σ is the fraction of the material that relates to the crystal interfaces, ${}^M\mathfrak{M}^c$ is the crystal response functional and ${}^M\mathfrak{M}^i$ the interfacial response functional. The first term in [4.2] is derivable from the energy potentials mentioned in section 2 and may formally be written as:

$${}^M\mathfrak{M}^c = \sum_{\alpha=1}^{N} \int_{{}^\alpha v} ({}^\alpha U^e + {}^\alpha\Phi^v)\,\varkappa({}^\alpha\underset{\approx}{Q})\,dv \quad [4.3]$$

where $\varkappa({}^\alpha\underset{\approx}{Q})$ indicates that the influence of the crystallographic axes orientations of each crystal is taken into account. On the other hand, using the distributions as discussed in section 3, ${}^M\mathfrak{M}^c$ can be written more realistically in the form of:

$${}^M\mathfrak{M}^c = \int_O f(O) \int_v f(v) \int_{{}^\alpha v} ({}^\alpha U^e + {}^\alpha\Phi^v)\, d^\alpha v\, dv\, dO \quad [4.4]$$

where "f" indicates density distributions.

The part due to the grain boundary effects can be expressed by:

$${}^M\mathfrak{M}^i = \tfrac{1}{2}\sum_{\alpha=1}^{N} \{\int_{{}^\alpha v}[\int_{{}^{\alpha\beta}\Psi_e} f({}^{\alpha\beta}\Psi_e)\,{}^{\alpha\beta}\Psi_e\,d\Psi_e + \int_{{}^{\alpha\beta}\Psi_v} f({}^{\alpha\beta}\Psi_v)\,{}^{\alpha\beta}\Psi_v\,d\Psi_v]\,dv\} \quad [4.5]$$

in which, as stated previously, the distributions $\pi({}^{\alpha\beta}\Psi_e)$ and $\pi({}^{\alpha\beta}\Psi_v)$ and their densities $f({}^{\alpha\beta}\Psi_e)$ and $f({}^{\alpha\beta}\Psi_v)$ are available from the knowledge of the material properties $\underset{\approx}{g}$, Ψ^e and Ψ^v as well as the distributions $\pi(\underset{\approx}{Q})$ and $\pi(v)$.

Combining [4.4] and [4.5] in [4.1] gives the total mesoscopic material functional as a function of the geometric, kinematic and the other physical parameters involved. The response characteristics of the polycrystalline solid can now be written on the assumption that the material functional is an energy operator between stress, strain and strain rate. In particular, assuming a constant stress input to the macroscopic body, a mesoscopic strain-time function based upon an after-effect approach can be formulated. First, a mesoscopic stress is defined in the following form:

$${}^M\underset{\approx}{\xi} = (1-\sigma)\langle{}^\alpha\underset{\approx}{\xi}\rangle_N + \sigma\langle{}^\alpha\underset{\approx}{\xi}^i\rangle_N$$

$$= \frac{(1-\sigma)}{N}\sum_{\alpha=1}^{N} {}^\alpha\underset{\approx}{Q}\,{}^\alpha\underset{\approx}{Q}\,{}^\alpha\underset{\approx}{\xi} + \frac{\sigma}{2N}\sum_{\alpha=1}^{N} {}^\alpha\underset{\approx}{Q}\,{}^\alpha\underset{\approx}{Q}\,{}^\alpha\underset{\approx}{\xi}^i \quad [4.6]$$

where ${}^\alpha\underset{\approx}{\xi}$ is defined in [2.18] and ${}^\alpha\underset{\approx}{\xi}^i$ in [2.25]. In a similar manner, a mesoscopic strain and strain rate may be expressed by:

$${}^M\underset{\approx}{\varepsilon} = (1-\sigma)\langle{}^\alpha\underset{\approx}{\varepsilon}\rangle_N + \frac{\sigma}{2}\langle{}^\alpha\underset{\approx}{\varepsilon}^i\rangle_N \quad [4.7]$$

and

$${}^M\dot{\underset{\approx}{\varepsilon}} = (1-\sigma)\langle{}^\alpha\dot{\underset{\approx}{\varepsilon}}\rangle_N + \frac{\sigma}{2}\langle{}^\alpha\dot{\underset{\approx}{\varepsilon}}^i\rangle_N. \quad [4.8]$$

With the above interpretation of the mesoscopic quantities, a memory integral including the stress history in the formulation can be written so that the total mesoscopic response becomes:

$${}^M\underset{\approx}{\varepsilon}(t) = \frac{(1-\sigma)}{N}\sum_{\alpha=1}^{N} {}^\alpha\underset{\approx}{Q}\,{}^\alpha\underset{\approx}{Q}\,\underline{\underset{\approx}{C}}\,{}^\alpha\underset{\approx}{\xi}(0) + \frac{\sigma}{2N}\sum_{\alpha=1}^{N}\sum_{\varrho=1}^{{}^\alpha p} {}^\alpha\underset{\approx}{Q}\,{}^\alpha\underset{\approx}{Q}\,{}^{\alpha\beta}\underline{\underset{\approx}{G}}^{\varrho\alpha}\underset{\approx}{\xi}^i(0) + \frac{(1-\sigma)}{N}\sum_{\alpha=1}^{N}\int_{0^+}^{t} {}^\alpha\underset{\approx}{Q}\,{}^\alpha\underset{\approx}{Q}\,\underline{\underset{\approx}{F}}(t-\tau)\,{}^\alpha\underset{\approx}{\xi}(\tau)\,d\tau + \frac{\sigma}{2N}\sum_{\alpha=1}^{N}\sum_{\varrho=1}^{{}^\alpha p}\int_{0^+}^{t} {}^\alpha\underset{\approx}{Q}\,{}^\alpha\underset{\approx}{Q}\,{}^{\alpha\beta}\underline{\underset{\approx}{H}}^{\varrho}(t-\tau)\,{}^\alpha\underset{\approx}{\xi}^i(\tau)\,d\tau. \quad [4.9]$$

In conclusion, it may be seen that the response of a polycrystalline solid to a constant stress input depends strongly upon the orientation of the crystals, the internal microstress, as well as the internal mesostress defined clearly in this paper, and finally, on the stress history.

References

1) *Axelrad, D. R.*, Stochastic Analysis of the Flow of Two-Phase Media. Proc. 5th Int. Cong. on Rheology **2**, 221–231 (Tokyo 1970).

2) *Axelrad, D. R.*, Arch. Mech. Stosowanej **23**, 131 to 140 (1971).

3) *Axelrad, D. R.* and *R. N. Yong*, Micro-Rheology of the Yielding of a Heterogeneous Medium. Proc. 5th Int. Cong. on Rheology **2**, 309–314 (1971).

4) *Axelrad, D. R.* and *J. W. Provan*, Rheol. Acta **10**, 330–335 (1971).

5) *Provan, J. W.*, Arch. Mech. Stosowanej **23**, 339 to 352 (1971).

6) *Goux, C.* et al., Theoretical and Experimental Determinations of Grain Boundary Structures and Energies. In: Grain Boundaries and Interfaces, Eds. *P. Chaudhari* and *J. W. Matthews*, North-Holland (1972).

7) *Bollmann, W.*, Crystal Defects and Crystalline Interfaces (Berlin-Heidelberg-New York 1970).

8) *Axelrad, D. R., D. Atack*, and *J. W. Provan*, Rheol. Acta **12**, 000–000 (1973).

9) *Ziegler, H.*, Some Extremum Principles in Irreversible Thermodynamics with Application to Continuum Mechanics. In: Progress in Solid Mechanics **4**, eds.: *I. N. Sneddon* and *R. Hill* (Amsterdam 1963).

10) *Yvon, J.*, Correlations and Entropy in Classical Statistical Mechanics (London 1969).

11) *Gel'Fand, I. M.* and *N. Ya. Vilenkin*, Generalized Functions Vol. 4 (London-New York 1964).

12) *Yaglom, A. M.*, An Introduction to the Theory of Stationary Random Functions (1962).

13) *Axelrad, D. R.* and *J. Kalousek*, Measurement of Microdeformations by Holographic X-Ray Diffraction. Paper 15, Int. Sym. on Experimental Mechanics, Univ. of Waterloo, Ontario, Canada, June 12–16, 1972.

Authors' address:

D. R. Axelrad
Professor of Mechanical Eng.
Micromechanics Lab., McGill University, P.O. Box 6070
Montreal (Canada)

and *J. W. Provan*
Associate Professor of Mechanical Eng.
Micromechanics Lab., McGill University, P.O. Box 6070
Montreal (Canada)

Rheol. Acta 12, 183–188 (1973)

From the Institute of Chemical Physics, USSR Academy of Sciences, Moscow (USSR)

Experimental and theoretical investigation of creep of rigid polymers on torsion

By I. N. Danilova and T. V. Sokolova

With 5 figures

(Received October 27, 1972)

Based on the results of investigations, an approximate method is proposed for determining the maximum tangential stresses in resin and for evaluating certain relaxation parameters contained in the generalized *Maxwell* equation, using short-time experiments on torsion of solid cylindrical specimens.

The behaviour of polymeric binders can be studied with the help of experiments on pure torsion of thin tubular specimens. Such specimens exhibit uniform stressed state and thus it is comparatively easy to determine both elastic and relaxation characteristics of polymers in pure shear (2). However, manufacturing of thin tubular specimens without internal defects and free of initial stresses is a fairly tedious and complicated process. But it is simpler to prepare solid cylindrical specimens. Therefore, the development of methods for studying the polymeric properties in shear, using volumetric specimens, is of practical value.

In theoretical consideration of the problem of torsion of circular polymeric rod on the basis of preliminary experimental observations it was assumed that the total deformation in polymeric rod consists of elastic deformation related by *Hooke* law to stress and of highly elastic deformation which is the sum of individual terms corresponding to different members of relaxation time spectrum.

The generalized nonlinear *Maxwell* equation describes the relation between rate of each component of highly elastic deformation respectively with the stress and highly elastic deformation itself. However, in contrast to the wellknown *Maxwell* equation in which the relaxation time is a constant quantity, relaxation time in generalized *Maxwell* equation is an exponential function of argument, which contains both stress and highly elastic deformation.

It is shown in many reports (3, 4, 5) that for describing the mechanical behaviour of certain rigid net-shaped polymers by generalized *Maxwell* equation it is sufficient to take into account only two terms of the time relaxation spectrum corresponding to minimum and maximum relaxation times. These terms are referred to as older and younger respectively.

The problem of torsion of solid circular rod, when using generalized *Maxwell* equation with older and younger terms of spectrum, leads to the following set of eqs. [6, 7]

$$\bar{\varrho}\bar{v} = \frac{\tau}{G} + \sum_{i=1}^{2} \varepsilon_i^* \qquad (i = 1, 2) \qquad [1]$$

where $\bar{\varrho} = \frac{\varrho}{R}$ – non-dimensional current radius, R – external radius of rod, $\tau \equiv \tau_{\theta z}$ – tangential stress in cylindrical coordinates, $\bar{v} = vR$ – non-dimensional twisting, $\varepsilon_i^* \equiv \varepsilon_{\theta z, i}^* (i = 1, 2)$ – components of highly elastic deformation, corresponding to older and younger terms of relaxation time spectrum, G – modulus of shear.

Eq. [1] shows that the total deformation $\varepsilon = \bar{v}\bar{\varrho}$ in rigid net-shaped polymer consists of elastic deformation $e = \tau/G$ and highly elastic deformation $\varepsilon^* = \sum_{i=1}^{2} \varepsilon_i^*$. Generalized *Maxwell* equation for each component of highly elastic deformation in pure shear is of the form

$$\frac{\partial \varepsilon_i^*}{\partial t} = \frac{\tau - G_{\infty,i}\varepsilon_i^*}{\eta_{0\ G,i}} \exp \frac{1}{m_{G,i}^*} |\tau - G_{\infty,i}\varepsilon_i^*| \quad (i = 1.2) \qquad [2]$$

where $G_{\infty,i}$ – moduli of high elasticity in shear, $\eta_{0G,i}^*$ – coefficients of initial relaxation viscosity in shear, $m_{G,i}^*$ – moduli of rate in shear.

The boundary conditions at the end faces of rod are satisfied in the integral form

$$M = 2\pi R^3 \int_0^1 \tau \bar{\varrho}^2 d\bar{\varrho} \qquad [3]$$

where M – torsional moment.

Two loading regimes were investigated: loading at constant rate of deformation, when

13*

$$\frac{d\bar{v}}{dt} = V_0 = \text{const}, \quad \bar{v} = \bar{V}_0 t$$

and in the case of creep, when $M = M_0 = \text{const.}$

In calculating the loading regime at constant deformation rate the eq. [2] is integrated numerically, using the initial values at $t = 0$, $\varepsilon_i^*(0, \bar{\varrho}) \equiv 0$ $(i = 1.2)$, and τ and M are determined from [1] and [3]. In the case of creep, simultaneously solving [1] and [3], we obtain

$$\bar{v} = \frac{2M_0}{\pi R^3 G} + 4\int_0^1 (\varepsilon_1^* + \varepsilon_2^*)\bar{\varrho}^2 d\bar{\varrho} \qquad [4]$$

$$\tau = G(\bar{v}\bar{\varrho} - \sum_{i=1}^{2} \varepsilon_i^*) \qquad (i = 1.2) \qquad [5]$$

Eq. [2] is integrated numerically. If the creep arises owing to elastic initial state, then at

$$\bar{t} = 0 \quad \varepsilon_i^*(o, \bar{\varrho}) \equiv 0 \ (i = 1.2) \quad \tau = \frac{2M_0}{\pi R^3}\bar{\varrho}.$$

If the initial state is not elastic, then at $t = 0$ stress field $\tau(o, \bar{\varrho})$ and field of highly elastic deformation $\varepsilon_i^*(o, \bar{\varrho})$ are given.

The calculations were made on the computer "Mir"; integration of eq. [2] was carried out at 10 points with respect to $\bar{\varrho}(\bar{\varrho} = 0.1, 0.2, \ldots \text{I})$.

Numerical integration was based on simple finite-differential approximation of derivatives, but time step was taken to be variable. Integral in [4] was calculated by *Simpson* formula.

Cylindrical specimens of 20 mm diameter and a working length of 120 mm were used in the experiments.

The specimens were prepared by mechanical treatment of semi-finished blanks cast from three epoxy compositions EDT-10, K-115 and LMT-6.

The base of compositions EDT-10 and K-115 was resin ED-5. The compound EDT-10 is a hot hardened composition but K-115 is a cold hardened composition.

The ZMI-6 compound on ED-6 resin base is a hot hardened composition.

After mechanical treatment the specimens were annealed for relieving the stress.

Torsional testing of specimens with constant rate of deformation was carried out on a standard testing machine KM-50 provided with lever-pendulum dynamometer and with special reduction gear which allowed to vary the angular speed of clamp displacement by 5 orders (9).

Creep testing was carried out on a machine KM-50 fitted with a contact type electro-mechanical servosystem which ensured automatic loading and on horizontal machines with steady load (5).

The angle of twist was measured at the middle section of the specimen with 50 mm base by a special device which allowed two different measurement methods: potentiometric method, using potentiometer having automatic recording, and mechanical method using indicator.

The experimental technique, equipment and instruments are described in detail in (9, 10).

Fig. 1 shows the experimental torsion diagrams, obtained at three different deformation rates, $d\bar{v}/dt = 1.3, 0.3, 0.06\%$ per min for EDT-10 [1, 2, 3] and K-115 [6, 7, 8] and at two deformation rates $d\bar{v}/dt = 0.2, 1.1\%/\text{min}$ for ZMI-6 specimen [4, 5].

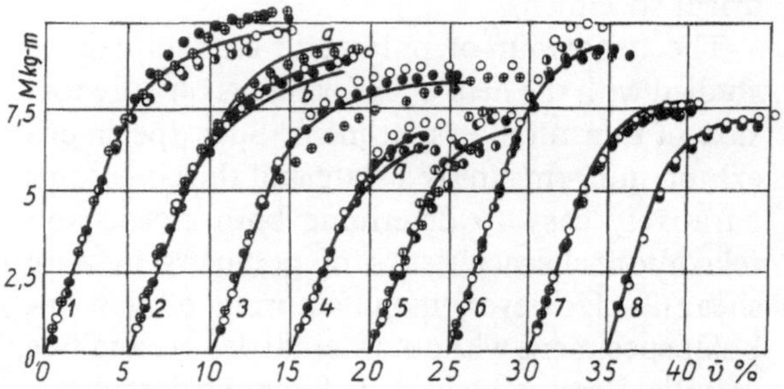

Fig. 1. Comparison of theoretical torsion diagram (solid lines), calculated with two members of relaxation time spectrum (dotted line), with experimental curves 1, 2, 3, and 6, 7, 8 – compositions EDT-10 and K-115, tested at deformation rates $d\bar{v}/dt = 1.3$, 0.3, 0.06%/min; 4, 5 – composition ZMI-6, tested at deformation rates $d\bar{v}/dt = 0.2$, 1.1%/min. Curves 2a, b and I-a calculated with maximum and minimum values of $\eta_{0G,2}^*$

Corresponding theoretical diagrams of creep were calculated preliminarily without allowance for the relaxation time spectrum, assuming that

$$\bar{\varrho}\bar{v} = \frac{\tau}{G} + \varepsilon^*.$$

The problem reduces to integration of differential equation relatively to function $\xi(\bar{\varrho}, \bar{v})$ (9).

$$\frac{\partial \xi}{\partial \bar{v}} = A\bar{\varrho}\left(1 - \frac{\xi \exp|3|}{B\bar{\varrho}}\right) \qquad [6]$$

where

$$A = \frac{G}{m_G^*}, \quad B = \frac{V_0 \cdot \eta_{0G}^*}{m_G^*(1 + G_\infty/G)}. \qquad [7]$$

Function $\xi(\bar{\varrho}, \bar{v})$ is related to stress τ and deformation $\bar{v}$ by the equation

$$\tau = \frac{2}{\pi R^3}\left(\frac{1}{4}C\xi + D\varepsilon\right) \qquad [8]$$

where $\varepsilon = \bar{\varrho}\bar{v} = V_0 t \bar{\varrho}$

$$C = 2\pi R^3 \frac{m_G^*}{1 + G_\infty/G},$$

$$D = \frac{\pi R^3}{2} \frac{G_\infty}{1 + G_\infty/G}. \qquad [9]$$

Function $\xi(\bar{\varrho}, \bar{v})$ with increasing of $\bar{v}$ tends to the limiting function $\xi_{max}(\bar{\varrho})$, moreover,

$$\xi_{max} \exp |\xi_{max}| = B \cdot \bar{\varrho}. \qquad [10]$$

Torsional moment M is expressed through function and deformation v, according to formula

$$M = CI + D\bar{v} \qquad [11]$$

where

$$I \equiv I(\bar{v}) = \int_0^1 \xi \bar{\varrho}^2 d\bar{\varrho} \qquad [12]$$

and

$$I_{max} = \int_0^1 \xi_{max} \bar{\varrho}^2 d\bar{\varrho}. \qquad [13]$$

The shape of functions $\xi(\bar{\varrho}, \bar{v})$ and $I(\bar{v})$ (for particular value of B) are shown in fig. 2. Eq. [6] is integrated numerically; stress τ and torsional moment M are calculated by formulae [8] and [11].

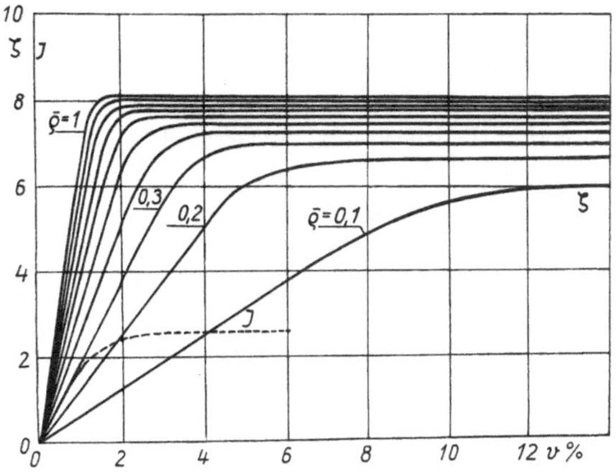

Fig. 2. Theoretical curves $\xi(\bar{\varrho}, \bar{v})$ and $I(\bar{v})$

Fig. 3 gives a comparison of theoretical diagram of shear calculated by eqs. [6–13] with experimental curves, obtained for organic glass EDT-10, K-115, ZMI-6 at different deformation rates. As a result of numerous computer-based calculations of torsion diagram it was found that the function $I(\bar{v})$ in the deformation region, corresponding to maximum torsional moment, differs from its maximum value by 1–2$^0/_0$ and almost a linear relation exists between I_{max} and $\xi_{max}(1)$:

$$I_{max} = K\xi_{max}(1). \qquad [14]$$

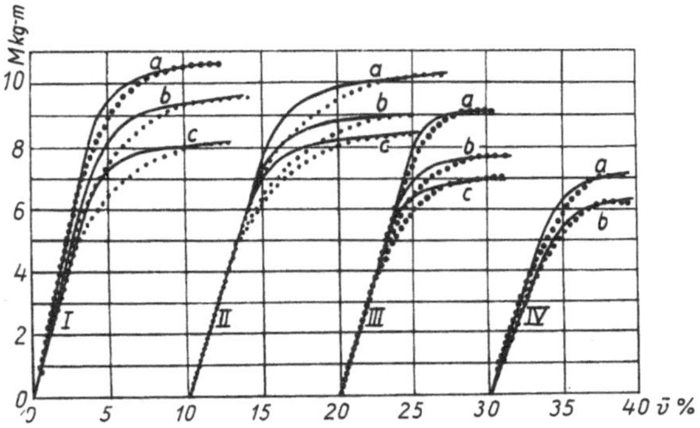

Fig. 3. Comparison of theoretical torsion diagram, calculated, with one member of relaxation time spectrum, with experimental curves for organic glass (I), (EDT-10) (II), K-115 (III), and ZMI-6 (IV) at different deformation rates

In our calculations the values of B were in the range $2.5 \cdot 10^4 \leqslant B \leqslant 4 \cdot 10^{11}$, and coefficient K varied in the interval $0.320 \leqslant K \leqslant 0.329$.

On the basis of observed regularities the following approximate method was proposed for determining the maximum tangential stresses, near the external surface of specimen at $\bar{\varrho} = 1$. Maximum tangential stresses are determined according to [8] by formula

$$\tau_{max}(1) = \frac{2}{\pi R^3}\left(\frac{1}{4} C\xi_{max}(1) + D\bar{v}_{max}\right) \qquad [15]$$

where $\bar{v}_{max}$ – value of $\bar{v}$, corresponding to $M = M_{max}$.

Experimental torsion diagrams obtained for three deformation rates (differing by one or more order) are approximated in the end regions by plotting straight lines whose equations according to [11], are of the form

$$M(\bar{v}) = CI_{max} + D\bar{v}. \qquad [16]$$

The initial linear regions of torsion diagram are given by straight lines

$$M(v) = \frac{\pi R^3}{2} G\bar{v}. \qquad [17]$$

The coordinates of intersection points of lines [16] and [17] – (M^*, $\bar{v}^*$) and slopes of lines [16], determing the coefficient D, are found from experimental torsion diagram.

According to [9] the coefficient D is expressed in terms of the modulus of high elasticity G_∞. The theoretical high elasticity modulus is independent of deformation rate and therefore the slopes of lines [16] must be practically constant at different rates of deformation.

The values of $\xi_{max}(1)$ determine the coordinates $(M^*, \bar{v}^*)$.

Simultaneously solving eq. [16] and [17] we find

$$I_{max} = \frac{M^*}{2\pi R^3 m_G^*}; \quad C = \frac{M^* - D\,\bar{v}^*}{I_{max}} \qquad [18]$$

and from [14]

$\xi_{max}(1) = 1/K I_{max}$.

Effective modulus of shear is determined, using initial linear regions of torsion diagram. This modulus, generally speaking, depends on the deformation rate because in polymers at rates of deformation, provided by standard testing machines, even at small stresses, highly elastic deformation corresponding to older members of time relaxation spectrum is developed.

In our experiments the deformation rates differed at most by two orders. The shear modulus G of EDT-10 and K115 practically did not change; deviations were within the scattering of experimental results, and it was noticed that modulus of shear of softer polymer ZMI-6 increased with increasing deformation rate.

From the formulae [15–18] for the determination of maximum tangential stress in addition to the data from the experimental torsion diagram we need the modulus of rate m_G^*.

Paper (3) describes the experimental technique for the determination of relaxation constants involved in the generalized *Maxwell* equation. The experimental studies described in paper (5) show that the constants, obtained from tension (compression) tests at single-axial stressed state, for the case of pure shear, employing the correlation existing between constants in shear and in tension can also be used.

$$m_G^* = \frac{2}{3} m^*, \quad G_\infty = \frac{1}{3} E_\infty,$$

$$\eta_{0G}^* = \frac{1}{3}\eta_0^*, \quad G = \frac{E}{2(1+\mu)}. \qquad [19]$$

Both elastic and relaxation constants are functions of temperature; constants $G_\infty(E_\infty)$ and $\eta_{0G}^*(\eta_0^*)$ may significantly differ for different components of spectrum. The constant $m_G^*(m^*)$ comparatively less dependent on polymer structure and in calculations its value may be assumed the same for all the members of spectrum.

It is recommended to determine the modulus of rate $m^*(m_G^*)$ from loading conditions (at constant deformation rate) in single-axial tension of small-size cylindrical specimens or in torsion of tubular specimens. The modulus of rate is determined, using the slope of line expressing the maximum stress as a function of the logarithm of deformation rate.

The modulus of high elasticity $E_\infty(G_\infty)$ and coefficient of initial relaxation viscosity $\eta_0^*(\eta_{0G}^*)$ for older component of relaxation time spectrum are determined with good accuracy from the diagrams of relaxation deformation.

The same constants for younger components of spectrum are determined from the diagram of creep (in single-axial tension of cylindrical specimens or in creep of tubular specimens).

It was shown that in each loading condition the different regions of experimental curves (while using the generalized *Maxwell* equation with one relaxation time) are best described with relaxation constants, corresponding to different members of relaxation time spectrum. For example, the slope of the tangent to end region of tension diagram (or shear) is determined, mainly, by $E_\infty(G_\infty)$, corresponding to younger member of spectrum.

In calculating the theoretical torsion diagrams (fig. 3), we used in [11] the coefficients B, C, D and modulus G, determined by formulae [10] at $\bar{\varrho} = I$ [17] and [18]. From these plots it is obvious that the theoretical curves well describe the end regions of experimental torsion diagram. According to above facts, one may suppose that G_∞ and η_{0G}^* calculated from formulae [7] and [9] using the values of B, C, D and G at known modulus m_G^* must be close to constants, corresponding to younger member of spectrum.

Fig. 1 shows the theoretical torsion diagrams, calculated with two members of relaxation time spectrum according to formulae [1–3]. The constants corresponding to younger member of spectrum were taken to be the values of $G_{\infty,2}$ and $\eta_{0G,2}^*$, calculated from formulae [7] and [9]. The constants $G_{\infty,1}$ and $\eta_{0G,1}^*$, corresponding to older member of spectrum, and the values of m^* were determined from [19], using the values m^*, E_∞, η_0^* given in (4).

Of three relaxation constants, the coefficient of initial relaxation viscosity η_G^* exhibits the most unstable character.

In calculating the coefficient η^*_{0G} by formula [7] for several specimens, tested at the same deformation rate we obtained scattering of η^*_{0G} in the following ranges: for EDT-10 – $5 \cdot 10^{12} \mathscr{F} \eta^*_{0G} \leqslant 5 \cdot 10^{13}$; for K-115 – $4 \cdot 10^{13} \leqslant \eta^*_{0G} \leqslant 4 \cdot 10^{15}$; for -6 $10^8 \leqslant \eta^*_{0G} \leqslant 10^9$. The average values of constants were used in calculating the theoretical curves (fig. 1); some theoretical curves, for example, were calculated with the maximum and the minimum (for this series of test) values of coefficient

$\eta^*_{0G,2}\,(2 - a, b, 4 - a)$.

From plots of fig. 1 it is evident that the theoretical curves, obtained with two members of spectrum, are in good agreement with the experimental curves.

A comparison of the maximum tangential stresses, calculated by the approximate method, with the values estimated with two members of spectrum has shown that these values practically coincided.

Fig. 4 shows the experimental curves of creep and a comparison of the theoretical creep curves with experimental curves.

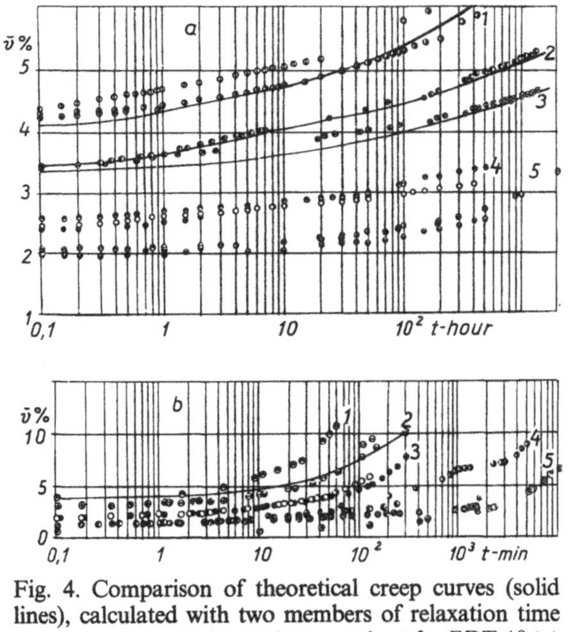

Fig. 4. Comparison of theoretical creep curves (solid lines), calculated with two members of relaxation time spectrum with experimental curves (dots) for EDT-10 (a) and ZMI-6 (b) for torsional moment: (a) M kg/m = 6 (I), 5 (2 and 3), 3, 5 (4), 3 (5) 6, (b) M kg/m = 6 (I), 5 (2), 3 (3), 2 (4), 1 (5). Theoretical curves 2 and 3 [for EDT-10 (a)], calculated, with maximum and minimum values of coefficient $\eta_{0G,1}$

The values of constants were taken to be the same as in calculating the theoretical torsion diagram with two members of spectrum. An example (fig. 4) of estimation of creep curves with minimum and maximum value of constant $\eta^*_{0G,1}$, is also shown in this figure.

Fig. 5 illustrates the variation of older and younger components of highly elastic deformation with logarithm of time. These plots show that the initial regions of creep curves are described, in general, with older component of spectrum; from certain moment of time the younger component begins to play a great role.

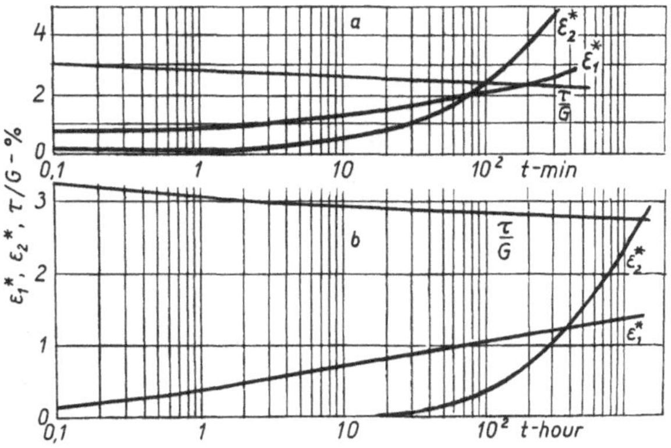

Fig. 5. Theoretical curves, showing variation in the process of creep of oldest (ε^*_1), and younger (ε^*_2) components of highly elastic deformation for EDT-10 (a) and ZMI-6 (b) at $M = 5$ kg/m. Curves τ/G show redistribution of stresses at creep

From plots τ/G it is evident that the stresses in volumetric specimens in creep are not constant and undergo redistribution during deformation.

The behaviour of rigid polymers (employed as a binder in reinforced systems) under pure shear has been studied. The properties of binder in shear such as deformation, strength and rheology is of great interest because of the special role that tangential stresses play in reinforced plastics. Because of the tangential stresses transmitted by the binder, all reinforced fibres and layers are involved in this process.

Weak resistance of reinforced materials to interlayer shear reduces the efficiency of the new high-modular fibres.

A study of deformation regularities of polymers (binder) is also necessary for developing the theory of deformation of reinforced plastics.

This paper reports the experimental and theoretical results on the creep on torsion and deformation (at constant rate) of circular polymeric rods of three different compositions, manufactured on the base of resins ED-5 and ED-6. It has been shown that the generalized *Maxwell* eq. (1) can be used to interconnect the stresses and deformations in describing the behaviour of rigid net-shaped polymers in the case of non-uniform stressed-deformation state.

Summary

1. Use of the generalized nonlinear *Maxwell* equation with due account for two members of relaxation time spectrum shows a good agreement between the theoretical and experimental torsion diagrams and creep curves over a large time range.

2. Calculation of maximum tangential stresses by an approximate method, based on experimental torsion diagram, obtained as a result of rod testing, gives such values that practically coincide with the values calculated with two members of relaxation time spectrum.

3. Constants of polymers, determined on the basis of experimental torsion diagrams, obtained as a result of rod testing, may be used for the evaluation in the first approximation of two relaxation constants of polymers ($G_{\infty,2}$ and $\eta^*_{0G.2}$) corresponding to younger component of spectrum.

References

1) *Gurevich, G. I.*, Tr. IFZ A.N. USSR No. **2** (1959).

2) *Goldman, A. Ya.* and *A. L. Rabinovich*, Mekhanika polimerov **2**, 214–228 (1966).

3) *Rabinovich, A. L.*, Vvedenie v meckhaniku armirovanykh polimerov (Izd. Nauka 1970).

4) *Babich, V. F.*, Kandidatskaia dissertatsiya (Moskva 1965).

5) *Goldman, A. Ya.*, Kandidatskaya dissertatsiya (Moskva 1966).

6) *Danilova, I. N.*, Mekhanika polimerov **5**, 906–914 (1967).

7) *Danilova, I. N., T. V. Sokolova*, Mekhanika polimerov **1972**, 4.

8) *Bernatsky, A. D., A. L. Rabinovich*, Standartizatsiya **1965**, 3.

9) *Danilova, I. N., Yu. I. Fedorova*, Sbornik, MISI **1970**, 64.

10) *Danilova, I. N., Yu. I. Fedorova*, Sbornik, MISI **1969**, 63.

Authors' address:

I. N. Danilova and *T. V. Sokolova*
Vorobievskge Chaussee 26
Institute of Chemical Physics, Academy of Sciences of USSR, Moscow V 334 (USSR)

Rheol. Acta 12, 189–193 (1973)

From the Department of Chemical Engineering and Materials Science, University of Minnesota, Minneapolis (U.S.A.)

Rheology of network forming systems

By F. G. Mussatti and C. W. Macosko

With 7 figures

(Received October 27, 1972)

Introduction

The following remarks made not so long ago by *Davies* and *Hill* (1) are still applicable today:

It is a surprising fact that despite their great technological importance many of the principal features of polycondensation and polyaddition reactions are unresolved by the published studies of these processes.

Although many new techniques have been developed since these remarks were made, studying thermosetting systems is not an easy task. The very properties sought in these materials – insolubility, strength, and high temperature stability – render standard solution techniques almost useless. Consequently, the chemist must resort to secondary means, such as ultraviolet or infrared absorption, conductivity, and calorimetry, to follow the reaction.

Continuous monitoring of rheological properties, specifically dynamic mechanical properties, can be used to follow thermosetting reactions. Changes in these properties are very sensitive to the molecular changes going on during the reaction. Furthermore, these same measurements are directly useful to the processing engineer and to the designer concerned with end product performance.

After the gel point in a thermosetting system, further reaction extends the network and increases the crosslink density. In this region the material can be treated as a rubber. Eventually, in most thermoset systems, enough crosslinks are formed to transform the material from a rubber to a glass. Any further reaction then proceeds in a solid, is diffusion limited, and is very slow.

We can monitor the reaction through the rubbery and into the glassy region with small amplitude sinusoidal oscillations. Such dynamic measurements have been used extensively to characterize solid polymers and to some extent for following rubber curing (2, 3, 4). Generally rubber studies have involved only the complex dynamic modulus, G^*, rather than its two components, G', the elastic or storage modulus, and G'', the loss or viscous modulus.

$$G^* = G' + iG''. \qquad [1]$$

A thermosetting system can be assumed to be an ideal rubber from the gel point to somewhere near the rubber to glass transition. From the theory of rubber elasticity, we can relate the modulus determined at small deformations directly to the crosslink density, X.

$$|G^*| = RTX. \qquad [2]$$

If the rate of formation of crosslinks is proportional to the number of crosslinks still to be formed, $X_\infty - X$, and if crosslinks are formed by an nth order irreversible reaction, then

$$\frac{dX}{dt} = k(X_\infty = X)^m. \qquad [3]$$

Furthermore, if the reaction is isothermal, eqs. [2] and [3] can be combined to relate G^* directly to the kinetics. For zero order reaction, if $|G_0^*| = RTX_0$ at the gel point, then

$$|G^*| = RTkt + |G_0^*|. \qquad [4]$$

Generally, $G'' \ll G'$ after the gel point. Thus,

$$G' \simeq RTkt + |G_0^*|. \qquad [5]$$

For a first order reaction,

$$G' \simeq |G_\infty^*| - (|G_\infty^*| - |G_0^*|)\,e^{-kt} \qquad [6]$$

where $|G_\infty^*|$ is the modulus at full cure. Thus, we should be able to extract reaction order and rate data directly from dynamic modulus measurements on curing thermoset systems. We can also find the activation energy from the inverse of the induction times of the dynamic storage modulus since $1/t_{ind} \alpha k_0$ (5). Then

$$\frac{1}{t_{ind}} = C e^{-E_A/RT}. \qquad [7]$$

Another application of this type of analysis is to the vulcanization of rubber systems. Since these systems are always in the rubbery region, testing of a vulcanization reaction should provide a good check on the theory presented here. In the case where reversion does not occur eqs. [5] and [6] are again applicable with the modification that $|G_0^*|$ is the initial modulus. When reversion is significant, terms must be included in eq. [3] which account for the destruction of crosslinks. In this case there will be a rate constant and an activation energy for crosslinking and for reversion.

Experimental

To test the relations derived above, an epoxy polymerization and a rubber vulcanization were used.

Epoxy polymerizations are a class of thermosetting reactions particularly well suited for this study since there are no by-products and there is very little volume change during the reaction. The epoxy system used in this work consists of Dow DER 332/LC (diglycidyl ether of bisphenol A), hexahydrophthalic anhydride (HHPA), and tris-2,4,6-dimethylaminomethyl phenol (DMP). The composition of the reaction system was 100 parts of epoxy, 80 parts of HHPA, and 1 part of DMP.

The rubber system was based on an ethylene-propylene-1,4 hexadiene rubber vulcanized by sulfur. The composition of this system is given in table 1.

Table 1

Component	Parts by weight
Nordel 1070 rubber (Dupont)	100
Zinc oxide	5
Stearic acid	1
Tetramethylthiuram disulfide	1.2
2-Mercaptobenzothiazole	0.5
Sulfur	1.29

The dynamic shear moduli of these systems were measured in the eccentric rotating disk geometry (6, 7) with a Rheometrics Mechanical Spectrometer (8). G' and G'' were measured continuously throughout the reaction at a fixed frequency, 1 rad/sec, and at a constant shear strain amplitude of 4%.

Results and discussion

Fig. 1 shows the dynamic storage modulus versus cure time data for the rubber system. The drop in the modulus during the initial part of the cure due to the transient heating of the material.

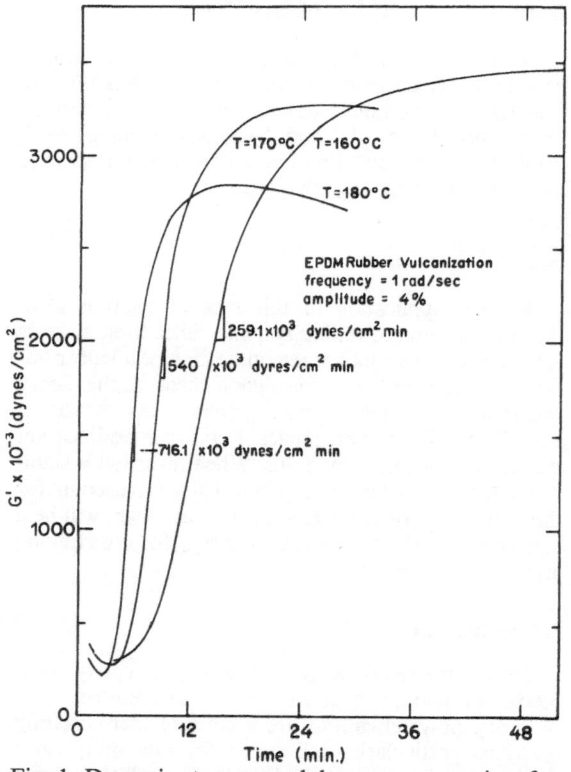

Fig. 1. Dynamic storage modulus versus cure time for rubber vulcanization at 160 °C, 170 °C, and 180 °C

In a study of the crosslink density from swelling measurements as a function of time for this system *Fujimoto* and *Nakade* (9) reached the following conclusions:

1. The crosslink density increases rapidly in the early stages of cure and then tapers off as the crosslinking agent is depleted.
2. There is a small amount of reversion which increases as the cure temperature increases.
3. The maximum crosslink density decreases as the cure temperature increases.

According to eq. [2], these same conclusions should hold for $|G^*|$ or for G' since G'' was found to be only a small fraction of G' after a short time. The curves shown in fig. 1 do show the same general trends as those reported in the literature. In addition, the values of $|G^*|$ computed from eq. [2] and from the crosslink densities reported in the literature (9) agree very well with the G' values measured here. A comparison is shown in table 2.

Table 2

T(°K)	Calculated $\|G^*\|$(dynes/cm²) [ref. (9)]	Experimental G'(dynes/cm²)
160	3350×10^3	3475×10^3
170	3282×10^3	3265×10^3
180	3200×10^3	2850×10^3

Since there is only a small amount of reversion, an attempt was made to analyze the data according to eq. [3]. Assuming a zero order reaction since the curves in fig. 1 are linear in cure time, calculated k values are shown on an *Arrhenius* plot in fig. 2. The activation energy is 20.1 kcal/mole. This value is somewhat lower than the literature value of 23.8 kcal/mole from the

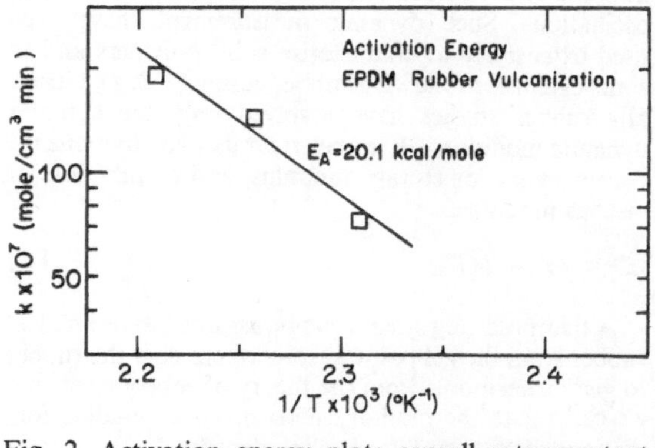

Fig. 2. Activation energy plot: overall rate constant versus reciprocal reaction temperature for rubber vulcanization

crosslinking rate constant which was calculated while taking reversion into account (7). The slightly lower activation energy found here is not unexpected since reversion is greater at the higher temperatures, thereby decreasing the k values more at the higher temperatures and decreasing the slope of the *Arrhenius* plot. Although no attempt has been made to model the reversion at this time, the data on this rubber system is considered to be a valid test of the theory.

Dynamic storage modulus data on the epoxy system was taken at 96.7 °C, 107.8 °C, and 118.4 °C. Fig. 3 shows a typical example of the type of data obtained. G' first rises gradually and then more rapidly as the network gels. The plot appears linear in cure time indicating a zero order reaction according to eq. [4]. Since G'' was found to be small, $G' = |G^*|$ and eq. [5] should hold. Calculated k values are shown on an *Arrhenius* plot in fig. 4. The activation energy is 20.0 kcal/mole. This value does not agree well with the values of 17.8 kcal/mole reported in the literature for the same system (10, 11). However, if the inverse of the induction times as determined in fig. are plotted according to eq. [7], an activation energy of 18.1 kcal/mole is obtained (see fig. 5). This value agrees favorably with that reported in the literature. Since the references cited for the activation energy used different methods of analysis, differential scanning calorimetry (10) and torsion pendulum (11), one must immediately ask what is the reason for the disagreement of the activation energy calculated from the k values. One point which may explain this discrepancy is the fact that, in all of the experiments on this epoxy system, small bubbles formed in the sample. The effect of these bubbles on the results is still uncertain. However, further work is being conducted with this system.

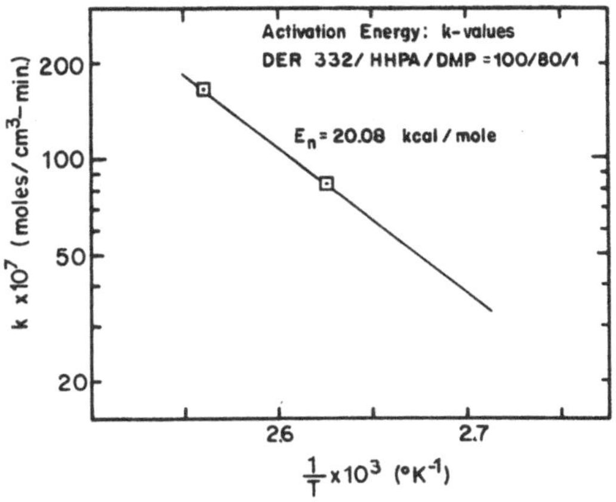

Fig. 4. Activation energy plot: overall rate constant versus reciprocal reaction temperature for epoxy cure

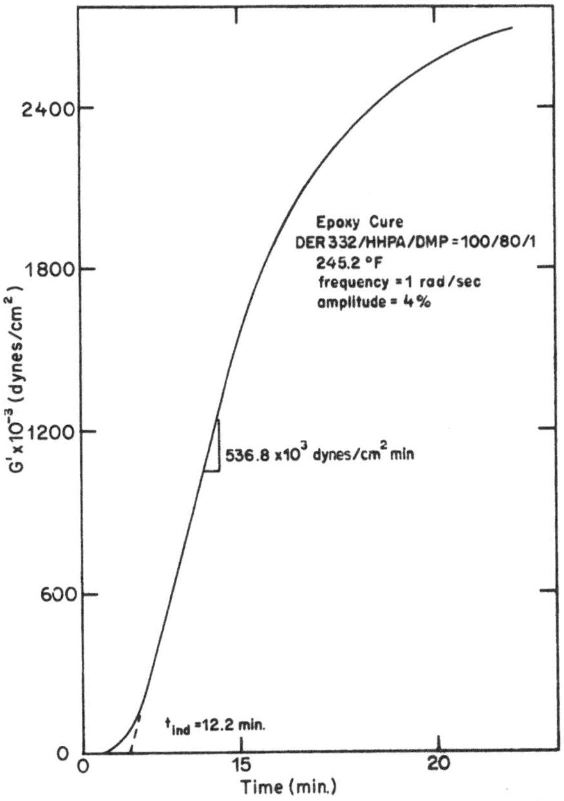

Fig. 3. Dynamic storage modulus versus cure time for epoxy cure at 108.4 °C

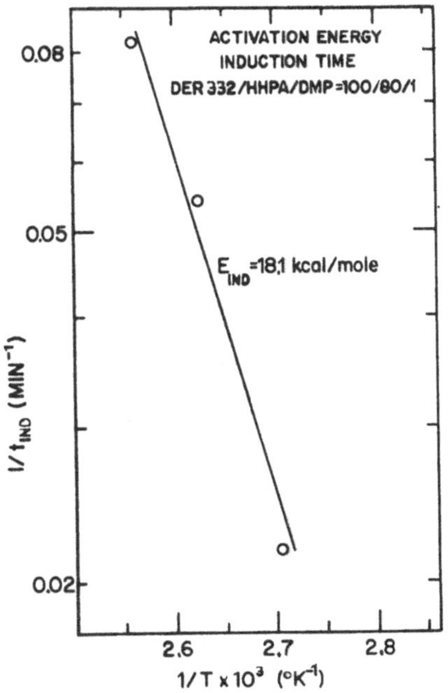

Fig. 5. Activation energy plot: reciprocal induction time versus reciprocal reaction temperatur for epoxy cure

Other thermosetting systems have also been tested according to the method described above. One such system consists of diallyl phthalate. Four samples of this material, each of which was known to behave differently from processing experience, were supplied to us for testing. In separate tests by the supplier using a differential scanning calorimeter no differences between the samples were detected. The results of tests in the eccentric rotating disks geometry are shown in fig. 6. The differences in the induction times, in the

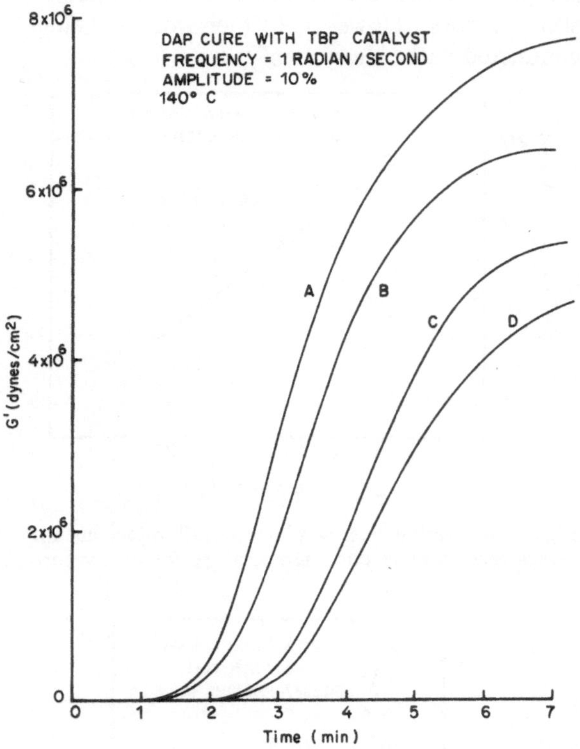

Fig. 6. Dynamic storage modulus versus time for diallyl phthalate at 140 °C

rates of increase of the storage modulus, and in the ultimate levels attained all correspond to behavior observed during processing. Fig. 7 shows similar results on two commercially used compounding formulas containing diallyl phthalate. The two samples differ in their useful life during processing which is clearly shown by the longer reaction time required for the soft flow sample. Fig. 7 also shows the repeatability attained on duplicate runs of the hard flow sample.

In summary we feel that this method of rheological analysis is useful for studying the reactions of network forming systems. The preliminary data presented here look promising and demon-

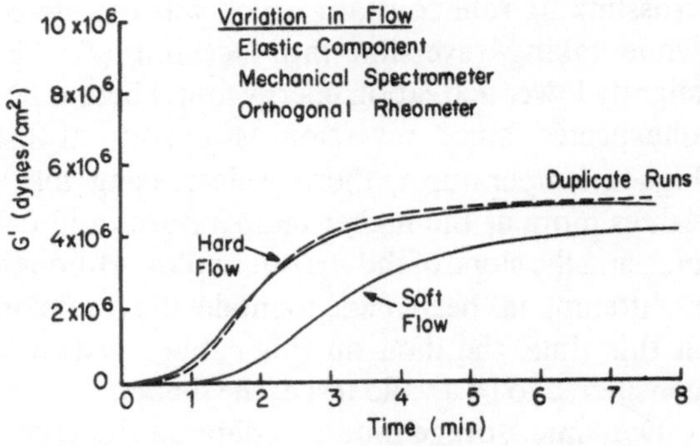

Fig. 7. Dynamic storage modulus versus cure time for two commercially used compounding formulas containing diallyl phthalate at 140 °C

strate the application of the method to commercially used materials. However, further work is required to improve our experimental technique and the theoretical considerations.

Acknowledgements

The authors wish to express their appreciation to Dr. *J. W. White* and Dr. *P. E. Willard* for supplying material and data on the epoxy and diallyl phthalate systems, respectively, and to Mr. *M. Tsuchimachi* for supplying the material for the rubber system.

Summary

The curing reactions of an epoxy-anhydride and a rubber vulcanization system were followed by rheological means. Dynamic shear measurements were used to follow the reaction. The rheological quantities were related to the reaction kinetics of the systems. From these relations, overall activation energies, reaction orders and rate constants were interpreted from the data. These compare favorably with results reported in the literature. The cures of several diallyl phthalate compounds were also examined with dynamic shear measurements. The results correlated well with thermoset molding experience.

References

1) *Davies, M.* and *D. R. J. Hill*, Trans. Far. Soc. **49**, 395 (1953).

2) *Robinson, R. W.* and *R. N. F. Pinfold*, Trans. Inst. Rub. Ind. **39**, T26 (1963).

3) *Lewis, A. F.* and *G. J. Pietsch*, SPE Tech. Papers **17**, 87 (1971).

4) ASTM D2705–68T.

5) *Tanaka, Y.* and *H. Kakiuchi*, J. Appl. Polym. Sci. **7**, 1063 (1963).

6) *Gent, A. N.*, Brit. J. Appl. Phys. **11**, 165 (1960). — *Maxwell, B.* and *R. P. Chartoff*, Trans. Soc. Rheol. **9**, 41 (1965).

7) *Macosko, C. W.* and *W. M. Davis*, Proc. 6th Int. Congr. Rheol., Lyon 1972 (to be published).

8) *Macosko, C. W.* and *J. M. Starita*, SPE Journal **27**, 38 (1971).

9) *Fijimoto, K.* and *S. Nakade*, J. Appl. Polymer Sci. **13**, 1509 (1969).

10) *Fava, R.*, Polymer **9**, 137 (1968).

11) *White, J.*, 3M Company, personal communication.

Authors' address:

Dr. *F. G. Mussati* and Dr. *C. W. Macosko*
Dept. of Chem. Engineering and Materials Science
University of Minnesota, 151 Chemical Engineering Building, Minneapolis, Minnesota 55455 (U.S.A.)

Rheol. Acta **12**, 194–199 (1973)

From the Department of Theoretical Mechanics, University of Nottingham

Plane mechanics and kinematics of compressible ideal granular materials

By *A. J. M. Spencer* and *M. R. Kingston**)

(Received October 27, 1972)

1. Introduction

The question of the correct formulation of constitutive equations to describe the mechanical and kinematic behaviour of granular materials such as soils and powders remains one of discussion and controversy. The purpose of this paper is to make some contributions to this discussion with particular reference to compressible materials. To this end we put forward some suggestions for consideration, while recognising the existence of alternative approaches to the problems discussed, and of differences of opinion which remain to be resolved.

In a previous paper *Spencer* (1) made proposals for the equations which govern the mechanical and kinematic behaviour in plane strain of an ideal incompressible granular material. The author subsequently learned that essentially the same equations had been formulated earlier by *Mandel* (2), although *Mandel* (3) has since raised objections to the theory which leads to these equations. The theory is based on material isotropy, incompressibility, the *Coulomb* yield condition (although it can readily be extended to any other yield function which is a function of the two invariants of the stress tensor in two dimensions) and, more controversially, an assumption that deformation takes place by simultaneous shears on the characteristic curves of the stress equations. The predictions of the resulting theory do not appear to conflict violently with observations of the behaviour of some granular materials which are to a reasonable approximation incompressible. However, there are alternative theories available, some of which were described in (1). Space does not permit discussion of more recent work. However, we mention papers by *Dais* (4), *Mandl* and *Fernandez Luque* (5) [see also the discussion by *Spencer* (6)], *de Josselin de Jong* (7), and *Goodman* and *Cowin* (8, 9), and take this opportunity to record our opinion that the best prospect for future progress lies in investigating the predictions of the various theories which have been proposed with a view to devising critical experiments to test them against one another.

Although some granular materials show only small volume changes in the range of stress states they encounter in practice, for others compressibility effects are certainly important. In this paper we attempt to extend to compressible materials some of the theory developed in (1). It is emphasized that these suggestions are put forward in a tentative manner, as a basis for discussion. They certainly do not amount to a complete theory. In the first place they deal only with plane deformations, and it has yet to be shown that the plane deformation theory can be embedded within a full three-dimensional theory. Second, we do not consider the relative motions and force interactions between the constituents which make up the material, and we judge these to be of importance in many situations. Third, the theory retains many features of the theory described in (1) which remain open to verification or otherwise. In spite of these uncertainties, we hope that this paper will make some contribution to the eventual formulation of a satisfactory and accepted theory to describe the mechanical behaviour of granular materials.

2. Densities and Specific Volumes

We envisage the granular material as comprising a mixture of two materials, although obviously more complex systems are possible. It is supposed that one of the phases is a solid phase, and the other is a gas. This approximates the situation in a dry powder or dry sand.

The density of the mixture (mass of mixture per unit volume of mixture) is denoted by ϱ. Its specific volume (volume of mixture per unit mass of mixture) is then ϱ^{-1}. The mass concentration of the solid phase (mass of solid per unit mass of mixture) is c, and that of the gas phase is then $1 - c$. The density of solid (mass of solid per unit volume of mixture) is $c\varrho$ and the density of gas is $(1 - c)\varrho$. The volume concentration of the solid phase (volume of solid per unit volume of mixture) is denoted v, and that of the gas phase is then $1 - v$. The specific volume of the solid phase (volume of solid per unit mass of mixture) is $v\varrho^{-1}$, and that of the gas phase is $(1 - v)\varrho^{-1}$. In practice we expect the density of the gas to be much less than that of the solid, so that

$$(1 - c) \ll 1, \quad c \simeq 1 \qquad [2.1]$$

but no similar relations need hold for v. In order to describe the density and composition of an element of the mixture at any time, it is necessary to specify all three of ϱ, c and v, or three equivalent quantities; any one of ϱ, c and v may in principle be varied independently of the other two.

*) Present address: Joseph Lucas Ltd., Solihull, Warwickshire (UK).

Although we have identified the two constitutents of the material, we are constructing a continuum theory and, as for example in the mathematical theory of mixtures, we do not distinguish the places occupied by the solid and gas. Each constituent is regarded as distributed continuously throughout the volume occupied by the material, and each point of the body as simultaneously occupied by both solid and gas particles. A more rigorous approach to this formulation is employed by *Goodman* and *Cowin* (9).

It is to be expected that normally the solid phase will be much less compressible than the gas phase. If we idealize this property by assuming that the mass of solid per unit volume of solid is a constant γ (which may be interpreted as the density of the granules in a real granular material) then

$$c\varrho = v\gamma. \qquad [2.2]$$

Thus in this case one of ϱ, c, v may be expressed in terms of the other two.

3. Freely-diffusing and non-diffusing materials

In order to investigate the relations between the variables ϱ, c and v and the forces acting in a body it would appear to be necessary to introduce separate velocity fields for each constituent, partial stress fields corresponding to each constituent, and diffusive forces arising from the relative motions of the constituents. There are, however, two cases which appear to be of interest in which these complications can to some extent be avoided.

Freely-diffusing Materials

We say the material is freely-diffusing if the gas phase can move freely relative to the solid phase, or, more exactly, if deformations of the material take place over times long compared to a characteristic time for gas to diffuse through the material. In such a situation it may be assumed that the pore pressure is constant in the material. If the pore pressure is also assumed to be a function of the pore density (defined as mass of gas per unit volume of gas) which has the value $(1 - c)\varrho/(1 - v)$, then in a freely diffusing mixture

$$(1 - c)\varrho/(1 - v) = \varrho_0 \qquad [3.1]$$

where the constant ϱ_0 may be taken to be the ambient density outside the material.

Non-diffusing Materials

The other extreme case of a non-diffusing material arises when there is no relative motion between the gas and solid constituents, or when the time over which deformation of the material takes place is short compared to a characteristic time for gas to diffuse through the material. In this case the mass concentrations at a particle of the mixture are constant, and so

$$c = c_0 \qquad [3.2]$$

where c_0 is constant at a particle. Also in this case the velocity field is unambiguously defined for the mixture, and there is no need to distinguish velocity fields for the individual constituents.

4. Flow condition

We consider two-dimensional states of stress and deformation which we refer to rectangular cartesian coordinates (x, y). Stress components in the (x, y) plane are denoted by t_{xx}, t_{yy}, t_{xy}, and tensile stress components are taken to be positive. The two principal stress components whose principal directions lie in the (x, y) plane are denoted by t_1, t_2 and without loss of generality we take $t_1 \geq t_2$. Define

$$p = -\tfrac{1}{2}(t_1 + t_2), \qquad q = \tfrac{1}{2}(t_1 - t_2). \qquad [4.1]$$

Then

$$\begin{aligned} t_{xx} &= -p + q\cos 2\psi \\ t_{yy} &= -p - q\cos 2\psi \\ t_{xy} &= q\sin 2\psi \end{aligned} \qquad [4.2]$$

$$\begin{aligned} p &= -\tfrac{1}{2}(t_{xx} + t_{yy}) \\ q &= \{\tfrac{1}{4}(t_{xx} - t_{yy})^2 + t_{xy}^2\}^{\frac{1}{2}} \end{aligned} \qquad [4.3]$$

where ψ is the angle of inclination, measured anticlockwise, of the t_1 principal stress direction to the x-axis. These stress components are all total stress components for the mixture. Partial stress components for the solid and gas will not be explicitly introduced.

We follow the usual procedure of metal and soil plasticity and postulate the existence of a function of the stress components which is equal to zero when the material is deforming and negative or zero otherwise. For isotropic material this function must be expressible as a function of the stress invariants p and q. Since we are considering a compressible material we also admit dependence on the density ϱ. It is assumed that [2.2] and either [3.1] or [3.2] hold, so that c and v can be expressed in terms of ϱ and need not be explicitly introduced. We therefore assume a relation of the form

$$F(p, q, \varrho) = 0. \qquad [4.4]$$

This relation is called the flow condition and the function F the flow function. These terms are used in preference to the terms yield condition and yield function because it seems that for granular materials the condition which determines the onset of yield may differ from that which holds during deformation. Probably the simplest flow condition is the *Coulomb* condition

$$F(p, q, \varrho) = q - p \sin \varphi_0 - k \cos \varphi_0 \qquad [4.5]$$

where φ_0 and k are constants called the angle of internal friction and cohesion respectively. A generalization of [4.5] which may be more appropriate for compressible granular materials is

$$F(p, q, \varrho) = q - \{p - p^*(\varrho)\} \sin \varphi_0 - k \cos \varphi_0 \qquad [4.6]$$

where the term $p^*(\varrho)$ is introduced to allow for the pore pressure of the gas phase. However, the general expression [4.4] will be used. Specializations from this can be made when required.

Eq. [4.4] represents a surface in p, q, ϱ space. Surfaces of this kind are frequently introduced in the soil and powder mechanics literature. References are given by *Ashton* et al. (10) who present data for a variety of dry powders.

No consideration has yet been given to the equations which determine the density ϱ. In general it is to be expected that the density of a material particle will depend in a complicated way on the stress and deformation history of that particle. We return to this in § 6 and for the present assume that at a given instant of time ϱ is known throughout the body considered.

5. Stress equations and characteristics

The equilibrium equations are

$$\frac{\partial t_{xx}}{\partial x} + \frac{\partial t_{xy}}{\partial y} + X = 0$$

$$\frac{\partial t_{xy}}{\partial x} + \frac{\partial t_{yy}}{\partial y} + Y = 0 \qquad [5.1]$$

where X and Y are body forces per unit volume in the x and y directions. The substitution of [4.2] into [5.1] yields the following two equations for p, q and ψ:

$$-p_x + q_x \cos 2\psi - 2q\psi_x \sin 2\psi + q_y \sin 2\psi + 2q\psi_y \cos 2\psi + X = 0$$

$$-p_y + q_x \sin 2\psi + 2q\psi_x \cos 2\psi - q_y \cos 2\psi - 2q\psi_y \sin 2\psi + Y = 0 \qquad [5.2]$$

where here the suffixes denote partial derivatives. In addition we have from [4.4], with suffixes again denoting partial differentiation,

$$F_p p_x + F_q q_x + F_\varrho \varrho_x = 0$$

$$F_p p_y + F_q q_y + F_\varrho \varrho_y = 0. \qquad [5.3]$$

These may be used to eliminate p_x and p_y from [5.2]. In the resulting equations, F_p, F_q, F_ϱ are specified functions of p, q and ϱ, and ϱ_x and ϱ_y are regarded as known.

Equations similar to [5.2] and [5.3] have been discussed frequently in soil plasticity literature, for example, for the case $F_\varrho = 0$, by *Kingston* and *Spencer* (11) who give references to other work. The only effect on the argument given in (11) of the introduction of the terms in F_ϱ is to replace X and Y by

$$X + (F_\varrho \varrho_x / F_p) \quad \text{and} \quad Y + (F_\varrho \varrho_y / F_p) \qquad [5.4)$$

respectively. By the standard methods used in (11) it follows that [5.2] and [5.3] form a system of hyperbolic equations provided that

$$|F_q / F_p| \geq 1 \qquad [5.5]$$

and we consider only this case. Then if

$$F_p / F_q = - \sin \varphi \qquad [5.6]$$

the equations of the characteristic curves are

$$\frac{dy}{dx} = \tan\left(\psi - \frac{1}{4}\pi - \frac{1}{2}\varphi\right)$$

$$\frac{dy}{dx} = \tan\left(\psi + \frac{1}{4}\pi + \frac{1}{2}\varphi\right) \qquad [5.7]$$

and these curves are called α- and β-curves respectively. They intersect at angles $\frac{1}{2}\pi \pm \varphi$, and are bisected by the principal axes of stress. The relations along the characteristics are readily obtained; they are given by eqs. [3.1] of (11) if the substitutions [5.4] are made.

When the *Coulomb* condition [4.5] or its generalization [4.6] is used, then $\varphi = \varphi_0$, so the angle φ can be regarded as a generalized angle of internal friction.

If $-\sigma$, τ are the normal and shear components of traction on the α- and β-curves, then

$$\sigma = p - q\sin\varphi, \quad \tau = \pm q\cos\varphi$$
$$p = \sigma \pm \tau\tan\varphi, \quad q = \pm\tau\sin\varphi \qquad [5.8]$$

and we can write

$$F(p, q, \varrho) = G(\sigma, \tau, \varrho). \qquad [5.9]$$

Then it was shown in (11) and elsewhere that

$$\tan\varphi = \pm G_\sigma/G_\tau. \qquad [5.10]$$

If deformation occurs by shear on the α- and β-curves, then σ and τ are the quantities measured directly in a shear box test. The characteristic relations in terms of σ and τ are also given in (11).

If c and v are not determined in terms of ϱ by [2.2] and [3.1] or [3.2], then we may replace [4.5] by a more general relation

$$F(p, q, \varrho, c, v) = 0. \qquad [5.11]$$

When the spatial distribution of ϱ, c and v is known, the effect of this generalization is only to introduce further terms on the right-hand sides of the characteristic relations. Additional equations are of course required for c and v; if diffusion is taking place one would expect these to involve the time derivatives of c and v.

6. Velocity equations and characteristics

The following discussion of velocity equations is largely independent of the stress equations described above. It is required only that the stress equations be hyperbolic with the characteristics [5.7], with the angle φ determined in some appropriate way.

The velocity components u, v satisfy first the continuity equation, which we take in the form

$$\frac{\partial u}{\partial x} + \frac{\partial v}{\partial y} + \frac{1}{\varrho}\dot{\varrho} = 0 \qquad [6.1]$$

where $\dot{\varrho}$ denotes the convected or material time derivative of ϱ. If the material is incompressible then $\dot{\varrho} = 0$ and [6.1] reduces to

$$\frac{\partial u}{\partial x} + \frac{\partial v}{\partial y} = 0. \qquad [6.2]$$

In (1) it was proposed that for incompressible material the other velocity equation should be

$$\sin 2\psi\left(\frac{\partial u}{\partial x} - \frac{\partial v}{\partial y}\right) - \cos 2\psi\left(\frac{\partial u}{\partial y} + \frac{\partial v}{\partial x}\right) + \sin\varphi\left(\frac{\partial v}{\partial x} - \frac{\partial u}{\partial y} - 2\Omega\right) = 0 \qquad [6.3]$$

where

$$\Omega = \dot{\psi} \qquad [6.4]$$

is the rate of rotation of the principal axes of stress at a particle. Eq. [6.4] can be interpreted physically in various ways. With [6.2] it expresses the condition that the material deforms by superposed shearing deformations on the α- and β-curves. As described in (5) and (6), in steady motions such that $\partial\varphi/\partial t = 0$ at each point (x, y), a shearing motion on say the α-curves is one in which each element undergoes a shear on the α-curve and a rotation as it follows the α-curve. A single such deformation is not consistent with the incompressibility condition, but two superposed deformations of this kind may be. If the motion is not steady, then [6.3] arises by assuming that the deformation is of this type relative to coordinates which are fixed locally relative to the principal stress axes through a point. Equivalently, [6.3] arises, as was shown in (1), by assuming that the deformation is two superposed shears on the α- and β-curves, without rotation, relative to axes which are fixed locally relative to the principal stress axes through a particle. Another interpretation of [6.2] and [6.3], also obtained in (1), is that the rate of extension of the α-curves and β-curves (as distinct from the material curves which instantaneously coincide with the α-curves and β-curves) is zero.

We note that although in (1) it was assumed that φ is constant, eq. [6.3] remains valid, under the same assumptions regarding the form of the deformation as those used in (1), when φ is variable, provided that the principal stress axes always bisect the angles between the α- and β-curves.

Eq. [6.3] can be regarded as a relation between the deformation-rate components d_{ij} and the spin component ω_{21}, where

$$d_{11} = \partial u/dx, \ d_{22} = dv/dy,$$
$$d_{12} = \tfrac{1}{2}(\partial u/\partial y + \partial v/\partial x),$$
$$\omega_{21} = \tfrac{1}{2}(\partial v/\partial x - \partial u/\partial y). \qquad [6.5]$$

However, it can equally well be interpreted as a relation between ω_{21} and the deviatoric deformation-rate components d'_{ij}, where

$$d'_{11} = d_{11} - \tfrac{1}{2}(d_{11} + d_{22})$$
$$d'_{22} = d_{22} - \tfrac{1}{2}(d_{11} + d_{22}), \quad d'_{12} = d_{12} \qquad [6.6]$$

and in fact [6.3] can be written

$$(d'_{11} - d'_{22})\sin 2\psi - 2d'_{12}\cos 2\psi + 2(\omega_{21} - \Omega)\sin\varphi = 0. \qquad [6.7]$$

This shows that [6.3] is unchanged by a superposed dilatation. This suggests that we adopt [6.3] for compressible as well as for incompressible material. Physically this amounts to the assumption that the deformation consists of the superposition of a dilatation on the superposed shearing deformations on the α- and β-curves described above. The governing equations for the velocity field thus become [6.1] and [6.3].

With a little straightforward manipulation, eqs. [6.1] and [6.3] can be expressed in characteristic form by referring them to the α- and β-curves as curvilinear coordinates. This gives

$$\begin{aligned} d(v_\alpha \cos\varphi) + (v_\alpha \sin\varphi - v_\beta)\, d(\psi + \tfrac{1}{2}\varphi) &= ds_\alpha\{\Omega \sin\varphi - \tfrac{1}{2}\varrho^{-1}\dot{\varrho}\cos\varphi\}, \\ &\quad \text{on an } \alpha\text{-curve}, \\ d(v_\beta \cos\varphi) + (v_\alpha - v_\beta \sin\varphi)\, d(\psi - \tfrac{1}{2}\varphi) &= ds_\beta\{-\Omega \sin\varphi - \tfrac{1}{2}\varrho^{-1}\dot{\varrho}\cos\varphi\}, \\ &\quad \text{on a } \beta\text{-curve} \end{aligned} \qquad [6.8]$$

where v_α and v_β are components of velocity in the directions of the α- and β-curves, and ds_α, ds_β denote elements of length along these curves.

Eqs. [6.1] and [6.3] or their equivalent [6.8] are independent of superposed rigid-body motions. They are from invariant under transformations of the rectangular coordinate system, and so satisfy conditions for material isotropy. They reduce to the equations given in (1) when ϱ is constant and to the velocity equations for an ideal compressible frictionless plastic material when $\varphi = 0$. They incorporate dependence of the velocity field on the stress-rate as well as on the stress, but only through the term Ω which represents the spin of the principal axes of stress. They do not, in general, require coincidence of the principal axes of stress and rate of deformation. This does not conflict with material isotropy, for reasons discussed in (1, 4, 5, 7) and elsewhere.

It is to be expected that, just as for instance in the theory of ideal metal plasticity, the constitutive equations should be supplemented by inequalities which ensure that the rate of working of the stress is positive. This point is made by *Dais* (4) and *de Josselin de Jong* (7). *De Josselin de Jong* makes the further plausible assumption that the rate of working associated with each shearing mechanism should be positive. Such conditions can readily be formulated in the context of the present theory, but we omit explicit details.

Eqs. [4.4], [5.2], [6.1] and [6.3] are five equations between the six dependent variables p, q, ψ, ϱ, u and v. The missing equation appears to be one which relates ϱ or $\dot{\varrho}$ to the stress variables. It was observed in § 4 that ϱ is likely to have complicated dependence on the stress and deformation history. It may be that a comparatively simple relation exists for $d\varrho/dp$. We suggest tentatively that this may have the form

$$\frac{d\varrho}{dp} = f(\varrho, p, k_R) \qquad R = 1, 2, \ldots, N \qquad [6.9]$$

where k_R are N parameters which depend on the stress and deformation history. As a very simple example, one of the k_R might be the maximum pressure to which a material particle has been subjected during its history. This question, however, requires further investigation.

Acknowledgement

This work was supported by the Ministry of Technology as part of the programme of the Warren Spring Laboratory (now of the Department of Trade and Industry) in the field of bulk materials storage and handling.

Summary

In a previous paper (1) proposals were made for the equations which govern the mechanical behaviour in plane strain of an ideal incompressible granular material. In this paper tentative suggestions are made for the extension of this theory to compressible granular materials. The material is envisaged as a mixture of an ideal gas and solid particles. The state of an element of the mixture is determined by its overall density and the mass and volume concentrations of the constituents. It is proposed that the material satisfies a flow condition which relates the two principal invariants of the stress (in two dimensions) and the density. The condition for the stress equations to be hyperbolic is obtained, and for the hyperbolic case a natural interpretation is obtained for the angle of internal friction.

For the kinematic behaviour it is proposed that the deformation consists of superposed shearing deformations on the stress characteristic surfaces, as described in (1), together with a superposed dilatation. The equations describing this behaviour are expressed in characteristic form. They reduce to the equations given in (1) when the density is constant.

References

1) *Spencer, A. J. M.*, J. Mech. Phys. Solids **12**, 337 (1964).

2) *Mandel, J.*, C. R. hebd. Séanc. Acad. Sci. **225**, 1272 (Paris 1947).

3) *Mandel, J.*, J. Mech. Phys. Solids **14**, 303 (1966).

4) *Dais, J. L.*, Int. J. Solids Struct. **6**, 1185 (1970).

5) *Mandl, G.* and *R. Fernandez Luque*, Géotechnique **20**, 277 (1970).

6) *Spencer, A. J. M.*, Géotechnique **21**, 190 (1971).
7) *de Josselin de Jong, G.*, Géotechnique **21**, 156 (1971).
8) *Goodman, M. A.* and *S. C. Cowin*, J. Fluid Mech. **45**, 321 (1971).
9) *Goodman, M. A.* and *S. C. Cowin*, Arch. Rational Mech. Anal. **44**, 249 (1972).
10) *Ashton, M. D., D. C.-H. Cheng, R. Farley*, and *F. H. H. Valentin*, Rheol. Acta **4**, 206 (1965).
11) *Kingston, M. R.* and *A. J. M. Spencer*, J. Mech. Phys. Solids **18**, 233 (1970).

Authors' address:

A. J. M. Spencer and *M. R. Kingston*
Dept. of Theoretical Mechanics
University of Nottingham
Nottingham NG7 2RD (England)

Rheol. Acta 12, 200–205 (1973)

From the Department of Engineering Mechanics University of Missouri-Rolla, Rolla (U.S.A.)

Application of large elastic deformation theory to the calculation of liquid drop shapes of some polymers

By *Xavier J. R. Avula*

With 8 figures

(Received October 27, 1972)

Introduction

Boundary tension is the best known property of liquid surfaces. This property is of outstanding importance in the phenomenon of adhesion which is an interdisciplinary subject involving surface chemistry, rheology, polymer physics, and fracture mechanics. Adhesion plays an important role in various industrial applications such as in packaging, construction and manufacturing. Also, the behavior of solder and printing inks in their respective practical applications is influenced by boundary tension.

The analogy of a liquid surface or interface to a stretched membrane is well known. Currently, the surface tensions of liquids are determined in terms of the shape factor parameters of sessile and pendent liquid drops in conjunction with the analytical solution of the *Young–Laplace* equation. In the investigations on boundary tension, so far reported, the drop shape parameters are not explicitly expressed as a function of pressure, but the pressure is carefully and conveniently eliminated in terms of curvature. However, in view of the advent of sophisticated pressure transducers, it is desirable to calculate the liquid drop shapes in terms of the crown pressure. It is expected that such calculations would simplify the experimental observations, because it is easier to measure pressure than to determine the curvature or shape parameters.

There are numerous methods available for the determination of boundary tension. The fundamental differential equation of a liquid drop shape is given in the work by *Bashforth* and *Adams* (1). The numerical solution of this equation and a table of shape factors versus boundary tension were presented by *Fordham* (2), *Staicopolus* (3) and also by *Niederhauser* and *Bartell* (4). The maximum bubble pressure method described by *Hypart* and *White* (5), the sessile bubble technique by *Sakai* (6), the ring method by *Schornhorn* and *Sharpe* (7), the plate method by *Dettre* and *Johnson* (8), the rotating drop method by *Patterson* et al. (9), and the pendent drop methods by *Roe* (10) and *Wu* (11) are among the various experimental methods of determining surface tension of polymer liquids.

In this paper the theory of large elastic deformations is used to determine the shapes and sizes of sessile and pendent liquid drops of some polymers by treating them as liquid-filled membranes subject to a variable hydrostatic pressure. In particular, the drop shapes of linear polyethylene (LPE) and polymethyl methylacrylate (PMMA) at 140 °C are calculated. A procedure for the estimation of an unknown surface tension is also given. The theory of large elastic deformations as applied to membranes has only recently been developed. The general theory of membranes is discussed in detail by *Green* and *Adkins* (12). Some problems of axisymmetric membranes were discussed by *Yu* and *Valanis* (13) and by *Avula* (14). The latter work is of particular interest in the formulation of the problem presented in this paper.

*) This work is an extension of the work performed under the National Science Foundation Grant No. GK-4842.

Governing equations of axisymmetric membranes

The geometries of undeformed and deformed middle surfaces are conveniently described by the cylindrical polar coordinates (ϱ, φ, x) and (r, φ, y), respectively. Following the notation used by *Green* and *Adkins* (12) the principal extension ratios λ_1, λ_2, and λ_3 for an incompressible material are given by

$$\lambda_1 = \frac{d\xi}{d\eta}, \quad \lambda_2 = \frac{r}{\varrho}, \quad \lambda_3 = \frac{1}{\lambda_1 \lambda_2} \qquad [1]$$

where, ξ and η are the meridional coordinates of the deformed and undeformed middle surfaces. The stress resultant equations for an incompressible material are

$$T_1 = 4\lambda_3 h_0 \{\lambda_1^2 - \lambda_3^2\} \left\{\frac{\partial W}{\partial I_1} + \lambda_2^2 \frac{\partial W}{\partial I_2}\right\} \qquad [2]$$

$$T_2 = 4\lambda_3 h_0 \{\lambda_2^2 - \lambda_3^2\} \left\{\frac{\partial W}{\partial I_1} + \lambda_1^2 \frac{\partial W}{\partial I_2}\right\} \qquad [3]$$

where, T_1 and T_2 are the physical components of the stress resultants measured per unit length of the membrane in the meridional and circumferential directions, respectively, $2h_0$ is the thickness of the undeformed membrane, I_1 and I_2 are the strain invariants, and W is the strain energy function. Note that the third strain invariant I_3 is equal to unity.

The strain energy function W for a homogeneous, isotropic and incompressible material is a function of I_1 and I_2. However, for simplicity a neo-*Hooke*an form of the strain energy function is assumed. Then

$$W = C_{10}(I_1 - 3) \quad [4]$$

where, C_{10} is a constant.

The equilibrium equations can be expressed in the form

$$\frac{d}{d\xi}(T_1 r) = T_2 \frac{dr}{d\xi} \quad [5]$$

$$K_1 T_1 + K_2 T_2 = p \quad [6]$$

where, K_1 and K_2 are the principal curvatures in the meridional and circumferential directions, and p is the resultant pressure in the direction of the outward normal to the deformed middle surface.

The compatibility equation is expressed in the form

$$\frac{d}{d\xi}(K_2 r) = K_1 \frac{dr}{d\xi}. \quad [7]$$

The curvature of a meridian curve in the deformed middle surface can be obtained from elementary differential geometry as

$$K_1 = -\frac{d^2r}{d\xi^2}\left\{1 - \left(\frac{dr}{d\xi}\right)^2\right\}^{-\frac{1}{2}}. \quad [8]$$

Introducing eq. [8] into eq. [7] and integrating yields

$$K_2 r = \left\{1 - \left(\frac{dr}{d\xi}\right)^2\right\}^{\frac{1}{2}}. \quad [9]$$

For a complete analysis of an axisymmetric membrane problem the system of differential eqs. [1] to [9] must be solved. However, in this paper the deformed shapes are only considered.

Substituting eq. [4] and λ_3 from eq. [1] into eqs. [2] and [3], and with the assumption of large deformations of the membrane in the circumferential and meridional directions such that λ_1 and $\lambda_2 \geq 2$, it has been shown (13, 14) that

$$T_1 = T_2 = 4h_0 C_{10} = \gamma \quad [10]$$

throughout the membrane, and

$$\lambda_1 = \lambda_2 \quad [11]$$

at every point on the deformed middle surface of the membrane. By analogy, the quantity γ in eq. [10] represents the surface tension of a liquid drop.

Now, consider axisymmetric liquid-filled membranes in the shapes of sessile and pendant liquid drops as shown in fig. 1. The pressure at a typical point on the deformed surface is given by

$$p = p_A \pm wy \quad [12]$$

where, p_A is the pressure at the tip of the drop, and w is the specific weight of the polymer liquid. Note that the "+" sign is for the sessile liquid drop and "−" sign for the pendant drop.

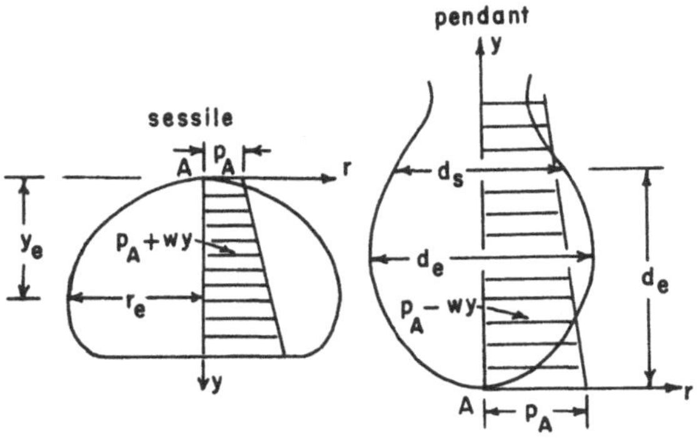

Fig. 1. Sessile and pendant liquid drop geometry

Introducing eqs. [9] and [10] into eq. [6] yields

$$-\gamma\frac{d^2r}{d\xi^2} + \frac{\gamma}{r}\left\{1 - \left(\frac{dr}{d\xi}\right)^2\right\} = \left\{1 - \left(\frac{dr}{d\xi}\right)^2\right\}^{\frac{1}{2}} p. \quad [13]$$

Since, in the deformed membrane,

$$(d\xi)^2 = (dy)^2 + (dr)^2 \quad [14]$$

one can write

$$y = \int_0^\xi \left\{1 - \left(\frac{dr}{d\xi}\right)^2\right\}^{\frac{1}{2}} d\xi. \quad [15]$$

A simultaneous solution of eqs. [13] and [15] determines the liquid drop shape which is of the form

$$r = r(y). \quad [16]$$

Obviously, the volume V of the drop from the tip ($r = y = 0$) to any height y is calculated by the equation

$$V = \pi \int_0^y \{r(y)\}^2 \, dy. \quad [17]$$

The total volume is obtained by computing V from the tip to the base in the case of a sessile drop, and from the tip to the neck in the case of a pendent drop.

Eq. [13] is a nonlinear second order ordinary differential equation and eq. [15] is a *Volterra* type integral equation. Due to their nonlinearity a numerical solution is sought on a digital computer employing a fourth-order *Runge-Kutta* integration process.

Deformation of a sessile liquid drop

For a sessile drop configuration the governing equations of deformation become

$$\frac{d^2r}{d\xi^2} - \frac{1}{r}\left\{1 - \left(\frac{dr}{d\xi}\right)^2\right\} + \frac{P_A + wy}{\gamma}\left\{1 - \left(\frac{dr}{d\xi}\right)^2\right\}^{\frac{1}{2}} = 0 \qquad [18]$$

$$y = \int_0^\xi \left\{1 - \left(\frac{dr}{d\xi}\right)^2\right\}^{\frac{1}{2}} d\xi. \qquad [19]$$

The pertinent boundary conditions are: At $r = 0$

$$\xi = 0, \quad y = 0, \quad \frac{dr}{d\xi} = 1. \qquad [20]$$

The configuration of the liquid drop depends upon the parameters p_A, w and γ. Numerical solutions of eqs. [18] and [19] with the boundary conditions [20] were obtained for LPE and PMMA at 140 °C, and the shapes of the corresponding sessile drops at various pressures are shown in fig. 2. Specific gravities of 0.811 and 1.19, and surface tensions of 28.8 and 32.0 dynes/cm were respectively used in the computations.

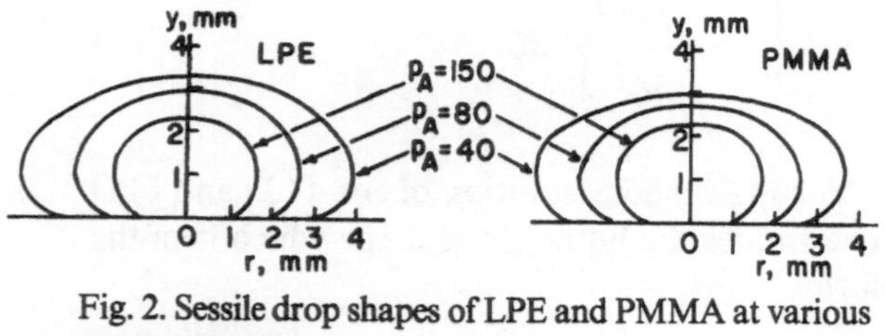

Fig. 2. Sessile drop shapes of LPE and PMMA at various crown pressures

A shape factor S is defined as the ratio of the equatorial radius r_e to the corresponding height y_e. Thus

$$S = \frac{r_e}{y_e}. \qquad [21]$$

The variation of S with respect to p_A is shown in fig. 3 for the two polymers in consideration. The values of r_e and y_e are picked in the computer output where $dr/d\xi = 0$. The slope $dr/d\xi = 0$ occurs naturally in the computations at the maximum radius of the liquid drop.

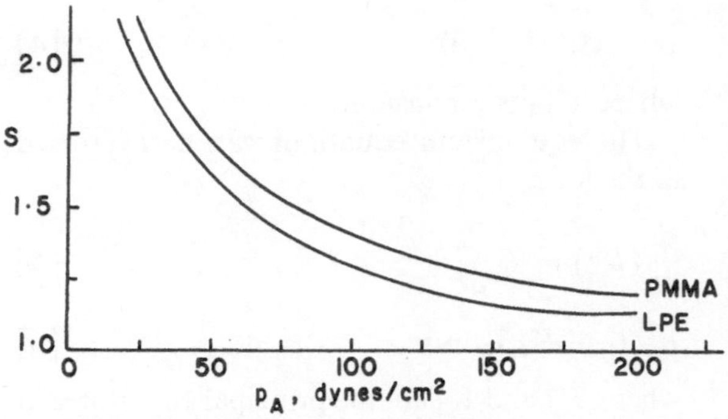

Fig. 3. Shape factor – crown pressure curves for LPE and PMMA

To estimate an unknown surface tension of a liquid whose specific weight w is known, one must solve eqs. [18], [19] and [20] for arbitrary values of p_A and γ, and plot S versus p_A curves. Since S and p_A are experimentally determinable for the given liquid the surface tension γ can be easily estimated by interpolation. An example of such a procedure is illustrated in fig. 4.

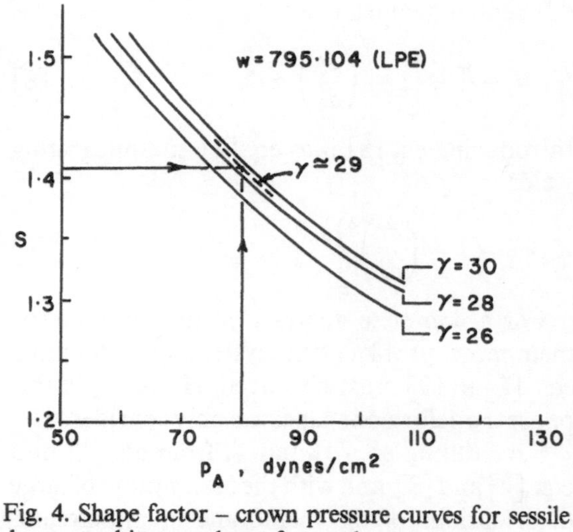

Fig. 4. Shape factor – crown pressure curves for sessile drops at arbitrary values of p_A and γ

The volume of a liquid drop is calculated by eq. [17]. The total volumes thus calculated at various pressure distributions for LPE and PMMA are shown in table 1. It can be observed here that the liquid drop volumes increase as the crown pressures decrease, and *vice versa*. This can be explained by the fact that, at the crown, it takes a small pressure to maintain a small curvature, or a flat configuration, caused by a large drop volume.

Table 1. Volume-pressure relationship for sessile liquid drops

Crown pressure dynes/cm²	Volume, cm³	
	LPE	PMMA
5	0.5397	0.4016
10	0.3648	0.2802
20	0.2205	0.1767
40	0.1139	0.0972
80	0.0465	0.0437
150	0.0155	0.0164
300	0.0032	0.0039

Deformation of a pendent liquid drop

The equations governing the deformation of a pendent liquid drop are

$$\frac{d^2r}{d\xi^2} - \frac{1}{r}\left\{1 - \left(\frac{dr}{d\xi}\right)^2\right\} + \frac{p_A - wy}{\gamma}\left\{1 - \left(\frac{dr}{d\xi}\right)^2\right\}^{\frac{1}{2}} = 0 \qquad [22]$$

$$y = \int_0^{\xi}\left\{1 - \left(\frac{dr}{d\xi}\right)^2\right\}^{\frac{1}{2}} d\xi. \qquad [23]$$

In eq. [21] p_A is the pressure at the bottom tip of the liquid drop. Observe the "$-$" sign in the pressure distribution term. The boundary conditions are: At $r = 0$

$$\xi = 0, \quad y = 0, \quad \frac{dr}{d\xi} = 1 \qquad [24]$$

The pendent drop shapes of LPE and PMMA were obtained from the numerical solutions of eqs. [22], [23] and [24], and are illustrated in fig. 5. A shape factor is described by the ratio

$$S_p = \frac{d_s}{d_e} \qquad [25]$$

where, d_e is the diameter at the equator, and d_s is the diameter at an arbitrarily selected plane at a distance from the tip of the drop equal to the equatorial diameter. The variation of S_p with respect to p_A for LPE and PMMA is shown in fig. 6.

The surface tension of a liquid whose specific weight is known can be obtained from S_p versus p_A curves computed for arbitrary values of p_A and γ in conjunction with the experimental pendent drop data by a procedure similar to the one described in the previous section.

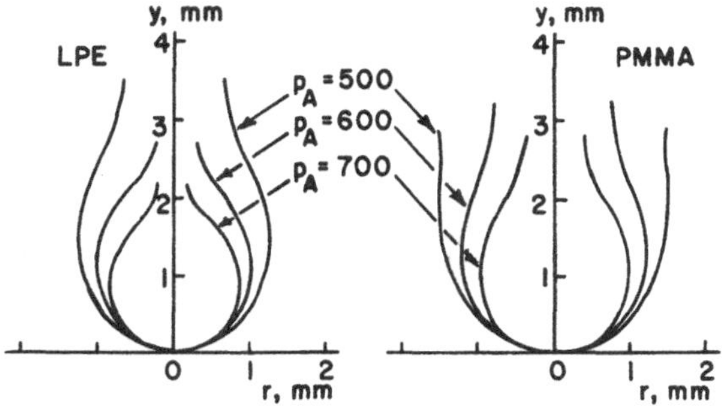

Fig. 5. Pendent drop shapes of LPE and PMMA at various crouch pressures

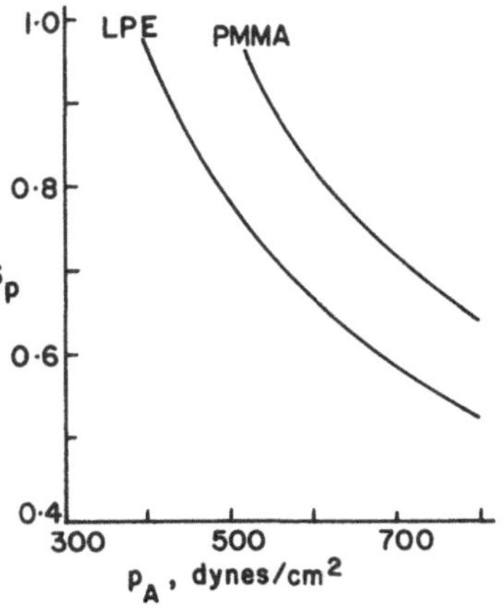

Fig. 6. Shape factor – crouch pressure curves for pendent drops

Table 2. Volume-pressure relationship for pendent liquid drops

Crown pressure dynes/cm²	Volume, cm³	
	LPE	PMMA
500	0.0111	0.0147
540	0.0079	0.0140
580	0.0059	0.0112
620	0.0045	0.0083
660	0.0036	0.0063
700	0.0029	0.0049

The volumes of pendent liquid drops of LPE and PMMA are calculated by eq. [17]. In the integral of eq. [17] the total height from the tip to the neck of the drop is taken as the upper limit. In this case too, it can be observed that the total drop volumes decrease with the increasing pressures at the tip. The explanation in the previous section for the sessile drops is also valid in the present case of pendent drops. The volume-pressure relationships for the pendent drops of LPE and PMMA are shown in table 2.

Separation of a pendent liquid drop

In order for a pendent drop to separate from a liquid column the weight of the drop must overcome the vertical component of the surface tension on a circumferential boundary and the intermolecular attraction force over the area enclosed by this boundary. A close observation of the pendent drop shapes in fig. 5 indicates that an increasing crouch pressure results in a rapid necking of the drop, and therefore contributes to the separation. However, the numerical solution of eqs. [22] and [23] exhibits that with increasing pressure, the weight of a drop decreases rapidly and then increases again displaying a minimum as shown in fig. 7. In fig. 7 a PMMA liquid drop is considered as an example. Using the pressure corresponding to a drop of minimum weight, the weight of a separated drop is calculated by the equation

$$W = 2\pi r \gamma \cos\theta + \pi r^2 \gamma_{11} \qquad [26]$$

where, W is the weight of the liquid drop in dynes, γ_{11} is the intramolecular attraction in dynes/cm^2, and θ is the slope of the meridional profile where the weight of the drop just balances the intermolecular and membrane forces (see fig. 8).

For the example considered, the profile of a drop which gives the minimum weight is shown on the right hand side in fig. 8. In calculating this profile the intramolecular attraction γ_{11} is not included. It must be noted that the drop profile will be altered as soon as complete separation occurs.

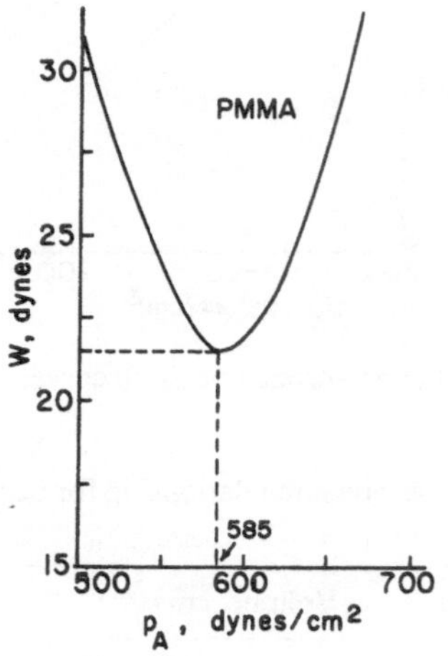

Fig. 7. Weight of separated drops at various p_A

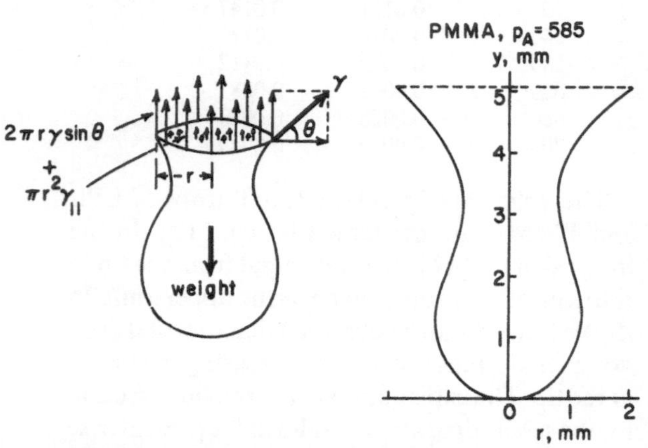

Fig. 8. Size of a separated pendent drop

Discussion

The theoretical results presented in this paper indicate that it is possible to predict the surface tension of a polymer liquid if the specific weight and the tip pressure under which the drop is formed are known. However, it is not possible to determine the various components of surface tension from these results. The difficulties involving the experimental determination of shape parameters and locating the inflexion points on liquid drop profiles are well acknowledged. Therefore, the formulation of the problem in terms of pressure which can be easily measured is certainly an advantage. At the present time, data on experimentally determined shape parameters of polymer liquid drops expressed as a function of tip pressure do not exist for verification of the solution presented in this paper. A comparison of the solution for water, with certain modifications, agreed well with the existing pendent drop data (15), and this establishes a degree of confidence in the method discussed herein.

Besides the absence of shape parameter data on polymer liquid drops accurate data on the specific weights of liquid polymers are not available. The theoretically calculated profiles of liquid drops are very sensitive to the changes in specific weight and pressure. Therefore, for a meaningful comparison the input parameters must be accurately determined under the given environmental conditions.

In the present formulation of the problem, the boundary conditions for a sessile drop at the maximum radius and at the circumference of contact, and for a pendent drop at the maximum radius and at the neck are obtained naturally as a part of the solution. These boundary conditions are used to fix the size of a liquid drop.

The governing equations can be easily modified for the determination of the interfacial tension between immiscible polymer systems. For interfacial tension the specific weight must be changed to the difference of specific weights, and the pressure to the pressure difference across the interface. Since, the theory presented in this paper is based on the kinematic and equilibrium conditions, it can be extended to other liquid systems.

Summary

Considering the analogy of a liquid surface to a stretched membrane, the theory of large elastic deformations is used to calculate the deformed shapes of sessile and pendent liquid drops of some polymers. In particular, the liquid drops are treated as highly deformable liquid-filled membranes that follow a neo-*Hooke*an constitutive equation. Since, with the advent of sophisticated pressure transducers, it is easier to measure pressure accurately than to determine the curvature of the liquid drop geometry, the problem is formulated in terms of the pressure prescribed at the tip of the liquid drop with a varying hydrostatic pressure. The governing equations consisting of a non-linear ordinary differential equation and a *Volterra* type integral equation are solved numerically on a digital computer.

Based on the solution of the governing equations, a procedure to estimate the surface tension of a polymer from the shape factor versus pressure curves is illustrated. Also, by increasing the pressure and by balancing the vertical component of the surface tension against the weight of a pendent liquid drop the size of a drop separated from a liquid column is predicted.

Zusammenfassung

Mit Rücksicht auf die Analogie zwischen einer flüssigen Oberfläche und einer gespannten Membrane wird die Theorie der großen elastischen Deformationen angewendet, um die deformierte Gestalt von sitzenden und hängenden flüssigen Tropfen einiger Polymere zu kalkulieren. Im besonderen werden die flüssigen Tropfen als sehr deformierbare liquid gefüllte Membranen behandelt, die einer neo-*Hooke*schen konstitutiven Gleichung folgen. Das Problem ist so formuliert, daß der Druck an der Spitze eines flüssigen Tropfens als ein veränderlicher hydrostatischer Druck ausgedrückt wird, weil mit den neuen hochentwickelten Druckvermittlern es leichter ist, den Druck direkt zu messen, als die Krümmung eines flüssigen Tropfens von seiner Geometrie zu bestimmen. Die Gleichungen, von denen die Geometrie eines Tropfens bestimmt werden, bestehen aus einer nichtlinearen Differentialgleichung und einer *Volterra*-Integralgleichung. Diese Gleichungen werden mit Hilfe eines Computers gelöst.

Aufgrund der Lösung dieser Gleichungen wird ein Verfahren entwickelt, das die Bestimmung der Oberflächenspannung eines Polymers von dem Formfaktor Gegendruckkurven ermöglicht. Die Größe eines von der flüssigen Säule getrennten Tropfens kann mit Hilfe von erhöhtem Druck durch Kräfteausgleich vorherbestimmt werden.

References

1) *Bashforth, F.* and *J. C. Adams*, An Attempt to Test the Theories of Capillary Action (Cambridge, England 1883).

2) *Fordham, S.*, Proc. Roy. Soc. **A 194**, 1 (London 1948).

3) *Staicopolus, D. N.*, J. Colloid Sci. **17**, 439 (1962).

4) *Niederhauser, D. O.* and *F. E. Bartell*, Report of Progress – Fundamental Research on Occurrence and Recovery of Petroleum, 1948–1949 (Baltimore, Maryland 1950).

5) *Hybart, F. J.* and *T. R. White*, J. Appl. Polymer Sci. **3**, 118 (1960).

6) *Sakai, T.*, Polymer **6**, 659 (1965).

7) *Schornhorn, H.* and *L. H. Sharpe*, J. Poly. Sci. **A 3**, 569 (1965).

8) *Dettre, R. H.* and *R. E. Johnson*, J. Colloid Interface Sci. **21**, 367 (1966).

9) *Patterson, H. T., K. H. Hu*, and *T. H. Grindstaff*, J. Poly. Sci. **C 34**, 31 (1971).

10) *Roe, R. J.*, J. Phys. Chem. **72**, 2013 (1968).

11) *Wu, S.*, J. Colloid Interface Sci. **31**, 153 (1969).

12) *Green, A. E.* and *J. E. Adkins*, Large Elastic Deformations and Nonlinear Continuum Mechanics (Oxford, England 1960).

13) *Yu, L. K.* and *K. C. Valanis*, Trans. Soc. Rheology **14**, 159 (1970).

14) *Avula, X. J. R.*, Proc. Sixth Southeastern Conference Theor. Appl. Mech. (Tampa, Florida 1972).

15) *Stauffer, C. E.*, J. Phys. Chem. **69**, 1933 (1965).

Author's address:

Dr. *Xavier J. R. Avula*
Dept. of Engineering Mechanics
University of Missouri-Rolla
Rolla, Missouri 65401 (U.S.A.)

Rheol. Acta 12, 206–211 (1973)

From the Institute of Science and Technology, The University of Manchester (England)

The shape of a liquid surface between rotating concentric cylinders

By A. Kaye

With 3 figures

(Received October 27, 1972)

I. Introduction

The purpose of this paper is to calculate the shape of the top free surface of a non-*Newton*ian liquid which is contained between axially rotating cylinders which have a common vertical axis. We shall consider the case in which the outer cylinder is stationary and the inner cylinder rotates with an angular velocity Ω. This is the flow situation described by *Weissenberg* (Nature **159**, 310, 1947) and for most non-*Newton*ian liquids, on rotation, the liquid rises near the inner cylinder and falls near the outer. This effect is now generally known as the *Weissenberg* effect. Inertial forces alone would, of course, produce a fall at the inner and a rise at the outer cylinder.

In our calculation we shall assume that the liquid is an incompressible liquid whose constitutive equation takes the form

$$p_{ij} + p\delta_{ij} = \nu\varrho A_{ij}^{(1)} + \alpha\varrho A_{ik}^{(1)} A_{kj}^{(1)} + \beta\varrho A_{ij}^{(2)} \qquad [1]$$

where p_{ij} are the components of the stress tensor referred to a set of rectangular *Cartes*ian co-ordinate axes, p is the magnitude of the isotropic pressure, ν, α, β are constants and ϱ is the density of the liquid. $A_{ij}^{(1)}$ is defined by

$$A_{ij}^{(1)} = \frac{\partial v_i}{\partial x_j} + \frac{\partial v_j}{\partial x_i} \qquad [2]$$

where v_i is the velocity of a typical particle of the liquid and

$$A_{ij}^{(2)} = \frac{\partial A_{ij}^{(1)}}{\partial t} + v_k \frac{\partial A_{ij}^{(1)}}{\partial x_k} + A_{ik}^{(1)} \frac{\partial v_k}{\partial x_j} + A_{kj}^{(1)} \frac{\partial v_k}{\partial x_i}. \qquad [3]$$

δ_{ij} has its usual meaning and the dummy suffix notation is employed.

The use of this constitutive equation is not unduly restrictive for, in the mathematical analysis, we shall use a perturbation procedure which assumes that the angular velocity is small and we have a situation described as a slow flow. It can be shown that the constitutive equation for a simple fluid may then be described by an equation for form [1] when the flow is slow enough (see e.g. *Colman* and *Noll*, p. 530, Second Order Effects in Elasticity, Plasticity and Fluid Dynamics, New York 1964, Macmillan).

II. Constitutive equation and equations of motion

We shall use cylindrical polar co-ordinates (r, θ, z) in which the origin is taken on the common axis of the vertical cylinders where it meets the horizontal, stationary, free liquid surface. The z axis is taken vertically upwards and the $\theta = 0$ plane is chosen arbitrarily since we shall assume that we have axial symmetry. The stress equations of motion are

$$\frac{\partial p_{rr}}{\partial r} + \frac{\partial p_{rz}}{\partial z} + \frac{1}{r}(p_{rr} - p_{zz}) = \varrho A \qquad [4]$$

$$\frac{\partial p_{r\theta}}{\partial r} + \frac{\partial p_{\theta z}}{\partial z} + \frac{2}{r} p_{r\theta} = \varrho B \qquad [5]$$

$$\frac{\partial p_{rz}}{\partial r} + \frac{\partial p_{zz}}{\partial z} + \frac{1}{r} p_{rz} = \varrho(C + g) \qquad [6]$$

where

$$A = u\frac{\partial u}{\partial r} + w\frac{\partial u}{\partial z} - \frac{v^2}{r} \qquad [7]$$

$$B = u\frac{\partial v}{\partial r} + w\frac{\partial v}{\partial z} + \frac{uv}{r} \qquad [8]$$

$$C = u\frac{\partial w}{\partial r} + w\frac{\partial w}{\partial z} \qquad [9]$$

with the usual notation for the physical components of the stress tensor and where the physical components of velocity in the r increasing, θ increasing and z increasing directions are u, v and w respectively. It will be noticed that the gravitational body force has been included.

The physical components of the $A_{(ij)}^{(1)}$ and $A_{(ij)}^{(2)}$ tensors may be computed and they are found to be

$(A_{ij}^{(1)})$

$$= \begin{pmatrix} 2\frac{\partial u}{\partial r} & \frac{\partial v}{\partial r} - \frac{v}{r} & \frac{\partial u}{\partial z} + \frac{\partial w}{\partial r} \\ \frac{\partial v}{\partial r} - \frac{v}{r} & 2\frac{u}{r} & \frac{\partial v}{\partial z} \\ \frac{\partial u}{\partial z} + \frac{\partial w}{\partial r} & \frac{\partial v}{\partial z} & 2\frac{\partial w}{\partial z} \end{pmatrix} \qquad [10]$$

and

$(A_{ij}^{(2)})$

$$= \begin{pmatrix} 2\frac{\partial A}{\partial r} & \frac{\partial B}{\partial r} - \frac{B}{r} & \frac{\partial A}{\partial z} + \frac{\partial C}{\partial r} \\ \frac{\partial B}{\partial r} - \frac{B}{r} & \frac{2A}{r} & \frac{\partial B}{\partial z} \\ \frac{\partial A}{\partial z} + \frac{\partial C}{\partial r} & \frac{\partial B}{\partial z} & 2\frac{\partial C}{\partial z} \end{pmatrix} + 2\tilde{Q}Q \qquad [11]$$

where

$$Q = \begin{pmatrix} \frac{\partial u}{\partial r} & -\frac{v}{r} & \frac{\partial u}{\partial z} \\ \frac{\partial v}{\partial r} & \frac{u}{r} & \frac{\partial v}{\partial z} \\ \frac{\partial w}{\partial r} & 0 & \frac{\partial w}{\partial z} \end{pmatrix} \qquad [12]$$

Substituting these expressions for $A_{ij}^{(1)}$, $A_{ij}^{(2)}$ into the constitutive equation to find p_{ij}, and then inserting these values of p_{ij} into the stress equations of motion, we find

$$-\frac{1}{\varrho}\frac{\partial p}{\partial r} + \nu\left\{\frac{\partial^2 u}{\partial r^2} + \frac{\partial^2 u}{\partial z^2} + \frac{1}{r}\frac{\partial u}{\partial r} - \frac{u}{r^2}\right\} = u\frac{\partial u}{\partial r} + w\frac{\partial u}{\partial z} - \frac{v^2}{r} - \alpha S_r - \beta K_r \qquad [13]$$

$$\nu\left\{\frac{\partial^2 v}{\partial r^2} + \frac{\partial^2 v}{\partial z^2} + \frac{1}{r}\frac{\partial v}{\partial r} - \frac{v}{r^2}\right\} = u\frac{\partial v}{\partial r} + w\frac{\partial v}{\partial z} + \frac{uv}{r} - \alpha S_\theta - \beta K_\theta \qquad [14]$$

$$-\frac{1}{\varrho}\frac{\partial p}{\partial z} + \nu\left\{\frac{\partial^2 w}{\partial r^2} + \frac{\partial^2 w}{\partial z^2} + \frac{1}{r}\frac{\partial w}{\partial r}\right\} = g + u\frac{\partial w}{\partial r} + w\frac{\partial w}{\partial z} - \alpha S_z - \beta K_z \qquad [15]$$

where (S_r, S_θ, S_z), (K_r, K_θ, K_z) are physical components in cylindrical polar co-ordinates of the vectors $\frac{\partial}{\partial x_i}(A_{ik}^{(1)}A_{kj}^{(1)})$ and $\frac{\partial}{\partial x_i}(A_{ij}^{(2)})$ respectively. The actual forms of the vectors $\underset{\sim}{S}$ and $\underset{\sim}{K}$ are not important at this stage. It is to be noted, however, that they are quadratic in velocity.

The equation of continuity is

$$\frac{\partial u}{\partial r} + \frac{\partial w}{\partial z} + \frac{u}{r} = 0. \qquad [16]$$

III. Boundary conditions

Let the inner and outer cylinders have radii a and b respectively. Then the boundary conditions on the cylinders are

$v = \Omega a$ on $r = a$ for all z

$v = 0$ on $r = b$ for all z

$u = w = 0$ on $r = a$ and $r = b$ for all z.

We now consider the boundary conditions on the free surface. Take a local set of rectangular cartesian co-ordinate axes $0'x_t$, $0'x_n$ and $0'x_\theta$ at some point $0'$ on the surface, where $0'x_t$ is tangential to the surface in a plane containing the z axis, $0'x_n$ is normal to the surface and $0'x_\theta$ is in the θ increasing direction; these axes are in the r increasing, z increasing and θ increasing directions respectively when the liquid is stationary. With an obvious notation the boundary conditions on the surface are

$$p_{nn} = p_{nt} = p_{n\theta} = 0$$

where we assume that pressures are measured relative to atmospheric pressure and we ignore surface tension forces. If $\frac{\pi}{2} - \psi$ is the angle between the $0'x_i$ and $0z$ axes then it can be shown that

$$p_{nn} = \cos^2\psi\, p_{zz} + p_{rr}\sin^2\psi - 2p_{rz}\sin\psi\cos\psi \qquad [17]$$

$$p_{tn} = p_{rz}(\cos^2\psi - \sin^2\psi) + (p_{zz} - p_{rr})\cos\psi\sin\psi \qquad [18]$$

$$p_{\theta n} = p_{z\theta}\cos\psi - p_{r\theta}\sin\psi. \qquad [19]$$

IV. The perturbation

When the inner cylinder is stationary the solution is trivial, namely

$$u = v = w = 0 \quad p = -\varrho g z \qquad [20]$$

and the free surface has the equation $z = f(r)$ where $f(r) = 0$.

Let us perturb this solution for small values of Ω. It is easily seen, by reversing the direction of rotation of the inner cylinder, that u, w, p, f are even functions of Ω and v is an odd function of Ω.

Let

$$v = \Omega v_1 + \Omega^3 v_3 + \cdots \qquad [21]$$

$$u = \Omega^2 u_2 + \Omega^4 u_4 + \cdots \qquad [22]$$

$$w = \Omega^2 w_2 + \Omega^4 w_4 + \cdots \qquad [23]$$

$$p = -\varrho g z + \Omega^2 p_2 + \Omega^4 p_4 + \cdots \qquad [24]$$

$$f(r) = \Omega^2 f_2 + \Omega^4 f_4 + \cdots \qquad [25]$$

If we now substitute in the equations of motion and equate coefficients of powers of Ω we find, from terms in Ω,

$$\frac{\partial^2 v_1}{\partial r^2} + \frac{\partial^2 v_1}{\partial z^2} + \frac{1}{r}\frac{\partial v_1}{\partial r} - \frac{v_1}{r^2} = 0 \qquad [26]$$

and from terms in Ω^2

$$-\frac{1}{\varrho}\frac{\partial p_2}{\partial r} + \nu\left\{\frac{\partial^2 u_2}{\partial r^2} + \frac{\partial^2 u_2}{\partial z^2} + \frac{1}{r}\frac{\partial u_2}{\partial r} - \frac{u_2}{r^2}\right\} = -\frac{v_1^2}{r} - \alpha S_r^* - \beta K_r^* \qquad [27]$$

$$-\frac{1}{\varrho}\frac{\partial p_2}{\partial z} + \nu\left\{\frac{\partial^2 w_2}{\partial r^2} + \frac{\partial^2 w_2}{\partial z^2} + \frac{1}{r}\frac{\partial w_2}{\partial r}\right\} = -\alpha S_z^* - \beta K_z^*. \qquad [28]$$

$\underline{S}^*$ and $\underline{K}^*$ are obtained from $\underline{S}$ and $\underline{K}$ by putting $u = w = 0$, $v = v_1$ because the only possible Ω^2 terms must come from v_1 since $\underline{S}$ and $\underline{K}$ are quadratic in velocity.

The equation of continuity gives

$$\frac{\partial u_2}{\partial r} + \frac{\partial w_2}{\partial z} + \frac{u_2}{r} = 0. \qquad [29]$$

The boundary conditions are

$v_1 = a \quad r = a \quad$ all z

$v_1 = 0 \quad r = b \quad$ all z

and

$w_2 = u_2 = 0$ on $r = a$ and $r = b$ all z.

The free boundary presents more difficulties. We deal with the problem by expressing all functions and their derivatives on the boundary in terms of functions and derivatives on the neighbouring fixed surface $z = 0$.

Consider, for example, the free boundary condition given by eq. [17]

$$0 = p_{zz} + p_{rr}\tan^2\psi - 2p_{rz}\tan\psi. \qquad [30]$$

From eqs. [1], [10], [11], [12] we find

$$p_{zz} = -p + 2\nu\varrho\frac{\partial w}{\partial z} + \varrho\alpha\left[\left(\frac{\partial w}{\partial r} + \frac{\partial u}{\partial z}\right)^2 + \left(\frac{\partial v}{\partial z}\right)^2 + 4\left(\frac{\partial w}{\partial z}\right)^2\right] + \varrho\beta\left[2\frac{\partial}{\partial z}\left(u\frac{\partial w}{\partial r} + w\frac{\partial w}{\partial z}\right) + 2\left(\left(\frac{\partial u}{\partial z}\right)^2 + \left(\frac{\partial v}{\partial z}\right)^2 + \left(\frac{\partial w}{\partial z}\right)^2\right)\right] \qquad [31]$$

Retaining terms only up to Ω^2 we find

$$p_{zz} = -\varrho g z + p_2\Omega^2 + 2\nu\varrho\frac{\partial w_2}{\partial z}\Omega^2 + \varrho\alpha\Omega^2\left(\frac{\partial v_1}{\partial z}\right)^2 + 2\varrho\beta\Omega^2\left(\frac{\partial v_1}{\partial z}\right)^2 \qquad [32]$$

p_{zz} is to be evaluated on the surface $z = f(r)$.

Thus, for example,

$$p_2 \text{ (on the boundary)} = p_2(r, f(r)) = p_2(r, 0) + \Omega^2 f_2\left(\frac{\partial p_2}{\partial z}\right)_{r,0} + \cdots \qquad [33]$$

We also note that

$$\tan\psi = \frac{df}{dr} = \Omega^2\frac{df_2}{dr} + \Omega^4\frac{df_4}{dr} + \cdots \qquad [34]$$

Thus the boundary condition given by eq. [17] gives, from terms in Ω^2,

$$0 = -\varrho g f_2(r) + p_2(r, 0) + 2\nu\varrho\left(\frac{\partial w_2}{\partial z}\right)_{r,0} \qquad [35]$$

Treating the other free boundary conditions in the same way we find that the only other condition obtained from Ω^2 terms is

$$\left(\frac{\partial w_2}{\partial r}\right)_{r,0} + \left(\frac{\partial u_2}{\partial z}\right)_{r,0} = 0. \qquad [36]$$

The only condition obtained from Ω terms is

$$\left(\frac{\partial v_1}{\partial z}\right)_{r,0} = 0. \qquad [37]$$

There are no problems in going to higher orders in Ω, other than algebraic ones.

V. The first order problem

We may now assemble the boundary value problems associated with each power in Ω in turn. From terms in Ω we find that we must solve

$$\frac{\partial^2 v_1}{\partial r^2} + \frac{\partial^2 v_1}{\partial z^2} + \frac{1}{r}\frac{\partial v_1}{\partial r} - \frac{v_1}{r^2} = 0 \qquad [38]$$

subject to the boundary conditions

$$v_1 = a \quad r = a \quad \text{all } z$$

$$v_1 = 0 \quad r = b \quad \text{all } z$$

$$\frac{\partial v_1}{\partial z} = 0 \quad z = 0 \quad \text{all } r.$$

The solution is

$$v_1 = \bar{A} r + \frac{\bar{B}}{r} \qquad [39]$$

where $\bar{A}$ and $\bar{B}$ are given by

$$a = \bar{A} a + \bar{B}/a \qquad [40]$$

$$0 = \bar{A} b + \bar{B}/b. \qquad [41]$$

VI. The second order problem

We must solve

$$-\frac{1}{\varrho}\frac{\partial p_2}{\partial r} + \nu\left\{\frac{\partial^2 u_2}{\partial r^2} + \frac{\partial^2 u_2}{\partial z^2} + \frac{1}{r}\frac{\partial u_2}{\partial r} - \frac{u_2}{r^2}\right\} = -\frac{v_1^2}{r} - \alpha S_r^* - \beta K_r^* \qquad [42]$$

$$\frac{1}{\varrho}\frac{\partial p_2}{\partial z} + \nu\left\{\frac{\partial^2 w_2}{\partial r^2} + \frac{\partial^2 w_2}{\partial z^2} + \frac{1}{r}\frac{\partial w_2}{\partial r}\right\} = -\alpha S_z^* - \beta K_z^* \qquad [43]$$

subject to the continuity equation

$$\frac{\partial u_2}{\partial r} + \frac{\partial w_2}{\partial z} + \frac{u_2}{r} = 0 \qquad [44]$$

and the boundary conditions. These are

$$g\varrho f_2(r) - p_2 + 2\nu\varrho\frac{\partial w_2}{\partial z} = 0$$

$$\text{on} \quad z = 0 \quad \text{for all } r \qquad [45]$$

$$\frac{\partial w_2}{\partial r} + \frac{\partial u_2}{\partial z} = 0 \quad \text{on} \quad z = 0 \quad \text{for all } r \qquad [46]$$

and

$$u_2 = w_2 = 0 \quad \text{on} \quad r = a \quad \text{and} \quad r = b \quad \text{all } z.$$

Notice that the first condition is not really a boundary condition at all; it is a means of finding f_2 when the problem, to find u_2, w_2 and p_2 is solved.

We may now calculate $\underline{S}^*$ and $\underline{K}^*$ since v_1 is known. Using eqs. [39], [1], [10], [11], [12], we find

$$S_\theta^* = S_z^* = K_\theta^* = K_z^* = 0 \qquad [47]$$

$$S_r^* = -\frac{16}{r^5}\bar{B}^2 \qquad [48]$$

$$K_r^* = -\frac{24}{r^5}\bar{B}^2. \qquad [49]$$

Fortunately there is a trivial solution of the problem: $u_2 = w_2 = 0$ with p_2 a function of r only.

By calculation we find

$$\frac{1}{\varrho} p_2 = \bar{A}^2\frac{r^2}{2} + 2\bar{A}\bar{B}\log r - \frac{\bar{B}^2}{2r^2} + \frac{\bar{B}^2}{r^4}\{4\alpha + 6\beta\} + K \qquad [50]$$

where K is a constant integration.

Using the boundary condition [45] we find the free boundary is

$$z = f(r) = \Omega^2 f_2(r)$$

$$= \frac{\Omega^2}{g}\left[\left[\frac{\bar{A}^2}{2}r^2 + 2\bar{A}\bar{B}\log r - \frac{\bar{B}^2}{2r^2}\right] + \frac{\bar{B}^2}{r^4}(4\alpha + 6\beta) + K\right]. \qquad [51]$$

The constant K may be found by using the fact that the volume of material is conserved and hence

$$\int_a^b f(r)\, r\, dr = 0. \qquad [52]$$

Thus, using this condition and the values of $\bar{A}$ and $\bar{B}$ obtained from eqs. [40], [41] we may calculate the shape of the surface.

VII. The shape of the surface

The shape of the surface is given by

$$z = f(r) = P + Q \qquad [53]$$

where

$$\frac{Pg}{\Omega^2}\frac{(b^2 - a^2)^2}{a^4} = \left(\frac{r^2}{2} - 2b^2\log r - \frac{b^4}{2r^2}\right) - \frac{2}{(b^2 - a^2)}\left[\frac{1}{8}(5b^2 + a^2)(b^2 - a^2) - 3/2\, b^4\log b + \frac{b^2}{2}(b^2 + 2a^2)\log a\right] \qquad [54]$$

$$\frac{g}{\Omega^2} Q = \left(\frac{a^2 b^2}{b^2 - a^2}\right)^2 \left(\frac{1}{r^4} - \frac{1}{a^2 b^2}\right)(4\alpha + 6\beta). \quad [55]$$

We see that P gives the shape of the surface for the inertia terms and Q for the non-*Newton*ian terms. We note that Q falls as r increases when $2\alpha + 3\beta > 0$: the behaviour which would explain the *Weissenberg* effect. It can be shown that, for $r < b$, P is an increasing function of r; thus the inertial forces tend to force the liquid up the wall of the outer cylinder. We also note that the function $\frac{1}{r^4} - \frac{1}{a^2 b^2}$ varies rapidly with r when r is small, explaining the observed fact that most of the *Weissenberg* rise takes place near the inner cylinder.

VIII. Experimental measurements

Some experimental measurements of the shape of the surface of the liquid were carried out and the results are given below. The inner cylinder was held in a milling machine and could be rotated about a vertical axis at speeds determined by the gear ratios of the machine. The outer cylinder was mounted on the table and the traverses could be used in order to ensure that the axes of the cylinders were accurately co-incident. The shape of the surface was determined by touching the surface of the liquid with a needle which was moved by the vertical and horizontal traverses of a travelling microscope. Hence its position could be determined accurately.

Two liquids were used: a silicone fluid, MS 200 series, whose viscosity of 9.7 poise was known to be constant over the range of shear rates of interest, and a non-*Newton*ian liquid, a solution of 6 g of B 100 polyisobutylene dissolved in 100 ml of decalin. The experiments were carried out at 25 °C in a temperature controlled laboratory.

For the *Newton*ian liquid, the height z was measured for various values of d, the distance measured radially from the inner cylinder wall for three different speeds of revolution. d is equal to $r - a$ where a is 1.27 cm and b is 3.3 cm. If the theory is to apply to this system, the height z should be proportional to the square of the angular velocity. Fig. 1 shows a plot of z against Ω^2 for a particular value of d. The theoretical value of z obtained from eqs. [53] and [54] is also plotted and we see that the higher angular velocities are too large for the theory to be applicable, but that there is good agreement at the smallest angular velocity. Fig. 2 shows a plot of z/Ω^2 for this angular velocity plotted against d. The full curve is the theoretical curve obtained from eq. [54]. There is good agreement.

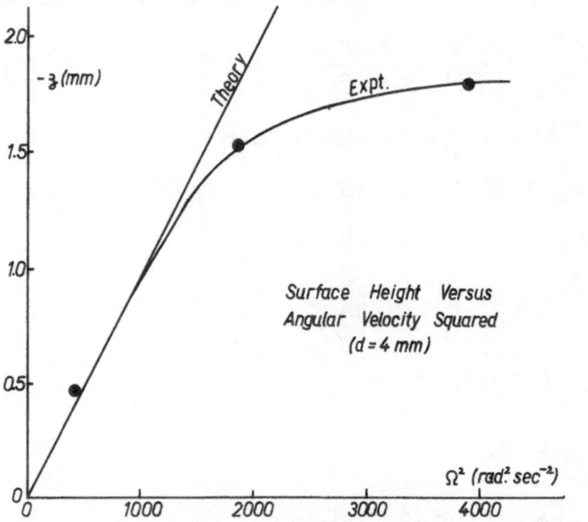

Fig. 1. A plot of the height rise of the surface, z, at a distance, d, of 4 mm from the wall of the inner cylinder as a function of the square of the angular velocity of the inner cylinder. The liquid used was the MS 200 silicone *Newton*ian liquid

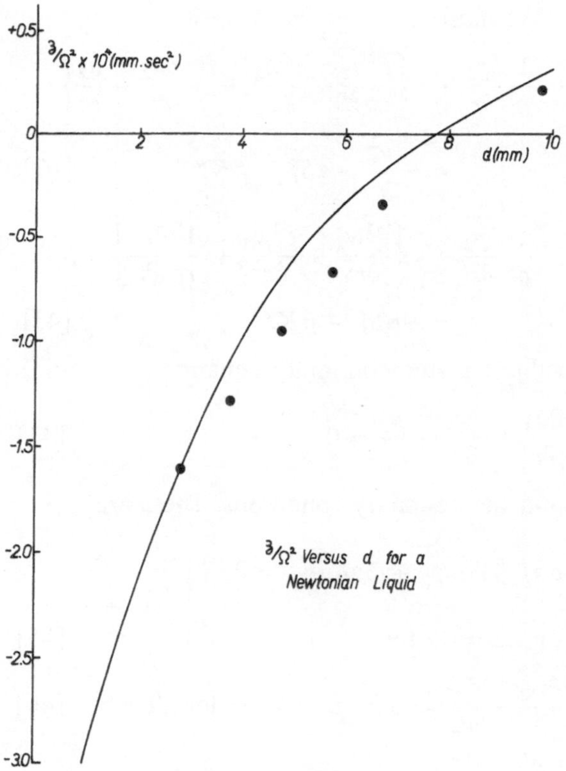

Fig. 2. A plot of z/Ω^2 against d for the *Newton*ian liquid in which the smallest value of angular velocity was used. The theoretical to curve was obtained from eqs. [53] and [54]

With the non-*Newton*ian liquid, a stable flow could only be obtained at the lowest angular velocity and thus it was not possible to test the variation of z with Ω^2. It would, however, be very unlikely for z to be proportional to Ω^2 since a measurement of the primary normal stress as a function of shear rate shows that the normal stress is not proportional to shear rate squared and hence the constitutive equation assumed in the theory does not hold in the shear rate range (0–10 $\sec^{-1}$) used. Thus, we cannot expect an exact agreement with theory.

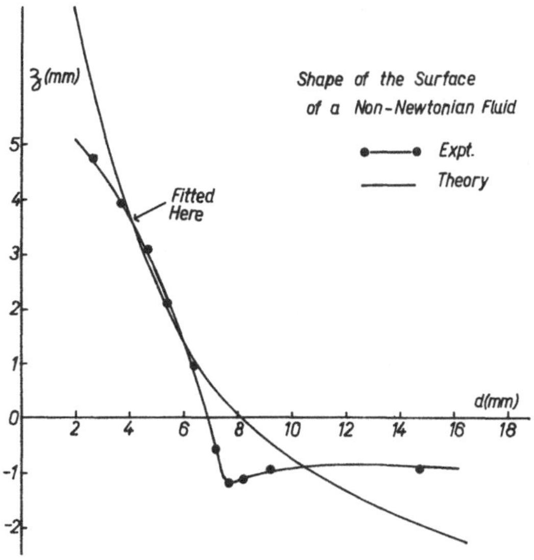

Fig. 3. A plot of z against d for the solution of 6 gr of B 100 polyisobutylene dissolved in 100 ml of decalin. The theoretical line was obtained from eq. [55] by fitting to the experimental data at the point shown

Fig. 3 shows a plot of z against d. Also shown is a theoretical curve obtained from eq. [55] which was made to agree with the experimental data when d equals 4 mm. We see that the curves are similar. We can use this fit to calculate $4\alpha + 6\beta$ from eq. [55]. Thus, assuming that the second normal stress difference is zero, and using the density of 0.90 g/cm³, we can show that the theory predicts that the primary normal stress difference, $p_{11} - p_{22}$, takes the form

$$p_{11} - p_{22} = 16.0\ \mathrm{G}^2\ \text{dynes/cm}^3.$$

This estimated value of the normal stress difference does not differ from the measured values by more than 50% over the range of shear rates used.

IX. Conclusion

The experimental work presented here is only intended to examine the order of magnitude of the effects predicted. More extensive measurements are needed to check the theoretical results with greater precision, and, in particular, measurements of non-*Newton*ian liquids using lower angular velocities will be needed in order to be in a region in which the second order approximations adequately explains the properties of the material. The height rise of the surface will then be too small and more accurate methods for measuring the shape of the surface will be needed. Perhaps optical interferometric techniques could be used. If such methods could be developed then the theory gives an indirect method of measuring normal stress differences at very much lower levels than measured previously. Since most microscopic theories of solution only make predictions of solution behaviours at small shear rates, such experimental measurements would be very useful.

The mathematical perturbation theory presented here can be extended to include higher orders of the angular velocity in the expansions. The constitutive equation for a simple fluid for slow flows can be given in the form of an expansion in terms of the *Rivlin-Ericksen* tensors and hence other constitutive equations can be included in the calculation. The calculation involves the solution of linear partial differential equations on fixed boundaries. The solutions may, perhaps, best be found by numerical computation.

Summary

The shape of the free surface of a second order fluid, enclosed between vertical concentric cylinders with the inner cylinder rotating with angular velocity Ω, is calculated. The calculation uses a perturbation procedure in which, for slow flows, the boundary conditions on the free surface are transformed to boundary conditions on the fixed horizontal surface obtained when Ω is zero. The results are tested experimentally.

Author's address:
R. Kaye
University of Manchester
Institute of Science and Technology
Sackville Street, P.O. Box 88, Manchester 1 (England)

Rheol. Acta **12**, 212–216 (1973)

From the New Mexico State University Mechanical Engineering Department Las Cruces, New Mexico, 88001 (U.S.A.)

Nonlinear motion equations for a Non-Newtonian incompressible fluid in an orthogonal coordinate system

By M. H. Cobble, P. R. Smith, and G. P. Mulholland

With 3 figures

(Received October 27, 1972)

Nomenclature

e_{ij}	strain rate tensor
$\bar{F}$	body force density, dynes/cm³
F_1, F_2, F_3	components of body force density, dynes/cm³
g	acceleration of gravity
H	function of time
h_1, h_2, h_3	metric coefficients
I_1, I_2, I_3	invariants
m	constant
P	pressure, dynes/cm²
r	radius, cm
t	time, sec
$\bar{v}$	velocity vector, cm/sec
v_1, v_2, v_3	velocities in the x_1, x_2, and x_3 directions, respectively, cm/sec
$v_n(t)$	velocity of the nth node, cm/sec
x_1, x_2, x_3	coordinate directions
z	coordinate, cm
δ	unit tensor
δ_{ij}	*Kronecker* delta
Δ_{ij}	$2\,e_{ij}$
∇	nabla
ε_{ijk}	alternating unit tensor
η	non-*Newton*ian viscosity, dynes/cm²
η_0, η_1	constant viscosities, dynes sec/cm², dynes sec^m/cm²
θ	angle, radians
ν_0, ν_1	constant kinematic viscosities, cm²/sec, cm² sec^{m-2}
ϱ	density, g/cm³
σ_{ij}	stress tensor
Φ	fluid dilation

Introduction

Much interest has developed in recent years in the flow of non-*Newton*ian fluids and many excellent articles and books have appeared upon this subject (1, 2, 3, 4). However, most of the solutions which have been published have been for steady flows. In this paper the nonlinear equations of motion in general orthogonal coordinates are developed for the non-steady flow of an incompressible non-*Newton*ian fluid. A solution is presented for the non-steady *Hagen-Poiseuille* flow through a circular pipe.

Theory

The equations of motion can be written in vector form, in terms of stress as

$$\frac{\varrho D\bar{v}}{Dt} = \bar{F} + \nabla \cdot \sigma \qquad [1]$$

where the term $\Delta \cdot \sigma$ represents the divergence of the stress tensor. Eq. [1] is written in orthogonal coordinates for the three components as (1)

x_1 *component:*

$$\frac{\varrho}{g}\left[\frac{\partial v_1}{\partial t} + \frac{v_1}{h_1}\frac{\partial v_1}{\partial x_1} + \frac{v_2}{h_2}\frac{\partial v_1}{\partial x_2} + \frac{v_3}{h_3}\frac{\partial v_1}{\partial x_3} - v_2\left(\frac{v_2}{h_2 h_1}\frac{\partial h_2}{\partial x_1} - \frac{v_1}{h_1 h_2}\frac{\partial h_1}{\partial x_2}\right) + v_3\left(\frac{v_1}{h_1 h_3}\frac{\partial h_1}{\partial x_3} - \frac{v_3}{h_3 h_1}\frac{\partial h_3}{\partial x_1}\right)\right]$$
$$= \frac{1}{h_1 h_2 h_3}\left[\frac{\partial(h_2 h_3 \sigma_{11})}{\partial x_1} + \frac{\partial(h_3 h_1 \sigma_{21})}{\partial x_2} + \frac{\partial(h_1 h_2 \sigma_{31})}{\partial x_3}\right] + \frac{\sigma_{12}}{h_1 h_2}\frac{\partial h_1}{\partial x_2} + \frac{\sigma_{31}}{h_1 h_3}\frac{\partial h_1}{\partial x_3} - \frac{\sigma_{22}}{h_1 h_2}\frac{\partial h_2}{\partial x_1} - \frac{\sigma_{33}}{h_1 h_3}\frac{\partial h_3}{\partial x_1} + F_1. \qquad [2a]$$

x_2 *component:*

$$\frac{\varrho}{g}\left[\frac{\partial v_2}{\partial t} + \frac{v_1}{h_1}\frac{\partial v_2}{\partial x_1} + \frac{v_2}{h_2}\frac{\partial v_2}{\partial x_2} + \frac{v_3}{h_3}\frac{\partial v_2}{\partial x_3} - v_3\left(\frac{v_3}{h_3 h_2}\frac{\partial h_3}{\partial x_2} - \frac{v_2}{h_2 h_3}\frac{\partial h_2}{\partial x_3}\right) + v_1\left(\frac{v_2}{h_2 h_1}\frac{\partial h_2}{\partial x_1} - \frac{v_1}{h_1 h_2}\frac{\partial h_1}{\partial x_2}\right)\right]$$
$$= \frac{1}{h_1 h_2 h_3}\left[\frac{\partial(h_2 h_3 \sigma_{12})}{\partial x_1} + \frac{\partial(h_3 h_1 \sigma_{22})}{\partial x_2}\right.$$

$$+\frac{\partial(h_1h_2\sigma_{32})}{\partial x_2}\Bigg]+\frac{\sigma_{23}}{h_1h_3}\frac{\partial h_2}{\partial x_3}+\frac{\sigma_{12}}{h_2h_1}\frac{\partial h_2}{\partial x_1} -\frac{\sigma_{33}}{h_2h_3}\frac{\partial h_3}{\partial x_2}-\frac{\sigma_{11}}{h_2h_1}\frac{\partial h_1}{\partial x_2}+F_2. \quad [2b]$$

x_3 component:

$$\frac{\partial}{g}\Bigg[\frac{\partial v_3}{\partial t}+\frac{v_1}{h_1}\frac{\partial v_3}{\partial x_1}+\frac{v_2}{h_2}\frac{\partial v_3}{\partial x_2}+\frac{v_3}{h_3}\frac{\partial v_3}{\partial x_3} -v_1\left(\frac{v_1}{h_1h_3}\frac{\partial h_1}{\partial x_3}-\frac{v_3}{h_3h_1}\frac{\partial h_3}{\partial x_1}\right) +v_2\left(\frac{v_3}{h_3h_2}\frac{\partial h_3}{\partial x_2}-\frac{v_2}{h_2h_3}\frac{\partial h_2}{\partial x_3}\right)\Bigg]$$
$$=\frac{1}{h_1h_2h_3}\Bigg[\frac{\partial(h_2h_3\sigma_{13})}{\partial x_1}+\frac{\partial(h_3h_1\sigma_{23})}{\partial x_2} +\frac{\partial(h_1h_2\sigma_{33})}{\partial x_3}\Bigg]+\frac{\sigma_{31}}{h_1h_3}\frac{\partial h_3}{\partial x_1}+\frac{\sigma_{23}}{h_3h_1}\frac{\partial h_3}{\partial x_2} -\frac{\sigma_{11}}{h_3h_1}\frac{\partial h_1}{\partial x_3}-\frac{\sigma_{22}}{h_3h_2}\frac{\partial h_2}{\partial x_3}+F_3. \quad [2c]$$

For certain types of non-*Newton*ian fluids

$$=\sigma_{ij}--P\delta_{ij}+2\eta e_{ij}=-P\delta_{ij}+\eta\Delta_{ij} \quad [3]$$

where

$$e_{ij}=\tfrac{1}{2}\Delta_{ij} \quad [4]$$

and for orthogonal curvilinear coordinate systems,

$$\sigma_{11}=-P+2\eta e_{11}$$
$$\sigma_{22}=-P+2\eta e_{22}$$
$$\sigma_{33}=-P+2\eta e_{33}$$
$$\sigma_{12}=\sigma_{21}=2\eta e_{12}=2\eta e_{21}$$
$$\sigma_{13}=\sigma_{31}=2\eta e_{13}=2\eta e_{31}$$
$$\sigma_{23}=\sigma_{32}=2\eta e_{23}=2\eta e_{32}$$

and in the above equations, η the non-*Newton*ian viscosity is a function of e (or Δ).

In order for η to be a scalar function of the tensor e (or Δ), η must depend on the invariants of e (or Δ) that transform as scalars with respect to orthogonal transformations of coordinates, and for this type of viscosity, we can write (2)

$$\eta=\eta(I_1,I_2,I_3) \quad [6a]$$

where I_1, I_2, and I_3 are the invariants, and further

$$I_1=(\Delta:\delta)=2\,(e:\delta) =2\sum_i e_{ii}=2(e_{11}+e_{22}+e_{33}) \quad [7]$$

$$I_2=(\Delta:\Delta)=4(e:e)=4\sum_i\sum_j e_{ij}e_{ji} =4[e_{11}^2+e_{22}^2+e_{33}^2+2(e_{12}^2+e_{13}^2+e_{23}^2)] \quad [8]$$

and

$$I_3=\det\Delta=8\det e=8\begin{vmatrix}e_{11}&e_{12}&e_{13}\\e_{21}&e_{22}&e_{23}\\e_{31}&e_{32}&e_{33}\end{vmatrix} =8\sum_i\sum_j\sum_k\varepsilon_{ijk}e_{1i}e_{2j}e_{3k}$$
$$=8[e_{11}(e_{22}e_{33}-e_{32}e_{23}) -e_{12}(e_{21}e_{33}-e_{31}e_{23}) +e_{13}(e_{21}e_{32}-e_{31}e_{22})]. \quad [9]$$

Additionally, the fluid dilation Φ, is defined as

$$\varphi=e_{11}+e_{22}+e_{33}=\nabla\cdot\bar{v} \quad [10]$$

and is equal to the divergence of the velocity. For an incompressible fluid $\Delta\cdot\bar{v}=0$, and so $I_1=0$. Thus we may write for a non-*Newton*ian fluid of this type,

$$\eta=\eta(I_2,I_3). \quad [6b]$$

Using the equations in [5] in eqs. [2a, b, c] for a non-*Newton*ian incompressible fluid of the type shown in eq. [6b], we get the following motion equations for a non-*Newton*ian fluid in an orthogonal coordinate system,

x_1 component:

$$\frac{\varrho}{g}\Bigg[\frac{\partial v_1}{\partial t}+\frac{v_1}{h_1}\frac{\partial v_1}{\partial x_1}+\frac{v_2}{h_2}\frac{\partial v_1}{\partial x_2}+\frac{v_3}{h_3}\frac{\partial v_1}{\partial x_3} -v_2\left(\frac{v_2}{h_2h_1}\frac{\partial h_2}{\partial x_1}-\frac{v_1}{h_1h_2}\frac{\partial h_1}{\partial x_2}\right) +v_3\left(\frac{v_1}{h_1h_3}\frac{\partial h_1}{\partial x_3}-\frac{v_3}{h_3h_1}\frac{\partial h_3}{\partial x_1}\right)\Bigg]$$
$$=F_1-\frac{1}{h_1}\frac{\partial P}{\partial x_1}+\frac{2}{h_1h_2h_3} \times\Bigg\{\eta\Bigg[\frac{h_2h_3\partial e_1}{\partial x_1}+\frac{h_3h_1\partial e_{21}}{\partial x_2} +\frac{h_1h_2\partial e_{31}}{\partial x_2}+\left(\frac{2h_3\partial h_1}{\partial x_2}+\frac{h_1\partial h_3}{\partial x_2}\right)e_{21} +\left(\frac{h_1\partial h_2}{\partial x_3}+\frac{2h_2\partial h_1}{\partial x_3}\right)e_{31} +\frac{h_2\partial h_3}{\partial x_1}(e_{11}-e_{33}) -\frac{h_3\partial h_2}{\partial x_1}(e_{11}-e_{22})\Bigg]+\frac{h_2h_3e_{11}\partial\eta}{\partial x_1} +\frac{h_3h_1e_{21}\partial\eta}{\partial x_2}+\frac{h_1h_2e_{31}\partial\eta}{\partial x_3}\Bigg\}. \quad [11a]$$

x_2 component:

$$\frac{\varrho}{g}\left[\frac{\partial v_2}{\partial t}+\frac{v_1}{h_1}\frac{\partial v_2}{\partial x_1}+\frac{v_2}{h_2}\frac{\partial v_2}{\partial x_2}+\frac{v_3}{h_3}\frac{\partial v_2}{\partial x_3}\right.$$
$$-v_3\left(\frac{v_3}{h_3h_2}\frac{\partial h_3}{\partial x_2}-\frac{v_2}{h_2h_3}\frac{\partial h_2}{\partial x_3}\right)$$
$$\left.+v_1\left(\frac{v_2}{h_2h_1}\frac{\partial h_2}{\partial x_1}-\frac{v_1}{h_1h_2}\frac{\partial h_1}{\partial x_2}\right)\right]$$
$$=F_2-\frac{1}{h_2}\frac{\partial P}{\partial x_2}+\frac{2}{h_1h_2h_3}$$
$$\times\left\{\eta\left[\frac{h_2h_3\partial e_{12}}{\partial x_1}+\frac{h_3h_1\partial e_{22}}{\partial x_2}\right.\right.$$
$$+\frac{h_1h_2\partial e_{32}}{\partial x_3}$$
$$+\left(\frac{h_2\partial h_3}{\partial x_1}+\frac{2h_3\partial h_2}{\partial x_1}\right)e_{12}$$
$$+\left(\frac{2h_1\partial h_2}{\partial x_3}+\frac{h_2\partial h_1}{\partial x_3}\right)e_{32}$$
$$\frac{h_3\partial h_1}{\partial x_2}(e_{22}-e_{11})$$
$$\left.+\frac{h_1\partial h_3}{\partial x_2}(e_{22}-e_{33})\right]$$
$$+\frac{h_2h_3e_{12}\partial\eta}{\partial x_1}$$
$$\left.+\frac{h_3h_1e_{22}\partial\eta}{\partial x_2}+\frac{h_1h_2e_{32}\partial\eta}{\partial x_3}\right\}. \quad [11b]$$

x_3 component:

$$\frac{\varrho}{g}\left[\frac{\partial v_3}{\partial t}+\frac{v_1}{h_1}\frac{\partial v_3}{\partial x_1}+\frac{v_2}{h_2}\frac{\partial v_3}{\partial x_2}+\frac{v_3}{h_3}\frac{\partial v_3}{\partial x_3}\right.$$
$$-v_1\left(\frac{v_1}{h_1h_3}\frac{\partial h_1}{\partial x_3}-\frac{v_3}{h_3h_1}\frac{\partial h_3}{\partial x_1}\right)$$
$$\left.+v_2\left(\frac{v_3}{h_3h_2}\frac{\partial h_3}{\partial x_2}-\frac{v_2}{h_2h_3}\frac{\partial h_2}{\partial x_3}\right)\right]$$
$$=F_3-\frac{1}{h_3}\frac{\partial P}{\partial x_3}+\frac{2}{h_1h_2h_3}$$
$$\times\left\{\eta\left[\frac{h_2h_3\partial e_{13}}{\partial x_1}+\frac{h_3h_1\partial e_{23}}{\partial x_2}\right.\right.$$
$$+\frac{h_1h_2\partial e_{33}}{\partial x_3}+\left(\frac{2h_2\partial h_3}{\partial x_1}+\frac{h_3\partial h_2}{\partial x_1}\right)e_{13}$$
$$+\left(\frac{h_3\partial h_1}{\partial x_2}+\frac{2h_1\partial h_3}{\partial x_2}\right)e_{23}$$
$$+\frac{h_2\partial h_1}{\partial x_3}(e_{33}-e_{22})$$
$$\left.+\frac{h_2\partial h_1}{\partial x_3}(e_{33}-e_{11})\right]+\frac{h_2h_3e_{13}\partial\eta}{\partial x_1}$$
$$\left.+\frac{h_3h_1e_{23}\partial\eta}{\partial x_2}+\frac{h_1h_2e_{33}\partial\eta}{\partial x_3}\right\}. \quad [11c]$$

In eqs. [7–11a, b, c], the strain rate relationships for an orthogonal coordinate system are

$$e_{11}=\frac{1}{h_1}\frac{\partial v_1}{\partial x_1}+\frac{v_2}{h_1h_2}\frac{\partial h_1}{\partial x_2}+\frac{v_3}{h_3h_1}\frac{\partial h_1}{\partial x_3}$$
$$e_{22}=\frac{1}{h_2}\frac{\partial v_2}{\partial x_2}+\frac{v_3}{h_2h_3}\frac{\partial h_2}{\partial x_3}+\frac{v_1}{h_1h_2}\frac{\partial h_2}{\partial x_1}$$
$$e_{33}=\frac{1}{h_3}\frac{\partial v_2}{\partial x_3}+\frac{v_1}{h_3h_1}\frac{\partial h_3}{\partial x_1}+\frac{v_2}{h_2h_3}\frac{\partial h_3}{\partial x_2}$$
$$e_{12}=e_{21}=\frac{1}{2}\left[\frac{h_2}{h_1}\frac{\partial}{\partial x_1}\left(\frac{v_2}{h_2}\right)+\frac{h_1}{h_2}\frac{\partial}{\partial x_2}\left(\frac{v_1}{h_1}\right)\right]$$
$$e_{13}=e_{31}=\frac{1}{2}\left[\frac{h_1}{h_3}\frac{\partial}{\partial x_3}\left(\frac{v_1}{h_1}\right)+\frac{h_3}{h_1}\frac{\partial}{\partial x_1}\left(\frac{v_3}{h_1}\right)\right]$$
$$e_{23}=e_{32}=\frac{1}{2}\left[\frac{h_3}{h_2}\frac{\partial}{\partial x_2}\left(\frac{v_3}{h_3}\right)+\frac{h_2}{h_3}\frac{\partial}{\partial x_3}\left(\frac{v_2}{h_2}\right)\right] \quad [12]$$

Using eq. [12], the motion equations for an incompressible, non-*Newton*ian fluid in any orthogonal coordinate system, are given in terms of the velocities, metric coefficients, body force, and pressure gradients, and the non-*Newton*ian viscosity, and their derivatives, and where the non-*Newton*ian viscosity η is a function of the invariants I_2 and I_3. The invariants I_2 and I_3 are found using eq. [12] in eq. [8] and [9] respectively for any orthogonal coordinate system. These non-*Newton*ian fluids are called time independent fluids. Time independent fluids are those for which the rates of shear at a given point is solely dependent upon the instantaneous shear stress at that point (3), and are members of the visco-inelastic fluids mentioned by *Rivlin* (4). Typical non-*Newton*ian fluids of this type are *Ostwald-de Waele*, *Herschel-Bulkley*, *Sisko*, *Reiner-Philippoff*.

Example

Commencement of *Hagen-Poiseuille* flow in a circular pipe for a non-*Newton*ian fluid.

If one assumes that

1. $x_1=r,\quad x_2=\theta,\quad x_3=z$

   (cylindrical coordinate system)

2. $v_r=v_\theta=0$
3. $F_r=F_\theta=F_z=0$
4. $v_3=v_z=v_z(r,t)=v(r,t)$

then the motion eqs. [11 a, b, c] reduce to the single equation,

$$\frac{\varrho}{g}\frac{\partial v}{\partial t} = -\frac{\partial P}{\partial z} + \eta\left[\frac{\partial^2 v}{\partial r^2} + \frac{1}{r}\frac{\partial v}{\partial r}\right] + \frac{\partial \eta}{\partial r}\frac{\partial v}{\partial r}. \quad [13]$$

The boundary and initial conditions for $v(r, t)$ are

1. $\dfrac{\partial v(0, t)}{\partial r} = 0$
2. $v(r_0, t) = 0$
3. $v(r, 0) = 0.$

It is further assumed that

$$\frac{-\partial P}{\partial z} = H(t) = \text{constant} \quad [14]$$

and the non-*Newton*ian fluid is to be of the *Ostwald-de Waele-Sisko* type where

$$\eta = \eta_0 + \eta_1 \left|\sqrt{\frac{I_2}{2}}\right|^{m-1}$$
$$= \eta_0 + \eta_1 \left|\sqrt{2(e : e)}\right|^{m-1}, \quad m \geq 1. \quad [15]$$

Now

$$2(e : e) = \left(\frac{\partial v}{\partial r}\right)^2 \quad [16]$$

so

$$\eta = \eta_0 + \eta_1 \left|\frac{\partial v}{\partial r}\right|^{m-1}. \quad [17]$$

The motion eq. [13], can now be written as

$$\frac{\partial v}{\partial t} = -\frac{g}{\varrho}\frac{\partial P}{\partial z} + \nu_0\left\{\left[1 + \frac{\nu_1}{\nu_0}\left|\frac{\partial v}{\partial r}\right|^{m-1}\right]\right.$$
$$\times\left[\frac{\partial^2 v}{\partial r^2} + \frac{1}{r}\frac{\partial v}{\partial r}\right]$$
$$\left. + (m-1)\frac{\nu_1}{\nu_0}\left|\frac{\partial v}{\partial r}\right|^{m-1}\frac{\partial^2 v}{\partial r^2}\right\}. \quad [18]$$

Using the discrete-space, continuous-time (DSCT) method of approximation, eq. [18] can be written for the internal nodes as,

$$\frac{dv_n(t)}{dt} = -\frac{g}{\varrho}\frac{\partial P}{\partial z} + \frac{\nu_0}{\Delta r^2}$$
$$\times\left\{\left[1 + \frac{\nu_1}{\nu_0}\left|\frac{v_{n+1}(t) - v_{n-1}(t)}{2\Delta r}\right|^{m-1}\right]\right.\left[\vphantom{\frac{1}{2}}\right.$$
$$\times\left[\left(1 + \frac{1}{2n}\right)v_{n+1}(t) + \left(1 - \frac{1}{2n}\right)v_{n-1}(t)\right.$$
$$\left. - 2v_n(t)\right]$$
$$+ (m-1)\frac{\nu_1}{\nu_0}\left|\frac{v_{n+1}(t) - v_{n-1}(t)}{2\Delta r}\right|^{m-1}$$
$$\left.\times\left[v_{n+1}(t) + v_{n-1}(t) - 2v_n(t)\right]\right\}$$
$$n = 1, 2, 3, \ldots, N-1. \quad [19]$$

From the boundary conditions, one obtains

$$\frac{dv_0(t)}{dt} = -\frac{g}{\varrho}\frac{\partial P}{\partial z} + \frac{4\nu_0}{\Delta r^2}\left[v_1(t) - v_0(t)\right] \quad [20]$$

and

$$v_N(t) = 0. \quad [21]$$

The initial conditions are,

$$v_n(0) = 0 \qquad n = 0, 1, 2, \ldots, N. \quad [22]$$

Eqs. [19] and [20] can now be solved on the analog computer, and the velocities $v_n(t)$ at the $N+1$ nodes can be determined for any time.

Fig. 1, 2, and 3 present the start-up velocity profiles for *Ostwald-de Waele-Sisko* fluids having $m = 3$, $\nu_0 = 1\ \text{cm}^2/\text{sec}$, and ν_1 equal to $1\ \text{cm}^1$ sec, $5\ \text{cm}^2$ sec, and $10\ \text{cm}^2$ sec, respectively. For each of these flows the negative pressure gradient over density H was $1\ \text{cm/sec}^2$. Notice that an

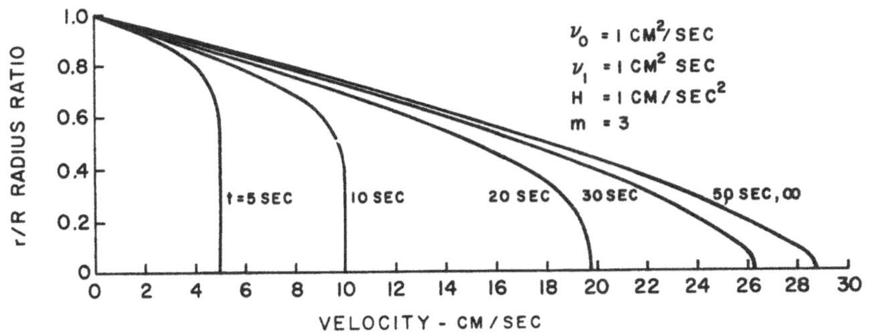

Fig. 1. Start-up flow velocity profiles, $\nu_1 = 1\ \text{cm}^2$ sec

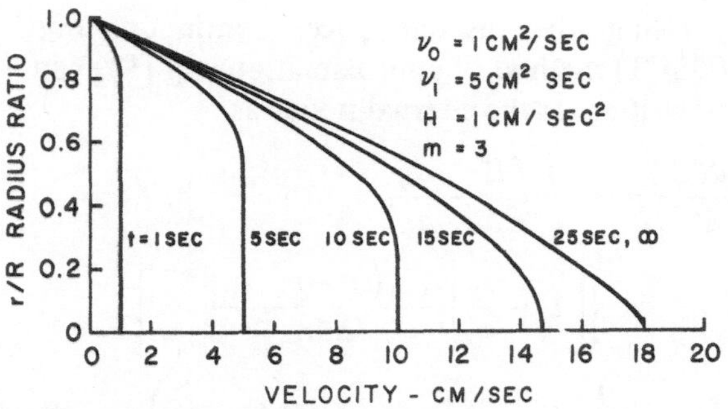

Fig. 2. Start-up flow velocity profiles, $\nu_1 = 5$ cm² sec

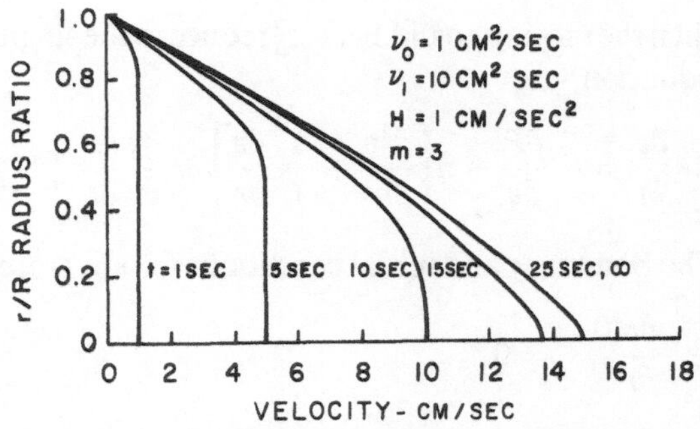

Fig. 3. Start-up flow velocity profiles, $\nu_1 = 10$ cm² sec

increase in the second coefficient of kinematic viscosity ν_1 causes a decrease in the total flow rate when all other parameters are held constant.

Conclusion

The equations of motion for the non-steady flow of incompressible, non-*Newton*ian fluids were derived and put in terms of the scalar velocities and the metric coefficients of the general orthogonal coordinate system. This form of the equations proves to be very useful as was demonstrated by solving an example problem, the commencement of flow through a cylindrical tube, using a discrete-continuous technique.

Summary

Starting with an assumed relationship between the stress tensor, the non-*Newton*ian viscosity, and the strain rate tensor, the nonlinear equations of motion are developed for use in any orthogonal coordinate system. The resulting equations are written in terms of the scalar velocities, the non-*Newton*ian viscosity, the metric coefficients, and their derivatives.

The non-*Newton*ian viscosity is assumed to be a scalar function of the strain rate tensor, and so depends upon the invariants of the strain rate tensor. For convenience, the necessary invariants are written out in complete form for use in any orthogonal coordinate system, in terms of the scalar velocities, the metric coefficients, and their derivatives.

Using the resulting motion equations and a model of this type of viscosity, the *Ostwald-de Waele* model, an example of time dependent flow is solved using a continuous time-discrete space method programmed on an analog computer.

References

1) *Hughes, W. F.* and *E. W. Gaylord*, Basic Equations of Engineering Science, p. 9 (1964).

2) *Bird, R. B., W. E. Stewart*, and *E. N. Lightfoot*, Transport Phenomena, Second Printing, pp. 101–103 (London-New York 1962).

3) *Skelland, A. H. P.*, Non-Newtonian Flow and Heat Transfer, p. 4 (London-New York 1966).

4) *Rivlin, R. S.*, Proc. Ray. Soc. A **193**, 261 (London 1948).

Authors' address:
M. H. Cobble, P. R. Smith, and *G. P. Mulholland*
New Mexico State University
Mechanical Engineering Department
Las Cruces, New Mexico 88001 (U.S.A.)

Rheol. Acta **12**, 217–223 (1973)

From the National Physical Laboratory Teddington, Middlesex (England)

Stability of plane Poiseuille flow of slightly viscoelastic fluids

By R. R. Cousins

With 6 figures

(Received October *27, 1972*)

A continuum approach to drag reduction by additives

In many practical situations involving fluid flow, turbulence or eddying occurs near solid surfaces, and resistance to motion is largely associated with this turbulence. Large reductions in turbulent frictional resistance, for example, of the pressure drop in pipe flows, can sometimes be achieved by dissolving small quantities of certain substances in the liquid. The first clear scientific description of the phenomenon which now bears his name was given by *Toms* (1949). The additive substances usually have very high molecular weights, of the order of 10^6, and are effective in concentrations of the order of ten to one hundred parts per million by weight. In these very low concentrations the solution has a viscosity, μ, which is independent of shear rate and indistinguishable from that of the solvent alone. Here, then, are fluids which are essentially the same as the solvent (usually water) in their values of density and viscosity, which are normally regarded as the relevant fluid parameters, yet they behave in a radically different way.

It is convenient to introduce a *Reynolds* number, $\varrho U L/\mu$, where ϱ is the density of the fluid and U and L are typical values of the speed and length of the flow situation. When a dilute polymer solution is made to pass through a sufficiently large pipe the curve of friction coefficient, a non-dimensional parameter proportional to the pressure drop down the pipe, against *Reynolds* number follows the curve for water until a certain threshold value of the shear stress at the wall is reached, after which drag reduction occurs (*Gadd*, 1966). A possible explanation of the threshold shear stress is provided by theories (*Ericksen*, 1962, *and* Tulin, 1966) which predict an elongation and orientation of the molecules for shear rates greater than some critical value. These theories imply that molecular elongation is the essential requirement for reduction of turbulent drag. The phenomenon of degradation, to which some polymer solutions are susceptible when subjected to continued shearing action, also supports this theory. For example, polyethylene oxide loses its effectiveness (*Gadd*, 1965), and it is observed that the molecular chains are broken up. On the other hand guar gum shows very little mechanical degradation, probably because any broken molecules rapidly reform, and the solution maintains its effectiveness.

If molecular elongation, or something equivalent to it, is the factor common to all fluids which reduce drag, *Gadd* (1966) has suggested that a primary mechanism may be a thickening on the laminar sublayer, a thin region close to the wall where the flow is essentially smooth or laminar. The presence of the surface suppresses turbulent eddies near to it. Any elongated molecular filaments would tend to become aligned in the flow direction due to the action of the high shear. We shall show that, in contrast to a *Newton*ian fluid, longitudinal vorticity may be established close to the wall in a non-*Newton*ian solution. The molecules will be "unwrapped" in a helical manner and have a resulting elongation in the direction of the main flow. The theory only relates to conditions at the breakdown of laminar flow, but it does suggest a mechanism which *may* result in a different turbulent structure in the important region close to the wall. Instead of a molecular description we could employ the continuum concept of normal stress differences to provide forces which would oppose transverse motions. Normal stress differences have been obserbed in drag reducing solutions by *Metzner* and *Park* (1964) who found that drag reduction increases with the ratio of a normal stress difference to shear stress. We shall use a continuum approach throughout this paper and work mainly with a constitutive equation that models fluids exhibiting normal stress effects.

Linear stability analysis for a second-order fluid

The second-order *Rivlin-Ericksen* fluid is a simple example of a fluid that exhibits normal stress effects. Its constitutive relation can be written, in non-dimensional form,

$$s = \frac{1}{R} A_1 + \lambda A_2 + \mu A_1^2, \qquad [1]$$

where s is the stress additional to that produced by a hydrostatic pressure p, R is the *Reynolds* number, λ and μ are viscoelastic parameters, and the kinematic matrices A_1 and A_2 are defined

$$A_1 = V + V^T,$$

$$A_2 = \frac{\partial A_1}{\partial t} + v_k \frac{\partial A_1}{\partial x_k} + A_1 V + (A_1 V)^T, \qquad [2]$$

where T denotes a transposed matrix,

$$V_{ij} = \partial v_i / \partial x_j \qquad [3]$$

and v_i are velocity components referred to *Cartes*ian coordinates x_j. The momentum equation is

$$-\frac{\partial p}{\partial x_i} + \frac{\partial s_{ij}}{\partial x_j} = \frac{\partial v_i}{\partial t} + v_k \frac{\partial v_i}{\partial x_k}, \qquad [4]$$

and the continuity equation for an incompressible fluid is

$$\partial v_k / \partial x_k = 0. \qquad [5]$$

We consider velocity fields of the form

$$v_1 = U(x_2) + u_1(x_2) \exp\{i(\alpha x_1 + \beta x_3 - \alpha\, \mathrm{ct})\},$$

$$v_2 = u_2(x_2) \exp\{i(\alpha x_1 + \beta x_3 - \alpha\, \mathrm{ct})\},$$

$$v_3 = u_3(x_2) \exp\{i(\alpha x_1 + \beta x_3 - \alpha\, \mathrm{ct})\},$$

which represents a steady parallel flow U and a disturbance of the form required in stability analyses. For *Poiseuille* flow between parallel planes $x_2 = \pm 1$

$$U(x_2) = 1 - x_2^2. \qquad [7]$$

When c_i, the imaginary part of c is negative, the disturbance will grow exponentially in time and the flow is unstable. When c_i is positive the disturbance decays and the flow is stable, Neutral conditions are given by $c_i = 0$. The least value of R for which $c_i = 0$ is called the critical *Reynolds* number. Flows with R less than this are stable.

We assume that the disturbance is small and neglect products of u_i when substituting [6] in [1] and [4]. If we introduce the transformation due to *Squire* (1933):

$$\tilde{\alpha}^2 = \alpha^2 + \beta^2, \quad \tilde{\alpha}\tilde{u}_1 = \alpha u_1 + \beta u_3,$$

$$\tilde{\alpha}\tilde{R} = \alpha R \qquad [8]$$

the first order terms from [4] and [5] give, after eliminating pressure, a modified *Orr-Sommerfeld* equation

$$\{(D^2 - \tilde{\alpha}^2)^2 - i\tilde{\alpha}\tilde{R}(U - c)(D^2 - \tilde{\alpha}^2) + i\tilde{\alpha}\tilde{R}U'')u_2\}$$

$$= \lambda i\tilde{\alpha}\tilde{R}\{U'''' - (U - c)(D^2 - \tilde{\alpha}^2)^2\} u_2$$

$$+ \frac{\beta^2}{\alpha^2}(2\lambda + \mu) i\tilde{\alpha}\tilde{R}\{U'(D^2 - \tilde{\alpha}^2) - U'''\}$$

$$(u_2' + i\tilde{\alpha}^2 u_3/\beta), \qquad [9]$$

where D or a prime denote differentiation with respect to x_2. For a *Newton*ian fluid, with $\lambda = \mu = 0$, *Squire* (1933) showed that eqs. [4] and [5] for three-dimensional motion take, under [8], the same form as those for two-dimensional disturbances. The equivalent two-dimensional motion takes place at a lower *Reynolds* number, and consequently critical conditions for three-dimensional flow are given by two-dimensional analysis. *Squire*s theorem does not hold for a second-order fluid unless $2\lambda + \mu = 0$ (*Lockett*, 1969a) and a full three-dimensional analysis is required. As viscoelasticity is assumed to be small for dilute polymer solutions a solution to [9] with boundary conditions $u_2 = u_2' = 0$ at the walls was obtained by perturbation about the solution for a *Newton*ian fluid (*Cousins*, 1970).

A result of particular interest (*Cousins*, 1970) is that $u_3' \neq 0$ at $x_2 = 1$. For a *Newton*ian fluid the first unstable disturbance to appear is in-plane with $u_3 = \beta = 0$, but *Lockett* (1969b) has shown that under certain conditions, which will be determined below, the first unstable disturbance in a viscoelastic fluid may be out-of-plane. The significance of this will be seen if we examine vorticity of the flow, whose components are given by

$$\omega_1 = (u_3' - i\beta u_2) \exp\{i(\alpha x_1 + \beta x_3 - \alpha\, \mathrm{ct})\},$$

$$\omega_2 = i(\beta u_1 - \alpha u_3) \exp\{i(\alpha x_1 + \beta x_3 - \alpha\, \mathrm{ct})\}. \qquad [10]$$

$$\omega_3 = U' + (i\alpha u_2 - u_1') \exp\{i(\alpha x_1 + \beta x_3 - \alpha\, \mathrm{ct})\},$$

For the first growing disturbance in a *Newton*ian fluid ω_1 and ω_2 vanish, whereas if the first perturbation in a viscoelastic fluid is out-of-plane ω_1 and ω_2 are non-zero, and, since u_3' is non-zero at the walls, ω_1 is also non-zero there. Slight viscoelasticity may therefore introduce a longitudinal component of vorticity which persists in the region close to the wall; conditions under which this occurs are discussed below. This vorticity component may play an important role in the subsequent development to turbulence, and the region near the wall is believed to play a major part in this transition. *Lockett* (1969 b) has suggested this mechanism as a possible starting point for a theory to explain the drag-reduction properties of certain long-chain polymer solutions.

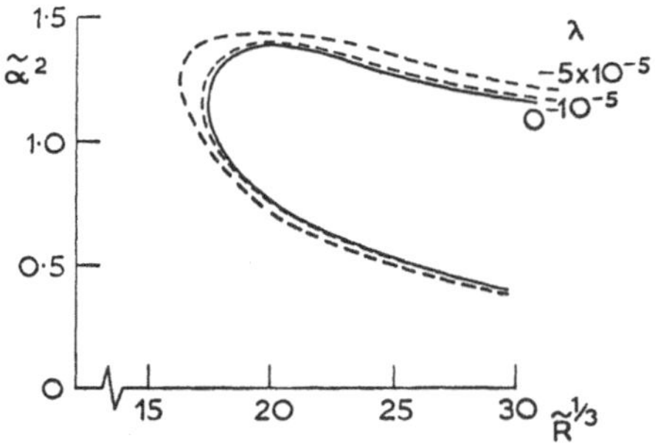

Fig. 1

Fig. 1 shows neutral stability curves ($c_i = 0$) for three different values of λ, when $\beta^2(2\lambda + \mu) = 0$. The solid line refers to a *Newton*ian fluid. It is known from thermodynamic considerations that λ is negative (see *Coleman* and *Markovitz*, 1964), and the limited experimental evidence that exists tends to confirm this result. The graphs show that for negative values of λ the presence of viscoelasticity in the fluid is destabilising in the sense that it reduces the critical *Reynolds* number, and these results (*Cousins*, 1970) agree with those of other workers (*Chan Man Fong* and *Walters*, 1965; *Mook*, 1967; *Jones* and *Walters*, 1968; *Jones*, 1967; *Schwarz* and *Chun*, 1968). The effects of perturbations corresponding to the term in $\beta^2(2\lambda + \mu)/\alpha^2$ are shown in fig. 2, where

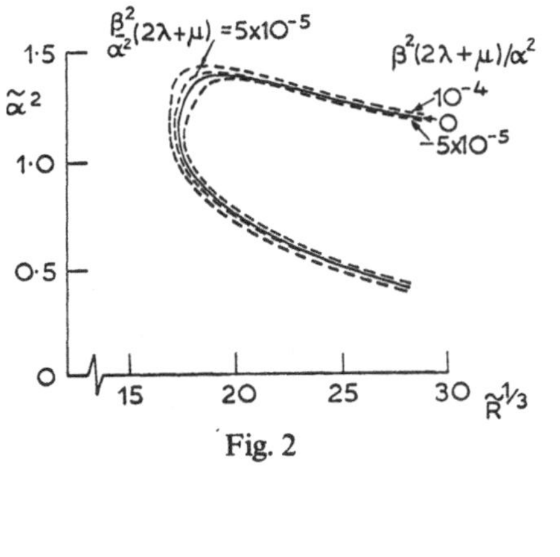

Fig. 2

the *Newton*ian case is again shown by a solid line. These curves are for constant values of β^2/α^2. A plot of neutral stability curves at constant β is obtained via [8] and shown in fig. 3. The parameter $2\lambda + \mu$ may be either positive or negative, the flow being destabilised for positive values or stabilised for negative values. An asymptotic method of solution was used to obtain the graphs in fig. 1 and 2 (*Cousins*, 1970). Although the approximations made in specifying u_2 are sufficiently good to determine whether the terms in λ and μ stabilise or destabilise the fluid, they do not enable us to determine the position of the neutral curve with any great accuracy. For instance, for $\lambda = \mu = 0$, the present results give a critical *Reynolds* number of about 5240, in contrast to the usually quoted figure of 5780 obtained by *Thomas* (1953). In the next section, where it is far more important to specify the eigenfunctions accurately, a direct numerical approach to [9] is used, and better results are obtained, namely a value of 5774.4 for R_c and a corresponding wave number of $\alpha = 1.02024$ (*Cousins*, 1972 b) which compare well with the results of *Porteous* and *Denn* (1971), who obtained $R_c = 5775, \alpha = 1.0206$.

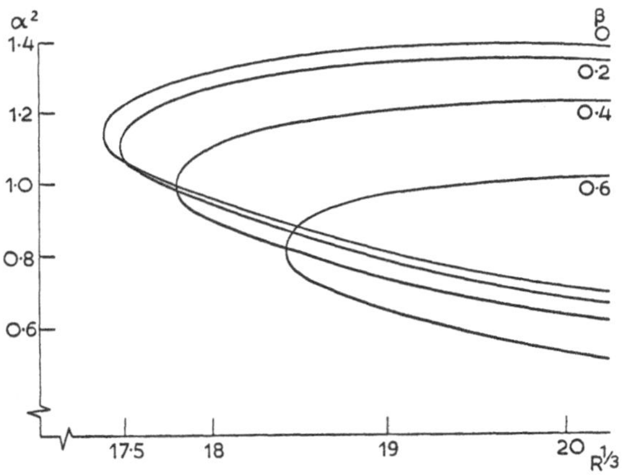

Fig. 3. Neutral stability curves for various values of β. $\lambda = 0$, $2\lambda + \mu = 5 \times 10^{-5}$

The investigation so far has been in terms of eigenvalues defined by transformations [8], but to determine how the value of β/α affects the critical *Reynolds* number we must work with α and R directly. The neutral stability curve becomes a surface if we use as axes α, β and R. We may cut this surface with a plane parallel to the αR-plane and obtain a neutral curve and consequently a critical *Reynolds* number corresponding to a particular value of β. Variation of R_c with β^2/α^2

is shown in fig. 4. The curve for a *Newton*ian fluid is labelled A, and the effect of terms corresponding to $\lambda = -10^{-6}$ is shown by B. We observe that R_c increases with β, showing that the least stable disturbance occurs when $\beta = 0$. We note also that the term in λ makes little difference to the slope of the curve, while perturbations corresponding to the $2\lambda + \mu$ term affect the gradient as indicated by C and D. For sufficiently large values of $2\lambda + \mu$ (>0.00106 approximately) the value of R_c decreases with β^2/α^2 as shown by E and F. The value 0.0106 is insensitive to variations in λ. Since we require the product $\beta^2(2\lambda + \mu)/\alpha^2$ rather than the combination of parameters, $2\lambda + \mu$, to be small for the method of perturbing the *Newton*ian solution to be valid, we may choose μ as large as necessary, but remaining consistent with the second-order model of a fluid, though its application will be restricted to small values of β^2/α^2. The quantity $2\lambda + \mu$ is a measure of the second normal stress difference.

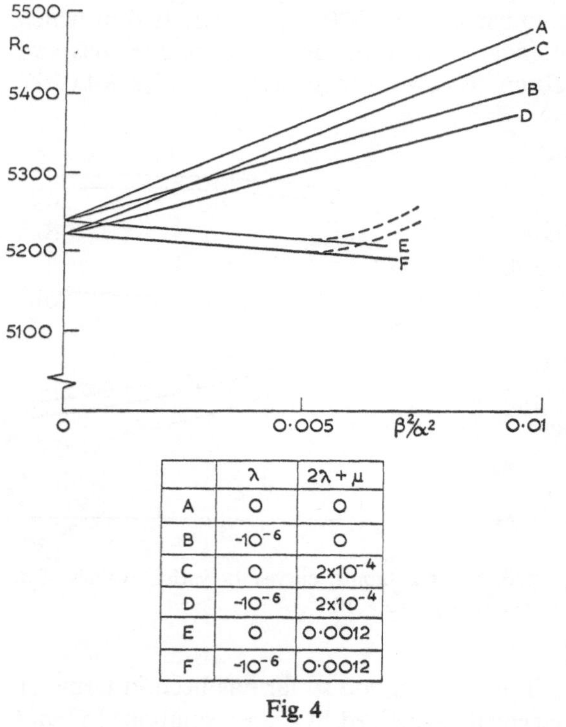

	λ	$2\lambda+\mu$
A	0	0
B	-10^{-6}	0
C	0	2×10^{-4}
D	-10^{-6}	2×10^{-4}
E	0	0·0012
F	-10^{-6}	0·0012

Fig. 4

We have assumed that slight viscoelasticity makes only a small change in R_c, and we expect curves E and F to remain close to A and B for every value of β/α. In particular $R_c \to \infty$ as $\beta^2/\alpha^2 \to \infty$ ($\alpha \to 0$) for both *Newton*ian and slightly viscoelastic fluids, so E and F will diverge from the straight lines shown in fig. 4 in a manner indicated by the broken curves, and there will be a minimum value of R_c at some value of $\beta \neq 0$ for viscoelastic fluids. This is in marked contrast to *Newton*ian fluids where the minimum value of R_c is given by $\beta = 0$. Linear perturbation theory leads to terms proportional to β^2, so R_c is linearly dependent on β^2/α^2 and hence there is no tendency for curves E and F to bend upwards. In the next section we shall examine non-linear effects, though the analysis does not lead to a critical direction along which the first (linearly) unstable wave would propagate. The first stage in performing a non-linear calculation is to solve the linear *Orr-Sommerfeld* equation.

Non-linear stability analysis for a second-order fluid

In the previous section conditions were obtained under which disturbances first become unstable. Linear theory predicts that such disturbances grow exponentially until they are sufficiently large for the transport of momentum by the finite fluctuations to be considerable, and the associated mean stress, the *Reynolds* stress, has an appreciable effect on the mean flow. This distortion of the mean flow modifies the rate of transfer of energy to the disturbance. As this energy transfer is the cause of the growth of the disturbance the rate of growth is itself modified. It is possible that there may exist an equilibrium state in which the rate of transfer of energy from the modified mean flow to the disturbance is exactly balanced by the rate of viscous dissipation of the energy of the disturbance. "Equilibrium" in this sense means that the oscillations have a steady finite amplitude.

The effect of the non-linear terms in the momentum equations is shown in three ways, [1] the generation of harmonics of the basic disturbance, [2] modification of the mean flow by the disturbance, an [3] modification of the fundamental. To see how these effects arise we shall consider for simplicity a disturbance whose x-dependence is $\exp(i\alpha x)$, where α is a wave number. In the non-linear terms the product of the fundamental with itself immediately introduces the harmonic proportional to $\exp(2i\alpha x)$, which in turn interacts with the fundamental and with itself to generate higher harmonics, and so on. In physical terms we are dealing with real quantities, so corresponding to $\exp(i\alpha x)$ we must introduce a term proportional to its complex conjugate $\exp(-i\alpha x)$.

The product of these expressions gives a term independent of x which represents the modification of the mean flow by the fundamental. The mean flow is modified by the harmonics and their conjugates in a similar way. The product of terms in $\exp(2i\alpha x)$ and $\exp(-i\alpha x)$ gives a term in $\exp(i\alpha x)$, and so represents a modification of the fundamental.

A *Fourier* analysis developed by *Stuart* (1960) and *Watson* (1960) has been extended to the interaction of three fundamental waveforms with each other and the main flow. The work is described fully elsewhere (*Cousins*, 1972b) and leads to a sequence of equations which can be solved successively. Computation of all the necessary functions is, however, prohibitively lengthy, and calculations have been restricted to the simpler case of a single two-dimensional disturbance. It is convenient to work with a stream function, θ, defined by

$$v_1 = \partial\theta/\partial x_2, \quad v_2 = -\partial\theta/\partial x_1, \qquad [11]$$

so that the continuity eq. [5] is satisfied identically. The momentum eq. [4] reduces to

$$\left(\frac{\partial}{\partial t} + \frac{\partial\theta}{\partial x_2}\frac{\partial}{\partial x_1} - \frac{\partial\theta}{\partial x_1}\frac{\partial}{\partial x_2}\right)(1 - \lambda\nabla^2)\nabla^2\theta - \frac{1}{R}\nabla^4\theta = 0, \qquad [12]$$

and we note that two-dimensional motions are independent of μ. We consider a disturbance proportional to $A(t)\exp(i\alpha x_1)$ and, following *Stuart* and *Watson*, expand the stream function in powers of $A(t)$:

$$\begin{aligned}\theta = \theta_0(x_2) &+ \theta_1(x_2)A(t)\exp(i\alpha x_1) \\ &+ \theta_2(x_2)\{A(t)\}^2\exp(2i\alpha x_1) \\ &+ f_1(x_2)|A(t)|^2 \\ &+ \theta_{11}(x_2)A(t)|A(t)|^2\exp(i\alpha x_1) \\ &+ \text{complex conjugates} + \cdots \qquad [13]\end{aligned}$$

where

$$\frac{dA}{dt} = A(a_0 + a_1|A|^2 + a_2|A|^4 + \cdots). \qquad [14]$$

For linear theory (first order in A) $a_0 = -i\alpha c$, leading to exponential growth or decay. If we substitute [13] and [14] in [12] and equate coefficients of powers of A we obtain equations to determine successively $\theta_0, \theta_1, \theta_2, f_1, \theta_{11}$, which represent the basic flow, the fundamental disturbance, the first harmonic to the disturbance, modification of the mean flow and modification of the fundamental. The process can in principle be continued indefinitely to determine further terms in the expansion [13], but *Watson* (1960) has shown for small, but nevertheless finite values of A, that it is self-consistent to ignore further terms in [13] and to truncate [16] after the term in a_1. The equation for θ_{11} involves the constant a_1, which may be determined using a function adjoint to θ_1. With a_1 determined [14] may be solved to give $|A(t)|$.

The numerical method and detailed results are given by *Cousins* (1972b). Some pertinent results are summarised in fig. 5 where curves on which a_{1r}, the real part of a_1, is zero are shown.

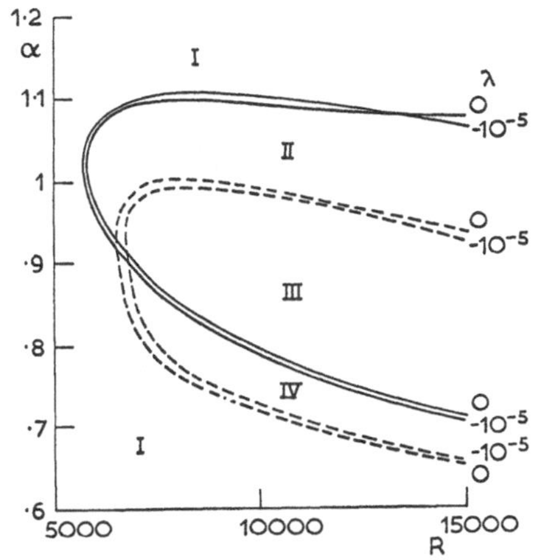

Fig. 5. ——— neutral stability curve ($c_i = 0$) for $\lambda = 0$, $-10^{-5}/a_{1r}$, ---- curve on which $= 0$ for $\lambda = 0, -10^{-5}$

Perturbations represented by points within the neutral stability curve (regions II and III) are unstable according to linear theory since $c_i > 0$ and the time dependence $\exp(i\alpha\,ct)$ indicates a growing disturbance. In region III, however, the non-linear theory of this chapter permits equilibrium states of finite amplitude given by

$$|A| = (\alpha c_i/-a_{1r})^{\frac{1}{2}}. \qquad [15]$$

In II no such equilibrium is reached, and disturbances remain unstable.

Points outside the neutral curve (I and IV) represent stable disturbances under linear theory, for c_i is negative, but non-linear effects result in instability in region I if the amplitude is greater than some finite value. We can estimate a reduction in the critical *Reynolds* number due to finite-amplitude disturbances. On the centre-line ($x_2 = 0$) the fluctuation intensity is, to order $|A|$

$$(\overline{V_2^2})^{\frac{1}{2}}/\bar{V}_1 = \sqrt{2}\,\alpha|A|, \qquad [16]$$

where the overbar denotes an average with respect to x_1. For a given value of the *Reynolds* number the minimum value of [16] is obtained. Disturbances with fluctuation intensities below this minimum decay, while those with greater intensities are unstable, so this minimum value, plotted in fig. 6, serves to define a relation between critical *Reynolds* number and the centre-line turbulence intensity present in a particular flow. Comparisons with results by other workers are shown.

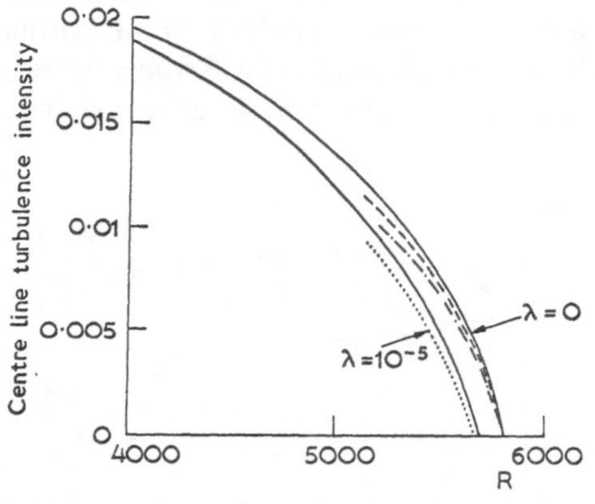

Fig. 6. Variation of critical *Reynolds* number with centre-line turbulence intensity for $\lambda = 0$ (—— this paper, ---- *Reynolds* and *Potter*, –·–·– *Porteous* and *Denn*) and $\lambda = -10^{-5}$ (—— this paper, ······· *Porteous* and *Denn*)

Effect of stress relaxation parameters on the stability of plane Poiseuille flow

Although the theory used previously does introduce non-*Newton*ian properties, especially normal stress effects, into the fluid, important features are neglected. The *Rivlin-Ericksen* description expresses stress in a fluid in terms of time derivatives of the rate of strain field evaluated at the time instant under consideration. The history of the motion is approximated, in a way essentially similar to a *Taylor* series expansion, by sufficient derivatives at one particular time, and there is no *explicit* use of the history in the final specification. Consequently properties such as stress relaxation are omitted. In this section some of the effects of stress relaxation are summarised (see *Cousins*, 1972a).

We take as a constitutive relation the form

$$s_{ij} = -p\delta_{ij} + \int_{-\infty}^{t} \varphi(t-\tau)\,[C_t(\tau)]_{ij}\,d\tau + \cdots, \qquad [17]$$

where C_t is the right *Cauchy-Green* tensor relative to time t. By taking the next term, a double integral, and using derivatives of the *Dirac* δ-function as kernels of the integrals the second-order fluid relation may be derived from [17]. However, we shall take

$$\varphi(s) = R^{-1}\delta'(s) + S\exp(-ks), \qquad [18]$$

the first term alone giving a *Newton*ian fluid and the second term introducing stress relaxation. For the flow field [6] we assume that the disturbance is small and determine $C_t(\tau)$ from the equation for particle paths. Assuming $k \gg 1$, *Squires* theorem is valid, and we obtain after much algebra the modified *Orr-Sommerfeld* equation

$$i\alpha(U-c)(D^2-\alpha^2-\beta^2)u_2 - i\alpha U'' u_2 - \left(\frac{1}{R}-\frac{S}{k^2}\right)(D^2-\alpha^2-\beta^2)^2 u_2 = 0. \qquad [19]$$

This is identical in form with the equation governing stability for a *Newton*ian fluid and we deduce that the critical *Reynolds* number R_c is given by

$$\frac{1}{R_c} - \frac{S}{k^2} = \frac{1}{5774}, \qquad [20]$$

the numerical value coming from the solution of the *Orr-Sommerfeld* equation for a *Newton*ian fluid. Since S is necessarily positive we see that $R_c < 5774$ and so the stress relaxation term in [18] is destabilising.

Summary

The title subject has been examined by the author in a series of papers (*Cousins*, 1970, 1972a, b), and the assumptions and principal results of those papers are discussed here. The work is motivated by the phenomenon evinced in fluid flow situations, of turbulent drag reduction by certain polymer additives. From a survey of experimental work it is clear that molecular elongation plays an important role in reducing drag by suppressing transverse motions. This effect may be interpreted as a normal stress effect in a continuum theory. A second-order fluid, which is a simple model exhibiting such a property, is used in a linear analysis of disturbances to plane *Poiseuille* flow. Unlike the *Newton*ian case Squire's theorem is not valid (*Lockett*, 1969a) and a three-dimensional analysis is required. The viscoelastic terms are in general destabilising. Under certain conditions the first growing disturbance will propagate at an angle to the basic flow, giving a longitudinal vortex structure close to the channel boundaries not present at the onset of instability in a *Newton*ian fluid. The analysis is extended to finite-amplitude disturbances by introducing a time-dependent amplitude, but calculations are here confined to the simpler two-dimensional case. Disturbances which would decay under linear theory may in fact grow provided the initial amplitude is sufficiently large. A threshold amplitude for instability is found as a function of *Reynolds* number. The viscoelastic terms are again found to be destabilising. Finally, a further viscoelastic property, that of stress relaxation, is introduced through an integral representation of the stress. A linear analysis is developed and stress relaxation is also shown to be a destabilising influence.

References

Chan Man Fong, C. F. and *K. Walters*, J. Mec. **4**, 439 (1965).

Coleman, B. D. and *H. Markovitz*, J. Appl. Phys. **35**, 1 (1964).

Cousins, R. R., Int. J. Eng. Sci. **8**, 595 (1970); Int. J. Eng. Sci. **10**, 301 (1972); Int. J. Eng. Sci. **10**, 511 (1972).

Ericksen, J. L., Trans. Soc. Rheol. **6**, 275 (1962).

Gadd, G. E., Nature **206**, 463 (1965); Nature **212**, 874 (1966).

Jones, D. T., Thesis, University of Wales (1967).

Jones, D. T. and *K. Walters*, Amer. Inst. Chem. Eng. J. **14**, 658 (1968).

Lockett, F. J., Int. J. Eng. Sci. **7**, 337 (1969); Nature **222**, 937 (1969).

Metzner, A. B. and *M. G. Park*, J. Fluid Mech. **20**, 291 (1964).

Mook, D. T., Univ. of Michigan Tech. Rep. 06505-2-T. (1967).

Porteous, K. C. and *M. M. Denn*, Univ. of Delaware Water Resources Center, No. 16 (1971).

Reynolds, W. C. and *M. C. Potter*, J. Fluid Mech. **27**, 465 (1967).

Schwarz, W. H. and *D. H. Chun*, Phys. Fluids **11**, 5 (1968).

Squire, H. B., Proc. Roy. Soc. **A 142**, 621 (1933).

Stuart, J. T., J. Fluid Mech. **9**, 353 (1960).

Toms, B. A., Proc. Second Int. Cong. Rheology, 135 (1949).

Tulin, M. P., Paper Presented at Sixth Symp. Naval Hydrodynamics, Washington D.C. (1966).

Watson, J., J. Fluid Mech. **9**, 371 (1960).

Author's address:

Dr. *R. R. Cousins*
National Physical Laboratory
Teddington, Mddx. (England)

Rheol. Acta 12, 224–227 (1973)

From the Department of Physics, Keio University, Hiyoshi, Yokohama (Japan)

Pressure development in a non-Newtonian flow through a tapered tube

By S. Oka

With 2 figures

(Received October 27, 1972)

1. General formula for the flow per unit time

The flow of viscous fluids through a tapered tube is very interesting from the standpoint of blood flow in blood vessels. The taper of the tube will be an important factor in the pressure development.

Before we calculate the pressure gradient in the flow through a tapered tube, we shall first give a brief summary of our theory of the steady convergent flow of non-*Newton*ian fluids characterized by an arbitrary time-independent flow curve through a slightly tapered tube. This theory was originally presented by Mr. *Murata* and the author (1). However, the following summary is given in a slightly modified form.

Following assumptions are made in order to simplify the problem: i) the fluid is incompressible; ii) the motion of the fluid is laminar; iii) the motion has an axial symmetry; iv) the motion is steady; v) no body force acts in the fluids; vi) the motion is so slow that the inertia term can be neglected; vii) the semiangle α of the cone is very small; and viii) there is no slip at the wall.

Let us take a cylindrical coordinate system (r, φ, z) whose z axis is the axis of the tapered tube.

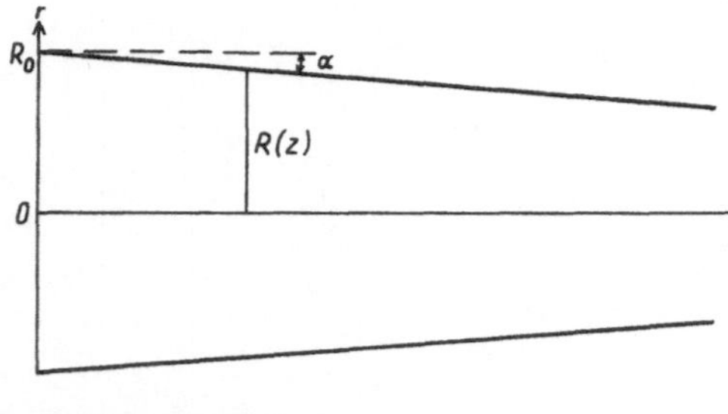

Fig. 1. Tapered tube

From fig. 1, the radius of the tube at a point z is approximately given by

$$R(z) = R_0 - \alpha z. \qquad [1]$$

For *Newton*ian fluids, the stress components are given by

$$\tau_{rr} = -p + 2\eta \frac{\partial v}{\partial r} \qquad [2]$$

$$\tau_{\varphi\varphi} = -p + 2\eta \frac{v}{r} \qquad [3]$$

$$\tau_{zz} = -p + 2\eta \frac{\partial u}{\partial z} \qquad [4]$$

$$\tau_{zr} = \eta \left(\frac{\partial u}{\partial r} + \frac{\partial v}{\partial z}\right) \qquad [5]$$

$$\tau_{r\varphi} = 0 \qquad [6]$$

$$\tau_{\varphi z} = 0 \qquad [7]$$

where u and v are z- and r-component of velocity, p is the pressure and η is the coefficient of viscosity.

For non-*Newton*ian fluids characterized by an arbitrary time-independent flow curve, however, we shall consider that η has the meaning of an apparent viscosity η_a, which is defined as the ratio between the shear stress and the shear rate. Hence η_a is not a constant but a function of the shear rate, that is, a function of the coordinates of the fluid particles.

On the basis of assumptions (iv–vi), the equations of motion become

$$\frac{1}{r}\frac{\partial}{\partial r}(r\tau_{rr}) - \frac{\tau_{\varphi\varphi}}{r} + \frac{\partial \tau_{zr}}{\partial z} = 0 \qquad [8]$$

$$\frac{1}{r}\frac{\partial}{\partial r}(r\tau_{zr}) + \frac{\partial \tau_{zz}}{\partial z} = 0. \qquad [9]$$

Substitution of eqs. [2–5] into the above equations yields

$$\frac{\partial p}{\partial r} = \frac{2}{r}\frac{\partial}{\partial r}\left(\eta_a r \frac{\partial v}{\partial r}\right) - 2\eta_a \frac{v}{r^2} + \frac{\partial}{\partial z}\left(\eta_a \frac{\partial u}{\partial r}\right) + \frac{\partial}{\partial z}\left(\eta_a \frac{\partial v}{\partial z}\right) \qquad [10]$$

$$\frac{\partial p}{\partial z} = \frac{1}{r}\frac{\partial}{\partial r}\left(\eta_a r \frac{\partial u}{\partial r}\right) + \frac{1}{r}\frac{\partial}{\partial r}\left(\eta_a r \frac{\partial v}{\partial z}\right) + 2\frac{\partial}{\partial z}\left(\eta_a \frac{\partial u}{\partial z}\right) \qquad [11]$$

where η_a is the apparent viscosity which is a function of $\partial u/\partial r + \partial v/\partial z$. Since α is very small, we may assume that u is of the order of 1 with respect to α **but both v and u are of the order of α, while $\partial v/\partial z$** is of the order of α^2. Further we may assume that η_a is of the order of 1 but $\partial \eta_a/\partial z$ is of the order of α. Then eqs. [5], [10] and [11] are reduced to

$$\tau_{zr} = \eta_a \frac{\partial u}{\partial r} \qquad [12]$$

$$\frac{\partial p}{\partial r} = \frac{2}{r}\frac{\partial}{\partial r}\left(\eta_a r \frac{dv}{\partial r}\right) - 2\eta_a \frac{v}{r^2} + \frac{\partial}{\partial z}\left(\eta_a \frac{\partial u}{\partial r}\right) \qquad [13]$$

$$\frac{\partial p}{\partial z} = \frac{1}{r}\frac{\partial}{\partial r}\left(\eta_a r \frac{\partial u}{\partial r}\right). \qquad [14]$$

Eliminating p we have

$$\frac{\partial}{\partial r}\left[\frac{1}{r}\frac{\partial}{\partial r}\left(\eta_a r \frac{\partial u}{\partial r}\right)\right] = \frac{\partial}{\partial z}\left[\frac{2}{r}\frac{\partial}{\partial r}\left(\eta_a r \frac{\partial v}{\partial r}\right) - 2\eta_a \frac{v}{r^2} + \frac{\partial}{\partial z}\left(\eta_a \frac{\partial u}{\partial r}\right)\right]. \qquad [15]$$

Since all terms on the right-hand side of this equation are of the order of α^2, we have

$$\frac{\partial}{\partial r}\left[\frac{1}{r}\frac{\partial}{\partial r}\left(\eta_a r \frac{\partial u}{\partial r}\right)\right]. \qquad [16]$$

Integration of this equation yields

$$\tau = -r\psi(z) \qquad [17]$$

where τ stands for $|\tau_{zr}| = -\tau_{zr}$. Then we obtain from eq. [14]

$$\psi(z) = \frac{1}{2}\frac{\partial p}{\partial z}. \qquad [18]$$

If we denote the shear stress τ at the wall by $\tau_w(z)$, then eq. [17] can be rewritten as

$$\tau = \frac{\tau_w(z)}{R(z)} r. \qquad [19]$$

If we denote the flow curve by $f(\tau)$, we have

$$f(\tau) = -\frac{\partial u}{\partial r} \qquad [20]$$

since $\partial v/\partial z$ is of the order of α^2. From the no slip condition at the wall we obtain

$$u(r, z) = \frac{R(z)}{\tau_w(z)} \int_{\tau}^{\tau_w(z)} f(\tau)\, d\tau \qquad [21]$$

by using eq. [19].

Integration of the equation of continuity

$$\frac{\partial}{\partial r}(rv) + \frac{\partial}{\partial z}(ru) = 0 \qquad [22]$$

yields

$$v(r, z) = \frac{R(z)}{\tau_w(z)} \frac{1}{\tau} \int_{\tau}^{\tau_w(z)} \tau \frac{\partial u}{\partial z}\, d\tau. \qquad [23]$$

The volume of the fluid flowing per unit time across a cross section is given by

$$Q = \int_0^{R(z)} 2\pi r u\, dr. \qquad [24]$$

Integration by parts yields

$$Q = \frac{\pi R^3(z)}{\tau_w^3(z)} \int_0^{\tau_w(z)} \tau^2 f(\tau)\, d\tau. \qquad [25]$$

This is the general formula for the flow per unit time of non-*Newton*ian fluids specified by an arbitrary time-independent flow curve $f(\tau)$ through a slightly tapered tube.

2. Distribution of pressure

In our general formula [25], Q is constant with regard to z, while $\tau_w(z)$ is given by

$$\tau_w(z) = -R(z)\,\psi(z) \qquad [26]$$

where $-\psi(z)$ is equal to half of the pressure gradient $-\partial p/\partial z$ as shown in eq. [18]. Consequently, eq. [25] will enable us to obtain the pressure gradient as a function of z, provided that the flow curve of a non-*Newton*ian fluid is known.

In the following, we shall give explicit expressions of $\psi(z)$ in several cases of non-*Newton*ian fluids specified by particular flow curves.

i) Power law fluid

The flow curve is given by

$$f(\tau) = k\tau^n \qquad [27]$$

k and n being positive constants. Substitution of eq. [27] into eq. [25] yields

$$Q = \frac{\pi R^3(z)\, k}{n+3} \tau_w^n(z) \qquad [28]$$

or

$$Q = \frac{\pi k}{n+3} R^{n+3}(z)\{-\psi(z)\}^n \qquad [29]$$

by using eq. [26].

Let us denote the value of $-\psi(z)$ at $z=0$ by $-\psi_0$. Then we have

$$R^{n+3}(z)\{-\psi(z)\}^n = R_0^{n+3}(-\psi_0)^n \qquad [30]$$

or

$$\frac{-\psi(z)}{-\psi_0} = \left\{\frac{R_0}{R(z)}\right\}^{(n+3)/n} = \left\{\frac{R_0}{R_0 - \alpha z}\right\}^{(n+3)/n}. \qquad [31]$$

It can be seen that the pressure gradient is not constant but increases with decrease in the radius of the tube.

Let us confine ourselves to the region $\alpha z/R_0 \ll 1$ and neglect small quantities of order $(\alpha z/R_0)^2$. Then eq. [31] is simplified to

$$\frac{-\psi(z)}{-\psi_0} = 1 + \frac{n+3}{n}\frac{\alpha z}{R_0}. \qquad [32]$$

Thus, the pressure gradient increases linearly with the distance z along the axis of the tube.

ii) Bingham body

The flow curve is given by

$$f(\tau) = 0 \qquad (\tau < f_B)$$
$$= \frac{1}{\eta_B}(\tau - f_B) \quad (\tau > f_B) \qquad [33]$$

where f_B is the *Bingham* yield value and η_B is the *Bingham* viscosity. We need only consider the case where $\tau_w(z) > f_B$, because Q vanishes when $\tau_w(z) < f_B$.

Substitution of eq. [33] into eq. [25] yields

$$Q = \frac{\pi R^3(z)\,\tau_w(z)}{4\eta_B} \times \left[1 - \frac{4}{3}\frac{f_B}{\tau_w(z)} + \frac{1}{3}\left\{\frac{f_B}{\tau_w(z)}\right\}^4\right]. \qquad [34]$$

We shall assume that the yield value is so small that the condition $f_B/\tau_w(z) \ll 1$ is satisfied. Neglecting the small term $\{f_B/\tau_w(z)\}^4$, we have

$$Q = \frac{\pi R^3(z)}{4\eta_B}\left[-R(z)\,\psi(z) - \frac{4}{3} f_B\right] \qquad [35]$$

or

$$R^3(z)\left[-R(z)\,\psi(z) - \tfrac{4}{3} f_B\right] = R_0^3\left[-R_0\psi_0 - \tfrac{4}{3} f_B\right]. \qquad [36]$$

It can easily be seen that the pressure gradient increases with decrease in the radius $R(z)$.

Let us again neglect small quantities of order $(\alpha z/R_0)^2$. Then we obtain

$$\frac{-\psi(z)}{-\psi_0} = 1 + \frac{4\alpha z}{R_0}\left(1 + \frac{f_B}{R_0\psi_0}\right). \qquad [37]$$

The factor $1 + (f_B/R_0\psi_0)$ is positive from the assumption $f_B/\tau_w(z) \ll 1$.

iii) Fluid obeying Casson's equation

It is generally accepted that blood obeys *Cassons* equation. In this case, the flow curve is given by

$$f(\tau) = 0 \qquad (\tau < f_c)$$
$$= \frac{1}{\eta_c}(\sqrt{\tau} - \sqrt{f_c})^2 \quad (\tau > f_c) \qquad [38]$$

where f_c is the *Casson* yield value and η_c is the *Casson* viscosity. We shall confine ourselves to the case $\tau_w(z) > f_c$, since Q vanishes when $\tau_w(z) < f_c$.

Substitution of eq. [38] into eq. [25] yields

$$Q = \frac{\pi R^3(z)\,\tau_w(z)}{4\eta_c}\left[1 - \frac{16}{7}\left\{\frac{f_c}{\tau_w(z)}\right\}^{1/2} + \frac{4}{3}\frac{f_c}{\tau_w(z)} - \frac{1}{21}\left\{\frac{f_c}{\tau_w(z)}\right\}^4\right]. \qquad [39]$$

We shall assume that the yield value f_c is so small that the condition $f_c/\tau_w(z) \ll 1$ is satisfied. Neglecting small terms $f_c/\tau_w(z)$ and $\{f_c/\tau_w(z)\}^4$, we have

$$Q = \frac{\pi R^3(z)}{4\eta_c} = \left[-R(z)\,\psi(z) - \frac{16}{7}\sqrt{-f_c R(z)\,\psi(z)}\right] \qquad [40]$$

or

$$R^3(z)\left[-R(z)\,\psi(z) - \frac{16}{7}\sqrt{-f_c R(z)\,\psi(z)}\right] = R_0^3\left[-R_0\psi_0 - \frac{16}{7}\sqrt{-f_c R_0\psi_0}\right.. \qquad [41]$$

Thus the pressure gradient increases with decrease in the radius $R(z)$.

In order to obtain an approximate solution of $\psi(z)$ we put

$$\psi(z) = \psi_0\{1 + \alpha\varphi(z)\}. \qquad [42]$$

Substituting eq. [42] into eq. [41] and neglecting small quantities of order α^2, we get

$$\varphi(z) = \frac{4z}{R_0}\left[1 + \frac{6}{7}\left\{\frac{f_c}{-R_0\psi_0}\right\}^{1/2}\right]. \qquad [43]$$

Hence we obtain

$$\frac{-\psi(z)}{-\psi_0} = 1 + \frac{4\alpha z}{R_0}\left[1 + \frac{6}{7}\left\{\frac{f_c}{-R_0\psi_0}\right\}^{1/2}\right]. \qquad [44]$$

The pressure gradient again increases linearly with the distance z.

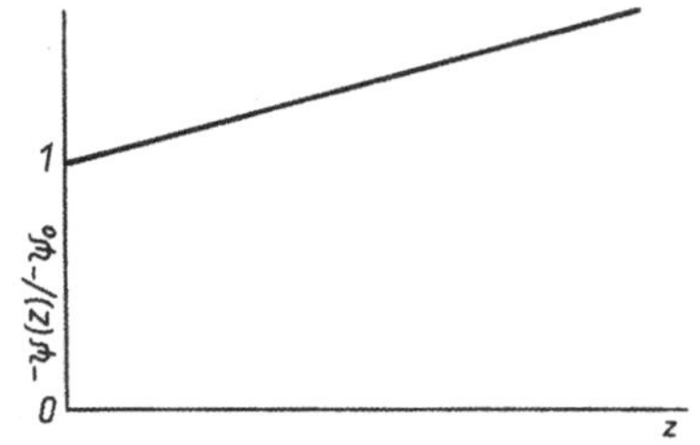

Fig. 2. Plot of $-\psi(z)/-\psi_0$ against z

Fig. 2 shows the plot of $-\psi(z)/(-\psi_0)$ against z. As is seen from eqs. [32] and [37], similar curves are obtained for the power law fluid as well as for the *Bingham* body. Comparison of our theory with experiment will be most desirable.

Summary

The flow of viscous fluids through a tapered tube is very interesting from the standpoint of blood flow in blood vessels. The taper of the tube is an important factor in the pressure development. In the first place, we have given a brief summary of our theory of the steady convergent flow of non-*Newton*ian fluids characterized by an arbitrary time-independent flow curve through a slightly tapered tube. Based on our general formula for the flow per unit time, explicit formulae of the pressure gradient are obtained in several cases of non-*Newton*ian fluids specified by particular flow curves: power law fluid, *Bingham* body, and the fluid obeying *Casson*s equation. In all these cases it is shown that the pressure gradient is not constant along the axis but increases with decrease in the radius of the tapered tube. If we neglect quantities of order α^2 (α: angle of taper), then the pressure gradient increases linearly with the distance along the axis of the tube.

Reference

1) *Oka, S.*, Japan. J. Appl. Phys. **8**, 5 (1969).

Author's address:
Dr. *Syoten Oka*
3-2-2 Eifuku Suginamiku, Tokyo (Japan)

Rheol. Acta **12**, 228–233 (1973)

From the Warren Spring Laboratory, Stevenage, Herts. (U.K.)

A differential form of constitutive relation for thixotropy

By D. C.-H. Cheng

(Received October 27, 1972)

Introduction

In 1965 we published a paper (1) on what may be called a structural theory of thixotropy. In this the constitutive relation is given by a pair of equations, viz, the equation of state

$$F = \eta(D, \lambda)\, D \quad [1]$$

and the rate equation

$$\frac{d\lambda}{dt} = g(D, \lambda) \quad [2]$$

in which the functions η and g are subjected to certain restrictions. (The symbols have the meanings given in the notations.) Experimental procedures were described which allow the equation of state to be mapped as constant-λ, i.e. constant structure curves and the rates g to be determined at different points along these curves or mapped as constant-g curves (2, 3). The theory formed the basis of the experimental investigation of *Joye* and *Poehlein* (4).

Parallel with this work a number of workers have proposed rate equations of the form

$$\frac{dx}{dt} = f(x). \quad [3]$$

Brodkey and co-workers (5) and *Pinder* et al. (6) equated x with the viscosity; *Fredrickson* (7) equated x with the fluidity whilst *Ritter* and *Govier* (8) equated x with the "structural stress component". Using what are termed chemical kinetic considerations all these authors put forward specific forms of the function f. *Harris* (9) used the concept of strain rate invariants and relaxation time spectrum, but his equations can be reduced to a rate equation involving viscosity, which depend on an "influence function" that is a function of time. Several specific forms of this are discussed. It can be seen that with suitable definitions of the structural parameter, λ, these models can be said to represent special cases of the generalized theory of *Cheng* and *Evans.*

Our structural theory of thixotropy is so designed that it can be used for material characterization as well as for the description of the complex flow behaviour of thixotropic fluids. It has been applied to the problem of the start-up flow of a pipeline containing a gelled thixotropic material (10, 11) such as clay slurries or waxy petroleum oils, and we are now studying the problem of the flow of thixotropic films of paints or other coating materials during levelling and sagging.

In the present paper an alternative constitutive relation for thixotropy is described. This takes the form of a differential equation involving two functions, $\alpha(F, D)$ and $\beta(F, D)$. The properties of these functions and the ways in which they can be determined experimentally are described. The relation between this formulation of thixotropy and the structural theory is also discussed.

The constitutive equation

The differential form of the constitutive relation for thixotropy is readily obtained by differentiating eq. [1] and substituting eq. [2]

$$\dot{F} = \alpha(F, D)\, \dot{D} + \beta(F, D) \quad [4]$$

where

$$\alpha = \left(\frac{\partial F}{\partial D}\right)_\lambda \quad [5]$$

and

$$\beta = \left(\frac{\partial F}{\partial \lambda}\right)_D g(\lambda, D). \quad [6]$$

But henceforth the dependence on λ need not be explicitly discussed.

Eqs. [4] to [6] are subjected to the following restrictions:

$$\frac{F}{D} > 0 \quad [7]$$

$$\alpha(F, D) > 0 \quad [8]$$

$$\beta(F, D) = 0, \qquad F = F_e(D) \quad [9]$$

$$\beta(F, D) < 0, \qquad F > F_e(D) \quad [10]$$

$$\beta(F, D) > 0, \qquad F < F_e(D). \quad [11]$$

The reasons for these restrictions are discussed in full in the previous paper (1). Briefly, eq. [7] ensures that viscosity is positive and F and D take the same sign and eq. [8] ensures that dF/dD is positive along constant structure curves. Eq. [9] defines the equilibrium flow curve which relates steady state, shear stress and shear rate. It divides the (F, D) field into two regions; one, described by eq. [10], represents conditions of structural break-down; the other, described by eq. [11] represents structural build-up. (Further restrictions imposed on the structural theory are not necessary here.)

It is also unnecessary to specify the functional dependence of α and β on F and D any further. As will be shown below, the two functions may be determined directly by experiment. However, to illustrate what they might look like the *Moore* (12) model may be taken as an example. In terms of the structural theory the constitutive equations are

$$F = (\eta_0 + c\lambda) D$$

$$g = a - (a + bD)\lambda. \qquad [12]$$

Therefore,

$$\alpha = \frac{F}{D}$$

$$\beta = (a + bD)(F_e - F) \qquad [13]$$

where

$$F_e = \left(\eta_0 + \frac{ca}{a + bD}\right) D. \qquad [14]$$

The functional dependence of α and β on F and D can be readily plotted. On $a(F, D)$-plot, the constant-α loci are straight lines through the origin, while the constant-β loci intercept the F-axis at $(-\beta/a)$ and gradually curve over to become asymptotic to the equilibrium flow curve as D becomes large.

The *Moore* model is not too realistic because of the absence of yield stress. This can be simply remedied by rewriting eq. [12] as follows:

$$F = F_Y + (\eta_0 + c\lambda) D$$

$$F_Y = F_{Y0} + \lambda F_{Y1}$$

$$g = a - (a + bD)\lambda. \qquad [15]$$

Then

$$\alpha = \frac{cF + \eta_0 F_{Y1} - cF_{Y0}}{cD + F_{Y1}}$$

$$\beta = (a + bD)(F_e - F) \qquad [16]$$

where

$$F_e = \eta_0 + \frac{ca}{a + bD} D + F_{Y0} + \frac{aF_{Y1}}{a + bD}. \qquad [17]$$

In this model the constant-α loci are still straight lines but the intercepts are non-zero; whilst the constant-β loci now intersect the F-axis at $(F_{Y0} + F_{Y1} - \beta/a)$ but become asymptotic to the equilibrium flow curve, eq. [17], as D becomes large.

Experimental determination of α and β

Traditionally the experimental determination of material properties involves assuming some specific functional relationship between the variables which, in the case of thixotropy, would be α, β and F, D. These relations would involve a number of material parameters. By solving the material equation, eq. [4], and the equations of motion for the experimental situation, the material parameters may be related to experimentally measurable quantities. Then, by suitably manipulating the experimental data or by comparing them with calculations based on assumed values of the material parameters, the numerical values of these parameters may be determined for the particular fluid being studied.

Bearing in mind the complex behaviours in thixotropy this traditional procedure is a hit-or-miss method. It is difficult to invent functional relations that closely approximate reality. Also it may not be easy to solve the equations involved. However it is not necessary to use this method. I shall show that it is possible to determine α and β directly from suitable experimental techniques and interpretation, and in such a form that the results may readily be used to solve industrial flow problems.

The functions α and β are the most easily determined using the *cone-and-plate viscometer*. It has been shown (13) that in this geometry F and D are effectively constant in the fluid sample. Most cone-and plate viscometers such as the *Weissenberg* Rheogoniometer, operate at constant-D. Eq. [4] shows that β can be immediately obtained from the experimental (F, t) relationship, by differentiation. If in addition experiments are made at constant-F conditions, the ratio $(-\beta/\alpha)$ can be determined from the (D, t) relationship and α can be calculated.

Alternatively the constant-D experiment may be carried out together with an experiment in which both F and D are allowed to vary with time. [This can be performed using the *Ferranti-Shirley* Viscometer; the details are given elsewhere (14).] From the latter experiment pairs of values of

$(\dot{F}, \dot{D})$ may be obtained for any F and D. And with β already known from the constant-D experiment α can be calculated using eq. [4].

The *co-axial cylinder viscometer* is less suitable for the determination of α and β, because of the appreciable variation of shear rate in the fluid sample, particularly at low rotational speeds and if the fluid possesses yield stress (1, 3). However, if experiments are performed using a range of different inner cylinder diameters α and β can again be directly measured. The same results can be obtained if the outer cylinder diameter is varied.

A combination of constant speed and constant torque experiments is required. For each experiment the fluid sample is preconditioned by shearing to equilibrium at (Ω_e, T_e). In the constant torque experiment the torque is suddenly changed to T and then maintained constant. The initial speed Ω_0 and rate of change of speed $\dot{\Omega}_0$ are measured. It is known from experimental experience (which can be taken as a defining property of thixotropy), or by appealing to the structural theory of thixotropy, that Ω_0 and Ω_0 depend on T_e and T. With the shear stress given (on neglecting inertia effects), by

$$F = \frac{T}{2\pi h r^2} \qquad [18]$$

the shear rate distribution must depend on T_e and T also. Thus on integrating the usual expression for shear rate,

$$\Omega_0 = \int_{r_a}^{r_b} D(T_e, T)\, d\ln r \qquad [19]$$

on differentiating and using eq. [4]

$$\dot{\Omega}_0 = \int_{r_a}^{r_b} \left[\frac{\dot{F}}{\alpha} + \gamma\right] d\ln r \qquad [20]$$

where

$$\gamma(F, D) = -\frac{\beta}{\alpha}. \qquad [21]$$

Because T is constant $\dot{F} = 0$, and so

$$\dot{\Omega}_0 = \int_{r_a}^{r_b} \gamma\, d\ln r. \qquad [22]$$

On repeating this constant torque experiment using different inner cylinders, eq. [22] may be differentiated,

$$\left(\frac{\partial \dot{\Omega}_0}{\partial \ln r_a}\right)_{T_e, T} = -\gamma(F_a, D_a). \qquad [23]$$

This equation allows γ to be evaluated at the corresponding shear stress of

$$F_a = \frac{T}{2\pi h r_a^2} \qquad [24]$$

and the corresponding shear rate, obtained from eq. [19] of

$$D_a = -\left(\frac{\partial \Omega_0}{\partial \ln r_a}\right)_{T_e, T}. \qquad [25]$$

In the constant speed experiment the preconditioned fluid sample is suddenly subjected to a speed change to Ω, which is maintained constant, and the initial torque T_0 and the rate of change of torque $\dot{T}_0$ are measured. As before,

$$F = \frac{T_0}{2\pi h r^2} \qquad [26]$$

and

$$\Omega = \int_{r_a}^{r_b} D(T_e, T_0)\, d\ln r. \qquad [27]$$

With Ω constant, $\dot{\Omega} = 0$, and eq. [27] differentiates to give

$$0 = \dot{T}_0 \int_{r_a}^{r_b} \frac{d\ln r}{\alpha 2\pi h r^2} + \int_{r_a}^{r_b} \gamma\, d\ln r. \qquad [28]$$

On repeating the constant speed experiment using different inner cylinders eq. [28] may be differentiated:

$$\left(\frac{\partial \dot{T}_0}{\partial \ln r_a}\right)_{T_e, T_{02}} \int_{r_a}^{r_b} \frac{d\ln r}{\alpha 2\pi h r^2} - \frac{\dot{T}_0}{\alpha_a 2\pi h r_a^2} - \gamma_a = 0. \qquad [29]$$

On using eq. [28], eq. [29] becomes

$$\left(\frac{\partial \ln \dot{T}_0}{\partial \ln r_a}\right)_{T_e, T_0} \int_{r_a}^{r_b} \gamma\, d\ln r + \frac{\dot{T}_0}{\alpha_a 2\pi h r_a^2} + \gamma_a = 0. \qquad [30]$$

This equation allows $\alpha_a = \alpha(F_a, D_a)$ to be evaluated at the shear stress of

$$F_a = \frac{T_0}{2\pi h r_a^2} \qquad [31]$$

and the shear rate obtained from eq. [27], of

$$D_a = -\left(\frac{\partial \Omega}{\partial \ln r_a}\right)_{T_e, T_0}. \qquad [32]$$

The way in which this is done is as follows.

From the variation of Ω with r_a under constant T_e and T_0, it is possible to determine the distribution of D with r over the range $r_a < r < r_b$. (In practice extrapolation outside the experimental range of r_a is necessary to reach r_b, unless supplementary experiments are made using a range of cylinders with radii greater than r_b and having r_b as the inner cylinder.) Then with eq. [26] and using the $\gamma(F, D)$ results obtained in the constant torque experiments, the integral in eq. [30] may be evaluated. Hence α_a may be calculated.

With α and γ thus determined, β may be found using eq. [21]. These methods measure the value of α and β at selected values of F, D. The constant-α and constant-β contour mapping on the (F, D) plot can be readily carried out graphically or otherwise.

There is the special locus represented by $\beta = 0$, viz, *the equilibrium flow curve*, whose determination is well known. Both the cone-and-plate and the coaxial cylinder viscometers or even the tube viscometer (with due allowance being made for entrance and exit effects) may be used. The relation between F_e and D is unique for any material, the interpretation of data follows the standard form for time-independent fluids (15).

The function $\beta(F, 0)$ describes *the rate of development of yield stress at rest*. (*See the Discussion Section.*) It can be determined directly using the *Ferranti-Shirley* Viscometer in a special mode. The details are given in a separate report (14).

Calculation of thixotropic behaviour

The behaviour of a thixotropic material can be predicted when the functions α and β are known. Because of the complex nature of thixotropy, generally it would not be possible to solve eq. [4] analytically. However, with the use of a high-speed computer it will be possible to carry out the calculations numerically using a step-by-step method. (In fact in similar applications of the structural theory the constitutive equations are used in the differential form, which constitutes eq. [4].) In a complicated flow situation the velocity field must be approximated by one of simple shear field. The velocity, shear rate and shear stress of an element of fluid are functions of position and time. By following the movement of the elements, calculating shear rate or shear stress changes (depending on the prescribed conditions) using eq. [4] and then integrating over the whole field, the overall response of the fluid may be predicted. We have applied this procedure to the problems of start-up flow of a pipeline containing a gelled material (10, 11) but it is outside the scope of the present paper to go into details. We are now using it on the problems of flow of thin fluid films or coatings on vertical or horizontal surfaces.

Discussion

The derivation of eq. [4] and discussion of the properties of the functions α and β have been done mainly using physical intuition. It is recognized that the formulation lacks mathematical rigor and it is possible that not all the necessary restrictions have been given to make it fully describe the thixotropic property. But this task and also the problem of generalizing eq. [4] into tensor form are outside the scope of the present paper.

Because this paper is given to a general discussion. I have not discussed the additional restrictions to which α and β are subjected if special cases of thixotropy are involved. These special cases include negative or anti-thixotropy, rheopexy (the distinction between anti-thixotropy and rheopexy is discussed by *Cheng* and *Evans*) and limiting values of α (corresponding to limits imposed by fully built-up and broken-down structural states). The possibility of the equilibrium curve showing maximum and minimum in F also requires special consideration.

It is worth noting that the present formulation represents an inelastic material in the sense that the restrictions on α and β preclude recoil phenomena and energy storage. If D is suddenly changed from one value to another, there is a correspondingly sudden change in F (i.e. with zero relaxation time), even though there is no sudden change in the deformation of the material. After the step-change in F, the subsequent variation of F is governed by eq. [4].

In this connection a special meaning must be attached to eq. [4] when $D = 0$ and is maintained at zero. Under this condition the shear stress on the material can be less than the yield stress but otherwise indeterminate. If the material is thixotropic the yield stress can increase as the build-up progresses: this is given by eq. [4]

$$\dot{F} = \beta(F, 0). \qquad [33]$$

which therefore does not describe the actual shear stress acting on the material at rest. On the other hand, if the material is anti-thixotropic the yield stress decreases with time and the residual shear stress on the material must relax simultaneously. Eq. [33] can then describe the actual shear stress at rest.

Despite the inelastic behaviour already noted, there are similarities between thixotropy and non-linear viscoelasticity which have been discussed and the question is asked whether the mathematical description of thixotropy can be formulated in an analogous way to that of viscoelasticity (1, 3). It is seen that the differential form of a constitutive equation, eq. [4], goes partly towards the answer. *Oldroyd* (16) proposed a general form of linear viscoelastic equation as

$$\sum_{n=0}^{n} a_n F^{(n)} = \sum_{n=0}^{n} b_n D^{(n)} \qquad [34]$$

in which a_n and b_n are constants and (^{n}) denotes the n-th time-derivative. It would seem that one way to render this non-linear would be to replace the constants by functions of F and D:

$$\sum_{n=0}^{n} \alpha_n(F, D)\, F^{(n)} = \sum_{n=0}^{n} \beta_n(F, D)\, D^{(n)}. \qquad [35]$$

Then it is at once obvious that eq. [4] is a special case of eq. [35], suggesting that thixotropy could be described in the mathematics of non-linear viscoelasticity. But it is outside the scope of the present paper to take the discussion further.

Turning now to the relationship between the structural theory of thixotropy and the differential equation, it is seen that both methods of formulating the constitutive relationships of thixotropy are equally useful when it comes to the application to industrial problems such as pipe and film flow. But there is a fundamental difference between them, viz, that the structural theory would provide more insight into the physical basis of thixotropy. The information that it gives on the shear stress/shear rate relationships and the rates of build-up and break-down for different levels of structure would be most suitable for correlating with the molecular and microscopic properties of a fluid and processes that go on in a fluid, and so to a better understanding of them.

With suitable forms of the functions α and β it is possible that eq. [4] might be sufficient to describe the thixotropic behaviour of viscoelastic fluids where it exists. But as discussed, one would expect a structural theory to be more informative. Following the inelastic theory, that for the viscoelastic fluid can be readily established. In terms of the spring and dash-pot model, the elastic modulus and the viscosity of the elements may be allowed to depend on some structural parameter which can vary according to some rate equation. Alternatively, the relaxation time spectrum may be similarly treated [*Yamamoto* (17) has described spectra that vary with time, but this treatment, paralleling the early treatment of inelastic thixotropy that discussed thixotropy in terms of (F, D, t) relations, (summarized by *Sherman* (18)) would not seem to go far enough in depth]. Again, it is outside the scope of the present paper to explore these suggestions but the comments are made to stimulate further work on the subject.

Conclusions

Using the structural theory of inelastic thixotropy discussed by *Cheng* and *Evans* (1, 2, 3), a differential form of the constitutive equation, eq. [4], is derived. In this formulation, thixotropy is characterized by two functions $\alpha(F, D)$ and $\beta(F, D)$ which are subjected to a number of restrictions, eq. [7] to [11]. Two specific models of these functions are described. One based on the *Moore* (12) model and the other is a modified *Moore* model showing yield stress.

It is shown that the functions α and β can be determined directly from experimental measurements without the need to make any a priori assumptions about their mathematical form. Both the cone-and-plate and the coaxial cylinder viscometers may be used. The determinations of the equilibrium flow curve, $\beta(F, D) = 0$, and of the rate of developed of the yield stress at rest, $\beta(F, 0)$, are indicated.

The way in which eq. [4] can be used to predict industrial problems such as pipeline flow and film or coating flow, is also indicated and reference is made to publications where details are given.

The relationship between the structural theory and the differential constitutive equation is discussed. Although both methods of describing thixotropy are equally useful in industrial applications, the structural theory is more informative about the molecular and microscopic nature of the fluid.

Ways in which the present theories of inelastic thixotropy may be developed further are indicated. The relation between the mathematics of thixotropy and nonlinear viscoelasticity is discussed. It is suggested that it may be possible to develop a theory to describe the thixotropic behaviour of viscoelastic fluids.

Notation

a, b	rate constants in the *Moore* model
c	viscosity constant in the *Moore* model
D	shear rate
F	shear stress
F_Y	yield stress
F_{Y0}, F_{Y1}	yield stress constants in the modified *Moore* model
g	rate of structural build-up
h	height of cylinder
r	radial co-ordinate
t	time
T	torque
T_0	initial torque
α, β	thixotropic functions
γ	$= -\beta/\alpha$
η	viscosity
η_0	viscosity constant in the *Moore* model
λ	structural parameter
Ω	rotational speed
Ω_0	initial rotational speed
subscripts	
a, b	pertaining to the inner, outer cylinder
e	pertaining to the equilibrium flow curve or equilibrium shearing condition

Summary

A differential form of constitutive equation for the thixotropy

$$\dot{F} = \alpha(F, D)\,\dot{D} + \beta(F, D)$$

(where F is the shear stress, D the shear rate, $\dot{F}$ and $\dot{D}$ the time derivatives, and α and β are material functions), is derived from the structural constitutive equations of *Cheng* and *Evans* (1965). The restrictions on the forms of α and β are discussed and the ways in which they may be determined using the cone-and-plate and coaxial cylinder viscometers are described. The use of the equation in the prediction of industrial problems such as pipeline flow and sagging and levelling of films and coatings are indicated.

The relationship between the structural theory and the differential constitution equation is discussed; and so is the relationship between these theories of thixotropy and non-linear viscoelasticity. The way in which the thixotropic theories may be extended to cover the thixotropy behaviour of viscoelastic fluids is suggested.

References

1) *Cheng, D. C.-H.* and *F. Evans*, Brit. J. Appl. Phys. **16**, 1599–1617 (1965).

2) *Cheng, D. C.-H.*, Nature **216**, 1099–1100 (London 1967).

3) *Cheng, D. C.-H.*, The characterization of thixotropic behaviour. Research Report No. LR **157** (MH) (Stevenage, U.K., Warren Spring Laboratory, 1971).

4) *Joye, D. D.* and *G. W. Poehlein*, Trans. Soc. Rheol. **15**, 51–61 (1971).

5) *Lee, K. H.* and *R. S. Brodkey*, Trans. Soc. Rheol. **15**, 627–646 (1971).

6) *Brown, J. P.* and *K. L. Pinder*, Canad. J. Chem. Eng. **49**, 38–43 (1971).

7) *Fredrickson, A. G.*, Amer. Inst. Chem. Eng. J. **16**, 436–441 (1970).

8) *Ritter, R. A.* and *G. W. Govier*, Canad. J. Chem. Eng. **48**, 505–513 (1970).

9) *Harris, J.*, Rheol. Acta **6**, 6–12 (1967).

10) *Cheng, D. C.-H.* and *W. Whittaker*, The start-up flow of thixotropic fluids in pipelines. Research Report No. LR **155** (MH) (Stevenage, U.K., Warren Spring Laboratory, 1972).

11) *Cheng, D. C.-H.*, and *W. Whittaker*, A method for assessing the thixotropic properties of fluids carried in pipelines. Paper to 2nd International Conference on the Hydraulic Transport of Solids in Pipes, British Hydromechanical Research Association, Coventry (England 1972).

12) *Moore, F.*, Trans. Br. Ceram. Soc. **58**, 470–492 (1959).

13) *Cheng, D. C.-H.*, Brit. J. Appl. Phys. **17**, 253–263 (1966).

14) *Cheng, D. C.-H.*, The interpretation of the Ferranti-Shirley viscometer data obtained on thixotropic fluids. Research Report No. LR **158** (MH) (Stevenage, U.K., Warren Spring Laboratory, 1970).

15) *Cheng, D. C.-H.*, in: Proceedings of a symposium on the physical properties of liquids and gases for plant and process design, March 1968, East Kilbridge, Glasgow, pp. C 48–C 71 (Edinburgh 1970).

16) *Oldroyd, J. G.*, in: Rheology Theory and Application, Vol. I, Chap. 16. Ed.: *F. R. Eirich* (New York 1956).

17) *Yamamoto, M.*, Trans. Soc. Rheol. **15**, 783–788 (1971).

18) *Sherman, P.*, Industrial Rheology, pp. 11–12 (London 1970).

Author's address:

Dr. *D. C.-H. Cheng*
Warren Spring Laboratory
Gunnels Wood Road
P.O. Box 20, Stevenage, Herts. (U.K.)

Rheol. Acta 12, 234–239 (1973)

Centre d'études Mathématiques, Beyrouth (Liban)

Comportement des matériaux plastiques parfaits, non visqueux

Par M. Hajal

Avec 3 figures

(Reçu p. p. le 27 octobre 1972)

Introduction

Différentes théories ont été proposées pour décrire l'écoulement d'un matériau plastique parfait, non visqueux. Citons la théorie du potentiel plastique (la vitesse de déformation est orthogonale à la surface d'écoulement), la théorie de *E. H. Brown*, reprise par *G. Gudehus* en ce qui concerne les sables (le déviateur de vitesse de déformation est orthogonal à la section de la surface d'écoulement par un plan déviatoire), et la théorie de la courbe intrinséque de *Mohr-Caquot* (l'écoulement s'effectue par glissement sur un plan paralléle à la contrainte intermédiaire). Chacune de ces théories ne semble valable que pour une classe de matériaux en écoulement plastique parfait.

A la lumiere de l'expérience, nous avons essayé de fixer le domaine d'application de chacune de ces théories; puis nous avons essayé de construire une théorie, valable pour tous les matériaux plastiques parfaits, et admettant comme cas particuliers les théories précédentes.

Notations

L'espace est rapporté à un système d'axes orthonormés notés 1, 2, 3. Nous utilisons la convention de l'indice muet.

σ_{ij} et ε_{ij}	tenseurs euleriens de contraintes et de vitesses de déformations
s_{ij} et e_{ij}	parties déviatoires de σ_{ij} et ε_{ij} respectivement

Le tenseur σ_{ij} possède trois invariants scalaires que nous choisirons comme suit:

$\sigma_m = \sigma_{ii/3}$	contrainte moyenne
$S_2 = (s_{ij}s_{ij})^{1/2}$	norme du déviateur de contraintes
$S_3 = (s_{ij}s_{jk}s_{ki})^{1/3}$	

De même le tenseur ε_{ij} possède trois invariants, soient:

$\varepsilon_m = \varepsilon_{ii/3}$	égal au tiers de la vitesse de dilatation cubique
$E_2 = (e_{ij}e_{ij})^{1/2}$	norme du déviateur de vitesses de déformations

Nous posons:

$p = \varepsilon_{m/E_2}$ et $q = E_{3/E_2}$.

La masse spécifique est désignée par:

ϱ

Equations de comportement – forme generale

Le matériau est supposé homogène, non vieillissant et plastique parfait. Sa loi de comportement relie donc les contraintes aux vitesses de déformations, à la masse spécifique, à la température et à leurs dérivées objectives successives.

Nous supposerons, de plus, le matériau simple et nous négligerons les effets de la température (matériau non thermique ou en écoulement isotherme). Avec ces hypothèses, la loi de comportement s'écrit:

$$\sigma_{ij} = F_{ij}(\varepsilon_{kl}, \varrho).$$

Nous allons introduire deux hypothèses supplémentaires.

Hypothèse 1:

Le matériau est supposé non visqueux.

Si nous remplaçons dans la loi de comportement ε_{kl} par $\lambda \varepsilon_{kl}$, λ étant une constante arbitraire, les contraintes σ_{ij} restent inchangées. Ceçi entraine que les fonctions F_{ij} sont homogènes de degré zéro par rapport aux composantes ε_{kl} du tenseur vitesses de déformations.

$$\forall \lambda, \quad F_{ij}(\varepsilon_{kl}, \varrho) = F_{ij}(\lambda \varepsilon_{kl}, \varrho).$$

A. Sawczuk et *P. Stutz* (ref: 1) ont montré que l'hypothèse précédente nécéssite que la loi de comportement soit de la forme:

$$\sigma_{ij} = G_{ij}(\varepsilon_{kl/E_2}, \varrho). \qquad [1]$$

De plus le principe d'isotropie de l'espace nécessite que si l'on transmue ε_{kl/E_2} par une transformation orthogonale arbitraire R, σ_{ij} soit transmué par R. D'après un résultat dû à *Rivlin* (ref. 2), ceçi entraîne que la fonction tensorielle G_{ij} est isotrope et s'écrit:

$$\sigma_{ij} = a\delta_{ij} + b\varepsilon_{ij/E_2} + c\varepsilon_{ik}\varepsilon_{kj/E_2^2}$$

a, b, c étant fonctions de ϱ et des invariants scalaires de ε_{ij/E_2}. [2]

ε_{ij/E_2} admet deux invariants scalaires indépendants, $p = \varepsilon_{m/E_2}$ et $q = E_{3/E_2}$ donc a, b, c sont fonctions de ϱ, p, q.

Décomposons [2] en parties isotropes et déviatoires, il vient:

$$\sigma_m = a + bp + c(1 + p^2)$$

$$s_{ij} = -c/3\delta_{ij} + (b + 2cp)\, e_{ij/E_2} + c e_{ik} e_{kj/E_2}$$

a, b, c: fonctions de ϱ, p, q. [3]

Hypothèse 2:

Nous supposons qu'entre la contrainte moyenne et la masse spécifique existe une relation biunivoque:

$$\sigma_m = A(\varrho) \quad \text{ou} \quad \varrho = A^{-1}(\sigma_m).$$

Pour les métaux, la compréssibilité volumétrique est généralement supposée élastique, et l'hypothèse précédente est vérifiée. Pour les milieux granulaires, essentiellement inélastiques, et pour de grandes déformations (palier de plasticité parfaite), la masse spécifique tend vers une masse spécifique dite critique, dépendant de la contrainte moyenne (*Lelong*, ref: 3) et probablement de l'expérience envisagée. A notre connaissance, ce dernier point n'est pas établi clairement, nous négligerons donc cette influence et supposerons qu'entre la masse spécifique et la contrainte moyenne éxiste une relation biunivoque.

En posant:

$$a = A(\varrho) - pb - c(p^2 + 1)$$

$$B = b + 2pc$$

les équations [3] s'écrivent:

$$\sigma_m = A(\varrho)$$

$$s_{ij} = -c/3\delta_{ij} + B e_{ij/E_2} + c e_{ik} e_{kj/E_2^2}$$

B, c: fonctions de ϱ, p, q. [4]

A l'aide de multiplications contractées, l'on peut calculer les invariants S_2 et S_3, et écrire les eq. [4] sous la forme invariante:

$$\sigma_m = A(\varrho)$$

$$S_2^2 = (B + cq^3)^2 + 1/6c^2(1 - 6q^6)$$

$$S_3^3 = (B + cq^3)^3 - 1/36c^3(1 - 6q^6)^2 + 1/2cB(B + cq^3)(1 - 6q^6)$$

B, c: fonctions de ϱ, p, q. [5]

Nous allons montrer qu'en fait ces équations sont indépendantes de p. En effet, l'équation de continuité s'écrit:

$$\frac{d\varrho}{dt} + 3\varrho\varepsilon_m = 0 \quad \text{ou} \quad p = -\frac{1}{3\varrho E_2}\frac{d\varrho}{dt}$$

or

$$\frac{d\varrho}{dt} = \frac{1}{\frac{dA(\varrho)}{d\varrho}}\frac{d\sigma_m}{dt}$$

d'où:

$$p = -\frac{1}{3\varrho E_2}\frac{1}{\frac{dA(\varrho)}{d\varrho}}\frac{d\sigma_m}{dt}.$$

Considérons un état de contraintes σ_{ij}, tel que le matériau soit en écoulement plastique, et appliquons un accroissement $d\sigma_{ij}$ pendant un temps dt, tel que le matériau reste en écoulement plastique. L'expression de p, montre que pour un même σ_{ij}, p peut prendre toute valeur selon l'accroissement $d\sigma_m$ résultant de $d\sigma_{ij}$. Donc à différentes valeurs de p correspond un même état de contraintes. Ceçi est incompatible avec les éq. [5] qui indiquent qu'à chaque triplet (ϱ, p, q) correspond une valeur de σ_m, S_2, S_3; à moins que ces équations ne soient indépendantes de p. Il en sera ainsi si les fonctions B et c sont indépendantes de p: $B(\varrho, q)$ et $c(\varrho, q)$.

Surface et loi d'ecoulement

Les invariants σ_m, S_2, S_3 n'étant fonctions que de ϱ et q, en éliminant ϱ et q entre les trois éq. [5], il vient une relation de la forme: $F(\sigma_m, S_2, S_3) = 0$ qui peut s'écrire en fonction des contraintes principales:

$$f(\sigma_1, \sigma_2, \sigma_3) = 0.$$

Ceçi prouve l'existence d'une surface d'écoulement ou critère d'écoulement.

Nous allons reécrire les lois de comportement sous une forme faisant apparaitre le critère d'écoulement, la loi de compréssibilité volumétrique et la relation entre déviateurs de vitesses de déformations et contraintes. Posons:

$$B' = B/S_2 \quad \text{et} \quad C = c/S_2$$

B' peut être exprimé en fonction de C en se servant de l'éq. [5] et en écrivant que $S_{2/S_2} = 1$. Il vient:

$$B' + Cq^3 = \pm[1 - 1/6C^2(1 - 6q^6)]^{1/2}.$$

Dans l'espace des contraintes principales, considérons un plan déviatoire. Les tenseurs s_{ij} et e_{ij} sont représentés dans ce plan par deux vecteurs $\vec{s}$ et $\vec{e}$ de modules S_2 et E_2 respectivement, et faisant un angle θ entre eux. L'on a:

$s_{ij}e_{ij} = \vec{s} \cdot \vec{e} = S_2 E_2 \cos\theta$

d'où:

$\cos\theta = s_{ij/S}\, e_{ij/E_2}.$

En remplaçant s_{ij} par son expression tirée de [4], il vient:

$\cos\theta = B' + Cq^3 = \pm [1 - 1/6\, C^2(1 - 6q^6)]^{1/2}.$

Pour choisir le signe, il suffit de remarquer que $s_{ij}e_{ij}$ n'est autre que l'énergie de distorsion qui doit être positive.

Finalement la loi de comportement prend la forme:

$F(\sigma_m, S_2, S_3) = f(\sigma_1, \sigma_2, \sigma_3) = 0$

surface d'écoulement

$\sigma_m = A(\varrho)$

compréssibilité volumétrique

$\cos\theta = [1 - 1/6\, C^2(1 - 6q^6)]^{1/2}$

relation entre déviateurs de vitesses de déformations et contraintes C: fonction de ϱ et q.

[6]

Recherche d'une loi d'écoulement

En ce qui concerne la loi de compressibilité volumétrique, il est illusoire de chercher à préciser la forme de la fonction $A(\varrho)$. En effet la nature physique de cette compréssibilité varie grandement selon la nature du matériau: pour les métaux, la compréssibilité est essentiellement élastique; pour les milieux granulaires elle est partiellement élastique (compréssibilité des grains) et partiellement irréversible (fractures de grains et glissements).

Par contre l'on peut espérer préciser la relation entre le déviateur de vitesses de déformations et les contraintes, et ceçi quel que soit le matériau plastique parfait. Il est généralement admis que le déviateur de déformations (changement de forme) est dû, quel que soit le matériau, à des glissements entre cristaux ou grains formant le matériau. Seul varie la loi régissant ces glissements (relation entre contraintes normales et tangentielles). La relation entre le déviateur de déformations et les contraintes résultera donc de l'orientation des glissements et de la loi régissant ces glissements. Or ces mêmes facteurs déterminent le critère d'écoulement. Il ne semble donc pas illogique de rechercher une relation entre déviateur de vitesses de déformations et surface d'écoulement. Par contre il ne semble pas exister de relation entre surface d'écoulement et compréssibilité volumétrique, qui font intervenir des phénomènes de nature physique différente.

Nous poserons comme hypothèse:

Hypothèse 3:

Pour tous les matériaux, en écoulement plastique parfait, il existe une même relation entre surface d'écoulement et déviateur de vitesses de déformations.

Toute relation proposée entre déviateur de vitesses de déformations et surface d'écoulement doit retrouver comme cas particuliers les quelques relations admises pour certaines classes de matériaux.

Rapportons l'espace aux axes principaux des contraintes de vecteurs unitaires $(\vec{i}_1, \vec{i}_2, \vec{i}_3)$. Le tenseur σ_{ij} est représenté par le vecteur $\vec{\sigma}(\sigma_1, \sigma_2, \sigma_3)$ et le déviateur e_{ij} par le vecteur $\vec{e}(e_1, e_2, e_3)$. La normale unitaire à la surface d'écoulement sera notée par $\vec{n}(n_1, n_2, n_3)$. Nous noterons par $\vec{n}_d$ le vecteur unitaire, projection de $\vec{n}$ sur le plan déviatoire ($\vec{n}_d$ est normal à la section de la surface d'écoulement par un plan déviatoire).

Citons les différentes relations proposées entre surface d'écoulement et déviateur de vitesses de déformations et essayons de fixer le domaine de validité de chacune d'elles.

a) Pour les matériaux possédant une courbe intrinséque, l'écoulement se fait par cissaillement sur un plan parallèle à la contrainte principale intermédiaire, soit σ. Dans ce cas la surface d'écoulement est indépendante de σ_2 et l'on a $e_2 = 0$. Pour ces matériaux l'on a:

$$\frac{\partial f}{\partial \sigma_2} = 0 \quad \text{entraine} \quad e_2 = 0$$

ou

$$n_2 = 0 \quad \text{entraine} \quad e_2 = 0.$$

b) La théorie du potentiel plastique s'écrit:

$$\varepsilon_{m/E_2} = p = 1/\sqrt{3} \times \frac{(n_1 + n_2 + n_3)}{[(n_1 - n_2)^2 + (n_1 - n_3)^2 + (n_2 - n_3)^2]^{1/2}}$$

$$\vec{e}_{/E_2} = \vec{n}_d.$$

Comme nous l'avons dit, la loi de compressibilité volumétrique semble difficilement admissible. De plus elle indiquerait qu'à un état de contraintes correspond une valeur de p, ce qui est en contradiction avec les éq. [6]. Cette théorie ne sera valable que pour les matériaux incompressibles. Il semble, de plus, qu'elle ne soit valable que pour les matériaux dont le critère d'écoulement est indépendant de la contrainte moyenne σ_m, c'est à dire si $\vec{n} = \vec{n}_d$. Nous admettrons donc:

matériaux incompressibles et $\vec{n} = \vec{n}_d$

entraine $\vec{e}_{/E_2} = \vec{n} = \vec{n}_d$.

c) La théorie de *Brown* et *Gudehus* (ref: 4) décompose la loi d'écoulement en une loi de compressibilité volumétrique et en une relation entre $\vec{e}$ et $\vec{n}_d$. Elle s'écrit:

$\sigma_m = A(\varrho)$ et $\vec{e}_{/E_2} = \vec{n}_d$.

De même que pour le potentiel plastique, il semble que cette théorie n'est valable que pour les matériaux dont le critère d'écoulement est indépendant de la contrainte moyenne; mais ces matériaux peuvent être compressibles. Nous admettrons donc:

matériaux compressibles et $\vec{n} = \vec{n}_d$

entraîne $\vec{e}_{/E_2} = \vec{n} = \vec{n}_d$.

A partir des cas particuliers précedemment cités nous avons essayé de trouver une relation plus générale entre surface d'écoulement et déviateur de vitesses de déformations.

Le cas particulier (a) indique que: $n_2 = 0$ entraine $e_{2/E_2} = 0$; donc nécessairement e_2 est de la forme:

$e_{2/E_2} = \lambda_2 n_2$.

Les indices 1, 2, 3 jouant des roles symétriques, il vient une relation de la forme:

$e_{i/E_2} = \lambda_i n_i$ (non sommation sur i).

Ceçi indique que la relation cherchée relie le déviateur de vitesses de déformations à la normale unitaire $\vec{n}$ à la surface d'écoulement (et non à $\vec{n}_d$). A titre de simplification, nous admettrons l'hypothèse suivante:

Hypothèse 4:

La relation entre $\vec{e}_{/E_2}$ et $\vec{n}$ est locale et du premier ordre.

La relation est locale, signifie que la donnée de la surface d'écoulement au voisinage d'un point détermine entierement $\vec{e}_{/E_2}$ en ce point. La relation est du premier ordre signifie que la donnée du plan tangent, ou de la normale $\vec{n}$ à la surface, en un point, détermine $\vec{e}_{/E}$ en ce point. D'où:

$\vec{e}_{/E_2} = g(\vec{n})$ $\quad g$ étant une fonction vectorielle du seul vecteur $\vec{n}$.

Afin de faciliter la formulation, adoptons en tout point de la surface d'écoulement un repère local orthonormé défini comme suit:

$\vec{u}$: vecteur unitaire de la trisectrice des directions principales des contraintes.

$\vec{n}_d$: normale, unitaire déviatoire, à la section de la surface d'écoulement par un plan déviatoire.

$\vec{t}$: tangente, unitaire déviatoire, à la section de la surface d'écoulement par un plan déviatoire (voir fig. 2).

$\vec{e}_{/E_2}$ est dans le plan déviatoire $(\vec{t} - \vec{n}_d)$ et $\vec{n}$ est dans le plan $(\vec{n}_d - \vec{u})$. Soient: ψ l'angle $(\vec{e}_{/E_2}, \vec{n}_d)$ et φ l'angle $(\vec{n}_d, \vec{n})$; il vient (voir fig. 3):

$$\begin{aligned} \vec{e}_{/E_2} &= \cos\psi\,\vec{n}_d + \sin\psi\,\vec{t} \\ \vec{n} &= \cos\varphi\,\vec{n}_d + \sin\varphi\,\vec{u}. \end{aligned} \qquad [7]$$

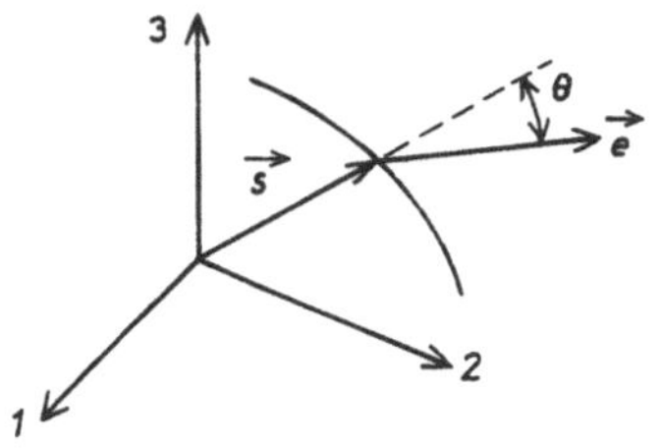

Fig. 1

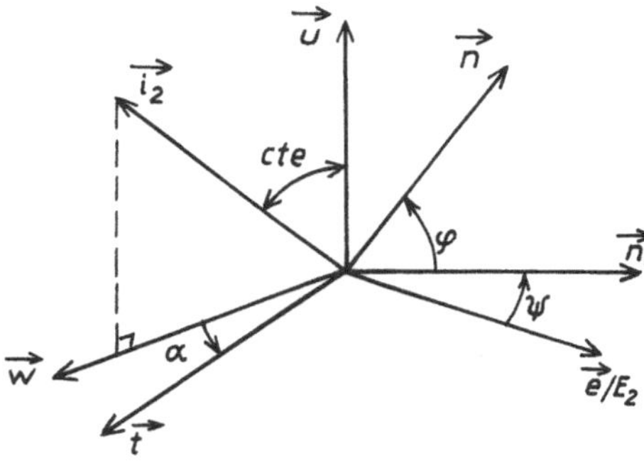

Fig. 2

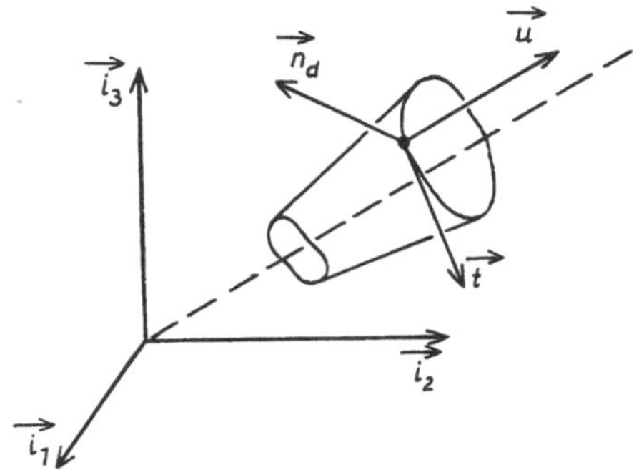

Fig. 3

Pour fixer la position de $\vec{n}$ par rapport aux axes principaux des contraintes $(\vec{i}_1, \vec{i}_2, \vec{i}_3)$, il suffit de fixer la position de $\vec{n}$ par rapport au repère local $(\vec{t}, \vec{n}_d, \vec{u})$ et la position de ce repère par rapport aux axes principaux des contraintes.

L'angle φ fixe $\vec{n}$ par rapport au repère local.

La position de $(\vec{t}, \vec{n}_d, \vec{u})$ est fixée par la donnée de $\vec{t}$ dans le plan déviatoire (car $\vec{u}$ est fixe). Notons par $\vec{w}$ le vecteur unitaire, projection de $\vec{i}_2$ sur le plan déviatoire; la position de $\vec{t}$ sera fixée par la donnée de l'angle α de $\vec{w}$ et $\vec{t}$:

$$\alpha = (\vec{w}, \vec{t}).$$

Donc la normale $\vec{n}$ est déterminée par le couple (φ, α).

D'après l'hypothèse [4], la relation entre $\vec{e}_{/E_2}$ et $\vec{n}$, en un point donné de la surface d'écoulement, s'écrit:

$$\psi = g(\varphi, \alpha) \qquad g\text{: fonction de } \varphi \text{ et } \alpha \qquad [8]$$

D'après l'hypothèse [3] la fonction g est la même pour tous les matériaux plastiques parfaits. De plus elle doit admettre les cas particuliers a, b et c précedemment cités.

D'après la théorie du potentiel plastique et celle de *Brown* et *Gudehus* si $\vec{n} = \vec{n}_d$ c'est à dire si $\varphi = 0$ (critère indépendant de la contrainte moyenne), alors $\vec{e}_{/E_2} = \vec{n} = \vec{n}_d$ c'est à dire $\psi = 0$. Donc $\varphi = 0$ entraine $\psi = 0$; d'où nécessairement ψ est de la forme:

$$\psi = \varphi h(\varphi, \alpha) \qquad h\text{: fonction de } \varphi \text{ et } \alpha \qquad [9]$$

Hypothèse 5:

Nous supposerons, pour raison de simplicité, que la relation cherchée est de la forme:

$$\psi = \varphi h(\alpha).$$

Ecrivons la condition imposée par le cas particulier des matériaux à courbe intrinséque. Si $n_2 = 0$ alors $e_{2/E_2} = 0$ ou encore:

$$\vec{n} \cdot \vec{i}_2 = 0 \quad \text{entraine} \quad \vec{e}_{/E_2} \cdot \vec{i}_2 = 0$$

$\vec{i}_2$ s'exprime dans le repère local $(\vec{t}, \vec{n}_d, \vec{u})$ par (voir fig. 3):

$$\vec{i}_2 = 1/\sqrt{3}\ \vec{u} + \sqrt{2/3}\,[\cos\alpha\,\vec{t} - \sin\alpha\,\vec{n}_d]$$

et $\vec{n} \cdot \vec{t}_2 = 0$ s'écrit: $\varphi = \mathrm{Arctg}\,[\sqrt{2}\sin\alpha]$.

Par ailleurs $\vec{e}_{/E_2} \cdot \vec{t}_2 = 0$ s'écrit: $\psi = \alpha$.

D'où nécessairement:

$$\varphi = \mathrm{Arctg}\,[\sqrt{2}\sin\alpha] \quad \text{entraine} \quad \psi = \alpha.$$

La fonction $h(\alpha)$ s'écrit alors:

$$h(\alpha) = \frac{\alpha}{\mathrm{Arctg}\,[\sqrt{2}\sin\alpha]}.$$

D'où la relation cherchée:

$$\psi = \frac{\alpha}{\mathrm{Arctg}\,[\sqrt{2}\sin\alpha]}\,\varphi. \qquad [10]$$

En définitive la loi de comportement d'un matériau plastique parfait s'écrit:

$$f(\sigma_1, \sigma_2, \sigma_3) = 0$$

critère ou surface d'écoulement

$$\sigma_m = A(\varrho)$$

loi de compréssibilité volumétrique

$$\psi = \frac{\alpha}{\mathrm{Arctg}\,[\sqrt{2}\sin\alpha]}\,\varphi$$

relation entre le déviateur de vitesses de déformations et les contraintes.

Conclusion

1. Un matériau plastique parfait, non visqueux, admet une surface d'écoulement.
2. Il est nécessaire de distinguer, dans la loi d'écoulement, entre la loi de compréssibilité volumétrique et celle reliant le déviateur de vitesses de déformations aux contraintes.
3. Cette derniere relie le déviateur de vitesses de déformations à la normale à la surface d'écoulement. Nous proposons une loi simple indiquant l'influence sur ψ (angle du déviateur de vitesses de déformations et de la normale dans le plan déviatoire) de l'inclinaison de la normale $\vec{n}$ sur la trisectrice (angle φ) et de la pente de la tangente $\vec{t}$ dans le plan déviatoire.

Résumé

Dans une première partie, partant d'une relation générale entre contraintes, vitesses de déformations et masse spécifique, nous démontrons l'existence d'une surface d'écoulement (critère d'écoulement en contraintes). Dans une deuxième partie, nous proposons une loi d'écoulement. Celle-çi admet comme cas particuliers: la théorie du potentiel plastique pour les matériaux incompressibles à critère d'écoulement indépendant de la contrainte moyenne, la théorie de *Brown* et *Gudehus* pour les matériaux compressibles à critère d'écoulement indépendant de la contrainte moyenne, et la théorie de la courbe intrinséque (cisaillement selon un plan) pour les matériaux à critère d'écoulement indépendant de la contrainte intermédiaire.

Littérature

Sawczuk, A. et *P. Stutz*, C.R.A.S. serie A (Paris), **1968**, p. 87–89.

Rivlin, R. S. et *J. L. Ericksen*, Rat. Mech. Anal. **4**, 323–425 (1955).

Lelong, Contribution à l'étude des propriétés mécaniques des sols sous fortes pressions. Th. Ing. Doct. (Grenoble 1968).

Brown, E. H., Proc. Congr. Appl. Mech. (Munich) **1964**, 183.

Adresse de l'auteur:
Prof. Dr. *Mounir Hajal*
Centre d'Etudes Mathématiques
B.P. 3855, Beyrouth (Liban)

Rheol. Acta **12**, 240–244 (1973)

From the Department of Chemical Engineering, University of Cambridge, Cambridge (U.K.),
and the Department of Chemical Engineering, University of Pittsburgh, Pittsburgh, Pennsylvania (U.S.A.)

On the stability of non-isothermal flow in channels

By J. R. A. Pearson and Y. T. Shah

With 4 figures

(Received October 27, 1972)

Notation

b rheological parameter of the fluid defined by eq. [4]
B dimensionless viscosity-temperature parameter defined by eq. [11]
C rheological parameter defined by eq. [4]
h distance between the two parallel plates, ft.
H a thermal transfer coefficient (1/h)
l length of the plates, ft.
p pressure
P inlet pressure
Gz *Graetz* number defined by eq. [11]
t time, h
T mean temperature as defined by eq. [2]
T_1 inlet temperature
$\boldsymbol{u}$ velocity vector with u_x, u_y, u_z as component velocities
v mean velocity vector as defined by eq. [1]
V mean steady state axial velocity
x, y, z *Cartes*ian coordinate system

Subscript

w refers to wall condition

Greek symbols

α thermal diffusivity, ft²/h
A effective thermal diffusivity tensor
ξ dimensionless x coordinate
λ wave number in y direction
Λ dimensionless wave number in y direction
μ_0 viscosity of fluid
ϱ density of fluid
ψ dimensionless velocity in x direction
ω growth rate of disturbances
Ω dimensionless growth rate
$\bar{K}$ proportionality constant for heat generation in eq. [5]

I. Introduction

When polymer melts flow through narrow channel or small-orifice dies, highly unsteady or irregular flow have sometimes been observed even when the imposed boundary conditions are steady and uniform (*Pearson*, 1969).

In this paper we briefly describe the departure from flow uniformity that would occur due to coupling between the energy equation, which describes the heat transfer mechanism between fluid and channel walls, and the flow equation, which includes the temperature dependence of viscosity. We first briefly describe a general model for the problem and then summarize some of the results published in a recent paper (*Pearson* and *Shah*, 1973). Finally we outline the future work that will be pursued on this problem.

II. Theoretical

We consider flow between two closely spaced parallel plates of infinite width and finite uniform length, maintained at constant temperature (see fig. 1). Fluid is introduced at constant pressure, and at a constant temperature above that of the plates, along one "edge" of the pair of plates and emerges at a lower, also constant, pressure at the other edge. The situation is such that a unidirectional flow, at right angles to the edges and parallel to the plates is to be expected. However it is found that for certain ranges of system parameters, the curve of volume flux against pressure difference (between the "edges") is not single valued, and it is usually assumed that a forbidden range of outputs therefore exists. If the flow is achieved by imposing a given pressure difference, then no difficulties in interpretation arise; if however a flow is established by imposing a fixed volume flux, within this forbidden range, then the predicted unidirectional flow would seem unstable.

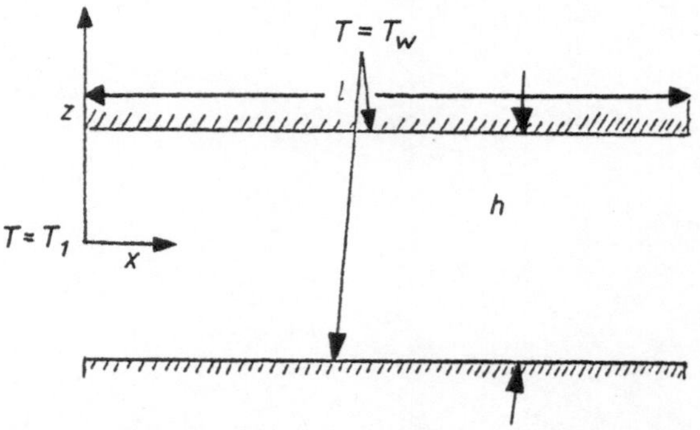

Fig. 1. Geometry of flow region and coordinate system

We introduce here the possibility of the flow adopting a less symmetric pattern, by allowing non-zero velocities in both directions parallel to the bounding plates. To simplify the analysis, we shall suppose the departures from the known steady unidirectional solution to be small and we shall linearize the resulting equations for these small disturbances. Unsteady disturbances are included. What we achieve is a linearized stability analysis.

The results obtained so far indicate that the unidirectional flow pattern is unstable to small disturbances for a range of system parameters wider than that given by the forbidden range of outputs derived from a purely unidirectional solution, thus indicating that a complex flow pattern would arise in practice for realizable inlet and wall conditions.

We use a *Cartesian* co-ordinate system. A fluid at inlet temperature, T_1, flows through a channel of constant depth h, length l and infinite width. The walls are maintained at constant temperature T_w. In general, the flow pattern would be described by a three-dimensional velocity vector $\mathbf{u}(\mathbf{x})$; however here we assume that the channel is narrow, i.e. $h \ll l$ and so $u_z \approx 0$. u_x and u_y will still be functions of x, y, and z, as will the local temperature T^*. Following the methods of lubrication theory (*Pearson*, 1966) we shall work in terms of a two-dimensional mean velocity vector (with x and y components only)

$$\boldsymbol{v}(x, y) = \frac{1}{h}\int_0^h \boldsymbol{u}(x, y, z)\, dz \qquad [1]$$

and a mean convected temperature

$$T(x, y) = \frac{\int_0^h T^*(x, y, z)\, |\boldsymbol{u}|\, dz}{|\boldsymbol{u}|}. \qquad [2]$$

Here we have implicitly assumed that u_x/u_y is independent of z.

In choosing flow and energy equations for the system, which we shall state but not derive here, we shall suppose that similarity profiles exist for u and T^* as functions of z at fixed (x, y); this assumption is the key postulate of lubrication theory. This means that the dependence on x and y in the integrals of [1] and [2] can be taken outside the integral sign. It also means that $\partial T^*/\partial z$ at the walls is linearly related to T, so that a fixed thermal transfer coefficient may be employed. Finally we shall suppose that the fluid is a linear viscous fluid, with viscosity an exponentially decreasing function of temperature, and that it is incompressible. The equations defining the flow then become conservation of mass

$$\nabla \cdot \boldsymbol{v} = 0. \qquad [3]$$

Stress equilibrium

$$\boldsymbol{v} = -C \exp[b(T - T_w)]\, \nabla p. \qquad [4]$$

Energy

$$\frac{\partial T}{\partial t} + (\boldsymbol{v} \cdot \nabla)\, T = H(T_w - T) + \nabla \cdot (\boldsymbol{A} \cdot \nabla T) + \bar{K} V^2 \exp[-b(T - T_w)]. \qquad [5]$$

Here C and b are essentially rheological parameters of the fluid, and H is a thermal transfer coefficient. C has the dimensions of h^2/μ_0 where μ_0 is the viscosity of the fluid at the wall temperature (for a fully developed parabolic flow, for example, $C = h^2/12\,\mu_0$). H has the dimensions of α/h^2 where α is the thermal diffusivity; the coefficient of proportionality is the local *Nusselt* number, which is taken to be constant.

The second and third terms on the right hand side in eq. [5] represent the contributions of conduction and heat generation respectively. $\boldsymbol{A}$ is the effective thermal diffusivity tensor; $\bar{K}$ is a proportionality constant for heat generation, which is numerically equal to $1/C\varrho C_p$, where ϱ is the density and C_p the specific heat. For convenience, a time-dependent term has been introduced in [5], to cover the case of unsteady flow; none of the arguments used above are altered by introducing the extra variable, time t. *Pearson* and *Shah* have reported the analysis of eq. [3]–[5] for the case when the contributions of the second and third terms on the right hand side of eq. [5] are negligible. For this case the boundary conditions, in the (x, y) plane, are taken to be

$$p(0, y) = P \qquad [6\text{i}]$$

$$p(l, y) = 0 \qquad [6\text{ii}]$$

$$T(0, y) = T_1. \qquad [6\text{iii}]$$

We will first describe briefly the important results of this analysis.

IV. Absence of conduction and heat generation

A. Steady Unidirectional Flow

A solution to eqs. [3–6] in terms of the x-coordinate only for this case is given by (*Pearson* and *Shah*, 1973)

$$\boldsymbol{v}^{(0)} = (V, 0)$$

$$T^{(0)} = T_w + (T_1 - T_w)\, e^{-Hx/V} \qquad [8]$$

and

$$P - p^{(0)}(x) = \int_0^x \exp[-b(T_1 - T_w)\, e^{-Hx/V}]\, \frac{V}{C}\, dx. \qquad [9]$$

An explicit relationship between P and V is given by

$$\frac{PC}{Vl} = \int_0^1 \exp[-B e^{-\xi/Gz}]\, d\xi \qquad [10]$$

where

$$\xi = x/l; \quad B = b(T_1 - T_w); \quad Gz = V/Hl. \qquad [11]$$

Thus, the dimensionless parameters important in this case are B, a viscosity-temperature parameter, and Gz, the *Graetz* number. From eq. [10] it is easy to note that for certain ranges of (B, Gz), $dP/dV < 0$, namely when

$$2\int_0^1 \exp[-B e^{-\xi/Gz}]\, d\xi \le \exp[-B e^{-Gz^{-1}}]. \qquad [12]$$

The curve in the (B, Gz)-plane representing equality in [12] is drawn in fig. 2, and the region of multivaluedness of V as a function of P is shown. As explained earlier, portions of the (P, V) curve which are such that $dP/dV \leq 0$, are not thought to correspond to physically realizable situations. As explained by *Pearson* and *Shah* (1973), this condition on dP/dV could cause a hysteresis effect.

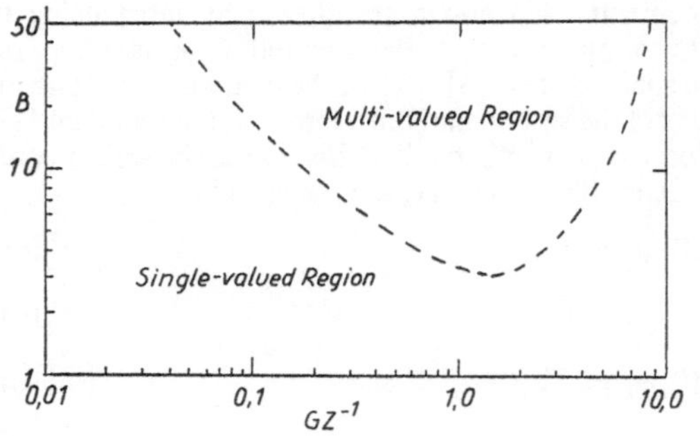

Fig. 2. Solution of

$$2 \int_{\exp(-Gz^{-1})}^{1} e^{-Bu} du = Gz^{-1} \exp(-Be^{-Gz})$$

B. Small Perturbation Analysis

For small disturbances from the unidirectional $\mathbf{v}^{(0)}, p^{(0)}, T^{(0)}$ flow, the solution can be written in the form

$$\left.\begin{aligned} \mathbf{v} &= \mathbf{v}^{(0)} + \mathbf{v}^{(1)}(x, y, t) \\ T &= T^{(0)}(x) + T^{(1)}(x, y, t) \\ p &= p^{(0)}(x) + p^{(1)}(x, y, t) \end{aligned}\right\} + \text{higher order terms} \qquad [13]$$

where $v^{(1)}$, $T^{(1)}$ and $p^{(1)}$ are small so that terms quadratic in them can be neglected. Considering typical solutions of the form

$$\mathbf{v}^{(1)}(x, y, t) = \mathbf{u}^{(1)}(x) \exp(-i\lambda y + \omega t) \qquad [14]$$

$$T^{(1)}(x, y, t) = s(x) \exp(-i\lambda y + \omega t)$$

and

$$p^{(1)}(x, y, t) = q(x) \exp(-i\lambda y + \omega t)$$

where λ is real positive and ω complex, *Pearson* and *Shah* (1973) have derived the resulting eigen-value equation. This third order eigenvalue equation can be expressed as

$$\begin{aligned} &\psi''' + \psi''(BGz^{-1}e^{-\xi/Gz} + Gz^{-1} + \Omega) \\ &+ \psi'(\Omega BGz^{-1}e^{-\xi/Gz} - \Lambda^2) \\ &+ \psi\Lambda^2(Gz^{-1}Be^{-\xi/Gz} - Gz^{-1} - \Omega) = 0 \end{aligned} \qquad [15]$$

where

$$\Omega = \omega l/V; \quad \Lambda = \lambda l; \quad \psi = u/V$$

$$\psi' = \frac{d\psi}{d\xi}.$$

Eq. [15] is subject to the boundary conditions

$$\psi'(0) = \psi'(1) = 0; \quad \psi''(0) = \Lambda^2\psi(0). \qquad [16]$$

Eq. [15] and conditions [16] are homogeneous in ψ. The physical parameters are represented by B and Gz. The range of solutions [14] is defined by the parameters Λ (a wave number) and Ω ($i\Omega$ is a frequency). For real Ω one can define the projections

$$\tilde{\Omega}(B, Gz) = max\,\Omega(\Lambda, B, Gz)$$

$$B_{min}(Gz) \text{ such that } \tilde{\Omega}(B_{min}, Gz) = 0.$$

From eq. [15] it can be seen that for fixed (B, Gz) any sum or integral of solutions of type [14] for eigen-value pairs (Λ, Ω) is itself a solution of the eigen-value problem. The long-time behavior of any general infinitesimal disturbance will be dominated by the largest value of Ω possible, i.e. $\tilde{\Omega}$. The curve representing $\tilde{\Omega} = 0$ in the (B, Gz)-plane is called the neutral stability curve (it may have branches). This will divide the (B, Gz)-plane into regions of stability and instability.

Results and discussion

A solution to eq. [15] with conditions [16] was obtained numerically on a computer using a fourth order *Runge-Kutta* method. The details of the procedure for obtaining solutions are described by *Pearson* and *Shah* (1972). We will briefly summarize here the important results.

First we note again that no complex eigen values Ω were obtained. Physically this means that periodic exchange of energy between fluid and walls is not possible.

Fig. 3 shows results that were obtained for $\Omega = 0$ in the ranges $10^{-7} < \Lambda^2 < 5 \times 10^4$; $0 < B < 50$; $0.02 < Gz < 100$ which are believed to include all those of practical interest. The regions in the (B, Gz)-plane marked y are those for which no eigenvalues were found; those marked Y the ones for which one or more eigen values Λ were obtained. Curve A is what we may expect, on physical grounds, to be a neutral stability curve separating stable from unstable regions; Region B is a more curious one on which $\Lambda^2 \to \infty$, for which the initial model breaks down, and which we do not fully understand. By examining positive as well as negative values of Ω *Pearson* and *Shah* (1973) have shown that the lower plot shown in fig. 3 is indeed a neutral stability plot.

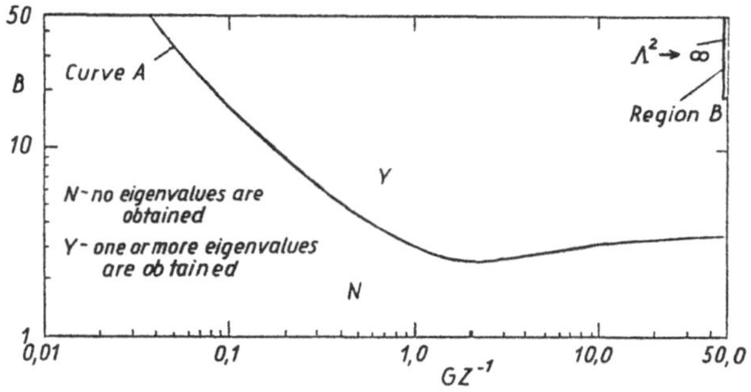

Fig. 3. Stability plot in (B, Gz) plane from eq. [15]–[16] $(\Omega = 0)$

Pearson and *Shah* (1973) have also shown that as $Gz \to \infty$, $B/Gz \to 1.55$; and as $Gz \to 0$, the curve A in fig. 3 tends to a limiting value of approximately 3.6.

The results shown in fig. 2 and 3 reveal some interesting aspects of the problem. We note that for $B < B_0 \approx 2.4$, the flow is always stable. Thus, in physical terms, whatever the flow rate, there will be no instability provided the viscosity ratio between inlet and wall temperatures is less than $e^{2.4} \approx 11$.

If we consider the case of some fixed flow rate, i.e. fixed Gz, and increase the temperature difference, i.e. B, from zero, we reach infinitesimal instability well before multivaluedness for $Gz < 1$.

Equally if we fix the temperature difference, B, and increase the flow rate, Gz, from zero, for B in the range $3.6 > B > B_0$, then again we reach infinitesimal instability before multivaluedness.

V. Entrance and exit pressure loss

By virtue of conditions (6i) and (6ii) *Pearson* and *Shah* (1973) neglected pressure loss due to entrance and exit effects. In the presence of entrance pressure losses, an alternative pressure boundary condition can be taken crudely to be

$$u(0) + Q_0 q(0) = 0; \quad Q_0 > 0 \qquad [17]$$

where Q_0 is an appropriate constant, because linear equations are being investigated.

Similarly,

$$u(1) + Q_1 q(1) = 0; \quad Q_1 < 0 \qquad [18]$$

for exit pressure loss.

For conditions [17] and [18], numerical results similar to the ones described in fig. 3 can be sought numerically as before. Using condition [17] in place of [6i], the effect of entrance pressure loss on the stability plot was examined. These results are shown in fig. 4. The results indicate, as one might expect, that the entrance pressure loss reduces the unstable region. Similar results can be obtained in the presence of the exit pressure loss.

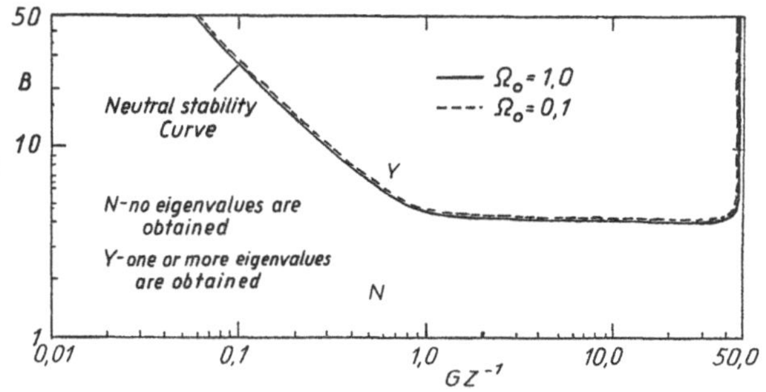

Fig. 4. Stability plot in (B, Gz) plane in the case of entrance pressure loss

VI. Future work

Presently, work is being pursued to evaluate the effects of the second and third terms on the right-hand side of eq. [5] on the stability plot described in fig. 3. Few remarks about this work can be made yet. The introduction of the conduction term will result in a fourth order eigenvalue equation. The relevant boundary conditions on temperature for this case are of the form

$$T - T_1 = \frac{\alpha}{V}\frac{\partial T}{\partial x} \quad \text{at} \quad x = 0 \qquad [19]$$

$$\frac{\partial T}{\partial x} = 0 \quad \text{at} \quad x = l \qquad [20]$$

The heat generation term in eq. [5] does not change the order of the resulting eigenvalue equation. However, the resulting eigenvalue equation is highly non linear in some of the parameters. The results should be of considerable importance because multivaluedness in the (P, V) relation has been shown to arise in very long channels (*Martin*, 1967).

We have not considered here the case when non-unidirectional disturbances [14] grow to finite size. However, we would expect the flow to settle down to some non uniform pattern, probably steady.

Finally, it should be noted that the present model also covers the case of isothermal flow of structured fluid, whose apparent viscosity increases exponentially with structure and whose structural state is governed by a first-order rate equation. Details are given in *Pearson* and *Shah* (1972).

Acknowledgement

The help of the University of Pittsburgh Computing Center and that of Mr. *Ron Schleppy* are gratefully acknowledged.

Summary

We study here stability of non-isothermal flow between two closely spaced, heat conducting, infinite parallel flat plates of length l and distance h apart. Fluid enters uniformly along $x = 0$ at temperature $T_1 > T_w$ the plate temperature. The flow non-uniformity is assumed to occur due to coupling between the energy equation, which describes the heat transfer mechanism between fluid and channel walls, and the flow equation which includes the temperature dependence of viscosity.

The model for the flow assumes that similarity profiles exist for velocity and temperature in the flow direction. The stability of the unidirectional flow by a linearized first order perturbation analysis of the proposed model is examined.

References

Pearson, J. R. A., Plastics and Polymers **37**, 285 (1969). — *Pearson, J. R. A.* and *Y. T. Shah*, Stability of Non-Isothermal Flow in Channels (Paper acepted by Chem. Eng. Sci.). – *Pearson, J. R. A.*, Mechanical Principles of Polymer Melt Processing, p. 60 (Oxford, England, 1966).

Martin, B., Intl. J. Non-Linear Mech. **2**, 285 (1967).

Authors' addresses:

Dr. *J. R. A. Pearson*
Department of Chemical Engineering
Chemical Technology
Imperial College, London (U.K.)

Dr. *Y. T. Shah*
Department of Chemical Enginnering
University of Pittsburgh, Pittsburgh Pa 15213 (U.S.A.)

Rheol. Acta **12**, 245–252 (1973)

Recherche au C.N.R.S., Institut de Mécanique de Grenoble, Laboratoire Associé au C.N.R.S. n° 6, Domaine Universitaire, Grenoble (France)

Sur la résolution numérique des problèmes à grandes déformations élastiques: cas de la symétrie sphérique

Par P. Guélin

Avec 1 figure

(Reçu p.p. le 27 octobre 1972)

Introduction

En examinant deux situations particulières, on montre qu'il est possible de développer une méthode de formulation et une méthode de résolution numérique des problèmes à grandes déformations pour une assez large gamme de corps: ceux qui obéissent au principe de superposition en repères rhéologiques (co-rotationnels) d'incréments de contrainte affectés ou non d'une mémoire.

La première partie de l'étude introduit la méthode de formulation grâce à l'examen du problème des grandes déformations de corps isothermes, homogènes, isotropes, compressibles, à comportement de solide élastique. Au moyen d'un traitement particulier des conditions aux limites (§ 1) et d'une linéarisation nouvelle de la théorie classique des petites déformations successives (§ 2), on rend la formulation du problème (§ 3), d'une part adaptée à la prise en compte de la gamme de corps évoqués ci-dessus (§ 4), et d'autre part spécialement compatible avec la méthode de résolution examinée dans la deuxième partie du mémoire.

Les bases en sont posées (§ 6) à propos du problème des coques à symétrie sphérique (§ 5); l'extension en est faite pour les problèmes axisymétriques (§ 7) et des résultats illustrent le rôle des non linéarités (§ 8). En vue d'une relaxation ultérieure, pratiquée avec ou sans pénalisation des opérateurs, on cherche à mettre sous une forme matricielle commode et précise une linéarisation du problème aux limites. On montre qu il est possible de construire une telle forme matricielle au moyen de développements de la solution cherchée en séries tronquées de fonctions orthogonales associées à une base ponctuelle quelconque. On présente les résultats des nombreux contrôles auxquels sont soumises les solutions obtenues.

1. Notations

L'espace est rapporté au système de coordonnées cartésiennes $(0, x^1, x^2, x^3)$. Le repère associé, d'origine 0, de vecteurs $\vec{\bar{e}}_i (i = 1, 2, 3)$, orthonormé, fixe, est noté (F). Le vecteur: $\overrightarrow{0M} = x^i \vec{\bar{e}}_i$ définit la position du point matériel M du domaine Ω, ouvert borné de R^3, simplement connexe, de frontière S régulière, occupé par un corps unique homogène isotrope isotherme élastique compressible en évolution quasi statique à partir d'un état initial neutre.

On envisage de déterminer une suite discrète finie d'états à l'équilibre de ce corps. Si t désigne un scalaire associé à une configuration d'équilibre, le problème est de déterminer la configuration à $t + \delta t$, où δt est petit mais fini, lorsque celle à t est connue.

Soit (N_t) le champ des repères naturels, d'origine $M(t)$, de vecteurs $\vec{e}_i(t)$, généralement non orthonormés, définis en considérant un système de coordonnées curvilignes $y^i (i = 1, 2, 3)$ matérielles, entraînées par le milieu continu. Pour assurer un traitement commode des conditions aux limites, nous choisissons le changement de coordonnées

de telle sorte que S soit formé de trois couples de nappes $S(i, \alpha) (i = 1, 2, 3; \alpha = 1, 2)$ de surface matérielles, régulières:

$$y^i = C(i, \alpha) \qquad [2]$$

constituant un hexaèdre curviligne. Chaque nappe est une surface à y^i constant. Les six constantes C sont donc déterminées par [1] (à $t = 0$) et par la géométrie initiale.

Soit $\overrightarrow{\delta U}$ le champ de vecteurs déplacement faisant passer de l'état t à l'état $t + \delta t$. Ce champ est de la forme $\overrightarrow{KW}$ avec $\overrightarrow{0M}(t + \delta t) = \overrightarrow{0M}(0) + \vec{U}(t) + \overrightarrow{KW} = \overrightarrow{0M}(t) + \overrightarrow{\delta U}$, formule où δt est choisi de telle sorte que les modules de $\vec{U}$ et $\vec{W}$ soient du même ordre et où K désigne un paramètre scalaire petit devant l'unité. Nous supposons que les champs tensoriels construits sur [1] sont continûment dérivables, de sorte qu'il est possible de calculer au premier ordre en K leurs variations entre t et $t + \delta t$. Rappelons que l'on a, d'une part:

$$\vec{e}_i \cdot \vec{e}_j = g_{ij} = \frac{\partial x^K}{\partial y^i} \frac{\partial x^K}{\partial y^j}$$

$$g^{ij} = \frac{1}{g} \operatorname{cof} g_{ji}; \qquad g = \det g_{ij}$$

$$2[ij, m] = \frac{\partial g_{mj}}{\partial y^i} + \frac{\partial g_{im}}{\partial y^j} - \frac{\partial g_{ij}}{\partial y^m}$$

$$\Gamma_{ij}^K = g^{Km}[ij, m]$$

$$\frac{Da_i}{Dy^j \,|\, (N_t)} = \frac{\partial a_i}{\partial y^j} - a_K \Gamma_{ij}^K$$

$$\frac{Da^{ij}}{Dy^j\,|\,(N_t)} = \frac{\partial a^{ij}}{\partial a^j} + a^{jr}\Gamma^i_{jr} + a^{ij}\Gamma^r_{jr} \qquad [3]$$

et d'autre part:

$$\delta g_{ij} = \frac{D\delta U_i}{Dy^j\,|\,(N_t)} + \frac{D\delta U_j}{Dy^i\,|\,(N_t)}$$

$$\delta g^{ij} = -g^{ir}g^{js}\delta g_{rs}$$

$$\delta g = g g^{ij}\delta g_{ij}$$

$$\delta\Gamma^K_{ij} = g^{Km}\delta\,[ij, m] + [ij, m]\,\delta g^{Km}$$

$$2\delta[ij, m] = \frac{\partial\delta g_{mj}}{\partial y^i} + \frac{\partial\delta g_{im}}{\partial y^j} - \frac{\partial\delta g_{ij}}{\partial y^m}$$

$$\delta\left[\frac{Da^{ij}}{Dy^j}\right] = \frac{D\delta a^{ij}}{Dy^j\,|\,(N_t)} + a^{jr}\delta\Gamma^i_{jr} + a^{ij}\delta\Gamma^r_{jr}. \qquad [4]$$

Dans les formules [3], $D/Dy\,|\,(N_t)$ note la dérivée covariante dans (N_t), et dans [4] tous les éléments non affectés du signe δ (indiquant la variation) ont pour valeurs celles de la configuration à t.

2. Expression du comportement élastique

On convient ici de noter G et g les composantes des tenseurs métriques à t et 0 respectivement.

Si W désigne l'énergie potentielle de déformation du corps par unité de volume de l'état initial neutre, on définit habituellement les composantes σ^{ij} du tenseur de contrainte σ à l'état t par:

$$\sigma = \sigma^{ij}\vec{e}_i \otimes \vec{e}_j = (\varphi g^{ij} + \psi B^{ij} + p G^{ij})\,\vec{e}_i \otimes \vec{e}_j$$

$$W = W(I_1, I_2, I_3)$$

$$\varphi = \frac{2}{\sqrt{I_3}}\frac{\partial W}{\partial I_1}; \qquad \Psi = \frac{2}{\sqrt{I_3}}\frac{\partial W}{\partial I_2}$$

$$p = 2\sqrt{I_3}\,\frac{\partial W}{\partial I_3}$$

$$B^{ij} = (g^{ij}g^{rs} - g^{ir}g^{js})\,G_{rs}$$

$$I_1 = g^{rs}G_{rs}; \quad I_2 = I_3 g_{rs}G^{rs}; \quad I_3 = G/g \qquad [5]$$

et on calcule les variations au premier ordre de ces quantités. Retenons les suivantes:

$$\delta B^{ij} = (g^{ij}g^{rs} - g^{ir}g^{js})\,\delta G_{rs}$$

$$\delta I_1 = g^{rs}\delta G_{rs}; \qquad \delta I_2 = g_{rs}(I_3\delta G^{rs} + \delta I_3 G^{rs})$$

$$\delta I_3 = \frac{\delta G}{G} = I_3 G^{rs}\delta G_{rs}$$

$$\delta G^{ij} = -G^{ir}G^{js}\delta G_{rs}. \qquad [6]$$

Au lieu de calculer $\delta\sigma$ par différenciation directe de σ à l'aide de [5], posons:

$$\frac{\partial W}{\partial I_i} = A_i = \alpha^i_0 + \alpha^i_1 I_1 + \alpha^i_2 \quad (i = 1, 2, 3) \qquad [7]$$

Dans ces expressions α^i_0 et α^i_1 dépendent linéairement des coefficients m_1, n_1, l_2, m_2 et n_2, coefficients qui définissent localement le comportement élastique. Par contre, les α^i_2 sont des polynomes en I, II et III. Compte denu de [5] et [7], nous écrivons:

$$W = m_1 I^2 + n_1\mathrm{II} + l_2 I^3 + m_2 I\,\mathrm{II} + n_2\mathrm{III} + \cdots$$

$$2\mathrm{I} = I_1 - 3; \quad 4\mathrm{II} = I_2 - 2I_1 + 3;$$

$$8\mathrm{III} = I_3 - I_2 + I_1 - 1$$

$$A_1 = \frac{1}{2}\frac{\partial W}{\partial \mathrm{I}} - \frac{1}{2}\frac{\partial W}{\partial \mathrm{II}} + \frac{1}{8}\frac{\partial W}{\partial \mathrm{III}}$$

$$A_2 = \frac{1}{4}\frac{\partial W}{\partial \mathrm{II}} - \frac{1}{8}\frac{\partial W}{\partial \mathrm{III}}$$

$$A_3 = \frac{1}{8}\frac{\partial W}{\partial \mathrm{III}}$$

$$\frac{\sigma^{ij}\sqrt{I_3}}{2} = A_1 g^{ij} + A_2 B^{ij} + I_3 A_3 G^{ij}. \qquad [8]$$

Portant [7] dans [8], effectuant la variation au premier ordre de l'expression de σ ainsi obtenue, et tenant compte de [5] et [6], on obtient:

$$\delta\sigma = [a^{irjs}\delta G_{rs} + \delta b^{ij}]\,\vec{e}_i(t) \otimes \vec{e}_j(t) \qquad [9]$$

forme fondamentale de l'incrément de contrainte (entre les états t et $t + \delta t$) qui s'explicite selon:

$$\frac{\sqrt{I_3}}{2}\delta b^{ij} = \delta\,[\alpha^1_2 g^{ij} + \alpha^2_2 B^{ij} + I_3\alpha^3_2 G^{ij}]$$

$$\frac{\sqrt{I_3}}{2}a^{irjs} = (\alpha^2_0 + \alpha^2_1 I_1)(g^{ij}g^{rs} - g^{ir}g^{js}) + I_3(\alpha^3_0 + \alpha^3_1 I_1)(G^{ij}G^{rs} - G^{ir}G^{js}) + \frac{\sqrt{I_3}}{2}\frac{1}{2}\sigma^{ij}G^{rs} + (\alpha^1_1 g^{ij} + \alpha^2_1 B^{ij} + \alpha^3_1 I_3 G^{ij})\,g^{rs}. \qquad [10]$$

Les relations [9] et [10] ne constituent qu'une forme particulière de l'expression classique:

$$\delta\sigma^{ij} = \delta\varphi g^{ij} + \delta\psi B^{ij} + \delta p G^{ij} + \psi\delta B^{ij} + p\delta G^{ij}$$

que l'on peut obtenir et expliciter à partir de [5] et [6] [cf. (1)].

Pour justifier l'utilité de cette transformation des relations classiques, il faut noter les points suivants. Tout d'abord on montrera (cf. §4) que la forme [9] est celle à laquelle on parvient lors de l'examen de la classe de corps non élastique qui nous intéresse. De sorte que la portée physique de la méthode de formulation s'en trouve étendue. De plus, la base tensorielle $\vec{e}_i \otimes \vec{e}_j$ est alors remplacée par une base orthonormée, de sorte qu'il est légitime de ne pas envisager à ce niveau de l'exposé la variation de la base $\vec{e}_i \otimes \vec{e}_j$ qu'il est classique de prendre en compte indirectement en appliquant la dernière formule [4] à l'équation d'équilibre indéfini [cf. (1)].

Ensuite, la linéarisation [7] permet de ménager une extension de la linéarisation infinitésimale qui est complète en ce sens qu'elle prend en compte une élasticité d'ordre quelconque.

Cette formulation contient évidemment le cas de l'élasticité infinitésimale. En effet, si on envisage l'évolution de 0 à δt avec δt tel que: $g = G, I_3 = 1$ et si on introduit les paramètres de *Lame* λ et μ en posant : $2m_1 = \lambda + 2\mu, n_1 = -2\mu$, on obtient: $2\alpha_1^1 = \frac{1}{2}(\lambda + 2\mu)$ et: $2\alpha_0^2 = -\mu$, seuls coefficients non nuls en dehors de α_0^1. D'après [10], on voit immédiatement que δb est nul et que l'expression de a^{irjs} se présente sous la forme classique d'élasticité infinitésimale.

Enfin, on verra que le fait de concentrer les non linéarités de la loi de comportement dans la partie δb de σ semble favorable au point de vue de la convergence des calculs numériques.

3. Formulation du problème aux limites

Il s'agit de déterminer la configuration à $t + \delta t$ connaissant celle à t considérée comme une condition initiale.

Compte tenu du choix [1], [2] des surfaces $S(i, \alpha)$ le vecteur contrainte associé à l'élément de $S(i, \alpha)$ dans sa configuration $S_t(i, \alpha)$ à t est $\vec{F}_{(i)}$ tel que

$$\sigma_{(t)}^{(i)j} \vec{e}_j(t) = \sqrt{g^{(ii)}(t)}\, \vec{F}_{(i)} \qquad [11]$$

sans sommation sur i. Dans la suite $\vec{F}_{(i)}$ est une donnée imposée sur $S_t(i, \alpha)$. Dans le cas général il est en principe commode de la définir dans (F) par ses composantes $\overset{*}{\vec{F}}{}^K_{(i)}$. Dans le cas particulier très important où elle correspond à une pression hydrostatique P on peut la définir dans (Nt) par:

$$\vec{F}_{(i)} = P \frac{\overset{*}{\vec{e}}{}^{(i)}}{\sqrt{g^{(ii)}}} = P \vec{n}_{(i)} \, (\varepsilon_{ijK} \overset{*}{\vec{e}}{}^K = \vec{e}_i \times \vec{e}_j)$$

où $\vec{n}(i)$ désigne la normale extérieure unitaire à $S(i, \alpha)$, de sorte que, avec [11]:

$$\sigma^{ij} = P g^{ij}. \qquad [12]$$

La variation au premier ordre de [12] donne:

$$\delta\sigma^{(i)j} + P g^{(i)r} g^{js} \left(\frac{D\delta U_r}{Dy^s} + \frac{D\delta U_A}{Dy^r} \right) = \delta P g^{(i)j}. \qquad [13]$$

Désignons par ϱ la masse du corps par unité de volume dans la configuration à t et par $\vec{\varphi}$ la densité de force volumique. Les bilans de masse et d'équilibre quasi statique des densités de force sont:

$$\varrho\sqrt{g} = \varrho(0)\sqrt{g(0)}; \qquad \frac{D\sigma^{ij}}{Dy^j \,|\, (N_t)} + \varrho\varphi^i = 0. \qquad [14]$$

Différenciant ces relations au premier ordre, tenant compte de [4], supposant $\delta\varphi^i$ nul et éliminant $\delta\varrho$, on obtient:

$$\frac{D\delta\sigma^{ij}}{Dy^j \,|\, (N_t)} + \sigma^{jr}\delta\Gamma^i_{jr} + \sigma^{ij}\delta\Gamma^r_{jr}$$
$$= \frac{1}{2} g^{rs}\delta g_{rs} \sqrt{\frac{g(0)}{g}}\, \varrho(0)\, \varphi^i \text{ dans } \Omega \qquad [15]$$

équations aux dérivées partielles non linéaires portant sur les δU_i. Avec [11] et [12] on obtient les équations et conditions aux limites du problème:

$$A_\Omega \delta U = \delta f_\Omega \quad \text{sur} \quad \Omega$$
$$A_S^{(i)} \delta\sigma = f_S^{(i)} \quad \text{sur} \quad S^i \qquad [16]$$

où A et A_s sont les opérateurs:

$$(A_\Omega \delta u)^i = \frac{D\delta\sigma^{ij}(\delta u)}{Dy^j} + \sigma^{jr}\delta\Gamma^i_{jr}(\delta u) + \sigma^{ij}\delta\Gamma^r_{jr}(\delta u)$$

$$(A_s^{(i)}\delta\sigma)^K_{(i)} = \delta\sigma^{(i)j} + P g^{ir} g^{js} \delta g_{rs}$$

où δf_Ω et $\delta f_S^{(i)}$ sont les vecteurs de R^3,

$$(\delta f_\Omega)^i = \varrho(0)\,\varphi^i \frac{1}{2} g^{rs}\delta g_{rs} \sqrt{\frac{g(0)}{g}}$$
$$(\delta f_S^{(i)})^K = \delta P g^{(i)K} \qquad [18]$$

et où les équations complémentaires sont [3] à [10].

Rappelons notamment que:

$$\delta\sigma^{ij} = a^{irjs}(\sigma, \delta u)\,\delta g_{rs} + \delta b^{ij}$$

$$\delta g_{rs} = \frac{D\delta u_r}{Dy^s\,|\,(N_t)} + \frac{D\delta u_s}{Dy^r\,|\,(N_t)}. \qquad [19]$$

Les données invariantes durant l'évolution sont: $x^i(y^1, y^2, y^3, 0)$, $m_1, n_1, l_2, n_2, \ldots \vec{F}_{(i)}(t)$ ou $P(t)$ c'est-à-dire la géométrie initiale, les paramètres elastiques et les forces appliquées aux-contours. Le problème aux limites défini par [16] à [19] et exprimé en fonction des δu_i, permet de définir les nouvelles positions du point matériel M:

$$\overline{0M}(t + \delta t) = \left[x^K(t) + g^{im}\delta u_m \frac{\partial x^K}{\partial y^i}\right]\mathring{\vec{e}}_K = x^K(t + \delta t)\,\mathring{\vec{e}}_K.$$

Ce problème est non linéaire dès que la déformation, de composantes:

$$\gamma_{ij} = \tfrac{1}{2}\,[g_{ij}(t) - g_{ij}(0)] \qquad [20]$$

n'est plus infinitésimale.

On constate que les non linéarités dues aux déformations interviennent de façon complexe. Nous verrons d'ailleurs, dans le cas de la symétrie sphérique, qu'elles jouent un rôle important de sorte que le problème d'identification des éventuels paramètres élastiques l_2, m_2, n_2, etc. peut exiger une résolution numérique précise.

4. Extension à certains comportements rhéologiques

La méthode de formulation exposée au paragraphe précédent met en œuvre d'une part une prise en compte particulière des conditions aux limites, d'autre part une linéarisation nouvelle de la théorie des petites déformations élastiques successives. Il s'agit à présent de montrer que la méthode peut être étendue de façon à traiter le cas d'une gamme de corps à comportement non élastique.

a) Le paramètre t a maintenant le sens d'un temps. Au point matériel M les composantes dans (F) des champs de vitesse, de vitesse de transformation linéaire tangente, de vitesse de déformation et de vitesse de rotation sont telles que:

$$v_i = \frac{dx^i}{dt\,|\,(F)} = \left.\frac{\partial x^i}{\partial t}\right|\,y^i = cste$$

$$\mathscr{C}_{ij}dx_j = \frac{d(dx^i)}{dt\,|\,(F)} = \frac{\partial v^i}{\partial x_j}dx_j = (\mathscr{D}_{ij} + R_{ij})\,dx_j$$

$$\mathscr{D}_{ij}dx_idx_j = \frac{d(ds^2)}{dt\,|\,(F)} = \frac{1}{2}\left(\frac{\partial v_i}{\partial x^j} + \frac{\partial v_j}{\partial x^i}\right)dx_idx_j$$

$$\mathscr{R}_{ij} = \frac{1}{2}\left(\frac{\partial v_i}{\partial x^j} - \frac{\partial v_j}{\partial x^i}\right). \qquad [21]$$

Soit (Rh) le champ des repères d'origine M de vecteurs $\vec{f}_i(t)$ $(i = 1, 2, 3)$, orthonormés, définis dans (F) à une rotation près par le champ de vecteur vitesse de rotation instantanée [cf. (2) et (3) § II. A].

$$\vec{\omega} = \omega^k\mathring{\vec{e}}_k = \varepsilon^{kji}\mathscr{R}_{ij}\vec{f}_k$$

$$\mathscr{R}_{ij} + \mathscr{R}_{ji} = 0; \qquad \frac{d\vec{f}_j}{dt\,|\,(F)} = \mathscr{R}^i_j\vec{f}_i. \qquad [22]$$

Ce champ (Rh) sera appelé co-rotationnel ou rhéologique. On désigne par $\mathring{r}$, r et $\bar{r}$ les changements de repères définis par:

$$\vec{f}_\alpha(t) = \mathring{r}^j_\alpha\mathring{\vec{e}}_j; \quad \vec{f}_\alpha(t) = \bar{r}^j_\alpha\vec{e}_j(t); \quad \overset{*}{\vec{e}}{}^i(t) = r^i_\alpha\overset{*}{\vec{f}}{}^\alpha(t). \qquad [23]$$

Compte tenu de [22] et [23], il vient:

$$\mathscr{R}^k_j\mathring{r}^i_k = d\mathring{r}^i_j/dt\,|\,(F). \qquad [24]$$

D'autre part on tire des définitions [23] les relations:

$$\mathring{r}^j_i = \bar{r}^k_i\frac{\partial x^j}{\partial y^k} \qquad [25]$$

et:

$$\bar{r}^i_\alpha g_{ij} r^j_\beta = 1 \quad \text{si} \quad \alpha = \beta 0 \quad \text{si} \quad \alpha \neq \beta. \qquad [26]$$

b) Considérons à nouveau la formulation [16] à [19] en supposant la forme [9] définie à présent dans (Rh) et non dans (Nt):

$$d\sigma = (a^{\alpha\varrho\beta\sigma}dg_{\varrho\sigma} + db^{\alpha\beta})\vec{f}_\alpha \otimes \vec{f}_\beta. \qquad [27]$$

Compte tenu de [23], [27] permet de remplacer [19] par:

$$d\sigma^{ij} = \bar{r}^i_\alpha\bar{r}^i_\beta(a^{\alpha\varrho\beta\sigma}r^l_\varrho r^m_\sigma dg_{lm} + db^{\alpha\beta})$$

$$dg_{rs} = \frac{D\delta u_r}{Dy^s\,|\,(N_t)} + \frac{D\delta du_s}{Dy^r\,|\,(N_t)}. \qquad [28]$$

Les valeurs de $\bar{r}^i_\alpha$ et r^i_α sont définies par [24] à [26], relations qui s'avèrent assez complexes. On constate cependant que [19] et [28] sont formellement peu différents alors que la portée en est, au point de vue physique, beaucoup plus large puisqu'il devient possible d'envisager des corps qui obéissent au principe de superposition dans les repères rhéologiques (co-rotationnels) d'incréments de contrainte affectés ou non d'une mémoire. En effet, considérons par exemple un corps pour lequel la contrainte à l'instant t est la somme dans (Rh) d'accroissement $d\Sigma(\tau)$ affecté d'une mémoire $\chi(t-\tau)$ [cf. (4)]:

$$\sigma_{\alpha\beta}(t)\vec{f}_\alpha(t) \otimes \vec{f}_\beta(t) = \int_{-\infty}^{t} \chi(t-\tau)\, d\Sigma_{\alpha\beta}(\tau) \times \vec{f}_\alpha(\tau) \otimes \vec{f}_\beta(\tau).$$

Le repère (Rh) étant orthonormé et la somme étant faite sur les composantes dans (Rh), cette relation équivaut à:

$$\sigma_{\alpha\beta}(t) = \int_{-\infty}^{t} \chi(t-\tau)\, d\Sigma_{\alpha\beta}(\tau). \qquad [29]$$

Munissons le corps d'une loi de comportement local (ou instantané) simple:

$$d\Sigma(\tau) = d\gamma_{\alpha\beta}(\tau)\vec{f}_\alpha(\tau) \otimes \vec{f}_\beta(\tau). \qquad [30]$$

En introduisant $\mathscr{D}$ dans [30] et en portant dans [29], on obtient:

$$d\sigma_{\alpha\beta}(t) = \int_{-\infty}^{t} \chi(t-\tau)\, \mathscr{D}_{\alpha\beta}(\tau)\, d\tau. \qquad [31]$$

Nous avons défini de façon objective (dans Rh) une loi de comportement de fluide viscoélastique. Notons que cette loi s'exprime dans (Nt) sous la forme non linéaire:

$$\sigma^{ij}(t) = \bar{r}^i_\alpha(t)\bar{r}^j_\beta(t) \int_{-\infty}^{t} \chi(t-\tau)\, r^k_\alpha(\tau)\, r^l_\beta(\tau) \times \mathscr{D}_{kl}(\tau)\, d\tau.$$

Dérivons [31] relativement à t. Il vient:

$$\frac{\partial\sigma_{\alpha\beta}}{\partial t} = \frac{d\sigma_{\alpha\beta}}{dt\,|\,(R_h)} = \chi(0)\,\mathscr{D}(t) + \int_{-\infty}^{t} \chi'_t(t-\tau)\, \mathscr{D}_{\alpha\beta}(\tau)\, d\tau.$$

Soit:

$$d\sigma_{\alpha\beta} = a^{\alpha\varrho\beta\sigma} dg_{\varrho\sigma} + db^{\alpha\beta}$$

avec:

$$a^{\alpha\varrho\beta\sigma} = \tfrac{1}{2}\,\chi(0)\, 1^{\alpha\varrho\beta\sigma}$$

$$db^{\alpha\beta} = \int_{-\infty}^{t} \chi'_t(t-\tau)\, \tfrac{1}{2}\, dg_{\alpha\beta}(\tau)\, d\tau.$$

Ces relations définissent a et db figurant dans [27] et montrent la possibilité d'étendre [9] à des comportements non élastiques.

5. Le problème des coques sphériques élastiques

On considère une coque comprise entre les sphères concentriques de centre 0 de rayon r_1 et $r_2 (0 < r_1 < r_2)$, soumise à une pression interne P et à une pression externe nulle. Le changement de coordonnées [1] est polaire de centre 0 et l'on opère le long de la droite: $y^2 = 0$, $y^3 = \pi/2$. Notons r et R les valeurs de la coordonnée x^1 du point matériel M dans ses positions à l'état neutre et à l'état déformé d'équilibre. Avec [3] et [20] on obtient les composantes non nulles de γ dans le repère N de l'état final:

$$2\gamma_{11} = 1 - \left(\frac{dr}{dR}\right)^2; \qquad 2\gamma_{22} = R^2 - r^2.$$

Avec les formules [5], on obtient:

$$I_1 = \left(\frac{dR}{dr}\right)^2 + 2\frac{R^2}{r^2}$$

$$I_2 = \frac{R^2}{r^2}\left[2\left(\frac{dR}{r^2}\right)^2 + \frac{R^2}{r^2}\right]$$

$$I_3 = \left(\frac{R}{r}\right)^4 \left(\frac{dR}{dr}\right)^2 \qquad [32]$$

et l'énergie élastique, l'équation d'équilibre indéfini et les composantes non nulles du tenseur contraintes, s'expriment par:

$$W = m_1 I^2 + n_1 \mathrm{II} + l_2 I^3 + m_2 \mathrm{I.II} + n_2 \mathrm{III}$$

$$\frac{d}{dr}(R^2\sigma^{11}) = 2R^3\frac{dR}{dr}\sigma^{22}$$

$$\sigma^{11} = 2\frac{dR}{dr}\frac{r^2}{R^2}\left(A_1 + 2A_2\frac{R^2}{r^2} + A_3\frac{R^4}{r^4}\right)$$

$$R^2\sigma^{22} = 2\frac{dr}{dR}\left\{A_1 + A_2\left[\frac{R^2}{r^2} + \left(\frac{dR}{dr}\right)^2\right] + A_3\frac{R^2}{r^2}\left(\frac{dR}{dr}\right)^2\right\}. \qquad [33]$$

On a négligé les termes d'ordre supérieur de W, ce qui semble être une approximation légitime dans l'état actuel de la question lorsque l'on envisage d'étudier des propriétés élastiques pour des déformations locales de l'ordre de 100%.

Compte tenu de [32], [33], il s'agit de déterminer $R(r)$ $(0 < r_1 \le r \le r_2)$ vérifiant

$$\frac{d}{dr}\left[2\frac{dR}{dr}r^2\left(A_1 + 2A_2\frac{R^2}{r^2} + A_3\frac{R^4}{r^4}\right)\right] = 4R\left\{A_1 + A_2\left[\frac{R^2}{r^2} + \left(\frac{dR}{dr}\right)^2\right] + A_3\frac{R^2}{r^2}\left(\frac{dR}{dr}\right)^2\right\}$$

$$2\frac{dR}{dr}r^2\left(A_1 + 2A_2\frac{R^2}{r^2} + A_3\frac{R^4}{r^4}\right) = -PR^2$$

$$\text{si } r = r_1, \quad 0 \text{ si } r = r_2 \qquad [34]$$

où $A_1 A_2 A_3$ sont des fonctions de R^2/r^2 et R'^2 définies compte tenu de [8].

Pour mettre en évidence les non linéarités, nous transformons [34] en employant [7] de sorte qu'il vient:

$$\frac{d}{dr}\left\{2\frac{dR}{dr}r^2\left[D_1 + D_2\left(\frac{dR}{dr}\right)^2 + D_3\frac{R^2}{r^2}\right]\right\}$$

$$= 4R\left[S_1 + S_2\left(\frac{dR}{dr}\right)^2 + S_3\frac{R^2}{r^2}\right]$$

$$2\frac{dR}{dr}r^2\left[D_1 + D_2\left(\frac{dR}{dr}\right)^2 + D_3\frac{R^2}{r^2}\right] = -PR^2$$

$$\text{si } r = r_1, \quad 0 \text{ si } r = r_2 \qquad [35]$$

où $D_1, D_2; \ldots, S_3$ sont des fonctions non linéaires lentement variables de R et dR/dr. Remarquons que l'on peut, dans le cas de l'élasticité du premier ordre ($l_2 = m_2 = n_2 = 0$) retrouver à partir de [35] ou de [34] les éqs. [2.12] et [3.9] de *Craine* dont l'écriture reflète moins que [35] (ou [34]) la structure physique explicitée par [32] et [33] [cf. (5)].

Pour résoudre numériquement [35] on opère par relaxation sur une forme linéaire:

$$\frac{d}{dr}\left(E_1 + E_2\frac{d\varrho}{dr} + E_3\varrho\right) = F_1 + F_2\frac{d\varrho}{dr} + F_3\varrho$$

$$E_1 + E_2\frac{d\varrho}{dr} + E_3\varrho = -P(r + \varrho)^2$$

$$\text{si } r = r_1, \quad 0 \text{ si } r = r_2 \qquad [36]$$

où $E_1, E_2, \ldots, F_3$ sont des fonctions non linéaires de r et ϱ calculées à partir de la solution $\varrho(r)$ obtenue lors du pas d'itération précédent.

Le problème aux limites [36] va permettre de décrire le principe de la méthode de résolution numérique.

6. Introduction à la méthode de résolution

Obtenir une solution numérique définissant la solution $\varrho(r)$ de [36] consiste à déterminer N valeurs $\varrho(r_i)$ $(1 \le i \le N)$ de ϱ en N points i de $[r_1, r_2]$ constituant une base ponctuelle de discrétisation à intervalle quelconque incluant les extrêmités de l'intervalle ($r_1 = r_1$, $r_N = r_2$). Soit φ_N le sous-espace des N premières fonctions d'un espace de *Hilbert* orthogonal et complet. On pose [cf. (6)]:

$$\varrho(r_i) = \sum_j c_j\varphi_j(r_i) \qquad 1 \le i, j \le N. \qquad [37]$$

Dérivant terme à terme:

$$\left.\frac{d\varrho}{dr}\right|_{r_i} = \sum_j c_j\left.\frac{d\varphi_j}{dr}\right|_{r_i} = \sum_j c_j\frac{d\varphi_j(r_i)}{dr}. \qquad [38]$$

Les fonctions φ_i étant orthogonales, la matrice de terme $\varphi_j(r_i)$ est bien conditionnée et l'élimination des c_j dans [37] et [38] conduit à:

$$\left.\frac{d\varrho}{dr}\right|_{r_i} = \left[\frac{d\varphi}{dr}\varphi^{-1}\right]\varrho = D_i^k\varrho(r_k) \qquad [39]$$

où D_i^k est la matrice de dérivation construite sur le sous-espace φ_N. Cette matrice permet de générer avec une très bonne précision une forme matricielle discrète des équations différentielles [36].

La première équation différentielle de [36] prend la forme matricielle:

$$D_i^k[E_{1k} + (E_2D)_k^j\varrho_j + (E_31)_k^j\varrho_j]$$
$$= F_{1i} + (F_2D)_i^j\varrho_j + (F_31)_i^j\varrho_j$$

avec:

$$(ED)_r^s = E_rD_r^s \quad \text{sans sommation sur} \quad r.$$

D'où:

$$A_i^j\varrho_j = B_i$$

avec

$$A_i^j = D_i^k[(E_2D)_k^j + (E_31)_k^j] - (F_2D)_i^j - (F_31)_i^j$$

$$B_i = F_{1i} - D_i^kE_{1k}.$$

Pour la deuxième équation différentielle de [36], il vient de même:

$$a_i^j = (E_2D)_i^j + (E_31)_i^j$$

$$b_i = \begin{cases} -E_{1i} & \text{si } i = 1 \\ -E_{1i} - [P(\varrho + r)^2]_i & \text{si } i = N. \end{cases}$$

La matrice représentant sous forme discrète le problème aux limites complet est donc définie par A_i^j et B_i si $1 < i < N$ et par a_i^j et b_i si $i = 1$ ou N. La discrétisation obtenue est précise dès que les solutions cherchées présentent une régularité suffisante. La programmation est commode même si le choix des bases ponctuelle r_i et fonctionnelle φ_N est laissé libre.

7. Extension au cas de l'axisymétrie

A présent $0x^3$ est axe de symétrie pour le corps et les lignes y^2 sont des cercles. Il suffit donc d'envisager Ω comme hexaèdre curviligne limité par les plans méridiens $y^2 = z_1^2$ et $y^2 = -z_1^2$. Le plan méridien de référence est : $y^2 = 0$ (ou : $x^2 = 0$). Il est plan de symétrie pour Ω. Un quadrilatère curviligne $A\,B\,C\,D$ de ce plan engendre par rotation autour de $0x^3$ la portion curviligne de S qui reste convenablement régulière pourvu que le quadrilatère soit écorné.

La discrétisation de Ω est définie par les trois familles de surfaces: $y^i = C(i, \alpha)$ ($\alpha = 1, \ldots, N_\alpha$), avec N_α entiers et $C(i, 1) = 1$, $C(i, N) = N$ par exemple.

Soit alors une fonction scalaire $a(y^1, y^3)$ continûment dérivable dans le domaine $ABCD$. Par exemple a représente x^3, g^{22}, δu_1, $\delta\sigma^{13}$, etc. Avec [39] appliquée respectivement le long des lignes $y^3 =$ cste précisée ou $y^1 =$ cste précisée, nous pouvons trouver:

$$\left.\frac{\partial a}{\partial y^1}\right|_i = D^j_{3i} a_j \qquad i, j = 1, N_3$$

$$\left.\frac{\partial a}{\partial y^3}\right|_i = D^j_{1i} a_j \qquad i, j = 1, N_1 \qquad [40]$$

où D_1 et D_3 sont analogues à D de [39].

Il en est de même pour calculer N_2 valeurs de $\partial a/\partial y^2$ le long d'une ligne $y^2 =$ cste précisée.

Les matrices D_1, D_2, D_3 restent invariantes durant l'évolution puisque les bases ponctuelles sont associées aux valeurs invariantes des y^i. D'autre part, le calcul des champs g_{ij}, g^{ij}, Γ^k_{ij}, etc. est immédiat: il suffit d'appliquer [40] à x^i, g_{ij}, etc. Enfin des tenseurs tels que Du_i/Dy^j ou $D\sigma^{ij}/Dy$, qui sont linéaires en les composantes de leurs arguments u_i ou σ^{ij}, peuvent être approchés sous forme matricielle discrète.

Pour illustrer ce dernier point considérons par exemple l'opérateur de divergence appliquée au tenseur contrainte. Si l'on note $\Gamma 1$ à $\Gamma 10$ les symboles $\Gamma^1_{11}, \Gamma^1_{33}, \Gamma^1_{13}, \Gamma^3_{11}, \ldots, \Gamma^1_{22}, \Gamma^3_{22}, \Gamma^2_{12}, \Gamma^2_{23}$, on a:

$$F^1 = \frac{\partial\sigma^{11}}{\partial y^1} + \frac{\partial\sigma^{13}}{\partial y^3} + \sigma^{11}(2\Gamma 1 + \Gamma 6 + \Gamma 9) + \sigma^{22}\Gamma 7 + \sigma^{33}\Gamma 2 + \sigma^{13}(3\Gamma 3 + \Gamma 5 + \Gamma 10)$$

$$F^3 = \frac{\partial\sigma^{13}}{\partial y^2} + \frac{\partial\sigma^{33}}{\partial y^3} + \sigma^{11}\Gamma 4 + \sigma^{22}\Gamma 8 + \sigma^{33}(2\Gamma 5 + \Gamma 3 + \Gamma 10) + \sigma^{13}(3\Gamma 6 + \Gamma 1 + \Gamma 9).$$

Soit, sous forme matricielle:

$$F_i = D^j_{vi}\,\sigma_j \qquad i = 1, 2\,N_1 \ldots N_3 \qquad j = 1, 4\,N_1 \ldots N_3. \qquad [41]$$

La matrice D_v est la somme de:

$$\left|\begin{matrix} D_1 & 0 & 0 & D_3 \\ 0 & 0 & D_3 & D_1 \end{matrix}\right|$$

et de:

$$\left|\begin{array}{ccc|c} 2\Gamma 1 + \Gamma 6 + \varepsilon 9 & \Gamma 7 & \Gamma 2 & \\ \hline \Gamma 4 & \Gamma 8 & 2\Gamma 5 + \Gamma 3 + \Gamma 10 & \end{array}\right. \quad \left.\begin{array}{c} 3\Gamma 3 + \Gamma 5 + \Gamma 10 \\ \hline 3\Gamma 6 + \Gamma 1 + \Gamma 9 \end{array}\right|$$

où chaque sous-matrice carrée est diagonale en les éléments (ou somme d'éléments) Γ indiqués.

Une mise en forme matricielle analogue à [41] pouvant être faite à propos de chacun des champs introduits, il est possible de construire une représentation matricielle discrète de [17] à [19] permettant d'exprimer [16] sous la forme:

$$(D_v \lambda D_g)\delta U = \delta f_\Omega \quad \text{sur} \quad \Omega$$

$$(\lambda' D'_g)\delta U = \delta f_s \quad \text{sur} \quad S$$

analogue à celle présentée au sujet du problème [36].

En pratique les valeurs des N peuvent rester petites (de l'ordre de 10) de sorte que le calcul ne dépasse pas la capacité de la mémoire centrale des grands ordinateurs [cf. (6) et (7)].

8. Quelques résultats numériques en symétrie sphérique

Diverses résolutions de [34] permettent de constater que les non linéarités d'origine géométrique jouent un rôle considérable même dans le cas où la loi élastique est linéaire ($l_2 = m_2 = n_2 = 0$). Ainsi, pour des déformations de l'ordre de 100%, nous allons voir que les courbes $\sigma^{11}(\gamma_{11})$ ont l'aspect classique du genre élastoplastique même si $l_1 = m_2 = n_2 = 0$. Dans ces conditions, on peut prévoir que des précisions relatives assez élevées ne sont pas forcément superflues si la résolution de [34] est destinée à interpréter des résultats expérimentaux en vue de distinguer une non linéarité éventuelle de la loi élastique et d'identifier les valeurs de l_2, m_2 et n_2 (cf. figure).

Faute d'avoir des critères de convergence sûrs, il importe à ce sujet de procéder à des tests de précision. Le contrôle des solutions numériques obtenues a été effectué grâce aux opérations suivantes: vérification de l'équilibre local des densités de forces dues aux tensions en portant la solution obtenue dans la deuxième équation [33], vérification de l'équilibre global entre forces exercées sur une demi-coque par la pression interne et par les tensions circonférencielles, vérification (locale) de l'égalité des valeurs du gradient d'énergie exprimé à partir de la première relation [33] et à partir des solutions obtenues portées dans la formule [cf. (7)]:

$$r^3 \frac{dW}{dr} = \frac{d}{dr}\left[\left(r\frac{dR}{dr} - R\right) R^2 \sigma^{11}\right] \qquad [42]$$

vérification de l'égalité de l'énergie fournie par la mise en pression et de la variation d'énergie potentielle élastique globale de la coque, vérification de la condition d'isotropie, vérification du caractère hydrostatique des contraintes calculées en imposant l'égalité des pressions interne et externe, vérification des solutions pour les petites déformations à l'aide de la solution classique de l'élasticité infinitésimale, vérification de l'identité des solutions numériques pour des discrétisations différentes (discrétisation à 4, 7 et 13 points par exemple), vérification de l'identité des solutions pour des processus de montées en charge différentes (P imposé d'un coup ou par fractions successives), vérification de l'identité des solutions avec ou sans pénalisation du problème aux limites.

Considérons, par exemple, une étude de coque épaisse ($r_1 = 50$ mm, $r_2 = 100$ mm) et une étude de membrane ($r_1 = 99{,}5$ mm, $r_2 = 100$ mm). Le matériau est tel que $l_2 = m_2 = n_2 = 0$ et que a et b correspondent soit à $E = 40$ kg/mm², $\nu = 1/3$, soit à $E = 4$ kg/mm², $\nu = 1/3$. Pour des pressions P valant respectivement $E/8$ et $6 \cdot 10^3 E$ les résultats obtenus permettent de tracer les courbes $\sigma_{11}(\gamma_{11})$ en $r = r_1$ jusqu'à des déformations de 100 et 300% respectivement. On constate une non linéarité telle que les courbes ont l'aspect classique du genre élastoplastique (cf. figure).

Les contrôles de précision relative ont été satisfaisants. Dans les cas examinés ci-dessus, les erreurs relatives atteintes à propos du contrôle basé sur [42] sont respectivement de 10^{-5} et 10^{-7}, alors que celles basées sur le premier contrôle cité sont respectivement de 10^{-9} et 10^{-7}. Pour les déformations extrêmes, ces chiffres passent respectivement à 10^{-4}, 10^{-6}, 10^{-8}, 10^{-6}. Compte tenu des majorations d'erreur introduites lors des contrôles cités, l'erreur relative doit varier entre 10^{-8} et 10^{-6}. Le fait de prendre en compte une élasticité du second ordre ($l_2 = -m_2 = n_2 = 15$ et 0,3 kg/mm² respectivement) affecte peu les précisions relatives qui sont du même ordre que celles obtenues dans le cas de l'axisymétrie [cf. (8)].

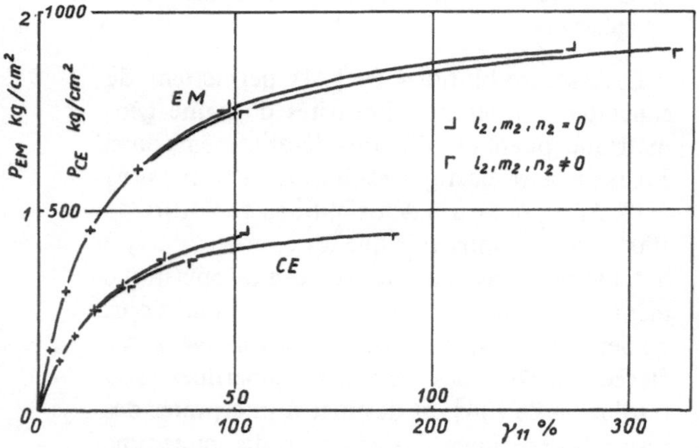

Fig. 1. Expansion d'enveloppes sphériques mince ou epaisse: Courbes de pression de gonflement P en fonction de la déformation radiale γ_{11} à la paroi interne.

Enveloppe mince (EM): $r_1 = 9{,}95$ cm; $r_2 = 10$ cm; $E = 400$ kg/cm²; $\nu = 1/3$; $l_2, m_2, n_2 = 0$ où + — + 30 kg/cm²

Coque épaisse (CE): $r_1 = 5$ cm; $r_2 = 10$ cm; $E = 4000$ kg/cm²; $\nu = 1/3$; $l_2, m_2, n_2 = 0$ où + — + 1500 kg/cm²

Littérature

1) *Green, A. E.* et *W. Zerna*, Theoretical Elasticity (Oxford 1954).

2) *Angles D'Auriac, P.*, Définitions et Principes en Rhéologie tensorielle. Symposium I.U.T.A.M., Grenoble 1964 (Berlin-Heidelberg-New York 1966).

3) *Le Roy, P.*, Contribution à l'étude de certains liquides non newtoniens. Thèse de Doctorat d'Etat (Grenoble 1968).

4) *Guelin, P., P. Le Roy* et *J. M. Pierrard*, Sur la détermination et l'utilisation de certaines lois de comportement de matériaux à mémoire. Colloque de Rhéologie et de Mécanique des Sols, Varsovie 1971.

5) *Craine, R. E.*, Quart. J. Mech. and Appl. Math. **21**, 279 (1968).

6) *Boisserie, J. M.*, Int. J. Num. Methods Eng. **3**, 327 (1971).

7) *Green, A. E.* et *J. E. Adkins*, Large Elastic Deformation and Non-Linear Continuum Mechanics (Oxford 1960).

8) *Guelin, P.*, C.R.A.S. **272**, 498–501 (Paris 1971).

Adresse de l'auteur:

Dr. *Pierre Guélin*
Laboratoire de Mécaniques des Sols Domaine Universitaire, Cédex 53 F-38 Grenoble (France)

Rheol. Acta 12, 253–262 (1973)

From the Department of Chemical Engineering, Princeton University, Princeton, U.S.A.

Nonlinear effects in the unsteady flow of viscoelastic fluids

By Ch. Goldstein and W. R. Schowalter

With 7 figures

(Received October 27, 1972)

1. Introduction

It has long been recognized that interesting features of viscoelastic materials can be revealed through study of the response of such materials to unsteady inputs. Indeed, most of the early work dealing with measurement of viscoelastic properties consisted of a study of the response of specimens to imposition of small-amplitude oscillatory stress or strain (1). Though much of the early work dealt with viscoelastic polymer solids, oscillatory testing has been extended to liquids, and there is now an extensive literature on the subject [see, for example, (2–5)]. However, by far the greatest interest to date has been in analyses and experiments which are restricted to a linear relation between input (for example, strain or strain rate) and output (for example, stress). This is the province of linear viscoelasticity, a subject which is now well developed and which has been quite successful for study of solid-like materials. Linear theory is insufficient, however, for characterization of many situations involving flow of liquid-like polymer melts, polymer solutions, and suspensions.

The mathematical complexity of nonlinear systems is of course a strong motivation for working within the regime of linearity. An inkling of the possibilities inherent in exploitation of the nonlinear attributes of rheologically complex liquids has been provided through a series of recent papers by *Walters* and coworkers. They have found, for example, that significant enhancement of pumping rates for a given mean pressure gradient can be attained through superposition of a pressure oscillation upon a steady flow. This flow enhancement has been demonstrated both theoretically and experimentally (6–8).

Though the analytical techniques which we have employed are similar to those of *Walters* and coworkers, the objectives are, for the most part, rather different. Our interests center on the potential of nonlinear analysis as a rheometrical tool. In this sense our objectives are similar to those of *MacDonald* et al. (3) but in the present analysis a more explicit recognition is given to the importance of nonlinear effects.

Our report is illustrative in nature and is offered with the hope that it will stimulate careful collection of data so that nonlinear effects can be properly identified and interpreted. We show first that retention of nonlinear terms in analysis of some classical flows leads to prediction of new effects which arise because of interaction of harmonics of the oscillatory input. Second, the nature of these interactions is identified with parameters of one or more specific constitutive models. No attempt is made in this paper to evaluate suitability of constitutive equations to describe real materials in oscillatory tests. It is hoped that the framework presented here will provide motivation for experiments which will allow such comparisons in the future.

The unsteady flows for which analyses are provided are oscillatory shearing, superposed steady and parallel oscillatory shearing, and superposed steady and transverse oscillatory shearing.

2. Background

2.1. Nonlinear Effects

Phenomena which are inherently dependent for their existence upon nonlinear behavior are familiar throughout science and technology. In fluid mechanics one can cite "acoustic streaming", the term used to describe a steady flow which is induced by an object, a cylinder for example, which has been placed in an initially quiescent fluid and is oscillated in a direction normal to the cylinder axis. Boundary-layer analysis (9, 10) has shown that the steady flow is a result of the nonlinear convective inertial terms in the equation of motion. The corresponding equation of motion for a rheologically complex fluid has two sources of nonlinearity. In addition to inertial effects, the constitutive equation relating stress to deformation rate is typically nonlinear in the deformation rate. Thus one expects, and finds, that the analogue to acoustic streaming which is exhibited by viscoelastic fluids is due to a combination of inertial and rheological properties.

We have already cited the work of *Walters* and his associates on the unsteady flow of non-*Newton*ian fluids in tubes. They have shown that the major qualitative features of such a pulsatile flow, i. e., a frequency dependent enhancement of pumping rate through unsteady operation, is predicted from an analysis in which one employs the viscous inelastic, but nevertheless nonlinear, nature of the fluid.

2.2. Choice of Constitutive Equation

Since the flows to be discussed here are all small perturbations about a viscometric base flow, including the limiting case of a fluid at rest, one would hope that a solution in terms of the general viscometric functions of a simple fluid (11) would be possible. One can in fact formulate the problem using the procedure of *Pipkin* and *Owen* (8, 12, 13) for nearly viscometric flows. However, in order to solve the resulting equations it is necessary to make a statement about the form of constitutive equation beyond specification that the material is a simple fluid. Hence it seemed desirable, for purposes of the present exposition, to carry out the computations for specific constitutive models which are simple enough to permit use of analytical techniques, but which are also complete enough to describe, in a qualitative way, the behavior of real viscoelastic fluids in a variety of flows. We deal primarily with the equation developed by *Bird* and *Carreau*. This model, an integral representation, is related to molecular theories of constitutive behavior and is appealing in that respect (14). The constitutive equation is

$$\tau = \int_{-\infty}^{t} m[(t-t'), \mathrm{II}(t')] \times \left[\left(1+\frac{\varepsilon}{2}\right)\bar{\Gamma} - \frac{\varepsilon}{2}\Gamma\right] dt'. \qquad [1]$$

Finite strain tensors are given by Γ and $\bar{\Gamma}$, where components in rectangular *Cartesian* coordinates are

$$\Gamma_{ij} = \delta_{ij} - (\partial x'_\alpha/\partial x_i)(\partial x'_\alpha/\partial x_j),$$

$$\bar{\Gamma}_{ij} = (\partial x_i/\partial x'_\alpha)(\partial x_j/\partial x'_\alpha) - \delta_{ij}.$$

Unless otherwise stated, summation over repeated indices is assumed. $\delta_{ij} = 1$ for $i = j$ and zero for $i \neq j$; x'_i and x_i are position coordinates at t' and t, respectively; and ε is a scalar constant; and τ is the stress with respect to an isotropic contribution, $-J$. When the material is in a completely relaxed state, P is identified with the hydrostatic pressure. The relaxation function is given by

$$m[(t-t'), \mathrm{II}(t')] = \sum_{p=1}^{\infty} \frac{\eta_p}{\lambda_{2p}^2} \frac{e^{-(t-t')/\lambda_{2p}}}{\left[1+\frac{1}{2}\lambda_{1p}^2 \mathrm{II}(t')\right]}$$

where λ_{1p}, λ_{2p}, and η_p are material constants. $\mathrm{II}(t')$ is the second invariant of the rate of strain tensor evaluated at time t',

$$\mathrm{II}(t') = 4\, e_{ij} e_{ij}$$

where $e_{ij} = \frac{1}{2}(\partial v_i/\partial x_j + \partial v_j/\partial x_i)$. *Bird* and *Carreau* state that the constants λ_{1p} and λ_{2p} can be related to rate of creation and loss of network junctions. The empirical constant ε allows for nonzero secondary normal stress difference (14).

In order to test the sensitivity of results to choice of constitutive model we have carried out a number of computations with an *Oldroyd* model (15). It is rather different from the *Bird-Carreau* model both in mathematical form and physical origin. *Oldroyd* has shown, for example, how this model is related to the behavior of a dilute suspension of elastic spheres (16). In *Cartesian* coordinates we write

$$\tau_{ij} + \lambda_1 \frac{\mathscr{D}\tau_{ij}}{\mathscr{D}t} + \mu_0 \tau_{\alpha\alpha} e_{ij} - \mu_1(\tau_{i\alpha} e_{\alpha j} + \tau_{\alpha j} e_{i\alpha}) + \nu_1 \tau_{\alpha\beta} e_{\alpha\beta} \delta_{ij} = 2\eta_0 \left[e_{ij} + \lambda_2 \frac{\mathscr{D}e_{ij}}{\mathscr{D}t} - 2\mu_2 e_{i\alpha} e_{\alpha j} + \nu_2 e_{\alpha\beta} e_{\alpha\beta} \delta_{ij}\right] \qquad [2]$$

where μ_i, λ_i, ν_i, and η_0 are material constants. The *Jaumann* derivative is

$$\frac{\mathscr{D}b_{ij}}{\mathscr{D}t} = \frac{\partial b_{ij}}{\partial t} + v_\alpha \frac{\partial b_{ij}}{\partial x_\alpha} + \omega_{i\alpha} b_{\alpha j} + \omega_{j\alpha} b_{i\alpha}$$

where

$$\omega_{ij} = \tfrac{1}{2}(\partial v_j/\partial x_i - \partial v_i/\partial x_j).$$

Booij (17, 18) has shown that the *Oldroyd* model equation provides qualitative agreement with both dynamic and steady shear viscosity measurements for several polymer solutions.

The form of the viscometric functions in steady laminar shear flow is well known for both models and will not be repeated here (14, 15, 18).

3. Unsteady flows

3.1. Oscillatory Shear Flow

There are numerous reports which deal with both theoretical and experimental aspects of oscillatory shearing of viscoelastic fluids. Most of these studies, however, contain one or more of the following constraints:

(i) a highly simplified (for example, linear) constitutive equation,

(ii) homogeneous rate of strain,

(iii) negligible effect of fluid density.

Presence of higher harmonics in oscillatory testing of rheologically complex fluids has been demonstrated by *Harris* and *Bogie* (19) and by *Walters* and *Jones* (20). *MacDonald* et al. (3) and *Onogi* et al. (21) have carried out finite-amplitude analyses and experiments with a cone-and-plate and with add a coaxial cylinder geometry, respectively. The effect of fluid density is ignored in all of these studies. *Dodge* and *Krieger* (22) have shown that the conventional analysis of flow between oscillating discs or cone and plate is not valid if fluid density is considered, because a secondary flow will be generated. This objection is not present for flow between parallel plates or coaxial cylinders, the latter geometry being the one chosen by *Dodge* and *Krieger* for their experiments (5).

In contrast to most of the above analyses, and similar to the development of *Barnes* et al. (8) for pulsatile pipe flow, we include the effect of fluid density in the time dependent term of the equation of motion but neglect nonlinearities due to inertial effects. Thus we restrict the analysis in this and subsequent sections to nonlinear effects which are due to material rather than flow properties.

Consider an idealized flow between two infinite parallel planes. The lower plane at $y = 0$ is stationary, the upper plane at $y = h$ oscillates with frequency ω and velocity amplitude V_0, and the pressure is uniform in the x- and z-directions. A velocity field of the form

$$v = \{u(y, t), 0, 0\} \qquad [3]$$

is assumed. The continuity equation for an incompressible fluid is identically satisfied, and the x-component of the equation of motion reduces to

$$\varrho \frac{\partial u}{\partial t} = \frac{\partial}{\partial y} \tau_{xy}. \qquad [4]$$

We first compute the velocity and stress fields according to the *Bird-Carreau* model, employing an expansion similar to that used by *Schlichting* (9) to analyze inertial effects in acoustic streaming, and by *Walters* and coworkers (7, 8) to assess nonlinear viscoelastic effects in pulsatile flows.

A dimensionless quantity $\alpha = V_0/(\omega h)$ is chosen as the expansion parameter. Note that α is the strain amplitude (scaled to h) applied to the upper plate in the case of a zero-density linear viscoelastic fluid. We assume

$$\begin{aligned} u &= \alpha u_1 + \alpha^3 u_3 + \cdots, \\ \tau_{xy} &= \alpha \tau_1 + \alpha^3 \tau_3 + \ldots, \\ \tau_{ii} &= \alpha^2 \tau_{ii2} + \alpha^4 \tau_{ii4} + \cdots \qquad (i = x, y, z; \text{ no sum}). \end{aligned} \qquad [5]$$

Terms in the expansions which can be shown to be zero have not been included. A simple calculation based upon the assumed form of the velocity profile, eq. [3], and the constitutive eq. [1] leads, for the *Bird-Carreau* model, to

$$\tau_{xy} = \int_0^\infty \sum_{p=1}^\infty \frac{\eta_p}{\lambda_{2p}^2} \exp(-s/\lambda_{2p}) \times \frac{\frac{\partial}{\partial y}\left[\int_{t-s}^{t} (\alpha u_1 + \alpha^3 u_3 + \cdots)\, dt''\right]}{1 + \lambda_{1p}^2 \left[\alpha \frac{\partial u_1}{\partial y} + \alpha^3 \frac{\partial u_3}{\partial y} + \cdots\right]^2} ds$$

where s is a dummy variable corresponding to $(t - t')$

$$\tfrac{1}{2}\, \mathrm{II}(t') = (\partial u/\partial y)^2_{t=t'}.$$

We next assume separable solutions

$$\tau_1 = \boldsymbol{R}(\tau_1^1(y)\, e^{i\omega t}); \quad u_1 = \boldsymbol{R}\,(u_1^1(y)\, e^{i\omega t})$$

where $\boldsymbol{R}$ signifies the real part. Note also that superscripts associate a term with the frequency of the time dependent portion while subscripts indicate the order of expansion in [5].

The first term in the expansion leads of course to known linear results for the velocity profile and the real and imaginary parts of the complex viscosity η^*. No normal stresses occur to first order. Terms proportional to α^2 have no effect upon shear stress or velocity components but normal stresses do appear. For convenience we introduce the following notation:

$$\tau_{xx} = p; \quad \tau_{yy} = q; \quad \tau_{zz} = r$$

$$\begin{aligned} p_2 &= \boldsymbol{R}[p_2^0 + p_2^2 e^{2i\omega t}] \\ q_2 &= \boldsymbol{R}[q_2^0 + q_2^2 e^{2i\omega t}] \\ r_2 &= \boldsymbol{R}[r_2^0 + r_2^2 e^{2i\omega t}]. \end{aligned} \qquad [6]$$

It is also convenient to define normal stress coefficients (23) to reflect both the steady and oscillatory portions which arise because of retention of nonlinear terms

$$\zeta^d = \boldsymbol{R}[(p_2^0 - q_2^0)/z_1 \tilde{z}_1] \qquad [7]$$

$$\zeta^* = (r_2^2 - q_2^2)/z_1^2 \qquad [8]$$

where $z_1 = du_1^1/dy$ and the tilde refers to complex conjugate. Secondary normal stress coefficients can be similarly defined.

$$\chi^d = (q_2^0 - r_2^0)/z_1 \tilde{z}_1 \qquad [9]$$

$$\chi^* = (q_2^2 - r_2^2)/z_1^2 \qquad [10]$$

from which it follows that

$$\chi^d = \frac{\varepsilon}{2} \zeta^d \qquad [11]$$

$$\chi^* = \frac{\varepsilon}{2} \zeta^*. \qquad [12]$$

The model employed here, as well as the *Oldroyd* model and many others, yield the familiar relations (23–25)

$$\zeta^* = \frac{\eta^*(\omega) - \eta^*(2\,\omega)}{i\omega} \qquad [13]$$

$$\zeta^d = \frac{\eta''}{\omega}. \qquad [14]$$

Though some oscillatory data exist with which one might compare these predictions (26), the significance of such a comparison is not clear since the data are not consistent with eq. [13].

The primary new contribution of this section results when one proceeds to terms of order α^3. We write

$$u_3 = \boldsymbol{R}\,[u_3^1 e^{i\omega t} + u_3^3 e^{3i\omega t}] \quad [15]$$

$$\tau_3 = \boldsymbol{R}\,[\tau_3^1 e^{i\omega t} + \tau_3^3 e^{3i\omega t}]. \quad [16]$$

Presence of both a fundamental and a third harmonic of the stress field is accounted for by defining new material functions η^{I} and η^{III}, where

$$\tau_3^1 = \eta^* du_3^1/dy - \eta^{\mathrm{I}}[z_1]^2\,\tilde{z}_1 \quad [17]$$

$$\tau_3^3 = \eta^*(3\,\omega)\,du_3^3/dy - \eta^{\mathrm{III}}[z_1]^3. \quad [18]$$

Expressions in terms of specific models are

(i) *Bird-Carreau*

$$\eta^* = \eta' - i\eta'' = \int_0^\infty m(s,0)\left[\frac{1-e^{-i\omega s}}{i\omega}\right]ds \quad [19]$$

$$\eta^{\mathrm{I}} = \int_0^\infty m_1(s,0)\left[\frac{1-e^{-i\omega s}}{2\,i\omega} - \frac{e^{-2i\omega s}-e^{-i\omega s}}{4\,i\omega}\right]ds \quad [20]$$

$$\eta^{\mathrm{III}} = \int_0^\infty m_1(s,0)\left[\frac{e^{-2i\omega s}-e^{-3i\omega s}}{4\,i\omega}\right]ds$$

$$m_1(s,0) = \sum_{p=1}^{\infty} \lambda_{1p}^2 \frac{\eta_p}{\lambda_{2p}^2}\, e^{-s/\lambda_{2p}}. \quad [21]$$

(ii) *Oldroyd* model

$$\eta^{\mathrm{I}} = \frac{\theta_1\eta^*\left[1+\dfrac{1}{(1+2\,i\omega\lambda_1)}\right] - \eta_0\theta_2\left[1+\dfrac{1}{(1+i\omega\lambda_1)}\right]}{2\,[1+i\omega\lambda_1]} \quad [22]$$

$$\eta^{\mathrm{III}} = \frac{\eta^*\theta_1 - \eta_0\theta_2}{2\,(1+2\,i\omega\lambda_1)\,(1+3\,i\omega\lambda_1)} \quad [23]$$

where $\eta^* = \eta_0(1+i\omega\lambda_2)/(1+i\omega\lambda_1)$

$$\theta_1 = \tfrac{1}{2}(\lambda_1-\mu_0+\mu_1)(\lambda_1-\mu_1+\nu_1) + \tfrac{1}{2}(\lambda_1+\mu_0-\mu_1)(\lambda_1+\mu_1-\nu_1) - \tfrac{1}{2}\mu_0\nu_1$$

$$\theta_2 = \tfrac{1}{2}(\lambda_1-\mu_0+\mu_1)(\lambda_2-\mu_2+\nu_2) + \tfrac{1}{2}(\lambda_1+\mu_0-\mu_1)(\lambda_2+\mu_2-\nu_2) - \tfrac{1}{2}\mu_0\nu_2.$$

Substitution of [17] and [18] into [4] and solution of the resulting ordinary differential equations in accord with the no-slip boundary conditions leads to equations for the contributions u_3^1 and u_3^3 to the velocity profile in terms of known quantities and the material parameters η^*, η^{I}, η^{III} (27). These results are given in the Appendix. Note that the third order stress contributions are readily obtained from these results.

An import result of this section is the indication that experiments, the purpose of which is to describe nonlinear viscoelastic response, should be designed so that one monitors both the fundamental frequency as well as effects due to higher harmonics. Though some experiments have been described in which nonlinear outputs are recorded (3, 19, 21, 28), an analysis of the data similar to that suggested here has not, to the authors' knowledge, been carried out.

It is perhaps useful to illustrate the behavior predicted by some of the present results. To do this we consider parameters which are approximately equal to those cited by *MacDonald* et al., (3) for a 5% polystyrene solution. They used only one term, and hence three parameters, in the expression for the relaxation function $\underline{m}$. We set $\lambda_{11} = 1$ sec, $\lambda_{21} = 1$ sec, and $\eta_1 = 10^3$ poise. Results obtained from eq. [7], showing the steady part of the normal stress difference and evaluated at $y = 0$, are shown in fig. 1. One notes that the normal stress difference can be an appreciable fraction of the shear stress amplitude. It is interesting to note that, for the parameters chosen, the effect of density is negligible. As one would expect, the normal stress difference increases with increasing frequency and/or strain amplitude.

The amplitude of the oscillatory component of the primary normal stress varies from one to one-half times the steady amplitude as the frequency increases. This result has also been shown for the *Oldroyd* model (18).

Since it has been shown that terms proportional to $e^{i\omega t}$ and to $e^{3i\omega t}$ arise when contributions to $0(\alpha^3)$ are considered, it is of interest to discuss the behavior of both terms for the *Bird-Carreau* parameters previously given. Again, the effect of density is negligible. To third order in α the shear stress τ_{xy} is of the form

$$\tau_{xy} = \boldsymbol{R}[(\alpha\eta^* z_1 - \alpha^3\eta^{\mathrm{I}} z_1^2\tilde{z}_1)\,e^{i\omega t} - \alpha^3\eta^{\mathrm{III}} z_1^3 e^{3i\omega t}].$$

We define

$$\boldsymbol{R}Y = -\,\alpha^2\,\frac{\boldsymbol{R}\eta^{\mathrm{I}}}{\eta'}\,z_1\tilde{z}_1$$

so that $\boldsymbol{R}Y$ represents the fractional increase in the amplitude of the $\cos\omega t$ component due to the inclusion of the $0(\alpha^3)$ terms. Fig. 2 shows $-\boldsymbol{R}Y$ plotted against strain amplitude for several values of ω. For the model parameters chosen, the effect of the nonlinearity becomes more pronounced at high frequencies and/or strains.

We emphasize that the results shown in figs. 1 and 2 are intended to illustrate the likely qualitative features of a nonlinear response. It is of course probable that, as the nonlinear effects grow in magnitude, higher-order terms not considered here will play an important role. The nature of the frequency dependence will also, no doubt, be model dependent. For example, one would expect a different frequency dependence if computations were performed for *Tanners* network rupture model (29), in which the nonlinearity occurs through the strain rather than, as in the present case, through the second invariant of the strain rate.

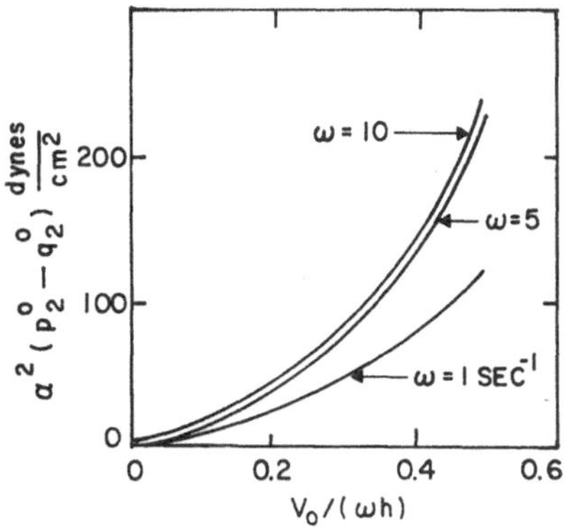

Fig. 1. Oscillatory shear: Steady portion of primary normal stress difference to $0(\alpha^2)$ as a function of strain amplitude

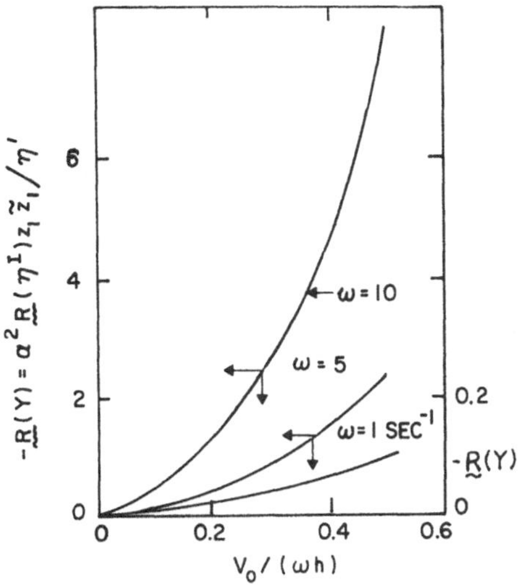

Fig. 2. Oscillatory shear: Fractional decrease of in-phase oscillatory stress to $0(\alpha^3)$

Further interesting results, for which space is insufficient here, can also be obtained from consideration of terms proportional to $\sin\omega t$ and $e^{3i\omega t}$. One finds, for example, that the term proportional to $\cos 3\omega t$ changes sign at a critical frequency.

It is worth noting that, as with viscometric flows, a general description of this flow can be provided which is independent of the model and involves a series of material parameters η^*, ζ^d, ζ^*, χ^d, χ^*, η^{I}, η^{III}. Each of these quantities is however related to the special parameters of the specific model, and the prediction of nonlinear behavior will be model dependent.

We close this section with a brief description of the contributions of elastic and inelastic properties to the phenomena described here. Using the *Bird-Carreau* model as a basis one generally associates fluid elasticity with λ_{2p} and non-*Newton*ian inelastic behavior with λ_{1p} (14). Since each term in expressions for ζ^d and ζ^* is proportional to λ_{2p}, we conclude that fluid elasticity is of primary importance to existence of normal stresses. It is interesting to note, however, that both η^{I} and η^{III} (see eqs. [17] and [18]) are proportional to the λ_{1p}^2. Hence the nonlinear *inelastic* nature of the fluid is reflected in the third-order material functions. The importance of inelastic nonlinearities has been pointed out previously in other contexts (8).

3.2. *Superposed Steady and Parallel Oscillatory Shear*

In recent years there has been considerable interest in the difference between fluid response in oscillatory flow and in a flow where the oscillation is superposed upon a steady flow. The importance of such a comparison is obvious if one thinks of constitutive equations which reflect the effect of flow upon formation and rupture of molecular entanglement points. The first experimental report of combined oscillatory and steady motion appears to be that of *Osaki* et al (30). They used a coaxial cylinder apparatus and described the effect of steady shear upon the relaxation time spectrum of the polymer molecules. The analysis of *Osaki* et al., as well as other experimental and theoretical treatments (17, 18, 20, 28, 31) have been concerned with situations where the amplitude of oscillation is assumed small enough so that the oscillatory portion of the flow is spatially homogeneous and oscillates with frequency ω.

The procedure which we follow in our analysis is analogous to that employed in Section 3.1. We emphasize, for illustrative purposes, the results that follow from the *Bird-Carreau* model.

The flow envisioned is similar to that of the previous section, but the boundary condition at the upper plate is now given by

$$u(y = h) = U_0 + V_0 \cos\omega t. \qquad [24]$$

Expansions [5] are used for velocity and stress components with $\alpha = (V_0/\omega h)$ again being used as the expansion parameter. However the summation now begins with α^0. Solution to order α^0 yields, of course, the well known viscometric results cited earlier. Results to order α^1 are significantly altered from those of the previous section because of the presence of the viscometric base flow. Thus if we write

$$\begin{aligned} u_1 &= \boldsymbol{R}[u_1^1 e^{i\omega t}] \\ \tau_{xy1} = \tau_1 &= \boldsymbol{R}[\tau_1^1 e^{i\omega t}] \\ \tau_{xx1} = p_1 &= \boldsymbol{R}[p_1^1 e^{i\omega t}] \\ \tau_{yy1} = q_1 &= \boldsymbol{R}[q_1^1 e^{i\omega t}] \\ \tau_{zz1} = r_1 &= \boldsymbol{R}[r_1^1 e^{i\omega t}]. \end{aligned} \qquad [25]$$

and

$$\tau_1^1 = \eta_{\varkappa_0}^* du_1^1/dy$$

where $\varkappa_0 = du_0/dy$, then the *Bird-Carreau* model leads to the prediction

$$\eta^* = \int_0^\infty \left[m(s, 2\varkappa_0^2)\left(\frac{1 - e^{-i\omega s}}{i\omega}\right) - m_1(s, 2\varkappa_0^2)\, 2\varkappa_0^2 s e^{-i\omega s}\right] ds \qquad [26]$$

with

$$m_1(s, 2\varkappa_0^2) = \sum_{p=1}^\infty \frac{\dfrac{\eta_p}{\lambda_{2p}} \lambda_{1p}^2 e^{-s/\lambda_{2p}}}{[1 + \lambda_{1p}^2 \varkappa_0^2]^2}.$$

Normal stress differences are

$$p_1^1 - q_1^1 = \varepsilon^* \varkappa_0 \frac{du_1^1}{dy} \qquad [27]$$

$$q_1^1 - r_1^1 = \frac{\varepsilon}{2} \varepsilon^* \varkappa_0 \frac{du_1^1}{dy} \qquad [28]$$

where

$$\varepsilon^* =: \int_0^\infty \left[m(s, 2\varkappa_0^2)\, 2s\left(\frac{1 - e^{-i\omega s}}{i\omega}\right) - m_1(s, 2\varkappa_0^2)\, 2s^2 \varkappa_0^2 e^{-i\omega s}\right] ds.$$

The contribution, to order α^1, to the velocity profile is

$$\alpha u_1^1 = \frac{V_0 \sinh b_p y}{\sinh b_p h} \qquad [29]$$

where

$$(b_p)^2 = i\omega\varrho/\eta_{\varkappa_0}^*.$$

Decomposing $\eta_{\varkappa_0}$ into real and imaginary parts one obtains

$$\eta_{\varkappa_0}^* = \eta'_{\varkappa_0} - i\eta''_{\varkappa_0} \qquad [30]$$

with

$$\eta'_{\varkappa_0} = \sum_{p=1}^\infty \eta_p \left\{\frac{(1 + \omega^2\lambda_{2p}^2) + \lambda_{1p}^2\varkappa_0^2(3\omega^2\lambda_{2p}^2 - 1)}{(1 + \lambda_{1p}^2\varkappa_0^2)^2(1 + \omega^2\lambda_{2p}^2)^2}\right\} \qquad [31]$$

$$\eta''_{\varkappa_0} = \sum_{p=1}^\infty \omega\lambda_{2p}\eta_p \left\{\frac{(1 + \omega^2\lambda_{2p}^2) + \lambda_{1p}^2\varkappa_0^2(\lambda_{2p}^2\omega^2 - 3)}{(1 + \lambda_{1p}^2\varkappa_0^2)^2(1 + \omega^2\lambda_{2p}^2)^2}\right\}. \qquad [32]$$

The exact forms of [31] and [32] will depend upon the rheological model, as can be seen by comparison with expressions obtained under different constitutive assumptions (13, 31).

The chief point of interest thus far is the fact that the addition of a steady flow component U^0 gives rise to *oscillatory* normal stresses which occur to $0(\alpha)$, in contrast to the results of the previous section.

Proceeding to terms of order α^2 we employ the notation

$$\begin{aligned} u_2 &= \boldsymbol{R}[u_2^0 + u_2^2 e^{2i\omega t}] \\ \tau_2 &= \boldsymbol{R}[\tau_2^0 + \tau_2^2 e^{2i\omega t}] \\ p_2 &= \boldsymbol{R}[p_2^0 + p_2^2 e^{2i\omega t}] \\ q_2 &= \boldsymbol{R}[q_2^0 + q_2^2 e^{2i\omega t}] \\ r_2 &= \boldsymbol{R}[r_2^0 + r_2^2 e^{2i\omega t}]. \end{aligned} \qquad [33]$$

Analogous to Section 3.1 we write

$$\tau_2^0 = \varphi_0 du_2^0/dy - \varphi_1 z_1 \bar{z}_1 \qquad [34]$$

$$\tau_2^2 = \varphi_2 du_2^2/dy - \varphi_3 z_1 z_1 \qquad [35]$$

where $z_1 = du_1^1/dy$.

The φ_i can be evaluated in terms of material functions and are given in the Appendix.

Solutions for u_2^0 and u_2^2 are then readily found by integration of the equation of motion. From the results, given in the Appendix, one sees that there is both a steady and an oscillating contribution to order α^2, the frequency of the latter being 2ω.

We present an illustration of the behavior which is predicted by considering the model parameters cited in Section 3.1. The behavior of the steady part of the shear stress that arises when terms $0(\alpha^2)$ are considered is readily found.

$$\alpha^2 \tau_2^0 \cong \frac{-\varphi_1 (V_0)^2 b_p \tilde{b}_p}{2 \sinh(b_p h) \sinh(\tilde{b}_p h)} \left[\frac{\sinh[(p_p + \tilde{b}_p) h]}{(b_p + \tilde{b}_p) h} + \frac{\sinh[(b_p - \tilde{b}_p) h]}{(b_p - \tilde{b}_p) h} \right].$$

Note that to $0(\alpha^2)$ the stress is constant and the direction is dependent on the sign of φ_1. Recall that, for the parameters chosen, effects of density are negligible. Then it is a simple matter to compute the ratio of steady shear stress due to terms $0(\alpha^2)$ to the shear stress caused by the basic flow $\varkappa_0$. This ratio, which we call B, is

$$B = -\left(\frac{V_0}{h}\right)^2 \frac{\varphi_1}{\tau_{\varkappa=\varkappa}}.$$

For single relaxation times λ_{11} and λ_{21} of the *Bird-Carreau* model it is possible for B to vanish at certain critical values of the frequency and shear rate $\varkappa_0$.

Fig. 3 shows the combinations of values $(\omega^2, \varkappa_0^2)$ for which B vanishes, given the indicated values of the parameters of the *Bird-Carreau* model. For the same parameters, figs. 4 and 5 illustrate the effect of strain amplitude on shear stress enhancement B for several frequencies and for two values of $\varkappa_0$. As noted previously, high frequencies and/or strains lead, in general, to larger contributions to the stress from the second order terms.

It is of interest to derive an approximate expression for the induced steady flow. If $(b_p)^2$ is small, the additional flow per unit width is

$$Q_A \cong -(\varphi_1/\varphi_0)\, V_0^2 (b_p^2 + \tilde{b}_p^2)\, h^2/24. \qquad [36]$$

It is interesting to note that, as in the previous section, the additional steady flow component is controlled by the terms governing *inelastic* character of the fluid. The reader is again reminded of the corresponding results of *Barnes* et al. (6, 8) for oscillatory pipe flow.

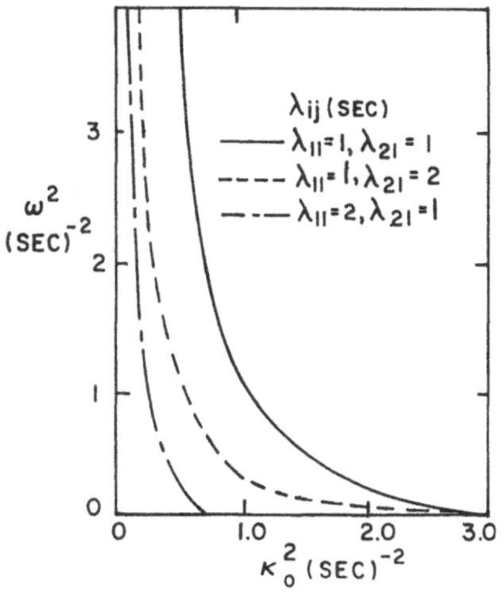

Fig. 3. Superposed steady and parallel oscillatory shear: Conditions for which steady part of second-order shear stress vanishes

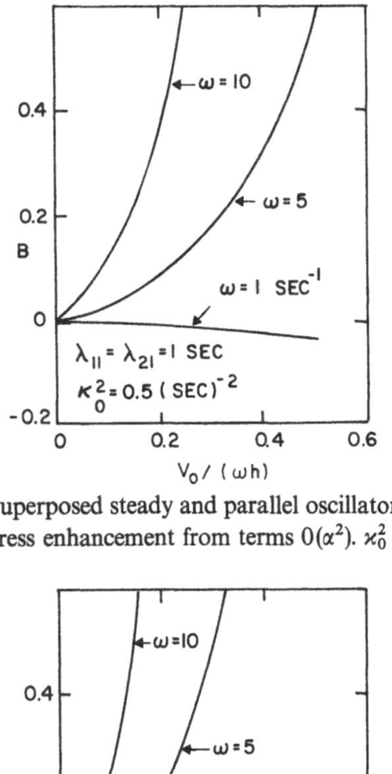

Fig. 4. Superposed steady and parallel oscillatory shear: Shear stress enhancement from terms $0(\alpha^2)$. $\varkappa_0^2 = \frac{1}{2}$

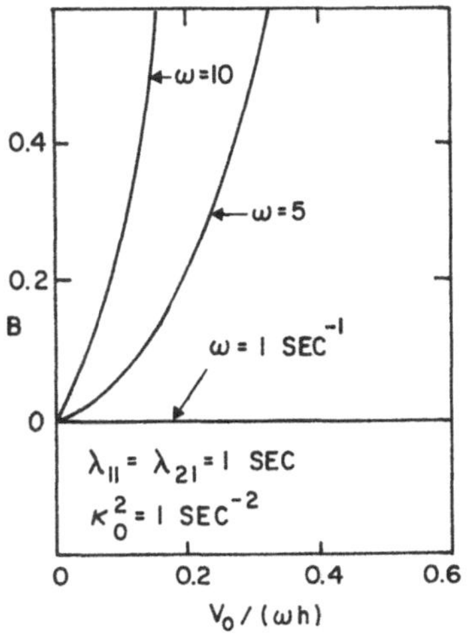

Fig. 5. Superposed steady and parallel oscillatory shear: Shear stress enhancement from terms $0(\alpha^2)$. $\varkappa_0^2 = 1$

3.3. Superposed Steady and Transverse Oscillatory Shear

Just as the previous sections have shown that new effects are predicted when nonlinear terms are retained in the analysis, so one finds new interactions when the oscillatory and steady shearing are orthogonal. Studies of this flow have, to date, been interpreted on the basis of a time dependent velocity profile which oscillates with frequency ω (2, 13, 32).

For this flow configuration the velocity of the upper plane is given by

$$v = \{U_0, 0, V_0 \cos\omega t\}. \qquad [37]$$

With the exception of the altered boundary condition, assumptions about the nature of the flow are similar to those employed in the solutions of the last two sections. We again assume validity of an expansion in powers of α for u, w, and τ_{ij}; and compute results according to the *Bird-Carreau* model. Results to order α are already available, but we include them here for completeness. Thus

$$w_1 = \boldsymbol{R}[w_1^1 e^{i\omega t}]$$

$$(\tau_{zy})_1 = \beta_1 = \boldsymbol{R}[\beta_1^1 e^{i\omega t}]$$

$$\beta_1^1 = \eta^*_{T_{\varkappa_0}} dw_1^1/dy \qquad [38]$$

with

$$\eta^*_{T_{\varkappa_0}} = \int_0^\infty m(s, 2\varkappa_0^2) \frac{1 - e^{-i\omega s}}{i\omega} ds \qquad [39]$$

and, to order α,

$$u_1 = (\tau_{xz})_1 = (\tau_{xy})_1 = (\tau_{ii})_1 = 0$$

$$(i = 1, 2, 3, \text{no sum})$$

$$\alpha w_1^1 = V_0 \sinh b_T y/(\sinh b_T h) \qquad [40]$$

where $(b_T)^2 = \dfrac{i\omega\varrho}{\eta^*_{T_{\varkappa_0}}}$. Comparison of $\eta^*_{T_{\varkappa_0}}$ with $\eta^*_{\varkappa_0}$ of the previous section reveals that even to first order in α there is a marked contrast between the prediction of parallel and transverse superposed shearing. This difference has been noted previously (13).

We proceed to terms of order α^2. Let

$$u_2 = \boldsymbol{R}[u_2^0 + u_2^2 e^{2i\omega t}]$$

$$(\tau_{xy})_2 = \Omega_2 = \boldsymbol{R}[\Omega_2^0 + \Omega_2^2 e^{2i\omega t}]$$

$$\Omega_2^0 = \varphi_0 du_2^0/dy - \varphi_4 z_1 \tilde{z}_1$$

$$\Omega_2^2 = \varphi_2 du_2^2/dy - \varphi_5 z_1^2$$

$$(\tau_{zy})_2 = w_2 = 0$$

where $z_1 = dw_1^1/dy$ and the φ_i are given in the Appendix. Solutions for the velocity fields u_2^0 and u_2^2 are also given in the Appendix. For small values of b_T and $(b_2)^2 = 2i\omega\varrho/\varphi_2$, the enhancement of the steady flow can be found.

$$Q_A \cong -(\varphi_4/\varphi_0)\, V_0^2 (b_T^2 + \bar{b}_T^2)\, V_0^2 h^2/24. \qquad [41]$$

Inspection of relevant terms in the Appendix again shows that the nonlinear behavior is dominated by the "inelastic" terms in the constitutive equation.

To provide some comparison with the parallel superposition of Section 3.2 we have plotted in figs. 6 and 7 $-B_T$, the fractional decrease in steady shear stress which is due to inclusion of the second order terms, for the same model parameters used earlier. It is easily shown that the

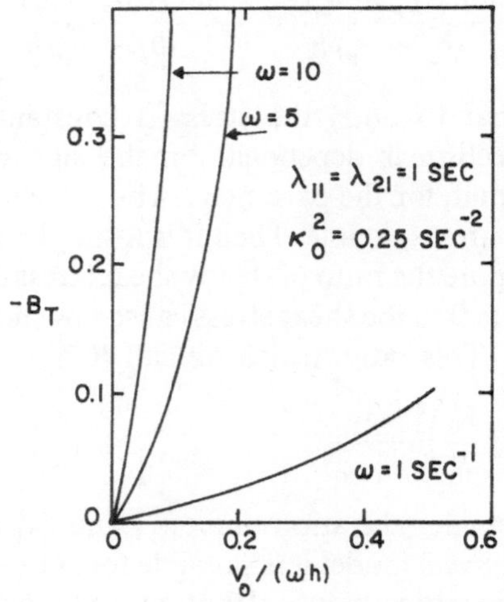

Fig. 6. Superposed steady and transverse oscillatory shear: Fractional decrease in steady shear stress due to terms $0(\alpha^2)$. $\varkappa_0^2 = \frac{1}{4}$

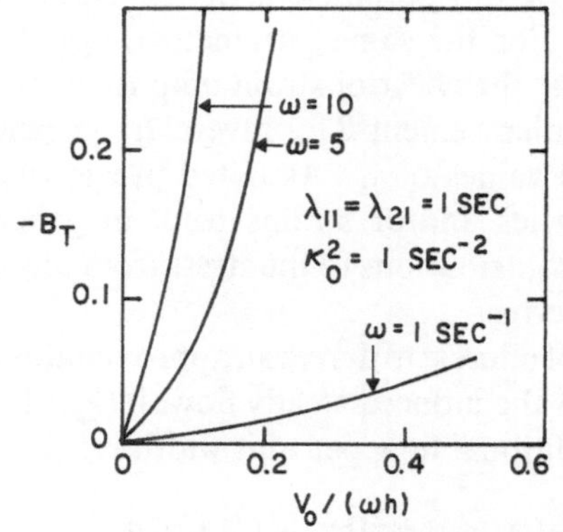

Fig. 7. Superposed steady and transverse oscillatory shear: Fractional decrease in steady shear stress due to terms $0(\alpha^2)$. $\varkappa_0^2 = 1$

steady-state contribution to the stress from second order terms is

$$\alpha^2 \Omega_2^0 = \frac{-\varphi_4 (V_0)^2 b_T \bar{b}_T}{2 \sinh(b_T h) \sinh(\bar{b}_T h)} \times \left[\frac{\sinh[(b_T + \bar{b}_T) h]}{(b_T + \bar{b}_T) h} + \frac{\sinh[(b_T - \bar{b}_T) h]}{(b_T - \bar{b}_T) h} \right].$$

Since φ_4 is positive, $-B_T$ is always positive. For the conditions chosen, the effect of density on $-B_T$ is negligible. One obtains

$$-B_T = \left(\frac{V_0}{h}\right)^2 \frac{\varphi_4}{\tau_{\varkappa=\varkappa_0}}.$$

4. Concluding remarks

Most of the interesting phenomena which one observes with non-*Newton*ian fluids are attributable to the nonlinear nature of fluid response to imposed forces. In the present paper we have illustrated how these nonlinearities can give rise to a number of unusual effects in time dependent flows. The exact nature and magnitude of these effects must be determined from properly designed experiments, the motivation for which we hope has been supplied.

Study of nonlinear behavior by expansion techniques is, at best, an incomplete approach and the present illustration is no exception. Two possible difficulties should be mentioned. First, one cannot be sure *a priori* of the range of practical validity of the expansion to the order used, simply because the necessary fluid parameters are not available for wide classes of fluids. Second, we have assumed throughout that the dimensionality of the flow is not altered by the nonlinear effects studied here. In an absolute sense this is not expected to be true. Spatially inhomogeneous normal stresses, for example, may induce new secondary flows. Whether in fact these flows will be sufficient to alter the results given here will again depend upon combinations of flow and material variables and must ultimately be answered by experiment.

Acknowledgment

The authors are grateful for partial support from the National Aeronautics and Space Administration through Grant NGR 31-001-025, and from the donors of the Petroleum Research Fund, administered by the American Chemical Society.

Appendix

I. Oscillatory Shear Flow

Contribution of terms $0(\alpha^3)$ to the velocity field.

$$\alpha^3 u_3^1 = \left(\frac{\eta^{\mathrm{I}}}{\eta^*}\right) \frac{V_0^3 b^2 \bar{b}}{4 \sinh^2(b h) \sinh(\bar{b} h)} \left\{ \left[\frac{2b + \bar{b}}{(2b + \bar{b})^2 - b^2} \right] \left[\sinh[(2b + \bar{b}) y] - \sinh[(2b + \bar{b}) h] \frac{\sinh(b y)}{\sinh(b h)} \right] + \left[\frac{2b - \bar{b}}{(2b - \bar{b})^2 - b^2} \right] \left[\sinh[(2b - \bar{b}) y] - \sinh[(2b - \bar{b}) h] \frac{\sinh(b y)}{\sinh(b h)} \right] + \left[\frac{2\bar{b}}{\bar{b}^2 - b^2} \right] \left[\sinh(\bar{b} y) - \sinh(\bar{b} h) \frac{\sinh(b y)}{\sinh(b h)} \right] \right\};$$

$$\alpha^3 u_3^3 = \left[\frac{\eta^{\mathrm{III}}}{\eta^*(3\omega)} \right] \left[\frac{3 V_0^3 b^4}{4 \sinh^2(b h)} \right] \left\{ \frac{\sinh(3 b y) - \sinh(3 b h) \frac{\sinh(b_3 y)}{\sinh(b_3 h)}}{9 b^2 - (b_3)^2} + \frac{\sinh(b y) - \sinh(b h) \frac{\sinh(b_3 y)}{\sinh(b_3 h)}}{b^2 - (b_3)^2} \right\}$$

where

$$(b_3)^2 = \frac{3 i \omega \varrho}{\eta^*(3\omega)}.$$

II. Superposed Steady and Parallel Oscillatory Shear

$$\varphi_0 = \int_0^\infty [m(s, 2\varkappa_0^2) s - m_1(s, 2\varkappa_0^2) 2\varkappa_0^2 s] \, ds$$

$$\varphi_1 = \int_0^\infty \left\{ m_1(s, 2\varkappa_0^2) \left[\frac{1}{2} \varkappa_0 s - \varkappa_0 \left(\frac{e^{-i\omega s} - 1}{i\omega} \right) \right] - m_2(s, 2\varkappa_0^2) 2\varkappa_0^3 s \right\} ds$$

$$\varphi_2 = \int_0^\infty \left\{ m(s, 2\varkappa_0^2) \left[\frac{1 - e^{-2i\omega s}}{2 i\omega} \right] - m_1(s, 2\varkappa_0^2) 2 s \varkappa_0^2 e^{-2i\omega s} \right\} ds$$

$$\varphi_3 = \int_0^\infty \left\{ m_1(s, 2\varkappa_0^2)\left[\frac{1}{2}\varkappa_0 s e^{-2i\omega s} + \varkappa_0\left(\frac{e^{-i\omega s} - e^{-2i\omega s}}{i\omega}\right)\right] - m_2(s^1, 2\varkappa_0^2)\, 2\varkappa_0^3 s e^{-2i\omega s}\right\} ds$$

where

$$m_2(s, 2\varkappa_0^2) = \sum_{p=1}^{\infty} \frac{\lambda_{1p}^4 \eta_p e^{-s/\lambda_{2p}}}{\lambda_{2p}^2(1 + \lambda_{1p}^2 \varkappa_0^2)^3}$$

$$\alpha^2 u_2^0 = \frac{\varphi_1}{\varphi_0}\left[\frac{V_0^2 b_p \tilde{b}_p}{2\sinh(b_p h)\sinh(\tilde{b}_p h)}\right]$$

$$\times \left\{\frac{\sinh[(b_p + \tilde{b}_p)y] - (y/h)\sinh[(b_p + \tilde{b}_p)h]}{(b_p + \tilde{b}_p)} + \frac{\sinh[(b_p - \tilde{b}_p)y] - (y/h)\sinh[(b_p - \tilde{b}_p)h]}{(b_p - \tilde{b}_p)}\right\}$$

$$\alpha^2 u_2^2 = \frac{\varphi_3}{\varphi_2}\left[\frac{V_0^2 b_p^3}{\sinh^2(b_p h)(4b_p^2 - b_2^2)}\right] \times \left[\sinh(2b_p y) - \frac{\sinh(b_2 y)}{\sinh(b_2 h)}\sinh(2b_p h)\right]$$

where

$$b_2^2 = 2i\omega\varrho/\varphi_2.$$

III. Superposed Steady and Transverse Oscillatory Shear

Equations for velocity are identical to those of paragraph II above, with b_p replaced by b_T, φ_1 by φ_4 and φ_3 by φ_5.

$$b_T^2 = i\omega\varrho/\eta_{T\varkappa_0}^*$$

$$\varphi_4 = \int_0^\infty m_1(s, 2\varkappa_0^2)\frac{\varkappa_0 s}{2}\, ds$$

$$\varphi_5 = \int_0^\infty m_1(s, 2\varkappa_0^2)\frac{\varkappa_0 s}{2} e^{-2i\omega s} ds.$$

Summary

Approximate solutions to the equations of motion for rheologically complex fluids have been obtained for the following unsteady flows between parallel boundaries: oscillatory shearing, superposed steady and parallel oscillatory shearing, and superposed steady and transverse oscillatory shearing. The chief objective is to elucidate contributions to these time-dependent flows which follow from interaction of harmonics of the flow and hence appear only when a nonlinear description of the flow is retained. It is shown that several new features arise. These may be helpful for discrimination between constitutive models.

References

1) *Ferry, J. D.*, Viscoelastic Properties of Polymers (New York 1961).
2) *Tanner, R. I.* and *J. M. Simmons*, Chem. Eng. Sci. **22**, 1803 (1967).
3) *Mac Donald, I. F., D. B. Marsh*, and *E. Ashare*, Chem. Eng. Sci. **24**, 1615 (1969).
4) *Jones, T. E. R.* and *K. Walters*, J. Phys. A: Gen. Phys. **4**, 85 (1971).
5) *Dodge, J. S.* and *I. M. Krieger*, Trans. Soc. Rheol. **15**, 589 (1971).
6) *Barnes, H. A., K. Walters*, and *P. Townsend*, Nature **224**, 585 (1969).
7) *Walters, K.* and *P. Townsend*, Proc. 5th Int. Cong. on Rheol. Ed.: *S. Onogi*, Vol. **4**, p. 471 (Tokyo 1970).
8) *Barnes, H. A., P. Townsend*, and *K. Walters*, Rheol. Acta **10**, 517 (1971).
9) *Schlichting, H.*, Boundary Layer Theory, 6th ed., p. 411 (New York 1968).
10) *Batchelor, G. K.*, An Introduction to Fluid Dynamics, p. 363 (Cambridge Univ. Press 1967).
11) *Coleman, B. D., H. Markovitz*, and *W. Noll*, Viscometric Flows of Non-Newtonian Fluids (New York 1966).
12) *Pipkin, A. C.* and *J. R. Owen*, Phys. Fluids **10**, 836 (1967).
13) *Markovitz, H.*, Proc. 5th Int. Cong. on Rheol. Ed.: *S. Onogi*, Vol. **1**, p. 499 (Tokyo 1969).
14) *Bird, R. B.* and *P. J. Carreau*, Chem. Eng. Sci. **23**, 427 (1968).
15) *Oldroyd, J. G.*, Proc. Roy. Soc. **A 245**, 278 (1958).
16) *Oldroyd, J. G.*, Proc. Roy. Soc. **A 200**, 523 (1950).
17) *Booij, H. C.*, Rheol. Acta **5**, 215 (1966).
18) *Booij, H. C.*, Rheol. Acta **5**, 222 (1966).
19) *Harris, J.* and *K. Bogie*, Rheol. Acta **6**, 3 (1967).
20) *Walters, K.* and *T. E. R. Jones*, Proc. 5th Int. Cong. on Rheol. Ed.: *S. Onogi*, Vol. **4**, p. 337 (Tokyo 1970).
21) *Onogi, S.*, Trans. Soc. Rheol. **14**, 275 (1970).
22) *Dodge, J. S.* and *I. M. Krieger*, Rheol. Acta **8**, 480 (1969).
23) *Williams, M. C.* and *R. B. Bird*, Ind. Eng. Chem. Fund. **3**, 42 (1964).
24) *Coleman, B. D.* and *W. Noll*, Rev. Mod. Phys. **33**, 239 (1961).
25) *Spriggs, T. F., J. D. Huppler*, and *R. B. Bird*, Trans. Soc. Rheol. **10**, 191 (1966).
26) *Endo, H.* and *M. Nagasawa*, J. Pol. Sci., Pt. A 2, **8**, 371 (1970).
27) *Goldstein, Charles*, Ph. D. Thesis, Princeton University, Princeton, New Jersey 1971.
28) *Philippoff, W.*, Trans. Soc. Rheol. **10**, 316 (1966).
29) *Tanner, R. I.*, Trans. Soc. Rheol. **12**, 155 (1968).
30) *Osaki, K., M. Tamura, M. Kurata*, and *T. Kotaka*, J. Phys. Chem. **69**, 4183 (1965).
31) *Mac Donald, I. F.* and *R. B. Bird*, J. Phys. Chem. **70**, 2068 (1966).
32) *Tanner, R. I.* and *G. Williams*, Rheol. Acta **10**, 528 (1971).

Authors' address:
Dr. *Charles Goldstein*
Research & Engineering Division Whirlpool Corporation
Benton Harbor, Michigan 49022 (U.S.A.)
Akron, Ohio 44316 (U.S.A.)

Rheol. Acta 12, 263–268 (1973)

From the Department of Chemical Engineering McGill University Montreal (Canada)

The effect of pressure on the shear viscosity of polymer melts

By M. R. Kamal and H. Nyun

With 5 figures

(Received October 27, 1972)

List of Symbols

C_v	specific heat at constant volume
L	total length of capillary
L_D	length of capillary over which developed flow exists
L_T	length of capillary over which hypothetical isothermal flow exists
P	pressure
ΔP_d, P	pressure drop over developed flow region
ΔP_{total}	total pressure drop
ΔP_{acc}	average temperature rise of the melt expressed as a pressure drop
ΔP_e	pressure drop in the entrance region
R	capillary radius
r	radial coordinate
T	temperature
T	average temperature of the melt
T_a	temperature of the barrel and capillary surface
Q	volumetric flow rate
q_r	radial heat conduction flux at position r
q_R	radial heat conduction flux at the capillary wall
τ_{rz}	shear stress
τ_w	shear stress at the capillary wall
$\dot{\gamma}$	shear rate
$\dot{\gamma}_w$	shear rate at the capillary wall
ϱ	density of the melt
η	viscosity of polymer melt

Introduction

The shear viscosity of fluids in general, and polymer melts in particular, depends on temperature, pressure, and shear rate. Although the effects of temperature and shear rate have been studied extensively for a large number of polymer types, only recently methods have been devised to study the effect of pressure. As a result of these studies, it has been found that the effect of pressure on the shear viscosity of polymer melts is quite significant. Furthermore, the effect of viscous heating on the determination of viscosity from capillary measurements has been recognized, and some attempts have been made to make corrections for this effect (1, 2).

Previous studies on the effect of pressure on the viscosity of polymer melts include *Westovers* (3) capillary viscometer, the rotating cylinder viscometer developed by *Semjownow* (4), and the falling cylinder viscometer employed by *Ramsteiner* (5). The above devices are limited due to complexities of operation and maintenance and the narrow range of shear rates that can be employed. Capillary devices are simpler and more versatile. Such devices have been used by *Duvdevani* and *Klein* (6), *Nakajima* and *Choi* (7), and *Penwell, Porter*, and *Middleman* (8).

In this paper we present a more general and quantitative method for obtaining the effect of pressure on the viscosity of polymer melts from capillary viscometer data. The proposed method employs less restrictive assumptions than previous techniques, and it allows corrections for the effect of viscous heating. Some of the results which show the general validity of the proposed method will be discussed.

Theoretical development

In determining the viscosity of a fluid, it is necessary to determine the shear stress and shear rate acting on the fluid at a point where the temperature and pressure are also known. The pressure acting on the fluid in a capillary tube changes with axial position, and the temperature of the fluid is nonuniform due to viscous heating even when the temperature of the wall of the capillary is constant. Furthermore, the shear rate varies at different points in the capillary. In measurements that have been carried out so far, the effects of pressure and viscous heating have been neglected. The shear rate is considered to be a function of radial position only, and it is calculated at the wall with the help of the *Weissenberg-Rabinowitch* eq. [14]. According to the proposed procedure, the effects of pressure and viscous heating are not neglected. The shear rate is determined at a point where both pressure and temperature are known.

The flow system in a capillary rheometer may be represented as in fig. 1. The piston, *P*, pushes the polymer melt from the barrel (reservoir) into the capillary at a constant speed. The origin of the cylindrical coordinate system is taken at the center of the capillary at the exit end. Developed flow is assumed to start at $Z = L_D$. The purpose of the following discussion is to determine the temperature, pressure, shear stress, and shear rate at the wall of the capillary at $Z = L_D$. The assumptions employed in the treatment of the problem are listed below.

1. The flow of the polymer melt is laminar, steady, and axisymmetric.
2. There is no slip at the wall of the capillary.
3. Heat conduction in the axial direction is negligible in comparison with heat convection.
4. The thermal conductivity and thermal diffusivity of the melt are constant and independent of temperature and pressure.
5. The effects of compressibility may be neglected.
6. Elastic effects are neglected.

18*

Pressure at $Z = L_D$

The pressure at $Z = L_D$ may be obtained from the following equation.

$$\frac{F(L)}{A} = \Delta P_{\text{total}} = \Delta P_e + \Delta P_d \cdot (L) \qquad [1]$$

where $F(L)$ is the total force on the plunger which has a cross-sectional area, A; ΔP_e is the entrance pressure drop which is assumed constant for a constant volumetric flow rate, diameter, and temperature; $\Delta P_d(L)$ is the pressure drop over the developed flow region. In this work, the quantity ΔP_e is obtained by fitting ΔP_{total} as a quadratic function of L. The constant term in the quadratic will give ΔP_e. Thus the pressure at $Z = L_D$ is obtained according to eq. [2].

$$(P)_{Z=L_D} = \Delta P_{\text{total}} - \Delta P_e . \qquad [2]$$

This method of finding ΔP_e is similar to the method proposed by *Bagley* (9), except for using a quadratic instead of a linear fit for ΔP_{total} as a function of capillary length. It has been found in this work that a linear fit is not possible over the range of variables of practical interest.

Temperature at $Z = L_D$

The pressure drop over the entry region is due to the generation of thermal, kinetic, and elastic energy in the fluid. It has been shown by *Bogue* (10) and *Collins* and *Schowalter* (11) that most of ΔP_e is due to the generation of kinetic energy, and that viscous heat generation contributes only a small fraction of the total entrance pressure drop. Since the total entrance pressure drop itself is rather small, viscous heat generation in the entry region may be neglected. The temperature at $Z = L_D$ may therefore be regarded equal to the temperature of the fluid in the barrel, which is T_0.

Shear rate at the wall at $Z = L_D$

The expression for the shear rate at the wall at $Z = L_D$ is obtained in the same manner as the *Weissenberg-Rabinowitch* eq. [14]. Thus, the true shear rate, $\dot{\gamma}_w$, at the point of interest is given by:

$$\dot{\gamma}(\tau_w, P, T)_{Z=L_D} = \frac{Q}{\pi R^3} \times \left[3 + \left(\frac{\partial \ln (Q/\pi R^3)}{\partial \ln \tau_w} \right)_{P,T} \right]_{Z=L_D} \qquad [3]$$

where τ_w, P, and T are the shear stress at the wall, the melt pressure, and the melt temperature, respectively, all evaluated at $Z = L_D$. The procedure to determine τ_w will be outlined in the next section.

It should be observed that the main difference between eq. [3] and the *Weissenberg-Rabinowitch* equation is that the former takes into consideration the possible variation of τ_w, P, and T in the axial direction. Therefore, the value of the shear rate is specified to correspond to the prevailing values of the above variables at $Z = L_D$.

Shear stress at the wall at $Z = L_D$

The temperature rise of the capillary wall depends on the rate of transfer of thermal energy from the fluid, the rate with which heat is conducted away from the surface, and on the time elapsed. The problem may be regarded as one of heat conduction in a region consisting of the capillary, the barrel, and the barrel jacket while a heat flux is applied at the inner wall of the capillary. *Carslaw* and *Jaeger* (12) give a solution to a similar problem for the case of an infinite medium. In their solution, the temperature rise of the inner wall, ΔT, is given by

$$\Delta T = \frac{q_R R}{2k} \ln \left(\frac{4 \alpha t}{C R^2} \right) \qquad [4]$$

where q_R is the heat flux at the inner wall, R is the inner radius of the cylinder, α is the thermal diffusivity of the fluid, t is the elapsed time, and C is a constant given as exp (0.57722).

Employing the above treatment to approximate the present experimental situation, the temperature rise at the capillary wall is found to be less than 1.5 °C. Therefore, the inner wall of the capillary may be considered isothermal.

In accordance with the assumptions that have been made, the equations of motion and energy may be written as follows:

$$\frac{1}{r} \frac{\partial}{\partial r} (r \tau_{rz}) = - \frac{\partial P}{\partial z} \qquad [5]$$

and

$$\varrho C_v v_z \frac{\partial T}{\partial z} = - \frac{1}{r} \left(\frac{\partial}{\partial r} (r q_r) \right) - \tau_{rz} \left(\frac{\partial v_z}{\partial r} \right) \qquad [6]$$

where r and z represent the radial and axial coordinates, respectively (in cylindrical coordinates), P is the pressure, T is the temperature, τ_{rz} is the viscous shear stress, q_r is the radial heat flux at position r, ϱ is the density of the melt, and C_v is the specific heat of the melt at constant volume. Eq. [3] yields the following expression for the shear stress, τ_{rz}:

$$\tau_{rz} = -\frac{r}{2}\frac{\partial P}{\partial z}. \qquad [7]$$

The magnitude of the pressure drop, ΔP, in the developed flow region, between $z = L_D$ and $z = 0$, is obtained by combining eqs. [5] and [6] and integration.

$$\Delta P = \varrho C_v[T_a(z=0) - T_a(z=L_D)] - \frac{2\pi R}{Q}\int_0^{L_D} q_R d_z \qquad [8]$$

where the average temperature, T_a, is defined by:

$$T_a = \frac{\int_0^R 2\pi r v_z T dr}{Q}. \qquad [9]$$

The volumetric flow rate, Q, is given by

$$Q = \int_0^R 2\pi r v_z dr \qquad [10]$$

and R is the radius of the capillary.

In order to obtain the wall shear stress at $z = L_D$ for the case of isothermal capillary wall, it is useful to consider the expressions for shear stress for isothermal and adiabatic flows.

Isothermal flow

When the fluid at all points in the capillary has the same temperature, the shear stress at any point in the capillary becomes independent of the capillary length, L. The developed-flow pressure drop, ΔP, is obtained by integrating eq. [7] as follows

$$\Delta P = -\int_0^{L_D} \frac{2}{R}\tau_w(z, Q, R, T_0)\, dz \qquad [11]$$

where T_0 is the melt temperature in the reservoir which also holds at all points in the capillary. Differentiation of eq. [11] with respect to L yields the value of τ_w at $z = L_D$.

$$(\tau_w)_{z=L_D} = -\frac{R}{2}\left(\frac{\partial \Delta P}{\partial L}\right)_{Q,R,T_0}. \qquad [12]$$

The derivative on the right-hand side of eq. [12] is evaluated for the capillary length, L, of interest in order to define τ_w at $z = L_D$ for the prevailing melt temperature and pressure at that point. Thus, eq. [12] in combination with eq. [3] may be used to obtain viscosity as a function of temperature, pressure, and shear rate when viscous heat generation is neglected. This is similar to the standard procedures for calculating viscosity, except that the effect of pressure on viscosity is included in the present treatment.

Adiabatic flow

For the case of adiabatic flow, the radial heat flux at the wall, Q_R, is zero, and eq. [3], rewritten in terms of T_a, takes the form:

$$\varrho C_v \frac{\partial T_a}{\partial z} = -\frac{\partial P}{\partial z}. \qquad [13]$$

Thus, combining eqs. [7] and [13]:

$$(\tau_w)_{Q,R,T_0,L} = \frac{\varrho C_v R}{2}\left(\frac{\partial T_a}{\partial z}\right)_{Q,R,T_0,L}. \qquad [14]$$

For the remainder of this discussion, Q and R will be held constant, and they are not included as variables in the functional dependence of T_a for simplicity.

Now, for $z = L_D$ we can write

$$\left(\frac{\partial T_a}{\partial z}\right)_{T_0,L} = \left(\frac{\partial T_a}{\partial L_D}\right)_{T_0,L}. \qquad [15]$$

Furthermore, it can be shown by the chain rule of partial derivatives that

$$\left(\frac{\partial T_a}{\partial L_D}\right)_{T_0,L} = -\left(\frac{\partial T_a}{\partial L}\right)_{T_0,z} \qquad [16]$$

since

$$L_D = L - e \qquad [17]$$

where e is an entry length independent of L and T_0. Remembering that

$$\left(\frac{\partial T_a(L_D, L, T_0)}{\partial T_0}\right)_{L,z} = 1 \qquad [18]$$

because

$$T_a = T_0 \quad \text{at} \quad z = L_D \qquad [19]$$

and recognizing that $T_a = T_a(L, T_0)$, we can write for $z = L_D$

$$\left.\frac{\partial T_0}{\partial L}\right)_{T_a,z} = -\frac{(\partial T_a/\partial L)_{T_0,z}}{(\partial T_a/\partial T_0)_{L,z}} = \left(\frac{\partial T_a}{\partial z}\right)_{T_0,L}. \qquad [20]$$

By combining eqs. [14], [15], [16], [18] and [20] and by differentiating eq. [8] with $q_R = 0$, we can write

$$(\tau_w)_{z=L_D} = -\frac{R}{2}\left(\frac{\left(\frac{\partial \Delta P}{\partial L}\right)_{Q,R,T_0}}{1 + \frac{1}{\varrho C_v}\left(\frac{\partial \Delta P}{\partial T_0}\right)_{Q,R,L}}\right) \qquad [21]$$

since for adiabatic flow

$$\left(\frac{\partial T_0}{\partial L}\right)_{T_a, z} = \left(\frac{\partial T_0}{\partial L}\right)_{T_a, z=0}. \qquad [22]$$

This is due to the fact that in adiabatic flow, if L and T_0 are changed so as to maintain a constant average temperature, T_a, at a fixed coordinate, z, then the average temperatures at other coordinates will not change, keeping in mind that Q and R are held constant.

Isothermal wall capillary

When the flow is neither adiabatic nor isothermal, the magnitude of the pressure drop in the developed flow region is given by eq. [8]. The real process in which the pressure is reduced by ΔP and the average temperature rises from $T_a(z = L_D)$ to $T_a(z = 0)$ may be replaced by a hypothetical combination of two processes. The first process is adiabatic where the fluid temperature is raised by the above amount, and the second process is isothermal where the balance of the pressure drop is achieved. The hypothetical combination leads to the definition of a new position coordinate at $z = L_T$. Isothermal flow is taken to occur between $z = 0$ and $z = L_T$, and adiabatic flow occurs between $z = L_T$ and $z = L_D$. Thus, the integrated energy equation may be rewritten as follows:

$$P = \varrho C_v[T_a(L_T, L, T_0) - T_a(L_D, L, T_0) - \frac{2}{R}\int_0^{L_T} \tau_w(z, T_0)\, dz. \qquad [23]$$

The variables Q and R are not included in eq. [23] because they are assumed to be held constant throughout this section.

By following mathematical procedures similar to those outlined above for adiabatic flow, the following relationship may be obtained:

$$(\tau_w)_{z=L_D} = -\frac{R}{2}\left[\frac{\left(\frac{\partial \Delta P}{\partial L}\right)_{Q,R,T_0}}{\left(1 + \frac{1}{\varrho C_v}\left(\frac{\partial \Delta P_{acc}}{\partial T_0}\right)_{Q,R,L}\right)}\right] \qquad [24]$$

where the accumulated pressure drop, ΔP_{acc}, represents the portion of the total pressure drop in the developed flow region that is dissipated as a result of viscous heating. The quantity ΔP_{acc} is defined as follows

$$\Delta P_{acc} = \varrho C_v[T_a(L_T, L, T_0) - T_a(L_D, L, T_0)]. \qquad [25]$$

The complex method for estimating ΔP_{acc} is based on the application of the procedures employed by *Bird* (13) and will be reported elsewhere.

It is interesting to note that eq. [12] for isothermal flow and eq. [21] for adiabatic flow may be derived as special cases from eq. [24].

Experimental

The Instron Capillary Rheometer was employed to obtain the experimental data reported in this study. Two sets of capillaries were used. One set had an inside diameter of 0.030 inch, and the following length/diameter ratios: 5, 10, 20, 40, 60, 80, and 100. The inside diameter of the other set was 0.052 inch, and the length/diameter ratios were: 5, 10, 20, 40, 60, and 80. Experiments were conducted at 180, 190, 200, 210, and 220 °C. The shear rate ranged from 15 sec^{-1} to 3000 sec^{-1}. Styron 683, obtained from Dow Chemical of Canada, was employed throughout this study.

Results and discussion

The common approach is to calculate the shear rate at the wall according to the following equation

$$\dot{\gamma}_w = \frac{Q}{\pi R^3}\left(3 + \left(\frac{\partial \ln(Q/\pi R^3)}{\partial \ln \tau_w}\right)_{T_0}\right). \qquad [26]$$

The shear stress at the wall is usually calculated from

$$\tau_w = \frac{R \Delta P}{2L}. \qquad [27]$$

In some instances, the overall pressure drop between the reservoir and the exit of the capillary is employed, especially when capillaries having large length-diameter ratio (L/D) are employed. In other instances, the pressure is corrected for the entrance region by linear extrapolation to zero length as indicated above. The end correction, ΔP_e, was calculated in this work by fitting ΔP_{total} as a quadratic function of L. This was found necessary because it was not possible to obtain a linear fit over the range of variables of interest.

Fig. 2 shows the results of the present work calculated without the end corrections. These results show that large L/D ratios, of the order of 40, are required before the effect of the end correction can be neglected. On the other hand, even when the end correction is applied, different viscosity values are obtained at the same shear rate when different capillaries are used as shown in fig. 3.

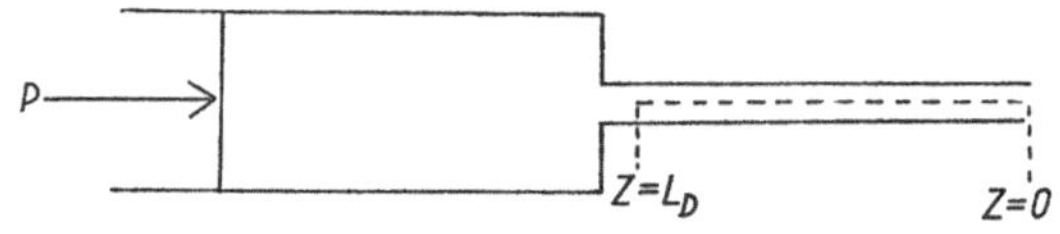

Fig. 1. Schematic diagramm of capillary rheometer

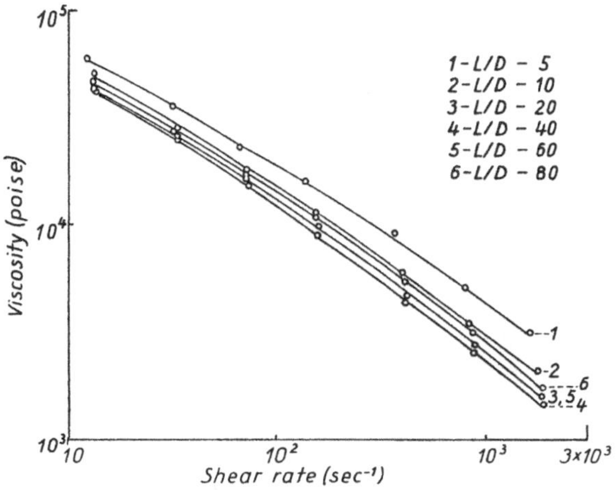

Fig. 2. Viscosity of Styron 683 at 180 °C obtained with 0.052 inch diameter capillaries without end correction

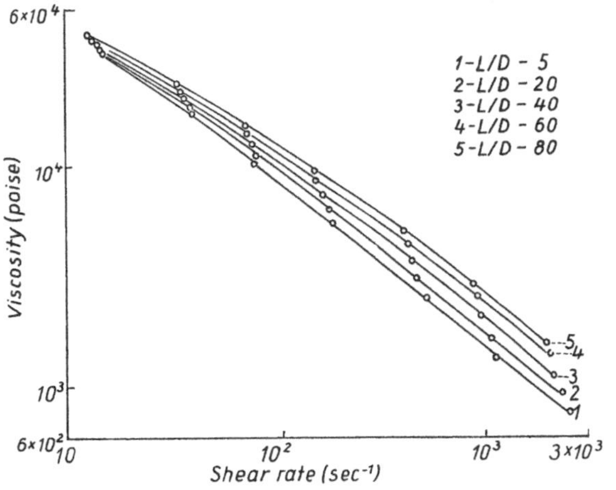

Fig. 3. Viscosity of Styron 683 at 180 °C obtained with 0.052 inch diameter capillaries with end correction

Fig. 4 shows the results that are obtained when the data are processed according to eqs. [3] and [12], i.e., when viscous heating effects are neglected. The data for the two sets of capillaries diverge at higher pressures, leading to higher viscosity values for the capillaries with the smaller diameters. This is probably because these capillaries offer a shorter path for the removal of heat by conduction, which results in lower melt temperatures.

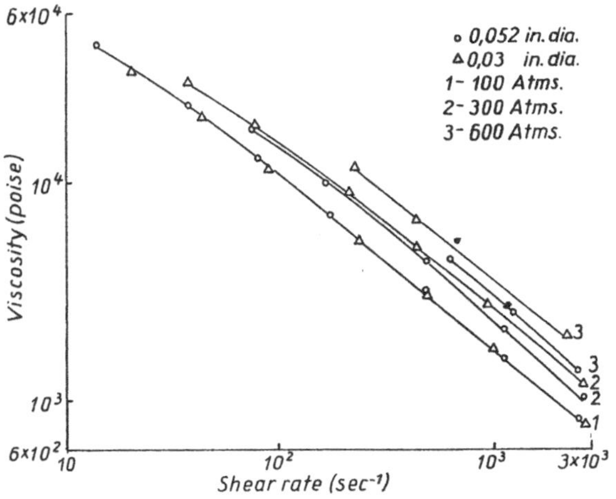

Fig. 4. Effect of pressure on viscosity of Styron 683 at 180 °C without correction for viscous heating

When both pressure and viscous heating effects are employed according to eqs. [3] and [24], the results shown in fig. 5 are obtained. The data from the two sets of capillaries, processed separately, show remarkable agreement with a maximum variation of 5%. Similar results are available for the other temperatures employed in this study.

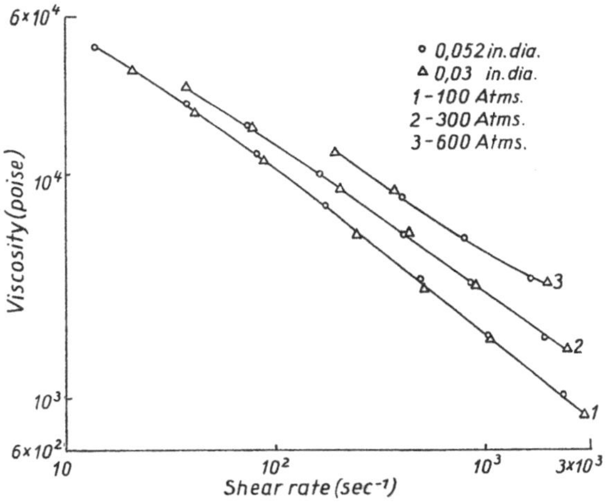

Fig. 5. Effect of pressure on viscosity of Styron 683 at 180 °C with correction for viscous heating

The results show that both pressure and viscous heating effects are important in extracting viscosity data on the basis of capillary measurements. Furthermore, they point out the significant effect of pressure on viscosity. The value of the pressure coefficient, defined by

$$b = \frac{d \ln \eta}{dP} \qquad [28]$$

where η is the viscosity of the melt, decreases with pressure. At atmospheric pressure and a shear rate of 2500 sec^{-1} we obtain $b = 2.07 \times 10^{-3}\,atm^{-1}$ which is in good agreement with the value of $2.13 \times 10^{-3}\,atm^{-1}$ obtained from the *Williams-Landel-Ferry* (W-L-F) eq. [15] according to the method reported by *Penwell, Porter*, and *Middleman* (8).

A method has been developed to obtain the dependence of viscosity on shear rate from capillary measurements with due consideration for pressure and viscous heating effects. The data obtained show the importance of including pressure and viscous heating effects in capillary measurements, and indicate the applicability of the proposed techniques. The results obtained on the effect of pressure on viscosity are in agreement with predictions based on the W-L-F-equation.

Acknowledgement

The authors wish to express their thanks to the National Research Council of Canada for financial support during the course of this work. Mr. *H. Nyun* holds a scholarship from the Canadian International Development Agency (CIDA).

References

1) *Garrard, J. E.* and *W. Philippoff*, Proc. Fourth Int. Congr. Rheol. **4**, 77 (1963).
2) *Cheng, D. C. H.*, Proc. Fifth Int. Congr. Rheol. **1**, 483 (1969).
3) *Westover, R. F.* and *B. Maxwell*, SPE Journal **13**, 27 (1957).
4) *Semjownow, von V.*, Rheol. Acta **2**, 138 (1962).
5) *Ramsteiner, von F.*, Rheol. Acta **9**, 374 (1970).
6) *Duvdevani, I. J.* and *I. Klein*, SPE Journal **23**, 41 (1967).
7) *Choi, S. Y.* and *N. Nakajima*, Fifth Int. Congr. Rheol. **4**, 287 (1970).
8) *Penwell, R. C., R. S. Porter*, and *S. Middleman*, J. Polym. Sci. **A-2**, 9, 731 (1971).
9) *Bagley, E. B.*, J. Appl. Phys. **28**, 624 (1957).
10) *Bogue, D. C.*, Ind. Eng. Chem. **51**, 874 (1959).
11) *Collins, M.* and *W. R. Schowalter*, Amer. Inst. Chem. Eng. J. **9**, 98 (1969).
12) *Carslaw, H. S.* and *J. C. Jaeger*, Conduction of Heat in Solids, p. 339 (London 1959).
13) *Bird, R. B.*, SPE Journal **11**, 35 (1955).
14) *Middleman, S.*, The Flow of High Polymers, p. 15 (London-New York 1968).
15) *Williams, M. R., R. F. Landel*, and *J. D. Ferry*, J. Amer. Chem. Soc. **77**, 3701 (1955).

Authors' address:

Dr. *M. R. Kamal*, Dr. *H. Nyun*
Department of Chemical Engineering
McGill University, Montreal/Canada

Rheol. Acta **12**, 269–275 (1973)

From the Cattedra di Principi d'ingegneria chimica, Università di Palermo (Italia) and the Istituto di Principi d'ingegneria chimica, Università di Napoli (Italia)

Testing of a constitutive equation for entangled networks by elongational and shear data of polymer melts

By G. Marrucci, G. Titomanlio, and G. C. Sarti

With 6 figures

(Received October 27, 1972)

Introduction

The flow behavior of melts or concentrated solutions of high molecular weight polymers seems to be dominated by the entanglement phenomenon. Although the detailed nature of the entanglements is not entirely understood, it is a common assumption to consider the polymer molecules in a melt or a concentrated solution organized in a network of which the entanglements are the junctions. These junctions are of course only temporary; they can slide along the chains, disappear and reform continuously.

Constitutive equations for entangled networks have been suggested by a number of authors (1–6). *Lodge* (7) reviewed these theories using a unified formalism for all of them and showing, among other things, the following facts: 1. Assuming linear elasticity of the molecular segments, all theories correspond to either simple or generalized contravariant *Maxwell* models. For a constant junction creation rate and loss probability, the parameters of the *Maxwell* model are constant. In this case, the *Maxwell* model is equivalent to *Lodge*s integral equation (8) with one or more exponential terms in the memory function i.e. one or more relaxation times. 2. Detailed predictions are very sensitive to assumptions concerning creation and loss of junctions. Constant rates or probabilities as assumed by *Green* and *Tobolsky* (1) and *Lodge* (2) yield a constant viscosity in shear flow. *Kaye* (6) and *Yamamoto* (3–5) assume a loss probability which depends on the stress or on the molecular segment extension, respectively. This allows for a non-*Newton*ian viscosity and for an elongational viscosity passing through a maximum (4). However, admittedly (4), all types of response can be easily accommodated in view of the arbitrary assumptions on the loss probability function which are necessarily made.

Following a different approach, *Graessley* (9, 10) has considered the friction in the entanglements associated with the sliding of the junctions. For the case of shearing flows only, he has calculated the viscosity, accounting for the fact that, with increasing the rate of shear, decreases the time allowed for entanglement formation between neighbouring molecules. The progressive loss of entanglements determines the smaller values of the viscosity which are experienced at higher shearing rates. Recently, *Ziabicki* and *Takserman-Krozer* (11) have undertaken a systematic re-examination of the network dynamics, but up to now no detailed results can be derived from the general theory for the case of entangled networks.

In this paper, a rather simple constitutive equation for entangled networks is proposed. For a large part, it is based on the theories referred to above as well as on phenomenological equations referred to in the following. It contains, however, physically significant novelties, especially with regards to what will be called the "balance of entanglements". Moreover, the equations of the model, in their final dimensionless formulation, do not contain arbitrary functions. The only two parameters which appear can be independently estimated, at least in the order of magnitude. It must be said immediately that the model is also simplicistic in some respects: For example, a single relaxation time is assumed. The extension to a set of relaxation times or, rather, to a continuous spectrum is being considered and will be presented in a later paper. However, most concepts are already outlined in the simple model discussed here and a qualitative comparison with existing data of shear and extensional viscosity of polymer melts is already promising.

Description of the model

a) The equation for the stress

The contravariant simple *Maxwell* equation is usually written in the form:

$$\tau + \lambda \frac{\mathfrak{d}\tau}{\mathfrak{d}t} = 2\eta \mathbf{D} \qquad [1]$$

where τ is the stress tensor, $\mathbf{D}$ the stretching tensor – the symmetric part of the velocity gradient, $\mathbf{L}$ – and $\mathfrak{d}/\mathfrak{d}t$ stands for the operator, defined by *Oldroyd* (12):

$$\frac{\mathfrak{d}\tau}{\mathfrak{d}t} = \frac{d\tau}{dt} - \mathbf{L}\cdot\tau - \tau\cdot\mathbf{L}^T. \qquad [2]$$

The scalar coefficients η and λ are easily shown to represent the shear viscosity and the relaxation time of the fluid. In terms of the familiar spring-dashpot uni-dimensional model, η is related to the friction in the dashpot and λ is the ratio η/G where G is the modulus of the spring.

In the phenomenological approach, eq. [1] is frequently used: in a cruder form, with constant values of λ and η; alternatively, with variable λ

and η depending upon the invariants of the stretching tensor, usually II_D, see e.g. (13).

We shall make here the following assumption: For a given material, the values of λ and η depend upon the average number of entanglements per macromolecule instantaneously present, n. The functions $\lambda(n)$ and $\eta(n)$ will be specified below.

Whenever the modulus

$$G = \eta\lambda \qquad [3]$$

is allowed to vary with time, as it is done here, eq. [1] requires a modification. This is easily understood by considering the uni-dimensional spring-dashpot model with a variable spring modulus. Eq. [1] becomes:

$$\frac{1}{G}\tau + \lambda\frac{\mathfrak{d}}{\mathfrak{d}t}\left(\frac{1}{G}\tau\right) = 2\lambda\mathbf{D} \qquad [4]$$

or

$$\tau\left(1 - \frac{\lambda}{G}\frac{dG}{dt}\right) + \lambda\frac{\mathfrak{d}\tau}{\mathfrak{d}t} = 2\eta\mathbf{D}. \qquad [4']$$

Eq. [4] is one of the equations of the model. The scalar quantities η, λ and G vary with the number of entanglements n; at all times they are related among themselves by eq. [3].

By a procedure first suggested by *Lodge* (8, 14) for the case of constant coefficients and extended by *Marrucci* (15) to the case of variable ones, eq. [4] can be transformed in the integral equation:

$$\tau(t) = G\int_0^\infty \frac{1}{\lambda}\exp\left(-\int_0^s \frac{ds'}{\lambda}\right)\left[\mathbf{C}_t^{-1}(s) - \mathbf{1}\right]ds \qquad [5]$$

where s is time measured backward from the present time, t, and $\mathbf{C}_t(s)$ is the history of the right *Cauchy-Green* tensor with the configuration at time t taken as reference. In eq. [5], G is the value of the modulus at time t, while λ, which appears within the integrals, depends in general on the integration variable. Eq. [5] can be compared with integral equations containing a relaxation function which does not depend only on time, see e.g. (16–18). In principle, in order to calculate τ from eq. [5] as it now stands, it is needed to know the history of both the deformation and the relaxation time, at least in the most recent past. Eqs. [4] and [5] are completely equivalent and the choice of one or the other in actual calculations will only be a matter of computational preference.

b) Dependence of G, η and λ on number of entanglements

In view of the network structure of the material, it is quite natural to assume that the modulus G can be calculated from the well known relationship of the rubber elasticity theory.

Assume that the material at equilibrium has on the average n_0 entanglements per molecule. The equilibrium modulus G_0 is given by:

$$G_0 = c(n_0 + 1)kT \qquad [6]$$

where c is the number of macromolecules per unit volume. Because the junctions are not permanent, the definition of G_0 requires some clarification. We must assume that a strain is applied sufficiently fast for the entanglements not to slip or be destroyed, yet not so fast as to get glassy rather than rubberlike response. G_0 is then the ratio between stress and strain at time zero in such an ideal experiment.

When the material is not at equilibrium, the average number of entanglements per molecule becomes n. The corresponding modulus becomes:

$$G = c(n + 1)kT. \qquad [7]$$

Now, G is the ratio between the stress to the strain increments in an ideal experiment such as the one above described, performed upon the material in the non equilibrium existing conditions. Should the previous deformation history have been such as to destroy all entanglements, the modulus would attain its minimum value, ckT.

The dependence of shear viscosity η on number of entanglements can be deduced from the experimental dependence of the zero shear viscosity on concentration in concentrated solutions. As shown by *Ferry* (19), all existing data (20–24) point to the relationship:

$$\eta_0 \propto c^{3.4}. \qquad [8]$$

The fact that η_0 is proportional to $c^{3.4}$ instead of c is related to the entanglement phenomenon. It is interpreted by considering that, with varying the concentration, also varies the number of entanglements and the friction associated with them (10). With the assumption that the average molecular weight spacing between entanglement points, M_e, is inversely proportional to c (19), one can interpret eq. [8] as saying that η_0 is separately proportional to c and to the 2.4 power of the number of strands per molecule, $n + 1$.

In the case considered here, the concentration does not vary while so does the number of entanglements, depending on the deformation history. We then assume:

$$\eta \propto (n K 1)^{2.4}. \qquad [9]$$

Finally, recalling eq. [3], we get:

$$u \propto (n K 1)^{1.4}. \qquad [10]$$

c) The balance of entanglements

It now remains to be specified how the number of entanglements varies as a consequence of the flow. When the material is at equilibrium, entanglements are continuously formed and destroyed by thermal motions. The average number of them is however constant in time at a value indicated as n_0. Correspondingly, the stress tensor is isotropic.

Generally, under non-equilibrium conditions, the number of entanglements will be less than n_0 and the stress will be anisotropic. We shall assume that a net rate of entanglement reformation exists, due to thermal motions, which is related to the "distance from equilibrium" $n_0 - n$. The rate of entanglement destruction will be related to the existence of a non-isotropic stress. The balance of entanglements then takes the form of a differential equation which, in words, is written as:

$$\begin{bmatrix}\text{rate of change}\\ \text{with time}\end{bmatrix} = \begin{bmatrix}\text{rate of formation due}\\ \text{to thermal motions}\end{bmatrix} - \begin{bmatrix}\text{rate of destruction}\\ \text{due to stress}\end{bmatrix}.$$

The net rate of formation due to thermal motions is explicitly assumed to be:

$$\left.\frac{dn}{dt}\right|_{\text{formation}} = \frac{n_0 - n}{\lambda}. \qquad [11]$$

Notice that eq. [11] does not imply first order kinetics because λ depends on n. The kinetic constant $1/\lambda$ is larger when there are less entanglements and, in the words of *Graessley* (9), the molecules are "freer of encumbrances". Of course, close to equilibrium, the rate of formation is indeed proportional to $n_0 - n$.

The explicit formulation of the destruction term requires a somewhat longer discussion. Consider first the case of a steady shearing flow with a rate of shear, γ.

As in the work of *Graessley* (9), define R as the radius of the sphere which, on the average, gives the space occupied by one macromolecule, and J_c^0 as the average number of entanglements that, under equilibrium conditions, join two macromolecules whose spheres overlap. Further, define J_c as the latter quantity in a general non-equilibrium situation and l_c as the average distance between successive entanglements of such molecular pairs.

In a shearing flow, considering one of the molecules as fixed in space, the distance that an other molecule entangled with the first must travel in order that one mutual entanglement be destroyed can be taken of order l_c. The average velocity of the second molecule is of order $R\gamma$ and thus the time needed for destruction of one entanglement of a molecular pair is $l_c/R\gamma$. Because each molecule is entangled with cR^3 other molecules, we get:

$$\left.\frac{dn}{dt}\right|_{\text{destruction}} \simeq \frac{cR^4\gamma}{l_c}. \qquad [12]$$

From statistical mechanics we know that, if $J_c \gg 1$:

$$l_c^2 J_c \simeq R^2. \qquad [13]$$

Eq. [13] together with the following relationships

$$n_0 \simeq cR^3 J_c^0 \qquad [14]$$

$$n \simeq cR^3 J_c \qquad [15]$$

allows one to write eq. [12] as:

$$\left.\frac{dn}{dt}\right|_{\text{destruction}} \simeq n_0 \frac{1}{\sqrt{J_c^0}} \sqrt{\frac{n}{n_0}}\,\gamma. \qquad [16]$$

The problem now arises on how to generalize eq. [16] to include situations different from the steady shear so far considered. We want the rate of destruction to depend on the stress in the material rather than on the rate of deformation. In fact, assume for example that a motion has gone on for some time and is suddenly brought to a halt. We expect that during the subsequent stress relaxation some structural rearrangment, destruction and reformation of entanglements, goes on and only stops when the stress is also vanished.

Of course, in eq. 16 γ can be written as τ_{12}/η. We shall assume that this term becomes in general $\sqrt{-\mathrm{II}_\tau}/\eta$, where II_τ is the second invariant of the stress tensor $\boldsymbol{\tau}$, such as calculated from eq. 4 or 5. Notice that $\boldsymbol{\tau}$ from those equations is not traceless. However, it is easily verified that $\sqrt{-\mathrm{II}_\tau} = \tau_{12}$ in a shear flow. Thus finally we write eq. 16 in the form:

$$\left.\frac{dn}{dt}\right|_{\text{destruction}} = n_0 a \sqrt{\frac{n}{n_0}} \frac{\sqrt{-\mathrm{II}_\tau}}{\eta} \qquad [17]$$

where the order of magnitude of the numerical parameter, a, can be estimated from:

$$a \simeq (J_c^0)^{-\frac{1}{2}}. \qquad [18]$$

The balance of entanglements is then written as:

$$\frac{dn}{dt} = \frac{n_0 - n}{\lambda} + n_0 a \sqrt{\frac{n}{n_0}} \frac{\sqrt{-\mathrm{II}_\tau}}{\eta}. \qquad [19]$$

Eq. 19 and the stress eqs. [4] or [5] are the two basic equations. These, together with eqs. [7], [9] and [10] for G, η and λ, fully determine the model. Of course, the details of these equations are by no means, "certain" and improvements can very probably be made as suggested by comparison with experiments or by more detailed analyses. In particular, the last term of eq.[19] has been obtained with several arbitrary assumptions. For example, the generalization from τ_{12} to $\sqrt{-\mathrm{II}_\tau}$ is certainly not unique; also the square root term $\sqrt{n/n_0}$ is related to the assumption that $J_c \gg 1$, which might not always be the case and which is certainly not true when many entanglements have been destroyed.

d) Dimensionless formulation

It is convenient to rewrite the equations of the model after introducing dimensionless variables based on values which are characteristic of equilibrium conditions. Thus, a dimensionless modulus, viscosity and relaxation time are defined as:

$$G = \frac{G}{G_0}; \quad \eta = \frac{\eta}{\eta_0}; \quad \lambda = \frac{\lambda}{\lambda_0}. \qquad [20]$$

Dimensionless G, η and λ are still related by eq. [3].

Stress, rate of stretch and time are made dimensionless in the following way:

$$\tau = \frac{1}{G_0}\tau; \quad \mathbf{D} = \lambda_0 \mathbf{D}; \quad t = \frac{t}{\lambda_0};$$

$$s = \frac{s}{\lambda_0}. \qquad [21]$$

In the rest of this paper, the symbols above will always represent dimensionless quantities.

Finally we define the fraction:

$$x = \frac{n}{n_0} \qquad [22]$$

and the parameter:

$$\alpha = \frac{1}{n_0}. \qquad [23]$$

Eqs. [7], [9], [10] and [19] become:

$$G = \frac{x + \alpha}{1 + \alpha} \qquad [24]$$

$$\eta = \left(\frac{x + \alpha}{1 + \alpha}\right)^{2.4} \qquad [25]$$

$$\lambda = \left(\frac{x + \alpha}{1 + \alpha}\right)^{1.4} \qquad [26]$$

$$\frac{dx}{dt} = \frac{1 - x}{\lambda} - a\sqrt{x}\,\frac{\sqrt{-\mathrm{II}_\tau}}{\eta}. \qquad [27]$$

The form of eqs. [4] and [5] is not modified when using dimensionless variables.

The solution of the set of eqs. [4] (or [5]) and [24–27], for any given kinematics, only depends on parameters a and α. The value of these parameters can be determined, at least approximately, from their definition, eqs. [18] and [23].

Comparison with existing steady elongational and shear data

For a constant stretch history flow such as a steady elongation or a steady shear, eqs. [5] and [27] take a much simpler form because x, as well as τ, does not depend on time. Accounting for eqs. [24–26], eqs. [5] and [27] become:

$$\tau = \left(\frac{1 + \alpha}{x + \alpha}\right)^{0.4} \int_0^\infty \exp\left[-s\left(\frac{1 + \alpha}{x + \alpha}\right)^{1.4}\right]$$

$$\times\, [\mathbf{C}_t^{-1}(s) - \mathbf{1}]\, ds \qquad [28]$$

$$0 = 1 - x - a\sqrt{x}\sqrt{-\mathrm{II}_\tau}\,\frac{1 + \alpha}{x + \alpha}. \qquad [29]$$

In a *Cartes*ian coordinate system, the matrix of $\mathbf{C}_t^{-1}(s) - \mathbf{1}$, for the two types of flow, is given by:

Shear

$$[\mathbf{C}_t^{-1}(s) - \mathbf{1}] = \begin{Vmatrix} \gamma^2 s^2 & \gamma s & 0 \\ \gamma s & 0 & 0 \\ 0 & 0 & 0 \end{Vmatrix}. \qquad [30]$$

Elongational

$$[\mathbf{C}_t^{-1}(s) - \mathbf{1}] = \begin{Vmatrix} e^{2\Gamma s} - 1 & 0 & 0 \\ 0 & e^{-\Gamma s} - 1 & 0 \\ 0 & 0 & e^{-\Gamma s} - 1 \end{Vmatrix} \quad [31]$$

where γ and Γ are dimensionless rate of shear and rate of elongation respectively.

The integrals of eq. [28] can then be calculated in closed form. As said above, $\sqrt{-\mathrm{II}_\tau}$ coincides with τ_{12} in a shear flow; in elongation we have:

$$\mathrm{II}_\tau = \tau_{11}\tau_{22} + \tau_{22}\tau_{33} + \tau_{33}\tau_{11} \, . \quad [32]$$

The system of eqs. 28 and 29 is now algebraic and can easily be solved numerically to obtain τ and x as a function of γ, or Γ, for any choice of the parameters a and α.

In fig. 1, the viscosity in shear is plotted vs. $a\gamma$ for different values of α. As expected, the shear viscosity decreases with increasing γ down to a minimum value η_∞ which is obtained when the entanglements are all destroyed. From eq. [25], η_∞ is given by:

$$\eta_\infty = \left(\frac{\alpha}{1+\alpha}\right)^{2.4} . \quad [33]$$

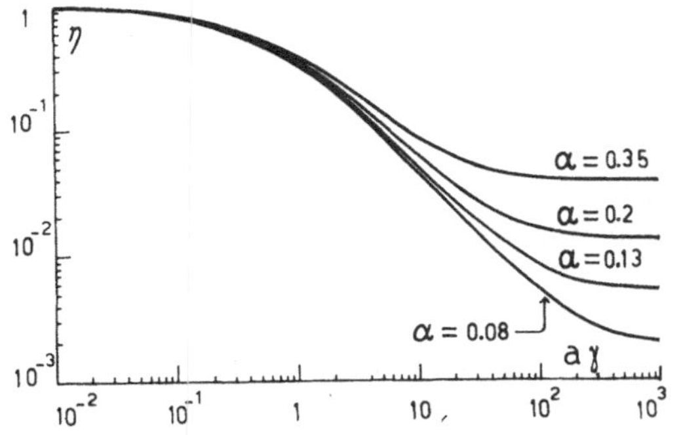

Fig. 1. Theoretical curves of shear viscosity

The influence of parameter *a* is that of shifting the curves $\log\eta$ vs. $\log\gamma$ horizontally.

Figs. 2 and 3 show some of the elongational results. The elongational viscosity is calculated from:

$$\eta_{el} = \frac{\tau_{11} - \tau_{22}}{\Gamma} \, . \quad [34]$$

As shown by figs. 2 and 3, a wide variety of behaviors is predicted, depending on *a* and α values. Starting from the *Trouton* value of 3 for low Γ values, η_{el} may always increase as Γ increases, or pass through a maximum or directly decrease, due to the conflicting influence of the molecule elongation (which increases η_{el}) and the entanglement destruction (which reduces it). In all cases, for sufficiently large Γ values, when all entanglements are destroyed, η_{el} starts increasing beyond any limit. This is due to the fact that, in adopting equations such as eq. 1, no upper limit to the elongation of the molecules is assumed.

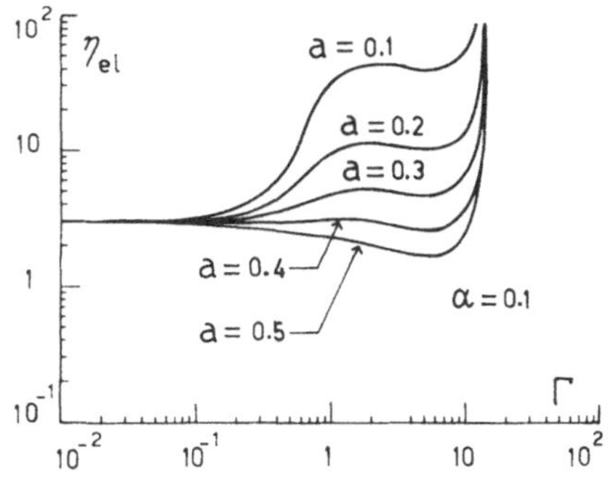

Fig. 2. Some predicted behaviors of elongational viscosity vs. stretching rate

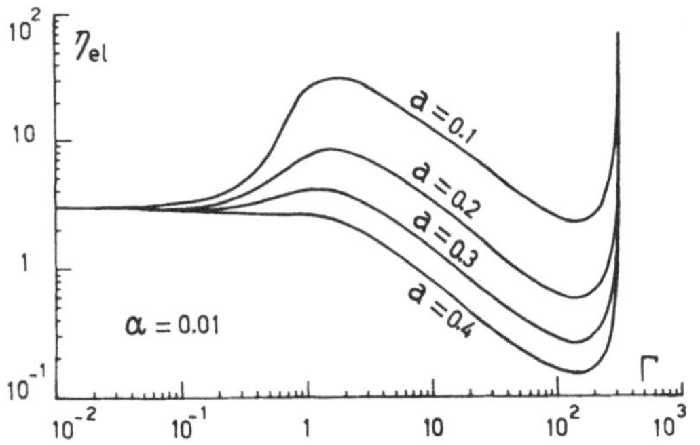

Fig. 3. Some predicted behaviors of elongational viscosity vs. stretching rate

Recently, a number of authors (25–32) has collected elongational viscosity data for a variety of polymer melts. Indeed, all types of behaviors such as those depicted in figs. 2 and 3 have been observed. We have then attempted a quantitative comparison between the experimental results and the predictions of the model. For the comparison to be meaningful, it has been limited to those cases where sufficient auxiliary information was available, i.e. shear curve, zero-shear viscosity and molecular weight.

The results are shown in figs. 4–6 in dimensionless coordinates. In order to make the abscissa of the data points dimensionless, the value of λ_0 was needed. This was available in one case (data of *Cogswell*, fig. 6) from a knowledge of the modulus G_0 obtained under conditions of small stresses. In the other two cases λ_0 was arbitrarily deter-

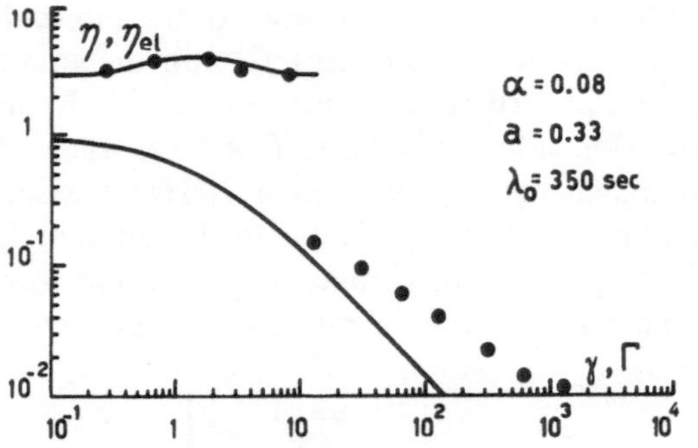

Fig. 4. Data of *Ballman* (25)

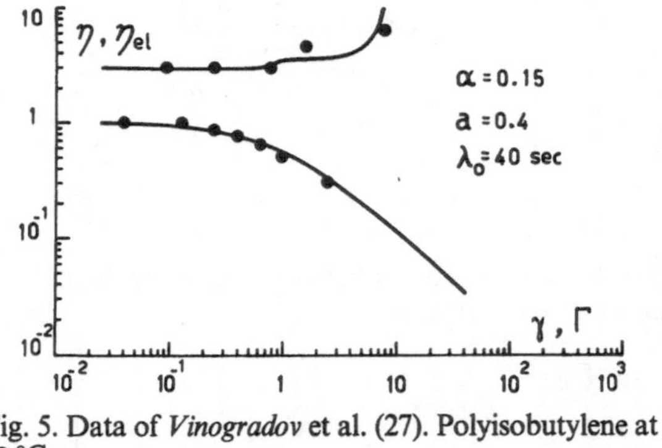

Fig. 5. Data of *Vinogradov* et al. (27). Polyisobutylene at 22 °C

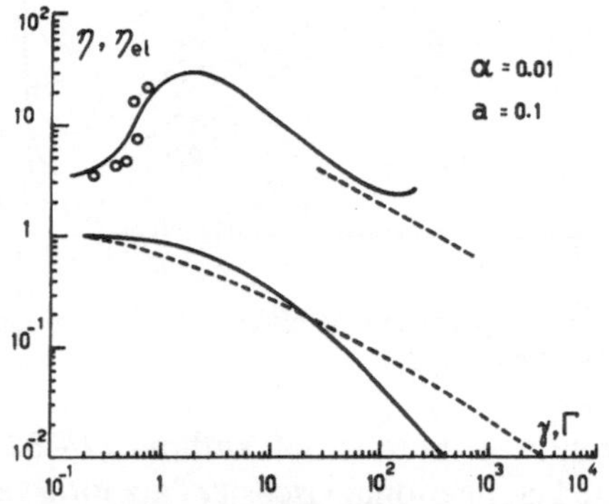

Fig. 6. Data of *Cogswell* (32). Blend of a L.D. and H.D. Polyethylene. η_0 is 1.4×10^5 N sec/m², G_0 is 3×10^3 N/m² and thus $\lambda_0 = 47$ sec. (Both the dots and the dashed curves are experimental)

mined by a horizontal shift of the data. In view of eq. [23], the value of the parameter α should be given by:

$$\alpha = \frac{M_e}{M} \qquad [35]$$

where M is the chain molecular weight and M_e that between subsequent entanglements along the chain under equilibrium conditions. Usually, it is assumed: $M_e = \frac{1}{2} M_c$, where M_c is the critical molecular weight for entanglement formation, as revealed by the change of slope in the η_0 vs. M curve. In the cases considered here, M_c was available but, because of polidispersity, the value of M to be used in eq. [35] was uncertain. However, the value of α could be approximately determined. The value of α could also be obtained from eq. [33]. In the cases considered here η_∞ was not reached, but the lowest data point of the shear curve poses an upper limit to the admissible values of α. The values of α actually used for the theoretical curves of figs. 4–6 are consistent with both the above observations. The prediction of the value of a from eqs. [18] and [14] is even more uncertain. However, we believe that the values adopted here are in the correct order of magnitude.

With regard to the elongational viscosity, figs. 4–6 show a fair agreement between theory and experiments, which on the whole can be considered satisfactory. In particular, for the data by *Cogswell*, we observe that a slight change in the value of α could have fitted the data for large Γ values even better. However, we hardly believe that those data, obtained in a converging flow apparatus, represent steady values of the elongational viscosity. They are most probably transient and are thus lower than the steady elongational viscosity which would be obtained for the same values of Γ.

The overall situation for the shear results is conversely quite bad, with the exception of the data by *Vinogradov* et al. which are limited to the leftmost part of the shear curve. The theory predicts a far too steep slope of the shear curve and thus the data for large γ values are much to the right of the theoretical curve. We believe this discrepancy to be due to the choice of a single relaxation time and/or to the effect of polydispersity which is also ignored in the model.

As a last comment, we briefly mention some preliminary results which also have been obtained from the present model. They are relative to the transient response which is obtained after a sudden application of a shear or elongational rate. Depending on values of the parameters, the well known phenomenon of the stress overshoot is found, particularly for the case of shear. From the physics of the model, the stress overshoot is easily understood. When a rate of strain is suddenly applied to the sample, the stress starts increasing like in a solid rubber with a value of the modulus which corresponds to the equilibrium number of entanglements. With some delay, the entanglement destruction phenomenon takes place and in the meantime the stress may well increase above the value which ultimately attains when the number of entanglements becomes stabilized to the value of the steady flow. Damped oscillatory behavior can also be obtained in some cases. Such behavior has also been experimentally observed (33).

Acknowledgment

Thanks are due to Dr. *D. Acierno*, whose experimental work prompted interest in the subject of this paper.

Summary

An entangled network such as a polymer melt or a concentrated solution is here described by a set of two simultaneous equations. One of them is a balance of entanglements, the other gives the stress in the classical form of a *Maxwell* equation.

The balance of entanglements contains both an entanglement generation term which depends on the "distance" from equilibrium and an entanglement destruction term which depends on the stress level. The parameters appearing in the *Maxwell* and the balance equations are made to depend in a specified way on the existing number of entanglements.

The model is here tested by comparison with existing data of steady-state elongational and shear viscosity of polymer melts.

References

1) *Green, M. S.* and *A. V. Tobolsky*, J. Chem. Phys. **14**, 80 (1946).
2) *Lodge, A. S.*, Trans. Faraday Soc. **52**, 120 (1956).
3) *Yamamoto, M.*, J. Phys. Soc. Japan **11**, 413 (1956).
4) *Yamamoto, M.*, ibidem **12**, 1148 (1957).
5) *Yamamoto, M.*, ibidem **13**, 1200 (1958).
6) *Kaye, A.*, Brit. J. Appl. Phys. **17**, 803 (1966).
7) *Lodge, A. S.*, Rheol. Acta **7**, 379 (1968).
8) *Lodge, A. S.*, Elastic Liquids (London 1964).
9) *Graessley, W. W.*, J. Chem. Phys. **43**, 2696 (1965).
10) *Graessley, W. W.*, ibidem **47**, 1942 (1967).
11) *Ziabicki, A.* and *R. Takserman-Krozer*, J. Polymer Sci. A-2, **7**, 2005 (1969); *Takserman-Krozer, R.* and *A. Ziabicki*, J. Polymer Sci. A-2, **8**, 321 (1970); *Ziabicki, A.*, J. Polymer Sci. A-2.
12) *Oldroyd, J. G.*, Proc. Roy. Soc. ≫ **200**, 523 (London 1950).
13) *White, J. L.* and *A. B. Metzner*, J. Polymer Sci. **7**, 1867 (1963).
14) *Lodge, A. S.*, Proc. 5th Int. Congr. Rheology, Vol. 4 (Baltimore 1970).
15) *Marrucci, G.*, Trans. Soc. Rheol. **16**, 321 (1972).
16) *Bogue, D. C.* and *J. O. Doughty*, Ind. Eng. Chem. Fund. **5**, 243 (1966); *Doughty, J. O.* and *D. C. Bogue*, ibidem **6**, 388 (1967).
17) *Bird, R. B.* and *P. J. Carreau*, Chem. Eng. Sci. **23**, 427 (1968).
18) *Tanner, R. I.* and *J. M. Simmons*, Chem. Eng. Sci. **22**, 1803 (1967).
19) *Ferry, J. D.*, Viscoelastic properties of polymers, p. 542 (New York 1970).
20) *Nakayasu, H.* and *T. G. Fox*, **Abstracts** 137th Meeting American Chem. Soc., p. 11-I (1960).
21) *Shultz, A. R.*, *W. R. Shultz*, and *T. G. Fox*, unpublished experiments.
22) *Ninomiya, K.*, *J. R. Richards*, and *J. D. Ferry*, J. Phys. Chem. **67**, 327 (1963).
23) *Oyanagi, Y.* and *J. D. Ferry*, J. Colloid Sci. **21**, 547 (1966).
24) *Berry, G. C.*, J. Phys. Chem. **70**, 1194 (1966).
25) *Ballman, R. L.*, Rheol. Acta **4**, 137 (1965).
26) *Meissner, J.*, Rheol. Acta **8**, 78 (1969).
27) *Vinogradov, G. V.*, *B. D. Radushkevich*, and *V. D. Fikhman*, J. Polymer Sci. A-2, **8**, 1 (1970).
28) *Vinogradov, G. V.*, *A. I. Leonov*, and *A. N. Prokunin*, Rheol. Acta **8**, 482 (1969).
29) *Vinogradov, G. V.*, *V. D. Fikhman*, *B. D. Radushkevich*, and *A. Ya. Malkin*, J. Polymer Sci. A-2, **8**, 657 (1970).
30) *Cogswell, F. N.*, Plastics & Polymers **36**, 109 (1968).
31) *Cogswell, F. N.*, Rheol. Acta **8**, 187 (1969).
32) *Cogswell, F. N.*, Measuring the extensional rheology of polymer melts, presented at American Soc. Rheol. Conference Knoxville 1971, to be published in Trans. Soc. Rheol.
33) *Chen, I-Jen* and *D. C. Bogue*, Trans. Soc. Rheol. **16**, 59 (1972).

Authors' addresses:

Dr. *G. Marrucci* and Dr. *G. Titomanlio*
Cattedra di Principi d'ingegneria chimica
Università di Palermo (Italia)

Dr. *G. C. Sarti*
Istituto di Principi d'ingegneria chimica
Università di Napoli (Italia)

Rheol. Acta 12, 276–278 (1973)

From the Department of Chemistry, University of Genoa (Italy)

Comments on recent results bearing on the rubber elasticity theory

By A. Ciferri

(Received October 27, 1972)

The elastic equation of state for a polymer network (1, 2) may be written in the form

$$\frac{\tau}{\alpha - \alpha^{-2}} = \mathcal{M} \qquad [1]$$

where τ is the force per unit area of the unstressed sample, $\mathcal{M}$ is the elasticity modulus, and α is the elongation ratio. The modulus $\mathcal{M}$ is often given in the form (3)

$$\mathcal{M} = \frac{\varrho RT}{M_c} \frac{\langle r^2 \rangle_i}{\langle r^2 \rangle_0} \qquad [2]$$

where ϱ is the polymer density, M_c is the molecular weight of a network chain, $\langle r^2 \rangle_i$ and $\langle r^2 \rangle_0$ being the mean-square end-to-end distance for an actual network chain and for a free chain, respectively. R and T have the usual meaning. Experimental data, particularly in a limited range of unidimensional elongation, are often well described by an equation containing two elastic constants C_1 and C_2 (1, 4)

$$\frac{\tau}{\alpha - \alpha^{-2}} = 2C_1 + 2C_2\alpha^{-1} \qquad [3]$$

which is based on purely phenomenological considerations (5).

A very large amount of work has been done to assess the significance of the C_2 term. Also discussed has been the possibility of identifying $2C_1$ with $\mathcal{M}$ (6–10) when $C_2 \neq 0$. The experimental data which have appeared, particularly during the past twenty years (11), offer a perhaps incomplete, but extensive description of the role, on C_2, of the following variables: 1. type and extent of deformation (1, 8, 10, 12), 2. temperature (6, 13), 3. degree of swelling and type of swelling liquid (6, 14), 4. time scale of experiment (6, 13, 15–17), 5. chemical nature of the polymer and of the cross-linkages (6, 13), 6. degree of cross-linking (13, 16), 7. network topology (9, 16, 18–24). When this body of results is considered, it is apparent that: i) the C_2 term may often, and in no-small measure, represent non-equilibrium effects, ii) equilibrium contributions to C_2 also appear to exist, iii) instances of networks obeying the elastic equation of state (i.e., $C_2 = 0$) are also not uncommon.

Several theoretical attempts to describe the C_2 term have been performed. Equilibrium contributions to C_2 have been described in terms of interchain packing (25–27), interchain obstructions and excluded volume effects (28, 29), short network chains (30), intra- and intermolecular energy effects (31, 32). These theoretical attempts appear to have met with only a limited success, as a widely accepted equilibrium description of C_2 has not emerged. On the other hand, it seems to be generally accepted that non-equilibrium effects should be associated to the occurrence of labile cross-linkages (entanglements or associations are generally considered). An approximate description, on the basis of a model containing a non-newtonian viscosity element, was presented (33) although a detailed theory of non-equilibrium effects has not been developed.

It seems important to comment on the implications of the occurrence of networks for which C_2 is effectively equal to zero. Of the seven variables considered above, let first consider the case in which only topology is altered. Differences due to changes in chemical structure, swelling, temperature, etc., are thus not complicating the analysis. Natural rubber (18, 19, 34) and polyethylacrylate (19) samples may be cross-linked in bulk, using a given amount of high energy radiation, both in the random, isotropic state or in the unidimensional oriented state. When both types of networks are analyzed, under similar conditions and in the unswollen state, the networks cross-linked in the isotropic state exhibit $C_2 > 0$, and the networks cross-linked in the oriented state may exhibit $C_2 = 0$, provided they are stretched in the direction of orientation. Biological polymers, where the cross-linking process occurs in the oriented (or native) state (35, 36), invariably exhibit $C_2 = 0$. Natural rubber (21, 22), polydimethylsiloxane [23], 1–4, cis polybutadiene (9, 24) may be cross-linked in the isotropic state, both in bulk and in solution. When both types of networks are analyzed, under similar experimental conditions and in the unswollen state [even at comparable C_1 values (9, 23)], the networks cross-linked in bulk exhibit $C_2 > 0$, and the networks cross-linked in solution may exhibit $C_2 = 0$. Which topological features may explain the above differences? Perhaps there is the possibility of a common rationalization of the results obtained using the above indicated methods for altering network topology. The network structure of both oriented and solution networks possesses after shrinkage or solvent removal, a "memory" of the rest (undeformed) state which is certainly not so pronunced as in the case of networks prepared directly in the random, bulk state. Such difficulties as chain connectivity patterns (37, 38), permanent (22, 23) or temporary (21, 33), entanglements hindering subsequent deformation should certainly be less severe in the case of networks possessing a topology "naturally compatible" with a three-dimensionally expanded state or with a preferential stretch direction, than in the case of conventional networks. These differences in topology might be directly responsible for the observed differences in the C_2 term, in line with suggestions already made in connection with the solution networks (9, 21, 22).

Admittedly, the shape of a stress-strain isotherm for a gaussian network should be independent of the state in which the cross-linkages are inserted, and a *Gaussian* network has to be an isotropic one (39, 40). Yet, rather than being unduly pessimistic about the validity and the applicability of the *Gaussian* theory, why not be more critical of the particular network structure which is obtained by cross-linking in the bulk state? The conventional networks could be regarded as "generally inappropriate" at least for some tests of a theory which virtually ignores the topological and viscoelastic complexities of the network. It seems that any attempt we make to modify the conventional bulk network, and in this respect are also particularly noteworthy the results obtained by transient swelling (6) and by cross-linking through the endgroups (16), brings about a closer approach to the *Gaussian* theory [generally also with respect to decreased stress relaxation (6, 9, 16), consistency between results obtained under different types of strain (1, 10), and perhaps thermoelasticity (24), although the thermoelastic behaviour is not usually affected by the deviations represented by the C_2 term (11, 19, 41)].

The occurrence of some networks and of some experimental conditions under which $C_2 = 0$, suggests (when also other features of the theory are verified) that the *Gaussian* theory is both valid and useful. For other networks, whenever $C_2 > 0$, a more complex theory is necessary, apparently to account for severe networks complexities (37, 38), and non-equilibrium effects (6, 33).

Summary

A short survey of the present status of the rubber elasticity theory, with specific regard to the controversial C_2 term of the *Mooney-Rivlin* equation, is given. The implications of recent results, particularly those concerned with alterations of the network topology, are discussed. The occurrence of the C_2 term in conventional networks, and deviations from theoretical expectation in the case of oriented networks, are attributed to complex topology and to non-equilibrium effects.

References

1) *Treloar, L. R. G.*, The Physics of Rubber Elasticity (London 1958).
2) *Flory, P. J.*, Principles of Polymer Chemistry (Ithaca, N.Y., 1953).
3) *Flory, P. J., C. A. J. Hoeve*, and *A. Ciferri*, J. Pol. Sci. **34**, 337 (1959).
4) *Mooney, M.*, J. Appl. Phys. **19**, 434 (1948).
5) *Rivlin, R. S.*, Phil. Trans. Roy. Soc. **A 240**, 459, 491, 509 (London 1948).
6) *Ciferri, A.* and *P. J. Flory*, J. Appl. Phys. **30**, 1498 (1959).
7) *Mullins, L.*, J. Appl. Pol. Sci. **2**, 1 (1959).
8) *Dusek, K.* and *W. Prins*, Fortschr. Hochpolym. Forschg. **6**, 1 (1969).
9) *Mark, J. E.*, J. Pol. Sci. **C 31**, 97 (1970).
10) *De Candia, F.* and *A. Ciferri*, Makrom. Chem. **134**, 535 (1970).
11) A more extensive list of references and detailed review shall be presented elsewhere.
12) *Rivlin, R. S.* and *D. W. Saunders*, Phil. Trans. Roy. Soc. **A 243**, 251 (London 1951).
13) *Smith, T. L.*, J. Pol. Sci. **C16**, 841 (1967).
14) *Mullins, L.*, J. Appl. Pol. Sci. **2**, 257 (1959).
15) *Roe, R. J.* and *W. R. Krigbaum*, J. Pol. Sci. **61**, 167 (1962); ibid., **A1**, 2049 (1963).
16) *Kraus, G.* and *G. A. Maczvgemba*, J. Pol. Sci. **A 2**, 277 (1964).
17) *Gent, A. N.* and *R. S. Rivlin*, Proc. Phys. Soc. **B65**, 487 (London 1952).
18) *Greene, A.* and *A. Ciferri*, Kolloid-Z. **186**, 1 (1962).
19) *Greene, A., K. J. Smith* jr., and *A. Ciferri*, Trans. Farad. Soc. **61**, 2772 (1965).
20) *Meissner, B., I. Klier*, and *S. Kucharik*, J. Pol. Sci. **C 16**, 793 (1967).
21) *Meissner, B.*, Paper presented at the 4th IUPAC Microsymposium, Prague 1969.
22) *Price, C., G. Allen, F. De Candia, M. C. Kirkham*, and *A. Subramaniam*, Polymer **11**, 468 (1970).
23) *Johnson, R. M.* and *J. E. Mark*, Macromolecules **5**, 41 (1972).

24) *De Candia, F., L. Amelino*, and *C. Price*, J. Pol. Sci. **A 2**, 10, 975 (1972).
25) *Volkstein, M. V., Yu. Ya. Gotlib*, and *O. B. Ptisyn*, Vysokomolekul. Soed. **1**, 10 (1959).
26) *Blockland, R.* and *W. Prins*, J. Pol. Sci. **A 2**, 7, 1595 (1969).
27) *Di Marzio, E. A.*, J. Chem. Phys. **36**, 1563 (1962).
28) *Jackson, J. L., P. C. Shen*, and *D. A. McQuire*, ibid. **44**, 2359 (1966).
29) *Gee, G.*, Polymer **7**, 373 (1966).
30) *Dobson, G. R.* and *M. Gordon*, J. Chem. Phys. **43**, 705 (1965).
31) *Krigbaum, W. R.* and *M. Kaneko*, J. Chem. Phys. **36**, 99 (1962).
32) *Guth, E.* and *H. M. James*, Proc. 3rd Rubber Techn. Conf., Heffer, London 1954.
33) *Ciferri, A.* and *J. J. Hermans*, J. Pol. Sci. **B 2**, 1089 (1964).
34) *Smith* jr., *K. J., A. Ciferri*, and *J. J. Hermans*, J. Pol. Sci. **A 2**, 1025 (1964).
35) *Puett, D., A. Ciferri*, and *L. V. Rajagh*, Biopolymers **3**, 438 (1965).
36) *Mistrali, F., D. Volpin, G. B. Garibaldo*, and *A. Ciferri*, J. Phys. Chem. **75**, 142 (1971).
37) *Mazur, J.* and *T. Alfrey*, Proc. Intl. Rubber Conf., Wash., D.C., 1959.
38) *Alfrey, T.* and *W. G. Lloyd*, J. Pol. Sci. **62**, 159 (1962).
39) *Berry, J. P., J. Scanlan*, and *W. F. Watson*, Trans. Farad. Soc. **52**, 1137 (1956).
40) *Flory, P. J.*, Trans. Farad. Soc. **57**, 829 (1961).
41) *Ciferri, A.*, J. Pol. Sci. **A 2**, 3089 (1964).

Author's address:
Dr. *A. Ciferri*
Istituto di Chimica Industriale
Università di Genova, Via Pastore 3
Genova/Italia

Rheol. Acta 12, 279–288 (1973)

Centre Technique des Tuiles et Briques, Paris (France)

Application à la viscoélasticité non linéaire du calcul symbolique à plusieurs variables

Par C. Huet

(Reçu p. p. le 27 octobre 1972)

1. Introduction

En viscoélasticité linéaire, le calcul symbolique (sous forme de transformation de *Laplace*) a souvent été employé pour ramener à la forme algébrique les équations intégro-différentielles du problème.

De plus, *Mandel* (1–3) a établi une correspondance complète de celles-ci avec les équations de l'élasticité linéaire par l'emploi de la transformation de *Carson* (celle de *Laplace* ne conduisant qu'à des correspondances incomplètes): par la méthode de *M. Mandel*, tous les problèmes de viscoélasticité linéaire, tant dynamiques que statiques, se ramènent à l'élastostatique.

Nous avons, à de nombreuses reprises, éprouvé personnellement la commodité et l'efficacité de la méthode de *M. Mandel* par l'usage que nous avons pu en faire dans des domaines variés couvrant aussi bien l'expérimentation (4–7), les applications techniques (8), que des travaux à caractère plus général (9–10) destinés à faciliter les premières.

Nous présentons ici une tentative pour étendre à la viscoélasticité non linéaire la méthode mise au point par *M. Mandel* pour le cas linéaire.

Nous verrons que, grâce aux progrès considérables réalisés par la viscoélasticité non linéaire [du fait notamment des travaux déterminants de *Green*, *Rivlin*, *Spencer*, *Winemann*, *Pipkin* (11–13)], il est effectivement possible d'aboutir à des résultats significatifs lorsque l'on utilise, après une première transformation que nous préciserons plus loin, la transformation de *Carson* à plusieurs variables, étudiée notamment par *Poli* et *Delerue* (14), et par *Melle Delavault* (15).

L'idée de départ sur la méthode que nous présentons et nos premiers résultats partiels remontent à 1968. Faute d'avoir pu à l'époque aboutir à des résultats vraiment utilisables, cette premiere tentative fût alors abandonnée, et nous ne l'avons reprise que tout dernièrement (1972).

Le calcul symbolique ne fournit de résultats intéressants que pour les comportements ne dépendant pas de l'âge du matériau. Nous laissons donc de côté le cas des matériaux présentant du vieillissement [à propos desquels nous avons eu l'occasion (17–18), d'appliquer une autre méthode de calcul opérationnel également due à *Mandel* (16)].

2. Rappels sur les principaux résultats de la viscoélasticité non linéaire

Rappelons tout d'abord les résultats généraux obtenus en ce qui concerne l'expression des lois de comportement en viscoélasticité non linéaire, expression dont la forme détermine directement la nature des résultats que nous obtiendrons et même, très probablement, la possibilité d'aboutir à un résultat.

2.1. Représentation fonctionnelle du comportement

Les travaux de *Noll*, *Coleman*, *Green*, *Rivlin*, etc. ont permis de montrer que la représentation par une fonctionnelle (notion introduite par *Volterra* en 1887 et appliquée dès cette époque à ce qu'il nommait «élasticité héréditaire» [cf. (19)]) est la seule représentation rigoureuse des équations de comportement [cf. *Mandel* (3)].

De façon à s'affranchir des complications d'ordre purement cinématique qui s'introduisent, dans la formulation de *Green* et *Rivlin* (11), par l'emploi de la contrainte d'*Euler*, nous utilisons la contrainte de *Kirchhoff* $\boldsymbol{\Pi}$ [cf. *Mandel* (3)]. On a alors en fonction de $\mathbf{E}(t)$, tenseur déformation de *Green* en variables de *Lagrange*:

$$\boldsymbol{\Pi}(t) = \mathscr{F}\left[\mathbf{E}(t_0 \overset{t}{\underset{\tau=0}{-}} \tau)\right]. \qquad [1]$$

Il en résulte que nos résultats ne pourront (sauf dans le cas des déformations infinitésimales), s'appliquer sans précautions à ceux obtenus par les expérimentateurs [cf. (22) à (28)] ayant utilisé ces dernières années la formulation de *Green*-*Rivlin* [cf. aussi (38)].

2.2. Formes possibles pour $\mathscr{F}$

En généralisant et précisant les premiers résultats de *Green* et *Rivlin* (11) d'une part, et *Spencer* et *Rivlin* d'autre part (12), *Winemann* et *Pipkin* (13) ont montré que toute fonctionnelle satisfaisant aux conditions imposées par l'invariance tensorielle et la symétrie matérielle (et par conséquent propre à entrer dans une équation de comportement), possède de façon nécessaire et suffisante, les propriétés suivantes:

– Elle est égale à la somme d'un nombre fini de fonctionnelles linéaires par rapport aux tenseurs de base de l'histoire de la sollicitation, et qui sont aussi des fonctionnelles par rapport aux invariants de base I_α de cette histoire. Le nombre de termes de la somme est égal au nombre (fini) de tenseurs de base, lui-même fonction du type de symétrie matérielle considéré.

– Les tenseurs de base f^β sont des fonctions polynômiales (en produits contractés) d'un nombre fini n de tenseurs $\psi_i = \psi(\xi_i)$ ($i = 1$ à n) correspondants aux valeurs prises par l'histoire tensorielle $\psi(\tau)$ à n instants différents; ces polynômes sont linéaires par rapport à chacun des $\psi(\xi_i)$.

Les coefficients de ces polynômes sont des fonctions des invariants de base I_α, dont les définitions et le nombre dépend du cas de symétrie matérielle envisagé, et du rang du tenseur formant l'histoire.

– Dans le cas isotrope, et pour un tenseur symétrique du 2e ordre, les invariants de base de l'histoire $\psi(\tau)$ sont en nombre égal à 6 et peuvent être pris sous la forme:

$I_k = \mathrm{tr}[\psi(\xi_1)\,\psi(\xi_2)\ldots\psi(\xi_k)]$ ($k = 1$ à 6); les tenseurs de base f^β sont alors également au nombre de 6 (dont $\mathbf{1}$, tenseur de *Kronecker*).

Dans le cas isotrope, la fonctionnelle s'exprime donc comme la somme de 6 termes, dont chacun est une fonctionnelle linéaire de l'un des tenseurs de base.

En supposant $\mathscr{F}$ continue, l'application à ces résultats du théorème d'*Hadamard* sur les fonctionnelles linéaires, et de la théorie de *Fréchet* (20) sur les fonctionnelles d'ordre entier fournit une expression sous forme d'intégrales multiples jusqu'à l'ordre 11*), que l'on peut dans notre cas exprimer par rapport aux vitesses de déformation $\mathring{\mathbf{E}}(t)$. On obtient une expression de la forme, en posant $\mathring{\mathbf{E}}_i = \mathring{\mathbf{E}}(\tau_i)$:

$$\boldsymbol{\Pi}(t) = \sum_{k=1}^{n} \int_0^t\int_0^t\int_0^t \ldots \int_0^t \sum_{j=1}^{m(k)} [r_{kj}(t-\tau_1,\ldots,t-\tau_k)\, \mathring{\mathbf{E}}_1\ldots\mathring{\mathbf{E}}_i\,\mathrm{tr}(\mathring{\mathbf{E}}_{i+1}\ldots\mathring{\mathbf{E}}_{i+p})\ldots \mathrm{tr}(\mathring{\mathbf{E}}_{i+p+1}\ldots\mathring{\mathbf{E}}_k)]\, d\tau_1\ldots d\tau_k. \quad [2]$$

Les noyaux $r_{kj}(\tau_1,\ldots,\tau_k)$ des intégrales sont indépendants de la sollicitation et leur trace (obtenue en faisant $\tau_i = t\ \forall i = 1, k$) fournit le terme d'ordre k de la réponse à une sollicitation constante. Ils peuvent donc être considérés comme des extensions des fonctions relaxations du cas linéaire.

*) Tout tenseur ou invariant d'ordre supérieur à 6 peut s'exprimer en fonction des tenseurs ou invariants de base, d'ordres au plus égal à 6, ce qui fournit un développement d'ordre 2 × 6 qui se réduit à 11 par application du théorème de *Cayley-Hamilton*.

Dans les divers cas d'anisotropie (orthotropie, etc.) l'ordre du developpement sera plus élevé.

La remarque de *Gradowczyk* à propos de l'influence de la précision expérimentale sur la détermination des r_{kj} permettra souvent dans la pratique, de diminuer l'ordre du développement (31).

S'appuyant sur les résultats de *Pipkin* (32), *Lai* et *Findley* (23) ont ainsi obtenu la forme explicite du développement d'ordre 3, forme que, pour plus de généralité dans la suite, nous considérons comme constituant les premiers termes d'un développement d'ordre n plus élevé. Ce développement, qui est bien de la forme générale mentionnée plus haut, s'écrit alors avec nos notations:

$$\begin{aligned}
\boldsymbol{\Pi}(t) = {} & \mathscr{F}[\mathbf{E}(t_0 \overset{t}{\underset{\tau=0}{-}} \tau)] \\
= {} & \int_0^t \{\mathbf{1} r_{11}(t-\tau_1)\,\mathrm{tr}[\mathring{\mathbf{E}}(\tau_1)] \\
& + r_{12}(t-\tau_1)\,\mathring{\mathbf{E}}(\tau_1)\}\, d\tau_1 \\
& + \int_0^t\int_0^t \{\mathbf{1} r_{21}(t-\tau_1, t-\tau_2)\,\mathrm{tr}[\mathring{\mathbf{E}}(\tau_1)] \\
& \quad \mathrm{tr}[\mathring{\mathbf{E}}(\tau_2)] \\
& + r_{22}(t-\tau_1, t-\tau_2)\,\mathrm{tr}[\mathring{\mathbf{E}}(\tau_1)\,\mathring{\mathbf{E}}(\tau_2)]\,\mathbf{1} \\
& + r_{23}(t-\tau_1, t-\tau_2)\,\mathring{\mathbf{E}}(\tau_1)\,\mathrm{tr}[\mathring{\mathbf{E}}(\tau_2)] \\
& + r_{24}(t-\tau_1, t-\tau_2)\,\mathring{\mathbf{E}}(\tau_1)\,\mathring{\mathbf{E}}(\tau_2)]\}\, d\tau_1\, d\tau_2 \\
& + \int_0^t\int_0^t\int_0^t \{r_{31}(t-\tau_1, t-\tau_2, t-\tau_3) \\
& \quad \mathrm{tr}[\mathring{\mathbf{E}}(\tau_1)]\,\mathrm{tr}[\mathring{\mathbf{E}}(\tau_2)]\,\mathrm{tr}[\mathring{\mathbf{E}}(\tau_3)]\,\mathbf{1} \\
& + \mathbf{1} r_{32}(t-\tau_1, t-\tau_2, t-\tau_3) \\
& \quad \mathrm{tr}[\mathring{\mathbf{E}}(\tau_1)]\,\mathrm{tr}[\mathring{\mathbf{E}}(\tau_2)\,\mathring{\mathbf{E}}(\tau_3)] \\
& + r_{33}(t-\tau_1, t-\tau_2, t-\tau_3)\,\mathring{\mathbf{E}}(\tau_1) \\
& \quad \mathrm{tr}[\mathring{\mathbf{E}}(\tau_2)]\,\mathrm{tr}[\mathring{\mathbf{E}}(\tau_3)] \\
& + r_{34}(t-\tau_1, t-\tau_2, t-\tau_3)\,\mathring{\mathbf{E}}(\tau_1) \\
& \quad \mathrm{tr}[\mathring{\mathbf{E}}(\tau_2)\,\mathring{\mathbf{E}}(\tau_3)] \\
& + r_{35}(t-\tau_1, t-\tau_2, t-\tau_3)\,\mathring{\mathbf{E}}(\tau_1)\,\mathring{\mathbf{E}}(\tau_2) \\
& \quad \mathrm{tr}[\mathring{\mathbf{E}}(\tau_3)] \\
& + r_{36}(t-\tau_1, t-\tau_2, t-\tau_3) \\
& \quad \mathring{\mathbf{E}}(\tau_1)\,\mathring{\mathbf{E}}(\tau_2)\,\mathring{\mathbf{E}}(\tau_3) \\
& + \cdots + \}\, d\tau_1\, d\tau_2\, d\tau_3 \\
& + \cdots \text{(jusqu'à l'ordre } n\text{)} \qquad [3]
\end{aligned}$$

où ne sont explicitement écrits que les termes obtenus dans un développement à l'ordre 3.

Bien qu'ils les supposent noyaux symétriques, *Lai* et *Findley* indiquent qu'ils ne le sont pas dans le cas général. *Volterra* montre par ailleurs (19) qu'on peut se ramener à des noyaux symétriques, mais ce n'est pas nécessaire à notre propos pour le moment.

On peut obtenir une expression de même forme pour la fonctionnelle $\mathscr{G} = \mathscr{F}^{-1}$ liant la déformation à l'histoire de la contrainte.

3. Remarques sur la forme du développement intégral de $\mathscr{F}$

La forme de ce développement présente des analogies avec la forme prise par $\mathscr{F}$ dans le cas linéaire (forme à laquelle se ramène le présent développement pour $n = 1$). En particulier (et cela apparaît encore mieux si l'on se contente d'examiner le cas unidimensionnel en remplaçant $\mathbf{E}$ par un scalaire e), les différents termes du développement apparaissent comme ayant sensiblement la forme de convolutions multiples [cf. *Delavault* (15) pour les principales propriétés de celles-ci] de deux fonctions à plusieurs variables (la deuxième fonction étant ici constituée par le produit de k fonctions à 1 variable).

On saît que dans le cas linéaire, le succès de la transformation de *Carson* courante (ou calcul symbolique) provient essentiellement de ce qu'elle transforme en un produit la convolution simple.

Par ailleurs, le calcul symbolique portant sur des fonctions d'une variable (ou CS_1) s'étant vu généraliser à des fonctions de plusieurs variables [ou CS_n; cf. *Poli* et *Delerue* (14), 1954; *Delavault* (15), 1961] son étude montre que la convolution multiple se transforme en produit des fonctions transformées (à un facteur près).

Par conséquent, on peut espérer obtenir une simplification importante des problèmes de viscoélasticité non linéaire si l'on parvient à tirer parti de cette remarque.

Si l'on cherche à appliquer directement le CS_n à [2], une difficulté d'inversion surgit, due au caractère symétrique en p du résultat obtenu.

La difficulté se trouve tournée si l'on fait précéder l'application du CS^n par une première transformation qui convertit en véritables convolutions multiples les différents termes du développement [2] et [4].

4. Description de la transformation proposée

Nous considérons ici la réponse $\boldsymbol{\Pi}(t)$ à une histoire $\mathbf{E}(t)$ étant entendu que, d'après la remarque faite à la fin de 2.2, tout ce qui sera dit maintenant sera transposable à la réponse de $\mathbf{E}(t)$ à une histoire $\boldsymbol{\Pi}(t)$.

Reprenons le développement [2]:

$$\boldsymbol{\Pi}(t) = \mathscr{F}_n[\mathbf{E}(t \underset{\tau=0}{\overset{t}{-}} \tau)]$$
$$= \sum_{k=1}^{n} \int_0^t \int_0^t \dots \int_0^t \{ \sum_{j=1}^{m(k)} r_{kj}(t-\tau_1, \dots, t-\tau_k)$$
$$\mathring{\mathbf{E}}_1 \dots \mathring{\mathbf{E}}_i \operatorname{tr}[\mathring{\mathbf{E}}_{i+1} \dots \mathring{\mathbf{E}}_{i+1}] \dots$$
$$\operatorname{tr}[\mathring{\mathbf{E}}_{1+i} - \mathring{\mathbf{E}}_k]\} \qquad d\tau_1 \dots d\tau_k \qquad [2]$$

où $\mathring{\mathbf{E}}_i$ désigne toujours $\mathring{\mathbf{E}}(\tau_i)$.

Nous appliquons à la fonctionnelle $\mathscr{F}_n = \boldsymbol{\Pi}(t)$ (où l'indice n indique qu'ils'agit d'un développement d'ordre n) la transformation suivante:

$$\mathscr{H}_n\{\mathscr{F}_n[\mathbf{E}(t \underset{\tau=0}{\overset{t}{-}} \tau)]\} = \mathscr{H}_n\{\boldsymbol{\Pi}(t)\}$$
$$= \mathscr{C}_n \mathscr{D}_n\{\mathscr{F}_n\}$$
$$= \mathscr{C}_n \mathscr{D}_n\{\boldsymbol{\Pi}\} \qquad [4]$$

produit d'une première transformation $\mathscr{D}_n$ que nous nommons «dissymétrisation» d'ordre n par la transformation de *Carson* à n variables $\mathscr{C}_n$.

4.1. *Première transformation: dissymétrisation $\mathscr{D}_n$*

La dissymétrisation $\mathscr{D}_n$ est, pour l'ordre n, définie par:

$$\mathscr{D}_n[\boldsymbol{\Pi}(t)] = \boldsymbol{\Phi}(t_1, \dots, t_k, \dots, t_n) = \mathscr{D}_n \mathscr{F} \qquad [5]$$

où $\boldsymbol{\Phi}$ est la fonctionnelle tensorielle à n variables obtenue en remplaçant, dans [2], chacun des t associé à une valeur τ_i donnée par une variable fictive t_i.

Ainsi,

$$\int_0^t d\tau_i \quad \text{devient} \quad \int_0^{t_i} d\tau_i \qquad \forall\, i = 1 \dots n;$$

et

$$r_{kj}(t-\tau_1, \dots, t-\tau_i, \dots, t-\tau_k)$$

devient

$$r_{kj}(t_1-\tau_1, \dots, t_i-\tau_i, \dots, t_k-\tau_k)$$

(en continuant à nommer r_{kj} la nouvelle fonction issue de ce changement de variables).

On aura donc:

$$\boldsymbol{\Phi}(t_1, \dots, t_k, \dots, t_n)$$
$$= \sum_{k=1}^{n} \int_0^{t_1} \int_0^{t_2} \dots \int_0^{t_k} \{ \sum_{j=0}^{m(k)} r_{kj}(t_1-\tau_1, \dots, t_k-\tau_k)$$
$$\mathring{\mathbf{E}}_1 \mathring{\mathbf{E}}_2 \dots \operatorname{tr}[\mathring{\mathbf{E}}_{l+s} \dots \mathring{\mathbf{E}}_k]\} \, d\tau_1 \dots d\tau_k. \qquad [6]$$

On remarque alors que la fonction de n variables Φ a bien la forme de la somme de convolutions multiples d'ordre $k (k = 1$ à $n)$ par rapport aux couples de fonctions à n variables.

$$r_{kj} = r_{kj}(t_1, \ldots, t_k) \qquad [k = 1 \text{ à } n; j = 1 \text{ à } m(k)]$$

$$\mathbf{S}_{kj} = \mathbf{S}_{kj}(t_1, \ldots, t_k) = \mathring{\mathbf{E}}(t_1) \cdot \mathring{\mathbf{E}}_2(t_2) \ldots \mathrm{tr}[\mathring{\mathbf{E}}_{l+s}(t_{l+s}) \ldots \mathring{\mathbf{E}}_k(t_k)] \qquad [7]$$

où $\mathbf{S}_{kj}$ a la forme particulière du produit de k fonctions à une seule variable.

Autrement dit, les convolutions multiples en question sont multilinéaires par rapport à chacune des fonctions d'une variable figurant dans la fonctions $\mathbf{S}_{kj}$.

La transformation $\mathscr{D}_n$ est évidemment linéaire:

$$\mathscr{D}_n(a\mathscr{F}_n + b\mathscr{G}_n) = a\mathscr{D}_n(\mathscr{F}_n) + b\mathscr{D}_n(\mathscr{G}_n) \qquad \forall\, a, b = \text{Ctes} \qquad [8]$$

si $\mathscr{F}_n$ et $\mathscr{G}_n$ ont la forme [2].

De plus elle est récurrente, i.e.:

$$\mathscr{D}_n(\mathscr{F}_k) = \mathscr{D}_k(\mathscr{F}_k) \qquad \forall\, 1 \leqslant k \leqslant n. \qquad [9]$$

Enfin elle est immédiatement inversible en prenant la trace de la fonction Φ, obtenue en faisant $t_i = t \;\forall\, i = 1, \ldots, n$.

$$\mathrm{Tr}\,\Phi(t_1, \ldots, t_n) = \Phi(t, \ldots, t) = \mathscr{D}_n^{-1}\,\Phi = \Pi(t) \qquad [10]$$

(la trace fonctionnelle Tr ainsi définie *a* ne étant pas confondre avec la trace tensorielle que nous avons notée tr).

4.2. Deuxième transformation: transformation de Carson à n variables $\mathscr{C}_n$.

Nous définissons la transformée de *Carson* à n variables de la fonction scalaire $f(t_1, \ldots, t_n)$, par

$$\mathscr{C}_n(f) = p_1 p_2 \ldots p_n \int_0^\infty \int_0^\infty \ldots \int_0^\infty e^{-(pt_1 + pt_2 + \cdots + p_n t_n)} f(t_1, \ldots, t_n)\, dt_1 \ldots dt_n. \qquad [11]$$

Nous désignerons l'application à f de cette transformation par la notation:

$$f(t_1, \ldots, t_n) \underset{n}{\supset} f^{n*}(p_1, \ldots, p_n). \qquad [12]$$

Pour l'étude de cette transformation, on pourra se reporter à *Poli* et *Delerue* (14). Des tables de transformées et de leurs inverses ont été établies par *Ditkin* et *Prudnikow* (33) pour les cas de $n = 1$ et 2 (calcul symbolique à 1 variable CS_1, et calcul symbolique à 2 variables CS_2).

Si l'extension du CS_1 au CS_2 présente certaines difficultés sur le plan mathématique, il n'en est plus ainsi, lorsqu'on les a résolues, pour l'extension du CS_2 au CS_n avec n entier fini quelconque [cf. (14) et (15)].

Nous utiliserons dans la suite, les propriétés suivantes du CS_n scalaire:

– linéarité:

$$a_1 f_1 + a_2 f_2 \underset{n}{\supset} a_1 f_1^{n*} + a_2 f_2^{n*} \quad \forall\, a_1, a_2 = \text{Ctes} \qquad [13]$$

– récurrence:

$$f(t_1, \ldots, t_{n-k}) \underset{n}{\supset} f^{(n-k)*}(p_1, \ldots, p_{n-k}) \qquad \forall\, k = 1, n \qquad [14]$$

– algébrisation de la convolution multiple:

$$\int_0^{t_1} \ldots \int_0^{t_k} f(t_1 - \tau_1, \ldots, t_k - \tau_k) g(\tau_1, \ldots, \tau_k)\, d\tau_1 \ldots d\tau_k \underset{n}{\supset} \frac{1}{p_1 \ldots p_k} f^{k*} \cdot g^{k*} \qquad [15]$$

– algébrisation de la dérivation:

$$\dot{f}(t_i) \underset{n}{\supset} p_i f^*(p_i) \qquad \forall\, f(0) = 0 \qquad [16]$$

(ce qu'il est toujours possible de supposer dans notre cas par choix de l'origine avant toute application des sollicitations).

De plus, la définition de l'intégrale d'un tenseur permet de définir la transformée de *Carson* par le tenseur dont les composantes sont les transformées du tenseur initial:

$$\mathbf{A}(t_1, \ldots, t_k) \underset{n}{\supset} \mathbf{A}^{n*}(p_1, \ldots, p_k)$$

$$\text{si} \quad a_{ij}(t_1, \ldots, t_k) \underset{n}{\supset} a_{ij}^{n*}(p_1, \ldots, p_k) \quad \forall\, i, j. \qquad [17]$$

On obtient alors facilement les propriétés suivantes du CS_n appliqué à un tenseur $\mathbf{A}$:

$$\mathbf{A}(t_1, \ldots, t_k) \underset{n}{\supset} \mathbf{A}^{k*}(p_1, \ldots, p_k) \qquad [18]$$

$$\mathbf{A}(t_i) \ldots \mathbf{A}(t_k) \underset{n}{\supset} \mathbf{A}^*(p_i) \ldots \mathbf{A}^*(p_k) \qquad [19]$$

$$\mathring{\mathbf{A}}(t_i) \ldots \mathring{\mathbf{A}}(t_k) \underset{n}{\supset} p_i \ldots p_k \mathbf{A}_i^* \ldots \mathbf{A}_k^* \qquad \forall\, i, k = 1 \text{ à } n \qquad \text{si} \quad \mathbf{A}_j(0) = 0 \;\forall\, j \qquad [20]$$

$$\mathrm{tr}\,\mathbf{A}(t_i) \underset{n}{\supset} \mathrm{tr}[\mathbf{A}_i^*] \qquad [21]$$

$$\mathrm{tr}[\mathbf{A}_i \ldots \mathbf{A}_k] \underset{n}{\supset} \mathrm{tr}[\mathbf{A}_i^* \ldots \mathbf{A}_k^*] \qquad \forall\, \mathbf{A}_j = \mathbf{A}(t_j) \qquad [22]$$

$$\mathrm{tr}[\mathring{\mathbf{A}}_i \ldots \mathring{\mathbf{A}}_k] \underset{n}{\supset} p_i \ldots p_k \,\mathrm{tr}[\mathbf{A}_i^* \ldots \mathbf{A}_k^*] \quad \text{si} \quad \mathbf{A}_j(0) = 0 \;\forall j \qquad [23]$$

$$\int_0^{t_1} \ldots \int_0^{t_k} \mathbf{A}(t_1 - \tau_1, \ldots, t_k - \tau_k)\, \mathbf{B}(\tau_1, \ldots, \tau_k) \cdot d\tau_1 \ldots d\tau_k \underset{n}{\supset} \frac{1}{p_1 \ldots p_k} \mathbf{A}^{k*}\, \mathbf{B}^{k*} \qquad [24]$$

$$\int_0^{t_1} \ldots \int_0^{t_k} f(t_1 - \tau_1, \ldots, t_k - \tau_k)\, \mathring{\mathbf{A}}(\tau_1) \cdot \mathring{\mathbf{A}}(\tau_2) \ldots \ldots \mathrm{tr}[\mathring{\mathbf{A}}_i \ldots \mathring{\mathbf{A}}_k]\, d\tau_1 \ldots d\tau_k \underset{n}{\supset} f^{k*}(p_1 \ldots p_k)\, \mathbf{A}_1^* \ldots \ldots \mathrm{tr}[\mathbf{A}_i^* \ldots \mathbf{A}_k^*]. \qquad [25]$$

Dans les formules [20] à [25], les **A** sont les transformées par CS_1 des $\mathbf{A}(t_i)$:

$$\mathbf{A}(t_i) \underset{1}{\supset} \mathbf{A}^*(p_i) = \mathbf{A}_i^*. \qquad [26]$$

On voit donc que le CS_n appliqué à un tenseur:
- possède la récurrence [14] du cas scalaire;
- transforme le produit contracté en le produit contracté des transformées (à cause du caractère multilinéaire de cette opération) si chacun des termes du produit ne contient pas deux fois des **A** fonction du même t_i;
- algébrise la dérivation, par rapport à chacune des variables t_i, du produit contracté;
- transforme la trace tensorielle en la trace tensorielle du transformé;
- algébrise l'expression de la trace du produit des dérivées, et l'exprime en fonction de la trace du produit des transformés;
- algébrise la convolution tensorielle multiple.

4.3. *Résultat final*

Si l'on applique ces propriétés, et particulièrement la propriété [25] qui se déduit des précédentes, à la fonction $\boldsymbol{\Phi}$ définie par la transformation $\mathscr{D}_n$ à partir du développement de $\mathscr{F}$, on voit que l'on obtient

$$\boldsymbol{\Phi}(t_1, \ldots, t_n) \underset{n}{\supset} \Phi^{n*}(p_1, \ldots, p_n) \qquad [27]$$

avec, après mise en facteur convenable:

$$\begin{aligned}\boldsymbol{\Phi}^{n*} = \mathbf{1}\{&r_{11}^{1*}\,\mathrm{tr}\,\mathbf{E}_1^* + r_{21}^{2*}\,\mathrm{tr}\,\mathbf{E}_1^*\,\mathrm{tr}\,\mathbf{E}_2^* \\ &+ r_{22}^{2*}\,\mathrm{tr}[\mathbf{E}_1^*\mathbf{E}_2^*] + r_{31}^{3*}\,\mathrm{tr}\,\mathbf{E}_1^*\,\mathrm{tr}\,\mathbf{E}_2^*\,\mathrm{tr}\,\mathbf{E}_3^* \\ &+ r_{32}^{3*}\,\mathrm{tr}\,\mathbf{E}_1^*\,\mathrm{tr}[\mathbf{E}_2^*\mathbf{E}_3^*] + \cdots\} \\ + \mathbf{E}_1^*\{&r_{12}^{1*} + r_{23}^{2*}\,\mathrm{tr}\,\mathbf{E}_2^* + r_{33}^{3*}\,\mathrm{tr}\,\mathbf{E}_2^*\,\mathrm{tr}\,\mathbf{E}_3^* \\ &+ r_{34}^{3*}\,\mathrm{tr}[\mathbf{E}_2^*\mathbf{E}_3^*] + \cdots\} \\ + \mathbf{E}_1^*\mathbf{E}_2^*\{&r_{24}^{2*} + r_{35}^{3*}\,\mathrm{tr}\,\mathbf{E}_3^* + \cdots\} \\ + \mathbf{E}_1^*\mathbf{E}_2^*\mathbf{E}_3^*\{&r_{36}^{3*} + \cdots\} + \cdots \\ + \mathbf{E}_1^* \ldots \mathbf{E}_n^*\, &r_{nm}^{n*}.\end{aligned} \qquad [28]$$

L'application de la transformation $\mathscr{H}_n$ transforme donc la fonctionnelle tensorielle $\mathscr{F}_n$ définissant la réponse de la contrainte $\boldsymbol{\Pi}(t)$ à une histoire quelconque $\mathbf{E}(t)$ de la déformation, en un polynôme tensoriel (en produits contractés), de degré n par rapport à l'ensemble des transformées $\mathbf{E}_i^* = \mathbf{E}^*(p_i)$, et multilinéaire par rapport à chacun d'entre eux.

De plus, les coefficients de ce polynôme sont eux-mêmes de degré n par rapport aux invariants de base des n tenseurs transformés $\mathbf{E}^*(p_i)$ constitués par les traces des divers produits des $\mathbf{E}^*(p_i)$ jusqu'à l'ordre n.

Ce résultat est valable en sollicitation multiaxiale quelconque, pour une histoire quelconque (définie sous forme de distribution pour permettre l'incorporation de l'influence des discontinuités pouvant survenir en des instants particuliers dont la réunion constitue un ensemble de mesure nulle).

On remarque en outre que ce développement a même forme qu'un développement polynômial très général donné par *Goldenblatt* (34) pour l'élasticité non linéaire. Cependant, ce développement ne peut se réduire, comme le fait *Goldenblatt* [cf. aussi *Mandel* (3)] à un développement limité à l'ordre 2 puisque la présence de n tenseurs différents (puisque fonction de t_i différents) impose, suivant un théorème mentionné en (13) par *Winemann* et *Pipkin*, la présence de 5 tenseurs de base[1]) et de 6 invariants (dans le cas isotrope).

La transformation $\mathscr{H}$ ramène donc bien, comme nous l'espérions, la viscoélasticité non linéaire à l'élasticité non linéaire, tout au moins au niveau des équations de comportement. Nous laissons de côté pour le moment la question de savoir si ce résultat peut s'étendre aux équations générales, et si oui, dans quelles conditions, et sous quelle forme.

4.4. *Retour à la réponse $\boldsymbol{\Pi}(t)$: inversion de la transformation $\mathscr{H}$*

Le retour à la réponse $\boldsymbol{\Pi}(t)$ de la contrainte à une histoire quelconque $\mathbf{E}(t)$ de la déformation se fait de la façon suivante:

– On inverse d'abord la transformation $\mathscr{C}_n$, soit par les tables, soit par la formule de *Mellin* étendue au cas multidimensionnel [cf. (14)], soit dans certains cas par applications successives de n inversions partielles d'ordre 1 (ou du nombre

---

[1]) outre le tenseur de *Kronecker* $\mathbf{1} = \|\delta_{ij}\|$

convenable d'inversions partielles d'ordre k inférieur à n).

Cette inversion fournit la fonction $\boldsymbol{\Phi}(t_1, \ldots, t_n)$.

– On inverse $\mathscr{D}_n$ en faisant, dans Φ, $t_i = t$ $\forall\, i = 1, n$, ce qui transforme $\boldsymbol{\Phi}$ en $\boldsymbol{\Pi}(t)$.

On voit que l'intervention de $\mathscr{D}_n$ ne constitue pas à proprement parler une complication.

On remarque que $\mathscr{H}_n$ contient la transformation de *Carson* habituelle $\mathscr{C}_1$ comme cas particulier, ce qui permet donc d'utiliser tous les résultats de la viscoélasticité linéaire pour les termes globalement linéaires dans $\boldsymbol{\Phi}$.

Bien entendu, tous ces résultats s'appliquent (mutatis mutandis) à la réponse $\mathbf{E}(t)$ de la déformation, à une histoire quelconque $\boldsymbol{\Pi}(t)$ de la contrainte (de *Kirchhoff*).

Les fonctions $r_{kj}^{k*}(p_1, \ldots, p_k)$ ne sont pas fonction des sollicitations, mais ne dépendent que des variables p_i. On peut les nommer modules opérationnels d'ordre k. De même, les fonctions $f_{kj}^{k*}(p_1, \ldots, p_k)$ pourront être nommées complaisances opérationnelles d'ordre k.

5. Exemples particuliers d'histoire de la sollicitation

5.1. *Réponses à un palier, à un escalier, à une rampe*

On vérifie facilement que l'application de la méthode proposée et des règles de transformation du CS_n fournit des résultats conformes au résultat direct dans les cas simples d'histoire de la sollicitation tels que sollicitation constante, en escalier, en rampe. Les principales correspondances utilisées dans ces cas sont:

$$f(t_1 - \tau_1, t_2 - \tau_2, \ldots, t_k - \tau_k) \underset{n}{\supset} e^{-(p_1\tau_1 + \cdots + p_k\tau_k)} \times f^{k*}(p_1, \ldots, p_k) \qquad [29]$$

$$t_1 \ldots t_n \underset{n}{\supset} \frac{1}{p_1 p_2 \ldots p_n} \qquad [30]$$

$$\int_0^{t_1} \ldots \int_0^{t_k} f(\tau_1, \ldots, \tau_k)\, d\tau_1 \ldots d\tau_k \underset{n}{\supset} \frac{1}{p_1 \ldots p_k} \times f^{k*}(p_1, \ldots, p_k). \qquad [31]$$

5.2. *Réponse à une histoire sinusoïdale*

La réponse à une histoire sinusoïdale correspond à un cas particulier important, tant au point de vue théorique qu'expérimental.

Pour une histoire sinusoïdale

$$\mathbf{E}(t) = \mathbf{E}_0 \sin \omega t = \mathscr{I}m[\mathbf{E}^+ e^{i\omega t}] \qquad [32]$$

on a encore en CS_1:

$$\mathbf{E}^*(p) = \frac{p}{p - i\omega} \mathbf{E}^+ \qquad [33]$$

on aura donc:

$$\mathbf{E}(t_1) \cdot \mathbf{E}(t_2) \ldots \mathbf{E}(t_i) \operatorname{tr}[\mathbf{E}_{i+1} \ldots \mathbf{E}_{i+1}] \ldots \ldots \operatorname{tr}[\mathbf{E}_{k-m} \ldots \mathbf{E}_k] \underset{n}{\supset} \frac{p_1}{p_1 - i\omega} \cdot \frac{p_2}{p_2 - i\omega} \ldots \frac{p_k}{p_k - i\omega} (\mathbf{E}^+)^i \ldots \ldots \operatorname{tr}[(\mathbf{E}^+)^k] \ldots \operatorname{tr}[(\mathbf{E}^+)^m]. \qquad [34]$$

Pour inverser un terme d'ordre k de l'expression donnant Φ^* on peut appliquer la formule de *Mellin-Bromwich* étendue à plusieurs variables, qui s'écrit:

$$f(t_1, \ldots, t_k) = \frac{1}{(2\pi_i)^k} \int_{a_1 - i\infty}^{a_1 + i\infty} \ldots \int_{a_k - i\infty}^{a_k + i\infty} \frac{e^{p_1 t_1 + p_2 t_2 + \cdots p_k t_k}}{p_1 p_2 \ldots p_k} \times f^{k*}(p_1, p_2, \ldots, p_k)\, dp_1 \ldots dp_k. \qquad [35]$$

Dans notre cas f^{k*} est égal au produit du 2ème membre de [35] par $r_{k1}^{*}(p_1, \ldots, p_k)$. L'ensemble des pôles de f^* est donc constitué par l'ensemble des pôles de r^* et de l'ensemble des pôles $p_i = i\omega$ $(i = 1, \ldots k)$.

Si dans l'expression de la transformée de r^{k*} on fixe tous les p_i sauf un (p_j), l'expression obtenue est une transformée de *Carson* à une variable p_j que nous notons r_j^*. Il en résulte que r_j^* est holomorphe dans presque tout plan parallèle au plan des p_j s'il est privé des pôles de r_j^*, lesquels sont à partie réelle négative. Par conséquent, r^{k*}, étant holomorphe par rapport à chacune de ses variables p_j, est holomorphe par rapport à l'ensemble dans l'espace des $p_1, \ldots, p_k$ privé au plus des pôles des r_j^* [cf. *Cartan* (35) pour les fonctions analytiques à plusieurs variables]. Ceux-ci étant à partie réelle négative les résidus correspondants tendent vers zéro lorsque $t \to \infty$ et, au bout d'un temps assez long, seuls subsistent les résidus correspondant aux k pôles $p_j = i\omega$ $(p_j = 1,$ à $\ldots, k)$.

Le α^e terme d'ordre k a donc pour original, par application du théorème des résidus:

$$\begin{aligned}\boldsymbol{\Phi}(t_1, \ldots, t_k) &= (\mathbf{E}^+)^i \cdot \operatorname{tr}[(\mathbf{E}^+)^1] \times r_{k\alpha}^{k*}(i\omega, i\omega, \ldots, i\omega) \times e^{i\omega t_1 + i\omega t_2 + \cdots + i\omega t_k} \\ &= (\mathbf{E}^+)^i \cdot \operatorname{tr}[(\mathbf{E}^+)^k]\, R_{k\alpha}^*(i\omega) \times e^{i\omega t_1 + \cdots + i\omega t_k}\end{aligned} \qquad [36]$$

d'où l'on tire le α^e terme d'ordre k de $\boldsymbol{\Pi}(t)$, en faisant $t_i = t \ \forall i = 1, \ldots k$:

$$\boldsymbol{\Pi}_{k\alpha}(t) = \mathscr{I}m\{(\mathbf{E}^+) \cdot \mathrm{tr}[(\mathbf{E}^+)^k] \, R^*_{k\alpha}(i\omega) \, e^{ik\omega t}\}$$
$$\forall \alpha = 1 \text{ à } m(k), \quad \forall k = 1 \text{ à } n. \qquad [37]$$

On voit donc immédiatement que l'emploi de $\mathscr{H}_n$ permet de montrer que, en viscoélasticité non linéaire, la réponse de régime permanent à une sollicitation sinusoïdale est constituée par la somme d'un terme à la fréquence d'excitation et de termes aux fréquences harmoniques de celle-ci. Ce résultat est conforme à ce que l'on connait des effets de la non linéarité sur la réponse des systèmes aux sollicitations sinusoïdales, et aussi au résultat de *Onogi* et al. (26) obtenu par intégration directe dans un cas particulier.

On pourra définir $R^*_{k\alpha}(i\omega)$ comme le module complexe d'ordre k.

Si le développement [2] a été mis au départ de façon telle que les $r_{k\alpha}(t_1, \ldots, t_k)$ soient symétriques (ce qui est toujours possible par des regroupements appropriés), **la détermination de $R^*_{k\alpha}(i\omega)$**, sur la gamme des fréquences permet de remonter à $r^{k*}_{k\alpha}(i\omega, i\omega, \ldots, i\omega)$, et donc à $r^{k*}_{k\alpha}(p_1, \ldots, p_k)$ en remplaçant ω par $-ip_j \ \forall j = 1$ à k.

On remarque qu'on **ne peut** mettre $\boldsymbol{\Pi}(t)$ sous la forme

$$\boldsymbol{\Pi}(t) = \mathscr{I}m[\boldsymbol{\Pi}^+ e^{i\omega t}]$$

et que l'on ne peut donc définir une réponse complexe $\boldsymbol{\Pi}^+$ à la sollicitation complexe, mais un spectre de réponses complexes $\boldsymbol{\Pi}^+_k$:

$$\boldsymbol{\Pi}(t) = \mathscr{I}m\{\sum_{k=1}^{n} \boldsymbol{\Pi}^+_k \, e^{ik\omega t}\}$$
$$= \sum_{k=1}^{n} \sum_{\alpha=1}^{m} \mathscr{I}m[\boldsymbol{\Pi}^+_{k\alpha} e^{ik\omega t}]. \qquad [39]$$

Si l'on exprime $R^*_{k\alpha}(i\omega)$ sous la forme:

$$R^*_{k\alpha}(i\omega) = R^*_{k\alpha} e^{i\theta k\alpha} \qquad [40]$$

expression que l'on explicite facilement en remontant à $r^{k*}_{k\alpha}$, on voit que les $\boldsymbol{\Pi}^+_k$ sont toujours en avance sur $\mathbf{E}^+$, les déphasages variant avec chaque k.

On pourra cependant (si utile) définir une contrainte complexe généralisée sous la forme $\mathbf{P}^+$ d'une matrice ligne (à éléments tensoriels) par l'expression:

$$\boldsymbol{\Pi}(t) = \mathscr{I}m\{[\boldsymbol{\Pi}^+_1 \ldots \boldsymbol{\Pi}^+_k] \, [e^{i\omega t} \ldots e^{ik\omega t}]^T\}$$
$$= \mathscr{I}m\{\mathbf{P}^+ \cdot \Omega^T\} \qquad [41]$$

où Ω^T est la matrice colonne transposée de la matrice ligne Ω.

On obtiendra facilement des expressions analytiques pour les $R^*_k(i\omega)$ en utilisant les résultats que nous avons obtenus dans le cas des modules complexes du cas linéaire par l'emploi des plans complexes de *Cole-Cole* et/ou de *Black* [cf. (9)]. A noter que les $\mathscr{R}e[R^{k*}(i\omega)]$ *et* $\mathscr{I}m[R^{k*}_k(i\omega)]$ se prêtent à des relations de *Kramers* généralisées.

6. Formes particulières des fonctions fluage ou relaxation

On tire le meilleur parti de la présente méthode en l'appliquant à des formes explicites des fonctions fluage ou relaxation d'ordre $k (k = 1$ à $n)$.

Nous examinons ici quelques formes déjà proposées dans la littérature [(22) à (28)] ou particulièrement intéressantes.

6.1. Noyaux produits

Si les noyaux (ou fonctions fluage/relaxation) peuvent se mettre sous forme de produit de fonctions d'une variable, nous avons vu qu'il y a séparation: on ramène ainsi le CS_n à n CS_1.

6.2. Noyaux paraboliques

Le modèle parabolique (d'exposant fractionnaire), simple ou généralisé, joue un rôle important en viscoélasticité linéaire [cf. (5), (9)]. Des formes paraboliques pour les noyaux d'ordre 1 à 3 ont été observés en viscoélasticité non linéaire [cf. par exemple (24)].

La méthode que nous proposons sera particulièrement bien adaptée à ces noyaux en tenant compte de la correspondance:

$$t_1^{m_1} \cdot t_2^{m_2} \ldots t_k^{m_k} \underset{n}{\supset} (m_1!\, m_2! \ldots m_k!) \, p_1^{-m_1} p_2^{-m_2} \ldots$$
$$\ldots p_k^{-m_k} \qquad [42]$$

(où les factorielles sont des fonctions Γ d'*Euler* $\forall m_i$ non entier).

En particulier, le module complexe d'ordre k correspondant est en

$$(i\omega)^{-(m_1 + \cdots + m_k)}.$$

6.3. Noyaux exponentiels

Si les noyaux de [2] ont la forme de polynômes de *Dirichlet* (ce qui sera toujours possible à la précision que l'on voudra) les modules ou complaisances opérationnels seront rationnels par rapport aux différents p_i. S'il en est de même de l'histoire (là aussi toujours possible avec une précision donnée à l'avance), l'expression de Φ^{n*} sera entièrement rationnelle et son inversion immédiate par décomposition en éléments simples.

Cette forme des noyaux correspond à une généralisation des modèles de *Maxwell* ou *Kelvin* généralisés du cas linéaire. On pourra définir les spectres des modèles d'ordre k correspondants.

En particulier, les modules complexes d'ordre k seront des fonctions rationnelles de $i\omega$.

7. Conclusion

Au niveau des équations de comportement, la transformation que nous avons définie sous forme du produit d'une dissymétrisation par une transformation de *Carson* à n variables fournit une généralisation de la plupart des résultats fournis dans le cas de la viscoélasticité linéaire par l'application de la transformation de *Carson* à une variable:
- algébrisation des équations de comportement;
- même forme que dans le cas élastique non linéaire;
- obtention commode de la réponse à une histoire quelconque;
- correspondance avec les modules complexes généralisés.

De plus, il y a cohérence avec le traitement de la viscoélasticité linéaire puisque notre transformation contient le CS_1 comme cas particulier (on a $\mathcal{D}_1 \equiv \mathcal{I}$ transformation identique).

Il ne semble pas faire de doute que ce faisceau de propriétés ne simplifie grandement l'étude et l'application de la viscoélasticité non linéaire (notamment à partir d'analogies, d'une part avec la viscoélasticité linéaire, d'autre part avec l'élasticité non linéaire).

Par exemple, on pourra chercher à en tirer les formes des réponses à des sollicitations complexes ou à transposer au cas viscoélastique des modèles non linéaires de l'élasticité.

On pourra également tenter de définir des modifications correctes du principe de superposition (éventuellement de validité limitée à certaines formes d'histoires) en examinant comment s'exprime la réponse opérationnelle à la somme de plusieurs histoires en fonction des réponses opérationnelles à chacune d'entre elles.

Dans tous les cas un peu compliqués, il est certain que l'application de cette méthode, dont la validité s'étend rappelons-le aux déformations finies, se montrera (notamment en appliquant 6.3) avantageuse sur le plan du calcul numérique. Dans beaucoup de cas, elle rendra possible d'éviter le recours à l'ordinateur.

Enfin, ces résultats, établis en coordonnées cartésiennes, pourront probablement être étendus au cas de coordonnées curvilignes quelconques en remplaçant le tenseur δ_{ij} par le tenseur métrique approprié g_{ij}, et les composantes cartésiennes des tenseurs, par les composantes covariantes ou mixtes également appropriées.

D'après ce qui précède, la formulation opérationnelle que nous présentons contient comme cas particuliers l'élasticité linéaire et non linéaire (en déformations finies ou non), et la viscoélasticité linéaire (idem).

Par ailleurs elle fournit un moyen commode de ramener au cas des fonctions tensorielles les considérations de *Green*, *Rivlin*, *Spencer*, *Winemann* et *Pipkin* (11), (12), (13) sur les fonctionnelles tensorielles devant respecter certaines formes d'invariance.

Dans les divers cas d'anisotropie, ces résultats se conservent dans l'ensemble· seul le nombre de tenseurs de base et d'invariants (et donc le degré n de l'expression de Φ^{n*}) sont changés.

Par exemple, d'après *Adkins* (36) [cité en (13)] on a $n = 9$ pour le matériau orthotrope, et le nombre d'invariants de base I_α est égal à 7.

Le cas orthotrope pourra donc se traiter, en introduisant les termes supplémentaires appropriés, sans changement de principe de la méthode ni de la forme des résultats.

La méthode que nous proposons pourra également s'appliquer à d'autres problèmes physiques à hérédité invariable: par exemple variations dimensionnelles lentes de matériaux sous l'effet de variations des conditions hygrométriques et/ou thermiques. L'étude des couplages entre ces derniers et les effets mécaniques sera particulièrement intéressante. La présente méthode devrait pouvoir s'appliquer par exemple à des problèmes du type de ceux étudiés par *Ting* (27) d'une part, *Creus* et *Onat* (37) d'autre part. Elle devrait fournir également une nouvelle façon d'aborder (en utilisant notamment les résultats d'*Adeyeri* et al.) (28) le problème que nous avons traité en (18).

Ce que nous venons de dire ne concerne que les équations de comportement. Les possibilités éventuelles d'application au niveau des équations générales pour permettre la résolution des problèmes de champ demandent des examens plus approfondis, si bien que, dans son état actuel tout au moins, la méthode que nous proposons reste en deçà de ce qui avait été obtenu en viscoélasticité linéaire. Cet inconvénient est amoindri par le fait que bien peu de problèmes connaissent une solution explicite en élasticité non linéaire. Par contre, l'emploi de notre méthode permettra peut-être d'utiliser en viscoélasticité non linéaire certaines des méthodes d'approximation utilisées en élasticité non linéaire (par exemple méthode des perturbations).

Elle fournit cependant directement la solution dans le cas où l'on connaît (ou suppose connaître) a priori la répartition du champ des déformations. On pourra par exemple, sur la base d'une hypothèse de conservation des sections planes, appliquer directement notre méthode à la flexion d'une tige ou d'une plaque par un poinçon de forme donnée (ce qui correspond donc à un problème de contraintes de formage), en fonction de l'histoire correspondant au programme d'avance du poinçon.

Résumé

Nous présentons une tentative pour étendre à la viscoélasticité non linéaire les méthodes du calcul symbolique, dont on sait qu'ils ramènent, dans le cas linéaire, tous les problèmes de viscoélasticité, tant dynamiques que statiques, à des problèmes d'élastostatique (*Mandel*, 1955).

Nous partons d'un développement de *Fréchet*, limité à l'ordre n, de la fonctionnelle non linéaire donnant l'expression du tenseur des contraintes de *Kirchhoff* $\boldsymbol{\Pi}$ en fonction de la représentation lagrangienne (en déformations finies) du tenseur de la déformation pure $\mathbf{E}(t)$.

Nous appliquons successivement deux transformations. La première dissymétrise $\boldsymbol{\Pi}(t)$ en $\Phi(t_1, \dots, t_k, \dots, t_n)$ ce qui transforme le développement de *Fréchet* limité en une somme des produits de convolution multiples, multilinéaires par rapport aux $\dot{\mathbf{E}}_i = \dot{\mathbf{E}}(\tau_i)$ et à leurs invariants simples $\mathrm{tr}\,\dot{\mathbf{E}}_i$, ou composés $\mathrm{tr}(\dot{\mathbf{E}}_j \dots \dot{\mathbf{E}}_k)$ (ces derniers étant eux-mêmes multilinéaires par rapport aux $\dot{\mathbf{E}}_i$).

Pour un choix convenable de l'origine des temps, l'application à Φ de la transformation de *Carson* à n variable (CS_n) transforme cette somme en un polynôme $\Phi^{n*}(p_1, \dots p_n)$ multilinéaire par rapport aux $\mathbf{E}_i^* = \mathbf{E}^*(p_i)$ (en CS_1) et à leurs invariants simples $\mathrm{tr}\,\mathbf{E}_i^*$ ou composés $\mathrm{tr}[\mathbf{E}_j^* \dots \mathbf{E}_k^*]$. Ses coefficients (scalaires) sont les transformés $r_{ki}^{k*}(p_1, \dots, p_k)$ des noyaux $r_{ki}(t_1, \dots, t_k)$.

Dans le cas d'un corps initialement isotrope, le résultat de *Spencer* et *Rivlin* (1960) permet de se limiter au CS_{11}. Dans la pratique, on peut même se limiter au CS_3 si l'on tient compte de la remarque de *Gradowczyk* (1969) sur le rôle de la précision expérimentale escomptable.

Dans tous les cas, Φ a même forme que l'expression obtenue en dissymétrisant le développement utilisé par *Goldenblatt* dans le cas de l'élasticité non linéaire en déformations finies, à ceci près que le nombre d'invariants de base étant supérieur à 3, il ne peut se réduire à un développement d'ordre 2 par rapport à l'ensemble des $\mathbf{E}_i^*$.

Le retour à $\boldsymbol{\Pi}(t)$ se fait en inversant (par des tables) la transformation de *Carson*, ce qui donne $\Phi(t_1, \dots t_n)$, puis en faisant $t_i = t \; \forall i = 1, \dots n$ dans Φ.

L'inversion se ramène à n inversions d'ordre 1 lorsque les noyaux K sont sous forme produit ou sous forme de polynômes de *Dirichlet* par rapport aux τ_i. Dans ce dernier cas, elle est immédiate (par décomposition en éléments simples) si la sollicitation fonction du temps est elle-même représentée par un polynôme de *Dirichlet*, propriété qui se transmet à la solution.

Bibliographie

1) *Mandel, J.*, C. R. Acad. Sci. **241**, 1910–1912 (1955). Cahier du Groupe Français d'Etudes de Rhéologie **3**, 4 (1958).

2) *Mandel, J.*, C. R. Acad. Sci. **245**, 2176–2178 (1957); C. R. Acad. Sci. **245**, 2004–2006 (1957). Vibrations des corps viscoélastiques linéaires. 10e Congrès Intern. de Mécanique Appliquée, Stresa, 1960.

3) *Mandel, J.*, Mécanique des Milieux Continus t. I et II (Paris 1966).

4) *Huet, C.*, C. R. Acad. Sci. **257**, 1438–1441 (1963).

5) *Huet, C.*, Thèse, Paris 1963; Ann. Ponts et Chaussées **6**, 373–429 (1965). Bull. Liaison des Laboratoires des Ponts et Chaussées **8**, 4.1–4.20 (1964).

6) *Huet, C.*, Détermination par une méthode de vibrations forcées des modules complexes de Young et de cisaillement des enrobés bitumineux routiers. Communication aux Journées d'étude du GAMI, Paris, Mai 1964.

7) *Huet, C., G. Sayegh, P. Solemani*, Ann. de l'ITBTP n° 219–220 (Mars-Avril 1966). Conférence prononcée le 12 Janvier 1965 au Centre d'Etudes Supérieures de l'ITBTP.

8) *Huet C.*, Contribution à l'étude des effets différés dans les constructions composites. CIB International Symposium on Bearing Walls, Warsaw, 9–12 June 1969.

9) *Huet, C.*, Cahiers du Groupe Franç. Rhéol. **1**, 5 (1967).

10) *Huet, C.*, C. R. Acad. Sci. **269**, série B, 869–872 (1969); J. Mécanique **10**, 1 (Paris 1971).

11) *Green, A. E.* et *R. S. Rivlin*, Arch. Rat. Mech. and Anal. **1**, 1–21 (1957). — *Green, A. E., R. S. Rivlin* et *A. J. M. Spencer*, Arch. Rat. Mech. and Anal. **3**, 82–90 (1959). — *Green, A. E.* et *R. S. Rivlin*, Arch. Rat. Mech. Anal. **4**, 26 (1960).

12) *Spencer, A. J. M.* et *R. S. Rivlin*, Arch. Rat. Mech. Anal. **4**, 214 (1960).

13) *Winemann, A. S.* et *A. C. Pipkin*, Arch. Rat. Anal. **17**, 184–214 (1964).

14) *Poli, L.* et *P. Delerue*, Le calcul symbolique à deux variables et ses applications. In: Mém. Sci. Math., Fasc. 127 (Paris 1954).

15) *Delavault, H.*, Les transformations intégrales à plusieurs variables et leurs applications. In: Mém. Sci. Math., Fasc. 148 (Paris 1961).

16) *Mandel, J.*, C. A. Acad. Sci. **247**, 175–198 (Paris 1958).

17) *Huet, C.*, C. R. Acad. Sci. **270**, 213–216 (Paris 1970).

18) *Huet, C.*, Bull. Groupe Français des Argiles **23**, 39–57 (1970).

19) *Volterra, V.*, Theory of functionals and of integral and integro differential equations (Madrid 1925; réédité New York 1959).

20) *Fréchet, M.*, Ann. Sci. **27**, 3, 193–216 (Paris 1910).

21) *Novacki, W.*, Théorie du fluage (Paris 1965).

22) *Neis, V. V.* et *J. L. Sackman*, Trans. Soc. Rheol. **11**, 3, 397–434 (1967).

23) *Findley, W. N.* et *J. S. Y. Lai*, Trans. Soc. Rheol. **11**, 3, 361–380 (1967).

24) *Findley, W. N.* et *K. Onaran*, Trans. Soc. Rheol. **12**, 2, 217–242 (1968).

25) *Lai, J. S. Y.* et *W. N. Findley*, Trans. Soc. Rheol. **12**, 2, 243–258 (1968); Trans. Soc. Rheol. **12**, 2, 259–280 (1968).

26) *Onogi, S.*, *T. Masuda* et *T. Matsumoto*, Trans. Soc. Rheol. **14**, 2, 275 (1970).

27) *Ting, E. C.*, Trans. Soc. Rheol. **14**, 3, 297–306 (1970).

28) *Adeyeri, J. B.*, *R. Y. Krizek*, *J. D. Achenbach*, Trans. Soc. Rheol. **14**, 3, 375–392 (1970).

29) *Spencer, A. J. M.* et *R. S. Rivlin*, Arch. Rat. Mech. Anal. **2**, 435 (1959).

30) *Spencer, A. J. M.* et *R. S. Rivlin*, Arch. Rat. Mech. Anal. **4**, 214 (1960).

31) *Gradowczyk, M. H.*, Int. J. Solids Structures **5**, 873–877 (1969).

32) *Pipkin, A. C.*, Rev. Mod. Phys. **36**, 4, 125–136 (1964).

33) *Ditkin, V. A.* et *A. P. Prudnikov*, Formulaire pour le calcul opérationnel (Paris 1967).

34) *Goldenblatt, I. I.*, Some problems of the mechanics of deformable media (Groningen, Hollande, 1963).

35) *Cartan, H.*, Théorie élémentaire des fonctions analytiques d'une ou plusieurs variables complexes (Paris 1963).

36) *Adkins, J. E.*, Arch. Rat. Mech. Anal. **4**, 193 (1960).

37) *Creus, G. J.* et *E. T. Onat*, Int. J. Eng. Sci. **10**, 8, 649–658 (1972).

38) *Bismuth, W.* et *M. Chezeaux*, Cahiers du Groupe Français de Rhéologie **1**, 6 (1968).

39) *Huet, C.*, C. R. Acad. Sci. **275**, 793–796 (1972).

Addresse de l'auteur:

Dr. *C. Huet*
Centre technique des tuiles et briques
2, Avenue Hoche
F-75008 Paris (France)

Rheol. Acta 12, 289–294 (1973)

From the University of Bucharest, Mathematical Institute, Bucharest (Romania)

Rheological aspects in plasticity

By N. Cristescu

With 4 figures

(Received October 27, 1972)

1. Introduction

As it is well known, the equations describing the "classical plastic" model are essentially time independent. This way to describe "plasticity" seems to go back to *Barré de Saint-Venant* in 1870. In the present paper one discusses the way in which with a rate-type constitutive equation one can describe various phenomena which have been observed to occur during plastic deformation, and which cannot be described with the classical model.

Generally when introducing the equations of plasticity theory one starts from a non-linear one-dimensional stress-strain relation written in finite form

$$\sigma = \varphi(\varepsilon) \qquad [1.1]$$

which represents the "quasi static stress-strain curve" and is obtained experimentally with standard testing machines. In [1.1] the function φ is generally a strictly increasing one, this assuring the description of "work-hardening". In addition, in order to describe "plasticity" one stipulates that the plastic strain is irreversible and takes place only if the stress surpasses a certain limit called yield stress. In subsequent loading-unloading tests, each producing plastic strains as well, the actual yield stress $\sigma_y(\varepsilon_p)$ is increasing. Stresses higher than the actual yield stress $[\sigma(\varepsilon_p) > \sigma_y(\varepsilon_p)]$ are impossible within the classical model. It is obvious that these types of phenomena as creep, relaxation, etc., are not described by such a model.

The basic three-dimensional generalization of the constitutive equation is written in the form

$$\dot{\varepsilon}^p_{ij} = \lambda \frac{\partial F}{\partial \sigma_{ij}} \qquad [1.2]$$

(*Prager*, 1949) where $F(\sigma, \varepsilon^p)$ is the smooth and unique yield function. It is important to observe that the proportionality coefficient λ is determined in the classical plasticity theory after introducing a work-hardening condition in such a way that time is excluded from the constitutive equation. For instance [1.2] can be written as

$$\dot{\varepsilon}^p_{ij} = G \frac{\partial F}{\partial \sigma_{ij}} \frac{\partial F}{\partial \sigma_{kl}} \dot{\sigma}_{kl} \qquad [1.3]$$

with the coefficient G depending on invariants.

Similarly as for the one-dimensional case, stress points on the actual yield surface represent "plastic" states, points inside the yield surface represent elastic states while exterior points have no physical meaning unless an additional work-hardening is further produced. Thus if σ^* and ε^{*p} satisfy $F(\sigma^*, \varepsilon^{*p}) = 0$, states σ satisfying $F(\sigma, \varepsilon^{*p}) > F(\sigma^*, \varepsilon^{*p})$ are impossible.

In plasticity theory there were mainly the dynamic problems which have suggested that phenomena as creep, relaxation, overstress, etc. have to be considered as well if a comprehensive description of dynamic plastic deformation of various materials is desired.

2. Rate-Type plasticity theory

In order to describe various rheological effects which may be present in dynamic testing of elastic-plastic materials, a rate-type constitutive equation has been introduced. For one-dimensional cases there are two basic concepts which have to be defined in order to make explicit a constitutive equation of the form

$$\dot{\varepsilon} = \frac{\dot{\sigma}}{E} + \Phi(\sigma, \varepsilon)\,\dot{\sigma} + \Psi(\sigma, \varepsilon). \qquad [2.1]$$

First is introduced the so called "relaxation boundary". This is a curve

$$\sigma = f(\varepsilon) \qquad [2.2]$$

in the stress-strain plane satisfying the following property: above this curve, i.e. for $\sigma > f(\varepsilon)$, the model may describe both creep and relaxation, while under this curve, i.e. for $\sigma \leqslant f(\varepsilon)$, the model is pure elastic. Generally it is no reason to assume that [2.2] always coincides with what is called the "quasi-static" stress-strain curve. Therefore the initial yield point (beginning of plastic deformation) may also differ from the quasi-static one.

Various expressions for $f(\varepsilon)$ have been used, all being some variants of a certain power law. For instance one has used [*Cristescu* (1972a)]

$$f(\varepsilon) = \begin{cases} \sigma_Y & \text{if } \varepsilon \leqslant \varepsilon_Y \\ \beta(\varepsilon + \varepsilon_0)^{1/\alpha} & \text{if } \varepsilon \geqslant \varepsilon_Y \end{cases} \qquad [2.3]$$

with α, β and ε_0 positive constants. Another used expression is

$$f(\varepsilon) = \begin{cases} \sigma_Y & \text{if} \quad \varepsilon \leqslant \varepsilon_Y \\ \sigma_Y + \dfrac{\beta}{2}\varepsilon_z^{-1/2}(\varepsilon - \varepsilon_Y) & \text{if} \quad \varepsilon_Y \leqslant \varepsilon \leqslant \varepsilon_z \\ \beta\varepsilon^{1/2} & \text{if} \quad \varepsilon_z \leqslant \varepsilon \end{cases} \quad [2.4]$$

where

$$\varepsilon_z = \left(\frac{\beta\varepsilon_Y}{\sigma_Y - \sqrt{\sigma_Y^2 - \varepsilon_Y\beta^2}}\right)^2. \quad [2.5]$$

After introducing the relaxation boundary it was further assumed that the function $\Psi(\sigma, \varepsilon)$ entering the constitutive eq. [2.1] depends on the so-called overstress $\sigma - f(\varepsilon)$ i.e. the difference between the actual (dynamic) stress σ and the stress $f(\varepsilon)$ at the point of the relaxation boundary corresponding to the same strain magnitude. Various laws of such dependence can be considered: linear, exponential, etc. For aluminium a linear dependence of the form

$$\Psi(\sigma, \varepsilon) = \begin{cases} \dfrac{k(\varepsilon)}{E}[\sigma - f(\varepsilon)] & \text{if} \quad \sigma > f(\varepsilon) \text{ and } \varepsilon \geqslant \varepsilon_Y \\ 0 & \text{if} \quad \sigma \leqslant f(\varepsilon) \text{ or } \varepsilon < \varepsilon_Y \end{cases} \quad [2.6]$$

seems quite reasonable [see *Cristescu* (1972a)].

The coefficient $E/k(\varepsilon)$ entering in [2.6] can be interpreted in a certain sense as a viscosity coefficient which has to be defined for $\varepsilon \geqslant \varepsilon_Y$. This definition in turn can be considered to define the position of the yield point. E is the *Youngs* modulus, while from a laborious comparison between various theoretical predictions and the experimental data it was found that while for higher strains k can be reasonable well assumed to be a constant, for small strains (just above the yield point) k has to be assumed an increasing with ε function. Here again various expressions describing the variation of k with ε have been used. Any such expression must be defined for the states $\sigma > f(\varepsilon)$ only. An exponential law of the form

$$k(\varepsilon) = k_0\left[1 - \exp\left(-\frac{\varepsilon}{\hat{\varepsilon}}\right)\right] \quad [2.7]$$

was found reasonable but not easy enough to handle in order to fit well the experimental data. Here k_0 and $\hat{\varepsilon}$ are material constants. A more convenient in application expression for $k(\varepsilon)$ is

$$k(\varepsilon) = \begin{cases} 0 & \text{if} \quad \varepsilon < \varepsilon_Y \\ k_1 & \text{if} \quad \varepsilon_Y \leqslant \varepsilon \leqslant \varepsilon_1 \\ k_2 + \dfrac{k_1 - k_2}{\varepsilon_1 - \varepsilon_2}(\varepsilon - \varepsilon_2) & \text{if} \quad \varepsilon_1 \leqslant \varepsilon \leqslant \varepsilon_2 \\ k_2 & \text{if} \quad \varepsilon_2 \leqslant \varepsilon \end{cases} \quad [2.8]$$

where k_1, k_2, ε_1 and ε_2 are constants. ε_Y is the strain at the yield point, or more precise the strain corresponding to the upper yield point, if the material under consideration possesses both upper and lower yield stresses. Therefore with a relationship of the form [2.8] one can describe analytically a possible presence of an upper and lower yield stress. What concerns the other constants involved in [2.8], as a rule k_1 is generally with several order of magnitude smaller than k_2; sometimes k_1 can simply be assumed to be zero. ε_1 and ε_2 have to be found from experiments.

In [2.1] the function Φ is the measure of the "instantaneous" nonlinear response of the material, that is $1/(\Phi + 1/E)$ is the slope of the dynamic instantaneous stress-strain curve. For various materials one can contemplate to use Φ in order to describe nonlinear reversible behaviour, however, in the previously mentioned paper it was used in order to describe irreversible (i.e. plastic) behaviour. Thus $\Phi\dot{\sigma}$ is in a certain sense the "time-independent" component of the rate of strain for which the term "instantaneous" will be used for convenience. In order to find an explicit expression for Φ one has made the additional assumption that the fastest possible response of the model is the elastic (*Hooke*) response. From here results a restrictive inequality, which for the expression [2.3] of the function f is of the form

$$1 \geqslant \frac{1}{E\Phi + 1} > \frac{1}{\alpha(\varepsilon + \varepsilon_0)}\left(\frac{\sigma}{E} - \frac{\Psi}{k}\right) \quad [2.9]$$

for any stress and strain states in the loading domain (i.e. there where $\Psi \neq 0$). Therefore these inequalities express the assumption that the slope of the instantaneous stress-strain curve is comprised between the elastic one and that furnished by the relaxation boundary for the same strain value. The inequality [2.9] was considered to be a "sufficient" one though one may contemplate to establish a weaker one if necessary.

The function $\Phi(\sigma, \varepsilon)$ is to be defined in the domain $\varepsilon > \sigma/E$, $\sigma > f(\varepsilon)$; one has assumed that $\Phi = 0$ in the domain $\sigma < f(\varepsilon)$ and $\varepsilon < \sigma/E$ while

$\Phi \neq 0$ possibly only in the domain $\sigma > f(\varepsilon)$, $\varepsilon > \sigma/E$. The coefficient function Φ was defined in the following way

$$\Phi = \chi(\sigma_m(x))\, \Omega(\sigma, \varepsilon) \qquad [2.10]$$

where the explicit form of $\Omega(\sigma, \varepsilon)$ will be given below and χ is a parameter playing an important role in the definition of Φ:

$$\chi = \begin{cases} 1 & \text{if} \quad \sigma = \sigma_m(x) \\ 0 & \text{if} \quad \sigma \leqslant f(\varepsilon) + \sigma^* \quad \text{or} \\ & \phantom{\text{if}} \quad \sigma < \sigma_m(x) \end{cases} \qquad [2.11]$$

with $\sigma_m(x)$ the maximum stress reached at the section x of the bar. In [2.11] there has also been introduced a threshold overstress σ^*, below which $\Phi = 0$, which was introduced according to

$$\frac{\sigma^*}{E} = \varepsilon^* = h - \lambda \sqrt{\left|\varepsilon - f^{-1}\left(\frac{\sigma_Y}{E} + h\right)\right|} \qquad [2.12]$$

with h and λ constants. In a first approximation one can well assume $\sigma^* = E\varepsilon^* = 0$.

The introduction of the parameter χ according to [2.11] in the definition of Φ is based on the assumption that $\Phi \neq 0$ only if $\dot{\sigma} > 0$. Satisfactory results have been obtained with this assumption though one can well contemplate to weaken it if necessary.

Since Φ is describing the instantaneous response of the material, the explicit expression for the function Φ could be determined from the speed of the acceleration waves, because Φ enters the expression for the characteristic speed

$$c = \frac{c_0}{\sqrt{1 + E\Phi}}. \qquad [2.13]$$

Since, however, experimental acceleration wave speed data for the whole loading range were not available, the explicit expression for Φ was determined from the arrival times of strains of various magnitudes at successive cross-sections of the longitudinally impacted bar. Thus various expressions for Φ have been determined. For aluminium one has used

$$\Phi(\sigma, \varepsilon) = \chi \frac{\gamma}{E}\left[3\left(\frac{E}{p + q\sqrt{\varepsilon}}\right)^3 \left(\frac{\sigma}{E}\right)^2 - 1\right] \qquad [2.14]$$

where γ, p and q are constants, or

$$\Phi(\sigma, \varepsilon) = \chi \left\{ \frac{3\left[\varepsilon - \frac{\sigma}{E} - \underset{\sim}{\varepsilon} + \left(\frac{a}{3E}\right)^{3/2}\right]^{2/3}}{a} - \frac{1}{E} \right\}$$

with [2.15]

$$a = m + n\sqrt{\varepsilon} \qquad [2.16]$$

m and n being constants. In a first approximation one can assume a = constant, $\underset{\sim}{\varepsilon}$ is also a constant, generally a very small one.

It is important to observe that the explicit form of Φ cannot be determined in a unique way by considering the strain-time curves at various cross-sections only. Since Φ is also responsible for the maximum stress reached in a dynamic loading at various cross-sections of the bar, one can uniquely determine Φ by knowing from experiment the variation of the stress in a single cross section (impacted end) of the bar.

We have to observe that with the model [2.1] the work-hardening concept is simply described by the function entering the expression of the relaxation boundary [2.2] and that it is not possible to define work-hardening using the sign of the ratio $d\sigma/d\varepsilon$. If for finite form theories for a given plastic stress-strain state the ratio $d\sigma/d\varepsilon$ is uniquely determined and work-hardening means simply that $d\sigma/d\varepsilon > 0$, for rate-type theories this ratio is not unique and depends on the loading history; now the material can well work-harden even if $d\sigma/d\varepsilon = 0$, or $d\sigma/d\varepsilon < 0$. In both kinds of theories work-hardening is associated with the increase of the size of $\sigma_Y(\varepsilon_p)$ of the elastic domains when the plastic strain is increasing. For rate-type theories "plastic strain" is to be understood in a more general sense: the total irreversible component of the strain.

What concerns the "initial" yield stress, i.e. the stress corresponding to the beginning of the plastic deformation, while it is a point on the curve [1.1], it is more difficult to define it for [2.1] by the expressions of both the relaxation boundary and $k(\varepsilon)$. Thus the description of an "upper" and "lower" yield stress is also possible.

The plastic deformation takes place for any state satisfying $\Psi > 0$. However, it is only if $\Psi > 0$ and $\Phi > 0$ that both mechanisms of plastic deformation are contributing to the plastic deformation.

3. An example

In order to make clear the way in which the previously mentioned model is describing "plasticity" in dynamic problems, an example will be given in which two identical aluminium bars collide longitudinally. The bars are cylindrical and the length of each of them is equal to ten diameters. This problem, where both loading and unloading processes have been studied, has been discussed elsewhere [*Cristescu* (1972a) (1972b)]. Here the reply of the model to such kind of impact will be shortly described. The model used corresponds to the formulas [2.4], [2.8] and [2.15]. Comparison with the experimental data are very satisfactory. For details see *Cristescu* (1972c).

Due to the sudden impact, the stress at the impacted end is increasing very fast towards what is called the "peak stress" (see fig. 1). This peak stress is at least of one order of magnitude higher than the initial yield stress. The peak is unstable and the stress relaxes in a few microseconds towards the so-called "stress plateau" (fig. 1).

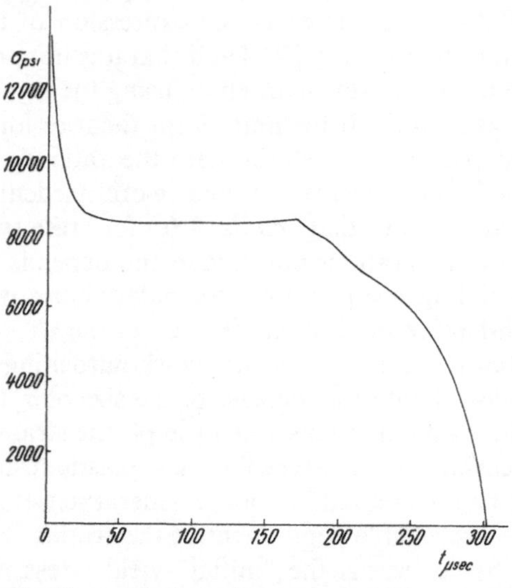

Fig. 1. The variation of stress at the impacted end showing the "peak stress" and the "plateau"

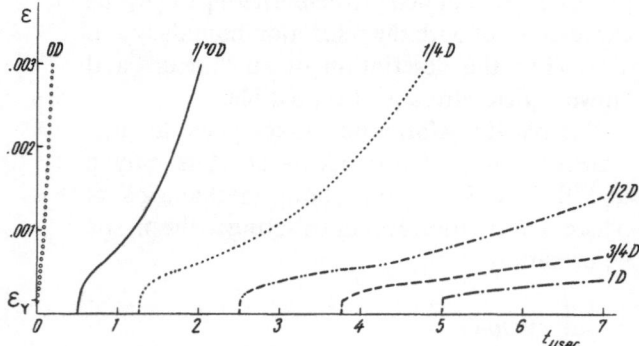

Fig. 2. The beginning of the strain-time curves showing the decay of the "strain bump"

During the period when stress is raising towards the peak stress, the strain is increasing in a nearly linear manner towards a value much higher than that of the yield strain. This initial increase of the strain due to the presence of the peak stress is further (during propagation in the bar) responsible for the formation of a "bump" on the strain-time curves (see fig. 2).

When it is propagating into the bar the peak stress is decaying quite fast together with the initial "bump" of the strain-time curve (see fig. 2) which during propagation in the first diameter is decreasing towards ε_Y. The presence of a peak stress, its rapid decay together with the initial bump of the strain-time curves, are experimental facts quite well described by constitutive equations of the form [2.1].

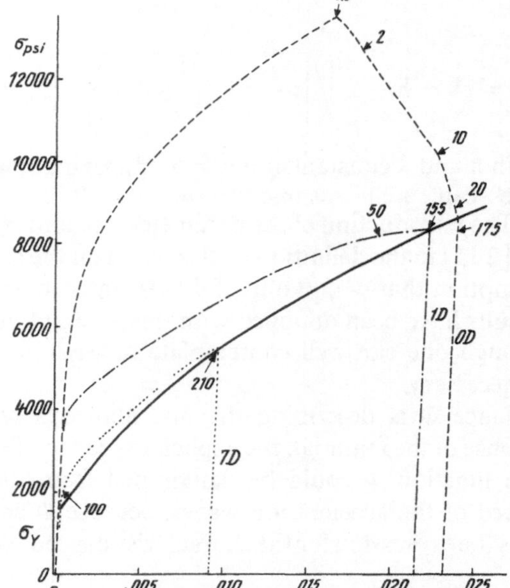

Fig. 3. The stress-strain curves at various cross-sections along the bar: the full line is the relaxation boundary. The numbers indicate how many microseconds have elapsed from the moment of impact

Fig. 3 shows the stress-strain curves at various cross-sections of the bar. The solid line is the relaxation boundary. The stress-strain curves at various cross-sections are certainly distinct, and some of them are situated well above the relaxation boundary. The decay of the peak stress during propagation can be seen here as well. The small numbers written along the curves are showing how many microseconds have elapsed from the moment of impact until this particular point of the stress-strain curve is reached. It is very important to observe that the states close to the relaxation boundary are quasi-stable states. This is the main property of the relaxation boundary which can be used to define it. In other words the values of stress and strain corresponding to the relaxation boundary are just those for which on both stress-time curves and strain-time curves are appearing some plateaus. Such strain plateaus are shown on fig. 4.

Another rheological phenomenon which can be described with a rate-type constitutive equation of the type [2.1] is the possible stress relaxation accompanied by a dynamic creep (increase of plastic strains is shown on fig. 3). It is also significant to observe that at the impacted end, the raising portion of the stress-strain curve (fig. 3) corresponds to the portion of the strain-time curve which is above the initial bump and which is concave upward. Therefore at the impacted end, the portion of the strain-time curve concave downwards corresponds to the decreasing portion of the stress-strain curve. Thus the "main" inflexion point of the strain-time curve (the one which lays somewhere in the middle of the rising portion of the strain-time curve) corresponds to the peak of the stress-strain curve [*Cristescu* (1972b)].

All the previously mentioned phenomena which are experimentally found, can reasonably well be described with a rate-type constitutive equation of the form presented in the previous section. At the same time most of these phenomena cannot be described within the framework of classical plasticity theory.

4. Remarks Concerning a Three-Dimensional Generalization

The experimental results available today are not sufficient for a full and comprehensive generalization for the three-dimensional case of the constitutive eq. [2.1]. It is however, easy to try an "isotropic" generalization following a procedure somehow similar to the one used in plasticity theory (when passing from [1.1]–[1.2] say).

While it is relatively easy to generalize the concept of the relaxation boundary into a relaxation surface if "isotropy" is assumed, it is more difficult to generalize the concept of instantaneous response [see *Cristescu* (1967) Ch. X].

The overstress above the relaxation boundary becomes now the overstress $\sqrt{3}\sqrt{II_s} - f(\bar{\varepsilon})$ above the relaxation surface. Here

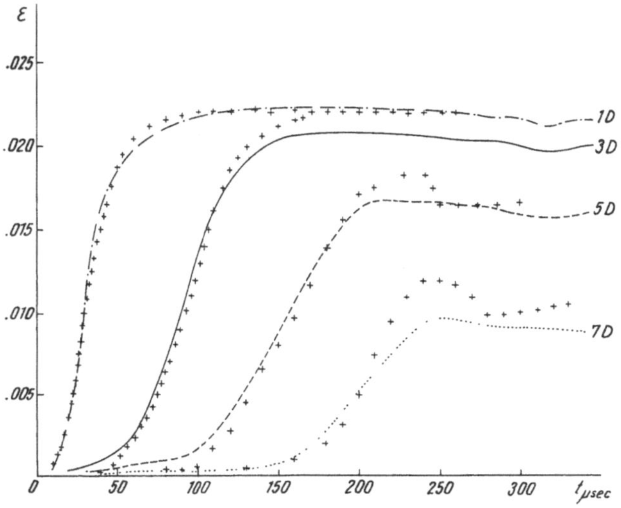

Fig. 4. Strain-time curves at various cross-sections of the bar; the experimental data are represented by crosses

$$II_s = \tfrac{1}{2} s_{ij} s_{ij}, \quad II_e = \tfrac{1}{2} e_{ij} e_{ij} \qquad [4.1]$$

are the invariants of the stress and strain deviators respectively and $\bar{\varepsilon} = (2/\sqrt{3})\sqrt{II_e}$, $\bar{\sigma} = \sqrt{3}\sqrt{II_s}$. It is also assumed that the coefficient k depends now on $\bar{\varepsilon}$ and is defined for stress-strain states above the relaxation surface.

Thus "work-hardening" is defined in the same way as for the one-dimensional model by the function f, while the initial yield surface by the definition of f and k.

What concerns the plastic instantaneous response of the material, it is assumed that if $s_{ij}\dot{s}_{ij} > 0$ the following relation is satisfied

$$s_{ij}\dot{e}^p_{ij} = F(II_s, II_e)\, s_{ij}\dot{s}_{ij} \qquad [4.2]$$

where the superscript p stands for the "plastic instantaneous" component and F is a strictly increasing function of its arguments. If the same notations as in §2 are used, [4.2] can be written finally as

$$\dot{e}^p_{ij} = \frac{3\Phi(\bar{\sigma}, \bar{\varepsilon})}{4 II_s} s_{kl}\dot{s}_{kl} s_{ij}. \qquad [4.3]$$

Therefore Φ is here the same function as furnished by [2.14] or [2.15] but where stress and strain have been replaced by $\bar{\sigma}$ and $\bar{\varepsilon}$ respectively.

It must be observed that though [4.2] is somehow remembering a work-hardening condition of classical plasticity theory, here the decrease of stress is not a pure unloading but starts by being a "time-dependent" phenomenon first. Generally, as long as the stress-state is above the relaxation boundary, the whole deformation process, during increasing or decreasing stresses ($\dot{\text{II}}_s > 0$ or $\dot{\text{II}}_s < 0$), is a time-dependent one, unless it is instantaneous.

The generalization of the constitutive eq. [2.1], if small strains and rotations are assumed can therefore be written as

$$\dot{e}_{ij} \frac{\dot{s}_{ij}}{2G} + \frac{3k(\bar{\varepsilon})}{2E}\left[1 - \frac{f(\bar{\varepsilon})}{\sqrt{3}\sqrt{\text{II}_s}}\right] s_{ij} + \frac{3\Phi(\bar{\varepsilon},\bar{\sigma})}{4\,\text{II}_s} s_{kl}\dot{s}_{kl}s_{ij} \qquad [4.4]$$

with an additional assumption for the behaviour of the volume. For metals an elastic behaviour

$$\sigma = 3K\varepsilon \qquad [4.5]$$

seems reasonable.

In the generalization [4.4] both instantaneous response curve and relaxation boundary were generalized in a global manner into an instantaneous response surface and a relaxation surface respectively. More general three-dimensional generalization can be contemplated, for instance by introducing the concept of normality

$$\dot{\varepsilon}^p_{ij} = 2\lambda \frac{\partial \mathscr{F}}{\partial \sigma_{ij}} \qquad [4.6]$$

to an instantaneous response surface $\mathscr{F}(\sigma_{ij}, \varepsilon^p_{ij}, \ldots) = 0$ and prescribing certain properties for the deformation and possible translation of this surface. It is possible that a certain kind of ***Koiter*** generalization be more reasonable than [4.2] in the sense that piece-wise nonglobal instantaneous reply relations are also allowed.

The previously described model in contrast with the classical plasticity model [1.3] is essentially a time-dependent one, describing phenomena as creep and relaxation as well. Two-dimensional experiments, both static and dynamic, are necessary in order to choose the way in which a model of the form [2.1] (which has proven to describe very well the experimental data) is to be generalized for the three-dimensional case.

Summary

One discusses several ways of improving the equations of plasticity theory by introducing rheological aspects as well. The main concepts whose meaning has been changed are yielding and workhardening. The modifications of these concepts have been suggested mainly by considering various aspects in problems of dynamic plasticity.

In order to be specific, a procedure is given to determine various functions and constants entering in a constitutive equation exhibiting both time-dependent and time-independent plasticity, starting from a set of experimental data. As typical experiments one has chosen the symmetric longitudinal impact of two identical bars, since for such kinds of problems a great deal of experimental data is available.

The unloading process is also considered since it is mainly during unloading that interesting aspects concerning "plasticity" may be put into evidence. A "relaxation boundary" which plays a main role in time-dependent plasticity is introduced. An example is given in order to show the influence of various functions or constants entering in the constitutive equation. Comparison with experimental data is excellent.

The discussion is then done for the three-dimensional problem though the experimental data are scarce. However, the one-dimensional approach is suggesting a way of defining yielding and workhardening for the three-dimensional case. Both these concepts can be introduced so that time-dependent and time-independent aspects would be present simultaneously.

References

Cristescu, N., Dynamic Plasticity (Amsterdam 1967). — Int. J. Solids Structures **8**, 511–531 (1972a). — Introduction to rate-dependent plasticity (a dynamical approach) (Berlin-Heidelberg-New York 1972b). — Rate-Type Constitutive Equations in Dynamic Plasticity. Int. Symposium on Foundations of Plasticity, Warsaw (1972c).

Prager, W., J. Appl. Phys. **20**, 235–241 (1949).

Author's address:
N. Cristescu
Mathematical Institute
Calea Grivitei 21, Bucharest 12 (Romania)

Rheol. Acta 12, 295–298 (1973)

From the Department of Mathematics, Illinois Institute of Technology, Chicago (U.S.A.)

On rheological relations

By Barry Bernstein

With 2 figures

(Received October 27, 1972)

Those who are serious about their desire to formulate constitutive equations which actually describe the behavior of visco-elastic media naturally wish to be able to have their equations subjected to experimental test. To this end, it is fruitful to seek to derive from a given theory relations among observable quantities. By an observable quantity in the context of our subject we mean a quantity which may be defined operationally in terms of a given flow or deformation history which may be carried out experimentally to within acceptable accuracy, i.e. a quantity which may be determined by rheological measurements. Steady state shearing stress, normal stress and complex dynamic shear modulus are examples of such quantities as are also elongational viscosity and the infinitesimal shear-relaxation modulus. We do not believe that there would be any objection if we referred to such quantities as "rheological quantities" and to relations among them as "rheological relations". But when we speak of rheological relations here, we understand that they are relations predicted by theory. Our main interest in such relations stems from their usefulness in studying non-linear behavior. However, for the purpose of illustration, we shall cite some relations which arise from well known and widely accepted theory, namely the theory of linear visco-elastic fluids. Some such examples are

$$G^*(\omega) = i\omega \int_0^\infty G(\xi)\, e^{-i\omega\xi} d\xi, \qquad [1]$$

$$\lim_{\omega\to 0} \eta'(\omega) = \eta_0, \qquad [2]$$

$$\eta_0 = \int_0^\infty G(\xi)\, d\xi. \qquad [3]$$

Here, as conventionally, $G(\xi)$ is the shear relaxation modulus, $G^*(\omega)$ is the complex dynamic modulus and η_0 is the steady shear viscosity. For these fluids the dynamic viscosity η' is defined by

$$\eta'(\omega) = \frac{Im\, G^*(\omega)}{\omega}. \qquad [4]$$

Let us note that [1] and [2] relate stress relaxation measurements ($G(\xi)$) to steady-state oscillation measurements ($G^*(\omega)$), whereas [3] says that the area under the stress-relaxation modulus is the viscosity to be observed in steady shear.

When [1], [2] and [3] they hold, they support the applicability of the theory of linear visco-elastic fluids. In circumstances in which they fail to hold the inadequacy of linear visco-elastic theory is indicated.

As we have already emphasized our principal concern is with non-linear behavior, for it is there, in the front echelons of our subject, that criteria to distinguish among theories can play a most vital role. In 1964 *Coleman* and *Markovitz* (1) obtained a relation from the theory of simple fluids, namely

$$\lim_{\omega\to 0} \frac{2\,G'(\omega)}{\omega^2} = \lim_{\varkappa\to 0} \frac{\nu(\varkappa)}{\varkappa^2} \qquad [5]$$

where $G^* = G' + iG''$ and $\nu(\varkappa)$ is the first normal stress difference, i.e. $\nu = \sigma_{11} - \sigma_{22}$ in a steady simple shearing flow with velocity V_i given by

$$V_1 = \varkappa \chi_2, \quad V_2 = V_3 = 0 \qquad [6]$$

where $\chi_i, i = 1, 2, 3$, are cartesian coordinates, σ_{ij} is the stress tensor and $\varkappa$ is a constant. We shall discuss the relation [5] in the context of another relation, namely [9] below.

In the past few years this author and some others have had a strong interest in a theory (2, 3) which has come to be known as the BKZ theory[1]). In its incompressible version the theory comprises a stress-strain relation of the form

$$\sigma_{ij} = -p\delta_{ij} + 2\int_{-\infty}^{t} \left[\frac{\partial U}{\partial I_1} B_{ij}(t,\tau) - \frac{\partial U}{\partial I_2} B_{ij}^{-1}(t,\tau)\right] d\tau \qquad [7]$$

[1]) This theory was first publically presented by the authors at the U.S. Soc. of Rheology in October 1962. Simultaneously and independently it was proposed by *A. Kaye* in England (4).

where $B_{ij}(t, \tau)$ is the left *Cauchy-Green* relative strain tensor and $U[I_1, I_2, t-\tau]$ is a function of the principal invariants I_1 and I_2 of $B_{ij}(t,\tau)$ and of the time interval $t-\tau$. Also, as is usual in many incompressible theories, the hydrostatic pressure is not assigned any dependence on other quantities a priori. *Zapas* and *Craft* (5) showed strong evidence for the applicability of this theory to bulk polyisobutylene and plasticized polyvinyl chloride through simple extension experiments. *Zapas* (6) and *Zapas* and *Phillips* (7) gave further evidence of the agreement of this theory with experiments on polymer solutions as well as on bulk polymer.

It is important to realize that in the relations to be discussed below, the theory is assumed in the generality of eq. [7]. No specific form of the function U is assumed. Furthermore we shall discuss here only relations which have been shown *not* to be obtainable for a general simple fluid. To this end, let us define a few quantities.

The in-line superposed complex dynamic modulus $G^*_{\text{in}}(\varkappa, \omega)$ is obtained by perturbing the flow [5] with a small in-line perturbation $A e^{i\omega t}$ in displacement, i.e.

$$V_1 = (\varkappa + iA\omega e^{i\omega t})\,\chi_2 .$$

If quantities are linearized about the steady shear flow [6], the ratios of the perturbations in the shear stress to the perturbed shear strain, both of which are sinusoidal, is $G^*_{\text{in}}[\varkappa, \omega] = G'_{\text{in}} + iG''_{\text{in}}$. Similarly the superposed complex dynamic normal modulus $N^*_{\text{in}}[\varkappa, \omega]$ is defined by using the perturbed values of $\sigma_{11} - \sigma_{22}$ instead of shear stress.

If the perturbations are transverse, i.e.

$$V_1 = \varkappa\chi_2, \quad V_2 = 0, \quad V_3 = i\omega A e^{i\omega t}\chi_2,$$

we obtain similarly the transverse superposed dynamic shear modulus

$$G^*_{\text{tr}}[\varkappa, \omega] = G'_{\text{tr}}[\varkappa, \omega] + iG''_{\text{tr}}[\varkappa, \omega].$$

(Here σ_{23} is divided by $A e^{i\omega t}$ to get $G^*_{\text{tr}}[\varkappa, \omega]$.) The dynamic viscosities are defined by

$$\eta'_{\text{in}} = G''_{\text{in}}/\omega, \quad \eta'_{\text{tr}} = G''_{\text{tr}}/\omega .$$

We shall discuss the following rheological relations.

$$\lim_{\omega\to 0} \frac{2G'_{\text{in}}[\varkappa, \omega]}{\omega^2} = \frac{d}{d\varkappa}\frac{\nu(\varkappa)}{\varkappa} = \frac{\nu}{\varkappa^2}\left[\frac{d\ln\nu}{d\ln\varkappa} - 1\right], \qquad [8]$$

$$\lim_{\omega\to 0} \frac{2G'_{\text{tr}}[\varkappa, \omega]}{\omega^2} = \frac{\nu(\varkappa)}{\varkappa^2}, \qquad [9]$$

$$\lim_{\omega\to 0} N'_{\text{in}}[\varkappa, \omega] = \frac{d}{d\varkappa}\varkappa^2\eta(\varkappa) \qquad [10]$$

where $\eta(\varkappa)$ is the viscosity corresponding to the steady shear flow [6].

Booij (8) claimed that data on N^* were not good enough to check [10]. However [8] and [9] have been the subject of some study.

Relations [8] and [10] were proposed by *Bernstein* (9) in 1968. In 1970 *Bernstein* and *Fosdick* (10) showed that [8] and [10] were not general simple fluid relations and could therefore be used to distinguish the BKZ theory from the general simple fluid theory. They also argued that the relation was not inconsistent with data of *Booij* (11). In particular the data quoted seemed not inconsistent with the prediction of [8] is that G'_{in}/ω^2 should approach a negative limit as $\omega \to 0$ for fixed $\varkappa$ when and only when the slope of $\ln\nu$ against $\ln\varkappa$ was less than unity. *Booij* (8) disputed this contention with other data. *Tanner* and *Williams* (12) countered with the contention that "the evidence is inconclusive and tests at lower frequencies are required to be certain one way or the other". Indeed, they pointed out that at frequencies $\leqq$ 3 Hz, the signal to noise ratio becomes low enough to cause serious experimental problems. They derived the relation [9] and then presented experimental evidence which they claim support the relation [9] and, consequently, the *Coleman-Markovitz* **relation**, which is merely the limiting case $\varkappa = 0$. Some of their data is presented in fig. 1.

Another set of rheological relations have been discussed by *Bernstein* and *Huilgol* (13) and by *Bernstein, Huilgol* and *Tanner* (14). These are

$$\lim_{\omega\to\infty} \omega^2\eta'_{\text{in}}[\varkappa, \omega] = \lim_{\omega\to\infty}\omega^2\,\eta'_{\text{tr}}[\varkappa, \omega] = \lim_{\omega\to\infty}\omega^2\eta'(\omega) = -\dot{G}(0) \qquad [11]$$

i.e. the dynamic viscosity curves become asymptotic to

$$-\frac{\dot{G}(0)}{\omega^2}$$

which is independent of $\varkappa$. Indeed, as predicted, evidence shows that if one plots against ω the superposed dynamic viscosity curves at fixed $\varkappa$, these curves come together at high enough values of ω. *Bernstein* and *Huilgol* (13) cited data of *Simmons* (15) to bear this out. An example of such data is given in fig. 2.

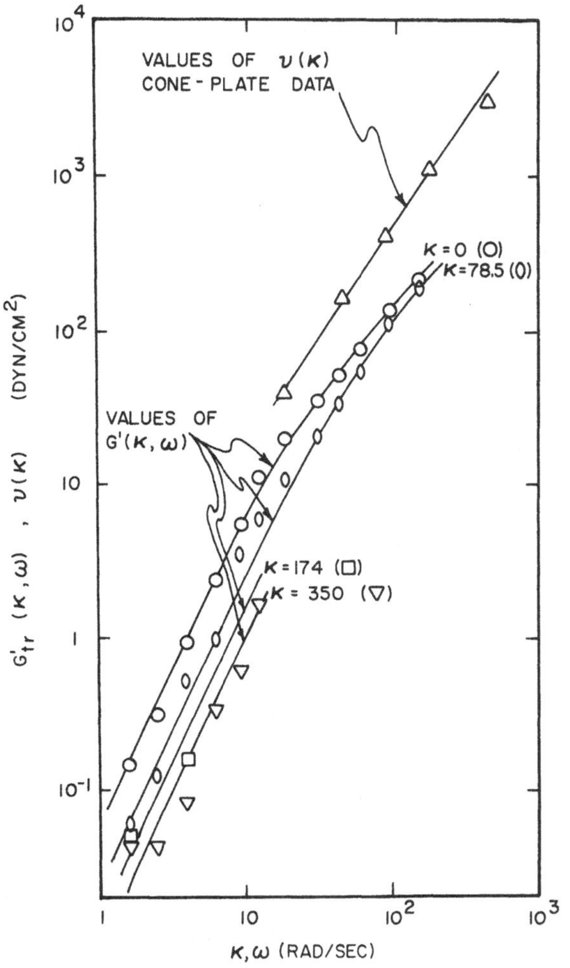

Fig. 1. First normal stress difference, $\nu(\varkappa)$ and $G'_{tr}[\varkappa, \omega]$ for 4.4% polyisobutylene solution as given by *Tanner* and *Williams* (12). The lines which extend to the lower left hand portions of the figure are plots of $\ln\left(\frac{\nu(\varkappa)}{2\varkappa^2}\right) - 2\ln\omega$, versus $\ln\omega$. According to relation [9] the values of $G'_{tr}[\varkappa, \omega]$ should approach these lines for $\omega \to 0$. The leftmost line, $\varkappa = 0$, pertains to the *Coleman-Markovitz* relation [5]

Bernstein, Huilgol and *Tanner* also showed that the BKZ theory implies

$$\lim_{\omega\to\infty} \omega^3 \frac{\partial}{\partial\omega} G'_{tr}[\varkappa, \omega] = \lim_{\omega\to\infty} \omega^3 \frac{\partial}{\partial\omega} G'_{in}[\varkappa, \omega]$$
$$= \lim_{\omega\to\infty} \omega^3 \frac{\partial G'(\omega)}{\partial\omega}$$
$$= 2\ddot{G}(0) \qquad [12]$$

which are again independent of the rate of shear of the base motion.

Experimental data of *Booij* (8, 11), *Osaki* et al. (16), *Kátaoka* and *Ueda* (17) and of *Walters* and *Jones* (18) showed that in some instances the graphs of $G'_{in}[\varkappa, \omega]$ versus ω for different fixed values of $\varkappa$ tend to approach each other at high ω and in other cases became parallel to each other at high enough ω. Both observations would be consistent with [12].

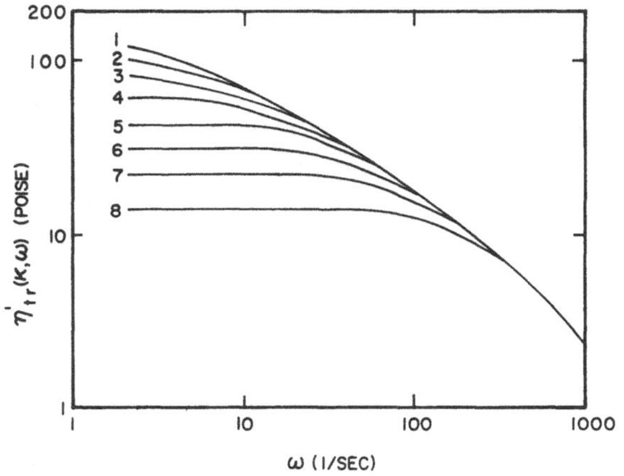

Fig. 2. Transverse Dynamic Viscosities of 8.54% polyisobutylene in cetane as obtained by *Simmons* (15). Each curve is for fixed $\varkappa$ as follows: 1: $\varkappa = 0.2$ sec^{-1}; 2: $\varkappa = 6.36$ sec^{-1}; 3: $\varkappa = 12.7$ sec^{-1}; 4: $\varkappa = 25.4$ sec^{-1}; 5: $\varkappa = 50.9$ sec^{-1}; 6: $\varkappa = 102$ sec^{-1}; 7: $\varkappa = 204$ sec^{-1}; 8: $\varkappa = 407$ sec^{-1}; Results appear consistent with [10]

We conclude with one more rheological relation which has not been properly tested, but which has a great advantage over the others. This is

$$G^*_{in}[\varkappa, \omega] = G^*_{tr}[\varkappa, \omega] + \varkappa \frac{\partial G^*_{tr}}{\partial\varkappa}$$

or

$$G^*_{in}[\varkappa, \omega] = \frac{\partial}{\partial\varkappa} \varkappa G^*_{tr}[\varkappa, \omega]$$

or

$$\frac{G^*_{in}}{G^*_{tr}} = 1 + \frac{\partial \ln G^*_{tr}}{\partial \ln\varkappa}.$$

This was shown by *Bernstein* (19) to be a BKZ relation but not a general simple fluid relation. It holds for the dynamic modulus or for the real or imaginary parts respectively. Also it does not involve limits at either high or low frequency. A test of it awaits the gathering of data on both in-line and transverse oscillations for the same fluid under the same conditions. Such data do not yet exist in the literature.

The foregoing, then, are relations intended to test the BKZ theory. No specific form of the function $U[I_1, I_2, t - \tau]$ of [7] is assumed.

Where the data are good and reliable we feel that they support the theory. Difficulties seem to occur experimentally at low rather than high frequency and it is at high frequency, where good data are easier to get, that consistency of these relations with experimental data is best. We feel that because of the theory has enjoyed very significant success in predicting non-linear rheological behavior, any relations which follow from it deserve serious consideration. For in testing them one is bound to obtain further insight into the extent and limits of validity and applicability of the theory and certainly further insight into rheological behavior. We hope that this presentation will encourage such work.

Acknowledgement

This work was supported by the U.S. National Science Foundation under grant number GP 32321.

Summary

A number of rheological relations are discussed with emphasis on non-linear visco-elastic behaviour and in particular the BKZ elastic fluid theory. Comparison with experimental data is discussed where possible and some examples of such comparison are given.

Résumé

Un nombre de relations rhéologiques sont analysées en mettant l'emphase sur le comportement viscoélastique non-linéaire et en particulier la théorie des fluides élastiques BKZ. Chaque fois qu'il est possible, une comparaison avec des résultats experimentaux est faite et discutée et quelques exemples de ces comparaisons sont donnés.

References

1) *Coleman, B. D.* and *H. Markovitz*, Arch. Rat. Mech. Anal. **17**, 1 (1964).

2) *Bernstein, B., E. A. Kearsley*, and *L. J. Zapus*, Trans. Soc. Rheology **7**, 391 (1963).

3) *Bernstein, B.*, Acta Mechanica **2**, 329 (1966).

4) *Kaye, A.*, College of Aeronautics, Cranfield Note No. 134 (1962).

5) *Zapas, L. J.* and *T. Craft*, J. Res. Nat. Bur. Standards U.S.A. **69A**, 541 (1965).

6) *Zapas, L. J.*, J. Res. Nat. Bur. Standards U.S.A. **70A**, 541 (1965).

7) *Zapas, L. J.* and *J. C. Phillips*, J. Res. Nat. Bur. Standards U.S.A. **75A**, 33 (1971).

8) *Booij, H. C.*, Effect of Superimposed Steady Shear Flow on the Dynamic Properties of Polymeric Fluids. Doctoral Thesis, Leiden (1970).

9) *Bernstein, B.*, Int. J. Nonlinear Mech. **4**, 183 (1969).

10) *Bernstein, B.* and *R. L. Fosdick*, Rheol. Acta **9**, 186 (1970).

11) *Booij, H. C.*, Rheol. Acta **5**, 216 (1966).

12) *Tanner, R. I.* and *G. W. Williams*, Rheol. Acta **10**, 528 (1971).

13) *Bernstein, B.* and *R. R. Huilgol*, Trans. Soc. Rheology **15**, 731 (1972).

14) *Bernstein, B., R. R. Huilgol*, and *R. I. Tanner*, Int. J. Eng. Sci. **10**, 263 (1972).

15) *Simmons, J. M.*, Rheol. Acta **7**, 184 (1968).

16) *Osaki, K., M. Tamura, M. Kurata*, and *T. Kotaka*, J. Phys. Chem. **69**, 4183 (1965).

17) *Kataoka, T.* and *S. Ueda*, J. Polymer Sci. Part A 2, **7**, 475 (1969).

18) *Walters, K.* and *T. E. R. Jones*, Proc. of Fifth Int. Cong. Rheology, edited by *S. Onogi*, **4**, 337 (Univ. Park Press 1970).

19) *Bernstein, B.*, Rheol. Acta **11**, 210 (1972).

Author's address:

Dr. *B. Bernstein*
Dept. of Mathematics
Illinois Institute of Technology
Chicago, Illinois 60616 (U.S.A.)

Rheol. Acta 12, 299–310 (1973)

De l'Institut National Polytechnique de Toulouse (France)

Panorama des milieux continus entre solides et liquides

Par D. Bellet

Avec 3 figures et 2 tableaux

(Reçu p. p. le 27 octobre 1972)

Les comportements mécaniques idéalisés, tels que le comportement «solide parfaitement élastique» introduit par *Hooke* ou le comportement «liquide parfaitement visqueux» défini par *Newton*, reflètent parfois assez bien la réalité dans le cas de certains matériaux soumis à certaines sollicitations, pour que l'on puisse les accepter sans grand risque d'erreur. Mais en fait, les corps manifestent, à des degrés plus ou moins importants, toutes les formes de réponses, depuis les réponses élastiques jusqu'aux réponses visqueuses en passant par toutes les formes intermédiaires telles que inélastiques, viscoélastiques, non newtoniennes ou plastiques.

Situer le comportement rhéologique d'un matériau au milieu de ce large spectre, revient à mettre en évidence sont caractère dominant. On doit se souvenir à ce sujet que ce n'est pas le matériau lui-même qui peut être caractérisé par une appreciation rhéologique, mais seulement son comportement dans certaines conditions bien définies.

Dans cet exposé, nous employons la procédure la plus directe généralement adoptée en Rheologie Theorique: elle consiste à développer les relations de simple proportionnalité entre tenseurs de contraintes, de déformations et de vitesses de déformations, établies en élasticité linéaire instantanée ou en hydrodynamique classique, et à les remplacer par des lois linéaires plus complexes, mais dont la valeur représentative est plus large. Mais la validité, concernant la précision et l'envergure de ces lois linéaires de comportement, est encore trop limitée, ce qui conduit à parler des théories non linéaires; de la sorte, il est possible de caractériser des milieux continus variés et d'expliquer simplement la nature et l'origine des comportements viscoélastiques et plastiques en faisant jouer à la variable temps un rôle prépondérant.

L'idéal serait bien sûr, de parvenir à la construction de lois régissant le comportement des milieux à partir des lois microscopiques de la Physique du discontinu et en les intégrant suivant un processus stochastique approprié. Malheureusement, les satisfactions obtenues dans cette voie sont encore bien rares.

I. Les fluides visqueux

Avant d'aborder les considérations relatives aux corps viscoélastiques et plastiques, nous allons effectuer quelques rappels concernant les schémas qu'il est d'usage d'adopter pour représenter le comportement visqueux des fluides.

En présentant ce panorama des relations représentatives des fluides visqueux, nous nous déplaçons en fait dans le sens inverse du chemin qui a été parcouru historiquement. Nous partons des relations les plus générales, relativement complexes, pour aboutir en définitive aux cas particuliers les plus simples. Les cas particuliers de ces relations générales ne représentent pas des comportements rares, bien au contraire. D'autre part, ces lois simplifiées constituent des formes asymptotiques des relations complètes et forment de très bonnes approximations de ces dernières, si la plage des variations des sollicitations reste faible.

A. Hypothèses générales relatives aux fluides visqueux

Les hypothèses les moins restrictives que les résultats expérimentaux permettent d'attribuer aux liquides et aux gaz visqueux sont les suivantes:

1. Au repos, le tenseur des contraintes Σ est sphérique, sa partie déviatrice S est nulle; la *tension* $\vec{t}$ dans une direction $\vec{n}$ se réduit à la contrainte normale $\vec{\sigma}(\vec{\tau} = 0)$, qui est la même quel que soit $\vec{n}$. De plus, cette contrainte normale, appelée pression hydrostatique p_0, est toujours orientée suivant $-\vec{n}$.

2. Appelons P le tenseur des contraintes de cisaillement, tel que:

$$P = \sum + pI \qquad [1]$$

expression dans laquelle p représente la valeur de la pression au sein du fluide. Lorsque le milieu est en mouvement, les composantes de P sont des fonctions des composantes du tenseur vitesses de déformation $\dot{\mathscr{E}}$ et de la température et d'elles seules (avec $\dot{\mathscr{E}}$ = Sym grad V).

3. Les fluides purement visqueux sont supposés *homogènes*, ce qui entraîne que la loi de

comportement adoptée est supposée être la même en tout point du milieu.

B. Hypothèses supplémentaires de Stokes

Stokes d'une part, et *Reiner-Rivlin* d'autre part, émirent deux hypothèses supplémentaires, suivies généralement par les fluides purement visqueux et qui en fait ne restreignent que très peu le vaste domaine défini précédemment.

1. Le milieu est supposé *isotrope*, il n'admet donc pas de direction privilégiée.

2. Lorsque le fluide purement visqueux est en mouvement, les composantes de P peuvent être écrites sous la forme polynomiale suivante:

$$p_{ij} = \sigma_{ij} + p_0 \delta_{ij} = A_{ij,mn} \dot{\varepsilon}^{mn} + B_{ij,mn} \dot{\varepsilon}^{mr} \dot{\varepsilon}_r^n + C_{ij,mn} \dot{\varepsilon}^{mr} \dot{\varepsilon}_r^s \dot{\varepsilon}_s^n + \cdots \qquad [2]$$

Les tenseurs $A_{ij,mn}$, $B_{ij,mn}$, $C_{ij,mn}, \ldots$ sont des tenseurs du quatrième ordre, caractéristiques des propriétés du fluide et ne dépendant que de l'état thermodynamique du système.

Le tenseur des contraintes Σ est symétrique (loi fondamentale de la dynamique), P l'est donc également, ainsi que $\dot{\mathscr{E}}$. Les tenseurs propriétés $A, B, C, \ldots$ sont des tenseurs isotropes. Une expression telle que $A_{ij,mn} \dot{\varepsilon}^{mn}$ peut être écrite:

$$A_{ij,mn} \dot{\varepsilon}^{mn} = a_1 \dot{e}_{\mathrm{I}} \delta_{ij} + 2 a_2 \dot{\varepsilon}_{ij} \qquad [3]$$

où a_1 et a_2 sont des coefficients scalaires fonction uniquement de la température T.

$\dot{e}_{\mathrm{I}}$ est le premier invariant principal de $\dot{\mathscr{E}}$, $\dot{e}_{\mathrm{II}}$ le second et $\dot{e}_{\mathrm{III}}$ le troisième.

L'application de la démarche précédente à l'ensemble de l'éq. [2] entraîne l'écriture de P suivante:

$$P = f(\dot{e}_{\mathrm{I}}, \dot{e}_{\mathrm{II}}, \dot{e}_{\mathrm{III}}, T) I + 2 a_2 \dot{\mathscr{E}} + 2 b_2 \dot{\mathscr{E}}^2 + 2 c_2 \dot{\mathscr{E}}^3 + \cdots \qquad [4]$$

La fonction $f(\dot{e}_{\mathrm{I}}, \dot{e}_{\mathrm{II}}, \dot{e}_{\mathrm{III}}, T)$ est une fonction scalaire des invariants principaux de $\dot{\mathscr{E}}$: $(\dot{e}_{\mathrm{I}}, \dot{e}_{\mathrm{II}}, \dot{e}_{\mathrm{III}})$ et de T; a_2, b_2, c_2 sont des constantes physiques du corps ne dépendant que de la température T.

La formule [4] peut être simplifiée en utilisant le théorème de *Cayley-Hamilton*. La forme plus simple obtenue est:

$$P = \alpha_0 (\dot{e}_{\mathrm{I}}, \dot{e}_{\mathrm{II}}, \dot{e}_{\mathrm{III}}, T) I + \alpha_1 (\dot{e}_{\mathrm{I}}, \dot{e}_{\mathrm{II}}, \dot{e}_{\mathrm{III}}, T) \dot{\mathscr{E}} + \alpha_2 (\dot{e}_{\mathrm{I}}, \dot{e}_{\mathrm{II}}, \dot{e}_{\mathrm{III}}, T) \dot{\mathscr{E}}^2. \qquad [5]$$

Expression dans laquelle $\alpha_0, \alpha_1, \alpha_2$ sont des polynômes en $\dot{e}_{\mathrm{I}}, \dot{e}_{\mathrm{II}}, \dot{e}_{\mathrm{III}}$ dont les coefficients sont caractéristiques des propriétés rhéologiques du corps qui ne dépendent que le la température T. Rappelons que $\alpha_0 = 0$ si $\dot{\mathscr{E}} = 0$.

Il est intéressant de remarquer à ce niveau qu'une théorie tout à fait analogue faisant intervenir $\mathscr{E}$ (et non $\dot{\mathscr{E}}$) a été proposée par *Prager* pour décrire les réponses observées sur des milieux purement élastiques.

C. Simplification de la loi rhéologique dans le cas de fluides incompressibles

Dans ce cas: div $\vec{V} = 0$, donc $\dot{e}_{\mathrm{I}} = 0$ et $\dot{e}_{\mathrm{II}} = -\frac{1}{2}$ trace $[(\dot{\mathscr{E}})^2]$. La pression p étant alors définie à une constante près, rien n'empêche d'inclure dans $-pI$ le tenseur sphérique $\alpha_0 I$. D'où:

$$\sum + pI = P = \alpha_1 (0, \dot{e}_{\mathrm{II}}, \dot{e}_{\mathrm{III}}, T) \dot{\mathscr{E}} + \alpha_2 (0, \dot{e}_{\mathrm{II}}, \dot{e}_{\mathrm{III}}, T) \dot{\mathscr{E}}^2. \qquad [6]$$

Dans la relation [6] les fonctions α_1 et α_2 doivent satisfaire à l'inégalité de *Clausius-Duhem*, à savoir que la puissance de dissipation π_{diss} doit être positive ou nulle, soit:

$$\alpha_1 \geqslant \frac{3}{2} \frac{\dot{e}_{\mathrm{III}}}{\dot{e}_{\mathrm{II}}} \alpha_2. \qquad [7]$$

De façon plus particulière, lorsqu $\dot{e}_{\mathrm{III}}$ est nul (cas du cisaillement pur par exemple) cette inégalité implique $\alpha_1 \geqslant 0$.

D. Théorie linéaire des fluides visqueux: fluides newtoniens et non newtoniens

Dans de très nombreux cas relatifs aux fluides visqueux satisfaisant à toutes les hypothèses que nous venons d'énoncer, la loi quadratique de comportement peut être réduite à une forme linéaire telle que:

$$\sum + pI = P = \alpha_0 I + \alpha_1 \dot{\mathscr{E}}. \qquad [8]$$

Cette simplification est en effet très souvent justifiée grâce à l'expérience. De plus, les coefficients α_0 et α_1 sont dans la théorie générale des polynômes en $\dot{e}_{\mathrm{I}}$, $\dot{e}_{\mathrm{II}}$, $\dot{e}_{\mathrm{III}}$ dont les coefficients dépendent de T. Ils se réduisent très fréquemment, dans la théorie linéaire, à des fonctions d'un seul ou deux invariants, et même parfois à des constantes; [8] s'écrit alors:

$$P = \beta_0 I + \beta_1 \dot{\mathscr{E}}. \qquad [9]$$

Pour chaque exemple que nous allons indiquer, la fonction β_0 adoptée prend la forme très simple:

$$\beta_0 = \lambda_L(T) \dot{e}_{\mathrm{I}}. \qquad [10]$$

Diverses formes ont été avancées pour la fonction β_1 en fonction des résultats expérimentaux obtenus, des corps considérés et des sollicitations imposées. Nous en donnons quatre exemples proposés par différents auteurs:

Comportements non newtoniens:

Loi d'*Ostwald-De Waele – Nutting* dite Loi « Puissance »:

$$\beta_1 = 2\varkappa\,[2\dot{e}_\mathrm{I}^2 - 4\dot{e}_\mathrm{II}] = 2\varkappa\,[\mathrm{trace}\,(\dot{\mathscr{E}}^2)]. \qquad [11]$$

Cette loi permet de rendre compte soit directement, soit sous une forme légèrement modifiée d'au moins 80% des recherches publiées dans le domaine des fluides non newtoniens purement visqueux.

Modéle d'*Ellis de Haven:*

$$\beta_1 = \frac{2a}{1 + b\,[2\dot{e}_\mathrm{I}^2 - 4\dot{e}_\mathrm{II}]^{(n-1)/2}}$$

$$\beta_1 = \frac{2a}{1 + b\,[2\,\mathrm{trace}\,(\dot{\mathscr{E}}^2)]^{(n-1)/2}}. \qquad [12]$$

Modèle de *Sisko:*

$$\beta_1 = 2A + 2B\,[2\dot{e}_\mathrm{I}^2 - 4\dot{e}_\mathrm{II}]^{\frac{n-1}{2}}$$

$$\beta_1 = 2A + 2B\,[2\,\mathrm{trace}\,(\dot{\mathscr{E}})^2]^{\frac{n-1}{2}}. \qquad [13]$$

Modèle de *Sutterby:*

$$\beta_1 = 2\eta_0 \left[\frac{\mathrm{Args}\,h(D\sqrt{2\dot{e}_\mathrm{I}^2 - 4\dot{e}_\mathrm{II}})}{D\sqrt{2\dot{e}_\mathrm{I}^2 - 4\dot{e}_\mathrm{II}}}\right]^E. \qquad [14]$$

Comportement newtonien:

Le comportement newtonien ou parfaitement visqueux est le plus simple qui puisse être envisagé pour les liquides et les gaz visqueux. Son modèle rhéologique est bien sûr linéaire et satisfait aux relations [9] et [10], mais de plus la fonction β_1 prend ici une forme très simple puisque:

$$\beta_1 = 2\mu_L = \text{constante}. \qquad [15]$$

La relation [9] devient:

$$P = \sum + pI = \lambda_L \dot{e}_\mathrm{I} I + 2\mu_L \dot{\mathscr{E}}. \qquad [16]$$

Bien que cette relation constitue une simplification à l'extrême du cas le plus général des fluides purement visqueux, elle caractérise très bien l'attitude et les résultats obtenus avec des liquides très répandus dans la nature (l'eau par exemple).

Rappelons que pour un fluide newtonien incompressible la relation [12] peut être écrite:

$$\left.\begin{aligned} &\textstyle\sum = -pI + 2\mu_L \dot{\mathscr{E}} \\ \text{et}\quad &\dot{e}_\mathrm{I} = \mathrm{div}\,\vec{V} = 0 \end{aligned}\right\}.$$

Signalons également qu'un fluide newtonien est considéré comme parfait lorsque les coefficients λ_L et μ_L sont négligeables devant la pression. On peut alors écrire en mouvement comme au repos: $\sum = -pI$.

De tels fluides sont appelés fluides de *Pascal* et constituent bien entendu le schéma de plus idéalisé qui puisse être fait sur les fluides visqueux.

Remarque: Nous proposons sur le tableau 1 différentes formes adoptées pour $\tau_{12} = \tau_{21} = p_{12} = p_{21}$. Les fonctions proposées sont en général le résultat de lissages de points expérimentaux obtenus par divers auteurs sur des produits variés.

II. La viscoélasticité

Tous les milieux manifestent à un degré plus ou moins important tous les comportements rhéologiques et dans de très nombreux cas, ils ne peuvent être identifiés à des schémas idéalisés tels que ceux des solide-élastiques ou des fluides visqueux, car ils sont à la fois élastiques et visqueux et ceci à un degré équivalent. Ce mode de comportement est appelé *viscoélastique.*

La définition de ce comportement, proposée par *Oldroyd*, à la fois simple et claire, peut être énoncée de la façon suivante:

«Il existe une relation entre l'histoire des contraintes, l'histoire des déformations et la température de chaque élément de matière d'un corps à comportement viscoélastique.» Le concept fondamental nouveau, appelé *principe de determinisme*, introduit par cette définition est celui d'«*histoire*».

La forme des équations de comportement d'un matériau n'est pas entièrement arbitraire. En plus de l'homogénéité dimensionelle (Invariance par changement des unités), ces relations doivent satisfaire aux principes de la thermodynamique, à des conditions nécessaires de stabilité des équilibres et à certaines conditions d'invariance par changement de repère de référence: Invariance tensorielle, Respect des symétries de la matière, Principe d'objectivité.

A. Théories linéaires de la viscoélasticité

1. Relaxation et fluage

Les théories de la Relaxation et du Fluage constituent une application directe du principe

Tableau 1. Classification des modèles rhéologiques

Auteurs	Modèles proposés	Mise sous la forme $\frac{du}{dy}=h(\tau_{yx}-\tau_0)\times(\tau_{yx}-\tau_0)$		Mise sous la forme τ_{yx}	$-\tau_0=g\left(\frac{du}{dy}\right)\frac{du}{dy}$
CAS où $\tau_0=0$		Expression $\frac{du}{dy}$	Fluidité équivalente $h(\tau_{yx}-\tau_0)$	Expression $(\tau_{yx}-\tau_0)$	Viscosite équivalente $g\left(\frac{du}{dy}\right)$
Newton	$\tau_{yx}=\mu\frac{du}{dy}$	$\frac{du}{dy}=\frac{1}{\mu}\tau_{yx}$	$\frac{1}{\mu}=\text{cte}$	$\tau_{yx}=\mu\frac{du}{dy}$	$\mu=\text{cte}$
Ostwald-De Waele	$\tau_{yx}=K\left(\frac{du}{dy}\right)^n$	$\frac{du}{dy}=\left(\frac{1}{K}\right)^{1/n}(\tau_{yx})^{(1-n)/n}\tau_{yx}$	$\left(\frac{1}{K}\right)^{1/n}(\tau_{yx})^{(1-n)/n}$	$\tau_{yx}=K\left(\frac{du}{dy}\right)^{n-1}\left(\frac{du}{dy}\right)$	$K\left(\frac{du}{dy}\right)^{n-1}$
Ellis-De Haven	$\tau_{yx}=\frac{\mu_0}{1+C\tau_{yx}^n}\left(\frac{du}{dy}\right)$	$\frac{du}{dy}=\frac{1}{\mu_0}(1+C\tau_{yx}^n)\tau_{yx}$	$\frac{1}{\mu_0}(1+C\tau_{yx}^n)$		
Reiner-Philippoff	$\tau_{yx}=\left[\mu_\infty+\frac{\mu_0-\mu_\infty}{1+(\tau_{yx}/A)^2}\right]\left(\frac{du}{dy}\right)$	$\frac{du}{dy}=\left[\frac{1+(\tau_{yx}/A)^2}{\mu_\infty(\tau_{yx}/A)^2+\mu_0}\right]\tau_{yx}$	$\frac{1+(\tau_{yx}/A)^2}{\mu_0+\mu_\infty(\tau_{yx}/A)^2}$		
Sisko	$\tau_{yx}=A\left(\frac{du}{dy}\right)+B\left(\frac{du}{dy}\right)^n$			$\tau_{yx}=\left[A+B\left(\frac{du}{dy}\right)^{n-1}\right]\left(\frac{du}{dy}\right)$	$A+B\left(\frac{du}{dy}\right)^{n-1}$
Prandt-Eyring	$\tau_{yx}=A\sinh^{-1}\left[\frac{1}{B}\left(\frac{du}{dy}\right)\right]$			$\tau_{yx}=\left[\frac{A\sinh^{-1}((1/B)du/dy)}{du/dy}\right]\frac{du}{dy}$	$\frac{A\sinh^{-1}((1/B)du/dy)}{du/dy}$
Powell-Eyring	$\tau_{yx}=C\left(\frac{du}{dy}\right)+D\sinh^{-1}\left[\frac{1}{E}\frac{du}{dy}\right]$			$\tau_{yx}=\left[C+\frac{D\sinh^{-1}((1/E)du/dy)}{du/dy}\right]\left(\frac{du}{dy}\right)$	$C+\frac{D\sinh^{-1}((1/E)du/dy)}{du/dy}$
CAS où $\tau_0\neq 0$					
Bingham	$\tau_{yx}=\tau_0+\eta\frac{du}{dy}$	$\frac{du}{dy}=\frac{1}{\eta}(\tau_{yx}-\tau_0)$	$\frac{1}{\eta}=\text{cte}$	$\tau_{yx}-\tau_0=\eta\frac{du}{dy}$	$\eta=\text{cte}$
Herschel-Bulkley n° 1	$\tau_{xy}=\tau_0+\left[\eta'\left(\frac{du}{dy}\right)\right]^{1/m}$	$\frac{du}{dy}=\left(\frac{1}{\eta}\right)^m(\tau_{yx}-\tau_0)^{m-1}(\tau_{xy}-\tau_0)$	$\left(\frac{1}{\eta}\right)(\tau_{yx}-\tau_0)^{m-1}$	$\tau_{yx}-\tau_0=(\eta')^{1/m}\left(\frac{du}{dy}\right)^{(1-m)/m}\frac{du}{dy}$	$(\eta')^{1/m}\left(\frac{du}{dy}\right)^{(1-m)/m}$
Herschel-Bulkley n° 2	$\tau_{yx}=\tau_0+\frac{\eta_0}{1+C(\tau_{yx}-\tau_0)^n}\left(\frac{du}{dy}\right)$	$\frac{du}{dy}=\frac{1}{\eta_0}[1+C(\tau_{yx}-\tau_0)^n](\tau_{yx}-\tau_0)$	$\frac{1}{\eta_0}[1+C(\tau_{yx}-\tau_0)^n]$		

de superposition de *Boltzmann*. Nous supposerons que *le milieu étudié est homogène et isotrope.*

a) Considérons tout d'abord E et S les parties déviatrices de $\mathscr{E}$ et de Σ

Dans le cas du phénomène de Relaxation, on écrit dans un système de coordonnées convectives:

$$\Delta S_c\,(\vec{\xi}, t) = 2\psi(t - t_1)\,\Delta E_c(\vec{\xi}, t_1). \qquad [15]$$

La cause agissant en un point ξ à l'instant t_1 est représentée ici par le tenseur de déformation ΔE_c. L'effet résultant de cette sollicitation est un ensemble de contraintes $\Delta S_c(\vec{\xi}, t)$ à un instant t ultérieur à t_1. La fonction ψ joue le rôle d'un module élasticité, qui est ici une fonction du temps. La forme de cette fonction dite «*fonction de relaxation*» toujours continue et décroissante, dépend du matériau étudié et de la température T à laquelle il est porté.

Si les déformations sont linéairement superposées, le principe de *Boltzmann* permet d'écrire en repère convectif:

$$S_c(\vec{\xi}, t) = 2\int_{-\infty}^{t} \psi(t - t')\frac{DE_c}{Dt'}(\vec{\xi}, t')\cdot dt'. \qquad [16]$$

$\frac{DE_c}{Dt'}$ représente la dérivée convective de E_c par rapport au temps.

b) Voyons maintenant le cas des parties sphériques de $\mathscr{E}$ et Σ

Il semblerait logique de faire intervenir pour des composantes une deuxième fonction de relaxation $\Psi(t)$. Cependant, les expériences montrent que si l'on se trouve dans un cas de contraintes et déformations purement sphériques, quelles que soient la température et les sollicitations, le comportement rhéologique du matériau est caractérisé par une relation de la forme:

$$s_1 = K_s[e_1 - \alpha\cdot(T - T_0)] + K_L\dot{e}_1 - 3p \qquad [17]$$

dans laquelle on a inclu le terme d'expansion thermique $\alpha\cdot(T - T_0)$, la pression hydrostatique p, la contribution élastique: $K_s = 3\lambda_s + 2\mu_s$ (K_s module de rigidité à la compression) et la contribution visqueuse: $K_L = 3\lambda_L + 2\mu_L$ (K_L module de viscosité à la compression).

Il est souvent utile, dans les théories linéaires de la viscoélasticité, de considérer les déformations comme des conséquences de l'application des contraintes. Les déformations E constituent alors les effets provoqués par les causes S. La relation déduite du principe de *Boltzmann* a alors la forme:

$$E_c(\vec{\xi}, t) = \frac{1}{2}\int_{-\infty}^{t} \varphi(t - t')\frac{D}{Dt'}S_c(\vec{\xi}, t')\,dt'. \qquad [18]$$

Le tenseur E pris pour référence est celui relatif à $t \to \infty$. $\varphi(t)$ est appelée «*fonction Fluage*» du matériau porté à la température T. Elle possède des propriétés analogues à celles de $\psi(t)$.

Il est possible de lier les fonctions fluage φ et relaxation ψ relatives à un même corps et à une même température T. Supposons que l'on adopte pour configurations de référence les configurations prises à l'instant $t = 0$ (et non pas $t \to -\infty$). Les transformées de *Laplace* de [16] et [18] sont respectivement:

$$\bar{S}_c(\vec{\xi}, p) = 2p\,\bar{\psi}(p)\cdot\bar{E}_c(\vec{\xi}, p), \qquad [19]$$

$$\bar{E}_c(\vec{\xi}, p) = \tfrac{1}{2}\,p\,\bar{\varphi}(p)\,\bar{S}_c(\vec{\xi}, p). \qquad [20]$$

On déduit de [19] et [20]:

$$\bar{\varphi}(p)\cdot\bar{\psi}(p) = \frac{1}{p^2}. \qquad [21]$$

2. Equations rhéologiques de comportement en viscoélasticité linéaire

La forme générale des équations rhéologiques, liant le tenseur déviateur des contraintes S au tenseur déviateur des déformations E pour les milieux à comportement viscoélastique linéaire, est:

$$Q[S\vec{x}, t)] = R[E(\vec{x}, t)] \qquad [22]$$

expression dans laquelle Q et R sont des opérateurs différentiels linéaires tels que:

$$Q = l_0 + l_1\frac{\delta}{\delta t} + l_2\frac{\delta^2}{\delta t^2} + \cdots + l_N\frac{\delta^N}{\delta t^N}, \qquad [23]$$

$$R = 2\left[m_0 + m_1\frac{\delta}{\delta t} + m_2\frac{\delta^2}{\delta t^2} + \cdots + m_P\frac{\delta^P}{\delta t^P}\right]. \qquad [24]$$

$\frac{\delta}{\delta t}$: représente l'opérateur de dérivation convective par rapport au temps; $l_0, l_1, \ldots, l_N, m_0, m_i, \ldots, m_P$ sont des coefficients ou modules viscoélastiques qui constituent des propriétés caractéristiques du matériau. Ces paramètres doivent être déterminés par voie expérimentale et dépendent de la température.

Quatre cas particuliers de la relation [22] sont très fréquemment utilisés en pratique pour représenter un comportement viscoélastique linéaire:

a) Le modèle rhéologique de *Maxwell*:

$$\frac{1}{2\mu}\dot{s}_{ij} + \frac{1}{2\eta}s_{ij} = \dot{e}_{ij}. \qquad [25]$$

b) Le modèle de *Kelvin-Voigt*:

$$s_{ij} = 2\eta\dot{e}_{ij} + 2\mu e_{ij}. \qquad [26]$$

c) Le modèle d'*Oldroyd*:

$$s_{ij} + \chi\dot{s}_{ij} = 2\eta\dot{e}_{ij} + 2\mu\ddot{e}_{ij}, \qquad [27]$$

d) Le modèle standard de *Zener*:

$$\frac{1}{2\eta}s_{ij} + \frac{1}{2\mu}\dot{s}_{ij} = \dot{e}_{ij} + \xi e_{ij}. \qquad [28]$$

Tous les coefficients intervenant dans ces relations dépendent de la température T.

Le comportement rhéologique complet est représenté en ajoutant aux relations précédentes la relation [17] liant les parties sphériques des tenseurs contraintes et déformations.

Ainsi, en faisant intervenir [17] dans [22] on obtient:

$$Q\left[\Sigma - \frac{s_I}{3}I\right] = R\left[\mathscr{E} - \frac{e_I}{3}I\right]. \qquad [29]$$

3. Relations liant équations de comportement et fonctions influence:

Il est possible de lier simplement l'équation de comportement aux relations faisant intervenir les fonctions influence ψ et φ. En repère convectif:

$$Q[S_c(\vec{\xi}, t)] = R[E_c(\vec{\xi}, t)]. \qquad [30]$$

Appliquons à [30] la transformation de *Laplace*, il vient:

$$\bar{S}_c(\vec{\xi}, p) = 2\frac{m_0 + m_1 p + m_2 p^2 + \cdots + m_P p^P}{l_0 + l_1 p + l_2 p^2 + \cdots + l_N p^N}\bar{E}_c \times (\vec{\xi}, p). \qquad [31]$$

Or:

$$\bar{S}_c(\vec{\xi}, p) = 2p\bar{\psi}(p)\,\bar{E}_c(\vec{\xi}, p) = 2\bar{\psi}(p)\cdot\bar{\dot{E}}_c(\vec{\xi}, p). \qquad [32]$$

On en déduit:

$$\psi(t) = \mathscr{L}^{-1}\left\{\frac{1}{p}\cdot\frac{m_0 + m_1 p + m_2 p^2 + \cdots + m_P p^P}{l_0 + l_1 p + l_2 p^2 + \cdots + l_N p^N}\right\}. \qquad [33]$$

$$\varphi(t) = \mathscr{L}^{-1}\left\{\frac{1}{p}\cdot\frac{l_0 + l_1 p + l_2 p^2 + \cdots + l_N p^N}{m_0 + m_1 p + m_2 p^2 + \cdots + m_P p^P}\right\}. \qquad [34]$$

4. Analogies mécaniques – Généralisation des modèles simples

On a pris l'habitude de comparer le comportement de chaque matériau à celui d'un assemblage plus ou moins compliqué d'éléments mécaniques tels que ressorts, amortisseurs... La connaissance d'un tel modèle analogique, conforme au comportement de la matière, conduit au même titre, mais mieux encore que l'équation rhéologique, à séparer nettement la part de la déformation ou de la contrainte qui revient à l'élasticité, à la viscosité ou à la viscoélasticité. Ce procédé facilite beaucoup la détermination des paramètres à adopter dans les équations de comportement à partir des résultats expérimentaux. Il convient cependant de considérer ces modèles comme un mode d'écriture idéographique plus condensé et plus caractéristique que les équations rhéologiques. Malheureusement, l'analogie mécanique ne peut être faite que pour une équation de comportement écrite dans un cas monodimensionnel.

4.1. Modèles analogiques simples:

L'analogie mécanique d'un solide de *Hooke* est constituée par un ressort et celle d'un fluide de *Newton* par un amortisseur et l'analogie mécanique des modèles viscoélastiques simples sera obtenue à l'aide d'un montage série ou d'un montage en parallèle d'un ressort et d'un amortisseur:

– Dans le cas d'un modèle de *Maxwell*, ressort et amortisseur sont disposés en série (tableau 2).

– Dans le cas d'un modèle de *Kelvin-Voigt*, ressort et amortisseur sont placés en parallèle (tableau 2).

Le tableau 2 rend compte d'un certain nombre d'autres modèles rhéologiques (et mécaniques) très souvent utilisés, pour représenter le comportement des corps viscoélastiques linéaires.

4.2. Généralisation des modèles simples

Lorsque l'on étudie le comportement des materiaux réels, on constate que le fait de leur attribuer un modèle du type *Maxwell*, ou *Kelvin*, ou autre, ne permet de représenter que de façon approchée les résultats obtenus expérimentalement. De façon à corriger cette imprécision, tout en conservant des modèles linéaires, on a l'habitude d'introduire des modèles dits «généralises» pour représenter le comportement des fluides

Tableau 2.

Nbre	Solides			Plastiques	Liquides		
	Modele	Fonction Fluage	Fonction Relaxation	Modele	Modele	Fonction Fluage	Fonction Relaxation
1	Hooke : H $2\mu_s$	$1/\mu_s$	μ_s	Saint-Venant StV	Newton : N N $2\eta_L$	t/η_L	$\eta_L \delta(t)$
2	Kelvin-Voigt : KV KV=H//N $2\mu_s$ $2\eta_s$	$\frac{1}{\mu_s}\left[1-\exp(-t\mu_s/\eta_s)\right]$	$\mu_s\left[1+\frac{\eta_s}{\mu_s}\delta(t)\right]$	Plastoelastique PE PE=StV-H	Maxwell : M M=N-H $2\mu_L$ $2\eta_L$	$\frac{1}{\mu_L}\left(1+\frac{\mu_L}{\eta_L}t\right)$	$\mu_L \exp\left(-\frac{\mu_L}{\eta_L}t\right)$
3	Zener : Z Z=H//M $2\mu_{s_1}$, η_{s_1} $2\mu_{s_2}$	—	$\mu_{s_1}+\mu_{s_2}\exp\left(-t\mu_{s_2}/\eta_{s_2}\right)$	Bingham B B=(N//StV)-H	Lethersich : L L=N-KV $2\mu_{L_2}$ $2\eta_{L_1}$ $2\eta_{L_2}$	$\frac{t}{\eta_{L_1}}+\frac{1}{\mu_{L_2}}\left[1-\exp\left(-\frac{t\mu_{L_2}}{\eta_{L_2}}\right)\right]$	—
3	Poynting-Thomson PTh=H-KV $2\mu_{s_2}$ $2\mu_{s_1}$ $2\eta_{s_2}$	$\frac{1}{\mu_{s_1}}+\frac{1}{\mu_{s_2}}\left[1-\exp\left(-t\mu_{s_2}/\eta_{s_2}\right)\right]$	—	Viscoplastique : VP VP=(H//StV)-N	Jeffreys : J J=N//M $2\eta_{L_1}$ $2\mu_{L_2}$ $2\eta_{L_2}$	—	$\eta_{L_1}\delta(t)+\mu_{L_2}e^{-\frac{t\mu_{L_2}}{\eta_{L_2}}}$
4	Solide A 4 Parametres S=KV-KV $2\mu_{s_1}$ $2\mu_{s_2}$ $2\eta_{s_1}$ $2\eta_{s_2}$	$\frac{1}{\mu_{s_1}}\left(1-e^{-\frac{t\mu_{s_1}}{\eta_{s_1}}}\right)+\frac{1}{\mu_{s_2}}\left(1-e^{-\frac{t\mu_{s_2}}{\eta_{s_2}}}\right)$	—	Schwedoff : Schw Schw=(M//StV)-H	Burgers : BU BU=M-KV-Z $2\mu_{L_1}$ $2\mu_{L_2}$ $2\eta_{L_1}$ $2\eta_{L_2}$	$\frac{1}{\mu_{L_1}}\left(1+\frac{t\mu_{L_1}}{\eta_{L_1}}\right)+\frac{1}{\mu_{L_2}}\left(1+\frac{t\mu_{L_2}}{\eta_{L_2}}\right)$	—
5				Schofield-Scottblair Sch Scb=(M//StV)-KV			

réels viscoélastiques linéaires. La représentation de ces généralisations peut être aisément faite à l'aide des modèles analogiques mécaniques. Elles consistent à connecter soit des éléments de *Kelvin-Voigt* en série, soit des éléments de *Maxwell* en parallèle. Ces généralisation sont justifiées car des éléments de *Kelvin-Voigt* en série ont des caractéristiques analogues à celles d'un élément de *Kelvin-Voigt* et de même des éléments de *Maxwell* en parallèle ont des caractéristiques analogues à celles d'un élément de *Maxwell*.

B. Théories non linéaires de la viscoélasticité

Les théories linéaires, d'un emploi très simple, ont le désavantage de ne pas avoir un domaine d'applicabilité très vaste: en particulier, les relations rhéologiques linéaires ne permettent pas une représentation correcte du comportement des corps soumis à des sollicitations importantes. Cette lacune est comblée grâce à l'introduction de théories non-linéaires. Nous nous proposons d'en présenter succintement quelques-unes des plus classiques.

1. Généralisation des théories linéaires: la théorie d'Oldroyd (1950)

Oldroyd a pensé à généraliser la relation [27] en la remplaçant par:

$$\left(1 + \varkappa \frac{\delta}{\delta t}\right) p_{ij} - 4\xi \dot{\varepsilon}_{ik} p_j^k = 2\left(\eta + \mu \frac{\delta}{\delta t}\right) \dot{\varepsilon}_{ij} - 8\eta\nu \dot{\varepsilon}_{ik} \dot{\varepsilon}_k^j . \qquad [35]$$

Il a récemment apporté une nouvelle amélioration à son modèle. Au lieu d'utiliser des dérivées convectives $\delta/\delta t$, il introduit, à la manière de *Prager*, des dérivées de *Jaumann* des sollicitations notées $\mathscr{D}/\mathscr{D}t$ (appelées également dérivées corotationnelles), forme de dérivation tout à fait en accord avec le principe d'invariance:

$$\left(1 + \varkappa \frac{\mathscr{D}}{\mathscr{D}t}\right) P - 4\xi P \dot{\mathscr{E}} = \left(2\eta + 2\mu \frac{\mathscr{D}}{\mathscr{D}t}\right) \dot{\mathscr{E}} - 8\eta\nu \dot{\mathscr{E}}^2 . \qquad [36]$$

2. Théorie du second ordre de Rivlin-Ericksen (1955)

Considérons maintenant la relation linéaire générale [22].

Tout d'abord, généralisons-la et remplaçons les déviateurs de $\mathscr{E}$ et Σ par $\mathscr{E}$ et P eux-mêmes. Dans certains cas, il est possible d'expliciter Σ en fonction de $\dot{\mathscr{E}}$ et de ses dérivées successives.

a) Théorie des fluides simples de Noll (1955)

$$\left[l_0 + l_1 \frac{\delta}{\delta t} + l_2 \frac{\delta^2}{\delta t^2} + \cdots + l_N \frac{\delta^N}{\delta t^N}\right] P$$
$$= 2\left[m_0 + m_1 \frac{\delta}{\delta t} + \cdots + m_P \frac{\delta^P}{\delta t^P}\right]. \qquad [37]$$

La théorie de *Noll* suppose que l'on peut écrire $m_0 = 0$ et:

$$P = 2\left[A_1 + A_2 \frac{\delta}{\delta t} + A_3 \frac{\delta^2}{\delta t^2} + \cdots\right] \dot{\mathcal{E}}. \qquad [38]$$

$A_1, A_2, A_3, \ldots$ sont des constantes s'exprimant en fonction des l_i et m_j

$$\begin{cases} i = 0, 1, \ldots, N \\ j = 0, 1, \ldots, P \end{cases}$$

b) Généralisation de Rivlin-Ericksen

Rivlin et *Ericksen* ajoutèrent les hypothèses suivantes:

+ P peut être exprimé explicitement comme une fonction polynomiale de $\dot{\mathcal{E}}$ et de ses $(N-1)$ premières dérivées convectives.

+ Le matériau est supposé isotrope et Σ est nul lorsque l'écoulement considéré est tel que:

$$\dot{\mathcal{E}} = \frac{\delta \dot{\mathcal{E}}}{\delta t} = \frac{\delta^2 \dot{\mathcal{E}}}{\delta t^2} = \cdots = 0. \qquad [39]$$

Si on se place dans le cas où l'écoulement satisfait à:

$$\frac{\delta^2 \dot{\mathcal{E}}}{\delta t^2} = \frac{\delta^3 \dot{\mathcal{E}}}{\delta t^3} = \cdots = 0. \qquad [40]$$

Il vient:

$$P = \alpha_0 I + 2\alpha_1 \dot{\mathcal{E}} + 4\alpha_2 \dot{\mathcal{E}}^2 + 2\alpha_3 \ddot{\mathcal{E}}$$
$$+ 4\alpha_4 \ddot{\mathcal{E}}^2 + 8\alpha_5 \dot{\mathcal{E}}\ddot{\mathcal{E}} + 16\alpha_6 \dot{\mathcal{E}}^2 \ddot{\mathcal{E}}$$
$$+ 16\alpha_7 \dot{\mathcal{E}}\ddot{\mathcal{E}}^2 + 32\alpha_8 \dot{\mathcal{E}}^2 \ddot{\mathcal{E}}^2 \qquad [41]$$

expression dans laquelle les coefficients α_0, $\alpha_1, \ldots, \alpha_8$ ne sont pas des constantes mais des fonctions polynomiales des 10 invariants suivants: trace$(\dot{\mathcal{E}})$, tr$(\dot{\mathcal{E}}^2)$, trace$(\dot{\mathcal{E}}^3)$, tr$(\ddot{\mathcal{E}})$, tr$(\ddot{\mathcal{E}}^2)$, tr$(\ddot{\mathcal{E}}^3)$, tr$(\dot{\mathcal{E}}\ddot{\mathcal{E}})$, tr$(\dot{\mathcal{E}}^2\ddot{\mathcal{E}})$, tr$(\dot{\mathcal{E}}\ddot{\mathcal{E}}^2)$, tr$(\dot{\mathcal{E}}^2\ddot{\mathcal{E}}^2)$.

Les travaux de *Noll* et de *Rivlin-Ericksen* ont eu une influence décisive sur les recherches dans le domaine de la Mécanique des Fluides Viscoélastiques. En plus de sa justification physique, l'équation [41] se prête particulièrement bien aux expériences de Mécanique des Fluides Viscoélastiques du type «héréditaire».

3. Autres théories non linéaires de la viscoélasticité

Voici très rapidement énumérées quelques théories très récentes de la viscoélasticité:

En 1958, *Noll* a proposé une théorie non linéaire de la viscoélasticité, généralisation des théories linéaires car elle met en œuvre un principe de superposition: le comportement rhéologique actuel d'un fluide au voisinage d'un point matériel est une fonction (non linéaire) de son histoire et en particulier une fonction des gradients de déformation passées en ce point matériel. Cette théorie matérialise le concept de l'influence héréditaire implicitement synonyme du caractère viscoélastique.

Coleman et *Noll* ont toutefois remarqué que cette théorie conduit aux mêmes informations que la théorie de *Rivilin-Ericksen* dans le cas d'un écoulement laminaire avec cisaillement pur.

Des théories semblables à celle de *Noll* ont été présentées vers 1960 par divers auteurs tels que *Bergen, Green* et *Rivlin*. Mais, à la même époque, *Ericksen* a effectué une approche tout à fait différente du problème de la viscoélasticité qui consiste à considérer les fluides étudiés comme des milieux continus anisotropes.

III. La plasticité

Le temps ne joue plus un rôle important en plasticité et il n'intervient pas dans les versions idéalisées des processus plastiques. D'autre part, un comportement viscoélastique est caractérisé par le fait que viscosité et élasticité ont lieu simultanément en un même point au sein d'un matériau, tandis que dans le cas d'un comportement plastique, ces deux phénomènes se manifestent successivement suivant le taux de sollicitation imposé au corps ou bien simultanément mais dans ce cas en des points distincts qui ne sont pas soumis aux mêmes sollicitations.

A. Etude phénoménologique de l'écoulement plastique – Ecrouissage

Considérons une barre initialement au repos dans un état non contraint qui, à partir de $t_0 = 0$, est soumise à une sollicitation simple augmentant régulièrement avec le temps (fig. 1).

× Si, en un point T, situé avant Y, on stoppe la mise en charge et si on décharge la barre, il y a parcours de $T0$ en sens inverse et retour à l'état initial lorsque $\sigma_1 = 0 (\varepsilon_1 = 0)$.

× Si, lors de la mise en charge, on dépasse le point Y appelé «limite élastique» (relativement à $0Y$) ou «seuil d'écoulement» (relativement à

YR) la pente de la courbe rhéologique change: il y a *Ecrouissage* du matériau.

× Si la charge atteint une valeur importante, il peut y avoir rupture (*R*) du matériau. Elle est en général précédée d'une striction de la barre qui accélère le processus.

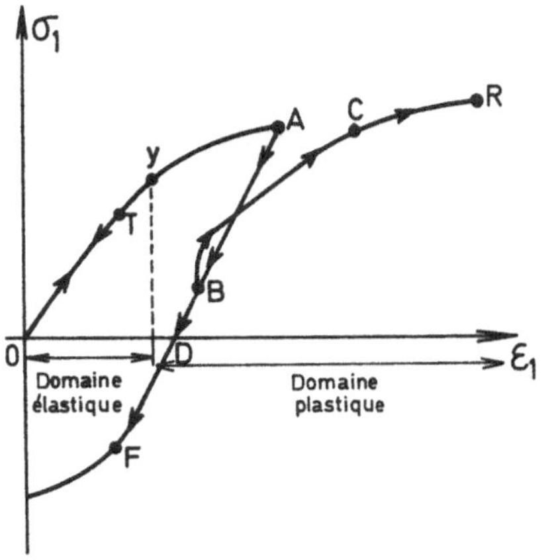

Fig. 1

A partir du point *Y*, le matériau est le siège d'un comportement plastique. Les déformations sont irréversibles.

On remarque expérimentalement que la déformation plastique en traction abaisse la limite élastique en compression: ce phénomène est appelé Effet *Bauschinger*.

B. Aspect théorique de la plasticité: théorie idéalisée; matériau écroui

1. Aspect unidimensionnel

L'idéalisation du phénomène précédent consiste à postuler que les pentes de la droite de charge initiale 0 *Y* et de la droite de décharge *AB* ont la même valeur. La déformation au point *A* peut être considérèe comme formé de deux composantes distinctes:
- la déformation élastique ε_1^{El}
- la déformation plastique ε_1^{Pl}

telles que:

$$\varepsilon_1 = \varepsilon_1^{Pl} + \varepsilon_1^{El},$$

$$\varepsilon_1^{El} = \frac{\sigma_1}{E}. \qquad [42]$$

Conditions d'irréversibilité: Les sens des déplacements différents sur *XZ* et *X'Z'* (fig. 2 et 3) sont imposés par une condition d'irréversibilité: lorsque l'on a une variation de ε_1^{El}, on postule que la déformation plastique élémentaire doit être telle que:

$$\sigma_1 \cdot d\varepsilon_1^{Pl} > 0 \qquad \forall\, d\varepsilon_1^{Pl} \neq 0. \qquad [43]$$

Cette inégalité découle de considérations thermodynamiques: la création d'une déformation plastique nouvelle (processus irréversible) doit être accompagnée d'une génération d'Entropie.

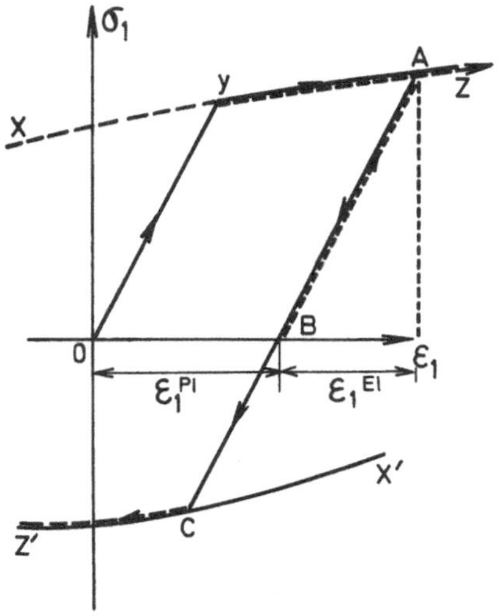

Fig. 2

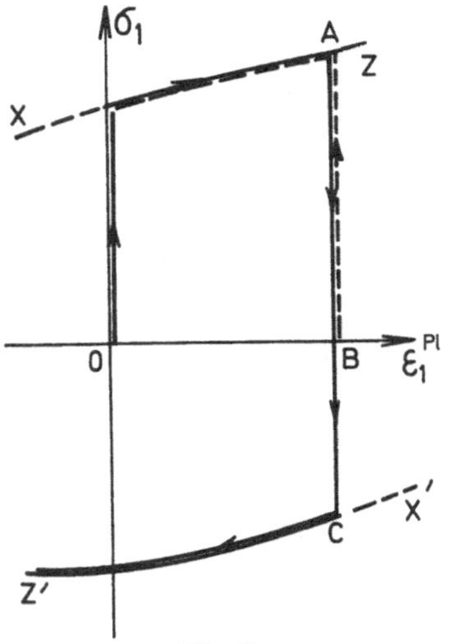

Fig. 3

Condition de stabilité: Sur la fig. 3, on peut remarquer que les courbes *XZ* et *X'Z'*, symétriques par rapport à 0 sont représentatives de fonctions croissantes de $\varepsilon_{1_{Pl}}$. Ceci vient du fait qu'un changement de $d\sigma_1$ de σ_1 et de $d\varepsilon_1^{Pl}$ de ε_1^{Pl} doit être tel que le point figuratif reste sur l'une des courbes *XZ* ou *X'Z'* et que l'on ait:

$$d\sigma_1 \cdot d\varepsilon_1^{Pl} > 0 \qquad \forall\, d\varepsilon_1^{Pl} \neq 0. \qquad [44]$$

2. Généralisation à l'étude tridimensionnelle

Ce qui a été dit précédemment peut être généralisé aux cas faisant intervenir la comportement des matériaux dans un espace d'étude à trois dimensions. Seules les relations liant les parties déviatrices des tenseurs $\sum(\vec{x}, t)$ et $\mathscr{E}(\vec{x}, t)$ peuvent mettre en évident le comportement plastique d'un matériau; la température $T(\vec{x}, t)$ intervient naturellement et d'elle dépend en particulier la forme de la condition limite.

Par analogie avec le cas unidimensionnel, posons:

$$E = E^{\text{El}} + E^{\text{Pl}}. \qquad [45]$$

La relation entre $S(\vec{x}, t)$ et E^{El}, dû au caractère élastique du matériau, s'écrit:

$$S = 2\mu E^{\text{El}}. \qquad [46]$$

Quant à la partie plastique E^{Pl} de la déformation, nous supposons qu'elle ne peut varier que si au début et durant le temps pendant lequel a lieu cette variation $S(\vec{x}, t)$, $E^{\text{Pl}}(\vec{x}, t)$ et $T(\vec{x}, t)$ sont liés par une relation tensorielle du type:

$$f(S, E^{\text{Pl}}, T) = 0. \qquad [47]$$

La fonction f est appelée «fonction limite» du matériau ou «*critère d'écoulement*». Elle est supposée continue, différentiable et avoir des dérivées partielles continues.

Quelles que soient les sollicitations imposées au corps étudié, on a obligatoirement:

$$f(S, E^{\text{Pl}}, T) = 0. \qquad [48]$$

Compte tenu des hypothèses précédemment citées, une variation de E^{Pl} avec t n'est possible que si le point figuratif demeure sur la surface limite définie par $f = 0$. *L'équation de Consistance* (ainsi nommée par *Prager*), traduisant cette condition, s'écrit:

$$\dot{f} = \frac{\partial f}{\partial s_{ij}} \dot{s}_{ij} + \frac{\partial f}{\partial T} \dot{T} + \frac{\partial f}{\partial e_{ij}^{\text{Pl}}} \dot{e}_{ij}^{\text{Pl}} = 0 \qquad [50]$$

La «fonction tensorielle limite» f doit satisfaire dans le cas isotherme aux relations [43] et [44]. On doit avoir dans tous les cas:

$$\dot{e}_{ij}^{\text{Pl}} = 0 \quad \text{si:} \begin{cases} f < 0 \\ \text{ou} \\ f = 0 \quad \text{et} \quad \dfrac{\partial f}{\partial s_{ij}} \dot{s}_{ij} + \dfrac{\partial f}{\partial T} \dot{T} < 0 \end{cases} \qquad [51]$$

$$\dot{e}_{ij}^{\text{Pl}} = \frac{1}{D} \frac{\partial f}{\partial s^{ij}} \left[\frac{\partial f}{\partial s_{kl}} \dot{s}_{kl} + \frac{\partial f}{\partial T} \dot{T} \right]$$

$$\text{si:} \quad f = 0 \quad \text{et} \quad \frac{\partial f}{\partial s_{ij}} \dot{s}_{ij} + \frac{\partial f}{\partial T} \dot{T} \geqslant 0. \qquad [52]$$

Relation dans laquelle:

$$D = -\frac{\partial f}{\partial s^{ij}} \cdot \frac{\partial f}{\partial e_{ij}^{\text{Pl}}}. \qquad [53]$$

A ces relations, il faut ajouter les relations élastiques liant S et E^{El} et les relations liant s_{I}, e_{I} et T.

C. Théorie idéalisée de la plasticité: matériau parfaitement plastique

De façon à pouvoir poursuivre une étude théorique systématique, nous allons atteindre une nouvelle étape dans l'idéalisation: nous ne considérerons que des corps parfaitement plastiques, c'est à dire tels que la «condition limite» $f = 0$ ne dépend pas de E^{Pl}, mais seulement de S et de T.

Il est intéressant de considérer alors la «fonction limite» f comme la limite de $f(S, cE^{\text{Pl}}, T)$ lorsque $c \to 0$.

Soit: $f_0(S, T) = f(S, \{0\}, T)$

et

$$\lambda = \left\{ \frac{[\partial f / \partial b_{kl}]\, \dot{e}_{kl}^{\text{Pl}}}{[\partial f / \partial s_{mn}]\, [\partial f / \partial b^{mn}]} \right\}_{c=0}. \qquad [53\text{a}]$$

$$\dot{e}_{ij}^{\text{Pl}} = 0 \qquad \text{si:} \begin{cases} f_0 < 0 \\ \text{ou} \\ f_0 = 0 \end{cases}$$

$$\text{et}$$

$$\frac{\partial f_0}{\partial s_{ij}} \dot{s}_{ij} + \frac{\partial f_0}{\partial T} \dot{T} < 0. \qquad [54]$$

$$\dot{e}_{ij}^{\text{Pl}} = \lambda \frac{\partial f_0}{\partial s^{ij}} \qquad \text{si:} \quad f_0 = 0$$

$$\text{et}$$

$$\frac{\partial f_0}{\partial s_{ij}} \dot{s}_{ij} + \frac{\partial f_0}{\partial T} \dot{T} = 0. \qquad [55]$$

D. Premier exemple de condition limite: le critère de von Mises

Ce critère consiste à supposer que la fonction limite (que nous rappellerons f dorénavant) prend la forme simple:

$$f = \tfrac{1}{2} s_{ij} s^{ij} - [k(T)]^2. \qquad [56]$$

Il est relatif aux matériaux isotropes dont le comportement est supposé parfaitement plastique. Le fonction $k(T)$ représente la valeur de la contrainte limite d'un matériau soumis à un cisaillement pur à T.

Dans le cas où f vérifie [56], les conditions [54] et [55] deviennent:

$$\dot{e}_{ij}^{\mathrm{Pl}} = 0 \quad \text{si:} \quad \begin{cases} f < 0 \quad \text{soit: } s_{ij} s^{ij} < k^2(T) \\ \quad\text{ou} \\ f = 0\text{: } s_{ij} s^{ij} = 2k^2(T) \\ \quad\text{et} \\ \quad s_{ij} \dot{s}^{ij} - 2kk'\dot{T} < 0, \end{cases} \qquad [57]$$

$$\dot{e}_{ij}^{\mathrm{Pl}} = \lambda s_{ij} \quad \text{si:} \quad s_{ij} s^{ij} = 2k^2(T) \quad \text{et} \quad s_{ij} \dot{s}^{ij} - 2kk'\dot{T} = 0. \qquad [58]$$

$\lambda > 0$

Valeur de λ compte tenu de [56]*:* en utilisant [58] on peut écrire:

$$\dot{e}_{ij} \dot{e}^{ij} = \lambda^2 s_{ij} s^{ij} = 2\lambda^2 k^2(T)$$

d'où:

$$\lambda = \frac{1}{k(T)} \sqrt{\frac{\dot{e}_{ij}^{\mathrm{Pl}} \dot{e}^{ij\mathrm{Pl}}}{2}}. \qquad [59]$$

Remarquons que le critère de *von Mises* représente parfaitement le comportement des métaux ductiles en particulier dans le cas où les essais sont isothermes [$k(T) = $ cte].

E. Deuxième exemple de fonction limite: le critère de Tresca

La condition limite est constituée de n fonctions limites indépendantes:

$$f_\alpha(S, T) \qquad \alpha = 1, 2, \ldots, n. \qquad [60]$$

Un ensemble de valeurs s_{ij}, et T, caractérise un état plastique au sens de *Tresca* si une (au moins) ou plusieurs fonctions $f(s_{ij}, T)$ appartenant à l'ensemble [60] prennent une valeur nulle tandis que toutes les autres prennent des valeurs négatives.

Le tenseur vitesse de déformation plastique $\dot{E}^{\mathrm{Pl}}$ est la somme des n composantes $\dot{E}_\alpha^{\mathrm{Pl}}$, soit:

$$\dot{E}^{\mathrm{Pl}} = \sum_{\alpha=1}^{n} \dot{E}_\alpha^{\mathrm{Pl}}. \qquad [61]$$

Chacune de ces n composantes vérifie les lois de l'écoulement [54] ou [55], c'est à dire:

$$\dot{E}_\alpha^{\mathrm{Pl}} = 0 \quad \text{si:} \quad \begin{cases} f_\alpha < 0 \\ \text{ou} \\ f_\alpha = 0 \end{cases} \quad \text{et} \quad \frac{\partial f}{\partial s_{ij}} \dot{s}_{ij} + \frac{\partial f}{\partial T} \dot{T} < 0, \qquad [62]$$

$$\dot{E}_\alpha^{\mathrm{Pl}} = \lambda_\alpha \frac{\partial f_\alpha}{\partial s_{ij}} \quad \text{si:} \quad f_\alpha = 0 \quad \text{et} \quad \frac{\partial f}{\partial s_{ij}} \dot{s}_{ij} + \frac{\partial f}{\partial T} \dot{T} \geqslant 0 \quad \text{avec } \lambda_\alpha > 0. \qquad [63]$$

Les fonctions limites étant indépendantes, il existe pour chacune d'elles, un processus expérimental pour lequel, seul, $\dot{E}_\alpha^{\mathrm{Pl}}$ est différent de zéro. La condition d'irréversibilité devant être vérifiée pour chacun de ces n processus, il s'ensuit que:

$$\lambda_\alpha > 0; \qquad \alpha = 1, 2, \ldots, n. \qquad [64]$$

Inversement, si les n conditions du type [64] sont vérifiées, il est clair que la condition d'irréversibilité est satisfaite quel que soit le comportement plastique envisagé pour lequel plusieurs quantités $\dot{E}_\alpha^{\mathrm{Pl}}$ sont différentes de zéro.

Conclusion

Le panorama que nous venons de brosser permet de ranger les milieux continus à l'aide de leur réponse et de leur comportement rhéologiques. Les cas d'étude présentés, bien que fondamentaux et d'un intérêt pratique considérable, n'en sont pas moins assez particuliers, ce qui rend cette classification quelque peu rigide. Rappelons que si l'on fait des mesures suffisamment précises et fines sur des temps très longs, aucun matériau ne manifeste un comportement idéal: les comportements rhéologiques réels ne sont en fait que le résultat de la superposition et de l'imbrication des comportements de base plus ou moins idéalisés, les uns prenant vis à vis des autres des places prépondérantes; ce phénomène est bien sûr une conséquence de la nature même des matériaux, mais également dépend intimement des sollicitations qu'il a subies. Les conclusions auxquelles on aboutit, bien que parfois complexes, restent malgré tout profondément subjectives. Mais, n'en est-il pas ainsi toutes les fois que l'on cherche à schématiser les processus Physiques et à modéliser grâce à la Mathématique les lois de la Nature?

Résumé

Les milieux continus, qui peuvent tous être considérés comme des fluides, occupent le vaste domaine s'étendant du solide indéformable d'*Euclide* au liquide incompressible et non visqueux de *Pascal* en passant par toutes les catégories intermédiaires de comportements: élastique, inélastique, viscoélastique, non-newtonien, newtonien, sans oublier l'état plastique.

Portant uniquement sur des corps isotropes, ce panorama est scindé en trois parties:

Etude des fluides purement visqueux

Les relations rhéologiques présentées ici vont des lois de comportement les plus complètes telles que celles de *Stokes* ou de *Reiner-Rivlin*, jusqu'aux modèles les plus simples de *Newton*. Une place importante est réservée aux **fluides *Ostwaldiens*.**

La Viscoélasticité

Les théories linéaires sont fondées sur l'emploi du principe de superposition de *Boltzmann* qui conduit à l'établissement des fonctions fluage et relaxation reliées de façon simple grâce à l'utilisation de la transformation de *Laplace*. L'emploi d'analogies mécaniques permet de déterminer commodément la contribution visqueuse ou élastique intervenant dans le modèle d'un corps soumis à **sollicitations**. Des analogies formelles sont également employées pour simplifier la résolution des problèmes de viscoélasticité.

Pour être suffisamment représentative, la viscoélasticité doit parfois faire intervenir des théories non linéaires de comportement qui sont présentées et analysées succintement. Elles aboutissent à la notion de fluides d'ordre *n*.

La Plasticité

Les corps plastiques sont le siège d'effets élastiques et visqueux successifs séparés par la limite élastique ou seuil d'écoulement. Les théories idéalisées considérées sont celles des matériaux écrouis et parfaitement plastiques. Les lois de l'écoulement sont établies puis illustrées grâce aux corps régis par le critère de *von Mises* et le critère de *Tresca*.

Dans chaque cas, des considérations relatives à l'aspect phénoménologique et les propriétés expérimentales descriptives précèdent et justifient l'établissement ou le rappel des lois rhéologiques de comportement adoptées, elles-mêmes déduites du développement des relations bien connues de l'Elasticité et de l'Hydrodynamique classique.

Littérature

Bland, D. R., The theory of linear viscoelasticity (London 1960).

Christensen, R. M., Theory of viscoelasticity – An introduction (New York 1971).

Coleman, B. D. et *W. Noll*, Arch. Rat. Mech. Anal. **3**, **(1959)**.

Eirich, F. R., Rheology – Tomes I, II, III (New York 1956).

Ericksen, J. L., Laminar shear flow of viscoelastic materials. Dans: Viscoelasticity, phenomenological aspects. Ed.: *Bergen* (New York 1960).

Ferry, J. D., Viscoelastic properties of polymers (New York 1961).

Lodge, A. S., Elastic liquids (New York 1964).

Metzner, A. B., Dans: Handbook of Fluid Dynamic (New York 1961). – *Metzner, R. B.* and *ϰ. C. Reed*, Amer. Inst. Chem. Eng. J. **1**, 434 (1955).

Oldroyd, J. G., Proc. Roy. Soc. **A 295**, 278 (1958).

Prager, W., Introduction to mechanics of continua (Chicago 1961).

Reiner, M., Deformation strain and flow (London 1960).

Rivlin, R. S., J. Rat. Mech. Anal. **5**, **(1956)**.

Sisko, A. W., Ind. Eng. Chem. **50**, 1789 (1958).

Slattery, J. C. and *R. B. Bird*, Chem. Eng. Sci. **16**, 231.

Tobolsky, A. V., Properties and structure of polymers (New York 1960).

Adresse de l'auteur:

D. Bellet
E.N.S.E.I.H.T.
2 rue Charles Camichel
31071 Toulouse-Cecex (France)

Rheol. Acta 12, 311–318 (1973)

From the Civil Engineering Department College of Engineering University of Utah Salt Lake City (U.S.A.)

On the general theory of Steklov-Aging materials

By *J. Edmund Fitzgerald*

With 14 figures

(Received October 27, 1972)

1. Introduction

The application of the linear and nonlinear theories of viscoelasticity has had wide application in the characterization and stress analysis of solid propellants, highly solids loaded polymers, and asphalt pavement during the past two decades.

So long as the characterization and verification of the selected constitutive equation was confined to the usual loading or straining conditions such as a constant strain rate, ramp relaxation, or constant stress creep the comparison of theory with experiment was generally satisfactory. This result was to be expected since the theory and experiment are often a curve fitting exercise over a narrow range of load types.

When viscoelastic theory is applied to the prediction of repeated loads such as a sawtooth input or multiple ramp input, the predictive results of the theory are often quite unsatisfactory. The above statement applies to linear viscoelasticity as well as to the several forms of nonlinear viscoelasticity utilizing the multiple integral approach.

Details of the previous statements are presented in *Farris* and *Fitzgerald* (1970), *Fitzgerald* and *Farris* (1970), and Chapter XI of *Fitzgerald* and *Hufferd* (1971). A most excellent comparative review is given in *Stafford* (1969).

Fig. 1 demonstrates the above problem based upon some of *Farris*' experiments, using a filled polyurethane propellant.

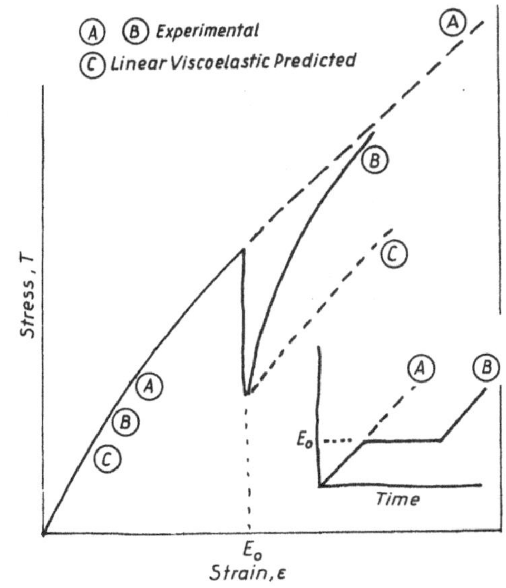

Fig. 1. Stress for an interrupted constant strain rate test

The use of a repeated sawtooth strain history is typified by the curves of fig. 2 for a highly solids loaded polybutadiene acrilonitrile propellant (*Bennett*, 1971).

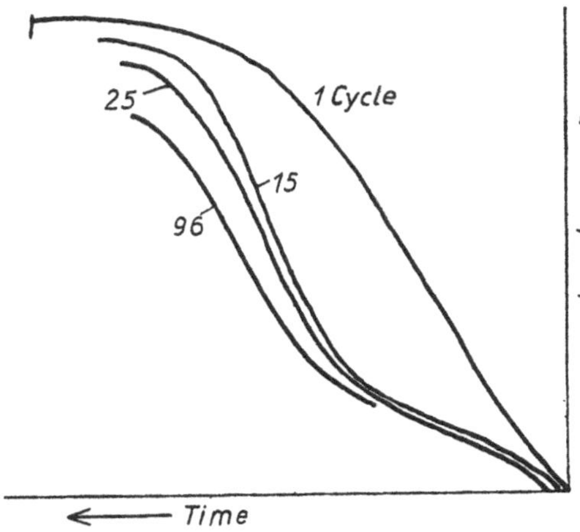

Fig. 2. Uniaxial constant rate cycling of PBAN propellant

Previous publications by the author and his co-workers employed the use of homogeneous functions of the strain history made specific through *Lebesgue* norms. The *Lebesgue* norm, $\|\varepsilon\|_{L_P}$ is essentially the weighted *P*-summable integral of the infinitesimal strain history, ε wherein

$$\|\varepsilon\|_{L_P} = [\int_{\text{meas } s} |\varepsilon|^P h_P(s)\, ds]^{1/P}. \qquad [1]$$

The term $h_P(s)$ is a positive decreasing function of s for fading memory theories where P is taken as unity. Certain boundedness restrictions apply to $h_P(s)$.

For $P \to \infty$, the *Chebyshev* norm results wherein

$$\|\varepsilon\|_{L_\infty} = \text{ess. sup}\, |\varepsilon|. \qquad [2]$$

Using a rather simple expression (*Farris*, 1970) for stress, T, versus strain ε, wherein

$$T_{11}(t) = 435 \left[\frac{\|\varepsilon_{11}\|_\infty}{\|\varepsilon_{11}\|_{21}}\right]^{2.25} \varepsilon_{11}(t) \qquad [3]$$

the results of fig. 3 were obtained (*Farris*, 1970).

It should be noted that the general form of eq. [3], which is the L_∞ norm divided by the L_P norm to the nth power (with $P \geqslant 1$), when multiplied by ε_{11} yields the following results:

21*

For a constant strain rate test, $\varepsilon_{11} = Rt$

$$\|\varepsilon_{11}\|_{\infty} = Rt; \quad \|\varepsilon_{11}\|_P = Rt - \frac{t^{1/P}}{(P+1)^{n/P}}$$

$$T_{11} = A(P+1)^{n/P} R t^{1-n/P}$$

or

$$T_{11} = A(P+1)^{n/P} \varepsilon_{11} t^{-n/P}. \qquad [4]$$

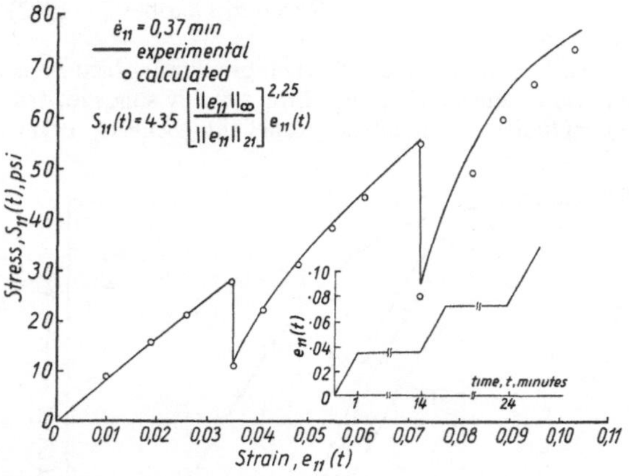

Fig. 3. Comparison of calculated and observed stress-strain output for an interrupted ramp strain input

For a step relaxation test, $\varepsilon_{11} = \varepsilon_0$

$$\|\varepsilon_{11}\|_{\infty} = \varepsilon_0; \quad \|\varepsilon_{11}\|_P = \varepsilon_0 t^{1/P}$$

$$T_{11} = A\varepsilon_0 t^{-n/P}. \qquad [5]$$

The last result predicts an inverse power law for the relaxation modulus, $At^{-n/P}$, so that the value of A and the slope, n/P, can be determined from a step relaxation test since $G(t)_{\text{relax.}} = At^{-n/P}$.

The constant strain rate test yields the secant modulus

$$G(t)_{\text{secant}} = (P+1)^{n/P} G(t)_{\text{relax.}} \qquad [6]$$

from which $P + 1$, hence P and n may be determined.

The predictions of fig. 3 were made using data obtained as described. A comparison of the above norm expression with linear viscoelasticity for a polyurethane propellant is given in fig. 4 (*Farris*, 1970).

A more general expansion of *Farris'* earlier expressions has been given by *Vakily* and *Fitzgerald* (1972) wherein the stress function involves a sum of *Lebesgue* norm ratios and the naturally occuring inverse, power law kernel in the viscoelastic integral as follows:

$$T_{11}(t) = \sum_{i=0}^{N} \sum_{J=0}^{N} A_{ij} \left(\frac{\|f\|\, q_i}{\|f\|_{P_i}} \right)^{n_i} \times \int_0^t (t-\tau)^{-m_j} \dot{e}_{11}(\tau)\, d\tau. \qquad [7]$$

The first term of the above expansion, $i = j = 0$ with $q_0 = \infty$ yields

$$T_{11}(t) = A_{00} \left[\frac{\|\varepsilon_{11}(t)\|_{\infty}}{\|\varepsilon_{11}(t)\|_{P_0}} \right]^{n_0} \varepsilon_{11}(t) \qquad [8]$$

which is the previous *Farris* expression, eq. [3].

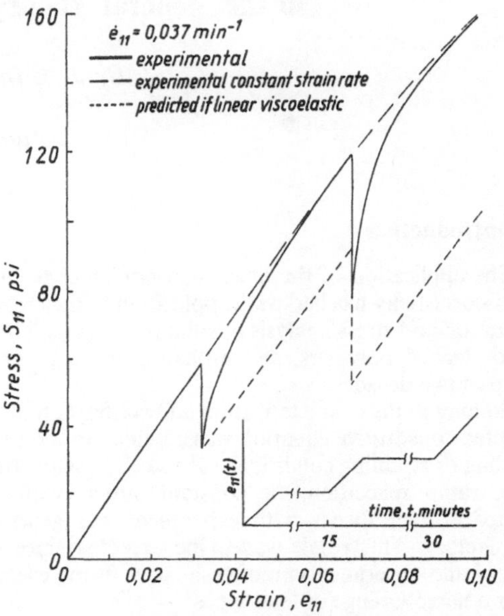

Fig. 4. Stress output for interrupted constant strain rate test

Including the current value of the strain, $|\varepsilon_{11}|$ and simplifying a three term expansion of eq. [7] results in

$$T_{11}(t) = A_1 \left[\frac{|\varepsilon_{11}|}{\|\varepsilon_{11}\|_{P_0}} \right]^{n_0} \varepsilon_{11}(t) + A_2 \left[1 - \frac{|\varepsilon_{11}|}{\|\varepsilon_{11}\|_{\infty}} \right]^{n_1} \times \int_0^t (t-\tau)^{-m_1} \dot{\varepsilon}_{11}(\tau)\, d\tau. \qquad [9]$$

where we have defined $\|\varepsilon\|_0 = |\varepsilon|$.

Various other specific forms of the above *Lebesgue* norms may be formulated.

Applying eq. [12] to a series of compression tests on a sand-asphalt mixture produced the following expression (*Vakily* and *Fitzgerald*, 1972) where the constants were evaluated from constant strain rate "step" relaxation tests:

$$T(t) = 310 \left\{ \left(\frac{|\varepsilon|}{\|\varepsilon\|_9} \right)^{4.32} \varepsilon(t) + \left[1 - \frac{|\varepsilon|}{\|\varepsilon\|_{\infty}} \right] \int_0^t (t-\tau)^{-0.8} \dot{\varepsilon}(\tau)\, d\tau \right\}. \qquad [10]$$

The moduli for a strain of 0.37 and 0.5% are shown in figs. 5 and 6.

The constant strain rate tests for two different rates are given in fig. 7.

The ramp relaxation test results are given in fig. 8.

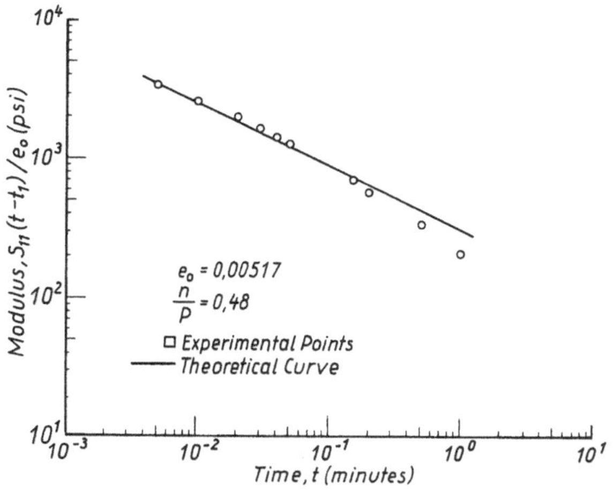

Fig. 5. Relaxation of sand-asphalt for first test

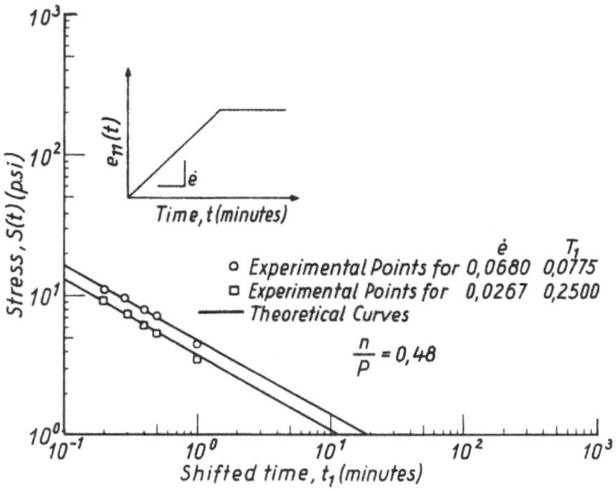

Fig. 8. Stress output for two different ramp relaxation tests

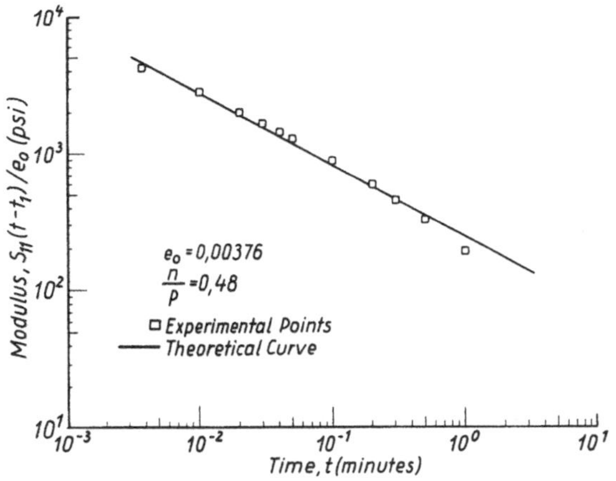

Fig. 6. Relaxation modulus of sand-asphalt for second test

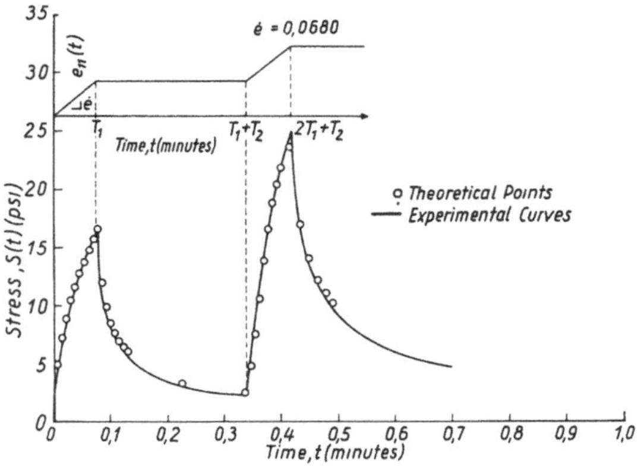

Fig. 9. Stress output for interrupted ramp strain input

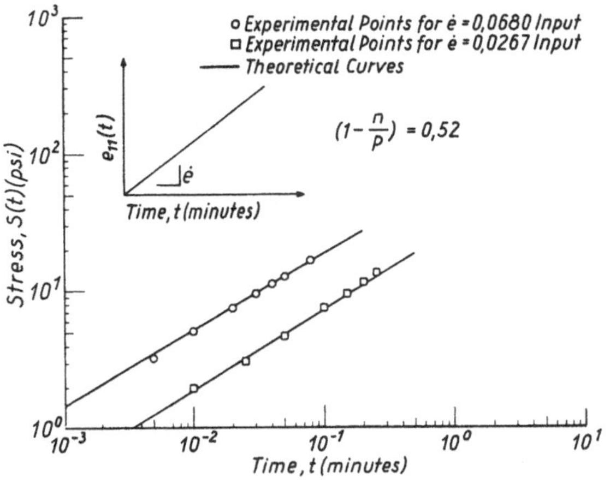

Fig. 7. Stress output for two different constant strain rate tests

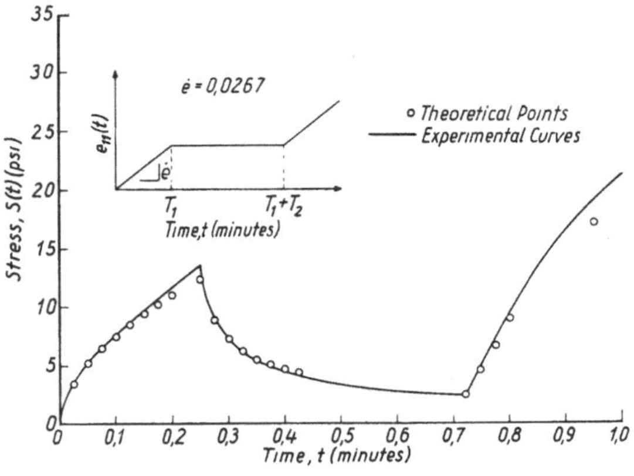

Fig. 10. Stress output for interrupted ramp strain input

Using the expression, eq. [10], derived from the above data, predictions shown as open circles and experiments, shown as solid lines, were conducted for:

1. An interrupted ramp strain input at two different strain rates, shown in figs. 9 and 10. A comparison of the above predictions with those of linear viscoelasticity is given in fig. 11.

2. A single constant strain rate "sawtooth" strain, shown in fig. 12.

3. A repeated constant strain rate test cycled between a set maximum strain and zero load, fig. 13.

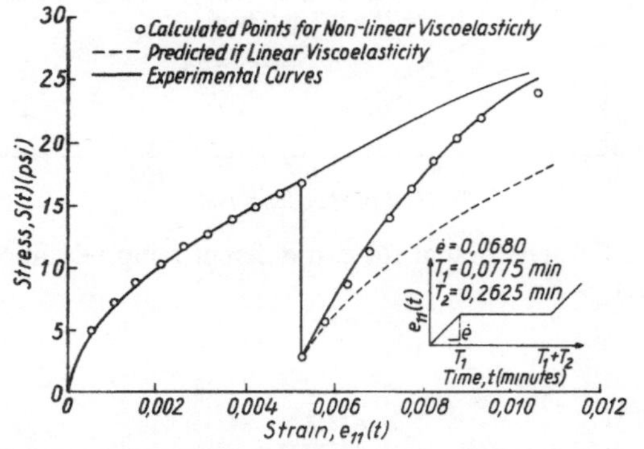

Fig. 11. Comparison of calculated and observed stress-strain output for an interrupted ramp strain input

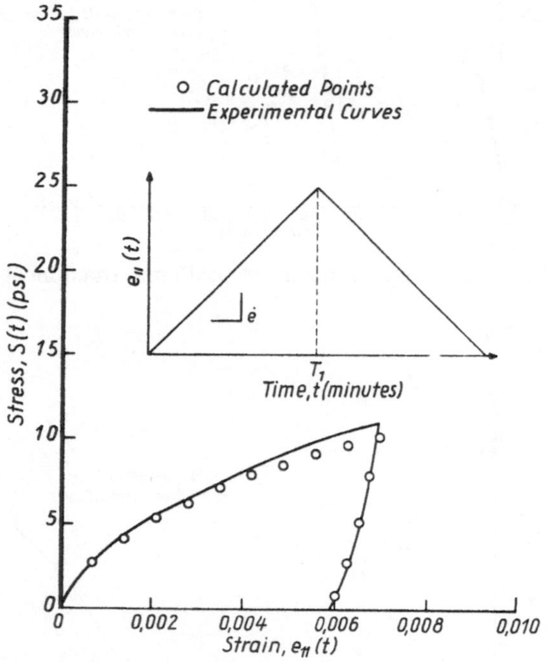

Fig. 12. Comparison of calculated and observed stress-strain behavior of sand-asphalt

Again, it is clear that the sand-asphalt material exhibits a degree of permanent memory similar to the filled polyurethane propellant and the PBAN propellant. Figs. 5 through 13 are from *Vakily* and *Fitzgerald* (1972).

An expression for the stress such as

$$T(t) = f\,[U(t), \|U(t)\|_{L_\infty}] \qquad [11]$$

produces a stress-strain relation with no relaxation but with the *Mullins* effect predominant (*Mullins*, 1947).

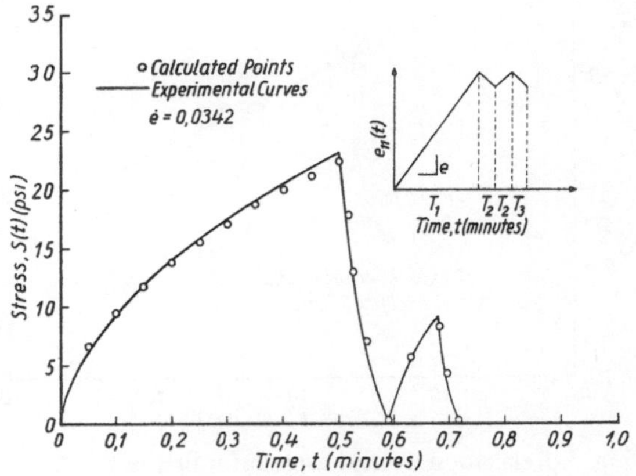

Fig. 13. Comparison of calculated and observed stress output for a reversed ramp strain test

A mathematical justification for the above norms, and their restrictions, is given in *Hufferd* and *Fitzgerald* (1972) including thermodynamic implications.

In general, one may express the *Cauchy* stress tensor, T, as a function of the norms, for example

$$T = f\,[U(t), \|U\|_{L_\infty}, \|U\|_{L_P}] \qquad [12]$$

where U is the positive square root of the right *Cauchy-Green* (finite strain) tensor. The resulting polynomial function following the well known *Rivlin-Spencer* expansion for initially isotropic materials will produce an expression up to the second power in the U and its norms with coefficients that are polynomials, or preferably here, rational fractions in the joint invariants

$$\operatorname{tr} U, \quad \operatorname{tr} U^2, \quad --, \quad \operatorname{tr}\|U\|_{L_P}, \quad \text{etc.}$$

Again, a three-dimensional expression similar to *Farris'* is derived for a single term, homogeneous form for T, namely

$$T(t) = A\left[\frac{\operatorname{tr}\|U\|_\infty}{\operatorname{tr}\|U\|_P}\right]^n (U(t) - 1). \qquad [13]$$

Application of the above tensorial equation to a uniaxial test produce results similar to those previously shown.

2. Aging Effects and the Steklov Average

The *Lebesgue* norms previously described are such that they produce the integral of the strain history on a P-summable basis. The use of a non-unit weighting function, $h_P(s)$, will provide for fading memory effects. The *Lebesgue* norms may have a basis in microscopic theory as presented by *Farris* in his doctoral thesis.

Nevertheless, the norms as used herein are simply continuum postulates whose use is justified by the results shown herein and in the various quoted references.

With the integrals as used, however, no aging effects are included.

Consider now a class of materials whose response is governed by certain weighted averages of the past strain history as well as by the present value of the strain and its several time derivatives.

A norm on such a space may be constructed as follows for a generalized input Λ

$$\|\Lambda^t(s)\|_{S_P^N} = \sum_{i=0}^{N} |\Lambda^{(i)}(o)|\, h_i(o) + \sum_{P=1}^{\infty}\left[\frac{1}{\text{mes}\, D}\int_D |\Lambda_r^t(s)|^P\, h_P^P(s)\, ds\right]^{1/P}$$

with

$$\left[\frac{1}{\text{mes}\, D}\int_D h_P^P(s)\, ds\right]^{1/P} < \infty \quad \text{and} \quad h_P^P(s) \geq 0 \qquad [14]$$

and $\Lambda^{(i)}(o)$ the present i-th rates of change. The above is actually a semi-norm unless meas D is finite.

Considering especially the second summand in eq. [14], we shall call it a *Steklov Average* since it is a generalization of the *Steklov-Lebesgue* average given in *Kantorovich* and *Akilov* (1964).

A slight variation in the above, which will be termed the *modified Steklov Average* and which satisfies all the requirements of a norm (or semi-norm) is given by

$$\|\Lambda\|_{S_P} = \left[\frac{1}{(1 + k\,\text{meas} D)}\int_D |\Lambda_r^t(s)|^P\, h_P^P(s)\, ds\right]^{1/P}$$

with $k \geq 0$. [15]

Dropping the fading memory factor $h_P(s)$, for simplicity, results then in

$$\|\Lambda\|_{S_P} = \left[\frac{1}{(1 + kt)}\int_{s=0}^{t} |\Lambda|^P\, ds\right]^{1/P} \qquad [16]$$

where we have taken meas $D = t$ for the usual rather smooth physical inputs, Λ.

Looking at the *Lebesgue* norm ratios of eqs. [7] or [8] for example and the definitions eqs. [1] and [16] yields the following relation between the L_P norms and the S_P norms

$$\left[\frac{\|\Lambda\|_{S_Q}}{\|\Lambda\|_{S_P}}\right]^n = \left[\frac{\|\Lambda\|_{L_Q}}{\|\Lambda\|_{L_P}}\right]^n (1 + kt)^{\left(\frac{Q-P}{QP}\right)n} \qquad [17]$$

and for $Q \to \infty$,

$$\left[\frac{\|\Lambda\|_{S_\infty}}{\|\Lambda\|_{S_P}}\right]^n = \left[\frac{\|\Lambda\|_{L_\infty}}{\|\Lambda\|_{L_P}}\right]^n (1 + kt)^{n/P}. \qquad [18]$$

It is readily shown that the ratio

$$\frac{\|\Lambda\|_{S_Q}}{\|\Lambda\|_{S_P}} \geq 1\, \forall Q > P, \quad P \geq 1 \qquad [19]$$

since the dominance of L_Q over L_P is well known when $Q > P$ (*Kantorovich* and *Akilov*, 1964) and for $k \geq 0$, the muliplier is also ≥ 1.

Defining T^L as the stress obtained from *Lebesgue* norms as in the previous section 1.0, the stress obtained by substituting modified *Steklov* norms is T^S where

$$T^S = T^L(1 + kt)^{\frac{Q-P}{QP}n} \qquad [20]$$

or with $Q \to \infty$

$$T^S = T^L(1 + kt)^{n/P}. \qquad [21]$$

Consider for example a material governed by a simple permanent memory norm relation such as eq. (3)

$$T_{11}^L = A\left[\frac{\|\varepsilon_{11}\|_\infty}{\|\varepsilon_{11}\|_{L_{21}}}\right]^n \varepsilon_{11}(t). \qquad [22]$$

For a step relaxation test with strain magnitude ε_0 [22] yields

$$T_{11}^L = A\varepsilon_0 t^{-n/P}. \qquad [23]$$

Applying instead the ratio of the modified *Steklov* norms S_P produces

$$T_{11}^s = A\varepsilon_0 t^{-n/P}(1 + kt^*)^{n/P}. \qquad [24]$$

For a material which is strained immediately upon being formed, $t = t^*$. Otherwise however, t is the time relative to the beginning of the step strain whereas t^* is the actual time from the initial creation of the material to the present. The difference in behavior of the L_P and S_P norms is shown in Fig. 14. A step load-unload input is used with the L_1, $\bar{S}_1$ and $L_\infty = S_\infty = \varepsilon$ max. curves plotted for an example. After the load occurs, L_1 is constant as is L_∞. However, the S_1 norm reduces with time since it is averaging over the life of the specimen.

The relation between the L_P and S_S norms is, from [1] and [16]

$$\|\Lambda\|_{S_P} = \|\Lambda\|_{L_P}(1 + kt)^{-1/P}. \qquad [25]$$

It will also be noticed from eq. [21] that the exponent of the "aging" term $1 + kt^*$ is equal to the positive numerical slope of the inverse power law relaxation modulus, n/P.

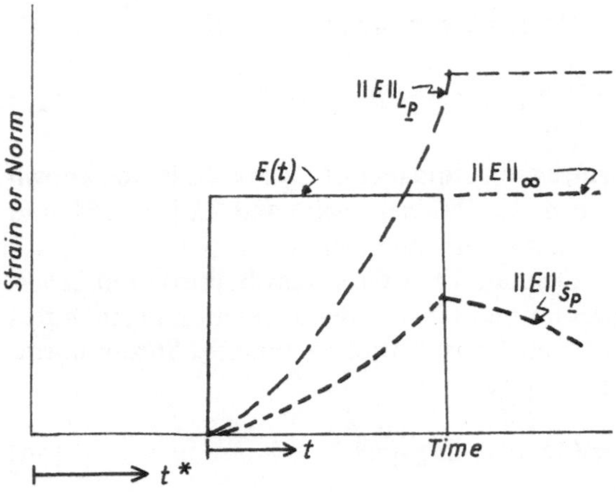

Fig. 14. Time vs. norms

Consider a typical solid propellant with $n/P = 0.2$, then

$$T_{11} = A\varepsilon_0 t^{-0.2}(1 + kt^*)^{0.2}. \qquad [26]$$

If the relaxation modulus increases by 20% over a one-year ambient aging then $k \cong 3 \times 10^{-8}$, $\sec^{-1}$.

A typical CTPB propellant increases its modulus by 20% in 100 days. With a slope $n/P = 0.2$, hence, the value of k is $15 \times 10^{-8}\ \sec^{-1}$.

It is also to be noted that for large values of kt^*, the aging is described by a straight line on a log-log scale. Further for $n/P = 0$, no relaxation and no strain rate effects, the aging is nonexistent.

Because of the small values of k, the use of the L_P rather than the S_P norms for short time loads on newly formed materials is fully equivalent. Further, for short loading times relative to the life of the material, $(1 + kt^*)^{n/P}$ is essentially a constant and *Steklov* aging reduces to the socalled homothetic aging process (*De Arriaga*, 1969).

For finite strain, the essential results also hold. Consider an incompressible material subject to a simple step elongation with a stretch ratio of λ where λ is the stretched length divided by the initial length. Using a simple constitutive equation of the form of [13] but with modified *Steklov* norms instead of *Lebesgue* norms as in [18] results in

$$T(t) = A\left[\frac{\lambda + 2}{\lambda + 2\lambda^{-1/2}}\right]^n t^{-n/P}(1 + kt^*)^{n/P} \times (\lambda - 1). \qquad [27]$$

For a material stretched when newly formed and then held at the stretch ratio, λ, with $t = t^*$

$$T(t) = A\left[\frac{\lambda + 2}{\lambda + 2\lambda^{-1/2}}\right]^n (t^{-1} + k)^{n/P}(\lambda - 1). \qquad [28]$$

For long times then as $t \to \infty$,

$$T(t) = A\left[\frac{\lambda + 2}{\lambda + 2\lambda^{-1/2}}\right]^n k^{n/P}(\lambda - 1). \qquad [29]$$

One could obviously extend the expansion to higher order terms in U, but the essential point to be made is that the material does not fully relax as $t \to \infty$. For the typical material previously mentioned, if the relaxation modulus at one second were 1000 psi, the final modulus would reduce to 60 psi. In addition, there can be considered an elastic component with no loss of generality.

3. Conclusions

It has been proposed that the characterization of relaxing, rate sensitive materials be based upon certain weighted averages of the past history which are semi-norms called modified *Steklov* averages.

The rationale is based upon previously mentioned guidance from molecular theory but is primarily based upon the postulate that the present response of a material is governed by its present deformation gradient and a selected weighted P-summable *Steklov* average [14].

Since in engineering practice, one seldom knows the detailed past of a structure's thermal and deformation history but usually knows the average and the maximums, no serious loss of applicability should result.

The justification for the use of the proposed norms is based upon the several examples given herein where the predictions using norms is shown to be superior in accuracy to the predictions based upon linear viscoelasticity. The use of the $\bar{S}_{\infty}$ or L_{∞} maximum value norms implies that the material also is sensitive to the maximum strain value it has ever been subjected to. The possible extension of the concept to viscoplastic materials is under study.

Farris has shown that for a selected class of solid propellants the use of the conventional time-temperature superposition integral is valid. The present author suggests that the time temperature shift integral

$$\xi = \int_{\tau=0}^{t} \frac{\tau}{\alpha_T} d\tau \qquad [30]$$

for reduced time ξ with the temperature shift factor α_T be used in the modified *Steklov* norm for aging effects. It is, of course, equally possible to use an absolute reaction rate correction for the kt term in [16].

The use of a rather simple homogeneous fraction involving only the max. norm and the P-norm of the modified *Steklov* average [22] produces four specific results for infinitesimal deformations

1. the relaxation modulus is described by an inverse power law, $t^{-n/P}$;
2. the constant strain rate curves are described by a power law whose exponent is unity plus the relaxation modulus exponent, $t^{1-n/P}$;
3. the relaxation and secant moduli are subject to an age hardening process n/P described by the factor $(1 + kt)$;
4. for very long term stretching and aging, the material stress response is nonvanishing as shown in [29].

It has been also shown that for short aging times the form of the equations reduces to ratios of *Lebesgue* norms only. For short term loading relative to the age of the material, a homothetic aging process is produced.

There does not exist a unique inverse for strain as a function of stress, i.e., a general creep inverse. Indeed, an inverse does not generally exist unless a nonconstant weighting function, $h_P(s)$ is used. Even then, there results a nonlinear integral equation of the type

$$\varphi(s)^m = \int |\varphi(s)|\, h_P(s)\, ds; \qquad \varphi(s) = \Lambda^P(s) \qquad [31]$$

with the exponent m a rational fraction whose value will generally be near unity.

Since the present value of stress is governed by certain averages of the past strain history, uniqueness in the inverse is not to be expected. It is therefore suggested that a creep law be formulated in the same fashion as [22], for example

$$\varepsilon(t) = f\,[\|T\|_{S_P}, \|T\|_{S_\infty}, T(t)] \qquad [32]$$

made specific as

$$\varepsilon(t) = B \left[\frac{\|T\|_{S_P}}{\|T\|_{S_\infty}}\right]^m T(t) \qquad [33]$$

which for a constant stress, T_0, results in

$$\varepsilon(t) = B T_0 t^{m/P} (1 + kt)^{-m/P}. \qquad [34]$$

The calculation time necessary to use the various norm equations given herein is much shorter for general inputs than the time needed for viscoelasticity. This computer time saving results from the fact that the norms are only a number at the present time whose value changes by the modified average at each step. If, however, nonconstant weighting functions, $h_P(s)$, are used, no computational advantage results.

Acknowledgements

The author wishes to acknowledge the past and continuing efforts of his colleagues *R. Farris*, *W. Hufferd*, *J. Vakily*, and *M. Quinlan* in pursuing and contributing to the ideas herein presented.

This work was sponsored in part by the Office of Naval Research under Contract No. 00014-67-A-0325-0001 and the Air Force Office of Scientific Research under Contract No. 72-2332.

Summary

A broad class of materials possessing both instantaneous nonlinear elasticity and dissipation in addition to fading memory with aging effects is described. The measure of the generalized input function, Λ, which is a multiplet in F, the deformation gradient; θ, the temperature; $g = \text{grad}\,\theta$, as well as various chemical affinities, A_k; is given by a semi-norm over a *Banach* space. With the definition of the history $\Lambda^t = \Lambda^t(s) \equiv \Lambda^t(t-s)$; $s\varepsilon(0, \infty)$ and with the restriction of Λ^t to past history given by $\Lambda^t_r = \Lambda^t_r(s) = \Lambda^t_r(t-s)$; $s\varepsilon(0, \infty)$ the semi-norm is:

$$\|\Lambda^t(s)\|_{S^N_P} = \sum_{i=0}^{N} |\Lambda^{(i)}(o)|\, h_i(o) + \sum_{P=1}^{\infty} \left[\frac{1}{1 + k \operatorname{meas} D} \int_D |\Lambda^t_r(s)|^P h^P_P(s)\, ds\right]^{1/P}$$

$$\text{with} \left[\frac{1}{1 + k \operatorname{meas} D} \int_D h^P_P(s)\, ds\right]^{1/P} < \infty$$

$$\text{and} \quad h^P_P(s) \geq 0. \qquad [35]$$

and $\Lambda^i(0)$ the present ith time rate of change.

The second summand is a modification of the *Steklov Average* (*Kantorovich*, 1964) to P – integrable *Lebesgue* functions.

It is assumed that the generalized response $\Omega(t)$ is a nonlinear function(al) of the present input $\Lambda(t)$ and a material property-history kernel determined by the *Steklov-Lebesgue* norm $\|\Lambda^t(s)\|_{S^N_P}$ such that

$$\Omega(t) = F\,[\|\Lambda^t(s)\|_{S^N_P}]\, \Lambda(t). \qquad [36]$$

Eq. [35] shows that as time increases and that the influence of the past history on present response decreases. *For a given finite duration input, its influence decreases both the longer in the past it occurred and the older the material is.* This latter effect is termed *Steklov-aging*. As the age of the material becomes very large the past history effects are obliterated and from eq. [35]

$$\lim_{\substack{t\to\infty\\ \mathrm{mes}\,D\to\infty}} \|\Lambda^t(s)\|_{S_P^N} = \sum_{i=0}^{N} |\Lambda^{(i)}(o)|\, h_i(o). \qquad [37]$$

Thus a *Steklov-aging material under long term aging approaches the behavior of a nonlinear viscoelastic or Markoffian type material.* For very small inputs, $\Lambda(0)$ and small time rates of change of inputs, $\Lambda^{(i)}(0)$, the material, after long term aging, becomes linearly viscoelastic.

Examples of the theory as applied to solid propellants and a sand-asphalt concrete are given.

Zusammenfassung

Es wird eine breite Klasse von Materialien beschrieben, die zusätzlich zu einem schwindenden Rückerinnerungsvermögen infolge Alterungserscheinungen sowohl ein nicht-lineares Elastizitätsverhalten als auch Dissipation aufweisen. Als Maß für die verallgemeinerte Eingangsfunktion, Λ, die ein Multiplet in F is, des Deformationsgradienten; der Temperatur, θ, $g = \mathrm{grad}\,\theta$ und verschiedener chemischer Affinitäten, A_k, wird eine Halb-Norm über einen *Banach*-Raum verwendet. Mit der Definition der Beanspruchungsgeschichte $\Lambda^t = \Lambda^t(s) = \Lambda^t(t-s)\, s\varepsilon\,[0,\infty]$ und mit der Einschränkung, daß die Gegenwart ausgeschlossen ist, d. h. $\Lambda_r^t = \Lambda_r^t(s) = \Lambda_r^t(t-s)$; $s\varepsilon(0,\infty)$ gibt:

$$\|\Lambda^t(s)\|_{S_P^N} = \sum_{i=0}^{N} |\Lambda^{(i)}(o)|\, h_i(o) + \sum_{P=1}^{\infty}\left[\frac{1}{1+k\,\mathrm{meas}\,D}\int_D |\Lambda_r^t(s)|^p\, h_p^P(s)\,ds\right]^{1/p}$$

$$\text{with} \left[\frac{1}{1+k\,\mathrm{meas}\,D}\int_D h_P^P(s)\,ds\right]^{1/P} < \infty$$

$$\text{and} \quad h_P^P(s) \geq 0. \qquad [35]$$

$\Lambda^i(0)$ die i-te Ableitung nach der Zeit (Gegenwart).

Der zweite Summand ist die Modifikation des *Steklov-Durchschnittswertes* (*Kantorovich*, 1964) zu P-integrierbaren *Lebesgue-Funktionen.*

Es wird angenommen, daß die verallgemeinerte Abhängigkeit $\Omega(t)$ von $\Lambda(t)$ eine nicht-lineare Funktion (Funktional) der gegenwärtigen Eingangsgrößen und eines Stoffeigenschaften-Vorgeschichte Kernel ist, das durch die *Steklov-Lebesgue*-Norm $\|\Lambda^t(s)\|_{S_P^N}$ vorgegeben ist. Somit ergibt sich:

$$\Omega(t) = F\,[\|\Lambda^t(s)\|_{S_P^N}]\,\Lambda(t). \qquad [36]$$

Gl. [35] besagt, daß der Einfluß der Vorgeschichte auf das Gegenwartsverhalten mit zunehmender Zeit geringer wird. *Der Einfluß eines Inputs endlicher Zeitdauer nimmt ab, je mehr Zeit verstreicht und je älter das Material ist.* Diese letztere Beobachtung wird als *Stecklov-Alterung* bezeichnet. Mit hohem Materialalter wird der Einfluß der Beanspruchungsvorgeschichte völlig ausgelöscht und Gl. [35] lautet dann

$$\lim_{\substack{t\to\infty\\ \mathrm{mes}\,D\to\infty}} \|\Lambda^t(s)\|_{S_P^N} = \sum_{i=0}^{N} |\Lambda^{(i)}(o)|\, h_i(o). \qquad [37]$$

Damit *nähert sich das Verhalten eines Steklov-alternden Materials mit zunehmendem Alter mehr und mehr dem eines nicht-linearen viscoelastischen Materials oder „Markoff"-Materials an.* Wenn die Eingangsgrößen und die zeitlichen Änderungen der Eingangsgrößen klein sind, dann wird das Material nach langer Alterung linear viscoelastisch.

Es werden Beispiele der Theorie in ihrer Anwendung auf Festkörper Treibstoffe und Sand-Asphalt-Beton beschrieben.

References

1) *Bennett, J.*, Private Communication, Thiokol Chemical Co. (1971).

2) *De Arriaga, F.*, Homothetic Aging Viscoelastic Materials. Int. Conf. on Structure and Materials, Southampton 1969.

3) *Farris, R. J.*, Homogeneous Constitutive Equations for Materials with Permanent Memory. UTEC TH 70-083, Project Themis Report AFOSR 70-1962 TR, University of Utah, June 1970.

4) *Farris, R. J.* and *J. E. Fitzgerald*, Deficiencies of Viscoelastic Theories as Applied to Solid Propellants. Bulletin JANNAF Mechanical Behavior Working Group, 8th Meeting 1969, CPIA Publication No. **193**, Vol. **1**, March 1970.

5) *Fitzgerald, J. E.* and *R. J. Farris*, Characterization and Analysis Methods for Nonlinear Viscoelastic Materials. Project Themis Report, UTEC TH 70-204, University of Utah, November 1970.

6) *Fitzgerald, J. E.* and *W. L. Hufferd*, Engineering Structural Analysis of Solid Propellants. CPIA Publication No. **214**, The Johns Hopkins University, July 1971.

7) *Hufferd, W. L.* and *J. E. Fitzgerald*, A Thermodynamic Theory of Materials with Permanent Memory. Proc. 1971 Int. Conf. on Mechanical Behavior of Materials, Vol. III, pp. 530–540, Soc. Mtr'l. Sci., Japan, 1972.

8) *Kantorovich, L. V.* and *G. P. Akilov*, Functional Analysis in Normed Spaces (New York 1964; translated from the 1959 Russian original).

9) *Mullins, L. J.*, J. Rubber Res. **16**, 275–289 (1947).

10) *Stafford, R. O.*, J. Mech. Phys. Solids **17**, 339–358 (1969).

11) *Fitzgerald, J. E.* and *J. Vakily*, Nonlinear Characterization of Sand Asphalt Concrete by Means of Permanent Memory Norms. UTEC CE 72-079, Univ. Utah, 1972. Soc. Exper. Stress Anat. papor No. 2105.

Author's address:

Prof. Dr. *J. Edmund Fitzgerald*
Civil Engineering Department, College of Engineering, University of Utah, Salt Lake City, Utah 84112 (U.S.A.)

Rheol. Acta 12, 319–320 (1973)

An extension of Reiner's "Deborah Number" concept to a Wide field of rheological investigations

By *G. W. Scott Blair* (*Iffley, Oxford*)

(Received October 27, 1972)

Metzner, *White* and *Denn* (1) have claimed that, under certain simplifying rheological conditions, "the *Reynolds*, *Weissenberg* and *Deborah* numbers become the sole dimensionless groups which must be considered". The *Reynolds* numbers define the relative importance of viscous and inertia forces; the *Weissenberg* numbers, the ratio of elastic to viscous behaviour; (2) and the *Deborah* number, (3) the ratio of a relaxation time (τ) to an experimental time (t). *Metzner* et al. point out that it would have been more in keeping with other work had *Reiner* used the reciprocal, t/τ; but this is not very important.

Reiners Number, which he writes as T ("printed dalet" for Hebrew "*Dvorah*") is, of course, a simplification. It assumes a single relaxation time ("a *Maxwell* model"); in practice, one generally has a continuous spectrum, which, in the simplest cases, is represented by a *Wiechert* distribution. But all such distributions include the term $\int_0^\infty F(\tau)(1 - e^{-t/\tau})\,d\tau$, so that it would be best to write, for an expanded *Deborah* Number, $\Sigma\tau_{rel}$ for relaxation in general.

There is also, of course, a corresponding Number for retardation, taking the *Kelvin-Voigt* model as the simplest case. This might be written T_{ret}. *Reiner's* and *Glücklich's* work (4) on concrete illustrates the importance of such Numbers. The purpose of the present paper is to show that there are many other rheological phenomena which are characterized by ratios of a time inherent in the system (for which we will now generalize τ) and a time chosen by the experimenter; t. (In fact there are many bees in the hive.)[1]

Some (but not all) types of thixotropy could possibly be treated in this way. Suppose that a thixotropic system is sheared for a fixed time t at a constant shear-rate and then allowed to recover: the time (τ) taken for the consistency to recover to $1/e$ of its initial value could give a $T_{thix} = \tau/t$ as a measure of thixotropic recovery rate. That is why almost all rheologists make the *arbitrary* distinction between thixotropy (when, under the experimental conditions, the recovery can be observed) and shear thinning, when it cannot. (To return to *Reiner's* original concept, we generally regard mountains as solid", however arbitrary this may be in God's sight!)

My own work has shown that for the rate of coagulation of all of the many blood[2]) samples that I have studied (bovine and human), the following equation holds: $G^* = G^*_\infty e^{-\tau/t}$ where G^* is the complex modulus, G^*_∞ is the value of G^* extrapolated on log-reciprocal paper, to infinity and τ is the time taken for G^* to reach $1/e$ of the value G^*_∞. After some years of struggle, I think I have at last found a simple explanation for this equation: earlier published attempts were not fully satisfactory.

The equation is, of course, an approximation, like all "basic" equations and the treatment is therefore oversimplified.

First, it is evident that a coagulation curve (G^* vs t) must be sinusoidal. Initially, there is an acceleration, partly because when two molecules link up at one point, they are increasingly likely to link up at other points; whereas in the later stages, the number of possible unattached junction points is gradually running out.

We start with the equation $dG^*/dn = f_1 G^*$ (where n is the number of linkages per unit volume). This is a function of G^* and not of n because one can imagine a situation in which large numbers of linkages join quite small numbers of molecules and also another situation in which almost all the molecules are linked together, but not in groups. These two conditions could have the same value of n, but the latter would have a higher G^*, especially since, except right at the start, almost all of the complex modulus is "real".

But dn/dt will also be an inverse function of t (time), or $dn/dt = f_2(1/t)$. (In the earlier papers, it was assumed that these functions are both exponential, but this need not necessarily be so.)

[1]) In Hebrew, the name "*Dvorah*" means a bee.

[2]) This equation holds for milk coagulation, but not quite so well.

Combining the two equations, we have: $dG^*/dt = f_3 G^*/t$. It is clear from the dimensions that f_3 must be a numeric; but it is also clear that it cannot be simply a single number. If it were, we should have a power equation which would imply acceleration or deceleration throughout, depending on whether the number were greater or less than one. We will suppose that the "power", starting very large, is diminishing progressively and (in the simplest case) linearly with time. We then have $dG^*/dt = \tau G^*/t^2$ where τ (needed to balance the dimensions) is a time characteristic of the material. Integration of the equation gives $G^* = G^*_\infty e^{-\tau/t}$.

It will be noticed that the case of coagulation is unlike relaxation and retardation in that the t/τ-ratio is inverted (as was *Reiner's* original T_{rel}). This is because the coagulation curve is a sigmoid, representing the interaction of two opposing factors.

As a final example, I would quote early work in the field of psycho-rheology. *Scott Blair* and *Coppen* (5) gave subjects two small cylinders, one in each hand. One was a steel spring, with rigid plastic disks top and bottom and surrounded by a cotton sheath (or, in earlier experiments a rubber cylinder). In the other hand was a cylinder of the same size and shape of a (*Newton*ian) Californian bitumen. Subjects were asked to squeeze the two samples with a steady pressure, change hands, squeeze again and report on which sample felt the softer. (Various other precautions were taken.) This was done, first with one spring against four bitumens of different viscosity ($\eta \sim 10^7$ p) each at four times ($^1/_4$, 1, 2 and 4 sec.) controlled by a metronome; and then with one bitumen against four springs of different "stiffness" ($\sim 10^7$ d/cm^2). All subjects, except sceptical physicists, found no difficulty in giving these judgments, which depended, in a very simple way on the viscosity of the bitumen (η), the stiffness of the rubber (n) and the time allowed for squeezing (t).

The double system could be characterized by the ratio $\eta/n(=\tau)$ which has the dimensions of time, and the judgment depended on the ratio, or τ/t.

Though not strictly a "*Deborah* Number", mention might be made of my "Springiness number" proposed some years ago (*Scott Blair*) (6) which is the ratio of τ_{rel}/τ_{ret} for a *Burgers* model. Again, this marks the first of a series for more complex models. I am slightly surprised that it has been overlooked, since the term "elasticity" often causes confusion among non-rheologists[3]). "Springiness" depends on a high relaxation and a low retardation time.

To conclude, I have time to mention only a few illustrations of what I think may well be found widely scattered throughout rheology: ratios of times characteristic of materials (τ) and times selected by the experimenter (t).

[3]) It is a pity that there is no word for "springiness" in French.

Summary

Reiner defined a numeric, which he called "the *Deborah* Number" to represent the ratio of a relaxation time to a "natural" (observation) time. This implies a *Maxwell* model but is readily extended to complete relaxation spectra. Similar Numbers are proposed for retardation times and also for some conditions of coagulation thixotropy and for data from certain psychophysical experiments.

References

1) *Metzner, A. B., J. L. White,* and *M. M. Denn*, Amer. Inst. Chem. Eng. J. **12**, 863 (1966).
2) *Weissenberg, K.*, in: The Principles of Rheological Measurement, p. 36 (London 1949).
3) *Reiner, M.*, Physics Today **17**, 62 (1964).
4) *Glücklich, J.*, Rheol. Acta **1**, 356 (1961).
5) *Scott Blair, G. W.* and *F. M. V. Coppen*, Nature **145**, 425 (1940).
6) *Scott Blair, G. W.*, Research **11**, 123 (1958).

Author's address:
Dr. *G. W. Scott Blair*
Grist Cottage, Iffley, Oxford (U.K.)

Rheol. Acta 12, 321–329 (1973)

From the University of Louvain, Louvain-la-Neuve (Belgium), and the University of California, Berkeley (U.S.A.)

On thermal effects in a special class of viscoelastic fluids

By M. J. Crochet and P. M. Naghdi

(Received October 27, 1972)

1. Introduction

It is known that certain polymeric materials, such as polymeric solutions and melts, flow like fluids while they strongly exhibit recovery usually associated with viscoelastic solids. Moreover, the behavior of these materials is highly influenced by their flow and temperature history. Some available experimental observations [see, e.g., *Ferry* (1)] indicate that the temperature dependence of these viscoelastic fluids may be described adequately by incorporating the notion of "temperature-time equivalence" in their constitutive relations. The "temperature-time equivalence" has been utilized previously in a thermodynamic study of a class of viscoelastic solids (2) and include as a special case results appropriate for "thermo-rheologically simple" solids (3)[1]).

In a recent paper (4), the present authors have considered a class of non-isothermal viscoelastic fluids which may be regarded as the counterpart of our earlier study for a restricted class of viscoelastic solids (2). Our starting in (4) is *Colemans* results (5) on thermodynamics of simple materials with fading memory. While the class of non-isothermal viscoelastic fluids in (4) is necessarily less general than that for simple fluids in (5), it is also less formidable and is partly determined from the knowledge of the corresponding isothermal theory.

After collecting some preliminaries on simple fluids in section 2, we summarize in section 3 the main results obtained in (4). First we examine the nature of the deformation history when, in the absence of stress, the material is subject to a variable temperature history. This leads to the definition of a constitutive functional which describes the thermal dilation of a simple fluid. The strain history is then decomposed into the product of a thermal dilatation and an "isothermal" strain. With the use of these results, the *Helmholtz* free energy in the restricted theory is expressed as the sum of two functionals: One of these depends on the temperature history only, while the other is determined from the knowledge of the isothermal free energy functional; the latter depends, however, on the "isothermal" strain history with a modified time scale.

In order to elaborate on the nature of thermal effects within the scope of the above restricted theory, in section 4 we examine the form of the material functionals of incompressible viscoelastic fluids when the temperature differs from its reference value. By considering a lineal flow and after recalling the definition of isothermal viscometric functions, for the special class of viscoelastic fluids under consideration, we find that a change of temperature in a lineal flow is equivalent to multiplying the shear rate by a factor depending on the temperature. This factor, which may be determined from the value of the viscosity of the fluid under vanishing shear rate, is identical to the so-called "shift function" often employed in studies of "thermo-rheologically simple" materials. Finally in section 5, we consider the example of a non-isothermal steady *Poiseuille* flow in which heat is generated by dissipation.

2. Preliminaries

Consider a body $\mathscr{B}$ with material points X and let $\boldsymbol{x}$ be the position in *Euclid*ean space $\mathscr{E}$ occupied by X at the present time t. We designate the position occupied by X in $\mathscr{E}$ at time $\tau \leqq t$ by

$$\boldsymbol{x}(\tau) = \boldsymbol{\chi}_t(\boldsymbol{x}, \tau)\,(-\infty < \tau \leqq t) \qquad [2.1]$$

where $\boldsymbol{\chi}_t$ is the relative deformation function. The relative deformation gradient at time τ with respect to $\boldsymbol{x}$ is given by

$$\boldsymbol{F}_t(\tau) = \operatorname{grad} \boldsymbol{\chi}_t(\boldsymbol{x}, \tau) \qquad [2.2]$$

where grad is the grad operator with respect to $\boldsymbol{x}$ keeping τ fixed. We define the right relative *Cauchy-Green* measure of deformation by

$$\boldsymbol{C}_t(\tau) = \boldsymbol{F}_t^T(\tau)\,\boldsymbol{F}_t(\tau) \qquad [2.3]$$

where f^T stands for the transpose of f. Also, we use the notation $f = f(t)$ so that [2.1]–[2.2] for $\tau = t$ reduce to $\boldsymbol{x} = \boldsymbol{x}(t) = \boldsymbol{\chi}_t(\boldsymbol{x}, t)$, $\boldsymbol{F}_t(t) = \boldsymbol{1}$ where $\boldsymbol{1}$ denotes the unit tensor.

For later reference, we recall here the energy equation in the form

$$-\varrho\,\dot{\varepsilon} + \operatorname{tr}\{\boldsymbol{TL}\} + \varrho r - \operatorname{div} \boldsymbol{q} = 0 \qquad [2.4]$$

where ϱ is the mass density at time t, ε is the internal energy per unit mass, r is the heat supply function per unit mass, $\boldsymbol{q}$ is the heat flux vector, $\boldsymbol{T}$ is the symmetric stress of *Cauchy*, $\boldsymbol{L} = \operatorname{grad} \boldsymbol{v}$ is the velocity gradient, $\boldsymbol{v}$ is the velocity, a superposed dot denotes the material time derivative, tr is the trace operator and div is the divergence operator with respect to $\boldsymbol{x}$ keeping t fixed. The above energy equation can also be expressed in terms of the *Helmholtz* free energy per unit mass defined by

[1]) The mechanical behavior of "thermo-rheologically simple" solids with infinitesimal deformation was first considered by *Schwarzl* and *Staverman* and rests on the "temperature-time equivalence" originally proposed by *Leaderman* and in a slightly different form by *Ferry*. For original references, see (1) or (2–3).

$$\psi = \varepsilon - \theta \eta \quad [2.5]$$

where $\theta (> 0)$ is the absolute temperature and η is the entropy per unit mass. We also recall the expression

$$\det \boldsymbol{F}_t(\tau) = [\det \boldsymbol{C}_t(\tau)]^{\frac{1}{2}} = \frac{v(\tau)}{v} \quad [2.6]$$

where $v(\tau)$ is the specific volume of the particle X at time τ.

It is convenient to introduce the notation $\tau = t - s$, $0 \leqq s < \infty$, and define

$$\boldsymbol{F}_t^t(s) = \boldsymbol{F}_t(t-s), \quad \boldsymbol{C}_t^t(s) = \boldsymbol{C}_t(t-s)$$

$$\theta^t(s) = \theta(t-s) \, (0 \leqq s < \infty). \quad [2.7]$$

The tensor functions $\boldsymbol{F}_t^t(s), \boldsymbol{C}_t^t(s)$ and the scalar function $\theta^t(s)$ are histories up to time t of the relative deformation gradient, deformation and the temperature, respectively. The restrictions $\boldsymbol{F}_{tr}^t(\cdot), \boldsymbol{C}_{tr}^t(\cdot), \theta_r^t(\cdot)$ of the functions [2.7] to the open interval $(0, \infty)$ are past histories while

$$\boldsymbol{F}_t^t(0) = \boldsymbol{F}_t(t) = \mathbf{1}, \quad \boldsymbol{C}_t^t(0) = \boldsymbol{C}_t(t) = \mathbf{1}$$

$$\theta = \theta^t(0) = \theta(t) \quad [2.8]$$

are the present values. We note that since $\boldsymbol{F}_t^t(s)$ and $\boldsymbol{C}_t^t(s)$ have always the value 1 at $s = 0$, it is not necessary to distinguish between $\boldsymbol{F}_{tr}^t(s)$ and $\boldsymbol{F}_t^t(s)$ and between $\boldsymbol{C}_{tr}^t(s)$ and $\boldsymbol{C}_t^t(s)$.

Within the scope of the thermo-mechanical theory of simple materials, a simple fluid may be defined by a set of constitutive equations in terms of response functionals for $\psi, \eta, \boldsymbol{T}$, and $\boldsymbol{q}$. Each of these response functionals may be regarded to depend on the present value of the specific volume v (or the mass density ϱ) and the histories $\boldsymbol{C}_t^t(s)$, $\theta^t(s)$, while the response functional for the heat flux depends in addition on the present value of the temperature gradient $\boldsymbol{g} = \operatorname{grad} \theta$. We record, in particular, the constitutive equation for the *Helmholtz* free energy in the form[2])

$$\psi = \mathop{\bar{\mathfrak{F}}}_{s=0}^{\infty} [\boldsymbol{C}_t^t(s), \theta_r^t(s); v, \theta], \quad [2.9]$$

where the domain of $\boldsymbol{C}_t^t(\cdot)$ and $\theta_r^t(\cdot)$ is the positive real line $0 < s < \infty$. Under suitable smoothness assumptions appropriate to simple materials with *fading memory*[3]), it follows from the results obtained by *Coleman* (5) that the response functionals for the entropy and the stress can be expressed in terms of the free energy functional $\mathfrak{F}$ by

[2]) The use of the overbar on the symbol designating the functional in [2.9] is for later convenience. The corresponding symbol without the overbar is used in section 3.

[3]) For details of the required smoothness properties for simple materials with fading memory, we refer the reader to (5) or (6). These smoothness properties are also stated in (2, 4).

$$\eta = -\frac{\partial}{\partial \theta} \{ \mathop{\mathfrak{F}}_{s=0}^{\infty} [\boldsymbol{C}_t^t(s), \theta_r^t(s); v, \theta] \}$$

$$\boldsymbol{T} = \frac{\partial}{\partial v} \{ \mathop{\mathfrak{F}}_{s=0}^{\infty} [\boldsymbol{C}_t^t(s), \theta_r^t(s); v, \theta] \} \mathbf{1}$$

$$- \varrho \mathop{\mathscr{T}}_{s=0}^{\infty} [\boldsymbol{C}_t^t(s), \theta_r^t(s); v, \theta : \boldsymbol{C}_t^t(s)] \quad [2.10]$$

$$\mathop{\mathscr{T}_{ij}}_{s=0}^{\infty} [\boldsymbol{C}_t^t(s), \theta_r^t(s); v, \theta : \boldsymbol{S}(s)]$$

$$= (\delta_{C_{ij}} + \delta_{C_{ji}}) \mathop{\mathfrak{F}}_{s=0}^{\infty} [\boldsymbol{C}_t^t(s), \theta_r^t(s); v, \theta \,|\, S_{ij}(s)] \quad [2.11]$$

where $\mathscr{T}_{ij}$ and S_{ij} designate the tensor components of $\mathscr{T}$ and an arbitrary tensor S in rectangular *Cartesian* coordinates, $\partial/\partial\theta$ and $\partial/\partial v$ denote the partial derivatives with respect to the present values θ and v, respectively, and $\delta_{C_{ij}}$ stands for the *Fréchet* derivative with respect to the rectangular *Cartesian* components $C_{t_{ij}}^t(s)$ of $\boldsymbol{C}_t^t(s)$.

In view of [2.10], it can be shown that the local equation of balance of energy reduces to the form

$$\varrho \theta \dot{\eta} = \varrho r - \operatorname{div} \boldsymbol{q} + \varrho \theta \sigma \quad [2.12]$$

where σ is the internal dissipation given by

$$\sigma = \frac{1}{2\theta} \operatorname{tr} \mathop{\mathscr{T}}_{s=0}^{\infty} \left[\boldsymbol{C}_t^t(s), \theta_r^t(s); v, \theta : \frac{d}{ds} \boldsymbol{C}_t^t(s) \right]$$

$$+ \frac{1}{\theta} \delta_\theta \mathop{\mathfrak{F}}_{s=0}^{\infty} \left[\boldsymbol{C}_t^t(s), \theta_r^t(s); v, \theta \,\middle|\, \frac{d}{ds} \theta_r^t(s) \right] \quad [2.13]$$

and the operator δ denotes the *Fréchet* derivative with respect to $\theta^t(s)$.

Let

$$\theta^t(s) = \theta_0 1^+(s) \quad [2.14]$$

be a constant temperature history at X, where θ_0 is a fixed reference temperature and $1^+(s)$ is a scalar function with constant value 1. When the temperature history is specified by [2.14], the free energy functional in [2.9] reduces to

$$\mathop{\mathfrak{F}}_{s=0}^{\infty} [\boldsymbol{C}_t^t(s), \theta_0 1^+(s); v, \theta_0] = \mathop{\mathfrak{F}^*}_{s=0}^{\infty} [\boldsymbol{C}_t^t(s); v] \quad [2.15]$$

where the starred functional on the right-hand side of [2.15] will be called an *isothermal functional*. For a restricted class of simple fluids, in the next section we elaborate on the determination of the free energy functional $\mathfrak{F}$ and the corresponding constitutive relations for the entropy and the stress from the knowledge of the isothermal functional $\mathfrak{F}^*$. In this connection, we introduce a *modified time scale* which depends on the past history of temperature as follows: For a given temperature history, to each value of $s \, (0 \leqq s < \infty)$ and for a fixed t, there corresponds a number $\xi^t(s)$ through the functional relation[4])

[4]) For further details regarding the modified time scale [2.16], we refer the reader to (2).

$$\xi^t(s) = \mathop{\mathfrak{Z}}_{u=0}^{\infty} (\theta_r^t(u); s). \qquad [2.16]$$

We also impose the conditions

$$\xi^t(0) = 0, \quad \frac{\partial \xi^t(s)}{\partial s} > 0 \qquad [2.17]$$

so that $\xi^t(s)$ is a monotone increasing function of s. The functional $\mathfrak{Z}$ in [2.16] has smoothness properties associated with *fading memory*. Moreover, we require that $\mathfrak{Z}$ reduce to the identity operator when the temperature has been kept at the constant reference value θ_0 for all past time, i.e.,

$$\mathop{\mathfrak{Z}}_{u=0}^{\infty} (\theta_0 1^+(u); s) = s. \qquad [2.18]$$

The functional relation between s and $\xi^t(s)$ may also be expressed through the inverse of $\mathfrak{Z}$, namely

$$s = \mathop{\mathfrak{U}}_{\sigma=0}^{\infty} (\theta_r^t(\sigma); \xi^t(s)). \qquad [2.19]$$

In the remainder of the paper, often we use the abbreviation ξ_s instead of $\xi^t(s)$.

If the temperature at a material point is held constant at some arbitrary value, say θ_1, for all past time, i.e., if $\theta^t(\cdot) = \theta_1 1^+(\cdot)$, then the functional in [2.16] reduces to a function of s. A special case of this latter results is when the function is linear in s so that, in view of [2.17] and [2.18],

$$\mathop{\mathfrak{Z}}_{u=0}^{\infty} [\theta_1 1^+(u); s] = \frac{s}{\Phi(\theta_1)}$$

$$\Phi(\theta_1) > 0, \quad \Phi(\theta_0) = 1. \qquad [2.20]$$

In the theory of "thermo-rheologically simple" solids, $\Phi(\theta)$ is called a "shift" function and the functional $\mathfrak{U}$ in [2.19] has the form

$$\mathop{\mathfrak{U}}_{\sigma=0}^{\infty} (\theta_r^t(\sigma); \xi^t(s)) = \int_0^{\xi} \Phi(\theta^t(\sigma))\, d\sigma \qquad [2.21]$$

which evidently satisfies [2.20].

3. A Special Class of Simple Fluids

Consider a deformation history at a material point X, which for all past time has been a *pure dilation*, i.e., a deformation at X for which the principal stretches (relative to the present configuration) have had a common value for all past time. For a pure dilation, the right relative *Cauchy-Green* tensor is spherical and by [2.6] takes the form

$$C_t^t(s) = c(s)\,\mathbf{1}, \quad c(s) = \left[\frac{v(t-s)}{v}\right]^{2/3}. \qquad [3.1]$$

When the strain history is specified by [3.1], it follows from invariance requirements under superposed rigid body motions that the stress tensor is spherical in the form

$$\boldsymbol{T} = -p\mathbf{1} = \mathop{\mathfrak{T}}_{s=0}^{\infty} \left[\left(\frac{v(t-s)}{v(t)}\right)^{2/3} \mathbf{1}, \theta_r^t(s); v, \theta\right]$$

$$= - \mathop{\not{p}}_{s=0}^{\infty} [v(t-s), \theta_r^t(s); v, \theta]\, \mathbf{1}. \qquad [3.2]$$

In what follows, corresponding to an arbitrary temperature history at a material point X, we shall be interested in dilation histories at X such that the pressure history is a constant history and maintained at the constant reference value p_0. We designate the value of the specific volume under these conditions by v_d. With the use of the specific *Gibbs* free energy function, it was shown in (4) that for dilation histories maintained at a constant pressure p_0, the specific volume $v_d(t)$ has a constitutive equation of the form

$$v_d = \mathop{\mathscr{V}}_{s=0}^{\infty} [\theta(t-s); \theta] \qquad [3.3]$$

and that the stress functional in [3.2] becomes

$$\mathop{\mathfrak{T}}_{s=0}^{\infty} \left[\left(\frac{v_d(t-s)}{v_d}\right)^{2/3} \mathbf{1}, \theta_r^t(s); v_d, \theta\right] = -p_0\mathbf{1}. \qquad [3.4]$$

When the fluid is maintained at a constant temperature history

$$\theta^t(s) = \theta 1^+(s) \qquad [3.5]$$

and at a constant reference pressure p_0 in a dilation history, the specific volume given by [3.3] reduces to a function of the temperature θ:

$$v_d = \mathop{\mathscr{V}}_{s=0}^{\infty} [\theta 1^+(s); \theta] = V(\theta). \qquad [3.6]$$

In particular, we denote by v_0 the value of the function V when θ is the reference temperature θ_0, i.e.

$$v_0 = V(\theta_0). \qquad [3.7]$$

Throughout the remainder of the paper, for simplicity we assume that $p_0 = 0$.

It is convenient to rewrite the constitutive eq. [2.9] in the equivalent form [5])

$$\psi = \mathop{\mathfrak{F}'}_{s=0}^{\infty} \left[C_t^t(s), \theta_r^t(s); \frac{v}{v_0}, \theta\right] \qquad [3.8]$$

[5]) Compare with [4.11] in (4).

where v_0 is defined by [3.7]. When the fluid is subject to a dilation history under vanishing pressure, with the help of [3.1] and [3.3], the free energy functional in [3.8] becomes

$$\underset{s=0}{\overset{\infty}{\mathfrak{F}'}}\left[\left(\frac{v_d(t-s)}{v_d}\right)^{2/3}\mathbf{1},\ \theta_r^t(s);\ \frac{v_d}{v_0},\ \theta\right] = \underset{s=0}{\overset{\infty}{\mathfrak{B}}}\,[\theta_r^t(s);\theta] \qquad [3.9]$$

where v_d is given by [3.6]. A new functional $\mathfrak{F}$ can then be defined by the relation

$$\underset{s=0}{\overset{\infty}{\mathfrak{F}'}}\left[\boldsymbol{C}_t^t(s),\ \theta_r^t(s);\ \frac{v}{v_0},\ \theta\right] = \underset{s=0}{\overset{\infty}{\mathfrak{F}}}\left[\boldsymbol{C}_t^t(s),\ \theta_r^t(s);\ \frac{v}{v_0},\ \theta\right] + \underset{s=0}{\overset{\infty}{\mathfrak{B}}}\,[\theta_r^t(s);\theta] \qquad [3.10]$$

such that

$$\underset{s=0}{\overset{\infty}{\mathfrak{F}}}\left[\left(\frac{v_d(t-s)}{v_d}\right)^{2/3}\mathbf{1},\ \theta_r^t(s);\ \frac{v_d}{v_0},\ \theta\right] = 0. \qquad [3.11]$$

When the fluid is maintained at the constant temperature history [2.14], the functional $\mathfrak{F}$ in [3.10] becomes

$$\underset{s=0}{\overset{\infty}{\mathfrak{F}}}\left[\boldsymbol{C}_t^t(s),\ \theta_0 1^+(s);\ \frac{v}{v_0},\ \theta_0\right] = \underset{s=0}{\overset{\infty}{\mathfrak{F}^*}}\left[\boldsymbol{C}_t^t(s);\ \frac{v}{v_0}\right]. \qquad [3.12]$$

As in [2.15], the functional $\mathfrak{F}^*$ is called an *isothermal functional.*

The functional $\mathfrak{F}$ in [3.10] depends upon the strain history relative to the present configuration at time t, and the ratio of the specific volume at time t and the fixed specific volume v_0 defined by [3.7]. In a dilation history under vanishing pressure and corresponding to the temperature history $\theta(t-s)$, the relative strain measure $\boldsymbol{C}_t^t(s)$ and the ratio v/v_0 would assume the values

$$\boldsymbol{C}_{td}^t(s) = \left[\frac{v_d(t-s)}{v_d}\right]^{2/3}\mathbf{1},\ \frac{v_d}{v_0}. \qquad [3.13]$$

The above values suggest that we decompose $\boldsymbol{C}_t^t(s)$ and v/v_0 in product forms as follows:

$$\boldsymbol{C}_t^t(s) = \left[\frac{v_d(t-s)}{v_d}\right]^{2/3} \boldsymbol{C}_t^{t*}(s)$$

$$\frac{v}{v_0} = \left(\frac{v_d}{v_0}\right)\frac{v}{v_d}. \qquad [3.14]$$

The first factor on the right-hand side of each of the quantities in [3.14] is due to *thermal stretch* defined by [3.13] while the second factor is associated with an *isothermal deformation.* The strain history $\boldsymbol{C}_t^{t*}(s)$ and the ratio v/v_d correspond to a deformation history resulting from a "contracted" motion by an amount which depends on the thermal expansion of the fluid.

The special class of simple fluids may now be specified as follows: For a class of simple fluids with fading memory, the free energy functional $\mathfrak{F}$ in [3.10] is equivalent to the isothermal functional $\mathfrak{F}^*$ in [3.12] provided that (i) the arguments $\boldsymbol{C}_t^t(\cdot)$ and v/v_0 of $\mathfrak{F}^*$ are scaled, respectively, by the values [3.13] and (ii) the variable s is replaced by the time variable [2.16]. By this assumption, we may write

$$\underset{s=0}{\overset{\infty}{\mathfrak{F}}}\left[\boldsymbol{C}_t^t(s),\ \theta_r^t(s);\ \frac{v}{v_0},\ \theta\right] = \underset{s=0}{\overset{\infty}{\mathfrak{F}^*}}\left[\left(\frac{v_d}{v_d(t-\xi_s)}\right)^{2/3}\boldsymbol{C}_t^t(\xi_s);\ \frac{v}{v_d}\right] \qquad [3.15]$$

where the notation ξ_s is a convenient abbreviation for $\xi^t(s)$ in [2.16]. With the use of [3.10] and [3.15], as shown in (4), the constitutive eq. [2.10] can be expressed in the forms

$$\eta = -\frac{\partial\mathfrak{B}}{\partial\theta} + \frac{1}{3}\,\mathrm{tr}\boldsymbol{T}\,\frac{v}{v_d}\frac{\partial\mathscr{V}}{\partial\theta}$$

$$\boldsymbol{T} = -\frac{1}{v}\left\{\mathscr{T}^*\left[\frac{\boldsymbol{C}_t^t(\xi_s)}{c_d(\xi_s)};\ \frac{v}{v_d}:\frac{\boldsymbol{C}_t^t(\xi_s)}{c_d(\xi_s)}\right] - \frac{v}{v_d}\frac{\partial\mathfrak{F}^*}{\partial(v/v_d)}\mathbf{1}\right\} \qquad [3.16]$$

where the operator $\mathscr{T}^*$ is defined similarly to $\mathscr{T}$ in [2.11] and we have introduced the notation

$$c_d(\xi_s) = \left[\frac{v_d}{v_d(t-\xi_s)}\right]^{2/3}. \qquad [3.17]$$

When the temperature history is given by [2.14], the stress constitutive relation in $[3.16]_2$ reduces to

$$T = -\frac{1}{v}\left\{\mathscr{T}^*\left[C_t^t(s);\ \frac{v}{v_0} : C_t^t(s)\right]\right.$$
$$\left. - \frac{v}{v_0}\frac{\partial \mathfrak{F}^*}{\partial(v/v_0)}\mathbf{1}\right\} \qquad [3.18]$$

which shows that the non-isothermal stress constitutive relation may be obtained from its isothermal counterpart through the use of the "contracted" motion with a modified time scale (4).

For later use, we record the reduction of our spěcial theory to the case of incompressible fluids for which the specifid volume remains a constant in all motions, so that

$$v(t) = v_d(t) = v_0, \quad c_d(s) = 1^+(s). \qquad [3.19]$$

We recall that for an incompressible viscoelastic fluid, the isothermal stress constitutive relation (at the reference temperature θ_0) has the form

$$T + \dot{p}\mathbf{1} = \mathop{\mathfrak{T}^*}_{s=0}^{\infty}[C_t^t(s)] \qquad [3.20]$$

where $\dot{p}$ is a hydrostatic pressure which does no work in motions satisfying the constraint $\det C_t^t(\cdot) = 1$. The corresponding nonisothermal stress constitutive relation is then given by

$$T + \dot{p}\mathbf{1} = \mathop{\mathfrak{T}^*}_{s=0}^{\infty}[C_t^t(\xi_s)]. \qquad [3.21]$$

The constitutive equation for the *Helmholtz* free energy assumes the form

$$\psi = \mathop{\tilde{\mathfrak{F}}^*}_{s=0}^{\infty}[C_t^t(\xi_s)] + \mathop{\mathfrak{B}}_{s=0}^{\infty}[\theta_r^t(s); \theta] \qquad [3.22]$$

and the constitutive eq. [3.16] becomes

$$\eta = -\frac{\partial \mathfrak{B}}{\partial \theta}$$

$$T + \dot{p}\mathbf{1} = -\frac{1}{v_C}\{\mathscr{T}^*[C_t^t(\xi_s);\ C_t^t(\xi_s)]\} \qquad [3.23]$$

where $\mathscr{T}^*$ is defined similarly to $\mathscr{T}^*$ in [3.16].

4. Viscometric Functions

Within the framework of the purely mechanical (or isothermal) theory of viscometric flows, viscometric functions are usually defined from the calculation of the stress components at a material point X and at time t in a *lineal flow*[6]). We discuss here the effect of the temperature history on the viscometric functions, using the results of the previous section for incompressible fluids. Thus, with reference to a rectangular *Cartes*ian coordinate system $\{x, y, z\}$, consider the motion of a material point in a fluid whose velocity components at time t in the x, y, z-directions are

$$u = 0, \quad v = v(x, t), \quad w = 0 \qquad [4.1]$$

respectively. The coordinates of the material point X at time $\tau = t - s$ are then given by

$$x(t - s) = x, \quad z(t - s) = z$$
$$y(t - s) = y + \int_t^{t-s} v(x, t')\,dt'. \qquad [4.2]$$

Using [4.2], the right-relative *Gauchy-Green* tensor is found to be

$$C_t^t(s) = \mathbf{1} + g(x, s)(N + N^T) + g^2(x, s)\,N^T N \qquad [4.3]$$

where N is a constant tensor with the component matrix

$$[N] = \begin{Vmatrix} 0 & 0 & 0 \\ 1 & 0 & 0 \\ 0 & 0 & 0 \end{Vmatrix} \qquad [4.4]$$

$g(x, s)$ is given by

$$g(x, s) = \int_t^{t-s} v'(x, \sigma)\,d\sigma \qquad [4.5]$$

v' denotes the partial derivative of v with respect to x and we emphasize that the spatial argument in g refers to the coordinate of X at time t.

When the fluid is at the spatially uniform reference temperature θ_0, the stress tensor at X and at time t is given by

$$T + \dot{p}\mathbf{1} = \mathop{\mathfrak{T}^*}_{s=0}^{\infty}[\ \mathbf{1} + g(x, s)(N + N^T)$$
$$+ g(x, s)^2\,N^T N]. \qquad [4.6]$$

By considering invariance of [4.6] under superposed rigid body motions, it can be shown that the stress components have the form (see *Truesdell* and *Noll* [6, *Sect.* 106]):

$$T_{12} = \mathop{\mathfrak{t}^*}_{s=0}^{\infty}[g(x, s)]$$
$$T_{11} - T_{33} = \mathop{\mathfrak{s}_1^*}_{s=0}^{\infty}[g(x, s)]$$
$$T_{22} - T_{33} = \mathop{\mathfrak{s}_2^*}_{s=0}^{\infty}[g(x, s)] \qquad [4.7]$$

[6]) We use the term *lineal flow* in the sense of *Truesdell* and *Noll* (6, p. 429).

where $\mathfrak{t}^*$, $\mathfrak{s}_1^*$, $\mathfrak{s}_2^*$ are material functionals. The functionals $\mathfrak{s}_1^*$ and $\mathfrak{s}_2^*$, which determine the normal stress differences, are even in $g(x, s)$ while $\mathfrak{t}^*$ is odd in $g(x, s)$.

When the velocity field is steady, instead of [4.1], we have

$$u = 0, \quad v = v(x), \quad w = 0 \tag{4.8}$$

and $g(x, s)$ is simply linear in s, i.e.,

$$g(x, s) = -v'(x)\, s. \tag{4.9}$$

The material functionals $\mathfrak{t}^*$, $\mathfrak{s}_1^*$, $\mathfrak{s}_2^*$ in [4.7] then reduce to

$$\begin{aligned}
\overset{\infty}{\underset{s=0}{\mathfrak{t}^*}}[-v'(x)\,s] &= \tau^*[v'(x)] \\
\overset{\infty}{\underset{s=0}{\mathfrak{s}_1^*}}[-v'(x)\,s] &= \sigma_1^*[v'(x)] \\
\underset{s=0}{\mathfrak{s}_2^*}[-v'(x)\,s] &= \sigma_2^*[v'(x)]
\end{aligned} \tag{4.10}$$

where τ^*, σ_1^*, σ_2^* are the viscometric functions of the fluid at the reference temperature θ_0.

Consider again the velocity field [4.1] and hence the deformation [4.2], but let the temperature history of the material point X which occupies the position $\{x, y, z\}$ at time t be specified by the function $\theta(x, y, z, t-s)$. In view of [4.3] and [3.21], the stress tensor at time t is given by

$$\begin{aligned}
\boldsymbol{T} = \dot{p}\boldsymbol{1} = \overset{\infty}{\underset{s=0}{\mathfrak{T}^*}}[\boldsymbol{1} + g(x, \xi_s)(\boldsymbol{N} + \boldsymbol{N}^T) \\
+ g(x, \xi_s)^2\, \boldsymbol{N}^T \boldsymbol{N}]
\end{aligned} \tag{4.11}$$

where by [2.13] the argument ξ_s is

$$\xi_s = \overset{\infty}{\underset{u=0}{\mathfrak{Z}}}[\theta(x, y, z, t-u); s]. \tag{4.12}$$

Since the stress tensor in the non-isothermal case is obtained simply by substituting ξ_s for s in the argument of the stress functional $\mathfrak{T}^*$ in [4.6], we recall the results [4.7] and conclude that

$$\begin{aligned}
T_{12} &= \overset{\infty}{\underset{s=0}{\mathfrak{t}^*}}[g(x, \xi_s)] \\
T_{11} - T_{33} &= \overset{\infty}{\underset{s=0}{\mathfrak{s}_1^*}}[g(x, \xi_s)] \\
T_{22} - T_{33} &= \overset{\infty}{\underset{s=0}{\mathfrak{s}_J^*}}[g(x, \xi_s)].
\end{aligned} \tag{4.13}$$

Similarly, when the velocity field is steady and given by [4.8], the expressions in [4.13] become

$$\begin{aligned}
T_{12} &= \overset{\infty}{\underset{s=0}{\mathfrak{t}^*}}[-v'(x)\,\xi_s] \\
T_{11} - T_{33} &= \overset{\infty}{\underset{s=0}{\mathfrak{s}_1^*}}[-v'(x)\,\xi_s] \\
T_{22} - T_{33} &= \overset{\infty}{\underset{s=0}{\mathfrak{s}_2^*}}[-v'(x)\,\xi_s].
\end{aligned} \tag{4.14}$$

The material functionals in [4.14] do not reduce to the viscometric functions since, in general, s does not appear linearly in the argument of $\mathfrak{t}^*$, $\mathfrak{s}_1^*$, $\mathfrak{s}_2^*$, as it was the case in [4.10].

A noteworthy simplification occurs, however, when the temperature at a material point may be spatially non-uniform but constant in time. Recalling [4.2], we assume that the temperature field is given by a function of the form $\theta(x, z)$. Then, the modified time scale is given by

$$\xi_s = \overset{\infty}{\underset{u=0}{\mathfrak{Z}}}[\theta(x, z)\,1^+(u); s] \tag{4.15}$$

and, in view of [2.20], we have

$$\xi_s = \frac{s}{\Phi[\theta(x, z)]}. \tag{4.16}$$

Introducing [4.16] into [4.14] and comparing the results with those in [4.10], we obtain

$$\begin{aligned}
T_{12} &= \tau^*\left\{\frac{v'(x)}{\Phi[\theta(x, z)]}\right\} \\
T_{11} - T_{33} &= \sigma_1^*\left\{\frac{v'(x)}{\Phi[\theta(x, z)]}\right\} \\
T_{22} - T_{33} &= \sigma_2^*\left\{\frac{v'(x)}{\Phi[\theta(x, z)]}\right\}.
\end{aligned} \tag{4.17}$$

Thus, when the temperature remains a constant along each path line, the effect of temperature variation (from the reference temperature θ_0) in a lineal flow is equivalent to a decrease (or an increase) of the shear rate in the isothermal viscometric functions depending on the "shift" function Φ.

To elaborate, we recall that the *viscosity function* μ is defined by the ratio of the shear stress T_{12} to the shear rate $v'(x)$ in a steady lineal flow. In particular, under isothermal conditions corresponding to the reference temperature θ_0, we have

$$\mu^*[v'(x)] = \frac{\tau^*[v'(x)]}{v'(x)}. \tag{4.18}$$

The zero-shear viscosity at reference temperature θ_0 is defined by

$$\mu_0^* = \lim_{v' \to 0} \mu^* [v'(x)]. \qquad [4.19]$$

When the temperature differs from the reference temperature θ_0 and the modified time scale is given by [4.16], from $[4.17]_1$ we obtain

$$\mu[v'(x), \theta(x, z)] = \frac{1}{v'(x)} \tau^* \left\{ \frac{v'(x)}{\Phi[\theta(x, z)]} \right\}$$

$$= \frac{1}{\Phi[\theta(x, z)]} \mu^* \left\{ \frac{v'(x)}{\Phi[\theta(x, z)]} \right\}. \qquad [4.20]$$

Also, from [4.19] and [4.20] follows the expression

$$\mu_0[\theta(x, z)] = \frac{\mu_0^*}{\Phi[\theta(x, z)]}. \qquad [4.21]$$

5. Poiseuille flow

In order to indicate explicitly the simplified nature of the results of the restricted theory discussed in sections 3 and 4, we consider now a simple example of *Poiseuille* flow while taking dissipation into account. With reference to a system of cylindrical polar coordinates $\{r, \varphi, z\}$, the steady flow of an incompressible fluid through a fixed infinite circular pipe of radius R is characterized by the velocity field

$$u = 0, \quad v = 0, \quad w = w(r), \quad (0 \leqq r \leqq R \qquad [5.1]$$

where $\{u, v, w\}$ are the velocity components in the $\{r, \varphi, z\}$-directions, respectively. The position at time $\tau = t - s$ of a material point which occupies the position $\{r, \varphi, z\}$ at time t is given by

$$r(t - s) = r, \quad \varphi(t - s) = \varphi$$

$$z(t - s) = z - w(r)s \qquad [5.2]$$

and the physical components $C_{t\langle ij \rangle}^{t}(s)$ of the relative *Cauchy-Green* tensor $C_t^t(s)$ are found to be

$$\|C \quad (s)\| = \begin{Vmatrix} 1 + w'^2 s^2 & 0 & -w's \\ 0 & 1 & 0 \\ -w's & 0 & 1 \end{Vmatrix}, \qquad [5.3]$$

where $w' = \dfrac{dw}{dr}$.

In a steady state, we assume that the temperature is a constant along each path line, i.e.,

$$\theta = \theta(r). \qquad [5.4]$$

Then, for a material point X at a distance r from the axis of the pipe, the modified time scale is given by

$$\xi_s = \mathop{\mathfrak{Z}}_{u=0}^{\infty} [\theta(r)\, 1^+(u); s] = \frac{s}{\Phi[\theta(r)]} \qquad [5.5]$$

and the stress tensor at X has the form

$$T + p\mathbf{1} = \mathop{\mathfrak{T}^*}_{s=0}^{\infty} \left\{ C_t^t \left[\frac{s}{\Phi[\theta(r)]} \right] \right\}. \qquad [5.6]$$

The values of the physical components of C_t^t in [5.3] are identical to those in [4.3], provided we associate a correspondence between $\{x, y, z\}$ and $\{r, z, \varphi\}$ components of the respective velocities fields and also substitute w' for v'. We may then conclude from [4.17] that

$$T_{rz} = \tau^* \left\{ \frac{w'(r)}{\Phi[\theta(r)]} \right\}$$

$$T_{rr} - T_{\varphi\varphi} = \sigma_1^* \left\{ \frac{w'(r)}{\Phi[\theta(r)]} \right\}$$

$$T_{zz} - T_{\varphi\varphi} = \sigma_2^* \left\{ \frac{w'(r)}{\Phi[\theta(r)]} \right\}$$

$$T_{r\varphi} = T_{\varphi z} = 0 \qquad [5.7]$$

where T_{rr}, $T_{\varphi\varphi}$, $T_{r\varphi}$, etc., in [5.7] are the physical components of the stress tensor $\boldsymbol{T}$ referred to the coordinates r, φ, z.

In terms of the physical components of stress and in the absence of the body forces, the equations of motion are:

$$\frac{\partial T_{rr}}{\partial r} + \frac{1}{r}(T_{rr} - T_{\varphi\varphi}) = 0$$

$$\frac{1}{r}\frac{\partial}{\partial r}(r T_{rz}) + \frac{\partial T_{zz}}{\partial z} = 0. \qquad [5.8]$$

From the elimination of T_{rr} between $[5.7]_2$ and [5.8] follows the expression

$$T_{\varphi\varphi} = -\int^{r} \frac{1}{\varrho} \frac{\partial}{\partial \varrho} \left[\varrho \sigma_1^* \left(\frac{w'}{\Phi(\theta)} \right) \right] d\varrho + f(z) \qquad [5.9]$$

where $f(z)$ is an arbitrary function of z. Next, by eliminating T_{zz} between $[5.7]_3$ and $[5.8]_2$ and making use of [5.9] and $[5.7]_1$, we obtain

$$\frac{1}{r}\frac{\partial}{\partial r}\left[r\tau^* \left(\frac{w'}{\Phi(\theta)} \right) \right] + f'(z) = 0 \qquad [5.10]$$

where $f'(z) = df/dz$. It follows from [5.10] that

$$f'(z) = A, \quad \frac{1}{r}\frac{\partial}{\partial r}\left[r\tau^* \left(\frac{w'}{\Phi(\theta)} \right) \right] = -A \qquad [5.11]$$

A being an arbitrary constant. But, since $\boldsymbol{T}$ is continuous at $r = 0$, from $[5.11]_2$ we obtain

$$\tau^*\left[-\frac{w'}{\Phi(\theta)}\right] = \frac{1}{2}A\,r \qquad [5.12]$$

where we have used the fact that τ^* is an odd function of its argument.

Let ζ^* be the inverse function of τ^*,

$$\zeta^* = (\tau^*)^{-1} \qquad [5.13]$$

and let us assume that the shear rate is confined to an interval where ζ^* exists. Then, from [5.12] and [5.13] we obtain

$$w'(r) = -\Phi[\theta(r)]\,\zeta^*(\tfrac{1}{2}A\,r) \qquad [5.14]$$

and, if the velocity vanishes on the boundary $r = R$, we have

$$w(r) = -\int_R^r \Phi[\theta(\varrho)]\,\zeta^*(\tfrac{1}{2}A\,\varrho)\,d\varrho\,. \qquad [5.15]$$

Once the temperature profile is known, for a given pressure gradient A, [5.15] determines the velocity field in the non-isothermal flow. We observe that in an isothermal flow at the reference temperature θ_0, since Φ reduces to $\Phi(\theta_0) = 1$, [5.15] gives the usual expression for the velocity $w(r)$.

The temperature field can be determined from the residual energy eq. [2.12]. However, in the present development, it is easier to return to the original balance of energy. For steady flow under consideration, it follows from [5.1–5.4] that along a path line the temperature and the strain history are time-independent. Also, from [3.22] and $[3.23]_2$ we conclude that ψ and η are constant along each path line, so that $\dot\psi, \dot\eta = 0$ along with $\dot\theta = 0$. We further assume that heat flows only through the wall of the pipe, in the absence of heat sources and sinks, and put $r = 0$ in the energy equation. With these observations and using [2.5], the energy eq. [2.4] becomes

$$\operatorname{div}\boldsymbol{q} - \operatorname{tr}(\boldsymbol{TL}) = 0 \qquad [5.16]$$

where

$$\operatorname{tr}(\boldsymbol{TL}) = T_{rz}w'(r) = w'(r)\,\tau^*\left[\frac{w'}{\Phi(\theta)}\right]$$

$$= \Phi[\theta(r)]\,\frac{A\,r}{2}\,\zeta^*\left(\frac{A\,r}{2}\right) \qquad [5.17]$$

is obtained with the help of [5.1], [5.12] and [5.14].

Using the result [5.17], the temperature profile can be determined from the energy eq. [5.16], which also requires a constitutive equation for the heat flux $\boldsymbol{q}$. The latter depends in general on the whole history of strain and temperature, in addition to the present value of the temperature gradient. In the remainder of the paper, we obtain a reduced form of the energy equation with the use of a special constitutive assumption for the heat flux, namely

$$q = \mathop{\mathfrak{Q}}_{s=0}^{\infty}[\theta_r^t(s);\,\theta,\boldsymbol{g}]\,. \qquad [5.18]$$

For the present steady flow, in view of [5.4], the functional $\mathfrak{Q}$ reduces to a function of θ and $\boldsymbol{g}$. Then, [5.18] becomes

$$\boldsymbol{q} = \mathfrak{Q}[\theta(r), \boldsymbol{g}] \qquad [5.19]$$

and since the only nonvanishing component of $\boldsymbol{g}$ is $g_r = \partial\theta/\partial r$, the components of the heat flux in the $\{r, \varphi, z\}$-coordinate directions are

$$q_r = -\varkappa\left[\left(\frac{d\theta}{dr}\right)^2, \theta\right]\frac{d\theta}{dr}, \qquad q_\varphi, q_z = 0. \qquad [5.20]$$

Finally, by inserting [5.20] and [5.17] into [5.16], we obtain

$$\frac{1}{r}\frac{d}{dr}\left\{r\varkappa\left[\left(\frac{d\theta}{dr}\right)^2, \theta\right]\frac{d\theta}{dr}\right\}$$

$$+\,\Phi(\theta)\,\frac{A\,r}{2}\,\zeta^*\left(\frac{A\,r}{2}\right) = 0 \qquad [5.21]$$

which for given values of the pressure gradient A is a nonlinear ordinary differential equation in r.

Acknowledgement

The results reported here were obtained in the course of research supported by the U.S. Office of Naval Research under Contract N00014-69-A-0200-1008 with the University of California, Berkeley (U.C.B.). One of us (M.J.C.) held a visiting appointment in U.C.B. during 1971 and the other (P.M.N.) would like to acknowledge an appointment during 1971–72 in U.C.B.'s Miller Institute for Basic Research in Science.

Summary

A brief account of a non-isothermal theory of a restricted class of viscoelastic fluids with fading memory followed by its application to some special viscometric flows.

References

1) *Ferry, J. D.*, Viscoelastic Properties of Polymers (2nd ed., London-New York 1970).

2) *Crochet, M. J.* and *P. M. Naghdi*, Int. J. Eng. Sci. 7, 1173 (1969).

3) *Crochet, M. J.* and *P. M. Naghdi*, Proc. IUTAM Symp. on Thermoinelasticity (East Kilbride, Scotland, 1968), p. 59 (Berlin-Heidelberg-New York 1970).

4) *Crochet, M. J.* and *P. M. Naghdi*, Int. J. Eng. Sci. **10** (1972, to appear).

5) *Coleman, B. D.*, Arch. Rational Mech. Anal. **17**, 1 (1964).

6) *Truesdell, C.* and *W. Noll*, The Non-Linear Field Theories of Mechanics, Handbuch der Physik, Vol. III/3. Edited by *S. Flügge* (Berlin-Heidelberg-New York 1965).

Authors' address:

Prof. *P. M. Naghdi*
6121 Etchevery Hall
Dept. of Mechanical Engineering
University of California
Berkeley, California 94720 (U.S.A.)

Rheol. Acta 12, 330–336 (1973)

From the Institute National du Verre, Charleroi (Belgium), and the University of Louvain, Louvain-La-Neuve (Belgium)

Thermal stresses during annealing of a glass ribbon

By G. Tackels *and* M. J. Crochet

With 5 figures

(Received October 27, 1972)

1. Introduction

In order to avoid fracture and to keep transient and residual stresses within acceptable bounds, it is necessary to include in the continuous process of glass forming an annealing phase, during which the temperature of the sheet is slowly brought from approximately 600 °C down to 25 °C. An adequate control of the annealing gallery requests a good understanding of the stress generation during cooling. The problem is quite complex, since the prediction of stresses is intimately related to the thermoviscoelastic character of glass. Moreover, stress relaxation is accompanied by structural relaxation, which is strongly dependent upon temperature history.

The first thermoviscoelastic analysis of stress generation during the cooling of flat glass was presentend by *Lee* et al. (6), who considered tempering at constant rate.

Their work is based upon the thermoviscoelastic behavior of glass, without allowing for structural relaxation. This last omission is not acceptable when the cooling rate is slower than the usual temperature rate of the tempering process. It is then necessary to consider a modified time-scale in the relaxation function which depends upon the past history of temperature. The theoretical background of such modified time-scales has been given recently by *Crochet* and *Naghdi* (1).

The importance of structural relaxation as regards the evaluation of residual stresses during annealing has been recognized by *Gardon* and *Narayanaswami* (5), who have proposed a mathematical model of annealing which considers both stress and structural relaxation. Their model requests however the determination of an adjustable parameter in order to fit the experimental data.

The physical properties of glass in the transformation range have been the object of an extensive study by *Debast* and *Gilard* (3, 4); in particular their measurements show that structural relaxation may be taken into account by making use of the fictive temperature concept developed by *Tool* (8) and *Ritland* (7). In a recent paper (2), we have used the experimental results of *Debast* and *Gilard* in order to calculate the transient and residual stresses in a flat sheet of glass during annealing. In particular, we have shown how it is possible to select an optimal process within a given class of temperature histories. The numerical programme has also been applied to the calculation of residual stresses in industrial processes. Calculated values of the residual stresses show good agreement with the measured values.

In section 2, we briefly review the basic constitutive equations to be used in our analysis. In particular, we will elaborate to some extent on the nature of the modified time-scale, and show how it is related to the viscosity. We also mention in section 2 the numerical values used in our calculations.

In our earlier paper (2), the calculations were limited to temperature fields which are symmetric with respect to the middle plane of the sheet. It is known however that, during annealing, the temperature may differ appreciably on both faces of the sheet. In section 3, we show how a non-symmetric temperature field may be taken into account in calculating the transient and residual stresses. The numerical programme is then applied to a typical temperature history in vertical drawing. The numerical results show that the transient stresses are affected by the lack of symmetry; its effect upon the residual stresses is hardly observable.

In the study of industrial processes and optimized temperature programmes, we have usually interpolated the temperature between points by linear segments. The resulting discontinuity in the rate of temperature causes peaks of the transient stresses. In section 4, we study the influence of smoothing on a typical annealing curve. The peaks are avoided; however, the residual stresses are affected very little by the smoothing process.

2. Constitutive relations

Experimental observations show that, for infinitesimal deformations, glass behaves as an isotropic linear viscoelastic material in a temperature range extending from 450–600 °C. At a fixed reference temperature T_0, let $\psi_1(t)$, $\psi_2(t)$, be the relaxation functions of the material in shear and pure dilatation, respectively, and let $\Phi_1(t)$, $\Phi_2(t)$ be the corresponding creep functions; the creep and relaxation functions are normalized such that

$$\Phi_1(0) = \Phi_2(0) = \psi_1(0) = \psi_2(0) = 1. \qquad [2.1]$$

The stress-strain relations are expressed as follows,

$$s_{ij}(t) = 2\mu_0 \int_{-\infty}^{t} \psi_1(t-t') \frac{\partial e_{ij}(t')}{\partial t'} dt'$$

$$s(t) = 3k_0 \int_{-\infty}^{t} \psi_2(t-t') \frac{\partial e(t')}{\partial t'} dt' \qquad [2.2]$$

or, equivalently,

$$e_{ij}(t) = \frac{1}{2\mu_0} \int_{-\infty}^{t} \Phi_1(t - t') \frac{\partial s_{ij}(t')}{\partial t'} dt',$$

$$e(t) = \frac{1}{3k_0} \int_{-\infty}^{t} \Phi_2(t - t') \frac{\partial s(t')}{\partial t'} dt'; \qquad [2.3]$$

in [2.2] and [2.3], s_{ij} and e_{ij} denote the deviatoric components of the stress and strain tensors, respectively, s and e denote the spherical components, μ_0 is the shear modulus and k_0 is the modulus of compression.

The mechanical behavior of glass at a given temperature T_0 is frequently characterized by its *viscosity* η_0, measured in a creep experiment, for which

$$t < 0, \quad \sigma(t) = 0; \quad t \geq 0, \quad \sigma(t) = \sigma_0,$$

$$\varepsilon(t) = \frac{\sigma_0}{E_0} \Phi(t), \qquad [2.4]$$

with σ_0 being a fixed constant and

$$E_0 = \frac{9k_0\mu_0}{3k_0 + \mu_0},$$

$$\Phi(t) = \frac{E_0}{3\mu_0} \Phi_1(t) + \frac{E_0}{9k_0} \Phi_2(t). \qquad [2.5]$$

It is observed experimentally that when t becomes large, the function $\Phi(t)$ shows a linear dependence on time, and the viscosity is defined by

$$\eta_0 = \lim_{t\to\infty} \frac{\sigma_0}{\dot{\varepsilon}(t)} = \frac{E_0}{\lim_{t\to\infty} \dot{\Phi}(t)}. \qquad [2.6]$$

We assume that, when time increases, $\dot{\Phi}(t)$ converges monotonically to the value E_0/η_0; thus, there exists a number t^* such that, within a given accuracy, we may write

$$t > t^*, \quad \dot{\Phi}(t) \cong \frac{E_0}{\eta_0}. \qquad [2.7]$$

Experimental observations (3) show that, in the viscoelastic domain, stabilized glass exhibits a thermorheologically simple response; at a constant temperature T_1 which differs from T_0, the constitutive relations [2.2] become

$$s_{ij}(t) = 2\mu_0 \int_{-\infty}^{t} \psi_1(\xi - \xi') \frac{\partial e_{ij}(t')}{\partial t'} dt',$$

$$s(t) = 3k_0 \int_{-\infty}^{t} \psi_2(\xi - \xi') \frac{\partial}{\partial t'} [e(t') - \alpha(T_1)]\, dt', \qquad [2.8]$$

where ξ, ξ' are modified times corresponding to t, t' through the relation

$$\xi = t\varphi(T_1), \qquad [2.9]$$

$\alpha(T)$ is the coefficient of thermal expansion, and $\varphi(T)$ is the shift function.

The shift function $\varphi(T_1)$ is easily related to the viscosity as a function of temperature; for a stress history given by [2.4], we obtain from [2.8] and [2.5],

$$\varepsilon(t) - \alpha(T_1) = \frac{\sigma_0}{E_0} \Phi[t\varphi(T_1)] \qquad [2.10]$$

and from [2.6],

$$\eta(T_1) = \lim_{t\to\infty} \frac{\sigma_0}{\dot{\varepsilon}(t)} = \frac{E_0}{\varphi(T_1)} \lim_{t\to\infty} \frac{1}{\dot{\Phi}[t\varphi(T_1)]} \qquad [2.11]$$

$$\varphi(T) = \frac{\eta_0}{\eta(T)}.$$

In view of [2.9] and [2.11], [2.8] become

$$s_{ij}(t) = 2\mu_0 \int_{-\infty}^{t} \Psi_1 \left[\frac{\eta_0}{\eta(T_1)}(t - t')\right] \frac{\partial e_{ij}(t')}{\partial t'} dt',$$

$$s(t) = 3k_0 \int_{-\infty}^{t} \Psi_2 \left[\frac{\eta_0}{\eta(T_1)}(t - t')\right] \frac{\partial}{\partial t'} \times [e(t') - \alpha(T_1)]\, dt'. \qquad [2.12]$$

When a thermorheologically simple material is not subject to structural modifications under a varying temperature field, the stress strain relations [2.8] remain valid for an arbitrary temperature history, provided $\alpha(T_1)$ in [2.8]$_2$ is replaced by $\alpha[T(t_1)]$, and the modified time ξ is defined by

$$\xi(t) = \int_0^t \varphi[T(\tau)]\, d\tau. \qquad [2.13]$$

In view of the strong dependence of glass properties upon the temperature history, [2.13] is in general not valid for glass. Experimental observations made by *Debast* and *Gilard* (4) on mechanical properties of glass in the transformation range indicate that the modified time must be determined by the history of the temperature and is given by a functional Z such that

$$\xi(t) = \underset{s=0}{\overset{\infty}{Z}} \{T(t - s)\}, \quad \xi(0) = 0, \quad \dot{\xi}(t) > 0. \qquad [2.14]$$

Let us consider a temperature history defined as follows,

$$t < 0, \quad T = T_1; \quad t \geq 0, \quad T = T(t), \qquad [2.15]$$

where $T(t)$ is selected arbitrarily and consider a time t^+ such that

$$t^+ = \frac{\eta(T_1)}{\eta_0} t^*, \qquad [2.16]$$

where t^* has been defined in [2.7]. In order to obtain the form of the functional Z in [2.14] for a temperature history defined by [2.15], let us consider a simple traction experiment [fig. 1] for which

$$t < -t^+, \quad \sigma(t) = 0;$$

$$-t^+ \leq t, \quad \sigma(t) = \sigma_0; \qquad [2.17]$$

and

$$\varepsilon(t) - \alpha(t) = \frac{\sigma_0}{E_0} \Phi[\xi(t) - \xi(-t^+)]. \qquad [2.18]$$

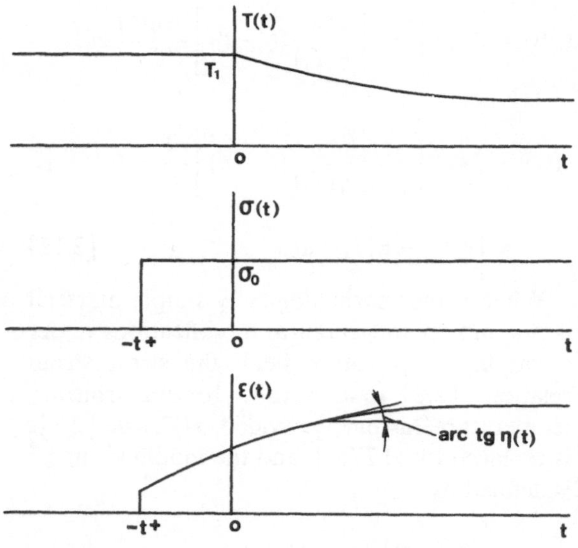

Fig. 1. Non isothermal simple traction experiment

In view of $[2.15]_1$, we have

$$\xi(-t^+) = -t^+ \varphi(T_1) = -t^*, \qquad [2.19]$$

and thus, from $[2.14]_3$,

$$t > 0, \quad \xi(t) - \xi(-t^+) = t^* + \xi(t) > t^*. \qquad [2.20]$$

From [2.7] and [2.20] we have

$$t > 0, \quad \frac{d\Phi}{d[\xi(t) - \xi(-t^+)]} \cong \frac{E_0}{\eta_0} \qquad [2.21]$$

and finally, from [2.18] and [2.21] we obtain

$$\frac{d}{dt}[\varepsilon(t) - \alpha(t)] = \frac{\sigma_0}{\eta_0} \frac{d\xi}{dt}. \qquad [2.22]$$

If in a simple traction experiment defined by [2.17], we introduce the notation,

$$t \geq 0, \quad \eta(t) = \frac{\sigma_0}{\frac{d}{dt}[\varepsilon(t) - \alpha(t)]}, \qquad [2.23]$$

we obtain from [2.22]

$$\frac{d\xi}{dt} = \frac{\eta_0}{\eta(t)}, \quad \xi(t) = \int_0^t \frac{\eta_0}{\eta(\tau)} d\tau. \qquad [2.24]$$

The modified time scale [2.14] is thus entirely defined by the viscosity function $\eta(t)$, which is itself determined by the history of the temperature.

In order to take the structural changes of glass into account, we will assume that its material properties at time t depend upon the temperature at time t and a fictive temperature T_f, introduced by *Tool* (8); the value of T_f depends upon the past history of temperature. In particular, the constitutive relations for the thermal expansion and the viscosity take the form,

$$\alpha(t) = \alpha[T(t), T_f(t)],$$

$$\eta(t) = \eta[T(t), T_f(t)] \qquad [2.25]$$

while the value of $T_f(t)$ is determined by a functional over the temperature history,

$$T_f(t) \underset{\tau \leq t}{\mathfrak{T}} [T(\tau)]. \qquad [2.26]$$

In later sections, we will consider temperature histories of the type exhibited in [2.15]; the constitutive relations are then given by

$$s_{ij}(t) = 2\mu_0 \int_{-\infty}^{t} \Psi_1(\xi - \xi') \frac{\partial}{\partial t'} e_{ij}(t')\, dt',$$

$$s(t) = 3k_0 \int_{-\infty}^{t} \Psi_2(\xi - \xi') \frac{\partial}{\partial t'}$$

$$\times \{e(t') - \alpha[T(t'), T_f(t')]\}\, dt',$$

$$\xi - \xi' = \int_{t'}^{t} \frac{\eta_0}{\eta[T(\tau), T_f(\tau)]} d\tau. \qquad [2.27]$$

The physical properties of a sodo-calcic window glass in the transformation range have been evaluated by *Debast* and *Gilard* (3). From their measurements, we may consider that the relaxation functions $\Psi_1(t)$, $\Psi_2(t)$ are identical and, at a constant temperature T_0, they are given by

$$\Psi_1(t) = \Psi_2(t) = \exp\left[-\left(K_R \frac{t}{\eta_0}\right)^{b_1}\right] \quad [2.28]$$

$K_R = 0.302\ 10^{12}$ g/cm sec^2, $b_1 = 0.5435$, where η_0 is the equilibrium value of the viscosity at temperature T_0.

The form [2.28] of the relaxation functions agrees with the change of time scale defined by [2.9] and [2.11]. *Young*s modulus and *Poisson*s ratio between 600 °C and 20 °C are given by the average values

$$E_0 = 6600\ \text{kg/mm}^2, \quad \nu_0 = 0.23. \quad [2.29]$$

The viscosity as a function of the temperature T and the fictive temperature T_f is given by

$$\log \frac{\eta(T, T_f)}{\eta_0} = -0.0592\,(T_f - T_0) - 0.0280\,(T - T_f), \quad [2.30]$$

where η_0 is the equilibrium viscosity corresponding to the reference temperature T_0; in the next section, we will use

$$T_0 = 500\ °\text{C}, \quad \eta_0 = 10^{15.87}\ \text{poises}. \quad [2.31]$$

The coefficient α of thermal expansion may be written as follows,

$$\alpha(T, T_f) = \alpha_0 + 3.47\ 10^{-5}(T_f - T_0) + 0.97 \times 10^{-5}(T - T_f); \quad [2.32]$$

there is however no need to specify the reference value α_0 since α appears only through its derivatives in the constitutive relations.

Finally, we must specify the form of the functional $\mathfrak{T}$ in [2.26]; we will assume that the fictive temperature T_f obeys a modified *Tools* equation. From experimental observations given in (4), we may write

$$\frac{dT_f}{dt} = \exp[0.1362\,(T - 500)] \times \exp[-0.0718(T - T_f)] \times [0.021(T - T_f) + 0.0002(T - T_f)^2], \quad [2.33]$$

where t is expressed in hours and T, T_f in °C.

3. Calculation of residual stresses

In our earlier paper [2], the fundamental equations of section 2 were used for calculating the residual stresses in a plane sheet of glass cooled from 575 °C to 20 °C; the temperature field was symmetric with respect to the middle plane of the plate.

The method has been applied to the analysis of industrial processes. The temperature history is measured on the faces of the glass sheet in the annealing gallery, and a numerical programme calculates the transient and residual stresses in the thickness of the sheet. The calculated value σ_c of the residual stress in the central plane has been compared to the corresponding measured stress σ_m; the average value of the ratio σ_c/σ_m over a series of 16 industrial curves was 0.90 (standard deviation: 0.16).

In the case of vertical drawing (Pittsburgh), it is known that the temperature differs on both faces of the sheet. We wish to analyze, on a typical temperature history in vertical drawing, the influence of the lack of symmetry of the temperature distribution upon the transient and residual stresses.

Let x, y, z be a *Cartes*ian coordinate system, and let the (x, y) plane coincide with the middle plane of the sheet. We shall assume that the temperature field does not depend upon the x and y coordinates; the temperature history is imposed on the boundaries $z = \pm l$, and we have

$$T(l, t) = T_1(t), \quad T(-l, t) = T_2(t), \quad [3.1]$$

where $T_1(t)$, $T_2(t)$ are known functions of time.

The temperature field within the sheet of glass is obtained by numerically integrating *Fourier*s law of heat conduction,

$$\varrho c \frac{\partial T(z, t)}{\partial t} = \frac{\partial}{\partial z}\left[k \frac{\partial}{\partial z} T(z, t)\right], \quad [3.2]$$

where ϱc is the specific heat per unit volume, and k is the coefficient of heat conduction; dissipation is not taken into account. We will adopt the value

$$k/\varrho c = 0.0045\ \text{cm}^2/\text{sec}. \quad [3.3]$$

We assume that the temperature field is initially uniform, together with the fictive temperature T_f which initially equals T. This last hypothesis is admissible provided the calculation is started at a high enough initial temperature. The fictive temperature $T_f(z, t)$ is calculated by a numerical integration of [2.33].

In order to obtain the stress distribution, we consider a sheet of infinite extent in a state of plane stress, for which the stress components are given by

$$\sigma_{xx} = \sigma_{yy} = \sigma(z, t),$$

$$\sigma_{zz} = \sigma_{xy} = \sigma_{yz} = \sigma_{zx} = 0, \quad [3.4]$$

which identically satisfy the equations of equilibrium. The strain components reduce to

$$\varepsilon_{xx} = \varepsilon_{yy} = \varepsilon_x(z, t),$$

$$\varepsilon_{zz} = \varepsilon_z(z, t),$$

$$\varepsilon_{xy} = \varepsilon_{yz} = \varepsilon_{zx} = 0, \quad [3.5]$$

while the only non trivial equation of compatibility is

$$\frac{\partial^2 \varepsilon_x(z, t)}{\partial z^2} = 0 \quad [3.6]$$

from which we obtain

$$\varepsilon_x(z, t) = A(t)\, z + B(t). \quad [3.7]$$

By inserting [3.4] and [3.5] into [2.27], eliminating $\varepsilon_z(z, t)$ between the resulting equations, and making use of [2.28], we obtain easily

$$\sigma(z, t) = \frac{E_0}{1 - \nu_0} \int_{-\infty}^{t} \Psi(\xi - \xi') \frac{\partial}{\partial t'} \times [\varepsilon_x(z, t') - \alpha(z, t')]\, dt', \quad [3.8]$$

where $(\xi - \xi')$ is given by $[2.27]_3$, and $\alpha(z, t')$ is obtained from [2.32]. In order to evaluate the functions $A(t)$ and $B(t)$ appearing in [3.7], we substitute the right hand side of [3.7] for $\varepsilon_x(z, t')$ in [3.8] and express that the resultant force and resultant moment on a plane normal to the middle surface vanish identically; we obtain

$$\int_{-l}^{l} dz \int_{-\infty}^{t} \Psi(\xi - \xi') \frac{\partial}{\partial t'} \times [z A(t') + B(t') - \alpha(z, t')]\, dt' = 0$$

$$\int_{-l}^{l} z dz \int_{-\infty}^{t} \Psi(\xi - \xi') \frac{\partial}{\partial t'} \times [z A(t') + B(t') - \alpha(z, t')]\, dt' = 0. \quad [3.9]$$

In particular, when the temperature distribution is symmetric with respect to the middle plane of the plate, $A(t)$ in [3.7] vanishes identically, and [3.9] reduces to

$$\int_{-l}^{l} z dz \int_{-\infty}^{t} \Psi(\xi - \xi') \frac{\partial}{\partial t'} \times [B(t') - \alpha(z, t')]\, dt' = 0. \quad [3.10]$$

A numerical algorithm for solving [3.10] has been proposed by *Lee* et al. (6), and applied in our earlier paper (2). A similar procedure applied to the system [3.9] results in a system of two linear algebraic equations in the increments $[A(t_{n+1}) - A(t_n)]$, $[B(t_{N+1}) - B(t_n)]$, where t_n, t_{n+1} ($n=1, N$) are the limits of the nth time interval. The residual stresses are obtained from [3.8] and [3.7], after the temperature has reached a uniform value throughout the sheet in the elastic range (say 25 °C). In general, however, it is possible to stop the calculation when the temperature is below 450 °C everywhere, since relaxation is then negligible.

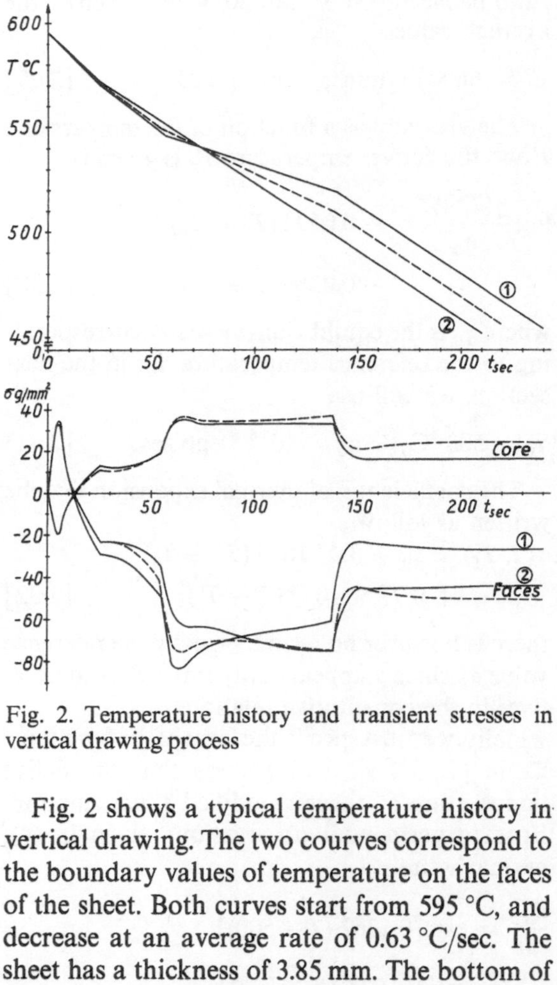

Fig. 2. Temperature history and transient stresses in vertical drawing process

Fig. 2 shows a typical temperature history in vertical drawing. The two courves correspond to the boundary values of temperature on the faces of the sheet. Both curves start from 595 °C, and decrease at an average rate of 0.63 °C/sec. The sheet has a thickness of 3.85 mm. The bottom of fig. 2 shows the transient stresses on the faces of the sheet and in the middle plane; the discontinuities of the time derivatives correspond to the changes of slope of the temperature history. After 160 sec, the transient stresses keep a constant value, because relaxation is then negligible. The residual stresses are obtained after uniformization of the temperature throughout the thickness of the sheet.

A second calculation has been performed by assuming that the temperature field is symmetric with respect to the middle plane of the sheet. We impose on both faces a temperature history which is the average of the temperatures on both faces in the actual process. Fig. 2 shows in dashed lines the transient stresses on the faces and in the middle plane of the sheet. It may be seen that, although the transient stresses differ appreciably in the symmetric and non-symmetric cases, due to the difference in the temperature gradients, the residual stresses are only slightly affected by the non-symmetric temperature field. We obtain, for the non-symmetric case

$\sigma\,(\text{face 1}) = -0.191, \quad \sigma\,(\text{middle}) = 0.092,$

$\sigma\,(\text{face 2}) = -0.172\ \text{kg/mm}^2.$

For the symmetric case, we have

$\sigma\,(\text{faces}) \quad = -0.197,$

$\sigma\,(\text{middle}) = 0.100\ \text{kg/mm}^2.$

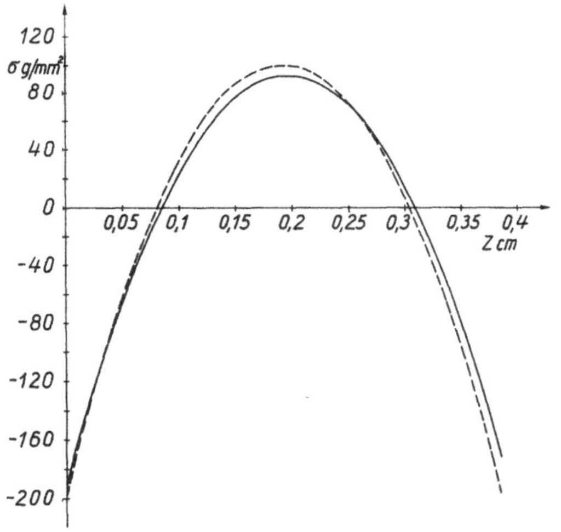

Fig. 3. Residual stresses for non-symmetric (—) and symmetric (---) temperature histories

Fig. 3 shows the stress profile in the thickness of the sheet for the symmetric and non-symmetric cases.

4. Influence of the changes of slope in temperature

Fig. 2 shows that changes of slope of the temperature imposed on the faces of the sheet result in high values of the transient stresses. The phenomenon is particularly noticeable when looking for optimal temperature histories with the purpose of minimizing the residual stresses.

In our earlier paper, we studied the annealing of a glass sheet with a thickness of 20 mm, by imposing on the faces temperature histories of the following kind: a temperature decrease at constant rate, followed by an interval at constant temperature, and again a decrease at constant rate. An optimal curve for the considered class is shown in fig. 4 (curve 1). The transient stresses on the faces are shown in fig. 5. Peaks in the transient stresses are observed after 210 and 950 sec, when the temperature rate on the faces is subject to a discontinuity.

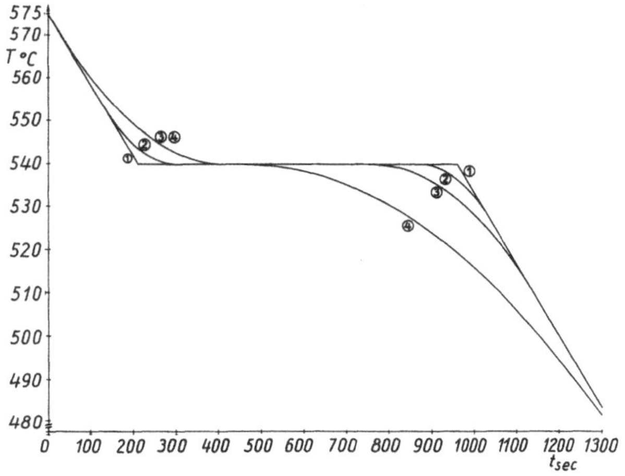

Fig. 4. Temperature histories

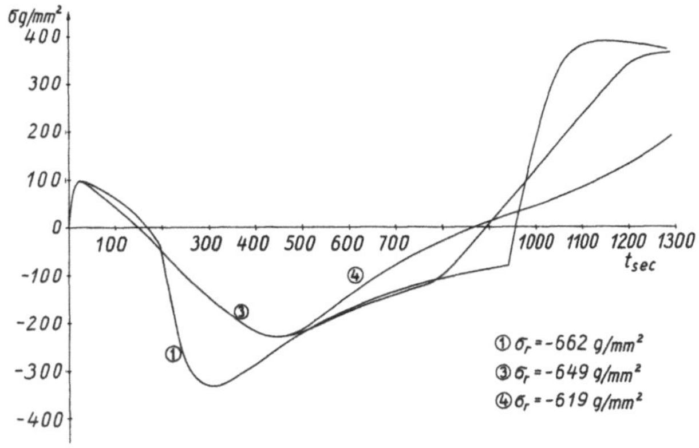

Fig. 5. Transient stresses corresponding to fig. 4

The numerical programme has been modified in order to smooth the corners of the temperature history; fig. 4 shows three typical temperature curves obtained after smoothing. Fig. 5 shows the transient stresses on the faces corresponding to the temperature curves 3 and 4. It is observed that the peaks in transient stress are lowered considerably. The influence of smoothing on the residuals stresses is a decrease to a limited extent.

The residual stresses corresponding to the temperature curves 1, 2, 3, 4, are, respectively,

	Faces kg/mm²	Core kg/mm²
1.	— 0.662	0.370
2.	— 0.659	0.365
3.	— 0.649	0.354
4.	— 0.619	0.323

In general, the actual temperature history in the annealing gallery will not follow closely the optimal curve, and smoothing will occur naturally. It may thus be expected that values of the residual stresses will be lowered.

Acknowledgements

The present paper has been prepared within the course of research conducted at the "Institut National du Verre", under the sponsorship of IRSIA (Institut pour l'Encouragement de la Recherche Scientifique dans l'Industrie et l'Agriculture). The collaboration of MM. *Debast* and *Gilard* is gratefully acknowledged.

References

1) *Crochet, M. J.* and *P. M. Naghdi*, Int. J. Eng. Sci. **7**, 1173–1198 (1969).

2) *Crochet, M. J.* and *G. Tackels*, Silic. Industr. **37**, 101–106 (1972).

3) *Debast, J.* and *P. Gilard*, IRSIA, Comptes-rendus de recherches **1965**, No. 32.

4) *Debast, J.* and *P. Gilard*, IRSIA, Comptes-rendus de recherches **1969**, No. 36.

5) *Gardon, R.* and *O. S. Narayanaswami*, J. Amer. Ceram. Soc. **53**, 380–385 (1970).

6) *Lee, E. H., T. G. Rogers*, and *T. C. Woo*, J. Amer. Ceram. Soc. **48**, 480–487 (1965).

7) *Ritland, H. N.*, J. Amer. Ceram. Soc. **37**, 370–378 (1959).

8) *Tool, A. Q.*, J. Amer. Ceram. Soc. **29**, 240–253 (1946).

Authors' addresses:

G. Tackels
Engineer, Institut National du Verre
B-6000 Charleroi (Belgium)

and *M. J. Crochet*
Professor, University of Louvain
B-1 348 Louvain-La-Neuve (Belgium)

Rheol. Acta 12, 337–344 (1973)

De C.N.R.S. Laboratoire des Verres, Paris (France)

Viscosité des verres présentant une séparation de phases

Par M. Prod'homme

Avec 5 figures and 3 tableaux

(Reçu p. p. le 27 octobre 1972)

1. Introduction

Parmi les verres à séparation de phases, les borosilicates de sodium ont une place privilégiée, étant donné les applications pratiques qu'ils ont reçues dans le domaine des verres «Pyrex» et «Vycor». On sait que lors de la démixtion, il se forme deux phases principales, l'une constituée surtout de silice, et l'autre enrichie en borate.

Après la séparation de phases, le verre n'est plus homogène et la présence de deux réseaux vitreux discontinus (une phase dispersée dans une matrice) ou continus (deux phases interconnectées) modifie ses propriétés.

La mesure de la viscosité qui exige l'emploi d'une méthode dynamique semble tout-à-fait appropriée à l'étude de ces verres et la variation isotherme de cette propriété permet de suivre l'évolution isotherme de la séparation de phases.

Ces dernières années, plusieurs chercheurs: *Li* et *Uhlmann* (1) et *Bernheim* et *Chaklader* (2) se sont intéressés aux effets de la séparation de phase et des traitements thermiques sur la viscosité des verres silico-sodiques et des verres de borosilicate. *Haller*, *Simmons* et *Napolitano* (3) ont proposé une méthode basée sur des mesures de viscosité pour mettre en évidence des séparations de phases dans les systèmes vitreux et déterminer la température critique d'immiscibilité liquide-liquide. Enfin *Mazurin* et ses collaborateurs (4, 5, 6) ont mis tout particulièrement l'accent sur la recherche de la nature de la distribution des différentes phases dans le verre et du changement qui intervient dans cette distribution après traitement termique. Ces auteurs pensent que la viscosité des verres à séparation de phases dépend en premier lieu de la façon selon laquelle la phase à viscosité élevée est distribuée dans ceux-ci.

En complément aux études déjà réalisées, le but de ce travail a été d'étudier en fonction du temps, la variation isotherme de la viscosité de plusieurs verres de borosilicate de sodium à phases séparées. Cette étude a été réalisée dans le domaine de transformation, c'est-à-dire pour des valeurs de viscosité variant de 10^{12} à 10^{15} poises. Ce domaine de température est particulièrement important car on sait qu'à l'intérieur de celui-ci, on arrive par un traitement thermique approprié à obtenir une structure en équilibre. A titre de comparaison, nous avons également étudié un verre d'optique, c'est-à-dire un verre très homogène qui n'a aucune tendance à la démixtion.

2. Méthode expérimentale

2.1. *Description de l'appareil*

Pour mesurer la déformation des échantillons, nous avons utilisé un dilatomètre à enregistrement électronique, spécialement adapté à notre étude par l'adjonction de ressorts engendrant la contrainte de compression uniaxiale à laquelle sont soumis les échantillons. Dans cet appareil déjà décrit (7), la baguette de verre à étudier est placée dans un tube de silice disposé dans l'axe d'un four. Sa déformation est transmise à un levier muni d'un miroir concave par l'intermédiaire d'un poussoir en silice et d'un coulisseau métallique. Un spot lumineux réfléchi par le miroir est focalisé sur une cellule photorésistante qui entraîne la plume enregistrant la déformation sur un tambour. La température de l'échantillon est mesurée par un thermocouple Pt/Pt-Rh dont la soudure est située juste au-dessus de l'échantillon.

2.2. Echantillons utilisés

Les échantillons sont taillés sous forme de baguettes de 50 mm de longueur et de section carrée de 2,5 × 2,5 mm. La composition des verres étudiés est la suivante (% mol.):

	SiO_2	B_2O_3	Na_2O	ZnO	K_2O	Al_2O_3	CaO	PbO
1209	60	30	10					
371	68	27	5					
299	73	19	8					
1435	77	13	3	7				
Pyrex	83	11	4		1	1		
Verre d'optique (crown)	68	8	8	6	6		2	2

2.3. Résultats expérimentaux

La viscosité des échantillons est calculée à partir de la vitesse de déformation de la baguette sous l'effet de la contrainte de compression:

$$\eta = \frac{\sigma}{3\dfrac{dl}{l \cdot dt}}$$

où $\sigma = mg/S$ désigne la contrainte; l et S sont respectivement la longueur et la section de la baguette; dl/dt est la vitesse de déformation de la baguette.

2.3.1. *Etude de la viscosité en fonction de la température*

On sait que les températures du domaine de transformation correspondent toujours aux mêmes valeurs de la viscosité quel que soit le verre étudié. En particulier, la majorité des auteurs attribuent à la température T_g déterminée d'après la courbe de dilatation, une valeur de viscosité voisine de 10^{13} à $10^{13,3}$ poises. Comme l'a fait remarquer *Mazurin* (4), dans un verre à deux phases, si la phase à viscosité élevée est continue, c'est sa viscosité même que l'on mesure, mais si elle est discontinue, on mesure alors la viscosité de la phase la plus fluide. Donc la comparaison de la température de transformation T_g à la température caractéristique $T_{13,3}$ d'une viscosité de $10^{13,3}$ poises renseigne, dans le cas d'un verre à deux phases sur la nature de la distribution de la phase la plus visqueuse. Pour faire cette comparaison, nous avons déterminé la viscosité des quatre borosilicates: 1209, 371, 299 et 1435 en fonction de la température (8). Sur le tableau 1 nous pouvons comparer les valeurs de le température correspondant à la viscosité de $10^{13,3}$ poises à la température de transformation T_g déterminée d'après la courbe de dilatation et sur les thermogrammes d'A.T.D. Nous pouvons ainsi constater que T_g ne coïncide avec $T_{13,3}$ que pour les verres 1209 et 299 dont les deux phases seraient alors discontinues.

Tableau 1

	$T_{13,3}$	$T_{g(\text{dil.})}$	$T_{g(\text{ATD})}$ 1° pic	$T_{g(\text{ATD})}$ 2° pic
1209	495	490	480	605
371	520	460	460	560
299	515	520	520	
1435	630	510 et 590	500	845

2.3.2. *Etude de la viscosité en fonction du temps*

2.3.2.1. *Verre d'optique crown*

Avant d'examiner le comportement rhéologique des verres à séparation de phases, au cours d'expériences de stabilisation, nous rappellerons celui d'un verre homogène, un verre d'optique en crown ordinaire. On sait qu'à l'intérieur du domaine de transformation, à une température donnée, la structure du verre évolue en fonction du temps pour atteindre la structure d'équilibre correspondant à cette température. Cet effet de stabilisation se manifeste sur toutes les propriétés. En particulier, la viscosité apparente d'un verre trempé augmente en fonction du temps jusqu'à ce qu'elle ait atteint sa valeur d'équilibre (fig. 1). Les temps de stabilisation sont de plus en

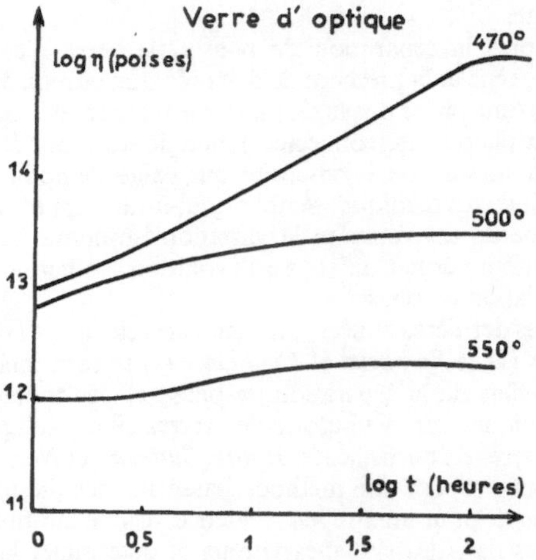

Fig. 1. Viscosité du verre d'optique crown en fonction du temps

plus longs et les variations totales de viscosité sont de plus en plus grandes lorsque la température diminue. Par contre, lorsqu'un verre est déjà stabilisé à la température de l'expérience, sa viscosité ne varie pas. Nous avons montré (7) que la variation isotherme de la viscosité apparente de ce verre dans le domaine de transformation provenait en fait de la variation isotherme de la densité. Sur la fig. 2, nous pouvons voir les déformations obtenues à 480 °C avec des échantillons trempés soumis à des charges de compression différentes (courbes a). Elles peuvent être décomposées en une contraction visqueuse linéaire (courbes b) et une contraction due à la stabilisation de la densité (courbe c). De la même façon nous avons porté sur la fig. 3, les déformations obtenues à 480 °C avec des échantillons recuits

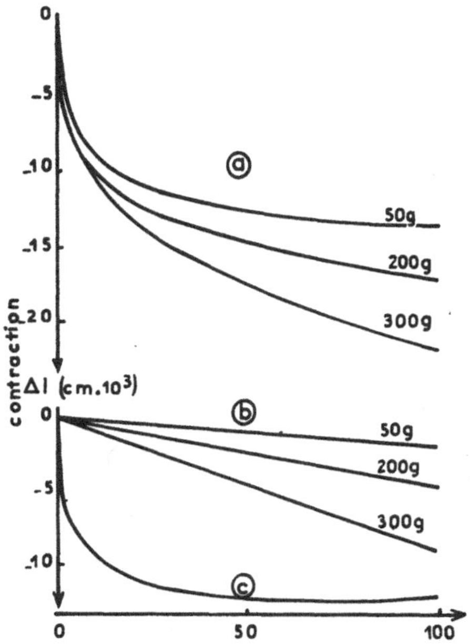

Fig. 2. Déformation à 480 °C des échantillons trempés de crown sous les charges de 50 g, 200 g et 300 g. a) déformation totale, b) déformation due à la contraction visqueuse, c) déformation due à la stabilisation de la densité (contraction)

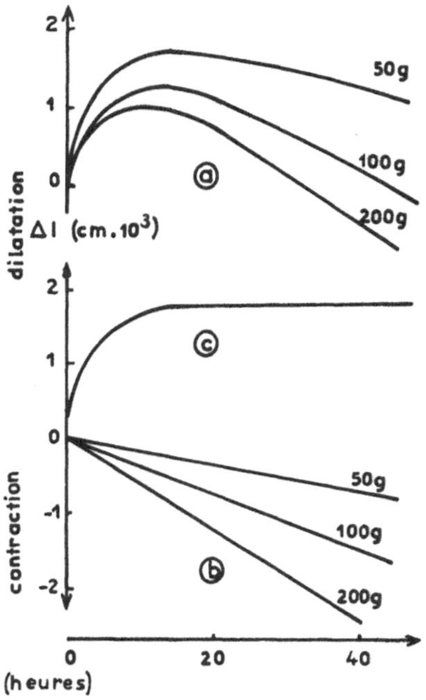

Fig. 3. Déformation à 480 °C des échantillons recuits de crown sous les charges de 50 g, 100 g et 200 g. a) déformation totale, b) déformation due à la contraction visqueuse, c) déformation due à la stabilisation de la densité (dilatation)

soumis à des charges de compression différentes (courbes a). Elles peuvent être décomposées en une contraction visqueuse linéaire (courbes b) et une dilatation due à la stabilisation de la densité (courbe c).

2.3.2.2. Borosilicate 1209

Les expériences ont été effectuées sur des échantillons brut d'usine[1]) et sur des échantillons traités à la température critique d'immiscibilité (710°). La fig. 4a montre l'influence du temps sur la viscosité à température constante pour les échantillons bruts. On peut voir que jusqu'à 500 °C, l'évolution totale de la viscosité et la durée de cette évolution diminuent lorsque la température augmente. Par exemple:

température	$\Delta(\log \eta)$	temps de stabilisation
450°	1,3	90 h
475°	0,5	50 h
490°	0	0

On peut penser que cette évolution est due au processus de stabilisation; à 490 °C, l'équilibre structural du verre est atteint et la viscosité ne varie plus.

[1]) C'est-à-dire pour ce verre assez bien recuit.

Au-dessus de 500 °C, il y a de nouveau une évolution très importante de la viscosité qui est de plus en plus grande et rapide lorsque la température augmente (fig. 4a); de sorte qu'à 520 °C, à partir de 40 h, la viscosité devient supérieure à 505 °C et à partir de 70 h supérieure à la viscosité à 490 °C.

Lorsque des expériences sont réalisées à des températures supérieures à 500 °C après avoir maintenu au préalable les échantillons à la température de l'expérience, on trouve que la viscosité initiale est alors beaucoup plus grande que celle des échantillons non traités (tableau 2).

Tableau 2. Borosilicate 1209, température 520 °C

temps	log viscosité du verre brut	log viscosité du verre traité 112 h à 520 °C
1 h	12,3	13,2
10 h	12,75	13,4
50 h	13,4	13,75
100 h	14,05	14,1
150 h	14,6	14,6

En fin d'expérience, la viscosité des échantillons traité et brut est identique et continue à croître en fonction du temps. Il est probable qu'aux températures supérieures à 500 °C, l'évolution de la visco-

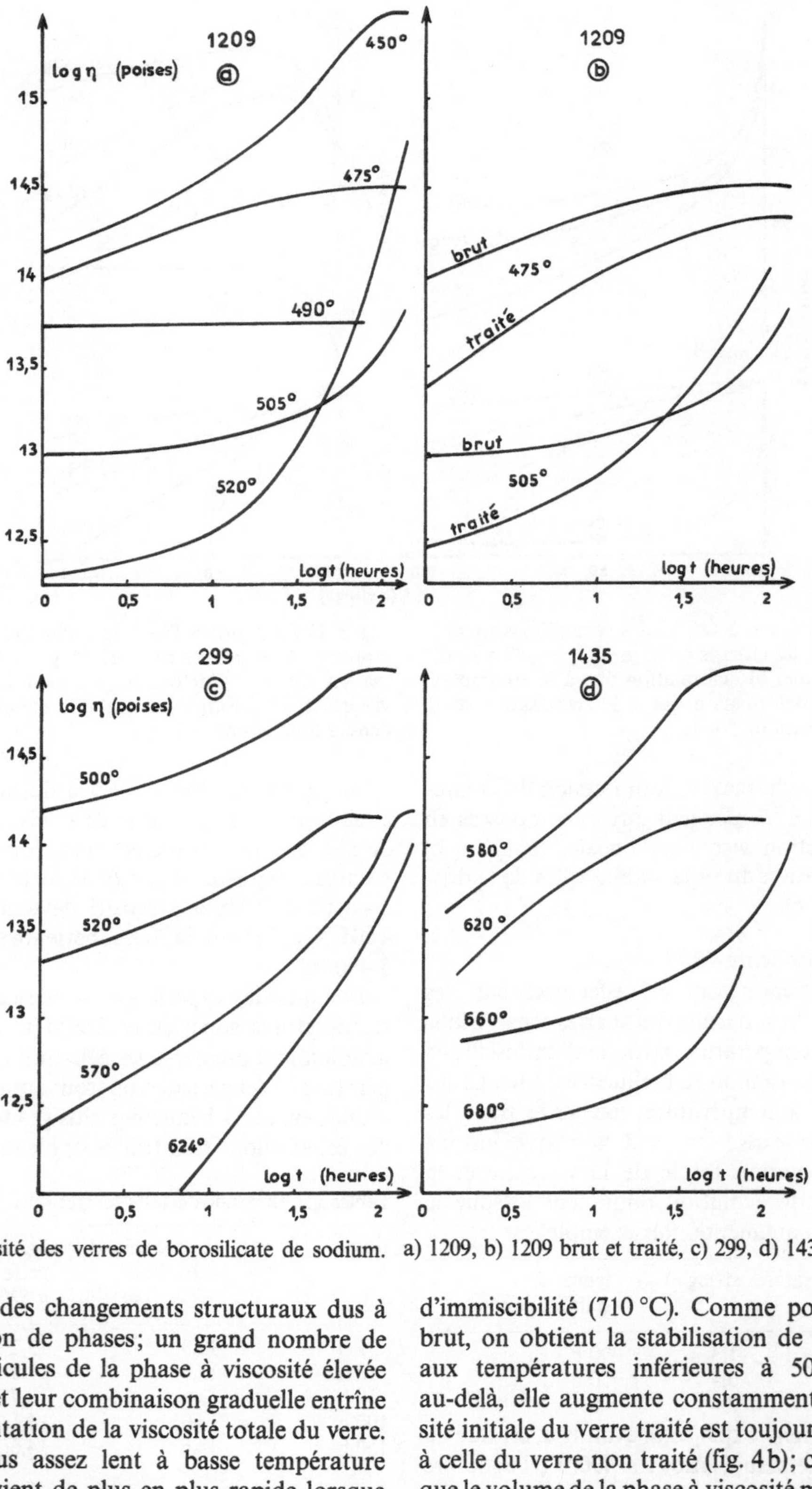

Fig. 4. Viscosité des verres de borosilicate de sodium. a) 1209, b) 1209 brut et traité, c) 299, d) 1435

sité résulte des changements structuraux dus à la séparation de phases; un grand nombre de petites particules de la phase à viscosité élevée se forment et leur combinaison graduelle entrîne une augmentation de la viscosité totale du verre. Ce processus assez lent à basse température (505 °C) devient de plus en plus rapide lorsque la température augmente.

Nous avons également étudié des échantillons de ce verre traités à la température critique d'immiscibilité (710 °C). Comme pour le verre brut, on obtient la stabilisation de la viscosité aux températures inférieures à 500 °C, mais au-delà, elle augmente constamment. La viscosité initiale du verre traité est toujours inférieure à celle du verre non traité (fig. 4b); cela indique que le volume de la phase à viscosité moins élevée est plus important dans le verre traité. En effet d'après l'étude réalisée au microscope électronique (9), on sait que ce verre après traitement à la température critique d'immiscibilité présente une structure discontinue[2]) formée de gouttelettes de silice dans une matrice boratée. La viscosité du verre est donc déterminée par celle de la phase borate qui est la plus fluide. L'augmentation de viscosité des échantillons traités à 710 °C, puis étudiés de nouveau à plus basse température qui est tout à fait comparable à celle des échantillons bruts est peut-être due à un processus de démixtion secondaire qui s'effectue au sein de la phase à viscosité faible. Comme cette phase n'est pas stable à ces basses températures, elle donne naissance à une partie de phase à viscosité élevée entraînant ainsi une augmentation de la viscosité totale.

2.3.2.3. *Borosilicate 371*

Les expériences ont été réalisées de 440 à 640 °C sur des échantillons bruts et sur des échantillons traités 72 h à 440 °C. Les résultats sont donnés dans le tableau 3. Nous constatons qu'à la température de 440 °C, la viscosité augmente de $10^{13,8}$ à $10^{15,6}$ poises en une durée d'environ 100 h et garde ensuite cette valeur. La variation totale de viscosité est à peu près identique ($\Delta \log\eta = 1{,}8$) pour les essais isothermes réalisés de 440 à 520 °C, puis elle diminue pour les températures inférieures. Les durées nécessaires pour obtenir une viscosité constante sont toujours de l'ordre de 100 h. On voit également que les valeurs de la viscosité finale varient peu dans le domaine de 440 à 520 °C ($\Delta \log\eta = 0{,}6$) et beaucoup dans le domaine de 520 à 600 °C ($\Delta \log\eta = 2{,}2$). Les essais effectués sur des baguettes traitées (tableau 3) montrent que la variation totale de viscosité et le temps de stabilisation diminuent beaucoup pour les expériences réalisées de 440 à 480 °C. Aux températures plus élevées, ces variations sont plus faibles, enfin à 560 °C, la variation totale de viscosité et le temps de stabilisation sont identiques pour le verre brut et pour le verre traité.

Tout ceci nous conduit à penser que la variation de viscosité au cours des expériences isothermes n'est pas due uniquement au processus de stabilisation. En effet, s'il en était ainsi, il faudrait de moins en moins de temps pour obtenir une viscosité constante et la variation totale de viscosité serait de moins en moins grande lorsque les expériences sont effectuées à des températures plus élevées. De plus, lors des expériences réalisées sur les baguettes traitées à 440 °C, ces deux facteurs devraient diminuer pour toutes les expériences. Pour ce verre, il est donc vraisemblable que l'évolution de la viscosité est due principalement au phénomène de la séparation de phases. La décomposition spinodale provoquerait une augmentation graduelle de la différence de composition entre les phases et entrînerait une augmentation de la viscosité de la phase à teneur élevée en silice si la concentration en silice augmente. Un changement de température de 440 à 520 °C s'accompagnant d'une faible variation de la composition de la phase silice alors qu'un changement de 520 à 600 °C provoque une variation appréciable.

Tableau 3. Borosilicate 371

tempé-rature	$\log\eta_i$	$\log\eta_f$	$\Delta(\log\eta)$	temps de stabilisation
Verre brut				
440	13,8	15,6	1,8	100 h
480	13,6	15,3	1,7	100 h
520	13,2	15,0	1,8	100 h
545	13,1	14,2	1,1	70 h
560	12,7	13,8	1,1	80 h
580	12,6	13,6	1,0	90 h
605	12,2	12,8	0,6	
Verre traité 72 h à 440 °C				
444	15,1	15,5	0,4	65 h
480	14,6	15,1	0,5	35 h
520	13,7	14,8	1,1	65 h
565	12,5	13,5	1,0	25 h
605	12,2	12,7	0,5	

2.3.2.4. *Borosilicate 299*

La viscosité de ce verre a été mesurée de 450 à 680 °C. Les mesures ont été effectuées sur le verre brut, sur le verre traité à 500 °C et sur le verre traité aux températures auxquelles les mesures sont réalisées. Les essais isothermes ont montré qu'à chaque température inférieure à 520 °C, la viscosité atteint une valeur constante qui décroît lorsque la température augmente (fig. 4c). A partir de 520 °C, quelle que soit la température, la viscosité se stabilise toujours approximativement à la même valeur située entre 10^{14} et $10^{14,2}$ poises et le temps pour arriver à ce palier est de plus en plus long au fur et à mesure que la température est plus élevée.

Nous pensons qu'au-dessus de 520 °C, la variation de viscosité au cours des expériences isothermes est due principalement au phénomène

[2]) C'est également ce que nous avions trouvé pour le verre brut en comparant T_g et $T_{13,3}$ (§ 2.3.1.).

de séparation de phases comme pour le verre étudié précédemment (borosilicate 371). Il y a une augmentation de la concentration de SiO_2 entraînant une augmentation de la viscosité jusqu'à une valeur constante correspondant à une concentration déterminée en silice de la phase de viscosité élevée.

L'influence de la séparation de phases se manifeste également pour les expériences réalisées sur des baguettes traitées. En effet la variation totale de viscosité et le temps de stabilisation sont réduits énormément lorsque leurs traitements thermiques sont effectués aux températures auxquelles les expériences sont réalisées, alors qu'un traitement à basse température a peu d'influence sur la viscosité comme pour le borosilicate 371.

2.3.2.5. Borosilicate 1435

Les expériences ont été effectuées de 560 à 680 °C sur des échantillons bruts et sur des échantillons traités à 560 et 600 °C.

Les essais isothermes réalisés aux températures inférieures à 650 °C ont tous conduit à une vitesse de déformation constante, donc à une viscosité constante (fig. 4d). La valeur de cette viscosité finale et le temps nécessaire pour l'obtenir diminuent lorsque la température augmente. Ce comportement est donc celui d'un verre au cours du processus de stabilisation. Les expériences sur les échantillons traités le confirment. On retrouve en effet les mêmes valeurs d'équilibre en partant de vitesses de déformation plus ou moins fortes suivant que le traitement est effectué à une température égale ou située endessous de celle de l'expérience isotherme de stabilisation.

Aux températures supérieures à 650 °C, la viscosité augmente constamment et il n'est pas possible d'obtenir la stabilisation, même pour des durées d'expériences supérieures à 100 h. A ces températures élevées, les traitements effectués à 560 et 600 °C n'ont pas d'influence sur la viscosité. Il est donc vraisemblable que la séparation de phases commence à se manifester à partir de 650 °C.

2.3.3. Nature de l'écoulement visqueux

On sait qu'en général l'écoulement visqueux des verres est newtonien (vitesses de déformation proportionnelle à la pression) bien que récemment des comportements non newtoniens aient été signalés dans la littérature (1, 10, 11, 12). Aussi avons-nous cherché à préciser la nature de l'écoulement visqueux des verres étudiés. Pour cela, les expériences de stabilisation isotherme de la viscosité ont été effectuées sous différentes charges de compression ($7 \cdot 10^5$ à 10^7 dynes/cm^2). En adoptant la représentation graphique de *Li* et *Uhlmann* (1) où l'on porte le logarithme de la vitesse de déformation ($\log dl/dt$) en fonction du logarithme de la contrainte ($\log \sigma$), un fluide newtonien sera caractérisé par une variation linéaire de pente égale à l'unité.

2.3.3.1. Verre d'optique crown

Ainsi que nous l'avons dit précédemment (§ 2.3.2.1), la viscosité apparente de ce verre dans le domaine de transformation provient en partie de la déformation due à la variation de la densité. Il en résulte que l'écoulement est newtonien uniquement lorsque le verre est stabilisé. C'est ce que nous constatons sur la fig. 5a pour une série d'expériences réalisées à 480 °C sur des échantillons trempés. Au début de la stabilisation, la vitesse de déformation est pratiquement indépendante de la charge appliquée car l'effet principal est dû à la variation de la densité. A la fin de la stabilisation, la déformation due à la densité est constante et la vitesse de déformation est proportionnelle à la charge appliquée.

2.3.3.2. Verre Pyrex

Les fig. 5b et 5c représentent les résultats obtenus avec le verre Pyrex à 550 °C. Les expériences ont été effectuées sur des baguettes non traitées et sur des baguettes traitées à la température de l'expérience. On peut voir que la viscosité varie beaucoup pendant les expériences isothermes: de $10^{13,4}$ poises pour $t = 1$ h à $10^{15,2}$ poises pour $t = 155$ h, mais que le fluage est newtonien dès le début. Un traitement préalable à 550 °C (fig. 5c) réduit la variation de la viscosité ($\Delta \log \eta = 0{,}9$) mais le fluage est newtonien. La variation isotherme de la viscosité de ce verre ne provient donc pas de la stabilisation de la densité mais des changements structuraux dus à la séparation de phases.

2.3.3.3. Borosilicate 1209

La vitesse de déformation des échantillons de ce verre brut a été mesurée aux températures de 475, 505 et 520 °C sous plusieurs charges de com-

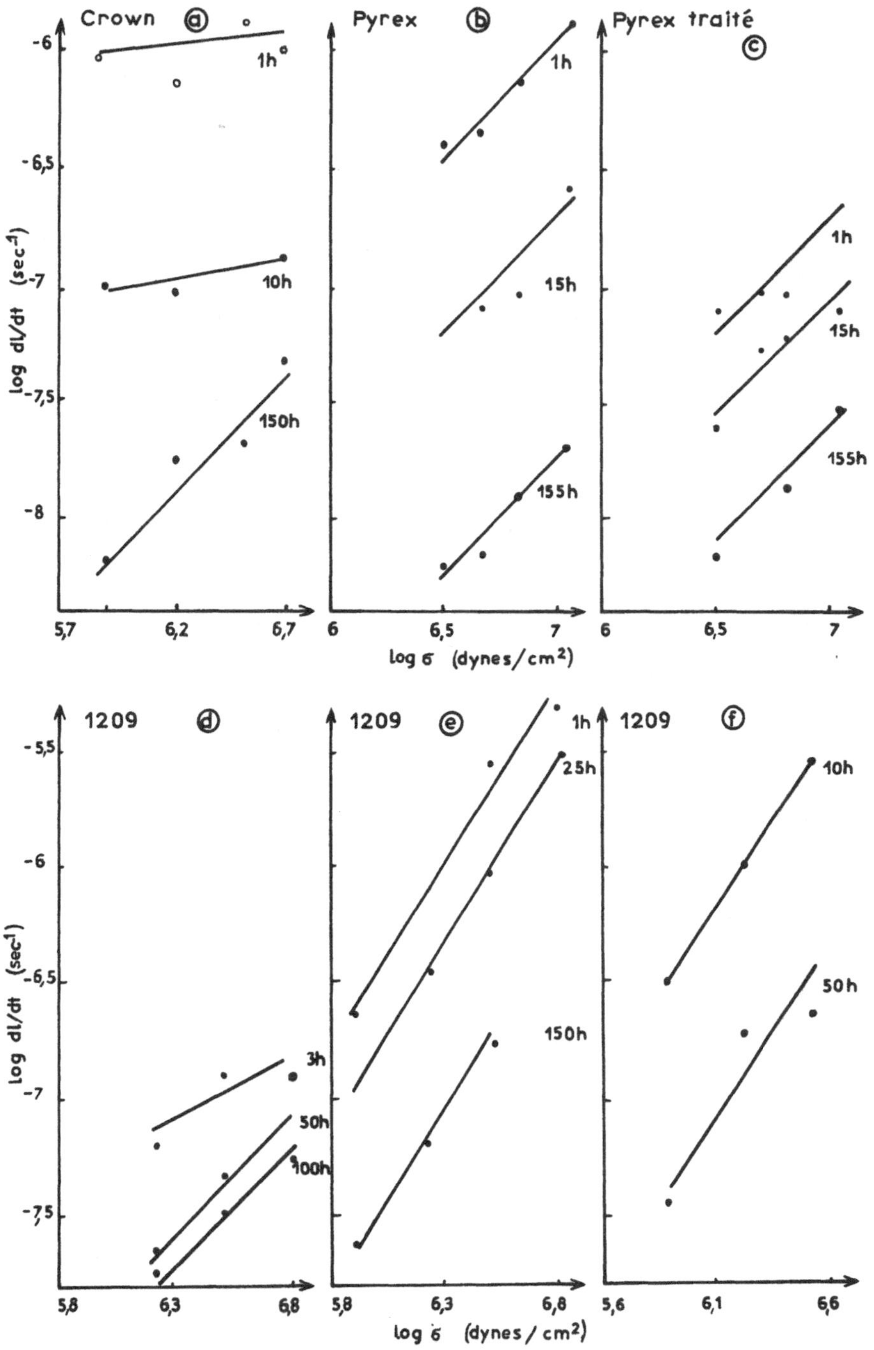

Fig. 5. Vitesse de déformation en fonction de la contrainte. a) verre d'optique crown à 480°C, b) Pyrex à 550°C, c) Pyrex à 550°C (traité auparavant à la température de l'expérience), d) 1209 à 475°C, e) 1209 à 505°C, f) 1209 à 520°C

pression. Malgré l'imprécision des mesures, on peut constater (fig. 5d) qu'à 475 °C, on obtient un fluage newtonien à partir d'une durée d'expériences d'environ 50 h. Pour les durées inférieures à 50 h la viscosité apparente $\sigma/(dl/dt)$ est une fonction croissante de la vitesse de déformation (pente inférieure à 1). C'est ce qui se produisait dans le cas d'un verre d'optique homogène en raison de la stabilisation de la densité. A 505 et 520 °C (fig. 5e et 5f), pour des durées d'expérience allant jusqu'à 100 h, on n'a pas obtenu de fluage newtonien. Il semble que dès le début des expériences, la viscosité apparente soit une fonction décroissante de la vitesse de déformation (pente supérieure à 1). On peut sans doute rapprocher cette viscosité anormale d'une viscosité de structure caractéristique du comportement des systèmes hétérogènes [suspensions, émulsions (13)].

3. Conclusion

En conclusion, nous remarquerons que les mesures de viscosité peuvent aboutir à de très nombreux résultats et à une meilleure compréhension des phénomènes très complexes qui ont lieu pendant les processus de séparation de phases. On peut seulement regretter que les mesures soient longues et imprécises.

Parmi les résultats, l'un des plus importants est peut-être que l'on ne peut pas bien souvent obtenir une structure en équilibre avec les verres à séparations de phases. Si dans certains cas, on arrive à un équilibre, celui-ci n'est réalisé qu'à des températures très inférieures à la température critique d'immiscibilité (viscosité supérieure à 10^{13} poises). L'étude de la nature de l'écoulement visqueux permet de distinguer plusieurs causes provoquant la variation isotherme de la viscosité:

1. Un déséquilibre structural sans démixtion à cette température. La viscosité apparente est une fonction croissante de la vitesse de déformation. L'écoulement n'est pas newtonien en raison de la variation de densité.

2. Une démixtion. L'écoulement est toujours newtonien. La démixtion entraîne une augmentation de la viscosité qui peut vraisemblablement s'expliquer par un accroissement de la teneur en silice de la phase principale (phases continues).

3. Une démixtion. La viscosité apparente est une fonction décroissante de la vitesse de déformation, l'écoulement n'est pas newtonien. On est en présence d'un système hétérogène (phases discontinues).

Littérature

1) *Li* et *Uhlmann*, J. non Cryst. Sol. **3**, 205 (1970).
2) *Bernheim* et *Chaklader*, J. non Cryst. Sol. **5**, 328 (1971).
3) *Haller*, *Simmons* et *Napolitano*, J. Amer. Ceram. Soc. **54**, 299 (1971).
4) *Mazurin*, *Streltsina* et *Totesh*, Phys. Chem. Gl. **10**, 63 (1969).
5) *Mazurin*, *Kluyev* et *Roskova*, Phys. Chem. Gl. **11**, 192 (1970).
6) *Mazurin*, *Roskova* et *Kluyev*, Disc. Far. Soc. **50**, 191 (1970).
7) *Prod'homme*, Verres et Réfract. **22**, 614 (1968).
8) *Prod'homme*, Verres et Réfract. **24**, 151 (1970).
9) *Cordelier*, Verres et Réfract. **24**, 113 (1970).
10) *Bartenev*, Phys. non Cryst. Sol. **1965**, 465.
11) *Stein* et *Stevels*, C. R. VII Cong. Int. Verre, 1965, n° 25.
12) *Watanabe* et *Koyama*, J. Soc. Gl. Tech. **41**, 137 (1957).
13) *de Vries*, Introduction à l'étude de la rhéologie (Dunod 1960).

Adresse de l'auteur:
Dr. *M. Prod'homme*
C.N.R.S. Laboratoire des Verres
5 Boulevard Pasteur
F-75015 Paris (France)

Rheol. Acta **12**, 345–348 (1973)

From the PPG Industries Inc., Harmarville, Pennsylvania (U.S.A.), and the University of Pittsburgh, Pittsburgh, Pennsylvania (U.S.A.)

Thermal stress analysis of glass with temperature dependent coefficient of expansion

By S. M. Ohlberg and T. C. Woo

With 4 figures

(Received October 27, 1972)

Introduction

In calculating transient and residual stresses in a glass plate during heat treatments, a theory based on the linear viscoelasticity has been used (1). The problem has been appropriately formulated with the initial and boundary conditions using the actual geometric and physical data and the measured material properties. Comparison between theoretical results and those observed in practice showed that such theory served as a good approximation in spite of highly idealized assumptions.

There are numerous aspects in which one could improve on the original theory. The experimental data on the viscoelastic properties and their temperature dependence, for instance, have been refined (2). Another possible modification would be a temperature dependent thermal expansion coefficient instead of the single room temperature value used in the original model. Here, one is concerned not only with the increase in expansion coefficient with temperature characteristic of a solid but even more so with the two- to three-fold increase that occurs when glass is heated through the transformation range.

Still another variation that would influence the results is the history effect embodied in a "fictive temperature" concept first proposed by *Tool* (3). The fictive temperature is the temperature at which the glass would be at structural equilibrium. A glass is undercooled or superheated depending on whether the fictive temperature and actual temperature become equal through heating or cooling, respectively. Since the fictive temperature effect is mainly on the volumetric strain, it seems that the thermal expansion could be split into the glassy state and liquid state and the viscoelastic stress analysis be properly modified. The combination of the fictive temperature evaluation and the thermal expansion coefficient change comprises the first method of approach in this paper.

A different and more phenomenological approach to take account of the change of the thermal expansion coefficient is to use the pseudo-temperature concept (4). This parallels the pseudo-time which was used to shift the relaxation curve along the logarithmic time scale for the pertinent temperature and therefore is quite consistent within the framework of linear viscoelasticity.

Determination of thermal expansion values

The coefficient of thermal expansion in the lower temperature range (25–500 °C) was measured with a conventional dilatometer. That in the higher range (870–1374 °C) was obtained from density measurements of the molten glass. The values in the intermediate range were obtained by linear extrapolation of both low temperature and high temperature measurements in order to get the values of α_{gl} and α_{liq} respectively for the fictive temperature method. For the pseudo temperature method, intermediate values were interpolated by drawing a smooth curve connecting the lower and upper regions. Further, a hyperbolic tangent function was taken as a good representation of this expansion coefficient versus temperature curve, thus enabling a closed-form expression for purposes of computation.

Fictive temperature method

The fictive temperature concept was proposed by *Tool* as one of the methods to discuss the annealing of glass (3). The relation of the fictive temperature to inelastic deformation expounded there is essentially implied by the shear components in the viscoelastic theory of the thermal stresses. The fictive temperature effect on the volumetric strains can therefore logically incorporate the variation of the thermal expansion coefficient in the temperature range considered.

Instead of using *Tools* original expression for the volume relaxation time we choose to use the more conventional *Arrhenius* form for the viscosity as

$$\bar{t} = t_0 \exp\left(\frac{K_1}{T}\right) \exp\left(\frac{K_2}{T_f}\right) \qquad [1]$$

where T and T_f are the actual and fictive temperatures respectively, and the constants K_1 and K_2 are related to the activation energy E through

$$K_1 + K_2 = \frac{E}{R} \qquad [2]$$

where R is the universal gas constant. The viscosity corresponding to the initial temperature T_0

is given by $t_0 \exp\left(\frac{K_1 + K_2}{T_0}\right)$. The magnitudes of K_1 and K_2 determine the relative influences of T and T_f on flow. Usually K_1 is roughly three-fold of K_2 for inorganic glasses.

Following *Tools* equation for the rate of approaching equilibrium we use

$$\frac{dT_f}{dt} = \frac{T - T_f}{\bar{t}} \qquad [3]$$

with t given by eq. [1]. In eq. [3] the actual temperature T is a known function of $\varkappa$ (the distance from the mid-plane) and time t. The fictive temperature T_f can be solved as a function of $\varkappa$ and t with the initial condition

$$(T_f)_{t=0} = (T)_{t=0} \quad \text{for all } \varkappa.$$

The differential eq. [3] was solved using the subroutine DHPCG in the IBM scientific Subroutine Package. This subroutine uses *Hammings* modified predictor-corrector method for the solution of general initial-value problems. The results have also been checked against the conventional *Runge-Kutta* method and found to be very accurate.

Once the fictive temperature is determined the volumetric strain e in the viscoelastic stress analysis is replaced by

$$e = \alpha_{\text{liq}}(T - T_0) + \alpha_{\text{gl}}(T - T_f) \qquad [4]$$

where α_{liq} and α_{gl} are the coefficients of thermal expansion at liquid and glassy states respectively. The shear response remains uneffected by the fictive temperature as mentioned before.

Pseudo-temperature method

The pseudo-temperature $\theta(\varkappa, t)$ is defined to be

$$\theta(\varkappa, t) = \frac{1}{\alpha_B} \int_{T}^{T(\varkappa, t)} \alpha(T') \, \alpha T' \qquad [5]$$

where T_B is the base temperature from which the relaxation function is shifted, α_B the coefficient of expansion at T_B, and $\alpha(T')$ is a given function of the coefficient of thermal expansion in terms of temperature. As an alternative to an abrupt change of the coefficient of expansion at some transformation temperature, one could assume a hyperbolic tangent for its representation. Thus

$$\alpha(T) = \alpha_m + \alpha_1 \tanh[\beta(T - T_m)] \qquad [6]$$

where β is a scale factor that controls how fast the α-curve approaches the asymptotes at both ends, α_1 is the half-range of the α variation, and α_m is the α value corresponding to T_m, the middle of the transformation range. With eq. [6] we can integrate [5] to give

$$\theta(\varkappa, t) = \frac{1}{\alpha_B}\left\{\alpha_m[T(\varkappa, t) - T_B] + \frac{\alpha_1}{\beta} \ln\left[\frac{\cosh\beta[T(\varkappa, t) - T_m]}{\cosh\beta[T_m - T_B]}\right]\right\}. \qquad [7]$$

With the introduction of the pseudo-temperature the viscoelastic stress analysis for heat treatment problems of glass is virtually unchanged except that in all the calculations the actual temperatures are to be replaced by the pseudo-temperatures. Modification of the temperature dependence of thermal expansion coefficient discussed above thus conforms to that of the viscoelastic properties and both are carried out in a similar manner, using all measured data for the input functions.

Calculations and discussion

In the fictive temperature method the value of the activation energy was taken to be 150 kcal/mol which is an average value for most of the common glasses [5]. The ratio K_1/K_2 was taken to be 3.5 and was not found to be too critical to the final results. The viscosity corresponding to the initial temperature (922 °K in the calculations) was extrapolated from the shift function presented in [2] to be 0.05 sec. The fictive temperatures for 1/8-inch and 1/4-inch thick glasses are given in figs. 1 and 2 using the same heat transfer coefficient. It is evident that in the mid-plane the fictive temperature lags further behind the actual temperature than that for the surfaces.

The residual stresses and the stress build-up were calculated for the above two thicknesses according to the modification on the volumetric strain given by eq. [4]. α_{liq} and α_{gl} were taken to be 23×10^{-6} and 9.2×10^{-6}/°C respectively. The results were given as in figs. 3 and 4. All computations were done on an IBM 360/40.

With the pseudo-temperature approach the results were obtained more directly. The case with 1/8-inch glass was taken as an example for comparison. The stress profile was superimposed on fig. 3. It is of interest to note that in the entirely two different approaches, with one somewhat microscopic and the other macroscopic, the results came out relatively close and the use of a temperature dependent thermal expansion co-

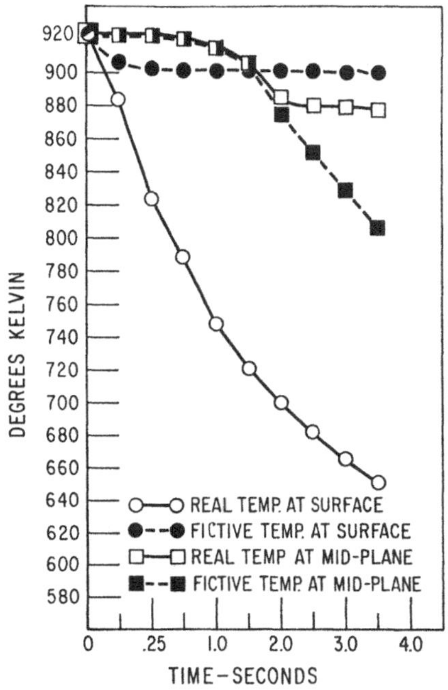

Fig. 1. Temperature vs. time; $^1/_8$″ glass (H = 80 BTU/HR–FT²–°F)

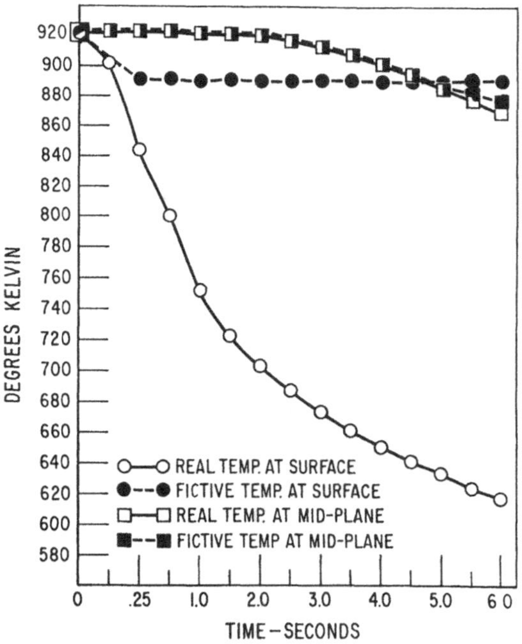

Fig. 2. Temperature vs. time; $^1/_4$″ glass (H = 80 BTU/HR–FT²–°F)

efficient significantly improves the agreement between experimental and calculated residual stresses.

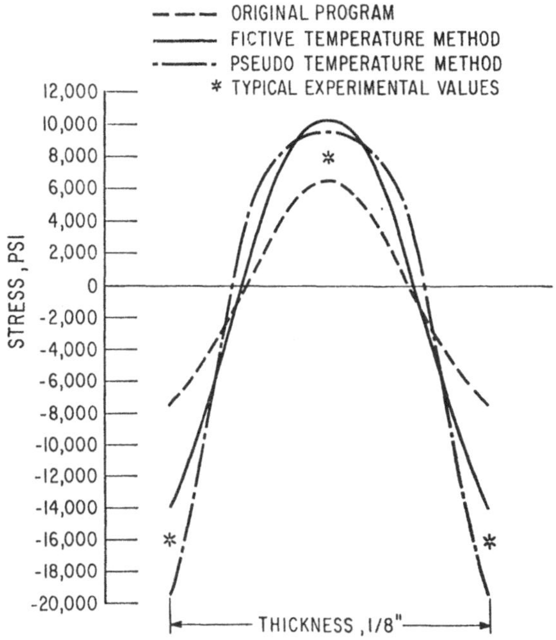

Fig. 3. Residual stress distribution; $^1/_8$″ float glass (H = 80 BTU/HR–FT²–°F, initial temp. —1200 °F)

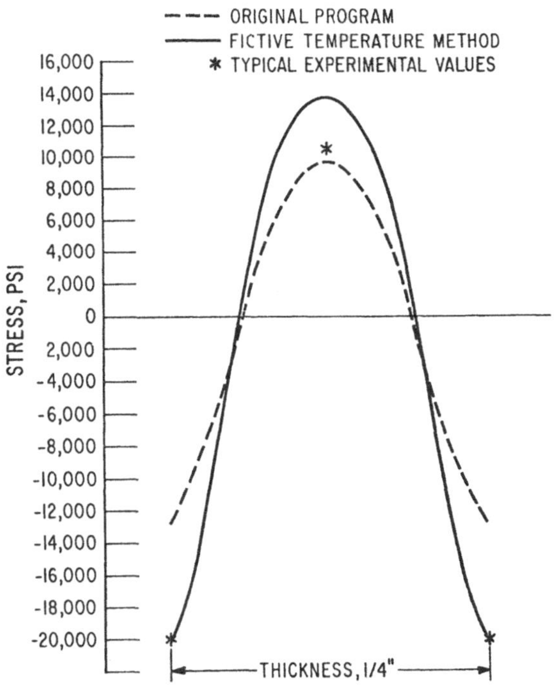

Fig. 4. Residual stress distribution; $^1/_4$″ float glass (H = 80 BTU/HR–FT²–°F, initial temp. —1200 °F)

Acknowledgement

Theoretical discussions with *M. Goldstein* of Yeshiva University were most helpful. *W. R. Krall, J. R. Gottschall,* and *K. M. Rovnak* developed our computational methods.

References

1) *Lee, E. H., T. G. Rogers*, and *T. C. Woo*, J. Amer. Ceram. Soc. **48**, 480–487 (1965).

2) *Woo, T. C., S. M. Ohlberg, E. R. Michalik*, and *R. A. Stewart*, Amorphous Materials. Edited by *R. W. Douglas* and *B. Ellis*, pp. 49–56 (London-New York 1972).

3) *Tool, A. Q.*, J. Amer. Ceram. Soc. **29**, 240 (1946).

4) *Muki, R.* and *E. Sternberg*, J. App. Mech. **28**, 193–207 (1961).

5) *Kurkjian, C. R.*, Physics Chem. Glasses **4**, 129 (1963).

Authors' addresses:

Dr. *S. M. Ohlberg*
P.P.G. Industries Inc.
Harmarville, Pennsylvania (U.S.A.)

Dr. *T. C. Woo*
University of Pittsburgh
Pittsburgh, Pa. 15213 (U.S.A.)

Rheol. Acta 12, 349–356 (1973)

From the Montecatini Edison S.p.A. Instituto Ricerche "G. Donegani" – Novara (Italia)

Rheology on the drawing zone in glass spinning

By G. *Manfrè*

With 10 figures and 2 tables

(Received October 27, 1972)

Symbols

- F_r *Froude* number $U_0^2/g R_0$ with g acceleration gravity (cm/sec^2)
- N_u *Nusselt'* number $2Rh/Ka$ with h heat transfer coefficient (cal/cm^2 sec °C) and K_a air thermal conductivity (cal/cm sec °C) around the forming fibre
- Q Volume rate of flow (cm^3/sec)
- r Radial distance from the central axis of the fibre (cm)
- R Cross section radius of the fibre (cm)
- R_0 Inside diameter of the nozzle (cm)
- t Quenching time (sec)
- $T_a T_s$ Temperature of fibre at the centre (°C)
- T_i Initial temperature at the distance $x = 0$ (°C)
- T_0 Mean value of temperature of air surrounding the forming fibre (°C)
- U_0 Mean value of velocity of glass at $x = 0$ (cm/sec)
- V Local velocity of fibre in the axial direction (cm/sec)
- x Axial distance of the fibre from the nozzle exit (cm/sec)
- W Weight rate of flow (g/minute)
- W_e *Weber* number $\varrho U_0^2 R_0/\alpha$
- α Glass surface tension (dynes/cm)
- φ Angle between the fibre axis and the tangent to the fibre surface in the r, x plane (radiant).
- ν Air kinematic viscosity (cm^2/sec)
- ϱ Glass density (g/cm^3)
- η Glass viscosity (poises)
- η_i Glass viscosity at T_i.
- τ *Maxwell* relaxation time η/G (sec) with G (dynes/cm^2) elastic shear modulus of glass

Introduction

Nearly all the textile synthetic fibres now available are industrially produced by bringing a spinnable material into liquid state, molten or concentrated solution, and forcing it through a small die to form a free liquid jet at the exit. This solidifies as it proceeds along the spinning path and the solid fibre is collected on a rotating drum. Solidification is due to cooling in the melt spinning, to evaporation of solvent in the dry spinning or to precipitation of polymer from solution in wet-spinning.

It is well known how the rheological behaviour of fiber forming polymer materials strongly affects the process of fibre formation and the resulting fibre structures and properties. It is not only the behaviour of spinning fluid or that of solidified fibre which is important in this connection but the rheological behaviour of all the intermediate states between the nozzle and the rotating drum.

As far as the glass is concerned, the fibre is made by a molten spinning process and the practical reasons of the recent increasing investigations of its behaviour in the forming fibre process, as has been shown for polymers (1, 2), are due to rapid growth in the production of continuous glass fibres, mainly for plastic reinforcement, and the spread out of their use and thus the necessity of increasing quality and reducing the cost. In fact it has been shown (3) that the main characteristics of continuous glass fibres, apart from glass, depend on the forming process which affects the diameter and its uniformity, the breaking strength and the spread of strength values and also the maximum output of the adopted equipment (4).

The forming glass fibre process can be fundamentally divided in three zones: extrusion, drawing and the solidified zone.

The extrusion zone involves the phenomena inside the reservoir as well as the nozzle and affects mainly the properties of the glass, such as its homogeneity and viscosity, and so the value of the rate of flow.

The solidified zone is confined between the solidification point and the rotating drum. Due to very rapid cooling, this zone seems to be subjected only to elastic deformations (5) and has a significant influence on the formation of the *Griffiths* flaws at the fibre surface in a very rapid tempering process for a tested glass (6).

The drawing zone (fig. 1) is typical of any kind of spinning process, whether from molten material or solution, flowing out from a nozzle; elsewhere it is called gob, tongue

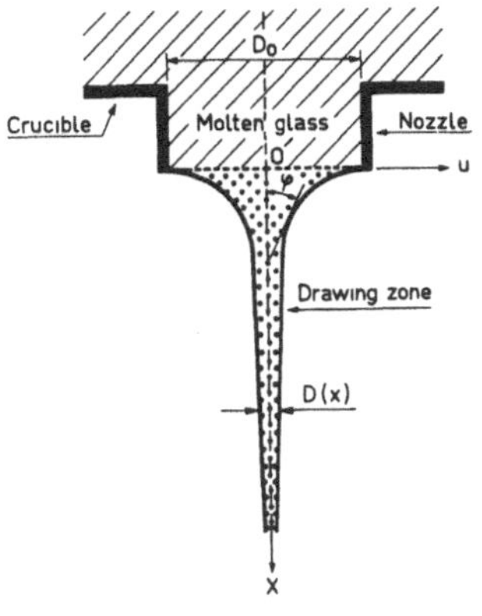

Fig. 1. Schematic drawing of a typical drawing zone of glass forming fibre

or jet; it is localized between the plane at the nozzle exit and the point at which the fibre starts being of constant diameter; elsewhere this point is called solidification point or "freezing in" point.

It has been shown that the quality and the characteristics of continuous glass fibre mainly depend on the performance of glass in this zone (7, 8, 9) and thus, the main purpose of the present paper will be to review the work carried out by other workers and the author on this part of the spinning glass process, and especially to see how the rheological point of view can bring new information on the mechanism of fibre formation.

Previous work

The earlier work on the glass spinning process was entirely dedicated to have quantitative information on the cooling time of the forming fibre in order to explain the experimentally found differences in the mechanical (5) and physical (6) properties between the bulk glass and the fibres, as well as among fibres differently drawn (10).

This whole work has been recently summarized by the author (11) and *Glicksman* (12) who pointed out, first, that only a general statement of the problem concerning the quenching time can give reliable results; then that any approach needs an actual experimental value of the heat transfer coefficient from the fibre surface; and finally that the successful approach of the problem jet applied to polymer molten spinning (2), cannot be used at the drawing zone of the glass spinning especially for the strong difference on boundary conditions (13) and temperature – velocity gradients between the two processes.

Further, the most more recent literature can be summarized in the following points:

– The governing equations and relative boundary conditions can be theoretically defined, but so far only a general one-dimensional approach has been worked out showing its validity only in a part of the drawing zone (12, 14).

– A general solution, also in a two-dimensional analysis, of the energy governing equation has been attempted in the whole spinning way (11) and inside the nozzle (15), and both can be tought reliable for the temperature distribution and thus for the quenching time calculation. Good results have been gained only after obtaining a realistic value of the heat transfer coefficient from the glass surface (16), including the upper part of the drawing zone (17), as shown in fig. 2.

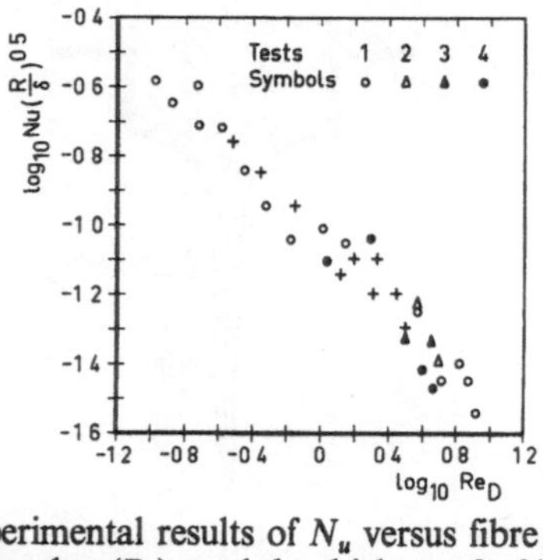

Fig. 2. Experimental results of N_u versus fibre diameter *Reynolds'* number $(R_e)_D$ and the thickness δ of boundary layer of air about the fibre. The spinning conditions are shown in table 1

– The experimental work in addition to the normal parameters of the glass and the process, can be based on photography or cinephotography of the diameter profile of the forming fibre (11, 12) the direct measurements of the temperature in the solidified part of the spinning path (13, 18), and on the measure of the tensile force performed by a rotating drum (19, 12).

The photographic experimental work showed that the drawing zone starts oscillating (9, 20), at least above a certain rate of flow, and more recently a description of three different shapes of the jet in function to the rate of flow has been published (21).

The present paper starts on the mentioned results, especially those obtained by *Glicksman* (12, 14) and the author (11, 13) concerning only the drawing zone with a concave shape which is the most common in ordinary production with a flow rate less than 0.7 g/minute per hole in the industrially used spinneret.

Table 1. Experimental spinning conditions, related to fig. 2, for direct temperature measurements on the forming fibre in the solidified zone. For the tests relating to *E* glass: $T_i = 1220$ °C, $T_0 = 50$ °C and $R_0 = 0.135$; for those relating to silica see (18)

Test	Glass	Q (cm³/sec)	D[1]) (micron)
1	*E*	0.0035	18
2	*E*	0.0035	660
3	*E*	0.0022	26
4	SiO_2	0.0032	48

[1]) *D* is the final fibre diameter.

For the sake of clearness, it has been thought useful to show in the present paper the figures earlier published by the author, starting from those of diameter profiles for *E* and *A* glasses with physical characteristics shown in table 2.

Table 2. Mean value in the temperature range of 1200 to 200 °C of the physical constants used for *E* and *A* glasses. *E* glass is the common borosilicate glass for the ordinary present production of the textile fibres. *A* glass is the common soda-lime glass for the windows. *S*, *C*, *K* are respectively the density (g/cm³), the heat capacity (cal/g °C) and the thermal conductivity (cal/cm sec °C) for the glasses

Glasses	*S*	*C*	*K*
E	2.35	0.26	2.65
A	2.42	0.27	2.50

The results mentioned, in addition to those showed in the *Glicksmans* papers, have driven the author to attempt a more general treatment in order to better know and define the phenomena occurring in the drawing zone with the following purposes:

– Experimental variation of the drawing zone with the spinning conditions.

– Possible solution in a two-dimensional analysis of the governing equations in order to have information about the influence of the surface tension as well as the temperature and the velocity gradients on the instability of the drawing zone. Furthermore, the knowledge of the normal stresses as well as of the possible shear stresses, above all in the lower part of the drawing zone, could bring information about the mechanism of fibre breaking occurring above a certain drawing ratio.

– Comparison between quenching and *Maxwell* relaxation time in function of spinning conditions.

The drawing zone

Practically, the drawing zone starts at the plane separating the flow inside the nozzle and the flow of the fluid jet with a free boundary surface, and ends at the point where any variation of diameter, along the spinning way, are not detectable with the photographic adopted equipment.

The glass flow inside the nozzle can be considered a developed *Poiseuille* flow (9, 22, 15) up to the vicinity of the exit plane, at least in the usual range of flow of the fibre production.

The photographic equipment used by the author (13) detected a variation of the forming fibre diameter up to 10^{-5} cm order of magnitude which is considered the lower limit of variation of the final fibre diameter involved in any production of glass fibre. The latter variation is usually detected by means of a good optical microscope.

The drawing zone in glass spinning mainly performs the following phenomena:

a) Absence of crystallization and molecular orientation along the spinning way which leads to neglect, in the governing equation, any complex consideration of the kinetics crystallization and orientation in a non steady state and non-isothermal conditions (23) certainly occurring in polymers.

b) Remarkable temperature gradient along the spinning way as well as in the cross section of the forming fibre as it can be seen in fig. 5 and 6 and fig. 7 and 8.

The latter, in addition to a cross-section velocity gradient, can be significant not only in the part of the drawing zone nearest the exit of the nozzle (13, 14, 15) but also in the lower part where a small variation of temperature corresponds to a high change in viscosity.

The very strong temperature gradient related to a very high viscosity-temperature coefficient of the glass practically leads to a very short length of the drawing zone in glass spinning, usually no more than 3 cm, in comparison with that of polymers whose length can be 60 cm (24).

c) The shape and the length of the drawing zone, in experiments concerning *A* and *E* glasses, varies with the spinning conditions as following:

1. In the range of weight rate of flow *W*, usually adopted in the industrial process for continuous glass fibres with final diameter from 3–20 μm, the shape of the jet can be considered concave. This means that the slope of the glass jet surface continuously decreases up to the solidification point. For *E* glass this range, in terms of flow *W*, can be defined from 0.2–0.7 g/minute and in terms of *Reynolds'* number R_e from 10^{-5} to 10^{-4}, as it was confirmed by *Burgman* experiments (21). For the *A* glass the upper limit of output is not more than 0.55 g/minute.

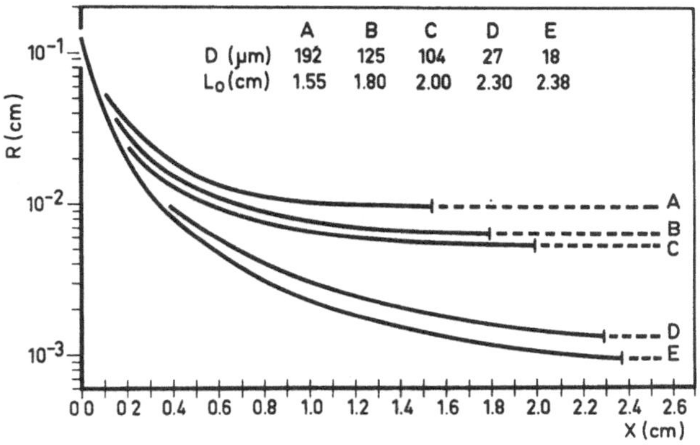

Fig. 3. Radius *R* versus *x*, for *E* glass fibres with different final diameter and spinning conditions $Q = 3.5 \cdot 10^{-3}$ cm³/sec, $R_0 = 0.135$ cm, $T_i = 1220$ °C, $T_0 = 50$ °C. The end of the solid lines shows the experimental length L_0 of the drawing zone

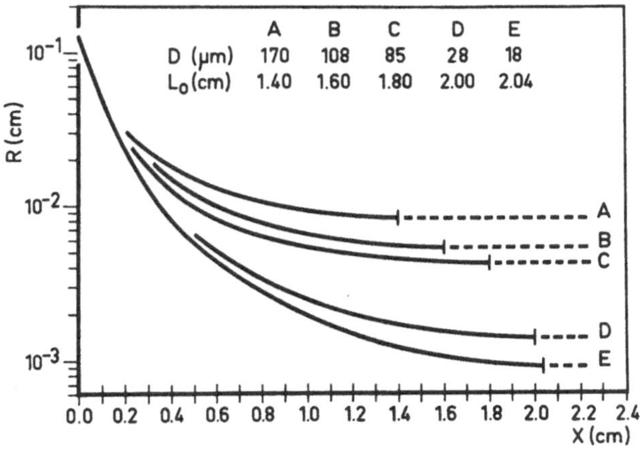

Fig. 4. Radius *R* versus *x* for *A* glass fibres with different final diameter and spinning conditions $Q = 2.2 \cdot 10^{-3}$ cm³/sec, $R_0 = 0.135$ cm, $T_i = 1212$ °C, $T_0 = 50$ °C. The end of solid lines shows the experimental length L_0 of the drawing zone

2. In the experimental range of spinning conditions with drawing speed V 100 ÷ 2500 m/minute, glass viscosity η $2 \times 10^2 \div 4 \times 10^3$ poises and nozzle inside diameter 0.1 ÷ 0.3 cm, for values W less than approximately 0.2 g/minute, the drawing zone shows that: its base diameter starts inside the nozzle, it is smaller than the exit nozzle diameter and its shape remains concave. At values W more than approximately 0.7 g/minute for E glass and 0.55 g/minute for A glass, the shape of the drawing zone shows a convex shape with an increasing slope at the plane of the exit nozzle up to a maximum diameter and than a decreasing slope up to a solidification point. In these spinning conditions the drawing zone shows unsteady jet flow, as was pointed out earlier (9, 20, 21).

Consequently, in this range of W, the glass fibres show nearly periodic variation of their final diameter and in the statement of the problem the flow in the jet should be considered unsteady and it was not included in our present investigation.

3. The length of the drawing zone, as shown in fig. 3 and 4, changes with the drawing speed, the other conditions taken as constant, and in particular it increases with the drawing speed. Simultaneously, the temperature of the solidification point, as shown in fig. 5 and 6, increases with drawing speed.

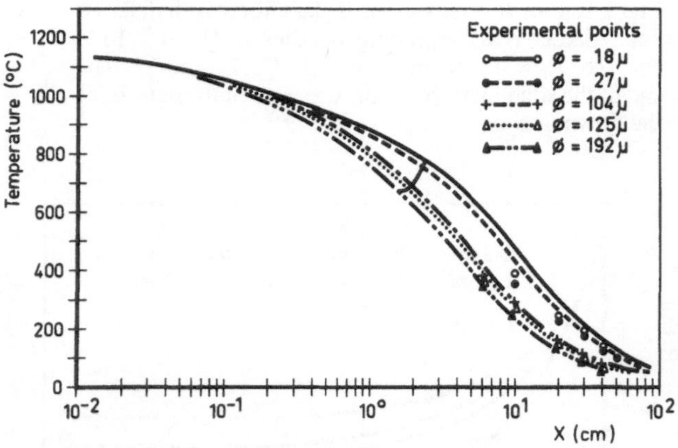

Fig. 5. Temperature T_s versus x for E glass fibres obtained by two-dimensional analysis. The solid crossing line shows the solidification temperature versus different final diameter of the forming fibre and the points are temperatures experimentally measured on the solidified part of the fibre. The spinning conditions are the same as in fig. 3

An important point, shown in fig. 3 and 4, is that the shape of the drawing zone can split in two parts: an upper part which in the range of spinning conditions mentioned in 2 appears to be independent of the drawing speed up to 2300 m/minute, and a lower part near the solidification point which varies consistently in shape as well as in length.

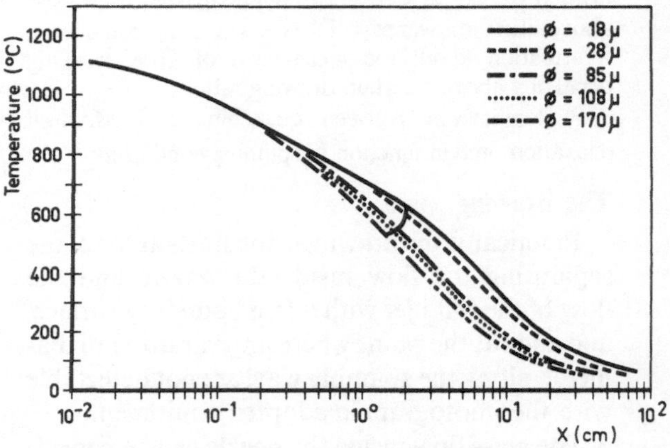

Fig. 6. Temperature T_s versus x for A glass fibres obtained by two-dimensional analysis. The solid crossing line shows the solidification temperature versus different final diameter of the forming fibre. The spinning conditions are the same as in fig. 4

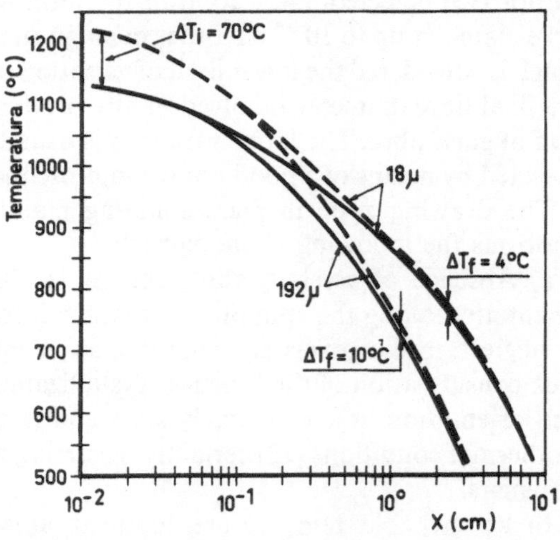

Fig. 7. Temperature T_s (solid line) and T_a (dashed lines) versus x for two E glass fibres of final diameter 192 and 18 micron. ΔT_f is the cross-sectional temperature difference at the solidification point. The spinning conditions are the same as in fig. 3

The upper part of the drawing zone

It has been shown that the one-dimensional analysis of *Glicksman* (12) mainly fails in the upper zone where, on the other hand, the sources of the possible jet instabilities are localized. Furthermore, the temperature gradients in this zone in both directions, shown in fig. 3, 4 and 6 and confirmed by *Sununu* and *Brown* (15), lead to state that the cross-sectional velocity gradient can be **neglected** and that it derives more from the viscosity gradient rather than from the shear rate of *Poiseuille* flow inside the nozzle.

Another important point to be tested should be the influence of the surface tension in this upper zone. In fact, for glasses it ranges from 200 up to 400 dynes/cm and, in comparison with that of polymer having nearly 20 dynes/cm, and so it plays an important role among the forces acting in glass spinning, as was shown earlier (19, 12).

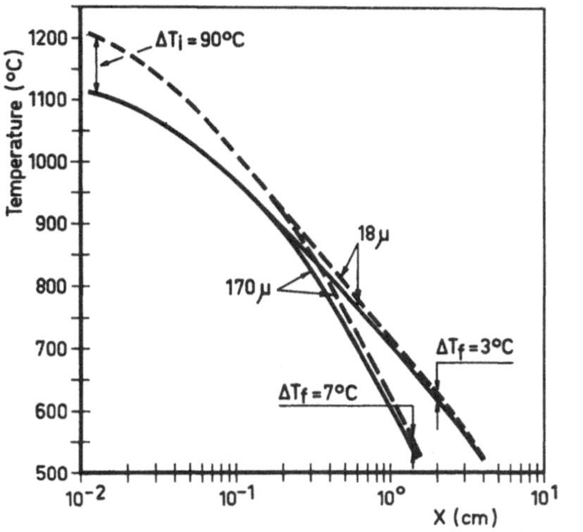

Fig. 8. Temperatures T_s (solid lines) and T_a (dashed lines) versus x for two *A* glass fibres of final diameters 170 and 18 micron. ΔT_f is the cross-sectional temperature difference at the solidification point. The spinning conditions are the same as in fig. 4

The presence of surface tension in the boundary conditions on the free surface of the jet considerably complicates the solution of the motion governing equations of the problem, as can be seen later, at least in a two-dimensional approach.

As the upper part of the jet can be responsible for instability of the jet, limiting the maximum output of glass per holes, we think that the lower part can be mainly responsible for the limitation of the glass spinning related to the breaking of the forming fibre beyond a certain value of the drawing ratio.

The scanty data of this kind of instability do not enable a quantitative conclusion to be drawn but the rheological point of view concerning the elongational shear stress involved, the knowledge of the relaxation time compared to quenching time and, if possible, some viscoelastic phenomena performing by the glass in this very particular zone, could bring important information. Concerning the shear stresses related to shear rate, it must take into account that, in the region near the solidification point, any small change in temperature along the cross section of the forming fibre could cause a significant change in viscosity and so to a telescopic deformation of the forming fibre, in addition to the elongational tensile stretching.

In regards to the rheological behaviour of the two tested glasses, the upper zone of the *A* glass appears to be longer than the corresponding zone of the *E* glass. This can be due to a viscosity – temperature curve, steeper in the case of *E* glass, and also to the surface tension lower for the case of *A* glass.

The viscosity can be related to temperature by

$$\eta = \eta_0 e^{\frac{B}{T - T_0}}. \qquad [1]$$

For *A* and *E* tested glasses η_0 is relatively 0.022 and 0.01 poises, B is 10,154 and 8020 and T_0 is 249 and 470 °C; the surface tension at the spinning temperature of 1212 °C for *A* glass is approximately 330 dynes/cm and for *E* glass at 1220 °C is nearly 380 dynes/cm.

To conclude this part we would like to point out that, to well define the upper part of drawing zone, it is necessary to carry out experiments in order to localize it between the exit of the nozzle and the point, or region, where the one-dimensional analysis can be applied successfully, following the *Glicksman*' work. For this purpose, the author stated the problem in a two-dimensional analysis considering the one-dimensional approach only an asymptotical solution of the two-dimensional approach, which might be reliable only downwards in a region after a certain distance from the nozzle.

Statement of the problem

The energy, the motion and the continuity equations are the governing equations which, in addition to the rheological constitutive equations of the tested material and the necessary boundary conditions, can completely state the problem.

We have considered the glass a *Newton*ian incomprensible flow and the jet in a steady state. Thus considering the glass spinning without any phase transformation, the solution of the governing equation could give the theoretical diameter profile of the jet to be compared with that experimentally obtained.

The boundary conditions of the energy equation contain the temperature variable and are not linear due to the heat loss by radiation as well as convection.

The convection heat loss involves the relation between the *Nusselt* and *Reynolds*' numbers, shown in fig. 2, which can be analytically stated by

$$N_u = 0.090(\delta/R)^{0.5}(R_e)^{-0.5} \qquad [2]$$

with

$$\delta/R = -1 + \sqrt{1 + f(x)}$$

$$f(x) = \frac{4\,v_m x}{Q}\left(\frac{1}{b} + \frac{1.57}{b^2} + \frac{1.84}{b^3}\right)$$

$$b = \log \frac{4\,v_m}{Q} x$$

δ is the thickness of boundary layer about the forming fibre and v_m is the mean value of the air kinematic viscosity assumed 0.5 cm^2/sec.

The eq. [2] has been shown to be very realistic since it takes into account the main spinning parameters and their variations along the spinning way.

The boundary conditions of the motion equations concerning the free surface of the jet are very complex and, due to the surface tension, they contain non-linearly the dR/dx derivatives.

Thus the problem can be solved only numerically, and in the two-dimensional analysis seems to be an elliptical problem rather than parabolic.

The details of the statement have recently been presented (25) in a recent congress and are to be published.

The first difficulty has been to find out only the necessary and realistic boundary conditions concerning the plane at the nozzle exit, the symmetry of the flow, which means the conditions at the fibre axis, and the actual forces acting on the free surface. It has been found that only the drag air force and the variation of surface tension with temperature can be actually neglected in the drawing zone.

The second difficulty was to present a system of the governing equations in a normal linear form in order to be numerically soluble not exhibiting any singularity in the matrix of the equation system.

The third problem was to present the equations and the relative boundary conditions in an adimensional form with dimensionless number having a physical meaning.

The dimensionless numbers, whose value seems to be determinant in the drawing zone of the spinning glass process, are:

$$R_e = \frac{\varrho R_0 U_0}{i}; \quad F_r = \frac{U_0^2}{g R_0}; \quad W_e = \frac{\varrho R_0 U_0^2}{\alpha}.$$

The ration R_e/W_e appears to be very important, confirming the important role of the ratio α/η (cm/sec) in the stability of the spinning process (9) and in the investigation of the spinnability of liquids (26).

So far we did not achieve a complete satisfactory solution due to instability of numerical computation but it can certainly conclude that the shape and the possible instability of the upper drawing zone does not depend on viscoelastic phenomena, as in polymers, and it is affected by other physical phenomena involving not only the *Reynolds*' number but also the *Weber* and the *Froude* numbers. The two latter can be neglected in polymer spinning.

Relaxation theory

The relaxation theory concerning the glass spinning rised to justify the difference in the mechanical and physical features between fibres and bulk glass and among the fibres with different diameter. It stated that the fibre cools fast enough for the elapsed time to be less than the *Maxwell* relaxation time for the shearing forces in the drawing zone of the forming fibre and this leads the fibre to reach an arbitrary temperature below the transformation temperature before rearrangement of the glass structure can take place.

Bateson (6) and *Blokh* (27) correlated the relationship between fibre strength and diameter with the ratio of the cooling time to the *Maxwell* relaxation $\tau = \eta/G$(sec), which is believed to be approximately of the order of $10^{-3} \div 10^{-4}$ sec.

This relationship was based on the calculation of the cooling time developed by *Deeg* and *Dietzel* (10) and *Anderson* (5). This has been shown unrealiable first by the experimental results of *Kutunov* (20) and *Arridge* (18) and later by the temperature profile in the drawing zone obtained by the author (13, 17, 11) and *Glicksman* (12).

These last results supported the conclusions, summarized by *Burgman* and *Hunia* (4), that there is no significant relation between fibre strength and total cooling time of forming fibre found to be of the order of 10^{-1} sec for *E* glass fibre of final diameter 7 ÷ 20 micron.

Thus other process parameters, like the environmental moisture (4) and the cross-sectional temperature gradient (28), have been appealed to justify the peculiar characteristics of the glass fibres.

Our opinion is that any parameter of the process might affect the features of the fibre but certainly, among the others, the temperature gradients in both directions, and thus the relative quenching time, in addition to the shearing stresses involved, have to play an important role.

The quantitative value of the shearing stresses can be actually obtained by a two-dimensional approach which has also to verify whether the *Newton*ian constitutive equation can be applied to the glass even in the region near the solidification point.

Before achieving results on it we would like to draw attention on the following points, shown in fig. 9 and 10:

– The cooling time can be defined as the time spending by the glass to go downwards from a certain temperature to the solidification point. In a particular region just before the latter the cooling time can be of the order of $10^{-3} \div 10^{-4}$ sec. The length of this region increases with the drawing speed.

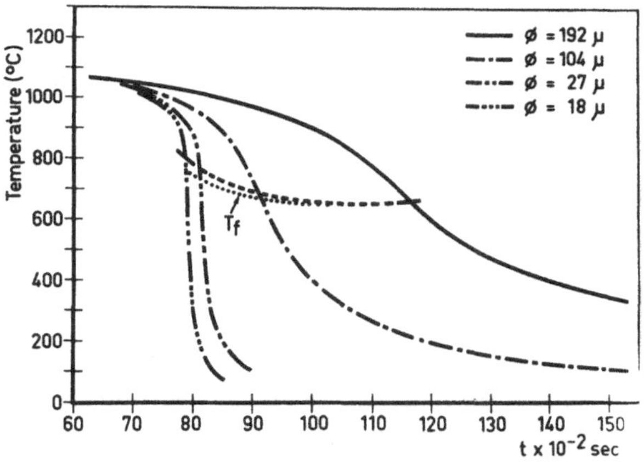

Fig. 9. Temperature T_s versus time t of the forming fibre for *E* glass with different final diameter. The solidification temperatures T_f are compared with temperatures (--- lines) corresponding to quenching time 10^{-3} sec in the zone just above the solidification point. The spinning conditions are the same as in fig. 3

For instance for the *E* glass fibres of diameter 104 micron in 10^{-3} sec the glass runs a distance of 0.03 cm with a difference in temperature along x of 4 °C; for the fiber of diameter 18 micron the relative values are 0.54 cm and 35 °C. On the other hand the quenching time of the upper part of the drawing zone is of the order of 10^{-1} sec and for the finer fibres is comparable with the total quenching time.

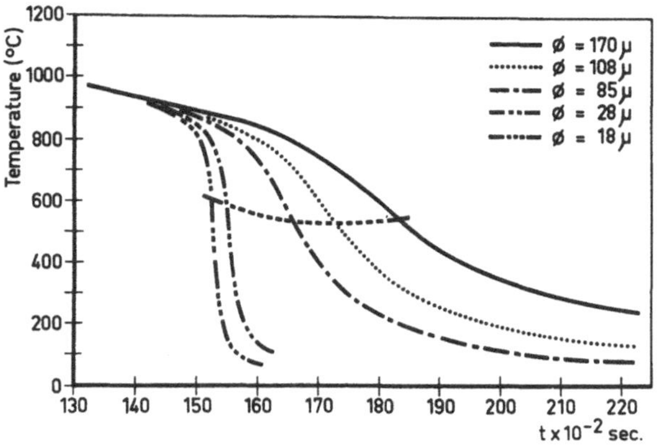

Fig. 10. Temperatures T_s versus time t for the forming fibre of *A* glass with different final diameters. The solidification temperatures are shown by the crossing curve. The spinning conditions are the same as in fig. 4

– The solidification point increases with the drawing speed.

These results lead to say that a relationship could actually exist between the features of the fibres and the cooling time of a lower part of the drawing zone and possibly to compare it not only with a single glass relaxation time but with a distribution of it, as in polymers.

Furthermore in addition to a rheological behaviour a simultaneous thermodynamical phenomena occur which can "freeze in" a structure, dislocations and strains corresponding to liquid glass structure. Higher is the $\Delta T/\Delta t$ ratio more should be the "frozen in" structure.

Conclusions

The two-dimensional approach of the glass liquid jet flowing from a nozzle has been shown necessary to give information on the instabilities of the process and in the possible relationship between the spinning conditions and the features of the fibres.

The investigation on the instability of the upper part of the drawing zone, which limits the maximum output of glass per nozzle, has to consider the temperature gradient, in both directions, the *Weber* and the *Froude* numbers involved in the problem in addition to the *Reynolds*' number.

The instability concerning the breaking of the forming fibre can be related to the rheological behaviour of glass and the shearing stresses involved in the region near the solidification point.

For the shearing stresses can be important the cross sectional temperature gradient which can cause a telescopic deformation in addition to the elongational tensile deformation.

The relaxation theory, concerning the relationship between the features of the fibres and the ratio of cooling time to *Maxwell* relaxation time, **could be appealed only whether is considered a** cooling time of the particular zone near the solidification point and not the total cooling time of the drawing zone.

A complete solution of our two-dimensional approach might bring new information on the rheological behaviour of glasses subjected to high shear stresses in the transformation region and also in the possible relationship between structure of glasses and cooling time.

Summary

Among investigations concerning the rheology of spinning materials from melt, or in other terms the problem of "spinnability", glasses perform an example of fibre forming without crystallization along the spinning way and with surface tension playing an important role. Furthermore glasses show a *Newton*ian behaviour at least in the upper part of the drawing zone.

As the absence of crystallization simplifies the formulation of the governing energy equation, on the other hand, the surface tension makes the applied motion equations quite complex to solve, above all in the two-dimensional analysis.

The present paper shows that only a two-dimensional approach can give reliable results on the temperature, velocity and stress distribution in the drawing zone by a comparison of the theoretical and the experimental diameter profile of the forming fibre.

The temperature profile has been obtained by a numerical solution of the energy equation, only after gaining experimentally the heat transfer coefficient. The results **shown in the one-dimensional analysis cannot be applied in the opper part of the drawing zone.**

The velocity and stress distribution can be obtained by very complex numerical solutions in the very upper part of the drawing zone where the one-dimensional approach is shown unreliable. This can be thought an asymptotic solution of two-dimensional approach, reliable only after a certain distance of the spinning way from the exit of the nozzle.

Furthermore, an analysis of the dimensionless numbers involved in the spinning phenomena brings up some information concerning the instability of the glass jet in comparison with that shown by materials as molten polymers or metals.

As far as the rheological behaviour of glasses in the elongational shear rate is concerned, some conclusions can be drawn.

References

1) *Ziabicki, A.*, Principles of spinning. Man Made Fibres, Science and Technology, Vol. 1, Edited by *Mark, Atlas, Cernia* (New York 1967).

2) *Acierno, D., J. N. Dalton, J. M. Rodriguez* and *J. L. White*, J. Appl. Poly. Sci. **15**, 2395–2415 (1971).

3) *Thomas, W. F.*, Phys. Chem. Glasses **1**, 4–18 (1960).

4) *Burgman, J. A.* and *E. M. Hunia*, Glass Technology **11**, 147–152 (1970).

5) *Anderson, O. L.*, J. Appl. Physics **29**, 9–12 (1958).

6) *Bateson, S.*, J. Appl. Physics, **29**, 13–21 (1958).

7) *Aslanova, M. S.* and *V. E. Khazanov*, Steklo Keram. **25**, 9, 1–4 (1968).

8) *Bruckner, R.*, The structure of glass fibres, in particular glass fibres. VII Congrès International du Verre, Paper n. 38, Bruxelles 1965.

9) *Manfrè, G.*, Some rheological aspects in spinning glass fibres. VII Congrès International du Verre, Paper n. 78, Bruxelles 1965.

10) *Deeg, E.* and *A. Dietzel*, Glastech. Ber. **28**, 221–232 (1955).

11) *Manfrè, G.*, Verres et Réfractaires **26**, 57–65 (1972).

12) *Glicksman, L. R.*, Trans. ASME, J. Basic Engineering **90**, 343–354 (1968).

13) *Manfrè, G.*, Temperature profile of drawing zone in spinning continuous glass fibres. Autumn Meeting Amer. Cer. Soc., Bedford Springs (USA), (1967).

14) *Krishman, S.* and *L. R. Glicksman*, Trans. ASME, J. Basic Engineering, Paper n. **70** – WA/FE – 3 (1970).

15) *Sununu, J. H.* and *G. A. Brown*, Trans. ASME, J. Basic Engineering, Paper n. **70** – WA/HT – 12 (1970).

16) *Glicksman, L. R.*, Glass Technology **9**, 131–138 (1968).

17) *Manfrè, G.*, Survey of cooling rate of glass fibres. Autumn Meeting Am. Cer. Soc., Bedford Springs (USA), (1967).

18) *Arridge, R. G. C.* and *K. Prior*, Nature **203**, 386–387 (1964).

19) *Manfrè, G.*, Glass Technology **10**, 99–106 (1969).

20) *Khodakovskii, M. D.* and *S. A. Kutunov*, Steklo Keramics **21**, 2, 3–10 (1964).

21) *Burgman, J. A.*, Glass Technology **11**, 110–116 (1970).

22) *Zawadcki, A.*, Szklo I Ceramic **18**, 40–44 (1967).

23) *Ziabicki, A.*, Kinetics of polymer crystallization and molecular orientation in the course of melt spinning. Applied Polymer Symposia n. 6, Fiber Spinning and Drawing, p. 1–18 (New York).

24) *Kase, S.* and *T. Matsuo*, J. Polymer Sci. Part **A 3**, 2541–2554 (1965); Part **A 11**, 251–287 (1966).

25) *Manfrè, G.*, Two dimensional analysis of the drawing zone in liquid molten spinning. Paper presented at the Euromech Colloquium 37 on Fluid Mechanics and Polymer Processing, Naples, 20–23 June (1972), (in press).

26) *Ziabicki, A.* and *R. Takserman-Krozer*, Roczniki Chemii, **37**, 1511–1520, (1963).

27) *Blokh, K. I.*, Relaxation theory of glass formation and strength of glass fibres. The structure of glass, Vol. 6, 222–224, Edited by *Porai-Koshits* (New York 1966).

28) *Dusollier, G.* and *J. Robredo*, Verres Réfract. **24**, 2, 63–70 (1970).

Author's address:

Dr. *G. Manfrè*, Montecatini Edison S.p.A.
Istituto Ricerche "G. Donegani"-Novara (Italia)

Für die Schriftleitung verantwortlich: Dr. W. Meskat, 5090 Leverkusen; Anzeigenverwaltung: Dr. Karl Niedermeyer Nachf., 6000 Frankfurt/M. 90, Georg-Speyer-Str. 76; Dr. Dietrich Steinkopff Verlag, 6100 Darmstadt, Saalbaustr. 12; Gesamtherstellung: Universitätsdruckerei Mainz GmbH

Rheol. Acta 12, 357–373 (1973)

From the Institute of Petrochemical Synthesis, Academy of Sciences of the USSR, Moscow (USSR)

Critical regimes of deformation of liquid polymeric systems

G. V. Vinogradov

With 18 figures and 2 tables

(Received October 27, 1972)

The principal peculiarity of typical polymers at temperatures above the glass-transition region and the melting point is their ability to exist in two physical states – fluid and high-elastic (rubbery). Accordingly, they manifest two main relaxation mechanisms. In the fluid state, irreversible displacement of the centres of gravity of macromolecules relative to each other is possible, and irreversible deformations may be infinitely large. In the high-elastic (rubbery) state, the determining role is played by rapid changes in the spatial arrangement of groups of atoms forming the macromolecular chain, and this enables it to unwind, thereby accumulating a considerable amount of reversible (high-elastic) deformation. In the fluid state polymers manifest both relaxation mechanisms, and during the flow they may accumulate large amounts of reversible deformations.

The transition from the fluid to the high-elastic (rubbery) state and back can be realized at isothermal conditions by a change of the deformation rate. Note that with an increase in the deformation rate when the high-elastic state is attained the development of large irreversible deformations may prove impossible, which excludes the possibility of realization of a steady flow. In short, they lose fluidity. This point provides a qualitative approach to the concept of critical deformation regimes connected with transition of polymeric systems from the fluid to the high-elastic state with increasing deformation rate. Such a transition, as a purely relaxation phenomenon has a general significance for polymers with sufficiently high molecular weights. It is manifested in the most distinct way for linear high-molecular-weight polymers characterized by narrow molecular weight distribution.

Amorphous and also crystallizing polymers, when passing to the high-elastic state at temperatures exceeding their T_g and melting points behave like cured polymers. In this connection the idea of the existence of entanglement networks in uncured polymers is of great importance.

Much experience has been accumulated in investigating the viscoelastic properties of melts of narrow-distribution linear polymers. They include data for polymer homologous series of polystyrenes, polybutadienes, polyisoprenes, polyvinyl acetates, polymethylmetacrylates and polydimethylsiloxanes. So far the greatest attention has been given to polystyrenes, which are obtained by anionic polymerization and are commercially available.

Using the concept of entanglement networks of macromolecules and proceeding from the results of investigations into the viscoelastic properties of narrow-distribution linear polymers, it is possible to give their qualitative classification, which is presented in table 1.

This classification is based on the normalization of molecular weights according to the value of M_e, the molecular weight of a chain section between entanglements of macromolecules. The value of M_e is determined by measuring various characteristics of the viscoelastic properties of polymers, in the first place of the dependence of the storage modulus that is modulus of elastic deformation on the frequency in regimes of small-amplitude cyclic deformation. It is customary to assume that M_e is in a simple relationship with the critical molecular weight M_c, which corresponds to a sharp change in the rate of dependence of the initial viscosity on the molecular weight.

Table 1. Classification of low-molecular weight substances, oligomers and polymers according to their viscoelastic properties

Low-viscosity liquids M_k-kinetic flow segment	Oligomers $M = M_e \simeq \lvert 5 + 10 \rvert M_k$ Appearance of entanglement network. M_e-kinetic unit in entanglement network.	
Polymers →		
$M = M_c \simeq \lvert 2 + 2.5 \rvert M_e$ Entanglement network. Increase in rate of viscosity-molecular weight dependence. Appearance of large reversible (high-elastic) deformations. Development of viscosity anomaly.	$M \simeq (4 + 5)\, M_e$ Appearance of signs of high-elastic (rubbery) state. Stabilization of glass-transition temperature.	$M \geqslant 20\, M_e$ Completion of development of signs of high-elastic (rubbery) state. Increase of uniformity of segmental density. Appearance of loss maximum and development of high-elasticity plateau. More abrupt transition from fluid to high-elastic state with increased deformation rate. Separation of relaxation mechanisms. Decrease in ability for viscosity anomaly. Loss of fluidity with increased deformation rate decrease and stabilization of zero shear modulus.

Let us consider (see fig. 1) the viscoelastic characteristics of monodisperse polystyrenes melt over a wide range of variation in the ratio of the molecular weight to M_e based on the measurement of storage and loss moduli at sinusoidal regimes of deformation of small-amplitude, when deformation does not effect the structure of the polymers. This figure includes data borrowed from the publication of *Sh. Onogi* with coauthors (1) *I. den Otters* dissertation (2) and also obtained in our laboratory by *Yu. G. Yanovsky*.

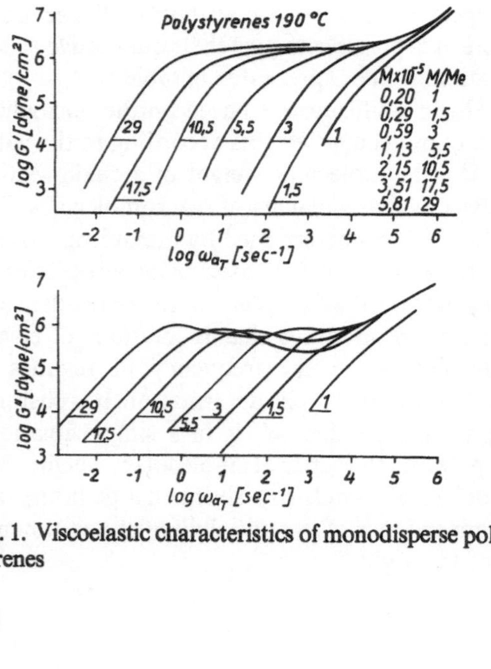

Fig. 1. Viscoelastic characteristics of monodisperse polystyrenes

The upper part of the figure shows the dependence of the storage modulus on frequency, the lower part, a similar dependence of the loss modulus. The numbers at the curves indicate the M/M_e ratio. It is seen from the figure that only when M/M_e ratio exceeds 10 distinct plateau of high-elasticity and maximum losses are displayed. The lower graph is of special interest, primarily in the region of maxima. They appear because with an increase in frequency the centres of gravity of macromolecules have no time to shift within each cycle of deformation and therefore the losses due to polymer fluidity diminish. Thus the maxima share the low-frequency region, where the greatest role is played by the relaxation mechanism due to the possibility of the displacement of the centres of gravity of macromolecules from the region of higher frequencies at which the determining role is played by the rapid variation in the conformations of the chains between the entanglements. It is important that the maxima appear long before the horizontal section of the high-elasticity plateau is achieved. In this way the maxima distinctly delimit the regions of transitions from the fluid to the high-elastic state and vice versa.

As has been indicated, the polymer homologous series of polystyrenes was studied in greatest detail. However, polystyrenes exhibit two peculiarities associated with the fact that polystyrene macromolecules do not show high flexibility. First, polystyrenes possess a rather high value of M_e, which is approximately equal to 20 thousands (1). Second, owing to the high T_g value investigations of their viscoelastic properties are usually carried out at temperatures not very far from the glass transition region.

The high value of M_e impedes detailed investigation of the viscoelastic properties of those polystyrene specimens which clearly exhibit all signs of the rubbery state, because they have extremely high molecular weights and a correspondingly tremendous viscosity.

On the other hand, polybutadienes obtained by anionic polymerization have a value of M_e six to seven times lower (3) than for polystyrenes and a glass transition point close to -100 °C. Being stabilized by antioxidants they possess a satisfactory thermal-oxidation stability up to temperatures of 120 °C. Thus it is possible, without great impedements to investigate polybutadienes in which the ratio of the molecular weight to M_e greatly exceeds twenty units (it may reach a value of about one hundred units) and the ratio of the experimental temperature to T_g exceeds two. Such conditions are hard to realize for pure polymers whose macromolecules do not exhibit high flexibility. Therefore in our laboratory much attention was devoted to investigating narrow-distribution polybutadienes. This permits systematizing the characteristics of the viscoelastic properties of narrow distribution linear polymers, using polybutadienes as an example, over a wide range of molecular weights.

Up to the present no theory has been developed which, at least qualitatively, would correctly describe the whole complex of viscoelastic properties of linear polymers with molecular weights exceeding twenty M_e. In the light of the aforesaid it is expedient to dwell on those theories which predict the appearance of a maximum on the curve showing the dependence of the loss modulus

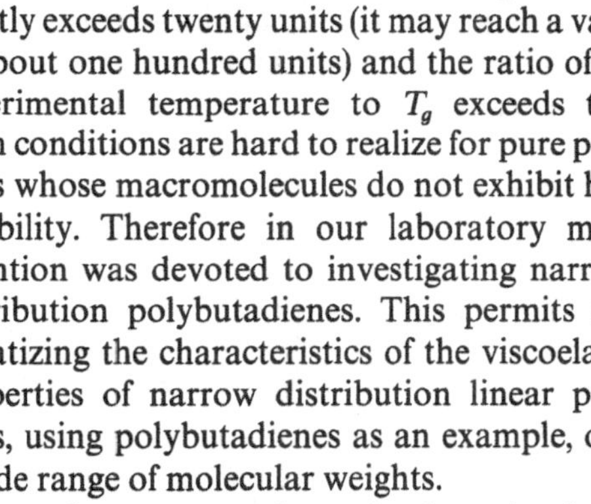

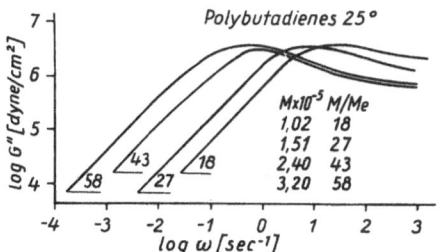

Fig. 2. Viscoelastic characteristics of narrow-distribution high-molecular-weight polybutadienes

on frequency and accordingly, a sharp transition from the fluid to the high-elastic state.

According to the theories of *S. Oser* and *R. Marvin* (4), *V. Pokrovsky* (5) and also *W. Graessly* (6) the maximum on the dependence of the loss modulus on frequency should appear at molecular weights close to ten M_e. When the ratio of the molecular weight to M_e exceeds twenty, the maximum values of the loss modulus should not depend on the molecular weight.

As is seen from the fig. 2 the experimental data confirm this statement.

The maximum value of the loss modulus is characteristic of each polymer homologous series, rising with diminishing M_e i.e. with increasing flexibility of the polymer chain. The maximum values of the loss modulus for various polymers lie within the range from 10^6 to 10^7 dyn/cm^2. Since the activation energy of the formation and disintegration of the entanglement network is very low, the values of M_e and the maximum values of the loss moduli are practically independent of temperature.

According to the theory of *W. Graessly*, it should be expected that this will also be in excellent agreement with our experimental information that the circular frequency corresponding to the maxima is inversely propertional to the initial viscosity and, consequently, inversely proportional to the molecular weight of the polymer to the power of 3.5. The dependence of the frequency at the point of maximum is determined to a first approximation by the activation energy of viscous flow.

The aforesaid is of great importance and reaches far beyond the limits of consideration of cyclical deformation of polymers at small-amplitude. The crux of the matter is as follows. There is an empirically well-established correlation between the regimes of polymers cyclical deformation at small-amplitude and their steady-flow (7). If we assume that the circular frequency is numerically equal to the shear rate, then in the cases at hand the loss modulus is equal to the shear stress. This means that the sections of the curves showing the dependence of the loss modulus on the frequency, which describe the fluid state below the maxima, at the same time are flow curves determining the viscous properties of polymers under conditions of steady flow. Since in narrow-distribution polymers the loss moduli are directly proportional to the frequency almost up to the maximum, the shear stress is directly proportional to the shear rate. This is the characteristic of a *Newton*ian fluid.

Why is it necessary to consider dependence of loss modulus on circular frequency only up to the maximum? Because at steady-regimes of flow the stress cannot reduce when the shear rate increases.

The maxima on dependence of loss moduli on the frequency should at the same time determine the critical values of shear stresses and rates, which cannot be exceeded at steady-flow regimes. This means that with regard to these critical values of the flow parameters all that has been said with regard to maxima of dependences of loss moduli on the frequency is correct. Consequently, for each polymer homologous series the critical shear stress should have a constant value, which is independent of temperature and lies between 10^6 and 10^7 dyn/cm^2. The critical shear rates should change in inverse proportion to the initial viscosity or to the molecular weight to the power of 3.5. Their temperature dependence should be determined, to a high approximation, by the activation energy of viscous flow. Of great importance is the fact that the above-considered predictions of the theory must be valid not only for shear but for uniaxial extension as well.

When discussing the classification of linear polymers attention was called to the fact that, with the development of the entanglement network, the manifestation of viscosity anomaly is enhanced. This is natural if the cause for the viscosity anomaly is considered to be the displacement of equilibrium in the processes of formation and disintegration of entanglements of macromolecular chains under the effect of deformation. Then the manifestation of this effect should be the sharper, the greater is the number of entanglements. This situation, however, can only exist until, with increasing number of entanglements per macromolecule, a high uniformity of segmental density is achieved. Then the stress acting in the flow should be distributed more and more uniformly over the segments of macromolecules and the entanglements binding them. Thus, a polymer melt modelled by an entanglement network becomes quasiisotropic. This inhibits a gradual reduction in the number of entanglements under the effect of the increasing shear stress, and hence the manifestation of the viscosity anomaly. With increasing deformation rate, when its critical value is achieved, a polymer melt turns into a quasicured polymer which is losing fluidity. How is this manifested in high-molecular-weight linear narrow distribution polymers?

Under the conditions of uniaxial extension attainment of the critical regime of deformation corresponds to the rupture of polymer samples. A much more complicated situation exists at simple shear. Therefore it is reasonable to focuss attention on simple shear which may be realized in a wide range of stresses and rates by the method of capillary viscosimetry.

It was shown in our laboratory (8) that in high-molecular-weight linear narrow-distribution amorphous polymers the critical flow regimes under shear correspond to a sharp transition from the *Newton*ian or a near-*Newton*ian regime of flow to a spurt, when the volume rate may increase stepwise tens, hundreds, and even thousands of times over. Besides it has also been established that in crystallizing polymers similar to linear polyethylene the critical parameters of flow at temperatures close to the melting point are associated with the beginning of their crystallization. What has just been said is illustrated schematically in fig. 3.

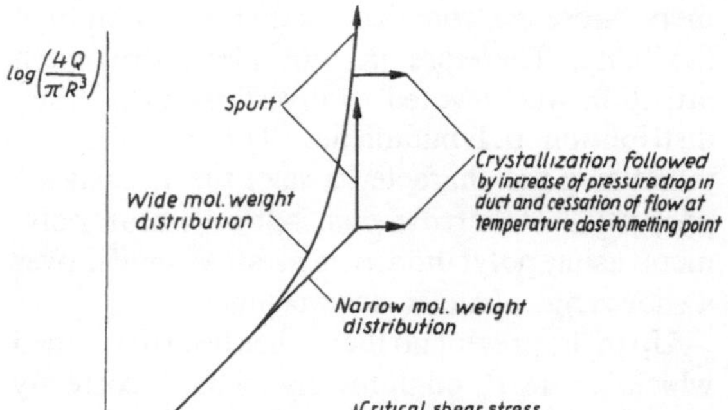

Fig. 3. Key diagram of development of crystallization and spurt processes in high-molecular-weight linear polymers on attainment of critical deformation regimes

Here the question arises immediately to what extent the critical flow regimes determined by the capillary viscosimetry method depend on the conditions of conducting such measurements, primarily on the size of the capillaries. This dependence is very slight. It does not effect the essence of the problem under discussion. As regards the nature and scale of the dependence, this will be considered below.

Concerning the spurt effect, experimental facts are systematized schematically in fig. 4.

Here only the lower part of the graph until the critical point is the proper flow curve. At regimes of deformation above the critical one it has nothing to do with the flow curve in the accepted sense of this notion. Therefore it is reasonable to use the more general concept of flow rate curve, that is the dependence of the volume rate on the pressure drop in the duct. Let us dwell on the figure for one minute.

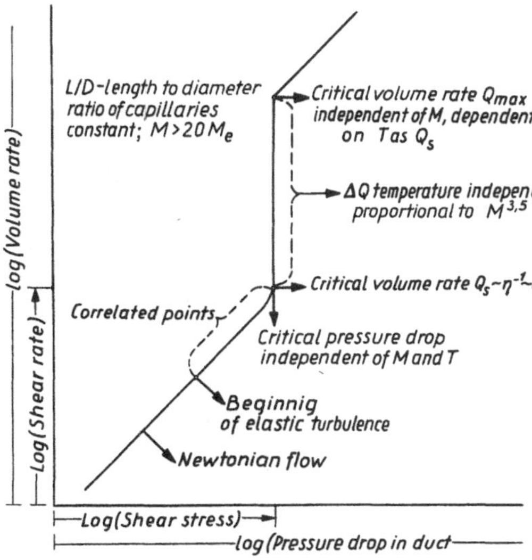

Fig. 4. Key diagram of phenomena observed during movement of narrow-distribution linear high-molecular-weight polymers in ducts and critical deformation parameters

Now let us turn to comparison of the results of investigation of polybutadienes at room temperature and at cyclic deformation regimes with capillary viscosimetry data. They have been provided by *Yu. G. Yanovsky* and *E. K. Borisenkova* in fig. 5. Here on the right-hand side are given the molecular weights of polybutadiene samples. The portions of the dashed lines with a slope tangent equal to unity represent flow curves of polybutadienes. The horizontal parts of the dashed lines demonstrate spurts. Monotonic thin solid lines give the dependence of the storage modulus on the circular frequency. The high-elastic plateaus are distinctly manifested. The curves with a maximum relate to the dependence of the loss modulus on the circular frequency. The beginning of elastic turbulence is indicated by arrows. The following facts stand out prominently here. The critical stress corresponding to the spurt is constant; the logarithm of its value is equal to 6.55. The critical stress is equal, to a fairly good approximation, to the maximum value of the loss modulus. The spurt occurs at shear rates which are numerically less than the frequency corresponding to the maximum at about three times and thirty times, respectively, less than the frequency at which the horizontal section of the high-elasticity plateau is achieved. Since the ratio between the critical shear rate and the frequency at the maximum point is constant, all the predictions of the theories with regard to the dependence of the frequency corresponding to the maximum of the loss modulus on the molecular weight or the initial viscosity of polymers are correct also with reference to the characteristic of the spurt regimes. Moreover, a simple relationship exists between the conditions of appearance of disturbances in the stream issuing from the ducts (so-called elastic turbulence) and the critical regime at which the spurt occurs. This directly suggests that these phenomena have the same nature. Further on this will be proved convincingly.

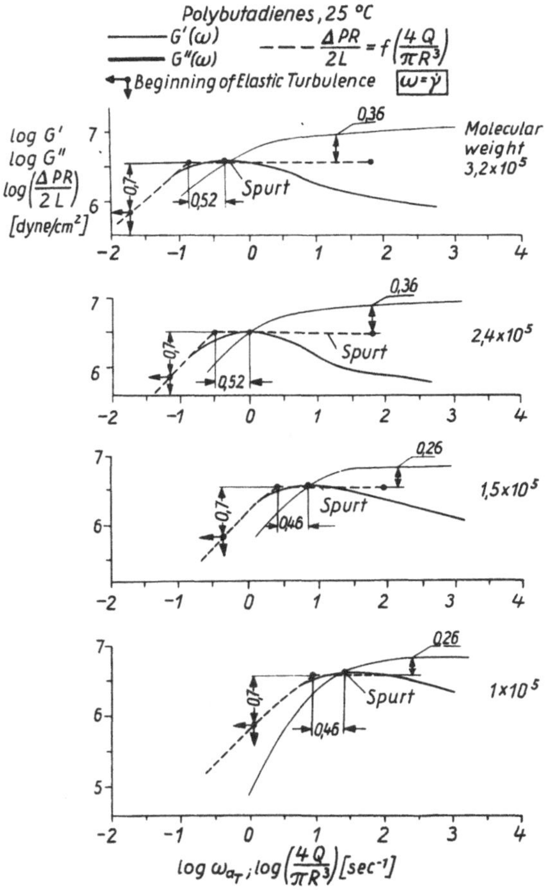

Fig. 5. Comparison of viscoelastic properties of narrow-distribution polybutadienes by capillary viscosimetry method and at regimes of small-amplitude cyclical deformation

Results similar to those shown in figure under consideration were obtained by *Yu. Yanovsky* and *E. Borisenkova* for polymer homologous series of polyisoprenes and polystyrenes. This points to the general significance of the above conclusion on the conditions of attainment of critical deformation regimes of polymer melts.

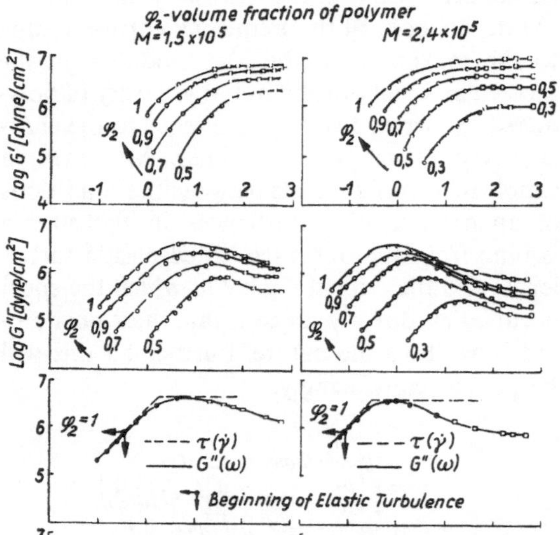

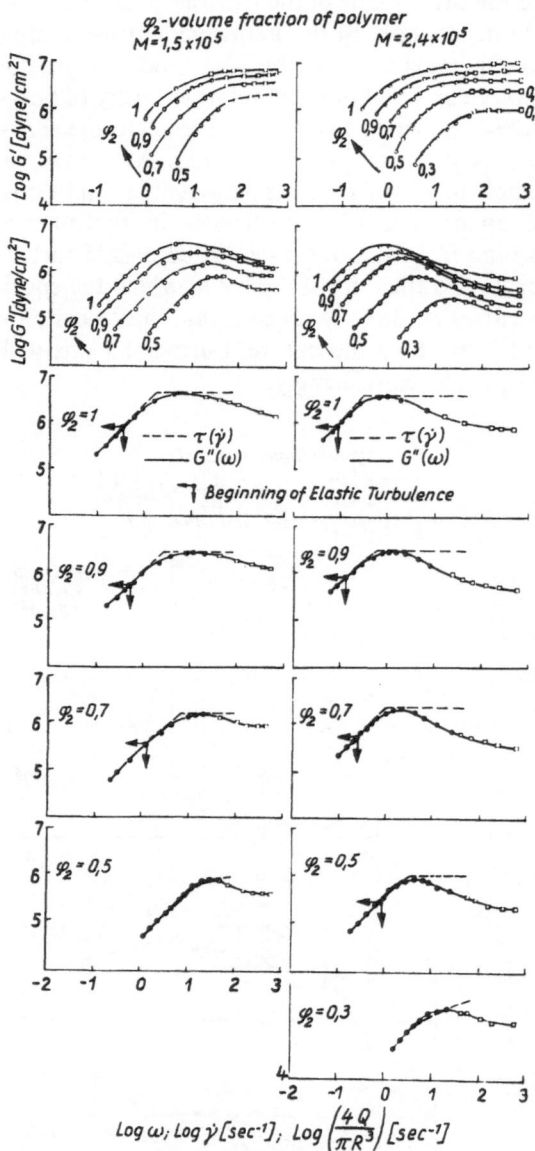

Fig. 6. Comparison of viscoelastic properties of narrow-distribution polybutadienes and of their solutions in α-methyl naphthalene by capillary viscosimetry method and at cyclical deformation regimes

The following must be emphasized. The existence of critical deformation regimes of bulk polymers corresponding to the above-formulated conditions – the presence of the spurt effect at shear stresses of the order of 10^6 dyn/cm^2, the independence of this stress from the molecular weight (provided it is sufficiently high) and from temperature, the variation in the critical shear rate with temperature according to the temperature dependence of the initial viscosity, and even the connection between the critical regimes and the frequency dependence, all this is known from many works in which the viscous properties of linear polyethylenes were investigated. And this is quite understandable, since in linear polyethylenes M_e is equal to about two thousands, i.e. almost one third less than in polybutadienes. However, these facts, which are known for linear polyethylenes, remained ungeneralized and their importance for the understanding of the peculiarities of the properties of high-molecular-weight linear polymers remained unrealized.

The manifestation of critical deformation regimes is of importance not only for bulk polymers but also for concentrated solutions of linear polymers. This can be demonstrated for solutions of polybutadienes in different solvents. The relevant data for solutions in α-methyl naphtalene at 25 °C are given in fig. 6. They are based on *Yu. Yanovskys and S. Sergeyenkovs* measurements at regimes of cyclic deformation and on the data of *A. Malkin* and *N. Blinova* obtained by the capillary viscosimetry method.

In the upper row of the graphs the dependences of the storage modulus on the circular frequency are given. The second row represents the dependences of the loss modulus on the circular frequency. All the other graphs give a comparison of the dependence of loss modulus on the frequency with the capillary viscometer data at volume fractions (denoted by φ_2) of the polymers diminishing from top to bottom.

It can be seen from the figure that in the region of highly concentrated solutions the spurt effect is manifested sharply. Where the spurt parameters are determined clearly, they are in a unique relationship with the parameters characterizing the maximum of the dependence of the loss modulus on the frequency and the beginning of elastic turbulence. All these parameters are connected by a quadratic dependence with the polymer concentration.

On considerable reduction in polymer concentration the spurt degenerates, and the region of the loss maximum corresponds to the viscosity anomaly. From this it follows that the presence of a loss maximum is a condition necessary but not sufficient for a clear-cut manifestation of the spurt. The sufficiency criterion has not been established so far.

Theories predict and experiments confirm the existence of critical deformation regimes. However, it is impossible to obtain from theories any information as to what happens to polymers during spurt and at above-critical regimes, and why at *Newton*ian flow regimes there appear distortions in the shape of the stream issuing from the capillaries, i.e. elastic turbulence is observed.

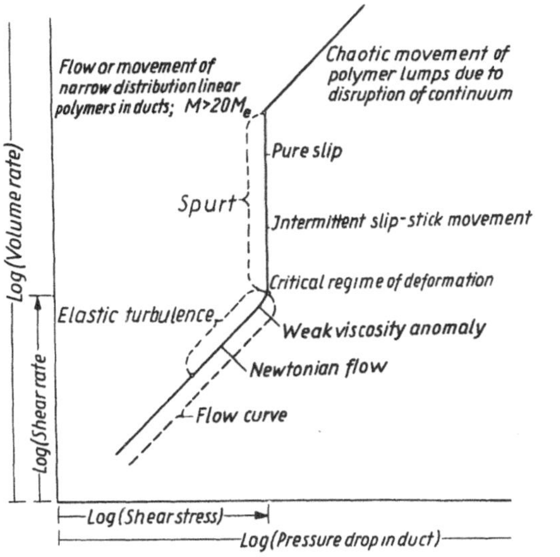

Fig. 7. Key diagram of important phenomena observed during movement of narrow-distribution high-molecular-weight linear polymers in ducts

The answer to these questions is supplied by the visualization of the polymer stream – direct observation of the polymer movement in the ducts. But before I show what the structure of the stream of a narrow-distribution polymer looks like, let us consider for one minute the key diagram of the phenomena observed (fig. 7).

What is achieved by the method of visualization of the flow of the polymers discussed can be illustrated by the data obtained by *N. I. Insarova* (10) in experiments on a flat slit with a narrow-distribution polybutadiene of molecular weight $1.5 \cdot 10^5$. The flow lines were recorded by the movement of 10–20 μ glass particles in the polymer. The stress distribution in the flow was determined by using circular-polarized monochromatic light. Fig. 8 shows that low shear rates are associated with a regular nature of flow lines in the duct and at its entrance.

Now let us see what can be achieved by observation of a polybutadiene flow in circular-polarized light (fig. 9).

Photos a and b depict, respectively, the zone of polymer entry into and exit from the duct and also the strictly regular distribution of interference bands inside the duct. The linear relationship between the shear stresses and the order of the band (near the wall) is fulfilled almost up to critical stress. The band value is, on the average, 9×10^4 dyn/cm^2. Note that polybutadienes are distinguished by very high polarization optical sensitivity.

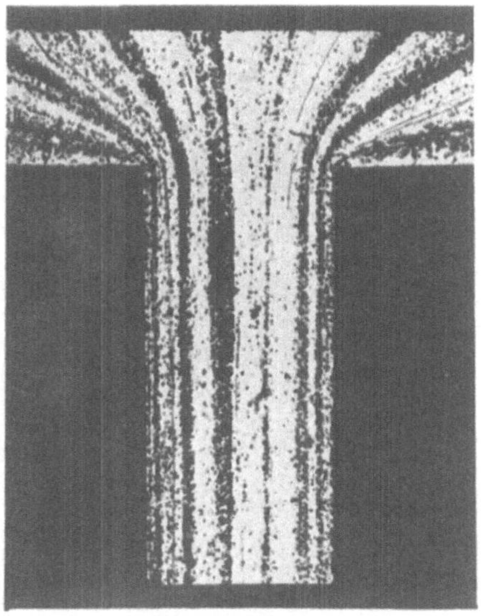

Fig. 8. Flow lines of polybutadiene at shear stress of 6.31×10^5 dyn/cm^2 at the walls of a rectangular duct

It should be pointed out that at the entrance into the duct and at its exit in the near-the-wall regions (in the zones adjacent to the edges) stress concentration is observed. The shear stress in these zones may achieve the critical value, whereas inside the duct the shear stress at the wall is substantially lower than critical. If at the duct entrance near the edge a critical stress is reached, the polymer passes into the high elastic state, its adhesion to the wall diminishes, and a local spurt occurs. As a result of detachment of the polymer from the walls the stress acting on it falls off, it relaxes and then begins to flow in the layer adjacent to the wall as a fluid. This continues as long as the shear stress inside the duct at the walls remains below critical.

On exit of the polymer from the duct the stress concentration up to critical values also leads to a local detachment of the polymer from the duct wall. As a result, a small portion of more or less relaxed polymer is ejected from the duct. It is followed by a portion of the polymer under a

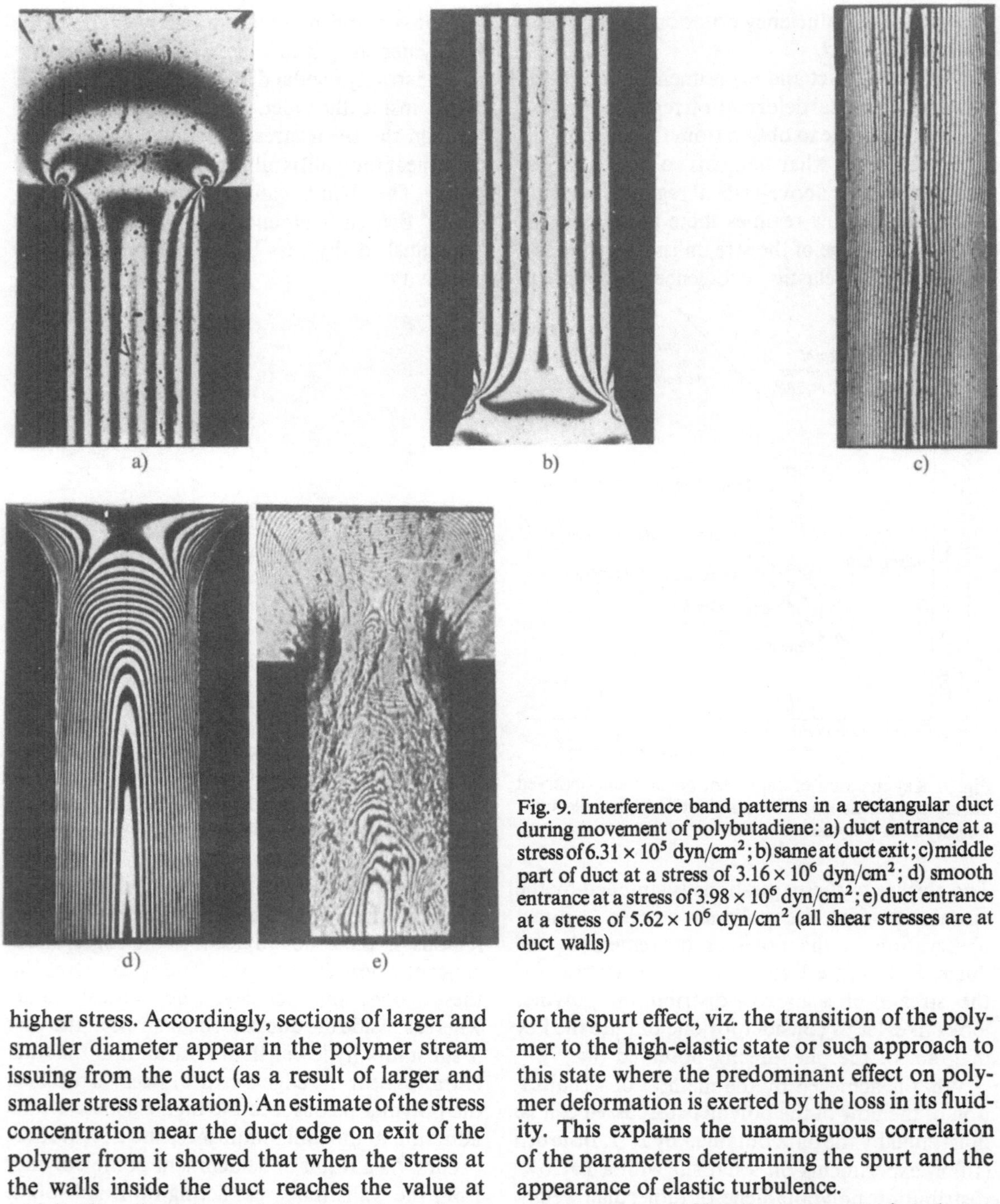

a) b) c)

d) e)

Fig. 9. Interference band patterns in a rectangular duct during movement of polybutadiene: a) duct entrance at a stress of 6.31×10^5 dyn/cm^2; b) same at duct exit; c) middle part of duct at a stress of 3.16×10^6 dyn/cm^2; d) smooth entrance at a stress of 3.98×10^6 dyn/cm^2; e) duct entrance at a stress of 5.62×10^6 dyn/cm^2 (all shear stresses are at duct walls)

higher stress. Accordingly, sections of larger and smaller diameter appear in the polymer stream issuing from the duct (as a result of larger and smaller stress relaxation). An estimate of the stress concentration near the duct edge on exit of the polymer from it showed that when the stress at the walls inside the duct reaches the value at which the distortion of the extrudate shape begins, it rises to the critical value in the near-the-edge zone. So this explains why small-scale periodic distortions in the extrudate shape may be observed at the duct exit under the *Newton*ian regime of PB flow in the duct. Another important conclusion is that the cause for the appearance of small-scale distortions in the extrudate shape and the enhancement of these distortions with increased shear stresses and rates is the same as for the spurt effect, viz. the transition of the polymer to the high-elastic state or such approach to this state where the predominant effect on polymer deformation is exerted by the loss in its fluidity. This explains the unambiguous correlation of the parameters determining the spurt and the appearance of elastic turbulence.

Let us now consider the movement of a polymer in a duct when critical values of pressure drops and flow rates are achieved inside it. When observing the movement of solid particles in the near-the wall zone, the stick-slip process is clearly registered. This process is very sharply reflected in the polarization-optical pattern, which is presented in photo c. One can see alternating narrowing and broadening of the interference bands along the duct walls. The narrow zones correspond to an increase in stress up to the critical value and the transition of the polymer to the high elastic state. This causes a reduction in the adhesion of the polymer to the wall and also results in slippage, which is accompanied by relaxation of the stress in the polymer, transition to the fluid state, sticking to the wall, a new increase in stress to the critical value, etc. Under these conditions the high stress concentration at the duct entrance, where the stress is above critical, is no longer attended by such stress relaxation as a result of which the polymer acquires a steady-state, fluidlike flow. The local spurts occurring along the length of the duct cause longitudinal fluctuations in the optical pattern at the duct entrance.

When the shear stress at the duct walls is equal to the critical, stress concentration occurs at the sharp edge at the duct entrance which leads to a rupture of the polymer – the deformed medium loses its continuity. This very important process will be discussed in more detail later on. But since it masks continuous slippage to some extent, it is convenient to use a smooth-entrance duct for observing slippage in its pure form. This is shown in photo d. The figure clearly demonstrates continuous slippage of a polymer and gradual relaxation of stresses along the duct walls. This is testified by the fact that the order of the bands decreases along the length of the duct from its entrance. Besides, in smooth-entrance ducts the disturbing influence of various factors affecting the asymmetry of the flow at the entrance diminishes, and the polarization-optical pattern of the polymer flow proves to be strictly symmetrical.

Already from the inspection of the interference band pattern in photos c and d it follows that at above-critical shear stresses and rates the routine methods for calculating the shear rate of a polymer at duct walls are useless. As for calculation of the shear stress from the pressure drop in ducts, the stress thus determined is actually a certain averaged value, and the nature of this averaging essentially depends on the velocity of polymeric system in the duct.

It is important to see what happens when the stress at the duct walls reaches or exceeds the critical one. It is obvious that in the zones of stress concentration near the edges at the duct entrance the stresses are still higher. Under the effect of high stresses the polymer breaks into irregularly shaped lumps, and the flow in the duct becomes chaotic. This is shown in photo e. The dark spots at the sharp edges appear as a result of light scattering in the non-uniform medium.

Of great interest is the observation of flow regimes corresponding to a pressure drop a little larger than the critical one. This is depicted in fig. 11 where the duct entrance is shown in the left part, and the duct proper in the right part; the lower edge of the photo corresponds to the middle part of the duct along its length. With pressure drops which create the flow pattern presented here the ruptures in the medium of the flowing polymer extend over a limited zone adjoining the

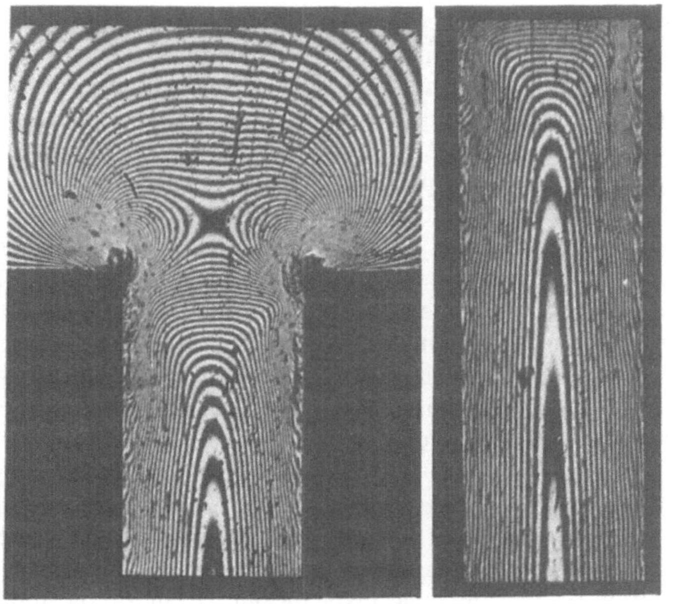

Fig. 10. Interference band pattern during flow of butadiene in rectangular duct at spurt regime

edges, and the chaotic movement of the polymer medium takes place only in layers adjacent to the wall. As polymer lumps advance along the duct walls the stress in them relaxes, and a distribution of interference bands of ever-improving regularity appears near the walls.

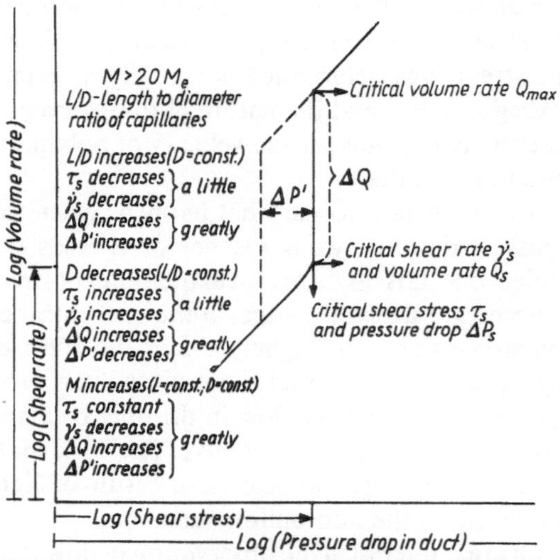

Fig. 11. Influence of size of cylindrical ducts and of molecular weights of linear high-molecular-weight polymers on spurt effect

The change in the slippage regime of the polymer medium – the movement of lumps of the polymer medium randomly shifting relative to one another – explains the increase in the pressure drop with increasing flow at above-critical regimes of polymer movement in the duct.

It should be emphasized that when sufficiently high pressure drops and flow rates are imparted, the continuity break in the polymer medium occurs also with a smooth entrance. The sharp edges at the duct entrance only facilitate this process but it is essentially determined by the fact that in the high elastic state the polymer undergoes a rupture when the deformation rates and stresses acting in it exceed certain critical values.

The photos displayed above show the actual nature of the break of the polymer as a continuous medium and indicate that in this state of the polymer breaking the movement of the medium has nothing to do with the usual laminar flow of polymer systems.

Thus, the visualization of the polymer movement in the ducts under spurt conditions, and all the more so at the above-critical regimes, shows that for such deformation regimes the concepts of viscosity and flow curves lose their meaning.

The visualization of the polymer deformation process in ducts makes it possible to explain the above-noted small (in absolute value) but systematic effect of the duct size on the critical parameters associated with the transition of polymers from the fluid to the high-elastic state. This is shown schematically in fig. 11 on the basis of *E. Borisenkovas* measurements with a capillary viscometer by the constant-pressure method.

If one approaches the assessment of the deformation process proceeding from assignment of a definite shear stress, this means that the pressure drop in the duct should be directly proportional to the length and inversely proportional to the diameter of the cylindrical duct. Then the increase of the duct length-to-diameter ratio corresponds to an increase in the pressure drop. At the regime of polymer deformation directly preceding the spurt the resistance in the duct decreases as a result of the development of near-the-wall slippage, and an ever-increasing share of the pressure drop falls on the entrance zone. This leads to an increase in stress concentration and also in polymer movement velocity in the entrance zone, which facilitates the manifestation of the spurt and is registered formally as a reduction in the critical values of shear stresses and rates (up to 35 and 40% respectively).

A reduction in the duct diameter at a constant length-to-diameter ratio entails the shortening of the duct. In this case the role of wall slippage in the duct diminishes, this leading to an overstatement of the critical parameters in the range of 15–25%.

After *E. Bagley* and cowerkers described (11) the presence of loops on the curves of the dependence of the flow rate on the pressure drop (see the dashed line parallelepiped in fig. 11), this effect has been repeatedly noted by other investigators, but until recently no systematic study of it was made. The nature of the effect under consideration, which in reality has nothing to do with hysteresis, was elucidated by *B. Yarlykov* and consists in the following. At the spurt regime, an intensive auto-oscillatory process of pulsations of the flow rate and pressure drop develops in the duct. In the course of the spurt, an increase in the flow rate results in a larger pulsation amplitude, which reaches a maximum and then decreases. When the chaotic movement of polymer lumps covers the entire entrance zone the pulsations die out. The pulsation amplitude is the higher, the greater is the pressure drop at the entrance.

Therefore, the width and height of the loop in fig. 11 are found to be interrelated. In order to damp the pulsations by lowering the pressure drop, it should be decreased to the minimum value corresponding to the vertical dashed line in fig. 11.

All that has been said with regard to the spurt effect and the attending phenomena is essential not only for high-molecular-weight narrow-distribution linear polymers. What is manifested in them with utmost clarity is important for high-molecular-weight linear polymers in general. It should be remembered that even a small broadening of the molecular-weight distribution and also approach to the glass transition temperature causes the appearance on the flow rate curves of a section of more or less clearly defined viscosity anomaly relating to the branches of *Newton*ian flow and the spurt, and this degenerates the transition to the spurt, masking this effect. The literature abounds with examples of curves for the dependence of the flow rate on the pressure drop with a degenerated (gradual) transition to the spurt. It is important to note that there are no criteria determining what section of the curves may be regarded as true flow curves.

In the light of the aforesaid, of great interest is modelling of polydisperse polymers by mixtures of very narrow fractions. Here, proceeding from the classification of high-molecular-weight compounds with which we began the consideration of the problem, it is desirable to elucidate the specific features of the variation in the properties of their mixtures if the components making up the mixtures belong to different classes. Taking into account the special characteristics of high-molecular-weight polymers, it is expedient to focus the attention on mixtures in which the molecular weight of one of the components exceeds twenty M_e. Naturally, the extreme cases are solutions of high-molecular-weight polymers in low-viscosity solvents. Therefore, we will consider the entire composition range, beginning with mixtures of high-molecular-weight polymers and ending with solutions of high-molecular-weight polymers in low-viscosity solvents. The relevant data from *N. Blinovas* experiments are presented in fig. 12. The experiments were made in a constant pressure viscometer. In the case of high-viscosity components the length-to-diameter ratio of capillaries was 22.5. For solutions of PB in low-viscosity solvents the above-mentioned ratio was at least 40; for dilute solutions it was 100 or more. It is important to note that the PB samples used in this work whose molecular weights were below 6.10^4 were not narrow-distribution polymers.

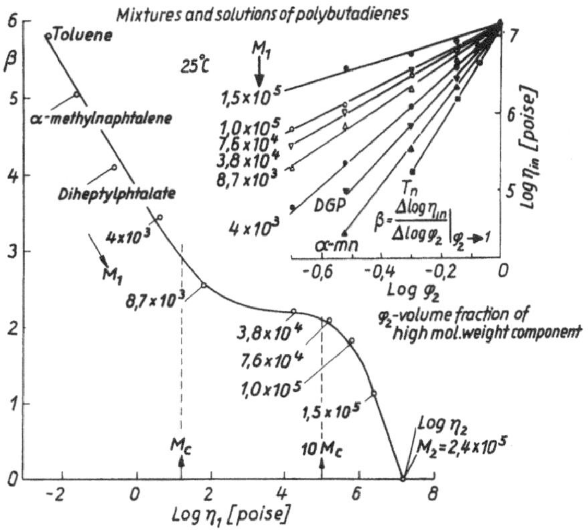

Fig. 12. Influence of additions of polybutadienes of various molecular weights and of low-molecular-weight solvents on initial viscosity of narrow-distribution high-molecular-weight polybutadiene (molecular weight 2.4×10^5)

The fig. 12 demonstrates how the initial viscosity of polybutadiene of molecular weight $2.4 \cdot 10^5$ is effected by additions of less viscous polybutadienes and low-viscosity solvents. In the right-hand top corner of the figure the dependences of initial viscosity of the mixtures and solutions on the concentration of the high-molecular-weight polybutadiene are presented. The molecular weights of the low-molecular-weight components are given in the left-hand column. The dependences under consideration for the solutions of high-molecular-weight polybutadiene PB in dipeptylphthalate (DGP), α-methyl-naphthalene (α-mn) and toluene (Tn) are also depicted. The dependences of the angular coefficients of the straight lines in the right-hand figure on the initial viscosity of low-molecular components are represented by a single curve in the left-hand side of fig. 12. Here, regardless of the thermodynamic quality of solvents their action fits into a unified system of facts embracing also oligomers and high-molecular-weight polybutadienes.

The introduction of a less viscous component reduces the viscosity of mixtures and solutions the greater, the lower its own viscosity. This, however, is nothing more than a general trend.

The effects of solvents, low-molecular-weight and high-molecular-weight components are clearly delimited. Two new circumstances draw one's attention. Low-viscosity solvents exert a relatively weaker effect on the viscosity of the high-molecular-weight component as compared with the effect of polybutadienes, whose molecular weights exceed 10 M_c. Quite an unexpectedly weak effect on the viscosity of high-molecular-weight polybutadiene is produced by components whose molecular weights range from M_c to 10 M_c. Note that both effects manifest themselves over a wide range of high concentrations of less viscous components.

It is impossible so far to explain the above-discussed set of new facts, but there is no doubt that they are of great importance for the understanding of the laws governing the viscous properties of polydisperse polymers, for locating the boundary of the polydisperse polymers and polymer solutions, it is essential for computing the viscosity of polymer mixtures, because the estimate of the viscosity from the weight average molecular weight shows the greater deviations from the expected values, the wider the difference between the molecular weights of the mixture components.

We will now consider the data obtained by *N. Blinova* over a wide range of flow rates, which yield a general system of the viscous properties of mixtures and solutions of narrow-distribution PB of molecular weight $2.4 \cdot 10^5$. The experiments were carried out in capillary viscometers at 25 °C. The size of capillaries was the same as was indicated above concerning the data of fig. 12. The results in the form of flow rate curves are given in fig. 13. As is clear from the figure, all the systems distinctly fall into two types of mixtures and solutions.

The first type relates to a set of graphs in the left-hand top corner of the figure. The second type is represented by the other sets of graphs in the upper row. The lower row of the graphs embraces different kinds of solutions.

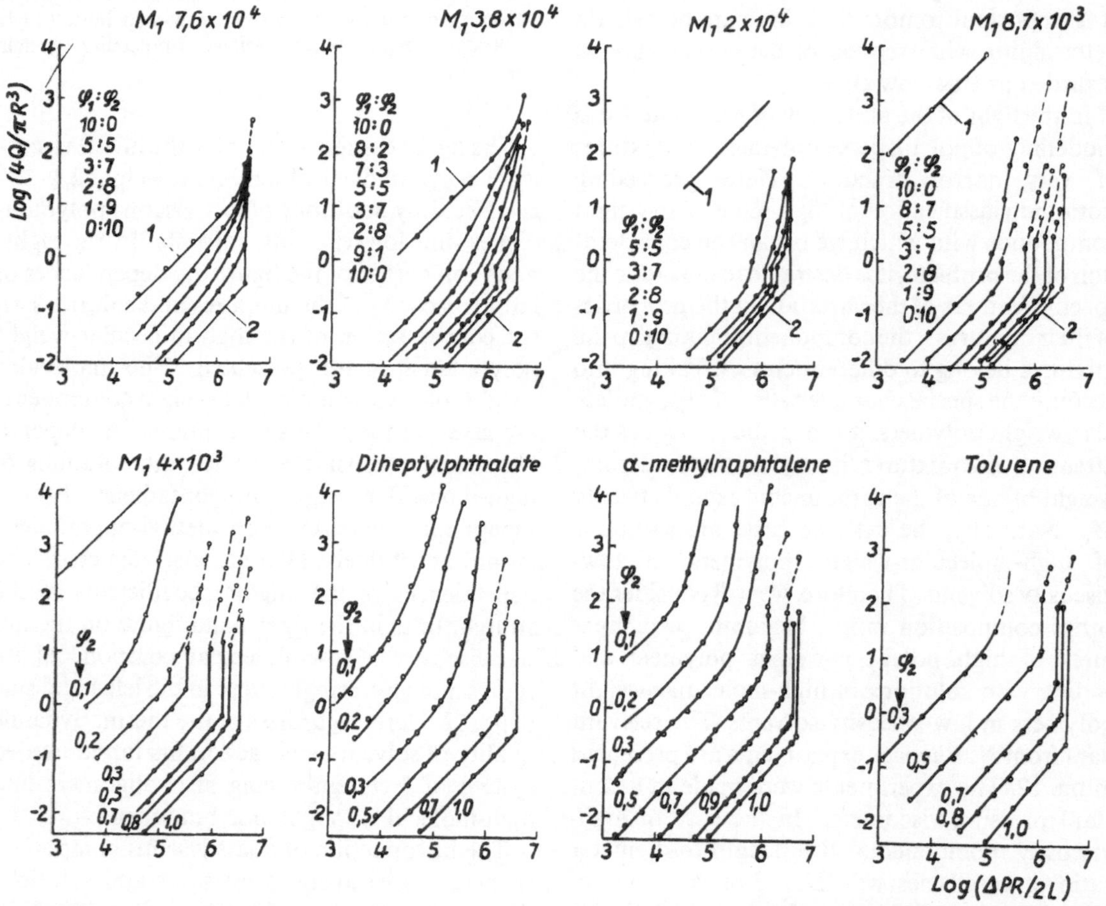

Fig. 13. Viscous properties of binary mixtures and solutions of polymers (molecular weight of high-molecular-weight component 2.4×10^5)

The common feature for all these systems is the fact that at volume fractions of the high-molecular-weight component exceeding 0.5 binary mixtures and solutions always show a spurt.

The first type of mixtures – molecular weight of both components exceeds 20 M_e

A spurt occurs at a critical stress characterizing the homologous series of monodisperse polybutadienes. The mixtures exhibit a viscosity anomaly – the flow rate increases faster than the pressure drop in that shear-stress region where both components behave as *Newton*ian fluids. This indicates unambiguously that the principal cause of the viscosity anomaly in high-molecular-weight linear polymers is their polydispersity. In this case the viscosity anomaly has a relaxation nature. The viscosity anomaly appears upon attainment of the critical shear rate, when the transition of the high-molecular-weight component to the high-elastic state begins and it ceases to behave as a fluid. A decrease in viscosity on further rise in shear rate is a consequence of decreasing losses on deformation of this component, since as a result of transition to the high-elastic state the process of the co-operative displacement of the kinetic segments in its macromolecules attenuates, and only fast conformation transitions and vibrations of segments about the centres of gravity of the macromolecules take place in them.

Thus, it is natural to make the following simple assumption: within the class of high-molecular-weight polymers the viscous properties of mixtures are additive, so that the dissipative losses are added up, while the losses of each of the components are determined by its relaxation state, which depends uniquely on the acting shear stresses and rates. In this respect the components do not exert a mutual effect on each other. Consequently, if, at a given deformation regime, the shear rate of i-th component is below critical, this component flows as a *Newton*ian liquid, but if it is above-critical, then its contribution to the dissipative losses of the system is determined by the losses in the polymer passing to or existing in the high-elastic state.

This approach enabled *A. Malkin* to suggest a simple method for computing the viscosity of multicomponent mixtures of linear polymers according to the arrangement shown in table 2. A comparison of calculation with experiment for a binary mixture is given in fig. 14.

Table 2. The formulae for calculation initial and apparent viscosities of monodisperse high molecular weight polymers and their mixtures

Initial viscosity

For a monodisperse polymer

$\eta_{in} = AM^{\alpha}; \quad M > M_c; \quad \alpha = 3.5$

For a mixture of monodisperse polymers

$\eta_{in} = A[\sum_i (\varphi_i M_i)]^{\alpha};$

φ_i-volume fraction of the polymer with molecular weight M_i.

Initial and apparent viscosity

$\eta = A\{\sum_i \varphi_i [|\eta^*|(\omega)]^{1/2}\}^{\alpha};$

$\dot{\gamma}$-shear rate, $|\eta^*|(\omega)$-absolute value of complex viscosity as a function of circular frequency ω.

$\eta^* = \dfrac{G^*}{\omega};$

G^*-complex shear modulus.

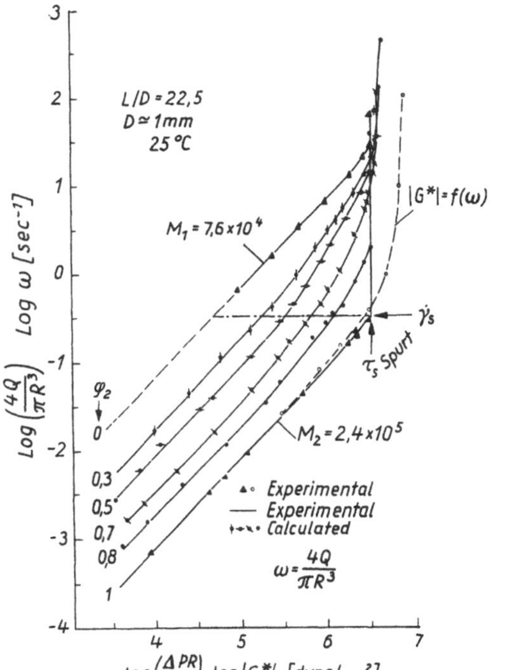

Fig. 14. Viscous properties of binary mixtures of high-molecular-weight polybutadienes

Equally good results are obtained for multicomponent mixtures and their concentrated solutions in low-viscosity solvents. This may serve as an indirect proof that from the standpoint of modelling the structure of uncured polymers by entanglement networks not a single network is formed in mixtures of the type discussed here because the high-molecular-weight component behaves independently from the lower-molecular-weight one.

The second type of mixtures – the low-molecular-weight component has a molecular weight from twenty to two – four M_e

At high concentrations of the high-molecular-weight component the flow rate curves consist of four sections (see fig. 13). Here, only the lower sections in which the viscosity is constant represent flow curves proper. Within the type of mixtures discussed, the first spurt begins at the critical shear rate of the high-molecular-weight component, but at stresses systematically reducing with decreasing concentration of the high-molecular-weight component. Previously, *O. Bartoš* (12) described a case which, using the notation adopted here, may be characterized as a flow rate curve with two spurts. Now we can assert that this is quite a common phenomenon. The nature of the first spurt and of the section of the flow rate curve connecting the two spurt sections is not quite clear. It can be assumed that the first spurt is due to the isolated position of the entanglement network of the high-molecular-weight component. This network, however, is weakened by the low-molecular-weight component. Disentanglement does not occur so sharply as in the high-molecular-weight component in its pure form, but since it does take place, it leads to a retarded increase in flow rates. On further increase in flow rates a critical stress is achieved which is characteristic of the homologous series of polybutadienes.

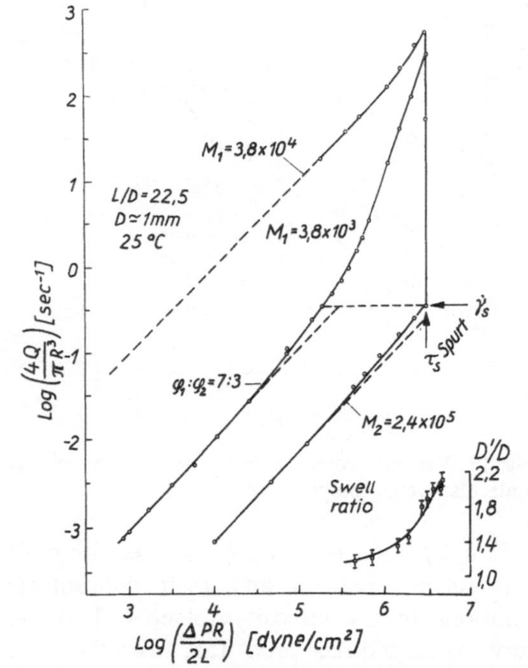

Fig. 15. Viscous properties of binary mixture of polybutadienes in which molecular weight of one of the components is higher and that of the other lower than 10 M_c

At volume fractions of the high-molecular-weight component below 0.5 the first spurt disappears and the flow rate curves take the form of typical curves of the flow of anomalously viscous fluids. Fig. 15 shows pictorially that in this case the sharply defined viscosity anomaly is manifested at shear rates exceeding the critical shear rate of the high-molecular-weight component. The viscosity anomaly at lower shear rates is due to the insufficiently narrow molecular-weight distribution of the high-viscosity component, i.e. to its own viscosity anomaly, which is enhanced in a mixture with the low-viscosity component.

Calculation of the viscosity of mixtures according to the above-considered scheme yields viscosity values below the measured ones. This shows that the contribution of the high-molecular-weight component to the dissipative losses is less than would correspond to its pure state at a given deformation regime of the system. Hence, there is no more additivity of losses, and the low-molecular-weight component reduces the losses of high-molecular-weight component. This is associated with the decrease in the level of the loss maximum and the high-elasticity plateau, which is typical of the effect of solvents on high-molecular-weight polymers. However, unlike solutions in low-viscosity solvents, polymer mixtures of molecular weights discussed here do not displace their zone of the loss maximum and the high-elasticity plateau towards smaller relaxation times. Precisely such is the specific nature of the effect of polymers whose molecular weights range from twenty to two-four M_e.

For the polymer mixture represented by the flow curve under consideration the die swelling was measured. Accordingly, the lower part of fig. 15 illustrates a graph showing the ratio of the stream diameter to the capillary diameter as the function of shear stress. This ratio is associated with the elasticity of the melt whose manifestation is enhanced when passing over to polymer mixtures, particularly in the region of their non-*Newton*ian flow. This poses the question as to the connection between the viscosity anomaly and the development of great high-elastic deformations in polymer systems. We will later revert to this question.

Solutions of a high-molecular-weight polymer M > 20 M_e in oligomers and low-viscosity solvents

Let us return to fig. 13, which gives the general systematization of the viscous properties of binary polymer mixtures. In the figure, solutions are associated with the lower set of graphs. In the region of concentrated solutions, when passing over from oligomers to typical solvents a decrease in polymer concentration results in ever-increasing critical shear rates corresponding to the spurt. This is the more pronounced, the lower the solvent viscosity is.

Now it is expedient to consider the flow rate curves of solutions of a high-molecular-weight PB in the low viscosity solvents over the entire composition range. Using the results obtained by *N. Blinova* in experiments on capillary viscometers at 25 °C, this is shown in fig. 16 where φ_2 denotes the volume fraction of the polymer. The experiments were carried out with capillaries whose length-to-diameter ratio was at least 40 for concentrated and 100 or more for dilute solutions. The dashed lines I–I and II–II in fig. 17 correspond, respectively, to the appearance of a spurt and of elastic turbulence. As is seen from the figure critical stresses relating to the appearance of distortions in the shape of the stream issuing from the capillary and to a spurt are in a constant relationship. This is quite understandable if one proceeds from the above-discussed mechanism of development of stream perturbations at the capillary exit.

A decrease in polymer concentration leads to the degradation of the spurt and then to transition to the typical flow curves of non-*Newton*ian fluids. With decreasing polymer concentration a change in the viscosity anomaly mechanisms takes place: (1) transition of the polymer to the high-elastic state and the reduction of its contribution to the dissipative losses; (2) displacement of the equilibrium, in the processes of formation and disintegration of entanglement, towards a decrease in their number; (3) orientation effect – the straightening of macromolecule chains in the stream and the consequent reduction in the resistance of the system to deformation. These mechanisms are superimposed, the geometrical factor (orientation effect) being of basic importance for dilute solutions.

Passing over to dilute solutions corresponds to the disappearance of the three-dimensional entanglement network and to a change in the nature of the concentration dependence of the parameters determing the viscoelastic behaviour of solutions. This is clearly seen from fig. 17. In the upper part of the figure are the data of *N. Blinova* for the concentration dependence of the critical value of the shear stress (τ_{nN}) at which the transition to a non-*Newton*ian flow takes place. The lower part of the figure shows the experi-

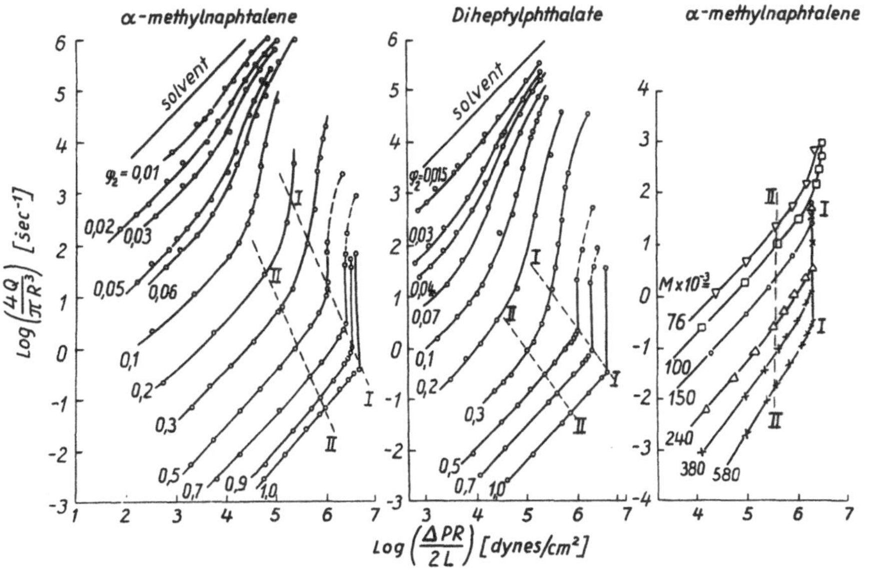

Fig. 16. Viscous properties of solutions of narrow-distribution high-molecular-weight polybutadiene (molecular weight 2.4×10^5) over entire concentration range

mental results of *G. Berezhnaya* obtained by measuring the initial values (extrapolated to zero shear stresses) of the high-elasticity modulus G_{in} (according to elastic recoil), viscosity η_{in} and normal stress coefficient ζ_{in}, which relates these stresses to the square of the shear rate.

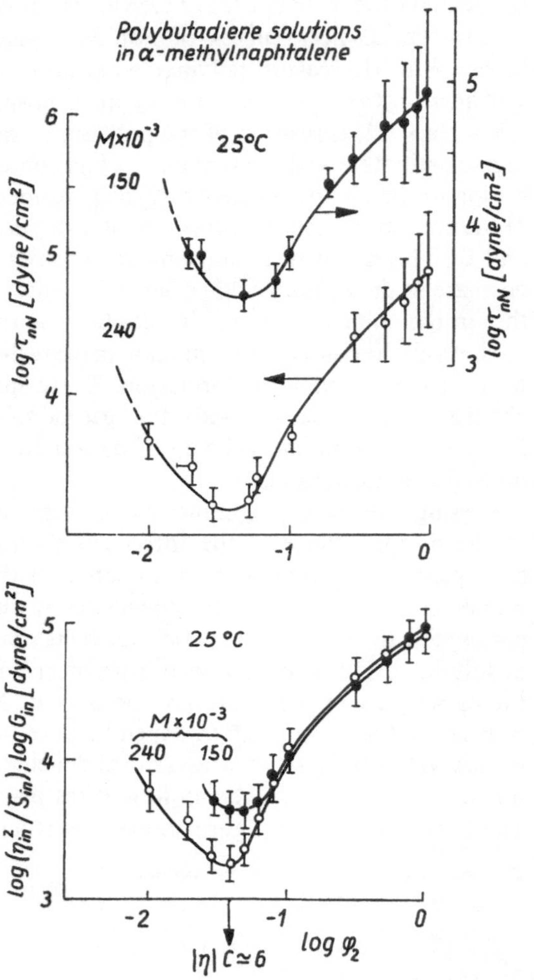

Fig. 17. Dependence of initial high-elasticity modulus, ratio of the square of initial viscosity to initial coefficient of normal stresses and critical shear stress corresponding to appearance of viscosity anomaly on concentration of narrow-distribution polybutadienes in solution

There is a good correlation of critical concentrations determining the variation in the structure of polymeric systems according to measurements of critical shear stresses and viscoelastic characteristics. This region of critical concentrations divides two kinds of states of the systems – when an entanglement network is present and absent in them.

When passing over from a bulk polymer to solutions the segment density in the entanglement network reduces. This facilitates reduction in the concentration of entanglements under the effect of deformation. As a result the critical shear stress at which the viscosity anomaly appears, decreases. On disappearance of the network the viscosity anomaly mechanism changes substantially. For it to manifest itself at the expense of straightening out of macromolecules (orientation effect) it is necessary that the shear stresses and rates rise considerably – the critical shear stresses increase.

The initial modulus of high elasticity and also the parameter equal to the ratio of the square of initial viscosity to the initial normal stress coefficient characterize the rigidity of polymer systems. With transition from a bulk polymer to solutions the rigidity of the systems decrease so long as the entanglement network exists. In the range of dilute solutions there is no entanglement network, and their viscoelastic properties are determined by the behaviour of the separate macromolecules. Here the macromolecule relaxation is facilitated, which is equivalent to an increase in the rigidity of the system with the polymer concentration reducing below the critical value.

It should be taken into account that measurements of the initial moduli relate only to the range of concentration where an entanglement network exists (the right-hand parts of the curves under consideration). In dilute solutions estimation of the normal stress coefficient was made predominantly by measurements of flow birefringence.

In the region of concentrated solutions of the high-molecular-weight polymers the effects of the spurt, the beginning of transition to the high-elastic state and the displacement of the equilibrium concentration of entanglements causing the appearance of viscosity anomaly are relaxation effects. They should obey one and the same law. This is confirmed by the data of *N. Blinova* and *S. Sergeyenkov* depicted in fig. 18. On the y-axis are given critical shear rates corresponding to the appearance of viscosity anomaly ($\dot{\gamma}_{nN}$), the spurt ($\dot{\gamma}_s$) and the frequency ($\omega_{G''_{max}}$) at which the loss modulus in the frequency dependence achieves the maximum value. On the x-axis the polymer concentrations are inicated.

Brief conclusion

Systematic study of the viscoelastic properties of high-molecular-weight narrow distribution polybutadienes has made it possible to establish a number of new facts important for the understanding of the peculiarities in the behaviour under deformation of linear polymers of narrow as well as wide molecular-weight distribution. Proceeding from the measurement of visco-elastic properties it is possible to create a classification of high-molecular-weight compounds, which will facilitate the understanding of the peculiarities in their behaviour depending on the molecular weight. It is especially significant that this enables the realization of a qualitatively unified approach to representatives of different polymer-homologous series.

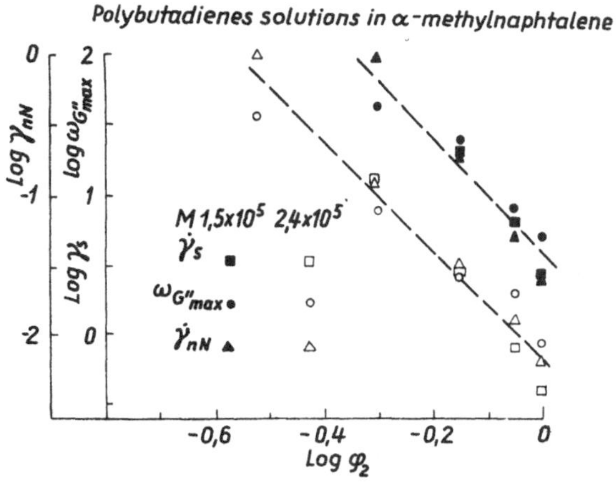

Fig. 18. Dependence of critical parameters of flow in capillary viscometers and at regimes of small-amplitude cyclical deformation on concentration of narrow-distribution polybutadiene in solutions

As shear rates and stresses increase, high-molecular-weight linear polymers pass to the high-elastic (rubbery) state, which changes radically their behaviour under deformation. This fact calls for a revision of traditional conceptions with respect to possible viscosity mechanism for many polymer systems. It has been proved experimentally that at rates and stresses exceeding definite critical values the viscosity concept cannot be used and any attempts at plotting flow curves are senseless.

The assessment of the behaviour of polymers at high shear stresses is of greatest technological importance and enables one to understand the causes for the difficulties involved in the processing of such polymers as linear polyethylenes, polybutadienes, polyisoprenes and other polymers.

The results discussed pose urgent problems. What are the criteria of a true viscous flow at high shear stresses in polydisperse polymers and in concentrated polymer solutions? How can one conjecture a network model of the structure of polymer systems in which the properties of polymers with molecular weights of the same order are manifested separately? What is the shear strength of polymer melts and solutions? How can adhesion of polymers to a solid be estimated in relation to the shear rate and stress? And so on, and so forth.

References

1) *Onogi, Sh., T. Masuda,* and *K. Kitagawa,* Macromolecules, **3**, 109 (1970).

2) *den Otter, I.,* Dynamic Properties of Some Polymeric Systems. Dissertation (Leiden 1967).

3) *Gruver, J. T.* and *G. Kraus,* J. Polymer Sci., **A-2**, 797 (1964).

4) *Oser, H.* and *R. S. Marvin,* J. Res. Nat. Bur. Stand. **67 B**, 87 (1963).

5) *Vinogradov, G. V., V. W. Pokrovsky* and *Yu. G. Yanovsky,* Rheol. Acta **11**, 258 (1972).

6) *Graessley, W.,* J. Chem. Phys., **54**, 5143 (1971).

7) *Vinogradov, G. V., A. Ya. Malkin, Yu. G. Yanovsky, E. A. Dzuyra, V. F. Schumsky* and *V. G. Kulichikhin,* Rheol. Acta, **8**, 490 (1969).

8. *Vinogradov, G. V.,* Pure and Appl. Chem., **26**, 423 (1971).

9) *Vinogradov, G. V., N. I. Insarova, B. B. Boiko,* and *E. K. Borisenkova,* Polymer Engng and Sci. **12**, 5, 323 (1972).

10) *Boiko, B. B., N. I. Insarova,* and *G. V. Vinogradov,* Doklady Akademii Nauk SSSR, 803, 159 (1972).

11) *Bagley, E. B., I. M. Cabbot,* and *D. C. West,* J. Appl. Phys., **29**, 109 (1958).

12) *Bartos, O.,* Polymer Letters, **3**, 1025 (1965).

Author's address:

Prof. *G. V. Vinogradov*
Leninski Prospekt 29, Institute of Petrochemical Synthesis, Academy of Sciences of the USSR
Moscow 117071 (USSR)

Rheol. Acta 12, 374–392 (1973)

From the Institute of Theoretical and Applied Physics, Stuttgart (Germany)

The rheological behaviour of metals

By E. Kröner

With 28 figures

(Received October 27, 1972)

1. Synopsis of the lecture

The rheological behaviour of metals and, more generally, of other crystalline materials, is completely determined by the imperfections of the crystal lattice which are also called lattice defects. This is why the contribution of physics to the rheology of these materials comes almost exclusively from the field of lattice defects.

In the first part of this lecture (sections 2–4) I shall try to give an idea of the physical research and its results pertinent to the rheology of crystalline matter. In this part 1 shall show what the relevant lattice defects are and explain their role in the process of plastic and viscoplastic deformation. I shall also give a visual impression of the defect state by a number of pictures most of which show direct observations of lattice defects by means of electron microscopes. These pictures have considerably raised the confidence in the theories which try to explain the rheological behaviour on a microscopic or atomic scale. The complexity of the rheology of metals is overwhelming. Therefore, it would not be sensible to try to cover the whole field in such a limited lecture.

Section 2 contains a brief introduction to the notion and concept of lattice imperfections. The most important defect properties are also explained there. In section 3 some aspects of transmission electron microscopy are discussed. I hope that this brief exposition will help the reader in his appraisal of the various electron micrographs shown in section 4. Using these pictures as illustrations I shall survey a number of results of the physical research on unidirectional and cyclic deformation of single crystals and from here proceed to polycrystal behaviour. In this section I concentrate on plasticity as the more specific deformation mode of crystalline matter, compared to their viscosity.

In the second part (section 5) of this lecture I shall investigate the implications of the results of the first part regarding a phenomenological or, at least, macroscopic description of the deformation of metals. An important question will be whether such implications exist at all.

This section is the most subjective and perhaps even controversial part of this lecture in which I summarize some thoughts which have occupied my mind for a long time. These remarks should only in the second place be considered as an evaluation of the present state of development of such theories. The primary aim is to exhibit those fundamental points a clear idea of which would be of considerable help in the development of phenomenological or macroscopic theories in the rheology of metals.

2. Classification of lattice defects and their role in the deformation of crystals

The lattice defects have been classified according to the number of dimensions into which they extend themselves. Most important are the zero dimensional or point defects, the one-dimensional or line defects and the two-dimensional or surface defects.

a) Point defects

Fig. 1 shows three point defects of particular interest, the vacancy, the substitutional foreign atom and the interstitial atom. It is clear that these defects cause a lattice distortion. In a more macroscopic manner they can be described as so-called elastic dipoles introduced by myself in 1956 (1). The elastic dipole represents a superposition of three *Boussinesq* double forces without moment which are oriented perpendicular to each other. The elastic dipole is a measure of the forces which the defect exerts on the neighbouring particles. Since another meaning of the elastic dipole is that of a singular impressed stress the dipole behaves like a tensor of second rank. According to elasticity theory which gives good results for larger distances r from the dipole one finds that the lattice distortion due to a point defect decays as $1/r^2$.

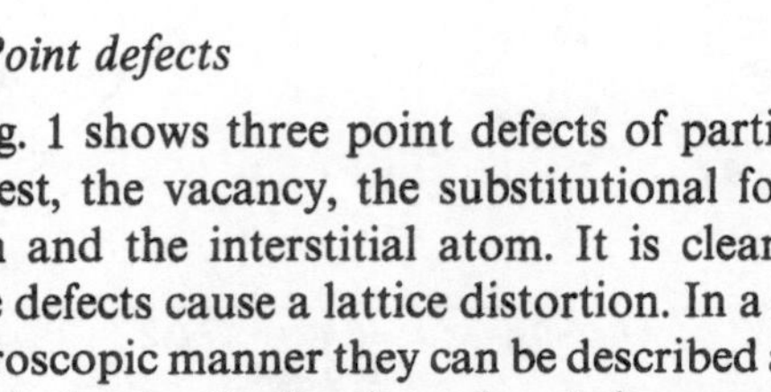

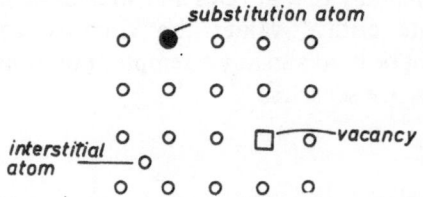

Fig. 1. Three important types of point-like lattice defects

Point defects possess a certain mobility which is strongly temperature dependent. This is due to the fact that the motion of point defects needs thermal activation. It is a diffusional type of motion which can become directional in a suitable stress and strain field. The observed temperature dependence and strain rate sensitivity of the motion of point defects are typical for viscous deformation rather than for plastic deformation. The importance of point defects for the viscous deformation is furthermore in agreement with the two intuitive expectations (i) that the temperature-induced non-directional motion of the point defects becomes directional under arbitrarily small stresses and strains – i.e. there is no yield limit – and (ii) that the internal state of the body is not changed by the motion of the point defects, because their number and arrangement is not changed in a macroscopic sense. The second statement is not true if the point defects interact with line defects and surface defects.

b) Line defects, in particular dislocations

Dislocations are by far the most important line defects. From the standpoint of rheology they are even the most important defects in general. Fig. 2 represents the original picture by which *G. I. Taylor* introduced the so-called edge dislocation in 1934. The second fundamental type of dislocation, the so-called screw dislocation, has been found by *J. M. Burgers* in 1939 and will not be discussed in this lecture. The fig. 2 shows one atomic plane of a three-dimensional primitive cubic lattice the third dimension of which is assumed to extend perpendicular to the picture as a repetition of this plane. Fig. 2b shows a topological peculiarity: Whereas all particles in the plane have four next neighbours and are therefore in the next neighbouring configuration (though distorted) of the regular lattice there is one atom (in the center) which has only three next neighbours the one below being absent. This can be considered as if one vertical atomic half-row were missing the insertion of which would make the crystal plane perfect again. In three dimensions one sees a topological line-peculiarity which is called an edge dislocation. I would like to emphasize here that the dislocation is fundamentally of a topological character. Hence it is not necessary to speak of the response of the body when one introduces the dislocation. Fig. 3 may underline this statement.

The fig. 2 gives the intuitive impression that such a dislocation can enter a volume element of a crystal, move through it and leave it so producing a relative glide motion of the two lattice planes adjacent to the plane in which the motion of the dislocation takes place. This plane is the so-called glide plane.

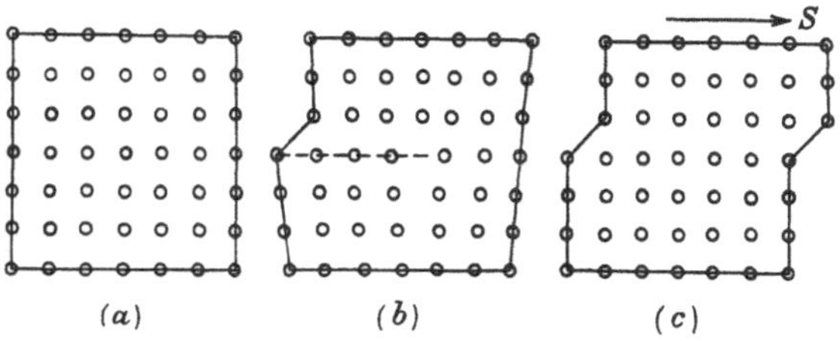

Fig. 2. The edge dislocation as introduced by *G. I. Taylor* in 1934 (2).

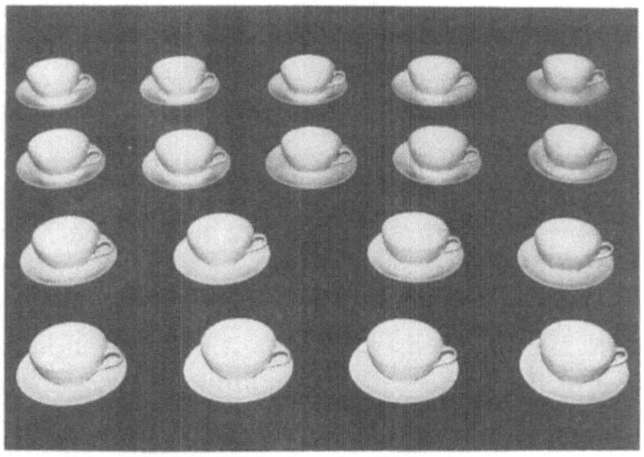

Fig. 3. The edge dislocation as a topological peculiarity.

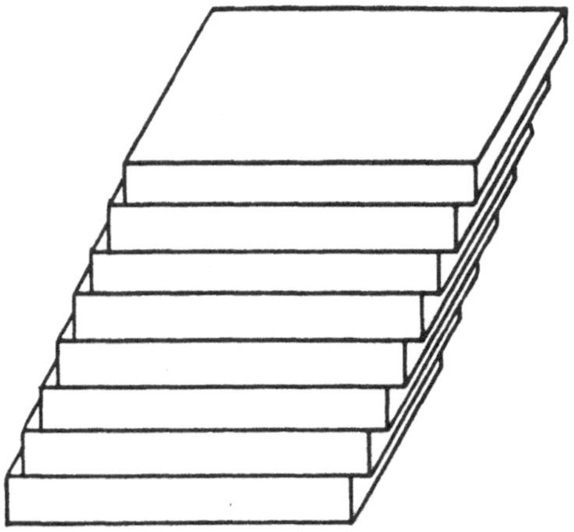

Fig. 4. The meaning of plastic shear strain

The glide step has the size of one lattice spacing. This is very small when viewed on the macroscopic scale. Nevertheless, if such a slip occurs on every hundredth lattice plane, one already obtains a permanent macroscopic shear strain of 1% (fig. 4). The conclusion is then that at least an important part of the macroscopically observed permanent deformation can be ascribed to the motion of dislocations. It is thus clear that under an applied shear stress dislocations feel a driving force in such a direction that their motion will lead to a shear strain of the right sign. Quantitatively this force is described by the famous formula of *Peach* and *Koehler* (1950).

Simultaneously with and independently of *Taylor* the role of the dislocation has been discovered also by *E. Orowan* and by *M. Polanyi* who explained the dislocation as the lattice defect which "can make a crystal plastic". In fact, in agreement with the phenomenological observation on plasticity one finds that dislocation motion is very little dependent on temperature and strain rate. This is true at least as long as the dislocation does not interact with other types of defects. The reason is that unlike the motion of point defects the motion of a dislocation occurs under stress alone. Thermal activation cannot assist because the motion of the dislocation implies a cooperative movement of many atoms, namely of all those along the line. Since high amplitudes of thermal stress waves occur only in very small volumes, of the order of a few lattice cells, thermal assistance cannot be large enough and of the correct direction along a dislocation segment which is longer than a few lattice spacings[1]).

Furthermore, we shall show in section 4 that a relevant amount of dislocation motion needs an applied stress which exceeds a certain limit, in agreement with the macroscopically observed phenomenon of the yield limit. We shall also establish in section 4 that the dislocation number and arrangement is changed drastically during the deformation and this results in an important change of the internal state of the deformed body. Thus the motion of the dislocations fulfills the basic criteria of plastic deformation as a counterpart to viscous deformation of solids.

c) Surface defects, in particular grain boundaries

The most important surface defect is the so-called grain boundary which separates two crystallites (grains) of different orientations in a polycrystalline aggregate[2]). This defect is so important because most, though by no means all, solid crystalline materials in practical use are polycrystals consisting of grains which are very small, for instance of diameter below 0.1 mm, which nevertheless possess more than 10^{15} particles for example.

The grain boundaries can be described to some extent as thin amorphous layers between the ordered crystallites. Due to the disorder in these layers it is possible that single particles jump into neighbouring positions in an uncorrelated way. Since such a process is very localized it can be thermally activated and becomes directional under an applied shear stress. Thus the situation is somewhat similar to the case of point defects. This means that the grain boundary slip is to be classified as a viscous deformation mode. Most metals exhibit this effect at elevated temperatures. This fact has been confirmed by means of a maximum of the mechanical damping in torsional vibrations of about 1 c. However, there are metals in which the grain boundary motion occurs already at room temperature to an observable degree. The low strength of lead found empirically is an example. The effect of grain boundary slip can become very pronounced at elevated temperatures if the grain size is so small (far below 1 μm) that no dislocation sources (see section 4) can be activated. This is the important phenomenon of superplasticity which is a viscous rather than a plastic deformation mode in our terminology.

There is another aspect of grain boundaries which perhaps is even more important than its viscous motion. For reasons of the crystal topology dislocations cannot pass through grain boundaries which therefore form impenetrable obstacles for dislocation motion. This effect makes the theoretical derivation of the deformation behaviour of polycrystals from the assumed-as-known behaviour of single crystals alone an unsolvable problem. I shall come back to this question in section 4.

3. Electron microscopy

The so-called direct observation of dislocations and other lattice defects forms the most important basis of the physical theories of plasticity and viscoplasticity. Among the various methods of direct observation transmission electron microscopy has been particularly useful. Most of the pictures shown in the following sections were obtained by this technique, which

[1]) The explanations of this paragraph are somewhat simplified. Also dislocations undergo thermally activated processes, for instance when two dislocations intersect during their motion. Such processes occur within very small volumes and are usually accompanied by the production of special point-like defects such as jogs, kinks, vacancies.

[2]) In multiphase materials the phase boundary plays a role which is similar to that of the grain boundary in pure metals.

was originally developed by *W. Bollmann* (3) and by *P. B. Hirsch* and his collaborators (4, 5) who first saw dislocation lines in the electron microscope (1956). For an excellent review of the older results see *S. Mader* (6).

There are serious theoretical and practical problems connected with the electron microscopy. In order to interpret the obtained pictures quantitatively, one needs a good theory of electron diffraction in crystals. A good theory, however, is complicated. We shall not go into the details of this theory but only state that the electrons are scattered by the gradients of the crystal lattice strain caused by the defects. In the case of straight dislocation lines these strains decay as $1/r$ with distance r from the line, a fact which establishes a far-reaching elastic interaction between dislocations. It is these strains which make the dislocations visible as lines on the electron micrograph although the resolution of the microscope is not high enough to see the particles of the lattice.

Since electrons suffer a strong absorption when passing through a crystal one is forced to use thin crystalline layers typically of order 0.1 μm. This requires particular skill in the preparation techniques. The question must be raised whether dislocation arrangements observed in thin layers are typical also for arrangements which occur in the bulk material. The critical point is that dislocations do not only interact with *each other* (via their elastic stress fields), but they also interact strongly with *free surfaces* if they are near enough. This can be explained as follows: The internal stress field connected with the internal elastic strain around a dislocation near a surface must satisfy the usual stress boundary conditions. In the case of a dislocation which extends parallel to a (plane) surface these conditions can be satisfied by assuming a fictive so-called image dislocation (or several of those) of opposite sign outside of the crystal. The real dislocation then feels strongly attracted to the surface by the stress field of the image dislocations. Since in thin layers all dislocations are near one of the surfaces one expects that all dislocations which are parallel to the surface are pulled out of the layer by their images whereas the dislocations which extend from one surface to the other try to do this along the shortest possible way. Hence the dislocation arrangement in thin layers is by no means a true image of the typical arrangement in the bulk material, at least as long as special precautions are not taken to prevent this type of motion of the dislocation.

In order to do this, *J. R. Low* and *A. M. Turkalo* (7) and also *U. Essmann* (8) have developed particularly successful techniques which rely on the elastic interaction of dislocations with intrinsic or extrinsic point defects[3]). This interaction is attractive due to the peculiarities of the strain field produced by the dislocation. As a consequence, point defects as well as clusters of them precipitated along the line pin the dislocation, i.e. they prevent it from moving away. In pure metals the best way to effect this pinning is to introduce point defect clusters in large numbers by neutron irradiation in the reactor (*U. Essmann* l.c.).

In performing this experiment one has to overcome two difficulties. (i) According to *Essmann* the irradiation should be carried out under the applied load because unloading causes some backflow of the dislocations. This should be avoided, since one is usually interested in the dislocation arrangement under the external load. (ii) The incoming neutron flux heats the sample up. This changes the resistance against plastic flow, i.e. against dislocation motion. One then expects some dislocation movements even if the applied load is not altered during the irradiation. In order to avoid this effect one is forced to cool the specimen to low temperatures during the irradiation. Irradiation at liquid helium temperature has been applied successfully. This means a considerable experimental effort [*H. Mughrabi* (10)]. The above procedure shows the extent of the troubles taken in order to observe realistic dislocation arrangements. Most pictures shown in the sequence of section 4 were obtained in this manner and therefore have a high degree of reality.

The close interplay of theory and experiment is very typical of the physical research on the deformation behaviour of crystals. In many cases theoretical predictions were given which were verified by experiment only much later. A typical example was the 1934 hypothesis of the existence of the dislocation which has been seen only in 1953. On the other hand, the electron micrographs revealed structures which had not been predicted and were investigated theoretically only after its experimental discovery. An example is seen in fig. 5.

[3]) Other methods which are especially useful in non-metallic crystals were introduced by *H. Alexander* (9).

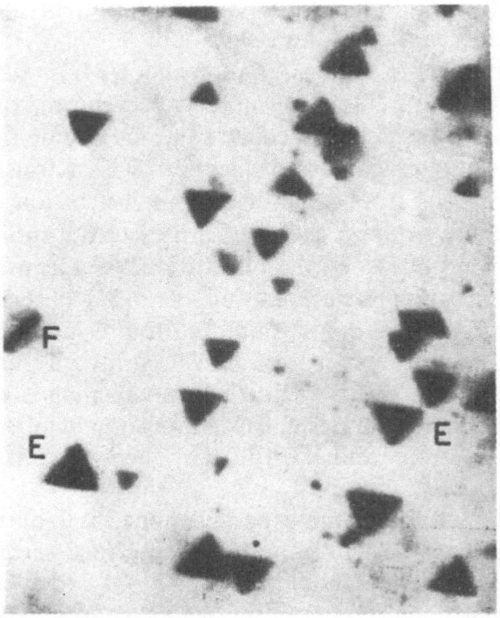

Fig. 5. The picture shows tetrahedry-like structural defects. These have been interpreted as a coagulation of vacancies in the tetrahedral planes. One speaks of stacking-fault tetrahedra (stacking-faults are surface defects which imply a disturbance in the natural sequence of close-packed atomic planes). The occurrence of these defects influences the work-hardening behaviour. [After *J. Silcox* and *P. B. Hirsch* (11)]

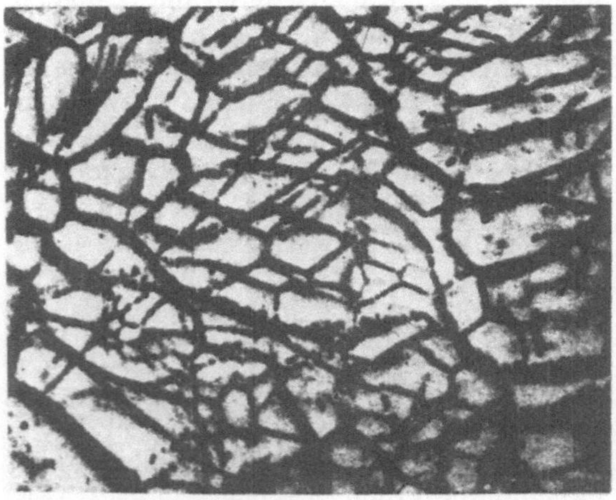

Fig. 6. A three-dimensional network of dislocations below the surface of a single crystal of AgBr. The network becomes visible in the microscope by the precipitation of silver along the dislocation lines. 1000×. [After *J. M. Hedges* and *J. E. Mitchell* (12)]

4. Plastic deformation of crystals

a) Preliminaries

The plastic deformation of crystals is always accompanied by a certain viscous component due to thermally activated processes. Although this time dependent behaviour is important in many applications, I shall concentrate on the plasticity because this is the most characteristic property of metals.

We first consider single crystals and only later give brief comments on polycrystalline aggregates. It is important to note that even undeformed crystals contain numerous dislocations which are usually arranged in the form of three-dimensional networks, as shown in fig. 6 due to *J. M. Hedges* and *J. W. Mitchell* (1953). Such networks have been predicted from the theory of crystal growth by *F. C. Frank* and others. The nodes of these networks cannot move because this would destroy the crystal lattice along the path of motion so leading to a very high localized increase of energy. This energy cannot be provided by applying forces from outside. As a result of the immobility of the nodes the dislocations bow out under an applied stress as shown schematically in fig. 7.

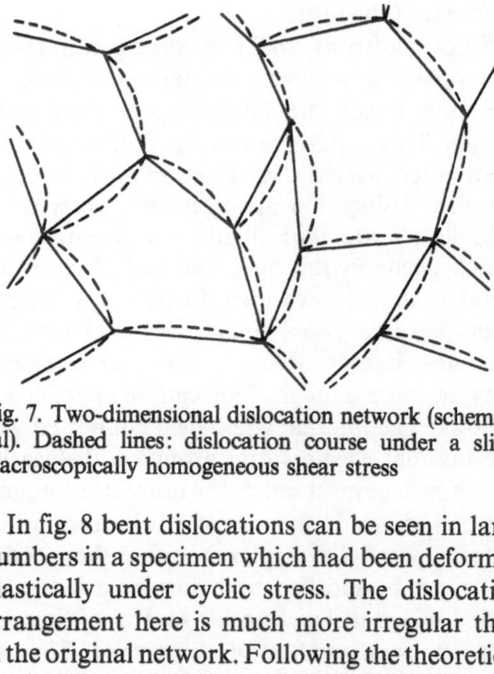

Fig. 7. Two-dimensional dislocation network (schematical). Dashed lines: dislocation course under a slight macroscopically homogeneous shear stress

In fig. 8 bent dislocations can be seen in large numbers in a specimen which had been deformed plastically under cyclic stress. The dislocation arrangement here is much more irregular than in the original network. Following the theoretical results of *A. de Wit* and *J. S. Koehler* (13) the curvature of the dislocations on this micrograph can be related directly to the local stress (applied plus internal) along the dislocation. Such calculations are performed with the help of the elasticity theory of dislocations originally developed by *G. I. Taylor*, *J. M. Burgers* and others. This fascinating possibility has been applied by *U. Essmann* (14), *J. C. Crump* and *F. W. Young* (15) and by *H. Mughrabi* (16). It is the most direct and most reliable way to measure the inter-

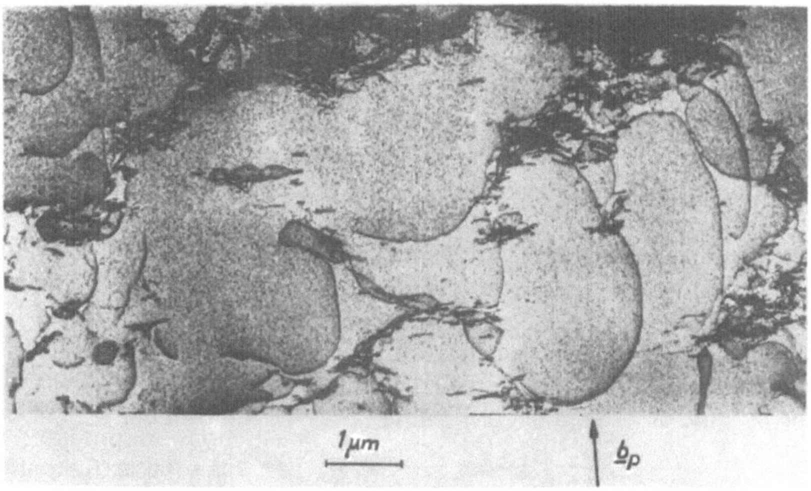

Fig. 8. Cu single crystal after 15 cycles of tension and compression. The radius of curvature of the dislocation lines is a measure of the internal stress. (After *H. Mughrabi*, unpublished)

nal fluctuating stresses in the interior of metals. Of course, it requires the destruction of the material.

In calculations of this kind the dislocation is treated as an elastic string which has a line tension and hence tends to become as straight as possible between two nodes or other fixpoints[4]).

b) Reversible plastic (anelastic) deformation

The bowing out of the dislocations implies gliding of atomic planes over each other. This takes place already at applied stresses far below the yield stress. This particular kind of plastic deformation is *reversible*. In fact, the dislocations will move back into the straight position if the applied loads are removed. It follows that a crystal containing dislocations has apparent elastic moduli which are smaller than they would be in a perfect crystal of the same particles. This effect can amount to several percents. It can be suppressed and in this way be verified by the pinning of the dislocations through point defects, for instance dissolved foreign atoms [*J. Friedel* (17), *G. Bradfield* and *H. Pursey* (18)]. The weakening effect is *anelastic* rather than *elastic* because the motion of the dislocations occurs in a finite time. Accordingly the effect dies out if the specimen is subjected to an alternating stress of high frequency so that the dislocations have no time to bow out. This picture suggests the existence of a resonance frequency which is the eigen-frequency of the dislocation strings. Such frequencies have been found in the megacycle range.

As is to be expected one observes a mechanical damping of vibrations of the specimen which is maximum if the specimen vibrates with the eigen-frequency of the dislocation strings. In a modified form, namely temperature dependent and at lower frequencies, this effect is also observed if the dislocations are pinned by few point defects which is often the case. Fig. 9 shows some steps in this motion. In order to bow out as before the dislocation has to overcome the attractive force of the point defects and this is done with the help of thermal activation. In typical cases the maximum of the damping at room temperature occurs in the low megacycle range. A theory of this effect was given by *A. Granato* and *K. Lücke*, see e.g. ref. (19).

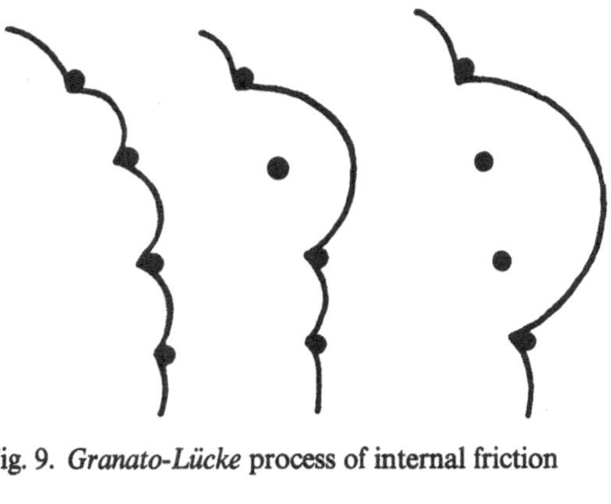

Fig. 9. *Granato-Lücke* process of internal friction

[4]) An exact calculation should include the fact that a moving dislocation feels a lattice-periodical potential, named after *E. Peierls* and *F. R. N. Nabarro*, of a very small height. However, except for extremely low temperatures (liquid air and below) there is always so much thermal motion of the particles that this potential is practically smeared out, i.e. may be considered to be constant. This applies to densely packed atomic lattices which are common to most metals.

c) Unidirectional glide on one glide system (stage I)

The simple picture of the bowing out of dislocations changes drastically if the stresses become larger. *F. C. Frank* and *W. T. Read* have explained in 1950 that an increasing applied stress on a bent segment of dislocations between two fixed points can lead to the production of one or several free dislocation loops. Fig. 10 shows schematically the sequence of dislocation shapes until the loop is separated from the fixpoints. This mechanism has often been compared with the production of soap bubbles from a straw. In metals one rarely observes a clean *Frank-Read* mechanism because the presence of other dislocations disturbs the effect. In materials which have much smaller numbers of dislocations than metals one has, in fact, found *Frank-Read* sources (fig. 11). Nevertheless the observation in metals shows that new dislocations are produced in great numbers by mechanisms which are similar to the *Frank-Read* one. In fact, the dislocation density increases from about 10^6 dislocations per cm^2 area in the undeformed crystal to about 10^{11} to 10^{12} in heavily deformed crystals.

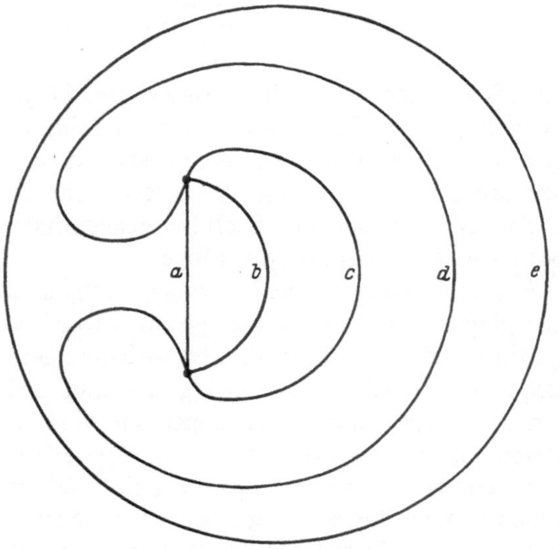

Fig. 10. Scheme of dislocation multiplication by the mechanism proposed by *F. C. Frank* and *W. T. Read* in 1950

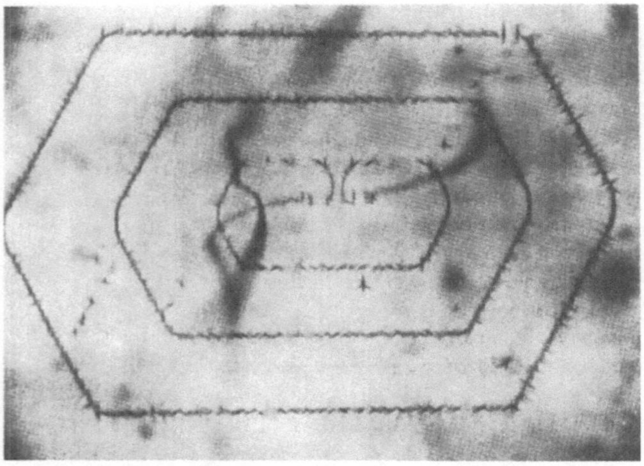

Fig. 11. A *Frank-Read* dislocation source in silicon. The dislocations have been made visible in the microscope by precipitations of solute copper atoms along the lines. [After *W. C. Dash* (20)]

The source mechanisms suggest that dislocations are produced and move in large numbers as soon as the critical source stress is reached. It seems, however, that this stress which can be calculated in some approximation by elasticity theory, is, in general, not identical with the yield stress since a higher stress is needed to drive dislocations through a crystal. This is the stress required to overcome the numerous obstacles. which are formed by the other dislocations through the already mentioned mutual elastic interaction of dislocations. Most important here is the interaction of two edge dislocations of opposite sign on parallel glide planes which run in opposite direction under the same applied shear stress. In fig. 12 the symbols ⊥ and ⊤ are used for these dislocations. Due to the special features of elasticity theory (bipotential instead of potential) the stress field around an edge dislocation has no cylindrical symmetry in contrast, for instance, to the magnetic field around a straight electric current line.

Since, except for the so-called climb at elevated temperatures, the edge dislocations cannot leave their glide plane it is meaningful to draw the

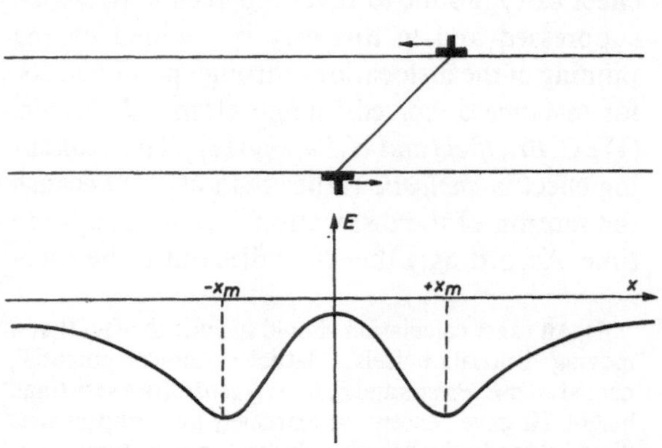

Fig. 12. Two edge dislocations of opposite sign in dipole configuration. Below: interaction energy E as a function of horizontal relative distance x of the dislocations (lower dislocation supposed to be fixed at $x = 0$)

interaction energy E as a function of the distance x in glide direction. Fig. 12 shows that besides $x = \pm\infty$ where the dislocations do not interact there are three distinct positions, namely the maximum at $x = 0$ and two minima at $x = \pm x_m$. This figure shows that the interaction assists the applied stress as long as the dislocations are at a distance larger than x_m. When, however, the dislocations have reached a distance x_m then the interaction is opposed to a further motion in the same direction, i.e. to a passing each other on the parallel glide planes. Since at this relatively small distance the interaction is rather large one needs a high increase of applied stress in order to shift the dislocations over the potential wall at $x = 0$.

As long as such a stress is not applied we expect that the two dislocations remain lying in the potential valleys and thus form what is called a dislocation dipole. This process of forming a dipole is not reversible because the crystal does not return to the initial state if the applied stress is removed: one observes a permanent change of the internal mechanical state.

The figs. 13–16 of a copper single crystal deformed in simple tension show what is revealed by the electron microscope at small plastic strains (resolved shear strain about 5%). Since the area seen in the microscope is very small, some μm in linear dimension, and the deformation is inhomogeneous on this scale, one sees different arrangements in different regions. Fig. 13 shows a region with only few dislocation lines which have not yet formed dipoles; hence they are able to move relatively freely. In fig. 14 we see lines which are thicker than those of fig. 13. A closer inspection which needs considerable experience with such pictures reveals that the bigger lines are groups of a few dislocations which form dislocation multipoles. There are also a few thinner pairs of lines which are to be considered as dipoles. Groups of such dipoles are shown in figs. 15 and 16.

The relatively ordered (parallel) appearance of the dislocations in these figures is due to the simplicity of the experimental situation: By choosing a favourable orientation between tensile axis and crystal orientation it was achieved that the resolved shear stress reached higher values only in one glide system so that the whole glide took place on parallel planes. This situation is typical of the so-called easy glide (or stage I) where only small work-hardening occurs. Obviously the changes of the internal mechanical state consist in the gradual building up of dislocation dipoles and multipoles during the deformation.

d) Cyclic deformation (fatigue)

The mentioned effect that irreversible changes of the internal state first occur already at applied stresses far below the yield stress is of fundamental importance to the phenomenon of fatigue. Also the cyclic loading causes a gradual build-up of dislocation groups (multipoles). The general feature of this process consists of the fact that by means of small repeatedly applied strains certain domains accumulate a great amount of internal strain energy in the form of dislocation multipole bundles. Such bundles interact with the mobile dislocations. They become more and more unstable as the stresses increase. Under special experimental conditions a correlated decomposition of the bundles can be observed in the form of so-called strain bursts: After a number of cycles during which not much seems to happen the instability point is exceeded at a certain stress level and a sudden and relatively large motion inside the crystals occurs during which the dislocations find new, more stable, arrangements. Of course, these motions are directed in such a way as to increase the plastic strain. Strain bursts of this kind have been observed in recent time. Fig. 17 due to *P. Neumann* shows an example.

In four of the further pictures (figs. 19–22) we see how the dislocation arrangements look in a copper crystal which has undergone 15 strain cycles of constant amplitude and fig. 18 shows how the operating stress varies with the number of cycles. A quasi-saturation is reached after some hundred cycles, and thereafter the change of the dislocation arrangement occurs very slowly. After 15 cycles we see rather large bundles of dislocation multipoles and a number of single dislocations which are bent under the effective (= applied plus internal) stress. Closer inspection shows that the bundles contain dislocations of secondary glide systems to a certain amount. The flow stress has reached a value much higher than the critical flow stress of the undeformed specimen. This is in agreement with the visual observation that the internal state has much more changed than in the previous pictures of small unidirectional deformation. Inspite of the apparent disorder seen in these pictures there is some order in the arrangement as can be seen from

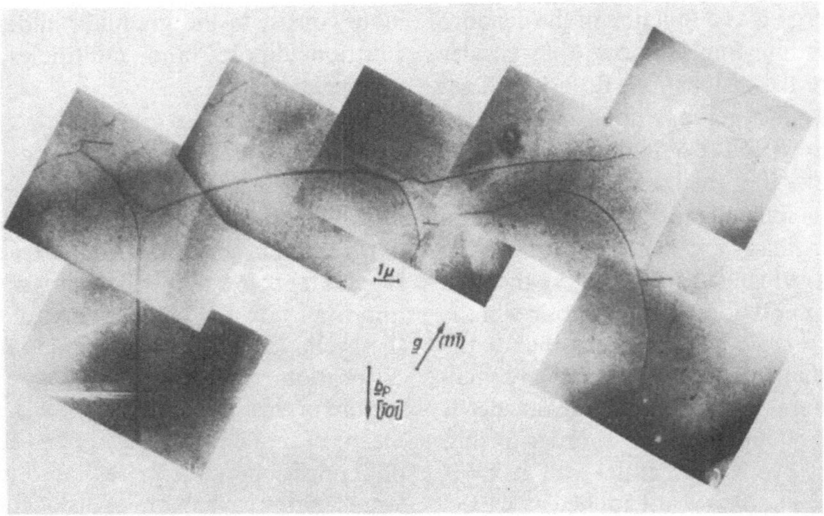

Fig. 13

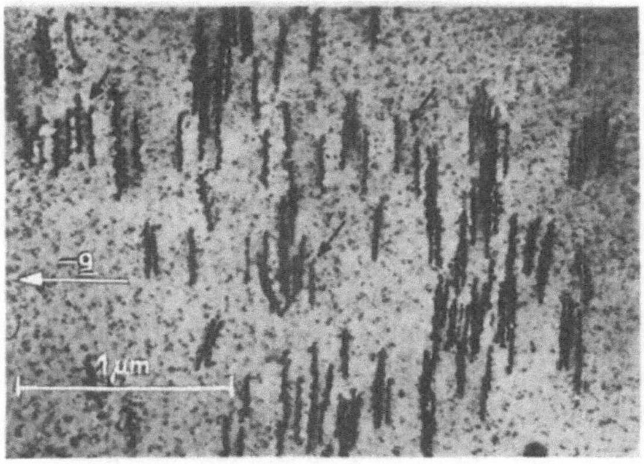

Fig. 14

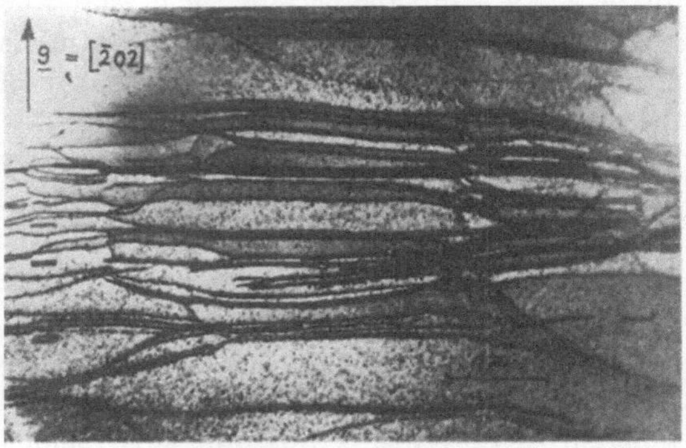

Fig. 16

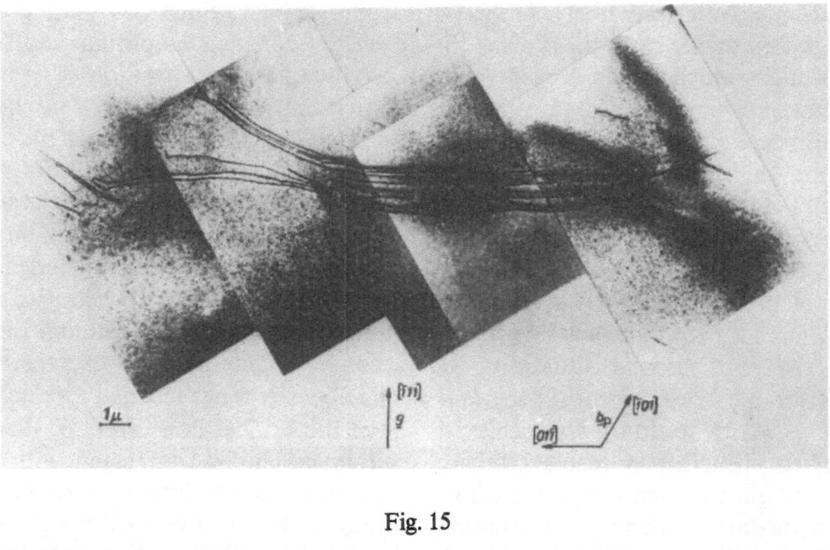

Fig. 15

Figs. 13–16. Dislocation arrangements in slightly deformed (5%) copper single crystals (stage I). Thin lines represent single dislocation lines, thicker lines correspond to bundles of a few dislocation lines. For details about figs. 13, 15 see *H. Mughrabi* (16), of figs. 14, 16 see *U. Essmann* (21) fig. 23 which shows the distribution of the radii of curvature of the bent single dislocations.

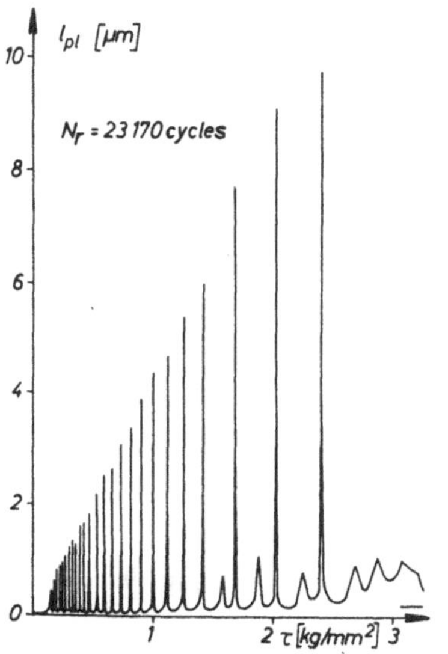

Fig. 17. Strain bursts during the linear build-up of the stress amplitude τ in a cyclic tension-compression test of a copper single crystal. The final stress amplitude of 3.2 kg/mm² was built up in 23 170 cycles. [After *P. Neumann* (19)]

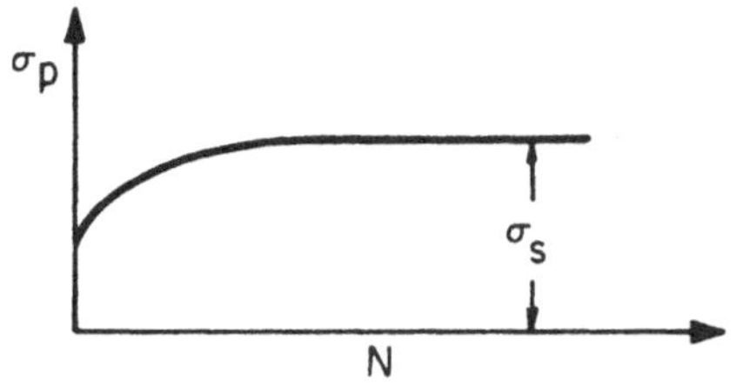

Fig. 18. Increase of stress amplitude σ_p in the constant strain fatigue test of figs. 19–24. σ_s = saturation stress, N = number of cycles. [After *J. C. Grosskreutz* (23)]

The micrograph of fig. 24 is taken after 300 cycles when the flow stress has become rather stationary. Already the visual impression shows a more stable situation. The dislocation multiple bundles have given way to rather distinct dislocation walls, forming a cell-like structure. From now on the change of the picture with increasing number of cycles is very slow, although important. The tendency is toward the formation of a more pronounced cell structure. Finally, we mention that in the present experiment fracture occurred after some 10 000 cycles.

These brief comments were supposed to give a typical example of the endeavour of the physicist to contribute to the solution of the extremely important problem of fatigue. It should be emphasized, however, that no satisfactory microscopic (nor macroscopic) theory exists for this problem[5]).

e) Unidirectional glide on two or more glide systems (stage II)

The experimental observation of tensile tests shows that the easy glide range develops into a situation of much larger workhardening (stage II) after a few percent of plastic strain. This is the case as soon as glide sets in in glide systems other than the primary one so far considered in the unidirectional experiments. This happens even if the resolved applied shear stresses in these secondary glide systems are far below the critical shear stress in these systems. In order to understand this let me refer to a basic theorem of plasticity which can be stated qualitatively as follows: If glide occurs locally inside a solid then internal stresses are built up which tend to hinder a further spreading out of the local glide and at the same time support glide tendencies at other places. This theorem which can be proved rigorously from elasticity theory has wide applications wherever discrete glide acts occur. Two examples may illustrate this.

(i) Earthquakes can be traced back to local glide acts. Hereby an internal stress which at the place of the glide has opposite direction of the original stress is produced so that a stress relaxation at the earthquake center occurs. On the other hand the new internal stresses increase the stresses already existing at neighbouring places. This effect leads to a correlation of earthquakes at neighbouring places, in agreement with numerous observations.

(ii) Creep in metals can, to some extent, be described as a sequence of "earthquakes", i.e. local glide acts produced by sudden motions of single dislocations. The analogy is rather far-reaching. The modern techniques even permit the observation of sound waves emitted from the local glide in a way similar to the emission of seismic waves. This so-called *acoustic emission* is observed in experiments which could be named *creep seismology*.

Returning to our stage II deformation we can explain the glide in secondary systems as follows: The local glidings in the primary system build

[5]) For a comprehensive and up-to-date review on the mechanisms of metal fatigue see *J. C. Grosskreutz* (23).

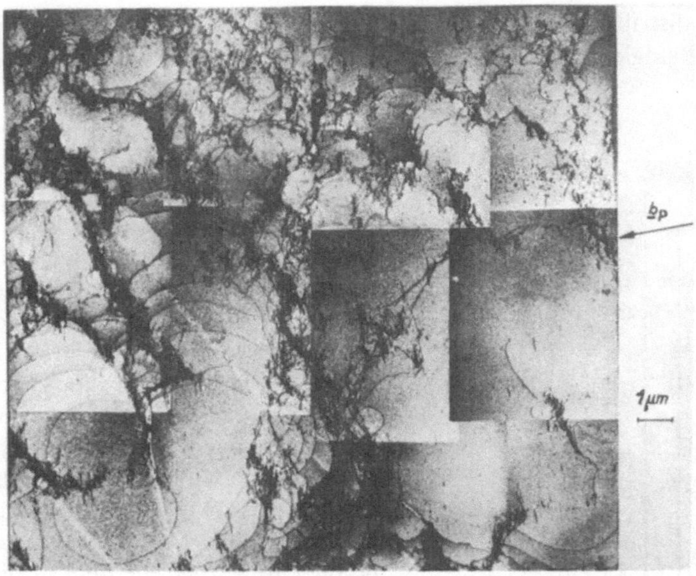

Fig. 19

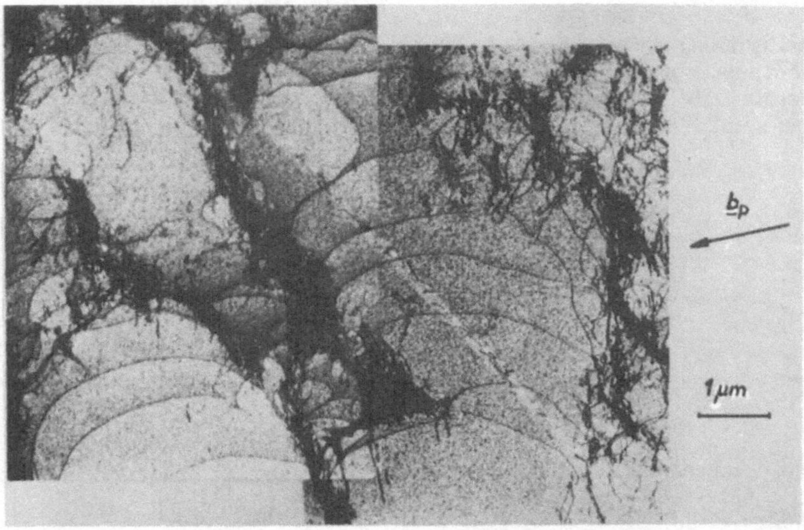

Fig. 20

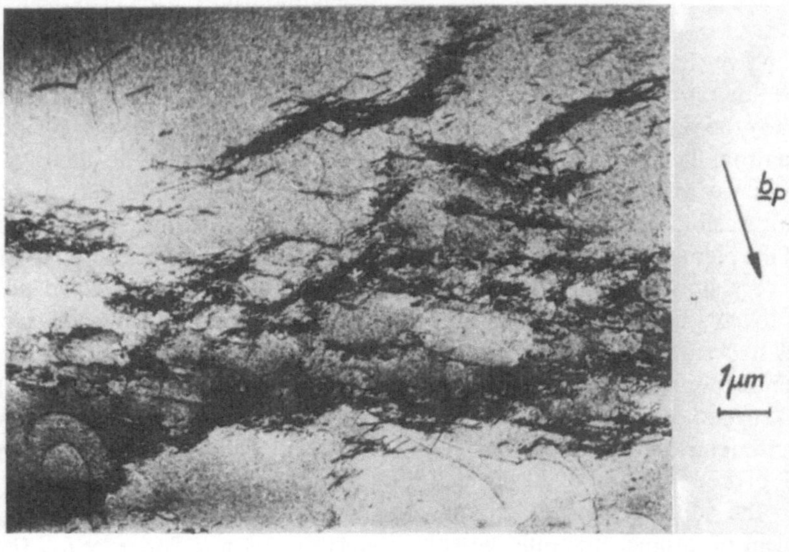

Fig. 21

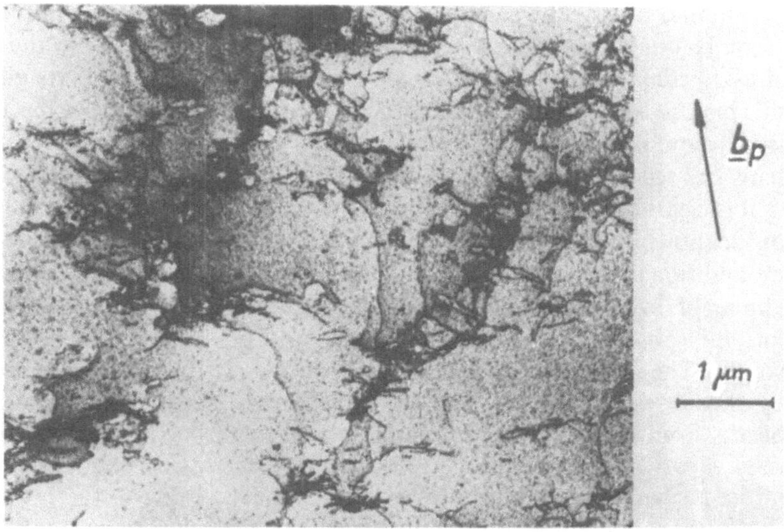

Fig. 22

Figs. 19–22. Dislocation arrangements in copper single crystals which have undergone 15 cycles in a tension-compression test of constant strain amplitude $\varepsilon = \pm 0.0065$ at room temperature. Typical is the formation of dislocation multipoles and the coexistence of free single dislocation lines. [After *H. Mughrabi* (24)]

Fig. 23. Distribution of radii of curvature of dislocations in figs. 19–22. The ordinate shows the number of dislocations with radius of curvature, drawn on the abscissa (in mμ). [After *H. Mughrabi* (24)]

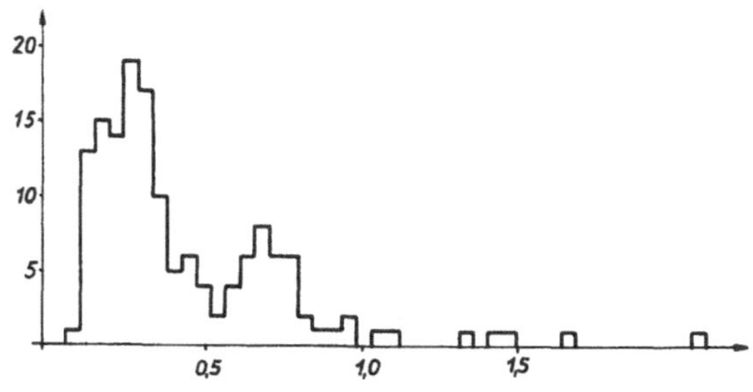

Fig. 24. Dislocation arrangement of the specimen of the preceeding figures after 300 cycles. [After *H. Mughrabi* (24)]

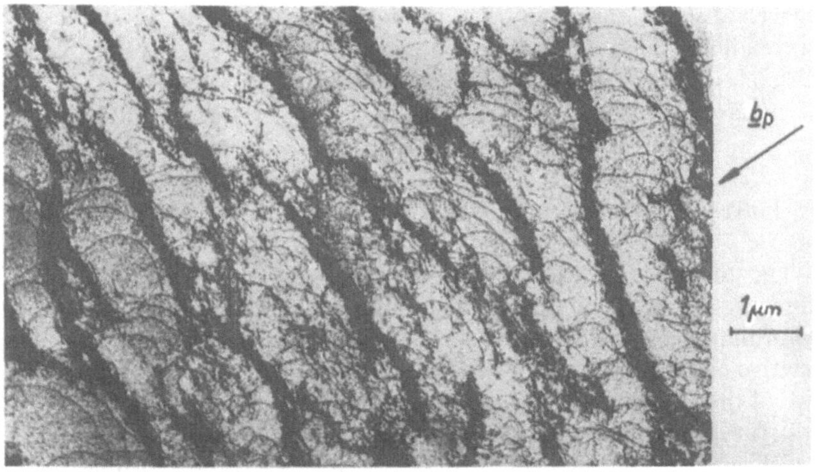

up internal stresses the resultants of which in the secondary systems assist the applied shear stresses already existent in these systems. If this picture is correct one expects secondary slip long before the applied shear stress in these systems has reached the critical value.

It is clear that the mutual elastic interaction of dislocations on different glide systems serves as an additional strong obstruction to the further gliding of dislocations and this effect is the cause for the observed strong work-hardening in stage II. The correctness of these ideas is confirmed by the observation that large work-hardening occurs from the beginning in tensile tests if the relative orientation of tensile axis and crystal lattice is such that the maximum resolved applied shear stress is about the same in two or more crystallographic glide systems.

Under simplifying assumptions *A. Seeger* and *H. Kronmüller* have developed a theory of the work-hardening in stage II which is in good agreement with the experiment. These and other interesting results are summarized in refs. (25, 26).

I mention in passing that at higher stresses the dislocations gain new freedom by the process of cross-slip which allows them to circumvent obstacles built up by the dislocations of the other glide systems. This new freedom demonstrates itself by a work-hardening which decreases with increasing deformation (stage III).

The micrographs of figs. 25–28 show how the dislocation arrangements look in stage II. The main features are: Due to the glide in several systems the arrangements are much more irregular than in stage I, in fact similar to those shown for the fatigue experiment. In addition we see nearly parallely extending groups of slightly curved dislocations which are produced in a correlated fashion and are pressed against the obstacles formed by interacting primary and secondary dislocations.

f) Polycrystal deformation

The theorem of the last paragraph is also fundamental for the understanding of the plastic deformation of polycrystals. Depending on the relative orientation between tensile axis and lattice of the crystallites the beginning of the glide will vary among the grains. In those grains in which glide occurs first, immediately a strong, usually triaxial, back stress is built up tending to stop glide. At the same time the internal stresses help to initiate glide in other grains earlier than would be expected under the applied stress alone.

Due to this effect the full plastic state in which all grains participate in further plastic deformation is reached at strains smaller than one percent. At even lower strains one observes that the deformation in the crystallites occurs by gliding on several glide systems. Also this effect is explained by the above theorem. As a consequence there is no easy glide in polycrystalline aggregates: From the beginning they exhibit a strong work-hardening.

For the reason just explained the work-hardening of the crystallites is more similar to that of crystals which glide on several glide systems from the beginning. It is therefore not surprising that one obtains very bad approximations if one tries to obtain theoretical work-hardening curves just by averaging the curves of the single crystals over the lattice orientations. *U. F. Kocks* (28) has shown that the agreement improves considerably if only curves belonging to multiple glide are taken into account.

More quantitative results have been obtained in a self-consistent calculation initially proposed by me in 1959 (29) and further developed by *B. Budjanski, T. T. Wu* (30), *J. Hutchinson* (31), *R. Hill* (32), *H. D. Bui* (33), *A. Zaoui* (34) and by *H. D. Bui, A. Zaoui* and *J. Zarka* (35). In this method the evolution of the mean internal stress and the active glide systems are calculated in dependence on the grain orientation in a self-consistent way. In a somewhat different calculation *T. H. Lin* has taken into account the fact that the internal stresses are not homogeneous within a grain. Comprehensive reviews on recent results in polycrystal plasticity are given by *T. H. Lin* (36) and *U. F. Kocks* (37).

A particular problem in deriving polycrystal work-hardening curves from those of single crystals is that the latter curves should be obtained from experiments with complex stress states rather than from those with homogeneous uniaxial applied stress. However, such curves have not been measured. Therefore, it would be highly desirable to have good theories of the hardening behaviour of single crystals under complex applied stress. Farthest in this direction seems to be the theory of *J. Zarka* (38), which I discuss briefly in section 5. Nevertheless, further work on this problem is needed.

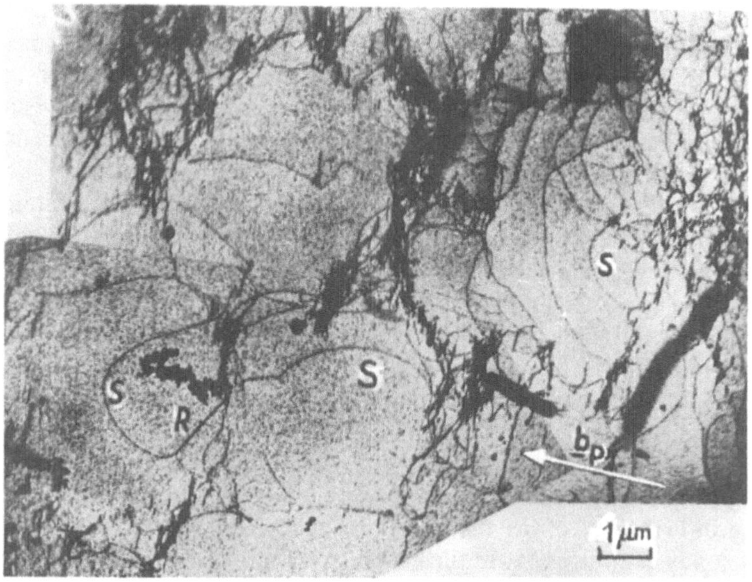

Fig. 25

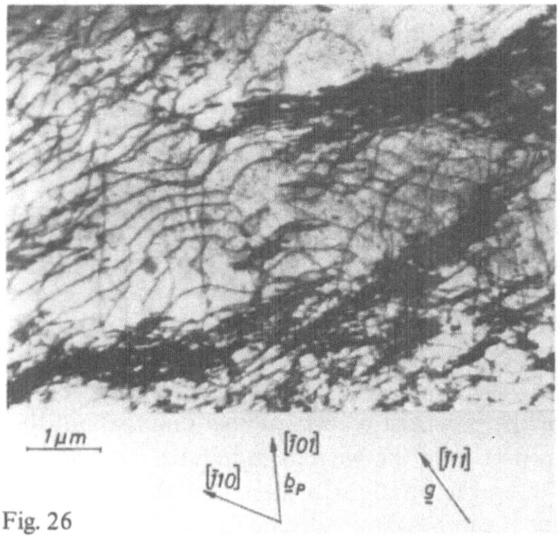

Fig. 26

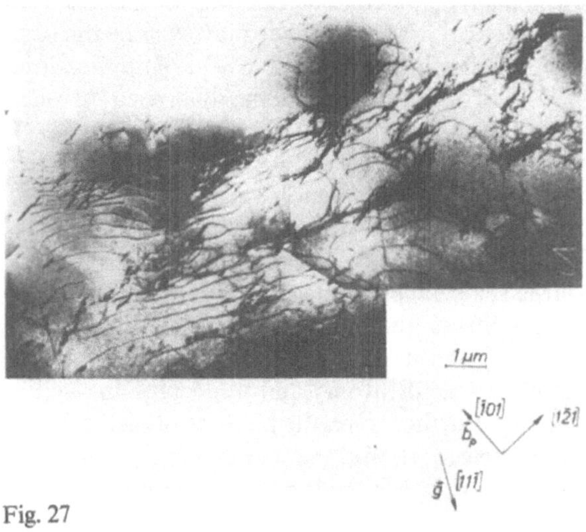

Fig. 27

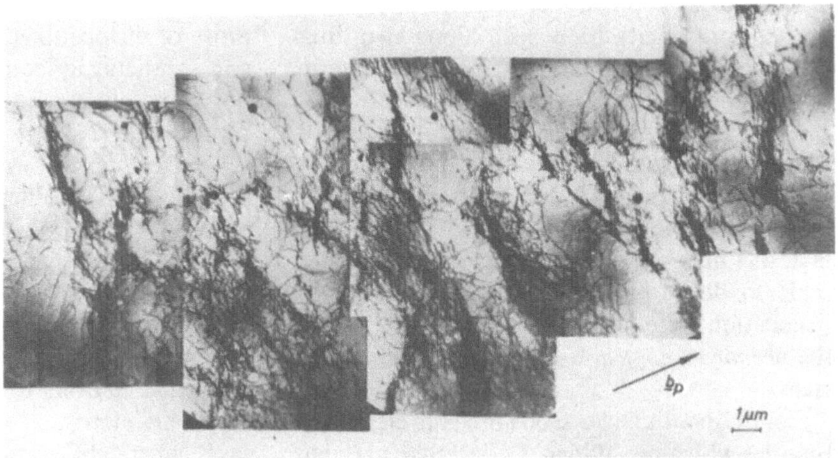

Fig. 28

Figs. 25–28. Dislocation arrangements in copper single crystals after a tensile strain of 0.17 (stage II). The pictures resemble those obtained after cyclic deformation. [After *U. Essmann* (14) and *H. Mughrabi* (16, 27)]

5. Phenomenological and macroscopic approaches

This section contains remarks on various approaches to phenomenological and macroscopic theories of the rheological behaviour of metals. In particular I would like to discuss possible consequences which the physical results reported in the previous sections may have for the development of such theories. The view adopted in this section is rather general. Since any well founded judgement of the numerous existing approaches within particular areas of our field of interest would require going into too many details I renounce any attempt of doing so in this lecture of limited length. I do this with some regret not only because many workers of great merits in the field of viscous and plastic behaviour of solids cannot be given the due reference, but also because there are important approaches which have proved to be valuable for the solution of practical problems. They may even be the best approaches in the particular field of application.

My first point concerns the difference between phenomenological and macroscopic theories. I classify a theory as phenomenological if it uses exclusively phenomenological quantities, i.e. quantities which can be measured in purely phenomenological experiments by which I mean large scale experiments (large scale in space and time). Stress and strain are the most important of such quantities for our problem. Clearly, there is no place for atoms and molecules in such a theory. Furthermore, the pictures of dislocations seen before should be convincing enough to show that neither single dislocations nor arrangements of them are phenomenological in this sense.

Recent developments of theories of plasticity and viscoplasticity have been along two directions which are characterized best by the terms *history* and *internal state*. As we have learned the internal state is the lattice defect state, in particular the dislocation state. But quantities describing the dislocation arrangements are *hidden* variables in the truest meaning of the word. In fact, one needs special devices on a microscopic scale to detect them. This is so although the dislocation state has profound influence also on the *phenomenological* behaviour of plastic materials.

For the mentioned reason I prefer to classify the theories which use hidden, i.e. internal, variables as macroscopic rather than as phenomenological although there is not much difference in the original meaning of the two words. It is intuitively clear that complete theories using only a few variables for the description of the rheological behaviour of metals must include explicitly the history of the material. This is usually done in the constitutive laws, which form the physically most interesting part of the theory.

This statement, namely that the constitutive law is the really relevant part of such theories from the point of view of physics, is another point of relevance. The constitutive equations are the place where physical insight is urgently needed. Phenomenological theories have led to frameworks for such equations and, in particular, have formulated restrictions on the form of these equations which stem from certain fundamental invariance laws of physics and from the 2nd law of thermodynamics.

It is rather clear that the contribution of solid state physics must be to fill the given framework of the constitutive laws with physical life. In fact, *specific* forms of these laws must be established which are sufficiently good approximations in the situations of interest.

To give an example: If in the history type theory of plastic or viscoplastic deformation it is stated that the stress is a time functional of the strain (which may also depend on temperature and perhaps on further parameters) then this statement does not help in the solution of actual problems as long as the explicit form of this time functional is not known. Here I feel an important task for the physicist who knows so much about the properties of the various materials.

I see essentially two ways how the time functional can be obtained. The first way is to use all the knowledge on the physical processes, mainly those of dislocations, on the microscopic scale, and synthetically construct the functional. I do not know of any successes of this method to be reported here.

The second way would be to perform a great number of phenomenological stress-strain experiments and in this manner obtain the information about the material response to operations from outside. The unpleasant feature of the latter procedure is that, in contrast to the elasticity theory, a large number of experiments have to be performed in order to obtain the full constitutive law. This statement certainly applies to those cases where dislocations are involved whereas the situation is simpler in the case of pure viscoelasticity in which the time functional approach has had its main successes. The reason for the complication in the case of plasticity and viscoplasticity is, of course, the complexity of the dislocation state which reflects itself in a most difficult stress-strain behaviour.

In the strict sense one has to perform all experiments with all possible strain histories in order to obtain the complete information on the material. The following example may explain this. Consider two specimens of materially identical properties. Apply two different loading programs in such a way that the two specimens are, macroscopically speaking, in the same state. We assume this to be possible but emphasize that it is not a trivial assumption. We expect that from now on the two specimens will respond to identical further loading in exactly the same way. Also this is a non-trivial assumption because of the stochastical nature of some of the fundamental processes. But within certain statistical fluctuations it may be considered as true.

Now we may argue that if we had an a priori information on the fact that the states of the two specimens are identical after the first part of the loading, then we could predict the behaviour of the second specimen during the second part of the loading from the experiment with the first specimen alone. However, the mentioned information can never be an *a priori* information in a purely phenomenological consideration since the state of the body has no direct place in such a theory. The fact that two loading histories lead to the same state can only be established *a posteriori*, namely by experiments which show that the two specimens behave identical with respect to any *further* loading program.

Generalizing this example to more than two loading histories we conclude that in the strict sense the response of a plastic or viscoplastic body to any loading program cannot be *predicted* without introducing considerations on a smaller scale. Such a proceeding would correspond to *J. Meixners higher level of observation* (40). *D. C. Druckers* remark (39) that the experiment itself must be run to get the answer applies to the purely phenomenological approach.

In a less strict sense one can, of course, make predictions, namely by means of interpolation. This procedure would be based on the assumption that the response to similar load programs is similar. It is questionable, however, whether such a procedure deserves the qualification of a physical theory.

One may conjecture that only experiments with homogeneous stress and strain are required in order to establish the plastic constitutive law. This would be true if plastic materials behaved like simple materials. Within the phenomenological theory the statement of simple behaviour is an a priori information which allows us to give genuine predictions on non-homogeneous experiments on the basis of homogeneous experiments only. However, the observation as well as the theory show that dislocation arrangements which never appear in the homogeneous tensile tests do occur in the non-homogeneous plastic bending experiments, for instance. These are arrangements which have excess dislocations of one sign on a macroscopic scale. Since the response to such a dislocation distribution cannot be found by simple tensile or shear experiments plastic materials are not simple in the usual sense. This seems to be an important feature of the plasticity of crystalline matter.

The observation just made does not exclude, of course, that the assumption of such materials as simple can be a reasonable or even excellent approximation in a certain situation and for the sake of simplicity one often will assume the material simplicity.

In the work mentioned earlier *D. C. Drucker* has also surmised that the number of variables in an internal state type theory of plasticity which we now consider would be infinite. That this idea is physically correct can be established in a convincing manner as follows. Leaving other types of lattice defects aside we are faced with the problem of describing the macroscopically effective part of the dislocation distribution. The pictures shown convey the clear impression that the distribution is disordered to a considerable degree. A full macroscopic description therefore requires the tools of probability theory and statistics, in particular the correlation functions of the dislocation distribution. The proof that an infinite number of variables is needed for a complete macroscopic description of the dislocation state is, as we shall see, already contained in a well-known theorem of probability theory which states that the complete macroscopic characterization of a random function requires an infinite number of n-point correlation functions ($n = 1, 2, \ldots, \infty$).

At this point it is useful to recall that in the so-called continuum theory of dislocations the dislocation arrangement is specified by the tensor

$\alpha_{ij}(x)$ of dislocation density which counts the number of dislocation lines of *Burgers* vector b_j piercing through a macroscopic area element dS_i at point x. However, in our pictures we have seen that dislocations do not form perfect line densities but are arranged in rather complex structures. Therefore, the continuum theory of dislocations gives a correct image of the situation in a crystal only if $\alpha_{ij}(x)$ has the form of δ-functions along the actual lines. In this way $\alpha_{ij}(x)$ becomes a random function of position.

The first term in the correlation description is then the average dislocation density $\langle\alpha_{ij}(x)\rangle$ taken in the sense of *Gibbs*. It is, in fact, this quantity which was introduced originally as the dislocation density tensor by *J. F. Nye* in 1953. This tensor is a very poor description of dislocation distributions, as can be seen best from the fact that $\langle\alpha_{ij}(x)\rangle = 0$ in a specimen which was deformed by uniaxial tension because in this case the macroscopic volume element of the crystal contains an equal number of positive and negative dislocations. All specimens of which we have seen electron micrographs had been deformed in uniaxial tension. They showed important changes of the internal state although having $\langle\alpha_{ij}(x)\rangle = 0$.

$\langle\alpha_{ij}(x)\rangle$ does not vanish in bending experiments. For this reason we had classified plastic materials as non-simple.

Of much more interest is the 2-point dislocation correlation function $\langle\alpha_{ij}(x_1)\,\alpha_{kl}(x_2)\rangle$, a 4th rank tensor, which contains such important information as the total length of all dislocations in a volume element, distributions of dislocation dipoles etc. (41). Therefore a theory which takes into account this correlation function can be expected to lead to interesting results.

A rigorous theory should include all correlations up to infinite order, i.e. an unlimited number of state variables. This again is an important point. Such a theory would be of statistical type such as the theory of turbulence. The role which *Navier-Stokes* equations play in the statistical theory of turbulence would be taken over by the equations of the continuum theory of dislocations. The considerations in this direction are still very preliminary (41). It may be worthwhile to mention that the statistical distributions of dislocations are accompanied by small scale fluctuations of stress and strain which obviously become random functions in this manner. Instead of saying that the change of the internal state is due to the elastic interactions between dislocations one can also say that this change is due to the fluctuating internal stresses and strains. These fluctuations do, in fact, determine the plastic behaviour of crystals to a large extent. Also this standpoint speaks for a close analogy between plasticity and turbulence.

As I have shown most physical experiments on dislocations aim at visualization and description of lattice defect states. For this reason internal state type theories can expect more assistance from solid state physics than history type theories. A number of macroscopic internal state theories have been proposed which seem to provide a good basis for a combination with dislocation models. Of recent work I mention that of *H. Ziegler, J. Mandel, K. C. Valanis, B. D. Coleman* and *M. E. Gurtin, E. Onat, M. A. Eisenberg, J. Kratochvil* and *O. W. Dillon, P. Perzyna, J. Rice, J. Lambermont* and myself (42–52). This list does not claim completeness. Based upon these and similar results dislocations have been introduced into the macroscopic theory of plasticity in a more direct form by *J. Zarka* (38), *C. Teodosiu* (53) and others. In all these approaches a rather limited number of internal parameters is used in order to specify dislocation arrangements. Hereby the statistical character of the theory is lost to some extent.

In *Zarka*s model, for example, the dislocations are described as straight segments which are distributed randomly as well in length as in space. This model misses certain features of real dislocation distributions, but nevertheless gives reasonable results in the construction of constitutive laws of plastic bodies. Some refinement also allows us to include thermally activated, i.e. viscoplastic, processes. For details I refer to the paper of *J. Zarka* (38). The appealing feature of this theory is that it leads to constitutive equations which also include more complex stress states than pure tension. It can, therefore, be readily applied also to the plasticity of polycrystals, for instance in the self-consistent method mentioned previously. This has been shown by *H. D. Bui, A. Zaoui* and *J. Zarka* (35).

It seems clear that theories of this kind are the simpler the smaller the number of internal state parameters is. The problem is then to find a set of such parameters which on one hand is as small as possible, and on the other hand as large as necessary in order to give a sufficiently good specification of the internal state. It is obvious that one and the same set of parameters will not cover all situations of plasticity and viscoplasticity except if one uses an infinite set. But such a rigorous theory which is desirable from a general standpoint of understanding, does not help much in practical applications. Here restrictive assumptions are indispensible and this means that one needs, and always will need, many theories each of which are adapted to a particular situation. For instance, the powerful theory of the perfectly plastic body will serve its purpose in many special applications. It will not help much in those situations of great practical importance where the main interest lies in the change of the internal mechanical state.

I conclude this lecture by giving my personal answer to the question raised in the introduction to this lecture. The insights gained in the physical research on lattice defects can, indeed, help in the search for realistic theories of the rheological behaviour of metals. These insights can provide a physical basis for an optimum selection of those internal variables which are needed for the description of the internal mechanical state. In addition, they may perhaps provide those physical laws according to which the internal state develops during the plastic or viscoplastic deformation. These two contributions from solid state physics are exactly the ones which cannot come from purely phenomenological considerations. This is an obvious fact which only confirms that the plasticity and viscoplasticity of metals cannot be understood by pure phenomenology.

Acknowledgment

I would like to thank Drs. *H. Kronmüller*, *W. Frank* and *H. Mughrabi* of the Max-Planck-Institut für Metallforschung in Stuttgart for valuable discussions during the preparation of the manuscript, Dr. *B. K. Datta-Gairola* for its thorough reading. In addition I am grateful to all authors the original figures of whom I have used to illustrate the present text.

References

1) *Kröner, E.*, Z. Naturforschg. **11a**, 969 (1956).
2) *Taylor, G. I.*, Proc. Roy. Soc. (Lond.) **A 145**, 362 (1934).
3) *Bollmann, W.*, Phys. Rev. **103**, 1588 (1956).
4) *Hirsch, P. B., R. W. Horne*, and *M. J. Whelan*, Phil. Mag. **1**, 677 (1956).
5) *Hirsch, P. B., A. Howie*, and *M. J. Whelan*, Phil. Trans. Roy. Soc. (Lond.) **A 252**, 499 (1960).
6) *Mader, S.*, Moderne Probleme der Metallphysik I, *A. Seeger* ed., p. 192 (Berlin-Heidelberg-New York 1965).
7) *Low, J. R.* and *A. M. Turkalo*, Acta Met. **10**, 215 (1962).
8) *Essmann, U.*, Phys. Stat. Sol. **3**, 932 (1963).
9) *Alexander, H.*, Phys. Stat. Sol. **26**, 75 (1968); **27**, 391 (1968).
10) *Mughrabi, H.*, J. Scient. Inst. **2**, 351 (1969).
11) *Silcox, J.* and *P. B. Hirsch*, Phil. Mag. **4**, 72 (1959).
12) *Hedges, J. M.* and *J. W. Mitchell*, Phil. Mag. **44**, 223 (1953).
13) *deWit, A.* and *J. S. Koehler*, Phys. Rev. **116**, 1113 (1959).
14) *Essmann, U.*, Phys. Stat. Sol. **12**, 723 (1965).
15) *Crump, J. C.* and *F. W. Young*, Phil. Mag. **17**, 381 (1968).
16) *Mughrabi, H.*, Phil. Mag. **23**, 869, 897 (1971).
17) *Friedel, J.*, Phil. Mag. **44**, 444 (1953).
18) *Bradfield, G.* and *H. Pursey*, Phil. Mag. **44**, 437 (1953).
19) *Granato, A. V.* and *K. Lücke*, J. Appl. Phys. **27**, 583 (1956).
20) *Dash*, W. C., Dislocations and the Mechanical Properties of Crystals, p. 57 (New York 1957).
21) *Essmann, U.*, Phys. Stat. Sol. **12**, 707 (1965); Acta Met. **12**, 1468 (1964).
22) *Neumann, P.*, Acta Met. **17**, 1219 (1969).
23) *Grosskreutz, J. C.*, Phys. Stat. Sol. (b) **47**, 11, 359 (1971).
24) *Mughrabi, H.*, Unpublished work, partly discussed in: *H. Mughrabi*, Proc. 3rd Int. Conf. an Strength of Metals and Alloys, Cambridge 1973.
25) *Seeger, A.*, Encyclopedia of Physics VII/2, p. 1 (Berlin-Heidelberg-New York).
26) *Kronmüller, H.*, Moderne Probleme der Metallphysik II, *A. Seeger* ed., p. 126 (Berlin-Heidelberg-New York 1965).
27) *Mughrabi, H.*, Phil. Mag. **18**, 1211 (1968).
28) Kocks, U. F., Acta Met. **6**, 85 (1958).
29) *Kröner, E.*, Teknika (Belgrade) **15**, 2187 (1960); Acta Met. **9**, 155 (1961).
30) *Budiansky, B.* and *Wu, T. T.*, Proc. IVth U.S. Congr. Appl. Mech., p. 1175 (1962).
31) *Hutchinson, J. W.*, Proc. Roy. Soc. Lond. **A 319**, 247 (1970).
32) *Hill, R.*, J. Mech. Phys. Solids **13**, 89 (1965).
33) *Bui, H. D.*, Mém. de l'Artill. Franç., Sciences et Techniques 1st fasc., 145 (1970) (Thesis Paris 1969).
34) *Zaoui, A.*, Mém. de l'Artill. Franç., Sciences et Techniques (1971), (Thesis Paris 1970).
35) *Bui, H. D., A. Zaoui*, and *J. Zarka*, Proc. Int. Symp. on Foundations of Plasticity, Warsaw 1972, Preprint ed., *A. Sawczuk* ed., p. 51 (Groningen 1972).
36) *Lin, T. H.*, Adv. Appl. Mech. **11**, 225 (1971).
37) *Kocks, U. F.*, Met. Trans. **1**, 1121 (1970).
38) *Zarka, J.*, Mém. de l'Artill. Franç., Sciences et Techniques 2nd fasc., 223 (1970) (Thesis Paris 1969).
39) *Drucker, D. C.*, Structural Mechanics, p. 407 (London 1960).
40) *Meixner*, J. in the Proceedings of this Conference.
41) *Kröner, E.*, Inelastic Behaviour of Solids, *M. F. Kanninen, W. F. Adler, A. R. Rosenfield, R. I. Jaffee* eds., p. 137 (London 1970).
42) *Ziegler, H.*, Progr. Sol. Mech. **4**, 91 (1963).
43) *Mandel, J.*, Proc. XIth Int. Congr. Appl. Mech., Munich 1964, p. 502 (Berlin-Heidelberg-New York 1966).
44) *Valanis, K. C.*, J. Math. and Phys. **45**, 197 (1966); Mechanical Behaviour of Materials under Dynamic Loads, *U. S. Lindholm* ed., p. 343 (Berlin-Heidelberg-New York 1967).

45) *Coleman, B. D.* and *M. E. Gurtin*, J. Chem. Phys. **47**, 597 (1967).

46) *Onat, E.*, Proc. IUTAM-Symp. on Irreversible Aspects of Continuum Mechanics, Vienna 1966, p. 292 (Berlin-Heidelberg-New York 1967).

47) *Eisenberg, M. A.*, Int. J. Engin. Sci. **8**, 261 (1970).

48) *Kratochvil, J.* and *O. W. Dillon*, J. Appl. Phys. **41**, 1470 (1970).

49) *Perzyna, P.*, Series of Lectures held at C.I.S.M. Udine, Italy, 1971.

50) *Rice, J.*, J. Appl. Mech. **37**, 728 (1970); J. Mech. Phys. Solids **19**, 433 (1971).

51) *Lambermont, J.*, Bull. Soc. Roy. Sci. Liège (Belgium) **41**, 572 (1972).

52) *Kröner, E.*, J. Math. and Phys. **42**, 27 (1963).

53) *Teodosiu, C.*, to appear in Proc. Int. Symp. on Foundations of Plasticity, Warsaw 1972 (Groningen 1973 or 1974).

Author's address:

Prof. Dr. *Ekkehart Kröner*
Institut für Theoretische und Angewandte Physik der Universität
D-7000 Stuttgart

Rheol. Acta 12, 393–397 (1973)

Laboratoire de Mécanique des Solides, Ecole Polytechnique Paris (France)

Essai de définition de quelques comportements rhéologiques

Par J. Mandel

Avec 3 figures

(Reçu p. p. le 27 octobre 1972)

Il est sans aucun doute extrêmement téméraire d'établir des classifications dans les sciences de la nature. Il est illusoire de chercher des définitions rigoureuses. Et le premier enseignement de la rhéologie est justement qu'il est difficile de distinguer fluide et solide, puisque toute matière flue.

Cependant pouvons-nous empêcher de nous demander quel sens précis il faut attacher à ces mots, que nous employons si souvent en rhéologie: solide, fluide, viscoélastique, plastique, etc.? Répondre à ces questions est un besoin de l'esprit; c'est aussi une condition du progrès de la science.

La réponse peut être recherchée à l'échelle de la micro-structure. Mais ce n'est pas la voie que nous suivrons ici, car on peut espérer atteindre une généralité plus grande par l'étude du comportement macroscopique. N'est-ce pas là ce qu'expriment les mots: comportement solide, comportement viscoélastique, etc.? Pour ce qui concerne la distinction entre fluide et solide, *W. Noll*[1]) a apporté en 1958 une contribution remarquable en se fondant sur les notions de réponse fonctionnelle et de groupe d'isotropie. Dans une large mesure notre exposé fait appel aux idées de *Noll*. Cependant notre point de vue est différent. Au lieu d'invoquer le groupe d'isotropie, nous ferons appel à la notion de *mémoire évanescente*, qui nous permettra d'établir notamment deux distinctions fondamentales; distinction entre fluide et solide, distinction entre viscoélasticité et plasticité.

Nous nous limiterons aux matériaux simples suivant la terminologie de *W. Noll*, ce qui signifie que le tenseur contrainte σ dépend du gradient γ de la transformation, mais ne dépend pas des dérivées spatiales de ce gradient. En outre nous supposerons qu'il n'y a pas de liaison interne telle que l'incompressibilité (invariance du volume quel que soit σ).

A partir d'une configuration $\varkappa$, dans laquelle le corps est en équilibre, on impose une transformation de gradient $\gamma(t) = \partial x/\partial X$, X désignant les coordonnées dans la configuration $\varkappa$ et x les coordonnées dans la configuration à l'instant t. *W. Noll* envisage la *réponse fonctionnelle*, i.e. la fonctionnelle:

$$\sigma(t) = \mathscr{F}_\varkappa[\gamma(\tau)] \qquad -\infty < \tau \leqq t$$

qui dépend de la configuration de référence $\varkappa$. C'est la nature de cette dépendance entre la fonctionnelle et la configuration de référence qui est la base de sa classification.

Ici, nous envisageons seulement deux types de fonctions que nous nommons échelon et créneau:

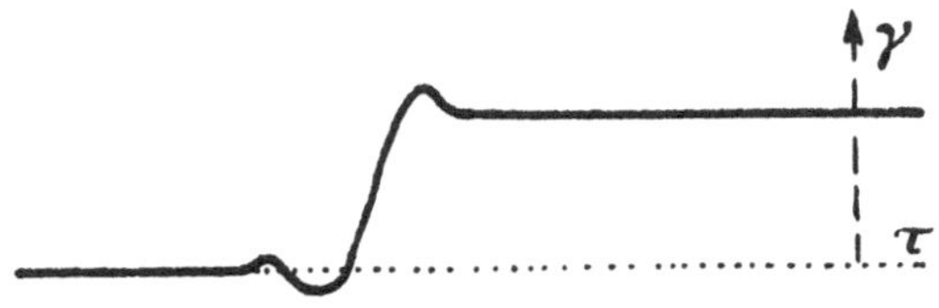

Fig. 1. *Echelon:* c'est le passage d'une valeur constante γ_0 à une valeur constante γ_1 en un temps fini, sans préciser le mode de passage. Le souvenir d'un échelon établit la distinction entre solide et fluide

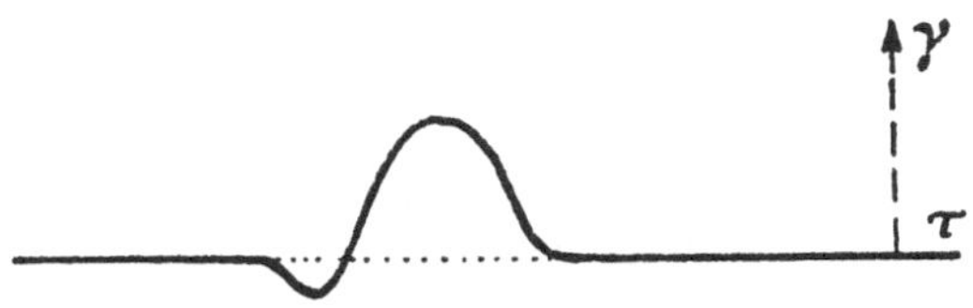

Fig. 2. *Créneau:* c'est un échelon d'amplitude nulle ($\gamma_1 = \gamma_0$). Le souvenir d'un créneau établit la distinction entre viscoélastique et plastique

[1]) *Noll, W.*, A mathematical theory of the mechanical behaviour of continuous media. (Arch. Rat. Mech. Anal. 2, pp. 197–226, 1958/59.)

Comportement fluide

Rappelons deux définitions usuelles, dont on peut montrer l'équivalence:

1. Un corps est fluide s'il reste en équilibre seulement sous une pression isotrope.

2. Un corps est un fluide si la relaxation est totale lorsque le volume n'a pas varié, i.e. lorsque la transformation imposée est un échelon «isochore».

Ces définitions simples suscitent diverses critiques. On sait par exemple qu'un métal flue même à la température ordinaire sous des contraintes très faibles. Existe-t-il un seuil pour le fluage? Certains en doutent. De toute manière il est très difficile à préciser. Or l'absence de seuil conduirait à dire que l'acier à la température ordinaire est un fluide.

W. Noll a donné une définition qui semble beaucoup plus satisfaisante: un corps est un fluide si la réponse fonctionnelle n'est pas modifiée par un changement isochore quelconque de la configuration $\varkappa$. De cette définition on déduit que i) le fluide est isotrope, ii) il n'existe dans son passé aucun état privilégié, iii) la contrainte actuelle $\sigma(t)$ ne dépend que de l'histoire des déformations relatives (i.e. rapportées à la configuration actuelle). Toutefois, et ce point semble être passé inaperçu, on ne retrouve la condition d'équilibre sous pression isotrope (ou de relaxation totale) que moyennant l'hypothèse suivante, implicitement admise: le fluide n'a aucun souvenir des transformations qui lui ont été imposées dans un passé très lointain. Cet énoncé implique notamment la relaxation totale, mais il est plus fort, parce qu'il concerne toutes les propriétés (mécaniques et thermiques) actuelles. En particulier la réponse fonctionnelle est indépendante d'un changement de configuration isochore effectué dans un passé infiniment lointain, ce qui redonne la définition de *W. Noll.* Nous adopterons donc la définition suivante:

Un corps a le comportement fluide dans une configuration $\varkappa$ s'il a une mémoire évanescente de toute transformation échelon isochore à partir de $\varkappa$ (à température constante)[2]).

L'ensemble $E_\varkappa$

Pour un corps quelconque nous définirons le comportement (fluide, solide, semi-fluide) dans une configuration $\varkappa$, grâce à l'ensemble $E_\varkappa$ des transformations par échelon, à partir de $\varkappa$, dont le souvenir est évanescent. Toute configuration déduite de $\varkappa$ par une des transformations de $E_\varkappa$ jouit par hypothèse des mêmes propriétés que $\varkappa$. Le comportement est donc le même pour toutes ces configurations.

Le comportement est fluide si l'ensemble $E_\varkappa$ est le groupe des transformations telles que dét $\gamma = +1$, appelées transformations unimodulaires *directes. Si le comportement est fluide dans une configuration, il l'est encore dans toutes les configurations de même densité* (à la même température). On peut donc parler de fluide sans préciser la configuration. $E_\varkappa$ contient notamment le sous-groupe des rotations, ce qui entraîne l'isotropie du fluide en équilibre.

Comportement solide

Suivant la terminologie usuelle un corps est solide s'il existe pour le déviateur des contraintes un seuil en-deçà duquel les déformations restent limitées. Comme on l'a déjà dit, cette notion de seuil soulève des difficultés. *W. Noll* a proposé la définition suivante: un corps est un solide s'il existe des configurations, dites non distordues, telles que la réponse fonctionnelle soit modifiée par tout changement non orthogonal, i.e. par tout changement de forme de la configuration de référence. Notre point de vue est à peine différent, mais nous considérons des changements de forme réellement effectués dans un passé lointain.

1. Cas d'une configuration d'équilibre sous contrainte isotrope. Il est possible que dans cette configuration le corps soit isotrope ou possède des directions de symétrie, i.e. conserve ses propriétés quand on lui impose, soit une rotation arbitraire, soit certaines rotations particulières. Par contre tout changement de forme modifie d'une manière permanente les propriétés d'un solide.

2. Cas d'une configuration d'équilibre sous une contrainte non isotrope. Supposons qu'on puisse se ramener par une transformation élastique de gradient F à une configuration du type précédent, dans laquelle une rotation Ω ne modifie pas le comportement, puis rétablir la contrainte initiale par la transformation F^{-1}. La transformation $T = F^{-1}\Omega F$ ne modifie pas les propriétés du corps. Une telle transformation sera dite orthogonalisable par F[3]). D'où la défini-

---

[2]) La densité est la même avant et après la transformation mais peut varier au cours de celle-ci.

[3]) Les transformations orthogonalisables jouissent des mêmes propriétés que les transformations orthogonales en ce qui concerne leurs valeurs propres. Car les valeurs propres de T sont les mêmes que celles de Ω (soit 1, $e^{i\varphi}$, $e^{-i\varphi}$, φ réel).

tion suivante (dont nous admettons la validité même si l'on ne peut pas se ramener à une configuration sous contrainte isotrope par une déformation élastique):

Le comportement est solide dans la configuration $\varkappa$ si l'ensemble $E_\varkappa$ ne contient, dans un voisinage de $\varkappa$, que des transformations orthogonalisables par une même transformation F (qu'on peut supposer symétrique).

La restriction à un voisinage de $\varkappa$ a pour but d'éliminer la possibilité de transformations (non orthogonalisables) comportant une déformation plastique sans écrouissage.

Comportement semi-fluide

Le comportement est dit semi-fluide dans une configuration $\varkappa$ si, dans n'importe quel voisinage de $\varkappa$, l'ensemble $E_\varkappa$ contient des transformations non orthogonalisables. Exemples:

1. Corps parfaitement plastique à la limite d'écoulement. La déformation plastique ne modifie pas dans ce cas les propriétés du corps, donc son comportement est alors semi-fluide. Il n'en est pas de même lorsqu'il y a écrouissage[4]).
2. Cristal-fluide. Si le comportement est semi-fluide dans une configuration d'équilibre sous contrainte isotrope on a un cristal fluide (qui ne peut rester en équilibre que lorsque deux des contraintes principales sont égales)[5]).
3. Corps hygrostériques de *W. Noll*, corps définis par une relation entre σ, $D\sigma/Dt$ et $\mathscr{D}$, tenseur vitesse de déformation.

Le fluide est évidemment un cas particulier de semi-fluide, plus précisément de cristal-fluide. En effet pour un cristal fluide l'ensemble $E_\varkappa$ est un sous-groupe du groupe unimodulaire direct. Si, dans une configuration $\varkappa$, il y a isotropie (non seulement du tenseur σ, mais de toutes les propriétés) alors d'après un théorème de *W. Noll*, l'ensemble $E_\varkappa$ est le groupe unimodulaire direct tout entier donc on a un fluide.

Remarques

Le type de comportement peut dépendre de la configuration $\varkappa$ (et de la contrainte d'équilibre et de la température). Nous ferons à ce sujet les remarques suivantes:

1. Le type de comportement est le même pour deux configurations déduites l'une de l'autre par rotation[6]).
2. Dans la plupart des cas il reste le même quelle que soit la configuration dans un «ouvert» de l'espace des configurations. On peut parler de fluide, de solide, de cristal-fluide, de semi-fluide hygrostérique, sans préciser la configuration. Il n'en est pas de même pour le comportement parfaitement plastique.
3. Dans le cas d'un solide il ne peut pas y avoir de relation entre les composantes σ_{ij} du tenseur des contraintes en équilibre. Sinon un même tenseur σ correspondrait (à $t = +\infty$) à une infinité de transformations échelons non orthogonalisables. Dans le cas d'un semi-fluide il y a une ou plusieurs relations entre les σ_{ij} (5 dans le cas d'un fluide).

Viscoélasticité

Ici nous utilisons la réponse de la contrainte σ à un créneau de γ, ou plus simplement à un créneau de déformation ε (car une même rotation appliquée puis retirée est sans effet sur σ). *Nous considérons les transformations créneaux de ε dont le souvenir est évanescent.* En particulier la contrainte revient à sa valeur initiale au bout d'un temps infini; nous dirons qu'il y a *effacement* total des contraintes.

Cette notion d'effacement n'ayant guère été jusqu'ici utilisée en rhéologie, quelques commentaires s'imposent avant de passer à la définition de la viscoélasticité. L'effacement est la propriété duale de la recouvrance (échange de σ et ε). Pour définir la recouvrance on considère un créneau de contrainte. Si, au bout d'un temps infini, la déformation revient à zéro, la recouvrance est totale. Mais la recouvrance totale ne permet pas de définir correctement la viscoélasticité, puisqu'un fluide viscoélastique n'a pas une recouvrance totale. La relation entre les deux propriétés est fournie par les deux propositions suivantes:

P.1. S'il n'y a pas de liaisons intérieures, recouvrance totale quel que soit le créneau de σ

[4]) Lorsque la résistance à la déformation augmente, on a le comportement solide, lorsqu'elle diminue on a un nouveau comportement qu'on peut appeler comportement liquéfiant.

[5]) Il s'agit des corps dits smectiques. Les autres corps mésomorphes (nématiques, cholestériques) ne sont pas des corps matériellement simples.

[6]) Soit $\varkappa'$ déduite de $\varkappa$ par la rotation $R(X' = RX)$. Si la transformation γ par rapport à $\varkappa$ est à souvenir évanescent, de même par rapport à $\varkappa'$ la transformation $\gamma' = R\gamma R^{-1}$.

implique effacement total (quel que soit le créneau de ε)[7]).

P. 2. S'il n'y a pas de relations entre les contraintes à l'équilibre, effacement total, quel que soit le créneau de ε, implique recouvrance totale (quel que soit le créneau de σ).

Définition du comportement viscoélastique

Le comportement est viscoélastique dans la configuration $\varkappa$ si le souvenir de n'importe quelle déformation créneau à partir et dans un voisinage de $\varkappa$ est évanescent. Cet énoncé implique notamment l'effacement total des contraintes, mais il est plus fort, parce qu'il concerne toutes les propriétés (mécaniques et thermiques) actuelles, en particulier la réponse fonctionnelle.

Un fluide est viscoélastique dans toute configuration. En effet un créneau est un cas particulier d'échelon isochore. D'après la proposition P. 2, pour un fluide, il n'y a pas recouvrance totale.

De même un semi-fluide peut être viscoélastique, mais il n'y a pas recouvrance totale.

Un solide peut être viscoélastique dans un certain domaine. Comme, dans ce domaine, il n'y a pas de relation entre les contraintes à l'équilibre, il y recouvrance totale.

Plasticité

Le comportement est plastique dans la configuration $\varkappa$ si, dans n'importe quel voisinage de $\varkappa$, il existe des transformations créneaux pour lesquelles l'effacement n'est pas total. D'après la proposition P.1, la recouvrance ne l'est pas non plus pour certains créneaux de contrainte (s'il n'y a pas de liaisons intérieures, ce qu'on a supposé au départ). Autrement dit, il y a des déformations permanentes.

On définit habituellement le comportement plastique par l'existence d'un seuil du déviateur des contraintes pour les déformations permanentes, ceci afin de distinguer le corps plastique du fluide pour lequel le seuil serait nul.

Notre définition a l'avantage d'être affranchie de la notion de seuil. Si le seuil est nul le domaine d'élasticité est réduit à zéro, sans que pour autant le comportement soit celui d'un fluide. La propriété commune avec le fluide est l'existence de déformations permanentes pour un déviateur de contraintes aussi petit que l'on veut. Mais pour le fluide, il y a effacement total, pour le solide plastique non.

L'absence de seuil ne signifie pas que les déformations permanentes soient illimitées, sauf lorsqu'il n'y a pas d'écrouissage. Lorsqu'il y a écrouissage, ou bien, après déformation le seuil cesse d'être nul (tel est le du cuivre électrolytique dont le seuil, à l'état naissant, est nul, mais ne l'est plus après déformation), ou bien on a un solide dont le domaine d'élasticité reste ponctuel, mais dépend des déformations permanentes subies. Cette conception étant peu courante en plasticité (parce que le modèle classique est celui du patin à frottement sec), nous donnons ici deux exemples:

1. Un sable lâche et sec subit des déformations permanentes sous des déviateurs de contrainte très faibles. Le domaine d'élasticité, s'il existe, est d'étendue très faible. Ce cas n'est pas exceptionnel. Pour la plupart des matériaux la limite d'élasticité ne peut être définie que d'une manière conventionnelle.

2. Considérons un modèle formé d'éléments placés en série. Chaque élément est constitué par un ressort et un patin à friction en parallèle. Les seuils des patins sont: $s_1 \geq s_2 \ldots \geq s_n$. L'étendue du domaine d'élasticité est donc $2s_n$.

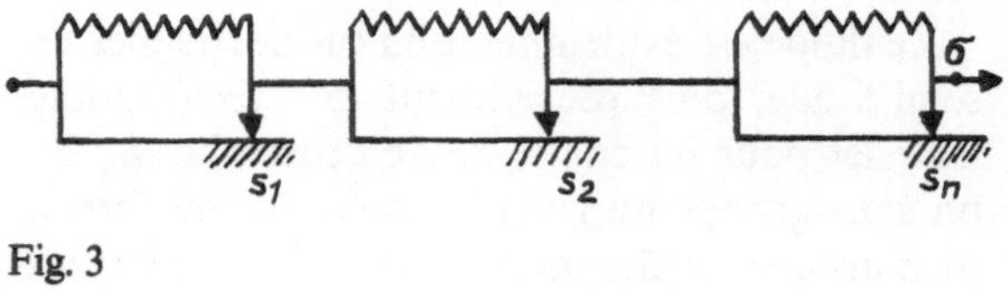

Fig. 3

Si nous prenons une suite infinie d'éléments de seuils décroissants, telle que s_n tend vers zéro quand n tend vers l'infini, nous obtenons un modèle qui reste en équilibre sous n'importe quelle contrainte (ce n'est donc pas un fluide) et dont cependant le domaine d'élasticité reste toujours nul.

Signalons sans insister que les phénomènes de radoucissement et de thixotropie sont aussi, évidemment, des effets de mémoire évanescente (d'un créneau de contrainte, le corps ne garde d'autre souvenir que des déformations permanentes). Nous conclurons que la faculté d'oubli (dont il faut préciser le contenu) est un élément essentiel du comportement rhéologique.

---

[7]) En effet, la recouvrance totale prouve que, en équilibre $\varepsilon = f(\sigma)$, relation inversible, à condition d'exclure les positions d'équilibre instable, lorsque les six composantes ε_{ij} sont indépendantes.

Traduction de quelques termes en anglais

tenseur contrainte	stress tensor
réponse fonctionnelle	response functional
mémoire évanescente	fading memory
matériau simple	simple material
liaison interne	internal constraint
non distordu	undistorted
vitesse de déformation	stretching
ouvert	open set
recouvrance	recovery

Adresse de l'auteur:
Prof. Dr. *J. Mandel*
Laboratoire de Mécanique des Solides
Ecole Polytechnique, Rue Descartes
F-75 Paris (France)

Rheol. Acta **12**, 398–403 (1973)

Centre de Recherches Physiques, Marseille (France)

L'acoustique en rhéologie et ses applications

Par W. Bismuth

Avec 4 figures

(Reçu p. p. le 27 Octobre 1972)

Introduction

Si l'on ne porte pas son attention sur les processus physico-chimiques, la Rhéologie constitue un chapitre de la mécanique des milieux continus; les relations contrainte-déformation qui peuvent être du type viscoélastique linéaire ou non, élastoplastique, viscoplastique, etc. ... forment la pierre angulaire de cette discipline. Or, ces corps sont compressibles et laissent par conséquent se propager des ondes acoustiques qui peuvent provenir d'une source aérienne aussi bien que d'un excitateur mécanique ou autre; d'où l'interdépendance de ces deux sciences Rhéologie et Acoustique qui semblaient à première vue si indépendantes.

I. L'acoustique en rhéologie

Nous n'étudierons ici que les milieux viscoélastiques linéaires.

A. Propagation d'onde dans un milieu viscoélastique linéaire à une dimension

On sait qu'on peut supposer, pour rendre compte du comportement viscoélastique d'un corps, que la relation entre σ et ε est différentielle, c'est-à-dire que l'équation de constitution qui le caractérise peut s'écrire:

$$f(\sigma, \sigma^{\cdot}, \sigma^{\cdot\cdot} \ldots \varepsilon \varepsilon^{\cdot} \varepsilon^{\cdot\cdot}) = 0$$

$$\sigma^{\cdot}, \sigma^{\cdot\cdot} \equiv \frac{d\sigma}{dt}, \quad \frac{d^2\sigma}{dt}. \qquad [1]$$

Cette hypothèse écarte les phénomènes tels que l'hystérésis, la capillarité, etc. ... Dans un certain domaine, la relation peut être considérée comme linéaire, d'où:

$$f(\sigma, \varepsilon) = \sum_{i=0}^{n} A_i \frac{\partial^i \sigma}{\partial t^i} - B_i \frac{\partial^i \varepsilon}{\partial t^i} \equiv 0. \qquad [2]$$

En excluant tout phénomène qui se traduit par une fonction à croissance plus rapide que celle d'une exponentielle, on peut appliquer à l'éq. [2] la transformation de *Laplace*. On aura:

$$A(p)\,\bar{\sigma} - B(p)\,\bar{\varepsilon} = 0$$

où $A(p)$ et $B(p)$ sont deux polynômes en p

$$\bar{\sigma} = Q(p)\,\bar{\varepsilon}$$

et

$$Q(p) = \frac{B(p)}{A(p)} = \frac{\sum_{i=0}^{n} B_i p^i}{\sum_{i=0}^{n} A_i p^i}. \qquad [3]$$

On remarque que l'équation de constitution d'un corps viscoélastique qui lie les transformées de la contrainte σ et de la déformation ε est formellement semblable à celle d'un corps élastique

$$\begin{cases} \bar{\sigma} = Q(p)\,\bar{\varepsilon} & \text{corps viscoélastique linéaire} \\ \sigma = E\varepsilon & \text{corps élastique linéaire} \end{cases}$$

Onde élastique

Considérons un long barreau de très petite section découpé dans le corps à étudier et supposons qu'on sache y exciter une onde longitudinale; à chaque instant t le point matériel qui avait pour abscisse x deviendra: $x + u(x, t)$.

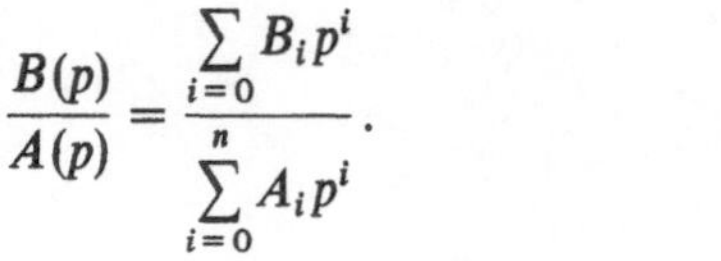

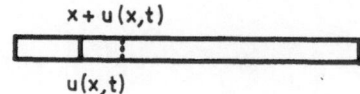

Fig. 1

De ce fait, le corps subira au point considéré une déformation

$$\varepsilon(x,t) = \frac{\partial}{\partial x} u(x,t) \qquad [4]$$

il en résultera une contrainte telle que:

$$\frac{\partial}{\partial x}\sigma(x,t) = \varrho \frac{\partial^2 u}{\partial t^2}(x,t) \qquad (\varrho\text{: densité du corps}). \qquad [5]$$

Pour un corps élastique, la relation contrainte déformation s'écrit: $\sigma = E\varepsilon$, E étant une constante on pourra écrire l'équation classique de propagation d'une onde longitudinale dans une tige élastique:

$$\boxed{\varrho \frac{\partial^2 u}{\partial t^2} = E \frac{\partial^2 u}{\partial x^2}}\,.$$

La propagation de l'onde longitudinale se fera avec une vitesse $c = \sqrt{\frac{E}{\varrho}}$.

Onde viscoélastique

Dans le cas viscoélastique, la relation contrainte déformation s'écrit:

$$\bar{\sigma} = Q(p)\,\bar{\varepsilon}.$$

En dynamique $p = i\omega$ (en supposant par exemple que la perturbation peut être représentée par une série de *Fourier*), il vient:

$$\sigma(i\omega) = E^*(i\omega)\,\varepsilon(i\omega). \qquad [6]$$

E^* s'appellera le module complexe. Les équations [4] et [5] deviennent:

$$\varepsilon(i\omega) = \partial_x u(i\omega)$$

$$\partial_x \sigma = \varrho\omega^2 u(i\omega).$$

En posant:

$$K^2 = -\frac{\varrho\omega^2}{E^*(i\omega)}; \qquad K = i\omega\sqrt{\frac{\varrho}{E^*(i\omega)}}$$

l'éq. [6] s'écrira:

$$\partial_{xx}\bar{u} - k^2\bar{u} = 0$$

$$\bar{u} = \bar{u}_0 e^{-kx}$$

$\bar{u}_0$ a pour image $u_0 e^{i\omega t}$ et $u = u_0 e^{-kx + i\omega t} = u_0 e^{i\omega}\left(t - \frac{kx}{i\omega}\right)$.

Si on pose $c^* = \sqrt{E^*(i\omega)/\varrho}$, c'est une vitesse (en élasticité $c = \sqrt{E/\varrho}$) qui varie avec la fréquence

$$u = u_0 e^{i\omega\left(t - \frac{x}{c^*}\right)}$$

E^* étant un module complexe, on peut écrire

$$E^* = |E_0|\,e^{i\delta} = E_0(\cos\delta + i\sin\delta)$$

$$E_1 = E_0\cos\delta$$

$$E_2 = E_0\sin\delta$$

or

$$c^{*-1} = \sqrt{\frac{\varrho}{E_0}}\,e^{-i\delta} = \sqrt{\frac{\varrho}{E_0}}\left(\cos\frac{\delta}{2} - i\sin\frac{\delta}{2}\right) = \frac{1}{c} - \frac{i\alpha}{\omega}$$

α et c réels, et la déformation u se propage de façon que:

$$u = u_0 e^{-\alpha x + i\omega\left(t - \frac{x}{c}\right)} \qquad [7]$$

avec une vitesse

$$c = \sqrt{\frac{|E_0|}{\varrho}}\,\frac{1}{\cos\frac{\delta}{2}} \qquad \operatorname{tg}\frac{\delta}{2} = \frac{\alpha c}{\omega}$$

et un amortissement

$$\alpha = \sqrt{\frac{\varrho}{E_0}}\sin\frac{\delta}{2} \qquad \operatorname{tg}\frac{\delta}{2} = \frac{\alpha c}{\omega}.$$

Par conséquent, si on connaît la vitesse de propagation c et l'amortissement α d'une onde progressive, on peut déterminer les parties réelles et imaginaires du module complexe E^* ainsi que les coefficients A_i et B_i de l'équation de constitution des corps viscoélastiques linéaires car on a,

$$[Q(p)]_{i\omega} = E^* = \frac{B_0 + i\omega B_1 + \cdots (i\omega)^n B_n + \cdots}{A_0 + i\omega A_1 + \cdots (i\omega)^n A_n + \cdots} = E_1 + iE_2.$$

Il suffit de connaître E_1 et E_2 en fonction de la fréquence pour avoir autant d'équations qu'on a de paramètres à déterminer, et de résoudre ces équations par une calculatrice électronique. Il existe plusieurs méthodes expérimentales qui permettent de déterminer la vitesse de propagation et l'amortissement d'une onde progressive. Nous n'en citerons ici que deux: la méthode de phase (1) utilisant des ondes entretenues, et la méthode d'ébranlement par choc (2).

a) Méthode de phase

La méthode ayant été longuement décrite, nous en dirons que quelques mots. Si on entretient dans une tige mince viscoélastique de longueur l des vibrations longitudinales à une extrémité et qu'on mesure le déphasage entre le point x_0 et le point d'abscisse x, on peut écrire en supposant que l'onde se propage comme une onde amortie [7]

$$\operatorname{tg}(\theta_0 - \theta_x) = \operatorname{tg}\frac{\omega x}{c}\,\operatorname{th}\alpha x. \qquad [8]$$

En faisant varier x, on obtient une courbe qui donne directement la valeur de c (par les zéro de la courbe) et on détermine α par une calculatrice électronique à partir des autres points expérimentaux. On vérifie que la courbe expérimentale est bien de la forme de l'éq. [8], ce qui montre que la propagation se fait bien suivant la loi [7].

b) Méthode d'ébranlement par choc

On produit à l'extrémité B d'une tige viscoélastique une percussion axiale qui engendre la propagation d'une impulsion de compression. Si la pression n'excède pas la limite d'élasticité du barreau la vitesse de propagation d'une onde plane sera donnée par $c = \sqrt{E/\varrho}$. Cette relation suppose que la pression est uniforme sur toute section transversale du barreau: ceci ne peut être vrai que pour des impulsions longues par rapport au diamètre de la barre. Pendant le temps dt où la force est appliquée, une longueur $c\,dt$ est comprimée et le reste est au repos. L'impulsion de contrainte se propagera le long de la barre avec la vitesse c. Lorsque cette impulsion atteint l'autre extrémité libre A du barreau, elle se réfléchit et on montre qu'elle donne en retour une impulsion de dilatation laquelle se réfléchit à son tour sur l'extrémité initiale B en donnant une nouvelle impulsion de compression et ainsi de suite.

En réalité, au cours de sa propagation dans la barre, la longueur de l'impulsion croît avec la distance parcourue comme conséquence du frottement interne; nous en tirerons un procédé de calcul de l'amortissement α. Les impulsions sont produites à une extrémité du barreau soit par un marteau pendulaire, soit par un canon pneumatique qui tire des billes d'acier. Les déplacements de l'autre extrémité sont décelés par un capteur capacitif. Le montage expérimental permet d'enregistrer en fonction du temps le déplacement de l'extrémité d'un barreau lorsque l'autre extrémité reçoit une percussion axiale. On en tire directement la vitesse de propagation de l'onde. Par différentiation de la fonction déplacement, on obtient la forme de l'impulsion de contrainte et l'on en déduit l'amortissement dans le matériau étudié.

Si on connaît c et α la détermination des A_i et B_i de l'équation de constitution se ramène au premier cas étudié.

B. Propagation d'onde dans un milieu viscoélastique à trois dimensions

La loi générale de *Hooke* pour un corps élastique s'écrit:

$$\sigma_{ij} = C_{ijkl}\,\varepsilon_{kl}.$$

Or, pour un corps viscoélastique à une dimension on peut écrire

$$\sigma(t) = \int_0^t E(t-\tau)\frac{d\varepsilon}{d\tau}\,d\tau.$$

On admettra que cette relation se généralise dans la viscoélasticité à trois dimensions en remplaçant σ et ε par des tenseurs σ_{ij}, ε_{ij} et E par un tenseur du 4e ordre fonction du temps C_{ijkl} qui est la généralisation du tenseur constant de *Hooke*

$$\sigma_{ij}(t) = \int_0^t C_{ijkl}(t-\tau)\frac{d\varepsilon_{kl}}{d\tau}\,d\tau = 0$$

$$= \int_0^t C_{ijkl}(\tau)\frac{d\varepsilon_{kl}}{d\tau}(t-\tau)\,d\tau.$$

Faisons subir à cette relation la transformation de *Laplace*

$$\int_0^\infty e^{-pt}\sigma_{ij}(t)$$

$$= \int_0^\infty\left[\int_0^t C_{ijkl}(\tau)\frac{d\varepsilon_{kl}(t-\tau)}{d\tau}\,d\tau\right]e^{-pt}\,dt.$$

C'est un produit de convolution qui donne:

$$\bar{\sigma}_{ij} = [p\,C_{ijkl}(p)]\,\varepsilon_{kl}.$$

Si on pose $\tilde{C}_{ijkl} = p\,C_{ijkl}(p)$, on aura: $\bar{\sigma}_{ij} = p\,C_{ijkl}\bar{\varepsilon}_{kl} = \tilde{C}_{ijkl}\bar{\varepsilon}_{kl}$

$$\bar{\sigma}_{ij} = \tilde{C}_{ijkl}\bar{\varepsilon}_{kl}.$$

Cette relation est formelle à la relation contrainte déformation des corps élastiques. Or, en élasticité, l'équation qui régit le mouvement est donnée par:

$$(\lambda + 2\mu)\,\mathrm{grad\,div}\,\vec{u} - \mu\,\mathrm{rot}\,(\mathrm{rot}\,\vec{u}) - \frac{\varrho\,\partial^2 \vec{u}}{\partial t^2} = 0\,.$$

Si l'on pose: $\vec{u} = \vec{u}_1 + \vec{u}_2$; $\vec{u}_1 = \mathrm{grad}\,\varphi$; $\vec{u}_2 = \mathrm{rot}\,\psi$ l'équation donne deux et seulement deux équations pour φ et ψ

$$(\lambda + 2\mu)\,\Delta\varphi = \varrho\,\frac{\partial^2\varphi}{\partial t^2}$$

$$\mu\,\Delta\psi = \varrho\,\frac{\partial^2\psi}{\partial t^2}$$

qui sont des équations d'ondes, l'une longitudinales et l'autre transversales se propageant avec une vitesse

$$C_L = \sqrt{\frac{\lambda + 2\mu}{\varrho}}$$

$$C_T = \sqrt{\frac{\mu}{\varrho}}\,.$$

La solution du problème est donc une combinaison linéaire des deux ondes des types L et T. Ce théorème est plus connu sous le nom de théorème de *Poisson*. En viscoélasticité, les équations de constitution étant formellement semblables, on peut écrire l'équation qui régit le mouvement:

$$(\tilde{\lambda} + 2\tilde{\mu})\,\mathrm{grad\,div}\,\vec{\bar{u}} - \tilde{\mu}\,\mathrm{rot}\,(\mathrm{rot}\,\vec{\bar{u}}) - \varrho p^2 \vec{\bar{u}} = 0\,.$$

Il y a donc bien deux vitesses de propagation

$$C_L = \sqrt{\frac{\tilde{\lambda} + 2\tilde{\mu}}{\varrho}} \quad \text{avec} \quad \tilde{\lambda} = p\lambda(p)$$

$$C_T = \sqrt{\frac{\tilde{\mu}}{\varrho}} \quad \text{avec} \quad \tilde{\mu} = p\mu(p)\,.$$

Le théorème de *Poisson* s'explique en viscoélasticité si on pose:

$$\tilde{\lambda} + 2\tilde{\mu} = E_0 e^{i\delta}\,; \quad \tilde{\mu} = G_0 e^{i\delta}\,.$$

Le même raisonnement que pour la propagation à une seule dimension montre que deux ondes se propagent avec des vitesses C_L et C_T telles que

$$u_L = u_0\, e^{-\alpha_L x + i\omega\left(t - \frac{x}{C_L}\right)}$$

$$u_T = u_0\, e^{-\alpha_T x + i\omega\left(t - \frac{x}{C_T}\right)}$$

avec:

$$C_L = \sqrt{\frac{E_0}{\varrho}}\,\frac{1}{\cos\frac{\delta}{2}} \qquad \alpha_L = \omega\sqrt{\frac{\varrho}{E_0}}\,\sin\frac{\delta}{2}$$

$$C_T = \sqrt{\frac{G_0}{\varrho}}\,\frac{1}{\cos\frac{\delta}{2}} \qquad \alpha_T = \omega\sqrt{\frac{\varrho}{G_0}}\,\sin\frac{\delta}{2}\,.$$

Des études tant théoriques qu'expérimentales ont été faites sur la propagation des ondes dans un barreau de PMM (3). On montre qu'on peut éliminer l'une des propagations et tout se passe comme si l'on avait affaire à une propagation uniaxiale.

Ainsi nous avons montré que l'on pouvait déterminer le comportement rhéologique d'un corps à partir d'équations purement acoustiques. Inversement, nous allons voir comment la rhéologie est utilisée en acoustique.

II. La rhéologie en acoustique

On a pu se rendre compte et surtout depuis la découverte des hauts polymères, de l'importance que les acousticiens accordaient aux matériaux qu'ils employaient.

A. Montage antivibratile

On sait que si une machine fait vibrer par l'intermédiaire de son socle le plancher sur lequel elle repose, on peut diminuer d'une façon notable ces vibrations par l'intermédiaire d'amortisseurs (silent-bloc). Or, ceux-ci ne sont pas autre chose que des élastomères armés ou non dont les caractéristiques mécaniques doivent être connues pour être employés judicieusement. Or, les méthodes acoustiques décrites au premier paragraphe permettent justement de déterminer d'une façon sûre ces caractéristiques. Le montage antivibratile le plus simple que l'on connaît peut être donné par le schéma ci-dessous:

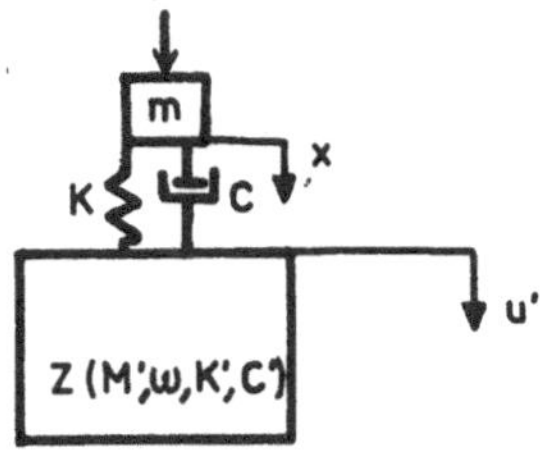

Fig. 2

Soit une fondation A sur laquelle repose un moteur m par l'intermédiaire d'un amortisseur dont le modèle mécanique peut être représenté par un ressort de raideur E et un «dashpot» de viscosité η disposé en parallèle (modèle de *Voigt*). Si u est l'amplitude des vibrations de la fondation A sans amortisseur et u' celle avec, on définit un rapport $R = u/u'$ appelé efficacité du montage. Or, pour ce montage, R est donné par (4):

$$R = \frac{u'}{u} = \sqrt{\frac{(E^2 + \eta^2\omega^2 - \gamma E)^2 + \gamma^2\eta^2\omega^2}{(E^2 + \eta^2\omega^2)^2}}$$

en posant:

$$\gamma = \frac{Zm\omega^2}{Z - m\omega^2}.$$

Par conséquent, pour connaître R à une fréquence donnée, il faut déterminer E et η de l'amortisseur. Or, la détermination de E et η revient à connaître les constantes de l'équation de constitution de ce modèle, soit:

$$\sigma - E\varepsilon - \eta\dot{\varepsilon} = 0.$$

C'est-à-dire de la forme [2]:

$$\sigma - B_0 E - B_1\dot{\varepsilon} = 0$$

avec:

$$A_0 = 1; \quad B_0 = E; \quad B_1 = \eta$$

et en les déterminant par les méthodes expérimentales décrites plus haut. Il est bien évident que les «silent-bloc» employés ne correspondent pas à ce modèle simple, mais le principe reste valable; il suffit de connaître les constantes A_i et B_i du matériau employé comme amortisseur pour connaître «l'efficacité» du montage. En outre, on connaît l'intérêt que représentent les revêtements viscoélastiques quand il s'agit d'atténuer les vibrations transversales de tôles. Les travaux de *M. Jullien* ont montré l'importance de la détermination des A_i et B_i de l'équation de constitution de ces revêtements et comment on peut les utiliser pour réduire les vibrations parasites; nous n'insisterons pas davantage sur ces travaux qui font intervenir la notion de non linéarité en viscoélasticité.

B. Transparence acoustique des parois

Une première approximation grossière de la transparence acoustique d'une cloison avait été obtenue en supposant que celle-ci se comportait comme un piston rigide. Ce résultat a pu d'abord être amélioré en considérant la cloison comme une plaque élastique fixée sur son pourtour. Enfin, on se rapproche encore plus des résultats de l'expérience, surtout aux hautes fréquences en prenant en considération l'amortissement interne du matériau composant la cloison. Les résultats obtenus par *F. Spronck* (5) (fig. 3), en considérant les vibrations transversales d'une plaque viscoélastique sont particulièrement démonstratifs à cet égard. De là, la nécessité de connaître l'amortissement global d'une cloison revêtue de matériaux viscoélastiques. C'est ainsi que parallèlement aux expérimentateurs les théoriciens ont obtenu ces dernières années des résultats très encourageants en ce qui concerne l'étude des propriétés élastiques et viscoélastiques des matériaux hétérogènes anisotropes et «composites .

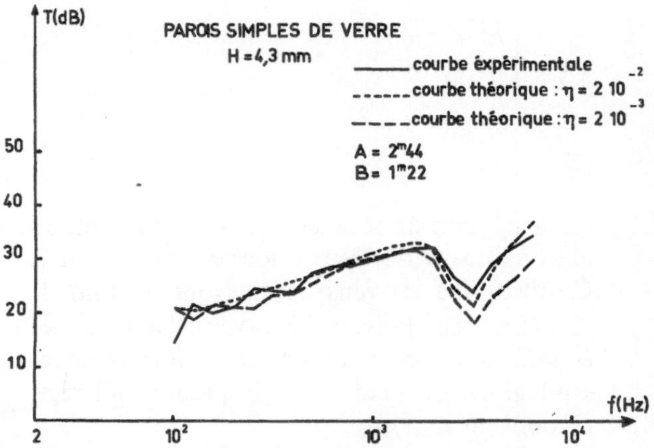

Fig. 3. Influence du coefficient d'amortissement η [cf. Thèse, *F. Spronck* (5)] sur la transparence en champs diffus de parois simples

C. Effet de filtrage des corps viscoélastiques

Dans une conférence générale présentée au 5e Congrès d'Acoustique (6) à Liège, *M. Vogel* a parlé de l'utilisation des cordes viscoélastiques comme vibrateurs filtrants. On sait, en effet, que les cordes élastiques ont des fréquences de résonance couvrant toute la gamme des fréquences au-delà de la fondamentale et qu'il n'en est plus de même pour une corde viscoélastique, car celle-ci a une relation de constitution comportant des dérivées temporelles d'ordre impair. Si on considère par exemple un élastomère dont la loi de constitution est de la forme:

$$A_0\sigma + A_1\dot{\sigma} - B_0\varepsilon - B_1\dot{\varepsilon} - B_2\ddot{\varepsilon} = 0$$

on montre que suivant les valeurs de certaines constantes B_i, il y aura des bandes de fréquence bien délimitées en dehors desquelles la corde ne pourra plus vibrer librement. En conclusion, en pouvant choisir convenablement les A_iB_i d'un corps viscoélastique, on pourra constituer des

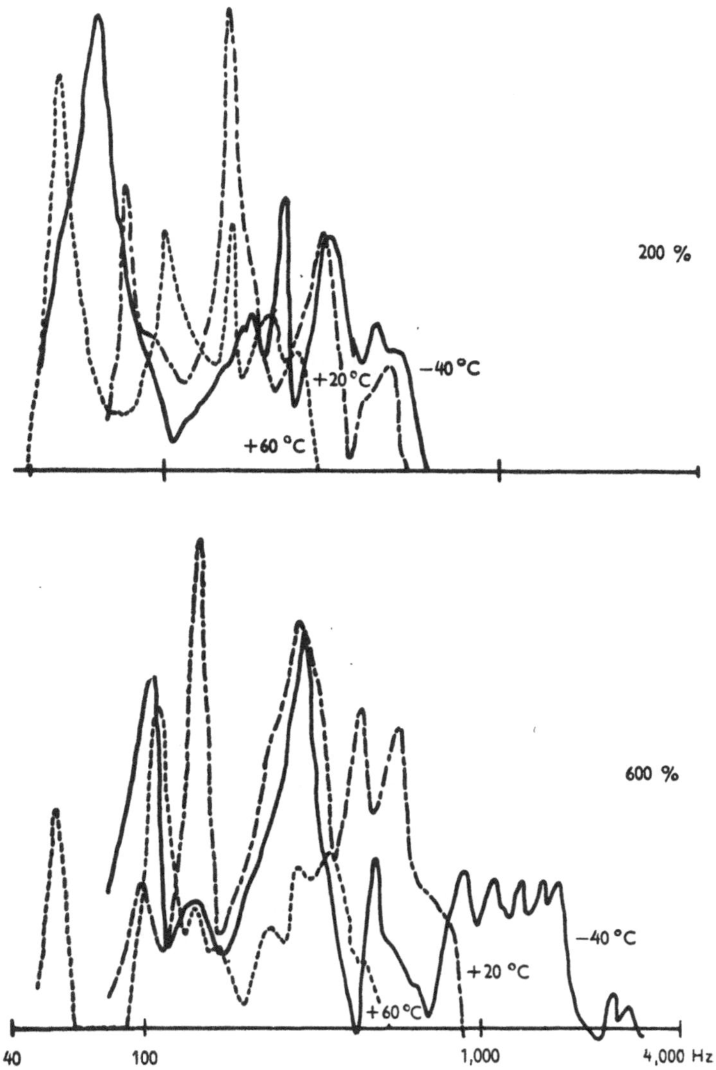

Fig. 4. Bandes passantes pour une même corde en caoutchouc, sous deux étirements et à trois températures différentes [cf. (6), *Vogel, Th.*]

filtres de fréquences. La fig. 4 montre l'effet de filtrage par une corde de caoutchouc dans différentes conditions d'étirement et de température.

Ces quelques exemples montrent les services que chacune de ces deux disciplines est susceptible de rendre à l'autre. Il serait facile de les multiplier si le temps ne nous était mesuré.

Littérature

1) *Bismuth, W.*, Etude du comportement des corps viscoélastiques par une méthode de phase. Thèse, Doct. ès-Sci. Phys., Fac. Sci. Marseille, 1962.

2) *Blanc, R. H.*, Détermination de l'équation de comportement des corps viscoélastiques linéaires par une méthode d'impulsion. Thèse, Doct. ès-Sci. Math., Fac. Sci. Marseille, 1971.

3) *Bonsignour, D.*, Contribution à l'étude des propriétés viscoélastiques d'un polymère solide. Thèse, Doct. ès-Sci. Phys., Fac. Sci. Marseille, 1970.

4) *Jullien, Y.*, Réduction des vibrations mécaniques dues aux machines. Colloque R.I.L.E.M., Budapest, Mesure et interprétation des effets dynamiques et vibrations des constructions, vol. 1, p. 65 (1963).

5) *Spronck, F.*, Transparence au son de parois minces viscoélastiques finies et infinies, simples et multiples. Thèse, Doct. spéc. d'Acoustique, Fac. Sci. Marseille, 1971.

6) *Vogel, Th.*, Vibrations de solides minces. Extrait des Rapports du 5e Congrès International d'Acoustique, Liège 1965, vol. II: Conférences Générales.

Adresse de l'auteur:
Dr. W. Bismuth
Centre de Recherches Physiques
31, Chemin Joséph Aiguier
F-13274 Marseille (France)

Rheol. Acta 12, 404–411 (1973)

From the School of Physics, The University, Newcastle upon Tyne (U.K.)

Isotropic composites with elastic or viscoelastic phases: General bounds for the moduli and solutions for special geometries

By R. Roscoe

With 4 figures

(Received October 27, 1972)

1. Introduction

This work is concerned with the mechanical properties of composite materials consisting of two phases which have linear isotropic elastic or viscoelastic properties, the geometry of the phases being such that the composites are themselves mechanically isotropic. The results for the viscoelastic case should have application to filled polymeric materials and composites formed from mixtures of two incompatible polymers.

2. Bounds on the moduli of elastic composites

Upper and lower bounds on the rigidity modulus μ of a composite of two phases, with rigidity and bulk moduli $\mu_1, \varkappa_1$ and $\mu_2, \varkappa_2$ respectively, have been obtained by *Hashin* and *Shtrikman* (1963). These are valid for any phase geometry which gives isotropic properties to the composite as a whole, provided the phase with the larger rigidity also has the larger bulk modulus. Taking $\mu_2 > \mu_1$ and $\varkappa_2 > \varkappa_1$ these bounds may be written

$$\mu_1 + \frac{(\mu_2 - \mu_1)\,c_2}{1 + \dfrac{(\mu_2 - \mu_1)\,c_1}{\mu_1 + \mu_g^*}} \geq \mu \geq \mu_1 + \frac{(\mu_2 - \mu_1)\,c_2}{1 + \dfrac{(\mu_2 - \mu_1)\,c_1}{\mu_1 + \mu_l^*}} \qquad [1]$$

where

$$\mu_g^* = \frac{\mu_2}{6}\,\frac{9\varkappa_2 + 8\mu_2}{\varkappa_2 + 2\mu_2}$$

$$\mu_l^* = \frac{\mu_1}{6}\,\frac{9\varkappa_1 + 8\mu_1}{\varkappa_1 + 2\mu_1} \qquad [2]$$

and c_1 is the volume concentration of phase 1 and $c_2 = 1 - c_1$ is that of phase 2. The general case has been treated by *Walpole* (1966) who has shown that, when $\mu_2 > \mu_1$, $\varkappa_2 < \varkappa_1$, eq. [2] only has to be modified by interchanging $\varkappa_1$ and $\varkappa_2$, leaving μ_1 and μ_2 unchanged. Either case gives the same result when both phases are incompressible. Somewhat similar bounds on the bulk modulus $\varkappa$ have been obtained by *Hashin* and *Shtrikman* (1963) for their case, and by *Hill* (1963) for the general case.

These bounds are close when the differences between the corresponding moduli of the phases are small. When the moduli are very different, however, the bounds become very wide so that it is important to ascertain whether more restrictive bounds can be obtained. In fact it has already been shown by *Hashin* and *Shtrikman* (1963) that their bounds on the bulk modulus, valid when $\mu_2 > \mu_1$, $\varkappa_2 > \varkappa_1$, are the best possible (i.e. most restrictive) bounds. This was demonstrated by consideration of a particular phase geometry originally proposed by *Hashin* (1962) in which one phase is in the form of spheres of different sizes, each surrounded by a spherical shell of the other phase with an outer radius bearing a constant ratio to the sphere radius, the whole space of the composite being filled with such spheres with enclosing shells. An exact expression for the bulk modulus was obtained which coincides with the lower bound when the spheres consist of the phase with the higher bulk modulus, and with the upper bound when they consist of the phase with the lower modulus. The corresponding result that [1] gives the best possible bounds on the rigidity (when $\mu_2 > \mu_1$, $\varkappa_2 > \varkappa_1$) will now be derived.

If a small concentration δc of very flat oblate spheroidal particles with moduli μ_1, $\varkappa_1$ are embedded at random in a homogeneous matrix with moduli μ, $\varkappa$ the moduli of the composite are $\mu + \delta\mu$, $\varkappa + \delta\varkappa$, where

$$\delta\mu = -(\mu - \mu_1)\left\{1 + \frac{6}{5}\frac{\varkappa_1 + 2\mu_1}{3\varkappa_1 + 4\mu_1}\frac{\mu - \mu_1}{\mu_1}\right\}\delta c + 0(\delta c)^2 \quad [3]$$

$$\delta\varkappa = -(\varkappa - \varkappa_1)\frac{3\varkappa + 4\mu_1}{3\varkappa_1 + 4\mu_1}\delta c + 0(\delta c)^2. \quad [4]$$

These results are obtainable from the general treatment of the ellipsoidal inhomogeneity given by *Eshelby* (1961). Now let μ_l be the best possible lower bound to the rigidity of a composite having concentrations c_1 of phase 1 and c_2 of phase 2 with $\mu_2 > \mu_1$. Then it must be possible to find such a composite which can be regarded as homogeneous on a sufficiently large scale with a rigidity equal to (or indefinitely close to) μ_l. If a concentration δc of sufficiently large flat oblate spheroidal particles of phase 1 is added to this, the change in rigidity is given by [3] with μ set equal to μ_l, while the increase δc_1 in concentration of phase 1 is

$$\delta c_1 = (1 - c_1)\,\delta c. \quad [5]$$

The reduction in rigidity cannot be greater than the reduction in μ_l for this increase in c_1, so

$$-\frac{d\mu_l}{dc_1} \geq \frac{\mu_l - \mu_1}{1 - c_1}\left\{1 + \frac{6}{5}\frac{\varkappa_1 + 2\mu_1}{3\varkappa_1 + 4\mu_1}\frac{\mu_l - \mu_1}{\mu_1}\right\}. \quad [6]$$

It follows from [1] that $\mu_l > \mu_1$, hence the expression on the right-hand side of [6] is positive, so the result may be re-written as

$$\frac{d}{dc_1}\log_e\left\{\frac{\mu_1}{\mu_l - \mu_1} + \frac{6}{5}\frac{\varkappa_1 + 2\mu_1}{3\varkappa_1 + 4\mu_1}\right\} \geq \frac{1}{1 - c_1}. \quad [7]$$

The term on the right-hand side of [6] is again positive, and $\mu_l = \mu_2$ when $c_1 = 0$, so a further result can be obtained by integrating both sides of [7]. On simplification, this gives the final result that μ_l is not greater than the expression on the right-hand side of [1], with μ_l^* given by [2], which is a lower bound to μ provided $\mu_2 > \mu_1$, $\varkappa_2 > \varkappa_1$. Hence that expression is actually the best possible lower bound in this case.

By considering the effect of adding a small concentration of large flat oblate spheroids of phase 2 to a composite of the two phases which has a rigidity equal to (or indefinitely close to) the best possible upper bound, it may be shown that the latter is given by the left-hand side of [1] with μ_g^* given by [2], provided $\mu_2 > \mu_1$, $\varkappa_2 > \varkappa_1$. The argument follows the same lines as in the case of the lower bound. A similar method may be employed using [4] to re-derive *Hashin* and *Shtrikmans* result that their bounds for $\varkappa$ are the best possible bounds in the case to which they apply.

It is an open question whether *Walpoles* bounds for the rigidity and *Hills* bounds for the bulk modulus are the best possible bounds for the case of $\mu_2 > \mu_1$, $\varkappa_2 < \varkappa_1$. This case does not often occur in actual composites, however, since large rigidity is usually associated with large bulk modulus.

3. Bounds on the complex moduli of viscoelastic composites

When a viscoelastic material is subjected to sinusoidally varying strain, the relation between complex stress and complex strain is formally the same as that between stress and strain in an elastic material, but the moduli are now complex quantities. Thus any algebraic expressions for the moduli of an elastic composite derived from the fundamental equations of elasticity also apply to the complex moduli of a viscoelastic composite with the same geometry.

The complex moduli of composites and their phases are expressed in terms of their real and imaginary parts, e.g.

$$\mu = \mu' + i\mu'', \qquad \varkappa = \varkappa' + i\varkappa''. \quad [8]$$

Bounds on these real and imaginary parts for a viscoelastic composite can be obtained directly from the known bounds on the moduli of elastic composites in the following special cases: when the composite consists of incompressible phases having the same loss tangent ($\mu_2''/\mu_2' = \mu_1''/\mu_1'$), when it consists of an incompressible matrix with rigid, incompressible inclusions and when it consists of an incompressible matrix surrounding voids. The more complicated general case has been treated by *Roscoe* (1972) who has shown that bounds may be set to the parts of the moduli of any viscoelastic composite in terms of the moduli of certain elastic composites with the same phase geometry. Since bounds are known for elastic composites of unspecified geometry, this result provides bounds on the parts of the moduli for a viscoelastic composite of unspecified geometry. These, like the bounds for elastic cases, become wide when the moduli of the phases are very different. In fact there is reason to suppose that they are not best possible bounds, even when based on best possible bounds for elastic cases, but it is doubtful whether an improved treatment would yield much closer results. It would seem, therefore, that further work should be directed to the study of special phase geometries which may approximate to those found in actual composites, and some of these are considered in the rest of this paper.

4. Solutions for special phase geometries

The problem of an elastic composite containing a small concentration of ellipsoidal particles of one phase embedded in a matrix of the other can be solved to give algebraic expressions for the moduli correct to the first power of the concentration (except in some special cases, such as that in which a finite proportion of the particles are in contact). The results for flat spheroids quoted in equations [3] and [4] provide an example. *Walpole* (1972) has obtained expressions correct to the second power of the concentration for the case of regularly spaced spherical inclusions of uniform size.

All these results are applicable when the moduli are complex but they are not of much practical value since they are restricted to small concentrations of the included phase. It is therefore necessary to look for solutions which are valid, if only to a good approximation, for finite concentrations. One such case appears in the literature: the exact result for the bulk modulus of a special composite, derived by *Hashin* (1962), which has already been mentioned. Unfortunately it is not possible to derive an exact expression for the rigidity with this phase geometry (which is not entirely specific). The following sections of this paper are therefore devoted to a consideration of certain other cases of spherical inclusions, for which algebraic expressions for the rigidity can be obtained.

5. Spherical inclusions with a wide range of sizes

Consider first a composite consisting of a matrix (μ_1, $\varkappa_1$) in which is embedded a small concentration δc of spheres (μ_2, $\varkappa_2$). If a further concentration δc of much larger spheres of the same material is then embedded in this composite, the latter can be regarded as a homogeneous material in calculating the moduli of the new composite. A finite concentration of spheres can be built up by adding successive sets of larger and larger spheres. At any stage in this process the next addition of spheres increases the moduli of the composite ($\mu, \varkappa$) by the amounts

$$\delta\mu = \frac{(\mu_2 - \mu)\,\delta c}{1 + \dfrac{6}{5}\dfrac{\varkappa + 2\mu}{3\varkappa + 4\mu}\dfrac{\mu_2 - \mu}{\mu}} + 0(\delta c)^2 \qquad [9]$$

$$\delta\varkappa = (\varkappa_2 - \varkappa)\frac{3\varkappa + 4\mu}{3\varkappa_2 + 4\mu}\,\delta c + 0(\delta c)^2 \qquad [10]$$

and the increase in the total concentration of spheres δc_2 is $(1 - c_2)\,\delta c$. If δc is made smaller and smaller, the rate of change of rigidity with concentration becomes

$$\frac{d\mu}{dc_2} = \frac{\mu_2 - \mu}{1 - c_2}\,\frac{1}{1 + \dfrac{6}{5}\dfrac{\varkappa + 2\mu}{3\varkappa + 4\mu}\dfrac{\mu_2 - \mu}{\mu}}. \qquad [11]$$

When both phases have the same bulk modulus, it follows from [10] that the composite also has the same bulk modulus, and [11] can then be integrated to obtain an algebraic relation for μ. A simpler case arises, however, when both phases are incompressible. Integration of [11] here gives the relation

$$\mu_2 - \mu = (\mu_2 - \mu_1)\left(\frac{\mu}{\mu_1}\right)^{2/5} c_1. \qquad [12]$$

For this result to be exact for a finite concentration c_2, the size range of the spheres must extend from an arbitrarily chosen maximum down to zero, as in the rather different geometry proposed by *Hashin* (1962), but its validity will not be seriously affected by introducing a sufficiently low cut-off in the size distribution. It must be remembered, however, that the formula has only been shown to hold for a particular type of size distribution and will not necessarily apply to all wide size distributions.

The method used here is a generalisation of that employed by *Brinkman* (1952) and by *Roscoe* (1952) in order to obtain an expression for the viscosity of a suspension of rigid spheres with this type of geometry.

6. Spherical inclusions of uniform size

As already mentioned, *Walpole* (1972) obtained an expression for the rigidity of a composite with regularly spaced spherical inclusions, which is correct to the second power of the concentration. Here it will be shown that an expression correct to the third power can be obtained in the case of a composite with still more specialised geometry. This composite is built up of a large number of domains, each containing many spherical inclusions with their centres situated at the lattice points of a face-centred cubic lattice, the lattices for the various domains being randomly oriented so that the composite as a whole is mechanically isotropic.

It is first necessary to determine the relation between stress and strain in a single domain, and for this purpose use is made of a number of general results obtained by *Eshelby* (1961). Thus it may be shown that a certain continuous displacement throughout a 2-phase composite gives rise to a strain field of the form

$$e'_{ij} = e^A_{ij} - 4C(1-\sigma)\,(\Phi_{,ik}e^A_{jk} + \Phi_{,jk}e^A_{ik}) + C\Psi_{,ijkl}e^A_{kl} \qquad [13]$$

which satisfies the equations for elastic equilibrium within each of the phases provided each has the same *Poissons* ratio σ. Here e^A_{ij} is a uniform deviatoric strain, Φ and Ψ are respectively the harmonic and biharmonic potentials which would result from uniform charge distributions within the two phases, and C is a constant. In this work the fictitious charge distribution is chosen to be a uniform distribution of density $-c_2/4\pi$ throughout the whole volume together with an additional uniform density of $1/4\pi$ within the spherical inclusions only, c_2 being the volume concentration of the inclusions. This choice is made so that the total "charge" in any large volume is zero, and this ensures the convergence of the summation process occurring later in the work.

In any actual strain field the surface tractions over the phase interfaces must be in balance, so it is necessary to find how closely this condition is obeyed by the strain field e'_{ij} with the best possible choice of the constant C. For this purpose, coordinate axes are chosen along the axes of the face-centred cubic lattice, and as each inclusion is surrounded by exactly the same arrangement of neighbours it is only necessary to consider the case of an inclusion centred on the origin. The surface tractions are obtainable from the strains in the phases just inside and just outside the interface, and in order to calculate these from [13] the "charge" distribution may be conveniently divided into two parts. The first part is a distribution of density $1/4\pi$ within the space of the inclusion considered (radius a) together with a density $-c_2/4\pi$ throughout the whole of the space within a concentric spherical surface of radius $a/c_2^{1/3}$ (this radius being chosen so that the total "charge" is zero). The second part consists of a distribution of density $-c_2/4\pi$ throughout the whole remaining space of the domain together with an additional density $1/4\pi$ within the other inclusions.

It may readily be shown that the part of e'_{ij} due to the first part of the "charge" distribution gives surface tractions which are balanced when the constant C is given the value

$$C = -\frac{1}{4(1-\sigma)}\,\frac{\mu_2-\mu_1}{\mu_1}\,\frac{1}{1+\dfrac{(\mu_2-\mu_1)}{\mu_1+\mu_1^*}c_1} \qquad [14]$$

where μ_1^* is given by eq. [2]. On the other hand, the second part of the distribution gives the same strain at points just inside and outside the spherical interface and thus gives rise to unbalanced surface tractions (since the elastic moduli of the phases are different). This strain may be evaluated from the parts of $\Phi_{,ij}$, $\Psi_{,ijkl}$ at the interface due to the second part of the distribution, using a summation process. Consideration is first given to the effect of the "charge" distribution of density $1/4\pi$ in nearest neighbours together with that of the distribution of density $-c_2/4\pi$ in a spherical annulus of inner radius $a/c_2^{1/3}$ having such an outer radius that the total charge now considered is zero. Then the effect of next nearest neighbours is considered together with that of a corresponding annulus, and so on. (In fact while the uniform "charge" in each annulus affects the values of Φ and Ψ, it leaves $\Phi_{,ij}$ and $\Psi_{,ijkl}$ unaltered).

For sufficiently small values of c_2, all effects of order $(a/r_1)^5$ may be neglected in this summation, a being the radius of the inclusions and r_1 the centre-to-centre distance of nearest neighbours. The total effect of the second part of the "charge" distribution is then found to consist of the addition of a strain δe_{ij} at all points adjacent to the interface with components given by

$$\delta e_{ij} = 3ke^A_{ij} \quad \text{when} \quad i = j \qquad [15]$$

$$\delta e_{ij} = -2ke^A_{ij} \quad \text{when} \quad i \neq j \qquad [16]$$

where

$$k = (1.21\ldots)\,C\,\frac{a^3}{r_1^3} = (.204\ldots)\,Cc_2\,. \qquad [17]$$

Thus if a uniform strain, equal and opposite to this is superposed upon e'_{ij} to give a strain

27*

$$e_{ij} = e'_{ij} - \delta e_{ij} \qquad [18]$$

throughout the domain, the surface tractions are balanced to the degree of approximation considered.

Now the volume average of e'_{ij} throughout the domain is equal to e^A_{ij}, so the volume average of e_{ij} is

$$\bar{e}_{ij} = e^A_{ij} - \delta e_{ij} \qquad [19]$$

and the volume average of the stress may be shown to be

$$\bar{p}_{ij} = 2\{\mu_l \bar{e}_{ij} + (\mu_l - \mu_1)\,\delta e_{ij}\} \qquad [20]$$

where μ_l is the expression on the right-hand side of [1] with μ_l^* given by [2]. (It may be recalled that e^A_{ij} is deviatoric, hence $\bar{e}_{ij}$ and δe_{ij} are also deviatoric.) The components of δe_{ij} can be expressed in terms of those of $\bar{e}_{ij}$ by elimination of e^A_{ij} between [15], [16] and [19]. The results may be put in the form

$$\delta e_{ij} = \frac{3k}{(1+2k)(1-3k)}\bar{e}_{ij} + \frac{6k^2}{(1+2k)(1-3k)}\bar{e}_{ij}, \quad i = j \qquad [21]$$

$$\delta e_{ij} = -\frac{2k}{(1+2k)(1-3k)}\bar{e}_{ij} + \frac{6k^2}{(1+2k)(1-3k)}\bar{e}_{ij}, \quad i \neq j. \qquad [22]$$

Evidently the material of the domain is somewhat anisotropic in its mechanical properties on account of the presence of the second term in the bracket in [20]. However, it is the average stress throughout the whole composite for a given average strain $\bar{e}_{ij}$ that is required, and this is obtained to a first approximation by averaging over all lattice orientations for a fixed $\bar{e}_{ij}$. For this purpose it is necessary to transform to a coordinate system in which the lattice axes have an arbitrary orientation, and then average over all such orientations. In the result, the first term in the bracket of [20] and the second part of δe_{ij} as given by [21] and [22] remain unchanged, but the first part of δe_{ij} contributes nothing to the averaged stress. Thus the final result for the average stress throughout the composite is

$$p_{ij} = 2\{\mu_l + (.250\ldots)(\mu_l - \mu_1)\,C^2 c_2^2 + \cdots\}\bar{e}_{ij} \qquad [23]$$

when the average strain is $\bar{e}_{ij}$. Higher powers of k or Cc_2 have been neglected in this expression, because the simple averaging process here used to obtain the average stress in the composite is only accurate when the anisotropy of the domains is small. The bracket represents the rigidity of the composite, and it may be noted that as $(\mu_l - \mu_1)$ is proportional to c_2 for small values of c_2, the second term in the bracket is of order c_2^3. The strain field in the domains given by [19] is, of course, only an approximation, because the corresponding surface tractions at phase interfaces are out of balance by fractional amounts of the order $(a/r_1^5)^5$. It may be shown, however, that improvement in the approximation gives an addition to the bracket term of [24] which is of order $c_2^{11/3}$ or even smaller.

In the case of a composite with *incompressible phases*, it is convenient to write

$$f = \frac{\mu_2 - \mu_1}{\mu_2 + 1.5\mu_1} \qquad [24]$$

and eq. [23] may then be put in the form

$$\frac{\mu}{\mu_1} = 1 + \frac{2.5\,f c_2}{1 - f c_2 - .390\,f^2 c_2^2 + \cdots}. \qquad [25]$$

The remainder in the denomination is of order $c_2^{8/3}$ or less, and for small values of f it is proportional to the second or some higher power of f. The reason for expressing μ in this way (with the series in a denominator term) is that the series terms appear to be converging much more rapidly than those in the direct expansion in powers of c_2. In the extreme case of $f = 1$ (rigid inclusions) and $c_2 = .741$ (close packing) the rigidity becomes infinite, so the remainder is here equal to $-.045$. In other cases, an outside estimate of the error in μ resulting from neglect of the remainder may be obtained by assuming the latter to vary as $f^2 c_2^{8/3}$. For equal concentrations of the phases ($c_1 = \frac{1}{2} = c_2$) this gives the error as 4% when $f = 1$ and .3% when $f = .5$. Evidently at this concentration the remainder is quite negligible except when f is close to unity.

In this composite the nearest-neighbour distance is the same for all inclusions (except for the infinitesimal proportion situated at the domain boundaries) and has the maximum value for such cases. An opposite extreme is represented by a composite in which the inclusions are bunched in large close-packed spherical aggregates. When these are sufficiently large, they may be treated as mechanically homogeneous, and for small values of f their rigidity will be given by an expression of the same form as [25] with $c_2 = .741$ and a slightly different coefficient of f^2 in the denominator. If these aggregates are taken to be distributed in the same way as the inclusions in the composite previously considered, the rigidity of the whole may be estimated using [25] with c_2 set equal to the volume concentration of the aggregates and f calculated from μ_1 and the rigidity of the aggregates, giving

$$\frac{\mu}{\mu_1} \simeq 1 + \frac{2.5 f c_2}{1 - f c_2 - .390 f^2 (c_2^2 + .549)} \qquad [26]$$

where c_2 now represents the concentration of the inclusions in the whole composite and f is given by [24]. This result only applies when f is small, and the figure .390 appearing in the denominator is only approximate. For this composite the nearest neighbour distance is the same for all inclusions and has the minimum value. It is therefore to be expected that the rigidity of actual composites in which the inclusions are distributed in some random way in space will lie between the values given by [25] and [26].

7. Application to viscoelastic composites

The algebraic expressions obtained in the last two sections may be applied to viscoelastic composites for the calculation of the complex rigidity. This quantity is usually expressed in terms of its real part μ' and the loss tangent $\tan \Phi = \mu''/\mu'$. For illustration, two hypothetical composites will be considered, both incompressible and containing equal volumes of the two phases ($c_1 = \frac{1}{2} = c_2$). For the first, the real parts of the rigidity and loss tangents for the two phases are supposed to vary with frequency as shown in figs. 1 and 2. The curves marked A give the values of μ' and $\tan \Phi$ for the composite as calculated from [12] for the case where the spherical inclusions (phase 2) have the wide size distribution. The curves marked B are calculated for the same case but with the inclusions consisting of phase 1, so as to illustrate the effect of phase inversion. Over most of the frequency range μ' for the composite lies between the corresponding values for the two phases, and it has the higher value when the continuous phase has the higher value of μ'. This effect is reversed, however, at low frequencies, and μ' for the composite can then be greater than its value for either phase. The loss tangent peak for the included phase is almost entirely suppressed in the composites.

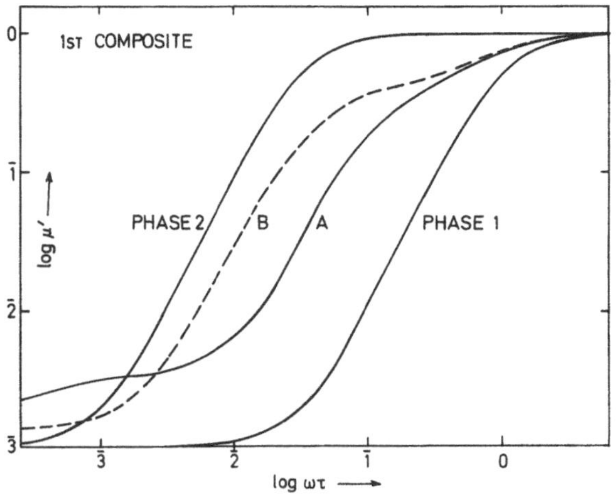

Fig. 1. Variation of the real part of the rigidity (arbitrary units) with $\omega\tau$ ($\omega = 2\pi \times$ frequency, τ an arbitrary constant). Curve A for the first composite as calculated from eq. [12], and B for the same case but with phase inversion. Also curves for the two phases

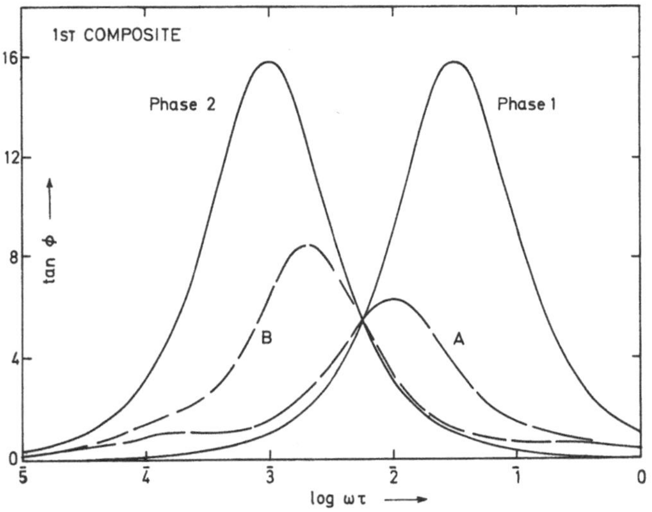

Fig. 2. Variation of the loss tangent with $\omega\tau$. Curves for the same cases as in fig. 1

Eq. [25] and [26] for spherical inclusions of equal size cannot be applied to this first composite, because there is too wide a difference in the complex rigidity of the two phases (i.e. f is close to unity) over much of the frequency range. A second composite is therefore considered with values of the real part of the rigidity and loss tangent for phases 1 and 2 as shown in figs. 3 and 4; and here μ_2' never exceeds $9\,\mu_1'$. Also shown in the figures are the curves A calculated for the wide size distribution, curves B for the same case but with phase inversion, and curves C and D calculated for spherical inclusions of equal size from [25] and [26]. Here the loss tangent peaks for both the phases are present in the curves of tan Φ for the composite. The curves *C* and *D* lie fairly closely on either side of *A*, indicating that the results for spherical inclusions of uniform size randomly distributed would be close to those for spherical inclusions with the wide size range. This suggests that variations in size and spatial distribution of spherical inclusions in an actual composite may not have very material effects on the viscoelastic properties, but the effect of phase inversion will always be appreciable. These conclusions have been drawn, however, from consideration of a particular case in which the two phases are present in equal volumes and have viscoelastic properties which are not very widely different.

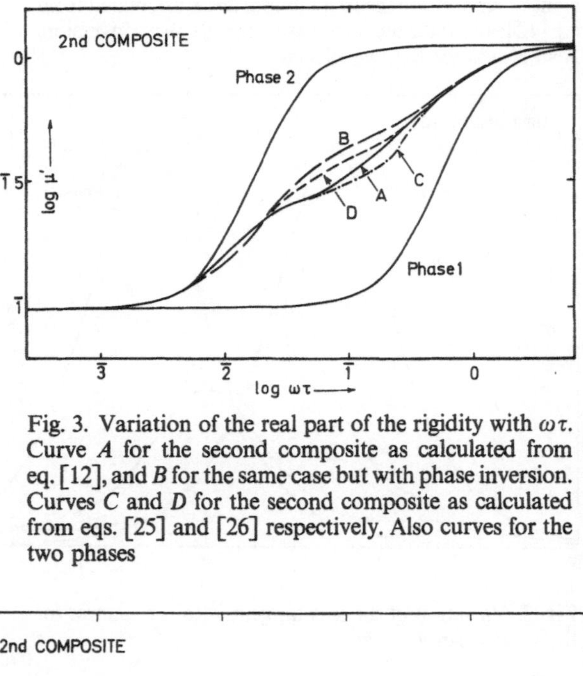

Fig. 3. Variation of the real part of the rigidity with $\omega\tau$. Curve *A* for the second composite as calculated from eq. [12], and *B* for the same case but with phase inversion. Curves *C* and *D* for the second composite as calculated from eqs. [25] and [26] respectively. Also curves for the two phases

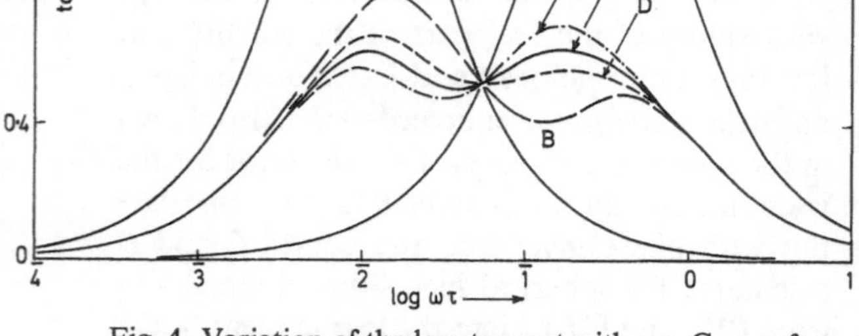

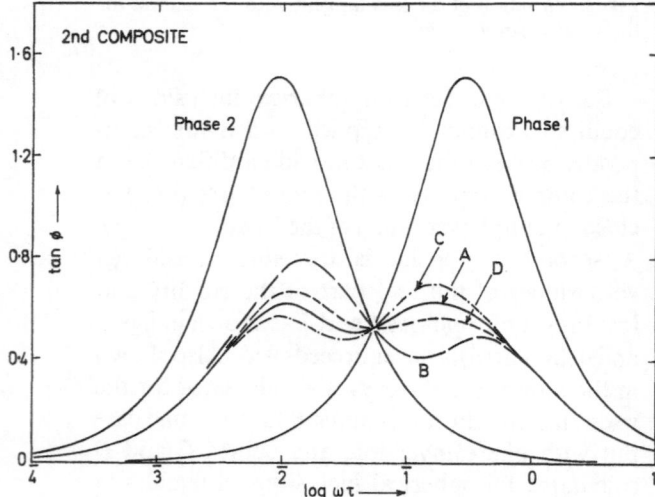

Fig. 4. Variation of the loss tangent with $\omega\tau$. Curves for the same cases as in fig. 3

Summary

General bounds on the moduli of isotropic composites of two isotropic and purely elastic phases have been obtained by *Hill, Hashin* and *Shtrikman*, and *Walpole*. It may be shown that, in most cases of importance, these are best possible bounds for arbitrary phase geometry. Bounds can also be found for the real and imaginary parts of the complex moduli of composites with viscoelastic phases. In all cases, however, the bounds are wide when the moduli of the phases are very different, so phase geometry must be taken into account in order to obtain results of practical value. Calculations of rigidity are therefore made for incompressible composites in which one phase forms spherical inclusions in the other. An exact formula is obtained for a case in which the size distribution of these inclusions is very wide, and approximate formulae for two cases of inclusions of uniform size (nearest-neighbour distance being a maximum in the one and a minimum in the other). These have been applied in a study of the frequency variation of the real part of the rigidity and the loss tangent of a composite of equal volumes of two viscoelastic phases, each having sharp loss-tangent peaks at different frequencies. The effect of variations in the spatial distribution and size distribution of the inclusions can be inferred from the results, as well as the effect of phase inversion.

Zusammenfassung

Allgemeine Schranken für die Moduln isotroper Stoffsysteme, bestehend aus zwei isotropen, rein elastischen Phasen, wurden von *Hill, Hashin* und *Shtrikman*, sowie *Walpole* angegeben. Es kann gezeigt werden, daß in den meisten wichtigen Fällen diese die besten möglichen Schranken darstellen, die für eine willkürliche Phasengeometrie erhalten werden können. Ebenfalls können Schranken für Real- und Imaginärteil der komplexen Moduln von zusammengesetzten viskoelastischen Phasen gewonnen werden. Indessen liegen diese Schranken weit auseinander, wenn die beiden Phasen sehr verschiedene Moduln haben. Daher muß in solchen Fällen die Phasengeometrie in die Betrachtung einbezogen werden, wenn man Ergebnisse von praktischer Bedeutung erhalten will. Es wird hier die Berechnung der Steifigkeit für inkompressible zusammengesetzte Stoffe durchgeführt, bei denen die eine Phase in Form von kugelförmigen Inklusionen in der andern vorliegt. Exakte Formeln werden für den Fall erhalten, daß die Größenverteilung der Inklusionen sehr weit ist, sowie Näherungsformeln für zwei Fälle von Inklusionen mit gleicher Größe – wobei der kleinste Abstand zwischen zwei Teilchen im ersten Fall maximal, im zweiten dagegen minimal gewählt wird. Diese Ergebnisse werden zur Untersuchung der Frequenzvariation des Realteils der Steifigkeit und der Dämpfung verwendet für eine Zusammensetzung gleicher Volumenanteile zweier viskoelastischer Phasen, die bei verschiedenen Frequenzen jeweils scharfe Dämpfungsmaxima besitzen. Die Auswirkung der Variation sowohl der räumlichen als auch der Größenverteilung der Inklusionen ebenso wie diejenige einer Phasenumkehr kann aus den Ergebnissen abgeleitet werden.

References

Brinkman, H. C., J. Chem. Phys. **20**, 571–573 (1952).

Eshelby, J. D., Progress in Solid Mechanics (ed. by *I. N. Sneddon and R. Hill*), Vol. 2, Chap. III (Amsterdam 1961).

Hashin, Z., J. Appl. Mech. **29**, 143–150 (1962).

Hashin, Z. and *S. Shtrikman*, J. Mech. Phys. Solids **11**, 127–140 (1963).

Hill, R., J. Mech. Phys. Solids **11**, 357–372 (1963).

Roscoe, R., Brit. J. Appl. Phys. **3**, 267–269 (1952); J. Mech. Phys. Solids **20**, 91–99 (1972).

Walpole, L. J., J. Mech. Phys. Solids **14**, 151–162 (1966); Q. J. Mech. & Appl. Maths. **25**, 153–160 (1972).

Author's address:
Dr. *R. Roscoe*
School of Physics, The University
Newcastel upon Tyne NE1 7RU (U.K.)

Rheol. Acta 12, 412–417 (1973)

Laboratoire de mécanique des solides de l'ecole Polytechnique Paris (France)

Comportements compares des sols et des roches

Par P. Habib

Avec 9 figures

(Reçu p. p. le 27 octobre 1972)

Les fondations des bâtiments, le génie civil, les travaux souterrains, les mines sont installés essentiellement dans deux types de matériaux: les terrains meubles, sables, limons, argiles, appelés généralement sols et les terrains rocheux. Le développement des activités de construction et de terrassement imposent une connaissance de plus en plus précise du comportement rhéologique de ces différents milieux.

I. Description elementaire des mecanismes de deformation

En simplifiant la description à l'extrême on peut dire que les sols sont des milieux constitués de grains en contact mutuel par des points ou par des surfaces petites. Les roches sont des milieux fracturés et fissurés et les contacts entre les blocs élémentaires (qui peuvent avoir quelques dm^3 ou quelques m^3) se font par des surfaces. Dans les deux cas la rupture se produit par des glissements accompagnés d'une désorganisation de l'architecture initiale. Dans les sables par exemple les contacts se polarisent vers la direction de la contrainte majeure; dans les massifs rocheux la transmission des forces se fait probablement par des mécanismes de bielles. En général, les particules élémentaires, grains de sol ou blocs rocheux, ne sont pas rompues; la différence des résistances entre les contacts et les grains est en effet très grande: les angles de frottement physique ψ ou de frottement interne φ sont parfois du même ordre de grandeur mais les cohésions sont dans des rapports de 1–100. Cependant les grains ne sont pas indestructibles; ainsi la granulométrie d'un sable moyen cisaillé sous une contrainte normale de 10 daN/cm^2 se dégrade vers les fines. Le décrochement d'un système de failles ou de diaclases nécessite le déverouillage d'un certain nombre d'obstacles accompagné de ruptures de la matrice rocheuse. Il ne faut pas négliger l'énergie mise en jeu dans de tels mouvements; l'effondrement sur 20 m d'un bloc de roches de 10 m d'arête libère une puissance de l'ordre de 100000 CV pendant 5–6 sec: il est bien évident qu'une telle puissance peut provoquer des ruptures en dehors des surfaces de glissement et casser de nombreux débris.

L'évaluation des caractéristiques mécaniques des matériaux en place pose des problèmes à peu près résolus pour les sols, mais particulièrement difficiles pour les massifs rocheux. Il existe alors un effet d'échelle qui se caractérise pour les roches par le fait que la résistance est une fonction décroissante de la dimension du volume considéré. L'effet d'échelle est moins marqué pour les sols, mais il a été observé notamment en laboratoire où différents auteurs (*Y. Tcheng*, 1966; *J. Weber*, 1971) ont constaté que les formules de la plasticité des matériaux de *Coulomb* étaient en défaut pour des poinçons de petites dimensions, les pressions de rupture étant alors plus grandes que la théorie le prévoit. L'origine de l'effet d'échelle est le plus généralement attribuée à la présence d'hétérogénéités; pour les roches il a été montré que cette hétérogénéité est celle des fissures. En particulier sous des contraintes moyennes de plus en plus grandes, lorsque la pression pince les lèvres des fissures et rétablit en quelque sorte la continuité de la transmission des contraintes, l'effet d'échelle et la dispersion des résultats disparaissent simultanément (*P. Habib, G. Vouille*, 1966). Pour les sables, considérés comme des corps statistiquement continus, nous allons voir que la discontinuité liée à la formation d'une ligne de glissement est responsable de l'effet d'échelle constaté sur les modèles réduits.

II. Description de la formation d'une ligne de glissement

a) Roches

Soit ψ l'angle de frottement physique de la roche. Les surfaces de contact dans une fissure ne sont pas planes; soit α l'angle moyen d'inclinaison des irrégularités par rapport à la direction générale de la fissure (fig. 1): cet angle α caractérise la direction du déplacement d'un bloc par rapport à l'autre, au début du mouvement. La résistance au glissement est définie par un coefficient de frottement apparent ($\psi + \alpha$) et il faut éventuellement ajouter une certaine cohésion c, s'il existe des forces d'adhésion entre les lèvres de la fissure.

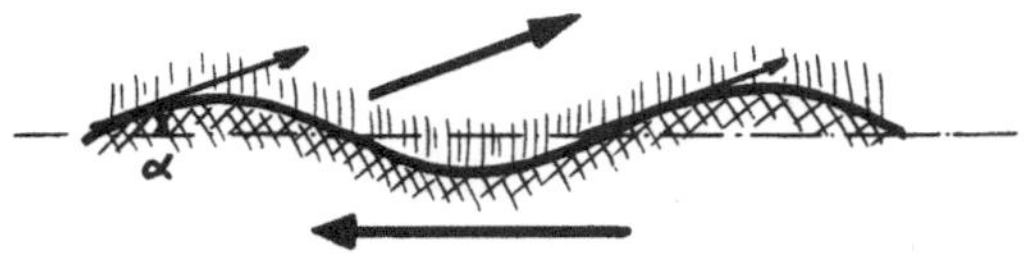

Fig. 1. Fissure avant déplacement

En définitive la résistance au cisaillement t d'un joint (*M. Goldstein*, 1966; *F. Patton*, 1966) est:

$$t = c + n\,\mathrm{tg}(\psi + \alpha) \qquad [1]$$

(n: contrainte normale au plan moyen de glissement).

On voit que la présence des irrégularités de la surface de glissement améliore la résistance au frottement et fait apparaitre simultanément une augmentation de volume en cours de cisaillement.

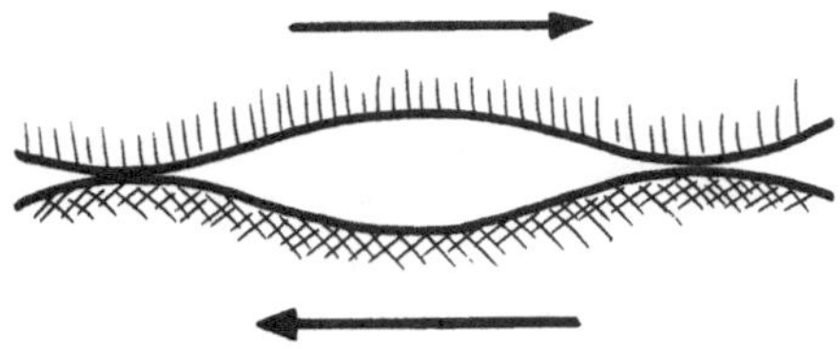

Fig. 2. Fissure après déplacement

Cette dilatance du joint est limitée par l'amplitude des irrégularités; la vitesse de la dilatance est maximale au début du mouvement et elle s'annule après le déplacement qui mène à un glissement sur les «pointes» (fig. 2) la résistance au cisaillement est alors à peu près constante et surtout influencée par ψ, encore que l'usure des surfaces ait pu faire varier le frottement physique jusqu'à la valeur ψ'. De toutes façons la courbe efforts-déformations présente un maximum (fig. 3), la pente initiale provenant des déformations pseudo-élastiques du jeu de la fissure.

La dilatance du joint varie avec la contrainte normale: en effet lorsque les contraintes augmentent il se produit un cisaillement des irrégularités; ce sont d'abord les sommets des pointes les plus aiguës, puis toutes les irrégularités sont progressivement affectées (fig. 4). Le critère de rupture n'est plus une droite, mais est composé de deux segments, le premier de pente $\psi + \alpha$, le second de pente ψ' (fig. 5).

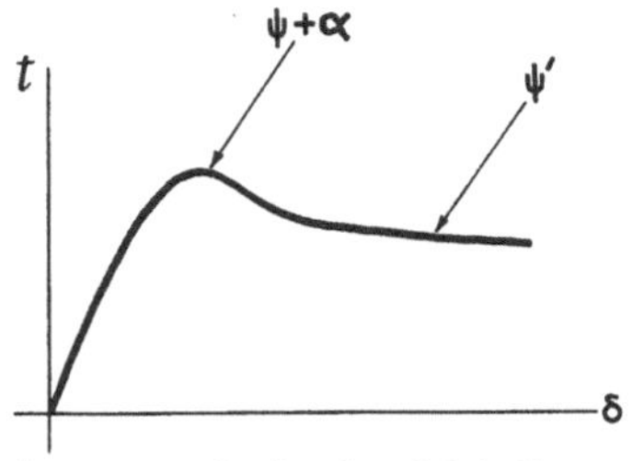

Fig. 3. Déplacement «δ» d'un bord de la fissure par rapport à l'autre en fonction d'un effort de cisaillement «t»

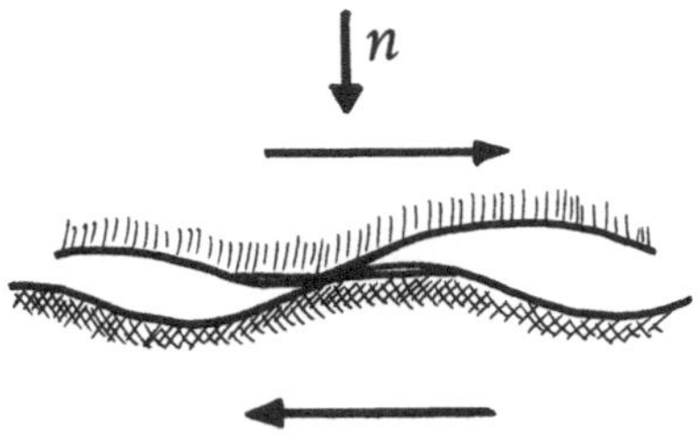

Fig. 4. Rupture des pointes

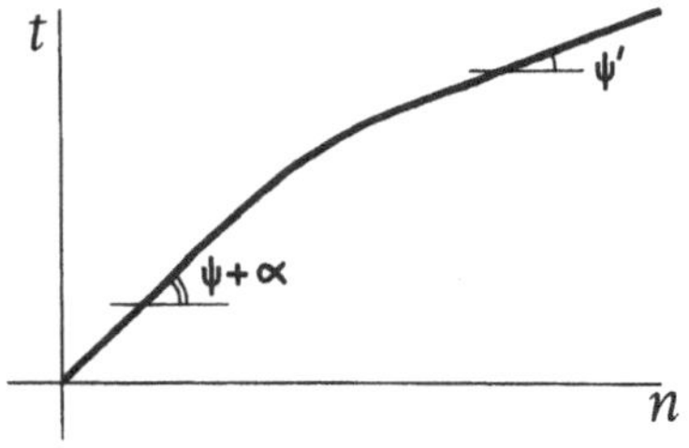

Fig. 5. Frottement d'une fissure

b) Sols

La résistance au cisaillement d'un sable, c'est à dire l'angle de frottement interne, croît avec le poids volumique γ. Sous une contrainte moyenne donnée il existe un poids volumique critique γ_c tel que la variation de volume globale soit nulle en cours de cisaillement. Si $\gamma < \gamma_c$ la pente de la courbe efforts-déformations est toujours positive: le matériau s'écrouit et la rupture se produit avec un écoulement plastique. Si $\gamma > \gamma_c$ le sable se dilate avant la rupture, la courbe efforts-déformations présente un maximum (fig. 6) et la rupture se produit avec apparition d'une ligne de glissement isolée. Dans un champ non homogène, un nombre limité de lignes de discontinuité se forment et par exemple dans le cas d'un poinçon en surface, une ligne de glissement extrême sépare le massif profond, pratiquement immobile, de la zone superficielle où règnent des grands mouvements par blocs (*Chazy-Habib*, 1961). La présence d'une ligne de glissement dans un sol n'est pas liée à la dilatance mais à la présence de l'anti-écrouissage, c'est à dire de la pente négative de la courbe efforts-déformations après le maximum; c'est d'ailleurs à partir du moment où la déformation atteint puis dépasse celle qui correspond au maximum de l'effort que la ligne de glissement apparait. On en verra une autre preuve en comparant avec la rupture des argiles. Pour ces matériaux, les trois types de comportement de la fig. 6 se rencontrent selon la nature ou l'état de consolidation. Les lignes de glissement n'apparaissent que lorsque le comportement du matériau présente un anti-écrouissage et ceci bien que la dilatance soit nulle puisque la déformation de l'argile saturée se fait à volume constant. L'anti-écrouissage de l'argile provient probablement d'une destruction de la structure minéralogique avec orientation privilégiée des particules dans la surface de glissement. La formation d'une ligne de discontinuité peut d'ailleurs trouver une explication à partir de la pente de la courbe efforts-déformations en admettant l'existence de défauts dans la matière, défauts qui sont susceptibles de se renforcer dans le cas de l'écrouissage, d'où une déformation plastique homogène, ou de se dégrader, dans le cas de l'anti-écrouissage, ce qui conduit à la concentration des déformations.

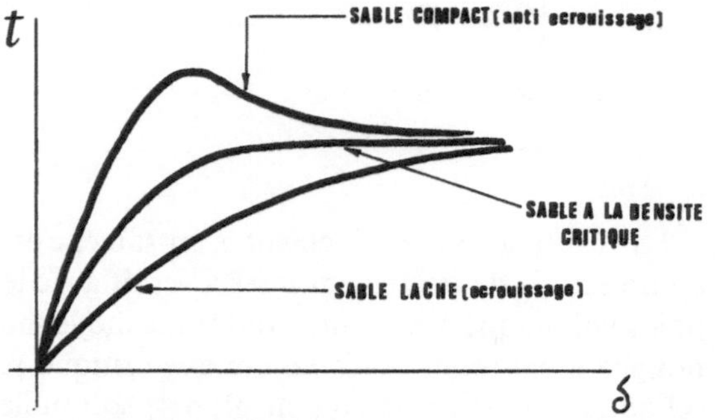

Fig. 6. Cisaillement d'un sable

La fig. 7 montre une représentation idéalisée de la formation d'une ligne de glissement dans un sable: il est probable que des réarrangements locaux se produisent autour des «creux» de la ligne de glissement créant une dilatation irréversible. D'ailleurs l'observation des déplacements des grains de sables, notamment par *Dantu*, sur des matériaux analogiques à deux dimensions, montre que la surface de glissement est en réalité un volume; la perturbation de l'architecture des grains s'étend sur un maximum de 3–4 couches de part et d'autre d'une surface plus ou moins irrégulière et c'est dans cette zone que se localise la variation de volume pendant l'anti-écrouissage (fig. 6). C'est dire que la dilatation dans la direction normale à un plan de glissement est fonction du diamètre des grains; elle varie probablement avec la densité initiale et comme pour la dilatance d'un joint rocheux elle est fonction décroissante de la contrainte moyenne.

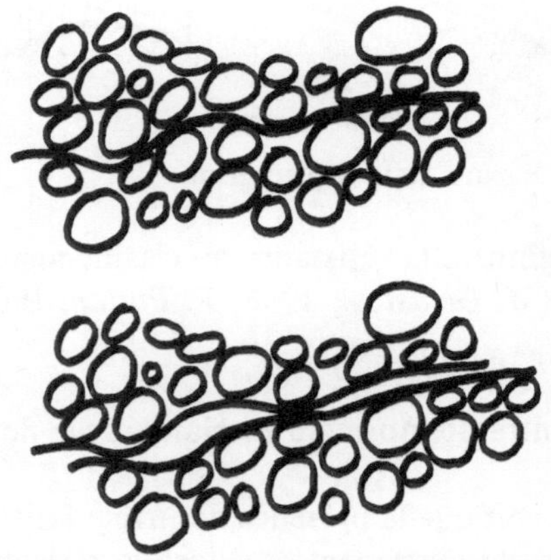

Fig. 7. Désenchevètrement d'un sable compact et formation d'une ligne de glissement

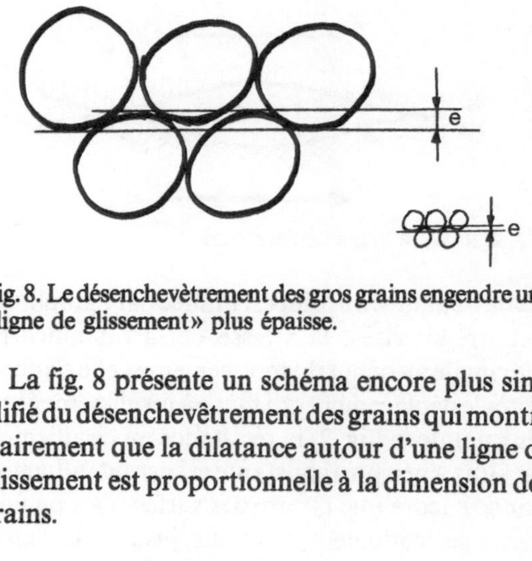

Fig. 8. Le désenchevètrement des gros grains engendre une «ligne de glissement» plus épaisse.

La fig. 8 présente un schéma encore plus simplifié du désenchevêtrement des grains qui montre clairement que la dilatance autour d'une ligne de glissement est proportionnelle à la dimension des grains.

III. Effet d'echelle en mecanique des sols

L'effet d'échelle en Mécanique des Sols n'a jamais été signalé pour un champ de contrainte uniforme, sauf pour les argiles fissurées, où l'on rejoint les problèmes de mécanique des roches, et pour les sols contenant de très grosses inclusions (argile à blocaux, sols de morraines) où les hétérogénéités engendrant dispersion des résultats et effet d'échelle sont évidentes. En particulier dans des essais de compression simple ou triaxiale, des éprouvettes de dimensions différentes d'argile (*Habib*, 1952), ou de sable, donnent les mêmes resistances: l'angle de frottement interne et la cohésion sont indépendants de la dimension des éprouvettes d'essais.

Par contre, avec des sables denses on á observé un effet d'échelle dans des expériences où le champ n'est pas uniforme comme par exemple le poinçonnement par une fondation superficielle.

On peut trouver une explication à ce phénomène dans l'apparition d'une ligne de glissement isolée. L'expérience montre que la formation des lignes de glissement est un phénomène discret; avec une fondation de largeur B la première discontinuité dans le milieu apparait pour un enfoncement de l'ordre de $e/B = 5\%$. La courbe effort-enfoncement passe alors par un maximum. Si on poursuit le poinçonnement, une deuxième puis une troisième ligne de glissement se produisent pour des enfoncements de plus en plus grands, accompagnées de nouveaux maximums. La portance de la fondation est liée à l'apparition des grandes déformations donc au premier maximum. Le travail dépensé par le poinçon pour engendrer la rupture du massif pulvérulent contient d'une part le travail du frottement entre les grains et d'autre part le travail dépensé contre la pesanteur pour soulever le centre de gravité du massif par l'effet de la dilatance.

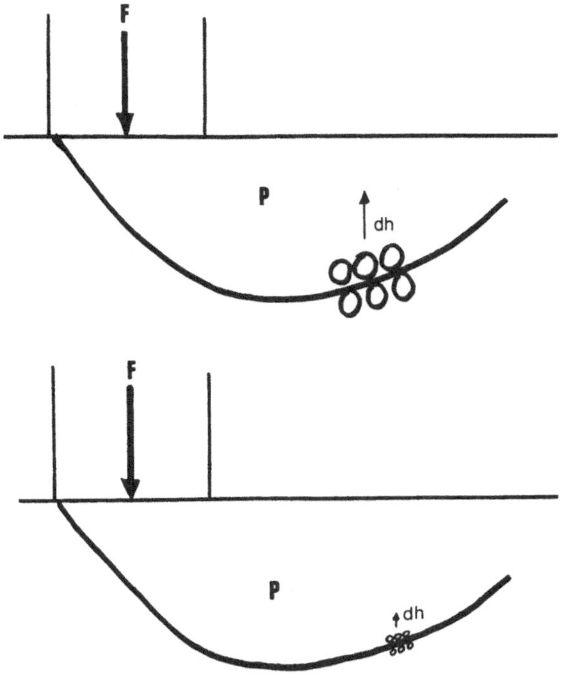

Fig. 9. Poinçonnement d'un milieu pulvérulent à gros grains ou à petits grains

Pour les deux poinçons de la fig. 9 qui ont même largeur mais reposent sur des massifs où les grains sont homothétiques, il est évident que le travail dû à la dilatance est plus grand pour les gros grains que pour les petits grains. Si maintenant nous réduisons les dimensions du poinçon sur les gros grains *et* les gros grains par une homothétie qui fasse correspondre les gros grains aux petits, les lois de similitude mécanique nous indiquent que les contraintes sont proportionnelles à l'échelle des longueurs puisque les grains pulvérulents sont pratiquement infiniment rigides. Il en résulte que le petit poinçon sur petits grains est homologue du grand poinçon sur gros grains, mais n'est pas homologue du grand poinçon sur petits grains et il donnera une portance plus grande que celle que l'on peut prévoir si l'on néglige la dimension des grains: c'est bien un effet d'échelle. On peut chercher à préciser ce résultat par un calcul sommaire destiné à donner un ordre de grandeur de la largeur de la fondation à partir de laquelle ce phénomène est sensible.

La force portante q_u à la surface d'un milieu pulvérulent est donnée par la relation classique:

$$q_u = \gamma \frac{B}{2} N_\gamma$$

où N_γ est une fonction connue de φ.

Sur un milieu composé de grains infiniment petits pour que la dilatance dans la ligne de glissement soit négligeable, le travail dépensé à la rupture, c'est à dire pour un enfoncement de l'ordre de 5%, est:

$$W = \frac{1}{2} \frac{5B}{100} \gamma \frac{B}{2} N_\gamma B.$$

Lorsque les grains ont une certaine dimension, il faut y ajouter le travail dû à la dilatance. La variation d'épaisseur de la ligne de glissement est de l'ordre de grandeur du diamètre Φ d'un grain. Cette constatation expérimentale, probablement approximative parce que ne tenant pas compte de la densité initiale, est en accord avec la remarque de *Dantu*, signalée plus haut, que l'architecture des grains est bousculée sur quelques couches de part et d'autre de la surface de glissement; or comme le désenchevêtrement d'une couche sur une autre correspond à un déplacement de l'ordre du $^1/_5$ ou du $^1/_4$ de grain, avec plusieurs couches en mouvement relatif on aboutit à un soulèvement de l'ordre de un diamètre. Le travail de la pesanteur est alors le produit du poids P du sable situé au-dessus de la ligne de glissement par le soulèvement. Une approximation de P est $P = 5B^2\gamma$. Si nous posons $\Phi = B/n$ le travail total pour le poinçon s'écrit:

$$W = \frac{1}{2}\,\frac{5B}{100}\,\gamma\,\frac{B}{2}\,N_\gamma B + 5B^2\gamma\,\frac{B}{n}.$$

On reviendra à la charge ultime par:

$$q_u = \frac{W}{\frac{1}{2}\,\frac{5B}{100}\,B} = \gamma\,\frac{B}{2}\,N_\gamma + \gamma\,\frac{B}{2}\,\frac{400}{n}$$

$$= \gamma\,\frac{B}{2}\left(N_\gamma + \frac{400}{n}\right).$$

Pour $\varphi = 35°$, $N_\gamma = 40$ et on voit que si $n = 10$ l'effet d'échelle correspond à un doublement de la force portante, ce qui est considérable. On peut dire encore que l'effet d'échelle est une majoration de 10% de la charge limite, pour une fondation de 10 cm de large reposant sur un sable de 35° d'angle de frottement interne et de 1 mm de grain moyen ($n = 100$). Une telle majoration est à la limite de ce que l'on peut déceler expérimentalement et l'effet d'échelle observé en laboratoire est surtout sensible pour des largeurs de poinçon inférieures à 10 cm.

Le calcul précédent n'a nullement la prétention de représenter la réalité; il a simplement pour but de donner une indication de l'ordre de grandeur de la largeur de fondation comparée à la dimension du grain moyen du milieu pulvérulent; il est basé sur l'hypothèse de la simultanéité du glissement et du soulèvement, ce qui est physiquement acceptable. La vérification précédente étant satisfaisante, on peut tenter de continuer le raisonnement. Ainsi, si l'effet d'échelle qui est observé sur les sables denses est lié à la dilatance dans la ligne de glissement, elle-même fonction de la dimension du grain, cet effet n'est pas lié à la présence d'un maximum, ni à l'anti-écrouissage qui le suit, ni même à la formation d'une ligne de glissement. Effectivement pour les argiles à fortes pressions de consolidation présentant de l'anti-écrouissage, l'interprétation des résultats d'essais de poinçonnements rapides conduit à la formule classique de la plasticité parfaite $p = (2 + \pi)\,c$ avec une bonne approximation et sans effet d'échelle pour un sol semi-infini homogène. Dans l'hypothèse précédente, ce résultat est évident puisque les grains d'argile sont infiniment petits par rapport à la largeur de la fondation et que la ligne de glissement se forme à volume constant. Une autre déduction peut être exprimée de la façon suivante: il n'y a pas d'effet d'échelle pour un poinçon sur un sable peu dense lorsqu'il n'y a pas formation d'une ligne de glissement isolée. Ce résultat est lui aussi conforme à l'expérience; il s'explique immédiatement puisque les variations de volume sont indépendantes de la dimension des grains dès lors que la déformation est continue.

IV. Conclusion

Les mécanismes de rupture des sols et des roches sont tout à fait différents. Pour les massifs rocheux, à de rares exceptions près en génie civil, les glissements se produisent sur des surfaces pré-existantes avec des phénomènes de dilatance des joints fonctions de la contrainte normale. Dans les sols la déformation est en général continue dans l'espace et peut être considérée comme homogène dans un petit domaine. Cependant lorsqu'il existe un anti-écrouissage ce qui conditionne l'existence d'une ligne de glissement, les mécanismes deviennent voisins et l'on voit la rupture se propager par le déplacement relatif de blocs presques rigides, séparés par des joints de résistance plus faible. Dans le cas des sables denses la formation d'une ligne de discontinuité s'accompagne d'une dilatance de la surface de glissement et les mécanismes de la déformation deviennent pratiquement identiques. Enfin dans ce dernier cas on peut interpréter l'effet d'échelle dans certains champs de contraintes hétérogènes par le non respect des conditions de similitude dans l'épaisseur du joint et l'on peut prévoir avec une bonne approximation la dimension des modèles réduits qui ne sont pas affectés par ce phénomène.

Résumé

Les sols et les roches sont des corps dont le comportement rhéologique est régi par les liaisons de contact entre les éléments qui les constituent. On peut schématiser ces structures en disant que dans les sols, les contacts sont ponctuels entre les grains alors qu'ils ont lieu sur les surfaces que forment les fissures entre les blocs pour les massifs rocheux. Au cours des déformations qui précèdent la rupture, des glissements se produisent le long de surfaces irrégulières et les mouvements sont accompagnés d'une variation de volume, qui est positive pour les roches et les sables denses, et négative pour les sables lâches et la plupart des argiles. Les roches présentent un effet d'échelle: les caractéristiques mécaniques de la rupture sont des fonctions décroissantes de la dimension de la zone soumise aux contraintes et ceci est lié à l'hétérogénéité de la fissuration. Cet effet n'apparaît pas dans les sols, mais dans un champ de contraintes hétérogène, comme celui qui existe sous un poinçon, un phénomène analogue se produit pour les sables denses. On peut l'expliquer par l'hétérogénéité introduite dans le milieu par la création d'une ligne de glissement isolée.

Summary

Rheological behaviour of soils and rocks is determined by contact forces between constitutive elements. For soils, the contacts are punctual. For rocks they occur along the surfaces of cracks or joints. During the deformation before the failure, sliding movements occur along irregularly shaped surfaces with a volume change, which is positive for rocks and dense sands, negative for loose sands and most of clays. Size effect in rocks is due to cracks heterogeneity: mechanical characteristics at failure decrease with increasing size of specimens. This effect does not exist in soils, but in a heterogeneous stress field a similar phenomenon appears in dense sands. It is connected with the heterogeneity induced by the creation of an isolated slip line.

Littérature

Chazy, C., P. Habib, Les piles du quai de la Floride. 5ème Congrès Int. de Mécanique des Sols, Paris 1961, Communication 3 A/26.

Goldstein, M. et al., Investigation of mechanical properties of cracked rock. Comptes-Rendus 1er Congrès de la Soc. Int. Mec. Roches, Lisboa 1966, T. I, thème 3 (1966).

Graham, J., J. G. Stuart, Proc. ASCE, Soil Mec. and Foundation Division, **1971**, p. 1533.

Habib, P., La résistance au cisaillement des sols. Annales I.T.B.T.P., **1953**, p. 17.

Habib, P., G. Vouille, Sur la disparition de l'effet d'échelle aux hautes pressions. Note aux C.R.Ac.Sc. Paris, **262**, 715–717 (1966).

Patton, F. D., Multiple modes of shear failure in rock. Comptes-Rendus 1er Congrès de la Soc. Int. Mec. Roches, Lisboa 1966, T. I, thème 3.

Tcheng, Y., J. Iseux, Nouvelles recherches sur le pouvoir portant des milieux pulvérulents: fondations superficielles et semi-profondes. Annales I.T.B.T.P., **1966**.

Weber, J. D., Les applications de la similitude physique aux problèmes de la Mécanique des Sols. pp. 22–26, (Paris 1971).

Adresse de l'auteur:

Dr. *P. Habib*
Laboratoire de Mécanique des Solides
Ecole Polytechnique
17, rue Descartes
F-75005 Paris (France)

Rheol. Acta 12, 418–424 (1973)

From the Purdue University, West Lafayette, Indiana (U.S.A.)

Theory of a second-order microfluid

By Severino L. Koh

With 2 figures

(Received October 27, 1972)

1. Introduction

To characterize the motion of fluids for which the *microstructure* of the given material is of the essence, *Eringen* (1) proposed a theory of "simple microfluids". Microstructure in the fluid may be due to the fact that the given body is a composite of two or more different types of material, e.g. a suspension, or that the body while constituting a single type of a material is characterized by nonhomogeneous local motions such as eddies in a turbulent flow. Since local "micromotion" is taken into account, the theory provides a mechanism whereby a wide variety of fluid flow problems traditionally treated in a statistical manner may be considered in a continuum formulation. Such an application is given by *Eringen* and *Chang* (2) in which a particular case of the theory is used to describe a simple problem in hydrodynamic turbulence.

Basic to the *Eringen* theory is the existence of a representative volume element (a *macroelement*) at the mass center of which averaged physical quantities are defined as continuous functions. Thus, the given body which is essentially heterogeneous is now represented by a hypothetical continuum composed "entirely" of mass centers. Motion of the continuum is described in terms of the motion of the mass center (*macromotion*) plus a *relative micromotion* that takes into account the deformation and motion of the given macroelement about its mass center. In the general case, this micromotion allows both the stretching and rigid rotation of the macroelement of the real body.

A fundamental assumption of the *Eringen* theory restricts the relative micromotion to an affine transformation. It can then be shown [*Eringen* and *Suhubi* (3)] that this assumption carries the mass center of a macroelement in one configuration into the mass center of that macroelement in all of its other configurations. Thus, a material particle in the real body occupying the mass center of a macroelement in one configuration of the body will continue to occupy the mass center of that macroelement no matter how the body deforms and for all times.

Clearly, this assumption of homogeneous micromotion is rather restrictive for certain cases of heterogeneous fluids and violent nonhomogeneous motions. For these cases, a more general description of the micromotion has to be considered. *Ahmadi*, *Koh* and *Goldschmidt* (4, 5) propose a more general theory of *nonsimple* microfluids. Particular results are presented for the case of a micromotion characterized by a quadratic relationship. Thus, the theory is laid for a *second-order microfluid.*

In these studies, thermal effects are neglected. However, due to the obvious pertinence of thermodynamical properties on fluid behavior, further consideration becomes necessary. This is attempted by *Cook* and *Koh* (6), who propose a theory for general microfluids in the presence of thermal inhomogeneities.

The present paper is concerned with second-order microfluids with microstructural thermal effects. In Section 2 we review briefly the kinematics of the theory and define the deformation tensors peculiar to the case of second-order microfluids. The concept of microtemperatures is discussed in Section 3. In Section 4, the derivation of the fundamental field equations is discussed. The development of a constitutive theory for second-order microfluids is presented in Section 5.

2. Kinematics

To study the motion of a given fluid, we consider two configurations of the given body as schematically shown in fig. 1. Configuration B is taken to be the *reference configuration* which the body assumes at some prescribed reference time t_0. At some subsequent time $t > t_0$, the body assumes a *deformed configuration b*.

Particles of the body for any instant are located with respect to some fixed reference frame Z^K. A material point is identified by its position in the reference configuration B in terms of its curvilinear coordinates X'^K. At time t, this same particle will occupy in configuration b the spatial point x'^k, in general different from X'^K. The *motion* of the given body may be formally expressed in the relation

$$x'^k = x'^k(X'^K, t). \qquad [1]$$

The *inverse motion* is taken to be

$$X'^K = X'^K(x'^k, t), \qquad [2]$$

where the axiom of continuity is implied.

A hypothetical continuum is constructed on the premise that its gross behavior is equivalent to that of the given fluid. We consider the average mass densities ϱ_0 and ϱ of some sufficiently small representative volume element (called *macroelement*), measured respectively at times t_0 and t, to be derivable from the mass densities ϱ'_0 and ϱ' of the *microelements* that constitute the macroelement. Explicitly, we have

$$\int_{dV} \varrho'_0 dV' = \varrho_0 dV \quad \text{at time } t_0 \qquad [3]$$

$$\int_{dv} \varrho' dv' = \varrho dv \quad \text{at time } t. \qquad [4]$$

In eqs. [3] and [4], dV and dv are the macroelements, while dV' and dv' are the microelements.

The mass center X^K of the reference macroelement dV is determined from the expression:

$$\int_{dV} \varrho'_0 X'^K dV' = \varrho_0 X^K dV. \qquad [5]$$

The motion [1] carries the material located at the mass center X^K to a spatial point x^k, which we shall call the *reference point*.

The position vector of a generic material point X'^K in a macroelement in the reference configuration is referred to the mass center X^K of the macroelement according to the relation

$$X'^K = X^K + \Xi^K \qquad [6]$$

where Ξ^K is the relative position vector of the material point (see fig. 2). Similarly, we define a relative position vector, ξ^k, in the deformed configuration. The motion may then be written in the form

$$x'^k = x^k(X^K, t) + \xi^k(X^K, \Xi^K, t) \qquad [7]$$

and the unique inverse motion

$$X'^K = X^K(x^k, t) + \Xi(x^k, \xi^k, t). \qquad [8]$$

Departing from the *Eringen-Suhubi* assumption of homogeneous micromotion, we consider the following quadratic relationship to hold between the relative position vectors:

$$\xi^k = \chi^k_K(X^M, t)\, \Xi^K + \eta^k_{KL}(X^M, t)\, \Xi^K \Xi^L. \qquad [9]$$

It must be remarked that one obtains the *Eringen* equations for simple microfluids by simply setting the second-order micromotion η^k_{KL} in [9] equal to zero.

Treating the constituent microelements of the macroelement as continua, we consider the mass of the microelement to be conserved, i.e.

$$\varrho'_0 dV' = \varrho' dv'. \qquad [10]$$

Eqs. [3] and [4] then imply

$$\varrho_0 dV = \varrho dv. \qquad [11]$$

We now proceed to prove an important distinction between simple microfluids and second-order microfluids embodied in the following theorem: For the second-order microfluid with its micromotion characterized by [9], the mass center X^K of dV is carried to the *reference point* x^k in dv, which is in general distinct from the *mass center* $\bar{x}^k$ of dv.

Proof: From eqs. [4], [7] and [9], we have the following calculations:

$$\begin{aligned} \int_{dv} \varrho' x'^k dv' &= \int_{dv} \varrho'(x^k + \chi^k_K \Xi^K + \eta^k_{KL} \Xi^K \Xi^L)\, dv' \\ &= \varrho x^k dv + \chi^k_K \int_{dv} \varrho' \Xi^K dv' \\ &\quad + \eta^k_{KL} \int_{dv} \varrho' \Xi^K \Xi^L dv' \\ &= \varrho x^k dv + \varrho \eta^k_{KL} I^{KL} dv, \end{aligned} \qquad [12]$$

where we have used the fact that $\int_{dV} \varrho'_0 \Xi^K dV' = 0$ since X^K is the mass center of dV. The second-order micro-inertia tensor, I^{KL}, has been defined as (cf. *Eringen* (1), eq. [12]).

$$\varrho_0 I^{KL} dV \simeq \varrho I^{KL} dv = \int \varrho'_0 \Xi^K \Xi^L dV'. \qquad [13]$$

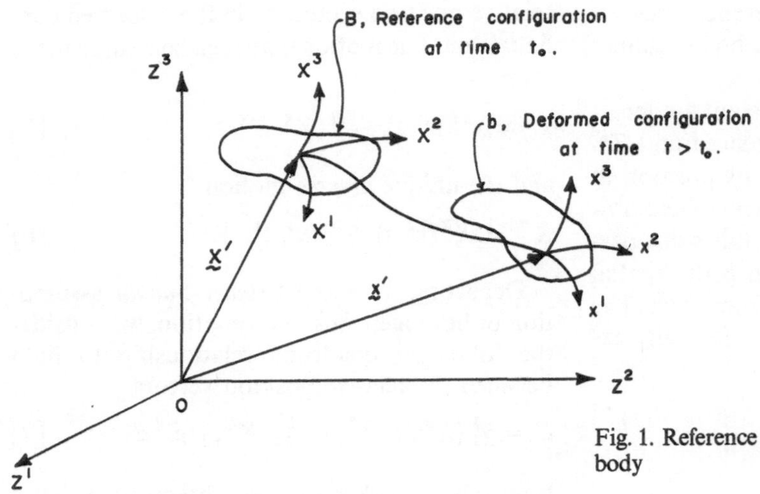

Fig. 1. Reference and deformed configurations of a given body

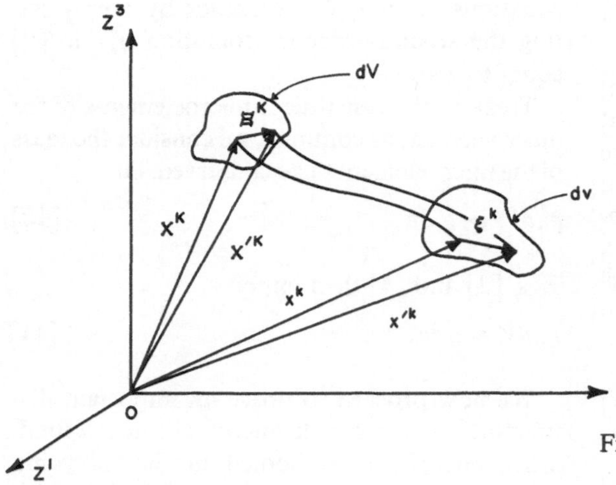

Fig. 2. Deformation of a macroelement

Let $\bar{x}^k$ be the mass center of dv, i.e.

$$\int_{dv} \varrho' x'^k dv' = \varrho \bar{x}^k dv. \qquad [14]$$

Comparing this equation with [12], we find that

$$\bar{x}^k = x^k + \beta^k, \qquad [15]$$

where β^k is the *eccentricity* vector defined according to

$$\beta^k = \eta^k_{KL} I^{KL}. \qquad [16]$$

This proves the theorem. It is to be noted from [15] and [16] that the reference point x^k coincides with the mass center $\bar{x}^k$ only if the eccentricity vector β^k vanishes, or equivalently if the second-order micromotion η^k_{KL} is zero. This is the case treated by *Eringen* (1).

For simplicity, it is assumed in subsequent calculations that the micromotion is only *slightly nonhomogeneous*. More precisely, η^k_{KL} is taken to be small such that powers of this tensor greater than one are ignorable.

The time rate of eq. [7] yields the velocity of the particle X'^K located at x'^k at instant t. That is

$$v'^k \equiv \dot{x}'^k = v^k + v^k_m \xi^m + \mu^k_{mn} \xi^m \xi^n, \qquad [17]$$

where v^k is the velocity of the reference point x^k, and v^k_m and μ^k_{mn} are the first and second *gyration tensors* defined as follows:

$$v^k_m \equiv \dot{\chi}^k_K \chi^K_m \qquad [18]$$

$$\mu^k_{mn} \equiv \chi^K_m \chi^L_n (\dot{\eta}^k_{KL} - v^k_p \eta^p_{KL}). \qquad [19]$$

It is again noted that in the theory of simple microfluids the second-order gyration tensor μ^k_{mn} is identically zero.

With the use of [15], the velocity v'^k may be expressed also in terms of the velocity of the mass center $\bar{v}^k \equiv \dot{\bar{x}}^k$. This gives rise to an expression involving the time rate of the eccentricity vector $\dot{\beta}^k \equiv \dot{\eta}^k_{KL} I^{KL}$. The acceleration a'^k of the material point X'^K is determined by taking the time rate of eq. [17], which gives a deviation from the acceleration of the reference point a^k. In turn, a'^k

may also be expressed in terms of the acceleration $\bar{a}^k \equiv \dot{\bar{v}}^k$ of the mass center of the macroelement. This last expressions give rise to the second time derivative of the eccentricity vector $\ddot{\beta}^k$. Details are presented by *Ahmadi* et al. (5).

In the formulation of the theory of second-order microfluids, certain kinematical variables are pertinent. These are naturally obtained by taking the material derivatives of elements of arc and surface. Dispensing with the lengthy calculations which are presented by *Ahmadi, Koh* and *Goldschmidt*, we simply enumerate below the deformation and microdeformation rate tensors needed in the analysis of second-order microfluids:

The *deformation rate tensor* d_{km} is defined as in classical theories according to:

$$2d_{km} = v_{k,m} + v_{m,k} \qquad [20]$$

where covariant differentiation is indicated by a comma preceding a subscript (k). The *micro-deformation rate tensors* b_{km} and a_{kmn} are expressible in terms of the gyration tensors ν_{km} and μ_{kmn} as follows:

$$b_{km} = \nu_{km} + v_{m,k} + \mu_{kmn}\xi^k \qquad [21]$$

$$a_{kmn} = \nu_{km,n} + (\mu_{kmp}\xi^p)_{,n}. \qquad [22]$$

The *Eringen* microdeformation rate tensors are obtained from eqs. [21] and [22] by equating the second gyration tensor μ_{kmn} to zero.

3. Concept of microtemperatures

Contrary to the assumption taken in most theories on materials with microstructure, thermal inhomogeneities in the macroelement cannot be disregarded. In general, each of the particles in the given body has a temperature θ'. A simple averaging of the temperature will not display the variation of temperature from microelement to microelement. *Wozniak* (7) introduced the concept of "microtemperature" through a vector θ_k in a theory characterizing the thermoelasticity of a solid with microstructure. *Grot* (8) reformulated the basic definition of microtemperature and incorporated this idea in the development of the fundamental field equations for the microelastic solid.

A more complicated temperature relation was also proposed by *Wozniak* (9), whereby the temperature θ' of the constituent particle is determined by the mean temperature θ of the macroelement, the microtemperature θ_k plus the local gradients of both θ and θ_k. We presume that this takes into account the case of a stronger variation of the temperature within the microelement than those accounted for in the earlier theories. However, it would seem that the gradient of the mean temperature is an unnecessary added complication since the microtemperature θ_k may, in a sense, be identified as the local temperature gradient in the microelement. On the other hand, the gradient of the microtemperature results in a second-order term that may be useful in characterizing a more complicated situation than those considered in the previous theories.

Cook and *Koh* (6) also considered higher order variations within the microelement. Following the approach considered in that work, we assume the following relation:

$$\frac{1}{\theta'} = \frac{1}{\theta} + \frac{1}{\theta^k}\xi^k + \frac{1}{\theta^{km}}\xi^k\xi^m \qquad [23]$$

where θ is the mean temperature of the macroelement. The variation of temperature in the macroelement is accounted for in the vector $1/\theta^k$, while a more severe variation within the microelements themselves is characterized through the second-order tensor $1/\theta^{km}$.

In the subsequent discussions, it is more convenient to express these vector and tensor in slightly different forms. For later usage, we define the following *first* and *second microtemperatures:*

$$\bar{\theta}_k \equiv \frac{\theta}{\theta^k} \qquad [24]$$

$$\bar{\theta}_{km} \equiv \frac{\theta}{\theta^{km}}. \qquad [25]$$

4. Fundamental field equations

The microelements that constitute the macroelement of the given body are treated as continuous media. Therefore, the basic field equations of classical continuum mechanics are applicable to these microcontinua. To obtain the fundamental field equations for the hypothetical continuum representing the given body, we follow the approach initially advanced by *Eringen* and *Suhubi* (3) and employ the extensions presented by *Cook* and *Koh* (6) explicitly for the case of a general fluid with microstructural effects in both mechanical and thermal behavior.

Briefly, this involves the application of the *global balance principles* to the given body. These are expressed in terms of averaged quantities over the macroelement. On the assumption that these expressions of global conservation are applicable to all parts of the hypothetical continuum, no matter how "small" that part might be, we immediately obtain the fundamental field equations and jump conditions.

After lengthy but rather straight-forward calculations, we obtain the following basic field equations for a second-order microfluid with microstructural effects.

Conservation of mass and microinertias:

$$\dot{\varrho} + \varrho v^k{}_{,m} = 0 \qquad [26]$$

$$\dot{i}^{km} - v^k_q i^{qm} - i^{qk} v^m_q = 0 \qquad [27]$$

$$\dot{i}^{kmn} - v^k_q i^{qmn} - v^m_q i^{kqn} - v^n_q i^{kmq} = 0 \qquad [28]$$

$$\dot{i}^{kmnp} - v^k_q i^{qmnp} - v^m_q i^{kqnp} - v^n_q i^{kmqp} - v^p_q i^{kmnq} = 0. \qquad [29]$$

Balance of linear momentum and moments of momentum:

$$t^{km}{}_{,k} + \varrho(f^m - \dot{\sigma}^m) = 0 \qquad [30]$$

$$t^{kmn}{}_{,k} + (t^{nm} - s^{nm}) + \varrho(f^{mn} - \dot{\sigma}^{mn}) = 0 \qquad [31]$$

$$t^{kmnp}{}_{,k} + (t^{nmp} + t^{pmn} - s^{nmp} - s^{pmn}) + \varrho(f^{mnp} - \dot{\sigma}^{mnp}) = 0 \qquad [32]$$

$$t^{kmnpq}{}_{,k} + (t^{nmpq} + t^{pmnq} + t^{qmnp} - s^{nmpq} - s^{pmnq} - s^{qmnp}) + \varrho(f^{mnpq} - \dot{\sigma}^{mnpq}) = 0. \qquad [33]$$

Balance of energy and moments of energy:

$$\varrho\dot{\varepsilon} = t^{km} v_{m,k} + t^{kmn} \nu_{mn,k} + t^{kmnp} \mu_{mnp,k} + \nu_{mn}(t^{nm} - s^{nm}) + \mu_{mnp}(t^{nmp} + t^{pmn} - s^{nmp} - s^{pmn}) + q^k{}_{,k} + \varrho h. \qquad [34]$$

$$\varrho(\dot{\varepsilon}^k - v^k_m \varepsilon^m) = t^{mnk} v_{n,m} + t^{mnpk} \nu_{np,m} + \nu_{nm}(t^{mnk} - s^{mnk}) + q^{mk}{}_{,m} + q^k - \gamma^k + \varrho h^k. \qquad [35]$$

$$\varrho(\dot{\varepsilon}^{kp} - v^k_m \varepsilon^{mp} - v^p_m \varepsilon^{km}) = t^{mnkp} v_{n,m} - t^{mnokp} \nu_{no,m} + \nu_{nm}(t^{mnkp} - s^{mnkp}) + q^{mkp}{}_{,m} + q^{kp} + q^{pk} - \gamma^{kp} - \gamma^{pk} + \varrho h^{kp}. \qquad [36]$$

In the above field equations i^{km}, i^{kmn} and i^{kmnp} are the *microinertias* of various orders defined according to the following relations:

$$\varrho\, i^{km} dv \equiv \int_{dv} \varrho' \xi^k \xi^m dv'$$

$$\varrho\, i^{kmn} dv \equiv \int_{dv} \varrho' \xi^k \xi^m \xi^n dv'$$

$$\varrho\, i^{kmnp} dv \equiv \int_{dv} \varrho' \xi^k \xi^m \xi^n \xi^p dv'.$$

These tensors are defined in the deformed configuration and are related to the corresponding microinertias I^{KM}, I^{KMN}, ... of the reference configuration in rather complicated expressions involving micromotions χ^k_K and η^k_{KL}. However, consistent with the initial assumption of only *slight nonhomogeneity*, it may be shown [*Ahmadi* et al. (5)] that

$$i^{km} = \chi^k_K \chi^m_M I^{KM} \qquad [37]$$

$$i^{kmn} = \chi^k_K \chi^m_M \chi^n_N I^{KMN} \qquad [38]$$

$$i^{kmnp} = \chi^k_K \chi^m_M \chi^n_N \chi^p_P I^{KMNP}. \qquad [39]$$

In eqs. [30]–[33], $\dot{\sigma}^m$, $\dot{\sigma}^{mn}$, $\dot{\sigma}^{mnp}$ and $\dot{\sigma}^{mnpq}$ are the *inertia spin tensors* defined according to the following:

$$\varrho\dot{\sigma}^m dv \equiv \int_{dv} \varrho' a'^m dv'$$

$$\varrho\dot{\sigma}^{mn} dv \equiv \int_{dv} \varrho' a'^m \xi^n dv'$$

$$\varrho\dot{\sigma}^{mnp} dv \equiv \int_{dv} \varrho' a'^m \xi^n \xi^p dv'$$

$$\varrho\dot{\sigma}^{mnpq} dv \equiv \int_{dv} \varrho' a'^m \xi^n \xi^p \xi^q dv'.$$

In the momenta eqs. [30]–[33], f^m is the average body force, while f^{mn}, f^{mnp} and f^{mnpq} are body moments of various orders. These are defined according to

$$\varrho f^m dv \equiv \int_{dv} \varrho' f'^m dv'$$

$$\varrho f^{mn} dv \equiv \int_{dv} \varrho' f'^m \xi^n dv'$$

$$\varrho f^{mnp} dv \equiv \int_{dv} \varrho' f'^m \xi^n \xi^p dv'$$

$$\varrho f^{mnpq} dv \equiv \int_{dv} \varrho' f'^m \xi^n \xi^p \xi^q dv'.$$

The asymmetric stress tensor t^{km} is averaged over the surface ds of the macroelement:

$$t^{km} da_k \equiv \int_{ds} t'^{km} da'_k.$$

The *stress moments* t^{kmn}, t^{kmnp}, t^{kmnpq} are similarly averaged over ds. For example,

$$t^{kmnp} da_k \equiv \int_{ds} t'^{km} \xi^n \xi^p da'_k.$$

Stress averages over the macroelement dv are called *microstresses*. The first of these, s^{mn}, is defined as

$$s^{mn}\,dv \equiv \int_{dv} t'^{mn}\,dv'.$$

Higher order microstresses, e.g. s^{mnpq}, are actually moments of stress calculated over the macroelement. For s^{mnpq}, we have

$$s^{mnpq}\,dv \equiv \int_{dv} t'^{mn}\,\xi^p\,\xi^q\,dv'.$$

In the energy eqs. [34]–[36], scalar ε is the average *internal energy density:*

$$\varrho\,\varepsilon\,dv \equiv \int_{dv} \varrho'\,\varepsilon'\,dv'.$$

The vector ε^k and the second order tensor ε^{km} are respectively the first and second *moments of internal energy*, defined as follows:

$$\varrho\,\varepsilon^k\,dv \equiv \int_{dv} \varrho'\,\varepsilon'\,\xi^k\,dv'$$

$$\varrho\,\varepsilon^{km}\,dv \equiv \int_{dv} \varrho'\,\varepsilon'\,\xi^k\,\xi^m\,dv'.$$

The *heat flux vector* q^k is averaged over the surface ds of the macroelement:

$$q^k\,da_k \equiv \int_{ds} q'^k\,da'_k.$$

Similarly, the *heat flux moments* q^{km} and q^{kmn} are surfaces averages:

$$q^{km}\,da_k \equiv \int_{ds} q'^k\,\xi^m\,da'_k$$

$$q^{kmn}\,da_k \equiv \int_{ds} q'^k\,\xi^m\,\xi^n\,da'_k.$$

On the other hand, the *microheat flux* γ^k and the *microheat-flux moment* γ^{km} are averages of the heat flux over the macroelement dv:

$$\gamma^k\,dv \equiv \int_{dv} q'^k\,dv'$$

$$\gamma^{km}\,dv \equiv \int_{dv} q'^k\,\xi^m\,dv'.$$

The average heat source h and the average body moments of heat source h^k and h^{km} are defined as follows:

$$\varrho\,h\,dv \equiv \int_{dv} \varrho'\,h'\,dv'$$

$$\varrho\,h^k\,dv \equiv \int_{dv} \varrho'\,h'\,\xi^k\,dv'$$

$$\varrho\,h^{km}\,dv \equiv \int_{dv} \varrho'\,h'\,\xi^k\,\xi^m\,dv'.$$

Finally, it must be emphasized that in the derivation of the above equations a consistent simplification of the results has been obtained with the assumption that the fluid motion is only slightly nonhomogeneous. For a less restrictive but more complicated theory, one may refer to *Cook* and *Koh* (6).

5. A constitutive theory

To complete the formulation of the theory of a second-order microfluid, a set of constitutive equations is required. Such a set may be derived through the axiomatic approach indicated in numerous modern works in continuum mechanics, e.g. *Eringen* (10) and *Koh* (11).

For our purpose, we start with the *Clausius-Duhem* inequality appropriate for the given microcontinua:

$$\varrho'\,\dot{\eta}' - \left(\frac{q'^k}{\theta'}\right)_{,k} - \frac{\varrho'\,h'}{\theta'} \geq 0. \qquad [40]$$

To obtain the corresponding inequality for the entire body b, we first integrate [40] over the macrovolume dv and then extend the integration over the entire volume V and surface S of the given body. Using eq. [23] in the process, we have the following inequality:

$$\frac{D}{Dt}\int_V \varrho\,\eta\,dv - \int_S\left(\frac{q^k}{\theta} + \frac{q^{km}}{\theta^m} + \frac{q^{kmn}}{\theta^{mn}}\right) da_k$$
$$- \int_V \varrho\left(\frac{h}{\theta} + \frac{k^k}{\theta^k} + \frac{h^{km}}{\theta^{km}}\right) dv \geq 0 \qquad [41]$$

where η is the average entropy density of the hypothetical continuum defined by

$$\varrho\,\eta\,dv \equiv \int_{dv} \varrho'\,\eta'\,dv'.$$

Again imposing the condition that [41] is suitable for any portion of the entire body b, we obtain the following result:

$$\varrho\,\dot{\eta} - \left(\frac{q^k}{\theta} + \frac{q^{km}}{\theta^m} + \frac{q^{kmn}}{\theta^{mn}}\right)_{,k}$$
$$- \varrho\left(\frac{h}{\theta} + \frac{h^k}{\theta^k} + \frac{h^{km}}{\theta^{km}}\right) \geq 0. \qquad [42]$$

Finally, eliminating h, h^k and h^{km} from [42] with the use of [34]–[36], we obtain the basic inequality needed for the formulation of the constitutive equations. Using the notations [24] and [25], we have

$$\begin{aligned}
&\frac{\varrho}{\theta}[\theta\dot{\eta} - \dot{\varepsilon} - \bar{\theta}_k(\dot{\varepsilon}^k - v^k_m \varepsilon^m) \\
&- \bar{\theta}_{km}(\dot{\varepsilon}^{km} - v^k_p \varepsilon^{pm} - v^m_p \varepsilon^{kp})] \\
&+ \frac{1}{\theta}\{[t^{km} v_{m,k} + t^{kmn} \nu_{mn,k} + t^{kmnp} \mu_{mnp,k} \\
&+ (t^{nm} - s^{nm})\nu_{mn} \\
&+ (t^{nmp} + t^{pmn} - s^{nmp} - s^{pmn})\mu_{mnp}] \\
&+ \bar{\theta}_k[t^{mnk} v_{n,m} + t^{mnpk}\nu_{np,m} + (t^{nmk} - s^{nmk})\nu_{mn} \\
&+ q^k - \gamma^k] \\
&+ \bar{\theta}_{kp}[t^{mnkp} v_{n,m} + t^{mnokp}\nu_{no,m} \\
&+ (t^{nmkp} - s^{nmkp})\nu_{mn} \\
&+ q^{kp} + q^{pk} - \gamma^{kp} - \gamma^{pk}] \\
&- \theta\left(\frac{1}{\theta}\right)_{,k}(q^k + q^{km}\bar{\theta}_m + q^{kmn}\bar{\theta}_{mn}) \\
&- (\bar{\theta}_m)_{,k}\, q^{km} - (\bar{\theta}_{mn})_{,k}\, q^{kmn}\} \geq 0. \qquad [43]
\end{aligned}$$

Applying the *axiom of causality*¹) we identify the following set of thermal and kinematical variables as the *independent variables* of the constitutive theory:

Set A:

$$\theta,\ \bar{\theta}_k,\ \bar{\theta}_{km},\ v_{m,k},\ \nu_{kp},\ \nu_{kp,m},\ \mu_{mnp},\ \mu_{mnp,k},\ \left(\frac{1}{\theta}\right)_{,k},\ (\bar{\theta}_m)_{,k},\ (\bar{\theta}_{mn})_{,k}.$$

The *dependent variables* are then the following:

Set B:

$$\eta,\ \varepsilon,\ \varepsilon^k,\ \varepsilon^{km},\ t^{km},\ t^{kmn},\ t^{kmnp},\ t^{kmnop},\ s^{nm},\ s^{nmp},\ q^k,\ q^{km},\ q^{kmn},\ \gamma^k,\ \gamma^{km}.$$

From the *axiom of equipresence*²), each of the dependent variables of Set B is initially considered to be a function of all of the independent variables enumerated in Set A. As an example, one of the constitutive equations is of the form:

$$q^{km} = q^{km}\left[\theta, \bar{\theta}_k, \bar{\theta}_{km}, v_{m,k}, \nu_{kp}, \nu_{kp,m}, \mu_{mnp}, \mu_{mnp,k}, \left(\frac{1}{\theta}\right)_{,k}, (\bar{\theta}_m)_{,k}, (\bar{\theta}_{mn})_{,k}\right]. \qquad [44]$$

¹) See, for instance, *Koh* (11).

²) See, for instance, *Koh* (11).

The set of constitutive equations are valid for nonlinear microfluids. However, in this generality the equations are extremely complicated. Simplification may be made by imposing other axioms of constitutive theories and also by perhaps diminishing the effects of certain terms. A suitable form for the free energy function ψ may also be defined as done by *Grot* [8], and *Cook* and *Koh* [6]. With the application of the *Clausius-Duhem* inequality [43] for arbitrary motions and thermal conditions, some simplification of the functional dependence of the free energy function is achieved. The remaining constitutive equations are also expressible in terms of this function ψ.

We dispense with the details of this expansion and simply refer the reader to *Cook* and *Koh* (6) for some explicit examples.

Acknowledgments

The author is indebted to the National Science Foundation for the support in undertaking some of the work reported here. The help and contributions made by his collaborators *G. Ahmadi, V. W. Goldschmidt* and *R. D. Cook* in various calculations are also gratefully acknowledged.

References

1) *Eringen, A. C.*, Int. J. Engng. Sci. **2**, 205 (1964).

2) *Eringen, A. C.* and *T. S. Chang*, Recent Adv. Engng. Sci. **5**, Part II, 1 (1970).

3) *Eringen, A. C.* and *E. S. Suhubi*, Int. J. Engng. Sci **2**, 189 (1964).

4) *Ahmadi, G., S. L. Koh* and *V. W. Goldschmidt*, Recent Adv. Engng. Sci. **5**, Part II, 9 (1970).

5) *Ahmadi, G., S. L. Koh,* and *V. W. Goldschmidt*, Iranian J. Sci. & Techn. **1**, 233 (1971).

6) *Cook, R. D.* and *S. L. Koh*, AA & ES Techn. Rept. No. 72-9-5, Purdue University (1972).

7) *Wozniak, C* , Bull. Acad. Polon. Sci., Ser. sci. techn. **14**, 573 [957] (1966).

8) *Grot, R. A.*, Int. J. Engng. Sci. **7**, 801 (1969).

9) *Wozniak, C.*, ibid. **5**, 605 (1967).

10) *Eringen, A. C.*, Mechanics of Continua (New York 1967).

11) *Koh, S. L.*, Mechanics 1971, *N. C. Lind*, ed., American Academy of Mechanics, 213 (1971).

Author's address:

Prof. *Severino L. Koh*
School of Mechanical Engineering
Purdue University
West Lafayette, Indiana 47907 (USA)

Rheol. Acta 12, 425–429 (1973)

Recoverable shear in rheology

By *W. Philippoff* (*Linden, N.J., USA*)

With 2 tables

(Received October 27, 1972)

The concept of elasticity or potential energy in the flow of polymer solutions as viscoelastic liquids was introduced by *K. Weissenberg* (1) in 1928 as a cause for the non-Newtonian behavior. In 1948 he correlated the "recoverable shear *s*" (now sometimes called the *Weissenberg* number) with the normal stresses $P_{11} - P_{22}$ and the shear stress τ:

$$s = (P_{11} - P_{22})/\tau \qquad [1]$$

Mooney (2) in 1951 introduced "a state of elastic stress" in the flow of viscoelastic liquids resulting from the continuous stressing and their simultaneous relaxation. The elastic strain s was treated as a finite shear deformation leading to eq. [1]. This is now termed the "solid or quasi-static" approach. The author (3) introduced the shear modulus G from the classical rubber theory to calculate s from τ according to *Hooke*s law in shear:

$$\tau = G \cdot s \qquad [2]$$

and after *A. S. Lodge*s suggestion (4) concerning the coaxiality of the stress- and optical tensor also the relation between s and the experimentally accessible "extinction angle χ" in flow-birefringence:

$$s = 2 \cot 2\chi \qquad [3]$$

s is a necessary property of viscoelastic models, such as *Maxwell* body or its modern development, the *Rouse* body. If a *Maxwell* body is loaded, it deforms instantly by an amount $s = \tau/G$ and then flows with a constant viscosity η. Upon unloading it "recoils" – changes its deformation opposite to the initially acting stress – by the same amount s. The amount of stored potential energy in flow per cc is $1/2\tau s$. Any number of *Maxwell* bodies in parallel (distribution of relaxation times) do not change this picture qualitatevely. A viscosity in parallel causes the initially instant response to become one with a finite speed, practically similar to an exponential approach to equilibrium.

The modern description of polymer molecules as freely jointed coils whose shape depends on the *Brownian* motion, leads to the "entropy-elasticity": G is determined by the presence of n particles/cc, in first approximation $G = nkT$, where k is the *Boltzmann* constant and T the absolute temperature in °K. A more elaborate statistical treatment by *Rouse* (5), *Zimm* (6) and *Williams* (7) added a multiplicative constant of 1.25 for the freely draining coil. The author prefers to use twice this constant as "A" to describe polymer solutions (8). For the rubber theory $A = 2$, for the *Rouse*-body $A = 2.5$, *Zimm* body $A = 5$, etc. Polydispersity can lower this value (9), say, to 1.5.

G_0 can be calculated from the rubber theory for monodisperse polymers and twice ratio of the experimental G to G_0 is A. With c_w the concentration in weight %, ϱ the density in g/cc and M the molecular weight in 10^6 the practical expression for G_0 is:

$$G_0 = 248 \cdot c_w \cdot \varrho/M \text{ in dynes/cm}^2. \qquad [4]$$

The concept of s leads to the existence of a third coaxial tensor in laminar flow: the elastic deformation tensor, one component of which is

s. Without further assumptions, finite shear causes $\chi < 45°$: the principal tensile stress follows the same direction and then normal stresses appear: s is the origin of the normal stresses (see eq. [1]). s requires the shear stress τ to be the cause of the observed effects, not the rate of shear. Therefore plotting of s or $P_{11} - P_{22}$ vs. τ in the non-*Newton*ian range result in simpler relations than vs. the rate of shear. This is indeed the case.

Further small variations in temperature T should not influence the s vs. τ relation. A change of viscosity with T changes the rate at constant τ, therefore the s vs. rate relation changes. Experimentally this is true: s vs. τ is nearly independent of T, not the more easily accessible $P_{11} - P_{22}$ vs. rate of shear.

One can use both steady-state and non-steady state methods to measure s. The steady-state methods most suitable are the normal stress measurements and flow-birefringence.

The reason flow-birefringence can be used to determine rheological properties (Rheo-optics) lies in the nature of the behavior of coiled molecules in flow. As introduced by *W. Kuhn* (10), the statistical element has a certain length. The force on the whole molecule in flow is calculated from the super-position of the action of these elements. The optical behavior is calculated by ascribing a difference in polarizabilities parallel to the element's length. The calculation of the observable effects is identical for the stresses and the birefringence independent of the particular mechanism chosen. This leads directly to the stress-optical law (proportionality of birefringence and difference of principal stresses) and the coaxiality of the tensors mentioned above. This law has been extensively proven for polymer solutions in very large ranges of variables, in a *Newton*ian or non-*Newton*ian range, for dilute and concentrated solution as well as monodisperse and polydisperse polymers. Besides χ gives an easily accessible new parameter, not directly measurable by mechanical means.

The coaxiality of stress- and optical tensor was first tested on polymer solutions (3, 11), recently also on polymer melts (12). χ was calculated by eq. [1] and [3] from normal stress (mechanical) measurements and compared with the directly measured optical χ. Both sets of values as a function of the rate of shear or τ coincide within 0, 5°, the experimental error. A better proof is to calculate s from the mechanical measurements by eq. [1] and [3], plotting them in a wide range of τ. Using the *Weissenberg* Rheogoniometer and flow-birefringence measurements these measurements have been made in a wide range of τ, concentrations and polymers, solutions as well as melts, polydisperse and monodisperse. The more carefully the measurements are done, the better is the check between both sets of values, even if performed in different laboratories for each set (13, 14).

This conclusively proves the coaxiality of the tensors and the applicability of optical measurements for determining normal stresses independent of the magnitude of s, and if the viscosity is constant or not, not only as a limiting law. Further, it is clear as expected that the principal stress is not oriented under 45° under the conditions used.

Any material law in mechanics is not valid for any amount of deformation: there is always a limit beyond which initially linear relations become non-linear. This is also the case with s. In general eq. [2] is valid in a wide range of τ. However, in more dilute solutions s tends to level off at higher τ, for some concentrated solutions on the contrary s increases slightly more than proportionally with τ. Such effects, without a thorough theoretical investigation of the statistics of a strongly deformed coil cannot be interpreted, but only stated as experimental facts.

The magnitude of s varies largely from the detectability limit of $\sim$0.01 shear units to a commonly occurring value of 20, sometimes up to 60. For some Al-soap gels even value of 400 have been found. This is considerably greater than 10, the ultimate elongation of gum-rubber.

Non-steady state measurements, especially recoil-measurements, are the most direct manifestation of s. To perform such tests one can use a rotational instrument, which must be unloaded practically instantly after a steady state flow and the motion of the moving part measured. This has also been defined as "creep-recovery" when very viscous materials, such as polymer melts are tested. A number of such devices have been used, the one in our laboratory is loaded by weights that are "picked up" by suitable means at a predetermined point. Any device has a finite inertia, therefore an estimate of its influence on the measurement has been made. An "instant" recoil is limited by the resonance frequency of the instrument, determined by its dimensions, inertia, the shear modulus of the liquid, as well as the degree of damping. The recoil may also be influenced by the simultaneous stress-relaxation during the measurement, which if comparable in rate to the recoil experiment, diminishes the measured value. In the non-Newtonian range there is an ambiguity as to which viscosity value to use: the initial viscosity or the viscosity "at rate". Estimates show that for most cases (using η_0) the recoil proceeds much faster than the stress-relaxation enabling valid results. Friction-free instruments are of course required, any friction as well as other errors tend to diminish the measured value of s.

Recoils, as well as steady-state value have been measured for a large number of polymer solutions and plotted according to eq. [2]: s vs. τ (13, 14). In all cases the slope of the resulting nearly straight line (G) was identical for all three tests. For the recoil measurements, it is mostly the case below $s = 0.35$, sometimes even $s = 1$, depending on the material. This proves, beyond the identity of eq. [1] and [3], that s has indeed a physical meaning: the recoil. It is not a "number". At higher s the recoil is increasingly less than the steady-state values, apparently reaching some constant fraction of the latter, maybe <0.5.

The reasons for a non linear s vs. τ relation is recoil are not completely known, one possibility is the mentioned interference of the stress-relaxation, which is known to become faster at

Table 1. Monodisperse Polystyrenes in Aroclor 1248 at 25 °C. (Flow-birefringence and Rheogoniometer results)

Code	$M \cdot 10^6$	c_w%	G_0 dynes/cm²	G dynes/cm²	A
S 111 (Dow Co)	0.220	JK 100	—	—	2.5
$M_w/M_n = 1.07$		20	30600	30400	2.00
		15	23500	21500	1.83
		10	15450	15500	2.01
		7	11250	10900	2.06
		4.10	6700	6700	2.00
		4.00	6520	6500	2.00
		3.00	4920	4800	1.95
		2.00	3310	3800	2.30
		1.00	1660	2300	2.77
		0.50	835	1180	2.83
		0.30	503	600	2.39
		0.15	250	350	2.80
		0.10	170	260	3.06
		0.05	85	130	3.04
Lot 14a	1.8	5.00	980	1150	2.35
(Pressure Chem. Co)		4.00	785	950	2.42
$M_w/M_n = 1.25$		3.25	642	720	2.24
		3.00	592	500	1.69
		2.00	397	280	1.41
		1.00	199	160	1.61
		0.50	100	95	1.90
NBS-2-35967	0.45	7.14	4770	4100	1.72
(Pressure Chem. Co)					

Last one in TCP/Aroclor
JK- melt at 196 °C by *Janeschitz-Kriegl*, Delft (Netherlands)

higher rates. The decrease in the relaxation time spectrum is presently being discussed, but its origin is far from being understood (15, 16).

Assuming a "memory function" as the origin of the viscoelastic behavior of "second order" or "rubberlike" fluids (17) only 1/2 as large an s has been calculated as from eq. [1]: "the factor of 2". Some measured values are between these limits. If the instruments were error-free, it is possible that the measurements were performed outside the mentioned limit of $s = 0.35$ or there was an influence of the mentioned stress-relaxation.

The recoil after stationary laminar flow can be more or less understood from linear viscoelasticity. However, if one measures s with increasing times, s can have a pronounced maximum (overshoot) or even maxima and minima at $\tau = \text{const.}$ (18). This is similar to the overshoot in the Rheogoniometer, when normal stresses at constant rate are measured. These maxima occur in the purely elastic deformation range (19), before any measurable irreversible flow had occurred. They are at present unexplainable without some very arbitrary assumptions or the introduction of new constants for the specific purpose. Similar overshoots are also observed in flow-birefringence: one can therefore state that they occur in any viscoelastic tests, given the right combination of material, instrument and conditions.

The constant A has been calculated, as mentioned above, for a large number of "monodisperse" polymer solutions. The results are listed in table 1. Some polymers (Lot 14a) are not as monodisperse as would be necessary, but are the best currently available. In a large range of M and c_w, not only for dilute solutions and the limit of small stresses, A is near to the value of 2.5 of *Rouse*. This proves that, using the reasoning

Table 1a. Monodisperse Polymethylstyrenes (Nagasawa, Nagoya) in Aroclor 1248 at 25 °C (only flow-birefringence)

Code	$M \cdot 10^6$	c_w%	G_0 dynes/cm²	G dynes/cm²	A
α-113	3.0	1.00	120	160	2.67
		0.50	60	62	2.07
		0.25	30	40	2.66
		0.15	18	28	3.10
α-112	1.8	1.00	199	220	2.21
		0.50	100	110	2.20
		0.25	50	70	2.80
		0.15	30	48	3.20
		0.10	20	36	3.60
α-317	0.513	0.50	350	520	2.97
		0.25	175	285	3.25
		0.15	105	160	3.04
		0.10	70	55	1.58
α-004	0.31	(N) 7	7170	8700	2.42
		1.00	1155	1600	2.77
		0.50	578	900	3.12
		0.25	289	440	3.05
		0.10	116	180	3.10
		0.05	58	115	3.97
α-5	0.058	2.58	15900	24000	3.02
		1.00	6200	14500	4.68
		0.50	3100	7400	4.77

(N) *Nagasawa* Rheogoniometer measurements.
$M_w/M_n < 1.05$

presented, one can predict the elastic behavior of monodisperse polymer solutions from non-rheological constants.

A direct conclusion from eq. [3] is following relation connecting steady-state and vibration measurements: (4, 20)

$$\delta = 2\chi \text{ or } 2G' = P_{11} - P_{22} \qquad [5]$$

where δ is the loss angle" and G' the shear storage modulus in vibrations for a rate of shear = circular frequency. The relation has been proved in numerous cases, mostly for lower frequencies or smaller s. This can be understood, as in steady state all partial mechanisms are exited, giving a higher s than the one at a single frequency, as only the mechanisms limited by it are exited. This relation, in the limit of frequency → 0 has also been derived from the "second order fluid" theory (21). Peculiarly G' is independent of the amplitude of vibration (linear viscoelasticity), whereas $P_{11} - P_{22}$ is a strongly non-linear effect.

A further conclusion, (3,20) following *Weissenberg*, is that the non-*Newton*ian viscosity depends on s through χ:

$$\eta = \eta_0 \cdot \sin 2\chi \qquad [6]$$

Indeed the principal values in birefringence give straighter lines when plotted over the rate of shear than the respective flowcurve, supporting idea.

Thus one sees that using the concept of s one can build a system that explains or predicts a number of rheological phenomena.

References

1) *Weissenberg, K.*, and *R.O. Herzog*, Kolloid-Z. **46**, 277 (1928).
2) *Mooney, M.*, J. Colloid Sci. **6**, 96 (1951).
3) *Philippoff, W.*, J. Appl. Physics **27**, 984 (1956); Trans. Soc. Rheol. **1**, 95 (1957).
4) *Lodge, A.S.*, Nature **176**, 838 (1955).
5) *Rouse, P.E.*, J. Chem. Physics **21**, 603 (1954).
6) *Zimm, B.*, J. Chem. Physics **24**, 269 (1956).
7) *Williams, M.C.*, J. Chem. Physics **42**, 2988 (1965).
8) *Philippoff, W.*, Trans. Soc. Rheology **10**,**1**, 1 (1966).
9) *Daum, U.*, J. Polymer Sci. **A2**, 141 (1968).
10) *Kuhn, W.* and *H. Kuhn*, Helv. Chim. Acta **26**, 1394 (1943).
11) *Philippoff, W.*, Nature **178**, 811 (1956).
12) TNO-Group, personal communication.
13) *Philippoff, W.* and *R.A. Stratton*, Trans. Soc. Rheology **10**, **2**, 467 (1966).
14) *Philippoff, W.* and *R.A. Stratton*, Proc. V Congr. Rheol. Vol. **4**, 13 (1970).
15) *Vinogradov, G. V.* and *A. Y. Malkin*, J. Polymer Sci. **A4**, 135 (1966). Earlier *G. V. Vinogradov* and *A.I. Leonov*, Doklady Akad. Nauk **155**, 406 (1964).
16) *Yamamoto, M.*, Trans. Soc. Rheology **15**, **2**, 331 (1971).
17) *Lodge, A.S.*, Elastic Liquids (New York 1964).
18) *Trapeznikov, A.A.* and *A.T. Pylaeva*, Vysokomol. Soed. **12**, 1294 (1970), and earlier papers, since 1956.
19) Unpublished measurements.
20) *Philippoff, W.*, J. Appl. Physics **36**, 3033 (1965) and Trans. Soc. Rheology **12**, **1**, 85 (1968).
21) *Coleman, B.D.* and *H. Markovitz*, J. Appl. Physics **35**, 1 (1964).

Author's address:

Prof. Dr. *W. Philippoff*,
Formerly with Esso Research & Engineering Company,
Products Research Division, P.O. Box 51.
Linden, N.J. 07036 (USA).

Rheol. Acta 12, 430–437 (1973)

Calcul de prévision des matériaux composites viscoélastiques renforcés de fibres unidirectionnelles

Par Yvon Chevalier et Vinh Tuong (Saint Ouen/France)

Avec 4 figures et 1 table

(Reçu p. p. 27 Octobre 1972)

1. Introduction

L'un des problèmes importants à résoudre pour les matériaux composites est le calcul de prévision de tels corps, calcul qui tient compte de la géométrie et des caractéristiques mécaniques de chaque phase. A travers une littérature scientifique extrêmement abondante quoique relativement récente, nous croyons distinguer les méthodes d'approche suivantes:

a) Utilisation de la technique des bornes; *Hill* (5) – Constantes élastiques. *Roscoe* (8) – Parties réelles et imaginaires des modules complexes en viscoélasticité.

b) Utilisation de la fonction d'*Airy*, calcul en élastostatique, méthodes numériques sur ordinateur; *Pickett* (7), *Leissa* et *Clausen* (6).

c) Calcul en élastodynamique par la méthode dite des ondes longues *Behrens* (1).

Il arrive fréquemment que l'un des deux composants du matériau orthotrope (ou les deux à la fois) est viscoélastique. Les méthodes numériques, valables en élasticité, s'avèrent alors difficilement exploitables car elles ne permettent pas de passer aux corps viscoélastiques par le principe dit de correspondance. Notre objectif, dans la première phase de notre étude, a donc été d'utiliser la mèthode b) la plus «rigoureuse» et concédant quelques approximations sur les conditions aux limites de tenter d'aboutir à une solution assez simple en élasticité. Dans une deuxième phase nous sommes passés au domaine viscoélastique. Des modèles simples nous ont permis de mettre en évidence des fonctions de relaxation du composite. Par contre des mesures sur les caractéristiques des composants nous ont permis de traiter numériquement le problème en régime harmonique, la connaissance des modèles rhéologiques de chaque composant n'étant plus nécessaire. Cette dernière approche pourrait être utile sur le plan pratique.

II. Calcul des paramètres élastiques d'un matériau composite à symétrie hexagonale

1. Hypothèses de travail

Pour que les expressions soient relativement maniables sans pour autant adopter des formules simplificatrices trop empiriques, nous sommes amenés à poser les hypothèses suivantes:

a) Constituants

Le matériau étudié sera constitué de fibres circulaires homogènes et isotropes de rayon r_2 (indice 2) noyées dans une matrice elle même homogène et isotrope (indice 1). Les fibres seront réparties hexagonalement (cf. fig. 1a). Le comportement de ces deux constituants sera supposé élastique et défini par les coefficients de *Lamé* (λ_1, μ_1), (λ_2, μ_2) ou éventuellement les modules d'*Young* et de *Poisson* (E_1, ν_1), (E_2, ν_2).

b) Composite

Nous admettons ici que le matériau composite est statistiquement homogène, nous postulons de plus que tout système de charge crée dans un échantillon de ce matériau des grandeurs mécaniques (σ_{ij} contraintes et U_i déplacement) périodiques, périodicité dans le plan normal à l'axe des fibres. Le composite est alors représenté par un volume élémentaire représentatif (V.E.R.) $\mathscr{V}$, cylindrique, de directrice hexagonale $\mathscr{B}$ (ou maille), et de génératrices parallèles à l'axe des fibres (cf. fig. 1b).

L'hypothèse d'homogénéité nous autorise à écrire le comportement du composite sous la forme:

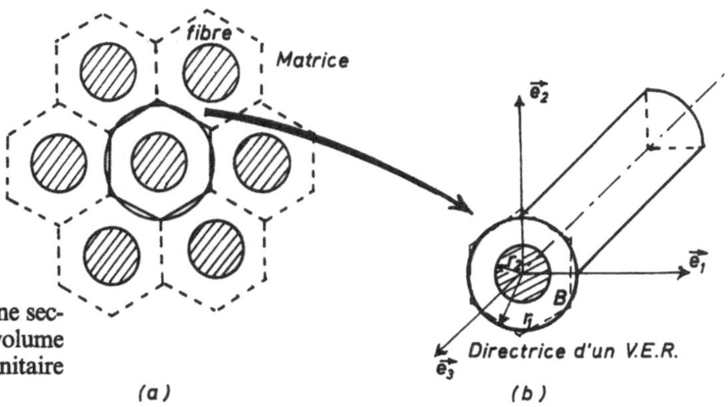

Fig. 1. a) Disposition des fibres circulaires dans une section transversale. b) La section hexagonale du volume élémentaire représentatif V.E.R. de longueur L unitaire est approchée par un cercle dans le plan $(\vec{e}_1, \vec{e}_2)$

$$\bar{\sigma}_{ij} = \bar{C}_{ijkl}\bar{\varepsilon}_{kl} \qquad [1]$$

où $\bar{C}_{ijkl}$ n'est fonction que des paramètres (λ_α, μ_α) ($\alpha = 1$ correspondant à la matrice tandis que $\alpha = 2$ représentant la fibre) et du pourcentage T de fibres.

La périodicité permet enfin d'interpréter $\bar{\sigma}_{ij}$ et $\bar{\varepsilon}_{kl}$ comme des valeurs moyennes des contraintes et des déformations sur le V.E.R. On notera que des déplacements $\vec{\hat{U}}$ de la forme:

$$\hat{U}_i = a_{ij}x_j; \quad i = 1, 2, 3; \quad a_{ij} = \text{constante} \qquad [2]$$

sur le bord $\partial\mathscr{V}$ du V.E.R. sont des conditions suffisantes d'existence des nombres $\bar{C}_{ijkl}$ [cf. (2)].

c) Conditions aux limites

Afin d'obtenir des solutions littérales qui permettent, à l'aide du principe de correspondance de passer au domaine viscoélastique, nous approchons l'hexagone régulier $\mathscr{B}$ par un cercle de rayon r_1 (cf. fig. 1b). Cette approximation devant respecter le volume relatif T de fibres ($T = 1/R^2 = r_2^2/r_1{}^2$) et les valeurs élastiques dans les cas extrêmes ($T = 0$, $T = 1$).

Les relations classiques relatives aux milieux orthotropes montrent qu'un tel matériau est défini par la connaissance de 5 coefficients $\bar{C}_{ijkl}$ [cf. (4) et (2)].

2. Calcul des constantes élastiques

Le champ des déformations planes $(\vec{e}_1, \vec{e}_2)$ dans la section transversale permet de calculer, sous forme de séries, les fonctions de contrainte qui satisfont, en coordonnées polaires, (r, θ) à l'équation biharmonique:

$$\sigma_{ij} = e_{ip}e_{jq}\frac{\partial^2\Phi}{\partial x_p \partial x_q}$$

$$e_{11} = e_{22}\ 0; \; e_{12} = -e_{21} = 1. \qquad [3]$$

$$\Phi(r, \theta) = \sum_{n=1}^{\infty} \alpha_{2n}(r)\cos 2n\theta + \tfrac{1}{2}\alpha_0(r) \qquad [4]$$

avec respectivement pour la matrice et les fibres:

$$2\alpha_0 = \mu_1(a_0 + b_0 \log r + c_0 r^2 + d_0 r^2 \log r)$$

$$\alpha_{2n} = \mu_1(a_{2n}r^{2n} + b_{2n}r^{-2n} + c_{2n}r^{2n+2} + d_{2n}r^{-2n+2}); \qquad [5]$$

$$2\alpha_0 = \mu_2(e_0 + f_0 r^2)$$

$$\alpha_{2n} = \mu_2(e_{2n}r^{2n} + f_{2n}r^{2n+2}). \qquad [6]$$

En imposant sur le bord $\mathscr{B}$ les déplacements: $U_1 = ux_1$, $U_2 = -vx_2$, $U_3 = 0$, on calcule les constantes $a_i, b_i, c_i, d_i, e_i, f_i$ en écrivant les conditions aux limites (contact parfait entre fibre et matrice).

Seuls les coefficients c_0, f_0, b_2, c_2 sont nécessaires aux calculs ultérieurs; $Q = \mu_2/\mu_1$ étant le rapport des deux modules de coulomb, on trouve:

$$c_0 = \frac{Q + 1 - 2\nu_2}{2[(1 - 2\nu_1)(1 - T)Q + (1 - 2\nu_2)(1 - 2\nu_1 + T)]}$$

$$f_0 = \frac{1 - \nu_1}{(1 - 2\nu_1)(1 - T)Q + (1 - 2\nu_2)(1 - 2\nu_1 + T)}$$

$$b_2 = \frac{3(1 - Q)}{2\Delta_1}\{[3 - 4\nu_2 + Q] \times [T^2 - T^3 + (1 - T^3)(3 - 4\nu_1)] + 4(\nu_1 - 1)T^2[(4\nu_2 - 3) + Q(1 - T)]\}$$

$$c_2 = \frac{3T^2(1 - Q)(1 - T)(3 - 4\nu_2 + Q)}{\Delta_1} \qquad [7]$$

avec

$$\Delta_1 = \{(4\nu_1 - 3)[T + Q(1 - T)] - 1\} \times \{3(3 - 4\nu_2 + Q)(3 - 4\nu_1)(1 - T^3) + 12T^2(\nu_1 - 1)(4\nu_2 - 3)\} - 9T(1 - Q)(3 - 4\nu_2 + Q)(1 - T)^2.$$

On peut montrer que les coefficients c_0, f_0, b_2, et c_2 sont positifs pour $(\nu_1, \nu_2, T) \in \Omega$ et $Q \geq 1$, Ω étant l'ouvert de R^3 limité par les plans $\nu_2 = 0$, $T = 0$, $T = 1$, $\nu_1 = 1/2$, $\nu_1 = \nu_2$.

A partir des tenseurs moyens des contraintes et des déformations, les moyennes envisagées étant arithmétiques [cf. (6) et (7)], on détermine finalement les modules élastiques du composite:

$$\bar{C}_{1111} = \mu_1(T_0 + T_2);$$

$$\bar{C}_{1122} = \mu_1(T_0 - T_2)$$

$$\bar{C}_{1133} = 4\mu_1[T\nu_2 Q f_0 + \nu_1 c_0(1 - T)]; \quad [8]$$

T_0 et T_2 valant respectivement:

$$T_0 = 4c_0(1 - \nu_1) - 1;$$

$$T_2 = 1 - 8(1 - \nu_1)[b_2 T^2 + c_2(1 - 2\nu_1)/T]. \quad [9]$$

Une traction simple dans la direction $\vec{e}_3$ et un cisaillement simple dans le plan $(\vec{e}_1, \vec{e}_2)$ permettent de calculer séparément les deux autres constantes:

$$\bar{C}_{2323} = \mu_1 \frac{(1 + Q) + (Q - 1)\,T}{(1 + Q) + (1 - Q)\,T},$$

$$\bar{C}_{3333} = T(E_2 - E_1) + E_1 + \bar{C}^2_{1133}/\mu_1 T_0. \quad [10]$$

Les relations [8] et [10] ont été confrontées d'une part avec les expressions obtenues par divers chercheurs, d'autre part avec des mesures. Cette comparaison tend à montrer que notre méthode de calcul constitue une approximation raisonnable (cf. fig. 2 et 3).

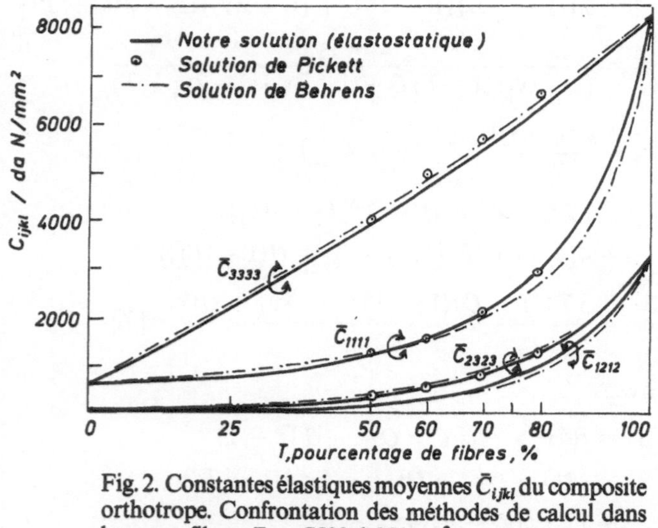

Fig. 2. Constantes élastiques moyennes $\bar{C}_{ijkl}$ du composite orthotrope. Confrontation des méthodes de calcul dans le cas: fibre $E_2 = 7500$ daN/mm², $\nu_2 = 0{,}2$; matrice $E_1 = 371$ daN/mm², $\nu_1 = 0{,}34$

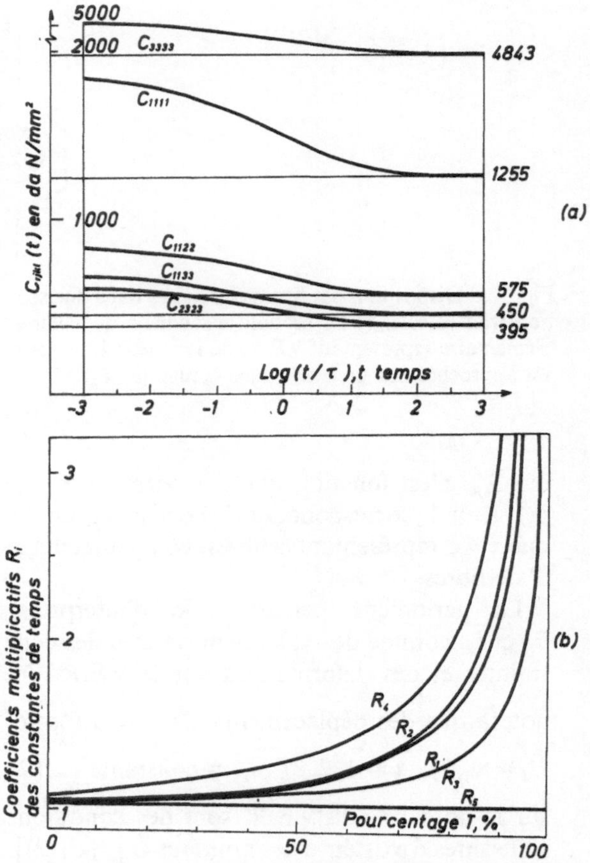

Fig. 3. Caractéristiques des composants: $E_2 = 7500$ daN/mm², $\nu_2 = 0{,}20$; $E_1 = 480$ daN/mm², $\nu_1 = 0{,}34$, $k = 0{,}625$, $\tau = 1{,}581$ s. a) *Modèle de Zener:* Variation en fonction du temps des fonctions de relaxation du composite. b) *Modèles de Maxwell et de Zener:* Variation en fonction du pourcentage des coefficients multiplicatifs R_i des constantes de temps

Table 1. Différentes valeurs des modules élastiques d'un composite silice-résine phénolique comportant 63% de fibres dans le cas: fibre $E_2 = 7500$ daN/mm², $\nu_2 = 0{,}2$; matrice $E_1 = 371$ daN/mm², $\nu_1 = 0{,}34$. (Unités: daN/mm²)

	Notre solution	Mesures *(T. Vinh)*	Solution *(E. Behrens)*	Solution *(G. Pickett)*
$\bar{C}_{1111}$	1616	1733	1651	1684
$\bar{C}_{1122}$	718	745	683	667
$\bar{C}_{1133}$	569	642	569	571
$\bar{C}_{3333}$	5140	4952	5142	5141
$\bar{C}_{2323}$	515	582	515	519
$\bar{C}_{1212}$	449	494	483	509

III. Calcul des paramètres viscoélastiques d'un materiau composite a symétrie hexagonale

Un milieu est viscoélastique si, au voisinage d'un état d'équilibre qui est un état neutre, à chaque instant l'état des contraintes est une fonctionnelle linéaire des déformations existant antérieurement. L'application de la transformation de *Carson*[1]) à ces lois intégrales permet d'écrire le comportement du matériau sous la forme:

$$\sigma^*_{ij}(p) = \lambda^*(p)\,\varepsilon^*_{kk}(p)\,\delta_{ij} + 2\mu^*(p)\,\varepsilon^*_{ij}(p) \qquad [11]$$

p désignant la variable complexe.

Adoptant les mêmes hypothèses de travail et la même méthode d'étude utilisée pour le composite élastique, nous pouvons écrire les lois reliant les contraintes moyennes aux déformations moyennes:

$$\bar{\sigma}^*_{ij}(p) = \bar{C}^*_{ijkl}(p)\,\bar{\varepsilon}^*_{kl}(p). \qquad [12]$$

Si de plus les hypothèses suivantes sont vérifiées,
– quasi stabilité;
– forces volumiques nulles;
– conditions aux limites écrites sous forme de produit d'une fonction temporelle par une fonction de la variable spatiale;
nous sommes en droit d'appliquer le principe de correspondance. Nous pouvons alors écrire la relation:

$$\bar{\sigma}^*_{ij}(p) = \bar{C}_{ijkl}[\mu^*_1(p), \nu^*_1(p), \mu^*_2(p), \nu^*_2(p)]\,\bar{\varepsilon}^*_{kl}(p) \qquad [13]$$

à partir des expressions littérales des constantes $\bar{C}_{ijkl}$ dans [8] et [10]. ν_i et μ_i, $i = 1, 2$, ont été simplement remplacés par leur transformée de *Carson*.

Des égalités [12] et [13] nous obtenons immédiatement les paramètres viscoélastiques du composite:

$$\bar{C}^*_{ijkl}(p) = \bar{C}_{ijkl}[\mu^*_1(p), \nu^*_1(p), \mu^*_2(p), \nu^*_2(p)]. \qquad [14]$$

L'étude que nous présentons ici est applicable aux problèmes dynamiques à basse fréquence telle que la longueur d'onde Λ est très supérieure à la plus grande dimension du V.E.R. Nous allons illustrer ce calcul par deux exemples simples où la fibre est élastique et la matrice caractérisée par une seule fonction de relaxation $\mu_1(t)$, $\nu_1(t)$ = constante. Cette hypothèse nous parait raisonnable pour un grand nombre de matériaux composites où l'on cherche à exploiter les «résistances mécaniques» élevées des fibres, la matrice plus souple servant à transmettre les efforts d'une fibre à l'autre.

1. Modèle de Maxwell

Adoptons pour $\mu_1(t)$ le modèle de *Maxwell*:

$$\mu_1(t) = \mu_1 e^{-t/\tau} \Leftrightarrow Q^*(p) = \frac{\mu_2}{\mu^*_1(p)} = Q_0\,\frac{(p + 1/\tau)}{p} \qquad [15]$$

avec

$$Q_0 = \frac{\mu_2}{\mu_1}.$$

En vertu des relation [8] et [10], $\bar{C}^*_{ijkl}(p)$ sera un rapport de deux polynômes en p de même degré. Le problème se ramène alors à une décomposition de fractions rationelles en éléments simples. L'inversion à partir des transformées donne:

$$\bar{C}_{ijkl}(t) = \bar{C}_{ijkl}\Big[K_{ijkl} + \sum_{s=0}^{s=5} X_{ijkl}(s)\exp(-t/R_s\tau)\Big], \qquad [16]$$

Les coefficients K_{ijkl} et X_{ijkl} sont des «constantes» dépendant des caractéristiques mécaniques des constituants et du pourcentage volumique de fibres tandis que $\bar{C}_{ijkl}$ désigne la valeur trouvée en élasticité (cf. relations [8] et [10]). $R_0 = 1$, on peut montrer que les coefficients multiplicatifs $R_s (s \geq 1)$ sont strictement supérieurs à 1 pour tout élément (ν_1, ν_2, T) de Ω et $Q_0 \geq 1$ [cf. fig. 3b et (2)]. La décroissance et la positivité des fonctions $\bar{C}_{ijkl}(t)$ peuvent être etablies sur des exemples numériques ou en valeur littérale pour $\bar{C}_{1133}(t)$ et $C_{2323}(t)$.

$$K_{1111} = K_{1122} = K_{1133} = K_{2323} = 0$$

$$K_{3333} = TE_2/\bar{C}_{3333}. \qquad [17]$$

Dans le tableau ci-dessous signalons les nombres X_{ijkl} non nuls dont les valeurs explicites seront données en annexe:

[1]) Si $f(t)$ est une fonction localement sommable, vérifiant en outre des conditions restrictives, on appelle transformée de *Carson* de f l'intégrale:

$$f^*(p) = p\int_0^\infty f(t)\,e^{-pt}\,dt.$$

ijkl \ s	0	1	2	3	4	5
1111	$X_{1111}(0)$	0	$X_{1111}(2)$	$X_{1111}(3)$	$X_{1111}(4)$	0
1122	$X_{1122}(0)$	0	$X_{1122}(2)$	$X_{1122}(3)$	$X_{1122}(4)$	0
1133	$X_{1133}(0)$	0	$X_{1133}(2)$	0	0	0
3333	$X_{3333}(0)$	0	$X_{3333}(2)$	0	0	$X_{3333}(5)$
2323	$X_{2323}(0)$	$X_{2323}(1)$	0	0	0	0

On remarquera que les relations [16] fournissent les caractéristiques de la matrice ou de la fibre pour les valeurs extrêmes du pourcentage T, les coefficients multiplicatifs R_s étant donnés par les relations [18] suivantes:

$$R_0 = 1$$

$$R_1 = \frac{Q_0 + 1 + (1 - Q_0)\,T}{Q_0(1 - T)}$$

$$R_2 = \frac{(1 - 2\nu_1)(1 - T)\,Q_0}{Q_0(1 - 2\nu_1)} \frac{+ (1 - 2\nu_2)(1 - 2\nu_1 + T)}{\times (1 - T)}$$

$-1/R_3\tau$, $-1/R_4\tau$ sont les zéros en p de $\Delta_1(Q^*)$, Q^* étant donné par [15].

$$R_5 = \frac{Q_0[T(1 - 2\nu_1) + 1] + (1 - 2\nu_2)(1 - T)}{Q_0[T(1 - 2\nu_1) + 1]} \quad [18]$$

2. Modèle de Zener

Ce modèle est plus voisin des corps réels que le précédent, il se traduit par les relations [19] suivantes:

$$\mu_1(t) = \mu_1[k + (1 - k)\,e^{-t/\tau}] \Leftrightarrow Q^*(p) = Q_0 \frac{(p + 1/\tau)}{(p + k/\tau)}, \quad 0 \le k \le 1. \quad [19]$$

Un changement de variables:

$$p' = p + k/\tau$$

$$\frac{1}{\tau'} = (1 - k)/\tau$$

permet de se ramener au modèle de *Maxwell*, p' ayant remplacé p et τ' ayant remplacé τ. Les résultats de [16] sont alors facilement transposables:

$$\bar{C}_{ijkl}(t) = \bar{C}_{ijkl}\Big[K'_{ijkl} + \sum_{s=0}^{s=5} X'_{ijkl}(s) \exp(-t\,\theta_s/R_s\tau)\Big] \quad [20]$$

avec

$$\theta_s = kR_s + 1 - k, \; s = 0, 1, 2, 3, 4, 5. \; \theta_s \ge 1.$$

Le calcul montre que les constantes K'_{ijkl} et X'_{ijkl} sont des combinaisons linéaires des nombres K_{ijkl} et X_{ijkl}, plus précisément:

$$K'_{ijkl} = K_{ijkl} + k \sum_{s=0}^{s=5} X_{ijkl}(s)\,R_s/\theta_s$$

$$X'_{ijkl}(s) = (1 - k)\,X_{ijkl}(s)/\theta_s. \quad [21]$$

La positivité et la non croissance des fonctions de relaxation données par les égalités [20] n'ont pas été établies littéralement mais numériquement pour un couple de matériaux donnés (cf. fig. 3a).

On peut généraliser à un modèle plus élaboré pour $\mu_1^*(p)$. Notons que n temps de relaxation de cette dernière donne lieu à $6n$ temps de relaxation pour les 5 paramètres $\bar{C}^*_{ijkl}(p)$. Faute de place nous n'aborderons pas le cas général de deux constituants viscoélastiques caractérisés chacun par deux fonctions de relaxation. On peut toutefois montrer que si s_1, s_2, n_1, n_2 sont respectivement les nombres de raies des spectres de μ_1^*, μ_2^*, ν_1^*, ν_2^*, le nombre total de raies pour tous les paramètres $\bar{C}^*_{ijkl}(p)$ sont:

$$N = s_1 + s_2 + n_1 + n_2 + 5(s_1 + s_2) + 4n_1 + 3n_2. \quad [22]$$

Les raies initiales représentées par les 4 premiers termes de [22] sont conservées.

3. Modules complexes

Dans la pratique, où l'on s'intéresse généralement aux phénomènes de propagation d'ondes et aux propriétés amortissantes du matériau, la notion de module complexe est de loin la plus importante.

Lorsqu'un matériau est excité sinusoïdalement par une sollicitation de pulsation ω, quand le régime permanent est établi nous savons que:

$$\sigma_{pq}(t) = \tilde{\sigma}_{pq}(\omega)\,e^{i\omega t}$$

$$\varepsilon_{pq}(t) = \tilde{\varepsilon}_{pq}(\omega)\,e^{i\omega t} \qquad i = \sqrt{-1} \quad [23]$$

$\tilde{\sigma}_{pq}(\omega) = \sigma^*_{pq}(i\omega)$, $\tilde{\varepsilon}_{pq}(\omega) = \varepsilon^*_{pq}(i\omega)$ étant des quantités complexes. Le matériau composite est alors caractérisé par ses modules complexes $\tilde{\bar{C}}_{ijkl}(\omega)$ dont les relations [12], [13] et [14] fournissent les expressions:

$$\tilde{\bar{C}}_{ijkl}(\omega) = \bar{C}^*_{ijkl}(i\omega) = \bar{C}_{ijkl}[\tilde{\mu}_1(\omega), \tilde{\nu}_1(\omega), \tilde{\mu}_2(\omega), \tilde{\nu}_2(\omega)]. \quad [24]$$

L'égalité [24] peut être exploitée sur le plan pratique ou sur le plan théorique

– Si l'on veut tenter une étude théorique systématique, il est nécessaire de connaitre sous forme littérale la loi de comportement de chaque composant:

$$\sigma_{ij} = \sigma_{ij}(\varepsilon_{kl}). \quad [25]$$

Sur le plan calcul c'est là que réside la difficulté car des expressions analytiques qui prétendent traduire les corps réels sont compliquées.

– Si l'on s'attache au côté pratique, sans tenir compte du fait qu'en régime harmonique il est très difficile de tracer les courbes de module viscoélastique pour une gamme étalée de fréquences, les relations [24] permettent de construire les courbes $\tilde{\bar{C}}_{ijkl}(\omega)$ sans adopter de modèle mathématique pour les constituants.

Nous nous sommes donc bornés dans ce paragraphe à une étude purement numérique en introduisant les modules des constituants par valeurs discrètes de ω. Nous avons ici encore considéré les fibres élastiques et le coefficient de *Poisson* ν_1 constant. δ_{ijkl} désignant l'angle de perte, nous avons construit les familles de courbes $\log|\tilde{\bar{C}}_{ijkl}|$ et $\operatorname{tg}\delta_{ijkl}$ en fonction de la pulsation ω, ceci pour des pourcentages de fibres variant entre 0 et 1 (cf. fig. 4a, a' et 4b, b').

On peut alors constater que:

– Le déphasage δ_{ijkl} diminue lorsque le pourcentage de fibres augmente.

– Les modules des quantités $\tilde{\bar{C}}_{ijkl}$ croissent avec le pourcentage de fibres.

– L'angle de perte δ_{ijkl} est compris entre 0 et $\pi/2$.

– Enfin le composite est très peu amortissant dans le sens des fibres.

IV. Conclusion

Notre méthode de calcul, basée sur une approximation géométrique des conditions aux limites, fournit des résultats concordant avec les diverses techniques de calcul. Ces résultats sont en effet voisins des nombres déduits des méthodes «plus rigoureuses» [cf. (7)] et satisfont à l'encadrement donné par le calcul variationnel, tant en élasticité [cf. (5)] qu'en viscoélasticité [cf. (8)]. On peut remarquer que dans ce dernier cas, notre solution est plus voisine de la borne inférieure que de la borne supérieure [cf. (2)].

L'étude que nous avons tenté en viscoélasticité, bien que nécessitant encore quelques précisions sur le plan théorique, présente, pensons nous, un intérêt dans les domaines suivants:

– Prévision sur le comportement en fluage (ou en relaxation) des composites viscoélastiques dans le domaine linéaire. La méthode que nous préconisons, bien qu'imparfaite, constitue d'ores et déjà une amélioration par rapport à la règle des mélanges qui à une portée pratique trop limitée.

– Dans notre étude nous avons admis certes l'hypothèse d'une adhésion parfaite entre la fibre et la matrice. Dans les problèmes réels on doit s'attendre cependant, surtout pour les composites viscoélastiques, à ce que le contact fibre-matrice n'est pas parfait et alors, pour les ondes viscoélastiques, la propagation s'accompagne d'une absorption d'énergie aux interfaces par frottements superficiels. Nous pouvons espérer toutefois que nos calculs de prévision fourniront des valeurs de référence pour les amortissements de diverses ondes à partir desquelles on peut étudier les amortissements réels.

Annexe

$$X_{1111}(0) = (1 - \nu_1) \times \left[8 b_2 T^2 \eta_0 + \frac{8 c_2}{T}(1 - 2\nu_1)\xi_0 + 4 c_0 \chi_0\right] \Big/ (T_0 + T_2)$$

$$X_{1111}(2) = (1 - \nu_1) 4 c_0 \chi_1 / (T_0 + T_2)$$

$$X_{1111}(3) = (1 - \nu_1) \times \left[8 b_2 T^2 \eta_1 + \frac{8 c_2}{T}(1 - 2\nu_1)\xi_1\right] \Big/ (T_0 + T_2)$$

$$X_{1111}(4) = (1 - \nu_1) \times \left[8 b_2 T^2 \eta_2 + \frac{8 c_2}{T}(1 - 2\nu_1)\xi_2\right] \Big/ (T_0 + T_2)$$

$$X_{1122}(0) = \left\{-2 + (1 - \nu_1)\left[4c_0\chi_0 - 8b_2T^2\eta_0 - \frac{8}{T}(1 - 2\nu_1)c_2\xi_0\right]\right\}\Big/(T_0 - T_2)$$

$$X_{1122}(2) = X_{1111}(2)(T_0 + T_2)/(T_0 - T_2)$$

$$X_{1122}(3) = X_{1111}(3)(T_0 + T_2)/(T_0 - T_2)$$

$$X_{1122}(4) = -X_{1111}(4)(T_0 + T_2)/(T_0 - T_2).$$

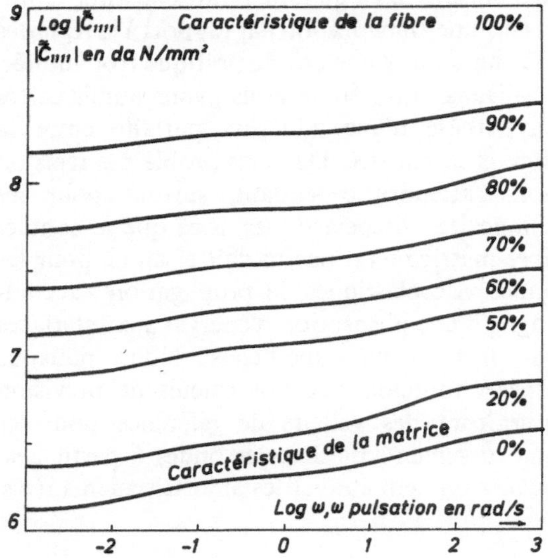

a

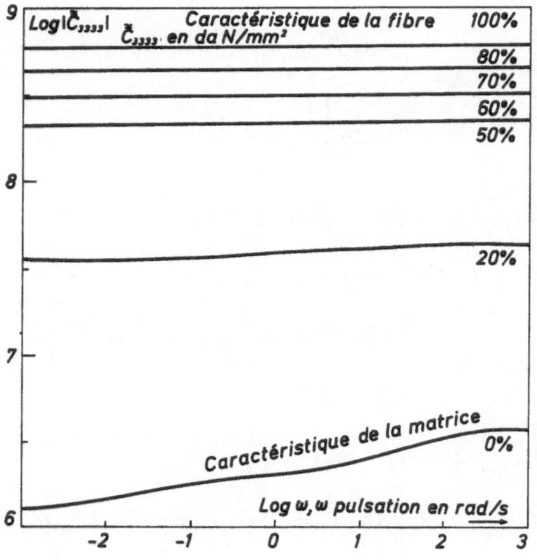

a′

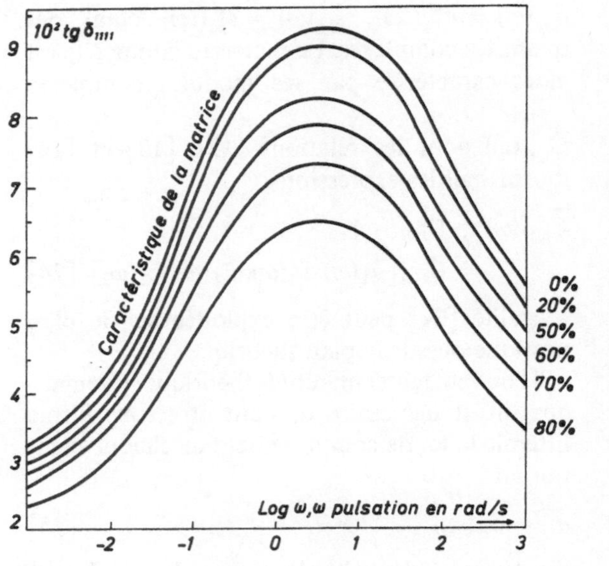

b

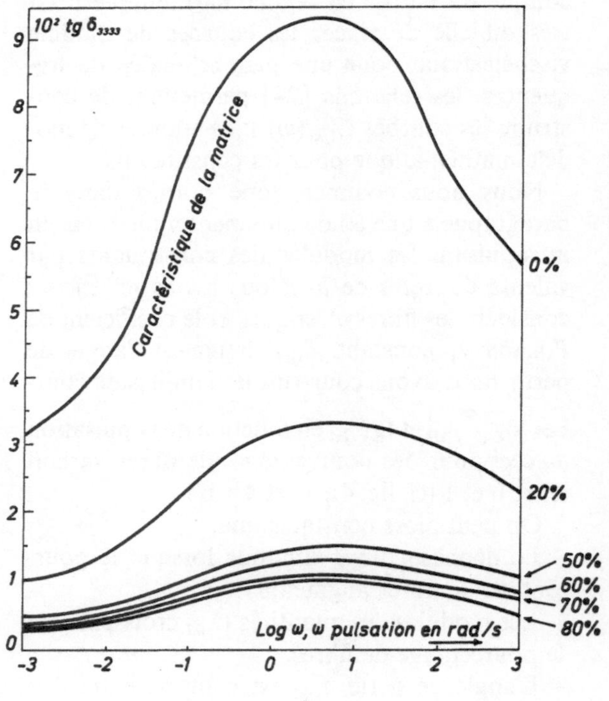

b′

Fig. 4. a) a′) Variation des modules complexes $|\tilde{\tilde{C}}_{1111}|$ et $|\tilde{\tilde{C}}_{3333}|$ en fonction de la pulsation ω pour des valeurs discrètes du pourcentage de fibres. b) b′) Variation de la tangente des angles de perte δ_{1111} et δ_{3333} en fonction de la pulsation ω pour des valeurs discrètes du pourcentage de fibres. *Caractéristiques des composants:* $\nu_1 = 0{,}34$; $\nu_2 = 0{,}2$; $E_2 = 7500$ daN/mm², $|\tilde{E}_1(0)| = 300$ daN/mm², $|\tilde{E}_1(\infty)| = 480$ daN/mm²

$$X_{1133}(0) = \frac{R_2 \nu_1 (1 - T)(1 - 2\nu_2)}{(R_2 - 1)\{[2T\nu_2(1 - \nu_1) + \nu_1(1 - T)]\, Q_0 + \nu_1(1 - T)(1 - 2\nu_2)\}}$$

$$X_{1133}(2) = 1 - X_{1133}(0).$$

$$X_{3333}(0) = \left[\frac{\bar{C}^2_{1133}}{\mu_1 T_0} Z_0 + E_1(1 - T)\right] \Big/ \bar{C}_{3333}$$

$$X_{3333}(2) = \frac{\bar{C}^2_{1133}}{\mu_1 T_0} Z_1 / \bar{C}_{3333}$$

$$X_{3333}(5) = \frac{\bar{C}^2_{1133}}{\mu_1 T_0} Z_2 / \bar{C}_{3333}.$$

$$X_{2323}(0) = \frac{(1 - T)[(1 - Q_0)T + 1 + Q_0]}{(1 + T)[(Q_0 - 1)T + 1 + Q_0]}$$

$$X_{2323}(1) = \frac{4 Q_0 T}{(1 - T)[1 + Q_0 + (Q_0 - 1)T]}.$$

Dans ces égalités T_0 et T_2 sont donnés par les relations [9]. η_i, ξ_i, Z_i $(i = 1, 2, 0)$, résultent respectivement des décompositions en éléments simples des quantités $\mu_1^* b_2^*/p$, $\mu_1^* c_2^*/p$ et $\bar{C}^*_{1133}/\mu_1^* T_0^* p$.

$$\frac{\mu_1^* b_2^*}{p} = \mu_1 b_2 \left(\frac{\eta_0}{p + 1/\tau} + \frac{\eta_1}{(p + 1/R_3 \tau)} + \frac{\eta_2}{(p + 1/R_4 \tau)} \right)$$

$$\frac{\mu_1^* c_2^*}{p} = \mu_1 c_2 \left(\frac{\xi_0}{p + 1/\tau} + \frac{\xi_1}{(p + 1/R_3 \tau)} + \frac{\xi_2}{(p + 1/R_4 \tau)} \right)$$

$$\frac{\bar{C}^*_{1133}}{\mu_1^* T_0^* p} = \bar{C}_{1133} \left(\frac{Z_0}{(p + 1/\tau)} + \frac{Z_1}{(p + 1/R_2 \tau)} + \frac{Z_2}{(p + 1/R_5 \tau)} \right).$$

χ_0 et χ_1, résultats de la décomposition de $\mu_1^* T_0^*$ sont enfin fournis par les expressions suivantes:

$$\chi_0 = \frac{(1 - 2\nu_2) R_2}{[Q_0 + (1 - 2\nu_2)](R_2 - 1)}$$

$$\chi_1 = 1\ N \chi_0.$$

N.B. Dans cette série de relations, R_i $(i = 0, 1, 2, 3, 4, 5)$ résultent des égalités [18].

Littérature

1) *Behrens, E.*, J. acoustical soc. Amer. **42**, 367–377 (1967).
2) *Chevalier, Y.*, Thèse de 3è cycle. Fac. Sci. de Paris (juin 1971).
3) *Chevalier, Y.* et *Vinh Tuong*, Comp. Rend. Acad. Sci. Paris **271**, 1268–1271 (1970); **273**, 735–738 (1971).
4) *Green* et *Zerna*, Theoretical Elasticity.
5) *Hill, R.*, J. mech. Phys. Solids. **12**, 199–212 (1964).
6) *Leissa, A. W.* et *Claussen*, Application of point matching to problems in micromechanics. Fundamental aspect of fiber reinforced plastic composites; Chap. 3 (New York).
7) *Pickett, G.*, Elastic moduli of fiber reinforced plastic composites. Fundamental aspect of fiber reinforced plastic composites. Chap. 2 (New York).
8) *Roscoe, R.*, J. Mech. Phys. solids. **17**, 17–23 (1969).

Adresse des auteurs:

Dr. *Yvon Chevalier*
I.S.M.C.M.
3 rue Fernand Hainaut
F-93 Saint Ouen (France)

Rheol. Acta 12, 438–448 (1973)

From the Department of Electronics and Electrical Engineering, The University of Glasgow (Scotland)

Mechanical retardation and relaxation in liquids

By J. Lamb

With 16 figures

(Received October 27, 1972)

1. Introduction

This research was initially motivated by interest in the bulk properties of lubricating fluids since little was known about the response of a liquid to a rapidly changing stress. In elasto-hydrodynamic lubrication, elastic deformation of the contact surfaces occurs with the accompanying development of a very high pressure which, in the centre of the contact zone, may reach 10,000 atm. The fact that this phenomenon is essentially non-linear does not detract from the need, in the first instance, to study the linear viscoelastic behaviour over wide ranges of temperature and pressure. Accordingly, the author and his colleagues have conducted a detailed study of the viscoelastic behaviour of liquids when these are subjected to sinusoidally oscillating shear stress, the amplitude of which is sufficiently small for the response to be restricted to the linear viscoelastic region.

Results now available permit analyses to be made, which lead at least to a phenomenological understanding and, to some extent, to explanations in terms of molecular behaviour. Investigations have naturally extended to polymer solutions and to polymer melts, the scientific study of which has a direct bearing on problems encountered in practice such as, for example, polymer extrusion or injection moulding. The co-ordinated study of the viscoelastic properties of both non-polymeric and polymeric liquids leads to questions of the difference between the flow behaviour of polymer molecules and that of shorter chain molecules. This, in turn, raises fundamental questions concerning the nature of a polymer: is there, for example, an abrupt or a progressive change in the mechanical properties as a function of increasing chain length in the progression from a non-polymeric liquid to a polymeric one of the same molecular species?

At the present stage, some tentative conclusions can be drawn, but further work is required to provide a satisfactory explanation. It is, therefore, appropriate to review the existing state of knowledge in this field, both from the point of view of consolidating that which is known and of emphasising the directions along which further effort might usefully be deployed.

2. Supercooled liquids

The simplest theoretical representation of viscoelastic response is the *Maxwell* model, according to which the complex compliance at an angular frequency $\omega\,(=2\pi f)$ is expressed by:

$$J^*(j\omega) = J_\infty[1 + 1/j\omega\tau_m] \quad [1]$$

where J_∞ is the limiting compliance at very high frequencies ($\omega\tau \gg 1$) and $\tau_m = \eta J_\infty$ is the *Maxwell* relaxation time. Measurements shown that the value of J_∞ is between 10^{-10} and 5×10^{-10} cm^2/dyne: taking a typical value of 2×10^{-10} cm^2/dyne for J_∞ gives the value of the characteristic frequency, f_c, corresponding to $\omega_c\tau_m = 1$, as $10^{10}/4\pi\eta$ Hz $\simeq 0.8/\eta$ GHz where η is in Poise. Thus for $\eta = 1$ cP, 10 cP, 1 P and 100 P, corresponding values of f_c are 80 GHz, 8 GHz, 800 MHz and 8 MHz, respectively. For experimental reasons, alternating shear wave measurements on liquids are sensibly confined to frequencies below 1 GHz and hence, in order, to determine the viscoelastic as distinct from the viscous response, it is necessary to work with supercooled liquids. In practice, many liquids do supercool readily, including those of particular interest which have been developed as synthetic lubricants or are used as plasticisers.

The limiting high frequency compliance, J_∞, is obtained by making measurements at a convenient frequency – generally 30 or 450 MHz – and by cooling the liquid to temperatures approaching the glass transition temperature T_g so that the condition $\omega\tau_m \gg 1$ is attained. Results obtained on some 50 or more liquids have all been fitted by a linear variation of J_∞ with T:

$$J_\infty = J_0 + C(T - T_0). \qquad [2]$$

Here T_0 is the reference temperature involved in the modified free-volume equation used to describe the dependence of steady-flow viscosity upon temperature in the supercooled region, namely:

$$\ln\eta = A + B/(T - T_0). \qquad [3]$$

For measurements made at 30 MHz the linear variation described by eq. [2] generally extends over some 20–30 °C with a somewhat longer range at the higher frequency of 450 MHz. However, as temperature increases, the liquid no longer behaves elastically and it is necessary to make an extrapolation, using eq. [2], in order to calculate the effective value of J_∞ at these higher temperatures. It is then possible to normalise experimental results and thereby compare the viscoelastic behaviour of different liquids.

In practice, the quantities which are measured experimentally using shear wave methods in the frequency range 20 kHz to 1 GHz are the resistive and reactive components of the shear mechanical impedance, $Z_L = R_L + jX_L$ and the complex compliance is then given by $J^*(j\omega) = \varrho/Z_L^2$, where ϱ is the density. It follows that the components of $J^*(j\omega) = J' - jJ''$ are determined from

$$J' = \frac{\varrho[R_L^2 - X_L^2]}{[R_L^2 + X_L^2]^2}, \quad J'' = \frac{2\varrho R_L X_L}{[R_L^2 + X_L^2]^2}. \qquad [4]$$

In the elastic region $X_L \simeq 0$: hence $J'' \simeq 0$ and $J' = J_\infty = \varrho/R_L^2$. In the purely viscous or *Newtonian* region, $R_L = X_L$, so that here $J' = 0$ and $J'' = 2\varrho/R_L^2 = 4/\omega\eta$.

Earlier work of the author and his colleagues (1–4) has shown that, at atmospheric pressure, results obtained over available ranges of temperature and frequency can be fitted within experimental error by the *Barlow*, *Erginsav* and *Lamb* (BEL) equation:

$$J^*(j\omega) = J_\infty[1 + 1/j\omega\tau_m] + J_\infty/(j\omega\tau_m)^{\frac{1}{2}}. \qquad [5]$$

Confirmation of the applicability of this equation is demonstrated by the results shown in fig. 1 for a number of liquids which are merely representative of a much larger group comprising liquids of widely differing molecular type. In fig. 1, the experimentally determined quantities, R_L and X_L, are plotted in normalised form, $R_L/(\varrho G_\infty)^{\frac{1}{2}}$ and $X_L/(\varrho G_\infty)^{\frac{1}{2}}$, where the limiting high frequency modulus $G_\infty = 1/J_\infty$. It has been established that frequency and temperature are interchangeable variables in this form of representation (1, 4).

Since $\tau_m = \eta J_\infty = \eta/G_\infty$, it follows that the behaviour in alternating shear can be predicted within the accuracy of the measurements, provided that the steady-flow viscosity, η, and the limiting high frequency compliance, J_∞, are known over the temperature range of interest.

Measurements have also been made as a function of pressure (5), the viscosity being determined using a falling-weight viscometer with automatic electrical recording (6). By increasing the pressure sufficiently, the viscoelastic relaxation region can be transposed to frequencies well below the frequencies of measurement (10 and 30 MHz), so that the limiting high frequency elastic modulus can be determined experimentally. Results obtained show that the shear modulus, G_∞, is a linear function of pressure, fig. 2. When the results are normalised in the manner previously indicated, it is found that the behaviour again conforms within experimental error to the predictions of the BEL eq. [5], as shown by fig. 3.

Despite the good agreement found between experimental results and calculated values based upon the BEL equation, it was realised that this equation is only an approximate representation. When transformed into the time domain, it leads to the prediction of unlimited creep strain, which is physically unacceptable.

Provided that experimental conditions are restricted to the amplitude region where linear viscoelastic theory applies, then following *Gross* (7), the complex compliance can be described by the equation

$$J^*(j\omega) = J_\infty + 1/j\omega\eta + J_r^*(j\omega)$$

$$= J_\infty + 1/j\omega\eta + J_r \int_0^\infty \frac{N(\tau)\,d\tau}{1 + j\omega\tau}$$

$$= J_\infty + 1/j\omega\eta + J_r\,\chi(j\omega) \qquad [6]$$

where $N(\tau)$ represents the spectrum of retardation times, normalised so that the integral of $N(\tau)\,d\tau$ is unity, and $J_r^*(j\omega)$ is the complex retardational compliance. The real part of J^* tends to a value $J_r + J_\infty$ at low frequencies and to J_∞ at high frequencies. For a *Maxwell* liquid, characterised by a single relaxation time, $\tau_m = \eta J_\infty$, the retardational compliance J_r is zero. The time variation of strain $\varepsilon(t)$ in response to a constant stress, γ_0, imposed at $t = 0$, is given by:

$$\frac{\varepsilon(t)}{\gamma_0} = J_\infty + t/\eta + J_r \int_0^\infty N(\tau)\left[1 - e^{t/\tau}\right] d\tau$$
$$= J_\infty + t/\eta + J_r \psi(t) \qquad [7]$$

where $\psi(t)$ is the normalised retardation function used, for example, by *Plazek* and *O'Rourke* (8).

Comparison of eqs. [5] and [6] shows that the amplitude $N(\tau)$ of the retardation spectrum appropriate to the BEL equation is proportional to $\tau^{\frac{1}{2}}$ and therefore, as mentioned previously, the corresponding recoverable creep strain increases without limit as $t^{\frac{1}{2}}$. This inconsistency has been

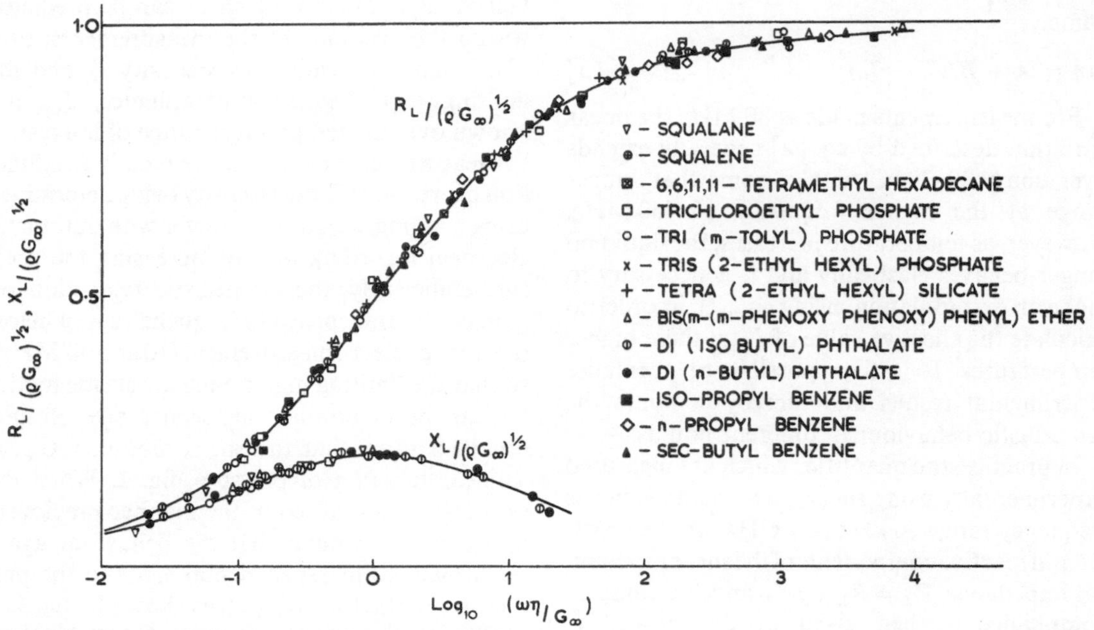

Fig. 1. Normalised plots of the resistive and reactive components of a number of liquids. The curves are calculated from the equation of *Barlow*, *Erginsav* and *Lamb*.

Fig. 2. Measured values of R_L^2/ϱ as a function of pressure for bis(m-(m-phenoxy phenoxy)phenyl) ether at 30 °C. ▽, 10 MHz; △, 30 MHz; ■, extrapolated value obtained at atmospheric pressure by variation of temperature. In the elastic high pressure region $R_L^2/\varrho \rightarrow G_\infty$. [100 MN/m^2 = 1000 bar] (By permission of the Royal Society)

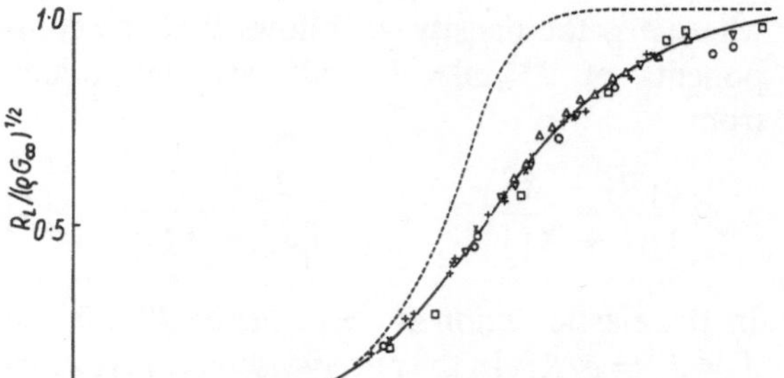

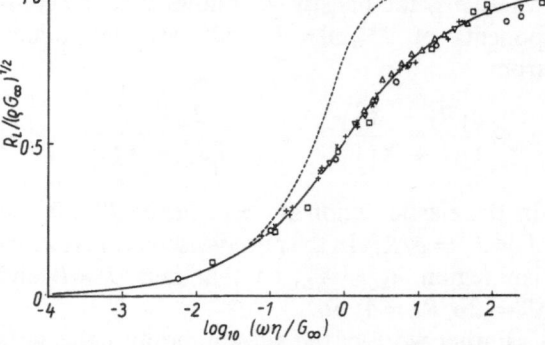

Fig. 3. Normalised values of the resistive component of the shear mechanical impedance including results obtained at 30 °C as a function of pressure and at atmospheric pressure as a function of temperature. Obtained at frequencies in the range 6–450 MHz at atmospheric pressure: +, bis(m-(m-phenoxy phenoxy)phenyl) ether. ×, di(2-ethyl hexyl) phthalate. Obtained at 30 °C as a function of pressure: bis(m-(m-phenoxy phenoxy)phenyl) ether. ▽, 10 MHz; △, 30 MHz. di(2-ethyl hexyl) phthalate: ○, 10 MHz; □, 30 MHz. The solid curve represents the behaviour calculated from the *Barlow*, *Erginsav* and *Lamb* equation. The broken curve represents the *Maxwell* behaviour for a single relaxation time. (By permission of the Royal Society)

resolved by *Barlow* and *Erginsav* (9) using a further development of the shear impedance technique, which provides an order of magnitude improvement in absolute accuracy over previous systems used in this laboratory. The new system operates at a single frequency of 30 MHz so that related measurements are made with temperature as the only experimental variable. The author's colleagues have shown that the complex retardational compliance can be fitted within experimental error by the form of empirical equation used by *Davidson* and *Cole* (10) to fit their dielectric data:

$$J_r^*(j\omega) = J_1(\omega) - jJ_2(\omega) = J_r/(1 + j\omega\tau_r)^\beta \qquad [8]$$

in which τ_r is a characteristic retardation time and $0 < \beta < 1$. For supercooled liquids, β is typically 0.5, although values ranging from 0.4 to 0.6 have been found at temperatures somewhat below the "*Arrhenius* temperature", T_A, above which the liquid can no longer be described as supercooled. The temperature, T_A, can be specified only approximately within a few degrees but the corresponding viscosity of organic non-polymeric liquids is almost invariably less than 0.5 P at this temperature.

In terms of the above representation, the total complex compliance is expressed by:

$$\frac{J^*(j\omega)}{J_\infty} = [1 + 1/j\omega\tau_m] + \frac{J_r/J_\infty}{(1 + j\omega\tau_r)^\beta}. \qquad [9]$$

Here the first term is recognised as the conventional *Maxwell* relaxation whilst the second term describes the retardational process. Fig. 4 shows the *Davidson-Cole* "skewed arc" plots of $J_2(\omega)/J_\infty$ versus $J_1(\omega)/J_\infty$ for tri(m-tolyl)phosphate and for tri(o-tolyl)phosphate, with $\beta = 0.5$ in each case. $J_1(\omega)/J_\infty$ and $J_2(\omega)/J_\infty$ are plotted as a function of $\log(\omega\tau_m)$ in fig. 5 for tri(m-tolyl)-phosphate, for which liquid $J_r/J_\infty = 7.6$. It is emphasised that these results are obtained by varying temperature at a fixed frequency of 30 MHz. Assuming that eq. [8] is valid, τ_r may be calculated at any given temperature in the retardation region and plotted, as in fig. 6, as $\log(\tau_r/\tau_m)$ versus $1/T$. Also shown on fig. 6 is the plot of $\log\eta$ versus $1/T$. At the *Arrhenius* temperature, T_A, it transpires that τ_r/τ_m is approximately equal to J_r/J_∞ and this is found to be the case for other liquids investigated. Since $\tau_m = \eta J_\infty$, this implies that, at T_A, $\tau_r \simeq \eta J_r$, so that, if we express the retardation time, τ_r, as the product of "a retardational viscosity", η_r, multiplied by the retardational compliance, J_r, ($\tau_r = \eta_r \cdot J_r$), then $\eta_r \simeq \eta$ at $T = T_A$.

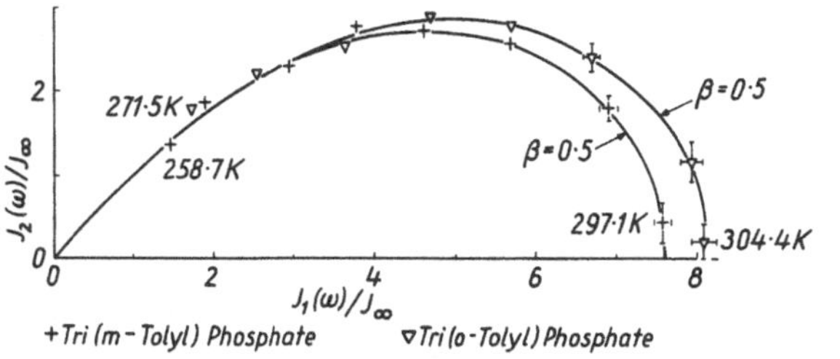

Fig. 4. The imaginary component of the retardational compliance plotted against the real part. +, tri(m-tolyl) phosphate. ▽, tri(o-tolyl) phosphate. The curves represent the *Davidson-Cole* eq. [8] with $\beta = 0.5$. Error bars define the estimated semi-interquartile range of error distribution. (By permission of the Royal Society)

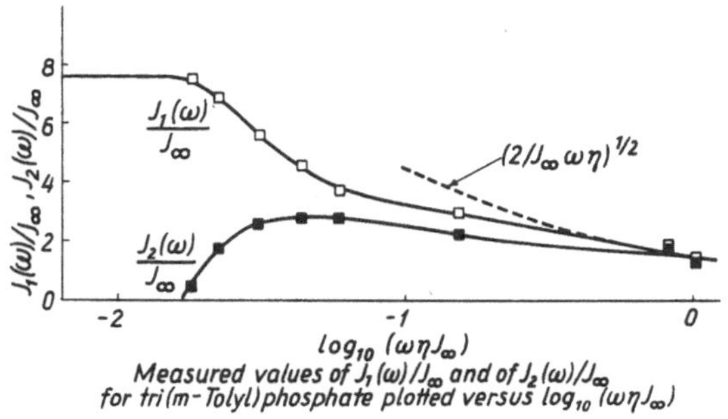

Fig. 5. Measured values of $J_1(\omega)/J_\infty$ and $J_2(\omega)/J_\infty$ for tri(m-tolyl) phosphate plotted as functions of $\log(\omega\eta J_\infty)$

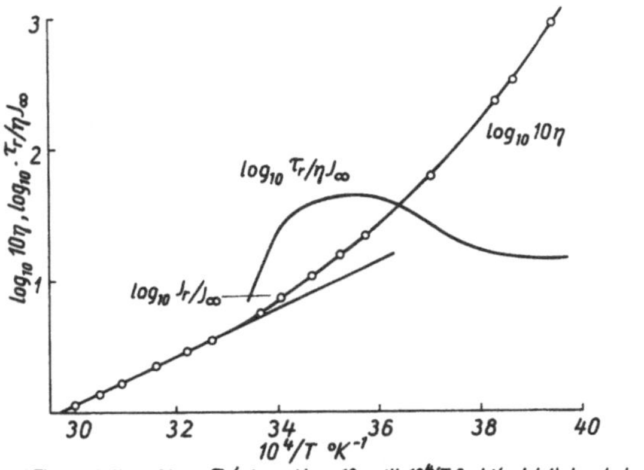

Fig. 6. The variation of $\log(\tau_r/\eta J_\infty)$ and $\log 10\eta$ with $10^4/T$ for tri(m-tolyl) phosphate

It is evident from fig. 6 that τ_r does not have the same dependence upon temperature as does τ_m so that, in terms of the representation given by eq. [10], frequency and temperature are not interchangeable variables. In the supercooled region, τ_r is greater than τ_m, the minimum value of the ratio τ_r/τ_m being equal to J_r/J_∞ at $T = T_A$. Below T_A this ratio increases rapidly as the temperature is reduced; it reaches a maximum and becomes constant at lower temperatures. Putting $\omega\tau_r \gg 1$ in eq. [8] gives $J_r^*(j\omega) \simeq J_r/(j\omega\tau_r)^\beta = J_r/(j\omega\tau_r)^{\frac{1}{2}}$, for $\beta = 0.5$. The BEL equation gives $J_r^*(j\omega) = 2J_\infty/(j\omega\tau_m)^{\frac{1}{2}}$; and comparing these results, we have:

$$\tau_r/\tau_m = [J_r/2J_\infty]^2 \quad \text{for} \quad \omega\tau_r \gg 1. \qquad [10]$$

In this temperature region τ_r/τ_m is constant (fig. 6): moreover, it has been assumed that J_r and J_∞ have the same linear dependence upon temperature and this would appear to be substantiated since the constant ratio found for τ_r/τ_m at low temperatures is precisely that calculated from eq. [10] for a liquid represented by $\beta = 0.5$. Thus, for the major part of the viscoelastic relaxation region, the predictions of the BEL equation and of eq. [9] are often indistinguishable within experimental error [fig. 7].

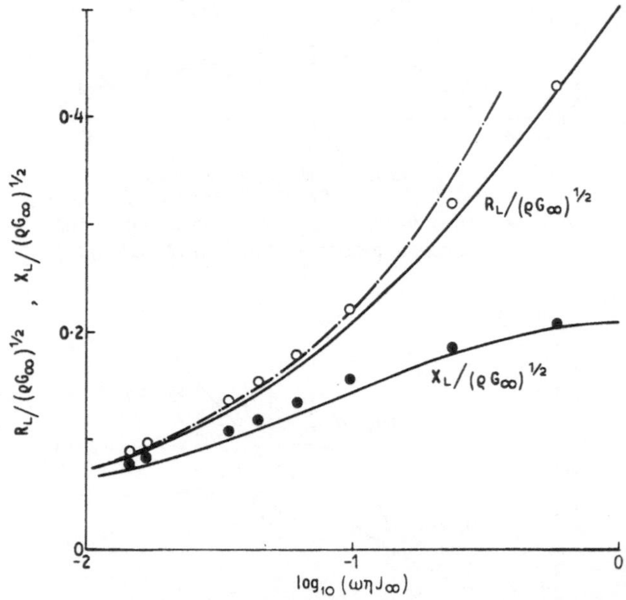

Fig. 7. Measured values of the normalised shear mechanical impedance components for tri(o-tolyl) phosphate. The solid curves show the variation calculated from the *Barlow, Erginsav* and *Lamb* equation. The broken line represents *Newton*ian behaviour, $R_L = X_L = (\pi f \eta \varrho)^{\frac{1}{2}}$. Divergence of the experimental points from the BEL curves in the region of $\log(\omega\eta J_\infty) = -1$ is a reflection of the influence of the retardational process. (By permission of the Royal Society)

However, if the values of the ratios J_r/J_∞ and τ_r/τ_m are not large, say, less than 10, then the effects of the retardation process can be significant. Computed plots of R_L and X_L are shown in fig. 8 and compared with the behaviour predicted from the BEL equation. Experimental results for a mixture (E 3) (4) of di(n-butyl)phthalate and di(iso-butyl)phthalate are given in fig. 9: these are obtained at constant frequency by variation of temperature. The computed curves of fig. 8 assume that the ratio τ_r/τ_m is constant, implying that the temperature is fixed and ω regarded as the variable, which is not so in practice. Nevertheless, there are similarities in the pattern of behaviour found experimentally and the predicted curve for an assumed set of parameters. This particular mixture (E 3) is chosen for this example, since it provides the smallest value of the ratio J_r/J_∞ of any of the liquids measured: hence the deviations from the BEL equation are here most pronounced.

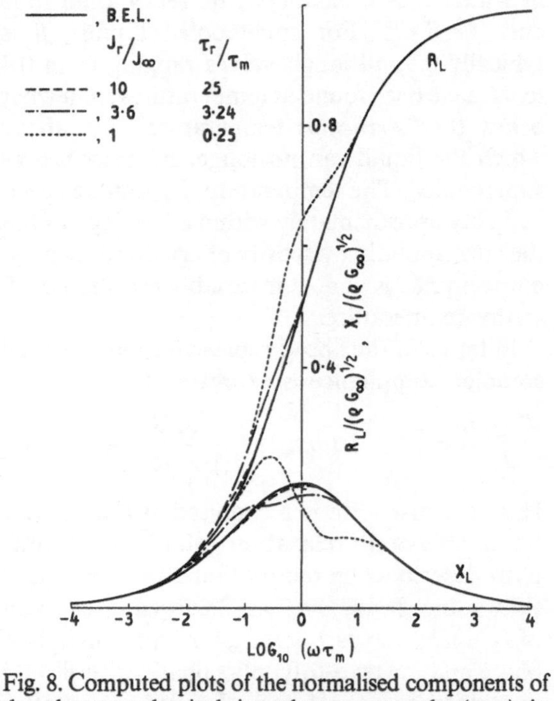

Fig. 8. Computed plots of the normalised components of the shear mechanical impedance versus $\log(\omega\tau_m)$ in accordance with eq. [10]: $J^*(j\omega) = J_\infty[1 + 1/j\omega\tau_m] + J_r/(1 + j\omega\tau_r)^\beta$. Values of τ_r/τ_m have been selected so that the high frequency behaviour conforms to the BEL equation, namely $\tau_r/\tau_m = [J_r/2J_\infty]^2$, in accordance with experimental observation. ------, $J_r/J_\infty = 1$, $\tau_r/\tau_m = 0.25$; –·–·–, $J_r/J_\infty = 3.6$, $\tau_r/\tau_m = 3.24$; ––––, $J_r/J_\infty = 10$, $\tau_r/\tau_m = 25$; ———, *Barlow, Erginsav* and *Lamb* equation

The retardational compliance, J_r, represents the delayed storage of elastic energy under shear stress: it is reasonable to presume that it is determined by local fluctuations in the neighbourhood of a molecule and hence by the intermolecular forces which control reorientation. Independent evidence for this point of view is

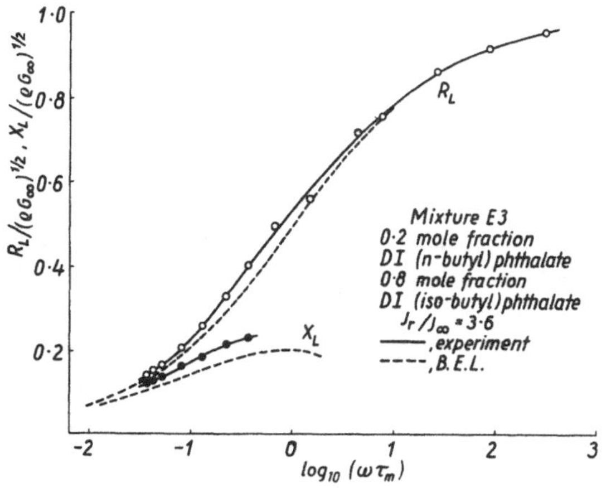

Fig. 9. Experimental results for mixture E 3 comprising 0.2 mole fraction di(n-butyl) phthalate and 0.8 mole fraction di(iso-butyl) phthalate. $J_r/J_\infty = 3.6$. ———, experimental curve; ----, BEL equation

provided by the results of experiments on dielectric relaxation and, following the helpful collaboration of Dr. *G. Williams*, it is now possible to compare the results of dielectric and viscoelastic experiments on the same liquids. Thus, *Shears* and *Williams* (11) have measured the dipole relaxation in some of the liquids investigated in our viscoelastic work. Because they worked at lower frequencies over the range 100–10^5 Hz their results were obtained at somewhat lower temperatures than those of the viscoelastic experiments. A comparison is shown in fig. 10 for tri(o-tolyl)phosphate and for di(n-butyl)phthalate. Both of these liquids show viscoelastic behaviour some 10 °C below the *Arrhenius* temperature corresponding to a value of $\beta = 0.5$ (eq. [8]). In the *Davidson-Cole* representation, the dielectric relaxation data are plotted in fig. 10 as $\varepsilon''/(\varepsilon_0 - \varepsilon_\infty)$ versus $(\varepsilon' - \varepsilon_\infty)/(\varepsilon_0 - \varepsilon_\infty)$: the corresponding viscoelastic data are plotted as J_2/J_r versus J_1/J_r, where J_r is the difference between the "zero" frequency compliance, $J_0(= J_r + J_\infty)$ and the high frequency value, J_∞. The agreement found between the results of these two quite independent studies is remarkable and reinforces the view expressed above that the retardational process involves local reorientational changes of the molecule. For this reason, it is to be expected that the value of J_r will be strongly influenced by the type of molecule. Thus, as the flexibility of the molecule is increased, larger values of J_r would be expected and, with the limited evidence available, this is found to be the case. For squalane, the value of J_r/J_∞ is 21.8 compared with 4.0 for di(iso-butyl)phthalate and for a six-ring polyphenyl ether the value of this ratio exceeds 50.

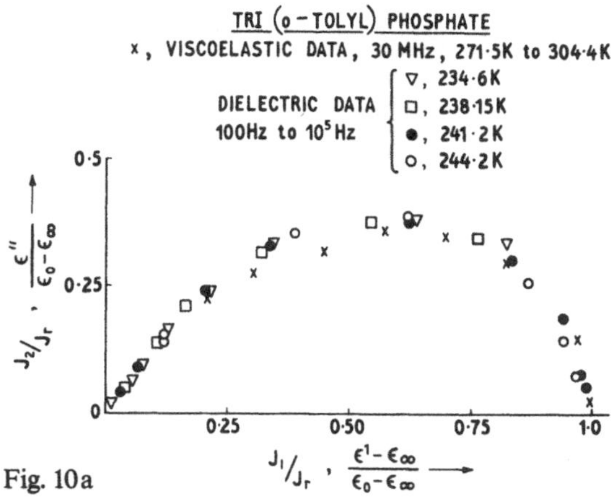

Fig. 10a

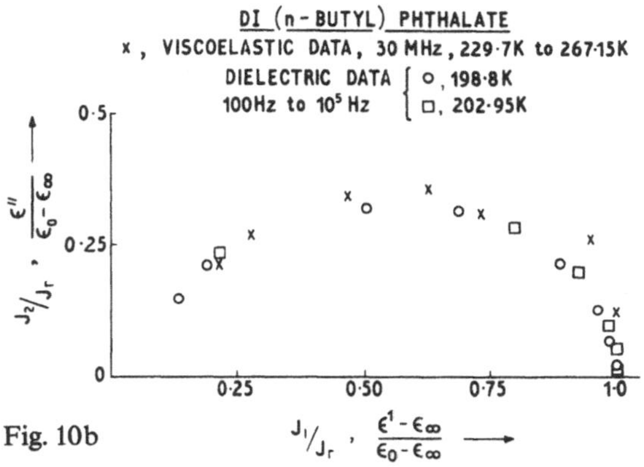

Fig. 10b

Fig. 10. Comparison of dielectric relaxation data of *Williams* and *Shears* and the retardational compliance. Dielectric results are plotted in the form of *Davidson* and *Cole* as $\varepsilon''/(\varepsilon_0 - \varepsilon_\infty)$ versus $(\varepsilon' - \varepsilon_\infty)/(\varepsilon_0 - \varepsilon_\infty)$. Viscoelastic retardational results are plotted on the same normalised scale as J_2/J_r versus J_1/J_r. a) tri(o-tolyl) phosphate: ×, viscoelastic data, 30 MHz, 271.5–304.4 K. Dielectric data, 100–10^5 Hz; ▽, 234.6 K; □, 238.15 K; ●, 241.2 K; ○, 244.2 K. b) di(n-butyl) phthalate: ×, viscoelastic data, 30 MHz, 229.7–267.15 K. Dielectric data, 100–10^5 Hz: ○, 198.8 K, □, 202.95 K

In principle, J_r can be determined either from linear creep experiments or by the use of alternating shear. To the author's knowledge, only a single non-polymeric liquid has been measured in shear creep and this is 1,3,5-tri(α-naphthyl)benzene studied by *Plazek* and *Magill* (12) at temperatures close to T_g. Their results for the total recoverable compliance $(J_r + J_\infty)$ of this liquid, reduced to 64.2 °C, are shown as a function of time by the curve of fig. 11. The curve tends to a value corresponding to $J_\infty = 0.81 \times 10^{-10}$ cm²/dyne at short times and to $J_0 = J_r + J_\infty = 2.58 \times 10^{-10}$ cm²/dyne at long times. The ratio $J_r/J_\infty = 2.2$ is less than the value for the liquids studied in this laboratory but is presumably due to the symmetry and lack of reorientation flexibility of this molecule. It is a simple matter to calculate the frequency-dependent complex compliance from the time-varying behaviour and vice versa and it is found that, within experimental error, the recoverable compliance curve of fig. 11 corresponds with a calculated curve based upon the *Davidson-Cole* eq. [8], taking $J_r = 1.77 \times 10^{-10}$ cm²/dyne, as found by *Plazek* and *Magill*, $\beta = 0.30$ and $\tau_r = J_r \cdot (3.55\,\eta)$, where η is the measured viscosity of $10^{12.35}$ P. The retardational viscosity, η_r, is therefore 3.55 times the steady-flow viscosity and the relatively low value of β is consistent with the fact that the results were obtained at temperatures close to T_g. There is ample evidence from dielectric studies that β decreases as the viscosity increases and values in the region of 0.3 are to be expected at temperatures near T_g.

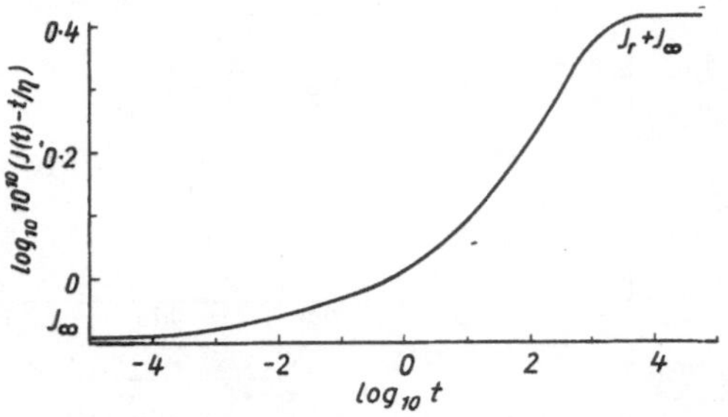

Fig. 11. Recoverable compliance of 1,3,5-tri(α-naphthyl) benzene at 64.2 °C. Temperature reduced curve of *Plazek* and *Magill* (ref. 12). J_∞=U 8.1 × 10^{-11} cm²/dyne, $J_r = 1.77 \times 10^{-10}$ cm²/dyne. The identical curve is calculated by transforming the frequency variation described by: $J^*(j\omega) = J_\infty + 1/j\omega\eta + J_r/(1 + j\omega\tau_r)^{0.3}$ with $\tau_r = J_r(3.55\,\eta)$ and $\eta = 10^{12.35}$ P., as measured by *Plazek* and *Magill*

The equivalent of recoverable creep compliance and the retardational compliance derived from shear wave measurements is predictable from linear viscoelastic theory. Experimentally, however, the two methods are complementary, since they yield results in different temperature ranges. The close correspondence between dielectric absorption due to dipole relaxation and viscoelastic retardation is new and interesting, since it promises to lead to an understanding of the viscoelastic mechanism on a molecular basis which has hitherto been absent. The BEL formulation is now seen as a first step on the route to a more detailed explanation. It remains, however, that, at sufficiently low temperatures and/or high frequencies, the measured behaviour tends to that described by the BEL equation. In phenomenological terms, this arises because in this region $\omega\tau_m \gg 1$ and τ_r/τ_m becomes constant independent of temperature and equal to $[J_r/2J_\infty]^2$. For $\beta = 0.5$ the right-hand side of eq. [8] then becomes simply $2J_\infty/(j\omega\tau_m)^{\frac{1}{2}}$ which is the second term in the BEL eq. [5]. In practice, this equation adequately describes the behaviour of most liquids in the region above $\omega\tau_m = 1$. It would, therefore, appear to have a basic significance which has been exemplified by its theoretical derivation from a defect-diffusion model (13). We are led to the conclusion that the essential features of the linear viscoelastic behaviour of non-polymeric liquids can be ascribed to a diffusion controlled process, the observed effects of which can be calculated from knowledge of the values of the two physical parameters, steady-flow viscosity, η, and limiting compliance, J_∞. Once these quantities are known at the relevant temperature and pressure, the cyclic shear properties are in the main predictable. There remains the additional factor of local readjustments, presumably on a molecular scale, which have been assumed to be responsible for the retardational process, evidence of which is generally found at values of $\omega\tau_m$ less than unity for supercooled liquids. Since the retardation time, τ_r, is found to be greater than τ_m, often by a factor of 10 or more, the effects of the retardational process are of greater importance in creep measurements than in those of alternating shear. In the latter case, the practical effect as regards the measurement of shear impedance or complex modulus is often a minor one and, at least for organic liquids, it is evident in the experimentally ill-defined region between *Newton*ian behaviour ($R_L = X_L$) and the onset of viscoelastic relaxation (fig. 7). In the region $\omega\tau_m \simeq 1$, where $\omega\tau_r > 1$, the retardational term merges with that of the diffusion controlled BEL behaviour and all liquids investigated exhibit this behaviour. The ratio τ_r/τ_m becomes sensibly independent of temperature as $\omega\tau_m$ increases further. This ratio differs for different liquids but, in each case, it is given by the value of $[J_r/2J_\infty]^2$. The effect is marked, but the explanation in molecular terms is elusive. Possibly this behaviour may be due to the fact that motions of translational diffusion take place on a time scale much shorter than that required for the establishment of local equilibrium around the site of the immediate molecular environment.

Although it has not yet been found practicable to obtain measurements of viscoelastic relaxation and/or retardation in the higher temperature region described by the *Arrhenius* equation of viscous flow, $\ln \eta = A' + B'/T$, it is interesting to speculate about the likely behaviour. Thus, as is evident from fig. 6, the retardation time, τ_r, decreases rapidly as the temperature of the supercooled liquid is increased to the approximate value of the *Arrhenius* temperature T_A. At T_A the ratio τ_r/τ_m is approximately equal to J_r/J_∞ and the retardational viscosity, η_r, is equal to the steady flow viscosity, η. It seems unlikely that an abrupt transition occurs at T_A and there is no evidence from present results that J_r should vanish above T_A. It is logical, therefore, to suggest that the retardation process persists at higher temperatures in the *Arrhenius* region with a retardational compliance J_r, not very different from its known value below T_A: this view has already been expressed by *Goldstein* (14). However, in the *Arrhenius* region the viscosity is energy controlled by the process of translational "jumps" of the molecules which occur more rapidly than the much slower diffusional motion below T_A. Local equilibrium arising from reorientational motions would also be established more rapidly at the higher temperatures and by analogy with dielectric relaxation, it might be expected that the viscoelastic behaviour would conform to the description of a *Maxwell* relaxation time and a single retardational time ($\beta = 1$). The possibility of testing these ideas experimentally would appear to be somewhat remote. It should, however, be noted that *Barlow* and *Erginsav* (9) have found evidence of a second retardational process in the temperature region above the transition temperature, T_K, found previously in the supercooled region for certain liquids (15). Their results for di(iso-butyl)phthalate are shown in figs. 12 and 13, in which the suffix 1 is used for the lower and suffix 2 to designate the higher temperature process. Moreover, the latter is approximately described by a single retardation time τ_{r_2} with $\beta \simeq 1$. The importance of this finding is that separate and distinct retardational mechanisms can apparently be associated with different regions of the viscosity-temperature characteristic. It is therefore not unreasonable, as suggested above, to suppose that a somewhat analogous situation involving a single retardation time might be found in the *Arrhenius* region above T_A.

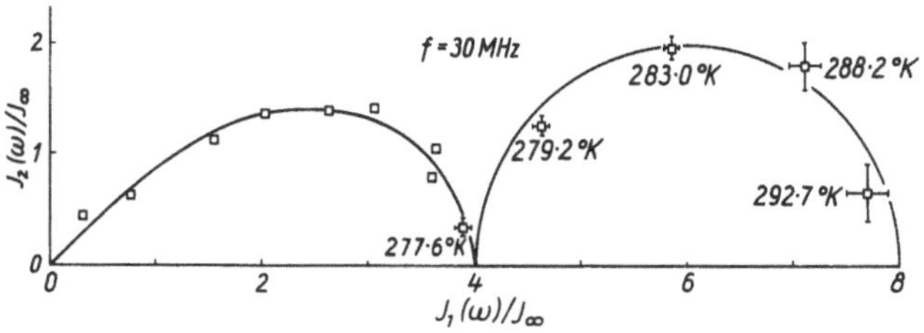

Fig. 12. Components of the retardational compliance of di(iso-butyl) phthalate. (By permission of the Royal Society)

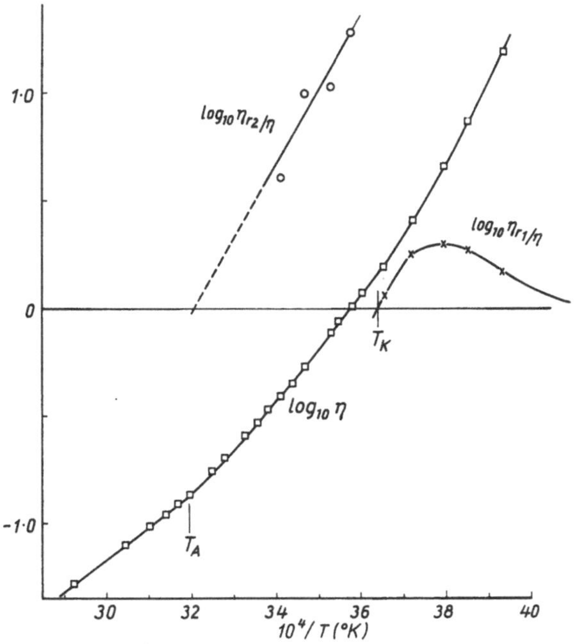

Fig. 13. The variation of $\log\eta$, $\log(\eta_{r_1}/\eta)$ and $\log(\eta_{r_2}/\eta)$ with $10^4/T$ for di(iso-butyl) phthalate. η_{r_1} represents the retardational viscosity below the transition temperature T_K and η_{r_2} the retardational viscosity above T_K but below T_A

3. Long chain molecules

Measurements have previously been made in this laboratory of the viscoelastic relaxation properties obtained in alternating shear on a range of polydimethyl siloxanes (16, 17), polybutenes (18), polyethylacrylates and poly-n-butyl acrylates (19). In analysing the results, use has been made of the *Rouse* theory involving a summation of modes of motion along the polymer chain. Each mode is characterised by a relaxation time, τ_p, and contributes an amount $(\varrho RT/\bar{M}_n)$. τ_p to the steady-flow viscosity and an amount $(\varrho RT/\bar{M}_n)$ to the limiting shear modulus, which is independent of the mode number. Good agreement has been found between calculated and experimental curves for the variation with frequency of the components of the complex modulus. Two different types of relaxation processes have been included in the analysis, namely: i) relaxation modes of the flexible polymer molecule mainly responsible for the viscoelastic behaviour at relatively low frequencies and ii) relaxation of small portions of the backbone chain and associated side group, as described by the BEL equation for supercooled liquids, which governs the high frequency behaviour.

Despite the agreement obtained between predicted and measured properties, there are fundamental objections to the use of the *Rouse* theory for undiluted polymer melts. The theory purports to deal with dilute solutions of monodisperse polymer molecules. Reasonable modifications can be made to deal with polydispersity and with entanglements due to polymer-polymer interactions. In order to clarify the position, our approach has been to study the behaviour in the melt of polystyrenes having narrow molecular weight distributions. These are available in the molecular weight range from 600 to over 10^6 and are particularly attractive, since the lower molecular weight samples might be expected to exhibit the behaviour found for non-polymeric supercooled liquids; thus the onset of polymer mode behaviour with increasing molecular weight may, in due course, be traced.

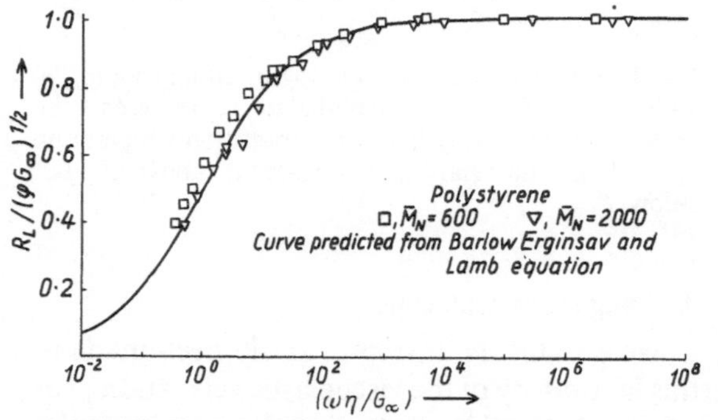

Fig. 14. The normalised resistive component of the shear mechanical impedance versus $\log(\omega\eta/G_\infty)$ for polystyrene samples of molecular weight 600 (□) and 2000 (▽). The curve is drawn in accordance with the BEL equation

The results shown in fig. 14 are for the two lowest molecular weight samples (600 and 2000). Measurements of the resistive component of the shear mechanical impedance are plotted in normalised form against $(\omega\tau_m)$ and these results are in close agreement with the prediction of the BEL curve. Such differences as exist between the experimental points and the calculated curve may be due either to the effect of the retardation process discussed in § 2 or to the fact that these polymer samples contain appreciable amounts of both shorter and longer molecules than the nominal 6 or 20 repeat units in the respective cases. Nevertheless, we conclude that the viscoelastic properties measured in cyclic shear are characteristic of a non-polymeric liquid.

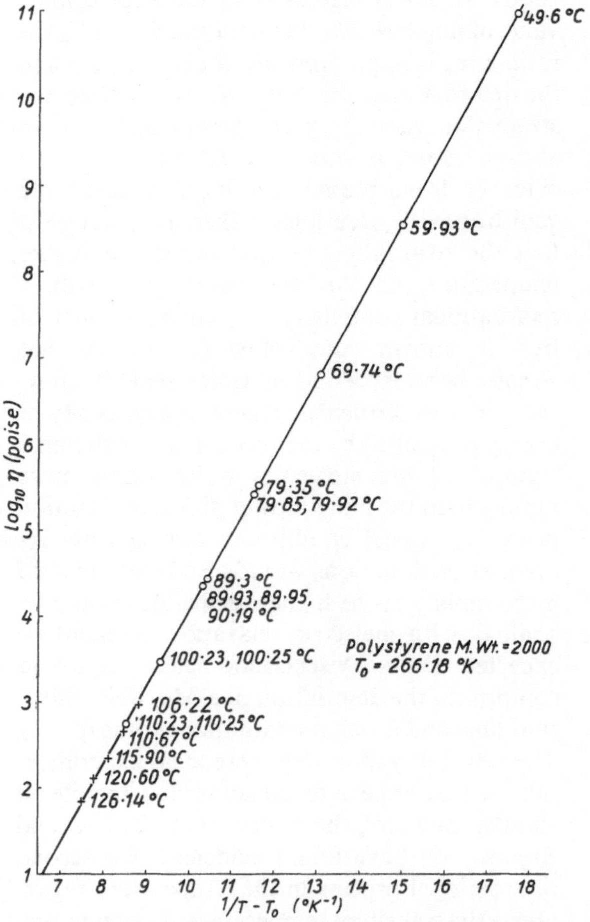

Fig. 15. $\log\eta$ versus $1/(T - T_0)$ for the 2000 molecular weight polystyrene showing the linear behaviour in accordance with the modified free volume eq. [3]

The steady-flow viscosity of the 2000 molecular weight sample has been measured using a shear creep apparatus based upon the design of *Plazek* (20). As shown by fig. 15, the results conform to the description of the modified free volume eq. [3] for viscosities up to 10^{11} P.

The limiting high frequency compliance of these polymers, J_∞, is of the order of 10^{-10} cm^2/dyne and depends linearly upon temperature as for non-polymeric liquids (eq. [2]). As the molecular weight increases, progressive changes occur, both in the magnitude of J_∞ at a given temperature above T_g and in the slope of J_∞ versus temperature until a critical molecular weight is reached. As shown by the results of fig. 16, values of J_∞ measured over a temperature range of some 40 °C above T_g are indistinguishable within experimental error for the two higher molecular weight samples, 20,400 and 97,000.

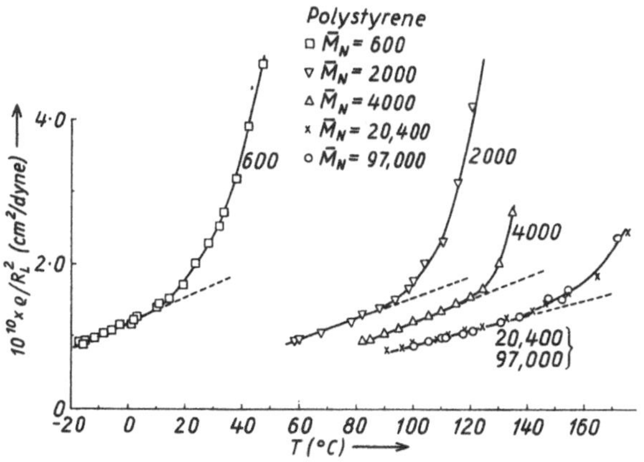

Fig. 16. Plots of ϱ/R_L^2 versus T for polystyrene samples of narrow molecular weight distribution. In the elastic region at sufficiently low temperatures $\varrho/R_L^2 \rightarrow J_\infty$. □, $M=600$; ▽, $M=2000$; △, $M=4000$; ×, $M=20{,}400$; ○, $M=97{,}000$

Extensive results are available from the work of *Plazek* and *O'Rourke* (8) on the creep behaviour of these polystyrenes. Completion of the present series of measurements should enable direct comparisons to be made with their earlier creep work, and it is particularly interesting to follow the influence of the retardational properties of non-polymeric liquids into the region where, with increasing molecular weight, evidence of polymer modes might be expected to be found in the viscoelastic properties.

In view of the wealth of results available in the literature dealing with the viscoelastic properties of polymer materials, it is perhaps surprising that a satisfactory treatment backed by reliable measurements has not yet emerged. In the author's view, the requirements for a scientific understanding are, firstly, to eliminate as far as possible in any analysis the uncertainties caused by a distribution of molecular weights and, secondly, to examine the effects of predetermined tacticity.

The author wishes to thank Dr. *G. Williams* and Mr. *M. Shears* for making their dielectric relaxation results available in advance of publication and for their willing collaboration. He is indebted to his colleagues, Drs. *A. J. Barlow, G. Harrison, A. Erginsav, S. Fewster, M. G. Kim* and *C. J. Steadman*, who are responsible for carrying out this programme of research. This work is supported financially by a Grant from the Science Research Council, which is most gratefully acknowledged.

Summary

According to linear viscoelastic theory, the expressions for the complex compliance in cyclic shear at angular frequency, ω, and for the creep function are:

$$J^*(j\omega) = J_\infty + 1/j\omega\eta + J_r\chi(j\omega) \qquad [1]$$

$$J(t) = J_\infty + t/\eta + J_r\psi(t). \qquad [2]$$

The normalised functions $\chi(j\omega)$ and $\psi(t)$ are related through the spectrum of retardation times.

Recent work in the author's laboratory has shown that for non-polymeric liquids $\chi(j\omega)$ can be closely represented by the empirical *Davidson-Cole* expression used extensively to describe dielectric relaxation:

$$\chi(j\omega) = \frac{1}{(1 + j\omega\tau_r)^\beta} \qquad [3]$$

τ_r is a characteristic retardation time parameter and $0 < \beta < 1$. For supercooled liquids, β is typically 0.5, in which case eq. [1] may be rewritten:

$$J^*(j\omega) = J_\infty \left[1 + \frac{1}{j\omega\tau_m}\right] + \frac{J_r}{(1 + j\omega\tau_r)^{\frac{1}{2}}} \qquad [4]$$

$\tau_m = \eta J_\infty$ is the *Maxwell* relaxation time.

Eq. [4] is fitted to experimental results for a number of liquids: the corresponding function $\psi(t)$ in the time domain is found to fit the creep data of *Plazek* for 1,3,5-tri-(α-naphthyl) benzene.

In practice $\tau_r > \tau_m$ below the "*Arrhenius* temperature" and for $\omega\tau_r \gg 1$ eq. [4] reduces to the form of the *Barlow, Erginsav* and *Lamb* equation used previously:

$$J^*(j\omega) = J_\infty \left[1 + \frac{1}{j\omega\tau_m}\right] + \frac{2J_\infty}{(j\omega\tau_m)^{\frac{1}{2}}} \qquad [5]$$

with $J_r/J_\infty = 2(\tau_r/\tau_m)^{\frac{1}{2}}$.

A review is given of available experimental results which confirm the applicability of eq. [4] or, in more approximate form, eq. [5]. It has also been shown that within experimental accuracy:

a) J_∞ is a linear function of temperature, and
b) $G_\infty (= 1/J_\infty)$ is a linear function of pressure.

In the case of polymer melts, the behaviour at high frequencies where molecular movements are restricted to small elements of the polymer chain is found to be similar to that observed in non-polymeric liquids, eq. [5]. Additional contributions are found at lower frequencies due to entanglement effects. Results are given of preliminary measurements on a range of polystyrenes of narrow molecular weight distribution.

References

1) *Barlow, A. J., J. Lamb, A. J. Matheson, P. R. K. L. Padmini,* and *J. Richter*, Proc. Roy. Soc. (London), **A 298**, 467–480 (1967).

2) *Barlow, A. J., A. Erginsav,* and *J. Lamb,* Proc. Roy. Soc. (London), **A 298**, 481–494 (1967).
3) *Barlow, A. J.* and *J. Lamb,* Disc. Faraday Soc. **43**, 223–230 (1967).
4) *Barlow, A. J., A. Erginsav,* and *J. Lamb,* Proc. Roy. Soc. (London), **A 309**, 473–496 (1969).
5) *Barlow, A. J., G. Harrison, J. B. Irving, M. G. Kim, J. Lamb,* and *W. C. Pursley,* Proc. Roy. Soc. (London), **A 327**, 403–412 (1972).
6) *Irving, J. B.* and *A. J. Barlow,* J. Physics E: Sci. Instruments, **4**, 232–236 (1971).
7) *Gross, B.,* Mathematical structure of the theories of viscoelasticity (Paris Hermann 1953).
8) *Plazek, D. J.* and *V. M. O'Rourke,* J. Pol. Sci., **9**, 209–243 (1971).
9) *Barlow, A. J.* and *A. Erginsav,* Proc. Roy. Soc. (London), **A 327**, 175–190 (1972).
10) *Davidson, D. W.* and *R. H. Cole,* J. Chem. Phys., **19**, 1484–1490 (1951).
11) *Williams, G.* and *M. Shears,* Private Communication.
12) *Plazek, D. J.* and *J. H. Magill,* J. Chem. Phys. **45**, 3038–3050 (1966); J. Chem. Phys. **49**, 3678–3682 (1968).
13) *Phillips, M. C., A. J. Barlow,* and *J. Lamb,* Proc. Roy. Soc., London **A 329**, 193–218 (1972).
14) *Goldstein, M.,* J. Chem. Phys. **51**, 3728–3739 (1969)
15) *Barlow, A. J., J. Lamb,* and *A. J. Matheson,* Proc. Roy. Soc., **A 292**, 322–342 (1966).
16) *Barlow, A. J., G. Harrison,* and *J. Lamb.,* Proc. Roy. Soc., **A 282**, 228–251 (1964).
17) *Lamb, J.* and *P. Lindon,* J. Acous. Soc. Amer. **41**, 1032–1042 (1967).
18) *Barlow, A. J., R. A. Dickie,* and *J. Lamb,* Proc. Roy. Soc., **A 300**, 356–372 (1967).
19) *Barlow, A. J., M. Day, G. Harrison, J. Lamb,* and *S. Subramanian,* Proc. Roy. Soc., **A 309**, 497–520 (1969).
20) *Plazek, D. J.,* J. Pol. Sci. **6**, 621–638 (1968).

Author's address:

Professor *John Lamb*
Dept. of Electronics and Electrical Engineering,
The University
Glasgow 912 8 QQ (Scotland)

Rheol. Acta 12, 449–454 (1973)

Université Scientifique et Médicale de Grenoble, Institut de Mécanique de Grenoble, Domaine Universitaire (France)

Fluides viscoélastiques non linéaires satisfaisant à un principe de superposition étude théorique et experimentale

Par Ph. Le Roy et J. M. Pierrard

Avec 6 figures

(Reçu p. p. 27 octobre 1972)

I. Loi de comportement de materiaux héréditaires

Nous présentons ici une formulation de lois de comportement de matériaux à mémoire obéissant à un principe de superposition des contraintes. Ce n'est, en fait, qu'une généralisation aux grandes déformations du principe de superposition de *Boltzmann* bien connu. Un tel principe traduit que la contrainte à un instant donné est la limite d'une somme infinie d'incréments de contrainte de tous les instants précédents, tous affectés d'un coefficient de mémoire.

Pour que la loi respecte le principe d'objectivité, il faut que cette sommation soit effectuée dans un repère rhéologique (Rh)[1]) qui n'est autre qu'un repère co-rotationnel de *Zaremba*. Cela s'écrit dans un tel repère:

$$\sigma^{ij} = \int_{-\infty}^{t} \chi_0(t-\tau)\, d\sigma_1^{ij} \qquad [1]$$

χ_0 est alors une fonction mémoire caractéristique du matériau.

Une telle loi (de comportement) peut aussi bien représenter un comportement de matériau viscoélastique *solide* ou *liquide*: en écrivant l'incrément de contrainte $d\sigma_1^{ij}$ sous la forme:

$$d\sigma_1^{ij} = f^{ij}\, dt$$

la fonction f^{ij} étant par exemple, suivant le cas, une fonction du tenseur déformation pure D, ou du tenseur vitesse de déformation $\mathscr{D}$.

II. Cas d'un fluide viscoélastique incompressible

Nous étudions dans ce qui suit un matériau incompressible: la relation [1] qui traduit le principe de superposition s'applique alors au deviateur de contrainte S:

$$S^{ij} = \int_{-\infty}^{t} \chi_0(t-\tau)\, dS_1^{ij}(\tau). \qquad [2]$$

Dans cette relation, l'incrément de déviateur de contrainte est donné alors par une loi visqueuse.

La loi visqueuse la plus générale pour un matériau incompressible s'écrit:

$$f^{ij} = A_1(B_2, B_3)\, \mathscr{D}^{ij} + A_2(B_2, B_3)\left[\mathscr{D}^{ih}\mathscr{D}^{hj} - \tfrac{1}{3} B_2 \delta^{ij}\right]$$

où A_1 et A_2 sont des fonctions des invariants:

$$B_1 = \mathscr{D}^{ii} = 0$$

$$B_2 = \mathscr{D}^{ih}\mathscr{D}^{hi}$$

$$B_3 = \mathscr{D}^{ij}\mathscr{D}^{jh}\mathscr{D}^{hi}.$$

Nous la mettons sous la forme:

$$f^{ij} = \frac{\Phi_1(B_2, B_3)}{B_2}\mathscr{D}^{ij} + \Phi_2(B_2, B_3) \times \left\{B_3\mathscr{D}^{ij} - B_2\mathscr{D}^{ih}\mathscr{D}^{hj} + \frac{(B_2)^2}{3}\delta^{ij}\right\} \qquad [3]$$

en posant

$$\Phi_1 = B_2 A_1 + B_3 A_2$$

$$\Phi_2 = -\frac{A_2}{B_2}.$$

Sous cette forme, on peut remarquer que l'énergie dissipée par unité de volume et par unité de temps est:

$$\mathscr{P} = \Phi_1(B_2, B_3) > 0.$$

Nous remarquons que le second terme de la loi [3] ne dissipe pas d'énergie.

Pour plus de généralité, on décrit le comportement héréditaire du matériau en appliquant le

[1]) Un repère (Rh) est un repère dans lequel la vitesse de la transformation est une vitesse de déformation pure (Ref. 1). C'est également un «Kinematically preferred coordinate system» de *Thomas* et autres.

principe de superposition des contraintes séparément à la partie dissipative et à celle non dissipative de la loi visqueuse [3].

On obtient alors une loi de comportement telle que l'on ait:

$S = S_{(1)} + S_{(2)}$.

Alors $S_{(1)}$ et $S_{(2)}$ sont données par leurs composants en repère Rh:

$$S_{(1)}^{ij}(t) = \int_{-\infty}^{t} \chi_1(t-\tau) \cdot \frac{\Phi_1(B_2, B_3)}{B_2} \mathscr{D}^{ij}(\tau)\, d\tau$$

$$S_{(2)}^{ij}(t) = \int_{-\infty}^{t} \chi_2(t-\tau) \cdot \Phi_2(B_2, B_3) \times \left\{ B_3 \mathscr{D}^{ij} - B_2 \mathscr{D}^{ih} \mathscr{D}^{hj} + \frac{(B_2)^2}{3} \delta^{ij} \right\} d\tau. \quad [4]$$

Cas simple

Le cas le plus simple pour cette loi est celui où la loi visqueuse incrémentale f^{ij} est de type *Newton*ien.

Alors, la loi de comportement du matériau se formule donc (en repère Rh):

$$S^{ij}(t) = \int_{-\infty}^{t} \chi_0(t-\tau)\, 2\mu \mathscr{D}^{ij}(\tau)\, d\tau$$

ou

$$S^{ij}(t) = \int_{-\infty}^{t} \chi(t-\tau)\, \mathscr{D}^{ij}(\tau)\, d\tau. \quad [5]$$

Ce cas correspond à:

$\Phi_1(B_2, B_3) = 2\mu B_2$ et $\Phi_2 = 0$.

Nous allons voir que de nombreux matériaux peuvent être représentés par une telle loi rhéologique. Pour cela, étudions des écoulements simples qui sont des essais viscosimétriques classiques (de la classe des écoulements visco-métriques de *Coleman-Noll*): Ecoulements à trajectoires circulaires.

III. Mise en équation d'écoulements à trajectoires circulaires

En un point de coordonnées cylindriques (ϱ, θ, z) la vitesse est supposée de la forme $\vec{V} = \varrho \cdot g(\varrho, z)\, \vec{h}_1$ et le repère orthonormé $[\vec{h}_i]$ est défini par:

$\vec{h}_2$, dans le plan méridien, normal aux lignes g = cte;

h_3, dans ce même plan, tangent aux lignes g = cte.

Dans ce repère, on a:

$$\mathscr{D} = \begin{Vmatrix} 0 & \Omega_1 & 0 \\ \Omega_1 & 0 & 0 \\ 0 & 0 & 0 \end{Vmatrix}$$

avec:

$$(\Omega_1)^2 = \varrho^2/4 \left[\left(\frac{\partial g}{\partial \varrho}\right)^2 + \left(\frac{\partial g}{\partial z}\right)^2 \right]$$

et:

$$S = \begin{Vmatrix} S_{(1)}^{11} + S_{(2)}^{11} & S^{12} & 0 \\ S^{12} & -S_{(1)}^{11} + S_{(2)}^{11} & 0 \\ 0 & 0 & -2S_{(2)}^{11} \end{Vmatrix} \quad [6]$$

Avec:

$$S_{(1)}^{11} = \frac{\Phi_1(2\Omega_1^2, 0)}{2\Omega_1} \int_0^\infty \chi_1(u) \sin(2\Omega_1 u)\, du$$

$$S^{12} = \frac{\Phi_1(2\Omega_1^2, 0)}{2\Omega_1} \int_0^\infty \chi_1(u) \cos(2\Omega_1 u)\, du$$

$$S_{(2)}^{11} = \tfrac{2}{3} \Omega_1^4 \cdot \Phi_2(2\Omega_1^2, 0) \int_0^\infty \chi_2(u)\, du. \quad [7]$$

En introduisant ce déviateur de contrainte dans l'équation de l'équilibre, on détermine la pression isotrope p (telle que $\sigma = -pI + S$).

Cette pression, définie par son gradient, n'existe que si l'écoulement satisfait à la condition:

$$\varrho^2 S^{12}(\Omega_1) = \frac{\partial}{\partial y^3}(H) \quad [8]$$

où H est telle que l'on ait:

$$\frac{\partial(H)}{\partial y^2} = 0.$$

Dans cette écriture $\frac{\partial}{\partial y^i}$ n'est pas la dérivée partielle par rapport à une variable, mais la dérivée dans la direction $\vec{h}_i$.

Les relations [7] permettent d'exprimer les trois fonctions viscométriques usuelles [Ref. (2) et (6)]:

$$\eta = \frac{S_1^{12}}{2\Omega_1}$$

$$N_1 = \sigma_2 - \sigma_1 = S^{11} - S^{22} = 2S_{(1)}^{11}$$

$$N_2 = \sigma_1 = S^{22} - S^{33} = -S_{(1)}^{11} + 3S_{(2)}^{11} \quad [9]$$

Dans l'hypothèse de la loi simple [5], les expressions [6] et [7] deviennent:

$$S = \begin{Vmatrix} S^{11} & S^{12} & 0 \\ S^{12} & -S^{11} & 0 \\ 0 & 0 & 0 \end{Vmatrix} \qquad [10]$$

avec:

$$S^{11} = \Omega_1 \int_0^\infty \chi(u) \sin(2\Omega_1 u)\, du$$

$$S^{12} = \Omega_1 \int_0^\infty \chi(u) \cos(2\Omega_1 u)\, du. \qquad [11]$$

IV. Cas particuliers d'écoulements

Dans les différents essais viscosimétriques usuels les surfaces $g =$ cte sont des plans, des cylindres, ou des cônes.

a) Surfaces g = constante cylindriques (écoulement de *Couette*).

g est fonction de ϱ seul. Pour une distribution de vitesses on peut tenir compte des forces d'inertie.

On a alors

$$\Omega_1 = -\frac{\varrho}{2}\frac{dg}{d\varrho}.$$

La condition [8] d'existence de la pression s'écrit $S^{12} = \dfrac{a}{\varrho 2}$ où a est une constante d'intégration. Cette relation définit par inversion $\dfrac{dg}{d\varrho}(\varrho)$, et donc g par intégration.

Les deux constantes introduites ci-dessus sont déterminées par les valeurs de g sur les cylindres qui limitent l'écoulement.

b) Surfaces g = constante coniques (écoulement *Plan-Cône*).

On pose $\varrho = r\cos\varphi$ et $z = r\sin\varphi$, g est alors fonction de φ seulement, et $\Omega_1 = \frac{1}{2}\cos\varphi\, dg/d\varphi$. Pour une telle distribution de vitesses il faut négliger les forces d'inertie. La condition [8] d'existence de la pression s'écrit $S^{12} = \dfrac{a}{\cos^2\varphi}$ où a est une constante d'intégration. Cette relation définit par inversion $\dfrac{dg}{d\varphi}(\varphi)$, et donc g par intégration.

Les deux constantes introduites ci-dessus sont déterminées par les valeurs de g sur les deux cônes qui limitent cet écoulement.

Cas d'impossibilité de tels écoulements

Dans les cas a) et b) évoqués ci-dessus, nous avons vu que Ω_1 est obtenu par inversion de la fonction $S^{12}(\Omega_1)$. Cette fonction a l'allure ci-dessous (fig. 1).

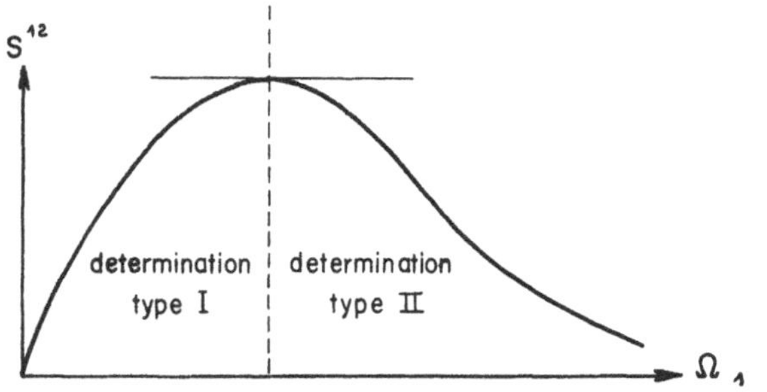

Fig. 1

L'inversion de cette fonction conduit à deux déterminations obtenues respectivement en utilisant la partie croissante et la partie décroissante de $S^{12}(\Omega_1)$.

Notons ω la différence de rotation entre les parois solides qui limitent l'écoulement. On démontre qu'il existe deux nombres B_1 et B_2, avec $B_1 < B_2$, ne dépendant que des données géométriques de l'écoulement, tels que:

$\omega < B_1$	l'écoulement est obtenu avec la détermination de type I;
$B_1 < \omega < B_2$	l'écoulement répondant aux hypothèses cinématiques envisagées ($\vec{V} = \varrho \cdot g(\varrho, z)\vec{h}_1$) est impossible;
$B_2 < \omega$	l'écoulement est obtenu avec la détermination de type II, mais il est instable et donc irréalisable expérimentalement.

Conclusion

Les expressions [7], qui donnent $S^{11}_{(1)}$ et S^{12} en fonction de Ω_1, dans un écoulement en régime permanent, se déduisent de la transformée de *Fourier* de $\chi_1(t)$. Mais la détermination de $\chi_1(t)$ à partir de ces relations [7] nécessite la connaissance de $S^{11}_{(1)}(\Omega_1)$ pour tout le domaine $0 < \Omega_1 < \infty$. Les expériences d'écoulements permanents, ne pouvant être faites que pour Ω_1 borné, ne permettent donc pas la détermination de $\chi_1(t)$.

V. Essais en régime transitoire

Par contre, des expériences de relaxation permettent d'obtenir $\chi_1(t)$ et $\chi_2(t)$.

Si, à $t = 0$, on suppose que l'on arrête instantanément l'écoulement, l'évolution des contraintes au cours du temps est donnée par:

$$S^{11}_{(1)} = -S^{22}_{(1)} = -\frac{\Phi_1(2\Omega_1^2)}{2\Omega_1} \times \int_t^\infty \chi_1(u)\sin[2\Omega_1(t-u)]\,du$$

$$S^{12}_{(1)} = \frac{\Phi_1(2\Omega_1^2)}{2\Omega_1}\int_t^\infty \chi_1(u)\cos[2\Omega_1(t-u)]\,du$$

$$S^{11}_{(2)} = -\frac{1}{2}S^{33}_{(2)} = -\frac{2}{3}\Omega_1^4\cdot\Phi_2(2\Omega_1^2)\int_t^\infty \chi_2(u)\,du \qquad [12]$$

où Ω_1 est celui de l'écoulement de rotation uniforme à partir duquel on effectue l'essai de relaxation.

Ces contraintes satisfont au système différentiel suivant:

$$\frac{dS^{12}}{dt} + 4\Omega_1^2\int_\infty^t S^{12}(\tau)\,d\tau = \frac{-\Phi_1(2\Omega_1^2)}{2\Omega_1}\cdot\chi_1(t)$$

$$\frac{dS^{22}}{dt} + 2\Omega_1 S^{12} = \frac{2}{3}\Omega_1^4\cdot\Phi_2(2\Omega_1^2)\cdot\chi_2(t). \qquad [13]$$

Ces relations permettent de déterminer les produits $\Phi_1\chi_1$ et $\Phi_2\chi_2$. Les résultats expérimentaux obtenus à partir de différentes valeurs de Ω_1 doivent donner des courbes en fonction de t qui sont affines; les rapports d'affinité déterminent $\Phi_1(\Omega_1)$ et $\Phi_2(\Omega_1)$.

VI. Résultats expérimentaux et comparaison avec les résultats théoriques

Les résultats dont nous rendons compte ici sont relatifs à une cinématique plan-cône (de type b) (fig. 2).

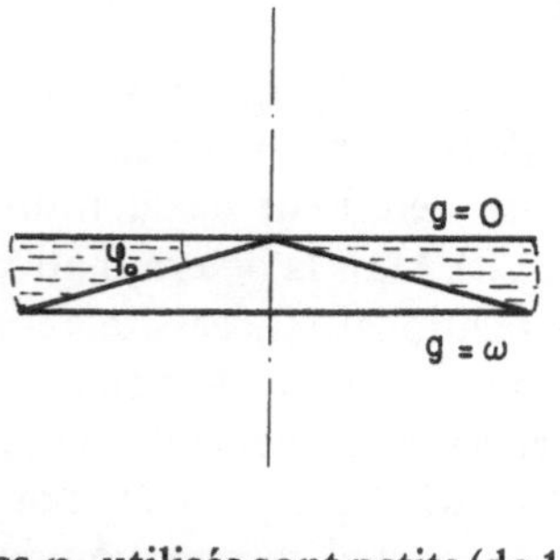

Fig. 2

Les angles φ_0 utilisés sont petits (de 1–4°); dans ce cas Ω_1 est constant dans tout l'échantillon et vaut:

$$\Omega_1 \neq \frac{1}{2}\frac{\omega}{\varphi_0}.$$

Le déviateur de contrainte est donc constant dans tout l'échantillon (cf. éq. [7]), ce qui rend cet essai particulièrement intéressant.

Le rhéogoniomètre de *Weissenberg* sur lequel nous avons effectué les essais ci-dessous permet de mesurer avec la précision souhaitable (moins de 1%) le couple T sur le plan supérieur et l'effort global normal N sur le cône.

On a alors:

$$T = A_1 S^{12}\left(\frac{\omega}{2\varphi_0}\right) \qquad A_1 = \int_{\text{plan}} \varrho\, ds$$

$$N = \int_{\text{plan}} (P - S^{22})\,ds = A\left[S^{11}_{(1)}\left(\frac{\omega}{2\varphi_0}\right) + 2S^{11}_{(2)}\left(\frac{\omega}{2\varphi_0}\right)\right];$$

$$A = \int_{\text{plan}} ds.$$

Les matériaux utilisés ont été des polymères (polyisobutylènes, silicones) en solution ou non, ainsi que des encres d'imprimerie, des argiles dans leur domaine de comportement liquide, ou des suspensions d'argiles de concentrations variées.

Dans des essais de relaxation, ce dispositif expérimental permet seulement la mesure de $T(t)$; la détermination des fonctions χ_1 a été faite au moyen de [13].

Sur les figs. 3 et 4 sont représentées les courbes $\frac{\Phi_1\chi_1}{2\Omega_1^2}(t)$ ainsi déterminées de deux matériaux très différents: une solution à 50% en poids de polyisobutylène vistanex LMMS (P.M ≠ 11000) et d'huile minérale mayoline, et une gomme rhodorsil (silicone chargé de type diméthyl-polysiloxamique).

On note sur ces figures que pour un très large domaine de valeur de Ω_1, les courbes coïncident et donc $\Phi_1(2\Omega_1^2) = 2\Omega_1^2$. Φ_1 et χ_1 sont donc bien alors des fonctions caractéristiques du matériau.

La mesure de $N(t)$ se fait au moyen d'un servomécanisme ramenant à déformation nulle la déflection d'un système élastique contraint par N et il y a alors couplage entre la réponse du servomécanisme et celle du matériau. La détermination de Φ_2 et χ_2 ne peut donc être faite que par des mesures ponctuelles de la contrainte normale au plan, par exemple avec des capteurs d'effort normal.

Pour confirmer plus complètement l'hypothèse de la loi choisie, nous avons recalculé par [11], et à partir des fonctions cidessus déterminées expérimentalement les valeurs du couple et de l'effort normal mesurées en régime permanent; pour l'effort normal nous avons négligé $S_{(2)}^{11}$ ce qui revient à utiliser la forme simplifiée [5] de la loi complète. Les calculs utilisant la transformée de *Fourier* de la fonction χ_1 ont été effectués numériquement sur ordinateur [1]). Les résultats comparés sont reportés sur les figs. 5 et 6.

La concordance des résultats est meilleure que la précision des mesures.

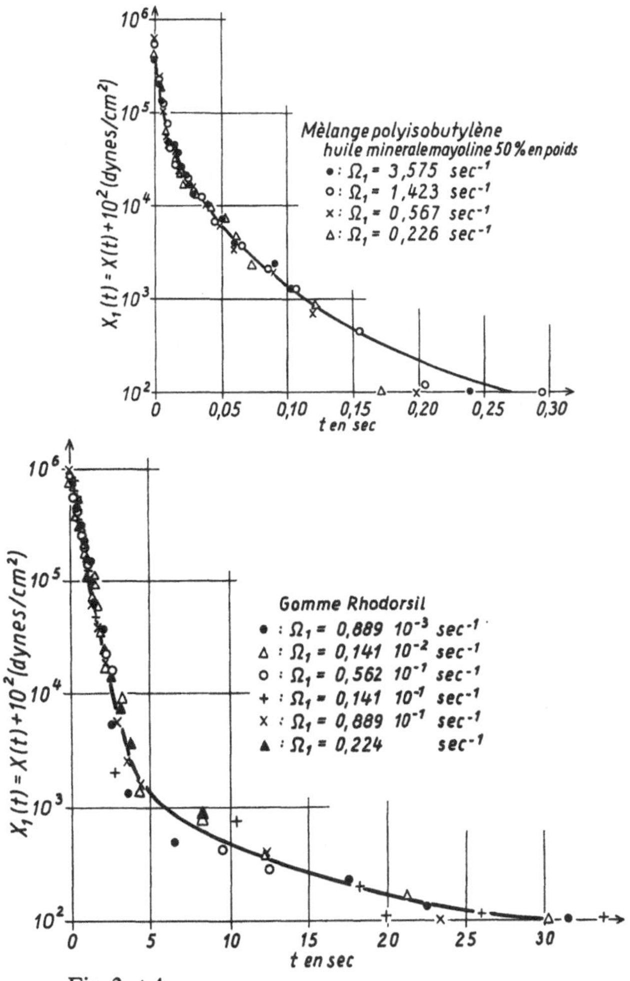

Fig. 3 et 4

Répartition de la contrainte normale au plan

Le déviateur de contrainte S est fonction de Ω_1 seul et donc ici indépendant de ϱ; par contre, la pression est donnée par la relation

$$p = Po - 6S_{(2)}^{11} \log\left(\frac{\varrho}{R}\right).$$

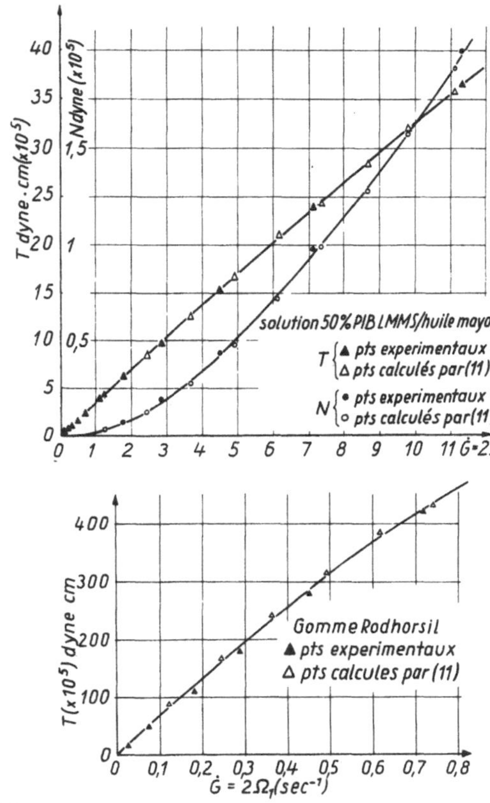

Fig. 5 et 6

La variation de contrainte normale est donc due à la seule variation de p. On observe expérimentalement que p est une fonction décroissante de ϱ, ce qui implique que $S_{(2)}^{11}$ est positif. D'après la relation [7] cela entraîne que $\Phi_2(2\Omega_1^2)$ est négatif.

Nous voyons d'après [9] que la fonction viscométrique N_1 est positive, et que N_2 peut être de signe quelconque, et petite en module par rapport à N_1 : $|N_2| \ll N_1$.

Nous retrouvons ici analytiquement des résultats connus expérimentalement (Ref. [6], [7]).

VII. Problèmes à résoudre

Pour compléter la détermination et le contrôle d'une telle loi de comportement, nous envisageons les études suivantes:

1. Détermination de Φ_2 et χ_2 à l'aide d'essais de relaxation d'efforts normaux mesurés par des micro-capteurs. Ces expériences sont actuellement en cours de réalisation.

[1]) Deux techniques de calcul ont été employées et ont donné toutes deux la même précision de résultats: transformée $\mathscr{F}(\chi)$ utilisant une extension de l'algorithme «Fast *Fourier* transform» de *Cooley-Tukey;* approximation de $\chi(t)$ par une série d'exponentielles à coefficients positifs (5 suffisaient ici) et inversion directe.

2. Etude des variations de Φ_1 et Φ_2 en fonction du troisième invariant B_3. Il faut noter que cette étude ne peut pas se faire par des mesures sur une paroi rigide car en point d'une telle paroi avec l'hypothèse d'adhérence du liquide on a toujours $B_3 = 0$ (cf. [8]). Il faudra donc utiliser un dispositif avec glissement sur la paroi, paroi déformable ou visualisation de l'écoulement.

Summary

This study is devoted to incompressible viscoelastic fluids in which a superposition principle for stress is assumed. In order to respect the material objectivity, this summation of incremental stress, with corresponding memory factor, is operated in a rheological (corotational) frame. Theoretical study for these fluids in viscometric flows was conducted up to explicit solutions. This study shows some impossibility ranges for such flows. The determination of the memory functions from this type of flow involves *Fourier* transform and so, would need the knowledge of all the kinematic range, and this is impossible.

The study in relaxation tests permits the determination of the memory functions. So, we performed these experiments, to get memory functions which were independent of the kinematics.

Using the function so obtained, we can compute and predict some experimental results in steady flows; and the comparison between these two results was quite good.

Zusammenfassung

Wir untersuchen inkompressible viskoelastische Fluide, die einem Superpositionsprinzip für die Spannungen genügen. Um dem Prinzip der Bezugsindifferenz zu entsprechen, wird die Summation der Spannungsinkremente in einem korotationalen Bezugssystem vorgenommen. Die theoretische Untersuchung wird für viskosimetrische Strömungen mit kreisförmigen oder geradlinigen Stromlinien für einfache Randwertprobleme bis zu expliziten Lösungen durchgeführt. Dabei findet man, daß es Bereiche gibt, für die solche Strömungen nicht existieren. Die Bestimmung der Gedächtnisfunktion aus diesen Strömungsformen würde eine *Fourier*-Transformation erfordern und darum die Kenntnis des gesamten kinematischen Bereichs voraussetzen, was nach vorstehendem Ergebnis unmöglich ist.

Mit Hilfe von Relaxationsversuchen kann man die Gedächtnisfunktion unmittelbar erschließen. Durch solche Experimente erhalten wir Gedächtnisfunktionen, die von der Kinematik unabhängig sind. Unter Verwendung dieser Funktionen lassen sich einige theoretische Aussagen über stationäre Strömungen erhalten, die mit dem Experiment verglichen werden können. Es ergibt sich eine recht gute Übereinstimmung.

Littérature

1) *Angles d'Auriac, P.*, Définitions et principes en rhéologie tensorielle, Symposium I.U.T.A.M. Rhéologie et Mécanique des Sols, Grenoble, 1964. *J. Kravtchenko* et *P. M. Sirieys* (Ed.) (Berlin-Heidelberg, New York 1966). Houille Blanche **5**, 427-432 (1970).

2) *Coleman-Markowitz-Noll*, Viscometric flows of non-Newtonian fluids (Berlin-Heidelberg-New York 1966).

3) *Le Roy, P.*, Etude de certains liquides non Newtoniens, Thèse, Grenoble (1968).

4) *Pierrard, J. M.*, Sur la détermination et l'utilisation de lois de comportement de fluides viscoélastiques, Thèse, Grenoble (1971).

5) *Guelin, P., J. M. Pierrard*, C.R.A.S., **269**, 1156–1159 (1969).

6) *Tanner, R. I.*, Trans. Soc. Rheol. **14**, 483–507 (1970).

7) *Kaye, A., A. S. Lodge, D. G. Vale*, Rheol. Acta **7**, 368 (1968).

8) *Clermont, J. R.*, Lois de comportement et classes d'écoulements de fluides non Newtoniens, Thèse de Doctorat de Spécialité, Grenoble (1972).

Adresse de les auteurs:

Ph. LeRoy et *J. M. Pierrard*
Université Scientifique et Médicale
Institut de Mécanique
Domaine Université
B. P. 53 Centre de Tri
F-38041 Grenoble (France)

Rheol. Acta 12, 455–464 (1973)

From the Division of Macromolecular Science, Case Western Reserve University, Cleveland (U.S.A.) and the Department of Chemical Engineering, Mc Gill University, Montreal (Canada)

The viscosity of concentrated polymer solutions: Corresponding states principles

By R. Simha and L. A. Utracki

With 7 figures and 1 table

(Received October 27, 1972)

The concern of this paper is the zero shear viscosity in the concentration region bridging the highly dilute system, where the properties of the isolated solute particle or molecule and its interaction with the solvent medium are the determining factors, and the dense fluid, formed by the polymer melt or closely packed suspension. In the former domain, *Einsteins* point of view, i.e., the consideration of difference effects, arising from a hydrodynamic perturbation of the solvent flow by suspended particles, has governed subsequent theoretical work for all types of macrosolutes.

In more concentrated systems, the quantitative description of hydrodynamic interactions becomes prohibitively difficult, except for the simplest particle geometries. Here the configurational thermodynamic aspects are still relatively simple. Section I deals briefly with such suspensions. For these systems, the theoretical and experimental results indicate the possibility of formulating the viscosity-concentration dependence in terms of reduced variables. The resulting master curve is a manifestation of corresponding states behavior.

It is of great interest and practical importance, to examine whether such corresponding states behavior can also be defined for polymer solutions. Such an approach does not involve a priori theory or development of empirical viscosity-concentration relations. Beyond the establishment of the validity of a corresponding state principle for the viscosity of polymer solutions, we investigate the range of its validity, define the characteristic reducing parameters and correlate these with polymer molecular weight, configurational parameters, temperature of measurement and quality of the solvent. These topics will be taken up in Section II.

Finally, in Section III, the limitations of the approach are discussed and further experiments are suggested.

I. Suspensions of spheres

A. General

At finite, but low concentrations, a consistent extension of *Einsteins* theory for spherical particles requires a consideration of hydrodynamic interactions. Such a theory to a first order approximation was first presented by *Guth* and *Simha* (1). With increasing concentration however, a shielding effect becomes increasingly significant (2), and in addition the non-random distribution of centers must be considered (2). In the limit of zero shear, this distribution is determined by thermodynamic equilibrium solely. Both factors have been treated in a simplified manner by *Simha*, who adopted a cell model to account for the positional and shielding effects. Thus each particle interacts with its nearest neighbors only, which are assumed to act as the wall of a spherical enclosure. In this formulation the hydrodynamic problem can be solved rigorously. The cell radius is a function of the volume fraction and of a packing parameter f which will approach in the limit of high volume fraction a value determined by the spatial arrangement of the spheres. Since the model is strictly appropriate to the high concentration region and thus tends to overemphasize the shielding effects at lower volume fractions, f should be expected to vary slowly with composition. A comparison of the theory with an empirical "master" curve for spherical suspensions (3) shows this indeed to be the case (4), with f increasing first more rapidly, but only by 3% between volume fractions of 0.20 and 0.56.

30*

The results for the viscosity may be written in the form (2):

$$\tilde{\eta} = \eta_{sp}/([\eta]\,\varphi) = \lambda(y)$$
$$\lambda(y) = 4(1 - y^7)/[4(1 + y^{10}) - 25\,y^3(1 + y^4) + 42\,y^5]$$
$$y = \varphi^{1/3}/(f - \varphi^{1/3}) \qquad [1]$$

where $[\eta] = 5/2$ represents the *Einstein* value for the intrinsic viscosity, based on volume fractions. The function λ has been tabulated for $0.1 \leq y \leq 0.9$ and increases sharply for $y > 0.5$ (2, 4). The appearance of fractional exponents is characteristic of the quasi-crystalline structure assumed. As φ approaches the maximum value φ_{max}, f tends to the limit $2\varphi_{max}^{1/3}$. Eq. [1] then reduces, disregarding the difference between specific and relative viscosity, to:

$$\lim_{\tilde{\varphi} \to 1} \tilde{\eta} = (27/50)\,\tilde{\varphi}/(1 - \tilde{\varphi})^3$$

with

$$\tilde{\varphi} = \varphi/\varphi_{max}\,. \qquad [1a]$$

Eq. [1a] is expressed in reduced variables and thus indicates a principle of corresponding states. Although eq. [1] could be similarly formulated, it has been left in the form shown, as a reminder of the variation of f with φ for moderate concentrations.

B. Some recent applications

Recently *Utracki* has examined aqueous (5a) and non-aqueous (5b) suspensions of poly(vinyl chloride). The main objective concerning the first was to study the shear coagulation of lattices, by varying the temperature and concentration as well as the type of polymer particles, of surfactant and of electrolyte. One of the parameters determined from the kinetics of coagulation by means of a newly derived theoretical relation is φ_{max}. Simultaneously, the zero-shear viscosities of these suspensions were measured, varying φ from 0.05–0.22. The results were analyzed by means of eq. [1], in which φ_{max} was assumed to be the single characteristic parameter of the system. Values of φ_{max} between 0.74 and 0.43 result and are in good agreement with the numerical values of φ_{max} as determined from the kinetics of coagulation.

Eq. [1] was also found to be very useful in predicting the zero shear viscosities of plastisols for numerous dispersion type resin-plasticizer pairs. Here again φ_{max} was determined by two methods: a) from low concentration viscosity data ($\varphi = 0.02$–0.05) using eq. [1], and b) by the centrifugation method. The differences between these two sets of data are within 5%. Next, using an average value of φ_{max}, for a desired viscosity of plastisol (10^2–10^4 Poises) the concentration φ was calculated from eq. [1]. The experimental values of viscosities agreed with the predicted ones within 50%.

It has to be stressed that for partially coagulated systems (5a) and for plastisols (5b) in which particles become highly swollen, the deviations from the viscosity behavior predicted by eq. [1] are severe. The particles are not spherical and moreover it becomes difficult if not meaningless to define a particle volume.

II. Solutions of flexible macromolecules

A. General considerations

Eq. [1a] illustrates two points, one a principle of corresponding states, the other the expected rapid increase of η, as the reduced concentration approaches unity. Particularly the first point is to be the subject of further discussion. As is well known, the representation of a single coiling molecule by an *Einstein* sphere is a fair approximation in the theory of intrinsic viscosity. When it comes to a consideration of packing effects and an analog to the parameter φ_{max} in eq. [1a], such an approximation breaks down, due to a) the compressibility of the coils at moderate concentrations, especially in good solvents (6, 7), and b) entanglements, which assume increasing importance at elevated concentrations. Nevertheless, one may define a concentration c_0 corresponding to incipient overlap of the average spherical coils, as they exist at infinite dilution. With the well known expression for the intrinsic viscosity of impenetrable coils, we find c_0 to be proportional to $[\eta]^{-1}$, where the proportionality factor is close to unity. For operational purposes one may then define moderate concentrations as a few multiples of c_0, i.e., a few percent for conventional molecular weights, and examine the role of c_0 as a reducing concentration factor (6, 7). Two facts emerge. First, no rapid increase of $\tilde{\eta}$ at $c/c_0 \simeq 1$ is observed, although the derivatives $[\partial\tilde{\eta}/\partial(c/c_0)]_M$ are larger in θ-systems, where the coils are more compact, than in a good solvent. Secondly, the product $c[\eta]$ does not represent a reduced variable.

Even for the moderate concentration region, a theory of viscosity requires a consideration of the long range hydrodynamic interactions discussed above for spherical suspensions, and most importantly, of the osmotic compression in good solvents, and of the formation and hydrodynamic contributions of aggregates (6, 7). We have attempted to describe the thermodynamic factors at least in a semiquantitative manner and been able to show that the compression effect accounts for a significant portion of the difference between solutions of a polystyrene fraction in toluene and cyclohexane (θ) respectively (7).

Even for *Newton*ian systems and relatively dilute solutions, the establishment of a theoretical rheological equation of state, which would describe the dependence of the viscosity on concentration, molecular weight, solvent conditions and temperature, represents an extremely difficult problem. We proceed therefore on a different basis and explore a corresponding states principle. If successful, this results in the establishment of master curves, which may be expressed in analytical form, if desired. In any case, the treatment leads to predictions to be shown in what follows and provides guidance for further theoretical approaches.

B. Corresponding states

Since no appropriate theoretical information is available for polymer solutions, it is necessary to establish first experimentally, whether viscosity-concentration curves for different degrees of polymerization are superimposable in a given polymer-solvent system. They are not in the $\tilde{\eta} - c[\eta]$ plane (6, 7). The concentration reducing parameter in a given solvent and at a given temperature is written as $\gamma(M)$, to distinguish it from the previous quantity $c_0 \propto [\eta]^{-1}$. Consider first moderate concentrations. Fig. 1 shows a successful superposition for a series of polystyrenes of various origin in toluene, in a range of number average molecular weights between 6×10^5 and 0.15×10^5 (8). A fraction in the middle of the range, $M_n = 1.46 \times 10^5$ is chosen as the reference compound and the concentrations do not exceed about $2/[\eta]$. Similarly, fig. 2 exhibits the behavior of the same fractions in a θ-solvent (8). Analogous superpositions have been demonstrated for other polymer-solvent systems, e.g. polystyrene-cyclohexane over a moderate range of temperatures around θ (9), and polyisobutylene in benzene (θ) and cyclohexane (9, 10).

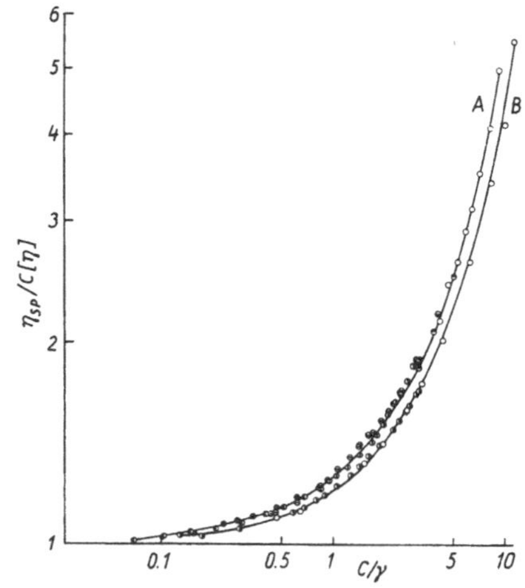

Fig. 1. Reduced viscosity-concentration curves, polystyrene in toluene; A. 30 °C, B. 48.2 °C. Solid line for $M_n = 1.46 \times 10^5$, points for different molecular weights

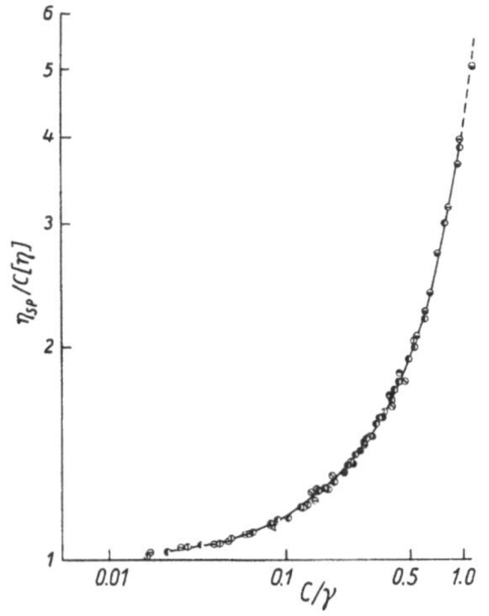

Fig. 2. Reduced viscosity-concentration curve, polystyrene in cyclohexane, 34 °C. Solid line for $M_n = 1.5 \times 10^5$, points for different molecular weights

These results encourage us to examine further systems at an extended range of concentrations (11), by making use of literature data. As an example, fig. 3 shows the behavior of poly(vinyl chloride)-cyclohexanone solutions over a considerable range of viscosity average molecular weight M_v and concentration extending from approximately 0.2–41 wt.-% (12). With the exception of the two lowest molecular weights in the upper range of concentrations, a gratifying master curve can be established over the whole range. Fig. 4 illustrates results on polyisobutene-isooctane mixtures (13). In the preparation of

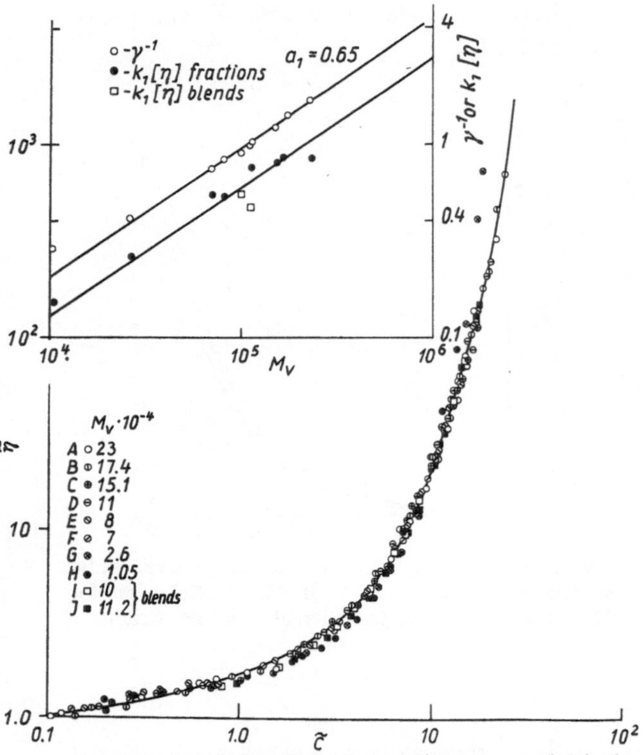

Fig. 3. Reduced viscosity-concentration curve, poly(vinyl chloride) in cyclohexanone. Solid line for $M_v = 1.1 \times 10^5$. Insert, characteristic concentration factor γ and $k_1[\eta]$, eq. [7′], vs. M_v.

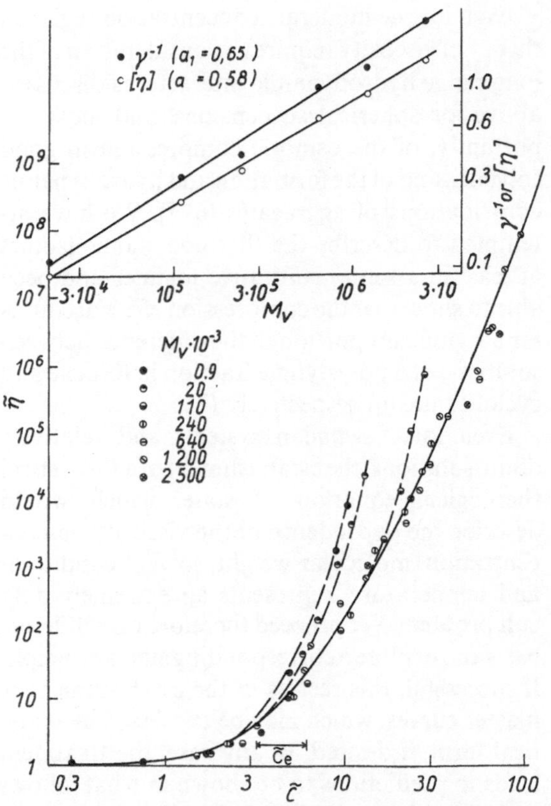

Fig. 4. Reduced viscosity-concentration curve, polyisobutene in isooctane. Range of reduced entanglement concentration $\tilde{c}_e$ computed from ref. 13 and γ. Insert, characteristic concentration factor γ and $k_1[\eta]$, eq. [7′], vs. M_v.

these graphs, double shifts on the logarithmic scales were required in the absence of information about $[\eta]$. Nevertheless: the following points are noteworthy. First, there is a departure of the two lowest molecular weights, viz. $M_v = 900$ and 20,000, and 110,000 for $\tilde{\eta}$-values exceeding 10^3. Secondly, the range of concentrations is very high, extending into the melt. Clearly these results are far from conclusive, but they do suggest the detailed examination of the high and melt viscosity ranges. Table 1, taken from ref. 11, contains a summary of the mixtures investigated by us.

Table 1. Characteristics of polymer-solvent systems

Polymer	Solvent	Temp. °C	Mol. weight $\bar{M}_v \times 10^{-4}$	Concentration range	η_{max} poise	No. of samples	a	a_1
Polystyrene	Cyclohexane	$34.4 = \theta$	1.5–60 [1])	0.22 g/dl	0.17	9	0.503	0.47
Polystyrene	Toluene	30	1.5–60 [1])	0–9 g/dl	0.5	9	~0.72	0.64
Polyisobutene	Benzene	$24 = \theta$	1.0–100	0–7 g/dl	0.14	4	0.50	0.47
Polyisobutene	Cyclohexane	25	1.0–100	0–5 g/dl	0.18	4	0.72	0.64
Polyisobutene	Xylene	25	32–400	5–40 wt.-%	210	5	0.64	0.68
Polyisobutene	Decalin	25	100–251	12–25 wt.-%	2.7×10^5	3	0.65	0.68
Polyisobutene	Isoctane	20	0.09–250	5–100 vol.-%	10^{11}	7	0.58 (?)	0.65
Poly(vinyl chloride)	Cyclohexanone	30	1.05–23	0–41 wt.-%	133	10	0.86	0.65
Poly(vinyl alcohol)	Water	30	2.44–32.2	4–20 g/dl	2×10^4	27	0.64	0.69
Ethyl cellulose	Toluene-ethanol 4:1	25	12–90	0.8–13 vol.-%	50	4	~1.0	0.73
Cellulose acetate	Acetone	20	17–90	0–4 vol.-%	0.35	6	~1.0	0.88
Polyisoprene	Decane	25	2.52–126	4.9–19.2 wt.-%	12	22	0.86	1.08

[1]) Number-average molecular weights.

C. The characteristic concentration; molecular weight and solvent effects

It can be anticipated that the parameter γ, although not proportional to $[\eta]$, will exhibit a similar behavior. That is, in analogy to the *Mark-Houwink* relation with an exponent a, characteristic of the solvent for a given polymer, an exponential relationship with analogous exponent a_1 is expected. We recapitulate the explicit relations for polystyrene in the two solvents, shown in figs. 1 and 2. The smoothed γ-values satisfy the following relations (8):

$$\text{Toluene:} \qquad \gamma(M) = 1.955 \times 10^3 M_n^{-0.640}$$

$$\text{Cyclohexane } (\theta)\text{:} \quad \gamma(M) = 2.596 \times 10^2 M_n^{-0.471}. \qquad [2]$$

The two expressions are normalized to slightly different values of M_n; $\gamma = 1$ for $M_n = 1.388$ and 1.334×10^5 respectively. We note that γ is independent of temperature between 30 and 48 °C in toluene, a very good solvent. Most noteworthy is the relationship between γ and $1/[\eta]$ which is proportional to the apparent overlap concentration c_0. Whereas in toluene $c_0/\gamma \propto M_n^{-0.082}$, the difference is considerably reduced in the θ-solvent and the ratio may be made practically equal to unity over a range of molecular weights, by suitable choice of a reference polymer.

Eq. [2] illustrates the parallel trends of the exponents a and a_1. A more complete listing appears in table 1. The values for the θ-solvent [which also apply to solutions at least some degrees above and below θ (9)], deserve particular comment. It could be demonstrated that γ is proportional to (or can be identified with) the critical concentration v_{crit} for phase separation (14), at least in these particular mixtures. Expressed in volume fractions, the relation is:

$$v_{crit} = 1.383 \times 10^3 M_n^{-0.471}. \qquad [3]$$

Furthermore, the intrinsic viscosities have also been measured as a function of temperature (15) and inter- or extrapolated to the respective critical temperatures T_c. The molecular weight range was such that these temperatures extended from about 6–28 °C. The results obey the equation

$$[\eta](T = T_c) = 1.51 \times v_{crit}^{-1}. \qquad [3']$$

Eqs. [2], [3] and [3′] indicate that in sufficiently poor solvents, γ is defined by the average volume of the encompassed coil, as measured at $T = T_c$. Under these conditions the average coil volume appears to play a role analogous to the particle volume in spherical suspensions. Moreover, the reduced $\tilde{\eta} - \tilde{c}$ curves tend in the direction of the theoretical $\tilde{\eta} - \tilde{\varphi}$ plot, discussed in Section I (see figs. 1 and 10 of ref. 11).

The importance of the critical point in connection with the viscosity at infinite dilution as well as at finite concentrations, makes further studies of precipitating systems important. We revert to the role of the critical temperature in the next section.

For a series of polymers dissolved in θ-solvents, the relative position of the reduced viscosity-concentration curves $\tilde{\eta} = \tilde{\eta}(c/v_{crit})$ is expected to be determined solely by the properties of the polymer chain, e.g. by its flexibility, polarity, and structure. Such a set of curves would be totally defined by the use of $[\eta]_\theta$ and v_{crit} as the reducing parameters for specific viscosity and concentration, respectively.

Some years ago we posed the question (16) whether a similar map can be constructed for a series of polymers in non-θ-solvents. The preliminary results on four vinyl aromatic polymers in the concentration range $c \leq 4/[\eta]$ indicate that if γ, identified with $1/k_1[\eta]$ (where k_1 is the *Huggins* parameter, see below), is compensated for differences in monomer molecular weight and polymer chain stiffness, then the relative position of the $\tilde{\eta}$ vs. $\tilde{c}$ plots, is defined by the magnitude of the thermodynamic long range interaction parameter B.

D. Temperature effects

So far we have operated primarily under isothermal conditions. The preceding results (8, 15) suggest that γ should be a slowly varying function of T. Furthermore, the role of the critical temperature T_c in connection with the intrinsic viscosity has been shown, see eq. [3′]. One may now ask whether characteristic reducing temperatures can be defined, such that $\gamma = \gamma(M, \tilde{T})$. We have recently explored this question by means of polystyrene solutions in 1-chlorodecane (17). The upper limit of concentration was about $4/[\eta]$. With the number average molecular weights of the anionically polymerized samples ranging from less than 10^4 to about 1.6×10^6, the maximum concentrations extended therefore from about 4–40 g/dl. A temperature change from 6.6–90 °C makes for a transition from approximate θ-conditions to a moderately good solvent, as is indicated by a *Mark-Houwink* coefficient a of 0.6. At each temperature superposition and the usual relationship

$$\gamma(M) = (M_0/M)^{a_1} = \tilde{M}^{-a_1} \quad [2']$$

are found to be valid, where M_0 and $\tilde{M}$ represent the reference and reduced molecular weight respectively. The only significant deviation again results from the lowest fraction. However, in this system $a_1 > a$ throughout. The generalization of eq. [2] for non-isothermal conditions is

$$\gamma(M, T) = \gamma(M_0, T) \times \tilde{M}^{-a_1(T)} \quad [2'']$$

with the condition $\gamma(M_0, T_0) = 1$, where T_0 represents a reference temperature. Fig. 5 exhibits the master curve with $T_0 = 60°$ and $M_0 = 2.4 \times 10^5$. Thus functions $\gamma(M_0, T)$ and $a_1(T)$ are definable and explicit expressions could be derived in terms of an arbitrarily chosen reference temperature. However, the results obtained for the polystyrene-cyclohexane system (8, 9, 15), and the involvement of the critical point, suggest an analogous approach here. Specifically, it is of interest to explore the proposition that the critical temperature is a corresponding temperature. Unfortunately, the requisite thermodynamic information, although available for cyclohexane (14), is lacking for 1-chlorodecane. To arrive at tentative conclusions at least, we consider the difference $T_\theta - T_{crit}$ for the two solvents and *assume* that it may be represented by the identical function of molecular weight for both. For cyclohexane an empirical relation has been proposed previously (18), which is in fair agreement with the results of *Debye* et al. (14). Fig. 6 shows $[\eta](T = T_c)$ as a function of M for cyclohexane, and the result is in accord with eqs. [3] and [3'].

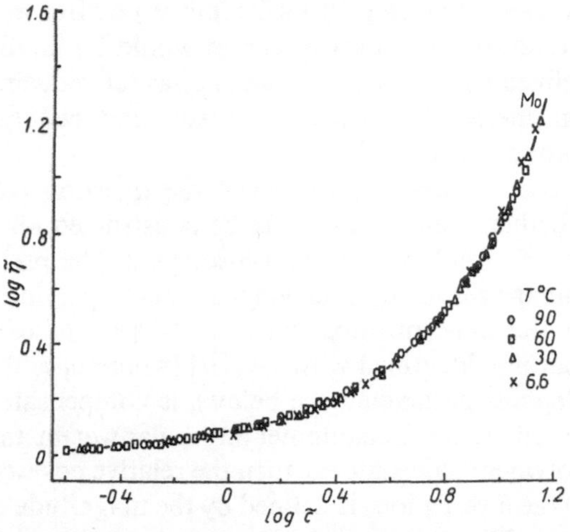

Fig. 5. Reduced viscosity-concentration curve, polystyrene in 1-chlorodecane, $M_n = 2.4 \times 10^5$. Solid line for 60 °C

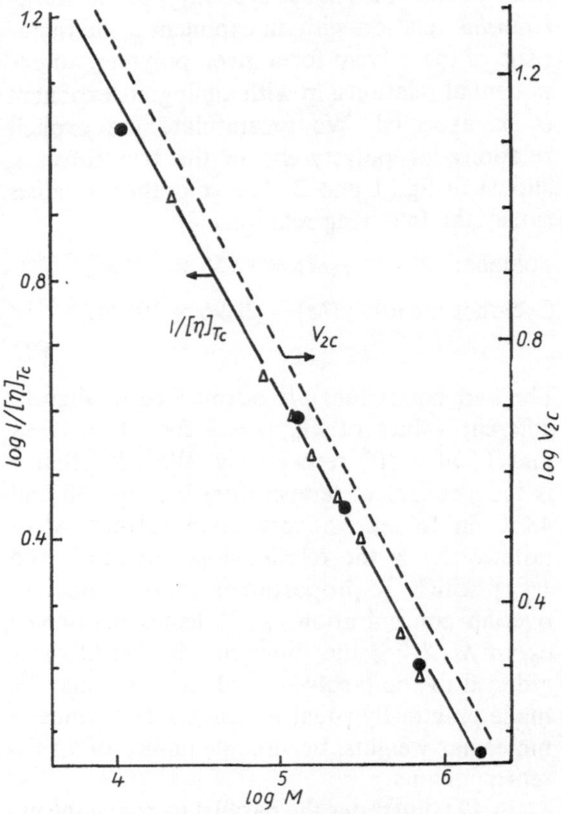

Fig. 6. Intrinsic viscosity at critical solution temperature and critical volume fraction as a function of molecular weight for polystyrene in cyclohexane (triangles) and estimates for 1-chlorodecane (circles)

On the same graph there appear the intrinsic viscosities estimated by extrapolation to the computed critical temperatures for 1-chlorodecane. The close correspondence is gratifying, although it does not constitute a proof.

We arrive finally at the following representation:

$$\gamma(M_0, \tilde{T}) = [(\tilde{T} - 1)/\tilde{T}_0 - 1)]^{0.1}$$

$$a_1 = 0.219 + \tilde{T}/3 \quad [4]$$

where the subscript 0 refers to the reference fraction and $\tilde{T} = T/T_c(M_0)$. The first expression represents the type of power relationship, frequently invoked in the neighborhood of a critical point. The second follows directly from experiment. Thus, we may consider $T_c(M_0)$ as a reducing temperature and identify it at least tentatively with a critical solution temperature.

E. Structural effects

So far we have been able to demonstrate (11, 16) that the corresponding states principle can be successfully applied to linear chain polymers of such widely different chemical and conformational characteristics as poly(isobutene), poly(vinyl chloride), poly(vinyl alcohol), poly(vinyl biphenyl), cellulose derivatives, and others. Recently, we have investigated (19) the viscoelastic properties of linear, 4- and 6-star branched polystyrenes under conditions of constant temperature (30.0 °C) and concentration ($c = 25.5$ g/dl) in di(ethyl) benzene, and also obtained the zero shear viscosity-molecular weight relations. Simultaneously, the mean square average radius of gyration $\langle s_0^2 \rangle$ and intrinsic viscosity in cyclohexane and toluene were determined (20). The results indicate the *Zimm-Kilb* relation (21)

$$[\eta]_{\theta,\,\text{branched}} = g^{1/2} [\eta]_{\theta,\,\text{linear}}$$

$$g = \langle s_0^2 \rangle_{\text{branched}} / \langle s_0^2 \rangle_{\text{linear}} \qquad [5]$$

to be valid.

All attempts to generate the viscosity-molecular weight dependence for the branched systems from the data on the linear polymer and the values of the parameter g through the conventional theoretical relations (22, 23) failed. However, in terms of the corresponding states principle very satisfactory results were obtained.

As indicated previously (8), the function $\tilde{\eta}(\tilde{c})$ can be expanded as (see also the following section)

$$\tilde{\eta} = 1 + \sum_i k_i([\eta]\,c)^i. \qquad [6]$$

The first term of the expansion yields *Huggins*' equation (24). At constant temperature and concentration eq. [6] can be cast in the form

$$(\eta - \eta_0)/[\eta] = f(c/\gamma) \qquad [6a]$$

provided

$$\gamma^{-i} \propto (k_i[\eta]^i)_{\text{linear}} = (k_i[\eta]^i)_{\text{branched}} \quad \text{for all } i \qquad [6b]$$

if superposition is to hold.

First, eq. [6b] was successfully tested for solutions in cyclohexane and toluene. Next, it was assumed that the highly concentrated di-(ethyl) benzene solutions of linear and branched polystyrene can be approximately treated as θ-systems. For this reason, the concentration reducing parameter γ was assumed to be proportional to $1/[\eta]_\theta$. Using the experimental values of $[\eta]_\theta$ in cyclohexane at 35.0 °C (19, 25), the data are plotted as $(\eta - \eta_0)/[\eta]_\theta$ vs. $[\eta]_\theta$ in fig. 7. An excellent superposition is observed. Moreover, if instead of $[\eta]_\theta$ the theoretical expression (26)

$$g = (3f - 2)/f^2$$

is used, with f the number of branches per molecule, the plot of $(\eta - \eta_0)/g^{1/2}$ vs. $(M_w g)$, based on relations [5] and [6a], also provides good superposition. Finally we note that the characteristic product $c[\eta]_\theta$ varies here from 3.8–32, 3.6–19 and 3.6–18 for the linear, 4- and 6-branched polymer, respectively, due to the different magnitudes of $[\eta]_\theta$.

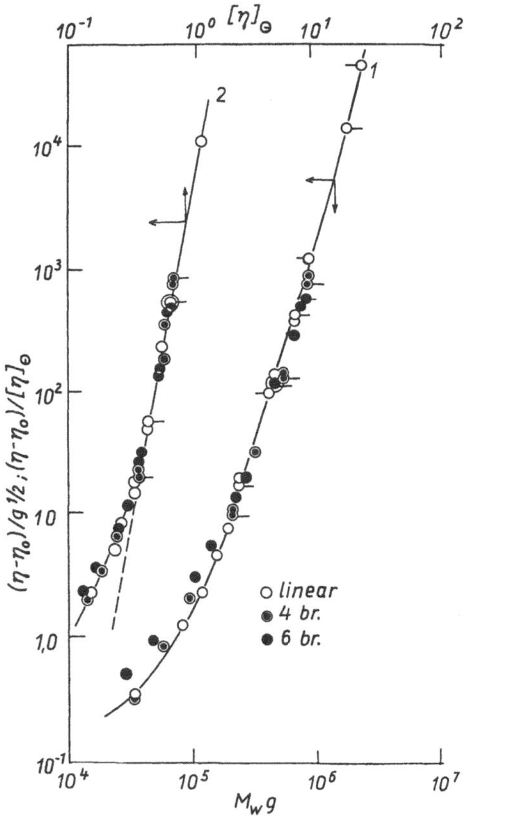

Fig. 7. Superposition of data for linear, 4- and 6-branched polystyrenes at constant concentration, 25.5 g/dl in diethyl benzene and temperature, 30.0 °C. Pip left-ref. 25, pip right-ref. 19, cone-and plate rheometer data, no pip-ref. 19, capillary viscometer data. Solid lines linear polymer. See text for choice of coordinates

F. Some consequences: Quantitative relationships

On the basis of the principle of corresponding states demonstrated in the preceding sections, one could express the relationship between the reduced variables $\tilde{\eta}$ and $\tilde{c}$ in analytical form. For the θ-mixture polystyrene-cyclohexane as well as other poor solvents, the empirical *Martin* relation (27) is applicable over a wider range of concentration than in a good solvent. On increasing the temperature from below to slightly above the θ-temperature, one finds that the departures from this relation occur in opposite directions, and the θ-temperature provides the dividing line (8, 11). This equation has the form

$$\ln \tilde{\eta} = A \cdot \tilde{c} \qquad [7]$$

where A represents a disposable parameter. A four-parameter extension of this equation to account for the curvature and a subsequent straight line portion for $\tilde{c} < 8$ has been proposed (8). Irrespective of such empirical representations of viscosity-concentration relations, there are a series of important predictions which may be derived.

In dilute solutions it has been customary to express experimental data in the form of *Huggins'* relation (24), valid for dilute solutions

$$\tilde{\eta} - 1 = k_1 [\eta] \gamma \cdot \tilde{c} + 0(\tilde{c}^2) \qquad [7']$$

which ascribes a definite form to the parameter A in eq. [7], and where k_1 is assumed to be characteristic of the polymer-solvent pair. Irrespective of the results of hydrodynamic treatments, one would intuitively expect k_1 to vary with molecular weight (or intrinsic viscosity). Various proposals have appeared in the literature, but the existence of a universal relationship between $\tilde{\eta}$ and $\tilde{c}$ makes this precise by requiring that the coefficient of $\tilde{c}$ be independent of molecular weight, or

$$k_1 \propto M^{a_1 - a} \qquad [8]$$

where the proportionality factor contains the preexponential factors in the expressions for $[\eta]$ and γ, and the first coefficient in the expansion of the left hand side (eq. 7'), in powers of $\tilde{c}$. A weak inverse relationship with molecular weight is thus predicted for good solvents. Accurate data for k_1 are not easy to come by, since both the intercept and the slope of the conventional viscosity-concentration curves are involved. However, from experimental data for molecular weights below 20,000, a trend towards increasing k_1 is discernible for polystyrene in toluene (7, 28). In poor solvents, not only is k_1 near unity, in comparison with values near 0.3 in good solvents, but also the variation with molecular weight is practically eliminated (15). The insert in fig. 3 is consistent with the proportionality of $k_1 [\eta]$ and γ. The insert in fig. 4 would predict an increase of k_1 with molecular weight, but no experimental $[\eta]$-data were available. We have disregarded in the whole discussion the question of polydispersity, which, not unexpectedly, appears to affect the magnitude of k_1 (8).

In the light of extensive studies (22) and practical interest, it is perhaps of greater interest to examine concentrated solutions. One important point emerges immediately from the observation, that the significance of the reduction parameter γ extends from moderate to elevated concentrations. That is, the thermodynamics of solvent-solute interactions continues to play a role.

Again, leaving aside explicit empirical expressions, we consider the isothermal concentration and molecular weight dependence of the relative viscosity in terms of the derivatives

$$\varepsilon_1 = (\partial \ln \eta_r / \partial \ln c)_{M,T}$$
$$\varepsilon_2 = (\partial \ln \eta_r / \partial \ln M)_{c,T} . \qquad [9]$$

In general, ε_1 and ε_2 should be functions of M and c, but may be treated as constants over limited ranges. The principle of corresponding states implies that the molecular weight dependence of the reduced viscosity variable $\tilde{\eta}$ is contained in $\tilde{c}$ and hence

$$\tilde{\varepsilon}_2 = (\partial \ln \tilde{\eta} / \partial \ln M)_{\tilde{c},T} = 0 . \qquad [10]$$

But with the *Mark-Houwink* relation and the prototype of eq. [2], we obtain by suitable partial differentiation:

$$\tilde{\varepsilon}_2 = (\tilde{\varepsilon}_1 + 1)(\varepsilon_2/\varepsilon_1) - (a_1 \tilde{\varepsilon}_1 + a)$$
$$(\tilde{\varepsilon}_1 + 1) = \varepsilon_1 \eta_r / (\eta_r - 1)$$

or

$$a_1 \varepsilon_1 = \varepsilon_2 + (a_1 - a)(\eta_r - 1)/\eta_r \simeq \varepsilon_2 + (a_1 - a). \qquad [11]$$

Now the difference $(a_1 - a)$ will not exceed a few tenths, see table 1. On the other hand, if ε_2 is of the order of 3–4 above the critical entanglement length (22), ε_1 and ε_2 are inversely related with a factor a_1, which increases with solvent power. In poor solvents, the derivative ε_1 should be larger, a qualitatively well known observation, which finds here as quantitative expression. Moreover, eq. [11] relates parameters obtainable in concentrated and moderately concentrated systems. It predicts that a rapid decrease of ε_2 in the entanglement region should be accompanied by a similar change in ε_1. However we also recall (8) that sufficiently low molecular weights, in polystyrene-toluene of the order of $M_n \simeq 1.5 \times 10^4$, i.e., below the entanglement region, depart from the master curves defined by higher degrees of polymerization. Similar departures will be noted at elevated concentrations in figs. 3 and 4 and for samples with degrees of polymerization below 340 and about 940 in benzene solutions ($c < 4/[\eta]$) of poly(1-vinylnaphtalene) and poly(4-binyl-biphenyl) respectively. A reason for this divergent behavior has been suggested (16).

We turn next to a consideration of the entanglement region itself. Independently of molecular interpretations offered (22, 23), this region may formally be characterized by the equations

$$\eta = \eta^{(1)}(c, M); \quad c \leq c_e$$
$$\eta = \eta^{(2)}(c, M); \quad c \geq c_e$$
$$\eta^{(1)} = \eta^{(2)}; \quad c = c_e$$

where the superscripts indicate two different functions. By considering the variation of η with variations in M and c, and eqs. [9] and [11], we arrive at the result [11]:

$$d \ln c_e / d \ln \mathrm{M} = -(\varepsilon_2^{(1)} - \varepsilon_2^{(2)})/(\varepsilon_1^{(1)} - \varepsilon_1^{(2)}) = -a_1$$

and

$$\varphi_e = (M_e/M)^{a_1}$$
$$(\eta_r)_e \simeq [\eta]\,\gamma \times \tilde{c}_e \cdot \tilde{\eta}(\tilde{c}_e) \propto M^{a-a_1} \qquad [12]$$

where φ_e is the entanglement volume fraction of the polymer in the solution and M_e the entanglement molecular weight of the melt. The following conclusions are drawn from eq. [12]: First, the entanglement concentration c_e is a reducing factor analogous to v_{crit} in poor solvents, see eqs. [2] and [3], as could be anticipated on dimensional grounds. Provided γ is defined in terms of the identical reference molecular weight, $\tilde{c}_e$ is constant and independent of the solvent. Secondly, the relationship between c_e and molecular weight depends on the solvent, characterized by the exponent a_1; c_e should be smaller for a given M in a good than in a poor solvent. Opinions in the literature vary between an inverse first and one half power dependence. It would be important to demonstrate by *direct* experimentation the validity of our conclusions regarding solvent effects, which are to persist into the region of elevated concentrations. In fig. 4 a spread of values for $\tilde{c}_e$, computed from γ will be noted, but this is based on a first power assumption for the molecular weight dependence of c_e by the original authors.

III. Outlook

In this concluding section we want to indicate the areas of future investigations which, in our opinion, are needed to complete the development of the proposed method of generalization of the polymer solution viscosity data.

Up to now, the method presented on the preceding pages, was found to be applicable to every system examined. Nevertheless, in certain cases we noted deviations from the general behavior by the lowest molecular weights. Two types of deviations are observed: a) the data points for $\tilde{\eta}$ as a function of $\tilde{c}$ depart from the master curve defined by the results for all other fractions of higher molecular weight (11, 16, 17), and b) the quantities γ or $k_1[\eta]$ do not obey eq. [2″] (11, 17, 19). The significance of these two types of deviations is quite different; whereas the first one indicates a true limitation of the proposed method, the second only points out the inadequacy of the employed empirical correlation. The onset of the first of these two was related (16) to chain stiffness, as measured by the conformational parameter σ. However, some other factors were also suspected and are not resolved. It would be desirable to analyze the viscosity of a series of low molecular weight polymers in a wide range of variables, in respect to type of solvent, temperature, concentration and molecular weight. The polymers should be made to differ systematically from each other, in regard to such structural characteristics as chain stiffness, polarity and configuration, in order to establish specific mechanisms for the deviations. It is pertinent to recall that dilute solution properties such as intrinsic viscosity or the linear expansion coefficient of the macromolecular coil also deviate from those of the higher molecular weight homologues (29, 30, 31).

It is becoming clear that θ-conditions can be uniquely defined only for nonpolar systems (32) to which the proposed method should be particularly well applicable. Studies of the viscosity-concentration behavior in the whole range of concentrations and at temperatures ranging from θ to the highest experimentally accessible, should permit an exhaustive and critical examination of the corresponding states principle in the present formulation. Furthermore, such investigations should provide means of evaluating the proposed reducing temperature parameters.

It is expected that, at least for a good solvent and in the high concentration region, the critical mixing temperature will loose its significance and either a different internal temperature scale should be employed or some mixing rule will have to be developed. Finally, these studies have a bearing on the question as to whether thermodynamic solvent effects extend to the viscosity of concentrated solutions, i.e., to the entanglement region and beyond.

After such model systems are tested, the careful examination of commercially interesting polymers is desirable, in which polar forces, inhomogeneity and branching often play an important role. It is of interest to note that the method proposed by us was found to be "quite successful" (33) in generating superposition of the solution viscosity data of nonpolar as well as polar (poly(methyl methacrylate)) polymers, at least within the range of variables investigated by the authors.

References

1) *Guth, E.* and *R. Simha*, Kolloid-Z. **74**, 266 (1936).

2) *Simha, R.*, J. Appl. Phys. **23**, 1020 (1952). For a review see *H. L. Frisch* and *R. Simha*, in: *F. R. Eirich*, Ed., Rheology, Vol. 1 (New York 1956).

3) *Thomas, D. G.*, J. Colloid Sci. **20**, 267 (1965).

4) *Simha, R.*, and *T. Somcynsky*, J. Colloid Sci. **20**, 278 (1965).

5a) *Utracki, L. A.*, J. Colloid, Interface Sci. **42**, 185 (1973).
b) Gulf Oil Canada, Research Laboratory Reports. The permission of Dr. *D. C. Downing* to publish this information, is gratefully acknowledged.

6) *Weissberg, S. G., R. Simha*, and *S. Rothman*, J. Res. Natl. Bur. Stands. **47**, 298 (1951). – *Simha, R.* and *J. L. Zakin*, J. Chem. Phys. **33**, 1791 (1960).

7) *Simha, R.* and *J. L. Zakin*, J. Colloid Sci. **17**, 270 (1962).

8) *Utracki, L.* and *R. Simha*, J. Polymer Sci. **1**, 1089 (1963).

9) *Utracki, L.*, Polimery, Warsaw, **9**, 50 (1964); ibid., **9**, 144 (1964).

10) *Chou, L.-Y.* and *J. L. Zakin*, J. Colloid Interface Sci. **25**, 547 (1967).

11) *Simha, R.*, and *L. Utracki*, J. Polymer Sci., A-2, **5**, 853 (1967).

12) *Pezzin, G.* and *N. Gligo*, J. Appl. Polymer Sci. **10**, 1 (1966).

13) *Tager, A. A., V. E. Dreval*, and *N. G. Trayanova*, Dokl. Akad. Nauk SSSR **151**, 140 (1963).

14) *Debye, P., H. Coll*, and *D. Woermann*, J. Chem. Phys. **33**, 1746 (1960). – *Debye, P., D. Woermann*, and *B. Chu*, ibid., **36**, 851 (1962).; see: "Note added in proof".

15) *Utracki, L.* and *R. Simha*, J. Phys. Chem. **67**, 1052 (1963).

16) *Utracki, L., R. Simha*, and *N. Eliezer*, Polymer **10**, 43 (1969).

17) *Simha, R.* and *F. S. Chan*, J. Phys. Chem. **75**, 256 (1971).

18) *Maron, S. H.* and *N. Nakajima*, J. Polymer Sci. **54**, 587 (1961).

19) *Utracki, L. A.* and *J. E. L. Roovers*, Paper presented at the 55th Annual Meeting of the Chemical Institute of Canada, Quebec (Canada), June 5–7, 1972; Macromols **6**, 366 (1973); Macromols **6**, 373 (1973).

20) *Roovers, J. E. L.* and *S. Bywater*, Macromols **5**, 384 (1972).

21) *Zimm, B. H.* and *R. W. Kilb*, J. Polymer Sci **37**, 19 (1959).

22) See the Discussion by *Berry G. C.* and *T. G. Fox*, Adv. Polymer Ser. **5**, 262 (1968).

23) *Imai, S.*, Rep. Prog. Polymer Phys. (Japan) **9**, 117 (1966).

24) *Huggins, M. L.*, J. Amer. Chem. Soc. **64**, 2716 (1942).

25) *Graessley, W. W.* and *L. Segal*, Macromols. **2**, 49 (1969). – *Graessley, W. W., R. L. Hazelton*, and *L. R. Lindeman*, Trans. Soc. Rheol. **11**, 267 (1967).

26) *Zimm, B. H.* and *W. H. Stockmayer*, J. Chem. Phys. **17**, 1301 (1949).

27) *Martin, A. F.*, Meeting of the Amer. Chem. Soc., Memphis, April 1942.

28) *McCormick, H. W.*, J. Colloid Sci **16**, 635 (1961).

29) *Utracki, L. A.* and *R. Simha*, J. Phys. Chem. **67**, 1056 (1963).

30) *Bianchi, U., M. Dalpior*, and *E. Patrone*, Makromol. Chem. **80**, 112 (1964).

31) *Cowie, J. M. G.*, Polymer **7**, 487 (1966).

32) *Utracki, L. A.*, J. Appl. Polymer Sci **16**, 1167 (1972); Polymer J. (Japan) **3**, 551 (1972).

33) *Gandhi, K. S.* and *M. C. Williams*, J. Polymer Sci., Part C, **35**, 211 (1971).

Note added in proof

It was brought to our attention that the maximum angular dissymmetry of light scattering, used to determine v_{crit} in ref. 14 occurs at lower concentration and higher temperature than the true critical phenomenon. In consequence v_{crit} used in this paper should be identified with the volume fraction at the maximum of the spinodal. E.g. see *W. Borchard* and *G. Rehage*, Adv. Chem. Ser. **99**, 42 (1971); *A. Vrij* and *M. W. van den Esker*, J. Chem. S., Faraday II, **68**, 513 (1972).

Authors' addresses:

Dr. *Robert Simha*
Div. of Macromolecular Science
Case Western Reserve University
Cleveland, Ohio 44106 (USA)

L. A. Utracki
Dept. of Chemical Engineering
McGill University
Montreal, P. Q. (Canada)

Rheol. Acta 12, 465–467 (1973)

From the Institute of Theoretical Physics, Rheinisch-Westfälische Technische Hochschule, Aachen (Germany)

The entropy problem in thermodynamics of processes

By J. Meixner

(Received October 27, 1972)

When building up thermodynamics of continuous matter, one of the main points is, of course, that the concepts to be used are clearly developed and generally accepted so that one knows what one is talking about. Such can be done only by using the available experience. The development of the concepts cannot be separated from laying down the basic laws which also can be inferred from experiments only.

As an example I mention the concept of a static or equilibrium entropy which has turned out to be very important and successful during the more than a hundred years since *Clausius* has established this concept.

The situation is, however, entirely different in thermodynamics, i.e. in the theory of irreversible processes. People use to believe that *Clausius* has also laid down the foundation for the concept of a non-equilibrium entropy. But this is not so, and this is quite obvious if one carefully studies his work. On irreversible processes he makes the following statement

$$S(B) - S(A) \geqq \int_A^B \delta Q/T \qquad [1]$$

where S is the equilibrium entropy of a system in the two consecutive equilibrium states A and B and the δQ are the heats transfered to the system at the exchange temperatures T during an irreversible process which connects A and B. He also states quite emphatically that A and B must be equilibrium states in order that [1] hold. But he contradicts himself when a few years later he writes [1] without argument or motivation in the "differential form"

$$\delta S \geqq \delta Q/T \qquad [2]$$

with the understanding that the states which are connected in this differential can be states within an irreversible process, i.e. non-equilibrium states and that the S is then an non-equilibrium entropy. In fact, the inequality, can be considered as the assertion that there exists a function of state S which, for a differential piece of an irreversible process, satisfies the inequality [2] and which in equilibrium becomes identical with the static entropy. In the same way the *Clausius-Duhem* inequality

$$\varrho \frac{ds}{dt} + \nabla \cdot \left(\frac{1}{T} q\right) \geqq 0 \qquad [3]$$

(ϱ = density, s = specific entropy, q = heat flow), which follows from [2], can be considered as the assertion that there exists a function of state s which for every irreversible process (without diffusion or electrical conduction) satisfies the inequality [3] and which in equilibrium reduces to the static specific entropy.

Thus there arises the question whether such a function of state S or s with the asserted properties exists at all; or, in case one has one such function, whether there are not many other such functions, i.e. whether S or s is a unique function of state.

One might try to get some answer to this question by looking at kinetic theories of matter. Indeed it seems that *Boltzmanns* kinetic theory of gases gives an excellent example for the existence of a non-equilibrium entropy; the H-theorem provides even an explicit expression for a non-equilibrium entropy with all the required properties. But there is no proof that there exists only one H-theorem for the *Boltzmann* equation, and consequently we do not know whether there is a unique non-equilibrium entropy. In the generalized kinetic theories for two-, three-etc. particle distribution functions (derived from the BBGKY-hierarchy) it is at the present time not even known whether H-theorems exist at all except for the N-particle distribution function of a gas with N particles which satisfies the *Liouville* equation.

There is another example which elucidates the situation and has, moreover, the advantages that it can be treated in purely macroscopic terms and that it is of utmost simplicity, namely electrical networks. Such electrical networks are composed, in the simplest case, of a finite number of ideal capacitances, inductances and resistances.

Electrical networks are thermodynamic systems of a special kind and they lend themselves to illustrate several aspects of continuum thermodynamics.

1. Lowest level of description

By this we understand in continuum thermodynamics constitutive equations which relate the values of external variables at time t (temperature, temperature gradient, stress tensor) to the histories up to time t of the conjugate external variables (internal energy, heat flow, deformation gradient). In electrical network theory this corresponds to a description which sets the currents at time t through the terminals in relation to the histories up to time t of the voltages applied across each pair of terminals. The electrical engineers call this the black box point of view because the mentioned relation can be established experimentally without looking into the internal structure of the electrical network. But also, the mentioned relation does not uniquely determine the internal structure of the network. A famous example is the voltage-current relation $u(t)=Ri(t)$ with a constant R. It can be realized by a pure resistance of magnitude R, but also by a parallel connection of 1. a resistance R in series with an arbitrary capacitance C and 2. a resistance R in series with an inductance $L=R^2C$. It can easily be seen that quite generally there exists even an infinite number of electrical networks which have the same voltage-current relation (provided the impedance is not of the reactance type).

This statement can be translated into continuum thermodynamics and in a cautious formulation we would expect that the thermodynamic constitutive equations of a material on the lowest level of description may be compatible with a multitude of possible internal structures.

Electrical networks with the same voltage-current relation are not equivalent with respect to their energy dissipations. This is quite obvious in the example of the voltage-current relation $u(t)=Ri(t)$. Therefore we should also expect that the thermodynamic constitutive equations of a material on the lowest level of description do not determine a unique entropy production. This would imply that we cannot assign a unique entropy to a material state on the lowest level of description.

2. The construction of non-equilibrium entropies

A long time ago I have been troubled how one could assign to an electrical network in a non-equilibrium state a distinguished loss function if it is taken as a black box, which means that only the impedance function is known. I arrived at the following proposal: In $-\infty<t<\tau$ a voltage $u(t)$ is applied to a one-port and produces a current $i(t)$: Then the energy delivered to the network is

$$W(\tau) = \int_{-\infty}^{\tau} u(t)\, i(t)\, dt \geqq 0. \quad [4]$$

Some of it is dissipated in the resistances and some is stored in the other elements. The function $W(\tau)$ is not necessarily an increasing function of τ, in other words, there may be time intervals during which $W(\tau)$ is decreasing and energy is returned from the network. This leads to the following problem: Assume τ to be a fixed time, consider all possible continuations of $u(t)$ beyond $t=\tau$ and determine the infimum $W_L(\tau)$ of the integral

$$W_1(\tau) = \left[\int_{-\infty}^{\tau} + \int_{\tau}^{\infty} \right] u(t)\, i(t)\, dt \quad [5]$$

over all these continuations. Then $W_L(\tau)$ is the energy at time which by no means can be recovered from the network if the application of voltage functions is the only permitted operation. $W_L(\tau)$ is called the non-recoverable or lost energy at time τ. Evidently $W_L(\tau)$ is monotonically increasing with τ.

Thus we have defined a function which in a well-defined manner characterizes the loss in an electrical network and it is also obvious that it depends only on the applied voltage $u(t)$ and on the black box-properties of the network, i.e. on the impedance. It is also clear that it is always greater than or equal to the energy which is dissipated in the resistances, because at most the stored energy can be recovered.

This problem can also be formulated for general vector-valued linear passive systems and has found a complete and explicit solution by *König* and *Tobergte* (1963) and *Tobergte* (1965). A special case which corresponds to RC-networks has been treated by *Breuer* and *Onat* (1964). Into the category of linear passive systems belong the after-effect equations of linear thermo-viscoelasticity (*Meixner*, 1969*a*, *b*); therefore one can define for them a loss function in the same manner and ultimately a non-equilibrium entropy which satisfies the *Clausius Duhem* inequality and reduces in equilibrium to the static entropy.

Essentially the same idea has led *Day* (1970) to the construction of a non-equilibrium entropy for thermoviscoelastic materials with large departures from equilibrium; but an explicit expression of this entropy in terms of the constitutive equations is not yet known.

In the case of electrical networks it is quite easy to construct other possible loss functions; if the impedance is known one chooses any one of the infinite number of network realizations of this impedance and defines as a loss function the time integral up to time τ of the *Joule*an heat produced in the resistances. It has all required properties and is in general less that $W_L(\tau)$.

As a counterpart we note that it is easily possible to derive an infinite number of other entropy functionals from *Days* entropy (*Kern*, 1973).

Although *Days* entropy is distinguished as being the infimum of all possible entropies for the same state, it has a very distinct disadvantage. Also this can be learned from the electrical networks. Consider an RC-network which contains by definition only resistances and capacitance. Then all other RC-networks with the same impedance coincide at all times in the *Joule*an heat provided the same voltage function is applied (*Staverman* and *Schwarzl*, 1953; *Tellegen*, 1952; *Meixner*, 1959). Consequently for such networks one would consider the time integral of the *Joule*an heat up to time τ as a reasonable loss function, because it conforms with the physical idea of losses in a network. But in general this function is different from $W_L(\tau)$.

Summing up we can say that there is no unique non-equilibrium entropy in thermo-viscoelastic materials, and then the same statement must hold when diffusion and electromagnetic phenomena are included. Furthermore, although the special entropy which is defined in analogy to the minimum loss in electrical networks, has a distinguished position, it does not seem to possess a deep physical significance.

3. Is the concept of a non-equilibrium entropy superfluous?

The irreversible behavior of an electric network on the lowest level of description is completely expressed by the property of the impedance function to be a so-called positive function of a parameter which is the complex frequency multiplied by minus the imaginary unit. There is no gain in introducing a loss function, be it $W_L(\tau)$ or any other loss function. We should expect that a similar conclusion holds for processes in thermoviscoelastic materials: the concept of a nonequilibrium entropy is not needed and thermodynamics can be developed from the conservation laws and, instead of [3], from the fundamental inequality [1]. That such an approach is possible has been demonstrated by the author (1969a, b).

Nevertheless one is so much accustomed to the concept of a non-equilibrium entropy that one would like to retain it as a quantity of physical significance. This is indeed possible on a higher level of description in which the state at time τ is not characterized by the histories of the independent external variables up to time τ, but instead by these external variables and by a set of internal variables, all of them being taken at the same time τ. The resulting theory is essentially the classical thermodynamics of irreversible processes and its entropy is the well-defined entropy of frozen equilibrium which is a function of the independent external and of the internal variables. In this description no after-effects or memory are present in the constitutive equations. It is a different matter, of course, whether in a given material the internal variables can always be properly identified with molecular processes. With electrical networks as an example of thermodynamic systems it is, however, quite easy to recognize the internal variables. One has just to open the black box and to choose appropriate charges on capacitances and currents through inductances as internal variables (*Meixner*, 1963, 1966).

References

Breuer, S. and *E. T. Onat*, J. Applied Math. Phys. (ZAMP) **15**, 12–21 (1964).

Day, W. A., Quart. J. Mech. Appl. Math. **23**, 1–15 (1970).

Kern, W., Zur Vieldeutigkeit der Nichtgleichgewichtsentropie in kontinuierlichen Medien. Doctoral Thesis, Aachen 1972.

König, H. and *J. Tobergte*, J. reine angew. Math. **212**, 104 bis 108 (1963).

Meixner, J., Z. Physik **156**, 200–210 (1959).

Meixner, J., J. Math. Phys. **4**, 154–159 (1963).

Meixner, J., Network Theory in its Relation to Thermodynamics, p. 13–25. In: Proceedings of the Symposium on Generalized Networks, New York 1966, Vol. XVI in the Microwave Research Institute Symposia Series (Polytech. Press Polytech. Inst. of Brooklyn, Brooklyn, N. Y., 1966).

Meixner, J., Z. Physik **219**, 79–104 (1969a).

Meixner, J., Arch. Rat. Mech. Anal. **33**, 33–53 (1969b).

Staverman, A. J. and *F. Schwarzl*, Proceed. Kon. Ned. Akad. Wetensch. B, **55**, 474–485 (1953).

Tellegen, B. D. H., Philips Research Reports **7**, 259–269 (1952).

Tobergte, J., Invariante Teilräume und die verlorene Energie linearer passiver Transformationen. Doctoral thesis, Cologne 1965.

Author's address:

Prof. Dr. *J. Meixner*
Institut für theoretische Physik der RWTH
D-5100 Aachen (Deutschland)
Templergraben 55

Rheol. Acta **12**, 468–478 (1973)

Vrije Universiteit Brussel (Belgique)

Analyse macro- et microrhéologique de certains effets observés en photoélasticité

Par R. E. Van Geen

(Reçu p. p. le 27. Octobre 1972)

1. Introduction

Un continu transparent, soumis à un champ triaxial de déformation et de contrainte présente en général un *effet photoélastique*, constitué par une modification de la célérité des ondes lumineuses. Le phénomène est dû à la genèse dans le milieu, d'un champ tensoriel de polarisabilités électroniques.

La *photoélasticimétrie* est une technique d'ingénieur, basée sur l'emploi de l'effet photoélastique à la détermination des contraintes *élastiques*, plus rarement des contraintes viscoélastiques linéaires.

Les techniques expérimentales actuelles (1) permettent de déterminer avec précision les propriétés optiques d'un milieu, et dès lors d'avoir accès au champ de polarisabilité. Les mesures sont alors transposées en termes de contraintes ou de déformation, à partir de la connaissance de la loi phénoménologique liant les composantes des tenseurs mécaniques et optiques.

Dans la théorie classique du phénomène, on admet:

1. La coïncidence des directions principales des tenseurs mécaniques et optiques, et par voie de conséquence, la coïncidence des cas de sphéricité de ces tenseurs.

2. Une relation élémentaire, en général *linéaire*, entre les composantes des tenseurs mécaniques et optiques.

Les lois classiques de *Maxwell-Neumann* (2) traduisent cette *linéarité* et *réversibilité* de l'effet photoélastique. Elles expriment en fait une simple proportionalité entre les différences des valeurs principales des tenseurs mécaniques et optiques, c'est-à-dire respectivement entre les quantités $(\sigma_1 - \sigma_2)$, $(\sigma_2 - \sigma_3)$, $(\sigma_1 - \sigma_3)$ [ou, alternativement $(\varepsilon_1 - \varepsilon_2)$, $(\varepsilon_2 - \varepsilon_3)$, $(\varepsilon_1 - \varepsilon_3)$] et $(n_1 - n_2)$, $(n_2 - n_3)$, $(n_1 - n_3)$.

($\sigma_{1,2,3}$; $\varepsilon_{1,2,3}$; $n_{1,2,3}$: valeurs principales respectivement du tenseur des contraintes, des déformations évanouissantes, des indices.) Hormis quelques rares matériaux sélectionnés en photoélasticimétrie, les corps réels présentent un comportement photoélastique beaucoup plus complexe. De nombreux travaux, répertoriés dans (3), se sont limités à généraliser la relation *scalaire* existant entre les différences des valeurs principales du tenseur des contraintes et du tenseur des indices. La généralisation consiste en l'introduction de relations non-linéaires, ou d'une dépendance au temps.

Elle maintient, sous-jacente, l'hypothèse d'identité des directions principales des quadriques associées aux tenseurs des contraintes, déformations et polarisabilités, et réduit, sans autre justification, la modification à la forme des relations scalaires liant les valeurs scalaires des axes des quadriques.

Cette manière de voir est en contradiction avec une analyse fine des résultats expérimentaux décrits au paragraphe suivant.

Dans la présente contribution nous examinons plus particulièrement l'un de ces résultats, la biréfringence dynamique en régime établi. Les effets observés ne sont pas essentiellement liés au caractère fini des déformations, ni à un comportement non-linéaire du matériau: nous admettons donc, par souci de simplicité, les hypothèses de déformations évanouissantes et de respect du principe de superposition de *Boltzmann*, sauf indication contraire.

Les matériaux étudiés sont isotropes dans l'état non déformé.

2. Faits expérimentaux en désaccord avec les lois classiques de la photoélasticité

Les résultats d'expériences effectuées aux Universités de Bruxelles permettent de dégager les conclusions suivantes, en contradiction avec les principes classiquement admis en la matière.

2.1.

Les directions principales du tenseur diélectrique ne coïncident, en général, ni avec celles du tenseur des contraintes, ni avec celles du tenseur des déformations. Il en résulte que les *isoclines* (lieu d'égale orientation des directions principales secondaires d'un tenseur) mécaniques et optiques *ne coïncident pas*, en règle générale.

2.2.

L'isotropie du tenseur des déformations *n'implique pas* l'isotropie du tenseur diélectrique.

Ces deux faits ont été analysés dans les publications antérieures (3), les résultats obtenus étant confirmés par l'analyse des nouveaux résultats suivants:

2.3. Comportement en régime sinusoïdal

Lorsqu'un matériau est soumis à une sollicitation périodique sinusoïdale, de pulsation ω, la déformation, *en régime établi*, est également périodique sinusoïdale, et *déphasée en retard* sur la sollicitation. L'effet photoélastique observé (7) est également périodique sinusoïdal; une polémique s'instaura naguère et se poursuit encore, en vue de déterminer si le signal photoélastique est en phase avec la déformation ou avec la contrainte. Les explications théoriques relatives à l'une ou l'autre option trouvaient des arguments dans les travaux expérimentaux disponibles, et dus, notamment, à *Read, Lagarde, Bourassa* et *Kolsky*.

En réalité, l'effet photoélastique n'est, en règle générale, en phase ni avec la sollicitation, ni avec la déformation. Il peut même être déphasé (en régime établi) en avance sur la sollicitation, alors que la déformation est nécessairement en retard par rapport à la sollicitation.

2.4. Dispersion chromatique de l'effet photoélastique

Nous avons déjà signalé précédemment (5, 6) que les directions principales du tenseur diélectrique *dépendaient de la longueur d'onde lumineuse utilisée lors de la mesure photoélastique.*

Nos expériences établissent également que le déphasage mentionné au § 2.3 dépend à son tour de la longueur d'onde d'observation. Le signal photoélastique peut apparaître retardé ou avancé par rapport à la sollicitation, suivant la fréquence de l'onde lumineuse. En particulier, pour une gamme de fréquences données, l'effet photoélastique peut apparaître, à la précision des mesures photoélastiques, en phase avec la contrainte.

Ces deux faits seront analysés dans la présente contribution.

2.5. Non bi-univocité des relations biréfringence-contrainte

A l'inverse de l'opinion répandue à cet égard, il n'existe pas de relation bi-univoque entre le tenseur des contraintes et celui des permittivités diélectriques (dont dérive l'indice de réfraction). Dans certaines circonstances, un état diélectrique peut correspondre à deux états de contrainte très distincts (7), ce qui pose des problèmes sérieux dans l'analyse photoélastique.

Par contre, la relation est univoque dans le sens contrainte-biréfringence. Le phénomène présente, de plus, une dispersion chromatique très prononcée. Ce dernier phénomène est lié, par certains aspects, à un comportement mécanique non-linéaire, et ne sera qu'évoqué dans le présent article.

3. Résultats expérimentaux obtenus en régime établi

3.1. Dispositifs expérimentaux

Les expériences ont été conduites au moyen d'échantillons viscoélastiques, soumis à des sollicitations sinusoïdales établies depuis très longtemps. Dans un premier montage, une éprouvette est soumise à une contrainte uniaxiale sinusoïdale, de fréquence variable, et se déforme en régime forcé permanent (8). Les résultats sont recoupés au moyen d'expériences en vibrations stationnaires à la première résonance du système d'ondes longitudinales (9).

Dans une seconde série d'expériences, un disque circulaire est soumis à une paire de forces diamétrales agissant dans son plan, et variant également sinusoïdalement en fonction du temps. Elles créent dans le disque un état plan de contraintes (les forces d'inertie sont ici négligées) (2, 7).

Les expériences se déroulent en environnement contrôlé, en particulier de température (de −15 °C à +250 °C).

Les élongations et la biréfringence sont mesurées par des méthodes extensométriques classiques (moirés, strain-gages, et scratches).

Une attention particulière est accordée à la mesure des effets photoélastiques en diverses longueurs d'ondes (de 3650–9000 Å) en vue de déterminer la dépendance chromatique du déphasage entre l'effet photoélastique et la sollicitation extérieure.

3.2. *Matériaux étudiés*

Cette étude porte sur quelques principes fondamentaux du comportement rhéooptique des matériaux et ne se propose pas d'étudier la structure macromoléculaire de corps déterminés.

L'expérimentation est dès lors limitée à trois types de polymères, présentant des comportements caractéristiques:

– Des résines epoxy, présentant un comportement viscoélastique classique, avec élasticité vitreuse à la température ambiante, température de transition vers 120 °C, et comportement *rubberlike* très net au dela de 150 °C. Leur comportement photoélastique est «simple» (2, 10, 11).

– Des polyesters non-saturés, des polystyrènes (à chaînes latérales ou non), présentant des groupements aromatiques. Ils présentent un comportement photoélastique inhabituel (7, 2, 12).

– Du polyurethane, à l'état *rubberlike* dès la température ambiante.

3.3. *Résultats expérimentaux*

La sollicitation est déterminée par la contrainte sinusoïdale:

$$\sigma(t) = \sigma_m \mathcal{R}(e^{j\omega t}) = \mathcal{R}\sigma^* \qquad [1]$$

où σ représente la contrainte de traction dans le cas d'expériences uniaxiales, et le déviateur $(\sigma_1 - \sigma_2)$ (fonction de t et des coordonnées spatiales) dans le cas d'expériences en état plan de contrainte. (σ_1, σ_2: contraintes principales *dans* le plan; la contrainte σ_3, parallèle à la direction du rayon lumineux lors d'une mesure photoélastique, est nulle.)

La déformation résultante est

$$\varepsilon(t) = \varepsilon_m \mathcal{R}(e^{j(\omega t + \psi)}) = \mathcal{R}\varepsilon^* \qquad [2]$$

où ε représente la déformation longitudinale dans le cas d'expériences uniaxiales, et la différence des déformations principales dans le plan, $(\varepsilon_1 - \varepsilon_2)$ dans l'état plan de contrainte.

$\varepsilon^*(t)$ est liée à la contrainte [1] par la complaisance complexe

$$J^* = J_A - jJ_B = \frac{\varepsilon^*}{\sigma^*}. \qquad [3]$$

La biréfringence photoélastique mesurée, $\delta = e(n_1 - n_2)$, devient, en régime établi

$$\delta(t) = \delta_m \mathcal{R}(e^{j(\omega t + \varphi)}) = \mathcal{R}\delta^* \qquad [4]$$

et permet de définir une complaisance photoélastique

$$C^* = W_1 - jW_2 = \frac{\delta^*}{\sigma^*}. \qquad [5]$$

On détermine ensuite, suivant la technique rhéologique classique, la variation des parties réelles et imaginaires des complaisances mécanique et photoélastique (pour une longueur d'onde λ donnée), en fonction de la période T de la sollicitation extérieure.

Les résultats relatifs au comportement mécanique ont la forme classique, caractéristique des polymères viscoélastiques linéaires (fig. 1).

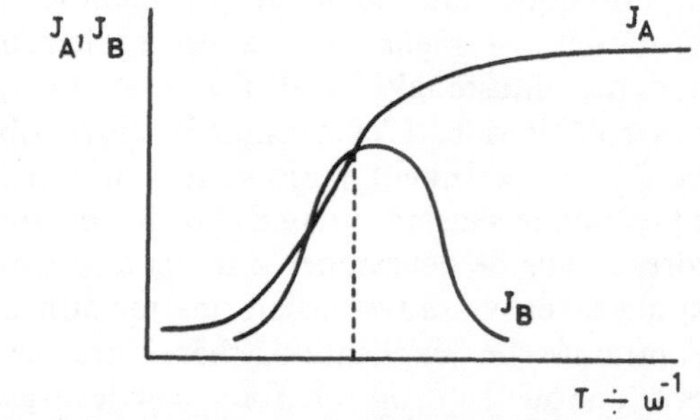

Fig. 1. Complaisances mécaniques, réelle et complexe

Par contre, les diagrammes relatifs au comportement photoélastique présentent, au contraire deux structures différentes pour la partie réelle de C^* (fig. 2, 3a et 3b).

Le diagramme donnant la valeur absolue $|W_2|$ présente une allure identique à celle du diagramme de J_B.

L'existence de ces deux types de comportement photoélastique est caractéristique de la complexité du phénomène photoélastique, fréquemment méconnue par les photoélasticimétristes.

Les comportements observés sont analogues à ceux observés par *Read* (13), nos expériences conduisant à l'observations supplémentaire qu'un changement de la *fréquence de l'onde lumineuse* utilisée dans le polariscope permet de passer, pour un même polymère, de l'un à l'autre de ces comportement.

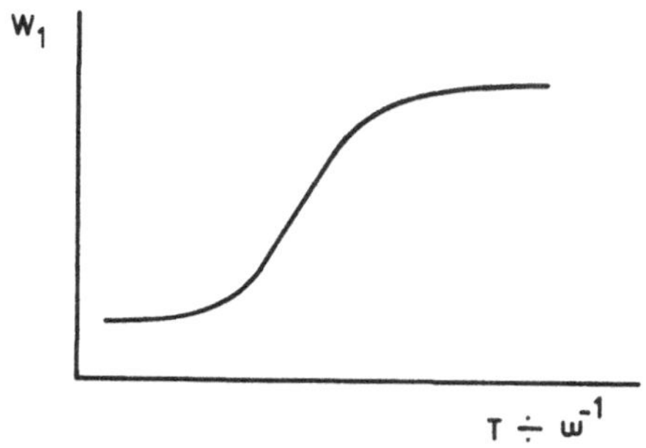

Fig. 2. Complaisance photoélastique réelle/comportement de type I

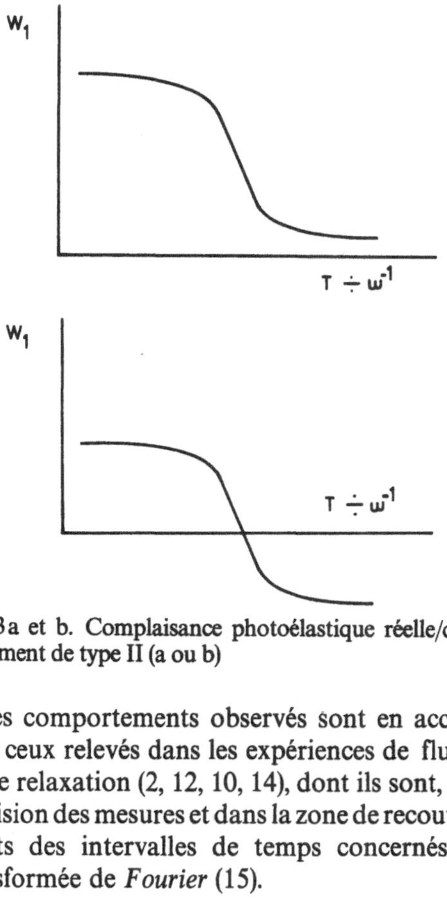

Fig. 3a et b. Complaisance photoélastique réelle/comportement de type II (a ou b)

Les comportements observés sont en accord avec ceux relevés dans les expériences de fluage ou de relaxation (2, 12, 10, 14), dont ils sont, à la précision des mesures et dans la zone de recouvrements des intervalles de temps concernés, la transformée de *Fourier* (15).

4. Interprétation macrorhéologique: fonction de transfert

4.1. Fonction de fluage

Le présent paragraphe s'inscrit dans le cadre théorique proposé par *Mindlin* (16) et *Dill* (17), mais analyse les conséquences de l'existence de biréfringence différée négative, au moyen d'un modèle simple de *Burgers*.

La comparaison des fig. 1, 2 et 3 permet de préciser les points de convergence d'une part, la divergence fondamentale, d'autre part, entre le comportement mécanique et photoélastique.

La forme sigmoïde du diagramme (J_A, T) (fig. 1) s'interprète classiquement, dans le domaine de la viscoélasticité linéaire, par l'existence de deux mécanismes, l'un dit «d'élasticité instantanée», l'autre «d'élasticité différée», ce dernier caractérisé par une fonction de retardement $\psi(t)$

$$J(t) = J_1 + J_2\psi(t) \qquad [6]$$

avec

$$\psi(0) = 0, \quad \psi(\infty) = 1 \quad \text{et } \exists \ \frac{d\psi}{dt}$$

d'où

$$J_A = J_1 + J_2 \int_0^\infty \frac{d\psi(u)}{du} \cos \omega u \, du \qquad [7]$$

$$J_B = -J_2 \int_0^\infty \frac{d\psi(u)}{du} \sin \omega u \, du. \qquad [8]$$

La forme du diagramme (W_1, T) rélève la même structure sigmoïde, de telle sorte qu'il est possible de représenter la complaisance photoélastique en fluage sous la forme

$$C(t) = M_1 + M_2 \Phi(t) \qquad [9]$$

dont on tirera des expressions pour W_1 et W_2, de manière identique à [7] et [8].

L'examen des résultats expérimentaux [p.ex. (2, 7, 10, 13)] indique que les fonctions $\psi(t)$ et $\Phi(t)$ sont, pour de nombreux matériaux, *peu différentes*, voire identiques. Ce fait traduit l'existence d'une relation étroite entre les mécanismes responsables de la biréfringence et de la déformation différées.

La différence essentielle réside dans les signes respectifs de J_2 et M_2. Alors que J_2 est (pour le cas d'un matériau non évolutif) nécessairement positif pour des raisons thermodynamiques, M_2 est indifféremment *positif* ou *négatif* (fig. 2 et 3).

4.2. Application: modèle à un seul temps de relaxation

L'adoption d'un modèle de *Kelvin-Voigt* constitue certes une approximation simpliste, mais permet de comprendre l'essentiel des phénomènes observés.

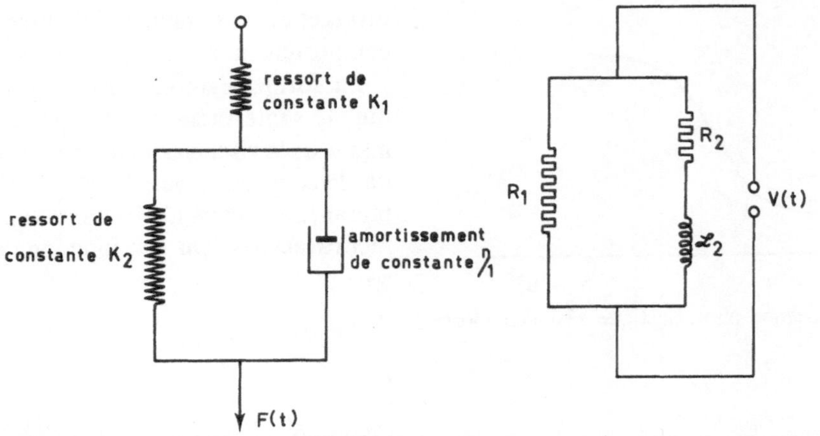

Fig. 4. Modèles analogiques. Pour la complaisance photoélastique des matériaux de comportement II, la résistance R_2 est *négative*

Soit

$$\psi(t) = \Phi(t) = [1 - e^{-\beta t}]. \quad [10]$$

Dans le cas d'une constante M_2 négative, le «ressort» de l'élément de *Kelvin-Voigt* présente la particularité d'être à *rigidité négative* [le modèle analogique électrique présente alors une résistance négative (12)] (fig. 4).

Examinons le cas des matériaux à $M_1 > 0$, $-M_2 > 0$, avec $|M_2| > |M_1|$.

L'emploi des transformées intégrales donne les expressions pour W_1 et W_2.

$$W_1 = (M_1 - |M_2|) + |M_2| \frac{\omega^2}{\beta^2 + \omega^2}$$

$$= M_1 - |M_2| \frac{K^2}{1 + K^2} \quad [11]$$

$$W_2 = |M_2| \frac{\omega\beta}{\beta^2 + \omega^2} = |M_2| \frac{K}{1 + K^2} \quad [12]$$

en introduisant la grandeur adimensionnelle $K = \beta/\omega \div T$ (fig. 5 et 6). On remarquera que la composante photoélastique en phase avec la sollicitation, s'annule pour $K = \sqrt{\frac{-M_1}{M_1 - |M_2|}}$, change de signe; et présente un point d'inflexion (indépendant des constantes physiques M_1 et M_2), au point $K = \frac{1}{\sqrt{3}}$, marquant la *transition* entre les deux mécanismes responsables du comportement photoélastique.

Ce comportement est expérimentalement vérifié pour les polyesters non-saturés, pour lesquels le point d'inflexion se situe à $K^2 = 1/3$. (Une valeur typique de β, à la température de 25 °C est $\beta = 10^{-4}\,s^{-1}$.) Il coïncide avec le point d'inflexion du diagramme (J_A, K) et avec l'inverse de l'abcisse du point d'inflexion du diagramme (J_B, K). Ceci confirme l'hypothèse que les mécanismes responsables du comportement mécanique et photoélastique sont étroitement liés.

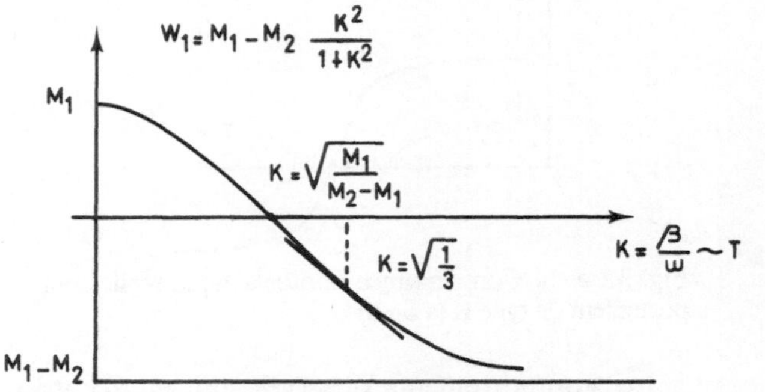

Fig. 5. Complaisance photoélastique (partie réelle) du modèle particulier

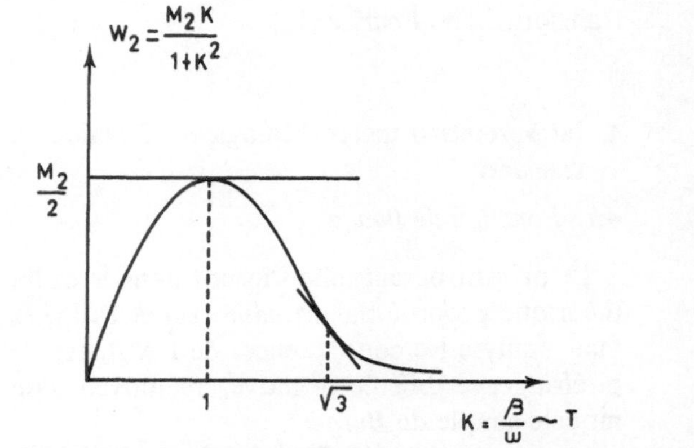

Fig. 6. Complaisance photoélastique (partie imaginaire) du modèle particulier

L'élimination de K entre W_1 et W_2 dans [11] et [12], et entre les équations similaires (aux signes de M_2 et J_2 près) pour la complaisance mécanique, fournit une double représentation de *Cole* et *Cole* (fig. 7). Avec la convention de signe adoptée, le demi-cercle de *Cole* et *Cole* (18) relatif au comportement mécanique se trouve dans le demi-plan inférieur, celui relatif au comportement photoélastique se trouve indifféremment du côté des ordonnées positives ou négatives. Sa localisation dépend de la *longueur d'onde d'observation.*

La comparaison, sur la fig. 7, du comportement de la déformation et de la biréfringence, permet immédiatement de dégager les résultats suivants:

a) Il n'y a, en général, pas de relation directe et encore moins de proportionnalité – entre la complaisance complexe et la constante photoélastique complexe.

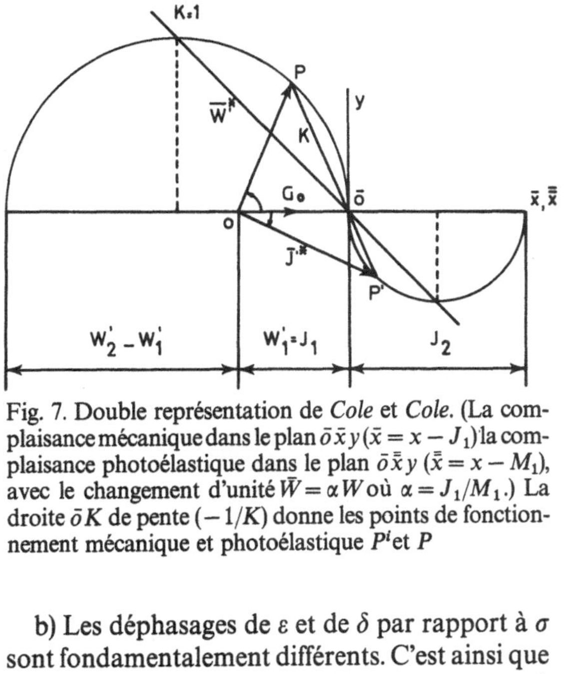

Fig. 7. Double représentation de *Cole* et *Cole*. (La complaisance mécanique dans le plan $\bar{o}\bar{x}y$ ($\bar{x} = x - J_1$) la complaisance photoélastique dans le plan $\bar{o}\bar{\bar{x}}y$ ($\bar{\bar{x}} = x - M_1$), avec le changement d'unité $\bar{W} = \alpha W$ où $\alpha = J_1/M_1$.) La droite $\bar{o}K$ de pente $(-1/K)$ donne les points de fonctionnement mécanique et photoélastique P' et P

b) Les déphasages de ε et de δ par rapport à σ sont fondamentalement différents. C'est ainsi que dans notre cas, δ est en avance sur σ et ε en retard.

c) Il n'y a pas de relation entre la résonance de phase de la biréfringence $\left(K^2 = \dfrac{M_1}{|M_2| - M_1}\right)$ et la résonance de phase de la déformation $\left(K^2 = \dfrac{J_1}{J_1 + J_2}\right)$.

d) Par contre, le maximum de la composante en quadrature (retard ou avance) est obtenu pour $K = 1$, aussi bien pour la biréfringence que pour la déformation. Ce résultat est prévisible, ce maximum ne dépendant que de la constante de temps β, dans la mesure où nous avons admis comme hypothèse que les deux phénomènes retardés s'établissaient suivant une loi en $(1 - e^{-\beta t})$. Ce fait constitue un critère de vérification de la validité du modèle. Il est aisément généralisé aux modèles à plusieurs temps de relaxation.

e) Pour des vibrations infiniment rapides, le matériau se comporte comme un matériau classique, avec constante positive $C = M_1$. Pour des vibrations infiniment lentes, le matériau se comporte comme un matériau à constante négative $C = M_1 - |M_2|$. Dans les deux cas, biréfringence, déformation et contraintes sont en phase.

4.3. Généralisation

La généralisation à un modèle présentant un nombre plus elevé de temps de relaxation conduit à deux spectres de raies $J_{2i}(\beta_i)$ et $M_{2i}(\beta_i)$, respectivement pour les complaisances mécanique et photoélastique (fig. 8), où les temps de relaxation sont *identiques*, et correspondent à des mécanismes moléculaires *communs*, les amplitudes J_{2i} et M_{2i} étant *distinctes* (J_{2i} positif, M_{2i} indifféremment positif ou négatif, voire nul, comme précisé au § 5).

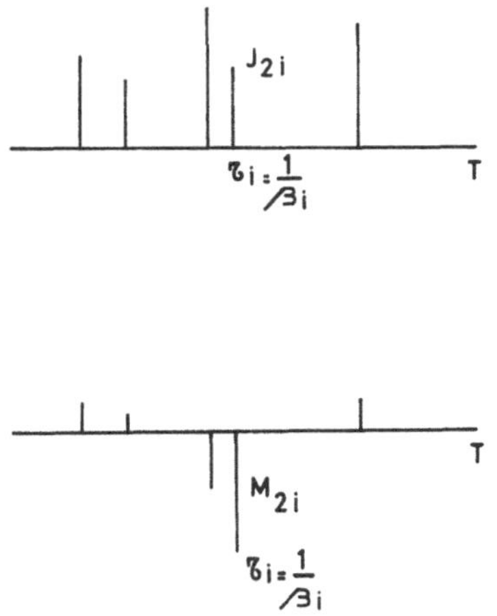

Fig. 8. Spectres de raies des temps de relaxation et des amplitudes correspondantes pour les comportements mécanique et photoélastique

$$J_A(\omega) = J_1 + \sum_i J_{2i} \frac{\beta_i^2}{\beta_i^2 + \omega^2} \qquad [13]$$

$$J_B(\omega) = \sum_i J_{2i} \frac{\omega \beta_i}{\beta_i^2 + \omega^2} \qquad [14]$$

$$W_1(\omega) = M_1 + \sum_i M_{2i} \frac{\beta_i^2}{\beta_i^2 + \omega^2} \qquad [15]$$

$$W_2(\omega) = \sum_i M_{2i} \frac{\omega \beta_i}{\beta_i^2 + \omega^2}. \quad [16]$$

Si l'on préfère représenter, par souci de commodité, le comportement d'un corps réel par une distribution spectrale continue de temps de retard, en échelle logarithmique, $L(\tau)$ (avec $\tau = 1/\beta$) pour la partie mécanique, et $M(\tau)$ pour la partie photoélastique, il vient:

$$J_A(\omega) = J_1 + \int_{-\infty}^{\infty} \frac{L(\tau)}{1 + \omega^2 \tau^2} d\log\tau \quad [17]$$

$$J_B(\omega) = \int_{-\infty}^{\infty} L(\tau) \frac{\omega\tau}{1 + \omega^2 \tau^2} d\log\tau \quad [18]$$

$$W_1(\omega) = M_1 + \int_{-\infty}^{\infty} \frac{M(\tau)}{1 + \omega^2 \tau^2} d\log\tau \quad [19]$$

$$W_2(\omega) = \int_{-\infty}^{\infty} M(\tau) \frac{\omega\tau}{1 + \omega^2 \tau^2} d\log\tau \quad [20]$$

avec comme différence essentielle entre ces deux groupes de relation le fait que $L(\tau)$ est essentiellement une fonction *définie positive*, alors que $M(\tau)$ est, en général, une fonction tantôt négative, tantôt positive, pouvant donc présenter des zéros, et en outre, fonction de la longueur d'onde d'observation.

5. Interprétation microrhéologique

L'existence de deux mécanismes responsables de la déformation et de la biréfringence peut être comprise, du point de vue thermodynamique, à partir de l'hypothèse de l'existence d'une énergie interne spécifique u et d'une équation calorique d'état (19).

$$u = \hat{u}(n_\alpha, S) \quad [21]$$

où S est l'entropie spécifique, et n_α le système de paramètres définissant l'état thermodynamique du système. Nous admettons que n_α se limite à $n_1 = V$, volume spécifique et $n_2 \dots n_{10}$ les 9 gradients de déformation $x^k_{,A}$ à partir d'un état de référence.

Par définition

$$\left(\frac{\partial \hat{u}}{\partial S}\right)_{n_\alpha} \equiv \theta$$

(température thermodynamique) [22]

$$\left(\frac{\partial \hat{u}}{\partial n_\alpha}\right)_{S, n_{\beta=\alpha}} \equiv f_\alpha$$

(contrainte thermodynamique conjugée à n_α). [23]

En particulier,

$f_1 = -p$ (pression thermodynamique)

$f_2 \dots f_{10}$ = composantes du tenseur des contraintes hyperélastiques.

Dans les limites de validité du principe de superposition de *Boltzmann*, nous pouvons nous limiter à une expérience de tractage où $n_{\alpha=2} \div l$ (longueur de l'éprouvette) et $f_{\alpha=2}$ est la force thermodynamiquement conjuguée, nécessaire à assurer la longueur l.

$$u = \hat{u}(n_1 = V, n_2 = l, S). \quad [24]$$

Les expériences de laboratoire peuvent être conduites dans les situations suivantes (à température constante)

a) $V = c^{ste}$; $f = c^{ste}$
(fluage à volume constant)

b) $V = c^{ste}$; $l = c^{ste}$
(relaxation à volume constant)

c) $p = c^{ste}$; $f = c^{ste}$
(fluage à pression thermodynamique constante)

d) $p = c^{ste}$; $l = c^{ste}$
(relaxation à pression thermodynamique constante)

Définissons les 5 potentiels

I $F = u - \theta S$
(énergie libre de *Helmholtz*) [25]

II $\xi = u - f_\alpha n_\alpha - \theta S$
(enthalpie libre de *Truesdell*) [26]

Qui se réduit ici à

$\chi = u + pV - f l - \theta S$

$\chi = u - f_\alpha n_\alpha = \xi + \theta S$
(enthalpie de *Truesdell*) [27]

III $R = u - f l$
(enthalpie avec exclusion de la pression) [28]

$X = R - \theta S$
(enthalpie libre avec exclusion de la pression) [29]

IV $H = u + pV$

(enthalpie de *Gibbs*) [30]

$G = H - \theta S$

(enthalpie libre de *Gibbs*) [31]

Le group I correspond aux variables (V, l), II à (p, f), III à (V, f), IV à (p, l).
Gee (20), *Elliott* et *Lippman* (21) et *Treloar* (22) ont démontré *à l'equilibre* les relations valables pour les groupes de variables I et IV.

I $$\left(\frac{\partial u}{\partial l}\right)_{V,\theta} = f - \theta\left(\frac{\partial f}{\partial \theta}\right)_{V,l} \qquad [32]$$

avec

$$S = -\left(\frac{\partial F}{\partial \theta}\right)_{V,l}$$

$$f = \left(\frac{\partial F}{\partial l}\right)_{V,\theta}$$

$$p = -\left(\frac{\partial F}{\partial V}\right)_{l,\theta}$$

IV $$\left(\frac{\partial F}{\partial T}\right)_{p,\theta} = f - \theta\left(\frac{\partial f}{\partial \theta}\right)_{p,l} \qquad [33]$$

avec

$$S = -\left(\frac{\partial G}{\partial \theta}\right)_{p,l}$$

$$f = \left(\frac{\partial G}{\partial l}\right)_{p,\theta}$$

$$V = \left(\frac{\partial G}{\partial p}\right)_{l,\theta}.$$

Nous établissons de même les relations correspondant aux autres choix de variables,

II $$\left(\frac{\partial \chi}{\partial f}\right)_{p,\theta} = l - \theta\left(\frac{\partial l}{\partial \theta}\right)_{p,f} \qquad [34]$$

avec

$$S = -\left(\frac{\partial \xi}{\partial \theta}\right)_{p,f}$$

$$l = \left(\frac{\partial \xi}{\partial f}\right)_{\theta,p}$$

$$V = \left(\frac{\partial \xi}{\partial p}\right)_{\theta,l}$$

III $$\left(\frac{\partial R}{\partial f}\right)_{V,\theta} = -l + \theta\left(\frac{\partial l}{\partial \theta}\right)_{V,f} \qquad [35]$$

avec

$$S = -\left(\frac{\partial X}{\partial \theta}\right)_{V,f}$$

$$l = -\left(\frac{\partial X}{\partial f}\right)_{V,\theta}$$

$$p = -\left(\frac{\partial X}{\partial V}\right)_{\theta,f}.$$

Les relations sont susceptibles d'une interprétation expérimentale immédiate: par exemple [35] s'interprète à partir du diagramme expérimental $l = l(\theta)$ à V et $f = c^{ste}$. En un point de fonctionnement donné, $\theta = \theta_0$, le coefficient angulaire de la tangente donne $\left(\frac{\partial l}{\partial \theta}\right)_{V,f}$, son *ordonnée à l'origine* donnant la dérivée partielle à déterminer $\left(\frac{\partial R}{\partial f}\right)_{V,\theta}$. Dans une expérience de fluage ([34], [35]), la déformation l apparait comme la somme de deux termes, l'un *hyperélastique*, dérivant d'un potentiel thermodynamique χ ou R, l'autre *entropique*.

En effet,

$$\left(\frac{\partial l}{\partial \theta}\right)_{V,f} = \left(\frac{\partial S}{\partial f}\right)_{V,\theta} \qquad [36]$$

et

$$\left(\frac{\partial l}{\partial \theta}\right)_{p,f} = -\left(\frac{\partial S}{\partial f}\right)_{p,\theta}. \qquad [37]$$

De même, dans une expérience de relaxation ([32], [33]), la force f est la somme d'un terme hyperélastique de potentiel H ou u, et d'un terme entropique, puisque

$$\left(\frac{\partial f}{\partial \theta}\right)_{V,l} = -\left(\frac{\partial S}{\partial l}\right)_{V,\theta} \qquad [38]$$

et

$$\left(\frac{\partial f}{\partial \theta}\right)_{p,l} = \left(\frac{\partial S}{\partial l}\right)_{p,\theta}. \qquad [39]$$

Les deux cas extrêmes sont désignés par *élasticité vitreuse* (dérivant d'un potentiel hyperélastique) et *élasticité rubberlike* ou *entropique*, correspondent respectivement au cas du verre et du caoutchouc, à la température ambiante.

L'élasticité entropique, entrevue par Joule, est thermodynamiquement irréversible, puisque dans le bilan d'énergie du continu

$$\varrho \dot{u} = \mathrm{tr}(\bar{\bar{\sigma}} \bar{\bar{V}}) + \mathrm{div}\vec{q} + \varrho s \qquad [40]$$

le terme $\dot{u}$ est nul par définition. Le travail des contraintes se traduit donc par un *nécessaire* échange de chaleur avec le monde extérieur.

Un effet photoélastique peut être associé à chacun de ces mécanismes. Cependant, le développement des relations ci-dessus implique l'adoption de modèles microscopiques particuliers, à la fois pour le mode de déformation mécanique et l'effet photoélastique.

Pour l'effet photoélastique associé à la composante hyperélastique, nous adopterons le modèle de polarisabilité de *Lorentz*, dont la rigidité devient anisotrope sous l'effet de la déformation (2). La biréfringence qui en résulte est «instantanée», au sens donné à ce terme dans la théorie classique de l'élasticité. La complaisance photoélastique qui en résulte est donnée par une relation de *Lind* (23), avec

$$M_1 \div \frac{(n_0^2 - 1)(n_0^2 + 2)}{n_0} \qquad [41]$$

(le facteur de proportionnalité étant $J_1/30$ pour un solide incompressible), n_0 est l'indice de réfraction en l'absence de déformation.

Pour l'effet photoélastique associé à la composante entropique nous adoptons le modèle de *Treloar* et *Saunders* (17), dans lequel le polymère est formé de chaînons moléculaires, optiquement anisotropes, et caractérisés par leurs polarisabilités transversale et longitudinale, au repos, α_T et α_L. La déformation tend à orienter ces segments moléculaires, modifiant ainsi l'entropie du système à l'équilibre. Cette entropie peut être calculée, en mécanique statistique, à partir de la probabilité de la configuration géométrique due à la déformation.

L'orientation des chaînes moléculaires n'est pas instantanée: elle s'opère suivant une cinétique extrêment complexe, liée à la formation et à la destruction des liaisons interchaînes. L'étude de cette cinétique (distinctes suivant que le matériau est fortement réticulé ou non) n'est pas essentielle dans la présente étude.

Il résulte du calcul

$$M_2 \div \frac{(n_0^2 + 2)^2}{n_0} (\alpha_L - \alpha_T). \qquad [42]$$

Si l'on se trouve en présence de mécanismes moléculaires multiples (orientation de chaînes latérales), on associera à chaque M_{2i} une différence de polarisabilités principales $(\alpha_L - \alpha_\theta)_i$.

Cette interprétation microrhéologique rend bien compte des observations expérimentales. En effet, la quantité M_1 est nécessairement positive, alors que M_2 voit son signe dépendre de celui de la différence $(\alpha_L - \alpha_T)$. Celle-ci peut être positive, nulle (chaînon isotrope) ou négative, comme le confirme le calcul quantique des polarisabilités calculées à partir de la nature des liaisons C—C, C—C, H—C.

La complaisance photoélastique totale peut donc évoluer de plusieurs manières, au fur et à mesure de l'établissement de la déformation d'origine entropique.

La *dispersion chromatique* du phènomène s'explique de manière similaire: la dispersion chromatique de M_1 est déterminée par celle de n_0, c'est-à-dire par la dispersion classique d'un oscillateur de *Lorentz*. Celle de M_2 est déterminée en autre par la dispersion de α_L et α_T, distincte de n_0, et dont la différence peut être de signe quelconque.

La dispersion chromatique de la complaisance photoélastique totale est donc essentiellement fonction de la fréquence de la sollicitation périodique en régime établi, comme vérifié par nos expériences. En particulier, l'effet photoélastique peut apparaître en phase avec la contrainte pour une longueur d'onde donnée, et en phase avec la déformation pour une autre longueur d'onde. La théorie explique aussi l'origine de la variation de la dispersion chromatique au cours du temps observé durant une expérience de fluage, en particulier par *Pindera* (24). Elle est par contre en dèsaccord avec l'interprétation photoplastique introduite par *Monch* et *Loreck* (25, 26).

6. Conclusions

La présente théorie n'est pas dynamique au sens défini dans (3). Elle peut cependant fournir des indications d'ordre expérimental pour la construction de modèles mathématiques dans le cadre de la mécanique rationnelle des milieux continus. Pour représenter valablement le comportement des matériaux investigués dans la présente contribution, un tel modèle devrait rendre compte de la dispersion chromatique de l'effet photoélastique, qui apparaît comme une donnée expérimentale essentielle dans ce problème. Les propriétés chromatiques dont un tel modèle devrait rendre compte peuvent être résumées comme suit:

1. En l'absence de contraintes, le matériau présente une dispersion chromatique classique de l'indice de réfraction, suivant les lois de *Lorentz-Drude-Sellmeyer*.

2. En présence de contraintes, le matériau présente une dispersion chromatique de ses propriétés photoélastiques (c'est-à-dire des *différences* des indices principaux) à priori quelconque, et pouvant changer de signe en fonction de la contrainte.

3. L'équation de comportement introduite pour l'énergie interne se décompose en une *somme* d'un potentiel hyperélastique (fonction quadratique des déformations dans le cas évanouissant), et d'un terme entropique ne dépendant pas du tenseur des déformations.

La présente contribution se proposait également de montrer, qu'à l'inverse des idées reçues en cette matière, le comportement photoélastique d'un matériau est *distinct* de son comportement mécanique, bien que ces deux comportements soient déterminés par les mêmes mécanismes.

En particulier, dans les modèles microrhéologiques simples adoptés, M_1 est fonction linéaire de J_1, M_2 étant une combinaison linéaire de J_{2i}, comme J_2. Mais les poids respectifs des divers mécanismes moléculaires en particulier par le truchement des $(\alpha_L - \alpha_T)_i$ (négatifs, positifs, ou nuls), sont très différents dans le comportement mécanique et photoélastique.

L'ensemble de ces conclusions peut être condensé dans l'observation suivante: alors que les complaisances mécaniques réelle et imaginaire satisfont aux relations de *Kramers-Kronig*, les complaisances photoélastiques n'y satisfont pas nécessairement, la partie imaginaire de la complaisance photoélastique puvant être *négative*, ce qui est contraire aux hypothèses nécessaires à l'établissement des relations de *Kramers-Kronig* (27).

Remerciements

Une partie des travaux théoriques et expérimentaux décrits dans cette contribution fut réalisée à l'occasion de séjours aux Universités de *Poitiers* (France), *Sherbrooke* et *Waterloo* (Canada). L'auteur remercie les Professeurs *A. Lagarde*, *P. Bourassa* et *J. Pindera* pour leur assistance dans la discussion des problèmes évoqués dans cette conférence.

Nous remerçions également le Fonds pour la Recherche Fondamental Collective (FKFO – Belgique) pour son assistance financière.

Résumé

Des mesures fines de l'effet photoélastique présenté par des matériaux soumis à une sollicitation sinuoïdale en régime établi, montrent un comportement inhabituel, et présentant une forte dépendance chromatique. En outre, la biréfringence peut être déphasée *en avance* sur la sollicitation. Une analyse macrorhéologique, basée sur les fonctions de transfert, et l'adoption d'un modèle microrhéologique distinguant une composante hyperélastique et une composante entropique, fournissent une explication.

Summary

Precise measurements on the photoelastic effect exhibited by materials harmonically loaded, show an unusual behaviour, with a strong chromatic dependance. The phase shift of the birefringence to the stress may also be positive, negative or zero. The use of transfer functions, at the macrorheological scale, and the assumption of a microrheological model with a hyperelastic and an entropic component, may explain the phenomenon.

Littérature

1) *Robert, A.*, Polarimétrie et photoélasticimétrie. Ecole Nationale Supérieure des Techniques Avancées, Laboratoire Matériaux Structures, Analyse Expérimentale des Contraintes (1970).

2) *Van Geen, R. E.*, Effet photoélastique et matériaux adéquats en photoélasticimétrie. Sciences et Techniques de l'Armement, Mémorial de l'Artillerie Française **2** (Paris 1971).

3) *Boulanger, P., G. Mayne, A. Hermanne, J. Kestens* et *R. Van Geen*, Cah. Groupe Franç. Rhéol. **2**, 5 (1971).

4) *Van Geen, R. E.*, C. R. Acad. Sci. (Paris) **258**, 5164 (1964).

5) *Van Geen, R. E.*, Strojnicky Casopis **4**, 19 (1968).

6) *Van Geen, R. E.*, Modern theories on the photoelastic effect. Proc. Symposium Experimental Mechanics, Waterloo, Canada (juin 1972) (à paraître).

7) *Van Geen*, R. E., Contribution à l'étude de l'effet photoélastique. Ph. D. Sc. Université de Bruxelles (1964).

8) *Staverman, A. J.* et *F. Schwarzl*, dans: *H. A. Stuart*, Die Physik der Hochpolymeren. Vol. IV, Ch. 1 (Berlin-Göttingen-Heidelberg 1956).

9) *Vinh Tuong, N. P.*, dans: *B. Persoz*, La Rheologie. Chap. V (Paris 1969).

10) *Theocaris, P. S.*, Experimental Mechanics 4. (1965).

11) *Brinson, H. F.*, Experimental Mechanics **8**, 12 (1968).

12) *Noblet, A., G. Sylin, M. Vandaele-Dossche* et *R. van Geen*, Bull. Cl. Sc. Acad. Royale de Belgique **50**, 8 (1964).

13) *Read, B. E.*, J. polymer Sci., Part C, Polymer Symposia **5** (1964).

14) *Williams, M. L.* et *R. J. Arenz*, Experimental Mechanics **9** (1964).

15) *Gross, B.*, J. Appl. Phys. **19**, 257 (1948).

16) *Mindlin, R. D.*, A mathematical theory of photoviscoelasticity. J. Appl. Mech. **1949**, 20.

17) *Dill, E. M.*, J. Polymer Sci., Part C, Polymer Symposia **5** (1964).

18) *Pierrard, J. M.*, Modèles viscoélastiques linéaires. Dans: *Persoz, B.*, La Rhéologie, cfr [9].

19) *Truesdell, C.* et *R. A. Toupin*, The Classical Field Theories, Chap. E, p. 607 (Berlin-Göttingen-Heidelberg 1960).

20) *Gee, G.*, Trans. Far. Soc. **42**, 585 (1946).

21) *Elliott, D. A.* et *S. A. Lippmann*, J. Appl. Physics **16**, 50 (1945).

22) *Treloar, L. R.*, dans: *Stuart, H. A.*, The structure and mechanical properties of rubberlike materials. Die Physik der Hochpolymeren. Vol. IV (Berlin-Göttingen-Heidelberg 1956).

23) *Lind, N.*, Experimental Mechanics **9**, 9 (1969).

24) *Pindera, J.* et *Peter Straka*, On physical measures of rheological responses of some materials in a wide temperature and spectral frequency range. Communication au Présent Congrès.

25) *Monch, E.*, Schweizer Arch. Angew. wiss. Tech. **25**, 5 (1959).

26) *Loreck, R.*, Untersuchung von Polyestergießharzen als photoplastiekens modelmaterial. Kunststoffe **52**, 3 (1962).

27) *Landau, L.* et *E. Lifchitz*, Electrodynamique des Milieux Continus (Moscou 1969).

Adresse de l'auteur:
R. E. Van Geen, Dr. Sc.
Professeur à la Vrije Universiteit Brussel
Dienst Toegepaste Mechanica van het Continuum
Ad. Buyllaan, 87
B-1050 Bruxelles (Belgique)

Rheol. Acta 12, 479–485 (1973)

From the Institute of Mathematics, Bucharest (Romania)

Mathematical models in rheology of the bodies with memory and ageing

By M. Predeleanu

(Received October 27, 1972)

1. Introduction

The changing in time of the capacity of materials of answering the inner and outer solicitations has become more obvious together with the quantitative observations on this rheological phenomenon, called briefly ageing.

Though its causes are of a structural nature, the experimental informations we have got nowadays are, generally, of a phenomenological order, lacking those resulted of studying its inner structure and changing during the time, under different conditions of solicitation.

It is obvious,that the attention of the researchers has been focussed to materials of technological interest which especially manifest this phenomenon and among these we must first of all enumerate the concrete and plastics. We shouldn't however omit certain soils or metals.

Thus, the instantaneous elastic deformation and that due to the creep of certain concrete piece diminish sensibly during the first year after processing. Also, for a kind of composite it has been established in (1) the increasing resistance to the creep deformation within the first two years of life.

Generally, as the experiments show, many materials exhibit rheological properties in the period immediately after the processing, *different* of those of the following service time. This fact complicates a great deal the design, the adopting of big safety coefficients being necessary.

In the construction of the mathematical model, besides kinematic and dynamic variables, the age of the material should be included. This one implies the characterization of the artificial materials by a "birth data", which on the scale of the newtonian time becomes an absolute constant.

The division of materials in classes of materials of the same nature (chemical composition, physical structure, etc.) simplifies for sure the experimental and theoretical research, otherwise insurmountable. And yet, the experimental data including the influence of ageing are now far of being satisfactory. If the phenomenon were set before with its fundamental qualitative aspects sufficient for elaborating mathematical models, the quantitative information concerning the rheological moduli, under their dependence on the kinematical and dynamical variables, are not sufficient. This situation makes late the rapid development of the theory of ageing materials, the only ones that could give the sentence when it is or not necessary to take into account in design the effects of ageing.

Nowadays, such a decision can be taken only in a few cases and with little certitude.

The elaboration of mathematical models has known a logical way: from the mere generalization of the constitutive equations in finite or differential form, considering the rheological moduli not constant but time variable (with the inherent inadventencies of such formal constructions), up to the nonlinear theory of the bodies with memory in which it has from the beginning included the natural assumption that materials are ageing during the time. The founders of this last mentioned theory have thus thought over the construction of the mathematical model.

Volterra (2) was the one who invented the term "closed cycle" for the simplificatory hypothesis of considering the rheological properties of the materials invariable during the time.

This hypothesis brings simplifications in the mathematical treatment but it doesn't justify the proliferation of a general theory on the deformable media ignoring the phenomenon of ageing even from the bases of constructing the mathematical model as it happened in the case of some important researches.

According to *Volterra*s studies, the mathematical models which were including the ageing effects have particularly referred to the linear-viscoelastic behaviour of the concrete (3–7). In the last two decades important progress has been made on the elaboration of the general theory for the linear as well as for the nonlinear case (8–12), permitting to find the solutions of the fundamental boundary-value problems in the linear case [see, for example, (13)], as well as for solving particular problems in the non linear case. Theorems of existence and unity have been found out in (14–16).

In the paper, on the occasion of discussing the linear model, represented by means of *Volterra-Stieltjes* operators, not commutative, the conditions of application of these operators were widened to include jump-discontinuous solicitations in time.

It also sets before the mathematical status of the linear theory as an approximation to some exact theory of visco-elastic behaviour.

2. The assumption of non-ageing and the principle of objectivity

Let us consider a deformable body, occupying a bounded region $\mathscr{B}$, with boundary S, in the three-dimensional euclidian space, referred to a fixed rectangular cartesian coordinate system x_i. Let be t, the time variable, denoting by t the present time moment and by τ the values of t for $\tau \leq t$. We assume, for shortness, that every particle of the body is at rest for $\tau \leq 0$ and $\mathscr{B}_0$ being the reference configuration, when $\tau = 0$. The particle X which at $\tau = 0$ occupies the point $X_i = x_i(0)$, at t will occupy the point $x_i = x_i(X_i, t)$.

Let us consider a sample which under a stimulus measured by $\sigma(t)$ gives a response measured by $\varepsilon(t)$.

For instance, in a simple uniaxial test, $\sigma(t)$ can be the tension or compression and $\varepsilon(t)$, the deformation (for which a certain measure was adopted). If two experiments on the same material are made, the moments of loading being different, $\tau_1 < \tau_2$, the response will be different.

Usually, the elastic instantaneous deformation as well as the creep deformation, measured after the same interval of time of the loading moment, will be less in the second experiment, than in the first one. The material hasn't got the same capacity of answering to the solicitations, it aged.

The hypothesis of non-ageing[1]) (time-translation invariance)

It is said that a body doesn't age or verifies the time-translation invariance hypothesis if and only if for two stimuli $\varepsilon_1(t)$ and $\varepsilon_2(t)$ with $\varepsilon_2(t) = \varepsilon_1(t-\lambda)$ the corresponding responses $\sigma_1(t)$ and $\sigma_2(t)$ verify the relation $\sigma_2(t) = \sigma_1(t-\lambda)$, for all t and λ arbitrary.

This assumption implies the following

Proposition 1

The operator from the set of the functions $\sigma(t)$ into the set of the functions $\varepsilon(t)$ cannot depend on t explicitely.

That is, for a constitutive equation in the form

$$\sigma(X,t) = \mathop{\mathscr{F}}_{\tau=0}^{t}[\varepsilon(\tau), X, \ldots] \qquad [1]$$

the functional $\mathscr{F}$ is independent of t. These functionals have been called hereditary functionals and respectively, hereditary materials. Obviously, [1] is the result of applying the principle of determinism[2]). If it is accepted the assumption of local action, $\varepsilon(t)$ is the deformation gradient, $x_{i,k}$ (simple materials). The form of the functional $\mathscr{F}$ cannot be arbitrary, but it must satisfy the principle of invariance under superposed rotations, which requires that should be form-invariant to the group of proper orthogonal transformations, defined by

$$\bar{x}_i = c_i(\tau) + q_{ij}(\tau)\,x_j \qquad [2]$$

with

$$q_{ij}q_{ik} = q_{ji}q_{ki} = \delta_{jk} \qquad [3]$$

$$|a_{ij}| = 1. \qquad [4]$$

Noll (17) arrives to the Proposition 1 not as a consequence of the non-ageing hypothesis but by applying the principle of objectivity[3]) which includes the time-translation, i.e.,

$$\bar{x}_i = c_i(\tau) + q_{ij}(\tau)\,x_j \qquad [5]$$

with

$$q_{ij}q_{ik} = q_{ji}q_{ki} = \delta_{jk} \qquad [6]$$

---

[1]) Named "closed cycle" by *V. Volterra.*

[2]) The version of this principle that requires that the past should determine the present is adopted.

[3]) This invariance principle is used under different names, that depending if it considered the proper group as [2]–[4] or full group as [5]–[7].

$$|a_{ij}| = \pm 1 \quad ^{4)} \tag{7}$$

and

$$t' = t + a. \tag{8}$$

The *Noll* conclusion is taken over in the treatise of wide spread (18, 19) and that has lead to a certain ambiguity [5]) concerning this subject-matter.

We should stress that in the theory of constitutive equations it is necessary that we should make the distinction between the general principles which must lie on the basis of the mathematical models in rheology (as for example the principle of determinism) by the constitutive hypothesis.

The hypothesis of non-ageing is a constitutive hypothesis and it cannot be present as a necessary principle to be verified by all the constitutive equations.

The fact that a corps modifies its properties during the time is a truth, verified by experiment and that is why it can be considered as a postulate, which should be initially accepted in the elaboration of a mathematical model in rheology.

3. Linear models

a) Integral operators

Assuming the physical and geometrical linearity of the rheological behaviour let us denote be ε, with components ε_{ij}, the infinitesimal strain tensor and by e, with components e_{ij} its deviator defined by $e_{ij} = \varepsilon_{ij} - 1/3\,\delta_{ij}\varepsilon_{hh}$. Let be σ, with components σ_{ij} the stress tensor and s with components s_{ij} its deviator: $s_{ij} = \sigma_{ij} - 1/3\,\delta_{ij}\sigma_{hh}$. For shortness, we shall ignore the presence of the thermal field or of other physical ones.

The linear mathematical model of the viscoelastic behaviour of a homogeneous ageing material can be defined by the following integral form of the constitutive equations

$$\sigma_{ij}(t) = \int_0^t E_{ijkl}(t, \tau)\, d\varepsilon_{kl}(\tau) \tag{9}$$

or

$$\varepsilon_{ij}(t) = \int_0^t E^*_{ijkl}(t, \tau)\, d\sigma_{kl}(\tau) \tag{10}$$

where $E_{ijkl}(t, \tau)$ and $E^*_{ijkl}(t, \tau)$ are the material functions. These functions change their form according to the point in time when the test is made, for including the ageing effects.

The integral representation given by [9] or [10] can be deduced [1]), assuming the following properties of the mapping $\mathscr{F}$ from ε_{ij} into σ_{ij}:

i. $\mathscr{F}$ is linear: $\mathscr{F}(a_1\varepsilon_1 + a_2\varepsilon_2) = a_1\mathscr{F}\varepsilon_1 + a_2\mathscr{F}\varepsilon_2$, for every real numbers a_1, a_2.

ii. $\mathscr{F}$ is not retroactive: for every fixed a, $\varepsilon = 0$ for all $t \le a$, implies $\sigma(t) = 0$ for all $t \le a$.

iii. $\mathscr{F}$ is continuous over a compact space of bounded variation functions (with respect to a suitable topology). For an isotropic medium the relations [9] became

$$s_{ij} = 2\int_0^t \chi(t, \tau)\, de_{ij} \tag{11a}$$

$$\sigma_{ii} = 3\int_0^t \Omega(t, \tau)\, d\varepsilon_{ii} \tag{11b}$$

where $\chi(t, \tau)$ and $\Omega(t, \tau)$ represent the relaxation function in shear and dilatation, respectively.

The dual stress-strain relations defined by the corresponding creep functions are the following

$$e_{ij} = \tfrac{1}{2}\int_0^t \chi^*(t, \tau)\, ds_{ij} \tag{12a}$$

$$\varepsilon_{ii} = \tfrac{1}{3}\int_0^t \Omega^*(t, \tau)\, d\sigma_{ii} \tag{12b}$$

where $\chi^*(t, \tau)$ and $\Omega^*(t, \tau)$ are the creep functions in shear and dilatation, respectively.

We assume also that instantaneous elastic response is time-variable, namely let $G(t) = \chi(t, t) = 1/\chi^*(t, t)$, $K(t) = \Omega(t, t) = 1/\Omega^*(t, t)$ be the instantaneous elastic ageing moduli.

If the *Poisson* ratio is time-invariable $\mu(t, \tau) = \mu = \text{const}$, then

$$\Omega(t, \tau) = c_0\,\chi(t, \tau) \tag{13}$$

or

$$\Omega^*(t, \tau) = \frac{1}{c_0}\chi^*(t, \tau) \tag{14}$$

with

$$c_0 = \frac{2(1 + \mu)}{3(1 - 2\mu)}. \tag{15}$$

It results also, reciprocally, that if the relaxation functions $\chi(t, \tau)$ and $\Omega(t, \tau)$ are proportional, then the *Poisson* ratio is time-invariable (9).

---

[4]) *Rivlin* (20) has discussed the use of the full group of orthogonal transformations with its unsuitable consequences.

[5]) Some reviewers have been surprised that the constitutive equations for ageing materials do not satisfy the principle of objectivity given in (17)!

The methods of invertibility of the *Volterra-Stieltjes* integral operators, used in [9]–[12], can be referred to those proposed by *Volterra* in the class of continuous functions, by *Trigomi* (25) in the space L_2 and *Hlavacek* and *Predoleanu* (15) in the *Soboleff* space $\mathring{W}_2^{(-1)}$.

It may be shown that within the physically admitted conditions for the material functions (for instance left-continuous functions) the integral operators used in [9]–[12] can be applied over the set of bounded variation functions and led to functions of bounded variation. Thus can be included the boundary data which are jump-discontinuous in time.

The model for the non-ageing materials is obtained from [9]–[12] putting $E_{ijkl}(t, \tau) = E_{ijke}(t - \tau)$ and so on.

It is easy to show that the non-ageing condition for the hereditary functionals is accomplished.

b) Differential operators

An immediate way to elaborate linear models consists in generalizing the differential constitutive equations of the viscoelasticity theory of the non-ageing materials, assuming the rheological moduli variable in time. Accordingly, we have

$$P_1(D)\, s_{ij} = P_2(D)\, e_{ij} \qquad [16a]$$

$$P_3(D)\, \sigma_{ii} = P_4(D)\, \varepsilon_{ii} \qquad [16b]$$

with

$$P_\alpha(D) = \sum_{k=0}^{N_\alpha} p_k^{(\alpha)}(t)\, D^k \qquad \alpha = 1, \ldots, 4, \quad D = \frac{d}{dt}.$$

The mathematical study of such models has not been developed, though the general theory of differential equations with variable coefficients gives many results which can be used.

Also, the general theory of systems with memory and ageing (not only rheological), offers many elements for rheological models [see, for instance, (24)].

Libliner and *Sackman* (10) analyse the model defined for a unidimensional state of stress, by the constitutive equation

$$\frac{d}{dt}\left[\frac{1}{E(t)}\frac{d\sigma}{dt}\right] + \alpha(t)\frac{d\sigma}{dt} + \beta(t)\,\sigma = \frac{d^2\varepsilon}{dt^2} + \gamma(t)\frac{d\varepsilon}{dt}. \qquad [17]$$

Thus it is generalized the *Burgers* model. It is shown that for a creep experiment, the model set before the creep functions used for concrete by *Iasin* (26):

$$C(t, \tau) = \frac{1}{E(\tau)} + f(\tau)\left[1 - e^{-c(t-\tau)}\right] + g(t) - g(\tau). \qquad [18]$$

For $g(t) = \text{const}$ it is obtained from [18] the *Arutinians* creep function (7) and if $g(t) = k \log t$ ($k = \text{const}$), it is deduced the creep function given by *Hansen* (27) *Bazant* (28) investigates the same model, but with $\beta(t) = 0$ and finds also the creep functions used in (7). It is shown that the time-varying *Maxwell* model, defined by

$$\frac{1}{E(t)}\frac{d\sigma}{dt} + \frac{1}{\eta(t)}\sigma = \frac{d\varepsilon}{dt} \qquad [19]$$

set before a creep function of *Dischinger-Whitney* type

$$C(t, \tau) = \mathscr{C}(t) - \mathscr{C}(\tau). \qquad [20]$$

4. Nonlinear models

The first study on the three-dimensional constitutive equation for non-linear materials with memory and ageing, might be considered that given by *Green* and *Rivlin* in 1957 (11). The authors admit from the beginning, that the constitutive functional depends explicitly on the time t, i.e. it is assumed, for an homogeneous simple material that

$$\sigma_{ij}(t) = \mathscr{F}_{ij}\left[\overset{t}{\underset{\tau=0}{x_{i,k}(\tau)}}, t\right]. \qquad [21]$$

By subjecting the functional to be form-invariant under the transformations [2]–[4] it is deduced

$$\sigma_{ij}(t) = x_{i,r}\, x_{j,s}\, \varphi_{rs}\left[\overset{t}{C_{pq}(\tau)}, t\right] \qquad [22]$$

where φ_{rs} is arbitrary, and $C_{pq}(\tau) = x_{i,p}(\tau)\, x_{i,q}(\tau)$ (*Cauchy* strain tensor).

The method of representing the functional φ_{rs} is based on the *Fréchet* theorem (29) and consists in approximating it by a polynomial $P_{rs}^{(N)}$, in linear functionals, which tends to φ_{rs} as $N \to \infty$. It is obtained

$$\varphi_{rs}\left[\underset{\tau=0}{\overset{t}{C_{pq}(\tau)}}, t\right] = K_{rs}(t) + \int_0^t K_{rsr_1s_1}(t, \tau)\, C_{r_1s_1}(\tau)\, d\tau + \int_0^t\int_0^t K_{rsr_1s_1r_2s_2}(t, \tau_1, \tau_2)\, C_{r_1s_1}(\tau_1) \times C_{r_2s_2}(\tau_2) + \cdots \qquad [23]$$

It is assumed that the functional is continuous over a compact space of continuous functions $C_{pq}(\tau)$ defined on $(0, t)$. These smoothness requirements can be relaxed, also for this theory, like it has been made for the hereditary theory.

The above representation allows to establish the status of the linear theory, formulated by [9]–[10] as an approximation to this exact theory.

The non-ageing condition is verified if the kernels have the following form

$$K_{rsr_1s,\ldots,r_Ns_N}(t, \tau_1, \ldots, \tau_N) = K_{rsr_1s_1,\ldots,r_Ns_N}(t-\tau_1, \ldots, t-\tau_N). \qquad [24]$$

The general theory of the ageing materials has not been further developed in this direction, as it happened with the hereditary materials; however certain results obtained can be used for:

i. Enlargement of the regularity conditions for the existence of the constitutive functionals and the use of a suitable topology necessary to the approximation theorems for these functionals.

ii. Representation theorems of the constitutive functionals for bodies with some material symmetry (including, the consequences of the invariance under the isotropy group).

iii. Experimental programs for the rheological functions characterizing any material.

iv. Formulation of finite linear viscoelasticity theory.

A greater attention has been given to the theory concerning the infinitesimal theory. Thus in (12) it is presented a non-linear theory of the bodies with memory and ageing, characterized by the following relations between stress tensor σ_{ij} and infinitesimal strain tensor ε_{ij}

$$\varepsilon_{ij}(t) = \int_0^t \Gamma^{(1)}_{iji_1j_1}(t,\tau_1)\,\sigma_{i_1j_1}(\tau_1)\,d\tau_1 + \int_0^t\int_0^t \Gamma^{(2)}_{iji_1j_1i_2j_2}(t,\tau_1,\tau_2)\,\sigma_{i_1j_1}(\tau_1) \times \sigma_{i_2j_2}(\tau_2)\,d\tau_1\,d\tau_2 + \int_0^t\ldots\int_0^t \Gamma^{(n)}_{iji_1j_1\ldots i_nj_n}(t,\tau_1,\ldots,\tau_n)\,\sigma_{i_1j_1}(\tau_1)\ldots \ldots\sigma_{i_nj_n}(\tau_n)\,d\tau_1\ldots d\tau_n + \cdots \qquad [25a]$$

or by invertibility

$$\sigma_{ij}(t) = \int_0^t K^{(1)}_{iji_1j_1}(t,\tau_1)\,\varepsilon_{i_1j_1}(\tau_1)\,d\tau_1 + \int_0^t\int_0^t K^{(2)}_{iji_1j_1i_2j_2}(t,\tau_1,\tau_2)\,\varepsilon_{i_1j_1}(\tau_1) \times \varepsilon_{i_2j_2}(\tau_2)\,d\tau_1\,d\tau_2 + \cdots + \int_0^t\ldots\int_0^t K^{(n)}_{iji_1j_1\ldots i_nj_n}(t,\tau_1,\ldots,\tau_n)\,\varepsilon_{i_1j_1}(\tau_1)\ldots \ldots\varepsilon_{i_nj_n}(\tau_n)\,d\tau_1\ldots d\tau_n + \cdots \qquad [25b]$$

where the kernels

$$\Gamma^{(n)}_{iji_1j_1\ldots i_nj_n}(t,\tau_1,\ldots,\tau_n)$$
$$K^{(n)}_{iji_1j_1\ldots i_nj_n}(t,\tau_1,\ldots,\tau_n)$$

are tensors [rank $2(n+1)$] symmetrical with respect to i_n and j_n.

A special version of this theory, called the quasi-linear theory of viscoelasticity has been proposed by *Iliusin* and *Oghibalov* [see (12)] within the assumptions:

i. Tensorial linearity.

ii. Symmetry of the kernel with respect to every pair of index i_kj_k and $i_lj_l (k, l = 1, 2, \ldots, n)$, i.e. the reciprocity relations are verified.

Retaining three terms of the series of multiple integrals, the relations between the deviators of stress and strain tensors for an isotropic material are

$$e_{ij} = \int_0^t \Gamma^{(1)}_2(t,\tau_1)\,s_{ij}(\tau_1)\,d\tau_1 + 4\int_0^t\int_0^t\int_0^t K^{(3)}_3(t,\tau_1,\tau_2,\tau_3)\,s(\tau_1,\tau_2)\,s_{ij}(\tau_3) \times d\tau_1\,d\tau_2\,d\tau_3 \qquad [26]$$

where

$$s(\tau_1,\tau_2) = s_{kl}(\tau_1)\,s_{kl}(\tau_2). \qquad [27]$$

Pobedria [see (12)] has obtained important results concerning the invertibility problem of the operators involved in the above theory.

Finally, we shall refer to some non-linear models, used especially in applications. The only mathematical reason of some of these models is the generalized superposition principles.

A class of models could be defined by the constitutive equation considered by *Babuska* and *Hlavacek* (16)

$$\varepsilon(t) = B(t)\,\sigma(t) - \int_0^t [A_1(\tau)\,\sigma(\tau) + A_2(t,\tau)(\sigma(\tau))]\,d\tau \qquad [28]$$

where B denotes the operator of the generalized *Hookes* law, i.e. $\{B(t)^{-1}\varepsilon(t)\}_{ij} = \sum_{l,m=1}^{3} c_{ijlm}\varepsilon_{lm}$: $A_1(\tau)$ is a linear ageing operator; A_2 is a nonlinear and noncommutative ageing operator defined by $A_2(t,\tau)(\sigma(t)) = \bar{A}_2(t,\tau)\,f[\sigma(\tau)]$ where $f[\sigma(\tau)]$ is a tensor with components $\{f[\sigma(\tau)]\}_{ij} = f(\sigma_{ij})$ $(i, j = 1, 2, 3)$.

There are given the conditions sufficient that the relation [28] could define an invertible mapping of L_2 to L_2.

Bychawski and *Fox* (30) have studied a special case of the model defined by [28].

The constitutive equations proposed by *Arutunian* (7)

$$E(t)\,\varepsilon(t) = \sigma(t) + \int_{t_0}^{t} \sigma(\tau)\,K_1(t,\tau)\,d\tau + \int_{t_0}^{t} f[\sigma(\tau)]\,K_2(t,\tau)\,d\tau \qquad [29]$$

can be also deduced from [28].

The equation proposed by *Rabotnov* (21)

$$E(t)\,\mathscr{C}[\varepsilon(\tau)] = \sigma(t) + \int_{t_0}^{t} \sigma(\tau)\,K(t,\tau)\,d\tau \qquad [30]$$

has also been used for concrete as well as for metals. We should stress the fact, that certain models above formulated, are equivalent in the way that they show the same aspects of viscoelastic behaviour. Such equivalence has been established in (32–34). Some of constitutive equations presented here could also be deduced from the general representation by multiple integrals, initially formulated by *Volterra*. For instance, *Rabotnovs* eq. [30] could be deduced from the equation used by *Nekada* (35).

5. Discussion

The methods for elaborating the mathematical models for rheological behaviour of the materials with memory and ageing, presented in this paper, do not exhaust this subject-matter. Thus, one of the direction possible to take into account the ageing effects is to introduce as arguments of the constitutive equation the histories of other variables, which could describe the changing of the rheological properties of the material during the time (for instance the temperature, chemical composition, structural parameters, etc.).

Secondly, the constitutive functional, as operator having a time-variable form, could be imagined so as to describe the rheological aspects of different nature during the time: for instance for $\tau_0 < \tau < \tau_1$ the physical linearity and for $\tau > \tau_1$, nonlinearity. The transition from linear to nonlinear behaviour has not been studied from point of view of time-influence.

The mathematical study of such operators has been insufficiently elaborated.

The development of the theory in this direction is of course conditioned by the progresses in thermodynamics of the materials with ageing. The restrictions on the constitutive equations and especially on the rheological moduli will be made much more precise than at the present moment.

Finally, so as it is deduced from this paper, the viscoelastic behaviour has received a greater attention. But, it remains also to solve the problem of the apparition of the plastic deformation in an ageing material, as a time-depending phenomenon.

References

1) *Martirosan, M. M.*, Mehanika Polimerov (Riga) **1**, 6, 20–29 (1965).

2) *Volterra, V.*, Leçons sur les fonctions des lignes (Paris 1913).

3) *Whitney, Ch. S.*, J. Amer. Concr. Inst. **28**, 479 (1932).

4) *Dischinger, F.*, Baningeniem **1937**, Heft 33/34, S. 487–520.

5) *Maslov, G. H.*, Izv. vses. n-i, in-ta ghidrotehn **28** (1940).

6) *McHenry, D.*, Proc. Am. Soc. Test. **43**, 1069 (1943).

7) *Arutuinian, N. K.*, Nekotorîe voprosî teorii polzucesti (Moskva 1952).

8) *Predeleanu, M.*, Contribuţii la studiul matematic al unei clase de corpuri cu proprietăţi reologice (Bucuresti 1961).

9) *Predeleanu, M.*, Bull. Math. de la Soc. Sci. Math. de la R.S.R. **9** (57), nr. 1–2, 115–127 (1965).

10) *Lubliner, J.* and *J. L. Sackman*, J. Mech. Phys. Solids **14**, 25–32 (1966).

11) *Green, A. E.* and *R. S. Rivlin*, Arch. Rat. Mech. Analysis **1**, 1–21 (1957).

12) *Iliuşin, A. A.* and *B. E. Pobedria*, Osnovî matematiceskoi teorii termoviazko-uprugosti (Moskva 1970).

13) *Predeleanu, M.*, Mathematical methods for stress analysis in viscoelastic ageing materials. The Proceedings of the Southampton 1969 Civil Engineering Materials Conference, edited by *M. Te'eni*, Part 2, 1219–1226 (London 1971).

14) *Predeleanu, M.*, Comptes Rendus, Acad. Sci. (Paris) **256**, 1, 71–74 (1963).

15) *Hlavacek, I.* and *M. Predeleanu*, Appl. Math. **9**, 321–327 (1964); Appl. Math. **11**, 199–210 (1965); Appl. Math. **11**, 199–210 (1965).

16) *Babuşka, I.* and *I. Hlavacek*, Arch. Mech. Stas **18**, 47–84 (1966).

17) *Noll, W.*, Arch. Rational Mech. Anal. **2**, 197–226 (1958).

18) *Eringen, A. C.*, Nonlinear theory of continuous media (New York 1962).

19) *Truesdell, C.* and *W. Noll*, The non-linear field theories of mechanics. Handbuch der Physik, Band III/3 (Berlin-Heidelberg-New York 1965).

20) *Rivlin, R. S.*, Technical Report No. CAM-**100**-9 (1969).

21) *Love, E. R.*, Australian J. Physics **9**, 1–12 (1956).

22) *König, H.* and *J. Meixner*, Math. Nachrichten **19**, 256–322 (1958).

23) *Gurtin, M. E.* and *E. Sternberg*, Arch. Rational Mech. Anal. **4**, 291–356 (1962).

24) *Stubberud, A. R.*, Analysis and synthesis of time-varying linear systems (Univ. California Press, 1964).

25) *Tricomi, F.*, Integral Equations (New York 1967).

26) *Yashin, A. V.*, Research on properties of concrete and reinforces concrete structures (in russian). Gos. Stroi. Izdat (Moscow 1959).

27) *Hansen, T. C.*, Creep and stress relaxation of concrete (Stockholm 1960).

28) *Bazant, Z. P.*, Acta Technica, CSAV **11**, 1, 82–109 (1966).

29) *Fréchet, M.*, Ann. Sci. de l'Ecole Norm. Sup. (3) **27**, 193–216 (1910).

30) *Bychawski, Z.* and *A. Fox*, Bull. Acad. Polonaise Sci. Série des Sci. tech. **15**, 453-460 (1967).

31) *Robotnov, Yu. N.*, Prikl. Mat. Mech. Mech. **12**, 53–62 (1948).

32) *Predeleanu, M.*, Colocviul Soc. Sci. Mat. R. S. România, Constanţa, 5–6 Dec. 1971.

33) *Trost, H.*, Beton und Stahlbetonbau **62**, 230–238, 261–269 (1967).

34) *Rjanitîn, A. R.*, Teoria polzucesti. Stroizdat (Moskvam 1968).

35) *Nakada, O.*, J. Phys. Soc. Japon **15**, 12 (1960).

Author's address:

Dr. *M. Predeleanu*
Institute of Mathematics
Caléa Grivitei 21
Bucuresti 12 (Romania)

---

[1]) Similar treatment has been given for the hereditary theory in (21–23).

Rheol. Acta 12, 486–495 (1973)

From the Institute of Petrochemical Synthesis of the USSR Academy of Sciences, Moscow (USSR)

General characteristics of the visco-elastic properties of flexible-chain polymer solutions over a wide range of concentrations

By A. Ya. Malkin

With 20 figures

(Received October 27, 1972)

Study of the rheological properties of polymer solutions offers a unique opportunity of examining the entire range of possible combinations of mechanical characteristics of visco-elastic liquids by gradually varying the composition of a system.

Roughly speaking, two principal trends can be distinguished in the flood of investigations on the rheology of polymer solutions. These are studies of the effect of a low-molecular-weight solvent on the properties of a melt and of the result obtained by incorporating a single macromolecule into a low-viscosity *Newton*ian liquid. In other words, what is basically meant are the extreme regions of the possible spectrum of compositions and, accordingly, of the properties of polymer solutions. But of fundamental importance are all-round measurements over an unlimitedly wide range of concentrations. What is to be clarified here is how the transition between the two limiting cases takes place. To what extent is the behaviour of concentrated solutions predetermined by the individual properties of a single macromolecular chain and by the simplest molecular interactions in dilute solutions? What is the relation between the changes in the viscosity, visco-elasticity and high-elasticity of polymer solutions? How does structure formation in a concentrated solution affect the entire range of its rheological properties? This list of interesting problems is far from being complete and it can be extended depending on the interests of the investigator.

The validity of the answers to these questions is largely dependent on the possibility of separating the various factors affecting the properties of systems under study. In this sense measurements on monodisperse samples (or those close to monodisperse ones) are of great value in studies of the rheology of polymer solutions, which has long been appreciated by many researchers working in the field of dilute solutions. What we are interested in here, however, is the entire range of concentrations.

Let us consider the results of measurements of the rheological properties of solutions of a number of samples of polybutadienes having a narrow molecular-weight distribution in methylnaphthalene, starting from a pure polymer. We shall consider here the following characteristics determined under steady flow conditions: the zero-shear viscosity (*Newton*ian viscosity) η_0, the initial coefficient of the primary normal stress difference ζ_0, and the initial (zero-shear) high-elasticity modulus G_0, the latter being measured from elastic recoil. The values of all these quantities have been found in the region of low shear rates, where they are independent of the shear rate $\dot{\gamma}$. It is not surprising and long known that a critical concentration can be singled out on the concentration-viscosity curve, at which the character of this dependence is changed. It is more interesting to note that at this concentration (fig. 1) the shape of the concentration dependence of the primary normal stress difference undergoes a similar change. Still more interesting is the fact that the same concentration is distinctly identified on the curve of elastic deformations versus concentration (fig. 2) as a point below which the high-elasticity of the solution disappears, though the solution apparently remains visco-elastic. This fact is evidently the most typical manifestation of the effect of the formation of a three-dimensional fluctuating network of entanglements permeating the whole of the solution at the critical concentration. At lower concentrations random chain contacts are of course possible, but they do not provide the formation in the solution of a quasi stable network when each chain is linked with other chains, at least at one site at any moment of time.

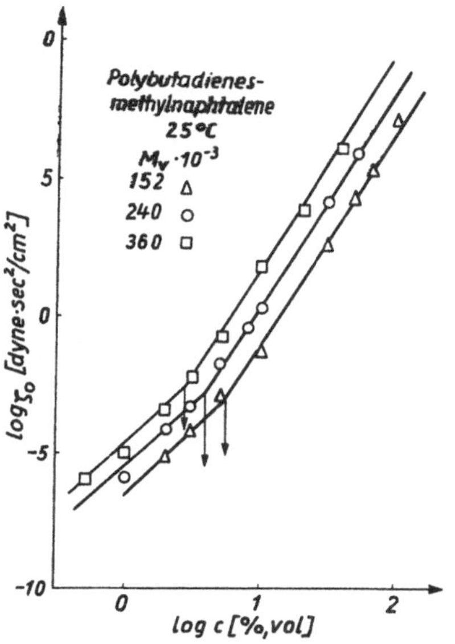

Fig. 1. Concentration dependences of the initial value of the normal stress coefficient for solutions of polybutadienes of different molecular weight in methylnapthalene

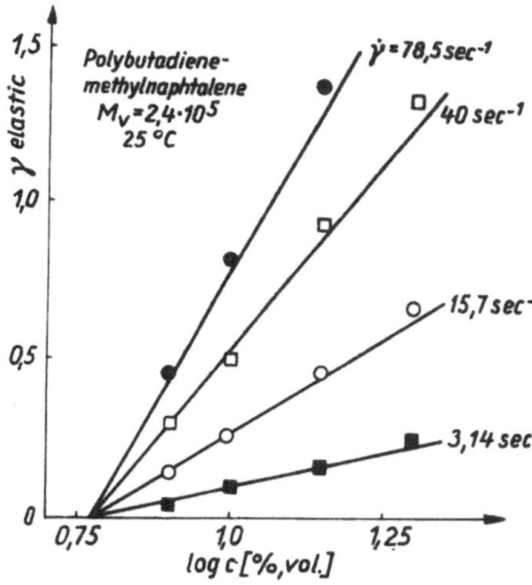

Fig. 2. Concentration dependences of elastic deformations developed at steady shear flow at different shear rates

The above considerations can be extended to a comparison of the ratio of the square of viscosity to the normal stress coefficient, η_0^2/ζ_0, with the measured modulus G_0 (fig. 3). Such a comparison can be made only in a concentration region above the critical concentration since solutions do not exhibit high-elasticity at lower concentrations. The behaviour of concentrated solutions is well described by *Lodges* formula, which is theoretically valid for any linear visco-elastic media (2, 3), namely $G_0 = \eta_0^2/\zeta_0$. But at low concentrations this correlation cannot be fulfilled because the solution is not high-elastic. To the critical concentration there corresponds rather exactly the minimum of the concentration dependence of the parameter concerned, that is, the ratio η_0^2/ζ_0, so that inversion of the effect of concentration on the visco-elasticity of the solution takes place. Analogous curves have been described in a number of works (4, 5), but they are reported to have been obtained by measurements of the dynamic modulus and not by stationary measurements. It may be supposed that this shape of the curve of the visco-elastic characteristics of polymer solutions against concentration is of a general nature, though for reliable identification of the minimum it is essential that this effect is not masked by the polydispersity of the polymer. It should be stressed here that not only dynamic measurements but also the existence of the normal stress effect unambiguously points to the presence of visco-elastic properties up to very dilute solutions.

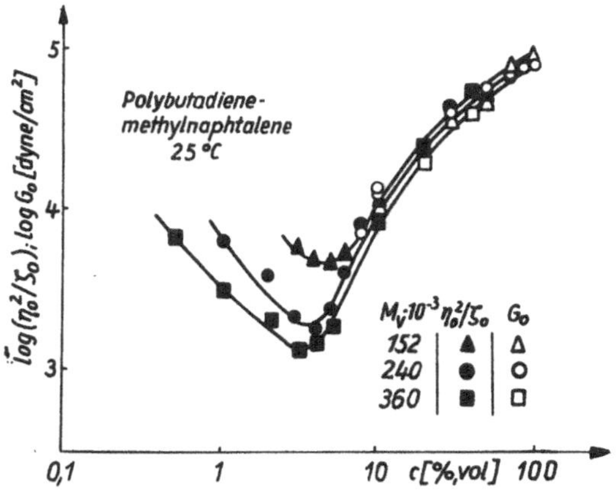

Fig. 3. The relation between the parameter η_0^2/ζ_0 and the equilibrium high-elastic modulus for solutions of polybutadienes of different molecular weight

Thus, below the critical concentration polymer solutions display visco-elastic properties, which become less pronounced (the parameter η_0^2/ζ_0 decreases) with increasing concentration, and high-elasticity is not observed. At concentrations higher than the critical, polymer solutions also exhibit visco-elastic properties supplemented by high-elasticity; in this case the modulus G_0 and the ratio of squared viscosity to the normal stress coefficient are equal, increasing with concentration. These general manifestations of the existence of a critical concentration in solutions are associated with a change in the mechanism of visco-elasticity. Indeed, in dilute solutions relaxation occurs at the level of the segmental motion of separate chains, and the observed effects are associated with local visco-elasticity. In concentrated solutions the whole fluctuating network is brought into motion and the solution assumes continuum visco-elastic and high-elastic properties.

Thus, the main physical cause of the difference in the behaviour of dilute and concentrated solutions is the formation of a three-dimensional fluctuating network at a certain critical concentration. But the individual properties of a chain are not changed only because it is incorporated into the network. The search for a correlation between the characteristics of a single chain, measured in the region of infinitely dilute solutions, and the behaviour of concentrated solutions is therefore of paramount significance. The general approach in the search for such correlations is associated with the principle of corresponding states or with the determination of main variables in dimensionless form. For concentration dependences of the visco-elastic characteristics of solutions such dimensionless variable is the product of the volume concentration c by the intrinsic viscosity $[\eta]$. This parameter was first introduced by *Debye* (6) as a measure of the volume filling of a solution by a polymer. This product ($c[\eta]$) has such clear meaning only for dilute solutions. Nevertheless, in the region of high concentrations too ($c[\eta] \gg 1$) the use of this parameter has meaning as a dimensionless measure of concentration.

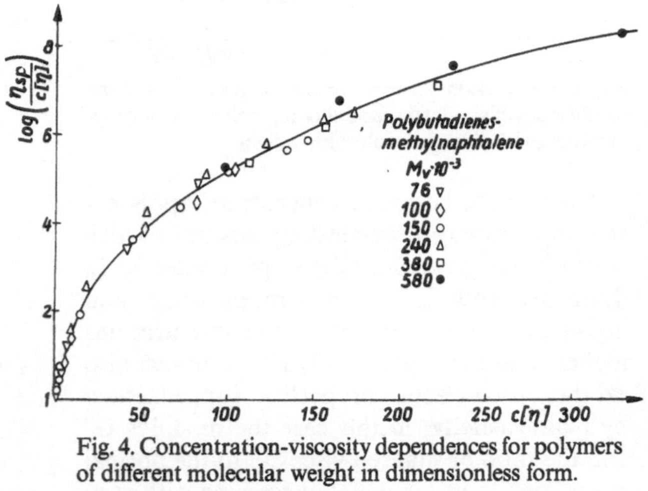

Fig. 4. Concentration-viscosity dependences for polymers of different molecular weight in dimensionless form.

As regards the viscosity function, following *Simha* and *Utracky* (7), we shall assume it to be the value of normalized viscosity $\tilde{\eta}$, which is equal to the ratio of specific viscosity to the product of concentration and intrinsic viscosity ($\eta_{sp}/c[\eta]$). This choice is justified by the fact that at $c \to 0$ this quantity is equal to unity, which gives a single point of origin common for all solutions. Coordinates of this type was successfully used by *Simha* and other authors, though not to very large values of dimensionless concentration. Fig. 4 shows that these coordinates are well applicable to solutions of polymers of different molecular weight in a single solvent over the entire range of concentrations up to polymer melts, which corresponds to the values of dimensionless concentration of the order of several hundreds.

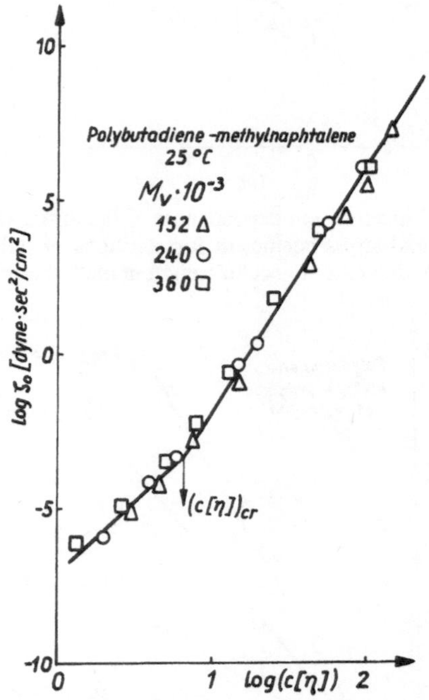

Fig. 5. The dependences of the normal stress coefficient on the dimensionless concentration for polybutadienes of different molecular weight

These results indicate that the effect of molecular weight on viscosity over the entire concentration range is associated with the influence of chain length on the size of a molecular coil characterized for an infinitely dilute solution.

The use of the dimensionless concentration ($c[\eta]$) enables one to construct a master curve of normal stresses versus concentration (fig. 5) and also the concentration dependence of the complex parameter $(\eta_0 - \eta_{sp})^2/\zeta_0$, which characterizes the visco-elastic properties of solutions over the entire concentration range (fig. 6). It is clearly seen that the minimum of this function and accordingly the transition from the region of dilute solutions to that of concentrated solutions correspond to the condition of the constancy of the critical value of the product of concentration by intrinsic viscosity, $(c[\eta]) = \text{const}$, which is close to 6 at this point.

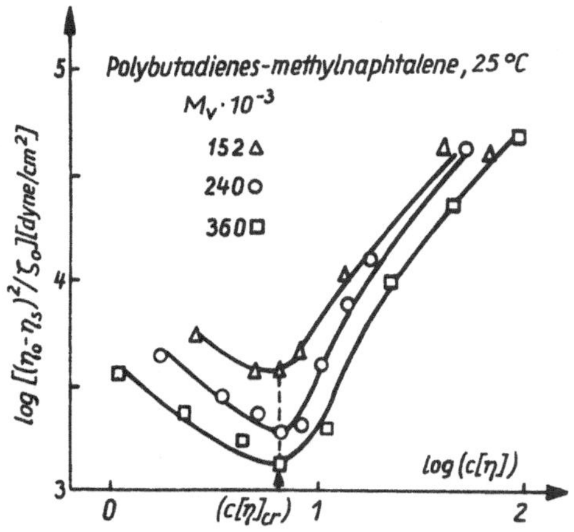

Fig. 6. The relation between the parameter $(\eta_0 - \eta_s)^2/\zeta_0$ as a measure of the visco-elasticity of the system and the dimensionless concentration

The extension of the method of reduced variables to the entire concentration range does not of course deny the existence and significance of the network of fluctuating chain contacts. We are speaking here of main variables but not of the shape of the dependences of visco-elastic characteristics on reduced concentration. The results obtained show that the properties of a chain determine not only its behaviour in dilute solutions, but also the conditions of the formation of a network, and still serve as a quantitative characteristic of the solution structure even when the network has already been formed. In this sense this result is close to the idea expressed by *Berry* and *Fox* (8) that the quantity determining the viscosity of a solution over the entire range of compositions is still their parameter *X*, which in turn depends on the size of an individual macromolecular coil.

The value of intrinsic viscosity varies from solvent to solvent. But is it the only factor that determines the role of the solvent nature? The answer is obviously negative, as seen from fig. 7, which shows concentration dependences of polymer viscosity for different solvents. Account should also be taken of the difference in molecular segmental interaction. But, as before, the question that is essential here is whether the nature of this interaction is the same for dilute and concentrated

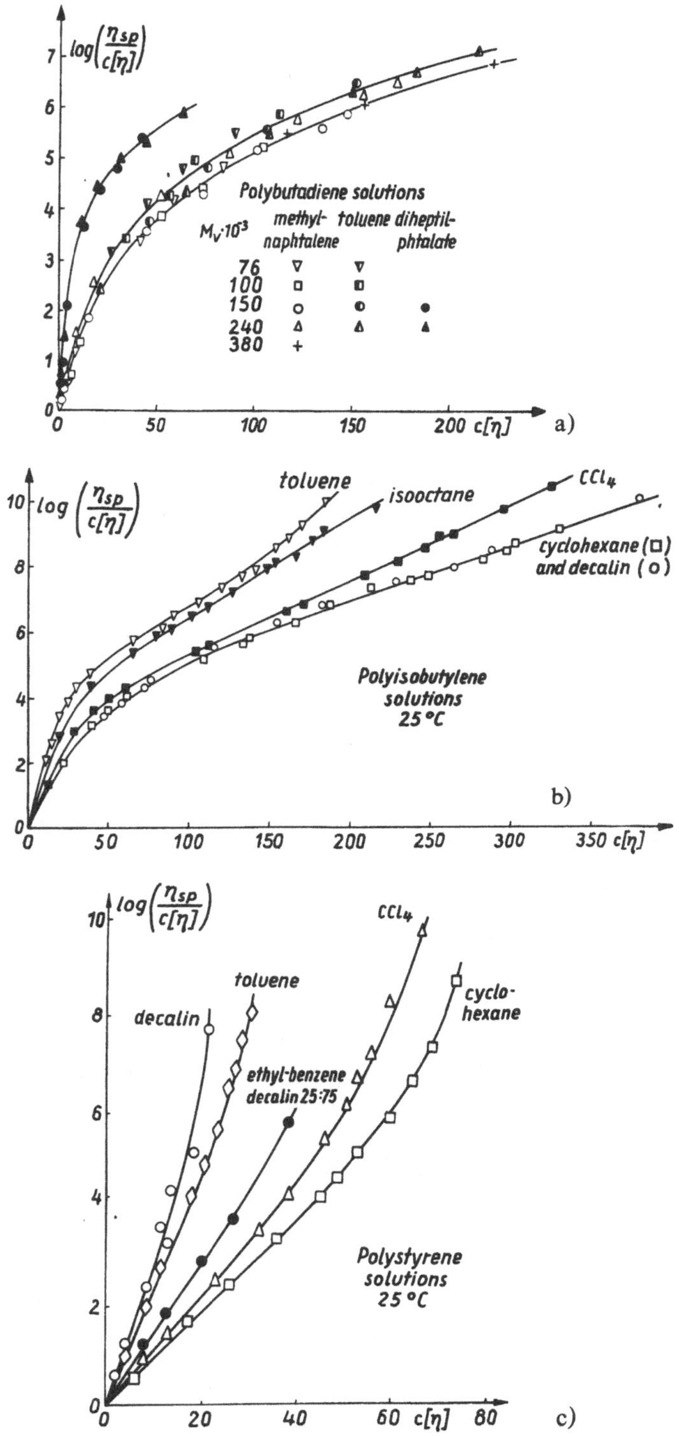

Fig. 7. Dependences of the viscosity on dimensionless concentration in different solvent (9). a) polybutadienes of different molecular weight; b) polyisobutylene; c) polystyrene

solutions. To answer this question, we are to find out what is the rheological measure of molecular interaction. This measure is a constant in any single-parameter equation describing the effect of concentration on the viscosity of the solution. At low concentrations the *Huggins* equation must always be fulfilled as a limiting case. Practically it is more reliable to approximate the initial region of the concentration-viscosity curve with the aid of the *Martin* exponential equation and to use *Martin* constant K_M as a rheological measure of molecular interaction. This is borne out by the linearity of the initial portions of the concentration-viscosity curve in semilogarithmic coordinates used in constructing the graphs described above (see fig. 7).

According to the statement of our problem, we are to find out whether the same constant characterizes the behaviour of solutions in the region of high concentrations. For this purpose we make use (9) of the reduced coordinates $\log \tilde{\eta}$ and $K_M c[\eta]$ and present data relating to different solvents in these coordinates. Some examples are shown in fig. 8. The result obtained by Dr. *Dreval'* and this author concerning the construction of master viscosity-concentration curves for different polymers are summarized in fig. 9. The conclusion made is that the constant K_M retains its meaning and value over the entire range of concentrations. Thus, the value of this constant, which is the rheological measure of the simplest interaction in dilute solutions determines not only the intrinsic viscosity but also the behaviour of solutions of a particular polymer over the entire range of concentrations.

Of fundamental importance is the elucidation of the internal causes giving rise to the effect of the solvent on the rheological properties of solutions. Two approaches differing in starting point are possible here – relaxational and thermodynamic. The concept of free volume as a measure of the rate of relaxation in a polymer system assumes that the properties must be the same at an equal distance from the glass temperature. Indeed (10), if we compare the isothermal concentration dependences of the viscosity of polystyrene in a good and a poor solvent (fig. 10), we shall see that the differences in the viscosity of equiconcentration solutions amount to several decimal orders. The right-hand part of fig. 10 shows the viscosity-concentration curves for temperatures equidistant from the glass temperature. In this case the solutions are found to be in corresponding states and the differences in their viscosities disappear. Thus, the corresponding states for these solutions can be constructed both from the side of infinitely dilute systems and from the side of concentrated systems. This emphasizes the genetic connection between the properties of an individual chain and the structure of a concentrated solution.

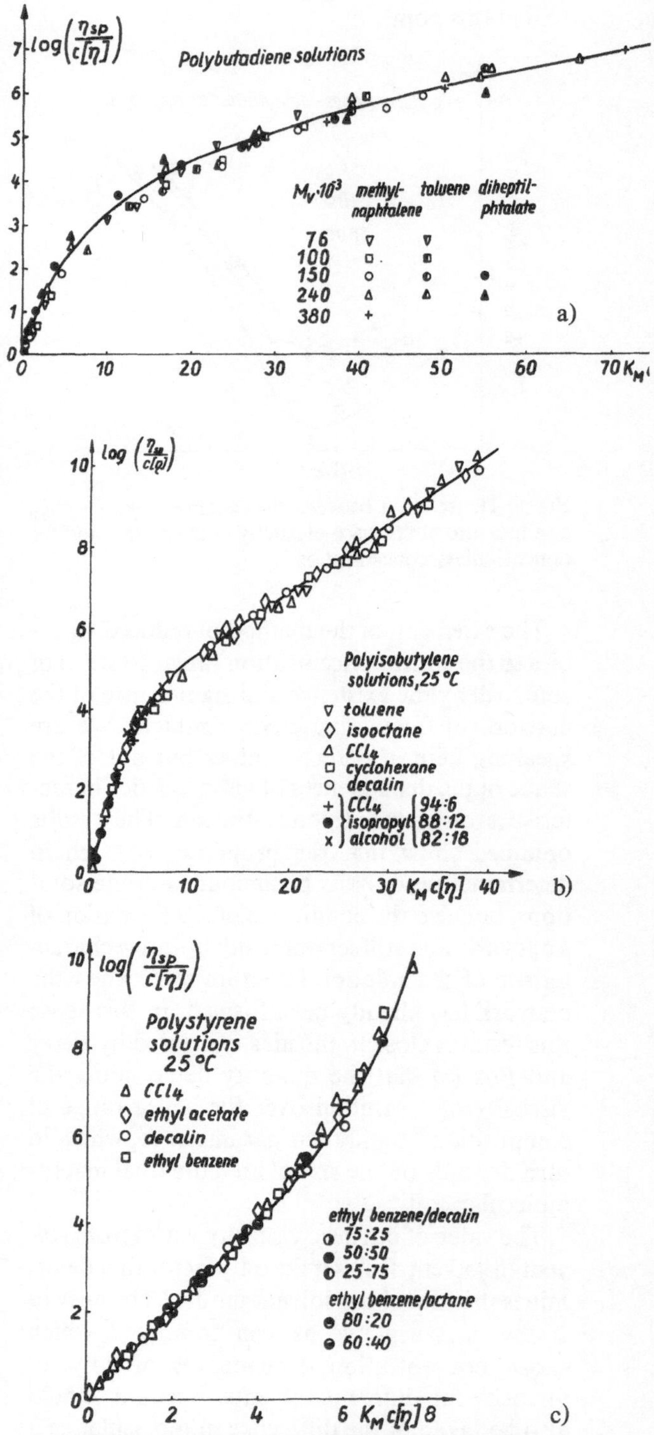

Fig. 8. Reduced concentration dependences of the viscosity of solutions in different solvents (9): a) polybutadienes of different molecular weight; b) polyisobutylenes; c) polystyrene.

There are also other systems in which this connection is disturbed. Indeed, an analogous approach to the determination of corresponding states for solutions of polymethyl methacrylate in different solvents (fig. 11), though lowering somewhat the difference in viscosities, does not eliminate it completely. The factor responsible for this is a strong intermolecular interaction in polar polymers, which depends on the solvent nature. This gives rise to a new factor, the association which operates along the path from a single macromolecule to a concentrated solution. The degree of association depends on the solvent nature. This is confirmed by the fact that the high-elasticity modulus is independent of the nature of the solvent for solutions of, say, polyisobutylene, and that there exists such dependence for solutions of polymethyl methacrylate (fig. 12). Besides, in the first case the modulus does not depend on temperature, while in the second it is strongly dependent on it (fig. 13). All this means that in the first case a homogeneous molecular network is formed, and in the second, owing to association, the network density varies depending on the solvent nature and temperature, which is what disturbs the unambiguity of the correspondence between the properties of a single chain and a concentrated solution. It is for this reason that the relaxational approach to the flow properties of solutions is not quite general.

Interesting and more general considerations concerning the behaviour of solutions can be put forward on the basis of a thermodynamic treatment. From the works of *Sakai* (11) and *Bohdanecký* (12) it is known that a correlation exists between the *Huggins* constant and the degree of swelling of a polymeric coil in different solvents. Similar results are presented in fig. 14 for *Martin* constant (9).

The interrelation between *Martin* constant K_M and the degree of swelling implies the existence of a dependence of this parameter on the energy and entropy of interaction between polymer and solvent. The separation of the effect of these factors is of interest. We shall make use of the *Flory* constant[1] $\chi = v_1(\delta_1 - \delta_2)^2/RT$ calculated for the difference in the solubility parameters of the components $(\delta_2 - \delta_1)$ in order to characterize the energy interactions. Fig. 15 shows the dependence of the reciprocal value of the *Martin* constant on the *Flory* constant χ; with these coordinates chosen, the dependence

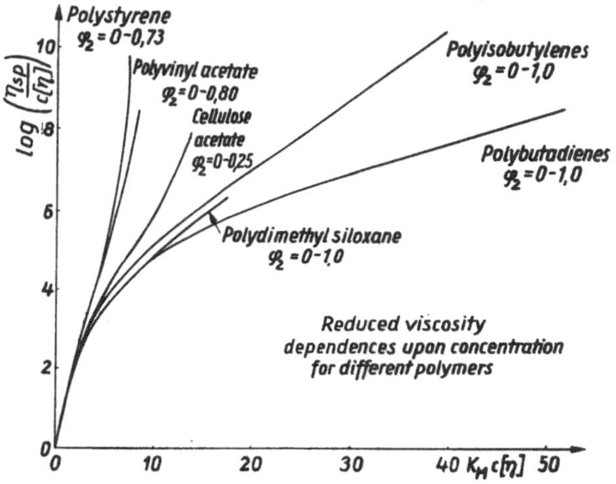

Fig. 9. Comparison of reduced concentration dependences of the viscosity for different polymer systems (9)

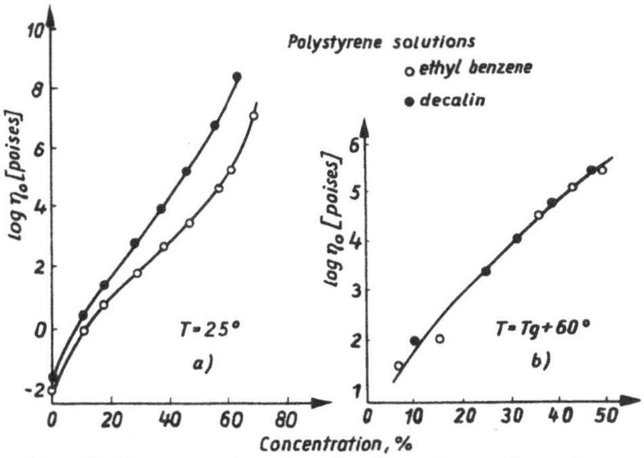

Fig. 10. Concentration-viscosity dependences for polystyrene solutions in different solvents under isothermal conditions (a) and in corresponding states at an equal distance from the glass temperature (b) (10)

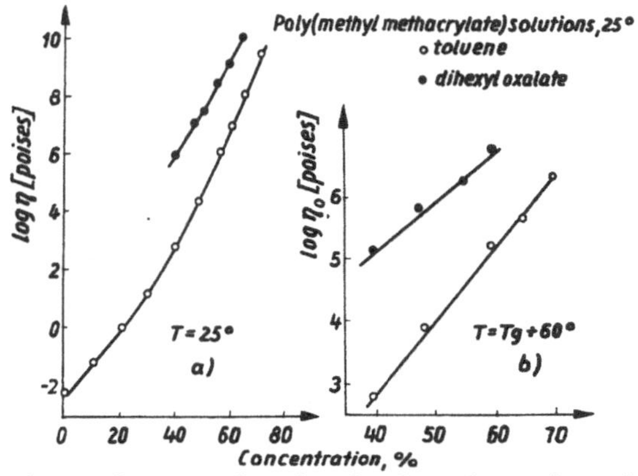

Fig. 11. Concentration-viscosity dependences for polymethyl methacrylate solutions in different solvents under isothermal conditions (a) and at an equal distance from the glass temperature (b) (10)

[1]) Here v_1 is the molar volume of the solvent and δ_1 and δ_2 are the square roots of the cohesive energy densities (or solubility parameters) of polymer and solvent.

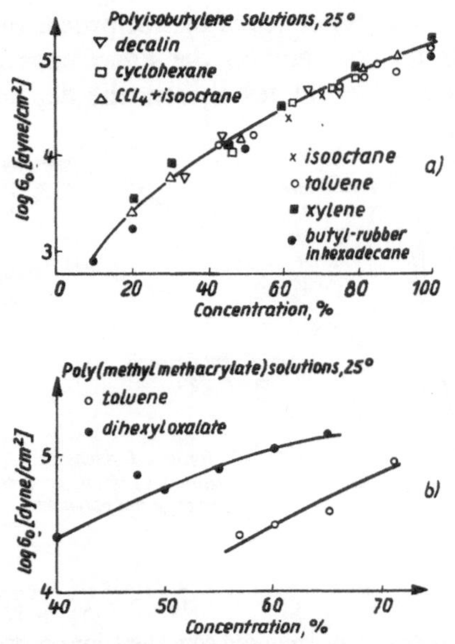

Fig. 12. Concentration dependences of the high-elastic modulus of solutions of polyisobutylene (a) and polymethyl methacrylate (b) in different solvents (10)

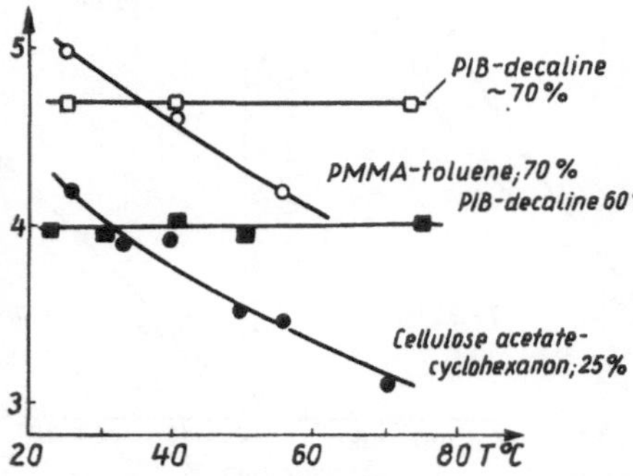

Fig. 13. Temperature dependences of the high-elastic modulus of solutions of some polymers (10)

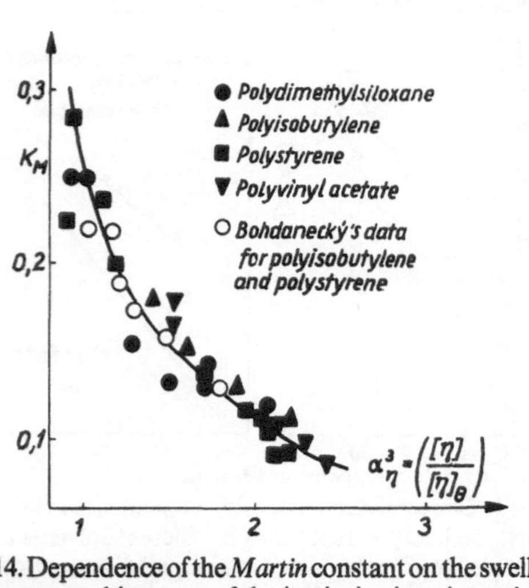

Fig. 14. Dependence of the *Martin* constant on the swelling ratio expressed in terms of the intrinsic viscosity ratio

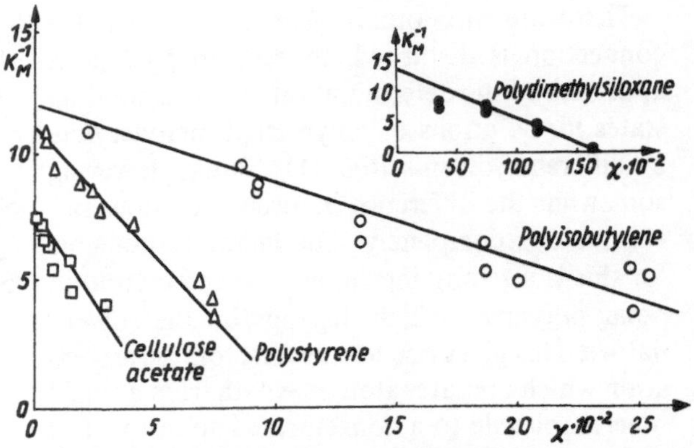

Fig. 15. Dependence of the inverse *Martin* constant on the parameter of energy interactions in solutions of different polymers

for each polymer becomes linear. The fact that the graphs plotted for different polymers do not coincide is associated with the entropy effects. If the energy interactions are excluded by extrapolating the curves of K_M^{-1} against χ to $\chi = 0$, it will be possible to establish a correspondence (fig. 16) between these limiting values of the *Martin* constant $K_{M,0}$ and the skeleton stiffness of the chain $(h_\theta^2/h_0^2)^{0.5}$, where h_θ^2 is the root-mean-square distance between the ends of the chain under theta conditions, and h_0^2 is the same quantity under the assumption of free rotation. The increase of the skeleton rigidity of the chain is accompanied by a decrease in the limiting values of the *Martin* constant $K_{M,0}$. This graph yields a correspondence between the structure factor and the contribution of the entropy effects, the energy interactions being excluded. An increase in energy interactions results in an increase in the *Martin* constant K_M as compared with its limiting low value $K_{M,0}$. Here the intensity of the influence of the *Flory* constant χ on the constant K_M correlates with the value of solubility parameter as seen from fig. 17, where the slopes of the K_M^{-1} versus χ curves are compared with the value of the cohesive energy density of the polymer, δ_2. Thus, the value of the rheological parameter of interactions K_M is the greater the stiffer the polymer chain, and its variation from solvent to solvent is the stronger the higher the energy of intermolecular interaction of the polymer itself. The effect of the solvent on viscosity is therefore the stronger the stiffer the macromolecular chain and the greater the cohesive energy density (9).

Let us now digress from the region of the limiting low rates of shear and discuss how the var-

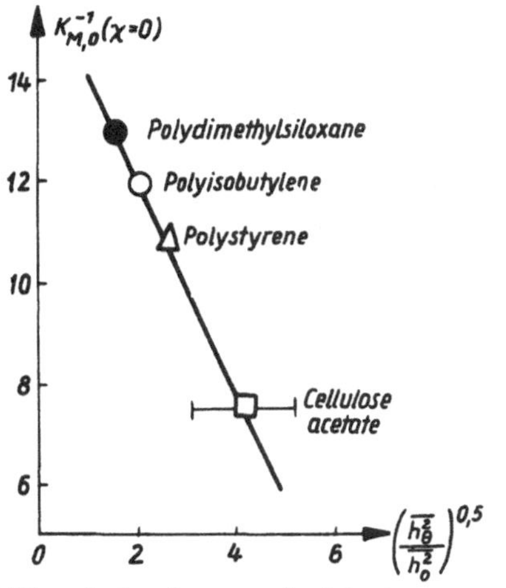

Fig. 16. The relation between the *Martin* constant in the absence of energy interactions and the skeleton rigidity of a macromolecular chain

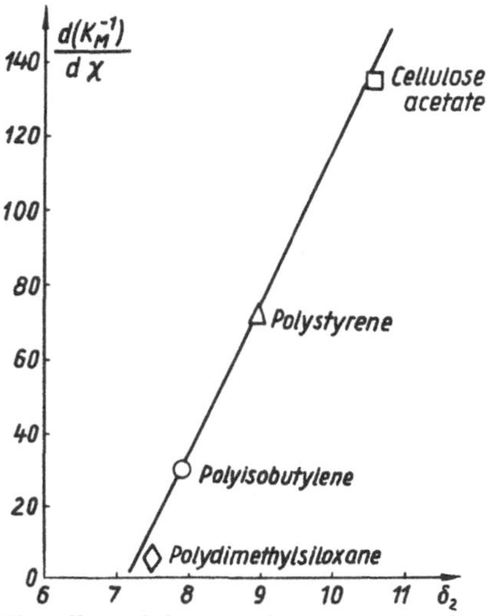

Fig. 17. The effect of the cohesive energy density of the polymer on the contribution of energy interactions to the value of *Martin* constant

iation of concentration affects the full flow curve (fig. 18). The phenomenon of anomalous viscosity is in general not characteristic of melts of monodisperse flexible-chain polymers far away from the glass temperature. As treated in detail in the lecture of Prof. *Vinogradov* at this Congress [see also ref. (13)], in the immediate vicinity of the *Newton*ian region, where the critical rate of deformation is reached, the polymer assumes a rubber-like state and loses the ability to flow. At high concentrations this picture remains in general unchanged. But the value of the critical shear stress decreases with dilution in just the same way as the visco-elastic properties of the solution (the storage modulus in the plateau region and the maximum of the loss modulus upon transition from the fluid to the rubbery state), which can be seen in fig. 19. This means that the criterion of the loss of fluidity is still the ratio of viscous and elastic forces in flow. But as the solution is diluted the region of anomalously viscous flow becomes gradually dominant since the presence of large amounts of solvent makes impossible such a pronounced manifestation of the transition to the high-elastic state as the complete loss of fluidity and the transition to slippage. Nevertheless, the continuation of the line connecting the critical values of shear stress τ_s to the region of relatively dilute solutions retains its meaning. Now this line delimits the region of elastic turbulence. This can be seen on photographs of extrudates. One can suppose that this picture will remain unchanged up to the critical concentration discussed in the first part of this communication. At still lower concentrations local perturbations

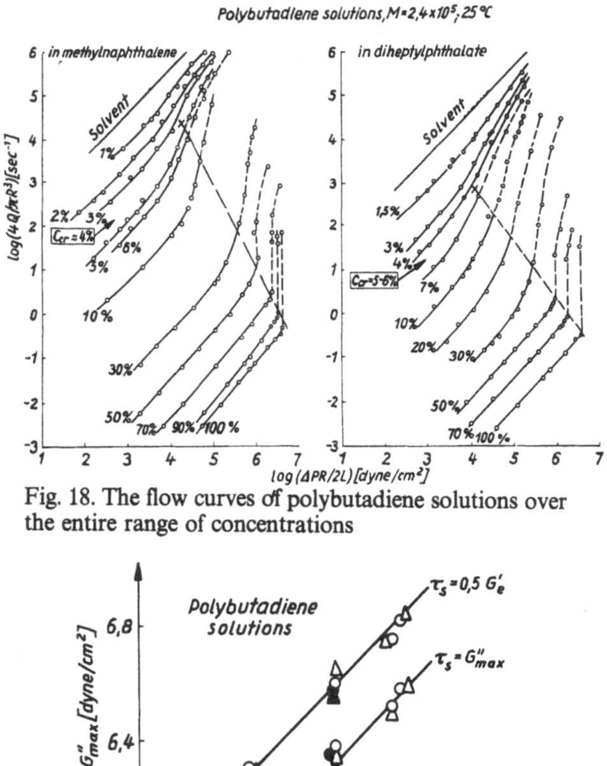

Fig. 18. The flow curves of polybutadiene solutions over the entire range of concentrations

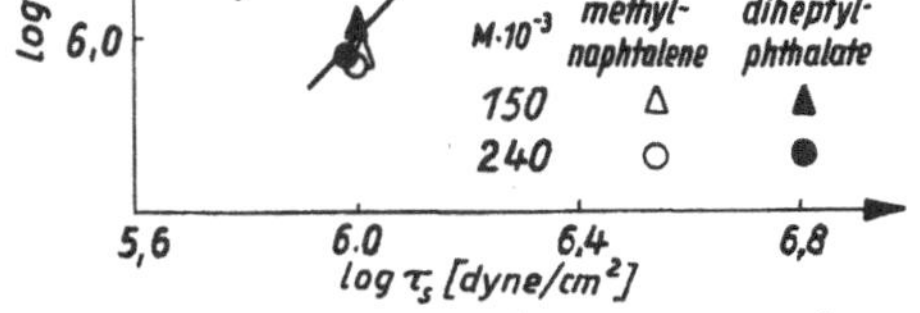

Fig. 19. The relation between the parameters determining the visco-elastic (dynamic) properties of polybutadiene solutions and the stresses corresponding to the loss of fluidity due to the transition to the high-elastic state.

can hardly lead to the occurrence of elastic turbulence characteristic of molten polymers and concentrated solutions. The local visco-elasticity of individual chains however is sufficient to be the cause for the retainment of the effect of viscosity anomaly up to very dilute solutions. For concentrated solutions, however, anomalous viscosity is found to be associated with the loss of fluidity of high-molecular-weight components rather than with the deformation of individual chains. This is clearly seen in fig. 20, which shows the flow curves for a polymer, oligomer and polymer solutions in an oligomer. Though the effect of anomalous viscosity is not observed in practice for initial components, it is strongly pronounced for solutions, the viscosity beginning to fall at the critical rate of shear of a high-molecular-weight component. In this sense the flow of a solution may be likened to the flow of a low-molecular-weight liquid with a visco-elastic filler. The viscosity of the system depends in this case both on the viscosity of the dispersion medium and on the dissipation losses when the filler is subjected to visco-elastic deformations, i.e., on its dynamic viscosity. The flow curves calculated for this model of the medium with a visco-elastic filler are shown in fig. 20 by a dashed line. They are in good agreement with experimental points.

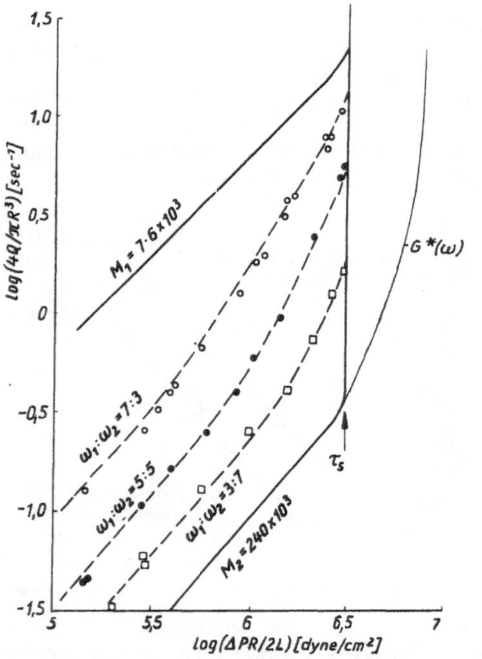

Fig. 20. The flow curves of a high-molecular-weight polybutadiene with a narrow molecular-weight distribution, an oligomer and a polymer solution in an oligomer

Generally it may be stated that the mechanism of viscosity anomaly in solutions and polydisperse polymers (which are also in a large measure nothing else but solutions) is associated with the gradual transition of high-molecular-weight fractions to the rubber-like nonfluid state at critical deformation rates. This results in a system being impoverished in visco-elastic components, which lose fluidity and become a visco-elastic filler. The losses during the visco-elastic vibrations of this filler are lower than would have been during viscous flow of the most viscous components at corresponding shear rates. The intensity of dissipation therefore decreases, which gives rise to the well-known effect of the so-called anomalous viscosity.

Preliminary calculations have shown that this qualitative model of viscosity anomaly enables one to estimate rather exactly the nonlinear characteristics of the system – the flow curve of a polydisperse polymer on the basis of its linear properties such as the *Newton*ian viscosity of samples of different molecular weight, the dynamic viscosity in the region of the high-elastic state and the value of the critical shear stress at which monodisperse fractions attain a visco-elastic state, this value being equal to the loss modulus at the point of transition.

In conclusion, I should like to say that some principal results described in this report have been obtained in collaboration with Dr. *Dreval'* and with the help of *Botvinnik*, *Blinova*, and *Berezhnaya*. The various aspects of this work have been discussed with Prof. *Vinogradov*. To them all I wish to express my deep gratitude.

Summary

The paper summarizes the ideas and results of investigations of viscous and high-elastic properties of polymer solutions in the whole interval of concentrations, chemical structure of the polymers used and the "goodness" of the solvents tried being varied in wide ranges.

Zusammenfassung

In diesem Beitrag sind die Ideen und experimentellen Resultate der Forschung von viscosen und hochelastischen Eigenschaften polymerer Lösungen im gesamten Konzentrationsbereich zusammengestellt. Dabei wurden die chemische Struktur und Lösungsmittelqualität in weiten Grenzen variiert.

References

1) *Lodge, A. S.*, Elastic Liquids (London-New York, 1964).

2) *Malkin, A. Ya.*, Rheol. Acta **7**, 335 (1968).

3) *Coleman, B. D.* and *H. Markovitz*, J. Appl. Phys. **39**, 1 (1964).

4) *Holmes, L. A., K. Ninomiya,* and *J. D. Ferry,* J. Phys. Chem. **70**, 2714 (1966). – *Holmes, L. A.* and *J. D. Ferry,* J. Polymer Sci., C, **23**, 291 (1968).

5) *Einaga, Y., K. Osaki, M. Kurata,* and *M. Tamura,* Macromolecules **4**, 87 (1971).

6) *Debye, P.,* J. Chem. Phys. **14**, 636 (1946).

7) *Utracky, L.* and *R. Simha,* J. Polymer Sci. **A 1**, 1089 (1963); *Simha, R.* and *L Utracky,* J. Polymer Sci. **A-2, 5**, 853 (1967).

8) *Berry, G. C.* and *T. G. Fox,* Adv. Polymer Sci. **5**, 262 (1968).

9) *Dreval', V. E., A. Ya. Malkin,* and *G. O. Botvinnik,* J. Polymer Sci.: Polymer Phys. Ed. **11**, 1055 (1973).

10) *Dreval', V. E., A. Ya. Malkin, G. V. Vinogradov,* and *A. A. Tager,* Europ. Polymer J. **9**, 85 (1973).

11) *Sakai, T.,* J. Polymer Sci. **A-2, 6**, 1535 (1968).

12) *Bohdanacký, M.,* Collect. Czechoslovak. Chem. Commun. **35**, 1972 (1970).

13) *Vinogradov, G. V., A. Ya. Malkin, Yu. G. Yanovskii, E. K. Borisenkova, B. V. Yarlykov,* and *G. V. Berezhnaya,* J. Polymer Sci. **A-2, 10**, 1061 (1972).

Author's address:

Dr. *A. Ya Malkin*
Leninski Prospekt 29
Institute of Petrochemical Synthesis
Academy of Sciences of the USSR
Moscow B 71 (USSR)

Rheol. Acta 12, 496–502 (1973)

From the Camille Dreyfus Laboratory, Research Triangle Institute, Research Triangle Park, North Carolina (U.S.A.)

Molecular model of internal viscosity

By A. Peterlin

With 4 figures and 1 table

(Received October 27, 1972)

Internal viscosity was introduced in the model of randomly coiled macromolecule by *Kuhn* and *Kuhn* (1) in order to explain the gradient dependence of intrinsic viscosity. In laminar flow with transverse gradient the molecule rotates with an angular velocity equal to half the gradient and during each rotation gets twice extended and twice compressed. Hence the amplitude and the frequency of shape change are increasing almost linearly, the rate of deformation quadratically with the gradient. The resistance of the macromolecular chain to rapid change of shape reduces the coil deformation below the value expected for a completely flexible elastic dumbbell or necklace model. As a consequence, the increase of end-to-end distance cannot compensate the decrease of viscosity contribution caused by chain orientation so that with increasing gradient the intrinsic viscosity drops below the initial value at zero gradient. The effect is extreme for perfectly rigid molecule and disappears for ideally flexible coil.

The resistance to rapid shape change is a consequence of the fact that such a change requires at least one conformational change from trans to gauche or gauche prime or from gauche to gauche prime or vice versa. Since these conformations are separated by energy barriers (fig. 1) the transition from one to another conformation requires a finite time which shows up as a viscous type resistance

$$\vec{F}_i = -(\varphi/Z)\,(\partial\vec{r}_{0Z}/\partial t)_{\mathrm{def}}\,. \qquad [1]$$

Here $\vec{r}_{0Z}$ is the end-to-end vector and φ, with the dimension g sec^{-1} = length times viscosity, is the coefficient of internal viscosity of the coil with Z joints where the conformational change may occur. The shape change resistance is the smaller the larger Z, i.e., the larger the molecular weight M or the degree of polymerization.

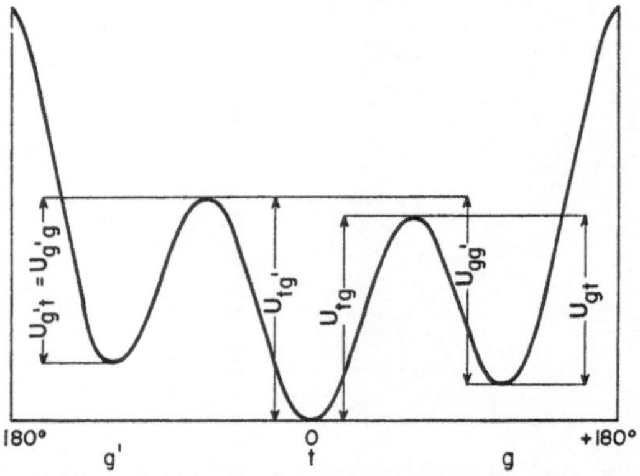

Fig. 1. Angular dependence of potential energy of a vinylidenic chain atom showing the energy barriers U_{tg}, $U_{tg'}$, $U_{gg'}$, $U_{g'g}$, U_{gt}, $U_{g't}$ separating the stable t, g, and g' conformations

Kuhn and *Kuhns* assumption of internal viscosity for the explanation of the gradient dependence of intrinsic viscosity is not more an absolute necessity because one knows now that the effect can be caused also by the anisotropy of hydrodynamic interaction (2–4) and the change of hydrodynamic interaction as a consequence of coil deformation in flow (5–7). But it turned out to be the only way to explain the dependence of the initial intrinsic orientation of streaming birefringence on the viscosity of the solvent and of the finite value of the second *Newton*ian dynamic intrinsic viscosity at very high frequency. Both effects are inexplicable by the ideally flexible coil model.

According to the elastic dumbbell or necklace model the initial change of extinction angle χ of the intrinsic streaming birefringence with the gradient Γ

$$(d\chi/d\Gamma)_{c=0,\,\Gamma=0} = [\omega]_{\Gamma=0} \qquad [2]$$

is expected to be proportional to the viscosity η_s of the solvent. Experiments on polystyrene (PS) (8), DNA (9) and polymethyl methacrylate

(PMMA) (10) have shown, however, that a proportionality exists only in a small range of η_s. The initial straight line of the plot of $[\omega]_{\Gamma=0}$ versus η_s very soon bends down rather sharply and continues as a straight line with a substantially smaller slope and a finite ordinate intercept. According to *Cerf* (11, 12) who introduced the internal viscosity concept into the elastic necklace model the initial intrinsic orientation can be expressed as (13)

$$\begin{aligned}[\omega]_{\Gamma=0} &= 0.67\,(M[\eta]/RT)\,\eta_s && \text{small } \eta_s\\ &= 0.1\ \ (M[\eta]/RT)\,\eta_s\\ &\quad + 0.0062\,h^2_{Z,0}\,\varphi/kT && \text{large } \eta_s \quad [3\text{a}]\end{aligned}$$

in the case of nearly free draining coil and as

$$\begin{aligned}[\omega]_{\Gamma=0} &= 0.51\,(M[\eta]/RT)\,\eta_s && \text{small } \eta_s\\ &= 0.1\ \ (M[\eta]/RT)\,\eta_s\\ &\quad + 0.0045\,h^2_{Z,0}\,\varphi/kT && \text{large } \eta_s \quad [3\text{b}]\end{aligned}$$

in the case of nearly impermeable coil. Here $h^2_{Z,0}$ is the average square of the end-to-end distance of the coil in solution at rest. All values refer to theta solutions with $h^2_{Z,0}$ proportional to M. The intercept of the asymptote of the orientational curve at high η_s yields the parameter φ of internal viscosity.

The proportionality of the intercept witn M was obtained by the special choice of the eigenvalues of the Φ tensor (11)

$$\varphi_p = \varphi p/Z \qquad p = 1, 2, \ldots, Z \qquad [4]$$

where Z is the number of elastic links of the necklace model. Such a choice means that each deformational eigenmode is resisted by the viscosity coefficient φ_p inversely proportional to the number of chain joints between two consecutive nodes. Since the p-th eigenmode has p nodes one has there Z/p links and hence a proportional number of chain atoms. The conformational chance required for the deformation described by the eigenmode can occur at any such atom. Therefore, the probability for its occurrence is proportional to Z/p and the resistance to deformational change proportional to its inverse value, i.e., to p/Z.

From the birefringence data of *Leray* on PS *Cerf* derives $\gamma = h^2_{Z,0}\,\varphi/MkT = 1.1 \times 10^{-8}$ sec/g. With the theta solvent value $h_{Z0}/M^{1/2} \sim 0.67$ Å one obtains $\varphi = 10^{-5}$ g sec^{-1}. *Janeschitz-Kriegl* (14) concludes from his measurements that this value is about ten times smaller, $\varphi \sim 10^{-6}$ g sec^{-1}, and even claims that the intrinsic orientation does not justifiy the introduction of internal viscosity.

The situation is less ambiguous in the frequency dependence of the dynamic intrinsic viscosity. In the second *Newton*ian range the experimentally observed $[\eta]_{\omega=\infty}$ has a finite value (15–18). This is in sharp contrast with the predictions of the perfectly soft necklace model which yields the complex intrinsic viscosity at zero gradient as function of circular frequency ω

$$\begin{aligned}[\eta]^*_\omega &= (RT/M\eta_s)\sum_1^Z \tau_p/(1 + i\omega\tau_p)\\ [\eta]_\infty &= 0\\ \tau_p &= 1/2\mu_0 D_0 \lambda_p\\ \mu_0 &= 3/2b_0^2\\ D_0 &= kT/f_0\\ f_0 &= 6\pi a_h \eta_s. \end{aligned} \qquad [5]$$

Here b_0 is the mean square equilibrium length of the link, a_h is the hydrodynamic radius of the bead of the elastic necklace model, λ_p is the p-th nonzero eigenvalue of the $\underset{\sim}{H}\underset{\sim}{A}$ tensor, $\underset{\sim}{H}$ is the tensor of hydrodynamic interaction and $\underset{\sim}{A}$ is the tensor of elastic forces of the links (5). By introducing the internal viscosity one obtains (19–21)

$$\begin{aligned}[\eta]_\omega &= (RT/M\eta_s)\sum_1^Z \tau_p \frac{1 + i\omega(\tau'_p - \tau_p)}{1 + i\omega\tau'_p}\\ [\eta]_\infty &= (RT/M\eta_s)\sum_1^Z \tau_p/(1 + f_0/\nu_p\varphi_p)\\ &= (RT/M\eta_s)\sum_1^Z \tau_p \cdot \frac{\nu_p\varphi_p/f_0}{1 + \nu_p\varphi_p/f_0}\\ \tau'_p &= \tau_p(1 + \nu_p\varphi_p/f_0) \end{aligned} \qquad [6]$$

in excellent agreement with experimental data of *Schrag* and coworkers (17, 18) if φ_p is chosen according to eq. [4]. Here ν_p is the p-th nonzero eigenvalue of $\underset{\sim}{H}$. It is certainly a comforting situation for a theory if supporting evidence can be derived from experimental data which were neither known nor considered at the time when the theory was established. Most of the data were collected on narrow fractions of PS in highly viscous Aroclor with η_s between 2 and 70 P where the second *Newton*ian range can be easily reached without the need for extremely high frequencies.

The experimentally determined limiting value $[\eta]_\infty$ turns out to be very little dependent on molecular weight of the solute with M ranging from 20 to 840 thousand (fig. 2) and viscosity of the solvent ranging from 2 to 70 P (17, 18) (fig. 3).

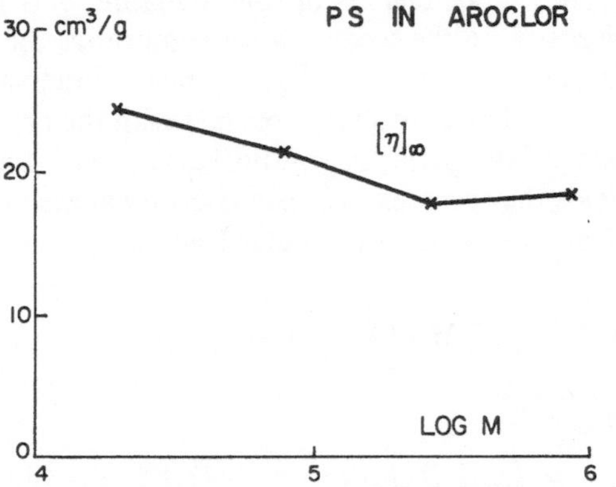

Fig. 2. Molecular weight dependence of $[\eta]_\infty$ (17)

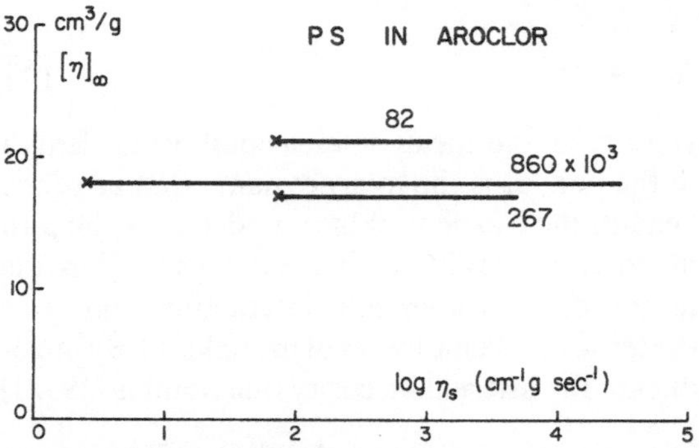

Fig. 3. Dependence of $[\eta]_\infty$ on the viscosity range of the solvents, Aroclor 1258 and 1254. The viscosity variation is a consequence of the temperature variation between -4.7 and $+35$ °C (17, 18)

The former fact is in perfect agreement with the predictions of the theory if the internal viscosity is introduced as formulated by *Cerf* (eq. [4]), i.e., connected with the deformation of the macromolecule as described by the eigenmodes and not with the displacement of a bead in the conventional space as a force depending on the deformation rate of the links. It is not possible to describe it by a simple dash-pot parallel to the representing the elastic link. This is a disadvantage of the model which is responsible for its rejection by many workers in the field.

A viscous element parallel to the elastic link yield in the force field a term dependent on the coordinates of the beads in a similar manner as the elastic force, i.e., a tensor like $\underset{\sim}{A}$. Such a tensor φ has eigenvalues proportional to p^2 (free draining coil) or $p^{3/2}$ (impermeable coil) instead of the required proportionality with p. That yields a proportionality of $[\eta]_\infty$ to M^{-1} instead of the experimentally observed independence of M. On the other hand, the simple correlation of φ with the eigenmodes and the frequency of conformational changes available for a shape change is equivalent with the existence of frictional resistance between each pair of beads of the necklace model. The resistance decreases as the inverse number of beads separating the pair. In the conventional dash-pot representation one has $Z(Z+1)/2$ of them instead of Z viscous elements as required in the case of one dash-pot per link. Such a model is indeed a straightforward extension of *Kuhns* original dumb-bell formulation with two beads to the necklace with $Z+1$ beads.

The fact, however, that $[\eta]_\infty$ is independent of the viscosity of the solvent is incompatible with the concept of internal viscosity based on the energy barriers separating the gauche and trans conformations. Since these barriers are a consequence of the potential field of the chemical bond and of the sterical hindrance of the substituents on the chain atoms they most likely do not depend very much on the solvent and particularly not on the viscosity of the solvent. Therefore, in first and probably very good approximation the coefficient φ of internal viscosity as derived from the energy barriers separating the gauche and trans conformations is independent of the solvent. According to eq. [6] the modification of intrinsic viscosity and the limiting value $[\eta]_\infty$ is proportional to the ratio φ/f_0 which is proportional to η_s^{-1} and hence vanishes with increasing η_s. As a consequence with solvent independent φ the macromolecular coil is becoming less rigid with increasing viscosity of the solvent so that in a very viscous solvent it is expected to be perfectly flexible with $[\eta]_\infty = 0$. Experiments just demonstrate that this is not so and that the basic concept of internal viscosity has to be modified.

An analysis of the actual chain displacement as a consequence of a conformational change indeed demonstrates the existence of an additional resistance force proportional to the viscosity of the solvent and independent of the energy barriers but not included in the conventional elastic necklace model. The model considers in the friction coefficient f_0 only the translational resistance of the subchain with all the chain elements having the same velocity. In particular $f_0(\vec{v} - \vec{v}_j)$ with $\vec{v}$ the unperturbed velocity of the liquid is the friction force the liquid exerts on the

j-th bead moving with the velocity $\vec{v}_j$. As soon as the motion pattern deviates from this uniformity, i.e., shows at any chain element a component perpendicular to the average translational velocity of the subchain additional work is performed in the viscous solvent and hence the resistance of the subchain is increased beyond f_0. This increase is not considered in the diffusion equation and has to be taken into account by the introduction of internal viscosity.

If one considers conformational changes leading to a shape change of the macromolecule one sees immediately that as a rule a single conformational change is practically impossible except at the very ends of the macromolecule because it requires a rotation by 120° of all the rest of the molecule. In such a case the work against the viscous friction forces of the solvent becomes prohibitive. Two simultaneous closeby conformational changes, however, may yield any desired displacement of the two adjacent beads without demanding an excessive translation or rotation of the rest of the molecule. A particularly simple example is shown in fig. 4. The *tgtg't* to *ttttt* transformation eliminating the *gtg'* kink in the chain involves a simultaneous *gt* and *g't* conformational change. The corresponding energy barrier is $U_{gt} + U_{g't}$. The energy barriers U_{gt} and $U_{g't}$ are equal in a vinylidenic but different in vinylic chain (fig. 1). The inverse transformation creating a kink has to overcome the 1.2 kcal/mole higher energy barrier $U_{tg} + U_{tg'}$.

The simultaneous jump *gt* and *g't* has an energy barrier $U = U_{gt} + U_{g't}$ of between 2 and 7 kcal/mole. At room temperature the ratio U/RT is between 3 and 12, $e^{U/RT}$ between 20 and 2×10^5.

On the basis of the activation energy dependence of rate processes one obtains a very rough estimate of the correlation between applied force F_i of internal viscosity and deformation rate $v_{\text{def}} = \Delta l/\Delta t$

$$F_i = (kT/2\nu l \Delta l)\, e^{U/RT} v_{\text{def}} = \varphi \cdot v_{\text{def}} \qquad [7]$$

where $\nu \sim 10^{10}$ sec^{-1} is the frequency of rotational oscillation of the chain element, $l \sim 10^{-8}$ cm the length of the path of the chain element between its energy minimum and potential barrier maximum position over which the external force is performing work and thus helping the element to cross the barrier, Δl the displacement in the direction of the force resulting from the conformational change. Depending on the height of the energy barrier one obtains φ between 10^{-3} and 10^{-7} g sec^{-1}. If there are n chain atoms in a subchain this value has to be divided by n in order to obtain the correct internal viscosity parameter φ. Since the subchain contains about 5 monomers, i.e., 10 chain atoms in vinylic polymers, φ is expected to be between 10^{-4} and 10^{-8} g sec^{-1}. The experimental data, 10^{-5} and 10^{-6} g sec^{-1}, are within these limits.

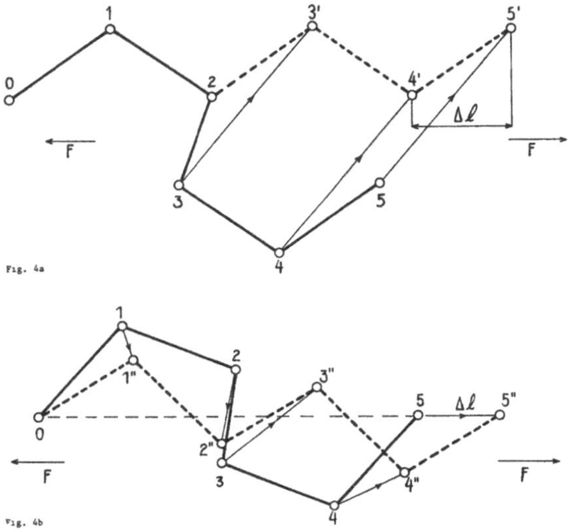

Fig. 4. Conformational change *tgtg't* to *ttttt* leading to an elongation $\Delta l = b \cos\beta/2$ in the direction of applied force and the transposition of 3, 4, 5 to 3', 4', 5' (a). By rotation of 05' in the direction 05 the chain is elongated by $\Delta l = 0.707\,b$ and the chain elements 1, 2, 3, 4, 5 are transposed to 1'', 2'', 3'', 4'', 5'' (b)

But the conformational change also involves a translation and rotation of chain elements which is not identical with the translation $1 = \Delta l$ of the end group 5 in the direction of the applied force. The chain displacement perpendicular to the applied external force is proportional to the displacement parallel to the force. The work against the viscous forces during this lateral displacement is equivalent to the work against a viscous type resistance which is proportional to the rate of chain deformation but is not considered by the frictional coefficient f_0 in the diffusion equation which takes into accout only the translation of the bead. Hence this contribution has to be accounted for independently as internal viscosity.

In the conformational change shown in fig. 4 the end point 5 moves to 5' with an elongation $\Delta l = b \cos\beta/2$ in the direction 02 of the applied force F and an effective translation of 3, 4, 5 to 3', 4', 5' by $2\Delta l$. The actual path of these three chain elements is still longer because the change from g to t at 2 involves a rotation of the link 23 by $\alpha = 120°$ and an identical motion of 4 and 5 on an arc of 120°. The ratio of the path on the circle and on the straight line connecting the two ends of the arc is $\alpha/2 \sin\alpha/2 = 1.21$. Hence the total path of 3, 4, 5 is $(2\pi b/3)\cos\beta/2 = 2\pi\Delta l/3$ and the friction work per second for each of them $(2\pi/3)^2\ (\Delta l/\Delta t)^2\, b$. In addition the chain element 2 rotates by 120° which involves a friction work per second $8\pi a_h^3 \eta_s (\Delta l/\Delta t)^2 = (4a_h^2 f/3)$ $(\Delta\alpha/\Delta t)^2$. All these contributions have to be compared with the friction work per second associated with a uniform extension of the chain leading to a displacement Δl of the end group 5 in the direction of the applied force. The translation for the m-th element is $m\Delta l/5$ yielding for the friction coefficient of the bead

$$f_0(\Delta l/\Delta t)^2 = \sum_{m=0}^{5} (m/5)^2 f(\Delta l/\Delta t)^2 = (11 f/5)(\Delta l/\Delta t)^2 \qquad [8]$$

which enters explicitly the diffusion equation. The ratio of the actual friction work and the friction work of the bead γ turns out to be

$$\gamma = (5/11)\left[3 \times (2\pi/3)^2 + (4a_h^2/3)\cdot(\Delta\alpha/\Delta l)^2\right] = 13.07 \qquad [9]$$

with the assumption $a_h = 2$ Å. This value γ is very large and yields for the ratio $\varphi/f_0 = \gamma - 1$ the much too large value 12.

A better procedure is to consider the total change in length $05\text{–}05' = \Delta l = 0.866 \times b\cos\beta/2 = 0.707\,b$ without any large lateral displacement. If one rotates 5′ to 5″ located on the continuation of the 05 direction the displacements 1 to 1″, 2 to 2″ and so on are much smaller than those shown in fig. 4a (table 1). If one considers only the work done along such shortest paths one obtains $\gamma = 2.02$ which yields $\varphi/f_0 = \gamma - 1 = 1.02$. This value is much smaller than the experimental data. It is also unrealistic because the procedure to obtain it completely neglects the actual paths the chain elements had to move in order to comply with the conformational conditions. That what matters, however, is the fact that even the shortest imaginable chain group displacement yields a nonvanishing φ/f_0, i.e., a finite contribution to internal viscosity proportional to the viscosity of the solvent. A more realistic consideration of the actual path of all chain atoms during a simultaneous conformational change yields a $\gamma - 1$ value above 1.02 and hence closer to the experimental value 2. A detailed investigation of the effect for all possible conformational changes is under way and will be published soon.

The total value of internal viscosity factor φ/f_0 is hence composed of two contributions. That originating from the energy barrier, φ'/f_0, is inversely proportional to the viscosity of the solvent which appears as a factor in f_0 because φ' is almost independent of η_s although it is certainly modified a little by the interaction of the macromolecule with the solvent. The contribution caused by the lateral displacement of chain atoms as a consequence of the conformational changes, φ''/f_0, is independent on η_s because φ'' is proportional to η_s in exactly the same manner as f_0. One hence has

$$\varphi/f_0 = \varphi'/f_0 + \varphi''/f_0 = A/\eta_s + \gamma - 1 \qquad [10]$$

with the limiting value $\gamma - 1$ at high η_s and a rapid increase as $1/\eta_s$ at small η_s. The former effect was observed by *Schrag* and coworkers (17, 18) who found $\varphi/f_0 \sim 2$ in the whole range of η_s between 2 and 70 P. The latter effect was prevailing in the streaming birefringence measurements of *Leray* (8) yielding $\varphi' \sim 10^{-5}$ g sec^{-1}. With 10 chain atoms in the link and an average hydrodynamic radius 2 Å for the chain atom, which seems to a reasonable value for PS, one obtains A $= 2$ cm^{-1} g sec^{-1}. That would yield a contribution 1 at $\eta_s = 2$ and 0.03 at $\eta_s = 70$ P so that the value φ/f_0 would drop from 3 to 2.03 in this viscosity range of the solvent. Since no such effect was observed one feels compelled to assume as smaller value for φ', e.g., 10^{-6} g sec^{-1} as suggested by *Janeschitz-Kriegl* (14). With such a value the variation of φ/f_0 would be from 2.1 to 2 which is perfectly compatible with the data of *Schrag* and coworkers.

Table 1. Coordinates and shortest displacements of chain elements (fig. 4) in multiples of $b\sin\beta/2$

1	(1, 0, 0.707)	1′	(1, 0, 0.707)	11′ = 0	1″	(1.134, −0.281, 0.368)	11″ = 0.461
2	(2, 0, 0)	2′	(2, 0, 0)	22′ = 0	2″	(1.847, −0.463, −0.611)	22″ = 0.781
3	(2, −1, −0.707)	3′	(3, 0, 0.707)	33′ = 2	3″	(2.981, −0.744, −0.243)	33″ = 1.115
4	(3, −1, −1.414)	4′	(4, 0, 0)	44′ = 2	4″	(3.695, −0.926, −1.222)	44″ = 0.725
5	(4, −1, −0.707)	5′	(5, 0, 0.707)	55′ = 2	5″	(4.828, −1.207, −0.854)	55″ = 0.866

Conclusions

The experimentally observed proportionality between the coefficient of internal viscosity of PS in Aroclor with the solvent viscosity η_s, at least in the range of η_s between 2 and 70 P and the molecular weight independence of the limiting intrinsic viscosity at very high frequency impose very specific conditions on the molecular model of internal viscosity. In purely mathematical terms the molecular weight independence of $[\eta]_\infty$ requires the proportionality of the eigenvalues φ_p of the internal viscosity to the index p and not to a higher power of p. That excludes the concept that the internal viscosity force is describable by a force proportional and parallel to the elastic link vector. Such a force would simply mean the introduction of a friction element parallel to the *Hooke*an spring representing the link in the conventional ideally flexible necklace model. The internal viscosity instead reflects the resistance against deformation as described by the eigenmodes. The resistance to each eigenmode is inversely proportional to the number of ways by which such a deformation can occur. Since simultaneous conformational changes can occur with equal probability at any chain atom the number of chain atoms between subsequent modes, $P/P \sim Z/P$, is inversely proportional to $\varphi_p = \varphi_p/Z$.

The constancy of internal viscosity parameter φ at low solvent viscosity and its proportionality to the solvent viscosity at high values of it requires a new term not considered in the original formulation of *Kuhn* and *Cerf*. The conventional term independent of the viscosity of the solvent takes into account the energy barriers separating the possible conformations at each chemical bond which oppose chain deformation by a viscous type resistance. Another term proportional to the viscosity of the solvent can be derived from the excess displacement of chain elements required at any link deformation as a consequence of conformational limitations by valency angles and rotational potential energy of the covalent bonds. The former contribution prevails at low and the latter at high viscosity of the solvent.

At high viscosity of the solvent the main contribution to internal viscosity is coming not from energy barriers between adjacent permitted chain conformations but from the excess friction resistance of the chain during a simultaneous conformational change involved in a deformation of the subchain represented by the elastic link between subsequent beads of the necklace model. The frictional resistance f_0 of the bead takes into consideration only the pure translation of the subchain. As a consequence of the conformational limitations of the molecular chain, however, the change of the link length involves not only the translation of the chain elements as showing up in the bead translation but also some transverse translation with a corresponding friction work performed against the solvent which is not considered by the conventional friction coefficient f_0 as it appears in the constitutive diffusion equation of the ideally flexible necklace model. This excess contribution proportional to the viscosity of the solvent needs a new term in the diffusion equation, i.e., the internal viscosity term.

Summary

Intrinsic viscosity measurement in highly viscous solvents at high frequency of the shear flow have demonstrated the molecular weight independence of limiting intrinsic viscosity and the constancy of the ratio of internal and solvent viscosity. The former fact proves the adequacy of the formulation of the internal viscosity concept in the normal coordinate space. The latter observation requires the introduction in the molecular model of a new term derived from the excess displacement of chain elements inherent to any deformation of the necklace. As a consequence of conformational limitations by the valency angles and rotational potential energy any deformation of the necklace as described by the displacement of the beads requires displacement of some elements of the real chain in a direction perpendicular to the deformation considered. The resulting frictional forces proportional to the viscosity of the solvent are not included in the frictional coefficient of the beads and hence have to be considered separately as a contribution to internal viscosity.

Zusammenfassung

Messungen des *Staudinger*schen Viskositätsindex in hochviskosen Lösungsmitteln im hochfrequenten Schergefälle haben die Unabhängigkeit des Grenzindex vom Molekulargewicht und die Konstanz des Verhältnisses zwischen der Innen- und Lösungsmittelviskosität erwiesen. Nach dem ersten Befund erweist sich die Formulierung der Innenviskosität im Normalkoordinatenraum als ausreichend. Das zweite Ergebnis verlangt die Einführung eines neuen Termes in der Innenviskosität, der von der Überschußverschiebung der Kettenelemente herrührt. Als Folge der Konformationseinschränkung durch die Valenzwinkel und Rotationspotentialenergie verlangt eine jede Deformation des Perlschnurmodels, wie sie durch die Perlverschiebung beschrieben wird, die Verschiebung von einigen Gliedern der wirklichen Molekülkette in einer dazu senkrechten Richtung. Die sich daraus ergebenden zusätzlichen Reibungskräfte sind proportional der Lösungsmittelviskosität. Sie sind jedoch nicht in dem Reibungskoeffizienten der Perlen einbegriffen und müssen deshalb gesondert als Beitrag zur Innenviskosität erfaßt werden.

References

1) *Kuhn, W.* and *H. Kuhn*, Helv. Chim. Acta **29**, 609, 830 (1946).
2) *Čopič, M.*, J. Chim. Phys. **54**, 348 (1956).
3) *Peterlin, A.* and *M. Čopič*, J. Appl. Phys. **27**, 434 (1956).
4) *Ikeda, E.*, Phys. Soc. Japan **12**, 378 (1957).
5) *Zimm, B. Z.*, J. Chem. Phys. **24**, 269 (1956).
6) *Peterlin, A.*, J. Chem. Phys. **33**, 1799 (1960).
7) *Fixmann, M.*, J. Chem. Phys. **5**, 793 (1966).
8) *Leray, J.*, Compt. Rend. Acad. Sci. (Paris) **241**, 1741 (1955); J. Polymer Sci. **23**, 167 (1957).
9) *Cerf, R.*, Compt. Rend. Acad. Sci. (Paris) **230**, 81 (1950); J. Chim. Phys. **48**, 85 (1951).
10) *Tsvetkov, V. N.* and *V. P. Budtov*, Vysokomol. Soedin **6**, 1209 (1964).
11) *Cerf, R.*, J. Phys. & Radium **19**, 122 (1958).
12) *Cerf, R.*, Adv. Polymer Sci. **1**, 382 (1959).
13) *Chaffey, C.*, J. Chem. Phys. **63**, 1385 (1966).
14) *Janeschitz-Kriegl, H.*, Adv. Polymer Sci. **6**, 170 (1969).
15) *Philippoff, W.*, Trans. Soc. Rheol. **8**, 117 (1964).
16) *Ferry, J. D., L. A. Holmes, J. Lamb*, and *A. J. Matheson*, J. Chem. Phys. **70**, 1685 (1966).
17) *Massa, D. J., J. L. Schrag*, and *J. D. Ferry*, Macromol. **4**, 210 (1961).
18) *Osaki, K.* and *J. L. Schrag*, Polymer J. Japan **2**, 541 (1971).
19) *Peterlin, A.*, Kolloid-Z. u. Z. Polymere **209**, 181 (1966).
20) *Peterlin, A.*, J. Polymer Sci. A-2, **5**, 179 (1967).
21) *Peterlin, A.* and *C. Reinhold*, Trans. Soc. Rheol. **11**:1, 15 (1967).

Author's address:

Prof. *A. Peterlin*
Camille Dreyfus Laboratory, Research Triangle Institute
P.O. Box 12194,
Research Triangle Park N.C. 27709 (USA)

Rheol. Acta 12, 503–515 (1973)

From the Laboratory of Inorganic Chemistry, Eindhoven University of Technology, Eindhoven (The Netherlands)

Rheological properties of alkali borate glasses

By J. M. Stevels

With 25 figures

(Received October 27, 1972)

1. Introduction

The question has been discussed very often, whether glasses show a real *Newton*ian behaviour, in other words if there is a proportionality between a given shear stress τ (g cm^{-1} sec^{-2}) and the velocity gradient $p = dv_x/dy$ (sec^{-1}) for the corresponding viscous flow of the material. If the glass behaves in such a way, the ratio between the two quantities is called the (coefficient of) viscosity η, which is defined by the equation $\tau = \eta p$, η being usually expressed in poises (g cm^{-1} sec^{-1}). In many materials, however, this equation is not fulfilled.

Deviations from *Newton*ian behaviour can often be described by an equation of the type

$$\tau = \tau_0 + \eta p \qquad [1]$$

and then it is said, that the rheological behaviour of the material is of the *Bingham* type. In fact is means that viscous flow only starts when the shear stress surpasses a certain threshold or yield value.

During the last few years systematic investigations have been carried out in the Inorganic Chemistry Department of the Eindhoven University of Technology on the rheological properties of inorganic vitreous systems (1) and the purpose of this paper is to present a number of the results.

So far only borate glasses have been considered. These systems are of interest, since they can easily be prepared and their structures are fairly well-known. The networks of alkali borate glasses are composed of three different types of structural units, which are shown in a schematic way in fig. 1.

The relative amounts of these structural units as a function of the composition of the glass are also well-known and shown in fig. 2. The first question that can be asked is, if there is a relation between the rheological properties and the relative amounts of these structural units.

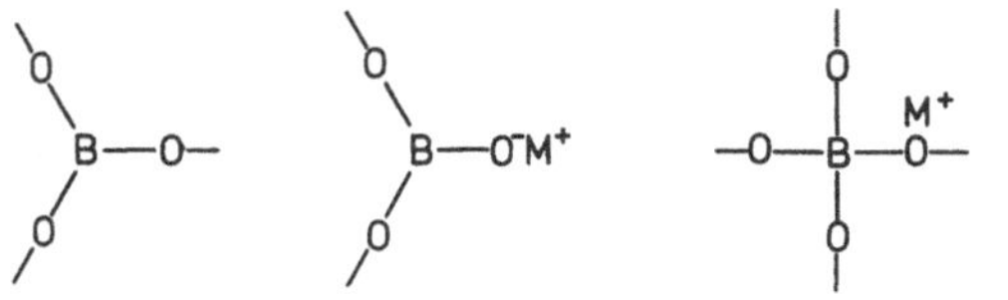

Fig. 1. Structural units in alkali borate glasses after *Beekenkamp* (1)

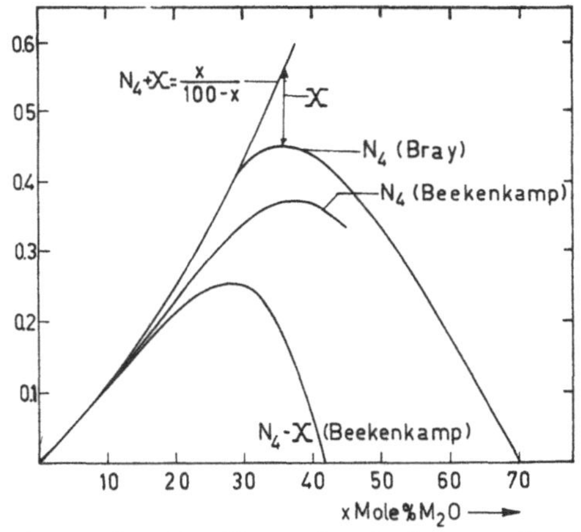

Fig. 2. Fraction N_4 of boron ions in four coordinations after *Bray* (2) and *Beekenkamp* (1). The curve $N_4 + X = \frac{x}{100 - x}$ is the limit of N_4, where X is the fraction of boron ions with a non-bridging oxygen ion

A second interesting question is if there is an influence of small amounts of water on the rheological properties of the glasses involved and if so, what conclusion can be drawn about the way in which OH groups or protons are taken up by the B–O network. Will water be taken up in a similar way as for instance the alkali oxides or in a different way?

A third question is whether *Newton*ian resp. *Bingham* behaviour of glasses is in direct relation with the presence or the absence of phase separation. Sometimes this idea is advocated in the literature (4), but our results seem to show, at least for the systems investigated, there is such no direct relation between the two. We shall not go into this matter in this paper, since this would lead us too far.

2. General remarks

At temperatures above the transition temperature inorganic glasses may sometimes exhibit delayed elastic deformation (5–9). *Taylor* (8) considered the elongation of a glass under stress as a summation of these effects (fig. 3), viz.

a) an instantaneous elastic deformation;

b) a time dependent delayed elastic deformation, and

c) a deformation, due to viscous flow, characterized by a linear time dependence.

Effects a) and b) are completely reversible on removal of the stress at time t', while the deformation caused by effect c) will remain, since viscous flow is an irreversible process.

To avoid long deformation times in order to obtain stationary viscous flow it is desirable to develop a method from which the stationary viscous flow can be determined, even when the delayed elasticity has not completely disappeared. For that purpose we are interested in the form of the curve representing the elongation versus the deformation time, when the stress is changed stepwise. In accordance with the results of *Taylor* an example of this kind of elongation is shown in fig. 4 for changes in the stress τ by a value $\Delta\tau$. If we determine from fig. 4 the corresponding elongation rate as a function of time, we obtain a curve as shown in fig. 5. We note that the stationary limit value of the elongation rate at a stress τ is approached from two sides:

i) from the higher side after a lower stress ($\tau - \Delta\tau \rightarrow \tau$), and

ii) from the lower side after a higher stress ($\tau + \Delta\tau \rightarrow \tau$).

The average of both velocity curves after some time may yield a good approximation of the stationary viscous flow at a stress τ.

3. Experimental

The viscosimeter employed was of the *Pochettino* type (12) described in more detail by *Stein* et al. (10) and *Cornelisse* at al. (11). It consists of two concentric cylinders, between which the glass to be examined is inserted by melting. A load is applied to the inner cylinder in the axial direction, while the outer cylinder is supported (fig. 6). The resulting axial motion of the inner cylinder is measured by an interferometrical method (10, 11). The lowest and highest viscosities that can be measured by this method are about 10^9 and 10^{14} poise.

If a flow behaviour of the *Bingham* type is considered, the mathematical relation between the load, the velocity of the inner cylinder, and the viscosity and the dimensions of the glass sample is:

$$\frac{(m - m_0)\,g}{2\pi h(R_2 - R_1)} \ln \frac{R_2}{R_1} = \eta \frac{v}{(R_2 - R_2)} \qquad [2]$$

where m is the sum of the masses of the inner cylinder and load, g the gravitation constant, h the height of the glass sample, R_2 and R_1 the inner radius of the outer cylinder and the radius of the inner cylinder respectively, η the viscosity of the glass, v the velocity of the inner cylinder and m_0 is related to the yield value. Flow behaviour of the *Bingham* type obeys the following relation

$$\tau - \tau_0 = p\eta \qquad [1]$$

where τ is a shear stress, τ_0 the yield value and p the velocity gradient. Comparing the two formulae gives:

$$\frac{(m - m_0)\,g}{2\pi h(R_2 - R_1)} \ln \frac{R_2}{R_1} = \tau - \tau_0 \qquad [3]$$

with m_0 related to the yield value as:

$$\tau_0 = \frac{m_0 g}{2\pi h(R_2 - R_1)} \ln \frac{R_2}{R_1} \qquad [4]$$

and

$$p = \frac{v}{(R_2 - R_1)}. \qquad [5]$$

4. Separation of delayed elasticity and viscous flow

The experiments consist in measuring the velocity gradient of the inner cylinder as a function of time at a constant temperature and constant stress. First of all, it is necessary to find a method to separate delayed elasticity and viscous flow. For the purpose the stress is applied after the glass has been stabilized for a long time at a stress $(\tau - \Delta\tau) < \tau$ and alternatively at a stress $(\tau + \Delta\tau)$

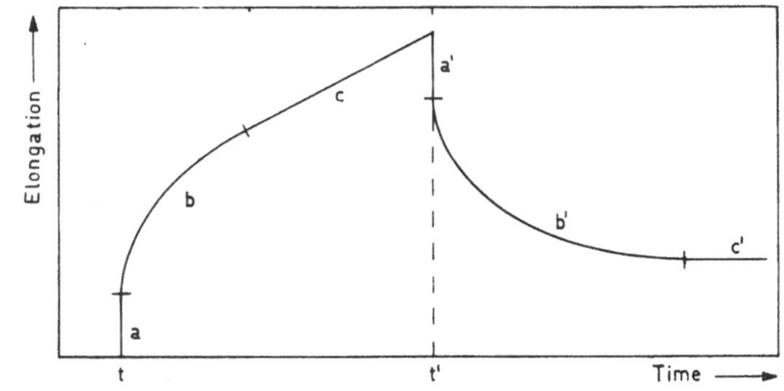

Fig. 3. Typical flow curve (for explanation see text)

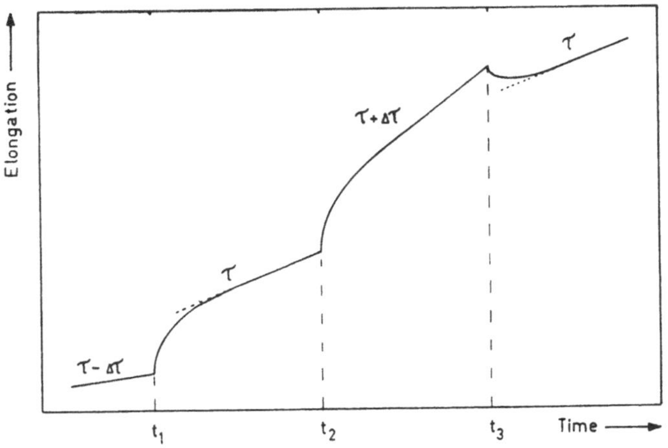

Fig. 4. Flow curve of changes in the stress τ by an amount $+\Delta\tau$ or $-\Delta\tau$.

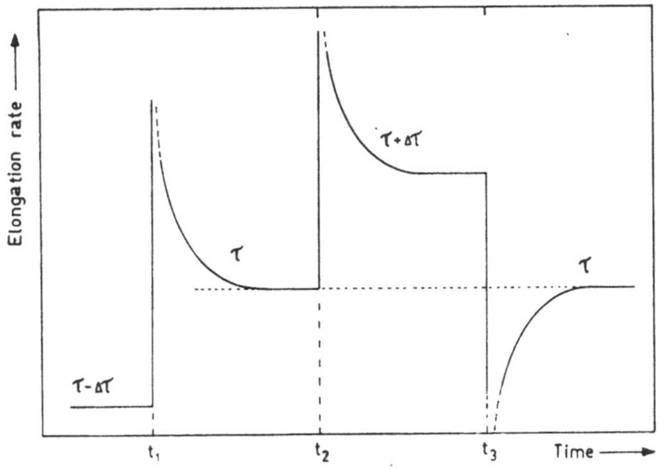

Fig. 5. Elongation rate corresponding with the case shown in fig. 4.

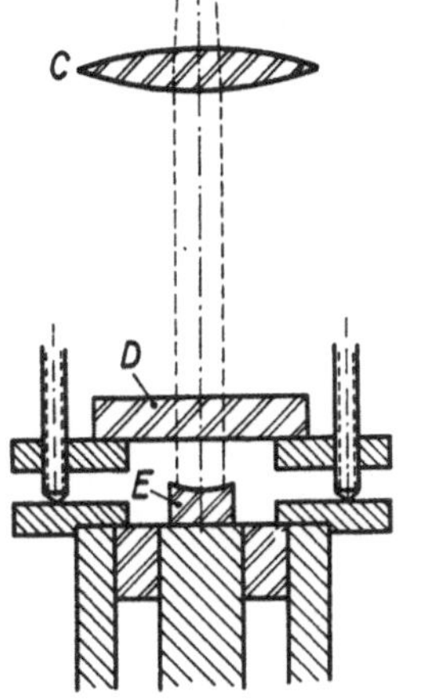

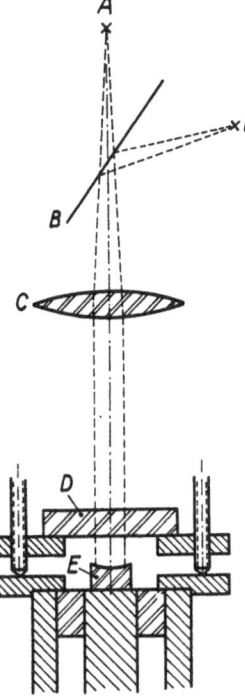

Fig. 6. Schematic representation of interferometric measuring device. A: light source ($\lambda = 543.1$ nm), B: half reflecting mirror, C: lense, D and E: quartz glass bodies effecting interference, and F: observer

$> \tau$ (the value of Δt is approx. $\frac{1}{2}\tau$ to $\frac{1}{4}\tau$) for vitreous B_2O_3, a lithium and a cesium borate glass only, since it was expected that the other alkali borate glasses would show a delayed elastic deformation of intermediate magnitude. Figs. 7 and 8 show the velocity gradients for these glasses at viscosities of about 10^{10} and 10^{12} poise, respectively. The average value of the two curves at one temperature and stress may indeed be taken as a good approximation of the velocity gradient at stationary viscous flow.

To support this conclusion further, more measurements were made in the viscosity region of 10^{12} poise in order to obtain information on the influence of the delayed elasticity on the determination of the rheological properties (η and τ_0). For the calculation of η and τ_0 it is necessary to determine at a constant temperature the velocity gradient versus time curves for different stresses. We restricted the experiments to two different stresses. Fig. 9 shows the results for a lithium and a cesium borate glass. From each pair of curves at one temperature η and yield value as a function of time at each temperature can be calculated from the average curves and the corresponding stresses. The values obtained are shown in fig. 10; the viscosity appears to be hardly dependent on time, while the yield value shows variations within the error expected (0.7×10^4 g cm^{-1} sec^{-2}), excluding the points at very low deformation times (some 25 min), where the results are less reliable. Summarizing

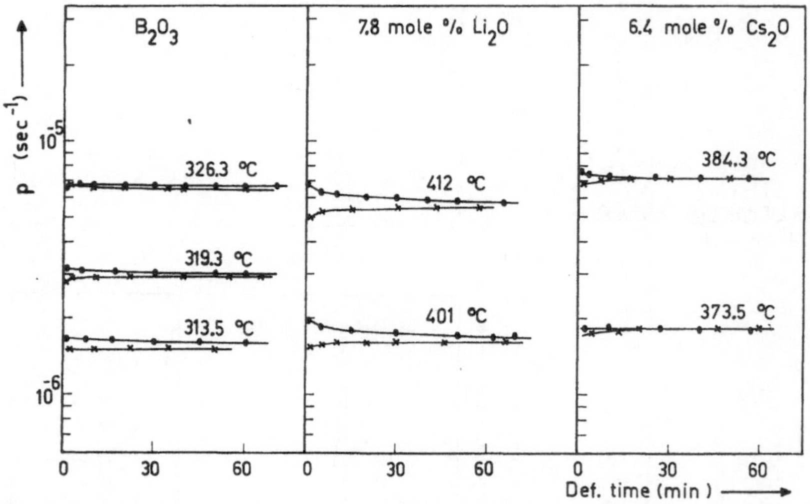

Fig. 7. Velocity gradient p versus time curves at various temperatures (● after a lower stress, and ∗ after a higher stress) for vitreous B_2O_3 and alkali borate glasses. Stress is about 3×10^4 g · cm^{-1} sec^{-2}

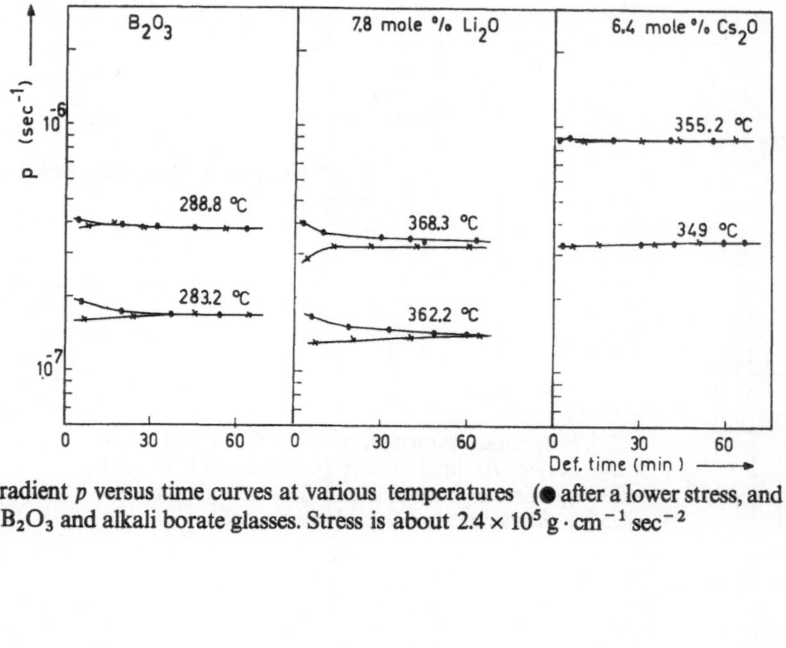

Fig. 8. Velocity gradient p versus time curves at various temperatures (● after a lower stress, and ∗ after a higher stress) for vitreous B_2O_3 and alkali borate glasses. Stress is about 2.4×10^5 g · cm^{-1} sec^{-2}

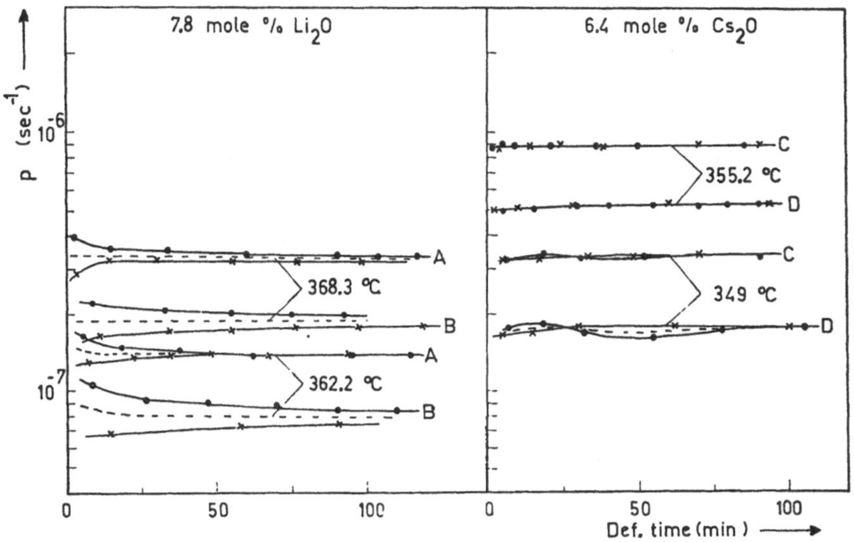

Fig. 9. Velocity gradient p versus time curves at various temperatures (● after a lower stress, and ∗ after a higher stress) for vitreous B_2O_3 and alkali borate glasses. Stress for A: 2.22×10^5 g · cm^{-1} sec^{-2}, for B: 1.36×10^5 g · cm^{-1} sec^{-2}; for C: 2.48×10^5 g · cm^{-1} sec^{-2}; for D: 1.53×10^5 g · cm^{-1} sec^{-2}

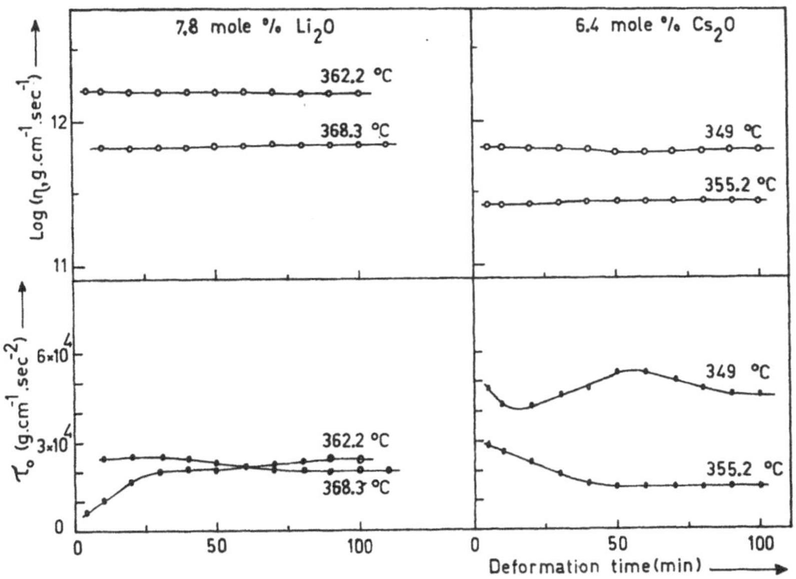

Fig. 10. Viscosity-time and yield value-time relations calculated from the average curves in fig. 9

we can say that to avoid long deformation times for obtaining stationary viscous flow, the measuring procedure is:

a) measuring the velocity gradient versus time curves after applying a lower and a higher stress alternatively, and

b) determining their average.

Rheological properties calculated from these average curves appear to be almost independent of time. It may be inferred, therefore that the applied procedure separates the delayed elastic deformation satisfactorily from the viscous flow, and that the average velocity gradient curves are sufficiently reliable to draw conclusions about the rheological properties.

5. Results

5.1. Boric oxide glass

For vitreous B_2O_3 the velocity gradient as a function of the stress is represented by straight lines intersecting the stress axis approximately at the origin (fig. 11). The lines have been calculated by the method of least squares, from which

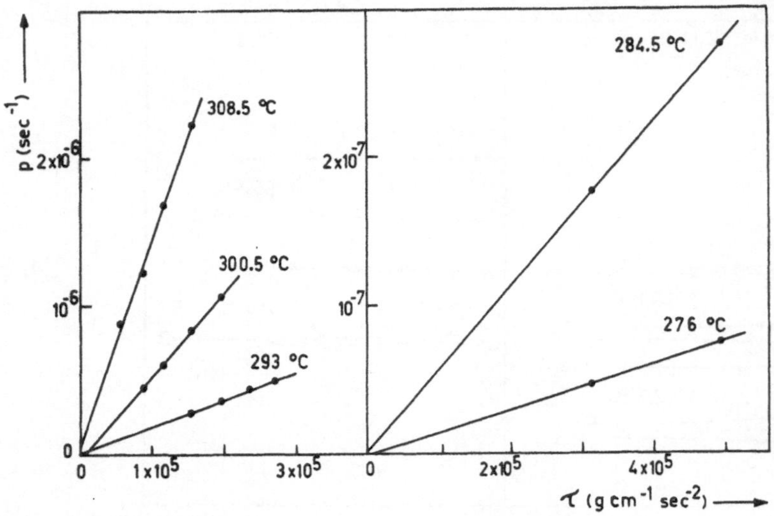

Fig. 11. Rheological behaviour of vitreous B_2O_3

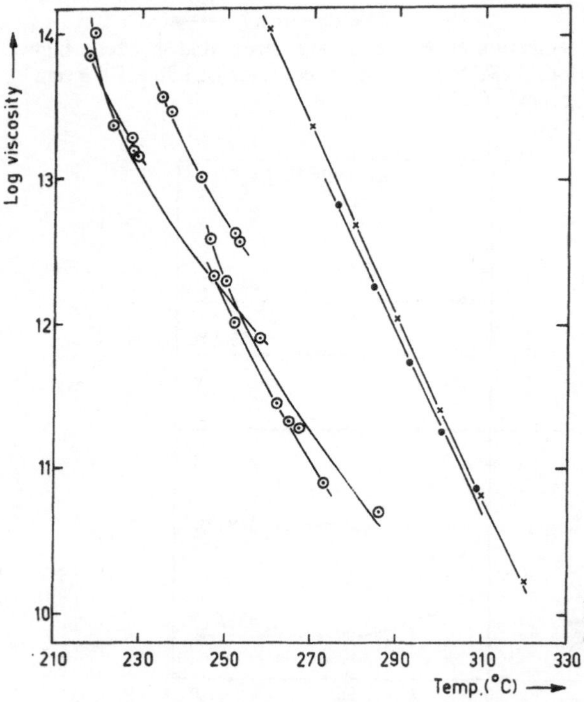

Fig. 12. Viscosity of vitreous B_2O_3: ⊙ after *Cornelisse* et al. (11); × after *Macedo* and *Napolitano*; ● present work

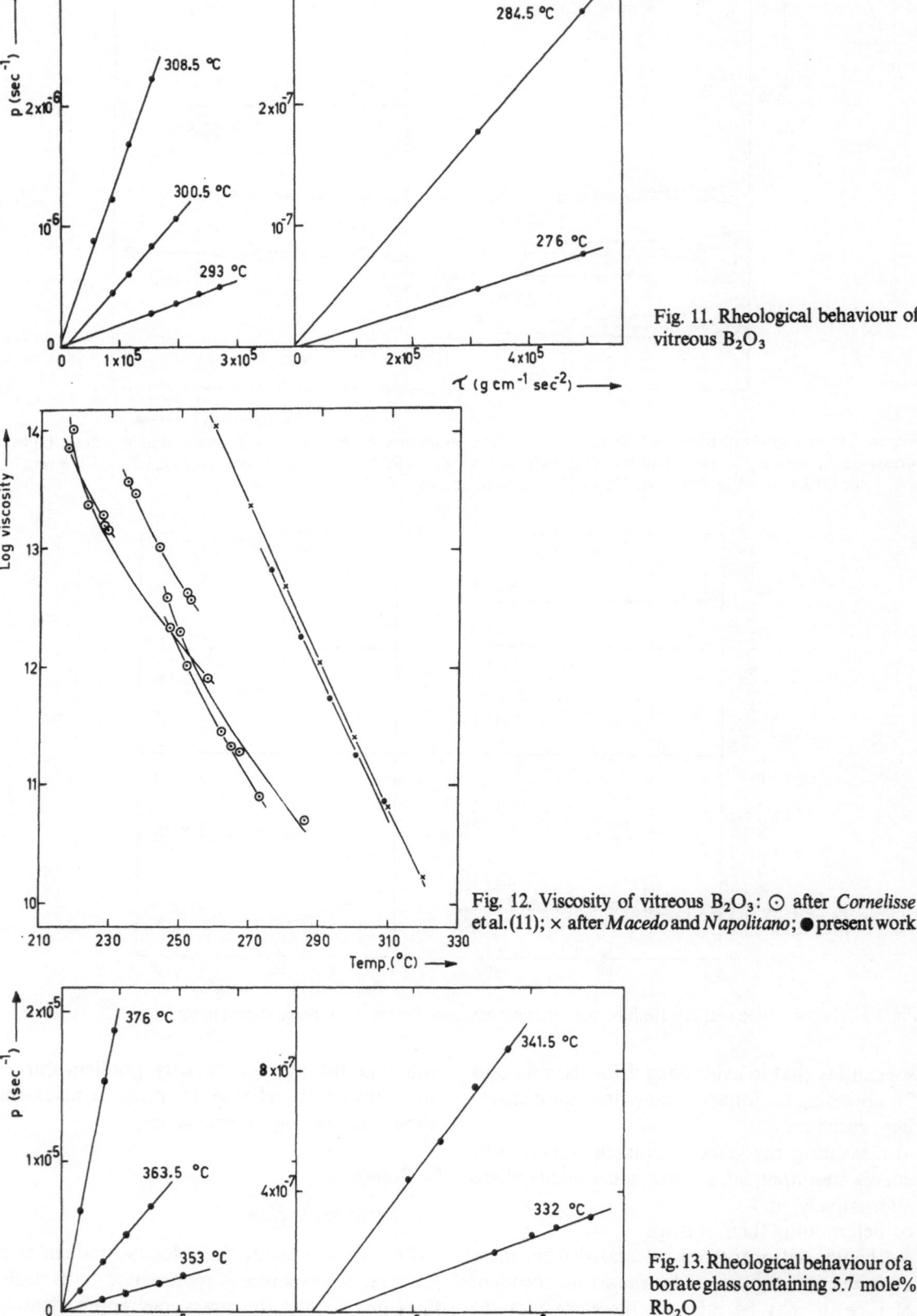

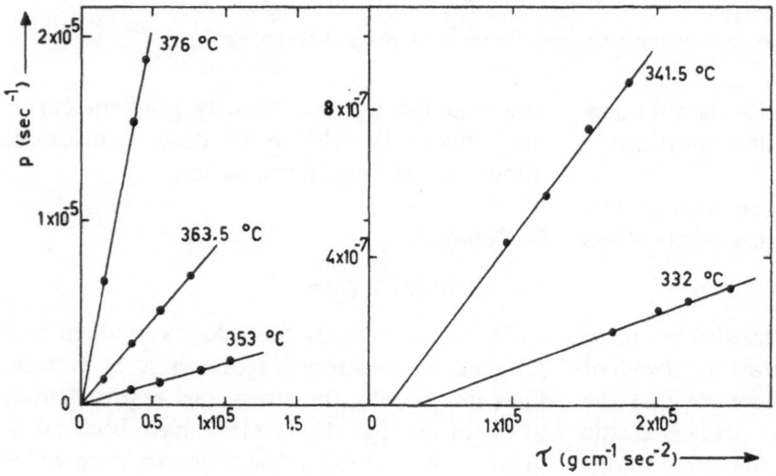

Fig. 13. Rheological behaviour of a borate glass containing 5.7 mole% Rb_2O

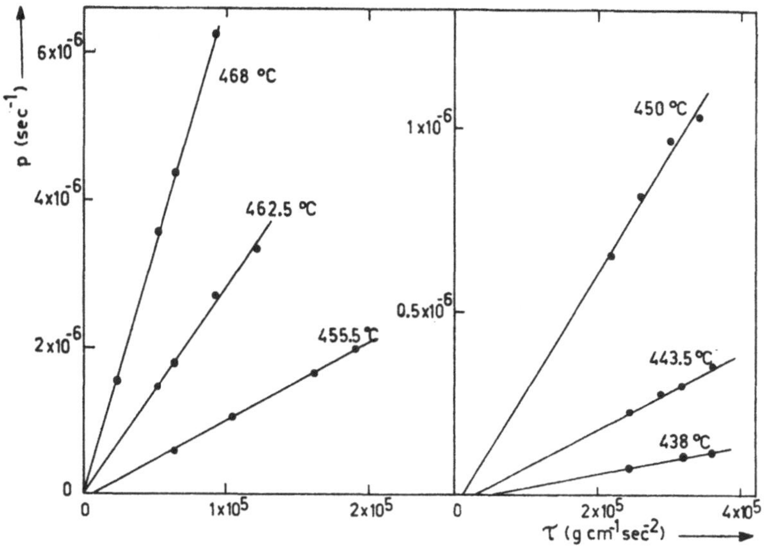

Fig. 14. Rheological behaviour of borate glass containing 23.8 mole% NaKO

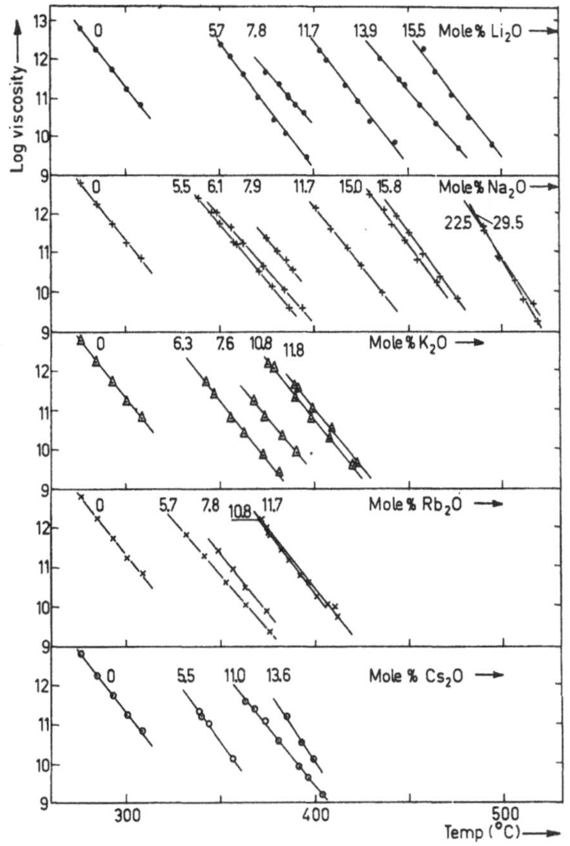

Fig. 15. Viscosities of alkali borate glasses as a function of temperature

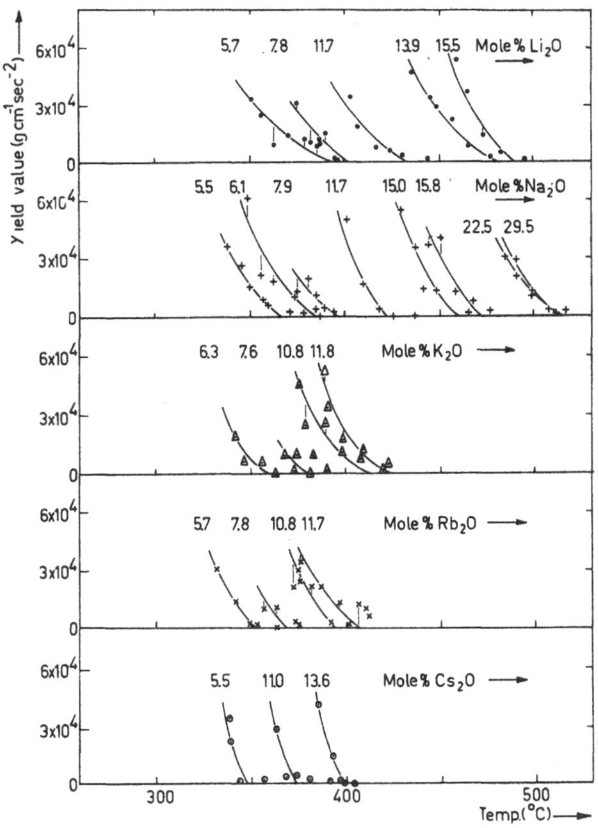

Fig. 16. Yield values of alkali borate glasses as a function of temperature

values for the viscosity and the yield value have been obtained. From the positive as well as negative deviations from *Newton*ian flow up to a viscosity of $10^{12.8}$ poise, it seems likely that they are not significant and only reflect the degree of inaccuracy of the measuring of the yield value.

In fig. 12 the viscosity of B_2O_3 glass is given as a function of the temperature as found by several investigators. The viscosity of boric oxide glass melted in vacuo (present work) is at a given temperature much higher than that obtained by *Cornelisse* et al. (11). However, the present values closely agree with the viscosity of B_2O_3 glass (bubbled with dry nitrogen at 1300 °C) as measured by *Macedo* and *Napolitano* (13). It is likely that the samples of *Cornelisse*, which were melted in a normal furnace at 1000 °C, still contain water, which reduces the viscosity considerably. The viscosity of boric oxide glass is found to obey the *Arrhenius* equation with an activation energy for viscous flow of 92 kcal/mole. According to *Macedo* and *Napolitano* this value is 94 kcal/mole.

5.2. Alkali borate glasses

The measurements in the alkali borate glasses can also be represented by straight lines, which pass through the origin of the velocity gradient versus stress diagrams at the higher temperatures, but when extrapolated intersect the stress axis at increasingly positive values with decreasing temperature. Figs. 13 and 14 represent some of the measurements obtained on the borate glasses examined. The viscosities and yield values calculated from the velocity gradient versus stress relations are all obtained by using the method of least squares. In figs. 15 and 16 the viscosities and the yield values, respectively, are given as a function of temperature for all the alkali borate glasses investigated. It is seen that the viscosity at constant temperature *increases with increasing alkali oxide content* and, at constant temperature and alkali oxide concentration, increases in the order *cesium to lithium*. The yield value appears only below a certain temperature and in all cases, increases with decreasing temperature.

The measurements in the sodium potassium borate glasses (figs. 17 and 18) give a more complete illustration of the relation between the rheological properties and the alkali oxide concentration up to 33 mole% NaKO. Fig. 18 gives some evidence that the change from *Newton*ian flow in the boric oxide glass into the *Bingham* behaviour as found *in the alkali borate glasses commences at concentrations lower than about 3mole% alkali oxide.*

6. Discussion

6.1. General remarks

It is generally accepted that the boron ions in boric oxide glass have a triangular coordination of oxygen ions (14, 15). In alkali borate glasses the excess oxygen ions introduced by the alkali oxide cause a certain number of boron ions to adopt a tetrahedral coordination or become non-bridging oxygen ions (2, 3, 16, 17). Therefore, 3 different types of structural units are pres-

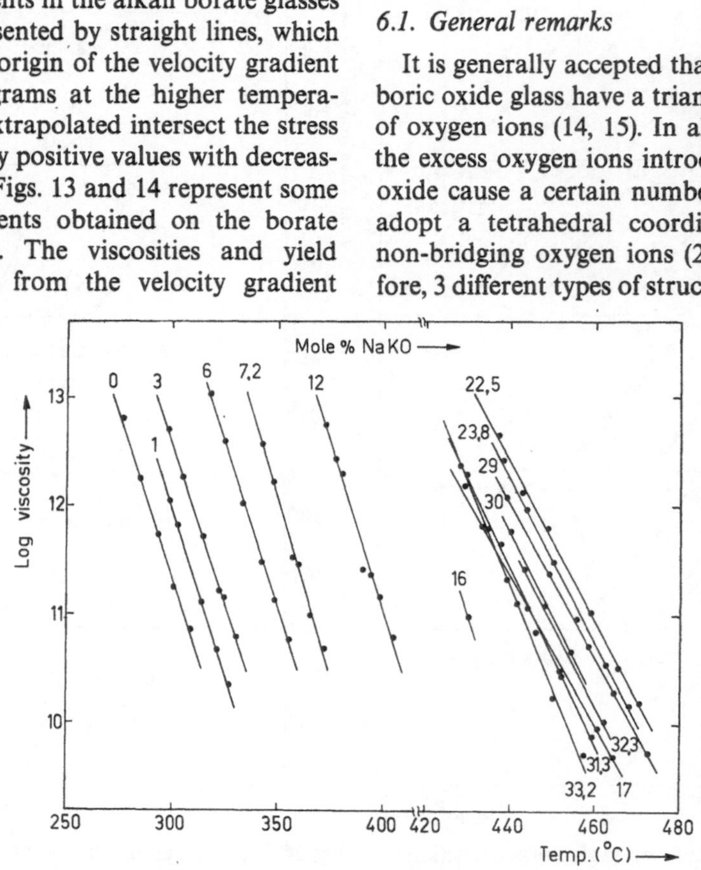

Fig. 17. Viscosities of NaKO borate glasses as a function of temperature

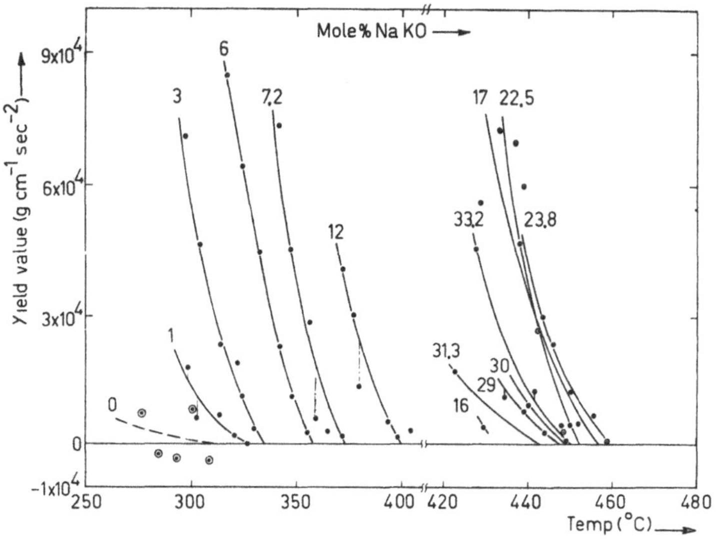

Fig. 18. Yield values of NaKO borate glasses as a function of temperature

ent in the B–O-network of borate glasses, viz.

a) BO_3 triangles with three bridging oxygen ions, further denoted BO_3 groups.

b) BO_3 triangles with two bridging and one non-bridging-oxygen ion, further denoted BO_3^- groups.

c) BO_4 tetrahedra with four bridging oxygen ions, further denoted BO_4^- groups.

In the glass literature, the relative amounts of these structural units are usually indicated by $(1 - N_4 - X)$, X and N_4. If the composition of the alkali borate glass is given by $xM_2O \cdot (100 - x) B_2O_3$, the following relation between N_4, X and x, must always exist:

$$N_4 + X = \frac{x}{1 - x}. \qquad [6]$$

Beekenkamp (3) and *Bray* (2) have shown by N.M.R. experiments that at room temperature, N_4 goes through a maximum as a function of x, as shown in fig. 2.

Thermodynamic considerations by *Beekenkamp*, have shown that the ratio $R = N_4/X$ is strongly temperature dependent. (One may consider an "equilibrium" between the BO_3^- groups and the BO_4^- groups; then the significance of R is a "reaction constant" which is given by

$$R = \exp\left(\frac{\Delta G}{kT}\right) \qquad [7]$$

in which ΔG is the difference in thermodynamic potential of the two different structural units. Since ΔG is positive R decreases with increasing temperature; in other words the amount by BO_3^- groups increases at higher temperatures at the cost of the amount of the BO_4^- groups. Since BO_4^- groups increase the rigidity (viscosity) of the network and BO_3^- groups decrease it. Isothermal changes of the viscosity of alkali borate glasses as a function of composition may be described by a quantity which is a function of N_4 and X:

$$\log\eta_{\text{alkali borate glass}} - \log\eta_{\text{boric oxide glass}}$$
$$= \Delta(\log\eta) = F(N_4, X, \ldots). \qquad [8]$$

A very simple function for this is:

$$\Delta(\log\eta) = B(N_4 - fX) \qquad [9]$$

where B and f are dimensionless constants. If the influence of one BO_4^- and one BO_3^- group is numerically equal, though opposed in sign, f will be equal to one.

$\Delta\log\eta$ is then proportional to $N_4 - X$. This expression is also shown as a function of x in fig. 2. The viscosity isotherms in fig. 19 may contribute to testing the function [8] for its validity. The strong increase at 320 °C will be mainly due to the formation of BO_4 groups. The assumption that no BO_3^- groups are formed in the low alkali oxide region ($x \sim 10$) enables the factor B to be calculated. It is found to be 50. The estimated smaller increase at 450 °C in the low alkali oxide region may be explained by some formation of BO_3^- groups at already small amounts of alkali oxide (see eq. [5]). It is not unlikely that one BO_3^- group will decrease the viscosity more than one BO_4^- group will increase it. This means that f

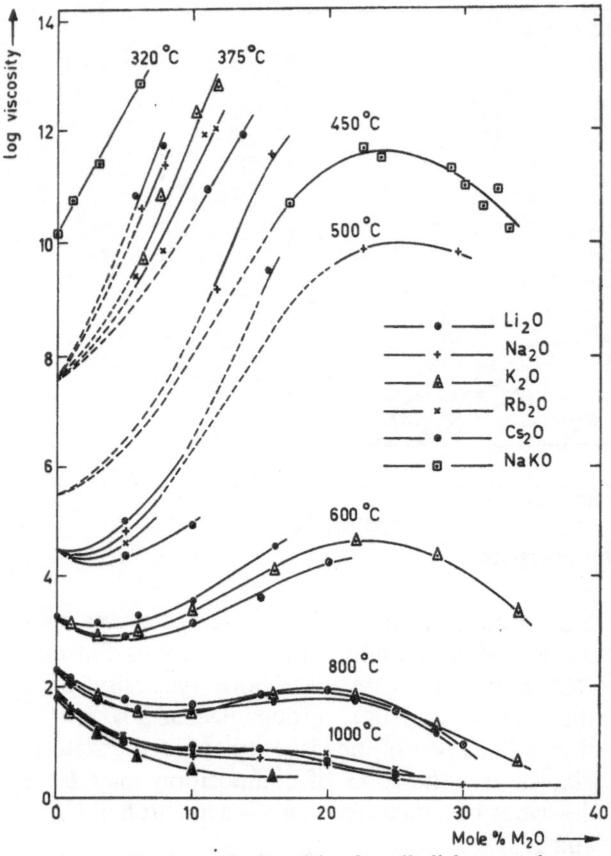

Fig. 19. Isothermal viscosities for alkali borate glasses after *Shartsis* et al. (17) and *Ti* et al.(4) ($\log \eta < 6$), and the present work ($\log \eta > 9$). The dashed parts of the curves are estimations

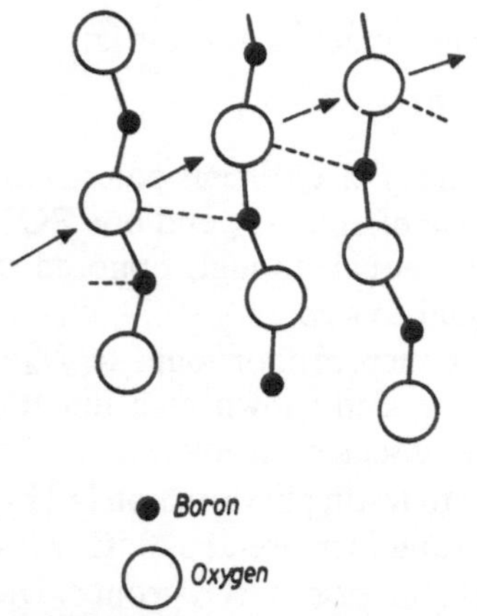

Fig. 20. Shear displacement in vitreous B_2O_3, effected through shifting in the network planes.

will be larger than one. This supposition finds some support in the fact that the maximum of the 450 °C isotherm lies at a lower composition ($x = 24$) than the one of the ($N_4 - X$) curve for room temperature ($x = 27$). This may result of a decrease in N_4 and an increase in X in consequence of the increase in temperature but, if the changes of N_4 and X as a function of temperature are only very small, the smaller increase may also be due to the fact that f is larger than one.

The isotherms at higher temperatures show only in a qualitative way that N_4 decreases and X increases with temperature. A preliminary conclusion may be that the function [9] is probably valid in the high viscosity region ($\eta > 10^{10}$ poise) and it may be assumed that B will be about 50 and f somewhat larger than one.

6.2. *Flow behaviour*

Boric oxide glass shows *Newton*ian flow for viscosities lower than 10^{13} poise, which is explained as follows. In boric oxide glass all BO_3 units are in approximately equally deep potential wells in the unloaded state. After applying a load the flow shows a purely viscous behaviour, implying a constant breaking and forming of B–O–B bonds between the flow units. Before as well as after the displacement the units are located in states of equally low energy. The local stress does not differ much from the average stress in the direction of flow and the required energetic situations will be readily realized by thermal fluctuations. This corresponds with *Newton*ian flow. This mechanism is visualized in fig. 20.

Alkali borate glass can show deviations from *Newton*ian flow. This can be explained as follows. Flow units bound to each other as in the boric oxide glass will show *Newton*ian flow. However, there will also be flow units which are bound to each other by BO_4 tetrahedra. The strong B–O bond in the tetrahedra, which has to be broken, will, however, experience an additional force in the direction of motion by stress concentration. If shearing takes place, a tetrahedrally surrounded boron ion will have to assume a triagonal coordination, and simultaneously another triagonally surrounded boron ion will have to assume a tetrahedral coordination. This leads to the occurrence of ions in highly excited energy states during such long times as cannot be accounted for by thermal fluctuations. This local stress concentration then provides the energy needed for motion. The occurrence of stress concentration leads to deviations from *Newton*ian behaviour (18) consistent with *Bingham* yield values. The mechanism is visualized in fig. 21.

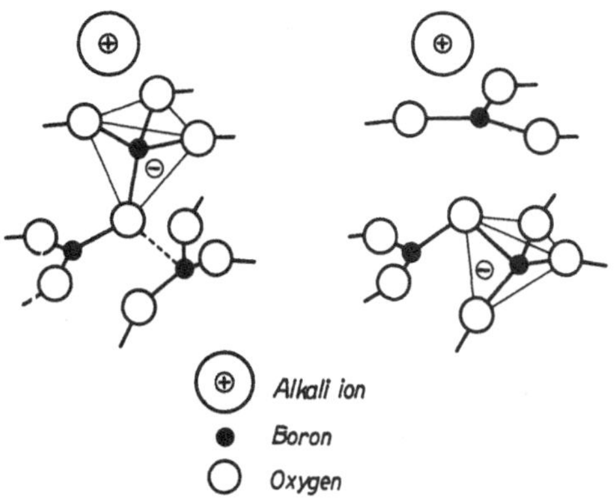

Fig. 21. Example of a situation in alkali borate glass, requiring stress concentration before displacement

7. Influence of water

In the course of our investigations we have been able to make samples containing small amounts of water. A reliable method for the analysis of these samples was developed (19), with these well-defined compositions the log viscosity and yield values could be measured as a function of temperature for vitreous B_2O_3 and two different alkali borate glasses. The results are shown in figs. 22 and 23.

7.1. Viscosity of water-containing borate glasses

Fig. 22 shows that at room temperature the introduction of H_2O has an effect different from that of the introduction of the alkali oxides. Whereas in the latter case BO_4^- groups are formed, giving rise to an increase of the viscosity, now the viscosity decreases, and the conclusions can only be that B–O–B bands are broken, B–OH groups formed and the coherence of the B–O network is diminished. This accounts for the considerable decrease in viscosity with small amounts of water (<0.3 mole%) as shown in fig. 24 (fig. 24 has been derived from fig. 22). However, it seems that a further increase of the water content (>0.5 mole%) gives rise to the

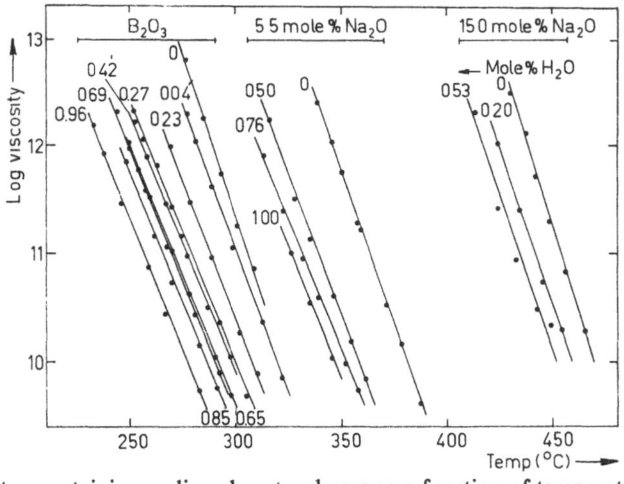

Fig. 22. Viscosities of water containing sodium borate glasses as a function of temperature

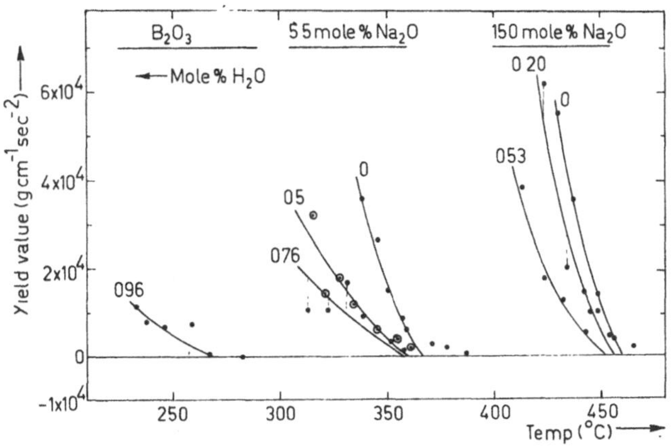

Fig. 23. Yield values of water containing sodium borate glasses as a function of temperature

formation of strong hydrogen bonds (20). These bonds oppose the decreasing coherence of the network and are probably responsible for the concave form of the viscosity curves shown in fig. 24. From the same figure it can be seen that at a constant water content (e.g. 0.5 mole%) the decrease in viscosity becomes smaller with increasing sodium oxide content. The increasing coherence and density of the network caused by the increasing amount of sodium oxide may account for this. The OH^- groups eventually give rise to still stronger hydrogen bonds, which oppose the decrease in viscosity. Strong hydrogen bonds will be formed likewise of non-bridging oxygen ions are present which associate with OH groups. This may be the case for the sodium borate glass with 15 mole% sodium oxide.

In section 6.1 we put forth a function which may describe the changes of the isothermal viscosity of boric oxide, as alkali oxide is introduced:

$$\Delta(\log\eta) = B(N_4 - fX) \qquad [9]$$

where N_4 and X are the fractions of BO_4^- groups and the fraction of BO_3^- groups, respectively, B and f are dimensionless constants. By assuming that a) water-containing borate glasses up to a content of 1 mole% H_2O X may be taken as the fraction of boron ions with an OH group, b) no non-bridging oxygens are formed when the sodium oxide concentration is lower than 15 mole%, and c) B has a value of 50 for the sodium borate system (5) it is possible to calculate f from fig. 24 as a function of the water concentration. The results are shown in fig. 25. It can be

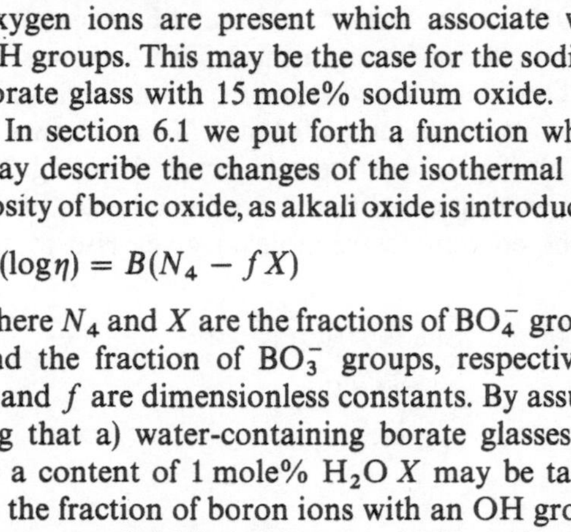

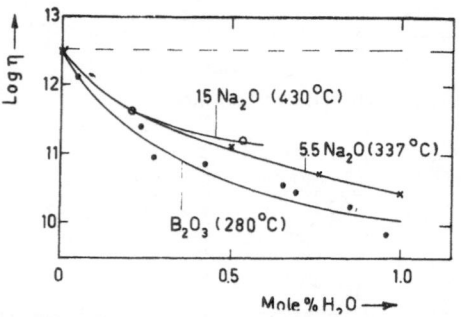

Fig. 24. Isothermal viscosities of sodium borate glasses at various temperatures as a function of the water concentration

seen that f approaches the value 5 at 1 mole% H_2O in the case of boric oxide. The value of f shows a slight dependence on the alkali oxide content, but this may be due to the increase in temperature.

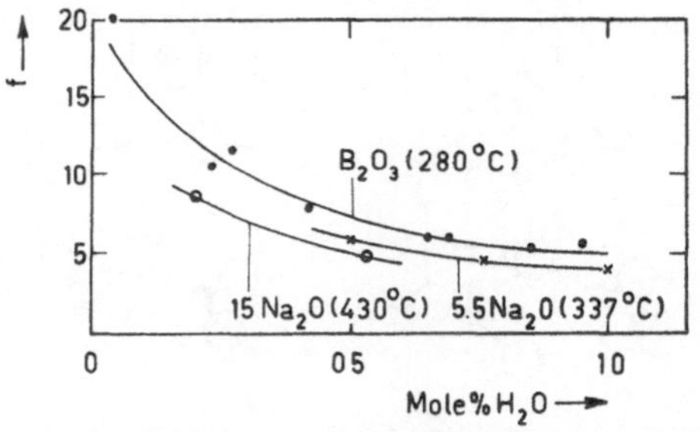

Fig. 25. Isothermal values of f for sodium borate glasses at various temperatures as a function of the water concentration

7.2. *Yield values for water-containing borate glasses*

The yield value in the sodium borate glasses also decreases with increasing water concentration (fig. 23) like the viscosity. This was to be expected, since it is likely that on account of the preferable flow along the OH groups less stress (18) will be required for viscous flow. However, it does not appear from the experiments that the yield value will disappear completely with further increase in the water content. An explanation for the appearance of the yield value in water-containing boric oxide glass at viscosities below 10^{13} poise may be that the strongest hydrogen bonds have caused the formation of BO_4 groups (21). If so, these groups will be responsible for the occurrence of local stress concentrations which differ from the average stress and so cause small yield values (18).

Summary

A survey is given of the recent work in the Department of Inorganic Chemistry of the Eindhoven University of Technology on the rheological properties of alkali borate glasses.

With the help of a specially adapted viscosimeter the rheological properties of systematically varied series of alkali borate glasses are measured. These include the coefficients of viscosity in the 10^9–10^{12} poise region and the corresponding yield values; they are discussed as a function of composition and of the water content of the samples.

Conclusions can be drawn about the structure of borate glasses in question and the manner in which the "water" is built into the network.

Whereas relative small amounts of alkali oxides built into the network of B_2O_3 give rise to boron atoms with a tetrahedral coordination, small amounts of water are taken up by the mechanism of breaking B–O–B bonds.

References

1) For many details Cf. *T. J. M. Visser*, Rheological properties of alkali borate glasses, Thesis Eindhoven, 1971.

2) *Bray, P. J.*, and *J. G. O'Keefe*, Phys. Chem. Glasses **4**, 37 (1963).
3) *Beekenkamp, P.*, in: Physics of Non-Crystalline Solids, Ed. *J. A. Prins*, p. 512 (Amsterdam 1965).
4) *Li, J. H.* and *D. R. Uhlmann*, J. Non-Crystalline Solids **3**, 127, 205 (1970).
5) *Lillie, H. R.*, J. Amer. Ceram. Soc. **16**, 619 (1933).
6) *Bair, G. J.*, J. Amer. Ceram. Soc. **19**, 347 (1936).
7) *Taylor, N. W., E. P. McNamara*, and *J. Sherman*, J. Soc. Glass Technol. **21**, 61 (1937).
8) *Taylor, N. W.* and *P. S. Dear*, J. Amer. Ceram. Soc. **20**, 296 (1937).
9) *Taylor, N. W.* and *R. F. Doran*, J. Amer. Ceram. Soc. **24**, 103 (1941).
10) *Stein, H. N.* and *J. M. Stevels*, in: 7th Intern. Congr. on Glass, paper 25, Brussels 1965.
11) *Cornelisse, Ch. W., T. J. M. Visser, H. N. Stein*, and *J. M. Stevels*, J. Non-Crystalline Solids **1**, 150 (1969).
12) *Pochettino, A.*, Nuovo Limento **8**, 77 (1914).
13) *Macedo, P. B.* and *A. Napolitano*, J. Chem. Phys. **49**, 1887 (1968).
14) *Boow, J.*, Phys. Chem. Glasses **8**, 45 (1967).
15) *Krogh-Moe, J.*, J. Non-Crystalline Solids **1**, 269 (1969).
16) *Krogh-Moe, J.*, Phys. Chem. Glasses **3**, 1 (1962).
17) *Shartsis, L., W. Capps*, and *S. Sprinner*, J. Amer. Ceram. Soc. **36**, 319 (1953).
18) *Stein, H. N., Ch. W. Cornelisse*, and *J. M. Stevels*, J. Non-Crystalline Solids **7**, 395 (1972).
19) *Visser, T. J. M.* and *J. M. Stevels*, J. Non-Crystalline Solids **1**, 347 (1969).
20) *Scholze, H.*, Glastech. Ber. **32**, 278 (1959).
21) *Uhlmann, D. R.* and *R. R. Shaw*, J. Non-Crystalline Solids **1**, 347 (1969).

Author's address:

Dr. *J. M. Stevels*
Laboratory of Inorganic Chemistry, Eindhoven
University of Technology, P. O. Box 513
Eindhoven (The Netherlands)

Rheol. Acta 12, 516–518 (1973)

From the Bundesanstalt für Materialprüfung (BAM) (Berlin, Federal Republic of Germany)

Documentation and information on rheology: Present state and development in its design and application

By K. Kirschke

With 1 table

(Received October 27, 1972)

The field of rheology and its boundaries

The edition of a documentation service is preconditioned by careful considerations as to a concise definition of the field to be covered. This is necessary to be able to decide which publications can be regarded as belonging to this specific field, where its boundary runs, and how a subdivision should be arranged for documentation purposes.

In the following the term "rheology" shall be subjected to close scrutiny, and scientific statements made about its importance. Rheology is the science which seeks to measure, describe, explain, and apply the phenomena of plastic deformation and flow occurring in bodies on being deformed.

According to *M. Reiner*[1]) rheology as a branch of science is closely related to mechanics as one of the physical disciplines. Classical mechanics on the one hand, and rheology on the other, pursue, however, essentially different aims. The theories of mechanics deal, above all, with statical and dynamical phenomena, while the reaction of materials is assumed to be ideal. Rheology, however, concentrates on materials and on the specification of their deformation and flow properties. Accordingly, interest is centered on the rheological constitutive equation, and processes are viewed above all with regard to properties of materials that could be derived. In the field of applied science the point of view is a different one, i.e. the properties of materials are viewed with respect to processes. This accounts for the importance of rheology to, for instance, chemical engineering. Here the influence of the properties of materials must be known with respect to processing as well as to the design of processing devices, transport equipment and the like.

As to materials, there are no restrictions to rheology. But the regular phenomena of water, air and solids are left to special disciplines as hydromechanics, aerodynamics and the theory of elasticity. Together with materials science rheology has gained great importance chiefly because new materials have been developed that have, due to technical progress, to endure greater stress. Moreover, rheology has gained access to other fields, as biology and medicine. As it turned out, a sufficiently detailed description and explanation of the deformation and flow properties is necessary for highly diverse materials.

[1]) *M. Reiner*, Rheologie, Carl Hanser Verlag, München 1968.

1. Available documentation services on rheology

Scientists concerned with rheology are offered two documentation services: one by the Deutsche Rheologische Gesellschaft (German Society of Rheology) and one by the British Society of Rheology. Both documentation services have been available for a considerable time, and they have been improved in the course of years. They are arranged in such a way as to make them suited for use on an international scale. The documentation by the Deutsche Rheologische Gesellschaft fairly completely covers the entire rheological literature from all over the world in the form of title references, whereas the British documentation called "Rheology Abstracts" consists of a collection of carefully prepared rheological abstracts. The societies have thus taken fundamentally different courses, which can be regarded as complementary.

2. The "Documentation Rheology" as edited by the Deutsche Rheologische Gesellschaft (German Society of Rheology)

Due to the fact that the Rheology Abstracts have been issued for a considerable time and have achieved a wide international circulation they can be regarded as internationally well-known. Actually this is not yet the case with the "Documentation Rheology" by the German Society of Rheology and therefore, more detailed information on it shall be given below. The German documentation has been published in the form of books, as part of the "Berichte der Deutschen Rheologischen Gesellschaft" (Reports of the German Society of Rheology). They are numbered consecutively and designated with the year of publication, with one or two numbers appearing every year. No. 37, 1971, is the latest issue and

Table 1. Subdivision of subject of the "Documentation Rheology" as edited by the German Society of Rheology

1. Fundamentals of rheology and rheology of materials
1.1. General rheology
  1.11. Bibliographies
  1.12. Conference proceedings, books, monographs
  1.13. Single papers and articles
1.2. Measuring methods (rheometry)
  1.21. Gases, liquids, multiphase liquids, colloids, fats, waxes, bituminous materials
  1.22. Solids
  1.23. Granular materials
1.3. Gases, vapors, plasmas
1.4. Homogeneous liquids and liquid solutions
  1.41. Inorganic melts
  1.42. Other inorganic liquids
  1.43. Polymer melts (incl. silicone oils)
  1.44. Polymer solutions
  1.45. Other organic liquids (incl. liquid crystals)
1.5. Homogeneous organic solids
  1.51. Simple organic solids
  1.52. Polymers (plastics and rubber)
  1.53. Natural and man-made fibres, films, foils
1.6. Homogeneous inorganic solids
  1.61. Metals
  1.62 Non-metallic materials (cf. 1.73)
  1.63. Glass
  1.64. Whisker, fibres, films, foils
1.7. Heterogeneous aggregates and materials of special type and use
  1.71. Multiphase liquids, colloids
  1.72. Granular materials, powders, soils
  1.73. Agglomerates, coal, rocks, cement, concrete, ceramics
  1.74. Composites
  1.75. Cellular materials
  1.76. Wood, paper, textile materials
  1.77. Materials of special type and use
    1.771. Pharmaceutical products, cosmetics
    1.772. Biological substances, blood
    1.773. Foods
    1.774. Fats, waxes
    1.775. Bituminous materials
  1.78. Miscellaneous
1.8. Interfaces

2. Rheology of processing
2.1. General
2.2. Forming
  2.21. Metals
  2.22. Polymers (plastics and rubber)
  2.23. Glass, ceramics, building materials
2.3. Cutting, machining
2.4. Adhesive bonding, coating
2.5. Flow and transport processes
2.6. Stirring, mixing
2.7. Miscellaneous

3. Rheology of friction and lubrication
3.1. Friction
3.2. Lubrication
3.3. Lubricants

It is desirable that, at home and abroad, as many scientists as possible should benefit from the considerable amount of work which is invested in a documentation service.

The German Society of Rheology, therefore, offers its documentation service, at a small charge, to all interested institutes, companies and individuals.

shall be referred to as an example. On the first 65 pages a survey on rheology symposia and other rheological activities is given, subdivided into countries. Added to this is information of general interest. The second part of the book – the main part – comprises 623 pages and covers the literature of rheology, classified into subject groups. On an average the number of references amounts to 2000–3000 per number. From No. 38 onwards, further subdivisions will be added to the present classification into subject groups. A carefully worked-out classification scheme covering the entire field under consideration is an essential prerequisite for a good documentation and will guarantee a smooth classification and retrieval of papers.

The classification scheme as from No. 38 is given below.

The "Documentation Rheology" may be ordered directly from the Deutsche Rheologische Gesellschaft, 1 Berlin 45, Unter den Eichen 87. Each number covers all subject groups and contains current rheological literature. The subscription fee for the "Documentation Rheology" (including cost of delivery) p.a. is:

Europe 40,— DM
overseas 14,— $

Members of the Deutsche Rheologische Gesellschaft pay, including membership fees, in

Europe 30,— DM
overseas 11,— $

Back numbers may be ordered if desired.

The "Documentation Rheology" provides reliable and continuous information on all publications of relevant current technical literature from all over the world. The documentation relieves the individual scientist of the pains of taking the precaution constantly to sort out and register the literature referring to his field of interest. He can concentrate on reading fewer journals to be informed about the current state of knowledge and can, when occasion arises, quickly select for evaluation purposes all those publications he deems relevant. It is the aim of the documentation to facilitate and shorten hereby the work of all scientists concerned with rheology.

The way titles are quoted makes the documentation suited for international use. Wherever the title is non-English, an English translation is given. In case complete English translations of, e.g. Russian or Japanese publications are available, it is indicated where they were published. Moreover, it is stated where abstracts in English can be found.

As the fields of rheology and tribology (wear, friction and lubrication) overlap to a certain degree, attention shall be drawn to the edition of another documentation service dealing with tribology. Both documentation services are arranged under similar aspects, yet they exist completely independent of each other.

Summary

As the number of rheological and other scientific publications increases considerably year by year, documentation services should be used more extensively. The Deutsche Rheologische Gesellschaft (German Society of Rheology), 1 Berlin 45, Unter den Eichen 87, publishes at regular intervals and for an international reader audience the "Documentation Rheology" which contains a fairly complete collection of relevant literature subdivided into subject groups. The papers listed cover fundamentals of rheology, rheometry, rheological behaviour of polymer melts and solutions, plastics and rubber, metals, pastes, colloids and suspensions, solids, ceramics, concrete, pharmaceutical products and cosmetics, food stuffs, chemical engineering, forming processes, flow processes, mixing processes and so on.

The documentation is available at a small charge. It has been published to make better use of the possibilities that lie in a systematic collection, arrangement, exploitation and application of findings achieved and published so far. Methodical and intensified research and development, as well as further rationalization and cost-saving measures in the field of practical application depend on a better utilization of existing findings. The above-mentioned documentation service covers basic research as well as technical application. With the aid of it literature search on general or more specific problems of research and practice can be done quicker and carried out much more easily. Literature search done this way guarantees that a research project will start from the current state of knowledge and that sufficient hints to solutions of practical problems are at hand. Moreover, the preparation and publication of literature reports will initiate a systematic evaluation of the numerous existing publications.

Author's address:

Prof. Dr. *K. Kirschke*
Bundesanstalt für Materialprüfung (BAM)
Fachgruppe 5-2
D-1000 Berlin 45
Unter den Eichen 87

Rheol. Acta 12, 519–523 (1973)

From the Shell Research Ltd., Thornton Research Centre, Chester (England) and Dept. of Pure and Applied Chemistry, The University of Salford (England)

Experience in making rheological measurements at high pressures

By G. D. Galvin, J. F. Hutton, B. Jones, H. Naylor, M. C. Phillips, G. Powell, and E. Wyn-Jones

With 3 figures

(Received October 27, 1972)

Introduction

In elastohydrodynamic lubrication (1) a film of liquid is entrained between two surfaces by rolling and is subjected to very high pressures and shear rates. Typical conditions in the film are:

pressure, $P = 1\,\mathrm{GN\,m^{-2}}$
$= 145{,}000\,\mathrm{lbf\,in^{-2}}$

shear rate, $\dot{\gamma} = 10^5\,\mathrm{s^{-1}}$

liquid transit time, $t_0 = 10^{-5}\,\mathrm{s}$.

The shear viscosity and shear modulus of non-polymeric lubricants at these high pressures may have values of the following magnitude (2, 3, 4, 5):

shear viscosity $\eta_s(P) = 10^5\,\mathrm{N\,s\,m^{-2}}$
$= 10^6$ poise

shear modulus $G_\infty(P) = 1\,\mathrm{GN\,m^{-2}}$.

The corresponding volume properties are of similar magnitude (6, 7):

bulk viscosity[1]), $\eta_b(P) \approx \eta_s(P)$

bulk modulus, $K_\infty(P) \approx G_\infty(P)$.

At atmospheric pressure the moduli are smaller by a factor of about three but the viscosities are several orders of magnitude smaller. The viscosity varies exponentially with pressure as follows (2):

$\eta_s(P) = \eta_s(0)\exp(\alpha P)$.

Typical values of the parameters are:

$n_s(0) = 10^{-1}\,\mathrm{N\,s\,m^{-2}} = 1$ poise

$\alpha^{-1} = 70\,\mathrm{MN\,m^{-2}}$.

[1]) Or volume viscosity.

Consider the behaviour of a liquid lubricant as it moves from a region at atmospheric pressure into the high-pressure region. Firstly, its volume decreases and viscosity increases. However, at high pressures volume retardation occurs with a time constant, τ_b, of the order of $\eta_b/K_\infty = 10^{-4}\,\mathrm{s}$. Since the liquid passes through the high-pressure region in a time, $t_0 \approx 10^{-5}\,\mathrm{s}$, it is obvious that volume retardation effects can be important and the shear and bulk viscosities may not reach the steady state values (8).

Secondly, the shear relaxation time, $\tau_s = \eta_s/G_\infty$, is of the order of the transit time, t_0, so that the shear viscosity will again be time dependent (9).

Thirdly, the shear stresses, which for a *Newtonian* liquid would equal $\eta_s(P)\dot{\gamma} \approx 10^{10}\,\mathrm{N\,m^{-2}}$, are of the same magnitude as the shear modulus, G_∞, so that large elastic strains will occur. Therefore, the shear stress can be expected to be non-linearly dependent on the shear rate; hence the viscosity will be shear-rate dependent.

Fourthly, the work done by the pressure rise and by the high shear forces induces a temperature rise in the liquid and this effect, like the three above, tends to reduce the apparent viscosity of the liquid. It is not surprising, therefore, that there have been reports (10, 11) of anomalously low viscosities and shear stresses in elastohydrodynamic systems.

We are attempting to investigate these four effects independently in our experiments. The types of rheological measurement that have been made are listed in the next section. The apparatus must be accurate and sensitive and sufficiently small to fit into the pressure vessel which has a cavity 25.5 mm in diameter and 240 mm long. The later sections deal with the difficulties encountered in the design and operation of the rheological apparatus. Problems of pressure generation and measurement are not discussed.

Methods of measurement

All the measurements can be made at pressures from atmospheric to about 900 MN m^{-2}.

At *low* shear rates the shear viscosity has been measured by the falling body method (2, 12) from 10^{-2} to 10^3 N s m^{-2} and 25–120 °C. A *Couette* viscometer is being used for shear viscosities from 10–10^5 N s m^{-2} and from −30 to 120 °C.

At *high* shear rates preliminary measurements are being made on another *Couette* viscometer designed to measure the shear-rate dependence of viscosity up to 10^6 s^{-1}: there is no provision for thermostating this apparatus.

Piezoelectric transducers are used to generate and detect shear (13) and longitudinal waves at frequencies from 6–78 MHz to measure the frequency dependence of the shear viscosity, bulk viscosity, shear elasticity and bulk elasticity. From these high-frequency measurements the time-dependent shear and bulk viscosities can be obtained (14). These measurements can be made at high pressure at temperatures from −30 to 120 °C, and also at atmospheric pressure down to −165 °C.

Static sealing

Both static and rotary seals have been used to maintain the high pressures. A rotary seal is a seal which acts on a rotating shaft. At the end of an experiment the seals must be broken to recover apparatus from inside the pressure vessel. It is desirable that the seals can be easily broken to avoid damage to delicate apparatus. For static sealing we have found that three rubber "O"-ring seals are extremely efficient and yet easily demountable (15). They are shown in fig. 1 on an end plug which incorporates the pressure inlet. However, at the highest temperatures (120 °C) and pressures, plastic flow of the rubber "O"-rings occurs, whilst at low temperatures (−30 °C) and high pressures *elastoplastic* flow can occur beyond the glass transition of the rubber. This latter phenomenon is particularly interesting. When an "O"-ring seal is compressed in its housing, shear strains of the order of 5% are introduced. Since the shear modulus of the rubber is around 10^5 N m^{-2}, the shear stress is 5.10^3 N m^{-2}. This is less than the yield stress of about 10^5 N m^{-2} [2]) and consequently no flow occurs. However, under pressure the rubber may pass into the glassy state when its modulus will rise to 10^9 N m^{-2}. If the shear stress is then greater than the yield stress in the glassy state (16, 17), about 3.10^7 N m^{-2}, flow occurs. When the "O"-ring is removed from the pressure vessel the initially smooth ring can be seen to have developed ridges due to this flow, as shown in fig. 2. When the ring is heated to about 150 °C at atmospheric pressure the ridges disappear, indicating that the flow was elastoplastic and hence recoverable, rather than plastic and irrecoverable.

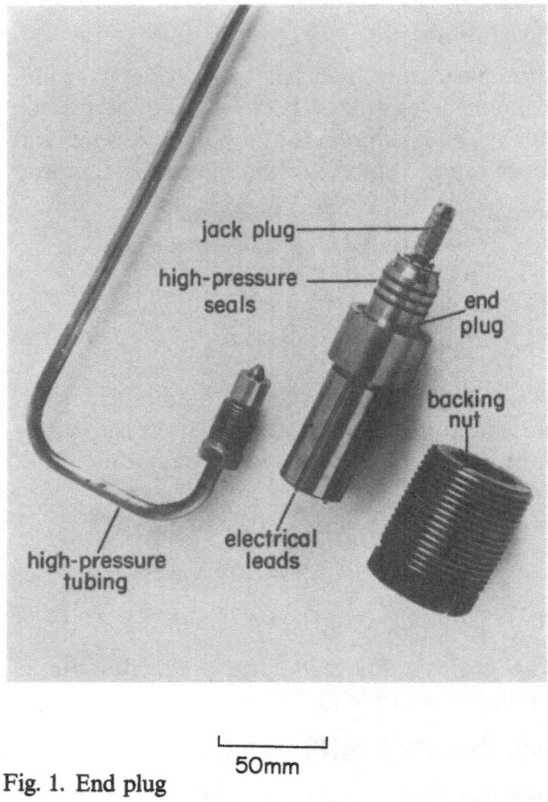

Fig. 1. End plug

Rotary seals

Morrison rotary seals (18) have been employed on the high shear rate viscometer and run at speeds up to 600 rev/min. These seals are a development of a simpler type of seal in which the unsupported area principle (19) is used to transmit an increased pressure to a rubber packing ring. The packing ring is forced directly onto the shaft and therefore the friction is high. In an

[2]) From this value and the modulus the yield strain is about 100%, in agreement with common experience.

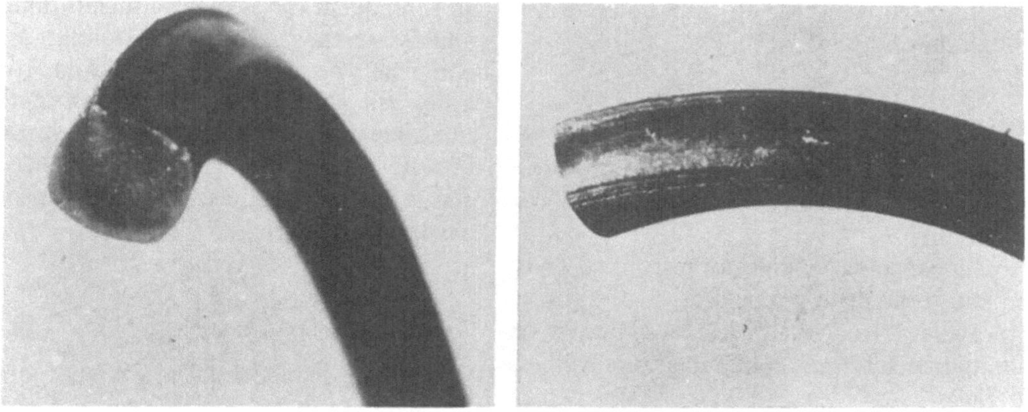

After elastoplastic flow at high pressures

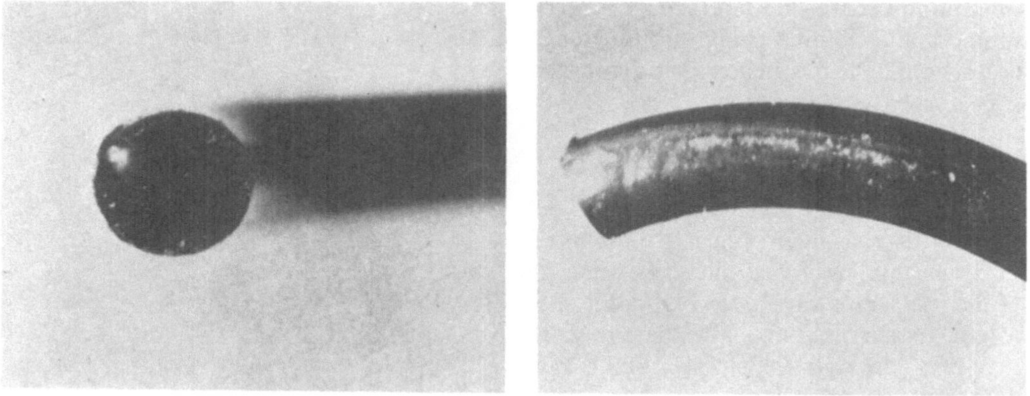

After subsequent heating, recovery occurs

Fig. 2. Elastoplastic flow of rubber "O" seals

attempt to reduce the friction *Morrison* introduced a metal bush into the seal design. The bush is positioned between the shaft and the packing ring. The clamping force is now partly opposed by the pressure of fluid leaking down the annulus between the shaft and bush. The seal is efficient up to 150 MN m^{-2} at which pressure it seizes. By adjusting the size of rubber packing other pressure ranges could be attained.

A simpler type of rotary seal has been developed for the low shear rate *Couette* viscometer and the acoustic absorptiometer described later. The seal consists of a shaft in a close-fitting hole of radius r. Very slow leakage of pressurizing fluid occurs along the annular clearance, h, since it is only 5 μm across but is 60 mm in length, L. The leakage rage, Q, is throttled by the exponential increase of viscosity with pressure, and for an incompressible liquid is:

$$Q = \pi r h^3 (1 - e^{-\alpha P})/6 \alpha L \eta_s(0).$$

The leakage thus reaches a maximum constant value at pressures of the order of α^{-1}, and for the fluid used is about 10^{-2} $cm^3\,s^{-1}$ at room temperature. At higher pressures the annulus, h, is reduced by the action of pressure on a sleeve that surrounds the shaft and projects inside the pressure vessel. This causes the leakage rate to drop to 10^{-5} $cm^3\,s^{-1}$ at 700 MN m^{-2}. Even under these conditions the shaft can be rotated by hand!

Electrical

Electrical leads are taken through each end plug as shown in fig. 1. They are supplied for use as thermocouples (20) and comprise pairs of wires embedded in a magnesia matrix in a stainless-steel tube of 1.5 mm diameter. The leads are dried to increase their leakage resistance to over 1 MΩ, and sealed at each end by epoxy resin. The stainless-steel tubes are bonded by epoxy resin or vacuum copper brazing into holes drilled through the end plug.

Separating the test and pressurizing fluids

It is useful to be able to separate the pressurizing fluid from the liquid under test for the following reasons:
a) to enable small volumes of test liquid to be used and to be readily replaced,
b) to avoid corrosive attack of test liquids on seals and metals and contamination by these materials,
c) to avoid degradation of polymeric materials by mechanical action in the pump.

Any container of the test liquid must be able to accommodate a volume change of up to 30% in the test liquid. Pistons, bellows and diaphragms have been used as separators.

Pressure changing

Transient pressure gradients can damage delicate apparatus, and hence must be minimized. Large transients are more likely to occur the faster the pressure is varied, the higher the shear and bulk viscosities, and the narrower the passageways through which the fluids must flow. Smooth increases in pressure require the use of a reciprocating pump which delivers only a small volume of liquid whilst, for smooth decreases in pressure, a sensitively controllable release tap is desirable. The convolutions in metal bellows are too narrow for the most viscous fluids studied in this work; hence pistons and flexible diaphragms are preferred.

Experimental technique

A dyed pressurizing fluid is always used so that leakage out of the system can be readily detected. A convenient test for leakage of pressurizing fluid into the test fluid can be made with a mock test fluid of high volatility such as isopentane (b.pt. 28 °C). This test fluid is evaporated after use to check for small quantities of the coloured pressurizing fluid.

A check on reversibility and the absence of hysteresis is made by performing experiments with increasing and decreasing pressure. Thus, if the pressure sequence used is $0 \rightarrow P_1 \rightarrow P_2 \rightarrow P_1 \rightarrow 0$ with $0 < P_1 < P_2$, and if the two experimental results at pressure P_1 agree, then there is no irreversibility between P_1 and P_2. In subsequent cycles of experiments different P_1 and P_2 are chosen so that reversibility over the complete pressure range is checked. Irreversibility can arise through leakage into the test fluid, creep of materials (e.g. the elastoplastic flow of seals mentioned earlier) or other ageing effects that may produce dimensional or physical changes in materials.

Acoustic absorptiometer

Many of the above ideas are incorporated in the acoustic absorptiometer which is shown in fig. 3. A rotational motion is given to the external

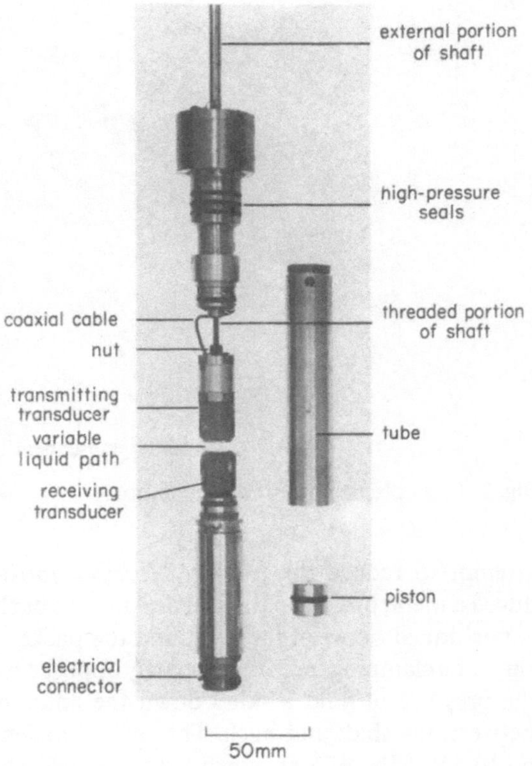

Fig. 3. Acoustic absorptiometer

portion of the shaft of the rotary seal. The part of the shaft inside the pressure vessel is threaded and carries a nut which is constrained to move along the shaft. The nut carries a piezoelectric transducer assembly. Sound waves are generated by the transducer, propagated through a length, l, of liquid and received at a second, fixed, transducer. From here the electrical signal, I, is taken to the electrical connector which fits over the jackplug, shown in fig. 1, when the apparatus is inserted in the pressure vessel.

The sound absorption coefficient, a, is obtained from:

$I \propto \exp(-al)$.

Differences in l are measured to 1 μm.

The sound absorption coefficient, a, arising from the frequency-dependent shear and volume viscosities is given by (16):

$$a = [2\pi^2 f^2/\varrho v(f)^3]\,[\eta_b(f) + (4/3)\,\eta_s(f)].$$

ϱ is the density and $v(f)$ is the sound velocity at frequency f, and both quantities are measured in separate experiments. Since $\eta_s(f)$ can be measured in a separate experiment using shear waves, $\eta_b(f)$ can be obtained.

Concluding remarks

In this paper we have described some of the special features of apparatus designed for high pressure rheological measurements on lubricating liquids. Many of the difficulties are mentioned as well as the procedures adopted for resolving them. Regarding the nature and accuracy of the results obtainable it is necessary to refer to other publications. Already, the low-shear viscosity by the falling body technique (2, 12), and the high-frequency shear modulus by the piezoelectric technique (5, 13) have been reported. Measurements by the other techniques have been made but not yet published.

As indicated in the introduction, a complete description of elastohydrodynamic lubrication requires a knowledge of many rheological parameters. Much work has already been reported by various authors on the more important of these parameters, such as the pressure coefficient of shear viscosity, but, for various reasons, information about the other parameters, such as the pressure coefficient of volume viscosity, is scanty. It is now possible, with the comprehensive range of apparatus mentioned in this paper, to evaluate lubricants completely for these properties.

Acknowledgements

This work was partly supported by an SRC studentship to G. P. and an SRC equipment grant to E. W-J.

Summary

We review the difficulties encountered in the design and operation of apparatus for rheological studies on liquids at pressures up to 900 MN m^{-2} and temperatures from -30 to 120 °C. Such rheological information is required in connection with elastohydrodynamic lubrication in which high pressures and shear rates are encountered.

References

1) *Dowson, D.* and *G. R. Higginson*, Elastohydrodynamic Lubrication: The Fundamentals of Roller and Gear Lubrication. (Oxford 1966.)

2) *Galvin, G. D., H. Naylor*, and *A. R. Wilson*, Proc. Inst. Mech. Engrs., **178** (Pt 3 N), 283 (1963-4).

3) *Slie, W. M.* and *W. M. Madigosky*, J. Chem. Phys., **48**, 2810 (1968).

4) *Barlow, A. J., G. Harrison, J. B. Irving, M. G. Kim, J. Lamb*, and *W. C. Pursley*, Proc. Roy. Soc. Lond., **A 327**, 403 (1972).

5) *Hutton, J. F.* and *M. C. Phillips*, Nature (in press).

6) *Dexter, A. R.* and *A. J. Matheson*, J. Chem. Phys., **54**, 3463 (1971).

7) *Barlow, A. J., J. Lamb*, and *N. S. Tasköprülü*, J. Acoust. Soc. Amer., **46**, 569 (1969).

8) *Harrison, G.* and *E. G. Trachman*, Trans. A.S.M.E., J. Lub. Tech. (in press).

9) *Dyson, A.*, Phil. Trans. Roy. Soc. Lond., **A 266**, 1 (1970).

10) *Johnson, K. L.* and *R. Cameron*, Proc. Inst. Mech. Engrs., **182** (1), 307 (1967-8).

11) *Plint, M. A.*, Proc. Inst. Mech. Engrs., **182** (1), 300 (1967-8).

12) *Galvin, G. D., B. Jones*, and *H. Naylor*, Proc. Inst. Mech. Engrs., **182** (36), 135 (1967-8).

13) *Hutton, J. F.* and *M. C. Phillips*, J. Chem. Phys., **51**, 1065 (1969).

14) *Herzfeld, K. F.* and *T. A. Litovitz*, Absorption and Dispersion of Ultrasonic Waves (New York 1959).

15) *Paterson, M. S.*, J. Sci. Instr., **39**, 173 (1962).

16) *Ueno, S., H. Yamazaki, T. Oue, K. Ito*, and *M. Tsutsui*, Trans. Soc. Rheol., **10**, 627 (1966).

17) *Bauwens, J. C., C. Bauwens-Crowet*, and *G. Homes*, J. Pol. Sci., A 2, **7**, 1745 (1969).

18) *Morrison, J. L. M., B. Crossland*, and *J. S. C. Parry*, Proc. Inst. Mech. Engrs., **170**, 21 (1956).

19) *Bridgman, P. W.*, The Physics of High Pressure (London 1958).

20) Ex. British Insulated Callender Cables Ltd.

Authors' addresses:

G. D. Galvin, J. F. Hutton, B. Jones, H. Naylor, and *M. C. Phillips*
Shell Research Ltd.
Thornton Research Centre
P.O. Box 1, Chester CH1 3SH (England)

G. Powell and *E. Wyn-Jones*
Dept. of Pure and Applied Chemistry
The University of Salford
Salford M5 4WT (England)

Rheol. Acta 12, 524–532 (1973)

From the Department of Applied Mathematics, University College of Wales, Aberystwyth (U.K.)

Nearly viscometric flows in the new rheometers

By D. G. Knight and K. Walters

With 4 figures

(Received October 27, 1972)

1. Introduction

In this paper, we consider the flow generated in rheometers which make use of a *steady* flow to determine the linear time-dependent behaviour of elastico-viscous liquids. The Orthogonal (1), Balance (2) and Eccentric-cylinder (3, 4) rheometers are perhaps the best known examples of this class of rheometers at the present time (see fig. 1). The basic idea is that the test fluid is contained between two instrument members, which rotate *with the same angular velocity* about axes which differ by a small linear displacement or a small angular displacement. The real and imaginary parts of the complex viscosity are then determined by measuring the components of the force or the components of the couple on one of the instrument members. Provided both instrument members rotate with the same angular velocity and the displacement between the axes of rotation is small enough, the fluid is subjected to a small-amplitude oscillatory shear and the results can be interpreted unambiguously in terms of the complex viscosity.

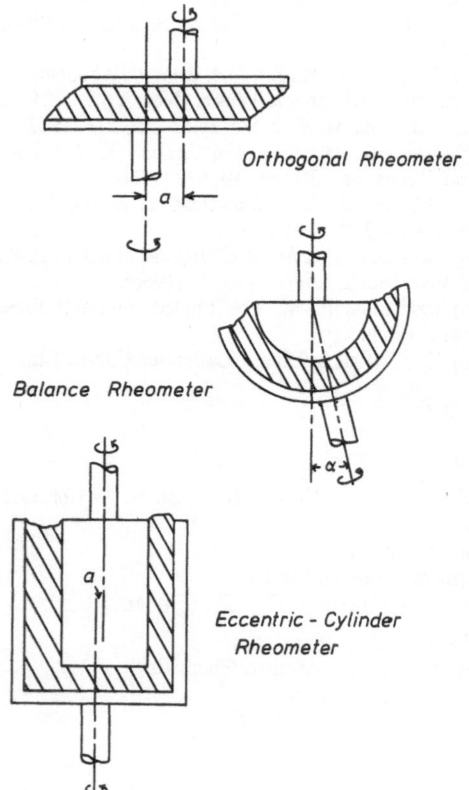

Fig. 1. The new rheometers

In the present paper, we give detailed theoretical and experimental consideration to the situation in which the instrument members rotate with *different* angular velocities. The work was motivated by the following considerations:

i) So far as we can determine, in all existing rheometers, commercial or otherwise, only one of the instrument members is driven mechanically and the other is driven by viscous forces. Bearing friction will therefore result in one of the members lagging the other, hopefully by a small amount. In a previous communication (4), we have conjectured that even a very small "lag" may have a significant effect on the interpretation of results in the case of elastico-viscous liquids (e.g. a lag of 1:500 can result in up to a 10% error in the interpretation of the experimental results). The present work will therefore throw light on the possible effects of "lag" and indicate how these effects may be accommodated in the new rheometers.

ii) When the instrument members are driven at different speeds in the Orthogonal and Eccentric-Cylinder rheometers, we have examples of the class of flows know as "nearly viscometric" (5), these being a type of combined steady and oscillatory shear. The more old-fashioned oscillatory instruments (e.g. the *Weissenberg* Rheogoniometer) have facilities for subjecting the test fluid to a combined steady and oscillatory shear of a *simple* type and this technique has been developed by numerous workers (6–10). Whether or not the new rheometers will supersede the more old fashioned instruments will therefore depend in part on the ease by which the rather complicated combined steady and oscillatory shear generated in the new rheometers will be amenable to mathematical analysis and on the predictive value that can be placed on such analyses.

2. The effect of lag on the interpretation of experimental results

The experiments described in this paper were carried out on the eccentric-cylinder rheometer adaptation to the *Weissenberg* Rheogoniometer described by *Broadbent* and *Walters* (4). This instrument consists essentially of a force-measuring cell, employing leaf springs and micrometer screw gauges. A null method is used to determine the forces on the inner cylinder in the eccentric-cylinder arrangement, the measured quantities at each rotational speed being the eccentricity "a" measured by means of a linear inductance transducer and the forces on the inner cylinder in the direction of the offset (the Y force) and perpendicular to this direction in a horizontal plane (the X force). If the inner and outer cylinders are rotating with the same angular velocity Ω and the offset a is sufficiently small, it can be shown that the relation between the forces X and Y and the dynamic viscosity η' and dynamic rigidity G' can be written in the form (4)

$$X = \beta \Omega \eta'$$
$$Y = \beta G' \qquad [1]$$

where β is a geometrical parameter involving the radii r_1 and r_2 of the cylinders ($r_2 > r_1$), the offset a and the length of column of fluid L.

In practice, the friction in the ball bearings associated with the free-running inner cylinder results in a small difference in the speeds of the two cylinders. In the experiments described in the present paper, the speed differential was increased by applying a "braking pad" to the inner cylinder. By this means, the angular velocity Ω_1 of the inner cylinder could be varied from zero to almost that of the outer cylinder Ω_2, the precise value of Ω_1 being determined by a simple photographic method.

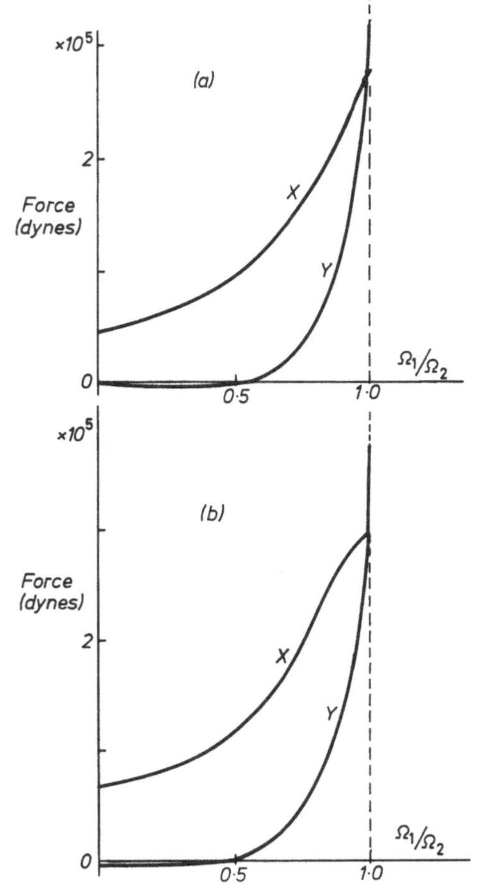

Fig. 2. Experimental results for a 5% aqueous solution of polyacrylamide. a) $\Omega_2 = 1.57\ \text{sec}^{-1}$, b) $\Omega_2 = 2.5\ \text{sec}^{-1}$. $r_1 = 2.65$ cm, $r_2 = 3$ cm, $L = 9.45$ cm, $a = 0.01$ cm

Fig. 2 contains experimental results for a 5% aqueous solution of polyacrylamide P 250. In the experiments, Ω_2 was fixed and the forces on the inner cylinder measured for various values of Ω_1. Of particular interest is the behaviour of the curves when $\Omega_2 - \Omega_1$ is small. Here, a small difference in the speeds of the cylinders is seen to make a significant difference to the forces measured, with the effect on the elastic force Y being greater than that on the viscous force X. This means that in existing instruments there will be a tendency to underestimate η' and G' due to inner-cylinder "lag", with the underestimation likely to be most severe in the case of G'. This supports the conclusion reached by *Broadbent* and *Walters* (4).

The above conclusions refer to the Eccentric-Cylinder Rheometer, but one would anticipate similar conclusions for the Orthogonal and Balance Rheometers.

If very accurate measurements are required with existing instruments which have no facilities for driving both instrument members, the present technique may supply them, since it is possible to measure X and Y for various small values of $\Omega_2 - \Omega_1$ and extrapolate to the case $\Omega_1 = \Omega_2$.

3. Combined steady and oscillatory shear

There are at least four commercial rheometers which make use of the "steady-flow" principle to determine the complex viscosity of elastico-viscous liquids. With this in mind, it is reasonable to seek further relevant flow situations which can be generated by simple modifications to these rheometers. Allowing both instrument members to rotate with different angular velocities would appear to be an obvious choice, especially in view of the fact that the resulting situations are likely to lead to types of combined steady and oscillatory shear flow, such flows being popular at the present time in characterization studies. In the present paper, we explore the feasibility of this suggestion for the three most popular rheometer geometries.

i) The Balance Rheometer

In the Balance Rheometer (see fig. 1), the test fluid is contained between two concentric spheres (hemispheres) which rotate with the same angular velocity about two axes which pass through the centre of the spheres, the angle between the axes being small. η' and G' are determined by measuring the components of the force or the components of the couple on the inner sphere (hemisphere) [see, for example (11, 12)]. If the angular velocities of the inner and outer spheres are now taken to be Ω_1 and Ω_2 respectively, with $\Omega_1 \neq \Omega_2$, the flow can be considered to be made up of a basic flow between spheres (hemispheres) rotating about the same diameter with different angular velocities, together with a perturbation to this flow brought about by tilting one of the axes of rotation through a small angle α. Unfortunately, the basic flow is not viscometric and its characterization is far from trivial [see, for example (13)]. The solution to the combined basic and perturbed flow is therefore sure to be extremely complicated and even if such a solution were forthcoming, it would certainly be of little predictive value. It must be concluded, therefore, that there will be little benefit in generalizing Balance Rheometer flow by allowing the instrument members to rotate with different angular velocities.

ii) The Orthogonal Rheometer

In the *Maxwell* Orthogonal Rheometer (see fig. 1), the test fluid is contained between two parallel plates which rotate with the same angular velocity about axes which are normal to the plates but not coincident. The distance "a" between the axes is small. The two components of the tangential force on one of the plates can be used to determine η' and G' [see, for example (1, 14)]. We now consider the situation when the plates rotate with *different* angular velocities.

We take a cylindrical polar coordinate system with the z axis along the bisection of the two axes of rotation. The plate at $z = 0$ rotates with angular velocity Ω_1 and that at $z = h$ with angular velocity Ω_2. The boundary conditions may therefore be taken to be

$$v_{(r)} = -\frac{\Omega_1 a}{2}\cos\theta, \quad v_{(\theta)} = \Omega_1\left[r + \frac{a}{2}\sin\theta\right]$$

$$v_{(z)} = 0 \quad \text{on} \quad z = 0$$

$$v_{(r)} = \frac{\Omega_2 a}{2}\cos\theta, \quad v_{(\theta)} = \Omega_2\left[r - \frac{a}{2}\sin\theta\right]$$

$$v_{(z)} = 0 \quad \text{on} \quad z = h \qquad [2]$$

where $v_{(r)}$, $v_{(\theta)}$ and $v_{(z)}$ are the physical components of the velocity vector. In the following, we shall work to first order in a.

In view of [2], we are led to take

$$v_{(r)} = a u(r, z)\, e^{i\theta}$$

$$v_{(\theta)} = r\Omega(z) + a v(r, z)\, e^{i\theta}$$

$$v_{(z)} = a w(r, z)\, e^{i\theta} \qquad [3]$$

where u, v and w may be complex and the real part is implied. The boundary conditions now imply

$$\Omega = \Omega_1 \quad \text{on} \quad z = 0$$

$$\Omega = \Omega_2 \quad \text{on} \quad z = h \qquad [4]$$

$$u = -\frac{\Omega_1}{2}, \quad v = -\frac{i\Omega_1}{2}$$

$$w = 0 \quad \text{on} \quad z = 0$$

$$u = \frac{\Omega_2}{2}, \quad v = \frac{i\Omega_2}{2}$$

$$w = 0 \quad \text{on} \quad z = h. \qquad [5]$$

We shall take as our deformation variable the right *Cauchy-Green* tensor C_{ik} which is essentially given by [cf. (15)]

$$C_{ik} = \frac{\partial x'^m}{\partial x^i} \frac{\partial x'^s}{\partial x^k} g_{ms}(\underline{x}') - g_{ik}(\underline{x}) \qquad [6]$$

where g_{ik} is the metric tensor of the cylindrical polar coordinate system and x'^i (which we shall write as r', θ', z') is the position at time t' of the element that is instantaneously at the point x^i at time t. The displacement functions corresponding to [3] can be shown to be

$$r' = r + \frac{iau}{\Omega} e^{i\theta}[1 - e^{-i\Omega(t-t')}]$$

$$\theta' = \theta - \Omega(z)(t - t') + \frac{iav}{r\Omega} e^{i\theta}\left[1 - e^{-i\Omega(t-t')}\right]$$

$$+ \frac{aw}{\Omega}\frac{d\Omega}{dz} e^{i\theta}$$

$$\times \left[-i(t - t') + \frac{1}{\Omega}[1 - e^{-i\Omega(t-t')}]\right]$$

$$z' = z + \frac{iaw}{\Omega} e^{i\theta}[1 - e^{-i\Omega(t-t')}]. \qquad [7]$$

Substituting [3] into the equation of continuity, given by

$$\frac{\partial}{\partial r}(r v_{(r)}) + \frac{\partial}{\partial \theta}(v_{(\theta)}) + \frac{\partial}{\partial z}(r v_{(z)}) = 0 \qquad [8]$$

we obtain

$$\frac{\partial}{\partial r}(ru) + iv + r\frac{\partial w}{\partial z} = 0. \qquad [9]$$

The boundary conditions [5] would strongly suggest that we try

$$u = u(z)$$

$$v = v(z) \qquad [10]$$

and from [9] we take

$$w = 0 \qquad [11]$$

to avoid singularities. Substituting [10] and [11] into [6] and [7], we obtain[1])

$$C_{(rz)} = \Delta C_{(rz)}$$

$$C_{(\theta z)} = -r\Omega' s + \Delta C_{(\theta z)}$$

$$C_{(zz)} = r^2(\Omega')^2 s^2 + \Delta C_{(zz)}$$

other

$$C_{(ik)} = 0 \qquad [12]$$

where $s = (t - t')$ and the dash refers to differentiation with respect to z. The perturbation parts of the deformation variable are given by

$$\Delta C_{(\theta z)} = \frac{a}{\Omega^2}[u\Omega' - u'\Omega] e^{i\theta}[1 - e^{-i\Omega s}]$$

$$- \frac{aui\Omega'}{\Omega} s e^{i\theta}$$

$$\Delta C_{(rz)} = -i\Delta C_{(\theta z)}$$

$$\Delta C_{(zz)} = -2r\Omega' s \Delta C_{(\theta z)}. \qquad [13]$$

The solution to the basic flow (i.e. with $a = 0$) in the absence of fluid inertia is easily shown to be the viscometric solution

$$\Omega = \frac{(\Omega_2 - \Omega_1)z}{h} + \Omega_1 \qquad [14]$$

and the corresponding shear rate γ is given by

$$\gamma = \frac{(\Omega_2 - \Omega_1)r}{h}. \qquad [15]$$

For the perturbed flow corresponding to ΔC_{ik}, the general equations of state can be written in the form [cf. (5)]

$$\Delta p'_{ij} = \int_{-\infty}^{t} \Psi_{ijkl}(\gamma, t - t')\Delta C_{kl}(t')\,dt' \qquad [16]$$

where $\Delta p'_{ij}$ is the extra stress due to the perturbed flow and various components of Ψ_{ijkl} obey certain consistency relations. For example, in the present situation $\Psi_{ijkl} = 0$ when there is an odd number of r indices, and

$$\frac{dp^{(0)'}_{(rr)}}{d\gamma} = \int_0^\infty [\Psi_{rrzz} 2\gamma s^2 - \Psi_{rr\theta z} 2s]\,ds$$

$$\frac{dp^{(0}_{(\theta\theta)}}{d\gamma} = \int_0^\infty [\Psi_{\theta\theta zz} 2\gamma s^2 - \Psi_{\theta\theta\theta z} 2s]\,ds$$

$$\frac{dp^{(0)'}_{(\theta z)}}{d\gamma} = \int_0^\infty [\Psi_{\theta zzz} 2\gamma s^2 - \Psi_{\theta z\theta z} 2s]\,ds$$

$$-p^{(0)'}_{(\theta z)} = \int_0^\infty \Psi_{rzrz} 2\gamma s\,ds$$

$$p^{(0)'}_{(rr)} - p^{(0)'}_{(\theta\theta)} = \int_0^\infty \Psi_{r\theta\theta z} 2\gamma s\,ds \qquad [17]$$

where the zero suffix refers to the basic viscometric flow.

If we write $\Delta C_{(\theta z)}$ as C, substitute [13] into [16] and hence into the stress equations of motion, we obtain, on eliminating the pressure,

---

[1]) Brackets placed round suffices are used to denote physical components of tensors.

$$0 = \int_0^\infty \left\{ \frac{\partial C}{\partial z} \left[-i\Psi_{rzrz} + ir\frac{\partial}{\partial r}(-\gamma s \Psi_{\theta zzz} + \Psi_{\theta z\theta z}) + i(-\gamma s \Psi_{\theta zzz} + \Psi_{\theta z\theta z}) \right] + C \left[r\frac{\partial^2}{\partial r^2}\Psi_{r\theta rz} + 3\frac{\partial}{\partial r}\Psi_{r\theta rz} + \frac{1}{r}\Psi_{r\theta rz} + \frac{\partial}{\partial r}(-\gamma s \Psi_{rrzz} + \Psi_{rr\theta z}) + \frac{1}{r}(-\gamma s \Psi_{rrzz} + \Psi_{rr\theta z}) - \frac{\partial}{\partial r}(-\gamma s \Psi_{\theta\theta zz} + \Psi_{\theta\theta\theta z}) - \frac{1}{r}(-\gamma s \Psi_{\theta\theta zz} + \Psi_{\theta\theta\theta z}) \right] \right\} ds \quad [18]$$

$$0 = \int_0^\infty \left\{ \frac{\partial^2 C}{\partial z^2} \left[-ir(-\gamma s \Psi_{\theta zzz} + \Psi_{\theta z\theta z}) \right] + \frac{\partial C}{\partial z} \left[r\frac{\partial}{\partial r}\Psi_{r\theta rz} + 2\Psi_{r\theta rz} + (\gamma s \Psi_{\theta\theta zz} - \Psi_{\theta\theta\theta z}) + (-\gamma s \Psi_{zzzz} + \Psi_{zz\theta z}) \right] + C \left[\frac{\partial}{\partial r}(-i\Psi_{rzrz}) + \frac{i}{r}(-\gamma s \Psi_{\theta zzz} + \Psi_{\theta z\theta z}) - \frac{i}{r}\Psi_{rzrz} \right] \right\} ds \quad [19]$$

where inertia effects have been ignored.

Eqs. [18] and [19] would appear to be incompatible in the general case [2]). This means that the simpifications assumed in [10] and [11] are not applicable and it would be necessary to take the velocity distribution [3] to obtain compatibility. If this were carried out, the resulting equations would be extremely complicated partial differential equations, and even if a solution were obtained, it would certainly be too complicated to have predictive value.

As a further conclusion, it may be assumed that generalized *Maxwell* Orthogonal Rheometer flow would be of limited use as a possible crucial test of simple approximate equations of state, since there would be no guarantee that the simplifications embodied in [10] and [11] would lead to compatible equations. There may, of course, be special circumstances when the general analysis is simplified and has predictive value. Clearly, the case $\Omega_1 = \Omega_2$ (i.e. $\Omega' = 0$) is such a one, this corresponding to conventional *Maxwell* Orthogonal Rheometer flow. Another corresponds to the case

$$u\Omega' - u'\Omega = 0 \quad [20]$$

which greatly simplifies [13]. The solution to [20] satisfying the boundary conditions corresponds to

$$\Omega_1 = 0 \quad \text{or} \quad \Omega_2 = 0. \quad [21]$$

Under these circumstances, it is not difficult to show, with the help of [17], and the viscometric flow equations, that eqs. [18] and [19] are compatible and that the resulting expressions for the forces X and Y, which are given by [3])

$$X = \int_0^R \int_0^{2\pi} [p_{(rz)} \cos\theta - p_{(\theta z)} \sin\theta]\, r\, d\theta\, dr \quad [22]$$

$$Y = \int_0^R \int_0^{2\pi} [p_{(rz)} \sin\theta + p_{(\theta z)} \cos\theta]\, r\, d\theta\, dr \quad [23]$$

reduce to

$$X = a\frac{\pi u \Omega'}{\Omega} \int_0^R \frac{r}{\gamma}\frac{d}{d\gamma}(\tau\gamma)\, dr \quad [24]$$

$$Y = 0 \quad [25]$$

where τ is the shear stress of the unperturbed flow. We see from [24] and [25] that measurement of the forces on one of the plates in this case yields

$\frac{1}{\gamma}\frac{d}{d\gamma}(\tau\gamma)$ [4]).

iii) The Eccentric-Cylinder Rheometer

The Eccentric-Cylinder Rheometer is shown in fig. 1 and has been briefly described in section 2. If the z axis of a suitable cylindrical polar coordinate system is taken to coincide with the axis of the inner cylinder, the relevant boundary conditions are

$$v_{(r)} = 0, \quad v_{(\theta)} = \Omega_1 r_1, \quad v_{(z)} = 0 \quad \text{on} \quad r = r_1$$

$$v_{(r)} = \Omega_2 a \cos\theta, \quad v_{(\theta)} = \Omega_2 [r - a \sin\theta] \quad v_{(z)} = 0 \quad \text{on} \quad r = r_2 + a \sin\theta \quad [26]$$

where the offset "a" is assumed small enough for second-order terms in a to be ignored.

[2]) The equations degenerate into compatible forms for some *simple* fluid models – e.g. for the *Newton*ian and finite linear viscoelasticity models.

[3]) R is the radius of the plate.

[4]) This result could also be obtained by noting that [21] corresponds to a viscometric flow, without any appeal to the general eq. [16].

The boundary conditions suggest a velocity distribution of the form [cf. (3, 4)]

$$v_{(r)} = a u(r, z)\, e^{i\theta},$$
$$v_{(\theta)} = r\Omega(r) + a v(r, z)\, e^{i\theta},$$
$$v_{(z)} = a w(r, z)\, e^{i\theta}, \qquad [27]$$

where u, v and w may be complex and the real part is implied. The corresponding displacement functions are

$$r' = r + \frac{iau}{\Omega} e^{i\theta}[1 - e^{-i\Omega(t-t')}]$$
$$\theta' = \theta - \Omega(r)(t - t') + \frac{iav}{r\Omega} e^{i\theta}[1 - e^{-i\Omega(t-t')}]$$
$$+ \frac{au}{\Omega}\frac{d\Omega}{dr} e^{i\theta}\left[-i(t - t') + \frac{1}{\Omega}[1 - e^{-i\Omega(t-t')}]\right]$$
$$z' = z + \frac{iaw}{\Omega} e^{i\theta}[1 - e^{-i\Omega(t-t')}]. \qquad [28]$$

In the present work, we limit consideration to the case in which end effects may be ignored. It is known that end effects can be severe when the cylinders rotate with the same angular velocity (4) but it is hoped that any end effect correction will be independent of Ω_1 and Ω_2 and may therefore be ignored within a suitable interpretation of the experimental results. This is a subject of current research and the relevant details are contained elsewhere (16). Writing

$$u = u(r), \quad v = i\frac{d}{dr}(ru), \quad w = 0 \qquad [29]$$

the relevant form for the perturbed part of the deformation tensor (ΔC_{ik}) is

$$\Delta C_{rr} = \left[2r(\Omega')^2 s^2 + 2\frac{\partial}{\partial r}\right]$$
$$\times \left[\frac{iau}{\Omega}(1 - e^{-i\Omega s})\, e^{i\theta}\right] + -2r\Omega' s \frac{\partial}{\partial r}$$
$$\times \left[-\frac{a(ru)'}{r\Omega}(1 - e^{-i\Omega s})\right.$$
$$\left. + \frac{au\Omega'}{\Omega}\left(-is + \frac{1}{\Omega}(1 - e^{-i\Omega s})\right)\right]$$

$$\Delta C_{\theta\theta} = 2\frac{iau}{\Omega} r(1 - e^{-i\Omega s})\, e^{i\theta}$$
$$+ 2iar^2\left[-\frac{(ru)'}{\Omega}(1 - e^{-i\Omega s})\right.$$
$$\left. + \frac{u\Omega'}{\Omega}\left(-is + \frac{1}{\Omega}(1 - e^{-i\Omega s})\right)\right] e^{i\theta}$$

$$\Delta C_{r\theta} = (i - 2r\Omega' s)\left[\frac{iau}{\Omega}(1 - e^{-i\Omega s})\, e^{i\theta}\right]$$
$$+ \left[-ir\Omega' s + r^2\frac{\partial}{\partial r}\right]\left[-\frac{a(ru)'}{\Omega}(1 - e^{-i\Omega s})\right.$$
$$\left. + \frac{au\Omega'}{\Omega}\left(-is + \frac{1}{\Omega}(1 - e^{-i\Omega s})\right)\right] e^{i\theta}. \qquad [30]$$

Where the dash now refers to differentiation with respect to r. Substituting [3] into the perturbed equations of state [16], we obtain

$$p'_{(rr)} = \int_0^\infty \{\Psi_{rrrr} C_{rr} + \Psi_{rr\theta\theta} C_{\theta\theta}$$
$$+ 2\Psi_{rrr\theta} C_{r\theta}\}\, ds$$
$$p'_{(\theta\theta)} = \int_0^\infty \{\Psi_{\theta\theta rr} C_{rr} + \Psi_{\theta\theta\theta\theta} C_{\theta\theta}$$
$$+ 2\Psi_{\theta\theta r\theta} C_{r\theta}\}\, ds$$
$$p'_{(r\theta)} = \int_0^\infty \{\Psi_{r\theta rr} C_{rr} + \Psi_{r\theta\theta\theta} C_{\theta\theta}$$
$$+ 2\Psi_{r\theta r\theta} C_{r\theta}\}\, ds$$
$$p'_{(zz)} = \int_0^\infty \{\Psi_{zzrr} C_{rr} + \Psi_{zz\theta\theta} C_{\theta\theta}\}\, ds$$
$$p'_{(rz)} = p'_{\theta z} = 0. \qquad [31]$$

If the stress components [31] are substituted into the stress equations of motion and the pressure eliminated, we obtain a complicated integro-differential equation for u involving a number of kernel functions. Although we do not now face any problem of "compatibility" as in the Orthogonal rheometer case, any hope of obtaining a solution which is simple enough to have predictive possibilities would appear to be out of the question. The flow must therefore be seen as a more complicated type of combined steady and oscillatory shear than has been considered hitherto in the more conventional instruments. The flow may nevertheless have attractions as providing a crucial experiment for testing proposed equations of state and it is to this important consideration we now turn.

We illustrate the basic theoretical ideas by considering a rather simple approximate equation of state, but the application of the theory to more complicated equations is fairly straightforward, in principle at least.

We consider on *Oldroyd* (17) model with equations of state

$$p_{ik} = -p g_{ik} + p'_{ik} \qquad [32]$$

$$p'^{ik} + \lambda_1 \frac{\vartheta}{\vartheta t} p'^{ik} + \mu_0 p'^{j}_{j} e^{(1)ik} = 2\eta_0 \left[1 + \lambda_2 \frac{\vartheta}{\vartheta t} \right] e^{(1)ik} \qquad [33]$$

where $e^{(1)}_{ik}$ is the rate of strain tensor, η_0, λ_1, λ_2 and μ_0 are material constants and $\vartheta/\vartheta t$ is the convected time derivative introduced by *Oldroyd* (15). We have found [32] and [33] to be a useful model to use in the qualitative prediction of elastico-viscous behaviour [see, for example (9)].

For the model [32] and [33] it can be shown after some tedious but routine calculation, that the equation for u (defined in eq. [29]) is

$$Au''' + Bu'' + Cu' + Du + E = 0 \qquad [34]$$

where

$$A = ir^2 G$$

$$B = ir^2(G' + H) + irG - rJ$$

$$C = ir^2(H' + I) + irH + r(M - K)$$

$$D = ir^2 I' + irI + r(N - L) \qquad [35]$$

and $G - N$ are given by

$$G = \left[\!\!\left[\{2i(\lambda_1 n - 2\eta_0 \lambda_2 \Omega')\} \left\{ \frac{\mu_0 \Omega'}{2} \right\} - \left\{ -\frac{i\mu_0 r^2 m}{2} + i\eta_0(1 + i\lambda_2 \Omega) \right\} \times \left\{ \frac{1 + i\lambda_1 \Omega}{r} \right\} \right]\!\!\right] \Bigg/ \left[-\frac{\mu_0 \lambda_1 (\Omega')^2}{r} - \frac{(1 + i\lambda_1 \Omega)}{r^3} \right]$$

$$H = \left[\!\!\left[\left\{ \frac{2i\lambda_1 n}{r} + m(\mu_0 - 2\lambda_1) - \frac{2\eta_0}{r_2}(1 + i\lambda_2(2r\Omega' + \Omega)) \right\} \left\{ \frac{\mu_0 \Omega'}{2} \right\} - \left\{ \frac{\Omega'(\lambda_1 - \mu_0/2)}{1 + i\lambda_1 \Omega} [2\eta_0(1 + i\lambda_2 \Omega) - \mu_0 r^2 m] - \frac{i\mu_0 r^2 m}{2} + \frac{i\eta_0}{r}(1 + i\lambda_2 \Omega) - 2\eta_0 \lambda_2 \Omega' \right\} \left\{ \frac{1 + i\lambda_1 \Omega}{r^2} \right\} \right]\!\!\right] \Bigg/ \left[-\frac{\mu_0 \lambda_1 (\Omega')^2}{r} - \frac{(1 + i\lambda_1 \Omega)^2}{r^3} \right]$$

$$I = \left[\!\!\left[\left\{ -\frac{\lambda_1}{r^2}\left(2in + \frac{d}{dr}(r^2 m)\right) + \frac{2\eta_0 i \lambda_2 \Omega'}{r^2} \right\} \times \left\{ \frac{\mu_0 \Omega'}{2} \right\} - \left\{ -\lambda_1 \left(\frac{1}{r}\frac{d}{dr}(rn) - im \right) + \frac{2i\Omega'}{(1 + i\lambda_1 \Omega)} \left(\lambda_1 - \frac{\mu_0}{2} \right)(\lambda_1 n - \eta_0 \lambda_2 \Omega') + \eta_0 \lambda_2 \left(\frac{\Omega'}{r} + \Omega'' \right) \right\} \left\{ \frac{1 + i\lambda_1 \Omega}{r^2} \right\} \right]\!\!\right] \Bigg/ \left[-\frac{\mu_0 \lambda_1 (\Omega')^2}{r} - \frac{(1 + i\lambda_1 \Omega)^2}{r^3} \right]$$

$$J = \left[\!\!\left[\left\{ -\frac{i\mu_0 r^2 m}{2} + i\eta_0(1 + i\lambda_2 \Omega) \right\} \times \left\{ -\frac{2\lambda_1 \Omega'}{r} \right\} - \{2i(\lambda_1 n - 2\eta_0 \lambda_2 \Omega')\} \times \left\{ \frac{1 + i\lambda_1 \Omega}{r} \right\} \right]\!\!\right] \Bigg/ \left[-\frac{\mu_0 \lambda_1 (\Omega')^2}{r} - \frac{(1 + i\lambda_1 \Omega)^2}{r^3} \right]$$

$$K = \left[\!\!\left[\left\{ \frac{\Omega'(\lambda_1 - \mu_0/2)}{(1 + i\lambda_1 \Omega)} [2\eta_0(1 + i\lambda_2 \Omega) - \mu_0 r^2 m] - \frac{i\mu_0 r^2 m}{2} + \frac{i\eta_0}{r}(1 + i\lambda_2 \Omega) - 2\eta_0 \lambda_2 \Omega' \right\} \times \left\{ -\frac{2\lambda_1 \Omega'}{r} \right\} - \left\{ \frac{2i\lambda_1 n}{r} + m(\mu_0 - 2\lambda_1) - \frac{2\eta_0}{r^2}(1 + i\lambda_2(2r\Omega' + \Omega)) \right\} \left\{ \frac{1 + i\lambda_1 \Omega}{r} \right\} \right]\!\!\right] \Bigg/ \left[-\frac{\mu_0 \lambda_1 (\Omega')^2}{r} - \frac{(1 + i\lambda_1 \Omega)^2}{r^3} \right]$$

$$M = \frac{2\eta_0(1 + i\lambda_2 \Omega) - \mu_0 r^2 m}{(1 + i\lambda_1 \Omega)}$$

$$N = \frac{2i}{(1 + i\lambda_1 \Omega)}(\lambda_1 n - \eta_0 \lambda_2 \Omega')$$

$$L = \left[\!\!\left[\left\{ -\lambda_1 \left(\frac{1}{r}\frac{d}{dr}(rn) - im \right) + \frac{2i\Omega'}{(1 + i\lambda_1 \Omega)} \left(\lambda_1 - \frac{\mu_0}{2} \right)(\lambda_1 n - \eta_0 \lambda_2 \Omega') + \eta_0 \lambda_2 \left(\frac{\Omega'}{r} + \Omega'' \right) \right\} \left\{ -\frac{2\lambda_1 \Omega'}{r} \right\}\right.$$

$$- \left\{ -\frac{\lambda_1}{r_2}\left(2in + \frac{d}{dr}(r^2 m)\right) + \frac{2\eta_0 i\lambda_2 \Omega'}{r^2} \right\}$$
$$\times \left\{ \frac{1 + i\lambda_1 \Omega}{r} \right\} \Bigg]\Bigg/$$
$$\left[-\frac{\mu_0 \lambda_1 (\Omega')^2}{r} - \frac{(1 + i\lambda_1 \Omega)^2}{r^3} \right]$$

and

$$m = \frac{2\eta_0 (\Omega')^2 (\lambda_1 - \lambda_2)}{(1 + \mu_0 \lambda_1 r^2 (\Omega')^2)}$$

$$n = \frac{\eta_0 \Omega' (1 + \mu_0 \lambda_2 r^2 (\Omega')^2)}{(1 + \mu_0 \lambda_1 r^2 (\Omega')^2)} . ^{5)}$$

An analytic solution of eq. [34] is clearly out of the question and we have resorted to a numerical procedure to obtain a solution, this being essentially a *Runge-Kutta* method used in conjunction with a "shooting technique". The relevant details are contained elsewhere (16).

The forces on the inner cylinder in the horizontal plane are given by

$$X = L r_1 \int_0^{2\pi} [p_{(rr)} \cos\theta - p_{(r\theta)} \sin\theta]\, d\theta \qquad [36]$$

$$Y = L r_1 \int_0^{2\pi} [p_{(rr)} \sin\theta + p_{(r\theta)} \cos\theta]\, d\theta . \qquad [37]$$

The relevant forms for X and Y for the *Oldroyd* model are obtained from [29]–[37].

We now have to consider the most convenient interpretation of the results. We associate a representative frequency of the motion with the rotational speed $\Omega_1 (\leq \Omega_2)$ and take q, given by

$$q = \frac{r_1(\Omega_2 - \Omega_2)}{d} \qquad [38]$$

to be the superimposed steady shear rate. In view of [1], we write

$$\eta'(\Omega_1, q) = \frac{2X}{\beta(\Omega_1 + \Omega_2)} \qquad [39]$$

$$G'(\Omega_1, q) = \frac{Y}{\beta} \qquad [40]$$

in which case, for a sufficiently small gap d, it can be shown that

$$\eta' = \text{const}$$
$$G' = 0 \qquad [41]$$

for a *Newtonian* liquid.

[5]) E is a constant of integration obtained from the boundary conditions.

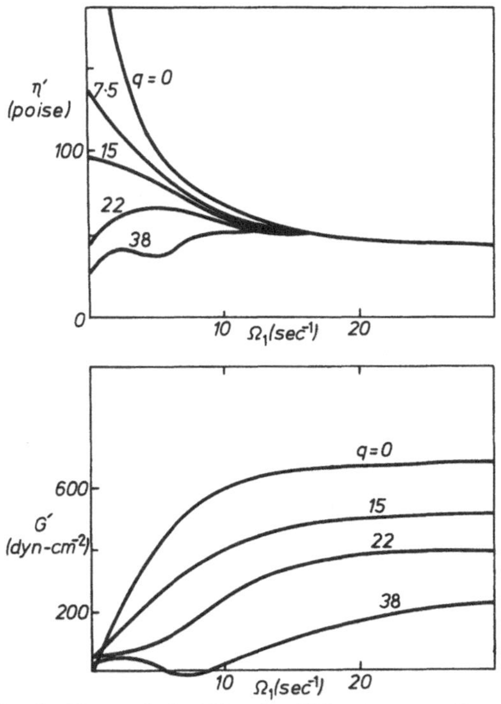

Fig. 3. Theoretical $\eta'(\Omega_1, q)$, $G'(\Omega_1, q)$ curves for the *Oldroyd* model. $\eta_0 = 200$, $\lambda_1 = 0.25$, $\lambda_2 = 0.05$, $\mu_0 = 0.005$, $r_1 = 2.65$, $r_2 = 3.0$

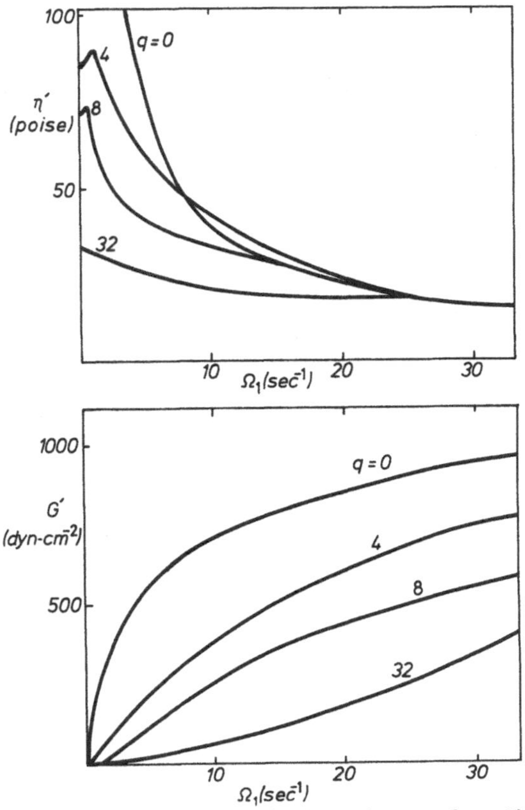

Fig. 4. Experimental $\eta'(\Omega_1, q)$, $G'(\Omega_1, q)$ curves for a 5% aqueous solution of polyacrylamide. $r_1 = 2.65$ cm, $r_2 = 3.0$ cm, $L = 9.45$ cm, $a = 0.01$ cm

Fig. 3 contains representative *theoretical* $\eta'(\Omega_1, q)$ and $G'(\Omega_1, q)$ results for the *Oldroyd* model. The curves show the same general features as the steady-and-oscillatory-shear data assocated with the more conventional instruments [see for example (9, 18)].

Fig. 4 contains *experimental* $\eta'(\Omega_1, q)$ and $G'(\Omega_1, q)$ results for a 5% aqueous solution of polyacrylamide. The curves have a qualitatively similar shape to the theoretical curves given in fig. 3. In the light of previous experience with such simple models as [32] and [33], we do not seek quantitative agreement between theory and experiment. We have used the model as a means of illustrating the basic ideas in the most convenient way. It is our future aim to generalize the theoretical analysis for more complicated fluid models in the hope that quantitative agreement between theory and experiment will be possible, i.e. we envisage generalized eccentric-cylinder flow as providing in the future a suitable (non-viscometric) test of continuum rheological equations of state.

Acknowledgements

We are happy to acknowledge substantial assistance given by Dr. *J. M. Broadbent* on various experimental aspects of the work. We have also benefitted from discussions with Mr. *J. M. Davies.*

Summary

The linear time-dependent behaviour of elastico-viscous liquids may be determined by conventional rheometers, or by any one of a number of new rheometers which have recently appeared in which the test fluid is contained between two surfaces which rotate with the same angular velocity about axes which are not coincident. (The Orthogonal, Balance and Eccentric-Cylinder rheometers are perhaps the most well known examples of such instruments at the present time.) In this communication, we consider the more general situation in the new rheometers in which the angular velocities of the two surfaces are different.

The first part of the paper is concerned with experimental results from the Eccentric-Cylinder Rheometer, with particular reference to the case in which the speeds of the two members are almost the same. The results throw light on the important problem of "lag" brought about by bearing friction.

The main body of work concerns a study of the potential use of the "combined steady and oscillatory shear" type of flow generated when the instrument members are driven at different speeds. The Balance rheometer is concluded to be of very little use in this connection, and most of the attention is confined to the "nearly-viscometric" flows generated in the Orthogonal and Eccentric-Cylinder rheometers. It is concluded that the analysis for the Orthogonal Rheometer is too complicated to have any predictive value, but that "generalized Eccentric-Cylinder flow" may be useful in the context of "model assessment".

References

1) *Maxwell, B.* and *R. P. Chartoff,* Trans. Soc. Rheol. **9**, 41 (1965).
2) *Kepes, A.*, Paper presented at 5th Int. Congress on Rheology, Kyoto (Japan) 1968.
3) *Abbott, T. N. G.* and *K. Walters*, J. Fluid Mechanics **43**, 257 (1970).
4) *Broadbent, J. M.* and *K. J. Walters*, Phys. D. **4**, 1863 (1971).
5) (a) *Pipkin, A. C.* and *D. R. Owen*, Phys. Fluids **10**, 836 (1967); (b) *Pipkin, A. C.*, Trans. Soc. Rheol. **12**, 397 (1968).
6) *Booij, H. C.*, Rheol. Acta **5**, 215 (1966).
7) *Macdonald, I. F.* and *R. B. Bird*, J. Chem. Phys. **70**, 2068 (1966).
8) *Osaki, K., M. Tamura, M. Kurata,* and *T. Kotaka*, J. Chem. Phys. **69**, 4183 (1965).
9) *Jones, T. E. R.* and *K. Walters*, J. Phys. A **4**, 85 (1971).
10) *Bernstein, B.* and *R. R. Huilgol*, Trans. Soc. Rheol. **15**, 731 (1971).
11) *Walters, K.* J. Fluid Mechanics **40**, 191 (1970).
12) *Abbott, T. N. G., G. W. Bowen,* and *K. Walters*, J. Phys. D. **4**, 190 (1971).
13) *Walters, K.* and *N. D. Waters*, Rheol. Acta **3**, 312 (1964).
14) *Abbott, T. N. G.* and *K. Walters*, J. Fluid Mechanics **40**, 205 (1970).
15) *Oldroyd, J. G.*, Proc. Roy. Soc. **A 200**, 523 (1950).
16) *Knight, D. G.*, Ph. D. thesis. (Univ. of Wales) 1973 (in prep.).
17) *Oldroyd, J. G.*, Proc. Roy. Soc. **A 245**, 278 (1958).
18) *Walters, K.* and *T. E. R. Jones*, Proc. 5th Int. Congress Rheol.. Vol. 4, p. 337 (Univ. Park Press 1970).

Authors' address:

D. G. Knight and *K. Walters*
Dept. of Applied Mathematics
University College of Wales
Aberystwyth (U.K.)

Rheol. Acta 12, 533–539 (1973)

From the Warren Spring Laboratory, Stevenage, Hertfordshire (U.K.)

A new torque and normal thrust measuring system for the Weissenberg rheogoniometer

By R. W. Higman

With 6 figures

(Received October 27, 1972)

At Warren Spring Laboratory a new torque and normal thrust measuring system, suitable for fitting to most models of the *Weissenberg* Rheogoniometer with the exception of the R 17, has been developed. A single piezoelectric crystal device completely replaces the air bearing torsion assembly, and therefore the need for a compressed air supply, and also replaces the servo assisted normal thrust measuring system.

Under steady rotational testing a wide range of torque and thrust are covered by the device. Without the need to change torsion bars, normal thrust springs and/or cone diameters similar ranges to those in the standard system are achieved. Any change in gap setting due to normal thrust, is virtually nil owing to the nature of the device and thus the servo assisted assembly is made redundant. This removes the difficulties associated with insensitivity and lag in the response of the assembly when used with high viscosity materials.

Under dynamic testing no corrections are required for the effect of the natural frequency of the device because it is far higher than the upper testing frequency of the instrument. When used for low viscosities corrections are required in the standard model. Higher strains are also achieved because of the torsional rigidity of the upper platen. With the fast response of the new device, the possibility of dynamic normal force measurements can be realized.

Some comparative data using the new and standard systems on a range of materials are given for both steady rotational and oscillatory testing.

Introduction

The Materials Handling Division of Warren Spring Laboratory has a section which carries out work in the field of non-*Newton*ian rheology. Its function is to assist industry to solve rheological problems covering a wide range of materials which have to be characterized. To provide data for the characterization of these materials the section operates R 15 and R 18 versions of the *Weissenberg* Rheogoniometer. This means that the instruments are frequently having one set of accessories, i.e., torsion bar, cone and normal-force spring, changed to another set in order to satisfy the requirements for testing. In spite of the fact that a nomograph (1) has been devised to assist in the selection of the accessories the time spent in setting up the instrument was becoming excessive. A simplified system of measurement (2) was therefore developed to overcome this difficulty while still retaining the versatility of the standard system. The new system consists of a combined torque and normal thrust piezoelectric transducer which replaces the air bearing/displacement transducer torque measuring unit and the servo/displacement transducer normal force measuring unit of the standard system. It is the purpose of this paper to describe the system and to indicate its advantages.

New system

Description

The basic construction and operation of the *Weissenberg* Rheogoniometer (Sangamo Weston Controls Ltd., Bognor Regis, U.K.) will not be described in detail because these points are well documented in the literature (3, 4).

The new crystal transducer system (fig. 1) is mounted on an L-shaped bracket in place of the standard air bearing/torsion bar arrangement. The upper platen is attached to the lower face of the transducer unit. The transducer consists of two piezoelectric crystals mounted as a single

35

Fig. 1. General view of new piezo-electric transducer system

unit (Type Z 3003 Kistler Instruments Ltd., Farnborough, U.K.). One crystal monitors the torque imposed on the cone by the fluid under steady or oscillatory shear whilst the second monitors any normal thrust that may be developed. The deformation of either crystal causes an electrostatic charge to be produced which is measured by a charge amplifier (Type 586 or 5001 Kistler Instruments Ltd.). The amplifier produces an output voltage directly proportional to the charge and proportional to the applied torque or thrust. The voltage output can be monitored in any convenient way; in the case of the present work it is done using ultra-violet (U.V.) recorders. The transducer is surrounded by a duralumin shield and the enclosed space is filled with foam sheet to provide adequate thermal stabilization for the system.

Advantages of the new system

Steady rotational shear

The main advantage of the new transducer system is that it allows the operator to use the instrument over the range of the stiffer torsion bars and the complete normal force range of the standard instrument without the need to interchange cones of different diameters and angles, torsion bars and normal force springs of different stiffnesses. In fact, by using one cone, e.g., 10 cm diameter 1° angle, it is possible to cover a range of viscosities from 10^0–10^{10} N/m^2 and shear rates between 2×10^{-3} and 2×10^3 S^{-1}. The normal thrust range is from 10^0–10^4 N/m^2. It removes the need for the special diaphragm mounting, the normal force spring, the micrometer, the servo motor, the servo amplifier and the associated transducer meter when taking normal thrust measurements. The variation in separation of the cone and plate is negligible during testing, being of the order 1 μm/5×10^3 N. The requirement for an air supply to the torsion head is also removed.

The response of the new system is improved, particularly during normal thrust measurement. Immediate response is obtained from an applied torque and/or normal thrust which is particularly useful in cases where viscous heating takes place in the material under test.

An overall simplification of the measuring system is obtained as seen from fig. 2. Each block in the figure represents a possible source of error in each system. The new system has fewer blocks and therefore the number of possible sources of error and the time required for fault searching are reduced.

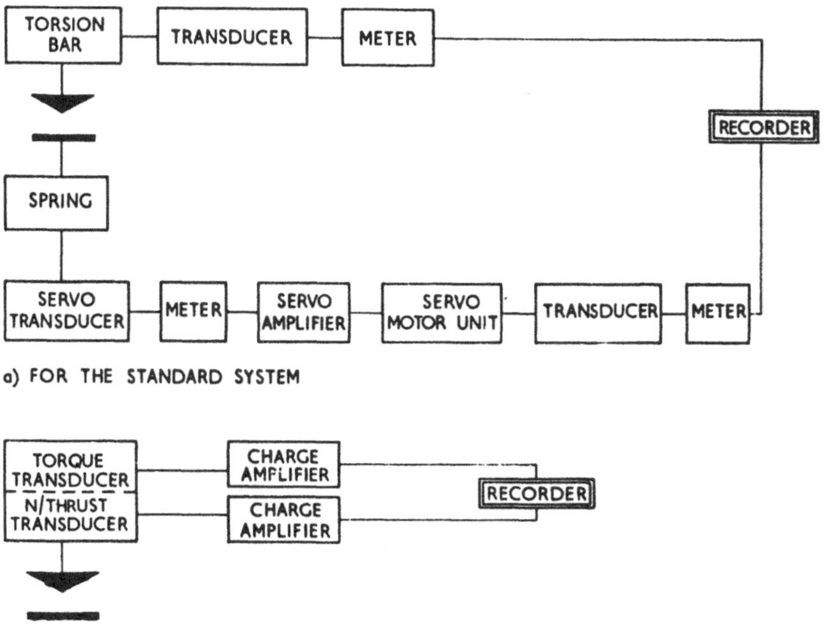

Fig. 2. Measurement flow paths

Oscillatory or alternating shear

The main advantage when testing in the oscillatory mode is the removal of the corrections for natural frequency of the torsion bar because the natural frequency of the crystal system is not less than 3 kHz, well in excess of the 50 or 60 Hz upper limit of the Rheogoniometer. Another factor is that the material under test can be subjected to greater strains than in the standard system due to the rigidity of the crystal allowing virtually no movement of the upper platen. All the advantages listed under steady shear also apply. In particular, the rapid response of the normal thrust system allows the possibility of measuring oscillatory normal force at the higher frequencies of the instruments.

Testing of the new system

So that the new system could be fully calibrated and compared with the standard system a number of tests were carried out on different samples. The standard system was fitted with a No. 8 torsion bar (1.311×10^{-2} Nm/μm) and the soft normal force spring (9.96×10^{-3} N/μm).

Samples were chosen primarily to determine the lowest practical values of torque and normal thrust that could be measured by the new system. Its behaviour is such that large torques or thrusts create no problems and improve the accuracy of measurement.

For steady shear testing two samples were used. They were (1) a 3% w/w polyisobutylene in dekaline solution and (2) a 4% w/w polyethylene oxide (WSR 301) in distilled water solution.

For oscillatory testing two samples were again used (1) the polyisobutylene in dekaline sample and (2) a 8% w/w polyethylene oxide (WSR 301) in distilled water solution.

Steady shear tests

Procedure

Tests were carried out in the normal way (4) on the two samples mentioned previously, over a range of shear rate from 10^{-1} to 10^3 S^{-1}. This was done using the R 15 version of the rheogoniometer with the standard system. The new system was fitted to the R 18 version and required a larger-angled cone to provide comparable data due to differences in the internal rotational gearing of the instruments.

Results and discussion

The results obtained from each system were treated in a similar way and according to standard practice (4), the only difference being that the new system provided a proportional voltage output rather than a measured displacement of the torsion bar or normal force spring. The results are presented in graphical form for shear and normal stress against shear rate in figs. 3 and 4.

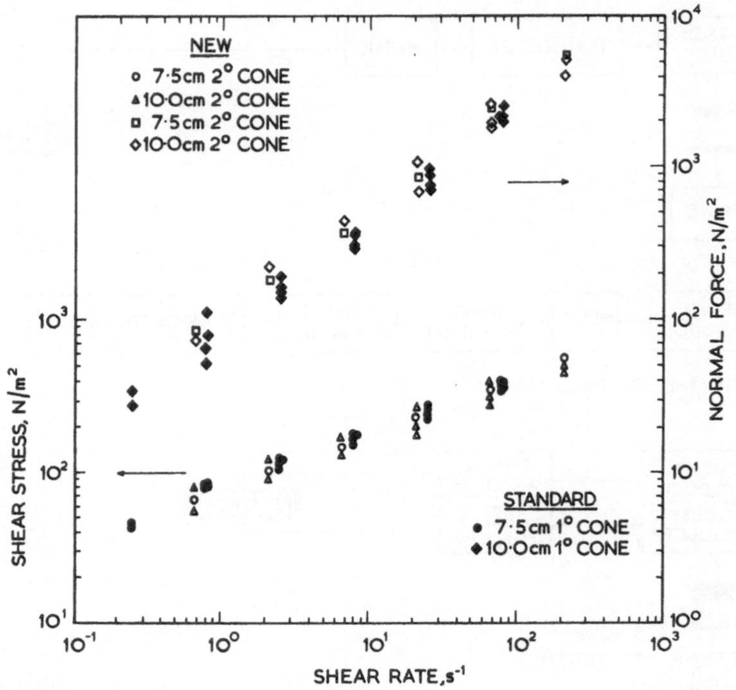

Fig. 3. 4% w/w polyethylene oxide in water

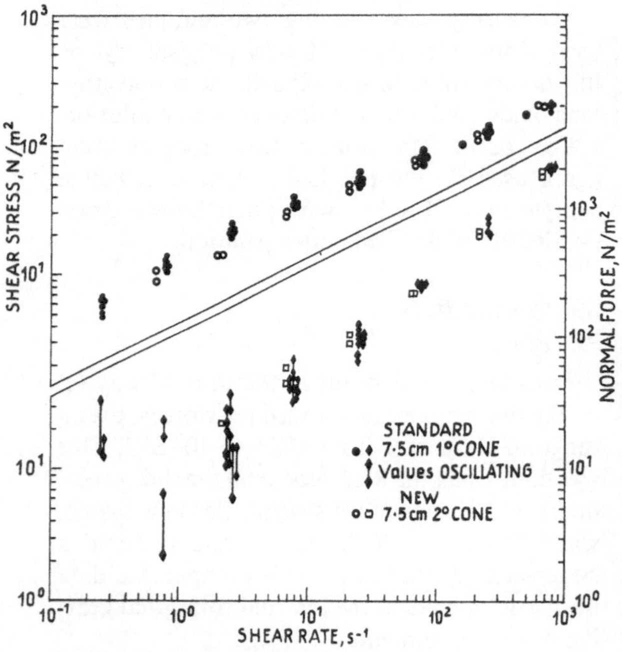

Fig. 4. 3% w/w polyisobutylene in dekalin

The graph (fig. 3) for the polyethylene oxide solution shows that the agreement between the two systems is good especially as a number of different cones were used.

The graph (fig. 4) for the polyisobutylene solution again shows good agreement between the two systems. It can be seen that the normal stress data for the standard system shows much scatter because it was operating at the lower limits of its range. The absence of oscillation in the results for the new system provides an indication of its reliability at its lower limit.

Oscillatory shear tests

Procedure

The R 15 version of the rheogoniometer was used with both measuring systems. The standard system was fitted first and calibrated by clamping the input and output platen supports together to determine the amplitude relationship and phase angle difference introduced by the electronic part of the measuring system at frequencies of 0.1, 0.5, 1.0, 5.0, 10.0, 20.0, and 40.0 Hz and at input amplitudes of 5, 10, 15, 20, and 25×10^{-3} inches displacement of the worm drive. Tests on the materials were made at the same frequencies and amplitudes.

When using the new crystal system it was not possible to calibrate it by the same clamping method because this would have broken the transducer. However, a similar method was used by placing a rod of *Tufnol*, having known elastic characteristics, between the two platen supports. This was sufficiently flexible to allow the crystal transducer to be calibrated over the same frequency and input amplitude range. The materials were retested with the new system.

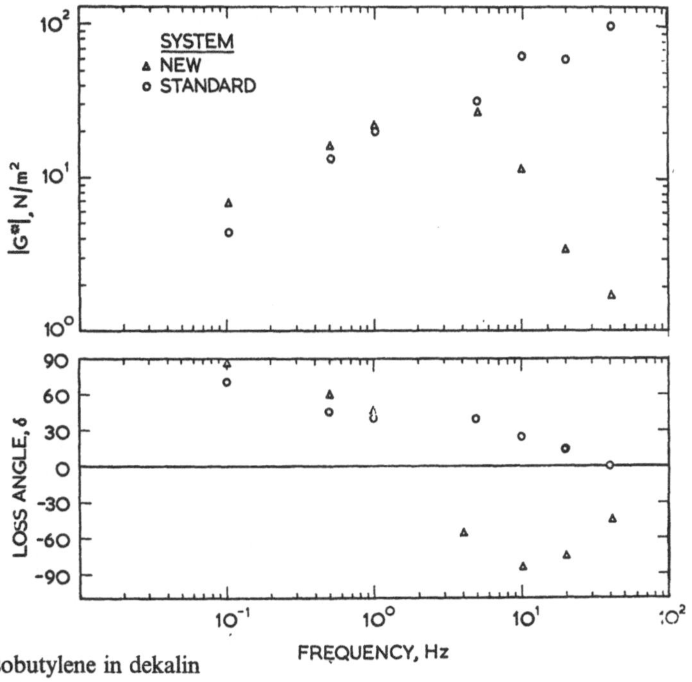

Fig. 5. 3% w/w polyisobutylene in dekalin

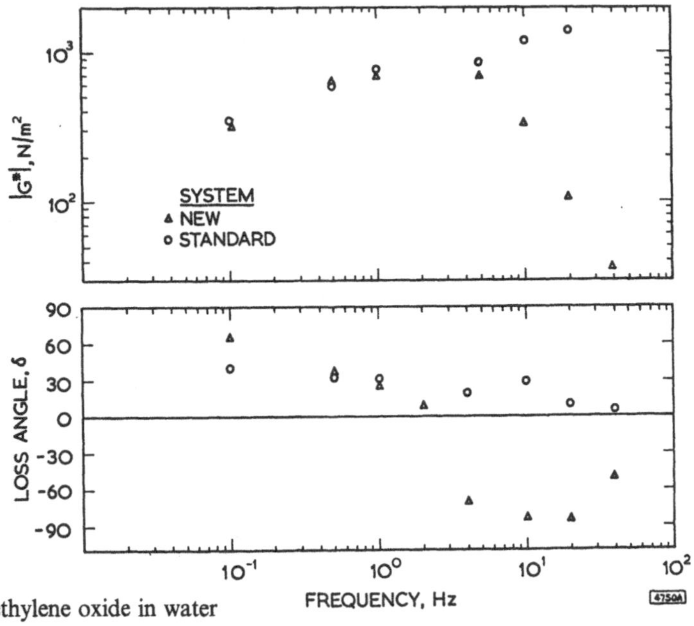

Fig. 6. 8% w/w polyethylene oxide in water

Results and discussion

The results were worked out according to standard formulae (5) with a correction being made for the fact that the natural frequency of the torsion bar (36 Hz) was within the range of testing frequencies.

The results are shown in graphical form for (G^*) and loss angle against frequency for the polyisobutylene and polyethylene samples in figs. 5 and 6. It can be seen that there is good agreement between the data of the new and the standard system at frequencies below 2 Hz. But at higher frequencies (G^*) and phase angle data from the new system fall abruptly. This cannot be natural frequency effect due to some unknown source because the phenomenon was not observed with the *tufnol* rod. Similar effects were observed with a range of other materials ranging from mineral oils and silicone liquid to latex gel, the fall-off being obtained at decreasing frequencies as the elasticity of the fluid increased. Because the tests with the new system were carried out at greater strains, it is thought that the fall-off in data was caused by shear fracture (6). This problem is still under investigation, but it seems that in addition to the increased ease in experimentation and data treatment, the new system would be useful in exploring the shear fracture properties of fluids.

Oscillatory normal thrust measurement

Using the standard system attempts were made to measure the oscillatory normal thrust that was present in both samples, but difficulty was experienced owing to its low levels and the slow response of the servo system. Thus no reliable data was obtained.

The attempt was repeated using the new system. However, any signal from the small normal forces present was masked by noise. The noise was found to be caused by the flexing of the back support member carrying the new system due to the vibrations created by the drive mechanism. Compression of the sample between the platens gave rise to apparent normal force signals at the same frequency as the drive and was detected by the sensitivity of the system. This noise will be insignificant when the normal force, such as obtained in polymer melts, is large but corrections to the instrument to remove the flexing are being made to check the lower limit of the system.

Conclusions

A new torque and normal thrust measuring system for the *Weissenberg* Rheogoniometer is described. Its design, method of working and advantages are explained. Some results on various test fluids are given to provide a comparison between the standard and new systems under conditions of steady and oscillatory shear.

The conclusions that may be drawn are as follows:

1. The new system is easier to set up, operate and maintain compared with the standard system.

2. The reproducibility of results is comparable with the standard system at the low stresses involved in the present work. The reproducibility is expected to be better with the new system when used for more viscous fluids.

3. The range of torque measurement is comparable with the three largest torsion bars on the standard system, i.e. it will measure between 10^{-3} and 25 Nm. This places the lower viscosity limit at about 10 P, but as there are other viscometers more suitable for low viscosity fluids, the new system does not impair the usefulness of the Rheogoniometer.

4. The range of normal thrust measurement is comparable with the standard system, i.e., it will measure between 10^{-1} and 10^4 N.

5. The response of both the torque and normal thrust components is extremely rapid, faster than the standard system under both rotational and oscillatory modes of operation.

6. The material under test can be subjected to greater strain than in the standard system due to virtually no movement of the upper platen taking place.

7. The means is available of measuring oscillatory normal force over the higher frequency range of the instrument.

From the above conclusions it may be seen that the original objectives of the investigation have been satisfied and the new system is now in normal use at Warren Spring Laboratory.

Zusammenfassung

Im Warren Spring Laboratory wurde ein neues Drehmoment- und Normaldruckmeßsystem entwickelt, das mit Ausnahme des R 17 zu den meisten Modellen des *Weissenberg*-Rheogoniometers paßt. Ein einziges piezoelektrisches Kristallgerät ersetzt die luftgelagerte Torsionsaufhängung und somit die Notwendigkeit einer Preßluftzufuhr, und es ersetzt außerdem das mit Servoeinrichtung versehene Normaldruckmeßsystem.

Bei gleichmäßiger Rotationsprüfung erfaßt das Gerät einen weiten Drehmoment- und Druckbereich. Ohne daß Torsionswellen, Normaldruckfedern und/oder Kegeldurchmesser ausgetauscht werden müssen, werden die gleichen Bereiche wie mit dem Standardsystem erreicht. Durch den Normaldruck verursachte etwaige Änderungen der Spaltweite sind dank der Natur dieses Gerätes praktisch bedeutungslos, und die mit Servoeinrichtung versehene Anordnung wird somit überflüssig. Damit entfallen die Schwierigkeiten durch Unempfindlichkeit und Ansprechverzögerung der Anordnung während des Einsatzes bei hochviskosen Materialien.

Beim dynamischen Testen sind keine Korrekturen aufgrund der Eigenfrequenz des Gerätes erforderlich, da diese wesentlich höher als die obere Testfrequenz des Instrumentes ist. Im niedrigen Viskositätsbereich sind beim Standardmodell Korrekturen erforderlich. Dank der Torsionsstabilität der Oberplatte werden außerdem höhere Spannungen ermöglicht. Durch die Ansprechgeschwindigkeit des neuen Gerätes werden dynamische Normalkraftmessungen möglich.

Es werden einige mit dem neuen und mit dem Standardsystem gewonnene vergleichende Daten einer Reihe von Materialien sowohl für gleichbleibende Rotations- als auch Oszillationstest angegeben.

References

1) *Higman, R. W.*, Bull. Brit. Soc. Rheol., **14** (1), 2–4, and **14** (3), 42 (1971).

2) *Higman, R. W.*, Instrument for measuring the characteristics of a sample. UK Pat. App. 43625/70.

3) *Van Wazer, J. R., J. W. Lyon, K. Y. Kim*, and *R. E. Colwell*, in: Viscosity and flow measurement, 113–116 (New York 1963).

4) Sangamo Weston Controls Ltd., The Weissenberg Rheogoniometer Instruction Manual Model R 15, Model R 18. (Bognor Regis: The Company, undated.)

5) *Walters, K.*, Basic concepts and formulae for the rheogoniometer. (Bognor Regis: Sangamo Weston Controls, 1968.)

6) *Hutton, J. F.*, Proc. Roy Soc., **A 287**; 222–239 (1965)

Author's address:
Mr. *R. W. Higman*
Warren Spring Laboratory
Dept. of Trade and Industry
P.O. Box 20
Gunnelswood Road
Stevenage, Herts. SG1 2BX (England)

Rheol. Acta **12**, 540–545 (1973)

From the Postgraduate School of Chemical Engineering, The University of Bradford (England)

Steady shear measurement of thixotropic fluid properties

By *J. C. Godfrey*

With 3 figures

(Received October 27, 1972)

Introduction

Techniques for the characterization of thixotropic fluids have been proposed and investigated for a long time, but as yet no quantitative technique has been widely adopted. The description of material properties in a quantitative manner still presents a problem and the application of constitutive equations to a specific flow situation is seldom possible. The most widely used procedure has been that of *Weltman* (1), and although the parameters obtained are not truly quantitative the technique still finds application in industrial and academic research.

Even the most characteristic property of thixotropic fluids, the decay of viscosity or stress under steady shear conditions, still presents difficulties in characterization. As a first approximation the decay is often represented as a first order rate process, relative to an equilibrium viscosity or stress value. This concept is usually presented in the form:

$$-\frac{d\mu}{dt} = k(\mu(t) - \mu_\infty) \qquad [1]$$

where $\mu(t)$ is the current viscosity at the time t, μ_∞ is the equilibrium viscosity attained after an "infinitely" long period of shearing and k is a rate constant; all values being for a particular shear rate.

However experimental observations have shown that the response of real materials is not as simple as this first order expression suggests and further hypotheses have been developed. A frequently proposed interpretation (2, 3, 4, 5) is that the change of viscosity is the net outcome of forward and backward rate processes; perhaps breakdown of viscosity relative to an equilibrium or minimum value and a recovery of viscosity relative to an initial or maximum value. A similar concept based on structural considerations has been successfully applied by *Ritter* and *Govier* (6) to the characterization of a thixotropic crude oil in the shear rate range 35–350 $\sec^{-1}$. *Slibar* and *Paslay* (7) have used a memory function model to analyse the results of both breakdown and recovery tests with some success, although the degree of fit of the transient sections of the data was not always very precise.

Examination of experimental data suggests that an alternative interpretation is possible as the reduction of viscosity with time appears to be a combination of short term and long term rate processes which could be described by a number of exponential terms. In contrast to many of the above mentioned techniques an exponential description requires no assumption as to the form of shear rate dependence of the process, which is an advantage when this aspect is not fully understood. *Joye* and *Poehlein* (8) have investigated the breakdown and recovery of a clay suspension in steady shear and were able to describe their results by a number of exponential terms. Three terms were required for the breakdown process and two for recovery, but the parameters determined could not be correlated with shear rate. The properties of the clay suspension tested were very complicated and it may be that a more "well behaved" fluid would provide a more logical response. An extension of the exponential description method has been considered by *Harris* (9) who suggests the use of a continuous spectrum of viscosity deficits and relaxation times to accommodate the complex behaviour of thixotropic materials, however, even for a simple block spectrum parameters can be difficult to evaluate.

In any discussion of models of thixotropy some attention should be paid to the dynamic response of test material and to any implication of this that appears in the models under examination. Given a successful model of the decay of viscosity with time that includes the effect of shear rate, it is important to observe that the shear rate dependency so expressed is not necessarily adequate to accommodate the effects of a change of shear rate during a test. *Cheng* and *Evans* (10) have discussed this point and recommend the use of a rate equation for the description of time dependency and an equation of state to describe the dynamic shear rate dependency of a material under conditions of constant structure.

In this work the decay of viscosity with time was measured under steady shear conditions and the experimental data interpreted as the sum of two exponential terms. The parameters of these terms were then investigated as functions of shear rate. Some additional investigations were made of constant structure properties and of the effect of shear history on the behaviour of the test material. The fuel oil examined proved to be "well behaved" in many respects and a satisfactory exponential description was obtained.

Experimental

The main requirement of the experimental program was to provide measurements of the decay of viscosity of a thixotropic fuel oil for a range of shear rates. These measurements were made with a Rheogoniometer model R 18 at a temperature of 20 °C and at shear rates of 0.90, 2.84, 9.0, 28.4 and 90 $\sec^{-1}$. A cone and plate geometry (59′16″ × 10 cm) was used to remove the problems of shear rate computation that accompany the use of co-axial cylinder viscometers. A number of shear rate computation techniques have been quoted for viscometric tests of thixotropic fluids in co-axial cylinder geometry (6, 10) but quantitative assessments of their applicability are not available. The Rheogoniometer also has the advantage of separate drive and measuring systems thus allowing short term transient measurements to be made. Modifications have been developed by some experimenters (11, 12) which use a very rigid torsion bar with a piezo-electric torque measuring system to improve transient response.

The fluid investigated in the main part of the program was a thixotropic fuel oil, crude oil being avoided because of the difficulty of controlling the loss of the more volatile constituents. With high viscosity fuel oils a limitation of the cone and plate geometry soon became apparent, as *Hutton* (13) has reported for lubricating oils, loss of sample can occur by secondary flow. The highest viscosity fuel oils could not be tested as the sample loss was both extensive and rapid. With all high viscosity fuel oils some sample loss was observed on occasions; experiments had to be abandoned when the loss became evident. Another problem that arose was the "damage" that was done during the loading of a thixotropic sample. Experiments showed that this effect was not very reproducible and the problem was overcome by shearing the material at a shear rate (0.09 $\sec^{-1}$) lower than those of the test program and then allowing the material to recover for a fixed period (5 h). This provided a reproducible initial condition for the decay tests, although a true maximum initial condition would only be obtained after a much longer recovery period.

In addition to the viscosity decay measurements some investigation was made of constant structure properties, shear history effects and the behaviour of other thixotropic fluids (crude oil, salad cream, yoghurt). These experiments were conducted with either the Rheogoniometer cone and plate geometry or the MVI co-axial cylinder geometry of the *Haake* Rotovisko. Although the co-axial cylinder geometry presents problems and the transient response is not particularly good the instrument offers good temperature control, reasonably rapid shear rate changes and does not suffer from sample loss. The response of the Rotovisko seemed adequate for much of the test program and the measurements were qualitatively the same as those obtained in the cone and plate geometry.

Results

The reduction of apparent viscosity with time was determined at shear rates between 0.90 and 90 $\sec^{-1}$ for shearing times of 5–3600 sec and the data extrapolated to estimate equilibrium viscosities. If the thixotropic process is represented by a number of exponential terms the relationship may be written:

$$\begin{aligned}\mu(t) = \mu_0 &- \Delta\mu_1(1 - e^{-t/\lambda_1}) \\ &- \Delta\mu_2(1 - e^{-t/\lambda_2}) - \cdots \\ &- \Delta\mu_n(1 - e^{-t/\lambda_n}) - \cdots \\ &- \Delta\mu_N(1 - e^{-t/\lambda_N})\end{aligned} \qquad [2]$$

or alternatively:

$$\begin{aligned}\mu(t) = \mu_\infty &+ \Delta\mu_1(e^{-t/\lambda_1}) \\ &+ \Delta\mu_2(e^{-t/\lambda_2}) + \cdots + \Delta\mu_n(e^{-t/\lambda_n}) + \cdots \\ &+ \Delta\mu_N(e^{-t/\lambda_N})\end{aligned} \qquad [3]$$

where there are N exponential terms and the $\Delta\mu_n$ are viscosity deficits associated with the decay time constants λ_n. The relationship between μ_0 and μ_∞ is given by:

$$\mu_0 = \mu_\infty + \Delta\mu_1 + \Delta\mu_2 + \cdots + \Delta\mu_n + \cdots + \Delta\mu_N \qquad [4]$$

the value of μ_∞ being more readily obtained by extrapolation than the value of μ_0; determination of the latter implying a high order of frequency response in the test instrument and an understanding of the short term behaviour of the test material. If the experimental results are interpreted according to eq. [3] the variation of $\Delta\mu_n$ and λ_n with shear rate may provide a useful characterization of the time dependency of the test fluid.

Examination of the results in the form of the variation of $\ln(\mu(t) - \mu_\infty)$ with time showed that the relationship is of a simple exponential kind for elapsed times greater than 300 sec. In this region the decay of viscosity can be represented by a viscosity deficit, $\Delta\mu_1$, and a time constant λ_1. For these experiments λ_1 was effectively independent of shear rate, while $\Delta\mu_1$ decreased with increasing shear rate. The data for elapsed times less than 300 sec was then re-examined in terms of the variation of $\ln(\mu(t) - \mu_\infty - \Delta\mu_1(e^{-t/\lambda_1}))$ with time and a second exponential term $\Delta\mu_2(e^{-t/\lambda})$ obtained, where λ_2 can be taken to be constant for the test shear rate range and $\Delta\mu_2$ decreases with increasing shear rate. With these two exponential terms and the equilibrium viscosity the experimentally observed variation of apparent viscosity with time can be described as:

$$\mu(t) = \mu_\infty + \Delta\mu_1(e^{-t/\lambda_1}) + \Delta\mu_2(e^{-t/\lambda_2}). \qquad [5]$$

For the elapsed times and shear rates of the data the description is within $\pm 8\%$, the comparison of eq. [5] and the experimental data is shown in fig. 1. The variation of $\Delta\mu_1$ and $\Delta\mu_2$ with shear rate is shown in fig. 2, and for shear rates of 2.84 to 90 sec^{-1} the variation could be approximated by a power-law equation:

$$\Delta\mu(\dot\gamma) = \Delta\mu^1 \dot\gamma^n \qquad [6]$$

where $\Delta\mu^1$ is the value of $\Delta\mu$ at $\dot\gamma = 1$ and the index, n, is the slope of the plot of fig. 2. A similar expression can be written for μ_0.

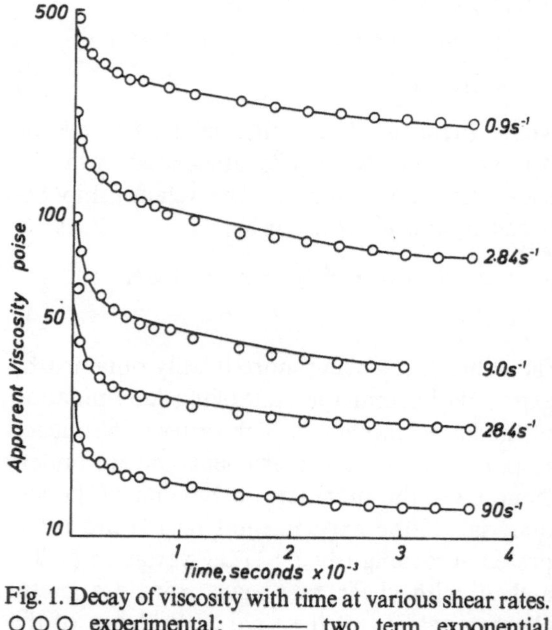

Fig. 1. Decay of viscosity with time at various shear rates. ○○○ experimental; —— two term exponential decay process

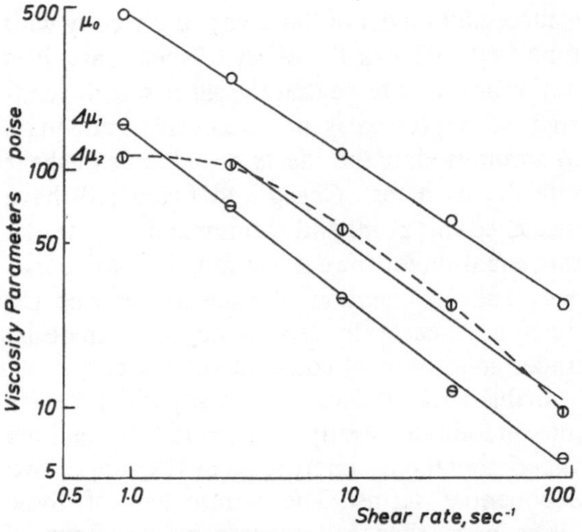

Fig. 2. Parameters for two term exponential decay process

While two exponential terms provide a reasonable fit of the experimental data a better fit could be achieved with three terms. The improvement resulting from the use of an extra term is to be expected, and it seems that, as for λ_1 and λ_2, the same value of a time constant λ_3 could be used at each shear rate. However, when so many parameters are used any experimental errors become more significant and there is some question as to whether the description is unique.

No detailed measurements were made of the recovery of viscosity of the test fluid at rest, but preliminary investigations showed that the rate of recovery was very much slower than the rate of breakdown. The recovery demonstrated on allowing the material to rest for a considerable time was very reproducible.

Measurements were also made of the constant structure shear stress-shear rate relationship by observing the change of stress accompanying a nominally rapid change of shear rate. The response of the test material to decreases in shear rate could be satisfactorily measured with the Rheogoniometer as the rate of recovery was very slow. In this case effects of instrument response and short rest periods between changes in shear rate were relatively unimportant. Similarly, if the material has already undergone considerable thixotropic decay the response to increases in shear rate was not difficult to measure. The measurements made are shown, together with isochronous "flow curves", in fig. 3 and illustrate the difference between the shear rate dependency of steady shearing and dynamic response experiments. The dynamic response measurements were made from a reference condition of near equilibrium stress, attained after shearing for 3600 sec. It is possible to conduct these experiments from any shear stress-shear rate combination as a starting point and, providing there are no problems of stress recovery or instrument response, the shear stress-shear rate relationships obtained are the same as those for the equilibrium reference conditions.

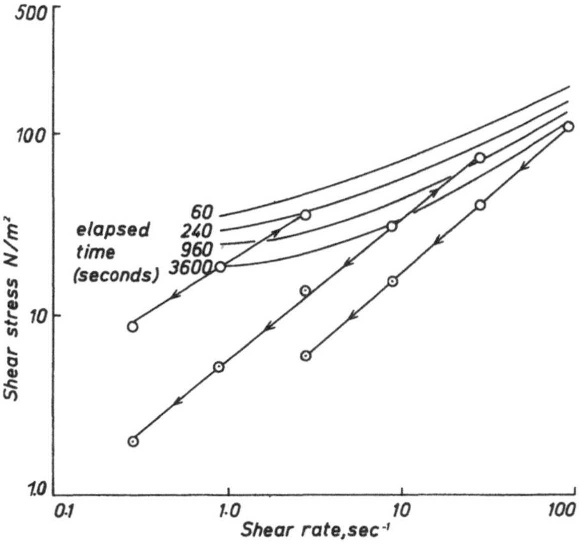

Fig. 3. Constant structure and isochronous stress-shear rate data

Discussion

When the first exponential term was determined from the long term decay data the time constant, λ_1, did not vary significantly with shear rate and one value could be used for the range 0.9–90 $(\text{sec})^{-1}$ with little error. Similarly in the determination of a second exponential term the assumption that the time constant, λ_2, was independent of shear rate provided a good fit for the data for the experimental time and shear rate range. This observation suggests that the time constants obtained could be regarded as fundamental properties of the test material, the long term behaviour being particularly reproducible. The phenomenum of shear rate independent time constants has been observed in other materials, crude oil, salad cream and yogurt all demonstrating this effect. Whether this behaviour is shown by a large number of thixotropic materials and over wide variations of shear rate remains to be seen. A two term exponential representation is not necessarily universally applicable to all thixotropic materials, it is most likely that the treatment used here only approximates the real situation. Measurements made with highly thixotropic samples of crude oil appeared to require three exponential terms for a reasonably accurate representation of the decay process. Although it seems that the three time constants required may be regarded as independent of shear rate there is the possibility that, given enough terms, a variety of parameters could be used to provide the same quality of description. *Joye* and *Poehlein* found that they could describe their time dependent data by three exponential terms, but that the parameters obtained were not logically related to shear rate. However, if the restraint of shear rate independent time constants proves to be realistic and widely applicable, the possibility of obtaining unique values for time constants and viscosity deficits is greatly increased.

For each shear rate and time constant there is a corresponding viscosity deficit and the relationship which exists between viscosity deficit and shear rate gives a good representation of the material properties. The thixotropic properties of the test fuel oil can be characterized by the three plots of fig. 2, showing the shear rate dependency of $\Delta\mu_1$, $\Delta\mu_2$ and μ_0. The only other information required is the value of the relevant time constants. For limited ranges of shear rate, in this case between 2.84 and 90 $(\text{sec})^{-1}$, shear rate dependencies can be expressed by the power law relationship of eq. [6] and the relationship between apparent viscosity and time at constant shear rate can be written as:

$$\mu(t) = \mu_0^1\dot{\gamma}^{n_0} - \Delta\mu_1^1\dot{\gamma}^{n_1}(1 - e^{-t/\lambda_1}) - \Delta\mu_2^1\dot{\gamma}^{n_2}(1 - e^{-t/\lambda_2})$$

where

μ_0^1	454 poise
n_0	−0.62
$\Delta\mu_1^1$	144 poise
n_1	−0.72
λ_1	1750 sec
$\Delta\mu_2^1$	226 poise
n_2	−0.68
λ_2	60 sec

The combination of time constants and the plots of fig. 2 provide a compact description of the experimental measurements with a terminology which has some physical significance. The initial viscosity μ_0, specifies a maximum value while the viscosity deficits specify the reduction associated with particular time constants. In the usual way the time constants define the time scales of the processes under examination. In this interpretation the value of μ_0 is defined by eq. [4] and is consequently an extrapolation and therefore not necessarily reliable. If the short term response of the material is the major interest consideration of instrument response and fluid elasticity may be necessary; as has been investigated by *Brodkey* (11).

Measurements of constant structure properties demonstrate the non-*Newton*ian shear rate-shear stress characteristics of a thixotropic fluid at an instant in time. When recovery of the test material is slow a decrease in shear rate gives a response which is easily measured. If the fluid is near an equilibrium stress condition it has already undergone considerable breakdown and the response to an increase in shear rate is not difficult to measure, if the increase is not too large. If the dynamic response appears to be a problem the reproducible nature of stress-time data provides a method of extrapolation to obtain a value appropriate to the onset of shearing at the new shear rate. In the results of fig. 3 the response to decreases or increases in shear rate is of an approximately power law nature. From the three sets of results it seems that as the value of the consistency index falls so the value of the power law index more closely approaches unity; as the material is broken down so it approaches more nearly *Newton*ian behaviour.

During the experimental work it was observed that the effect of shear history always appeared to be completely described by the current shear rate-shear stress combination. Given a particular combination any subsequent viscosity (or stress) decay seemed completely predictable from previous experiments at the same shear rate. Similarly, if recovery was allowed for a set time from a certain shear rate-shear stress combination the extent of recovery was independent of all else. This applied even if heating and cooling cycles of an undefined nature were included in test procedures prior to the final shear rate-shear stress combination. Also the response of a fluid to shear rate changes in the determination of constant structure properties was apparently a function only of the shear stress-shear rate combination at the commencement of the test. Thus for fuel oil it seems that the effects of past shear (and perhaps thermal) history are completely expressed at any time by the operating shear rate-shear stress condition. As yet the effect of shear history on other materials have not been checked, but other thixotropic materials have been investigated (8) which have more complicated behaviour.

Conclusions

The thixotropic fuel oil tested in these experiments produced complicated but well behaved responses. Two exponential terms provide a characterization of the steady shear viscosity decay to an accuracy better than $\pm 8\%$. The time constants λ_1 and λ_2 were independent of shear rate and the viscosity deficits $\Delta\mu_1$ and $\Delta\mu_2$ both show a simple relation to shear rate. Similar behaviour has been observed in measurements of the thixotropic behaviour of crude oil, salad cream and yogurt and suggests that the observations recorded for fuel oil may be applicable to other materials. However samples of very thixotropic material appear to required three exponential terms for an accurate description, and this may bring problems.

Examination of the dynamic response of the fluid to changes in shear rate in constant structure tests showed a simple power law relationship between shear rate and shear stress.

The overall effect of shear history on the test material was very simple. It seems that the effects of the stress history are entirely represented at any time by the current shear stress-shear rate condition and that the material response is a function of this condition only.

Summary

Measurements of the viscometric properties of a thixotropic fuel oil at constant shear rate have shown a reduction of viscosity that has the characteristics of combined long term and short term exponential decay processes. It is possible to evaluate parameters from experimental data for decay processes which combine to represent the observed time dependence of viscosity.

At a particular shear rate the time dependence can be represented as:

$$\mu(t) = \mu_0 - \Delta\mu_1(1 - e^{-t/\lambda_1}) - \Delta\mu_2(1 - e^{-t/\lambda_2}).$$

When measurements are made for a range of shear rates it is found that the time constants, λ_1 and λ_2, are relatively unchanged while the viscosity deficits, $\Delta\mu_1$ and $\Delta\mu_2$, and the initial viscosity are shear rate dependent. For a limited shear rate range the nature of this dependency can be expressed as:

$$\Delta\mu(\dot{\gamma}) = \Delta\mu^1 \dot{\gamma}^n$$

$$\mu_0(\dot{\gamma}) = \mu_0^1 \dot{\gamma}^{n_0}$$

where $\Delta\mu^1$ and μ_0^1 are evaluated at $\dot{\gamma} = 1$ and the various indices all lie between 0 and -1. The time dependence of viscosity measured at constant shear rate can then be represented as:

$$\mu(t) = \mu_0^1 \dot{\gamma}^{n_0} - \Delta\mu_1^1 \dot{\gamma}^{n_1}(1 - e^{-t/\lambda_1}) - \Delta\mu_2^1 \dot{\gamma}^{n_2}(1 - e^{-t/\lambda_2}).$$

With this characterization method long term, short term, time independent and shear rate dependent characteristics of a material can be individually identified.

References

1) *Weltman, R. N.*, Rheology, **3**, 205 (New York 1960).
2) *Hahn, S. J., T. Ree*, and *H. Eyring*, Ind. Eng. Chem., **51**, 856 (1959).
3) *Kim, H. T.* and *R. S. Brodkey*, Amer. Inst. Chem. Eng. J., **14**, 61 (1968).
4) *Fredrickson, A. G.*, Amer. Inst. Chem. Eng. J., **16**, 436 (1970).
5) *Pinder, K. L.*, Canad. J. Chem. Eng., **42**, 132 (1964).
6) *Ritter, R. A.* and *G. W. Govier*, Canad. J. Chem. Eng., **48**, 505 (1970).
7) *Slibar, A.* and *P. R. Paslay*, Proc. IUTAM – Symposium on Second Order Effects in Elasticity, Plasticity and Fluid Dynamics, 314 (Oxford 1962).
8) *Joye, D. D.* and *G. W. Poehlein*, Trans. Soc. Rheol., **15**, 51 (1971).
9) *Harris, J.*, Rheol. Acta., **6**, 6 (1967).
10) *Cheng, D. C-H.* and *F. Evans*, Brit. J. Appl. Phys., **16**, 1599 (1965).
11) *Lee, K. H.* and *R. S. Brodkey*, Trans. Soc. Rheol., **15**, 627 (1971).
12) *Higman, R. W.*, B.S.R. Autumn Conference, Exeter University, 20–23 September 1971.
13) *Hutton, J. F.*, Conference on the Rheology of Lubricants, Trent Polytechnic, Nottingham, 6–7 July 1972.

Author's address:

J. C. Godfrey
School of Chemical Engineering
University of Bradford
Bradford BD7 1DP (England)

Rheol. Acta **12**, 546–549 (1973)

From the National Bureau of Standards Washington, D. C. (U.S.A.)

A test sample to standardize measurements of normal stress

By E. A. Kearsley

With 3 figures and 1 table

(Received October 27, 1972)

In the early 1940's, *Garner, Nissan* and *Wood* (1)[1]) and *Weissenberg* (2) made the first attempts to measure normal stress[2]) in viscoelastic fluids undergoing shear. Fifteen years later, by the late 50's, a considerable body of data on various fluids and techniques had been generated. *Markovitz* and his colleagues, recognizing the importance of collecting different measurements on one sample, completed an extensive program on a solution of polyisobutylene. In surveying this work in 1958, *Markovitz* (3) said ... "The data lead us to the conclusion that two contradictory types of theory are needed to correlate different pairs of experiments. Obviously, there is something wrong. We need more and better experiments to decide among the various possibilities." The statement could well be made at this Congress, although the 14 years intervening have been filled with activity, devising new methods, collecting data and increasing our understanding.

Presently, it is easy to list more than a dozen distinct techniques which have actually been used to measure normal stresses, although there is considerable uncertainty about the accuracy of any of them and probably no consensus among rheologists on which of them is most reliable.

In view of the many methods available for measuring different combinations of normal stress, why not resolve these questions by testing the various methods against each other? Such comparisons have been done in the past, but, by and large, in a piecemeal fashion. Each laboratory has only a limited number of methods available; moreover, there has been insufficient exchange of samples among laboratories. It is, perhaps, significant that the most extensive set of different measurements on a common sample, that of *Kaye, Lodge* and *Vale* (4), revealed a fundamental error in many of the previously accepted methods of measuring normal stress.

In 1970, at the annual meeting of the Society of Rheology (USA), held at Princeton, N. J., there was a panel discussion on the measurement of normal stress. I took that occasion first to propose a cooperative program among laboratories measuring a common test sample of viscoelastic fluid and second to solicit comments and suggestions. Researchers representing over twenty laboratories expressed an interest in joining such a program and subsequently sent me requirements for their participation and many suggestions for a suitable test fluid. That program is now underway and this paper constitutes a report on the selection of a test sample.

There are four particular goals this program is designed to achieve, beyond the general goal of intercomparing measurements among the participating laboratories, viz:

1. A comparison of the limiting value of the dynamic modulus to the limiting value of the first normal stress difference

Coleman and *Markovitz* (5) show that for any simple fluid (6) the following condition must hold

$$\lim_{\omega \to 0} \frac{2G'(\omega)}{\omega^2} = \lim_{\varkappa \to 0} \frac{N_1}{\varkappa^2}$$

where $G'(\omega)$ is the dynamic shear modulus, N_1 is the first normal stress difference in a viscometric flow and ω and $\varkappa$ are frequency and rate of shear respectively. Since many of the methods of measuring normal stress are meaningless if this relation does not hold, it is important that it be

[1]) Numbers in brackets refer to the references listed at the end of this paper.

[2]) More correctly, I should say normal stress differences but will often use this abbreviation throughout this paper.

established for the test sample. For that to be practical, the sample must have normal stress of measurable magnitude at rates of shear sufficiently low that the extrapolation to zero shear rate is valid.

2. A complete test of the linear relation between the index of refraction tensor and the stress tensor in streaming birefringence

Although this relationship has been partially checked in the past for polymer solutions, the proportionality of the second normal stress and the corresponding index of refraction difference has never been established. All the streaming birefringence methods of measuring normal stress depend on this linear relation. It is known that for the relation to hold for a polymer solution, the solvent must be matching to suppress form birefringence. Further, the sample must be clear and free of spurious light scattering particles. Detailed explanations of these matters are to be found in the treatise of *Janeschitz-Kriegl* (7).

3. A comparison of the various methods of measuring normal stress in viscometric flows and an evaluation of their validity

Table 1 gives a list of methods which have actually been used to measure normal stress in viscometric flow. Almost all of these have been used at some time by at least one of the participants in the program. Their recommendations suggest that a highly elastic fluid with a viscosity of the order of 1000 poise would be most generally convenient. Some of the techniques require several liters of sample so that a large batch is called for.

4. Measurement of rheological properties in non-viscometric flows

Various devices make rheological measurements in non-viscometric flow, for example the *Maxwell-Chartoff* rheometer (8), *Képès* rheometer (9), apparatus for measuring extensional flow and dynamic measurement devices. Some of these measurements relate to rheological relations which characterize whole classes of materials (10) and are thus of particular interest. I hope to include a full range of such measurements in the program.

These goals determined the characteristics required of the test sample. Particularly stringent is the requirement that the normal stress be measurably large at rates of shear within the region describable by a second order fluid model. It is possible to adjust the normal stress of a polymer solution at a given rate of shear by selecting average molecular weight, concentration and solvent viscosity. The rate of shear limiting the second order fluid region also changes with these variables and in such a way that the normal stress at the limit is insensitive to these changes. A narrowing of molecular weight distribution, however, can produce a substantial change in the normal stress at the limiting rate of shear. Fig. 1, for example, shows some data on two solutions of polystyrene, a broad and a narrow molecular weight distribution. Although the average molecular weights are not the same (540,000 and 680,000) the data have been reduced with the limiting value of the dynamic viscosity in the manner of *Ferry* (11). The value of $2G'(\omega)$ gives a lower bound on the first normal stress and indicates that the second order fluid region should occur at higher stresses for the narrow distribution than for the broad distribution.

Table 1. Methods used to measure normal stress differences in viscometric flows

Method	Normal stress	References on limitations
Cone and Plate (total thrust)	N_1	(12)
Cone and Plate (pressure distribution)	$N_1 + 2N_2$	(4)
Cone and Plate (pressure at rim)	N_2	(12)
Cone and Plate with separation	N_2	(4)
Parallel plates (pres. dist.)	$N_1 - N_2 + k\frac{\partial N_2}{\partial k}$	(4)
Concentric cylinders (axial flow)	N_2	(13, 14)
Concentric cylinders (couette flow)	N_1	(4, 14)
Gravity flow across slit	N_1	(13)
Gravity flow along slit	N_2	(14)
Gravity flow, surface shape	N_2	(12)
Concentric cylinder (birefringence)	N_1	(16, 7)
Slit capillary (birefringence)	N_2	(17, 7)
Stability of couette flow	N_2	(18)
Jet expansion	N_1	(19)

General remarks on errors in measuring N_2 by various methods are found in (12) and (15).

From these considerations, I was led to the decision to form the test sample of a solution of polymer of narrow molecular weight distribution. Polystyrene was the natural choice since very narrow distributions of molecular weight are available in batches of several pounds.

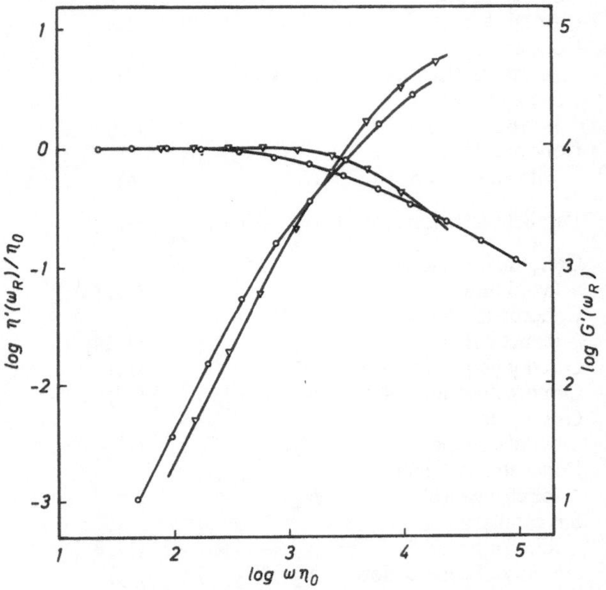

Fig. 1. A comparison of the mechanical properties of solutions of narrow (Δ) and broad (0) molecular weight distributions of polystyrene. Both solutions were 10% by weight

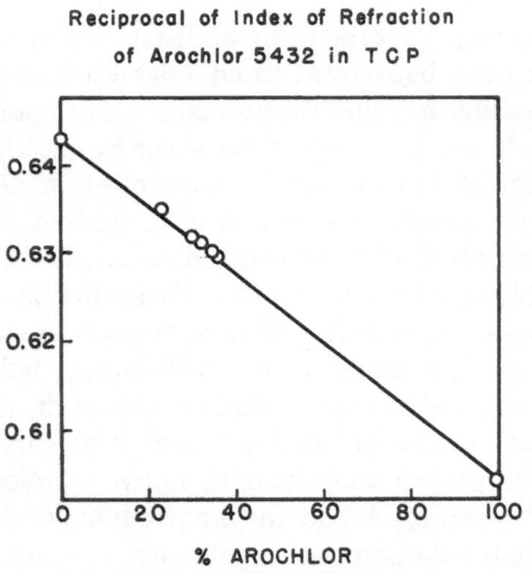

Fig. 2. The plot of reciprocal of index of refraction versus percent Arochlor is very nearly linear. The matching index was found to be 1.5880 at 25 °C with 36.1% Arochlor

Having chosen the polymer, it was necessary to find a suitable matching solvent to suppress the effects of form birefringence. Customarily, flow birefringence of polystyrene solutions has been done with Arochlor 1248 (a chlorinated biphenyl) or bromoform as a solvent. The latter is too unpleasant to handle in large batches in the laboratory and slowly decomposes in air and light. Arochlor 1248 is no longer available in quantity. Consequently, I decided to use a mixture of tri-cresyl-phosphate and Arochlor 5432 (which was then still available) as the solvent. Fig. 2 shows the reciprocal of the index of refraction as a function of percent Arochlor for such mixtures. The plot is linear to the accuracy of the data. The mixture selected had exactly the same index of refraction at 25 °C as the polystyrene. Strictly speaking, a matching solvent for birefringence is one for which the change in index with concentration is zero, but when the solvent and polymer have almost the same index of refraction, this criterion is usually well approximated. As it turns out, Dr. *Philippoff* has run some birefringence measurements on the test sample and reports that it seems satisfactory without further filtration, although the mixed solvent did produce some interesting and unexpected results. It is best I leave it to him to report the details. The solvent is very stable chemically, and exhibits almost no evaporation at room temperature even under vacuum for 48h[3]). The properties and toxicity of chlorinated biphenyls and tri-cresyl-phosphate are well known and must be taken into account when handling the sample.

The technique of dissolving the polymer in the solvent was carefully worked out by experiment on laboratory batches. It is necessary to produce a homogeneous viscous mixture without violent agitation or strong heating (which would alter the molecular weight distribution). I tried the method of using a preliminary solution of polystyrene in benzene and ultimately evaporating off the benzene in a vacuum. The method was abandoned because it proved too difficult to remove all traces of the benzene and the properties of the resulting sample were not constant. The method finally adopted involved heating to 65 °C and gentle mixing at intervals by hand operated paddle in a process which took about two weeks. Tests of samples from the top and bottom of the final mix showed no differences in viscosity or index of refraction.

---

[3]) There is evidence that the viscosity will change as much as 5% in six months if the sample is not kept sealed and in the dark.

The final test sample consists of a solution (7.14% by weight) of polystyrene with molecular weight average of about 440,000. The sample has a nominal zero-shear viscosity of about 680 poise at 25 °C and a very high temperature coefficient of viscosity, possibly as high as 12%/°C. Fig. 3 shows some preliminary data on the sample, viscosity as a function of rate of shear, normal stress as measured by the total thrust method on a *Weissenberg* rheogoniometer and some dynamic data.

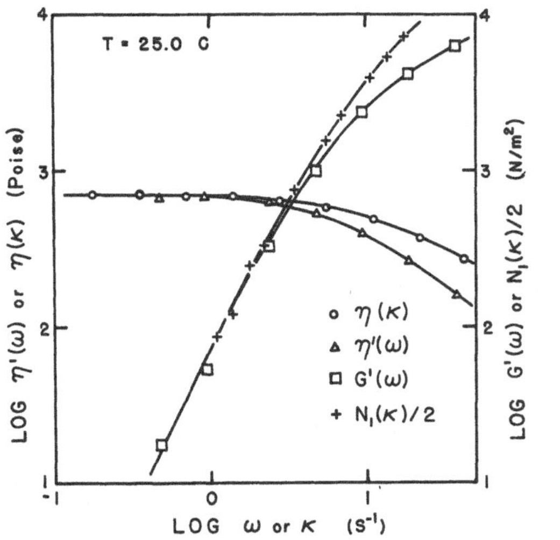

Fig. 3. Measurements of the properties of NBS non-linear test sample no. 1

Participants are requested to return their results to NBS by January 1, 1973. New participants will be most welcome. Quite probably further measurements will be suggested by the first results. I hope this program will settle some of the open questions of measurement of normal stress differences. Further, I expect to find the program valuable in evaluating the usefulness and practicality of issuing an NBS normal stress standard in the future. I invite comments from rheologists and particularly from industrial laboratories on this matter.

Summary

This paper is an account of the design of a test sample of polystyrene solution which will be distributed to research laboratories interested in participating in a comparison of techniques of measurement of normal stresses through a common sample. Questions of concentration, molecular weight distribution and solvent properties are considered.

References

1) *Garner, F. H., A. H. Nissan,* and *G. F. Wood,* Phil. Trans. Roy. Soc. (London) **A 243**, 31–66 (1950).

2) *Weissenberg, K.,* Specification of rheological phenomena by means of a rheogoniometer, Proc. Intl. Congr. Rheol. 1948, II, 114–118.

3) *Markovitz, H.,* Viscoelasticity, Phenomenological Aspects (*J. T. Bergen,* Ed.), p. 142 (New York 1960).

4) *Kaye, A., A. S. Lodge,* and *D. G. Vale,* Rheol. Acta 7, 368–379 (1968).

5) *Coleman, B. D.* and *H. Markovitz,* J. Appl. Phys. **35**, 1–9.

6) *Noll, W.,* Arch. Rational Mech. Anal. **2**, 197–226 (1958).

7) *Janeschitz-Kriegl, H.,* Adv. Polymer Sci. **6**, 170–318 (1969).

Maxwell, B. and *R. P. Chartoff,* Trans. Soc. Rheol. **9**, 41 (1965).

9) *Jones, P. E. R.* and *K. Walters,* Brit. Appl. Phys. Ser. 2, **2**, 815–820 (1969).

10) *Bernstein, B.,* Rheol. Acta **12** (1973).

11) *Ferry, J. D.,* Viscoelastic Properties of Polymers (second ed.) p. 556 (New York 1970).

12) *Tanner, R. I.,* Trans. Soc. Rheol. **14**, 483–507 (1970).

13) *Tanner, R. I.* and *A. C. Pipkin,* Trans. Soc. Rheol. **13**, 471–484 (1969).

14) *Kearsley, E. A.,* Trans. Soc. Rheol. **14**, 419–424 (1970).

15) *Ginn, R. F.* and *A. B. Metzner,* Trans. Soc. Rheol. **13**, 429–453 (1969).

16) *Philippoff, W.,* Studies of flow birefringence of polystyrene solutions. Proc. of the Fourth Internat. Congress on Rheol. part 2 (*E. H. Lee,* Ed.), p. 343–372 (New York 1963).

17) *Philipoff, W.,* Polymer Letters; **8**, 107–108 (1970).

18) *Denn, M. M.* and *J. J. Roisman,* Amer. Inst. Chem. Eng. J. **15**, 454–459 (1969).

19) *Harris, J.,* Nature. **190**, 993 (1961).

Author's address:

Dr. *E. A. Kearsley*
National Bureau of Standards
Washington, D.C. 20234 (U.S.A.)

Rheol. Acta 12, 550–558 (1973)

From the Department of Chemical Engineering, McGill University Montreal (Canada)

A concentric-cylinder rheometer for polymer melts

By J. M. Dealy, J. F. Petersen) and T.-T. Tee*

With 3 figures

(Received October 27, 1972)

Nomenclature

$\underline{C}$ *Cauchy-Green* strain tensor
$\underline{C}^{-1}$ *Finger* strain tensor
h gap spacing
k thermal conductivity
ΔT difference between maximum and minimum temperatures in sheared fluid
$\underline{II}$ second invariant of rate-of-strain tensor; $\underline{\Delta} : \underline{\Delta}$

Greek Letters

β material constant of second-order fluid
γ material constant of second-order fluid
γ_0 shear strain amplitude
$\dot{\gamma}$ shear rate in simple shear
$\dot{\gamma}_0$ maximum shear rate; equal to $\gamma_0 \omega$
$\underline{\Delta}$ rate-of-deformation tensor; grad velocity plus its transpose
ε material constant in eq. [8]
η viscosity
η^* complex viscosity
η' dynamic viscosity
λ characteristic time of fluid
μ material constant with units of viscosity
ϱ fluid density
$\underline{\tau}$ viscous or extra stress tensor
ω frequency (radians/sec.)

Introduction

Polymer melt testing devices now commercially available are adequate for studies of the viscometric functions up to moderately high shear rates and the complex viscosity over a wide range of frequencies. However, these devices have characteristics which limit their effectiveness at large shear rates or amplitudes of oscillation. For example in the cone-plate system, material is thrown out of the gap at large rates of rotation or large amplitudes of oscillation. In the case of the capillary viscometer, pressure and temperature are highly nonuniform at large shear rates. This paper describes a rheometer design for measurement of the shear viscosity of melts at high shear rates and also the response to large amplitude oscillatory shear. But only applications to the study of oscillatory shear will be discussed in detail.

In the study of nonlinear viscoelasticity, several types of dynamic test have been employed. *Marsh* (1) employed hysteresis loops generated by arbitrary but carefully controlled deformation histories in which the starting and ending states were isotropic. Stress growth after the sudden imposition of a steady shear has been employed by a number of people, including *Chen* and *Bogue* (2) who studied melts. Combined high shear rate steady shear and small-amplitude oscillatory shear (codirectional or orthogonal) have been employed by *Jones* and *Walters* (3) among others. Other possibilities are stress relaxation after cessation of steady-shear, large-stress creep, and large amplitude oscillatory shear. The last mentioned was chosen for use in the work described here because it does not involve the sudden starting and stopping of flows and because the authors feel it yields results which stand the best chance of being directly interpreted without the need for lengthy calculations.

In selecting a geometry for the shear-generating equipment the prime consideration was the need to operate at high shear rates or large amplitudes while at the same time limiting the errors associated with flow instabilities and secondary flows, nonuniformity in temperature, pressure and strain, and any tendency of the fluid to flow out of the high-shear-rate region. For dynamic testing of liquids, rotational flows are the only practical alternative, but several geometries are possible including cone-plate, parallel discs and concentric cylinders. *Dodge* and *Krieger* (4) have recommended concentric cylinders for dynamic testing on the basis of a detailed analysis of the governing equations of motion. In any event, it is very difficult to control the outflow of material

*) Veba-Chemie AG, Gelsenkirchen-Buer (Germany).

at high shear rates in the cone-plate and parallel disc systems. Moreover, the temperature can be made more uniform in the concentric cylinder system than in the cone-plate system.

Once the concentric-cylinder geometry had been selected it was necessary to study the effects of the errors mentioned above. First mentioned was stability and secondary flows. Except for end effects, flow in concentric cylinders, unlike cone-plate flow, is a true viscometric flow. However, the viscometric flow between concentric cylinders is not stable under all conditions and secondary flows and turbulence have been observed. For the case in which the inner cylinder rotates, theoretical stability limits were first worked out by *G. I. Taylor*, and the observed secondary flows are commonly associated with his name. The case in which the inner cylinder is stationary and the outer cylinder rotates was first studied by *Couette* and this flow is considerably more stable. For this and other reasons, it was decided to rotate the outer cylinder in the new melt rheometer. The flow even for this case eventually becomes unstable, although no satisfactory theory for this instability has been worked out. In the present case the high viscosity of melts together with the small gap spacing made necessary by other considerations might be expected to contribute to the stability of the flow.

The extent to which intertia or fluid acceleration interferes with the maintenance of a uniform shear rate has been examined for two special cases. The oscillatory shear flow of a second-order fluid has been analyzed for both the cases of parallel plates (5) and concentric cylinders (6). The second-order slow-flow approximation for the "simple fluid" can be written as:

$$\underset{\sim}{\tau} = \mu \underset{\sim}{A} + \beta \underset{\sim}{A}^2 + \gamma \underset{\sim}{A}^{(2)}. \qquad [1]$$

If (h/R) is small, the two solutions converge, and the criterion for a uniform strain rate is:

$$\frac{h^2 \omega \varrho}{2\mu}\left[\frac{(1+\omega^2\lambda^2)^{1/2}+\omega\lambda}{(1+\omega^2\lambda^2)}\right] \ll 1 \qquad [2]$$

where $\lambda = \gamma/\mu$ (a characteristic time).

MacDonald et al. (7) carried out an approximate analysis of oscillatory rectilinear flow for the case of linear viscoelasticity. They presented two criteria which must be satisfied in order to limit the effect of nonlinearity in the velocity profile on the apparent complex viscosity. The error in the dynamic viscosity (and loss modulus) will be negligible when

$$\frac{h^2 \omega \varrho \eta''}{6|\eta^*|^2} \ll 1. \qquad [3]$$

The error in the storage modulus (and η'') will be negligible when

$$\frac{h^2 \omega \varrho \eta''}{6|\eta^*|^2}\left[2 + \frac{\eta'^2}{\eta''^2}\right] \ll 1. \qquad [4]$$

The second criterion is the more severe and is thus the controlling factor. For a *Maxwell* fluid with relaxation time λ, this criterion becomes:

$$\frac{h^2 \varrho \omega^2 \lambda}{6\mu}\left[2 + \frac{1}{(\omega\lambda)^2}\right] \ll 1. \qquad [5]$$

It is apparent that, as in the case of stability considerations, a small gap and a high viscosity are favourable factors, and this seems likely to be true even in the case of highly nonlinear elastic behavior.

Temperature nonuniformity is governed by the balance between viscous heat dissipation and heat conduction. If both the inner and outer cylinders are cooled, there must be a temperature maximum somewhere in the sheared layer of fluid at all finite shear rates. For steady shear, energy considerations indicate that the difference between the maximum and wall temperatures will be of the following order of magnitude:

$$\Delta T = 0\left[\frac{\eta(\dot\gamma)\,\dot\gamma^2 h^2}{2k}\right]. \qquad [6]$$

In the case of oscillatory shear, the average rate of energy dissipation per unit volume is:

$$\tfrac{1}{2}\eta' \omega^2 \gamma_0^2 .$$

The difference between the maximum and minimum temperatures in the sheared fluid will vary periodically with an amplitude of the following order:

$$\Delta T = 0\left[\frac{\eta'(\omega)\,\gamma_0^2 \omega^2 h^2}{4k}\right]. \qquad [7]$$

Since η' decreases with ω, effects of increasing ω are partially offset. Still, in the case of both steady and oscillatory shear, there will always be a temperature variation within the test fluid. For polymers, η and μ' have rather high values while k is quite low relative to other types of materials. Furthermore, there are practical limits on how small h can be made as the ratio of machining tolerance to h must be kept small. For these reasons, it can be assumed that in general, the generation of temperature gradients as a result of viscous dissipation will be the limiting factor in the operation of such equipment.

End effects are an ever present problem in experimental rheology as it is virtually impossible to generate a shearing pattern which is perfectly uniform throughout the volume of fluid under test. In the case of *Newton*ian fluids it is normally possible to design equipment which can be calibrated to give a reliable measurement of the viscosity. Furthermore, the flow geometry can be so arranged that in the region where the shear rate is not uniform it is very small so that for approximate results the end effects can be neglected. In the case of non-*Newton*ian fluids, however, the calibration becomes fluid dependent. In addition, for pseudoplastic fluids, the effect on local shear stress of lower shear rate is offset by the higher viscosity at lower shear rate. *Highgate* and *Whorlow* (8) have reported experimental results which illustrate these effects and their importance in viscometry. They conclude that there is no completely reliable way of eliminating the importance of, or compensating for, end effects in the case of pseudoplastic fluids. Thus, end effects are a source of error which one must live with especially when one is studying non-*Newton*ian materials and wishes to avoid equipment which is complicated and difficult to clean.

Design details

The drive system is based on a 5 hp motor equipped with an SCR speed control unit. The output shaft speed is adjustable between 50 and 3470 rpm. This shaft is connected through a flexible coupling to a miter box. The standard 1-to-1 miter box is replaced by a 10-to-1 gear reduction system for low speed or low frequency operation. The vertical shaft of the miter box is connected to one of two optional transmission systems. One is simply a straight shaft connected to the outer cylinder of the rheometer and is employed for steady shear measurements. The other is a combination of gears, an eccentric and a rack and pinion for generating oscillatory motion. A pressurized mist lubrication system serves all the bearings for high speed operation.

In any system employing a rack and pinion to generate oscillatory rotation some backlash is unavoidable. It is not simply that the peaks of the wave are flattened; there is a periodic alternation between two waveforms displaced vertically with respect to each other. This is clearly undesirable so that gear backlash must be carefully controlled if meaningful results are to be obtained.

The speed of rotation of the miter box output shaft is monitored by counting the pulses coming from a photocell which is lighted through a plate with 100 slots, which rotates with the drive shaft. In the oscillatory mode of operation, this speed measurement gives the frequency, while the actual instantaneous strain is monitored by means of a linear displacement transducer whose shaft is fastened to one end of the rack of the oscillating mechanism.

Fig. 1 shows the essential features of the drive system while fig. 2 is a sketch of the concentric cylinders which are the heart of the instrument. The cylinder walls are slightly tapered to facilitate assembly and disassembly. The 50-to-1 typer also makes it possible to vary the gap spacing by simply raising or lowering the inner cylinder. While it has been found possible to operate with a gap of only 0.1 mm with viscous calibrating oils, with polymer melts precise alignment is more difficult, and a gap of 0.25 mm must be allowed. This gap spacing would allow the generation of shear rates as high as 5000 sec^{-1}. There is an overflow reservoir above the gap and a conical layer below it. It has been estimated that for a *Newton*ian fluid the end effects make a contribution of only 0.1% to the measured torque. But for a typical polyethylene melt in the power-law region of shear rates this contribution can rise to 1% or more. Bubble-free test samples are prepared in a mold having the same shape as the outer cylinder.

The temperature of the sheared fluid is controlled by circulating oil from a thermostated reservoir both inside the inner cylinder and outside the outer cylinder. Cooling water circulates through the coupling between the inner cylinder and the torquemeter tube to prevent heating of the latter as the magnetic properties of the torque tube are somewhat temperature dependent.

The torquemeter which monitors the torque on the inner, stationary cylinder operates on the basis of magnetic stress anisotropy. The use of this type of torquemeter with rotating machinery has been discussed by *Barton* et al. (9, 10). The inner cylinder transmits the torque generated by the shearing in the fluid under test to the rigidly fixed support through a ferromagnetic steel tube. The magnetic properties of this tube are modified by very small amounts of torsional strain. Two coils surround the tube; one is energized by a

400 Hz alternating current, and the other is a detector winding. Careful design can produce a characteristic which is remarkably linear. This torquemeter has a precision comparable to that of good strain-gauge types but with much greater output and overload capacity. It is thus quite robust and relatively immune to noise. It is not difficult or expensive to make, and requires only a few inches of shaft length. Robustness was considered to be a primary requirement because of the high stresses anticipated in the study of molten polymers at high shear rates.

Because the effects of thermal expansion on the cylinders cannot be calculated with a high degree of certainty, it was necessary to calibrate the instrument over a range of temperatures. From the results of the calibration runs a chart was made up showing gap spacing as a function of temperature.

Signal conditioning

The output from the torque tube coils is a 400 Hz AC signal which must be first demodulated and then filtered. When the rheometer is being used to measure viscosity in steady shear, it is sufficient to use a simple L-C filter and an ammeter to obtain a reading of the torque. For oscillatory testing, however, it is necessary to eliminate the 400 Hz component in the demodulator output without introducing significant phase shift or amplitude attentuation in the range of frequencies over which the instrument is to be used. The use of a 400 Hz energizing signal was based on a selection of 10 Hz as the maximum frequency of operation, and the assumption that the third and fifth harmonics should be recoverable without substantial phase shift or attenuation. A variable active filter is employed in this application. It has a 48 DB per octave attenuation slope and the other advantages of "Butterworth" circuit characteristics. The output of the active filter can be passed to an oscillograph for recording of the waveform or to a dualtrace storage oscilloscope equipped with Polaroid camera. The oscilloscope can be used to observe the waveform by use of a sweep generator on the horizontal axis or it can be used to observe closed curves of stress (torque) versus strain by feeding the displacement transducer output to the X axis input. In addition, for reasons to be discussed in the next section, it was desired to observe also the stress versus the rate of strain during oscillatory shear. Differentiation of a signal electronically is not recommended except when the signal has an extraordinarily low noise level, because the differentiation process amplifies noise. However, for a simple harmonic signal we note that integration as well as differentiation leads to the desired 90 degree phase shift. Therefore, a variable-gain integrator and back-biasing circuit was patched together and employed to condition the strain signal.

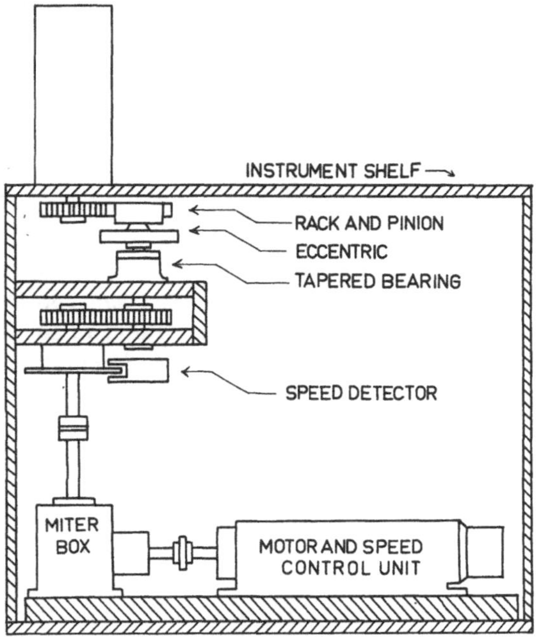

Fig. 1. Rheometer drive system

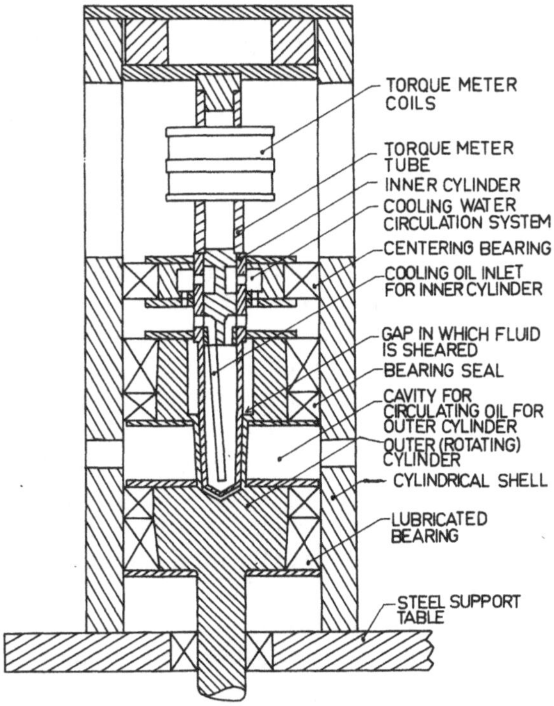

Fig. 2. Cross section of rheometer

Methods of data interpretation

It has been observed that there is a small but finite range of amplitudes over which real fluids in oscillatory shear exhibit linear viscoelasticity. That is to say, the output waveform is simple harmonic and the amplitude ratio and phase shift are independent of amplitude. A number of observers have reported that linearity is observed when the strain amplitude is below some critical value which has been variously reported to be between 0.2 and 1.0. The way in which the nonlinearity appears when this critical amplitude is exceeded is of great importance in the extension of this test method to large deformations. Although a dependence of phase shift and amplitude ratio on amplitude must be accompanied by the appearance of higher harmonics in the response characteristics, these have been reported to be quite small in a significant range of amplitudes so that several researchers have employed some of the concepts of linear viscoelasticity in the presentation of results on nonlinear tests. That is, they have computed the components of the complex viscosity or modulus as for linear tests, assuming a sinusoidal output, but have given the results as functions of amplitude as well as frequency.

MacDonald et al. (7) used a rheogoniometer to study the effect of amplitude on measurements of complex viscosity. They studied several polymer solutions and concluded that the optimum range of amplitudes for the rheogoniometer is from 0.1–0.3. They found that, even up to the largest amplitude which they used ($\gamma_0 = 1.4$), anharmonicity was not apparent in the stress. However, they admitted that their measurements were subject to large errors. *Philippoff* (11) also reported results of large amplitude oscillatory shear tests of polymer melts. The dynamic viscosity was independent of amplitude up to amplitudes of about 1.0. However, the dynamic modulus, G', varied with amplitude over the entire range of amplitudes studied, that is, down to $\gamma_0 = 0.25$. The substantial difference between the dependency of η' and G' on γ_0 was surprising and has not been explained. For the largest amplitudes (up to 7.0) anharmonicity was apparent, but the amplitude of the third harmonic was estimated to be less than 5% of the fundamental in the most nonlinear case.

Vinogradov et al. interpreted their results on large amplitude oscillatory shear of polypropylene (12), polyisobutene (13) and a monodisperse polybutadiene (14) in terms of a truncation of the long-time part of the relaxation spectrum and suggested that the information obtained from this type of experiment is equivalent to that obtained in the case of small amplitude oscillatory shear superposed on high shear rate steady simple shear. They observed that the maximum amplitude for linear behavior was constant at about 0.2 up to frequencies of about 10 Hz. Above this frequency, however, the maximum amplitude decreased slowly.

An obvious way of interpreting nonlinear test results is to compute the material constants of an assumed constitutive equation. All of the models which have been proposed for isotropic fluids are examples of "simple fluids". That is, they are based on the simple assumption that the stress can be represented as some functional of the history of the deformation gradient. If this functional is written as an infinite series of multiple integrals by use of the *Frechet* expansion, any truncated portion of this series can be considered to be an approximation valid up to some particular level of deviation from linearity. Such an expansion could, in principle, be employed to interpret the results of nonlinear tests by determining the kernels of the integrals. This approach has been employed to analyze "weakly nonlinear" behavior of solid polymers in relaxation (15, 16) and creep (17, 18) experiments. *Yanas* and *Haskell* (15) found in their studies of stress relaxation in amorphous glassy polymers that for elongational strain below 1% linear viscoelasticity was exhibited and that for strains up to 5% only the first two kernels were necessary to describe the behavior. However, they observed that the second kernel varies in a complex way with temperature and has no obvious physical significance. *Onogi* et al. (19) have presented relationships between various integrals of the kernel functions and measurable quantities in oscillatory shear. However, these involve sums of multiple integrals and it would be practically impossible to make use of them for routine data analysis.

A more promising possibility is the use of explicit empirical constitutive equations containing a reasonable number of material constants. One type of equation which has been much studied is of the form:

$$\underset{\sim}{\tau} = \int_{-\infty}^{t} m[(t-t'), \Pi(t')] \times [(1+\varepsilon)\, \underset{\sim}{C}^{-1} + \varepsilon \underset{\sim}{C}]\, dt'. \qquad [8]$$

The *Bird-Carreau* model (20) is an example of a constitutive equation of this type. We note, however, that the nonlinearity enters in through the occurrence of $\Pi(t')$ in the relaxation function. In oscillatory shear this quantity is:

$$\Pi(t') = 2\dot{\gamma}^2(t') = 2(\gamma_0 \omega \cos \omega t')^2. \qquad [9]$$

We note that the frequency and amplitude occur as a product and are thus predicted to have an equivalent influence on nonlinearity. As *Tanner* and *Simmons* (21) have pointed out, this is inconsistent with the observation discussed earlier that at low but finite amplitudes, oscillatory shear results are independent of amplitude. They proposed a "network rupture model" which avoids some of the problems of the simpler models and does predict the correct approach to linear viscoelasticity for small amplitude oscillatory shear. *Tanner* (22) has computed the stress predicted by this model for large amplitude oscillatory shear for the case of one combination of fluid, frequency and amplitude. However, *Carreau* (23) has pointed out that the simple network-rupture hypothesis predicts that when the amplitude reaches some critical value, the dynamic viscosity and modulus decrease suddenly, especially at high frequencies, and this is inconsistent with observations. *Carreau* (24) and *Chen* and *Bogue* (2) have recently discussed the problem of fitting the results of nonlinear dynamic tests with constitutive equations of various types. Their conclusions have contributed to the growing weight of evidence that integral constitutive equations can predict quantitatively correct results for nonlinear deformations only if the relaxation times of the model are made dependent on the strain history. Unfortunately this makes the model substantially more complex. Thus the use of a specific constitutive equation for routine analysis of nonlinear viscoelasticity data is not presently attractive.

A method of analyzing the stress response which does not rely on any specific model is the use of harmonic analysis. Because of the physical limitations on the type of response which could occur, the *Dirichlet* conditions guarantee that the shear stress can be represented by a convergent *Fourier* series and that the *Euler* coefficients can be computed if the shear stress as a function of time is known with sufficient accuracy. If the fluid of interest has no yield stress and is isotropic in its rest state, then even harmonics must be absent after a sufficient length of time.

There are several ways of resolving the first few *Fourier* components. The straightforward application of the *Euler* formulae to numerical data was employed by *Onogi* et al. (19), but this is a tedious procedure. *Philippoff* (11) employed a procedure in which electronically synthesized sums of the first and third harmonics were compared with the experimental stress output, and a trial-and-error procedure used to determine the *Euler* coefficients for the first two terms. This procedure, of course, relies heavily on the judgement of the experimenter. A spectrum analyzer could be used to produce a completely electronic characterization of the output. *Ibrahim* (25) has described a correlator which could be used for this purpose, and *Watson* (26) and *Warburton* and *Davis* (27) have discussed the use of electronic instruments in oscillatory shear rheometry.

To make practical use of the results of a harmonic analysis, it is desirable to relate the *Euler* coefficients to fundamental material constants whose dependence on temperature and molecular structure could be studied. *Onogi* et al. (19) have presented relationships between the *Euler* coefficients and certain combinations of definite integrals of the kernel functions of the *Green-Rivlin* (28) integral expansion representation of nonlinear viscoelasticity. However, these relationships are very complex, and operational techniques for their use are not available. Furthermore, as has already been pointed out, these kernel functions do not provide a practical basis for characterizing nonlinear viscoelastic properties of materials. *Dodge* and *Krieger* (29) have successfully employed harmonic analysis in their technique for characterizing inelastic fluids by use of oscillatory shear. But in general, harmonic analysis does not appear to provide a useful basis for the characterization of rheologically complex materials. Only the first few terms of the *Fourier* series can be resolved because of experimental limitation on precision and elimination of noise, and the coefficients of these terms cannot be related to physically meaningful constants. *Walters* and *Jones* (30) on the basis of their detailed studies of the behavior of materials in the *Weissenberg* rheogoniometer, concluded that harmonic analysis is not a useful tool in viscoelasticity studies.

Closed loop displays

All of the methods for the interpretation of nonlinear viscoelastic responses which have been

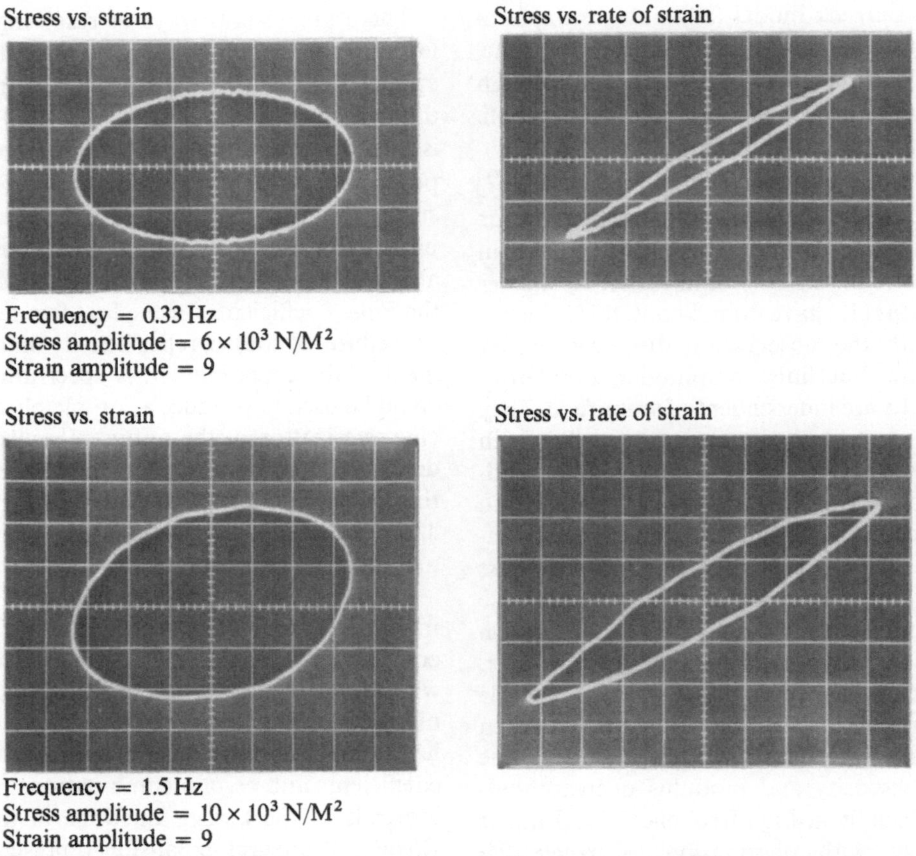

Fig. 3. Response loops for Dupont of Canada type 2511-4V medium density polyethylene resin at 175 °C

discussed above require substantial calculation, and none of these methods gives a unique characterization of the fluid under test. Thus, it is very desirable to establish some means for classifying and comparing materials by direct visual inspection of the response data. A direct inspection of the output wave form, i.e. stress versus time, is not very fruitful as the eye can detect only gross nonlinearities from this type of representation. The authors feel that a more discriminating interpretation can be made by inspection of stationary closed curves produced by plotting stress versus some other periodic function such as strain, rate of strain or the derivative of the stress.

The third type of representation mentioned, stress vs. its derivative, is used to study the properties of nonlinear differential equations where it is called a phase plane representation. However, the use of this type of representation would require the differentiation of the stress, and the instrumental generation of the derivative of a signal is not feasible unless the signal is virtually noise-free. The first two examples have more obvious physical significance, and produce a type of display which has been extensively employed in the science of acoustics where the curves produced are called *Lissajou* figures. In acoustics the response is generally linear, but the term "*Lissajou* figure" has been carried over to the general field of nonlinear system response. If the fluid characteristics are constant with time and the response is linear, the *Lissajou* figure will be an ellipse. *Philippoff* (11) used a *Lissajou*-type display in his study of nonlinear oscillatory flow of polymer solutions. *Davis* (31) has used stress-strain curves to characterize pharmaceutical semisolids, and *Payne* and *Whittaker* (32) and *Krizek* (33) have used this technique to study the rheological properties of clay-water systems.

Fig. 3 shows examples of experimental response loops for a medium density polyethylene at 175 °C. The zero-shear viscosity of this resin at 175 °C is 4×10^3 poises. The strain amplitude was 9 for these tests, and results are shown for 2 frequencies. This material exhibits a clearly defined nonlinearity, but is not as elastic as a high-density polyethylene which was studied under similar conditions. Nonlinearity is indicated by deviation from ellipsoidal form, and elasticity manifests itself in an opening-up of the stress-rate of strain loop. The noise which can be seen in the stress output at the lower frequency is a result of vibrations transmitted directly from the drive motor to the torquemeter through the frame of the rheometer. It can be controlled by mounting the motor on rubber pads or on a separate frame. The irregularities in the higher frequency loops result partly from gear backlash, and partly from a nonuniform distribution of polymer melt near the top of the gap. The equipment is currently being modified to reduce the contribution of these factors to the output noise level.

Acknowledgment

The development of the new melt rheometer was supported by the National Research Council of Canada.

Summary

If one wishes to measure the viscosity of a polymer melt at high shear rates there are substantial fluid dynamical and heat transfer difficulties. Cone-plate instruments are limited because of secondary flows and because the fluid tends to leave the gap. In capillary-flow instruments, there are substantial radial temperature gradients and the possibility of flow irregularities.

Similar difficulties are met in trying to study the response of melts to large-amplitude oscillatory shear, and fluid inertia must be added to the list. However, large-amplitude oscillatory shear is a test which is useful for studying non-linear viscoelasticity, and many flows of practical importance involve deformations outside the range of validity of the assumptions of linear viscoelasticity theory.

A heavy duty rheometer has been designed in which shear viscosity can be measured at high shear rates, and which can also be used for large-amplitude oscillatory shear tests. The melt is sheared between concentric cylinders; the torque on the inner, stationary cylinder is monitored while the outer cylinder either rotates at steady speed or oscillates. The shear rate of frequency and amplitude are continuously variable over wide ranges.

Careful consideration was given to the problems posed by hydrodynamic stability, fluid acceleration, heat generation and end effects, and the final design represents what the authors feel is a reasonable compromise between minimizing the influence of these factors and the basic practical requirement that the instrument have a reasonable cost and uncomplicated operating procedure.

In large-amplitude oscillatory shear, the interpretation of the experimental results poses special problems. The stress response is periodic, but not sinusoidal so that it is not possible to apply the methods of linear viscoelasticity. A number of possibilities suggest themselves, but it has been concluded that the best method of representation involves the plotting of stress versus strain or rate-of-strain. This results in closed curves which have distinctive shapes depending on the basic nature of the fluid response.

Zusammenfassung

Bei der Viskositätsmessung an Polymerschmelzen unter hohen Schergeschwindigkeiten treten strömungsdynamische und Wärmeübergangsprobleme in Erscheinung. Kegel-Platte-Rheometer sind nur bei relativ geringen Deformationen brauchbar, weil bei höherer Scherbeanspruchung die Meßsubstanz infolge Sekundärströmungserscheinungen bzw. Trägheitskräften aus dem Meßspalt austritt. Bei Kapillarrheometern ergeben sich unter hohen Scherbedingungen gravierende radiale Temperaturgradienten und Fließinstabilitäten.

Die gleichen Probleme finden sich, wenn das Verhalten von Schmelzen bei oszillatorischer Beanspruchung mit großer Scheramplitude untersucht werden soll. Allerdings ist diese Beanspruchungsart besonders geeignet, wenn die nicht-lineare Viskoelastizität studiert werden soll – und viele praktische Strömungsverhältnisse lassen sich durchaus nicht im Rahmen der linearen Viskoelastizitätstheorie behandeln.

Es wurde ein robustes Rheometer vom Couettetyp entworfen, mit dem die Scherviskosität bei hohen Schergeschwindigkeiten gemessen werden kann und das ebenso für Oszillationsversuche mit großer Scheramplitude geeignet ist. Der stetige oder oszillatorische Antrieb erfolgt auf den äußeren Zylinder, während am inneren, ruhenden Zylinder das übertragene Drehmoment gemessen wird.

Besonderer Wert wurde bei dem Entwurf des Rheometers den Problemen beigemessen, die sich durch hydrodynamische Stabilität, Beschleunigungsvorgänge, Wärmeentwicklung sowie Endeffekte ergeben. Die endgültige Form des Rheometers ist in den Augen der Autoren ein vernünftiger Kompromiß zwischen der Minimierung der genannten Einflüsse und den praktischen Voraussetzungen, daß ein Rheometer nicht zu teuer ist, aber einfach zu bedienen sein muß.

Die Interpretation der experimentellen Ergebnisse solcher Oszillationsversuche ist einigermaßen problematisch. Der resultierende Spannungsverlauf ist zwar periodisch, aber nicht sinusförmig, so daß die Methoden der linearen Viskoelastizität nicht anwendbar sind. Obwohl sich mehrere Möglichkeiten anbieten, dürfte die beste Darstellungsmethode der Ergebnisse durch die Auftragung der Spannung in Abhängigkeit von der Scherung bzw. der Schergeschwindigkeit gegeben sein. Dieses ergibt geschlossene Kurven, deren besonderes Aussehen durch die Eigenschaften des Testmaterials gekennzeichnet ist.

References

1) *Marsh, B. D.*, Trans. Soc. Rheol., **12**, 489 (1968).

2) *Jen Chen, I.* and *D. C. Bogue*, Trans. Soc. Rheol., **16**, 59 (1972).

3) *Jones, T. E.* and *K. Walters*, J. Phys. B, **4**, 85 (1971).

4) *Dodge, J. S.* and *I. M. Krieger*, Rheol. Acta, **8**, 480 (1969).

5) *Markovitz, H.* and *B. D. Coleman*, Adv. Appl. Mech., **8**, 69 (1964).

6) *Markovitz, H.* and *B. D. Coleman*, Phys. Fluids, **7**, 833 (1964).

7) *MacDonald, I. F., B. C. Marsh*, and *E. Ashare*, Chem. Eng. Sci., **24**, 1615 (1969).

8) *Highgate, D. J.* and *R. W. Whorlow*, Rheol. Acta, **8**, 142 (1969).

9) *Barton, T. H.* and *R. J. Ionides*, IEEE Trans., PAS-85, 152 (1966).

10) *Barton, T. H.* and *L. Solar*, IEEE Trans., IGA-3, 310 (1967).

11) *Philippoff, W.*, Trans. Soc. Rheol., **10**, 317 (1966).

12) *Vinogradov, G. V., Y. G. Yanovsky*, and *A. I. Isaev*, Int. Chem. Eng., **11**, 309 (1971).

13) *Vinogradov, G. V., Y. G. Yanovsky*, and *A. I. Isaev*, J. Polym. Sci., **A-2**, **8**, 1239 (1970).

14) *Vinogradov, G. V., Y. G. Yanovsky, A. I. Isaev, V. P. Shatalov*, and *V. G. Shalgonova*, Int. J. Polym. Mater., **1**, 17 (1971).

15) *Yanas, I. V.* and *V. C. Haskell*, J. Appl. Phys., **42**, 610 (1971).

16) *Yanas, I. V., N. H. Sung*, and *A. C. Lunn*, J. Macromol. Sci., Phys., **5**, 487 (1971).

17) *Yanas, I. V.* and *A. C. Lunn*, J. Macromol. Sci., Phys., **4**, 603 (1970).

18) *Lai, J. S. Y.* and *W. N. Findley*, Polym. Eng. Sci., **9**, 378 (1969).

19) *Onogi, S., T. Masuda*, and *T. Matsumoto*, Trans. Soc. Rheol., **14**, 275 (1970).

20) *Bird, R. B.* and *P. J. Carreau*, Chem. Eng. Sci., **23**, 427 (1968).

21) *Tanner, R. I.* and *J. M. Simmons*, Chem. Eng. Sci., **22**, 1803 (1967).

22) *Tanner, R. I.*, Trans. Soc. Rheol., **12**, 155 (1968).

23) *Carreau, P. J.*, Ph. D. Dissertation, Univ. of Wisconsin, 1968.

24) *Carreau, P. J.*, Trans. Soc. Rheol., **16**, 99 (1972).

25) *Ibrahim, O. E.*, Rheol. Acta, **8**, 214 (1969).

26) *Watson, J. D.*, Rheol. Acta, **8**, 201 (1969).

27) *Warburton, B.* and *S. S. Davis*, Rheol. Acta, **8**, 205 (1969).

28) *Green, A. E.* and *R. S. Rivlin*, Arch. Rat. Mech. Anal., **1**, 1 (1957); **3**, 82 (1959); **4**, 387 (1960).

29) *Dodge, J. S.* and *I. M. Krieger*, Trans. Soc. Rheol., **15**, 589 (1971).

30) *Walters, K.* and *T. E. R. Jones*, Proc. 5th Int. Cong. Rheol., **4**, 337 (Baltimore 1970).

31) *Davis, S. S.*, J. Pharmac. Sci., **60**, 1356 (1971).

32) *Payne, A. R.* and *R. E. Whittaker*, Rheol. Acta, **9**, 91 (1970).

33) *Krizek, R. J.*, Trans. Soc. Rheol., **15**, 433 (1971).

Authors' addresses:

Dr. *J. M. Dealy* and *T. T. Tee*
Dept. of Chemical Engineering
McGill University, Montreal (Canada)

Dipl.-Ing. *J. F. Petersen*
Veba-Chemie AG
D-4660 Gelsenkirchen-Buer (Germany)
Postfach 45

Rheol. Acta **12**, 559–562 (1973)

Laboratoire de Rhéologie, I.S.M.C.M., Saint Ouen (France)

Sur la mesure des constantes viscoélastiques des plaques de matériaux composites par les ultrasons

Par N. P. Vinh Tuong)*

Avec 3 figures

(Reçu p. p. le 27 Octobre 1972)

Introduction

Les techniques ultrasonores constituent un instrument de choix pour les études expérimentales en rhéologie relatives aux corps isotropes. Divers chercheurs les ont utilisés (*McSkimin*, *Kono*, *Buvet*, *Smadja* etc.). Une revue des méthodes principales a été faite notamment par *Ferry* (1). Récemment, on cherche à étudier les matériaux composites anisotropes par les ultrasons, *Markham* (2), *Vinh* (3), *Tauchert* et *Guzelsu* (4). Toutefois les travaux sont encore rares dans ce domaine. Dans cet exposé, nous nous attachons à faire ressortir les avantages et les inconvénients des techniques ultrasonores pour l'étude des plaques anisotropes de composite.

Principe des méthodes utilisées

On suppose que les ondes ultrasonores crées dans l'éprouvette soient de faible amplitude afin que le matériau anisotrope étudié soit linéairement viscoélastique. En appelant σ^*_{mn} et ε^*_{op} les contraintes et les déformations, l'étoile désignant les variables en régime harmonique, la loi de comportement est:

$$\sigma^*_{mn}(\omega) = C^*_{mnop}\varepsilon^*_{op}(\omega) \qquad [1]$$

ω est la pulsation, C^*_{mnop} les modules complexes:

$$C^*_{mnop} = C'_{mnop} + i\,C''_{mnop}. \qquad [2]$$

Faute de place, nous ne considérons ici que les matériaux à symétrie hexagonale caractérisés par cinq modules complexes rassemblés dans la matrice suivante:

$$\begin{matrix} C^*_{1111} & C^*_{1122} & C^*_{1133} & 0 & 0 & 0 \\ C^*_{1122} & C^*_{1111} & C^*_{2233} & 0 & 0 & 0 \\ C^*_{1133} & C^*_{2233} & C^*_{3333} & 0 & 0 & 0 \\ 0 & 0 & 0 & C^*_{2323} & 0 & 0 \\ 0 & 0 & 0 & 0 & C^*_{2323} & 0 \\ 0 & 0 & 0 & 0 & 0 & C^*_{1212} \end{matrix} = C^*_{mnop}\,. \qquad [3]$$

La fig. 1 montre les axes de symétrie, x_3 coïncidant avec la direction des fibres. Dans [3] les constantes viscoélastiques figurant sur la diagonale peuvent être évaluées à l'aide des ondes longitudinales ou transversales découplées.

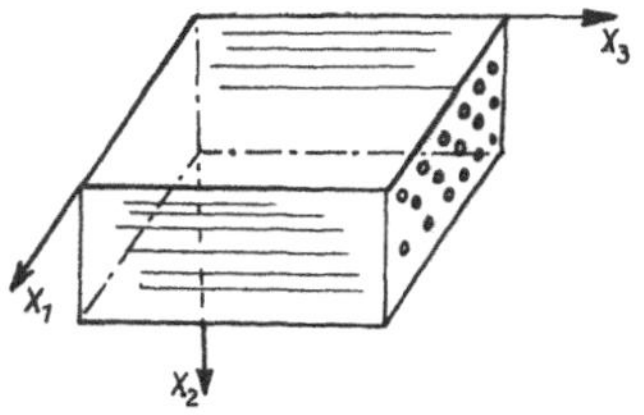

Fig. 1. Axes de symétrie du matériau à isotropie transverse (plan d'isotropie x_1, x_2) x_3 est la direction des fibres

C^*_{1111} et $C^*_{2222} = C_{1111}$ correspondent aux ondes longitudinales perpendiculaires à x_3 dans les directions x_1 et x_2.

Les ondes longitudinales dans le sens x_3 des fibres permettent d'atteindre C^*_{3333}. Toutefois, si la plaque est mince ce qui se produit fréquemment en pratique, il ne faut pas confondre les ondes longitudinales de plaque avec les ondes longitudinales (de dilatation) d'un milieu supposé indéfini.

Les ondes transversales polarisées permettent d'atteindre les modules de cisaillement C^*_{2323}, $C^*_{3131} = C^*_{2323}$, $C^*_{1212} = \frac{1}{2}C^*_{1111} - C^*_{1122}$. Pour une plaque mince, la mesure directe de C^*_{1212}

*) Docteur ès-science, maître de conférence, Laboratoire de Rhéologie I.S.M.C.M., 3, Rue Fernand Hainaut, F-93 Saint-Ouen (France).

peut présenter des difficultés comme celles de C^*_{3333} signalées plus haut[1]).

Les termes non diagonaux dans [3] ne peuvent être évalués qu'avec des ondes ultrasonores se propageant dans une direction non confondue avec les axes de symétrie.

En Annexe (I) nous présentons les détails sur la vitesse de propagation v obtenue à l'aide des raideurs de *Christoffel* Γ_{ij}.

On a affaire alors à des ondes dites quasi-longitudinales ou quasi-transversales.

Ondes découplées et mesure de l'attenuation de l'onde propagée

Si les ondes sont découplées et si l'amortissement $\operatorname{tg}\delta_{mnop}$ n'est pas trop important,

$$\operatorname{tg}\delta_{mnop} = \frac{\mathcal{I}_m(C^*_{mnop})}{\mathcal{R}_e(C^*_{mnop})}. \qquad [4]$$

L'onde garde pratiquement leur forme en se propageant alors il est possible d'évaluer [4] en mesurant l'atténuation α de l'onde après le parcours d'une distance d

$$\alpha = \frac{1}{d}\left[\ln\frac{A_0}{A_d}\right]. \qquad [5]$$

A_0, A_d étant les amplitudes

$$\operatorname{tg}\left(\frac{\delta_{mnop}}{2}\right) = \frac{\alpha v}{\omega} \qquad [6]$$

v vitesse de l'onde, ω pulsation

$$|C_{mnop}| = \frac{v^2\varrho}{1 + \frac{\alpha^2 v^2}{\omega^2}} \qquad [7]$$

ϱ masse volumique, α atténuation donnée par [5].

Les amortissements relatifs aux termes non diagonaux dans [3] sont difficiles à évaleur[2]).

Description sommaire de l'appareillage

a)

Nous ne faisons pas ici une étude comparative de nombreuses techniques ultrasonores utilisées (voir par exemple *Truell, Elbaum* et *Chick* (5). Nous ne présentons que la technique des ondes progressives par paquets d'ondes sinusoïdales. Deux bancs ultrasonores ont été utilisés. La fig. 2 représente un banc par transmission. On y distingue une paire de transducteurs (à quartz ou au zirconate-titanate de plomb). Dont l'un sert d'émetteur et l'autre, de récepteur. Le générateur de puissance délivre des trains d'impulsions électriques à l'émetteur.

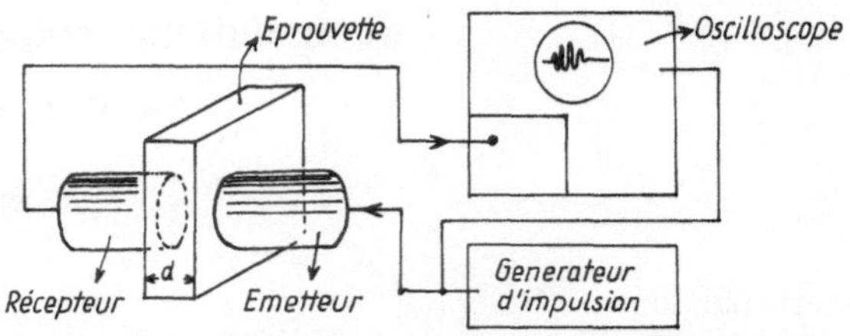

Fig. 2. Banc ultrasonore par couplage direct des transducteurs piézoélectriques sur l'éprouvette

On mesure le temps de propagation t d'un paquet d'ondes. On en déduit la vitesse, d étant la distance de parcours

$$v = \frac{d}{t}. \qquad [8]$$

Le couplage des transducteurs sur l'éprouvette s'effectue soit par collage soit par application d'une résine spéciale (Aroclor Monsanto) qui laisse passer aisément les ondes transversales et est facile d'usage. Il faut s'assurer que les couches de résine soient très minces par rapport à l'épaisseur de l'éprouvette. Les dépouillements des résultats expérimentaux sont faits par [4], [5], [6] et [7].

b) Banc ultrasonore par immersion dans l'eau

Les détails de ce banc peuvent être trouvés ailleur (6). Ici nous voulons montrer la facilité avec laquelle on crée des ondes de direction quelconque dans l'éprouvette par inclinaison de cette dernière, fig. 3.

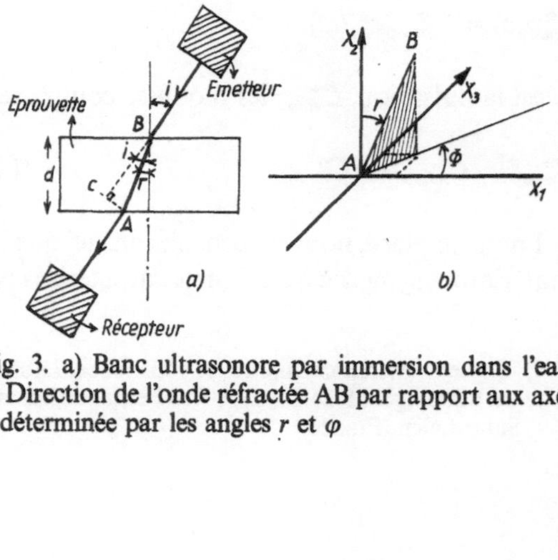

Fig. 3. a) Banc ultrasonore par immersion dans l'eau. b) Direction de l'onde réfractée AB par rapport aux axes x_i déterminée par les angles r et φ

[1]) Notons que la mesure indirecte de C_{1212} par l'intermédiaire de C_{1111} et C_{1122} supposent que le matériau soit isotrope transverse.

[2]) L'évaluation indirecte de C^*_{1122} à partir de C^*_{1111} et C^*_{1212} peut donner lieu à des erreurs importantes pour l'amortissement.

Appelons i et r l'angle d'incidence et l'angle de réfraction, on a les cosinus directeurs suivants du rayon réfracté dans l'éprouvette.

$$n_r = \cos r$$

$$n_r = \sin r \cdot \cos\varphi, \qquad n_r = \sin r \cdot \sin\varphi$$

avec

$$\frac{\sin i}{\sin r} = \frac{v_{\text{éprouvette}}}{v_{\text{eau}}}$$

$$\operatorname{tg} r = \frac{\sin i}{\cos i + \frac{\tau v_{\text{eau}}}{d}} \qquad [8]$$

τ est la différence de temps de parcours dans l'éprouvette et sur la distance équivalente en eau

$$\tau = \frac{AB}{v} - \frac{CB}{v_{\text{eau}}}. \qquad [9]$$

On peut donc évaluer r à l'aide de [8] et [9] et calculer la vitesse v en fonction des constantes C_{mnop}, voir Annexe (I).

Les mesures d'intervalles de temps sont faites à l'oscillographe cathodique. Les mesures d'amplitude, à l'aide de tiroir comparateur spéciaux sont précises au $\frac{1}{1000}\,e$.

Une force de compression appliquée aux transducteurs est nécessaire pour réduire au minimum les couches de résine.

Etudes possibles avec les ondes ultrasonores

a)

L'un des aspects intéressants du comportement dynamique est constitué par les propriétés dispersives des ondes planes (4). Les ultrasons permettent de vérifier les récentes études théoriques, *Sun*, *Achenbach* et *Hermann* (7) notamment.

b)

Les atténuations des ondes découplées sont attribuables aux propriétés viscoélastiques de chaque phase du composite certes mais aussi à l'adhésion interfaciale. Or les études expérimentales de cette dernière peuvent être effectuées par les ultrasons qui est une technique essentiellement non destructive.

Une onde de cisaillement (transverale) perpendiculaire aux fibres est fortement influencée au point de vue atténuation par les interfaces elles mêmes.

Ainsi, moyennant une étude systématique, nous pensons qu'on a affaire à une méthode intéressante qui permet de vérifier les calculs de prévision théoriques.

c)

Si les longueurs d'onde sont faibles, le phénomène de diffraction peut être prédominant [3]). Dans ce cas, une étude théorique de diffraction des interfaces, disposées périodiquement dans l'espace, serait intéressante. C'est une voie possible d'étude de l'adhésion interfaciale.

Conclusion

Les ultrasons permettent d'évaluer aisément les constantes viscoélastiques d'un matériau anisotrope. L'intérêt de la méthode est que les mesures intéressent le même élément de volume. Les amortissements peuvent être évalués. Toutefois il s'agit de mesures en haute fréquence ($f > 500$ kHz). Avec certaines précautions on peut utiliser les ultrasons pour évaluer les modules complexes des plaques de composite orthotropes.

Annexe

Ondes planes dans les composites orthotropes

L'équation du mouvement est μ_m étant le déplacement

$$C_{mnop} \frac{\partial^2 \mu_p}{\partial x_n \partial x_0} = \varrho \frac{\partial^2 \mu_m}{\partial t^2} \qquad (m, n, o, p) = 1, 2, 3 \qquad [10]$$

avec

$$\mu_m = U_m \exp i(\omega t - k_n x_n)$$

$$i = \sqrt{-1}. \qquad [11]$$

Portant [11] dans [10] et en développant on aboutit au déterminant suivant, δ_{ij} étant la notation de *Kronecker*

$$|\Gamma^*_{mn} - p(v^*)^2 \delta_{mn}| = 0 \quad (m, n) = 1, 2, 3 \qquad [12]$$

$$\Gamma^*_{11} = n_1^2 C^*_{1111} + n_2^2 C^*_{1212} + n_3^2 C^*_{3131}$$

$$\Gamma^*_{22} = n_1^2 C^*_{1212} + n_2^2 C^*_{2222} + n_3^2 C^*_{2323}$$

$$\Gamma^*_{33} = n_1^2 C^*_{3131} + n_2^2 C^*_{2323} + n_3^2 C^*_{3333}$$

$$\Gamma^*_{23} = n_2 n_3 (C^*_{2233} + C^*_{2323})$$

---

[3]) Jusqu'ici nous n'envisageons que des ondes longues vis à vis desquelles le matériau peut être considéré comme homogène.

$$\Gamma^*_{13} = n_1 n_3 (C^*_{1133} + C^*_{3131})$$

$$\Gamma^*_{12} = n_1 n_2 (C^*_{1122} + C^*_{1212}) \qquad [13]$$

n_i sont les cosinus directeurs de la normale au plan d'onde.

Si $n_1 = 0$, $n_2 = n_3 = 1/\sqrt{2}$, on peut écrire à l'aide de [13] et [12]:

$$C_{2233} = [C_{2222} + C_{2323}\varrho v^2 C_{3333} + C_{2323} - v^2]^{1/2} - C_{2323}. \qquad [14]$$

Littérature

1) *Ferry, J. D.*, Viscoelastic properties of polymers (New York 1970).

2) *Markham, M. F.*, The N.P.L. Ultrasonic tank, its uses in polymer and fibre composite – Agard conference proceedings n° 63 on composite materials – Paris (1970).

3) *Vinh, T.*, Compt Rendus Acad. Sci. (Paris) **271**, 1268–1271 (1970).

4) *Tauchert, T. R.* and *A. N. Guzelsu*, Applied mech. **1972**, 98–102.

5) *Meeker, T. R.* et *A. H. Meitzler*, Guided wave propagation in elongated cylinders and plates. Article dans: Physical Acoustics, Vol. I, Part A, pp. 117–167, Edit. par *Mason* (New York 1960).

6) *Truell, Elbaum* et *Chick*, Ultrasonic methods in solid state physics (New York 1969).

7) *Sun, C. T., J. D. Achenbach* et *G. Hermann*, Appl. Mech **35**; Trans A.S.M.E. **90**, Séries E, 467–475 (1968).

8) *Bourbion, M.*, Détermination des constantes viscoélastiques des milieux anisotropes. Thèse de doctorat de 3ème cycle – Université de Paris VI (1969).

Adresse de l'auteur:

Dr. *N. P. Vinh Tuong*
Laboratoire de Rhéologie, I.S.M.C.M.
3, Rue Fernand Hainaut
F-93 Saint Ouen (France)

Rheol. Acta **12**, 563–566 (1973)

Société des Usines Chimiques Rhône-Poulenc, Centre de Recherches des Carrières, Service de Physique, Saint-Fons (France)

Pendule de torsion automatique

Par Michel Ferrara

Avec 4 figures

(Reçu p. p. le 27 Octobre 1972)

1. Introduction

Le pendule de torsion est un appareil connu et utilisé depuis longtemps pour étudier les propriétés mécaniques des matériaux (métaux d'abord, matières plastiques ensuite); les mesures sont longues et nécessitent la présence permanente d'un opérateur; c'est ce qui explique que divers laboratoires aient essayé d'automatiser au maximum cet appareil.

Si cela fut relativement simple avec les métaux, ce fut plus délicat pour les matières plastiques à cause de l'importante chute du module de torsion dans la zone de transition (on atteint des rapports de l'ordre de 1000 entre les modules avant et après la transition) et aussi à cause du phénomène de dérive, c'est à dire de déformation de l'éprouvette sous la seule action d'une élévation de température.

Un pendule de torsion automatique a été étudié, réalisé, mis au point et utilisé dans le Service de Physique du Centre de Recherches des Carrières de la Société Rhône-Poulenc à Saint-Fons (Rhône).

2. Description

L'appareil se compose de deux parties:

– la partie mécanique constituant le pendule proprement dit;

– l'armoire électronique de pilotage.

Cette dernière comprend, de haut en bas:

– le programmateur de température à montée linéaire avec affichage analogique de la température de régulation du four;

– l'imprimante avec affichage numérique de la température de l'échantillon témoin;

– les racks contenant les circuits réalisant les divers automatismes;

– les alimentations.

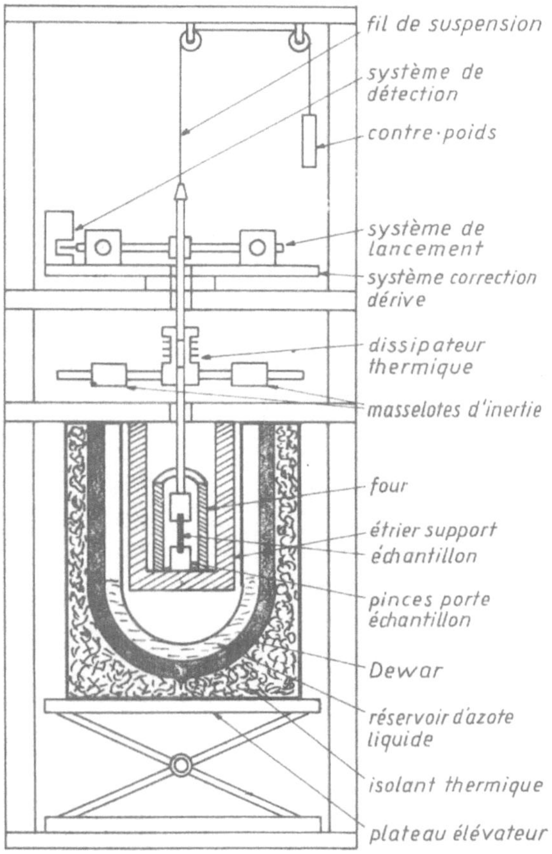

Fig. 1. Vue générale de l'appareil

Le schéma d'une coupe de la partie mécanique montre les différentes parties:

– le fil de suspension et le contrepoids: ce dernier équilibre le poids de tout l'équipage mobile de façon à ce qu'il ne s'exerce aucune traction sur l'éprouvette;

– le système de lancement constitué de deux paires de bobines agissant alternativement sur des noyaux de fer doux solidaires du bras mobile et disposés symétriquement par rapport à l'axe de l'appareil de façon à minimiser les risques d'oscillations parasites;

– le système de détection du mouvement constitué d'une palette solidaire de l'équipage mobile et coupant plus ou moins le faisceau lumineux tombant sur deux cellules solaires;

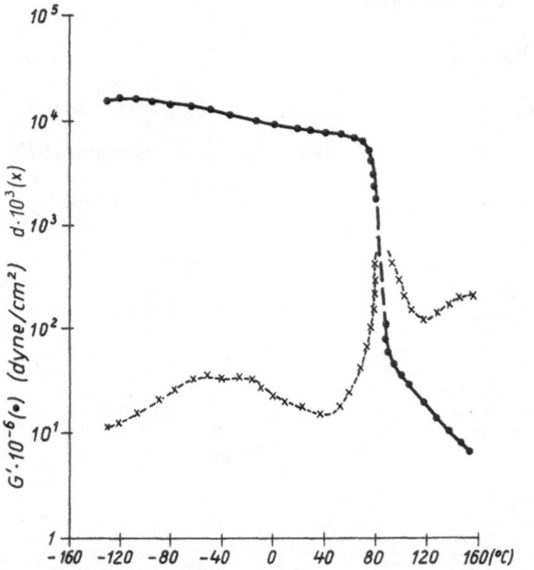

Fig. 2. Coupe de la partie mécanique

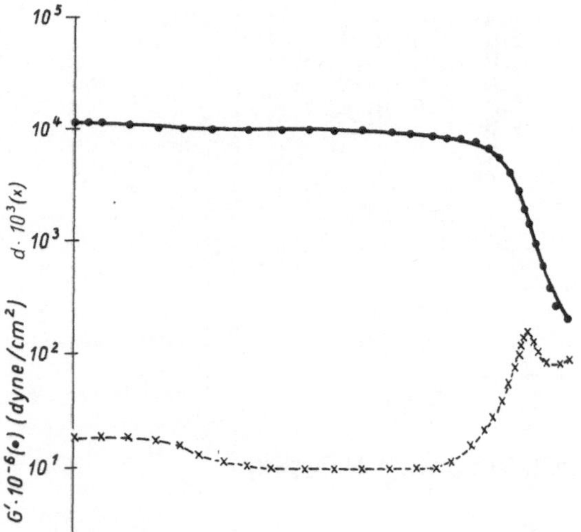

Fig. 3. Chlorure de polyvinyle

– le système de correction de dérive constitué par un plateau mobile supportant les organes de lancement et de détection du mouvement;

– le dissipateur et le bras à inertie variable;

– la potence qui supporte la pince inférieure de fixation de l'échantillon; cette potence a été réalisée en acier spécial traité puis usinée de façon symétrique afin de pouvoir résister aux très fortes variations de températures sans déformations. L'appareil est réalisé de façon telle qu'il n'y ait aucun contact métallique entre le bâti et les pièces portées aux températures extrêmes. Tous les fils situés près de l'éprouvette (thermocouples de mesure et de régulation, alimentation du four) sont isolés par de la gaine de verre. Le four coulisse le long de la potence afin de permettre la mise en place de l'échantillon;

– l'enceinte thermique est montée sur un plateau mobile; elle isole l'échantillon et le four de l'extérieur et permet, grâce à une double paroi, le remplissage du fluide réfrigérant (azote liquide) sans contact entre ce dernier et la potence. Un léger balayage d'azote gazeux évite toute condensation et tout givrage sans nuire à la régulation de température. L'isolement thermique est assuré par de la mousse polyuréthanne.

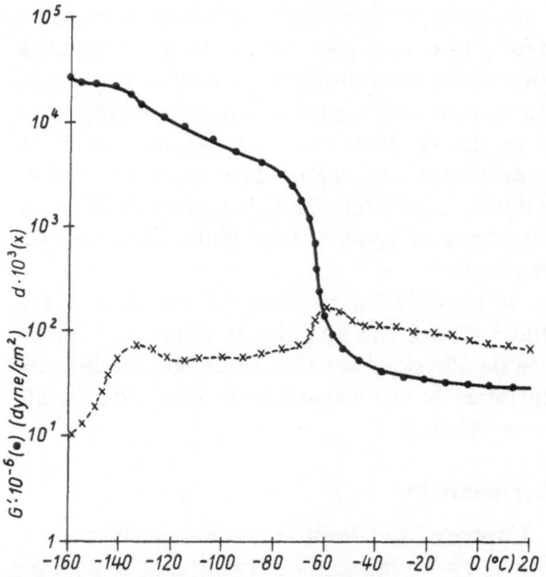

Fig. 4. Thermostable

3. Fonctionnement

Nous avons adopté le fonctionnement en oscillations amorties, cela par souci de simplicité et également afin de pouvoir comparer directement nos résultats avec ceux obtenus à l'aide d'un appareil conventionnel.

Les mesures se reproduisent au bout d'un temps préréglé; une mesure comprend:

– la correction éventuelle de dérive, c'est à dire la mise en coincidence du zéro électrique de l'appareil avec le zéro mécanique imposé par l'échantillon à l'équipage mobile, cela sans imposer de contraintes à l'éprouvette;

– le lancement c'est à dire l'envoi alternativement aux deux paires de bobines des impulsions synchronisées avec le passage à zéro de l'ensemble mobile jusqu'à ce que l'amplitude du mouvement atteigne le seuil préréglé; l'excitation est alors arrêtée et les différentes mesures sont effectuées:

– mesure de la période des oscillations par un chronomètre électronique;

– mesure du décrément logarithmique des oscillations par un circuit analogique piloté par un ensemble logique et suivi d'un convertisseur analogique numérique;

– mesure de la température;

– avance d'un pas du compteur «numéro de la mesure»;

– impression des résultats.

Deux particularités améliorent de façon notable le fonctionnement de l'appareil:

– la tension d'excitation des bobines est, à chaque impulsion, asservie au module instantané de l'échantillon; c'est à dire que la tension de l'impulsion de rang $n+1$ dépend du résultat obtenu par les n premières impulsions de lancement;

– la vitesse de scrutation est automatiquement modulée, pour les échantillons présentant une forte zone de transition, c'est à dire que, quelle que soit la vitesse de mesure choisie au départ, l'appareil passe sur une mesure par minute dans la zone de transition et retourne sur la vitesse préréglée une fois la transition passée.

4. Mise en œuvre et utilisation

On met en place l'éprouvette ainsi qu'un échantillon témoin (dans lequel se fait la mesure de température) et on positionne l'enceinte thermique. L'éprouvette et le témoin sont placés côte à côte et sont issus du même matériau.

On règle la vitesse de montée en température et la température maximum que l'on veut atteindre; deux possibilités sont alors présélectionnables: soit la mise en palier, soit l'arrêt automatique quand on atteint la température maximum.

On règle la vitesse de scrutation.

Puis, on passe sur la position «automatique». Les mesures se déroulent alors sans intervention de l'opérateur; les résultats sont imprimés à chaque cycle; sur la bande de papier, on a, de gauche à droite: le numéro de la mesure, la période des oscillations, le décrément logarithmique et la température à laquelle a été faite la mesure.

Le dépouillement des résultats bruts se fait sur calculateur Hewlett Packard; un programme est stocké sur carte magnétique et, après entrée des résultats bruts lus sur la bande de l'imprimante, le calculateur imprime:

– les parties réelle G' et imaginaire G'' du module complexe de torsion G^*;

– l'amortissement d;

– la température correspondante.

5. Performances

– La gamme de températures couvre de -190 à $+400$ et éventuellement $+500$ °C;

– la vitesse de montée peut être choisie entre 0,1 et 10 °C/mn;

– la vitesse de scrutation va de 1 mesure/2 mn à 1 mesure/6 mn par bonds de 1 mn (passage automatique sur 1 mes/mn lors des fortes zones de transition);

– la température de l'échantillon témoin est affichée en permanence ainsi que la température de régulation du four (on adopte une vitesse de montée telle que l'écart entre les deux ne soit pas trop grand);

– un dispositif de sécurité met l'appareil hors tension en cas de rupture d'un thermocouple ou de flash électrique important à proximité; il interdit également la remise en route après une interruption de la tension secteur;

– l'appareil possède une position de fonctionnement manuel grâce à laquelle on peut déclencher des cycles de mesures indépendamment de la base de temps;

– la reproductibilité des résultats étudiée sur 3 échantillons maintenus à température constante et portant sur 3 fois 200 mesures est meilleure que 1% pour la mesure de période et de décrément logarithmique.

6. Résultats obtenus

Les échantillons les plus divers ont été étudiés avec cet appareil; par exemple les chlorures de polyvinyle, les thermostables, les polyuréthannes, les polyéthylènes, les organosiliciques, etc.

Les courbes représentant le module et l'amortissement en fonction de la température pour trois produits sont représentées ci-joint.

7. Conclusion

Pour étudier un échantillon sur une large gamme de températures il fallait une journée à un opérateur bien entraîné; avec cet appareil, l'opérateur se borne à mettre l'éprouvette en place et à dépouiller sur calculateur les résultats bruts lus sur la bande de l'imprimante; entre ces deux opérations l'appareil effectue seul toutes les mesures.

Les essais demandent donc moins d'heures de travail, les résultats sont plus précis et les mesures plus reproductibles.

Malgré l'intérêt théorique et pratique des mesures faites au pendule de torsion, son utilisation n'était pas générale du fait du travail long et fastidieux que cela représentait. Dans sa version automatique, cet appareil devrait permettre l'étude systématique des propriétés mécaniques de tous les nouveaux polymères.

Adresse de l'auteur:

Dr. *Michel Ferrara*
Centre de Recherches des Carrières
Service de Physique
Société des Usines Chimiques Rhône-Poulenc
F-69 Saint Fons (France)

Rheol. Acta 12, 567–571 (1973)

From the Departments of Chemistry and Macromolecular Science, Case Western Reserve University, Cleveland (U.S.A.)

A rheometer for oscillatory studies of nonlinear fluids

By I. M. Krieger and Tyan-Faung Niu

With 6 figures

(Received October 27, 1972)

Introduction

In order to study their linear viscoelasticity, fluids and solids are frequently subjected to small-amplitude oscillatory shear in rotational rheometers. Steady rotational shear is employed to characterize nonlinear (non-*Newton*ian) viscosity, while the shear rate is sometimes programmed to generate a "thixotropic loop" or a step-jump in order to follow the effect on viscosity of the sample's prior shear history. Recently, however, a number of authors (1–4) have turned to oscillatory shear at large amplitudes, where the response frequently shows a harmonic content resulting from the nonlinear nature of the sample. Further development of this technique is hampered both by difficulties in analyzing and interpreting the data and by the limited capabilities of existing rheometers. The purpose of the present work is to remove some of the difficulties in measurement and analysis, by constructing a new rheometer specifically designed for oscillatory studies of nonlinear fluids. The resultant rheometer-computer system has proven to be suitable also for small-amplitude oscillations and for steady or time-programmed shear measurements.

Rheometer design

Dodge and *Krieger* (5) used a modified *Weissenberg* Rheogoniometer (Sangamo-Weston Ltd., Bognor Regis, U.K.) in their preliminary investigation of oscillatory shear of non-linear fluids. In discussing the limitations of their apparatus, they pointed to the need for: (a) continuous variation of frequency; (b) larger shear amplitudes, particularly at low frequencies; (c) upward extension of the frequency range, and (d) better methods of data acquisition and analysis, so that signal averaging could be conveniently employed. They suggested a servo-motor drive system, and an online computer.

The first stage of the present work consisted in coupling the modified Rheogoniometer to a LAB-8 Signal Averaging Computer (Digital Equipment Corp., Maynard, Massachusetts, U.S.A.). A schematic diagram of this coupling is shown in fig. 1. In operation, the stress and strain output signals (from the Rheogoniometer's transducer meters) were sampled and stored in the computer's memory 32 times per cycle of oscillation; sampling was triggered by pulses from a timing disk mounted on the oscillatory drive shaft. After 8 successive cycles were sampled, average cycles for both stress and strain were computed, and displayed on the oscilloscope. Upon command, the computer *Fourier*-analyzed both signals, giving amplitudes and phases for harmonics up to the fifth.

The use of the computer effectively removed previous limitations in data acquisition and analysis. However, a completely new drive system is needed in order to solve the other problems mentioned above. The availability of direct current torque motors which are capable of speed ranges from 5000 rpm down almost to zero makes possible a gearless drive unit. With a coupled tachometer and a differential servo amplifier, the motor speed can be made accurately proportional to the voltage of a control signal. Low-inertia "printed circuit" rotors, formed by stamping the metal winding onto the surfaces of a fiberglass disk, make it possible for the motor speeds to follow rapid changes in the control voltage. One such system tested in our laboratory gave sinusoidal oscillation up to 2 kHz when controlled by an audio signal generator.

Concentric cylinder rheometer geometry, shown by *Dodge* and *Krieger* (6) to be preferable

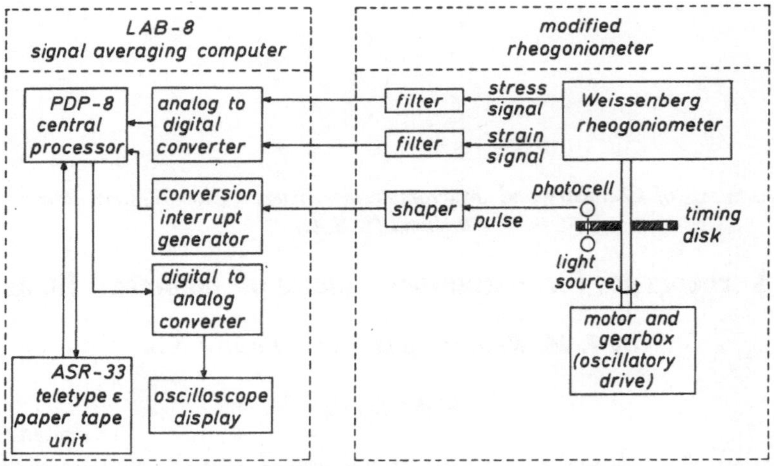

Fig. 1. Schematic diagram showing coupling of *Weissenberg* Rheogoniometer to LAB-8/L signal-averaging minicomputer

to cone-plate or parallel disk for nonsteady shear of nonlinear fluids, was adopted for the new rheometer. A stainless steel outer cylinder or cup was attached directly to the shaft of a printed circuit motor-tachometer (PMI Photocircuits Div., Glen Cove, N.Y., model U12M4T). A mercury seal was mounted on the shaft between motor and cup, to permit use of a water bath for temperature control. The inner cylinder was suspended through an air bearing from a transducer-equipped torsion bar, exactly as in the Rheogoniometer. Both the motor-cup assembly and the torsion head were mounted on a vertical dovetail slide; a lead screw permitted the torsion head to be raised and lowered. The weight of the water bath was borne by the base of the instrument. Fig. 2 is a drawing of the instrument.

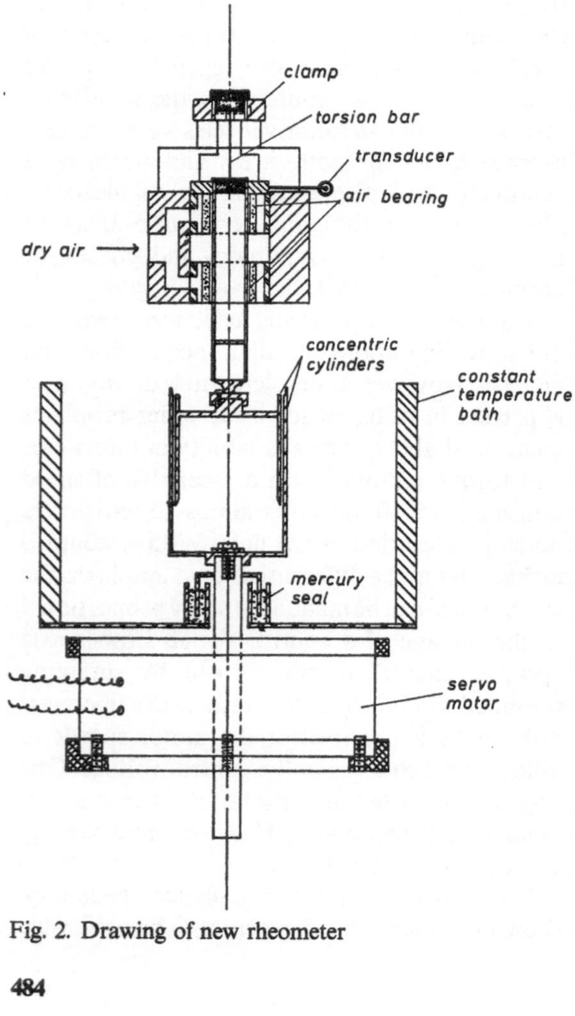

Fig. 2. Drawing of new rheometer

Control of motor speed

Fig. 3 shows a schematic diagram of the control and data acquisition system of the new rheometer. A differential amplifier (Control Systems Research, Inc., Pittsburgh, Pa., Model 500) compares the tachometer output with the control voltage, adjusting the motor speed to null the difference. At present, the control voltage is taken from the difference between outputs of the two digital-to-analog converters of the LAB-8 computer. Since these converters were originally designed to provide control of the oscilloscope display axes, they have a precision of only 8 bits, and hence must be operated near their full output at ± 5 volts to give precision better than 1%. Whenever smaller control voltages are needed, the output is reduced by an attenuator. (Because of adjustment problems inherent in the converters provided in the LAB-8 computer, a separate 12-bit analog-to digital converter is being constructed; this will provide more precise control signals.)

In steady rotation, a speed of ca. 1000 rpm will produce a tachometer output of ca. 5 volts;

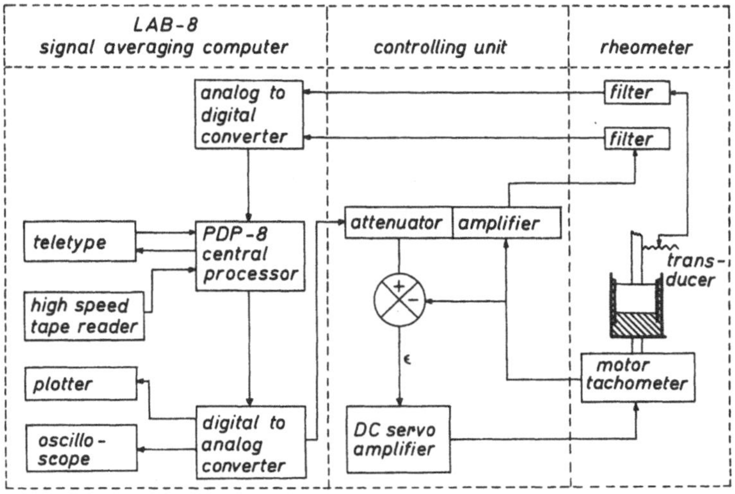

Fig. 3. Schematic diagram showing coupling of new rheometer to LAB-8/L mini-computer

hence a 5-volt control voltage is needed to produce a rotor speed of 1000 rpm. Since the 8-bit analog-to-digital converter accepts voltages ranging from -1 to $+1$ volts, an amplifier-attenuator is needed to provide input voltages which can be converted with adequate precision.

For oscillatory shear, the computer is used to generate a sinusoidally varying control signal of desired frequency and amplitude, as was done by *Birnboim* (7). Sixty-four equally spaced sine values are calculated for the first quadrant (0–90°); these values are multiplied by the amplitude factor and stored in 256 consecutive memory locations, with proper order and sign to provide a full sine wave. They are called out sequentially on pulses from the computer's crystal clock at intervals appropriate to the desired frequency, and presented to the digital-to-analog converter. The output is thus a 256-step approximation to a sine wave. At frequencies above 19.5 Hz, the number of points per cycle is reduced, to allow the computer time to sample data between output steps. The upper frequency limit with sampling of 32 points per cycle is 156 Hz, although higher drive frequencies are possible if the data acquisition steps are omitted. Fourier analyses of the tachometer output through the first five harmonics show that the oscillation is sinusoidal, with negligible distortion (see fig. 5). Superposition of oscillation on steady rotation is accomplished by displacing the control signal by a fixed amount.

To produce thixotropic loops, it is necessary to provide one cycle of a triangular wave control signal, of the desired amplitude and frequency. The desired "ramp function" is readily approximated by adding constant increments or decrements to the control signal at regular intervals. A step-jump in speed is even more easily achieved, by a corresponding jump in the control signal.

Data acquisition and analysis

In the new rheometer, acquisition and analysis of data are accomplished in the same way as with the computer-coupled Rheogoniometer. The torque and tachometer signals are both sampled 32 times per cycle; i.e., with 256 output steps per cycle, samples are taken every eighth output step. Sampling begins after a few cycles of oscillation, so as not to include starting transients, and continues for 8 cycles, after which an average cycle is calculated. This is *Fourier*-analyzed for the first 5 harmonics. Output of the average waveform is on an oscilloscope (fig. 4) or an X-Y plotter; amplitudes and phases of the *Fourier* components are presented in alphanumeric form either on the oscilloscope face or on the teletypewriter. A bar-graph spectrum of the tachometer output (fig. 5) shows negligible harmonic content, while the spectrum of the response torque (fig. 6) has a substantial content of odd harmonics.

Steady rotations are produced by calling forth steady control signals via the teletype keyboard. Torque and tachometer signals are acquired repetitively at specified intervals, and an average value valculated. These are presented in alphanumeric form on the oscilloscope or on the teletypewriter, after conversion to shear stress and apparent shear rate by multiplication by appropriate instrumental constants. The apparent viscosity may also be presented. The thixotropic loop program, described below, is used when it is desired to scan a range of shear rates automatically.

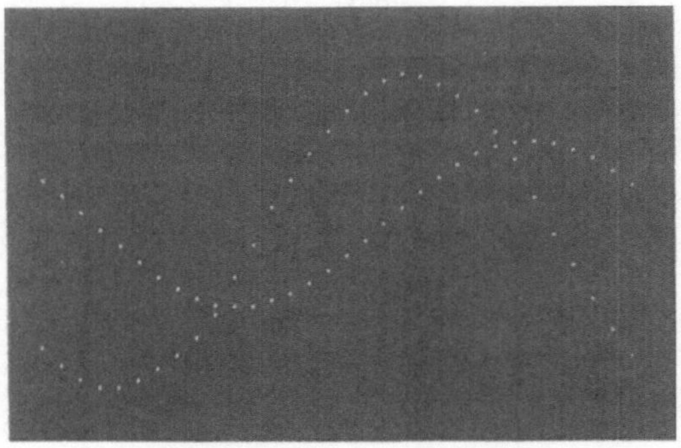

Fig. 4. Average cycle of stress signal and strain rate signal, as presented on oscilloscope

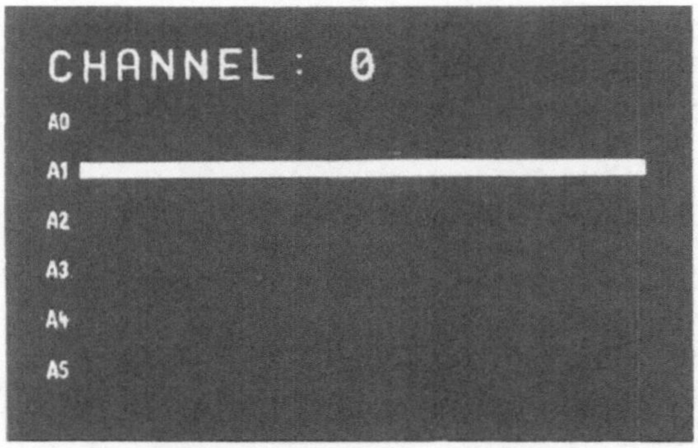

Fig. 5. Bar graph of spectrum of strain rate signal, as seen on oscilloscope. Absence of harmonics indicates a pure sine wave

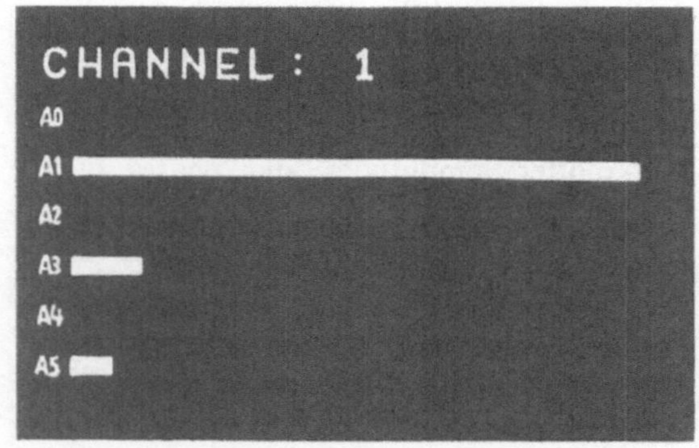

Fig. 6. Bar graph of spectrum of stress signal, showing significant content of odd harmonics

For thixotropic loop generation, the desired range of rotational speeds, number of steps, and interval between steps are keyed in on the teletypewriter, as well as the number of data samplings at each speed level. Average values for speed and torque are calculated at each level, and stored. After the speed has returned to zero, the data, converted to shear stress and apparent shear rate, can be presented on the X-Y plotter. They can also be presented numerically on the teletypewriter, either in toto or, as is usually preferred, a subset of equally spaced points can be called out, together with the apparent viscosities. To characterize steady-state flow behavior, the desired shear rate range is scanned in discrete steps using the thixotropic loop mode, at a rate sufficiently slow that the ascending and descending data coincide.

For studies of linear viscoelasticity, the oscillatory mode is used, with the amplitude reduced as necessary to produce negligible harmonic content in the response torque. The phases and amplitudes needed for linear analysis are those at the fundamental frequency.

Discussion

The new instrument satisfies most of the needs of oscillatory rheometry, providing a wide frequency range, large shear amplitudes at low frequencies, signal-to-noise ratio, and immediate data analysis. Using the computer to generate the control signal, the frequency is varied in small steps with an upper limit of ca. 78 Hz. Continuous variation can be provided by using a separate signal generator, however, with concommitant upward extension of the frequency range toward 1 kHz (7). Characterization of linear viscoelasticity, of non-*Newton*ian behavior under steady shear, and of time-dependent thixotropy by means of thixotropic loops and step-jump in shear rate, are also accomplished with accuracy, speed and convenience in the new apparatus.

The principal advantages of the new rheometer over previous instruments stem from the use of an on-line computer for instrument control, data acquisition and analysis. The other major source of improvement is the printed-circuit motor, which gives a continuous range of rotational speeds without use of gears, allows virtually unlimited shear amplitudes in low-frequency oscillation, and extends the high-frequency limit into the lower sonic range.

It is possible to utilize the new rheometer without the on-line computer. One needs only an inexpensive signal generator and a meter or chart recorder to acquire data. Analysis of the data can be done either manually or by use of an off-line computer. The added speed and convenience of the on-line computer are frequently indispensable, however. Signal averaging can add significantly to experimental accuracy, especially in determining phase relationships between stress and strain rate signals. The prompt presentation of results allow the experimenter to decide whether the experiment need be repeated, and allows him to modify his experimental strategy in the light of the results of prior experiments. This is especially important when the drive frequency or one of its lower harmonics falls close to the resonance frequency of the torsion head assembly. Finally, the speed with which rheological characterization can be completed is of vital importance when dealing with biological or other fluids whose properties change with time.

In its present version, the apparatus has some disadvantages vis-a-vis the *Weissenberg* Rheogoniometer. There is no provision for normal stress measurement, nor will this be readily accomplished without providing platens for cone-plate or parallel disk modes. The use of concentric cylinders with the outer cylinder immersed in a water bath and driven from below, while providing excellent thermostatic control at temperatures not too far from ambient, is a source of some difficulties. Sample sizes are larger than those required in the other rotational geometries. Replacement of the outer cylinder entails realignment, so that changing of sample must be accomplished with the cup in place. Also, the mercury seal and the bath water add to the inertia of the rotating member, thereby reducing the upper frequency limit in oscillation and making step-jumps less precise.

The study reported here shows the many advantages occurring from the marriage of a minicomputer with an appropriately designed rheometer. Recommendations for further improvement include:

1. Provide cone-plate and parallel disk platens.
2. Introduce a normal stress measurement capability.
3. Construct a miniaturized instrument, to use smaller samples and attain higher frequencies.
4. Provide an external signal generator syncronized to the computer's clock, for continuous speed variation.
5. Introduce auto-ranging of the amplifier-attenuator; this would reduce the chance of operator errors, and would permit the computer to control the complete experimental characterization of the sample.

Acknowledgments

The authors wish to acknowledge the help of *John Meszaros* and *Thomas Seitz* in formulating the computer program for the Rheogoniometer, and of *Larry Rogers* for the ramp function program. The instrument maker who did the exacting machining of the new rheometer is *Robert DiSantis*; *Andrew Tanos* designed and constructed the amplifier-attenuator. The research was supported by the Public Health Service, and Mr. *Niu* held a Paint Research Institute Fellowship during the course of the work. The dovetail slide was contributed by the Warner and Swasey Company.

Summary

This paper describes the coupling of a minicomputer with two rotational rheometers. In the first instance, the computer acquires output data from a *Weissenberg* Rheogoniometer in oscillatory shear, and *Fourier*-analyzes stress and strain signals. The second instance utilizes a rotational rheometer specifically designed to be controlled by the computer, employing a servo-driven torque motor. The computer-rheometer system is used to analyze steady-state non-*Newton*ian viscosity and linear viscoelasticity, as well as non-linear dynamic behavior.

Zusammenfassung

Diese Arbeit beschreibt die Kopplung eines Minicomputers mit zwei Rotationsrheometern. Im ersten Fall erhält der Computer die Daten von einem oszillierenden *Weissenberg*-Rheogoniometer und macht eine *Fourier*-Analyse der Spannungs- und Schergefällsignale. Im zweiten Fall benützt er ein speziell entworfenes Rotationsrheometer, welches vom Computer kontrolliert wird und einem Torsionsmotor mit Servomechanismus. Das Computer-Rheometer-System wird dazu verwendet, sowohl zeitunabhängige nicht *Newton*sche Viskosität und lineare Viskoelastizität als auch nichtlineares dynamisches Verhalten zu analysieren.

References

1) *Harris, J.*, Nature, **207**, 744 (1965).
2) *Harris, J.* and *K. Bogie*, Rheol. Acta, **6**, 3 (1967).
3) *Onogi, S., T. Masuda*, and *T. Matsumoto*, Trans. Soc. Rheology, **14**, 275 (1970).
4) *Dodge, J. S.*, Oscillatory Shear of Non-Linear Fluids, Ph. D. Thesis, Case Western Reserve University (1969).
5) *Dodge, J. S.* and *I. M. Krieger*, Trans. Soc. Rheology, **15**, 589 (1971).
6) *Dodge, J. S.* and *I. M. Krieger*, Rheol. Acta, **8**, 480 (1969).
7) *Birnboim, M. H., J. S. Burke*, and *R. L. Anderson*, Proc. 5th Intl. Cong. on Rheology, Vol. 1, p. 409 (*S. Onogi*, Ed.), (Univ. of Tokyo Press 1969).

Authors' address:

I. M. Krieger and *Tyan-Faung Niu*
Dept. of Chemistry and Div. of Macromolecular Science
Case Western Reserve University
Cleveland, Ohio 44106 (U.S.A.)

Rheol. Acta 12, 572–577 (1973)

Mitteilung aus dem Institut für Mechanische Verfahrenstechnik der Universität Karlsruhe (Bundesrepublik Deutschland)

Kurzzeit-Rotations-Rheometer zur Messung von hohen Schub- und Normalspannungen bei großen Schergefällen

Meßmethode und Prinzip der technischen Realisierung [1])

Von W. Gleissle und H. Reichert

Mit 3 Abbildungen und 1 Tabelle

(Eingegangen am 27. Oktober 1972)

Am Institut für Mechanische Verfahrenstechnik der Universität Karlsruhe wird seit langem die Beanspruchung von Feststoffteilchen untersucht, die sich in einer hochzähen Scherströmung befinden (1, 2, 3, 4). Die von der Strömung auf die Teilchenoberfläche übertragenen Spannungen bewirken eine Beanspruchung des Feststoffes, die bei Materialien niedriger Festigkeit, wie z. B. Agglomeraten, ausreichen kann, um einen Bruch der Teilchen auszulösen. Nach *Raasch* (1) ergibt die Beanspruchung von kugelförmigen starren Teilchen in der ebenen Scherströmung einer *Newton*schen Flüssigkeit eine dem hydrostatischen Druck überlagerte reine Schubbeanspruchung

$$T_{12} = 2{,}5\eta\varkappa, \quad T_{13} = 0, \quad T_{23} = 0. \qquad [1]$$

Hierbei ist η die dynamische Viskosität und $\varkappa$ das Schergefälle der ungestörten Scherströmung. Das Koordinatensystem ist so gewählt, daß die Orientierung 1 in die Scherrichtung und 2 in Richtung des Schergradienten zeigt, während 3 die indifferente Richtung darstellt.

Will man in Teilchen durch Scherung große Spannungen erzeugen, so muß man sie in Flüssigkeiten mit großer Zähigkeit und bei großen Schergeschwindigkeiten beanspruchen. Stoffe mit großer Zähigkeit, wie z. B. Hochpolymere, sind aber in der Regel bei hohen Schergefällen nicht-*Newton*isch und weisen häufig Normalspannungseffekte auf. Zur experimentellen Untersuchung des Einflusses der Flüssigkeitseigenschaften auf den Bruchvorgang von Teilchen in Scherströmungen ist daher die Kenntnis des Fließverhaltens der zur Scherung verwendeten Flüssigkeit erforderlich. Neben dieser sehr speziellen Aufgabenstellung ist die Messung der Zähigkeit und der Normalspannungsfunktionen von hochzähen Flüssigkeiten, gerade im Bereich sehr großer Schergefälle, im Hinblick auf die Probleme der Kunststoffverarbeitung von allgemeinem Interesse.

Die Ermittlung rheologischer Größen bei hohen Schergefällen bereitet aber wegen der großen Dissipation und der damit verbundenen bedeutenden Temperaturunterschiede im Scherspalt erhebliche Schwierigkeiten, da die zu messenden Stoffeigenschaften von der Temperatur abhängig sind.

Meßprinzip

Für die Temperaturdifferenz zwischen Spaltmitte und den begrenzenden isothermen Wänden mit dem Abstand b gilt für die thermisch stationäre ebene Scherströmung einer *Newton*schen Flüssigkeit mit der Wärmeleitzahl λ (5)

$$\Delta\vartheta_{st} = \eta\varkappa^2 b^2/(8\lambda). \qquad [2]$$

Betrachtet man die zeitliche Entwicklung des Temperaturprofils nach dem plötzlichen Einschalten einer homogenen Scherung (5), so zeigen vergleichende Rechnungen, daß für kleine Zeiten die Wärmeleitung bei den für Polymere typischen Wärmeleitzahlen vernachlässigt werden kann. Der Meßspalt kann als adiabates System behandelt werden. Als adiabate Temperaturerhöhung im Scherspalt erhält man

$$\Delta\vartheta_{ad} = \eta\varkappa^2 t/(c_w\varrho) \qquad [3]$$

---

[1]) Die konstruktiven Details des Rheometers sind in Rheol. Acta 12, 77–81 (1973) beschrieben.

wobei t die Scherdauer, c_w die spezifische Wärme und ϱ die Dichte der Meßzubstanz bedeuten. Die Temperaturerhöhung ist der Dauer der Scherung in diesem Beanspruchungsfall direkt proportional. Durch Einschränkung der Versuchsdauer kann die Temperaturdifferenz $\Delta\vartheta_{ad}$ in den gewünschten Grenzen gehalten werden. Vergleicht man für eine zulässige Temperaturerhöhung $\Delta\vartheta_{zul} = \Delta\vartheta_{st} = \Delta\vartheta_{ad}$ die bei gleichem Schergefälle $\varkappa$ noch meßbaren Zähigkeiten $\eta_{max\,ad}$ und $\eta_{max\,st}$, so ergibt sich aus den Gl. [2] und [3]

$$\eta_{max\,ad}/\eta_{max\,st} = \varrho\, c_w \cdot b^2/(8\lambda t_m). \qquad [4]$$

Die Zeit t_m bedeutet die für das vorgegebene Schergefälle $\varkappa$ in einem Viskosimeter noch sinnvoll verwirklichbare Versuchsdauer.

Für ein Dimethylpolysiloxan (z. B. Siliconöl AK 2000000 der Firma Wacker, München), das bei 20 °C und der Schergeschwindigkeit $\varkappa = 10^3\,s^{-1}$ etwa die Zähigkeit $\eta = 10^3$ P hat (6) und dessen Stoffwerte $\lambda = 1{,}68 \cdot 10^4$ dyn/s °C, $c_w = 1{,}47 \cdot 10^7$ erg/g °C und $\varrho = 0{,}98$ g/cm³ sind, ergibt sich bei einer Spaltbreite $b = 0{,}05$ cm – wesentlich kleinere Spaltweiten sind aus bearbeitungstechnischen Gründen kaum realisierbar – für die Temperaturdifferenz zwischen Spaltmitte und den begrenzenden Wänden im thermisch stationären Fall aus Gl. [2] $\Delta\vartheta_{st} = 18{,}7$ °C. Wenn eine Versuchsdauer $t_m = 2 \cdot 10^{-2}$ s erreicht werden kann, so folgt aus Gl. [3] für diese Substanz eine adiabate Temperaturerhöhung $\Delta\vartheta_{ad} = 1{,}37$ °C. Berücksichtigt man die Wärmeleitung (5), so hat sich die Flüssigkeit um $\Delta\vartheta = 1{,}36$ °C erwärmt. Die Abweichung vom adiabaten Fall ist also vernachlässigbar klein. Gl. [4] ergibt für Siliconöle mit den oben angegebenen thermischen Stoffkonstanten das Verhältnis $\eta_{max\,ad}/\eta_{max\,st} = 13{,}7$ und für ein Polyäthylen (Lupolen der BASF) $\eta_{max\,ad}/\eta_{max\,st} = 17{,}2$. Das bedeutet eine Erweiterung des Meßbereiches um mehr als eine Zehnerpotenz der Zähigkeit durch die Kurzzeitmeßmethode oder bei gleicher Zähigkeit eine Verminderung des Temperaturanstieges um die Faktoren 13,7 bzw. 17,2.

Da beim Kegel-Platte-Viskosimeter die Scherspaltweite dem Radius direkt proportional ist mit $b = r \cdot \alpha$ (α = Öffnungswinkel des Scherspaltes), ist auch die Temperaturdifferenz zwischen Spaltmitte und den Begrenzungsflächen eine vom Radius r abhängige Größe. In grober Abschätzung gilt

$$\Delta\vartheta_{st}(r) \approx \eta\varkappa^2(r\alpha)^2/(8\lambda) \qquad [2a]$$

wobei

$$\varkappa = \omega/\alpha \qquad [5]$$

das Schergefälle und ω die Winkelgeschwindigkeitsdifferenz zwischen Kegel und Platte ist. Die größte Temperaturdifferenz entsteht außen bei $r = R$, und die Gl. [4] wird für das Kegel-Platte-Viskosimeter

$$\eta_{max\,ad}/\eta_{max\,st} \approx \varrho\, c_w (r\alpha)^2/(8\lambda t_m). \qquad [4a]$$

Mit dem Winkel $\alpha = 1{,}5°$, $r = R = 4$ cm und den Stoffwerten von Polyäthylen wird

$$\eta_{max\,ad}/\eta_{max\,st} \approx 100.$$

Wie die angestellten Überlegungen zeigen, ist die Kurzzeitmessung im Bereich hoher Zähigkeiten und hoher Schergefälle die einzige Methode, um die störende Erwärmung der Meßflüssigkeit in erträglichen Grenzen zu halten. Die Methode beruht auf der Tatsache, daß wegen der großen Zähigkeit bei plötzlichem Anlauf die hydrodynamisch stationäre Scherströmung wesentlich schneller erreicht ist als die thermisch stationäre. Das gilt für alle Materialien, die keine langsam verlaufenden Thixotropie- oder Relaxationserscheinungen zeigen. Für eine *Newton*sche Flüssigkeit läßt sich die nach dem Anlauf zur Ausbildung der stationären Scherströmung erforderliche Zeit leicht angeben (11). Sie ist nach der Theorie unendlich, es weicht jedoch das Geschwindigkeitsprofil im Spalt schon nach sehr kurzen Zeiten nur noch äußerst wenig von dem der stationären Scherströmung ab. Für das zuvor angegebene Siliconöl beispielsweise beträgt die praktische Anlaufzeit zur Ausbildung der hydrodynamisch stationären Scherströmung selbst bei Schergefällen bis $10^4\,s^{-1}$ weit weniger als 1 ms.

Zur Bestimmung der Schubspannungsfunktion $\tau(\varkappa)$ oder der Viskositätsfunktion $\eta(\varkappa)$ im Kegel-Platte-System wird das durch die Flüssigkeit übertragene Drehmoment Md gemessen

$$\mathrm{Md} = (2/3)\,\pi R^3 \cdot \tau(\varkappa) = (2/3)\,\pi R^3 \eta(\varkappa)\,\varkappa. \qquad [6]$$

Für die Untersuchung der Normalspannungseffekte in der stationären viskosimetrischen Scherströmung kann man die Theorie der inkompressiblen einfachen Flüssigkeit von *Coleman*, *Markovitz* und *Noll* (7) benutzen. Hiernach sind zur Beschreibung des Spannungszustandes der Flüssigkeit in stationären viskosimetrischen Strömungen zusätzlich zur Viskositätsfunktion $\eta(\varkappa)$ noch die zwei Normalspannungsfunktionen $\sigma_1(\varkappa)$ und $\sigma_2(\varkappa)$ notwendig. Es ist

$$\sigma_1(\varkappa) = T_{11} - T_{33}$$

und

$$\sigma_2(\varkappa) = T_{22} - T_{33}. \qquad [7]$$

Zur Messung der beiden Normalspannungsfunktionen ist die Kegel-Platte-Scheranordnung am besten geeignet. Für die Normalspannung auf Kegel oder Platte in Abhängigkeit vom Radius r ergibt sich

$$T_{\theta\theta}(r) = \sigma_2(\varkappa) - \ln(R/r)\left[\sigma_2(\varkappa) + \sigma_1(\varkappa)\right] - p_0 \qquad [8]$$

wobei p_0 der Umgebungsdruck ist. Die Normalspannungsdifferenzen in Kugelkoordinaten sind

$$\sigma_1(\varkappa) = T_{\varphi\varphi} - T_{rr}$$

und

$$\sigma_2(\varkappa) = T_{\theta\theta} - T_{rr}. \qquad [9]$$

Durch Messung des Druckes auf den Kegel oder die Platte in Abhängigkeit vom Radius kann man beide Normalspannungsdifferenzen bestimmen (7). Der Druckverlauf, aufgezeichnet im einfachlogarithmischen Netz, ergibt eine Gerade mit Steigung $\sigma_2(\varkappa) + \sigma_1(\varkappa)$. An der Stelle $r = R$ reduziert sich Gl. [8] zu

$$T_{\theta\theta}(R) = \sigma_2(\varkappa) - p_0. \qquad [10]$$

Durch Integration der Gl. [8] von $r = 0$ bis $r = R$ erhält man die Gesamtkraft auf den Kegel und die Platte

$$N_{KP} = (\pi/2)\,R^2\left[\sigma_2(\varkappa) - \sigma_1(\varkappa)\right] - \pi R^2 p_0. \qquad [11]$$

Für Materialien, die nur eine Normalspannungsdifferenz haben (z.B. *Weissenberg*-Flüssigkeiten) bzw. deren zweite Normalspannungsdifferenz genügend klein ist, genügt die Messung der Gesamtkraft N_{KP}, um die Normalspannungseffekte zu ermitteln. Die Bestimmung von N_{KP} ist aber immer nützlich, um die aus Gl. [8] für $\sigma_1(\varkappa)$ und $\sigma_2(\varkappa)$ gefundenen Werte zu kontrollieren.

Zur Trennung der beiden Normalspannungsdifferenzen mit Hilfe von Gesamtkraftmessungen wurde bisher in zwei Scherversuchen mit verschiedenen Geometrien die Gesamtkraft gemessen (8), z. B. mit der Kegel-Platte-Geometrie und einer Platte-Platte-Anordnung, wobei für die Gesamtkraft bei Platte-Platte-Scherung gilt:

$$N_{PP} = \frac{\pi R^2}{\varkappa^{*2}} \int_0^{\varkappa^*} \varkappa\left[2\sigma_2(\varkappa) - \sigma_1(\varkappa)\right] d\varkappa - \pi R^2 p_0 \qquad [12]$$

$$\varkappa^* = \frac{R \cdot \omega}{d} \qquad [13]$$

wobei R den Außenradius und d den Plattenabstand bezeichnen. Wie man Gl. [12] nach $(2\sigma_2 - \sigma_1)$ auflösen kann, wurde von *Kotaka*, *Kurata* und *Tamura* (9) gezeigt. Eine andere Methode zur Bestimmung der Normalspannungsfunktionen mit dem Kegel-Platte-System allein besteht darin, die Axialkraft in zwei Anteile zu zerlegen und diese gleichzeitig zu messen. *Pollett* (10) benutzte dieses Verfahren, indem er die Platte in eine innere Kreisfläche und eine äußere Ringfläche unterteilte.

Gerätebeschreibung

Das neu entwickelte Rheometer ist ein Rotationsgerät, das so konzipiert wurde, daß der Meßbereich gegenüber der thermisch stationären Methode optimal erweitert ist. Dies erfordert, wie aus Gl. [4] hervorgeht, die Realisierung sehr kleiner Versuchszeiten t_m. Um die Erwärmung der Meßsubstanz schon während des Hochlaufes auf Nenndrehzahl so gering wie möglich zu halten, muß das Anfahren sehr schnell erfolgen. Die Abbremsung der Rheometerwelle nach der Zeit t_m sollte ebenfalls sehr rasch geschehen, so daß das Materialverhalten auch unter diesen Bedingungen untersucht werden kann. Das schnelle Anhalten begrenzt zudem die Erwärmung während des Bremsens. Es kann also unmittelbar auf einen Scherversuch ein zweiter folgen, ohne daß dieser durch die Temperaturerhöhung während des ersten stark beeinflußt wird. Dadurch können Erholungsvorgänge in Materialien, die durch die Deformation bei hohen Schergefällen Strukturzusammenbrüche erfahren, auch nach sehr kurzen Erholungszeiten studiert werden. Wie schon bei anderen Viskosimeterausführungen (12, 13, 14) wurde somit ein möglichst rechteckförmiger Drehzahlverlauf angestrebt.

Abb. 1 zeigt schematisch den Aufbau des Rheometers. Das Antriebsaggregat ist ein Gleichstrommotor, der vor Versuchsbeginn zusammen mit einem Schwungrad auf Solldrehzahl gebracht wird. Hiernach wird zunächst die Kupplung eingeschaltet, worauf die Rheometerwelle nach der Kuppelzeit t_K die Motordrehzahl erreicht. Am Ende der Versuchszeit t_m wird die Bremse ein- und die Kupplung ausgeschaltet. Nach Ablauf der Bremszeit t_B steht die Rheometerwelle wieder still. Der Motor ist mit einer Drehzahlregelung und -steuerung ausgestattet, weswegen auf ein schaltbares mechanisches Übersetzungsgetriebe verzichtet werden konnte. Das Schwungrad ist zur Verhinderung von Einschwingvorgängen des Motors in Folge der Drehmomentstöße beim Ein- und Auskuppeln zwischen Motor und Antriebswelle geschaltet.

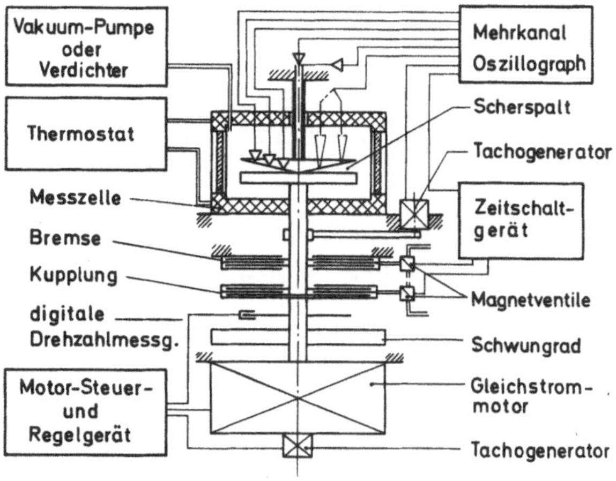

Abb. 1. Aufbauschema des Rheometers

Zur Realisierung des angestrebten rechteckförmigen Drehzahlverlaufs konnte auf die Erfahrungen aufgebaut werden, die an zwei Scherapparaturen zur Beanspruchung von Feststoffteilchen gesammelt worden waren (3, 6). An diese Apparaturen sind bezüglich des Drehzahlverlaufes ähnliche Anforderungen gestellt. Es eignen sich solche Kupplungen und Bremsen, die bei geringen eigenen Schwungmassen hohe Drehmomente übertragen können. Als besonders günstig erwies sich die Kombination einer starken pneumatischen Lamellenkupplung mit der gleichen Einheit als Bremse. Durch Variation des Luftdruckes kann die Kuppelzeit t_K und die Bremszeit t_B in gewissen Grenzen verändert werden. Der plötzliche Übergang vom Schergefälle $\varkappa = 0$ auf $\varkappa =$ konstant ist technisch natürlich nicht realisierbar. Um die genaue Berechnung der Scherdeformation der Flüssigkeit auch während des Hochlaufes und der Bremsung zu ermöglichen, wird die Drehzahl der Viskosimeterwelle während des ganzen Versuches gemessen und aufgezeichnet.

Mit dem neuen Rheometer sollen schnell verlaufende Vorgänge beim Beginn und am Ende der Scherung zeitlich auflösbar sein. Die Meßwertgeber müssen daher eine hohe Eigenfrequenz besitzen. Das von der Flüssigkeit übertragene Drehmoment wird mit einer drehsteifen Torsionsfeder bestimmt. Zur Bestimmung der Gesamtkraft N_{KP} in Gl. [11] wird die Stauchung der Torsionsfeder gemessen. Diese Einrichtung erlaubt zudem eine Kontrolle der Scherspaltweite während des Versuches.

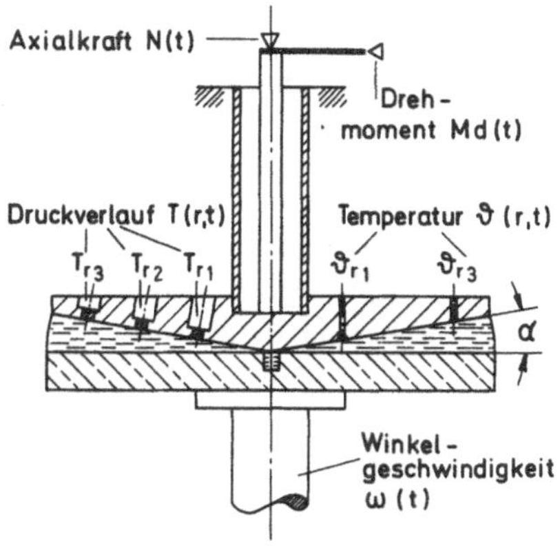

Abb. 2. Meßstelle des Rheometers

Neueste Entwicklungen auf dem Gebiet der Miniaturdruckaufnehmer erlauben es, die Druckverteilung $T_{\theta\theta}(r)$ direkt an den Spaltwänden festzustellen. Es werden drei druckempfindliche Halbleiterelemente (Abb. 2) verwendet, deren Membranen in der Plattenebene bzw. auf dem Kegelmantel liegen. Vor kurzem wurde von *Miller* und *Christiansen* (15) über Versuche berichtet, bei denen die radiale Druckverteilung auf dieselbe Weise gemessen wurde. Zur experimentellen Untersuchung des Einflusses von Bohrungen (16) auf die Messung der Normalspannungen werden auch Kegel eingesetzt, bei denen die Druckgeber über enge Bohrungen mit dem Scherspalt verbunden sind.

In der Meßzelle wird die Temperatur durch Thermoelemente gemessen, die an verschiedenen Stellen des Kegelmantels angebracht sind (Abb. 2). Die Temperaturänderung während der Scherung wird aufgezeichnet. Zusätzlich zur Temperatur kann in der Meßzelle auch der Absolutdruck variiert werden. Der Druck ist wählbar zwischen 10^3 dyn/cm^2 $< p_0 <$ 10^7 dyn/cm^2. Die Einstellung eines Unterdruckes gegenüber dem Atmosphärendruck erleichtert die Präparation der Proben, da eingeschlossene Gasblasen bei Unterdruck aus der Flüssigkeit schneller entweichen. Strömungsinstabilitäten, die durch die Normalspannungsdifferenzen verursacht werden, können durch die Erhöhung des Umgebungsdruckes p_0 beeinflußt werden.

Tab. 1. Meßbereich des Rheometers

Schergefälle	$10\,s^{-1} < \varkappa < 10^4\,s^{-1}$
Schubspannung	$2 \cdot 10^3$ dyn/cm$^2 < \tau < 10^7$ dyn/cm^2
Normaldruck	10^4 dyn/cm$^2 < T_{\theta\theta} < 10^7$ dyn/cm^2
Gesamtkraft	$2 \cdot 10^5$ dyn $< N < 10^8$ dyn
Temperaturbereich	$20\,°C < \vartheta < 260\,°C$
Kürzeste Kupplungs- und Bremszeit bei $\varkappa = 10^4\,s^{-1}$	$t_K \approx t \approx 5 \cdot 10^{-3}$ s
Versuchszeit	$10^{-2}\,s < t_m < 10^2\,s$

Der Meßbereich des Rheometers ist in Tab. 1 angegeben. Die aufgeführten Daten beziehen sich auf den derzeitigen Stand der meßtechnischen Einrichtungen. Die Meßbereiche von Schubspannung τ, Gesamtkraft N und der Normaldrücke $T_{\theta\theta}$ können durch Austauschen der Torsionsfeder bzw. der Druckgeber nach oben und unten weiter ausgedehnt werden. Auch kann der Schergefällebereich nach unten durch Übergang vom geregelten zum ungeregelten Motorbetrieb erweitert werden. Folgende Stoffeigenschaften können gemessen werden:

1. Viskositätsfunktion $\eta(\varkappa)$
2. Normalspannungsfunktionen $\sigma_1(\varkappa)$ und $\sigma_2(\varkappa)$
3. Die elastischen Parameter Schermodul und Relaxationszeit
4. Temperatureinfluß bis 260 °C

Erste Ergebnisse

Scherversuche, die mit einer *Couette*-Scherapparatur, deren Antrieb ähnlich wie der des Rheometers aufgebaut ist, ausgeführt wurden, zeigen erste interessante Ergebnisse. In Abb. 3

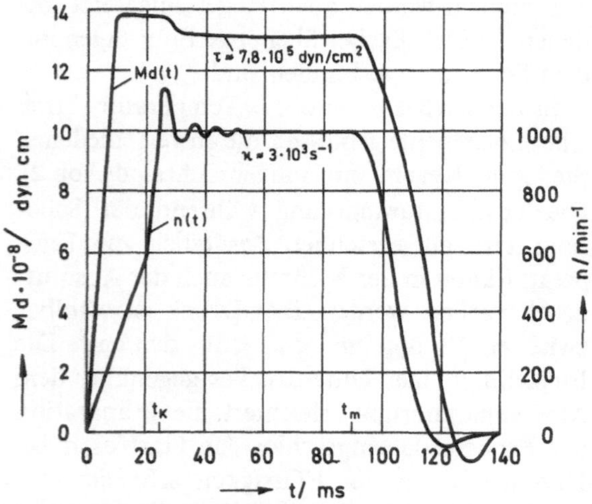

Abb. 3. Drehmoment- und Drehzahlverlauf eines Scherversuches mit Siliconöl ($\eta_0 = 2 \cdot 10^4$ P, $\vartheta = 20\,°C$)

ist der Drehmomentverlauf $Md(t)$ und der Drehzahlverlauf des Außenzylinders $n(t)$ aufgezeichnet. Versuchsflüssigkeit ist das Dimethylpolysiloxan AK 2000000 der Firma Wacker mit der Zähigkeit $\eta_0 = 2 \cdot 10^4$ P für $\varkappa \to 0$. Die Drehzahl hat nach einer Kuppelzeit $t_K = 25$ ms den stationären Wert von 1000 1/min erreicht, was einem Schergefälle von $\varkappa = 3 \cdot 10^3\,s^{-1}$ entspricht. Die rasch abklingenden Schwingungen haben in der Drehzahlmeßeinrichtung ihren Ursprung. Der Hochlauf ist in zwei Abschnitten sehr gut linear, woraus sich die Scherdeformation der Flüssigkeit leicht zu jedem Zeitpunkt bestimmen läßt. Die Versuchsdauer t_m beträgt 90 ms. Danach wird die Scherung in der Bremszeit $t_B = 25$ ms gestoppt. Durch ein geringes Spiel der Bremse bedingt treten am Ende kleine Verdrehungen der Viskosimeterwelle in negativer Richtung auf. Die Drehmoment-Zeit-Kurve $Md(t)$ zeigt vier charakteristische Bereiche:

Im Bereich I ($0 < t < 10$ ms) erfolgt ein sehr steiler Anstieg des Drehmomentes mit linear steigender Drehzahl bzw. linearem Schergefälle.

Im Bereich II ($10\,ms < t < 25\,ms$) tritt eine plötzliche Änderung des Materialverhaltens ein. Trotz weiterer Steigerung des Schergefälles um den Faktor 3 bleibt die Schubspannung τ konstant.

Im Bereich III ($25\,ms < t < 90\,ms$) fällt, nachdem die Enddrehzahl erreicht ist, die Schubspannung zunächst leicht ab. Schon 30 ms nach Beginn des Scherversuches hat sich ein stationärer Schubspannungszustand eingestellt.

Im Bereich IV ($90\,ms < t < 140\,ms$) zur Versuchszeit t_m beginnt gleichzeitig mit dem Drehzahlabfall ein rasches Abklingen des Drehmomentes. Der Drehmomentabfall hat gegenüber der Drehzahl nur eine geringe Zeitverzögerung. Der Nulldurchgang der Momentenkurve, verursacht durch die zeitweilige negative Drehrichtung, erfolgt etwa 7 ms nach der Drehrichtungsänderung.

Die Aufzeichnung des Drehzahlverlaufes verdeutlicht, daß die angestrebte Kurzzeitmessung technisch realisierbar ist, aber auch, daß der in der Literatur oft beschriebene Sprung des Schergefälles von Null auf einen stationären Wert $\varkappa$ = konstant trotz Verwendung extrem starker Kupplungen nur annähernd erreicht werden kann. Dies kann zum Beispiel bei der Berechnung der Deformation, bei der zu Beginn des Bereiches II der Strukturzusammenbruch erfolgt, zu erheblichen Fehlern führen. Würde man beispielsweise aus Unkenntnis des genauen Drehzahlverlaufes mit $\varkappa = 3 \cdot 10^3\,s^{-1}$ rechnen, so ergäbe sich für diese Deformation bei $t = 10$ ms der Wert $\gamma = \varkappa \cdot t = 30$. Der genaue Drehzahlverlauf nach Abb. 3 liefert aber nur den Wert 5. Die gemessene Drehmomentkurve zeigt, daß bei dieser Beanspruchungsart das stationäre Fließen, das zur Bestimmung der Schubspannungsfunktion $\tau(\varkappa)$ notwendig ist, in den kurzen Zeiten, in denen die Temperaturänderung im Meßspalt noch vernachlässigbar klein ist, tatsächlich auch erreicht werden kann.

Zusammenfassung

Die Schwierigkeiten, die sich bei der Ermittlung rheologischer Stoffeigenschaften bei großen Zähigkeiten und hohen Schergefällen durch die starke Wärmeentwicklung in der Strömung ergeben, können in gewissen Grenzen durch sehr kurze Scherzeiten behoben werden. Es wurde ein Rotationsrheometer gebaut, dessen Antrieb und Meßeinrichtungen extrem kurze Versuchszeiten im Bereich von 10 bzw. 100 Millisekunden zulassen. Das Rheometer ist zur Messung der Schubspannungsfunktion und der beiden Normalspannungsdifferenzen bei Schergefällen bis $10^4\,s^{-1}$ und Temperaturen bis 260 °C eingerichtet. Aus dem Anlauf- und Bremsverhalten der Flüssigkeit können ferner elastische Parameter bestimmt werden. Es werden erste Ergebnisse mitgeteilt, die an einer *Couette*-Apparatur mit ähnlichem Antrieb für hochviskoses Siliconöl erhalten wurden. Das Material zeigt während des Anlaufes einen ausgeprägten Strukturzusammenbruch.

Summary

The difficulties in determining the rheological properties of highly viscous liquids at high shear rates, caused by the considerable heat production in the flow, can be resolved within certain limits by drastically reducing the shearing time. A rotational rheometer has been designed, drive and measuring equipment of which allow to do the tests within 10 to 100 milliseconds. The rheometer is suited for measuring the shear stress function and the two normal stress functions at shear rates up to $10^4\,sec^{-1}$ and at temperatures up to 260 °C. Moreover, from the behaviour of the liquid during acceleration and deceleration elastic parameters can be determined. First results are presented which have been obtained with highly viscous silicon-oil using a *Couette* apparatus instead of a cone and plate arrangement. The material showed a pronounced structure break-down during acceleratio

Literatur

1) *Raasch, J.*, Beanspruchung und Verhalten suspendierter Feststoffteilchen in Scherströmungen hoher Zähigkeit, Dissertation Karlsruhe 1961.

2) *Rumpf, H.* und *J. Raasch*, Desagglomeration in Strömungen, Symposion Zerkleinern 1962 Frankfurt/M., S. 151–154 (Weinheim-Düsseldorf 1962).

3) *Krekel, J.*, Herstellung und Messung von Scherströmungen mit extrem großer Schubspannung und ihr Einfluß auf die Zerkleinerung, Dissertation Karlsruhe 1964, Kurzfassung in: Chem.-Ing.-Techn. **38**, 229–234 (1966).

4) *Reichert, H.*, Chem.-Ing.-Techn. **45**, 391-395 (1973).

5) *Carslaw, H. S.* und *J. C. Jaeger*, Conduction of heat in solids (Oxford 1959).

6) *Reichert, H.*, Desagglomeration organischer Farbpigmente in Couetteströmungen hochzäher viskoelastischer Flüssigkeiten, Dissertation Karlsruhe 1973, erscheint teilweise demnächst in Rheol. Acta.

7) *Coleman, B. D.*, *H. Markovitz* und *W. Noll.*, Viscometric flow of non-Newtonian fluids (Berlin-Heidelberg-New York 1966).

8) *Meißner, J.*, Kunststoffe **57**, 702–710 (1967).

9) *Kotaka, T.*, *M. Kurata* und *M. Tamura*, J. Appl. Physics **30**, 1705–1712 (1959).

10) *Pollett, W. F.*, Brit. J. Appl. Phys. **6**, 199–206 (1955).

11) *Schlichting, H.*, Grenzschichttheorie (Karlsruhe 1965).

12) *Lederer, K.*, *J. Schurz* und *F. Königshofer*, Rheol. Acta **8**, 456–465 (1969).

13) *Vinogradov, G. V.* und *J. M. Belkin*, J. Polym. Sci. Part **A**, **3**, 917–932 (1965).

14) *Vinogradov, G. V.* und *A. Ya. Malkin*, Rheol. Acta **5**, 188–193 (1966).

15) *Miller, M. J.* und *E. B. Christiansen*, Amer. Inst. Chem. Eng. J. **18**, 600–608 (1972).

16) *Pritchard, W. G.*, Rheol. Acta **9**, 200–207 (1970).

Anschrift der Verfasser:

W. Gleissle und *H. Reichert*
Institut für Mechanische Verfahrenstechnik
D-75 Karlsruhe, Richard-Willstätter-Allee

Rheol. Acta 12, 578–587 (1973)

From the Department of Chemical Engineering, Queen's University, Kingston (Canada)

Rates of shear in coaxial cylinder viscometers

By R. K. Code and J. D. Raal

With 11 figures

(Received October 27, 1972)

Nomenclature

C_R	correction factor for non-*Newtonian* effects
$C_{R\alpha}$	correction factor based on constant power-law departure factor α
C_{RP}	correction factor based on power-law approximation
C_{RK_3}	correction factor based on three-term series of *Krieger* and *Elrod*
C_{RK_4}	correction factor based on four-term series of *Krieger* and *Elrod*
$f(\tau)$	rate of shear $= dv/dr$
K	consistency index
m	reciprocal of flow-behavior index $= 1/n = d\ln\Omega/d\ln\tau$
n	flow-behavior index $= d\ln\tau/d\ln\Omega$
N	number of terms
p	index variable
r	radial direction
s	radius ratio $= r_1/r_2$
v	linear velocity
α	power-law departure factor $= d\ln m/d\ln\tau$
τ	shear stress
ω or Ω	angular velocity

Subscripts

r	radial direction
θ	angular direction
1	outer surface of inner cylinder
2	inner surface of outer cylinder

A common geometry used in rotational viscometry consists of two concentric cylinders in which the test fluid is sheared in the annulus between the cylinders. In this type of instrument the calculation of the shear stress from the measured torque is straightforward but the determination of the corresponding rate of shear is a difficult problem unless the type of fluid behavior, e.g. power law, is known *a priori*. The equation of motion may be solved (1) analytically for the special case of power-law behavior but a completely general analysis results in an expression for the difference in the rates of shear at the two surfaces (2).

$$\frac{d\Omega}{d\tau_1} = \frac{1}{2\tau_1}[f(\tau_1) - f(\tau_2)]. \qquad [1]$$

A rigorous infinite series solution (3, 4) to this difference equation is available

$$f(\tau_1) = 2\sum_{p=0}^{\infty} \Omega|_{s^{2p}\tau_1} \cdot m|_{s^{2p}\tau_1} \qquad [2]$$

but in practice, the series is slowly convergent for narrow-gap instruments requiring the laborious graphical evaluation of a large number of m's for the determination of the one value of the rate of shear corresponding to the shear stress τ_1. An approximate series solution has been developed by *Krieger* and *Elrod* (5).

In this paper the infinite series solution is reexamined and an analysis is proposed in terms of a power-law departure factor

$$\alpha = \frac{d\ln m}{d\ln\tau} = \text{constant}. \qquad [3]$$

The results of machine computations, covering a range of radius ratios, flow-behavior indexes, and power-law departure factors, are used to test the adequacy of the power-law approximation, and the approximate series solution of *Krieger* and *Elrod* in describing this postulated behavior.

Analysis

In this analysis conditions of simple shear are assumed for the test fluid in the annular gap between the rotating outer cylinder and the stationary inner cylinder. Under these conditions the rate of shear may be expressed as a function of the shear stress

$$r\frac{d\omega}{dr} = f(\tau) \qquad [4]$$

and

$$r^2\tau = \text{constant} \quad [5]$$

$$r^2 d\tau = 2r\tau dr = 0 \quad [6]$$

$$dr = -\frac{r}{2}\frac{d\tau}{r}. \quad [7]$$

Use may be made of this transformation in eq. [4] to obtain

$$d\omega = -\frac{1}{2}\frac{f(\tau)}{\tau}d\tau. \quad [8]$$

At r_2, $\omega = \Omega$ and $\tau = \tau_2$ and at r_1, $\omega = 0$ and $\tau = \tau_1$.

$$\Omega = \frac{1}{2}\int_{\tau_2}^{\tau_1}\frac{f(\tau)}{\tau}d\tau. \quad [9]$$

Differentiating,

$$\frac{d\Omega}{d\tau_1} = \frac{1}{2}\left[\frac{f(\tau_1)}{\tau_1} - \frac{f(\tau_2)}{\tau_2}\frac{d\tau_2}{d\tau_1}\right]. \quad [10]$$

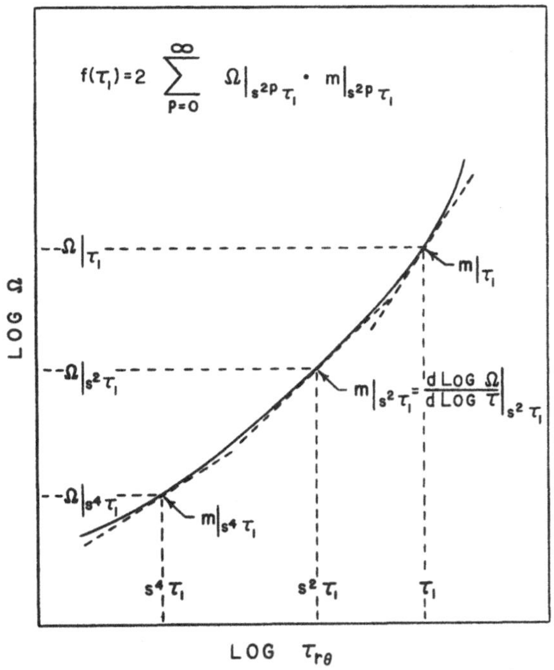

Fig. 1. Evaluation of Ω and m at successive values of shear stress equal to $s^{2p}\tau_1$

From the condition of constancy of torque

$$\tau_1 r_1^2 = \tau_2 r_2^2 \quad [11]$$

$$\frac{d\tau_2}{d\tau_1} = \left(\frac{r_1}{r_2}\right)^2 = s^2 \quad [12]$$

$$\frac{d\Omega}{d\tau_1} = \frac{1}{2}\left[\frac{f(\tau_1)}{\tau_1} - \frac{f(s^2\tau_1)}{s^2\tau_1}s^2\right] \quad [13]$$

$$2\tau_1\frac{d\Omega}{d\tau_1} = f(\tau_1) - f(s^2\tau_1) \quad [14]$$

In general,

$$2\tau_{r\theta}\frac{d\Omega}{d\tau_{r\theta}} = f(\tau_{r\theta}) - f(s^2\tau_{r\theta}). \quad [15]$$

Approximate solutions to this difference equation have been presented by a number of authors (5, 6. 7). In this development the treatment of *Middleman* (3) is followed. Since eq. [15] is valid for all values of shear stress, $s^2\tau_{r\theta}$ may be substituted for $\tau_{r\theta}$. In general,

$$2s^2\tau_{r\theta}\left.\frac{d\Omega}{d\tau_{r\theta}}\right|_{s^2\tau_{r\theta}} = f(s^2\tau_{r\theta}) - f(s^4\tau_{r\theta})$$

$$2s^4\tau_{r\theta}\left.\frac{d\Omega}{d\tau_{r\theta}}\right|_{s^4\tau_{r\theta}} = f(s^4\tau_{r\theta}) - f(s^6\tau_{r\theta})$$

$$\vdots$$

$$2s^{2(N-1)}\tau_{r\theta}\left.\frac{d\Omega}{d\tau_{r\theta}}\right|_{s^{2(N-1)}\tau_{r\theta}} = f(s^{2(N-1)}\tau_{r\theta}) - f(s^{2N}\tau_{r\theta}). \quad [16]$$

Summing the complete set of equations

$$f(\tau_{r\theta}) - f(s^{2N}\tau_{r\theta}) = \sum_{p=0}^{N-1} 2s^{2p}\tau_{r\theta}\left.\frac{d\Omega}{d\tau_{r\theta}}\right|_{s^{2p}\tau_{r\theta}}. \quad [17]$$

Since s is less than unity s^{2N} vanishes in the limit as N becomes large 4), and

$$f(\tau_{r\theta}) = \frac{dv_\theta}{dr} = 2\sum_{p=0}^{\infty} s^{2p}\tau_{r\theta}\left.\frac{d\Omega}{d\tau_{r\theta}}\right|_{s^{2p}\tau_{r\theta}}. \quad [18]$$

For convenience $d\Omega/d\tau_{r\theta}$ may be expressed in terms of the flow-behavior index n or its reciprocal m.

$$m = \frac{d\log\Omega}{d\log\tau_{r\theta}} \quad [19]$$

$$f(\tau_{r\theta}) = \frac{dv_\theta}{dr} = 2\sum_{p=0}^{\infty}\Omega|_{s^{2p}\tau_{r\theta}}\cdot m|_{s^{2p}\tau_{r\theta}}. \quad [20]$$

The rate of shear corresponding to the shear stress, $\tau_{r\theta}$, may be evaluated to the required accuracy by summing sufficient terms in the series. The summation requires the evaluation of corresponding values of Ω and m at successive values of shear stress equal to $s^{2p}\tau_{r\theta}$ as illustrated in fig. 1. In principle the procedure is straight-

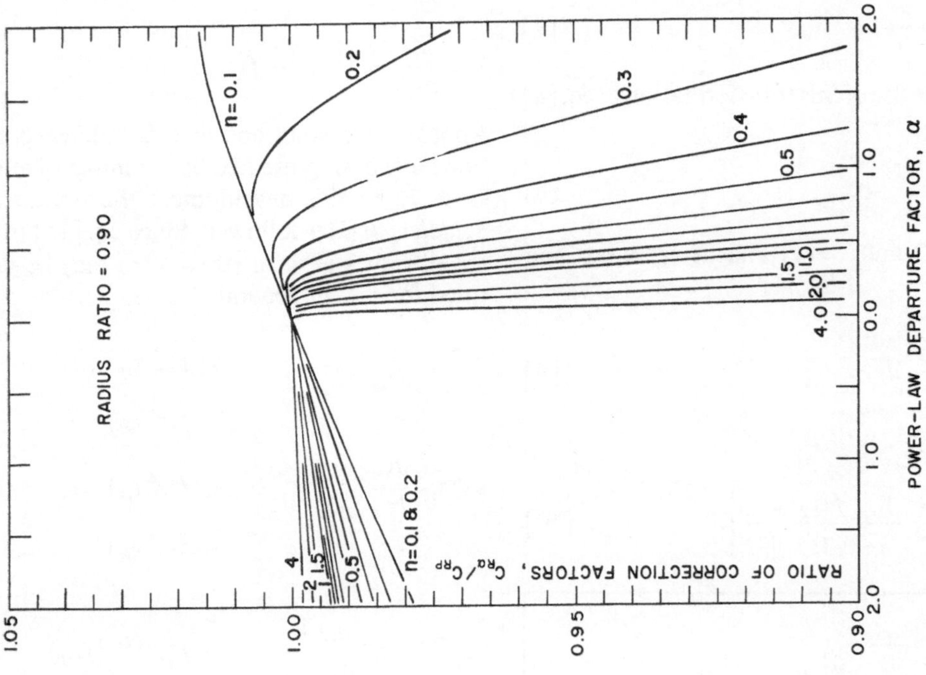
RADIUS RATIO = 0.90
RATIO OF CORRECTION FACTORS, $C_{R\alpha}/C_{RP}$
POWER-LAW DEPARTURE FACTOR, α
n = 0.1
0.2
0.3
0.4
0.5
1.0
1.5
2.0
4.0
n = 0.1 & 0.2
0.5
1.5
1
2
4
1.05
1.00
0.95
0.90
2.0
1.0
0.0
1.0
2.0

Fig. 3

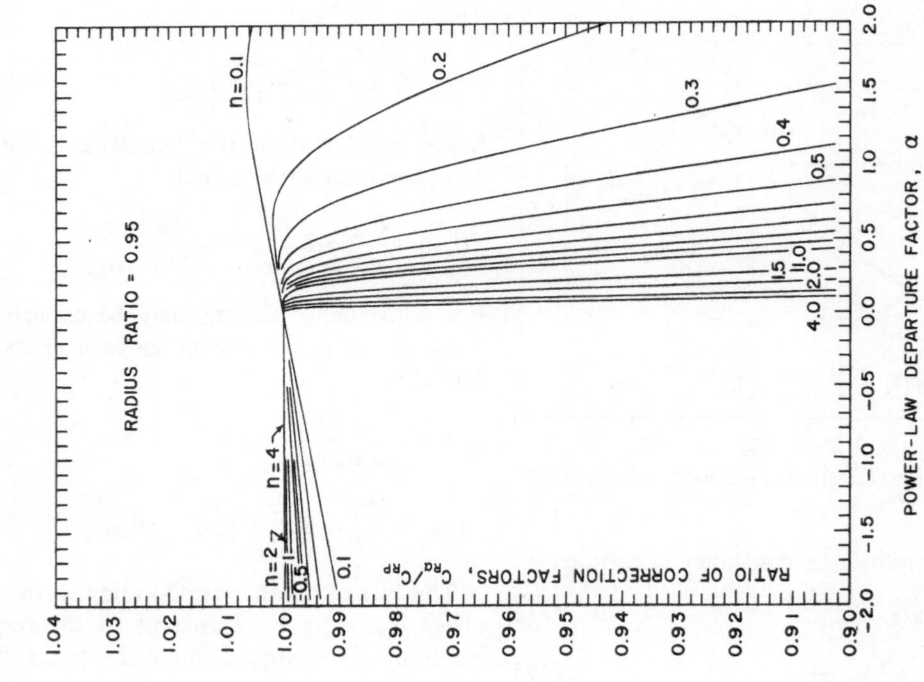
RADIUS RATIO = 0.95
RATIO OF CORRECTION FACTORS, $C_{R\alpha}/C_{RP}$
POWER-LAW DEPARTURE FACTOR, α
n = 0.1
0.2
0.3
0.4
0.5
1.0
1.5
2.0
4.0
n = 4
n = 2
1
0.5
0.1
1.04
1.03
1.02
1.01
1.00
0.99
0.98
0.97
0.96
0.95
0.94
0.93
0.92
0.91
0.90
-2.0
-1.5
-1.0
-0.5
0.0
0.5
1.0
1.5
2.0

Fig. 2

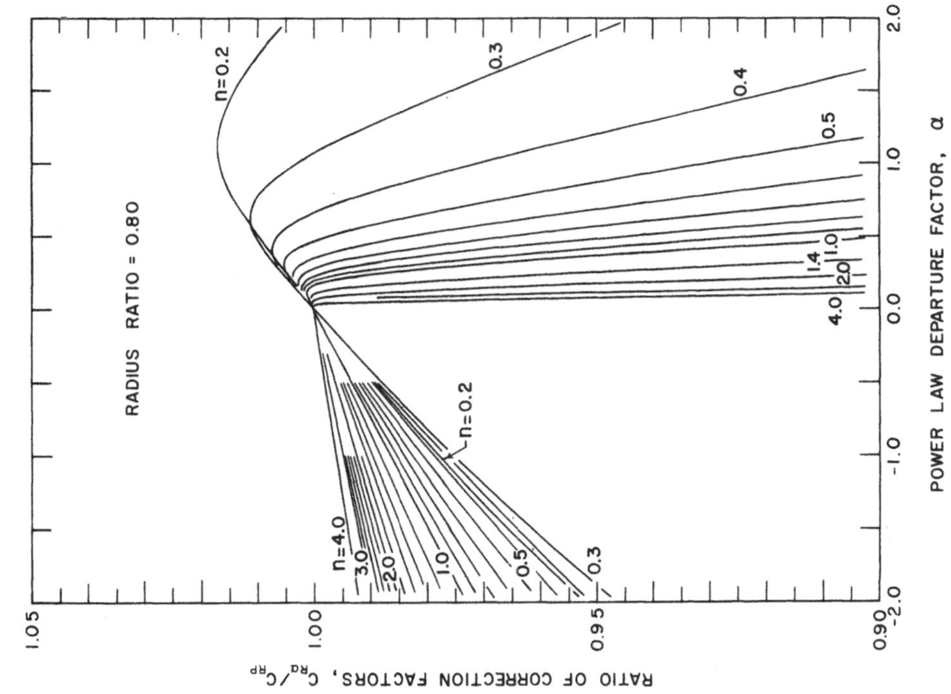

Fig. 4

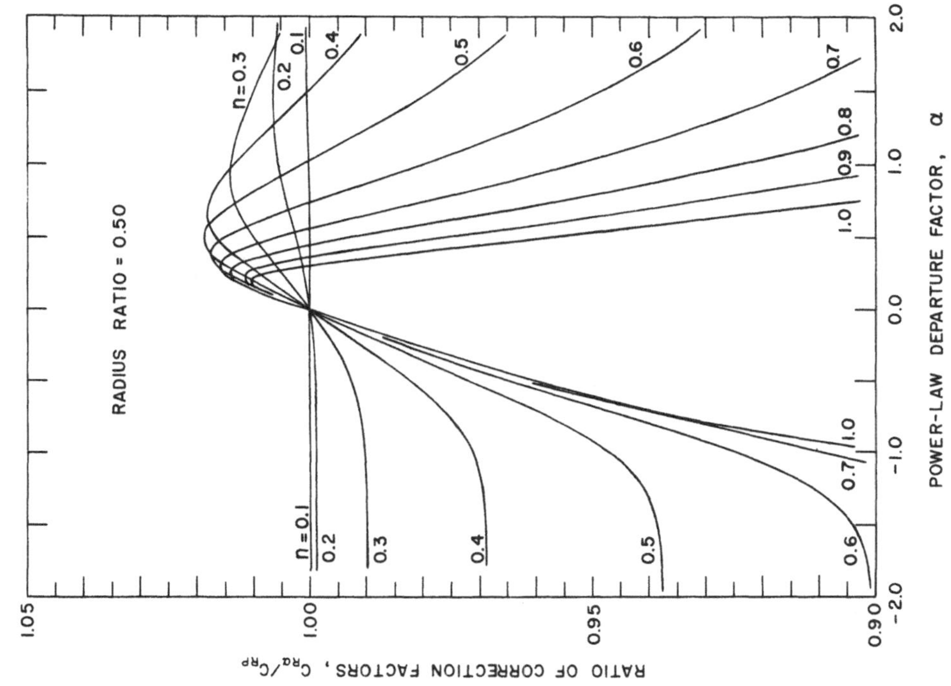

Fig. 5

Figs. 2–5. Test of adequacy of the power-law approximation

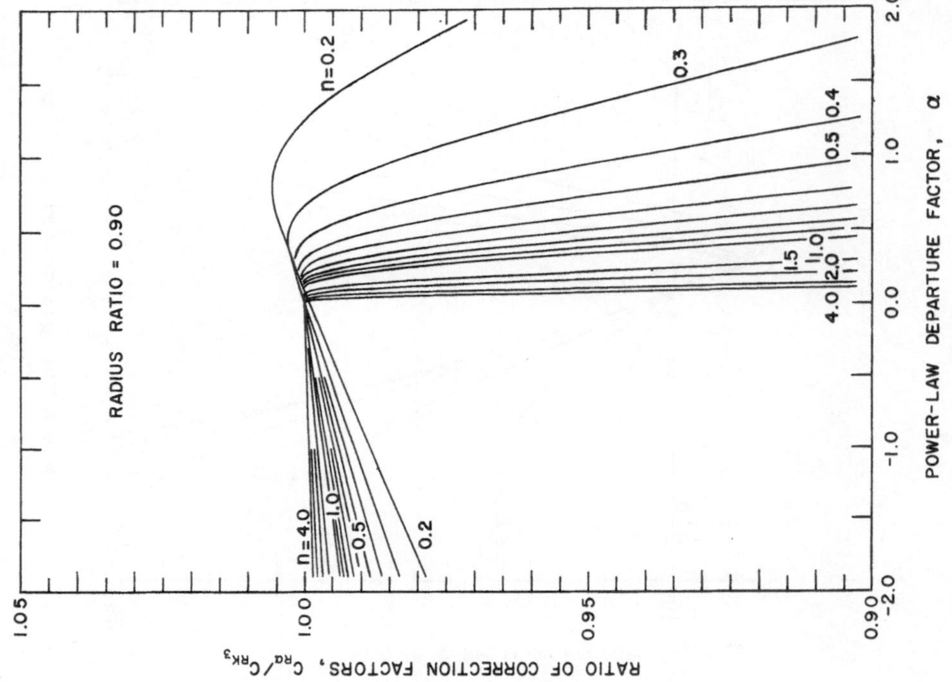
RADIUS RATIO = 0.90
n=0.2
0.3
0.4
0.5
1.0
1.5
2.0
4.0
n=4.0
1.0
0.5
0.2
POWER-LAW DEPARTURE FACTOR, α
RATIO OF CORRECTION FACTORS, $C_{R\alpha}/C_{RK_3}$

Fig. 7

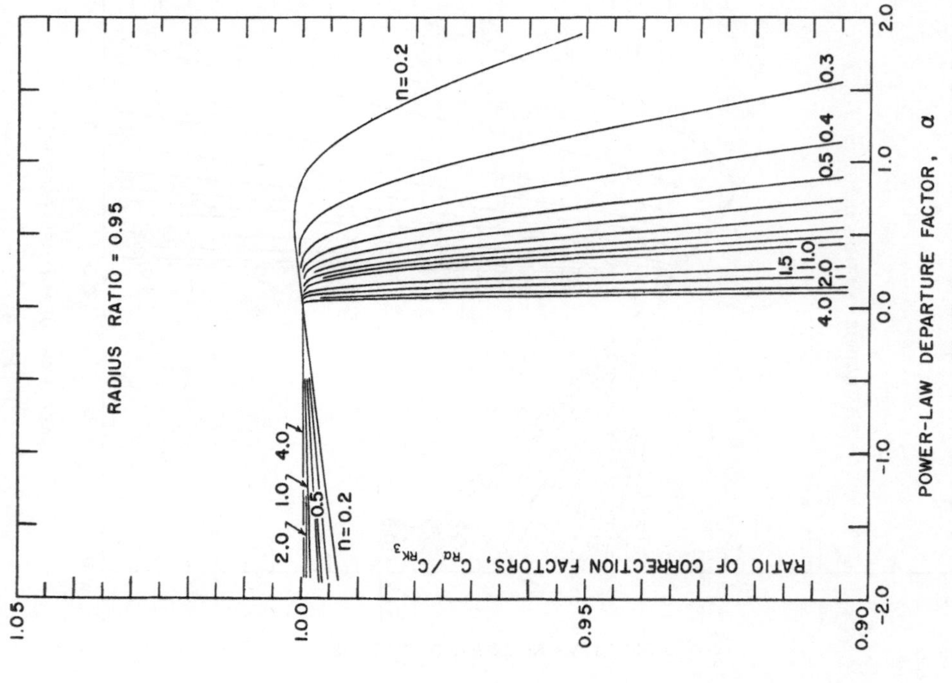
RADIUS RATIO = 0.95
n=0.2
0.3
0.4
0.5
1.0
1.5
2.0
4.0
2.0
1.0
4.0
0.5
n=0.2
POWER-LAW DEPARTURE FACTOR, α
RATIO OF CORRECTION FACTORS, $C_{R\alpha}/C_{RK_3}$

Fig. 6

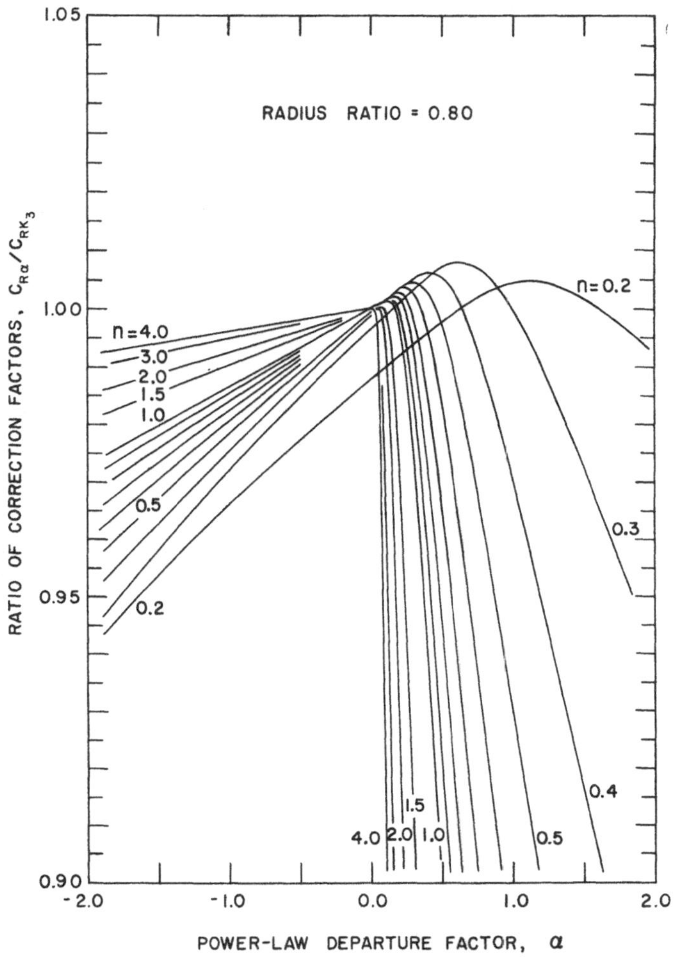

Fig. 8

Figs. 6–8. Test of adequacy of the three-term series of *Krieger* and *Elrod*

forward but in application two difficulties are encountered: 1. the series is slowly convergent for narrow-gap instruments, i.e., for $s \to 1.0$, and 2. values of m and Ω may be required for a range of several decades of shear stress below the region of interest.

For power-law behavior

$$f(\tau) = K\tau^m \qquad [21]$$

where

$$m = \frac{d\log\Omega}{d\log\tau} = \text{constant} \qquad [22]$$

and

$$\Omega|_{s^{2p}\tau} = \Omega_1\left[\frac{s^{2p}\tau_1}{\tau_1}\right]^m = \Omega_1 s^{2pm}. \qquad [23]$$

Substituting in eq. [20]

$$f(\tau_1) = 2\sum_{p=0}^{\infty}\Omega|_{s^{2p}\tau_1}\cdot m|_{s^{2p}\tau} = 2\Omega_1 m\sum_{p=0}^{\infty}s^{2pm} \qquad [24]$$

$$f(\tau_1) = 2\Omega_1 m[1 + s^{2m} + s^{4m} + s^{6m} + \cdots] = \frac{2\Omega_1 m}{1 - s^{2m}}. \qquad [25]$$

This result, which agrees with the analytical solution obtained by *Brodkey* (1), may be used to correct for non-*Newton*ian effects provided the assumption of power-law behavior is a good approximation or the gap width is sufficiently small that the variation in the flow-behavior index is negligible. The geometric dependency of the correction, once the flow-behavior index departs from unity, may be demonstrated by expressing the above result in terms of a correction factor to pseudo-shear rates based on *Newton*ian behavior.

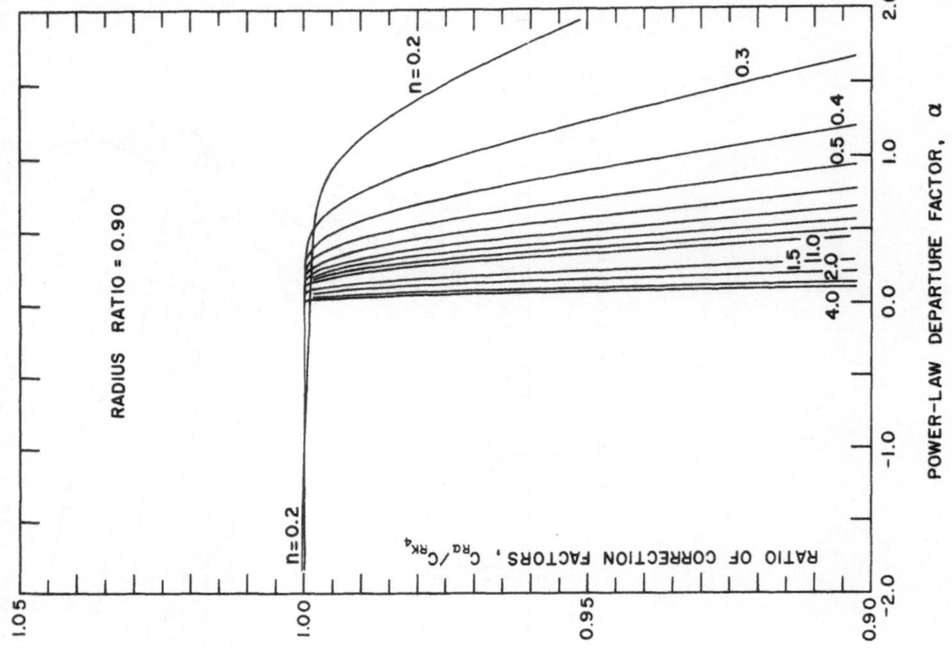
RADIUS RATIO = 0.90
RATIO OF CORRECTION FACTORS, $C_{R\alpha}/C_{RK_4}$
POWER-LAW DEPARTURE FACTOR, α
n=0.2
n=0.2
0.3
0.4
0.5
1.0
1.5
2.0
4.0
1.05
1.00
0.95
0.90
-2.0
-1.0
0.0
1.0
2.0

Fig. 10

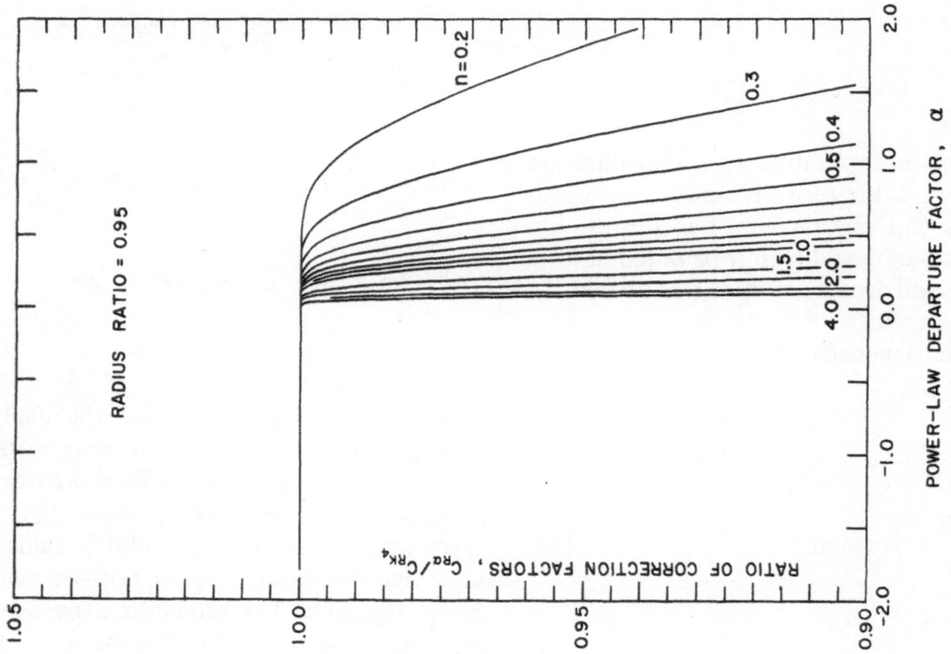
RADIUS RATIO = 0.95
RATIO OF CORRECTION FACTORS, $C_{R\alpha}/C_{RK_4}$
POWER-LAW DEPARTURE FACTOR, α
n=0.2
0.3
0.4
0.5
1.0
1.5
2.0
4.0
1.05
1.00
0.95
0.90
-2.0
-1.0
0.0
1.0
2.0

Fig. 9

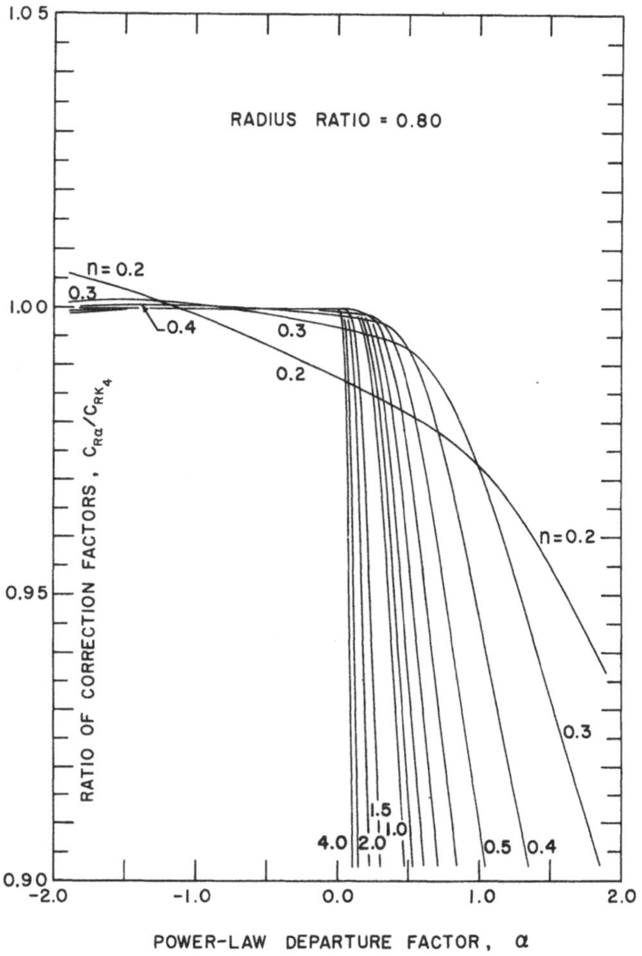

Figs. 9–11. Test of adequacy of the four-term series of *Krieger* and *Elrod*

Fig. 11

$$f(\tau)\big|_{\text{non-}Newtonian} = f(\tau)\big|_{Newtonian} \cdot C_R = \frac{2\Omega}{1 - s^2} \cdot C_R . \qquad [26]$$

For power-law behavior

$$C_{RP} = \frac{m(1 - s^2)}{1 - s^{2m}} . \qquad [27]$$

For systems where the assumption of power-law behavior is not an adequate approximation the following treatment is proposed. Define a power-law departure factor, α, as

$$\alpha = \frac{d \ln m}{d \ln \tau} = \text{constant}. \qquad [3]$$

The proposed function includes power-law behavior as a limiting case ($\alpha = 0$). The expressions for m and Ω successive values of shear stress equal to $s^{2p}\tau_1$ required for substitution in eq. [20] may be developed as follows.

From the definition of m and the assumed functional variation in m

$$\frac{d \ln \Omega}{d \ln \tau} = m \qquad [19]$$

and

$$m = m_1 \left(\frac{\tau}{\tau_1}\right)^{\alpha} \qquad [28]$$

$$m\big|_{s^{2p}\tau} = m_1 s^{2p\alpha} \qquad [29]$$

$$\ln\left[\frac{\Omega}{\Omega_1}\right] = \frac{m_1}{\tau_1^{\alpha}} \int_{\tau}^{s^{2p}\tau_1} \tau^{\alpha} d \ln \tau = \frac{m_1}{\alpha}[s^{2p\alpha} - 1] \qquad [30]$$

$$\Omega\big|_{s^{2p}\tau_1} = \Omega\big|_{\tau_1} \cdot e^{\frac{m}{\alpha}[s^{2p} - 1]} \qquad [31]$$

Substituting for $\Omega|_{s^{2p}\tau_1}$ and $m|_{s^{2p}\tau}$ in eq. [20]

$$f(\tau_1) = 2\Omega_1 \cdot m_1 \sum_{p=0}^{\infty} e^{\frac{m}{\alpha}[s^{2p_\alpha} - 1]} \cdot s^{2p\alpha} \qquad [32]$$

or in terms of a correction factor

$$C_{R\alpha} = m_1(1 - s^2) \sum_{p=0}^{\infty} e^{\frac{m_1}{\alpha}[s^{2p\alpha} - 1]} \cdot s^{2p\alpha}. \qquad [33]$$

Although the result is exact only for the assumed condition of constant power-law departure factor, in practice it is necessary only that $(d \ln m / d \ln \tau) = \text{constant}$ give an adequate description of the variation in the flow-behavior index over the range of shear stress required for convergence.

As expected the result reduces to the expression for power-law behavior in the limit as $\alpha \to 0$. Applying *Lhopitals* rule

$$\lim_{\alpha \to 0} \left[\frac{s^{2p\alpha} - 1}{\alpha}\right] = \lim_{\alpha \to 0} (s^{2p\alpha} 2p \ln s) = 2p \ln s \quad [34]$$

and

$$C_{R\alpha}\big|_{\alpha=0} = C_{RP} = m(1 - s^2) \sum_{p=0}^{\infty} e^{2pm \ln s} \qquad [35]$$

$$C_{RP} = m(1 - s^2) \sum_{p=0}^{\infty} s^{2pm} = \frac{m(1 - s^2)}{1 - s^{2m}}. \qquad [36]$$

The series result of *Krieger* and *Elrod* (5)

$$f(\tau_1) = \frac{\Omega}{-\ln s}\left[1 - m \ln s + \frac{1}{3}(m \ln s)^2 + \frac{(\ln s)^2}{3} \frac{dm}{d \ln \tau_1}\right] \qquad [37]$$

may be modified to include the factor α.

$$f(\tau_1) = \frac{\Omega}{-\ln s}\left[1 - m \ln s + \frac{1}{3}(m \ln s)^2 + \frac{\alpha m (\ln s)^2}{3}\right] \qquad [38]$$

or

$$C_{RK_4} = \frac{1 - s^2}{-2 \ln s}\left[1 - m \ln s + \frac{1}{3}(m \ln s)^2 + \frac{\alpha m (\ln s)^2}{3}\right]. \qquad [39]$$

The authors (5) state that the fourth term may be required for systems where $-m \ln s > 1$. *Middleman* (3) recommends the use of the three-term series for $-m \ln s < 0.5$.

$$C_{RK_3} = \frac{1 - s^2}{-2 \ln s}\left[1 - m \ln s + \frac{1}{3}(m \ln s)^2\right]. \quad [40]$$

Machine computations

A computer program was written for the generation of correction factors $C_{R\alpha}$, C_{RP}, C_{RK}, and C_{RK} for a range of flow-behavior indexes, radius ratios and power-law departure factors in order to compare the proposed analysis with the power-law approximation and the series solution of *Krieger* and *Elrod* (5). Since eq. [33] is exact for constant departure factor and the series may be summed to any specified accuracy, a comparison of correction factors provides a test of the adequacy of the simpler approximations in describing the postulated behavior as the gap width increases and the flow-behavior index departs from a constant value.

Convergence criteria for eq. [33] were developed using

$$\sum_{p=0}^{N} s^{2p|\alpha|} \qquad [41]$$

as a bounding series. Since s is less than unity, the exponential portion of eq. [33] is less than unity for α positive and $p > 0$ and each term in the series is less than the corresponding term in the bounding series. For positive departure factors convergence was based on the residual error in the bounding series.

$$\sum_{p=0}^{\infty} s^{2p|\alpha|} - \sum_{p=0}^{N} s^{2p|\alpha|} = \frac{1}{1 - s^{2\alpha}} - \sum_{p=0}^{N} s^{2p\alpha} < 10^{-5}. \qquad [42]$$

For negative departure factors the problem is not as straightforward. Again examining eq. [33] as p increases, the series increases above the bounding series initially, passes through a maximum and decreases to a value below the bounding series. In this region the series decreases more rapidly and remains bounded by $s^{2p|\alpha|}$ as p continues to increase. This behavior was checked numerically at extreme values of the parameters studies but as a precaution the computer was instructed to compare corresponding terms in the two series and to determine and print out the number of terms required for the series to become bounded by $s^{2p|\alpha|}$. For negative departure factors convergence was based on the relative error

$$\frac{\dfrac{1}{1 - s^{2|\alpha|}} - \sum_{p=0}^{N} s^{2p|\alpha|}}{\sum_{p=0}^{N} e^{-\frac{m}{|\alpha|}[s^{-2p|\alpha|} - 1]} \cdot s^{-2p|\alpha|}} < 10^{-5}. \qquad [43]$$

Results

The results as copies of computer printouts are available from the authors for the following range of parameters:

s 0.5–0.90 ($\Delta = 0.1$); 0.95
n 0.1–1.0 ($\Delta = 0.1$); 1.0–4.0 ($\Delta = 0.2$)
α -2.0 to $+2.0$ ($\Delta = 0.1$)

Selected results are summarized as correction factor ratios $C_{R\alpha}/C_{RP}$, $C_{R\alpha}/C_{RK}$, $C_{R\alpha}/C_{RK_4}$ versus power-law departure factor α in fig. 2 through 11. An estimate of the error involved in applying the power-law approximation or the *Krieger* and *Elrod* series may be obtained from the charts from an estimate of the departure factor α obtained from the pseudo-shear curve (assumed *Newton*ian behavior). An iterative estimation of α may be required if the true shear curve is not approximately parallel to the pseudo-shear curve.

As a final point it should be noted that conditions of simple shear are implied for the corrections for non-*Newton*ian effects discussed in this paper. Depending upon the viscometer design and type of fluid additional corrections may be required for end effects (8) and/or effective slip (9, 10).

Acknowledgement

This work was supported in part by a grant from the National Research Council of Canada, Grant No. A-952.

Summary

The general difference equation for coaxial cylinder geometry

$$\frac{d\Omega}{d\tau_1} = \frac{1}{2\tau_1}[f(\tau_1) - f(\tau_2)]$$

has been solved as an infinite series to obtain an expression for the rate of shear at one of the surfaces in terms of the radius ratio ($s = r_1/r_2$), the reciprocal of the flow-behavior index ($m = 1/n$), and a power-law departure factor ($\alpha = d\ln m/d\ln\tau$). The result expressed as a correction factor to pseudo-shear rates based on *Newton*ian behavior is:

$$f(\tau_1) = \frac{2\Omega}{1 - s^2} \cdot C_R$$

where

$$C_R = m(1 - s^2) \sum_{p=0} e^{\frac{m}{\alpha}[s^{2p\alpha} - 1]} \cdot s^{2p\alpha}.$$

Although the result is exact only for the assumed condition of constant power-law departure factor, in practice it is necessary only that $(d\ln m/d\ln\tau)$ = constant give an adequate description of the variation in the flow-behavior index over the range of shear stress required for convergence. The results of machine computations, covering a range of radius ratios, flow-behavior indexes, and power-law departure factors, are used to test the adequacy of the power-law approximation, and the approximate series solution of *Krieger* and *Elrod* in describing the postulated behavior.

References

1) *Brodkey, R. S.*, Ind. Eng. Chem. **54**, No. 9, 44 (1962).
2) *Brodkey, R. S.*, The Phenomena of Fluid Motions, p. 422 (Reading, Massachusetts, 1967).
3) *Middleman, S.*, The Flow of High Polymers, p. 21 (New York 1968).
4) *Coleman, B. D.* and *W. Noll*, Arch. Ratl. Mech. Anal. **3**, 289 (1959).
5) *Krieger, I. M.* and *H. Elrod*, J. Appl. Phys. **24**, 134 (1953).
6) *Krieger, I. M.* and *S. H. Maron*, J. Appl. Phys. **25**, 72 (1954).
7) *Savins, J. G., G. C. Wallick*, and *W. R. Foster*, Soc. Pet. Engrs. J. **2**, 211 (1962); **3**, 14, 177 (1963).
8) *Van Wazer, J. R.*, et al., Viscosity and Flow-Measurement, p. 68 (New York 1963).
9) *Mooney, M.*, J. Rheol. **2**, 210 (1931).
10) *Metzner, A. B.*, in: Handbook of Fluid Dynamics, Chapt. 7, p. 13 (New York 1961).

Authors' address:

Dept. of Chemical Engineering
Queen's University
Kingston (Canada)

Rheol. Acta 12, 588–592 (1973)

From the VEB-MLW Prüfgeräte-Medingen, Sitz Freital (DDR)

Rotating viscosimeters for process control

By S. Putzker

With 3 figures

(Received October 27, 1972)

For a great number of processes in chemical engineering viscosity is an essential factor for the valuation of the quality of primary, intermediate or final products. This is why the measurement of viscosity is frequently an integrating component of the measures taken for the purpose of monitoring, controlling and regulating production processes. In this case there is to be solved a great number of various measurement problems.

1. Demands on process viscosimeters

Compared to the viscosimeters destined for application in laboratories viscosimeters for process control are additionally required to effect the viscosity measurement continuously, i.e. without any manual taking of samples. In this connection in process viscosimetry there must be solved 2 fundamentally different problems dependent on the way in which the viscosimeters are applied:

– viscosity measurement in reaction and storage tanks and

– measuring the viscosity of liquids flowing through pipe lines.

Moreover process viscosimeters must be characterized by the following properties with respect to their value in use:

– The viscosity value of the substance to be measured must be measured as an instantaneous value. Discontinuous periodical viscosity measurements present the disadvantage of considerable dead-time periods.

– For the viscosity control in production processes with a non-*Newton*ian flow behaviour of the substance to be measured well-defined conditions of measurement with respect to the rheological parameters velocity gradient and shearing stress must be aspired to.

– Because of the diversified problems in the field of application technique there are required viscosity measuring ranges covering several decimal powers of viscosity. In particular cases of application, however, it is frequently necessary to determine only a limited variation of viscosity, but with a high sensitivity.

– There are differentiated requirements with respect to the temperature of the substance to be measured and to the working pressure. Some of these requirements are extreme ones.

– As a rule a constant reference temperature for the viscosity value measured is required in case of the continuous viscosity measurement. However, it is only in rare cases that the conditions involved by the given process will ensure a constant temperature of measurement. Consequently, there must exist devices destined for regulating the temperature of the substance to be measured or for compensating the influence of the temperature variation on the viscosity value measured.

– The components of the process viscosimeter coming into contact with the substance to be measured must be made of high-alloyed chromium-nickel steels or, respectively, of high-temperature resistant materials coming up to the corrosion, temperature, and pressure stresses exerted by the substance to be measured.

– The measurement signal "viscosity" must be transformed by the process viscosimeter into an electrical or pneumatical signal permitting the connection of secondary instruments intended for displaying, recording or data gathering purposes. The same necessity is in the combination of control units destined for regulating the mixing and the temperature of the substance to be measured.

– Viscosity measurement under industrial conditions requires a high technical stressability of the measuring instrument which is reflected in special demands concerning shock protection, water, explosion and climatic protection as well as resistance to vibration.

2. Measuring method

Among the known viscosity measuring instruments there are especially the rotational-type viscosimeters with a coaxial cylinder system type *Couette* or *Searle* which owing to their aptitude for modification lend themselves to the assembly of process viscosimeters.

As you know, the flow of the substance exposed to a shearing stress in the annular clearance between 2 cylinders arranged coaxially with respect to each other obeys an equation in which the torque being proportional to the flow resistance is a function of the angular velocity of the rotating cylinder, of the geometrical dimensions of the cylinder system, and of the viscosity of the substance to be measured. This measuring method fulfills the requirement necessary for process viscosimetry, i. e. that the instantaneous value of viscosity is to be determined. With a cylinder system of suitable design there can be realized specified measurement conditions regarding the measurement parameters "shearing gradient" and "shearing stress". In that way a correct control of the apparent viscosity of non-*Newton*ian liquids is also possible.

The fact that for the continuous exchange of the substance to be measured in the annular clearance an axial flow is superimposed on the *Couette* flow will result in an additional error of the measurement result in case of non-*Newton*ian liquids being measured provided that there do not exist any specified correlations of the above-mentioned flows.

Though the systems with rotating external cylinder present some advantages as to flow tech-

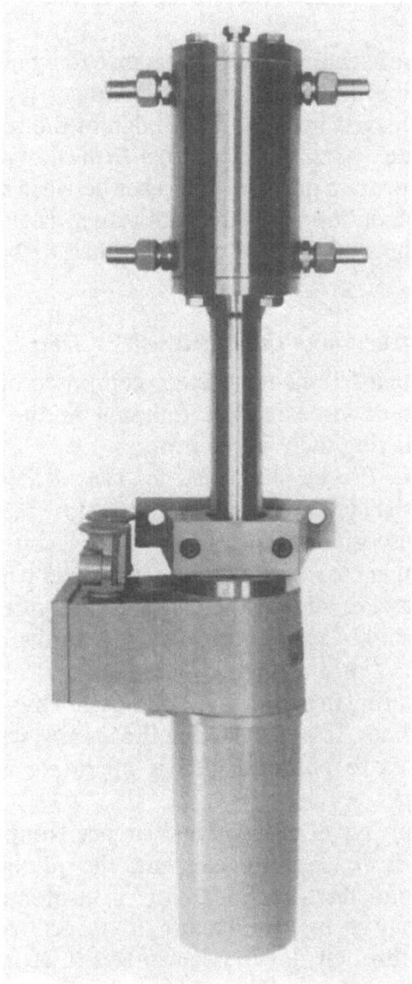

Fig. 1. Flow-through type viscosimeter

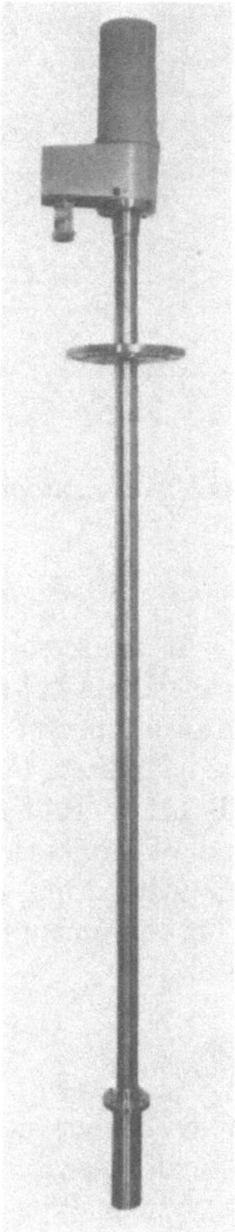

Fig. 2. Plunger-type viscosimeter

nique, the process viscosimeters with rotating internal cylinder are frequently given preference owing to their less complicated construction. In that case the torque measurement is accomplished in the rotating system.

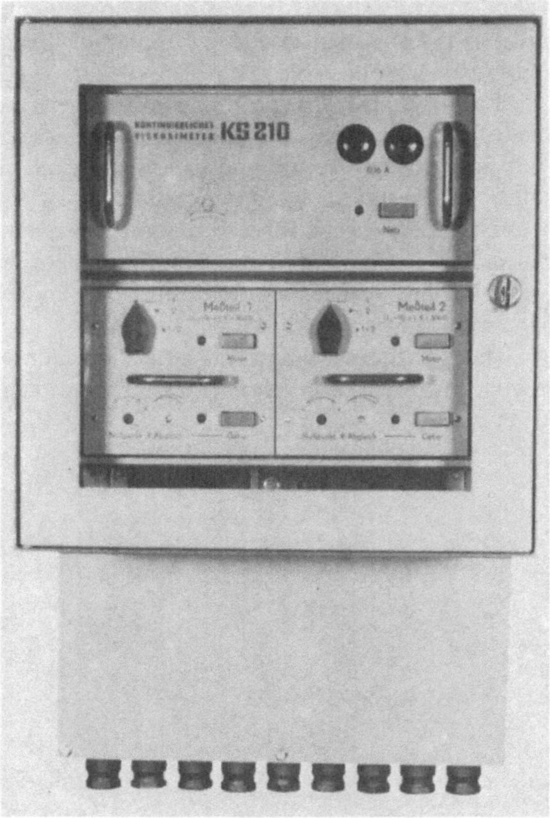

Fig. 3. Measuring unit with electronic sub-assemblies

3. Instrument system "process viscosimeters"

The advantages of the remarkable possibility of modification of the rotational-type viscosimeter are made use of in an instrument system of process viscosimeters manufactured by the nationally owned firm VEB MLW Prüfgeräte-Werk Medingen, Sitz Freital. With only a few basic modular units this system will permit the assembly of viscosity measuring equipment for differentiated cases of application.

Principle of measurement

The process viscosimeters of the instrument system are rotational-type viscosimeters with a coaxial cylinder system type *Searle-Couette*. During measurement the substance to be measured is in a double annular clearance formed by a fixed internal cylinder, a rotor having the shape of a hollow cylinder, and a fixed external cylinder. A constant angular velocity is given to the hollow cylinder accomplishing the function of a rotor which is located in the hermetically encapsulated measurement space. This is done by a synchronous motor with intermediate transmission by way of a permanent-magnet coupling. The drive of the rotor and the measurement of the torque are effected by means of the same shaft. A spring member whose deflection is a measure of the torque acting on the shaft serves for measuring the torque. This deflection is sensed by an instrument potentiometer situated in the branch of a bridge circuit. The diagonal voltage of the bridge is directly proportional to the viscosity of the measured substance.

An alternative solution for viscosity measurements of liquids which are not under pressure provides for a single annular clearance with rotating internal cylinder and fixed external cylinder.

The exchange of the substance to be measured in the cylinder system of the plunger-type probe is achieved by a specific design of the rotor and of the external cylinder (slots). In the flow-through type probe a pressure difference between inlet and outlet of the measured substance is necessary for the exchange of the substance to be measured.

Construction of the instrument system

The instrument system is composed of 2 basic types of viscosimeters differing in the way of employing them in the process.

The *Process Viscosimeter Type KS* is an industrial viscosimeter intended for measuring the viscosity of liquids in receptacles. It is made use of in case of production processes performed by batches and for the continuous control of the viscosity of substances in storage tanks.

The *Process Viscosimeter Type KD* serves for measuring the viscosity of liquids flowing through pipe lines. It is preferred in those cases where the process to be controlled is operating continuously.

Both types of viscosimeters are composed of a viscosity sensing element, the plunger-type viscosimeter (intended for measurements of substances in receptacles) or, respectively, the flow-through type viscosimeter (destined for measurements of substances flowing through), and an electronic measuring unit.

The plunger-type viscosimeter and the flow-through type viscosimeter differ only in the type of measuring probe used. Both viscosity sensing elements are provided with a measuring head in which there are located the devices necessary for driving the rotor and for measuring the torque. The synchronous rotational speed being independent of load and voltage can be varied by changing gears in 8 steps from 1 to 250 r.p.m. In that way there can be set 8 different viscosity measuring ranges.

Already as a separate unit the measuring head is made explosion-proof according to the protective system "pressure-proof encapsulation". This is why it can also be employed for solving other measuring problems of industrial practice which are in no connection with viscosity measurements, e.g. as measuring stirrers.

The synchronous rotational speed is given to the rotor of the plunger-type probe of the plunger-type viscosimeter from the measuring head by way of a connecting shaft. For the transmission of the constant angular velocity a magnetic coupling is inserted into the pressure space. With the aid of 2 variants of the plunger-type probe differing by the dimensions of the annular clearance there can be obtained viscosity measuring ranges from 2–10^6 cP. The different conditions of installation are taken into consideration by variable lengths of the plunger-type probes up to 1500 mm. The filling level in the receptacle receiving the substance to be measured is of no influence on the measurement result. However, one must not fall below a minimum filling level depending on the dimensions of the cylinder system. The inclination of the plunger-type viscosimeter by 30 angular degrees with respect to the vertical line does not effect any additional measuring errors. Furthermore the plunger-type viscosimeter of pressure-tight construction can be employed even in extreme positions of installation, e.g. horizontally and in case of overhead installation.

The flow-through type viscosimeter differs from the plunger-type viscosimeter only by the design of the external cylinder as a pressure-proof jacket surrounded by a chamber for the temperature regulation by means of a liquid circulation thermostat. In case of flow rates between 0 and 50 l/h the flow-through type viscosimeter is installed in the plant directly in the main flow.

Greater flow rates require the insertion of the flow-through type probe in a by-pass with respect to the main line. A supply of substance to be measured which is independent of the changing technological conditions of the process can be ensured by a proportionating pump. By means of the flow-through type viscosimeter, too, viscosity measurements from 2–10^6 cP can be effected. The permissible measurement temperatures go as far as +300 °C, the nominal pressures as far as ND 25.

The measuring unit of the instrument system process viscosimeter accommodates the electronic sub-assemblies and controls destined for starting and balancing the viscosity measuring equipment. In this basic modular unit the signal "torque" is transformed into an electrically analogous output signal, optionally from 0 to 50 mV, 0–5 mA or 4–20 mA, intended for the connection of secondary instruments serving for displaying, recording, data gathering, and regulation purposes. The indicating circuits of the measuring unit are intrinsically safe.

For the flow-through type viscosimeter it is possible to fulfil the requirement of a constant measurement temperature by means of a temperature regulation of the substance to be measured in the sensing element. For this purpose it is sometimes necessary, in case of great temperature differences, to provide for additional temperature-regulating stretches on the inlet of the substance to be measured. In case of the application of the plunger-type viscosimeter, on the other hand, only the electric temperature compensation of the process viscosimeter can be made use of for eliminating temperature variations affecting the result of the viscosity measurement. The variation of the output signal resulting from the difference between the temperature of the measured substance and the reference temperature is compensated by mixing the output signal with a directional electric signal. This method requires the knowledge of the viscosity-temperature function.

The instrument system "process viscosimeters" has a wide field of application, e.g.

the viscosity control of the reaction progress in case of polymerization and condensation processes,

the measurement and regulation of the viscosity of spinning solutions in the synthetic fibres industry,

the regulation of ink viscosity on rotary gravure printing presses, the viscosity regulation in oil mixing plants,

the viscosity regulation of emulsions for films and photographic papers in order to ensure a uniform coating,

the viscosity control in the glue production as well as,

the viscosity measurement and regulation in the pharmaceutical and food industry.

Summary

Compared to the viscosity measurement under laboratory conditions there are additional demands on the measuring instrument when viscosity-dependent processes are to be monitored, controlled or regulated. Rotational-type viscosimeters with a coaxial cylinder system lend themselves for solving this problem. The various problems of viscosity measurement in production processes make high demands on the universal applicability of the viscosimeters. An instrument-technical solution on the basis of standardized basic modular units is of advantage. An instrument system "process viscosimeters" for viscosity measurements in receptacles and pipe lines is described.

Zusammenfassung

Gegenüber der Viskositätsmessung unter Laboratoriumsbedingungen ergeben sich zusätzliche Forderungen an das Meßgerät, wenn viskositätsabhängige Produktionsprozesse überwacht, gesteuert oder geregelt werden müssen. Rotationsviskosimeter mit koaxialem Zylindersystem sind zur Lösung dieser Aufgabe geeignet. Die unterschiedlichen Viskositätsmeßprobleme in Produktionsprozessen stellen hohe Anforderungen an die universelle Einsetzbarkeit der Viskosimeter. Vorteilhaft ist eine gerätetechnische Lösung auf der Basis von typisierten Grundbausteinen. Ein Gerätesystem Prozeßviskosimeter für Viskositätsmessungen in Behältern und Rohrleitungen wird beschrieben.

References

1) *Süß, R.*, ATM Lief. **324**, 9–12 (1963).
2) *Putzker, S.*, Labortechnik **4**, 10–15 (1971).

Author's address:
Dipl.-Ing. *S. Putzker*
VEB MLW Prüfgeräte-Werk Medingen, Sitz Freital
DDR-821 Freital, Lesskestr. 10

RHEOLOGICA ACTA

AN INTERNATIONAL JOURNAL OF RHEOLOGY

Vol. 13 February 1974 No. 1

From the Nippon Oil Seal Industry Co., Ltd., Tokyo (Japan)

Development and application of micro pressure-pickups to measure point-pressure of elastomer flow

By Shintaro Miyakawa, Takao Ishii

With 6 figures

(Received October 27, 1972)

Introduction

We know three kinds of processes to mold rubber products: compression molding, transfer molding, and injection molding, among them there not being essential difference in view of considering only the basic relation between pressure and flow of rubber, or shearing stress and rate of shearing strain. The factors which govern the flow are polymer pressure, mold temperature, condition of mold surfaces, geometrical shape and dimension of runner, flow characteristics and curing characteristics of compound, etc. Molding engineers usually obtain the practically optimum values of these variables primarily after their own long experience through trial and error methods. It is, therefore, necessary for molding engineers to have more basically satisfactory data with which they can surmise the effect of each variable on the result of molds on purpose of making better design of molding process.

In some limited cases, a theoretical calculation of pressure loss of flowing polymer is possible (1). Such is not the case of most of practical process. Whereas measurement of pressure is $\dot{a} \rightarrow a$ feasible alternative means to see the state of real flow.

One of the difficulties which we have to overcome before trying to measure the compound pressure in relatively small cavities is to develop a small pressure pick-up of which sensing head has a sufficiently small diameter so as not to give any serious disturbance to the stream line of the flow. This requirement of small diameter becomes much more pressing when one needs to measure the pressure distribution along the cavity surface. Another difficulty to be overcome is that reading signal from the pressure pick-up should not be subjected to the influence of big temperature change of the cavity.

We have developed simple micro-pick-ups of which sensing head has a diameter of 1–2 mm in order to satisfy the requirements mentioned above.

Mechanism of micro-pick-ups

Fig. 1 shows one of simple micro-pick-ups we developed. Water cooling jacket is fitted around a portion of the stainless steel pipe. The air inside the pipe was discharged before the pipe was filled up with pure dry mercury. One end of the pipe was then plugged completely. A small foil strain gauge is glued at the surface of the pipe between the water jacket and the seal plug so as the strain gauge is not affected by high temperature of the mold.

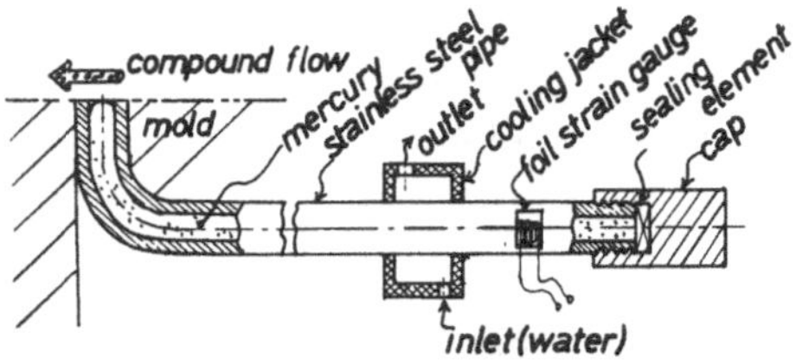

Fig. 1. Mechanism of micro pressure pick-up

A hole is bored by drilling at the cavity surface whereat pressure measurement is required. The open end of the pipe is carefully pressed into the hole so as the end section of the pipe is flushed with the cavity surface. The strong surface tension force of mercury can hold itself within the pipe against shock or vibration, if not excessively violent one, which the pressure pick-up may encounter in the handling or installation process.

At the measuring stage flowing polymer presses directly the surface of mercury and then caused pressure in mercury is transmitted all along inside of the pipe. The scraped volume of mercury by flowing polymer is usually very small amount; moreover, the scraped cavity can be filled up immediately with polymer. The scraping effect, therefore, has actually no significant obstruct for the reliability of measured value.

If necessary, the open end of the pipe may be filled up with some suitable relatively hard material.

Calibrated test data, pressure vs. strain of pipe, shows exactly linear relation. The measured maximum pressure with our micro-pick-ups is 1200 kg/cm^2 at 220 °C of polymer temperature. Measurement of much higher pressure is possible, if required.

Example of pressure distribution measurement along polymer flow

For the purpose of confirming the function of the micro pressure-pick-ups in practice, we take a way to compare measured pressure values with those of theoretical calculation in two cases of simple NBR compound flow, such as shown in fig. 2 and fig. 4, for which analytical pressure distribution is obtainable as follows.

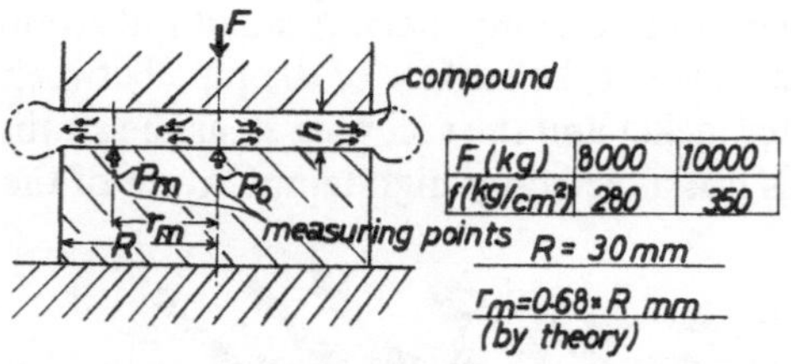

Fig. 2. Compound flow between compressing two disks

i) Pressure distribution of radial flow between compressing to parallel disks (fig. 2)

It is known of non-*Newton*ian polymer that the rate of shearing strain is proportional to the n-th power of the shearing stress within some limits of stress level (2), that is

$$\dot{\gamma} = k(\tau - \tau_y)^n \qquad [1]$$

where

$\dot{\gamma}$ the rate of shearing strain
k constant
τ shearing stress
τ_y the stress at yield point of material
n number of the structure viscosity power

As the stress at yield point of rubber compound is usually very small comparing with the working stress of shearing, τ_y is negligible in the eq. [1]. We can obtain the following expression from the eq. [1] wich is applied to the flow shown at fig. 2.

$$\frac{p(\bar{r})}{f} = \frac{3n+1}{n+1} \cdot \left(1 - \bar{r}^{\frac{n+1}{n}}\right) \qquad [2]$$

where

f $F/\pi R^2$ (see fig. 2)
$\bar{r}$ r/R (see fig. 2)
$p(\bar{r})$ pressure at the point of radius $\bar{r}$

As one can see from eq. [2], the nondimensional pressure $p(\bar{r})/f$ can be regarded a function of nondimensional radius $\bar{r}$ only, provided "n" is constant. It has been indicated as a result of the foregoing measurement for our compound that "n" is 4.6 approximately within the pressure level of our experiment. We can get the pressure value at the center of disk, $\bar{r} = 0$, from eq. [2];

$$\left(\frac{p}{f}\right)_{\bar{r}=0} = 2.6.$$

By solving $p(\bar{r})/f = 1$, we can see that a pressure which is equivalent to the mean pressure presents at a point of $\bar{r} = 0.68$.

Measured pressure values p_0 at $\bar{r} = 0$ and p_m at $\bar{r} = 0.68$ are shown in fig. 3. Fig. 3 shows that measured values agree fairly well with those of the theoretically predicted and also shows that the pressure value is independent of the distance between two disks, h, as eq. [2] implies.

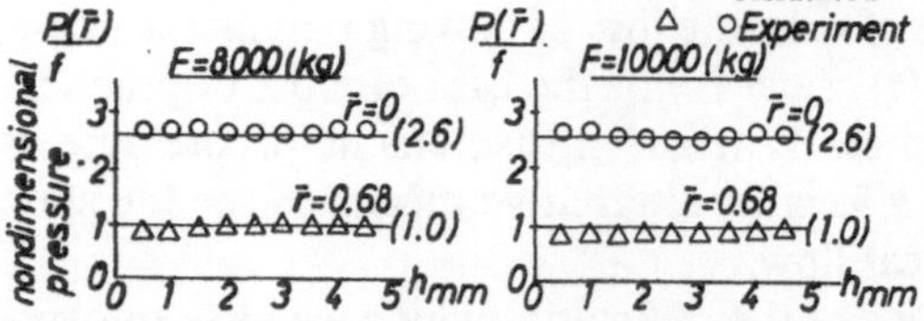

Fig. 3. Comparison of measured pressure and calculated pressure for a model in fig. 2

ii) Transitional pressure distribution caused in a cylindrical mold cavity (fig. 4)

By combining two solutions for a compressing radial flow and a cylindrical parallel flow we can obtain an expression of the pressure distribution along the radius of the base disk in fig. 4.

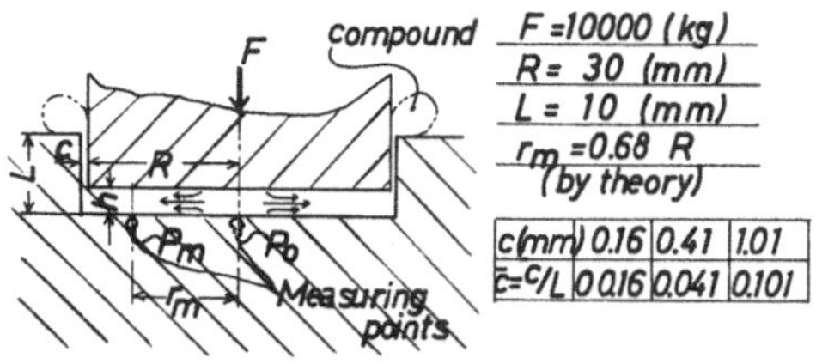

Fig. 4. Compound flow between cylindrical mold

The result is as follows:

$$\frac{p(\bar{r})}{f} = \frac{3n+1}{n+1} \times \frac{\left(1 - \bar{r}^{\frac{n+1}{n}}\right) A \bar{c}^{\frac{n+2}{n}} + \frac{n+1}{n}(1-\bar{h})\bar{h}^{\frac{n+2}{n}}}{A\bar{c}^{\frac{n+2}{n}} + \frac{3n+1}{n}(1-\bar{h})\bar{h}^{\frac{n+2}{n}}} \qquad [3]$$

where

the definitions of P, f, r, R and h are same with those of the previous case,

- L the depth of the outer mold
- c the clearance between inner and outer mold
- $\bar{h}$ h/L
- A R/L
- $\bar{c}$ c/L

Calculated pressure with eq. [3] at the center of base, $\bar{r} = 0$, and the theoretical mean pressure position, $\bar{r}_m = 0.68$, is shown in fig. 5.

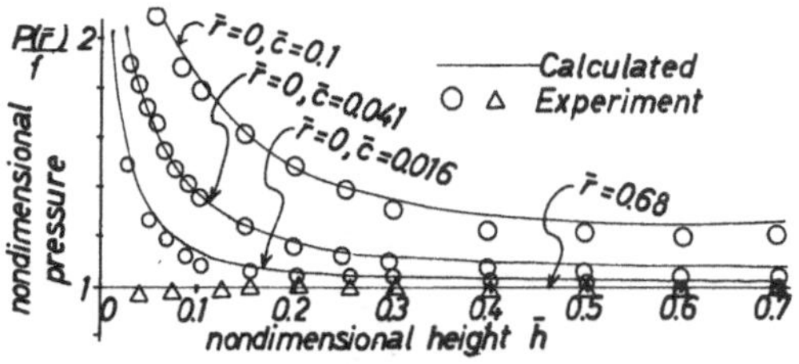

Fig. 5. Comparison of measured pressure and calculated pressure for a model in fig. 4

The corresponding pressure value obtained by measurement is also shown in fig. 5. These results show that the measured pressure distributions agree satisfactorily well with the theoretical ones.

An example of micro pressure pick-up application

We have obtained many fruitful results by using our micro pick-ups for the technical works of rubber molding. Here we outline one of problems concerning the reasonable determination of a injection pressure, which one may meet frequently especially when one tries to mold a thin and big size of rubber product like a big diaphragm.

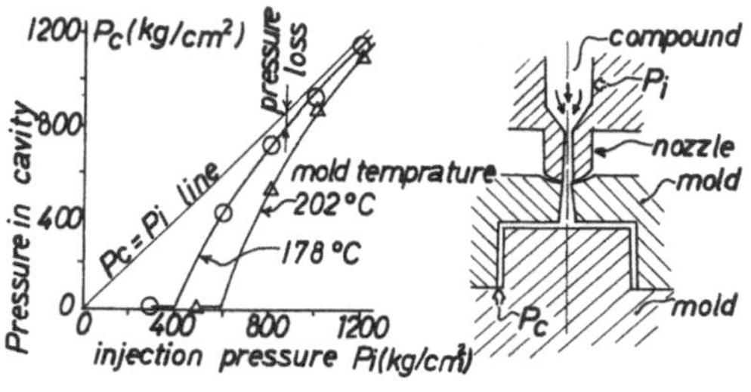

Fig. 6. Injection pressure and pressure in cavity

Fig. 6 shows the measured relation between the injection pressure and the rubber pressure in cavity at two different levels of mold temperature. The cavity pressure *Pc* is measured with a micro pick-up which is inserted at the dead end of the cavity, and the injection pressure *Pi* is with another one in the barrel. Rubber is injected into the cavity through nozzle first then through runner and gate. During this process the "flowability" of rubber decreases so much because of *scorching* by heat built in and heat transfered from the mold furface. Whereas the rubber pressure in cavity undergoes a considerable decrease even after the time when the cavity is filled up with rubber; the smaller the injection pressure, the larger the pressure loss, because the scorch becomes more considerable as the injection time increases due to the lower injection pressure.

One cán see this circumstance from fig. 6 and then can make use of these results to make a reasonable selection of the injection pressure and the mold temperature in order to avoid molding failure.

References

1) *Urabe, N.*, Properties and Processing of Rubbers (in Japanese). In: A Series of Monographs on the Science and Engineering of High Polymers; Vol. VII (Tokyo 1966).

2) *Shiromatsu, T.*, Processing of Plastics (in Japanese). In: A Series of Monographs on the Science and Engineering of High Polymers; Vol. VI (Tokyo 1966).

Authors' address:
Shintaro Miyakawa, Takao Ishii
Nippon Oil Seal Industry Co. Ltd.
Tokyo (Japan)

Rheol. Acta 13, 4–11 (1974)

From the Instron Corporation, Canton, Mass. (U.S.A.)

A new versatile rheological instrument: Design, testing and data analysis

By C. J. Drislane, J. P. DeNicola, W. M. Wareham, and R. I. Tanner*)

With 3 figures and 1 table

(Received October 27, 1972)

1. Introduction

Recent work in rheology has indicated the need to explore methods of testing viscoelastic fluids beyond the simple measurement of steady shearing viscosity; we do not imply that there is no longer any need to measure steady shearing viscosity, but there are already many instruments that are capable of making such measurements (1, 2). We present in this paper a short account of the design and proof testing of a new instrument capable of testing fluids and solids in many different modes of deformation derived from *rotary* motion complementing existing instruments for viscosity measurement (2) and tensile testing (3) which are based on linear motion. If a rotary instrument is used, then the basic type of flow in the sample will naturally be a shearing motion (steady or unsteady) or some deviation from a solid-body rotation. In table 1 we show a classification of shearing motions based on the *Pipkin* (4) flow diagnosis diagram. Note that the sample response depends on the sample mean relaxation time (T) and on the shearing amplitude (A). These parameters lead to a dimensionless classification in terms of ωT and A where ω is a characteristic rate or frequency describing the rate of change of stress at a particle. In a time-sinusoidal motion ω is the frequency of the input, for example, and in steady shearing ω is zero. The shear amplitude A is easy to define in a sinusoidal motion; in steady shear it may be thought of as the amount of shearing taking place in one relaxation time, so that here we may set $A = \gamma T$, where γ is the steady shear rate.

In order to cover all categories in table 1, an instrument must be capable of (i) measuring the three viscometric functions η, N_1 and N_2 in a steady shearing flow (4). This takes care of the top row in table 1. (ii) Measuring $G'(\omega)$ and $G''(\omega)$, the components of the complex modulus, as a function of frequency. This corresponds to the small strain column in table 1. (iii) Applying suitable rapid deformations which are required to take care of the row of boxes where ωT is large in table 1. Sudden starting and stopping of shearing, measuring not only the shear component but also the normal force components of stress, are methods of accomplishing this.

Table 1. Flow classification table

ωT	$A(\gamma T)$ Small	Medium	Large
Small	*Navier-Stokes* fluid (viscosity)	Viscometric flows (normal stress effects)	Viscometric flows (normal stress effects)
Medium	Linear viscoelastic body (G', G'')	General non-linear viscoelastic behavior	General non-linear viscoelastic behavior
Large	Linear elastic body (G_∞)	Rubber-like elasticity (recoil)	Rubber-like elasticity (recoil)

Certain combinations of tests – for example, superposition of small-strain oscillations and steady shearing – are also of interest. In considering the problems of measurement, the categories set out above present an increasing order of difficulty; in the case of G' and G'' the principal difficulties lie in phase angle measurement. This may be avoided by using as an alternative the eccentric disc (5) or eccentric cylinder (6) arrangements. By this means only steady forces need be

*) Permanent Address: Division of Engineering, Brown University, Providence, Rhode Island 02912 (U.S.A.).

measured in finding G' and G''. Some of our experiences with these relatively new methods are discussed below.

The transient measurements are much more difficult to make accurately, and yet they are of great interest. It is true, especially for the normal force components, that the natural compliance of any instrument tends to frustrate accurate measurements of the torque and force components; and it seems questionable whether any truly reliable results are currently available. The anticipation of this problem has set some limits on our design, which are discussed below.

2. The instrument

In order to allow valid testing in these regimes, a rheological instrument must have the following characteristics:

(i) Stable, controlled speeds over a wide range.

(ii) Accurate, sensitive, high response measurement of all sample reactions.

(iii) Precise, rigid test geometry.

(iv) A variety of well controlled test conditions (e.g., temperature, atmosphere).

These requirements are perhaps obvious but, historically, have not been fully achieved. Current technology, properly applied, can allow these to be realized in an instrument for rheological testing. The following is a description of such as instrument.

This Instron Rotary Rheometer incorporates these characteristics in a system that is very simple (as shown in fig. 1). It consists of two rotary test platens: the upper is directly driven from a motor-tachometer through a radial/thrust air bearing, and the lower platen is supported on a rigid transducer which senses the material response. These elements are supported in a frame providing complete rigidity and alignment. A separate control console provides the drive controls, transducer conditioning, and data readout.

a) Drive control, stability and range

The drive consists of a D. C. torque motor with an integral tachometer which provides the velocity feedback signal for closed-loop con-

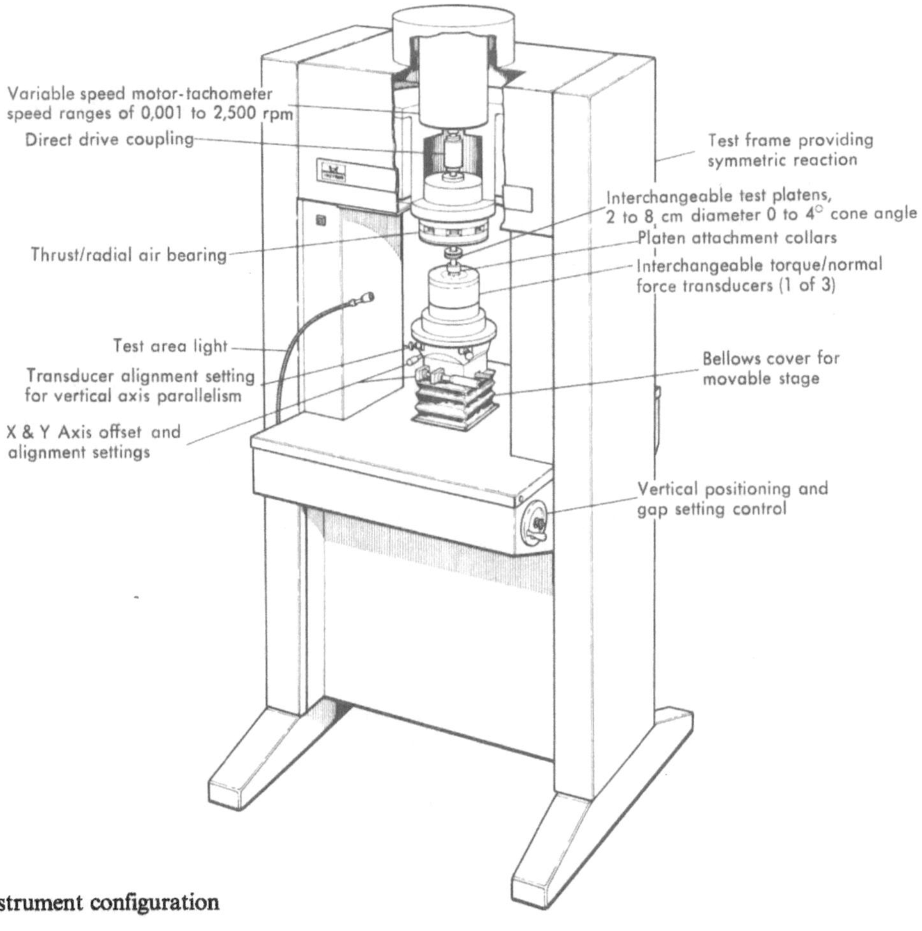

Fig. 1. Instrument configuration

trol. This system provides the widest speed range currently available in a direct drive system with the needed accuracy. Speed ranges of 10,000 and 25,000 to 1 are provided at accuracies of 1% of set speed. Speeds beyond these ranges are achievable but only at reduced accuracy.

Use of such a drive allows the rheometer to cover the above speed range directly without gearing or other intermediate reductions. The motor-tachometer is directly coupled to the main drive air bearing to which the upper test platen is attached. Not only are transmission errors associated with gearing avoided, but also speed control can be much more versatile. It is possible without stopping to change speed anywhere within the speed range. Also, direction of rotation can be quickly reversed. The system can also initiate or terminate a constant shear rate at acceleration rates of 1000 radians per sec^2 for transient testing. This allows the drive system to bring the air bearing rotor and upper test platen to a constant speed of 100 rpm in 10 milliseconds. Additional features such as adjustable acceleration rate and oscillatory motion can be directly attached. An electronic oscillator can be included for oscillatory testing over a frequency range of 0.003 to 100 Hz. The system also accommodates signals from external function generators for operation with any generated velocity waveform.

Although a drive system of this type can provide smooth, accurately controlled speeds over a 10,000 or 25,000 to 1 range, a further extension is required to satisfy the needs of all fields of rheological testing. With some materials in the cone-plate mode, speeds of 0.001 rpm will be desired while other materials in eccentric disc testing require speeds of 2500 rpm. Since no one direct drive motor-tachometer can cover this 2.5×10^6 speed range with adequate accuracy, interchangeable drive modules are provided. Three motor tachometers having ranges of 10,000, 15,000 and 25,000 to 1 cover this overall range with accuracies of 1–2% of set speed at any load anywhere within the speed range. These motor tachometer modules are interchanged in the frame using the same air bearing and are controlled with the same solid-state controller.

b) *Measurement accuracy, sensitivity, response time*

Both cone and plate and eccentric disc testing involve two primary reactions. These reactions can best be measured by transducers attached to the lower platen. A single dual channel transducer and conditioning system is used for simultaneous measurement of the two reactions associated with each type of testing. In the cone-plate mode a torque-normal force transducer operates with the conditioning system to measure these reactions. In eccentric testing an X-Y force transducer is interchanged with a torque-normal force transducer, and it operates in conjunction with the same conditioning equipment.

In the cone-plate mode the lower test platen is fixed in this transducer which, in turn, is mounted to the frame. The transducer for eccentric disc testing incorporates a central air bearing to which the lower test platen can be attached while allowing free rotation in synchronism with the upper drive test platen.

The transducers use strain gage sensing elements with extremely high stiffness to minimize deflection under sample load reactions. Each channel uses a strain gage bridge which is zero and span temperature compensated. The conditioning on each channel includes an amplifier with zero, balance and calibrate controls, and range selection providing a choice of full scale ranges from 100% to 2% of capacity of that transducer channel. Force and torque measurements may be made over this range with 1% accuracy through the linearity and drift-free operation of the transducer and conditioning electronics. Since accurate measurements over wider ranges are required for the many materials anticipated, a selection of three torque-normal force transducers is provided. These easily interchanged transducers extend the range of measurement to approximately 10,000 to 1 with the necessary linearity and stability. Although similar modularity is possible with the orthogonal force transducer for eccentric disc testing, it has not been provided here due to the wide variation of test geometry (offset, gap and platen radius) which can be easily achieved to extend the range of testing.

Readout requirements will vary with the type of testing being done. A multiple pen strip chart recorder can be provided for constant shear and low frequency oscillatory testing in the cone-plate mode. This is also an appropriate readout device for the constant reactions produced in eccentric geometry testing. An oscillograph recorder is available for transient and high frequency oscillatory testing. Additional readout devices which can also be used include oscilloscopes, digital voltmeters, phase measuring instruments, and data processing equipment.

c) Geometric precision and rigidity

Test geometry, one of the important instrument characteristics, becomes critical when one realizes the reliance that is placed on it with no feedback during test verifying that it is, in fact, correct. Proper design can assure that instrument alignment will not be easily lost through inadvertant shifting of parts or excessive deflection of the system from sample reaction forces.

The overall frame configuration differs from traditional rotary instruments in that the upper platen support is a closed bridge rather than an overhung cantilever arrangement. This provides two major benefits: vertical deflection resulting from developed normal force is significantly reduced; and those resulting deflections do not cause skewing of the rotational axes of the platens as occurs in cantilevered structures. Further stability is provided through thermal insulation of the frame to minimize thermal distortion.

The use of strain gage elements in the transducers results in measurement of sample reactions with greatly reduced deflections. These features in the frame and transducers along with very high stiffness in the air bearing have resulted in system vertical stiffness of 1.6×10^4 kg/cm between opposing test platens. The vertical stiffness of the frame alone is 1.8×10^5 kg/cm; the air bearing, 3.6×10^4 kg/cm and the high capacity transducer, 3.6×10^4 kg/cm^2.

Accuracy of the test geometry is afforded by precision machining of the system elements and through simplicity of design so that a minimum of separate mechanical parts are involved. In addition, means are provided to verify and, if necessary, adjust to achieve complete alignment. Dial gages and micrometer tables are provided to position the lower platen directly below the upper platen. This position can be checked by attaching the dial gage to the upper platen shaft and rotating it while indicating the outer surface or shaft of the lower test platen. Upper and lower axis parallelism can be set with adjustments on the lower support and verified by rotating the dial gage attached to the upper test platen while indicating the upper surface of the lower platen near its periphery. Vertical alignment and verification is accomplished by bringing the upper and lower platens to zero gap as indicated by normal force measurement (without rotation) and appropriate adjustment downward to the desired gap setting. All conical platens are slightly truncated so that at the proper gap there is no actual cone-plate contact, thus avoiding frictional torques and platen wear.

d) Controlled test conditions

In addition to test conditions such as velocity and geometry which are controlled in the manner previously discussed, other test conditions which may be prescribed and controlled are temperature and atmosphere. A temperature chamber is available providing the ability to test at temperatures from ambient to 400 °C. The system has a control accuracy of 1 °C with the ability in the cone-plate mode to monitor actual platen temperatures from a calibrated thermocouple imbedded in the non-rotating lower platen. This allows temperature to be monitored directly without the inherent problems associated with slip rings. Chamber heating is accomplished by radiation with heat being radiated to the platens and sample from heated chamber walls. This approach was chosen so that flowing gases, necessary for convective heating, could be avoided. A secondary benefit is realized at higher temperatures. In convective chambers, radiation loss to chamber walls which may be lower in temperature becomes significant. When the wall is used as the control medium in a radiation system, the error due to this thermal loss is eliminated. Platens have been designed to limit heat flow out through the shaft so that temperature gradients across the platens are minimized.

The chamber has been configured to provide different atmospheres at ambient or elevated temperatures. The chamber has inlets to introduce gases in circulatory flow in the test region. This can be done as a non-pressurized system with continuous purging and the gas escaping around the shaft openings.

3. Proof testing

a) Silicone fluids in steady shear

Using the high capacity cone-plate transducer, samples of silicone fluids of two different viscosities were tested. Fig. 2 shows the results for the viscosity and first normal stress functions taken with a 4-cm diameter, 2-degree cone at 25.6 °C. As expected, the viscosity drops slowly at higher shear rates; the first normal stress difference rises as the square of the shear rate at first and then starts to deviate from this line. This behavior is in accordance with the *Rivlin-Ericksen* (4) expansion scheme; and one could, in fact, fit the data shown using a fourth-order theory.

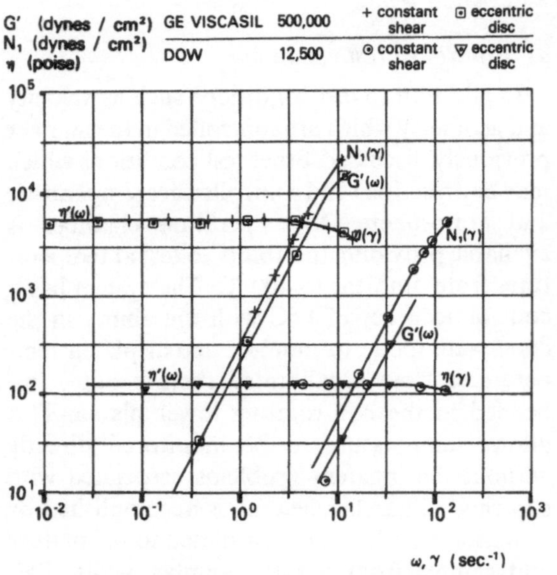

Fig. 2. Results for silicone fluids at 25.6 °C. Showing (i) viscosity and first normal stress difference versus shear rate for two fluids in constant shear with cone and plate of 4 cm diameter and 2° cone angle and (ii) viscosity and storage modulus versus shear rate in eccentric disc testing with 6 cm diameter platens with a .127 cm gap and .0508 cm offset

Below a normal stress of about 300 dyn/cm² we see errors (~10%) appearing in the square-law behavior, and these represent probable limits of measurement with this transducer. This corresponds to a load of less than 2 gm (or a probable error in load of about $\frac{1}{4}$ gm). When it is recalled that the maximum permitted load with this transducer is 10 kg, we see that an accurate working load range of nearly 3 decades is possible; by using different size cones this may be extended to nearly 4 decades of normal stress. From fig. 2 we also see that the corresponding torque readings are of comparable accuracy and that one can approach 4 decades of accurate shear stress measurement if required.

The second silicone fluid shown has a viscosity of about 122 poises (at 25.6 °C); once again, the viscosity test is successful in the expected limits and we also see large normal force errors for $N_1 \lesssim 300$ dyn/cm². In both cases failure to reach higher shear rates is due to the sample failing at the edges in the manner described by *Hutton* (7).

Switching from a 2-degree to a 4-degree cone angle was also tried, but no significant difference in results was noted. We also obtained the normal stress results from a 5.0% polyisobutylene (Vistanex L-140) in cetane sample; due to the less steeply rising response with shear rate, it is easier to cover a given shear rate range than with the silicone fluids.

b) Oscillatory testing

Space does not allow reporting of our data on oscillatory testing, but tests may be made out to 100 Hz (628 rad/sec) with an accuracy determined largely by one's ability to measure phase angle. In some tests on a polyisobutylene solution of about 1000 poises zero-shear viscosity, a commerical phasemeter was used to measure the phase angle between the velocity signal from the tachometer and the torque signal. It would be much more accurate to use a correlation technique, but even so, the accurate measurement of G' and G'' over the entire frequency range is not a trivial operation.

To check on the basic rheometer response, a *Newton*ian calibration fluid was loaded into the instrument between parallel 6-cm diameter plates spaced about .02 mm apart, and tests for phase shift were made up to 100 Hz. Unfortunately, the resultant torque level at these frequencies was too small; and the associated signal level was too low for angle measurement by the phasemeter. However, no phase shift was detected using an oscilloscope. This means that the phase shift is no more than a few degrees at 100 Hz. In concluding this section we note that many of the problems involved in finding G' and G'' in oscillatory test modes (e.g., difficulty of phase measurements at low frequencies) may be partly circumvented by using eccentric disc testing (5) as discussed below.

c) Some aspects of eccentric disc testing

Although there are good reasons to believe that the eccentric disc test measures G' and G'' in accordance with the account (5), there are some aspects of this test which need further consideration. For example, one strictly has to maintain a very small offset of the discs in order to be sure that G' and G'' are being measured correctly. One method is simply to take readings at various offsets and check that there is a linear relation between force and offset.

Another observation which has been made is that slip, or relative rotation, can occur between the discs. Sometimes this has been attributed to friction at the spindle of the lower disc (6), but we can show that this is an intrinsic effect not connected with spindle friction. To do this, consider an ideal machine with upper and lower discs driven at the same speed so that $\Omega_U = \Omega_L = \Omega$; here the subscripts U and L refer to the upper and lower discs, respectively. Each is offset $\delta/2$ from the origin of the coordinate axis. The force on each plate (radius R) transverse to the line of offset is $\pi R^2 G''(\Omega)\,\delta/h$, and thus the moments on the upper and lower platens, M_U and M_L respectively are

$$M_U = M_L = \pi R^2 G'' \delta^2/h \qquad [1]$$

and the total rate of energy input is

$$E = \pi R^2 G'' \delta^2 \Omega/h. \qquad [2]$$

We can also compute the rate of energy dissipation and obtain the same result by regarding this flow as a perturbation of a steady rotation. Now suppose that there is no torque applied through the lower shaft; in short, there is an ideal frictionless bearing there. To maintain the flow, we still need to apply the same net torque to the lower disc and now this must be done via the fluid with a slip rate between the discs. Regarding the slip motion as another perturbation of solid-body rotation in the form of a torsional flow, we have

$$M_L = \int_0^R 2\pi r^3 \frac{\eta}{h}(\Omega_U - \Omega_L)\, dr + F\delta = 0 \qquad [3]$$

where η is the relevant viscosity which is supposed constant in the small perturbation. Then we can equate the expression for M_L to zero, finding

$$\frac{\pi\eta}{2h}(\Omega_U - \Omega_L) R^4 = \pi R^2 G'' \frac{\delta^2}{h} \qquad [4]$$

where G'' ist the value relevant to the frequency Ω_U. Thus we find the slip rate $(\Omega_U - \Omega_L)$ to be

$$\Omega_U - \Omega_L = 2\frac{G''}{\eta}\left(\frac{\delta}{R}\right)^2. \qquad [5]$$

In case the fluid is *Newtonian*, $G'' = \Omega\eta$ and we have the relative slip as

$$\Omega_U - \Omega_L/\Omega_U = 2\left(\frac{\delta}{R}\right)^2. \qquad [6]$$

This formula is in good agreement with measurements on *Newtonian* fluids (*P. Payvar*, private communication). From this analysis we see that the slip is intrinsic and is not necessarily caused by design faults; to be sure, friction in the lower platen will aggravate slip and, if excessive, will make the measurements useless. In the present design an air bearing is used to obviate friction; also an acceleration control is provided to allow the lower (free) disc to accelerate slowly. In most tests it is easy to make slip negligible. Our experience with this mode of testing shows that it is much more convenient than oscillatory testing due to the ability to amplify the (steady) X and Y force components by different amounts as required.

Results of eccentric disc tests on silicone fluids are shown in fig. 2 along with the constant shear data. Good agreement is found between the two test modes in viscosity measurement of the two fluids. Results show that the storage modulus rises as the square of the shear rate consistent with expectations. The eccentric disc data was done at 21.6 °C and was adjusted with the manufacturers viscosity-temperature coefficient and the time-temperature shift factor (4). This shows valid G' and G'' measurements can be obtained over 2–3 decades of shear rate in one setup. By varying gap and offset and interchanging platens of other diameters, valid results should be obtainable over 4–5 decades.

d) Aspects of transient testing

Recently a great deal of interest has been generated in the measurement of the transient normal stresses and shear stresses when there is a sudden change in shear rate. It is much more difficult to design an adequate machine for such measurements, as we shall show. We shall ignore the mass of the machine in our response calculation this is a valid approximation provided the lowest natural frequency is very much higher than any rate associated with the polymer which is of interest.

The normal stresses in the cone-plate configuration tend to separate the cone and the plate and this leads to some negligible errors in the steady state. The complication in the transient case is that, as the plates start rotating, the normal stresses build up and start to separate the plates; this requires that fluid be drawn into the gap to fill the expanding volume. This flow can cause a severe reduction in the measured thrust. Although it is clear that a *Newton*ian fluid will behave worse in this respect, truly *Newton*ian fluids do not have normal stress differences and no plate separation occurs. A more realistic analysis uses a viscoelastic fluid of the *Maxwell* type giving an idea of the magnitude of the problem. An analysis by one of the present authors suggests that the parameter θ (time constant) must be much smaller than the fluid response time or the transient start-up time. The parameter θ is for a cone-plate device of radius R, and cone angle β is:

$$\theta = \frac{6\pi\eta R}{K\beta^3} \qquad [7]$$

where η is the fluid viscosity and K the vertical machine stiffness. Trial calculations show that with more viscous fluids one must use a very stiff machine and wide cone angles (large β) in order to measure correctly the transient normal stress behavior. (The situation is not as critical with transient shear stress testing because one has the option there of using a non-lifting *Couette* geometry.) Therefore, this instrument has been designed with a frame giving great overall stiffness. Fig. 3a shows the characteristics of the

a)
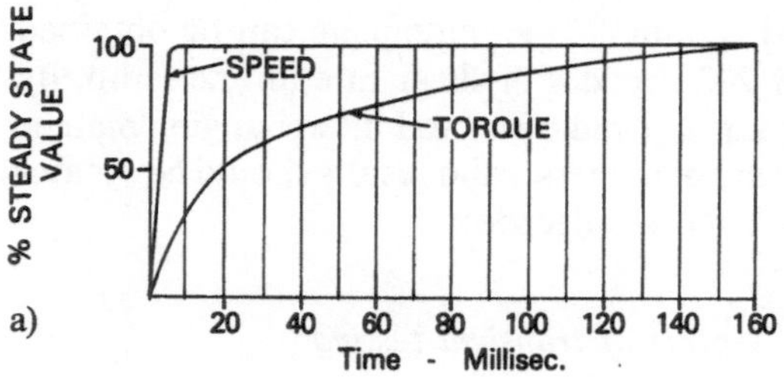

b)
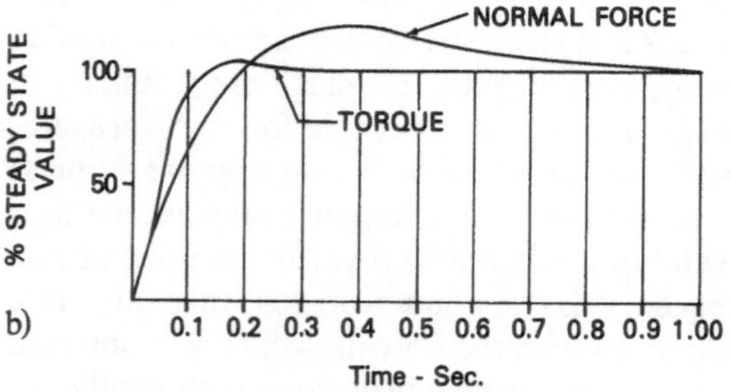

Fig. 3. Transient testing. With polyisobutylene fluid of approximately 1000 poise zero-shear viscosity. a) Showing start-up time at 6.7 sec^{-1}; b) Showing shear and normal force transients at 15 sec^{-1}

start-up process taken from a storage-screen oscilloscope. The fluid is a 1000-poise polyisobutylene solution (at 23 °C) and the steady shear rate is 6.7 sec^{-1}; a 6-degree cone was used. The start-up time is about 10 ms and here the fluid response time is at least $\frac{1}{6}$ sec. At a higher shear rate an overshoot appears on both torque and normal force traces (fig. 3b); here the final shear rate was 15 sec^{-1}.

4. Conclusion

We have discussed the new rheological instrument and have begun to show its ability to fulfill the basic requirements shown in table 1. It is adequate except that we have not discussed measurements of the second normal stress differences (N_2). This is a difficult operation, but it has been shown by *Ginn* and *Metzner* (8) that very careful total thrust measurements are capable of giving a reasonably accurate definition of N_2. If this is accepted, then the machine is capable of covering the boxes in table 1 for a very wide range of materials.

It has been found that the use of eccentric disc and sinusoidal testing are complementary in that the eccentric disc test is difficult to control at very high frequencies due to sample loss from centrifugal action, while the oscillatory testing is difficult at low frequencies due to signal noise and associated problems of small phase-angle measurements. A combination of the two can be used to cover a wide frequency range. The slip that must occur in eccentric disc testing has been shown to be negligible when suitable air bearings are provided for the lower platen. Finally, the stiffness of the machine has been made sufficient to make dynamic normal force testing feasible. Further proof testing is underway to fully define the capabilities of the instrument.

Summary

Recent work in rheology has indicated a need to explore, on a routine basis, methods of testing viscoelastic fluids beyond the measurement of steady shearing viscosity. In particular, measurements in transient flows (starting and stopping of shear) and the investigation of the flow between eccentric discs and cylinders as an alternative to time-oscillatory shear testing are current topics of great interest. The present paper discusses the design philosophy and proof testing of a new rotational rheometer capable of making (i) accurate measurements of steady shearing viscosity and first normal stress difference; (ii) eccentric disc mode testing for G' and G''; (iii) transient measurements of shear and first normal stress difference; (iv) sinusoidal time-shearing or other time-varying input measurements; a wide variety of these may be programmed. Tests illustrating the machine capabilities are shown for several fluids.

The machine is designed to work accurately over a wide range of materials from polymer melts to dilute solutions and over adequate shear rate and temperature ranges. The desire to measure transients inevitably leads one to consider a stiff machine. In the measurement of transient normal thrusts, difficulties arise in a cone-plate device because of the "squeeze-film" effect in a real machine with finite stiffness. We conclude that the use of rotating eccentric discs is an attractive alternative to sinusoidal testing and, with proper safeguards, gives accurate results for G' and G''.

References

1) *Van Wazer, J. R., J. W. Lyons, K. Y. Kim,* and *R. E. Colwell,* Viscosity & Flow Measurement (New York 1963).

2) *Mendelson, R. A.,* Melt Viscosity, Encyclopedia of Polymer Science & Technology, p. 587 (New York 1970).

3) *Morrow, JoDean,* Cyclic Plastic Strain Energy and Fatigue of Model, Internal Friction, Damping and Cyclic Plasticity (Philadelphia 1965).

4) *Pipkin, A. C.,* Lectures on Viscoelastic Theory (Berlin-Heidelberg-New York 1972).

5) *Maxwell, B.,* Polymer Eng. Sci., **7,** 145 (1967).

6) *Broadbent, J. M.* and *K. Walters,* J. Phys. D: Appl. Phys., **4,** 1863 (1971).

7) *Hutton, T. F.,* Rheol. Acta **8,** 54 (1969).

8) *Ginn, R. F.* and *A. B. Metzner,* Transactions Soc. Rheol., **13,** 429 (1969).

Authors' address:

Dr. *R. I. Tanner* et al.
Division of Engineering
Brown University
Providence, Rhode Island 02912 (U.S.A.)

Rheol. Acta 13, 12–21 (1974)

From the Cementation Research Ltd., Rickmansworth, Herts (England)

A concrete rheometer and its application to a rheological study of concrete mixes

By Osondu J. Uzomaka

With 15 figures and 1 table

(Received October 27, 1972)

1. Introduction

Various civil engineering constructions below ground level such as diaphragm walling stabilize the excavation with bentonite mud. The structural wall is then formed by tremmieing fluid concrete mixes into the bottom of the excavation thereby displacing the resident mud. Our Group of Companies makes extensive use of this type of system and Cementation Research Ltd. has carried out research into various aspects of this subject.

In one aspect of the programme the author investigated, through theoretical analysis and model tests, the efficiency of bentonite displacement by tremmied concrete. The solution of the governing equation derived in this theoretical analysis demanded the definition of some inherent rheological properties of concrete in fundamental units. However, no commercially available instrument was known to satisfy the above requirement. On the other hand, the relevant rheological quantities could not be extrapolated from previous rheological studies of cement-based materials (e.g. ref. 1). These studies have either employed cement pastes, mortars or microconcrete, rather than full scale concrete mixes. Neither have they defined concrete mix properties in quantities directly amenable to classical solution of fluid dynamic problems. It was therefore decided to develop an instrument which could measure the classical flow properties of fluid concrete mixes.

The design of the instrument was therefore aimed at measuring the *Bingham* yield value and plastic viscosity of the mix. Careful consideration of the principles of viscometry and the structure of a concrete mix led to the choice of the rotational coaxial cylinder viscometer. The chief advantage of this type of viscometer lies in the possibility of continuous measurements at a given shear rate for extended periods. Subsequent measurements at other conditions in the sample may also be made. It is possible therefore to study time-effects.

This decision was made with full awareness of the difficulties which could arise from limitations in the essential geometry of the instrument when applied to a rheological study of very coarse suspensions, such as full scale concrete mixes which contain solid particles of size up to 20 mm. These possible limitations were accepted in view of the immediate demand for the information and the results of the rheological measurements are discussed with these limitations in mind.

The experimental work is being extended to investigate the effect of concrete composition and the effect of instrument geometry. However, the results so far available furnish a working knowledge of the flow properties of concrete mixes and are therefore considered to warrant a discussion. A description of the instrument is given with a discussion of the results from a rheological standpoint. The use of the instrument data for site quality control purposes is also indicated.

The primary contribution of the present study is the extension of the application of rotational type rheometers to the study of full scale concrete mixes, and thus to furnish data on some basic aspects of the rheological behaviour of full scale concrete mixes.

2. Specifications and design of the present rheometer

2.1. The concentric cylinders

The two major design considerations are the geometry of the coaxial cylinders and their surface characteristics. The interaction of the instrument geometry and its power rating determines the extent in the gap to which flow of the material occurs. Results presented by *Van Wazer* et al. (2) show that the power required for flow across the entire gap increases exponentially with increasing ratio of cup to bob radius. They have therefore suggested a ratio of 1.02 as desirable. A gap size 10–100 times the largest particle size is also recommended to minimize the effect of particle interference; end effects will be minimized provided the bob height is not less than its diameter (2). The tendency of suspensoid fluids to separate into their basic phases necessitates profiling the surfaces. This treatment also prevents slippage between the instrument and the test material, an assumption made in the theory of rotational viscometers.

The following specifications based on a compromise between the above ideal requirements, and practical and economic considerations, were found suitable; while a vane arrangement was used to check phase separation and slippage.

Bob radius (R_b)	=	125 mm
Cup radius (R_c)	=	158 mm
R_c/R_b	=	1.26
Gap size	=	33 mm
$\frac{\text{Gap size}}{\text{max. particle size}}$	=	1.65
Height of bob (H_b)	=	93 mm
H_b/R_b	=	0.75
Depth of cup (D_c)	=	125 mm
Bob vane protrusion (P_b)	=	25 mm
Cup vane protrusion (P_c)	=	18 mm
$(R_b + P_b)/R_b$	=	1.20
$R_c/(R_c - P_c)$	=	1.13
No. of bob vanes	=	12
No. of cup vanes	=	15
Root-spacing of bob vanes (S_b)	=	25 mm
S_b/max. particle size	=	2.67
Root-spacing of cup vanes (S_c)	=	73 mm
S_c/max. particle size	=	3.91

2.2. General description of the instrument

Plate I shows the prototype of the instrument in use, while a detailed diagrammatic description of its layout is given in fig. 1. The rotor (cup) is driven by a 0.75 HP 3-phase A.C. induction motor through an infinitely variable speed gear box and a worm gear box (50 : 1 speed ratio).

The stator (bob) is suspended from a frame through a bearing housing which is fixed to the suspension frame, and is restrained from moving by a torque arm which itself is held by a spring loaded latch (fig. 2). The torque arm consists of a steel bar (307 mm long, 12.5 mm wide and 15 mm deep) with two electrical strain gauges mounted on each side of the bar to give maximum sensitivity and temperature compensation (fig. 2).

The structure of the suspension frame was designed to ensure that it suffered virtually no deflection under the under-hung weight. All structural members were designed to have a permissible stress, either in bending or shear, of 60.0 MN/m^2. The maximum deflection of the torque arm is limited to 1.25 mm under an ultimate bending moment 6.0 kNm, i.e., twice the working value. This limit on deflection (0.5% of the effective gauge length of the arm) ensured a linear calibration curve.

The torque arm was protected from being strained to permanent set by a torque limiting device (fig. 2), which operated on a lever system. A micro-switch protects the motor from overload.

The speed of the system is too low for commercial tachometers. A device employing a cam and D2 series DC-LVDT transducer system was designed to measure the instrument rotation. The signals from both the strain gauges and the transducer were fed into a U-V recorder which produced a continuous signature trace of torque and speed on a photographic paper.

3. Experimental details

3.1. Calibration of torque arm and speed indicator

The torque arm was calibrated statically with dead weights, readings being taken for both increasing and decreasing weights. All calibration curves are linear.

The speed selector on the infinitely variable speed gear box was calibrated with the instrument empty, by counting the number of revolutions per minute at five speed settings. Again the calibration is linear.

3.2. Calibration for end effect

In calibrating for the end effect the extreme water/cement ratios were used (fig. 3). The maximum end correction was found to be 10% of the actual immersed depth of the stator while no end correction was necessary for the wettest mix. The end correction for other mixes would therefore fall within these bounds, and were thus estimated since the probable error in estimation would be less than 10%.

3.3. Material specification and mix proportions

The aggregates were *Thames Valley* aggregates, the coarse aggregates being classified as irregular. All the aggregates were delivered in single sizes and in an air dry state. They were combined to correspond exactly to the

Plate 1. Prototype of the concrete rheometer

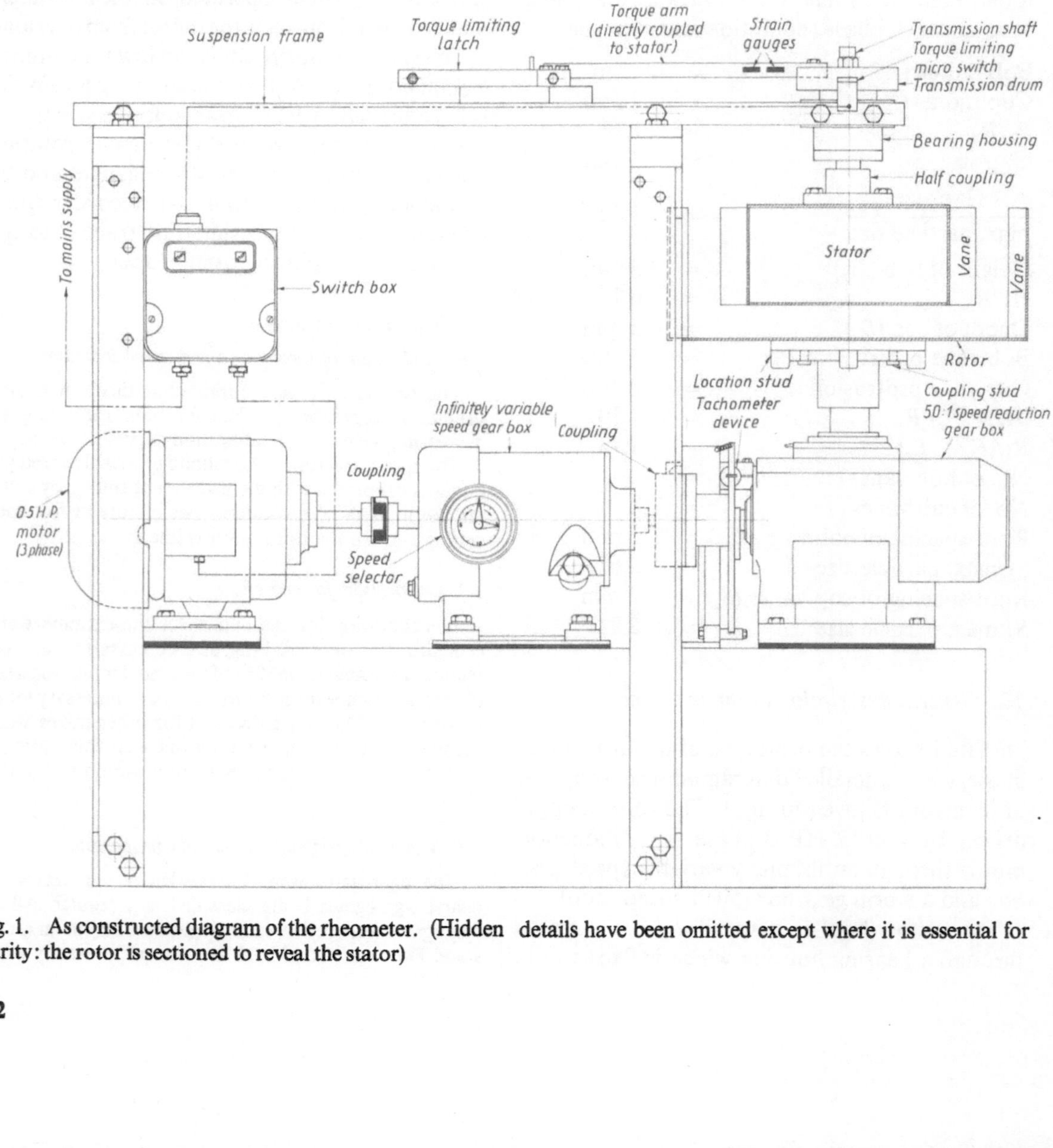

Fig. 1. As constructed diagram of the rheometer. (Hidden details have been omitted except where it is essential for clarity: the rotor is sectioned to reveal the stator)

required grading curve and stored in waterproof bags until required.

Ordinary Portland Cement was used throughout, and was obtained from the same source.

Fresh clean tap water was used in all cases.

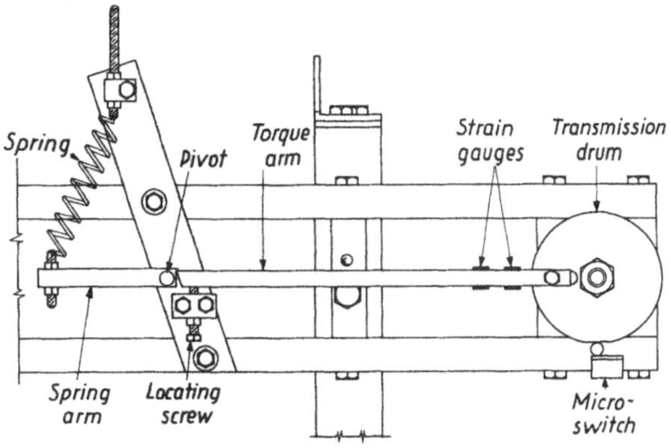

Fig. 2. Details of the torque limiting latch (plan view)

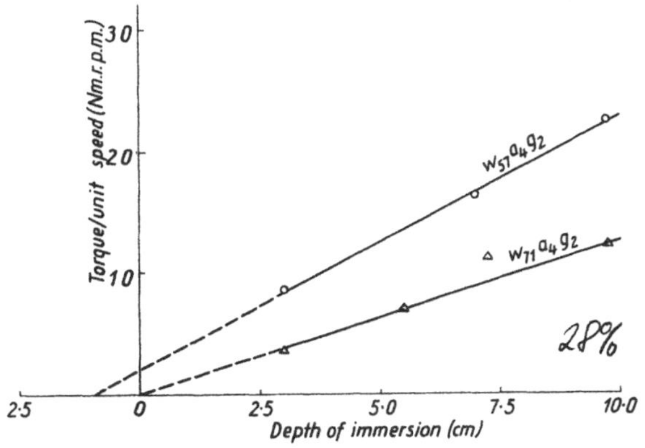

Fig. 3. Calibration for end effect

It was intended to test a range of aggregate/cement ratios but it was found that with aggregate/cement ratios above 4.0 the maximum permissible applied torque was frequently exceeded. This value of aggregate/cement ratio was therefore chosen as the limit for this investigation; in piling practice, for example, the value is around 5.5. Five different water/cement ratios were chosen to give a range of slump values from 170–250 mm, which covers the usual piling site mixes. Aggregate grading and aggregate/cement ratio were alternately changed to obtain an indication of their effects on the test results.

The following system is used to describe the mixes: $w_x a_y g_z$ implies that

a) water/cement ratio = $0.x$,

b) aggregate/cement ratio = $y.0$,

c) aggregate grading curve No. z of Road Note 4 (3).

3.4. Test procedure

The combined aggregate and cement was mixed dry for one minute. The required amount of mixing water was then added and mixing continued for a further 2 minutes, using a Cumflow pan mixer.

A slump test was carried out at the end of mixing. It had been found from pilot tests that at the normal rate of filling the cone, the test could be completed (i.e. the cone lifted) in exactly 3 minutes from end of mixing.

The specimen used for the slump test was returned into the mixer and the concrete was remixed by hand for another minute.

The annulus of the rheometer was next filled with concrete. During this process the mix was slightly tamped with a standard rod to distribute it within the annulus and to reduce the tendency for vertical flow during test as assumed in the theory. Preliminary tests had also shown that the sample preparation took 6 minutes to complete. Consequently, shearing of all specimens was commenced 10 minutes after mixing and was continued for between 30–60 seconds. In all tests therefore, the *time factor was held constant.*

4. Results and discussion

4.1. Stress decay trace

The instrument trace gives instantaneous records of torque with time. The general features of each trace which are apparent in fig. 4 and 5, are as follows:

a) a build up torque (or stress) to a peak within 1 ± 0.5 seconds from commencement of rotation;

b) a period, between 5 and 30 seconds from peak point of rapid drop in torque reading;

c) a prolonged period of stable torque reading following the period of rapid drop.

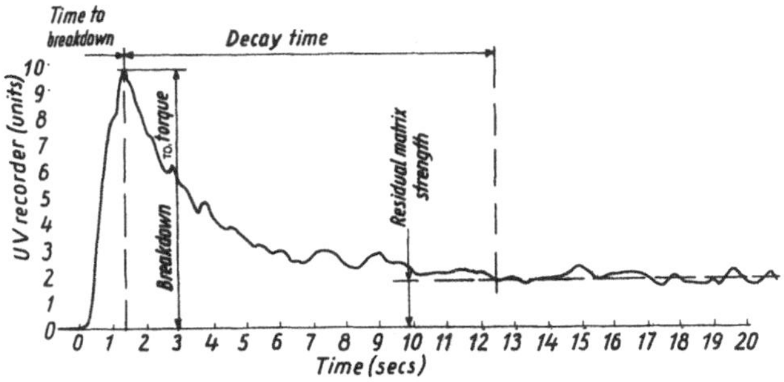

Fig. 4. Torque trace from rheometer

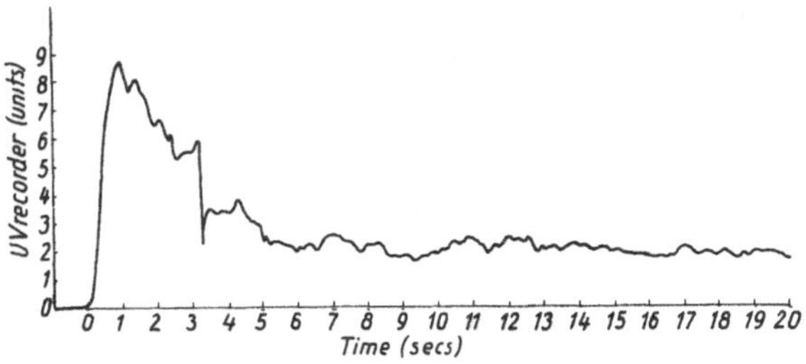

Fig. 5. Torque trace from rheometer

The stress decay, following the peak, is attributed to thixotropic breakdwon and it is hoped that experiments will be carried out in future to confirm thixotropy in accordance with *Greens* definition (4) – i.e., establishing hysteresis loops by plotting the up and down curves. However, such loops have already been obtained for cement pastes (5 and 6) which constitute the matrix within which the granular particles are interspersed to form a concrete mix. Thus, it is reasonable to presume the stress decay to be due to thixotropic breakdown.

Following earlier research investigations (5 and 6) an exponential curve was fitted to the thixotropic breakdown curve (fig. 6). It was found, however, that two different curves were required to describe the full range of each breakdown curve (fig. 6 and 7). *Nessim* and *Wajdar* (6) also found this with cement paste.

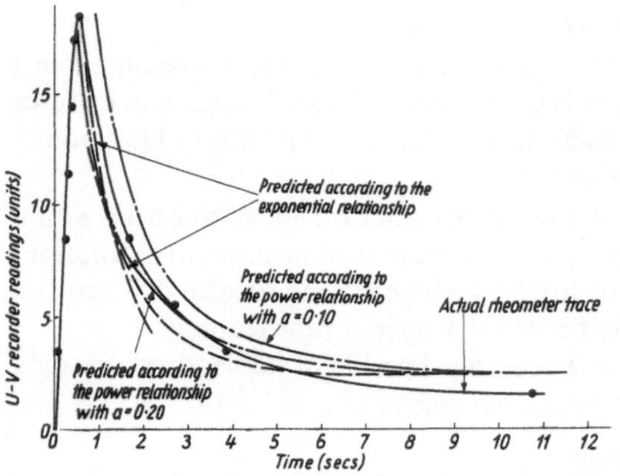

Fig. 6. Curve fitting to the thixotropic breakdown trace

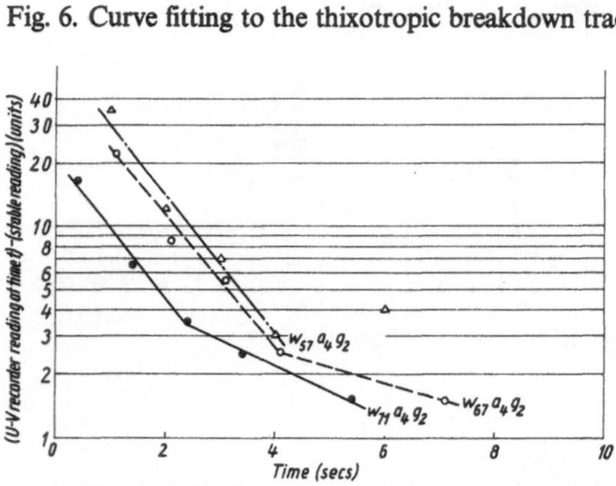

Fig. 7. Decay of U-V recorder reading with time based on exponential relationship

The departure of the exponential curve from the actual trace would suggest that the assumptions made in the theoretical derivation of the exponential relationship (5) may not be strictly valid for concrete mixes. An alternative theoretical expression based on a different assumption was therefore derived.

Tattersall (5) had assumed that the torque $(T - T_E)$ in excess of the equilibrium torque (T_E) at any time is related to the number of "linkages" yet unbroken at that time, thus

$$(T - T_E) = Z(n_0 - n) \qquad [1]$$

where Z is a function of the angular speed, n_0 is the number of linkages at the onset of structural breakdown; n is the total number of linkages broken, up to the material time.

A concrete mix is a more complex rheological body than cement paste, on which *Tattersalls* work was based, and the structural breakdwon of concrete mixes also includes the disentanglement of the mechanical interlock of particles. The breakdown should therefore require a much greater force initially. Consequently, the assumption of eq. [1] was modified thus:

$$(T - T_E) = Z(n_0 - n)^{\frac{1}{a}} \qquad [2]$$

where "a" is a positive fractional constant.

The assumption of eq. [2] marks the departure of the present analysis from *Tattersalls* approach. The calculus involved in the derivation of the final expression is similar to *Tattersalls* and is consequently omitted. The final expression obtained for the breakdown curve is

$$(t - t_0) = \frac{a\varphi}{2\pi Z^a \omega(1 - a)} \times \{(T - T_E)^{-(1-a)} - (T_0 - T_E)^{-(1-a)}\} \qquad [3]$$

where t is any time when the torque is T, t_0 is time corresponding to peak torque (T_0), φ is the work done in breaking a linkage.

Eq. [3] is a power relationship, and can be rewritten by taking the logarithm of both sides:

$$\log_e(t - t_0) = \log_e A + \log_e\{(T - T_E)^{-(1-a)} - B\} \qquad [4]$$

where

$$A = \frac{a\varphi}{2\pi Z^a \omega(1 - a)} \qquad [5]$$

and

$$B = (T_0 - T_E)^{-(1-a)}. \qquad [6]$$

Eq. [4] describes a linear curve with slope equal to unity. Numerical solutions of the equation have been obtained for various assumed values of "a" from 0.50–0.05 (fig. 8). The slopes of the curves of fig. 8 have been plotted in fig. 9 from which the values of "a" corresponding to a slope of unity may be read off as the most probable value of "*a*". This value lies between 0.1 and 0.2. Using these two values, the predicted curves were computed and also plotted in fig. 6. It is apparent that the power relationship predicts a *single* curve to cover the whole range of the trace and also makes a closer fit, than the exponential curve, to the actual trace.

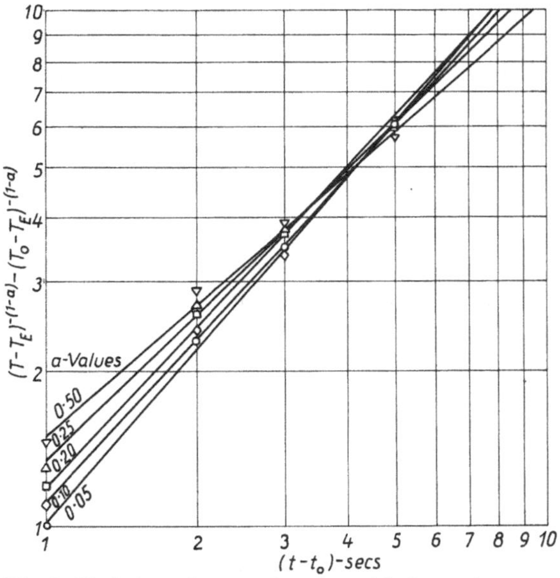

Fig. 8. Variation of torque function with decay time

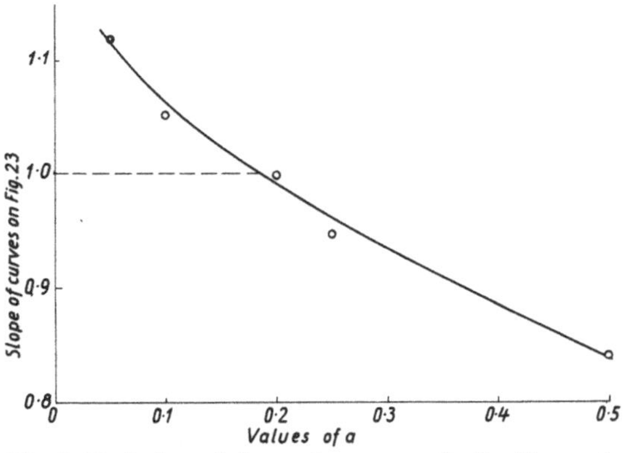

Fig. 9. Variation of slope of curves on fig. 8 with *a*-values

Further graphical analyses of eq. [4] for various water/cement ratios showed that the probable value of "*a*" lies between 0.1 and 0.2 for all the mixes tested. Consequently "*a*" may be considered independent of water/cement ratio but dependent on rheometer speed. The scope of the present investigation does not permit the effect of other mix parameters to be investigated.

4.2. Consistency curves

The r.p.m. torque curves for various mixes, based on average values, are presented in fig. 10. The angular speeds used in the plot are the preset values because preliminary tests showed that the instrument attained full speed before the peak torque was reached.

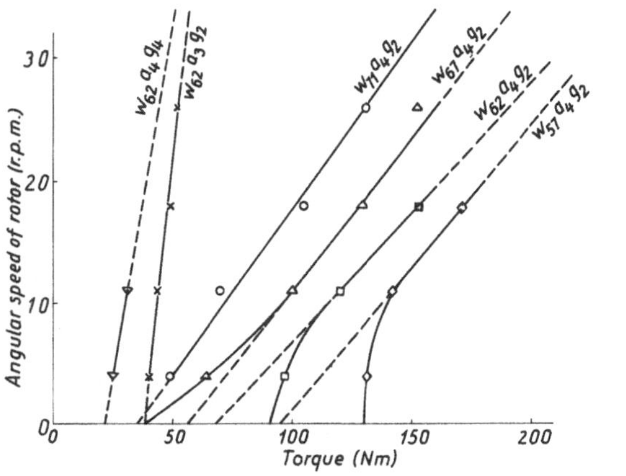

Fig. 10. Variation of torque with angular speed of rotor. ($w_{71}a_4g_2$ represents a mix with a water/cement ratio of 0.71, aggregate/cement ratio of 4.0 and aggregate grading-curve No. 2 of road note 4)

The true shear surface is strictly indeterminate as there could be local eddies at each vane location and the influence of the vane will also give an irregular profile. However, the vane protrusion is small compared to the bob radius, and observations of the surface of the concrete during test showed no evidence of eddies, and the rupture surface which should follow the configuration of the fluidified zone could be seen to approximate the cylindrical boundary of the vanes of the bob. A cylindrical shear surface of radius equal to that of the bob plus vane protrusion was therefore used in all the computations.

The major consequence of the geometry is its influence on the extent of plug flow within the gap. The occurrence of substantial plug flow at all the operating speeds of the instrument will result in the measurement being too localized to give any really meaningful assessment of the flow properties of the mix. Therefore it is most desirable that the concrete mix across the entire annular gap should be subjected to flow. It is only under such condition that the instrument can be regarded as reflecting the flow behaviour of the full mix within the annulus.

To examine this problem, let it be assumed that flow is occurring only in the immediate vicinity of the bob so that S_b/ϑ is very close to unity. Assuming also a *Bingham* body, it can be shown that the flow equation is

$$S_b = \vartheta + \left(\frac{8\pi\eta_p\vartheta}{60}\right)^{1/2} \Omega^{1/2} \quad [7]$$

where η_p is the plastic viscosity, S_b is the stress at the bob, Ω is the angular velocity (r.p.m.), ϑ is the yield value.

The relationship between S_b and $\Omega^{1/2}$ should be linear over the range of Ω where the assumption, S_b/ϑ is close to unity, is valid. A plot of the relationship is shown in fig. 11, in which straight lines, which should be appropriate to the very initial stage have been extended over the whole range of points (i.e. implying that plug flow predominates at all speeds). However, the distribution of the points about the lines suggests that, the lines should, strictly, be curvilinear. Thus, the presupposition that plug flow exists at all speeds is not corroborated by the empirical result.

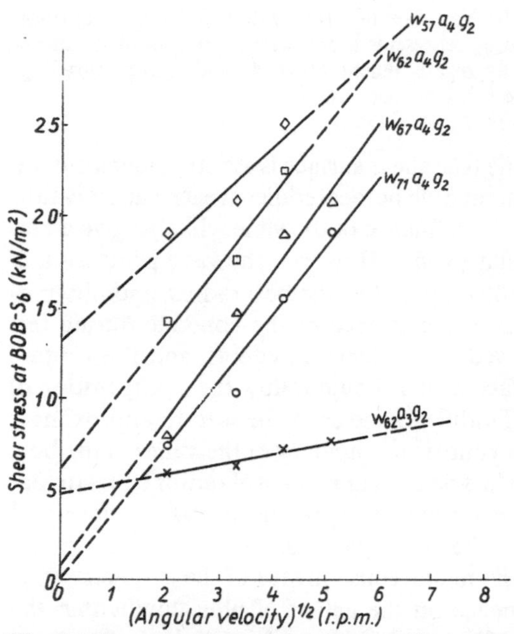

Fig. 11. Flow curves assuming existence of large region of plug flow

Furthermore, the critical radius (R') – i.e. the extent to which flow of the material occurs – at any speed can be calculated from the well established equation:

$$R' = \left(\frac{T}{2\pi\vartheta l}\right)^{1/2} \quad [8]$$

where T is the applied torque, l is the immersed depth of bob.

Using the yield values obtained from fig. 10, this calculation shows that at the lowest test speed of 4 r.p.m., flow occurs within 50 and 70% of the cross-sectional area of the annulus for mixes $w_{57}a_4g_2$ and $w_{62}a_3g_2$ respectively. Further calculation showed that flow should occur within the entire annulus at 15 and 13 r.p.m. for mixes $w_{57}a_4g_2$ and $w_{62}a_3g_2$ respectively. The rheometer tests were run up to a speed of 26 r.p.m., and, consequently, the flow curves of the mixes can legitimately be said to contain an upper linear portion and a lower curved portion. This characteristic is typical of flow curves obtained for a *Bingham* body when tested in a coaxial-cylinder viscometer (2 and 7).

Two deductions are made from this established nature of the flow curves of concrete mixes.

a) A significant proportion of the mix in the gap was always subjected to flow; at higher speeds, flow occurred across the entire gap (i.e. the linear range of the flow curve). Thus, it is argued that the instrument reflected the flow behaviour of the full mix.

b) The mixes tested behaved as *Bingham* plastics up to a speed of 26 r.p.m., their empirical flow curves being characterized by a lower curved portion and an upper linear portion.

Taken together, the foregoing deductions imply that the plastic viscosity and *Bingham* yield value of fluid concrete mixes can be determined from tests in the rheometer.

4.3. Plastic viscosity and yield value of the mixes

Attempts were made to simulate a static test with the view to determining the true yield values. However, the instrument was either unstable or sometimes would not run evenly at speeds less than 4 r.p.m. Intercepts of the linear portions of the curves on the torque-axis were therefore taken for computing the yield values; this is strictly the apparent, or *Bingham*, yield value. These yield values were then used to compute the shear rate, after correcting for the end effect, using the *Reiner-Riwlin* eq. [2]. The computed shear rate values have been plotted against their corresponding shear stresses in fig. 12.

The yield values and plastic viscosities derived from fig. 12 were correlated with water/cement ratio in fig. 13, using a semi-log plot to obtain linear relationships. Thus both quantities decrease exponentially with increasing water/cement ratio.

Using mix $w_{62}a_4g_2$ as a "standard", the effect of altering a mix variable other than water/cement ratio was investigated. Mix $w_{62}a_3g_2$

which has a lower aggregate/cement ratio was tested for aggregate/cement ratio effect, and a finer grading – mix $w_{62}a_4g_2$ – was tested for the effect of aggregate grading. The results are tabulated below:

Table 1. Variation of plastic viscosity and yield value with aggregate/cement ratio and aggregate grading

Mix	Plastic viscosity ($kN sec m^{-2}$)	Yield value (kN/m^2)	Standard slump (cm)
$w_{62}a_4g_2$ (h)(Standard)	4.8	8.0	21.5
$w_{62}a_3g_2$ (agg/cement ratio)	0.5	4.2	22.5
$w_{62}a_4g_4$ (agg. grad)	0.8	2.5	20.5

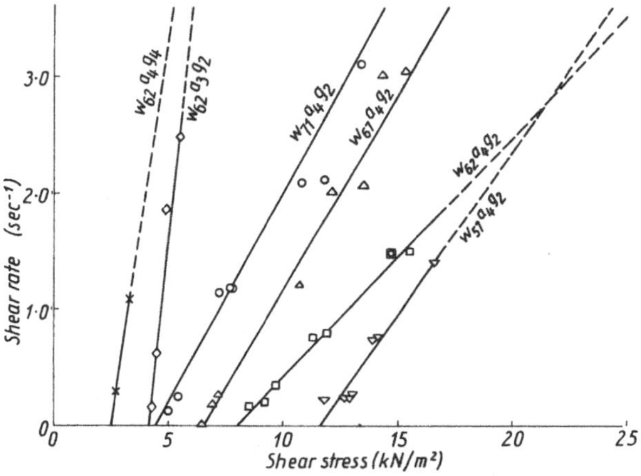

Fig. 12. Normalized flow curves for concrete mixes

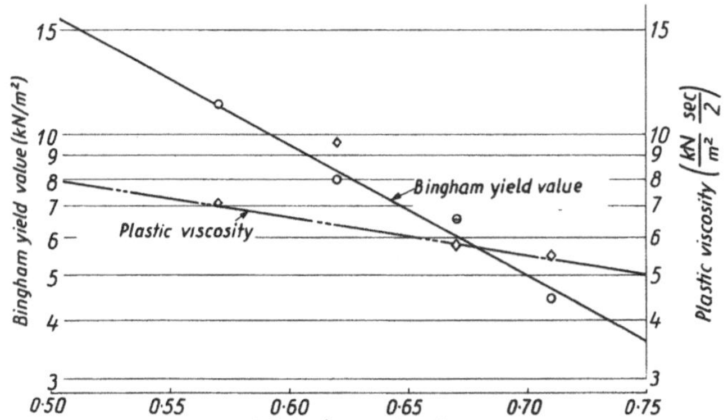

Fig. 13. Correlation of flow parameters of mix with water/cement ratio. (Aggregate/cement ratio –4; aggregate grading: curve No. 2)

In all these tests the water/cement ratio was maintained constant at 0.62. It is seen that a reduction in aggregate/cement ratio reduces the plastic viscosity by 89%, the yield value by 47% and increases the slump value by 5%. A change in aggregate grading from a coarse grading-curve No. 2 of Road Note 4 (3) – to a finer grading-curve 4 – reduced the plastic viscosity by 84%, the yield value by 69% and the slump by 7%. The marked changes in the rheological quantities, in contrast to the slump values, is noteworthy.

A plot of the apparent viscosity against the shear rate, on double logarithmic scales, is given in fig. 14. The slopes of the lines to the shear rate axis are all greater than zero but less than 45°. This observation is consistent with results presented by *Van Wazer* et al. (2) for pseudoplastics in general, a classification to which *Bingham* plastics belong. Thus this result is consistent with the earlier conclusion that the mixes tested behave as *Bingham* bodies.

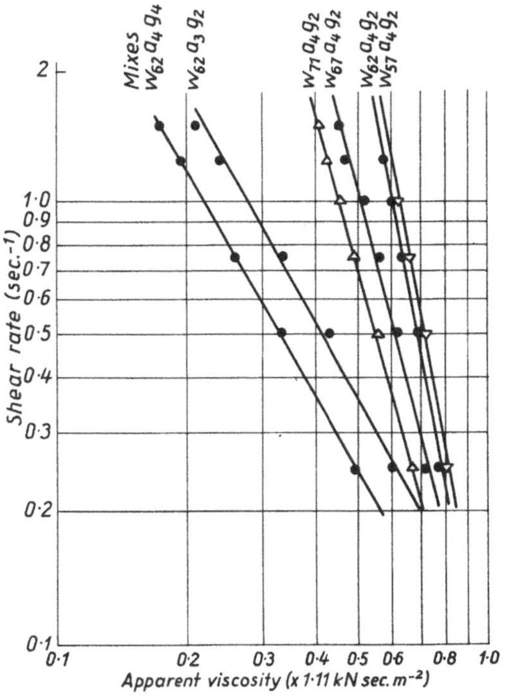

Fig. 14. Variation of apparent viscosity with shear rate

5. Potentials of rheometer data for site quality control

An analysis of the features of the rheometer trace suggests that some of them are significant in relation to the performance characteristics of the mix. The various features of the traces have also been found to correlate with variations in mix parameters. In general, these instrument data were found to be both more sensitive to variations in the mix composition (see fig. 15) and more consistent for any given mix than the slump test. Theoretical considerations and analysis of the test results suggest that the standard slump test (8) is in fact not a suitable descriptive test for fluid concrete mixes, as employed in geotechnical processes. Tamping of the sample as required in the standard test simply produces a solid cone at the base of the slump cone, which relates more to the tamping effort than to the mix itself. A modified slump test, consisting simply of pouring the concrete into the cone without tamping is a more relevant test. The rheometer has also been used for site purposes. Details of this aspect of the investigation will, however, be reported later.

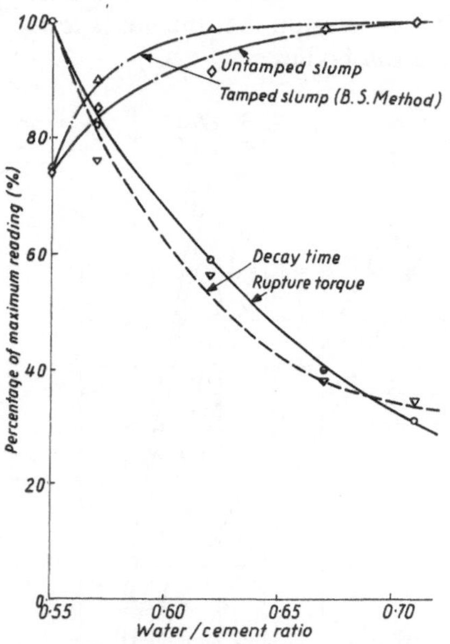

Fig. 15. Correlation of various mix parameters with water/cement ratio

6. Conclusions

1. When a fluid concrete mix is subjected to shear in a coaxial cylinder rheometer at constant angular speed, thixotropic breakdown of the material structure occurs after a peak torque (or stress) has been attained.

2. The thixotropic breakdown curve obeys a power law better than an exponential one.

3. When a full scale fluid concrete mix (containing up to 20 mm particle size) is tested in the instrument flow does occur over a significant proportion of the gap to justify the instrument being considered to reflect the flow properties of the full mix.

4. Fluid concrete mixes behave like *Bingham* plastics, and thus their flow characteristics can be defined by their yield value and plastic viscosity.

5. Both the yield value and plastic viscosity, as would be expected, are affected by variations in mix parameters; both quantities decrease exponentially with increasing water/cement ratio.

6. The yield value of the mixes tested is of the order of 2.5–12.0 kN/m^2 (or 25000–120000 dynes/cm^2).

7. Their plastic viscosity is of the order of 5000 to 40000 poises (i.e. 5–40 kP or 0.5–4 kN sec cm^{-2}).

8. Plots of the logarithms of apparent viscosity and shear rate are linear.

9. The concrete rheometer described in this paper can also be used for site quality control and available empirical results suggest that it is a potentially superior instrument, in this context, to the slump cone.

Summary

The description of a *Couette* type rheometer which has been developed for measuring some rheological properties of fluid concrete mixes is given with a discussion of experimental results.

The instrument produces a recorded trace which is essentially a stress decay curve attributed to thixotropic breakdown of the material structure. This breakdown curve is predicted better by an expression of the power law type than of the exponential type. Analysis of the empirical flow curves shows that, despite some limitations in its essential geometry, the instrument reflects the flow behaviour of the full concrete mix. Up to rotational speeds of 26 r.p.m., the flow curves are typical of *Bingham* plastics.

The yield values and plastic viscosities of the mixes which are very fluid and typical of piling concrete are of the order of 2.5–12.0 kN/m^2 (or 25000–120000 dynes/cm^2) and 5–40 kilopoises (5000–40000 poises) respectively. These quantities vary with concrete mix parameters, and increase exponentially with decrease in water/cement ratio.

References

1) *Powers, T. C.* and *E. M. Wiier*, Proc. Amer. Soc. Test. Mater. **41**, 1003–1015 (1941).

2) *Van Wazer, J. E.*, et al., Viscosity and flow measurement. (New York 1963.)

3) *Dsir*, Road Res. Laboratory, Road Note No. 4, Second Edn. (1962).

4) *Green, H.*, Industrial Rheology and Rheological Structures. (New York 1949.)

5) *Tattersall, G. H.*, Brit. Appl. Physics. **6**, 165–167 (1955).

6) *Nessim, A. A.* and *Wajdar*, Mag. of Conc. Res. **17**, No. 51, 59–68 (1965).
7) *Scott Blair, G. W.*, An introduction to industrial rheology. (Philadelphia 1938.)
8) *British Standard Institution*, Methods of testing concrete. B.S. 1881 (Part II) (1970).

Author's address:

Osundo J. Uzomaka
Dept. of Civil Engineering
University of Nigeria
Nsukka, E.C.S., Nigeria, W. Africa

Rheol. Acta 13, 22–27 (1974)

Mitteilung aus der Anwendungstechnischen Abteilung der Brabender OHG, Duisburg (Deutschland)

Meßsystem für verarbeitungstechnische Eigenschaften von Kunststoffen

Von A. Rothenpieler

Mit 9 Abbildungen

(Eingegangen am 27. Oktober 1972)

Einleitung

Verarbeitungstechnische Eigenschaftswerte werden von Kunststoffherstellern, Verarbeitern und Maschinenherstellern benötigt. Die Rohstoffhersteller ermitteln diese bei neuen Produkten, um ihren Kunden Verarbeitungshinweise zu geben und überwachen damit laufend ihre Produktion. Maschinenhersteller verwenden solche Werte zur Konstruktion und Entwicklung neuer oder zur Weiterentwicklung vorhandener Verarbeitungsverfahren. Kunststoffverarbeiter führen mit den entsprechenden Geräten die Wareneingangskontrolle durch und können vor allen Dingen im Labormaßstab neue Produkte entwickeln.

Meßaufgaben

Aufbereiten und Verarbeiten von Kunststoffen erfolgen mit Energiezufuhr durch Erwärmen und mechanische Energie. Während bei älteren Verfahren z. B. Pressen und Kolben-Spritzgießen der Einfluß der äußeren Heizung überwiegt, wird bei den meisten modernen Aufbereitungs- und Verarbeitungsverfahren wie Schnellmischen, Kalandrieren, Extrudieren, Hohlkörperblasen und Spritzgießen der größte Energieanteil durch mechanische Einleitung aufgebracht. Daher interessiert bei den verarbeitungstechnischen Eigenschaften besonders das Verhalten unter Scherbeanspruchung. Hierfür hat sich seit Jahrzehnten ein Meßsystem bewährt, mit dem das vom Kunststoff aufgenommene Drehmoment gemessen werden kann. Das in verschiedenen Meßvorsätzen in den Kunststoff einzubringende Drehmoment wird von einem pendelnd gelagerten Antrieb übertragen. Dieser wirkt als Dynamometer, dessen Drehmoment mit einer präzisen mechanischen Waage gemessen wird.

Für die verschiedenen Meßaufgaben stehen Antriebseinheiten unterschiedlicher Bauarten und Leistungen zur Verfügung. Die Geräte geringerer Leistung haben Drehstromantrieb mit polumschaltbaren Motoren für zwei verschiedene Drehzahlen oder ein mechanisches Verstellgetriebe für die stufenlose Drehzahleinstellung. Die Geräte höherer Leistung haben thyristor-gesteuerte Gleichstrommotoren. Diese messenden Antriebseinheiten werden mit verschiedenen Verarbeitungseinrichtungen kombiniert, die die Vorgänge in Produktionsmaschinen möglichst genau nachbilden.

Plastifizieren

Das Plastifizieren von Kunststoffen zum Extrudieren, Hohlkörperblasen und Spritzgießen erfolgt heute fast ausschließlich mit Schneckenmaschinen. Das Material wird eingezogen, komprimiert, durch Energiezufuhr in einen formfähigen, plastischen Zustand gebracht und den Formgebungswerkzeugen zugeführt. Beim Extrudieren muß auch der Formgebungsdruck durch die Schnecke aufgebracht werden.

Diese Vorgänge laufen zeitlich hintereinander an verschiedenen Stellen längs der Schnecke ab. Druck und Temperatur in der Masse können an verschiedenen Stellen gemessen werden. Da die Zylinder trotzdem geschlossene Systeme sind, lassen sich die im Kunststoff ablaufenden Vorgänge nur schwer trennen und beobachten.

In Meßknetern werden diese Vorgänge an einer Stelle, aber zeitlich hintereinander, untersucht und können registriert werden. Das Versuchsmaterial wird bei Raumtemperatur eingefüllt und durch die Heizung des Kneters und die von den Knetschaufeln abgegebene Scherung erwärmt. Die Temperatur des Kneters kann in einem weiten Bereich eingestellt werden und ist ein Maß für die an Schneckenmaschinen einzustellende Zylindertemperatur. Die Scherung kann durch Verstellen der Schaufeldrehzahl und durch Verwenden verschiedener Schaufelformen mit unterschiedlicher Scherwirkung variiert werden. Die Kneterdrehzahl ist ein Maß für die anzuwendende Schneckendrehzahl, während die Schaufelform verschiedene Schneckengeometrien nachbilden kann.

Mit solchen Messungen läßt sich das Verhalten der verschiedenen Kunststoffgruppen beim Plastifizieren durch den zeitlichen Verlauf des Drehmomentes charakterisieren. Materialschädigende Zersetzungen oder Vernetzungen von Thermoplasten durch zu hohe oder zu lange Scher- und Wärmebeanspruchung werden im Drehmomentverlauf sichtbar. Daraus lassen sich die Grenzen des Verarbeitungsbereiches feststellen. Bei vernetzenden Materialien, d. h. Duroplasten und Kautschuken, beginnt nach der plastischen Phase die Aushärtung oder Vulkanisation. Der dabei zu beobachtende Drehmomentanstieg kennzeichnet die Vernetzungsgeschwindigkeit.

Extrudieren

Als Ergänzung zu den differenzierenden Meßmethoden mit Knetern werden Meßextruder eingesetzt, die den tatsächlichen Produktionsmaschinen entsprechen und die Vorgänge längs der Schneckenachse meß- und registrierbar machen (Abb. 1). Für die unterschiedlichen Aufgaben stehen Extruder mit Schneckendurchmessern von 19 und 30 mm und *L/D*-Verhältnissen von 10/1 bis 25/1 in Stufen von 5 *D* zur Verfügung. Die Ermittlung des Drehmomentes erlaubt auch hier Rückschlüsse auf den Energiebedarf der Verarbeitungsmaschinen. Durch Druck- und Temperaturmessungen in der Masse können die Vorgänge des Aufschmelzens und Plastifizierens im Extruder erfaßt werden. Die für die entsprechenden Kunststoffe entsprechende optimale Schneckengeometrie kann unter praxisnahen Bedingungen ausgesucht werden. Die kühlbare Einzugsbuchse ist austauschbar, um für jedes Versuchsmaterial die förderwirksamste Ausführung einsetzen zu können.

Mit diesen Extrudern lassen sich Produktionsanlagen im Labormaßstab zusammenstellen. Damit können Profile, Rohre, Schläuche, Rundstränge, Monofile, Flach- und Blasfolien extrudiert sowie Kabel ummantelt werden. So können mit kleinen Mengen und geringem Aufwand die geeigneten Kunststoffe oder Mischungen für neue oder laufende Produkte ausgewählt werden. An den Erzeugnissen können die notwendigen Eigenschaften geprüft werden. Besonders aufschlußreich ist das Herstellen von Flach- und Blasfolien auch bei Kunststoffen, die beispielsweise für Rohre oder Kabelummantelungen eingesetzt werden. Die Folien stellen praktisch Dünnschnitte dar, an denen Gleichmäßigkeit, Mischgüte, Farbverteilung und Fehlstellen, wie Stippen oder Fischaugen, deutlich zu erkennen sind.

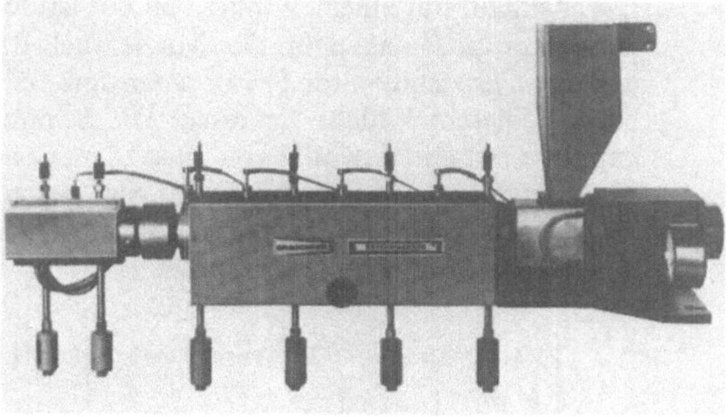

Abb. 1. Vgl. Text

In der Kunststoffprüfung werden Meßextruder vorteilhaft zu Untersuchungen der Materialeigenschaften in der plastischen Phase eingesetzt. Bei normalen Kapillarviskosimetern erfolgt die Plastifizierung durch äußere Beheizung in einem Zylinder und ist wegen der schlechten Wärmeleitfähigkeit der Kunststoffe entsprechend ungleichmäßig. Die Plastifizierung entspricht der früher üblichen Methode des Kolbenspritzgießens. Wenn die Erwärmungszeit zum Erreichen einer möglichst gleichmäßigen Temperatur verlängert wird, besteht die Gefahr von Veränderungen des Versuchsmaterials. Meßextruder dagegen plastifizieren die zu untersuchenden Kunststoffe in gleicher Weise wie die heute fast ausschließlich eingesetzten Schneckenspritzgießmaschinen und Extruder und haben außerdem den Vorteil einer kontinuierlichen Arbeitsweise. Vor der Auswertung von Meßdaten wird jeweils deren Konstanz abgewartet.

Versuchsdurchführung

Als Beispiel werden einige Versuchsergebnisse mit einem Polyäthylen mittlerer Dichte, Lupolen 2435K der BASF, gezeigt. Das Material wurde mit einem Extrusiograph mit einem Schneckendurchmesser von 19 mm und einer Schnekkenlänge von 25 D extrudiert, der mit verschiedenen Meßstellen für Massetemperaturen und -drücke ausgerüstet ist (Abb. 2). Die eingesetzte Kurz-Kompressionsschnecke hat ein Gangtiefenverhältnis von 4 : 1 bei einer Ausstoßzonenlänge von 10 D. Der Extruder wurde mit verschiedenen Rund- und einer Schlitzkapillare ausgerüstet. Die Runddüsen haben Bohrungsdurchmesser von 1, 2 und 3 mm bei einer Kapillarlänge von 30 mm sowie Kapillarlängen von 10 und 20 mm mit einem Durchmesser von 1 mm. Der Düseneinlauf mit einem Winkel von 60° ist der Schneckenspitze angepaßt. Der Massedruck P5 und die Massetemperatur TM5 werden unmittelbar vor diesem Einlauf gemessen. Die Schlitzkapillare mit einem Kanal von 20 × 0,8 mm hat eine Länge von 155 mm. Im Kanal sind gegenüberliegend Druckgeber und Thermoelemente im Abstand von 75 mm angeordnet.

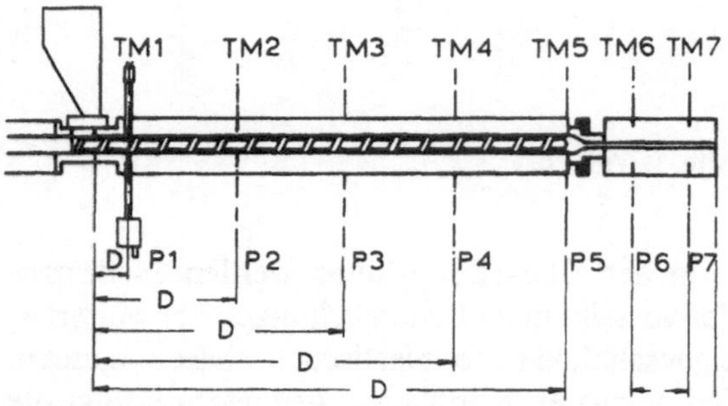

Abb. 2. Vgl. Text

Die Versuche wurden mit 5 Schneckendrehzahlen von 30–150 min^{-1} durchgeführt. Die Zylindertemperaturen wurden in 3 Programmen jeweils so eingestellt, daß die Massetemperatur bei der geringsten Schneckendrehzahl 180, 200 bzw. 220 °C an der Meßstelle TM5, d. h. direkt vor dem Eintritt in die Kapillare betrug. Mit steigender Schneckendrehzahl erhöhte sich diese Temperatur durch zusätzliche Reibungswärme um 3–6 °C. Die im Kanal der Schlitzkapillare gemessene Temperatur lag 2 bis 8 °C über der Eintrittstemperatur, was auf eine weitere Reibungserwärmung schließen läßt. Die Unterschiede zwischen den Temperaturen an den beiden Meßstellen in der Kapillare betrugen maximal 2 °C und waren im Bereich der Meßgenauigkeit.

Versuchsergebnisse

In Abb. 3 ist der Mengenstrom Q abhängig von der Schneckendrehzahl n bei einigen Düsen aufgetragen. Die Temperatur TM5 stieg mit der Drehzahl von 180 bis 186 °C an; bei den anderen Temperaturen zeigte sich der gleiche Verlauf, so daß auf eine zusätzliche Erläuterung dieser Ergebnisse verzichtet werden kann. Die Menge steigt mit der Drehzahl linear an. Die Rundkapillare mit dem kleinsten Durchmesser und der größten Länge und die Schlitzkapillare haben die niedrigste und praktisch gleiche Menge. Bei Verkürzung der Rundkapillare und Vergrößerung des Durchmessers ist jeweils ein Ansteigen der Menge festzustellen. Die höchsten Werte ergeben sich bei 3 mm Düsendurchmesser, da diese Düse den geringsten Widerstand hat.

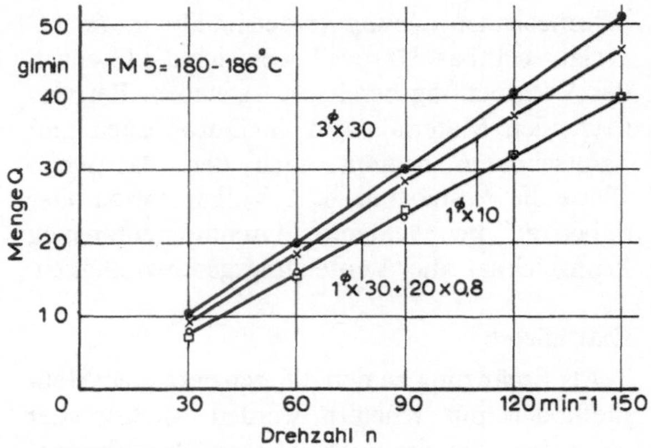

Abb. 3. Vgl. Text

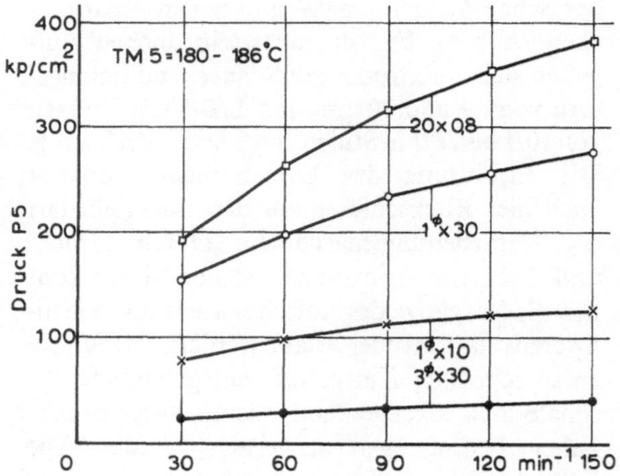

Abb. 4. Vgl. Text

Der Druck an der Meßstelle P5 ist in Abb. 4 abhängig von der Drehzahl dargestellt. Hier ist der Anstieg nicht mehr linear, sondern verringert sich mit steigender Drehzahl. Neben der geringfügig höheren Temperatur liegt der Grund vor allem in der mit steigender Schergeschwindigkeit kleiner werdenden Viskosität des Materials. Dieser Einfluß erklärt auch den Unterschied im Druckaufbau durch die Rundkapillare mit 1 mm Durchmesser und 30 mm Länge sowie die Schlitzkapillare. Bei gleicher Menge ist die Schergeschwindigkeit in den Rundkapillaren erheblich höher und senkt dadurch die Viskosität. Durch Verkürzen der Kapillare oder Vergrößern des Durchmessers wird der Druckaufbau verringert. Einen zum Druck vor dem Düseneintritt parallelen Verlauf hat die Schneckenrückdruckkraft PR (Abb. 5). Der auf den Schneckenquerschnitt bezogene Rückdruck ist jedoch immer höher als der vor der Schnecke gemessene Druck P5. Der von den Schneckenflanken aufgebaute Förderdruck muß sich daher ebenfalls auf die Rückdruckkraft auswirken.

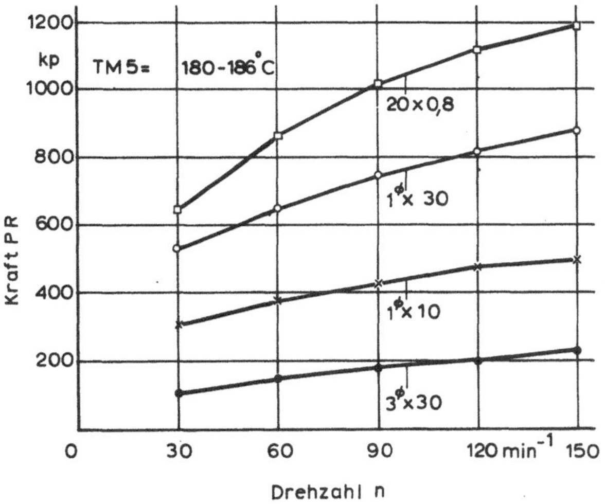

Abb. 5. Vgl. Text

Versuchsauswertung

Zur Auswertung der Versuchsergebnisse wird das exakt nur für *Newton*sche Flüssigkeiten geltende *Hagen-Poiseuille*sche Gesetz verwendet. Die Einlaufkorrekturen entfallen bei der Schlitzkapillare wegen der Druckmessung in der Kapillare selbst und bei den Rundkapillaren müssen sie aus den Ergebnissen ermittelt werden. Die Wandschubspannung ist bei einer Schlitzkapillare

$$\tau = \frac{\Delta p H}{2L}$$

und bei einer Rundkapillare

$$\tau = \frac{\Delta p R}{2L}$$

H ist die Höhe des Schlitzes, *R* der Radius der Bohrung, *L* die Länge der Kapillare und Δp der zwischen Eingang und Ausgang wirksame Druckunterschied. Diese Druckdifferenz wird bei der Schlitzdüse aus dem Unterschied zwischen den Werten an den Meßstellen P6 und P7 in der Kapillare gebildet, während bei den Runddüsen der Druck an der Meßstelle P5 vor dem Düseneinlauf ausgewertet wird.

Das Schergeschwindigkeitsgefälle ist bei einer Schlitzkapillare

$$\vartheta' = \frac{6Q}{BH^2}$$

mit *B* als Breite des Schlitzes und bei einer Rundkapillare

$$\vartheta' = \frac{4Q}{\pi R^3}.$$

Der Mengenstrom *Q* wird hier als das durchgesetzte Volumen/Zeit eingesetzt. Dazu muß das zunächst ermittelte Gewicht mit dem spezifischen Volumen bei der entsprechenden Temperatur multipliziert werden. Der Einfluß des Druckes auf das spezifische Volumen wurde hierbei nicht berücksichtigt, zumal in den Kapillaren kein konstanter Druck vorliegt.

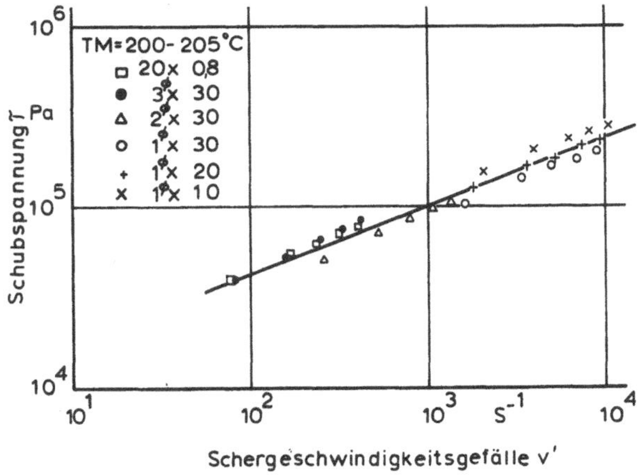

Abb. 6. Vgl. Text

Abb. 6 zeigt im doppelt-logarithmischen System die Abhängigkeit der Schubspannung τ vom Schergeschwindigkeitsgefälle ϑ' für das mittlere Temperaturprogramm mit einer Massetemperatur von 200–205 °C beim Düseneintritt. Mit den eingesetzten Kapillaren läßt sich der beim Extrudieren übliche Schergeschwindigkeitsbereich von 10^2–$10^4\,s^{-1}$ ausfahren. Die Schlitzkapillare und die Rundkapillare mit 3 mm Bohrung ergeben fast identische Werte. Eine rechnerisch mögliche Einlaufkorrektur für die Rundkapillare würde zu geringeren Schubspannungen führen, was durch die Übereinstimmung mit den Ergebnissen der Schlitzkapillare allerdings nicht gerechtfertigt ist. Die gute Plastifizierung im Schneckenzylinder und die Druckmessung direkt vor dem Düseneintritt führen bei der Rundkapillare mit dem größten Durchmesser zu ausreichend genauen Ergebnissen. Die höheren Schergeschwindigkeiten werden mit den Kapillaren mit 1 mm Durchmesser erreicht. Dabei ist die Schubspannung bei den kürzeren Düsen höher. Daraus läßt sich auf grafischem Wege die Einlaufkorrektur E ermitteln. Bei dem untersuchten Polyäthylen beträgt dieser Faktor zwischen 5 und 6. Die korrigierte Kapillarenlänge $L1$ läßt sich dann mit der Gleichung

$$L1 = (L/R + E)\,R$$

ermitteln. Eine solche Korrektur führt zu geringeren Schubspannungen, die etwas unter den Werten für die Düse mit 30 mm Kapillarlänge liegen. Der grundlegende Verlauf der Versuchsergebnisse wird dadurch jedoch nicht verändert. Zusammenfassend läßt sich die Abhängigkeit der Schubspannung vom Schergeschwindigkeitsgefälle im doppelt-logarithmischen System als eine Gerade darstellen. Die Möglichkeiten zu einer genaueren Auswertung sind jedoch vorhanden.

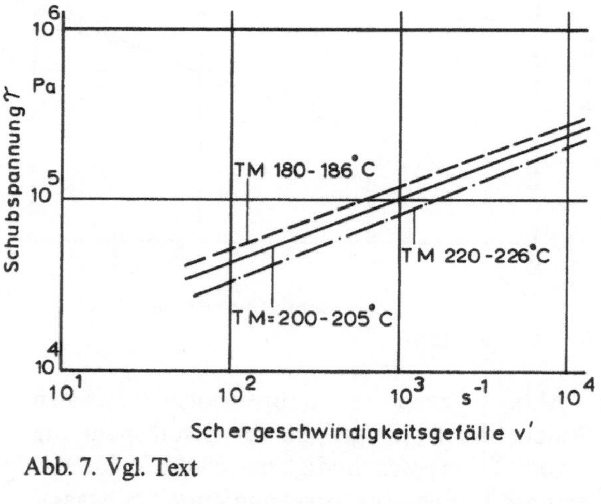

Abb. 7. Vgl. Text

Der Einfluß der Massetemperatur auf den Zusammenhang zwischen Schubspannung und Schergeschwindigkeitsgefälle ist in Abb. 7 dargestellt. Für die 3 untersuchten Temperaturbereiche ist jeweils die zusammenfassende Gerade eingezeichnet. Die Schubspannung sinkt erwartungsgemäß mit höherer Temperatur. Mit steigendem Schergeschwindigkeitsgefälle nimmt der Einfluß der Temperatur jedoch erheblich ab. Das exakte Einhalten bestimmter Temperaturen hat also bei praxisgerechten Schergeschwindigkeiten keine so große Bedeutung wie bei Viskositätsmessungen mit niedrigen Geschwindigkeiten.

Aus der Wandschubspannung τ und dem Schergeschwindigkeitsgefälle ϑ' wird die kinematische Viskosität μ ermittelt.

$$\mu = \tau/\vartheta'.$$

Diese bei plastifizierten Kunststoffen als scheinbare Viskosität bezeichnete Größe kennzeichnet das Verformungsverhalten der Werkstoffe bei der Kunststoffverarbeitung. In Abb. 8 sind die

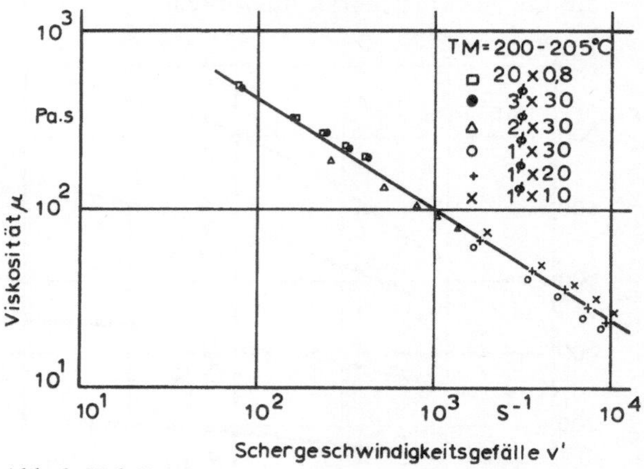

Abb. 8. Vgl. Text

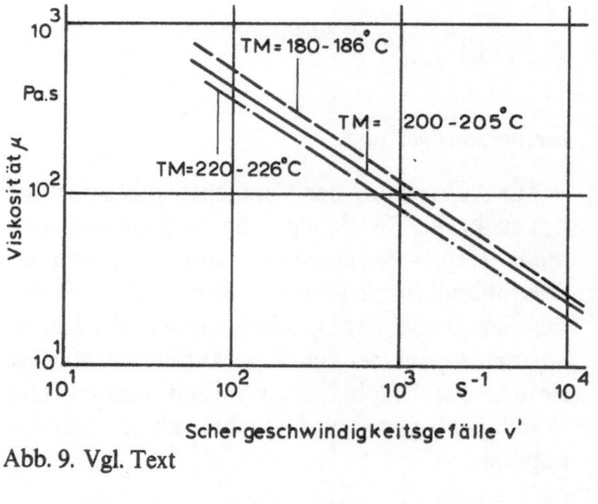

Abb. 9. Vgl. Text

Versuchsergebnisse, abhängig vom Schergeschwindigkeitsgefälle, wieder für das mittlere Temperaturprogramm, aufgetragen. Für die Einlaufkorrektur der Runddüsen gilt das bei der Erläuterung der Schubspannung Gesagte. Im oberen Schergeschwindigkeitsbereich verschiebt sich dann die scheinbare Viskosität etwas unter die Werte für die Kapillare mit 1 mm Durchmesser und 30 mm Länge. Im Schergeschwindigkeitsbereich von 10^2–$10^4\,s^{-1}$ sinkt die scheinbare Viskosität von 500 auf 25 Pa s, d. h. um den Faktor 20. Im doppelt-logarithmischen System ist dieser Zusammenhang praktisch linear. Den Viskositätsverlauf bei verschiedenen Massetemperaturen zeigt Abb. 9. Auch hier nimmt der Einfluß der Temperatur mit steigender Geschwindigkeit ab. Mit der Schergeschwindigkeit läßt sich die scheinbare Viskosität erheblich stärker verändern als mit der Temperatur.

Zusammenfassung

Ein vielseitiges Meßsystem ermöglicht die differenzierende Prüfung des Verhaltens von Kunststoffen beim Plastifizieren, die kontinuierliche Ermittlung der Eigenschaften des plastifizierten Materials und Prüfungen an Produkten, die durch Extrudieren hergestellt werden können. Sämtliche Meßeinrichtungen lassen sich aus einem Baukastensystem zusammenstellen, das aus messenden Antriebseinheiten verschiedener Leistungsstufen und den austauschbaren Plastifizieraggregaten und Nachfolgeeinrichtungen besteht. Eine Laborausrüstung mit diesen Geräten ist für vielfältige Untersuchungen der verarbeitungstechnischen Eigenschaften von Thermoplasten, Duroplasten, Elastomeren und Kautschuk einzusetzen.

Anschrift des Verfassers:

Dr.-Ing. *Arne Rothenpieler*
Anwendungstechnische Abteilung der Brabender OHG
D-4100 Duisburg (Deutschland)

Rheol. Acta **13**, 28–32 (1974)

From the Department of Electronics and Electrical Engineering, The University Glasgow, G12 8LT. (U.K.)

A low frequency torsional rheometer

By G. Harrison

With 3 figures

(Received October 27, 1972)

Introduction

Previous work in the author's laboratory has been largely concerned with the study of the viscoelastic properties of supercooled liquids and polymers of relatively low molecular weight (1, 2). Measurements on these materials can conveniently be made in the frequency range 10^4 to 10^9 Hz. In extending the scope of the research to include the study of high molecular weight polymers it has been necessary to develop apparatus to enable results to be obtained at lower frequencies. This paper describes an instrument which has been designed to measure the components of the complex shear modulus in the frequency range 10^{-3}–50 Hz.

The rheometer is similar in principle to those described by *Duiser* (3) and *Den Otter* (4) but instead of holding the sample between two coaxial cylinders it uses either a cone and plate or a parallel plate system. This arrangement was chosen to simplify the sample preparation and handling procedure. The upper plate of the system is fixed, the cone (or lower plate) being driven in sinusoidal torsional oscillation via a torsion wire, thus subjecting the sample between the plate and the cone to sinusoidal shear. The viscoelastic properties of the sample, the inertia of the moving system and the stiffness of the torsion wire result in a difference in the amplitude and phase of the motions at the ends of the torsion wire. Measurement of the amplitude of the angular displacements of the cone and the driven end of the torsion wire, and of the phase difference between them, enables the components of the shear modulus to be determined.

Theory

The motion of this type of system has been analysed by *Dekking* (5). If the cone oscillates sinusoidally with amplitude θ_2 and angular frequency ω, the torque exerted by a sample having a complex shear modulus $G^*(j\omega) = G'(\omega) + jG''(\omega)$ is given by $G^*(j\omega)\,D_2\,\theta_2 \exp j\omega t$. The geometry factor D_2 is given for the cone and plate system by

$$D_2 = 2\pi R^3/3\psi \qquad [1]$$

where R is the radius of the cone and the semi-vertical angle of the cone is equal to $(\pi/2 - \psi)$. The torque combined with the effect of the inertia of the moving system equals the torque exerted by the torsion wire. Denoting the phase angle between the displacements at the ends of the wire by φ, the angular displacement of the driven end of the torsion wire is given by $\theta_1 \exp j(\omega t + \varphi)$. The torque exerted on the cone is thus $D_1(\theta_2 \exp j(\omega t + \varphi) - \theta_2 \exp j\omega t)$ where D_1 is the torsional stiffness coefficient of the torsion wire. The equation of motion of the cone is then given by eq. [2]

$$(-\omega^2 I + D_2 G^*(j\omega))\,\theta_2 \exp j\omega t = D_1(\theta_1 \exp j(\omega t + \varphi) - \theta_2 \exp j\omega t). \qquad [2]$$

Then

$$-\omega^2 I + D_2 G'(\omega) + jD_2 G''(\omega) = D_1\,[\theta_1/\theta_2\,(\cos\varphi + j\sin\varphi) - 1].$$

Equating the real and imaginary parts, the components of the shear modulus are given by eqs. [3] and [4]

$$G'(\omega) = \frac{D_1}{D_2}\left[\frac{\theta_1}{\theta_2}\cos\varphi - 1\right] + \omega^2\frac{I}{D_2} \qquad [3]$$

$$G''(\omega) = \frac{D_1}{D_2}\frac{\theta_1}{\theta_2}\sin\varphi. \qquad [4]$$

The determination of the shear modulus therefore requires a knowledge of the instrument parameters D_1, D_2 and I, for the particular cone and torsion wire used, and the measurement of the amplitude ratio θ_1/θ_2 and the phase angle φ.

Description of apparatus

A schematic diagram of the apparatus is shown in fig. 1. The sample to be studied is placed between the fixed plate and the cone. Alternatively, the cone may be replaced by a flat plate to give a parallel plate system. The geometry of the sample may be controlled by adjusting the vertical position of the fixed plate with a screw. The cone is mounted on a rotor which is supported by an air bearing. This allows friction-free rotation while controlling the vertical and lateral position of the rotor. Both the cone, and the mounting block for the fixed plate, are fitted with adjusting screws, to permit alignment to the axis of rotation of the rotor.

The system by which the lower end of the torsion wire is driven consists of a quadrant, to which the torsion wire is attached, linked to two vibrators through flexible steel strips. This arrangement is shown in fig. 2. The quadrant is supported and constrained to move in an arc by four crossed leaf springs. This system allows friction-free rotation of the quadrant for several degrees about an axis at the intersection of the planes containing the leaf springs. The sinusoidal drive voltage for the vibrators is obtained from a Test Waveform Generator (Feedback Ltd., type TWG 300) followed by a d.c. amplifier. A low-pass filter is used to reduce the harmonic distortion of the output of the generator to less than 0.2%. The frequency range of the driving system is from 10^{-3} to above 10^2 Hz.

The angular displacements at the ends of the torsion wire are determined by using diffraction gratings. Each measuring assembly consists of a section of grating mounted on an arm at a radius of 4″ from the torsion wire, together with a reading head, containing a reference grating, which is mounted on the frame of the instrument. The two gratings are adjusted so that the rulings are parallel and lie along a radius, and the gratings are separated by a distance of approximately 0.1 mm. Twisting of the torsion wire then results in a cyclic variation of the intensity of light passing through the two gratings as one moves relative to the other.

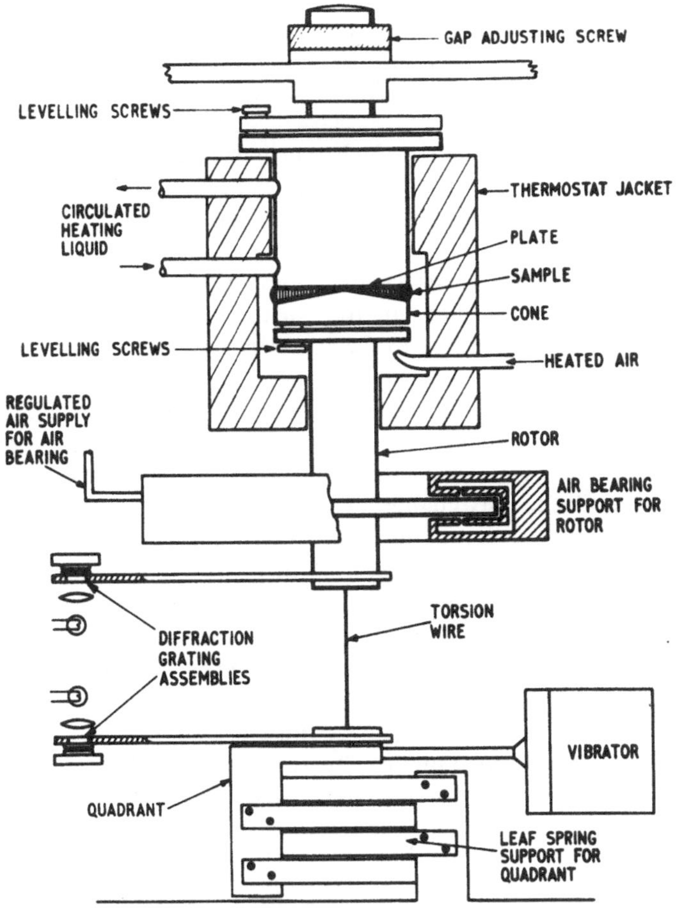

Fig. 1. Schematic diagram of rheometer

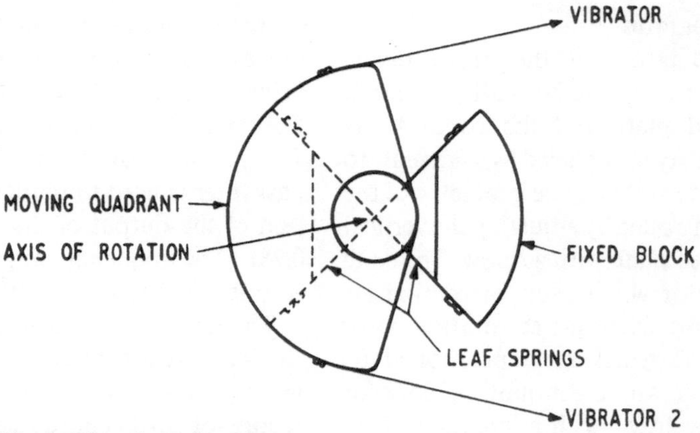

Fig. 2. Plan view of quadrant

The reading head contains a lamp and focussing lens together with the reference grating. This consists of four small sections of grating about 5 mm square, each successive section having the rulings displaced from the preceding section by a quarter of the line spacing. Photodiodes mounted beneath these sections detect the light transmitted through the moving grating and the reference grating. By suitably combining the outputs of the diodes, two signals are obtained which each vary sinusoidally but differ in phase by $\pm 90°$, the sign of the angle depending on the direction of travel of the moving grating.

The gratings and reading heads were supplied by Whitwell Electronics Ltd., Glasgow, to a design by the National Engineering Laboratory, East Kilbridge, Glasgow.

The two signals are amplified and then shaped in *Schmitt* trigger circuits. The resulting square waves may be regarded as corresponding to the two digits of a binary number having 4 values. The sequence of counting – 1234 or 1432 – then depends on the direction of the moving grating, and will change at the extremes of the oscillatory motion of the moving grating. This change in the counting sequence is used to generate signals which open and close the input gate of a digital counter. Pulses produced at each change of the binary number are counted while the grating is moving from one extremity of its travel to the other. The counter reading is thus proportional to the peak to peak amplitude of the motion. The gratings are ruled with 2000 lines per inch and four pulses are generated for a movement equal to one line spacing. The resolution of the system is therefore 1.25×10^{-4} in. The counter reading is also proportional to the angular displacement if the motion is restricted to a few degrees. With the gratings mounted at a radius of 4″ the angular resolution is 3×10^{-5} radian.

The phase angle φ between the motions at the ends of the torsion wire is determined from a comparison of the signals derived from the two diffraction grating assemblies. The signals are either displayed on an oscilloscope or recorded on a high speed ultra-violet oscillograph. With both methods the phase angle can be determined to within $\pm 1°$, but oscillograph recording of the signals is considerably more convenient at the lower frequencies.

The temperature of the sample is controlled primarily by heating the block supporting the upper plate. Fluid from a thermostatted bath is circulated through the hollow block. The cone is separately heated by blowing air at a controlled temperature onto the lower part of the cone. An insulating jacket surrounds the cone and plate assembly. This arrangement enables the temperature of the plate and cone to be controlled independently. The temperature of the sample can be controlled to within ± 0.01 °C with a differential across the sample of the same order, over the range 20–90 °C. An extension of this temperature range to 250 °C is proposed, using foil heaters cemented to the plate mounting block.

The temperatures of the plate and cone are measured using a platinum resistance thermometer in conjunction with a type VLF 51 A resistance comparator bridge (Rosemount Engineering Co. Ltd.). The platinum element has an overall diameter of 0.094 in and can be inserted into holes drilled close to the surfaces of the plate and cone.

Results and discussion

To determine the moment of inertia I and the torsional stiffness coefficient D_1 the quadrant is clamped, to prevent movement of the lower end of the torsion wire, and the cone allowed to vibrate with no sample present. Rings having different moments of inertia are added to the moving system and the period of free oscillation measured in each case. The period of oscillation τ is given eq. [5],

$$\tau^2 = \frac{4\pi^2}{D_1}(I + \Delta I) \qquad [5]$$

where ΔI is the moment of inertia of the added ring, calculated from its mass and dimensions. The measured values of τ are fitted to eq. [5] and the coefficients D_1 and I determined using the method of least squares. The values obtained are estimated to be accurate to better than $\pm 0.1\%$.

The coefficient D_2 is calculated from the sample geometry using eq. [1]. The cone angle ψ is determined by measuring the profile of the cone across a diameter, using a dial gauge mounted on a travelling microscope, and is estimated to be accurate to $\pm 0.5\%$. It is difficult to ensure that the sample completely fills the gap between the cone and the plate, as drainage of the sample over the edge of the cone usually results in the free surface of the sample being concave. In these cases the minimum diameter of the sample is measured using a micrometer eyepiece fitted to a travelling microscope in order to determine the radius R. The radius can be determined to an accuracy of $\pm 0.1\%$, giving an overall error in the coefficient D_2 of less than $\pm 1\%$.

To avoid friction between the cone and the plate, the tip of the cone is truncated by approximately 20 μm. After placing the sample on the cone, the plate is lowered to contact the sample. When thermal equilibrium has been attained the plate is lowered further until electrical contact is made between the cone and the plate. The plate is then raised by an amount equal to the truncation. The motion of the plate is monitored by a dial gauge mounted on the frame of the instrument.

The frequency range of the instrument is limited at frequencies below resonance by the decreasing value of the phase angle φ, and at frequencies above resonance by the reduction in the amplitude of the cone oscillation. The phase angle can be measured to an accuracy of $\pm 1°$, and the oscillation amplitudes to ± 1 count ($\pm 3 \times 10^{-5}$ rad), giving practical limits to the range of the instrument of a minimum phase angle of about 15°, and a minimum oscillation amplitude of 30 counts (10^{-2} rad). Measurements were made on a standard oil (Cannon Instrument Co. No. S-30000) with a nominal viscosity of 50 N s m^{-2} (500

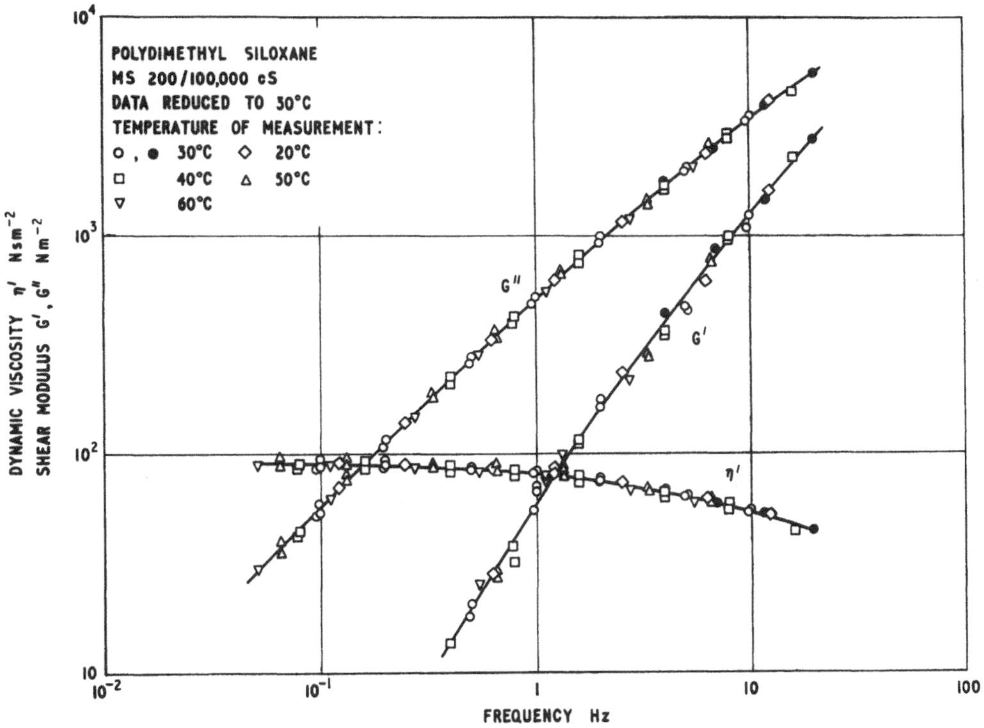

Fig. 3. Variation of G', G'' and η' with frequency for polydimethyl siloxane MS200/100000cS at 30 °C

Poise) at 30 °C using a 50 mm diameter cone with an angle ψ of 2°, the resonant frequency of the system was 2.54 Hz. The results obtained at frequencies between 0.2 Hz and 10 Hz gave values of dynamic viscosity $\eta'(\omega)$ $(= G''(\omega)/\omega)$ which were independent of frequency and which were within $\pm 3\%$ of the viscosity value determined using a suspended level viscometer. The error in the values of $G'(\omega)$ depends on the relative magnitudes of the terms in eq. [3] but tends to increase with frequency. The maximum value found in the measurements on the standard oil, which should not exhibit viscoelastic behaviour in this frequency range, was 52 Nm^{-2} at a frequency of 10 Hz.

Results obtained on a polydimethyl siloxane of nominal kinematic viscosity 0.1 $m^2 sec^{-1}$ (MS 200/100000) are shown in fig. 3. Data obtained at temperatures in the range 20–60 °C have been reduced to 30 °C following the method of reduced variables (6). The values shown with open symbols were obtained with a torsion wire giving a resonant frequency of 2.54 Hz, those shown with a solid symbol were obtained using a torsion wire giving a resonant frequency of 11.4 Hz.

The low frequency limiting value of the dynamic viscosity is equal to the steady flow viscosity determined in a capillary tube viscometer (94 $N s m^{-2}$) within experimental error.

The instrument has been found convenient and simple to operate, and the amplitude measuring system has proved entirely satisfactory. No calibration is required provided that both gratings are mounted at the same radius. The angular resolution enables amplitude measurements to be made with an accuracy of better than $\pm 1\%$ while keeping the amplitude of shear in the sample below 0.1 so that a linear stress-strain relation is maintained.

Support for this work has been provided by a grant from the Science Research Council, and this assistance is gratefully acknowledged.

Summary

An instrument has been developed for measuring the viscoelastic behaviour of polymer melts at low frequencies, in the range 10^{-3} to 50 Hz. The sample is contained between a cone and a fixed plate, or between parallel plates. The moving member is driven in torsional oscillation through a torsion wire. The amplitude of the resulting oscillation is compared in amplitude and phase with the driven end of the torsion wire. The amplitudes are measured digitally using optical diffraction gratings, and either an oscilloscope or a high-speed ultra-violet recorder is used to determine the phase angle between the two signals. The moving member is supported on an air bearing, which provides a very low friction support with a high degree of positional control thus giving a well defined sample geometry. The torsion wire is driven using a vibrator with a d.c. drive amplifier fed from a very low frequency oscillator. The sample temperature is controlled to better than 0.01 °C, with temperature gradients across the sample of a similar order of magnitude. The temperature range of the instrument is from —50 °C to +200 °C.

The angular resolution of the measuring system is 3×10^{-5} radius, so that an accuracy of better than $\pm 1\%$ in the amplitude measurements can be obtained with the amplitude of shear in the sample kept sufficiently low that a linear stress-strain relation is maintained.

References

1) *Barlow, A. J.* and *A. Erginsav*, Proc. Roy. Soc. **A 327**, 175–190 (1972).
2) *Barlow, A. J., M. Day, G. Harrison, J. Lamb*, and *S. Subramanian*, Proc. Roy. Soc. **A 309**, 497–520 (1969).
3) *Duiser, J. A.*, Thesis, Leiden (1965).
4) *Den Otter, J. L.*, Rheologica Acta **8**, 355–363 (1969).
5) *Dekking, P.*, Thesis, Leiden (1961).
6) *Ferry, J. D.*, Viscoelastic properties of polymers. (New York 1970).

Author's address:

Dr. *G. Harrison*
Dept. of Electronics and Electrical Engineering
The University
Glasgow, G12 8LT (UK)

Rheol. Acta 13, 33–39 (1974)

From the Department of Chemistry, Northwestern University, Evanston (U.S.A.) and Naval Research Laboratory, Chemistry Division, Washington, D.C. (U.S.A.)

Some graphical methods for the analysis of mechanical and dielectric relaxation data

By D. L. Hunston

With 3 figures and 2 tables

(Received October 27, 1972)

Introduction

A fundamental characteristic of viscoelastic materials is that their mechanical properties are time dependent or, in the case of sinusoidal deformation, frequency dependent. Moreover, it is often the primary objective of an experiment to analyse this time or frequency dependence and fit it with a mathematical model having one or more parameters. A number of methods have been developed for this kind of analysis. They range from very simple techniques (1, 2) to more complicated computerized procedures (3). In this paper we will be concerned only with simple graphical methods. Such methods find their greatest application in the initial stages of an analysis where they can provide a quick and easy way to evaluate parameters as well as a visual indication of how reliable these evaluations are.

A number of graphical techniques have been suggested previously for the analysis of specific kinds of relaxation data. The two most widely used are those proposed by *Mikhailov* (1) and by *Lamb* and *Huddart* (2). These techniques are designed to evaluate longitudinal wave propagation data when only one relaxation time is observed. In this paper we will show that these methods are actually special cases of a more general technique which can be applied to many different types of relaxation data. Furthermore, this general method can be extended to situations where more than one relaxation time is present, and thus relaxation spectra can be studied without assuming a continuous distribution.

Longitudinal wave propagation

The analysis of longitudinal wave propagation begins with the equations given by *Herzfeld* and *Litovitz* (4) for velocity, V, and excess attenuation, α'

$$V^2 - V_0^2 = \frac{(V_\infty^2 - V_0^2)(\tau_a^2)(\omega^2)}{1 + (\tau_a^2)(\omega^2)} \qquad [1]$$

$$(2V^3\alpha') = \frac{[(V_\infty^2 - V_0^2)/\tau_a](\tau_a^2)(\omega^2)}{1 + (\tau_a^2)(\omega^2)} \qquad [2]$$

where V_0 and V_∞ are the low frequency and high frequency limiting velocities, and ω is the angular frequency. The apparent relaxation time, τ_a, is related to the actual relaxation time, τ, by

$$\tau_a = \frac{C_V - C'}{C_V}\tau \qquad [3]$$

where C_V is the specific heat at constant volume and C' is the relaxing specific heat. The terms in eqs. [1] and [2] are grouped so as to indicate the mathematical similarity between these equations and the following general equation (corresponding variables are listed in table 1).

$$y = \frac{abx}{1 + bx}. \qquad [4]$$

Here x and y represent experimentally measured quantities while a and b correspond to the parameters we would like to determine.

Expressions similar to eq. [4] are found in many areas of science and have received a great deal of study (5–10). They are often evaluated by graphical methods (5–7) using one of the three "linear" forms for eq. [4].

$$\frac{1}{y} = \frac{1}{ab}\frac{1}{x} + \frac{1}{b} \qquad [5]$$

$$\frac{x}{y} = \frac{1}{b}x + \frac{1}{ab} \qquad [6]$$

$$\frac{y}{x} = -by + ab \qquad [7]$$

Table 1. Correspondence between variables

General equation	Shear modulus, G^*		Velocity and attenuation	
	G'	G''	V	α'
y	G'	$\omega G''$	$(V^2 - V_0^2)$	$2V^3\alpha'$
a	G_∞	G_∞/τ	$(V_\infty^2 - V_0^2)$	$(V_\infty^2 - V_0^2)/\tau_a$
b	τ^2	τ^2	τ_a^2	τ_a^2
x	ω^2	ω^2	ω^2	ω^2
a_i	$G_{\infty i}$	$G_{\infty i}/\tau_i$	$(V_{\infty i}^2 - V_{0i}^2)$	$(V_{\infty i}^2 - V_{0i}^2)/\tau_{ai}$
b_i	τ_i^2	τ_i^2	τ_{ai}^2	τ_{ai}^2

It is clear that a graph using as coordinates $(1/y)$ vs. $(1/x)$, (x/y) vs. x, or (y/x) vs. y will be a straight line whose slope and intercepts can be used to calculate a and b.

Using the analogy between the variables in eqs. [1], [2], and [4] we can determine the corresponding "linear" forms of eqs. [1] and [2]. This gives us six coordinate systems in which data can be graphically evaluated: $1/(V^2 - V_0^2)$ vs. $1/\omega^2$, $\omega^2/(V^2 - V_0^2)$ vs. ω^2, $(V^2 - V_0^2)/\omega^2$ vs. $(V^2 - V_0^2)$, $1/2V^3\alpha'$ vs. $1/\omega^2$, $\omega^2/2V^3\alpha'$ vs. ω^2, and $2V^3\alpha'/\omega^2$ vs. $2V^3\alpha'$. This list includes several new coordinate systems as well as those proposed by *Mikhailov*, $\omega^2/2V^3\alpha'$ vs. ω^2, and by *Lamb* and *Huddart*, $2V^3\alpha'/\omega^2$ vs. $2V^3\alpha'$. When experimental results are plotted in one of these coordinate systems, a straight line indicates that the data are consistent with a single relaxation time. Furthermore the slope and intercepts of this line can be used to calculate V_∞ and τ_a.

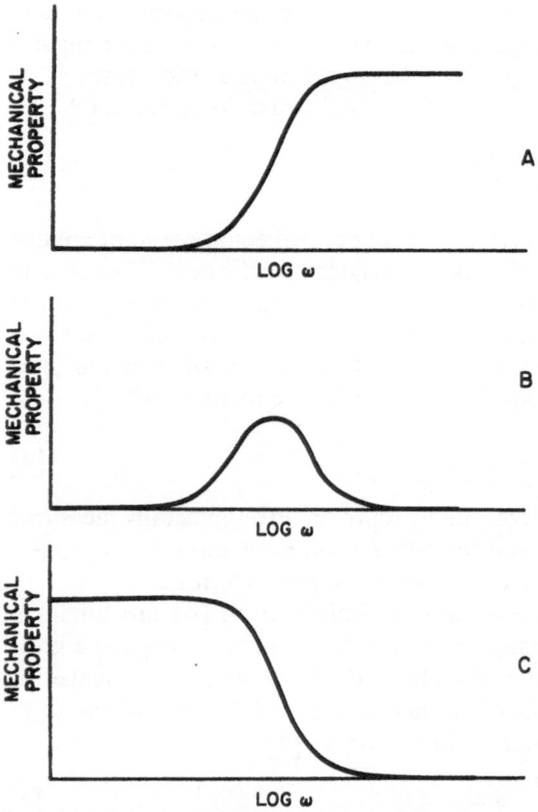

Fig. 1. Many mechanical properties exhibit a frequency dependence similar to one of those shown here. The mechanical properties illustrated in parts *A* and *B*, for example, might be the storage and loss components of the complex shear modulus while part *C* might correspond to the dynamic viscosity

Shear modulus

In addition to longitudinal wave propagation, this analogy with eq. [4] can be used to evaluate many other types of data; in fact, any property which exhibits the well known "S" shaped or "Bell' shaped curve in a semi-logarithmic plot (fig. 1) is a possible candidate. This includes many of the common mechanical (11) and dielectric (12) properties. As an example consider the complex shear modulus, G^*. The in phase component, G', and the out of phase component, G'', can be written for a simple *Maxwell* model as

$$G' = \frac{G_\infty(\tau^2)(\omega^2)}{1 + (\tau^2)(\omega^2)} \qquad [8]$$

and

$$(\omega G'') = \frac{(G_\infty/\tau)(\tau^2)(\omega^2)}{1 + (\tau^2)(\omega^2)} \qquad [9]$$

where G_∞ is the high frequency limiting modulus. The correspondence with eq. [4] is clear (table 1) and can be used to determine six graphical coordinate systems. The resulting graphs are shown in fig. 2 along with the values for the slopes and intercepts. Such graphs not only let us evaluate G_∞ and τ but also the limiting dynamic viscosity, η.

$$\eta = G_\infty \tau . \qquad [10]$$

One factor which must be considered when applying these techniques is the bias that may be given to some data points by virtue of the coordinate system used (3). The selection of the straight line that best fits the data depends on the spacing of the points and this, in turn, is a function of the coordinate system. For a given type of experiment, then, one coordinate system may provide a much better display of the data than another. This fact has often been overlooked in previous work where the *Mikhailov* or *Lamb-Huddart* coordinates were used (3). Fortunately, detailed statistical studies of the weighting effect have been made in connection with the use of eqs. [5] to [7] in other areas [9, 10] and the results of those studies are equally valid here. While specific recommendations depend on the nature of the data being considered, it can be concluded that a coordinate system based on eq. [7] usually provides the best presentation of the data (fig. 2C and 2F). Coordinate systems based on eq. [6] are also good (fig. 2B and 2E) but those based on eq. [5] are less desirable (fig. 2A and 2D).

Multiple relaxation processes

For many mechanical properties more than one relaxation time is required to adequately describe the data. In these cases the graphical coordinate systems will give curves (fig. 3). These curves can still be used to estimate relaxation parameters, however, if the equations involved are mathematically similar to this generalized form of equation (4).

$$y = \sum_{i=1}^{m} \frac{a_i b_i x}{1 + b_i x}. \qquad [11]$$

A recent paper (13) considers graphical representations for this type of equation and the results can be applied here.

To illustrate this type of graphical evaluation we will again take G^* as an example. For a system with m *Maxwell* relaxation times, G' and G''

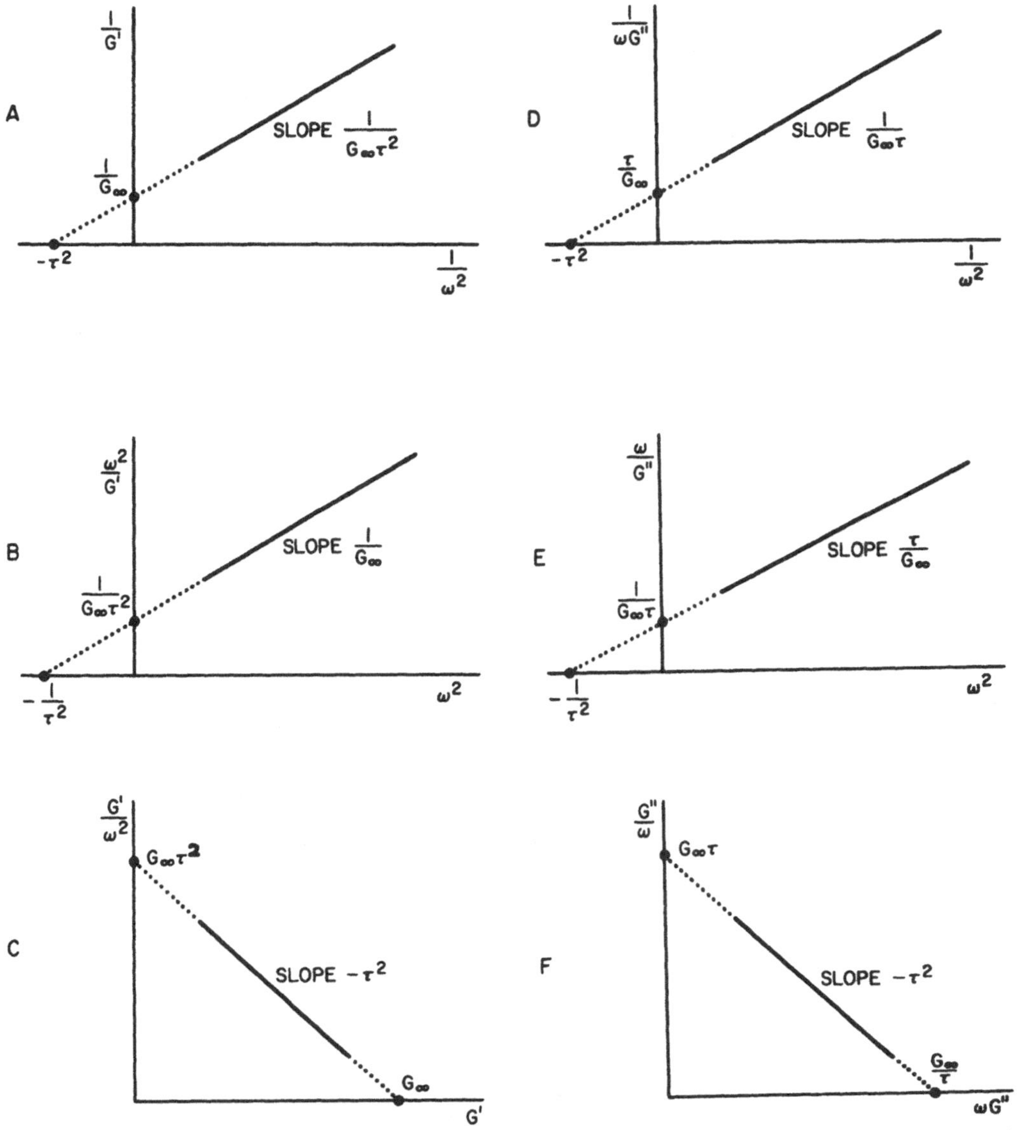

Fig. 2. These graphs show the six linear forms of the equations for the in phase component, G', and the out of phase component, G'', of the complex shear modulus. The slopes and intercepts are indicated

3*

can be written as follows (11)

$$G' = \sum_{i=1}^{m} \frac{G_{\infty i}(\tau_i^2)(\omega^2)}{1 + (\tau_i^2)(\omega^2)} \qquad [12]$$

and

$$(\omega G'') = \sum_{i=1}^{m} \frac{(G_{\infty i}/\tau_i)(\tau_i^2)(\omega^2)}{1 + (\tau_i^2)(\omega^2)} \qquad [13]$$

where $G_{\infty i}$ and τ_i are the ith component of the limiting modulus and the ith relaxation time. The similarity between these expressions and eq. [11] is obvious (table 1). Fig. 3 shows the graphical representations for such a system. When the data is measured over a sufficiently large range of frequencies, the limiting slopes and intercepts shown in this figure can be determined. These graphical quantities are related to the limiting moduli and certain "average" relaxation times. Consequently such graphs can be used to evaluate relaxation parameters from the experimental data. The limiting moduli and "p^{th} order average" relaxation time are defined as follows:

$$G_\infty = \sum_{i=1}^{m} G_{\infty i} \qquad [14]$$

$$\eta = \sum_{i=1}^{m} G_{\infty i}\tau_i \qquad [15]$$

and

$$\langle\tau\rangle_p = \frac{1}{G_\infty} \sum_{i=1}^{m} G_{\infty i}\tau_i^p. \qquad [16]$$

In a similar manner many other relaxation phenomena involve relationships that are mathematically similar to eq. [11]. By using table 2 as an example we can determine what the limiting slopes and intercepts would be in graphical representations of these properties. For example,

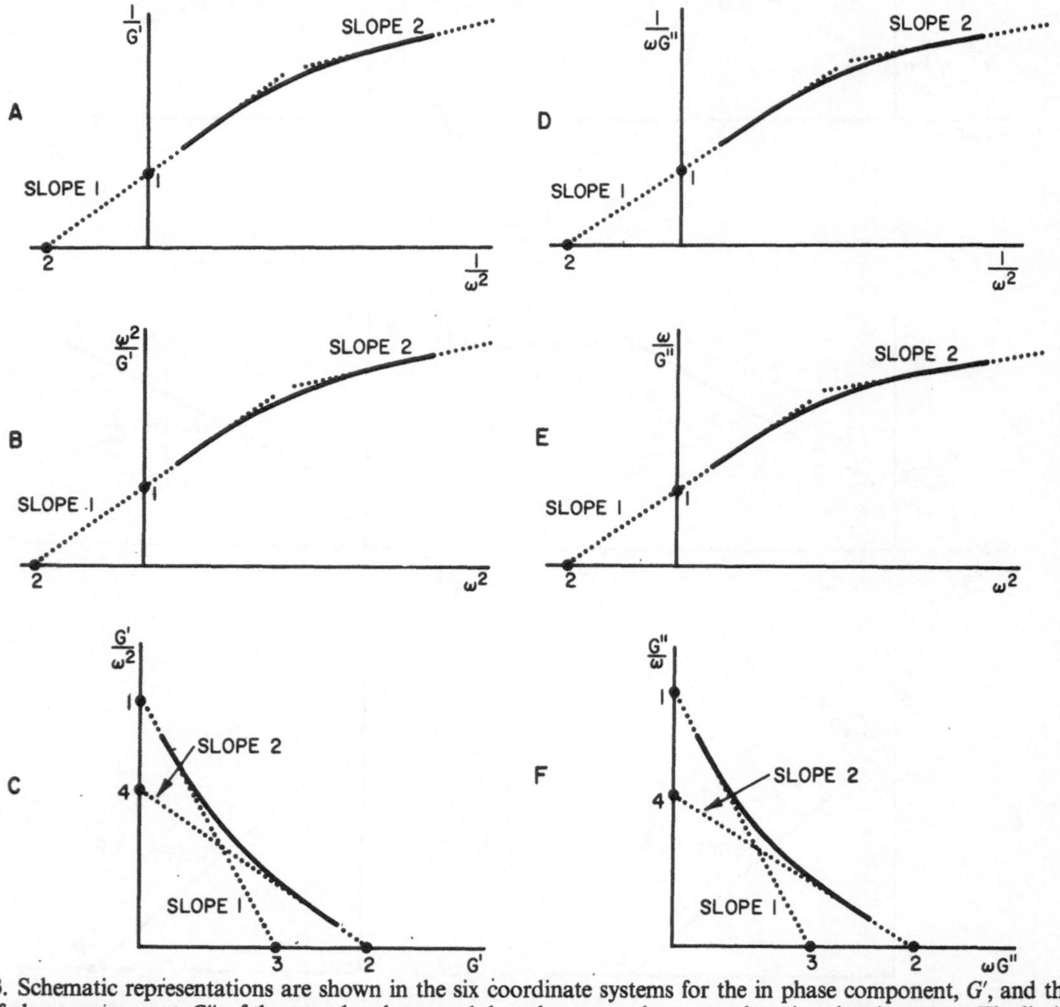

Fig. 3. Schematic representations are shown in the six coordinate systems for the in phase component, G', and the out of phase component, G'', of the complex shear modulus when more than one relaxation time is present. The limiting slopes and intercepts are indicated

table 1 lists the correspondence between the variables for longitudinal wave propagation and those for the complex shear modulus. From this correspondence we can conclude that the limiting slopes and intercepts for longitudinal wave propagation data would be the same as those given in table 2 except that G_∞ and τ would be replaced by $(V_\infty^2 - V_0^2)$ and τ_a respectively.

Table 2. Slopes and intercepts of fig. 3

	Fig. 3A	Fig. 3D
Intercept 1	$1/G_\infty$	$1/G_\infty\langle\tau\rangle_{-1}$
Intercept 2	$-1/\langle\tau\rangle_{-2}$	$-\langle\tau\rangle_{-1}/\langle\tau\rangle_{-3}$
Slope 1	$\langle\tau\rangle_{-2}/G_\infty$	$\langle\tau\rangle_{-3}/G_\infty(\langle\tau\rangle_{-1})^2$
Slope 2	$1/G_\infty\langle\tau\rangle_2$	$1/G_\infty\langle\tau\rangle$
	Fig. 3B	**Fig. 3E**
Intercept 1	$1/G_\infty\langle\tau\rangle_2$	$1/G_\infty\langle\tau\rangle$
Intercept 2	$-\langle\tau\rangle_2/\langle\tau\rangle_4$	$-\langle\tau\rangle/\langle\tau\rangle_3$
Slope 1	$\langle\tau\rangle_4/G_\infty(\langle\tau\rangle_2)^2$	$\langle\tau\rangle_3/G_\infty(\langle\tau\rangle^2$
Slope 2	$1/G_\infty$	$1/G_\infty\langle\tau\rangle_{-1}$
	Fig. 3C	**Fig. 3F**
Intercept 1	$G_\infty\langle\tau\rangle_2$	$G_\infty\langle\tau\rangle$
Intercept 2	G_∞	$G_\infty\langle\tau\rangle_{-1}$
Intercept 3	$G_\infty(\langle\tau\rangle_2)^2/\langle\tau\rangle_4$	$G_\infty(\langle\tau\rangle)^2/\langle\tau\rangle_3$
Intercept 4	$G_\infty/\langle\tau\rangle_{-2}$	$G_\infty(\langle\tau\rangle_{-1})^2/\langle\tau\rangle_{-3}$
Slope 1	$-\langle\tau\rangle_4/\langle\tau\rangle_2$	$-\langle\tau\rangle_3/\langle\tau\rangle$
Slope 2	$-1/\langle\tau\rangle_{-2}$	$-\langle\tau\rangle_{-1}/\langle\tau\rangle_{-3}$

Analysis of average relaxation times

The average τ's are useful in that they provide a means to describe the distribution of relaxation times. This was illustrated recently by *Montrose* and *Litovitz* (14) when they used the expression

$$J_\infty\left[\frac{\langle\tau\rangle_2}{(\langle\tau\rangle)^2} - 1\right] \qquad [17]$$

as a measure of the width of the distribution for shear relaxation times. J_∞ is the high frequency limiting compliance. Actually their expression is closely related to the standard deviation, s.

$$s^2 = \langle\tau\rangle_2 - (\langle\tau\rangle)^2. \qquad [18]$$

In other recent papers the averages themselves are given theoretical significance. *Gerhard Schwarz* (15), for example, has used three average relaxation times ($p = 1$, -1, and -2) in his discussion of "Kinetic Analysis by Chemical Relaxation Methods".

An interesting special case is obtained when only two relaxation times are present. For such a system we can obtain the parameters for both relaxations from the slopes and intercepts of the graphical representations. When the relaxation times are close together, it is obvious that the exact solutions for the slopes and intercepts must be used for the analysis. When the relaxation times are considerably different, however, one might expect the limiting slope and intercepts at low frequencies to be characteristic of the slower relaxation while the high frequency limits would describe the faster relaxation.

To test the accuracy of this intuitive supposition, consider fig. 2C when

$$m = 2 \qquad [19]$$

and

$$G_{\infty 1} = G_{\infty 2} = 1 \qquad [20]$$

The exact values are

Intercept 1: $\tau_1^2 + \tau_2^2$ [21]

Intercept 2: 2 [22]

Intercept 3: $(\tau_1^2 + \tau_2^2)^2/(\tau_1^4 + \tau_2^4)$ [23]

Intercept 4: $4\tau_1^2\tau_2^2/(\tau_1^2 + \tau_2^2)$ [24]

Slope 1: $-(\tau_1^4 + \tau_2^4)/(\tau_1^2 + \tau_2^2)$ [25]

Slope 2: $-2\tau_1^2\tau_2^2/(\tau_1^2 + \tau_2^2)$. [26]

If $\tau_1 > 10\tau_2$, the slope and intercepts of the low frequency limiting line are what one would expect for τ_1 only

Intercept 1: τ_1^2 [27]

Intercept 3: 1 [28]

Slope 1: $-\tau_1^2$. [29]

Furthermore, the values for the high frequency limiting line are independent of τ_1 (although they involve $G_{\infty 1}$).

Intercept 4: $4\tau_2^2$ [30]

Intercept 2: 2 [31]

Slope 2: $-2\tau_2^2$. [32]

They are not, however, what would be expected for τ_2 alone. A similar situation exists with the other graphical representations shown in fig. 3.

We can see then that these representations can be used to determine the relaxation parameters when there are two relaxation times. It is advisable however to begin with the exact solutions for the slopes and intercepts even when the relaxation times are widely separated.

Conclusions

Six graphical coordinate systems for evaluating many kinds of relaxation data are proposed. For material properties with only one relaxation time, the data points will fall on straight lines. The slopes and intercepts of these lines can be used to determine the relaxation time and other parameters describing the relaxation process. The advantages and disadvantages for using each of these coordinate systems to represent data are known and can be applied to their use in analyzing relaxation phenomena. When a material property has more than one relaxation time, the graphical representations are curves. The limiting slopes and intercepts in these graphs can be used to estimate relaxation parameters and to characterize the distribution of relaxation times in terms of certain average values of τ.

Acknowledgements

The author wishes to thank the Office of Naval Research whose support made presentation of this work possible. Additional support was provided by a Postdoctoral Fellowship from the National Institute of General Medical Science (1 F02 GM46385-01 A1) and grant (GB7122) from the National Science Foundation to Dr. *I. M. Klotz*.

Summary

Three types of coordinate systems for the analysis of relaxation data are examined. When the data can be described by a single relaxation time, the graphs are straight lines. The slopes and intercepts of these lines can be used to evaluate the parameters which characterize the process, such as relaxation time. The advantages and disadvantages of displaying data with each of these types of coordinate systems are known and can be applied here.

When more than one relaxation time is present, the graphs are curves. Nevertheless, the limiting slopes and intercepts can still be used to estimate the relaxation parameters. In this case various average relaxation times, $\langle\tau\rangle_p$, are obtained.

$$\langle\tau\rangle_p = \sum_{i=1}^{m} \Delta_i \tau_i^p \Big/ \sum_{i=1}^{m} \Delta_i$$

where m is the number of relaxation processes, τ_i and Δ_i are the time and magnitude of the i-th process and p is any integer between -3 and $+4$. These averages are useful both as a means of characterizing the distribution of relaxation times and determining other parameters describing the relaxation process.

The application of these graphical methods is illustrated in two specific areas: the frequency dependence of the shear modulus and ultrasonic longitudinal wave propagation. For these two examples the expressions for the intercepts and limiting slopes are evaluated. With these expressions the graphs of experimental data can be used to estimate the relevant parameters.

A section discussing the special case of a system with two relaxation times is also included. This section illustrates how intuitive generalization from a case with one relaxation time can be misleading.

Zusammenfassung

Drei Typen von Koordinatensystemen für die Analyse von Relaxationsdaten werden untersucht. Sind die Daten mittels einer einzigen Relaxationszeit beschreibbar, dann besteht die graphische Darstellung aus geraden Linien. Die Steigungen und Achsenschnittpunkte dieser Linien können dazu benützt werden, um die den Prozeß charakterisierenden Parameter, z. B. die Relaxationszeit, abzuschätzen. Die Vor- und Nachteile der Datendarstellung mit jedem dieser drei Koordinatensystemtypen sind wohlbekannt und können hier angewandt werden.

Gibt es mehr als eine Relaxationszeit, dann bestehen die graphischen Darstellungen aus Kurven. Jedoch können die begrenzenden Steigungen und Schnittpunkte immer noch benützt werden, um die Relaxationsparameter abzuschätzen. In diesem Fall werden verschiedene mittlere Relaxationszeiten $\langle\tau\rangle_p$ erhalten:

$$\langle\tau\rangle_p = \sum_{i=1}^{m} \Delta_i \tau_i^p \Big/ \sum_{i=1}^{m} \Delta_i$$

wobei m die Zahl der Relaxationsprozesse ist, τ_i und Δ_i die Zeit und Größe des i-ten Prozesses sind, und p irgendeine ganze Zahl zwischen -3 und $+4$ ist. Diese Mittelwerte sind zur Charakterisierung der Verteilung von Relaxationszeiten sowie zur Bestimmung anderer Parameter, die den Relaxationsprozeß beschrieben, nützlich.

Die Anwendung dieser graphischen Methode wird in zwei bestimmten Gebieten erläutert, und zwar für die Frequenzabhängigkeit des Schermoduls und der Ausbreitung von longitudinalen Ultraschallwellen. Für diese zwei Beispiele werden die Ausdrücke für Schnittpunkte und begrenzenden Steigungen abgeschätzt. Mit diesen Ausdrücken können die graphischen Darstellungen von experimentellen Daten zur Abschätzung der entsprechenden Parameter benützt werden.

Der spezielle Fall eines Systems mit zwei Relaxationszeiten wird auch diskutiert. Dieser Abschnitt zeigt, inwieweit intuitive Verallgemeinerung aus dem Fall einer einzigen Relaxationszeit irreführend sein kann.

References

1) *Mikhailov, I. G.*, Dokl. Akad. Nauk SSSR, **89**, 991 (1953).

2) *Lamb, J.* and *D. H. A. Huddart*, Trans. Faraday Soc. **46**, 540 (1950).

3) *Jackopin, L. G.* and *E. Yeager*, J. Phys. Chem. **74**, 3766 (1970).

4) *Herzfeld, K. F.* and *T. A. Litovitz*, Absorption and Dispersion of Ultrasonic Waves, p. 98 (New York 1959).

5) *Klotz, I. M.*, Arch. Biochem. **9**, 109 (1946).

6) *Scatchard, G.*, Ann. N. Y. Acad. Sci. **51**, 660 (1949).

7) *Klotz, I. M.*, in: *H. Neurath* and *K. Bailey* (Eds.), The Proteins, 1st ed., Vol. II, Chapter 8 (New York 1953).

8) *Mahler, H. R.* and *E. H. Cordes*, Biological Chemistry, pp. 227–230 (New York 1966).

9) *Dowd, J. E.* and *D. S. Riggs*, J. Biol. Chem. **240**, 863 (1965).
10) *Johansen, G.* and *R. Lumry*, Comptes Rendus Des Travaux Du Laboratoire Carlsberg. **32**, 185 (1961).
11) *Ferry, J. D.*, Viscoelastic Properties of Polymers, 2nd ed. (New York 1970).
12) *Daniel, V. V.*, Dielectric Relaxation (New York 1967).
13) *Klotz, I. M.* and *D. L. Hunston*, Biochemistry **10**, 3065 (1971).
14) *Montrose, C. J.* and *T. A. Litovitz*, J. Acoust. Soc. **47**, 1250 (1970).
15) *Schwarz, G.*, Rev. Mod. Phys. **40**, 206 (1968).

Author's address:

Dr. *Donald L. Hunston*
Naval Research Laboratory
Chemistry Division
Washington, D.C. 20375 (USA)

Rheol. Acta 13, 40–48 (1974)

From the Department of Astronautics & Engineering Sciences, Purdue University, Lafayette, Indiana U.S.A.

Experimental investigation of wave propagation in nonlinear viscoelastic materials

By *Hsiu-lin Yuan*, and *G. Lianis*

With 10 figures

(Received October 27, 1972)

1. Introduction

By establishing a similarity between the short-time elastic response and the long-time equilibrium stress for materials with fading memory, *Lianis* (1) proposed an approximate form of constitutive equation for finite linear viscoelasticity [Ref. (2)]. This constitutive equation was successful in the study of uniaxial deformation histories (3–5) in the static region. This investigation aims at measuring the material functions of viscoelastic materials by a series of wave propagation tests, in order to verify the validity of *Lianis*' constitutive equation at very short times.

If a relaxed static finite deformation is taken as the reference state in a problem of small dynamic strains superposed on the finite static deformation, the material appears as anisotropic. This anisotropy displays the nonlinear viscoelastic behavior even when the superposed dynamic strains are very small and the mechanical equations are linearized accordingly. Thus, performing tests on small amplitude waves, we can obtain the material functions entering into the nonlinear constitutive equation.

The subject of wave propagation is of practical importance both to the engineer, who needs to know the dynamic response of viscoelastic materials used in structural applications and to the physicist, who desires to known the dynamic properties of polymers.

Most experimental studies of wave propagation in linear viscoelastic solids have been aiming at the determination of the complex elastic modulus for a given range of frequencies.

However, in general boundary-value problems involving the propagation of waves in linear viscoelastic solids, it is often more convenient to deal directly with the creep and relaxation functions rather than with the complex modulus of the medium. Having this in mind, *Sackman* and *Kaya* (6) developed four methods for direct experimental determination of the creep and relaxation functions of linear viscoelastic media for very short times. They also proved, Ref. (7), that at least one method, namely method III in Ref. (6), was sufficiently accurate and simple. This method is also the one used in the present investigation.

The sample material, Geon 2042, was selected on the basis of its relaxation response and supplied in sheet stock form by the *B. F. Goodrich* Chemical Company. For the static region, its material functions and constants which appeared in *Lianis*' constitutive equation, were experimentally determined in the same way as in Ref. (3–5).

For the dynamic region, the same material functions were obtained in this investigation by propagating a pulse into specimens strained by a finite extension.

In Section 2, the problem and results of small dynamic strain superposed on large static deformation are presented briefly. The theory of linear longitudinal waves, *Sackman* and *Kayas* method, and numerical techniques are given respectively in Sections 3 through 5. In Section 6 we describe the equipment and test procedure which were designed for this investigation. Finally in Section 7, we present the experimental results which we compare with the theory of Ref. (1).

2. Small dynamic strain superposed on large static deformation

Let us consider a small displacement vector $\varepsilon u_m(y_k, \tau)$ superimposed on a static deformed state y_i. The scale factor $\varepsilon \ll 1$ is inserted here to show that the superposed deformation is small. *Lianis* (8) derived the stress equation for materials with fading memory as follows:

$$\underset{\sim}{\sigma}(t) = \underset{\sim}{\sigma}_r + \underset{\sim}{W}(t)\,\underset{\sim}{\sigma}_r - \underset{\sim}{\sigma}_r\,\underset{\sim}{W}(t) + \underset{\sim}{\Omega}(\underset{\sim}{B}^0)\,\{\underset{\sim}{E}(t)\}$$
$$+ 2\,\underset{\sim}{\Phi}(0,\underset{\sim}{B}^0)\,\{\underset{\sim}{E}(t)\}$$
$$+ 2\int_0^\infty \underset{\sim}{\dot\Phi}(s,\underset{\sim}{B}^0)\,\{\underset{\sim}{E}(t-s)\}\,ds \qquad [1]$$

where $\underset{\sim}{\sigma}(t)$ is the stress tensor at time t, $\underset{\sim}{\sigma}_r$ the residual stress tensor at the steady state, and $\underset{\sim}{B}^0$ the left *Cauchy-Green* deformation tensor at the steady state. $\underset{\sim}{\Omega}(\underset{\sim}{B}^0)\,\{\underset{\sim}{E}(t)\}$ is a linear transformation of $\underset{\sim}{E}$ and $\underset{\sim}{\Phi}(s,\underset{\sim}{B}^0)\,\{\underset{\sim}{E}(t-s)\}$ a linear transformation of $\underset{\sim}{E}(t-s)$. The infinitesimal rotation tensor $\underset{\sim}{W}$ and strain tensor $\underset{\sim}{E}$ referred to the deformed steady state y_i are given by:

$$\underset{\sim}{W}(t) = \frac{\varepsilon}{2}\,(u_{i,j} - u_{j,i})$$
$$\underset{\sim}{E}(t) = \frac{\varepsilon}{2}\,(u_{i,j} + u_{j,i}). \qquad [2]$$

If the initial static deformation is zero, i.e., $\underset{\sim}{B}^0 = \underset{\sim}{I}$, then $\underset{\sim}{\sigma}_r = 0$ and

$$2\,\underset{\sim}{\Phi}(s,\underset{\sim}{I})\,\{\underset{\sim}{E}(t-s)\} = G_1(s)\,\underset{\sim}{E}(t-s)$$
$$+ G_2(s)\,\underset{\sim}{I}\,\mathrm{tr}\,\underset{\sim}{E}(t-s). \qquad [4]$$

In this case, eq. [1] is reduced to the constitutive equation of linear infinitesimal viscoelasticity for isotropic bodies in terms of two relaxation functions G_1 and G_2 of time. In the presence of initial large deformation, twelve relaxation functions are needed, according to eq. [1]. This is so because the configuration at the static deformation, taken as the reference configuration, introduces effectively anisotropy, and the resulting configuration behaves as a linearly anisotropic viscoelastic body under small deformations. The relaxation functions depend, of course, on the state of initial deformation $\underset{\sim}{B}^0$.

By using *Lianis* approximate constitutive theory, Ref. (1), eq. [1] is simplified furthermore as:

$$\underset{\sim}{\sigma}(t) = \underset{\sim}{\sigma}_r + \underset{\sim}{W}(t)\,\underset{\sim}{\sigma}_r - \underset{\sim}{\sigma}_r\,\underset{\sim}{W}(t) + \underset{\sim}{\Omega}(\underset{\sim}{B}^0)\,\{\underset{\sim}{E}(t)\}$$
$$+ 4\int_{0-}^t \varphi_0(t-\tau)\,\underset{\sim}{\dot E}(\tau)\,d\tau$$
$$+ 2\underset{\sim}{B}^0 \int_{0-}^t \varphi_1(t-\tau)\,\underset{\sim}{\dot E}(\tau)\,d\tau$$
$$+ 2\int_{0-}^t \varphi_1(t-\tau)\,\underset{\sim}{\dot E}(\tau)\,d\tau\,\underset{\sim}{B}^0$$

$$+ 2\underset{\sim}{B}^{02} \int_{0-}^t \varphi_2(t-\tau)\,\underset{\sim}{\dot E}(\tau)\,d\tau$$
$$+ 2\int_{0-}^t \varphi_2(t-\tau)\,\underset{\sim}{\dot E}(\tau)\,d\tau\,\underset{\sim}{B}^{02}$$
$$+ 2\underset{\sim}{B}^0 \int_{0-}^t \varphi_3(t-\tau)\,\mathrm{tr}\,\underset{\sim}{B}^0\,\underset{\sim}{\dot E}(\tau)\,d\tau \qquad [4]$$

$\underset{\sim}{\Omega}(\underset{\sim}{B}^0)\,\{\underset{\sim}{E}(t)\}$ denotes the linear transformation:

$$\underset{\sim}{\Omega}(\underset{\sim}{B}^0)\,\{\underset{\sim}{E}(t)\} = q(t)\,I$$
$$+ [(a-3c) + (I_1 - 3)(b-c)]$$
$$\times [\underset{\sim}{B}^0\,\underset{\sim}{E}(t) + \underset{\sim}{E}(t)\,\underset{\sim}{B}^0]$$
$$+ c[\underset{\sim}{B}^{02}\,\underset{\sim}{E}(t) + \underset{\sim}{E}(t)\,\underset{\sim}{B}^{02}]$$
$$+ 2cI_2\,\underset{\sim}{E}(t) + 2b\underset{\sim}{B}^0\,\mathrm{tr}\,\underset{\sim}{B}^0\,\underset{\sim}{E}(t) \qquad [5]$$

where $q(t)$ is an indeterminate hydrostatic pressure and I_1 and I_2 are evaluated at the steady state.

Suppose that a small axial strain $\varepsilon(\tau)$ is superposed on a finite extension in a bar with the stretch ratio λ, and both $\varepsilon(\tau)$ and the initial finite extension are in the x_1 direction. Then, considering that the bar is incompressible, we have:

$$\underset{\sim}{B}^0 = \mathrm{diag}\left\{\lambda^2, \frac{1}{\lambda}, \frac{1}{\lambda}\right\}$$
$$\underset{\sim}{E}(\tau) = \mathrm{diag}\left\{\varepsilon(\tau), -\frac{\varepsilon(\tau)}{2}, -\frac{\varepsilon(\tau)}{2}\right\}, \underset{\sim}{W}(\tau) = 0 \qquad [6]$$

and therefore

$$I_1 = \lambda^2 + \frac{2}{\lambda},\ I_2 = \left(2\lambda + \frac{1}{\lambda^2}\right),$$
$$\mathrm{tr}\,[\underset{\sim}{B}^0\,\underset{\sim}{\dot E}(\tau)] = \dot\varepsilon(\tau)\left(\lambda^2 - \frac{1}{\lambda}\right). \qquad [7]$$

If we call $\underset{\sim}{\sigma}'(t)$ the increment of the stress, i.e.,

$$\underset{\sim}{\sigma}'(t) = \underset{\sim}{\sigma}(t) - \underset{\sim}{\sigma}_r \qquad [8]$$

and calculate the hydrostatic pressure $q(t)$ from the condition of zero stresses on the lateral surface of the bar i.e.

$$\sigma'_{22}(t) = \sigma'_{33}(t) = 0 \qquad [9]$$

we obtain the expression:

$$\sigma(t) = \sigma'_{11}(t)$$
$$= \left\{(a-3b)\left(2\lambda^2 + \frac{1}{\lambda}\right) + c\left(\lambda + \frac{2}{\lambda^2}\right)\right.$$
$$\left. + b\left(4\lambda^4 + \lambda + \frac{4}{\lambda^2}\right)\right\}\varepsilon(t)$$

$$+ 2 \int_{0-}^{t} \left[3\varphi_0(t-\tau) + \left(2\lambda^2 + \frac{1}{\lambda}\right)\varphi_1(t-\tau) + \left(2\lambda^4 + \frac{1}{\lambda^2}\right)\varphi_2(t-\tau) + \left(\lambda^2 - \frac{1}{\lambda}\right)^2 \varphi_3(t-\tau)\right] \dot{\varepsilon}(\tau)\, d\tau. \qquad [10]$$

Eq. [10] is of the form:

$$\sigma(t) = G_0 \varepsilon(t) + \int_{0-}^{t} G(t-\tau)\,\dot{\varepsilon}(\tau)\, d\tau \qquad [11]$$

where G_0 and $G(t)$ depend on the stretch ratio. Since eq. [11] is a linear functional of the history of $\sigma(t)$ and $\varepsilon(t)$, it can be inverted by using, say, a *Laplace* transform procedure. Thus we will have a relation:

$$\varepsilon(t) = J_0 \sigma(t) + \int_{0-}^{t} J(t-\tau)\,\dot{\sigma}(\tau)\, d\tau. \qquad [12]$$

For the static region, *Lianis'* relaxation functions and material constants for Geon 2042, which was used in this investigation, were determined by single-step and two-step uniaxial relaxation tests. Since the straight line approximation was used on uniaxial isochrones [1], the following relations were deduced

$$b = 0, \quad \varphi_3(t) = -2\varphi_2(t).$$

The other material constants are

$$a = 172.13 \text{ psi} \qquad c = 133.44 \text{ psi}.$$

The relaxation functions for Geon 2042 are shown in fig. 1.

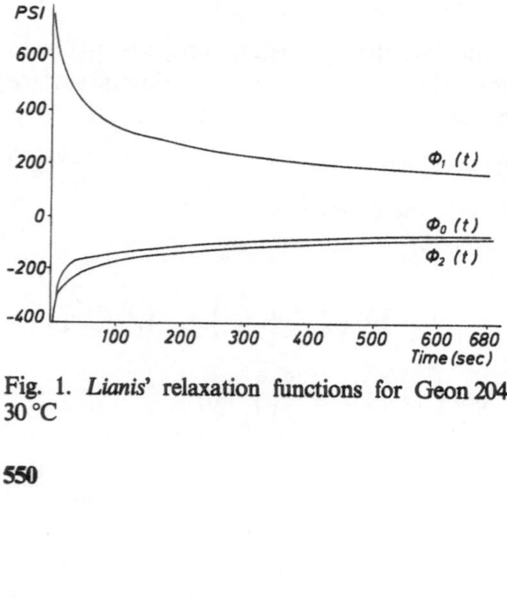

Fig. 1. *Lianis'* relaxation functions for Geon 2042 at 30 °C

3. Theory of linear longitudinal waves

The mathematical formulation of the problem of wave propagation in a viscoelastic material follows closely the analogous elastic problem and is defined in terms of the equation of motion, the strain-displacement relationship, the stress-strain law and the appropriate boundary condition. Since we are interested in the direct determination of the creep and relaxation functions, we choose eq. [11] as our stress-strain relationship.

For simplicity, we will consider only one-dimensional wave propagation. According to the so-called "Thin Rod" theory, the one-dimensional wave propagation theory applies well when the characteristic length of the specimen is much smaller than the wave lengths.

Let σ and ε be the uniaxial stress and strain respectively, v the particle velocity and ϱ the mass density. Then the governing equations are

$$\frac{\partial \sigma}{\partial x} = \varrho \frac{\partial v}{\partial t} \qquad [15]$$

$$\frac{\partial \varepsilon}{\partial t} = \frac{\partial v}{\partial x} \qquad [16]$$

$$\sigma(x, t) = G_0 \varepsilon(t) + \int_{0-}^{t} G(t-\tau) \frac{\partial \varepsilon}{\partial \tau}\, d\tau \qquad [11]$$

where G_0 and $G(t)$ are instantaneous modulus and relaxation function respectively. If the specimen is pre-extended by an extention ratio λ, then G_0 and $G(t)$ will contain λ as a parameter.

Defining a new relaxation function $g(t)$ and a new creep function $j(t)$ such that

$$g(t) = G_0 + G(t) \quad (0 \le t < \infty)$$

$$\lim_{t \to \infty} g(t) = G_0$$

and

$$j(t) = J_0 + J(t) \quad (0 \le t < \infty)$$

$$j(0) = J_0$$

we can write eqs. [11] and [12] as follows:

$$\sigma(x, t) = \int_{-\infty}^{t} g(t-\tau) \frac{\partial}{\partial \tau} \varepsilon(x, \tau)\, d\tau \qquad [17]$$

and

$$\varepsilon(x, t) = \int_{-\infty}^{t} j(t-\tau) \frac{\partial}{\partial \tau} \sigma(x, \tau)\, d\tau. \qquad [18]$$

Differentiating [17] with respect to x and combining it with eqs. [15] and [16], we obtain

$$\varrho \frac{\partial v}{\partial t} = \int_{-\infty}^{t} g(t-\tau) \frac{\partial^2 v}{\partial x^2} d\tau \qquad [19]$$

whose *Laplace* transform is

$$\varrho p \bar{v} = \bar{g}(p) \frac{d^2 \bar{v}}{dx^2}. \qquad [20]$$

The solution of eq. [20] is obviously:

$$\bar{v}(x,p) = A e^{\sqrt{(\varrho p/\bar{g}(p))}\, x} + B e^{-\sqrt{(\varrho p/\bar{g}(p))}\, x} \qquad [21]$$

where A and B are arbitrary constants.

When x approaches infinity, the velocity v is finite and then its *Laplace* transform $\bar{v}$ is also finite; therefore from [21] we obtain $A = 0$. At $x = 0$, the initial condition $v(0, t) = f(t)$ implies

$$B = \bar{v}(0, p) = \bar{f}(p). \qquad [22]$$

Therefore:

$$\bar{v}(x, p) = \bar{f}(p)\, e^{-px\bar{a}(p)} \qquad [23]$$

where

$$\bar{a}(p) = \sqrt{\frac{\varrho}{p\bar{g}(p)}}. \qquad [24]$$

4. Sackman – Kaya's method

Sackman and *Kaya* (6) proposed to record two particle velocity histories at station 1 and 2, along the axis of the rod for evaluating the relaxation function $g(t)$ numerically. If the origin of time is taken as the instant the front of the disturbance arrives at the station x_i, the velocity histories $q_i(t)$ have the following form

$$q_i(t) = v(x_i, t + x_i/V_0)\, H(t) \qquad [25]$$

where V_0 is the velocity of propagation of the wave front and $H(t)$ the unit step function.

Having solved the wave propagation problem in the *Laplace* domain and inverting it in the real time domain, the authors of Ref. [6] obtained the following equations:

$$q_1(t) = S_0 q_2(t) + \int_{0+}^{t} q_2(t)\, S_R(t-\tau)\, d\tau \qquad [26]$$

$$\frac{1}{\Delta x} t S_R(t) = S_0 t \dot{m}(t) + \int_{0+}^{t} \tau \dot{m}(\tau)\, S_R(t-\tau)\, d\tau \qquad [27]$$

$$\dot{m}(0^+) = \frac{S_R(0^+)}{\Delta x S_0} \qquad [28]$$

$$m(t) = m(0^+) + \int_{0+}^{t} \dot{m}(\tau)\, d\tau \qquad [29]$$

$$m(0^+) = \frac{1}{\Delta x} \ln S_0 \qquad [30]$$

$$j(t) = \frac{1}{\varrho V_0^2} + \frac{1}{\varrho} \int_{0+}^{t} \left[\frac{2}{V_0} m(\tau) + \int_{0+}^{\tau} m(\xi)\, m(\tau - \xi)\, d\xi \right] d\tau \qquad [31]$$

$$g(0^+) j(0^+) + \int_{0+}^{t} g(t-\tau)\, \dot{j}(\tau)\, d\tau = 1 \qquad t > 0 \qquad [32]$$

where $S_R(t)$ and $m(t)$ are auxiliary functions introduced for mathematical convenience, and S_0 a constant which should be predetermined. If $q_1(t)$ and $q_2(t)$ are known, one can solve eqs. [26] through [32] to find $S_R(t)$, $\dot{m}(t)$, $m(t)$, $j(t)$ and $g(t)$.

In principle, the method proposed allows one to obtain $j(t)$ and $g(t)$ over the entire range of time. However, in practice, accurate data for $q_1(t)$ and $q_2(t)$, which are used in [26] for the calculation of S_R, will exist only for some finite range of time, during which the pulse is passing through x_1 or x_2. Beyond this range, the values of $q_1(t)$ and $q_2(t)$ would generally be either too small to be recorded or will be influenced by the reflecting wave.

5. Numerical techniques

Eqs. [26] and [27] are *Volterra* integral equations. The general type of the *Volterra* integral equation is of the form:

$$a g(x) - \int_{0}^{x} W(x-\xi)\, g(\xi)\, d\xi = f(x) \qquad [33]$$

where a is a constant, and f, W are given functions of ξ.

For $a \neq 0$, eq. [33] can be solved numerically and step by step by the following form:

$$\left\{a - \frac{\delta}{2} W(0)\right\} g^*(m\delta) = f(m\delta) + \delta \left\{ \frac{1}{2} W(m\delta)\, g(0) + \sum_{i=1}^{m-1} W([m-i]\,\delta)\, g^*(i\delta) \right\} \qquad [34]$$

where δ is the interval of equal length along x-axis and an asterisk denotes an approximate value.

There are two different types of error inherent in this procedure due to:
a) replacing the integral by a quadrature formula,
b) the fact that all the values of the solution of the first m steps enter in the evaluation of the solution at the $(m + 1)$ th step. Thus their errors are carried in the $(m + 1)$ th step.

J. G. Jones (9) proved that for a fixed value of x, the errors counted above tend to zero as the step length tends to zero.

When $W(x)$ and $f(x)$ are given in tabular form (data collected from experiment), the procedure discussed above yielded a very unsatisfactory result. The significant figures of the experimental data are not sufficient for the solution of the equation under investigation, with any degree of accuracy. This was also pointed out in Ref. (10). It is necessary to approximate the tabulated data by analytic functions then carry out the numerical integration with these functions.

Because the first derivative of creep function $j(t)$ appears in the integrand of eq. [32], the following technique should be used. Divide t into equal intervals of length δ. As a mean value of $g(t)$ is used at each time interval eq. [32] reduces to:

$$g(0^+)j(0^+) + \sum_{i=1}^{m} \frac{1}{2} \{g([m-i-1]\delta) + g([m-i]\delta)\} \int_{(i-1)\delta}^{i\delta} \dot{j}(\tau)\, d\tau = 1 \qquad [35]$$

and thus

$$1 = g(0^+)j(0^+) + \sum_{i=1}^{m} \frac{1}{2} \{g[(m-i-1)\delta] + g[(m-i)\delta]\} \times [j(i\delta) - j(i\delta - \delta)]. \qquad [36]$$

Solving eq. [36], we obtain the relaxation function $g(t)$ step by step.

6. Equipment and test procedure

The apparatus used in this experimental program consists of one wave-producing assembly, one oscilloscope, two *Dana, D. C.*, amplifiers, two permanent magnets and a cathetometer. A block diagram is shown in fig. 2. The wave-producing assembly sketched in fig. 2 was specifically designed for our program. It consists of small dynamic strains superimposed on an initial large static deformation. Near the left end of the specimen two small aluminum plates fastened together by two screws were used to grip the specimen firmly. By hitting the plates with the pendulum, a tension wave was produced in the specimen. The pendulum could be adjusted so that its center line coincided with the center line of the specimen.

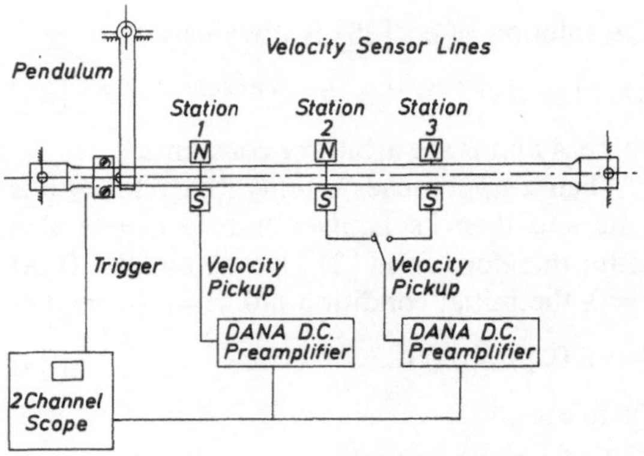

Fig. 2. Block diagram of particle velocity measurement

The strength of the two permanent magnets used are 4680 gauss and 4500 gauss when the gaps are 0.265″ and 0.327″ respectively. The entire wave-producing unit and the magnets were covered by a wood box in which two 100 watts electrical bulbs were installed to give the light and to supply heat for the purpose of temperature control. To record the temperature two thermometers were used, one at the end of the box and the other very near to the test position. The reading 30 °C of the second thermometer was taken as the testing temperature.

The material Geon 2042 was provided by the *B. F. Goodrich* Chemical Company in sheet form. It was selected on account of its relaxation response. The configuration of the specimen of the long dog-bone type.

The set-up for particle velocity measurements is based on Ref. (11). According to the *Faraday* principle, one could measure the particle velocity in a rod by cementing a short piece of wire to the surface of the rod and measuring the voltage developed across the wire as it moves through a magnetic field, in response to the passage of a stress wave. In our tests conducting lines were painted with a conductive paint on the surface and lateral sides of the specimen. The former acted as a piece of wire which cut the magnetic flux to generate a voltage. The latter acted as lead wires and bridged to connecting posts by the use of transparent scotch tape. This scotch tape is a superior adhesive on one hand and it is easy to write on it.

It also served as a cushion element between the wave-propagating specimen and the fixed connecting posts which were connected to the *Dana* amplifiers permanently.

The signals (which had an average intensity of 0.4 mv) were fed into two *Dana, D. C.*, amplifiers of 0–20 KC frequency range. These signals were recorded on a 555 Tektronix Oscilloscope. A *Dumont-Bolsey* oscillograph-record camera was used to photograph the generated voltage with ordinary black and white film. Another *Dumont* oscillograph-record camera was also used to record the generated voltage with polaroid black and white films.

The accuracy of both *Dana* amplifier and the 555 Tektronix oscilloscope were found satisfactory. The field strength of the magnets was measured repeatedly with a Scalmp Fluxmeter, and their relative values were found with satisfactory accuracy. The overall recording accuracy of the particle velocities is within one or two percent.

For large values of the stretch ratio, new conducting lines were well arranged and re-painted as velocity sensor lines before each test. Their relative positions were measured optically before and after the specimen was prestrained. From these measurements the stretched ratio λ was evaluated.

After all adjustments were made, the wood box was closed and the lights were adjusted suitably to maintain the temperature inside the box at 30 °C. About an hour later, the specimen was almost completely relaxed from the prestrain and was steadily and uniformly at 30 °C. At the moment the pendulum hit the plates H to produce a wave pulse, the oscilloscope was triggered by a special simple circuit. The induced voltage of the sensor lines was amplified and finally displayed on the oscilloscope screen and photographed.

7. Experimental results and comparison with the static test

After the photographs were enlarged and printed, the profiles of the particle velocities for different prestrains and stations were read out with the use of the cathetometer. The velocity of propagation V_0 was also calculated. The curve V_0 versus λ is shown in fig. 3.

To obtain $q_i(t)$, the analytic form of particle velocity at different stations, the polynomial approximation was used. Since the velocity signals start from zero, the approximations have no constant term. They do not have first order term either since the strain rate signals which are a measure of the slope of the velocity also started from zero. Thus

$$q_i t) = A_{i2}t^2 + A_{i3}t^3 + \cdots \qquad i = 1, 2. \qquad [37]$$

Different numbers of terms of $q_i(t)$ were tried. It was found that six terms gave the best fitting for the first station, and five terms for the second. More than six or five terms for velocity signals might give a negative value on S_0 and cause trouble in calculating $m(0^+)$ through eq. [30].

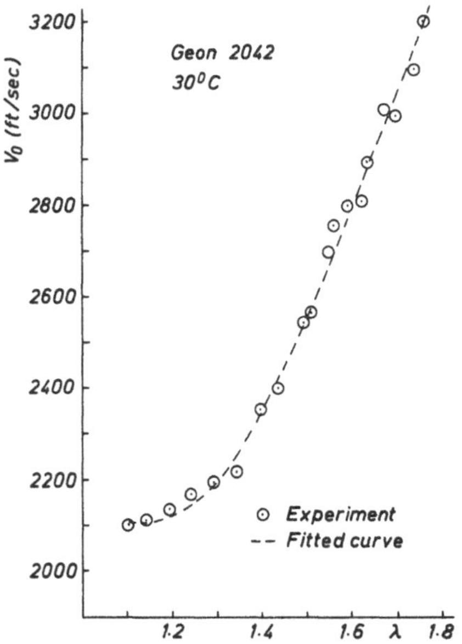

Fig. 3. Wave propagating velocity V_0 vs. λ

For evaluating S_0 the following approximation was used:

$$q_1(\delta t) \approx S_0 q_2(\delta t) \qquad [38]$$

where δt was chosen to be 0.001 m sec which is one-tenth of the time interval used for calculating $S_R(t)$ step by step. With the chosen δt, $S_R(0^+)$ is determined as

$$S_R(0^+) = \frac{2[q_1(2\delta t) - S_0 q_2(2\delta t)]}{q_2(\delta t) \cdot \delta t}. \qquad [39]$$

Eq. [39] has been derived from the general eq. [34] by retaining the first two terms only. A few calculations showed that the value of S_0 and $S_R(0^+)$ did not affect $S_R(t)$ appreciably. Knowing S_0, $S_R(0^+)$ and $q_i(t)$, we can calculate $S_R(t)$, $m(t)$, $j(t)$ and $g(t)$ through eqs. [26] through [32] with the numerical techniques discussed in Section 5.

Two of the calculated creep functions $j(\lambda, t)$ are shown in fig. 4 and the corresponding relaxation functions $g(\lambda, t)$ are shown in fig. 5.

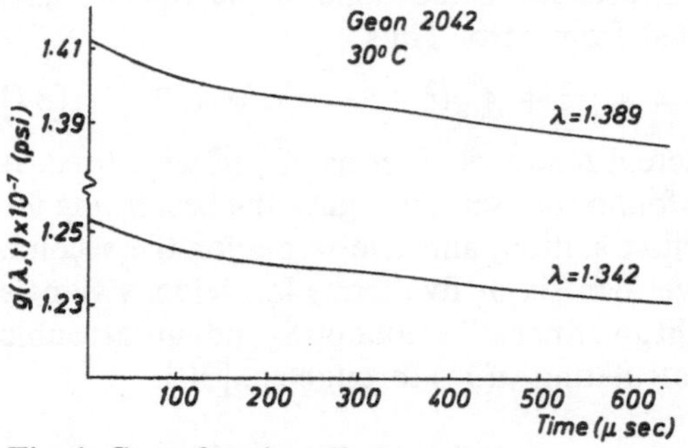

Fig. 4. Creep function $j(\lambda, t)$ vs. time

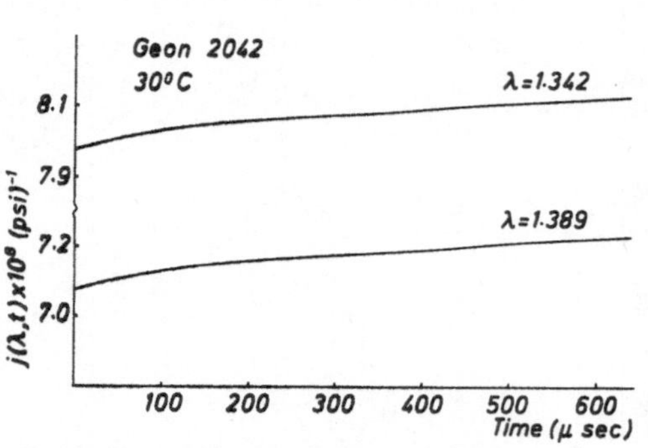

Fig. 5. Relaxation function $g(\lambda, t)$ vs. time

The range of prestrains covered in this investigation is from $\lambda = 1.1$ to $\lambda = 1.8$. Four relaxation functions $g(\lambda, t)$ whose corresponding λ's are 1.142, 1.290, 1.539, and 1.731 were fitted with four terms polynomial series through the method of least squares. The approximated curves and the original calculated data are shown in fig. 6.

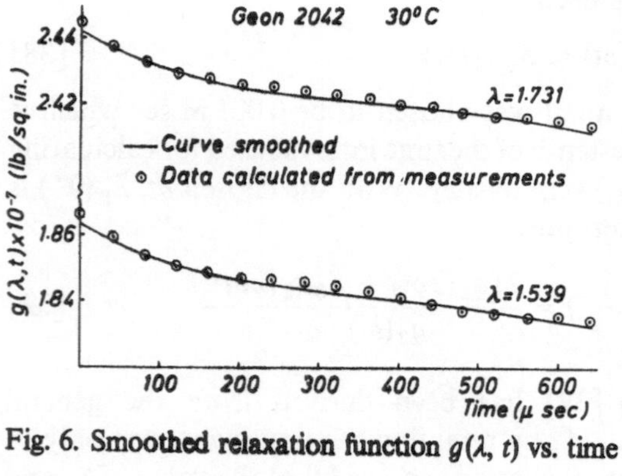

Fig. 6. Smoothed relaxation function $g(\lambda, t)$ vs. time

With these four smoothed functions $g(\lambda, t)$ *Lianis*' relaxation functions $\varphi_0(t)$, $\varphi_1(t)$, $\varphi_2(t)$, and $\varphi_3(t)$ were calculated through the following equation:

$$g(\lambda, t) = (a - 3b)\left(2\lambda^2 + \frac{1}{\lambda}\right) + c\left(\lambda + \frac{2}{\lambda^2}\right) + b\left(4\lambda^4 + \lambda + \frac{4}{\lambda^2}\right) + 2\left[3\varphi_0(t) + \left(2\lambda^2 + \frac{1}{\lambda}\right)\varphi_1(t) + \left(2\lambda^4 + \frac{1}{\lambda^2}\right)\varphi_2(t) + \left(\lambda^2 - \frac{1}{\lambda}\right)^2\varphi_3(t)\right] \qquad [40]$$

which is obtained directly from the definition of $g(\lambda, t)$. From the static test, we know that viscoelastic materials whose isochronic plots of $\sigma(t)/(\lambda^2 - 1/\lambda)$ versus $1/\lambda$ can be approximated with straight lines have the additional property:

$$\varphi_3(t) = -2\varphi_2(t),\ b = 0 \qquad [41]$$

The relation $b = 0$ is derived from the long-time material behavior and related to long-time property only. Thus it would be true for both static and dynamic tests. But, the relation [41] might not appear in the dynamic response of the material, although it is true for the static responses. Therefore, we kept relation $b = 0$ only in eq. [40]. Thus:

$$g(\lambda, t) = a\left(2\lambda^2 + \frac{1}{\lambda}\right) + c\left(\lambda + \frac{2}{\lambda^2}\right) + 2\left[3\varphi_0(t) + \left(2\lambda^2 + \frac{1}{\lambda}\right)\varphi_1(t) + \left(2\lambda^4 + \frac{1}{\lambda^2}\right)\varphi_2(t) + \left(\lambda^2 - \frac{1}{\lambda}\right)^2\varphi_3(t)\right] \qquad [42]$$

Lianis' relaxation functions for Geon 2042, which were calculated in the way discussed above are shown in fig. 7 and 8.

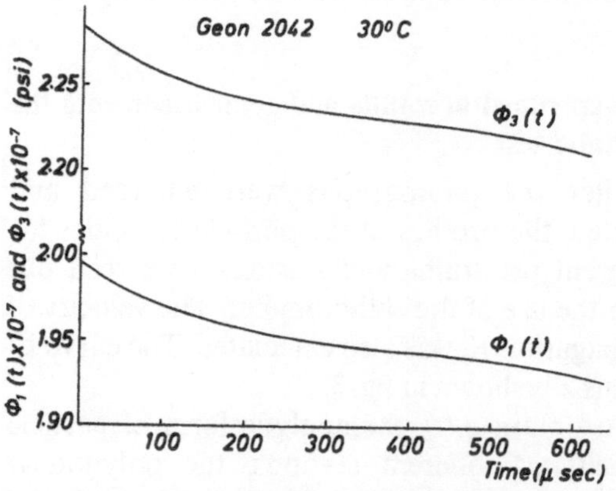

Fig. 7. *Lianis*' relaxation functions for Geon 2042 vs. time

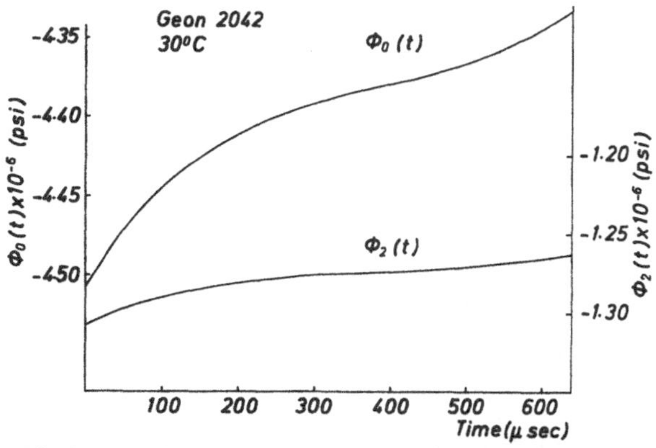

Fig. 8. *Lianis*' relaxation functions for Geon 2042 vs. time

Figs. 9 and 10 show *Lianis*' relaxation functions versus time ranging from 10^{-5} to 10^{3} seconds. Dotted lines are used again to connect the dynamic and quasistatic regions.

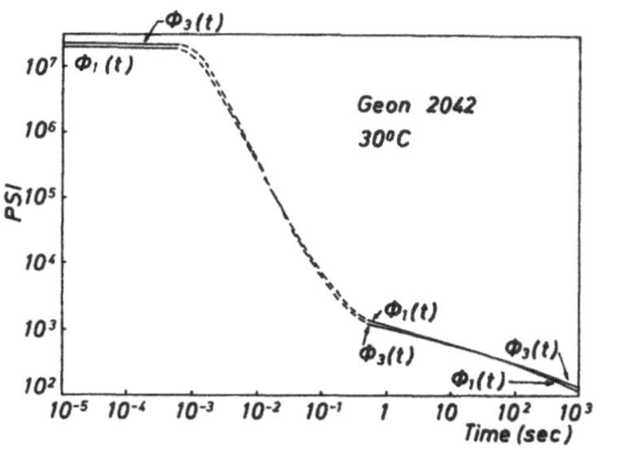

Fig. 9. *Lianis*' relaxation functions φ_0, φ_2 vs. time

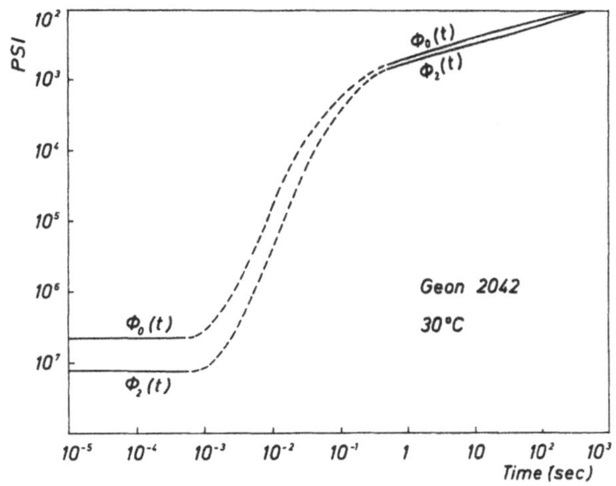

Fig. 10. *Lianis*' relaxation functions φ_1, φ_3 vs. time

The straight isochromes of $\sigma(t)/\lambda^2 - 1/\lambda$ versus $1/\lambda$ of Geon 2042 have an approximate intercept with the vertical axis when the relaxation time is over 10 seconds. This property implies that $\varphi_1(t) = \varphi_3(t)$. This is why the curves of $\varphi_1(t)$ and $\varphi_3(t)$ almost coincide in the static region. It appears that this property appears again in some region between the dynamic and static regions, and the difference between $\varphi_1(t)$ and $\varphi_3(t)$ is very small even in the dynamic region. Instead of $\varphi_3 = -2\varphi_2$ relation of the static region, we have the relation $\varphi_3 \simeq -1.74\varphi_2$ in the dynamic region.

Using *Lianis*' theory we can predict some $g(\lambda, t)$'s for given values of λ. However, because $g(\lambda, t)$ changed only within $2 \div 3\%$ during the whole time interval, 0.6 m sec, and we could not measure V_0 without error, we shifted the calculated data up or down with a constant number to get a comparison with the predicted quantities. For example, the data corresponding to $\lambda = 1.098$ have been shifted by 0.054×10^7 psi, which is the largest shift and corresponds to an error of less than 4.6% in $g(\lambda, t)$ and an error of less than 2.5% in the measured V_0. From this we conclude that *Lianis*' theory predicts satisfactory results on one hand and on the other hand, the very **earlytime material functions, $j(\lambda, t)$ and $g(\lambda, t)$, have a shift property with respect to λ in the dynamic region.**

The functions $\varphi_0(t)$, $\varphi_1(t)$, $\varphi_2(t)$, and $\varphi_3(t)$, which were calculated in the dynamic region, include all the properties of the static region and the dotted lines in fig. 9 and 10 connect reasonably well dynamic and static regions. We can conclude therefore that they belong to the same set of material functions. Thus we see that *Lianis*' assumptions made in Ref. [1] are accurate and his theory applies not only in the static but in the dynamic region as well.

8. Conclusions

A dynamic test unit was designed, and the earlytime material functions of Geon 2042 were calculated from recorded particle velocities. *Lianis*' relaxation functions of Geon 2042 for the dynamic region were thus evaluated. For the *Lianis*' relaxation functions, the comparison between the quantities which were measured in dynamic and static regions shows that they have the same characteristics in both regions and they can be correlated in a satisfactory way. This proves that *Lianis*' theory applies to most viscoelastic materials not only in the static region but in the dynamic region as well.

Acknowledgement

The results presented in this paper were obtained in the course of research sponsored by National Science

Foundation under Grant GK-1969. The authors wish to express their appreciation to their sponsor.

Summary

In this paper we present an experimental technique and a numerical procedure to measure the short-time relaxation functions of nonlinear viscoelastic media. The experimental technique consists of a longitudinal stress wave propagated through a pretensioned bar, while the numerical procedure incorporates experimental data into the solution of a *Volterra* type integral equation.

References

1) *Lianis, G.*, Constitutive Equations of Viscoelastic Solids under Finite Deformation. Purdue University Report AA & ES 63-11 (1963).

2) *Coleman, B. D.* and *W. Noll*, Rev. Mod. Phys. **33**, 239 (1961).

3) *Goldberg, W., B. Bernstein*, and *G. Lianis*, Int. J. Non-linear Mechanics, Vol. **4**, 277–300 (1969).

4) *McGuirt, C. W.* and *G. Lianis*, The Constant Stretch Rate History. Comparison of Theory with Experiment. Purdue University Report AA & ES 67-2 (1967).

5) *Goldberg, W.* and *G. Lianis*, J. Applied Mech. **35**, No. 3, Trans. ASME **90**, Series E (1968).

6) *Sackman, J. L.* and *I. Kaya*, J. Mech. Phys. Solids **16**, 121–132 (1968).

7) *Sackman, J. L.* and *I. Kaya*, J. Mech. Phys. Solids **16**, 349–356 (1968).

8) *Lianis, G.*, Small Deformations Superposed on an Initial Large Deformation in Viscoelastic Bodies, Proceedings, Fourth International Congress on Rheology, Part 2, p. 109 (1965).

9) *Jones, J. G.*, Math. of Computation **15**, No. 74, p. 131 (1961).

10) *Kaya, I.*, Ph.D. Thesis, Univ. of California, Berkeley (1968).

11) *Repperger, E. A.* and *L. M. Yeakley*, Experimental Mechanics, Feb. (1963).

Authors' addresses:
Dr. *Hsiu-lin Yuan*
Chungshan Institute of Science and Technology, Taiwan

Prof. Dr. *G. Lianis*
Dept. of Astronautics and Engineering Sciences, Purdue University
Lafayette, Indiana (USA)

Rheol. Acta 13, 49–53 (1974)

From the Industrial Chemistry Department of the Polytechnic Institute of Milan, Milan (Italy)

Mechanical-dynamic properties of some polymers grafted on polypropylene

By M. Pegoraro, L. Szilágyi, *and* A. Penati

With 6 figures and 1 table

(Received October 27, 1972)

Introduction

With a view to studying the relationships between properties and structure of grafted polymers, a series of dynamic resonance measurements was done in order to determine the values of the complex modulus vs. temperature at frequencies of the order of 1000 cps.

The materials under examination consisted of polypropylene grafted with acrylic acid, acrylonitrile, acrylamide. As known (1), by dynamic measurements it is possible to detect whether the copolymers consist of one or more phases through the determination of the existence of one or more glass transition temperatures.

This work is connected with the characterization of these materials, which exhibit the properties of both constituting polymers; hence materials with intermediate characteristics may be obtained.

Experimental part

Preparation of the materials

Runs were carried out with isotactic polypropylene (PP) having a viscosimetric molecular weight of 400000, supplied by *Montedison.*

Polyacrylic acid (PAA) was obtained by polymerization in boiling benzene of freshly distilled acrylic acid with benzoyl peroxide as an initiator (2).

Polyacrylonitrile (PAN) was obtained by polymerization in water at 35 °C of freshly distilled acrylonitrile, in the presence of ammonium persulfate and sodium metabisulfite as a redox initiator (3).

Polyacrylamide (PAAm) was obtained by polymerization of acrylamide in boiling water in the presence of hydrogen peroxide as an initiator (4).

Grafted polymers were obtained by a method of synthesis (5, 6), which consisted in causing to react PP dissolved in orthodichlorobenzene at 120 °C with the monomer in the presence of benzoylperoxide as an initiator for the time required (a few hrs). The solution containing the reaction crude product was poured into acetone, thus obtaining by precipitation the grafted polymer and a homopolymer that forms during the reaction. The reaction crude products were extracted with boiling methanol in the case of grafting with PAA, with dimethyl formamide in the case of grafting with PAN, and finally with boiling water in the case of grafting with PAAm.

Dried residues consisting of the pure grafted product, and possibly of unreacted PP, were subsequently moulded.

Product compositions shown in table 1 were evaluated by elemental analysis.

Table 1. Moulding conditions, densities and compositions of the materials used. Moulding pressure 50 kg/cm^2

Material	Moulding temperature (°C)	Moulding time (min)	g/cm^3	Composition (%)
PP	170	5	0.907	
PAA	170	5	1.440	
PP-PAA 1	170	5	0.950	12.3 PAA
PP-PAA 2	170	5	1.110	57 PAA
PAN	150	10	1.184	
PP-PAN	150	10	0.951	16 PAN
PAAm	130	20	1.294	
PP-PAAm	130	20	1.025	25 PAAm

Densities reported in table 1 were determined by a *Westphal* balance using anhydrous test-pieces of moulded polymer, immersed in benzene at 20 °C. Being this method of grafting done in solution, the samples obtained appeared to be grafted uniformly.

Mechanical-dynamic measurements

The previous homopolymer-free polymers were moulded using a pressure of 50 kg/cm^2 by a simple compression press between two aluminium sheets. Moulding conditions, densities and composition of the various polymers are described in table 1.

From the plates obtained, we drew prismatic test-pieces with the following dimensions: length: 26 mm; thickness: 2 mm; width: 10 mm.

Materials were stuck with a very thin layer of an adhesive consisting of an epoxy resin (CIBA, Araldite AV 138) and a hardening agent (HV 998) on a brass support with the same dimensions except for thickness (0.5 mm).

Both frequencies and resonance curves of the composite test-pieces obtained in this way as well as of the support

4

alone, were measured at different temperatures under vacuum (0.5 Tor.).

E', E'', Q^{-1} (where $Q^{-1} \simeq \text{tg}\,\delta$ and $E'' \simeq E' Q^{-1}$) of the materials were measured according to the calculation program described in a previous paper (7).

Discussion of the results

As previously mentioned (8) the first problem arisen when performing measurements on hydrophilic polymers was the removal of water from the samples. Actually, the absorbed water acts as plastifier which causes different elastic properties according its amount. Measurements were therefore repeated at different times until obtaining dry samples and constant values of the resonance curve. Instead, no similar problem has arisen with PP, PAN, and grafted PAN.

Homopolymers

The values of E', E'' *and* Q^{-1} of all homopolymers studied (PP, PAA, PAN, PAAm) may be detected in figs. 1 to 6.

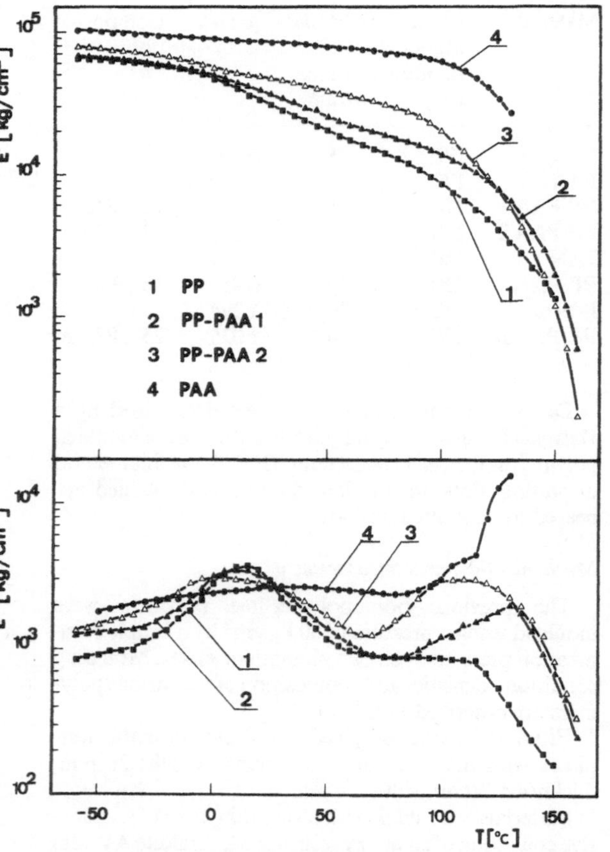

Fig. 1. Real (E') and imaginary (E'') parts of the elastic modulus vs. temperature for PP, PAA, PP–PAA 1 (12.3% of grafted PAA) and PP-PAA 2 (57% of grafted PAA)

In the case of crystalline isotactic PP, the behavior of E' and E'' shown for comparison in figs. 1, 3, 5, is well known: relaxation takes place in correspondence of the glass transition of the amorphous part (9) with a maximum of E'' at about 15 °C. A relaxation α with a maximum of E'' is also observed; its position depends on the thermal history of the sample: in the case of figs. 1, 3, 5, it occurs at 115 °C. As known (10), the α transition is attributed to molecular motions within the crystals or to a slipping of the crystalline lamellae: i. e. it depends on the morphology of the crystals.

Anhydrous PAA is a rigid glass material (figs. 1, 2), which in the investigated range of temperatures shows its glass transition (11) beginning at ca. 100 °C.

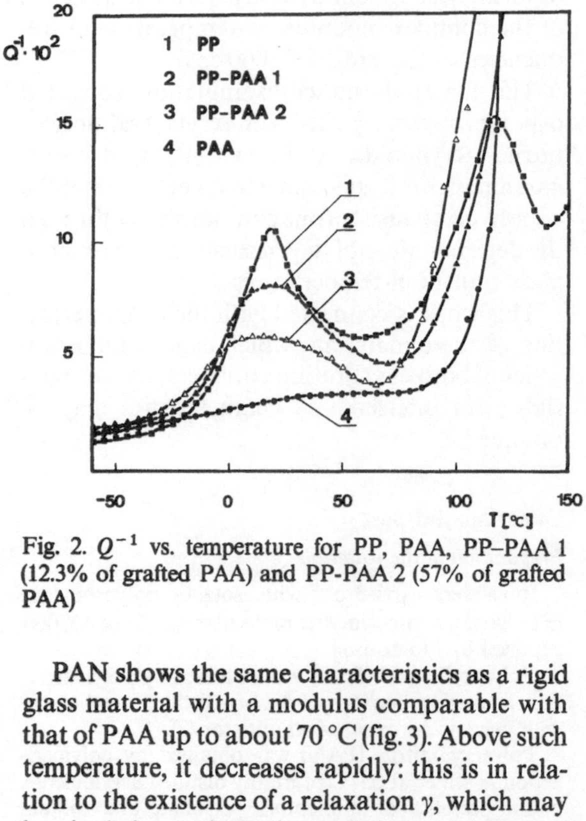

Fig. 2. Q^{-1} vs. temperature for PP, PAA, PP–PAA 1 (12.3% of grafted PAA) and PP-PAA 2 (57% of grafted PAA)

PAN shows the same characteristics as a rigid glass material with a modulus comparable with that of PAA up to about 70 °C (fig. 3). Above such temperature, it decreases rapidly: this is in relation to the existence of a relaxation γ, which may be singled out through a relative maximum of curve E'' (or of Q^{-1} in fig. 4), which occurs at about 45 °C.

Transition γ was found by *Saito* (12) by dielectric measurements; however, this author did not discuss it. The glass transition of PAN with a maximum of E'' at 105 °C may also be clearly detected.

PAAm is an amorphous material; its elastic modulus at room temperature is one of the highest

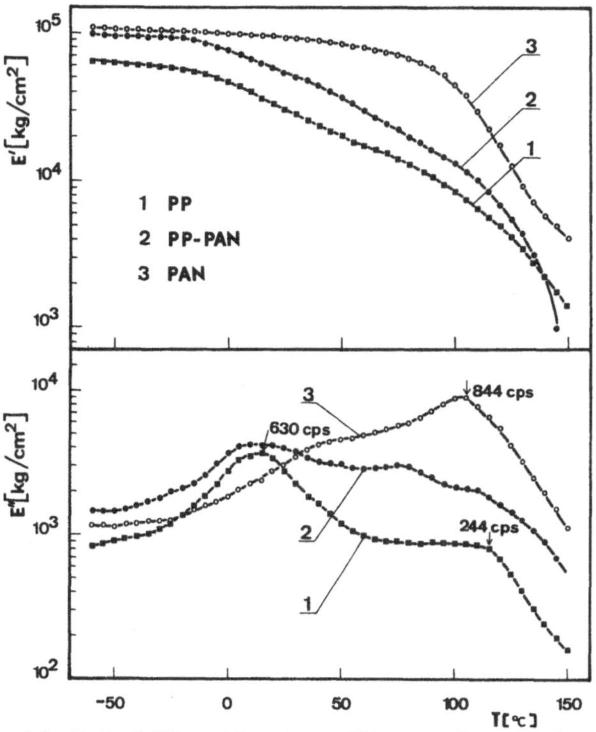

Fig. 3. Real (E') and imaginary (E'') parts of the elastic modulus vs. temperature for PP, PAN and PP-PAN (16% of grafted PAN)

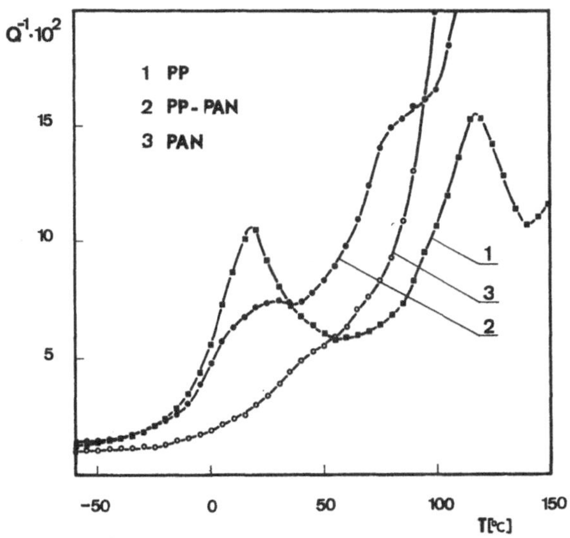

Fig. 4. Q^{-1} vs. temperature for PP, PAN and PP-PAN (16% of grafted PAN)

known for plastomers (fig. 5). This has to be attributed to the existence of frequent hydrogen bonds that reduce the mobility of the chain segments. The modulus value slowly decreases with temperature until 90 °C; then it sharply falls in correspondence of glass transition. The plot of E'' vs. temperature indicates three transitions β, γ, and δ; β is the glass transition; the other two correspond to the freezing of secondary molecular motions, which cannot be attributed so far. The macromolecule, which initially at low temperatures is in the glassy state with the greatest stiffness, acquires energy on increasing temperature; segments acquire mobility; hydrogen bonds should first break, then it would become possible the rotations of the pending groups —$CONH_2$; finally, the rotations around the C—C bonds of the main chain would occur. Transitions may be detected also from the Q^{-1}, T curve (fig. 6).

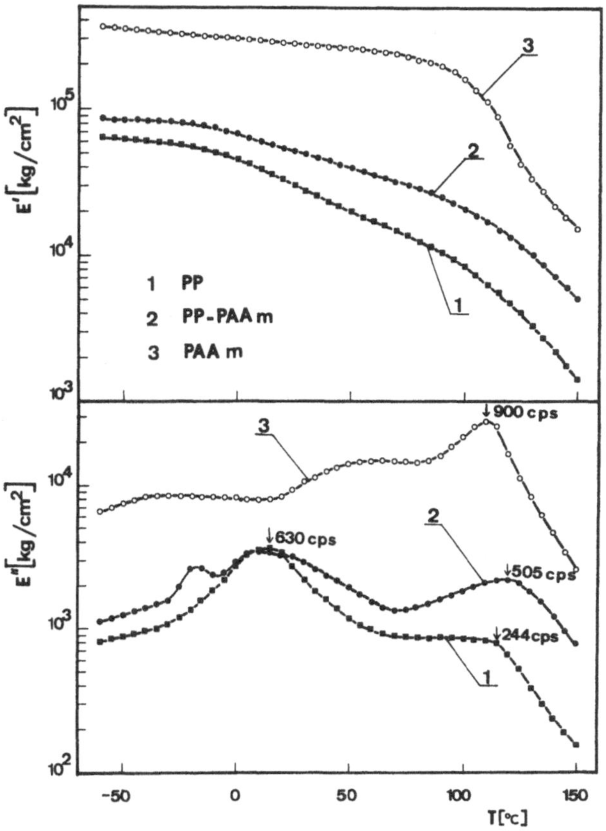

Fig. 5. Real (E') and imaginary (E'') parts of the elastic modulus vs. temperature for PP, PAAm and PP-PAAm (25% of grafted PAAm)

Grafted polymers

The grafted polymers under examination, described in figs. 1 to 6, are free from homopolymers produced by the monomers used. The peculiar transitions of the polymeric materials constituting the grafted polymers appear from our experience to be superposed.

PAA grafted on PP

In the case of PAA grafted on PP, our investigations also concerned different grafting percentages: in all cases, the PP glass transition may be clearly detected (figs. 1–2). The peak height markedly decreases on increasing the percent amount of PAA (effect of dilution) and the position of the maximum is slightly shifted toward low temperatures, revealing an easier relaxation motion of the amorphous regions of PP. This is caused by grafting, which might influence the interactions between the PP chains. For example we noticed that the specific volumes of grafted products are slightly higher than those calculated by the additivity rule.

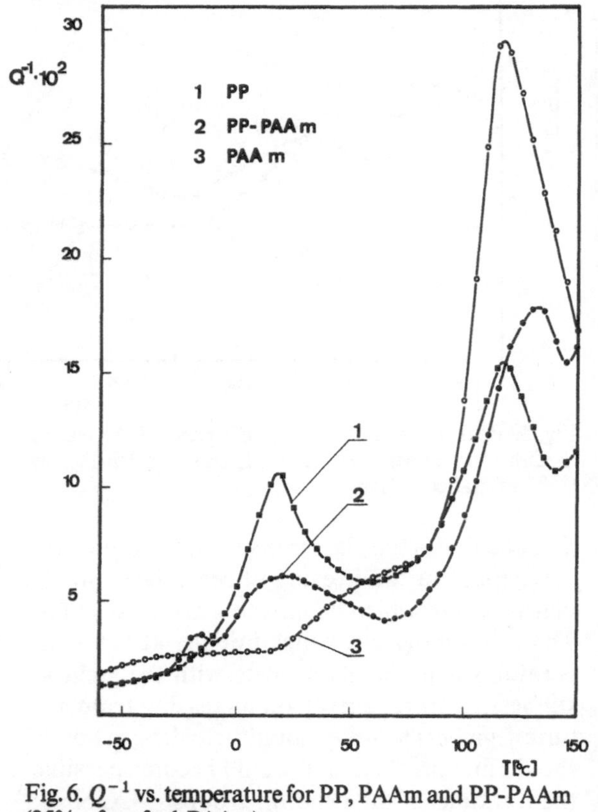

Fig. 6. Q^{-1} vs. temperature for PP, PAAm and PP-PAAm (25% of grafted PAAm)

Transitions β of PAA and α of PP are superposed; however, in that range of temperatures, the grafted polymers show much higher Q^{-1} values than PP. This agrees with the existence of PAA homogeneous domains also found by electron microscope (8). High values of Q^{-1} are reached at temperatures that are the lower the higher is the percentage of PAA. Hence also the glass transition of PAA is influenced by grafting.

PAN grafted on PP

The behavior of E'' vs. temperature indicates three relaxation processes (fig. 3); the most evident has a maximum at 15 °C attributable to relaxation of polypropylene. A relative maximum of E'' is also observed at 75 °C: such a peak cannot be attributed to the γ transition of PAN, the percent amount of grafted PAN being too low (16%) to evidence a process that may be hardly detected in PAN homopolymer. Instead, this is probably the α transition of PP.

As known (13) thermal annealing on quenched PP samples shift maximum α toward high temperatures even by 38 °C; this might be in connection with the fact that crystalline superstructures of PP may find an ever better arrangement on increasing annealing time. Lower temperatures of peak α involve rather disordered structures. Grafting might cause the decrease of the degree of order of such superstructures. Melting points of PP also decrease, although slightly, by the introduction of incompatible grafted chains (14).

The last relative maximum of E'' at 110 °C corresponds to the β transition of PAN. The previous statements may be deduced even by considering the behavior of Q^{-1} vs. temperature (fig. 4).

The value of the elastic modulus E' (fig. 3) of the grafted polymer is intermediate between the corresponding values of the two homopolymers; it behaves like the modulus of PP, as it has to be because of the low concentration of PAN.

PAAm grafted on PP

The value of the elastic modulus of PAAm grafted on PP (fig. 5) is intermediate between the corresponding values of the two homopolymers and the E' slope variations take place, in correspondence with the modulus variations of PP: this is in agreement with the fact that the percent amount of PP is predominant. Plot E'' vs. temperature (fig. 5) indicates the existence of the relative maxima. That occurring at 120 °C falls in the region of maximum β of PPAm and of maximum α of PP; the two relaxation processes are superposed.

The maximum at 15 °C must be attributed exclusively to relaxation β of PP. At the present state of knowledge, the maximum at −18 °C cannot be attributed to determinate relaxation processes, but is a consequence of the presence of PAAm.

Conclusions

The grafting of polymers with an energy of cohesion that noticeably differs from that of the matrix polymer is responsible for the separation of the two polymers into distinct domains (15, 16).

The properties of the grafted polymer are the result of the composition of the properties of the two homopolymers. In particular, all transitions characteristic of the two homopolymers are usually maintained.

Summary

Homopolymer-free grafted polymers consisting of chains of polyacrylic acid, polyacrylonitrile and polyacrylamide chemically bound to isotactic polypropylene have been prepared by fractionation of raw materials obtained by radical polymerization.

The real and the imaginary parts of the elastic modulus both of grafted polymers and of homopolymers have been measured vs. temperature.

The grafted polymers show the β transition of polypropylene; instead, the polypropylene α transition is superposed on the relaxations β of the grafted chains. Results are in agreement with the existence of a biphasic structure, as confirmed by the analysis by electron microscope.

References

1) *Nielsen, L. E.*, Mechanical Properties of Polymers, p. 132 (New York 1962).

2) *Miller, M. L.*, in: Encyclopedia of Polymer Science, and Technology, Vol. 1, p. 204 (New York 1964).

3) *Bamford, C. H.* and *G. C. Eastmond*, in: Encyclopedia of Polymer Science and Technology, Vol. 1, p. 299 (New York 1964).

4) *Schulz, P., G. Renner, A. Henglein,* and *W. Kern*, Makromol. Chem. **12**, 20 (1954).

5) *Natta, G., M. Pegoraro,* and *A. Penati*, Italian Pat. 865199 (May 22, 1969).

6) *Pegoraro, M., A. Penati,* and *G. Natta*, Chim. Ind. **53**, 235 (1971).

7) *Szilagyi, L., G. Locati,* and *M. Pegoraro*, Kolloid-Z. u. Z. Polymere **223**, 94 (1968).

8) *Pegoraro, M., L. Szilagyi, A. Penati,* and *G. Alessandrini*, European Polymer J. **7**, 1709 (1971).

9) *McCrum, N. G., B. E. Read,* and *G. Williams*, Anelastic and Dielectric Effects in Polymeric Solids, p. 377 (London 1967).

10) Ref. (9), p. 384.

11) *Hughes, L. J. F.* and *D. B. Fordyce*, J. Polymer Sci. **22**, 509 (1956).

12) *Saito, S.* and *T. Nakajima*, J. Appl. Polymer Sci. **2**, 93 (1959).

13) *McCrum, N. G.*, Polymer Letters **2**, 495 (1964).

14) *Pegoraro, M., A. Penati, G. Gianotti,* and *A. Capizzi*, Chemické Zvesti (Bratislava) **26**, 242 (1972).

15) *Bianchi, U., E. Pedemonte,* and *A. Turturro*, J. Polymer Sci. (B) **7**, 785 (1969).

16) *Bianchi, U., E. Pedemonte,* and *A. Turturro*, Polymer, Lond. **11**, 268 (1970).

Authors' address:

Istituto di Chimica Industriale
Polytechnico di Milano
32 Piazza Leonardo da Vinci
I-20133 Milano (Italia)

Rheol. Acta 13, 54–59 (1974)

From the Faculty of Engineering, Kyushu University, Fukuoka (Japan)

General equation of stress-strain behavior in crystalline polymers

By M. Takayanagi, K. Imada, S. Maruyama), and K. Nakamura**)*

With 7 figures and 3 tables

(Received October 27, 1972)

Introduction

Crystalline polymers show a remarkable strain hardening in its plastic deformation compared with metals. In the case of metals, the power law holds between true stress and true strain, the power law index of which is lower than one (1). True stress tends to saturate with increasing true strain. Polymer chain shows very high modulus and high bonding energy along molecular axis, whereas intermolecular force is weak and the modulus perpendicular to the molecular axis is about one hundredth of that along molecular axis. With the progress of unfolding of molecular chains from folded structure in lamellar crystals, the deformation stress increases rapidly. To describe such a deformation process of crystalline polymers, it is especially necessary to employ true stress and true strain instead of nominal stress and nominal strain. In our previous paper (2), we have presented the equation to describe the true stress-strain relationship in a very simplified form, which was verified for linear polyethylene (HDPE) and nylon 6. In this paper, we have examined the applicability of our equation to isotactic polypropylene (PP), polyoxymethylene (POM) and polytetrafluoroethylene (PTFE).

Experimental

The specimen for tensile testing was cut out of the quenched sheet, its dimensions being 20 mm long, 5 mm wide and 0.3–0.6 mm thick. Tensile testing was conducted using Tensilon UTM-III, equipped with an air-oven (Toyo Baldwin Co., Ltd.). True stress σ and true strain ε are defined as follows.

$$\sigma = F/A = (F/A_0)(l/l_0) \qquad [1]$$

$$\varepsilon = \int_{l_0}^{l} \frac{dl}{l} = \ln(l/l_0) = \ln(A_0/A) \qquad [2]$$

where A_0 and l_0 is the cross-sectional area and length of the original sample, respectively, and A and l are those of the deformed sample. Before necking takes place, it was sufficient to measure the sample length, l, but once necking takes place, it was necessary to remove the sample from the tensile tester and measure the cross-sectional area, A, to determine the true strain according to eq. [2]. More detailed description of experimental technique is given in our previous paper (2).

The samples used were listed in table 1.

Results

Experimental data of true stress and true strain during elongation of HDPE and nylon 6 were given in our previous paper (2). Fig. 1 (a) and (b) show $\sigma - \varepsilon$ relationships for PP and PTFE at various temperatures, respectively. The solid lines are the calculated ones according to our generalized eq. [3]. Similar results were obtained for POM.

Fig. 2 shows the composite curve of PP together with those of HDPE and nylon 6. Fig. 3 shows the composite curves of POM and PTFE. These composite curves were prepared by shifting the doubly logarithmic plots of true stress and true strain curves parallel to the abscissa and the ordinate by $\log\varepsilon^*$ and $\log\sigma^*$, respectively. All these curves are again represented by the same equation as given in the previous paper (2).

$$\log(\sigma/\sigma^*) \cdot \log(\varepsilon/\varepsilon^*) = -c \qquad [3]$$

where c is the constant characteristic of chemical species of polymers, and σ^* and ε^* are normalizing factors of σ and ε, respectively.

Table 2 lists the drawing conditions and the values of c, σ^* and ε^* for HDPE and nylon 6. Table 3 lists those for PP, POM and PTFE.

*) Present address: Visiting research fellow from the Mitsubishi Chemical Ind. Co., Ltd. (Japan).

**) Visiting research fellow from the Sumitomo Chemical Ind. Co., Ltd. (Japan).

Table 1. Samples used

Sample	Polymer	Grade (company)
PE-1	High density polyethylene	Hizex 1200J (Mitsui Petrochemical Co.)
PE-2		Sholex 4002B (Nihon Olefin Co.)
PE-3		Novatec JVO40 (Mitsubishi Chemical Co.)
PE-3a	Blend of PE and paraffin	Novatec JVO40 (90%) + paraffin A (mp 50 °C)
PE-3b		Novatec JVO40 (90%) + paraffin B (mp 70 °C)
PE-3c		Novatec JVO40 (90%) + paraffin C (mp 108 °C)
PE-3d		Novatec JVO40 (80%) + paraffin C
PE-3e		Novatec JVO40 (67%) + paraffin C
POM	Polyoxymethylene	Delrin (DuPont Co.)
NYL	Nylon 6	Amilan CM1031 (Toray Co.)
PTFE	Polytetrafluoro ethylene	Commercial product
PP-1	Isotactic polypropylene	Noblen W101 (Sumitomo Chemical Co.)
PP-2		Noblen D501 (Sumitomo Chemical Co.)

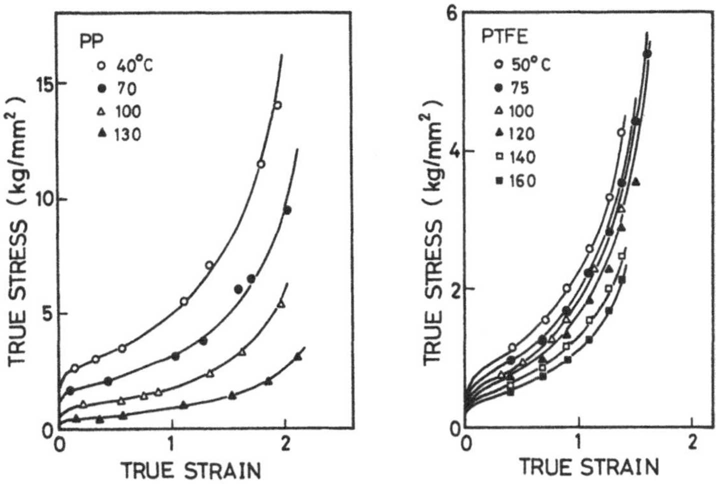

Fig. 1. True stress (σ) vs. true strain (ε) for PP (a) and PTFE (b) at various drawing temperatures

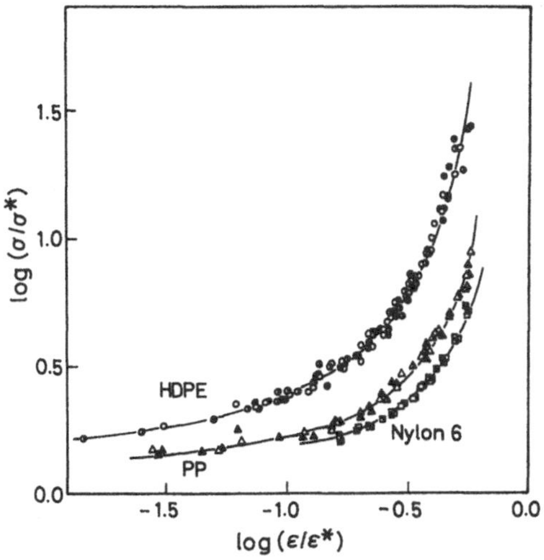

Fig. 2. Composite curves of $\log(\sigma/\sigma^*)$ vs. $\log(\varepsilon/\varepsilon^*)$ for HDPE, PP and nylon 6. Solid lines are calculated by eq. [3]

Tables 2 and 3 show that the values of c are characteristic of only polymer species, being independent of molecular weight, strain rate, blend ratio and drawing temperature. The values of c are 0.384 for HDPE, 0.175 for nylon 6, 0.30 for POM, 0.384 for PTFE and 0.23 for PP.

On the other hand, σ^* and ε^* are dependent on drawing conditions, as seen in fig. 4 for HDPE and fig. 5 for PP. Especially the temperature dependence of σ^* is always remarkable, whereas that of ε^* is not so sensitive but shows a somewhat complicated curve having a minimum against drawing temperatures which corresponds with the range where the σ^* vs. temperature curve changes its slope. The effect of elongation rate is also found in the case of HDPE as shown in fig. 4. The values of σ^* and ε^* increase with increasing elongation rate.

Table 2. Drawing conditions and some parameters characterizing the true stress–true strain curves for PE, the blend of PE and paraffin, and nylon 6

Sample	Drawing temperature (°C)	Elongation rate¹) (mm/min)	c (kg/mm²)	σ¹) (kg/mm²)	ε¹) (kg/mm²)
PE-1	30	5	0.384	1.28	5.75
	50	5	0.384	0.84	5.33
	70	5	0.384	0.63	5.38
	90	5	0.384	0.46	5.55
	110	5	0.384	0.27	5.58
	90	20	0.384	0.50	5.75
	50	50	0.384	1.26	6.31
	70	50	0.384	0.82	5.91
	90	50	0.384	0.56	5.93
PE-2	90	5	0.384	0.41	4.82
PE-3	50	5	0.384	0.80	5.55
PE-3a	50	5	0.384	0.58	5.55
PE-3b	50	5	0.384	0.68	5.55
PE-3c	50	5	0.384	0.76	5.55
PE-3d	50	5	0.384	0.69	5.55
PE-3e	50	5	0.384	0.62	5.55
NYL	80	5	0.175	1.37	1.70
	120	5	0.175	1.26	1.78
	160	5	0.175	0.89	1.76

¹) Original length of the specimen is 20 mm.

Table 3. Drawing conditions and some parameters characterizing the true stress–true strain curves for POM, PTFE, and PP

Sample	Drawing temperature (°C)	Elongation rate¹) (mm/min)	c (kg/mm²)	σ¹) (kg/mm²)	ε¹) (kg/mm²)
POM	130	5	0.30	1.23	3.98
PTFE	50	5	0.384	0.43	3.40
	75	5	0.384	0.36	3.33
	100	5	0.384	0.32	3.27
	120	5	0.384	0.28	3.23
	140	5	0.384	0.24	3.32
	160	5	0.384	0.20	3.26
PP-1	40	5	0.23	1.82	3.45
	70	5	0.23	1.20	3.60
	100	5	0.23	0.65	3.45
	130	5	0.23	0.34	3.65
PP-2	40	5	0.23	1.55	3.10
	70	5	0.23	1.00	3.50
	100	5	0.23	0.55	3.35
	130	5	0.23	0.30	3.55

¹) Original length of the specimen is 20 mm.

Discussion

The very simplified representation of true stress and true strain relationship is given by eq. [3], which has already been proved to hold for PE and nylon 6. Here again the same equation holds for POM, PP, and PTFE. We may say, therefore, that eq. [3] is a general equation for uniaxial deformation of crystalline polymers.

The physical meaning of eq. [3] has already been mentioned in our previous paper (2). Fig. 6 shows the schematic representation of eq. [3], in which the upper two figures (a) correspond with the case of $c = 0$ and the lower two ones (b) are for the actual case. Mathematically, for $c > 0$, σ should be larger than σ^* and ε smaller than ε^*.

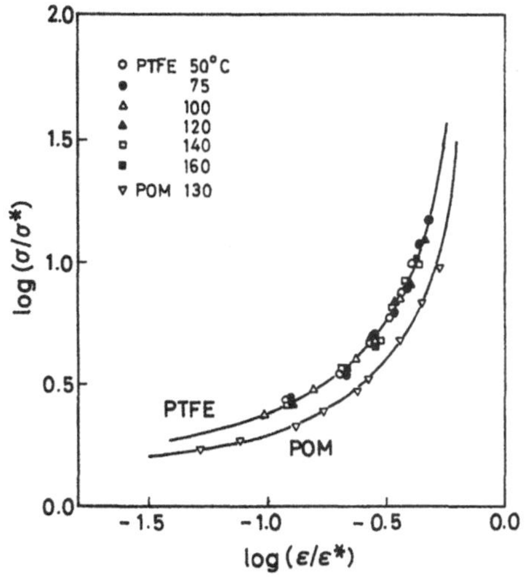

Fig. 3. Composite curves of $\log(\sigma/\sigma^*)$ vs. $\log(\varepsilon/\varepsilon^*)$ for PTFE and POM. Solid lines are calculated by eq. [3]

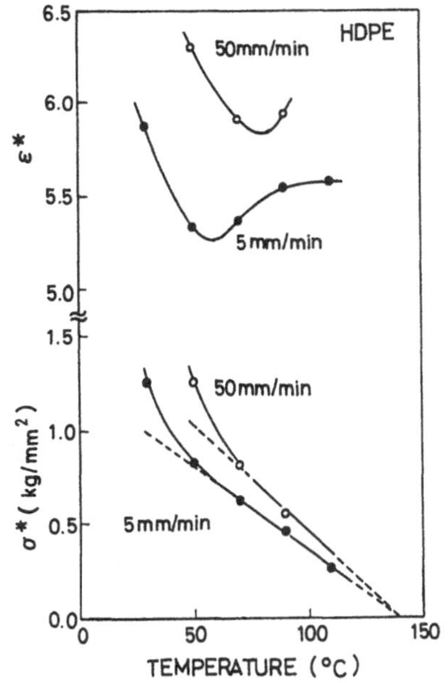

Fig. 4. ε^* and σ^* vs. drawing temperatures for HDPE at different rates of elongation (original length 20 mm)

In an idealized case of $c = 0$, the sample starts to be deformed when σ reaches σ^* and continues to be deformed until $\varepsilon = \varepsilon^*$, where the stress increases to infinite value. Such a deformation behavior is typical of purely plastic body, but the only difference of idealized crystalline polymer from the purely plastic body is in its displaying a rapid stress increase at $\varepsilon = \varepsilon^*$. The corresponding molecular process will be mentioned later.

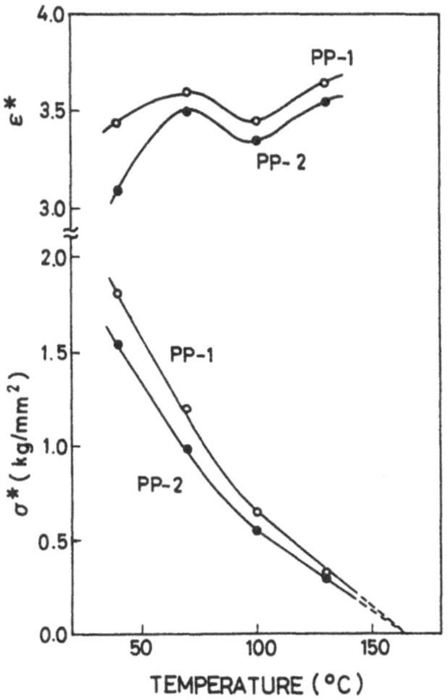

Fig. 5. ε^* and σ^* vs. drawing temperatures for PP-1 (Mol.wt. 27×10^3) and PP-2 (Mol.wt. 47×10^3)

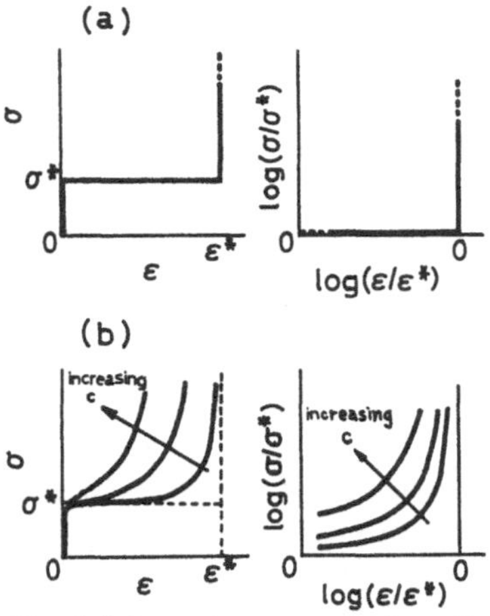

Fig. 6. Schematic representation of true stress-true strain relationship according to the general equation for (a) idealized case of $c = 0$ and (b) actual systems of crystalline polymers

Next we consider the actual deformation process represented by fig. 6. With increasing c value, the deviation from the ideal behavior represented by fig. 6 (a) is more remarkable. But, if we recognize the fact that the value of c increases in the order of nylon 6 (0.175), PP (0.23), POM (0.30), PTFE (0.384), and HDPE (0.384), we can not say that the higher c value means the larger deviation from the ideal behavior of crystalline polymer, because PE, PTFE, POM are believed to be more highly crystalline than PP or nylon 6. The degree of crystallinity of the latter polymers is about

60%, whereas that of the former ones is about 80%. The polymers with higher crystallinity are expected to show a more typical deformation pattern of crystalline polymers. Before we will discuss this subject, it would be more appropriate to recede to the ideal case of $c = 0$.

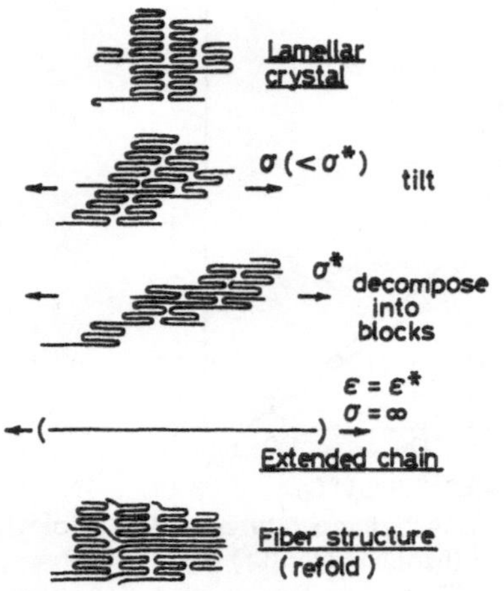

Fig. 7. Schematic representation of molecular process of plastic deformation of lamellar crystals

Fig. 7 shows a schematic representation of the molecular process of plastic deformation of lamellar crystals in the bulk crystallized sample, which is drawn corresponding with the ideal case of $c = 0$. That is, by applying stress, lamellar crystals first become thin by tilting of molecules and then decompose into blocks by slippage of intermosaic region. The critical stress σ^* will be necessary to initiate the slip of the mosaic blocks with each other. By continuing such an unfolding process, the molecular chain takes an extended chain conformation at $\varepsilon = \varepsilon^*$ to generate the infinite stress as mentioned above. In an actual case, for $c > 0$, such an extended chain conformation can not be realized as the intermolecular force restraining the both ends of the chain will not be sufficiently strong to sustain an isolated extended chain. In the midway to be fully extended, the chain will fold back to form a folded chain crystal at the stage of shortage in the restraining force. At that time, other chains may be still in a stage of being extended. In such a case, the latter extended chains are buried by the lamellar crystals generated by folding back, and remain as tie molecules. Such structures conform with the fiber structure composed of stacked lamellar crystals and crystal bridges or tie-link spanned between neighboring lamellae.

By taking into account of such a molecular process, the meaning of constant c in eq. [3] will be explained. Since strain hardening effect is more remarkable with increasing c value, it might be appropriate to call the constant c the "strain hardening parameter". Such a strain hardening process will be caused in the process of unfolding of lamellar crystals by increasing contribution from the additional mechanism that the crystal bridges or tie-molecules with high modulus partially sustain the applied stress. In order to make further deformation, it is necessary to overcome the increased resistance caused by the tie-molecules. Such an explanation of strain hardening mechanism predicts that comparatively large values of c such as 0.384 for HDPE and PTFE, and 0.30 for POM will be due to the nature of these polymers in their easiness of unfolding and refolding during deformation process. Crystal bridge will be more easily formed during the deformation process of these polymers. On the other hand, the comparatively lower values of c are found in 0.175 of nylon 6 and 0.23 of PP. Nylon 6 has a theoretically high value of modulus along its molecular axis corresponding to its all trans conformation as found in PE, but nylon 6 can not crystallize to a larger extent due to its capability of formation of hydrogen bond, which disturbs the long range migration of molecular chains necessary to form well developed crystals. Owing to the same reason, well developed crystal bridge will not be formed between neighboring crystal blocks aligned along the deformation direction. The modulus value of nylon 6 fiber is in the order of 10^{10} dyne/cm^2, which is smaller by one decade compared with those of other commercial fibers which have the moduli in the order of 10^{11} dyne/cm^2. Isotactic polypropylene molecule has 3_1 helix conformation in the crystal and its theoretical modulus is as low as 10^{11} dyne/cm^2 (3) and the actual fiber modulus is below 10^{11} dyne/cm^2. Crystallinity of PP is usually 60%. Both factors of low modulus and low crystallinity will be closely related to decrease in the value of strain-hardening parameter c. That is, if the crystal bridge develops and crystallinity increases, the modulus and the c value of nylon 6 should be large. These are not realized. In conclusion, polymers with high degree of crystallinity and high modulus anisotropy seem to have an increased tendency of strain-hardening and the high value of parameter c.

Next, the relationship between critical strain ε^* and polymer structure will be discussed. As discussed on figs. 4 and 5, ε^* does not vary so much with drawing temperature. We will take the average value of ε^*. Referring to the molecular process as illustrated in fig. 7, ε^* is the fictitious true strain where the extended chain conformation is assumed after completion of unfolding from lamellar crystals. If such an explanation is adopted, the lamellar crystals well developed over an extended area domain will give rise to the increased ε^* value. The lamellar crystals of the bulk crystallized samples and the single crystals for PE and POM are widely observed with electronmicroscope. They show comparatively large values of ε^* such as 5.5 for PE and 4.0 for POM. On the other hand, ε^* of nylon 6 is 1.8 and that of PP is 3.4. The latter two polymers are also known to form single crystals (4, 5). But the preparation method is not so easy and electronmicroscopically the crystals can not so beautifully develop as represented by the term of lamellar crystal. At present we assume that the last two polymers can not form the lamellar crystals which are suitable to be smoothly unfolded by drawing. Thus the value of ε^* is considered as the measure of easiness of unfolding of lamellar crystals, which reflects at the same time the degree of performance of crystallization of the corresponding molecular chains. To make these predictions more decisive, a more extended work will be desired. Especially to inspect the effect of regulation of crystalline texture of the starting materials on ε^* value will be effective to check our prediction.

σ^* is also expected to be in a close relation to the molecular structure. But σ^* is very sensitive to drawing temperature and at present we can not directly correlate them to molecular structure of polymers. σ^* is conceivable to be related to the viscoelastic properties of crystalline polymers. We have reported in the previous paper (2) the strain rate dependence of σ^* as a function of drawing temperature (refer to fig. 4) and evaluated the activation energy for deformation to be 27 kcal/mole for HDPE below 60 °C, which corresponds well with those evaluated for viscoelastic relaxation associated with the intermosaic block region (6).

Peterlin and his group (7) have reported a very extensive study on plastic deformation of crystalline polymers. They assumed that the lamellar crystals are sheared to destruct into small blocks which correspond with the mosaic blocks of lamellar crystals, which are remelted by generated heat and form the fiber structure composed of reformed crystallites and tie molecules. His prediction does not contradict essentially our view. We replaced the remelting and recrystallization of crystal blocks during deformation in *Peterlins* mechanisms by the process of unfolding and refolding of mosaic blocks.

References

1) Ref. to *Dieter, G.*, Mechanical Metallurgy, p. 247 (New York 1961).

2) *Maruyama, S., K. Imada,* and *M. Takayanagi,* Intern. J. Polymeric Mater. **1**, 211 (1972).

3) *Shimanouchi, T., M. Ashida,* and *S. Enomoto,* J. Polymer Sci. **59**, 93 (1962).

4) *Geil, P. H.*, J. Polymer Sci. **44**, 449 (1960).

5) *Kojima, M.*, J. Polymer Sci. A-2, **5**, 597 (1967).

6) *Kajiyama, T., T. Okada, A. Sakoda,* and *M. Takayanagi,* J. Macromol. Sci-Phys. B 7 (3), 583 (1973).

7) *Peterlin, A.*, a) J. Polymer Sci., C (9), 61 (1965); b) ibid., C (15), 427 (1967); c) ibid., C (18), 123 (1967); d) Kolloid-Z. u. Z. Polymere **216/217**, 129 (1967); e) ibid., **233**, 857 (1969); f) Polymer Eng. & Sci. **8**, 172 (1969); g) Man made fibers (eds. *Mark, Atlas,* and *Cernia*), Vol. 1, p. 283 (New York 1967).

Authors' address:

M. Takayanagi, K. Imada, S. Maruyama, and *K. Nakamura*
Faculty of Engineering
Kyushu University
Fukuoka (Japan)

Rheol. Acta 13, 60–65 (1974)

Laboratoire de Chimie Macromoléculaire – U.E.R. de Chimie – Biochimie – Université Claude Bernard «Lyon 1», Villeurbanne (France)

Influence du polymorphisme sur les propriétés mecaniques de quelques polyoléfines

Par J.-Y. Decroix, J.-F. May et G. Vallet

Avec 9 figures and 3 tableaux

(Reçu p. p. le 27 Octobre 1972)

Introduction

Depuis la découverte des catalyseurs du type *Ziegler-Natta*, un nombre important de travaux a été consacré aux propriétés mécaniques des polyoléfines en particulier du polyéthylène et du polypropylène. Dans le cadre plus général de la recherche de la relation entre la morphologie et les propriétés mécaniques d'un polymère, nous nous sommes intéressés à l'étude de l'influence de la variété allotropique sur ces propriétés. Dans la famille des polyoléfines, nous avons choisi le polybutène 1 qui présente deux formes dont l'une, instable, se transforme au cours du temps en variété stable et le polypentène 1 pour lequel les deux formes sont stables dans notre domaine d'expérimentation.

I. Appareillage et produits de départ

A. Appareillage

Cette étude a été réalisée à l'aide d'un viscoélasticimètre Rhéovibron DDVII commercialisé par la Société TMI (Toyo Measuring Instrument).

Cet appareil permet la détermination du module complexe d'élasticité E^* (traction alternée).

Une contrainte de traction sinusoïdale de fréquence et d'amplitude connues est appliquée à l'une des extrémités de l'éprouvette; la déformation résultante est mesurée à l'autre extrémité. L'angle de déphasage entre cette contrainte et la déformation, correspond à l'angle de perte δ; il est lu directement sur l'appareil de mesure. La mesure de la contrainte et de la déformation se fait par capteurs à jauges de part et d'autre des dispositifs d'amarrage de l'éprouvette. Leurs signaux de sortie sont appliqués à un comparateur de phase et la valeur de l'angle de déphasage est lu directement sur celui-ci.

Les mesures s'effectuent aux quatre fréquences fixes 3,5; 11; 35 et 110 Hz dans un domaine de températures comprises entre – 150 et 200 °C.

B. Obtention et caractéristiques des produits de départ

Les films de polybutène 1 (forme II) sont obtenus par fusion du polymère à 150 °C puis compression et refroidissement jusqu'à 80 °C et démoulage à cette température. La forme II ainsi obtenue se transforme spontanément au cours du temps en variété I stable. L'origine de l'échelle des temps est prise une demi-heure après le démoulage, aucune évolution détectable n'étant constatée durant cette période. Ceci a été vérifié en suivant la cinétique de transformation par spectroscopie infrarouge et par diffraction des rayons X. Les temps de demitransformation sont de 22 et 23 h et la transformation est pratiquement totale après 40 h. Nous considérerons donc que les propriétés obtenues après des durées de vieillissement de cet ordre, sont pratiquement celles de la forme I.

Le polypentène 1 utilisé a été étudié sous ses deux formes au moyen d'un procédé original mis au point au laboratoire par *D. Convard* (1). L'obtention des deux variétés cristallines est délicate car la cinétique de cristallisation est très lente. Les caractéristiques essentielles de ces deux formes ainsi que celles du polybutène 1 sont regroupées dans le tableau 1.

Tableau 1

Polyoléfine	Variété cristalline	T (°C) fusion	Masse moléculaire	$\{\eta\}$ (cm³/g) [1]
Polybutène P.B.	I II	135 126	$\overline{Mv} = 1{,}15 \cdot 10^6$ [2]	249
Polypentène P.Pe.	I II	130 80	$\overline{Mn} = 1{,}5 \cdot 10^5$ [3]	215

[1]) Mesures effectuées dans la décaline à 115 °C pour le P.B., dans l'heptane à 20 °C pour le P. Pe.
[2]) Loi de *Mark-Houwink*: $\{\eta\} = 9{,}49 \cdot 10^{-3}\, \overline{Mv}^{0{,}73}$ (2).
[3]) Mesures effectuées par osmométrie à 37 °C en solution dans le chloroforme.

Tableau 2 (d'après 4)

	Maille	Paramètres cristallins (Å)			Type d'hélice	Distance entre 2 unités monomères (Å)
		a	b	c		
Variété I	Rhomboédrique	17,5	17,7	6,5	3_1	2,17
Variété II	Tétragonale	14,85	14,85	20,6	11_3	1,87

II. Etude du polybutène 1

Depuis la mise en évidence du polymorphisme du polybutène 1 par *G. Natta* et coll (3), on a pu caractériser trois variétés cristallines: la forme III obtenue par précipitation dans de nombreux solvants; la forme II in stable obtenue par fusion du polymère, qui se transforme en variété I stable de caractéristiques cristallines différentes (cf. tableau 2).

Nous avons suivi l'évolution des parties réelle E' et imaginaire E'' du module complexe d'élasticité en fonction du temps et réalisé une étude en température sur des échantillons pris à des temps $t = 0$, $t = 17$ h et $t = 41$ h, ce qui a mis en évidence l'existence de deux pics d'absorption, l'un situé à 273 °K (pic β), l'autre vers 130 °K (pic γ).

A. Cinétique de transformation suivie en mesures mécaniques

L'évolution de E' et E'' en fonction du temps à température fixe est représentée sur la fig. 1 à différentes fréquences. Les courbes représentatives de E' présentent un point d'inflexion (correspondant au minimum de E'') vers 23 h soit le temps de demi-transformation de la forme II en forme I.

B. Etude en fonction de la température

B.1. Pic β

L'évolution de tangente δ avec la température est représentée sur les fig. 2 (forme II) et 3 (forme I). On notera que l'amplitude du pic diminue au cours du temps mais que sa position reste inchangée. En supposant que la relation d'*Arrhénius* est applicable dans ce domaine étroit de températures et de fréquences, nous pouvons calculer une énergie d'activation de 32 kcal/élément cinétique pour les deux formes. Les deux courbes maîtresses d'absorption obtenues en utilisant le principe d'équivalence temps – température (5) indiquent un temps de relaxation identique pour les deux formes, soit 10^{-8} secondes à 25 °C (fig. 4).

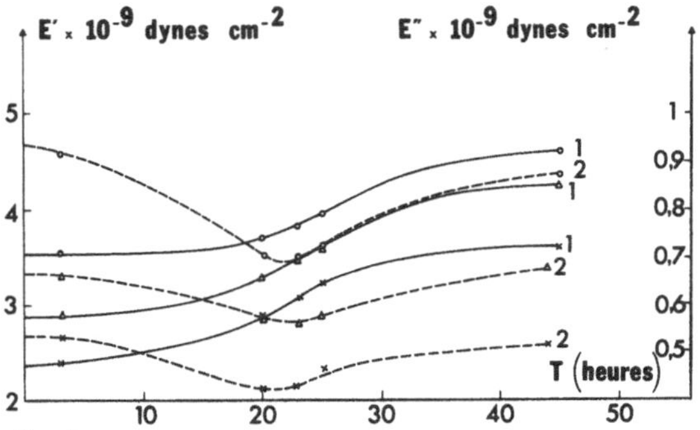

Fig. 1

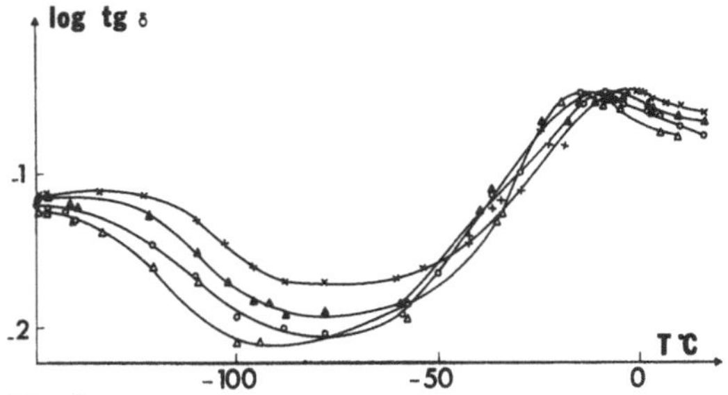

Fig. 2

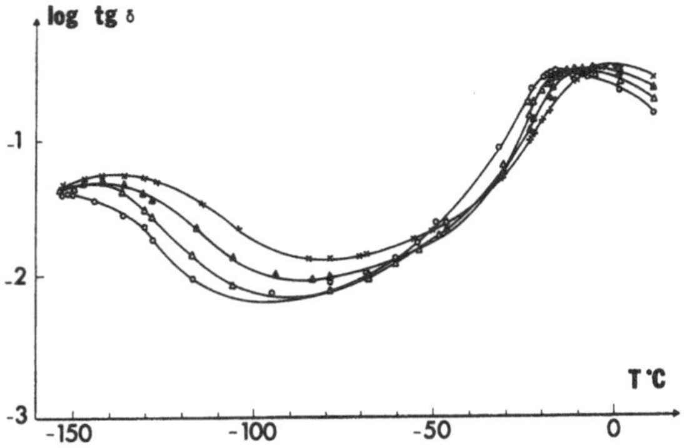

Fig. 3

D'après les travaux effectués au laboratoire sur les polyéthylènes (6), le polypropylène (7), le polybutène 1 (8) et le polyhexène 1 (9), il semble possible d'associer ce mécanisme de relaxation à un mouvement des chaînes latérales situées dans la partie amorphe, ce qui entraînerait un mouvement généralisé de la chaîne principale. Dans cette hypothèse, les ramifications étant

Tableau 3 (d'après 4)

Forme	Maille	Paramètres cristallins (Å)			Type d'hélice	Distance entre deux unités monomères (Å)
		a	b	c		
I	monoclinique	11,2	20,85	6,49	3_1	2,16
II	monoclinique	19,3	16,9	7,08	4_1	1,77

identiques lors de la transformation de la forme II en forme I, il ne doit pas y avoir déplacement du maximum d'absorption au cours du temps ce qui a bien été constaté expérimentalement. La diminution d'amplitude peut être reliée à l'accroissement de cristallinité qui a lieu lors de la transformation, phénomène constaté par *J. C. Boor* et *J. C. Mitchell* (10), par *A. J. Kovacs* (11) et qui serait dû à la poursuite de la cristallisation secondaire de la forme II.

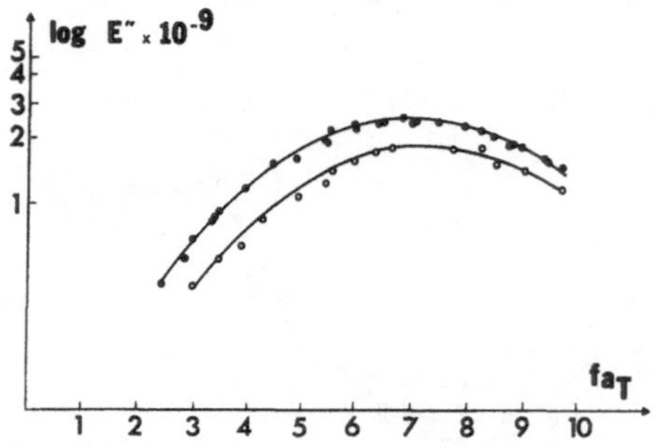

Fig. 4

B.2. Pic de relaxation γ

A fréquence fixe, on peut noter une diminution de l'amplitude du pic au cours du temps. Le maximum de tangente δ évolue de $7{,}8 \cdot 10^{-2}$ à $t = 0$, à $7{,}0 \cdot 10^{-2}$ à $t = 17$ h et à $5{,}4 \cdot 10^{-2}$ à $t = 41$ h. Un décalage a lieu vers les plus basses températures lors du passage de la forme II en forme I. Le mécanisme responsable de la relaxation pourrait être une oscillation localisée des groupements latéraux n'entraînant pas une mise en mouvement de la chaîne principale. Etant donnée la largeur du pic, on peut supposer que celui-ci est la somme de deux composantes γ_A (attribuable aux groupements latéraux situés dans la zone amorphe) et γ_C (groupements rattachés à la zone organisée). Lors de la transformation de II → I, les groupements latéraux s'éloignent (tableau 2): la distance passe de 1,87 Å à 2,17 Å, donc la facilité de mouvement augmente (car l'empêchement stérique diminue) et la mise en mouvement nécessite moins d'énergie. Comme γ_A diminue car la cristallinité augmente, l'enveloppe de ces deux pics verra son intensité diminuer et il y aura déplacement vers les basses températures. La corrélation entre l'amplitude du pic et la transformation a été faite à la suite de l'étude en frottement interne de copolymères stabilisant l'une ou l'autre des deux formes (12).

III. Etude du polypentène 1

Contrairement au cas du polybutène 1, il n'existe pas de transformation monotropique entre les deux formes, chacune ayant son propre domaine de stabilité. Les caractéristiques cristallines sont représentées dans le tableau 3.

Nous avons réalisé une étude en fonction de la température qui a mis en évidence l'existence de deux pics de relaxation β et γ.

A. Pic β

Les courbes de variation de tangente δ en fonction de la température sont représentées sur les figs. 5 (forme I) et 6 (forme II). On notera qu'à une fréquence donnée, le maximum d'absorption se trouve à une température plus élevée pour la forme II que pour la forme I. Par ailleurs la dissymétrie des courbes de variation de tangente δ est plus accentuée pour la forme II, les énergies d'activation étant respectivement de 35 kcal/élément cinétique (variété I) et de 39 kcal/élément cinétique (forme II). Les temps de relaxation obtenus à partir des courbes maîtresses d'absorption ont pour valeurs respectives $5 \cdot 10^{-6}$ secondes et $16 \cdot 10^{-6}$ secondes à 278 °K (fig. 7).

Ces résultats nous ont conduit à compléter (13) l'hypothèse émise sur l'origine de la relaxation β dans la première partie. En effet, il est possible de distinguer deux types principaux de ramifications dans un polymère: celles de la partie totalement amorphe et celles, qui, reliées à la partie cristalline sont rejetées dans la zone amorphe. Quand les groupements latéraux sont petits (CH_3, C_2H_5) seules les chaînes se trouvant entièrement dans la zone amorphe se mettent en mouvement.

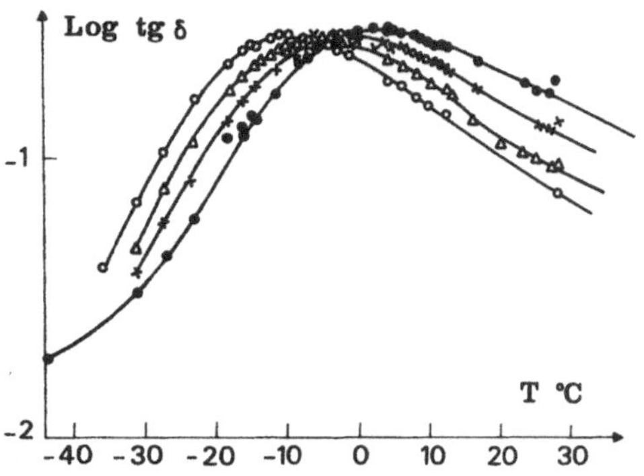

Fig. 5

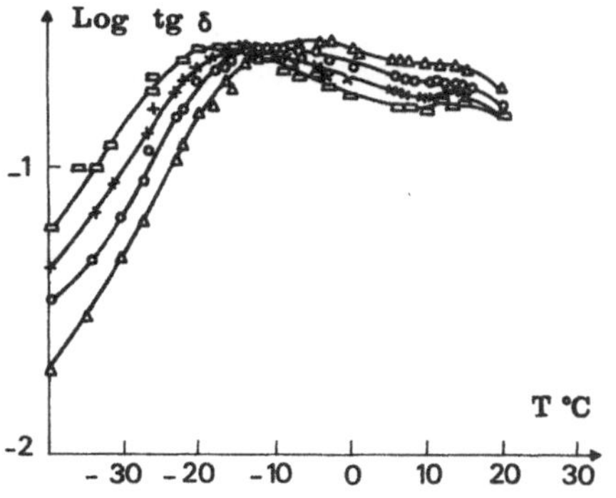

Fig. 6

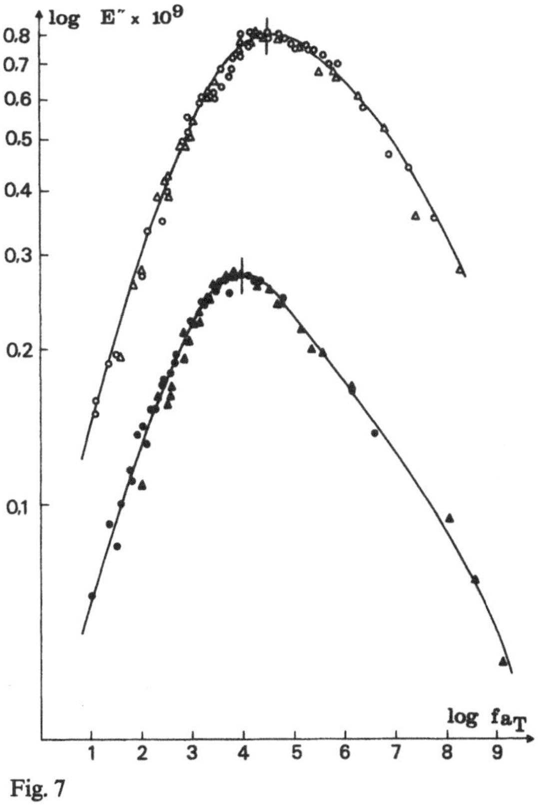

Fig. 7

A partir d'une certaine taille (en l'occurrence trois carbones) il y a superposition du phénomène de mise en mouvement généralisé des chaînes amorphes et de la mise en vibration des groupements R rejetés à l'extérieur des cristallites, ce qui explique la dissymétrie des courbes de variations de tangente δ. Par ailleurs, dans le cas du polypentène amorphe (14) et du polyhexène, on n'a pas observé de dissymétrie. En effet, il n'y a pas de cristal et il ne saurait donc y avoir de vibration de chaînes rejetées à l'extérieur des cristallites. Cette dissymétrie est accentuée dans le cas de la forme II du polypentène car la distance entre deux unités monomères étant plus faible (tableau 1) l'énergie à fournir pour atteindre la résonance est plus grande (39 kcal au lieu de 35) et le temps de relaxation plus long (l'empêchement stérique augmente).

Il nous a été possible de montrer (9), en calculant les temps de relaxation à partir des courbes maîtresses du polypropylène, du polybutène 1 (I et II), du polypentène 1 (I et II) et du polyhexène 1, à différentes températures, que le polypentène 1 présentait effectivement une anomalie (fig. 8 et 9). Celle-ci disparaît quand la température diminue car les chaînes latérales reliées à la partie cristalline sont figées, alors que celles situées dans la zone amorphe sont encore susceptibles de se mettre en mouvement.

B. *Pic γ*

Les premiers résultats semblent indiquer l'existence de deux pics non résolus pour la forme I et bien séparés pour la forme II, le pic le plus élevé en température de la variété II étant situé plus haut que le pic correspondant de la forme I.

Il semblerait possible, comme dans le cas du polybutène, de les attribuer aux mouvements de rotation empêchée des groupements situés dans la zone amorphe (γ_A) (pic basse température) et à ceux des groupements rattachés à la zone cristalline (γ_C) (pic haute température). Dans cette hypothèse, le pic γ_C serait donc bien situé plus haut dans le cas de la forme II que dans le cas de la forme I car l'empêchement stérique augmente. Néanmoins, il nous est actuellement difficile de vérifier ce mécanisme: il est, en effet, délicat de comparer ces deux formes qui n'ont pas le même type de préparation et des organisations cristallines différentes. Nous réalisons actuellement de nouvelles manipulations sur la variété II dont nous chercherons à faire varier la cristallinité,

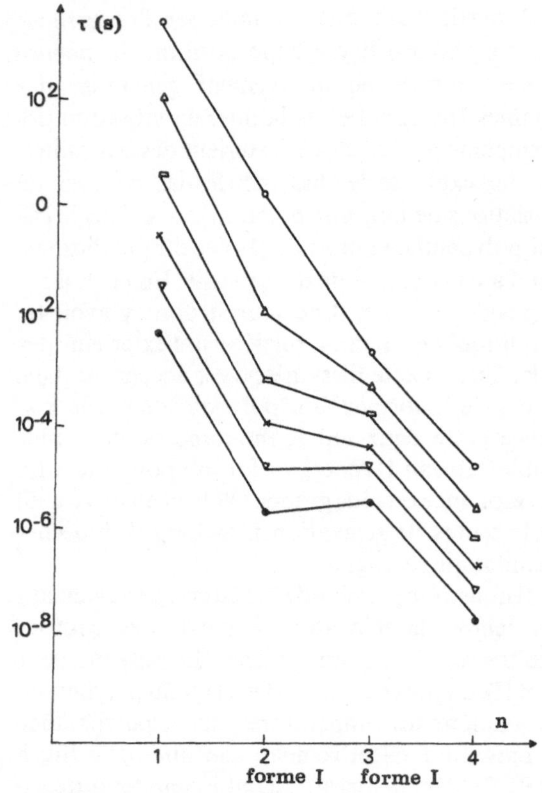

Fig. 8

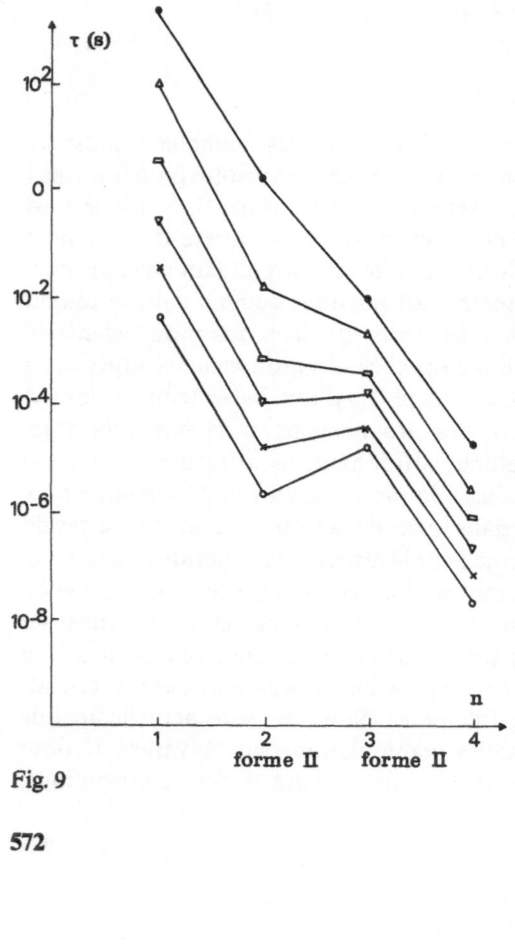

Fig. 9

paramètre que nous ne pouvons que très difficilement maîtriser lors de la préparation.

Conclusions

Dans le cadre d'une étude plus générale de la relation entre structure et morphologie des polymères et de leurs caractéristiques rhéologiques, nous avons recherché l'influence des ramifications sur les mécanismes de relaxation dans les polyoléfines de type $(CH_2—CHR)_n$. Nous avons pu mettre en évidence l'importance de la variété cristalline dans le cas de deux polymères présentant des variétés allotropiques.

Pour le polybutène 1, le pic β serait dû à la mise en mouvement des groupements latéraux situés dans la partie amorphe et entraînant une mise en mouvement généralisée de la chaîne principale. Les temps de relaxation sont identiques pour les deux formes, les ramifications étant les mêmes. Au niveau du pic γ a lieu un déplacement vers les basses températures et une diminution d'amplitude attribuée à la transformation II→I, la distance entre deux groupements latéraux jouant ici un rôle important.

Pour le polypentène 1, au niveau du pic β viendrait se superposer aux mouvements situés dans la zone amorphe, la vibration des groupements latéraux reliés à la phase organisée et rejetés à son extérieur. Le décalage entre les courbes de variation de tangente δ des formes I et II pouvant s'interpréter en faisant intervenir la distance entre deux groupes latéraux. Au niveau du pic γ, les premiers résultats obtenus semblent compatibles avec les hypothèses émises.

Remerciements

Les auteurs remercient la Société *Ethylène Plastique* d'avoir assuré le support matériel de ce travail.

Littérature

1) *Convard, D.*, Thèse de Doctorat de Spécialité (Lyon 1972).
2) *Stivala, S. S., R. J. Valles* et *D. W. Levi*, J. Applied Polym. Sci **7**, 97 (1963).
3) *Natta, G., P. Corradini* et *I. W. Bassi*, Nuovo Cimento **15**, 52 (1960).
4) *Turner Jones, A.*, Polymer **7**, 23 (1966).
5) *Williams, M. L., R. F. Landel* et *J. D. Ferry*, J. Applied Phys. **26**, 359 (1955).
6) *Lissac, P.*, Thèse d'Etat (Lyon 1970).
7) *Smadja, C.*, Thèse de Docteur-Ingénieur (Lyon 1967).
8) *Decroix, J. Y., J. F. May* et *G. Vallet*, Makromol. Chem. **163**, 295 (1973).

9) *Decroix, J. Y., P. Lissac, J. F. May* et *G. Vallet*, European Polym. J. **9**, 137 (1973).

10) *Boor, J. C.* et *J. C. Mitchell*, J. P. S. **62**, 70 (1962).

11) *Vidotto, G.* et *A. J. Kovacs*, Kolloid-Z. u. Z. Polymere **220**, 1 (1967).

12) *Pineri, M.*, Communication privée (Grenoble 1969).

13) *Decroix, J. Y., G. Nemoz, J. F. May* et *G. Vallet*, C. R. Acad. Sci. **C 275**, 605 (1972).

14) *Woodward, A. E., J. A. Sauer* et *R. A. Wall*, J. P. S. **50**, 117 (1961).

Adresse de l'auteur

Laboratoire de Chimie Macromoléculaire
U.E.R. de Chimie–Biochimie
Université Claude Bernard «Lyon 1»
43, Boulevard du 11 Novembre 1918
F-69100 Villeurbanne (France)

Rheol. Acta 13, 66–71 (1974)

Laboratoire de Chimie Macromoléculaire – U.E.R. de Chimie – Biochimie – Université Claude Bernard «Lyon 1», Villeurbanne (France)

Propriétés mécaniques dynamiques de quelques polytéréphthalates de polymèthylèneglycol*)

Par G. Nemoz, J.-F. May et G. Vallet

Avec 11 figures et 4 tableaux

(Reçu p. p. le 27. Octobre 1972)

Introduction

Différents polytéréphtalates de polyméthylèneglycol de formule générale:

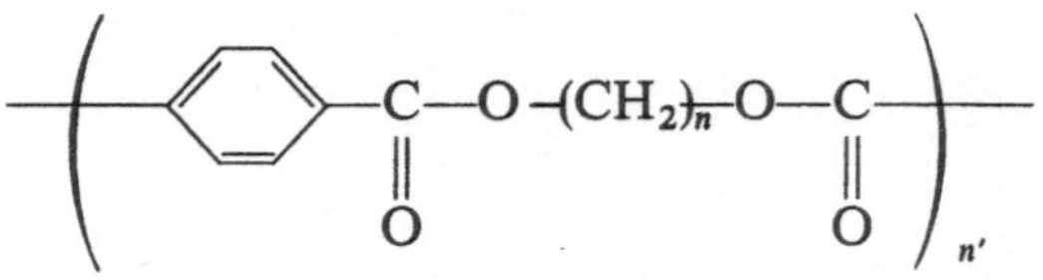

($n = 2$, 4, 5, 6 nommés respectivement 2 GT, 4 GT, 5 GT et 6 GT)

ont été étudiés dans notre laboratoire. Nous décrirons plus particulièrement certaines de leurs propriétés mécaniques dynamiques.

Deux pics de relaxation mécanique importants se manifestent quand on augmente la température; le pic basse température correspond à la relaxation γ et le pic haute température est attribué à la transition vitreuse.

De nombreux auteurs ont proposé différentes hypothèses pour expliquer ces transitions: nous nous sommes intéressés plus particulièrement à la relaxation γ, et les résultats obtenus liés à une étude antérieure de la structure et de la morphologie nous ont suggéré de nouvelles hypothèses permettant de mieux interpréter l'ensemble des données expérimentales.

A. Elaboration et caractérisation des échantillons

I. Elaboration des échantillons

Les échantillons sont moulés par fusion sous vide (10^{-2} torr), puis compression, sous forme de films, à partir de grains de polymères placés entre deux feuilles d'aluminium; les films cristallisent ensuite par refroidissement lent (0,5 °C/mn) dans le moule. Les échantillons ont une épaisseur de l'ordre de $7 \cdot 10^{-2}$ mm.

*) Le texte de cette communication reprend l'essentiel d'un article publié dans European Polymer J. (1973).

Différents taux de cristallinité du film de polyester 6 GT sont obtenus par les traitements thermiques décrits dans le tableau 1.

Tableau 1

Echantillons	Traitements thermiques
1	Fondu à 170 °C et trempé à −80 °C
2	Fondu à 170 °C et trempé à +15 °C
3 a, b, c, d	Fondu à 170 °C et refroidi dans le moule
4	Fondu à 170 °C, refroidi dans le moule et recuit à 146 °C pendant 48 h

II. Caractéristiques structurales et morphologiques des échantillons

Les viscosités intrinsèques $\{\eta\}$ déterminées sur des solutions de polymère dans l'orthochlorophénol à 25 °C, la masse spécifique μ_{cr} de la phase cristalline et les conformations des chaînes dans la maille cristalline ont été déterminées par *A. M. Joly* (1) (tableau 2).

Tableau 2

Polyester	$\{\eta\}$ $cm^3 \cdot g^{-1}$	μ_{cr} $g \cdot cm^{-3}$	Conformations des CH_2 [1])
2 GT	–	1,455 (3)	T
4 GT	79	1,394	GTG
5 GT	55	1,340	TGGT
6 GT	72	1,335	TTTTT

[1]) G = conformation gauche, T = conformation trans.

1. Cristallinité

La détermination de la cristallinité est réalisée soit par mesure de l'enthalpie de fusion ΔH_f, soit par diffraction des rayons X. Les valeurs de la cristallinité sont donc calculées en utilisant les relations classiques:

$$X' = \frac{\Delta H_f}{\Delta H_\infty} \cdot 100$$

[les valeurs de ΔH_∞ que nous emploierons sont celles données par *Kirshenbaum* (2)].

$$X = \frac{A_c}{A_a + A_c} \cdot 100.$$

A_c et A_a sont les aires correspondant à la diffraction des rayons X par les phases cristalline et amorphe.

Les résultats de nos mesures sont reportés dans le tableau 3.

Tableau 3

Echantillon	X' (%)	X (%)	Echantillon	X (%)
2 GT	46	–	6 GT n° a	12
4 GT	27	–	6 GT n° b	16,5
5 GT	21	–	6 GT n° c	17,5
6 GT n° 1	–	0	6 GT n° d	20
6 GT n° 2	–	16		
6 GT n° 3	23	22		
6 GT n° 4	30	27,5		

2. *Analyse des conformations trans et gauche du polyester 6 GT par spectrométrie infrarouge*

a) *Comparaison des spectres des polyesters 2 GT et 6 GT*

Les spectres de la fig. 1 nous montrent qualitativement l'évolution des bandes d'absorption caractéristiques des formes trans et gauche des groupements méthylènes; on peut attribuer la signification suivante aux bandes d'absorption du polyester 2 GT (4) (tableau 4). Ces bandes se retrouvent dans le spectre du polyester 6 GT.

Tableau 4

ν cm^{-1}	Nature des mouvements
1171	Torsion des CH_2 en forme gauche
896	Rotation des CH_2 en forme gauche
871	Vibration hors du plan des CH du noyau benzénique
847	Rotation des CH_2 en forme trans

On peut comparer qualitativement l'évolution de ces formes, en mesurant l'absorption relative $A = \log I_0/I$ d'une bande par rapport à la bande 871 cm^{-1}; les lignes de base explicitées sur la fig. 1 sont construites comme l'indique la plupart des auteurs (5, 6).

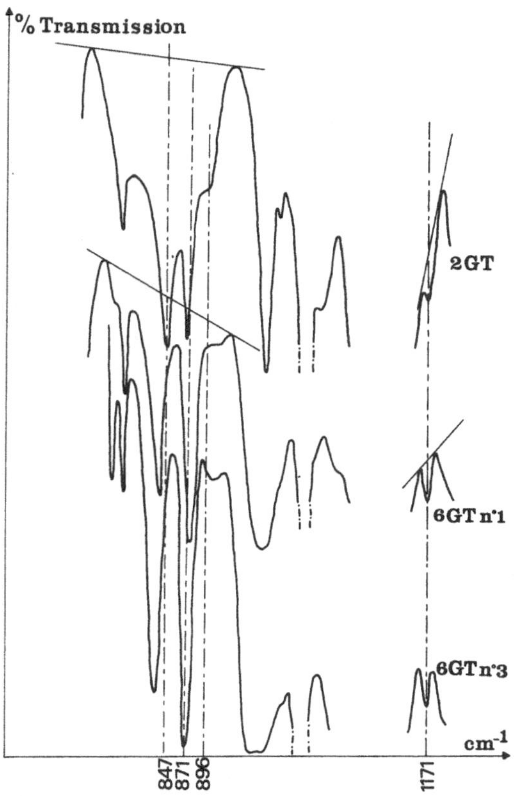

Fig. 1. Spectres infrarouges des polyesters

b) *Influence de la cristallinité du polyester 6 GT*

P. G. Schmidt (5) a montré, pour le polyester 2 GT, que, d'une part, la cristallinité augmentait linéairement avec la quantité de forme trans pour un échantillon non orienté et, d'autre part, qu'un échantillon amorphe contient une certaine proportion déterminée de forme trans et de forme gauche.

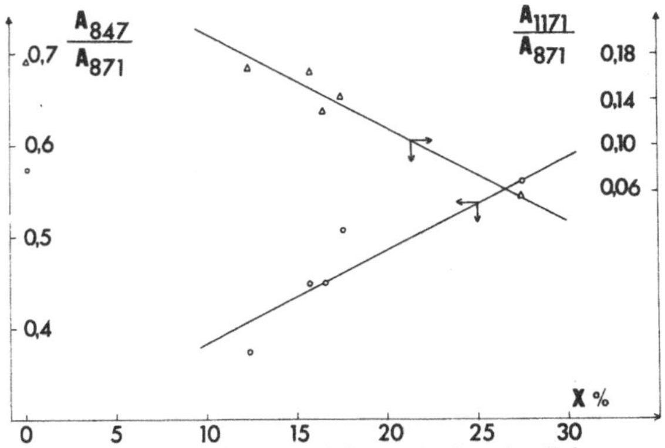

Fig. 2. Influence de la cristallinité sur les bandes d'absorption infrarouge du polyester 6 GT. △ forme gauche, ○ forme trans

La fig. 2 montre une évolution analogue des formes trans et gauche en fonction de la cristallinité pour le polyester 6 GT; il y a une exception à cette évolution: la trempe de l'échantillon amorphe a dû provoquer des contraintes, et il possède sans doute, pour cette raison, plus de forme trans.

A partir de ces mesures, il est possible de proposer un schéma de principe reporté sur la fig. 3. Si on représente l'évolution de la forme trans T en fonction de la cristallinité, on peut séparer la forme trans dans la phase cristalline (T_c) et dans la phase amorphe (T_a), sachant que pour $X = 100\%$ il n'y a que de la forme trans et qu'à $X = 0\%$ il y a plus de forme trans que de forme gauche d'après les spectres d'absorption infrarouge. Etant donné que, quelle que soit la cristallinité le nombre d'enchaînements trans + gauches est constant, par différence, nous pouvons suivre l'évolution du nombre G d'enchaînements gauches en fonction de la cristallinité.

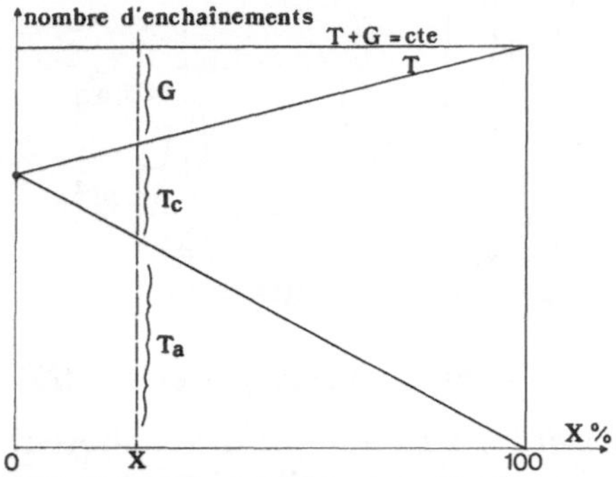

Fig. 3. Schéma de principe de l'évolution des bandes d'absorption infrarouge. G: forme gauche, T_c: forme trans de la phase cristalline, T_a: forme trans de la phase amorphe

B. Propriétés mécaniques dynamiques

I. Principe de mesure

Les mesures de la tangente de l'angle de perte mécanique tgδ ont été réalisées à l'aide d'un viscoélasticimètre rhéovibron DDVII (Toyo Measuring Instrument). Il est possible d'apprécier des valeurs de tgδ de l'ordre de $2 \cdot 10^{-3}$; on effectue les mesures en fonction de la température de -180 °C à $+200$ °C, à quatre fréquences fixes: 3,5, 11, 35 et 110 Hz.

L'appareil est construit sur le principe des essais de tractions alternées. L'échantillon, placé sous tension, est soumis à des vibrations longitudinales de faibles amplitudes. Etant données les caractéristiques mécaniques des échantillons, la contrainte statique qui leur est imposée, est très nettement en-dessous de la limite élastique; la microdéformation dynamique est de l'ordre de $6 \cdot 10^{-4}$.

La mesure de tgδ est effectuée par comparaison entre le signal émis par le capteur de force qui enregistre la vibration transmise par l'échantillon et le signal émis par le capteur de déplacement qui mesure la vibration envoyée à l'échantillon.

II. Résultats expérimentaux

L'évolution du pic de relaxation γ, est analysée en fonction du nombre de méthylènes du motif de répétition de chaque polyester et de la cristallinité.

1. Influence du nombre n de méthylènes

D'après les figs. 4 et 5 nous constatons que les pics de relaxation se déplacent vers les basses températures quand le nombre de méthylènes augmente et se rapprochent du pic de relaxation γ du polyéthylène (7): son maximum étant vers -110 °C pour une fréquence de 110 Hz.

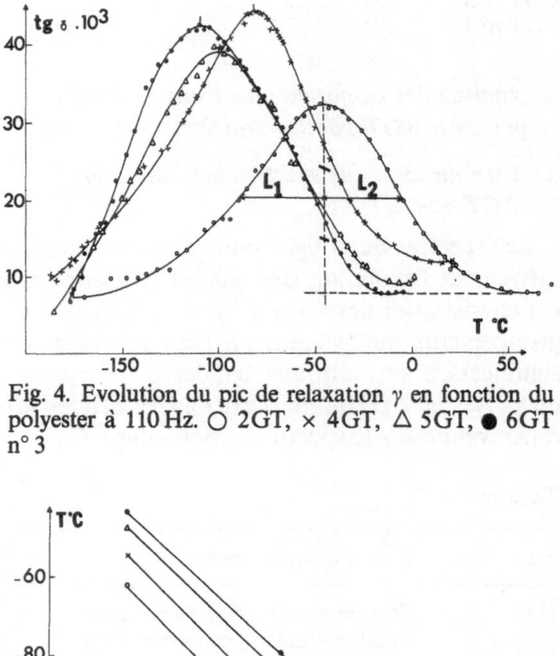

Fig. 4. Evolution du pic de relaxation γ en fonction du polyester à 110 Hz. ○ 2GT, × 4GT, △ 5GT, ● 6GT n° 3

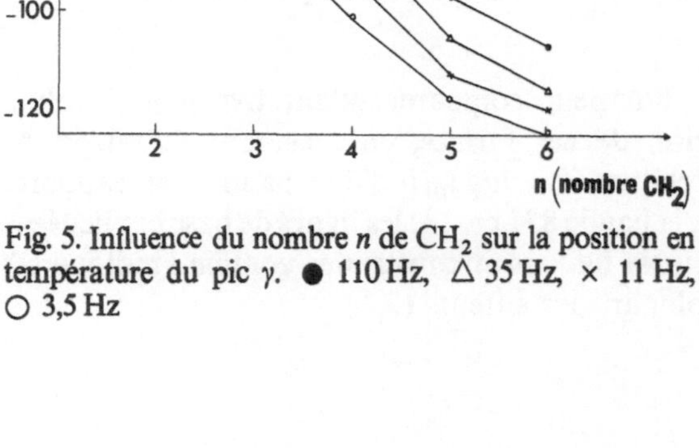

Fig. 5. Influence du nombre n de CH_2 sur la position en température du pic γ. ● 110 Hz, △ 35 Hz, × 11 Hz, ○ 3,5 Hz

L'énergie d'activation E_γ, calculée par la loi d'*Arrhénius*, diminue rapidement quand le nombre de méthylènes augmente (fig. 6), ce qui traduit l'accroissement de la mobilité des chaînons de la chaîne quand le nombre de méthylènes entre les noyaux benzéniques devient plus grand. Ces résultats sont en accord avec ceux de plusieurs auteurs (8–12).

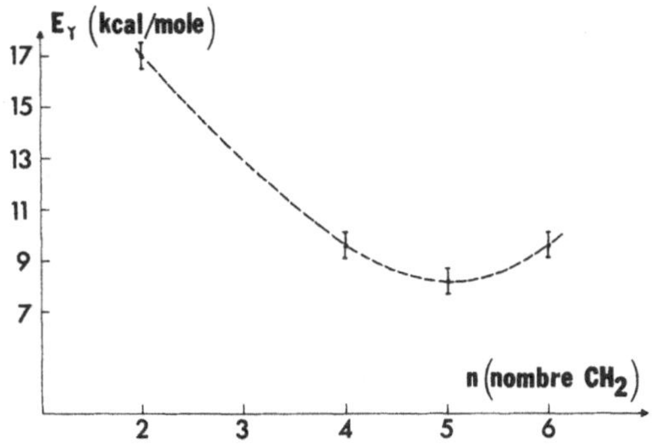

Fig. 6. Influence du nombre n de CH_2 sur l'énergie d'activation E_γ de la relaxation γ

Cette valeur de l'énergie d'activation n'est cependant qu'apparente, car nous constatons que les pics de relaxation ne sont pas symétriques et que leur coefficient de symétrie $R = L_1/L_2$, défini sur la fig. 4 comme étant le rapport des largeurs à mi-hauteur, diminue quand le nombre de méthylènes augmente (fig. 7).

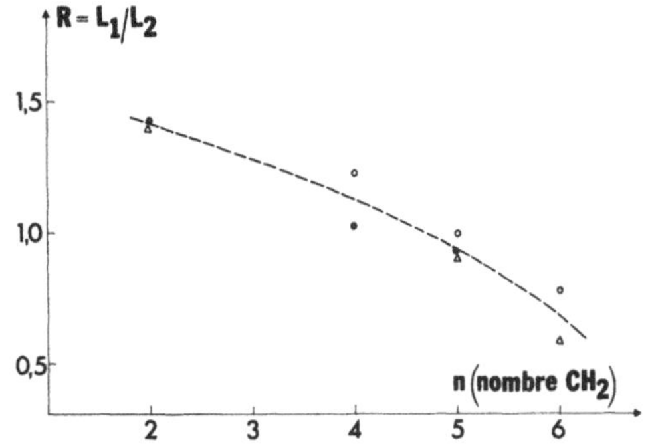

Fig. 7. Evolution du rapport de symétrie $R = L_1L_2$ en fonction de n (nombre de CH_2). ● 3,5 Hz, △ 11 Hz, ○ 110 Hz

Ainsi, nous sommes amenés à penser que ces pics peuvent se décomposer en plusieurs composantes, comme l'on suggéré divers auteurs (8, 9, 13).

2. *Influence de la cristallinité sur le pic de relaxation γ du polyester 6 GT*

Les courbes de la fig. 8 montrent que lorsque la cristallinité diminue, l'intensité du pic de relaxation augmente et qu'un épaulement apparaît vers les hautes températures par rapport au maximum initial. Cet épaulement pour $X = 0$ devient très important puisque la totalité du pic se déplace vers les plus hautes températures. Ceci suggère donc une décomposition de la courbe en 2 composantes principales γ_1 et γ_2.

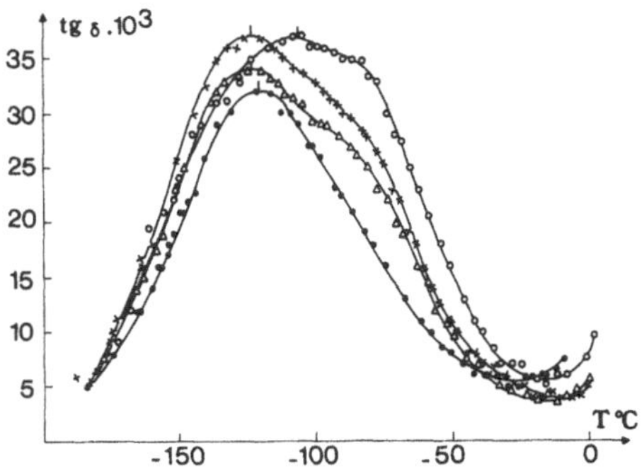

Fig. 8. Pic de relaxation γ du polyester 6 GT à 11 Hz – Influence de la cristallinité. ○ 1 ($X \simeq 0\%$), × 2 ($X \simeq 16\%$), △ 3 ($X \simeq 22\%$), ● 4 ($X \simeq 27{,}5\%$)

3. *Décomposition des courbes de relaxation*

Notre principe de décomposition est représenté sur la fig. 9; nous utilisons deux composantes gaussiennes dont nous fixons la largeur à mi-hauteur et la position relative, pour une même fréquence. Les seules variables seront alors la hauteur de chacune d'elles et un paramètre morphologique.

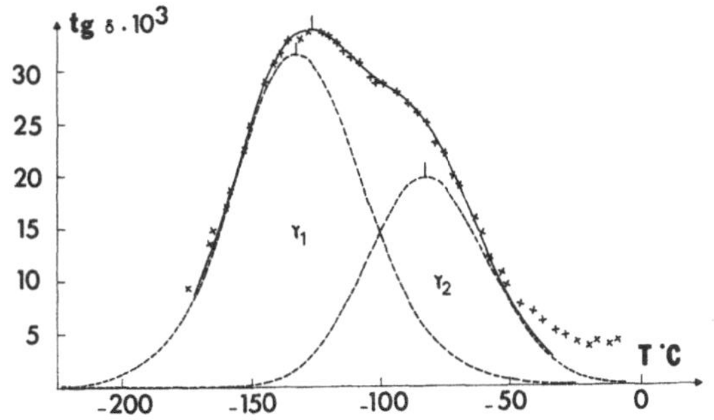

Fig. 9. Décomposition du pic γ du polyester 6 GT à 11 Hz en deux composantes γ_1 et γ_2

Ainsi, il nous est possible de suivre l'évolution des aires relatives S_1/S_2 et du maximum $\operatorname{tg}\delta_1$ et $\operatorname{tg}\delta_2$ de chacune de ces composantes, quand la cristallinité du polyester 6 GT varie (fig. 10, 11).

Nous constatons que la composante basse température γ_1 évolue peu et demeure plus importante que la composante γ_2 qui diminue très vite quand la cristallinité augmente, le rapport des aires ne demeurant pas constant.

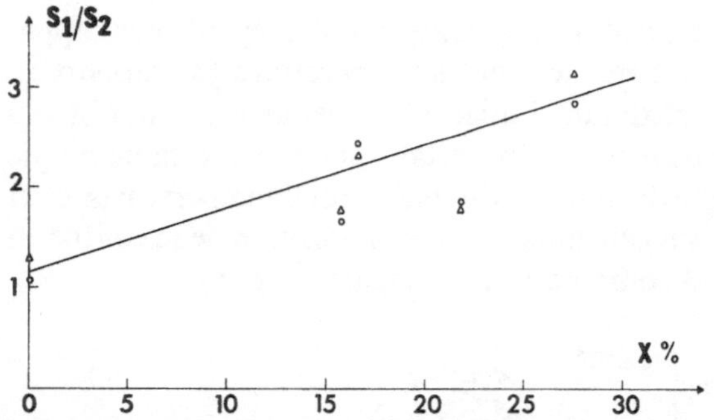

Fig. 10. Evolution du rapport des aires des composantes 1 et 2 du pic γ du polyester 6 GT en fonction de la cristallinité. ○ 110 Hz, △ 11 Hz

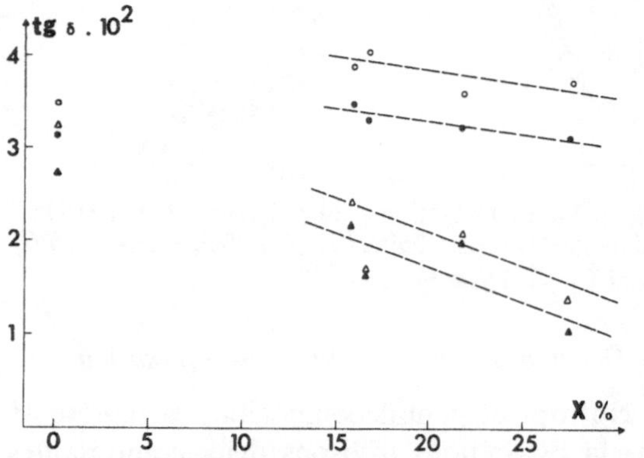

Fig. 11. Evolution des maximums des composantes γ_1 et γ_2 du polyester 6 GT avec la cristallinité. ○ $tg\delta_1$ et △ $tg\delta_2$ = 110 Hz; ● $tg\delta_1$ et ▲ $tg\delta_2$ = 11 Hz

Discussion

De nombreux auteurs (8, 10) admettent que l'on peut attribuer à la composante basse température, des mouvements empêchés de rotation des groupements méthylènes et à la composante haute température des vibrations dues au groupement carboxyle. Si cette hypothèse devait se vérifier, il nous semble que le rapport des aires des composantes γ_1 et γ_2 ne devrait pas varier quand on augmente la cristallinité d'un même polyester, ce qui n'est pas vérifié (cf. fig. 10).

Récemment *E. Sacher* (14), à partir de mesures diélectriques et de données de spectrométrie infrarouge sur un échantillon de polyester 2 GT, a émis l'hypothèse suivante: les mouvements localisés entraînant la relaxation γ sont des mouvements coopératifs impliquant plusieurs motifs élémentaires, engendrés par les dipôles du groupement carboxyle soumis à l'action du champ électrique; des différences de conformation pourraient être à l'origine des deux composantes du pic: le pic γ_2 serait associé à des mouvements dus aux conformations trans et le pic γ_1 aux conformations gauches.

Cette hypothèse est intéressante. Cependant, elle fait intervenir une mesure globale du rapport trans/gauche sans différencier la phase amorphe de la phase cristalline, impliquant par la même une intensité analogue des mouvements dans les deux phases.

Il nous semble fondamental de distinguer si les mouvements ont lieu dans l'une des phases ou dans les deux à la fois; en opposition avec l'hypothèse de *E. Sacher*, nous avons constaté sur le polyester 6 GT que, lorsque la cristallinité augmente, le nombre total d'enchaînements trans augmente alors que l'intensité du pic γ_2 diminue rapidement (fig. 11).

Ainsi, il nous semble plausible d'envisager les hypothèses suivantes: le pic γ_1 serait dû à des mouvements des groupements méthylènes en conformation gauche qui n'existent que dans la phase amorphe et le pic γ_2 à des mouvements des groupements méthylènes, en conformation trans, de la phase amorphe.

En effet, le pic γ_1 correspond à des mouvements plus faciles que le pic γ_2; pour le polyester 6 GT, le pic γ_1 est plus intense que le pic γ_2 car les enchaînements gauches, bien que moins nombreux que les enchaînements trans, ont plus de liberté pour vibrer. Comme le montre le schéma de principe (fig. 3), le nombre des enchaînements trans dans la phase amorphe décroît plus vite que le nombre des enchaînements gauches quand la cristallinité augmente. Cette évolution est à rapprocher de celles des maximums $tg\,\delta_1$ et $tg\,\delta_2$ des composantes γ_1 et γ_2 (fig. 11).

Ces hypothèses permettent d'expliquer que l'échantillon amorphe de polyester 6 GT présente un pic γ_2 relativement important car il a beaucoup d'enchaînements trans dus à une certaine orientation au cours de son élaboration.

Nos hypothèses sont applicables également aux polyesters 2 GT, 4 GT, 5 GT car nous avons constaté le rôle prépondérant que jouait le nombre de méthylènes dans la flexibilité de la chaîne.

Nous tenons à exprimer nos remerciements à Monsieur *Rochas*, Directeur du Centre de Recherches de la Soierie et des Industries Textiles de l'aide qu'il a bien voulu nous apporter.

Résumé

L'étude par spectrophotométrie infrarouge de la conformation des méthylènes dans les phases amorphe et cristalline en fonction du taux de cristallinité du polytéréphtalate d'hexaméthylèneglycol nous a permis de proposer un mécanisme de relaxation γ pour divers polytéréphtalates.

Deux composantes γ_1 et γ_2 sont associées à ce mécanisme. Elles font intervenir respectivement des mouvements, dans la phase amorphe, des méthylènes en conformation gauche et trans.

Summary

The conformation of methylenes in amorphous and crystalline phases as a function of cristallinity of polyhexamethylene terephtalate is investigated by infrared spectrophotometry.

The γ relaxation mechanism of several polyterephtalates can be explained using two components γ_1 and γ_2. They involve respective motions of gauche and trans methylenes in the amorphous phase.

Littérature

1) *Joly, A. M.*, Thèse Faculté des Sciences (Lyon 1970).

2) *Kirshenbaum, I.*, J. Polymer. Sci. **A**, 1869 (1965).

3) *Hellwege, K. H., J. Hennig* et *W. Knappe*, Kolloid-Z. u. Z. Polymere **186**, 29 (1962).

4) *Hummel/Scholl*, Infra-red Analysis of Polymers.

5) *Schmidt, P. G.*, J. Polymer. Sci. **A1**, 1271 (1963).

6) *Koenig, J. L.* et *M. J. Hannon*, J. Macromol. Sci. Phys. **B1**, 119 (1967).

7) *Takayanagi, M.*, Proc. 4th Inter. Congress on Rheology, Part 1, 161 (1965).

8) *Armeniades, C. D.*, J. Polymer Sci. A2, **9**, 1345 (1971).

9) *Sacher, E.*, J. Polymer Sci. A2, **6**, 1935 (1968).

10) *Farrow, G., J. McIntosh* et *I. M. Ward*, Makromol. Chem. **38**, 147 (1960).

11) *Armeniades, C. D.*, J. Macromol. Sci. Phys. **B1**, 777 (1967).

12) *Illers, K. H.* et *H. Breuer*, J. Colloid Sci. **18**, 1 (1963).

13) *Tagayanagi, M.*, J. Macromol. Sci. Phys. **B1**, 741 (1967).

14) *Sacher, E.*, J. Macromol. Sci. Phys. **B5**, 739 (1971).

Adresse de l'auteur:

Laboratoire de Chimie Macromoléculaire
U.E.R. de Chimie-Biochimie
Université Claude Bernard «Lyon 1»
43, Boulevard du 11. Novembre 1918
F-69100 Villeurbanne (France)

Rheol. Acta 13, 72–77 (1974)

From the Department of Fibre Science, University of Strathclyde, Glasgow (Scotland)

Influence of chemical structure on the rheological properties of polyurethane fibres with varying hard segment concentrations

By J. Ferguson and D. Patsavoudis

With 9 figures and 1 table

(Received October 27, 1972)

Polyurethanes containing alternating soft and hard segments along their chain length can be regarded as one type of molecular species that form polymers containing multiphase systems. Others include the styrene-butadiene block copolymers and segmented polycarbonates. The examination of such polymers offer unique opportunities for studying the interactions between chemical and physical structures and their influence on the rheological properties of the polymer. For example a considerable amount of research has been carried out to find the influence of segment concentration on such properties as tensile strength elongation modulus and melting point (1–7). For commercial reasons most of the work in this field has been carried out on elastomeric systems in an attempt to obtain polymers with high strength, elongation and elastic recovery, and low stress relaxation. Little work has been carried out on those multiphase systems which contain high hard segment concentrations and show low elastic recovery.

In a previous study, a series of polyurethanes were synthesised from Adiprene L 100 (a polyether-based macrodiisocyanate) of molecular weight of approximately 2000, trimethylene diamine and 4,4′ diphenylmethane diisocyanate (MDI). These were wet spun into fibres and their tensile (8) and dynamic mechanical properties (9) measured. It was found that their specific (actual) breaking stress, elongation, elastic recovery and stress relaxation characteristics were all dependent on soft segment (polyether) concentration. A transition region appeared at approximately 50% soft segment/50% hard segment concentration. This was particularly obvious in the stress relaxation index which went through a maximum in the same concentration region. Morphological examination also showed that a phase change occurred at approximately this concentration. At lower hard segment concentrations the soft segments became the continuous phase and vice versa. Several points arise as a result of this work. For example what is the effect of soft segment molecular weight, and do domains continue to be formed even at extremely high soft segment concentrations? To answer these questions a further series of polyurethanes were synthesised using Adiprene L 42 instead of Adiprene L 100. The former has a higher molecular weight (Ω 3000) than Adiprene L 100. The effect is to produce a series of polyurethanes with higher molecular weight soft segments.

Experimental

A total of seven polymers were synthesised. Fig. 1 gives typical examples of their chemical structure. Six of these had almost exactly the same concentration of hard and soft segments as polymers in the first series. The polymer containing 100% hard segment can be regarded as being common to both series. Table 1 shows the concentrations of soft segments in both series.

Polymer synthesis

Adiprene L 42 was supplied by Du Pont de Nemours Ltd and the isocyanate content determined according to ASTM D. 1638-67 T.

A solution polymerisation technique was used. The solvent was dimethylacetamide and the reaction was carried out under nitrogen. Care was taken in the preparation to avoid the formation of side branches in the polymer. The ratios of soft to hard segment was varied by altering the ratios of MDI, trimethylene diamine and Adiprene L 42 in the reaction. The total amount of diisocyanate always equaled the amount of diamine.

Wet spinning

The filaments were wet spun as previously described (7), in an aqueous coagulating bath at 50 °C. When coagulation was complete the filaments were washed in distilled

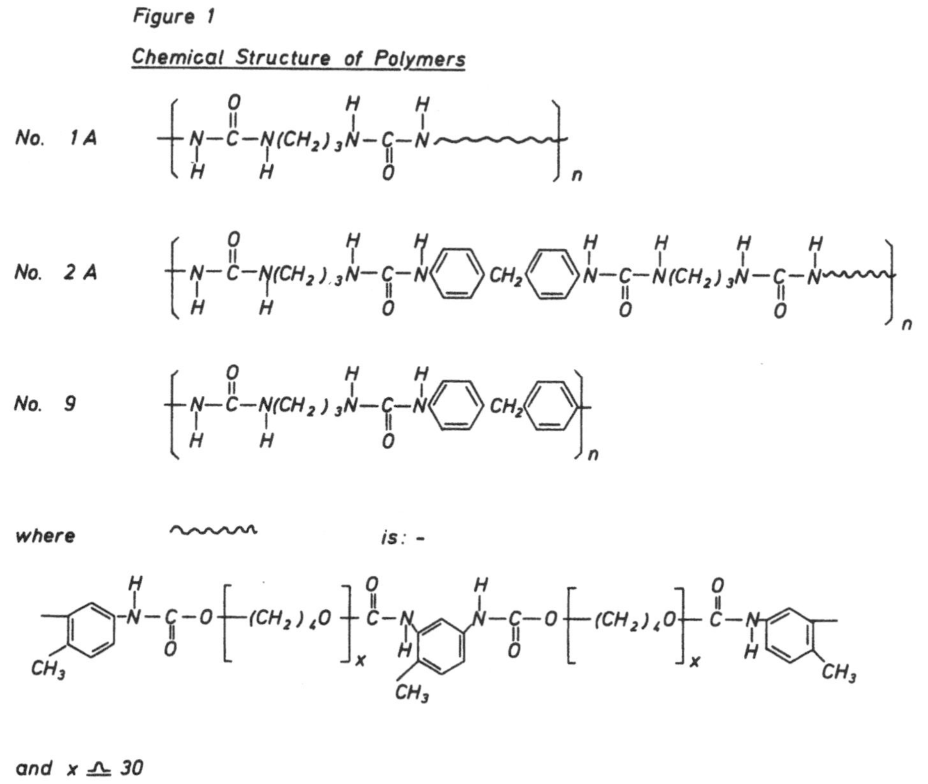

Fig. 1. Chemical structures of some typical polyurethanes

Table 1. Comparison of soft segment concentration in first and second series

Polymer	1st series		1	2	3	4	5	6	7	8	9
	2nd series	1A	2A	–	–	3A	4A	5A	6A	7	
Soft segment concentration (%)	1st series	–	70.8	69.1	67.4	65.8	61.6	54.9	46.1	32.1	0
	2nd series	80.8	70.9	–	–	66.0	61.7	54.9	46.1	32.1	

water at 50 °C, dried, then conditioned at 20 °C and 65% relative humidity for 48 h.

Measurement of rheological properties

A table model Instron tensile tester was used. For stress-strain measurements a rate of extension of 1000%/min was employed with a one inch gauge length. A correction was applied to allow for slippage at the jaws. Elastic recovery was determined from 100 and 200% extension and stress relaxation at 50, 200, and 400% extension over a period of 2 h.

Results and discussion

The count of the filaments ranged from 2.8 to 10.6 tex. Their physical appearence varied widely rangeing from those with a densely wrinkled surface and irregular cross-section to those with an even, glossy surface and nearly circular cross-section. Unlike the previous series no correlation between physical appearance and chemical structure seemed to exist. However, inspection under the microscope revealed a more or less gradual transition of the filaments from translucence to opacity with decreasing soft segment concentration.

Under the circumstances of changing morphology it is extremely important to establish that there are no exceptional structural changes such as voids, which might create errors in subsequent measurements. The density of the filaments was, therefore, measured in a density gradient column and the results are shown in fig. 2. The continuous line gives the correlation between density and soft segment concentration obtained for the first series. The open circles are the results for the second series. All the points fall fairly close to the line indicating that in none of the fibres was the presence of voids significant. It also shows quite clearly that the change in length of the soft segment had no effect on the density/soft segment concentration relationship. This suggests that the morphology is substantially the same in both series.

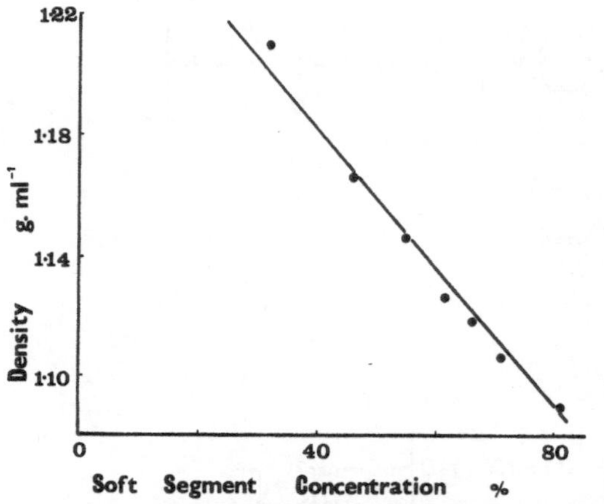

Fig. 2. Influence of soft segment concentration on density

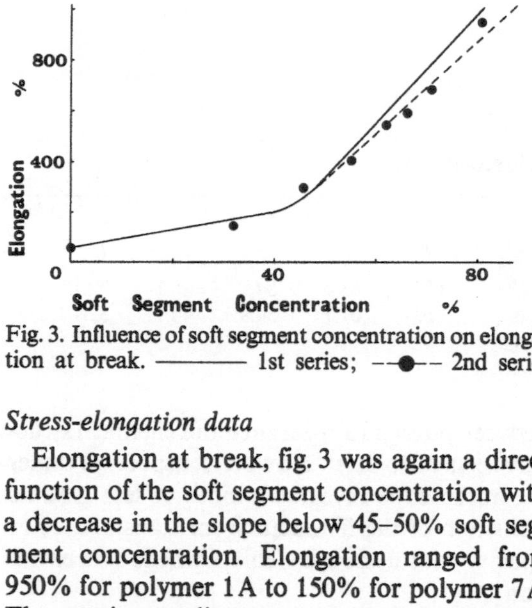

Fig. 3. Influence of soft segment concentration on elongation at break. ——— 1st series; --●-- 2nd series

Stress-elongation data

Elongation at break, fig. 3 was again a direct function of the soft segment concentration with a decrease in the slope below 45–50% soft segment concentration. Elongation ranged from 950% for polymer 1 A to 150% for polymer 7 A. The continuous line once again shows the best curve through the data for the first series. Close examination of fig. 3 reveals some interesting points. Firstly, as in the first series the rate of increase in elongation with increase in soft segment concentration is lower at low than at high soft segment values. It has already been established that below about 50% soft segment concentration the hard phase is continuous and the soft segment phase discontinuous. The reverse is true above about 50% soft segment concentration. Secondly the data for both series would seem to fall on the same line below the transition. However at high soft segment concentrations the presence of a higher molecular weight soft segment in the polyurethane causes a small but definite fall in extension. The explanation is almost certainly that where the hard segment is the continuous phase changes in chemical structure in the discontinuous phase are relatively unimportant since the hard segment is controlling the elongation. Once the phase change occurs and the soft segment becomes continuous the soft segment chemical structure becomes the controlling factor.

The reason why increasing the length of the soft segment causes a fall in elongation is not immediately obvious. This point must be considered in conjunction with other tensile data.

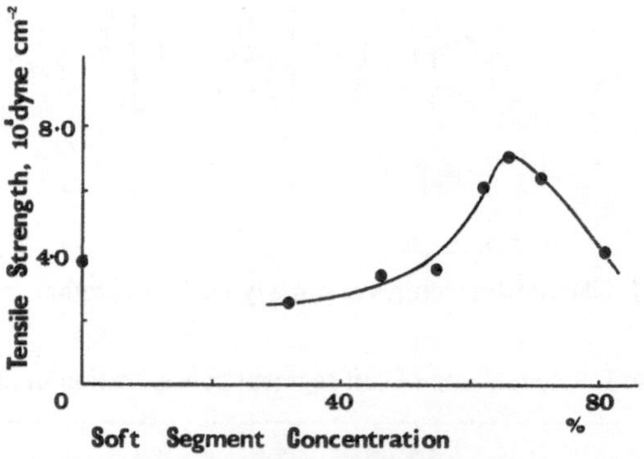

Fig. 4. Influence of soft segment concentration on tensile strength

Tensile strength, fig. 4, is normally calculated using the diametre of the fibre before elongation. Considerable errors can be introduced, however, when, as in the present case, the elongations and as a result the diametres at break, vary so widely. The actual breaking stress, fig. 5, is a quantity which takes into consideration the cross-section of the filament at break. As a result it gives a much more accurate determination of the true variation of breaking stress with soft segment concentration. Two main features are evident in the variation of actual breaking stress with increase in soft segment concentration. Firstly an increase in slope between 40 and 55% soft segment concentration and secondly a maximum at approximately 70% soft segment concentration. The increase in slope is clearly associated with the phase change. The maximum can be regarded as occurring at the concentration at which the reinforcing effect of the hard domains is a maximum.

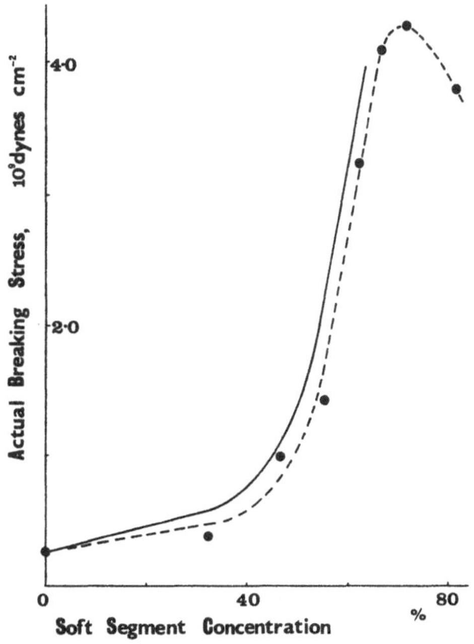

Fig. 5. Influence of soft segment concentration on actual breaking stress. ——— 1st series; --●-- 2nd series

As is generally accepted the hard segment domains act as pseudo cross-links. Since the easily deformable soft segments are linked by covalent bonds to the hard domains viscous flow of the molecule as a whole is prevented. The result is a high elastic recovery and high breaking stress. As soft segment concentration increases a concentration must be reached at which the hard segments become ineffective. This could be due to the domains being too few in number to prevent viscous flow and easy rupture. An alternative explanation is that at very high soft segment concentrations domain formation is less efficient. It is necessary for a number of hard segments in each chain to come in contact before domain formation can occur. The probability of this occurring will fall drastically as their concentration decreases. Thus at extremely high soft segment concentrations there must be a relatively large percentage of the hard segments which are not included in domains. Their reinforcing action would, as a result, be largely lost. Since the polyether itself (polytetrahydrofuran) is of very low tensile strength a maximum must therefore occur in the tensile strength/soft segment concentration relationship.

Fig. 5 also gives the variation in actual breaking stress found in the first series. This shows that the effect of increasing the molecular weight of the soft segment is to cause a fall in actual breaking stress at any given soft segment concentration. This is not only the case when the soft segment is the continuous phase, but also where the hard segment is the continuous phase. As *Huh* and *Cooper* (10) have pointed out at a given soft segment concentration the polymers with the higher soft segment molecular weight must also have higher molecular weights in their hard segments, and will tend to have fewer but larger domains. There is also the possibility that domain formation is less complete. Domain formation must be by a flow or diffusion process through the viscous coagulating solution of the discontinuous phase. The higher the molecular weight of, for example the diffusing hard segment, the less perfect will domain formation be. As a result elongation and tensile stress would both fall.

In view of the above hypothesis it is of interest to examine elasticity and stress relaxation.

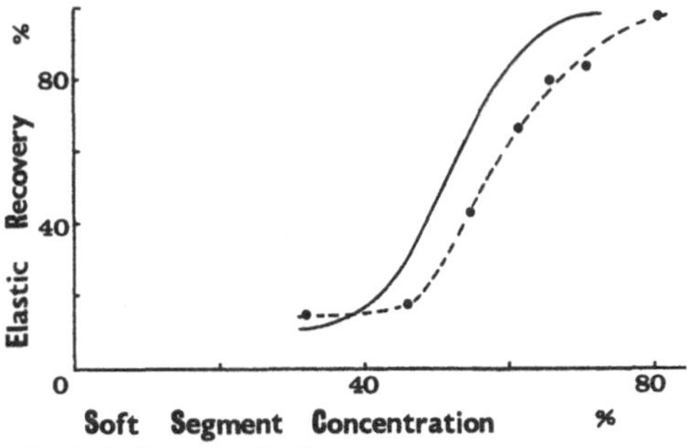

Fig. 6. Influence of soft segment concentration on elastic recovery from 100% extension. ——— 1st series; --●-- 2nd series

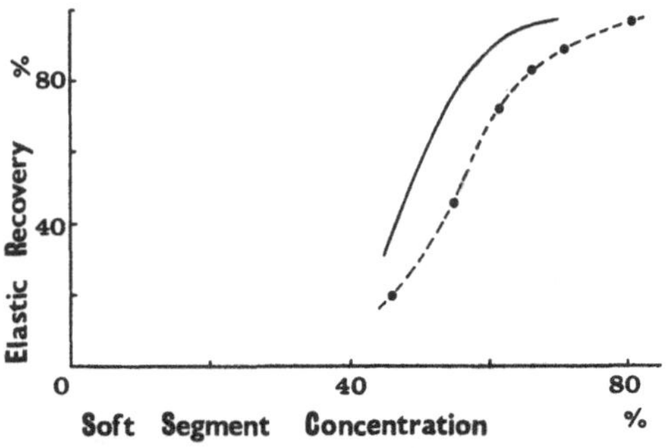

Fig. 7. Influence of soft segment concentration on elastic recovery from 200% extension. ——— 1st series; --●-- 2nd series

Elastic recovery

The instantaneous elastic recovery from 100 and 200% extension are shown in fig. 6 and 7 respectively. Elastic recovery at both extensions increased dramatically with increase in soft segment concentration. At high soft segment concentrations the elastic recovery was very similar to that normally found in commercial elastomers. At high hard segment concentrations the very low elastic recoveries are to be expected since the hard segment is the continuous phase.

The effect of increase in soft segment molecular weight is to produce a fall in elastic recovery at all but the lowest soft segment concentration. This result would tend to support the theory that domain formation is less complete when the higher molecular weight soft segment was used. The same explanation could possibly be used to explain the reversal of the effect at very low soft segment concentrations. It seems reasonable to assume that contamination of the hard segment by soft segment would increase the elastic recovery of the continuous hard phase.

Stress relaxation

Stress relaxation was calculated from the equation

$$f_t = f_{t_0} \exp - 2.3 K \log_{10}(t/t_0)$$

where f_t = stress at time t, f_{t_0} = stress at time t_0, and K = constant.

The stress relaxation indices $(-d \log f_t/d \log t_{t_0})$ at 50 and 200% extension are plotted against soft segment concentration in fig. 8 and 9 respectively. Also shown are the same plots for the first series which contained a low molecular weight soft segment. At low soft segment concentrations the effect of increase in the molecular weight of the soft segment is to decrease the rate of stress relaxation. At high soft segment concentration the reverse is the case. Furthermore the maximum is less sharp with the higher molecular weight soft segment. It is perhaps pertinent to note that the same was true of the phase change. This was very clearly discernable visually when Adiprene L 100 was used in the first series. The phase change, as has already been pointed out, was much less clear with the higher molecular weight soft segment.

The less sharply defined maximum in the stress relaxation index also suggests that each phase is contaminated with the other. The lower stress relaxation at low soft segment concentrations is rather more difficult to explain although the effect might be associated with the higher elastic recovery noted at very low soft segment concentrations.

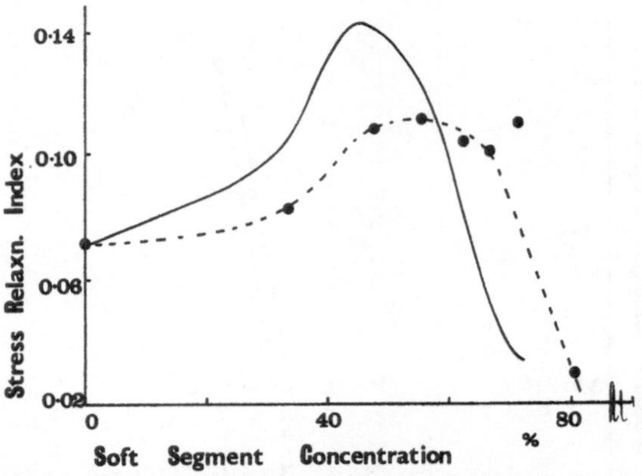

Fig. 8. Influence of soft segment concentration on stress relaxation at 50% extension. ——— 1st series; 2nd series

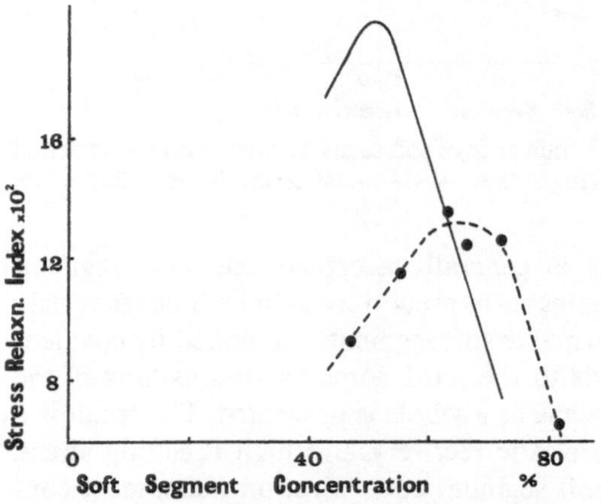

Fig. 9. Influence of soft segment concentration on stress relaxation at 200% extension. ——— 1st series; --●-- 2nd series

The stress relaxation data supports the hypothesis, therefore, that the effect of using a higher molecular weight soft segment is to produce domains which are contaminated with the opposite phase. *Huh* and *Cooper* drew the opposite conclusion from their results on the influence of soft segment molecular weight on the dynamic mechanical properties of polyurethane elastomers. Based on changes in the glass transition temperature they concluded that the longer chain segments would yield a more ordered and well defined domain structure. The apparent discrepancy could have arisen from the method of sample preparation. *Huh* and *Coopers* polyurethanes (10) were either solvent cast or compression moulded – both processes which allow ample time during solidification for complete domain formation to occur. The fibres described were wet spun. The coagulation process was extremely rapid giving much less time for complete phase separation.

References

1) *Heis, H. L.*, et al., Ind. Eng. Chem. **46**, 1498 (1954).
2) *Axelrood, S. L.* and *K. C. Frisch*, Rubber Age **88**, 465 (1960).
3) *Axelrood, S. L.*, et al., Ind. Eng. Chem. **53**, 889 (1961).
4) *Saunders, J. H.* and *K. C. Frisch*, Polyurethanes, Chemistry and Technology Part 1, Chemistry, p. 288 (New York 1962).
5) *Clough, S. B.* and *N. S. Schneider*, J. Macromol. Sci. Phys. **B2** (**4**), 553 (1968).
6) *Trappe, G.*, Advances in Polyurethane Technology, p. 112 (London 1968).
7) *Ferguson, J.* and *S. M. Al Khayatt*, Proceedings of the 5th International Congress on Rheology, Vol. 3, p. 373 (1971).
8) *Ferguson, J.* and *D. Patsavoudis*, European Polymer J. **8**, 385 (1972).
9) *Ferguson, J., D. J. Hourston, R. Meredith*, and *D. Patsavoudis*, European Polymer J. **8**, 369 (1972).
10) *Huh, Dong S.* and *S. L. Cooper*, Polymer Eng. Sci **11**, 369 (1971).

Authors' address:

Dr. *J. Ferguson* and Dr. *D. Patsavoudis*
Dept. of Fibre Science, University of Strathclyde
George Street, Glasgow (Scotland)

Rheol. Acta 13, 78–85 (1974)

From the Department of Mechanical Engineering, Chalmers University of Technology, Gothenburg (Sweden)

Comparison of stress-activated models and linear spectral models for visco-elasticity

By Bertil Olofsson

With 6 figures and 1 table

(Received October 27, 1972)

Introduction

The purpose of the present investigation is a phenomenological comparison of two models as their mode of deformation is varied. It was initiated by the approximately linear relationship between stress σ and logarithmic time t observed in relaxation of metals, polymers, paper etc., and especially the relationship

$$-(d\sigma/d\ln t)_{\text{infl.}}/(\sigma_0 - \sigma_\infty) = \text{Ku} \approx 0.1 \qquad [1]$$

first observed and extensively studied by *Kubát* (11, 12).

Here σ_0 and σ_∞ are the values of σ at times $t = 0$ and $t = \infty$, and $d\sigma/d\ln t$ is measured at the inflexion point of the relaxation-curve. Stress-activated models of the *Eyring* type as well linear spectral models of the "box" type are known to account for such effects (10, 11). But going from relaxation to creep, creep-recovery, constant strain rate extension and periodic deformation, new aspects on the applicability of either model are revealed which is further demonstrated in a discussion of the physical explanation for the phenomena.

Models and parameters

Fig. 1 shows the stress-activated model of *Eyring* type with the elastic parameters E_1 and E_2 and the viscous parameters α and $\varkappa$. The corresponding constitutive equation is

$$E_1 \varkappa \sinh \alpha(\sigma - \varepsilon E_2) = d[\varepsilon(E_1 + E_2) - \sigma]/dt. \qquad [2]$$

For a straightforward analysis of more complicated modes of deformation of a spectral model, it is appropriate to start with discrete *Maxwell* elements in parallel (fig. 2) with parameters E_n, η_n, $\tau_n = \eta_n/E_n$ and E_∞ and integrate between states 1 and 2.

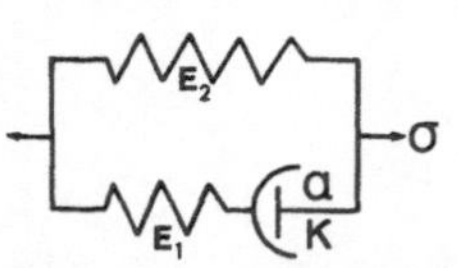

Fig. 1. Model of stress-activation (*Eyring*)

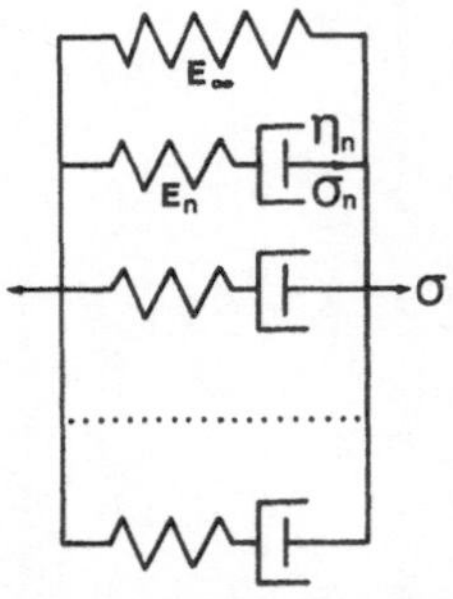

Fig. 2. Spectral model of parallel *Maxwell* units

$$\varepsilon_2 \exp(t_2/\tau_n) - \varepsilon_1 \exp(t_1/\tau_n) - (1/\tau_n)\int_{t_1}^{t_2} \varepsilon(t)\exp(t/\tau_n)\,dt = (1/E_n) \times [\sigma_{n2}\exp(t_2/\tau_n) - \sigma_{n1}\exp(t_1/\tau_n)] \qquad [3']$$

$$\sigma(t) = E_\infty \cdot \varepsilon(t) + \sum_n \sigma_n(t). \qquad [3'']$$

The transition to a continuous spectrum is made by putting

$$\tau_n \to \tau; \quad E_n \to E(\tau); \quad \eta_n \to \tau \cdot E(\tau) = H(\tau)$$

$$\sum_n E_n = \int_{\tau_L}^{\tau_M} E(\tau)\,d\tau = \int_{\tau_L}^{\tau_M} H(\tau)\,d(\ln\tau) \quad \textit{etc.} \qquad [4]$$

where τ_M and τ_L are upper and lower limits of τ. For the specific "box" the independent parameters derived are

E_∞, τ_M, τ_L and $H_0 = H(\tau) = \text{const.}$

Relaxation

The condition for relaxation is

$$\varepsilon(t) = \text{const} = \varepsilon_0 \qquad [5]$$

which is introduced in [2] and [3]. This gives for the "box"-model

$$\sigma(t)/\varepsilon_0 = E_\infty + H_0[Ei(-t/\tau_L) - Ei(-t/\tau_M)] \qquad [6']$$

$$\sigma_0/\varepsilon_0 = E_\infty + H_0 \ln(\tau_M/\tau_L) \qquad [6'']$$

$$\sigma_\infty/\varepsilon_0 = E_\infty \qquad [6''']$$

which corresponds to the smooth curve in fig. 3.

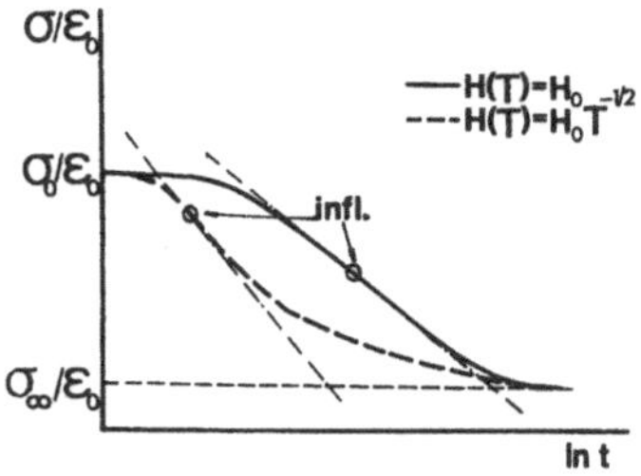

Fig. 3. Stress relaxation of "box"- and "wedge"-models

Because of the properties of the *Ei*-function (9)

$$Ei(\pm x) = -\int_{\mp x} [\exp(-y)/y]\, dy = \gamma + \ln x + \sum_{v=1}^{\infty} \frac{x^v(1 \pm 1)^v}{v!\, v}$$

where

$$x \geq 0;\quad Ei(+x) = \overline{Ei}(x);\quad \gamma = \textit{Eulers}\ \text{constant} \approx 0.5772 \qquad [7]$$

[6′] may be substituted for

$$\sigma(t)/\varepsilon_0 = E_\infty + H_0(\ln \tau_M - \gamma) - H_0 \ln t \qquad [8]$$

and the range for this equation $\tau_M \gg t \gg \tau_L$ may be extended to the lower limit $t = \tau_L \exp(-\gamma)$ and upper limit $t = \tau_M \exp(-\gamma)$, where [8] takes the values σ_0/ε_0 and $\sigma_0/\varepsilon_\infty$ in order.

By a corresponding derivation for the *Eyring* model with approximations $\sigma_0 \gg \sigma \gg \sigma_\infty$, it is demonstrated, that the relaxation curve is described in an almost identical way, provided there is a one-to-one correspondence of the four parameters of either model

$$E_2 = E_\infty;\quad E_1 = H_0 \ln(\tau_M/\tau_L);\quad \alpha\varepsilon_0 = 1/H_0;$$

$$\ln(\alpha\varkappa E_1/2) = \gamma - \ln\tau_M;\quad (1/E_e = 1/E_1 + 1/E_2). \qquad [9]$$

Also we get for the alternative choice of parameters from eq. [1]

$$\mathrm{Ku} = H_0\varepsilon_0/H_0\varepsilon_0 \ln(\tau_M/\tau_L) = \ln^{-1}(\tau_M/\tau_L) = (\varepsilon_0\alpha E_1)^{-1}. \qquad [10]$$

One important observation is made here: if the parameter H_0 is independent of ε_0, then from [9] the parameter α must be inversely proportional to ε_0, i.e. not constant. Alternatively the constancy of α means, that H_0 is in inverse proportion to ε_0. But experiments support the first alternative rather then the second, which is thus a point for the spectral model.

Creep

The condition for creep is

$$\sigma(t) = \text{const} = \sigma_0. \qquad [11]$$

For a spectral model the conventional way of calculating $\varepsilon(t)$ is by *Laplace* transformations ⟨as $\varepsilon(t)$ appears in the integrand in eq. [3′]⟩ according to

$$\bar{\varepsilon}(p) = \int_0^\infty \varepsilon(t) \exp(-pt)\, dt \qquad [12']$$

$$\varepsilon(t) = (1/2\pi i) \int_{c-i\infty}^{c+i\infty} \bar{\varepsilon}(p) \exp(pt)\, dp. \qquad [12'']$$

The equations for creep are well known but for the inverse transformation according to [12″] a rather specific method is generally cited in the literature (7). However, the more complicated relaxation-creep behaviour is best calculated by the conventional method of contour integration in the complex plane (8) and it is informative to derive the simple creep behaviour by the same method. Thus application of eq. [12′] to eq. [3] for the creep case and introduction of the box spectrum gives

$$\bar{\varepsilon}(p) = (\sigma_0/p)/\{E_\infty + H_0 \ln[(1 + p\tau_M)/(1 + p\tau_L)]\}. \qquad [13]$$

The contour to be used for the transformation is drawn in fig. 4, giving

$$\varepsilon(t) = \lim_{\substack{r \to 0 \\ R \to \infty}} (1/2\pi i) \int_A^{A'} \bar{\varepsilon}(p) \exp(pt)\, dp \qquad [14]$$

where

$$\int_A^{A'} = -\left[\int_{A'}^{B'} + \int_{B'}^{C'} + \int_{C'}^{D'} + \int_{D'}^{E'} + \int_{E'}^{E} + \int_{E}^{D} + \int_{D}^{C} + \int_{C}^{B} + \int_{B}^{A}\right]$$

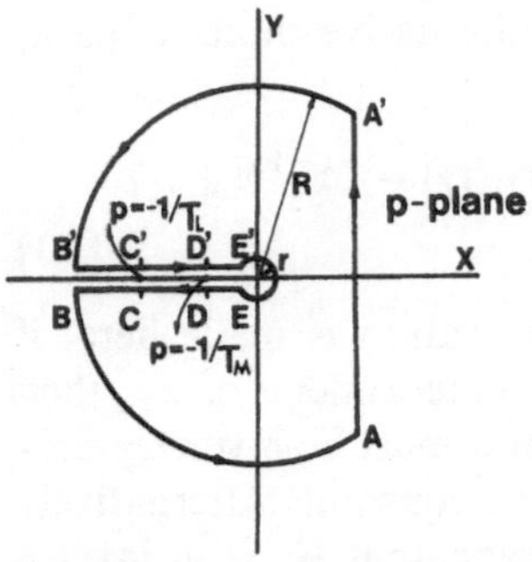

Fig. 4. Contour for inverse *Laplace* transformation

Here C and C' cut off $p = -1/\tau_L$ and D and D' cut off $p = -1/\tau_M$, which are branch points for $\ln(1 + p\tau_L)$ and $\ln(1 + p\tau_M)$. Thus $\ln(1 + p\tau_L)$ has the value $\ln(x\tau_L - 1) + \pi i$ on $B'C'$, $\ln(1 - x\tau_L)$ on $C'E'$ and EC, $\ln(x\tau_L - 1) - \pi i$ on CB and $\ln(1 + p\tau_M)$ varies correspondingly in passing D' and D. The branching gives contributions to the final integral along $C'D'$ and DC. From [13], $p_0 = 0$ and p_1 satisfying $E_\infty + H_0 \ln[(1 + p\tau_M)/(1 + p\tau_L)] = 0$ are poles. The former pole gives a contribution along the circle $E'E$, as its radius $r \to 0$. The pole p_1 however is so close to $p = -1/\tau_M$ that its contribution is negligible. The final result is written

$$\frac{\varepsilon(t)}{\sigma_0} = \frac{1}{E_\infty}\left[1 - \frac{E_\infty}{H_0}\int_{\tau_L}^{\tau_M}\frac{\exp(-t/\tau)\,d(\ln\tau)}{\{E_\infty/H_0 + \ln[(\tau_M - \tau)/(\tau - \tau_L)]\}^2 + \pi^2}\right] \qquad [15]$$

whose value for $t = \infty$ is

$$\varepsilon_\infty/\sigma_0 = 1/E_\infty \qquad [15']$$

and whose value for $t = 0$ from [15] should be equivalent to

$$\varepsilon_0/\sigma_0 = \lim_{p\to\infty} p\,\bar\varepsilon(p) = 1/[E_\infty + H_0\ln(\tau_M/\tau_L)]. \qquad [15'']$$

For the stress activated model, condition [11] is introduced in [2] and direct integration performed with the approximations used for relaxation to get the linear $\varepsilon - \ln t$-curve

$$\varepsilon(t)/\sigma_0 = 1/E_2 + (1/\alpha E_2\sigma_0)\,[\ln t + \ln\alpha\varkappa E_e/2]. \qquad [16]$$

For some experimental material of acetate and viscose filaments, whose behaviour was tested by a coupled plasto-visco-elastic model (16), calculations were performed by computer according to [15] and compared with results from [16]. The agreement seemed to be just as good as in the relaxation case, meaning that eq. [15] also gives a linear $\varepsilon - \ln t$-curve with parameters related to those of the stress-activated model according to eq. [9].

Relaxation-creep recovery

By relaxation for $t = t_R$ at extension $\varepsilon(t) = \text{const} = \varepsilon_R$ the stress has decreased from $\sigma = \sigma_0$ to $\sigma = \sigma(t_R)$, derived as before. Then this stress is released at the rate of initial extension, which is assumed to be instantaneous in the model description. This gives for the spectral model an extension ε_{0_c}

$$\varepsilon_{0_c} = \varepsilon_R - \sigma(t_R)/(E_\infty + \sum_n E_n) \qquad [17]$$

which is then considered to be the initial extension for creep recovery. This recovery is then calculated from eq. [3], introducing

$$\sigma(t) = \text{const} = 0. \qquad [18]$$

The further calculation follows the same scheme as for simple creep. Making the transformation [12'] and introducing the box-distribution gives

$$\bar\varepsilon_C(p) = (H_0\varepsilon_R/p)/$$
$$\{E_\infty + H_0\ln[(1 + p\tau_M)/(1 + p\tau_L)]\}$$
$$\times \left\{\ln\frac{1 + p\tau_M}{1 + p\tau_L} + Ei\left(-\frac{t_R}{\tau_M}\right)\right.$$
$$- Ei\left(-\frac{t_R}{\tau_L}\right) + \exp(pt_R)$$
$$\times \left[Ei\left(-\frac{t_R}{\tau_L} - pt_R\right)\right.$$
$$\left.\left. - Ei\left(-\frac{t_R}{\tau_M} - pt_R\right)\right]\right\}. \qquad [19]$$

The inverse transformation according to [12''] is performed with the same contour and the same consideration of poles and branch points as for the creep case (fig. 4). However, in the present case $p = 0$ is no pole and the points $p = -1/\tau_L$ and $p = -1/\tau_M$ are branch points also for $Ei[-t_R(1 + p\tau_L)/\tau_L]$ and $Ei[-t_R(1 + p\tau_M)/\tau_M]$. Thus the value of $Ei[-t_R(1 + p\tau_L)/\tau_L]$ is $\overline{Ei}[t_R(x\tau_L - 1)/\tau_L] + \pi i$ on $B'C'$, $Ei[-t_R(1 - x\tau_L)/$

$\tau_L]$ on $C'E'$ and EC and $\overline{Ei}[t_R(x\tau_L - 1)/\tau_L] - \pi i$ on CB. The final value is

$$\frac{\varepsilon_c(t)}{\varepsilon_R} = \int_{\tau_L}^{\tau_M} \frac{\exp(-t/\tau)\, d(\ln \tau)}{\{E_\infty/H_0 + \ln[(\tau_M - \tau)/(\tau - \tau_L)]\}^2 + \pi^2}$$
$$\times \left\{\frac{E_\infty}{H_0}[1 - \exp(-t_R/\tau)] - \exp\left(-\frac{t_R}{\tau}\right)\right.$$
$$\times \left[\ln\frac{\tau_M - \tau}{\tau - \tau_L} + Ei\left(-\frac{t_R}{\tau_L} + \frac{t_R}{\tau}\right)\right.$$
$$\left.\left. - \overline{Ei}\left(\frac{t_R}{\tau} - \frac{t_R}{\tau_M}\right) - Ei\left(-\frac{t_R}{\tau_M}\right) + Ei\left(-\frac{t_R}{\tau_L}\right)\right]\right\} \quad [20]$$

Putting $t = \infty$ gives

$$(\varepsilon_c/\varepsilon_R)_\infty \equiv 0 \quad [20']$$

and $t = 0$ in [20] is equivalent to

$$\left(\frac{\varepsilon_c}{\varepsilon_R}\right)_0 = \lim_{p\to\infty} p\bar{\varepsilon}_c(p) = \frac{\ln(\tau_M/\tau_L) - Ei(-t_R/\tau_L) + Ei(-t_R/\tau_M)}{E_\infty/H_0 + \ln(\tau_M/\tau_L)} \quad [20'']$$

For the stress-activated model we get

$$\varepsilon_c(t) = -(1/\alpha E_2)[\ln t + \ln \alpha\varkappa E_e/2]. \quad [21]$$

In the experiments mentioned (16), filaments were relaxed after 5 or 10% extension for $t_R = 0$, 5, 30, 60, 300 sec and then released and recovered in creep for $t = 0, 5, 30, 60, 300$ sec, all the 25 combinations of t and t_R studied. The results got the form

$$\varepsilon_c = A + B \ln t_R - C \ln t \quad [22]$$

where A, B, C are independent of t_R and t but affected by the initial extension. Combinations with t_R, t or both t_R and t equal zero have of course B, C or both B and C equal zero. For the box-model, parameters where chosen to fit the relaxation part $A + B \ln t_R$ of [22] and then C calculated according to [20] by computer and compared to C of [22]. Table 1 gives representative results for $E_\infty/H_0 = 1.862$, $\tau_M = 500$; $\tau_L = 0{,}02$.

Thus the box-model gives results whose magnitude is in the experimental range but increases somewhat with t_R in contradiction to the experiment. The *Eyring*-model has a different behaviour, its creep rate being independent of t_R but of a completely different magnitude than the experimental. The reason for this is that the α-value for the creep recovery rate ($=1/\alpha E_2$) is assumed to be that of the previous relaxation.

Table 1

t_R(sec)	$C = -d\varepsilon_c/d(\ln t)$ Experiment	Box-model	*Eyring*-model
5	0.0012	0.0008	0.0269
30	0.0012	0.0026	0.0269
60	0.0012	0.0034	0.0269
300	0.0012	0.0045	0.0269

Putting α (creep)/α (relaxation) = 269/12 = 22,4 removes the difference. And it was further demonstrated that eq. [1] is consistent with such a large value of α creep, where [1] gives for the creep recovery case $(\alpha E_2 \varepsilon_c)^{-1} \approx 0.1$. However, a further condition is that the initial extension is small (less than 10% where appreciable plastic flow takes place) and that t_R is small, i.e. that the previous structural distorsion is negligible. If the dependence on t_R in table 1 is neglected, the box again has the same H_0 as for relaxation and simple creep.

Constant strain-rate extension

The stress-time relationship for relaxation seems to be independent of the extent of original deformation, which might be of a non-*Hooke*an type (11) as well as accompanied by definite yielding (16). We might ask how much of this initial deformation in the straining process is compatible with the parameters of the succeeding relaxation and again compare the two models.

Putting

$$d\varepsilon/dt = \text{const} = K_0 \quad [23]$$

in [2] and [3], stress-strain curves are calculated (no approximations introduced).

Thus we get for the box-model

$$\sigma_s = \varepsilon \cdot E_\infty + K_0 H_0 \times \{\tau_M[1 - \exp(-\varepsilon/K_0\tau_M)] - \tau_L[1 - \exp(-\varepsilon/K_0\tau_L)] + (\varepsilon/K_0) \cdot Ei(-\varepsilon/K_0\tau_L) - (\varepsilon/K_0) \cdot Ei(-\varepsilon/K_0\tau_M)\} \quad [24]$$

and for the *Eyring*-model

$$\sigma_s = \varepsilon E_2 + (1/\alpha)\ln\{K_0/\varkappa + S\tanh[\alpha\varkappa E_1 S\varepsilon/2K_0 + \tanh^{-1}(1 - K_0/\varkappa)/S]\} \qquad [25]$$

where $S^2 = 1 + (K_0/\varkappa)^2$.

In fig. 5 (right part) a relaxation curve for polycrystalline Cd studied by *Kubát* (11) is reproduced and from this curve parameters for the alternative models are derived as before and introduced in [24] and [25] for calculation of the stress-strain curves, which are then drawn in fig. 5 together with the experimental curve.

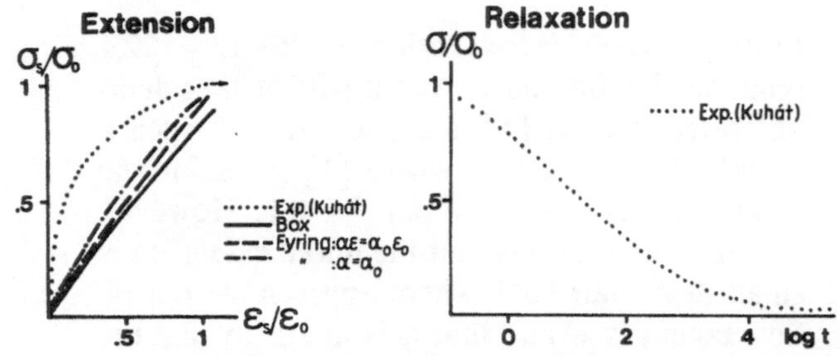

Fig. 5. Comparison of stress-strain curves from experiments and calculated from relaxation curve [polycrystalline Cd, ref. (11)]

Evidently the box-model predicts a rather uniform creep producing an almost linear stress-strain curve, while the stress activated model predicts a significant yield by a creep contribution, which increases with extension. But if the condition α = const is exchanged for the condition $\alpha \cdot \varepsilon$ = const in the stress-activated model [25] the nonlinear yield is also decreased. However, the experimental non-linearity is much larger than that predicted from the relaxation and not time dependent to the same extent. Similar results are obtained for the fibers discussed in the preceeding part (16).

The principal result for the metal as well as the fiber is the existence of mechanisms of a plastic type with small times of relaxation responsible for yield or non-linear elasticity, which are not included in the further long time relaxation.

Periodic strain

In most work where the stress-activated model has been applied for analysis of periodic stress or strain, it has been assumed, that a linear small order approximation is acceptable. However, although σ and ε are small, the product $\alpha(\sigma - \varepsilon E_2)$ is not necessarily small, if $\sigma - \varepsilon E_2$ deviates from its equilibrium value (=0). Thus the approximation $\sinh x = x$ might be erraneous. For a spectral model the calculation is straightforward. It is assumed that the strain has a conventional form, i.e.

$$\varepsilon = \varepsilon_0 \exp(i\omega t) \qquad [26]$$

and that the complex modulus may be written

$$\sigma(t)/\varepsilon(t) = (\sigma/\varepsilon_0)\exp(-i\omega t) = G'(\omega) + iG''(\omega). \qquad [27]$$

Introducing [26] and [27] in [3], then introducing the box-symbols and removing the transient effect ($t \to \infty$) yields for the storage modulus G' and the loss modulus G''

$$G'(\omega) = E_\infty + \frac{H_0}{2}\ln\frac{1 + \omega^2\tau_M^2}{1 + \omega^2\tau_L^2} \qquad [28']$$

$$G''(\omega) = H_0 \tan^{-1}\frac{\omega(\tau_M - \tau_L)}{\omega^2\tau_M\tau_L + 1}. \qquad [28'']$$

The dependence on ω is of a conventional type: $G'(\omega)$ increases with ω from the value E_∞ at $\omega = 0$ to $E_\infty + H_0\ln(\tau_M/\tau_L)$ at $\omega \to \infty$, while $G''(\omega)$ increases from 0 at $\omega = 0$ to a maximum at $\omega^2 = 1/(\tau_M\tau_L)$ and then decreases to 0 at $\omega \to \infty$.

If the same procedure is applied for the stress activated model, a conventional form of $G'(\omega)$ and $G''(\omega)$ is obtained if and only if the approximation $\sinh\alpha(\sigma - \varepsilon E_2) \equiv \alpha(\sigma - \varepsilon E_2)$ is introduced in [2]. However, assuming that there exists a non-transient modulus, which is a function not only of ω but also of $\omega t = \varphi$ we might write, c.f. [27]

$$(\sigma/\varepsilon_0)\exp(-i\omega t) = G_1(\omega, \varphi) + iG_2(\omega, \varphi) \qquad [29]$$

where $\varphi = \omega t$ and introduce [29] together with [26] in [2]. After simplification and rearrangements we get for $G_1(\omega, \varphi) = G_1$ and $G_2(\omega, \varphi) = G_2$

$$\frac{dG_1}{d\varphi} - G_2 + \frac{E_1\alpha\varkappa}{\omega}(G_1 - E_2) = -\frac{E_1\varkappa}{\varepsilon_0\omega}[\cos\varphi(\cos\theta\sinh\xi - \xi) + \sin\varphi(\sin\theta\cosh\xi - \theta)] \qquad [30']$$

$$\frac{dG_2}{d\varphi} + G_1 - (E_1 + E_2) + \frac{E_1\alpha\varkappa}{\omega}G_2 = \frac{E_1\varkappa}{\varepsilon_0\omega}[\sin\varphi(\cos\theta\sinh\xi - \xi) - \cos\varphi(\sin\theta\cosh\xi - \theta)] \qquad [30'']$$

where the functions ξ and θ are obtained from

$$\xi = \alpha\varepsilon_0[(G_1 - E_2)\cos\varphi - G_2\sin\varphi] \qquad [31']$$

$$\theta = \alpha\varepsilon_0[(G_1 - E_2)\sin\varphi + G_2\cos\varphi]. \qquad [31'']$$

The quantities $\cos\theta\sinh\xi$ and $\sin\theta\cosh\xi$ are critical in magnitude for the approximations applicable and it is appropriate to make series developments

$$\cos\theta\sinh\xi - \xi = \xi\left[\left(\frac{\xi^2}{3!} - \frac{\theta^2}{2!}\right) + \left(\frac{\xi^4}{5!} - \frac{\xi^2\theta^2}{3!\,2!} + \frac{\theta^4}{4!}\right) + \cdots\right] \qquad [32']$$

$$\sin\theta\cosh\xi - \theta = \theta\left[\left(\frac{\xi^2}{2!} - \frac{\theta^2}{3!}\right) + \left(\frac{\xi^4}{4!} - \frac{\xi^2\theta^2}{2!\,3!} + \frac{\theta^4}{5!}\right) + \cdots\right] \qquad [32'']$$

Introducing [32] in right side members of [30] we obtain combinations of $\sin\varphi$ and $\cos\varphi$ with factors $\xi^v\theta^\mu$ of different degree in v and μ. And if [31] is applied these terms are reduced to a series of terms with factors $\sin m\varphi$ and $\cos m\varphi$, where m takes only even integer values. The first order approximation is thus obtained by putting right hand members of [30] = 0 and then solving the partial differential equations for the non-transient part. For second order approximations the first order results are introduced in the right hand members of [30] and terms $\xi^v\theta^\mu$ for $v+\mu \leq 3$ kept. In the next approximation terms of $v+\mu \leq 5$ are kept etc. From this structural study it is concluded that the solutions of [30] can be obtained simply by putting

$$G_1 - E_2 = G_1^0 - E_2 + G_1^1\sin 2\varphi + G_2^1\cos 2\varphi + G_1^2\sin 4\varphi + G_2^2\cos 4\varphi + \cdots \qquad [33']$$

$$G_2 = G_2^0 + G_3^1\sin 2\varphi + G_4^1\cos 2\varphi + G_3^2\sin 4\varphi + G_4^2\cos 4\varphi + \cdots \qquad [33'']$$

then introducing these expressions to the left in [30′] and [30″] and identifying coefficients for $\sin 2\varphi$ etc. We have performed the calculations until third order terms and the results may be written as follows

$$G_1(\omega,\varphi)/E_1 = G_1(\mu,\omega t)/E_1 = E_2/E_1 + 1/(\mu^2+1) + (\alpha\varepsilon_0E_1)^2 \times [F_1^1(\mu)\cos 2\omega t - F_2^1(\mu)\sin 2\omega t] - (\alpha\varepsilon_0E_1)^4 \times [F_1^2(\mu)\cos 4\omega t - F_2^2(\mu)\sin 4\omega t] \qquad [34']$$

$$G_2(\omega,\varphi)/E_1 = G_2(\mu,\omega t)/E_1 = \mu/(\mu^2+1) + (\alpha\varepsilon_0E_1)^2 \times [F_2^1(\mu)\cos 2\omega t + F_1^1(\mu)\sin 2\omega t] - (\alpha\varepsilon_0E_1)^4 \times [F_2^2(\mu)\cos 4\omega t + F_1^2(\mu)\sin 4\omega t] \qquad [34'']$$

where

$$\mu = \frac{\alpha\varkappa E_1}{\omega}$$

$$F_1^1(\mu) = \frac{\mu^2(3\mu^2-5)}{3(\mu^2+1)^3(\mu^2+9)}$$

$$F_2^1(\mu) = \frac{\mu(\mu^4-12\mu^2+3)}{6(\mu^2+1)^3(\mu^2+9)}$$

$$F_1^2(\mu) = \frac{\mu^2(5\mu^4-30\mu^2+13)}{60(\mu^2+1)^5(\mu^2+25)}$$

$$F_2^2(\mu) = \frac{\mu(\mu^6-35\mu^4+55\mu^2-5)}{120(\mu^2+1)^5(\mu^2+25)}$$

It means that the simple harmonic contributions to G_1 and G_2, i.e. $E_2 + E_1/(\mu^2+1)$ and $E_1\mu/(\mu^2+1)$ are superposed by time dependent periodic contributions, which vary with μ or ω in a very complicated way. Thus $F_1^1(\mu)$ has two maxima at $\mu^2 \approx 4.0$ and -5.8 and one minimum at $\mu^2 = 0.35$. The contributions to G_1 and G_2 are further maxima if the corresponding trigonometric factors are maxima. Regarding $F_1^1(\mu)$ in G_1, we get $|\cos 2\omega t| = 1$ for $t = nT/4$, where T is the period time and $n = 0, 1, 2, 3$.

The coefficients $F_j^i(\mu)$ are all limited and $\lim F_j^i(\mu) = 0$ for $\mu \to 0$ as well as $\mu \to \infty$. Also for $\mu \to \infty\; F_j^{i+1}(\mu)/F_j^i(\mu) = 0(\mu^{-2}) \to 0$, i.e. the series is convergent. But for $\mu \to 0\; F_j^{i+1}(\mu)/F_j^i(\mu) = 0(\mu^0)$ and thus the factor $(E_1\varepsilon_0\alpha)^{2i}$ is significant for the convergency. Now $E_1\varepsilon_0\alpha$ is just $= \mathrm{Ku}^{-1} \approx 10$ from eq. [10], meaning that the series gets conditionally convergent for $\mu \to 0$ or $\omega \to \infty$. Thus the stress-activated model might be consistent with failure at high frequency deformation. And the range for such failure is extended to lower frequencies, a critical limit being $\mu \approx 10$ or $\omega = \alpha\varkappa E_1/10$. In every case two suggestions present themselves: further studies of periodic stress-strain deformation as a function of the time t within the period and further analysis of tenacity, fatigue etc. in terms of stress activation [cf. (1)].

Physics of the models

It is evident that the two models discussed are complementary in their phenomenological description although different incitaments might favor one or the other. It is of special interest to consider relationship [1] with different parameters introduced [cf. (10)].

$$\mathrm{Ku}^{-1} = \sigma_0 \alpha = \sigma_0 v/kT = \alpha \varepsilon_0 E_1 = E_1/H_0 = \ln(\tau_M/\tau_L) \quad [35]$$

where the quantitity $v = \alpha k T$ known as activation volume is also introduced. For a material of *Hooke*an elasticity, E_1 is constant and [35] is thus consistent with a box-spectrum of constant $H_0 = E(\tau) \cdot \tau = \eta_0$, the large variation in τ explained as variation in $E(\tau)$. The numerical value $\mathrm{Ku}^{-1} \approx 10$ would give $\tau_M/\tau_L \approx 2.2 \cdot 10^4$. Now this τ-range is very much affected by deviations from the box-model. Thus a wedge-type spectrum with $H = H_0 \cdot \tau^{-1/2}$ (fig. 3) satisfies eq. [1], whatever are its τ-limits, because the point of inflection gets close to τ_L, where the quantity in [1] just has the numeric value 0.1. For this spectrum however, there is a continuous decrease of the parameter from [1] with increasing time of relaxation. Another observation is that substitution of linear elements for stress-activated elements with uniform distribution in space predicts, that the box-spectrum is restricted to $\tau_M/\tau_L \approx 400$ for $\mathrm{Ku}^{-1} \approx 10$ because of anisotropic effects from σ in the exponential position (17). As relationship [35] is not restricted to the *Hooke*an range, the constancy of H_0 or $\varepsilon_0 \sim 1/\alpha$ is a rough approximation, while the relationship $\sigma_0 \sim 1/\alpha$ seems very significant (12).

One reason for a non-*Hooke*an and yet elastic behaviour is, that the anharmonic contribution to the molecular force-displacement equation is significant. This means a decrease of the elastic modulus with increasing stress and increasing space for higher modes of vibration which is an alternative way of describing stress activation (14, 15). But according to classical bead-spring models for molecular segment movement, higher modes of vibration gives modes of closer spacing, which might be identified with smaller activation volume (4, 18). This is a qualitative explanation for the relationship $v \sim 1/\sigma_0$, although a questionable point is its compatibility with the effect of T on v.

In the case of metals, dislocation considerations are the basis of all non-elastic deformations. And dislocations might be significant for polymers in a less regular semi-crystalline state too, if dislocation is generalized to mean misfittings of the molecules along some curve in the structure, the degree and direction given by a generalized *Burgers* vector $\underline{b}$ (6). In terms of dislocation theory the quantity $\sigma_0 v$ gets $\sigma_0 \underline{b}^2 l$, i.e. the force $\sigma_0 \underline{b} l$ acting on the dislocation of length l displaces it the distance $\underline{b}$ (2). There is further a close similarity between the molecular segment motion (fig. 6) and the "elastic string" behaviour of the dislocation, which might be actived to higher frequency by increased node density (5). Fig. 6 illustrates primarily an edge dislocation, while a screw dislocation would give longitudinal vibrations, provided the dislocation follows the segment. There are other possibilities to explain the $\sigma_0 - 1/v$-proportionality, although theories of jog-formation by glide through a dislocation forest etc. are not compatible with the dominating elastic nature at small strain. However, irregularities in the structure, especially due to density gradients, will give concentration of stresses. As the external force increases, the simultaneous decrease in width of already existing "dislocations" of this type and activation of new dislocations might result in the decrease of volume of the units for flow, which are diffusion controlled with high activation energy (parameter $\varkappa$). Such stress distributions would be of the *Peierl-Nabarro*-type and compatible with the rather arbitrary model characteristics applicable, e.g. the displacement energy being little dependent upon the exact stress-strain-relationship used (3, 13).

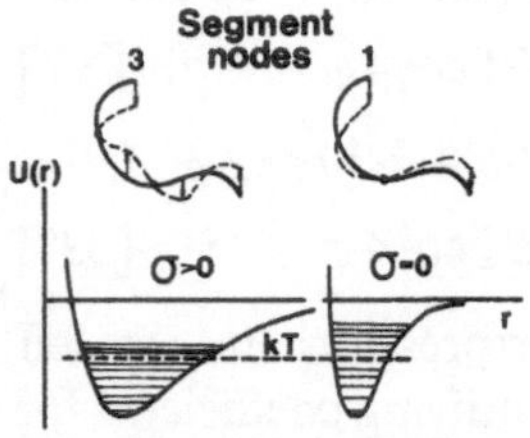

Fig. 6. Relationship between stress-activation and molecular segment motion [refs. (15 and 18)]

A third – and most elementary and concise – explanation of the constancy of $\sigma_0 v$ is the identification of v with the width of the potential barrier (fig. 6) $r_m - r$, where r is attained by an elastic deformation, $\sigma_0 = dU(r)/dr$, and $r = r_m$ is the limit (peak) value. However, the solution of the differential equation $(dU(r)/dr)(r_m - r) = \mathrm{const}$ – although qualitatively acceptable in a mean range $0 < r < r_m$ – must be modified to fit existing barrier models.

The existence of $\lim_{\sigma_0 \to 0} \sigma_0 \cdot v(\sigma_0) = v_0$ might have important technical consequences as indicated by the previous analysis of periodic deformations.

Acknowledgments

My best thanks to Prof. *J. Kubát* for stimulating discussions and to Dr. *A. Johnson* for effective help with the computer work.

Summary

An analytical comparison is made of a stress-activated nonlinear model (*E*) and a linear spectral model (*M*) for visco-elastic bodies in relaxation, creep, relaxation-creep recovery, constant strain-rate extension and periodic strain. A special reason for this is an observed "invariance" in relaxation (eq. [1]). Significant parameters are the frequency function *H* for (*M*) and the rate constant α for (*E*). *H* should be roughly constant in relaxation and in still rougher approximation also in creep recovery, while α should be inversely proportional to the stress and in addition α (creep-recovery)/α (relaxation) $\approx$ 22. For strain extension both models predict similar results but none takes the plastic creep of large rate in consideration. Regarding periodic strain model (*M*) gives a conventional behaviour, while model (*E*) predicts significant time dependent frequency effects.

Possible physical mechanisms for explanation of the results obtained are discussed.

Zusammenfassung

Ein analytischer Vergleich von einem spannungsaktivierten nichtlinearen Modell (*E*) und einem linearen Spektralmodell (*M*) für visco-elastische Körper in Relaxation, Kriechen, Relaxationkriecherholung, Dehnung mit konstanter Geschwindigkeit und periodische Dehnung wird gemacht. Eine spezielle Ursache dafür ist eine „Invarianz" in der Relaxation [1]. Signifikante Parameter sind die Frequenzfunktion *H* für (*M*) und die Geschwindigkeitskonstante α für (*E*). *H* sollte in grober Näherung in der Relaxation konstant sein und in noch gröberer Näherung auch in Kriecherholung. α sollte aber im inversen Verhältnis zur Spannung stehen, und weiter sollte α (Kriecherholung)/α (Relaxation) $\approx$ 22 sein. Für Dehnung geben beide Modelle ungefähr dieselben Resultate, keines nimmt aber das plastische Kriechen von größerer Geschwindigkeit in Betracht. Für periodische Dehnung sieht (*M*) ein konventionelles Verhalten voraus, für (*E*) findet man aber signifikante zeitabhängige Frequenzeffekte.

Mögliche physikalische Mechanismen zur Erklärung der Resultate werden diskutiert.

References

1) *Bartenev, G. M.* and *Yu. S. Zuyev*, Strength and Failure of Visco-Elastic Materials, p. 189 (Oxford 1968).

2) *Cherry, B. W.* and *P. L. McGinley*, Appl. Polymer Symposia No. **17**, 59 (1971).

3) *Cottrell, A. H.*, Theory of Crystal Dislocations, p. 57 (1962).

4) *Ferry, J. D., R. F. Landel* and *M. L. Williams*, J. Appl. Physics **26**, 359 (1955).

5) *Friedel, J.*, Dislocations, p. 67 (Oxford 1964).

6) *Gilman, J. J.*, Physics of Strength and Plasticity, p. 3 (1969).

7) *Gross, B.* and *H. Pelzer*, J. Appl. Physics **22**, 1035 (1951).

8) *Irving, J.* and *N. Mullineux*, Mathematics in Physics and Engineering, p. 599 (New York 1959).

9) *Jahnke* and *Emde*, Tables of Higher Functions (1948).

10) *Joshi, V. S.* and *N. R. Kothari*, Text. Res. J. **40**, 764 (1970).

11) *Kubát, J.*, A Similarity in the Stress Relaxation Behaviour of High Polymers and Metals. (Diss.) (Stockholm 1965).

12) *Kubát, J.*, IV. Int. Congr. Rheology, Part 1, 281 (1963).

13) *Kuhlmann-Wilsdorf, D.*, Phys. Review **120**, 773 (1960).

14) *Leibfried, G.* and *W. Ludwig*, Solid State Physics **12**, 365 (1961).

15) *Müller, F. H.*, Rheology V, p. 447 (New York 1969).

16) *Olofsson, B.*, Deformation and Recovery Properties of Textile Materials. (Diss.) (Gothenburg 1968).

17) *Olofsson, B.*, To be published.

18) *Zimm, B. H.*, Rheology III, p. 1 (New York 1960).

Author's address:

Bertil Olofsson
Department of Mechanical Engineering
Chalmers University of Technology
Fack 40220, Gothenburg 5 (Sweden)

Rheol. Acta 13, 86–92 (1974)

From the Monsanto Company, St. Louis, Missouri (U.S.A.)

Morphology and the elastic modulus of block polymers and polyblends

By L. E. Nielsen*)

With 7 figures and 1 table

(Received October 27, 1972)

Introduction

The elastic moduli and morphology of block polymers and polyblends can be modified by changing the relative concentration of the components, by heat and solvent treatments, and by the intensity of mechanical mixing. In these two-phase systems the dispersed phase can be spheres, aggregated spheres, or cylindrical or plate-like in shape. Over certain composition ranges, both phases can be continuous, or the phases can become inverted- the dispersed phase can become the continuous phase. These different morphologies have widely different moduli and other mechanical properties. It will be shown that if the morphology of a given block polymer or polyblend is known, then reasonable predictions of the elastic moduli can be made by using the theoretical equations developed for composite materials.

Theory

Earlier attempts to relate the concentration of the two components and the morphology of two phase systems such as polyblends and block polymers with their elastic moduli have been made by *Takayanagi* (1, 2) and by *Kaelble* (3, 4). These workers used series and parallel combinations of the components in models which give the highest upper bound and the lowest lower bound to the modulus. It is very difficult to relate the modulus of such a model with the actual morphology of the real system.

A better method which imposes much narrower limits on the moduli and which is capable of incorporating the morphology of the two-phase system in a less ambiguous manner is to use the recent theory of the moduli of composite materials. *Kerner* (5) developed a theory in which either of the phases can be a dispersion of spheres in a matrix of the other component. More recently *Halpin* and *Tsai* (6, 7) have developed equations which are general enough to cover the complete range of moduli from the lowest lower bound (series models) to the highest upper bound (parallel models). They also showed how the moduli can be calculated for many systems of widely different morphologies including dispersions of spheres, fiber-filled materials, etc. *Nielsen* (8) extended the *Halpin-Tsai* equations by using the concept of a generalized *Einstein* coefficient to cover still other morphologies, including dispersions of aggregated spheres and short fibers randomly oriented. *Lewis* and *Nielsen* (9, 10) were able to narrow the limits on the upper and lower bounds on the moduli by taking into account the maximum packing fraction of the filler phase.

The highest upper bound of the modulus is given by the rule of mixtures:

$$M = M_1 \varphi_1 + M_2 \varphi_2 \qquad [1]$$

where M is the modulus of the composite, M_1 is the modulus of component 1, and φ_1 is the volume fraction of component 1. This equation holds for models in which the components are arranged parallel to one another so that an applied stress elongates each component the same amount. The lowest lower bound to the modulus is found in models in which the components are arranged in series with the applied stress; the equation for this case is:

$$\frac{1}{M} = \frac{\varphi_1}{M_1} + \frac{\varphi_2}{M_2}. \qquad [2]$$

Curves numbered 1 and 2 of fig. 1 for the modulus of the composite divided by the modulus of the rubber correspond to eqs. [1] and [2] for the case of $M_2/M_1 = 1000$. This modulus ratio is approximately the correct value for composites made up of a rigid polymer and an elastomer.

*) HPC 71-144 from the Monsanto/Washington University Association sponsored by the Advanced Research Projects Agency, Department of Defense, under Office of Naval Research Contract N00014-67-C-0218.

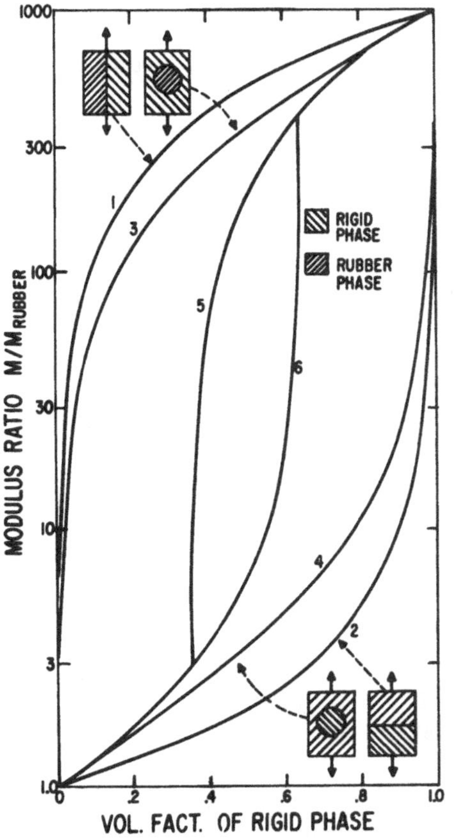

Fig. 1. Modulus ratio of composites. Curve 1: parallel element model; curve 2: series model; curve 3: *Halpin-Tsai* (or *Kerner*) shear modulus for elastomeric spheres dispersed in a rigid polymer; curve 4: rigid spheres dispersed in elastomeric phase; curve 5: elastomeric spheres in rigid matrix for $\varphi_m' = 0.64$ (random packing); and curve 6: rigid spheres dispersed in an elastomeric matrix for $\varphi_m = 0.64$

The *Halpin-Tsai* equations are (6, 7):

$$\frac{M}{M_1} = \frac{1 + AB\varphi_2}{1 - B\varphi_2} \qquad [3]$$

$$B = \frac{M_2/M_1 - 1}{M_2/M_1 + A} \qquad [4]$$

where the subscripts 1 and 2 refer to the continuous phase and the dispersed phase, respectively. The constant A is determined by the morphology of the system; for dispersed spheres in an elastomeric matrix, $A = 1.5$, for instance. The extension of these equations is given by (8–10):

$$\frac{M}{M} = \frac{1 + AB\varphi_2}{1 - B\psi\varphi_2} \qquad [5]$$

$$A = k - 1 \qquad [6]$$

$$\psi \doteq 1 + \frac{(1 - \varphi_m)}{\varphi_m^2}\varphi_2 \qquad [7]$$

where k is a generalized *Einstein* coefficient, and ψ is a function which takes into account the maximum packing fraction φ_m of the dispersed phase. The maximum volumetric packing fraction φ_m is indirectly related to morphology, and it generally has a value between 0.5 and 0.9. It has a value of 1.0 in the original *Halpin-Tsai* equations. The constants A and k are strongly dependent upon the morphology of the composite.

For inverted systems in which the continuous phase is the more rigid one, it is convenient to rewrite equations 4–6 as

$$\frac{M_1}{M} = \frac{1 + A_i B_i \varphi_2}{1 - B_i \psi \varphi_2} \qquad [8]$$

$$B_i = \frac{M_1/M_2 - 1}{M_1/M_2 + A_i} \qquad [9]$$

and

$$A_i = \frac{1}{A}. \qquad [10]$$

In the inverted system, the subscript 2 still refers to the dispersed phase, which is now the low modulus phase.

Curves 3 and 4 of fig. 1 illustrate the *Halpin-Tsai* equations for dispersed spheres. Curves 5 and 6 of fig. 1 illustrate the modified eqs. [5]–[10] for dispersed spheres with $\varphi_m = 0.64$ (random close packing), a *Poissons* ratio of 0.5 for the elastomer phase, and a *Poissons* ratio of 0.35 for the rigid phase. The *Halpin-Tsai* equations and their modification put much narrower limits on the moduli than the series or parallel models used in the past. Morphologies other than dispersed spheres generally have similar spreads between the upper and lower modulus curves.

In real systems of polyblends and block polymers, both phases may be continuous, or there may be an inversion of the phases as the composition ratio is changed. In this situation the equations giving the upper and lower bounds to the modulus must be combined in some manner. Empirically it has been known for a long time, and recent calculations on crystalline polymers (11) indicate, that a combination of the two equations is approximately given by

$$\log M = \varphi_u \log M_u + \varphi_L \log M_L . \qquad [11]$$

M_u and M_L are the upper and lower bounds to the modulus, respectively, at a given composition. In this equation φ_u is the fraction of the low modulus material that is in a continuous phase in the overlap region where both phases are essentially continuous, while φ_L is the fraction of the rigid material that is in the overlap region (10).

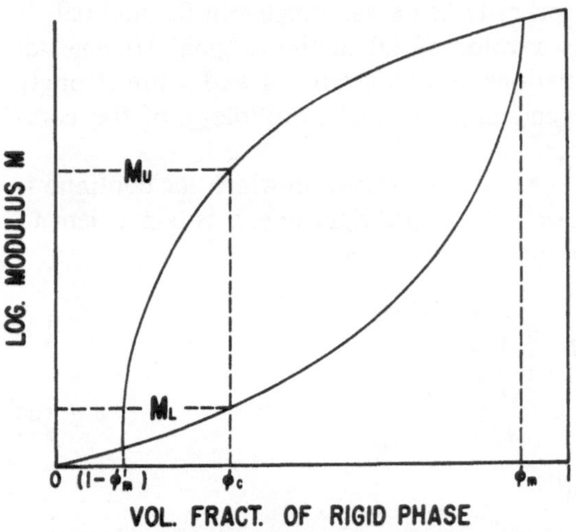

Fig. 2. Notation used in calculation of modulus when there are two continuous phases over part of the composition range

Fig. 2 may illustrate the point more clearly for some chosen over-all composition φ_c.

$$\varphi_L = \frac{\varphi_c - (1 - \varphi_m')}{\varphi_m - (1 - \varphi_m')} \qquad [12]$$

$$\varphi_u = 1 - \varphi_L = \frac{\varphi_m - \varphi_c}{\varphi_m - (1 - \varphi_m')}. \qquad [13]$$

φ_m' is the packing fraction of the low modulus material in the inverted system. For the unmodified *Halpin-Tsai* equations, φ_L is the volume fraction of the rigid phase, and φ_u is the volume fraction of the elastomeric phase for any value of φ_c.

The *Einstein* coefficient k (or A), which is very sensitive to the morphology of the system, is proportional to the initial slope of the modulus-concentration curve near $\varphi = 0$ and $\varphi = 1$ for dispersed systems in which both phases are not continuous. *Einstein* coefficients have been published for many systems (6, 12, 13). Some of these are listed in table 1. The listed values of k [or of $(A + 1)$] are for rigid particles in a matrix of lower modulus. The values for inverted systems, in which the matrix has a higher modulus than the dispersed phase, can be determined from eqs. [6] and [10]. If *Poissons* ratio of the matrix is not 0.5, a correction should be applied to the *Einstein* coefficient (8).

Table 1. *Einstein* coefficients for composites

Filler phase	Modulus	*Einstein* coefficient k
Spheres	G	$1 + (7 - 5v_1)/(8 - 10v_1)$
Large aggregates of spheres	G	$2.50/\varphi_a$
Aggregates of 2 spheres	G	2.58 +
Rods-axial ratio 4	G	3.08
Rods-axial ratio 6	G	3.84
Rods-axial ratio 8	G	4.80
Rods-axial ratio 10	G	5.93
Rods-axial ratio 15	G	9.4
Uniaxial fiber-filled	E_L	$1 + 2L/D$
Uniaxial fiber-filled	E_T	1.5
Uniaxial fiber-filled	G_{LT}	2.0
Uniaxial fiber-filled	G_T	1.5
Ribbon-filled $(w/t) \to \infty$	E_L	∞
Ribbon-filled	E_T	$1 + 2w/t$
Ribbon-filled	E_{TT}	1.0
Ribbon-filled	G_{LT}	$1 + (w/t)$ (13)
Ribbon-filled $(w/t) \to \infty$	$G_{LT'}$	1.0
Ribbon-filled $(w/t) \to \infty$	G_{TT}	1.0
$M = M_1\varphi_1 + M_2\varphi_2$	M	∞
$\frac{1}{M} = \frac{\varphi_1}{M_1} + \frac{\varphi_2}{M_2}$	M	1.0

v_1 *Poissons* ratio of matrix
φ_a Packing fraction of spheres in aggregate
L Length of rod
D Diameter of rod
w Width of ribbon
t Thickness of ribbon
T Transverse to fibers or ribbons
Subscript L Longitudinal direction

The values of the *Einstein* coefficient or of A do not uniquely define the morphology of a system; more than one kind of morphology can have the same *Einstein* coefficient. For instance, aggregates of spheres can have the same values as short fibers or rods. In addition, some phases may appear to be continuous when they actually are not. For instance, very long, but discontinuous, fibers, ribbons, and oriented flakes may give high moduli characteristic of a continuous phase. Thus, if one or more of the dimensions of a particle are very large compared to the other dimensions, such fillers appear to the matrix phase to be continuous. Therefore, moduli do not completely describe a system; additional morphological information is required. However, on the other hand, if the morphology of a system is known, in principle the moduli can be accurately calculated. In the overlap region where both phases are essentially continuous, the exact morphology does not appear to be important.

The important factor is then how much of each phase is present, and this is determined by φ_m and φ'_m. Changes in morphology which occur as the concentration changes are also largely compensated for by the values of φ_m and φ'_m since they depend upon the packing and morphology at high concentrations while the value of the *Einstein* coefficient is determined by the morphology at very low concentrations.

Experimental

Four examples from the literature will be used to illustrate the use of the theory. The first example is a series of styrene-butadiene-styrene block polymers reported by *Holden* et al. (14) and discussed by *Kaelble* (3, 4). Fig. 3 shows the experimental data for *Youngs* modulus and a calculated curve which fits the data quite well.

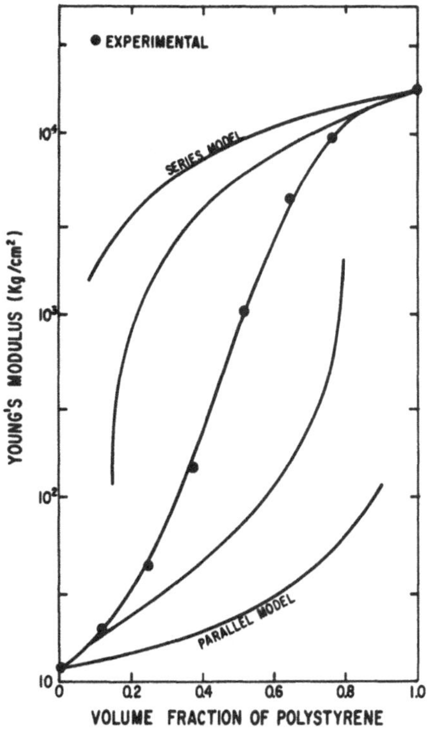

Fig. 3. Modulus of S-BD-S block polymers as a function of composition. Center curve calculated using $A = 3.0$, $\varphi_m = 0.80$, $v_1 = 0.5$, $A_i = 0.86$, $\varphi'_m = 0.85$, $v_1 = 0.35$

The calculated curve results from two curves using the following values: 1. Polystyrene dispersed in polybutadiene with $A = 3.0$ and $\varphi_m = 0.8$. 2. Polybutadiene dispersed in polystyrene with $A_i = 0.86$, $\varphi'_m = 0.85$, and a polystyrene *Poissons* ratio of 0.35. (A better fit to the data might be obtained by changing these values somewhat, but these illustrate the point.) Unfortunately, the morphology of these samples has not been published, but the above values strongly suggest the following changes in morphology as the concentration of polystyrene increases: At low concentrations of polystyrene, the polystyrene appears to be either aggregates of about six spheres (13) or rods with an aspect ratio of about 6 to 1.0 (12). Already at about 15% polystyrene, both phases tend to be continuous. The region of phase inversion where both phases are more or less continuous covers the range from 15–80% polystyrene. From 80–100% polystyrene, the polybutadiene is dispersed as spheres in the polystyrene. This type of information would be hard to deduce from series or parallel models, the limits of which are also shown in fig. 3. However, as pointed out by *Kaelble*, such models do also predict phase inversion for this series of block polymers.

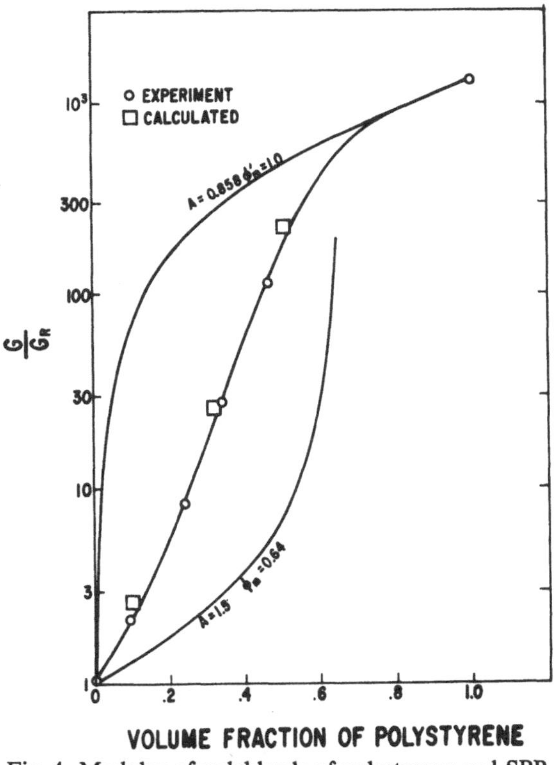

Fig. 4. Modulus of polyblends of polystyrene and SBR

A second example is a series of polyblends of polystyrene in styrene-butadiene rubber as reported by *Kraus* et al. (15). Fig. 4 shows the experimental values along with calculated values using the following constants: $A = 1.5$, $\varphi_m = 0.64$, $A_i = 0.86$, $\varphi'_m = 1.0$, and *Poissons* ratios of 0.5 and 0.35 for the rubber and the polystyrene, respectively. Again, the detailed morphology of these materials has not been published, but the above constants suggest the following behavior on addition of polystyrene to rubber: Almost immediately at very low concentrations of polystyrene, it tends to become a continuous phase, possibly by forming fibrous strings with an aspect ratio of about 15 to 1.0. The region of two continuous phases continues until a volume fraction of about 0.64 is reached. At higher concentrations of polystyrene, the rubber appears to be dispersed as spheres, although there are no experimental points except the point for pure polystyrene to really make this conclusion valid. In any case, the morphology is such as to make the experimental curve unsymmetrical about the mid composition point of $\varphi = 0.5$. On the other hand, the block polymer case is nearly symmetrical about $\varphi = 0.5$.

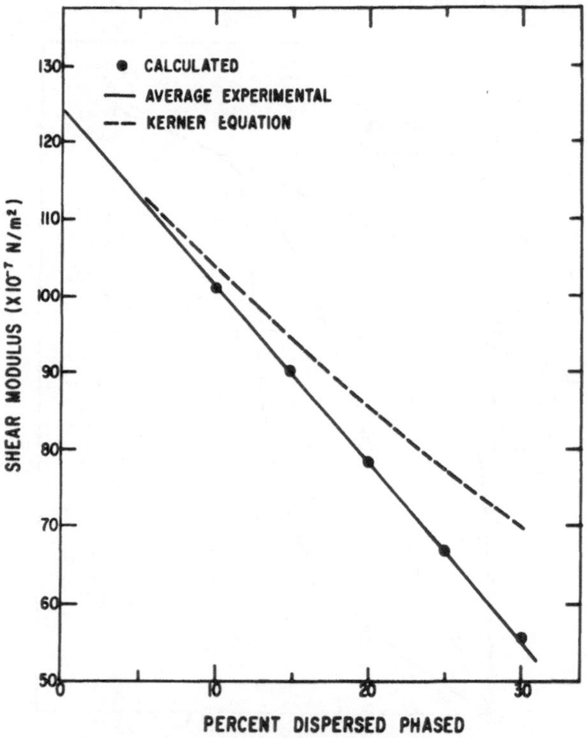

Fig. 5. Modulus of polystyrene containing a dispersed elastomeric phase. The solid line is the average experimental modulus line of *Cigna*. The points were calculated using $A_i = 0.86$ (spheres), $\varphi_m' = 0.55$, and $v_1 = 0.35$

The third example is a series of polystyrene-elastomer blends studied by *Cigna* (16). Electron microscopy showed these materials to be essentially spheres of the elastomers in a polystyrene matrix with little, if any, tendency for phase inversion to occur at concentrations of elastomers below 30%. Many polyblends were studied, undoubtedly with small changes in morphology, so at a given concentration of elastomer there was some scatter in the values of the shear moduli. An average experimental curve was drawn by *Cigna* through the experimental points as shown in fig. 5. The *Kerner* equation (or the *Halpin-Tsai* equation for spheres) gave values considerably higher than the experimental values. However, the modified equations using $\varphi_m' = 0.55$ and $A_i = 0.86$ gave good agreement with the experimental values. A value of $A_i = 0.86$ is the expected value for spheres in a matrix of *Poissons* ratio 0.35. A value of $\varphi_m' = 0.55$ is between the values for random loose packing (0.60) and simple cubic packing (0.524) of spheres. Thus, in this case the behavior predicted from the morphology is in very good agreement with the actually observed experimental results.

An example where the discontinuous phase does not appear to be dispersed spheres at either end of the composition range is the series of polyblends of polybutadiene and styrene-butadiene (SBR) rubber reported by *Fujimoto* et al. (17). The SBR copolymer contained 57.3% styrene. A comparison of the experimental data at 25 °C and the calculated results are given in fig. 6. An

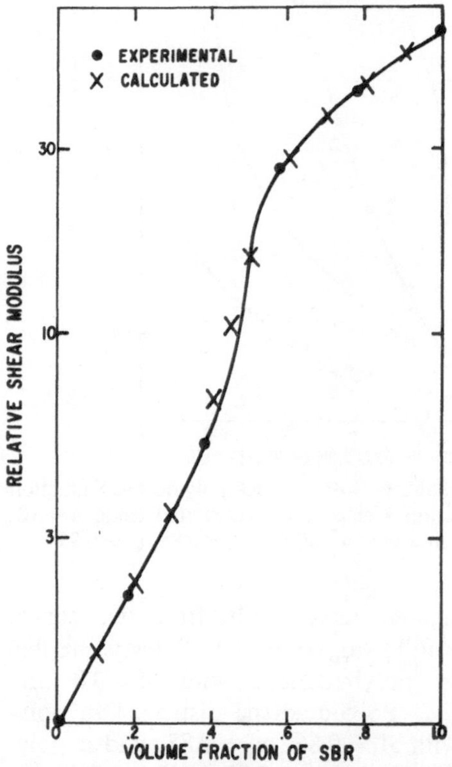

Fig. 6. Modulus of polybutadiene – SBR rubber blends at −25 °C

excellent match of the experimental results is achieved by using the values: $A = 4.35, B = 0.92, \varphi_m = 0.60, A_i = 0.3, B_i = 0.97, \varphi'_m = 0.70$. The *Einstein* coefficients (or A values) indicate that the dispersed phases near $\varphi \simeq 0$ and $\varphi \simeq 1$ are not dispersed spheres but are either fibers with an average aspect ratio of about 8 to 1 or very large aggregates. This system also has a very narrow range of compositions in which two continuous phases simultaneously exist; phase inversion is complete between a volume fraction of 0.3–0.6 of SBR copolymer.

Where the necessary morphological information is available in the four examples discussed, there is agreement between the experimental and the calculated results. In the other cases, reasonable morphologies have been deduced by fitting the theoretical equations to the experimental data. However, additional information on the morphology is needed to test the complete validity of this approach.

Acknowledgment

This research was supported by the Advanced Research Projects Agency of the Department of Defense and was monitored by the Office of Naval Research under Contract No. N 00014-67-C-0218.

Appendix

Justification for the logarithmic rule of mixtures in the overlap region

Empirically, it has been known that the logarithmic rule of mixtures (eq. [11]) is useful for predicting the modulus of crystalline polymers as a function of the degree of crystallinity or for predicting the modulus of polyblends in the region of phase inversion. Very little theoretical justification for its use has been presented, however. There are two intuitive arguments for its use:

1. There are two general equations which are capable of covering all values of the moduli M of composite systems from the lowest lower bound to the greatest upper bound. These are the *Halpin-Tsai* equations (eqs. [3]–[5]), and

$$M^n = M_1^n \varphi_1 + M_2^n \varphi_2; \quad -1 \le n \le +1. \qquad [a]$$

The logarithmic rule of mixtures is the limiting case of eq. [a] as $n \to 0$. The *Halpin-Tsai* equations and eq. [a] are symmetrical with respect to all upper and lower bounds when $\log M$ is plotted as a function of φ_2. They are not symmetrical on a M versus φ_2 plot.

2. The second argument involves giving equal relative weighting factors to both low and high moduli. When the upper and lower limits to the moduli are greatly different, equal weighting is achieved on a logarithmic scale but not on a linear scale.

The most important argument for using a logarithmic rule of mixtures can be visualized by writing eq. [11] in exponential form

$$M = M_U^{\varphi} \cdot M_L^{\varphi_L} \qquad [b]$$

where subscripts U and L refer to equations for the upper and lower bounds of the moduli. The quantities φ_L and φ_U refer to the relative amount of each phase participating as a continuous phase – not the total amount of each phase present. Thus, φ_L can be considered the "connectivity" of the rigid phase. If $\varphi_L = 0$, none of the rigid phase is continuous, while if $\varphi_L = 1$, all of the rigid phase is continuous in the region of overlap between $(1 - \varphi'_m)$ and φ_m (see fig. 2). Likewise, φ_U can be considered the "connectivity" of the low modulus component in the overlap region. The concept of connectivity can be illustrated by the overly simplified sketches in fig. 7. In fig. 7A and 7 C, the rubber phase is dispersed (its connectivity is zero), and the modulus of the composite is high. In fig. 7B and 7D the connectivity φ_U has increased, and the modulus of the composite decreases even though the quantity of rubber has not changed and the appearance of the rubber phase is nearly the same. A similar, but inverse situation, holds if $G_1 > G_2$ so that the connectivity φ_L of the rigid component increases in going from fig. 7A to 7B. As the connectivity of one phase increases, the connectivity of the other phase decreases in accordance with the equation

$$\varphi_L + \varphi_U = 1. \qquad [c]$$

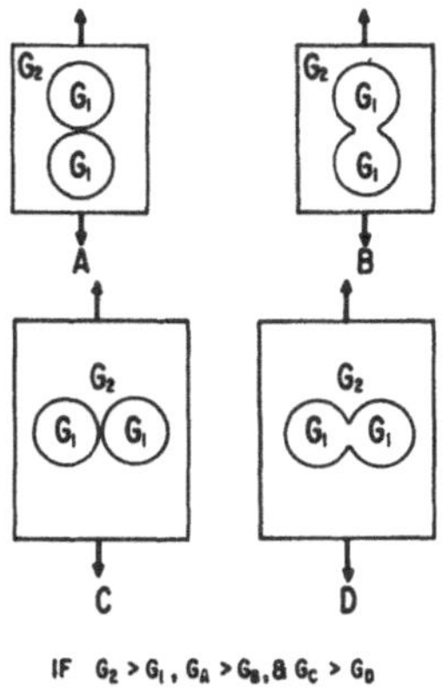

Fig. 7. Schematic diagrams illustrating the concept of connectivity of phases. Phase 1 has zero connectivity in *A* and *C*

In the overlap region where phase inversion is taking place, the connectivities φ_L and φ_U change more rapidly than the volume fractions of the components φ_1 and φ_2.

The connectivity must be random in nature for the logarithmic rule of mixtures to hold. If the connectivity of the particles becomes oriented primarily in one direction, anisotropy develops. In the extreme case of orientation parallel to the direction of stress, a parallel type of model holds in which the ordinary rule of mixtures is obeyed:

$$M = M_1 \varphi_1 + M_2 \varphi_2 . \qquad [d]$$

The other extreme is orientation perpendicular to the direction of applied tensile stress; in this case a series type of model system develops in which the inverse rule of mixtures holds:

$$\frac{1}{M} = \frac{\varphi_1}{M_1} + \frac{\varphi_2}{M_2} . \qquad [e]$$

On a log$M$ versus φ_2 plot, eqs. [d] and [e] are symmetrical about the line for the logarithmic rule of mixtures. Rotation of such plots 180° corresponds to rotating the specimen 90°.

Summary

The theory of the elastic moduli of composite materials in which an inversion of the phases can occur is reviewed. The morphology of the system and the packing fraction of the dispersed phase are important in determining the moduli. The applicability of the theoretical equations is illustrated for four systems of block polymers and polyblends. In three of the systems, phase inversion occurs. Agreement between theory and experiment is good, and where the morphology of the composites is known, the moduli agree with the values expected for that morphology.

Zusammenfassung

Die Elastizitätsmodul-Theorie der zusammengesetzten Stoffe, in welchen eine Phaseninversion vorkommen kann, wird untersucht. Die Systemmorphologie und die Packungsfraktion der dispersen Phase sind für die Modulbestimmung wichtig. Die Anwendbarkeit der theoretischen Gleichungen ist für vier Systeme von Blockpolymeren und Polygemischen veranschaulicht. Eine Phaseninversion kommt in drei von den Systemen vor. Die Theorie und Praxis sind in einer guten Übereinstimmung, und da, wo die Morphologie der zusammengesetzten Stoffe bekannt ist, stimmen die Moduli mit den für die Morphologie erwarteten Werten überein.

References

1) *Takayanagi, M.*, Proc. 4th Internat. Congr. Rheol., Part 1, p. 161, *Lee* and *Copley* (Ed.) (New York 1965).

2) *Manabe, S., R. Murakami*, and *M. Takayanagi*, Memoirs Faculty of Eng. Kyushu Univ. **28**, 295 (1969).

3) *Kaelble, D. H.*, Physical Chemistry of Adhesion, p. 415 (New York 1971).

4) *Kaelble, D. H.*, Trans. Soc. Rheol. **15**, 235 (1971).

5) *Kerner, E. H.*, Proc. Phys. Soc. **B 69**, 808 (1956).

6) *Ashton, J. E., J. C. Halpin*, and *P. H. Petit*, Primer on Composite Analysis, Chap. 5 (Stamford, Conn., (1969)

7) *Halpin, J. C.*, J. Compos. Mater. **3**, 732 (1969).

8) *Nielsen, L. E.*, J. Appl. Phys. **41**, 4626 (1970).

9) *Lewis, T. B.* and *L. E. Nielsen*, J. Appl. Polym. Sci. **14**, 1449 (1970).

10) *Nielsen, L. E.*, Appl. Polym. Symp., No. 12, 249 (1969).

11) *Gray, R. W.* and *N. G. McCrum*, J. Pol. Sci., Part A 2, **7**, 1329 (1969).

12) *Burgers, J. M.*, Second Report on Viscosity and Plasticity, p. 113 (New York 1938).

13) *Lewis, T. B.* and *L. E. Nielsen*, Trans. Soc. Rheol. **12**, 421 (1968).

14) *Holden, G., E. T. Bishop*, and *N. R. Legge*, J. Polym. Sci. **C 26**, 37 (1969).

15) *Kraus, G., K. W. Rollmann*, and *J. T. Gruver*, Macromol. 3, 92 (1970).

16) *Cigna, G.*, J. Appl. Polym. Sci. **14**, 1781 (1970).

17) *Fujimoto, K.* and *N. Yoshimura*, Rubber Chem. Tech. **41**, 1109 (1968).

Author's address:

Dr. *Lawrence E. Nielsen*
Monsanto Company
800 N Lindbergh
St. Louis, Missouri, 63166 (USA)

Rheol. Acta 13, 93–98 (1974)

Institut des Matériaux, Université libre de Bruxelles (Belgique)

Déformation plastique du polyéthylène haute densité soumis des essais de traction-torsion

Par J. C. Bauwens

Avec 5 figures

(Reçu p. p. le 27 Octobre 1972)

Introduction

Nous avons proposé, dans des publications antérieures (1, 2), un critère de plasticité applicable aux hauts polymères et basé sur la théorie d'activation d'*Eyring* (3). Selon ce critère, l'état de contrainte pour lequel la déformation plastique débute satisfait à équation:

$$\tau_{\text{oct}} + \mu p = f(\dot{\varepsilon}, T) \qquad [1]$$

où τ_{oct} est la composante tangentielle de la contrainte octaédrale, p la contrainte hydrostatique, μ une constante et $f(\dot{\varepsilon}, T)$ une fonction de la vitesse de déformation $\dot{\varepsilon}$ et de la température T.

Nous insistons sur le fait que cette relation [1] se rapporte à une déformation homogène et purement plastique correspondant à un endroit de la courbe charge-déformation où la contrainte est stationnaire. Nous avons jusqu'à présent vérifié la validité de ce critère en mesurant la contrainte relative au sommet du crochet à la limite élastique (upper yield point) que présentent les courbes charge-déformation des hauts polymères étudiés, tant en traction qu'en compression ou en torsion; c'est cette contrainte (upper yield stress) que nous désignerons par limite élastique dans la suite de l'exposé.

Dans les considérations qui ont permis d'exprimer ce critère, nous n'avons pas envisagé l'influence d'une déformation élastique ou viscoélastique préalable, sur la valeur de la contrainte à la limite élastique; ce traitement concernait donc un solide rigide-plastique et nous avons établi qu'il décrivait convenablement le comportement du chlorure de polyvinyle rigide dont la déformation à la limite élastique est faible. Nous allons considérer ici un solide élastique-plastique et calculer les corrections impliquées par l'existence d'une déformation préalable à la déformation purement plastique.

Pour confronter la théorie et l'expérience, nous avons choisi d'effectuer des essais de traction-torsion sur le polyéthylène haute densité, car ce haut polymère présente un grand allongement et un crochet à la limite élastique.

Considérations théoriques

Plusieurs auteurs (4, 5) ont proposé d'une façon purement formelle des critères de plasticité applicables aux hauts polymères. Les contraintes qui interviennent dans leurs expressions sont les contraintes vraies (6, 7, 8).

Le calcul de l'équilibre des solides soumis à des sollicitations, nécessite l'évaluation des contraintes à l'état déformé et il en découle la justification et l'emploi des contraintes vraies. Transposer cet usage à la formulation d'un critère de plasticité nous paraît sans fondement; nous pensons qu'il convient d'envisager, non pas le formalisme mathématique, mais la signification physique du critère, pour connaître l'influence d'une déformation préalable sur les grandeurs physiques qui en sont à la base.

Le critère que nous avons proposé est lié à la valeur de l'énergie que requiert le déplacement d'une unité rhéologique d'une position d'équilibre à une autre voisine. Cette énergie doit être constante, pour des essais effectués à même température et ramenés à des vitesses de déformation et des déformations plastiques équivalentes, et ce, quelle que soit la géométrie des contraintes. Nous avons établi (1) que cette condition s'exprimait à l'échelle macroscopique par la relation énergétique suivante dont découle l'expression [1]:

$$W_\tau + W_p = \text{constante} \qquad [2]$$

où W_τ désigne le travail absorbé par la déformation plastique et W_p est une énergie proportionnelle à la contrainte hydrostatique.

La relation [2] doit donc être indépendante de la valeur et de la nature (élastique, visco-élastique, plastique, etc.) de la déformation déjà subie par le corps au moment où il est soumis à l'action d'une contrainte satisfaisant au critère.

Evaluons le travail requis pour conférer, à la limite élastique, des déformations plastiques identiques à un solide rigide-plastique et à un solide élastique-plastique, ainsi que nous l'avions déjà fait dans les cas particuliers de la traction et de la compression (9).

Soit un solide ayant subi une déformation plastique définie par son tenseur de déplacement:

$$D'_{ij} = \delta_{ij} + u'_{i,j}$$

δ_{ij} étant le symbole de *Kronecker* et $u'_{i,j}$ le gradient de déplacement.

Appliquons-lui une seconde déformation, élastique par exemple:

$$D''_{ij} = \delta_{ij} + u'_{i,j}\,. \qquad [4]$$

La déformation totale produit un déplacement dont le tenseur vaut:

$$D = D'' \cdot D' = \delta_{ij} + u_{i,j}\,. \qquad [5]$$

Lorsque les composantes $u'_{i,j}$ tendent vers 0, ce qui correspond à une déformation plastique infiniment petite, les composantes $u_{i,j}$ sont les déplacements résultant de la déformation à la limite élastique.

Le travail par unité de volume qu'il faut fournir pour produire un accroissement de déformation plastique $\delta D'_{ij}$ vaut:

$$\delta W = \sum_{rs} \frac{\partial \sum_{ij} u_{i,j}\tau_{ij}}{\partial u'_{r,s}}\,\delta u'_{r,s} \qquad [6]$$

τ_{ij} étant la contrainte nominale.

Pour que ce travail soit constant, il faut que:

$$\frac{\partial \sum_{ij} u_{i,j}\tau_{ij}}{\partial u'_{r,s}}\,\delta u'_{r,s} = \tau^*_{rs}\delta u'_{r,s} = \text{constante} \qquad [7]$$

où $\tau^*_{rs}\delta u'_{r,s}$ se rapporte au travail de déformation d'un solide rigide-plastique sous l'action de la contrainte τ^*_{rs}. On a ainsi:

$$\tau^*_{rs} = \frac{\partial \sum_{ij} \tau_{ij}u_{i,j}}{\partial u'_{r,s}} \qquad [8]$$

et par conséquent:

$$\tau^*_{rs} = \left(\frac{\partial}{\partial u'_{r,s}} \sum_{ij} \frac{u_{i,j}\tau_{ij}}{\tau_{rs}}\right)\tau_{rs} = C_{rs}\tau_{rs} \qquad [9]$$

C_{rs} est donc le facteur de correction qui intervient lorsqu'il existe une déformation D''_{ij} à la limite élastique.

Si le critère qui s'applique à un solide rigide-plastique répond à la relation:

$$f(\tau^*_{rs}) = 0 \qquad [10]$$

l'équation du critère convenant à un solide élastique-plastique est alors:

$$f(C_{rs}\tau_{rs}) = 0. \qquad [11]$$

Dans le cas particulier d'un essai de traction-torsion dont les composantes respectives sont σ_x et τ_{yx}, on obtient:

$$u_{x,x} = (1 + u'_{x,x})(1 + u''_{x,x}) \qquad [12]$$

$$u_{y,x} = u'_{y,x}(1 + u''_{y,y}) + u''_{y,x}(1 + u'_{x,x}) \qquad [13]$$

et en vertu des eqs. [8] et [9], il vient:

$$\frac{\partial}{\partial u_{x,x}} \sum_{ij} \tau_{ij}u_{i,j} = \sigma_x(1 + u''_{x,x}) + \tau_{yx}u''_{y,x} = C_{xx}\sigma_x \qquad [14]$$

$$\frac{\partial}{\partial u_{y,x}} \sum_{ij} \tau_{ij}u_{i,j} = \tau_{yx}(1 + u''_{y,y}) = C_{yx}\tau_{yx} \qquad [15]$$

les coefficients de correction valent donc:

$$C_{xx} = 1 + u''_{x,x} + \frac{\tau_{yx}}{\sigma_x}u''_{y,x} = C_\sigma \qquad [16]$$

$$C_{yx} = 1 + u''_{y,y} = 1 - \nu u''_{x,x} = C_\tau \qquad [17]$$

ν étant le coefficient de *Poisson*.

En posant

$$u''_{x,x} = \varepsilon_e \quad \text{et} \quad u''_{y,x} = \gamma_e$$

ε_e et γ_e étant donc les déformations de traction et de torsion qui existent à la limite élastique, on obtient finalement, en supprimant les indices x et y devenus superflus:

$$C_\sigma = 1 + \varepsilon_e + \frac{\tau}{\sigma}\gamma_e \qquad [18]$$

$$C_\tau = 1 - \nu\varepsilon_e\,. \qquad [19]$$

Or, les déformations ε_e et γ_e sont liées par l'équation:

$$\gamma_e = 2(1 + \nu)\frac{\tau}{\sigma}\varepsilon_e \qquad [20]$$

les relations [18] et [19] deviennent ainsi:

$$C_\sigma = 1 + \varepsilon_e\left[1 + 2(1 + \nu)\frac{\tau^2}{\sigma^2}\right] \qquad [21]$$

$$C_\tau = 1 - \nu \varepsilon_e. \qquad [22]$$

Nous avons vu antérieurement (2) que le critère appliqué à un solide rigide-plastique soumis à un essai de traction-torsion avait la forme:

$$\sqrt{6\tau^2 + 2\sigma^2} + 3\mu p = C. \qquad [23]$$

Pour un solide élastique-plastique, ce critère s'exprime par l'équation

$$\sqrt{6\tau^2(1 - \nu\varepsilon_e)^2 + 2\sigma^2 \times \left\{1 + \varepsilon_e\left[1 + 2(1+\nu)\frac{\tau^2}{\sigma^2}\right]\right\}^2} + \frac{\mu\sigma}{(1 - \nu\varepsilon_e)^2} = C. \qquad [24]$$

Si l'on avait considéré, ainsi que l'ont fait les autres auteurs, que la correction qu'il convenait de faire pour tenir compte de la déformation à la limite élastique consistait à introduire dans le critère les contraintes vraies, on aurait obtenu:

$$\sqrt{6\tau^2 + 2\sigma^2} + \mu\sigma = (1 - \nu\varepsilon_e)^2\, C. \qquad [25]$$

La correction de déformation que nous proposons diffère donc de l'usage de la contrainte vraie.

Il existe une autre relation où les différences sont encore plus marquées; il s'agit de l'équation de *Levy-von Mises* qui découle d'ailleurs directement de la théorie d'*Eyring*, ainsi que nous l'avons déjà fait remarquer (1).

Cette équation s'écrit, pour un solide rigide-plastique soumis à une épreuve de traction-torsion (2):

$$\frac{\tau}{\gamma} = \frac{\sigma}{3\varepsilon}. \qquad [26]$$

Compte tenu des relations [12], [13], [18] et [19], elle devient, en vertu des corrections qui s'appliquent à σ, τ, ε et γ:

$$\frac{\sigma\left(1 + \varepsilon_e + \gamma_e \frac{\tau}{\sigma}\right)(1 + \varepsilon_e)}{3\varepsilon} = \frac{\tau(1 - \nu\varepsilon_e)^2}{\gamma\left(1 - \gamma_e \frac{\varepsilon}{\gamma}\right)}. \qquad [27]$$

Cette relation peut se mettre sous la forme:

$$\tau\varepsilon = \frac{\sigma\gamma}{3} f(\sigma, \tau) \qquad [28]$$

$f(\sigma, \tau)$ est le facteur de correction qu'il est nécessaire d'introduire lorsqu'il existe une déformation ε_e, γ_e à la limite élastique.

En tenant compte de la relation [21], en négligeant les corrections du second ordre et en prenant:

$$\varepsilon_e = k\sigma \qquad [29]$$

on obtient:

$$f(\sigma, \tau) = \left\{1 + k\sigma\left[1 + 2\nu + 2(1+\nu)\frac{\tau^2}{\sigma^2}\right]\right\}. \qquad [30]$$

Ce facteur devient très important pour les valeurs élevées du rapport τ/σ.

L'usage de la contrainte vraie, au contraire, ne modifie en rien l'éq. [26] *de Levy-von Mises* car les corrections qui affectent σ et τ sont identiques, ainsi que celles de ε et γ; elles s'annulent donc dans cette équation.

Méthode expérimentale et discussion des résultats

Les mesures ont été faites sur un polyéthylène haute densité dont le nom commercial est ELTEX 54001 (Solvay et Cie). Nous avons employé un dispositif classique de traction-torsion sur tubes à paroi mince (fig. 1) qui a été décrit dans un article antérieur (2). Rappelons seulement que les grandeurs imposées sont la vitesse de déformation en traction et la con-

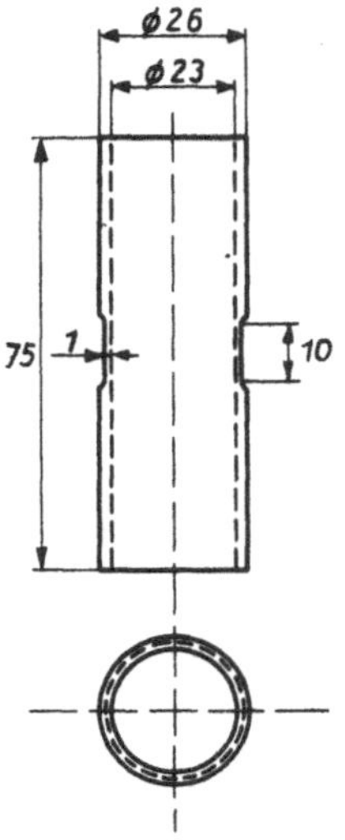

Fig. 1. Eprouvette de traction-torsion

trainte de torsion. Il en découle que la vitesse de déformation résultante n'est pas identique à chaque essai, elle croît avec la contrainte de torsion. Or, pour comparer les différents essais, il convient de les rapporter non seulement à une même température, mais également à des vitesses équivalentes et nous avons vu (2) que celles-ci sont liées par l'équation:

$$\dot{\gamma}^2 + 3\dot{\varepsilon}^2 = \text{constante}. \quad [31]$$

La mesure de $\dot{\gamma}$ est aisée, car l'enregistreur de la courbe de traction est équipé d'un dispositif de cochage. Il suffit de tracer sur cette courbe (fig. 2) des traits dont l'écartement correspond à une variation d'angle de torsion donnée.

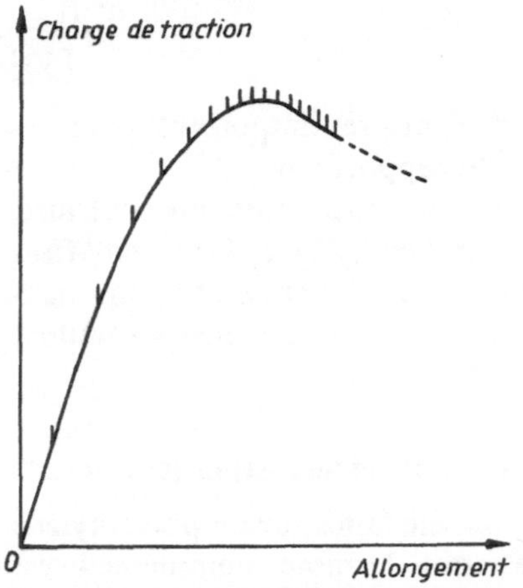

Fig. 2. Courbe charge de traction-allongement relevée lors d'un essai de traction-torsion. Le cochage de la courbe permet de mesurer la déformation en torsion

Nous avons fait des essais de traction à différentes températures et vitesses pour connaître les corrections que les variations de ces paramètres nécessitent. La limite élastique dépend linéairement de la température ainsi que du logarithme de la vitesse de déformation; nos mesures ont donné:

$$\frac{\partial \sigma}{\partial T} = 0{,}035\ \text{kg} \cdot \text{mm}^{-2} \cdot \text{K}^{-1} \quad [32]$$

$$\frac{\partial \sigma}{\partial \log \dot{\varepsilon}_t} = 0{,}36\ \text{kg} \cdot \text{mm}^{-2}. \quad [33]$$

Tous les essais ont ainsi pu être rapportés à $T = 21\ ^\circ\text{C}$ et $\dot{\varepsilon} = 0{,}1\ \text{min}^{-1}$.

La correction que nous proposons pour tenir compte de la déformation à la limite élastique dépend de la composante de traction ε_e. Nous avons pris:

$$\varepsilon_e = k\sigma \quad [29]$$

il nous faut donc déterminer k et pour ce faire, connaître l'allongement ε_{et} à la limite élastique en traction pure σ_t.

Les mesures extensométriques sur des éprouvettes de traction en forme d'haltères donnent:

$$\varepsilon_{et} = 0{,}12 \quad [34]$$

et

$$\sigma_t = 2{,}12\ \text{kg} \cdot \text{mm}^{-2} \quad [35]$$

il vient donc:

$$k = 0{,}057\ \text{kg}^{-1} \cdot \text{mm}^2. \quad [36]$$

Lorsque la contrainte de torsion est élevée, le tube s'écrase avant que la limite élastique ne soit atteinte. Pour éviter cet inconvénient, nous avons introduit dans le tube un cylindre lubrifié ayant le diamètre intérieur que prend le tube à la limite élastique. On prévient ainsi l'écrasement de la paroi tout en évitant que le tube ne serre le cylindre et, de la sorte, perturbe les contraintes.

Pour déterminer tous les paramètres de l'équation exprimant le critère de plasticité, il nous faut en outre connaître l'influence de la pression hydrostatique, c'est-à-dire la valeur de μ. Une évaluation directe de ce paramètre peut être obtenue en effectuant des essais de traction sous pression hydrostatique. De tels essais ont été réalisés sur le polyéthylène par *Mears*, *Pae* et *Sauer* (10) qui ont obtenu:

$$\mu = 0{,}049. \quad [37]$$

Ils ont en outre mesuré le coefficient de *Poisson* qui vaut:

$$\nu = 0{,}38. \quad [38]$$

Nous disposons ainsi de toutes les données qui sont nécessaires pour calculer le paramètre C des eqs. [24] et [25]. En supposant que les deux équations donnent la même limite élastique en traction définie par [35], les relations [24] et [25] donnent respectivement:

$$C = 3{,}46 \quad [39]$$

$$C' = 3{,}39. \quad [40]$$

Remarquons que si l'on néglige les corrections du second ordre, l'eq. [25] peut s'écrire:

$$6\tau^2 + 2\sigma^2 + (\mu + 2\nu k C)\sigma = C \quad [41]$$

et l'on obtient les mêmes paramètres que pour un corps rigide-plastique dont le coefficient de la contrainte hydrostatique vaudrait:

$$\mu' = \mu + 2\nu k C. \quad [42]$$

Nous avons tracé sur le graphique de la fig. 3 les courbes répondant aux eqs. [24] et [25] qui sont respectivement représentées en traits continus et discontinus, nous avons aussi porté les points expérimentaux provenant des essais de traction-torsion.

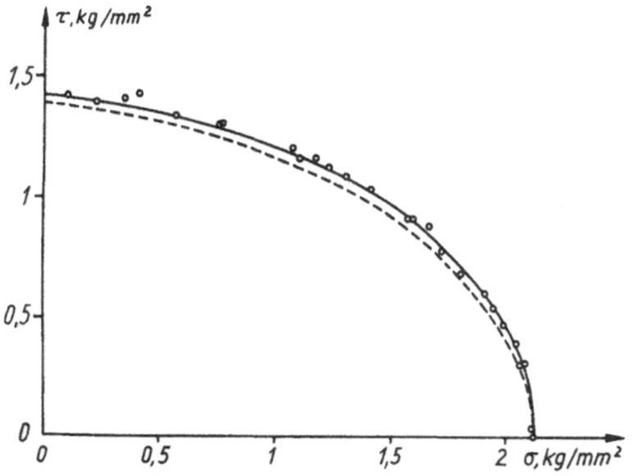

Fig. 3. Relation entre τ et σ à la limite élastique. La courbe en trait plein correspond à la relation [24] que nous proposons; la courbe en pointillé se rapporte à l'eq. [25] où l'on utilise les contraintes vraies. Les cercles désignent les mesures expérimentales

Ce graphique montre que la correction que nous proposons concorde fort convenablement avec les résultats expérimentaux, alors que l'usage de la contrainte vraie conduit à une courbe qui s'en écarte légèrement et par défaut.

Nous avons également comparé à l'expérience l'équation de *Levy-von Mises*; si l'on dérive les déformations par rapport au temps, elle s'écrit:

$$\tau \dot{\varepsilon} = \frac{\sigma \dot{\gamma}}{3} \qquad [43]$$

et nous avons vu que l'usage de la contrainte vraie n'introduisait pas de correction.

Au contraire, notre traitement implique une correction et l'eq. [43] devient:

$$\tau \dot{\varepsilon} = f(\sigma, \tau) \frac{\sigma \dot{\gamma}}{3} \qquad [44]$$

avec:

$$f(\sigma, \tau) = \left\{1 + 0{,}057\,\sigma \times \left[1 + 2 \cdot 0{,}38 + 2(1 + 0{,}38) \frac{\tau^2}{\sigma^2}\right]\right\}. \qquad [45]$$

Les graphiques des figs. 4 et 5 comparent respectivement les eqs. [43] et [44] aux valeurs expérimentales.

L'écart entre les deux traitements est grand et il est manifeste que la correction que nous proposons conduit à un bon accord avec l'expérience, tandis que l'usage de la contrainte vraie donne un rapport $\tau \dot{\varepsilon}/\sigma \dot{\gamma}$ qui est systématiquement trop élevé.

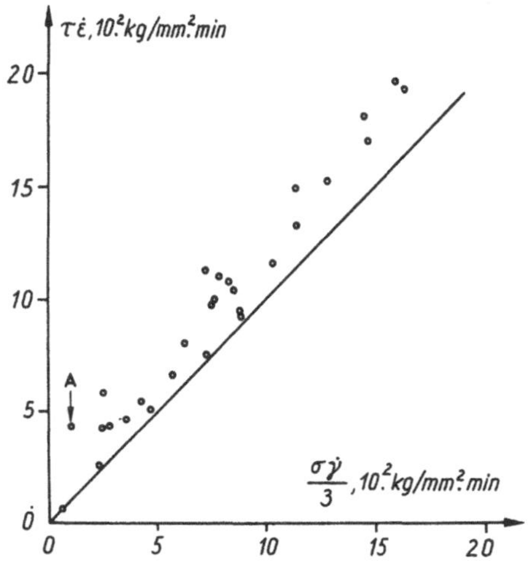

Fig. 4. Relation de *Levy-von Mises* non corrigée, telle que l'implique l'usage de la contrainte et des déformations vraies. Les cercles désignent les mesures expérimentales

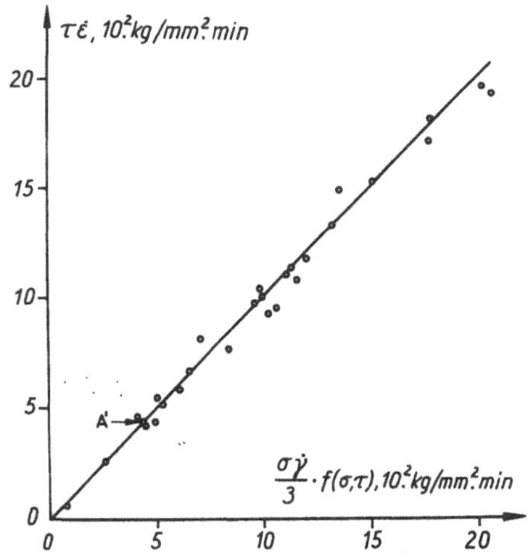

Fig. 5. Relation de *Levy-von Mises* corrigée ainsi que l'exige l'invariance du travail spécifique de déformation plastique. Les cercles désignent les mesures expérimentales

Le facteur de correction $f(\sigma, \tau)$ devient très important pour les grandes valeurs du rapport τ/σ. Par exemple, le point A de la fig. 4 est fort éloigné de la droite théorique, alors qu'après correction (fig. 5), il vient en A' sur la droite theorique; pour ce point, $f(\sigma, \tau)$ atteint la valeur 4.

Conclusions

Si l'on fait abstraction de l'influence de la contrainte hydrostatique, l'application de la théorie d'*Eyring* à la déformation plastique d'un solide rigide-plastique conduit à une loi dont le formalisme est identique à celui du critère de *von Mises*. La signification physique en est pourtant toute différente; le premier concept implique l'invariance du travail spécifique de déformation plastique, alors que le second est basé sur l'invariance de l'énergie élastique de distorsion. Nous avons montré que les deux formalismes ne sont cependant plus identiques lorsqu'on envisage le cas d'un solide élastique-plastique, car les deux théories conduisent à des corrections de déformation différentes.

Les mesures sur le polyéthylène haute densité dont la déformation à la limite élastique est relativement grande sont en bon accord avec les corrections qui découlent de l'invariance du travail de déformation plastique, tant pour l'expression du critère que pour l'équation de *Levy-von Mises*. L'usage de la contrainte vraie conduit à des écarts entre la théorie et l'expérience; ils sont faibles pour l'expression du critère, nettement plus grands pour la relation de *Levy-von Mises*.

La correction que nous proposons est donc mieux vérifiée, pour ce qui concerne le polyéthylène, et le travail spécifique de déformation plastique semble donc indépendant de la déformation préalable à la déformation purement plastique.

Nous pouvons remarquer de surcroît que le traitement proposé ne présume en aucune facon de la nature de cette déformation préalable, ni de la relation qui la lie aux contraintes.

Remerciements

Qu'il nous soit permis de remercier Monsieur le Professeur *Georges Homès*, directeur de l'Institut des Matériaux de l'Université libre de Bruxelles. Nous remercions aussi Monsieur *Léon Hoeymans* de la Société Solvay et Cie, qui nous a fourni le polyéthylène et a participé aux essais.

Littérature

1) *Bauwens, J. C.*, J. Polymer Sci. A 2, **5**, 1145 (1967).
2) *Bauwens, J. C.*, J. Polymer Sci. A 2, **8**, 893 (1970).
3) *Eyring, J.*, J. Chem. Phys. **4**, 283 (1936).
4) *Bowden, P. B.* et *J. A. Jukes*, J. Mater. Sci. **3**, 183 (1968).
5) *Sternstein, S. S.* et *L. Ongchin*, Amer. Chem. Soc. (Polymer Preprints) **10**, 117 (1969).
6) *Bowden, P. B.* et *J. A. Jukes*, J. Mater. Sci. **7**, 52 (1972).
7) *Sternstein, S. S., L. Ongchin* et *A. Silverman*, Applied Polymer Symposia **7**, 175 (1968).
8) *Christiansen, A. W., E. Baer* et *S. V. Radcliffe*, Phil. Mag. **24**, 451 (1971).
9) *Bauwens-Crowet, C., J. C. Bauwens* et *G. Homès*, J. Mater. Sci. **7**, 176 (1972).
10) *Mears, D. R., K. D. Pae* et *J. A. Sauer*, J. Applied Phys. **40**, 4229 (1969).

Adresse de l'auteur:
J. C. Bauwens
Institut des Matériaux
Université libre de Bruxelles
87 Avenue A. Buyl
B-1050 Bruxelles (Belgique)

Rheol. Acta 13, 99–102 (1974)

From the Manufacturing Development Laboratory, Mitsubishi Electric Corporation, Amagasaki Hyogo (Japan)

Dynamic mechanical properties of cured epoxy resin filled with mica flake

By K. Shibayama, K. Iisaka, and T. Kitagawa

With 3 figures and 2 tables

(Received October 27, 1972)

Introduction

Mechanical properties of filled polymeric system are affected by various factors, e. g. size, size distribution, shape, oriention and agglomeration of filler particles, and strength of the adhesive bond on filler surface. It is difficult, in general, to investigate the effects of each factor separately. Most of the theory so far developed in explaining the properties of filled polymers has been based on spherical particles.

Many fillers used practically are, however, irregular, platelet, or fibrous in shape. With nonspherical fillers, shape and degree of orientation could greatly modify mechanical behavior of the system. Except for the case of fibrous fillers, very little has been published on behavior of systems containing oriented filler (1).

The purpose of this study is to investigate the effect of orientation of a two dimensional filler on the mechanical properties of resinous composite system.

Experimental

Small muscovite mica flake about 1 μ in thickness and 700 μ in width was used as the filler. Mica flake was obtained by crushing mechanically a reconstituted mica paper supplied from 3 M Co. and sifted out (28 $\sim$ 40 mesh portion was used).

Unfilled batches of epoxy were made with composition ratio Epikote 828 (Shell Chemical Co.): methyl tetrahydrophthalic anhydride: benzyl dimethyl amine = 100:90:1 and last into sheet specimen between two glass plates. Filled batches of less than 16.4% volume fraction of filler was prepared by mixing filler and resin with the same composition ratio as unfilled ones and casted. Filled batches of more than 26.6% volume fraction of filler was made by impregnating same resin to mica mat prepared by a sheet macine in wet method. These batches were cured at 150 °C for 16 h followed by 200 °C for 3 h. Table 1 lists volume fraction of filler in each sample determined from the residue by heating a specimen at 500 °C for 3 h in air.

Measurement of dynamic mechanical properties were carried out over a temperature range from room temperature to 200 °C at 1 Hz by a viscoelastic spectrometer. Both tensile and shear moduli were determined in parallel direction to specimen surface (mica flake surface).

Table 1. Sample

No.	Volume fraction of filler (%)
1	0
2	16.4
3	26.6
4	38.7
5	65.6

Results and discussion

Temperature dependence of tensile storage modulus E' and mechanical loss $\tan\delta$ was shown in fig. 1. Two mechanical loss peaks were observed for mica flake filled samples on $\tan\delta$ curves, one around 80 °C and the other around 150 °C. Values of E' at rubbery state for filled specimens are extraordinarily high, comparing with values for particulate filled system reported elsewhere (2). Higher temperature peak (α) can be ascribed to the onset of microbrownian motion of chain segments, corresponding to glass to rubber transition. The position of α peak for filled samples is about ten degrees higher than that of unfilled resin. This may suggest the existence of certain interaction between polymer molecule and mica surface.

7*

Appearance of lower temperature loss peak (β) in filled specimens would suggest specific effects of mica flake on composite structure, since β peak is usually observed below room temperature for unfilled resin. Possible explanation of this behavior will be found in chemical structure of cured resin, boundary layer on mica surface, or geometry of the composite structure.

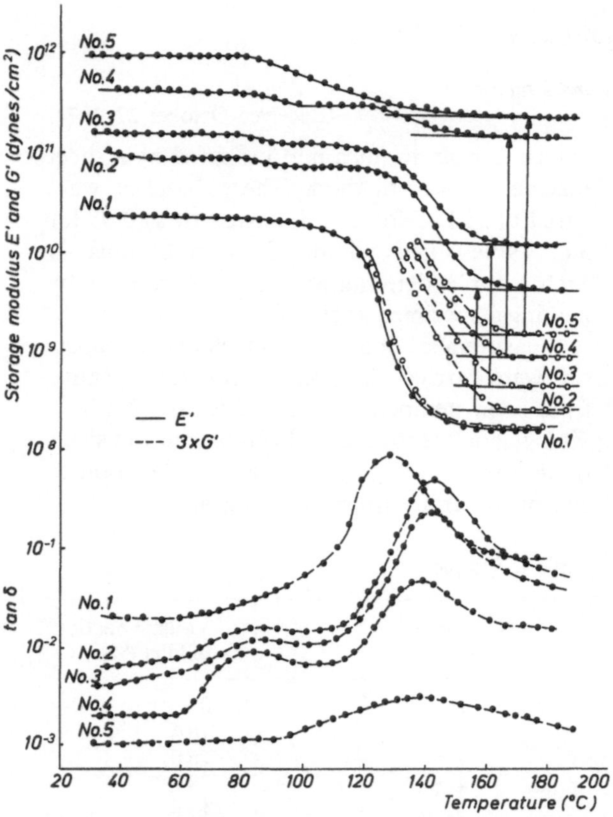

Fig. 1. Temperature dependence of tensile and shear modulus and mechanical loss tan δ in each specimen

In this paper, mechanism of originating extraordinary high modulus value in rubbery region is pursued. Relative modulus $E'c/E'r$ ($E'c$ and $E'r$ are modulus of filled and unfilled specimens, respectively) is plotted against volume fraction of filler in fig. 2, where calculated results using *Kerners* equation (3) are also shown. Increase in relative modulus with filler content is far more marked for rubbery state than in glassy state, though observed relative modulus is higher than the calculation both in rubbery and glassy regions.

Three ways are apparently possible to explain this behavior, 1. difference in *Poissons* ratio of resin between glassy and rubbery state, 2. strong adsorption of chain segments on filler surface, and 3. mode of deformation of polymeric matrix. As for the first way, numerical calculation according to *Kerners* equation predicts only twenty percent increase in relative modulus if *Poissons* ratio increases from 0.2–0.5 going from glassy to rubbery region. The second way is considered to be insufficient to explain the observed high modulus from usual knowledge of inert surface property of mica though the interfacial attraction may exist to some extent as revealed by the increase of tan δ peak temperature.

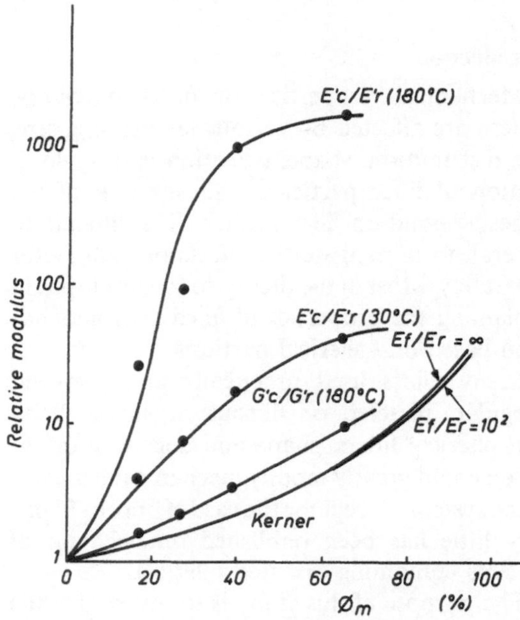

Fig. 2. Relative modulus vs. volume fraction of filler

Along with the third way, it will be plausible to consider that a part of the tension applied to bulk specimen is converted to shear stress in thin polymeric layer sandwitched between two mica flakes when the mica content of the specimen is increased and orientation of mica flakes in the direction of bulk deformation is enhanced.

An experimental evidence of this schema is given by results of shear measurement, shown in fig. 1 and 2. From fig. 1, one can see values of shear modulus for filled specimens full in a range which is similar to one usually met with particulate filled systems. Relative modulus is fairly close to the prediction by *Kerners* equation, as seen in fig. 2.

An equivalent mechanical model can be constructed to prove the validity of the above scheme more quantitatively, as shown in fig. 3. Mechanically, the model is composed of two kinds of

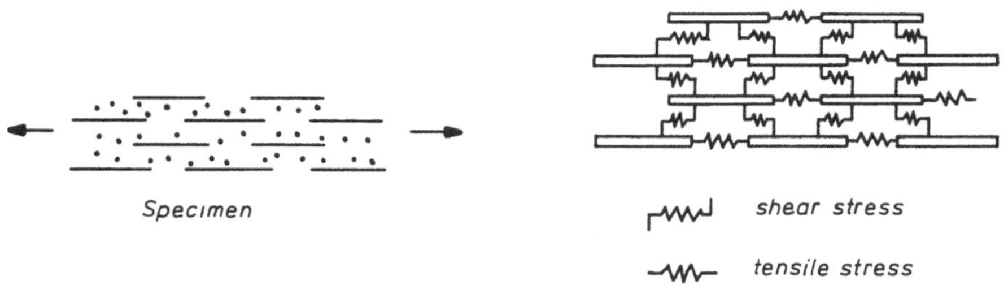

Fig. 3. Model explanation for high rubbery modulus

elastic elements; one corresponds to shear stress between two mica flakes neighbouring each other in direction perpendicular to mica flake surface and the other corresponds to tensile stress between two mica flakes neighbouring each other in one plane. Modulus of mica (10^{12} dyne/cm^2) is high enough comparing to modulus of rubbery resin (10^8 dyne/cm^2) and is assumed to be perfectly rigid. Responce of the model under an applied tensile force F in direction parallel to mica flake surface can be expressed as eq. [1],

$$\frac{F}{n \cdot w} = Gr \cdot \frac{\Delta l}{dr} \cdot l^2 + Er \cdot \frac{\Delta l}{\alpha \cdot l} \cdot dml \qquad [1]$$

where Gr and Er are shear and tensile moduli of resin, respectively; n and w are number of mica flakes aligned in thickness and length directions of specimen, respectively; $\Delta l = \Delta L/N$ is deformation of resin per one mica flake, ΔL and N being macroscopic deformation of specimen and the number of mica flakes in a plane parallel to specimen length, respectively; l^2 is surface area of mica flake, assuming its square shape; dr and dm are thickness of one resin layer and mica flake, respectively; α is a factor which takes spacing and geometry of tensile stress field between neighbouring mica flakes in an oriented plane into account.

Eq. [1] is rewritten as eq. [2], assuming $Er = 3Gr$

$$\frac{F}{n \cdot w} = Gr \cdot \frac{\Delta l}{dr} \cdot l^2 \left(1 + \frac{3\,dm\,dr}{\alpha \cdot l^2}\right). \qquad [2]$$

The second term in the right hand side of eq. [2] can be neglected because it hardly exceeds 10^{-1} when dr is less than 100 μ.

To examine the validity of eq. [2], the thickness of resin layer dr was computed by the equation observed stress and strain. It is necessary to determine number of layers n in order to carry out the computation. Thickness of specimen D is given by $n(dr + dm)$, and it can be assumed that the ratio dr/dm is proportional to the ratio of volume fraction of resin to mica $\varphi r/\varphi m$, hence one get

$$n = D/dr(l + \varphi m/\varphi r). \qquad [3]$$

Calculated results are listed in table 2. Resin thickness dr decreases with φm and its magnitude is comparable to direct microscopic data. Thickness of mica flake dm thus computed shows a minimum at $\varphi m = 0.4$. Decreasing tendency in lower loading region may come from overestimation of φ, since φm should be a volume fraction of mica flakes which are oriented in parallel to force direction. Increasing tendency in higher loading region is considered to be due to incomplete mixing.

Table 2. Calculated results of d_r and d_m according to eq. [2]

No.	φ_m (%)	$d_r (\mu)$	$d_m (\mu)$
2	16.4	76	15
3	26.6	33	12
4	38.7	7.8	4.9
5	65.6	5.8	11

Modulus behavior for whole temperature region including glassy state can be analyzed by eq. [4] which takes the tensile deformation of mica flake into account

$$\frac{F}{n \cdot w} = Gr \frac{\Delta l}{dr} l^2 \left(1 + \frac{Gr}{Em} \cdot \frac{l^2}{dm \cdot dr}\right)^{-1} \qquad [4]$$

where Em is tensile modulus of mica.

Though eq. [4] gives somewhat smaller value of dr for highly loaded system than eq. [2] does, the difference is not essential.

In conclusion, it was found that hard fillers tabular in shape could produce an extraordinary temperature insensitive property when fillers are oriented in one direction by a mechanism of converting tensile external force to internal shear stress. Similar mechanisms may play a role in more complex heterogeneous system composed of phases having appreciable difference of modulus.

Summary

Dynamic mechanical properties of cured epoxy resins filled with small mica flake were investigated.

A secondary absorption peak was observed for filled specimens around 80 °C, while unfilled specimen showed only a primary absorption peak around 140 °C. Modulus of filled specimens at high temperature rubbery region showed extraordinary high value. As an explanation to this behavior, it was proposed that the tensile stress applied to specimen was converted to shear stress in a thin resinous layer sandwiched by two mica flakes.

Zusammenfassung

Es werden die dynamisch-mechanischen Eigenschaften von ausgehärteten Epoxydharzen untersucht, die mit kleinen Glimmerplättchen gefüllt sind.

Bei den gefüllten Proben wird ein Nebenmaximum der Absorption bei etwa 80 °C beobachtet, wohingegen die ungefüllte Probe nur ein Hauptmaximum bei ungefähr 140 °C aufweist. Der Modul der gefüllten Proben zeigt in dem bei hohen Temperaturen liegenden kautschukelastischen Bereich außergewöhnlich hohe Werte. Zur Erklärung dieses Verhaltens wird angenommen, daß die auf die Probe im Ganzen wirkende Zugspannung in den dünnen Harzschichten, die jeweils zwischen zwei benachbarten Glimmerplättchen vorhanden sind, in eine Schubspannung verwandelt wird.

References

1) *Nielsen, L. E.*, J. Composite Materials **1**, 100 (1967).

2) *Nielsen, L. E.*, Appl. Polym. Symp. No. 12, 249 (1969).

3) *Kerner, E. H.*, Proc. Phys. Soc. **69 B**, 808 (1956).

Authors' address:

Dr. *Kyoichi Shibayama* et al.
Manufacturing Development Laboratory
Mitsubishi Electric Corporation
Amagasaki Hyago (Japan)

Rheol. Acta 13, 103–112 (1974)

From the Procter Department of Food and Leather Science, University of Leeds (England)

The mechanical properties of leather

By A. G. Ward

With 3 figures and 2 tables

(Received October 27, 1972)

1. Introduction

The mechanical properties of leather are encountered by members of the general public many times each day, although they are usually accepted without thought. The scientific study of these properties was, until recently, limited to investigations carried out by leather scientists and also by scientists concerned with the uses of leather, e.g. in footwear. Recently, however, the behaviour of leather and the structural reasons which permit that behaviour have become important to some polymer technologists, whose objective has been to make materials able to replace leather in some of its main uses. Unfortunately these latter researches are rarely published, although they would valuably supplement papers originating from leather sources.

There are good reasons why leather has not appealed more generally as a subject for the academic study of rheological behaviour. The term "leather" is used to include materials as different as the stiff sole leather of a heavy boot, the soft and yet tenacious leather of ladies' fashion gloves, the flexible membrane which undergoes repeated distortion for perhaps forty years in a gas meter, mouldable leathers used as oil seals and pump washers, wash (chamois) leather and many others. The reasons for the very large differences in mechanical behaviour characteristic of these leathers originate in part from the natural structure of the skin and in part from the choice of manufacturing process. With so wide a range of variability there cannot be any clearly defined and measured properties of "leather" as such, only properties of particular types of leather. Perhaps the most striking common feature of many types of leather is the ability to withstand repeated flexing without failure. Shoe upper leather in the vamp region provides an appropriate illustration of this point.

Even for one category of leather, however, very substantial differences in properties between samples are observed in practice, so that inter-laboratory comparisons are not readily made. As is discussed in some detail in section 3 iii many properties vary with the location of the material of the test samples on an individual skin. So any attempt to establish physical relations and laws requires statistical planning and extensive replication of measurements.

Unlike most textile materials, the fibre structure of skin and leather exhibits little regularity (fig. 1). Pronounced differences in fibre structure are observed as one proceeds from the outer, epidermal surface, through the grain layer to the main fibrous structure of the remainder of the skin. So theoretical studies relating properties to structure are made difficult and the extensive work on the structure and properties of textile materials carried out by textile physicists and engineers can only rarely be used as a basis for work with leather (*Hearle, Grosberg* and *Backer*, 1969).

The present paper cannot attempt a comprehensive review of the whole subject of leather mechanical properties. Some aspects of current interest, with which the author has been associated, will be used to illustrate the behaviour of this group of materials, in which biological origin and manufacture combine to give the properties required in use. Some references to early work will also be included.

It should perhaps be made clear what the group of materials is which is given the generic name "leather". As noted by *H. R. Procter* (1903), the object of tanning is to render animal skin imputrescible and pliable. Even these criteria are an imperfect description since leather is by no means permanently imputrescible and, as has already been emphasised, widely differing levels of pliability are encountered. However, it clearly indicates the objectives of leather manufacture.

Fig. 1. Photomicrograph of limed hide side showing grain and fibre structure

2. Leather structure

Only the briefest account of the structures of the various leathers can be given here, but further information is available in the literature. The essential structural feature of all leather is the network of tanned fibres of collagen, in which non-collagenous material plays only a secondary role. At the molecular level, the unique polypeptide chains which constitute, for a particular species and tissue, the protein collagen, occur each in helical form. Three such helices coil round each other to give a fairly rigid rodlike basic unit of nearly 300000 molecular weight, of length 300 nm and diameter 1.5 nm. Although the intra-molecular forces in these "molecules" are mainly hydrogen bonds and *van der Waals* forces, a small, variable number of covalent inter-chain links are formed at specific regions.

The distribution of polar regions on the surface of the rod-like particles causes them to associate in a regular way into long fibrils, which show the characteristic highly organised banded structure, with a main repeat of c. 64 nm, so clearly revealed by the electron microscope and confirmed by X-ray diffraction studies. The fibrils vary in diameter from 5–200 nm but 80 to 100 nm might be regarded as typical for the diameter of the fibrils of mature mammalian skin. The fibrils are of very considerable length. Covalent cross links are formed between the rod-like particles making up the fibrils, so insolubilising them. In skin the fibrils associate together to give the fibres and fibre bundles which constitute the appearance of a skin section (fig. 1) in the light microscope.

As fig. 1 shows, the fibres of the grain layer are much finer than those in the central portion of the skin. The grain also contains a much higher proportion of non-collagenous components than the remainder of the skin, corresponding to a number of specialised biological functions.

Fig. 1 illustrates very clearly the irregular three dimensional woven structure of skin. Some fibres can be seen to have been cut through in sectioning while others lie at various angles to each other, in the plane of the section. This fibre weave is retained in its essential features in the finished leather although fibres may be split or otherwise modified. The fibres may be oriented by the various processes to lie more parallel to the skin surface (low angle of weave) or more perpendicular to that surface (high angle of weave). The angle of weave is a significant parameter in relating fibre structure to mechanical behaviour. The special structure of the grain layer and the proportion the grain constitutes of the whole thickness, together with the form of transitional layer between the grain and the main fibre structure each also contribute to the leather properties.

In converting skin to leather, the removal of hair and of various impurities (polysaccharides, fats, the epidermal layer, etc.) is followed by stabilisation ("tanning") of the structure either by a few per cent of a strong cross linking agent (e.g. chrome complexes, aldehydes) or larger quantities of weak cross linking agents (e.g. vegetable tanning agents, syntans), which may also be deposited as discrete particles among the fibres. On drying the tanned skin, the tendency for fibres to stick too rigidly together is counteracted by the use of emulsified fatty materials which act as lubricants. The nearly dry leather may be subjected to complex flexing mechanical actions ("staking") which also help to soften the leather. Today synthetic impregnants applied at a late stage in leather making may be used to modify the leather for some distance below the grain surface. Various forms of polymeric finish layer are applied finally to the grain surface to improve appearance and to protect the leather. Other processes involved in leather making, such as dyeing, which are not designed to modify mechanical properties, may incidentally bring about some minor changes.

The production of each type of leather is a complex set of operations, involving chemical and physical changes in the skin, swelling, the deposition of materials in or on the structure, removal of impurities and finally of water and also all the varied mechanical actions which occur during the stages of processing.

The original ordered structure at the molecular level within the fibrils is largely retained in the finished leather, and is stabilised by the cross linking action of the tanning agents. So the conditions for the occurrence of rubberlike elasticity are not present. If bovine raw skin is heated in water a point is reached, at about 65 °C, at which the fibres contract longitudinally and become readily deformable in a rubberlike manner. This change from an ordered to a more disordered state, known as thermal shrinkage, is made more difficult by the crosslinking reactions of tannage. In chrome leather the shrinkage temperature may be raised to well over 100 °C. Once thermal shrinkage has occurred, cooling only partially restores the previous state of order, although stretching the leather or skin during cooling facilitates the reappearance of an ordered structure.

The chemical substances present in leather and skin contain many groupings capable of forming hydrogen bonds. The open network of the structure allows ready access for air and for water vapour. The moisture content therefore varies with the humidity of the surrounding air. It may be deliberately varied in making articles from leather. The mechanical behaviour of leather is markedly affected by its water content. Temperature also has an effect on properties, although this is rather limited, except in extreme conditions, and especially at or above the thermal shrinkage temperature. So studies of mechanical properties, as well as the use of leather for the fabrication of leather articles, have to be carried out with due precautions concerning the moisture content and its effect on properties and, to a lesser extent, the effect of temperature change. In the presence of a thermal gradient, moisture movement will occur also.

3. Linear extension

i) General

The availability of tensile test machines and the importance of tensile strength and extension at rupture for some leather uses early resulted in investigations of linear extension. *Wilson* (1929) summarised his own extensive researches. Since the structure varies through the thickness of the leather, a calculation of tensile strength as load per unit area is meaningless except as an overall average figure. *Wilson* (1929) demonstrated, by splitting typical chrome tanned and vegetable tanned calf leathers into two layers parallel to the surface, that the grain layer was much weaker than the remainder of the section. As would be expected, splitting reduces the combined strength of the two split portions, since the fibres are cut and less firmly held (see below) in each portion.

Most leathers require to retain their shape reasonably in use. This is certainly true of shoe upper leathers, with which much of the following sections is concerned. But for gloving and chamois leathers different considerations apply. The action of putting on a glove and securing a perfect fit to the fingers and hand involves stretching in some directions at the expense of others, with little residual stress remaining after deformation to produce elastic recovery. Similarly chamois leather may be pulled in one direction, causing contraction on the perpendicular direction in the plane of the leather and, in this instance, some change in thickness also. Pulling in the perpendicular direction to the original easily reverses the deformations. This property of these two types of leather is known as "run". It is secured by selecting a raw material (e.g. hair sheepskin) with a fine network of collagen fibres which remain well separated in the form in which they are left after processing. Deformation is then able to occur, as in some textile materials, by change in angle between fibres in the network, without yield occurring at points where fibres are linked together. The deformation occurring in run must be characterised as plastic since it is not restored on release of the external force. It does not, however, involve any rupture of the main bonds giving coherence to the material. While attempts have been made to develop test methods for the assessment of "run", little has so far been achieved in a basic study of the rheological behaviour of gloving and chamois leathers.

ii) Upper and similar leathers at normal room temperatures

As has been stated above, most leathers show, when stretched, a mainly elastic response, although varying degrees of delayed elastic and plastic response occur. Caution must be exercised in generalising results since the findings for a particular type of leather may not apply to other, closely similar, types.

The load-extension graph at a constant rate of extension does not exhibit the same shape for all leathers (*Upstone* and *Ward*, 1969). While for some leathers it is possible to take the simple *Hooke*an relation between percent extension y and load x (or load/width or load/original cross section), $y = bx$, for others a curve concave to the load axis is observed. It is also possible to obtain leathers giving load/extension curves concave to the extension axis. It would be expected, for a linked fibrous network, that as extension proceeds the load would increase more than proportionately to the extension i.e. a graph concave to the load axis, but this is not always observed.

Maeser (1965) has shown that, for an upper leather sample, a four hundred fold increase in the rate of extension left the breaking load unchanged but reduced the extension at rupture by about 20%. So the load extension behaviour shows only limited variation with rate of strain. A constant tensile load gives rise to significant creep and removal of the load is followed by delayed elastic recovery, often extending over long times. Addition of moisture at this stage may release further elastic recovery mechanisms, reducing the apparent plastic flow.

Further examination of the detail of the load extension curves, *Upstone* and *Ward* (1969), reveals that so far from the graph becoming linear at low strain (below 1% extension), there is an additional region of curvature, concave to the load axis, which may be very well marked for some leathers. Fig. 2 shows three forms of curve which have been observed for different leathers. It is clear that some mechanism of deformation is operating in the low strain region which achieves its maximum contribution to the extension by the time 1% strain is reached. The effect is extremely variable even between neighbouring strips of the same leather and appears little related to the general stiffness or softness of the particular leather. By extrapolating the linear portion in type *A* curves, $y = bx + c$, fig. 2 or the quadratic portion of type *B* curves which can be fitted for higher strains with the relation $y = ax^2 + bx + c$, the additional extension provided by the low strain curvature can be determined. For type *C* curves only an approximation to this can be estimated. It is likely that the extent of easy small strain deformation for a particular leather is determined by one or other of the processes in leather manufacture which involve passing the leather between rolls, or slicking it out onto a flat surface or drying it under tension. An attempt, *Upstone* and *Ward* (1969), to remove the low strain curvature by heat setting the leather in a stretched state was, however, unsuccessful. The property is not unique to leather since most of the poromeric leather substitutes, which are based mainly on non-woven fibrous structures, also show a substantial curvature in the low strain region.

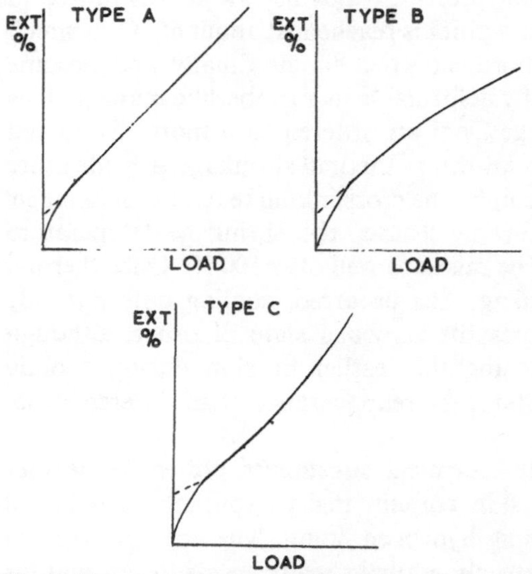

Fig. 2. Forms of the load extension curve for strips of upper leather

Popplewell and *Ward* (1963) have investigated stress relaxation at constant extension, for upper leathers, at 20 °C. The apparatus used secured rapid extension to a constant value by means of a piston driven by nitrogen pressure, which allowed the load to be applied in c. 0.1 sec. The load was measured by the deflection of a stiff spring steel bar and recorded by ciné camera.

The strips used measured 100 mm between the clamps, with a width of 15 mm. For times from a few tenths of a second to several days after stretching to constant extension, the load (or stress) fell linearly with log (time). This result was confirmed for many specimens. After c. 2 days the relaxation became discontinuous. The linear relation is in agreement with the results of *Shestokova* and *Kalinina* (1969) who give their results in the equivalent form $\sigma_t = \sigma_0 t^{-m}$, where σ_0 and σ_t are the stresses at time zero and t respectively. These relations indicate a very extended process of stress relaxation, some part of which is certainly a delayed elastic effect and not plastic flow. In the older forms of shoe making process where, after lasting, the shoes were retained on the last for a minimum of several days, the fit obtained depended in part on true plastic deformation but in part on the very slow elastic recovery process. The recovery time is much longer than the time for which the leather is stretched.

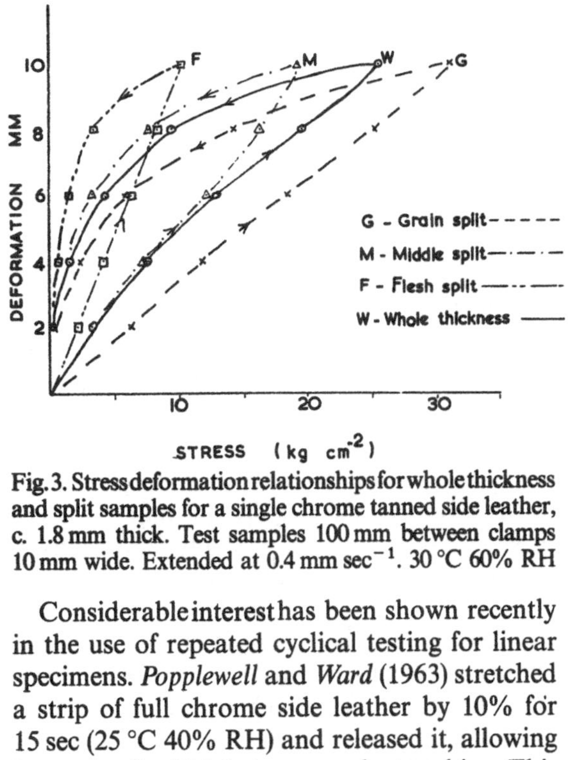

Fig. 3. Stress deformation relationships for whole thickness and split samples for a single chrome tanned side leather, c. 1.8 mm thick. Test samples 100 mm between clamps 10 mm wide. Extended at 0.4 mm sec^{-1}. 30 °C 60% RH

Considerable interest has been shown recently in the use of repeated cyclical testing for linear specimens. *Popplewell* and *Ward* (1963) stretched a strip of full chrome side leather by 10% for 15 sec (25 °C 40% RH) and released it, allowing it to relax for 24 h between each stretching. This procedure was repeated 17 times. By the tenth cycle the stress required to stretch the leather 10% was some 20% higher than the original value and the recovered length after 24 h had become almost constant. Further cycling raised the stress needed but had almost no effect on the recovered length. Further work is reported by *Guy* (1972). His results illustrate the difficulties of interpretation of cyclical testing in the absence of a theoretical model.

iii) Variation with location on a hide or skin

Maeser (1960) has carefully examined the variation in the tensile properties of leather over the area of chrome tanned paste-dried kip leathers. He determined the maximum load (at rupture), the extension at rupture, and what are effectively differential elastic moduli at high and low strains. He has compared his results with those of *Wilson* (1929), *Mann* (1951) and *Kanagy* (1952). The tensile strength variation shows a common pattern in the results of all workers, which it is presumed reflects the structure of the skin as it comes from the animal rather than the consequences of the leather making process. *Maeser* (1960) has also shown that the high strain differential modulus follows a similar pattern, sharing with the tensile strength good agreement between samples in the same location for right and left sides of a single hide. In contrast the low strain differential modulus showed much greater differences, for the same location, between samples from right and left sides and appeared much more influenced by the variable factors of the processing. This is consistent with the erratic nature of the small strain behaviour observed by *Upstone* and *Ward* (1969).

Maeser (1960) made measurements on strips cut parallel, perpendicular and at 45° to the backbone for the various locations on the hide. These results well illustrated the anisotropy of leather properties. From the results ellipses were calculated to represent the variation of properties at each location with angle.

Popplewell and *Ward* (1963) and *Popplewell* (1971) have further confirmed the effect of hide structure by measurements of the stress 5 sec after applying a 10% extension to leather strips from full chrome side leather. The extensions were applied for 15 sec and the recovered length measured after 24 h recovery. Percentage sets were calculated as the percentage ratio of the residual deformation to the imposed strain and were of the order of 5–10%. While there was some evidence of systematic variation of percentage set with hide location, the differences were much less marked than for the stress at constant strain. The extreme values of the 5 sec stress for different locations gave a 6 fold ratio. Table 1 (*Popplewell* and *Ward*, 1963) shows that these differences do not arise because of gross chemical differences between the samples and so must be related to the fibrous structure.

Table 1. Chemical analyses of leather strips showing widely differing stresses for 10% strain

	Sample 1	Sample 2
5 sec stress (kg cm^{-2})	12.80	70.70
24 h set	8.80%	6.90%
Grease content	1.36%	1.36%
Chrome content (Cr_2O_3)	5.20%	5.25%
Moisture	12.80%	12.80%

Poulter and *Ward* (1966) applied similar techniques to those of *Popplewell* and *Ward* to chrome tanned sheepskin. In addition to revealing a distribution of extension properties broadly similar to that already observed for bovine hide leather, it was possible to interpret to some extent the mechanical properties in terms of the fibre structure as shown by the examination of sections in the microscope. Where fibres run largely parallel with the surface, deformation must occur by stretching these fibres so giving rise to high forces. More ready extension can occur when the fibres are inclined to the surface. Fibre orientation is not the sole factor determining the extension properties, but it is clearly one of the most important.

iv) Set and its dependence on moisture and temperature

It was early shown by *Butlin* (1963) that leather behaved differently in relation to set, when subjected to two dimensional strains, from when a simple extension is applied. Two dimensional strains are considered later in this paper.

Popplewell (1971) showed that for linear extension of already equilibriated chrome side leather strips, no very marked effect of variation of equilibrium temperature on set was observed in the range 25–55 °C, nor of equilibrium moisture content in the range 11–30%. Although the higher temperatures gave higher initial sets (after 5 min recovery), if the samples were then left for 24 h, elastic after-effects sharply reduced the apparent set, especially if the sample was exposed during this period to both high temperature (55 °C) and high moisture content.

Butlin (1963) showed conclusively that either by the use of sufficiently intense heat alone, or of heat and moisture, on leather strips which had been already extended, very marked improvements in the retention of set could be obtained. In practical terms, his techniques allowed a reduction of the time required to give acceptable shape retention in shoe uppers from days and weeks to a few minutes. The combination of a first stage of treatment with a steam/hot air mixture and a subsequent stage of hot air alone (the whole operation is known commercially as moist heat setting) has proved particularly effective.

Popplewell (1971) has suggested that even where heat alone is used, consequent migration of water and water vapour within the specimen is a main agent in causing accelerated stress decay and increased set. Temperature in the absence of moisture movement appears not to be very effective. In an attempt to elucidate the mechanism of the action of condensing moisture, *Popplewell* (1971) also studied the effect of exposure of stretched leather samples to toluene vapour and to methyl, ethyl and isopropyl alcohol vapours. He showed that each was to some degree effective in causing both stress relaxation and increased set. There was, however, no unique quantitative relationship between either setting or stress relaxation and such quantities as the heat liberated by condensation, the volume of liquid condensed or the gram moles of liquid condensed.

v) Variation in properties through the leather thickness

Reference has already been made to the work of *Wilson* (1929) in which he conclusively demonstrated that the grain layers of vegetable tanned and chrome tanned calf leather differed in mechanical behaviour from the remaining portions of the tanned skin. Despite the importance of these results, the subject was only reinvestigated nearly forty years later by *Ward* and *Brooks* (1965 and 1967a, b). In the intervening period many studies had been carried out on the distribution of various chemical components (chromium, grease, etc.) through the thickness of the leather, using layer analysis, but no attempt was made to verify whether the chemical differences observed had any influence on physical properties. Extensive studies had also been made of the fibre structure as revealed in the microscope, using sections perpendicular to the leather surface. Here again, the appearance was reported without relation to measurable mechanical properties.

It is possible to make some deductions concerning the non-uniformity of leather through the thickness by comparison of mechanical behaviour in simple extension and in bending. It is, however, only the preparation of test samples formed by cutting the leather parallel to the grain surface at various positions through the thickness which enables the distribution of behaviour to be studied. As *Wilson* (1929) demonstrated, this procedure causes some degree of artefact through releasing new fibre ends where the fibres are cut. This is especially serious in reducing the tensile strength, but it is rather less important when more limited extensions are used. *Ward* and *Brooks* (1965) examined the effect of sectioning by preparing three splits, grain *G* containing in the main the grain layer, middle *M* and *F* the third split towards the flesh side, from unfinished full chrome side leather. Since the hide from which the split samples were prepared had already had a substantial split taken from the flesh side at the tannery, to give the appropriate thickness for upper leather, the split labelled *F* is actually derived from the centre of the whole skin. The sections were extended, as strips 10 mm wide 100 mm free length, by 10% at 0.4 mm sec^{-1}, with measurement of the load. The motion was then reversed to give the complete cycle. The samples were conditioned and measured at 30 °C 60% RH and the results compared with those for neighbouring strips similarly tested which had not been sectioned (fig. 2). The sectioning caused the combined load required to extend the three sectioned strips, compiled from the individual results, to drop by about 27% compared with the intact whole thickness strip. In a more thorough study of this effect a specially tanned cow hide 3.2 mm thickness had sets of neighbouring samples cut into 2, 3, 4 and 5 layers. The comparison of the recombined stresses for the various sectionings is given in table 2. The reduction in stress averages just over 6% for each cut, rather less than for the three sections of the full chrome side leather. It is evident that the magnitude of the effect will depend on the thickness of the splits taken and perhaps also on the orientation of the fibres relative to the direction of cutting. The reduction in loads where 3 sections are used, compared to those needed to extend the layers in whole thickness samples is, for upper leathers, insufficient to affect appreciably the relative values for the different layers or sections. So it is possible to represent the distribution of resistance to extension through the thickness.

Table 2. Average stress for neighbouring samples split into 2–5 layers parallel to surface

Number of layers	Combined average stress ($kg\ cm^{-2}$)
1	26.7
2	24.1
3	24.0
4	21.4
5	20.0

Ward and *Brooks* (1967a) studied the variation of stress strain behaviour through the thickness by this technique in relation to location on the hide, for two large chrome tanned calfskins (veals). No consistent pattern of distribution of stress strain properties through the thickness emerged. For some samples the grain layer gave the greatest resistance to stretching, for others the flesh side and for yet others the middle layer. Examination of the fibre structure enabled the results to be explained, at least in part, for the middle and flesh side layers, in terms of the average angles the fibres made with the direction of extension. The very different fibre structure of the grain layer (fig. 1) does not allow a valid comparison to be made in which it is involved. No relation was found between stress strain behaviour and any of the usually estimated chemical components of the leather, so confirming the information given in table 1.

Leather is valued not only for its flexibility, strength and endurance, but also for certain aspects of its appearance. One of these aspects is the way in which it behaves when folded or flexed to give a sharp curve with the grain surface inwards. In these conditions the grain of the leather assumes a wrinkled appearance known as "break". If the wrinkles are small and regular the leather has a "fine" break, if large, coarse and irregular it has a "coarse" break. When the leather is bent in this way, the grain is put into compression parallel to the surface and so can be regarded as forming the wrinkles as a result of instability in compression. While semi-quantitative methods of evaluating break have been developed (*Hole* and *Popplewell*, 1966; *Upstone*, 1969), subjective evaluation by trained observers can also be used reliably to rank leathers for break properties. *Ward* and *Brooks* (1967a, b) examined whether the pattern of resistance to deformation through the leather thickness could be related, as might seem plausible, to break characteristics. No such relation was detectable. It was observed that leathers with fibres, especially near the grain layer, making low angles with the grain surface and having poor interweaving or even some open spaces, gave coarse breaks. It may be that the grain layer in these leathers is poorly constrained by the layer below it. It therefore readily deflects into a coarse ripple pattern when compressed along its length as a consequence of bending. In their final paper, *Ward* and *Brooks* (1967b) tried to examine the relative importance of the original hide structure and of the leather manufacturing processes to the layerwise distribution of mechanical properties. Two experimental tannages were carried out, using conventional chrome tanning techniques. While the absolute values of stress at 10% extension varied with the stage of the process from raw skin to final leather, the pattern through the thickness did not. These results can only be regarded as a first attempt at detecting the origin of the differences observed.

4. Two dimensional studies

i) General

In most uses of leather, stresses are not confined to a single direction and complex strains are produced. This is very clearly shown in the lasting of shoe uppers where the strains can be followed by printing a grid on the leather before it is deformed. The behaviour of leather when subjected to two dimensional distortion in the plane parallel to the leather surface cannot safely be predicted from its behaviour in simple extension. A simple method for studying two dimensional set under a variety of conditions was developed at SATRA (*Butlin*, 1963). In this a disc of leather is held at its periphery while a spherical surface in the form of a dome constrains it, grain outside, to a spherical shape. The leather can then be given any required treatment with hot air, steam, etc. On removal of dome, recovery occurs and the extent of the permanent set can be calculated after suitable time periods. Much useful work has been done with this equipment but it is more empirical than is apparent, since friction on the dome surface (*Holmes*, 1969) makes the deformation far from uniform over the dome. In addition it lacks any means of measuring the stresses involved. So it should be regarded as only a means of partially matching the conditions involved in the actual lasting of shoe uppers.

ii) Elastic two dimensional behaviour

Ward and *Chinn* (1971) have constructed a two dimensional extension apparatus in which independent constant strains may be rapidly applied in two directions at right angles and the corresponding loads measured as a function of time. The principle of the method was based on the work of *Treloar* (1948) for rubber. Full details of the apparatus are given in *Chinn* (1967). A square grid 100 mm square is printed on the test area of the 140 mm square leather sample and the extensions are applied to five lugs on each side. Predetermined strains are applied in 0.1 sec by means of nitrogen pressure loaded pistons and these strains are checked by photography of the printed grid in conjunction with a reference grid. The total loads in each direction are obtained by means of two proving rings, the gauges of which can be recorded as a function of time by the use of small ciné cameras. Each lug of the specimen is connected to Mechanite GD strips equipped with resistance strain gauges. In this way the uniformity of the stress can be determined. The constancy over the 100 mm square was satisfactory for uniform materials, such as some leather substitutes. The variability of properties even over the 140 mm × 140 mm leather samples used was the origin of the observed stress differences for the leather samples. The whole measurements could be completed at constant temperature and humidity, the specimens having been conditioned in the same atmosphere. The first application of the apparatus was to evaluate the relation between two dimensional elastic strain and stress for upper leathers. In the conditions used, plastic components were small and readily subtracted to give the elastic strain. Using stepwise extensions in the two directions it was shown that, to a first approximation, for up to 10% strain in each direction:

$$P_x = K_A x + K_B y$$

$$P_y = K_D x + K_C y$$

where x and y are the elastic extensions and P_x, P_y the loads in the x and y directions. The values of P_x, P_y can be converted into average stresses if required. In a typical experiment $K_A = 12.9$ kg, $K_B = 1.8$ kg, $K_C = 6.9$ kg and $K_D = 1.8$ kg, showing the much greater dependence of the stress on the deformation in the same direction. A more refined study would allow non-linearity to be investigated and also permit the low strain region to be examined. It is of interest to note that the linear behaviour characterised by the above equations was also shown by the synthetic material "Corfam".

Chinn and *Ward* (1971) have reported the variation of the four constants K_A, K_B, K_C, K_D over the area of a matched pair (i.e. from one hide) of sides of full chrome upper leather. As would be expected these constants show closely comparable variations to those already derived from linear extension measurements. Each reflect the effect of the internal structure on the mechanical behaviour.

iii) Plastic two dimensional behaviour

Chinn (1967) modified the two dimensional apparatus to allow hot air or steam and hot air to be applied to the leather specimen after straining. This work has been refined and extended by *Holmes* and *Ward* (1971) and applied to a systematic study of heat setting of leather under two dimensional stress. The effect of area strain on area set has been shown to be significant. The main investigations were designed to show the relations, firstly between set and duration, temperature and steam output when steam/air mixtures were used to set the leather and secondly between set and duration, temperature and air flow rate for dry heat setting with hot air alone. Experimental work has also been completed on the combination of the two processes.

5. The behaviour of single fibres

It is plausible to seek an explanation for the complex mechanical behaviour of leather at least in part in terms of the individual fibres. *Mitton* (1945) demonstrated approximately *Hooke*an behaviour for single leather fibres. *Conabere* and *Hall* (1946) further showed that on first extending fibres of leather, substantial permanent sets were obtained (c. 20% for an extension of 10%) but that subsequent extensions were almost wholly elastic. This parallels the behaviour of some leather samples when stretched for the first time. In a lengthy series of experiments by *Mitton* and *Morgan*, which have been summarised by *Morgan* (1960), the properties of raw fibres teased from wet salted cowhide and of vegetable and chrome tanned fibres have been examined. The large random variations occurring with such fibres necessitated the use of statistical techniques, involving measurements on over 2000 fibres. Variations in tanning procedure had almost no effect on the rupture properties although tanned fibres gave somewhat lower breaking loads and extensions than raw fibres. Thin fibres are relatively stronger than thicker fibres. At low humidities the fibres are, as would be expected, more brittle. At very high humidities the breaking load is reduced.

Variations in tanning practice, comparable to those applied by *Mitton* and *Morgan* to single fibres, produce very large changes in the mechanical properties of leathers. So the behaviour of the individual fibre cannot be used adequately to explain the properties of the different sorts of leather. Clearly the interaction between the fibres is a very significant contributory factor. The fibres which could be studied in isolation are also not representative of those in the grain layer, as can be seen in fig. 1.

6. Conclusion

This paper has concentrated on those studies of leather mechanical properties where an attempt has been made to explain the origins of the properties. There has also been extensive work on the development of mechanical methods of test of value to users of leather. By enabling the tanner to test his products without the delays and uncertainties of assessing them directly in their practical uses, it is made easier to select and adapt the tannage to give leather of the required properties. Mechanical properties are not the sole criteria – colour, fastness, waterproofness and the many facets of aesthetic appeal are all significant – but they still constitute a major area for future study. Further basic work requires to have two main objectives, firstly to relate properties to their structural causes and secondly to show how, in the leather manufacturing processes, the appropriate structures can be deliberately created so providing the properties that are required.

Acknowledgments

I am indebted to the research students who have, during the last twelve years, worked with me on various facets of the mechanical behaviour of leather. This paper is mainly based on their contributions to the study of leather.

Summary

The many uses of leather have largely relied on the range of mechanical properties which it can provide, according to the raw material employed in its manufacture and the manufacturing processes themselves. The contrasting behaviour of a stiff sole leather and of a fine gloving leather exemplify this point. The last twenty years have seen intensive investigation of many mechanical properties of leather and the design of test methods now accepted internationally.

At ambient temperatures and humidities most types of leather show mainly elastic behaviour, although delayed elastic effects may give the semblance of plasticity. The stress relaxation-time relation for constant linear strain shows the stress decaying linearly with log (time). The stress decay becomes discontinuous after sufficient time. The stress-strain relation for extension of leather strips is often markedly non-linear even at low strains ($<2\%$). Two dimensional extension of leather has been analysed using an instrument allowing independent extension in two perpendicular directions. To a first approximation each stress component is linearly related to the two elastic strain components in the perpendicular directions.

As with other materials of biological origin, the mechanical behaviour of leather varies from place to place in the skin, not only over its area, but also through its thickness. The extent of variation is briefly discussed and related to the underlying fibre structure.

Leather which has been strained and then subjected to either heat alone or heat and moisture, shows much more extensive plasticity than occurs at lower temperatures. This behaviour has been used to enable leather to be given appropriate shapes, as in the heat setting of upper leathers. Quantitative studies of heat setting are reported and the influence of such variables as temperature, moisture content of the applied air stream, the air stream velocity and the duration of treatment are discussed. The plastic deformation obtained in this way is contrasted with "run" in gloving leather.

References

Butlin, J., J. Soc. Leather Trades' Chem. **47**, 3 (1963).

Chinn, S. J., Thesis The Two Dimensional Deformation of Upper Leather (University of Leeds 1967).

Conabere, G. O. and *R. H. Hall*, J. Int. Soc. Leather Trades' Chem. **30**, 214 (1946).

Guy, R., J. Soc. Leather Trades' Chem. **56**, 246 (1972).

Hearle, J. W. S., P. Grosberg and *S. Backer*, Structural Mechanics of Fibers, Yarns and Fabrics, Vol. 1 (New York 1969).

Hole, L. G. and *D. Popplewell*, SATRA TM 1348 (Satra, Kettering (England) 1966).

Holmes, C. M., Private communication (1968).

Holmes, C. M. and *A. G. Ward*, J. Soc. Leather Trades' Chem. **55**, 242 (1971).

Kanagy, J. R., J. Amer. Leather Chem. Assn. **47**, 726 (1952).

Maeser, M., J. Amer. Leather Chem. Assn. **55**, 501 (1960).

Maeser, M., The Chemistry and Technology of Leather, Vol. 4, pp. 310–332 (New York 1965). Ed. *O'Flaherty, F., W. T. Roddy* and *R. M. Lollar*.

Mann, C. W., J. Amer. Leather Chem. Assn. **46**, 228 (1951).

Mitton, R. G., J. Int. Soc. Leather Trades' Chem. **29**, 169 (1946).

Morgan, F. R., J. Amer. Leather Chem. Assn. **55**, 4 (1960).

Popplewell, D. and *A. G. Ward*, J. Soc. Leather Trades' Chem. **47**, 502 (1963).

Popplewell, D., Thesis Some Stress Relaxation and Set Properties of Full Chrome Side Leather (University of Leeds 1971).

Procter, H. R., The Principles of Leather Manufacture, p. 7 (London 1903).

Shestokova and *Kalinina*, Nauch. Tr. Mosk. Teknol. Inst. Legk. prom 1, **36**, 109 (1969).

Upstone, P. J., Thesis Subjective Assessment and Objective Measurement of the Physical Properties of Leather (University of Leeds 1969a).

Upstone, P. J. and *A. G. Ward*, J. Soc. Leather Trades' Chem. **53**, 361 (1969b).

Ward, A. G. and *F. W. Brooks*, J. Soc. Leather Trades' Chem. **49**, 312 (1965).

Ward, A. G. and *F. W. Brooks*, J. Soc. Leather Trades' Chem. **51**, 199 (1967a).

Ward, A. G. and *F. W. Brooks*, J. Soc. Leather Trades' Chem. **51**, 211 (1967b).

Ward, A. G. and *S. J. Chinn*, J. Soc. Leather Trades' Chem. **55**, 221 (1971).

Wilson, J. A., The Chemistry of Leather Manufacture, pp. 1054–1086 (New York 1929).

Author's address:

A. G. Ward
Procter Department of Food and Leather Science
University of Leeds (England)

Für die Schriftleitung verantwortlich: Dr. W. Meskat, 5090 Leverkusen
Anzeigenverwaltung: Dr. Karl Niedermeyer Nachf., 6000 Frankfurt/M. 90, Georg-Speyer-Straße 76
Dr. Dietrich Steinkopff Verlag, 6100 Darmstadt, Saalbaustraße 12
Gesamtherstellung: Universitätsdruckerei Mainz GmbH

From the Institute of Space and Aeronautical Science, University of Tokyo, Komaba, Meguro-ku, Tokyo (Japan)

Dynamic mechanical and thermo-mechanical properties of stretched polypyromellitimide films

By H. Kambe and T. Kato

With 3 figures

(Received October 27, 1972)

Introduction

The enhanced thermal stability of polypyromellitimide (PI) is due to the incorporation of rigid aromatic and heterocyclic rings in the backbone chain of the polymer. In the present paper, the relation of chain rigidity with the orientation of molecules was looked for in the dynamic mechanical and thermal shrinkage behaviors of the stretched PI films.

In the temperature dependence of the dynamic loss modulus of commercial PI film, a few peaks have been found below 100 °C (1). We have measured the loss of the film at the higher temperature range and found a broad peak around 300 °C at 0.1 Hz (2–4). This peak was magnified by stretching the film up to 40%, and disappeared after heating. We also observed a marked thermal shrinkage of the stretched film by thermomechanical analysis (TMA). It is concluded here the loss peak at high temperature range is caused by the thermal shrinkage of the film.

Experimental

Materials

A commercial film of polypyromellitimide, du Pont Kapon H, has been used as received. Its basic structure is:

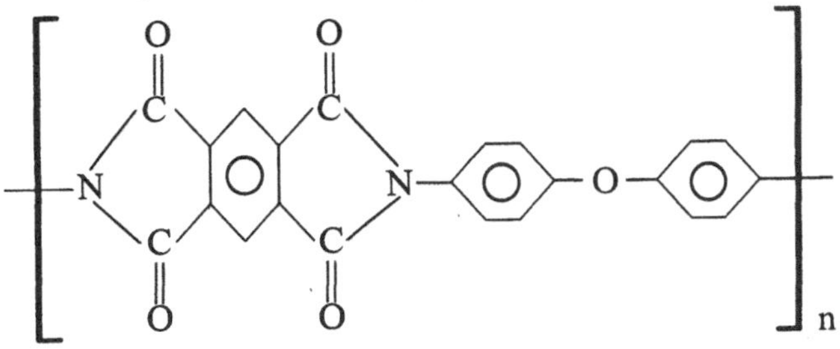

The thickness of the sample is 75 μ.

Apparatus

A torsion pendulum of the invert type was designed and constructed for the use at high temperatures. The free oscillation of the pendulum was measured at a frequency range of 0.1 ~ 0.2 Hz from the ambient temperature up to 500 °C.

The sample film was stretched by 40%, by a Shimadzu tensile tester with a very low constant rate of stretching at the room temperature.

The TMA apparatus is the modified linear expansion instrument supplied by Rigaku Denki Co. It is essentially constituted by a balance type deflection detector and a recorder. The change of length of the specimen under load can be recorded against temperature at a uniform rate of heating.

Procedure

The dynamic mechanical measurements were carried out for a thin strip of sample film, 5 × 20 ~ 30 mm. The recorded curve of free oscillation was analyzed by a usual method, and the temperature dependence of the dynamic storage and loss moduli were determined.

The thermal shrinkage of the stretched film, 5 × 2 or 20 mm, was measured in vacuum thermomechanically at a uniform heating rate of 10 °C/min under the tensile load of 10 g. The creep of film was negligible throughout the work.

Results and discussion

Dynamic properties

In fig. 1, the temperature dependence of dynamic storage modulus and loss modulus of the Kapton H film are shown. In the figure is also shown the result for the film stretched by 40% at room temperature. The specimen strip was cut out along the direction of stretching and measurements were repeated by heating and cooling cycles. A broad peak appeared in the loss modulus curve around 300 °C is magnified for the stretched film and disappeared for the annealed film at 500 °C. This peak seems to be dependent on the orientation of rigid chain molecules.

Cold-stretching

In fig. 2 are shown the stress-strain curves for PI film obtained at various temperatures. The

8

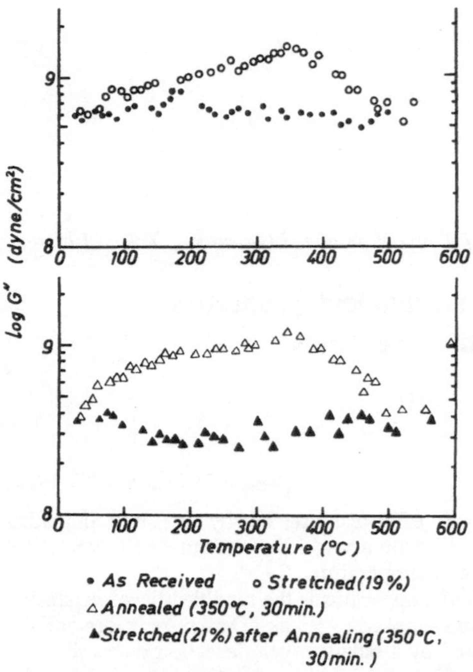

Fig. 1. Temperature dependence of dynamic loss modulus of polypyromellitimide films

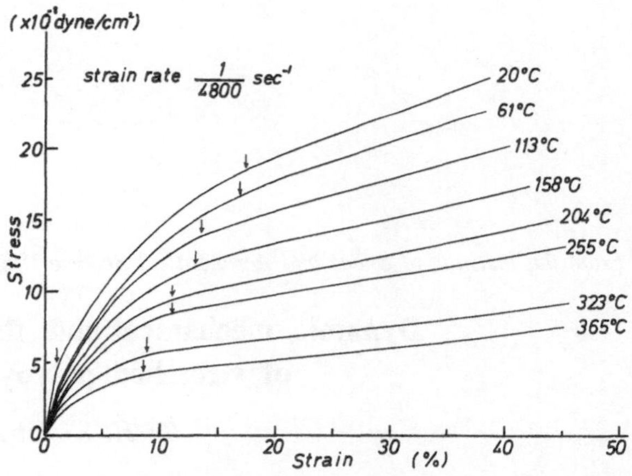

Fig. 2. Stress-strain curves of polypyromellitimide film at various temperatures

stress-strain curve may be divided into three regions as shown by arrows in the figure. Range I is the *Hooke*an elastic region up to a few percent strain, and range II is also elastic. Above range III the plastic flow commences markedly. The elastic modulus of the sample decreases at the higher temperatures and the plastic flow begins at lower strain for higher temperatures.

The cold-stretched films recovered somewhat in the cold retardedly. However, it should be stressed that these films shrink markedly on heating.

Thermal shrinkage

The thermal shrinkage was investigated kinetically by TMA method. The TMA curves obtained for the cold-drawn specimens with different degree of stretching are shown in fig. 3. The ordinate shows the percentage contraction of the stretched films in reference to the original length before stretching. The unstretched specimen shows only a uniform thermal expansion up to 400 °C. The stretched films show a shrinkage at the temperature range of 100–400 °C.

The ultimate degree of shrinkage increases with the degree of stretching, and the temperature range in which the shrinkage occurs is broadened with the degree of stretching. The arrow in the figure indicates the temperature at the optimum rate of shrinkage. This temperature estimated 350 °C for the 37.0% stretched film is identified to the peak temperature in the dynamic loss curve.

This polymer has a molecular structure of rigid planar rings connected by ether linkages. The molecular chains would be oriented within the film plane when they were cast from the solution. If these films are coldstretched, the internal rotation around the ether bond causes the

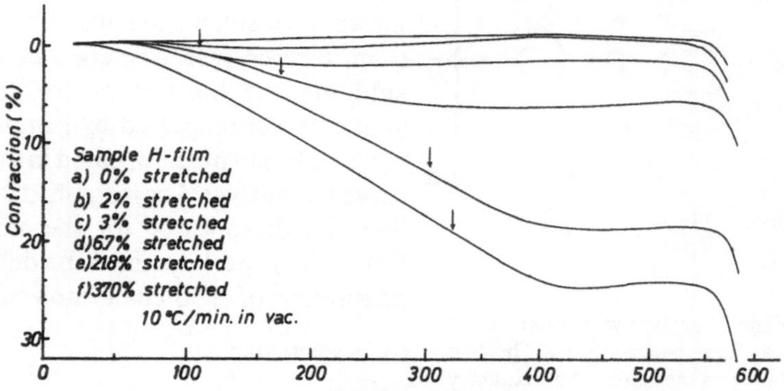

Fig. 3. Thermomechanical shrinkage curves of stretched polypyromellitimide films

expansion of free volume between chains and the planar orientation of the chains is disturbed. Chains would be frozen in a metastable state in the cold. On heating these frozen chain molecules would be back to the initial state. The film shows an overall shrinkage of the length. The absorption peak of dynamic loss modulus around 300 °C would be caused by the same recovery of metastable chains to the more stable state by the internal rotation released at higher temperatures.

Recently, *Gillham* and *Roller* (5) have found the same high-temperature loss peak for PI film by torsional pendulum method, and pointed out this peak disappeared after heating. They interpreted this peak as the α dispersion by the glass transition of the polymer and it disappears by heating due to the degradation of polymer. They don't touch the effect of stretching on this peak. We found, however, the peak appears again after re-stretching of the annealed film. The interpretation by them is apparently not valid for this case. The rate dependent shrinkage of the stretched films with a kinetic method of TMA may be analyzed by a two-state model of the oriented chain molecules. The theoretical considerations have been published elsewhere (6).

Summary

The relation of chain rigidity of thermally stable polypyromellitimide (PI) with the orientation of molecules in the stretched films was looked for in the temperature dependence of the dynamic mechanical properties and also in the thermal shrinkage of the film.

In the temperature dependence of the dynamic loss modulus of the commercial PI film, du Pont, Kapton H, a broad peak was found around 300 °C at 0.1 ~ 0.2 Hz. This peak was magnified by stretching the film by 40%, and disappeared after heating up to 500 °C. Thermal shrinkage of the stretched film was measured by thermomechanical analysis (TMA) at a uniform heating rate, and a significant shrinkage was found over the range 100 to 400 °C for the stretched films.

When stretched, the rigid polymer chains are extended and frozen in a metastable state. The extended chains recoil back to the more stable state when heated in TMA and also in dynamic measurements.

References

1) *Ikeda, R. M.*, J. Polymer Sci. B **4**, 353 (1966).

2) *Kambe, H.* and *T. Kato* et al., Rept. Progr. Polymer Phys. Japan **13**, 277 (1970).

3) *Kato, T.* and *H. Kambe* et al., Zairyo **20**, 582 (1971).

4) *Kambe, H.* and *T. Kato* et al., ibid. **21**, 405 (1972).

5) *Gillham, J. K.* and *M. B. Roller*, Polymer Eng. and Sci. **11**, 295 (1971).

6) *Kambe, H.* and *T. Kato*, Appl. Polym. Symp. **20**, 365 (1973).

Authors' address:

Prof. *Hirotaro Kambe* and Dr. *Teijo Kato*
Institute of Space and Aeronautical Science
University of Tokyo
Komaba, Meguro-ku, Tokyo (Japan)

Rheol. Acta **13**, 116–126 (1974)

From the Institute of Macromolecular Chemistry, Czechoslovak Academy of Sciences, Prague (Czechoslovakia)

Comparison of the time-temperature superposition of the relaxation modulus and time-to-break of preswollen gels

By J. Janáček, M. Raab, and M. Štol

With 11 figures and 5 tables

(Received October 27, 1972)

It is well known that the viscoelastic functions measured at different temperatures can be superimposed by means of the WLF equation, thus yielding generalized curves which extend the experimental time scale. It has been found (1) that the WLF equation holds well for the reduction of the viscoelastic data of a great majority of polymers measured in the main transition region, approximately over the temperature range from T_g to $T_g + 100°$. The moduli of crosslinked polymers approach with increasing temperature and time a certain limiting value predicted by the kinetic theory of elasticity. In this region, the viscoelastic curves determined at different temperatures are almost parallel, and the superimposed data are often subjected to a considerable error and sensitive to the use of various correction factors.

It has been shown that the ultimate properties (2, 3) can also be superimposed similarly to the viscoelastic data. The viscoelastic conception of strength (4) assumes that the temperature and time dependence of the ultimate behaviour is controlled by the viscoelastic processes in the roots of the growing cracks which proceed faster by several orders of magnitude that macroscopic creep or macroscopic relaxation.

There are some cases when the time-temperature superposition cannot be used within the whole field of interest owing to the fact that temperature changes lead to changes in the structure of the polymers under investigation. This happens in the case of the crystalline polymers in the first place, where temperature changes cause also changes in the content of the crystalline phase; a similar situation may also be observed if in a system polymer–low-molecular-weight substance another separated phase is formed with decreasing temperature. An example can be seen in hydrophilic polymers or porous systems swollen with water. In the case of poly(2-hydroxyethyl methacrylate) swollen with water to equilibrium (5) ($v_2 = 0.55$), ice is formed at 0°, which prevents viscoelastic measurements of the above system in the main transition region and over the range of reasonably attainable times. In this case, information on the temperature dependence of the viscoelastic processes near the main transition region can be obtained by ultimate experiments. Here, the time dependence of the breaking stress corresponds to the time or frequency dependence of the modulus shifted toward higher temperatures by several tens of degrees.

The work reported here has been devoted to a comparison of the relaxation curves and time-to-break curves of preswollen poly(2-hydroxyethyl methacrylate) gels (PHEMA), both homogeneous and having various heterogeneous structures, measured approximately over the same time and temperature interval. Owing to the fact that the region involved here is the rubber-like region, the changes in the breaking stress within the given temperature and time interval correspond to approximately ten times smaller changes in the relaxation modulus. A comparison of both types of experiments allows the determination of limits of the agreement between the viscoelastic and ultimate superposition; moreover, it allows a characterization of the analogies and differences between the mechanism of the stress relaxation in the linear viscoelastic region and the mechanism of failure.

Experimental

Polymers used

Two series of samples of the swollen PHEMA were used in the work. The samples of the first series were prepared by crosslinking polymerization of HEMA in the presence of various concentrations of water (volume fraction of the polymer $v_0 = 0.8$, 0.5, 0.4, 0.3, and 0.2).

Ethylene glycol dimethacrylate in concentrations 0.21 and 0.65×10^{-4} mol cm^{-3} was used as the crosslinking agent. At a volume fraction of the polymer $v_0 = 0.5$ and lower, water separates in the form of an independent phase, and heterogeneous sponges of different structure and porosity are formed. The procedure of preparation of the samples and their morphological characteristics have been described elsewhere (5). In the other series, preswollen samples were prepared in the presence of various concentrations of ethylene glycol during the polymerization ($v_0 = 0.6$, 0.5, 0.4, and 0.3). Here, too, the two concentrations of the crosslinking agent mentioned above were used. The equilibrium degree of swelling with water, v_2, of samples of the first series was determined by weighing twice the dry samples and samples swollen to equilibrium at different temperatures within the range from 20 to 80°; the volume degree of swelling was calculated on the basis of assumed additive weight changes (table 1). The density of dry PHEMA was (6) $\varrho = 1.288$ g cm^{-3}, the volume expansion coefficient in the rubberlike state was $\alpha_l = 4.6 \times 10^{-4}$.

Measuring procedure

Determination of the viscoelastic characteristics. Samples in the form of strips, approximately $120 \times 10 \times 1$ mm in size, were deformed at a strain rate of 100 cm min^{-1} on an Instron apparatus (table model) to a strain $\varepsilon = 0.2$. The course of the stress relaxation was measured within the time interval from 3–300 s; the stress values thus obtained were recalculated to tensile moduli $E(t)$ related to a dry nondeformed cross-section from the relationship $E(t) = f v_2^{1/3}/A_0 \varepsilon$, where f is the force, A_0 is the cross-section of the swollen samples, v_2 is the volume fraction of the polymer in the swollen (or swollen porous) system, $\varepsilon = \Delta l/l_0$ is the strain. The measurements were carried out at 6–17 temperatures within the interval from −3 to 80 °C, and the v_2 values were corrected to the degree of swelling determined independently for the given temperatures. The elongation of the samples was returned to zero position between the individual experiments, and the samples were left for thermostating at the nearest higher temperature used for 30 min. The working length of the samples between two clamps was measured optically with a cathetometer at each temperature, and so was the magnitude of deformation. Prior to the measurements, the samples prepared in the presence of water were swollen with water to equilibrium during the polymerization and measured in water; the samples preswollen in ethylene glycol were measured in paraffin oil so as to preclude their further swelling by air moisture, or on the other hand their deswelling (table 1).

v_0 is the volume fraction of the polymer in polymer-diluent system during polymerization; v_2 volume fraction of the dry polymer in the swollen homogeneous or heterogeneous system; v_x volume fraction of the swollen homogeneous polymer in the swollen heterogeneous system.

Ultimate measurements

The time-to-break of the test pieces at a constant load was measured with a tensile apparatus using a procedure described elsewhere (5). The dumbbell test pieces had a working length of 22.5 mm and width 4.1 mm. The stress varied from 0.5–40 kp cm^{-2}, the temperature within the range from −20–80 °C; the time-to-break of the samples was recorded automatically and varied within an approximate range from 1–10^3 s. Both the $E(t)$ moduli and the stress employed for the time-to-break measurements were reduced by the factor T_0/T required by the kinetic theory of elasticity.

Table 1. Swelling degress v_2 and porosities v_x of PHEMA gels

v_0	Preswollen in water; Equilibrium swollen in water; Concentration of the crosslinking agent $c \times 10^4$ mol cm^{-3}: 0.21		0.65		Preswollen in ethylene glycol; Swollen in ethylene glycol: 0.21	0.65
	v_2	v_x	v_2	v_x	v_2	v_2
0.8	0.529	1.0	0.541	1.0	–	–
0.6	–	–	–	–	0.6	0.6
0.5	0.518	0.95	0.523	0.81	0.5	0.5
0.4	0.487	0.90	0.487	0.72	0.4	0.4
0.3	0.356	0.73	0.384	0.62	0.3	0.3
0.2	0.198	0.49	0.274	0.54	–	–

Results and discussion

Temperature dependence of the E(t) values

The temperature dependences of the $\log E(t)(T_0/T)$ values (where $t = 300$ s) of the polymer networks under investigation are given in fig. 1. In all cases the temperature-reduced moduli decrease with increasing temperature, both for networks prepared in the presence of water (fig. 1 a) and in the presence of ethylene glycol (fig. 1 b), which can with good accuracy be approximated by straight lines. The slopes of the decrease of the reduced moduli with temperature are given in table 2. In most cases, higher slopes pertain to networks containing a lower concentration of the crosslinking agent; it can be deduced therefrom that the decrease in the moduli with temperature is in the first place affected by the distance of systems under investigation from the equilibrium.

In the case of lightly-crosslinked water-preswollen networks prepared in the presence of the crosslinking agent concentration $c = 0.21 \times 10^{-4}$ mol cm^{-3}, the largest decrease in the moduli with temperature takes place in the region $v_0 = 0.5 - 0.4$. For a system corresponding to $v_0 = 0.8$ the decrease in the moduli is slower. On

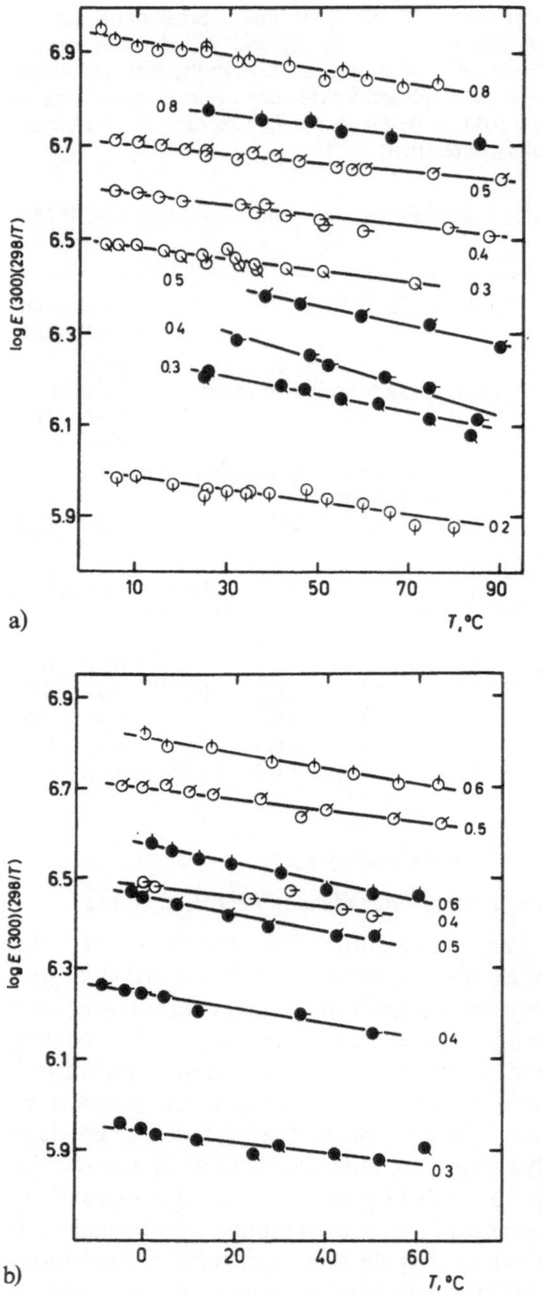

Fig. 1. $\log E(t)(298/T)$ values plotted against temperature for preswollen PHEMA samples at $t = 300$ s. Concentration of the crosslinking agent: black points: $c = 0.21 \times 10^{-4}$ mol cm^{-3}; white points: $c = 0.65 \times 10^{-4}$ mol cm^{-3}. The numbers correspond to the volume fractions of the polymer at the moment of polymerization, v_0. a) Samples prepared in the presence of water. b) Samples prepared in the presence of ethylene glycol

the contrary for a network prepared in the presence of a higher concentration of the crosslinking agent and $v_0 = 0.8$, the largest decrease in the moduli with temperature is observed. For samples preswollen in ethylene glycol the slopes decrease with decreasing v_0 in both cases. These tendencies reflect changes in the polymer microstructure and in the topology of the polymer network due to the different type and degree of preswelling.

Table 2. $d[\log E(t)\, T_0/T]/dT$ and T_s values of preswollen PHEMA networks

v_0	Concentration of the crosslinking agent $c \times 10^4$ mol cm^{-3}			
	0.21	0.65	0.21	0.65
	$-d[\log E(t)\, T_0/T]/dT \times 10^4$		T_s values (ultimate data) (°C)	
Samples preswollen in water				
0.8	13	15	36	1
0.5	21	10	− 2	− 2
0.4	31	11	28	12
0.3	19	12	13	18
0.2	–	13	−30	−14
Samples preswollen in ethylene glycol				
0.6	20	21	28	33
0.5	20	13	4	8
0.4	16	13	–	− 4
0.3	12	–	5	–

Even in the case of a dense network no temperature-independent constant values of the reduced moduli were found. It is known [cf. (7)] that the temperature dependence of the end-to-end distance of the polymethacrylate chains is comparatively large; for PHEMA a value of $d\log E_e/dT = -7.3 \times 10^{-4}$ grad^{-1} has been found previously (8). It has been demonstrated (9) that for the time-temperature superposition of viscoelastic data in the proximity of the equilibrium a correction factor of the temperature dependence of the internal energy of chains should not be neglected and an experimental verification was given for dry poly(butyl methacrylate) (10). If we take into account the correction factor for the temperature changes in density, $\log(\varrho_0/\varrho)$, which as a rule can be neglected and which for our systems does not exceed 0.0002 grad^{-1}, we obtain the total correction factor $d[\log E(t)\, T_0/T]/dT = -1 \times 10^{-3}$ grad^{-1} which corresponds to the really lowest experimentally determined value given in table 2. It seems therefore reasonable to use this correction for superposition of the viscoelastic data measured in the immediate proximity to the equilibrium.

However, the decrease in moduli with temperature can also be affected by another effect. This has been found for uncrosslinked polymethacrylates [cf. (11)], and also for very lightly crosslinked poly(butyl methacrylate) networks at temperatures above 80 °C [using both dynamic (12) and stress relaxation measurements (10)] and explained (11) by the influence of temperature on network spacing of these polymers. In this case simple WLF equation derived from the main transition region cannot be used for superposition of the data obtained at higher temperatures, but still below $T_g + 100$ °C; to attain the superposition of the reduced compliance data over the whole range, $\log J'' f$ vs. $\log \omega a_T f^2$ and $\log J' f$ vs. $\log \omega a_T f^2$ plots were introduced [cf. (11)], where f is an additional shift factor. In the flow region a shift on the vertical axis, $2.4 \log f$, was suggested. It was shown (12) that the necessity of a correction disappears with increasing concentration of the crosslinking agent.

Temperature dependence of the breaking stress in the creep rupture experiment

As has been shown earlier (5), the time-to-break of the PHEMA gels decrease in some cases monotonically with increasing stress and temperature, while in other cases characteristic maxima appear on the $\log t_b$ vs. $\log \sigma$ curves. At low stresses the curves approach asymptotically the safe stress; at stresses lower than the safe stress the survival of the test pieces is virtually unlimited. An example of the dependence of time-to-break values on stress for various temperatures for a sample prepared in the presence of ethylene glycol is shown in fig. 2. The temperature dependences of breaking stress values reduced to 25 °C, $\log \sigma (298/T)$ and corresponding to time-to-break $t_b = 10$ s are shown in fig. 3.

The decrease of the breaking stress with temperature is much steeper than the corresponding decrease in the modulus. This difference between the temperature sensitivity of deformation and ultimate behaviour has been explained by the viscoelastic concept of strength as mentioned above. The effect of preswelling by means of ethylene glycol and by means of water is different. In the case of ethylene glycol the given amount of a diluent remains in the gel and the curves corresponding to a higher swelling are shifted both towards lower stresses and towards lower temperatures, the shapes of the curves being virtually the same. Moreover a lower network density is reflected in a higher temperature sensitivity of the breaking stress. On the other hand, the preswelling with water influences after all the microstructure of the gel, the equilibrium amount of water in the polymer phase being only a little changed. According to this, the steepness of the curves and their position on the temperature scale is similar, the curves corresponding to higher heterogeneity are shifted only in the direction of lower stresses. The exception is made by the sample having the highest heterogeneity and interconnected pores (5) ($v_0 = 0.2$); which has a small temperature sensitivity of the breaking stress in comparison to the other samples. The crosslinking agent concentration has very little influence in most cases of water preswollen systems.

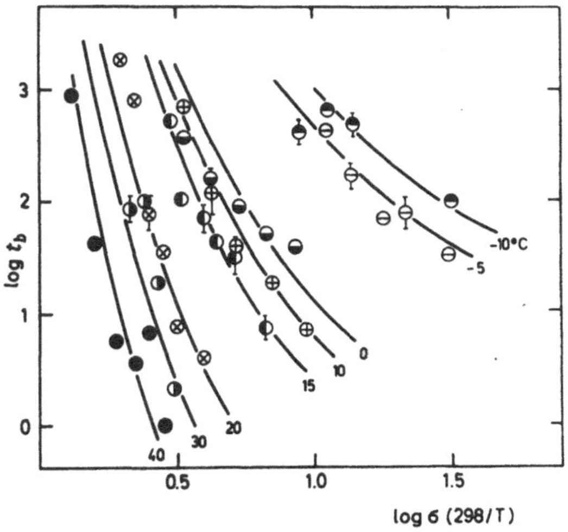

Fig. 2. Example of $\log t_b$ vs. $\log \sigma (298/T)$ plots obtained at different temperatures for PHEMA samples prepared in the presence of ethylene glycol ($v_0 = 0.6$, $c = 0.21 \times 10^{-4}$ mol cm^{-3}). Temperatures of the measurements are shown in the figure, solid lines are recalculated to the given temperatures from the reduced curve

Time-temperature superposition in viscoelastic and ultimate behaviour

It was possible to realize time-temperature superposition of time-to-break, t_b, vs. stress σ curves obtained at temperatures $T < 30$ °C for all samples under investigation; with respect to comparatively large changes in t_b with σ, the shift factors $\log a_T$ were obtained with a relatively good reliability. An example of the superimposed curves of PHEMA samples preswollen in ethy-

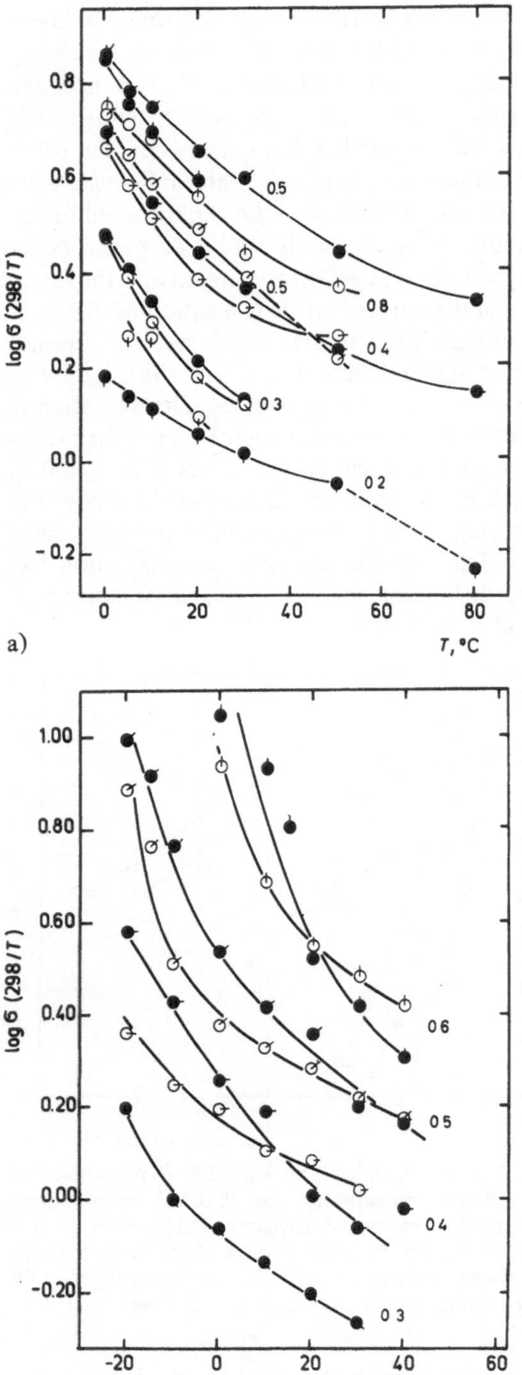

Fig. 3. Log$\sigma(t)(298/T)$ plotted against temperature for preswollen PHEMA samples at $t_b = 10$ s. The values are related to the swollen cross-section of the samples, the significance of points is the same as in fig. 1. a) Samples prepared in the presence of water. b) Samples prepared in the presence of ethylene glycol

lene glycol and obtained in the presence of two concentrations of the crosslinking agent is given in fig. 4. For temperatures 30 °C and higher, the

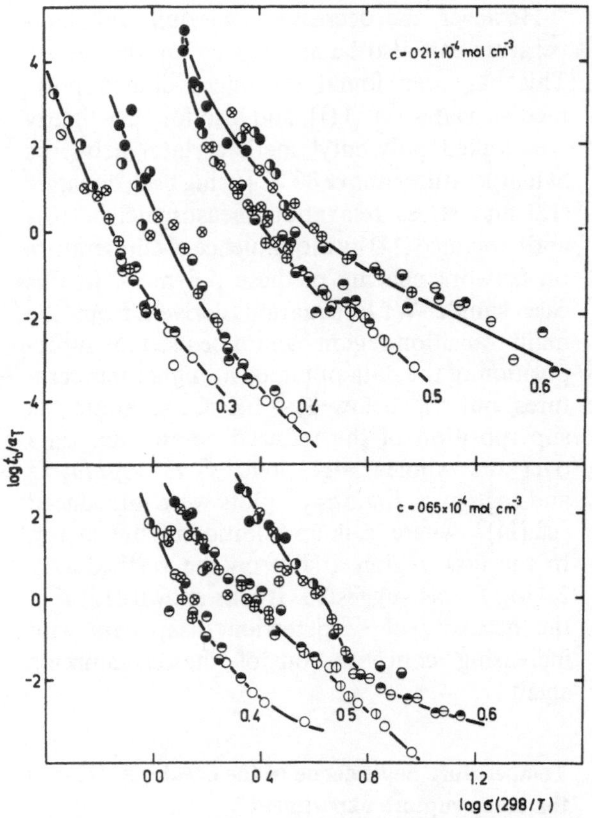

Fig. 4. Reduced stress-at-break as a function of temperature for ethylene glycol preswollen PHEMA gels. Concentration of the crosslinking agent: upper curves 0.21×10^{-4} mol cm^{-3}, lower curves 0.65×10^{-4} mol cm^{-3}. The values correspond to the volume fractions of the polymer, v_0, in the preswollen system. Temperatures: (○) −20 °C, (⦶) −15 °C, (◓) −10 °C, (⊖) −5 °C, (◒) 0 °C, (⊕) 10 °C, (◐) 15 °C, (⊗) 20 °C, (◑) 30 °C, and (●) 40 °C

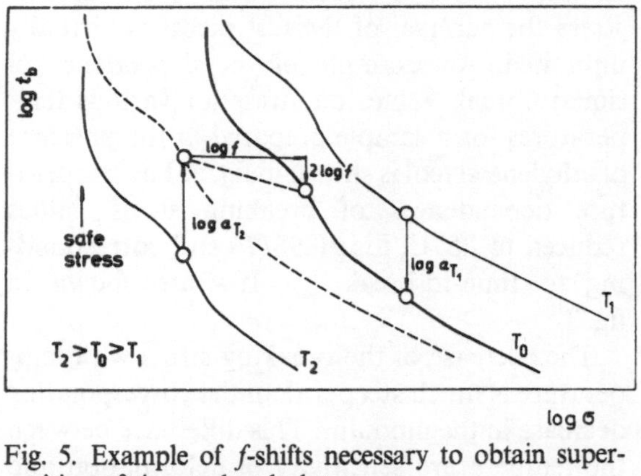

Fig. 5. Example of f-shifts necessary to obtain superposition of time-to-break data

superposition was in most cases not successful by using the same temperature dependence of the shift factors and it was evident (cf. fig. 5) that additional shifts along the log axis are necessary. In such cases we found the double shift mentioned above in the form $\log t_b f^2/a_T$ vs. $\log \sigma(298/T) f^{-1}$

to be very suitable. It was possible to determine these shifts objectively, especially if the t_b vs. σ curves exhibited characteristic maxima or if the curves were approaching the limits of safe stress. Table 3 gives f values found for these cases.

Table 3. Absolute values of f-shifts used for the superposition of time-to-break data of water preswollen PHEMA networks

$c \times 10^4$ mol cm^{-3}	v_0	Temperature °C 30	50	80
0.21	0.8	a	a	a
0.21	0.5	0	0.097	0.210
0.21	0.4	0	0.072	0.128
0.65	0.8	0.077	0.113	a
0.65	0.5	0.073	0.210	a
0.65	0.4	0	0	a

a = not measured

After carrying our corrections described in the Experimental, it was possible to determine reliably the shift factors also from viscoelastic measurements and to construct superimposed curves for all samples measured. An example of reduced curves obtained for systems preswollen in water is shown in fig. 6a ($c = 0.65 \times 10^{-4}$ mol cm^{-3}; only lower temperatures are given). Since at temperatures $T > 30$ °C the plots of $\log E(t)\,(298/T)$ vs. $\log t/a_T$ gave nearly parallel lines with the same slopes, it was very difficult to decide from the experimental results, whether also in the case of viscoelastic data f-shifts should be used, similarly to the case of time-to-break

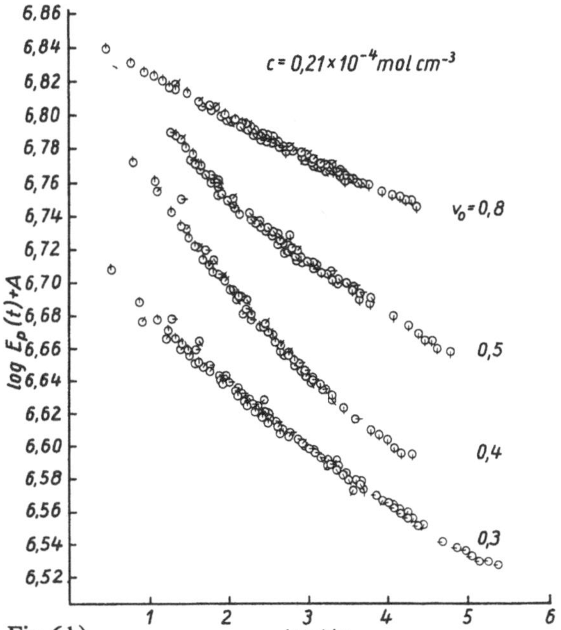

Fig. 6 b)

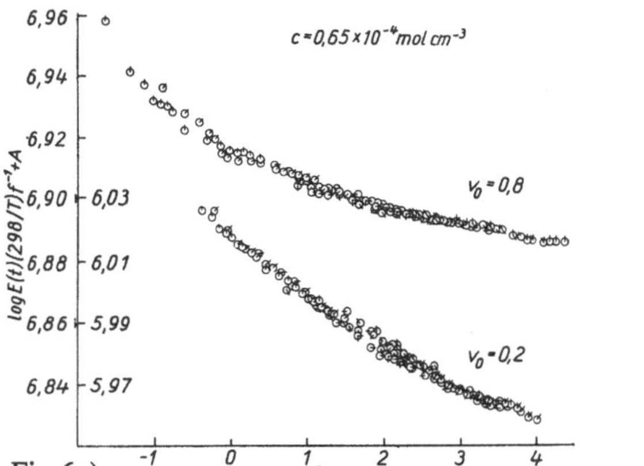

Fig. 6 c)

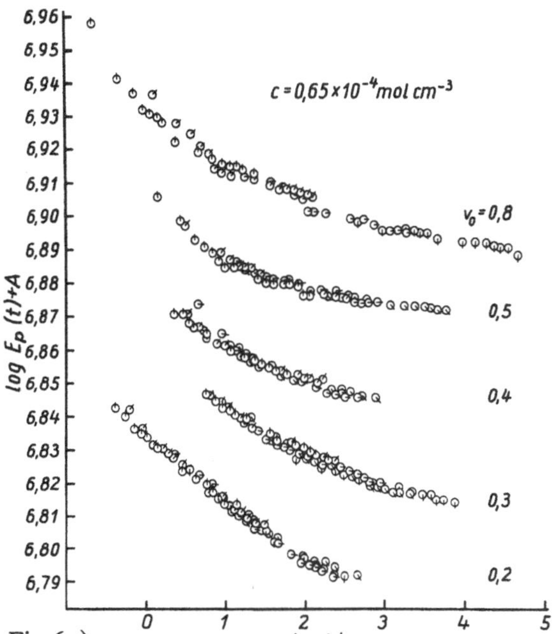

Fig. 6 a)

Fig. 6. $\log E(t)\,(298/T) = \log E_p(t)$ values plotted against $\log t/a_T$ for PHEMA samples prepared in the presence of water. The numbers of individual curves correspond to v_0 values. Systems prepared in the presence of ethylene dimethacrylate: a) 0.65×10^{-4} mol cm^{-3}; b) 0.21×10^{-4} mol cm^{-3}. In each series pip-up denotes the lowest temperature and successive 45° rotations clockwise denote successively higher temperatures listed in text: a) $v_0 = 0.8$, $A = 0$, temperatures from 5–32 °C (four steps); $v_0 = 0.5$, $A = -0.18$, temperatures from 5–21 °C (four steps); $v_0 = 0.4$, $A = -0.265$, temperatures from 6–20 °C (four steps); $v_0 = 0.3$, $A = -0.308$, temperatures from 3.5–36 °C (five steps); $v_0 = 0.2$, $A = -0.825$, temperatures from 3–26 °C (five steps). b) $v_0 = 0.8$, $A = 0$, temperatures from 26–76 °C (five steps); $v_0 = 0.5$, $A = -0.32$, temperatures from 38–88.5 °C (five steps); $v_0 = 0.4$, $A = -0.36$, temperatures from 32–74 °C (five steps); $v_0 = 0.3$, $A = -0.316$, temperatures from 26 to 83 °C (seven steps). c) Example of superimposed curves with additional f-shifts; $c = 0.65 \times 10^{-4}$ mol cm^{-3}, f-shifts used for 4 additional temperatures in the range of 34–68 °C ($v_0 = 0.8$, $A = 0$); 5 additional temperatures in the range 48–72 °C ($v_0 = 0.2$, $A = -0.87$)

data. Superimposed curves corresponding to higher temperatures of the same system ($c = 0.21 \times 10^{-4}$ mol cm^{-3}) are given in fig. 6b.

Temperature dependence of the shift factors

The temperature dependences of the shift factors of the time-to-break curves satisfied well the WLF equation with the characteristic temperature T_s which it was possible to determine in most cases with an accuracy of approximately ± 2 °C (table 2). The shift factors, $\log a_T$, referred to the respective temperatures T_s plotted against $(T - T_s)$ gave a good agreement (for water preswollen samples cf. fig. 7).

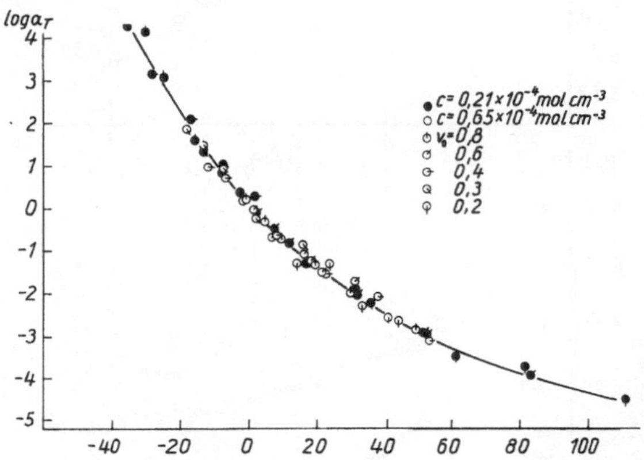

Fig. 7. Loga_T corresponding to time-to-break data plotted against $(T - T_0)$. PHEMA gels preswollen in water, a_T values related to corresponding temperatures T_s of table 2. v_0: (ȸ) 0.8, (ø) 0.5, (⊖) 0.4, (Q) 0.3, (φ) 0.2. Concentration of the crosslinking agent: $c = 0.21 \times 10^{-4}$ mol cm^{-3} (black points), 0.65×10^{-4} mol cm^{-3} (empty points)

A comparison of the temperature dependences of the shift factors corresponding to the ultimate and viscoelastic measurements (referred to the reference temperature 25 °C) for the individual systems and some selected temperatures is given in fig. 8. For water preswollen systems prepared in the presence of $c = 0.65 \times 10^{-4}$ mol cm^{-3} of ethylene dimethacrylate, an agreement between the temperature dependences of the ultimate and viscoelastic shift factors was found up to 30 °C. As has been mentioned the temperature dependence of $\log a_T$ of the f-shifted ultimate data fitted the WLF relationship over the whole range of the temperatures measured. On the contrary, in the case of the viscoelastic characteristics deviations from the common WLF function can be seen above 30 °C which increase with increasing temperature. Here, the temperature dependence of the shift factors satisfies better *Arrhenius*' relationship with the activation energy $\Delta H = 62.3$ kcal/mol (fig. 9) irrespective of the degree of dilution during the polymerization. Assuming the validity of the same a_T function over the whole range of temperatures, similarly to the case of ultimate measurements, an agree-

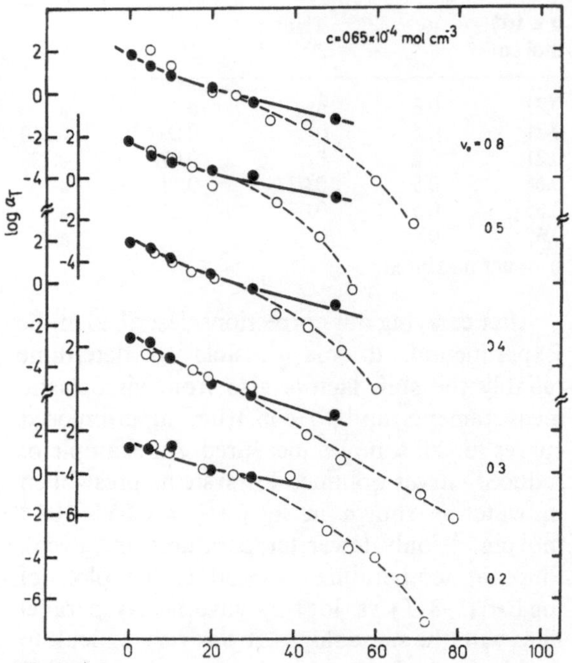

Fig. 8 a)

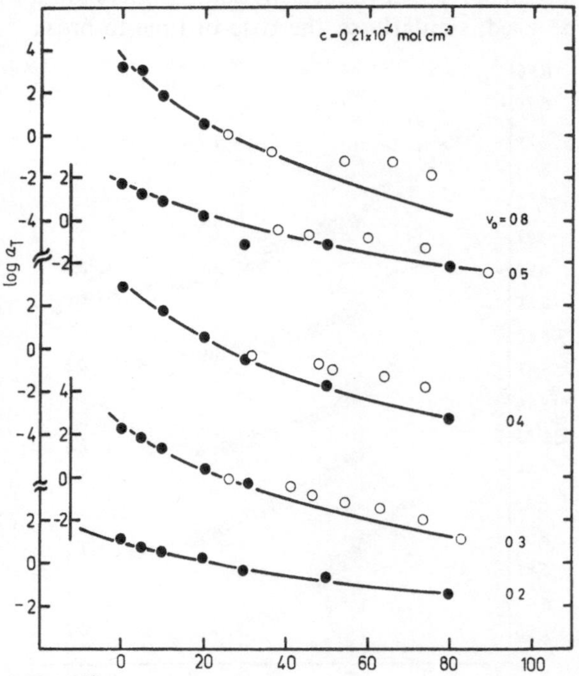

Fig. 8b)

ment of reduced data was obtained by plotting $\log E(t)\,(298/T)\,f^{-1}$ vs. $\log t f^2/a_T$ (cf. fig. 6c). The average value of $\Delta H'$ calculated from f-shifts using an equation of the *Arrhenius* form (11) has been compared with the same values obtained from the f-shifts of ultimate data as well as with the $\Delta H'$ values calculated for polybutyl methacrylate previously (12) (table 4). The values of $\Delta H'$ for PHEMA, calculated from the stress relaxation and time-to break data and the same values, calculated for the poly(butyl methacrylate) networks from the shifts of J' and J'' values are of the same order of magnitude. They support the conclusion that the origin of the effect is for both polymers the same and that the effect is typical not only of the viscoelastic but also of the ultimate data. Correction for the temperature dependence of the internal energy of chains used in the case of viscoelastic measurements is probably responsible for a lower $\Delta H'$ value in comparison to that obtained from time-at-break data.

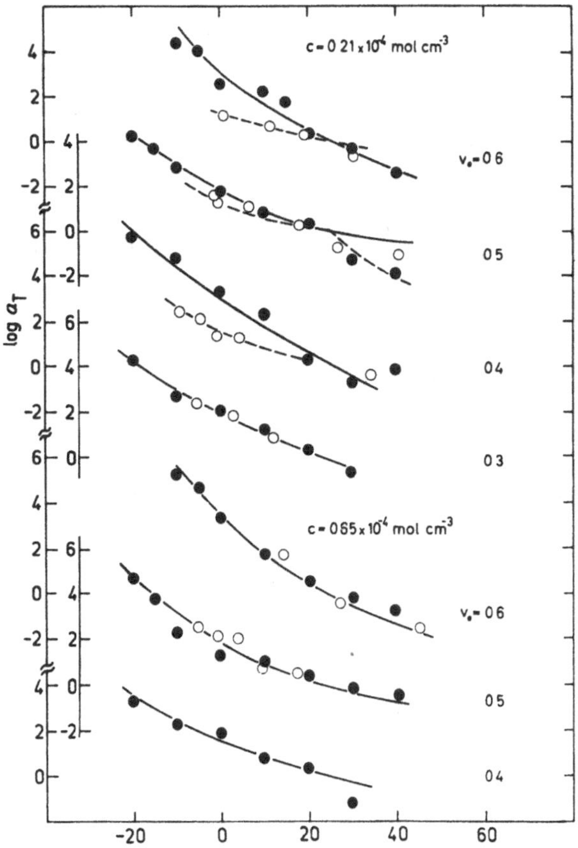

Fig. 8c) PHEMA - EG

Fig. 8. Log a_T vs. temperature plots of preswollen PHEMA gels. White points – viscoelastic data, black points – time-to-break data. The numbers indicate volume fractions of polymer in polymer-diluent systems, v_0, all values reduced to 25 °C. a) Water preswollen systems, $c = 0.21 \times 10^{-4}$ mol cm^{-3}. b) Water preswollen systems, $c = 0.65 \times 10^{-4}$ mol cm^{-3}. c) Ethylene glycol preswollen systems, $c = 0.21 \times 10^{-4}$ mol cm^{-3} (upper curves), $c = 0.65 \times 10^{-4}$ mol cm^{-3} (lower curves)

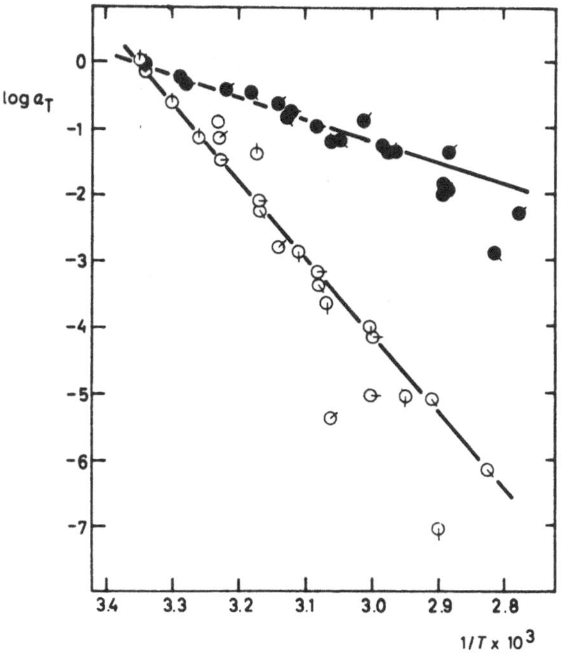

Fig. 9. *Arrhenius'* plot of shift factors obtained from viscoelastic measurements of water preswollen systems at temperatures $T > 25$ °C. Significance of points as in fig. 7

Table 4. $\Delta H'$ values found using *Arrhenius*-type equation from f-shifts of viscoelastic and time-to-break data of polymethacrylates. Measured in the rubberlike region

Polymer	$c \times 10^4$ mol cm^{-3}	$\Delta H'$	Note
PBuMA (12)	0	1.6	$\log J_p'$ vs. $\log \omega a_T$ data
PBuMA (12)	0.012	1.4	$\log J_p'$ vs. $\log \omega a_T$ data
PBuMA (12)	0.047	0.9	$\log J_p'$ vs. $\log \omega a_T$ data
PHEMA-water	0.21	1.5[1])	$\log t_b/a_T$ vs. $\log \sigma$ data
PHEMA-water	0.65	3.0[1])	$\log t_b/a_T$ vs. $\log \sigma$ data
PHEMA-water	0.65	0.2[1])	$\log E(t)$ vs. t/a_T data
PBuMA (12)	1.04	0	$\log J_p'$ vs. $\log \omega a_T$ data

[1]) Mean value.

A comparison of the temperature dependences of the shift factors of water preswollen networks prepared in the presence of 0.21×10^{-4} mol cm^{-3} of the crosslinking agent is shown in fig. 8b. The viscoelastic measurements were performed mostly in the region of temperatures $T > 30$ °C, therefore no definitive conclusions could be obtained from the comparison of both sets of the data. However it seems that the relaxation modulus was less temperature-sensitive than the time-to-break for all samples having different v_0 values. The strain-at-break of lightly-crosslinked networks subjected to constant load is much larger than deformations used when measuring the relaxation modulus. Therefore the discordance between the shift factors of both types of measurements is probably due to the different contributions of long relaxation times of lightly-crosslinked networks measured at small and large deformations. Here too was it possible to use *Arrhenius*' relationship (fig. 9) with the activation energy $\Delta H = 10\,\mathrm{kcal\,mol^{-1}}$ for the temperature dependence of the viscoelastic shift factors.

The above conclusions concerning the relationship between the temperature dependences of the shift factors ensuing from both types of measurements hold for both homogeneous and heterogeneous networks (fig. 8c), which follows from fig. 8c where homogeneous systems preswollen with ethyleneglycol are compared with each other.

T_s temperatures

The T_s temperatures of the samples under investigation depend on the degree of swelling, on the one hand, and are connected with the differences in the topology of networks prepared in the presence of different concentrations of the diluent, on the other. In the case of the ultimate characteristics, the high degree of orientation in the roots of the primary defects also plays its role. It can be expected, therefore, that the T_s temperatures will be a complicated function of a number of variables and that the effect of a different degree of swelling (and also of the presence of water in the pores) will affect the T_s temperatures of heterogeneous systems in a very complex manner. In fact, table 2 demonstrates that the T_s values of heterogeneous systems were not a simple function of v_0 for both concentrations used of the crosslinking agent. On the other hand, for ethylene glycol preswollen systems T_s decreased as expected with increasing concentration of ethylene glycol.

Effect of preswelling on the relaxation modulus and ultimate behaviour

It is usually assumed that the end-to-end distance of the polymer chains in the moment of the crosslinking polymerization corresponds to *Gaussian* statistics, even if there is a diluent present during the polymerization. The modulus of the preswollen system must therefore be corrected by the factor $v_0^{2/3}$ with respect to the system without the diluent, if it is related to a dry cross-section of the sample. The actual experimental values of moduli are often lower, because the presence of a low-molecular weight compound reduces the effectiveness of the crosslinking agent. The effect of the diluent on the viscoelastic behaviour of our systems was investigated using the value of the reduced relaxation modulus $E(t)(298/T)$ for $\log t = 2.5$ and temperature 25° (table 5). The dependence of this characteristic on dilution during the polymerization, v_0, in a bilogarithmic plot is linear (fig. 10). For the majority of water preswollen systems the slope is -1, in the case of preswelling with ethylene glycol the modulus falls off still faster. A less pronounced influence of water can be explained by phase separation during the polymerization, so that after polymerization the polymer network of heterogeneous systems contains an approximately constant amount of the diluent regardless of the composition of the polymerization mixture.

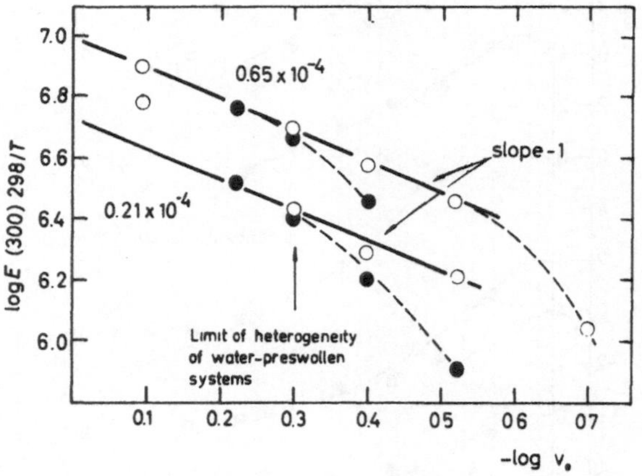

Fig. 10. $\log E(300)\,298/T$ plotted vs. $-\log v_0$. White points, water preswollen, black points ethylene glycol preswollen system. Concentration of crosslinking agent is given in the figure

The effect of preswelling on the reduced breaking stress corresponding to different values of time-at-break is given in table 5. Depending on dilution during the polymerization, the isochronous values of breaking stress pass through a maximum for systems swollen with water; the maximum disappears for longer times (fig. 11).

Table 5. Log$E(t)$ (298/T) and log$\sigma(T)$ (293/T) values of preswollen. PHEMA gels

$c \times 10^{-4}$ mol cm^{-3}	log$E(t)$ (298/T) 0.21	0.65	log$\sigma(t)$ (293/T) 0.21			0.65		
Samples preswollen in water								
log$t$ or logt_b	2.5	2.5	−1.0	0	2.5	−1.0	0	2.5
v_0								
0.8	6.780	6.900	0.77	0.68	0.50	0.82	0.71	0.45
0.5	6.428	6.695	0.90	0.80	0.52	0.93	0.66	0.35
0.4	6.287	6.580	0.62	0.52	0.37	0.74	0.56	0.26
0.3	6.213	6.460	0.51	0.37	0.11	0.49	0.32	0.21
0.2	–	6.041	0.37	0.20	0.12	0.23	0.19	–
Samples preswollen in ethylene glycol								
log$t$ or logt_b	2.5	2.5	−2.5	0	2.5	−2.5	0	2.5
v_0								
0.6	6.520	6.770	1.20	0.60	0.27	0.98	0.63	0.35
0.5	6.405	6.665	0.85	0.43	0.18	0.75	0.37	0.06
0.4	6.200	6.455	0.27	0.05	−0.11	0.42	0.13	−0.04
0.3	5.910		0.10	−0.14	−0.31	–	–	

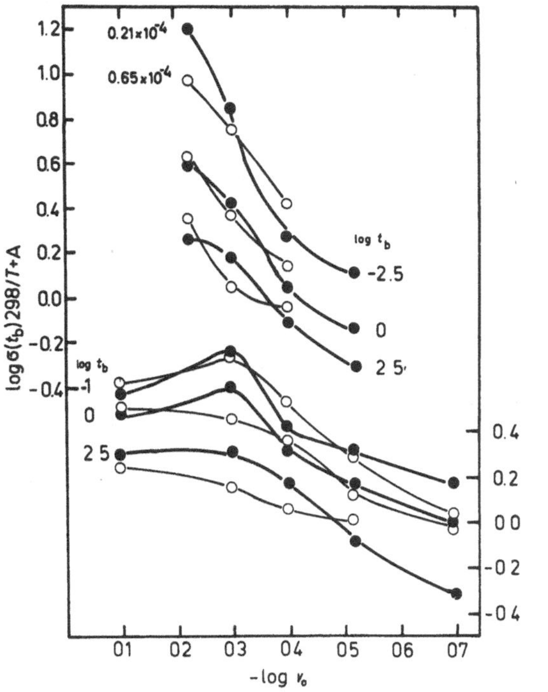

Fig. 11. Log$\sigma(t_b)$(298/T) plotted vs. – logv_0. Upper curves – ethylene glycol preswollen systems, lower curves – water preswollen systems. Logt_b and concentration of the crosslinking agent given in the figure

The breaking stress decreases pronouncedly with increasing temperature; at the same time, the order of samples having a higher and a lower network density also changes (fig. 3). These results document the importance of non-equilibrium effects during the breaking process which are also responsible for the formation of the maximum on the curve of dependence of tensile strength on network density (14). Although this maximum has not yet been proved experimentally for 2-hydroxyethyl methacrylate polymers, earlier results indicate (3) that it lies in the region of very low network densities ($c < 10^{-5}$ mol cm^{-3}). This means that no straightforward analogy between the viscoelastic and ultimate behaviour is possible in this region, since the structural factors influence both types of behaviour in a different way. Here too, however, the time and temperature dependence of breaking stress is controlled by viscoelastic processes in micro-volumes.

Careful technical assistance of Mrs. *E. Ledabylová* and Mrs. *B. Maršíková* is gratefully acknowledged.

Summary

Both stress relaxation behaviour and time-to-break t_b as a function of stress σ were studied in identical time and temperature intervals using poly(2-hydroxyethyl-methacrylate) gels in the swollen state, prepared both in the homogeneous and in the heterogeneous form. Homogeneous samples were prepared by means of crosslinking polymerization (concentration of ethylene dimethacrylate as a crosslinking agent $c = 0.21$ and 0.65×10^{-4} mol cm^{-3}) in the presence of ethylene glycol ($v_0 = 0.6 - 0.3$); heterogeneous samples with various degrees of porosity were prepared in the presence of water ($v_0 = 0.8 - 0.2$). Irrespective of v_0, an agreement was found between the temperature shift factors, log a_T, obtained from the viscoelastic and ultimate measurements for the systems prepared with a higher concentration of the crosslinking agent; for the lightly crosslinked systems the a_T values from the viscoelastic data were less sensitive to temperature. The temperature dependence of loga_T was not the same at temperatures $T > 30$ °C and additional horizontal-vertical shifts had to be used; this effect is very similar to that observed previously for the other polymethacrylates by *Ferry* and coworkers.

In the water preswollen systems the isochronous breaking stress vs. v_0 curves exhibit a maximum which disappears with time. The moduli $E(t)$ of the same systems decrease approximately with v_0^{-1} irrespective of the degree of preswelling. The effect of crosslinking and preswelling on the shape of the superimposed $\log E(t)$ vs. $\log t/a_T$ and $\log t_b/a_T$ vs. $\log \sigma$ curves was ascribed to the different role of viscoelastic mechanism in the relaxation and in the ultimate process and to different topological features of the networks prepared.

Zusammenfassung

Die Spannungsrelaxation und die Bruchzeit t_b als Funktion der Spannung σ wurden am gequollenen homogenen und heterogenen Poly(2-Hydroxyäthylmethacrylat) im gleichen Zeit- und Temperaturintervall untersucht. Die homogenen Proben wurden durch Vernetzungspolymerisation von 2-Hydroxyäthylmethacrylat mit Äthylendimethacrylat als Vernetzungsmittel (c = 0,21 und 0,65 10^4 mol cm^{-3}) und Äthylenglykol als Verdünnungsmittel hergestellt; die heterogenen Proben mit verschiedener Porosität wurden mit Wasser als Verdünnungsmittel hergestellt. Die Verschiebungsfaktoren $\log a_T$ aus Relaxations- und Bruchversuch stimmen nur bei Proben mit einem höheren Vernetzungsgrad überein. Für Systeme mit einem niedrigeren Vernetzungsgrad sind die viskoelastischen Kenngrößen weniger temperaturempfindlich. Bei höheren Temperaturen ($T > 30$ °C) ist die Konstruktion der generalisierten Kurven nur mit zusätzlichen Verschiebungsfaktoren möglich; dieses Verhalten wurde schon früher an anderen Polymethacrylaten von *Ferry* und Mitarb. beobachtet.

Der Einfluß von Vernetzungsdichte und von Quellung auf die generalisierten Kurven $\log E(t)$ vs. $\log t/a_T$ und $\log t_b/a_T$ vs. $\log \sigma$ wurde den Differenzen in dem viskoelastischen Mechanismus der Spannungsrelaxation und des Bruches sowie den Differenzen in der Netzwerk-Topologie zugeschrieben.

References

1) *Ferry, J. D.*, Viscoelastic Properties of Polymers, IInd. Ed. (New York 1970).

2) *Smith, T. L.*, Rheology, Vol. V, *F. R. Eirich* (Ed.) (New York 1969).

3) *Raab, M.* and *J. Janáček*, Rheol. Acta **10**, 280 (1971).

4) *Bueche, F.* and *J. C. Halpin*, J. Appl. Phys. **35**, 36 (1964); *Halpin, J. C.*, J. Appl. Phys. **35**, 3133 (1964).

5) *Raab, M., Z. Pelzbauer, J. Janáček*, and *M. Štol*, J. Polymer Sci.-Phys. **C38**, 221 (1972).

Ilavský, M. and *J. Hasa*, Collection Czechoslov. Chem. Commun. **33**, 2142 (1968).

7) *Ciferri, A.*, J. Polymer Sci. **A2**, 3089 (1964).

8) *Ilavský M.* and *J. Hasa*, Collection Czechoslov. Chem. Commun. **34**, 2205 (1969).

9) *Ilavský, M., J. Hasa*, and *I. Havliček*, J. Polymer Sci.-Phys. **A2**, **10**, 1775 (1972).

10) *Ilavský* and *J. Hasa*, J. Polymer Sci. Polymer Phys. Ed. **11**, 539 (1973).

11) Ref. 1, p. 335.

12) *Hrouz, J.* and *J. Janáček*, J. Polymer Sci. **A2**, **10**, 1383 (1972).

13) *Kolařik, J.* and *J. Janáček*, J. Polymer Sci. **A2**, 10, 11 (1972).

14) *Bueche, F.* and *T. J. Dudek*, Rubber Chem. Technol. **36**, 1 (1963).

Authors' address:

Doz. Dr. *J. Janáček*, Dr. *M. Raab*, and Dipl.-Ing. *M. Štol*
Institute of Macromolecular Chemistry
Czechoslovak Academy of Sciences
Praha (Czechoslowakia)

Rheol. Acta 13, 127–134 (1974)

From the Tohoku University, Tokyo (Japan)

Characterization of polymers in large deformation by repeated chemical stress relaxation of vulcanized rubbers

By K. Murakami and T. Kusano

With 16 figures and 1 table

(Received October 27, 1972)

Introduction

Many reports on chemorheological studies of rubbers have been presented but those for large deformations were very few. At large deformation, polymer chains would be oriented and accordingly statistical treatment become impossible to be applied, and also the reactivity of those polymers (1) will be changed.

In this paper, our investigations were carried out for chemorheological behavior of rubbers being treated mechanically, that is, what kind of effects on the cured polymer structures and the relation of chemical degradation would be expected under such mechanical and chemorheological actions.

These problems were discussed by the repeated stress relaxation test under oxidation reaction at high temperature.

Experiments

Unfilled natural rubber vulcanizates were used for these experiments.

A special apparatus was deviced to perform the intermittent stress relaxation measurement by changing the rotating motion for the reciprocation motion. The sample was maintained in a relaxed, unstretched condition between upper and lower clamps. At given time intervals, the sample was rapidly stretched to a fixed elongation, together the equilibrium stress was rapidly recorded, and then was immediately returned to its original condition. The interval of frequency is changeable from 0.7–225th per min and its extensive strain is from 0–80 mm.

Figures of clamping parts and detecting parts of stress were shown in fig. 1 (2). Experimental materials used were prepared under the following condition as shown in table 1.

Results and discussion

Similar to the continuous stress relaxation, intermittent repeated stress relaxation curves are also affected by temperature, time repeated cycling number and strain etc. From the results obtained, the factor of temperature was found to be the most sensitive among them and that of elongation came to the next.

Studies on repeated stress measurement in small deformation have been reported (3–5) but they dealt with the relation between repeated number and stress decay in the system of filled vulcanizates.

Oxidative degradation will usually occur above approximately 70 °C in air for natural rubber as shown in fig. 2, because the value of k equals to zero means no degradation occurs. Therefore, it is considered that the degradation observed lower than 70 °C if exists will involve only mechanical degradation in the repeated stress relaxation test.

The relaxation behavior at 88 °C was shown in fig. 3. These were obtained from the experiments with variable cycle rates, from 0.8–17th per min under the constant extension ratio α. These facts indicate that the effect on stress relaxation curves by the new crosslinkage yielded can be neglected. From these results, the time observed in repeated stress relaxation measurement will

Table 1. Mix detail and curing condition

	NR	ZnO	St-acid	DM	Sulfur	PβNA	TMTD	Curring condition
Standard material	100	5.0	3.0	1.0	3.0	0	0	145 °C 20 min
PβNA	100	5.0	3.0	1.0	3.0	1.0	0	145 °C 20 min
TMTD-vulcanizate	100	5.0	3.0	1.0	0	0	3.0	145 °C 45 min

St-acid Stearic acid; DM Dibenzothiazyldisulfide
TMTD Tetramethylthiuramdisulfide; PβNA N-phenyl-β-naphthylamine

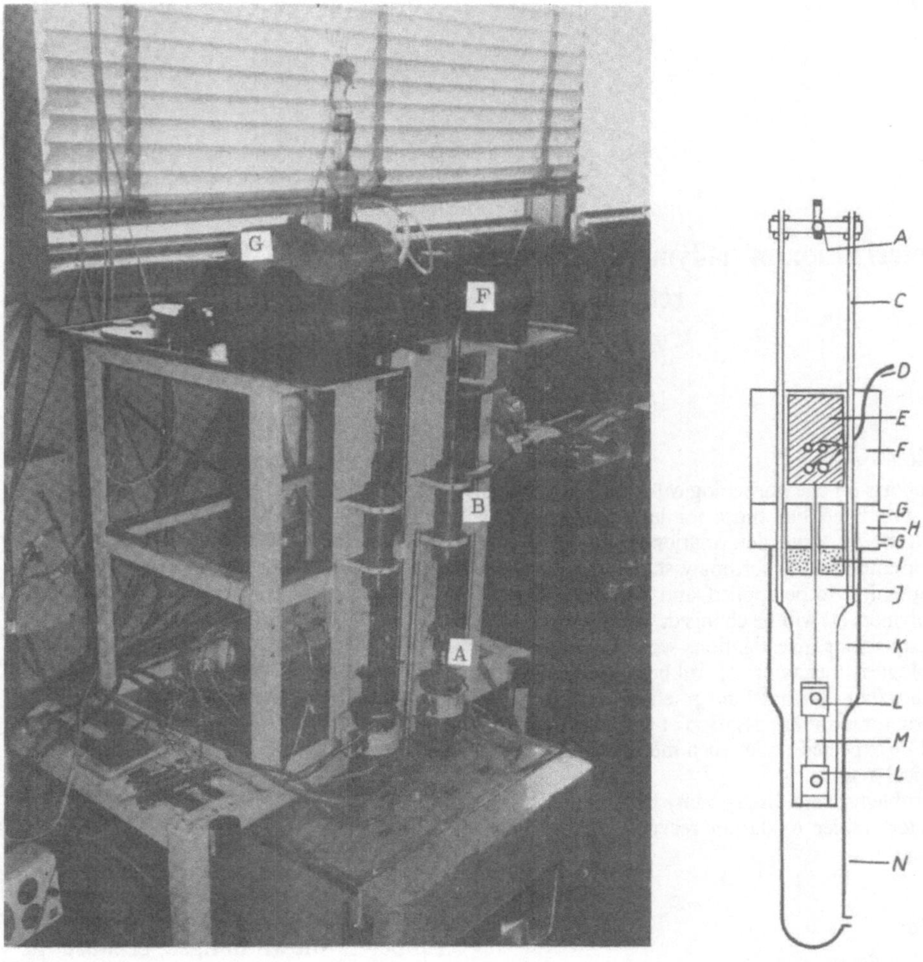

Fig. 1. Apparatus of repeated stress relaxation test. A. For reciprocating motion; B. Piston & Guide; C. K. Sliding bar; D., E. J. Strain gage; F. Setting part of frequency and elongation; G. Gear motor; H. Cooler; I. Cork; L. Clamps; M. Sample; N. Glass tube (changeable atmosphere)

be expressed by a formula $t = n \cdot \Delta t_{(\alpha\text{-max})}$, where n is the cycle number and $\Delta t_{(\alpha\text{-max})}$ is the summation of time at the nearness of maximum strain in cycle strains as shown in fig. 4. Consequently, both the total times t of low and high cycle rates should be equal. The causes of independence of stress decay on cycle rate are not clear.

In fig. 5 and 6, those intermittent relaxation curves of natural rubber in air were shown, in which the intermittent stress relaxation curves were consistent with for the range of small deformation, and when in larger deformation above the values of $\alpha = 1.5$ the rate of relaxation, that is, k value (the exponential value in *Maxwell*ian decay) will increase.

Similarly to the continuous stress relaxation behaviors, regions of linear behavior will exist under a given value of deformation. These behaviors are also shown at each temperature as in fig. 5 and 6. The relation between the k value obtained from the above data and the extension ratio α was depicted in fig. 7. The relation between $\log k$ and $1/T$ becomes straight lines as *Arrhenius* plot. From these results, the relation between activation energy of repeated stress relaxation and α will be shown in fig. 8. As is seen from fig. 8, the activation energy E_{act} of repeated stress relaxation becomes larger as the values of α increased, and at last the activation energy will be constant when α is more than approximately 3. This will be based upon the orientation of the polymer network in large deformation. Oxidative activation energy of natural rubber is about 30 kcal/mol while the activation energy of repeated stress relaxation appeared as 11.4 kcal/mol in small deformation. It is considered that the activity is accelerated by repeating strain. *Frenkel* (7) has reported

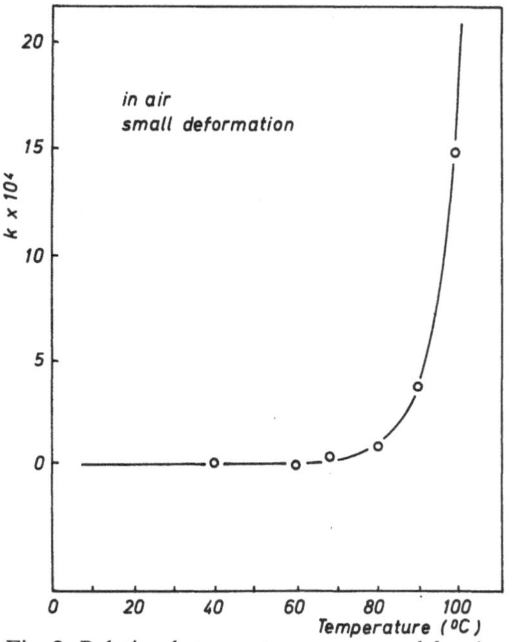

Fig. 2. Relation between temperature and k value (the exponential value in *Maxwellian* decay) which were measured by continuous relaxation test in small deformation

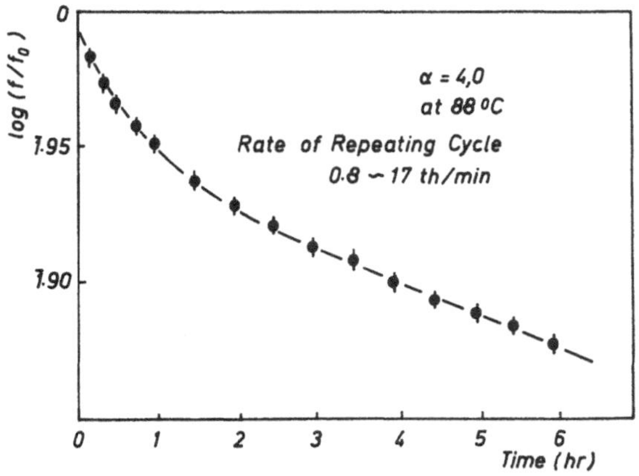

Fig. 3. Relation between stress relaxation and rate of repeating cycle (0.8 th/min ~ 17 th/min) at 88 °C in air

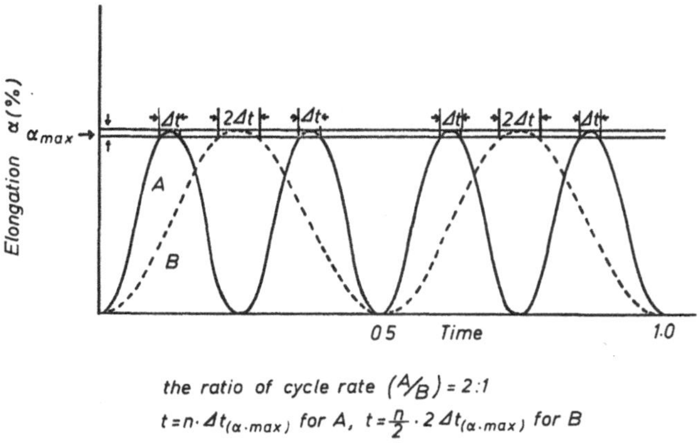

Fig. 4. Relation between the summation of time at the nearness of maximum strain and cycle rate. $t = n \cdot \Delta t_{(\alpha\text{-max})}$

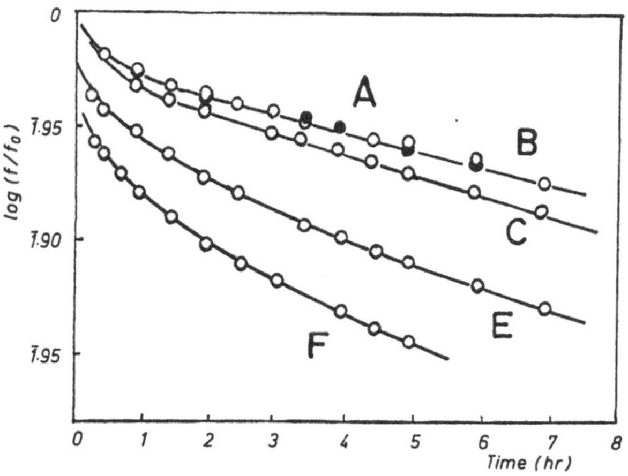

Fig. 5. Intermittent stress relaxation curves with variable strains at 88 °C in air, rate of repeating cycle (RRC) = 14 th/min. A. $\alpha = 1.25$; B. $\alpha = 1.5$; C. $\alpha = 2.5$; E. $\alpha = 4.0$; F. $\alpha = 5.0$

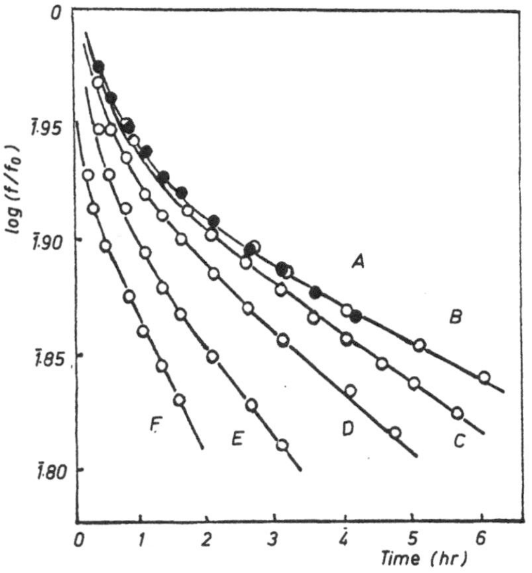

Fig. 6. Intermittent stress relaxation curves with variable strains at 100 °C; in air, RRC = 14 th/min. A. $\alpha = 1.25$; B. $\alpha = 1.5$; C. $\alpha = 2.5$; D. $\alpha = 3.25$; E. $\alpha = 4.0$; F. $\alpha = 5.0$

the existence of similar effects under such repeated conditions. And recently *Portter* (8) has suggested the mechanism of chain scission under mechanical stress.

For the purpose of observation between orientation and α, fig. 9 shows the well known *Mooney-Rivlin* plots of the same sample as in fig. 8. It was seen that the samples which would give the equal E_{act} in fig. 9 would be included in the region from the minimum point to the large deformation of the curves in fig. 9, thus chemical relaxation of large deformation will depend greatly upon the orientation.

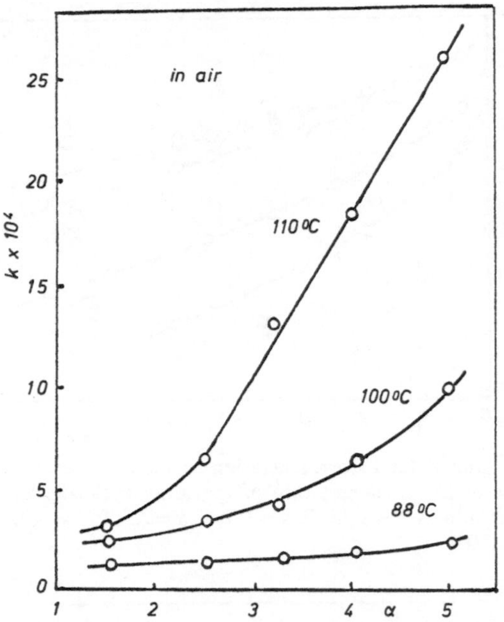

Fig. 7. Relation between relaxation rate and strains. RRC = 14 th/min

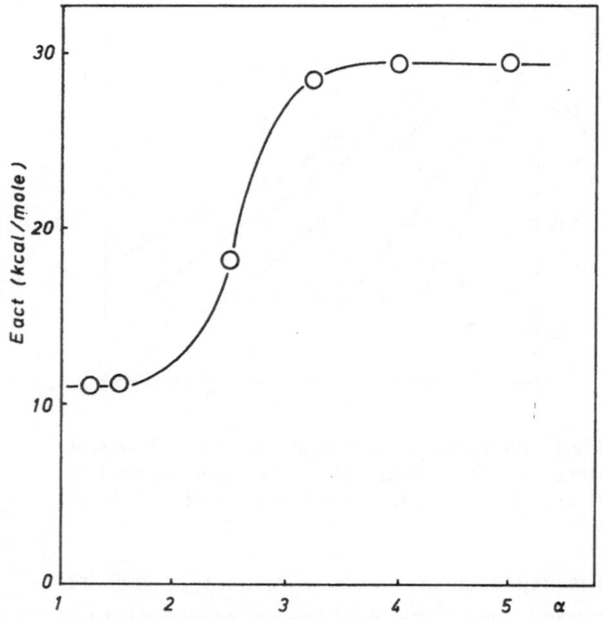

Fig. 8. Dependence between apparent activation energy and repeated strains.

In fig. 5 and 6 the relaxation rate k becomes increased when the extension ratio α was larger in large deformation, this was analogous to the change of relaxation rate when temperature was high. And it will be suggested that the increase of α will progress the relaxation degraded mechanically. From this point of view, assuming that mechanical stimulus was equivalent to temperature one, α could be referred to T in plotting *Arrhenius* type as shown in fig. 10.

The reason why k becomes increased under larger extension of α is considered to be based upon the more active mechano-chemical action of the larger extension of α. The mechano-chemical chain scission occurs predominantly on those network which are subject to large strain and isotropical temperature in this case. It is worthy of notice in fig. 10 that the straight lines paralleled to the abscissa indicate the linear regions in the lower extension ratio α, while the non-linear regions expressed by higher extension α were shown by the straight lines of slope equal to -3.0, independent of temperatures.

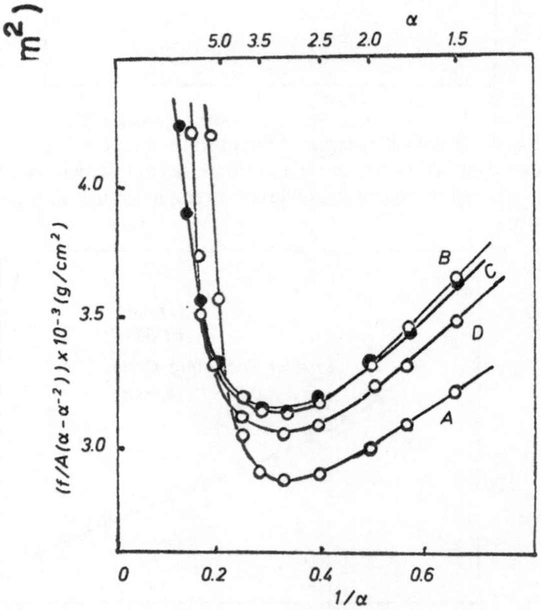

Fig. 9. *Mooney-Rivlin* plot of unfilled vulcanizates. A. 25 °C; B. 88 °C; C. 100 °C; D. 110 °C

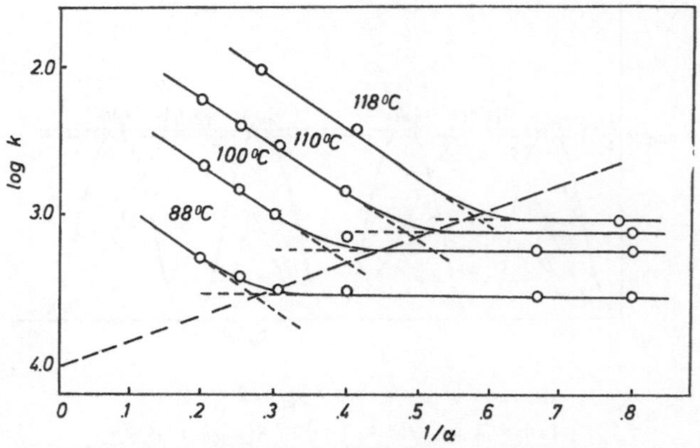

Fig. 10. Relation between $\log k$ and $1/\alpha$ at various temperatures. RRC = 14 th/min

The meaning of the straight lines of slope equal to -3.0 in the higher extension ratio α will be discussed as follows.

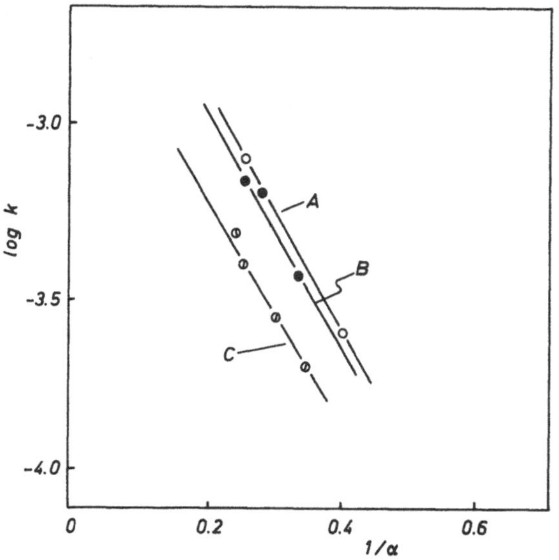

Fig. 11. *Arrhenius* type plot of three materials of natural rubber vulcanizates. RRC = 14 th/min. A. Sulfur vulcanizate with PβNA 1 phr, in air, at 80 °C; B. TMTD vulcanizate, in air, at 88 °C; C. Sulfur vulcanizate in N_2 at 80 °C

The samples shown in fig. 11 are natural rubber vulcanizates with the antioxidant N-phenyl-β-naphthylamine (PβNA) 1 phr and those cured with tetramethylthiuramdisulfide (TMTD) and natural rubber sulfur vulcanizates. If the mixed effects of oxidative reaction and mechanical degradation on these samples in fig. 11 exist, there must appear the change of slope. But the constant value of slopes was obtained as -3.0 and this would consist with the results of fig. 10 and again of sample C in fig. 11 with -3.2 in N_2. In conclusion, those three samples were consistent with the value of -3.0 ± 0.2. From these results it was surely seen that oxidative reaction could not affect on the slopes of straight lines at large α.

In general, the continuous chemical stress relaxation in small deformation can be expressed by eq. [1] as suggested by *Tobolsky* (9). The k value is independent on extension ratio α in such a linear region, while it will depend upon

$$f(t)/f(0) = e^{-kt} \qquad [1]$$

the strain in large deformation. It was found that the intermittent repeated chemical stress relaxation curves are also indicated by *Maxwellian* decay curves such as eq. [1] both in small deformation and in large one, where in the former, the k value is independent of α, in the latter, that is dependent of α as shown in fig. 5 and 6.

The values of slope of the straight lines in the region of large deformation in fig. 11 were identified at each temperature. From the above results, we can try to determine the relation between α and temperatures.

Those straight lines at higher α in fig. 10 and 11 can be expressed by eq. [2],

$$\ln k = -\frac{E_f}{\alpha} + \ln A, \quad k = A \cdot e^{-\frac{E_f}{\alpha}} \qquad [2]$$

where A and E_f are the constants. By substituting eq. [2] into eq. [1], eq. [3] can be obtained as an intermittent chemical relaxation formula in large deformation,

$$f(t)/f(0) = \exp \cdot \left(A \cdot e^{-\frac{E_f}{\alpha}} \cdot t\right) \qquad [3]$$

here, E_f indicates the slopes which will be different with the kind of polymers in higher α. The intermittent stress relaxation curves in large deformation depend upon the α keeping exponential linearity also depending upon temperature similar to that continuous stress relaxation.

From the results in fig. 10, the slopes are independent of temperature in higher α and *Arrhenius* formula can be expressed as eq. [4].

$$k_T = A \cdot e^{-\frac{E}{RT}}. \qquad [4]$$

Eq. [4] can also be substituted as eq. [5], where T_0 is adequate temperature.

$$k_{T_0} = A \cdot e^{-\frac{E}{RT_0}}. \qquad [5]$$

From eqs. [4] and [5], eq. [6] can be obtained.

$$\ln \frac{k_T}{k_{T_0}} = \frac{E}{R}\left(\frac{1}{T_0} - \frac{1}{T}\right). \qquad [6]$$

By substituting eq. [2] into eq. [6], eq. [7] can be obtained.

$$\ln k_T = \frac{E}{R}\left(\frac{1}{T_0} - \frac{1}{T}\right) + \ln A - \frac{E_f}{\alpha_T}. \qquad [7]$$

On the other hand, changing eq. [1] and rearranging for T, eq. [8] will be shown as follows.

$$k_T = [\ln(f(0)/f(t))]/t. \qquad [8]$$

By substituting eq. [8] into eq. [7], the following formula [9] is obtained.

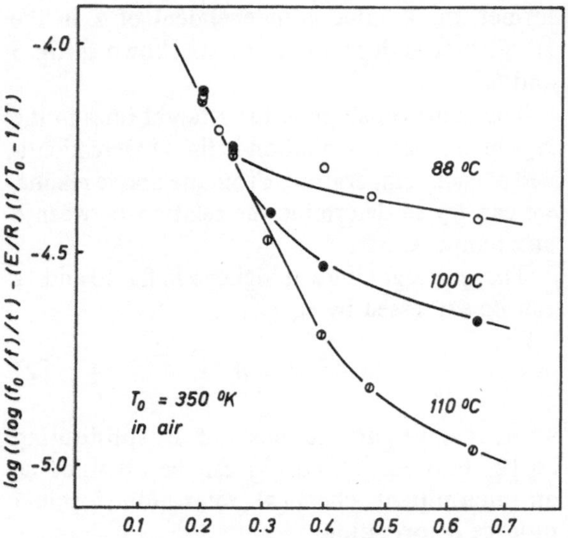

Fig. 12. Relation between the experimental data and eq. [9],
$\ln\{\ln(f(0)/f(t))/t\} - \frac{E}{R}\left(\frac{1}{T_0} - \frac{1}{T}\right) = \ln A - \frac{E_f}{\alpha_T}$ for various temperatures

$$\ln[\ln(f(0)/f(t))/t] - \frac{E}{R}\left(\frac{1}{T_0} - \frac{1}{T}\right) = \ln A - \frac{E_f}{\alpha_T}. \qquad [9]$$

If the left hand side term will be plotted against $1/\alpha$, the universal straight line which is independent of relative stress $f(t)/f(0)$, time t, temperature T, and extension ratio α can be obtained in the region of large deformation.

This was shown to be right by the universal straight line indicated in the region of the left upper side of fig. 12, using the experimental results.

Fig. 13 shows the relation between $\log k$ and $1/T$, or *Arrhenius* plot for different extension ratio α. In small deformation, as the value of k is independent of α, just one straight line can be obtained as shown in fig. 13. In large deformation, the straight lines which slopes are dependent on α can be obtained and in the region of approximately $\alpha > 4.0$, the slopes of the straight lines become constant. It is interesting that when

Fig. 13. Relation between $\log k$ and $1/T$ for various extension ratio α

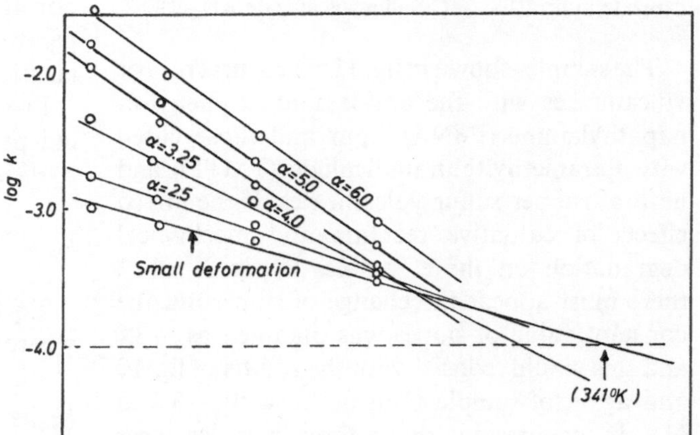

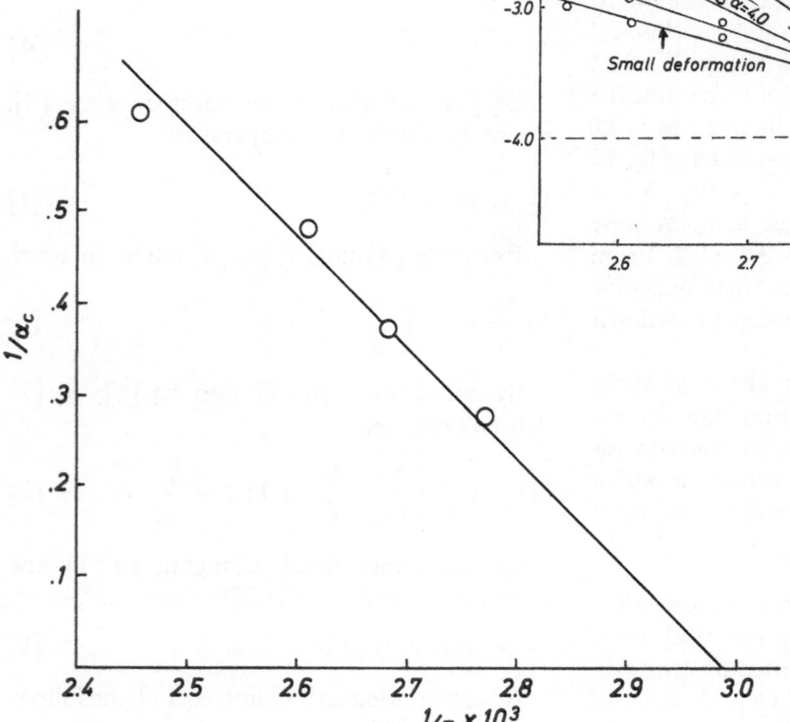

Fig. 14. Relation between $1/\alpha_c$ and $1/T$

$\log k$ is plotted against $1/T$ in fig. 13, the tendency of change of the slopes is consistent with that between E_{act} and α in fig. 8. On the other hand, no nonlinear regions are seen to exist in the range of temperatures where $\log k \leqq -0.4$ from fig. 10. This critical temperature is easily calculated from fig. 13 by obtaining the crossing point between the straight line for small deformation and the dotted line of $\log k = -0.4$, and the critical temperature is found to be 341 °K (68 °C). Similarly the relation between the critical extension ratio α_c and the temperature measured is obtained from fig. 10 as shown in fig. 14. By extrapolating the straight line in fig. 14, approximately 340 °K was found on the abscissa where $1/\alpha_c = 0$.

This result is consistent with those in fig. 13, and it was reaffirmed that no chemical and mechanical degradations occur theoretically lower than the approximate value of 68 °C indifferent of α.

From usual continuous chemical stress relaxation curves as indicated in fig. 15, it is clearly seen that chemical reaction will begin to occur from approximately 68 °C.

Then it will be concluded that the mechanochemical degradation will be stimulated and accelerated by the oxidative chemical reaction in the repeated chemical stress relaxation.

It will be noticed that $k = 0$ for any extension ratio of α lower than 68 °C in fig. 16 are confined to natural rubber vulcanizates without fillers. So it is necessary to study on the other materials whether the same tendency will be obtained or not.

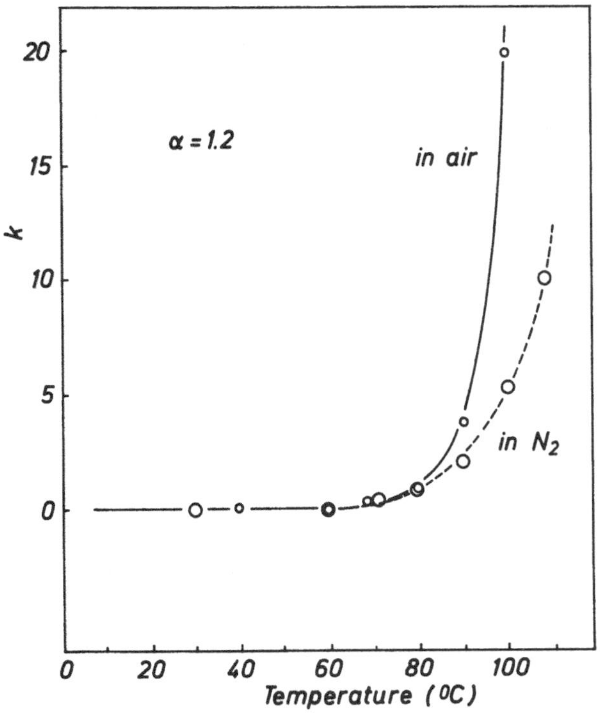

Fig. 15. Relation between k and temperature for continuous chemical stress relaxation

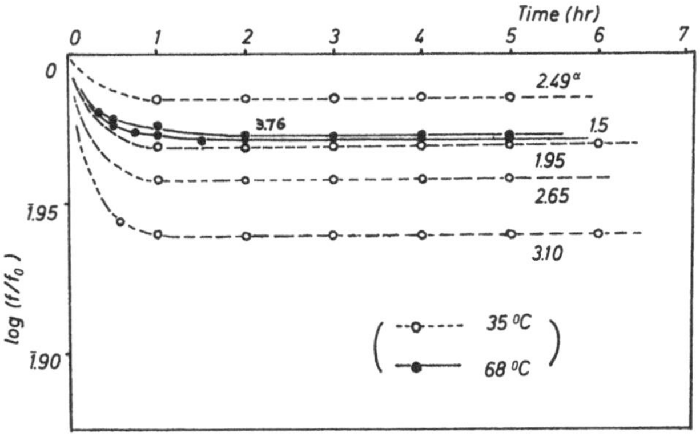

Fig. 16. $k = 0$ (no degradation occurs) for any extension ratio α at temperatures lower than 68 °C

Summary

In this investigation, the repeated chemical stress relaxation measurements were carried out to observe the relaxation behaviors at large deformation.

It was found that the repeated chemical stress relaxation curves were affected by both temperature measured and extension ratio of rubber. It was suggested from the results obtained that temperature and mechanical stimulus have a similar effect on the stress relaxation curves. Thus we propose the following *Arrhenius* type eq. [1] for high extension ratios.

$$f(t)/f(0) = \exp\left(A \cdot e^{\frac{E_f}{\alpha}} \cdot t\right). \qquad [1]$$

Where, α is the extension ratio, and A, and E_f are the constants determined experimentally. On the other hand, from eq. [1] and usual *Arrhenius* equation, the universal eq. [2] for the extension ratio and the temperature in large deformation was derived as follows,

$$\ln\{(\ln f(0)/f(t))/t\} - E/R(1/T_0 - 1/T) = \ln A - E_f/\alpha \qquad [2]$$

where, T_0 is adequate temperature. The curves obtained for different temperatures and extensions were very well consistent with those by eq. [2] in large deformation.

References

1) *Kusano, T., Y. Sutoh,* and *K. Murakami,* Paper presented at 21th Annual Meeting of the Society of Polymer Science of Japan, Tokyo (1972).

2) *Kusano, T.* and *K. Murakami*, Bull. Chem. Res. Ins. Nonaqueous Solu., Tohoku Univ. (Japan), **20**, 199 (1970).

3) *Sawaragi, G.* and *M. Fukuda*, Testing Materials **9**, 293 (1960).

4) *Fujimoto, K.*, Kobunshi **12**, 232 (1963).

5) *Bartenev, G. M.* and *F. A. Galil-Ogly*, Rubber Chem. Technol. **29**, 2504 (1956).

6) *Kusano, T.*, *Y. Suzuki*, and *K. Murakami*, J. Appl. Polymer Sci. **15**, 2453 (1971).

7) *Frenkel, J.*, Acta Physicochem.(URSS) **19**, 51 (1944).

8) *Portter, W. D.* and *G. D. Scott*, European Polymer J. **7**, 489 (1971).

9) *Tobolsky, A. V.*, *I. B. Prettyman*, and *J. H. Dillon*, J. Appl. Phys. **15**, 380 (1944).

Authors' address:

K. Murakami and *T. Kusano*
Tohoku University
4-14-17 Nishi Azaku
Minato-Ku, Tokyo (Japan)

Rheol. Acta **13**, 135–138 (1974)

From the Department of Polymer Chemistry, Kyoto University, Kyoto (Japan)

Measurements of viscoelasticity of polymer films and fabrics by the vibrating-reed method at very low frequencies

By S. Onogi, T. Kondo*), *and* Y. Tabata

With 3 figures and 3 tables

(Received October 27, 1972)

Introduction

The vibrating-reed method is one of the simplest methods for measuring the elasticity or viscoelasticity of solid materials; it has been applied to various kinds of materials including fibers, films, papers, and fabrics (1–13). This method is usually used in the audiofrequency range, but in many cases measurements at very low frequencies are desired. In making such measurements very long, and hence very heavy, sample reeds must be employed to obtain the resonance frequencies or resonance curves from which the viscoelastic functions, such as the storage and loss moduli, dynamic viscosity, and mechanical loss tangent can be evaluated.

One of the very important assumptions upon which the vibrating-reed method is based is that the mass of the reed is negligibly small as compared with that of the vibrator and that hence the effect of gravity can be ignored. However, the use of long and heavy reeds at very low frequencies tends to conflict with this assumption. In such cases, the effect of gravity should be taken into consideration.

The present paper is intended to offer some experimental evidence for the existence of the effect of gravity and to present a new theory of the vibrating-reed method when it is employed at very low frequencies, in which the effect of gravity is considered.

Theoretical

The transverse vibration of a viscoelastic beam or reed having a transverse dimension which is small in comparison with both its length and the wave length can be expressed by

$$\varrho \frac{\partial^2 y}{\partial t^2} + E' \varkappa^2 \frac{\partial^4 y}{\partial x^4} + \eta' \varkappa^2 \frac{\partial^5 y}{\partial x^4 \partial t} = 0 \qquad [1]$$

where the axis of the beam in the position of equilibrium is presented as x and the axis perpendicular to it and parallel to the plane of symmetry as y (14, 15). Here t = time, ϱ = density, E' = modulus of elasticity (bending modulus), η' = coefficient of viscosity, and $\varkappa^2$ = moment of inertia of the cross section around the neutral axis per cross-sectional area. This equation was solved by *Horio* and *Onogi* (16) for the case of forced vibration to give

$$E' = \frac{4\pi^2 \varrho l^4}{a_0^4 \varkappa^2} \left[\nu_r^2 + \frac{1}{8} (\Delta\nu)^2 \right] \qquad [2]$$

and

$$E'' = \frac{4\pi^2 \varrho l^4}{a_0^4 \varkappa^2} \nu_r \cdot \Delta\nu \qquad [3]$$

or

$$\eta' = \frac{2\pi \varrho l^4}{a_0^4 \varkappa^2} \cdot \Delta\nu \qquad [4]$$

where E'' is the loss modulus or the imaginary part of the complex modulus, ν_r is the resonance frequency, $\Delta\nu$ is the band width, l is the effective length, and

$$a_0 = 1.875,\ 4.694,\ 7.855,\ \ldots \qquad [5]$$

When such a viscoelastic beam is held vertically and affected by gravity, eq. [1] should contain another term for the effect of gravity; in that case it becomes too complex to be solved even in an approximate manner. Therefore, we consider an elastic beam affected by gravity. The equation of motion for the beam is given by

$$E' \varkappa^2 \frac{\partial^4 y}{\partial x^4} - \frac{\partial}{\partial x}\left(T \frac{\partial y}{\partial x}\right) + \varrho \frac{\partial^2 y}{\partial t^2} = 0 \qquad [6]$$

where

$$T = \varrho g (l - x) \qquad [7]$$

and g is the acceleration of gravity.

*) Present address: Central Research Laboratories, Unitika Ltd., Uji, Kyoto Pref. (Japan).

Assuming that y is presented by $Ye^{i\omega t}$ in the stationary state of motion, i is being $(-1)^{1/2}$, we obtain

$$\frac{d^4Y}{dx^4} - \frac{\varrho g}{E'\varkappa^2}(l-x)\frac{d^2Y}{dx^2} + \frac{\varrho g}{E'\varkappa^2}\frac{dY}{dx} - \frac{\varrho\omega^2}{E'\varkappa^2}Y = 0. \qquad [8]$$

This equation cannot be solved rigorously. Therefore, we employ the following approximation:

$$\frac{d^4Y}{dx^4} = \frac{\alpha_0^4}{l^4}Y \qquad [9]$$

which is the case where no effect of gravity exists. Then, eq. [8] becomes

$$g(l-x)\frac{d^2Y}{dx^2} - g\frac{dY}{dx} + \left(\omega^2 - \frac{a_0^4 E'\varkappa^2}{\varrho l^4}\right)Y = 0. \qquad [10]$$

When we put

$$z = l - x$$

and

$$u = 2\left(\frac{z}{g}\right)^{1/2}\left(\omega^2 - \frac{a_0^4 E'\varkappa^2}{\varrho l^4}\right)^{1/2}$$

eq. [10] becomes

$$\frac{d^2Y}{du^2} + \frac{1}{u}\frac{dY}{du} + Y = 0.$$

This equation is *Bessel*s differential equation and its solution can be given by a *Bessel* function

$$Y = \sum_{m=0}^{\infty}\frac{(-1)^m}{m!^2}\left(\frac{l}{g}\right)^m\left(\omega^2 - \frac{a_0^4 E'\varkappa^2}{\varrho l^4}\right)^m = 0.$$

The numerical calculation of this function gives

$$\frac{l}{g}\left(\omega^2 - \frac{a_0^4 E'\varkappa^2}{\varrho l^4}\right) \cong 1.45. \qquad [11]$$

And hence we obtain the following equation for the resonance frequency ν_r

$$\nu_r^2 l \cong E'\frac{a_0^4\varkappa^2}{4\pi^2\varrho l^3} + 36\ [\text{cm/sec}^2] \qquad [12]$$

or

$$E' \cong \frac{4\pi^2\varrho l^3}{a_0^4\varkappa^2}(\nu_r^2 l - 36). \qquad [13]$$

Eq. [12] is valid only when l is large. When l is smaller, we obtain

$$E' \cong \frac{4\pi^2\varrho l^3}{a_0^4\varkappa^2}(\nu_r^2 l - 50). \qquad [14]$$

Experimental

By means of a vibrating-reed apparatus, resonance curves for various materials were measured in a frequency range between about 1.0 and 20 Hz. The measuring temperature and humidity were kept constant at 25 °C and 65% R. H.

In tables 1–3 are listed the thicknesses and densities of the samples used in this study.

Table 1. The thickness d and density ϱ of aluminium foils, Al-1 and Al-2, and nichrome wire, Ni—Cr

Sample	d × 10² (cm)	ϱ (g/cm³)
Al-1	0.215	2.69
Al-2	0.976	2.69
Ni—Cr	1.730¹)	8.40

¹) d = radius

Table 2. The thickness d and density ϱ of polyvinyl chloride (PVC) and polyethylene (PE-1 and PE-2) films

Sample	$d \times 10^2$ (cm)	ϱ (g/cm³)
PVC	1.960	1.260
PE-1	0.526	0.899
PE-2	0.740	0.915

Table 3. The thickness d and apparent density ϱ of fabric samples. The S series are blends of wool and acrylic fibers, and the A series are blends of polyester and acrylic fibers

Sample	Blending ratio, Acrylic (wt.%)	$d \times 10^2$ (cm)	ϱ (g/cm³)
S_1	0	3.95	0.621
S_2	25	3.85	0.567
S_3	50	3.91	0.555
S_4	75	3.93	0.541
S_5	100	4.04	0.520
A_1	0	2.95	0.575
A_2	50	3.50	0.515
A_3	100	3.57	0.498

Results and discussion

When the band width in eq. [2] is negligibly small and the modulus E' is constant independent of frequency in the frequency range covered, the resonance frequency ν_r plotted against $1/l^2$ should give a straight line passing through the origin. This type of plot is usually made in order to determine the modulus from the slope m of the straight line by the following equation (11):

$$E' = \frac{m^2 \varrho}{(0.162 d)^2}$$

for the fundamental oscillation,
where $a_0 = 1.875$ [15]

where d is the thickness of a beam having rectangular cross section.

In the usual measurements in the audio-frequency range, ν_r vs. $1/l^2$ curves for various materials including fibers, papers, fabrics, films, and sheets are linear. But eq. [2] is no longer valid at frequencies as low as 1–20 Hz, and ν_r plotted against $1/l^2$ cannot give a straight line passing through the origin, as is shown in fig. 1 for the fabric samples A_1, A_2, and A_3, as an example.

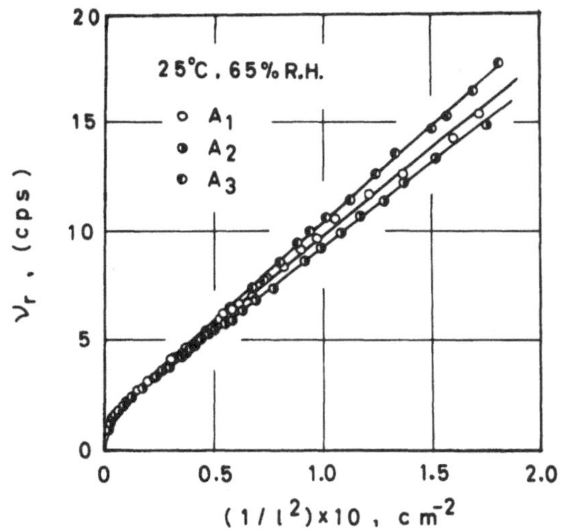

Fig. 1. ν_r vs. $1/l^2$ for the fabric samples A_1, A_2, and A_3.

On the other hand when eq. [13], including the effect of gravity, is valid $\nu_r^2 l$ plotted against $1/l^3$ should give a straight line having the constant intercept of 36 independent of the nature of the material. Fig. 2 shows such plots for the various kinds materials mentioned above. Although it is not clear from this figure, the plots are not completely linear, indicating that eq. [13] holds only approximately. Nevertheless, the plots have the intercept of 36 as is required by eq. [13]. Furthermore, the extrapolation of a linear portion of the plot at higher $1/l^3$ or smaller l values gives the intercept of about 50, which is equal to 50 in eq. [14].

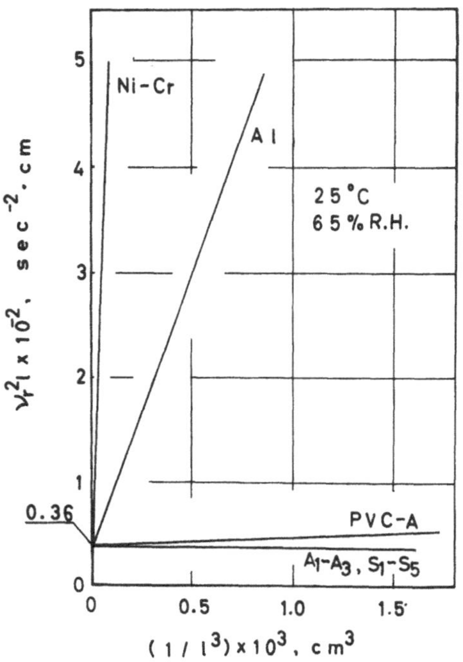

Fig. 2. $\nu_r^2 l$ vs. $1/l^3$ for various samples

Some results were obtained by experiments in a vacuum which gave resonance frequencies somewhat higher than those obtained in air, because of the absence of the frictional resistance of air. Furthermore, we notice that some curves of $\nu_r^2 l$ vs. $1/l^3$ are almost linear. It is noteworthy that the intercepts of all the curves are 36, which is consistent with eq. [13]. It is also evident from fig. 2 that the slope of the curves is different from material to material, because the moduli of elasticity for these materials are quite different. From the slopes m' of these curves, the moduli E' can be evaluated using

$$E' = \frac{4\pi^2 \varrho}{a_0^4 \varkappa^2} m'. \quad [16]$$

An example of E' thus determined for the sample S_1 is shown by the open circles and solid line in fig. 3. As is clear from this figure, E' is almost constant, independent of the sample length l or the measuring frequency. But E' evaluated from the slope m of the ν_r vs. $1/l^2$ plot using eq. [15] increase very rapidly with increasing l, as is shown by half closed circles and solid line in the same figure. The modulus determined by eq. [13] seems to be more reasonable, because the rapid increase of the modulus in a narrow frequency range is impossible.

Next, the determination of the loss modulus E'' may be discussed. When we employ eq. [3] to determine E'', the evaluated value of E'' increases very unreasonably with increasing l, as can be seen from the typical example for the sample S_1 shown in fig. 3 (half closed circles and broken line). This indicates that the effect of gravity should be considered here again.

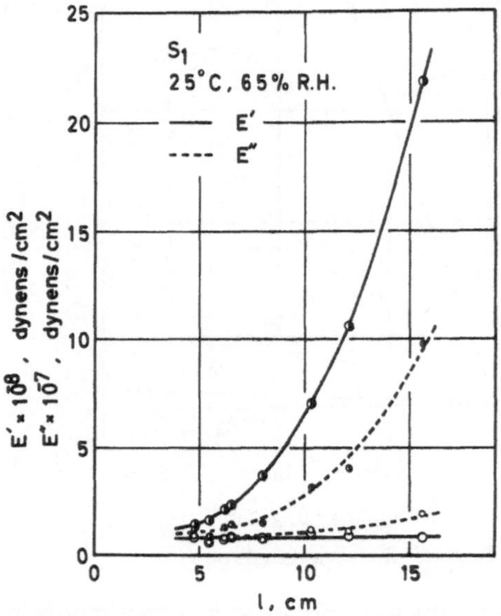

Fig. 3. E' and E'' of S_1 determined from eq. [2] (half closed circles and solid line), eq. [13] (open circles and solid line), *eq.* [3] (half closed circles and broken line), and eq. [17] (open circles and broken line)

The only difference between eqs. [2] and [13] for E' was that v^2 in eq. [2] was replaced with $(v_r^2 - 36/l)$ in eq. [13], neglecting Δv in eq. [2]. Therefore, when we replace v_r in eq. [3] with $(v_r^2 - 36/l)^{1/2}$, we obtain

$$E'' = \frac{4\pi^2 \varrho l^4}{a_0^4 \varkappa^2} (v_r^2 - 36/l)^{1/2} \Delta v. \qquad [17]$$

Using this equation, E'' for the sample S_1 was evaluated and compared with that from eq. [3] in fig. 3 (open circles and broken line). As is evident from this figure, E'' from eq. [17] increases only slightly with increasing l or decreasing frequency, indicating that eq. [17] is more reasonable than eq. [3].

Summarizing all these results, it can be concluded that eqs. [13] and [17] rather eqs. [2] and [3] should be employed to evaluate E' and E'' from v_r and Δv measured at very low frequencies, though eq. [13] and [17] are not completely rigorous.

Acknowledgement

This study was supported by a grant from the Japan Research Association for Textile End-Uses and the Japan Chemical Fibres Association, for which the authors are grateful.

Summary

Measurements of the elasticity or viscoelasticity of various materials by the vibrating-reed method at very low frequencies require the use of long and heavy samples. The effect of gravity on these samples has been considered theoretically, and new equations for the storage and loss moduli have been derived. Vibrating-reed measurements were carried out with several kinds of materials at frequencies ranging from about 1–20 Hz. The experimental results indicate that the above equations are satisfactory in practice.

References

1) *Lochner, J. P. A.*, J. Text. Inst. **40**, T220 (1949).
2) *Ballou, J. W.* and *J. C. Smith*, J. Appl. Phys. **20**, 493 (1949).
3) *Hillier, K. W.*, Proc. Phys. Soc. (London), **B64** (1951).
4) *Horio, M., S. Onogi, C. Nakayama,* and *K. Yamamoto*, J. Appl. Phys. **22**, 966 (1951).
5) *Onogi, S., S. Ando,* and *I. Suginaka*, Sen-i Gakkaishi **9**, 617 (1953); **10**, 32 (1954); **10**, 390 (1954).
6) *Kärrholm, E. M.* and *B. Schröder*, Text. Res. J. **23**, 207 (1953).
7) *Guthrie, J. C., D. H. Morton*, and *P. H. Oliver*, J. Text. Inst. **45**, T912 (1954).
8) *Onogi, S.* and *K. Ui*, J. Colloid Sci. **11**, 214 (1956).
9) *Merz, E., L. Nielsen*, and *R. Buchdahl*, J. Polymer Sci. **4**, 605 (1949).
10) *Nolle, A. W.*, J. Polymer Sci. **5**, 1 (1950).
11) *Oliver, D. A.*, Phil. Mag. **14**, 318 (1932).
12) *Horio, M.* and *S. Onogi*, J. Appl. Phys. **22**, 971 (1951).
13) *Onogi, S., S. Ando,* and *T. Yamamoto*, Kogyo Kagaku Zasshi, **57**, 251, 253 (1954).
14) *Sezawa, K.*, Bull. Earthquak Researches Inst. (Japan) **3**, 50 (1927).
15) *Suyehiro, K.*, Proc. Imp. Academy (Tokyo) **4**, 263 (1928).
16) *Horio, M.* and *S. Onogi*, J. Appl. Phys. **22**, 977 (1951).

Authors' address:

Dr. *Shigeharu Onogi* and Dr. *Yoshishige Tabata*
Dept. of Polymer Chemistry
Kyoto University
Kyoto (Japan)

Rheol. Acta 13, 139–148 (1974)

From the Industrial Research Laboratories, Kao Soap Company, Wakayama (Japan)

Effects of pseudo-network on physical properties of polyurethane

By *Y. Kazama* and *Y. Suzuki*

With 11 figures and 1 table

(Received October 27, 1972)

1. Introduction

In the styrene-isoprene type of block copolymer, the glassy phase of agglomerated polystyrene block is incompatible with the polyisoprene block, causing the bulk polymer to show rubber-like elasticity without incorporation of chemical crosslinking (1–4).

The rubber-like elasticity of polyurethane elastomers has been elucidated according to the concept of segmented block copolymer, in which highly hydrogen-bonded hard segment is dispersed in an amorphous matrix of the soft segment (5–9).

Although an aromatic association was accounted for the unusually high modulus and long rubbery plateau in the Estane type of block copolyurethane because of no evidence of crystallinity from X-ray diffraction study, little has been known in detail to substantiate this association (10).

Meanwhile, an emphasis was made on the intersegmental affinity in the investigation of the softening temperature of segmented polyurethanes (11, 12).

This was quite suggestive and instructive to understand the nature of the rubber-like elasticity of thermoplastic elastomers in conjunction with the concept of segmented block copolymer.

However, the essential feature should be disclosed in terms of chemical structure before a reliable interpretation of this nature can be given.

The present investigation is concerned with the effects of pseudo-networks on the physical properties of polyurethanes through the study of the temperature dependency of the infra-red spectral behavior, especially in the far infrared (FIR) region, which enables us to visualize the chemical nature of these pseudo-networks since a relatively small vibrational energy allows the FIR stretching to be more sensitive against the physical condition of samples and their molecular orientation than in the mid infra-red (MIR) region.

Supplemental discussions of results are attempted in combination with dynamic mechanical and X-ray diffraction measurements.

2. Experimental

Materials

Poly-1,4-butyleneglycol-adipate with two terminal hydroxyl groups (PBA) or polytetramethyleneglycol (PTMG) were addition-polymerized with diisocyanates for soft polyurethanes. Number average molecular weight of diols was about 1000.

Butanediol (BD), and the equimolar mixture of polymer-diol and BD were used for hard-and co-polyurethane, respectively.

Diisocyanates used were 4,4′-diphenylmethane diisocyanate (MDI), 2,2/2,6-toluene diisocyanate (TDI), 1,6-hexamethylene diisocyanate (HMDI) and 1,3/1,4-xylylene diisocyanate (XDI).

IR measurement

A dry powdery mixture containing 0.5% by weight of polyurethane and KBr, prepared after removal of solvent from a mixture of DMF solution of the polyurethane and KBr powder, was molded into a tablet, which was annealed at 150 °C for 30 min under nitrogen stream, cooled and kept standing over-night at room temperature.

After confirming the disappearance of IR absorption at 1660 cm^{-1} for the complete removal of DMF solvent, the IR measurement was carried out on this tablet with modified *Hitachi-Perkin-Elmer* Grating Infra-red Spectrophotometer type 125, equipped with a specially designed thermostat and a temperature program controller, which allowed the measurement in the temperature range between −150 and 250 °C.

Supplemental measurements

Dynamic mechanical measurements were carried out on a strip specimen with Iwamoto Viscoelastic Spectrometer. X-ray diffraction measurements were made with X-ray diffractometer, Geigerflex D-1.

3. Results and discussion

3.1. Assignment of FIR absorption

The FIR spectra of four copolyurethanes are shown in fig. 1.

Polyurethanes with —NH—CO—O— structure should exhibit spectral similarity to amides. Thus, in the FIR, absorption band which appears between 640 and 660 cm^{-1} has been assigned to the amide band V of secondary amides arising from out-of-plane deformation vibration of the associated NH-group, $\gamma(NH)_{assoc.}$ (13, 14). Analogically, absorption band between 580 and 590 cm^{-1} may be regarded as similar to the amide band VI arising from out-of-plane deformation vibration of the associated C=O group, $\gamma(C{=}O)_{assoc.}$.

Absorption band appearing between 700 and 760 cm^{-1} arises from a rocking mode vibration of the $(CH_2)_n$ group, $r(CH_2)_n$, and there has been a suggestion of a gradual shift towards higher frequency as the number of group is reduced (15). According to this suggestion, assignments are attempted using the data in fig. 1. In the fig. 2 is plotted the relation between wave numbers observed and possible numbers of the methylene groups in the polyurethanes. By the number of methylene group of unity is meant the methylene group in between two phenyl groups for MDI series or the one in between phenyl group and carbamoyl group for XDI series. Attempted are also the assignments of *Corish*s data (16) for the soft polyurethanes of 1,5-naphthalene diisocyanate with poly-ethylene-glycol-adipate and -sebacate, shown by open stars in this figure.

These polyurethanes give a linear relationship so far in contrast to *Shimanouchi*s data for octane isomers referred as a dashed curve (17).

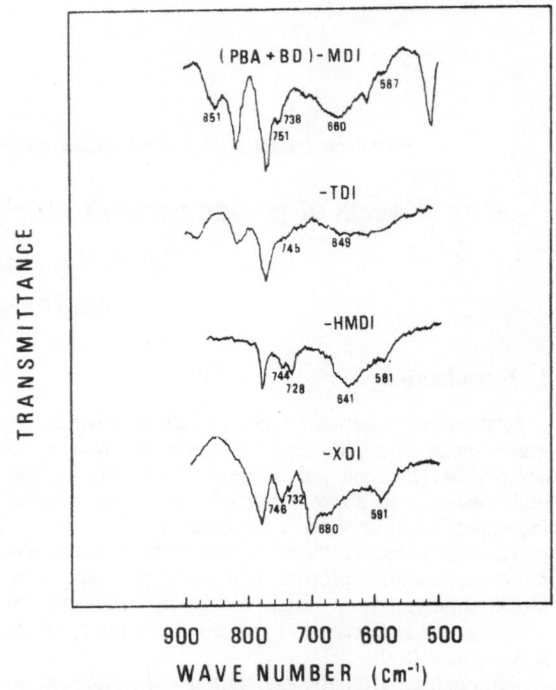

Fig. 1. FIR spectra of polyurethanes based on various diisocyanates and equimolar mixture of poly-1,4-butylene-glycoladipate and 1,4-butanediol, at $37 \pm 2\,°C$

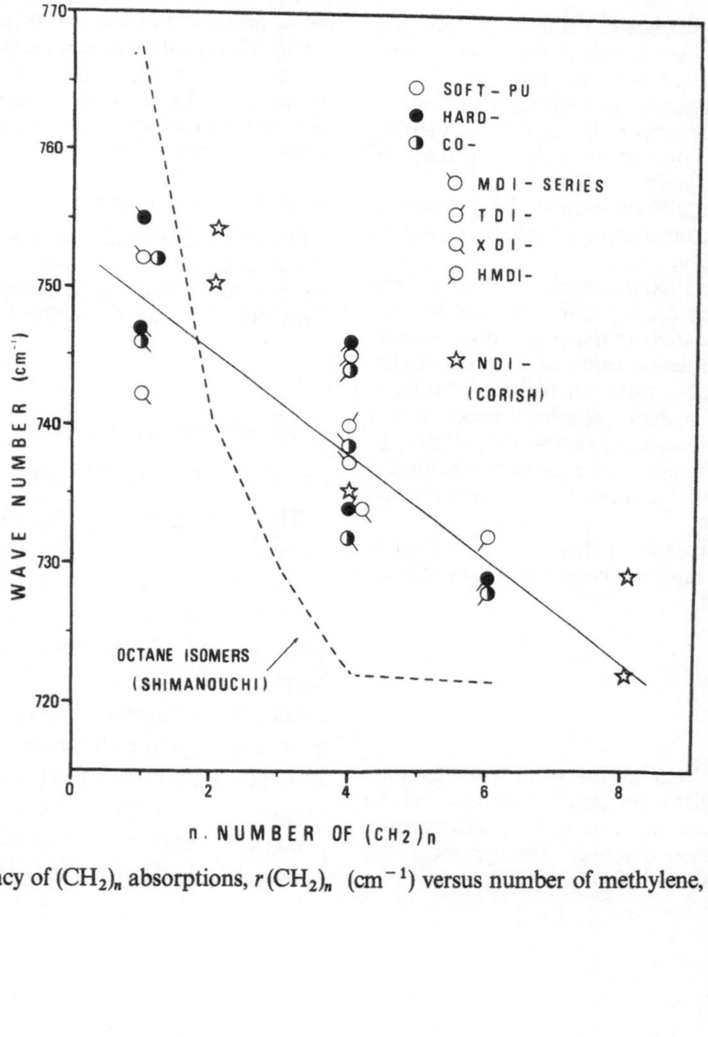

Fig. 2. Plot of frequency of $(CH_2)_n$ absorptions, $r(CH_2)_n$ (cm^{-1}) versus number of methylene, n

In the polyurethane of MDI series, characteristic absorption bands at 850 and 860 cm^{-1} give rise to a temperature dependency, which may well be ascribed to out-of-plane bending vibration $\gamma(CH)_{arom.ring}$ of two adjacent hydrogen atoms on the benzene ring of diphenylmethane.

FIR wave numbers and assignments are listed in table 1.

According to X-ray diffraction study presented by *Kilian* and *Jenckel* (18), BD–HMDI has two crystalline features; one is hydrogen bond crystal with (200) interference, and the other is *van der Waals* crystal with (002) interference transverse to (200) plane. As is seen from fig. 3, cited from the reference (18), these crystals can be characterised by their temperature behavior, i.e., the crystallinity of hydrogen bond crystal, α_H, stays almost unchanged until rapid decrease at the carbonamide dissociation temperature region at 170 up to 187 °C, while that of *van der Waals* crystal, α_V, decreases rather simply with increasing temperature, seemingly extrapolating to zero crystallinity near 187 °C.

From the mode of infrared vibration concerning hydrogen atom in associated NH— group, $\gamma(NH)_{assoc.}$ and $\nu(NH)_{assoc.}$ should be correspond to α_V and α_H, respectively, since the direction of out-of-plane vibration coincides with the direction transverse to the symmetrical lattice-plane of hydrogen bond in which stretching vibration oscillates perpendicular to the polymer chain axis. For this reason, plots of optical density versus temperature for $\gamma(NH)_{assoc.}$ and $\nu(NH)_{assoc.}$ resemble temperature plots of α_V and α_H, respectively. This is illustrated in figs. 3 and 4, for BD–HMDI. The optical density of $\nu(NH)_{assoc.}$ and crystallinity, α_H, of BD–HMDI stay almost unchanged until 170 °C, irrespective of gradual weakening of hydrogen bond until 162 °C, as will be shown in fig. 10. While, optical density of $\gamma(NH)_{assoc.}$ decreases as simply as α_V did with increasing temperature and zero optical density is reached at 191 °C. Optical density of $\gamma(NH)_{assoc.}$ may consequently be regarded as a measure of the amount of *van der Waals* interaction arising from the associated carbonamide group.

3.2. Temperature dependency of pseudo-networks

General feature

Four series of temperature dependent FIR spectra of soft- and hard-, as well as co-poly-

Table 1. Wave numbers and assignments of FIR-absorptions of various Polyurethanes

Diisocyanate	MDI				TDI				HMDI					XDI			
Diol / Mode of vibration	PBA	BD	PBA + BD	PTMG + BD	PBA	BD	PBA + BD	PTMG + BD	PBA	BD	PBA + BD	PTMG	PTMG + BD	PBA	BD	PBA + BD	PTMG + BD
$\gamma(C{=}O)_{assoc.}$	586	586	587	584	586	–	–	–	583	583	581	–	584	590	588	591	592
$\gamma(N{-}H)_{assoc.}$	660	661	661	667	–	–	650	–	631	639	641	667	646	–	–	680	659
$r(CH_2)_1$	752	755	751	750	–	–	–	–	–	–	–	–	–	742	747	746	750
$r(CH_2)_4$	737	–	738	–	740	734	745	748	745	746	744	752	744	734	–	732	–
$r(CH_2)_6$	–	–	–	–	–	–	–	–	732	729	728	–	729	–	–	–	–
$\gamma(CH)_{arom.ring}$	852	854	851	850	–	–	–	–	–	–	–	–	–	–	–	–	–
	–	861	–	–	–	–	–	–	–	–	–	–	–	–	–	–	–

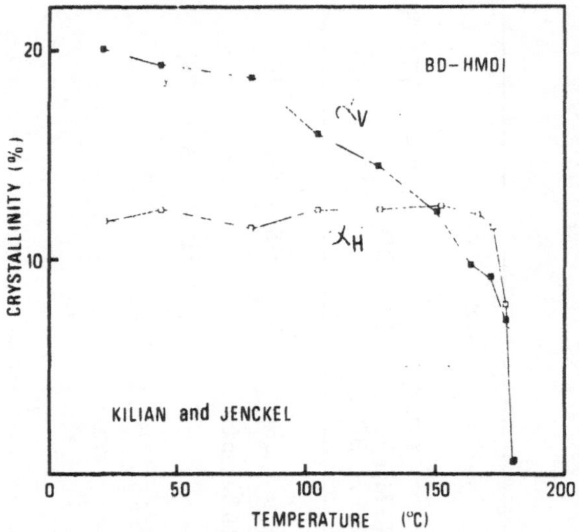

Fig. 3. Temperature-dependency of crystallinity of hydrogen bond-, (α_H) and *van der Waals*-crystal, (α_V), from X-ray scattering measurements by *H. G. Kilian* and *E. Jenckel* (18)

esterurethanes are shown in figs. 5 and 6. In fig. 7 are given the same data for co-polyether-urethanes from PTMG.

The wave numbers of absorption bands with marked temperature dependency appear on the figures. The areal concentrations of polyurethane in KBr tablet are also shown in the figures using the unit of mole-NH group per irradiated section.

Dissociations originated from associated carbonamide group, and from its neighboring methylene groups are seen, in general, to complete around the melting point of polyurethanes.

In MDI-series, a unique aromatic association can be observed through γ(CH) for aromatic ring at 850 and 860 cm^{-1} and r(CH$_2$) for CH$_2$-group in between two phenyl groups; this can not be seen in other phenyl-group-containing polyurethanes.

The soft polyurethanes are characterized by strong $\gamma(C=O)_{assoc.}$ and $r(CH_2)_4$, while strong $\gamma(NH)_{assoc.}$ is the characteristic of the hard polyurethanes.

As for the copolyurethanes, the patterns of the spectra are strongly affected by the hard segments rather than the soft. The spectral intensity increases in the order of the series, TDI, XDI$\approx$HMDI, MDI. The replacement of polyester (PBA) soft segment by polyether (PTMG) markedly increases the intensity with MDI and HMDI series; whereas, it decreases with XDI series and no difference is found with TDI series.

HMDI series

Polyurethanes of HMDI series show two aliphatic associations, one of which originates from $(CH_2)_6$ group for the absorption at 728 to 732 cm^{-1}, and the other originates from $(CH_2)_4$ group for 744–746 cm^{-1}. Temperature dependent behavior of $(CH_2)_6$ group for PBA–HMDI, BD–HMDI and (PBA + BD)–HMDI are quite different from each other, and it reflects spatial

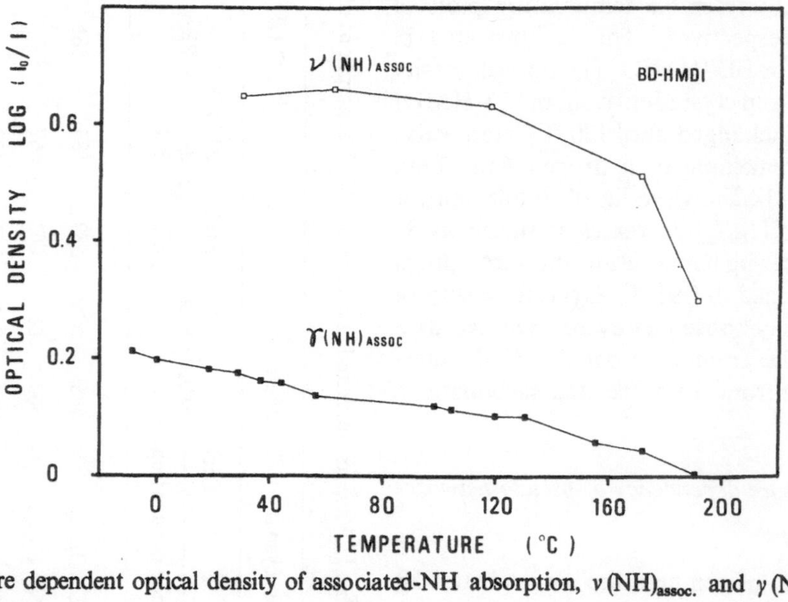

Fig. 4. Temperature dependent optical density of associated-NH absorption, $\nu(NH)_{assoc.}$ and $\gamma(NH)_{assoc.}$ for BD–HMDI

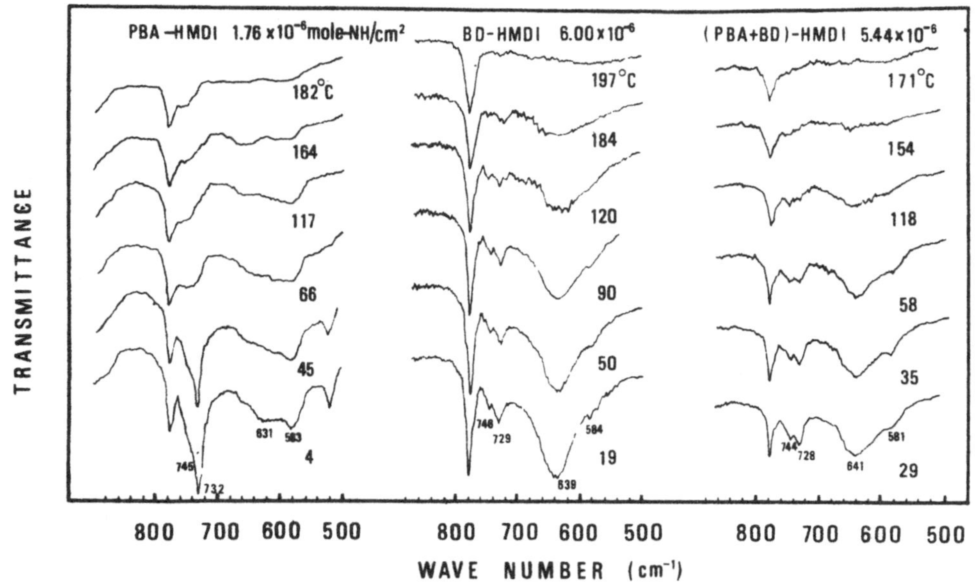

Fig. 5. Temperature dependent FIR spectra for soft-, hard- and co-polyurethanes of HMDI series

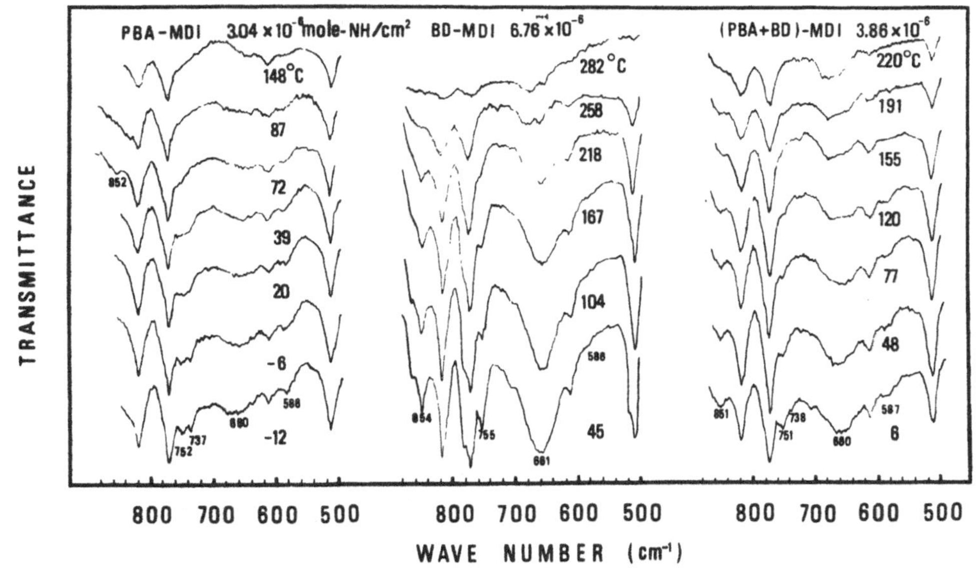

Fig. 6. Temperature dependent FIR spectra for soft-, hard- and co-polyurethanes of MDI series

molecular arrangements in the associated domains.

In PBA–HMDI, absorption band for $r(CH_2)_6$ at 732 cm^{-1} vanishes at between 45–66 °C. While, a slight indication of absorption for $r(CH_2)_4$ observed on the shoulder at around 745 cm^{-1} vanishes together at this temperature. Around this temperature region, sample became transparent from opaque, and DSC thermogram gave an endotherm peak for melting (at 36 °C), while viscoelastic modulus curve showing a steep decay in $\log E'$ at 50 °C, characteristic of the melting of crystalline polymers. As can be seen from the absorption bands for $\gamma(NH)_{assoc}$ and $\gamma(C{=}O)_{assoc.}$ at 634 and 583 cm^{-1} in fig. 5, carbonamide association still remains above 164 °C. In PBA–HMDI, therefore, $(CH_2)_6$ group from diisocyanate contribute predominantly to a hydrogen bond lattice-cell, and $(CH_2)_4$ group of polyester sequence, probably adjacent to a urethane linkage is suspected to be merely involved in the lattice-cell to a noticeable extent. Between 45–66 °C, predominantly associated $(CH_2)_6$ group, including small amount of $(CH_2)_4$

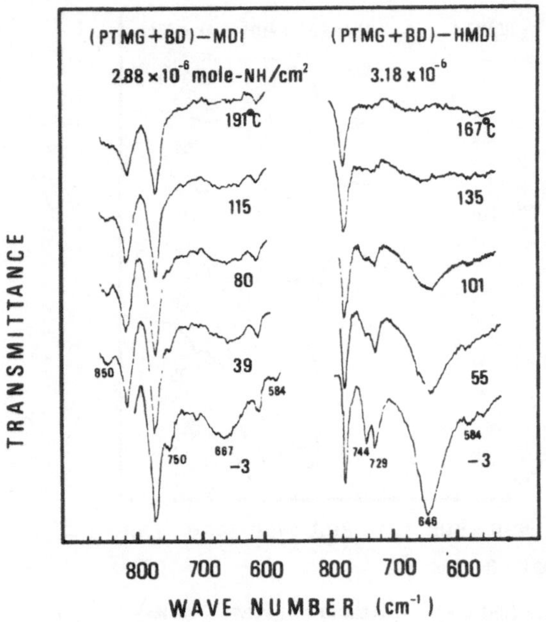

Fig. 7. Temperature dependent FIR spectra for copolyurethanes based on polytetramethylene glycol (PTMG)

group, in the lattice-cell dissociates to form an amorphous matrix containing hydrogen-bonded "bare" carbonamide group until its dissociation at above 164 °C.

In BD–HMDI, associated $(CH_2)_4$ and associated $(CH_2)_6$ go parallel with associated carbonamide group until its dissociation above 184 °C, and the FIR spectra become quite clear-cut. Therefore, in BD–HMDI, both of $(CH_2)_4$ and $(CH_2)_6$ groups are involved in the carbonamide association to form a highly ordered, rigid crystalline structure.

In the co-polyurethane, (PBA + BD)–HMDI, the dissociation of $(CH_2)_6$ at between 118 and 154 °C is followed by the dissociation of carbonamide group at around 171 °C which accompanies $(CH_2)_4$ group. Because of the presence of associated $(CH_2)_6$ absorption even in the higher temperature region than the melting point of associated $(CH_2)_6$ in the soft segment, the $(CH_2)_4$ and $(CH_2)_6$ groups should necessarily be assigned to the hard segment in the copolyurethane, and —$(CH_2)_6$NHCOO— group in the hard segment may be regarded as being solvated preferentially by the same group in the soft segment due to the similarity in local polarity and local high symmetry of each segment. The viscoelastic spectra of unique thermoplasticity for a polyurethane may be related to a plasticized matrix thus obtained through the relatively strong interaction between soft and hard segments in (PBA + BD)–HMDI. The complemental plot of logarithm of viscoelastic shift factor versus reciprocal temperature for (PBA + BD)–HMDI is not seriously affected by segregated segment and gives almost straight line with very large value of activation energy for segmental flow, as shown in fig. 8. Though time-temperature superposition was fairly difficult in this series, it seems to be sufficient enough to support the view that a highly internally-plasticized network exists in (PBA + BD)–HMDI. In contrast to HMDI series, (PBA + BD)–MDI, as shown in fig. 9, clearly shows first the contribution from segregated PBA–MDI at lower temperature region and the one from segregated BD-MDI at higher temperature region. It also shows the low activation energy of flow in comparison with (PBA + BD)–HMDI probably because of the segregation.

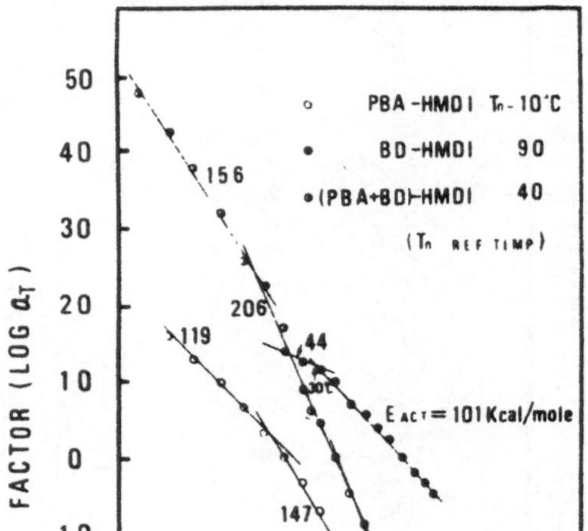

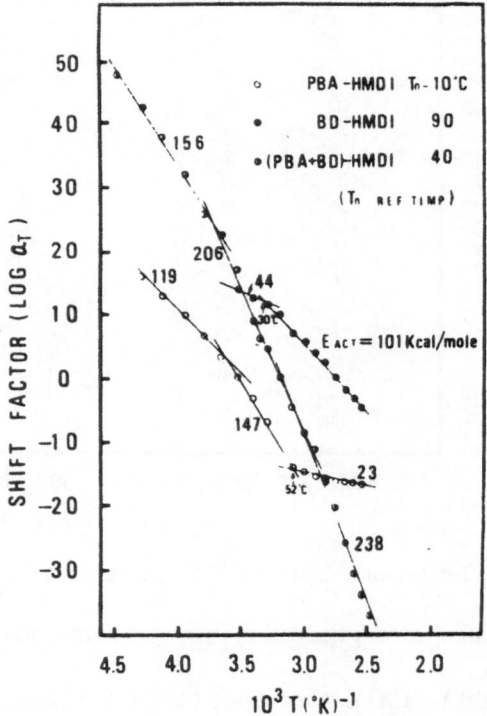

Fig. 8. *Arrhenius* plots of logarithm of viscoelastic shift factor for soft-, hard- and co-polyurethanes of HMDI series

Replacement of polyester soft segment by less polar polyether gives rise to much more clear-cut FIR spectra, in which associated $(CH_2)_6$ group of hard segment is not affected as was in the previous case. In consequence of the

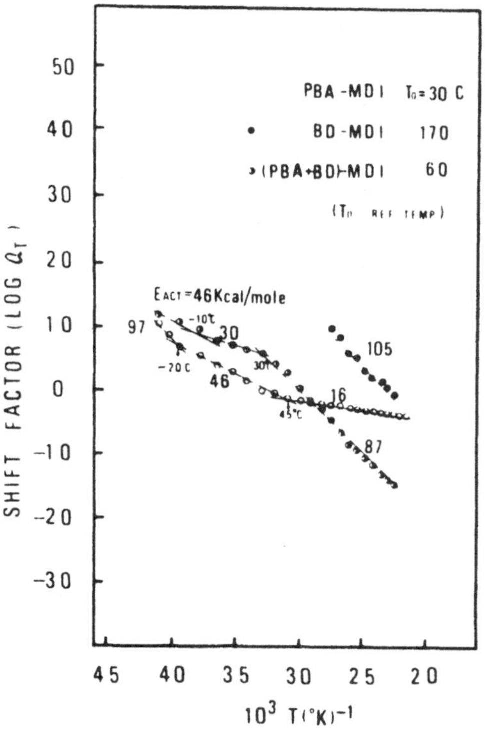

Fig. 9. *Arrhenius* plots of logarithm of viscoelastic shift factor for soft-, hard- and co-polyurethanes of MDI series

reduction of intersegmental interaction, a viscoelastic spectra becomes typical to thermoelastic materials.

Although an increase in heat distortion temperature associated perhaps with the onset of carbonamide dissociation has been reported by *Bonart* et al. (12) in the same replacement for segmented polyurethanes chain-extended with diamines, no remarkable difference in the depression from the dissociation temperature of BD–HMDI to that of copolyurethane can be observed in FIR spectra as follows: between 184 and 197 °C (174 °C by DSC) for BD–HMDI, between 154 and 171 °C (125 °C) for (PBA + BD)–HMDI and around 167 °C (127 °C) for (PTMG + BD)–HMDI.

As having noticed already with PBA–HMDI, the melting points observed by DSC do not coincide with the dissociation temperatures observed in FIR spectra.

Though PBA–HMDI, BD–HMDI and (PBA + BD)–HMDI have $\tan\delta$ maxima at their glass temperature, −50, 35 and −40 °C, respectively, none of which has been detectable directly by means of FIR spectra so far.

MDI series

Polyurethanes of MDI series exhibit unique aromatic association.

In PBA–MDI, two rocking mode vibrations for CH_2— group belonging to diphenylmethane at 751–755 cm^{-1}, and for $(CH_2)_4$ group at 737 cm^{-1} can be observed, which vanish at between 39 and 72 °C. This shows that diphenylmethane group and $(CH_2)_4$ group, probably adjacent to a urethane group, are involved in the associated phase of hydrogen bond lattice-cell.

In BD–MDI, only a rocking mode vibration of CH_2— group in between two phenyl groups, along with out-of-plane deformation vibration of CH group in aromatic ring at 854 and 860 cm^{-1} exhibit temperature dependent spectra. A $(CH_2)_4$ group seems not capable of entering into the area of association probably because of a steric hindrance due to two bulky diphenylmethane rings. This association, melting above 218 °C forms a very rigid hard segment especially when the sample is annealed. Only an amorphous pattern of FIR is seen on the contrary when the sample is quenched from molten state. This effect on hard segment may also affect most likely the crystallinity of copolyurethane. The FIR spectra of annealed (PBA + BD)–MDI shows a relatively strong contribution of hard segment by absorption at 751 cm^{-1}, and a slight indication of the presence of an isolated soft segment by absorption at 736 cm^{-1}.

No large modification except for the relative intensities is observed in FIR spectra by replacement of PBA by PTMG, but a lower half of modulus-temperature curve shifts towards lower temperature, subsequently a broadening of the plateau taking place in the viscoelastic spectrum. Although it had been recognized that Estane, (PBA + BD)–MDI, showed a completely amorphous X-ray diffraction scattering diagram (10), FIR spectral data presented here are capable of revealing the existence of an associated phase whose structure is quite sensible to thermal treatment.

XDI and TDI series

No aromatic association such as in MDI is observed in other phenyl-group-containing polyurethanes from XDI and TDI series.

In XDI series, associated $(CH_2)_4$ is seen to dissociate at the same temperature region between 49 and 67 °C in PBA–XDI and (PBA

+ BD)–XDI. BD–XDI does not show this association and only a rocking mode vibration of CH_2— group from diisocyanate is observable until carbonamide dissociation instead. Low optical density for the carbonamide association is observed. Viscoelastic spectrum shows only perceptible narrow plateau, rather similar to thermoplastics in spite of the segmental segregation indicated by FIR spectrum.

In TDI series, the spectral change is not so striking in comparison with other three series already mentioned.

As for the hard segments in XDI and TDI series, small change in FIR spectra and low optical densities of absorptions for $\gamma(NH)_{assoc.}$ and $\gamma(C{=}O)_{assoc.}$ should correlate with the low melting points compared to HMDI and MDI series.

3.3. Strength of networks

In mid IR region, absorption takes place at ca. 3330 cm^{-1} for the stretching vibration of associated NH group, $\nu(NH)_{assoc.}$, which diffuses and shifts towards higher frequency as it dissociates. This has been conventionally adopted to estimate the degree of hydrogen bond in polyurethanes or polyamides (19).

In figs. 10 and 11, the frequency shift of $\nu(NH)_{assoc.}$ for the copolyurethanes of HMDI and MDI series are plotted against temperature.

Polyurethanes of HMDI series tend to shift more easily compared with MDI series. BD–HMDI shows gradual dissociation of hydrogen bond, which turns steep around 170 °C. From the qualitative point of view, (PBA + BD)–HMDI possesses as high hydrogen bonding energy as BD–HMDI does, irrespective of participation of weakly hydrogen-bonded PBA–HMDI. The hard segment, no doubt, is internally-plasticized by the soft segment and a new hydrogen-bonded network is introduced into (PBA + BD)–HMDI. This is consistent with the views from the FIR and viscoelastic experiments, as having been illustrated in figs. 5 and 8.

The patterns of the frequency shift for polyurethanes of MDI series indicate much higher melting points compared with HMDI series. Of these, BD–MDI shows very rigid hydrogen bond network. While, in (PBA + BD)–MDI, as shown in fig. 11, it is relaxed to more than average bond strength in the co-existence of the soft segment. The frequency progressively shifts towards its dissociation temperature. Obviously, both of the hard and soft segments interact together to form a pseudo-network to which contributes the hard segment with hydrogen bond weakened, perhaps by an aromatic cohesion between both segments.

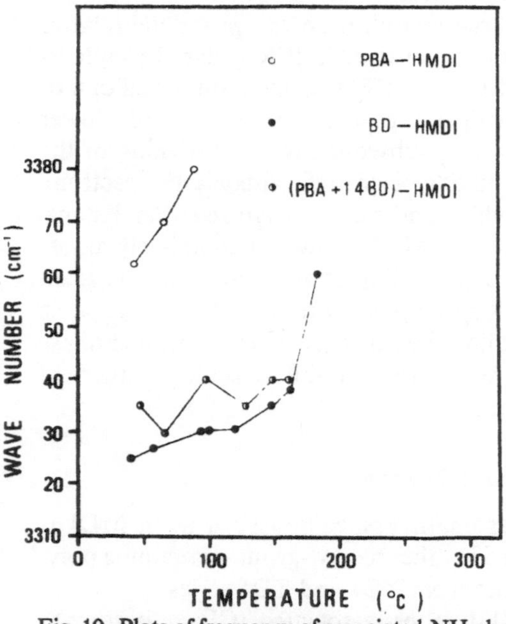

Fig. 10. Plots of frequency of associated-NH absorption, $\nu(NH)_{assoc.}$ (cm^{-1}) versus temperature, for soft-, hard- and co-polyurethanes of HMDI series

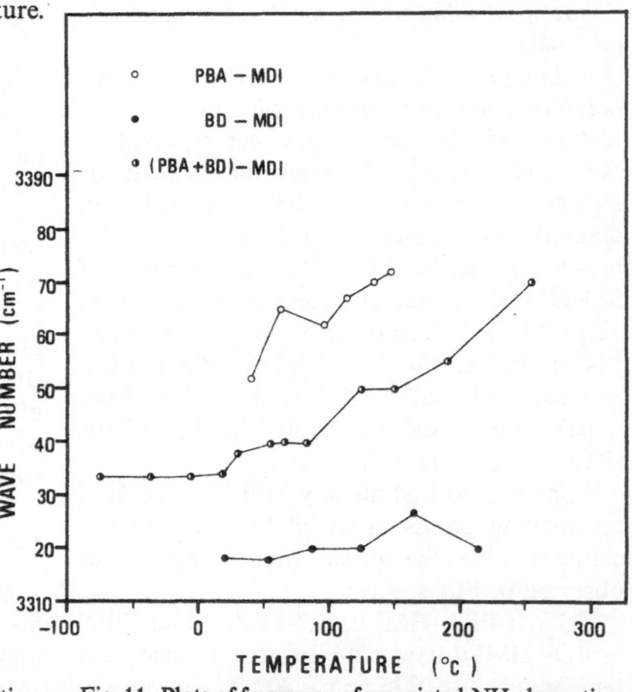

Fig. 11. Plots of frequency of associated-NH absorption, $\nu(NH)_{assoc.}$ (cm^{-1}) versus temperature, for soft-, hard- and co-polyurethanes of MDI series

The weakening of hydrogen bond is also observable in (PTMG + BD)-HMDI, which showed a typical thermoelastic response optically and mechanically, as mentioned in the preceding chapter.

4. Conclusion

In (PBA + BD)-HMDI, a strong intersegmental carbonamide association was found to occur through a preferential "solvation" of $-(CH_2)_6NHCOO-$ group in weakly hydrogen-bonded soft segment to the same group of hard segment with highly ordered association of $-(CH_2)_6NHCOO(CH_2)_4-$ group, thus yielding a thermoplastic network.

In BD-MDI a strong aromatic-association of diphenylmethane group was found to intensify the carbonamide association of hard segment. Soft segment was seen to interact with and weaken the rigidly associated hard segment in (PBA + BD)-MDI, giving rise to a thermoelastic response typical to block-copolymers.

The same weakening of hydrogen bond and a subsequent thermoelasticity was observed by the replacement of polyester segment of (PBA + BD)-HMDI by less polar polyether, PTMG.

Synopsis

Intersegmental affinities of various segmented copolyurethanes, derived from butanediol, poly-butyleneglycol-adipate and their mixture were analyzed by means of infrared measurements. A preferential solvation between $-(CH_2)_6NHCOO-$ groups of soft- and hard-segment were found to form an internally-plasticized aliphatic copolyurethane from 1,6-hexamethylene diisocyanate. A rigid aromatic association was found with hard segment polyurethane from butanediol and 4,4'-diphenylmethane diisocyanate (MDI). In the copolyurethane from MDI, the intensified hydrogen bond of hard segment was observed to interact with and become weakened by soft segment with an aromatic cohesion.

The infrared spectral behaviors were compared with subsequent dynamic viscoelastic responses.

Acknowledgments

Appreciation is expressed to the Kao Soap Company, Ltd. for the permission to publish this paper. We are also grateful to Mr. *Yasuo Miura*, Miss *Hiroko Ogawa* and Mr. *Tatsuo Hiraoka* who carried out various measurements of this report. We also wish to thank Prof. *Rodney D. Andrews* of Stevens Institute of Technology and Prof. *Junji Furukawa* of Kyoto University for helpful discussions.

Summary

Temperature dependent properties of some segmented copolyurethanes were compared with those of soft- and hard-segment polyurethanes.

To understand the chemical nature of associated phase more explicitly, far-infrared spectral behaviors were precisely analyzed.

Mid-infrared frequency shift of NH-stretching vibration of associated carbonamide group was taken as a measure of the strength of hydrogen bond.

Dynamic viscoelastic response as well as flow property estimated from viscoelastic shift factor were compared with the infrared spectral behaviors.

A strong intersegmental carbonamide association was found through a preferential "solvation" of $-(CH_2)_6NHCOO-$ group in the soft segment derived from poly-butyleneglycol-adipate (PBA) and hexamethylene diisocyanate (HMDI), to the same group in the hard segment derived from butanediol (BD) and HMDI, yielding a thermoplastic copolyurethane, (PBA + BD) —HMDI, with high activation energy of segmental flow.

While, a unique aromatic association was observed in copolyurethane derived from diphenylmethane diisocyanate (MDI). A weakening of hydrogen bond in hard segment by aromatic cohesion of soft segment was observed from mid-infrared measurements, which gave a typical thermoelastic copolyurethane with moderate activation energy of flow. The same effect of weakening of hydrogen bond was observed by the replacement of polyester soft segment by less polar polyether, which gave also thermoelasticity.

Polyurethanes derived from toluene diisocyanate (TDI) and xylylene diisocyanate (XDI) did not show the aromatic association of MDI type.

The mid-infrared stretching vibration and far-infrared out-of-plane deformation vibration of associated NH-group were found to correlate with hydrogen bond- and *van der Waals*-crystals having been assigned from (200) and (002) interferences, respectively, in X-ray scattering measurements by *Kilian* et al.

References

1) *Rembaum, A., F. R. Ells, R. C. Morrow*, and *A. V. Tobolsky*, J. Polym. Sci. **61**, 155 (1962).

2) *Angelo, R. J., R. M. Ikeda*, and *M. L. Wallach*, Polymer (London) **6**, 141 (1965).

3) *Murakami, K.*, Bull. Chem. Inst. of Non-Aq. Soln. Tohoku Univ. **17**, 145 (1967).

4) *Murakami, K., K. Shiina, Y. Itoh*, and *T. Ueono*, Reports on Progress in Polymer Physics in Japan **12**, 289 (1969).

5) *Bonart, R.*, J. Macromol. Sci. **B2**, 115 (1968).

6) *Clough, S. B., N. S. Schneider*, and *A. O. King*, J. Macromol. Sci. **B2**, 641 (1968).

7) *Bonart, R., L. Morbitzer*, and *G. Hentze*, J. Macromol. Sci. **B3**, 337 (1969).

8) *Koutsky, J. A., N. V. Hien*, and *S. L. Cooper*, Polym. Letters **8**, 353 (1970).

9) *Huh, D. S.* and *S. L. Cooper*, Polym. Eng. Sci. **11**, 369 (1971).

10) *Cooper, S. L.* and *A. V. Tobolsky*, J. Appl. Polym. Sci. **10**, 1837 (1966).

11) *Bonart, R.* and *L. Morbitzer*, Kolloid-Z. u. Z. Polymere **241**, 909 (1970).

12) *Bonart, R., L. Morbitzer*, and *H. Rinke*, Kolloid-Z. u. Z. Polymere **240**, 807 (1970).

13) *Hummel, D. O.*, Infrared Spectra of Polymers (New York 1966).

14) *Hummel, D. O.*, Atlas der Kunststoff-Analyse, Bd. I (München 1968).

15) *Bellamy, L. J.*, The Infra-red Spectra of Complex Molecules (London 1954).

16) *Corish, P. J.*, Anal. Chem. **31**, 1298 (1959).

17) *Shimanouchi, T.*, Analysis of Infrared Spectra (Tokyo 1960).

18) *Kilian, H. G.* and *E. Jenckel*, Kolloid-Z. **165**, 25 (1959).

19) *Nakayama, K., T. Ino* and *I. Matsubara*, J. Macromol. Sci.-Chem. **A3**, 1005 (1969).

Authors' address:

Y. Kazama and *Y. Suzuki*
Industrial Research Laboratories
Kao Soap Company
Wakayama (Japan)

Rheol. Acta 13, 149–156 (1974)

From the Research Department Plastics and Additives Division Ciba-Geigy Corporation Ardsley, N. Y. (U.S.A.)

The use of torsional braid analysis to predict rheological properties of polyimides*)

By Robert E. Coulehan

With 12 figures and 2 tables

(Received October 27, 1972)

Introduction

In recent years a number of new heterocyclic type polymers have been synthesized in an attempt to obtain polymers with improved thermal stability at high temperatures. The environmental factors that must be met by any material at high temperatures demand improved processing and characterization to insure that the material will meet these conditions. As in many polymer systems, the way the polymer is synthesized determines much of the approach to its characterization. For the most part the polyimide is prepared as a prepolymer that is first imidized and then cures to an intractable polymer at some high temperature. While one can relate these general concepts of processing in rather simple terms, there are numerous complexities in arriving at a polyimide that will perform to the requirements needed for high temperature application. A basis of these complexities is in the imidization step. While soluble as a prepolymer in a variety of solvents, in most cases it becomes insoluble after imidization. Such a situation precludes many traditional ways of characterization of the polymer, particularly in terms of molecular weights. One technique that has been used to characterize these polymers is the Torsional Braid Analysis (TBA). This approach has the advantage following the polymer from its prepolymer form through the imidization step to the curing and in most cases a post curing step. Once this has been done the same sample can be exposed to a number of additional thermal histories and these effects measured.

Experimental

The Torsional Braid Analysis (TBA) has been described by several authors (1–3). However, much of the intent of previous work has been over wide ranges of temperatures with little emphasis on the problems of determining processing data for the polymer. In this technique a loosely woven glass braid is used to support the initially liquid prepolymer. The braid is suspended as a pendulum in a thermal jacket that can be programmed over a temperature range of −190 to 500 °C. Attached to the free end of the pendulum is a glass disc with an optical coating that varies linearly over a 270 °C arc. The motion of the pendulum is one of torsion giving rise to shear forces in the sample. The oscillation of the disc is detected by a photocell and this output is used to determine the apparent rigidity and damping factor. It can be shown that for a freely oscillating pendulum the apparent rigidity is proportional to the inverse of the square of the period and that the damping factor is proportional to

$$\Delta = \ln \frac{A_i}{A_{i+1}} \qquad [1]$$

where A_i is the amplitude of the ith wave. These data are plotted with respect to temperature or time to determine the rheological properties of the polymer braid system. The instrument used for this work was built by Chemical Instruments Company who also built a data compiler (4) that prints out the temperature, ratio of successive double peak amplitudes and the period of oscillations. These data were used to plot the curves.

Four polyimides were used in this study as representative of the types of materials that can be investigated with the TBA. These were P 13 N and a modified P 13 N which are commercial resins of Ciba-Geigy and polymers of 3,3′,4,4′-benzophenonetetracarboxylic dianhydride (BTDA) and 4,4′-methylenedianiline (MDA), and BTDA plus 4,4′-oxidianiline (ODA). The structures of these four polymers are given in fig. 1.

Since thermal history determines much of the ultimate properties of these materials, three different thermal profiles were used for the commercial resins and one for the remaining

*) Presented at the Sixth International Congress on Rheology Lyon (France), september 4–8, 1972.

two resins. These heating schedules are given in table 1. For schedule C, the sample was imidized at 200 °C, cooled to room temperature

Table 1

Schedule	Imidization temperature (°C)	Curing temperature (°C)
A	–	350
B	200	350
C	200	300, 350

and removed from the heating chamber. The chamber was then heated to 300 °C and the braid assembly inserted. Curing was followed as a function of time. After curing, the sample was cooled and then reheated to 350 °C for 30 min. All the measurements made for this work were done in an atmosphere of argon.

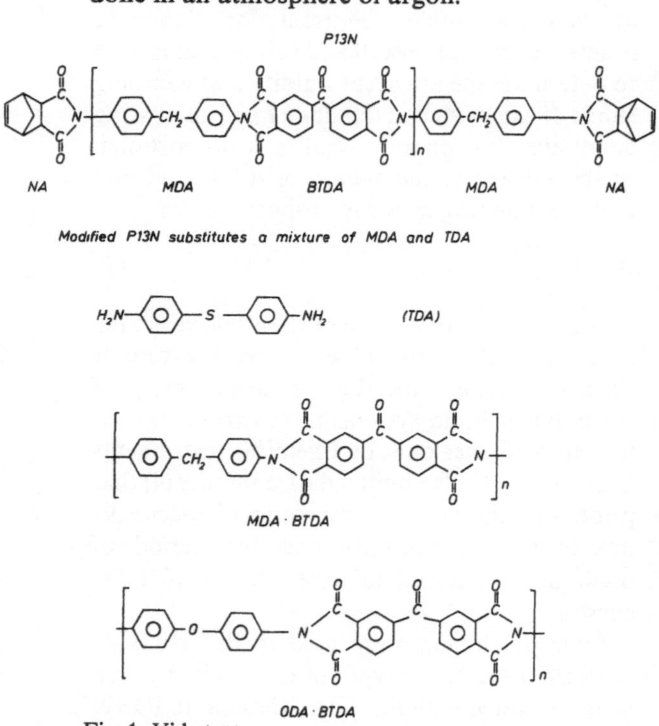

Fig. 1. Vide text

The P 13 N polyimide resin used for this study is currently being marketed by Ciba-Geigy. The polymer is unique in that it exhibits low voids when processed in laminate form. It is also available as a molding powder. A sample of this material was coated on a braid from a prepolymer solution and heated in the TBA according to schedule A. The heating rate was 3–5 °C per minute. A plot of the results is given in fig. 2. The onset of imidization is seen at 140 °C during the first heating. This continues past 200 °C. At the same time the solvent dimethylformamide (DMF) is being driven off. Some softening of the imidized polymer takes place between 250 and 300 °C. At 300 °C the polymer begins to rapidly build up crosslinks and by the time the sample had reached 340 °C it is fully cured for temperatures below 350 °C. The sample was then cycled in temperature to − 170 °C and up to 350 °C. As is seen in fig. 2, a secondary dispersion peak lies between 50 and 200 °C. No attempt was made in this measurement to go above the glass temperature for this sample since it was known from previous work (5) that the glass temperature would change to a higher temperature.

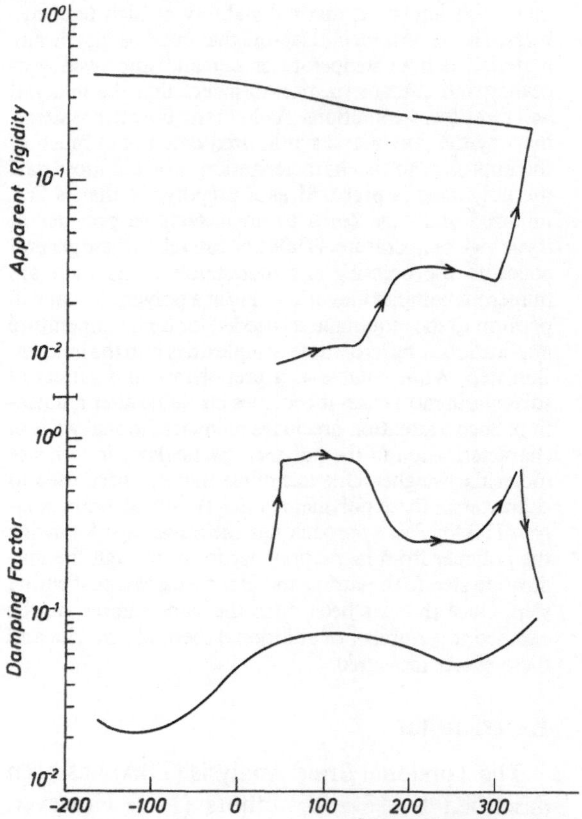

Fig. 2. Vide text

Using schedule B a second braid of P 13 N was heated over the same temperature range. These results are shown in figs. 3 and 4. The initial heating to 200 °C is identical to that of fig. 2. The sample was held at this temperature for 90 min and then cooled to − 170 °C and then heated to 350 °C and cured for 30 min. A glass temperature of 194 °C was found for the imidized

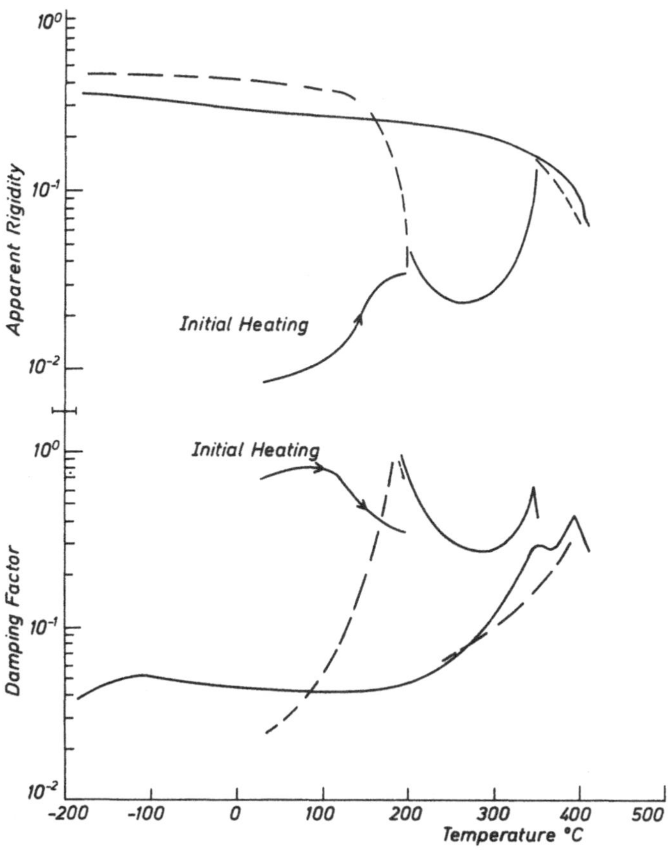

Fig. 3. Vide text

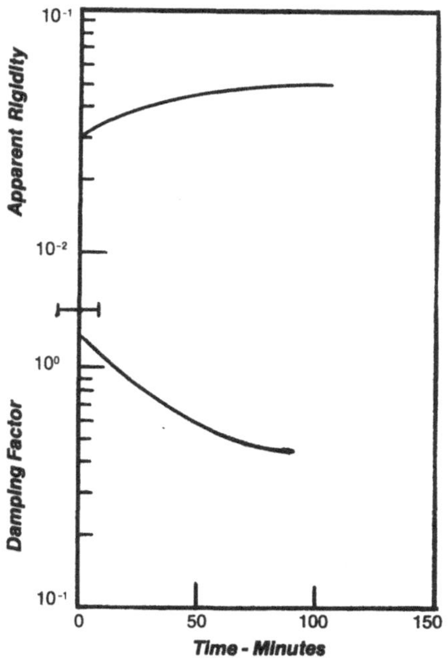

Fig. 4. Vide text

polymer. From this result it is possible to estimate the time required to imidize this material. Fig. 4 is a plot of the TBA output as a function of time at 200 °C. The apparent rigidity increases with time while the damping peak does not have a pronounced peak but indicates a loss region on the same level throughout the imidization. The apparent rigidity levels off after about 75 min indicating the completion of imidization. Further heating following the imidizing again shows softening about 250 °C with curing beginning as before at 300 °C. After curing at 350 °C the sample was heated through the glass temperature at 380 °C and on cooling to room temperature it was found that the glass temperature had been displaced higher. There was no pronounced dispersion region between 50 and 200°C as in the case of fig. 2.

Using schedule C another profile was run on a new sample of the P13N resin. These results are shown in figs. 5 and 6. Examining the curing at 300 °C in fig. 6 reveals that the glass temperature reached 300 °C after 30–35 min of the cure. This is shown in the peak of the damping curve. The Tg increased in temperature with increasing time until the rigidity curve reached a plateau value for this cure temperature. This occurs about 120 min after the start of the cure. If one compares these results with those of fig. 2 where the sample was heated directly to 350 °C then the time of 2 h agrees very well with first order rate processes. Following the cure, fig. 5 shows a secondary dispersion region seen between −100 and 100 °C on scanning from −170 to 400 °C.

The above data are confirmed in the laminating and molding of the resin. The cure in some instances was found to be too rapid leading to a resin viscosity buildup at temperatures below 300 °C that precluded such applications as autoclaving. In the autoclave process the total pressure available for consolidation is not sufficient to overcome the viscosity buildup in the P13N resin. To overcome this a modification of the resin structure was made to slow down the cure rate to improve flow at lower temperatures of 260 °C. The structure of this resin is given in fig. 1. This modified P13N resin was investigated using the heating schedule of table 1. The results

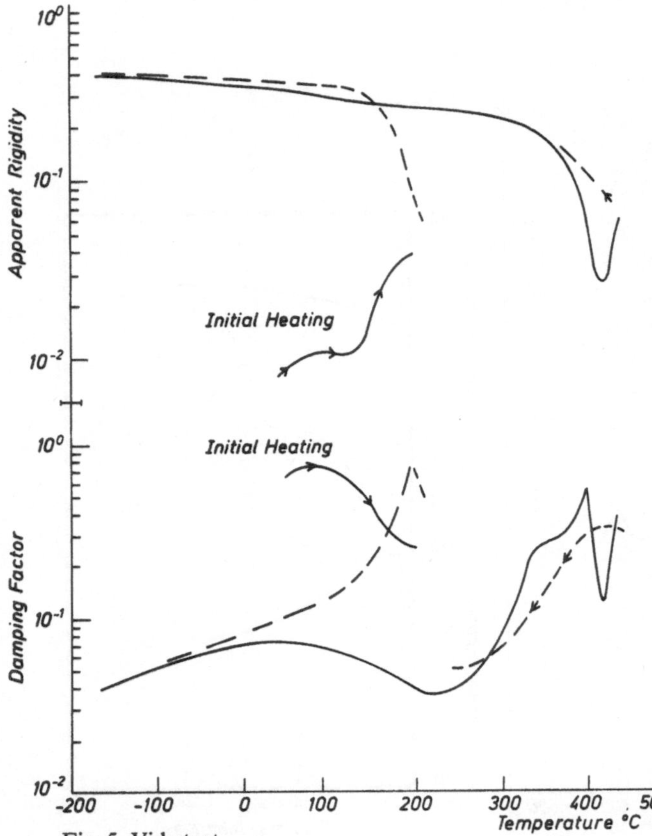

Fig. 5. Vide text

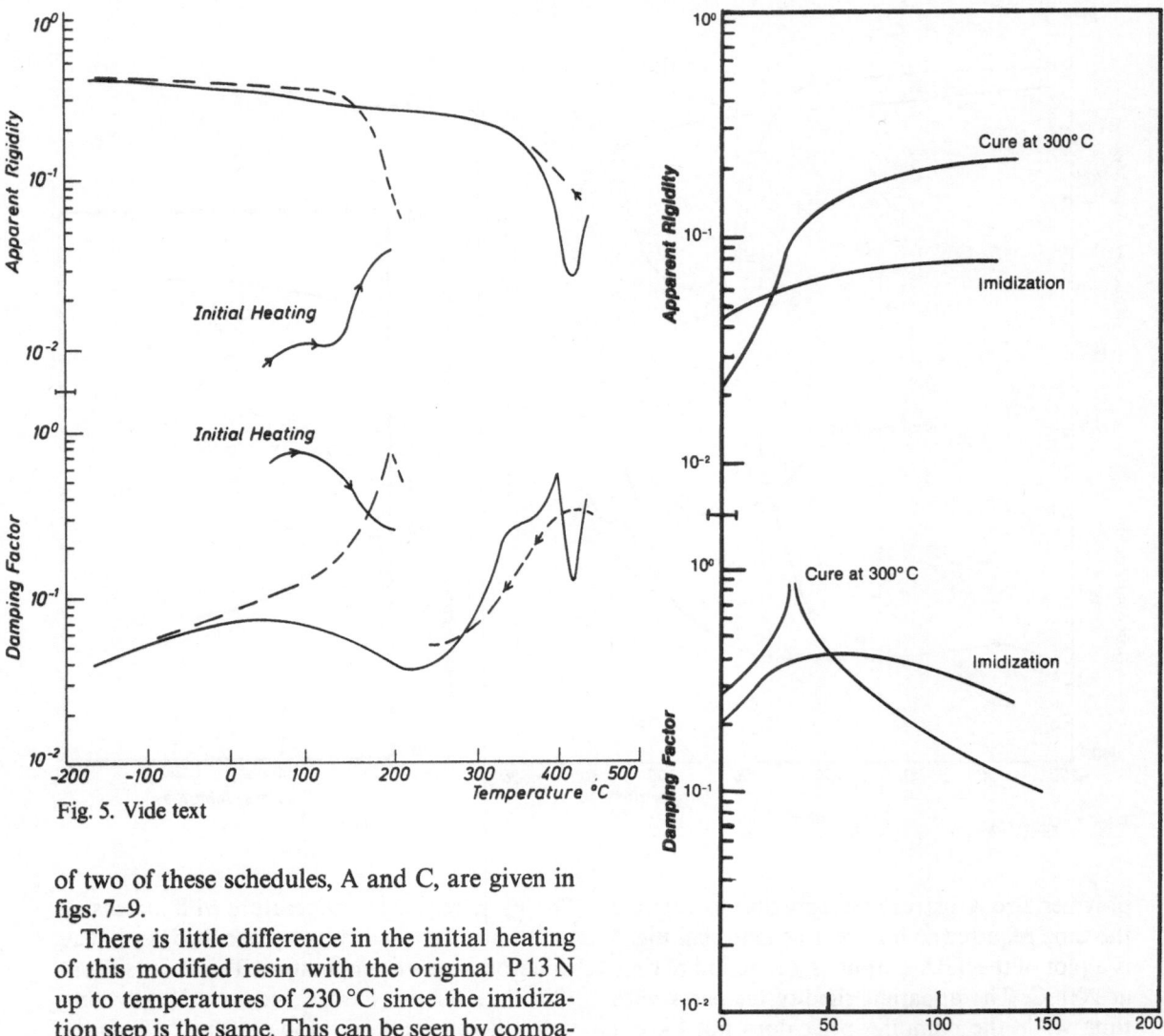

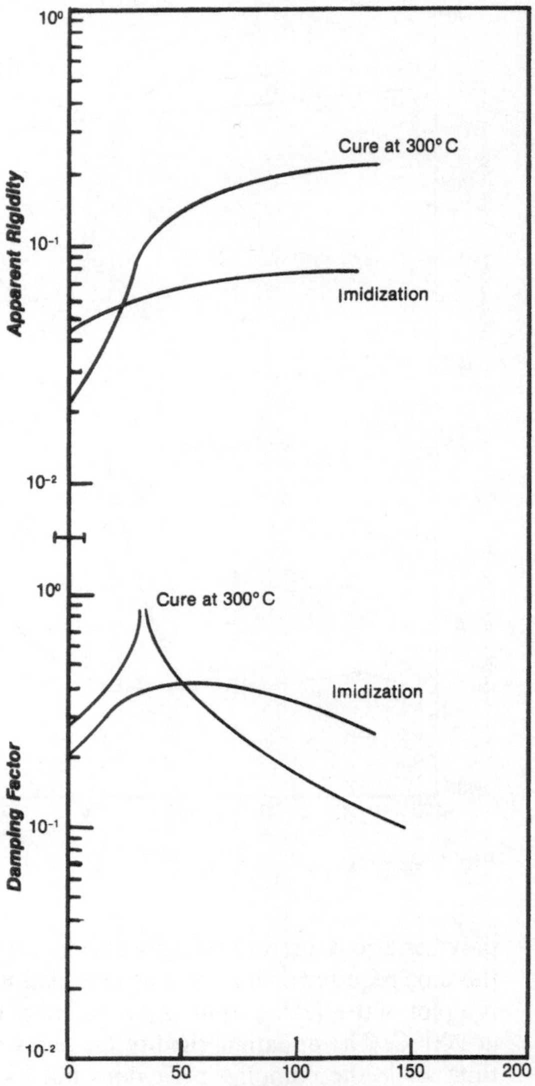

Fig. 6. Vide text

of two of these schedules, A and C, are given in figs. 7–9.

There is little difference in the initial heating of this modified resin with the original P13N up to temperatures of 230 °C since the imidization step is the same. This can be seen by comparing figs. 7 and 2. The material softens more than P13N above 250 °C and the onset of cure is now at 320 °C rather than 300 °C. The sample in fig. 7 gave a Tg of 400 °C on heating through glass temperature. However, on cooling both the apparent rigidity and damping curves show a sharp drop in the glass temperature to 350 °C. A dispersion region between 0 and 150 °C for the material is also seen for this thermal history. However, after cooling from 430 °C this region appears to have been changed which would tend to indicate poorer thermal stability of a material given a thermal treatment according to schedule A. The broadening of the damping peak and shift to lower temperature indicates a substantial change in molecular weight.

A second sample of the modified P13N resin was run according to schedule C. These results are given in figs. 8 and 9. Following the imidization at 200 °C one finds a substantially lower glass temperature of 167 °C. This difference reflects a difference in molecular weight.

The curing at 300 °C is given in fig. 9. Both the apparent rigidity and the damping curves reflect a slower cure than P13N. The glass temperature passes 300 °C 80 min after the start of the cure. This is twice as long as for P13N. The rigidity levels off after 140 min. The sample was then cooled to −170 °C and heated to 440 °C. Significant differences were found. The dispersion region was now shifted to between −125 °C and −25 °C. The glass temperature of the sample as cured at 300 is 325 °C. However, this can be called only an apparent glass temperature for the material with a thermal history limited to 300 °C maximum. As long as the tem-

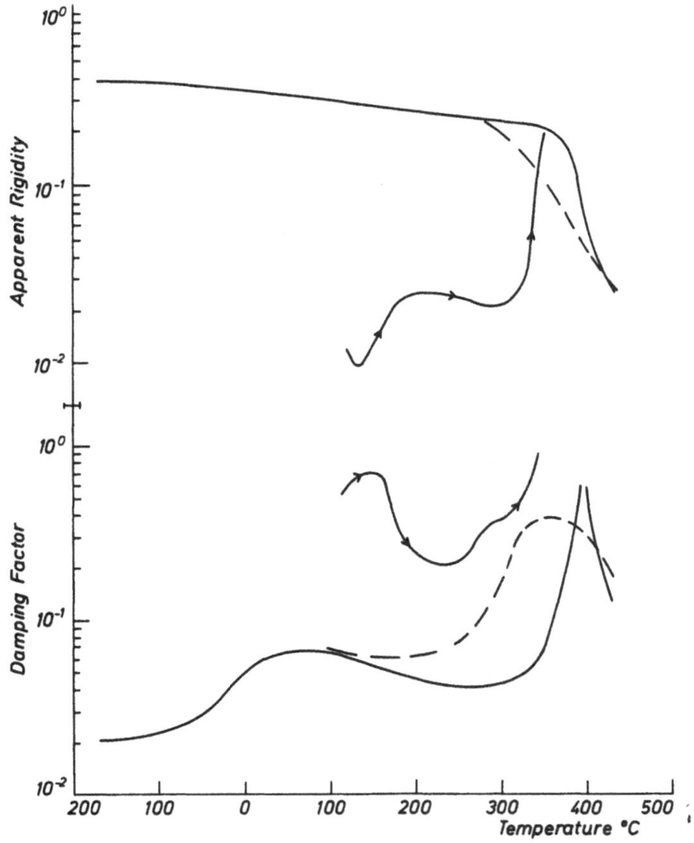

Fig. 7. Vide text

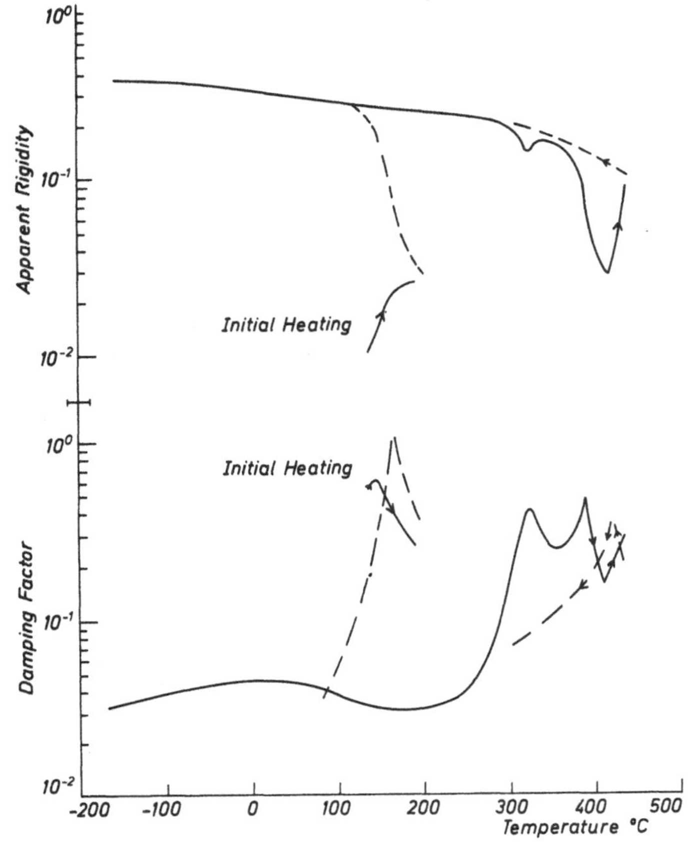

Fig. 8. Vide text

perature remains below 300 °C, this will be its glass temperature. As is seen in fig. 8, additional curing takes place as soon as the temperature reaches 320 °C. The material again passes through a glass temperature at 390 °C where now more crosslinks are formed pushing the Tg higher. This time on cooling no displacement of the Tg to a lower temperature is seen indicating improved thermal stability for this resin when cured under schedule C. The apparent rigidity shows no change on cooling to 25 °C.

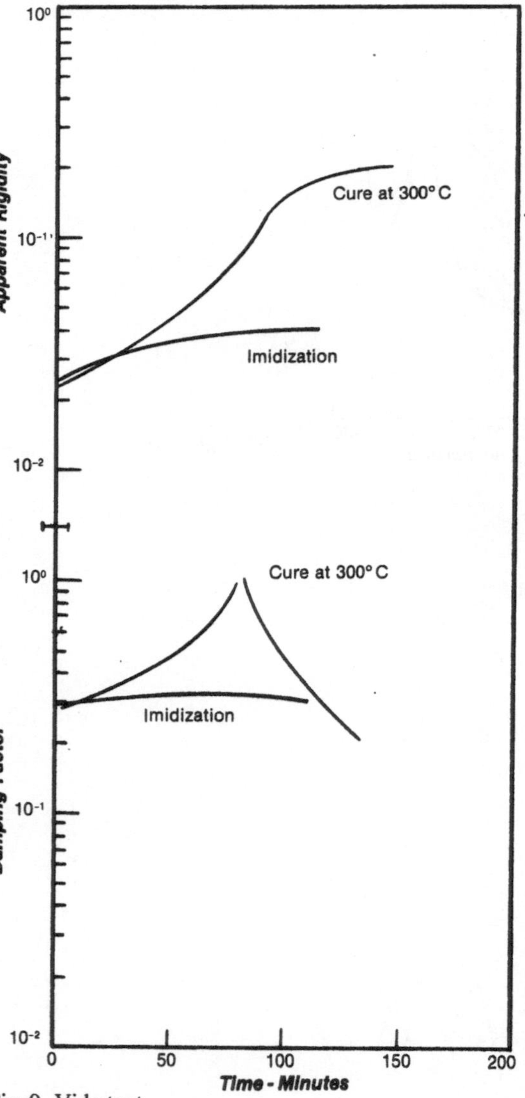

Fig. 9. Vide text

Two other samples were investigated using schedule B. These were both 1:1 ratios of a dianhydride and a diamine shown in fig. 1. TBA data on these two materials was previously reported by *Gillham* (1). Fig. 10 is a plot of results for BTDA · MDA. After imidization the sample was cooled to −170 °C and then heated to 350 °C. A Tg of 285 °C was found for this heating. After 30 min at 350 °C, the Tg was 300 °C. Two secondary dispersion regions were found, one centering around −100 °C and the other at 80 °C. These agree with those found by *Gillham* (1).

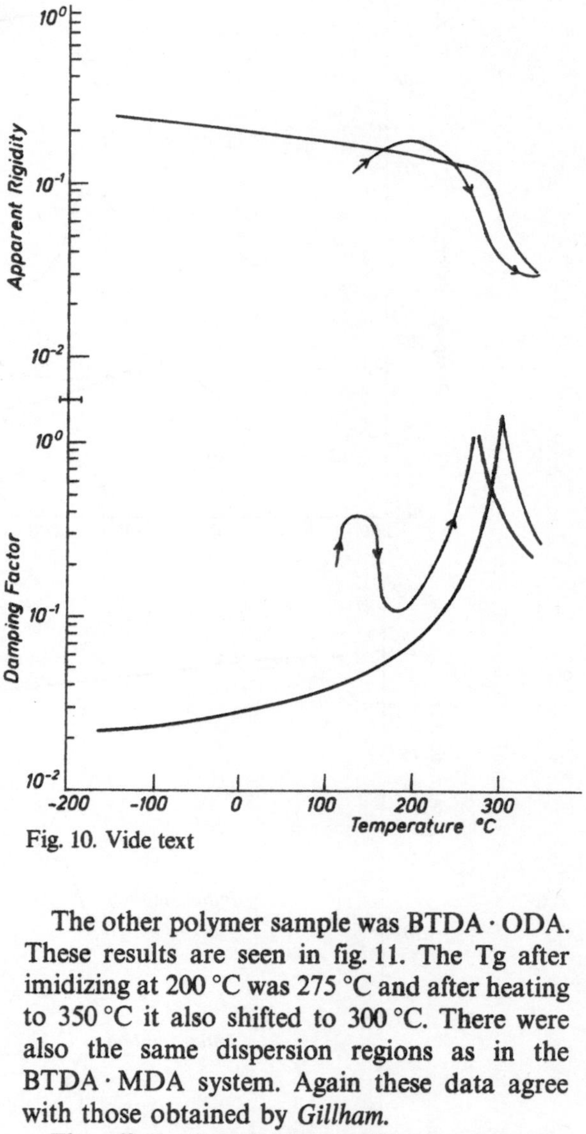

Fig. 10. Vide text

The other polymer sample was BTDA · ODA. These results are seen in fig. 11. The Tg after imidizing at 200 °C was 275 °C and after heating to 350 °C it also shifted to 300 °C. There were also the same dispersion regions as in the BTDA · MDA system. Again these data agree with those obtained by *Gillham.*

The effect of molecular weight on the glass temperature of polyimides was previously reported for the MDA · BTDA polyimide (5). The molecular weight refers to that of the amic acid prepolymer. These results showed that the higher the prepolymer molecular weight, the lower the final glass temperature of the polyimide. It was felt that similar results would hold for the P13N polyimide. Two P13N polymers of 3000 and 5000 molecular weight were cured on the TBA and their final Tg determined. Table 2 lists these results along with the P13N and the modified P13N. A plot of these results is seen in fig. 12.

Table 2

Polymer	Mw	Tg °K	1/Tg × 10³
P13N	5000	598	1.672
P13N	3000	628	1.592
P13N	1300	663	1.508
Modified P13N	1050	673	1.486

Discussion and conclusions

From the preceding results we have shown that considerable information is obtained from the TBA on the thermomechanical behavior of polyimides that can lead to insight into the processing of the polymer. It is important in any development program that this information be translatable into processing parameters. There are a variety of methods of molding and laminating these types of resins. Since the prime objective of work on polyimides is to provide a thermally stable material it is necessary to determine a thermal history that will give the optimum physical properties. As pointed out previously, the imidization step greatly influences subsequent behavior.

The process of removing the water of imidization is the key to ultimate thermal stability of the material. It is possible to chemically imidize some polyimides but there will still remain a solvent removal problem. It has been noticed that in many polyimides after imidization at 200 °C the apparent rigidity has a higher value on cooling than the same sample has following curing and post curing. This can be seen in figs. 3 and 8 for both the P13N and the modified P13N. Possible explanation for this behavior must be in the fact these ring structures in the intermediate form randomly orient themselves to give an encumbered structure. The process of curing and post curing the polymer gives much more order and compactness to the polymer chains and will actually give a lower modulus, or, phrasing it another way, a more flexible configuration than before curing. There are some manifestations of this concept in the handling of the individual glass laminate layers. Follow-

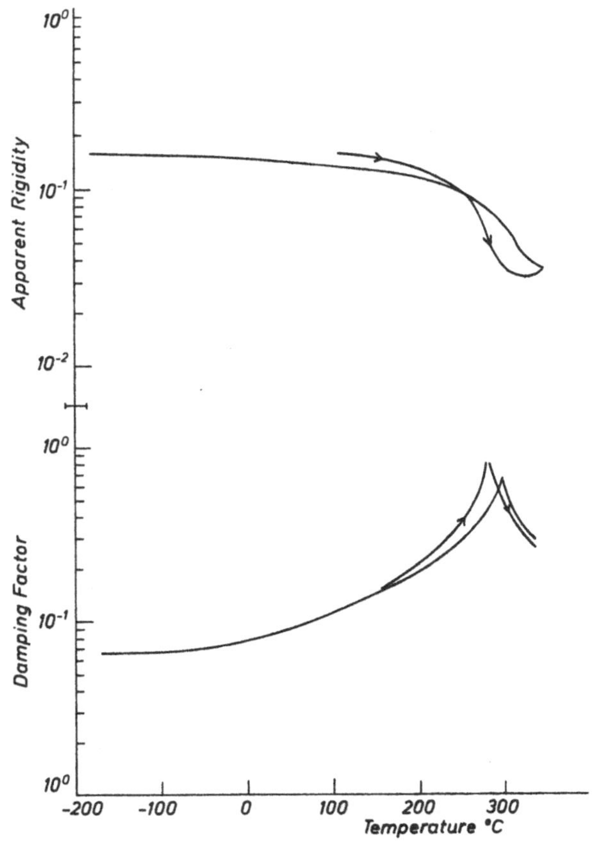

Fig. 11. Vide text

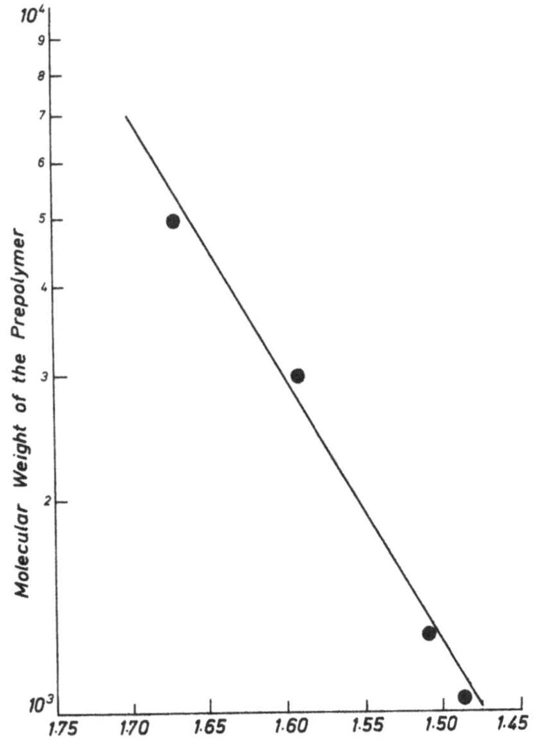

Fig. 12. Vide text

ing the "B" staging or imidization step these pieces become stiff and not very flexible. If one cures these individual pieces then they become more flexible.

The temperature and time of curing and post curing are very important parameters for processing. The cure temperature should be as low as possible and the time short enough so as to process as many items as possible. For the P13N resin we see that at a temperature of 300 °C the time to reach steady value of apparent rigidity is about 120 min. The cure can be shortened if it is possible to have an extended post cure cycle. In fig. 2 where the P13N resin was heated directly to 350 °C we found that the sample was cured by the time it had reached 340 °C. Hence a 4-h post cure at 305 °C would be sufficient to develop the ultimate properties that were found by heating to 350 °C. Hence it has been found that this resin can be processed at 288 °C for the cure step and post cured at 305 °C.

An equally important aspect of the thermophysical curve is the softening to the point of flow under a moderate amount of pressure between the imidization temperature and the cure temperature. Where both the P13N and the modified P13N show this softening character neither the ODA · BTDA or MDA · BTDA polymer showed this behavior. It would be extremely difficult to process either of these in a manner similar to the P13N system.

Finally some comments should be made about the fact that it is very difficult to obtain the absolute modulus value from TBA data. *Nielsen* (6) has shown that one can determine the modulus based on knowing the volume fraction of resin present. However, the real value of this supported braid technique lies in the ability to determine changes in rigidity and damping from the glassy region through the Tg to the rubbery plateau. This is the region where processing and buildup of ultimate properties of a polymer take place. The process development of a polymer system can use the TBA data to fix the temperature and time to process the polymer to give the required behavior. Ultimate thermal stability can also be determined by exposing the individual braid to thermal conditions outside the TBA instrument and then determining the thermomechanical behavior after various exposure times. Future work on the polyimides will report on this aspect of predicting service life at various temperatures.

Acknowledgements

The author wishes to acknowledge the assistance of Mr. *J. Velten* and Miss *C. Dunn* for operating the instrument and reducing the data, and Mrs. *M. DiBenedetto* for the MDA · BTDA and ODA · BTDA samples. The P13N resins were invented by TRW systems Group under contract from NASA-Lewis Research Center. Ciba-Geigy holds an exclusive license for the manufacture and sale of these resins.

Summary

The processing of polyimides gives rise to a number of rheological problems that cannot be solved using normal flow measurements. In particular, the effect of molecular structure on curing, flow and molecular weight leads to different ultimate properties depending on the thermal history. One technique that has been used to study these systems is the Torsional Braid Analysis (TBA). It is shown that with the TBA one obtains time-temperature relationship for cyclization of the prepolymer; curing and post cure parameters needed for processing. Thermal stability can also be determined. In addition, it is possible to relate the glass temperature of the cured system to the molecular weight of the prepolymer.

Zusammenfassung

Bei der Verarbeitung von Polyimiden ergeben sich zahlreiche rheologische Probleme, für deren Untersuchung übliche Fließmessungen unzureichend sind. So sind die Endeigenschaften solcher Werkstoffe nicht nur von chemischer Struktur und dem Fließverhalten während der Verarbeitung, sondern auch besonders von der thermischen Vorgeschichte abhängig. Torsionsschwingungsmessungen an imprägnierten Glasfasergeweben (Torsional Braid Analysis, Abkürzung: TBA) eignen sich vorzüglich für die Untersuchung solcher Systeme. Dieses Verfahren liefert Aussagen über Zeit-Temperaturabhängigkeit von Ringschluß-, Härtungs- und Nachhärtungsreaktionen sowie thermische Stabilität und Glasumwandlungstemperatur als Funktion von Struktur und thermischer Vorgeschichte.

References

1) *Gillham, J. K.*, ACS Polymer Preprints **13**, 221–226 (1972).

2) *Gillham, J. K., G. F. Pezdirtz*, and *L. Epps*, J. Macromol. Sci.-Chem. **A3 6**, 1183–1195 (1969).

3) *Myers, R. R., C. J. Krauss*, and *R. N. Schroff*, Proceeding of the Fifth International Congress on Rheology, Vol. 1, 473–482.

4) *Coulehan, R. E.*, SPE ANTEC **17**, 86 (1971).

5) *Coulehan, R. E.* and *T. L. Pickering*, ACS Polymer Preprints **12**, 1, 305–310 (1971).

6) *Nielsen, L.*, private communication.

Author's address:

Robert E. Coulehan
Research Department
Plastics and Additives Division
Ciba-Geigy Corp.
Ardsley, N.Y., 10502 (USA)

Rheol. Acta 13, 157–167 (1974)

Études statistiques sur l'hérédité en matière de rupture

Par *M. Davin (Royat/France)*

Avec 5 figures et 4 tableaux

(Reçu p. p. le 27 octobre 1972)

Un corps étant à l'instant t_1 soumis à des forces extérieures $F(t_1)$, son état mécanique est, au même instant, défini par un certain nombre de grandeurs G (en premier lieu celles qui définissent sa déformation). Dans certaines théories classiques comme celle de l'élasticité les valeurs $G(t_1)$ des grandeurs G, à l'instant t_1, ne sont fonction que des $F(t_1)$. Mais ces théories ne sont pas applicables à tous les corps, ni à toutes les valeurs des F. En général «l'histoire» antérieure des forces appliquées intervient, et $G(t_1)$ est une *fonctionnelle* des fonctions $F(t)$ représentant les forces en fonction de t depuis un instant antérieur t_0 jusqu'à t_1. Tel est le phénomène dénommé «hérédité».

Une de nos publications antérieures[1]) présentait des réflexions sur ces fonctionnelles et sur les modèles mécaniques qui y correspond. Nous avions signalé l'intérêt de l'hypothèse de leur linéarité, mais aussi les objections qu'elle soulève, surtout lorsque l'on veut étudier *la rupture* des corps doués d'hérédité.

Nous avions alors émis l'opinion que, si des éprouvettes d'une «population» sont soumises à une charge F croissant par «montées» de durée T_m et de vitesse uniforme dF/dt, alternant avec des «*paliers*» de durée T_p la statistique de répartition des instants de rupture dans les montées et les paliers pourrait donner des informations utiles sur ces fonctionnelles. A l'appui nous avions présenté un petit travail expérimental; mais le matériau essayé (conglomérat obtenu par compression de gravillon) avait été choisi bien plus pour sa commodité (résistance faible permettant une expérimentation sans construction de matériel spécial) que pour son intérêt pratique, et la taille de l'échantillon (103 unités) était notoirement insuffisante.

[1]) *M. Davin*, «Etudes statistiques sur la rupture en vue de l'ajustement d'un modèle non linéaire. Revue de l'industrie minérale, Cahier du Groupe Français de Rhéologie T II, page 143.

Nous avons voulu reprendre ces études en employant un liant hydraulique, et en opérant sur plusieurs milliers d'éprouvettes.

Éprouvettes et collages

Une telle «taille n'était conciliable avec les moyens limites dont nous disposions, que par l'adoption d'un type d'éprouvette très petit. Nous avons fixé les caractéristiques suivantes: composition: 450 g de ciment CPA pour 550 g de filler calcaire. Forme et dimensions: cylindrique, diamètre 12 mm, hauteur 21,5–23 mm. L'essai choisi est la traction directe, exercée par l'intermédiaire de boulons 6 pans ∅ 8 mm, collés sur les bases de l'éprouvette par la face plane de leur tête héxagonale (largeur 13 mm entre pans opposés).

Ce collage a posé le problème le plus difficile de toute notre recherche

Nous avions d'abord utilisé la formule suivante: Araldite AW 108 + durcisseur HY 997 (80% ± 10%).

Cette formule, à laquelle nous demandions plus que la résistance spécifiée par le fabricant,

Fig. 1

a répondu au début à nos exigences, mais pas, ensuite, de façon constante.

La formule suivante, utilisée ensuite, s'est révélée plus «fiable»: Araldite GY 257 + durcisseur X 157 (19% ± 1%) + filler «Plenamix».

La faible marge acceptable pour la proportion du durcisseur nous avait fait hésiter à l'employer. Comme nous ne pouvions mettre en œuvre pendant la «durée de vie» que quelques grammes de mélange, il faillait doser le durcisseur à $^1/_2$ décigramme près. L'emploi d'une balance de précision soulevait des difficultés. Nous avons finalement opté pour un dosage volumétrique par seringue graduée.

Matériel utilisé

Il comprend:
1. la batterie de traction à 10 postes;
2. le coffret électronique avec les circuits;
3. l'enregistreur «Mémotop rapide» à 6 voies.

Batterie

Elle comprend:

Un bâti de 60 cm de longueur environ, comportant 10 alvéoles affectés chacun à un poste de traction, et renfermant dans sa partie supérieure la vis sans fin V qui commande tous ces postes.

Sur la face arrière se trouvent:
- le moteur (puissance 100 W vitesse 1500 t/min);
- le carter contenant embrayage et frein.

Le premier réducteur, formant boite de vitesses à 6 combinaisons.

Nous l'avons fixé, une fois pour toutes, à celle qui donne, au niveau des postes, *1 mm d e montée en 20 sec.*

Un 2ème réducteur assurant par engrenages coniques le renvoi du mouvement à la vis V.

Chaque poste comporte, *de haut en bas* (en ordre de marche).

Une tige verticale T actionnée par V (translation verticale suivant son axe) terminée par un étrier E_1.

L'étrier de traction E_2 rattaché à E_1 par une broche B_r.

Le boulon supérieur b_1 de l'éprouvette, vissé dans la base de E_2, puis l'éprouvette, puis son boulon inférieur b_2.

L'embout supérieur e_1, vissé dans b_2, d'un ressort à boudin taré R (raideur 21,8 newtons par mm, longueur au repos 9 cm, diamètre d'encombrement 24)[2]) le ressort lui-même et son embout inférieur e_2 dont l'embase débordante *em* a 34 mm de diamètre.

Et, *de bas en haut*, encerclant la partie inférieure du ressort, les pièces suivantes qui transmettent en compression l'effort que les pièces précédentes transmettent en traction.

Une couronne en acier C_1 s'appuyant sur *em*.

Une couronne en Afcodur C_2 s'appuyant sur C_1.

Une bague B vissée sur C_2, portant latéralement 2 ergots qui s'engagent dans des créneaux Cr pratiqués à la base des parois limitant latéralement l'alvéole. Sur la face supérieure de B s'appuient les pointes de deux vis v liées au bâti, pour éviter tout basculement (la transmission de l'effort par C_1, C_2, B, se faisant en compression pourrait en effet amplifier un basculement amorcé). Sur cette face de B s'applique également le poussoir du microcontact m_1 commandant le fonctionnement de l'enregistreur.

Lors de la rupture de l'éprouvette, sa partie inférieure avec les pièces b_2, e_1, R, e_2 avec *em*, C_1, C_2, B, tombe *librement* dans le panier disposé sous la batterie. Nous n'avons jamais observé le moindre accrochage. Le microcontact m_1 est immédiatement coupé et la rupture enregistrée dans un délai de quelques millisecondes au plus.

Coffret électronique

Comprend une partie fonctionnant sur le courant secteur (triphasé 220–280 V) et une fonctionnant sur courant redressé 24 V, à peu près continu, obtenu dans le coffret lui-même, à partir du courant secteur, par un transformateur suivi d'un redresseur à 4 diodes.

Partie 220–380 V

Après les protections (interrupteur général et fusibles) le courant alternatif alimente d'un côte le moteur par l'intermédiaire d'un inverseur, d'un autre (en monophasé) le transformateur et les circuites 220 V des deux minuteries, M_m affectée à la phase «montée» et M_t affectée à la phase «palier». Caractéristiques: M_m = temps

[2]) Les 10 ressorts on été choisis par le Service des Etalonnages du L.C.P.C., comme ayant donné les meilleurs résultats au point de vue fidélité et linéarité, dans un lot de plus de 20 fourni par un constructeur agréé pour le matériel de précision. Plusieurs d'entre eux, retardés après 1 an $^1/_2$ de service, n'ont accusé aucune variation.

de marche maximal 77″, *réglage effectué à 40″*; M_t = temps de marche maximal 6′5, *réglage effectué à 1′5* (division *centésimale* de la minute).

La réglage du temps de marche effectif se réalise par simple déplacement d'une aiguille rouge; mais nous ne l'avons jamais modifié. On peut, sur le cadran de chaque minuterie, suivre la marche de l'aiguille noire, depuis l'aiguille rouge jusqu'au zéro où elle actionne, en fin de course, le microcontact qui déclenche: le retour de l'aiguille noire vers l'aiguille rouge, la mise en marche de l'autre minuterie, et la commande du frein (en fin de marche de M_m) ou de l'embrayage (en fin de marche de M_t). Grâce à la possibilité de suivre l'aiguille, nous avons pu, dans les cas d'indisponibilité du «Mémotop» relever directement les instants de rupture avec une précision de $\pm 1''$ en montée et $\pm 2''$ en palier (contre $\pm {}^1/_2''$ avec le Mémotop).

Partie 24 V

Elle consiste essentiellement dans le circuit de commande, par l'intermédiaire des circuits 24 V des minuteries, de l'embrayage et du frein. Elle est protégée par des diodes «Zener».

L'alternance du fonctionnement des deux minuteries est assurée du seul fait de la fermeture de l'interrupteur général, quelle que soit la position de l'inverseur. Si ce dernier est au point neutre, le moteur est arrêté. S'il est à la position «marche avant», l'alternance des minuteries entraine l'alternance des période d'embrayage (montées) et de freinage (paliers); s'il est à la position «marche arrière», le moteur fonctionne en marche arrière et l'embrayage est constamment en prise, quelle que soit la situation des minuteries. Cela permet d'économiser du temps dans le retour du mécanisme à sa position de départ en vue de l'essai suivant.

Enregistreur

En tant que «Mémotop rapide» il assure une vitesse de déroulement de 1 mm/″. Sur ses 6 voies, une est utilisée pour l'enregistrement des «instants de référence», passage du palier à la montée et de la montée au palier. Comme il n'en reste que 5 pour 10 postes de traction, ces derniers sont couplés 2 par 2. Dans chaque couple les microcontacts m_1 sont en série: mais des circuits de shuntage permettent, par la manœuvre d'un interrupteur I placé au-dessus de chaque poste, sur le bâti, de rétablir le courant lorsqu'il a été coupé par la rupture de l'éprouvette correspondante.

Fig. 2

Le Mémotop est alimenté en courant secteur, tant pour le moteur de déroulement que pour les électros actionnant les «plumes» sur les 6 voies. Toutefois, pour celui de la voie affecté aux «instants de référence», comme ces derniers ne sont matérialisés que par des coupures ou rétablissements *sur des circuits 24 V* (commandes de l'embrayage et du frein) il a été nécessaire d'adjoindre un relais, qui ferme le circuit 220 V quand le frein est alimenté et l'ouvre quand l'embrayage est en prise.

Fig. 3

Déroulement d'une opération

Préparation pour chaque poste

Replacer sur le ressort *R* (muni à demeure de ses embouts) les pièces C_1 (reposant sur l'embase em) puis C_2 muni de la bague *B* vissée.

Visser sur l'embout e_1 l'un des boulons collés à l'éprouvette. Sur l'autre (b_2) visser l'étrier E_2.

Mettre en place l'ensemble ainsi constitué dans son alvéole, en liant l'étrier E_2 à l'étrier E_1

par la broche B_r tandis que les ergots de B sont engagés dans les créneaux C_r.

Faire tourner C_2 vers la gauche: par l'effet de vissage sur B, les ergots prennent contact avec les fonds des créneaux, *en même temps* que la face supérieure de B avec les vis v (si celles-ci sont bien réglées) et que le microcontact m_1 s'enclenche. Continuer la rotation de C_2 pour repousser vers bas l'ensemble C_1, C_2, e_2 et précontraindre ainsi le ressort R (30–40 newtons).

Durée de la préparation: 10 min.

Essai proprement dit

Fermer l'interrupteur général, mettre l'inverseur en «marche avant». Les cycles se déroulent automatiquement jusqu'à ce que l'inverseur soit remis au point neutre, manœuvre que l'opérateur exécute quand toutes les éprouvettes sont rompues, et en tout cas, au plus tard, vers la fin du palier du 15ème cycle (en cas d'omission, un microcontact «fin de course» y pourvoit peu après).

A chaque cycle le ressort s'allongeant de 40×0 m, $05 = 2$ mm, se tend de $2 \times 21{,}8 = 43{,}6$ newtons. Sa tension maximale $40 + (14 \times 43{,}6) = 650$ N correspond pour l'éprouvette (section $1{,}13$ m^2) à 57,5 bars.

Pendant la montée, l'opérateur assume les tâches suivantes:

a) Après chaque rupture, dans un délai de quelques secondes, manœuvre de l'interrupteur I correspondant (rétablissement du circuit d'enregistrement)

Si dans ce délai, l'autre éprouvette de la même paire s'est rompue (cas très rare), sa rupture n'est pas enregistrée: mais l'opérateur note, au jugé, l'intervalle des 2 ruptures (précision 1 sec).

b)

Relevé, pour chaque rupture, du n° d'ordre du cycle et de l'indication «montée» ou «palier». Ces informations sont fournies par le Mémotop, mais leur connaissance préalable facilite beaucoup le dépouillement de la bande.

c) Surveillance générale de l'installation

Durée de la période: Maximale $14 \times (90 + 40) + 90 = 1{,}910$ sec. Moyenne = environ 1800 sec soit 30 min.

Retour à la position de départ

L'opérateur met l'inverseur en «marche arrière» et le Mémotop hors circuit, rouvre tous les interrupteurs I qui ont été fermés, récupère les pièces tombées dans le panier, dévisse les moignons d'éprouvettes restés chacun collé à son boulon, et commence le dépouillement de la bande pendant que le moteur ramène le mécanisme à sa position de départ en $14 \times 40 = 560$ sec seulement, car en marche arrière il n'y a pas de palier. Cette marche est arrêtée automatiquement par une butée de «fin de course». Mais le dépouillement de la bande n'est généralement terminé que 5 min plus tard (15 min après la fin de la 2ème période).

Si l'on compte 5 min pour la mise au net des résultats, l'ensemble de l'épreuve dure $10 + 30 + 15 + 5 = 60$ min soit 1 h.

Résultats obtenus

Nous avons vite ressenti la nécessité de distingueur plusieurs «populations» obéissant à des lois statistiques différentes. Nous en avons retenu 3, suivant l'aspect de la rupture:

I (941épr)

Ruptures se produisant franchement dans le ciment.

II (598épr)

Ruptures paraissant amorcées au contact araldite-ciment. Dans ces ruptures, l'aspect est granuleux des deux côtés, comme dans le ler cas. Mais du côté araldite, l'épaisseur du ciment est au plus d'un ou quelques dixièmes de mm et en quelques coups de lime on voit apparaître une surface polie d'araldite.

III (328épr)

Ruptures paraissant amorcées au contact araldite-acier (une partie de la surface d'acier est mise à nu).

A l'intérieur de chacune de ces «populations» subsiste une très grande dispersion des résistances, bien que les fortes résistances se rencontrent plutôt dans la lère, les faibles plutôt dans la 3ème. Chacune peut cependant être considérée comme à peu près statistiquement homogène.

Avant de donner des résultats, nous devons nous demander s'ils ne sont pas faussées par certains effets liés au mode opératoire et notam-

Tableau 1. Ruptures par couples et chaînes (intervalle ≤ 3 sec)

N° du postes	*M* (sec)	*P* (min)	N° du postes	*M* (sec)	*P* (min)	N° du postes	*M* (sec)	*P* (min)	N° du postes	*M* (sec)	*P* (min)
6 7 8		0,25	2 3		0,30	1 3	36		9 8	30	
1 10		0,01	2 5		0,78	8 9	38		4 7	31	
8 10	32,5		9 8		0,07	2 6	37		6 2 8		0,18
1 2 9	30,5		8 10		1,16	2 5	33		8 2	34,5	
6 1	27,5		2 5		0,33	5 10	33		4 10	29	
6 9	21,5		2 5	36		4 5	33		9 4		0,63
1 4		0,20	8 7 5	34,5		1 9 6	37		7 6	23,5	
1 8	28		3 10		0,07	3 6	33		4 8	27,5	
2 7	34,5		1 3	23		2 7	35		3 5		0,18
1 7	16		6 5	39		9 5		0,55	6 2		0,14
6 5	33,5		6 1	18,5		3 8	36		4 3		0,02
3 4		0,03	1 3		0,28	9 1	25		1 2	25	
2 3	38,5		7 8		0,32	4 5	35		7 6	35,5	
9 10	46,5		7 4		0,01	4 3	35				
1 4		0,05	5 6	28		2 5	18,5				

Couples simples: 53. Chaînes de 3: 5 $53+(2\times 5)=63$

Tableau 1 bis. Dépouillement du tableau 1 par groupes de postes

1 2 2	2 3 2	3 4 3	4 5 2	5 6 3	6 7 3	7 8 3	8 9 3	9 10 1
1 3 3	2 4	3 5 1	4 6	5 7 1	6 8	7 9	8 10 2	
1 4 2	2 5 5	3 6 1	4 7 2	5 8	6 9 2	7 10		
1 5	2 6 3	3 7	4 8 1	5 9 1	6 10			
1 6 2	2 7 2	3 8 1	4 9 1	5 10 1				
1 7 1	2 8 2	3 9	4 10 1					
1 8 1	2 9 1	2 10 1						
1 9 2	2 10							
1 10 1								

L'ordre des 2 ruptures n'est pas pris en considération: ainsi la rupture 7–4 est ici comptée dans la case 4–7. Pour les chaînes on compte 2 couples *p* ex la chaîne 1–9–6 est considérée comme formée du couple 1–9 et du couple 9–6 compté dans la case 6–9. Total général 63. Total 1ᵉ ligne 22.

ment au matériel. Du fait de notre conception d'essais par «batterie» de 10 postes, nous acceptions le risque d'un ébranlement qui, se produisant à chaque rupture d'éprouvette, pouvait se transmettre aux voisines et hâter leur rupture. Nous avons donc (tableaux 1, I bis, I ter) relevé toutes les ruptures séparées par un intervalle de temps inférieur à 3 sec, et les postes où elles s'étaient produites. Nous avons trouvé 63 couples (en comptant pour 2 couples un groupe de 3 ruptures séparées par moins de 3 sec). Ce nombre est plutôt faible par rapport à ce qui résulterait du hasard, d'après des calculs grossièrement approchés tenant compte de la dispersion des instants de rupture sur plusieurs cycles. Par contre, la proportion des couples *sur postes contigus* est de:

Tableau 1 ter

N: Nombre de couples ayant eu leur 1e rupture dans la tranche de temps considérée
$P = N/\Sigma N\,(\Sigma N = 63)$
p: probabilité de rupture dans la tranche considérée (evaluée d'après les courbes lissées) pour *l'ensemble* des éprouvettes.
$R = p^2/\Sigma p^2$
P_c: cumul des *P*; R_c: cumul des *R*

	Montée (40 sec)				Palier (1,5 min = 90 sec)								
	0^3–10	10–20	20–30	30–40	0–10	10–20	20–30	30–40	40–50	50–60	60–70	70–80	80–90
N		3	10	28	8	10		2	1		1		
P		0,05	0,16	0,445	0,125	0,16		0,03	0,015		0,015		
p	0,03	0,10	0,16	0,24	0,15	0,08	0,05	0,04	0,03	0,03	0,03	0,03	0,03
R	0,007	0,076	0,194	0,437	0,171	0,049	0,019	0,012	0,007	0,007	0,007	0,007	0,007

P_c	0	0	0,05	0,21	0,655	0,78	0,94	0,94	0,97	0,985	0,985	1,00	1,00	1,00
R_c	0	0,007	0,083	0,277	0,714	0,885	0,934	0,953	0,965	0,972	0,979	0,986	0,993	1,00

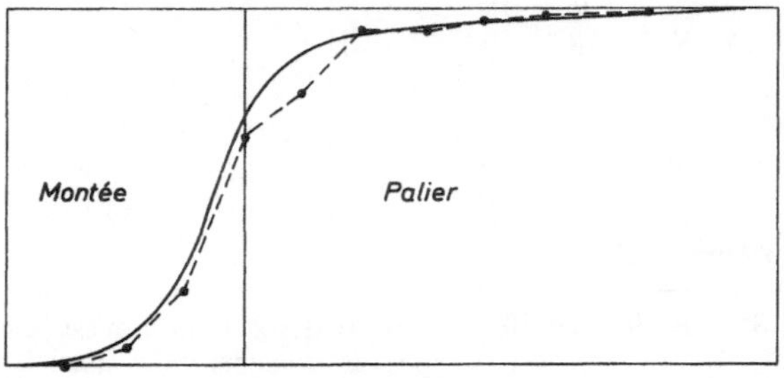

Courbes de P_c (en tireté) et de R_c (en trait plein)

35% (22 couples) contre 20% résultant du hasard pur. Cette différence, bien qu'importante, est peu significative vu la faible taille de l'échantillon. La répartition des instants de début des «couples» cadre fort bien, conformément à la théorie, avec la répartition des carrés des densités de probabilité des ruptures au cours du cycle montée-palier.

Par ailleurs, depuis la série 115, nous avons sur 67 séries, relevé l'ordre chronologique des ruptures et la proportion des séquences de ruptures sur postes contigus. Nous l'avons trouvée légèrement (et non significativement) supérieure à celle qui résulterait du hasard pur, soit 20% (108 séquences sur postes contigus, sur 514 séquences).

Enfin, nous avons cherché s'il y avait corrélation entre le n° du poste et la proportion des ruptures en montée dans le total des ruptures obtenues à ce poste. Le poste n° 9 a présenté un écart un peu fort, 58% de ruptures en montée au lieu de 53% en moyenne, écart qui n'aurait eu que 5% de probabilité à priori d'être dépassé. Mais un tel résultat est normal pour le poste présentant le plus grand écart sur un ensemble de 10 postes. Nous donnons la courbe classant les postes par probabilité croissante d'écart algébriquement inférieur à celui constaté. Elle est remarquablement voisine de la diagonale théorique. De plus, sur cette courbe, les postes 9 et 10 sont aux 2 extrémités (tableau 2 et courbe) alors que, dans la batterie, ils sont voisins de l'entrée de la vis V.

En conclusion, le 1er de ces 3 tests exclut toute influence notable de l'ébranlement dû à une rupture, tout au plus cet ébranlement peut-il hâter légèrement, *sur les postes contigus*, une rupture qui était *imminente*. Le 2ème test exclut une influence à échéance d'une ou plusieurs minutes, temps qui sépare le plus souvent deux ruptures consécutives. Le 3ème exclut l'influence des vibrations transmises par le mécanisme et des particularités des 10 postes.

D'autres tests ont été faits pour déceler en bloc les influences parasites qui pourraient provenir de la confection des éprouvettes, de l'âge du ciment et du collage lors de l'essai, des petites variations de température et de l'hygrométrie du laboratoire, en un mot de tout ce qui est identique pour toutes les éprouvettes d'une même série, mais varie d'une série à l'autre. A cet effet, nous avons, pour des raisons de commodité de dépouillement, classé nos éprouvettes, non pas suivant les trois «populations» retenues pour la statistique finale, mais suivant la taille des séries *de ruptures* après élimination des éprouvettes non rompues, rompues à la descente, ou des ruptures

Fig. 4

VE vers microcontacts de rupture et enregistreur
MO moteur
fc microcontacts fin de course
em commande de l'embrayage
fr commande du frein
Lt_1 lampe témoin néon 220 V
Lt_2 lampe témoin 28 V
z diodes *Zener*
T transformateur 220–240 V
R redresseur
rv résistance variable 100 Ω 5 W
co condensateur 470 μF 40 V
M_m minuterie 77 sec, affectée aux montées
M_p minuterie 6,5 min affectée aux palier
En trait fort: triphasé 220/380 V 50~ (secteur)
En trait fin: 24 V, continu après redressement

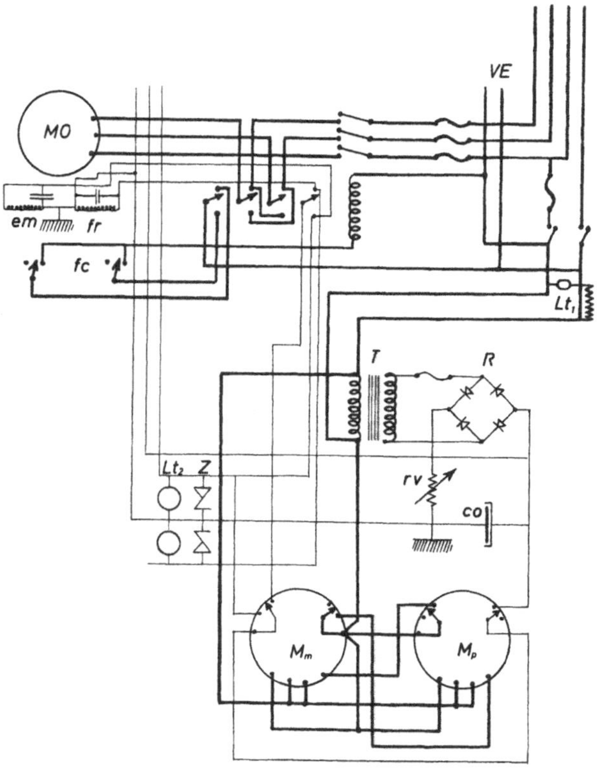

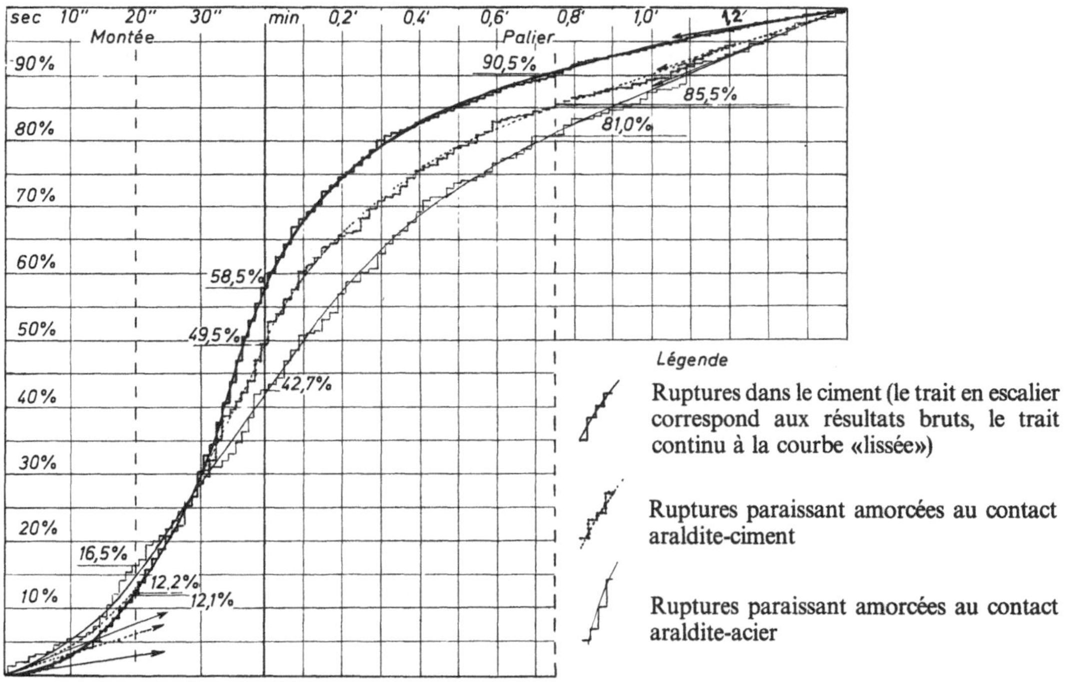

Fig. 5

non relevéees en raison d'une défaillance momentanée (Nous rappelons que la taille des séries *essayées* n'est qu'exceptionnellement inférieure à 10). En gros, les séries complètes comprennent la plupart des éprouvettes des «populations» II et III (car une série dont le collage n'est pas de

Tableau 2. Proportion des *ruptures en monté* en fonction du *N° du poste*

N° du poste	10	4	6	7	1	2	3	8	5	9
Proportion des ruptures en montée	48,5%	49%	51,5%	51,5%	52%	53%	54,5%	54,5%	57,5%	58%
	0,09	0,13	0,32	0,32	0,38	0,50	0,68	0,68	0,91	0,95

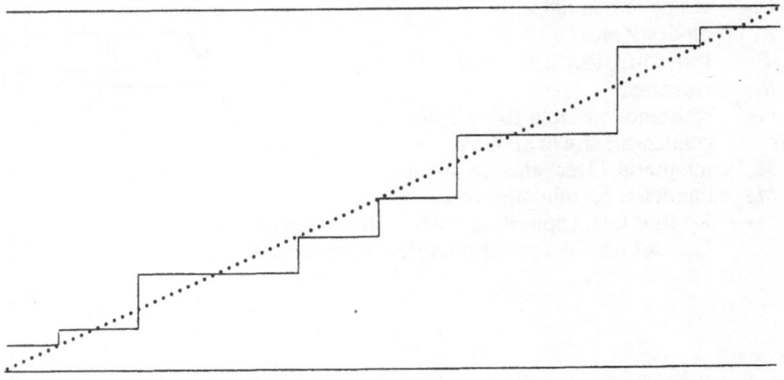

Courbe des (p)
(en tireté la diagonale théorique)

première qualité ne peut guère comprendre des éprouvettes capables de subir près de 60 bars sans amorce de rupture), et plus la série est incomplète, plus la proportion de la population I est forte. Le tableau 3 donne, en fonction de la taille, la répartition des séries d'après le nombre d'éprouvettes rompues en palier, et d'après le nombre d'éprouvettes rompues au total dans la 1ère moitié de la montée et la 2ème moitié du palier.

La répartition théorique pour une population homogène de même moyenne et de même taille est mise en regard, et les variances sont comparées. Une seule variance expérimentale dépasse significativement la variance théorique correspondante, celle du nombre de ruptures en palier dans les séries complètes (3,29 contre 2,50). Mais dans ces séries chacune de 3 «populations» (qui diffèrent notablement entre elles à ce point de vue) est largement représentée.

En définitive, les résultats bruts, que nous joignons dans le tableau 4 et les résultats du lissage graphique, paraissent traduire intrinséquement le phénomène étudié. On se doutait bien, avant toute expérimentation, qu'en gros la fréquence des ruptures augmente au cours de la montée et diminue au cours du palier. Mais la présente étude fait ressortir plusieurs particularités.

D'abord, les ruptures dans le collage ou au voisinage immédiat sont plus fréquentes en palier que les ruptures en plein ciment, leur proportion pouvant même, avec 58%, se rapprocher de la proportion de la durée du palier à la durée du cycle (69%).

D'autre part, pour les ruptures en plein ciment le phénomène «d'hérédité», moins marqué en ce qui concerne les ruptures en palier (42%) est très marqué en ce qui concerne le retard de l'augmentation de la fréquence dans le début de la montée. Au milieu de la montée, les ruptures cumulées représentent 21% de celles de la montée entière.

Mais le résultat le plus intéressant paraît être la mise en évidence d'une légère décroissance, ou au moins d'un plafonnement, de la fréquence des ruptures avant la fin de la montée, pour les ruptures dans le ciment tout au moins. La taille de nos «populations» n'est malheureusement pas encore suffisante pour que le phénomène puisse être affirmé avec une quasi-certitude, mais nous avons de fortes raisons de croire à sa réalité.

1.

La fréquence moyenne observée est, au début, du palier, inférieure de plus de 39% (pour la tranche 0–6 sec) à la fréquence moyenne observée en fin de montee pour la tranche de même durée (34–40 sec). Or, pour des raisons théoriques, la courbe des fréquences f ne doit pas présenter de discontinuité mais seulement un point anguleux, à la fin de la montée. Une valeur largement positive de df/dt avant la fin de la montée ne serait compatible avec les moyennes observées que si, après le passage du point anguleux, f subissait une chute invraisemblablement rapide.

Tableau 3. Répartition des séries suivant le nombre d'éprouvettes rompues dans certaines portions du cycle

Taille *N* de la série	Nombre de séries	Moyenne *M* de ruptures par série	Nature de la répartition	Nombre de *séries* présentant, dans la phase considérée, le nombre de *ruptures* inscrit dans la bande ci-dessous											Variance
				0	1	2	3	4	5	6	7	8	9	10	
I. Ruptures en palier															
10	122	4,877	obs^{vée}	1	2	11	12	23	32	13	21	6	"	1	3,29
			théor^{e}	0,15	1,42	6,20	15,73	26,21	29,94	23,75	12,92	4,61	0,98	0,09	2,50
9	32	4,062^5	obs^{vée}	"	"	5	9	5	8	2	3	"			2,24^5
			théor^{e}	0,145	1,065	3,51	6,74	8,32	6,845	3,755	1,325	0,27	0,025		2,23
8	16	4,187^5	obs^{vée}			2	3	4	5	1	1	"			1,81
			théor^{e}	0,04	0,38	1,44	3,16	4,33	3,81	2,09	0,66	0,09			1,99^5
7	22	2,818	obs^{vée}	1	3	3	10	2	3	"	"				1,69
			théor^{e}	0,59	2,82	5,70	6,405	4,315	1,74	0,395	0,03				1,68^5
<7	15														
II. Ruptures dans la 1e moitié de la montée et la 2e moitié du palier															
10	122	2,738	obs^{vée}	6	18	29	32	26	9	1	1	"	"	"	1,92
			théor^{e}	4,97	18,77	31,84	32,01	21,12	9,55	3,00	0,65	0,09	0,01	0,00	1,99
9	32	2,156	obs^{vée}	3	9	8	5	6	1	"	"	"	"		1,82
			théor^{e}	2,72	7,71	9,72	7,15	3,38	1,065	0,22	0,03	0,00			1,64
8	16	1,50	obs^{vée}	3	6	4	2	1	"	"	"	"			1,25
			théor^{e}	3,04	5,615	4,53	2,09	0,60	0,11	0,015	0,00				1,22
7	22	1,41	obs^{vée}	3	12	4	1	2	"	"	"				1,17
			théor^{e}	4,55	8,045	6,09	2,56	0,645	0,10	0,01	0,00				1,12
<7	15														

Tableau 4. Résults bruts généraux. (Pour chacune des 3 populations, la 6ème colonne est le total par ligne des 5 premières, la 7ème le cumul de la 6ème, la 8ème la proportion)

	Population I								Population II								Population III							
Montée							0	0							0	0							0	0
0– 5 sec	1	1	2	3	1	8	8	0,008^{5}	4	1	2	1	1	9	9	0,015	5	1	"	3	1	10	10	0,030^{5}
5–10 sec	4	3	6	4	4	21	29	0,031	3	2	8	4	3	20	29	0,048	"	2	4	1	2	9	19	0,058
10–15 sec	5	4	2	8	16	35	64	0,068	2	6	1	4	7	20	49	0,082	1	"	2	3	5	11	30	0,091^{5}
15–20 sec	5	10	17	8	10	50	114	0,121	4	3	8	5	4	24	73	0,122	5	6	6	4	3	24	54	0,165
20–25 sec	20	16	15	11	20	82	196	0,208	11	13	8	8	13	53	126	0,211	3	7	2	3	3	18	72	0,219^{5}
25–30 sec	9	7	28	14	32	90	286	0,303^{5}	5	6	13	11	11	46	172	0,288	4	4	5	4	5	22	94	0,286^{5}
30–35 sec	25	16	31	27	27	126	412	0,438	11	12	17	21	5	66	238	0,398	3	4	"	3	4	14	108	0,329
35–40 sec	30	31	21	28	24	134	**546**	**0,580**	10	9	11	16	12	58	**296**	**0,495**	6	5	8	11	2	32	**140**	**0,427**
Palier																								
0 –0,10 min	23	22	19	21	12	97	643	0,683	20	7	13	10	15	65	361	0,604	3	3	8	5	7	26	166	0,506
0,10–0,20 min	9	8	22	9	10	58	701	0,745	6	7	10	2	7	32	393	0,657	1	1	6	4	9	21	187	0,570
0,20–0,30 min	8	9	15	9	14	55	756	0,803	2	1	13	9	6	31	424	0,709	5	"	5	2	8	20	207	0,631
0,30–0,40 min	3	5	6	3	6	23	779	0,828	5	3	8	7	7	30	454	0,758^{5}	5	4	2	4	6	21	228	0,695
0,40–0,50 min	2	4	10	5	5	26	805	0,856	3	1	6	4	5	19	473	0,791	5	2	"	5	2	14	242	0,737
0,50–0,60 min	1	5	4	6	4	20	825	0,877	1	4	7	3	8	23	496	0,829^{5}	2	"	2	2	4	10	252	0,768
0,60–0,70 min	4	4	3	3	1	15	840	0,893	1	1	1	2	4	9	505	0,845	2	"	2	3	6	13	265	0,808
0,70–0,80 min	2	3	8	3	5	21	861	0,915	2	1	3	3	3	12	517	0,865	"	"	1	4	3	8	273	0,832
0,80–0,90 min	4	3	4	"	3	14	875	0,930	1	"	3	1	3	8	525	0,878	"	"	2	1	2	5	278	0,847^{5}
0,90–1 min	3	1	3	1	3	11	886	0,941^{5}	1	"	3	1	4	9	534	0,893	"	2	2	2	3	9	287	0,875
1 –1,10 min	3	1	4	1	4	13	899	0,955	1	"	7	4	3	15	549	0,918	1	"	1	2	9	13	300	0,915
1,10–1,20 min	4	1	3	1	1	10	909	0,965^{5}	4	2	6	1	1	14	563	0,942	1	1	"	1	2	5	305	0,930
1,20–1,30 min	1	2	3	6	1	13	922	0,980	3	1	3	2	"	9	572	0,957	1	"	5	"	1	7	312	0,951
1,30–1,40 min	1	2	3	"	3	9	931	0,989^{5}	3	2	5	3	5	18	590	0,987	3	"	3	3	"	9	321	0,979
1,40–1,50 min	1	2	1	4	2	10	**941**	**1,000**	1	"	2	2	3	8	**598**	**1,000**	1	1	1	2	2	7	**328**	**1,000**

Pour chacune des 3 populations: 5 premières colonnes correspondent chacune à $^1/_5$ de l'intervalle, soit 1 sec en montée et 0 min, 02 = 1 sec, 2 en palier. Ex. Popul. I, ligne palier 0 min, 10–0,20 3ème colonne: 22 signifie que 22 éprouvettes se sont rompues dans le ciment, dans l'intervalle 0 min, 14–0 min, 16 du palier.

2.

L'étude des «couples», qui malheureusement ne peut porter que sur des «populations» de taille réduite, mais qui accentue les variations de fréquence (voir tableau) donne des indications de même sens.

3.

Notre étude référencée ci-dessus, portant sur un phénomène de rupture assez différent, mais dans lequel certaines caractéristiques (par exemple le rapport des fréquences moyennes en palier et en montée) étaient du même ordre de grandeur, paraissait mettre en évidence le même phénomène.

Il semble que le maximum de la fréquence se situerait 5–6 sec avant la fin de la montée et dépasserait d'environ 20–30% la fréquence à l'instant fin de montée – début de palier.

Nous tenons à remercier le Service du Matériel du L.C.P.C. qui a étudié et réalisé le matériel spécialisé nécessaire au présent travail (sauf l'enregistrement) et le Laboratoire Régional de Clermont-Ferrand, notamment Monsieur *Patier*, précédent Directeur, grâce auquel nous avons pu disposer d'un emplacement bien adapté, et Monsieur *Laporte*, Chef du Service Electronique qui a étudié et réalisé les circuits d'enregistrement et dépanné le matériel à diverses reprises.

Résumé

La charge de rupture que peut subir une éprouvette dépend beaucoup des variations antérieures de la charge appliqueé, en fonction du temps. On observe couramment des ruptures au cours d'un palier de charge.

Nous nous sommes proposé de déterminer, lors d'un processus de charge par montées et paliers alternés, la répartition des instants de rupture par rapport à la «période» constituée par une montée et un palier (c'est la vitesse d'augmentation de la charge qui est périodique, constante pendant la montée et nulle pendant le palier). Opérant sur prés de 2000 éprouvettes, nous avons obtenu une forte augmentation de la fréquence des ruptures entre le début et la fin de la montée, une forte diminution entre le début et la fin du palier. (Il ne senible pas qu'il y ait une discontinuité de cette fréquence entre montée et palier.) Le maximum de la fréquence se situerait toutefois, non lors du passage de la montée au palier, mais un peu avant.

Zusammenfassung

Die Bruchlast eines Probekörpers hängt in hohem Maße von den vorhergehenden, zeitlich abhängigen Belastungsveränderungen ab. Oft werden Brüche bereits während einer Belastungsstufe festgestellt.
Daher wollten wir bei einem Belastungsvorgang mit regelmäßig abwechselnden Belastungssteigerungen und Belastungsstufen die Verteilung von den Bruchzeitpunkten in bezug auf die aus einer Belastungssteigerung und einer Belastungsstufe bestehenden Periode feststellen (dabei ist die Geschwindigkeit der Belastungserhöhung periodisch: sie ist konstant während der Belastungssteigerung und gleich Null während der Belastungsstufe).

Versuche an 2000 Probekörpern zeigten eine starke Zunahme der Bruchhäufigkeit zwischen dem Anfang und dem Ende der Belastungssteigerung und eine starke Abnahme dieser Häufigkeit zwischen dem Anfang und dem Ende der Belastungsstufe. (Anscheinend gibt es keinen Häufigkeitssprung zwischen der Belastungssteigerung und der Belastungsstufe.)

Die häufigsten Brüche wurden jedoch nicht während des Überganges der Belastungssteigerung zur Belastungsstufe festgestellt, sondern etwas früher.

Summary

The failure load which a sample can withstand depends largely on previous variations in the load applied, in function of time. Failures are commonly observed in the course of a given level of loading.

The author sets out to determine, in the course of a loading process by alternate increases and levels, the distribution of moments of failure in relation to the "period" (that is to say that the rate of increase in load is periodic, constant during the time of increase and zero while the load remains at a given level).

Working on more than 2000 samples, the author obtained a marked increase in the frequency of failures between the beginning and the end of the load increase, and a marked reduction between the beginning and the end of the time during which the load remained at a given level. (There does not seem to be a discontinuity of this frequency between load increase and load level.) The maximum frequency lies, however, not during the transition between load increase and load level, but slightly before that stage.

Adresse de l'auteur:

Dr. *M. Davin*
18, avenue Jocelyn Bargoin
F-63130 Royat (France)

Rheol. Acta 13, 168–172 (1974)

Mitteilung aus dem Laboratorium der Firma Brabender Messtechnik KG, Duisburg (Deutschland)

Ein neuartiges Prozeßviskosimeter

Von W. Heinz

Mit 6 Abbildungen

1. Einführung

Ein Prozeßviskosimeter soll dazu dienen, während eines Herstellungsprozesses kontinuierlich die Viskosität einer Flüssigkeit oder pastösen Substanz zu messen, und zwar in einer Rohrleitung, einem Kessel oder in einem sonstigen Produktionsgefäß. Rotationsviskosimeter, bei denen das Drehmoment gemessen wird, eignen sich hierfür deswegen in besonderer Weise, da bei diesem Gerätetyp das Meßsignal kontinuierlich ansteht. Entsprechend sind eine Anzahl von Konstruktionen bekannt geworden, bei denen ein Meßkörper, der meistens rotationssymmetrische Gestalt hat, an einer Meßwelle befestigt ist. Die Meßwelle wird mit konstanter Geschwindigkeit angetrieben; das aus dem viskosen Fließwiderstand resultierende Drehmoment wird gemessen. Eine besondere Schwierigkeit besteht nun darin, daß der Meßraum, also die Rohrleitung oder der Produktionskessel, vielfach unter Über- oder Unterdruck steht. Damit muß die Meßwelle also gegen den Meßraum abgedichtet werden. Übliche Dichtungen, wie man sie sonst von der Technik kennt, scheiden jedoch als Dichtungselement aus, da dadurch ja zusätzliche Reibung an der Meßwelle entsteht, die eine Meßwertverfälschung zur Folge hat. Es sind daher Lösungen bekannt geworden, bei denen der Antrieb des Meßkörpers über eine Magnetkupplung erfolgt. Es ist dann jedoch erforderlich, den Meßkörper innerhalb der Meßsubstanz zu lagern, da anders eine konstante Positionierung des Körpers nicht zu erreichen ist. Die Lager werden in den meisten Fällen als Zapfenlager ausgebildet, sie werden von der Meßsubstanz benetzt. Diese Lösung kann daher dann nicht befriedigen, wenn die Substanz körnige oder faserige Partikel enthält oder nach Beendigung des Herstellungsprozesses zur Bildung aushärtender Filme neigt. Weiterhin verlangt die Magnetkupplung eine verhältnismäßig eng dimensionierte Übergangszone für die magnetischen Feldlinien. In diese enge Übergangszone dringt aber Meßsubstanz ein, die dort meistens verbleibt, sich also mit der Umgebung nur sehr ungenügend austauscht. Auch hier muß es zu unerwünschten Meßeffekten kommen, wenn sich die Viskosität der Substanz schnell ändert, oder wenn die Substanz aushärtet.

2. Prinzip und Konstruktion

Hier soll über ein neuartiges Prozeßviskosimeter berichtet werden, das zunächst ebenfalls ein Rotationsviskosimeter ist. Der Handelsname des Gerätes ist Convimeter[1]). Zur Lösung der vorstehend skizzierten Problematik wurde von der Erkenntnis ausgegangen, daß sich eine Drehbewegung durch Superposition von zwei zeitlich und räumlich um 90° versetzten Hin- und Herbewegungen erzeugen läßt. Eine Welle oder ein Stab, die in einen druckerfüllten Meßraum hineinragen und dort Hin- und Herbewegungen vollführen, lassen sich aber bequem und zuverlässig und vor allen Dingen ohne Reibung abdichten durch einen flexiblen Hohlkörper, der mit seinem einen Ende an dem Stab und mit seinem anderen Ende an der Wand des Meßraumes befestigt ist. Als flexibler Hohlkörper dient zweckmäßigerweise ein Metallbalg bzw. ein gewelltes Metallrohr, und zwar aus korrosionsfestem Edelstahl.

Abb. 1 zeigt zunächst ganz allgemein den Grundgedanken, auf dem das Prozeßviskosimeter basiert. Die Welle hat im Meßraum einen Knick. Bei ihrer Drehung führt das im Meßraum gelegene Wellenende eine Drehbewegung um eine körperfremde Achse, nämlich um die

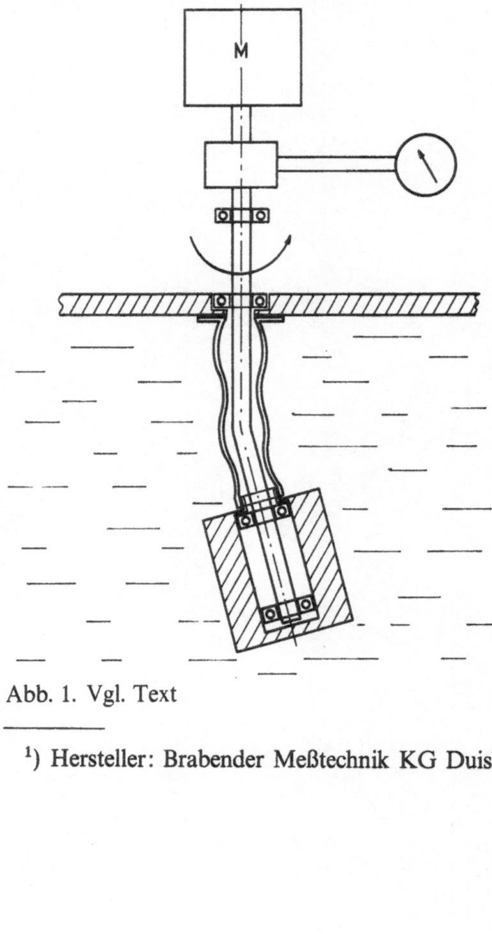

Abb. 1. Vgl. Text

[1]) Hersteller: Brabender Meßtechnik KG Duisburg.

Seele des geraden Teiles der Welle, aus. Eine solche Bewegung nennt man auch Taumelbewegung. Der Metallbalg und im Anschluß daran der Meßkörper werden über die beiden Lager von dem die Taumelbewegung vollführenden Wellenende mitgenommen und führen dementsprechend ebenfalls die Taumelbewegung aus. Diese Taumelbewegung ist nichts anderes als eine Superposition von zwei zeitlich und räumlich um 90° zueinander versetzten Hin- und Herbewegungen. Durch den viskosen Fließwiderstand entsteht ein kontinuierliches, d. h. dauernd anliegendes und diesem Fließwiderstand proportionales Drehmoment an dem Teil der Welle, der sich außerhalb des Meßraumes befindet. Die Lager werden von der Meßsubstanz nicht benetzt. Totzonen mit geringem Substanzaustausch sind, wie die Abbildung erkennen läßt, nicht vorhanden.

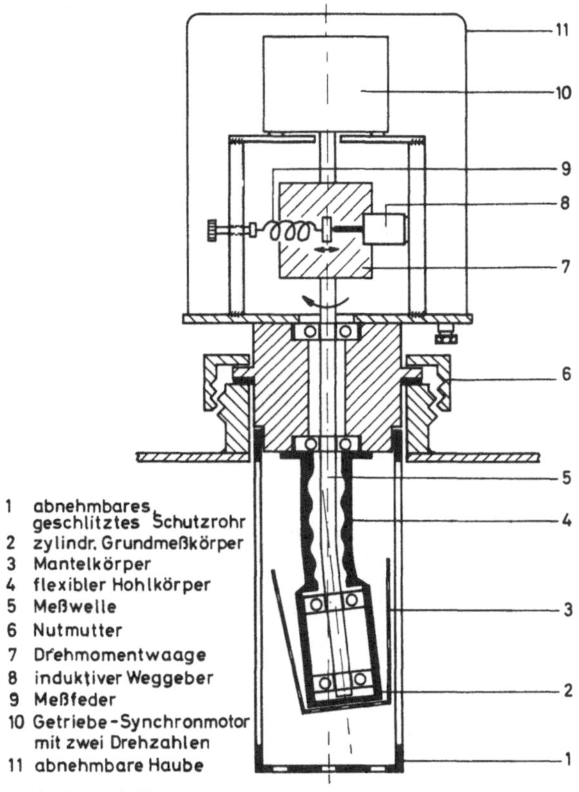

Abb. 2. Vgl. Text

Abb. 2 zeigt nun die technische Ausführung des Gerätes. Als Antrieb dient ein Synchronmotor (10), der zwei Drehzahlen abgibt. Die Drehmomentmessung erfolgt über eine sogenannte Drehmomentwaage (7), die gegen eine Meßfeder (9) arbeitet. Der Federweg, der dem Drehmoment proportional ist, wird über den induktiven Geber (8) einem Meßgehäuse übermittelt. Bemerkenswert ist noch die Formgebung des Meßkörpers (3). Dieser hat eine glockenartige, konische Gestalt, wobei der Winkel des Konus identisch ist mit dem des abgewinkelten Teiles der Welle. Auf diese Weise wird eine geringe Spaltbreite zu dem umgebenden Schutzrohr (1) erzeugt, was für die Messung sehr niedrigviskoser Substanzen wichtig ist. Bekanntlich besteht ja die grundsätzliche Schwierigkeit bei der Messung niedrigviskoser Substanzen darin, ein genügend großes Drehmoment zu erzeugen. Mit sinkender Spaltbreite wächst aber das Drehmoment bei gleichbleibender Viskosität. Das Schutzrohr (1) ist aus mehreren Gründen vorhanden. Zunächst dient es dazu, den Einfluß der Strömung der Meßsubstanz, z. B. in einer Rohrleitung oder durch die Wirkung eines Rührwerkes, auf den Meßwert zu eliminieren. Weiterhin vergrößert das Schutzrohr das entstehende Drehmoment. Um einen Substanzaustausch zwischen dem Inneren des Schutzrohres, der sogenannten Meßzone, und der Umgebung zu ermöglichen, ist das Schutzrohr mit mehreren Längsschlitzen versehen. Der Substanzaustausch zwischen der Meßzone und der Umgebung erfolgt nun durch einen Pumpeffekt, den der Meßkörper bei seiner Taumelbewegung erzeugt. Vor dem Meßkörper herrscht nämlich ein kleiner Überdruck und hinter ihm herrscht durch die Taumelbewegung ein Unterdruck. Entsprechend wird vor dem Meßkörper ständig Substanz aus der Meßzone nach außen gedrückt, und sie wird hinter ihm aus der Umgebung in das Schutzrohr hineingesogen. Dieser Zwangsaustausch der Substanz ist außerordentlich wichtig für ein schnelles Ansprechen des Gerätes auch bei sehr schnellen Viskositätsänderungen. Die Ausgleichszeit des Gerätes, d. h. die Zeit, in der das Gerät einer plötzlichen Viskositätsänderung der Umgebung bis auf 90% gefolgt ist, beträgt dadurch nur ca. 30 s, was als ein weiterer Vorteil dieses Konstruktionsprinzips zu werten ist.

3. Störgrößen

Als Störgröße ist zunächst der Druck im Meßraum zu nennen. Durch den Über- oder Unterdruck im Meßraum resultiert auf die Lager des Gerätes eine Axialkraft. Ansteigende Axialkraft bedeutet ansteigende Blindreibung der Kugellager. Die Größe der Axialkraft ergibt sich aus dem Druck und dem wirksamen Querschnitt des Metallbalges, der 1,2 cm^2 beträgt. Weiterhin wird bei steigendem Druck der Metallbalg seitlich gegen die Welle gedrückt, wobei der Anlagepunkt mit der Drehbewegung auf der Welle herumwandert, so daß außer kleinen Versatzbewegungen keine nennenswerte Friktion zwischen Metallbalg und Welle entsteht. Die Welle ist übrigens mit einem Teflonrohr überzogen. Die Seitenkraft auf den Balg beträgt etwa $^1/_{10}$–$^1/_{20}$ der Axialkraft auf die Lager. Das durch diese Einflüsse auftretende Drehmoment hängt nur vom Druck ab und ist nicht viskositätsabhängig. Demzufolge beobachtet man also durch steigenden Druck eine Nullpunktwanderung des Gerätes. Die Abb. 3 zeigt die Abhängigkeit der Blindreibung vom Druck. Beim Normalgerät ergibt sich so die Abhängigkeit des Nullpunktes vom Druck zu einem Skalenteil pro atü, wobei das Gerät eine 100teilige Skala aufweist. Durch entsprechende Korrektur des Nullpunktes läßt sich der Druckeinfluß eliminieren[2]).

[2]) Automatische Druckkompensation ist vorgesehen.

Man könnte auf den Gedanken kommen, daß der Metallbalg eine Störquelle darstellt. Daher hierzu eine kurze Betrachtung. Bei Normaldruck im Meßraum berührt der Metallbalg die Meßwelle praktisch nicht. Durch die Bewegung des Balges entsteht kein meßbares Drehmoment, denn er ist weitgehend frei von innerer Reibung.

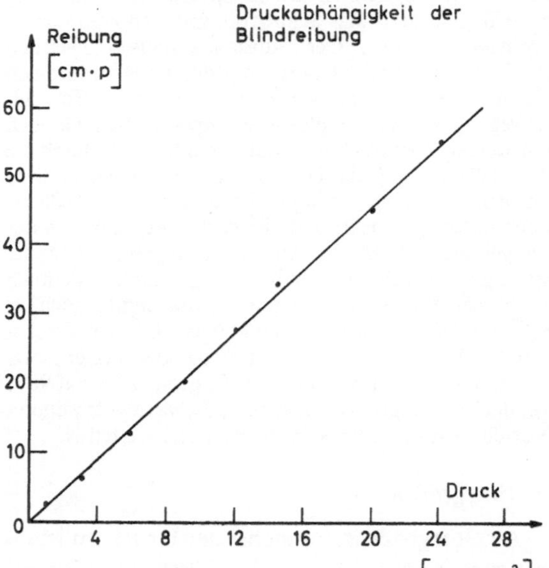

Abb. 3. Vgl. Text

Der Balg hat eine gewellte Form. Die Wellen können sich mit Meßstubstanz zusetzen, besonders bei aushärtenden Substanzen. Dieser Vorgang bewirkt zunächst grundsätzlich ein Ansteigen der inneren Reibung des Balges. Nicht nur praktische Versuche, sondern auch eine theoretische Überlegung zeigen jedoch, daß das daraus resultierende Stördrehmoment sehr klein ist: bei Bewegung des Balges führen die Wellen Abstandsänderungen aus. Rein rechnerisch kann man daher den Balg mit den bekannten Formeln behandeln, die für ein Parallelplatten-Kompressions-Viskosimeter gelten. Die Rechnung zeigt dann, daß bei den gewählten Balgabmessungen eine Substanz mit einer Viskosität von 10^6 cP, die sich zwischen den Wellen befindet, ein sehr kleines Drehmoment von 0,6 cmp an der Meßwelle erzeugt. Das sind nur zwei Promille des Nenndrehmomentes des Gerätes. Es kann also auch eine im Laufe der Zeit aushärtende und sehr hochviskos werdende Substanz sich in den Wellen des Balges befinden, ohne daß eine ins Gewicht fallende Fehlmessung daraus resultiert, oder mit anderen Worten, es ist keinesfalls erforderlich, den Balg von anhaftender Substanz zu säubern.

4. Temperaturkompensation

Ein Prozeßviskosimeter soll in vielen Fällen nicht die bei irgendeiner Betriebssituation tatsächlich vorhandene Viskosität messen, sondern es soll die gemessene Viskosität auf eine in der Nähe der Betriebstemperatur liegende Referenz- oder Bezugstemperatur beziehen. Das folgt daraus, daß die Viskositätsmessung sehr oft nicht um ihrer selbst willen erfolgt, sondern um ein Maß für den chemischen Zustand der Substanz zu erhalten. Bei dem hier beschriebenen Prozeßviskosimeter ist eine automatische Temperaturkompensation vorgesehen. Die Temperatur wird mit einem Widerstandsthermometer Pt 100 in der Nähe des Meßortes erfaßt. Um die Wirkungsweise der Temperaturkompensation zu verstehen, blicken wir auf Abb. 4. Die Viskosität

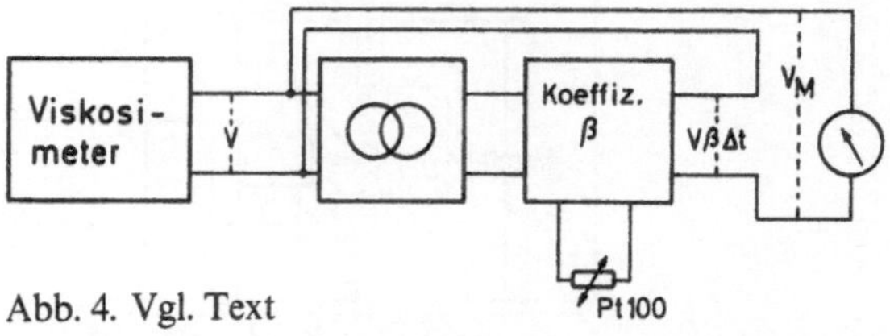

Abb. 4. Vgl. Text

einer Substanz hängt in den meisten Fällen von der Temperatur nach der Gleichung 1 ab, die

$$\eta_{(T)} = \eta_\infty e^{\frac{a}{T}} [P] \qquad [1]$$

von *Vogel* angegeben wurde. Diese Gleichung läßt sich in der Umgebung der Referenztemperatur T_0 nach der Gleichung 2 in eine Potenzreihe

η_∞, a = Stoffkonstanten, T = abs. Temperatur

$$\eta_{(T_0+\Delta t)} = \eta_{(T_0)}[1 - \alpha_1 \Delta t + \alpha_2 \Delta t^2 - \alpha_3 \Delta t^3 + \cdots]$$

$$\alpha_1 = \frac{a}{T_0^2}; \quad \alpha_2 = \frac{a}{T_0^3} + \frac{a^2}{2T_0^4}; \quad \alpha_3 = \frac{a}{T_0^4} + \frac{a^2}{T_0^5} + \frac{a^3}{6T_0^6} \qquad [2]$$

entwickeln mit den Koeffizienten α_1, α_2, α_3, wobei die Koeffizienten α in der Weise von der Stoffkonstanten a und der Referenztemperatur T_0 abhängen, wie es die Gleichung zeigt. Wenn man mit Zahlen in die Gleichung für die Koeffizienten eingeht, so sieht man, daß die Gleichung 2 das Viskositäts-Temperaturverhalten sehr gut wiedergibt bis zu Temperaturabweichungen Δt von etwa $\pm 10°$. Im Bild ist der Schaltungsgedanke dargestellt. Das Viskosimeter liefert zunächst eine Meßspannung V, die proportional ist der tatsächlich vorhandenen Viskosität. Zu dieser Spannung wird elektrisch eine Spannung addiert, die proportional ist der Meßspannung V, der Temperaturabweichung Δt und einem einstellbaren Koeffizienten β. Wenn man für die viskositätsabhängige Meßspannung V_M (siehe Gleichung 3)

T_0 = Referenztemperatur

V = proport. $\eta_{(T_0+\Delta t)}$

$$V_M = V + V\beta\Delta t = V[1 + \beta\Delta t] \qquad [3]$$

nun die Gleichung 2 für die temperaturabhängige Viskosität eingehenläßt, so erhält man die Gleichung 4. Von der Kompensation wird nun verlangt, daß das durch eine Klammer zusammengefaßte Glied der Gleichung 4 der Zahl 1

$$V_M \simeq \eta_{(T_0)} \underbrace{[1 - \alpha_1\Delta t + \alpha_2\Delta t^2 - \alpha_3\Delta t^3](1 + \beta\Delta t)}_{\approx 1} \qquad [4]$$

möglichst nahe kommt, denn dann ist die vom Gerät angezeigte Meßspannung V_M proportional der Viskosität bei der Referenztemperatur T_0.

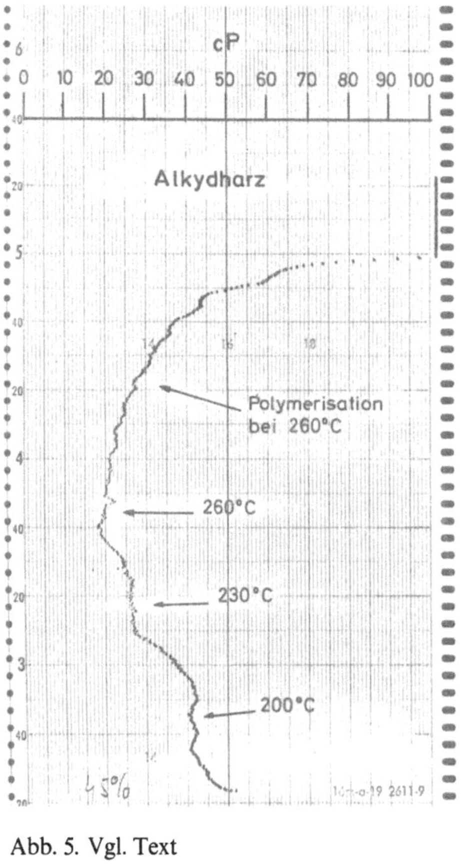

Abb. 5. Vgl. Text

Wenn man in die Gleichung 4 mit Zahlen eingeht, so zeigt sich, daß das Klammerglied der Zahl 1 tatsächlich sehr nahe kommt, sofern β etwa 5% kleiner gewählt wird als der Koeffizient α_1. Der Kompensationsfehler bleibt dann innerhalb 2%, wenn die Abweichungen der Temperatur am Meßort von der eingestellten Referenztemperatur kleiner als 3° bleiben, bei einem Temperaturkoeffizienten der Substanz von 10% pro °C. Oder mit anderen Worten: Die Kompensation funktioniert befriedigend bei durch Temperaturunterschiede bewirkten Viskositäts-

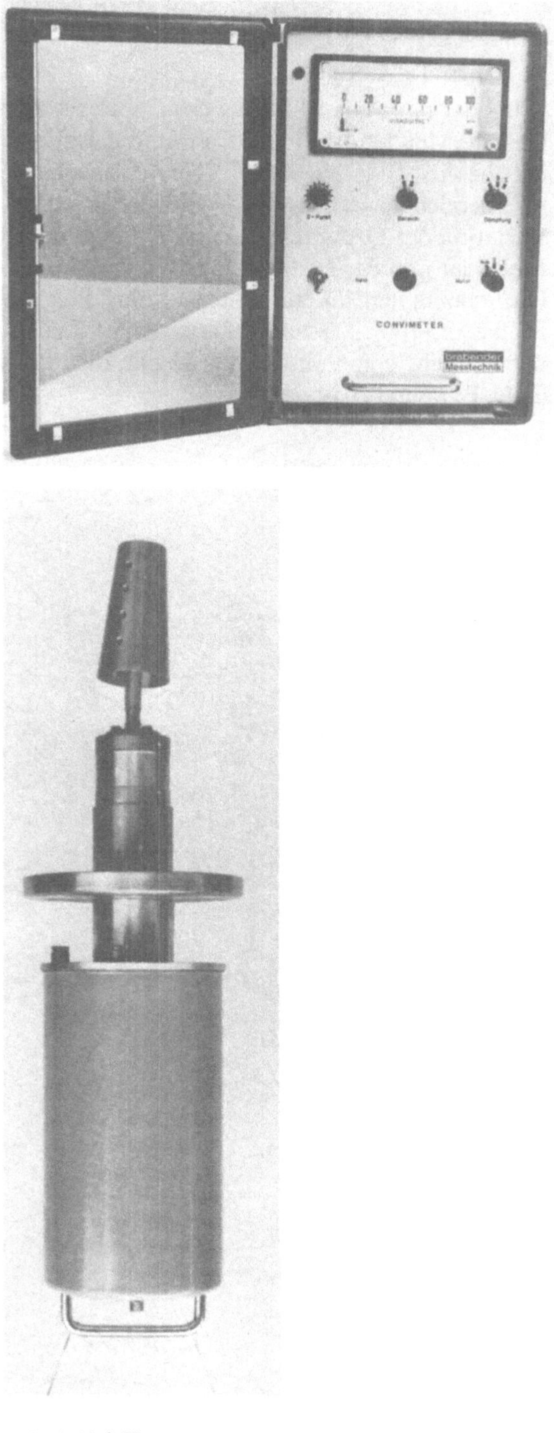

Abb. 6. Vgl. Text

gängen bis zu 30%. Bei kleineren Temperaturkoeffizienten der Substanz können die Temperaturabweichungen also größer sein. In der Praxis wird der an der Temperaturkompensationseinrichtung einzustellende Koeffizient dadurch bestimmt, daß die Viskosität der Substanz bei zwei dicht benachbarten Temperaturen gemessen und der Differenzquotient gebildet wird.

5. Daten des Gerätes

Bei der Normalausführung des Gerätes ist der kleinste Meßbereich 0 ... 100 cP und der größte Meßbereich 0 ... 250000 cP. Die verschiedenen Meßbereiche werden vom Anwender durch Variation der Drehzahl (120 und 15 U/min), der Meßfeder und der Meßkörpergröße vorgewählt. Die maximalen Drehmomente sind, je nach Meßfeder, 100, 300 und 900 cmp. Der Druckbereich geht von Vakuum bis 25 atü, die maximale Temperatur ist 280 °C.

In Sonderausführung wird das Gerät bis zu Viskositäten von 10^7 cP und bis zu Drucken von 75 atü geliefert. Der Meßkopf weist Nennweiten von 50 und 80 mm auf (wählbar); er wird bis zu Eintauchtiefen von 2000 mm gebaut. Das Gerät wird mit *Newton*schen Eichflüssigkeiten eingeeicht. Abb. 5 zeigt den Viskositätsverlauf eines Herstellprozesses von Alkydharz in einem Reaktionskessel, gemessen mit dem Convimeter. Die Meßwerte wurden von einem Punktschreiber registriert.

Abb. 6 zeigt eine Ansicht des Prozeßviskosimeters, bestehend aus dem Meßkopf und dem Meßgehäuse.

Anschrift des Verfassers:
Dipl.-Phys. *W. Heinz*
Brabender Messtechnik KG
4100 Duisburg (Deutschland)

Rheol. Acta **13**, 173–176 (1974)

Évolution des paramètres rhéologiques de polymères en solution avec application à l'hydrodynamique

Par L.-A. Sackmann, C. Gebel, H. Reitzer et O. Scrivener

Avec 9 figures

(Reçu p. p. le 27 octobre 1972)

Les travaux présentés dans cette communication sont les résultats de recherches entreprises à l'Institut de Mécanique de Fluides de Strasbourg sur les solutions acqueuses de corps macromoléculaires. Dans ces études la rhéologie sert de support à des recherches sur la modification des caractéristiques hydrodynamiques des écoulements tant internes qu'externes en présence de corps macromoléculaires. Parmi les solutions étudiées nous nous limiterons dans le cadre de cette communication à deux polymères: la Gomme de Guar et le Polyox 301.

1. Evolution des caractéristiques rhéologiques des solutions

Plusieurs paramètres sont responsables de l'évolution des caractéristiques hydrodynamiques des écoulements de fluide non newtonien, à savoir principalement: la concentration en polymère, la température et l'âge de la solution. L'étude de l'évolution des caractéristiques rhéologiques des solutions de Gomme de Guar et de Polyox en fonction de ces différents paramètres a été réalisée au viscosimètre rotatif et au viscosimètre d'écoulement.

a) Concentration

En ce qui concerne la Gomme de Guar les rhéogrammes, tension de cisaillement τ en fonction du gradient de vitesse D font apparaître en représentation logarithmique une évolution linéaire de τ pour les différentes concentrations (fig. 1). La solution acqueuse de ce polymère répond donc assez bien au modèle d'*Ostwald de Waele* dont la représentation logarithmique $\log\tau = n \log D + K$ permet la détermination de l'indice n du fluide et de sa consistance K. Les graphes de n et K font apparaître une faible variation de n pour les concentrations inférieures à 500 ppm, le caractère non newtonien de la solution étant peu marqué. Pour les fortes concentrations il semble que n tende vers une limite. L'augmentation de K avec la concentration est d'autant plus important que la concentration est élevée (fig. 2).

Contrairement aux précédentes, les solutions de Polyox 301 ne répondent pas parfaitement au modèle d'*Ostwald* (fig. 1), en particulier pour les gradients de vitesse élevés. La représentation de l'évolution de la viscosité apparente η_a en fonction de la concentration fait apparaître qu'à concentration égale la viscosité est une fonction du gradient de vitesse, confirmant ainsi le caractère non newtonien de la solution (fig. 3).

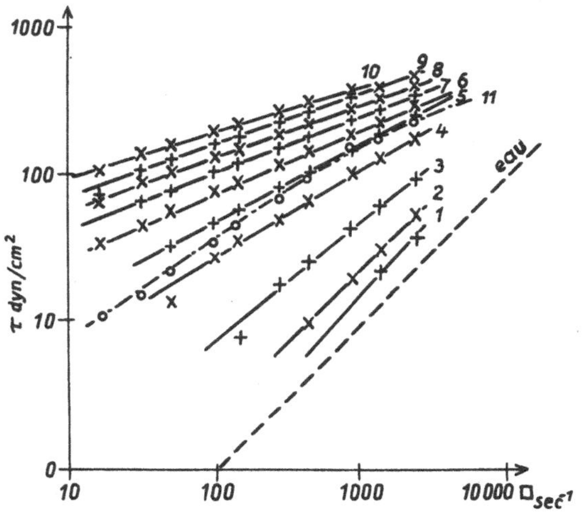

Fig. 1. $\log\tau = f(\log D)$. Gomme de Guar: $\theta = 18°$, $\Delta T = 24$ h. 1. $C = 0{,}05\%$; 2. $C = 0{,}1\%$; 3. $C = 0{,}2\%$; 4. $C = 0{,}4\%$; 5. $C = 0{,}5\%$; 6. $C = 0{,}6\%$; 7. $C = 0{,}7\%$; 8. $C = 0{,}8\%$; 9. $C = 0{,}9\%$; 10. $C = 1\%$. Polyox 301: $\theta = 20$ °C. 11. $C = 0{,}5\%$

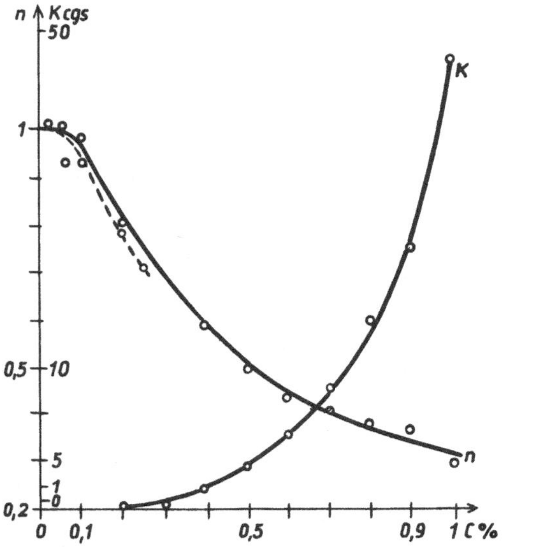

Fig. 2. Evolution de n et K en fonction de la concentration. Gomme de Guar: $\theta = 18°$, $\Delta T = 24$ h

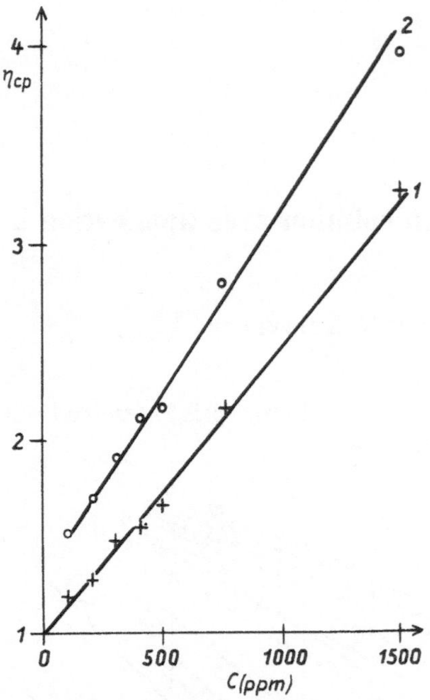

Fig. 3. Evolution de la viscosité en fonction de la concentration. Polyox 301: $\theta = 20\,°C$, $\Delta T = 20$ h. 1. $D = 2620$ sec^{-1}; 2. $D = 873{,}33$ sec^{-1}

b) Influence de la température

Un deuxième paramètre, la température, intervient dans l'étude de l'évolution des caractéristiques rhéologiques des solutions de fluides non newtoniens. L'essentiel de nos résultats concerne des solutions fortement concentrées (5000 ppm) de Polyox 301. A faible température (15 °C) l'influence de gradient de vitesse est plus importante qu'aux températures plus élevées (fig. 4). La viscosité varie peu en fonction de la température aux gradients de vitesse élevés. Inversement aux faibles gradients de vitesse la viscosité décroît rapidement avec la température.

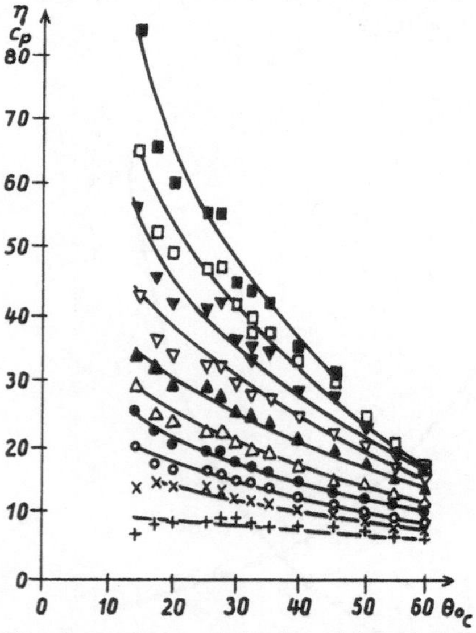

Fig. 4. Evolution de la viscosité en fonction de la température. Polyox 301: $C = 0{,}5\%$. 1. $D = 2620\,sec^{-1}$; 2. $D = 1310\,sec^{-1}$; 3. $D = 873{,}3\,sec^{-1}$; 4. $D = 476{,}6\,sec^{-1}$; 5. $D = 291{,}1\,sec^{-1}$; 6. $D = 145{,}5\,sec^{-1}$; 7. $D = 97{,}1\,sec^{-1}$; 8. $D = 48{,}5\,sec^{-1}$; 9. $D = 32{,}3\,sec^{-1}$; 10. $D = 16{,}2\,sec^{-1}$

c) Vieillissement des solutions

Certaines solutions de polymères sont très sensibles à une dégradation naturelle provoquant une évolution des caractéristiques rhéologiques dans le temps. C'est le cas en particulier pour les solutions de Gomme de Guar dont l'évolution est très marquée dans les premières heures qui suivent leur préparation. Des mesures de perte de charge en écoulement laminaire (fig. 5) montrent une augmentation de la perte de charge au cours des premières heures qui suivent la préparation puis une diminution liée à la dégradation de la solution. Ceci se traduit par une diminution de la consistance qui atteint rapidement une valeur voisine de la viscosité de l'eau, et par une valeur maximum pour l'indice du fluide.

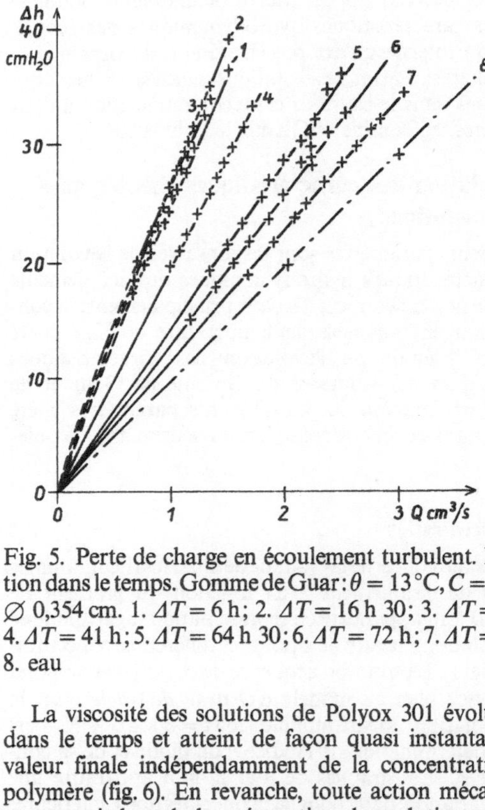

Fig. 5. Perte de charge en écoulement turbulent. Evolution dans le temps. Gomme de Guar: $\theta = 13\,°C$, $C = 0{,}1\,\%$, ∅ 0,354 cm. 1. $\Delta T = 6$ h; 2. $\Delta T = 16$ h 30; 3. $\Delta T = 18$ h; 4. $\Delta T = 41$ h; 5. $\Delta T = 64$ h 30; 6. $\Delta T = 72$ h; 7. $\Delta T = 89$ h; 8. eau

La viscosité des solutions de Polyox 301 évolue peu dans le temps et atteint de façon quasi instantanée sa valeur finale indépendamment de la concentration en polymère (fig. 6). En revanche, toute action mécanique, que ce soit lors de la préparation des solutions ou par l'action de gradients de vitesse ou de pression importants, provoque une dégradation importante de la solution difficilement reproductible et chiffrable et due vraisemblablement à la rupture des chaînes macromoléculaires.

2. Ecoulement interne de fluide non newtonien (dans une conduite)

Les résultats précédents ont abouti à l'étude des modifications des caractéristiques hydrodynamiques de l'écoulement turbulent de solutions de polyméres et en particulier de l'effet *Toms* (1).

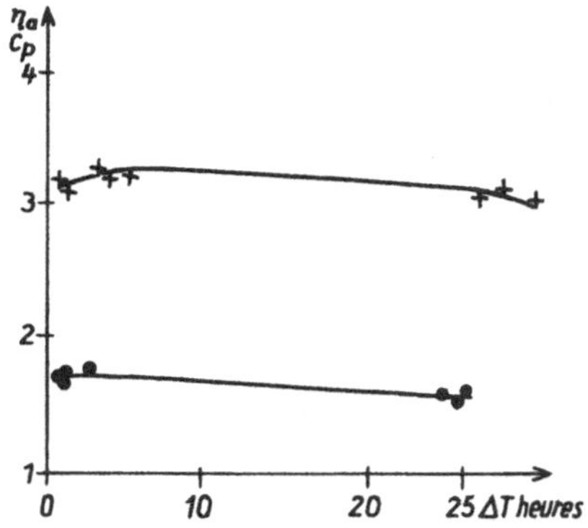

Fig. 6. Evolution de la viscosité dans le temps. Polyox 301: 1. $C = 300$ ppm, $\theta = 20$ °C, $D = 2620$ sec^{-1}; 2. $C = 1500$ ppm

a) Ecoulement interne en milieu diffus

Des réductions de perte de charge importantes ont été obtenues avec des solutions de Gomme de Guar à faible concentration (fig. 7). Cette réduction $\overline{\Delta H} = \frac{\Delta h - \Delta h\,\mathrm{eau}}{\Delta h\,\mathrm{eau}}$

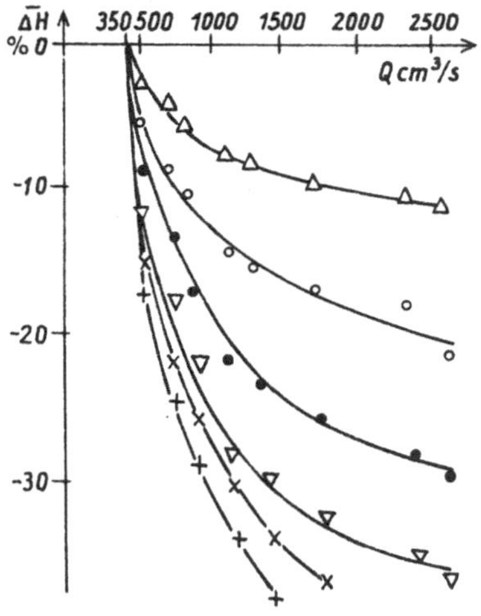

Fig. 7. Evolution de la réduction de perte de charge en fonction du débit et de la concentration. ∅ 2,198 cm. 1. $C = 1{,}5 \times 10^{-4}$; 2. $C = 1{,}0 \times 10^{-4}$; 3. $C = 7{,}5 \times 10^{-5}$; 4. $C = 5{,}0 \times 10^{-5}$; 5. $C = 2{,}5 \times 10^{-5}$; 6. $C = 1{,}0 \times 10^{-5}$

est une fonction directe de la concentration. A concentration égale, la réduction est d'autant plus importante que le débit est élevé, de sorte qu'à faible concentration (50 ppm) il est encore possible d'atteindre une réduction voisine de 30% pour un débit important. Il apparaît d'autre part que la réduction n'intervient qu'au delà d'un débit critique indépendant de la concentration, ce débit correspond à un régime turbulent de nombre de *Reynolds* supérieur à la transition laminaire – turbulent pour un écoulement d'eau. La réduction de perte de charge semble d'autre part tendre vers une limite supérieure pour les débits croissants, cette limite étant fonction de la concentration et donc de la viscosité ou de la consistance. Dans le cas du Polyox cette réduction peut atteindre 65%.

b) Ecoulement interne avec injection pariétale

Pour la Gomme de Guar l'injection pariétale annulaire à faible débit de solutions concentrées (5000 ppm) conduit à des résultats comparables, avec toutefois des performances légèrement inférieures dans le cas des débits élevés. Dans ce dernier cas, la modification des caractéristiques hydrodynamiques de l'écoulement due à l'injection elle même peut en être la cause. Ici encore un nombre de *Reynolds* critique de début de réduction a été mis en évidence.

Une solution de Polyox 301 injectée annulairement provoque une réduction de perte de charge qui augmente d'une part en fonction du débit d'injection et d'autre part en fonction de la concentration. Nos conditions expérimentales ne nous ont pas permis de montrer d'une manière formelle l'existence d'un maximum de réduction. Néanmoins l'ensemble des graphes accuse des différences de plus en plus faibles au fur et à mesure que l'on augmente soit le débit d'injection, soit la concentration (fig. 8). Ces effets sont dus à une réduction partielle de la turbulence.

En effet, et d'une façon générale, l'injection annulaire, en écoulement turbulent, se traduit par une diffusion progressive du fluide secondaire injecté, laissant au centre un cône de non mélange du fluide primaire. Dans le cas particulier de l'injection d'un fluide non newtonien la longueur de ce cône de non mélange va en croissant, ce qui correspond bien à une réduction de la turbulence.

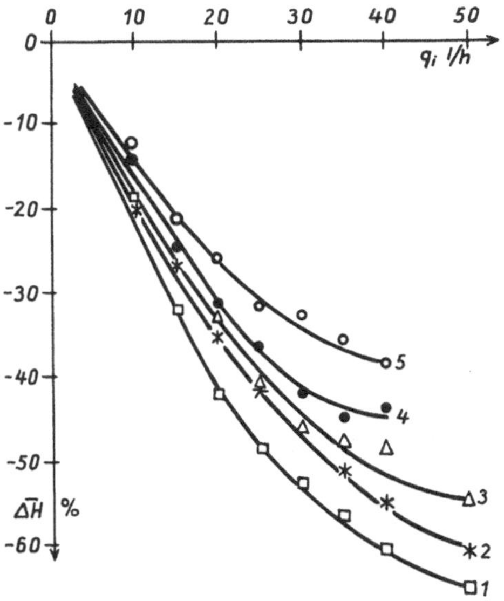

Fig. 8. Evolution de la réduction de perte de charge sous l'influence d'une injection annulaire. Polyox 301: ∅ 5 cm. 1. $C = 900$ ppm; 2. $C = 600$ ppm; 3. $C = 400$ ppm; 4. $C = 300$ ppm; 5. $C = 200$ ppm

3. Ecoulement externe en présence de fluides non newtoniens (plaque plane)

De même que dans les conduites, la présence de macromolécules dans la couche limite sur une plaque réduit la traînée de frottement dans les écoulements externes. Les études théoriques de *Wells* (2), faites à partir des résultats obtenus dans les conduites, montrent que des réductions de traînée de frottement de 80% sont possibles avec le

Polyox. Les résultats expérimentaux ont permis d'approcher ces précisions théoriques.

C'est ainsi que *Jin Wu* (3) a obtenu une réduction de 60% de la traînée de frottement d'une plaque en injectant localement en amont de la plaque par une fente des solutions macromoléculaires.

A Strasbourg, des essais en tunnel hydrodynamique ont permis d'atteindre une réduction de la traînée de frottement de plus de 70% sous l'influence d'une injection de Polyox 301. La fig. 9 montre un exemple de résultats

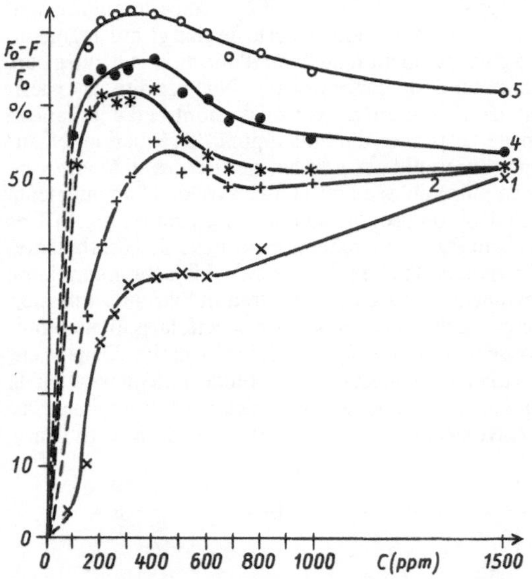

Fig. 9. Evolution de la traînée de frottement d'une plaque plane sous l'influence d'une injection de Polyox 301 dans la couche limite. U = 5,7 m/s. 1. $q_i = 0{,}5\ m^3/h$; 2. $q_i = 1\ m^3/h$; 3. $q_i = 2\ m^3/h$; 4. $q_i = 2{,}5\ m^3/h$; 5. $q_i = 3{,}5\ m^3/h$

obtenus avec le Polyox 301 pour différentes concentrations et débits d'injection. Il en résulte, pour une vitesse extérieure donnée, l'existence d'une injection optimale en débit et concentration d'une solution macromoléculaire. Remarquons que dans nos expériences les débits d'injection sont supérieurs aux débits utilisés par *Jin Wu*, notre plaque plane ayant également des dimensions plus élevées (400 × 1000 mm).

La géométrie des injecteurs a une influence sur la réduction du frottement relativement moins importante que celle des paramètres rhéologiques.

Il apparaît en effet que les produits tels que la Gomme de Guar, le CMC, le Polyox 3000, utilisés dans les mêmes conditions expérimentales, fournissent des réductions de frottement qui ne dépassent pas 30%.

Ce résultat est à comparer aux réductions de perte de charge dans les conduites sous l'influence d'une injection localisée annulaire ou le Polyox 301 donne seul des réduction importantes, alors qu'en milieu diffus ces produits non newtoniens fournissent des réductions de perte de charge supérieures à 40%.

Ces effets spectaculaires sont les conséquences des modifications des propriétés des paramètres hydrodynamiques de l'écoulement tels que le gradient de vitesse et le degré de turbulence aux différents niveaux de la couche limite. Plusieurs auteurs situent l'ensemble du mécanisme de réduction de la traînée de frottement au niveau de la sous-couche transitoire entre la sous-couche visqueuse et la sous-couche turbulente interne. Néanmoins c'est l'ensemble des vitesses qui est affecté par la présence des macromolécules.

Des mesures de vitesses au moyen de traceurs nous ont en effet montré que très près de la paroi les vitesses sont diminuées alors que dans le restant de la couche limite elles subissent une augmentation très nette faisant ainsi apparaître un τ_0 plus faible.

L'exploitation de ces travaux expérimentaux, actuellement en cours, de même que leur poursuite, est orientée d'une part vers une meilleure connaissance des mécanismes de réduction de frottement due à la présence de solutions macromoléculaires, et d'autre part vers l'étude cinématique de la diffusion des macromolécules dans la couche limite hydrodynamique.

Littérature

1) *Toms, B. A.*, Some observations on the flow of linear polymer solutions through straight tubes at large *Reynolds* numbers. Proc. Int. Congress on Rheol. 1948.

2) *Wells*, The flow of a dilute polymer solution in a turbulent boundary-layer. Symp. Dallas 1968.

3) *Jin Wu*, Drag Reduction in external flows of additive solutions. Symp. Dallas 1968.

4) *Kilian, F. P.*, Widerstandsverminderung durch Fadenmoleküle in der Grenzschicht. Schiffbautechn. Gesell. 63 (1969).

5) *Sackmann, L.-A., C. Gebel*, and *H. Reitzer*, Influence de l'injection de fluide secondaire dans la couche limite hydrodynamique. C.R.A.S. 1969.

Adresse des auteurs:

D. *L. A. Sackmann*
Institut de Mécanique des Fluides
2, rue Boussingault
F-67 Strasbourg (France)

Rheol. Acta 13, 177–179 (1974)

From the Department of Mathematics, Indian Institute of Technology, Bombay, Powai (India)

Stokes problem for elastico-viscous fluid

By V. M. Soundalgekar

With 1 figure and 1 table

(Received October 27, 1972)

1. Introduction

One of the earliest exact solutions of the well-known *Navier-Stokes* equation was given by *Stokes* (1) in the case of the flow past an impulsively started infinite plate. In the literature, it is known as *Rayleigh*'s problem. Because of its practical importance, the problem has received the attention of many research workers who solved it in case of the impulsively started bodies of different shapes. These studies are confined to *Newton*ian fluids.

In the present day technology, the Non-*Newton*ian fluids play an important part. A number of such fluids have been defined by different types of the constitutive equations. One such fluid, with slight elastic effects in shear flow, has been proposed by *Walters* (2) and is known as *Walters* liquid *B'*. The constitutive equation for this type of fluid is given as

$$p_{ik} = -p g_{ik} + p'_{ik} \qquad [1]$$

$$p'^{ik}(x, t) = 2 \int_{-\infty}^{t} \psi(t - t') \frac{\partial x^i}{\partial x'^m} \frac{\partial x^k}{\partial x'^r} \times e^{(1)mr}(x', t')\, dt' \qquad [2]$$

where p_{ik} is the stress tensor, p an arbitrary isotropic pressure, g_{ik} the metric tensor of a fixed co-ordinate system x^i, x'^i the position at time t' of the element which is instantaneously at the point x at time t, $e^{(1)}_{ik}$ the rate of strain tensor and

$$\psi(t - t') = \int_0^{\infty} \frac{N(\tau)}{\tau} \exp(-(t - t')/\tau)\, d\tau .$$

$N(\tau)$ being the distribution function of relaxation times τ. *Walters* (3) has shown that in the case of liquids with short memories (i.e. short relaxation times), the equation of state can be written in a simplified form

$$p'^{ik} = 2\eta_0 e^{(1)ik} - 2k_0 \frac{\delta}{\delta t} e^{(1)ik} \qquad [3]$$

where

$$\eta_0 = \int_0^{\infty} N(\tau)\, d\tau$$

is the limiting viscosity at small rates of shear

$$k_0 = \int_0^{\infty} \tau N(\tau)\, d\tau$$

and $\delta/\delta t$ denotes the convected differentiation of a tensor quantity, which for any contravarient tensor b^{ik} is given by

$$\frac{\delta b^{ik}}{\delta t} = \frac{\partial b^{ik}}{\partial t} + v^m \frac{\partial b^{ik}}{\partial x^m} - b^{im} \frac{\partial v^k}{\partial x^m} - b^{mk} \frac{\partial v^i}{\partial x^m} \qquad [4]$$

where v^i is the velocity vector.

2. Mathematical analysis

We now consider a plane, infinite plate moving impulsively in its own plane. The x'-axis is taken along the plate in the direction of the flow and the y'-axis is taken normal to it. Under these conditions, the flow is independent of x'. Hence, from [1], [3], and [4], the unsteady flow of an incompressible elastico-viscous fluid is governed by the following equations of motion

$$\varrho' \frac{\partial u'}{\partial t'} = \eta_0 \frac{\partial^2 u'}{\partial y'^2} - k_0 \frac{\partial^3 u'}{\partial y'^2 \partial t'} . \qquad [5]$$

The boundary conditions are

$$u' = U_0 \text{ at } y' = 0, \quad u' = 0 \text{ as } y' \to \infty. \qquad [6]$$

Here ϱ' is the density. On introducing the following non-dimensional quantities

12

$$y = y'/\sqrt{\nu\tau}, \quad T = t'/\tau, \quad u = u'/U_0$$
$$k = k_0/\eta_0\tau \qquad [7]$$

in [5] and [6], we get

$$\frac{\partial u}{\partial T} = \frac{\partial^2 u}{\partial y^2} - k\frac{\partial^3 u}{\partial y^2 \partial T} \qquad [8]$$

$$u = 1 \text{ at } y = 0, \quad u = 0 \text{ as } y \to \infty. \qquad [9]$$

Eq. [1] is the third-order differential equation when $k \neq 0$ and for $k = 0$, it reduces to an equation governing the *Newton*ian fluid. This is due to the presence of k, the elastic property parameter. Mathematically, we need three boundary conditions for a unique solution. So we follow *Beard* and *Walters* (4) and assume the solution in the form as follows:

$$u = u_0 + ku_1. \qquad [10]$$

This is a valid expansion as the values of k are $\ll 1$ and [3] is obtained from [2] on the same reasoning.

Substituting [10] in [8] and [9] and applying the usual *Laplace*-transform technique, we have the following solution:

$$u = \operatorname{erfc}(y/2\sqrt{T}) + \frac{ky}{4\sqrt{\pi T^3}}\left[1 - \frac{3}{4}\frac{y^2}{T} + \frac{5}{32}\frac{y^4}{T^2} - \frac{7}{384}\frac{y^6}{T^3}\cdots\right]. \qquad [11]$$

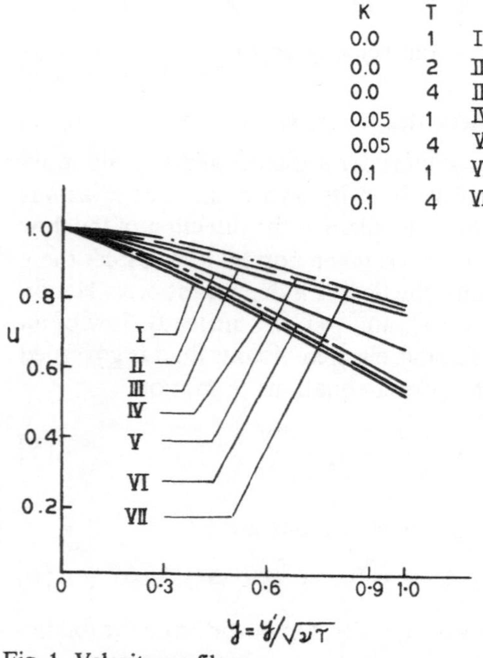

Fig. 1. Velocity profiles

Here erfc is the complimentary error function. The velocity profiles are shown on fig. 1.

We observe from this fig. 1 that due to the presence of the elastic property of the fluid, the velocity increases. It also increases with increasing time.

Knowing the velocity field, we can now calculate the expression for shearing stress. It is given by

$$P'_{x'y'}|_{y'=0} = -\left(\eta_0\frac{\partial u'}{\partial y'} - k_0\left(\frac{\partial^2 u'}{\partial y'\partial t'}\right)\right) \qquad [12]$$

and in virtue of [7], it reduces to

$$\tau_{xy} = -\frac{\sqrt{\nu\tau}}{\eta_0 U_0}P'_{x'y'}$$
$$= -\left(\frac{\partial u}{\partial y} - k\frac{\partial^2 u}{\partial y \partial T}\right)_{y=0} \qquad [13]$$

Substituting for u from [11] and putting $y=0$, we get

$$\tau_{xy} = \frac{1}{\sqrt{\pi T}} - \frac{k}{4\sqrt{\pi T^3}} - \frac{3k}{8\sqrt{\pi T^5}}. \qquad [14]$$

The numerical values of τ_{xy} are entered in table 1.

Table 1. Values of τ_{xy}

T/k	0	0.05	0.1
1	0.5641	0.5465	0.5289
2	0.3989	0.3945	0.3902
3	0.3257	0.3237	0.3216
4	0.2820	0.2808	0.2796

We observe from this table that in general, the shearing stress decreases with time. An increase in k also leads to a decrease in the shearing stress.

Acknowledgement

I am grateful to Atomic Energy of India, for the grant of a research grant to carry out this research.

Summary

An approximate solution to the flow past an impulsively started infinite plate in an elastico-viscous fluid is derived for the velocity and shearing stress. It is observed that the velocity increases with increasing the elastic parameter k and the shearing stress decreases with increasing k.

Zusammenfassung

Für den Fall der Strömung längs einer ruckartig in Bewegung gesetzten unendlichen Platte in einer viscoelastischen Flüssigkeit ist eine Näherungslösung für Geschwindigkeit und Scherspannung abgeleitet. Dabei wurde beobachtet, daß die Geschwindigkeit mit Zunahme des

Elastizitätsparameters *k* ansteigt und die Scherspannung mit zunehmendem *k* abnimmt.

References

1) *Stokes, G. G.*, Camb. Phil. Trans **9**, 8 (1851).

2) *Walters, K.*, IUTAM Int. Symp. on Second Order Effects in Elasticity, Plasticity and Fluid Dynamics, p. 507 (Eds. *M. Reiner* and *D. Abir*) (New York 1964).

3) *Walters, K.*, J. Mechanique **1**, 474 (1964).

4) *Beard, D. W.* and *Walters, K.*, Proc. Camb. Phil. Soc. **60**, 667 (1964).

Author's address:

V. M. Soundalgekar
Dept. of Mathematics
Indian Institute of Technology
Powai, Bombay (400076) (India)

Present address:

Dept. of Mech. Eng.
UWIST
Cardiff CF1 3NU (U.K.)

Rheol. Acta 13, 180–184 (1974)

From the Food Science Laboratories, Department of Applied Biochemistry and Nutrition, University of Nottingham School of Agriculture, Sutton Bonington, Loughborough, Leics. (England)

Viscoelastic behaviour of alginate gels

By J. R. Mitchell and J. M. V. Blanshard

With 6 figures and 1 table

(Received October 27, 1972)

Introduction

A number of polysaccharides are employed in food products partly because of their ability to gel under various conditions. In recent years considerable interest has been shown in these materials and a certain amount of progress has been made in elucidating the chemical nature of the inter-chain linkages which form on gelation (1).

The rheological properties of these gels have been investigated mainly by empirical methods which often involve the subjection of the material to very high stresses. Although work of this sort has proved useful in predicting the performance of the gel for a particular application, it is difficult to obtain detailed information about the gel structure from such methods. It is considered that a systematic study of the viscoelastic behaviour of polysaccharide gels in the small deformation region could provide valuable results which would complement the information on gel structure obtained from other techniques.

In this paper we report some preliminary data obtained from creep measurements on calcium alginate gels. The influence of the polysaccharide concentration, calcium ion concentration and pH on the rheological properties of gels formed from one alginate sample are considered. Recently *Smidsrod* and *Haug* have reported the effect of other variables, in particular the polysaccharide composition and the molecular weight on the stiffness of calcium alginate gels (2).

Experimental

1. Formation of gels

A sample of sodium alginate containing a relatively high proportion of D-mannuronate residues was generously supplied by Alginate Industries[1]). Calcium alginate gels were formed from this sample by two different methods both of which involved the slow release of calcium ions from a calcium salt dispersed in a solution of sodium alginate (3).

Method 1

Gels with a pH of approximately 3.5 were obtained when a solution of glucono delta lactone was added to a dispersion of calcium hydrogen orthophosphate ($CaHPO_4$) in a sodium alginate solution. The glucono delta lactone hydrolyses to give gluconic acid and the resultant lowering of the pH increases the solubility of the $CaHPO_4$. A gel is formed as the calcium ions released with the soluble alginate. In this work the concentration of sodium alginate and $CaHPO_4$ was varied but that of glucono delta lactone was maintained at 2% w/v in the final mixture.

Method 2

Gels with a pH of approximately 6.0 were prepared by mixing an aqueous slurry of calcium citrate with a sodium alginate solution. In certain cases the pH of gels of this type was varied either by the addition of different amounts of glucono delta lactone to the calcium citrate slurry or by adding sodium hydroxide to the alginate solution.

The gelling mixtures were allowed to set in rectangular moulds.

2. Measurement of creep compliance

Creep measurements were made using the parallel plate viscoelastometer described by *Shama* and *Sherman* (4). The movement of the centre plate after the application of the shear stress was followed continuously by recording the signal from a displacement transducer. The humidity of the air surrounding the gels was kept high by enclosing the viscoelastometer in a perspex box, the walls of which were coated with moist filter paper. The temperature of the gels was maintained at 24 ± 1 °C by pumping water through tanks in contact with the two samples. The measurement of the creep compliance at long times was complicated by the fact that some of the gels prepared exhibited considerable syneresis. Although the rate of loss of water from an unstressed gel which had been aged for several hours was very small, a significant quantity of water was exuded when the aged gel was compressed. To minimise errors caused by changes in the volume of the material during the course of an experiment the following procedure was adopted. The gels were aged for 24 h after formation at room temperature in a high humidity atmosphere. The samples were then placed in the viscoelastometer, compressed by approximately 5% and aged for a further hour before the stress was applied. During this time the temperature of the gel equilibrated, the stress exerted by the gel on the plates relaxed and the samples prone to syneresis exuded some

[1]) Alginate Industries, 22 Henrietta Street, London, WC2E 8NB.

water. The shear stress was then applied and measurements were taken for a period of 1 h. The creep recovery was generally followed for a few minutes after the stress was removed. It was found that the largest decrease in weight during the course of the creep experiment amounted to 2–3%.

The magnitude of the stress applied depended on the strength of the gel under investigation and was in the range 1000–3000 dyne cm^{-2}. Preliminary measurements showed that for strains up to at least 0.1 the viscoelastic response was linear within experimental error. The stress was normally chosen so that the maximum strain observed was less than this value. It was not possible to obtain satisfactory measurements of the creep compliance at the longer times for the weak gels formed at neutral pH. In such cases the only result reported is the compliance measured ten seconds after the application of the stress.

Results

Gels formed at pH 6.0 by method 2 exhibited little syneresis and were relatively weak in comparison with gels formed at lower pH's. Fig. 1 illustrates the compliance measured ten seconds after the application of stress for a gel at pH 6.0 which contained 2% sodium alginate and varying quantities of calcium citrate. As with all the data presented here the results were obtained after the gel had been aged for 25 h following formation. The properties of the gel changed only very slowly after this period. It was clear from the results shown in fig. 1 that the strength of the gel was independent of the calcium citrate concentration provided this was sufficiently high. Fig. 2 shows that an approximately linear relationship was observed between the reciprocal of the ten second compliance and the alginate concentration for the gels at pH 6.0. The calcium citrate level was maintained at 0.035 M.

Gels with a final pH of approximately 3.5 formed by method 1 exhibited considerable syneresis. Because gels of this type were appreciably stronger than those formed at higher pH's it proved easier to make satisfactory creep measurements for long times.

Fig. 3 illustrates the influence of $CaHPO_4$ concentration on the resultant creep compliance curve obtained from a gel which on formation contained 1% sodium alginate. The quoted concentrations of $CaHPO_4$ refer to the amount of the salt in the gel on formation. Table 1 displays the quantity of water lost from the system during the 25 h ageing period before the creep measurements were made. The extent of syneresis increased with the calcium ion concentration.

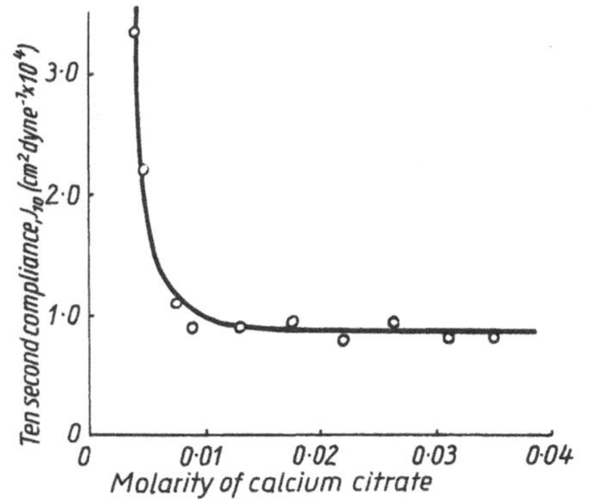

Fig. 1. The influence of calcium citrate concentration on the ten second compliance observed for gels formed at pH 6.0. (Alginate concentration 2 g/100 ml)

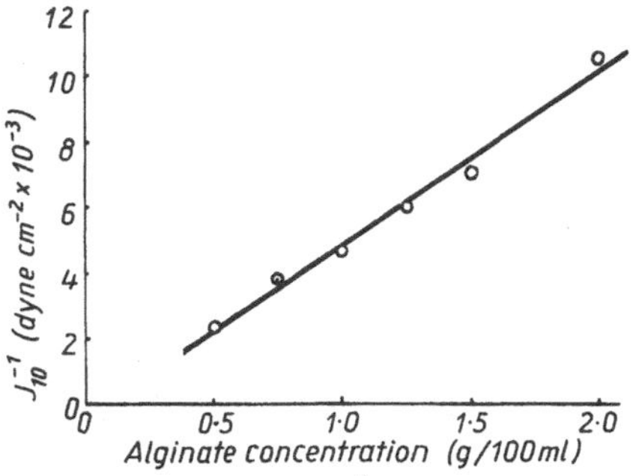

Fig. 2. The relationship between alginate concentration and the reciprocal of the ten second compliance for gels formed at pH 6.0. (Calcium citrate concentration 0.035 M)

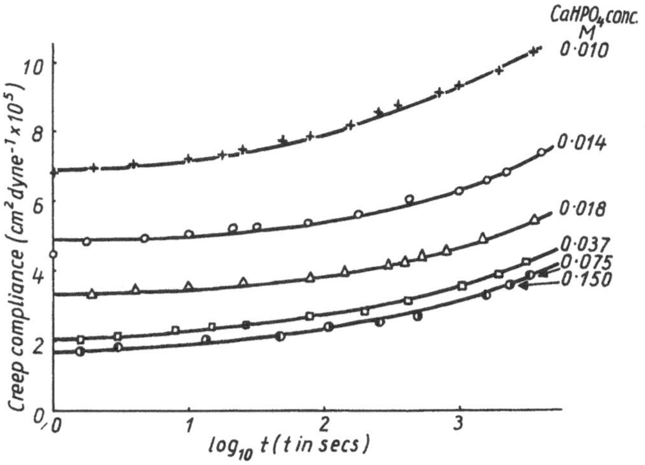

Fig. 3. The effect of the $CaHPO_4$ concentration on formation on the creep compliance-time curves observed for gels formed at pH 3.5. (Alginate concentration on formation 1 g/100 ml, the alginate concentration at the time the creep curve was measured can be calculated from the data displayed in table 1)

The influence of polysaccharide concentration on the resultant creep curve is shown in fig. 4. The concentration of $CaHPO_4$ in these systems on formation of the gel was 0.15 M. In fig. 4 the quoted polysaccharide concentration refers to the quantity of alginate in the gel at the time the creep curve was obtained. The extent of syneresis increased with decreasing polysaccharide concentration as is shown by the data displayed in table 1.

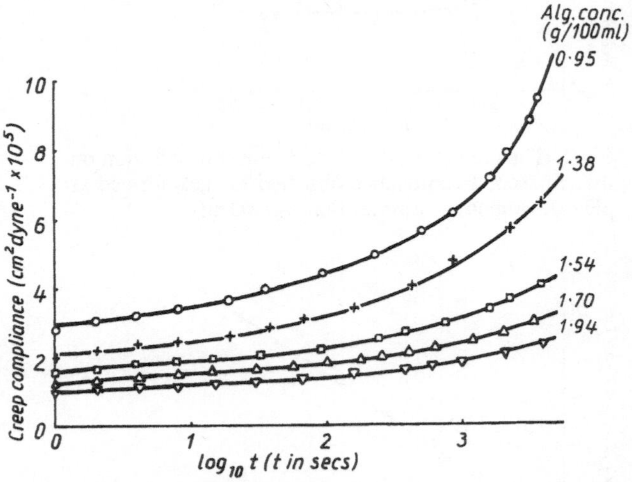

Fig. 4. The effect of alginate concentration on the creep compliance-time curves observed for gels formed at pH 3.5. ($CaHPO_4$ concentration on formation 0.15 M)

Table 1. Effect of $CaHPO_4$ and alginate concentration on the degree of syneresis of gels formed by method 1

$CaHPO_4$ concentration on formation	Alginate concentration on formation (%)	Percent water remaining in gel after 25 h ageing.
0.010 M	1	96
0.014 M	1	88
0.018 M	1	85
0.037 M	1	77
0.075 M	1	66
0.150 M	1	66
0.150 M	0.5	53
0.150 M	1.5	79
0.150 M	2	87

The striking influence of pH on the behaviour of these alginate gels is confirmed by the results obtained for calcium citrate gels where the pH was varied by the addition of glucono delta lactone or sodium hydroxide. Fig. 5 displays the compliance measured ten seconds after the application of stress for a gel which on formation contained 1% w/v sodium alginate and 0.035 M calcium citrate. Also shown in fig. 5 is the concentration of polysaccharide at the time when the compliance was measured. This latter data confirms that the amount of water lost by syneresis during the ageing period is greater for acid gels than for gels formed at neutral pH. Fig. 6 displays the complete creep curves obtained for the above gels at a series of representative pH's. Good agreement was obtained at low pH between the results for these calcium citrate gels and the gels formed using $CaHPO_4$.

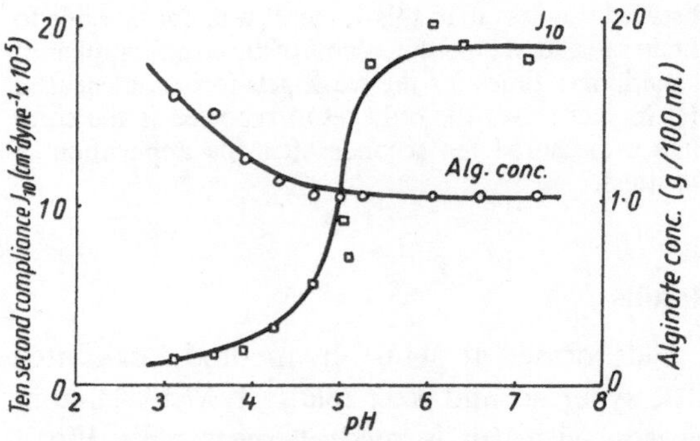

Fig. 5. The relationship between the ten second compliance and pH for gels which on formation contained 0.035 M calcium citrate and 1 g/100 ml alginate. Also shown is the alginate concentration at the time the creep curve was measured

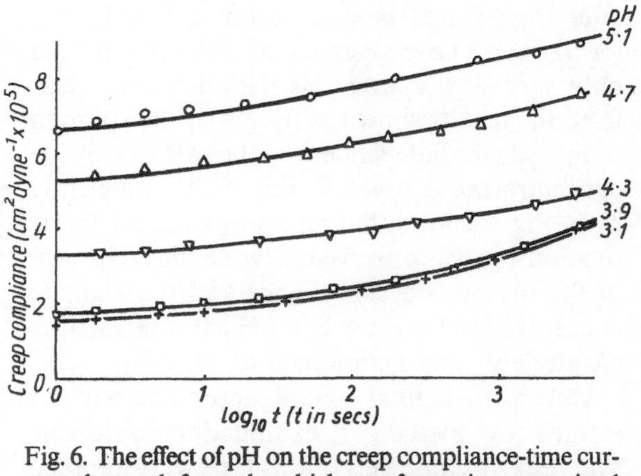

Fig. 6. The effect of pH on the creep compliance-time curves observed for gels which on formation contained 0.35 M calcium citrate and 1 g/100 ml alginate. (The alginate concentration at the time the creep curve was measured can be obtained from the data given in fig. 5)

Discussion

The pK_s of alginic acid depends slightly on the ratio of mannuronic to guluronic residues in the polymer but is approximately 4.0 (5). The increase in gel strength as the pH is lowered could therefore be interpreted in terms of the formation of additional interchain linkages involving undissociated alginic acid residues. The structures formed might possibly correspond to those suggested from the results of X-ray investigations on alginic acid fibres (6). In these structures the chains were held together by hydrogen bonds.

It has been shown that the degree of swelling of a polymer network should decrease with an increase in the number of cross-links or a reduction in the average distance between cross-links (7). The observation that for the gels at low pH syneresis increases with increasing calcium ion concentration would suggest that some of the additional bonds formed involve calcium ions. It is possible that interactions between undissociated alginic acid residues mainly occur near the original junction zones where the chains are already close together.

The shape of the creep compliance curve resembles that observed for viscoelastic liquids as during the second half hour of the creep measurement the compliance appears to be linearly related to time. However when the corresponding recovery experiment was carried out the strain did not reach a constant value even after one hour. Therefore it is clear that retarded elastic elements are still present at the longest times investigated and the magnitude of the viscous component cannot be determined with confidence. It is unfortunate that because of syneresis it is not possible to obtain reliable data for times longer than about one hour. The origin of the long time elements is not clear. Since they appear to depend primarily on the alginate concentration and are not affected by a change in the density of crosslinks involving calcium ions it seems unlikely that they can be associated with a breakage of the bonds between the calcium ions and the polysaccharide chain. It is possible that the long time elements could involve the slippage of entanglement couplings not involving calcium ions. If a viscous component is present this may indicate that some depolymerisation is occurring as the network is sheared.

For both the gels at pH 6.0 and the gels at low pH the minimum concentration of calcium citrate or $CaHPO_4$ which gives the optimum gel strength would correspond to approximately one calcium ion for every four polysaccharide residues. Since some of the calcium salt remains undissolved the true value will be somewhat lower than this and as it is not known how many calcium ions are involved in each junction it is impossible to estimate even approximately the number of cross-links from this result. It has been shown from a study of the effect of the degree of polymerisation on the stiffness of alginate gels that cross-links must occur more frequently along the chain than one every sixty five residues (2).

The linear relationship observed between the reciprocal of the ten second compliance and the alginate concentration found for the gels at pH 6.0, contrasts with the stronger concentration dependence observed for the gels at pH 3.5 and the proportionality to the square of the concentration observed for the shear modulus of gelatin gels (8). The weaker concentration dependence found for the alginate system at pH 6.0 may reflect a low degree of entanglement coupling at this pH. Rheological investigations of solutions of sodium alginate have shown that the frequency of entanglement couplings in this system is low. This has been attributed to the polyelectrolyte nature of the polymer (9).

Acknowledgements

We are particularly grateful to J. Lyons & Co. Ltd. for a research grant to support this project.

Summary

The rheological behaviour of alginate gels with a high D-mannuronate content has been investigated at strains of less than 0.1 and under stresses of 1000–3000 dyne cm^{-2}. Under these conditions linear viscoelastic behaviour was observed. The effects of varying pH, calcium ion and alginate concentrations were examined in gels which had been aged for 25 h. At pH 6.0 the reciprocal of the ten second compliance was directly proportional to the alginate concentration. The compliance of the gels exhibited a pronounced dependence on pH being larger at pH 6.0 than at lower pHs. Syneresis was greater for acid gels than those formed at neutral pH and increased with decreasing polysaccharide concentration. In general an enhanced calcium ion concentration resulted in a decrease in the compliance of the gels. The results are discussed in terms of polymer network theory.

References

1) *Rees, D. A.*, Advan. Carbohydrate Chem. Biochem. **24**, 267 (1969).

2. *Smidsrod, O.* and *A. Haug*, Acta Chem. Scand. **26**, 79 (1972).

3) "Gel formation with alginates" issued by Alginate Industries Ltd.

4) *Shama, F.* and *P. Sherman*, S.C.I. Monograph No. 27, "Rheology and Texture of Foodstuffs", p. 77 (1968).

5) *Haug, A.*, Acta. Chem. Scand. **15**, 950 (1951).

6) *Atkins, E. D. T., W. Mackie, K. D. Parker*, and *E. E. Smolko*, Polymer Letters **9**, 311 (1971).

7) *Flory, P. J.*, in: Principles of Polymer Chemistry, pp. 577ff. (New York 1953).

8) *Saunders, P. R.* and *A. G. Ward*. Proc. 2nd Intern. Congr. Rheol., p. 284 (London 1954).

9) *Amari, T.* and *M. Nakamura*, Kogyo Kagaku Zasshi **74**, 2140 (1971).

Authors' address:

J. R. Mitchell and *J. M. V. Blanshard*
Food Science Laboratories
Dept. of Applied Biochemistry and Nutrition
University of Nottingham, School of Agriculture
Sutton Bonington, Loughborough, Leics. (England)

Rheol. Acta 13, 185–190 (1974)

From the Department of Physical Metallurgy, University of Birmingham, Birmingham 15 (England)

The role of entanglement networks in the fracture of polysulfone

By N. J. *Mills*

With 7 figures and 2 tables

(Received October 27, 1972)

Introduction

The structure of the polysulfone from bisphenol *A* and dichlorodiphenylsulfone is shown in fig. 1. Study (1) of the shape of isolated polysulfone chains in solution has shown that the unperturbed mean square chain end-to-end distance $\langle \Gamma_0^2 \rangle$ is approximately the same as that of a model chain having free rotation about linear bonds of length ~5.8 Å through the phenylene groups. The best way to compare the flexibility of isolated polymer chains is to calculate the length of the segment of a freely jointed model chain that has the same root mean square length and fully extended length Γ_{max} as the real chain. The segment length is given by

$$l_s = <\Gamma_0^2> / \Gamma_{max}$$

and the values are 14.8 Å for polysulfone and 18.4 Å for polystyrene. Polysulfone is thus more flexible in these terms.

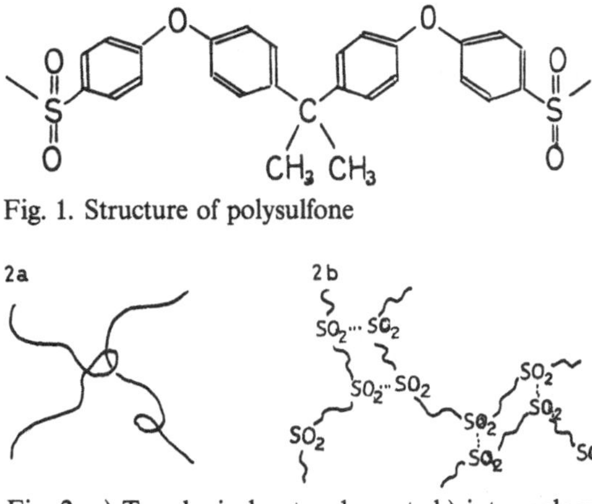

Fig. 1. Structure of polysulfone

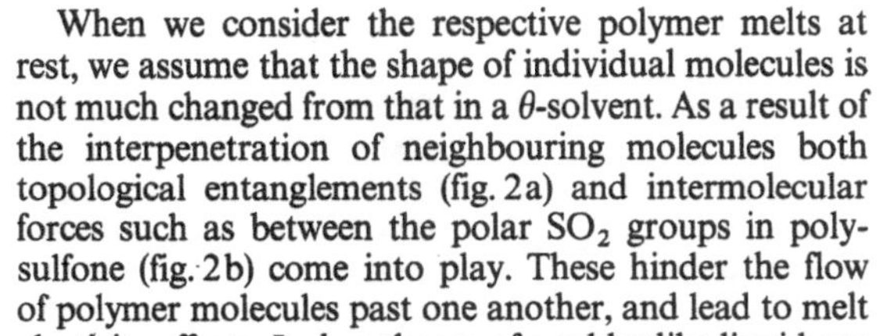

Fig. 2. a) Topological entanglement; b) intermolecular interactions

When we consider the respective polymer melts at rest, we assume that the shape of individual molecules is not much changed from that in a θ-solvent. As a result of the interpenetration of neighbouring molecules both topological entanglements (fig. 2a) and intermolecular forces such as between the polar SO_2 groups in polysulfone (fig. 2b) come into play. These hinder the flow of polymer molecules past one another, and lead to melt elasticity effects. *Lodges* theory of a rubberlike liquid can be interpreted as a network theory of polymer solutions and used to describe elasticity effects in polymer melts. In this theory "entanglement points" are continually being lost and reforming, and the liquid can be characterised by the number of "entanglements" present (or alternatively the average chain molecular weight between "entanglements" M_e) and their lifetime function. The term "entanglements" has been used since either form of interaction of fig. 2 could be effective.

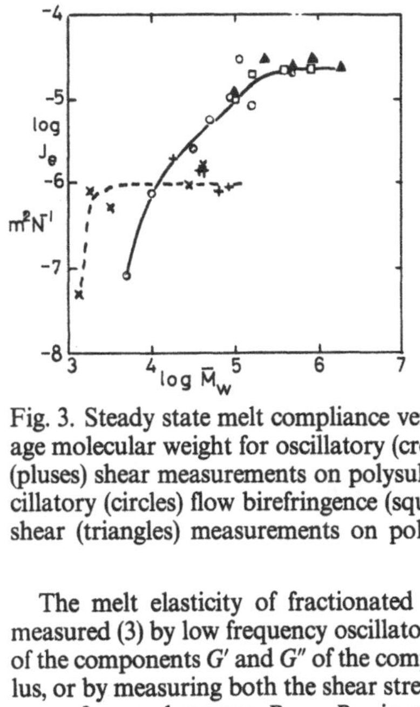

Fig. 3. Steady state melt compliance versus weight average molecular weight for oscillatory (crosses) and steady (pluses) shear measurements on polysulfone, and for oscillatory (circles) flow birefringence (squares) and steady shear (triangles) measurements on polystyrene. Ref. (3)

The melt elasticity of fractionated polysulfone was measured (3) by low frequency oscillatory measurements of the components G' and G'' of the complex shear modulus, or by measuring both the shear stress P_{12} and difference of normal stresses $P_{11} - P_{22}$ in steady shear flow. From this the steady shear compliance J_e, defined by

$$J_e = \underset{w \to 0}{Lt} \frac{G'}{(G'')^2} = \underset{\dot{\gamma} \to 0}{Lt} \frac{P_{11} - P_{22}}{2\,P_{12}^2}$$

where w is the frequency and $\dot{\gamma}$ is the shear rate, was calculated. Fig. 3 shows that for narrow molecular weight distribution polysulfone J_e reaches a maximum value for $\bar{M}_w > 2000$ whereas for polystyrene (4) a maximum value is reached for $\bar{M}_w > 200000$. For the purposes of comparison it is assumed that the average molecular weight between entanglements can be calculated from the high molecular weight compliance (corrected for polydispersity) by using the rubber elasticity relation

$$M_e = J_e \varrho RT$$

where ϱ is the density, R the gas constant, and T the absolute temperature. There may be a numerical constant in this equation, but the values obtained for polystyrene (37000) agree with values obtained for polystyrene by other methods (5). For polysulfone $2500 < M_e < 5000$ so it has much more densely "entangled" melt. This is provisionally explained by the slightly greater flexibility of polysulfone, and its greater intermolecular forces between the SO_2 groups.

When a polymer melt is cooled into the glassy state there is no reason to believe that the chain shapes alter, so presumably an "entanglement" network exists in glassy polymers. When it is considered which mechanical properties of polymeric glasses may be affected by a network structure the following tentative conclusions can be drawn.

1.

A network structure is unlikely to influence the small strain elastic modulus, since this has been shown to depend on the magnitude of the *Van der Waals* forces between chains. Nor should it influence the yield stress, since the yield strains in glassy polymers are not far short of the theoretical values calculated on the basic of short range intermolecular forces.

2.

By analogy with the relation between the maximum extensibility of rubbers and the network chain length, orientation hardening at high strains in polymeric glasses may depend on the network structure.

3.

If brittle fracture occurs, there will be highly strained regions in the neighbourhood of the fracture surface. Thus the magnitude of some of the fracture parameters may depend on the network structure.

Experimental materials

50 gm of Union Carbide polysulfone P3500 was fractionated (1) by cooling a 1% solution in dimethylsulphoxide in stages from 80–20 °C. The fractions were characterised by Gel Permeation Chromatography using tetrahydrofuran as a solvent. Table 1 gives the molecular weight averages computed from the traces a) as if the polymer were polystyrene using the polystyrene calibration of the columns, b) assuming that the product $[\eta]M$ of the intrinsic viscosity and the molecular weight gives a universal GPC calibration against elution volume, and using the known $[\eta] - M$ relationships of polysulfone and polystyrene to recalculate the molecular weight averages.

The amount of the fractions varied from 1–10 gm so samples for mechanical tests were necessarily limited in size. Discs of 28 mm diameter and 3 mm thickness were compression moulded at 230 °C from the fractions using a mould designed to make Infra Red KBr discs that could be evacuated. From the discs, strips 5 mm wide were cut out. These were notched if necessary, then annealed at 160 °C for 15 min in a silicone oil bath to relieve internal stresses and give a standard thermal history. The disc of fraction A2 shattered on cooling, under the influence of thermal contraction stresses.

Mechanical tests

Plane strain compression tests were made on samples of the fractions 10 mm long by 5 mm high by 3.1 mm thick by compressing them between parallel steel anvils lubricated with silicone grease in a channel 3.1 mm wide that restricted lateral strain to one direction only. These were compressed in an Instron at a crosshead rate of 0.005 cm/min at 20 °C until a load maximum was just passed. The samples were subsequently examined by transmitted light microscopy. Further samples of the unfractionated polysulfone and a commercial general purpose polystyrene were compressed to failure at a crosshead rate of 0.05 cm/min, unloading several times during the test to check the sample condition.

Strips 20 mm long by 5 mm wide were edge notched at the mid point to a depth between 1 and 2.5 mm, the notch being cut perpendicular to the tensile direction with a saw with a *V* tooth profile and sharpened with a fresh razor blade.

Table 1

Polymer	$\bar{M}_n(a)$	$\bar{M}_w(a)$	$\bar{M}_w/\bar{M}_n$	$\bar{M}_n(b)$	$\bar{M}_w(b)$
P3500	5730	27200	4.74	[1])	21000
A1	23000	35100	1.53	17000	27000
A2	2740	7750	2.82	[1])	[1])
B1	85700	116000	1.36	81000	112000
B2	34000	45500	1.34	26000	36000
B3	25400	32900	1.29	19000	25000
B4	17800	24500	1.38	13000	18000
B5	13700	18600	1.36	10000	13500

[1]) Uncertain since $[\eta]$ not known for polysulfone for $M < 10000$.

These strips were then pulled in an Instron at a crosshead speed of 0.05 cm/min to failure. The fracture surfaces were observed by reflected light microscopy and by scanning electron microscopy.

Results

Compression tests

Fig. 4 shows a plot of the compressive stress versus the shear strain for the plane strain compression of P3500 polysulfone. Once the maximum or yield stress has been passed at a shear strain $\gamma = 0.15$ the deformation becomes inhomogeneous due to the formation of shear bands at 45° to the compressive direction (fig. 7a). In cases where surface markings have been deformed by the shear band the amount of shear strain can be calculated, and results are given for this in table 2. When compression is continued the shear bands widen and rotate towards the perpendicular to the compression direction. Eventually at a shear strain $\gamma \sim 0.9$ the deformation becomes nearly homogeneous again and the stress rises sharply. The yielded polymer is highly birefringent showing that orientation of the monomer units has occurred.

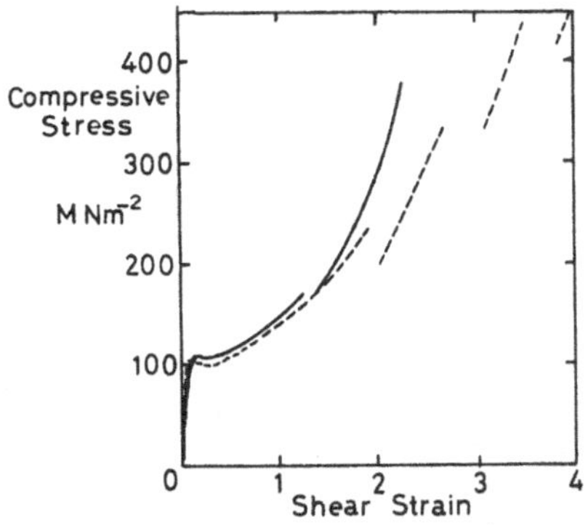

Fig. 4. Compressive stress versus shear strain in the plane strain compression of polysulfone (solid line) and polystyrene (dashes) at 23 °C. The curves are broken at points where unloading occurred

Fig. 5 shows the dependence of the yield stress of polysulfone on molecular weight. Literature data (6) showing the variations in the tensile strength of polystyrene with molecular weight is also included. By comparing figs. 3 and 5 it can be seen that for both polymers the strength is independent of molecular weight at high molecular weights and decreases rapidly to zero in roughly the same molecular weight range as does the steady shear melt compliance. The absolute value of the yield stress (the compressive yield stress of polystyrene is not known for a range of molecular weights) does not appear to depend on the network structure in the glass. A value of 93 MNm^{-2} has been reported (10) for the plain strain compressive yield stress of polystyrene.

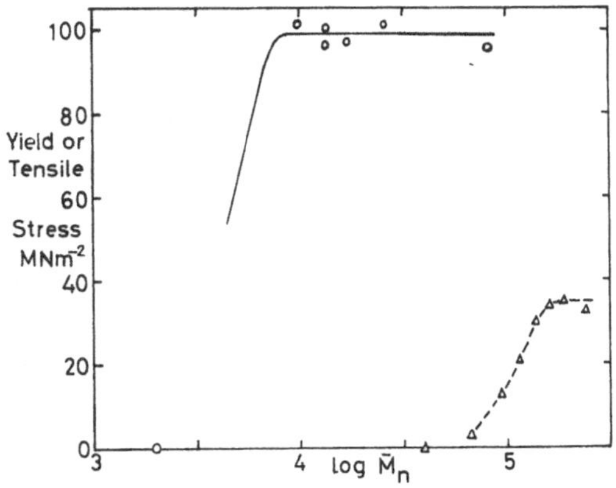

Fig. 5. Compressive yield stress for polysulfone (circles) and tensile strength for compression moulded anionic polystyrene (6) (triangles) versus number average molecular weight

Table 2. Maximum shear strains found in shear bands

Sample	Maximum shear strain
P3500	0.86
A1	0.34
B2	0.65
B3	0.72
B4	0.57
B5	0.56

However the shear strain at which orientation hardening occurs in polystyrene (fig. 4) is greater than for polysulfone. There is at present no adequate molecular treatment of orientation hardening in polymeric glasses. An approximate treatment (7), uses the rubber elasticity theory result that the maximum extension ratio of the network is equal to the square root of the number of freely jointed segments between crosslinks. If this were the case then using the chain flexibility parameters and entanglement molecular weights quoted earlier the maximum extension ratio of a polysulfone network would be 3–4 and a polystyrene one 7. In fact these are both about twice the observed values. Thus there is evidence that the "entanglement" network density controls orientation hardening in glassy polymers, but no exact theory. N. B. Polysulfone fails suddenly at high shear strains, whereas polystyrene develops internal and surface cracks in the compression direction that eventually cause failure.

Fracture tests

Previous work (8) has established that in P3500 polysulfone it was possible to make plain strain fracture toughness measurements with 3 mm thick samples (the maximum plastic zone size around the crack must be much less than the sheet thickness). From measurements of the failure stress σ, the initial crack length a and the specimen width w, the stress field intensity around the crack tip at failure K_c was calculated

$$K_c = Y\sigma\sqrt{a}$$

where

$$Y = 1.99 - 0.41\,(a/w) + 18.70\,(a/w)^2 - 38.48\,(a/w)^3 + \cdots$$

is a theoretical correction (9) for the finite width of the single edge notched test piece. It has been shown that for commercial polysulfone containing cracks of lengths between 1 and 5 mm, fracture occurred in tension when the stress field around the crack tip reached a certain level, characterised by the parameter K_c. Fig. 6 shows the variation of the experimental K_c values with $\bar{M}_n$. There is a great deal of scatter as is common with these measurements, but there appears to be little change in K_c with molecular weight, apart from a possible decrease at high molecular weight.

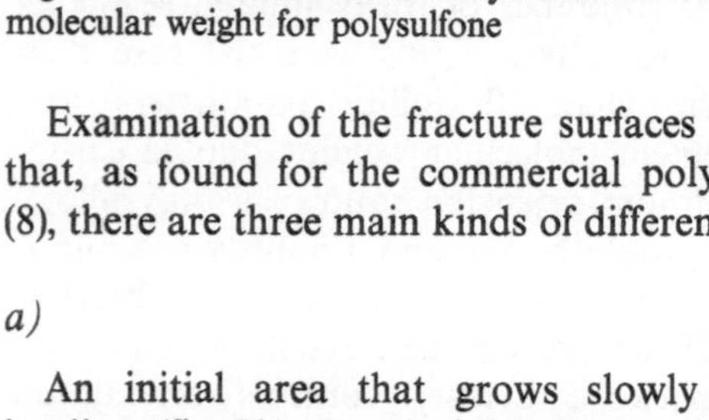

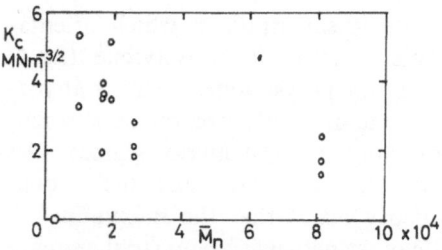

Fig. 6. Critical stress field intensity versus number average molecular weight for polysulfone

Examination of the fracture surfaces showed that, as found for the commercial polysulfone (8), there are three main kinds of difference.

a)

An initial area that grows slowly during loading (fig. 7b). It consists of many smooth surfaced parallel planes (i.e. some are below the actual fracture surface) the edges of which when they appear show evidence of marked plastic deformation (fig. 7c). Photoelastic measurements show that this area is initially load bearing, so it appears likely that it consists of many parallel crazes, some of which have eventually broken when the crack propagates.

b)

An area of parabolic markings (fig. 7d), similar to those found on polymethylmethacrylate. When the crack accelerates during fracture it must overtake the slow craze growth, and its stress field must initiate new disc shaped cracks in its path. There are no subsidiary crazes below the fracture surface in this area. The edges of the parabola are visible because the cracks form on levels that differ by ~0.5 μm.

c)

If the crack accelerates still faster the cracks bifurcate and a rougher area results. This consists of many roughly parallel planar crazes plus highly deformed projections of polymer (fig. 7e). The appearance of the fracture surfaces was as follows:

For all the high molecular weight fractions B 1, B 2, B 3 and A 1 the areas a) and b) were present.

For fraction B 5 of molecular weight ~10000, region b) was absent and the appearance changed directly to c).

For the fraction A2 of molecular weight ~5000 the fracture surface of disc that cracked up after moulding showed crack bifurcation but very much less evidence of plastic deformation in the process (fig. 7f).

Discussion

In comparing the mechanical properties of polystyrene and polysulfone we have two glassy polymers that differ marked in the density of the "entanglement" network structure present. When it is found that the mechanical strength of polysulfone reaches its maximum level for molecular weights greater than 10000, whereas for polystyrene the limit is 150000, it seems that the strength of glassy polymers must depend on the molecules being sufficiently "entangled" together. However there is less direct evidence of the effect

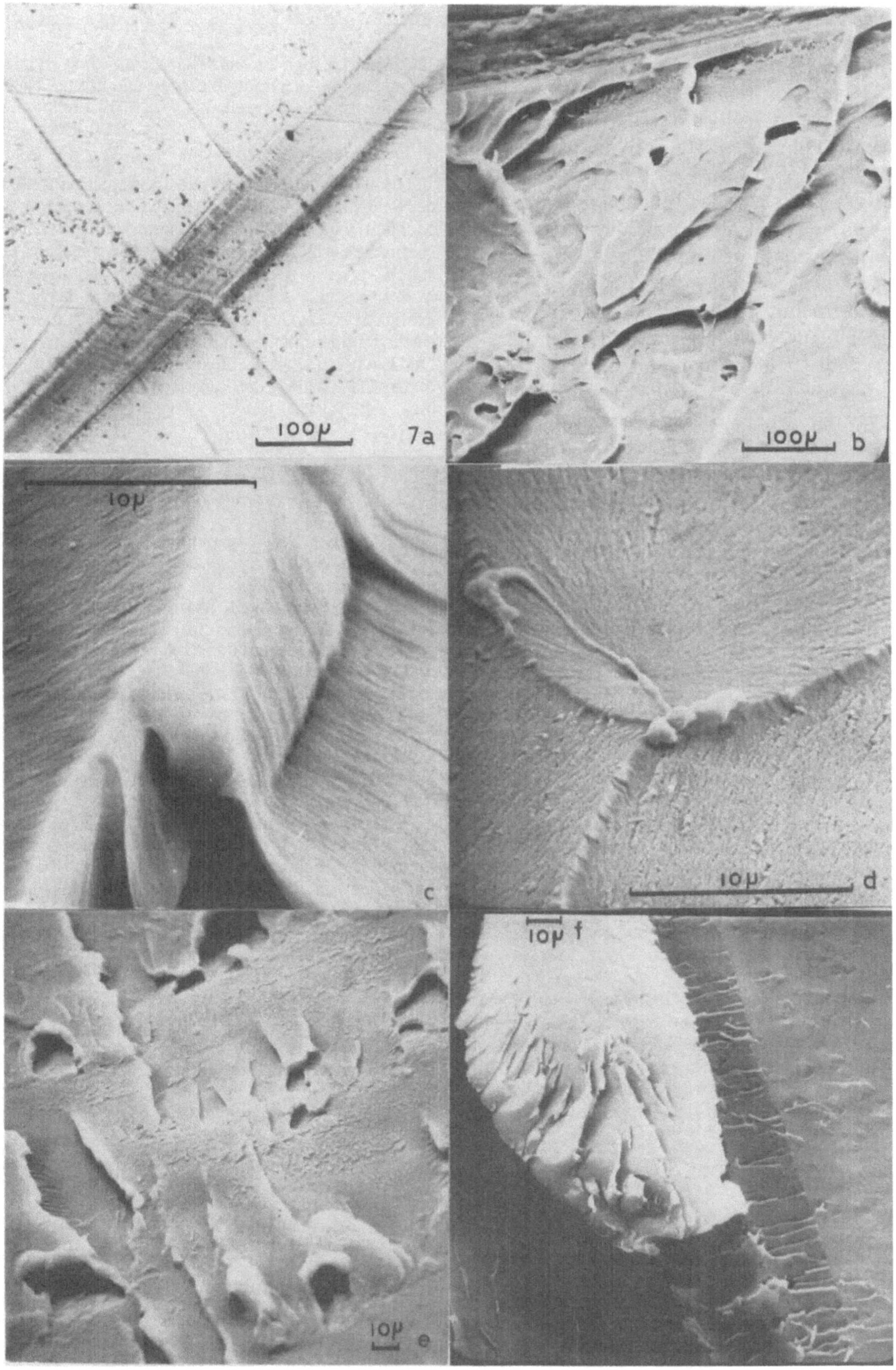

Fig. 7. a) Optical micrograph of shear band in polysulfone; b) to f) Scanning Electron Micrographs of b) crack edge and slow growth area in B3; c) detail of edge of planar region in B2; d) parabola edges in B2; e) rough fast fracture region in B5; f) surface of A2

of the degree of "entanglement". Orientation hardening measurements give some clue, but there are the difficulties of having no adequate molecular model, and the obvious strain rate dependence of the results. Polysulfone forms relatively wide shear bands in compression from which the shear strain can be measured, but in polystyrene these are much finer (10) and no measurements of strains are possible.

When fracture behaviour is considered, there is the difficulty that in a tensile test on an unnotched specimen polysulfone yields and forms a neck, whereas polystyrene develops crazes under stress which become cracks and cause failure. Even when the specimens are deliberately notched, the critical stress field parameter for crack growth does not differentiate between the polymers. (K_c for polystyrene is 1.1 $MNm^{-3/2}$ (11) and there is no data for its molecular weight dependence.) Since glassy polymers always form crazes ahead of the crack in plane strain fracture it may be that some detail of the craze fracture process, such as its critical opening displacement before failure, depends on the network structure. Further work is in progress to investigate this.

Acknowledgements

Thanks are due to the Science Research Council for supporting this research, to Dr. *F. W. Peaker* of the Chemistry Department, University of Birmingham for Gel Permeation Chromatography measurements, and to Mr. *J. C. Moore* for experimental assistance.

Summary

Study of the melt elasticity of many high polymers has indicated that an "entanglement" network exists between molecules. By comparing the mechanical properties of glassy polysulfone and polystyrene it appears that this network must be present to give adequate mechanical strength, and that it affects the degree of orientation hardening at high strains. The phenomenon of crazing is at the root of the fracture process and it is not clear yet how this is affected by the network structure.

Zusammenfassung

Die Untersuchung der Gleichgewichtsnachgiebigkeit der Schmelzen von Hochpolymeren hat gezeigt, daß die Moleküle ein verflochtenes Netzwerk bilden. Das mechanische Verhalten von amorphem Polystyrol und Polysulfon wurde verglichen, und es scheint, daß dieses Netzwerk vorhanden sein muß, um eine hohe mechanische Festigkeit zu erzielen. Außerdem kontrolliert es den Zuwachs im Spannungsertrag bei hohen Bruchdehnungen. Der Einfluß des verflochtenen Netzwerkes auf die Bildung von Rissen und den folgenden Bruchprozeß ist noch nicht geklärt.

References

1) *Allen, G., J. McAinsh*, and *C. Strazielle*, European Polymer J. **5**, 319 (1969).

2) *Lodge, A. S.*, Elastic Liquids, p. 118 (London 1964).

3) *Mills, N. J., A. Nevin*, and *J. McAinsh*, J. Macromol. Sci. **B4**, 863 (1970).

4) *Mills, N. J.* and *A. Nevin*, J. Polymer Sci. A2, **9**, 267 (1971).

5) *Porter, R. S.* and *J. F. Johnson*, Chem. Rev. **66**, 1 (1966).

6) *McCormick, H. W., F. M. Bower*, and *L. Kin*, J. Polymer Sci. **39**, 87 (1959).

7) *Haward, R. N.* and *G. Thackray*, Proc. Roy. Soc. **A 302**, 453 (1968).

8) *Gales, R. D. R.* and *N. J. Mills*, Eng. Fract. Mech. (1974).

9) *Brown, W. F.* and *J. E. Srawley*, ASTM STP 410 (1966) "Plane strain crack toughness testing of high strength metallic materials".

10) *Bowden, P. B.* and *S. Raha*, Phil. Mag. **22**, 463 (1970).

11) *Marshall, G. P., L. E. Culver*, and *J. G. Williams*, Int. J. Fract. (1973).

Author's address:

N. J. Mills
Dept of Physical Metallurgy and Science of Materials
University of Birmingham
Birmingham 15 (England)

Rheol. Acta 13, 191–198 (1974)

From the Department of Civil Engineering, University of Strathclyde, Glasgow (U.K.)

Strain predictions in granular soils

By *W. M. Kirkpatrick*

With 4 figures

(Received October 27, 1972)

Introduction

Strains in soils are predominantly non-recoverable and it is therefore appropriate to look towards plastic theory to provide a description of stress strain behaviour. A condition favourable to this approach involving a stress-strain law relating stress to strain increment for granular (non-cohesive) soils was illustrated by *Roscoe, Basset,* and *Cole* (1967). They demonstrated in simple shear tests with monotonically increasing loads on Leighton Buzzard sand that the principal axes of strain increment coincided closely with the principal axes of stress over all but the initial (low stress) stages of the tests.

Kirkpatrick in 1957 and since then others have shown that the *Mohr-Coulomb* failure theory describes with reasonable accuracy failure stress states in granular soils. As is well known however the associated flow rule provided by plastic theory is unsuccessful in estimating strains since it overpredicts volume dilations and also, due to the straight sides and corners of the six sided pyramid representing the theoretical failure surface, ambiguity exists in defining the strain increment vector.

The weight of experimental evidence on the failure of granular soils indicates that the experimental failure surface is continuously curved coinciding approximately with the theoretical *Mohr-Coulomb* surface at its corners. It has been suggested by *Kirkpatrick* (1970) that this surface in principal stress space can be represented with acceptable accuracy by

$$\Phi^2 = (\sigma_1 - \sigma_2)^2 + (\sigma_2 - \sigma_3)^2 + (\sigma_3 - \sigma_1)^2 - \left[\frac{2(1-k)(\sigma_1 + \sigma_2 + \sigma_3)}{(\sigma_1 + 3\sigma_2 - 4\sigma_3) + 2k(\sigma_1 - \sigma_3)}\right]^2 \quad [1]$$

in segments in which the stresses are ordered $\sigma_1 \geq \sigma_2 \geq \sigma_3$, $k = (1 - \sin\Phi)/(1 + \sin\Phi)$ where Φ = angle of shearing resistance of the soil at peak stress conditions.

Kirkpatrick (1970) pointed out that the curved sided curve of the experimental yield surface has certain desirable features as a plastic potential in that it provides uniqueness of the strain increment vector and also that it provides an explanation of some of the differences that are seen in the behaviour of the material under different failure stress regimes such as triaxial compression ($\sigma_1 > \sigma_2 = \sigma_3$), triaxial extension ($\sigma_1 = \sigma_2 > \sigma_3$) and in plane strain ($\sigma_1 > \sigma_2 > \sigma_3$, $\varepsilon_2 = 0$). The over prediction of the volume dilatations using this plastic potential however remains.

A flow rule of the type which is associated with either the theoretical *Mohr-Coulomb* yield surface or the experimental yield surface is thus not appropriate to sands. As a result advantage cannot be taken of the many convenient implied conditions of plastic theory when dealing with such materials. The problem of finding a meaningful stress-strain law for sands remains however and the approach described in the following is to examine further the parts of plastic theory which appear most fruitful in providing a more realistic description of behaviour.

Other possible plastic potentials

The use of the experimental yield surface as a plastic potential although not entirely appropriate has the features described above in its favour. These features are inferred from the geometric condition of normality of the strain increment vector to the yield surface. The condition of normality of the strain increment vector provides a very powerful tool since from purely geometric considerations it is possible to predict strain increments from any possible stress system once the plastic potential is known. Other potential surfaces are now examined to determine if they can be applied successfully to predict strains in sands. The appropriate plastic potential will not however be the yield surface so that the flow rule provided will be non-associated and the condition assumed of the normality of the strain increment vector merely becomes an expedient for determining values of strain increment.

An examination of existing data on soil behaviour will be of assistance in selecting suitable plastic potentials. In an analysis of a fairly large number of triaxial compression and extension test results *Kirkpatrick* (1961) pointed out that the volume strain–axial strain curves had the same form irrespective of the initial porosity of the samples. This was an initial reduction in volume followed by a volume increase as the major principal stress ratio σ_1/σ_3 was increased as illustrated by the characteristic stress-strain curve in fig. 1a. It was further noted that the point at which the volume strain reached its maximum (compressive) value coincided with a stress ratio σ_1/σ_{3F} which was approximately a constant for the compression and extension tests and independent of the initial porosity (or density) of sand. A simplified model of behaviour was constructed which postulated that a general state of shear must be accompanied by a volume dilation and the stress ratio at the threshold of this, σ_1/σ_{3F}, defines an angle of friction φ_F of the material such that $\sin\varphi_F = (\sigma_1/\sigma_{3F} - 1)/(\sigma_1/\sigma_{3F} + 1)$.

This simplified model does not contradict the more sophisticated stress dilatancy theory of *Rowe* (1962), although arguments in this latter theory would suggest the value of φ_F should increase slightly with increasing porosity for a given sand.

Difficulties in estimating σ_1/σ_{3F}, and hence φ_F from the stress strain curves were discussed by *Kirkpatrick* (1961) and it was concluded that even with refined volume strain observations the values are likely to be over-estimated. In an effort to reduce the masking effect of the volume reductions due to elastic strain which continue beyond the point at which dilations due to shear initiate, a number of triaxial compression tests were performed on cubes and cylinders of Leighton Buzzard Sand in which the stress ratio was cycled a number of times from unity to the point at which the volume changed from positive to zero values. This removed the residual volume reduction and allowed a modulus of elastic volume change to be determined. Deducting the estimated elastic volume reduction from the next continuous reloading curve it was possible to obtain an adjusted point of initial volume increase and hence a more accurate estimate of σ_1/σ_{3F} and hence of φ_F. The adjusted values of φ_F at an average of 26.3° are somewhat smaller than the initial values and although some scatter was present there appeared to be little variation with initial porosity. These values of φ_F would be expected to occur under compression and extension conditions where $\sigma_r = \sigma_\theta$ but for stress systems in which the intermediate principal stress is truly intermediate such as in plane strain a higher value of φ_F might be expected. In these cases as pointed out by *Cornforth* (1964) the material is no longer free to deform in the direction of minimum strength as in the triaxial test but is constrained to deform parallel to the σ_2 plane and is likely to be somewhat stronger as a result.

Stress-strain hypothesis

Inevitably in any strain regime both elastic and plastic strains are likely to be present. If the components of each are significant a two stage elasto-plastic treatment based on separate stress-strain laws will be required. The present treatment is an attempt to find a simple stress-strain law and hence only stages where elastic components can be ignored will be considered. From present knowledge of soil mechanics it can be stated that plastic strains are dominant at all stress states in which the hydrostatic component of the stress change is small compared to the change in deviatoric component. Only in situations in which the deviatoric component is small can elastic strains be considered in any way significant. In many of these situations also however elastic strains may be ignored with little loss of accuracy. Due to the almost exclusive concern with plastic strains the symbol ε will be taken to represent plastic strain in this paper.

As a sign convention compressive stress and compressive strain are considered positive. In view of this and the previous discussion the following conditions can be formulated to assist initially at least in the selection of the plastic potential. (The assumption that the material is homogeneous and isotropic is made in the analysis.)

For a stress state in which $\sigma_1 \geq \sigma_2 \geq \sigma_3$:

1.

The state of plasticity exists for stress ratios $\sigma_1/\sigma_3 \geq \sigma_1/\sigma_{3F}$ and that the plastic potential will apply to these stress states. Stress states prior to the development of φ_F are excluded from consideration at this stage.

2.

The plastic potential must be such that with normality of the strain increment vector it will predict zero volume strain increments at $\sigma_1/\sigma_3 = \sigma_1/\sigma_{3F}$ but volume dilations at $\sigma_1/\sigma_3 > \sigma_1/\sigma_{3F}$.

Considering the geometry of the plastic potential in principal stress space condition 2 indicates that:

i)

At a stress ratio $\sigma_1/\sigma_3 = \sigma_1/\sigma_{3F}$ the plastic potential must coincide or lie parallel to the hydrostatic line and

ii)

at stress ratios $\sigma_1/\sigma_3 > \sigma_1/\sigma_{3F}$ where dilations occur the plastic potential must be inclined outwards from the hydrostatic line.

Although the plastic potential g in the flow rule provided by plastic theory having the usual form

$$d\varepsilon_{ij} = \lambda \frac{\partial g}{\partial \sigma_{ij}} \qquad [2]$$

(where $g = f(\sigma_{ij})$ and λ is an instantaneous constant)

cannot thus be taken as the yield function Φ but must be some function of the hydrostatic stress component such that sections of it made by octahedral planes increase in magnitude as the hydrostatic stress increases.

The *Mohr Coulomb* failure theory defines that a failure state will be reached when

$$\sigma_3 = k\varphi\sigma_1 \quad [3]$$

(where $k\varphi = (1 - \sin\varphi)/(1 + \sin\varphi)$)

and $k\varphi$ can be used to define the shape of the theoretical failure surface. $k\varphi$ if obtained from triaxial compression or extension tests can also be used to define the shape of the experimental yield surface through for example eq. [1].

The previous discussion has shown that the shape of plastic potential such as suggested by the experimental yield surface has a number of desirable features and might be appropriate to the special conditions for sands. The above observations indicate however that such a plastic potential cannot be defined by peak value of $k = k\varphi$ but by some revised value of $k = k\psi$ where:

$$k\psi = \frac{1 - \sin\psi}{1 + \sin\psi} \quad [4a]$$

and

$$\psi = \varphi - \varphi_F \quad [4b]$$

where

$$\varphi = \sin^{-1}(\sigma_1/\sigma_3 - 1)/(\sigma_1/\sigma_3 + 1)$$

and

$$\varphi_F = \sin^{-1}(\sigma_1/\sigma_{3F} - 1)/(\sigma_1/\sigma_{3F} + 1).$$

A plastic potential of this form is therefore here proposed for sands. The appropriateness of the assumed plastic potential can be examined when strain predictions are compared with experimental data.

The plastic potential surface being a straight sides cone with non circular octahedral sections will have a relatively complex equation (e.g. eq. [1]) and will be cumbersome to handle mathematically. If however the experimental yield surface has been defined and the basic data for φ_F are available the application of the theory to certain stress states of practical interest should be straightforward.

It was intially contemplated using this concept of the plastic potential to estimate strains at peak or constant stress conditions in which case the maximum value of φ would be substituted in eq. [4b]. Considering the characteristic behaviour of the material however the possibility suggests itself of extending the hypothesis to include all dilating deformations that is to cover the entire stress strain curve beyong σ_1/σ_{3F} including the work hardening the work softening situations i.e. a–d and d–f respectively of the characteristic stress strain curve shown in fig. 1a. This may be done by defining a new parameter $k = k\psi i$ such that

$$k\psi i = (1 - \sin\psi i)/(1 + \sin\psi i) \quad [5a]$$

and

$$\psi i = \varphi i - \varphi_F \quad [5b]$$

where

$$\varphi i = \sin^{-1}(\sigma_1/\sigma_{3i} - 1)/(\sigma_1/\sigma_{3i} + 1).$$

The suffice i refers to the instantaneous values of the parameters, and σ_1/σ_{3i} is the instantaneous stress ratio such that $\sigma_1/\sigma_{3F} \leq \sigma_1/\sigma_{3i} \leq \sigma_1/\sigma_3$ peak where $\sigma_1 \geq \sigma_2 \geq \sigma_3$. Defining $\sigma_1/\sigma_{3i} = Ri$ then $k\psi i$ is given by expression 6.

$$k\psi i = \frac{Ri(1 - \cos\varphi_F) + (1 + \cos\varphi_F) + 2Ri\sin\varphi_F}{Ri(1 + \cos\varphi_F) + (1 - \cos\varphi_F) - 2Ri\sin\varphi_F}. \quad [6]$$

The inclusion of work softening curves represents a further departure from plastic theory. The condition of positive plastic work requires a stable inelastic material as postulated by *Drucker* (1959). This is satisfied by perfectly plastic or work hardening stress strain curves but is exclusive of strain softening. The further limitations that this implies in the application of the mathematical theory of plasticity to soils will have to be accepted in the interests of providing a realistic description of behaviour.

Eq. [5] can be used to define a family of plastic potentials which are curved lines in octahedral planes such as shown in fig. 1b. These curved plastic potentials circumscribe the hexagons whose sides are represented by the equation

$$\sigma_{\text{minor}} - k\psi_{\text{major}} = 0$$

and coincide with them at their apices. Fig. 1c shows the section of these plastic potentials made by the plane containing σ_1 and the hydrostatic axis. As a result of the implied uniqueness of the curved surfaces the conditions for triaxial compression and extension will lie in the plane of fig. 1c. Figs. 1b and 1c are drawn for a material which has $(\sigma_{\text{maj}}/\sigma_{\text{min}})_F = 3.0$ and $(\sigma_{\text{maj}}/\sigma_{\text{min}})_{\max} = 6.0$ ($\varphi = 46°$). The plastic potential for $(\sigma_{\text{maj}}/\sigma_{\text{min}})_F$ must coincide with the hydrostatic axis in fig. 1c and is thus the point at 0 the centre

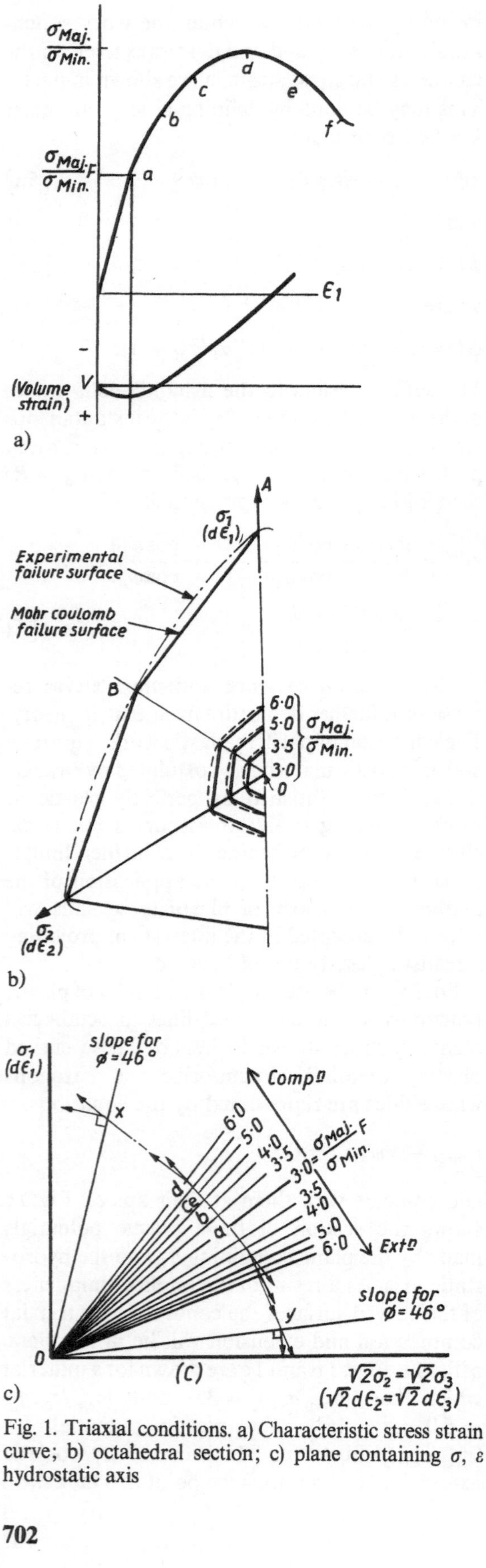

Fig. 1. Triaxial conditions. a) Characteristic stress strain curve; b) octahedral section; c) plane containing σ, ε hydrostatic axis

of the diagram in fig. 1b. The strain increment vector on this plastic potential will of course lead to zero volume strain. Families for the various principal stress ratios as they increase will lie above the hydrostatic axis in fig. 1c for the compression test and below for the extension test, corresponding families in octahedral planes are shown in fig. 1b. Assuming the stress strain curve shown in fig. 1a represents a compression test for this material the plastic potentials for the various points abcde on the curve would correspond to those shown for the same points in fig. 1c. The normal vectors on these plastic potentials are shown and the influence of strain hardening and strain softening on the strain increment ratios and the volume strain predictions can be anticipated. The difference in the slope of the strain increment vector at maximum stress conditions predicted by the present theory and that of the associated flow rule is shown at point x for the compression conditions and at y for the extension conditions. From the geometry of fig. 1b and 1c the following strain predictions can be made for the cases

i) Triaxial compression in which $\sigma_a > \sigma_\theta = \sigma_r$ (cylindrical coordinate system in which a, θ and r refer to axial, circumferential and radial directions respectively).

$$d\varepsilon_a/d\varepsilon_r = -2k\psi i \qquad [7a]$$

$$(d\varepsilon_r = d\varepsilon_\theta)$$

$$d\varepsilon_v = d\varepsilon_a(1 - 1/k\psi i) \qquad [7b]$$

ii) Triaxial extension in which $\sigma_r = \sigma_\theta > \sigma_a$

$$d\varepsilon_a/d\varepsilon_r = -2/k\psi i \qquad [8a]$$

$$(d\varepsilon_r = d\varepsilon_\theta)$$

$$d\varepsilon_v = d\varepsilon_a(1 - k\psi i). \qquad [8b]$$

The conditions representing plane strain are indicated in fig. 2 for $\varepsilon_2 = 0$. The sections of the surface shown in fig. 2 are made by the plane, $\sigma_2 = M =$ constant. The theoretical and experimental failure surfaces are illustrated. Plane strain conditions can only exist in sectors paq and ras. The only points which have zero component of strain in the 2 direction are the points at which the surfaces are tangent to lines emanating from the stress origin. Uniqueness for plane strain conditions can be obtained by assuming curved plastic potential surfaces. The plastic potential family in this section and the locations of the strain increment vectors for plane strain are illustrated in fig. 2.

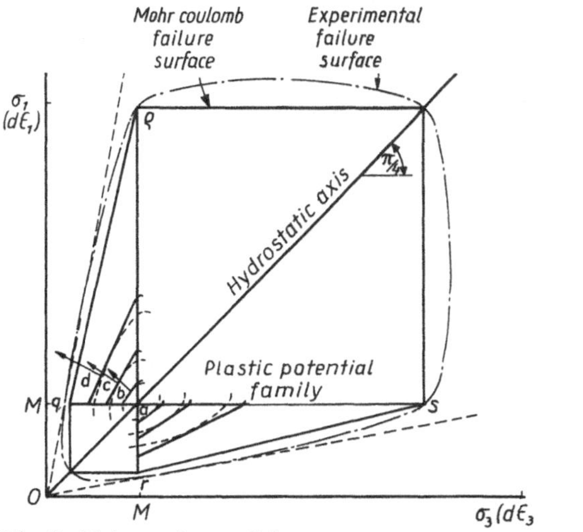

Fig. 2. Plain strain conditions

The strain predictions of the theory for plane strain conditions in which $\sigma_1 > \sigma_2 > \sigma_3 (\varepsilon_2 = 0)$ are thus

$$d\varepsilon_1/d\varepsilon_3 = -k\psi i \qquad [9a]$$

$$d\varepsilon_v = d\varepsilon_1(1 - 1/k\psi i). \qquad [9b]$$

Strain ratios for other stress states can be determined in the same way from the geometry of the plastic potential curves.

The eqs. [5]–[9] were first proposed by *Kirkpatrick* (1970) when it was shown that they were successful in predicting strains ratios during dilating deformation. It would be desirable if the equations could be used to predict strains in the initial or non-dilating stages of the stress-strain curve also. This extension of the theory is now proposed merely as an expedient so that the entire stress-strain behaviour can be estimated using a single process. The proposal is made for simplicity despite the lack of evidence of coincidence of the axes of stress and strain increment at low stress ratios. In practice the extension merely involves the application of eqs. [5]–[9] to all states of deformation. The conception of the shapes of the plastic potentials for nondilating deformation is however more obscure. In practical cases deformation in normally consolidated sands will be caused by a loading or an unloading from the in situ (at rest) condition at which $\sigma_3/\sigma_1 = k_0$. Since $\sigma_1/\sigma_{3F} \doteqdot 1/k_0$ errors as a result of ignoring plastic strains might be expected to be minimal in situations where σ_1/σ_3 is increasing. Deformations as a result of loading from a hydrostatic condition are more of academic interest but for these cases also *Barden, Khayatt* and *Wightman* (1969) have argued that for first loading and reloading the recoverable strains are negligible compared with the non recoverable strains.

This together with the simplifications involved might be considered justification for applying the theory to the entire stress-strain curve for loading situations.

The stress strain hypothesis presented here might be considered to be an extension of a concept proposed by *Hansen* (1959) and used later by *James* and *Bransby* (1970) to calculate strain increments under plane strain conditions for sands which were assumed to have perfectly plastic types of stress-strain curves.

Experimental verification

The applicability of the theory can be checked over a wide range of possible stress conditions by comparing experimental strain data for the common test cases of triaxial compression, plane strain and triaxial extension with the prediction of eqs. [7], [8] and [9]. In order that the comparison be meaningful however it is necessary that the assumptions made in the theory regarding the conditions in these tests are developed in practice.

The main assumption in strength tests of this nature is that a homogeneous state of stress and strain is developed throughout the test samples. In addition in the cases of triaxial compression and extension it has been tacitly assumed that equality exists between the lateral principal stresses and between the lateral principal strains. *Kirkpatrick* and *Belshaw* (1968) and *Kirkpatrick* and *Younger* (1970) have shown from X-ray measurements of internal displacement that the assumptions with regard to strains are valid in the compression test only if effectively frictionless end platens such as that described by *Rowe* and *Barden* (1962) are used. Similar studies on the extension test and the plane strain test by *Kirkpatrick* and *Younger* (1971) indicate in these cases also that effective platen lubrication results in the development of the desired strain conditions. The presence of small amounts of end friction however produces non-homogeneous strain states and in the case of the triaxial cylinder results in situations in which lateral principal strains are not equal.

Less can be said of the assumptions made in the tests with regard to stress. *Kirkpatrick, Seals,* and *Newman* (1974) however report data which indicates that platen lubrication produces uniform distributions of stress across the platens in the compression test and imply that a homogeneous stress state develops throughout the sample with $\sigma_r = \sigma_\theta$. The analysis of the data which is now presented has been made on the basis of the validity of these stress and strain assumptions.

The requirement concerning platen lubrication excludes the use of much of the test data in the literature in any attempt to appraise theory. The data presented here have been gathered from a number of sources but in all cases refer to normally consolidated sand tested using platen lubrication and refined methods of measuring stress, displacement and volume strain. In some cases lateral strains have been measured directly and in others they have been arrived at through a knowledge of the axial and volume strains.

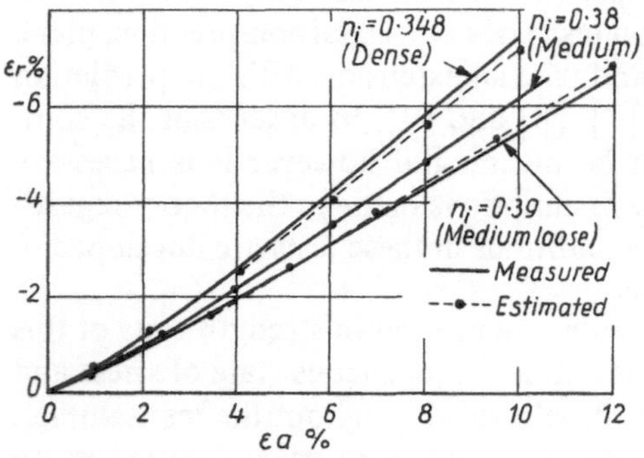

Fig. 3. Triaxial compression tests on Leighton Buzzard Sand

Fig. 3 refer to triaxial compression tests on Leighton Buzzard Sand performed by *Khataniar* (1968). This material is a quartz sand having rounded grains and the size fraction tested (passing no. 25 and retaining on no. 52 British Standard Sieve) had a porosity (ratio of void volume to total volume) range of 34.5–42%. Fig. 3 shows plots of the measured radial strain against axial strain (shown as full lines) for tests at dense, medium and medium loose porosities, n_i. The stress ratio – axial strain curves for the tests have been used in conjunction with eqs. [7] and [5] taking φ_F as the adjusted value of 26.3° to predict radial strains and allow theoretical curves relating radial and axial strain to be constructed. The predictions are shown as dots beside the appropriate curves and indicate close agreement with measured values over an extensive range of strains including both dilating and non dilating deformations for a fairly wide range of porosities. The agreement is best for the test at medium porosity there being a tendency for the theory to underestimate ε_r at low porosities while at high porosities ε_r is over-estimated.

The slight lack of agreement might arise if φ_F is not a constant as assumed but varies with porosity. To check this possibility values of φ_F were calculated from the theory by applying eqs. [7a] and [5] to the strains measured over the peak of the stress strain curve. The values estimated in this way from a number of compression tests over a range of porosities showed a tendency for φ_F to increase slightly with increasing porosity but the mean value of φ_F remained close to the value obtained experimentally by the method described previously. Using these calculated values of φ_F in the equations the estimated curves relating ε_r to ε_a were found to fit very closely with the measured curves. Similar observations can be made when the theory is checked against the results of tests on very loose sands in which the non-dilating deformation continues up to high strain values.

For most purposes sufficient accuracy of predictions will be obtained by assuming φ_F is a constant at the average value for the range of porosities for the material. For greater accuracy the actual value of φ_F for the porosity may be taken. Although this may be less convenient it does not represent a serious drawback in the use of the method since φ_F is considered basic data for the material which can readily be acquired.

A check on the accuracy of the predictions when applied fo a different material can be obtained from the data published on River Welland Sand. This is described as an angular or subangular material with a porosity range between 38.5 and 48.5%.

The data analysed comes from compression and extension tests by *Khayatt* (1967) and from plane strain by *Tong* (1970). The special experimental techniques described in this paper to obtain adjusted values of φ_F were not undertaken in these tests and the values of φ_F used in this analysis were estimated from eqs. [5], [7], [8], and [9] by the method described above. In the light of the findings for Leighton Buzzard sand this approach seems justified. Data from three compression tests at porosities covering the full range gave φ_F values varying between

29.5 and 31° (increasing as the porosity increased) with an average value of just below 30°. Plane strain tests over the same porosity range gave values of φ_F between 30.5 and 31.5° with an average value just below 31°. For this sand also the calculated values of φ_F tended to increase with increasing porosity. A value of φ_F was not determined for the extension tests since it is predicted that this should be the same as in the compression test.

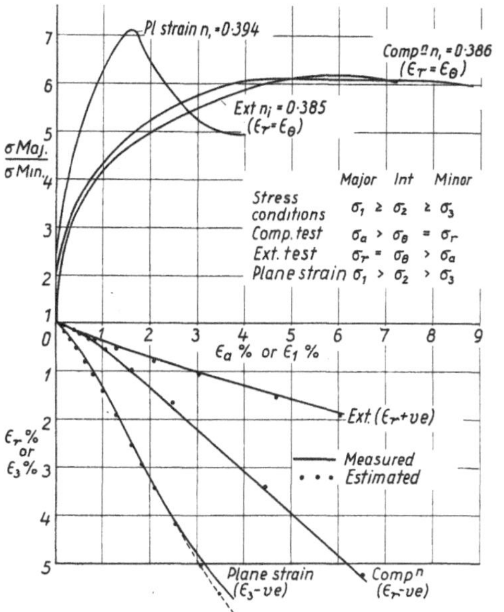

Fig. 4. Triaxial compression, triaxial extrnsion and plain strain tests on dense River Welland sand

The theoretical strain predictions shown in fig. 4 were made using the appropriate average φ_F values (compression or plane strain). The experimentally obtained results including the various stress ratio-strain curves are shown as full lines in fig. 4 while the strain predictions are indicated in the lower part of the diagram by the dots. Tests at approximately identical porosities were selected to aid the comparison. The actual φ_F values for these tests are by coincidence close to the average values used in the analysis.

Again very close agreement is obtained between the measured strains and the theoretical predictions. The agreement is particularly noteworthy in the case of the plane strain test where the theory is able to predict the strains successfully throughout both the hardening and the marked softening parts of the stress-strain curve.

Conclusions

A stress-strain theory is presented for predicting plastic strain increments in drained sands at medium stress levels at which grain crushing is an unimportant aspect of behaviour. The theory specifies the relationship between strain increments in the various directions with stress ratio. The theory is general and has been demonstrated to apply to normally consolidated materials of widely differing character over a wide range of porosities and a comprehensive range of loading stress conditions.

The stress-strain law which can be described as a non-associated flow rule is based on the assumption of normality of the strain increment vector to families of plastic potentials. The plastic potential surfaces in principal stress space for dilating deformation are straight sided cones with curved sections similar in shape to the experimental yield surface but defined by a reduced parameter ψ rather than the peak angle of shearing resistance φ.

The equation which describes the plastic potentials accurately is likely to be complex (such as suggested in eq. [1]) and will involve complexity in mathematical manipulation. The success of the method however lends encouragement in the search for a simplified but appropriate form for this equation.

The problem of finding a description of the stress-strain curve remains, plastic theory is however considered a fruitful basis for providing a complete description of stress-strain behaviour for cohesionless granular materials.

Summary

The paper presents a stress-strain theory for predicting incremental strain ratios in sands deforming under conditions of full drainage. The theory is developed from concepts of plastic theory involving normality of the strain increment vector to families of plastic potentials.

Conditions of uniqueness of strain demand that octahedral sections of the plastic potential surfaces be curved with shapes similar to that of the experimental yield surface in principal stress space but defined by an instantaneous constant ψ which is always less than the peak angle of shearing resistance φ. Values of ψ are determined from a single parameter φ_F which can be obtained from standard strength tests.

The theory is general and experimental evidence is provided to show that it is successful in predicting strain increments for a wide range of normally consolidated sands over a wide range of porosities and covering a comprehensive range of loading stress conditions.

References

1) *Barden, L., A. J. Kayatt,* and *A. Wightman,* Canad. Geotech. J. **6**, 227–240 (1969).

2) *Cornforth, D. H.,* Geotechnique **14**, 143–167 (1964).

3) *Drucker, D. C.*, Trans. ASME Jour. App. Mech., pp. 101–106 (1959).

4) *Hansen, J. B.*, Proc. Conf. Earth Press. Probs. Brussels **1**, 39–48 (1958).

5) *James, R. G.* and *P. L. Bransby*, Geotechnique **20**, 17–37 (1970).

6) *Khataniar, B. C.*, An experimental investigation of assumptions made in the interpretation of triaxial test data. Thesis presented in partial fulfillment of the requirement for M. Sc. University of Strathclyde, Glasgow (Scotland).

7) *Khayatt, A. J.*, Segme incremental stress strain relations for sand. Ph. D. thesis University of Manchester, Manchester (England) (1967).

8) *Kirkpatrick, W. M.*, Proc. IVth Conf. Soil Mech. and Found. Eng. **1**, 172–178 (1957).

9) *Kirkpatrick, W. M.*, Proc. Vth Int. Conf. Soil Mech. and Found. Eng. **3**, 131–133 (1961).

10) *Kirkpatrick, W. M.*, Civil Engineering Department Internal report No. SM 70/01 (April 1970), University of Strathclyde.

11) *Kirkpatrick, W. M.* and *D. J. Belshaw*, Geotechnique **18**, 336–350 (1968).

12) *Kirkpatrick, W. M.* and *J. S. Younger*, Proc. ASCE vol. 96 SM 5, pp. 1683–1695 (1970).

13) *Kirkpatrick, W. M.* and *J. S. Younger*, Proc. 1st Aust. New Zeal. Conf. Geomechanics Melbourne (1971).

14) *Rowe, P. W.*, Proc. Royal Soc. Serial **A 269**, 500 to 527 (1962).

15) *Rowe, P. W.* and *L. Barden*, Proc. ASCE **90** SMI, 1–27 (1964).

16) *Roscoe, K. H.*, *R. H. Basset*, and *E. R. L. Cole*, Proc. Geotech. Conf. Oslo **1**, 231–237 (1967).

17) *Kirkpatrick M. W.*, *R. K. Seals*, and *B. Newman*, Proc ASCE, vol. 100, SM 1.

18) *Tong, P.*, Private communication to author (1970).

Author's address:

W. M. Kirkpatrick
Department of Civil Engineering
University of Strathclyde
Glasgow (U.K.)

VIe CONGRÈS INTERNATIONAL DE RHÉOLOGIE LYON 1972
VIth INTERNATIONAL CONGRESS OF RHEOLOGY LYON 1972
VI. INTERNATIONALER KONGRESS FÜR RHEOLOGIE LYON 1972

Editors:
G. VALLET (Lyon) · W. MESKAT (Leverkusen)

Rheological Theories · Measuring Techniques in Rheology
Test Methods in Rheology · Fractures
Rheological Properties of Materials · Rheo-Optics · Biorheology

Organizing Committee: G. Vallet (President), B. Persoz and M. Joly (Vice-Presidents), C. Smadja (Secretary), A. Larcan, Ph. Comte, J. F. May, and Ph. Berticat (Members)

(Special Issue from Rheologica Acta, Vols. 12 and 13)

With 1203 figures, 6 diagrams and 162 tables

SPRINGER-VERLAG BERLIN HEIDELBERG GMBH 1975

ISBN 978-3-7985-0424-0 ISBN 978-3-662-41458-3 (eBook)
DOI 10.1007/978-3-662-41458-3

Originally published by Dr. Dietrich Steinkopff Verlag · Darmstadt in 1975

Manufactured by Universitätsdruckerei Mainz GmbH, Mainz, and
Dr. Alexander Krebs, Hemsbach/Bergstr. und Bad Homburg v. d. H.

CONTENTS · INHALT

---

*) Die Seitenzahlen beziehen sich auf die fetten Zahlen am Fuße.
The page numbers quoted in these contents are that to be found **bold typed** *at the bottom of each page!*

Rheol. Acta **13**, 199–208 (1974)

From the Department of Civil Engineering, Oxford, Miss. (U.S.A.)

Rheological behavior of soil-cement slab subjected to drying

By K. P. George) and M. Haroon**)*

With 11 figures

(Received October 27, 1972)

Introduction

A new aspect of deformation or warping has evolved from the laboratory investigations which were designed to study the shrinkage cracking of soil-cement base. A one-inch thick soil-cement slab (fine-grained soil), upon drying from the top face, first deforms with a concave surface at the top and gradually assumes an opposite curvature with convex surface at the top (fig. 1).

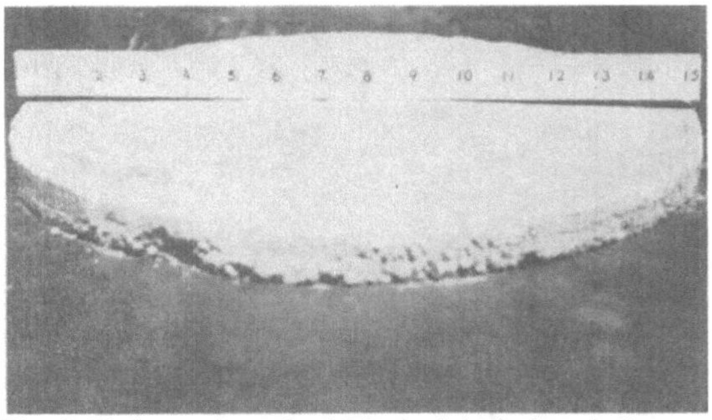

Fig. 1. Deformation of a 15-inch slab during drying

A classical approach to ascertain stresses and displacements in these panels, and for that matter in the prototype, is to use the expressions derived by *Pickett* (13) or modify equations derived by *Westergaard* (16). Fig. 2a depicts the maximum deflection of a long slab drying from one face (top) only, in which case the slab deflects with concave face upwards. On the contrary, fig. 2b, which is a sectional view of fig. 1, shows a different deformed shape. The phenomenon is contradictory to previous predictions. To the writers' knowledge, no mention has ever been made of such a phenomenon of warping with the convex surface at the top. This report, therefore, attempts to explain the paradoxial behavior of drying soil-cement panels.

In order to substantiate the doming of panels, we propose the following analyses. Moisture migration will be governed by the unsaturated capillary flow theory, modified to account for the fact that soil-cement, like natural soils, will exhibit hysterisis in the relation between pressure head (capillary tension) and its moisture content. By postulating that moisture tension (stress) results in deformation or shrinkage, a relation between shrinkage and moisture tension can be obtained from experimental data. Imposing the shrinkage strain in a free finite plate, we present a quasi-static analysis of transient shrinkage-stress and deflection in the linear theory of viscoelastic solids with time-dependent modulus. In order to take into account the time- and humidity-dependent modulus, however, a plane problem with prescribed shrinkage distribution is solved by the finite element method.

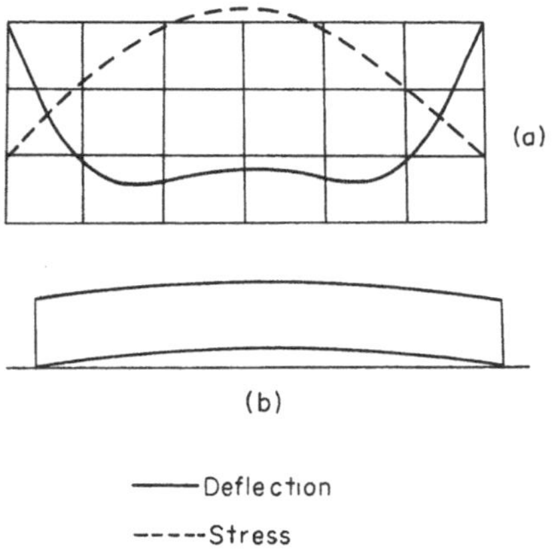

Fig. 2. (a) Stress and deflection in a strip slab of width "b" [after ref. (16)]. (b) Sectional view of the warped slab in fig. 1

Drying shrinkage of soil-cement slab

Conditions in a drying soil (or any medium for that matter) are never static, since the soil is always either losing or gaining moisture. The transient conditions associated with the moisture movement are believed to be a key factor in the problem. That the new aspect of doming corresponds to the direction of moisture movement lends support to this belief. Among the various theories governing moisture migration in soils, the diffusion theory and the capillary flow theory have received widespread recognition.

---

*) Professor of Civil Engineering, The University of Mississippi, USA.

**) Graduate Assistant, The University of Mississippi, USA.

Diffusion theory of moisture movement

Childs (3) and many others explicitly proposed the movement of moisture by diffusion as the principal flow mechanism. In systems possessing appreciable amounts of internal surface, when water is able to form structures and shapes that affect the transmission process, the diffusion equation is inexact (7). The diffusion theory, therefore, is abandoned in favor of the capillary flow theory.

Capillary flow equation

The distribution of soil moisture, θ, with depth, z, can be computed by solving the flow equation:

$$\frac{\partial \theta}{\partial t} = \frac{\partial}{\partial z}\left(K_{(\theta)} \frac{\partial \Psi}{\partial z}\right) + \frac{\partial K}{\partial z} \qquad [1]$$

with the following boundary and initial conditions.

$$\frac{\partial \theta}{\partial z} = 0, \quad t > 0, \quad z = h \qquad [2]$$

$$D \frac{\partial \theta}{\partial z} = f \log \frac{\theta}{\theta_e}, \quad t = t, \quad z = 0 \qquad [3]$$

$$\theta = \theta_i, \quad t = 0, \quad 0 \leq z \leq h \qquad [4]$$

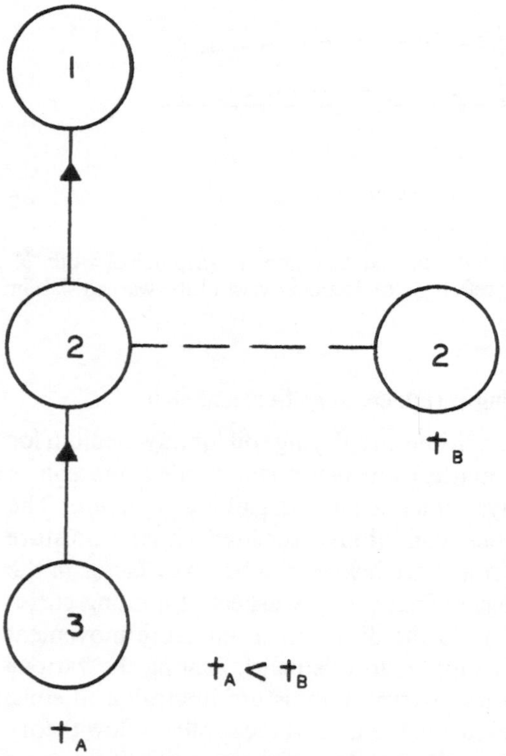

Fig. 3. Nodal arrangements to derive eqs. [5] and [6]

where θ = volumetric moisture content (cm^3/cm^3); $K_{(\theta)}$ = hydraulic conductivity of soil (cm/sec), which is assumed to be a single valued function of θ (fig. 4); $\Psi_{(\theta)}$ = soil water pressure head (cm of water), negative for unsaturated flow (fig. 5); θ_i = initial moisture content; θ_e = moisture content in equilibrium with the vapor pressure in atmosphere; D = diffusivity coefficient (cm^2/sec) $= K \frac{d\Psi}{d\theta}$; and f = surface factor (cm/sec).

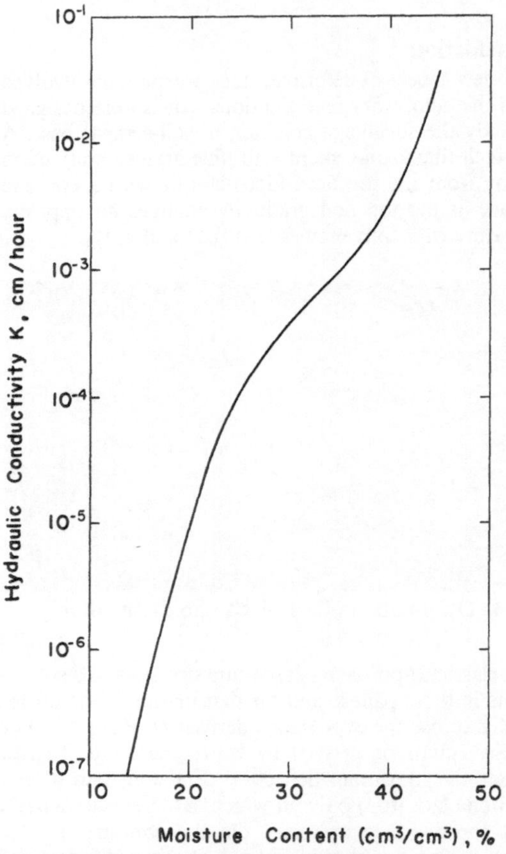

Fig. 4. Hydraulic conductivity related to moisture content

Referring to the nodal arrangements shown in fig. 3, the explicit finite difference form of eq. [1] is

$$\theta_{2B} = \theta_{2A} + \frac{t_B - t_A}{2(\Delta z)^2}\,[(K_3 + K_2)(\Psi_3 - \Psi_2) + (K_1 + K_2)(\Psi_1 - \Psi_2) + (K_3 - K_1)\,\Delta z]. \qquad [5]$$

For the evaporative boundary, eq. [3] in conjunction with eq. [1] gives

$$\theta_{2B} = \theta_{2A} + \frac{t_B - t_A}{2(\Delta z)^2}\left[(K_3 + K_2)(\Psi_3 - \Psi_2) + K_2\left\{\Psi_3 - \frac{4\Delta z\, f \log\left(\frac{\theta_2}{\theta_e}\right)}{K_3}\right\} - K_2\Psi_2 + K_3\Delta z\right] \quad [6]$$

where θ_{2A}, θ_{2B} = moisture contents at node 2 at respective times t_A and t_B; Ψ_1, Ψ_2, Ψ_3 and K_1, K_2, K_3 are, respectively, capillary potentials and hydraulic conductivity at nodes 1, 2, and 3 at time t_A.

Hysterisis associated with capillary flow

One peculiarity of capillary flow analysis arises from the complex dependence of liquid-filled space upon soil moisture tension. As early as 1926 *Haines* (6), and others subsequently, noted that the tension depends upon the previous history of moisture content. Consequently, Ψ is a multivalued function of θ, one relation for dessication and another for remoistening (fig. 5). In other words, soil exhibits hysterisis in relation between Ψ and θ.

Under conditions wherein every thin soil section is actually in the drying stage, the capillary potential corresponding to the measured moisture content is truly the desorption potential, Ψ_d. In slabs of finite thickness, however, each thin soil section transmits moisture from the adjacent element below to that at the top. It could be that the moisture content in each element, except that at the bottom of the slab, fluctuates in small excursions, and the potential follows a drying and remoistening cycle. This possibility is schematically represented in fig. 5. Accordingly, as the top layers undergo drying and rewetting, the capillary potential is no longer the dessication potential Ψ_d but approaches the remoistening potential $\Psi_h(\Psi_w \leq \Psi_h < \Psi_d)$; where Ψ_w = sorption potential. The reduction in potential in each layer is proportional to the amount of moisture transmitted. This latter result is incorporated in eqs. [5] and [6], from which the moisture content and potential are computed during the course of drying.

Drying shrinkage

That shrinkage is proportional to moisture loss, a tacit assumption made in previous investigations, can be shown to be (13, 15) inexact. According to *Hrennikoff* (10), the drier the material the greater the shrinkage per unit volume of moisture evaporated.

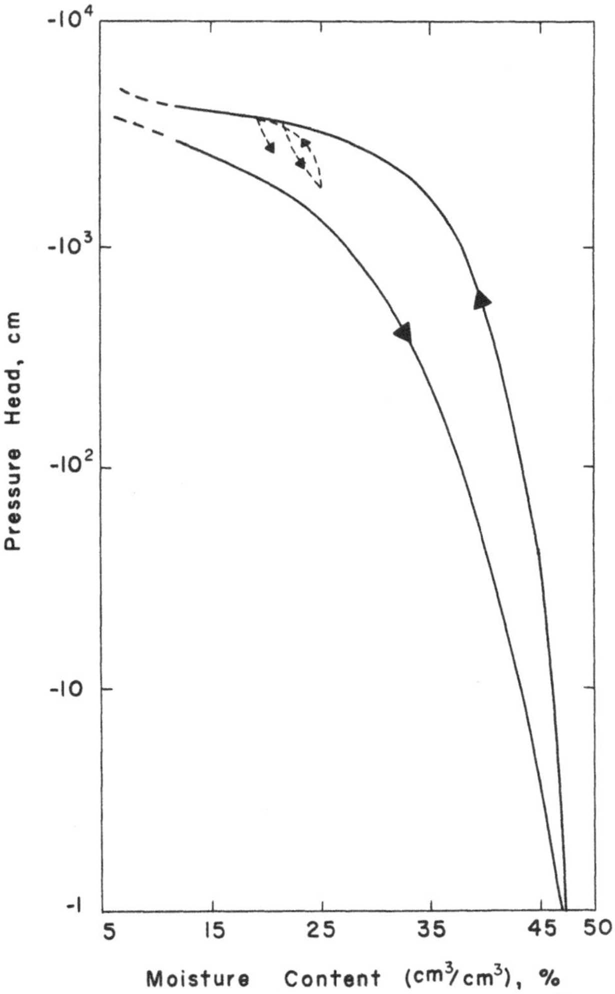

Fig. 5. The hysterisis-affected relations between soil moisture content and pressure head in silty clay soil-cement; drying curve on the right and wetting curve on the left.

In the present study, however, we propose a semiempirical theory which asserts that the volume change behavior in partly saturated soil conforms to the principle of effective stress. *Blight* (1) experimentally proved that in producing changes of effective stresses, the moisture potential is equivalent to mechanical stress. Stated differently, the stress from the pore water induces shrinkage deformation of the solid macro-structure in the same manner as though this stress was applied from a superimposed load.

An analytical relation between moisture tension and the volume change that arises therefrom is unduly complex; therefore, an empirical relation is obtained between the experimental values of Ψ and percent shrinkage S. The relation for the silty clay soil investigated here is

$$S = \exp(9.7858 - 31.8647\,x + 22.4525\,x^2 - 6.2153\,x^3 + 0.6175\,x^4) \qquad [7]$$

where $x = \log_{10}\Psi$; and Ψ expressed in cm of water.

From the moisture tension, as solved for from eqs. [5] and [6] and allowing for hysterisis, free shrinkage of a one-inch slab is computed with the aid of eq. [7]. Typical distributions are shown in fig. 6a.

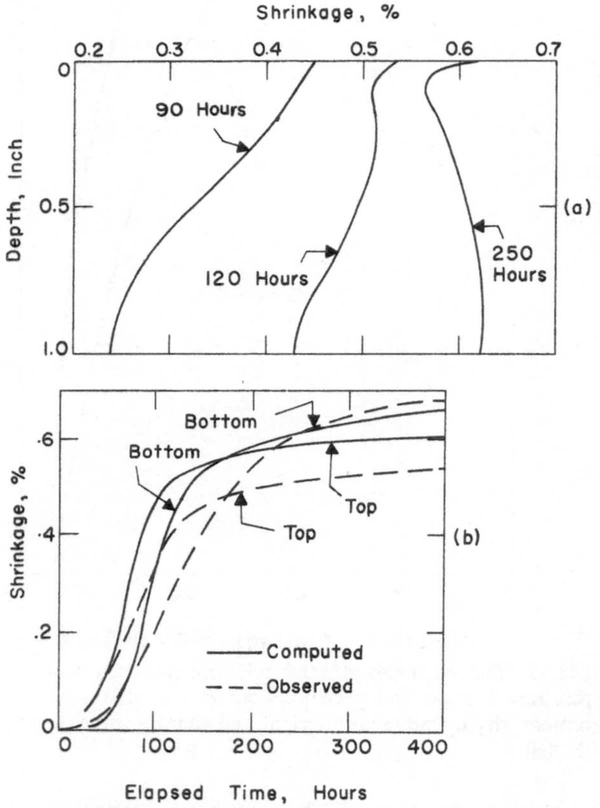

Fig. 6. (a) Computed shrinkage distribution with depth. (b) Comparison between observed and computed shrinkages; top and bottom only

The computed shrinkage variation is compared with the experimentally determined shrinkages at the top and bottom (fig. 6b). The shrinkage strain measured by strain gages may be open to question as being modified by creep; this influence may be less serious, however, if only the difference between the top and bottom gages is of primary concern. Excluding the top crest, where the state of shrinkage is postulated to be less than that predicted by the moisture tension theory (1), the computed strain variation is in satisfactory agreement with the measured one.

Transient shrinkage stresses and deflections in an elastic slab

Consider a plate of arbitrary planform and of constant thickness h (fig. 7). The plate is completely free of surface traction and the shrinkage varies through the thickness only, that is $S = S(z)$.

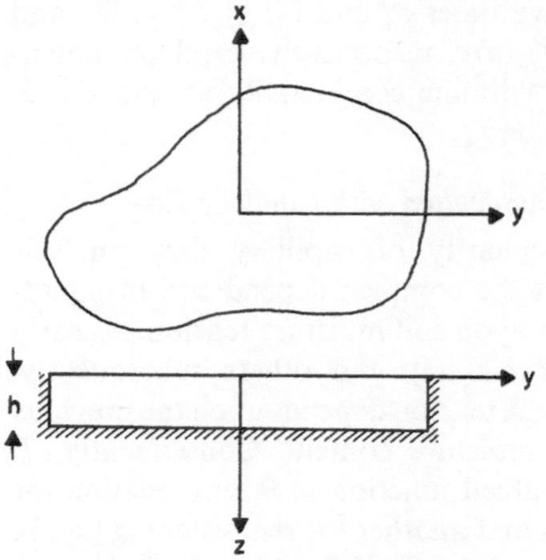

Fig. 7. Slab drying from top face only

Under these conditions the nonzero stress components will be

$$\sigma_{xx} = \sigma_{yy} = f(z) \qquad [8]$$

and

$$\sigma_{zz} = \sigma_{xz} = \sigma_{yz} = \sigma_{zy} = 0.$$

With the boundary conditions that the resultant force and moment (per unit length) produced by σ_{xx} and σ_{yy} are zero over the edges of the plate, *Boley* and *Weiner* (2) solved for stresses given by

$$\sigma_{xx} = \sigma_{yy} = \frac{ES(z)}{1-\nu} + \frac{6M_T}{h^2(1-\nu)}\left(\frac{2z}{h} - 1\right) - \frac{N_T}{h(1-\nu)} \qquad [9]$$

where

$$M_T = \int_0^h ES(z)\left(\frac{h}{2} - z\right) dz$$

and

$$N_T = \int_0^h ES(z)\, dz$$

where E = *Youngs* modulus; and ν = *Poissons* ratio.

The displacement in the Z-direction corresponding to stresses expressed by eq. [9] is (2),

$$U_z = -\frac{6M_T}{h^3 E}(x^2 + y^2) + \frac{1}{(1-\nu)E} \times \left[E(1+\nu)\int_0^z S(z)\,dz - \frac{\nu z N_T}{h} - \frac{12\nu M_T}{h^2} \times \left(\frac{z^2}{h} - z\right)\right]. \qquad [10]$$

Analysis of a viscoelastic slab with moisture-dependent material properties

Consider the slab in fig. 7 subjected to one dimensional drying. We assume that the drying problem is uncoupled with the mechanical one, and that the transient shrinkage $S(z, t)$ is known (fig. 6).

Constitutive relation in linear theory of viscoelasticity

In the linear theory of viscoelasticity, for a medium with a finite and discrete spectrum of relaxation and retardation times, the linear constitutive law admits the differential-operator representation.

$$P\sigma = Q\varepsilon \qquad [11]$$

where σ and ε represent stress and strain, respectively. P and Q are linear operators of the form $\sum_0^m a_n D^n$ and D is the time derivative $\partial/\partial t$. In this analysis, however, we adopt a form that corresponds to the hereditary integral method of specifying viscoelastic behavior, which gives a continuous spectrum, in preference to the discrete spectrum of relaxation times as represented by eq. [11].

In a general state of stress and strain in three dimensions, and for an isotropic medium, two operators are required to describe the stress-strain relationships as shown by *Lee* (11). Usually chosen in viscoelastic analysis are the relations between the dilative and deviatoric components of stress and strain (12). *Zienkiewicz* (17), however, showed that two operators, corresponding to E and ν in elastic representation, will describe the constitutive relation in concrete. In evaluating thermal stresses in concrete, he further simplified the problem by a postulation, supported by experimental results (17), that *Poissons* ratio in the material remains constant during creep. Following the same reasoning, we assume the following integral law for stress in terms of strain:

$$\sigma(x_k, t) = \int_{-\infty}^{t} G(t-\tau)\frac{d}{d\tau}[\varepsilon(x_k, \tau)]\,d\tau$$

$$k = 1, 2, 3 \qquad [12]$$

where $G(t)$ is the relaxation modulus in uniaxial load.

Laboratory investigation of relaxation modulus

The stiffness characteristics should, ideally, be determined by means of relaxation tests. Such tests are extremely difficult to perform, and the usual procedure is to conduct creep tests and derive the relaxation modulus from the results. Assuming that the material is linearly viscoelastic in response, the following superposition relations, which are due to *Boltzmann*, are applicable:

$$\int_0^t G(t)\,J(t-\tau)\,d\tau = \int_0^t G(t-\tau)\,J(t)\,d\tau = t \qquad [13]$$

where $J(t)$ is the creep compliance. *Hopkins* and *Hamming* (8) showed that it is possible to solve eq. [13] for one function (e.g. the modulus) if the other (i.e., the compliance) is defined on the time scale. *Pretorius* and *Monismith* (14) successfully applied this technique to test data on asphalt concrete and soil-cement.

Relaxation moduli for soil-cement as a function of time and relative humidity, computed using eq. [13] by finite difference method, are presented in fig. 8.

Deflections and stresses due to imposition of shrinkage: $G = G(t)$

Employing the second version of correspondence principle, *Zienkiewicz* (17) asserted that, due to imposition of initial shrinkage strains, the displacement of a viscoelastic solid, could be obtained by conventional elastic analysis, and the stresses by considering an equivalent elastic structure with initial modulus E_0 on which shrinkage changes of a magnitude

$$\frac{1}{E_0}\int_0^t G(t-\tau)\frac{\partial S(\tau)}{\partial \tau}\,d\tau \qquad [14]$$

are applied. The expression given by eq. [14] is denoted as equivalent shrinkage, $S_E(t)$. Accordingly, the stress in the viscoelastic slab obtained from eq. [9] is

$$\sigma_{xx}(t) = \sigma_{yy}(t) = \frac{E_0}{1-\nu} \times \left[S_E(t) + \frac{6}{h^2}\left(\frac{2z}{h}-1\right) \int_0^h S_E(t)\left(\frac{h}{2}-z\right) dz - \frac{1}{h}\int_0^h S_E(t)\, dz \right]. \quad [15]$$

The viscoelastic deflection, however, is given by eq. [10].

In making numerical computations, it is tacitly assumed that the relaxation modulus corresponding to 90% R. H. is applicable at any time, $t > 0$.

Shrinkage stresses and deflections:
Influence of humidity: $G = G(t, H)$

Time-humidity equivalence

To include the dependence of the modulus upon a variable humidity (moisture) field the constitutive law must be modified. The basis of this modification is to assume that the material during shrinkage exhibits "hydrorheologically simple" behavior. The distinction between the thermorheological and hydrorheological behaviors should be noted. Thermorheological behavior would imply that an increase in temperature corresponds to an increase in time. In a "hydrorheologically simple" material, however, a decrease in humidity would correspond to an increase in time. Based on the work of *Muki* and *Sternberg* (12), if $G(t, H)$ is the relaxation modulus function in tension at a constant humidity H, then

$$G(t, H_0) = G'(t) = G''(\log t) \quad [16]$$

where H_0 is the reference humidity.

Since the material is assumed hydrorheologically simple, the relaxation modulus can be expressed as

$$G(t, H) = G''[\log t + \log \varphi(H)] = G''[\log(t \cdot \varphi(H))] \quad [17]$$

where $\varphi(H)$ = "shift function".
Also by eq. [16] we have

$$G(t, H) = G'(\zeta) \quad [18]$$

where the "reduced" time ζ is given by

$$\zeta = t \cdot \varphi(H). \quad [19]$$

Eqs. [18] and [19] enable us to pass from eq. [12], which holds for a constant humidity H_0, to the corresponding relaxation integral law applicable at any constant humidity H. Thus, knowing $\varphi(H)$ gives the entire family of $G(t, H)$ for uniform humidity changes.

Fig. 8 pictures graphically the development of a time-humidity relaxation modulus; the step by step procedures for developing the composite curve are given by *Ferry* (4) and will not be discussed here. It should be noted, however, that the humidity-influenced vertical shifting is significant; therefore, the experimental values at low humidities are increased by a factor H_0/H_i to obtain $G(t)$ at 100% R. H. (fig. 8).

Stress and deflection analysis

In the present problem, because the medium is under the influence of a variable humidity field, $H(z, t)$, the instantaneous stresses at each material point of the body are, in general, functionals of the entire preceding local strain and humidity history. Since an analytical approach is unduly complicated, the problem is solved by the finite element method.

The problem of an infinitely long strip of finite width (plane strain) subjected to prescribed initial strains, typically as shown in fig. 6, is solved for here. A constant strain triangular element program has been used.

The incremental procedure that is followed and the method by which the shrinkage strains are handled are outlined in Appendix II. By this approach the nodal points of each element are locked, and the stresses developed in the elements due to shrinkage, are converted to nodal point forces by eq. [A 2]. These forces are reversed, applied to the system, and the structure solved for nodal point displacements by eq. [A 6], and therefrom stresses are calculated by eqs. [A 4] and [A 5]. This cycle is completed for every strain input over a small time increment Δt (10 h in this problem) and the resulting stresses and strains added to the previous values. For each time increment the relaxation modulus corresponding to the difference between the final time and time of increment is used.

Results

Soil

A silty clay soil (20% two micron clay and 75% silt) treated with 10% cement and compacted

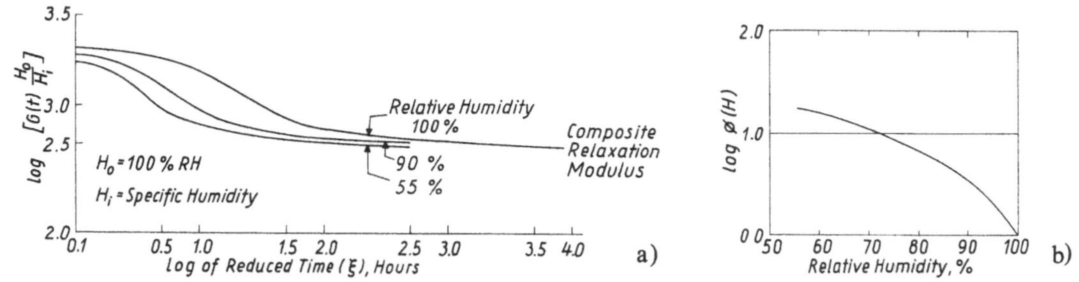

Fig. 8. (a) Composite relaxation modulus curve at a reference humidity of 100%. (b) Shift function vs. humidity

to 110 pcf at 17% (weight basis) was used in this study. The derived equations and the finite element formulation were solved for deflections and stresses using the following properties for the soil:

$E_0 = 280000$ psi (initial modulus)

$\nu \;= 0.25$ and $f = 0.095$ (cm/sec)

Deflection

The maximum deflections of a 15-inch slab, computed by elastic theory (eq. [10]) and by viscoelastic theory [plain strain finite element method using a general $G(t, H)$], are shown in fig. 9. We observe that during the entire drying the two curves almost coalesce, the response of the material being nearly elastic (and hence humidity independent).

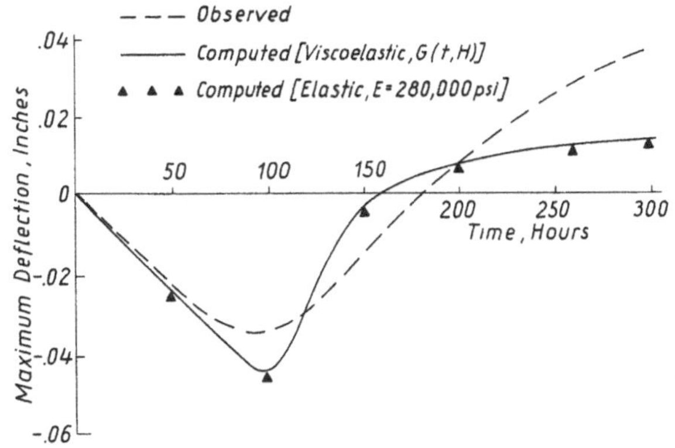

Fig. 9. Comparison between observed and computed (both viscoelastic and elastic) deflections

The deflections of two 15-inch diameter 1-inch thick slabs were measured. In the first, drying took place from the top, and in the second, from the bottom face. Shown by dotted curve in fig. 9 is the average of the two measurements.

Qualitatively at least, the computed and the experimental results compare in a reasonable way. The viscoelastic analysis predicts the extent of curling of the slab and forecasts the elapsed time for the slab to assume rectilinear position (160 h cf. 180 h). The theory, however, underpredicts the extent of doming. The large difference may be attributed partly to the finding that capillary theory ceases to apply in extremely dessicated cemented materials. *Feldman* and *Sereda* (4) showed that when dessication continues until all capillary pore spaces have emptied, the *Kelvin* equation ceases to apply. Then either contraction ceases or, in some materials, a spontaneous expansion occurs. The theory of shrinkage presented here does not account for this relaxation of moisture stresses; therefore, it may be conjectured that the theory will overpredict the shrinkage on the top face. Accordingly, the extent of doming will be underestimated.

Shrinkage stress

The stress variation with time in the slab, computed by the finite element viscoelastic analysis (curve 1), is compared with the elastic stresses (curve 2) in fig. 10. The viscoelastic stress falls below their elastic counterpart. Both of the stresses increase rather abruptly, attain their

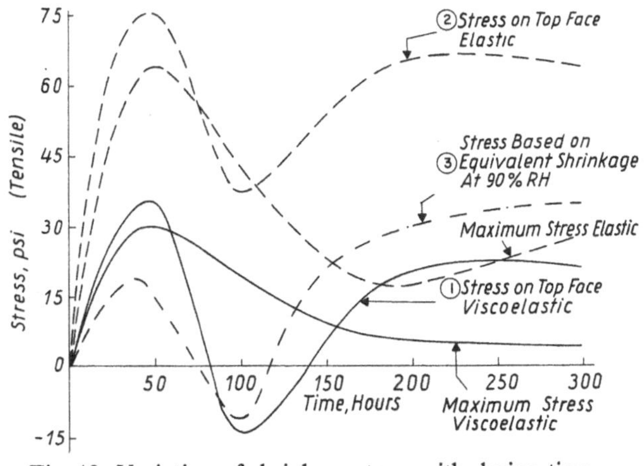

Fig. 10. Variation of shrinkage-stress with drying time. Stresses according to elastic, viscoelastic and equivalent shrinkage concepts are pictured

maximum values in about 50 h, and then decrease. Furthermore, the stresses pass through a minimum value (saddle point). Judging from the steep initial rise in stress, it should be concluded that differential shrinkage causes much severe shrinkage stress.

The elastic stress distribution is compared with the viscoelastic in fig. 11. The elastic shrinkage stress on the surface is always tensile and decreases with time, whereas the viscoelastic counterpart not only decreases but becomes compressive. In other word, a complete reversal of the stress picture predicted by the elastic theory occurs as soon as 100 h.

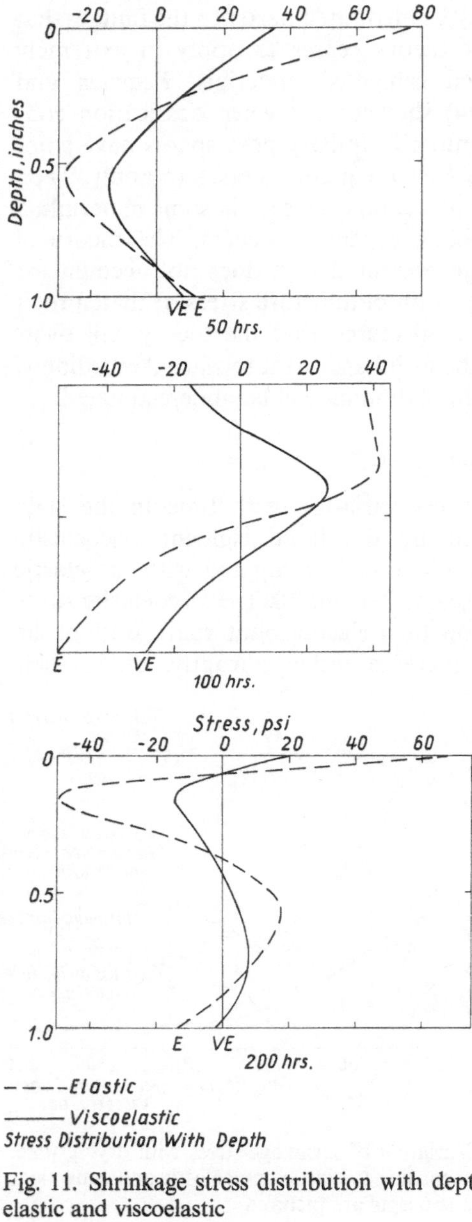

Fig. 11. Shrinkage stress distribution with depth; both elastic and viscoelastic

Such stresses at the exposed face of the slab, computed by the principle of equivalent shrinkage concept (curve 3), assuming 90% relative humidity throughout, are also compared in fig. 10. We note that during the initial period of about 80 h (curve 3) underpredicts the "actual" viscoelastic stress values (curve 1); later on, however, the former curve progressively overestimates the latter. As is evident from calculations, the humidity during the early stage is greater than the average humidity of 90% on which modulus $G(t)$ underlying curve 3 is based. Therefore, the stress relaxation, predicted by the modulus at 90% humidity, will be far more rapid than that implied by the $G(t, H)$ corresponding to the "actual" humidity; accordingly the stresses will be lower. This conclusion confirms the unrealistic character of any viscoelastic shrinkage-stress analysis which neglects the humidity dependence of the viscoelastic properties.

Conclusion

Contrary to the results obtained by *Pickett* and others, a one-inch thick soil-cement slab (fine-grained soil), upon drying from the top face, deforms with the convex surface at the top. In this paper a theory is proposed whereby the doming can be predicted.

The shrinkage results, computed by the capillary tension theory providing for hysterisis, show that the sealed bottom face shrinks more than the exposed top face. Qualitatively, at least, the computed values agree with the experimentally determined values.

Assuming *Poissons* ratio to be constant, deflections and stresses are computed from two points of view: 1. $G = G(t)$ and 2. $G = G(t, H)$. Comparison of the deflections of these numerical calculations shows that the two results compare so well that a general order-of-magnitude estimate can be made by the first method – that is, the conventional elastic analysis. The predicted and measured deformations are in reasonable agreement and thus lend support to the methods of analysis utilized.

The viscoelastic stresses fall below their elastic counterparts. With time both of the stresses increase somewhat abruptly, attain their maximum value in about 50 h, and then decrease. The results also show that for realistic shrinkage-stress computations humidity dependent modulus must be used.

In summary, moisture movement is the key factor in the problem; and the fact that the top shrinkage trails the bottom shrinkage appears to be the mechanism responsible for the reversed deformations.

Appendix

Finite element displacement formulation for shrinkage analysis

In the absence of body forces and surface tractions, the equilibrium relations at the nodes of an assemblage of elements may be represented by the matrix equation (18)

$$\{F\}^e = [K]^e \{u\}^e + \{F\}^e_{\varepsilon 0} \qquad [\text{A}1]$$

where $\{F\}^e$ is the equivalent concentrated nodal force vector; $[K]^e$ is the stiffness matrix; $\{u\}^e$ is the vector of nodal displacements; and $\{F\}^e_{\varepsilon 0}$ is the vector of nodal forces corresponding to initial strain

$$\{F\}^e_{\varepsilon 0} = -\int_V [B]^T [D] \{\varepsilon_0\}\, dv \qquad [\text{A}2]$$

and

$$[K]^e = \int_V [B]^T [D] [B]\, dv \qquad [\text{A}3]$$

in which $[B]$ is given by the relation between element strain and element nodal displacements.

$$\{\varepsilon\} = [B] \{u\}^e \qquad [\text{A}4]$$

$[D]$ is the elasticity matrix relating stress and strain

$$\{\sigma\} = [D] (\{\varepsilon\} - \{\varepsilon_0\}). \qquad [\text{A}5]$$

With $\{F\}^e = 0$ in eq. [A1], for the entire continuum

$$[K] \{u\} = -\{F\}_{\varepsilon_0}. \qquad [\text{A}6]$$

Eq. [A6] can be solved for $\{u\}$, therefrom with the aid of eq. [A4], the stress-induced strains $\{\varepsilon\}$ and finally stresses by eq. [A5].

Shrinkage and/or thermal effects are included in eq. [A5] for the plane strain analysis as:

$$\begin{Bmatrix} \sigma_{xx} \\ \sigma_{yy} \\ \sigma_{xy} \end{Bmatrix} = \frac{E}{(1-\nu)(1-2\nu)} \begin{bmatrix} 1-\nu & \nu & 0 \\ \nu & 1-\nu & 0 \\ 0 & 0 & \dfrac{1-2\nu}{2} \end{bmatrix} \begin{Bmatrix} \varepsilon_{xx} - \varepsilon_{0x} \\ \varepsilon_{yy} - \varepsilon_{0y} \\ \varepsilon_{xy} \end{Bmatrix} \qquad [\text{A}7]$$

where ε_{0x}, ε_{0y} = initial strain, respectively, in x- and y-directions.

The shrinkage stresses for the restraint conditions are transformed to nodal point forces by eq. [A2].

For a one dimensional visco-elastic material subjected to a time dependent strain, the stresses after an elapsed time t is given by

$$\sigma(t) = \int_0^t G(t-\tau) \frac{\partial \varepsilon(\tau)}{\partial \tau}\, d\tau$$

$$\sigma(t) = \sum_{i=0}^{n-1} \int_{t_i}^{t_{i+1}} G(t-\tau) \frac{\partial \varepsilon(\tau)}{\partial \tau}\, d\tau.$$

By making the time increments sufficiently small, the trapezoidal rule can be applied:

$$\sigma(t_n) = \sum_{i=0}^{n-1} \frac{1}{2} [G(t_n - t_{i+1}) + G(t_n - t_i)] \times [\varepsilon(t_{i+1}) - \varepsilon(t_i)]$$

$$= \sum_{i=0}^{n-1} [G(t_n - t_{i+1/2})] [\varepsilon(t_{i+1}) - \varepsilon(t_i)].$$

For a plane strain problem this equation becomes

$$\begin{Bmatrix} \sigma_{xx} \\ \sigma_{yy} \\ \sigma_{xy} \end{Bmatrix} = \sum_{i=0}^{n-1} \frac{G(t_n - t_{i+1/2})}{(1+\nu)(1-2\nu)} \begin{bmatrix} (1-\nu) & \nu & 0 \\ \nu & (1-\nu) & 0 \\ 0 & 0 & \dfrac{(1-2\nu)}{2} \end{bmatrix} \begin{Bmatrix} \varepsilon_{xx}(t_{i+1}) - \varepsilon_{xx}(t_i) \\ \varepsilon_{yy}(t_{i+1}) - \varepsilon_{yy}(t_i) \\ 0 \end{Bmatrix}.$$

Acknowledgement

The studies described in this report were financed through NSF Grant GK-25604. This support is gratefully acknowledged. The writers also wish to thank Mr. *S. S. Kong*, graduate student, for modifying the finite element program to suit the present problem.

Summary

It has been experimentally observed that one-inch thick soil-cement slab (fine-grained soil), upon drying from the top face, first deforms with a concave surface at the top and gradually assumes an opposite curvature with convex surface at the top. The latter shape contradicts the classical analysis of *Pickett*. A theory whereby the doming can be predicted is proposed in this paper.

Moisture migration in the drying slab is determined by the unsaturated capillary flow theory, modified to account for the fact that soil-cement will exhibit hysterisis in the relation between pressure head (capillary tension) and its moisture content. By postulating that moisture tension (stress) results in deformation or shrinkage, a relation between shrinkage and moisture tension is obtained from experimental data. The results seem to violate an intuitive feeling that the exposed top face should shrink more than the sealed bottom face.

The creep compliance of soil-cement specimens is determined for a range of humidities from 55–100%. Finite difference techniques have been employed to compute the relaxation modulus from creep compliance which are related through *Boltzmann* integral. To include the dependence of the modulus upon a variable humidity field, constitutive law is modified on the basis of time-humidity equivalence hypothesis.

Imposing on the theoretical shrinkage distribution, computations for stresses and deflections are made from two points of view: 1. Using the simplifying assumption that the relaxation modulus $G(t)$ does not vary with humidity and 2. allowing for the humidity variation in the transient problem and using a general $G(t, H)$. The solution from the first point of view is obtained by equivalent shrinkage concept and from the latter by finite element method. The computed deflections not only predict doming but also show good agreement with the observed values.

It should be noted that the current theory predicts deflections opposite to those predicted by conventional analysis when the moisture movement is assumed to be governed by diffusion law and shrinkage to be proportional to moisture loss.

References

1) *Blight, G. E.*, J. Soil Mech. and Found. Div., Proc. ASCE **92**, No. SM 6 (1966).

2) *Boley, B. A.* and *J. H. Weiner*, Theory of Thermal Stresses (New York 1967).

3) *Childs, E. C.*, J. Agricultural Science **26** (1936).

4) *Feldman, R. F.* and *P. J. Serda*, J. App. Chemistry (London) **14** (1964).

5) *Ferry, J. D.*, Viscoelastic Properties of Polymers. (New York 1961.)

6) *Haines, W. B.*, J. Agr. Res. **17** (1927).

7) *Hallaire, M.*, Soil Water Movement in the Film and Vapor Phase Under the Influence of Evapotranspiration. Water and Its Conduction in Soil. Special Report 40, Highway Research Board, 1958.

8) *Hopkins, I. L.* and *R. W. Hamming*, J. App. Physics **28**, 906–909 (1957).

9) *Hrennikoff, A.*, J. Engr. Mech. Div., Proc. ASCE, **55** (1958).

10) *Hrennikoff, A.*, J. Engr. Mech. Div., Proc. ASCE **96** (1959).

11) *Lee, E. H.*, Quart. App. Math. **8**, 183 (1955).

12) *Muki, R.* and *E. Sternberg*, J. App. Mech. **28** (1961).

13) *Picket, G.*, ACI J. Proc. **42**, 165–204 (1946).

14) *Pretorius, P. C.* and *C. L. Monismith*, Prediction of Shrinkage Stresses in Pavements Containing Soil-Cement Bases. Record No. 362, Highway Research Board, 1971.

15) *Sanan, B. K.* and *K. P. George*, J. Soil Mech. and Found. Div., Proc. ASCE (1972).

16) *Westergaard, H. M.*, Analysis of Stresses in Concrete Pavements Due to Variations of Temperature. Proc. 6th Annual Meeting, Highway Research Board, 1927.

17) *Zienkiewicz, O. C.*, ACI J., Proc. **58**, 383 (1961).

18) *Zienkiewicz, O. C.*, The Finite Element Method in Structural and Continuum Mechanics (New York 1967).

Authors' address:
Dr. *K. P. George* and Dr. *M. Haroon*
Dept. of Civil Engineering
University of Mississippi
P.O. Box 525, Oxford

Rheol. Acta 13, 209–215 (1974)

From the Centraal Laboratorium TNO Delft (The Netherlands)

Extrusion rheology of rigid PVC

By *J. L. den Otter, J. Schijf, J. L. S. Wales,* and *F. R. Schwarzl*

With 7 figures and 2 tables

(Received October 27, 1972)

1. Introduction

PVC is a very versatile thermoplastic. Unfortunately, its potentialities often cannot be realized in practice because of the difficulties encountered in processing. For instance, successful extrusion is possible only in a limited range of conditions and with the use of processing aids. If processing temperatures and shear stresses are loo low, or residence times in the screw are too short, one obtains what is sometimes called "insufficient plastification". This results in products with poor mechanical properties. If, on the other hand, processing temperatures are too high or residence times are too long, degradation of the PVC occurs which again leads to an unsatisfactory product. These restrictions only leave a small range of conditions where rigid PVC can be processed successfully. Even then, it is necessary to add processing aids such as stabilizers to retard the degradation process, and lubricants which are thought to facilitate powder transport and melt flow. In this situation a better insight into the influence of processing conditions and processing aids on the rheological properties of PVC melts and on the mechanical properties of PVC products seems highly desirable.

The flow properties of PVC melt differ in various respects from those of other commercially important thermoplastics. First, PVC shows a tendency to crystallize in a wide temperature region between its glass transition and its degradation (1). The degree of crystallinity remains small, but it changes with temperature, and recrystallization processes may also occur (2). Therefore, PVC behaves as an amorphous thermoplastic with some crystalline regions. Its flow properties are simultaneously determined by the presence of entanglements and crystallites (3).

Second, the flow of PVC melts seems to depend not only on the molecular but also on the microscopic structure of the material. The morphology of PVC powder is different for products polymerized in different ways. Emulsion (polymerized) PVC consists of small, dense resin particles of spherical shape with diameters between 0.5 and 1 μm (4). Suspension (polymerized) PVC consists of irregular, porous granules of the order of 100 μm in diameter (4). Investigations by *Berens* and *Folt* (4, 5, 6) indicated that the powder particles of emulsion PVC are only partly fused under processing conditions. Electron micrographs of fracture surfaces of emulsion PVC products showed that the resin particles had retained their identity and shape during extrusion up to high temperatures. From this evidence it was concluded that flow of emulsion PVC, under unfavourable circumstances, involves slippage of resin particles past one another rather than a homogeneous flow of the melt. This phenomenon was called particulate flow. Evidence for the occurrence of particulate flow in the melts of suspension PVC is less convincing.

It is to be expected that the occurrence of particulate flow during processing will change simultaneously the flow properties of the melt and the mechanical properties of the final product. The melt might become less viscous and less elastic, and the products might become inferior.

We have recently started a research program to investigate the influence of extrusion conditions and processing aids on the properties of the processed melt and on the quality of the product. Some preliminary results are reported here.

2. The research extruder

Investigation of the rheological properties of the melt and of the quality of the extrudate should be performed under well-controlled and stable conditions. We have therefore constructed a small highly instrumented research extruder, in order to meet these requirements (7). Pressures and temperatures may, moreover, be monitored continuously during operation and may be changed systematically.

Fig. 1 shows the instrumentation of the research extruder, with radial temperature profile monitor (*F*) and slit viscometer (*E*) attached. The extruder is a modified 30-mm Troester machine, extensively instrumented for the measurement of temperatures and pressures. Measured are temperatures in the barrel close to the inner wall (at positions indicated by triangles), temperatures inside the hollow screw (at positions indicated by circles), the temperature at the screw tip (circle), temperatures in the viscometer near the slit (indicated by squares). Pressures are determined by means of electronic pressure gauges of the piston type at four places in the cylinder wall, ① to ④, in the temperature profile

14

monitor ⑤, just before the entrance of the slit viscometer ⑥, and at three places within the rectangular slit of the viscometer, ⑦ to ○.

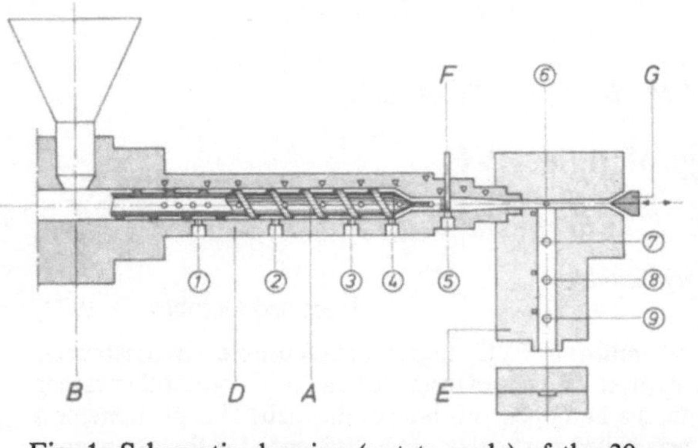

Fig. 1. Schematic drawing (not to scale) of the 30 mm research extruder. *A* screw, *B* hopper, *D* extruder cylinder, *E* slit viscometer, *F* temperature profile monitor, *G* by-pass; ①, ②; ... pressure transducers; thermocouples: ○ in screw, ▽ in cylinder, □ in viscometer, — in monitor

A temperature profile monitor (*F*) is inserted between the extruder cylinder and the slit viscometer. This monitor consists of a cylindrical rod on which seven temperature sensing elements are mounted in upstream direction at various fixed radial positions. These sensors indicate the radial temperature distribution of the melt just beyond the screw tip (8, 9).

A slit viscometer (*E*) may be attached to the exit of the extruder. The extruder is operated with a by-pass (*G*), which regulates that part of the output which passes through the viscometer. By this means, both, the pressure before the entrance of the viscometer, and the ratio of the outputs through by-pass and viscometer are regulated. In the viscometer the melt is forced through a rectangular slit channel. The axial pressure distribution in the channel is measured with gauges inserted in the flat wall of the channel. In this way, it is possible to investigate the influence of screw speed and pressure on the properties of the melt.

Different types of slit viscometers are available. The slit exit of some of these is provided with a regulating valve of the slit which permits viscosity measurements to be made at higher pressures. Other instruments have slits provided with windows, so that velocity profiles or flow birefringence can be measured in translucent melts. The slit viscometer may be replaced by a capillary viscometer of similar construction. Here, only the pressure before the entrance to the capillary is measured. However, length and diameter of the capillaries may be varied.

The screw used throughout was a 15 D screw for PVC extrusion with a compression ratio 1:2.4. Its dimensions are given in fig. 2.

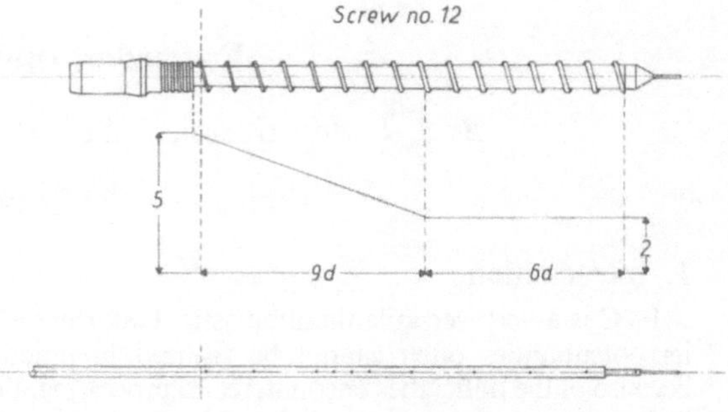

Fig. 2. Details of screw used in experimental work: Upper drawing shows general features; the central drawing shows the profile in channel depth; the lower drawing shows the hollow rod inserted for temperature measurements; the dots indicate the positions of the thermocouples

All measuring points can be recorded with a data logger with 50 channels, which produces punched tape. Only the readings of the outputs through viscometer and by-pass and the values of die swell are taken by hand.

3. Experimental

We investigated the extrusion behaviour of two compounds *A* and *B*, which were respectively based on an emulsion type PVC and a suspension type PVC, both of commercial origin. The compounds were prepared at the Plastics and Rubber Research Institute TNO by dry blending of polymer in the form of a powder. The composition of the compounds and the processing conditions are given in tables 1 and 2.

Table 1. Composition and processing conditions of PVC compound *A*.

Emulsion type PVC					p.b.w.
Polymer	Vestolit P 1328/K 60				100
Stabilizer	Advastab 17 M				2
Internal lubricant	Loxiol G 10				1
External lubricant	Advawa 360				0,5
Processing temp. °C	160	170	180	190	
Screw speed r.p.m.	15	20	16	20	

Table 2. Composition and processing conditions of PVC compound *B*

Suspension type PVC					p.b.w.
Polymer	Solvic 229				100
Stabilizer	Advastab 17M				2
Internal lubricant	Loxiol G10				1
External lubricant	Advawax 360				0,5
Processing temp. °C	170	180	190	200	
Screw speed r.p.m.	5	10	16	17	

We tried to maintain the screw speed in the range of 15–20 r.p.m., and to process the compounds in the temperature region between 160 and 220 °C. This could not be completely realized. With compound *B* at the temperature of 160 °C, reasonable outputs could not be obtained, as the torque on the screw became dangerously high. On the other hand, compound *A* showed the first sign of distortion of the extrudate at 190 °C; therefore, we did not process it at higher temperatures. Under the circumstances listed no serious distortion or discolouration of the extrudate was observed.

The machine was regulated so that the temperature of the centrally placed thermocouple in the monitor (*F*) was as close as possible to the desired working temperature. Whenever possible, the settings of the heating elements of the barrel were further adjusted to obtain uniform radial and axial temperature distributions in the melt (see, however, under 4.). Viscometers or dies were regulated to the desired working temperature. For each set of processing conditions the following experiments were performed:

a) viscosity measurements with the slit viscometer at various outputs;

b) production of extrudate in the shape of a flat strip by extrusion through the slit viscometer with closed by-pass;

c) viscosity measurements with the capillary viscometer at various outputs;

d) production of extrudate in the shape of a round bar by extrusion through cylindrical dies with diameters of 10 and 15 mm.

Ad. a) There was some interaction between the pressure built up and the temperature distribution in the extruder, especially at the lower working temperatures. Therefore, the temperature of the extruder had to be regulated again after each new adjustment of the by-pass of the viscometer. When the temperature of the centrally placed thermocouple was at the desired value and when the pressures in the viscometer were stationary, all data points were recorded with the logger. Simultaneously, two measurements of the outputs of the viscometer and the by-pass were taken by weighing. Die swell was measured on the cold extrudate.

Ad. b) No forced cooling or calibration was applied to the extrudate. The extrudate was used for the preparation of specimens for the acetone and the tensile test.

Ad. d) The extrudate was used for the preparation of specimens for the acetone test, for the measurement of dynamic shear modulus, damping and dynamic viscosities, and for impact testing.

4. Temperature distribution in the melt

Axial temperature distributions in the extruder during the processing of compound *A* are shown in fig. 3 for two processing temperatures and for two positions of the by-pass. The upper part of the figure shows the situation at a working temperature of 180 °C, the lower part the situation at a working temperature of 160 °C. Profiles indicated by drawn lines relate to the situation with open by-pass (corresponding to the minimum die head pressure and the minimum output through the slit), profiles indicated by broken lines relate to the situation with closed by-pass (corresponding to the maximum die head pressure and the maximum output through the slit).

At the working temperature of 180 °C we have an example of a well controlled thermal condition in the extruder. Starting from a point 25 cm from the hopper, the temperature within the screw was nearly uniform and about equal to the uniform temperature in the extruder cylinder. Adjustment of the by-pass had little influence on the temperature profiles. We conclude that the temperature in the melt was also uniform over the greater part of the screw channel, and we expect this to hold in the radial direction too. Indeed, the radial temperature distribution monitored in the same experiment (fig. 4) was uniform. The temperature of the screw tip, also indicated in the same picture, was only 1 °C higher than the melt temperature.

At the working temperature of 160 °C, the thermal conditions are much less favourable. It is seen from the lower part of fig. 3 that temperatures in screw and cylinder were much lower than the working temperature. The heat

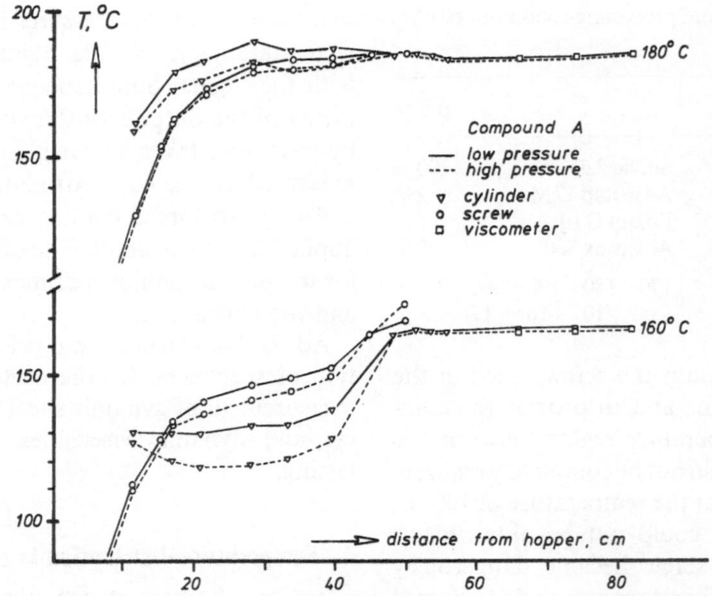

Fig. 3. Axial temperature distribution in the extruder during processing of compound *A* at working temperatures of 160 and 180 °C for two positions of the by-pass (——— open, – – – – closed)

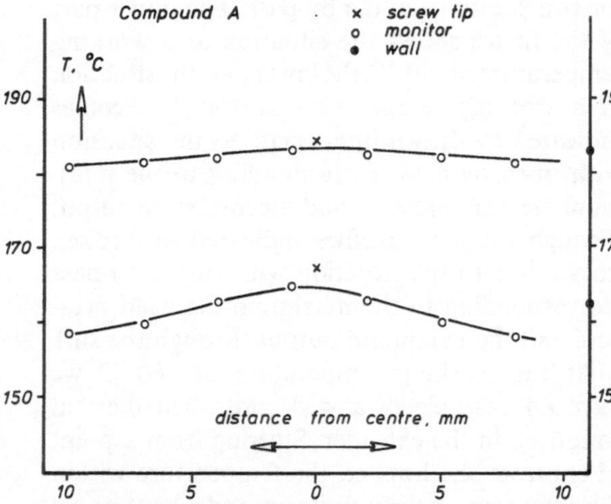

Fig. 4. Radial temperature distribution in the melt before screw tip for processing of compound *A* with open by-pass at working temperatures of 160 and 180 °C

production in the processed powder or melt was so high that strong cooling of the cylinder was necessary to keep the melt temperature in the monitor near the prescribed value. It therefore seems likely that the true temperature of the melt in the screw channel was higher than the temperatures indicated by the thermocouples in screw and cylinder. The true axial profile of the melt temperature is unknown. It is further seen that the temperature distribution in the cylinder wall is much more influenced by the position of the by-pass. A closed by-pass corresponds to a higher pressure and a larger heat production in the screw channel, which again requires stronger cooling. The cylinder temperatures are therefore lower with closed by-pass. Radial temperature profiles in front of the screw (fig. 4) are also less uniform at the lower processing temperature. The temperature of the melt is highest near the screw and decreases by about 8 °C towards the cylinder wall. The temperature in the screw tip was 2 °C higher than the temperature maximum in the melt.

The trend described here for two cases, was generally found for both compounds: At the higher processing temperatures, the homogeneity of the melt temperature was good, at the lower processing temperatures, it was poor. This is an unavoidable result of a high frictional heat production in the screw channel.

5. Viscometry

Results of measurements with the slit viscometer were briefly reported in the previous paper[1]). A more extensive discussion will be given elsewhere[2]).

6. Evaluation of the extrudate

The quality of the extrudate was investigated by different methods. One of these was the

[1]) *J. L. S. Wales*, these proceedings.
[2]) In preparation.

measurement of dynamic mechanical properties, as functions of temperature and frequency, of specimens prepared from the extrudates. Especially the measurement of dynamic viscosity and dynamic elasticity of the remelted extrudate gave significant results that will be discussed separately[3]).

It was further found that materials extruded at different temperatures gave marked differences when subjected to the acetone test. For this purpose, the extrudate as obtained by extrusion through the slit die was cut into shorter pieces and immersed in dry acetone for periods up to 48 h, during which the extrudates continued to swell and to lose cohesion. The rate of disintegration depended strongly on the processing temperature of the material. The extrudates processed at the lowest temperature (160 °C for compound *A* and 170 °C for compound *B*) fell into pieces during the immersion period. The other extrudates also showed cracks which clearly decreased in number and size with increasing processing temperature. Extrudates processed at the highest temperature showed at most one or two single cracks after the immersion period. An illustration is given for both compounds in fig. 5.

This test was repeated several times and proved to be reproducible. Similar conclusions could also be drawn from the acetone test as performed on slices cut from bars extruded through cylindrical dies. The results of the acetone tests showed the most pronounced differences found so far between materials extruded at different temperatures.

We tried to demonstrate differences in the strength properties of materials processed at various temperatures. To this end, dumbbell shaped tensile specimens were machined from the extrudates of compound *A* as obtained by slit die extrusion. These specimens were tested for their tensile behaviour in an Instron testing machine with a constant cross-head speed of 20 cm/min at various temperatures between −40 and +30 °C.

In all cases, the stress strain diagrams showed a pronounced yield point. At the test temperatures of −20, 25 and 29 °C, the stress decreased again after the yield point had been passed, and the specimen broke in the decreasing part of the stress strain curve. In these cases, the values of yield stress, σ_y, yield strain, ε_y, rupture stress, σ_b, and rupture strain, ε_b, were determined. In duplicate measurements, the scatter in the values of σ_y and ε_y was usually small, but that in the values of σ_b and ε_b was much larger. At the lowest testing temperature, −40 °C, the specimen broke at or just after the yield point. For this case, only the values of σ_y and ε_y were obtained.

Results of the measurements are shown in fig. 6, where the yield stress at the test temperatures of −40, −20, and +25 °C is plotted vs. the processing temperature of the specimen. The same figure shows the values of rupture stress and rupture strain at 25 °C.

Though the yield stress increased strongly with test temperature, it was only weakly dependent on the processing temperature of the specimen. A slight increase in yield stress with increasing processing temperature was observed at the lower test temperatures. With specimens extruded at a higher temperature, rupture in the decreasing part of the stress-strain curve occurred later. This resulted in a slight decrease in rupture stress and an appreciable increase in deformation at rupture with increasing processing temperature.

So far, these investigations could not be extended into a temperature region low enough for brittle rupture to occur far below yielding. To investigate possible differences in brittle rupture behaviour, we determined the impact strength properties of two extrudates of compound *B*, extruded at 170 and 200 °C.

From extruded rods, V-notched specimens for the Charpy impact test were machined. The notch had a radius of 2 mm at the apex. Specimens were slowly conditioned to the test temperature in a gas thermostat of special construction[4]), and pushed out of the thermostat just before the test procedure. This ensures an accurate temperature control without build-up of thermal stresses.

The notched Charpy impact strength is shown in fig. 7 vs. temperature for two extrudates of compound *B*. At each temperature, at least five specimens of the two materials were tested. Indicated are the mean values, and the standard deviations of the mean values. Significant though rather small differences were obtained between the two materials. The material extruded at 200 °C has an impact strength 5–10% higher than that of the material extruded at 170 °C.

[3]) *J. L. den Otter*, these proceedings.

[4]) *A. G. M. Tak*, private communication.

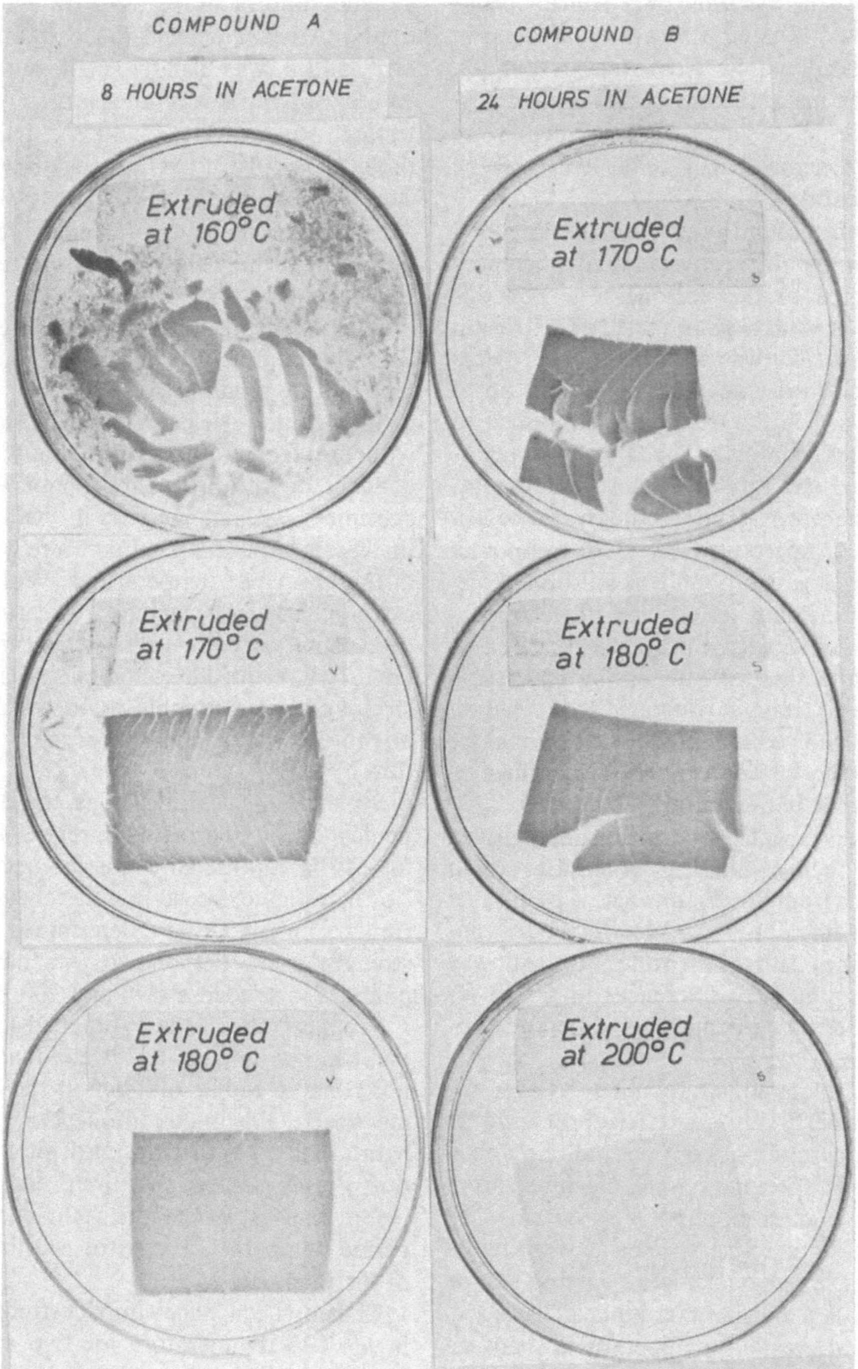

Fig. 5. Disintegration behaviour in acetone immersion test of compounds *A* and *B* extruded at various pressing temperatures

The experimental material presented here is of course very incomplete. If we were to draw a provisional conclusion at this stage of the investigation, it would be that the large differences observed in the acetone test for both compounds between extrudates processed at various temperatures, have no counterpart in the ultimate properties determined so far. Neither the impact strength, nor the yield strength shows differences large enough to indicate a superior quality of products processed at higher temperatures. Differences may show up in the long

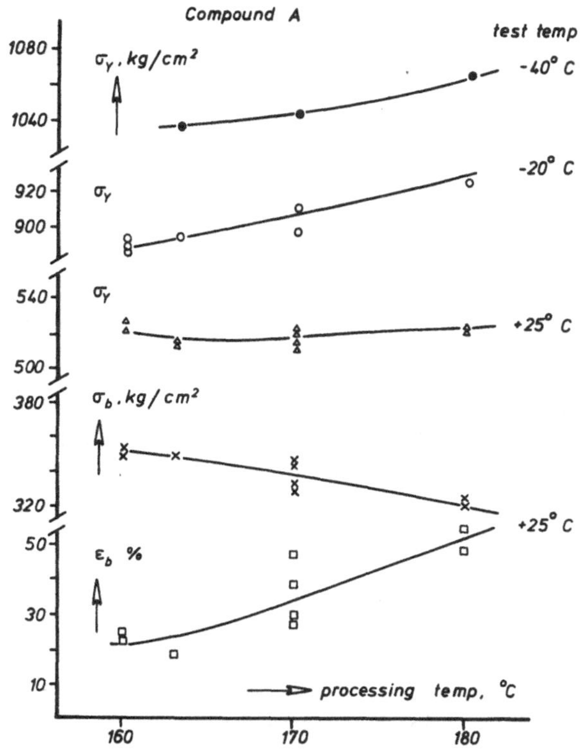

Fig. 6. Ultimate properties of extrudates of compound *A* at various test temperatures, vs. processing temperature

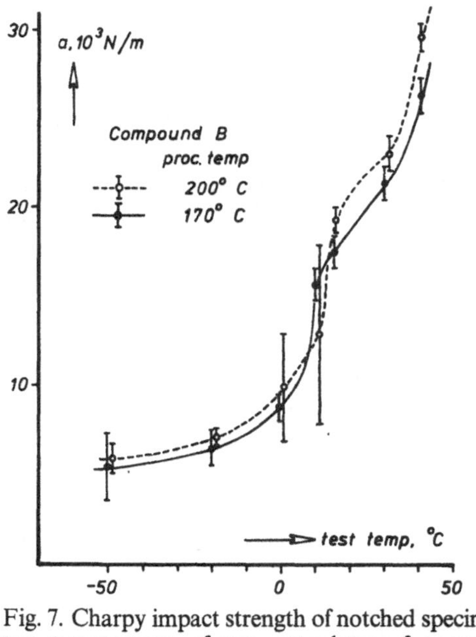

Fig. 7. Charpy impact strength of notched specimens, vs. test temperature, of two extrudates of compound *B* processed at 170 and 200 °C, respectively

term fatigue behaviour, for which tests are under preparation.

Acknowledgements

The authors are indebted to Mr. *H. W. Bree*, for performing the tensile tests, Mr. *A. G. M. Tak*, for performing the impact test, and to Prof. *H. Janeschitz-Kriegl*, Delft University of Technology, for helpful discussions.

Summary

Two compounds or rigid PVC, one of the emulsion type, the other of the suspension type, were processed at different temperatures by means of a highly instrumented 30-mm research extruder. During processing, measurements were made of the axial distribution of pressures and temperatures at the inner wall of the barrel, of the axial temperature distribution in the hollow screw and of the radial temperature distribution of the melt close to the screw tip.

The melts were forced through capillaries and slit dies and their viscosity behaviour was measured. Extrudates were subjected to an immersion test in acetone, in which they showed very significant differences in behaviour that depended on the temperatures at which they were processed. These differences did not show up in impact strength or yield strength in stress strain testing.

References

1) *Pezzin, G.*, Pure and Appl. Chem. **25**, 241 (1971).
2) *Juijn, J. A.*, Crystallinity in atactic polyvinyl chloride. (Thesis TH Delft 1972.)
3) *te Nijenhuis, K.* and *H. Dijkstra*, Rheol. Acta (in press 1974).
4) *Berens, A. R.* and *V. L. Folt*, Polymer Eng. Sci. **8**, 5 (1968).
5) *Berens, A. R.* and *V. L. Folt*, Trans Soc. Rheology **11**, 27 (1969).
6) *Berens, A. R.* and *V. L. Folt*, Polymer Eng. Sci. **9**, 27 (1969).
7) *Janeschitz-Kriegl, H.*, Proc. 4th Internat. Congr. on Rheology, Part I, 140 (New York 1965).
8) *Schläffer, W., J. Schijf*, and *H. Janeschitz-Kriegl*, Plastics and Polymers **39**, 193 (1971).
9) *v. Leeuwen, J.*, Poly. Eng. Sci. **7**, 98 (1969).

Authors' address:

J. L. den Otter, J. Schijf, J. L. S. Wales and *F. R. Schwarzl*
Centraal Laboratorium TNO
Delft (The Netherlands)

Rheol. Acta **13**, 216–222 (1974)

From the Department of Chemical Engineering, University of Salford, Salford 5, Lancs. (England)

Equations for the interpretation of normal stress differences in polymer solutions and melts

By *M. Soylu, R. A. Mashelkar, and J. Ulbrecht*

With 4 tables

(Received October 27, 1972)

Polymer solutions and melts are known to exhibit two types of anomalies, viz. viscous (resulting in shear dependent viscosity under steady flow conditions) and elastic (manifesting in normal stresses under steady flow conditions). An industrial rheologist is faced with the pressing problem of describing such experimentally observed viscous and elastic anomalies in mathematical terms of general utility or specifying in turn explicit constitutive equations with a small number of constants so as to be able to correlate the phenomena satisfactorily. Once such equations are found one hopes to apply these equations directly to the problem of interest. If the problems under consideration are too complicated then one may be able to at least suggest relevant dimensionless groups based on these equations.

Although most processing equipment does not operate in viscometric flow, the simplicity of the flow field makes viscometric flow a logical starting point for any characterisation. It has been shown (1) that three material functions, namely a shear stress function and two independent normal stress combinations, are required to completely specify the rheological properties of a viscoelastic material in viscometric flows. For a simple shear flow described by the velocity field $v_1 = v_1(x_2)$ the three material functions are determined as

$$S_{12} = \tau(\dot{\gamma}) \quad [1]$$

$$S_{11} - S_{22} = F_1(\dot{\gamma}) \quad [2]$$

$$S_{22} - S_{33} = F_2(\dot{\gamma}) \quad [3]$$

where $\tau(\dot{\gamma})$ is the shear stress function and $F_1(\dot{\gamma})$ and $F_2(\dot{\gamma})$ are primary and secondary normal stress difference functions, respectively. The three material functions τ, F_1 and F_2 are related to the viscosity (η) and the normal stress coefficients, F, by

$$\tau = \dot{\gamma}\eta(\dot{\gamma}) \quad [4]$$

$$F_1 = \dot{\gamma}^2 \sigma_1(\dot{\gamma}) \quad [5]$$

$$F_2 = \dot{\gamma}^2 \sigma_2(\dot{\gamma}). \quad [6]$$

The theories which have been proposed on the basis of molecular considerations are able to interpret both the shear dependent viscosity (2, 3, 4) as well as the presence of normal stresses (3, 4). The description is, however, successful only partially in that the functional dependence of these material functions on the rate of shear is far from satisfactorily described, particularly in the range shear rates of practical interest.

In the frame work of non-linear continuum mechanics, several constitutive equations have been proposed which depict the flow behaviour of liquids. A useful summary of these equations is made by *Bogue* and *Doughty* (5). The minimum desirable properties of any of these constitutive equations are as follows (5): 1. They should be coordinate invariant. 2. They should be capable of predicting shear dependent viscosities and normal stresses in steady shear. 3. They should be able to reconcile with the dynamic experiments of linear viscoelasticity. 4. They should express the stress relaxation phenomena. 5. They should be explicit with definite constants to be fitted to experimental data.

To satisfy all these imposed restrictions is a formidable task for any constitutive equation. As a net result of this the constitutive equations somewhat manage to describe all the relevant phenomena qualitatively but do only a mediocre job when dealing with one of the phenomena in an exact quantitative manner. Thus, for instance, the prediction of the viscosity function $\eta(\dot{\gamma})$

made by *Oldroyd*-3 constant model (6) or a *Coleman-Noll* third order fluid (7) is found to be completely unsatisfactory except for correlating either dilute solution data or data obtained at very small rates of deformation. It is of course possible to increase the complexity of the model (e.g. using *Oldroyd*-8 constant model or using a higher order approximation for a *Coleman-Noll* fluid), but the undetermined constants involved in the function $\tau(\dot{\gamma})$ also increase considerably and the equations become devoid of any practical utility.

The attempts to circumvent these difficulties in the case of viscosity function are well known. Two or more parameter Generalised *Newton*ian Fluid models (GNF models) have been used for a number of years (8). These empirical or semi-empirical models prescribe arbitrarily a functional form for the viscosity function as either $\eta(\mathrm{II}_D)$ or $\eta(\dot{\gamma})$. These equations have found enormous use in correlating rheological data as well as in treating engineering flow situations.

This is, however, not the case with the two other material functions related to the normal stress differences. It is well recognised that the second order approximation predicting a linear dependence between F_1 and $\dot{\gamma}^2$ is far from satisfied in practical situations. It is also agreed that F_1 and F_2 are functions of the second invariant of the rate of deformation tensor, i.e. $F(\mathrm{II}_D)$ or in simple shear flow, $F(\dot{\gamma})$. However, the exact functional forms of $F(\dot{\gamma})$ have not been well established so far. Suitable empirical models devised to describe these functions are likely to have a considerable importance for a rheologist as well as an engineer. Some of the areas of their utility have been discussed later.

In view of this, in this work we have examined the possibility of devising suitable empirical models capable of interpreting normal stress differences in liquids. Some data from this laboratory and extensive data from the literature have been examined to find out the best models which describe these functional relationships. Based on this study recommendations are made which will help in selecting suitable empirical models for the correlation of normal stress data.

Selection of empirical models

The selection of the empirical models to be tested was based on three primary considerations:

1.

The chosen model should correlate the normal stress data for sufficient number of liquids in a large range of shear rates.

2.

The model should have minimum number of unknown parameters.

3.

The mathematical expression should be simple so that it can be handled easily in engineering applications.

We have sought functional forms for both the primary normal stress difference function, $F_1(\dot{\gamma})$ as well as the secondary normal stress difference function, $F_2(\dot{\gamma})$. There is some dispute in the literature about the magnitude of $F_2(\dot{\gamma})$ in comparison to $F_1(\dot{\gamma})$ as well as about the sign of $F_2(\dot{\gamma})$ [see ref. (21)]. However, it has been generally agreed that both F_1 and F_2 are directly proportional to each other. In view of this both the functions are expected to have the same functional forms. Further in the ensuing discussion we have used F to signify either F_1 or F_2.

The first functional forms chosen were analogous formally to the GNF models used for describing the functional dependence of $\tau(\dot{\gamma})$. The models resembling a power-law model, *Ellis* model, *Sisko* model and a *Prandtl-Eyring* model respectively, were given by

$$F = K(\dot{\gamma})^n \quad [7]$$

$$\dot{\gamma} = AF^{\alpha} + BF \quad [8]$$

$$F = A_1\dot{\gamma}^{\alpha_1} + B_1\dot{\gamma} \quad [9]$$

$$F = A_2 \sinh^{-1}(\dot{\gamma}/B_2). \quad [10]$$

An examination of eqs. [7]–[10] indicates that these equations, although possibly useful in correlating the normal stress data, do not bring out the even character of the normal stress difference functions. Further, all of them are bound to fail in the lower region of shear rates where the theoretical considerations predict that F varies directly as $\dot{\gamma}^2$, whereas eqs. [7] to [10] predict that F varies directly as $\dot{\gamma}$.

In view of this we thought of examining another set of models which could overcome these difficulties. A simple superposition of the behaviour in the limit of lower shear rate range

($F \propto \dot{\gamma}^2$) and a power-law dependence in the higher shear rate range ($F \propto \dot{\gamma}^n$) gives on rearrangement,

$$\frac{1}{F/\dot{\gamma}^2} = 1/G(1 + (F/F_0)^{Q-1}). \qquad [11]$$

This is an analog of the *Ellis* model for the viscosity function. It is seen that the value of G is the zero shear normal stress coefficient. The constant Q in the equation gives an indication of the nonlinearity of the normal stress difference-shear rate curve as compared to a second order fluid. The term G/F_0 describes the scale by which different normal stress difference-shear rate curves will be shifted for different materials.

Evidently several other superpositions could be attempted to fit the $F - \dot{\gamma}$ data. Some of them are listed below,

$$F/\dot{\gamma}^2 = G[1 - (\dot{\gamma}/\gamma_0)^{n-1}] \qquad [12]$$

$$F/\dot{\gamma}^2 = G\left(\frac{\sinh^{-1}\beta\gamma}{\beta\gamma}\right)^Z. \qquad [13]$$

Similar significance could be attached to the parameters involved in the above equations as well. Some constitutive equations [e.g. *Oldroyd* 5 or 8 constant models (9), *Pao* model (10)] predict the presence of a lower limiting value of normal stress coefficient as well as an upper limiting value in the region of very high shear rates. This behaviour is analogous to the presence of a lower limiting viscosity (η_0) and an upper limiting viscosity (η_∞) in the case of the viscosity function, $\eta(\dot{\gamma})$. Empirical models to take into account this type of behaviour could also be devised on similar lines, e.g. eqs. [14] and [15] below

$$\frac{F/\dot{\gamma}^2 - H}{G - H} = \frac{\sinh^{-1}\beta\dot{\gamma}}{\beta\dot{\gamma}} \qquad [14]$$

$$\frac{F/\dot{\gamma}^2 - H}{G - H} = \frac{1}{1 + m_1\dot{\gamma}^{m_2}} \qquad [15]$$

where G and H are the values of the lower and upper limiting values of the normal stress coefficient, respectively.

The suitability of the above models to correlate the data was examined by using several data points obtained in this work as well as from the literature. It was thought desirable to choose data on rather a wide variety of liquids obtained in a rather large range of shear rates. The data obtained in this laboratory have been collected on a *Weissenberg* rheogoniometer (Model R-18). The list of the liquids used, the range of shear rates, the range of normal stress differences and the relevant source of data have been listed in table 1. It is seen that an impressive range of all the above variables was covered.

Two procedures were used for examining the suitability of various equations. The first procedure involved the use of a least square procedure. The equations were first suitably linearised. This required an accurate differentiation of the data. This was done graphically. The linearised equations were fitted by the least square technique on an ICL-KDF 9 digital computer.

Another procedure involving an optimisation technique was also used. For this minimisation procedure a programme XPOWMIN available at the University of Salford computing library was used. The details of the programme could be found in ref. (17).

Results and discussion

Considering the vast extent of the data which were analysed it is not possible to list the parameters of each model. The suitability of different models in describing the data is best examined by reporting the percentage absolute deviations in the case of each model tested. The average percentage deviation corresponds to the mean of all the individual percentage deviations obtained while testing a given empirical model. The minimum and the maximum percentage deviations have also been listed for the sake of comparison. In the first class of models (eqs. [7], [8], [9] and [10]) all the models were tested. Eqs. [14] and [15] could not be tested for the data analysed in this work, because no suitable set of data were reported in the literature which showed the variation of F as required by eqs. [14] and [15]. Among the second class of models eq. [11] was found to be a representative superposition and has been most extensively tested in the case of all the solutions used. Some typical values of the parameters in the case of some of the solutions used are listed in table 2.

An examination of table 3 indicates that eq. [11] is found to be far more satisfactory in comparison to eqs. [7], [8], [9] and [10]. The power law form of eq. [7] appears to be the next best form. However, it should be emphasized that this equation best describes the data in rather a narrow range of shear rates and never more than for about decades of shear rates. Models presented by eqs. [7], [8], [9] and [10] fail particularly

Table 1. Normal stress data analysed in this work

No.	Solutions or melts	Concentration (wt.-%)	Approximate shear rate range, sec^{-1}	Technique	Source of data
1	Polyacrylamide (ET-597)	0.01–0.05	$10^4 - 5\times10^6$	jet thrust	11
2	Polyhall (295)	0.01–0.05	0.1 – 500	cone and plate	12
3	Polyethylene oxide (WSR-301)	0.01–0.05	$10^4 - 5\times10^6$	jet thrust	11
4	Polyethylene oxide (WSR-301)	1–2	$2\times10^2 - 10^6$	cone and plate	13
5	Polyethylene oxide (WSR-205)	2–5.63	$1 - 5\times10^3$	cone and plate	13
6	Polyethylene oxide (WSR-301)	0.7–1.1	$10 - 10^3$	axial annular flow, cone and plate	14
7	Polyacrylamide (Separan AP-30)	0.0025–0.5	$0.5–10^3$	cone and plate	15
8	Polyacrylamide (Separan AP-30)	0.5–2	$1 - 10^3$	cone and plate	16, 17
9	Polyacrylamide (Separan MGL)	0.05–1.19	$1 - 10^3$	cone and plate	15
10	Hydroxy-ethyl cellulose	0.5–0.9	$10 - 10^3$	cone and plate	13
11	Hydroxy-ethyl cellulose	0.9	$2\times10^2 - 10^3$	separated cone and plate	18
12	Carboxy methyl cellulose (P-75-XH)	0.4–0.5	$10 - 10^3$	axial annular flow, cone and plate	14
13	Carboxy methyl cellulose (P-75-XH) (med)	1.1–1.5	$10 - 10^3$	axial annular flow, cone and plate	14
14	Carboxy methyl cellulose (P-75-XH) (low)	2.6	$10 - 10^3$	axial annular flow, cone and plate	14
15	Carboxy methyl cellulose (P-75-XH) (Hercules, High)	1–3	$5 - 10^3$	parallel plate	19
16	Jaguar (A-20-D)	0.3–0.6	$10 - 10^3$	axial annular flow, cone and plate	14
17	Natrosol (250-H)	0.5–0.9	$10 - 10^3$	axial annular flow, cone and plate	14
18	Polyisobutylene in decalin	5	$4 - 5\times10^2$	cone and plate, parallel plate	20
19	Polyisobutylene in decalin	6.5–10.5	0.4 – 100	cone and plate, parallel plate	21
20	Polyisobutylene in decalin	0.25–9	$10^{-1} - 500$	cone and plate parallel plate	22
21	Polyisobutylene in cetane	3.9–8.54	$1 - 5\times10^2$	cone and plate, parallel plate	23
22	Polystyrene in toluene	5–35	$1 - 10^2$	parallel plate	24
23	Polystyrene in decalin	12–31	$8 - 10^2$	parallel plate	19
24	Polystyrene in dioctylphthalate	0.8–2	$10 - 10^2$	parallel plate	25
25	Polystyrene in chlorinated diphenyl	0.3–2	$1 - 10^2$	parallel plate	26
26	Polystyrene in Aroclor (1232)	2–5	$1 - 10^2$	separated cone and plate	18
27	Silicone oil (MS 200–60000 cs)	–	$1 - 10^2$	cone and plate	27
28	Molten polyethylene	–	$10^{-1} - 10$	parallel plate	28
29	Molten polyethylene	–	$10^{-1} - 30$	parallel plate	29

Table 2. Some typical values of the parameters of the empirical model based on eq. [11]

Solution or melt	Concentration (wt.-%)	G (gm/cm)	F_0 (dyne/cm²)	Q	Source
Polyhall (295)	0.01	3.93	0.0047	1.258	12
Polyhall (295)	0.025	24.87	0.758	3.268	12
Polyhall (295)	0.05	165.36	0.765	3.227	12
Polyacrylamide (Separan-AP-30)	0.5	70	50.5	3.31	16, 17
Polyacrylamide (Separan-AP-30)	1	200	41.5	2.933	16, 17
Polyacrylamide (Separan-AP-30)	1.5	445	230	3.165	16, 17
Polyacrylamide (Separan-AP-30)	2	1000	216	3	16, 17
Polyethylene melt (PE I)	–	1.37×10^5	3.91×10^4	3.90	28
Polyethylene melt (PE II)	–	1.1×10^6	1.31×10^5	1.66	28

Table 3. Suitability of different empirical models for the correlation of normal stress functions

Empirical model	The mean deviation (%)	The maximum deviation (%)	The minimum deviation (%)
Eq. [7]	11.02	45.88	0.24
Eq. [8]	12.89	37.40	2.076
Eq. [9]	18.93	72.59	1.43
Eq. [10]	36.61	89.54	8.60
Eq. [11]	2.14	7.40	0.16

in the region of low shear rates. The model given by eq. [11], however, was able to correlate the data in about three to four decades of shear rates and could hence be recommended as a satisfactory equation for the correlation of normal stress difference-shear rate data.

An attempt towards molecular interpretation of the parameters

It should be emphasized that the above mentioned models are simply *empirical* functions designed to describe experimentally observed normal stress differences under steady state conditions in simple flow systems. The parameters appearing in these models may not be directly related to the structure or concentration of the fluid by means of the molecular theory. However, an attempt has been made to analyse the empirical functions derived in terms of the molecular theories.

Most of the analyses agree (9) that the fluid relaxation time λ_f is related to the zero shear primary normal stress coefficient by an equation of the type

$$\lambda_f = \frac{(F_1/\gamma^2)_{\gamma \to 0}}{\eta_0} \quad [16]$$

where η_0 is the zero shear viscosity.

For the case of the model proposed in eq. [11], $(F_1/\dot{\gamma}^2)_{\gamma \to 0} = G$. Hence the fluid relaxation time would be given by

$$\lambda_f = \frac{G}{\eta_0}. \quad [17]$$

Based on the molecular considerations, the molecular relaxation time λ_m has been obtained by several workers, e.g. *Bueche* (2), *Rouse* (30), *Zimm* (31) etc. A comparison of the molecular relaxation time λ_m obtained on the basis of molecular theories and the fluid relaxation time λ_f obtained on the basis of rheogoniometric measurements may reveal some relevant information. The molecular relaxation time, λ_m obtained from the *Bueche* theory (2) is given by

$$\lambda_m = \frac{12\eta_0 M}{\pi^2 c R T}. \quad [18]$$

For an estimation of the value of λ_f and λ_m we require accurate values of the zero shear viscosity, η_0 the weight average molecular weight, M, etc. Many of the systems examined in this work were not completely characterised so as to yield all the necessary information. Only those data in table 1 could be used for comparison where such information was available. The value of η_0, however, was not directly available since the shear stress-shear rate data were usually reported in the proper non-*Newton*ian regime. The value of η_0 was hence estimated by making use of the extrapolation principles developed by *Subbaraman* et al. (32) and *Mashelkar* et al. (7) for non-viscometric translational and rotational flows. The viscometric data were plotted as logarithm of apparent viscosity versus the shear stress. The plots were found to be linear. The value of η_0 was obtained by reading the value of η when the shear stress was extrapolated to zero.

Table 4 lists the values of G, η_0, λ_f and λ_m estimated by using the above procedures. Generally the values of λ_f and λ_m are found to agree reasonably well, particularly for dilute solutions in high viscosity solvents. It must be emphasized that none of the materials tested in this work had well established molecular weight distribution and some of the discrepancy observed may be really due to the rather wide molecular weight distribution in some of the systems.

The success observed above prompted us to correlate the normal stress data for different systems by using dimensionless plots of reduced normal stress difference (normal stress coefficient divided by the zero shear normal stress coefficient) against the reduced shear rate $\dot{\gamma}\lambda$. This attempt, however, was not very successful and hence has not been dealt with any further.

In conclusion, the empirical models tested in this work may be of significance in a number of situations. For instance, they may be used to correlate the experimental data on normal stresses over a rather wide range of shear rates. The mathematical forms of these models are far simpler as compared to the other equations based

Table 4. Comparison of the fluid relaxation time ($\lambda_f = G/\eta_0$) with the molecular relaxation time obtained from *Bueche* theory (λ_m)

No.	Solution or melt	Molecular weight	Concentration (wt.-%)	G (gm/cm)	η_0 poise	Fluid relaxation time $\lambda_f = G/\eta_0$	Molecular relaxation time λ_m	Source
1	Polystyrene in chlorinated diphenyl	1.2×10^6	0.3	0.6091	5.9	0.1032	0.1153	26
2	,,	1.2×10^6	0.5	0.7454	9.4	0.0956	0.1006	26
3	,,	1.2×10^6	1	2.5521	19.5	0.1308	0.1131	26
4	,,	5×10^6	0.5	2.0239	8.99	0.2251	0.44	26
5	,,	0.66×10^6	0.5	0.1296	8	0.0162	0.0517	26
6	,,	0.54×10^6	2	1.8769	43.5	0.0431	0.055	26
7	,,	0.27×10^6	2	0.4925	29.5	0.0166	0.0158	26
8	,,	0.239×10^5	2	0.0304	17.5	0.0017	0.0102	26
9	Polystyrene in dioctyl phthalate	0.66×10^6	2	0.0253	3.78	0.0066	0.0062	25
10	,,	1.2×10^6	2	0.1031	4.75	0.0217	0.0142	25
11	,,	5×10^6	0.8	0.0248	2.07	0.0118	0.0646	25
12	,,	5×10^6	1	0.07869	2.55	0.0308	0.0635	25
13	,,	5×10^6	1.5	0.4695	5.08	0.0924	0.083	25
14	,,	5×10^6	2	1.6893	10	0.1689	0.1224	25
15	Polystyrene in toluene	1.82×10^6	12.5	568	400	1.4	0.29	24
16	Polyisobutylene in cetane	1.2×10^6	3.9	0.1885	4.79	0.039	0.0072	23
17	,,		5.39	2.0846	18.5	0.1126	0.0201	23
18	,,		6.86	9.8141	60.2	0.1630	0.0515	23
19	Polyethylene melt	0.79×10^6	–	1.36×10^5	7.5×10^4	1.822	0.2059	28
20	Polyethylene melt	1.31×10^6	–	1.09×10^6	4.2×10^5	2.61	2.26	28

on the formal constitutive equations [e.g. *Spriggs* four constant model (33)]. The latter are too complex to have any practical utility and the former may find preference in these instances. If the flow behaviour of the liquids being processed is characterised by these models, then they may also be helpful in formulating relevant dimensionless groups. This will help in the proper correlation of the transport processes involving viscoelastic liquids. *Kelkar* et al. (27) have recently used the empirical model (eq. [11]) for correlating their primary normal stress data and have made use of the resulting dimensionless groups for correlating the dynamics of the agitation of viscoelastic liquids. It is conceivable that such a utility may be found in many other engineering situations as well. Lastly, some of the techniques of normal stress measurements [e.g. measurement of radial pressure drop in axial annular flow (14)] require that a model be fitted for $F(\dot{\gamma})$ curve. The parameters of the model could be obtained by curvefitting the model to the data. The form of such models could be judiciously selected based on the results of this work.

References

1) *Coleman, B. D., H. Markowitz*, and *W. Noll*, Viscometric flows of non-Newtonian fluids. (New York 1966.)

2) *Bueche, F.*, J. Chem. Phys. **22**, 1570 (1954).

3) *Yamamoto, M.*, J. Phys. Soc. Japan **11**, 413 (1956); **12**, 1148 (1957); **13**, 1200 (1958).

4) *Pao, Y. H.*, J. Appl. Phys. **28**, 591 (1957).

5) *Bogue, D. C.* and *J. O. Doughty*, Ind. Eng. Chem. Fundls. **2**, 243 (1966).

6) *Williams, M. C.* and *R. B. Bird*, Ind. Eng. Chem. Fundls. **3**, 42 (1964).

7) *Mashelkar, R. A., D. D. Kale, J. V. Kelkar*, and *J. Ulbrecht*, Chem. Eng. Sci. **27**, 973 (1972).

8) *Bird, R. B., W. E. Stewart*, and *E. W. Lightfoot*, Transport Phenomena. (New York 1960.)

9) *Middleman, S.*, The flow of high polymers. (New York 1968.)

10) *Pao, Y. H.*, J. Polymer Sci. **61**, 413 (1962).

11) *Oliver, D. R.*, Can. Jl. Chem. Engg. **44**, 100 (1966).

12) *Darby, R.*, Trans. Soc. Rheol. **14**, 185 (1970).

13) *Williams, M. C.*, Ph. D. Thesis (University of Wisconsin, 1964).

14) *Huppler, J. D.*, Ph. D. Thesis (University of Wisconsin, 1965).

15) *Bruce, C.* and *W. H. Schwarz*, J. Polymer Sci. **7**, 909 (1969).

16) *Kale, D. D.*, Ph. D. Thesis (University of Salford, 1973).

17) *Soylu, M.*, M. Sc. Thesis (University of Salford, 1971).

18) *Marsh, B. D.* and *J. R. A. Pearson*, Rheol. Acta **7**, 326 (1968).
19) *Kotaka, T., M. Kurata*, and *M. Tamura*, Jl. Appl. Phys. **30**, 1705 (1959).
20) *Ginn, R. F.* and *A. B. Metzner*, Proc. 4th International Rheological Congress, Part 2, 1965, p. 583.
21) *Ginn, R. F.* and *A. B. Metzner*, Trans. Soc. Rheol. **13**, 429 (1969).
22) *Brodnyan, J. G., F. H. Gaskins*, and *W. Philippoff*, Trans. Soc. Rheol. **1**, 109 (1957).
23) *Markowitz, H.* and *D. R. Brown*, Trans. Soc. Rheol. **7**, 137 (1963).
24) *Kotaka, T., M. Kurata*, and *M. Tamura*, Rheol. Acta **2**, 179 (1962).
25) *Tamura, M., M. Kurata, K. Osaki*, and *K. Tanaka*, J. Phys. Chem **70**, 516 (1966).
26) *Osaki, K., K. Tanaka, M. Kurata*, and *M. Tamura*, J. Phys. Chem. **70**, 2271 (1966).
27) *Kelkar, J. V., R. A. Mashelkar*, and *J. Ulbrecht*, Trans. Inst. Chem. Engrs. **50**, 343 (1972).
28) *Blyler, L. L.*, Trans. Soc. Rheol. **13**, 39 (1969).
29) *Sakamoto, K., N. Ishida*, and *Y. Fukasawa*, J. Polymer Sci. **6**, 1999 (1968).
30) *Rouse, P. E.*, J. Chem. Phys. **21**, 1272 (1953).
31) *Zimm, B. H.*, J. Chem. Phys. **24**, 269 (1956).
32) *Subbaraman, V., R. A. Mashelkar*, and *J. Ulbrecht*, Rheol. Acta **10**, 429 (1970).
33) *Spriggs, T. W.*, Chem. Eng. Sci. **20**, 931 (1965).

Authors' address:

M. Soylu, R. A. Mashelkar, and *J. Ulbrecht*
Dept. of Chemical Engineering
University of Salford
Salford 5, Lancs. (England)

Rheol. Acta 13, 223–227 (1974)

From the Department of Chemical Engineering, McMaster University, Hamilton, Ontario (Canada)

Die swell and melt fracture: Effects of molecular weight distribution

By J. Vlachopoulos

With 3 figures

(Received October 27, 1972)

When polymers are extruded at high stresses it is often observed that the surface or shape of the extrudate is impaired by the presence of certain flow defects. These flow defects are commonly known as swelling, mattness, sharkskin and melt fracture. Because of the great importance of these phenomena in polymer processing, many investigations have been carried out with the objective of determining the mechanism involved and the influence of various processing variables and material properties.

Extrudate swelling is observed when a polymer emerges from an orifice or tube with a diameter larger than the ejection diameter. This phenomenon is sometimes called the "*Barus* effect" and is usually referred to industrially as "die swell". This effect has variously been attributed to polymer memory conditions before the die, to shear recovery, to normal stresses and to velocity profile changes at the tube exit.

Mattness consists of a loss of surface gloss of the extrudate under certain processing conditions. Sharkskin is the flow defect that gives finely spaced, sharp, circumferential ridges in the extrudate. Both mattness and sharkskin can often be tolerated because they are not very catastrophic effects.

Melt fracture exhibits itself as a gross distortion of the extrudate when molten polymers are extruded at shear stresses of the order of 10^6 dynes/cm^2. A number of mechanisms have been proposed for this phenomenon, which is often referred to as "melt flow instability", the most important of which are (1): *Reynolds* turbulence, outlet phenomena, viscous heating, fracture hypothesis, stick-slip flow, elastic energy hypothesis.

Despite the many investigations that have been carried out on the flow defects, these phenomena are still not fully understood and there is often contradiction for various aspects of the problem in the literature. The objective of the present paper is to present some new aspects of die swell and melt fracture, especially on the effects of the distribution of molecular weights in the extruded polymer.

The die swell equations

Die swell is usually described quantitatively in terms of the swelling ratio (d/D), which is the ratio of extrudate diameter over the die diameter. Several expressions have been proposed for the prediction of die swell, centered on the concept of elastic recovery of the swelling process. These expressions relate the swelling ratio (d/D) to the recoverable shear (σ).

The recoverable shear is usually defined from *Hookes* law, which seems to hold for most polymers:

$$\sigma = \tau_{12} J_0 \qquad [1]$$

where τ_{12} is the shear stress and J_0 the shear compliance.

Coleman and *Markovitz* (2) have shown that:

$$\tau_{11} - \tau_{22} = 2 J_0 \tau_{12}^2 \qquad [2]$$

where τ_{11} and τ_{22} are the first and second normal stress respectively. Several questions have been raised in the past as to the numerical factor 2, but presently it is generally accepted for viscoelastic fluids. From eqs. [1] and [2]:

$$\frac{\tau_{11} - \tau_{22}}{2\tau_{12}} = \sigma . \qquad [3]$$

This expression led some authors to the definition of σ as "Stress Ratio".

Nakajima and *Shida* (3) used the concepts of rubber elasticity to calculate the tensile stress for the elastic deformation of the extrudate until its diameter becomes that of the capillary. They arrived at the equation:

$$\bar{\sigma} = \left(\frac{d}{D}\right)^2 - \left(\frac{d}{D}\right)^{-4} \qquad [4]$$

where $\bar{\sigma}$ is the recoverable shear averaged over the cross sectional area. *Bagley* and *Duffey* (4) following the analysis of *Nakajima* (3) and using a one-constant stored energy function for a *Mooney* material (5) obtained the relation:

$$\bar{\sigma}^2 = \left(\frac{d}{D}\right)^4 - \left(\frac{d}{D}\right)^{-2}. \qquad [5]$$

Bagley and *Duffey* also carried out an energy balance analysis along *Graessleys* (6) lines and arrived at the expression:

$$\bar{\sigma}^2 = \left(\frac{d}{D}\right)^4 + 2\left(\frac{d}{D}\right)^{-2} - 3. \qquad [6]$$

Graessley, *Glasscock*, and *Crawley* (6) assumed that the increase in the diameter of the extrudate is due to the release of elastic energy stored in the fluid inside the capillary. They obtained the equation:

$$\left(\frac{d}{D}\right)^4 + 2\left(\frac{d}{D}\right)^{-2} - 3 = \frac{J_0 \tau_w^2}{2G_s} \frac{\int_0^1 \xi^5 \, d\xi/(\eta/\eta_0)}{\int_0^1 \xi^3 \, d\xi/(\eta/\eta_0)} \qquad [7]$$

where $\xi = r/R$ is the dimensionless radial position, J_0 is the experimental shear compliance and G_s is the modulus of elasticity estimated from the theory of rubber elasticity in terms of the average molecular weight between entanglements:

$$G_s = \varrho RT/M_e \qquad [8]$$

τ_w is the shear stress at the capillary wall. The factor

$$\frac{1}{2} \frac{\int_0^1 \xi^5 \, d\xi/(\eta/\eta_0)}{\int_0^1 \xi^3 \, d\xi/(\eta/\eta_0)}$$

accounts for the velocity and stress distribution across the capillary.

By assuming a parabolic velocity profile inside the capillary (*Newton*ian fluid), we have:

$$\frac{1}{2} \frac{\int_0^1 \xi^5 \, d\xi/(\eta/\eta_0)}{\int_0^1 \xi^3 \, d\xi/(\eta/\eta_0)} = \frac{1}{3}. \qquad [9]$$

It can, then, be easily shown that

$$\sigma_w = \sqrt{3}\,\bar{\sigma}. \qquad [10]$$

This equation relates the recoverable shear at the wall to the average recoverable shear. The numerical factor $\sqrt{3}$ accounts for the velocity and stress distribution.

Another expression relating recoverable shear at the wall to the swelling ratio has been proposed by *Tanner* (7), based on the elastic recovery concept of a *Poiseuille* flow of a BKZ fluid.

$$\sigma_w^2 = 2\left(\left(\frac{d}{D}\right)^6 - 1\right). \qquad [11]$$

An experimental evaluation of the die swell equations as to their predictive power has been performed recently by *Vlachopoulos* et al. (8). The results seem to favor *Tanners* expression but definite conclusions cannot be made at present.

Critical conditions for melt fracture

The onset of melt fracture is detected by the appearance of gross distortions in the extrudate. Several observations and conjectures for the critical conditions have appeared in the polymer literature.

Spencer and *Dillons* (9) results for polystyrene show that the product of the shear stress (critical) at the onset of melt fracture and the weight average molecular weight is constant, i.e.

$$\tau_c \cdot \bar{M}_w \approx 3 \times 10^{11}. \qquad [12]$$

Barnett (10) obtained that for polypropylene, the critical stress rather than the product $\tau_c \cdot \bar{M}_w$ is constant, i.e.

$$\tau_c = \text{const} \quad (\text{of the order of } 10^6 \text{ dynes/cm}^2). \qquad [13]$$

Bagley (11, 12) and *White* (13) conjectured that melt fracture occurs at a constant value of recoverable shear (critical), i.e.

$$\bar{\sigma}_c = 7. \qquad [14]$$

This criterion can be easily derived from the definition of recoverable shear (eq. [1]) and the expression for shear compliance $J = \bar{M}_w/\varrho RT$.

Bartos (14) suggested that at the onset of melt fracture the ratio of apparent viscosity to zero shear viscosity is constant

$$\eta_c/\eta_0 = 0.025. \qquad [15]$$

Recently, *Vlachopoulos* and *Alam* (15) concluded that the critical recoverable shear at the wall, as determined from die swell, is proportional to the factor $(\bar{M}_z \bar{M}_{z+1}/\bar{M}_w^2)$ which represents the distribution of molecular weights, i.e.

$$\sigma_{wc} = \text{const}\left(\frac{\bar{M}_z \bar{M}_{z+1}}{\bar{M}_w^2}\right). \qquad [16]$$

This result was also interpreted in terms of the equality of shear wave velocity and friction velocity by *Boger* and *Williams* (16).

The importance of molecular weight distribution

The die swell equations relate the swelling index to the recoverable shear

$$\frac{d}{D} = f(\sigma). \qquad [17]$$

The recoverable shear has been defined as the product of shear stress (τ_{12}) and shear compliance (J_0). Therefore,

$$\frac{d}{D} = f(\tau_{12} J_0). \qquad [18]$$

The shear compliance for a polydisperse collection of *Rouse* (17) molecules is

$$J_R = \frac{2}{5}\left(\frac{\bar{M}_z \bar{M}_{z+1}}{\bar{M}_w^2}\right)\frac{\bar{M}_w}{\varrho R T}. \qquad [19]$$

By substituting τ_{12} with the shear stress at the wall (τ_w) and J_0 with J_R, we have

$$\left(\frac{d}{D}\right) = f(\tau_w J_R). \qquad [20]$$

This implies that for the same shear stress at the wall (τ_w) and the same molecular weight $\bar{M}_w$ the amount of swelling will be influenced by the distribution of molecular weights as represented by the factor $\bar{M}_z \bar{M}_{z+1}/\bar{M}_w^2$. *Vlachopoulos* et al. (8) results confirm this, as shown in fig. 1. No attempt was made to check how well the experimental data fit the theoretical predictions because of the difficulty in choosing an accurate expression for functional dependence of d/D on σ, and the approximate nature of the *Rouse* shear compliance. However, the conclusion has been reached that the broader the molecular weight distribution the larger the swelling of the extrudates. Dependence on the factor $\bar{M}_z \bar{M}_{z+1}/\bar{M}_w^2$ implies that the high molecular weight tail influences greatly the swelling process.

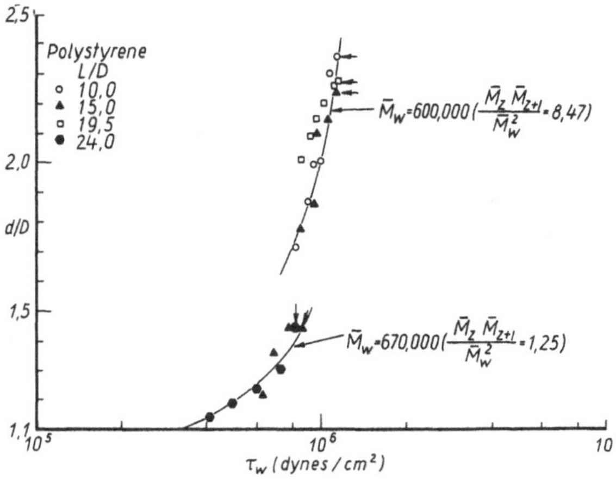

Fig. 1. Influence of molecular weight distribution on die swell for several L/D ratios (capillary length/capillary diameter). Arrows indicate extrudate distortion

By measuring the die swell at the onset of melt fracture for a large number of polystyrene samples it was found (15) that the critical recoverable shear at the wall was nearly proportional to the factor $\bar{M}_z \bar{M}_{z+1}/\bar{M}_w^2$, as stated in the previous chapter and shown in fig. 2. This proportionality

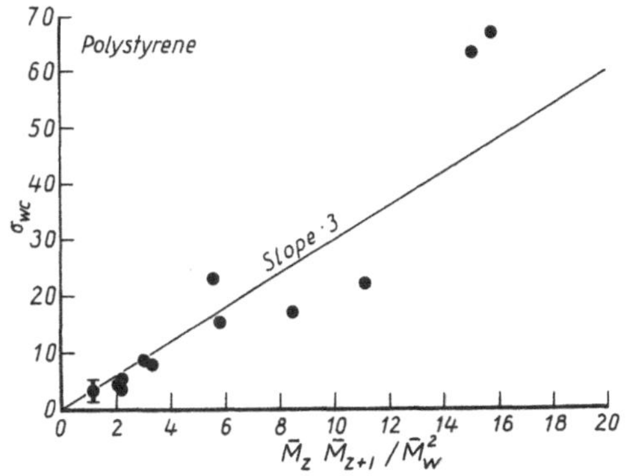

Fig. 2. Dependence of critical recoverable shear at the wall on molecular weight distribution

was confirmed by using the critical swelling ratios and any one of the die swell equations. The numerical values of the constant, however, varied with the different die swell equations. Using *Tanners* equation the numerical constant was estimated to be about 3, for polystyrene, i.e.

$$\sigma_{wc} = 3\left(\frac{\bar{M}_z \bar{M}_{z+1}}{\bar{M}_w^2}\right). \qquad [21]$$

This result can be used in *Hookes* law to explain the polymer behavior at the onset of melt fracture. Eq. [1] can be written for the shear stress at the wall

$$\sigma_w = \tau_w J_0 . \quad [22]$$

For the shear compliance J_0 we use *Graessleys* correlation (18) between experimental and *Rouse* compliance, which is

$$\frac{J_0}{J_R} = \frac{A}{1 + B\left(\frac{E}{E_0}\right)E_0} \quad [23]$$

where A, B are experimental constants

$$E_0 = 2\bar{M}_w/M_c$$

is the entanglement density at zero shear and E is the entanglement density at finite shear. By combining eqs. [19], [22], and [23] we have:

$$\tau_w = \sigma_w \left(\frac{\bar{M}_z \bar{M}_{z+1}}{\bar{M}_w^2}\right)^{-1} \left(\frac{5}{2}\right) \frac{\varrho R T}{A} \times \left[\frac{B}{M_c}\left(\frac{E}{E_0}\right) + \frac{1}{\bar{M}_w}\right]. \quad [24]$$

For polystyrene $A = 2.2$ and $B/M_c = 2.1 \times 10^{-5}$ and since the quantity

$$\sigma_w \left(\frac{\bar{M}_z \bar{M}_{z+1}}{\bar{M}_w^2}\right)^{-1}$$

has a constant value of about 3 at the onset of melt fracture we have:

$$\tau_c = 3.41 \varrho R T \left(2.1 \times 10^{-5}\left(\frac{E}{E_0}\right) + \frac{1}{\bar{M}_w}\right). \quad [25]$$

For $\varrho \approx 0.976$, $R = 8.314 \times 10^7$, we obtain:

$$\tau_c/T = 5.8 \times 10^3 \left(\frac{E}{E_0}\right) + 2.76 \times 10^8/\bar{M}_w . \quad [26]$$

Experimental measurements for several linear polymers shown in fig. 3 resulted in the expression

$$\tau_c/T = 1.717 \times 10^3 + 2.67 \times 10^8/\bar{M}_w . \quad [27]$$

Comparison to eq. [26] shows that the analysis predicts the correct slope of the straight line. From the intercept at the τ_c/T axis we can calculate the value of the entanglement density ratio

$$E/E_0 = 0.34 . \quad [28]$$

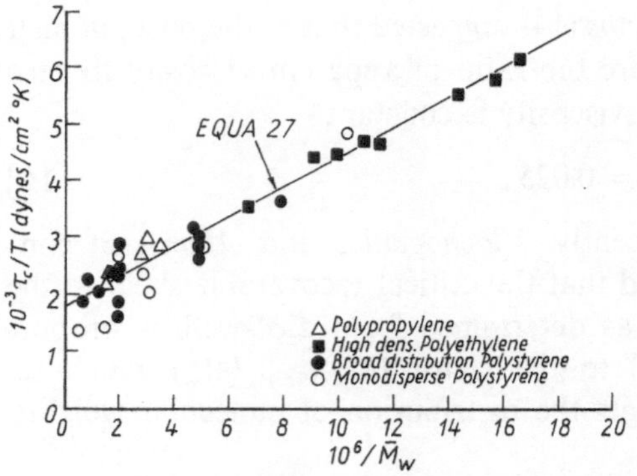

Fig. 3. Dependence of critical shear stress on weight average molecular weight

From *Graessleys* molecular entanglement theory (15, 18) we can calculate the corresponding viscosity ratio for a monodisperse sample

$$\eta_c/\eta_0 = 0.0065 . \quad [29]$$

This result is equivalent to *Bartos*' criterion for melt fracture (eq. [15]). The numerical value is much smaller than *Bartos*' average (actual values ranged from 0.006 to 0.0359). These discrepancies are by no means disconcerting because of the sensitivity of η/η_0 to changes in E/E_0 for example $E/E_0 = 0.486$ corresponds to $\eta_c/\eta_0 = 0.0302$.

It is further interesting to observe that if $E \ll 1$, from eq. [23], $J = A J_R$ and therefore $\tau_c \cdot \bar{M}_w = \text{const.}$ If $E \gg 1$, from eq. [23], $J = (A/B) J_R/E$ and therefore $\tau_c = \text{const.}$ This suggests that no sound conclusions can be made about the constancy of τ_c or $\tau_c \cdot \bar{M}_w$ unless large molecular weight ranges are examined. However, the use of *Graessleys* correlation for J_0/J_R seems to be very helpful in explaining the seemingly contradictory results which are available in the literature.

It is also interesting to note that for the most probable distribution $\bar{M}_z \bar{M}_{z+1}/\bar{M}_w^2 = 3$ which gives $\sigma_{wc} \approx 9$. It is therefore not surprising that many investigators who used commercial samples obtained a constant value of recoverable shear of about 7 units. In the present investigation however special samples of narrow molecular weight distribution ($\bar{M}_z \bar{M}_{z+1}/\bar{M}_w^2 = 1.25$) were used and blends were prepared with $\bar{M}_z \bar{M}_{z+1}/\bar{M}_w^2$ up to 16. Consequently, the characterization of polymer samples as to the distribution of molecular weights is very important both in melt fracture and die swell studies.

Acknowledgements

Financial assistance from the National Research Council of Canada is gratefully acknowledged.

Summary

The equations for determining die swell of molten polymers are discussed. The critical conditions for melt fracture and die swell results are combined in order to explain the flow behavior at the onset of instability. It is shown that the factor $\bar{M}_z \bar{M}_{z+1}/\bar{M}_w^2$ is the most important parameter in die swell and melt fracture studies.

Zusammenfassung

Die Formeln für die Bestimmung der Schwellrate von Kunststoffschmelzen wurden behandelt. Die kritischen Bedingungen für die Ergebnisse von Schmelzbruch und Schwellrate sind kombiniert, um das Flußverhalten beim Einsetzen der Instabilität zu erklären. Es wurde dargestellt, daß der Faktor $\bar{M}_z \bar{M}_{z+1}/\bar{M}_w^2$ in der Schwellrate sowie im Schmelzbruch der wichtigste Parameter ist.

References

1) *Tordella, J. P.*, Unstable Flow of Molten Polymers. Chapter 2, in: Rheology, vol. 5 (Ed. *Eirich*) (New York 1969).

2) *Coleman, B. D.* and *H. Markovitz*, J. Appl. Phys. **35**, 1 (1964).

3) *Nakajima, N.* and *M. Shida*, Trans. Soc. Rheol. **10**, 299 (1966).

4) *Bagley, E. B.* and *H. J. Duffey*, Trans. Soc. Rheol. **14**, 545 (1970).

5) *Treloar, L. R. G.*, The Physics of Rubber Elasticity (Oxford Univ. Press 1967).

6) *Graessley, W. W., S. D. Glasscock*, and *R. L. Crawley*, Trans. Soc. Rheol. **14**, 519 (1970).

7) *Tanner, R. I.*, J. Poly. Sci. (A2), **8**, 2067 (1970).

8) *Vlachopoulos, J., M. Horie*, and *S. Lidorikis*, Trans. Soc. Rheol. **16**, 669 (1972).

9) *Spencer, R. S.* and *R. E. Dillon*, J. Coll. Sci. **4**, 241 (1949).

10) *Barnett, S. M.*, Polym. Eng. Sci. **7**, 168 (1967).

11) *Bagley, E. B.*, J. Appl. Phys. **31**, 1126 (1960).

12) *Bagley, E. B.*, Trans. Soc. Rheol. **5**, 355 (1961).

13) *White, J. L.*, J. Appl. Sci. **8**, 2339 (1964).

14) *Bartos, O.*, J. Appl. Phys. **35**, 2767 (1964).

15) *Vlachopoulos, J.* and *M. Alam*, Polym. Eng. Sci. **12**, 184 (1972).

16) *Boger, D. V.* and *H. L. Williams*, Polym. Eng. Sci. **12**, 309 (1972).

17) *Ferry, J. D.*, Viscoelastic Properties of Polymers. (New York 1970.)

18) *Graessley, W. W.* and *L. Segal*, Macromol. **2**, 49 (1969).

Dr. *John Vlachopoulos*
Ass. Professor
Dept. of Chemical Engineering
McMaster University
Hamilton, Ontario L8S 4L7 (Canada)

Rheol. Acta 13, 228–232 (1974)

Centre de Recherches Physiques, Marseille (France)

Spectre instantané d'une impulsion dans un barreau viscoélastique

Par *R. H. Blanc*

Avec 4 figures

(Reçu p. p. le 27 octobre 1972)

Introduction

Nous avons déjà abordé l'étude d'un milieu viscoélastique linéaire par la détermination de la vitesse de propagation $c(\omega)$ et l'amortissement $\alpha(\omega)$ des ondes longitudinales planes se propageant dans ce matériau. Pour connaître ces fonctions dans un intervalle de fréquence donné, on peut exciter le milieu étudié au moyen d'un unique transitoire contenant le spectre de fréquence considéré, et chercher à déduire $c(\omega)$ et $\alpha(\omega)$ de l'évolution de la forme de l'onde au cours de sa propagation. Pour cela, l'impulsion peut être représentée par une intégrale de *Fourier* faisant intervenir, outre les fonctions c et α, l'impulsion initiale. Plusieurs voies s'ouvrent pour tirer de cette intégrale la vitesse de propagation et l'amortissement au sein du milieu considéré. On peut dans certains cas chercher à l'intégrer exactement, au moyen d'hypothèses formelles simplificatrices sur les fonctions $c(\omega)$ et $\alpha(\omega)$ ainsi que sur l'impulsion initiale (1, 2, 3). Il est possible de développer cette intégrale en série en se limitant aux n premiers termes (4). Enfin, nous avons établi les expressions générales de $c(\omega)$ et $\alpha(\omega)$ en fonction respectivement de l'argument et du module de la transformée de *Fourier* de l'impulsion (3); mais l'application numérique de ces relations exige pratiquement l'emploi de l'ordinateur pour le calcul des transformées.

L'objet de la présente communication est: de proposer une méthode qui conserve le principe de l'excitation par impulsion et permette le calcul direct de $c(\omega)$ et $\alpha(\omega)$ dans le cas général; de décrire un montage expérimental conforme; enfin, de discuter cette méthode.

Principe de la méthode

Considérons un barreau cylindrique du matériau étudié, limité par deux sections droites distantes de l. On produit contre une extrémité de ce barreau une percussion axiale qui engendre la propagation d'une impulsion. Celle-ci se réfléchit sur l'extrémité opposée et revient vers l'extrémité origine où elle se réfléchit à nouveau et le processus recommence (5, 1, 3). Appelons vitesse de particule $v(x, t)$, la vitesse de déplacement d'un point matériel du barreau d'abscisse x, associée au passage de l'onde. Le montage expérimental décrit plus loin permet de relever directement en fonction du temps la vitesse de particule associée aux passages successifs de l'impulsion au bout de la barre; passages qui correspondent à des distances de parcours égales à l, $3l, \ldots, (2k+1)l, \ldots$ le long d'une barre semi-infinie. Le changement dans la forme de l'onde dépend des fonctions $c(\omega)$ et $\alpha(\omega)$.

Appliquons ces signaux à un système dont la réponse dépende de la fréquence du signal d'entrée. Le signal à chaque fois délivré en réponse sera fonction non seulement de l'action mais encore des caractéristiques propres du système. Nous allons appliquer ici à un même filtre une même impulsion relevée après deux distances de parcours, x_1 et x_2 et comparer les réponses obtenues.

Théorie

A la distance x et à l'instant t, l'impulsion de vitesse de particule revêt la forme de l'intégrale (6):

$$v(x, t) = \frac{1}{2\pi} \int_{-\infty}^{+\infty} \hat{v}(0, \omega)\, e^{-\alpha(\omega)x + i\omega\left[t - \frac{x}{c(\omega)}\right]} d\omega \quad [1]$$

dans laquelle $\hat{v}(0, \omega)$ représente la transformée de *Fourier* de l'impulsion à l'origine des espaces. Considérons un système linéaire dont la réponse indicielle (7) est $h(t)$. Sa réponse à l'action $v(x, t)$ sera donnée par l'intégrale de *Duhamel*:

$$y(x, t) = \int_0^t v'_\tau(x, \tau)\, h(t - \tau)\, d\tau \qquad [2]$$

qui devient ici:

$$y(x, t) = \frac{i}{2\pi} \int_{-\infty}^{+\infty} \omega \hat{v}(0, \omega)\, e^{-\alpha(\omega) x - i \frac{\omega x}{c(\omega)}} d\omega \times \int_0^t h(t - \tau)\, e^{i\omega\tau} d\tau. \qquad [3]$$

Dans le second membre intervient l'intégrale

$$I = \int_0^t h(t - \tau)\, e^{i\omega\tau} d\tau. \qquad [4]$$

Choisissons, comme système, un circuit oscillant dont la réponse indicielle est de la forme:

$$h(t) = A e^{-\beta t} \sin \Omega t. \qquad [5]$$

Dans ces conditions, le calcul de l'intégrale [4] donne:

$$I = \frac{A}{(\beta + i\omega) 2 + \Omega^2} \times \{\Omega e^{i\omega t} - e^{-\beta t}[(\beta + i\omega) \sin \Omega t + \Omega \cos \Omega t]\}. \qquad [6]$$

Au bout d'un temps t suffisamment grand, I s'écarte aussi peu que l'on veut de sa limite:

$$I_l = \frac{A \Omega e^{i\omega t}}{(\beta + i\omega)^2 + \Omega^2}. \qquad [7]$$

Supposons cet instant atteint et portons l'expression ci-dessus dans [3], la relation obtenue revêt la forme d'une integrale de *Fourier*. Inversons cette transformation, on obtient:

$$\frac{i A \Omega \omega \hat{v}(0, \omega)}{(\beta + i\omega)^2 + \Omega^2} e^{-\alpha(\omega) x - i \frac{\omega x}{c(\omega)}} = \int_{-\infty}^{+\infty} y(x, t)\, e^{-i\omega t} dt. \qquad [8]$$

Dans le premier membre, intervient en facteur la fonction de ω:

$$K = \frac{\omega}{(\beta + i\omega)^2 + \Omega^2}. \qquad [9]$$

Le module de cette expression admet un unique maximum sur le demi-axe positif, atteint pour:

$$\omega = \sqrt{\Omega^2 + \beta^2} \qquad [10]$$

et qui a pour valeur:

$$M = \frac{1}{2\beta}. \qquad [11]$$

Ainsi, à la condition d'un amortissement β suffisamment faible, la relation [8] ne subsiste que pour la fréquence de résonance Ω et devient:

$$\frac{A \Omega \hat{v}(0, \Omega)}{2\beta} e^{-\alpha(\Omega) x - i \frac{\Omega x}{c(\Omega)}} = \int_{-\infty}^{+\infty} y(x, t)\, e^{-i\Omega t} dt. \qquad [12]$$

Supposons notre système attaqué par un transitoire de durée finie, c'est-à-dire variable dans l'intervalle $[0, T]$ et constant, mais non nécessairement nul, en dehors de cet intervalle. La réponse du système est donnée par la relation [2] qui prend la forme suivante, pour $t > T$:

$$y(x, t) = A e^{-\beta t} \varrho(\Omega) \sin [\Omega t + \theta(\Omega)] \qquad [13]$$

avec

$$\theta(0) = 0. \qquad [14]$$

Portons [13] dans la relation [12] en supposant de plus β négligeable devant Ω on est conduit au résultat:

$$i \Omega \hat{v}(0, \Omega)\, e^{-\alpha(\Omega) x - i \frac{\Omega x}{c(\Omega)}} = \varrho(\Omega)\, e^{i\theta(\Omega)}. \qquad [15]$$

Ecrivons cette relation aux deux distances x_1 et x_2, divisions membre à membre et décomposons la relation complexe obtenue en deux, respectivement entre les arguments et les modules, on en tire finalement:

$$c(\Omega) = -\Omega \frac{x_2 - x_1}{\theta_2(\Omega) - \theta_1(\Omega)} \qquad [16]$$

$$\alpha(\Omega) = -\frac{1}{x_2 - x_1} \log \frac{\varrho_2(\Omega)}{\varrho_1(\Omega)}. \qquad [17]$$

Pour les deux impulsions $v(x_1, t)$ et $v(x_2, t)$, la relation [13] montre qu'au bout d'un temps supérieur à la durée T de la plus longue impulsion, la différence des phases et le rapport des amplitudes des deux réponses sont des grandeurs toutes deux indépendantes du temps. Les relations [16] et [17] sont donc applicables dès l'instant T.

Pour le calcul de c à différentes fréquences, les différences $\theta_2 - \theta_1$ ne sont connues qu'à $2k\pi$ près. Toutefois, on peut faire les remarques suivantes. Tout d'abord, d'après [14]:

$$\theta_2(0) = \theta_1(0) = 0 \qquad [18]$$

d'où:

$$\Omega = 0 \Rightarrow k = 0. \qquad [19]$$

Enfin, on peut toujours préciser la détermination de c si on en connaît la valeur pour une fréquence quelconque par une mesure indépendante.

Examinons l'influence d'une distorsion éventuelle d'amplitude et de phase introduite par le transducteur dans la mesure de la grandeur mécanique considérée. Cette grandeur est ici la vitesse de particule, mais la présente théorie serait la même pour la contrainte ou la déformation, grandeurs représentables au moyen d'expressions formellement identiques à [1], (6). On sait qu'il existe une grande quantité de transducteurs électromécaniques mais beaucoup d'entr'eux ne sont utilisables que dans un domaine trop limité vers les hautes fréquences, surtout les capteurs électromagnétiques, sensibles à la vitesse de particule, ce qui les rend généralement impropres à des mesures sur les ondes de contrainte (8). De même, une jauge de déformation dans un pont d'extensométrie peut introduire un déphasage inconnu important croissant avec la fréquence (9). D'une façon générale, si pour chaque pulsation ω le transducteur multiplie l'amplitude par $\mu(\omega)$ et fait tourner la phase de $\varphi(\omega)$, on montre (10) que sa réponse indicielle est:

$$h_1(t) = \frac{E_0}{2\pi i} \int_{-\infty}^{+\infty} \frac{\mu(\omega)\, e^{i[\omega t + \varphi(\omega)]}}{\omega} d\omega. \qquad [20]$$

De sorte que la grandeur délivrée $g(x, t)$ est donnée en fonction de la grandeur vraie $v(x, t)$, par une relation analogue à [2]. Si $v(x, t)$ est une impulsion de durée finie T, pour $t > T$, $g(x, t)$ prend la forme:

$$g(x, t) = \frac{E_0}{2\pi i} \int_{-\infty}^{+\infty} \frac{S(\omega)\, \mu(\omega)\, e^{i[\omega t + \varphi(\omega)]}}{\omega} d\omega \qquad [21]$$

où

$$S(\omega) = i\omega \hat{v}(0, \omega)\, e^{-\alpha(\omega)x - i\frac{\omega x}{c(\omega)}} \qquad [22]$$

transformée de *Fourier* de $v_t'(x, t)$, est maintenant à remplacer dans la suite de notre théorie par la même quantité multipliée par le facteur $E_0 \mu(\omega)\, e^{i\varphi(\omega)}/i\omega$ lequel disparaît dans la division membre à membre des deux égalités déduites de [15]; finalement les relations [16] et [17] restent inchangées. Ainsi, une distorsion arbitraire introduite par le transducteur ne change pas les résultats obtenus. Il en résulte que le domaine d'utilisation des capteurs se trouve étendu, avec cette méthode, de leur bande passante à tout le domaine où leur sensibilité reste appréciable. Nous en donnerons un exemple dans notre montage expérimental. Considérons les fonctions v et v_t' et leurs transformées de *Fourier* $\hat{v}$ et $i\omega\hat{v}$; en regardant le facteur $i\omega$ comme la distorsion $\omega e^{i \cdot \pi/2}$, on voit qu'on pourrait utiliser comme capteur un accéléromètre.

Montage expérimental

Le choc qui produit l'impulsion est également utilisé pour déclencher, par rupture de contact, un balayage de l'écran d'un oscillographe cathodique bicourbe muni d'un appareil photographique.

Un transducteur électromécanique permet de relever la vitesse de particule $v(x, t)$ de l'extrémité opposée de la barre. La tension électrique obtenue est appliquée à la fois à une voie d'entrée de l'oscillographe et au filtre. Celui-ci délivre en réponse un signal $y(x, t)$ qui est appliqué sur la deuxième voie de l'oscillographe.

Nous avons essayé divers capteurs électromagnétiques au-delà de leur bande passante; par exemple le capteur Philips, type PR 9262 (1–2000 Hz), ou le capteur Brüel & Kjaer, type MM 0002.

Dans notre théorie, nous avons étudié le cas du circuit oscillant. Il est extrêmement simple de réaliser un tel circuit, toutefois, pour pouvoir atteindre de très faibles valeurs de β et pour sa commodité d'emploi, nous avons utilisé un appareil étalonné, le filtre A.I.M., type TFO 129. On démontre en effet (11), et nous avons vérifié expérimentalement, qu'il se comporte comme un circuit oscillant. Ce filtre est à amortissement β variable, ce qui permet de choisir cette grandeur à la fois, en vue d'obtenir une sensibilité maximum, mais aussi en sorte que chaque nouvelle impulsion trouve le filtre au repos.

Résultats expérimentaux

A titre d'illustration, nous avons appliqué cette méthode au lucoflex, variété par ailleurs connue de chlorure de polyvinyle. Les mesures ont porté sur deux barreaux de 2 cm de diamètre. L'un, de 2 m de long, était excité par le choc coaxial d'une barre d'acier de 2,5 cm de long. Pour atteindre des fréquences plus élevées, on a utilisé le choc d'une petite bille d'acier de 6 mm de diamètre lancée par un canon à air comprimé contre l'extrémité d'un deuxième barreau, d'une longueur de 80 cm, du même matériau.

On a calculé c et α au moyen des formules [16] et [17] pour un certain nombre de fréquences. L'ensemble est représenté graphiquement sur les figs. 3 et 4. A titre de comparaison, ces résultats sont rapprochés de ceux obtenus sur le même matériau par la méthode décrite en (3).

Conclusion

L'interposition d'un filtre sélectif permet d'obtenir directement la vitesse de propagation c et l'amortissement α.

La méthode fournit une mesure sûre et précise de α. Par contre, le choix de la détermination de c peut nécessiter une discussion ou une mesure indépendante à une fréquence quelconque.

Une distorsion arbitraire d'amplitude et de phase, introduite par le transducteur, n'affecte pas les résultats. Cette méthode accroît le nombre et la variété des capteurs propres aux mesures sur les ondes de contrainte. Le domaine d'utilisation de ces capteurs se trouve étendu à tout l'intervalle de fréquence où leur sensibilité reste appréciable.

Résumé

On produit un choc contre une extrémité d'un barreau du milieu viscoélastique étudié et on relève en fonction du temps, la forme de l'onde à l'extrêmité opposée. Un transducteur électromécanique fournit une tension proportionnelle. Ce signal est appliqué à un filtre sélectif. On étudie le cas d'un circuit oscillant. On considère les réponses d'un même filtre à une même impulsion après deux distances de parcours. De leur comparaison, d'une part on élimine les caractéristiques du filtre moyennant une condition de sélectivité que l'on discute; cette condition est remplie en pratique par des circuits très simples. D'autre part, on en déduit la vitesse de propagation $c(\omega)$ et l'amortissement $\alpha(\omega)$ dans le matériau étudié. α est déterminé sans ambiguité. On discute la détermination de c. On montre qu'une distorsion quelconque introduite par le transducteur n'altère pas les résultats; on peut ainsi utiliser un capteur électromagnétique au delà de sa bande passante pour mesurer la vitesse de particule. Plus généralement, le domaine d'utilisation des transducteurs électromécaniques est étendu à tout le domaine de fréquence où leur sensibilité reste suffisante. A titre d'illustration, la méthode est appliquée à deux barreaux de chlorure de polyvinyle. Les résultats sont rapprochés de ceux obtenus par une méthode d'analyse de *Fourier* nécessitant une calculatrice numérique. La comparaison est satisfaisante.

Zusammenfassung

Es wird an dem Ende einer aus dem zu untersuchenden viskoelastischen Material gefertigten Stange ein Stoß

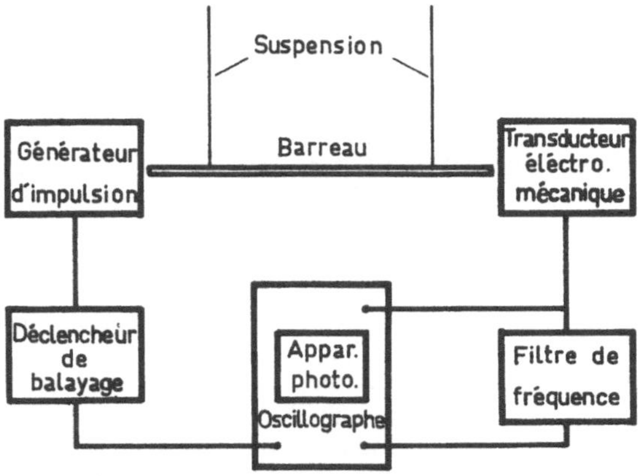

Fig. 1. Schéma synoptique

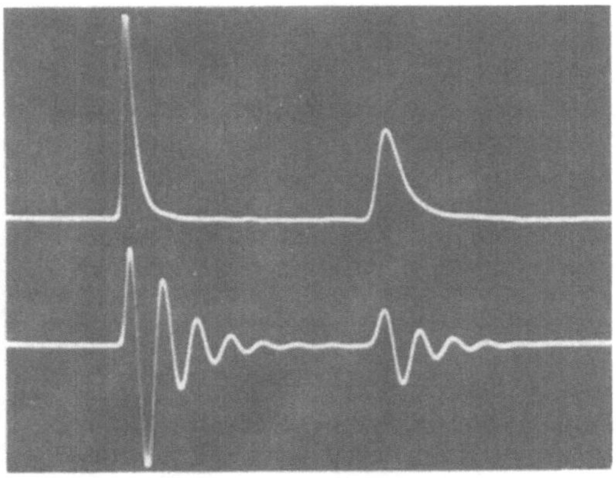

Fig. 2. En haut: enregistrement, en fonction du temps, de la vitesse de particule associée à la propagation d'une impulsion dans un barreau de lucoflex, après 2 et 6 m de parcours. En bas: réponse à ces transitoires d'un circuit oscillant sur la fréquence de 3 kHz.

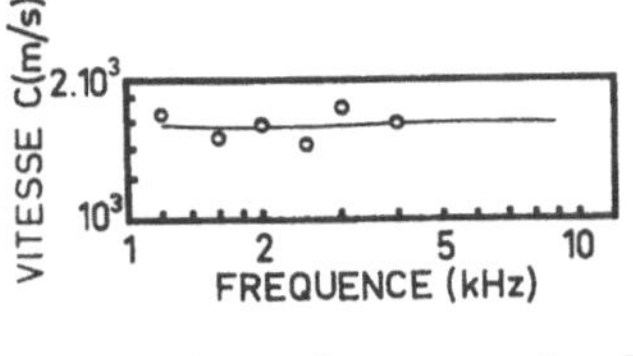

Fig. 3. Vitesse de propagation [en trait continu: valeur calculée par la méthode (3)]

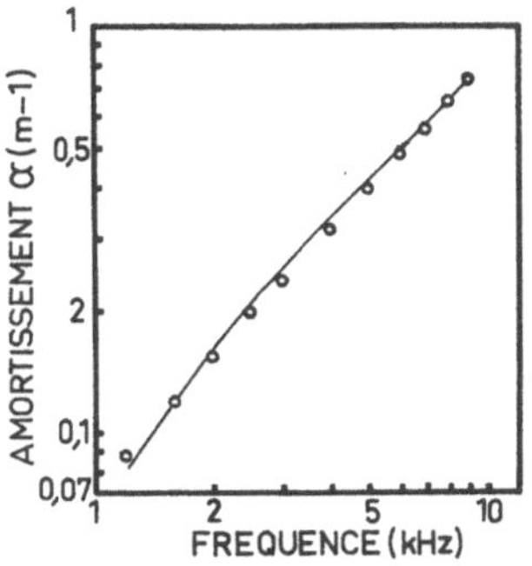

Fig. 4. Amortissement [en trait continu: valeur calculée par la méthode (3)]

gegeben und die Wellenform an dem Gegenende als Funktion der Zeit aufgezeichnet. Eine proportionale Spannung wird von einem elektromechanischen Wandler geliefert. Dieses Signal wird an ein selektives Filter übertragen. Es wird der Fall eines Schwingkreises untersucht. Das Ansprechen eines selbigen Filters einem selbigen Stoß entsprechend wird an zwei Punkten auf dem Stoßweg betrachtet. Durch den Vergleich dieser zwei Angaben miteinander können einerseits die Filtereigenschaften, unter Voraussetzung einer gewissen Filterselektivität, die hier diskutiert wird, eliminiert werden; diese Voraussetzung ist in der Praxis mittels sehr einfacher Kreise leicht erfüllt. Andererseits werden die Phasengeschwindigkeit $c(\omega)$ und die Dämpfung $\alpha(\omega)$ für das untersuchte Material dabei ohne Ungewißheit abgeleitet. Die Bestimmung von c wird diskutiert. Es wird gezeigt, daß eine etwaige durch den Wandler eingeführte Deformation keinen Einfluß auf die Ergebnisse hat; ein elektromagnetischer Wandler läßt sich daher über den Durchlaßbereich hinaus zum Messen der Teilchengeschwindigkeit verwenden. Im allgemeinen wird der Anwendungsbereich elektromechanischer Wandler zum ganzen Frequenzbereich ausgedehnt, in dem sie eine genügende Empfindlichkeit erweisen. Beispielsweise wird die Methode auf zwei Polyvinylchloridstangen angewandt. Die Ergebnisse, im Vergleich mit denen, die mittels einer *Fourier*analysenmethode, die einen Digitalrechner erfordert, erzielt werden, erweisen sich als befriedigend.

Summary

A shock is produced at one end of a bar of the viscoelastic material studied, and the wave shape at the opposite end is recorded in relation to time. A proportional voltage is delivered by an electromechanical transducer. This signal is transmitted to a selective filter. The case of an oscillating circuit is studied. The responses of one same filter to one same impulse are considered at two distances on the impulse course. By comparing these two responses, we are able on the one hand to eliminate the characteristics of the filter, given a certain filter selectivity condition which is discussed; this condition is fulfilled in practice with very simple circuits. On the other hand, the phase velocity $c(\omega)$ and the damping $\alpha(\omega)$ are deduced for the material studies. α is determined without ambiguity. The determination of c is discussed. It is shown that any distortion introduced by the transducer does not affect the results; so, an electromagnetic transducer can be used beyond its normal band-pass to measure the particle velocity. On a more general level, the application range for electromagnetic transducers is extended to the whole frequency range within which there is sufficient sensitivity. As an example, this method is applied to two bars of polyvinyl chloride. The results are compared with those obtained by a *Fourier* analysis method requiring the use of a computer. The comparison is satisfactory.

Littérature

1) *Kolsky, H.*, Phil. Mag. **1**, n° 8, 693–710, pl. 35–36 (1956).

2) *Bismuth, W.* et *R. H. Blanc*, Cah. Gr. Fr. Rhéo. **1**, 259–265 (1967).

3) *Blanc, R.*, Détermination de l'équation de comportement des corps viscoélastiques linéaires par une méthode d'impulsion. Thèse, Fac. Sci. (Marseille 1971).

4) *Kolsky, H.* et *S. S. Lee*, Off. Naval Res. 1962, Contr. Nonr 562 (30)/5, Brown Univ., Providence.

5) *Davies, R. M.*, Phil. Trans. Roy. Soc. London **A 240**, 375–457 (1948).

6) *Hunter, S. C.*, Viscoelastic waves. In: Progress in Solid Mechanics **1**, 1–57 (Amsterdam 1960).

7) *Naslin, P.*, Les régimes variables dans les systèmes linéaires et non linéaires (Paris 1962).

8) *Kolsky, H.*, The detection and measurement of stress waves. In: Experimental techniques in shock and vibration, 11–24 (New York 1962).

9) *Lagarde, A.*, Etude du comportement mécanique et photoélastique d'un barreau biréfringent soumis à des vibrations, Thèse, Fac. Sci. (Paris 1962).

10) *Rocard, Y.*, Dynamique générale des vibrations (Paris 1960).

11) *Edge, G. M.*, Instrument Review (G. B.), **1967**, 279–283.

Adresse de l'auteur:

Dr. *R. H. Blanc*
Centre de Recherches Physiques
31, Chemin Joseph-Aiguier (9° arr.)
F-13274 Marseille Cedex 2 (France)

Rheol. Acta 13, 233–235 (1974)

Centre de Recherches Physiques, Marseille (France)

Vérification par des méthodes statiques et dynamiques de la théorie de l'adhésion basée sur des phénomènes de rupture

Par J.-M. Tatraux-Paro et W. Bismuth

Avec 4 figures et 1 tableau

(Reçu p. p. le 27 octobre 1972)

Nous nous proposons de relier le comportement du frottement d'adhésion du caoutchouc en fonction de la vitesse de glissement aux caractéristiques viscoélastiques du caoutchouc. On verra par la suite qu'il est possible expérimentalement d'isoler la composante adhésion du frottement des composantes d'abrasion et d'hystérésis (1). Des études tant expérimentales que théoriques (3, 4, 5) ont traité de ce type de problème; on admet généralement que le frottement d'adhésion est dû à l'étirement des agrégats moléculaires caoutchoutiques qui sont à l'état lié avec le substrat pendant le glissement.

Nous schématiserons la mise en équation du phénomène en supposant que la rupture de la liaison entre l'agrégat et le substrat est provoquée lorsque l'énergie emmagasinée pendant l'élongation atteint un niveau critique W_0; la force globale de frottement en un instant donné se calcule alors comme la somme de toutes les liaisons. Soit un élément de la surface de caoutchouc au contact avec le substrat; on considère dans cet élément les n agrégats moléculaires susceptibles de liaisons avec le substrat; dans la suite, on supposera tous les agrégats identiques et obéissant à la même loi de relaxation

$$r(t) = E_0\left(1 + b e^{-\frac{t}{\tau}}\right). \qquad [1]$$

Parmi les n agrégats moléculaires susceptibles de liaisons, n_0 sont à l'état lié, n_1 à l'état libre; soit t_1 le temps qu'un agrégat demeure à l'état libre, t_0 le temps qu'il demeure à l'état lié; on considère que temps et nombres sont liés par la relation:

$$\frac{n_0}{t_0} = \frac{n_1}{t_1} = \frac{n}{t_0 + t_1}. \qquad [2]$$

On suppose par ailleurs que le temps pendant lequel un agrégat demeure à l'état libre est proportionnel au temps de relaxation τ

$$t_1 = a\tau. \qquad [3]$$

Le déplacement étant proportionnel à la vitesse de glissement V que l'on considère constante pour le moment, la force de liaison f des agrégats s'obtient par l'intégrale de *Boltzmann*

$$f(t) = V \int_0^t r(t - t')\, dt. \qquad [4]$$

En supposant qu'il y a rupture de liaison lorsque l'énergie emmagasinée par l'agrégat lors du frottement atteint le niveau critique W_0, on détermine le temps t_0 de durée d'une liaison par la relation

$$W_0 = V \int_0^{t_0} f(t)\, dt. \qquad [5]$$

La force de frottement F qui est la valeur moyenne des forces de liaison des agrégats s'obtient par la formule

$$F = \frac{n_0}{t_0} \int_0^{t_0} f(t)\, dt \qquad [6]$$

soit

$$F = \frac{n}{t_0 + a\tau} \frac{W_0}{V}. \qquad [7]$$

Les relations [1, 4, 5, 7] permettent de déterminer le comportement de F en fonction de V, il suffit d'éliminer α_0 entre les relations [8, 9]

$$L = \sqrt{\frac{W_0/E}{\frac{\alpha_0^2}{2} + b(\alpha_0 + e^{-\alpha_0} - 1)}} \qquad [8]$$

$$F = \frac{n W_0}{L(\alpha_0 + a)} \qquad [9]$$

avec

$$\alpha = \frac{t}{\tau}; \quad L = V\tau. \qquad [10]$$

On remarque que le résultat que nous venons d'obtenir est compatible avec le principe d'équivalence vitesse-température, car vitesse et temps de relaxation interviennent simultanément sous la forme du produit $L = V\tau$ (6, 7).

L'exploitation des formules [8, 9] permet d'obtenir les résultats ci-après (fig. 1). On a porté,

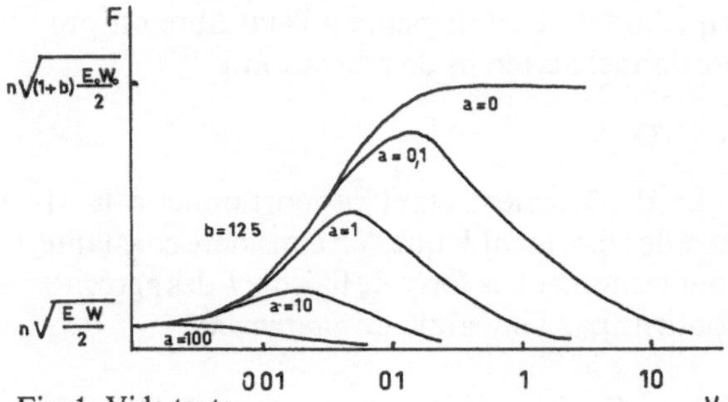

Fig. 1. Vide texte

sur la fig. 1, F en fonction de $\log(V)$; le frottement tend vers $F_0 = n\sqrt{E_0 W_0/2}$ si a est nul, ou vers 0 si a est non nul, lorsque V tend vers l'infini; une variation de E_0 ou W_0 a un effet de translation sur les abscisses et de changement d'unité sur les ordonnées. On remarque que la valeur initiale et la valeur maximale possible sont dans le rapport $\sqrt{1+b}$. La comparaison de ces résultats avec les expériences de *Grosch* (2) permettent de déterminer la valeur de b et celle de a dans le cas du contact néoprène-verre cathédrale (fig. 2).

Nous avons cherché à vérifier sur plusieurs types de caoutchoucs la corrélation entre la valeur de b et le rapport des valeurs maximums aux valeurs initiales du frottement; nous avons utilisé du caoutchouc naturel vulcanisé à divers pourcentages de soufre (1, 3,5, 8, 12%). Les expériences de frottement ont été faites à l'aide d'un tribomètre constitué d'un cylindre creux tournant autour de son axe normalement à la surface du caoutchouc (fig. 3). Le procédé permet de conserver identiques à elles-mêmes au

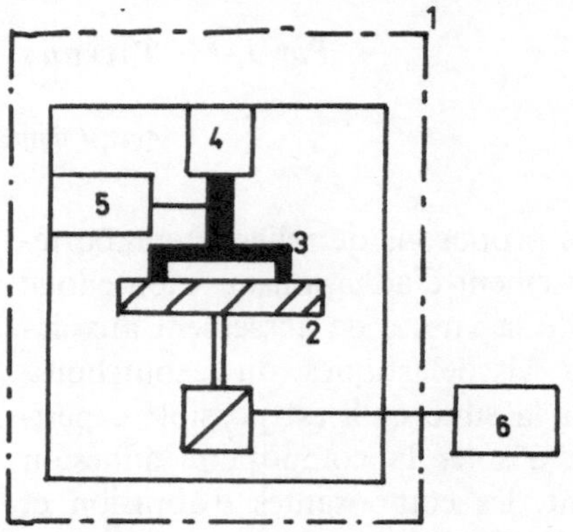

Fig. 3. Tribomètre. 1 Enceinte thermique, 2 caoutchouc, 3 piste, 4 effort normal, 5 couple, 6 entrainement

cours de l'expérience l'aire de contact et la distribution des pressions de contact, ce qui élimine la possibilité de frottement par hystérésis. On réduit d'autre part suffisamment la pression de contact (32 g/cm^2) de manière à pouvoir considérer l'abrasion négligeable. La rotation du cylindre permet d'obtenir des vitesses de glissement variant de manière géometrique entre 0,006 et 1,7 mm/s; le maximum du frottement est déterminé par le *Stick-Slip* (8). L'expérience est maintenue constante à la température de 20 °C; l'ensemble des résultats obtenus sur le frottement est reporté fig. 4.

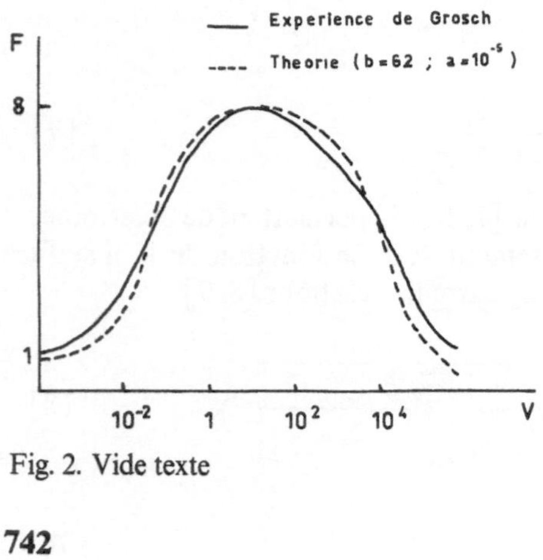

Fig. 2. Vide texte

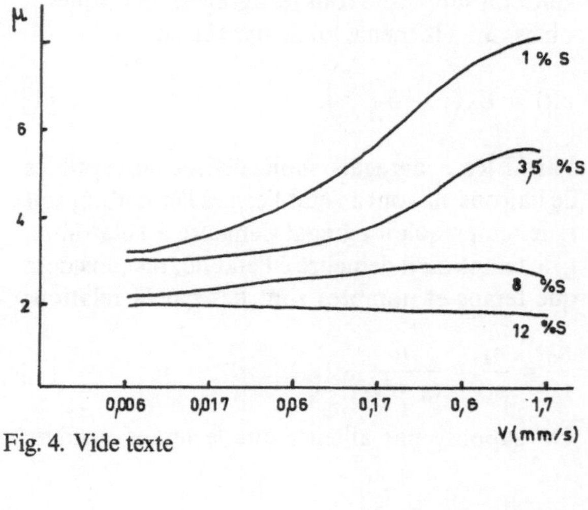

Fig. 4. Vide texte

Par ailleurs, on détermine b à partir d'une expérience de relaxation par la formule

$$b = \frac{E(0) - E(\infty)}{E(\infty)}. \quad [11]$$

Les résultats théoriques se rapprochent d'autant plus des résultats expérimentaux que b est mesuré pour un allongement voisin de celui de la rupture; la comparaison du rapport entre frottement maximum et frottement initial, et de $\sqrt{1 + b}$ est donné par le tableau ci-dessous pour les quatre types de caoutchoucs étudiés.

Tableau 1

%S	1%	3,5%	8%	12%
b	2,2	1,5	0,4	0,05
$\sqrt{1+b}$	1,8	1,6	1,2	1
$F_{max}/F_{init.}$	2,5	1,8	1,3	0,9

On observe que les résultats théoriques sont inférieurs aux résultats expérimentaux lorsque le degré de réticulation du caoutchouc est faible, ce qu'on attribue, d'une part, à la difficulté de déterminer b avec précision au voisinage de la rupture, et d'autre part, à la simplification excessive du modèle rhéologique choisi.

Nous pouvons cependant considérer pour les types de caoutchoucs étudiés qu'il existe une certaine corrélation entre le coefficient b que l'on peut déterminer par une expérience statique (relaxation par exemple), et les propriétés du frottement d'adhésion en fonction de la vitesse de glissement.

Littérature

1) *Kummer, H. W.*, Unified theory of rubber and tire friction (1964).
2) *Grosch, K. A.*, Nature **197**, 856 (1963).
3) *Schallamach, A.*, Wear **6**, 375 (1963).
4) *Savkoor, A. R.*, Wear **8**, 222 (1965).
5) *Thirion, P.* et *R. Chasset*, Le frottement du caoutchouc. Rapport bibliographique, mai 1964.
6) *Williams, M. L., R. F. Landel*, and *J. D. Ferry*, J. amer. Chem. Soc. **77**, 3701 (1955).
7) *Huet, C.*, Etude, par une méthode d'impédance du comportement viscoélastique des matériaux hydrocarbonés. Thèse, Ing-doc., Fac. Sci. (Paris 1963).
8) *Tatraux-Paro, J.-M., J.-C. Chezeaux* et *W. Bismuth*, Acustica **28**, 272–278 (1973).

Adresse des auteurs:

J.-M. Tatraux-Paro et *W. Bismuth*
Centre de Recherches Physiques
31, Chemin Joseph-Aiguier (9° arr.)
F-13274 Marseille Cedex 2 (France)

Rheol. Acta 13, 236–240 (1974)

I. U. T. Belfort (France)

Étude de la brossabilité des peintures

Par C. Oiknine et P. Jacquet

Avec 6 figures, 1 schéma et 3 tableaux

(Reçu p. p. le 27 octobre 1972)

L'application correcte d'une peinture sur une surface doit obéir à de nombreuses conditions; bien sûr les efforts d'application ne doivent pas être trop importants mais la peinture ne doit ni couler ni laisser apparaître les reprises par exemple.

Afin d'étudier la brossabilité des peintures nous avons à la demande des Industries des Peintures Associées, réalisé et mis au point un appareil permettant de mesurer les efforts d'application d'une peinture.

1. Dispositif expérimental

Pour mesurer les efforts appliqués par le rouleau du peintre nous avons choisi le principe de l'extensométrie.

L'appareil (fig. 1) est constitué par une planche à peindre verticale d'un mètre de large sur un mètre cinquante de haut dont les angles sont liés à un cadre intermédiaire par quatre lames de flexion à section circulaire; les lames sont encastrées dans la planche à peindre et articulées au moyen de cardans sur le cadre intermédiaire. Ce cadre intermédiaire est lié à un bâti fixe par quatre lames de flexion à section rectangulaire encastrées aux quatre coins du bâti fixe et articulées sur le cadre intermédiaire.

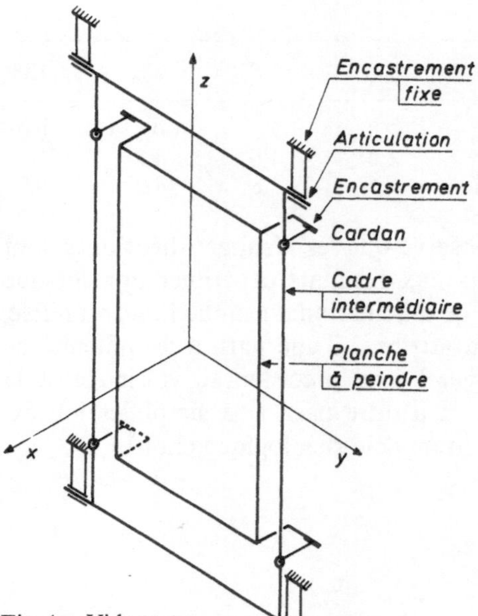

Fig. 1a. Vide texte

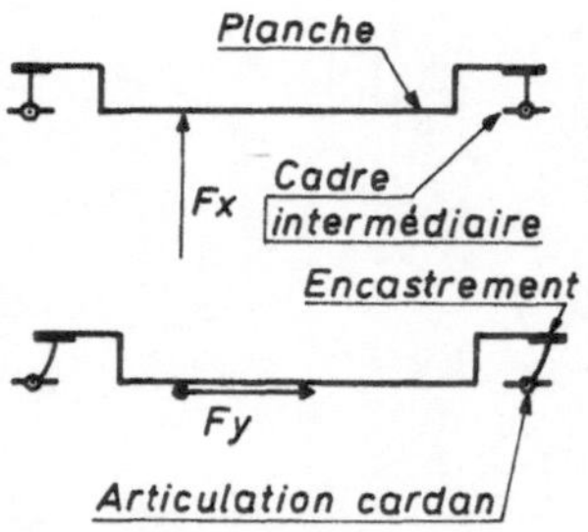

Fig. 1b. Vide texte

1.1. *Mesure de l'effort tangentiel*

Lorsque la planche est soumise à une force quelconque $\vec{F}$, les lames sont sollicitées en traction par la composante normale Fx, en flexion dans le plan x, y par la composante tangentielle horizontale Fy, et en flexion également dans le plan x, z par la composante tangentielle verticale Fz.

Principe de la mesure de Fy (fig. 2)

Deux jauges de traction identiques sont collées diamétralement opposées dans le plan de flexion x, y et à proximité de l'encastrement sur les quatre lames. Appliquons un effort Fy dirigé de gauche à droite; les jauges 1′, 1″, 3′ et 3″ sont en extension; les jauges 4′, 4″, 2′ et 2″ sont en compression. Ces huit jauges sont cablées en pont complet (fig. 3) et les jauges diamétralement opposées sont montées adjacentes dans ce pont. Ainsi les effets de traction dûs aux variations de température sont éliminés. La flexion dans le plan x, z due à l'effort tangentiel Fz n'intervient pas puisque les jauges sont collées sur les fibres neutres.

Le signal débité par le pont d'extensométrie dynamique est proportionnel à la somme des composantes des actions tangentielles horizontales sur les quatre lames élastiques donc proportionnel à la composante tangentielle Fy de l'effort quelconque appliqué en un point quelconque de la planche.

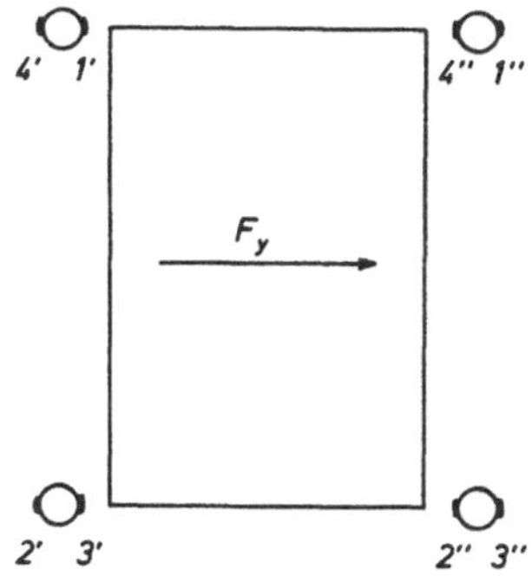

Fig. 2. Vide texte

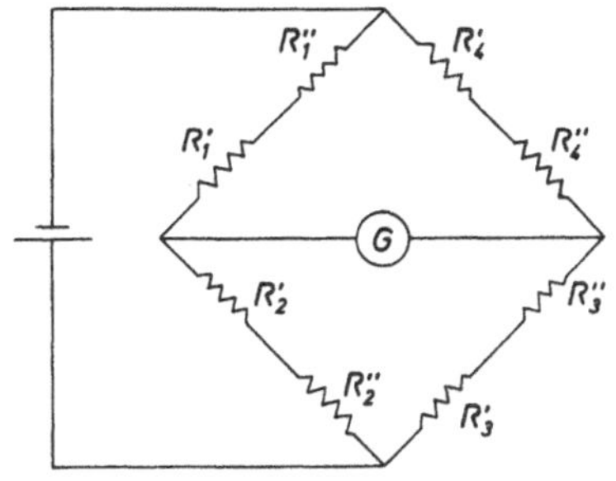

Fig. 3. Vide texte

Principe de la mesure de Fz

La mesure de Fz est rigoureusement identique à celle de Fy. Il suffit de faire une rotation de 90° dans le plan y, z pour trouver les huit jauges qui permettent la mesure de Fz.

Sur cette direction nous avons compensé la pesanteur par des contrepoids.

En résumé, pour la mesure de l'effort tangentiel, quatre lames élastiques à section circulaire supportent chacune quatre jauges collées tous les 90° sur les axes y et z.

1.2. Mesure de l'effort normal Fx (fig. 4)

Lorsque la planche est soumise à un effort quelconque les lames à section rectangulaire sont sollicitées en traction par le poids du cadre intermédiaire et la composante Fz, en flexion par Fy et par Fx.

La flexion due à Fy est négligeable car la largeur des lames est dix fois plus grande que leur épaisseur; de plus les jauges sont collées sur la fibre neutre au milieu des grandes faces et à proximité de l'encastrement.

Grâce à un montage analogue aux précédents, le pont débite un signal proportionnel à l'effort normal Fx.

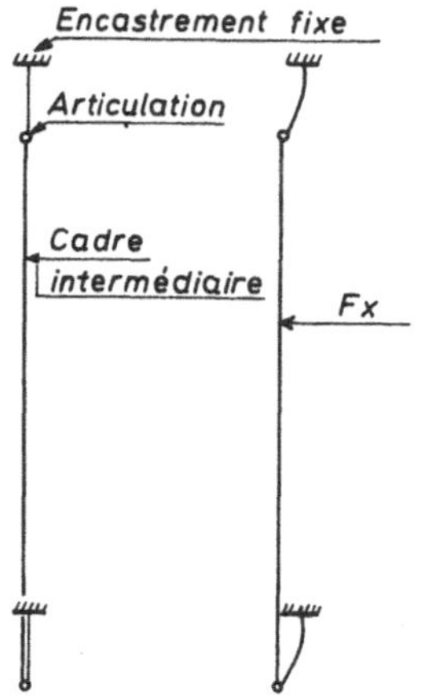

Fig. 4. Vide texte

1.3. Chaine de mesure

Elle se compose de trois ponts d'extensométrie dynamiques, trois filtres passe bas, trois amplificateurs et un enregistreur ultra violet sur lequel nous utilisons trois voies.

Lors de l'application de la peinture nous enregistrons simultanément les efforts Fx, Fy et Fz en fonction tu temps.

2. Processus expérimental

Dans un premier stade nous avons voulu mesurer les ordres de grandeur des efforts normaux et tangentiels développés lors de l'application d'une peinture.

Aussi le processus expérimental retenu est-il très voisin du processus réel d'application; il comprend trois phases:

– Une phase de dépose: pendant laquelle une certaine quantité de peinture est déposée sur la planche.

– Une phase d'étalement pendant laquelle la peinture déposée est répartie sur la planche.

– Une phase de lissage qui permet d'obtenir un bon état de surface.

Afin d'avoir des résultats les plus comparables possibles, ces trois phases ont toujours été effectuées dans les mêmes conditions.

La dépose se fait en trois coups verticaux en commençant par le haut de la planche.

L'étalement en huit coups horizontaux de gauche à droite.

Enfin le lissage en trois coups verticaux de haut en bas (schéma 1)

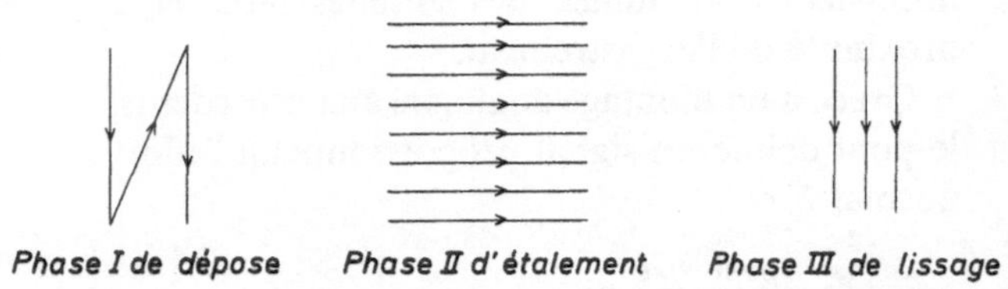

Nous avons fait réaliser 5 essais par un peintre professionnel et 7 essais par 5 amateurs différents avec un rouleau standard de 18 cm de longueur à poil demi long synthétique.

La peinture utilisée est une peinture en phase acqueuse ayant une viscosité apparente, mesurée à l'aide d'un viscomètre rotatif type Drage, de 43 poises à 20 °C.

Nous donnons (fig. 5) un exemple d'enregistrement obtenu; nous voyons que les coups composant la phase de dépose nécessitent des efforts très différents alors que les phases d'étalement et de lissage semblent plus reproductibles.

Nous avons, dans toute la suite, examiné les efforts maxima; nous indiquons dans le tableau 1 les résultats obtenus en désignant par X l'effort normal, Y l'effort tangentiel horizontal, Z l'effort tangentiel vertical et par les indices I, II, III les phases d'application de la peinture.

Ce tableau montre qu'il existe une grande dispersion d'un essai à l'autre même pour un peintre professionnel; toutefois il est possible de tirer certaines conclusions générales.

L'effort normal est environ 4 fois plus grand que l'effort tangentiel dans toutes les phases;

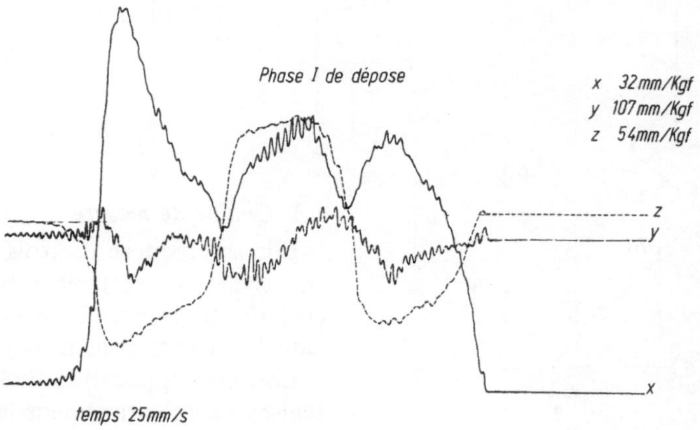

Fig. 5a. Vide texte

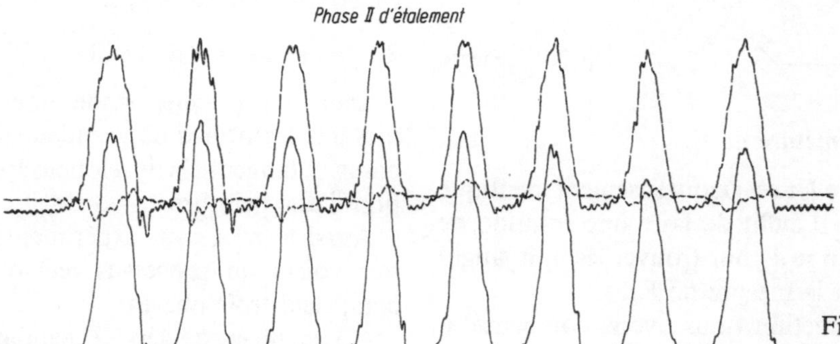

Fig. 5b. Vide texte

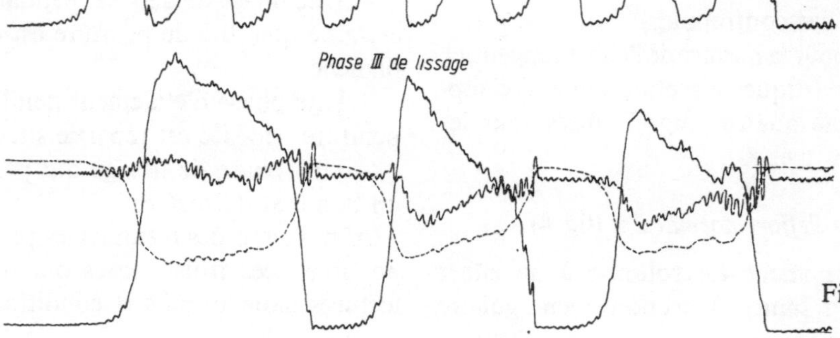

Fig. 5c. Vide texte

l'ordre de grandeur de l'effort normal este de 2 kgf alors que celui de l'effort tangentiel est de 0,5 kgf pour un professionnel, avec la peinture utilisée.

Pour des amateurs les efforts développés sont d'environ 20% supérieurs à ceux d'un professionnel.

Par la suite tous les essais ont été effectués par un peintre professionnel.

Tableau 1 a

	X_I	X_{II}	X_{III}	Z_I	Y_{II}	Z_{III}
Professionnel essai n° 1	3,379	2,433	2,679	0,740	0,543	0,576
Professionnel essai n° 2	2,820	1,266	1,322	0,694	0,347	0,405
Professionnel essai n° 3	1,727	1,453	1,486	0,499	0,419	0,472
Professionnel essai n° 4	3,333	2,418	2,025	0,840	0,558	0,449
Professionnel essai n° 5	2,600	1,781	2,183	0,651	0,488	0,504
Moyenne professionnel	2,771	1,870	1,939	0,684	0,471	0,481

Tableau 1 b

	X_I	X_{II}	X_{III}	Z_I	Y_{II}	Z_{III}
Amateur n° 1	3,640	2,624	2,727	0,913	0,643	0,670
Amateur n° 2	3,907	2,611	2,584	0,749	0,569	0,602
Amateur n° 3	3,061	2,244	2,800	0,740	0,546	0,653
Amateur	3,394	2,314	3,086	0,739	0,552	0,657
n° 4	2,466	2,457	3,245	0,587	0,585	0,694
Amateur	3,168	2,207	2,589	0,639	0,505	0,572
n° 5	3,887	2,080	2,266	0,742	0,477	0,301
Moyenne amateur	3,360	2,362	2,757	0,730	0,554	0,593

3. Influence du matériel d'application

Nous avons utilisé pour appliquer la même peinture trois rouleaux différents mais de même longueur; les différences obtenues ne sont pas significatives, les ordres de grandeurs des efforts moyens restant les mêmes.

Par contre (tableau 2) les résultats moyens obtenus avec une brosse d'une largeur de 4,5 cm montrent que l'effort normal est de l'ordre de 0,4 kgf, l'effort tangentiel de 0,160 kgf. L'effort tangentiel par cm, sensiblement le même au rouleau et à la brosse, est donc d'environ 30 à 40 gf.

Tableau 2

	X_I	X_{II}	X_{III}	Z_I	Y_{II}	Z_{III}
Rouleau n° 1	2,771	1,870	1,939	0,684	0,471	0,481
Rouleau n° 2	2,573	1,991	2,255	0,797	0,547	0,597
Rouleau n° 3	2,850	2,176	3,271	0,560	0,423	0,523
Moyenne brosse	0,420	0,505	0,353	0,150	0,164	0,160

4. Influence de la peinture

Vu la variation des efforts pendant les différentes phases de l'application d'une peinture, nous avons choisi pour comparer les résultats obtenus avec plusieurs peintures le processus expérimental suivant; nous peignons correctement environ un demi mètre carré puis immédiatement nous enregistrons sept coups de rouleaux appliqués horizontalement sur cette surface fraîchement peinte.

Les résultats du tableau 3 ont été obtenus avec un rouleau à poil demi long synthétique et sur un support non absorbant; nous avons indiqué la moyenne des efforts maxima obtenus avec quatre peintures de composition et de viscosité apparente très différentes. Nous voyons que la grandeur des efforts obtenus ne semble pas directement liée à la viscosité apparente de la peinture; par contre le rapport de l'effort normal sur l'effort tangentiel semble caractériser chaque type de peinture.

Tableau 3

Peinture	Viscosité en poises	X	Y	X/Y
Skinroc	104	0,92	0,39	2,3
Swip	43	1,69	0,47	3,6
Zinolit ée	6	0,95	0,21	4,4
Semi-gloss	5,7	1,27	0,41	3,1

5. Temps de reprise

Pour mesurer le temps de reprise nous avons peint une bande verticale d'environ 40 cm de largeur et repassé horizontalement deux coups de rouleau toutes les dix minutes sur cette bande; pendant toute l'expérience le rouleau est enveloppé dans un linge humide pour maintenir la peinture dans son état initial.

Nous avons tracé fig. 6 lamoyenne des efforts normaux et tangentiels en fonction du temps.

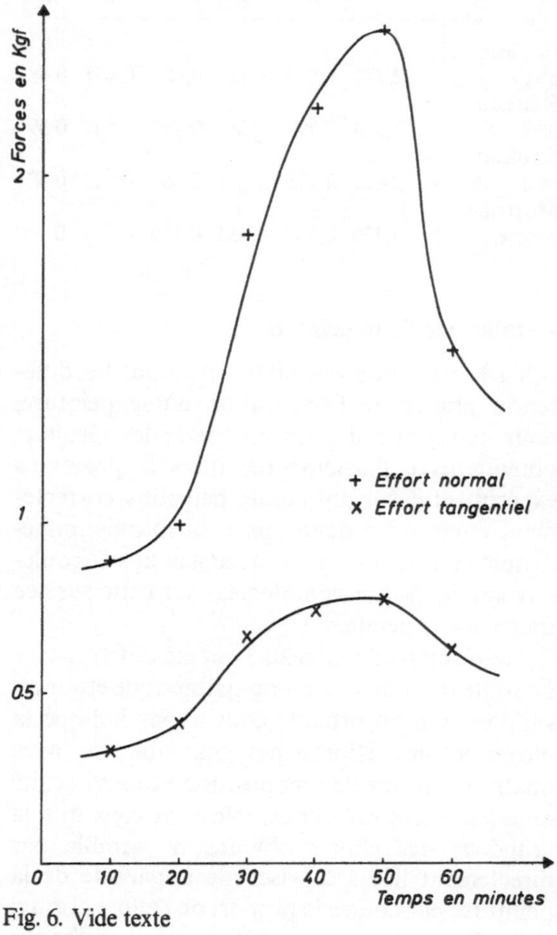

Fig. 6. Vide texte

Les courbes obtenues montrent une brusque augmentation des efforts au bout d'environ 30 min suivie d'une décroissance plus lente correspondant au séchage de la peinture.

L'essai est effectué sur une peinture oléoglycérophthalique dont la viscosité apparente, mesurée à l'aide d'un viscosimètre rotatif type *Ferranti*, est de 5,7 poises. Le temps de reprise trouvé est d'environ 30 min; ce temps est confirmé par les indications obtenues au banc *Erichsen*.

Résumé

Les premiers essais réalisés sur la planche à peindre nous ont permis de chiffrer l'ordre de grandeur des efforts d'application et de mettre en évidence l'importance pour une peinture donnée, du rapport de l'effort normal sur l'effort tangentiel.

Cette technique simple peut permettre un contrôle rapide, en cours de fabrication, de la brossabilité d'une peinture.

Toutefois la mise au point d'un applicateur automatique éliminant sur les résultats la dispersion due au peintre semble indispensable si l'on veut lier la brossabilité aux caractéristiques rhéologiques d'une peinture.

Adresse des auteurs:

Dr. *C. Oiknine*
I.U.T., Rue Engel-Gros
F-90 Belfort

Rheol. Acta 13, 241–246 (1974)

Laboratoire de Mécanique des Contacts, Institut National des Sciences Appliquées de Lyon, Villeurbanne (France)

Tribomécanique

Par M. Godet, D. Play, D. Berthe et J. Frene

Avec 5 figures

(Reçu p. p. le 27 octobre 1972)

I. Introduction

1. La tribologie

La tribologie (1–2) rassemble toutes les sciences qui touchent de près ou de loin à l'étude des contacts. Elle comporte des aspects chimiques, physiques, métallurgiques et des aspects mécaniques. Les sujets qu'elle traite sont mieux connus sous le nom de lubrification et de frottement. Nous n'aborderons ici que l'aspect mécanique de la tribologie, aspect souvent négligé et comme nous le verrons quelquefois en contradiction avec les vues des physiciens. Nous insisterons sur la recherche des éléments nécessaires pour le calcul de la force de frottement et proposerons une approche basée sur la mécanique des milieux continus.

2. Le contact

Tous les mécanismes (engrenages, roulements, articulations, etc.) comportent des liaisons qui dans la plupart des cas comprennent deux solides chargés en mouvement relatif. Les deux solides qui forment le contact sont le plus souvent bordés de films ou de revêtements créés soit:

– Artificiellement, c'est le cas des revêtements de surface (3, 4, 5).

– Par une action mécanique, c'est le cas du film d'huile qui est entraîné au contact et dont les lois de comportement sont relativement bien connues (6, 7).

– Par une action physico-chimique, c'est le cas de films d'oxydes (8), ou de films «extrême-pression» (9) nés de l'intéraction de l'ambiance et des surfaces solides.

Selon les conditions au contact, ces films peuvent soit complètement, soit seulement partiellement recouvrir les surfaces des solides. Ce cas a été étudié pour les oxydes (10) et l'huile (11). Ils peuvent également se former et s'éliminer au cours du temps (12).

Un contact réel peut enfin comprendre un nombre élevé (13) de films superposés, le plus souvent de nature et d'épaisseur aujourd'hui encore peu connues.

II. Le problème mécanique

1. La force de frottement

La solution générale au problème mécanique de la tribologie consiste à trouver l'opérateur $\mathscr{L}$ qui agissant sur les paramètres p_i définissant les aspects mécaniques du contact, permettra de calculer le torseur action de contact $\mathscr{C}$. Nous avons $\mathscr{C} = \mathscr{L}\{p_i\}$.

Nous retiendrons parmi ces paramètres:

a) La charge normale N;

b) la cinématique U des solides;

c) la forme géométrique G du contact;

d) le comportement mécanique (ou rhéologique) R_1, $R_2 \ldots R_i \ldots R_n$ des n corps formant le contact;

e) les conditions de déformation K_D et de rupture K_R des n matériaux;

f) les conditions d'adhérence entre deux matériaux en contact.

Précisons qu'il ne s'agit pas d'étudier ici une avarie particulière, telle que le grippage (14), ou l'écaillage (15) d'un système, mais uniquement d'obtenir une expression de la force de frottement qui entre pour une part dans l'apparition de ces avaries.

2. Analyse des paramètres de l'étude

Dans les faits, les paramètres, ci-dessus, ne sont pas nécessairement constants. On sait par exemple que les propriétés rhéologiques des matériaux ainsi que leurs limites sous contraintes dépendent de la température et que, comme nous l'avons vu précédemment, l'épaisseur de certains films et donc la géométrie du contact varie dans le temps. Ces deux phénomènes se produisent même dans des essais menés à charge et à vitesse constantes. Une première approche se devra toutefois de les considérer comme invariables. Une seconde pourrait ne traiter que les aspects transitoires inhérents à la formulation du problème avec paramètres fixes et d'ignorer les effets transitoires introduits par la variation de ces paramètres. Pour illustrer cette distinction, nous considérerons, par exemple, des retards de phase introduits par la nature visqueuse ou viscoélastique d'un matériau, mais non des différences dans ces retards provoquées par une modification des constantes, qui caractérisent le comportement visqueux ou visco-élastique du matériau.

3. Le modèle mécanique

Une première approche n'abordera également que le problème plan du contact à trois corps décrit dans la fig. 1. Dans cette schématisation, se trouvent deux solides A et B et un film intercalaire F. Tous trois sont susceptibles de supporter l'action de contact normale $\bar{N}$ dans les conditions de fonctionnement imposées. En dehors de cette condition, leurs comportements rhéologiques sont quelconques. Les méthodes utilisées dans l'étude du contact à trois corps pourront ultérieurement être élargies à l'étude du contact à n corps.

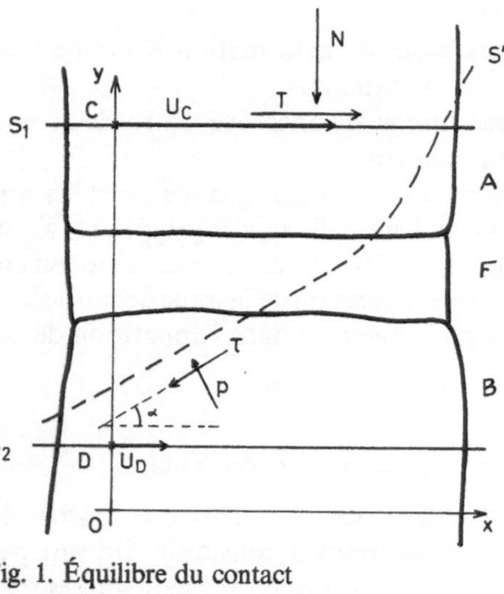

Fig. 1. Équilibre du contact

Considérons les points C et D, d'abcisse $x=0$ appartenant aux surfaces S_1 et S_2 en translation, de longueur unitaire dans le sens des Z et animés de vitesse $\bar{U}_c$ et $\bar{U}_D$ constantes, que dans cette première approche nous choisirons parallèles à l'axe des x. Cette condition est représentative d'un état de glissement pur. Une charge normale N est appliquée au système et soit T l'effort tangentiel – que l'on cherche à calculer – nécessaire pour maintenir la différence de vitesse $\bar{U}_c - \bar{U}_D$. Considérons une ligne S courbe le long de laquelle une déformation se produit et qui définira une frontière à l'intérieur du corps «composite» A, B et F. En négligeant les forces d'accélération, nous avons à tout moment et quelque soit S:

$$N - \int_S p \cdot \cos\alpha \cdot dS + \int_S \tau \cdot \sin\alpha \cdot dS = 0$$

$$T - \int_S p \cdot \sin\alpha \cdot dS - \int_S \tau \cdot \cos\alpha \cdot dS = 0 \qquad [1]$$

où p et τ, les pressions et les contraintes de cisaillement qui agissent sur S et α, l'angle de ces contraintes avec le repère $x\,y$, sont variables le long de S. La description complète de l'état d'équilibre exige l'écriture de l'équation du moment qui n'apporte pas ici d'information supplémentaire.

Les relations [1] montrent que l'action de contact $\bar{N} + \bar{T}$ ou force extérieure se calcule à partir du champ de contraintes à l'intérieur du corps. A son tour, ce champ de contraintes dépend du champ de déformations des corps. Enfin ce champ de déformations doit être compatible avec la définition cinématique imposée par les vitesses $\bar{U}_C$ et $\bar{U}_D$ des points C et D. Le problème revient donc à trouver le champ de déformation. Nous distinguerons deux cas: dans le premier, la déformation se produira entièrement dans les matériaux du contact, dans le second, la déformation se produira en partie à la frontière du film F.

III. Le champ de déformation

1. Les déformations se produisent dans les matériaux

Il n'y a pas de glissement à la frontière du film, la courbe de vitesse ou fonction $U(y)$ est continue, mais sa dérivée ne le sera qu'exceptionnellement. La fig. 2 montre quelques exemples donnant des courbes de $U(y)$ obéissant à la condition d'adhérence parfaite à la paroi, condition que l'on écrit en mécanique des films minces visqueux et en mécanique des fluides en général. Dans la courbe 1, la déformation est entièrement contenue dans le film. Dans la courbe 2, la déformation est répartie dans les trois corps. Dans la courbe 3, une rupture amorcée dans le film accommode l'essentiel des déformations.

L'établissement des champs de déformations de la fig. 2 suppose connues les conditions de déformation des trois matériaux du contact. Un matériau peut changer de configuration à la suite de déformations élastiques, plastiques ou visqueuses ou d'une combinaison de ces déformations. Il peut également se rompre. L'amorce de ces déformations se produit quand une condition de la forme:

$$K = f(\sigma_{ij}) \quad [2]$$

où σ_{ij} représente les contraintes en un point, est atteinte. Pour généraliser le problème, nous dirons qu'il existe pour les corps *A*, *B* et *F* trois constantes K_A, K_B et K_F correspondant respectivement à une condition de la forme donnée dans la relation [2] – c'est-à-dire:

$$K_A = f_A(\sigma_{ij}); \quad K_B = f_B(\sigma_{ij}); \quad K_F = f_C(\sigma_{ij})$$

pour lesquelles les matériaux formant les corps de *A*, *B* et *F* peuvent se déformer suffisamment pour accommoder la totalité ou une partie non négligeable de la différence de vitesse $\bar{U}_C - \bar{U}_D$, quelqu'en soit le mode de déformation.

Le problème revient donc à calculer en tous points du contact les fonctions $f_A(\sigma_{ij})$, $f_B(\sigma_{ij})$ et $f_C(\sigma_{ij})$ pour une action de contact et une cinématique *U* données. Lorsque les valeurs trouvées sont supérieures à la valeur *K* limite, on aura isolé des zones de déformations possibles. Ces zones de déformation transformeront dans certains cas, la répartition de contraintes et ce n'est qu'au terme d'un calcul itératif qu'il sera possible d'obtenir par la relation [1] la valeur de la composante *T* de l'action de contact.

Considérons la section verticale obtenue en un point $x = x_0$ du contact (fig. 3). On retrouve les trois corps *A*, *B* et *F* et on porte les trois limites K_A, K_B et K_F des matériaux, soit f_A, f_B et f_F les trois courbes donnant l'allure des fonctions $f(\sigma_{ij})$ de la relation [2] pour les trois matériaux. On voit que, dans l'exemple choisi, la déformation ne se produira pas dans le corps *A*, mais qu'elle se produira dans tout le film *F* et dans une partie du corps *B*.

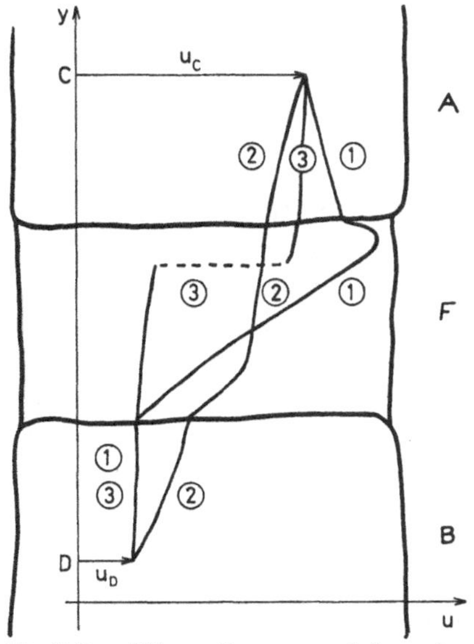

Fig. 2. Les déformations se produisent dans les matériaux

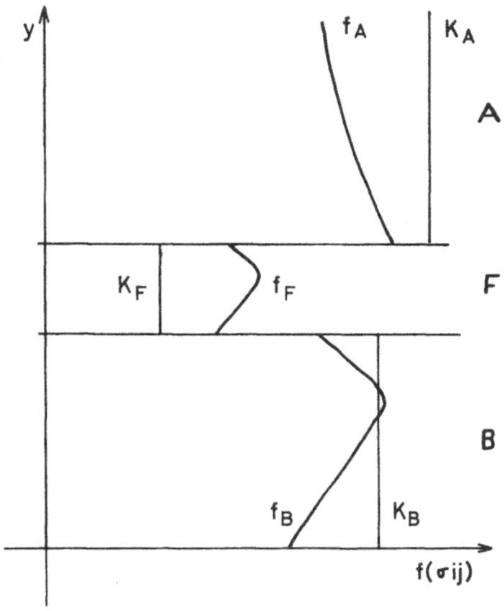

Fig. 3. Détermination des zones de déformation

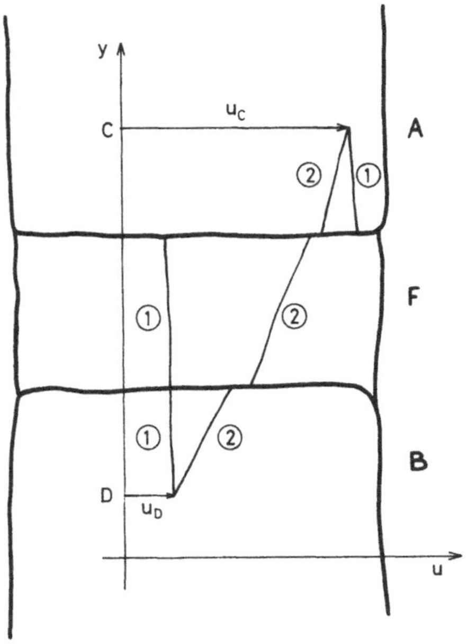

Fig. 4. Les déformations se produisent à l'intérieur et à la frontière des matériaux

2. Les déformations se produisent en partie à la surface du film

Deux exemples de champ de vitesse qui respectent cette condition sont donnés dans la fig. 4. Dans le premier cas, la différence de vitesse est entièrement localisée à la frontière, dans le second

16*

cette différence est répartie entre les matériaux et les deux frontières.

La condition d'égalité de déplacement à la frontière utilisée dans le cas précédent est remplacée par une valeur maximale du taux de cisaillement supportable à la paroi au-delà de laquelle le glissement intervient. On retrouve ici la condition souvent utilisée en mécanique des solides.

L'existence de contacts tribo-rugueux pour lesquels les surfaces réelles sont faibles devant les surfaces apparentes du contact complique l'analyse du problème, mais ne modifie pas l'approche utilisée ci-dessus. On substitue à la fonction τ_{xy} continue des valeurs discrètes de τ statistiquement distribuées sur une frontière moyenne.

IV. Les conditions aux limites du problème

Considérons un point *P* situé sur la courbe *S* de la fig. 1 que nous avons retracée dans la fig. 5 pour qu'elle puisse correspondre à une surface de déformation réelle. Dans les faits, cette surface peut être unique et donc parfaitement définie – c'est par exemple le cas dans les matériaux non ductiles – ou multiple et indéfinie – c'est le cas des matériaux visqueux. Dans le cas d'une surface unique la relation [1] est appliquée sur cette surface. Dans le cas de surface multiple la relation [1] est appliquée sur une ou plusieurs surfaces arbitrairement choisie. C'est ainsi qu'en mécanique des films minces visqueux et pour des raisons techniques évidentes, on choisit pour *S* les surfaces S_1 et S_2 qui bordent ce film visqueux. Les contraintes au point *P* peuvent être calculées dans les cas simples si les contraintes et les déplacements qui agissent sur les bords d'un domaine contenant le point *P* et les caractéristiques rhéologiques du matériau du domaine sont connus. Isolons le domaine *ABCD*. Nous voyons que les contraintes en *P* dépendent en particulier des conditions sur *AB* et *CD*.

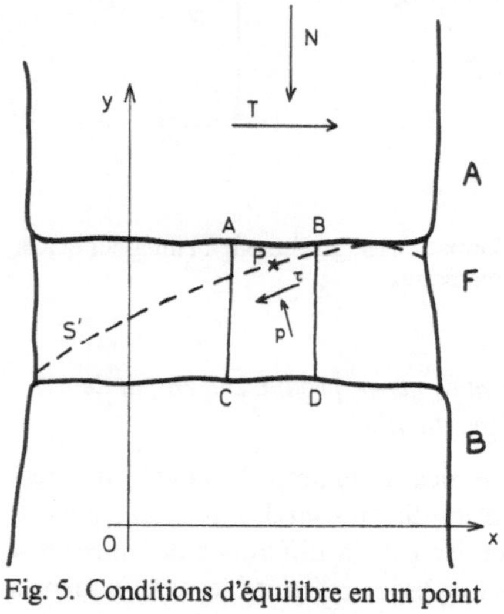

Fig. 5. Conditions d'équilibre en un point

Ces conditions qui dépendent des conditions d'adhérence entre le film *F* et les solides *A* et *B* sont gouvernées par des phénomènes physiques. Mais comme nous l'avons vu plus haut, elles sont traduites en termes mécaniques soit par la condition de non-glissement à la paroi, soit par la condition de taux de cisaillement maximal. Dans une direction normale à la paroi, la résistance à la rupture de la liaison film *F* – solide *A* ou *B* n'entre pas en ligne de compte en glissement pur, car la contrainte est de compression. Cette condition par contre intervient dans le roulement.

L'essentiel est donc de comprendre que les contraintes τ et p au point *P* situées sur la ou l'une des surfaces *S* sur laquelle on calcule la force tangentielle *T* par la relation [1] dépendent du champ de contraintes sur les frontières *AB* et *CD* du domaine qui se présentent comme des conditions aux limites du problème.

Il n'est donc plus possible, comme le veulent les théories courantes, de considérer la force de frottement comme étant constituée de deux termes, un terme d'adhérence et un de déformation. Par contre, les contraintes en *P* et donc la force de frottement dépendent des conditions d'adhérence des matériaux, comme la valeur d'une fonction dépend des conditions aux limites qu'on lui impose.

V. Applications et conclusions

1. *Le modèle de trois corps*

L'utilisation du modèle des trois corps présenté dans la fig. 1 peut surprendre quand il s'agit d'aborder le problème particulier du frottement (16) ou même le problème général de la tribologie. Il est toutefois utilisé depuis longtemps en mécanique des films minces visqueux, c'est-à-dire dans la théorie de la lubrification. Dans la lubrification classique, on considère en effet deux corps *A* et *B* parfaitement rigides séparés par un film *F* visqueux. En lubrification élastohydrodynamique (18) on considère deux corps *A* et *B* élastiques donc déformables séparés par un film *F* visqueux. En lubrification plastohydrodynamique (19) on considère deux corps *A* et *B* l'un rigide l'autre plastique séparés par un film visqueux, etc. ... S'il est clair qu'un gros effort a été fait pour calculer la portance hydrodynamique – une charge permise pour une épaisseur de film donnée – dans chacun des cas cités ci-dessus on s'oriente de plus en plus vers le calcul de l'effort de frottement ou effort tangentiel T.

2. Conditions pour qu'un matériau soit ou non lubrifiant

L'analyse précédente nous permet de ne plus juger globalement de la performance d'un lubrifiant mais de préciser un certain nombre de conditions auxquelles ce lubrifiant doit satisfaire pour être utilisable quand des matériaux de géométrie, de rhéologie et d'activité physico-chimique donnés, sont animés de vitesses et soumis à des charges connues.

Première condition:
Un film de lubrifiant doit pouvoir être formé

Cette condition est évidemment satisfaite pour les lubrifiants artificiels tels que les revêtements de surface disposés sur les surfaces avant tout contact. Elle l'est souvent quand le film est formé par action mécanique comme dans le cas des films hydrodynamiques ou films minces visqueux, toutefois l'état d'avancement de la théorie de la mécanique des films minces visqueux ne permet pas encore de garantir la formation de ces films. Elle l'est enfin si un film lubrifiant peut être engendré par action physico-chimique entre les surfaces solides et l'ambiance.

Deuxième condition:
Le film formé doit être maintenu au contact

Cette condition suppose que le film artificiel résiste à l'usure. Elle suppose que le film visqueux puisse résister aux pressions de contact. Elle suppose que le taux de formation du film créé par interaction avec l'ambiance soit plus élevé que son taux d'élimination ou de destruction.

Les conditions dans lesquelles un film visqueux formé peut être maintenu sont relativement bien décrites en mécanique des films minces visqueux. Les films artificiels et les films formés par interaction, dont le comportement rhéologique est le plus souvent mal connu, ne peuvent se maintenir sans se rompre que s'ils sont susceptibles de supporter les champs de déformation que subissent en surface les matériaux supports et les contraintes thermiques qui naissent de l'élévation de température au contact.

Troisième condition:
Le film formé et maintenu doit présenter ou créer une zone ou un plan de faible résistance au cisaillement

Dans la formule [1] l'angle α étant généralement petit le terme τ doit être faible pour que la composante N de l'action de contact ou force de frottement soit faible. Comme nous l'avons vu, ce taux de cisaillement se produit soit à la frontière soit dans la masse du film soit aux deux à la fois. Le film doit être de composition telle qu'au moins une de ces deux conditions soit remplie.

3. Les lubrifiants liquides et solides

Il est intéressant de constater, qu'un film d'huile se forme relativement facilement, qu'une fois formé, il est susceptible de resister à tous les champs de déformation que lui imposent ses supports et qu'il offre une faible résistance au cisaillement. Il se détruit toutefois à haute température. L'huile répond donc aux 3 conditions qu'un lubrifiant doit remplir à l'intérieur d'un intervalle de température donnée. Une approche semblable appliquée aux lubrifiants solides qu'ils soient naturels (oxydes) ou artificiels (revêtements) se heurte à des problèmes beaucoup plus considérables. En effet dans la mesure où la part solide dans le comportement de ces films est importante sinon prépondérante on se heurte immédiatement à toutes les limites des solides. Dans un corps composite tel qu'il s'en trouve dans les contacts, ces limites sont rapidement atteintes à cause des différences dans les modules de déformation et les coefficients de dilatation thermique qui existent dans le film et ses supports.

Littérature

1) *Godet, M.*, Les fondements mécaniques de la tribologie. Journées d'étude du G.A.M.I. sur l'usure (Paris 8–10 septembre 1970).

2) *Godet, M.*, Sur le regroupement des problèmes qui traitent de la lubrification et du frottement. C. R. Ac. Sciences (Paris) **273**, série A, 999–1002 (1971).

3) *Caubet, J. J.*, Théorie et pratique industrielle du frottement (Paris 1964).

4) *Caubet, J. J.*, Hydromécanique et frottement les traitement de surface contre l'usure (Paris 1967).

5) *Schaeffer, G.* et *M. Godet*, Etude à haute température des paliers lisses fortement chargés. Journées sur le frottement et l'usure du G.A.M.I. (Paris septembre 1968).

6) *Fantino, B., J. Frene* et *M. Godet*, C. R. Ac. Sciences (Paris) **272**, 691–693 (1971).

7) *Berthe, D.* et *M. Godet*, C. R. Ac. Sciences (Paris) **272**, 1010–1013 (1971).

8) *Tenwick, N.* and *S. W. E. Earles*, Wear **17**, 381–389 (1971).

9) *Deyber, P.*, Contribution à l'étude du coefficient de frottement et de la température en frottement mixte. Thèse Fac. Sciences (Lyon 28 septembre 1970).

10) *Schaeffer, G.*, Etude du frottement sec et lubrifié à haute température. Thèse Fac. Sciences (Lyon 6 juin

11) *Godet, M.*, L'apport hydrodynamique dans le frottement lubrifié. Thèse Fac. Sciences (Paris 20 février 1967).

12) Référence 9 – page 57.

13) *Groszek, A. J.*, Role of absorption in liquid lubrication. Interdisciplinary approach to liquid lubricant technology NASA Symposium Cleveland, U.S.A. (Janv. 1972).

14) *Rozeanu, L.*, A model of seizure. Conference on the limits in lubrication (Londres juillet 1971).

15) *Scott, D.* and *J. Blackwell*, Wear **17**, 323–333 (1971).

16) *Courtel, R.*, Cahiers du groupe français de rhéologie **2**, n° 3 (1970).

17) *Tipei, N.*, Theory of lubrication (New York 1962).

18) *Dowson, D.* et *G. R. Higginson*, Elastohydrodynamic lubrication (Oxford 1966).

19) *Cheng, H. S.*, Plastohydrodynamic lubrication. Friction and lubrication in metal processes. A.S.M.E., p. 69 (1966).

Adresse des auteurs:

M. Godet et al.
Laboratoire de Mécanique des Contacts
Institut National des Sciences Appliquées de Lyon
F-69621 Villeurbanne (France)

Rheol. Acta 13, 247–251 (1974)

From the Nezu Chemical Institute, Musashi University, Tokyo (Japan)

Flow properties of smectic liquid crystals*)

By B. *Tamamushi*

With 12 figures and 2 tables

(Received October 27, 1972)

Introduction

Liquid crystals or mesomorphic systems are the subject of interest from the viewpoint of rheology. Since the discovery by *Lehmann* (1) in the end of the last century, pretty many investigations concerning flow properties of liquid crystals, especially viscosity studies on these substances have been carried out (2). However, relatively few data on flow properties of smectic liquid crystals, i.e. soap-like substances, in their molten states have been reported. In the present paper the flow properties of some ammonium salts of higher normal carboxylic acids in their mesomorphic states will be described and discussed.

Experimental and result

Ammonium salts of higher normal carboxylic acids (myristic, palmitic, stearic and oleic acids) were prepared by passing dry ammonia through ethanol or methanol solution of the respective acid of high purity (over 95%). The salt precipitate formed was recrystallized from acetone and the product was kept in a deciccator filled with ammonia.

These salts have commonly two melting points, namely, one is the temperature for the phase change of crystals → smectic melt (C–S point) and another is the temperature for the phase change of smectic melt → isotropic liquid (S–L point). Table 1 and 2 indicate the data for the free acids and their salts applied in our study (3).

Flow properties of the melts of these salts were measured by applying a rotational viscometer (L-II type, Iwamoto Co.) in the range of shear rate of 10–500 sec^{-1} and also in the range of temperature from below to above

Table 1

Substance	Melting point (°C)
Myristic acid	56.5 ± 1
Palmitic acid	62.6 ± 1
Stearic acid	69.4 ± 1
Oleic acid	14

*) Another version of this paper appeared in Biorheology, **10**, 239 (1973) in memory of *Aharon Katzir-Katchalsky* by invitation.

Table 2

Substance	C–S point (°C)	S–L point (°C)
Ammonium myristate	80.7 ± 0.2	116.5 ± 0.3
Ammonium palmitate	87.8 ± 1	113 ± 1
Ammonium stearate	91.5 ± 1	115.5 ± 0.5
Ammonium oleate	41.6 ± 1	69.4 ± 1

the S–L point for each substance. The ammonium salts of carboxylic acids are liable to be decomposed in ordinary atmosphere, especially at higher temperatures, and therefore the care was taken to avoid this effect as possible by carrying out the measurement in a short time and with a fresh sample in each series of measurement. The result obtained is illustrated in figs. 1–12.

Figs. 1–3 show the relation between shear rate and shear force at various temperatures, the relation between viscosity and shear force at various temperatures, and the relation between viscosity and temperature at various shear rates for ammonium myristate, respectively.

Figs. 4–6 show the relation between shear rate and shear force at various temperatures, the relation between viscosity and shear force at various temperatures, and the relation between viscosity and temperature at various shear rates for ammonium palmitate, respectively.

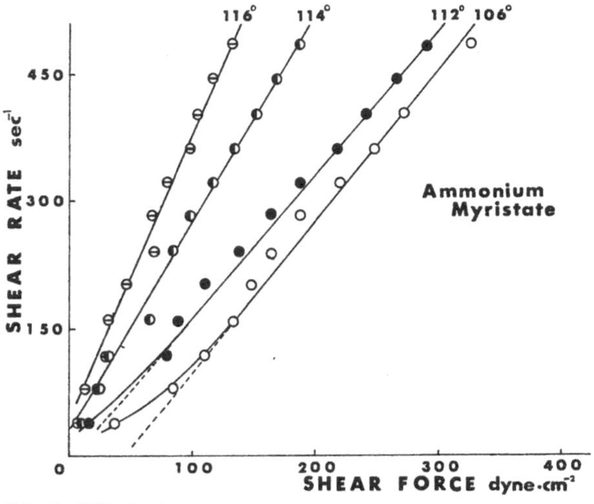

Fig. 1. Vide text

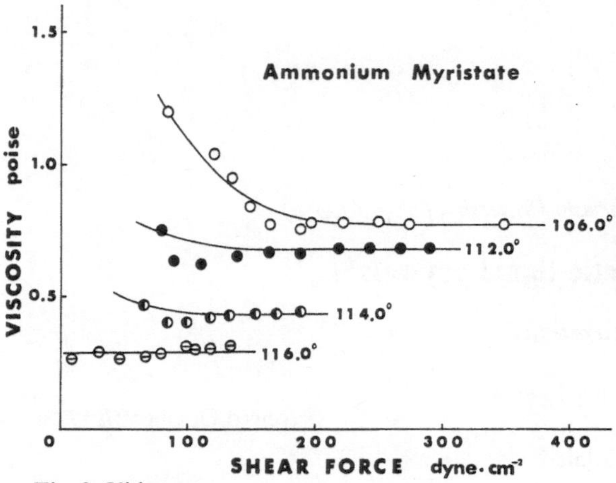

Fig. 2. Vide text

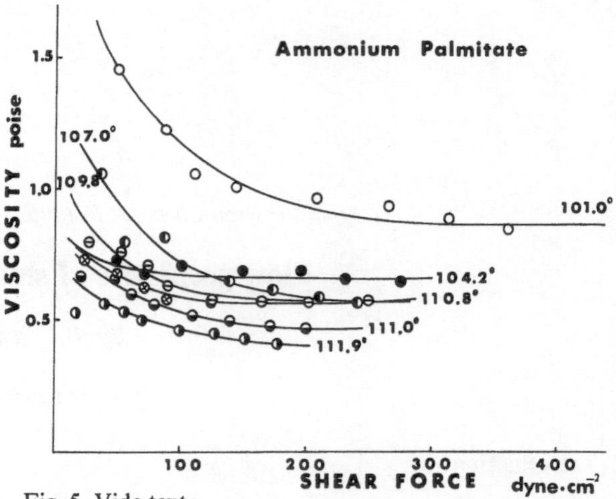

Fig. 5. Vide text

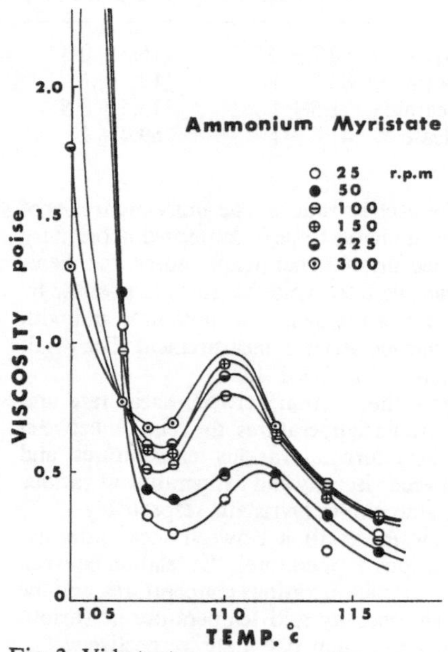

Fig. 3. Vide text

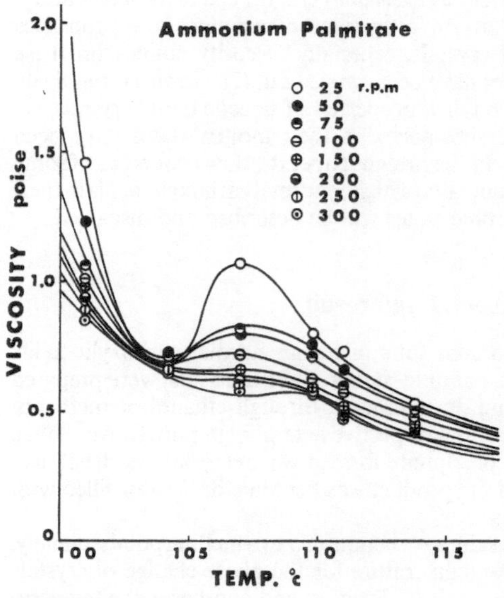

Fig. 6. Vide text

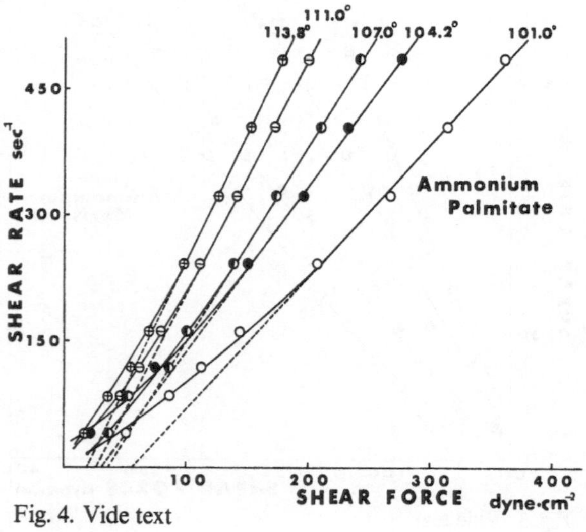

Fig. 4. Vide text

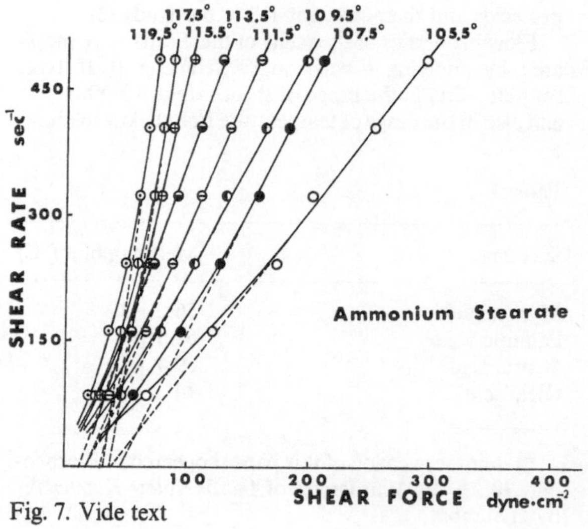

Fig. 7. Vide text

Figs. 7–9 show the relation between shear rate and shear force at various temperatures, the relation between viscosity and shear force at various temperatures, and the relation between viscosity and temperature at various shear rates for ammonium stearate, respectively.

Figs 10–12 show the relation between shear rate and shear force at various temperatures, the relation between viscosity and shear force at various temperatures, and the relation between viscosity and temperature at various shear rates for ammonium oleate, respectively.

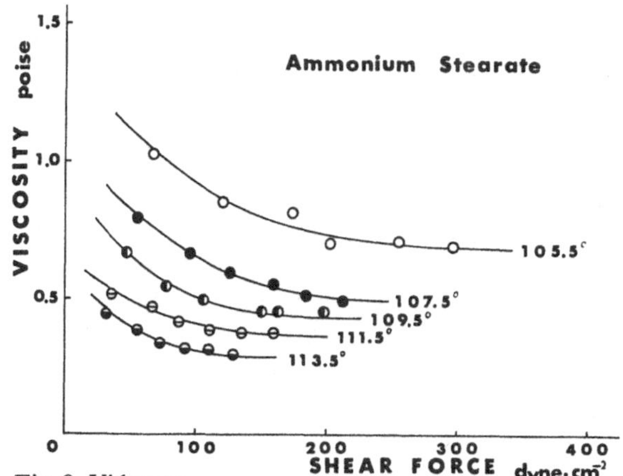

Fig. 8. Vide text

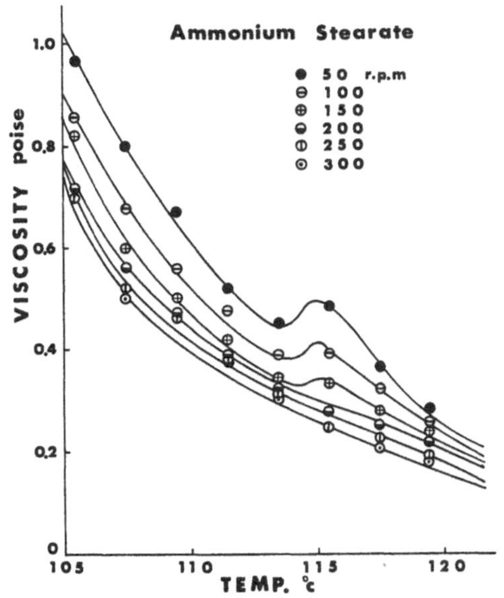

Fig. 9. Vide text

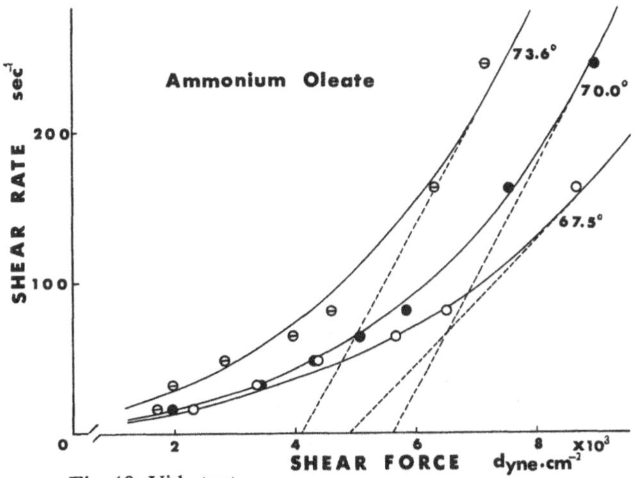

Fig. 10. Vide text

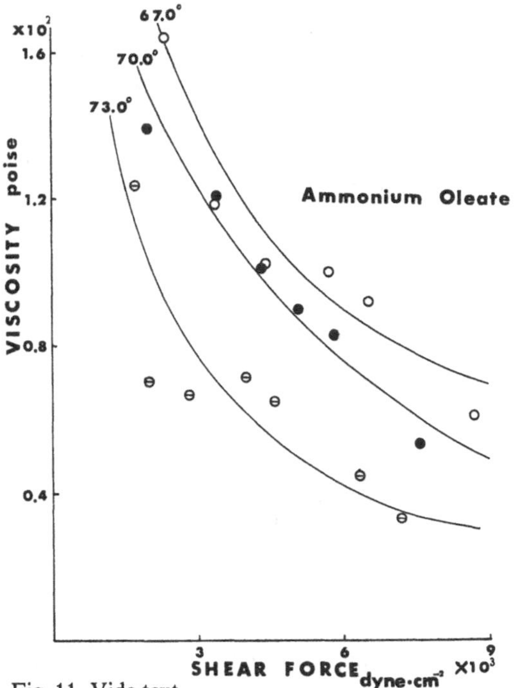

Fig. 11. Vide text

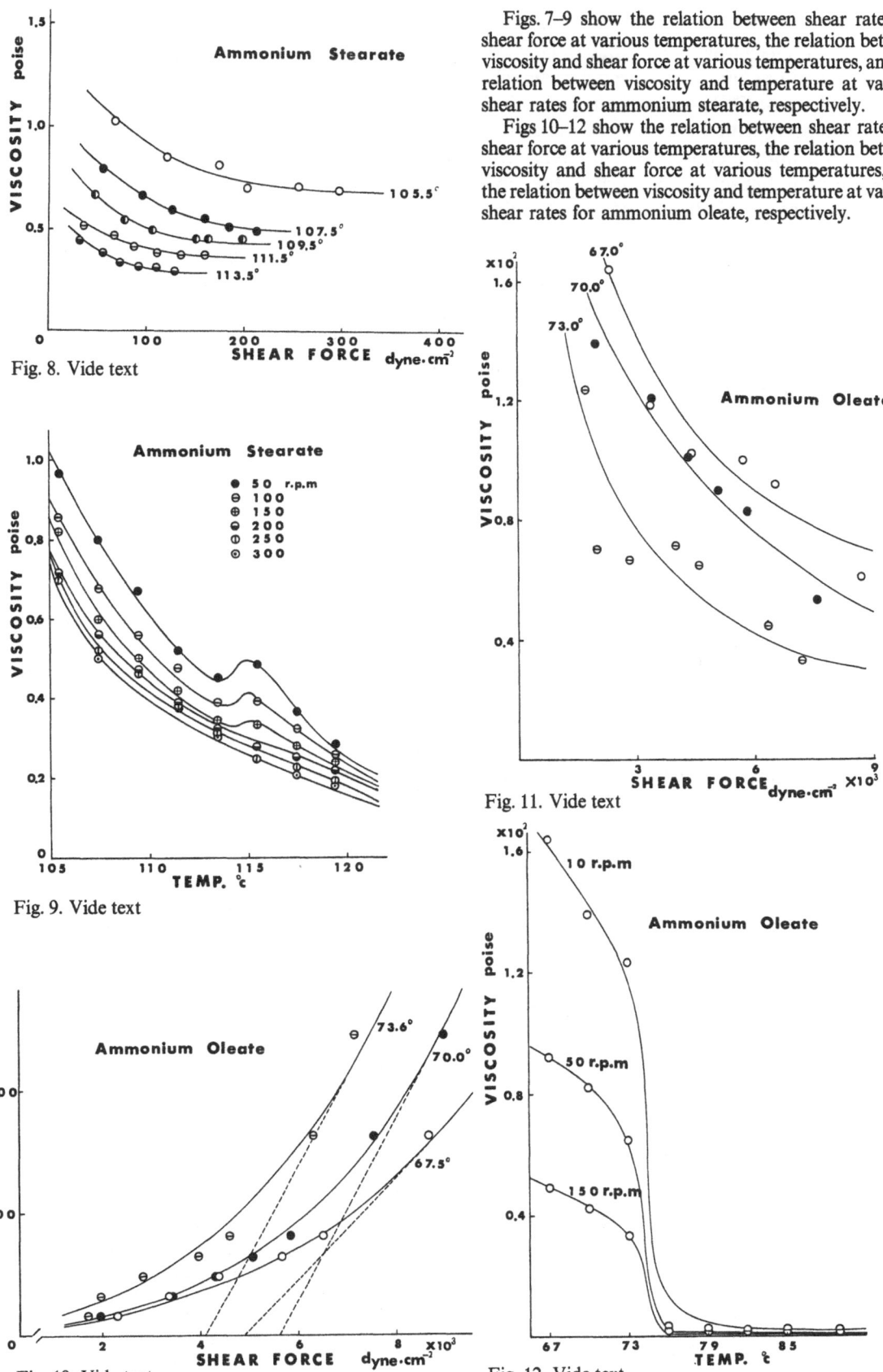

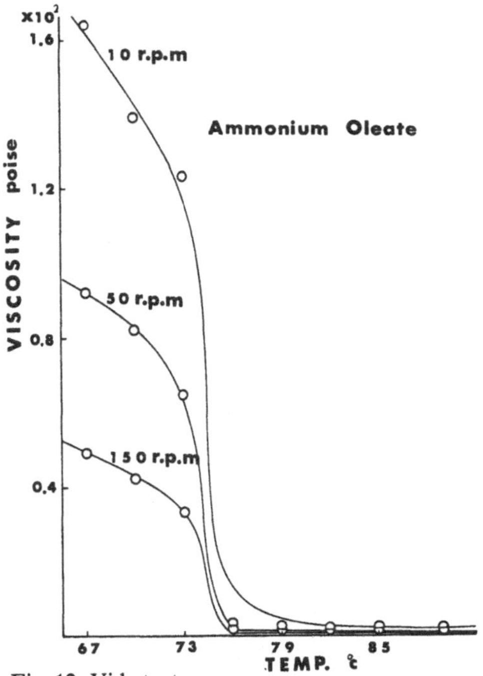

Fig. 12. Vide text

Discussion

It is to be noticed from table 1 and 2 that while the melting point of normal saturated carboxylic acids as well as the C–S point of their ammonium salts rise stepwise with the increase of the carbon number in a homologous series, the S–L point of their ammonium salts falls nearly in the same range of temperature, independent of the number of carbon atoms. This may be due to the partial decomposition of these ammonium salts in the temperature range of C–S and S–L points.

From fig. 1, 4, 7 and 10 we can notice that the flow of the mesomorphic melts are generally non-*Newton*ian and plastic or pseudoplastic, as far as the measuring temperature is lower than the S–L point. However, the *Bingham*s yield value becomes smaller with the rise of temperature and at temperatures above the S–L point, the flow of isotropic liquids becomes normal, i.e. *Newton*ian.

In figs. 2, 5, 8, and 11 it is shown that viscosity decreases with increasing shear force and that the rate of its decrease becomes more gradual with the rise of temperature and finally viscosity becomes almost constant. It is here to be remarked that the similar relationship has been found with ethyl p-azoxybenzoate, another typical smectic liquid crystal, and also with other kinds (nematic and cholesteric) of liquid crystals (4).

In figs. 3, 6, and 9 it is remarkable that in the viscosity – temperature curve a maximum appears at a certain temperature which nearly coincides with the S–L point of the given substance and also that the height of the maximum is dependent on the shear rate. In fig. 12, owing to the smaller temperature range studied, no maximum appears but a sharp decrease in viscosity is noticed.

Such abrupt rise and fall in viscosity near the transition temperature of the anisotropic → isotropic phase change as here illustrated is not only characteristic for smectic mesophases but also found in other kinds of liquid crystals, especially cholesteric ones (5).

However, as far as smectic liquid crystals concern, they are characterized by such a structure as that long-shaped molecules are not only parallel oriented to each other but their end groups lie in the same plane and so that they produce a layer structure. In this structure molecules have more restricted freedoms of motion – two in translational and one in rotational – than other types of liquid crystals. The facts, that smectic liquid crystals have in general relatively high viscosities and that they show such plastic or pseudoplastic flow as above found, are to be attributed to their higher ordered structures.

Under high shear forces, single molecules or molecular aggregates (swarms) may orient in such a manner as the axes of long-shaped molecules lie parallel to each other in the direction of the shear force, so that internal friction, i.e. viscosity becomes smaller. As temperature rises, the freedom of translational motion in the direction of molecular axis will increase, which will also cause the decrease in viscosity. However, when temperature attains the S–L point, the rotational motion in the direction perpendicular to the molecular axis may suddenly take place and accordingly there will be strong interactions among neighboring molecules which will cause a sudden increase in viscosity. Such an effect may however be lessened by higher shear rates.

On the basis of such considerations as above stated, which remain still qualitative, the experimental facts here found for smectic liquid crystals can be explained. A more or less quantitative theory concerning the viscosity of liquid crystals as a function of molecular rotations and translations was given by *Herzog* and *Kudar* (6). But, a more precise theoretical treatment of the problem should be worked out in future.

Lastly, it is to be remarked that, as far as ammonium salts of normal carboxylic acids concern, there is a possibility that the materials under experiment are not quite pure, but they are rather the mixtures of salts and free acids, the latter of which being produced by the decomposition of the former. Nevertheless, it is supposed that the characteristic smectic structure should remain even in such mixtures, and therefore the discussion above developed may not be essentially altered by this experimental condition.

Zusammenfassung

Es wurden die Fließeigenschaften der mesomorphen Schmelze von Ammonium-alkylcarboxylaten gemessen. Es wurde gefunden, daß diese mesomorphen Schmelzen im allgemeinen nicht-*Newton*sche Fließtypen zeigen und auch daß sie anomale Viskositätstemperaturkurven geben. Diese Kurven haben ein scharfes Maximum in der Nähe der Umwandlungstemperatur (anisotrop → isotrop), wenn die verwendete Scherspannung nicht so groß ist. Solche charakteristischen Eigenschaften wurden aufgrund der molekularen Bewegungen in den mesomorphen Schmelzen diskutiert.

References

1) *Lehmann, O.*, Z. physik. Chem. **4**, 462 (1889); **5**, 427 (1890).

2) *Brown, G. H.* and *W. G. Shaw*, Chem. Rev. **57**, 1049 (1957). – *Porter, R. S.* and *J. F. Johnson*, The Rheology of Liquid Crystals. In: Rheology IV, Chapt. 5, ed. by *F. Eirich* (New York 1968).

3) cf. *Lawrence, A. C. S.*, Trans. Farad. Soc. **34**, 660 (1938); The Faraday Society's General Discussion on Liquid Crystals and Anisotropic Melts, p. 1008 (1933).

4) *Ostwald, Wo.*, The Faraday Society's General Discussion on Liquid Crystals and Anisotropic Melts, p. 1002 (1933). – *Porter, R. S.* and *J. F. Johnson*, loc. cit.

5) *Sakamoto, K.*, *R. S. Porter*, and *J. F. Johnson*, Proc. 2nd Intern. Conf. on Liquid Crystals, Part 1, 237 (1968). – *Fukada, E.*, unpublished, private communication.

6) *Herzog, R. O.* and *H. Kudar*, The Faraday Society's General Discussion on Liquid Crystals and Anisotropic Melts, p. 1006 (1933).

Author's address:
Prof. Dr. *B. Tamamushi*
Nezu Chemical Institute
Musashi University
1–26 Toyotama, Nerimaku
Tokyo (Japan)

Rheol. Acta 13, 252–259 (1974)

From the Department of Polymer Chemistry, Faculty of Engineering, Kyoto University, Kyoto (Japan)

Frequency dispersion of strain-induced crystallization coefficient of natural rubber vulcanizates in sub-sonic range

By H. Hiratsuka, M. Hashiyama, S. Tomita, and H. Kawai

With 3 figures and 1 table

(Received October 27, 1972)

Introduction

The orientation crystallization behavior of linear polymers is significant not only from basic view point of crystallization kinetics but also from engineering view point, such as fiber spinning and its resulted properties. Since early works by *Singer* et al. (1, 2) and *Thiessen* (3), the investigation of the rate of orientation crystallization of natural rubber vulcanizates has been performed by several authors using different techniques over considerably wide ranges of crystallization temperature and degree of stretching of bulk specimens (4–8).

When the temperature is low enough to slow down the crystallization rate, the usual dilatometric technique may be useful for following the crystallization behavior (6–8). However, for example, when the vulcanized rubber is highly stretched at room temperature, the strain-induced orientation crystallization seems to be completed within a period of less than one second, as seen from some experimental results of early works by *Singer* et al. (1, 2) using a dynamic X-ray photographic technique and of recent works by *Stein* and *Yau* (9, 10), *Mitchell* and *Meier* (11), and *Dunning* and *Pennells* (12), using a high-speed light scattering technique, a thermal analysis technique, and an X-ray diffraction technique for continuously stretched endless belts of the specimens, respectively.

In the previous paper of this series of study (13), the orientation crystallization behavior of natural and synthetic isoprene rubber vulcanizates was investigated qualitatively by some of the present authors using a dynamic X-ray diffraction technique (14) utilizing a narrow sector which synchronizes with tensile dynamic strain of specimen and enables the measurements of X-ray diffraction intensity distribution at a given phase angle of the dynamic strain of the specimen as function of temperature, frequency, tensile static strain, and dynamic strain amplitude. Some critical frequencies around 10 cps at 30 °C and 1 cps at 38 °C, beyond which the rubber molecules can not crystallize appreciably even under good enough strain condition for the orientation crystallization at lower frequencies, as well as anisotropy in the crystallization behavior with respect to the principal crystallographic axes were found.

In this paper, the above behavior of orientation crystallization of natural rubber vulcanizates will be investigated more quantitatively by replacing the narrow sector with four sets of semi-circular sectors. The semi-circular sectors are arranged so as to shift from each other by phase angle of $\pi/2$ radians and to measure two components of the dynamic X-ray diffraction intensity distribution from a given crystal plane, in-phase and out-of-phase components with respect to the dynamic strain of the specimen (14).

Preparation of test specimens

Table 1 shows the characteristics of five kinds of vulcanized specimens of natural rubber (RSS No. 3) differing in degree of crosslink, including the compound formula, vulcanization condition, and number average molecular weight of the network chains estimated from the thermostress behavior. Every specimen was cast into sheet from of 1 mm thickness by using a laboratory press during the vulcanization process under a condition given in table 1.

Experimental procedures

General principle of dynamic X-ray diffraction technique

Let us assume that there are N diffraction peaks from N kinds of crystal planes, which do not necessarily separate from each other but overlap in a linear manner. Let each of the diffraction peaks be represented by either *Lorentz*ian or *Gauss*ian, or any other function such as that the peaks are expressed in intensity distribution as function of $(2\theta - 2\theta^0_{B,j})/\beta^0_j$ and $(\varphi - \varphi^0_{c,j})/\gamma^0_j$ to give a symmetric function with respect to $2\theta^0_{B,j}$ and $\varphi^0_{c,j}$, respectively, where $2\theta^0_{B,j}$ and $\varphi^0_{c,j}$ are diffraction angle (twice the *Bragg* angle) and azimuthal angle, respectively, both giving a maximum intensity of the jth peak $I^0_{m,j}$, and β^0_j and γ^0_j are parameters representing the broadness of the diffraction peak and are the so-called half-width of the peak when the peak is given by *Lorentz*ian function. The superscription 0 means hereafter the static state with extension ratio of the specimen $\lambda = \lambda^0$.

The specimen is subjected to a forced-periodic strain

$$\lambda = \lambda^0 + \Delta\lambda e^{i\omega t} \quad [1]$$

where λ^0 is the static extension ratio as mentioned just before, $\Delta\lambda$ the amplitude of the dynamic strain, and ω the angular frequency. Let us further assume that the sinusoidal strain produces sinusoidal variations of $I_{m,j}$, $2\theta_{B,j}$, $\varphi_{c,j}$, β_j and γ_j with phase delayed angle, $\delta_0, \delta_1, \delta_2, \delta_3$, and δ_4, respectively, all with respect to the sinusoidal strain.

These variation may be given by

$$I_{m,j} = I^0_{m,j} + \Delta I_{m,j} e^{i(\omega t - \delta_0)} \quad [2]$$

$$2\theta_{B,j} = 2\theta^0_{B,j} + \Delta 2\theta_{B,j} e^{i(\omega t - \delta_1)} \quad [3]$$

and so on.

The dynamic diffraction intensity from the jth crystal planes at given diffraction and azimuthal angles, $I_j(2\theta, \varphi)$ may be given by

$$I_j(2\theta, \varphi) = \chi(U_j)\,\zeta(V_j)\,I_{m,j} \quad [4]$$

where χ and ζ are symmetric functions to characterize the shape of the jth peak with respect to the diffraction and azimuthal angles, respectively, and U_j and V_j are variables defined by

$$U_j = (2\theta - 2\theta_{B,j})/\beta_j$$

$$V_j = (\varphi - \varphi_{c,j})/\gamma_j \quad . \quad [5]$$

Taking the total differential of eq. [4], as fully discussed elsewhere (14), the dynamic X-ray diffraction intensity $I_j(2\theta, \varphi)$ may be approximated as follows:

$$I_j(2\theta, \varphi) = a^0_j I^0_{m,j} + a^0_j \Delta I_{m,j} e^{i(\omega t - \delta_0)} + b^0_j \Delta 2\theta_{B,j} e^{i(\omega t - \delta_1)} + c^0_j \Delta\varphi_{c,j} e^{i(\omega t - \delta_2)} + d^0_j \Delta\beta_j e^{i(\omega t - \delta_3)} + e^0_j \Delta\gamma_j e^{i(\omega t - \delta_4)} \quad [6]$$

where

$$a^0_j = \chi(U^0_j)\,\zeta(V^0_j)$$

$$b^0_j = I^0_{m,j}\zeta(V^0_j)\,k^0_j$$

$$c^0_j = I^0_{m,j}\chi(U^0_j)\,l^0_j$$

$$d^0_j = I^0_{m,j}\zeta(V^0_j)\,k^0_j\,U^0_j$$

$$e^0_j = I^0_{m,j}\chi(U^0_j)\,l^0_j\,V^0_j \quad [7]$$

and where

$$k^0_j = -(1/\beta^0_j)\,[\partial\chi(U_j)/\partial U_j]_{U_j = U^0_j}$$

$$l^0_j = -(1/\gamma^0_j)\,[\partial\zeta(V_j)/\partial V_j]_{V_j = V^0_j}$$

$$U^0_j = (2\theta - 2\theta^0_{B,j})/\beta^0_j$$

$$V^0_j = (\varphi - \varphi^0_{c,j})/\gamma^0_j \,. \quad [8]$$

Consequently, the diffraction intensity at given diffraction and azimuthal angles, $I(2\theta, \varphi)$ can be written as follows:

$$I(2\theta, \varphi) = \sum_{j=1}^{N} I_j(2\theta, \varphi) = I^0(2\theta, \varphi) + [\Delta I'(2\theta, \varphi) - i\Delta I''(2\theta, \varphi)]\,e^{i\omega t} = I^0(2\theta, \varphi) + \Delta I(2\theta, \varphi)\,e^{i(\omega t - \delta)} \quad [9]$$

where

$$I^0(2\theta, \varphi) = \sum_{j=1}^{N} a^0_j I^0_{m,j} \quad [10]$$

$$\Delta I(2\theta, \varphi) = [\Delta I'(2\theta, \varphi)^2 + \Delta I''(2\theta, \varphi)^2]^{1/2} \quad [11]$$

$$\Delta I'(2\theta, \varphi) = \sum_{j=1}^{N} [a^0_j \Delta I_{m,j} \cos\delta_0 + b^0_j \Delta 2\theta_{B,j} \cos\delta_1 + c^0_j \Delta\varphi_{c,j} \cos\delta_2 + d^0_j \Delta\beta_j \cos\delta_3 + e^0_j \Delta\gamma_j \cos\delta_4] \quad [12]$$

Table 1. Characterization of test specimens

Specimen code	NR-1	NR-2	NR-3	NR-4	NR-5
	Compound formula by weight				
RSS No. 3	100	100	100	100	100
Dicumylperoxide D-98 R	0.5	0.75	1.0	1.5	2.0
	Number average molecular weight of the network chains				
	20200	13500	12800	8400	7700
	Vulcanization condition				
	150 °C and 150 kg/cm² for 40 min				

$$\Delta I''(2\theta,\varphi) = \sum_{j=1}^{N} [a_j^0 \Delta I_{m,j} \sin\delta_0 + b_j^0 \Delta 2\theta_{b,j} \sin\delta_1 + c_j^0 \Delta\varphi_{c,j} \sin\delta_2 + d_j^0 \Delta\beta_j \sin\delta_3 + e_j^0 \Delta\gamma_j \sin\delta_4]. \quad [13]$$

The parameters having the superscription 0, such as a_j^0 through e_j^0, can be determined from the static measurement of X-ray diffraction intensity distribution at $\lambda = \lambda^0$, and the unknown quantities, $\Delta I_{m,j}$, $\Delta 2\theta_{B,j}$, $\Delta\varphi_{c,j}$, $\Delta\beta_j$, and $\Delta\gamma_j$, and subsequently, δ_0, δ_1, δ_2, δ_3, and δ_4, can be determined from the dynamic measurements of $I^0(2\theta,\varphi)$, $\Delta I'(2\theta,\varphi)$, and $\Delta I''(2\theta,\varphi)$ by utilizing the semi-circular sector technique, which will be discussed later, at $5N$ different positions of $(2\theta,\varphi)$.

Semi-circular sector technique

Fig. 1 shows a schematic diagram of the dynamic X-ray diffractometer (14) which can measure the diffraction intensity $I(2\theta,\varphi)$ for four sets of any given phase intervals at any given phase angles of the sinusoidal strain of test specimen as function of angular frequency of the sinusoidal strain, temperature, static strain λ^0, and dynamic strain amplitude $\Delta\lambda$. In practice, however, the mechanical gating device in fig. 1a has been modified by replacing with a photoelectronic gating device composing of four couples of photo-switches, as illustrated in fig. 1b, whose detail was described elsewhere (14).

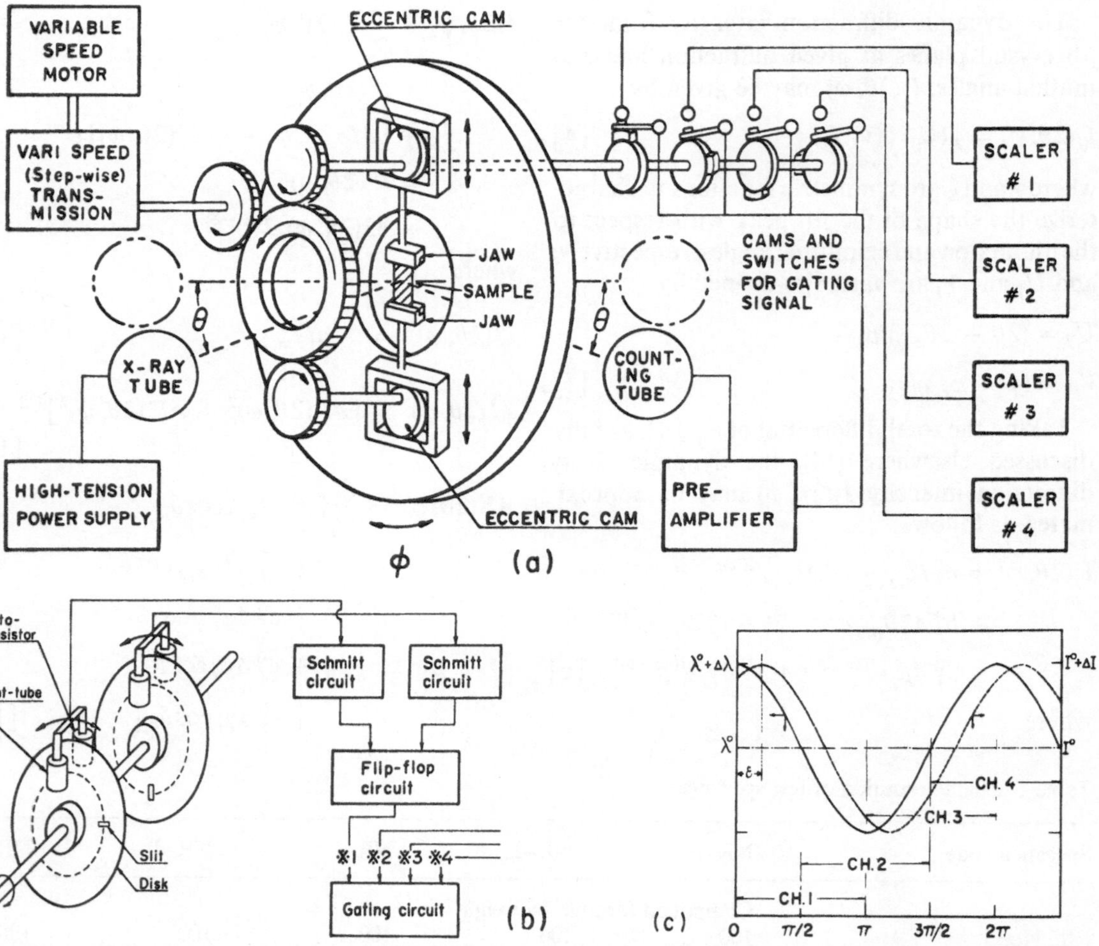

Fig. 1. a) Schematic diagram showing the principle of the dynamic X-ray diffraction technique and the construction of the diffractometer. b) Photoelectric switch set used instead of the mechanical gating device in fig. 1a. Two sets of the photoelectric switches provide gate signal for activating and deactivating each scaler. c) Schematic diagram illustrating the semi-circular sector technique measuring dynamic X-ray diffraction intensities for four π-radians-phase-intervals shifted from each other by $\pi/2$ radians with respect to the sinusoidal strain of test specimen

Repeating n cycles of the simusoidal strain with a given frequency of $f = {}^{\omega}/2\pi$, the X-ray diffraction intensity at $(2\theta, \varphi)$, accumulated for a given phase interval of the sinusoidal strain from $\alpha_1\pi$ to $\alpha_2\pi$, is

$$N(\alpha_1\pi \sim \alpha_2\pi) = \frac{n}{(\lambda^0)^{1/2} h K^0} \times \left[\frac{I^0(\alpha_2 - \alpha_1)}{\omega} - \frac{\Delta R I^0 - \Delta I'}{\omega} \times (\sin\alpha_2\pi - \sin\alpha_1\pi) - \frac{\Delta I''}{\omega}(\cos\alpha_2\pi - \cos\alpha_1\pi)\right] \quad [14]$$

where h is the correction factor for polarization, K^0 is that for absorption due to the change of thickness of the specimen during sinusoidal deformation, and ΔR is given by

$$\Delta R = (\Delta\lambda/2\lambda^0)[1 - \mu D_0 \sec\theta/(\lambda^0)^{1/2}] \quad [15]$$

and where μ is the linear absorption coefficient, D_0 is the thickness of undeformed specimen at $\lambda = 1$, and the above corrections are based on the incompressibility hypothesis and are normalized at D_0 thickness.

Measuring the diffraction intensity for four different phase intervals, $[0 \sim \pi]$, $[\pi/2 \sim 3\pi/2]$, $[\pi \sim 2\pi]$, and $[-\pi/2 \sim \pi/2]$, as illustrated in fig. 1c in terms of CH 1, CH 2, CH 3, and CH 4, the respective accumulated intensities can be given by

$$N_1 = \frac{(n/\omega)}{(\lambda^0)^{1/2} h K^0}(\pi I^0 + 2\Delta I'') \quad [16]$$

$$N_2 = \frac{(n/\omega)}{(\lambda^0)^{1/2} h K^0}(\pi I^0 + 2\Delta R I^0 - 2\Delta I') \quad [17]$$

$$N_3 = \frac{(n/\omega)}{(\lambda^0)^{1/2} h K^0}(\pi I^0 - 2\Delta I'') \quad [18]$$

$$N_4 = \frac{(n/\omega)}{(\lambda^0)^{1/2} h K^0}(\pi I^0 - 2\Delta R I^0 + 2\Delta I'). \quad [19]$$

From these accumulated intensities, the necessary quantities, $I^0(2\theta, \varphi)$, $\Delta I'(2\theta, \varphi)$, and $\Delta I''(2\theta, \varphi)$ can be determined as follows:

$$I^0(2\theta, \varphi) = \frac{(\omega/2\pi)(\lambda^0)^{1/2} h K^0}{n}(N_1 + N_3) = \frac{(\omega/2\pi)(\lambda^0)^{1/2} h K^0}{n}(N_2 + N_4) \quad [20]$$

$$\Delta I'(2\theta, \varphi) = \frac{(\omega/2\pi)(\lambda^0)^{1/2} h K^0}{(2n/\pi)}(N_4 - N_2) + \Delta R I^0 \quad [21]$$

$$\Delta I''(2\theta, \varphi) = \frac{(\omega/2\pi)(\lambda^0)^{1/2} h K^0}{(2n/\pi)}(N_1 - N_3). \quad [22]$$

Simplified semi-circular sector technique for measuring the orientation

Crystallization rates of vulcanized rubbers

As demonstrated in the previous paper of orientation crystallization behavior of natural and synthetic isoprene rubber vulcanizates (13), the diffraction intensity distributions from respective crystal planes, such as (200), (120), and (002) crystal planes, separate from each other without any overlapping, and the intensity distributions seem to be identical and are very sharp in such extent as giving one of the second order orientation factors (15) of reciprocal lattice vector of (002) crystal plane, $F_{20}{}^{(002)}$ as highly as 0.95, both irrespective of the magnitude of extension ratio. Actually, as illustrated in figs. 2a and 2b quantitatively for the dynamic diffraction intensity distributions of the (200) and (002) crystal planes, which were obtained for the specimen of NR-3 by means of the narrow sector technique (13) with a given dynamic strain amplitude of 0.5 at a given phase angle, as function of the azimuthal angle and diffraction angle, respectively, the intensity distributions are hardly affected by either of the static extension ratio λ^0, dynamic frequency f, or measuring temperature, giving always common sharp and symmetric distributions when the distributions are normalized.

These facts suggest that for the orientation crystallization of natural rubber vulcanizates the dynamic X-ray diffraction intensity $I_j(2\theta, \varphi)$ given by eq. [4] is mostly affected from $I_{m,j}$ but hardly from $\chi(U_j)\zeta(V_j)$ term. In other words, $\Delta 2\theta_{B,j}$, $\Delta\varphi_{c,j}$, $\Delta\beta_j$, and $\Delta\gamma_j$ in eq. [6] are negligibly small, so that the dynamic measurements can be performed at N position of $(2\theta^0_{B,j}, \varphi^0_{c,j})$ for the respective diffraction peaks. Unter these conditions, $U^0_j = 0$, $\chi(U^0_j) = 1$, $k^0_j = 0$, $V^0_j = 0$, $\zeta(V^0_j) = 1$, and $l^0_j = 0$, which make b^0_j through e^0_j in eq. [7] all zero and further reduce eq. [6] to such simple one as given by

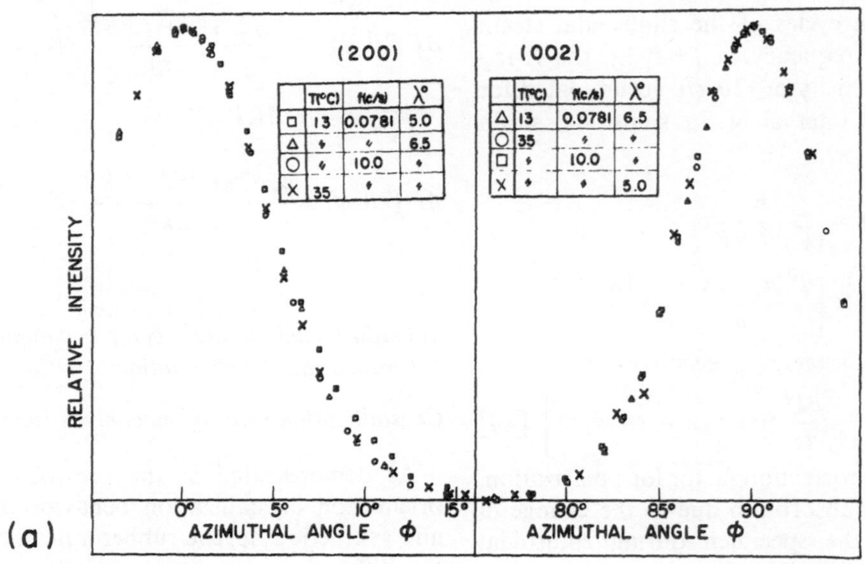

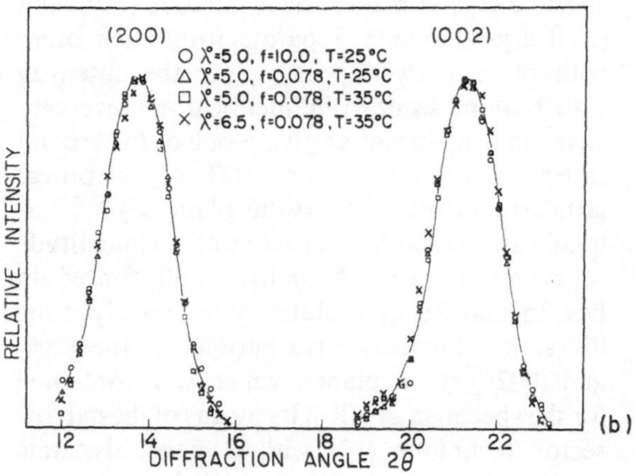

Fig. 2. Dynamic X-ray diffraction intensity distribution from (200) and (002) crystal planes with respect to a) azimuthal angle and b) diffraction angle, both obtained by using a narrow sector technique (13) with $\Delta\lambda = 0.5$, demonstrating that the distributions are very sharp and symmetric and are given by common curves, when normalized, irrespective of static extension ratio, temperature, and frequency

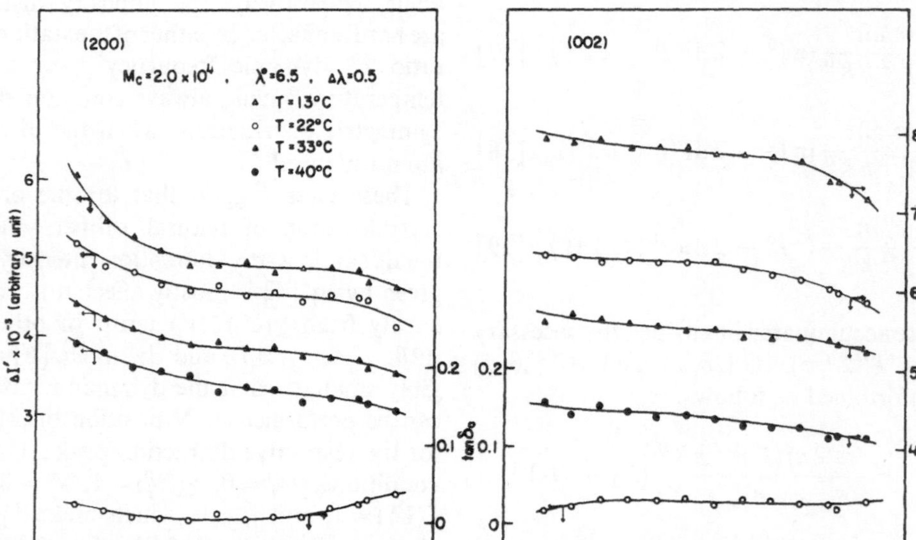

Fig. 3 a)

$$I_j(2\theta^0_{B,j}, \varphi^0_{c,j}) = I^0_{m,j} + \Delta I_{m,j}\, e^{i(\omega t - \delta_0)} = I^0_{m,j} + (\Delta I'_{m,j} - i\Delta I''_{m,j})\, e^{i\omega t}. \quad [23]$$

Similarly with the determination of $I^0(2\theta, \varphi)$, $\Delta I'(2\theta, \varphi)$, and $\Delta I''(2\theta, \varphi)$ in eqs. [20] through [22], $I^0_{m,j}$, $\Delta I'_{m,j}$, and $\Delta I''_{m,j}$, and subsequently, $\delta_0 = \tan^{-1}(\Delta I''_{m,j}/\Delta I'_{m,j})$ can be determined by using the semi-circular technique. Strictly speaking, however, the accumulated intensities, N_1 through N_4 thus obtained, include the contributions of the air-scattering, incoherent scattering, and noncrystalline scattering as well. These additional contributions can be eliminated, for example, by measuring the dynamic diffraction intensity at a particular position, $I_j(2\theta^0_{B,j}, \varphi^0_{c,j} \pm 15°)$ and by subtracting this intensity from $I_j(2\theta^0_{B,j}, \varphi^0_{c,j})$, as recognized from fig. 2a.

Furthermore, the degree of crystallinity is an almost linear function of extension ratio over a

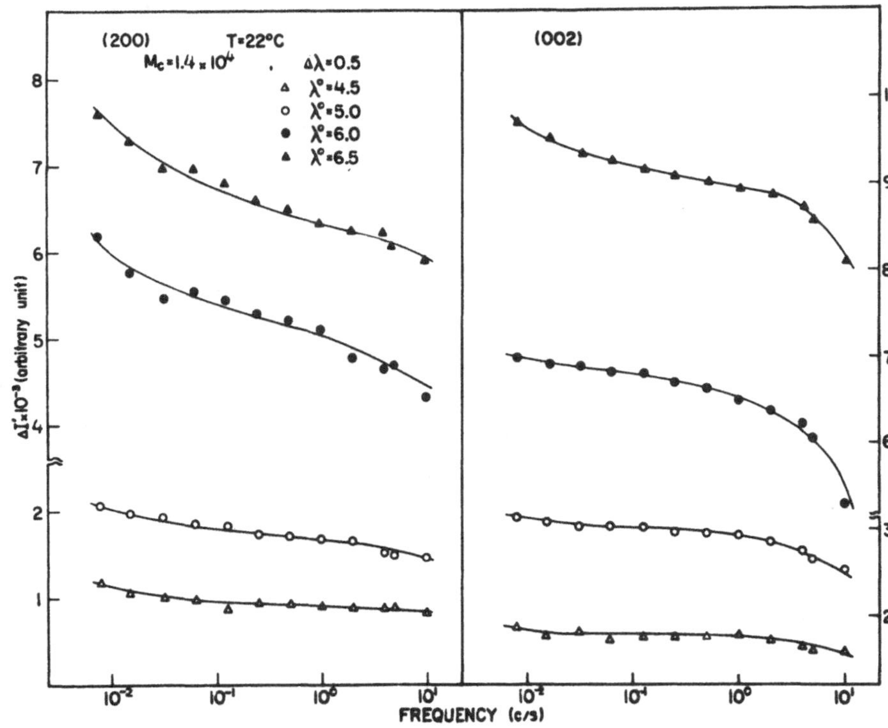

Fig. 3b)

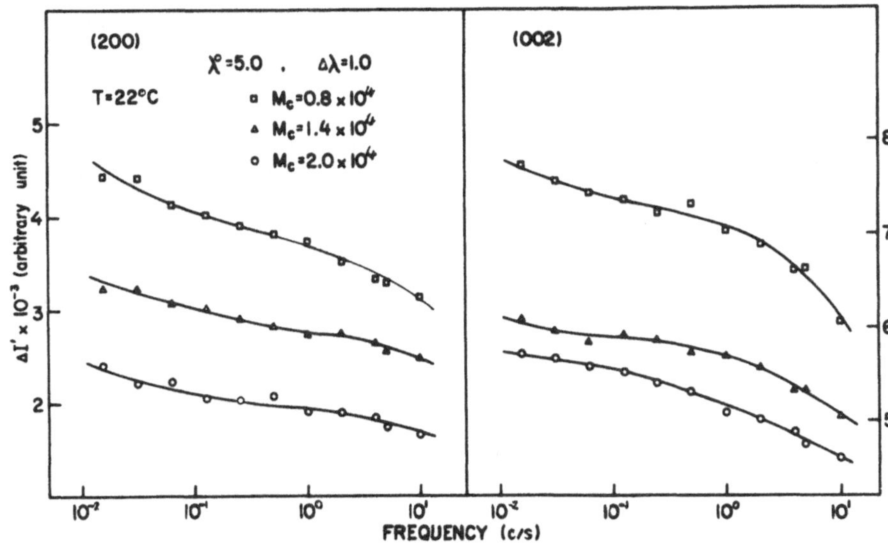

Fig. 3c)

Fig. 3. Effects of a) measuring temperature, b) static extension ratio, c) degree of crosslink, all upon the frequency dispersions of $\Delta I'_{m,j}$ of (200) and (002) crystal planes considerably wide range when the ratio is larger than a critical value, as demonstrated in the previous paper (13), and the dynamic strain-induced crystallization coefficient of the jth crystal plane $Q_j^*(i\omega)$ can be defined by

$$\Delta I_{m,j}^* / \Delta\lambda |_{\lambda=\lambda^0} \equiv Q_j^*(i\omega) = Q_j'(\omega) - i Q_j''(\omega). \quad [24]$$

The quantity $Q_j^*(i\omega)$ must be further a function of temperature and be considered as one of the most basic functions for characterizing the orientation crystallization kinetics of linear polymers.

Results and discussions

As already pointed out qualitatively in the previous paper (13), the anisotropy of crystallization, especially in the directions of paratropic and diatropic crystal planes, is noticeable, and the dynamic measurements are mostly carried out for the (002) and (200) crystal diffractions as function of static extension ratio, dynamic strain amplitude, temperature, and degree of crosslink.

Fig. 3 shows the results of the measurements in terms of the frequency dependence of $\Delta I'_{m,j}$ and, some times, $\tan\delta_0$ for a given dynamic strain amplitude, mostly $\Delta\lambda = 0.5$. As seen in the figures, the most characteristic is the appearance of two frequency dispersions, in general, at relatively low frequency region around 10^{-2} cps and high frequency region around 10^1 cps. The former dispersion is rather obvious for the (200) crystal plane, while the latter dispersion is for the (002) crystal plane.

As mentioned previously, the orientation crystallization of natural rubber gives an extreme orientation of resulted crystal, almost perfect orientation of crystal c-axis with respect to the stretching direction, irrespective of the magnitude of the extension ratio. In other words, this extreme behavior of crystal orientation suggests that the noncrystalline chains oriented extremely highly in the stretching direction can only form the crystal nuclei [the so-called γ-filament by *Andrews* (16, 17)].

Taking these postulations into consideration, the γ-filament must be formed almost instantaneously in association with forced orientation of rubber molecules while the α-filament must grow with considerably low rate of crystallization due to self-diffusion of rubber molecules, and the frequency dispersion of $\Delta I'_{m,j}$ at the low frequency region must be assigned as arising from the crystallization process of the α-filament while the dispersion at the high frequency region must be assigned as from the crystallization process of the γ-filament.

In addition to the above characteristics, it must be noted that the value of $\tan\delta_0$ is relatively small ranging around 0.02 but that its sign is definitely positive, as illustrated in fig. 3a just for one example for avoiding the complexity of the figure. In other words, the dynamic orientation crystallization of natural rubber vulcanizates occurs in advance of the dynamic strain of the specimen, probably rather close in phase with the dynamic stress of the specimen.

Fig. 3a demonstrates the effect of measuring temperature upon the frequency dispersions of $\Delta I'_{m,j}$ for a given specimen of NR–1 with keeping the static extension ratio λ^0 at a relatively high value of 6.5 and the dynamic strain amplitude at 0.5. As seen in the figure, the magnitude of $\Delta I'_{m,(002)}$ is larger than that of $\Delta I'_{m,(200)}$ at relatively low temperatures, but with increase of the temperature $\Delta I'_{m,(002)}$ decreases more rapidly and becomes probably smaller than $\Delta I'_{m,(200)}$ at higher temperatures than 40 °C. Although the frequency dispersions at the high and low frequency regions, which are obvious for the (200) and (002) crystal planes, respectively, become less obvious with increase of the temperature, the dispersion at the low frequency region seems to shift to higher frequencies while that at the high frequency region seems not to shift at all, both with increase of the temperature.

Fig. 3b shows the effect of static extension ratio upon the frequency dispersions of $\Delta I'_{m,j}$ for a given specimen of NR–2 with keeping the measuring temperature at 22 °C and the dynamic strain amplitude at 0.5, and fig. 3c shows the effect of degree of crosslink upon the frequency dispersions of $\Delta I'_{m,j}$ with keeping the static extension ratio at 5.0 and dynamic strain amplitude at relatively large magnitude of 1.0. As seen in the figures, the effects of static strain and degree of crosslink both upon the frequency dispersions seem to be identical, at least qualitatively, with that of temperature; i.e., the lower the static strain and degree of crosslink, the dispersions become less obvious and the magnitudes of $\Delta I'_{m,j}$ decrease as likely as the case of temperature increase in fig. 3a. The shift of the frequency dispersion to higher frequencies, which is observed in some extent for $\Delta I'_{j,(200)}$ at the low frequency region, may be interpreted in terms of the increase of mobility of the self-diffusion of rubber molecules to form the α-filament with decreases of the static strain and degree of crosslink.

Acknowledgements

The authors are indebted to the Toyo Rayon Science Foundation, Japan (Kenkyu Josei-kin, 1964), which enabled them to construct the dynamic X-ray diffractometer used in this experiment. Appreciation is also expressed for the financial support provided by a grant from the Japan Synthetic Rubber Co. Ltd., Tokyo, Japan, and the Bridgestone Tire Co. Ltd., Tokyo, Japan.

Summary

The orientation crystallization behavior of natural rubber vulcanizates was investigated by means of a dynamic X-ray diffraction technique utilizing the semi-circular sector technique, and the frequency dependence of dynamic X-ray diffraction intensities from diatropic and paratropic crystal planes, (002) and (200) crystal planes, was observed over a frequency range from 10^{-3} to 10^{1} cps as function of temperature, degree of crosslink, static extension ratio, and dynamic strain amplitude.

The frequency dependence of the dynamic X-ray diffraction gives two dispersion regimes around 10^{-2} and 10^{1} cps, and the dynamic orientation crystallization occurs in advance of the dynamic strain of the specimen. The frequency dispersion at low frequency side around 10^{-2} cps is much obvious for the (200) crystal plane and shifted in some extent to higher frequencies by several factors increasing the mobility of self-diffusion of rubber molecules, such as increase of temperature and decreases of extension ratio and degree of crosslink, while the frequency dispersion at high frequency side around 10^{1} cps is rather obvious for the (002) crystal plane and not shifted by the factors, appreciably, suggesting that these frequencies must be considered as critical ones beyond which the crystallizations in the forms of the so-called α- and γ-filaments can not appreciably proceed, respectively.

References

1) *Acken, M. F.* and *W. E. Singer*, Ind. Eng. Chem. **24**, 54 (1932).

2) *Long, J. D.* and *W. E. Singer*, Ind. Eng. Chem. **26**, 543 (1934).

3) *Thiesen*, Z. Phys. Chemie **29**, 359 (1935).

4) *Nyburg, S. C.*, Brit. J. Appl. Phys. **5**, 321 (1954).

5) *Goppel, J. M.*, Rubber Chem. Technol. **21**, 773 (1948); Appl. Sci. Res. **A1**, 3 (1947).

6) *Wood, L. A.* and *N. Bekkedahl*, J. Appl. Phys. **17**, 362 (1946).

7) *Gent, A. N.*, Trans. Farad. Soc. **50**, 521 (1954).

8) *Kim, H. G.* and *Mandelkern, L.*, J. Polymer Sci. A2, **6**, 181 (1968).

9) *Yau, W.* and *R. S. Stein*, J. Polymer Sci. B**2**, 231 (1964).

10) *Yau, W.* and *R. S. Stein*, J. Polymer Sci. A 2, **6**, 1 (1968).

11) *Mitchell, J. C.* and *D. J. Meier*, J. Polymer Sci. A2, **6**, 1689 (1968).

12) *Dunning, D. J.* and *P. J. Pennells*, Rubber Chem. Technol. **41**, 1381 (1968).

13) *Kawai, H., T. Oda, S. Tomita,* and *I. Furuta*, Proc. 5th Intern. Congr. Rheology **4**, 51 (1970).

14) *Ito, T., T. Oda, H. Kawai, T. Kawaguchi, D. A. Keedy*, and *R. S. Stein*, Rev. Sci. Instr. **39**, 1847 (1968).

15) *Nomura, S., H. Kawai, I. Kimura*, and *M. Kagiyama*, J. Polymer. Sci. A2, **8**, 383 (1970).

16) *Andrews, E. H.*, Proc. Royal Soc. **A270**, 232 (1963).

17) *Andrews, E. H.*, Proc. Royal. Soc. **A277**, 562 (1964).

Authors' addresses:

Dr. *Hiroaki Hiratsuka*
Ibaraki Electrical Communication Laboratory
Nippon Telegraph and Telephone Public Cooperation
Tokai-mura, Ibaraki-ken (Japan)

Dr. *Mitsuaki Hashiyama*
Polymer Research Institute
University of Massachusetts
Amherst, Mass. 01002 (U.S.A.)

Dr. *Seisuke Tomita*
Tire Research and Development Department
Bridgestone Tire Co. Ltd.
Kodaira, Tokyo (Japan)

and Dr. *Hiromichi Kawai*
Dept. of Polymer Chemistry
Kyoto University
Kyoto (Japan)

Rheol. Acta 13, 260–264 (1974)

Laboratoire de Physique des Solides, Université Paris-Sud, Centre d'Orsay, Orsay (France)

Biréfringence et dichroïsme linéaire des ferrofluides sous champ magnétique

Par A. Martinet

Avec 9 figures

(Reçu p. p. le 27 octobre 1972)

I. Introduction

Les ferrofluides sont des colloïdes magnétiques très stables qui ont été mis au point ces dernières années (1, 2). Leur stabilité dans le temps s'étend sur des années et en champ magnétique jusqu'à 10 kG au moins. De plus, ils conservent leurs propriétés fluides en champ fort, la viscosité n'étant que peu affectée. Ceci les distingue radicalement des liquides développés pour les embrayages magnétiques. Les aimantations à saturation des ces fluides sont élevées. On a couramment des $4\pi M$ de 100 G/cm^3 et même de 600 G/cm^3, c'est-à-dire des aimantations inférieures à celle du nickel d'à peine un ordre de grandeur.

Il est donc possible de transmettre directement des forces importantes au liquide par l'intermédiaire d'un gradient de champ magnétique. On peut aussi, de la même façon, modifier la densité apparente du liquide et ceci sélectivement par rapport à un autre corps (solide ou liquide) non magnétique d'où la possibilité de faire de la lévitation et des mesures directes de densité (3). Il faut remarquer que la qualité essentielle qui est demandée à un tel fluide est de s'opposer énergiquement à des gradients importants de concentration des grains sous l'action d'un champ magnétique inhomogène[1]). Le domaine d'application des ferrofluides est vaste et sort du cadre de cette communication. On trouvera de nombreuses suggestions dans les références (1 et 3).

Nous nous proposons ici de discuter les propriétés optiques de ces liquides. L'application d'un champ magnétique transversal crée une direction d'anisotropie qui se traduit par de la biréfringence et par du dichroïsme linéaire. Ces effets sont importants en valeur absolue (la différence des indices $n_{\parallel}$ et $n_{\perp}$ correspondant à des directions $\parallel$ et $\perp$ au champ magnétique est de l'ordre de quelques 10^{-3}). Nous avons également étudié les propriétés magnétiques de ces fluides et leur anisotropie magnétique. Ceci donne des informations sur l'état d'association des grains dans le colloïde. Au préalable, nous décrirons plus en détail les caractéristiques des matériaux utilisés.

II. Matériaux

Nous avons étudié trois ferrofluides désignés ci-après par *T*, *K* et *A*. *T* est un colloïde de cobalt dans du toluène, *K* de Fe_3O_4 dans de l'heptane et *A* de Fe_3O_4 dans de l'eau.

1. Ferrofluide T

Ce ferrofluide à base de toluène a été préparé à notre laboratoire par *Liebert* et *Strzelecki* suivant la méthode de *Hess* et *Parker* (4). On a $4\pi M_S \simeq 80$ G/cm^3. Les grains (Co) ont un diamètre d'environ 100 Å. Une couche de polymères adsorbés à la surface des grains stabilise le colloïde. Elle diminue l'interaction magnétique entre grains donnée par μ^2/a^3 où μ est le moment magnétique porté par un grain et a la distance minimum d'approche. Pour ce ferrofluide, le paramètre $\lambda = \mu^2/a^3 kT$ qui compare cette énergie magnétique à l'agitation thermique est de l'ordre de 5. On a donc une légère tendance à l'association, particulièrement en présence d'un champ magnétique.

2. Ferrofluide K

Il s'agit d'un fluide disponible dans le commerce (5) pour lequel $4\pi M_S \simeq 120$ G/cm^3. Il est constitué de grains de Fe_3O_4 d'un diamètre moyen de 80 Å dispersés dans de l'heptane. Pour ce fluide $\lambda \simeq 0{,}2$. Ce fluide ne présente jamais de tendance à l'association en chaînes, même en champ fort.

3. Ferrofluide A

C'est un fluide manufacturé par la même compagnie que le précédent avec $4\pi M_S \simeq 120$ G/cm^3. Il est également constitué de grains de Fe_3O_4 d'un diamètre légèrement supérieur pour l'échantillon particulier dont nous dispositions ($\varnothing \sim 120$ Å) mais en suspension dans de l'eau. $\lambda \sim 1{,}5$. On observe des chaînes même en champ faible.

Des études de microscopie électronique ont été faites sur ces fluides. Afin d'obtenir des informations sur l'état d'association entre grains, nous avons utilisé une méthode de polymérisation pour les ferrofluides *T* et *K* (6): par remplacement du liquide porteur par un mélange de styrène et de divinylbenzène et adjonction d'un catalyseur, on a pu «figer» par polymérisation l'état du liquide dans un champ donné et l'étudier ensuite sur des coupes obtenues à l'ultramicrotome. Cette méthode a été utilisée aussi pour déterminer l'anisotropie magnétique induite dans le fluide par des mesures de susceptibilité sur des échantillons solides polymérisés sous champ.

III. Propriétés optiques

1. Dispositif expérimental

Nous avons mesuré la biréfringence et le dichroïsme au microscope polarisant avec un

[1]) En ce sens un bon ferrofluide fait une mauvaise solution de *Bitter* pour l'observation des domaines magnétiques. On a d'ailleurs là un moyen qualitatif très simple pour tester un ferrofluide.

analyseur à pénombre schématisé sur la fig. 1. Le faisceau de lumière passe par le polariseur P, traverse l'échantillon de ferrofluide qui est contenu entre deux lames de verre distantes de e (typiquement $10 < e < 100\,\mu$ pour les concentrations d'origine) et soumis à un champ magnétique $H (0 < H < 2\,\text{kG})$ placé à 45° de P. Après traversée d'une lame quart d'onde, le faisceau tombe sur l'analyseur à pénombre à quatre plages (*Macé de Lépinay* avec et sans $\lambda/4$ + analyseur tournant). Les angles θ_1 entre P et la $\lambda/4$ et θ_2 entre la $\lambda/4$ et A, pour la position qui donne l'égalité d'éclairement des quatre plages, sont mesurés avec une précision d'environ 10'. θ_1 et $2\theta_2$ donnent, après les corrections habituelles, le dichroïsme et la biréfringence, respectivement.

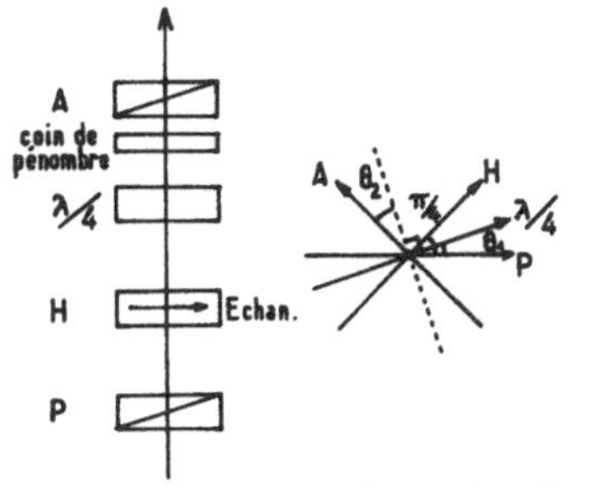

Fig. 1. Schéma de principe du dispositif expérimental. A droite: disposition angulaire des différents éléments, vus par-dessus.

2. Résultats

Sur la fig. 2, sont portées les valeurs de la différence d'indice $\Delta_n = n_{||} - n_{\perp}$ et de la différence d'absorption $K_{||} - K_{\perp}$ pour de la lumière polarisée parallèlement et perpendiculairement au champ magnétique. On a porté également les valeurs en champ fort pour des échantillons dilués à 1%. On en déduit que les effets sont proportionnels à la concentration, ce qui montre qu'il ne s'agit pas d'effets de surface mais bien de volume. Les trois ferrofluides étudiés nous ont donné des résultats analogues.

On notera que ces effets sont déjà appréciables en champs faibles. La fig. 3 donne les résultats près de l'origine pour un échantillon de A dilué à 1%. Des champs d'une fraction de *Gauss* donnent déjà des effets détectables.

IV. Propriétés magnétiques

1. Superparamagnétisme des grains

La courbe d'aimantation du liquide est donnée fig. 4. On remarque l'absence d'hystérésis et le fait qu'elle est voisine d'une courbe de *Langevin*. Nous avons également vérifié qu'à différentes températures les courbes d'aimantation se superposent quand on les porte en fonction de H/T. Ceci est caractéristique d'un comportement superparamagnétique. Ce superparamagnétisme est intrinsèque aux grains: en effet, les courbes d'aimantation sont essentiellement les mêmes sur le liquide et sur l'échantillon solide polymérisé, en particulier le solide ne présente pas non plus d'hystérésis.

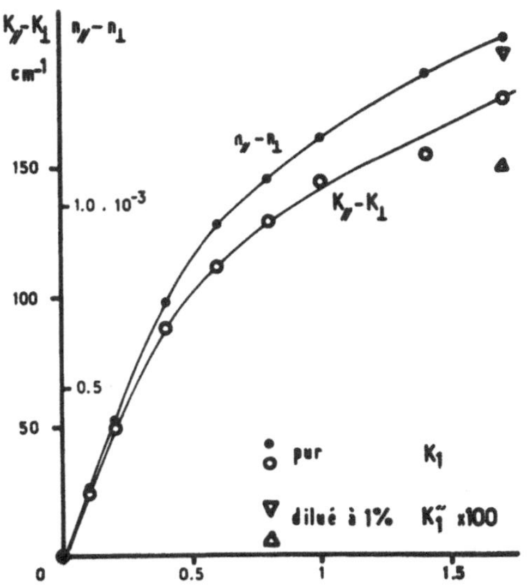

Fig. 2. Biréfringence (échelle de droite) et dichroïsme linéaire (échelle de gauche) en fonction du champ magnétique, pour le ferrofluide K

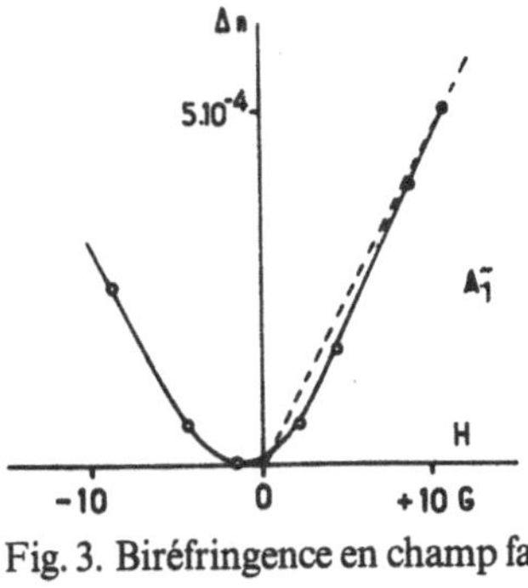

Fig. 3. Biréfringence en champ faible pour le ferrofluide A

2. Anisotropie magnétique

Une série d'échantillons a été polymérisée dans des champs magnétiques allant de 0 à 5 kG. Nous avons mesuré la susceptibilité initiale χ de ces échantillons à différents angles θ par rapport à la direction du champ H_p appliqué pendant la polymérisation (fig. 5). L'anisotropie peut être caractérisée par le rapport $\chi_{||}/\chi_{\perp}$ correspondant à $\theta = 0$ et $\theta = \pi/2$. Cette anisotropie croît avec le champ (figs. 6 et 7).

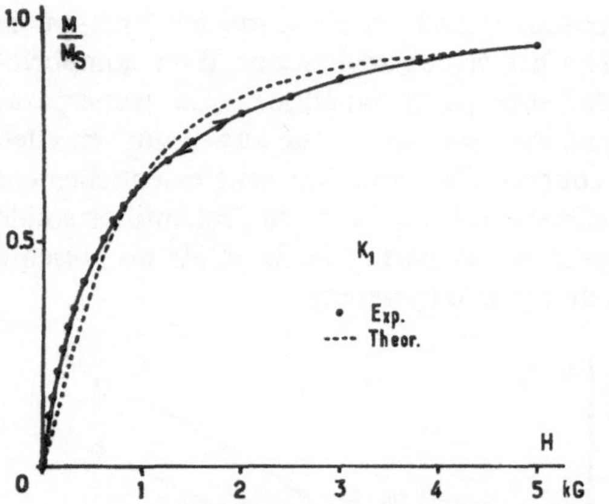

Fig. 4. Courbe d'aimantation du ferrofluide *K* liquide. La courbe théorique est une courbe de *Langevin*. La courbe expérimentale a été normalisée à 5 kG

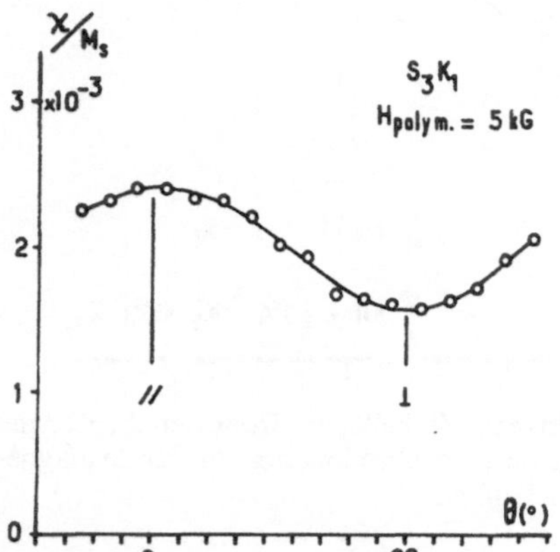

Fig. 5. Variation de la susceptibilité magnétique initiale pour des échantillons de *K* polymérisés dans un champ H_p de 5 kG, en fonction de l'angle par rapport à H_p

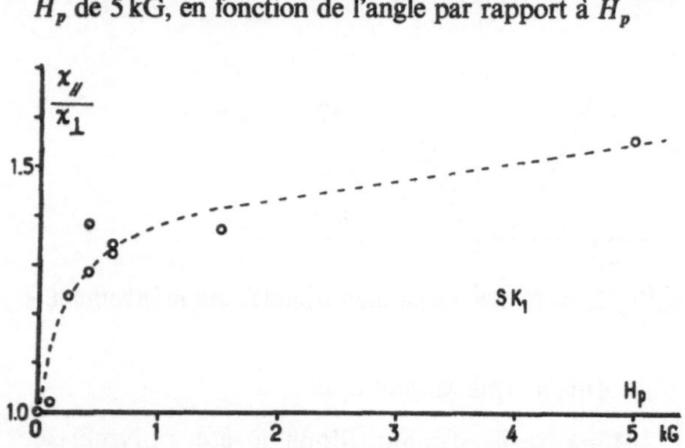

Fig. 6. Variation de l'anisotropie magnétique en fonction du champ de polymérisation H_p, pour le ferrofluide *K*

V. Discussion

On remarque sur la fig. 2 que les courbes ont un départ linéaire près de l'origine. Si l'on interprète les anisotropies observées comme résultant d'un effet *Cotton-Mouton* dû à l'orientation des grains par le champ, on est conduit au résultat classique d'un départ en H^2. Dans notre cas, ce domaine parabolique, s'il existe (fig. 3), est très petit. On a porté fig. 8 la variation normalisée à 1,5 kG que l'on attendrait pour des grains indépendants en prenant pour moment moyen μ par grain celui qui donne une courbe de *Langevin* en meilleur accord avec la courbe d'aimantation. Il semble donc que ces effets ne puissent pas s'interpréter simplement par des effets d'orientation des grains. De plus, les grains étant superparamagnétiques, l'aimantation n'est pas rigidement couplée aux axes cristallographiques du grain et le fait que l'aimantation soit orientée par le champ n'implique pas l'orientation du grain lui-même.

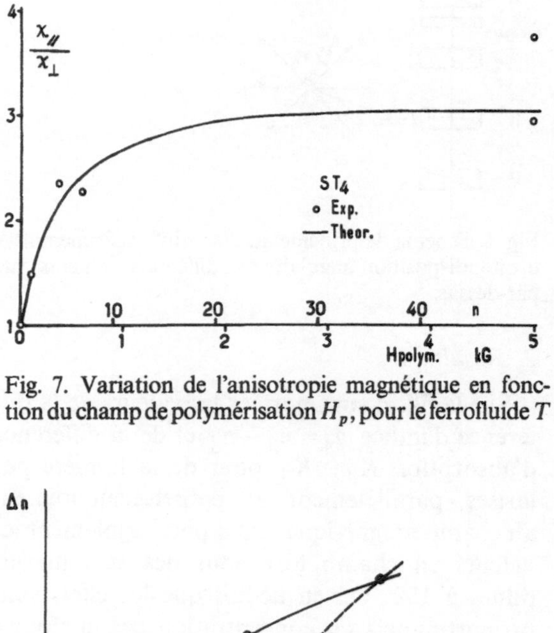

Fig. 7. Variation de l'anisotropie magnétique en fonction du champ de polymérisation H_p, pour le ferrofluide *T*

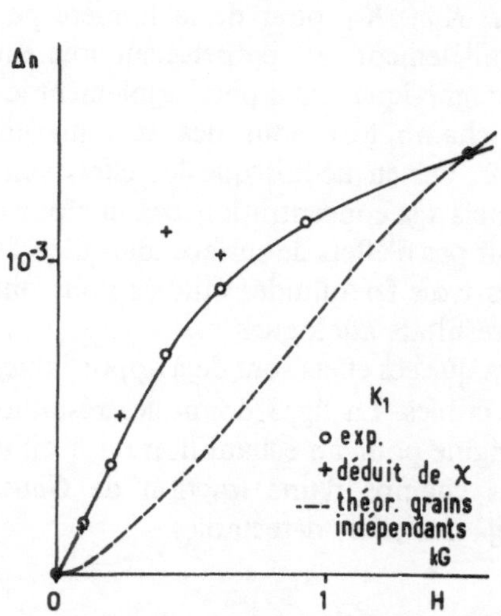

Fig. 8. Biréfringence en fonction du champ magnétique pour le ferrofluide *K*. ○ Points expérimentaux (mesures optiques); + valeurs déduites des mesures de susceptibilité; --- variation théorique pour des grains indépendants

Nous avons cherché à relier ces effets à des corrélations entre grains. A titre d'essai, on considère la suspension comme étant constituée de grains sphériques qui s'alignent dans le champ en formant des chaînes plus ou moins jointives. *Jacobs* et *Bean* (7) ont élaboré un modèle pour décrire un milieu magnétique constitué de chaînes de sphères en contact. *Burger* (8) en a déduit la susceptibilité initiale $\parallel$ et $\perp$, pour le cas où les sphères de diamètre d_0 ne sont plus forcément en contact mais séparées de d, on a:

$$\chi_{\parallel} = \frac{\chi_0}{1 - \frac{2\pi}{3}\varkappa(n)\chi_0\left(\frac{d_0}{d}\right)^3} \qquad [1]$$

$$\chi_{\perp} = \frac{\chi_0}{1 + \frac{\pi}{3}\varkappa(\eta)\chi_0\left(\frac{d_0}{d}\right)^3} \qquad [2]$$

χ_0 susceptibilité d'une sphère seule

$\varkappa(n)$ coefficient numérique donné par $\sum_{i=1}^{n} \frac{n-i}{n i^3}$ qui tend vers 1,2 pour $n \to \infty$

n nombre de grains par chaîne

– Dans le cas du ferrofluide $T(\lambda > 1)$, il s'agit réellement de chaînes de sphères en contact comme l'ont montré les observations de microscopie électroniques. L'application des formules [1] et [2] aux valeurs expérimentales de $\chi_{\parallel}/\chi_{\perp}$ permet de déduire une valeur du nombre moyen de grains n par chaîne en fonction du champ magnétique. On voit sur la fig. 7 que n est approximativement donné par:

$$n - 1 = \alpha H \qquad [3]$$

avec $\alpha \simeq 10^{-2}\,\underline{G}^{-1}$.

– Dans le cas du ferrofluide K, pour lequel $\lambda < 1$, on n'observe jamais de chaînes proprement dites. L'effet du champ est plutôt de diminuer la distance moyenne entre grains dans la direction parallèle ($d_{\parallel}$) et de l'augmenter dans la direction perpendiculaire ($d_{\perp}$). En appliquant l'analogue diélectrique des formules [1] et [2], on déduit une constante diélectrique $\parallel$ et $\perp$ effective des grains dont on tire les indices $n_{\parallel}$ et $n_{\perp}$ et la biréfringence Δn. On obtient dans le cas du ferrofluide K:

$$\Delta n = 18 \cdot 10^{-3}\,\Delta\left(\frac{d_0}{d}\right)^3$$

avec

$$\Delta\left(\frac{d_0}{d}\right)^3 = \left(\frac{d_0}{d_{\parallel}}\right)^3 - \left(\frac{d_0}{d_{\perp}}\right)^3 \qquad [4]$$

$\Delta(d_0/d)^3$ se déduit des mesures de susceptibilité initiale par les formules [1] et [2] en prenant $n = \infty$. On a d'ailleurs deux déterminations différentes à partir de $\chi_0/\chi_{\parallel}$ et de $\chi_0/\chi_{\perp}$ qui se recoupent assez bien (les variations de susceptibilité avec H_p étant faibles, la précision des mesures est limitée). Les valeurs de $d_{\parallel}$ et $d_{\perp}$ que l'on déduit ainsi des mesures de susceptibilité sont portées sur la fig. 9 en fonction de H_p. Les Δn correspondants sont portés sur la fig. 8, après normalisation à 1,5 kG. Ils donnent une variation en bon accord avec les mesures optiques directes. Les valeurs absolues déduites de la formule [4] sont cependant trop grandes par un facteur de l'ordre de 5.

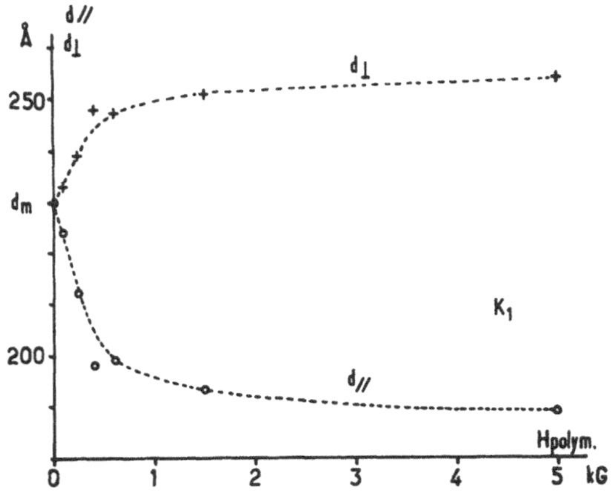

Fig. 9. Variation de la distance moyenne entre grains en fonction du champ de polymérisation H_p, déduite des mesures de susceptibilité, initiale pour le ferrofluide K

VI. Conclusion

Un ferrofluide soumis à l'action d'un champ magnétique présente une anisotropie. Cette anisotropie est révélée par les propriétés optiques (biréfringence et dichroïsme) et magnétiques (susceptibilité $\parallel$ et $\perp$ différentes). Nous avons relié les déterminations expérimentales de ces deux formes d'anisotropie et montré que l'une comme l'autre étaient reliées à des corrélations entre grains soit sous forme d'association en chaînes, soit sous forme d'un tassement des grains dans la direction parallèle au champ magnétique.

Résumé

On étudie l'anisotropie qu'un ferrofluide (colloïde magnétique) présente en champ magnétique taut du point de vue optique que magnétique. Cette anisotropie s'interprète par des conélations entre grains.

Littérature

1) *Rosensweig, R. E.*, Intern. Sci. Technol., p. 48 (1966).

2) *Kaiser, R.* and *G. Miskolczy*, J. Appl. Phys. **41**, 1064 (1970).

3) *Kaiser, R.* and *G. Miskolczy*, I.E.E.E. Trans. Magn. Vol. 6, 694 (1970).

4) *Hess, P. H.* and *P. H. Parker jr.*, J. Appl. Polymer Sci. **10**, 1915 (1966).

5) Ferrofluidics Corp. 144 Middlesex Turnpike. Burlington, Massachusetts 01803.

6) *Liebert, L., A. Martinet*, and *L. Strzelecky*, J. Coll. Interface Sci. **41**, 391 (1972).

7) *Jacobs, I. S.* and *C. P. Bean*, Phys. Rev. **100**, 1060 (1955).

8) *Burger, J. P.* (à paraître).

Adresse de l'auteur:
Dr. *A. Martinet*
Laboratoire de Physique des Solides
Université Paris Sud
Centre d'Orsay
F-91 Orsay (France)

Rheol. Acta **13**, 265–273 (1974)

From the Sandia Laboratories, Albuquerque, New Mexico (U.S.A.)

The dynamic mechanical behavior of polymethyl methacrylate*)

By K. W. Schuler and J. W. Nunziato

With 5 figures and 1 table

(Received October 27, 1972)

1. Introduction

At very low stress levels (< 1 kbar), the short time response of polymeric materials is described as "glassy" and they are often modeled using linear elasticity. However, at higher stress levels, such simple models may not be applicable. The purpose of this paper is to present experimental observations of wave propagation in polymethyl methacrylate (PMMA) up to 60 kbar which indicate that such materials exhibit considerable nonlinearities as well as significant rate-dependence. We compare numerical calculations based on both nonlinear elastic and nonlinear viscoelastic models with these observations.

The wave propagation experiments were conducted using the plate impact technique. In this technique a flat-nosed projectile, launched by either a gas or powder gun (1, 2), impacts a target plate, thus producing compressive plane waves which propagate into both the target and projectile. Depending on the details of the projectile's construction, the compressive wave propagating in the projectile may be reflected back into the target either as an unloading wave or as a wave which produces further loading (reloading wave). The evolutionary behavior of this series of waves in the target was observed by employing a velocity interferometer instrumentation technique (3). This technique permits an accurate determination of the particle velocity history of a point within the target. The particle velocity histories, along with wave velocity measurements, can be used to determine the loading and unloading stress-strain paths followed by the material. A detailed description of the experiments and the results are given in Section 2.

In Section 3, we propose a nonlinear viscoelastic model for PMMA and give a procedure for evaluating the relevant material functions from experimental data. In Section 4, numerical calculations obtained using this constitutive model in a finite-difference computer code are compared with the experimentally observed wave profiles. In addition, calculations based on a nonlinear elastic model are also included. The comparisons show that below ≈ 7 kbar and above ≈ 40 kbar the rate-effects are of secondary importance and that the nonlinear elastic model agrees quite well with the observations. However, between ≈ 7 and ≈ 40 kbar there is considerable stress relaxation observed and the comparisons show that a viscoelastic model must be employed.

2. Experimental technique and observations

The experimental arrangement which was employed to generate and to observe the propagation of loading and unloading waves in PMMA is shown schematically in fig. 1. Basically, a projectile launched by either a 10 cm bore gas gun (1) or a 8.9 cm powder gun (2) was impacted onto a PMMA target[1]). For times prior to the arrival of unloading waves from the lateral surfaces of the target and projectile plates, a uniaxial strain state is produced at the center of the plates which permits a one-dimensional treatment to be used in the reduction of the data and the analysis of the results.

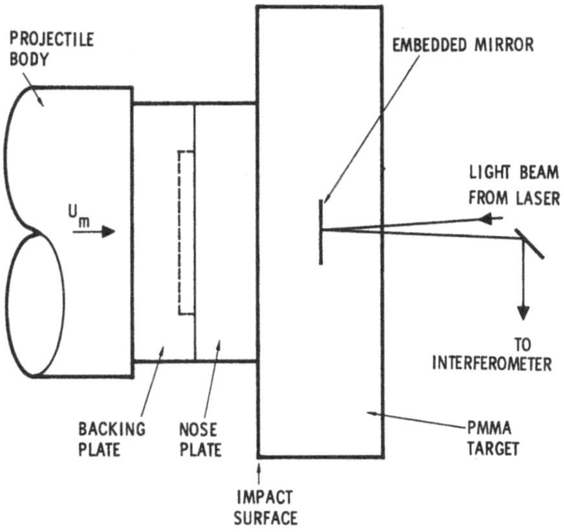

Fig. 1. Schematic of experiment configuration

*) This work was supported by the U.S. Atomic Energy Commission.

Presented at The Sixth International Congress of Rheology, Lyon (France).

[1]) All of the PMMA used in the experiments was Röhm & Haas Type II UVA Plexiglas sheet stock.

Table 1 summarizes the nine experiments conducted using the powder gun and the one reloading experiment conducted using the gas gun. The plate assembly which is mounted on the nose of the projectile determines the type of stress history to which the target is subjected. For shots 12101-13, a cavity machined in the backing plate (indicated by dashed lines in fig. 1) provided a free surface on the projectile nose plate from which an unloading wave originated. For shots 12115-19, the high impedance fused silica nose plate had a backing plate of low impedance PMMA. This arrangement produces a partial unloading. For the reloading experiment, 2131, the nose plate was PMMA and the backing plate was aluminum.

The target was fabricated from two pieces of PMMA joined by a thin (<0.02 mm thick) layer of epoxy cement. Prior to fabrication, a thin aluminum mirror, typically 100 nm thick, was vapor deposited on the surface of one of the plates. Because of its thinness, this mirror has negligible inertia and reached equilibrium with its surroundings in a fraction of a nanosecond. Thus, measurement of the motion of this mirror is essentially a direct measurement of the particle velocity in the sample. The velocity interferometer described in Ref. (3) was used to monitor the motion of the mirror.

The target was mounted on the gun using an optical alignment technique to minimize the tilt between the impact surfaces of the projectile and target (1). Also, to prevent an air cushion from forming between the projectile and the target prior to impact, the barrel of the gun and the target chamber were evacuated to less than 100 microns. An electrically charged pin technique was used to measure the projectile velocity. The resulting uncertainty in projectile velocity was about 0.5%. The tilt of the shock wave and the time at which the centers of the plates impacted were determined by electrically charged pins set flush with the target's impact surface. From knowledge of the time of impact, the time at which the shock wave arrives at the mirror, and the distance from the impact surface to the mirror, an average shock wave velocity may be calculated with an uncertainty of about 1%. For impact stresses between 6 and 12 kbar, Ref. (4) and (5) reported a slight decrease (≈2%) of wave velocity with increased propagation distance and, coincidently, a large amount of structure in the wave fronts. For impact stresses above 20 kbar, very little structure is evident in the compressive wave fronts; and hence, little change of wave velocity with propagation distance would be expected. The measured shock wave velocity is plotted as a function of the maximum particle velocity behind the shock front (fig. 2a) and the maximum or equilibrium strain (fig. 2b).

Also shown in fig. 2b is the variation of the velocity of the unloading wave front with equilibrium strain. In order to determine this velocity, the time at which the unloading wave arrives at the impact surface was calculated from the known properties of the fused silica (4). Subtraction of this calculated time from the time of arrival of this wave front at the mirror provides

Table 1. Summary of PMMA shots

Shot No.	Projectile velocity (mm/μsec)	Projectile thickness[1]) (mm)	Target thickness (mm)	Shock velocity (mm/μsec)	Particle velocities (mm/μsec)		Unloading wave velocity (mm/μsec)	Equilibrium state	
					v_0	v_{max}		Stress (kbar)	Strain
2131	.456	6.298	6.59	3.22	.175	.228	–	–	–
12101	.926	6.367	12.060	3.59	.618	.686	5.90	29.0	.190
12104	1.000	9.650	12.001	3.32	.318	.500	5.06	19.4	.154
12106	1.243	9.499	12.395	3.83	–	.897	6.79	40.7	.234
12111	1.242	9.606	12.639	3.88	.850	.895	6.88	41.2	.230
12113	1.453	9.525	12.682	4.08	.985	1.024	6.30	49.5	.250
12115	1.669	6.375	13.010	4.35	1.120	1.155	8.00	59.5	.266
12116	1.624	6.218	12.652	4.33	1.085	1.122	8.04	57.6	.259
12118	1.206	6.320	13.322	3.87	.821	.870	6.79	39.9	.225
12119	1.224	6.387	12.883	3.85	.817	.873	6.73	39.8	.227

[1]) The projectile nose plate was fused silica except for shot 12104 which was PMMA and shot 2131 which was PMMA and aluminum.

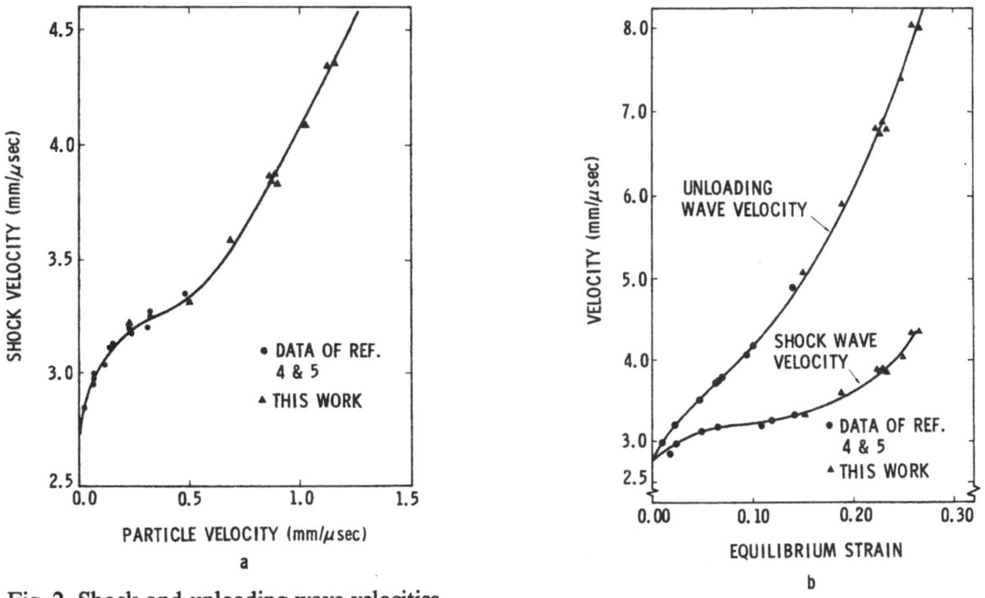

Fig. 2. Shock and unloading wave velocities

a transit time from which the unloading wave velocity may be calculated. Figs. 2a and 2b illustrate the large nonlinearities which are displayed by PMMA. We note that at a strain of 26% the unloading wave velocity has almost tripled from its zero strain value. This corresponds to the slope of the stress-strain curve increasing by a factor of almost nine.

Because the laser beam monitoring the motion of the mirror passes through shock compressed material, a slight correction must be made to the particle velocity measurements to account for the shock induced index of refraction change. For particle velocities below 0.32 mm/μsec, Ref. (4) reports this correction to be less than 1%. For higher particle velocities, the correction factor has not yet been accurately measured. However, by employing an impedance mismatch analysis, the value of the peak particle velocity may be obtained. The observed particle velocity history can then be normalized to this value. This analysis is credible because of the simple nature of the observed compressive waves at the higher particle velocities. As can be seen from fig. 3b, the compressive wave has very little structure and for all practical purposes is a shock wave[2]). For a shock wave propagating into an unstressed region which is at rest, the stress behind the wave, σ, is related to the particle velocity behind the wave, v, by[3])

$$\sigma = \varrho_0 U v \qquad [2.1]$$

where ϱ_0 is the reference density of the material (for PMMA, $\varrho_0 = 1.185$ gm/cc) and U is the shock velocity. Similarly, the known nonlinear elastic properties of fused silica given in Ref. (4) determine the relation between the stress, σ_1, and particle velocity, v_1, in the projectile nose plate[4]).

$$\sigma_1 = a v_1 + b v_1^2 + c v_1^3 + d v_1^4 . \qquad [2.2]$$

Since the stress and particle velocity must be continuous at the impact surface, $\sigma = \sigma_1$ and $v = v_m - v_1$, where v_m is the projectile velocity. Using these relations, [2.1] and [2.2] may be solved for v in terms of the measured quantities U and v_m. The resulting values for v are tabulated in table 1 under the heading v_{max}. Some of the observed particle velocity histories are shown in fig. 3 and 4.

The history observed in shot 12115, fig. 3b is typical of shots 12101 and 12106-19. Again we note the lack of structure in the transition region between the shock and the maximum particle velocity. Indeed, in shot 12115 this transition region accounted for only about 3% of the total

[2]) For shot 12104, fig. 4a, significant structure was present; however, in this experiment the projectile nosepiece was PMMA. For this arrangement symmetry considerations require the peak particle velocity to be one half of the projectile velocity.

[3]) Throughout this paper compressive stresses and strains are considered positive.

[4]) The coefficients for stresses in kbar and particle velocities in mm/μsec are: $a = 131.7$, $b = -73.61$, $c = 99.47$, and $d = -41.63$.

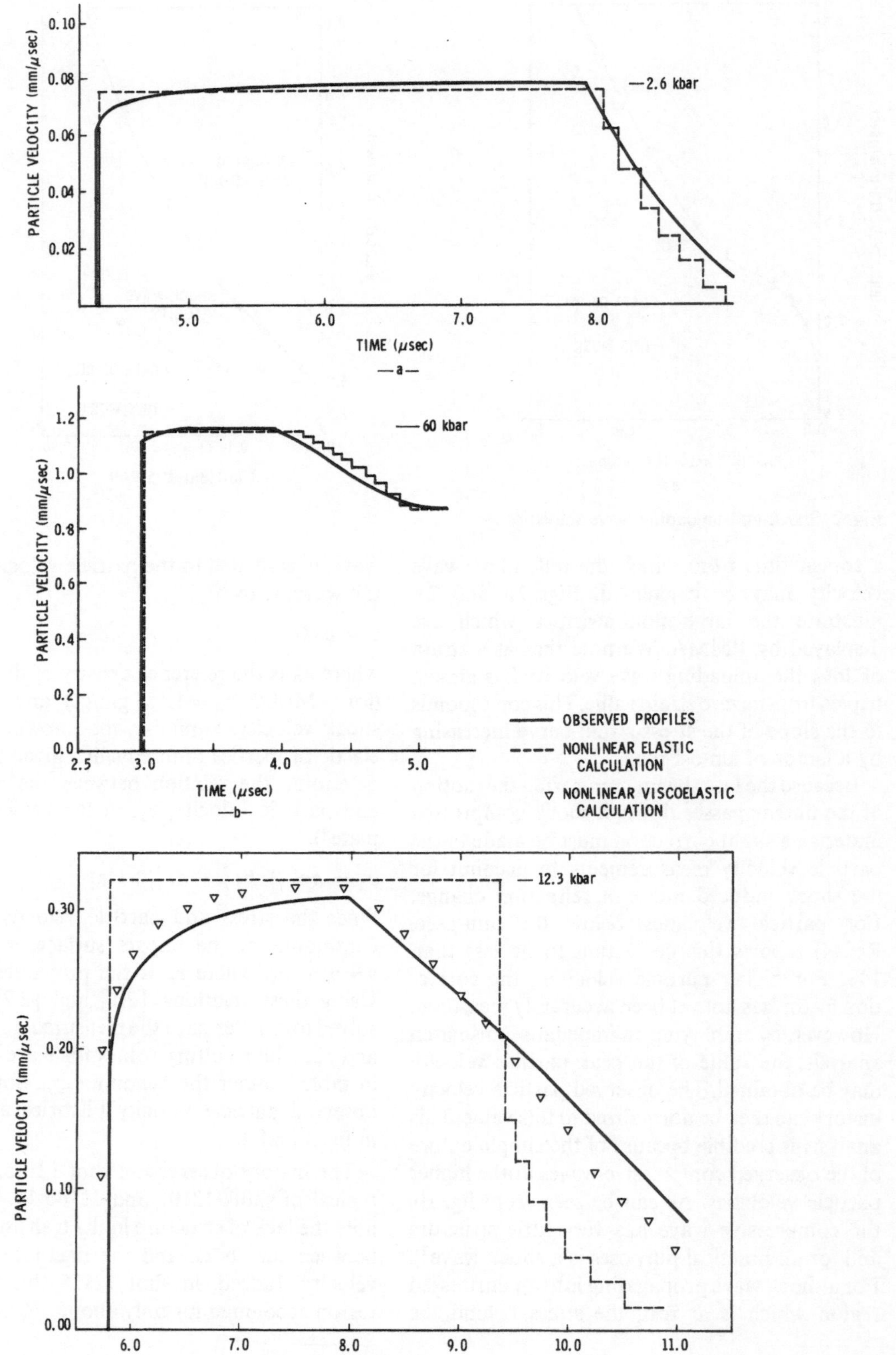

Fig. 3. Comparison of calculated and observed wave profiles. a) Shot 2110 [Ref. (4, 5)], b) Shot 12115, c) Shot 2104 [Ref. (4, 5)]

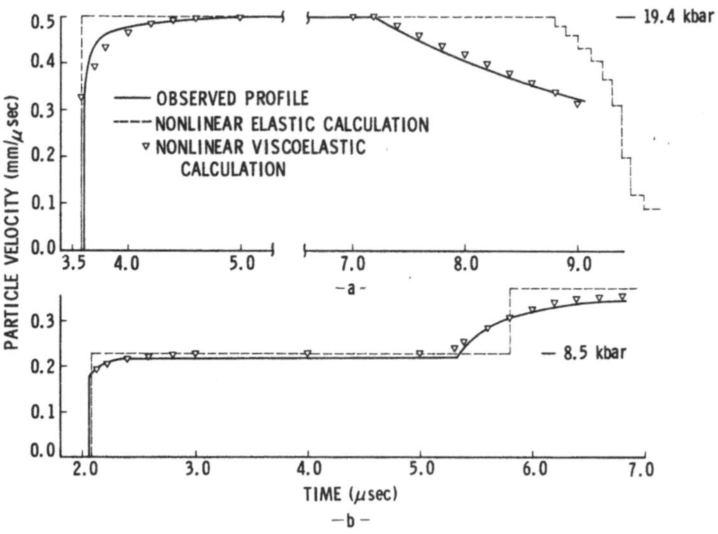

Fig. 4. Comparison of calculated and observed wave profiles. a) Shot 12104, b) Shot 2131

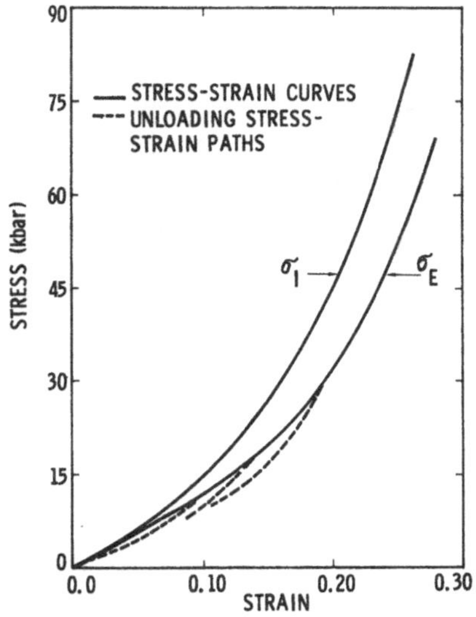

Fig. 5. Stress-strain curves and unloading paths

particle velocity change. The column in table 1 labeled v_0 indicates the particle velocity level at which this transition region begins. For shots 12106, 12111, 12118, and 12119, the transition region accounts for about 5% of the total particle velocity change. For these experiments, the stress and strain values shown in table 1 were calculated from the observed shock and particle velocities using [2.1] with the strain given by $\varepsilon = v/U$.

Fig. 4a shows the particle velocity history observed in shot 12104. Here a significant amount of structure is evident in the compressive wave. In order to calculate the stress-strain path followed by the material in this experiment, the procedure discussed in Ref. (4) was used. Basically this procedure treats the wave as a simple wave and applies $\Delta\sigma = \varrho_0 C(v)\,\Delta v$ and $\Delta\varepsilon = \Delta v/C(v)$ incrementally through the wave. Here $C(v)$ is the speed (assumed constant) at which a given level of particle velocity propagates. The same procedure was employed to obtain the stress-strain paths followed by the material during the unloading. Three of these unloading stress-strain paths are shown in fig. 5 as dashed lines. These unloading stress-strain paths illustrate the marked hysteresis present upon unloading from higher stress levels (>7 kbar). For unloadings from stress levels below 7 kbar, very little hysteresis is observed. This behavior suggests that a yield phenomenon occurs at about 7 kbar.

3. Constitutive models

In a number of recent studies (5–8), the dynamic mechanical response of PMMA below 7 kbar has been characterized by a nonlinear constitutive equation proposed by *Coleman* and *Noll* (9) which they termed *finite linear viscoelasticity.* The one dimensional motion of the material is represented by the function $u = u(X, t)$ which gives the displacement u of the material point, X, for all time, t. The constitutive assumption relates the present value of the stress, $\sigma(t)$, at any material point to the strain history, $\varepsilon(t-s) = -\partial_X u(X, t-s)$; $0 < s < \infty$, and its present value, $\varepsilon = \varepsilon(t)$:

$$\sigma(t) = \sigma_E(\varepsilon) + \int_0^\infty K(\varepsilon; s) \times \left\{\left(\frac{1-\varepsilon(t-s)}{1-\varepsilon}\right)^2 - 1\right\} ds. \qquad [3.1]$$

Here $\sigma_E(\varepsilon)$, the equilibrium response function, represents the response of the material when the strain has been held constant for all time and the stress relaxation function, $K(\varepsilon; s)$, is usually assumed to have the form (5, 6)

$$K(\varepsilon; s) = K(\varepsilon; 0) \exp\left(-\frac{s}{\tau}\right) \quad [3.2]$$

where $\tau > 0$ is the relaxation time and

$$K(\varepsilon; 0) = \frac{(1-\varepsilon)^2 [\sigma_I(\varepsilon) - \sigma_E(\varepsilon)]}{\tau\varepsilon(2-\varepsilon)}.$$

The function $\sigma_I(\varepsilon)$, the instantaneous response function, gives the response of the material to a sudden jump in strain at time, t, from an unstrained state. Because the experimental observations span only four decades of time ($10^{-9} - 10^{-5}$ sec), we feel that a single exponential relaxation function should suffice. The validity of this assumption at low stress levels has been established by the analysis of acoustic dispersion data (10).

By differentiating the constitutive eq. [3.1] with respect to time and using [3.2], we obtain (11):

$$\dot{\sigma} = E(\sigma, \varepsilon)\,\dot{\varepsilon} + G(\sigma, \varepsilon) \quad [3.3]$$

where

$$\dot{\sigma} = \frac{\partial \sigma}{\partial t}$$

$$\dot{\varepsilon} = \frac{\partial \varepsilon}{\partial t}$$

$$E(\sigma, \varepsilon) = \frac{d\sigma_I(\varepsilon)}{d\varepsilon} + \frac{d(\ln K(\varepsilon; 0))}{d\varepsilon}[\sigma - \sigma_I(\varepsilon)] \quad [3.4]$$

and

$$G(\sigma, \varepsilon) = -\frac{1}{\tau}[\sigma - \sigma_E(\varepsilon)]. \quad [3.5]$$

Materials which can be characterized by a general rate-equation of the type [3.3] are called *Maxwellian*. If we identify the tangent modulus, E, and the relaxation function, G, as those given in [3.4] and [3.5], then [3.3] and [3.1] are exactly equivalent only when a single relaxation time is used. However, more general forms for E and G may be used successfully to describe the behavior of viscoelastic materials (12). Furthermore, constitutive equations of the type [3.3] have been successfully applied to dynamic problems involving the viscoplastic response of metals (13).

In this latter case, the modulus, E, is taken to be a function of strain only and the relaxation function, G, is taken to be proportional to the plastic strain rate.

The experimental observation of a yield phenomenon at ≈ 7 kbar indicates that the constitutive model of finite linear viscoelasticity [3.1], is not appropriate for stress levels above ≈ 7 kbar. Motivated by the above considerations of viscoplastic constitutive equations, we propose to characterize PMMA over the entire range of stresses considered by a model of the type [3.3] which differs from that of finite linear viscoelasticity [3.1], in that the tangent modulus is a function of strain only:

$$E(\varepsilon) = \frac{d\sigma_I(\varepsilon)}{d\varepsilon}. \quad [3.6]$$

Furthermore, we generalize the relaxation function of finite linear viscoelasticity [3.5], and take the relaxation time to be a function of stress and strain:

$$G(\sigma, \varepsilon) = -\frac{\sigma - \sigma_E(\varepsilon)}{\tau(\sigma, \varepsilon)}. \quad [3.7]$$

This generalization was required in order to model the observed hysteresis upon unloading.

A least squares fit to the equilibrium stress-strain points which were determined from the compressive wave profiles (see table 1) yields the following σ_E function[5]):

$$\sigma_E(\varepsilon) = (k_E + l_E\varepsilon + m_E\varepsilon^2 + n_E\varepsilon^3)\,\varepsilon. \quad [3.8]$$

The coefficient $k_E = \varrho_0 C_E^2$ gives an equilibrium longitudinal sound speed, C_E, of 2.7526 mm/μsec which is slightly greater than the corresponding speed of 2.73 mm/μsec determined from acoustic data (10).

It has been shown (7) that unloading wave fronts can be modeled as acceleration waves which, in the context of singular surfaces, are fronts across which the particle acceleration undergoes a jump discontinuity. The calculation of the acceleration wave speed, C, based on the rate eq. [3.3] shows that $\varrho_0 C^2(\varepsilon) = E(\varepsilon)$. Thus, the measured unloading wave speed was fit as a least-square polynomial of the equilibrium strain to give[6]):

---

[5]) The coefficients of this fit are: $k_E = 89.79$ kbar, $l_E = 700$ kbar, $m_E = -5869$ kbar, and $n_E = 19652$ kbar.

[6]) The coefficients of this fit are: $k_I = 90.31$ kbar, $l_I = 1414$ kbar, $m_I = -6779$ kbar, and $n_I = 41600$ kbar.

$$E(\varepsilon) = k_I + l_I\varepsilon + m_I\varepsilon^2 + n_I\varepsilon^3. \quad [3.9]$$

This fit determined the solid line labeled unloading wave velocity shown in fig. 2b. The coefficient $k_I = \varrho_0 C_I^2$ gives an instantaneous longitudinal sound speed, C_I, of 2.7606 mm/μsec which correlates well with the speed of 2.763 mm/μsec determined from acoustic data (14). Using [3.9] in [3.6], the instantaneous curve, σ_I, can be determined by a simple integration. The curves σ_I and σ_E are shown in fig. 5.

At present no direct means is available to determine the dependence of the relaxation time on stress and strain. Examination of the relaxation functions employed in viscoplastic studies of metals (13) and recent work on thermally activated flow processes in polymers (17) indicate that the relaxation function and, hence, the relaxation time should depend on the shear stress and, in the case of polymers, on the hydrostatic pressure. We had attempted to resolve the uniaxial stress observed in our experiments into pressure and shear components by using available hydrostatic data. Unfortunately, all of the available data were obtained at low strain rate, making comparison of these data with our high strain rate data questionable. For this reason, we will consider the relaxation time to be a function of the over-stress measured from the equilibrium stress-strain curve. Comparing the numerical calculations with the compressive wave profiles and the reloading wave profile indicated that the relaxation time should decrease with increasing over-stress. The function

$$\tau = \tau_0 \exp\left[-\left(\frac{\sigma - \sigma_e}{k}\right)\right], \quad \sigma - \sigma_E > 0, \quad [3.10]$$

where $\tau_0 = 0.25$ μsec and $k = 0.8$ kbar, was found to give reasonable fits to the loading and reloading wave profiles. Upon unloading ($\sigma - \sigma_E < 0$), we originally attempted to follow the conventional practice of viscoplasticity theory and set the relaxation function to zero, i.e., letting $\tau \to \infty$. However, this resulted in unloading waves which had the correct arrival times but were much steeper than experimentally observed. Thus, it appears that even on unloading appreciable relaxation occurs. In order to model this relaxation, the following function was used for values of $\sigma - \sigma_E < 0$:

$$\tau = \tau_0 \left[2 - \exp\left(-\left(\frac{|\sigma - \sigma_E|}{k}\right)\right)\right], \quad \sigma - \sigma_E < 0. \quad [3.11]$$

At $\sigma - \sigma_E = 0$, the values and derivatives of [3.10] and [3.11] are continuous.

The nonlinear elastic model which is compared to the wave profiles was determined using the conventional procedure of obtaining a close fit to the shock velocity-particle velocity data. Due to the complicated variation of shock velocity with particle velocity (fig. 2a), it was found convenient to calculate stress and strain values from the shock and particle velocity points and fit these points. The resulting stress-strain curve is[7]):

$$\sigma_s = k_s\varepsilon + l_s\varepsilon^2 + m_s\varepsilon^3 + n_s\varepsilon^4.$$

The shock and particle velocity corresponding to this curve are shown as the solid line in fig. 2a.

4. Comparison of numerical and experimental results

In this section, comparisons are made between the experimentally observed wave profiles and profiles calculated numerically using the viscoelastic and elastic models described in Section 3. The calculations based on the viscoelastic model were obtained using the finite-difference computer code, WONDY IIIA (15); while those involving the elastic model were obtained using the characteristic code, SWAP-9 (16).

In fig. 3, we show the comparison for three different impact conditions, in which the peak stress attained varies from 2.6–60 kbar. Fig. 3a compares a wave profile with a peak stress of 2.6 kbar and the elastic calculation. There is fairly good agreement both in the compressive part of the wave and in the unloading. The slight rounding behind the shock and the shallower unloading wave which was experimentally observed indicate that even at this low stress level rate effects are present, although of secondary importance. Fig. 3b shows a wave profile with a peak stress of 60 kbar and again the elastic calculation compares fairly well with the observed profile; however, the calculated arrival time of the unloading wave is somewhat later than that observed[8]). While the predictions of the visco-

[7]) Here $k_s = 87$ kbar, $l_s = 859$ kbar, $m_s = -7063$ kbar, and $n_s = 22040$ kbar.

[8]) It should be noted that while the discrepancies between the elastic calculation and the experimental observations are small for these experiments, there may be situations, such as the attenuation of thin stress pulses (6) in which these discrepancies are significant and a viscoelastic description would be required.

elastic model are not shown in fig. 3a and b; this model predicts wave forms which are also in excellent agreement with the experimental observations. In fig. 3c, we show the comparison for a wave with a peak stress of 11 kbar. Here, we see that the elastic calculation does a rather poor job of reproducing the observed profile. There is significant rounding behind the shock front in the observed profile but not in the calculated profile. Furthermore, the calculated unloading wave arrives late and is considerably steeper than that observed. The much better agreement between the calculated wave profile based on the viscoelastic model and the observed profile indicates the importance of rate-effects at this stress level. It should be noted that in each of the profiles shown there is a slight discrepancy between the maximum particle velocity observed and that calculated. This is due to the fact that the calculated value is based on the fit of all the data and hence it will not in general agree exactly with that observed in an individual experiment.

As a further illustration of the significance of rate-effects in the range of 7–40 kbar, we show, in fig. 4a, a loading-unloading wave with a peak stress of 19 kbar and, in fig. 4b, a loading-reloading wave with the loading to 8 kbar and reloading to 14 kbar. For the loading-unloading wave, it was seen that the elastic calculation is not satisfactory, particularly with regard to the unloading wave. However, the viscoelastic calculation gives excellent agreement. For the recompression profile in fig. 4b, it is seen that the elastic calculation is reasonable for the initial part of the wave; however, the arrival of the recompression wave is late and the wave is predicted to be a shock. On the other hand, the viscoelastic calculation gives good agreement except in the vicinity of the foot of the reloading wave where there has been some smoothing of the profile.

As we have already mentioned, reloading of PMMA is accomplished by a smooth transition to a higher particle velocity. Predicting such transitions should be extremely sensitive to the relaxation function chosen and this suggests that such experiments may prove useful in determining the appropriate function. In view of the results, particularly those in fig. 3c, we feel that such further refinements in the viscoelastic model are warranted. A more detailed study of reloading of PMMA will therefore be considered in a future paper.

5. Conclusion

In this paper, we have attempted to characterize the dynamic (short time) response of PMMA which, it is felt, should be representative of many solid polymers. Using a high resolution laser interferometer, the detailed structures of shock waves produced in plate impact experiments were observed. Stress levels up to 60 kbar were achieved in these experiments and the results show that the stress-strain behavior of PMMA exhibits large strain nonlinearities. Moreover, between ≈ 7 and ≈ 40 kbar, significant rate-dependence was observed as evidenced by greater hysteresis upon unloading and marked dispersion upon reloading. Numerical calculations based on an elastic model were compared with the experimental profiles. It was found that outside the range of ≈ 7–40 kbar a nonlinear elastic description of the material behavior is sufficient for most practical purposes. For the stresses between ≈ 7–40 kbar; the elastic model is simply not appropriate.

To model observed behavior over the entire stress range, we proposed a viscoelastic constitutive relation and gave a procedure for evaluating the relevant material functions from the experimental data. Using this model, numerical calculations of wave profiles, even in the range of ≈ 7–40 kbar, were found to agree favorably with those observed.

Acknowledgment

The authors wish to thank *C. Konrad*, *B. Hardy*, and *M. Gonzales* for their able assistance in fabricating and conducting the experiments.

Summary

In order to investigate the short time response of solid polymers, a high resolution laser interferometer has been employed to observe the detailed structure of stress waves produced by the impact of polymeric plates. From the observed stress waves the loading and unloading stress-strain paths followed by the material can be determined. For the polymer, polymethyl methacrylate, over the stress range from 0–60 kbar such observations have disclosed a stress-strain behavior which exhibits large strain nonlinearities and significant rate-dependence. For stress levels below ≈ 7 kbar and above ≈ 40 kbar, rate-effects are of secondary importance and the observed wave profiles may be modeled using a nonlinear elastic description of the material behavior. However, at about 7 kbar, a transition is observed in the dynamic stress-strain behavior which is characterized by greater stress relaxation and significant hysteresis upon unloading. This transition, which we attribute to a yield phenomenon, requires a rate-dependent description of the material behavior to be used in the stress range of approximately 7–40 kbar.

References

1) *Barker, L. M.* and *R. E. Hollenbach*, Rev. Sci. Instr. **35**, 742 (1964).

2) *May, R. P.*, Description of the Sandia Laboratories Research Multi-Purpose, Hypervelocity Gun Facility, Sandia Laboratories Rept. DR-72-0180 (1972), Albuquerque, New Mexico.

3) *Barker, L. M.*, Fine Structure of Compressive and Release Wave Shapes in Aluminum Measured by the Velocity Interferometer Technique. In: Proc. IUTAM Symp. on High Dynamic Pressure, p. 483 (1968).

4) *Barker, L. M.* and *R. E. Hollenbach*, J. Appl. Phys. **41**, 4208 (1970).

5) *Schuler, K. W.*, J. Mech. Phys. Solids **18**, 277 (1970).

6) *Nunziato, J. W.* and *K. W. Schuler*, J. Mech. Phys. Solids (in press).

7) *Schuler, K. W.*, The Speed of Propagation of Release Waves in Polymethyl Methacrylate. In: Proc. Fifth Internat. Symp. on Detonation (1970).

8) *Nunziato, J. W., K. W. Schuler*, and *E. K. Walsh*, Trans. Soc. Rheol. **16**, 15 (1972).

9) *Coleman, B. D.* and *W. Noll*, Rev. Mod. Phys. **33**, 239 (1961); erratum ibid, **36**, 1103 (1964).

10) *Nunziato, J. W.* and *H. J. Sutherland*, J. Appl. Phys. **1973**, 184

11) *Herrmann, W.* and *J. W. Nunziato*, Nonlinear Constitutive Equations. In: Dynamic Response of Materials Under Impulsive Loading, p. 123 (1973).

12) *Greenberg, J. M.*, Quart. Appl. Math. **26**, 27 (1968).

13) *Herrmann, W.*, Nonlinear Stress Waves in Metals. In: Proc. ASME Symp. on Wave Propagation in Solids, p. 129 (1970).

14) *Asay, J. R., D. L. Lamberson*, and *A. H. Guenther*, J. Appl. Phys. **40**, 1768 (1969).

15) *Lawrence, R. J.*, A General Viscoelastic Constitutive Relation for Use in Wave Propagation Calculations, Sandia Laboratories Rept. RR-72-0114 (1972), Albuquerque, New Mexico.

16) *Barker, L. M.*, SWAP-9: An Improved Stress Wave Analyzing Program. Sandia Laboratories Rept. RR-69-233 (1969), Albuquerque, New Mexico.

17) *Davis, L. A.* and *C. A. Pampillo*, J. Appl. Phys. **42**, 4659 (1971).

Authors' address:

K. W. Schuler and *J. W. Nunziato*
Albuquerque, New Mexico (USA)

Rheol. Acta 13, 274–277 (1974)

Laboratoire de Rhéologie – Institut de mécanique technique, Sofia (Bulgarie)

Sur le comportement rhéologique des résines thermoréactives durcies et chargées

Par J. Simeonov et J. Christova

Avec 7 figures

(Reçu p. p. le 27 octobre 1972)

Notes générales

Des résines thermoréactives, remplies de charges macrogranulées d'origine rocheuse, forment des compositions qui ont été appelées bétons-polymères. A l'instar de tous les systèmes polymères, également les bétons polymères réunissent en soi des propriétés caractéristiques pour les liquides et pour les corps solides, c'est-à-dire qu'ils sont des matériaux viscoélastiques. Les charges élémentaires forment le squelette élastique dur du corps viscoélastique, alors que le liant polymère (la résine thermoréactive) emplit les espaces entre les éléments durs. Les modèles rhéologiques qui imitent le comportement des matériaux réels sont utilisées lors de la description du comportement rhéologique des corps viscoélastiques. Cependant, les modèles rhéologiques n'imitent que le comportement macroscopique et n'élucident pas la base physico-chimique et moléculaire des propriétés viscoélastiques. Dans le présent travail l'étude du comportement rhéologique des corps viscoélastiques est différente. Les propriétés rhéologiques des bétons-polymères sont étudiées en fonction de leur structure: elles sont examinées en tant que systèmes dispersés durs, constitués par deux phases solides. La phase dispersante, où la matrice est le liant (la résine thermoréactive), alors que les charges minérales sont la phase dispersée. Ces systèmes se caractérisent par une structure nettement formée et par une différence sensible dans les propriétés rhéologiques des deux phases solides. Ils peuvent être divisés en trois groupes structuraux selon que la charge est ségrégée ou agrégée, et que le système est compact ou poreux (1, 2, 3). Dans le premier groupe se trouvent des systèmes compacts chez lesquels les grains (les particules) de la charge sont espacés (charge ségrégée), alors que la matrice est compacte. On appelle aussi de tels systèmes: liant rempli. Au fur et à mesure de l'augmentation de la quantité de la charge ses grains se rapprochent progressivement jusqu'à se toucher (charge agrégée). Les systèmes à charge agrégée et à matrice compacte s'avèrent être une limite entre les systèmes relevant des groupes premier et second. Par l'augmentation ultérieure de la quantité de la charge, l'agrégat formé reste dans le même état (sans variations), mais dans la matrice apparaissent des pores isolées, à cause de l'insuffisance en liant. De tels systèmes sont classées dans le second groupe qu'on appelle aussi «charge liée». Le troisième groupe se caractérise par une charge agrégée et une matrice discontinue, dont les pores et les creux sont liés et occupés par une troisième phase, une phase fluide. Les systèmes du troisième groupe présentent une charge cimentée avec des ponts de liants.

Les propriétés rhéologiques du système général dépendent des propriétés rhéologiques du liant polymère, de la corrélation quantitative entre les deux phases solides, du volume intérieur et de la surface intérieure du système, des efforts interphases au contact et des influences exercées par le milieu ambiant.

Sur les propriétés rhéologiques et le comportement des systèmes influent encore les contraintes intérieures préalables, apparues lors de la formation du système dispersé (par exemple, par suite de la variation du volume de la matrice au cours du processus de durcissement), ainsi que la redistribution des champs des contraintes avec le temps, par suite de la différence dans les propriétés rhéologiques des deux phases solides. Indépendamment de la structure ces facteurs ne peuvent être séparés et pris en considération. Lors de nos recherches, leur influence sur le comportement des systèmes avait été évaluée en même temps que la structure, pour autant qu'elles se manifestent de manière différente dans des systèmes relevant des trois groupes.

Matériaux étudiés

Les bétons-polymères étudiés sont obtenus sur la base de la résine polyester non saturée bulgare «Vinalkide 550 P». La phase dispersante ou la matrice des systèmes n'est pas un liant pur (résine polyester non saturée à durcisseurs), mais elle constitue le liant et les éléments fins de la charge, c'est-à-dire qu'elle représente aussi un système dispersé du type «liant chargé». L'utilisation d'une matrice «liant chargé» est dictée par des considérations quant à l'augmentation de la dureté de la matrice et des systèmes visant à la diminution de la quantité du liant pur, nécessaires à l'obtention des systèmes compacts. Le microsquelette représente une farine de diabase avec une superficie spécifique considérable (5000 cm^2/g environ) et des hauts rapports d'adhésion avec la résine.

La phase dispersée ou la charge, constitue un mélange de diabase granulé. Le nombre des fractions est prédéterminé par le grain maximum de la plus grosse fraction et la règle de la granulométrie discontinue.

Là seront indiqués des résultats obtenus lors de l'étude des systèmes avec une matrice à composition et quantité égales (respectivement à quantité en liant pur égale), mais à charge différente de granulométrie et de composition fractionnaire. On procédera à une comparaison entre les propriétés rhéologiques des systèmes compacts à charge agrégée (cas limite entre systèmes à «liant rempli» et systèmes à «charge liée») et systèmes poreux à pores liées.

Résultats expérimentaux et discussions

La viscoélasticité des matériaux réels trouve une expression qualitative dans les processus de fluage et de relaxation (4). Pour l'étude du comportement des bétons-polymères nous nous basions sur les résultats expérimentaux obtenus lors de l'étude du fluage. On étudiait la grandeur et l'évolution de la déformation avec le temps sous l'action des contraintes de longue durée à indice et à intensité variés.

Le liant polymère qui forme les éléments fins de la matrice ou la «microstructure» des systèmes, est porteur des propriétés de viscosité des bétons-polymères. Toujours est-il que sa quantité ne détermine pas un comportement rhéologique univoque de ces matériaux. Ainsi que le montrent les recherches, leur comportement varie selon que le squelette de la matrice est continu ou non, c'est-à-dire si le système est compact ou poreux, et dans les systèmes poreux – selon que les pores sont isolées ou liées. Cela signifie que la quantité du liant polymère influe sur le comportement rhéologique des bétons-polymères au moyen de leur structure, formée dans des conditions données. La granulométrie et la composition granulométrique de la charge qui déterminent le volume intérieur et la surface interphase, exercent une influence substantielle sur la formation de la structure et, de là, également sur les propriétés rhéologiques des systèmes dispersés de cette espèce. L'augmentation de la dispersion et de la surface interphase du système peut avoir une répercussion différente sur ses caractéristiques en fonction des conditions. Quand la quantité de la matrice reste constante, la plus grande dispersion de la charge influe d'une façon défavorable sur le comportement rhéologique des bétons-polymères.

Ainsi qu'on le verra de l'analyse des résultats exposés, des systèmes ayant la même quantité de liant, mais une charge à granulométrie différente et une compacité de la matrice variée, accusent des déformations à évolution et à grandeur différentes, pour une intensité des contraites égale.

Aux figures suivantes seront comparées les déformations lors du fluage de systèmes à charge agrégée et à squelette compact (système S_1), ainsi que de systèmes à squelette poreux à pores liées (systèmes S_2 et S_3).

Au moyen de la détermination de la porosité des systèmes, à l'aide des méthodes physiques, ont peut obtenir des données sur l'existence des pores, tandis que pour le caractère des pores (isolées ou liées) on peut juger d'après la variation avec le temps de la masse d'éprouvettes, immergées dans l'eau. Concernant les systèmes étudiés, tout cela est montré à la fig. 1.

Comme on le voit, le système S_1 est compact, alors que les systèmes S_2 et S_3, sont poreux et renferment une quantité variée de pores liées. Cela donne une répercussion sur le comportement rhéologique de ces systèmes. Les courbes illustrées à la fig. 2 montrent que les déformations lors du fluage, sous l'effet de la contrainte à la compression $\sigma_c = 0.4 R_c$, accusent une évolu-

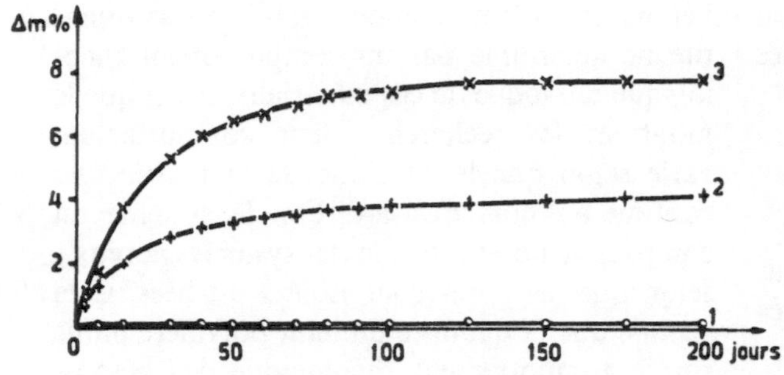

Fig. 1. Variations avec le temps de la masse d'éprouvettes immergées dans l'eau. 1. Système S_1; 2. Système S_2; 3. Système S_3

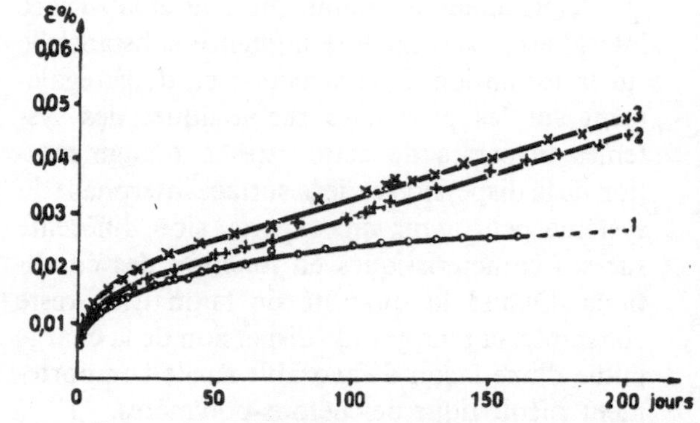

Fig. 2. Déformations au fluage sous l'effet d'une contrainte axiale à la compression $\sigma_c = 0{,}4 R_c$. 1. Système S_1; 2. Système S_2; 3. Système S_3

tion et une grandeur variées pour des systèmes compacts et des systèmes poreux à charge agrégée et une quantité de matrice égale (respectivement – un liant pur). On voit la même chose également à la fig. 3, où sont présentées les déformations, provoquées par unité de tension à une intensité de $0{,}4 R_c$.

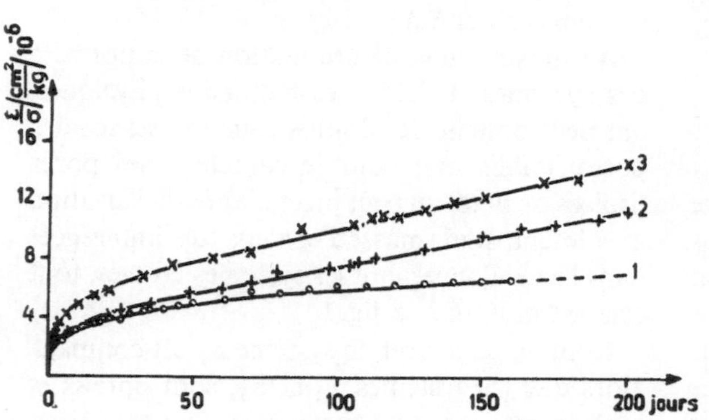

Fig. 3. Déformations au fluage sous l'effet d'une contrainte axial à la compression

$$\sigma_c = 1 \left(\frac{\text{kg}}{\text{cm}^2}\right).$$

1. Système S_1; 2. Système S_2; 3. Système S_3

La fig. 4 montre les déformations pour les systèmes S_1 et S_2, résultats de l'effect de traction et d'une tension à intensité variée ($\sigma_t = 0{,}1 R_t$; $0{,}2 R_t$; $0{,}3 R_t$). Ici l'influence de la structure se manifeste dans variation de la grandeur des déformations qui, dans les systèmes poreux, est sensiblement plus grande. Ce qui est montré également à la fig. 5, où sont illustrées les déformations par unité de tension.

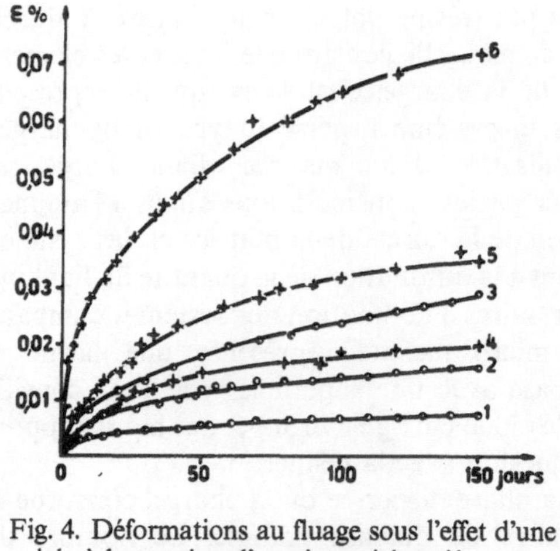

Fig. 4. Déformations au fluage sous l'effet d'une tension axiale à la traction d'une intensité variée.

$$1 \text{ et } 4 = \frac{\sigma_t}{R_t} = 0{,}1; \quad 2 \text{ et } 5 = \frac{\sigma_t}{R_t} = 0{,}2;$$

$$3 \text{ et } 6 = \frac{\sigma_t}{R_t} = 0{,}3;$$

○ = Système S_1; + = Système S_2

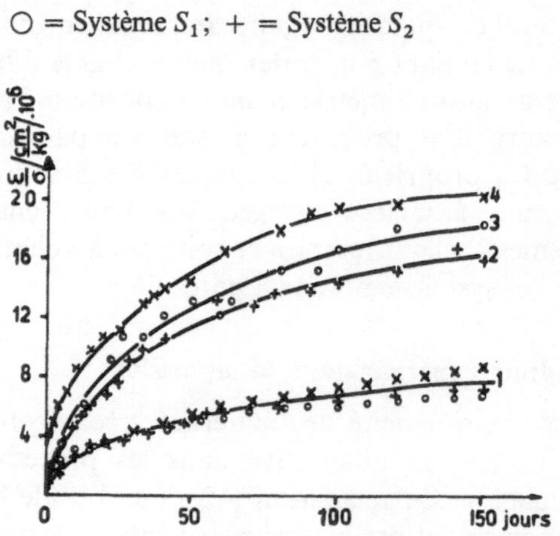

Fig. 5. Déformations au fluage sous l'effet d'une tension axiale à la traction

$$\sigma_t = 1 \left(\frac{\text{kg}}{\text{cm}^2}\right)$$

à une intensité

$$\bigcirc = \frac{\sigma_t}{R_t} = 0{,}1; \quad + = \frac{\sigma_t}{R_t} = 0{,}2; \quad x = \frac{\sigma_t}{R_t} = 0{,}3.$$

1. Système S_1; 2., 3 et 4. Système S_2

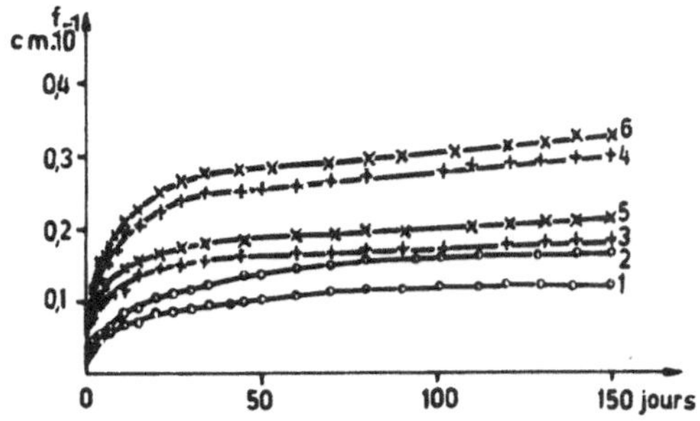

Fig. 6. Variations avec le temps du fléchissement des poutres soumises à la fléxion sous l'effet des tensions à intensité variée

○ – $\frac{\sigma_f}{R_f} = 0{,}2$; + – $\frac{\sigma_f}{R_f} = 0{,}3$.

1., 2. Système S_1; 3. 4. Système S_2; 5., 6. Système S_3

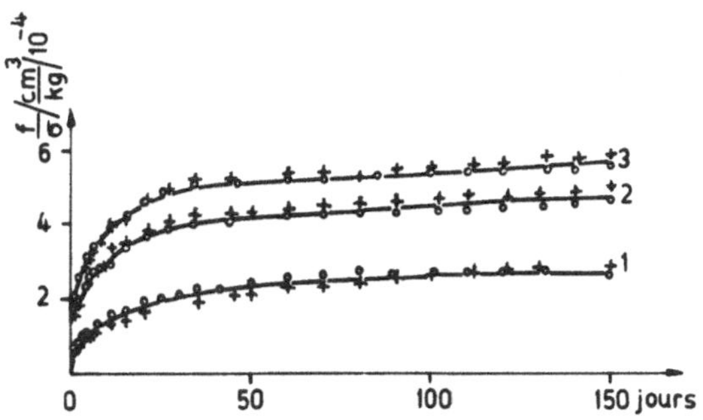

Fig. 7. Variations avec le temps du fléchissement des poutres soumises à la flexion sous l'effet d'une tension

$\sigma_f - 1 \left(\frac{kg}{cm^2}\right)$

à une intensité

○ – $\frac{\sigma_f}{R_f} = 0{,}2$; + – $\frac{\sigma_f}{R_f} = 0{,}3$.

1. Système S_1; 2. Système S_2; 3. Système S_3

L'étude du fluage sous l'effet des contraintes de flexion, aboutit également à des conclusions analogues. En ce qui concerne les systèmes S_1, S_2 et S_3, on voit à la fig. 6 une augmentation avec le temps du fléchissement des poutres, soumises à la flexion, tandis que la fig. 7 nous montre l'augmentation du fléchissement sous l'effet de l'unité de contrainte de flexion. La grandeur des déformations lors de la flexion, est de même différente pour les systèmes compacts et les systèmes poreux, à intensité égale des contraintes.

Conclusions

1. Par une matrice de composition et de quantité égales, et respectivement à liant pur, de quantité et de genre égaux, peuvent être obtenus des bétons-polymères à structure et, de là, avec des propriétés rhéologiques variées. Cela dépend de la granulométrie et de la composition fractionaire de la charge.

2. Le fluage des bétons-polymères, à charge agrégée et à liant polymère en quantité égale, représentant des systèmes poreux à pores liées, s'avère plus intensif, en comparaison avec les systèmes compacts.

Littérature

1) *Bares, R.*, Bulletin RILEM **1965**, N° 28.
2) *Bares, R.*, Bulletin RILEM **1965**, N° 28.

Adresse des auteurs:

J. Simeonov et *J. Christova*
Laboratoire de Rhéologie
Institut de Mécanique Technique
Bul. Lenin 4, «A»
Sofia 13 (Bulgaria)

Rheol. Acta 13, 278–282 (1974)

From the Snam Progetti-Laboratori Studi e Ricerche, S. Donato Milanese (Italy)

Molecular weight distribution from rheological measurements

By *G. Locati, L. Gargani, and A. De Chirico*

With 2 figures and 2 tables

(Received October 27, 1972)

Introduction

The determination of the molecular weight distribution of polymers can be performed by a variety of methods, both absolute and relative. The absolute methods (fractionation, light scattering, osmometry) are used only when strictly necessary, due to lenghty and/or experimentally difficult procedures; the relative methods (GPC, ultracentrifugation, intrinsic viscosity) need a calibration by absolute methods. All of these methods utilize the polymer dissolved in solvents.

Some attempts have been made to obtain the molecular weight distribution from measurements on polymer melts, by elaborating stress relaxation curves [for instance *Peticolas* (1), *Conti* (2), *Giuliani* (3)], or rheological curves [for instance *Graessley* (4), *Conti* (5)]. These methods, however, are quite complex and have not become routine methods for evaluating molecular parameters. Besides, they rest on theories whose validity has not completely proved.

We propose here a simple relative method for characterizing the molecular weight distribution by two parameters: the weight-average molecular weight and an index of polydispersity. The method is based on the measurement of two rheological properties, one measured in solution, the other on the melt and it is limited to polymers which are known, *a priori*, to follow a log-normal molecular weight distribution.

Description of the method

It is generally accepted that the zero-shear viscosity, η_0, of monodisperse polymers is proportional to the 3.4 (or 3.5) power of the molecular weight, at constant temperature

$$\log\eta_0 = \log k + 3.4 \log M \quad (M > M_c). \qquad [1]$$

The extension of this relationship to polydisperse polymers, however, is not universally accepted and has aroused some discussion.

Some polymers, like polyisobutylene and polystyrene were shown to follow a law of proportionality to the 3.4 power of the weight-average molecular weight, $\bar{M}_w$ (6, 7, 8, 9); for other polymers, e.g. polyethylene, the zero-shear viscosity was found experimentally to be related to some average molecular weight, $\bar{M}_t$, by a relationship (10)

$$\eta_0 = k_1 \bar{M}_t^{3.4} \qquad [2]$$

where k_1 is a new constant and $\bar{M}_w < \bar{M}_t < \bar{M}_z \cdot \bar{M}_t$ seems to be much closer to $\bar{M}_w$ when the molecular weight distribution is narrow and approaches $\bar{M}_z$ as the molecular weight distribution broadens.

We shall discuss soon elsewhere the apparent discrepancy between results obtained on different polymer species, which seems arising from their peculiar kind of molecular weight distribution.

Let us consider now polymers with log-normal molecular weight distribution (11)

$$w(M)\, dM = \frac{1}{\beta\sqrt{\pi} M} \exp\left[-\left(\frac{1}{\beta} \ln \frac{M}{M_0}\right)^2\right] dM \qquad [3]$$

$w(M)\, dM$ is the weight fraction of polymer with a molecular weight between M and $M + dM$; M_0 and β are respectively the average value and the breadth of the distribution. They are related to the weight-average molecular weight, $\bar{M}_w$, and to the index of polydispersity, $Q = \bar{M}_w/\bar{M}_n$, by eqs. [4] and [5]

$$M_0 = \bar{M}_w \exp(-\beta^2/4) \qquad [4]$$

$$\beta^2 = 2 \ln Q. \qquad [5]$$

We have shown that the zero shear viscosity of polymers having log-normal molecular weight distribution, e.g. polyethylene, can be written (12)

$$\log\eta_0 = \log k + \alpha \log \bar{M}_w + \beta \log Q. \qquad [6]$$

(Here β is not to be confused with β of eq. [3].) For reasons of consistency, the values of the coefficients $\log k$ and α

must reduce to the values found for monodisperse polymers, eq. [1].

The specific values of the coefficients of eq. [6] can be obtained either from experiments, as we did (12), or from structural theories on viscosity, for instance by the *Graessleys* theory (4). In our work the determination of the coefficients was made for polyethylene by multiple regression analysis. Both the experimental and the theorical methods were shown to satisfy the consistency condition. As for the coefficient β, its experimental value turned out to be about one third (or one half) of the theoretical value calculated by the *Graessleys* theory. The experimental value seems to be more acceptable, as discussed. Table 1 shows the values of the coefficients obtained from data taken from the literature (10).

Table 1

a) Coefficients of eqs. [6] and [11] obtained from data taken from ref. (10).

$$\log\eta_0 = -13.19 + 3.498\log\bar{M}_w + 0.8883\log Q \quad (T = 190\,°\mathrm{C})$$

$$\log\{\eta\} = -3.590 + 0.7658\log\bar{M}_w - 0.08797\log Q \quad (T = 130\,°\mathrm{C}, \text{tetralin})$$

b) Correlation coefficients

	$\log\bar{M}_w$	$\log Q$	Total
$\log\eta_0$	0.985	0.936	0.991
$\log\{\eta\}$	0.990	0.858	0.993

A procedure similar to that just used for deriving the dependence of the zero-shear viscosity from the weight-average molecular weight and an index of polydispersity can be applied to the intrinsic viscosity, as it is well known (13). The procedure goes as follows. The intrinsic viscosity of monodisperse polymers is proportional to the molecular weight, the exponent being a certain value "a" ($0.5 < a < 0.1$)

$$[\eta] = k' M^a. \quad [7]$$

This relationship can be extended to polydisperse polymers by defining an appropriate average molecular weight $\bar{M}_v$

$$\bar{M}_v = \left[\frac{\int_0^\infty w(M)\,M^a\,dM}{\int_0^\infty w(M)\,dM}\right]^{1/a} \quad [8]$$

$$[\eta] = k' \bar{M}_v^a. \quad [9]$$

By introducing the log-normal molecular weight distribution, eq. [3] into eq. [8], and integrating we obtain

$$\bar{M}_v = \bar{M}_w Q^{(a-1)/2} \quad [10]$$

so that

$$\log[\eta] = \log k' + \gamma\log\bar{M}_w + \delta\log Q \quad [11]$$

where

$$\gamma = a; \quad \delta = -\frac{a(1-a)}{2}.$$

Again the coefficients of eq. [11] must satisfy the consistency condition which holds for the generalized equation for the zero-shear viscosity, eq. [6].

The $\log Q$ coefficient, δ, cannot be positive and also is a function of a only. This last statement amounts to say that the determination of the coefficient δ requires only measurements on monodisperse polymers.

Eq. [6] and [11], linear in the variables $\log\bar{M}_w$ and $\log Q$, can now be used to evaluate $\bar{M}_w$ and Q. For this purpose we need first a precise determination of the coefficients of the two equations for a given polymer species (calibration). $\bar{M}_w$ and Q for a polymer of the some species are then obtained by introducing the values η_0 and $[\eta]$, measured at the same temperature used for calibration, and solving the system of eq. [6] and [11].

Error analysis

We estimate now the errors which affect the weight-average molecular weight and the index of polydispersity calculated by the method we have suggested.

For this purpose we assume that the coefficients of eqs. [6] and [11] were determined with high precision. Possible errors of calibration, however, would affect the absolute value of the parameters characterizing the molecular weight distribution, not their relative values.

For sufficiently small errors of measurement of η_0 and $[\eta]$, a linearized error analysis is applicable. Eqs. [6] and [11] can be written:

$$\alpha\ln\bar{M}_w + \beta\ln Q = \ln\eta_0 + \text{const} \quad [12a]$$

$$\gamma\ln\bar{M}_w + \delta\ln Q = \ln[\eta] + \text{const}. \quad [12b]$$

Using differentials and logarithmic differentiation, the relative changes of the molecular parameters are

$$\frac{\Delta\bar{M}_w}{\bar{M}_w} = \frac{\beta}{\beta\gamma - \alpha\delta}\frac{\Delta[\eta]}{[\eta]} - \frac{\delta}{\beta\gamma - \alpha\delta}\frac{\Delta\eta_0}{\eta_0} \quad [13a]$$

$$\frac{\Delta Q}{Q} = \frac{-\alpha}{\beta\gamma - \alpha\delta}\frac{\Delta[\eta]}{[\eta]} + \frac{\gamma}{\beta\gamma - \alpha\delta}\frac{\Delta\eta_0}{\eta_0}. \quad [13b]$$

Calling ε the maximum percentage error on the measurements of $[\eta]$ and η_0, we can estimate an upper limit for the error on the molecular parameters $\bar{M}_w$ and Q. We obtain

$$\left(\frac{\Delta\bar{M}_w}{\bar{M}_w}\right)_{\max} = \pm\varepsilon\frac{|\beta| + |\delta|}{|\beta\gamma - \alpha\delta|} \quad [14a]$$

$$\left(\frac{\Delta Q}{Q}\right)_{\max} = +\varepsilon\frac{|\alpha| + |\gamma|}{|\beta\gamma - \alpha\delta|}. \quad [14b]$$

For polyethylene we have found (12)

$\alpha = 3.36 \qquad \beta = +0.51$

$\gamma = 0.70 \qquad \delta = -0.105$

so that

$$\left(\frac{\Delta M_w}{\bar{M}_w}\right)_{max} = \pm 0.86\,\varepsilon \qquad [15a]$$

$$\left(\frac{\Delta Q}{Q}\right)_{max} = \pm 5.69\,\varepsilon. \qquad [15b]$$

These equations show that the error on the determination of the weight-average molecular weight is lower than the experimental errors; on the contrary the error on the determination of the index of polydispersity is higher than the experimental errors. The error amplification factor on Q is relatively high.

For the practical case where $[\eta]$ and η_0 are experimentally measured within a $\pm 5\%$ error we get

$$\left(\frac{\Delta \bar{M}_w}{\bar{M}_w}\right)_{max} = \pm 4.3\% \qquad [16a]$$

$$\left(\frac{\Delta Q}{Q}\right)_{max} = \pm 28.5\%. \qquad [16b]$$

Then it is apparent that the weight-average molecular weight is estimated quite accurately. The tolerance on the polydispersity index, which looks to be quite high, is comparable to the precision obtained by using other common methods.

The total errors on $\bar{M}_w$ and Q can be traced more precisely to the experimental errors on $[\eta]$ and η_0 by using eqs. [13a] and [13b]. We get

$$\frac{\Delta \bar{M}_w}{\bar{M}} = 0.72\,\frac{\Delta[\eta]}{[\eta]} + 0.15\,\frac{\Delta\eta_0}{\eta_0} \qquad [17a]$$

$$\frac{\Delta Q}{Q} = -4.73\,\frac{\Delta[\eta]}{[\eta]} + 0.98\,\frac{\Delta\eta_0}{\eta_0}. \qquad [17b]$$

The experimental error which is subjected to amplification appears to be the error on the intrinsic viscosity, in the case of the evaluation of the polydispersity index. A careful determination of $[\eta]$ would greatly restrict the error limits indicated.

Example of calibration: high density polyethylene

To perform the calibration we have used a series of published data concerning high density polyethylene (10). The weight average molecular weights and the indices of polydispersity of ten polydisperse samples were measured by fractionation. The zero-shear viscosity was extrapolated from the rheological curve by making use of an equation given by *Sabia* (14). The rheological curve was measured at 190 °C by a cone and plate instrument.

The choice of the instrument and of the method of extrapolation appears to be important since η_0 is found experimentally to be very sensitive to the instrument and to the procedure of extrapolation. A unification of methods would appear necessary.

The reported solution viscosity was the inherent viscosity in place of the intrinsic viscosity. Obviously the substitution is not crucial for the application of the method. The inherent viscosity was measured at 130 °C in tetralin, 0.1% concentration.

Table 1 show the eqs. [6] and [11] with completely determined coefficients. They were calculated by multiple regression analysis. It is also shown the matrix of the coefficients of correlation.

The correlations appear to be very good since the total correlation coefficients are very near to 1. The partial correlation coefficients between the couples zero-shear viscosity $-\bar{M}_w$ and inherent viscosity $-\bar{M}_w$ are very high, which indicates, as expected, that the variable $\bar{M}_w$ is by far the most important variable. Moreover, it can be seen that both correlations improve by introducing the index of polydispersity.

The absolute values of the coefficients of both equations are to be taken cautiously since they depend on the instruments used for measuring the rheological properties and on the method of extrapolation, as said before.

For our purposes it is only useful to note that the coefficients of eqs. [6] and [11] reduce, for $Q \to 1$, to the values which are generally obtained for monodisperse polymers. Once obtained the coefficients of eqs. [6] and [11], we have calculated $\bar{M}_w$ and Q of the samples used for calibration. The calculated values are expected to coincide with the reported data, the differences being due to the averaging process. Figs. 1 and 2 show the calculated versus the original $\bar{M}_w$ and Q.

The agreement for $\bar{M}_w$ looks very good; the deviations occurring in fig. 2 are well within the range of error which was expected from error analysis.

A further checking was performed by computing $\bar{M}_w$ and Q of polyethylene fractions for

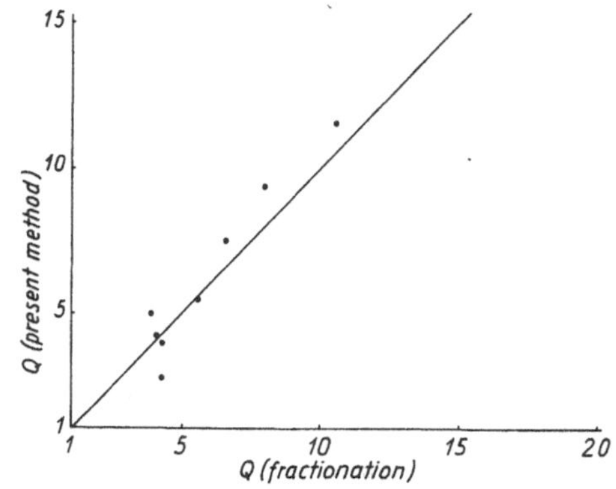

Fig. 1. HDPE. $\bar{M}_w$ calculated with the present method vs. published data, ref. (10)

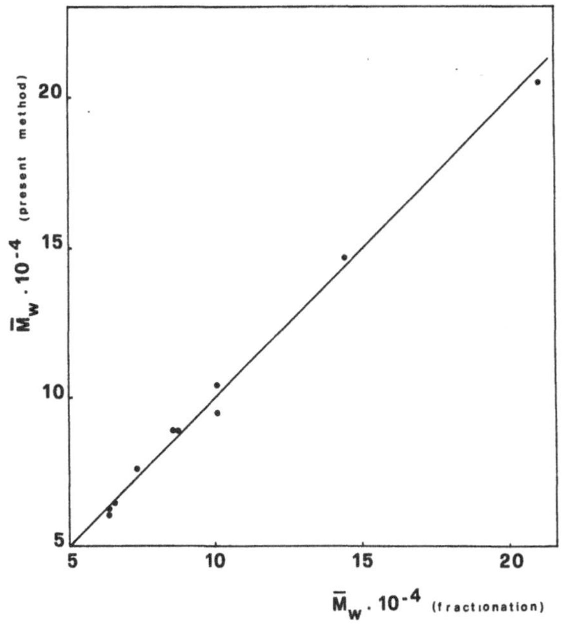

Fig. 2. HDPE. Index of polydispersity, Q, calculated with the present method vs. published data, ref. (10)

Table 2. Weight-average molecular weight and index of polydispersity of fractions calculated according to the method proposed in this work. Comparison data taken from ref. (10)

Sample	$\bar{M}_w$ calc.	$\bar{M}_w$ published	Q calc.	Q publ.
F 1	21500	19200	1.28	–
F 2	56900	53800	1.47	–
F 3	97500	97300	1.72	–
F 4	624000	724000	2.34	–
F 5	7200	6040	1.16	1.22
F 6	19100	17400	1.13	1.24
F 7	48700	44700	1.88	1.25
F 8	77700	75100	1.76	1.29
F 9	173000	182000	1.47	1.29
F10	430000	440000	3.16	–

which zero-shear viscosity and inherent viscosity was published by the same authors (10), table 2. The calculated molecular parameters can be seen to approach those evaluated by classical methods, expecially the molecular weights. Possibly our calculated Q are systematically a little too high than should be for fractions, perhaps because of poor calibration. Finally, we note that the Q of the last fractions, F 4 and F 10, turned out to be significantly higher than the Q values of the other fractions. This would mean that F 4 and F 10 were fractionation tails.

Conclusions

We have described a method for evaluating the weight-average molecular weight and an index of polydispersity, which applies to polymer species whose molecular weight distribution can be reasonably approximated by a log-normal molecular weight distribution, e.g. linear polyethylene. The method requires the knowledge of two data, η_0, measured on polymer melt and $[\eta]$, measured on polymer solution. Both data can be obtained with high precision with the available laboratory techniques.

Polyethylene was chosen as an example of calibration. It is warned that the calibration coefficients derived on the basis of the data of a single author, should be regarded as tentative. In order to apply routinely the method, a better determination of the coefficients of eqs. [6] and [11], a standardization of the measurement apparatus and of the extrapolation of the rheological curve are required. The method herein shown applies to linear polymer. Work is in progress to extend it to branched polymers, too.

Summary

A simple and reliable relative method to derive the molecular weight distribution of linear polymers is proposed.

It is shown that both the zero-shear viscosity, η_0, and the intrinsic viscosity, $[\eta]$, have a logarithmic dependence on the weight average molecular weight, $\bar{M}_w$, and the polydispersity, $Q = \bar{M}_w/\bar{M}_n$. The coefficients of these relationships can be determined by applying a multiple regression analysis to a series of samples for which $\bar{M}_w$ and Q are known.

By making use of the two established relationships, the determination of $\bar{M}_w$ and Q for a given polymer sample reduces to the experimental measurement of its η_0 and $[\eta]$.

An analysis has been performed to estimate to what extent experimental errors on η_0 and $[\eta]$ affect the calculated molecular weight distribution.

It has been found that only the experimental error on $[\eta]$ contributes heavily to the final error on the polydispersity.

Zusammenfassung

Es wird eine einfache und zuverlässige Relativmethode vorgeschlagen, um die Uneinheitlichkeit linearer Polymere abzuleiten.

Es wird gezeigt, daß alle beide, Nullschergradientviskosität η_0, und Grenzviskositätszahl $[\eta]$, einfach logarithmisch vom Gewichtsmittel des Molekulargewichts $\bar{M}_w$, und vom Polymolekularitätsindex $Q = \bar{M}_w/\bar{M}_n$, abhängig sind.

Die Koeffizienten dieser Beziehungen können mit statistischer Analyse festgesetzt werden, wenn $\bar{M}_w$ und Q einer Probenreihe bekannt sind.

Mit den zwei vorher festgesetzten Beziehungen besteht die Bestimmung von $\bar{M}_w$ und Q einer gegebenen Polymersprobe nur aus den experimentellen Massen seiner η_0- und $[\eta]$-Werte.

Eine Analyse wurde ausgeführt, um die Bedeutung des experimentellen Irrtums über die berechnete Uneinheitlichkeit zu wissen.

Es wurde gefunden, daß ein experimenteller Irrtum betreffs $[\eta]$ schwer an endlichem Irrtum der Uneinheitlichkeit teilnimmt.

References

1) *Peticolas, W. L.*, J. Chem. Phys. **39**, 3392 (1963).

2) *Conti, W.* and *E. Sorta*, paper presented at the I Meeting, Società Italiana di Reologia, Siena, 1971.

3) *Giuliani, G.* and *A. De Chirico*, J. Macrom. Sci. **B5**, 429 (1971).

4) *Graessley, W. W.*, J. Chem. Phys. **47**, 1942 (1967).

5) *Conti, W.* and *I. Gigli*, J. Pol. Sci. A-1, **4**, 1093 (1966); Makromol. Chem. **107**, 16 (1967).

6) *Fox, T. G.* and *P. J. Flory*, J. Amer. Chem. Soc. **70**, 2348 (1948).

7) *Fox, T. G.* and *V. R. Allen*, J. Chem. Phys. **41**, 337 (1964).

8) *Casale, A.* and *R. S. Porter*, J. Macromol. Sci. **C5 (2)**, 387 (1971).

9) *Zosel, A.*, Rheol. Acta **18**, 215 (1971).

10) *Saeda, S.*, *J. Yotsuyanagi*, and *K. Yamaguchi*, J. Appl. Pol. Sci. **15**, 277 (1971).

11) *Wesslau, W.*, Makromol. Chem. **20**, 111 (1956).

12) *Locati, G.* and *L. Gargani*, Journal of Polym. Science, Polym. Letter Ed. **11** (2), 95 (1973).

13) *Onyon, P. F.*, in Polymer Characterization. *P. W. Allen* Editor (London 1959).

14) *Sabia, R.*, J. Appl. Pol. Sci. **7**, 347 (1963).

Authors' address:

G. Locati, *L. Gargani*, and *A. De Chirico*
SNAM Progetti
Direzione Ricerca e Sviluppo
I-20097 S. Donato Milanese (Italia)

Rheol. Acta 13, 283–288 (1974)

Institut National Polytechnique, Toulouse (France) et Univérsité Fédérale, Rio de Janeiro (Brésil)

Rhéologie particulière des solutions de polymères en milieu poreux

Par C. Thirriot et G. Massaran

Avec 5 figures et 1 tableau

(Reçu p. p. le 27 October 1972)

I. Introduction

Quand on parle rhéologie, il faut d'abord bien préciser les systèmes qui sont considérés. Lorsqu'il s'agit d'un fluide, on peut s'intéresser à la loi rhéologique intrinsèque qui caractérise l'action des molécules les unes sur les autres: c'est un examen à *l'échelle microscopique* d'un milieu supposé homogène continu. Le graphe rhéologique sera alors donné par la relation entre tenseur de contrainte Σ et tenseurs des déformations $\mathscr{E}$ et leurs dérivées par rapport au temps. Mais on peut aussi prendre *un point de vue macroscopique* et considérer ensemble les molécules fluides et les canaux dans lesquelles elles circulent. C'est ce que l'on admet lorsqu'on parle d'écoulement en milieu poreux pour lesquels le graphe rhéologique sera une information globale constituée par la relation entre vecteur débit spécifique $\vec{q}$ et gradient de la grandeur caractérisant le niveau énergétique dans le fluide (par exemple la charge hydraulique, $\mathscr{H}$, si le fluide est considéré comme incompressible).

Pour les écoulements newtoniens en milieu poreux, la première forme explicitée de graphe rhéologique est la loi de *Darcy* présentée en 1856

$$\vec{q} = -\frac{k}{\mu}\,\overrightarrow{\mathrm{grad}}\,\varrho g \mathscr{H} = -\frac{k}{\mu}\,(\overrightarrow{\mathrm{grad}}\,p - \varrho\vec{g}). \quad [1]$$

Cette loi est la présentation rationnelle de résultats d'expériences de filtration réalisés par *Darcy*. Pour pallier le caractère empirique original de cette loi, de nombreuses tentatives théoriques furent amorcées. En gros, la plupart consiste à montrer qu'il existe un opérateur linéaire reliant vecteur débit spécifique $\vec{q}$ et gradient énergétique $\overrightarrow{\mathrm{grad}}\,\mathscr{H}$ pour les fluides newtoniens. Certaines démonstrations partent directement des équations de *Stokes* pour les écoulements lents d'autres, plus imagées utilisent l'analogie de forme entre loi de *Poiseuille* et loi de *Darcy* et proposent de simuler le milieu poreux comme un ensemble de capillaires cylindriques.

Compte tenu de la simplicité de ce modèle théorique, on est tenté de l'exploiter aussi pour prévoir les écoulements en milieu poreux de fluides non newtoniens tels que par exemple de nombreuses solutions de polymères qui ont un comportement pseudoplastique marqué.

Dans ce qui suit, nous essaierons de faire le point sur cette question afin d'aboutir à une généralisation de la loi de *Darcy*. Mais nous tenons au préalable à signaler que si les études sur ce thème sont récentes, elles fourmillent déjà compte tenu de leur intérêt aux divers points de vue rhéologique, mécanique des fluides et industriel. Sans prétendre n'omettre personne, nous citerons *Bird* et ses collaborateurs (1960), *Sadowsky* (1963), *Christopher* et *Middleman* (1965), *Gregory* (1966), *McKinley* (1966), *Dauben* et *Menzle* (1967), *Gaitonde* et *Middleman* (1967), *Marshall* et *Metzner* (1967), *Siskovic* (1969).

Lorsqu'ils utilisent l'analogie avec les écoulements de *Stokes* dans les capillaires, la plupart des auteurs font l'hypothèse que les fluides obéissent au modèle rhéologique d'*Ostwald-de Waele*.

Pour notre part, nous étendrons l'analogie à tout fluide simple de *Noll*.

II. Analogie entre filtration en milieu poreux et écoulement dans un ensemble de tubes capillaires

Par souci de simplicité dans la présentation, nous supposerons d'abord que les successions de pores du milieu poreux puissent être grossièrement représentées par des tubes capillaires cylindriques de même diamètre sans aucune interconnexion. Il est évidemment inutile d'insister sur la naiveté de l'image et la rudesse des approximations que l'on peut atténuer en considérant un faisceau de tubes à diamètres moyens différents et formés de successions de convergents et divergents (fig. 1).

Considérons donc l'écoulement laminaire d'un fluide simple incompressible de *Noll* dans un tube rectiligne de section circulaire.

Soit le graphe rhéologique simplifié intervenant dans le glissement relatif des particules à filets parallèles (fig. 2)

$$\frac{dv}{dr} = -\frac{1}{T^*}\,h\left(\frac{\tau}{\tau^*}\right) \quad [2]$$

avec T^* et τ^* respectivement temps et contrainte de référence τ contrainte de cisaillement.

L'application de la méthode de *Rabinovitsch-Mooney* pour déterminer le débit conduit à l'expression suivante de la vitesse moyenne.

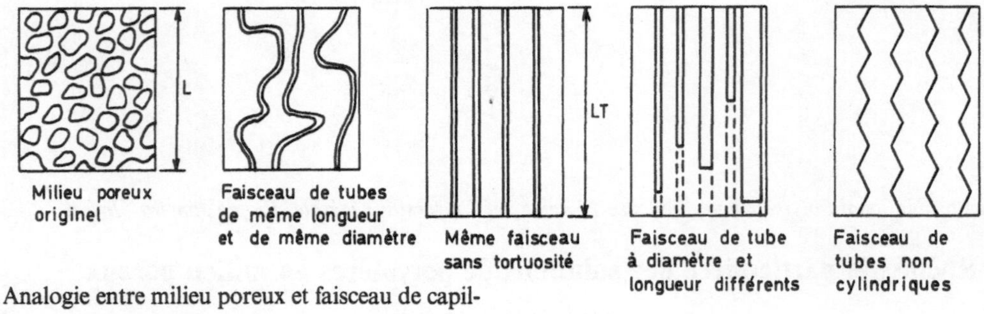

Fig. 1. Analogie entre milieu poreux et faisceau de capillaires

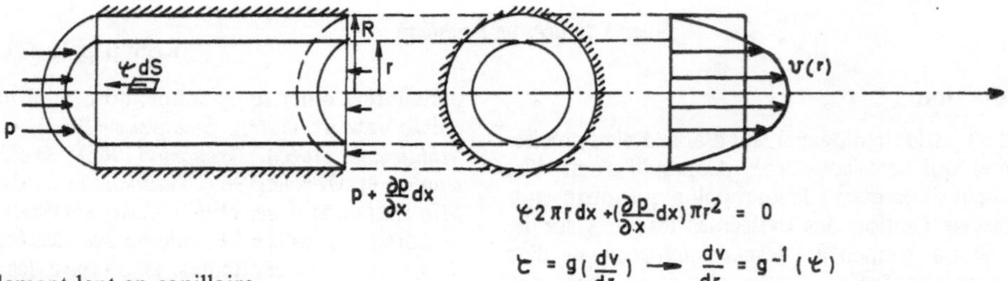

Fig. 2. Ecoulement lent en capillaire

$$v_m = \frac{R}{T^*} \int_0^1 y^2 h(\alpha y)\, dy \qquad [3]$$

avec

$$\alpha = \frac{R \Delta p^*}{2 \Delta \varkappa \tau^*} = \frac{\tau_p}{\tau^*} \qquad [4]$$

avec τ_p contrainte à la paroi; $p^* = p + \varrho g z$; p^* pression piézométrique.

Pour l'écoulement unidirectionnel à travers un milieu poreux isotrope, nous écrirons par analogie:

$$\frac{q}{P} = \frac{R^*}{T^*} \int_0^1 y^2 h(\alpha^* y)\, dy \qquad [5]$$

avec P porosité; R^* rayon équivalent

$$\alpha^* = \frac{R^* \Delta P^*}{2 \Delta x \tau^*} \frac{1}{t} \qquad [6]$$

t tortuosité.

Pour définir le rayon équivalent, nous imposerons que l'éq. [5] coïncide avec la loi de *Darcy* (1) lorsque le fluide a un comportement newtonien c'est-à-dire lorsque $h(y) = y$ et $\tau^* T^* = \mu$. D'où:

$$R^* = 2 \sqrt{\frac{2kt}{P}}. \qquad [7]$$

L'expression du débit spécifique devient alors

$$q = \frac{2\sqrt{2Pkt}}{T^*} \int_0^1 y^2 h \left[\frac{y}{\tau^*} \sqrt{\frac{2k}{Pt}} \left| \frac{dp^*}{dx} \right| \right] dy. \qquad [8]$$

Il est important de noter que cette présentation n'exige pas d'expliciter la forme mathématique du modèle rhéologique pour le fluide. Avec l'hypothèse de l'analogie, la fonction rhéologique peut être obtenue à partir de l'expérimentation directe sur des tubes capillaires.

2. *Faisceau de tubes non cylindriques mais identiques*

Considérons des capillaires constitués de divergents et convergents successifs, nous avons cherché la loi de perte de charge $v_m(dp^*/dx)$ en suivant un cheminement analogue à celui présenté ci-dessus en raccourci, et fondé essentiellement sur l'hypothèse de variation longitudinale très progressive de section.

Les résultats essentiels sont les suivants:

a) Pour un fluide suivant le modèle rhéologique d'*Ostwald* la forme du graphe est conservée lorsqu'on passe du comportement microscopique $(dv/dr, \tau)$ au comportement macroscopique $(v_m, dp/dx)$. Donc, l'analogie faisceau de capillaires/milieux poreux est encore satisfaisante.

Si

$$\frac{dv}{dr} = \frac{1}{T^*} \left(\frac{\tau}{\tau^*} \right)^{1/n}$$

$$v_m \sim \left(\frac{\Delta p^*}{\Delta^x} \right)^{1/n} \sim \frac{q}{P}.$$

Cependant, un effet d'échelle apparait qui dépend du contraste entre section minimale et section maximale.

b) Pour un comportement rhéologique différent, il apparait une légère distorsion: le graphe global a une allure légèrement différente du graphe intrinsèque mais l'écart est rarement significatif compte tenu de l'arbitraire du choix du modèle rhéologique et de la précision de l'ajustement.

c) Dans tous les cas, la valeur du rayon du capillaire cylindrique équivalent au point de vue perte de charge dépend (mais modérément) du graphe rhéologique (par exemple pour les fluides d'*Ostwald* ce rayon dépend de l'indice n).

En conclusion, on peut donc extrapoler l'image des faisceaux de capillaires lorsqu'on examine des écoulements non newtoniens en milieu poreux.

III. Généralisation de la loi de Darcy

Bien entendu, il n'est pas possible de déduire directement de l'éq. [8], l'expression vectorielle généralisée représentant la loi d'écoulement de fluides non newtoniens en milieu poreux. Mais en s'inspirant de la théorie d'*Onsager*, on peut proposer comme hypothèse l'expression:

$$\vec{q} = \vec{f}(\overrightarrow{\text{grad}}\, p^*), \quad \text{avec } \vec{f}(0) = \vec{0} \qquad [9]$$

où f est une fonction dérivalle à l'origine.

La vérification du principe de l'invariance lors de changement de système de coordonnées conduit pour un milieu isotrope à:

$$\vec{q} = -\frac{k}{\mu^*}\,\psi(|\overrightarrow{\text{grad}}\, p^*|)\cdot\overrightarrow{\text{grad}}\, p^*. \qquad [10]$$

D'après l'analogie présentée plus haut, il vient:

$$\psi = \frac{4\mu^* \int\limits_0^1 y^2 h\left(\frac{y}{\tau^*}\sqrt{\frac{2k}{Pt}}\,|\overrightarrow{\text{grad}}\, p^*|\right) dy}{T^*\cdot\sqrt{\frac{2k}{Pt}}\,|\overrightarrow{\text{grad}}\, p^*|}$$

$$= \frac{4\mu^*}{T^*\tau^*}\cdot\frac{H^*(\alpha^*)}{\alpha^*} \qquad [11]$$

avec

$$H^*(\alpha^*) = \int\limits_0^1 y^2 h(\alpha^* y)\, dy$$

fonction rhéologique qui d'après l'hypothèse de l'analogie peut être déterminée par essai sur des tubes capillaires par l'application $\alpha^* = \alpha$, $H^* = H$.

La détermination expérimentale de la fonction rhéologique $H(\alpha)$ est aisée puisque $H = \frac{QT^*}{\pi R^3}$. On peut ensuité déduire aisément la fonction rhéologique intrinsèque $h(\alpha)$ (mais ceci n'est pas du tout nécessaire pour l'étude d'écoulement en milieu poreux).

$$H^*(\alpha^*) = \int\limits_0^\alpha u^2 h(u)\, du.$$

Posons

$$\alpha^* y = u, \quad \alpha^{*3} H^*(\alpha^*) = \int\limits_0^{\alpha^*} u^2 h(u)\, du$$

et

$$h(\alpha^*) = \alpha^* \frac{dH^*}{d\alpha^*} + 3H^*. \qquad [13]$$

Pour les fluides dont le comportement est décrit par un modèle d'*Ostwald* (14) $h(\alpha^*) = C\alpha^{*m}$, la fonction rhéologique globale est aussi une fonction puissance avec le même exposant

$$H^*(\alpha^*) = \frac{C\alpha^{*m}}{3+m} = \frac{1}{3+m}\, h(\alpha^*). \qquad [15]$$

Ceci explique la faveur qu'a eu très tôt l'analogie capillaire pour les solutions de polymères nombreuses qui suivent convenablement la loi d'*Ostwald*. Mais cette coïncidence entre graphe rhéologique intrinsèque $(dv/dr, \tau)$ et graphe global $(v_m, |\overrightarrow{\text{grad}}\, p|)$ reste un cas particulier qui ne doit pas masquer les possibilités beaucoup plus larges de l'analogie et de la relation (8).

IV. Etude expérimentale

Afin d'éprouver les hypothèses et prévisions précédentes, une étude systématique a été menée, sur une vingtaine de milieux poreux et avec huit fluides différents.

Pour éviter les conséquences du vieillissement du produit dans l'essai de vérification de l'analogie entre milieu poreux et faisceau de tube capillaire, dans l'installation expérimentale, milieu poreux et capillaire étaient disposés en série.

Les fonctions rhéologiques déduites des mesures effectuées sur les tubes capillaires sont réprésentées sur la fig. 3. A première vue, compte tenu du caractère approximativement rectiligne des graphes, on pourrait penser que le comporte-

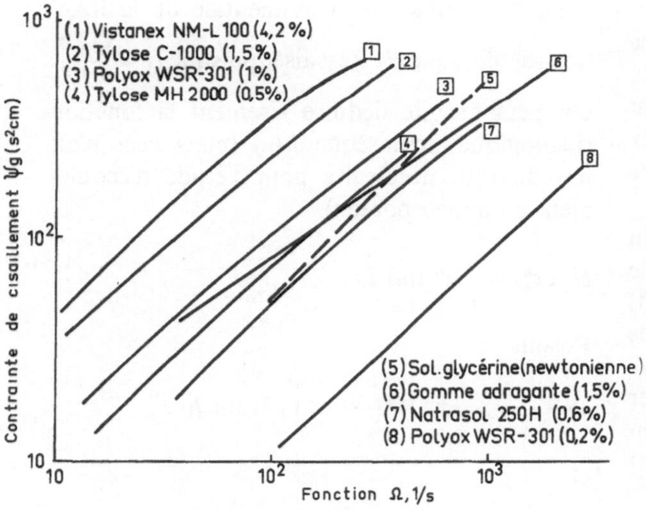

Fig. 3. Fonction rhéologique

ment rhéologique est ostwaldien. En fait, Vistanex (1) et Polyox WSR 301 (3) accusent une déviation assez nette par rapport à la loi monome puissance.

Les milieux perméables utilisés sont artificiels. Certains non consolidés (sable, bille de verre), d'autres consolidés à l'araldite suivant un procédé original. Les perméabilités varient de 5000 microns carrés (soit 5000 darcys) à $9\,\mu^2$ (ou 9 darcys). Sur l'ensemble des échantillons, la porosité varie de 20–46%.

A titre de synthèse de l'ensemble des essais, sur la fig. 4, nous présentons la corrélation entre débit spécifique mesuré et débit spécifique prévu par application de l'éq. [8]: l'accord, est assez satisfaisant puisqu'on ne note pas d'écart extrême supérieur à 10% sauf dans le cas de la gomme adragante et du produit Tylose C 1000 en solution à 1,5%. Pour ces solutions, apparait une réduction énorme du débit de filtration expérimental par rapport à la prévision théorique. Le même phénomène, mais tout de même moins accusé, est mentionné; par *Dauben* dans sa thèse. A notre connaissance, aujourd'hui, encore aucune explication décisive n'a été donnée. On peut avancer quelques hypothèses soit d'ordre cinématique soit d'ordre morphologique.

Au point de vue cinématique, tout au long de leur pérégrination à l'intérieur du milieu poreux les particules de fluides non newtoniens vont subir accélération et décélération successives. Or, de nombreux essais à l'aide du viscosimètre *Ferranti* ont montré que le graphe rhéologique était profondément altéré par des fluctuations fréquentes de vitesse. Une hystérésis très nette apparait qui peut dans certain cas à fort débit spécifique modifier nettement le graphe de perte de charge. Ce phénomène apparait déjà sans ambiguité pour des fluides sans mémoire. Il est évidemment beaucoup plus marqué pour les fluides thixotropes. Cependant, d'après *Dwight Dauben*, les constantes de temps de relaxation des fluides habituels seraient faibles par rapport aux durées de passage.

Du point de vue morphologique, la traversée des cols entre pores sera peut être très influente sur l'écoulement de solution contenant des macromolécules à très grande masse moléculaire supérieure à un million pour les molécules formées de longues chaines. Celles-ci s'enroulent

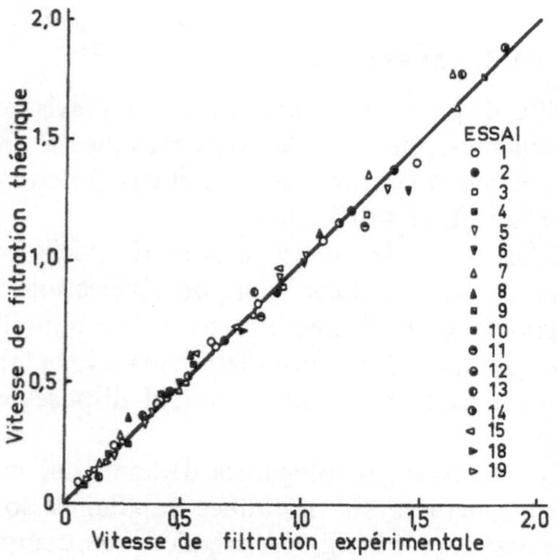

Fig. 4. Ecoulement de fluides non newtoniens en milieu poreux: vitesses de filtration expérimentales et théoriques

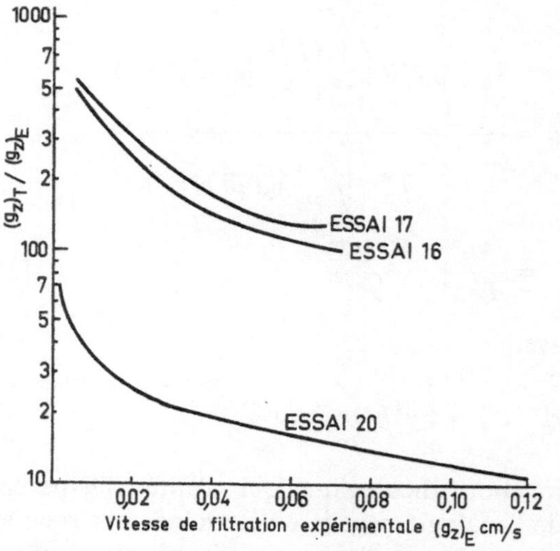

Fig. 5. Ecoulement non newtonien en milieu poreux: effets de reduction apparente de permeabilité

en pelote et retiennent à l'intérieur beaucoup de solvant comme une éponge. La dimension de ces macromolécules est difficile à évaluer car la physionomie change constamment, longiforme si la chaine est déroulée, plutôt sphérique lorsque le rouleau est concentré. La dimension caractéristique L serait donnée par la relation:

$$L = a\sqrt{N}$$

où N est le nombre de liaison; a la longeur d'une liaison.

A l'aide de telle formule, des appréciations évidemment très imprécises ont pu être réalisées

Tableau 1. Valeurs caractéristiques pour des expériences avec de fluides non newtoniens

Essais n°	Milieu poreux	Perméabilité k, cm²	Porosité ε, %	Polymère	Concentration sol, aqueuse (%)	Nombre de *Reynolds*, Re_k [1])
1	sable (hom. consolidé)	$5{,}9 \times 10^{-6}$	44,0	Tylose MH-2000K	1,0	$6{,}2 \times 10^{-4} - 80 \times 10^{-4}$
2	sable (hom. consolidé)	$4{,}2 \times 10^{-5}$	45,3	Tylose MH-2000K	1,0	$1{,}7 \times 10^{-3} - 15 \times 10^{-3}$
3	bille verre (hom. non consolidé)	$2{,}0 \times 10^{-5}$	35,5	Tylose MH-2000K	1,0	$7{,}5 \times 10^{-4} - 110 \times 10^{-4}$
4	bille verre (hom. non consolidé)	$3{,}5 \times 10^{-5}$	36,7	Tylose C-1000	1,5	$5{,}9 \times 10^{-4} - 224 \times 10^{-4}$
5	sable (hom. consolidé)	$2{,}2 \times 10^{-5}$	46,3	Tylose C-1000	1,5	$5{,}6 \times 10^{-4} - 96 \times 10^{-4}$
6	sable (hom. consolidé)	$7{,}7 \times 10^{-6}$	31,6	Tylose C-1000	1,5	$3{,}4 \times 10^{-4} - 124 \times 10^{-4}$
7	sable (hom. consolidé)	$7{,}9 \times 10^{-8}$	38,0	Polyox WSR-301	1,0 [2])	$1{,}2 \times 10^{-5} - 23 \times 10^{-5}$
8	sable (hom. consolidé)	$1{,}4 \times 10^{-6}$	38,0	Natrosol 250H	0,6	$1{,}2 \times 10^{-3} - 13 \times 10^{-3}$
9	sable (hom. consolidé)	$2{,}2 \times 10^{-7}$	20,2	Natrosol 250H	0,6	$7{,}2 \times 10^{-4} - 45 \times 10^{-4}$
10	sable (hom. consolidé)	$9{,}0 \times 10^{-8}$	38,0	Natrosol 250H	0,6	$5{,}8 \times 10^{-5} - 73 \times 10^{-5}$
11	sable (hom. consolidé)	$4{,}2 \times 10^{-5}$	45,3	Natrosol 250H	0,6	$1{,}4 \times 10^{-2} - 4{,}8 \times 10^{-2}$
12	sable (hom. consolidé	$3{,}4 \times 10^{-7}$	36,5	Natrosol 250H	0,6	$1{,}6 \times 10^{-4} - 58 \times 10^{-4}$
13	sable (stratifié)	$1{,}8 \times 10^{-5}$	45,0	Tylose MH-2000K	0,5	$2{,}7 \times 10^{-3} - 39 \times 10^{-3}$
14	sable (hétérogène)	$2{,}09 \times 10^{-5}$	37,0	Tylose MH-2000K	0,5	$2{,}0 \times 10^{-2} - 7{,}6 \times 10^{-2}$
15	sable (hom. consolidé)	$9{,}0 \times 10^{-8}$	38,0	Polyox 301	0,2	$2{,}5 \times 10^{-3} - 18 \times 10^{-3}$
16	sable (hom. consolidé)	$1{,}4 \times 10^{-6}$	38,0	Gomme adragante	1,5	$1{,}8 \times 10^{-7} - 130 \times 10^{-7}$
17	sable (hom. consolidé)	$6{,}6 \times 10^{-7}$	37,0	Gomme adragante	1,5	$1{,}0 \times 10^{-7} - 82 \times 10^{-7}$
18	sable (hom. consolidé)	$4{,}2 \times 10^{-5}$	45,0	Vistanex NM-L 100	4,23 [2]) (xylène)	$3{,}8 \times 10^{-4} - 12{,}3 \times 10^{-4}$
19	sable (hom. consolidé)	$3{,}1 \times 10^{-6}$	37,0	Vistanex NM-L 100	4,23 [2]) (xylène)	$9{,}5 \times 10^{-5} - 38{,}6 \times 10^{-5}$
20	sable (hom. consolidé)	$1{,}2 \times 10^{-6}$	36,0	Tylose C-1000	1,5	$4{,}5 \times 10^{-6} - 41 \times 10^{-6}$
21	sable (hom. consolidé)	$4{,}2 \times 10^{-5}$	45,0	Natrosol 250H	0,2	$5{,}5 \times 10^{-2} - 210 \times 10^{-2}$

[1]) $Re_k = \dfrac{k^{1/2} q_z \varrho}{\mu_{ef}}, \quad \mu_{ef} = \dfrac{\psi_0}{4\Omega}, \quad \psi_0 = (2k)^{1/2} \left|\dfrac{\Delta \mathbb{P}}{L}\right|.$

[2]) Présence des tensions normales constatée par visualisation du phénomène de «gonflement» apparaissant lorsque le fluide sort du tube capillaire.

qui donnent des longueurs de 0,1 μ pour les pelotes et 50 μ pour la chaine complètement déroulée. Ces dimensions sont de l'ordre de celle rencontrée dans les pores de matrices pétrolifères et dans les interstices entre particules d'argiles.

Ce phénomène de blocage partiel serait dû aussi selon certains à la gélification du fluide autour des particules du milieu poreux où à l'adsorption du fluide à la surface solide.

V. Conclusion

Après ce rapide panorama bien incomplet des écoulements non newtoniens en milieu poreux, la première constatation est celle de la complexité de la question. Déjà, tout problème d'écoulement en milieu poreux est déroutant car il est difficile de tenir compte des labyrinthes emmélés constitués par les tubes de courant. L'image des tubes capillaires bien que rassurante est insuffisante. Cependant, elle explique ce que l'expérience affirme comme résultat assez simple: la similitude du graphe rhéologique d'une part, et du graphe perte de charge/débit d'autre part, pour les fluides *Ostwald*iens. L'affinité qui apparait sur ce graphe et provenant à la fois de la courbe porométrique et de la variation continuelle de section est un paramètre important qui pourrait être relié à la structure du milieu poreux. Peut être est-ce là, une voie intéressante de recherche parmi les problèmes nombreux qui restent à éclaircir ou approfondir. Mais la question rhéologique la plus importante à notre avis, est l'explication de la forte réduction de perméabilité observée au passage de quelques solutions de polymères.

Dans la nature des phénomènes aussi bien que dans l'affinement de la métrologie, ce domaine des écoulements non newtoniens en milieu poreux constitue un Marché Commun Scientifique réunissant l'hydrogéologue et le physicien, le biologiste et le mécanicien des fluides, le chimiste et le mathématicien.

Littérature

Almeida, N. H., Escoamento nao linear em meio poroso. Tese de M. Sci., COPPE/UFRJ, Rio de Janeiro, Juin 1970.

Christopher, R. H. et *S. Middelman*, IEC Fundamentals **4**, 422 (1965).

Coleman, B. D., H. Markovitz et *W. Noll*, Viscosimetric flow of non newtonian fluids (Berlin-Heidelberg-New York 1966).

Dauben, D. L. et *D. E. Menzie*, Petroleum Transactions Soc. Petroleum Eng. **1967**, 1065.

Gaitonde, N. Y. et *S. Middleman*, IEC Fundamentals **6**, 145 (1967).

Gregory, D. R., Non newtonian polymeric flow through porous media. Ph. D. Thesis, Virginia Polytechnic Institute, 1966.

Marshall, R. J. et *A. B. Metzner*, IEC Fundamentals **6**, 393 (1967).

McKinley, R. M., H. O. Jahns, W. W. Harris et *R. A. Greenkorn*, Amer. Inst. Chem. Eng. J. **12**, 17 (1966).

Sadowski, T. J., Non newtonian flow through porous media. Ph. D. Thesis, University of Wiscosin, 1963.

Siskovic, N., R. Griskey et *D. R. Gregory*, Viscoelastic behavior of molten polymers in porous media. AICHE sixty-second annual meeting, Washington, D. C., Novembre 1969.

Slattery, J. C., Amer. Inst. Chem. Eng. J. **13**, 1066 (1967).

Adresses de les auteurs:

Claude Thirriot
Institut National Polytechnique
de Toulouse (France)

Giulio Massarani
Université Fédérale
de Rio de Janeiro (Brésil)

Rheol. Acta 13, 289–295 (1974)

Institut des Recherches sur les Eaux, Rome, Université de l'Aquila (Italie)

Rhéologie des mélanges d'eau et d'argile

Par M. Benedini et G. Margaritora

Avec 6 figures et 1 tableau

(Reçu p. p. le 27 octobre 1972)

1. Comment poser le problème

Pour traiter les problèmes de transport solide par l'eau on peut procéder fondamentalement de deux façons: considérer le mouvement du fluide comme étant newtonien en étudiant l'influence que les particules solides ont sur lui; considérer le fluide résultant non plus comme eau mais comme un nouveau liquide à comportement non newtonien.

La tendance moderne est d'enquêter sur cette seconde hypothèse, spécialement quand il s'agit de matériel argileux.

Dans le travail présent on a voulu enquêter sur les propriétés rhéologiques des mélanges eau et argile avec basses concentrations d'argile. Pour le choix du champs de concentration sur lequel enquêter nous nous sommes référés aux turbidités maxima des fleuves italiens qui ont un grand transport solide.

Des «Annales» du «Service hydrographique italien» on peut déduire que les turbidités maxima autour de 30 à 40 g/l sont très fréquentes et qu'on peut avoir des maxima supérieures aux 100 g/l.

Le rhéomètre que nous avons choisi nous a permis d'enquêter sur un domaine de 5 à 80 g/l. Quatre types d'argile standard ont été utilisées pour expérimenter les mélanges, à savoir: Chlorite California; Illite 48W-1535; Kaolinite 48W-0290; Vermiculite Africa flakes.

Le matériel a été fourni par le Ward's Natural Science Establishment, Rochester, New York.

Des nombreuses expériences conduites jusqu'à maintenant dans ce secteur et rapportées par écrit on déduit que les mélanges eau et argile n'ont pas un comportement newtonien. *Daido* (1) en expérimentant avec des mélanges ayant des concentrations de 23,4–28,5% en volume et en employant différentes argiles dans lesquelles les minéraux composants sont pyreux et kaolineux, déduit que les mélanges à la plus basse concentration se comportent comme des fluides *Bingham*, tandis que ceci n'a pas lieu pour les concentrations les plus hautes. Selon *Yano* et *Daido* (2) les mélanges avec plus de 3% en volume d'argile se comportent comme des pseudo-plastiques et ceux avec plus de 5% comme *Bingham*.

Du Plessis et *Ansley* (3) trouvent avec des mélanges d'eau et argile de 8,1–17,1% en volume un comportement rhéologique très proche de celui du modèle *Bingham*, caractérisé cependant par un comportement non linéaire du rhéogramme relatif.

Howard (4) observe qu'une suspension d'eau et argile avec 13% en volume ne peut être cataloguée ni avec le modèle *Bingham* ni avec un modèle à loi de puissance, mais avec une combinaison des deux.

En effet l'argile mélangée à l'eau forme une suspension dont les caractéristiques rhéologiques dépendent de la concentration et distribution des granules, des dimensions et des réactions réciproques entre eux. C'est donc dans le but d'avoir des données le plus possible compatibles que dans cette étude on a aussi employé non seulement toujours la même méthode pour faire les mélanges et le même type d'eau de même que la même température d'essai, mais aussi des argiles facilement identifiables.

2. Choix du rhéomètre

Afin de procéder à une expérience correcte, est déterminant le choix de l'instrument et de la technique de mesure pour les grandeurs rhéologiques. Comme on sait, parmi les différents types d'instruments employés, la préférence est donnée aux deux instruments suivants:

– rhéomètre capillaire, basé sur le mouvement dans un tube de petit diamètre (et par conséquent conforme à la loi de *Poiseuille*);

– rhéomètre tournant, dans lequel la vitesse de cisaillement est réalisée dans le liquide disposé entre une paroi fixe et une paroi mobile (principe de *Couette*).

Les deux instruments présentent des avantages et des défauts qui concourent à établir auquel des deux doit aller la préférence, selon le problème examiné et le liquide devant être expérimenté. Le rhéomètre capillaire est basé sur un principe de fonctionnement très simple et ne requiert pas l'intervention de parties mécaniques. Il permet donc des lectures nettement facilitées et en définitive très précises. En outre, étant donné qu'il s'agit d'un mouvement laminaire dans un tube, les conditions dynamiques sont les mêmes qui se présentent dans la pratique, quand on veut étudier le phénomène de mouvement non newtonien dans une conduite à basses valeurs du numéro de *Reynolds*.

Par contre la détermination des constantes rhéologiques d'un liquide par un viscosimètre capillaire demande de pouvoir disposer d'une série d'instruments de diamètre différent, chose qui peut difficilement être réalisée pour des liquides ayant une basse viscosité, où il est nécessaire que le diamètre reste au-dessous de quelque fraction de millimètre. L'opération avec un rhéomètre capillaire reste en outre plus difficile en présence de liquides non homogènes et de suspensions.

Un instrument tournant a, au contraire, l'avantage de pouvoir donner une gamme étendue de valeurs de dv/dr en variant simplement la vitesse relative entre la paroi fixe et celle mobile: il est donc possible d'expérimenter dans un champs étendu du diagramme rhéologique de la substance étudiée et mettre en évidence des nombreuses particularités de comportement du fluide (5). Le rhéomètre tournant présente cependant quelques inconvénients, comme en premier lieu, la présence de composantes centrifugues dans le fluide, notamment à des valeurs élevées de la vitesse de rotation. Ceci conduit à une situation hydrodynamique sensiblement différente de celle qui a lieu dans la plupart des cas pratiques.

On ajoute encore que les opérations de mesure résultent plus complexes, par la présence de parties mécaniques en mouvement et en raison de la nécessité de devoir déterminer quelques grandeurs au moyen d'un opportun étalonnage.

Dans le cas qui nous intéresse était nécessaire un rhéomètre de haute sensibilité, étant donné que les expériences concernaient des fluides de basse viscosité; en outre il s'agissait de liquides non homogènes et il était donc nécessaire de ne pas avoir de dépôts pendant les expériences. L'instrument choisi est du *type tournant* et il est doté d'un équipement de mesure «à double interspace», c'est-à-dire qu'il dispose deux zones destinées au liquide examiné, obtenues, ainsi qu'il est démontré dans la fig. 1 entre une «coupe renversée» tournante et une cannelure annulaire fixe. De la sorte, les actions dynamiques liquide-paroi sont accrues par l'effet de l'important développement superficiel des parois fixes et tournantes. Il est ainsi possible d'obtenir même avec des liquides à basses viscosité, un remarquable couple résultant dû à la tension tangentielle liquide – paroi, à opposer, pendant le mouvement, au couple-moteur appliqué à l'équipage par effet d'un moteur électrique à vitesse variable.

L'équipage de mesure est rélié au moteur électrique au moyen d'un axe rigide vertical; le moteur est à son tour suspendu au carter de l'instrument au moyen d'un fil de torsion qui a le devoir d'emmagasiner sous forme d'énergie élastique le travail conséquent à un déséquilibre entre couple moteur et couple résistant. Un index, coaxial au fil, indique la tension obtenue et donne par conséquent une indication (en «unités dynamiques») proportionnelle à la tension de cisaillement liquide paroi dans les interspaces de l'équipage.

Le moteur électrique à vitesse variable (du type synchroné) permet de réaliser un nombre de quinze valeurs différentes de la vitesse angulaire, d'un minimum de $0{,}5869\ \mathrm{sec}^{-1}$ à un maximum de $36{,}96\ \mathrm{sec}^{-1}$.

L'instrument possède aussi un bain thermostatique réglé électriquement à la température constante de $15 \pm 0{,}1$ °C et est disposé dans une chambre avec température constante équivalente à 25 °C.

C'est avec un soin particulier qu'on a rempli l'instrument avec le liquide devant être examiné, dans le but d'éviter que le liquide, préparé en partie avec le pourcentage établi d'argile, soit placé dans l'instrument avec des caractéristiques différentes de celles originales. Le remplissage a été effectué au moyen de l'emploi d'une burette graduée adéquate.

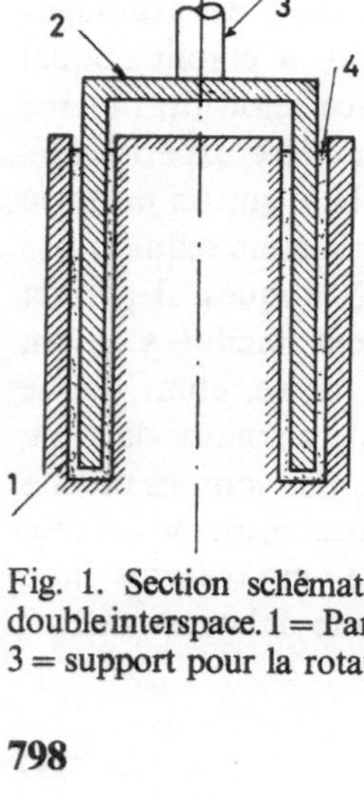

Fig. 1. Section schématique verticale du rhéomètre à double interspace. 1 = Partie fixe; 2 = «coupe» tournante; 3 = support pour la rotation; 4 = liquide étudié

3. Equations du rhéomètre

Pour la déduction des équations fondamentales du rhéomètre on a procédé de la façon suivante (fig. 2).

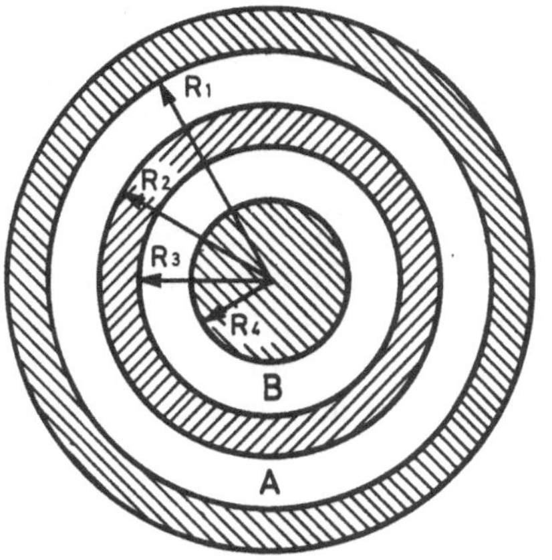

Fig. 2. Section horizontale schématique du rhéomètre avec indication des éléments géométriques pour la déduction des équations caractéristiques. A et B = Champs fluides.

Dans le cas d'un *fluide newtonien*, c'est-à-dire pour un fluide dont l'équation rhéologique est

$$\tau = \mu \frac{dv}{dR},$$

nous rapellons que

$v = \omega R$ (ω = vitesse angulaire du fluide)

et par conséquent

$$\tau = \mu \left(R \frac{d\omega}{dR} + \omega \right). \qquad [1]$$

Pour une «chemise» cylindrique fluide de rayon R et de hauteur l vaut la relation d'equilibre suivante

$$2\pi l R^2 \tau = -M. \qquad [2]$$

M étant le moment du couple appliqué à l'équipage par le moteur électrique. L'instrument employé ayant deux zones liquides, individuées par les valeurs du rayon R_1 et R_2 (zone «A») et respectivement R_3 et R_4 (zone «B»), il est légitime de penser que le moment externe se répartit en deux parties distinctes, M_A et M_B appliquées aux deux différentes zones fluides et définies selon la relation d'équilibre

$$M_A + M_B = M.$$

En combinant les deux relations [1] et [2] on obtient les deux équations différentielles distinctes

$$\frac{d\omega}{dR} + \frac{\omega}{R} + \frac{M_A}{2\pi\mu l R^2} = 0 \qquad [3']$$

et

$$\frac{d\omega}{dR} + \frac{\omega}{R} + \frac{M_B}{2\pi\mu l R^2} = 0. \qquad [3'']$$

En limitant l'étude à ce qui a lieu dans la zone «A», l'eq. [3'] admet comme intégrale l'expression:

$$\omega = e^{-\int \frac{dR}{R}} \left[C + \frac{M_A}{2\pi\mu l} \int e^{\int \frac{dR}{R}} dR \right]$$

$$= \frac{1}{R} \left[C + \frac{M_A}{2\pi\mu l R} \right]. \qquad [4]$$

Déterminant la constante C avec les conditions au contour

$\omega = 0$ pour $R = R_1$

$\omega = \Omega$ pour $R = R_2$

(la vitesse de rotation de la partie rigide de l'équipage étant Ω) on arrive à la relation finale

$$M_A = \frac{2\pi\mu l \Omega R_2}{\frac{1}{R_2} + \frac{1}{R_1}}. \qquad [5]$$

On veut maintenant connaître l'effet du fluide sur la paroi fixe (déterminée par le rayon R_1), effet qu'on peut porter dans la forme

$$2\pi l R^2 \tau = -M_A. \qquad [6]$$

Combinant la [5] et la [6] on obtient:

$$\tau_1 = -\frac{\mu \Omega R_2^2}{R_1(R_1 - R_2)} \qquad [7]$$

(le «−» indique que τ s'oppose au mouvement du fluide). L'expression [7] permet de calculer la valeur de τ_1 étant connues les grandeurs géométriques et connue la viscosité d'un fluide déterminé pour chaque valeur de Ω.

On peut aussi parvenir au calcul de τ_1 par d'autres voies, c'est-à-dire par l'expression

$$\tau_1 = \mu \left(\frac{dv}{dR} \right)_{R=R_1}.$$

Etant

$$v = \omega R$$

rapellant la [4]

$$v = C + \frac{M_A}{2\pi\mu l R}$$

et par conséquent, dérivant et posant $R = R_1$

$$\tau_1 = -\frac{M_A}{2\pi\mu l R_1^2}$$

qui peut servir de comparaison à la [7], étant connue la valeur de M_A pour chaque Ω et pour un μ déterminé, au moyen de [5].

Le rhéomètre utilisé maintenant était pourvu d'un tableau spécial d'étalonnage, fourni par la maison, qui donnait pour chacune des quinze valeurs de Ω la valeur correspondante de dv/dR.

Avec ces valeurs et utilisant des liquides à viscosité connue et à comportement newtonien (solutions aqueuses de glycérine à 5% et benzène) a été calculée la valeur de τ_1, qui a ensuite été trouvée bien d'acord avec la valeur correspondante calculée au moyen de la [7], obtenant une confirmation satisfaisante.

4. Interprétation des données expérimentales

Avec les données expérimentales a été tracé, pour chaque substance essayée, le rhéogramme correspondant, sur l'échelle logarithmique (fig. 3).

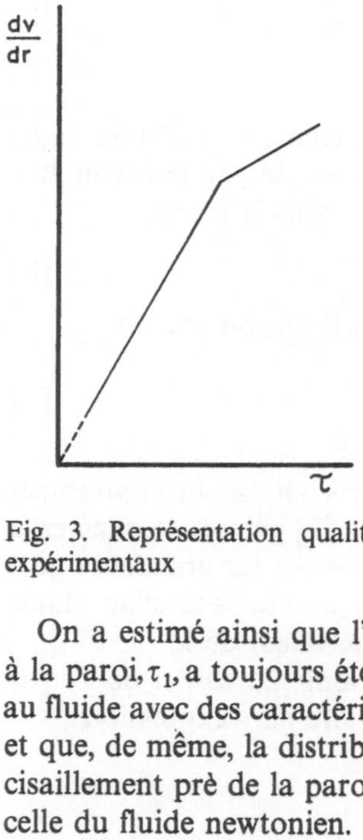

Fig. 3. Représentation qualitative des rhéogrammes expérimentaux

On a estimé ainsi que l'effort de cisaillement à la paroi, τ_1, a toujours été celui correspondant au fluide avec des caractéristiques newtoniennes et que, de même, la distribution de la vitesse de cisaillement prè de la paroi a toujours été aussi celle du fluide newtonien.

En réalité il faut tenir compte du fait que, même près de la paroi, toutes les grandeurs caractéristiques doivent obéir à l'équation rhéologique du fluide examiné.

En ce qui concerne l'effort à la paroi, τ_1, il est aisé de démontrer qu'il est indépendant de l'équation rhéologique; il suffit en effet de considérer la relation d'équilibre

$$\tau_1 = \frac{M_A}{2\pi l R_1^2}.$$

Il est par conséquent correct d'employer les mêmes valeurs déterminées sur le rhéomètre au moyen d'étalonnage avec fluide newtonien.

Quant à la vitesse de cisaillement, pour une détermination correcte des entités numériques qui doivent comparaitre dans le rhéogramme, il faut au contraire connaitre au préalable l'équation rhéologique du fluide, fait qui n'est pas toujours possible. On peut toutefois lui donner une valeur *conventionnelle* équivalente à la valeur newtonienne.

Les premiers rhéogrammes obtenus pour étalonner l'appareil (comme nous avons vu, des solutions aqueuses de glycérine 5% et benzène) mettaient en évidence une anomalie: aux vitesses de rotation les plus élevées du rhéomètre, le rhéogramme ne suivait plus la même loi rectiligne, mais s'inclinait dans le sens des τ croissantes. D'autres rhéogrammes ont alors été éxécutés avec des fluides sûrement newtoniens à viscosité connue: essence super, essence normale, eau distillée, eau potable, solution aqueuse de glycérine à 2,5%, cychloéxane, saumure à 10% et à 20% de NaCl, tridécan.

On a pu constater que le point de discontinuité où le rhéogramme commençait à ne plus être rectiligne commençait par τ toujours plus petite au moment de la diminution de la viscosité des fluides expérimentés (fig. 4). On a alors calculé

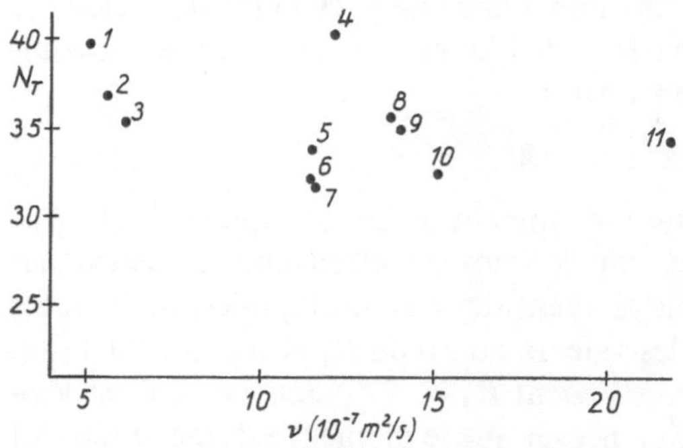

Fig. 4. Nombres de *Taylor* N_T limites pour différents liquides newtoniens en fonction de la viscosité cinématique. 1 = Essence super; 2 = essence normale; 3 = benzène; 4 = eau et glycérine 2%; 5 = saumure 20%; 6 = eau distillée; 7 = eau potable; 8 = cychloéxane; 9 = saumure 10%; 10 = glycérine; 11 = tridécan

les numéros de *Taylor* (6), N_T, qui sont liés au numéro de *Reynolds*, N_R, par la relation suivante:

$$N_T = N_R \sqrt{\frac{\delta}{R}} = \frac{\Omega R \delta}{\nu} \sqrt{\frac{\delta}{R}}$$

où Ω est la vitesse angulaire de l'instrument et ν est la viscosité cinématique du mélange; δ a été placé égal à la largeur de l'interspace du rhéomètre et R au rayon R_1 (voir fig. 2). Il en résulte que la discontinuité (voir fig. 4) se vérifie autour d'une valeur critique de N_T comprise entre 30 et 40.

Il est clair que le numéro de *Taylor* reporté n'est pas à exactement celui critique, étant donné que le rhéomètre ne dispose pas d'un champs de vitesse continu mais, comme on l'a rapellé plus haut, la vitesse varie dans des valeurs discrètes dans des intervalles préétablis. Il n'a d'ailleurs pas été possible de construire une courbe au-delà du point de discontinuité telle qu'elle s'interpolant à la ligne précédente, étant donné qu'on n'a pu obtenir qu'une ou deux valeurs de N_T plus grandes que celui critique. Cette élaboration des données expérimentales nous a permis d'interprêter la discontinuité dans le rhéogramme comme index du début de ce mouvement hélicoidal qui se produit dans les interspaces du rhéomètre et qui précède immédiatement le mouvement turbulent proprement dit. La présence de la turbulence contrastant avec les hypothèses de fonctionement du rhéomètre, il a été nécessaire de limiter les expériences aux points pour lesquels le numéro de *Taylor* soit au maximum égal à 30.

En plus des points supérieurs du rhéogramme, n'ont pas été pris en considération même ceux inférieurs, qui ne se disposaient pas selon la ligne correspondante au comportement newtonien du fluide qui devrait passer par l'origine des coordonnées, étant donné que pour les vitesses les plus petites la substance est partiellement cisaillée. Le phénomène se vérifie aussi quand il s'agit de plastiques *Bingham* (7). Les mélanges vers lesquels se dirige notre analyse ayant démontré un comportement qui en augmentant le pourcentage d'argile tend au plastique *Bingham*, le champs de validité des rhéogrammes pour les mélanges eau-argile à la suite expérimentés a été restreint par la vitesse angulaire pour laquelle on a dans le début du diagramme rectiligne pour les fluides newtoniens à celles pour lesquelles le numéro de *Taylor* est ≤ 30.

Le champs de validité des épreuves rhéométriques ayant été établi ainsi, des expériences avec les mélanges eau et argile ont été éxécutés. Comme on peut voir des fig. 3, les caractéristiques rhéologiques des mélanges sont semblables a celles des fluides pseudo-plastiques. Pour déterminer le modèle qui s'approche le mieux du comportement rhéologique ont été pris en considération les modèles suivants (8):

Ostwald-De Waele $\tau = K\left(\frac{dv}{dR}\right)^n$

Bingham $\tau = \tau_0 + \eta\frac{dv}{dR}$

Sisko $\tau = A\frac{dv}{dR} + B\left(\frac{dv}{dR}\right)^n$

Herschel-Bulkley $\tau = \tau_0 + \left(\eta\frac{dv}{dR}\right)^n$

à deux paramètres et
à trois paramètres

Une fois tirés des rhéogrammes les $(\tau_1)_m$ et les dv/dR correspondantes, ont été déterminées les constantes de façon à rendre minima la déviation standard[1]) σ entre les valeurs de τ_1 calculées au moyen des équations du modèle, $(\tau_1)_c$ et les valeurs mesurées, $(\tau_1)_m$. Ceci a été obtenu en posant $d\sigma/dK = 0$, représentant K les différentes constantes qui apparaissent dans les relations rhéologiques.

La valeur de déviation standard est assumée comme index de qualité pour le choix du modèle qui peut représenter le mieux les résultats expérimentaux.

On a d'abord constaté que les lois à deux paramètres ne permettent pas une approximation suffisante, concordant donc avec tous ces auteurs pour lesquelles les mélanges eau et argile ont un comportement intérmédiaire entre le modèle *Bingham* et le modèle *Ostwald*.

Des deux lois à trois paramètres, celle qui a donné les résultats les plus satisfaisants est celle de *Sisko*: en effet avec la loi de *Herschel-Bulkley* on obtient une déviation standard σ de l'ordre de 10^{-2}, tandis qu'avec la loi de *Sisko* on a σ de l'ordre de 10^{-4}.

Il semble opportun de rapeller que le modèle de *Sisko* (9) fut proposé par son auteur pour 5 types de gras lubrifiants sous la forme

$$\mu = \frac{\tau}{\frac{dv}{dR}} = A + B\left(\frac{dv}{dR}\right)^{n-1}$$

avec μ viscosité du fluide.

[1]) Pour la déviation standard on a choisi l'expression

$$\sigma = \sqrt{\frac{\Sigma\left[(\tau_1)_c - (\tau_1)_m\right]^2}{N}}$$

N étant le numbre d'observations.

Pour n qui tend à 1 le fluide tend au comportement newtonien

$$\tau = (A + B)\frac{dv}{dR}$$

pour n qui tend à zéro le fluide tend au comportement *Bingham*

$$\tau = B + A\frac{dv}{dR}.$$

Dans le tableau 1 nous avons reporté les valeurs de A, B et n tirées pour les quatre types d'argile d'abord précisés pour des mélanges qui vont de 5 à 80 g/l.

Comme on voit de la fig. 6 les n par chlorite, vermiculite et kaolinite vont en diminuant avec l'augmentation de la concentration de façon pas très régulière, toutefois pouvant être posée interpolée par les courbes que nous avons tracées. La tendance au passage à des lois toujours plus proches de celles de *Bingham* apparaît ainsi clairement. En ce qui concerne l'illite les valeurs de n se disposent à l'intérieur d'un filet horizontal au moins pour les mélanges de 20 à 80 g/l, pour lesquelles nous avons réussi à obtenir des rhéogrammes répétables et donc sûrs.

En ce qui concerne A et B on voit d'aprés le tableau 1 il y a une tendance générale à l'augmentation en croissant la concentration pour les quatre types de mélanges expérimentés.

Le ci-dessus exposé permet les déductions suivantes générales en ce qui concerne le mélange eau et argile avec concentration de cette derniére de 5–80 g/l:

a)

Les mélanges ont décidément un comportement non newtonien qu'on ne peut identifier correctement ni avec la loi de *Ostwald* ni avec celle de *Bingham*.

b)

Avec la loi de *Sisko* on peut interpoler les graphiques expérimentaux de façon satisfaisante.

c)

Dans le cadre de la loi de *Sisko* le comportement des mélanges avec les argiles standard prises séparément et sur lesquelles s'est basée l'enquête, est différent: on note notamment que tandis que pour les mélanges avec kaolinite, vermiculite et chlorite en augmentant la concentration d'argile on a une nette tendance vers le comportement selon le plastique *Bingham*, pour les mélanges avec illite, cette tendance n'est pas évidente, du moins jusqu' aux concentrations expérimentées.

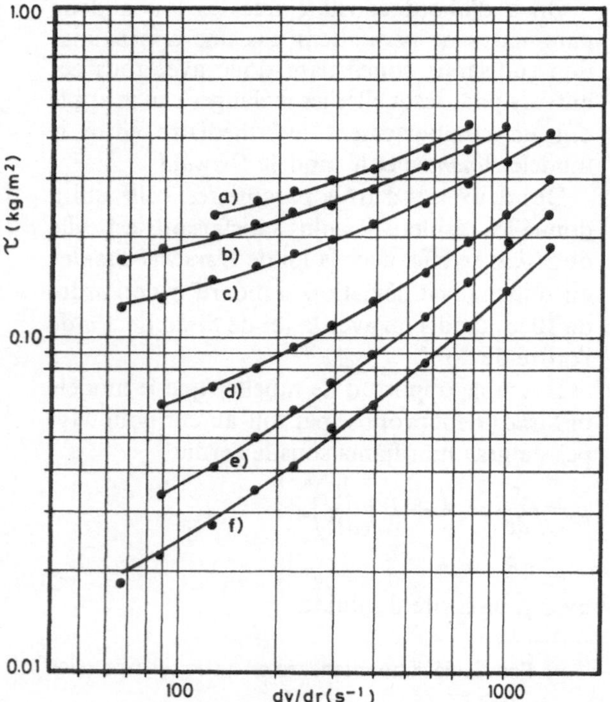

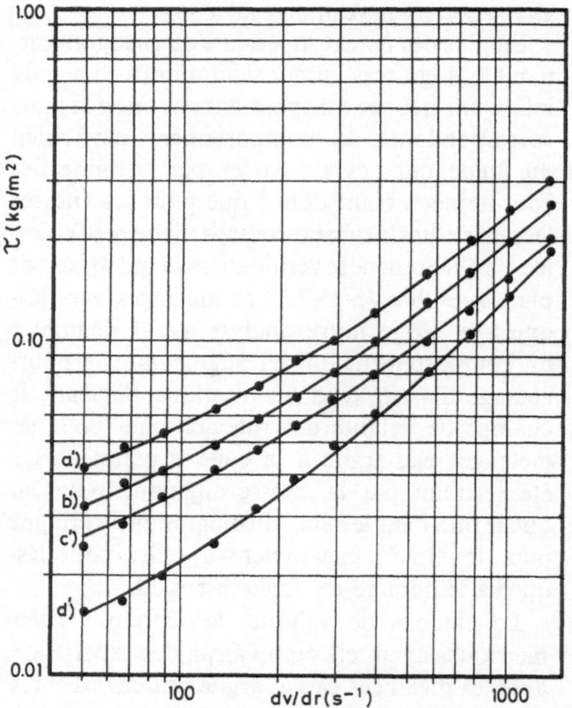

Fig. 5. Confrontation entre les données expérimentales et les courbes représentatives de l'équation de *Sisko*. a = Kaolinite 75 g/l; b = Chlorite 60 g/l; c = Chlorite 50 g/l; d = Chlorite 25 g/l; e = Chlorite 15 g/l; f = Illite 25 g/l. – a' = Illite 70 g/l; b' = Illite 60 g/l; c' = Illite 40 g/l; d' = Illite 20 g/l

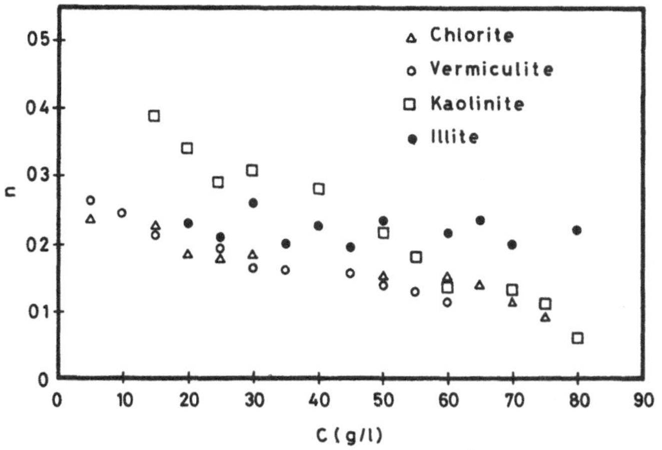

Fig. 6. Marche de l'exposant n de l'équation de *Sisko* en fonction de la concentration C d'argile

Tableau 1. Constants de l'équation de *Sisko* pour suspensions eau et argile en differente concentration

Argile	Conc. (g/l)	A $\left(\frac{kg \cdot sec}{m^2}\right)$ $\times 10^{-3}$	B $\left(\frac{kg \cdot sec^n}{m^2}\right)$	n	Argile	Conc. (g/l)	A $\left(\frac{kg \cdot sec}{m^2}\right)$ $\times 10^{-3}$	B $\left(\frac{kg \cdot sec^n}{m^2}\right)$	n
Chlorite	5	0,1176	0,00266	0,233	Kaolinite	15	0,12528	0,000272	0,39032
Chlorite	15	0,1478	0,00749	0,228	Kaolinite	20	0,11258	0,000945	0,34241
Chlorite	20	0,15815	0,01223	0,189	Kaolinite	25	0,13542	0,0017484	0,29144
Chlorite	25	0,1573	0,02119	0,187	Kaolinite	30	0,12438	0,0018345	0,31724
Chlorite	30	0,1452	0,02703	0,182	Kaolinite	40	0,15648	0,020647	0,283274
Chlorite	50	0,168	0,0611	0,149	Kaolinite	50	0,13580	0,014831	0,22315
Chlorite	60	0,183	0,0852	0,150	Kaolinite	55	0,14132	0,011832	0,18231
Chlorite	65	0,194	0,1041	0,138	Kaolinite	60	0,11946	0,015843	0,14553
Chlorite	70	0,237	0,1175	0,115	Kaolinite	70	0,12423	0,011636	0,13429
Chlorite	75	0,248	0,1285	0,0931	Kaolinite	75	0,195	0,011688	0,11406
					Kaolinite	80	0,2767	0,14923	0,060964
Illite	20	0,1174	0,003366	0,23503					
Illite	25	0,1188	0,004827	0,2144	Vermiculite	7	0,1224	0,003623	0,2647
Illite	30	0,108	0,0050009	0,26903	Vermiculite	10	0,13697	0,004687	0,24477
Illite	35	0,164	0,01176	0,2013	Vermiculite	15	0,14519	0,010542	0,22231
Illite	40	0,148	0,010434	0,22604	Vermiculite	25	0,1642	0,022674	0,19580
Illite	45	0,163	0,014781	0,19683	Vermiculite	30	0,16586	0,032394	0,16476
Illite	50	0,17416	0,013586	0,2302	Vermiculite	35	0,17732	0,050215	0,16049
Illite	60	0,14842	0,01092	0,217227	Vermiculite	45	0,1670	0,05368	0,15648
Illite	65	0,16096	0,010479	0,23750	Vermiculite	50	0,1375	0,01073	0,13854
Illite	70	0,17957	0,015344	0,20296	Vermiculite	55	0,19421	0,06874	0,129768
Illite	80	0,187563	0,01938	0,2217	Vermiculite	60	0,19932	0,06781	0,11537

Littérature

1) *Daido, A.*, Studi fondamentali riguardanti il moto degli ammassi fangosi (Dosekiryu ni Kan suru Kisoteki Kenkyu). (Osaka Pref. Tech. Coll. Publ., 1970.)

2) *Yano, K.* et *A. Daido*, Bull. Disaster Prev. Res. Institute **14**, Part 2 (1965).

3) *du Plessis, M. P.* et *R. W. Ansley*, J. Pipel. Div., Proc. A.S.C.E. **93**, n° PL 2 (1967).

4) *Howard, C. D. D.*, Hydr. Div., Proc. A.S.C.E. **89**, n° HY 5 (1963).

5) *Behn, V. C.*, J. Water Poll. Control Federation (1960).

6) *Schlichting, H.*, Boundary Layer Theory. 6th Edit. (London 1968).

7) *Parzonka, W.*, La Houille Blanche, n° 8 (1964).

8) *Bellet, D.* et *C. Thirriot*, La Houille Blanche, n° 5 (1970).

9) *Sisko, A. W.*, Ind. Eng. Chem. **50** (1958).

Adresse des auteurs:

Dr. *Marcello Benedini*
Institut des Recherches sur les Eaux, Rome
Rome (Italie)

Dr. *Gianmarco Margaritora*
Université de l'Aquila
L'Aquila (Italie)

Rheol. Acta 13, 296–304 (1974)

Aus dem Wissenschaftlichen Forschungsinstitut für Straßenverkehr, Budapest (Ungarn)

Untersuchung und Anwendung der dynamisch belasteten Konstruktionsteile mit Polyamid-Überzug

Von E. Vadász

Mit 14 Abbildungen

(Eingegangen am 27. Oktober 1972)

Die sich aufeinander verschiebenden Maschinenbauteile nützen sich ab. Die Abnutzung bedeutet die ansteigende und schädliche Abtrennung kleiner Materialteilchen von der Oberfläche. Infolge der Abtrennung der Materialteilchen tritt an den Bauteilen eine Maß- und Gewichtsverminderung auf. Abhängig vom Grade der Abnutzung müssen die Bauteile entweder ausgeschieden oder auf irgendwelche Art erneuert werden.

Die Materialflächen sind niemals so glatt, daß sie – mikrogeometrisch betrachtet – vollkommen aufeinander aufliegen könnten. Die Berührung zwischen bearbeiteten Flächen kann stets als eine Paarung von „Mikrobergen" aufgefaßt werden.

Bei Spielsitz brechen auf Einwirkung der spezifischen Belastung infolge der gegenseitigen „Kammwirkung" der gepaarten Mikroberge die hervorstehenden Spitzen ab. Im Falle eines Schmierungsausfalles oxydieren diese sich in Anwesenheit von Sauerstoff und anderen chemischen Wirkstoffen infolge der erhöhten Reibungswärme, reißen sich von den verhärteten Flächen ab und fressen sich im Gegenstück ein.

Bei dynamisch stark in Anspruch genommenen Fahrzeugteilen sind an zahlreichen Stellen mit fester Anpassung Wellen, Keilbahnen, Sternnaben und hauptsächlich Einbaustellen für Walzlager zu finden. Bei diesen sich aufeinander verschiebenden Gleitflächen (Wellen, Buchsen, Federplatten, Bauteilen von Wasserpumpen) entstehen Schadhaftwerdungen, die teils auf mechanische Abreibung, teils auf Korrosion zurückgeführt werden können.

Die Reibung (Haftreibung), die während des relativen Verschubs aneinander aufliegender Gleitflächen auftritt, stellt ein bisher noch immer nicht vollständig geklärtes wissenschaftliches Forschungsproblem dar.

Der Verschleiß entsteht als Folge der Korrosion und wegen des Abriebs des durch Dauerbeanspruchung ermüdeten Werkstoffes; damit begründet *Bowdens* neueste Theorie (1957) den Verschleißvorgang. In dieser Theorie wird besonders die Rolle der Adhesionskräfte betont, die an den am weitesten herausragenden einander berührenden Oberflächenpartien wirken. Während der Reibung entstehen an den Berührungspunkten der Metalloberflächen im weicheren Metall Kratzer, die die herausragenden Spitzen des härteren Metalls durch die von ihnen örtlich verursachten hohen spezifischen Belastungen hinterlassen. Dadurch bedingt wird die Höhe der Reibungskraft sehr wesentlich vom sich unter der Einwirkung der Adhesionskräfte abspielenden Verschleißungs- und Abreißprozeß beeinflußt.

Wenn zwei herausragende Spitzen der Reibflächen aneinanderprallen, werden Partikelchen vom Metall abgeschert, dessen Scherfestigkeit geringer ist.

Die abbröckelnden Partikelchen bleiben an der Oberfläche des härteren Metalls haften.

Prallen herausragende Spitzen gleichharter Metallflächen aneinander, dann brechen nicht die äußersten Spitzen ab, sondern die Masse der aneinander aufliegenden Teile wird u. U. herausgerissen.

Der Verschleiß der Gleitflächen spielt sich üblicherweise in zwei Stufen ab. Zuerst lagert sich auf der harten, meistens aufgekohlten und geschliffenen Zapfenoberfläche ein dünner Film des weichen Metalls ab, und in der zweiten Stufe verschweißen dann die Metallpartikelchen an den Berührungsflächen: infolgedessen werden bereits größere Teile aus der Oberfläche herausgerissen.

Es wurden an verschiedenen Plasten, Gußeisen, Stahl, Stahllegierungen, Buntmetall und Aluminium Vergleichsuntersuchungen durchgeführt, um die Verschleißfestigkeit zu bestimmen. Die Untersuchungen wurden auf einer Amsler-Verschleißprüfmaschine vollzogen. Die spezifische Belastung wurde nach einem Anfangswert von 20 kp/cm^2 nach je 10000 Umdrehungen um 20–20 kp/cm^2 gesteigert. Die Umlaufgeschwindigkeit betrug 0,4 m/sec. Der aus dem Versuchsmaterial gefertigte Probedrehkörper hatte eine Abmessung von ∅ 40 mal 10 mm, das fixe Gegenstück wurde in jedem Fall aus Chromstahl mit einer Härte von 60 HRc hergestellt.

Mit der angewendeten Untersuchung sollte man an den Fall eines betrieblichen Schmierungsausfalles herankommen. Die Schmierung spielt außer in der Verminderung der Abreibung auch eine Rolle in der Wärmeabfuhr. Schon bei einem geringerem Schmierungsausfall wird infolge der plötzlichen Erwärmung die ohnehin kleinere Menge des Schmiermittels zwischen den Flächen beseitigt und die Reibungsflächen fressen sich ein.

Die Untersuchungen laut Punkt 1–3 wurden vollendet, die letzten aber nur teilweise durchgeführt, weil diese mit dem Erscheinen neuerer Werkstoffe laufend fortgesetzt werden.

Die Mikrospitzen werden an der Metallfläche infolge der durch die Reibung hervorgerufenen Wärmewirkung abgeschmolzen, infolge der durch die dynamische Beanspruchung entstandenen Vibrationsschwingungen abgebrochen. Die abgetrennten Teilchen verursachen während ihrer Fortbewegung auf der Oberfläche tiefe Einfressungen. Infolge der guten Wärmeleitfähigkeit der Metalle stellt sich der völlige Verschleiß erst nach längerer Zeit ein.

Die Flächen der thermoplastischen Werkstoffe aus Spritzguß werden infolge der Reibungswärme und der sich mit der Wandstärke proportional vermindernden Wärmeabfuhr erweicht und danach in kurzer Zeit geschmolzen. Infolge der schlechten Wärmeabfuhr tritt die Erwärmung nicht im ganzen Querschnitt auf, der Verschleiß zeigt nicht die Erscheinung eines „Lagerauslaufes", sondern mit der Verminderung der mechanischen Festigkeit werden die bis zum Schmelzpunkt erhitzten Oberflächenteilchen abgeschmolzen.

Bei Hartgeweben werden die Oberflächen auf Einwirkung der Reibungswärme verkohlt.

Die Untersuchungen wurden in vier Gruppen aufgeteilt

1. *Verschiedene mit Spritzguß hergestellte Plaste mit kristalliner Struktur.*
2. *Metalle (Gußeisen, unlegierte und legierte Stähle).*
3. *Buntmetall, Aluminium und Hartgewebe (Textilbakelit).*
4. *Auf Metallflächen angeschmolzene Plastüberzüge.*

1. Untersuchung verschiedener mit Spritzguß hergestellten Plasten mit kristalliner Struktur

Auf Abb. 1 sind die Vergleichswerte der massiven Probekörper im Maß von ∅ 40 × 10 mm und die der im selben Maß gefertigten Stahlscheiben mit einer 0,5–0,8 mm starken Schicht von Polyamidüberzug angeführt. Es ist ersichtlich, daß mit Ausnahme der Polyformaldehyd- und Polyamidüberzüge die massiven Probekörper bis zu einer spezifischen Belastung von 60

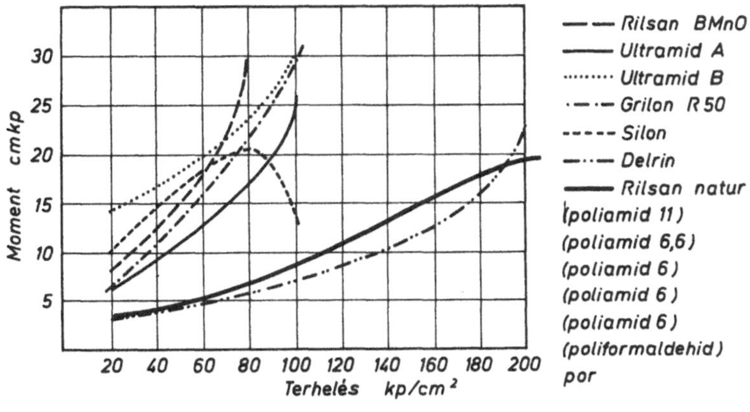

Abb. 1

bis 80 kp/cm^2 ein Reibungsmoment von 20 bis 25 cmkp aufweisen. Bei Steigerung der spezifischen Belastung über 80 kp/cm^2 erhöht sich plötzlich das Reibungsmoment, und die Materialoberfläche schmilzt ab. Diese Abnutzung stellt sich nach 40000–50000 Umdrehungen ein.

Das Verhalten der Polyformaldehyd- und Polyamidüberzüge weist im Vergleich zu den anderen Werkstoffen eine wesentliche Abweichung auf. Das Reibungsmoment dieser zwei Werkstoffe ist selbst noch bei einer spezifischen Belastung von 200 kp/cm^2 weniger als 20 cmkp. Die Abnutzung des Polyformaldehyds erfolgt ähnlich wie bei den anderen Werkstoffen, aber wesentlich später – nach 100000 Umdrehungen – bei einer spezifischen Belastung von 200 kp/cm^2. Der angeschmolzene Polyamidüberzug zeigte auch noch nach 400000 Umdrehungen keine Abnutzung und das Reibungsmoment verblieb dennoch unter 20 cmkp.

2. Untersuchungen von Metallen (Gußeisen, unlegierte und legierte Stähle)

Die Verschleißfestigkeit war bei den untersuchten Metallen sehr unterschiedlich. Die Zusammenstellung im Diagramm auf Abb. 2 veranschaulicht, um wieviel höher die Verschleißfestigkeitswerte von Gußeisen 18, Gußeisen 22, der Stahlarten mit der Qualitätsbezeichnung C 60 und MS 135, als die der anderen untersuchten Metalle stehen. Der Kohlenstoffgehalt der Werkstoffe mit vorteilhafteren Eigenschaften schwankte zwischen 0,4–3,5%. Ihre Oberflächen sind infolge der groben Kristallstruktur gute Schmiermittelspeicher. Dies alles trägt zur Erhöhung der Verschleißfestigkeit beträchtlich bei. Das spezifische Belastungsvermögen beträgt, ohne einer Einfressung Gefahr zu laufen, das zweifache der spezifischen Belastungsfähigkeit anderer Metalle.

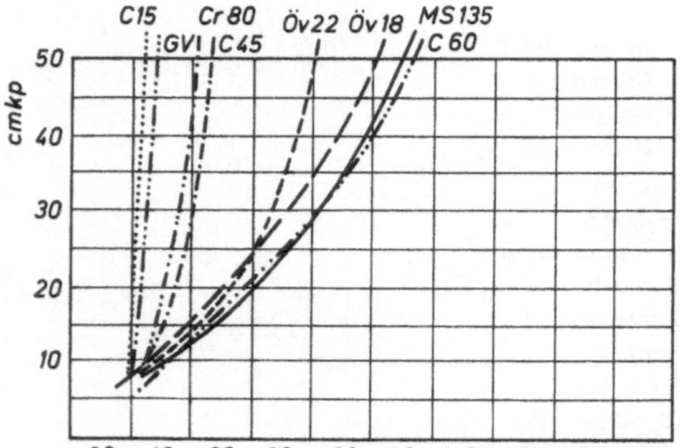

Abb. 2

Aufgrund der abgeschlossenen Etappen der bisherigen Vergleichsuntersuchungen kann schon jetzt festgesetzt werden, daß Metalle im Falle eines Schmierungsausfalles – wegen des darauffolgenden Trockenlaufs – nur mit niedriger 10–15 kp/cm^2 spezifischen Belastung und nur für eine beschränkte Zeitdauer in Anspruch genommen werden dürfen, wegen des beträchtlichen Verschleißes.

Aus dem Bild der abgeriebenen Flächen der Metallprobekörper sowie aus der Gestaltung des Verschleißdiagrammes können vergleichende Folgerungen gezogen werden.

3. Untersuchungen von Buntmetall, Aluminium und Hartgewebe

Die Untersuchungsresultate dieser Gruppe sind auch deshalb wichtig, weil diese Werkstoffe und ihre Legierungen auch heute noch als traditionelles Buntmetallmaterial bei Gleitlagern in Verwendung stehen. Aus dem vereinenden Diagramm (Abb. 3) kann festgestellt werden, daß Aluminium, Messing, Rotlegierung schon bei einer spezifischen Belastung von 40 kp/cm^2 vollständig abgenutzt wird (sich einfrißt). Die gesinterte (pulvermetallurgische) Bronze und die Bleibronze werden bei einer spezifischen Belastung von 60 kp/cm^2, das Hartgewebe bei 100 kp/cm^2 abgenutzt. Während die Metalle nach 10000–20000 Umdrehungen bei 20 bis 40 kp/cm^2 Belastung abgenutzt werden, tritt das bei Hartgeweben erst nach 40000–50000 Umdrehungen bei einer Belastung von 80-100 kp/cm^2 ein.

4. Untersuchung von auf Metalloberfläche angeschmolzenen Plastüberzügen

Das auf Stahlflächen angeschmolzene Plast war Naturpolyamid oder Polyamid 11 mit Mo-

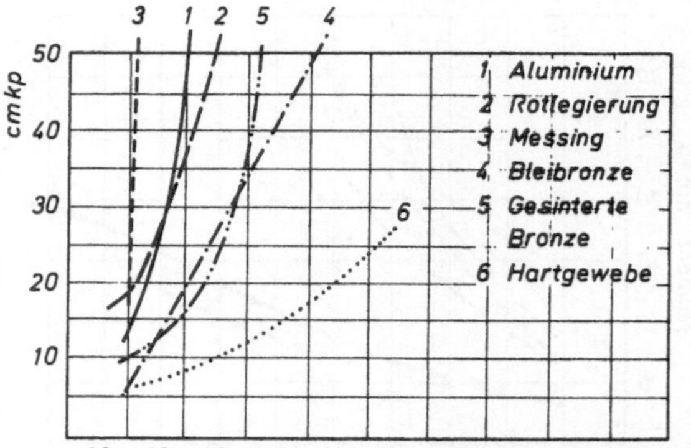

Abb. 3

lybdändisulfidzugaben Typ Rilsan und Polyamid 6 Typ Bonamid.

Die Untersuchungen wurden wie bei den vorhergehenden auf derselben Maschine und unter gleichen Umständen durchgeführt. Laut der früheren Untersuchungen erlitt der auf Metallflächen angeschmolzene Polyamidüberzug selbst bei einer 200 kp/cm² spez. Flächenbelastung keine Schäden, im Gegensatz zu den in vorgehendem Abschnitt erwähnten Werkstoffen. Andere Werkstoffe wurden bei 60–80 kp/cm² spez. Belastung nach 40000–60000 Umdrehungen abgenutzt. Der angeschmolzene Polyamidüberzug mit dem Molybdändisulfidgehalt wurde bei 200 kp/cm² spez. Belastung auch noch nach 400000 Umdrehungen nicht abgenutzt.

Unsere Untersuchungen erstreckten sich außer der Prüfung des Verhaltens verschiedener Metalle anderen Metallen bzw. Kunststoffen gegenüber (im letzteren Fall war der stillstehende Prüfling aus Stahl der Härte HR_c 60 der rotierende aus Metall mit einem angeschmolzenen Plastbelag) auch auf das Verhalten der Werkstoffpaarungen Kunststoff–Kunststoff und Metall + Kunststoff–Metall + Kunststoff.

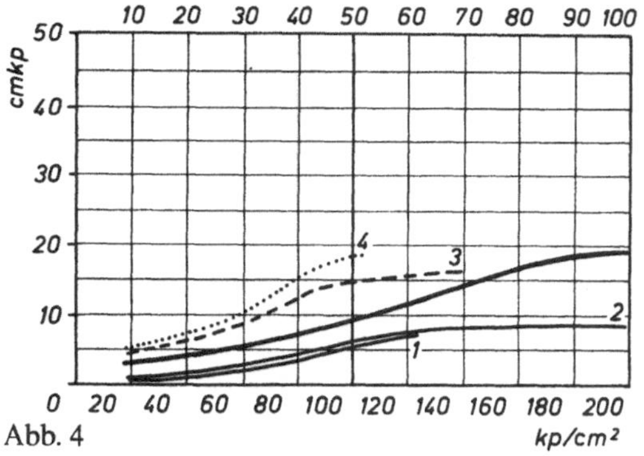

Abb. 4

Auf der Abb. 4 sind die mit angeschmolzenem Plastbelag bei der Paarung Metall + Kunststoff–Metall + Kunststoff erzielten Versuchsergebnisse eingetragen. Unter einem Prüfling aus Metall + Kunststoff ist ein 0,5 mm starker Kunststoffbelag zu verstehen, der auf die Oberfläche des rotierenden Metallprüflings angeschmolzen worden ist. Angewendet wurde dabei Polyamid 11 (das Rilsan französischer Herkunft) und Polyamid 6 (pulverförmiger Kunststoff, ungarisches Erzeugnis). Dieser Fall entspricht dem eines Gleitlager-Gegenstücks, einer Zapfenoberfläche. Der ruhende Prüfling ist ein Gleitlager aus M-DU, M-DX oder Speraflon mit einer Buchse aus Stahl + Kunststoff. Mit einer auf den ruhenden Prüfling geklebten Fläche von 1 cm² wollten wir die bei der Lagerbuchse vorherrschenden Verhältnisse simulieren. Durch diese Prüfung haben wir über die spezifische Belastbarkeit der Gleitflächen Kunststoff auf Metallgrundlage–Kunststoff bei einmaliger Schmierung, und über die Reibungsmomente und als Funktion der spezifischen Belastung Aufschluß erhalten.

Das R-Diagramm dient als Referenz (rotierender Teil aus Rilsan, stillstehender Teil aus Stahl). Kurve 1 zeigt die mit der Paarung Rilsan–DX, Kurve 2 die mit der Paarung Polyamid–DX erhaltenen Werte. Die im Diagramm 2 über der spezifischen Belastung eingetragenen Reibungsmomente zeigen einen sehr langsam ansteigenden Charakter und bleiben stets unter der roten Linie. Selbst bei einer spezifischen Belastung von 200 kp/cm² beträgt das Reibungsmoment 10 cmkp. Dieser recht günstige Wert ist dem Schmierstoffaufnahmevermögen des Lagerwerkstoffs M-DX und der höheren Belastbarkeit der Werkstoffe Typ Polyamid 6 zuzuschreiben.

Diagramm 3 zeigt die mit der Werkstoff-Paarung Rilsan und M-DU, Diagramm 4 die mit Rilsan und Sprelaflon erzielten Ergebnisse. In beiden Fällen übersteigt das Reibungsmoment die im Referenz-Diagramm angeführten Werte, und der Verschleiß erfolgt bei etwa 100 bis 140 kp/cm².

Die Prüfergebnisse beziehen sich auf eine rotierende Bewegung. Man könnte wohl annehmen, daß bei geringerem Verschub eine längere Lebensdauer zu erzielen wäre, und hauptsächlich könnte man durch das Anschmelzen von Plastbelag auf die Metalloberflächen den Verschleiß der bloß schwer oder überhaupt nicht schmierbaren Flächen herabsetzen.

Wendet man die Lagerwerkstoffe Glacier, Sprelaflon, M-DU oder M-DX im feuchten Betrieb an, muß man für den Korrosionsschutz der Zapfenflächen Sorge tragen (sie sollen daher entweder aus rostfreiem Stahl gefertigt oder verchromt werden). Bei unseren Vorprüfungen gewannen wir die Erfahrung, daß die Korrosion durch den angeschmolzenen Kunststoffbelag wirtschaftlich beseitigt werden kann.

Nach mehreren Kontrolluntersuchungen wurde mit der Bestimmung der höchstwirtschaftlichen Oberflächenfeinheit bei den mit Plast überzogenen Bauteilen begonnen. Die Rolle der Oberflächenbearbeitung – die der Oberflächengüte – kann mit der von Metallen nicht in Vergleich gezogen werden. Die Härte, Glätte der

Metallflächen erhöht die Verschleißfestigkeit, fördert die Gestaltung des nötigen Schmierfilms und bringt die hydrodynamische Schmierungstheorie zur Geltung. Infolge der Einwirkung der auf die Stahlflächen angeschmolzenen Plastschichten ergeben sich solche Umstände, auf welche die Voraussetzungen der traditionellen Schmierungstheorie nicht bezogen werden können. Die Erforschung dieser Erscheinung ist heute noch im Gange. Die ausführliche Erläuterung dieser ist gegenwärtig noch nicht aktuell, weil die Laborversuche noch mit Betriebsangaben bestätigt werden müssen. Dagegen kann schon heute festgestellt werden, daß die Oberflächenbearbeitung der Plaste keine so primäre Bedeutung hat, wie die der Metalle. Die Einfachheit der Plastbearbeitung ist ein derartiger technologischer Vorteil, der die Verarbeitungs- und Bearbeitungskosten in bedeutendem Maße herabsetzt.

Untersuchungen der Materialstruktur von auf Metallflächen angeschmolzenen Plastüberzügen

Nachfolgend soll anhand einiger Abbildungen die eigenartige Struktur der auf Metallflächen angeschmolzenen und mittels der Auftragetechnologie gestalteten Plastüberzüge vorgeführt werden.

Die rasche betriebsmäßige Verbreitung der neuen Technologie wurde durch die polarisationsmikroskopischen Untersuchungen beschleunigt und mit Hilfe dieser Untersuchungen wurde betreffs der betriebsgemäßen der Verwendung eine zuverlässige Kontrolle gewährleistet.

Auf Abb. 5 ist ein aus Normal-Polyamid 11 gefertigter Anschmelzungsüberzug in 60facher Vergrößerung zu sehen. Die in ihrem ganzen Querschnitt gleichmäßig feinkristalline Struktur ist die Grundstruktur, welche mit Hilfe einer für das Material charakteristischen und genauen Technologie ausgebildet werden kann.

Auf Abb. 6 ist die Struktur von Polyamid 11 mit Molybdänsulfidzugaben in 60facher Vergrößerung dargestellt. Das zwecks Verbesserung der Gleiteigenschaften beigemischte Molybdändisulfid ist durch seine charakteristische fächerförmige Verteilung leicht zu erkennen. Die vollkommene Vermischung wird durch die feinverteilten Fächerflügel angezeigt, wodurch die Materialstruktur sehr homogen gestaltet wird. Der sehr große Verschleißwiderstand des so mit Zugaben gebildeten Überzuges wird nicht nur durch die beigemischten Zugaben, sondern auch durch deren gleichmäßige Struktur beeinflußt. Bei der Montage der sich aufeinander verschiebenden, beweglichen Bauteile wird ein Schmierfett auf Lithium-Molybdänbasis verwendet. Das im Schmierfett befindliche Molybdänsulfid wird auf der Metallfläche zwischen die Mikrospitzen eingebaut, wozu auch die Lithiumzugabe mit ihrer hohen Wärmebeständigkeit beiträgt.

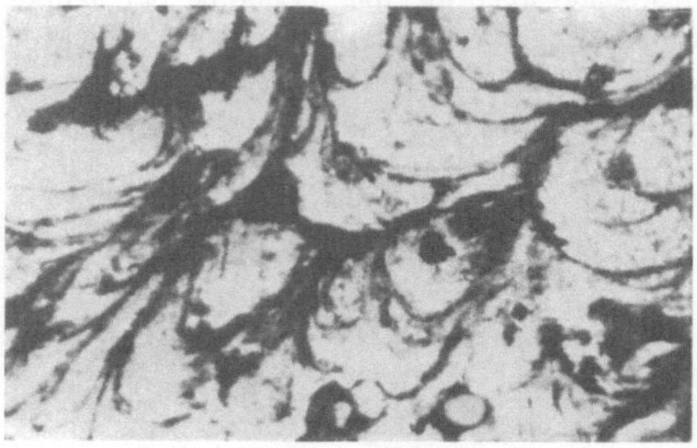

Abb. 6

Auf Abb. 7 ist ein aus Polyamid 11, mit einer Molybdänzugabe von großem Feuchtigkeitsgehalt hergestellter, bei hoher Temperatur angeschmolzener Überzug dargestellt. Die im

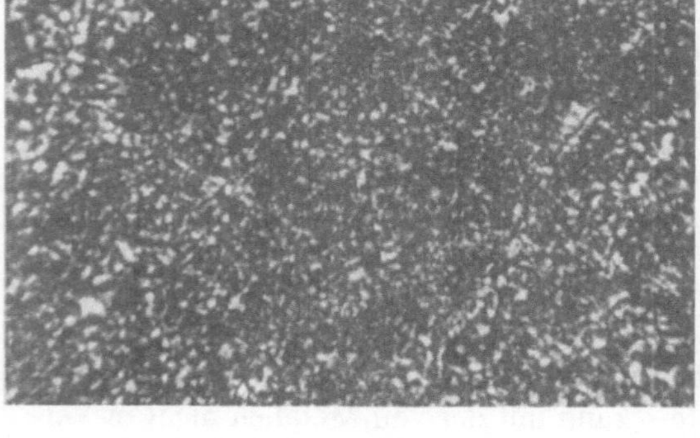

Abb. 5

Abb. 7

Überzug befindlichen kleinen Blasen verursachen in der Struktur Kontinuitätsmängel und eine geringere Haftfähigkeit. Infolge der hohen Verarbeitungstemperatur und des ungleich vermischten Molybdänsulfides nahm die Inhomogenität der Kristallstruktur zu.

Es wäre erfreulich, wenn es durch die Darstellung dieser Beispiele gelungen wäre, die Notwendigkeit und die Bedeutung der polarisationsmikroskopischen Materialuntersuchungen bei der Anschmelzung von Plastüberzügen auf Metallflächen hervorzuheben. Die polarisationsoptische Methode ergibt bei Untersuchungen von Plaststrukturen einen praktischen, handgreiflichen Nutzen. Jedes Eindringen in das Innere des Materials und dessen Erforschung stellen sehr interessante Aufgaben, die in bezug auf Plaste über die Identifizierung der Materialgrundstruktur hinweg zur Entwicklung einer zuverlässigen Anwendungstechnologie beitragen.

Nach Abschluß der vorgehend bekanntgegebenen, sehr ausführlichen und zahlreiche Materialarten umfassenden Untersuchungen, wurde mit der betriebsmäßigen Anwendung verschiedener Bauteile begonnen.

Den Anwendungen entsprechend, haben wir die Maschinenelemente in folgende Gruppen eingeteilt:

a) *Festsitzende Elemente* (z. B. Achsschenkel, Radnabe und andere Gleitlagerplätze usw.).

b) *Bauteile mit bestimmter Bogenbewegung* (z. B. Bremsschlüssel, Ausheber, Schwingebock, Kugelgelenk, Bolzen, Hängezapfen usw.).

c) *Bauteile mit Gleitbewegung* (z. B. Ventilhebel, Deckel, Schieber, Würfelzapfen usw.).

d) *Bauteile mit Drehbewegung* (z. B. Gleitlager).

Diese Gliederung wurde auf der Grundlage der mechanischen Beanspruchung vorgenommen und berücksichtigt nicht den Korrosionsschutz, obwohl in vielen Fällen die Korrosion die Ursache der Schädigung sein kann (z. B. bei Pumpen, Kühlern usw.).

Schutz gegen Reibungskorrosion durch angeschmolzene Polyamid-Überzüge

Reibungskorrosion kann zwischen Metallen, zwischen Nichtmetallen, aber auch zwischen Metall- und Nichtmetall-Flächen auftreten. Bei Metallen kommt es auch dann zu einer Reibungskorrosion, wenn bei den abgescheuerten, abgeriebenen Partikelchen keine chemische Umwandlung erfolgt. Tritt aber auch noch eine chemische Umwandlung auf, dann verläuft die Zerstörung natürlich erheblich schneller.

Außer der Reibungskorrosion können natürlich auch noch andere Korrosionserscheinungen vorkommen. Eine der wichtigsten ist Lochfraß (Pitting), der an Maschinenteilen, besonders unter Dauerbelastung, an der der Belastung gegenüberliegenden Seite der einen bogenförmigen Weg hinterlegenden Teile vorzukommen pflegt.

Aufgrund unserer mehr als siebenjährigen Erfahrungen können wir heute getrost die Behauptung wagen, daß durch Anschmelzen eines Polyamidbelags auf die Metalloberfläche die gegenseitige Berührung der Mikrospitzen der Metallflächen vermieden werden kann. Infolgedessen spielt sich die Reibungskorrosion zwischen Kunststoff und Metall bzw. zwischen Kunststoff und Kunststoff ab, und dadurch können gegenüber der Haftreibung zwischen Metallflächen völlig andere, viel günstigere Ergebnisse erzielt werden.

Die sich nicht gegeneinander bewegenden Oberflächen von Maschinenelementen sowie von zahlreichen Anlagen und Geräten werden z. T. aus ästhetischen Gründen in erster Linie jedoch zum Korrosionsschutz mit unterschiedlichen Anstrichsystemen versehen. Es gibt aber auch viele Beispiele, bei denen sich periodisch oder ständig gegeneinander bewegende Flächen durch einen Überzug gegen Korrosion geschützt werden müssen. Diese Schichten müssen auch den Beanspruchungen ohne Schaden widerstehen, die den Verschleiß hervorrufen.

Im folgenden werden in einer dem Charakter und den Anwendungsgebieten entsprechenden Gliederung einige Teile behandelt, die mit einer Polyamidschicht versehen wurden.

Das Problem des Oberflächenschutzes von Federn in Fahrzeugen oder anderen Geräten oder Anlagen ist bis heute nicht gelöst. Bei derartigen Federn wirkt eine Schicht nicht nur als ausgesprochener Korrosionsschutz, sie verhindert auch die Rostbildung auf Blattfedern und gewährleistet, daß die dadurch verursachte Haftung der einzelnen Federblätter aufeinander aufgehoben wird. Bei jedem Federblatt ist die gesamte Last gleichmäßig auf die ganze Oberfläche verteilt, wodurch auch die Lebensdauer der Feder wesentlich erhöht wird. Wir untersuchten in Betriebsversuchen unter Berücksichtigung der Wirtschaftlichkeit und der Lebensdauer die vollständige und die einseitige Beschichtung von Federblättern. Auch Spiralfedern, bei denen die einzelnen Windungen nicht aufeinanderliegen, können Plastschutzschichten erhalten. Dieses Verfahren hat sich zum Schutz der Federn von Lebensmittelabfüllautomaten besonders bewährt (Abb. 8 und 9).

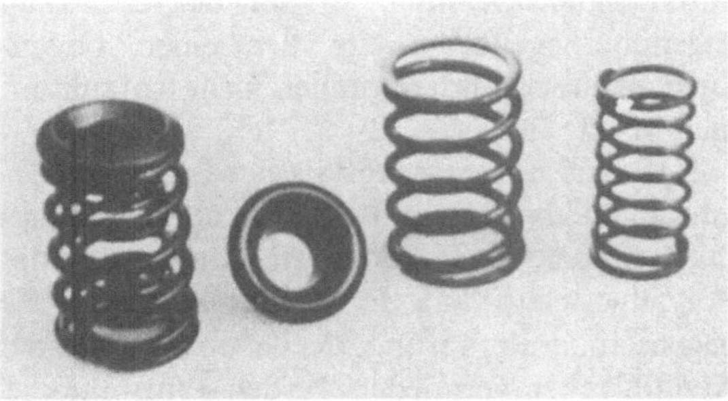

Abb. 8

Abb. 9

Bei lösbaren Verbindungen entstehen durch Rost oft Schwierigkeiten und ein erhöhter Arbeitsaufwand beim Lösen der Verbindung; oft werden die Verbindungselemente dabei sogar zerstört. Sehr ungünstig sind ungeschützte Verbindungselemente an Stellen, die der Witterung bzw. Wasserdampf oder anderen korrodierend wirkenden Chemikaliendämpfen ausgesetzt sind. Dabei wird die Korrosion durch die Potentialdifferenz noch gefördert, die sich aus den unterschiedlichen Metallen bzw. Legierungen ergibt, aus denen die Verbindungselemente bzw. die zu verbindenden Teile bestehen. Der von den Verbindungselementen ablaufende Rost zerstört außerdem die umliegenden mit einem Anstrich versehenen oder emailierten Oberflächen.

Die Anstrichschicht, die vor dem Verbinden aufgebracht wurde, wird bei der Montage verletzt, und bei einem Neuanstrich dringt der Anstrichstoff nicht zwischen die Teile, so daß die Fläche zwischen dem Verbindungselement und den verbundenen Teilen vollkommen ungeschützt ist.

Eine auf die Schraube aufgeschmolzene dünne, vollkommen homogene, Plastschicht bildet eine elastische Zwischenschicht zwischen den Metallen und verhindert, daß die beiden Metalloberflächen zusammenrosten. Gleichzeitig kann die Verbindung mehrmals ohne Schwierigkeit gelöst und wiederhergestellt werden (Abb. 10).

Abb. 10

Spezialschrauben, die in Sonderanfertigung gefertigt werden, können wirtschaftlicher mit einer Plastschicht erneuert, als neu hergestellt werden. In solchen wie auch in anderen Fällen sind Zweck und Wirtschaftlichkeit sorgfältig gegeneinander abzuwägen.

Bei manchen Elementen mit Gewinde ist es für die Anwendung entscheidend, ob sie ausgetauscht oder verstellt werden können, z. B. bei Kraftfahrzeugen die Hilfsführung, ein Element mit Innen- oder Außengewinde, das eine sichere Auflage auf den erhöhten Flächen der Gewinde gewährleistet. Der Verschleiß tritt bei diesem Element nicht an einer Fläche, sondern verteilt auf die Oberflächen der Gewindegänge auf. Zahlreiche derartige Teile wurden bereits durch Aufbringen einer Polyamidschicht repariert und befinden sich seitdem ohne Beanstandung in Betrieb. Ein Nachstellen ist leicht möglich.

Tragbolzen aus Gußeisen sind oft undicht und rosten häufig ein. Durch Aufschmelzen eines Plastes kann man ohne Nachbearbeitung eine dünne, zusammenhängende Schicht erzielen, die durch ihre Elastizität gleichzeitig auch als Dichtung zwischen den beiden Gewinden wirkt.

Selbständige Konstruktionseinheiten und Kräfte übertragende Verbindungen müssen im Interesse der Betriebssicherheit oft montiert und geschmiert werden. Diese Teile aus Metall werden zunehmend durch solche aus Plasten ersetzt oder mit einer Plastschicht vergütet; dadurch wird der Instandhaltungszyklus auf 4 bis 6 Monate bzw. 20000–30000 km vergrößert. Bei einer entsprechenden Auswahl des aufzubringenden Plastgemisches kann man die Zeitspanne sogar noch verlängern.

Bei den herkömmlichen Gleitpaaren aus Metallen werden zwei unterschiedliche Metalle oder Metalle verschiedener Härte gepaart. In beiden Fällen können die Metalle kein Schmiermittel aufnehmen (z. B. Hängebolzen); das hat eine kurze Lebensdauer der Gleitfläche und damit die Notwendigkeit häufiger Instandhaltungsarbeiten zur Folge (Abb. 11).

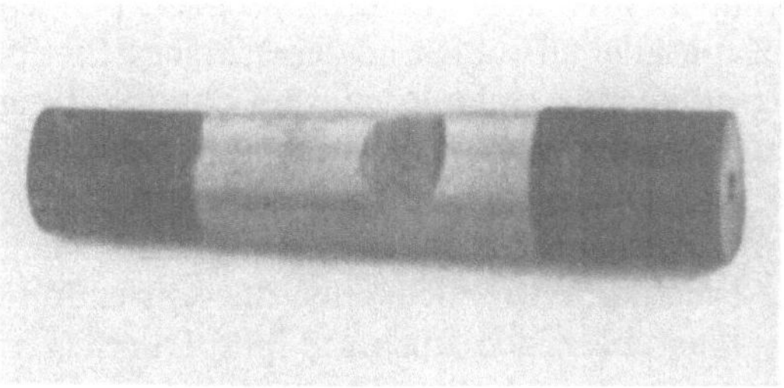

Abb. 11

Der Hängebolzen ist sowohl hinsichtlich der persönlichen als auch der materiellen Gefährdung ein äußerst wichtiges Element. Im allgemeinen werden als Gegenstück Mäntel aus Bronze, Stahl oder Hartgewebeschichtstoff eingesetzt. Wir fanden auch Vorschläge, spritzgegossene Mäntel aus Polyamid zu verwenden. Ein spanend oder im Spritzgießverfahren hergestellter Mantel aus Polyamid mit einer Wanddicke von 3 bis 4 mm kann aber ausbrechen und aus der Bohrung fallen. Das Ausbrechen kann ein Verklemmen zwischen Bolzen und Mantel und damit eine wesentliche Veränderung der Radneigung zur Folge haben, was die Lenkung stören oder gar unmöglich werden lassen kann.

Bei der Beschädigung einer 0,4–0,5 mm dikken Plastbeschichtung kann auch im Durchmesser ein Verschleiß von maximal 1 mm auftreten, der weder zu einem Verklemmen noch dadurch zu einem Unfall führt. Anstelle des Bronzemantels kann man einen solchen aus Gußeisen verwenden; die darauf befindliche Plastschicht kann man stets auf die erforderlichen Maße bearbeiten.

Kugelgelenke müssen oft ausgewechselt bzw. erneuert werden. Ihre Herstellung ist kompliziert und erfordert mehrere technologische Stufen. Die Kugelschalen und Kugelbolzen werden geschliffen. Im allgemeinen sind Kugelgelenke nach 1–2 Wochen zu schmieren.

Man kann zahlreiche technologische Prozesse einsparen, wenn man auf die gleitenden Teile von Kugelgelenken eine Plastschicht aufbringt, und zwar können das Zementieren oder Flammenhärten des Kugelbolzens sowie das Zementieren und Schleifen des unteren Teils der oberen Schalenhälfte entfallen. Unverändert müssen unbedingt die Grundfestigkeit des zur Herstellung eines neuen Kugelbolzens verwendeten Werkstoffs sowie dessen Wärmebehandlung zur Vergütung bleiben (Abb. 12).

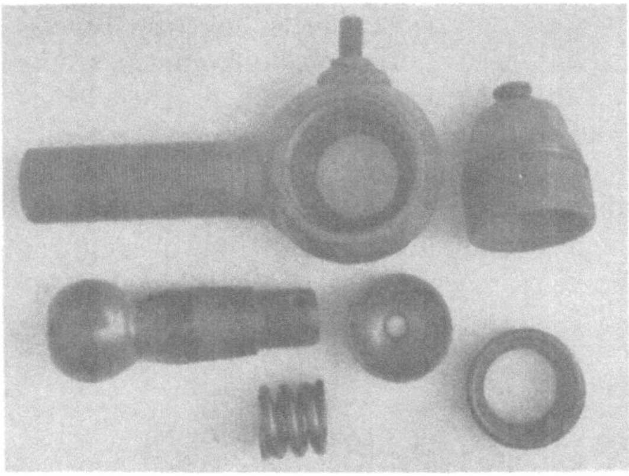

Abb. 12

Die Wagen der Budapester Vorortbahnen (HEV) sind über jeweils eine Drehtellerhälfte mit dem Drehschemel verbunden. Wenn der Drehteller beschädigt ist und ausgewechselt werden muß, fällt der Wagen länger als einen Tag aus. Die mit Plast beschichteten Drehteller wurden ohne Nachbearbeitung eingebaut. Das erste Paar befindet sich jetzt bereits mehr als 100000 km in Betrieb. In etwa 6 Monaten wurden 120 Paar eingebaut (Abb. 13).

Abb. 13

Die Bremsscheibe und der Bremsbackenbolzen können nur schwer gegen Korrosion geschützt werden. Stark korrodierend wirkt die zwischen Bremstrommel und Bremsscheibe eindringende Feuchtigkeit. Ein Schmieren erfordert viel Umsicht und kann nur bei der Montage oder der Durchsicht vorgenommen werden. Überschüssiges Schmiermittel kann von der Bremsscheibe auf die Bremsbacken gelangen und damit zum Rutschen der Bremse führen. Zusammengerostete Bremsbackenbolzen behindern die Bewegung der Elemente, verklemmen und können leicht einen Unfall verursachen.

Durch Aufbringen von Plastüberzügen kann man die Bremsbackenbolzen und die Bremsscheiben gegen Korrosion vollkommen schützen. Verwendet man beim Einbau einen Plastmantel als Lagerschale, so können diese Elemente schmierungsfrei betrieben werden. In ähnlicher Weise kann man auch bei den Elementen des Handbremssystems mit vollständiger Sicherheit aufgeschmolzene Plastüberzüge anwenden.

Abb. 14

Bei Kraftfahrzeugen mit Luftbremse ist es schwierig, die Radbremszylinder gegen Korrosion zu schützen. Die in den Bremszylinder eindringende Feuchtigkeit und die Kondensatbildung bewirken eine Korrosion der inneren Flächen, durch die die Funktion des Bremskolbens aus Gummi behindert wird. Bei Bremszylindern, die durch eine Plastschicht geschützt sind, ist die Korrosion vollkommen ausgeschlossen, außerdem werden durch die Plastschicht die Gleiteigenschaften verbessert. Beim Budapester Autobusbetrieb laufen die mit einem Plastüberzug erneuerten Bremszylinder nunmehr 50000 bis 60000 km (Abb. 14).

Eine Beschichtung von Kettenrädern erscheint dadurch gerechtfertigt, daß der Verschleiß durch die Plastschicht vermindert und die Korrosion verhindert wird. Durch die Plastschicht können die Zahnräder an Stellen, an denen keine Schmierung möglich ist, schmierungsfrei ohne Störung betrieben werden. Die Versuche auf diesem Gebiet sind noch nicht abgeschlossen.

Mit den hier aufgezählten Beispielen sind noch längst nicht alle Möglichkeiten der Anwendung von Plastüberzügen ausgeschöpft. Durch Laboratoriums- und Betriebsversuche wird es möglich sein, weitere ökonomisch und technisch effektive Anwendungen zu finden.

Die Betriebserfahrungen beweisen, daß bei den genannten Teilen eine Plastschicht aufgebracht werden kann. Die dem Verschleiß am stärksten ausgesetzten Flächen können nach diesem Verfahren schnell wieder auf die geforderten Maße bearbeitet werden; das eröffnet große Möglichkeiten für Reparaturen und Instandhaltungen.

Anschrift des Verfassers:

E. Vadász
Wissenschaftl. Forschungsinstitut für Straßenverkehr
Than Kalroly ut 3–5
Budapest XI (Ungarn)

Rheol. Acta 13, 305–317 (1974)

From the Sun Research and Development Company, Marcus Hook, Pennsylvania (U.S.A.)

High shear viscometry of concentrated solutions of poly (alkylmethacrylate) in a petroleum lubricating oil

By A. F. Talbot

With 11 figures and 3 tables

(Received October 27, 1972)

Nomenclature

η Apparent viscosity at experimental conditions, centipoise
$[\eta_M]$ Intrinsic viscosity via extrapolation of *Martin* equation, dl/g
$\dot{\gamma}$ Shear rate, sec^{-1}
τ Stress level, $dynes/cm^2$
$\hat{E}$ Energy level at inflection point ($\dot{\gamma}\tau$), $dynes/cm^2$ sec

Subscripts

0 Limiting low shear rate (1st *Newton*ian) condition.
∞ Limiting high shear rate (2nd *Newton*ian) condition
s Refers to solvent

Introduction

The addition of high molecular weight polymers to various types of petroleum lubricants and fluids has become accepted practice, justified by the improved performance of the fluid composition over a broader range of operating temperature. Increasingly, these polymer-modified oils are encountered in service, including automatic transmission fluids and multi-graded engine oils in automotive applications, multi-graded gear oils and hydraulic fluids. Despite the widespread use of these fluids, there have been published relatively few quantitative descriptions of their flow characteristics under the high shear conditions encountered in use.

Much of the data establishing non-*Newton*ian flow behavior of solutions of high polymers has been obtained in low viscosity, pure solvents, often at low concentrations. From a utilitarian view, there is considerable incentive to derive reliable generalizations for high shear flow of concentrated polymer solutions in higher viscosity media, such as petroleum lubricating oil fractions. The investigator concerned primarily with the dynamics of a hydraulic or lubrication system would welcome such a general scheme, whether or not it evolved from a fully developed flow theory.

Mathematical modeling of hydraulic or lubricating systems can be highly informative, from the standpoint of mechanical design of the system hardware, or with the view of developing improved lubricant formulations. A component of the model which expresses the viscosity of the fluid under wide ranges of temperature, pressure and shear rate is required. The response of the model to variations in the composition of the polymer-oil solution, including the inevitable reduction in polymer molecular weight during the useful life of the fluid, might also be examined. Recent work combining the analytical treatment of hydrodynamic bearings with pseudo-plastic lubricant characteristics has been reported. Substantially increased seal leakage (1) and reduced journal bearing capacity (2) are among the effects attributed to the shear-thinning of lubricants.

This paper describes the high shear steady-flow behavior of one type of polymer-modified lubricating oil which was examined in this laboratory. The parametric relationships presented are empirical in nature; however, for the particular polymer-oil system studied, they demonstrate an encouraging degree of consistency.

Background

Isolated descriptions of the high shear rate steady flow of solutions of polymers in petroleum lubricating oil are to be found; however, they offer few opportunities to derive inter-relationship among the many variables that can be imposed on the system. *Appeldoorn* and *Philippoff* (3) showed, for polyisobutylenes, that elastic and non-*Newton*ian effects are encountered at lower shear rates in higher viscosity solvent (decalin vs. white oil). In their summary on non-*Newton*ian flow in concentrated solutions (low viscosity solvent), *Tager* and *Dreval* (4) indicate relatively rigid polymer molecules and poor polymer solvents tend to produce more definitive 2nd *Newton*ian plateaus. *Johnson* and *Wright* (5, 6) showed that shear-degraded engine oil and automatic transmission fluid containing VI improvers have substantially different viscosity-shear rate-temperature relationship than the same fluids in new condition. Polymer type and concentration were not given. *Novak* (7) presented pseudoplastic flow patterns for several polymer-oil systems at pressures up to 3.45 kbar (50000 psig), although the effect of pressure on

20

the shape of the flow curve, and the quantitative effect of the polymer parameters were not extracted.

Horowitz (8) used shear stress, with varying success, to correlate the effect of temperature on pseudoplastic flow of several polymer types in oil solution. Normalization of concentration and molecular weight effects via the appropriate reduced variables were partially effective in producing representative master curves.

Ram and *Siegman* (9) investigated the high shear rate viscosities of a series of polyisobutylenes in the molecular weight range from 1×10^6 to 7×10^6 in both pure solvents and light petroleum fractions. Throughout a 6-fold increase in solvent viscosity, they observed a regular decrease in the *Mark-Houwink* exponent for the high shear rate polymer intrinsic viscosity.

Thus, for typical commercial fluids incorporating a polymeric additive in a petroleum lubricating oil, there is little information available to indicate how the apparent viscosity may change with wide variations in the more accessible solution parameters, when subjected to high shear conditions. The purpose of this work is to describe the combined effects of shear rate, polymer molecular weight, and polymer concentration within a single polymer-solvent system closely related to those commonly found in the petroleum lubricants.

Experimental

Petroleum oil

The oil was a solvent refined petroleum distillate oil, with a kinematic viscosity of 5.08 centistokes at 98.9 °C (210 °F) and 32.05 centistokes at 37.8 °C (100 °F). Since the polymeric additives were supplied in a carrier oil, the base oil "solvent" was chosen to match closely the properties and constitution of the carrier oil. Thus, variations in polymer concentrations would not be confounded with variations in the type of solvent, or polymer-solvent interaction.

The petroleum oil may be considered typical of the paraffinic blending stocks used in formulating many modern automative and industrial lubricants. It is obtained by fractional distillation, solvent extraction and dewaxing of a West Texas crude mix. Additional physical properties of this oil are summarized in table 1, which also includes estimates of the distribution of carbon atoms according to structure type by the $n-d-M$ method of *Van Nes* and *Van Weston* (10).

Polymers

The four polymers used in this study were selected from a series of commercial polyalkylmethacrylate viscosity index improvers supplied as solutions in a petroleum carrier oil for ease in handling. Solids content and polymer molecular weight are normally controlled during manufacture to maintain uniform viscosity of the polymer-carrier blend.

Table 1. Properties of petroleum lubricating oil

Property	Method	Value
Kinematic viscosity		
@ 100 °F (37.8 °C)	ASTM D-445	32.05 cSt
@ 210 °F (98.8 °C)	ASTM D-445	5.08 cSt
Viscosity index	ASTM D-2270	92
Specific Gravity @ 60 °F/60 °F (14.4 °C)	ASTM D-1298	0.8687
Flash point (Cleveland open cup)	ASTM D-92	390 °F (198 °C)
Aniline point	ASTM D-611	210 °F (99 °C)
Pour point	ASTM D-97	0 °F (— 18 °C)
Refractive index ($n_d^{20°C}$)	ASTM D-1218	1.4772
Molecular weight (est.)	[1])	383
Carbon type composition	[2])	
% Ca		5.9
% Cn		27.2
% Cp		66.9
Ra		0.27
Rn		1.23
Rt		1.50

[1]) *Hirschler, A. E.*, J. Inst. Pet. **32**, 133 (1946).

[2]) By *n–d–M* method of structural group analysis by *Van Nes* and *Van Weston* (10).

% Ca, Cn, and Cp represent percentage of carbon atoms in aromatic rings, naphthenic rings, and paraffinic chains, respectively.

Ra, Rn, and Rt represent average number of aromatic rings, naphthenic rings, and total rings per molecule.

Additional descriptive information for these polymers is presented in table 2, which shows that the viscosity average molecular weight covered the range from 355000 to 1650000. The polymers are not homopolymers, but are prepared from a mixture of alkylmethacrylates of various alkyl chain lengths, ranging from about C_4 to about C_{18}. At least three alkyl chain lengths are present in major proportions; several others, in minor proportions. The mean alkyl chain length is calculated to be about C_9. Molecular weight distribution ($\bar{M}w/\bar{M}n$) is typically in the range of 4–6.

Polymer solutions

Oil solutions of each of the four polymers were prepared at three concentration levels, to give comparable values of relative viscosity ($\eta_{rel} = 2.5$, 6.8, and 26) independent of polymer molecular weight. The range of concentrations used (2–21 wt.-%) encompassed more than the typical polymer dosage found in many automative and industrial lubricants, in order to obtain better definition of some of the phenomena being examined.

Table 2. Description of polymers

Type: Poly (alkylmethacrylate)
Form: Solution in petroleum carrier oil
Monomer composition: 3 major constituents ranging from C_4- to C_{18}-methacrylate undefined number of minor constituents
Monomer mixture approximates C_9-methacrylate

Molecular weight:				
Code	1	2	3	4
Nominal solids, wt. %	42	35	28	19.5
$\bar{M}_v$*)	3.5×10^5	5.6×10^5	8.33×10^5	1.65×10^6
Molecular weight dist'n.: $\bar{M}_w / \bar{M}_n$ 4–6				

*) By $[\eta] = 3.4 \times 10^{-5} \bar{M}_v^{0.72}$ in *n*-heptane at 30 °C.

Low shear rate viscosity measurements were obtained periodically on the solutions after a series of high shear experiments to insure that no significant amount of permanent viscosity loss had occurred through mechanical degradation of the polymer.

Equipment

Low shear rate viscosities of the base oil and the polymer-oil solutions were determined in modified *Ostwald* type glass capillary viscometers. For these measurements, the shear rate was estimated to be from 10–50 reciprocal seconds. These shear rates are considered sufficiently low that the solutions would not exhibit significant non-*Newtonian* flow. This assumption was generally confirmed, in that viscosities obtained at the lowest shear rates in the high shear apparatus agreed with those obtained in the glass viscometers.

The high shear apparatus consists of two fluid reservoirs connected by a length of precision bore stainless steel capillary mounted in a heavy walled stainless steel holder. Several interchangeable capillary-holder combinations, encompassing a range of capillary diameters (0.1–1 mm) are available, giving access to shear rates in the range of 10^2–10^6 sec^{-1}. Capillary L/R ratios ranged from 100–400. The test fluid is driven from one cell to the other by nitrogen pressure, which is recorded from either a mercury manometer or a series of precision pressure gauges. The flow time for a measured volume of test fluid to pass through the capillary is recorded. Consecutive passes in each direction are averaged, for each pressure level studied. The test apparatus is contained in a constant temperature bath maintained at the test temperature. The bath, in turn, is installed in a convected oven maintained at slightly below the test temperature. All data reported in this paper have been collected at a test temperature of 37.8 °C (100 °F).

Flow calculations

Raw data from the high shear viscometer are processed by computer. In the first program, appropriate corrections are made, including kinetic energy terms, pressure effects and the effect of viscous heating, before apparent viscosity, shear rate, and stress level are calculated. Elastic energy and viscous end effects are considered negligible,

Table 3a. Polymer solution flow data at 37.8 °C (100 °F)

Code	925	924	923	916	915	914
Polymer $\bar{M}_v$	355,000			560,000		
Polymer conc'n., g/dl	3.62	9.10	18.4	3.01	7.57	15.3
η_0, cps	63.3	177	750	66.8	188	742
η_0/η_s	2.33	6.51	27.6	2.46	6.92	27.3
$(\eta_0/\eta_s - 1)/C$	0.366	0.604	1.44	0.484	0.781	1.72
From approximating flow curve						
η_∞, cps	47.8	118	456	47.4	123	454
$(\eta_\infty/\eta_s - 1)/C$	0.209	0.367	0.856	0.247	0.465	1.03
$(\dot{\gamma}\tau)$ @ Inflection point, dyne sec/cm²	4.0×10^8	2.3×10^8	1.4×10^8	3.2×10^8	7.1×10^7	3.1×10^7
Slop of $\log E - f(\Phi)$ line	−1.15	−1.09	−1.21	−1.42	−1.11	−1.11

Table 3b. Polymer solution flow data at 37.8 °C (100 °F)

Code	922	921	920	919	918	917
Polymer $\bar{M}_v$	833,000			1,650,000		
Polymer conc'n., g/dl	2.41	6.04	12.2	1.68	4.20	8.44
η_0, cps	66.9	182	688	69.9	188	643
η_0/η_s	2.46	6.72	25.3	2.57	6.90	23.7
$(\eta_0/\eta_s - 1)/C$	0.606	0.946	2.00	0.934	1.40	2.68
From approximating flow curve						
η_∞, cps	47.6	114	398	48.6	108	313
$(\eta_\infty/\eta_s - 1)/C$	0.312	0.530	1.12	0.468	0.705	1.25
$(\dot{\gamma}\tau)$ @ Inflection point, dyne sec/cm²	5.6×10^7	2.8×10^7	9.6×10^6	4.1×10^6	3.6×10^6	2.1×10^6
Slope of $\log E - f(\Phi)$ line	−1.51	−1.38	−1.40	−1.54	−1.54	−1.39

since capillary L/R ratios ≥ 100. Attempts to estimate an overall end effect via the *Bagley* method confirmed this assumption. The *Rabinowitsch* correction for non-parabolic velocity distribution is also not made, as its effect was found to be quite small. *Reynolds* numbers, based on the apparent viscosity of the polymer solution at the experimental shear rate, ranged from 0.1–100, well below the conventional criterion for the transition from laminar to turbulent flow. In a second program, the entire series of points for a single fluid composition are fit by a curve which expresses the variation in apparent viscosity of the polymer solution over the entire shear rate range in which valid data were collected. This procedure will be discussed in a subsequent section.

Data treatment

Characteristics of flow curves

The raw data, suitably corrected, show a characteristic variation in apparent viscosity of the polymer solution with shear rate. A typical plot, for a solution of 10.5 weight % of the 355000 molecular weight polyalkylmethacrylate in the lubricating oil, is shown in fig. 1. A gradual loss in apparent viscosity occurs over several decades of shear rate. At both very low and very high shear rates, the apparent viscosity approaches, asymptotically, viscosity levels which are significantly higher than the viscosity of the base oil. These asymptotes are the "first *Newton*ian" and "second *Newton*ian" levels, respectively. Between these two plateaus, the apparent viscosity decreases – first slowly, then rapidly until the rate of change is a maximum. Beyond this inflection point, the rate of viscosity loss diminishes again, to approach the upper *Newton*ian viscosity level.

The data points of fig. 1 illustrate the typically good precision obtained with this series of capillaries. Within the smooth portion of the flow curve, consecutive duplicate measurements for a single capillary seldom vary by more than 1%. In the upper shear rate regime, however, a flow instability is apparently encountered, wherein the calculated viscosity falls sharply with shear rate and duplicate measurements can vary by as much as 10–50%. This unstable flow region is shown by the dotted lines of fig. 1. These indicate the complete 2nd *Newton*ian level is often inaccessible, experimentally. It is believed the discontinuity arises from a departure from a streamline flow pattern brought about by one or a combination of elastic effects and self-heating effects. These data points are omitted when the approximating function is sought.

Fitting the flow curve

Mathematical expressions of this variation in apparent viscosity may derive from a strictly empirical approach, or from fluid theory. *Cramer* (11) has evaluated a number of these expressions, concluding that two models, an extended *Williamson* model:

$$\eta = \eta_\infty + \frac{(\eta_0 - \eta_\infty)}{1 + \left(\frac{\dot{\gamma}}{\alpha_1}\right)^{\alpha_2}} \qquad [1]$$

and a five parameter *Powell-Eyring* model:

$$\eta = \eta_\infty + \frac{\alpha_1}{\dot{\gamma}} \sinh^{-1}(\alpha_2 \dot{\gamma}) + \frac{\alpha_3}{\dot{\gamma}} \sinh^{-1}(\alpha_4 \dot{\gamma}) \qquad [2]$$

where: α_i = model parameters

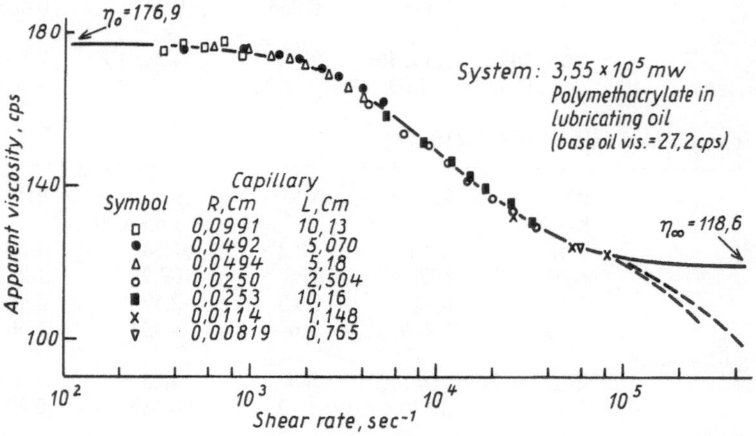

Fig. 1. Typical viscosity-shear rate data from capillary viscometer (solid line: calculated from approximating function; dotted lines: unstable flow area)

gave the best representations of the flow curves of a number of polymer melts and polymer solutions. The extended *Williamson* model is an empirical extension of an interacting sphere model, while the *Powell-Eyring* model derives from thermodynamic treatment of an activated state model. The latter may be desirable where a sound thermodynamic basis is important. Mean errors of about 5% were indicated.

Another approach to an approximating function derives from the high degree of symmetry of the transition region between the first *Newton*ian region and the second *Newton*ian region, particularly when the curves are drawn as functions of the product of the shear rate and the stress level, i.e., the energy level. If this distribution can be assumed *Gauss*ian, the probability function may be used to represent the flow curve. Although not included in *Cramer*s analysis, the usefulness of this or related approximations in treating pseudoplastic fluid flow has been established [12]–[16]. No fundamental derivation has yet been advanced, however.

Details on the manner in which the probability function was used to fit the experimental data acquired in this experimental program are contained in the Appendix. It should be noted that the value (η_∞) is obtained as a result of this optimization, rather than being fully defined experimentally. In fig. 1, the solid line was calculated from the approximating function; the fit to the experimental points is quite good. The fit of the probability function has not been compared to that of either of eqs. [1] or [2]. In this study, representation of the original data was generally within 2%. In using the probability function, we have assumed that the logarithm of apparent viscosity is symmetrically and normally distributed between η_0 and η_∞, according to the experimental variable – the log of the rate of viscous energy dissipation ($\dot{\gamma}\tau$). The function is thus smooth and continuous. It can conveniently be located by the inflection point. Fig. 2 is a plot of the approximating function that was fit to the experimental data of fig. 1, and presents the distribution of $\log\eta$ vs. $\log\dot{\gamma}\tau$. The energy level at the inflection point ($\hat{E}$) corresponds to a value of flow parameter (Φ) of 0.5 where:

Flow parameter

$$(\Phi) = \log(\eta/\eta_\infty)/\log(\eta_0/\eta_\infty). \qquad [3]$$

The fourth parameter necessary to define the approximating function is the "breadth" of the probability curve used to express the variation of Φ with $\log(\dot{\gamma}\tau)$. This is more readily visualized if Φ is transformed, via the probability function and plotted as a straight line against $\log(\dot{\gamma}\tau)$. This produces a series of straight lines for each concentration series. Fig. 3 shows such a series for the 560000 molecular weight polymer in the oil. Each polymer concentration shows a unique $\log\hat{E}$, or intercept at $\Phi = 0.5$, and a unique slope (m). The approximating function is now fully defined by η_0, η_∞, $\hat{E}$ and m.

The general method of obtaining the approximating function for any flow curve, utilizing the

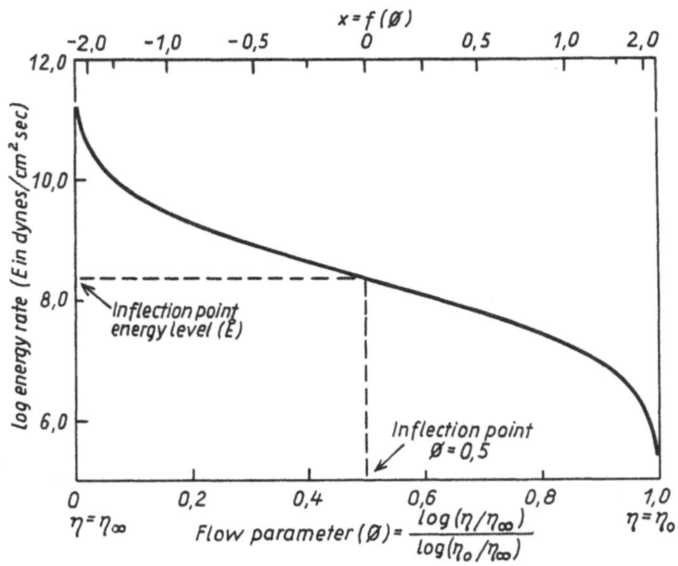

Fig. 2. The approximating probability function for the experimental data of fig. 1

probability function, has been described (16). The current refinement obtains the best fit by regression of the linearized flow parameter (Φ) of eq. [3], so that:

$$y = mx + b$$

where

$$y = \log(\dot{\gamma}\tau)$$

and

$$x = f(\Phi).$$

The transformation $x = f(\Phi)$ is equivalent to plotting Φ values on a probability scale; both the probability and rectilinear scales are shown in figs. 2 and 3, for clarity.

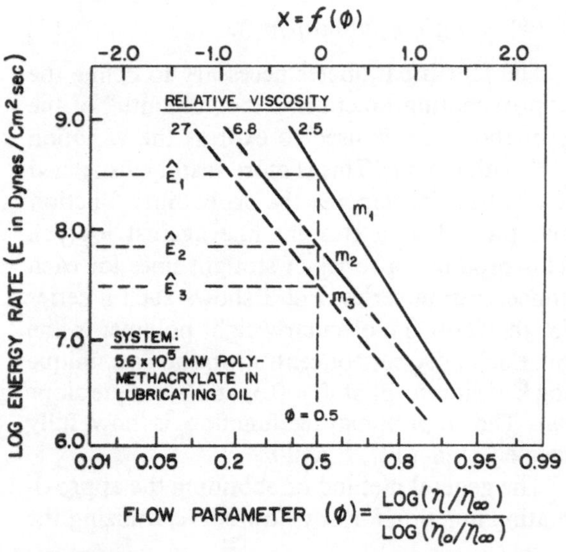

Fig. 3. Flow curve approximations after linearization by probability function

Concentration effects

The high polymer concentrations used in this study negated use of the *Huggins* (17) equation:

$$\frac{\eta_{sp}}{c} = [\eta] + K'[\eta]^2 c \qquad [4]$$

due to excessive curvature. Unter these circumstances, extrapolation to infinite dilution is not reliable. An empirical relationship attributed to *Martin* (18) has been used successfully for several concentrated polymer solutions (19, 20), and was successfully applied to all four molecular weight species used in this program:

$$\log\frac{(\eta_{sp})}{c} = \log[\eta] + k[\eta]c. \qquad [5]$$

The *Martin* equation may be considered a more universal version of the *Huggins* equation, since the latter can be obtained from the *Martin* equation:

$$\frac{\eta_{sp}}{c} = [\eta]\exp(K[\eta]c) \qquad [6]$$

upon expansion of the exponential term, in which terms higher than the third order are neglected (17, 21). Fig. 4 illustrates the quality of fit of the zero shear viscosity data to this relationship, permitting extrapolation to infinite dilution. The intercept at infinite dilution, or intrinsic viscosity, is denoted by the symbol $[\eta_M]$ to distinguish from the intrinsic viscosity conventionally obtained from the *Huggins* equation.

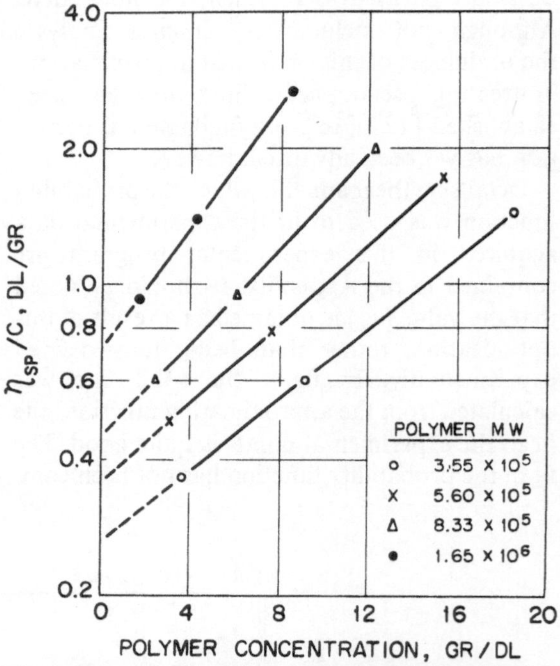

Fig. 4. Extrapolation of low shear reduced viscosities by *Martin* equation

Notwithstanding the success of the *Martin* extrapolation for concentrated solutions, further justification for this type of treatment is available from the work of *Ram* and *Siegman* (9) with polyisobutylene solution. In spite of direct evidence of shear thinning, *Ram* and *Siegman* consistently observed $[\eta]_\infty > [\eta]_0$ when zero shear intrinsic viscosities were extrapolated from dilute solution data. Only when the extrapolations of both low shear and high shear data were made at *comparable* concentration levels, did they observe that $[\eta]_\infty < [\eta]_0$. A corrollary requirement would seem to be that the same extrapolation technique be used. We have been able to adhere to these rules in the present work.

Experimental results

Viscosity relationships

Fig. 4 presents the low shear rate *Martin*-type relationship for the four polymers; fig. 5, the high shear or 2nd *Newton*ian results. While only three points are available for each polymer molecular weight, the relationship appears valid for both the low shear and high shear conditions at all molecular weights, permitting extrapolation to infinite dilution to obtain intrinsic viscosity.

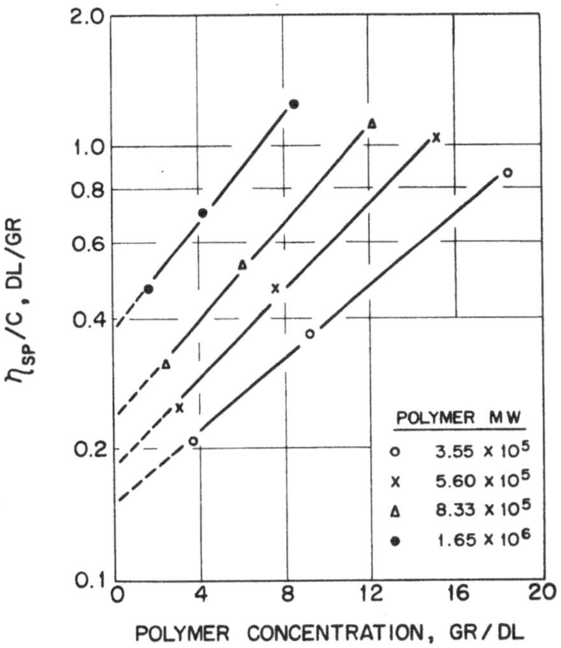

Fig. 5. Extrapolation of high shear reduced viscosities by *Martin* equation

The values of both $[\eta_M]_0$ and $[\eta_M]_\infty$ so obtained increase regularly with polymer molecular weight. Fig. 6 illustrates the bi-logarithmic relationship between either $[\eta_M]_0$ or $[\eta_M]_\infty$ and molecular weight, both of which can be expressed in *Mark-Houwink* form. For the low shear results:

$$[\eta_M]_0 = 5.668 \times 10^{-5} M_v^{0.660}. \quad [7]$$

For the high shear condition, the values for the three highest molecular weights result in:

$$[\eta_M]_\infty = 2.574 \times 10^{-5} M_v^{0.660}. \quad [8]$$

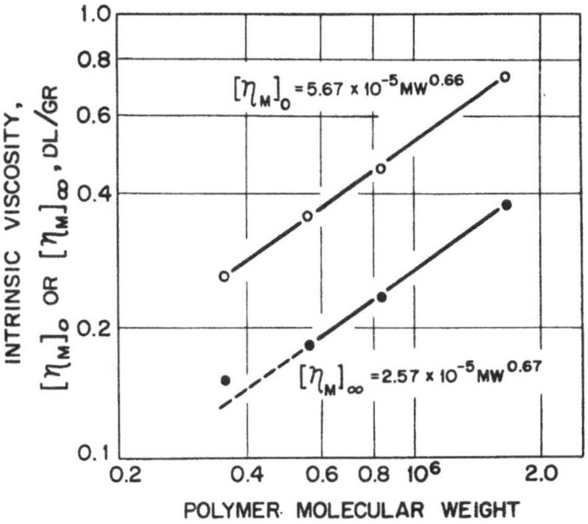

Fig. 6. Intrinsic viscosity-molecular weight relationships at low shear and high shear

As shown in fig. 6, the value for the lowest molecular weight is only slightly removed from this relationship. The exponents are comparable for these two conditions, indicating that both the first *Newton*ian and second *Newton*ian intrinsic viscosities are similarly affected by molecular weight changes in the polymer solution. The ratio of K_0 to K_∞ is 2.2. Thus, $[\eta_M]_\infty$ is approximately one-half $[\eta_M]_0$, over the molecular weight range studed.

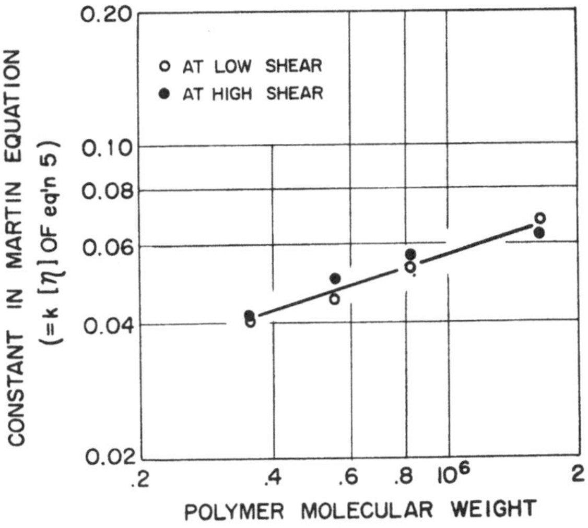

Fig. 7. Effect of polymer molecular weight on *Martin* equation slope

The slopes ($m_M = k[\eta]$ of eq. [5]) of the *Martin* plots (figs. 4 and 5) gradually increase with polymer molecular weight. Fig. 7 indicates there is some scatter about a power law relationship between the *Martin* slope and polymer molecular weight:

$m_M = 8.158 \times 10^{-4} M_v^{0.307}$. [9]

Definition of a specific relationship for each of the low shear and the high shear groups does not appear justified. From eq. [7]–[9], we note that the *Martin k* is not single-valued as in *Martins* (18a) polyhomologous series, but shows a slight M dependency – to about $M^{-1/3}$.

The above equations provide a means for defining both the low shear and high shear viscosity limits for the entire family of polymers over the very broad concentration range examined. The relationships described above are quite unique, compared to the limited data published to date.

Energy relationships

Fig. 3 is a plot of the log of the energy rate versus the transformed flow parameter – $f(\Phi)$ – for the three concentrations of an intermediate molecular weight polymer. The displacement of each line with concentration is quite evident. Similar plots constructed for the other three molecular weight polymers confirm the trend toward lower energy levels with increasing polymer content, at a given level of the flow parameter.

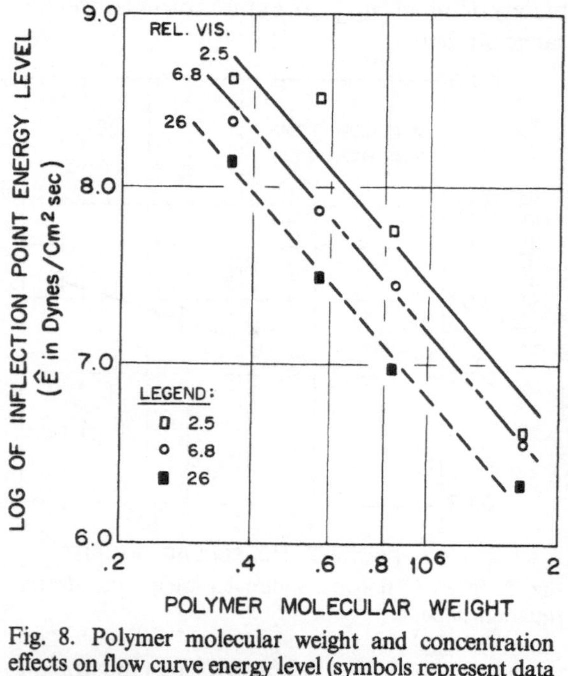

Fig. 8. Polymer molecular weight and concentration effects on flow curve energy level (symbols represent data points; lines are best fit)

Using the energy rate at the inflection point ($\hat{E}$) as a correlating basis, a regular trend with both molecular weight and polymer level (expressed as relative viscosity) emerges. This relationship is shown in fig. 8. The regression is of the form:

$$\log_{10}\hat{E} = 24.79 - 0.618 \log_{10}\eta_{rel} - 2.85 \log_{10}\bar{M}_v \quad [10]$$

where the units of E are dynes/cm² sec. Although only the energy level at the inflection point was considered in the above relationship, comparable trends would be evident were some other position on the flow curve considered, i.e., the energy level required decreases rapidly with increasing molecular weight, less drastically with increasing relative viscosity.

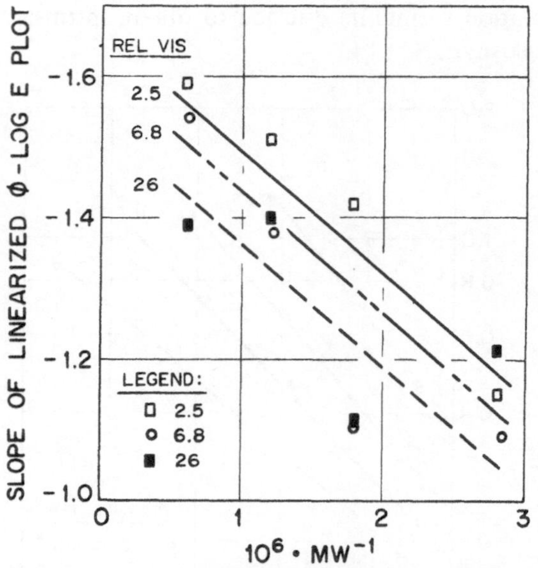

Fig. 9. Polymer molecular weight and concentration effects on slope of flow function (symbols represent data points; lines are best fit)

The plots of $\log E$ versus the transformed flow parameter also suggest a variation in the slopes of the straight lines with the two principle variables – molecular weight and relative viscosity. Although considerable scattering is evident, the trend toward larger values for the slope with decreasing molecular weight and decreasing polymer level (again, expressed as relative viscosity) is noticeable in fig. 9. The resultant linear regression is:

$$m = -1.712 + 0.171 \times 10^6 \bar{M}_v^{-1} + 0.128 \log_{10}\eta_{rel} \quad [11]$$

for which the correlation coefficient is 0.86. This is a considerably poorer fit than any of the preceding relationships. The lack-of-fit in this relationship is more likely due to random variations in m, rather than existence of a higher order relationship, for which there are too few data to derive. We note, in fact, that in the region of near-convergency of the fitting routine, the rate of change of m with η_∞ is substantially greater than the rate of change of $\log \hat{E}$. This relationship, alone, would tend to produce much more scatter in the values of m.

Discussion

Until recently, emphasis on non-*Newton*ian flow has been directed toward entangled systems, even to the extent of precluding non-*Newton*ian flow in non-entangled systems. Thus, *Sell* and *Forsman* (22) present strong evidence that disentanglement is the principal mechanism producing non-*Newton*ian flow in entangled systems, while *Sikri* (23) maintains that *Newton*ian and non-*Newton*ian solutions are distinguished by a critical M_v. However, the end-use requirements of commercial viscosity index improvers require a compromise in molecular weight between thickening ability and resistance to permanent shear degradation. It was anticipated that this factor would result in polymer molecular weights of these materials that were less than the critical level for entanglement (M_c).

Early studies (5, 6, 8) of lubricant systems containing polymeric viscosity index improvers have demonstrated that pseudoplastic flow is a reality, although these have not offered analyses of the mechanisms involved. Recently, *Stratton* (24) has shown that non-*Newton*ian flow could, indeed, be encountered at $M < M_c$, and that the estimated characteristic relaxation time was proportional to M/c, consistent with the *Pao-Rouse* theory. The exponential molecular weight dependency was 0.75 for entangled systems, implying different mechanisms were involved for pseudoplastic flow above or below M_c.

The molecular weights of our polyalkylmethacrylates would appear to be well into the entanglement region; however, much of the weight consists of side chain substitution. For main chain length (Z) estimates of 3000 to 14000 atoms, vZ (v = volume fraction polymer) ranged from about 120 to 1400. These may be considered to be at, or somewhat less than, the traditional critical vZ for entanglement. Slopes of bilogarithmic η vs. M plots, following interpolation for concentration variations, are generally ≤ 1.0, rather than the 3.5 relationship for entangled systems. Thus, by the traditional criteria, it is concluded that disentanglement is not a significant factor in the non-*Newton*ian flow of these solutions.

The polymer-solvent interaction, through its effect on the relative contributions of localized polymer segment-segment and polymer segment-solvent effects might be expected to exert an influence on both the low shear and high shear concentration effects. While not directly accessible or previously reported, the solvent quality of the lubricating oil for the poly (alkylmethacrylate) can be estimated (25) by use of the power-law correlation:

$$\eta_{\rm rel} = f(cM^b) \qquad [12]$$

where c is the concentration in g/dl, and the exponent b has been shown to reflect the degree of solvent power for the polymer. Fig. 10 indicates the $M^{0.68}$ relationship for a good solvent is unsuccessful in correlating the solutions in this

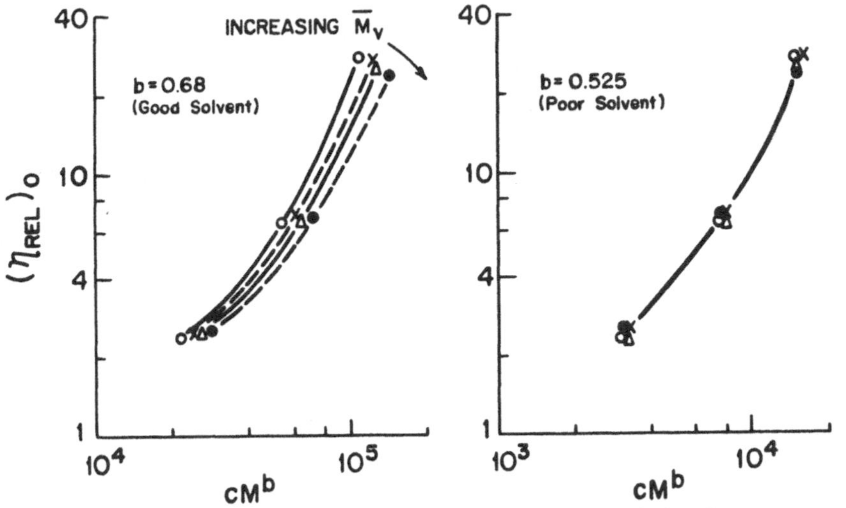

Fig. 10. Poor solvent exponent improves power-law correlation of low shear solution data

series, but a poor solvent $M^{0.525}$ relationship does succeed. Comparable results are obtained if the 2nd *Newton*ian viscosities are similarly treated. As a result of the relatively poor solvency for the polymer, this solvent system may be expected to produce a high degree of polymer aggregation. Apparently, the polar nature of the methacrylate function remains operable, in spite of any possible "shielding" effect of the long chain alkyl substituents.

Several aspects of the present work contrast notably with earlier observations on pseudoplastic flow of polymer solutions, although some disparities are inevitable due to wide variations in experimental conditions and methods of observation and analysis.

In their analysis of the data of *Brodnyan* et al. (26) for polyisobutylene in decalin and the data of *Philippoff* et al. (27) for nitrocellulose in *n*-butyl acetate, *Wright* and *Crouse* (16) noted the concentration-independence of the transformed flow curves. The polyisobutylene-in-decalin data of *Brodnyan* et al. however, might justifiably be presented as a series of straight lines of approximately constant intercept (i.e., $\log \hat{E}$) but of increasingly steeper slope with higher polymer concentration. In the present work on the polyalkyl methacrylates in petroleum oil, separate lines with unique slopes and intercepts were observed for each concentration and molecular weight (cf. fig. 3).

The data treatment by *Rodriguez* and *Goettler* (14) and by *Rodriguez* (15) is similar to that employed here, with important distinctions. To reduce the number of parameters, *Rodriguez* and *Goettler* proposed that the log of the apparent viscosity was symmetrical and normally distributed between the limits η_0 and η_s, rather than between the limits η_0 and η_∞. This approach may be acceptable in cases where no inflection point is in evidence. However, with the assumption $\eta_\infty = \eta_s$, the high shear reduced viscosity:

$$(\eta_{\text{red}})_\infty = \frac{\eta_\infty - \eta_s}{\eta_s C}$$

assumes a zero value. Numerous investigations of the upper *Newton*ian region have indicated a finite value for the high shear reduced viscosity.

With their model, *Rodriguez* et al. indicated that, in many aqueous and non-aqueous polymer solvent systems, the inflection point energy level was exponentially related to polymer chain length:

$$E = A \cdot Z^C$$

where Z = number of chain-atoms; A and C are experimentally determined constants specific to the polymer-solvent system under investigation; $C = -3$ to -4.

A similar relationship has been observed in the present work (eq. [10]).

Ram and associates (9, 28–30) explored the upper *Newton*ian region at great length. *Ram* (30) has recently summarized these results. For polyisobutylene of $\text{MW} > 5 \times 10^6$, $(\eta_{\text{red}})_\infty$ increased with decreasing concentration in a nonlinear fashion, preventing reasonable extrapolation to $[\eta]_\infty$. At $\text{MW} = 10^6$, $(\eta_{\text{red}})_\infty$ was independent of concentration, so that $[\eta]_\infty > [\eta]_0$, while at $\text{MW} < 10^6$, $(\eta_{\text{red}})_\infty$ increased linearly with concentration whence $[\eta]_\infty \approx [\eta]_0$.

Ram et al. (9) included solvent variations – toluene, kerosine, decalin and gas, oil – to provide a 6-fold increase in viscosity. The irregular $(\eta_{\text{red}})_\infty$ concentration relationships appeared at successively lower molecular weights as solvent viscosity increased. These were quite non-linear, and frequently showed increasing $(\eta_{\text{red}})_\infty$ with decreasing concentration. Although pseudoplastic flow of these polymer solutions was observed, the second *Newton*ian intrinsic viscosities consistently exceeded the first *Newton*ian intrinsic viscosities. A regular decrease in the *Mark-Houwink* exponent with increasing solvent viscosity, from 0.6 to 0.3, was attributed to a greater elongation of the polymer coils in the more viscous solvents. Energy relationships could not be extracted from their data.

The authors attributed the $[\eta]_\infty \geq [\eta]_0$ anomaly to differences in the polymer concentrations used for the low and high shear studies. When the low shear extrapolations were made at the same concentration range as for the high shear data, a new value of $[\eta]_0$ was obtained such that $[\eta]_\infty < [\eta]_0$. This, of course, would not alter the very unique $(\eta_{\text{red}})_\infty$ concentration relationships of the high shear regime; these were ascribed to the chain expansion effect predicted and verified by *Peterlin*.

In contrast to the above observations, our work has been relatively untroubled by these discrepancies and/or anomalies. Extrapolations of $(\eta_{\text{red}})_0$ and $(\eta_{\text{red}})_\infty$ with concentration to $[\eta_M]_0$ and $[\eta_M]_\infty$ were straight-forward, via a relationship (the *Martin* equation) previously found to be quite suitable for the concentration range of interest. Intrinsic viscosities increased regularly with polymer molecular weight, while the high shear intrinsic viscosities were found lower – by half – than the low shear intrinsic viscosities, in the expected direction. Further, the molecular weight dependency of $[\eta_M]$ was the same at both the low and high shear *Newton*ian levels.

In contrast to the work by *Ram* et al. on non-polar polymers (PIB, PS) *Tager* and *Dreval* (4) have shown that concentrated solutions of a relatively flexible, non-polar macromolecule, e.g., PIB in decalin, are characterized by incomplete flow curves whereas solutions of rigid and/or polar polymer molecules demonstrate complete flow curves (i.e., curves include both an inflection point and approach the second *Newton*ian viscosity). The authors attribute this phenomenon to the influence of two flow mechanisms: the partial or complete disruption of structure within the system upon the action of either solvent, heat or stress; and the orientation of the resultant smaller flow units in the flow field. We also have observed this phenomenon, in examining solutions of PIB viscosity-index improvers in the same lubricating oil as used for the polyalkylmethacrylates. Several flow curves for the former system did not include a distinct inflection point. In these cases, the computer routine for determining the approximating function cannot converge on a reasonable solution. In this instance, the technique of *Rodriguez* and *Goettler* might be used to advantage.

The various correlations just cited indicate that the high shear behavior of concentrated solutions of even these relatively simple polymer structures is not readily predicted, let alone generalized. Certain characteristics of the respective systems may be of value in understanding some of the flow phenomena involved.

When differences in weight of the repeat unit and molecular weight are considered, the chain lengths of *Rams* polystyrene and polyisobutylene are about an order of magnitude greater than the polyalkylmethacrylates used in this study. Chain disentanglement undoubtedly constitutes a significant portion of their high shear flow effects, while elongation and orientation in the flow field are implied through the decreasing molecular weight dependency with increasing solvent viscosity. The progressive orientation within the shear field of highly flexible polymer chain coils, such as *Rams* PIB, may be a consequence of a less-than-complete chain disentanglement. Such an interaction seems more plausible in explaining the reduced molecular weight dependency normally considered a characteristic of rigid polymer structures. *Ram* suggests that polymer expansion under high stress may also be occurring as dilution increases.

It has been shown that the present system can be characterized as a solution of non-entangled, slightly polar polymer chains in a relatively poor solvent. Thus, the distortion and/or disruption of polymer "aggregates" may constitute one physical factor contributing to pseudoplastic flow in the present study. In addition, both deformation of polymer coils and orientation of the coils in the velocity field may be contributing phenomena, although in the latter case this has not been sufficient to alter the molecular weight dependency of $[\eta_M]_\infty$. This conclusion is consistent with the image of a relatively flexible polymer coil. By the criteria of *Peterlin* (i.e., very high molecular weight flexible polymer coils, very high viscosity solvent, high dilution) polymer coil expansion under shear stress is not considered to be a factor.

The principal molecular motions expected to operate in the case of *Rams* high MW PIB, then (chain disentanglement, chain orientation, coil expansion), are precisely those which might not be found in the present system. Instead the polyalkylmethacrylate-oil solutions reported here are believed more susceptible to disruption of polymer clusters and coil deformation. Whether these phenomena are in fact operative in each situation, and thus account for the markedly different observations between the two systems, can only be determined with considerably more experimental evidence.

Conclusions

1. An approximating function based on the probability function provides a satisfactory fit of the flow curve of these pseudoplastic polymer solutions between the limits η_0 and η_∞, when correlated against the experimental variable $\dot{\gamma}\tau$), the rate of viscous energy dissipation.

2. For this series of alkylmethacrylate polymers in petroleum oil at 100 °F, the parameters of the approximating function vary in a regular manner with polymer molecular weight and with polymer concentration. Thus, the flow properties of polymer solutions falling within the broad limits of the above polymer-solvent system can be represented with reasonable accuracy, by application of the relationships presented here.

3. It appears that a successful generalization of steady shear flow of concentrated polymer solutions must account for at least the effects of solvent power, polymer molecular weight and concentration, factors which would tend to strongly influence the relative contributions of the micro- and/or macro-mechanisms in which non-*Newton*ian flow originates. The pseudoplastic flow of the polymer solutions of this study is believed to be caused mainly by the disruption of aggregates of non-entangled, somewhat polar polymer chains in a relatively poor solvent.

Acknowledgement

The author appreciates the opportunity to investigate these phenomena and the permission to publish the results of this work, granted by the Sun Research and Development Company, Subsidiary of Sun Oil Company. The painstaking efforts of Messrs. *William W. Crouse* and *Robert H. Johnson* in the calibration of the equipment and collection of data is acknowledged. Many stimulating discussions and valuable contributions were provided by Mr. *W. A. Wright*.

Appendix

Fitting the flow curve

A computer program was developed to optimize the fit of the approximating function to the experimental flow curve, thereby eliminating the variability of graphical fitting on logarithmic probability paper. To obtain a linear relationship in a rectilinear coordinate system, the flow function (Φ) has been transformed via the error function, permitting a least squares regression of the entire data set according to:

$$y = mx + b$$

$$y = \log(\dot{\gamma}\tau) \quad \text{or} \quad \log E$$

where

$$x = f(\Phi)$$

$$\Phi = \log(\eta/\eta_\infty)/\log(\eta_0/\eta_\infty) \qquad [3]$$

and

$$m \text{ and } b = \text{const}$$

employing the transformation:

$$\Phi = \frac{1}{2} + \int_0^x \frac{1}{\sqrt{2\pi}} e^{-x^2/2} dx.$$

Thus, at $x = 0$, $\Phi = 0.5$, which is the inflection point of the flow curve. The constant b represents the log of the energy rate at this point, and m is the rate of change of $\log E$ with x. This relationship, illustrated by the twofold ordinate scale of fig. 3, is equivalent to the probit transformation (31), but without the usual displacement of the ordinate by 5σ.

The experimental observations for each polymer solution provide a sequence of $\log\eta$ vs. $\log E$, while $\log\eta_0$ is determined by kinematic viscosity and density measurements. Although the asymptotic approach to $\log\eta_\infty$ is evident in the flow curves, an unequivocal definition of $\log\eta_\infty$ is seldom realized. The value of $\log\eta_\infty$ is, therefore, determined by an iterative procedure in which various values of η_∞ are substituted into eq. [3], taking successively smaller increments, until the η_∞ value that produces the minimum error is defined with sufficient precision. Inasmuch as neither the Φ values nor the E values are considered independently controlled experimental values, the regression is performed by minimizing the sum of squares of the normal distances between the points and the straight line, rather than by minimizing the sum of squares of the distances between the points and the straight line in either the x or the y direction. Fig. 11 is an example of the variation of the sum of squares of error terms as a function of η_∞ for the flow curve presented in fig. 1. The convergence is seen to be fairly sharp, although the error relationship is not symmetrical. The symmetry, and thus the reliability of the definition of η_∞, increases as more data points are obtained beyond the inflection point.

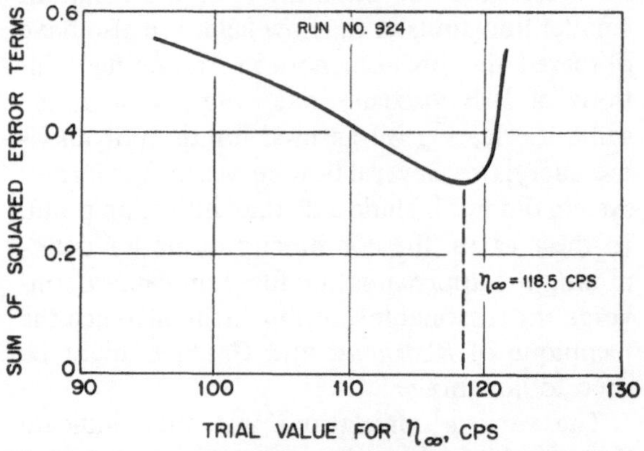

Fig. 11. Convergence of flow curve approximation to η_∞

Conversely, if there are not sufficient points to define an inflection, the routine will not converge, and no value for η_∞ can be defined.

Summary

High shear capillary viscometry at 37.8 °C (100 °F) of concentrated solutions of a series of polyalkylmethacrylate viscosity index improvers in a petroleum lubricating oil is reported. Viscosity average molecular weights of the four polymers varied from 355000–1650000 and solution concentrations varied from 2–20 wt.-%. An approximating function based on the error function was computer-fit to the complete flow curves by correlating the distribution of apparent viscosity with the product ($\dot{\gamma}\tau$), the rate of viscous energy dissipation. This gave an estimate of the second *Newton*ian viscosity (η_∞) and two parameters of the approximating function. The fourth quantity required to completely define the flow curve is the low shear or first *Newton*ian viscosity (η_0). Representation of the original data was within 2%, by this technique.

The parameters of the flow function – the energy level at the inflection point and the slope of the transformed flow function – were found to vary in a regular manner with both molecular weight of the polymer and polymer concentration, expressed as relative viscosity (η_{rel}). The limiting asymptotes of the approximating function – η_0 and η_∞ – could not be treated according to the conventional *Huggins* equation, but they were fit adequately by the *Martin* equation: $\log(\eta_{sp}/C) = \log[\eta] + K[\eta]C$. The intrinsic viscosities thus determined for both low shear ($[\eta_M]_0$) and high shear ($[\eta_M]_0$) demonstrate a *Mark-Houwink* relationship, i.e., $[\eta_M]_0 = 5.668 \times 10^{-5} M_v^{0.660}$ and $[\eta_M]_\infty = 2.574 \times 10^{-3} M_v^{0.0669}$ so that ($[\eta_M]_\infty \approx [\eta_M]_0/^2$) over the range studied. The relationship of these results to other reports of high shear viscometry of polymer solutions is discussed.

References

1) *Hodgson, J. W.*, A.S.L.E. Trans. **14**, 318 (1971).

2) *Wada, S.* and *H. Hayashi*, Bull. Japan Soc. Mech. Eng. **14**, 279 (1971).

3) *Appeldoorn, J. K.* and *W. Philippoff*, Amer. Chem. Soc. Preprints, Div. Petr. Chem., Atl. City, Sept. (1962).

4) *Tager, A. A.* and *V. E. Dreval*, Rheol. Acta **9**, 517 (1970).

5) *Johnson, R. H.* and *W. A. Wright*, Soc. Auto. Eng. Paper n° 680072, S.A.E. Annual Meeting, Detroit, Mich. (1968).

6) *Wright, W. A.* and *R. H. Johnson*, Soc. Auto. Eng. Paper n° 680437, S.A.E. Mid-Year Meeting, Detroit, Mich. (1968).

7) *Novak, J. D.*, Ph. D. Thesis, Univ. of Michigan (1968).

8) *Horowitz, H. H.*, Ind. Eng. Chem. **80**, 1089 (1958).

9) *Ram, A.* and *A. Siegman*, J. Appl. Poly. Sci. **12**, 59 (1968).

10) *Van Nes, K.* and *H. Van Weston*, Aspects of the Constitution of Mineral Oils (New York, 1951).

11) *Cramer, S. D.*, Ph. D. Thesis, Univ. of Maryland (1968).

12) *Umstatter, H.*, Proc. Second Int'l. Congr. Rheol., Pt. 5 (New York 1954).

13) *Kirschke, K.*, Rheol. Acta **2**, 147 (1962).

14) *Rodriguez, F.* and *L. A. Goettler*, Trans. Soc. Rheol. **8**, 3 (1964).

15) *Rodriguez, F.*, Trans. Soc. Rheol. **10**, 169 (1966).

16) *Wright, W. A.* and *W. W. Crouse*, A.S.L.E. Trans. **8**, 184 (1965).

17) *Huggins, M. L.*, J. Amer. Chem. Soc. **64**, 2716 (1942).

18a) *Martin, A. F.*, Amer. Chem. Soc. Abstracts, Div. Cellulose Chem., Memphis, April (1942).

18b) Discussed by *M. L. Huggins*, in: Cellulose and Cellulose Derivatives – High Polymers, Vol. 5 (New York 1943).

19) *Spurlin, H. M.*, *A. F. Martin*, and *H. G. Tennent*, J. Poly. Sci. **1**, 63 (1946).

20) *Sakai, T.*, J. Poly. Sci. A2, **6**, 1659 (1968).

21) *Hirose, M.*, *E. O'Shima*, and *H. Inoue*, J. Appl. Poly. Sci. **12**, 9 (1968).

22) *Sell, J. W.* and *W. C. Forsman*, Macromolecules **5**, 23 (1972).

23) *Sikri, A. P.*, Ph. D. Thesis, Univ. of Pennsylvania (1971).

24) *Stratton, R. A.*, Macromolecules **5**, 304 (1972).

25) *Gandhi, K. S.* and *M. C. Williams*, J. Poly. Sci.-Pt. C **35**, 211 (1971).

26) *Brodnyan, J. G.*, *F. H. Gaskins*, and *W. Philippoff*, Trans. Soc. Rheol. **1**, 109 (1957).

27) *Philippoff, W.*, *F. H. Gaskins*, and *J. G. Brodnyan*, J. Appl. Phys. **28**, 1118 (1957).

28) *Merrill, E. W.*, *H. S. Mickley*, *A. Ram*, and *G. Perkinson*, Trans. Soc. Rheol. **5**, 237 (1961).

29) *Merrill, E. W.*, *A. Ram*, *H. S. Mickley*, and *W. H. Stockmayer*, J. Poly. Sci.-Pt. A **1**, 1201 (1963).

30) *Ram, A.*, in: Rheology. Ed. *F. R. Eirich*, Vol. IV Ch. 3 (New York 1967).

31) Documenta Geigy-Scientific Tables. Ed. *K. Diem*, 6th Edition (New York 1962).

Author's address:

A. F. Talbot

Sun Research and Development Company
Marcus Hook, Pennsylvania (USA)

Rheol. Acta 13, 318–322 (1974)

Commisariat à l'Energie Atomique, Sevrans (France)

Tensions de surface et adhésivité

Par R. Mevrel, S. Poulard, J. Dumiel et M. Vignollet

Avec 6 tables

(Reçu p. p. le 27 octobre 1972)

1. Introduction

Cet exposé se rapporte à la recherche de polymères thermoplastiques compatibles avec un cristal moléculaire organique, l'octogène, du point de vue de l'adhésion.

La méthode que nous avons employée peut également s'appliquer au cas de l'adhésivité de deux polymères, et de façon plus générale, de deux matériaux susceptibles d'interagir par forces de *Van de Waals.*

Rappelons qu'il est habituel de distinguer, en première approximation, d'après un développement en perturbation, les forces de dispersion (dipole instantane–dipole instantané), les forces polaires (dipole–dipole) et les forces d'induction (dipole–dipole induit).

Ces forces étant à portée relativement courte (elles varient en $1/r^6$), il est nécessaire d'assurer le meilleur contact entre adhésif et support.

De ce fait, nous avons d'abord étudié la mouillabilité des surfaces à l'aide du concept de tension critique.

Schématiquement, la formation d'un joint adhésif entre deux matériaux 1 et 2 s'accompagne d'une variation d'énergie libre:

$$\Delta F = -W_A = \gamma_{12} - \gamma_1 - \gamma_2$$

γ_1, γ_2 sont les tensions de surface de 1 et 2, γ_{12} est la tension interfaciale et W_A l'énergie d'adhésion.

Ce processus est thermodynamiquement favorable si γ_{12} est minimale.

Toutefois, il est difficile d'atteindre directement ces grandeurs.

L'étude de l'interface solide-liquide montre qu'à l'équilibre (3) (équation d'*Young*):

$$\gamma_{lo} \cos\theta = \gamma_{sv} - \gamma_{sl}$$

où $\gamma_{lo}, \gamma_{so}, \gamma_{sl}$ sont les tensions liquide/gaz, solide/gaz et solide/liquide.

Du fait que les solides qui nous intéressent ont des faibles tensions de surface ($\gamma_s < 50$ dyne/cm²), il est possible de négliger les phénomènes d'adsorption et ainsi d'écrire:

$$\gamma_{so} \simeq \gamma_s .$$

Seuls θ et γ_{lo} sont accessibles à l'expérience. Il est donc nécessaire de formuler des hypothèses pour atteindre γ_s et γ_{sl}.

2. Tensions critiques

Dans une première approche, *Zisman* (4), d'après de nombreuses mesures d'angles de contact, a introduit le concept de tension critique de mouillage. Remarquant que la variation de $\cos\theta$ en fonction de la tension superficielle pour une série de liquides homologues (ex. *Alcanes*) est approximativement linéaire, il définit la tension critique de mouillage γ_c par extrapolation à $\cos\theta = 1$.

Cependant, il ne faudrait pas assimiler tension critique et tension superficielle (ce qui supposerait $\gamma_{sl} = 0$ d'après l'équation d'*Young*, ce qui est faux en général).

La tension critique n'est qu'une mesure empirique de la mouillabilité d'une surface.

Le calcul de la tension critique de nos matériaux a été effectué selon une méthode de régression linéaire en ne tenant compte que des points appartenant au domaine linéaire dans le graphique $\cos\theta = f(\gamma_L)$, comme le font *Kaelble* (5) ou *Davidson* (6). Les liquides qui attaquent le support ont été éliminés (micrographie).

L'estimation de la variance de γ_c a été calculée d'après la relation (12):

$$\sigma^2 = \frac{1}{N-2} s_{y,x}^2 \left[1 + \frac{(1-\bar{x})^2}{s_x^2}\right].$$

où $y = \gamma$ et $x = \cos\theta$, $s_{y,x}^2$ est la covariance de x et y, s_x^2 la variance de x, $\bar{x} = \frac{\Sigma x}{N}$ est la moyenne des x et N le nombre de points.

Ces valeurs de tensions critiques ne sont qu'indicatives et ne permettent qu'une comparaison de la mouillabilité des matériaux. Contrairement à *Kaelble* (5), à *Sharpe* et *Schonhorn* (7) et à *P. Weiss* (16) il ne nous a pas été possible d'établir une corrélation satisfaisante entre tensions critiques et propriétés adhésives. Remarquons simplement que des polymères de tensions critiques faibles, tels que le *Kelf* et le copoly (butadiène-styrène) n'adhèrent pas à l'octogène qui a une tension critique élevée.

D'autre part, la tension critique dépend fortement de la série de liquides employée et n'est pas intrinsèque au matériau. Ainsi, la tension critique de l'octogène (011) calculée avec une série de mélanges éthylèneglycol-2 éthoxyéthanol est de l'ordre de 29 dyne/cm, alors que la valeur calculée avec les liquides purs est de 42 dyne/cm. Ce fait a été souligné récemment par *Murphy* (17) pour des polymères.

Tableau 1. Tensions superficielles critiques de mouillabilité à 20 °C

Solide	γ_c dyne/cm	σ dyne/cm	γ_c litt dyne/cm
Octogène (101)	41,7	0,7	
Nitrate de polyvinyle	41,5	1,1	
Géon 222	36,8	2,0	39 (13)
Rhodopas	37,5	1,9	37 (14)
Kelf	30,6	0,8	31 (13)
Copoly (butadiène styrène)	26,6	1,1	
Esthane	42,5	0,3	
Polysulfone	38,5	1,5	39 (6)

3. Modèles de Fowkes-Kaelble

Selon *Fowkes* (8), il est possible de distinguer dans la tension superficielle γ, les contributions des forces polaires γ^p et des forces de dispersion γ^d, les autres contributions étant en général négligeables:

$$\gamma = \gamma^d + \gamma^p.$$

Suite à une approche de *Good* et *Girifalco* (2) et d'après l'expression de l'énergie entre deux molécules non polaires, il déduit pour la tension interfaciale entre deux corps interagissant par forces de dispersion uniquement:

$$\gamma_{12} = \gamma_1 + \gamma_2 - 2\sqrt{\gamma_1^d \gamma_2^d}.$$

Nous avons employé une expression due à *Kaelble* (9), généralisant la précédente au cas de l'intéraction de deux corps polaires:

$$\gamma_{12} = \gamma_1 + \gamma_2 - 2\left[\sqrt{\gamma_1^d \gamma_2^d} + \sqrt{\gamma_1^p \gamma_2^p}\right]$$

soit:

$$\gamma_{12} = (\sqrt{\gamma_1^d} - \sqrt{\gamma_2^d})^2 + (\sqrt{\gamma_1^p} - \sqrt{\gamma_2^p})^2.$$

Cette dernière relation montre qu'adhésif et support sont compatibles si leurs composantes de dispersion sont égales, ainsi que leurs composantes polaires; dans ce cas, la tension interfaciale est minimale.

Cette condition est analogue à la condition de compatibilité entre un polymère et un solvant à l'aide du paramètre de solubilité.

Cette relation et l'équation d'*Young* nous ont permis de calculer les composantes γ^d, γ^p de la tension superficielle de l'octogène et de plusieurs polymères à partir de mesures d'angles de contact et de tensions superficielles de liquides.

Ce moyen d'atteindre γ^d n'est pas unique. *Zettlemoyer* a proposé une méthode basée sur les mesures de chaleurs d'immersion. Cependant, les difficultés expérimentales restent assez grandes (10).

4. Partie expérimentale

Les tensions superficielles des liquides ont été mesurées à l'aide d'un tensiomètre de *Thibaud* équipé d'un étrier de platine.

Les angles de contact (angles à l'avancée) ont été mesurés par la méthode de la goutte sessile à l'aide d'un oculaire micrométrique muni d'un réticule. Les résultats présentés sont les moyennes de mesures effectuées sur une vingtaine de gouttes à 20 °C.

En général, la reproductibilité des valeurs d'angles de contact est de l'ordre de $\pm 3°$.

L'octogène (forme β) se présente sous la forme de cristaux de 10–20 mm obtenus par recristallisation. Seules les faces (011), les plus développées, ont été étudiées.

Les polymères employés sont: un nitrate de polyvinyle, un polysulfone, un kel – *F* (polychloro trifluoroéthylène), un géon 222 (préparé à base d'un copolymère de chlorure de polyvinyle) un copolymère butadiène-styrène, un esthane (polyuréthane) un rhodopas (acétate de polyvinyle).

Ils ont été préparés sous forme de film à partir de solutions.

Toutes les surfaces ont été nettoyées avec un détergent puis rincées à l'eau distillée sous ultrason et conservées sous dessicateur.

5. Interprétations

– Composantes de la tension superficielle des liquides:

Les composantes γ_l^d et γ_l^p de la tension superficielle γ_l d'un liquide ont été calculées d'après la mesure d'angles de contact sur un corps de

Tableau 2. Angles de contact (20 °C) (degrés)

Liquides	γ dyne/cm	Octogène (011)	Nitrate de polyvinyle	Polysulfone	Kel F	Géon 222	Copoly butadiène Styrène	Esthane	Rhodopas	Paraffine
Eau	67,4	69	68	74	66	75	86	63	63	98
Glycérol	61,8	62	64	69	89	66	81	64	66	96
Formamide	56,2	50	69	60	74	63	70	47	27	89
Iodure de méthylène	46,9	30	39	21	63	25	64	26	40	54
Ethylène-glycol	47,3	51	48	56		56		45	47	82
Bromo-naphtalène	43,1		18	4	50	14		5	12	29
Butane-diol	43,3	48	49	47		49				78
Chloro-naphtalène	39,1			4		17		22	10	31
Bromo-benzène	35,7			14	31	20			15	16
Phospate de tricrésyle	42,2				46					55
Phtalate de dibutyle	32,7				24					42
1-2 dichloro-éthane	29,9				11					26
Heptanol	26,3									32
2 Ethoxy-éthanol	27,4									38
Polyéthylène glycol 400	43,3	32								70
Ethylène glycol/ 2 éthoxy éthanol p. en p.										
80/20	39,0	43								
60/40	33,3	31								
40/60	31,0	21								

référence, une paraffine, et à l'aide du système suivant:

$$\gamma_i = \gamma_i^d + \gamma_i^p$$

$$\gamma_i(1 + \cos\theta_i) = 2\,(\sqrt{\gamma_i^d\gamma^d} + \sqrt{\gamma_i^p\gamma^p}).$$

Cette dernière relation exprime l'énergie d'adhésion du liquide i sur la paraffine (équation d'*Young* et de *Kaelble*).

La composante de dispersion de la paraffine $\gamma^d = 25{,}4$ dyne/cm a été déterminée d'après l'interaction alcanes-paraffine, en accord avec des résultats déjà publiés (1, 11).

La composante polaire $\gamma^p = 0{,}5$ dyne/cm qui traduit en fait des effets d'induction a été relevée dans un article de *Kaelble* (9).

Nous avons vérifié par le calcul qu'une variation de ces paramètres n'altère pas les valeurs finales.

Tableau 3. Composantes de la tension superficielle des liquides (20 °C)

Liquides	γ_L dyne/cm	γ_L^p dyne/cm	$\gamma_L^d(\sigma)$ dyne/cm
Eau	67,4	44,6	22,8 (5,5)
Glycérol	61,8	40,6	21,2 (6,4)
Formamide	56,2	32,1	24,1 (5,3)
Iodure de méthylène	46,9	1*)	45,9 (4)
Ethylène-glycol	47,3	25,9	21,4 (2,5)
Butane-diol 1–4	43,3	23,2	20,1 (2,9)
2-Ethoxy-éthanol	27,4	7,3	20,1 (1,4)
Polyéthylène-glycol 400	43,3	16,3	27,0 (5,5)
Thiodiglycol	55,6	21,2	34,4 (2,3)

*) Réf. (11) *Owens, D. K.* and *R. C. Wendt*, J. Appl. Poly. Sci. **13**, 174 (1969).

γ_L^p, γ_L^d: composantes polaires et de dispersion de la tension superficielle γ_L.

Les écarts types élevés sont dus à la difficulté d'obtenir des surfaces homogènes et reproductibles.

Des liquides tels que le chloronaphtalène, le bromobenzène, le phosphate de tricrésyle, le phtalate de butyle, le 1–2 dichloroéthane ont dû être éliminés car ils attaquent la paraffine.

– Composantes de la tension superficielle des solides:

L'intéraction d'un solide s et d'un liquide i se traduit par la relation:

$$\gamma_i(1 + \cos\theta_i) = 2\left(\sqrt{\gamma_i^d \gamma_s^d} + \sqrt{\gamma_i^p \gamma_s^p}\right).$$

Il suffit d'une autre mesure avec un liquide j pour obtenir les solutions $(\gamma_s^d(i-j),\ \gamma_s^p(i-j)$ d'un système linéaire de 2 équations à 2 inconnues. La cohérence du modèle doit se vérifier d'après la dispersion des valeurs $(\gamma_s^d(i-j),\ \gamma_s^p(i-j)$ et $\gamma_s(i-j)$ obtenues à partir de couples $(i-j)$ différentes, compte tenu de l'erreur expérimentale.

Pour que ces solutions aient une signification physique, il faut que le déterminant Δ du système ne soit pas trop faible, pratiquement $|\Delta| \geq 10$ c.g.s. (condition analogue à $\Delta \neq 0$ pour un système mathématique). Cette condition élimine un certain nombre de couples de liquides.

Les couples $(i-j)$ utilisables sont ainsi les suivants:

Tableau 4

Couples $(i-j)$	$\|\Delta\|$ c.g.s.	Couples $(i-j)$	$\|\Delta\|$ c.g.s.
1–3	38,5	3–4	33,5
1–7	16,1	3–5	29,8
1–8	14,5	3–6	28,2
2–3	40,5	3–8	22,2
2–7	17,1	4–7	12,1
2–8	15,5	5–7	10,3

1 Glycérol
2 Eau
3 Iodure de méthylène
4 Formamide
5 Ethylène Glycol
6 Butane Diol 1–4
7 2 Ethoxy éthanol
8 Polyéthylène Glycol 400

Les moyennes $\bar{\gamma}_s^d$, $\bar{\gamma}_s^p$, $\bar{\gamma}_s$ des solutions $\gamma_s^d(i-j)$, $\gamma_s^p(i-j)$, $\gamma_s(i-j)$ et leurs écarts types, calculés d'après N couples sont reportés dans le tableau 5.

D'après ces résultats, nous pouvons prévoir que le N.P.V., le Rhodopas et l'Esthane sont compatibles avec l'octogène au point de vue de l'adhésivité, leurs composantes de tensions superficielles étant de même ordre de grandeur.

Tableau 5. Tensions superficielles des solides à 20 °C (dyne/cm)

Solides	$\bar{\gamma}_s^d$	σ	$\bar{\gamma}_s^p$	σ	$\bar{\gamma}_s$	σ	N
Octogène (011)	34,8	1,8	6,8	1,1	41,6	1,0	8
Nitrate de polyvinyle	33,4	0,4	5,8	1,1	39,2	0,7	5
Polysulfone	41,2	0,2	3,2	0,4	44,4	0,2	5
Géon 222	40,2	0,3	3,1	0,5	43,3	0,3	5
Kel F	22,4	0,3	3,7	0,9	26,1	0,6	3
Copoly-butadiène-styrène	21,7	1,4	5,7	1,2	27,4	0,2	4
Esthane	38,1	0,4	7,3	1,0	45,4	0,7	4
Rhodopas	32,3	0,4	8,2	1,3	40,5	1,0	4

$\bar{\gamma}_s^d$ moyenne de la composante de dispersion
$\bar{\gamma}_s^p$ moyenne de la composante polaire
$\bar{\gamma}_s$ moyenne de la tension superficielle
σ écart type de la moyenne calculée d'après N couples

En revanche, il ne faut pas s'attendre à une bonne adhésion du Kel-F et du Copoly-Butadiène-Styrène, leurs composantes étant très différentes.

Il est difficile de prévoir le comportement des cas intermédiaires, tels que le Géon 222 et le Polysulfone.

Afin de vérifier la validité de ces prévisions, nous avons testé la résistance mécanique de joints réalisés à partir des matériaux précédents. Deux types d'essai ont été employés:
- test de traction,
- test de peeling.

Les essais ont été effectués sur monocristaux d'octogène; le polymère est appliqué sous forme de solution. Le test a lieu, après évaporation du solvant, sur machine Instron.

Les propriétés mécaniques (résistance à la rupture) d'un joint adhésif dépendent essentiellement des défauts situés dans l'adhésif, le support et la zone interfaciale.

Lors d'une bonne adhésion, la rupture a lieu dans l'un des matériaux, le plus faible pour le mode de sollicitation envisagé.

Si l'adhésion est mauvaise, la rupture a lieu près de l'interface. La rupture purement interfaciale est rarement réalisable, comme l'a souligné *Bikerman* (15).

Les valeurs de contrainte à la rupture F et d'énergie de peeling P reportées dans le tableau 6 ne sont qu'indicatives, du fait de la faible re-

Tableau 6. Propriétés de surface – Essai mécanique

Matériau	$\bar{\gamma}_s$ (dyne/cm)	$\bar{\gamma}_s^d$ (dyne/cm)	$\bar{\gamma}_s^p$ (dyne/cm)	Traction F (bars)		Peeling p (g/cm)	
Octogène face (011)	43,8	36,4	7,4				
Nitrate de polyvinyle	39,2	33,4	5,8	16	C_{cr}	200	C_f
Géon 222	43,3	40,2	3,1	13	C_{cr}	350	C_f
Rhodopas	40,5	32,3	8,2	12	C_{cr}	230	+ A C_{cr}
Esthane	45,4	38,1	7,3	10	C_{cr}	80	+ A C_{cr}
Kel *F* polychlorotrifluoro-éthylène	26,1	22,4	3,7	3,5	A		
Butadiène-Styrène	27,4	21,7	5,7	6	A + C_{cr}		
Polysulfone	44,4	41,2	3,2	3,5	A		

A rupture adhésive
C_{cr} rupture dans le cristal
C_f rupture dans le film de polymère

productibilité de telles expériences. (Coefficient de variation ~25%.) En revanche, le caractère de la rupture (cohésive ou adhésive) est plus significatif.

Ce tableau montre une bonne adhésion des polymères déjà définis comme compatibles avec l'octogène.

Nous n'avons pu expliquer le comportement du Géon 222 et du polysulfone d'après leurs propriétés de surface. Le premier adhère et non le second, bien que leurs composantes de tension superficielle soient comparables.

Par la suite, nous nous proposons d'appliquer ce modèle au contrôle de modifications de surface destinées à rendre compatibles le cristal et un polymère et d'essayer de le développer afin de prendre en forces des contributions différentes des forces de *Van der Waals*.

Il doit être possible, par ailleurs de l'appliquer à l'étude de l'adhésivité de deux polymères.

6. Conclusion

Rappelons tout d'abord que ce modèle ne s'applique qu'au cas de matériaux de faibles tensions de surface (polymères, cristaux, moléculaires) susceptibles d'intéragir par forces de *Van der Waals*.

Nous avons déterminé les composantes des tensions superficielles de l'octogène et de plusieurs polymères thermoplastiques. Lorsque ces composantes étaient de même ordre de grandeur, il a été possible de prévoir la compatibilité de certains polymères et de l'octogène du point de vue de l'adhésivité.

La vérification de ces prévisions a été faites par des tests mécaniques (peeling, traction).

Littérature

1) *Fowkes, F. M.*, Ind. Chem. Eng. **56**, 40 (1964); Chemistry and physics of interfaces, p. 1 (Amer. Chem. Soc. 1965).
2) *Good, R. J.* and *A. Girifalco*, J. Phys. Chem. **61**, 904 (1957); J. Phys. Chem. **62**, 1418 (1958).
3) *Johnson, R. E.*, J. Phys. Chem. **63**, 1655 (1959).
4) *Zisman, W. A.*, Adv. Chem. Series **43**, 1 (1964).
5) *Kaelble, D. H.*, J. Adhesion **1**, 102 (1969).
6) *Davidson, E. B.* and *G. Lei*, Poly. Letters **9**, 5490 (1971).
7) *Sharpe, L. H.* and *H. Schonhorn*, Adv. Chem. Series **43**, 189 (1964).
8) *Fowkes, F. M.*, Treatise on adhesion and adhesives, vol. 1, p. 325 (1967).
9) *Kaelble, D. H.*, J. Adhesion **2**, 66 (1970); Physical Chemistry of adhesion (New York 1971).
10) *Zettlemoyer, A. C.*, J. Coll. Interf. Sci. **28**, 343 (1968).
11) *Owens, D. K.* and *R. C. Wendt*, J. Appl. Poly. Sci. **13**, 1741 (1969).
12) *Spiegel, M. R.*, Statistics, p. 247 (London 1961).
13) *Shafrin, E. G.*, Polymer Handbook, p. III 113 (1966).
14) *Lee, L. H.*, J. Appl. Poly. Sci. **12**, 719 (1968).
15) *Bikerman, J. J.*, The Science of adhesive joints (New York 1961).
16) *Weiss, P.*, J. Poly. Sci. C n° **12**, p. 169 (1966).
17) *Murphy, W. J.*, *M. W. Roberts*, and *J. R. H. Ross*, Farad. Transactions I **68**, 1190 (1972).

Adresse des auteurs:
Commisariat à l'Energie Atomique, Etablissement T
B. P. No. 7, F-93 Sevrans (France)

Rheol. Acta **13**, 323–332 (2974)

From the Ural State University, Sverdlovsk (USSR)

Effect of solvent quality on the viscosity of flexible-chain and rigid-chain polymers in a wide range of concentrations

By A. A. Tager

With 10 figures and 1 table

(Received October 27, 1972)

The effect of the nature of a solvent on the viscosity of a polymer solution is a most important problem but one on which opinions are divided. The general view is that the intrinsic viscosity of a polymer is lower in a poor solvent than in a good solvent, but we have shown that the initial *Newton*ian viscosity of concentrated polymer solutions of polystyrene (PS), cellulose acetate (CA), and polymethylmethacrylate (PMMA) is much greater in a poor solvent than in a good one (1). As will be seen from fig. 1, the absolute viscosity of solutions in various solvents may differ by a factor of 10^2–10^4. The minimal viscosities of concentrated solutions of cellulose esters in good solvents is a fact well known to technologists.

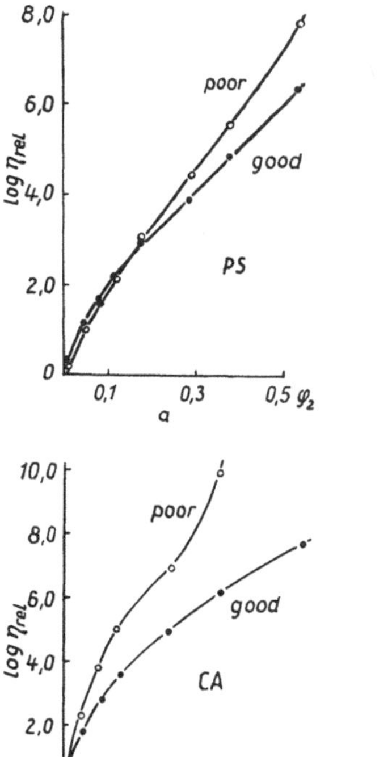

Fig. 1. The dependence of relative solution viscosity on volume fraction of polymers. $t = 25\,°C$. a) Polystyrene in ethylbenzene (●) and decalin (○); b) cellulose acetate in dimethylformamide (●) and cyclohexanone (○)

The different rheological behaviour of dilute and concentrated solutions is due to differences in their structure, which depends on the flexibility of the chain and the energy of the intermolecular interaction.

By the structure of a polymer we mean both the spatial arrangement of its components and their interdependence. Thus the spatial arrangement of the chemically bonded atoms forms the structure or configuration of the molecule while the mutual arrangement of the molecules forms the supermolecular structure, an example of which is the molecular crystal. But a variety of supermolecular structures has been found in the amorphous state as well as the crystalline. These structures may be either non-fluctuational or fluctuational, the former being typical of crystalline and amorphous solids and the latter of liquids. Short-range order has been detected in liquids by X-ray analysis.

Kargin, *Kitaigorodsky* and *Slonimsky* suggested that the macromolecules in a real amorphous polymer were arranged either in bundles or in globules, and supposed that the bundles existed in the vitreous, elastic, and molten states (2).

Their view, which was subsequently confirmed by experiment, influenced other investigators. Thus *Jeil* wrote that there must be local regions in molten polymers where the segments lay parallel (3), and termed these groups "bundles" or "clusters", pointing out the absence of direct evidence of the existence of order in molten polymers. *Kargin*, *Ovchinnikov*, and *Markova*, however, showed by electron diffraction that melts of polyethylene and other polymers consisted of various reference regions in which the segments of chain molecules lay parallel to one another, creating orientation and short-range order near the center of gravity (4). Thus the presence of structure in both the liquid and solid states of polymers has been proved experimentally.

When two liquids are mixed the structure of each is altered, as has been demonstrated for water-alcohol solutions by X-ray analysis and infrared spectroscopy (5); in solutions with less water, the structure of the alcohol is preserved, while the structure of the water is retained in solutions with less alcohol. In solutions with intermediate concentration and in dynamic equilibrium, there are associates of like molecules and solvates of unlike molecules.

When polymers are dissolved the structure of both components should also change, but attention has hitherto been directed mainly to the alteration of polymer structure under the influence of the solvent. Terms like "interstructural" and "intrastructural" swelling have become current in polymer chemistry. But the structure of the solvent also changes on dissolving a polymer, and its molecules are oriented along the polymer chain, which is expressed in a fall in its partial entropy (6), a change in the compressibility of the orientated layers of the solvate (7) and the dielectric properties of the solutions (8), and in the sign of their birefringence (9), which correlates with the heats of solution of the polymers (10). *Borisova, Cherkov,* and *Shemelov* have shown that the orientation of the solvent's molecules in a polymer solution leads to a drop in its relaxation time (11), which is quite analogous to the wellknown change in the order and relaxation time of water and other organic liquids when an electrolyte is added to them, as expressed in positive and negative solvation.

The existence of structure in solutions of crystalline polymers has been shown by *Frenkel* and *Baranov* by means of the scattering of polarized light (12), while *Berry* has recently demonstrated the fact of supermolecular structure in concentrated solutions of amorphous polymers, using low-angle X-ray scattering. Their observations suggest that there is an increase in supermolecular structure as concentration increases (13). Hence it is now evident that isolated macromolecules exist only in infinitely dilute solutions.

As concentration rises the macromolecules begin to interact, i.e. to form associates; and in interacting their molecular coils may either remain unaltered or be straightened out an interact as more or less straightened chains (14–16). The structures formed in solutions are not the result of purely geometrical entanglements as proposed by *Bueche* (17). His concept of "entanglement" is a reflection of earlier ideas of chain entanglement that have now been rejected. As *Fox* wrote: "It may well be asked whether it is valid to consider that molecular chains behave like ropes; whether 'entanglement' occurs if and only if two chains are looped over one another, or whether the entanglements reflect, instead, or in addition, intermolecular association through hydrogen bonding or some other specific chemical interaction." (18)

Ferry et al. have also shown that there are apparently several types of long-range intermolecular coupling in concentrated polymer solutions that differ from the entanglement coupling described by *Bueche.* The coupling points between macromolecules are not cross-links, but slip and permit viscous flow (19), a point of view in accord with the ideas already formulated by *Kargin* in 1939 (20).

With a further rise in concentration the interaction between these aggregates results in a three-dimensional network. Thus the fluctuational networks contained in moderately concentrated solutions consist of aggregates rather than of individual macromolecular. As concentration rises the orderliness of the aggregates apparently increases, as is indicated by the decrease in the scattering of light by the solution (21). At a definite, fairly high concentration, well ordered formations appear in the solution, which are preserved in the polymer formed from it. This is in accordance with the view of *Prins* (22), who writes: "Amorphous polymer networks are usually considered as assemblies of cross-linked coiled chains. The real structure however, may very well be more complicated. In the field of rubber elasticity, for example, experimental as well as theoretical evidence has in recent years emerged pointing to the existence of intermolecular effects which may perhaps be related to some ordering in the network. The crosslinking of polymers in solution may lead also to more complicated structures than random networks." (22)

Concentrated polymer solutions contain more or less broken structures of the polymers themselves. Depending on the structure of the macromolecules and the functionality of their constituents, these structures possess different strengths, so that whether they can be destroyed depends on the ratio of the energy of the structure to that of the polymer-solvent interaction. Poor solvents cannot destroy strong structures, but the molecules of a good solvent can penetrate a structure and destroy it. For that reason structure should be looser in polymer solutions in good solvents and larger and less mobile in solutions with poor solvents. And in fact the relaxation time of a polymer solution in a poor solvent is longer than the relaxation time of a solution in a good solvent (fig. 2).

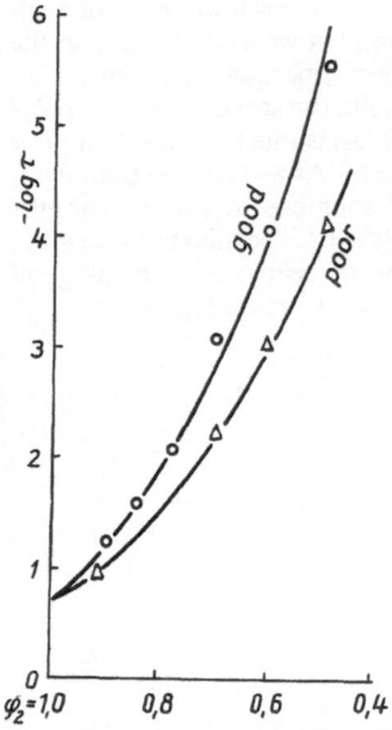

Fig. 2. The dependence of relaxation time of polymethylmethacrilate solutions on the volume fraction of polymers in good (△) and in poor (○) solvents. $t = 25$ °C

The formation of structure in liquids and liquid mixtures affects their viscosity (23); and the mutual dependence of structure and viscosity has been demonstrated for low-molecular mixtures (24). So the more compact arrangement of the molecules of cyclic alkynes than of linear ones leads to a stronger intermolecular interaction and therefore to great viscosity and lower compressibility. Isoviscous substances have the same short-range order and compressibility, which is understandable since both properties depend on the free volume, the parameter that is the most important element of liquid structure. But this term has different meanings. *Howard* (25) recently defined it as follows:

The following volumes should be distinguished:

1.

Empty volume, i.e. the difference between the observed volume of a substance at a given temperature V_T and the volume per mol of the substance as calculated from the *van der Waals* dimensions obtained by X-ray diffraction.

2.

Expansion volume, i.e. the difference between V_T and the volume occupied by its molecules at 0 °K in a closely packed crystalline state – $V_0 (V_f = V_T - V_0)$.

3.

Fluctuation volume, i.e. $V_f = N_A \cdot V_0$, where V_0 is the volume swept out by the centre of gravity of one molecule as a result of its thermal vibration, and N_A is *Avogadros* number.

Viscosity is linked with expansion volume V_f (26), which is always less than the empty volume. Owing to their difference in molecular size a polymer has a smaller free volume than a low-molecular-mass liquid (27). On dissolving a polymer passes into a medium with a greater free volume and, as it were, expands, whereas the solvent passes into a medium with a smaller free volume and therefore contracts. The net effect will depend on the ratio of the two processes, so that contraction or expansion can be expected when a polymer dissolves (28). Volume changes in the condensed state are small and used usually to be neglected; but it has now been shown that they play an important role and affect all the thermodynamic properties of the solution (29). For that reason the idea that the free volume of a solution is additive is not correct.

Thus the structure of a concentrated polymer solution consists of two elements, the fluctuation network and the free (expansion) volume. The solvent mainly affects the density of the fluctuation network. In the medium of a good solvent, the molecules of which can penetrate its structure, the network is looser, but in a poor solvent it is denser and more ordered, which is precisely why there is an enormous difference between the viscosities of concentrated polymer solutions in good and poor solvents (fig. 1). This view, advanced by us in 1962 (1) has since been confirmed by many facts.

1. *Vinogradov* and *Titkova*, by subjecting concentrated polystyrene solutions to sublimation drying, showed that the specific surface area of the PS aerogels obtained from solutions in a poor solvent was 15–20 times less than that of aerogels obtained from solutions of the same concentration in a good solvent (30).

2. The total pore volume of films and fibres from solutions prepared in good solvents is greater than that of films and fibres from similar solutions prepared with poor solvents (31).

A huge difference in the viscosity of equiconcentrated solutions has been observed with PS and CA solutions, but the viscosity of concentrated solutions of polyisobutylene (PI) does not depend on the quality of the solvent (1a, 32).

To elucidate the role of chain flexibility and intermolecular interactions, the initial *Newtonian* viscosities of solutions of five polymers with different chain flexibilities and energies of intermolecular interaction was made (33). The polymers studied are listed in table 1 together with their molecular weights, the ratios of the root-mean-square end-to-end distances $\left(\frac{\overline{h_\theta^2}}{\overline{h_0^2}}\right)^{1/2}$ the sizes of *Kuhns* segments, and their cohesive energy density (c.e.d.) δ. The value of δ for polyarylate (PA) was calculated by adding together the increments of the c.e.d. of the chemical groups of the monomeric unit, i.e. the so-called molecular cohesive constants (34). It will be seen from the table that polydimethylsiloxane (PDMS) had the most flexible chains, followed by PI, PS, PA, and CA.

The objects of the investigation and the techniques employed to prepare the solutions used have been described in detail (33), and a description of the viscometers used to measure viscosity has been given (35).

The quality of the solvents was varied by adding a poor solvent or non-solvent to the good one in amounts that did not make the solution opalescent. The viscosities of the solutions of different concentration, and those of the individual solvents and their mixtures, were measured and relative viscosities η_{rel} calculated. The dependence of η_{rel} and of intrinsic viscosity $[\eta]$ on the volume fraction of the non-solvent in the solution is shown in fig. 3. It will be seen that, irrespective of the nature of the polymer investigated, the intrinsic viscosities fell as the quality of the solvent deteriorated. The same dependence was observed for η_{rel} for PDMS solutions over the whole range of concentrations.

The relative viscosity of PI solutions having a polymer volume fraction $\varphi_2 < 0.15$ also fell as solvent quality deteriorated, and at $\varphi_2 > 0.15$ did not depend on solvent quality, which agrees with the conclusions drawn (32). The relative viscosity of PS solutions with a concentration corresponding to a polymer volume fraction of $\varphi_2 < 0.15$ fell as the quality of the solvent deteriorated. With increasing concentration the difference between the viscosities of solutions in good and poor solvents became smaller, and reached zero at $\varphi_2 = 0.15$–0.18. The same thing has been observed by *Onogi* (36). Above that figure, however, the higher the concentration of the solution the greater was the increase of η_{rel} as the solvent deteriorated. The relative viscosity of CA and PA solutions fell as the solvent deteriorated only in the region of very high dilutions, and at $\varphi_2 \cong 0.004$ increased when a poor solvent was added. These findings clearly demon-

Table 1. Characteristic of polymers

Polymer	$\left(\frac{\overline{h_\theta^2}}{\overline{h_0^2}}\right)^{1/2}$	$\left(\frac{\overline{h_\theta^2}}{\varrho}\right)^{1/2} 10^8$	$A \cdot 10^8$ cm	$\delta \left(\frac{\text{Kal}}{\text{cm}^3}\right)^{0,5}$	$\bar{M}_v \cdot 10^5$
PDMS	1.4–1.6	–	14.0	7.3– 7.6	9.35
PJ	2.2	–	18.3	7.7– 8.8	1.42
PS	2.4	7.1	20.0	6.9– 9.1	2.24
PA	–	18.0–19.0	–	10.3	1.10
CA	3.0–5.0	10.0–20.0	200.0	10.4–11.3	1.10

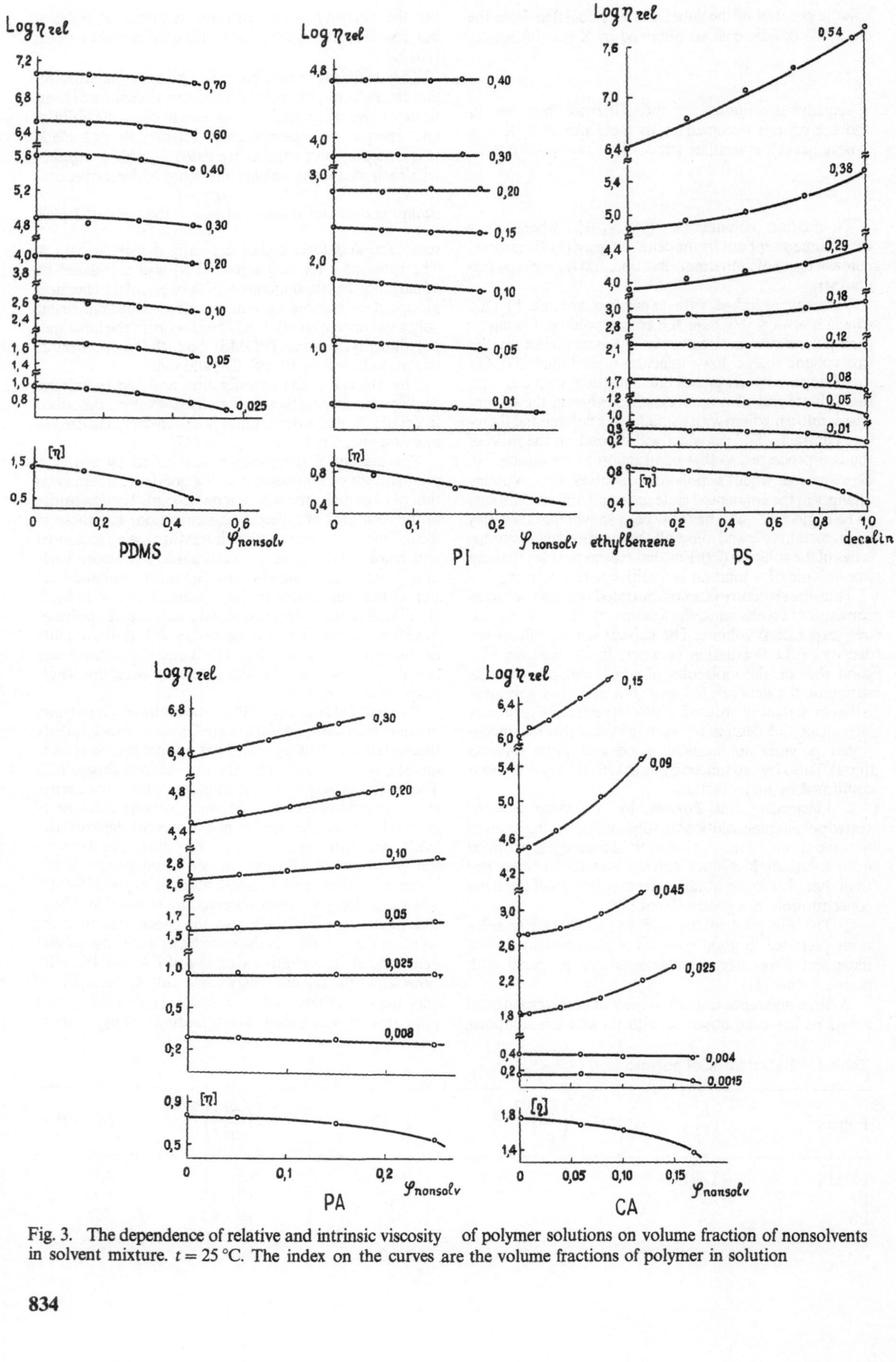

Fig. 3. The dependence of relative and intrinsic viscosity of polymer solutions on volume fraction of nonsolvents in solvent mixture. $t = 25\,°C$. The index on the curves are the volume fractions of polymer in solution

strate that there are no general laws governing the flow processes of flexible and rigid chain polymers.

The strength of the structures formed by flexible non-polar chains is not very great so that they should therefore be broken up by the action of poor solvents as well as by good ones. The solvent mainly affects chain flexibility and conformation, which is why its quality is manifest in same manner in both concentrated and dilute solutions. As chain rigidity and cohesive energy increase the super-molecular structure, i.e. the more or less ordered bundles or fluctuation network, begins to play a considerable role, and the solvent affects the degree of aggregation as well as chain conformation; and in that case, of course, the correlation between its effect on intrinsic viscosity and the viscosity of concentrated polymer solutions disappears.

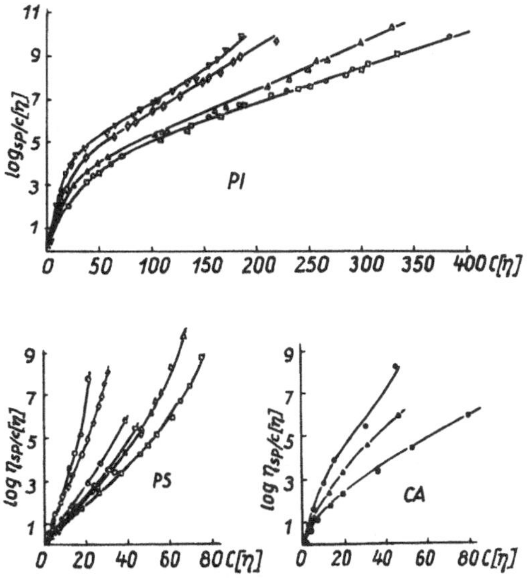

Fig. 4. The solution viscosity in different solvents in reduced coordinates. $t = 25$ °C. a) PJ in decalin (○), cyclohexane (□), CCl_4 (△), isooctane (◊), toluene (▽); b) PS in ethylbenzene (□), decalin (○), CCl_4 (△), ethylacetate (◊), ethylbenzene-decalin: (75:25) (ϙ), (50:50) (O-); c) CA in dimethylformamide (□), dioxan (▲), cyclohexanone (●)

It follows, therefore, that the flow of a concentrated polymer solution cannot be described by a parameter like intrinsic viscosity, as is evident from fig. 4 (37) where the same data are presented in reduced co-ordinates, as proposed by *Simha* et al. (38). The curves for solutions in different solvents are not brought into juxtaposition. Obviously, the reduction parameter will be a function not only of the chain conformation but also of the degree of aggregation. The constant K_M in *Martins* equation is such a parameter (37). It is thought that the constants in *Huggins* and *Martins* equations reflect the interaction of polymer and solvent, but that opinion is not quite correct. These constants reflect how far viscosity increased with concentration, which is, of course, due to the interaction of the segments of neightbouring chains, i.e. to aggregation. Aggregation, however, depends on the interaction of polymer and solvent, as is evident from fig. 5, where the dependence of the reciprocal K_M on the parameters of the thermodynamic quality of the solvent is shown. As *Huggins* constant χ increases, the value of K_M^{-1} decreases, i.e. K_M increases. That means that the degree of aggregation rises as the quality of a solvent deteriorates, and that viscosity increases more sharply with concentration. But that happens differently with different polymers. The curve showing the relationship between I/K_M and χ for PDMS solutions is not steep,

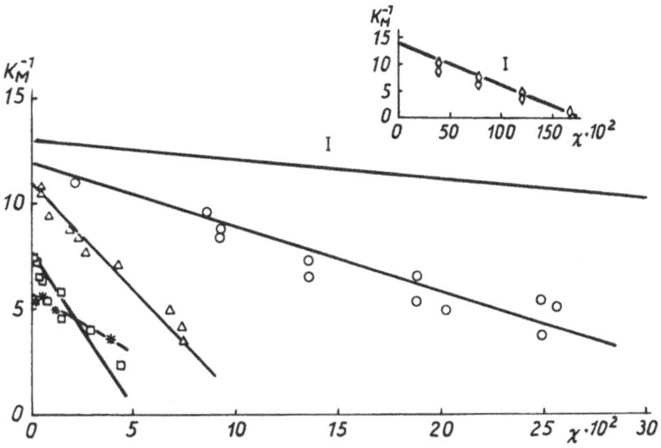

Fig. 5. The dependence of reciprocal value of K_M on the interaction parameter χ_1. PDMS (◊), PJ (○), PS (△), CA (□), PA (✻)

which indicates that the effect of the solvent's quality on viscosity is very weak. But as chain rigidity increases the effect of solvent quality becomes more and more pronounced, and is manifested in the various slopes of the lines and values of the segments I/K_0 intersected by them on the ordinate axis. As will be seen from fig. 6, the de-pendences $\frac{I}{K_0} - vs\left(\frac{\bar{h}_\theta^2}{\bar{h}_0^2}\right)^{1/2}$ and $tg\gamma - vs\delta$ are straight lines, which permits the following empirical formula to be deduced (39)

$$K_M^{-1} = 16 - 3\left(\frac{\bar{h}_\theta^2}{\bar{h}_0^2}\right)^{1/2} + (280 - 40\delta_2) \quad . \qquad [1]$$

Formula [1] shows that the greater the rigidity of the chain and the cohesive forces, and the poorer the solvent, the lower will be the value of I/K_M, and hence the greater the value of K_M, i.e. the higher the viscosity of the solution. The equation also enables ones to calculate the value of K_M from the tabulated values of $\left(\frac{\bar{h}_\theta^0}{\bar{h}_0^2}\right)^{1/2}$, δ, and χ for any polymer-solvent system. Using this value as a reduced parameter it is possible to obtain a single curve for a polymer in all solutions (fig. 7). Intrinsic viscosity takes into account the effect of solvent quality on chain conformation and the constant K_M on the degree of aggregation.

The method of reduction proposed by *Dreval, Botvinnik*, and *Malkin* (37) enables the viscosity of a polymer solution of any concentration in any solvent to be calculated we know the dependence of viscosity on concentration for solutions of the same polymer in any other solvent, the intrinsic viscosity for all solvents, and the parameters $\left(\frac{\bar{h}_\theta^2}{\bar{h}_0^2}\right)^{1/2}$, δ and χ.

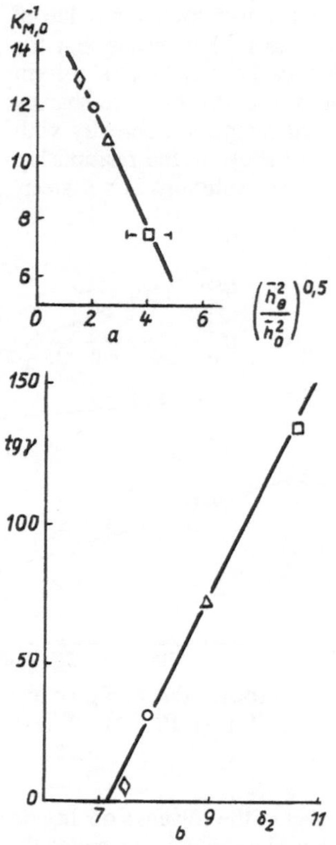

Fig. 6. The dependence of a) $K_{M(0)}^{-1}$ on $\left(\frac{\overline{h_\theta^2}}{\overline{h_0^2}}\right)^{1/2}$, b) tg$\gamma$ on δ_2, PDM (◊), PJ (○), PS (△), CA (□)

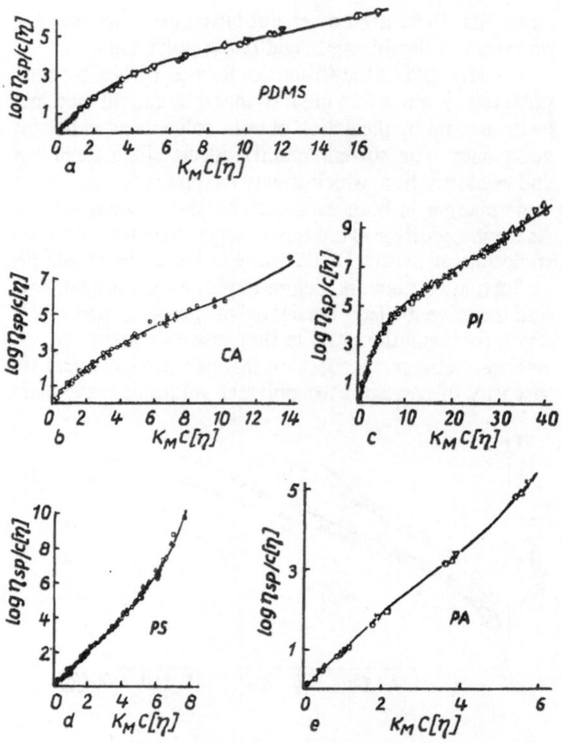

Fig. 7. The solution viscosity in different solvents in new reduced coordinates. $t = 25$ °C. a) PDMS in ethylbenzene (○), ethylbenzenebutanol: (80:20) (△), (60:40) (▽), (40:60) (□); b) CA in dimethylformamide (■), dioxan (▲), cyclohexanone (●), tetrachlorethane (○), tetrachlorethane-octane: (95:5) (♀), (90:10) (○–), (85:15) (♁); tetraclorethane-butanol: (95:5) (●), (90:10) (◑), (85:15) (◒), (75:25) (◐); c) PJ in toluene (▽), isooctane (◊), cyclohexane (□), decalin (○), CClu (△), CCl-isopropanol: (94:6) (⊖), (88:12) (⏀), (82:18) (×); d) PS in ethylbenzene (□), CCl_4 (△), ethylacetate (◊), decalin (○), ethylbenzene-octane: (80:20) (×), (60:40) (▽), ethylbenzene-decalin: (75:25) (♀), (50:50) (–○–), (25:75) (♁); e) PA in tetrachlorethane (○), tetraclorethane-octane: (95:5) (△), (85:15) (▽), (75:25) (□)

The effect of temperature on the viscosity of polymer solutions

The initial *Newton*ian viscosity of polymer solutions diminishes as temperature rises. The curve logη versus I/T is a straight line for small temperature ranges but is curvilinear for wide ones. Curvilinear character of this dependence reflects the nature of the liquid state (23). As temperature rises, the following changes in liquid structure occur: (i) the free volume of the liquid or solution increases, facilitating jumping of molecules or chain segments; (ii) the frequency of jumps itself increases; and (iii) the structure of the liquid is destroyed by the thermal motion.

The change in solvent quality with temperature must also be taken into account for polymer solutions. The change depends on the type of critical solution (or consolute) temperature. When a system has an upper critical solution temperature (UCST), the quality of the solvent improves as temperature rises, the solvent itself becoming better and breaking structure down better at high temperatures. When, however, a system has a lower critical solution temperature (LCST), the solvent deteriorates as temperature rises and breaks structure up less well at high temperatures (40). The fall in viscosity with rise of temperature in the first case is less than in the second. The classical examples are PS–ethylbenzene (41) and PS–decalin (42) respectively. The dependence of viscosity on temperature in these two systems is shown in fig. 8; at room temperature their viscosities differ by a factor of 10^2, but at 130 °C they are equal.

The type of critical solution temperature affects the temperature coefficient of the intrinsic viscosity (43–44). In systems with UCST the molecular coil swells more as temperature rises, leading to an increase in $[\eta]$, while in systems with LCST it swells less, which lowers solvent quality. The value of $[\eta]$, therefore, falls with rise of temperature.

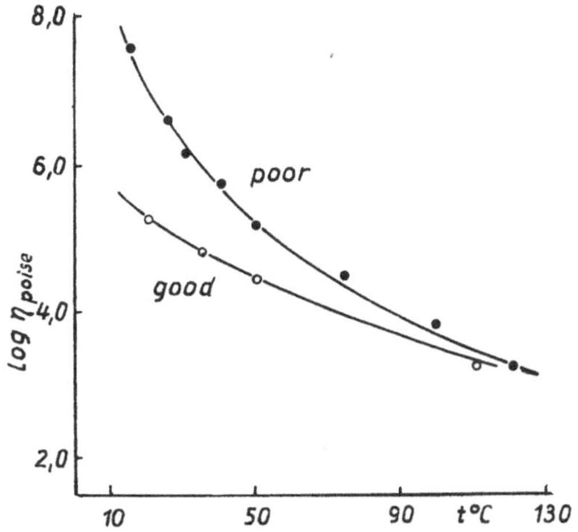

Fig. 8. The dependence of viscosity on temperature of polystyrene solution in ethylbenzene (○) and decalin (●). The volume fraction of polymer in solution is 0.57

Because of the dependence of viscosity on temperature, the activation parameters of viscous flow can be determined from the viscosity-temperature curve. These parameters are sensitive to the structure formation of solutions; to calculate them we used the *Frenkel-Eyring* equation

$$\eta = A_0 e^{\Delta G_{\text{visc}}/RT} \equiv A_0 e^{\frac{\Delta H_{\text{visc}} - T\Delta S_{\text{visc}}}{RT}} \qquad [2]$$

where ΔG_{visc} is the free activation energy of viscous flow; ΔH_{visc} is the heat of activation of viscous flow;

ΔS_{visc} is the activation entropy of viscous flow.

There are different views on the possibility of calculating A_0. With the *Eyring* equation, A_0 depends on the molar volume of the substance, i.e. has a different value for different substances. *Kobekos* view (23), that "only the same value of A_0 for different substances, and its correct order of magnitude, give confidence that the theoretical premises are right", seems more correct. By analysing the viscosity-temperature curves of most liquids, *Kobeko* concluded that A_0 should be 10^{-2} to 10^{-3} ps for all substances, which is confirmed when the viscosity of a substance at $T \to \infty$ is examined. At $T \to \infty$ the substance is converted into a gas, but its viscosity remains within the same limits.

Thus both the experimental data and the physical considerations are persuasive that $A_0 = 10^{-3}$ ps for all substances, in which case the potential barrier is completely permeable, i.e. the activation energy $\Delta U = 0$ and the translational frequency of molecules $\nu = 10^{13}$ period/sec.

Taking $A_0 = 10^{-3}$ we have calculated the free energy, heat of activation, and activation entropy for various systems (39, 45). The values obtained are given in fig. 9, which shows that

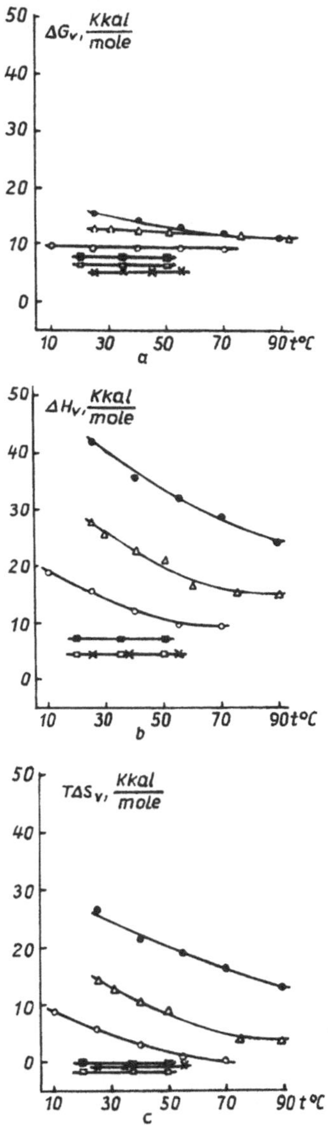

Fig. 9. The dependence of a) free energy; b) heat; and c) entropy of flow activation on temperature. PJ–CCl_4 (×), $\varphi_2 = 0.25$; PS–ethylbenzene (□), $\varphi_2 = 0.25$; PS–decalin (■), $\varphi_2 = 0.25$; PS–decalin (△), $\varphi_2 = 0.57$; CA–dimethylformamide (○), $\varphi_2 = 0.25$; CA–cyclohexane (●), $\varphi_2 = 0.25$

the values of ΔH_{visc} and ΔS_{visc} are very small for PY and PS solutions in a good solvent like ethylbenzene, and are of the same order as the activation parameters of low-molecular liquids. They do not depend on temperature within the temperature range investigated. For solutions in a poor solvent like decalin, and especially for CA solutions, the values of ΔH_{visc} and ΔS_{visc} are very large and diminish with rise of temperature.

These anomalous high values of ΔH_{visc} are due to the "structure activation energy" expended on degradation of the system's structure during flow. As temperature rises the value of ΔH_{visc} diminishes and the structure is broken down by thermal motion.

Entropy is a function directly linked with the degree of order in a system and its diminution points to a loss of order. The large positive values of ΔS_{visc} are evidence that the initial state is more ordered than the activated one.

Thus the high values of ΔH_{visc} and ΔS_{visc}, and their sharp diminution with rise of temperature, indicate that the structure of the initial system is very strong and ordered, as is observed in AC and PS solutions in a poor solvent. On the other hand, PI and PS solutions in a good solvent are less structured. The compensation effect of ΔH_{visc} and ΔS_{visc}, which is met in chemical reactions (46–47), is of great interest. The values of ΔG_{visc} are not large for any polymer and have practically no dependence on temperature.

Free activation energy changes with increase in the concentration and molecular mass of the polymer, as is shown in fig. 10a. The heat of activation, as previously shown (46, 48–49), (see fig. 10b) does not depend on molecular mass, but the latter does affect the activation entropy of viscous flow, altering both its value and sign.

The negative activation entropy observed in certain chemical reactions is explained by the fact that a definite mutual orientation of the molecules is important as well as molecular energy, for the realization of the reaction. The same processes of degradation of structure and orientation of macromolecules occur during flow, and both also require expenditure of energy, so that ΔH_{visc} is always positive. ΔS_{visc} can have either sign, depending on which of the two processes predominates. In dilute high-molecular PI solutions, the uncoiling of the macromolecular coils and their mutual orientation during flow take precedence, so that $\Delta S_{visc} < 0$. As concentration increases during flow the solution becomes more and more structured. When degradation of structure predominates, ΔS_{visc} is less negative.

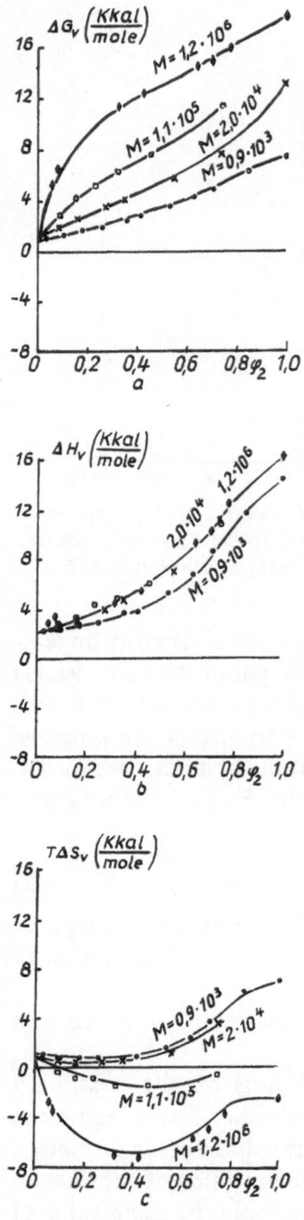

Fig. 10. The dependence of a) free energy; b) heat; and c) entropy of flow activation on volume fraction of PJ in isooctane. $t = 20\,°C$

The ability of the polymer chain to coil up diminishes with loss of molecular mass, so that the orientation process plays a less important role here.

The results presented in fig. 10 thus show that increase of viscosity with the molecular mass of a polymer is due to increase in the activation entropy of viscous flow.

Conclusion

From analysis of a large volume of experimental data it is shown that there are no general laws governing the processes of flow of flexible-chain and rigid-chain polymers. This is primarily evident from the effect exerted on the viscosity of solutions by the thermodynamic affinity of the solvent for the polymer, i. e. by solvent quality, which affects the flexibility of the polymer chain and the degree of aggregation of chains, i. e. the structure of the solution.

For solutions of non-polar flexible-chain polymers, the chain interaction energy of which is very low, the structures formed in solution are extremely fragile and collapse under the action of many solvents. Solvent quality primarily affects the flexibility of chains and their conformational transformations. Hence, for solutions of polymers with maximally flexible chains (polydimethyl siloxanes), deterioration of solvent quality leads to diminution of both the intrinsic and the relative viscosity of highly concentrated solutions.

Concentrated solutions of polar polymers, particularly those with specific chain interactions, are characterized by the formation of strong supermolecular structures, which only collapse under the action of good solvents. In poor solvent media many polymer structures remain intact, which leads to huge differences in the viscosity of equiconcentrated solutions of one and the same polymer in various solvents (which may be as high as 10^4). It is noted that solution viscosity is always higher in poor solvents, than in good ones i. e. the position is diametrically opposite to that which is characteristic of dilute solutions of the same polymers. The curves for the dependence of the viscosity of polymer solutions on concentration in good and poor solvents consequently intersect.

The parameters extremely sensitive to solution structure are heat of activation and activation entropy of viscous flow, as calculated from the temperature dependence of viscosity.

References

1a) *Tager, A. A.* und *V. E. Dreval*, Dokl. Ak. Nauk USSR **145**, 136 (1962). – b) *Tager, A. A., V. E. Dreval* and *F. A. Hasina*, Vysokomolek. soed **5**, 432 (1963). – c) *Tager, A. A., V. E. Drevel* and *A. S. Fomina*, ibid **5**, 1404 (1963). – d) *Tager, A. A., V. E. Dreval, M. Kurbanaliev, M. S. Lutski, N. E. Berkovitz, J. M. Granovskaya*, and *T. A. Charikova*, ibid **10 A**, 2044 (1968). – e) *Tager, A. A., V. E. Dreval*, Rheol. Acta **9**, 517 (1970).

2) *Kargin, V. A., A. J. Kitaigorodsky*, and *G. L. Slonimsky*, Kolloid. Zhur **19**, 131 (1957).

3) *Jell, B. Y. P. H.*, Polymer Single Cristals. (New York-London-Sidney 1965).

4) *Ovchinnikov, Ju. K., G. S. Markova*, and *V. A. Kargin*, Vysokomolek. soed. **A 11**, 329 (1969).

5) *Belousov, V.* and *A. Morachevski*, Coll. Khimiya i termodinamika rastvorov, p. 145 (Leningrad 1964).

6) *Tager, A. A.*, Physical Chemistry of Polymers. "Khimiya", 1968, "Mir", 1972.

7) *Pasynski, A. G.*, J. Phys. Chim. (USSR) **20**, 981 (1946), Kolloidn. J. **8**, 53 (1946).

8) *Marinesko*, Kolloid-Z. **58**, 285 (1932).

9) *Frisman, E. V.* and *A. K. Dadibanyan*, Dynzhev, Dokl. Ak. Nauk USSR **153**, 1062 (1963).

10) *Borisova, T. J., V. M. Chirkov*, and *V. A. Shevelev*, Vysokomolek. soed. **A 14**, 1240 (1972).

11) *Tager, A. A., A. J. Suvorova, Ju. S. Bessonov, A. J. Podlesnjak, J. A. Koroleva, L. V. Agamola*, and *M. V. Tsilipotkina*, Visokomolek. soed. **A 13**, 2454 (1971).

12) *Baranov, V. T.*, Disc. Faraday Soc. **49**, 137 (1970).

13) *Berry, G. C.*, Discuss. Faraday Soc. **49**, 121 (1970).

14) *Tager, A. A.* and *V. M. Andreeva*, J. Polymer Sci. **C 16**, 1145 (1967).

15) *Flory, P. J.*, Proc. Roy Soc. **A 234**, 73 (1956).

16a) *Panov, Ju. N., K. S. Nordbek*, and *S. Ya. Frenkel*, Vysokomolek. Soed. **6**, 47 (1964). – b) *Panov, Ju. N.* and *S. Ya. Frenkel*, ibid, **A.G**, 937 (1967). – c) *Frenkel, S. Ya.* and *Ju. N. Panov*, J. Polymer Sci. **C 3(** , 503 (1970). – d) *Frenkel, S. Ya., G. K. El'yashevich*, and *Ju. N. Panov*, Coll. Uspehi Khimii i fisiki polymerov, p. 87 (Khimiya 1970).

17) *Bueche, F.*, Physical Properties of Polymers (New York 1962).

18) *Fox, T.*, Unsolved Problems in Polymer Science, p. 145 (Nation Acad. of Sci. 1962).

19a) *Saunders, P. R., D. M. Stern, S. F. Kurat, C. Sakoonkin*, and *J. D. Ferry*, J. Colloid Sci. **14**, 222 (1959). – b) *Newlin, T. E., S. E. Jovell, P. R. Saunders*, and *J. D. Ferry*, ibid **17**, 10 (1962).

20) *Kargin, V. A., S. P. Papkov*, and *Z. A. Rogovin*, J. Phys. Chem. (USSR) **13**, 206 (1939).

21) *Boedtker, H.* and *P. Doty*, J. Phys. Chem. **58**, 1968 (1954).

22) *Donkersloot, M. C. A., J. H. Gouda, J. J. van Aarsten*, and *W. Prins*, Rec. trav. Chim. **86**, 321 (1967).

23) *Kobeko, P. P.*, Amorfnye veshchestva (Publ. Ak. Nauk USSR, 1952).

24a) *Golik, A. Z.*, Ukrain Fis. Zh **7**, 806 (1962), **10**, 444 (1965). – b) *A. F. Skryshevski*, Strukturnyi analis zhidkostei (Vysshaya shkola, 1971).

25) *Howard, P. H.*, J. Macrom. Sci, Rev. Macrom. Chem. **C 4** (2), 191 (1970).

26) *Frenkel, Ya. J.*, Kineticheskaya teoriya zhidkostey (Publ. Ak. Nauk USSR, 1954).

27) *Klimenkov, V. S., V. A. Kargin*, and *A. J. Kitaygorodsky*, J. Phys. Chem. (USSR) **27**, 1217 (1953).

28) *Patterson, D. D.* and *A. A. Tager*, Visokomolek. Soed. **A 9**, 1814, 1967. – *Tager, A. A.*, ibid. **A 13**, 467 (1970).

29) *Rowlinson, J. S.*, Liquids and Linguids Mixture (London 1959).

30) *Vinogradov, G. V.* and *L. V. Titkova*, Vysokomolek. soed. **A 11**, 951 (1969), Rheol. Acta **7**, 297 (1968).

31) *Tsilipotkina, M. V., A. A. Tager, E. B. Makovskaya*, and *V. Partina*, Vysokomolek. soed. **A 12**, 1082 (1970).

32) *Berry, G. C.* and *T. G. Fox*, Adv. Polymer Sci. **5**, 261 (1968).

33) *Tager, A. A., V. E. Dreval, G. O. Botvinnik, S. B. Kenina, V. J. Novitskaya, L. K. Sidorova*, and *T. A. Usoltseva*, Vysokomolek. soed. **A14**, 1381 (1972).

34a) *Gardon, J. L.*, Cohesive Energy Density. In: Enciclopedia of Polymer Science and Technology. Ed. *H. F. Mark, N. Gaylord*, and *N. M. Bikales*, p. 833 (New York 1966). – b) *Small, P. A.*, J. Appl. Chem. **3**, 71 (1953). – c) *Bunn, C. W.*, J. Polymer Sci. **16**, 232 (1955).

35) *Tager, A. A., A. I. Suvorova, V. E. Dreval, N. P. Gakova*, and *S. P. Lutskaya*, Vysokomolek. soed. **A10**, 2278 (1968).

36) *Onogi, S., S. Kimura, T. Kato* and *T. Masuda*, J. Polymer Sci. **C15**, 381 (1966).

37a) *Dreval, V. E., G. O. Botvinnik, A. Ja. Malkin*, and *A. A. Tager*, Mehanika polymerov (in press). – b) *Dreval, V. E., A. Ja. Malkin*, and *G. O. Botvinnik*, J. Polymer Sci. A2 (in press).

38a) *Weissberg, G. G., R. Simaha*, and *S. Rothan*, J. Res. Natl. Bur. Stand. **47**, 298 (1951). – b) *Simha, R.* and *J. L. Zakin*, J. Colloid Sci. **17**, 270 (1962).

39) *Botvinnik, G. O.*, These (Sverdlovsk 1972).

40) *Delmas, G.* and *D. Patterson*, Polymer **7**, 513 (1966).

41) *Andreeva, V. M., A. A. Anikeera, S. A. Vshivkov*, and *A. A. Tager*, Visokomolek. Soed. **11**, 1 (1970).

42) *Schulz, G. V.* and *H. Bauman*, Makromolek. Chem. **60**, 120 (1963).

43) *Okada, R.* and *H. Tanzawa*, J. Polymer Sci. A**3**, 4294 (1965).

44) *Tager, A. A., A. A. Anikeeva, V. M. Andreeva*, and *S. A. Vshivkov*, Vysokomolek. soed. B**14**, 231 (1972).

45) *Tager, A. A.* and *G. O. Botvinnik*, Vysokomolek. soed. (in press).

46) *Glasstone, S., K. Laidler*, and *H. Eyring*, The theory of Rate Processes (New York 1941).

47) *Hinshelwood, C. N.*, J. Chem. Soc., 694 (1947).

48) *Hirai, N.*, J. Polymer Sci. **39**, 435 (1959); **40**, 255 (1959).

49) *Tager, A. A., V. E. Dreval*, and *N. G. Trayanova*, Dokl. Ak. Nauk USSR **151**, 140 (1963).

Author's address:
A. A. Tager
Ural State University
Sverdlovsk (UdSSR)

Für die Schriftleitung verantwortlich: Dr. W. Meskat, 5090 Leverkusen
Anzeigenverwaltung: Dr. Karl Niedermeyer Nachf., 6000 Frankfurt/M. 90, Georg-Speyer-Straße 76
Dr. Dietrich Steinkopff Verlag, 6100 Darmstadt, Saalbaustraße 12
Gesamtherstellung: Universitätsdruckerei Mainz GmbH

// RHEOLOGICA ACTA

AN INTERNATIONAL JOURNAL OF RHEOLOGY

Vol. 13 June 1974 No. 3

Institut National Polytechnique de Toulouse (ENSEEIHT, Laboratoire Associé au C.N.R.S.), Toulouse (France)

Expériences utilisant la biréfringence des fluides en mouvement

Par J. J. Barrau, Ch. Truchasson

Avec 6 figures

(Reçu p. p. le 27 octobre 1972)

I. Biréfringence d'écoulement

L'exploitation de la biréfringence des solides pour la détermination des contraintes est utilisée depuis longtemps dans le domaine élastique: c'est la photoélasticité. Cette biréfringence se manifeste aussi dans d'autres comportements, et même dans certains liquides en mouvement, comme le signala *Maxwell* (1) dès 1873.

Par exemple, si l'on observe certains liquides entre polariseurs croisés dans une cellule de *Couette*, constituée de deux cylindres coaxiaux de révolution, une certaine biréfringence apparaît lorsqu'un cylindre est mis en rotation. En un point $M(x, y, z)$ de l'écoulement permanent, on peut alors parler de lignes neutres et de biréfringence du fluide, par exemple en relevant les axes et le retard du biréfringent équivalent. Les expérimentateurs (2) qui ont utilisé des liquides purs (cinnamate d'éthyle; aldéhyde cinnamique) ou des huiles (huile de sesame) ont dû réaliser des gradients de vitesse importants de l'ordre de 5000–20000 s^{-1} pour obtenir des valeurs mesurables de la biréfringence. Les axes du biréfringent équivalent font alors un angle $\varphi = 45°$ avec la vitesse v et le retard R est proportionnel au gradient de vitesse dv/dn. Les nombreuses études (3) portant sur les solutions colloïdales (de bentonite ou de Milling Yellow) ont montré la proportionnalité entre r et dv/dn en écoulement permanent uniforme.

Lodge (4) attribue la biréfringence des solutions comportant des macromolécules à l'état de contrainte σ_{ij} à l'intérieur du liquide.

Parce que la biréfringence d'écoulement est un outil de recherche concernant d'une part les champs de vitesse, d'autre part la structure et les propriétés d'un fluide, nous avons entrepris des expériences dans les laboratoires de l'Institut de Mécanique des Fluides, dépendant de l'ENSEEIHT (Institut National Polytechnique de Toulouse).

II. Expériences mises en œuvre

Le liquide est une solution aqueuse de Carboxyméthylcellulose (en abrégé C.M.C.). Ce liquide intervient dans la fabrication des colles, des papiers, des produits pharmaceutiques et des détergents.

Le C.M.C. en solution est un liquide non newtonien que l'on peut considérer comme *Ostwald*ien. La loi donnant la valeur de la contrainte tangentielle τ_{xy} en fonction du gradient de vitesse $\partial u/\partial y$ est de la forme:

$$\tau_{xy} = K\left(\frac{\partial u}{\partial y}\right)^n$$

où K et n sont des fonctions de la concentration C.

Les expériences concernent l'écoulement dans un canal à faces parallèles: nous voulions effectuer nos mesures pour des gradients de vitesse ayant des valeurs courantes (fig. 1).

Le canal de mesure est en plexiglas; toutefois, les deux faces permettant l'étude, sont réalisées en verre spécial et exempt de biréfringence (moins de $^1/_{100}$ de frange). Il mesure 23 cm de long, 1 cm de haut, 3 cm de large. Des cales amovibles permettent de réaliser plusieurs configurations d'écoulement.

Les mesures de biréfringence sont effectuées avec l'appareil de *Robert* (6). Il presente le double avantage de réaliser des mesures ponctuelles (faible diamètre du pinceau laser) et de déterminer facilement des biréfringences relative-

22

ment faibles (sa précision étant de l'ordre du centième de frange).

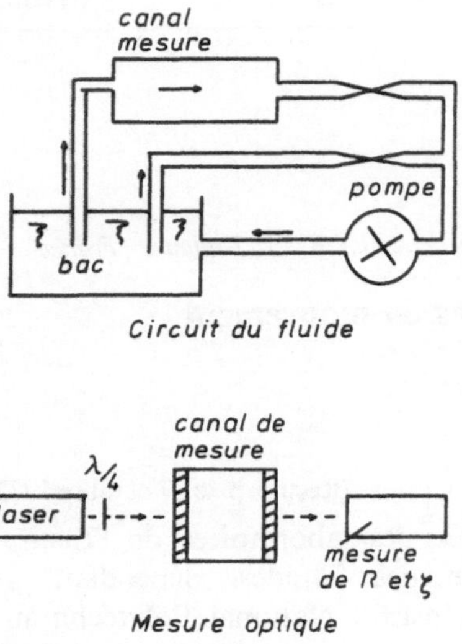

Fig. 1

Dans une première étape, nous avons déterminé la vitesse et le gradient de vitesse à l'intérieur de l'écoulement, afin de relier biréfringence, vitesse et gradient de vitesse.

Grâce aux études de Monsieur *Bellet* (7), Maître-Assistant à l'ENSEEIHT, on établit par le calcul la forme théorique de l'écoulement dans un canal à plaque parallèle, en introduisant dans les équations universelles en contrainte, la valeur de la contrainte tangentielle:

$$\tau_{xy} = K \left(\frac{\partial u}{\partial y} \right)^n .$$

On trouve alors:

$$\frac{u}{U} = \frac{1 - (y/a)^2 - 4 \sum_{n=1,3,5}^{\infty} N_n^{-3} (-1)^{(n/2-1/2)} \operatorname{ch}(N_n z/a) \cos(N_n y/a) \operatorname{ch}^{-1}(N_n b/a)}{\frac{2}{3} - 4 \frac{a}{b} \sum_n N_n^{-5} \operatorname{th}(N_n b/a)}$$

où

$$N_n = \frac{n\pi}{2}$$

2a, 2b, hauteur et largeur du canal.

Nous déterminons ainsi numériquement les valeurs de la vitesse et donc du gradient.

Par ailleurs, nous vérifions la bonne concordance de ces résultats avec l'expérience.

Nous opérons par chronophotographie de bulles d'air injectées dans le canal.

Nous définissons alors, d'une part la vitesse à l'intérieur du canal, mais aussi le gradient de vitesse en chaque point.

III. Résultats

Etude en écoulement permanent uniforme

Nous avons d'abord étudié la variation de la biréfringence R exprimée en degré (360° représente un retard d'une frange) en fonction du temps, à température sensiblement constante (fig. 2).

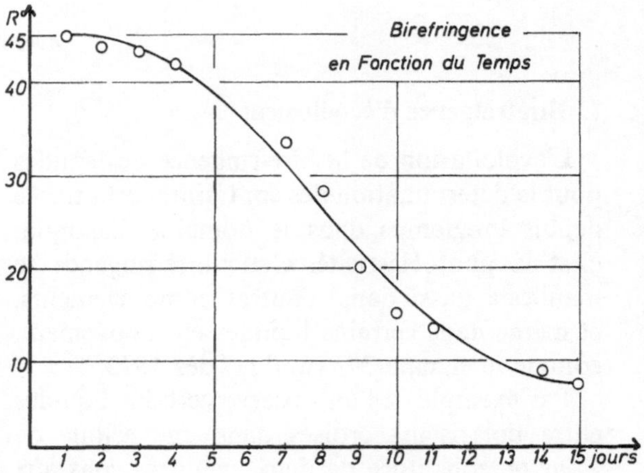

Fig. 2

On remarque, à partir du troisième jour, une décroissance rapide presque linéaire, puis une stabilisation à partir du quatorzième jour à une valeur assez faible.

D'après les travaux de *E. H. Buth*, *H. A. Hudy* et *H. H. Elliot* (5) la cellulose qui permet la préparation du C.M.C. est constituée à la fois de régions cristallines et amorphes, de sorte qu'un faible pourcentage de C.M.C. est réuni par des liens cristallins provenant du cellulose initial. Ces zones faiblement solubles, agissent comme des centres d'attraction pour le C.M.C. soluble. Il existe alors, à l'intérieur de la solution, une masse asez importante de C.M.C. soluble, dont les éléments sont attirés entre eux par des forces de *Van der Waals*, et des forces électrostatiques, isolant des gels insolubles ayant une structure du genre cristallin.

Quand la solution s'écoule, les gels d'une part s'orientent, d'autre part sont soumis à des efforts. La biréfringence peut alors s'expliquer:
- par un effort d'orientation,
- par un effort de déformation.

A mesure que le temps s'écoule, les gels doivent se dissoudre progressivement, et ainsi la biréfringence de la solution diminue. (Grâce à des mesures optiques, on pourrait peut être déterminer l'âge d'une solution.)

Pour ces raisons, notre étude porte sur des préparations ayant le même âge (en général 24 h d'existence).

Dans une deuxième étape, nous avons cherché les relations entre la valeur de la biréfringence R, de l'angle d'extinction φ (fig. 3) et les caractéristiques de l'écoulement.

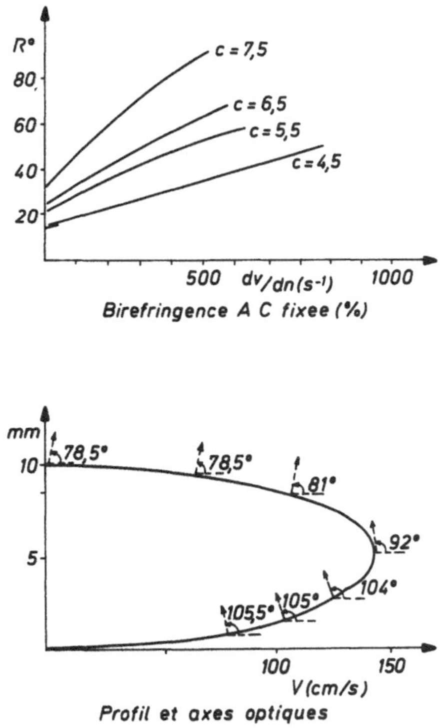

Fig. 3

On remarque une certaine linéarité entre R et le gradient pour une concentration c assez faible, mais lorsque c augmente, la biréfringence croît moins vite.

La valeur de la biréfringence varie dans le même sens que la concentration tandis que l'angle d'extinction reste à peu près constant.

Par raison de symétrie, l'axe du canal doit être une isocline: on trouve bien que sur cette ligne l'angle d'extinction φ est nul.

A partir de ces résultats, on pourrait admettre comme beaucoup d'investigateurs, que la biréfringence dépend de la vitesse de distorsion maximale E.

$$(E)\, E = \frac{1}{2}\sqrt{\frac{4v^2}{r'^2} + \left(\frac{\partial v}{\partial n} - \frac{v}{r}\right)^2}$$

où

V est le module de la vitesse;

r le rayon de courbure de la trajectoire au point considéré;

r' rayon de courbure de l'orthogonale à la trajectoire.

Cette biréfringence s'explique en considérant que les particules s'orientent sous l'effet du gradient de vitesse; en raison de leurs dimensions, on peut penser que l'effet du mouvement brownien est négligeable. *Mason* et *Manley* (8) ont montré que pour des particules rigides cylindriques, la distribution devient:

$$p(\alpha) = \frac{r_e}{2\pi(r_e \cos^2\alpha + \sin^2\alpha)}$$

r_e rapport des dimensions de la particule.

On voit que $p(\alpha)$ est maximum quand $\alpha = \pi/2$ (fig. 4). Les particules ont une orientation préférentielle dans la direction de l'écoulement. Comme de plus, ces particules sont flexibles, à l'effet d'orientation se superpose un effet de déformation. *Lodge* (4) a étudié des solutions de polymères, et il suppose que la biréfringence est liée au contraintes principales. Mais les formules qu'il donne supposent que la contrainte tangentielle est proportionnelle aux gradients, ce qui n'est pas vrai pour le C.M.C.

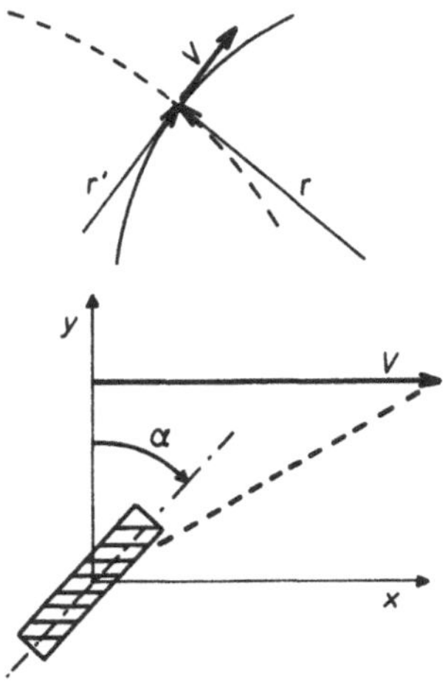

Fig. 4

En admettant que la molécule était représentée par une haltère élastique, *Kramer* (9) arrive malheureusement toujours au fait que $\tau_{x\dot{y}}$ est proportionnel au gradient.

Dans le cas de notre expérience, il doit cependant également se superposer un phénomène d'orientation et un effet de déformation, mais plus complexe.

L'orientation explique que l'angle d'extinction soit nul pour un gradient très faible, et la déformation explique sa croissance avec le gradient de vitesse.

Ecoulement permanent non uniforme

En étudiant une solution à 4,5 g/l où la biréfringence est liée de façon linéaire au gradient de vitesse, il semblerait possible de déterminer dans un convergent la valeur du vecteur vitesse en chaque point, à partir de résultats provenant de l'anisotropie du milieu.

En effet, en chaque point d'un tel écoulement, nous connaissons le rayon de courbure aux lignes de courant et aux orthogonales. A partir de la relation (E) il est alors possible d'obtenir dv/dn. Sachant que sur un bord la vitesse est nulle, on peut facilement intégrer pas à pas. Nous obtenons alors des vitesses largement supérieures aux vitesses réelles. La biréfringence ne suivrait donc pas les mêmes lois dans un écoulement permanent et uniforme que dans un écoulement permanent accéléré.

L'étude de la biréfringence sur l'axe du canal permet de fournir une réponse (fig. 5).

Entre les points A et B, la vitesse de dilatation maximale est égale à:

$$E = \frac{1}{2}\sqrt{\frac{4v^2}{r'^2} + \left(\frac{\partial v}{\partial n} - \frac{v}{r}\right)^2}.$$

Or

$$r = \infty$$

et

$$\frac{dv}{dn} = 0.$$

Soit

$$E = \frac{v}{r'}.$$

Traçons, pour la partie AB la courbe donnant la biréfringence en fonction de $v/r' = E$.

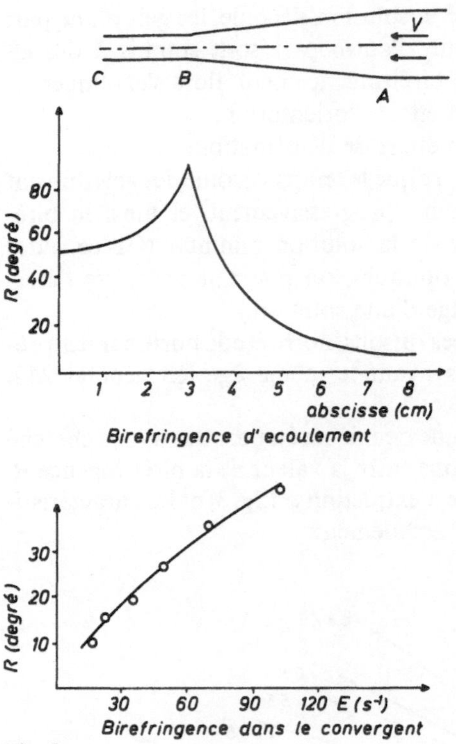

Fig. 5

Il existe toujours une loi linéaire mais dont la pente est totalement différente de celle trouvée pour un écoulement uniforme.

Dans cet écoulement, les particules sont toujours orientées suivant le vecteur vitesse, mais cette fois la particule subit une élongation dans le sens de son orientation.

Ceci suggère que la particule est anisotrope si bien qu'un gradient de vitesse n'a pas le même effet suivant sa direction par rapport à l'axe de la particule.

Pour étudier l'effet de biréfringence, supposons les molécules anisotropes.

Entre B et C, la biréfringence diminue exponentiellement; on l'explique avec ces particules viscoélastiques (leurs déformations dépendent du temps à sollicitation constante).

Un canal comportant deux convergents–divergents successifs (fig. 6) montre en outre que les propriétés de la solution sont fonction de l'histoire de celle-ci.

La valeur de la biréfringence R, le long de l'axe de symétrie, présente comme prévu, un maximum au voisinage des points A, C et E (gradient de vitesse maximum) et un minimum près de B et de D. Il est à remarquer que la valeur des extremums décroît du point A vers E. Il est possible que les solutions présentent un caractère thixo-

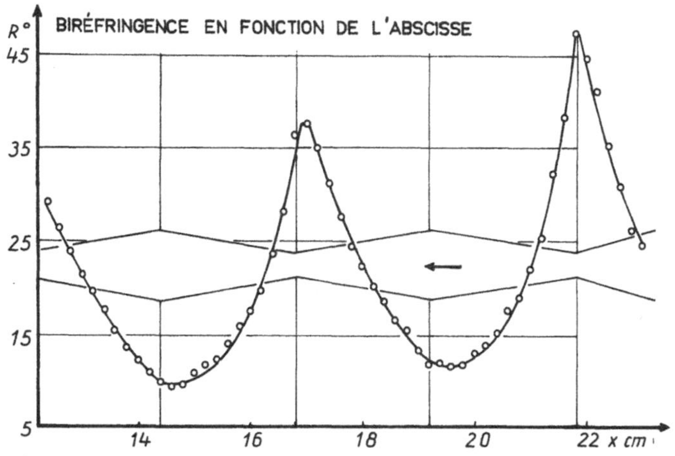

Fig. 6

trope. Sous l'effet d'un gradient de vitesse les gels peuvent être dissociés ce qui entraînerait une baisse de la biréfringence. L'histoire récentre du matériau, semble avoir un rôle important.

Ceci nous conduit à penser qu'il sera difficile avec le C.M.C. de déterminer des vitesses à partir des mesures de biréfringence puisque la déformation de la molécule dépend de l'histoire des sollicitations. On voit également le danger qu'il y aurait à extrapoler les résultats provenant d'écoulements simples (cylindre de *Couette*) à des écoulements quelconques.

IV. Conclusion

Les études entreprises montrent que l'exploitation de la biréfringence d'écoulement n'est pas simple lorsque l'écoulement permanent n'est pas rectiligne uniforme.

Il est toutefois possible de l'expliquer dans ces premières expériences sans abandonner l'hypothèse d'une origine due aux seules contraintes. Il s'agit là, dans des conditions d'expérience bien précises, d'un moyen expérimental d'investigation de la structure du fluide à champ de vitesses connu ou réciproquement.

D'autres expériences en cours permettront de confronter hypothèses et réalité, grâce à la sensibilité du photoélasticimètre à deux dimensions type *Robert* et à la connaissance du comportement rhéooptique du fluide.

Littérature

1) *Maxwell, J. C.*, Proc. Roy. Soc. **146**, 46–47 (1873).

2) *Sadron, G.*, Compt. Rend. Acad. Sci. **197**, 1293 (1933). – *Raman, C. V.* and *K. S. Krishnan*, Phil. Hag. **5**, 769 (1928).

3) *Prodos, J. W.* and *F. N. Peebles*, Amer. Ind. Eng. J. **5**, 225 (1959).

4) *Lodge, A. S.*, Proc. 2nd Intern. Congr. Rheol., p. 229. Edited by *V. L. W. Harrison* (London 1954); Trans. Faraday Soc. **52**, 120 (1951).

5) *De Butts, E. H.*, *J. A. Hudy*, and *J. H. Elliott*, Ind. Eng. Chem. **49**, 94–98 (1957).

6) *Robert, A.*, Methodes nouvelle de photoelasticimetrie (*Jobin* et *Yvon*), p. 92 (Arcueil 1966).

7) *Bellet, D. P.*, Thèse de doctorat ès sciences (à paraitre Toulouse 1972).

8) *Mason, S. G.* and *R. Manley*, Proc. Roy. Soc. A 238, 117 (1956).

9) *Kramers, H. A.*, Physica **11**, 1 (1944) translated in I. Chem. Phys. **14**, 415 (1946).

Adresse des auteurs:

J. J. Barrau et *Ch. Truchasson*
Institut National Polytechnique de Toulouse
(ENSEEIHT, Laboratoire Associé au C.N.R.S.)
2, rue Charles Camichel
F-31 Toulouse (France)

Rheol. Acta **13**, 338–351 (1974)

From the Experimental Mechanics Laboratory, Department of Civil Engineering, University of Waterloo (Canada)

On physical measures of rheological responses of some materials in wide ranges of temperature and spectral frequency

By *Jerzy T. Pindera*, and *Peter Straka*

With 7 figures

(Received October 27, 1972)

Introduction – frame of considerations

The major problem in applied sciences, with regard to the response of materials to loads, is the degree of correlation between actual material response and the predictions of corresponding mathematical model or models presented in a form of constitutive equations.

This problem becomes especially interesting when responses of materials to loads depend on time.

It is customary to investigate the correlation between the actual and the analytically predicted material response with regard to the accuracy of analytical prediction within a determined range of variation of major parameters.

It is less customary to analyse this correlation with regard to the reliability of a mathematical model, defined as a degree of correspondence between the actual mechanism of rheological process and the structure of the mathematical model chosen as the foundation of constitutive equation.

Let us denote, for convenience, the first approach as a phenomenological approach, and the second approach as a physical approach or structural approach.

In both approaches the main problem is how to choose a suitable set of measures of material response which could be taken as representative set of rheological responses of material.

In the case of the rheo-mechanical and the rheo-optical response of materials the typical set of independent and the dependent parameters consists of stress, σ, strain, ε, birefringence, Δn, time, t, and temperature, T; the influence of the time and of the temperature is coupled by means of the time-temperature equivalent hypothesis.

In a phenomenological approach, the solid can be considered as a simple system in a form of a "black box", or as an operator. Consequently the rheo-mechanical, $\varepsilon(t)$, and the rheo-optical, $\Delta n(t)$, responses to stress, $\sigma(t)$, in a chosen observation time interval $[0, t]$ are presented often in a general form:

$$\varepsilon(t) = F'[\sigma(t)] \qquad [1.1]$$

$$\Delta n(t) = F''[\sigma(t)] = F'''[\varepsilon(t)] \qquad [1.2]$$

where operator F is continuous.

The response of the "black box", represented by the operator F is often characterized by a particular response to a particular load, or by a more or less comprehensive set of material parameters, P_m, and system parameters, P_s, for instance in a general form, [1]:

$$(O_k/I_l) = f(P_{m1}, P_{m2}, \ldots, P_{mi}, \ldots; P_{s1}, \ldots, P_{s2}, \ldots, P_{sj}, \ldots) \qquad [1.3]$$

where O and I, denote an output signal and an input signal, respectively.

Obviously, eq. [1.3] can represent any physical event. The question arises what is the minimum set of material parameters, i, and system parameters, j, which can satisfactorily characterize particular responses of a material to particular loads:

$$(O_k/I_l) \approx (O_k/I_l)' = f'(P_{m1}, P_{m2}, \ldots, P_{mi}; P_{s1}, P_{s2}, \ldots, P_{sj}). \qquad [1.4]$$

As a result of simplifications, eq. [1.4] can be misleading without a suitable set of conditions and constraints.

A second question arises which sets of pairs (O_k/I_l) are most representative for the body under observation.

Obviously, all patterns of response (O_k/I_l), are more or less closely associated:

$$(O_k/I_l) = G[(O_m/I_n)]. \qquad [1.5]$$

Determination of operators G is often of a paramount interest in practical engineering application, and is usually of major interest in analysis of the mechanism of phenomenon.

Very often some parameters of the response of a body are closely related, and their interaction represents so-called clustered phenomena, f'. In such case the response of a body can be represented as:

$$(O_k/I_l)' = g'(f'_1, f'_2, \ldots, f'_j). \quad [1.6]$$

In a physical, or structural approach, the main problem is to establish relations between the structure of body and functions G and g. Such an approach is not compatible with the concept of a simple homogeneous body.

In the physical approach a body is considered as a kind of system which – in a general case – consists of interrelated subsystems (clustered parameters). In this presentation one of the main points is which material and system parameters are closely interrelated and which kinds of response are closely interrelated. Consequently, the problem arises which responses should be considered as characteristic responses.

The answer to this question can represent the basis for formulation of conditions for such a class of mathematical models of rheological response of materials which is able not only to describe numerically the response of bodies in a particular limited ranges, but also explain the mechanism of response, give a physical justification to constraints and make it possible to predict response of materials beyond the known range.

Distinction is made between accuracy of prediction of the mathematical model of material response, and the reliability of the mathematical model, as related to the correspondence between the structure of the model and the major physical events.

It follows that it is desirable that the functions which describe the rheological response of bodies to various signals, in a general case consist of terms which represent particular clustered parameter or particular subsystems. The problem is reduced to the determination of this particular phenomena and to the determination of major independent parameters which may influence the response.

In this paper an attempt is made to present some of the responses considered as main responses, to present related constraints, and to present some interrelations between particular responses.

The phenomenological theories of the rheo-mechanical and the rheo-optical response of assumed materials, in the linear and the nonlinear ranges are well developed (2, 4, 5, 6, 7, 8, 9, 10, 11, 12).

The question arises to what extent the available mathematical models can be applied for description and prediction of response of real rheological bodies, and what information should be included in the models. An attempt is made in this paper to contribute to this problem.

Suggested measures of rheological response of rheo-optical materials

An analysis of experimental data (3, 4, 13–24), shows that the typical measures of rheological response of materials as the mechanical creep compliance function, $D(t)$, the mechanical relaxation modulus function, $E(t)$, the photoelastic creep compliance functions $C_\sigma(t)$ and $C_\varepsilon(t)$, and the time-temperature equivalence functions are insufficient for an adequate and commensurate description of actual rheological responses.

On basis of information presented in pertinent bibliography, illustrated by fig. 1a, b, c, the following quantities appear to be of major importance either as quantities complementary to the set of rheological functions mentioned above, or quantities necessary for a satisfactory design of these functions.

1. The creep recovery and relaxation recovery functions, which represent an extension of the creep compliance functions and the relaxation functions into the creep recovery and the relaxation recovery regions. Fig. 1c (15).

2. The quantities describing ranges of linear rheological response, presented as the linear limit stresses, σ_{ll}, and the corresponding linear limit strains, ε_{ll}, fig. 2a (17).

3. The quantity describing dependence of magnitude of relative retardation on spectral frequency of radiation, presented as the normalized spectral birefringence, fig. 2b (18).

4. The quantity describing the rate of change of the spectral birefringence with the radiation frequency, presented as the normalized spectral dispersion of birefringence, fig. 2c (18, 22).

5. The quantity describing dependence of dielectric coefficient on field frequency, presented as normalized dielectric coefficient, fig. 2d (19).

6. The quantity describing dependence of the angle between chosen mechanical principal directions and optical principal directions, on frequency of radiation, presented as the spectral dispersion of optical axes, fig. 2e (22).

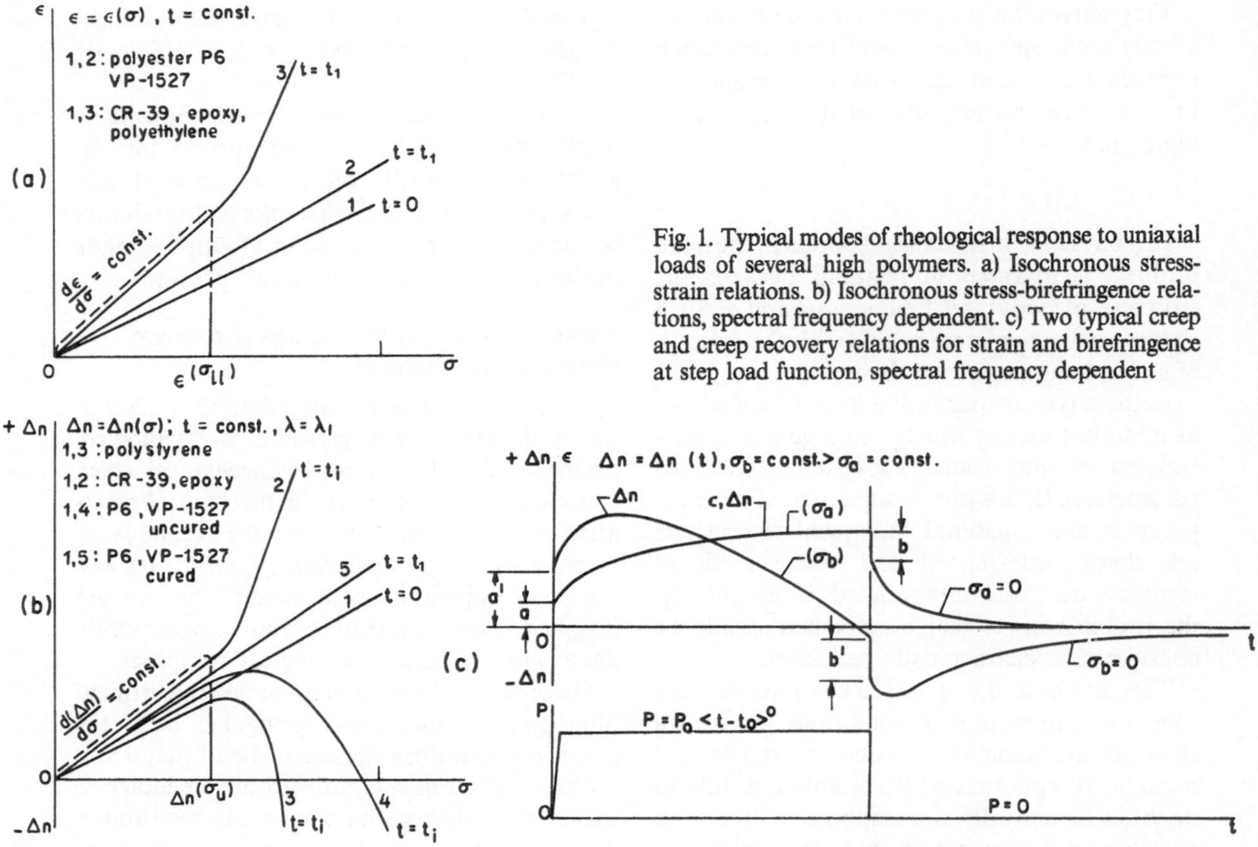

Fig. 1. Typical modes of rheological response to uniaxial loads of several high polymers. a) Isochronous stress-strain relations. b) Isochronous stress-birefringence relations, spectral frequency dependent. c) Two typical creep and creep recovery relations for strain and birefringence at step load function, spectral frequency dependent

7. The quantity describing related interaction between radiant energy and matter, presented as the spectral transmittance, fig. 3a.

8. The quantities describing character of influence of temperature on response of materials, presented as the variation of the thermal expansion coefficient with the temperature and the related variation of mechanical damping, fig. 3b, c (19).

Reasons for the design or the choice of the above quantities follow from the available experimental evidence (3, 4, 13–24). Samples of characteristic responses are given in figs. 4, 5, and 6.

Recovery functions

The mechanical and optical creep recovery functions carry information that is not only supplementary to the information provided by the creep function, but also supply essential information on mechanism of deformation. Obviously only creep and creep recovery functions which are produced by a step load function are amemable to simple analysis without additional assumptions.

Experimental results shown on fig. 1b, c and fig. 6a, b, c, can be explained satisfactorily by assuming that the rheo-birefringence is produced by a set of partial effect. While the particular shape of the creep function can be explained as a result of two partial effects, one distortional and one orientational, the creep recovery function indicates that there exists at least two orientational birefringence effects, produced at different stress levels and related to different time intervals.

Since the creep recovery function carries information pertinent to the understanding of the creep function, it seems necessary to present both these components of the creep response as two components of one function. Experimental results show that the simplest presentation of the optical creep for step loading and unloading in a form:

$$\Delta n(t) = [C(t - t_1)]\, \sigma_0(\langle t - t_0\rangle - \langle t - t_1\rangle^0) \qquad [2.1]$$

where t_0 denotes time of loading and t_1 denotes the time of unloading – is not satisfactory, neither quantitatively nor qualitatively.

Mathematical models of rheological response of several high polymers such as represented by eq. [2.1] are not commensurate with the real response of several major groups of polymers, such as polyester resins or epoxy resins, in both linear and nonlinear range. The creep compliance functions designed to represent the creep part of creep response, in a form

$$\varepsilon(t),\ \Delta n(t) = \int_0^t D(t-\tau)\,\dot{\sigma}\,d\tau \qquad [2.2]$$

for the linear range, and

$$\varepsilon(t),\ \Delta n(t) = \int_0^t D_1(t-\tau_1)\,\dot{\sigma}\,d\tau + \int_0^t\int_0^t D_2(t-\tau_1, t-\tau_2) + \dot{\sigma}(\tau_1)\,\dot{\sigma}(\tau_2)\,d\tau_1 d\tau_2 + \cdots \qquad [2.3]$$

for the nonlinear range,

can not describe satisfactorily response of several materials, figs. 1b and 6b, c.

The creep recovery function seems to be of primary importance for design of a general creep function, describing both the creep and the creep recovery part of creep response, and for understanding of the mechanism of coupling between the mechanical and the optical effects.

Experimental results indicate that the birefringence creep response of several materials could be approximated by a creep compliance function of a character:

$$D(t) = D_0(t) + D_1(t-t_1) + D_2(t-t_2) + \cdots \qquad [2.4]$$

where the time intervals $[t - t_i]$ depend on strain and strain rate,

$$t_i = t_i(\varepsilon, \dot{\varepsilon}). \qquad [2.5]$$

Creep recovery curves show that for several materials the mechanical and the birefingence response to the step load in the time interval $[t_0, t']$, where t_0 and t' denote time of loading and of unloading, respectively, can be presented in a form:

$$\Delta n(t),\ \varepsilon(t) = [D(t_0) + D(t)] \times \sigma(\langle t-t_0\rangle^0 - \langle t-t'\rangle^0) \qquad [2.6]$$

for

$$t_0 < t < (t')^+.$$

The statement represented by eq. [2.6], that the mechanical instantaneous and the birefringence instantaneous effects of the step function loading and unloading are equal, characterises response of several high polymers. However it can occur that the instantaneous mechanical effects are equal, while the instantaneous birefringence recovery effect is greater than the instantaneous load effect, fig. 6b.

Linear limit stresses

The isochronous relations between stress, strain and birefringence are linear in limited range or ranges. Response of materials can be linear in one or more ranges, as presented schematically on fig. 2a.

It is useful to introduce measures for the linear ranges of the rheological response of materials in form of linear limit stresses, σ_{ll}, and linear limite strains, ε_{ll} (15, 17, 21), fig. 2a. Two particular measures have been introduced (17), the nominal linear limit stress, $(\sigma_{ll})_{\text{nom}}$ or σ_{ll}, and the percentual linear limit stress, $(\sigma_{ll})_{\%}$. Both measures can be related to the primary linear range of response, σ'_{ll}, or to the secondary linear range of response, σ''_{ll} (21).

The concept of the nominal limit stress is of interest with regard to the mathematical models of material response. The percentual linear limit stress, which often describes the width of the transition range between two linear ranges, is also of engineering interest.

The linear limit stresses and strains are time-dependent and may be different for the strain and the birefringence, fig. 5.

The linear limit stresses depend on temperature and decrease to zero in ranges of transition temperatures (19).

The linear limit stresses for birefringence may depend also on the wavelength of radiation (22), hence the relation should be presented as follows:

for strain: $\varepsilon(\sigma_{ll}) = \sigma_{ll}(t, T)$

for birefringence: $\Delta n(\sigma_{ll}) = \sigma_{ll}(t, T, \lambda)$. [2.7]

The dependence of the linear limit stress on time, temperature and wavelength of radiation supplies information on the mechanism of deformation.

In accordance with interpretation of experimental results, presented above and subsequently, the multi-linear response of materials presented schematically on fig. 1, can be approximated (23) by the relation:

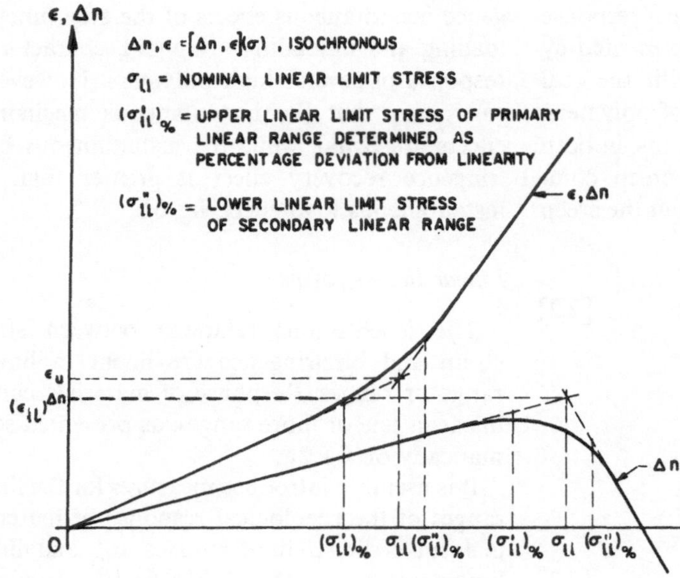

Fig. 2a

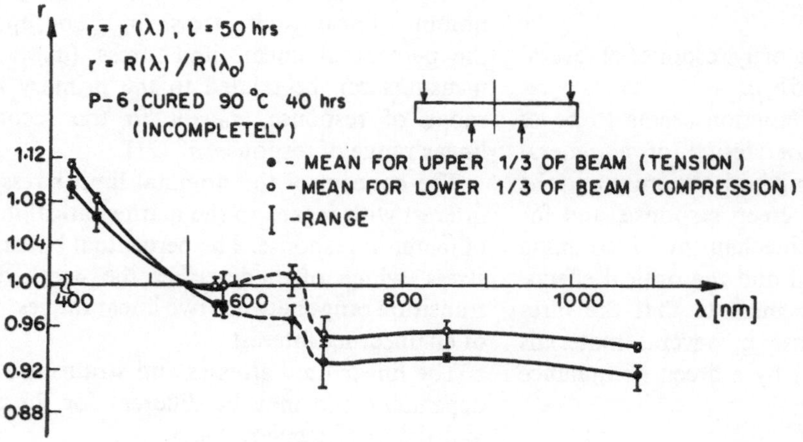

Fig. 2b

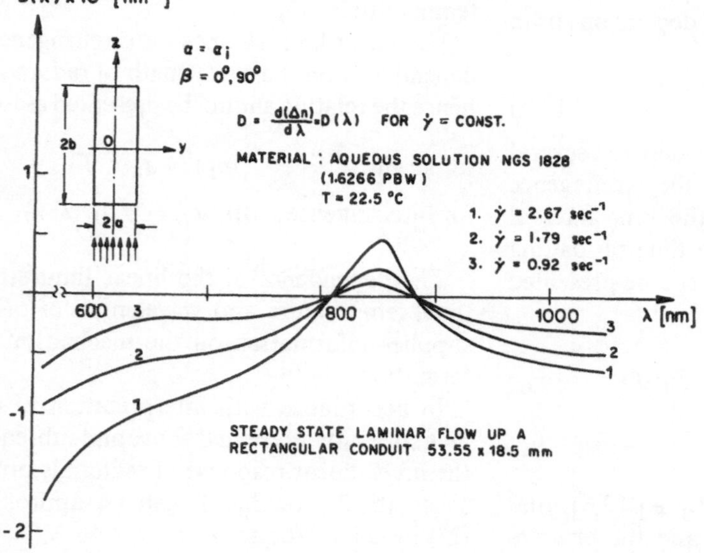

Fig. 2c

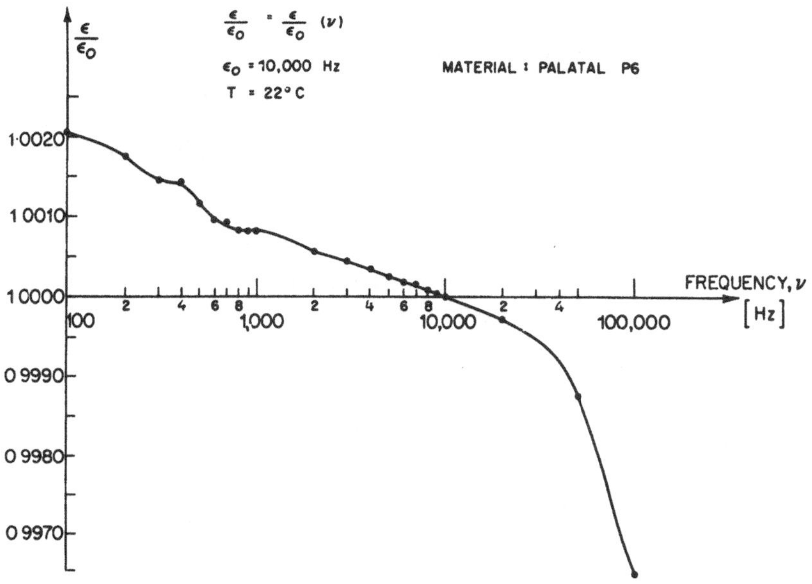

Fig. 2d

Fig. 2. Suggested measures of complex rheological response of viscoelastic bodies such as high polymers, glasses, birefringent liquids. a) Linear-limit stress definition. b) Normalized spectral birefringence. Sample: spectral birefringence of a polyester resin Palatal P6. c) Spectral dispersion of birefringence. Sample: Dispersion of birefringence of Milling Yellow colloid suspension. d) Frequency dependence of the dielectric coefficient. Sample. e) Spectral dispersion of optical axes. Sample: spectral dispersion of optical axes of Milling Yellow, for various shear strain rates

Fig. 2e

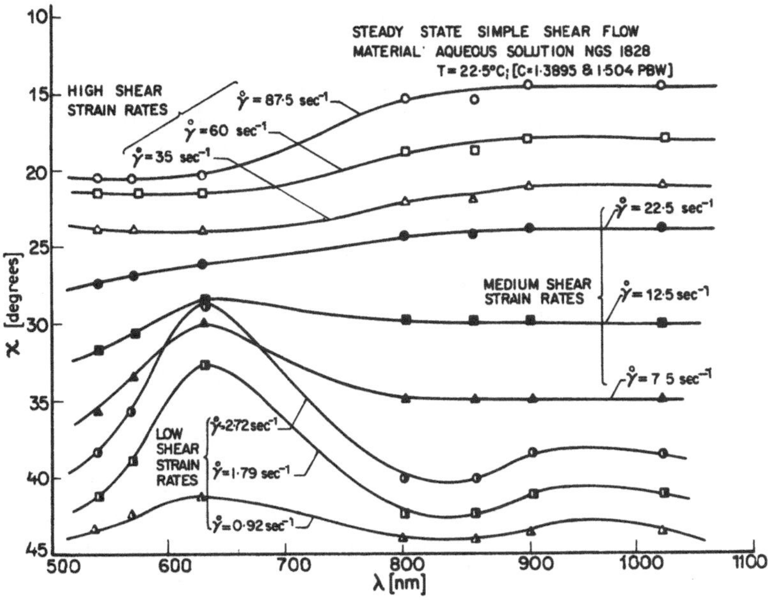

$$\varepsilon(t), \Delta n(t) = D_0 \sigma + \sum_k D_k(t) W_k \langle \sigma - \sigma_{ll}(t)_k \rangle^1 \quad [2.8]$$

where W denotes statistical weight temperature dependent, D_0 denotes instantaneous value of the creep compliance function, D_k denotes partial creep compliance functions representing partial effects, and $(\sigma_{ll})_k$ denotes linear limit stress of the k-th partial effect.

Normalized spectral birefringence

The dependence of the components of the index tensor on the frequency of electromagnetic radiation is one of the major features of the interaction between radiation and matter (25). Consequently the absolute stress-induced birefringence and all derived quantities must depend on spectral frequency. Experimental results show (15, 18, 19, 20, 22) that also the relative birefingence, and the derived quantities depend on spectral frequency.

It is convenient to represent the spectral frequency dependence of the birefringence by introducing the normalized spectral birefringence, r, defined in the following manner:

$$r = \frac{R(\lambda)}{R(\lambda_0)} = \frac{\Delta n(\lambda)}{\Delta n(\lambda_0)} = r(\lambda) = k r(v) \quad [2.9]$$

where R denotes relative retardation in terms of distance between two retarded wave fronts, Δn denotes relative retardation in terms of the difference between the principal indices of refraction related to both wave fronts, λ and v denote wave length or spectral frequency of radiation, respectively, and k denotes transformation factor.

In the linear range of response for $t = \text{const}$

$$r(\lambda) = \frac{C_\sigma(\lambda)}{C_\sigma(\lambda_0)} = \frac{C_\varepsilon(\lambda)}{C_\varepsilon(\lambda_0)}. \quad [2.10]$$

Since the dependence of the indices of refraction on the spectral frequency is a fundamental feature of the interaction between radiation and matter, depending directly on the structure of matter and the spectral frequency, the spectral birefringence should be considered as a basic measure of the mechanism of birefringence and of mechanism of rheological processes.

Experimental results confirm the above interpretation. Fig. 4a presents the influence of the manner of producing stress birefringence on the spectral birefringence. The birefringence produced in glass by residual stresses induced by a suitable rheological process is different from the birefringence produced by the force-induced stresses. Difference between both spectral birefringence functions is qualitative in the near infrared band of radiation. In this case the spectral birefringence is a suitable measure of the rheological past.

Similarly the relations presented on fig. 4b show that the spectral birefringence is related to the different mechanisms of deformation below and above transition temperature.

G. L. Cloud (24) has shown that for some polymers the spectral birefringence at constant temperature, normalized with regard to the time, does not depend on the spectral frequency in a wide spectral band, including infrared.

Normalized spectral dispersion of birefringence

In accordance with common practice, the spectral dispersion of birefringence is defined as (18):

$$D = \frac{dr}{d\lambda} = \frac{1}{\Delta n(\lambda_0)} \frac{d(\Delta n(\lambda))}{d\lambda} = D(\lambda). \quad [2.11]$$

In the linear range of response, for $t = \text{const}$

$$D(\lambda) = \frac{1}{C_\sigma(\lambda_0)} \frac{dC_\sigma(\lambda)}{d\lambda} = \frac{1}{C_\varepsilon(\lambda_0)} \frac{dC_\varepsilon(\lambda)}{d\lambda}. \quad [2.12]$$

It is convenient to normalize this quantity in a form

$$D_n(\lambda) = \frac{D(\lambda)}{D(\lambda_0)}. \quad [2.13]$$

The normalized spectral dispersion of birefringence is a convenient manner of presenting the rate of change of spectral birefringence with respect to wavelength, time and temperature.

According to experimental evidence (18, 22), this quantity is closely related to the character of rheological deformation, linear or non linear.

A sample of the spectral dispersion of birefringence in a birefringent liquid is presented in fig. 2c (22). The spectral range where the spectral dispersion of birefringence assumes minimum values, coincides with the spectral range of the maximum range of the linear birefringence response with the shear strain rate.

Normalized dielectric coefficient

The fundamental mathematical model of the photoelastic effect relates changes of components of dielectric tensor to values of components of stress or strain tensors in a known manner (25).

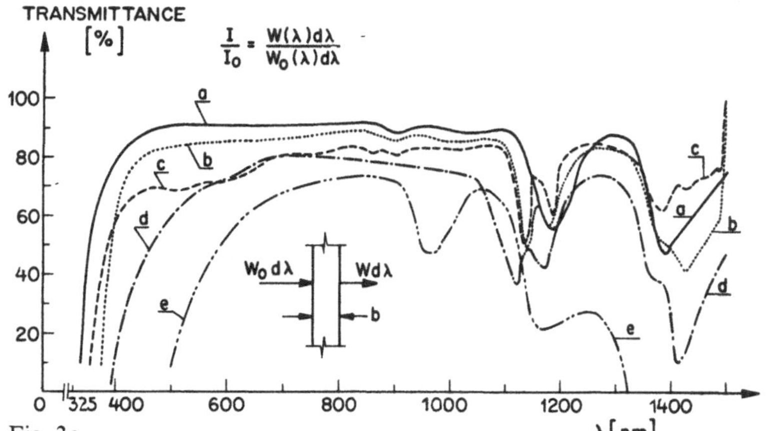

a - Allyl-polyester resin CR-39
b - Polyester resin Palatal P6
c - Polycarbonate Zelux
d - Epoxy resin Araldite 6060
e - Milling yellow aqueous solution

Fig. 3a

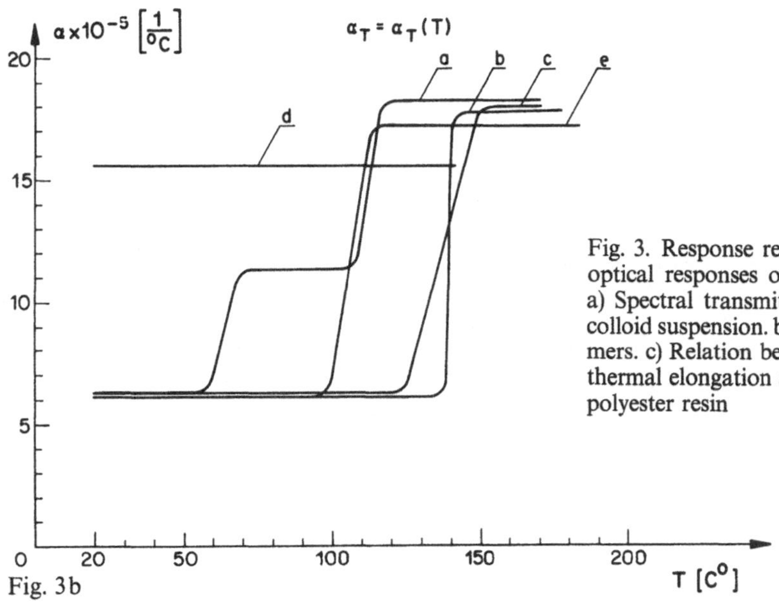

Fig. 3b

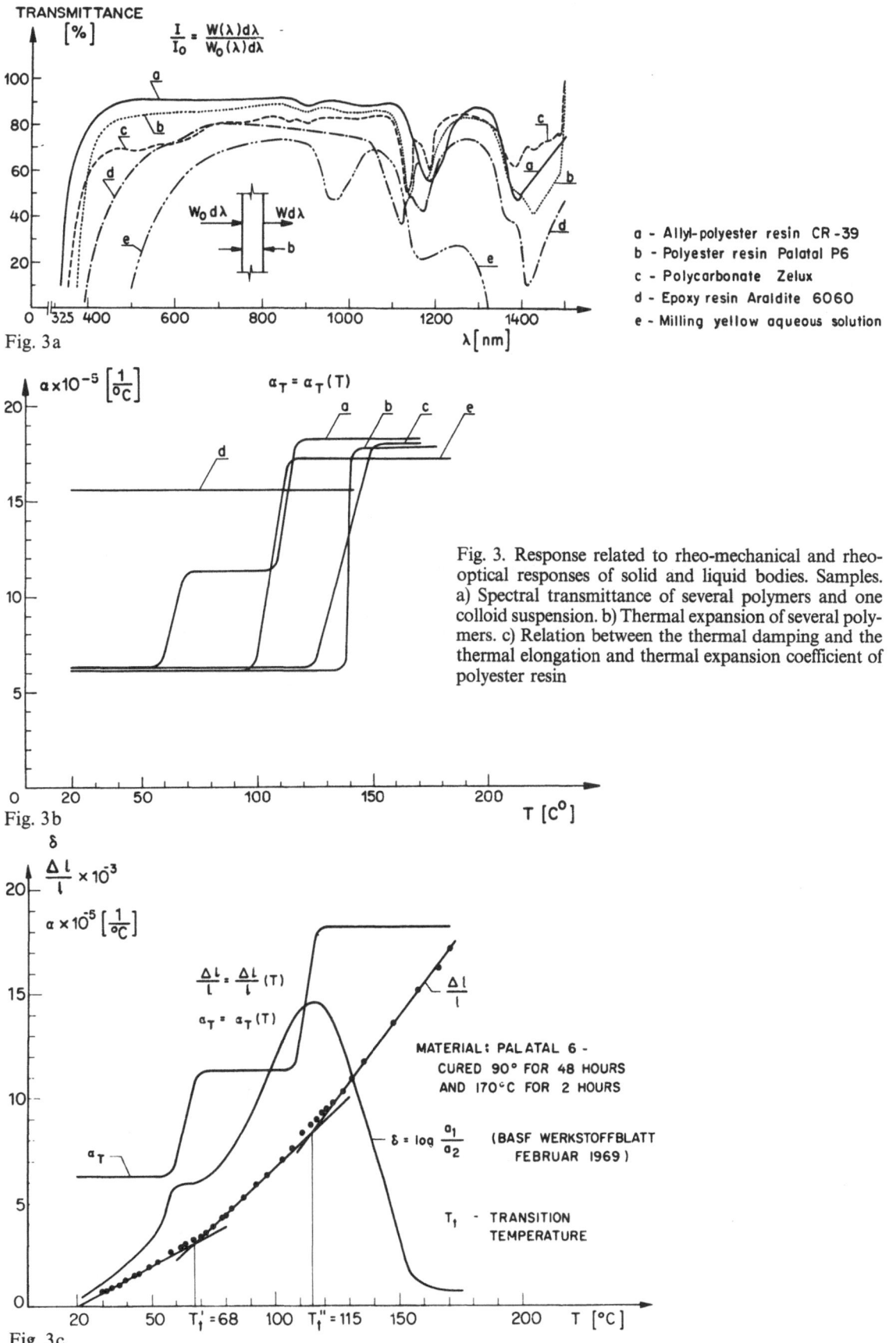

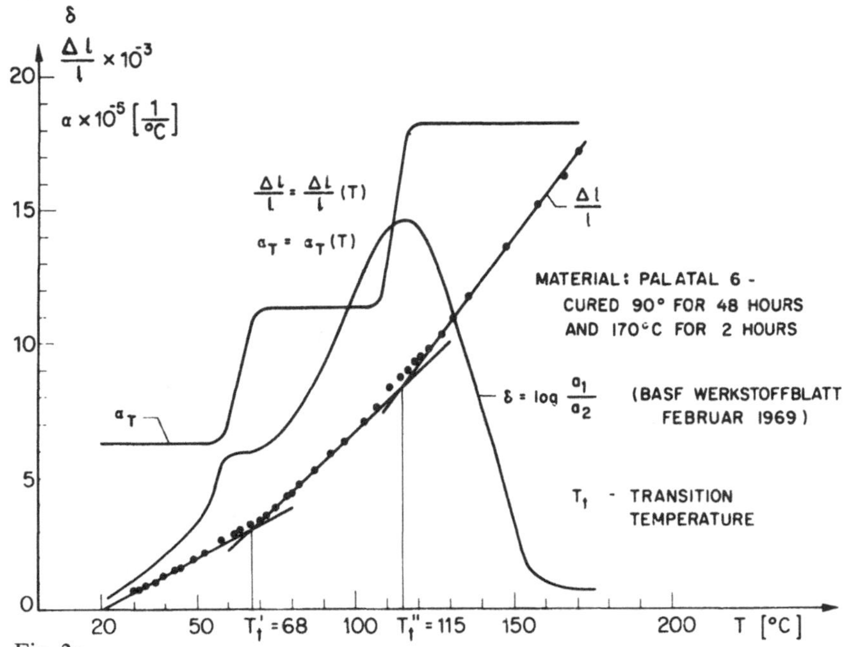

Fig. 3c

Fig. 3. Response related to rheo-mechanical and rheo-optical responses of solid and liquid bodies. Samples. a) Spectral transmittance of several polymers and one colloid suspension. b) Thermal expansion of several polymers. c) Relation between the thermal damping and the thermal elongation and thermal expansion coefficient of polyester resin

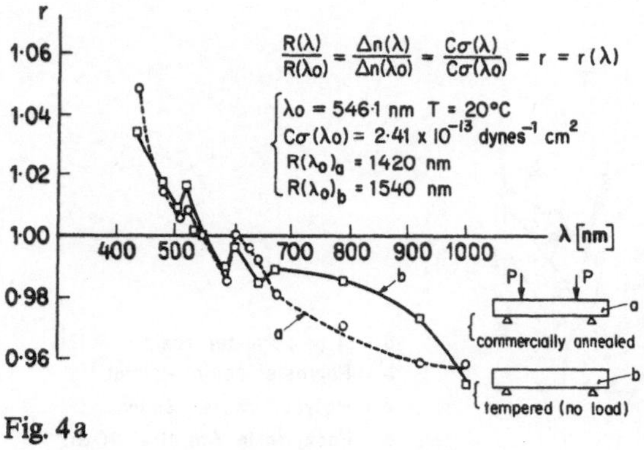

Fig. 4a

Fig. 4. Dependence of normalized spectral birefringence on various factors. a) Dependence on rheological past; soda-lime-silica plate glass. b) Dependence on temperature: epoxy resin Araldite 6010

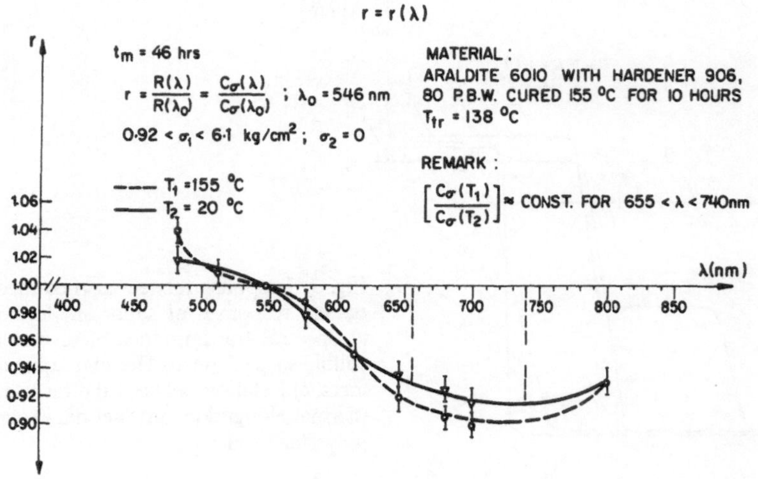

Fig. 4b

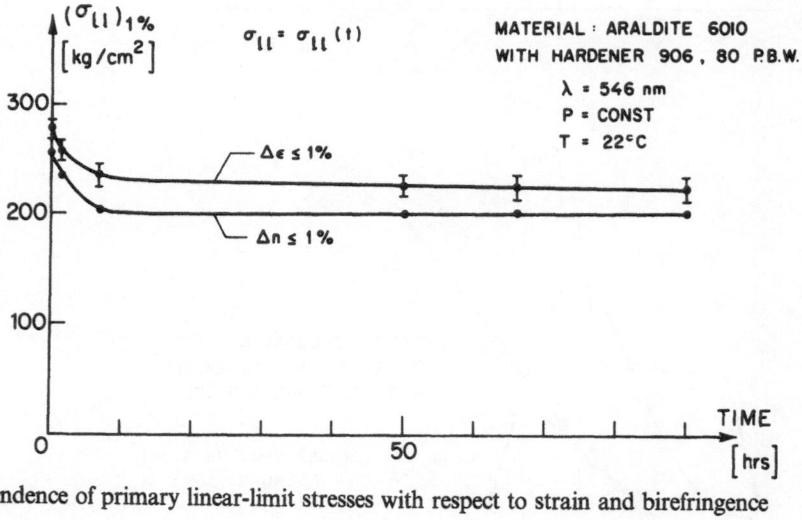

Fig. 5. Time dependence of primary linear-limit stresses with respect to strain and birefringence

Since *Maxwells* formula relates index of refraction to the dielectric coefficient, the latter should depend on the stress and strain state, and similar measures as those developed for birefringence should be applicable.

At the present time only spectral dependence of the dielectric coefficients for some polymers can be reported, in frequency band 100 to 1000000 Hz. A sample of this dependence is given in fig. 2d.

Spectral dispersion of optical axes

It is assumed in elementary photoelasticity that the principal axes of the index tensor coincide with the principal axes of the stress tensor or strain tensor, in solids, or of the strain rate tensor in liquids, regardless of the spectral frequency.

Experimental evidence shows that the validity of this model of response is limited. Results presented on fig. 2e illustrate changes in orientation of optical axes with the spectral frequency and with the shear strain rate in a birefringent liquid.

This element of the interaction between radiation and matter is of paramount interest in experimental work based on the effect of birefringence. Since it has been already proven that spectral dependence of the optical axes does exist,

$$\alpha = \alpha(\lambda) \qquad [2.14]$$

it appears desirable to consider the spectral dispersion of optical axes as a basic parameter of rheological material response.

Spectral transmittance

It is well known that the spectral transmittance of material is a function of the structure of material and therefore could be related to the rheological response of real materials.

One of the simplest measures for the spectral transmissivity of material is the spectral transmittance, defined, according to fig. 3a, as

$$Tr = \frac{I}{I_0} = \frac{W(\lambda)\,d\lambda}{W_0(\lambda)\,d\lambda} \qquad [2.15]$$

where $W(\lambda)$ denotes radiant power per unit crossection and per unit wave length.

The spectral frequency gradient of spectral transmittance appears to be of an interest in experimental rheology:

$$t_r(\lambda) = \frac{d}{d\lambda}(T_r). \qquad [2.16]$$

Some samples of the spectral transmittance of several materials used in Experimental Mechanics are presented in fig. 3a.

There seems to exist a correlation between the spectral transmittance and its spectral frequency gradient, the spectral dispersion of birefringence and the spectral band of optimum linear response with regard to birefringence.

From the point of view of optimizing a rheo-birefringence experiment it seems desirable to chose the spectral band of radiation sufficiently remote from the absorbtion bands.

Thermal expansion coefficient

The concept of a rheologically simple material, thermal response of which is satisfactorily described by the time-temperature equivalence principle is often applied for real materials, without a sufficient justification.

In reality, the application of elementary relations based on this principle is limited by the structural changes related to the changes of temperature.

Several high polymers exhibit rapid changes of the rheological responses in particular temperature ranges, called transition temperature ranges which are often reduced to the transition temperatures, T_r.

Among many interrelated temperature responses two responses seem to be very convenient for determination of transition ranges: the thermal expansion described by the thermal linear expansion coefficient and the change of the mechanical damping with the temperature, described by the logarithmic decrement of damping:

$$\alpha = \alpha(T) \qquad [2.17]$$

and

$$\delta = \delta(T) \qquad [2.18]$$

where α denotes the thermal linear expansion coefficient and δ denotes the logarithmic decrement of damping.

Some typical relations are presented on fig. 3b. Correlation between two different responses, the temperature dependence of the thermal expansion coefficient and the logarithmic decrement of damping for a polyester resin is illustrated by fig. 3c.

The shape of the function $\alpha = \alpha(T)$ is one of the measures for the suitability of material for the so-called photoelastic stress freezing technique.

The thermal expansion appears to be one of the most fundamental measures of the transition temperatures, from the point of view of Experimental Mechanics. It is necessary to know the temperature depence of the thermal expansion coefficient to optimize the stress freezing experi-

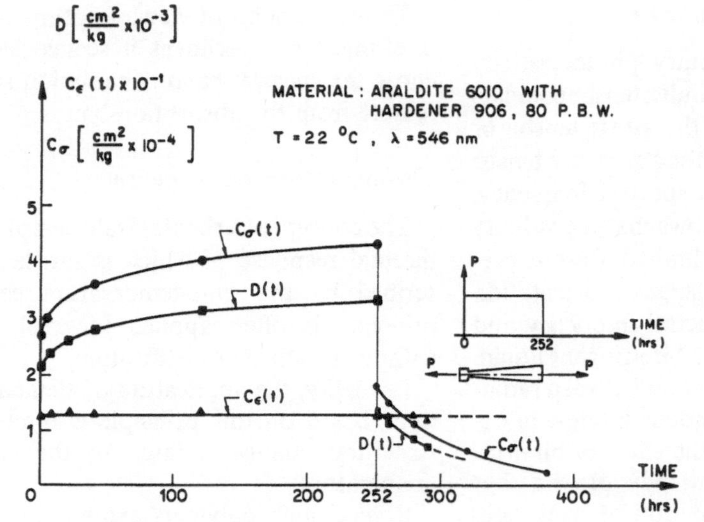

Fig. 6a

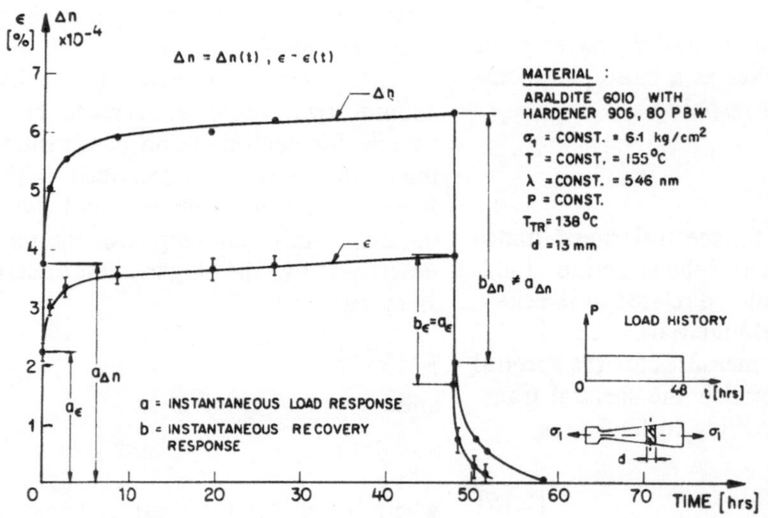

Fig. 6b

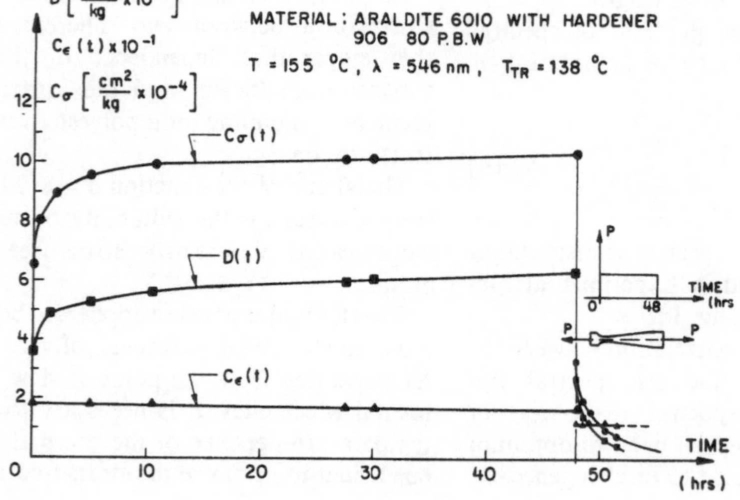

Fig. 6c

Fig. 6d

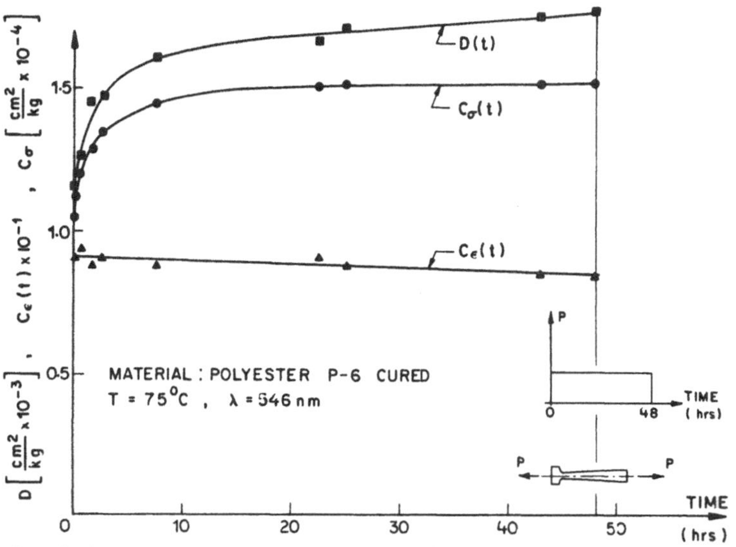

Fig. 6. Mechanical and optical creep and creep recovery functions. a) Epoxy resin at 20 °C (below transition): Creep compliance functions. b) Epoxy resin at 155 °C (above transition): Mechanical and optical creep and creep recovery response. c) Epoxy resin at 155 °C (above transition): Creep compliance functions. d) Polyester resin Palatal P6 at 75 °C: Creep compliance functions

ments which represent a major technique of the three dimensional stress analysis.

Discussion

The first problem which needs clarification is the meaning of the presented relations, parameters and coefficients. Obviously, this problem is a part of a fundamental problem of science– whether our mathematical models represent discoveries or represent inventions, and – to be more specific – whether parameters and coefficients of rheological equations represent "Properties" of materials or "Parameters of Response" of materials to specified inputs. Clearly, the term "Property" connotes two particular assumptions, namely that such quantity really exist and that it can be determined with a desired accuracy.

The first assumption is a fundamental philosophical assumption that seems to be redundant regarding development of a commensurate and adequate description of material behaviour, presented in an analogous or an analytical form. The second assumption neglects the fact that it is not possible to perform any observation without disturbing process under observation, and consequently neglects the influence of the interactions between components of the measuring system and the influence of the energy flow thru the system on the results of experiment.

The terms "response" of materials, "parameters of response" and "coefficients" of response, used in this paper connote that all kinds of response are inherently interrelated and that form of energy used for the determination of particular responses can be of a paramount importance.

One of the practical consequences of such an approach is requirement that the mathematical models of the material response should take in the account the temporal variation of energy flow, regardless of the forms of energy. It is obvious that such a requirement necessitates a physical approach to the design of a mathematical model of material response. For instance, it is possible to chose coefficients of a suitable function to match the spectral frequency dependence of birefringence. Such a mathematical model would appear artificial in comparison with a model based on relations between the resonant frequencies of particular groups of atoms, the frequency of electromagnetic radiation and the resulting spectral birefringence.

According to the presented results it is advantageous to consider the rheological response of a body as an inherently interrelated set of responses of one system. Consequently the mathematical models describing the various kinds of responses of the same body should be compatible and complementary.

For practical purposes it could be sufficient to apply even the simplest relations provided that the suitable set of conditions and constraint is

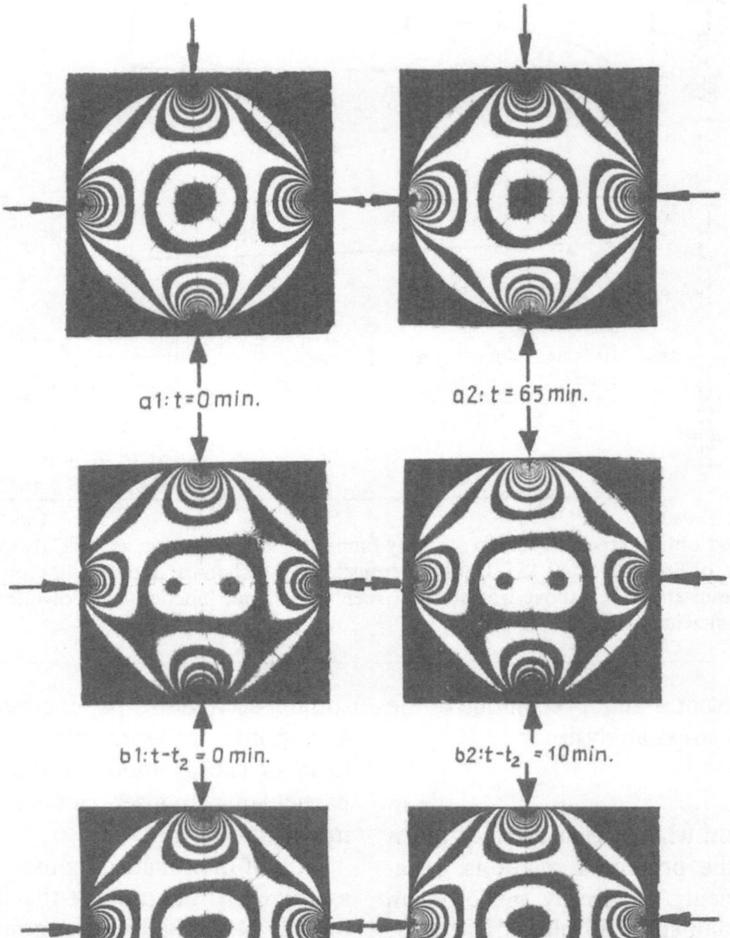

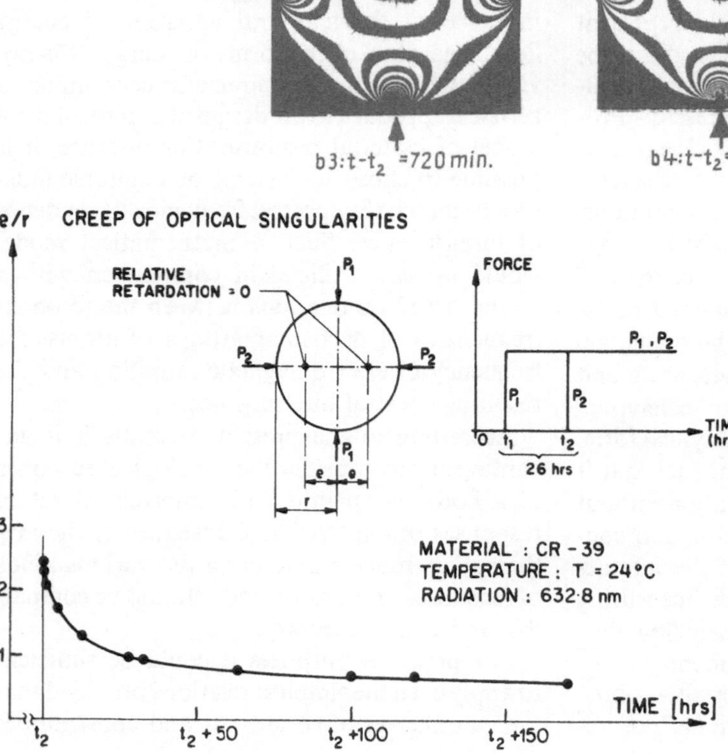

Fig. 7. Illustration of time-dependent birefringence in circular disc related to two dimensional stress-state caused by two consecutively applied loads along perpendicular diameters. (Example of a linear viscoelastic response.) a) Both loads are applied at the same time, a_1: $t = 0$ min, a_2: $t = 65$ min. b) Load P_2 was applied 26 h after load P_1. b_1: $t - t_2 = 0$ min, b_2: $t - t_2 = 10$ min, b_3: $t = t_2 = 720$ min = 12 h, b_4: $t - t_2 = 10\,200$ min = 170 h. c) Creep of optical singularities ($\Delta R = 0$), illustrated by isochromatics pattern, related to fig. 7b

given. Typical fundamental relations in the form

$$\Delta n(t) = \int_0^t C_\sigma(t - \tau)\,\dot{\sigma}\,d\tau \qquad [3.1]$$

can be succesfully applied when the following conditions are satisfied.

$$-(\sigma_{ll})_{\Delta n} < \sigma < (\sigma_{ll})_{\Delta n} \qquad [3.2]$$

and

$$T < T_r \quad \text{or} \quad T > T_r.$$

Experiment can yield incorrect results when the spectral dependence of the photoelastic coefficients $C = C(\lambda)$ is neglected.

The time-temperature equivalent function can be applied only in the temperature range determined by:

$$T_{r1} < T < T_{r2}. \qquad [3.4]$$

Results of rheo-optical experiments performed within such a set of constraints consciously chosen can be successfully interpreted in terms of generalized relation [3.1], as illustrated by fig. 7a, b, and c.

Conclusion

The presented experimental evidence supports the conclusion that it is advantageous – from the point of view of the degree of correlation between the real rheological response and the chosen mathematical model of response – to consider a rheological body as a system, characterized by major responses which are inherently interrelated. Design and choise of the parameters of response should be correlated with the transfer function of the measuring system.

This approach results in the condition that the mathematical models developed for description and prediction of particular responses of real bodies be compatible with each other and complementary.

Acknowledgement

Authors gratefully acknowledge the support given by the National Research Council of Canada under research grant No. A2939. The assistance of *Wolfgang Hickenfelder* during preparation of the experiments is greatly appreciated.

References

1) *Tiller, W. A.*, Science **165**, 469–475 (1965).
2) *Alfrey, T.*, Mechanical Behaviour of High Polymers (New York 1948).
3) *Stuart, H. A.*, Die Physik der Hochpolymeren (Berlin-Heidelberg-New York 1956).
4) *Persoz, B.*, Introduction a l'etude de la rhéologie (Paris 1960).
5) *Flügge, W.*, Viscoelasticity (Verlagsort 1967).
6) *Mindlin, R. D.*, Appl. Phys. **20**, 206–216 (1949).
7) *Nowacki, W.*, Theorie du Fluage (Paris 1965).
8) *Dill, E. H.*, Polymer Sci. Part C 5, 67–74 (1964).
9) *Green, A. E.* and *R. S. Rivlin*, Arch. ration Mech. Analysis **1**, 1–27 (1957).
10) *Green, A. E., R. S. Rivlin*, and *A. J. M. Spencer*, Archs. ration Mech. Analysis **3**, 82–90 (1959).
11) *Green, A. E.* and *R. S. Rivlin*, Archs. ration Mech. Analysis **4**, 387–404 (1960).
12) *Drescher, A.* and *K. Kwaszczyńska*, J. Non-Linear Mech. **5**, 11–22 (1970).
13) *Kolsky, H.*, in: *H. A. Stuart*, Die Physik der Hochpolymeren, Bd. 4 (Berlin-Heidelberg-New York 1956).
14) *Hiltscher, R.*, VDI-Z **23** (1953).
15) *Pindera, J. T.*, Rozprawy Inżynierskie **1959**, No. 3.
16) *Read, B. E.*, J. Polymer Sci., Part C **5**, 87–100 (1964).
17) *Pindera, J. T.* and *E. W. Kiesling*, On the Linear Range of Behavior of Photoelastic and Model Materials. Proceedings of the Third International Conference on Stress Analysis, March 1966, Berlin, VDI-Berichte, No. 102, 1966.
18) *Pindera, J. T.* and *G. L. Cloud*, Exp. Mechanics **6**, 470–480 (1966).
19) *Pindera, J. T.* and *P. Straka*, Mechanical and Dielectric Rheological Responses of Some High Polymers in a Wide Temperature and Spectral Frequency Ranges. Proceedings, Int. Symposium on Exp. Mechanics, Univ. of Waterloo, 1972; Experimental Mechanics in Research and Development. Study No. 9. Solid Mechanics Division, 1973.
20) *Pindera, J. T.* and *N. K. Sinha*, On the Studies of Residual Stresses in Glass Plates. Exp. Mechanics, March 1971.
21) *Pindera, J. T.* and *Y. Sze*, Trans. CSME **1**, No. 2 (1972).
22) *Pindera, J. T.* and *A. R. Krishnamurthy*, Foundation of Flow Birefringence in some Liquids. Proceedings of International Symposium on Exp. Mechanics, Univ. of Waterloo 1972: Experimental Mechanics in Research and Development. Study No. 9. Solid Mechanics Division, 1973.
23) *Pindera, J. T.*, On the Physical Basis of Modern Photoelasticity Techniques. Beiträge zur Spannungs- und Dehnungsanalyse, Vol. V (Berlin 1968).
24) *Cloud, G. L.*, Exp. Mechanics, November **1969**.
25) *Ramachandra, G. N.*, Crystal Optics. In: Encyclopedia of Physics. Edited by *S. Flügge*, Vol. XXV/1 (Berlin-Heidelberg-New York 1961).

Authors' address:

Jerzy T. Pindera and *Peter Straka*
Experimental Mechanics Laboratory
Department of Civil Engineering
University of Waterloo
Waterloo, Ontario, Canada

Rheol. Acta 13, 352–366 (1974)

From the College Engineering, University of Utah, Salt Lake City, Utah (U.S.A.)

A reaction rate model for deformation and fracture in polymeric fibers*)

By *B. A. Lloyd*, *K. L. DeVries*, and *M. L. Williams*

With 8 figures and 2 tables

(Received October 27, 1973)

Introduction

Fracture in polymers is a complex phenomenon of considerable recent interest. Here we shall discuss an important subclass, i.e. highly-oriented polymer fibers. Several models have been proposed to describe the microstructure in fibrous polymers. These range from the early fringed micelle structures (1) to more recently proposed folded chain models (2, 3). Although models proposed by *Hearle* (4); *Bosley* (5); *Keller* (6); *Fischer* et al. (7); *Peterlin* (2); *Dismore* and *Statton* (8); *Gubanov* and *Chevychelov* (9); *Takayangai* et al. (10); and *Bonart* and *Hosemann* (3) differ in some very important details, they have a basic common element in their microstructure; that is, they view polymers as alternating between highly-ordered regions and less ordered or "tie molecule" regions. For the particular melt spun, hot-drawn Nylon 6 fibers used in this research, *Joon B. Park* (11) has recently proposed a paracrystalline, folded chain structure as representative of the fiber microstructure.

Many researchers have proposed theories and molecular models to explain fracture from a microscopic viewpoint (12–15). However, it was only as experimental techniques which could be directly related to atomic occurrence were developed that theoretical predictions of failure on a microscopic scale could be compared with experiments. *S. N. Zhurkov* and his associates were the first to demonstrate that electron paramagnetic resonance (EPR) could be used to measure chain scission resulting from stress during fracture in polymers (16). Since that time the authors and a number of other researchers have extended these studies with the goal of establishing a satisfactory failure model for fibers.

In retrospect is appears that many molecular models of fracture have a common oversimplification. They neglect the difference in stress among the polymer chains. Reaction rate theory is well established and any model of failure should include its features. The probability, P_b, of bond rupture would then be given by

$$P_b = w_0 \exp[-(U_0 - \gamma\sigma)/kT] \qquad [1]$$

where w_0, U_0, and γ are kinetic parameters commonly known as the collision parameter, the activation energy, and activation volume, respectively. It should be noted that σ in this equation is the atomic stress and how this is related to applied or local macroscopic stress will be left as an open question for the time being. This relationship has been a subject of discussion in several recent investigations by *Roylance* et al. (17); *Kausch* et al. (18); *Peterlin* (2); and *Verma* and *Peterlin* (19). Further development of the statistical mathematical approach to the problem was presented by *Kausch* (20, 21). A more recent study has been conducted by the authors and the results published (22). In this paper model parameter limits are investigated and defined in detail. Results from theoretical and experimental stress, strain, and bond rupture comparisons reported previously (22) for Nylon 6 fibers are extended to other loading, particularly fatigue, creep, and stress relaxation. The effects of temperature are mentioned [for a detailed presentation see (23)] and preliminary tests on other Nylon 6 and on Nylon 66 fibers are presented.

*) Portions of this work were sponsored by the National Science Foundation and the National Aeronautics and Space Administration.

Test facilities and procedures

The experimental equipment, sample preparation, and experimental procedures were as previously reported (22, 23).

Results

Several researchers (19, 24) have concluded from their studies that fracture initiates in the quasi-amorphous regions in semi-crystalline fibers. An important point to make here is that even though chain scission is considered to occur almost wholly in the more disordered or tie molecule regions of the polymer fibers, there still remains the question: Does failure occur in every tie molecular region or are there specific critical tie molecule or flaw regions distributed throughout the fiber in which failure predominates? Observation of the number of free radicals generated during fracture has substantiated the concept of a failure process occurring throughout the fiber at many critical flaw regions (25) and analysis of this number in terms of the total chain scissions that could occur should all tie regions be active seems to indicate that only a fraction of the number possible are generated and/or observed by EPR (26). The concept of a critical flaw region in which failure predominates, i.e. a region which would perhaps contain the fewest and/or the most poorly distributed load-bearing tie molecules, is in basic agreement with previous observations in this laboratory and elsewhere (17, 25, 27, 28, 29). The first experimental verification of such a distribution was reported by *Kausch* and *Becht* (18) where bond scission data from EPR step-strain tests was plotted in the form of a histogram giving the number of chain scissions versus the strain increment (chain length). Similar tests by the authors indicate that not only does there seem to exist an "apparent or effective distribution" of chain lengths in the critical regions but this distribution is temperature sensitive and seems to comply with a normal or *Gauss*ian distribution criterion (22, 23). Fig. 1 shows the results of step-strain room temperature tests of two different Nylon 6 samples plotted as histograms.

It should be noted that due to the fact that the sample fails rather catastrophically, the tailing side of the distribution is missing. These experimental studies, therefore, give no information on the distribution in this region. Plotting these histograms on probability paper indicated that the distribution was *Gauss*ian and for Nylon 6, No. 1, at room temperature the distribution of one standard deviation value was observed to be approximately ten per cent of the distribution mean strain value.

The current study points out several features that a "correct" model must contain. These include:

1. There must be weak "flaw" regions reasonably uniformly distributed throughout the material.

2. Crack arrestors (densely packed regions) must be distributed about the flaw regions.

3. There must be a distribution in either the effective length of the chains in a given flaw region or in the size (criticality) of the flaw regions. The proposed model does not distinguish between these two possibilities.

One other possibility is perhaps worthy of note. One can envision mechanisms where some polymeric chains might be broken without the development of observable free radicals. Careful studies of free radical decay conducted both in this laboratory and elsewhere have indicated that up to room temperature and above free radicals in nylon are fairly stable. Some mechanisms might, however, not be expected to show up in such decay tests. Even if such is not the case and there are more bonds broken than free radicals formed, one would logically expect the free radical concentration to be proprotional to the total number of chains ruptured. As a consequence this number would not deter from the proposed basic model but might modify some of the parameters.

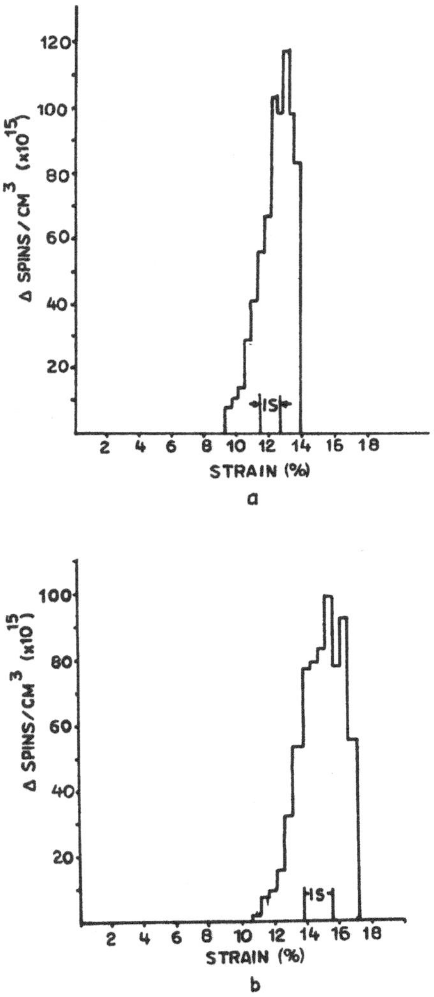

Figs. 1a and b. Histograms of Nylon 6 samples from step strain EPR data at room temperature; a) Nylon 6, No. 1, and b) Nylon 6, No. 2, annealed

It should be made clear that the experimentally determined distribution could be either of two phenomena or a combination of them. First, there is the tie-chain distribution in a given flaw and secondly, there is the distribution in size or criticality of flaws.

The assumed mathematical model incorporates a basic alternating sandwich structure of crystalline blocks and critical flaw regions. These regions have no predetermined size and are related to each other by ratio values only. A distribution in the critical flaw regions is included and, hence, different chains are under different stresses. The distribution in stress among the tie chains (or critical regions) changes as bond rupture occurs and the load carried must be redistributed among the remaining intact chains. Fracture is assumed to be controlled by a *Tobolsky-Eyring* (12) type rate process or reaction rate theory, i.e. a stress-aided activation energy process. The model takes one such crystalline block and flaw region as representative of the average of all critical regions of the fiber. The ordered or "crystalline block" region is represented by a number of aligned polymer chains, each with an average bond area, A_b. The disordered or "critical flaw region" is assumed composed of sets of tie chains. Each of these "n" sets has a different effective length and number of chains according to a prescribed distribution. The original length of the flaw region is taken as l_{0a} and some of the tie chains are initially considered to be taut whereas all other sets may be kinked, disoriented, folded, etc. and do not become taut until the region is sufficiently strained to remove the slack for a particular set. The derivation for these equations has been previously outlined (22) with the basic assumptions:

1. The shear mode of failure, i.e. chain slippage or flow due to secondary bond failure in the average of the critical regions, is negligible compared to backbone chain scission.
2. Chain scission in the "crystalline block" region is negligible.
3. *Hookes* law is applicable on an atomic scale.
4. A chain assumes no load-carrying ability until the strain in the flaw region is such that the chain will be taut and after it is taut, it is assumed to act elastically.
5. Chain scission is predicted by kinetic reaction rate theory.

The governing equations for chain rupture as a function of time are:

$$C_{BT} = \sum_{i=1}^{n} C_{Bi}$$

where

$C_{Bi} = C_{u0i} - C_{ui}$

C_{BT} the total broken polymer chain in the average of the critical regions of the fiber at time t

C_{Bi} the number of broken polymer chains in i^{th} set at time t

C_{u0i} the original number of polymer chains in i^{th} set

C_{ui} the number of unbroken chains in i^{th} set at time t

C_{ui} was determined by the numerical solution of the set of equations:

$$\frac{dC_{ui}}{dt} = -C_{ui} A \exp \beta \sigma_i \qquad \text{for} \quad i = 1, n$$

where

n the total number of chain sets in critical region

A $w_0 \exp(-U_0/kT)$

β γ/kT

σ_i the stress on each bond in the ith set in the critical region

k *Boltzmanns* constant

T the absolute temperature

w_0 the collision parameter

U_0 the activation energy

γ the activation volume

The stress σ_i on each bond in the critical region was related to applied macroscopic stress by a force balance. A detailed outline of the mathematical derivation is given in (22 and 30).

The input parameters necessary for a theoretical solution using the mathematical model of the polymer fiber are:

1. S = the standard deviation of the distribution of chains in the critical region.
2. RC = the ratio of the number of polymer chains in the "cry-blk" region to number of tie chains in the flaw region.
3. RL = the ratio of the original length of the "cry-blk" region to the original length of the flaw region.
4. E_b = the modulus of elasticity of a single polymer chain.
5. w_0 = the collision parameter.
6. U_0 = the activation energy.
7. γ = the activation volume.
8. σ_T or ε_T = the applied macroscopic stress (or strain) as a function of time.

9. T = the absolute temperature.

10. W = the total chains/cm^2 in a flaw region at time zero.

While no absolute values of the theoretical parameter are known physically they must fall within certain limits which can normally be fairly well defined. Briefly these are as follows:

1. The distribution parameter "S"

A detailed discussion of this parameter and how it varies with treatment, temperature, etc. is given in (23). For our purposes here it suffices that it can effectively be experimentally determined by methods previously outlined (22). It is worth noting, however, that it appears to be the most important parameter effecting the strength of a given polymer. Physically "S" probably includes either of two effects – the distribution of chains within a flaw region and/or the distribution or criticality of regions.

2. The chain ratio parameter "RC"

The ratio of the number of chains in the critical region to that in a crystallite can be thought of as a stress concentration factor. For a perfect crystal the effective cross-sectional area of a single Nylon 6 chain has been calculated as 17.7 A^2 (31). This corresponds to 5.65×10^{14} chains/cm^2 for perfect orientation. This represents the maximum number of chains which could occur in a square centimeter of polymer cross-section. The density of 100% crystalline Nylon 6 (calculated from single crystal data) is reported as 1.23 gm/cm^3 where 100% amorphous (extrapolated from bulk polymer data) is listed as 1.10 gm/cm^3 (32). X-ray evidence has also been interpreted as indicating that no clear-cut, two-phase system exists in highly-drawn fibers but rather only degrees of order and disorder or a paracrystalline-type structure (8, 11, 33). Even so, the existence of less-ordered or flaw regions between the highly-ordered regions seems necessary to explain strength and fracture data. Also, it is reasonable to assume that these regions consist of chain segments that are incorporated in the two adjacent "crystalline-like" regions (tie molecules), chain loops (polymer chains re-entering the same "crystalline-like" region) and chain ends. Therefore, the tie molecule or load-bearing chain portion is pictured as only some fraction of the total number of chains in the critical regions. *Meinel* and *Peterlin* conducted studies with fuming nitric acid treatments of highly-drawn polyethylene which they interpreted to indicate what fraction of polymer chains were tie chains (2, 34). From the relative values of the molecular weight species, the fraction of tie molecules was estimated. These results indicated ratios of the number of crystalline polymer chains to amorphous tie chains between three and ten. In light of the a priori possibility of the fracture process occurring in the "weak link" or "critical flaw" regions, one can speculate that there most likely exists a distribution of disordered or amorphous regions throughout the fiber. In fact, only the weakest or most critical regions would contribute to the fracture process. The proposed model for fracture considers only those regions which contribute significantly to the fracture process. These we have called "critical" regions whereas all other disordered regions are combined with the ordered regions and designated "crystalline-block" regions. Other experiments indicate that this effective ratio (*RC*) may be significantly smaller than the average ratio in some materials (19, 28). Part of the explanation for the apparently low strength of polymers compared to theories based on chain strength is that cracks can progress selectively through the weakest regions of the polymer.

3. The chain length parameter "RL"

Because of a lack of concrete experimental evidence, the ratio of the original length of the "crystalline-block" region to the original length of the "critical" region can at best only be speculated. Some evidence does exist, however, from small angle X-ray studies as to the length of the ordered or crystalline regions. Small angle X-ray patterns result only where there occurs an inhomogeneity in the microstructure of the polymer. Interpretation of long period values obtained from such measurements in highly-oriented fibers has varied. Early interpretation of this data was used as evidence for the length between crystalline and amorphous regions in the fringed micelle model (36, 37). Later analysis of the data by *Dismore* and *Statton* (8) indicated that long period data was evidence for chain-folded regions. For Nylon 6 and Nylon 66 fibers the long period has been reported to be approximately 100 Å in length (8, 11). Length of the tie chains or amorphous region has been speculated mostly from crystallinity percentage although *Peterlin* (2) has interpreted results from fuming nitric acid treatments in polyethylene to give a crystalline region of 120 Å and an amorphous tie chain region of 50 Å. If we assume the "crystalline block or ordered" regions are 100% crystalline and all the less-ordered regions are active, then the total sample crystallinity could be approximated by the original length of the regions. As these assumptions are probably not valid, i.e. the "cry-blk" regions are most likely *not* 100% crystalline and the active "critical" regions are probably only the least dense of all less-ordered regions, estimates of crystallinity from such a relationship are probably high. Furthermore, the meaning of crystallinity is not clear (8, 11, 33). Results from EPR bond rupture data also add to the speculation. Most researchers report the number of free radicals observed at fracture in Nylon 6 as approximately 5×10^{17} spins/cm^3 or 2.5×10^{17} chains/cm^3 (17, 24, 38, 39). If we assume EPR results give total tie-chain rupture and fracture occurs in every amorphous tie-chain region with these regions occurring every 100 Å, then we could conclude that there was approximately 2.5×10^{11} chains/cm^2 in each amorphous region. This figure is two or three orders of magnitude lower than any of the above estimates and, therefore, the assumptions become suspect.

They key to such a discrepancy most likely can be explained through either of two mechanisms, *i*, the number of free radicals observed by EPR, is representative of more broken bonds than the simple conversion of two free radicals per bond would indicate; or *ii*, all of the less-ordered regions are not active in the fracture process, i.e. these regions differ to such an extent that only a fraction are critical and contribute to the accumulation of broken bonds. Although the authors favor this latter concept, the truth most likely lies in a combination of the two concepts. In light of the above discussion, the ratio *RL* could easily range between two and one hundred depending on the exact mechanisms of rupture.

4. *The modulus of elasticity of a single Nylon 6 polymer chain, "E_b"*

Treloar (40) has calculated the modulus of elasticity of a single Nylon 66 chain based on a single chain area of 17.6 Å^2 as 2.84×10^7 lb/in.^{2} (2×10^4 kg/mm^2). The chemical composition of Nylon 6 is the same as Nylon 66 with a slightly different arrangement of the CH_2 groups and the single crystal area is 17.7 Å^2. These similarities make use of the Nylon 66 modulus value for Nylon 6 reasonable.

5. *The collision parameter, "w_0"*

This constant in a rate process failure criterion should theoretically have a value which approaches the thermal vibration frequency of the atom. *Zhurkov* (38) reports this value as 10^{12} to 10^{13} vibrations per second.

6. *The activation energy, "U_0"*

The parameter U_0 in the model corresponds to the height of the activation barrier which must be overcome for breakage of the chemical bonds. The activation energy for bond dissociation is not necessarily the same as the bond dissociation energy and in general one might not always expect a very close relationship between the two energies. However, in reactions of this type, i.e. separation or fracture of chemical bonds, the activation energy will be equal to or larger than the heat of reaction (41). The bond dissociation energy for a carbon-carbon bond in a paraffin hydrocarbon chain has been determined by *Gëro* to be 82.59 kcal/mole (42). The weakest bond in the Nylon 6 chain has been theoretically calculated to have a bond dissociation energy of approximately 15 kcal/mole less than an ordinary C—C bond (43). Significantly lower values of the activation energy have been obtained from studying the effect of temperature on fracture from macroscopic observations of thermal or mechanical degradation. *Zhurkov* (44) has obtained activation energy values for mechanical degradation from polymer liefetime data. The energy of activation was then obtained from a plot of the natural logarithm of time-to-failure versus one over the temperature for a constant applied macroscopic stress. From this data he reported a value of U_0 as 45 kcal/mole for Nylon 6. A recent publication by *Chevychelov* (29) reports work incorporating a distribution of stresses with rate process theory and indicates an activation energy of 57 kcal/mole for Nylon 6 necessary to fit experimental data. One might expect that activation energy values obtained from studying the effects of temperature on the rate of bond rupture on a macroscopic basis either by thermal or mechanical degradation may *not* agree with the U_0 values for atomic or microscopic reactions due to the inhomogeneity of the stresses on the bonds. Various reactions in solid state polymers might lower the activation energy for bond rupture.

7. *The activation volume "γ"*

This constant is often referred to as the "activation area" though it has the dimensions of volume. Theoretically one might think of this value as the area of a chemical bond times the distance necessary to separate the adjoining atoms to cause rupture. *Perepelkin* (45) states that while the value of the collision parameter w_0 is practically independent of the fiber structure, the activation volume "γ" *as determined from macromeasurements* is a coefficient which must take into account the structure of the material and the nonuniformity of stress distribution in molecular chains. From this basis he calculates "γ" values for various structural considerations, stating that the closer to an ideal structure the smaller "γ" becomes approaching a value on the order of a molecular volume. Table 1 below summarizes these results for Nylon 6 fibers.

Table 1. Activation volume "γ" calculated by *Perepelkin* (45) for Nylon 6 fibers

(cm³)		
Calculated for real real fibers	Calculated for ideally ordered fibers	Calculated for ideally ordered fibers with molecular interaction
$107–219 \times 10^{-24}$	$17–28.9 \times 10^{-24}$	$1.02–17.1 \times 10^{-24}$

Experimental values of "γ" for Nylon 6 were determined by *Zhurkov* (38, 44, 45) where the stress was varied at constant temperature and the results plotted in terms of natural logarithm of the time-to-failure versus applied macroscopic stress. From this linear plot the slope "γ" was evaluated and reported by *Perepelkin* (45) as 206×10^{-24} cm³. *Zhurkov* reported that an equivalent "γ" value was obtained from a similar linear plot of rate of chain scission data for quasi-load rate tests (38). Since then, "γ" has been determined for a variety of other loadings based on the macro-stress (25, 30). Comparison of the activation values obtained from these macromeasurements indicate a wide variation (30 to 300×10^{-24} cm³) and are generally much larger than those theoretical values reported by *Perepelkin* in table 1 for ideal fibers. Calculation of the distance necessary to rupture adjoining atoms in Nylon 6 chains, assuming a bond area of 17.7×10^{-16} cm² (31) and these "γ" values give values ranging from 2 to 45 Å. *Peterlin* reports a similar value of 8 Å from macromeasurement data (2) and states that this cannot possibly be considered an actual displacement. He further indicates that such a value should be more on the order of 0.3 Å. Assuming a bond area of 17.7×10^{-16} cm² and a displacement of 0.3×10^{-8} cm necessary to cause fracture, one obtains an activation volume value "γ" of 5.2 $\times 10^{-24}$ cm³. In essence, the activation volume term as determined from macromeasurement data appears to have little, if any, meaning and does not even appear to be constant. Once again, the key to obtaining reasonable values for "γ" appears to lie in an understanding of the distribution in stress in the polymer.

8. *The parameter "W"*

The model predicts failure in terms of the number of tie chains per unit area while the experimental values are in terms of the number of free radicals per unit volume of material in the EPR spectrometer cavity. In order to compare the two, a knowledge of the amount of microcrack or flaw area per unit volume is required. Such information is currently not completely available and is a subject of current investigation. Some interesting speculation is, however, possible from the present study. First, it will be noted that all the tie chains in the model were not ruptured at the corresponding experimental catastrophic fracture point. Estimates of the total number of tie chains can be made from both experimental and theoretical observations of tests completed in this study. From the experimental histogram plots, assuming a symmetrical *Gaussian* distribution, the magnitude of the missing portion or tailing side of the distribution can be made. From such analysis it is estimated that between fifty and ninety per cent of the total number of tie molecules are observed at fracture by EPR techniques (depending on the temperature, distribution, effect of plasticizers, etc.). *Peterlin* (46), however, has indicated that the tailing side of the distribution may be somewhat artifact due to a selective failure process proceeding through only the "most critical" of the flaws after the point of maximum load capacity has been exceeded. However, if such a process was taking place, we might expect the tailing side of the distribution plots to deviate somewhat from a normal or *Gauss*ian distribution. Indeed such a deviation has been noted in probability plots of the experimental histogram data especially at high temperatures where more of the tailing side is present. From such observations, we might then expect the number of tie molecules observed at fracture to be somewhat less than the fifty to ninety per cent estimated value.

From these observations of the model parameters one senses that the values which could be selected encompass a fairly broad spectrum. There is, of course, some interaction between the various model parameters and where one is incorrectly chosen, other parameters must also be chosen in error to help compensate or obtain a reasonable fit. In fact, it was found for any *one* type of experimental loading, a broad range of parameters could be chosen and adjusted such that a reasonable theoretical fit could be obtained. However, when these same parameters were used to predict other loadings, an extremely poor or inadequate fit was obtained for the majority of parameter choices. From these observations it was concluded that the only valid parameters would be those which successfully predicted the experimental results under several types of loadings. As a result of this observation the basic approach for determination of the parameters was as follows. Those parameters which were best known, either from experiment or theory, would be selected as base values. This included the distribution standard deviation value (S) obtained from EPR data, the modulus of elasticity (E_b) for a single Nylon 6 chain, and the rate theory collision parameter (w_0). The other four parameters, RC, RL, U_0, and γ, were then selected by comparison of the theoretical solution, using various combinations of the parameters with experimentally observed data from constant strain rate and constant stress (creep) tests of Nylon 6, No. 1, at room temperature. Four average experimental curves (two strain rate and two creep) were selected as the comparison standards. With these guidelines parameter values at room temperatures were then selected and the theoretical results plotted and compared with the experimental standards. To obtain the "best fit" of the experimental standards over 150 different combinations of the parameters, RL, RC, U_0, and γ, were computed and plotted as well as some variations of the parameters, S, E_b, and w_0. The ratio of number of chains, RC, was varied from ten to fifty; the ratio or original lengths RL was varied from one to ten; the activation energy, U_0, was varied from 45 to 82 kcal/mole; and the activation volume, γ, was varied from 0.5×10^{-24} to 200×10^{-24} cm^3. In order to more easily compare experimental free radicals versus time curves with theoretical free radical or chain scission versus time data, values between 0.5 to 2.5×10^{13} chains/cm^2 were selected as the total number of chains/cm^2 in the "critical flaw" region. After considerable analysis a group of parameters was selected as the "best fit" with the knowledge that moderate variations would not substantially degrade and possibly would improve the comparison with the experimental standards of Nylon 6, No. 1, at room temperature. Once these room temperature parameters were determined for Nylon 6, No. 1, they were used to predict behavior at other strain rates, creep, constant stress rate, and one frequenty (0.1 cycles/second) of cyclic stress fatigue. A very satisfactory correlation between experimental results and theoretical predictions from the model was found and has been presented (22). Further results with stress fatigue at different frequencies, constant strain (stress relaxation), and step strain tests have been conducted on Nylon 6, No. 1, at room temperature as shown in figs. 2, 3, 4, and 5. The original results from the creep tests are shown in fig. 6 for further discussion. Additional tests at room temperature and constant strain rate loadings were conducted on a second Nylon 6 material and also on Nylon 66. Best fit parameters from the model were determined for these materials also. Fig. 7 is a composite of the experimental observations and model predictions of these three types of nylon fibers, i.e. Nylon 6, No. 1, Nylon 6, No. 2, and Nylon 66, at room temperature and at two different strain rate loadings.

Table 2 summarizes the "best fit" parameters for these materials and the limiting range which resulted in a reasonably compatible fit of the model data to the experimental data for Nylon 6, No. 1.

The parameter range values were obtained by selectively varying the parameters and comparing

Table 2

Model Parameter	Nylon 6 No. 1 "best fit" value	Nylon 6 No. 1 parameter range	Parameter "potential" range	Nylon 6 No. 2 "best fit" value	Nylon 66 "best fit" value
S^1) (% strain)	1.25	1.25	no data available	1.44	1.74
RC	30	20–40	3–50	30	40
RL	5	4–10	2–100	5	4.5
E_b (lb/in.^{2})	3×10^7	3×10^7	2.84×10^7	3×10^7	3×10^7
w_0 (sec^{-1})	10^{13}	10^{13}	10^{12}–10^{13}	10^{13}	10^{13}
U_0 (kcal/mole)	67.5	65–75	43–82	67.5	67.5
γ (Å^3)	5	3–15	1–300	5	5
W^2) (chains/cm^2)	1×10^{13}	0.5–2.5×10^{13}	1×10^{11}–5×10^{14}	1×10^{13}	0.6×10^{13}

[1]) Obtained from EPR data and varies with temperature and material.
[2]) Arbitrary in theoretical model.

the variations with experimental data. The limiting values chosen were those range of values which the author considered from this analysis could be used to predict the experimental results without unreasonable degradation of the comparison. The "best fit" parameters, however, were considered to be the most compatible with the experimental results. An additional column in table 2 is noted as parameter "potential" range. The values for this column are tabulated from the previous parameter discussion and include the "potential" values that have been reported or discussed by other authors. It is interesting and important to note that the range of data presented by the authors is much more narrow than previous reported "potential" values.

Discussion

Some important observations between theoretical predictions and experimental tests are worthy of comment.

1. Fatigue tests (figs. 2 and 3)

Fatigue failure in materials have been the subject of much research and discussion. Most fatigue failure data contains a large amount of scatter which makes prediction difficult. Analysis of figs. 2 and 3 in the authors' opinion adds knowledge to this failure process for these high-strength polymer fibers. First, main backbone chain scission does not begin until approximately sixty per cent of the maximum stress is reached. Second, chain rupture occurs in cycles corresponding to the cyclic loading but slightly out of phase due to the time effects of the rate process failure theory. Third, the length of time and the magnitude of the peak stress half of the cycle is the most important factor in determining bond rupture. Fourth, actual failure of the material is determined by the "most critical" of the flaw regions and is extremely sensitive to time and magnitude of loading resulting in the scatter observed during fatigue failure.

The ability of the model to predict failure within a few cycles of experimentally observed failures was most gratifying considering that the failure parameters were obtained from experimental constant stress and constrain strain rate tests.

2. Stress relaxations (figs. 4 and 5)

The role of chain scission with regard to stress relaxation has been a subject of much debate. Probably in many materials, stress relaxation can be attributed in a large portion to viscoelastic time effects, i.e. chain slippage (secondary bond failure), unkinking, uncoiling or unfolding, but what portion, if any, and at what levels does chain scission contribute to such effects? Results from experimental and theoretical constant strain tests for strain levels above six per cent show that for these highly-oriented fibers, stress relaxation can almost wholly be accounted for by chain scission. Fig. 4 is an example of such tests for strain values of $11^1/_2$ and 13%. Fig. 5 is a step-strain test showing the increase in stress relaxation with the increase in bond rupture at high strain levels. Viscoelastic stress relaxation effects are observed in experimental tests at strain levels below six per cent where negligible chain scission and no stress relaxation is predicted by this model. The fact that some slight but

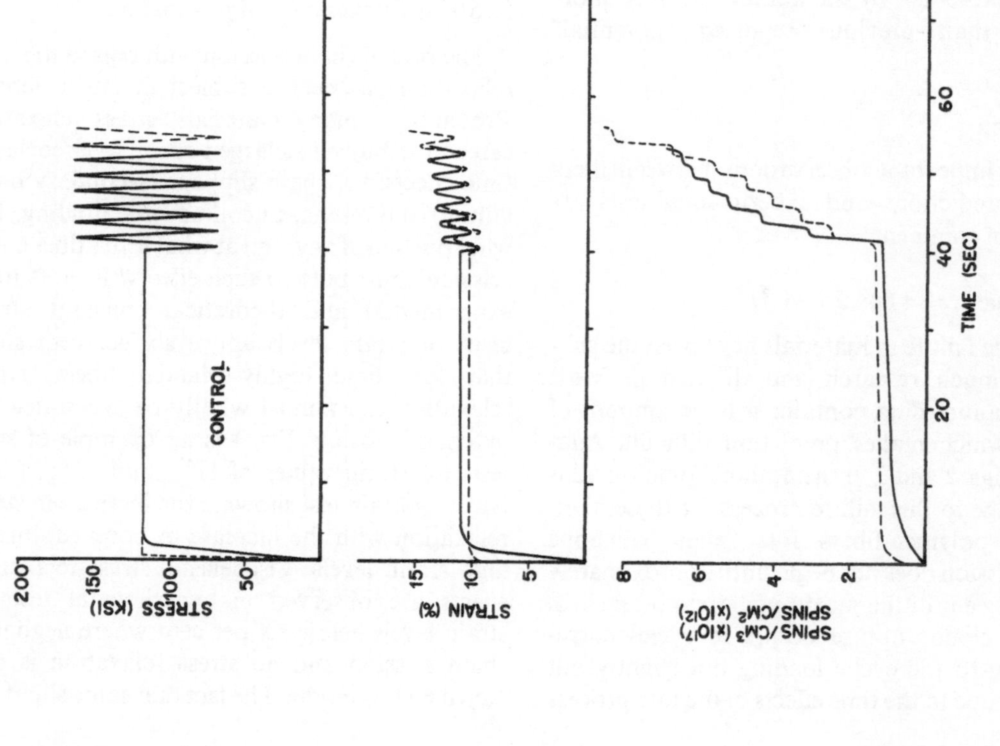

Fig. 2. Cyclic stress fatigue (0.05 cycle/sec), Nylon 6, No. 1, at room temperature; dashed line: theory (spins/cm^2); solid line: experimental (spins/cm^3)

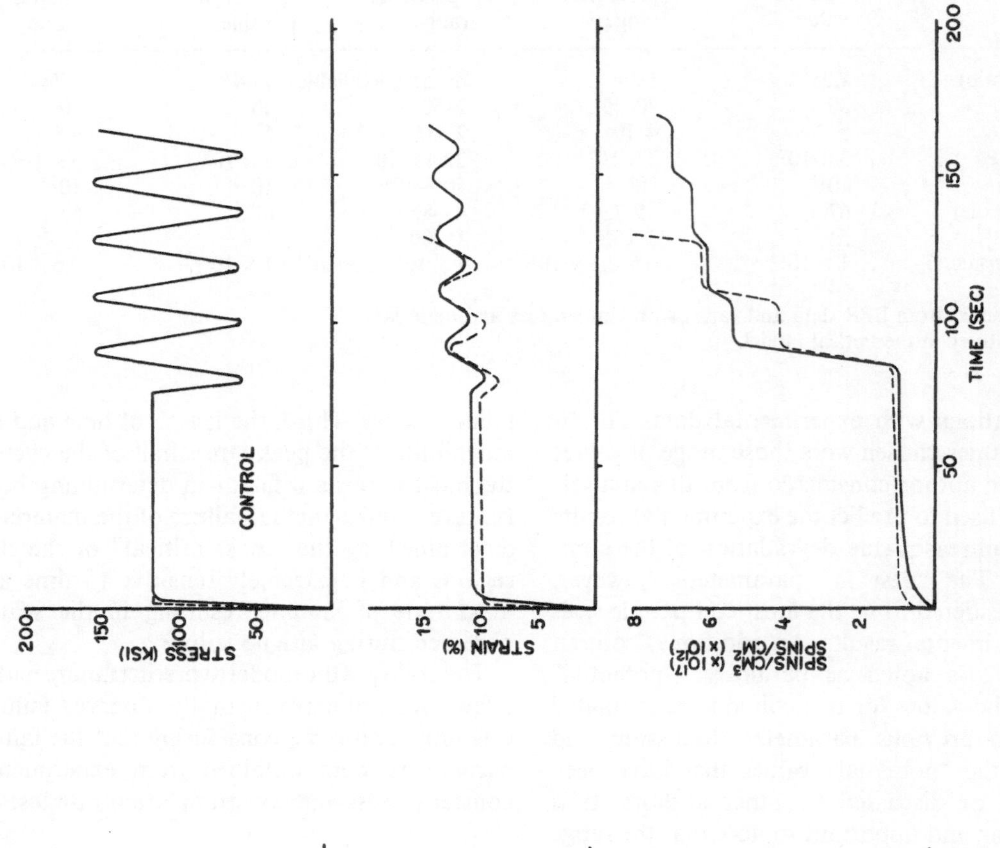

Fig. 3. Cyclic stress fatigue (0.5 cycle/sec), Nylon 6, No. 1, at room temperature; dashed line: theory (spins/cm^2); solid line: experimental (spins/cm^3)

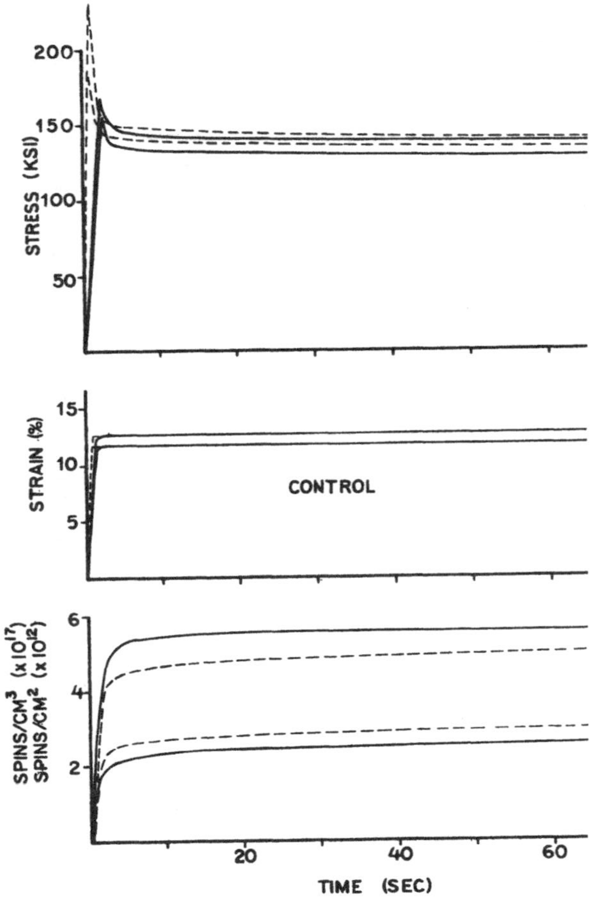

Fig. 4. Constant strain (stress relaxation), Nylon 6, No. 1, at room temperature; dashed line: theory (spins/cm^2); solid line: experimental (spins/cm^3)

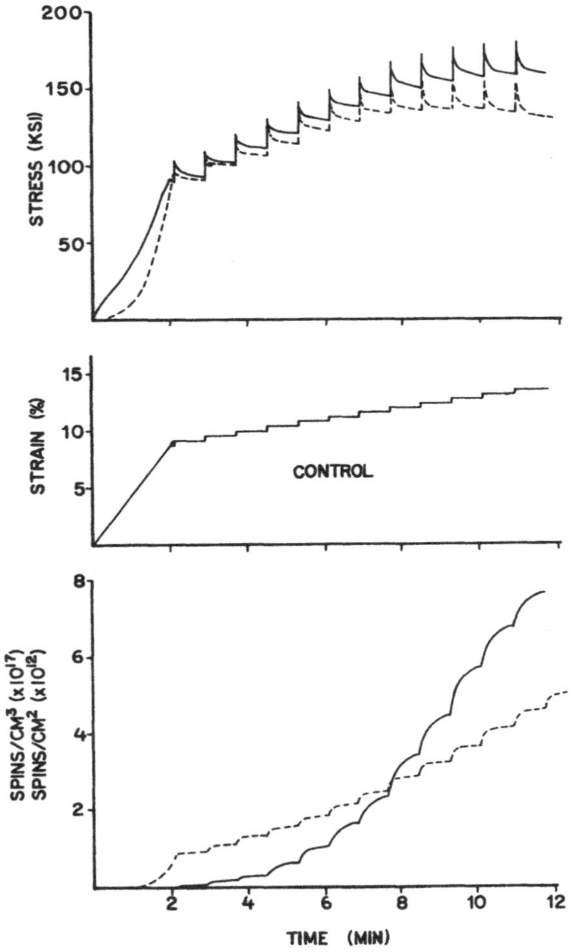

Fig. 5. Step-strain, Nylon 6, No. 1, at room temperature; dashed line: theory (spins/cm^2); solid line: experimental (spins/cm^3)

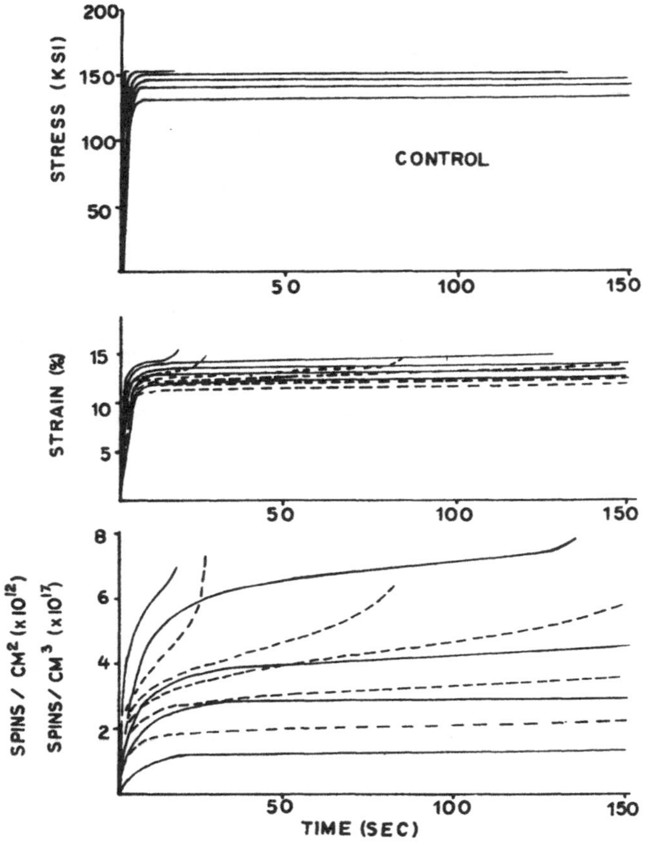

Fig. 6. Constant stress (creep), Nylon 6, No. 1, at room temperature; dashed line, theory (spins/cm^2); solid line, experimental (spins/cm^3)

measurable stress relaxation takes place at strains below that where measurable bond rupture occurs probably indicates that secondary bond rupture or "slippage" is occurring. Such effects must affect the stress distribution among the polymeric chains. Neglecting them in the analysis and in the subsequent curve fit to determine the model parameters in effect lumps viscoelasticity and slip in the kinetic parameters. The goodness of fit and "representative" values experimentally determined for these parameters leads us to conclude these secondary effects are comparatively rather small. The authors find the close correspondence between bond rupture and stress relaxation for the step displacement tests of figs. 4 and 5 particularly significant in this respect. The difference between the experimental and theoretical bond rupture observed in fig. 5 was the largest variation observed in any of the tests conducted. It was found that this correlation could be improved by slight adjustments in the rate parameters and the distribution function.

3. Creep (fig. 6)

The role of chain scission in creep is another interesting phenomenon. One might expect stress-relaxation due to viscoelastic effects to play a large part in creep. However, at stress levels near the fracture stress, chain rupture to relieve the stress can be a significant factor as shown in fig. 6. Both mechanisms, i.e. bond rupture and viscoelastic relaxation, most likely occur simultaneously. However, the model is based only on bond rupture; therefore, the model parameters have been chosen such as to compensate for both processes. In reality two sets of kinetic parameters exist, i.e. one for primary bond rupture and one for secondary bond rupture. Some preliminary work has been completed including viscoelastic effects with some success, especially in improving the initial growth comparison of bond rupture results in creep fatigue tests and stress relaxation tests.

4. Other materials (fig. 7)

The success of the model in predicting failure at other temperatures and with other types of Nylon 6 have been presented in detail in (22 and 23). In fig. 7, curves 1 and 2, are comparisons of experimental and theoretical results from Nylon 6, No. 1, and Nylon 6, No. 2, respectively, at two different strain rates. The basic similarity of these two fibers is evident from table 2 where only the apparent chain distribution values "S" parameter had to be altered in the model to obtain good agreement. The more narrow distribution of Nylon 6, No. 1, than that of Nylon 6, No. 2, makes it a stronger (greater maximum stress) but less tough material (other factors such as total number of tie chains, temperature, etc. being equal) than Nylon 6, No. 2. Such information about the relationship between the microstructure and physical properties is extremely valuable in fiber development and methods for changing this distribution have been investigated in this laboratory (11).

An important inquiry about the proposed model and failure criterion is whether the concepts can be extended to materials other than Nylon 6. Nylon 66 being similar in chemical composition but exhibiting several different physical properties was chosen for preliminary testing and evaluation. The apparent distribution width was measured by step strain tests at room temperature and experimental constant strain rate tests conducted at two different rates and also at room temperature (fig. 7). As noted in table 2 in order to make these correlations it was *not* necessary to modify the kinetic rate constants, U_0, w_0, and γ. As the basic chemical chain for Nylon 6 and Nylon 66 is similar and the EPR spectra is the same, we would assume that bond rupture is occurring in the same portion of the basic chain. Therefore, if reaction rate theory is a correct failure criterion, the rate constants should not have to be altered to obtain correspondence and only structural changes should be responsible for the difference in macroscopic properties. To obtain and understanding of the difference between the Nylon 6, No. 1, and Nylon 66 fibers, a theoretical curve for constant strain rate (curve 1 in fig. 8) was obtained using the Nylon 6 "best fit" parameters with the Nylon 66 effective distribution. Two important differences were noted between the theoretical and experimental curves. First, the theoretical strength was too high and, second, the number of spins or chain scission at fracture was too great. From these observations and other supporting evidence it was concluded that the flaw regions in this particular Nylon 66 sample material contained slightly fewer tie molecules than the corresponding Nylon 6 materials. Also the distribution for Nylon 66 appeared to be shifted toward higher strain values than that observed for the Nylon 6

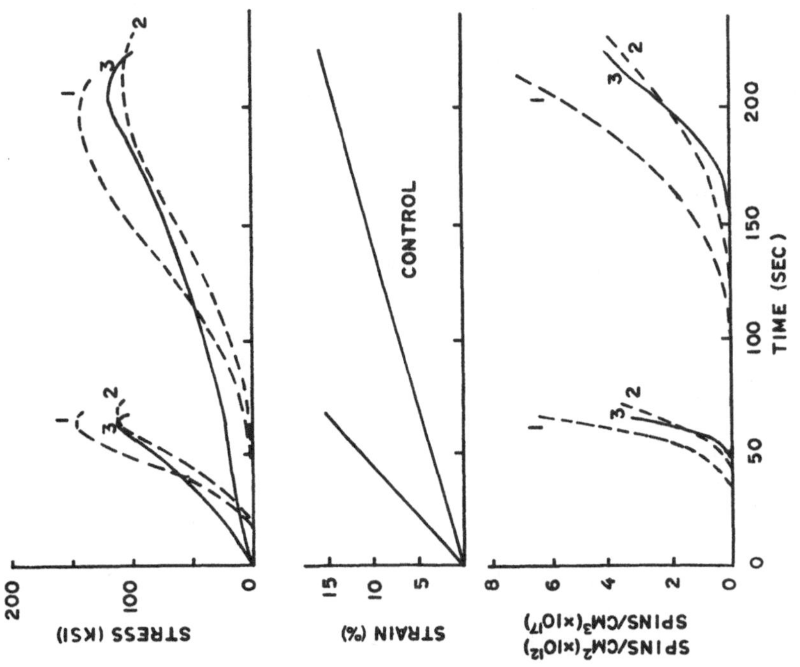

Fig. 8. Nylon 66 constant strain rate room temperature tests; curve 1, theory using Nylon 66 distribution and Nylon 6, No. 1 best fit parameters; curve 2, theory using Nylon 66 distribution and modified structural parameters; curve 3, experimental observations; theory (spins/cm^2); experiment (spins/cm^3)

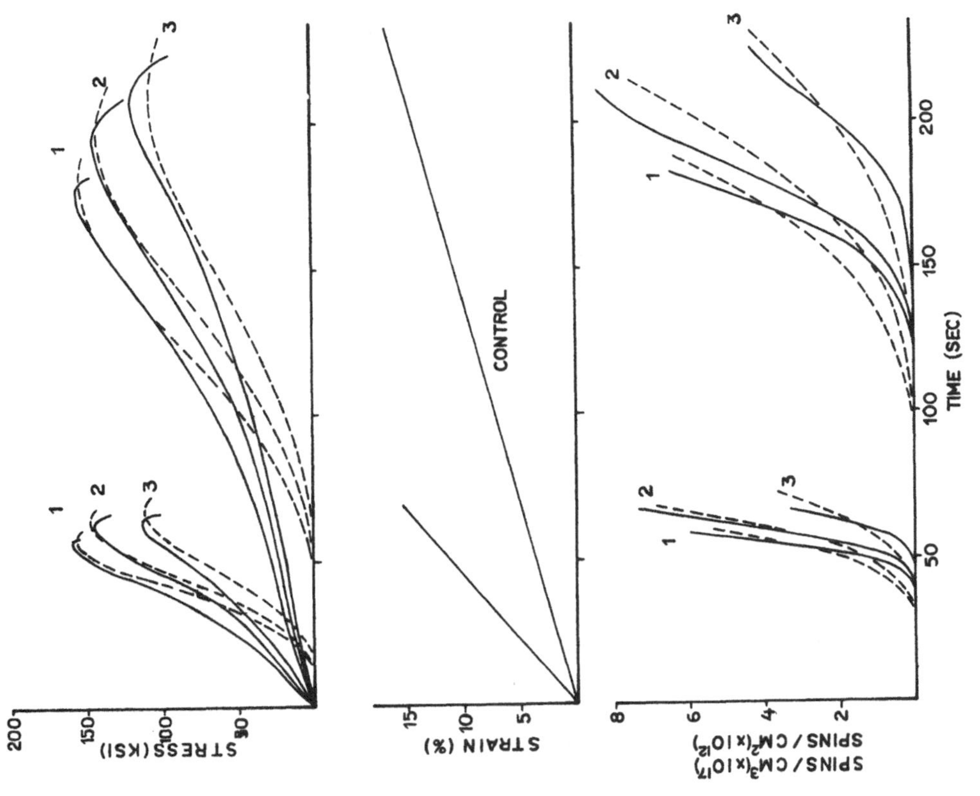

Fig. 7. Constant strain rate room temperature tests; 1 Nylon 6, No. 1, 2 Nylon 6, No. 2, 3 Nylon 66, dashed line, theory (spins/cm^2); solid line, experimental (spins/cm^3)

samples. One other comment is worthy of note. Even though the Nylon 6 and Nylon 66 are studied at the same temperatures, differences might be expected since they have different thermodynamic scales, i.e. do not have the same glass transition temperatures. Compensation for such effects was made in the model parameters, i.e. the total number of tie molecules in the flaw region was set at 0.6×10^{13} chains/cm^2 and the ratio of number of chains in the "cry-blk" to the critical region was increased to a *RC* value of forty. The relative position of the Nylon 66 distribution with respect to the Nylon 6 value was noted and the chain length ratio *RL* changed to 4.5. As already noted, the other model paramaters, i.e. the kinetic rate constants U_0, w_0, and γ along with the modulus of elasticity E_b, remain unchanged. These values would not be expected to change as the EPR spectra for Nylon 66 is the same as Nylon 6. With the above parameters the predicted behavior (curve 2 of fig. 8) compared very well with the experimental data (curve 3). A further interesting observation is that X-ray results from these same two fiber types (11) indicate that more chain folding occurs in the Nylon 66 samples. Such folding may account for the larger *RC* ratio value and slightly smaller total number of tie molecules in the flaw regions. These factors in turn result in a weaker fiber and less chain scission at fracture.

Conclusions

The goal of this study was to establish a microscopic failure criterion for highly-ordered semicrystalline polymers in fiber form which would relate microscopic occurrences to macroscopic behavior. The study was limited to a narrow and yet very important class of materials, the highly-oriented semicrystalline polymer fibers, specifically Nylon 6 and Nylon 66, where failure is suspected to be dominated by polymer chain scission. The mathematical model is an idealized model of the polymer fiber microstructure, representing mainly how the structure acts during the failure process and undoubtedly neglects many morphological intricacies.

Uniaxial tensile loadings were the only modes used in testing the fibers. Loading rates were relatively high and time-to-failure short (less than ten minutes) in order to minimize viscoelastic and free radical decay effects.

Viscoelastic time effects and entropic forces were not included in the mathematical model; however, portions of the curves attributed to this latter effect (22) have been treated with some success by other authors (47). Specifically, the major results are:

1. Reaction rate theory with a stress-aided activation energy is a successful failure criterion for predicting covalent bond rupture under mechanically induced stress. The key to the success of the theory lies in the propert interpretation and application in the stress σ. The stress, σ, is the stress on an individual bond and must be propertly related to the macroscopic applied loadings by the micro-macro structure relationships.

2. The mathematical model provides direct verification of the need for a distribution in length of tie molecules to explain experimentally observed strength and other fracture-related properties of these nylon fibers. EPR provided a method of experimentally characterizing the "apparent or effective" distribution and observing changes of differences in the distribution due to temperature, annealing treatments, or manufacturing processes. This apparent distribution was characterized from experimental EPR step-strain data and found to approximate a *Gaussian* or normal distribution criterion.

3. In addition to the effect of the apparent distribution on the strength and other failure characteristics of these fibers, other important microstructure concepts verified by the mathematical model were: a) chain scission is the main mode of failure; b) chain scission occurs almost exclusively in "critical" or "weak-link" regions distributed throughout the fiber; and c) failure properties and fiber strength at a given temperature depend on two microstructure properties – one, the existence of "weak-link" or "critical" regions that contain load-bearing tie molecules and two, the existence of an effective distribution in length of these tie molecules. Alternately, one might envision a distribution in the criticality of the region.

In conclusion it appears that the key to successful use of absolute reaction rate theory in predicting failure is an understanding of the changing distribution in stress among the polymer chains. A factor deserving further study, incidentally, is the size and distribution of these critical regions and means by which they might be controlled and/or modified.

The key to improving the properties of polymeric fibers, therefore, lies in modifying the alignment and effective length and/or number of tie chains. *Park, Statton*, and *DeVries* (48) have been rather successful in developing stretch-annealing treatments for use in bringing about such modifications. The techniques and models outlined herein provide a systematic means of analyzing the effect of such treatments.

After completion of the above outlined study the authors have become aware of another paper in the excellent series of papers by *S. N. Zhurkov* and his associates relating macroscopic and molecular fracture (35). In this extension of their earlier EPR studies they have made an "end group analysis" of oriented polymeric materials preceding and subsequent to fracture. In this article they report that the number of chains ruptured substantially exceeds the number of free radicals produced during fracture. The authors are currently trying to analyze the ramifications this would make in the proposed model and also are attempting to confirm the numbers by various other experimental methods. It appears that as long as the number of free radicals is proportional to the number of molecular chains ruptured, this would not seriously deter from the mathematical model.

Acknowledgments

Grateful acknowlegment is given to the National Science Foundation for their portion of the financial support of this work. Use was made of facilities purchased under a National Aeronautics and Space Administration grant. Thanks are also expressed to Drs. *J. B. Park, W. O. Statton, A. Peterlin*, and *H. H. Kausch* for their interest and suggestions. Gratitude is also expressed to Messrs. *Stephan Nichols* and *William H. Hassell* for their help in accumulating and reducing the data.

Summary

A model to describe bond rupture and fracture in polymeric fibers is described. An experimentally determined distribution in stresses is incorporated with absolute reaction rate theory in the model to predict bond rupture. Model predictions are compared with experimentally determined fracture and free radical concentrations for various loadings. The experimental parameters in the model are discussed and their "best fit" values given. In general, these values are found to compare quite satisfactorily with accepted values from theory or other tests in the literature.

References

1) *Hearle, J. W. S.* and *R. H. Peters*, Fiber Structure (Manchester 1963).

2) *Peterlin, A.*, Polymer Sci. **A2**, 7, 1151 (1969).

3) *Bonart, R.* and *R. Hosemann*, Z. Elektrochem. **64**, 314 (1960).

4) *Hearle, J. W. S.*, Polymer Sci. **C20**, 215 (1967).

5) *Boseley, D. E.*, Polymer Sci. **C20**, 77 (1967).

6) *Keller, A.*, Philosophical Magazine **2**, 1171 (1957).

7) *Fischer, E. W., H. Goddar*, and *G. F. Schmidt*, Makromolecular Chemistry **117**, 170 (1968).

8) *Dismore, P. F.* and *W. O. Statton*, J. Polymer Sci. **B2**, 1113 (1964) and **C13**, 133 (1966).

9) *Gubanov, A. I.* and *A. D. Chevychelov*, Soviet Physics-Solid State **4**, 4 (1962).

10) *Takayanagi, M., K. Imada*, and *T. Kajiyama*, Polymer Sci. **C15**, 263 (1966).

11) *Park, J. B.*, Fracture Morphology of Nylon 6 Fibers. Unpublished Ph. D. Dissertation, Division of Materials Science and Engineering, Department of Mechanical Engineering, University of Utah (Salt Lake City, Utah, 1971).

12) *Tobolsky, A.* and *H. Eyring*, J. Chem. Physics **11**, 125 (1943).

13) *Bueche, F.*, Physical Properties of Polymers (New York 1961).

14) *Gubanov, A. I.* and *A. D. Chevychelov*, Soviet Physics-Solid State **5**, 62 (1963).

15) *Rosen, B.*, ed., Fracture Processes in Polymer Solids (New York 1964).

16) *Zhurkov, S. N., A. Y. Savostin*, and *E. E. Tomashevskii*, Soviet Physics-Doklaky **9**, 968 (1964).

17) *Roylance, D. K., K. L. De Vries*, and *M. L. Williams*, Fracture. Ed. *P. Pratt* (London 1969).

18) *Kausch-Blecken von Schmelling, H. H.* and *J. Becht*, Rheol. Acta **9**, 137 (1970).

19) *Verma, G. S. P.* and *A. Peterlin*, Polymer Preprints **10**, 2, 1051 (1969).

20) *Kausch-Blecken von Schmelling, H. H.*, Internat. J. Fracture Mechanics **6**, 301 (1970).

21) *Kausch-Blecken von Schmelling, H. H.*, J. Macromol. Sci. – Rev. Macromol. Chem. **C4 (2)**, 243 (1970).

22) *De Vries, K. L., B. A. Lloyd*, and *M. L. Williams*, J. Appl. Phys. **42**, 4633 (1971).

23) *Lloyd, B. A., K. L. De Vries*, and *M. L. Williams*, J. Polymer Sci. **A2**, 10, 1415 (1972).

24) *Becht, J.* and *H. Fischer*, Kolloid-Z. **229**, 167 (1969).

25) *Roylance, D. K.*, An EPR Investigation of Polymer Fracture. An unpublished Ph. D. dissertation, Department of Mechanical Engineering, University of Utah (Salt Lake City, Utah, 1968).

26) *Becht, J., K. L. De Vries*, and *H. H. Kausch-Blecken von Schmelling*, European Polymer J. **7**, 105 (1971).

27) *Backman, D. K.*, An EPR Investigation of Polymer Fracture Surfaces. An unpublished Ph. D. dissertation, Department of Mechanical Engineering (University of Utah 1969).

28) *Backman, D. K.* and *K. L. De Vries*, J. Polymer Sci. **A1**, 7, 2125 (1969).

29) *Chevychelov, A. D.*, Polymer Mechanics **3**, 5 (1970).

30) *Lloyd, B. A.*, Fracture Behavior in Nylon 6 Fibers. Unpublished Ph. D. dissertation, Mechanical Engineering Department, University of Utah (Salt Lake City, Utah, 1972).

31) *Holmes, D. R., C. W. Bunn*, and *D. J. Smith*, J. Polymer Sci. **17**, 159 (1955).

32) *Geil, P. H.*, Polymer Single Crystals (New York 1963).
33) *Statton, W. O.*, Polymer Sci. **C18**, 33 (1967).
34) *Meinel, G.* and *A. Peterlin*, J. Polymer Sci. **A2**, 6, 587 (1968).
35) *Zhurkov, S. N., V. A. Zakrevskii, V. E. Korsukov,* and *V. S. Kuksenko*, Soviet Physics-Solid State **13**, 7, 1680 (1972).
36) *Hess, K.* and *H. Kiessig*, Z. Physik Chem. (Leipzig) **193**, 196 (1944).
37) *Statton, W. O.*, Polymer Sci. **41**, 143 (1959).
38) *Zhurkov, S. N.* and *E. E. Tomoshevskii*, Proceedings of the Conference on the Physical Basis of Yield and Fracture (Oxford/England, Oxford Univ. Press. 1966).
39) *Campbell, D.* and *A. Peterlin*, Polymer Letters **6**, 481 (1968).
40) *Treloar, L. R. G.*, Polymer 1. **5**, 95 (1960).
41) *Roberts, J. D.* and *M. C. Caserio*, Basic Principles of Organic Chemistry (New York and Amsterdam 1965).
42) *Gerö, L.*, Journal of Chemical Physics **16**, 1011 (1948).
43) *Zakrevskii, V. A., E. E. Tomashevskii,* and *V. V. Baptizmanskii*, Soviet Physics-Solid State **9**, 1118 (1967).
44) *Zhurkov, S. N.*, International Journal of Fracture Mechanics **1**, 311 (1965).
45) *Perepelkin, K. E.*, Polymer Mechanics **2**, 536 (1969).
46) *Peterlin, A.*, Personal communication.
47) *Pechhold, W.*, J. Polymer Sci. **C32**, 123 (1971).
48) *Park, J. B., W. O. Statton,* and *K. L. De Vries*, UTEC ME 71–122.

Authors' addresses:

Dr. *B. A. Lloyd*
Research Associate
Department of Mechanical Engineering
University of Utah
Salt Lake City, Utah (U.S.A.)

Prof. Dr. *K. L. De Vries* and Prof. Dr. *M. L. Williams*
College of Engineering
University of Utah
Salt Lake City, Utah (U.S.A.)

Rheol. Acta **13**, 367–376 (1974)

From the Department of Metallurgical and Materials Engineering, University of Pittsburgh, Pittsburgh (U.S.A.), and the Mellon Institute, Pittsburgh (U.S.A.)

The creep behavior of ideally atactic and commercial polymethylmethacrylate

By D. J. Plazek, V. Tan), and V. M. O'Rourke*

With 12 figures and 1 table

(Received October 27, 1972)

Introduction

It is perhaps surprising that the study of the time dependence of the mechanical behavior of such a common commercial polymer as polymethylmethacrylate, PMMA, at temperatures above its glass temperature, T_g, has been quite limited (1–4). Many investigations, too numerous to cite here, have been carried out on its mechanical behavior and other physical properties, mostly below its T_g. The general features of its stress relaxation behavior, including the great sensitivity of the rate of relaxation on the presence of absorbed moisture, have been described by *McLoughlin* and *Tobolsky*. It is not widely appreciated that the effect of absorbed water is present in most if not all polymers. The shift to shorter times or higher frequencies, of course, is far more significant in polar polymers. Scientific interest in the methacrylates of late has been spurred by the availability of samples with a wide range of stereochemical structures and the detailed knowledge of these structures afforded by the nuclear magnetic resonance, NMR, technique. The strong dependence of their T_g's on tacticity added immensely to their overall interest.

Our initial interest was birthed by what still may be the only polymerization which yielded an "ideally atactic" PMMA. At the Mellon Institute Dr. *Thomas G Fox* and his coworkers were studying polymerization mechanisms using NMR as their primary analytical tool (5, 6). The totally random PMMA was prepared by *Shiao-Ping Yen* in a two stage anionic polymerization. She also found that the thermal stability of PMMA is greatly dependent on the nature of the chain end groups. In contrast to commercial PMMA, which is about 76% syndiotactic, the "ideally atactic" material was shown by its NMR spectrum to have 50% syndiotactic placements, $p(S) = 0.50$ with 25% syndiotactic pairs, $p(SS) = 0.25$, 50% heterotactic pairs, $p(SI_v IS) = 0.50$ and a persistence ratio $\varrho = 1.00$. The stereochemical designations are given according to the statistical analysis scheme of *Coleman* and *Fox* (7). The results of creep studies on a sample of commercial or conventional PMMA and the atactic material are presented here as the initial phase of an investigation of the influence of stereochemical structure on viscoelastic behavior.

*) Data on commercial PMMA is incorporated in a thesis which has been submitted in partial fulfillment of the requirements for the degree of Master of Science in Materials Engineering, University of Pittsburgh, 1970.

Experimental

Torsional creep and creep recovery measurements were carried out *in vacuo* with a frictionless instrument employing a levitation magnetic bearing (8). Two fractions of the atactic sample were studied; the second cut, $Y[2]$, $\bar{M}_v = 1.99 \times 10^5$, and the third cut, $Y[3]$, $\bar{M}_v = 1.28 \times 10^5$ and $\bar{M}_n = 0.60 \times 10^5$. The commercial PMMA, which was precipitated from THF, was sent to us by *Robert E. Coulehan*. He reported that its $\bar{M}_v = 7.56 \times 10^5$ and that its $T_g = 121$ °C. The latter was a DSC determination made at a rate of heating of 20 °C/min. Simultaneous DTA determinations (1 °C/min) were made for us by *Lyle Chandler* of the Goodrich Chemical Co. of the $Y[3]$, the commercial PMMA and an anionic polystyrene, $M = 1.8 \times 10^6$ (Pressure Chemicals Co. Lot 14a). Repeated measurements of T_g were made in vacuo with different thermal histories. The initial determinations were 108, 119.5, and 104 °C for the atactic PMMA, the commercial PMMA and the polystyrene respectively. After holding the samples under vacuum at 170 °C for 68 h to remove absorbed moisture, T_g's of 110, 122, and 104.5 °C were obtained. Conventional dilatometric determinations of T_g carried out at Mellon Institute (5) yielded a value of 105.5 ± 1 °C for $Y[3]$. The cooling rate in the neighborhood of T_g was about 1°/h. From similar measurements on a series of conventional PMMA's, with molecular weights between 2.3×10^3 and 26×10^3 and $p(S) = 0.76$, a value of 117 °C is found to be appropriate for our commercial sample. The most significant fact to be noted from the above is that the commercial PMMA has a T_g about 11 °C higher than that of sample $Y[3]$.

Results

Creep and recoverable creep compliances

The plasticizing effect of water was found to be present after molding specimens *in vacuo*. Even after days of evacuation near T_g specimens with high surface to volume ratios, such as freeze-dried samples, which were subsequently molded at 180 °C, were still found to exhibit creep rates that decreased appreciably during prolonged *in situo* evacuation well above T_g. Only after

24*

three weeks of measurement with the specimen continually under vacuum conditions (c. 5×10^{-3} torr) at temperatures above T_g did the creep response become reproducible. Measurements at a series of temperatures from the neighborhood of T_g to 200 °C could then be made which yielded dependable temperature shift factors. Decreasing creep rates as a function of the time of residence above T_g are illustrated in fig. 1 with four creep curves obtained

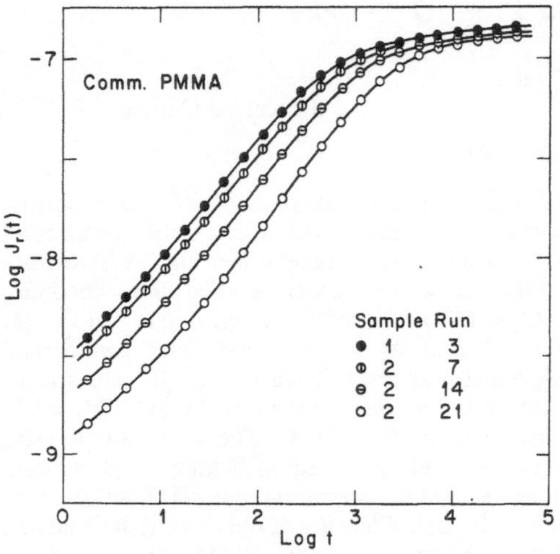

Fig. 1. Recoverable compliance data, $J_r(t)$, for commercial PMMA at 120 °C. Residence times of samples above T_g: Run 1–3, 3 days; 2–7, 7 days; 2–14, 12 days; 2–21, 21 days

at 120 °C from the commercial PMMA. Run 3 was made on our first sample after it was held above T_g for about three days; runs 7, 14 and 21 were made on our second sample after it was above T_g for 5, 12 and 26 days respectively. The curves indicate a decrease in creep rates of five fold between residence times of 3 and 26 days. We have no idea how fast a PMMA would creep were its water content in equilibrium with ambient humidity. The slow painful process of removing dissolved water as described above should be evidence for the hydrogen bonded condition of the last trace amounts. When measurements were made at temperatures below the established glass temperatures, repetitive measurements were made over a period of several days up to a week until creep rates were obtained that reflected the establishment of an equilibrium density.

The recoverable shear creep compliance, $J_r(t)$, cm^2/dyne, of the commercial PMMA is shown in fig. 2 as observed at seven temperatures between 114 and 189 °C. The $J_r(t)$ curve at 188.9 °C was obtained from a creep compliance curve, $J(t) = J_r(t) + t/\eta$, where the small viscous contribution reflected in the t/η term was subtracted. The viscosity value used was 2.8×10^{10} poise (dyne sec/cm^2). At the other temperatures either no appreciable viscous deformation was present or the compliance curves were obtained directly from recovery measurements.

Plots of each compliance curve which extends into the rubbery plateau between 10^{-7} and 10^{-6} cm^2/dyne, against the cube root of time, $t^{1/3}$, reveal linear sections that suggest that the dominant contribution to the deformation in the plateau region is *Andrade* creep (9–11), $J(t) = J_A + \beta t^{1/3}$. J_A and β are characterizing constants. In addition, the short time behavior at 114 °C exhibits the same functional form at compliances that are two orders of magnitude smaller. Retardation spectra shown later clearly indicate where *Adrade* creep dominates.

The $J_r(t)$ curves were shifted along the time axis to form a reduced curve representing extended behavior at the reference temperature, $T_0 = 120$ °C, see fig. 3. The quality of the reduction is surprisingly good in the light of the anomalous behavior exhibited by other methacrylates as observed by *Ferry* and his coworkers (12–15). The rubbery behavior in this case is well developed and extensive. Its level corresponds to a molecular weight per entanglement, M_{eN}, of 4700 or 94 chain backbone atoms per entanglement. These values are calculated from the rubbery plateau compliance, $J_{eN} = M_{eN}/\rho R T$. J_{eN} should be determined by 1. integrating over the peak of the retardation spectrum which is associated with the primary softening dispersion, 2. integrating over the corresponding peak in $J''(\omega)$, the dynamic loss compliance (16), or 3. by an extrapolation procedure utilizing viscoelastic data from a well-developed rubbery plateau (10). We have found that the *Andrade* intercept, J_A, yields values which are within 10% of those obtained by procedure 1. The rather high compliance seen in the glassy region of response coupled with this rather low rubbery level reveals a glass-like to rubber-like dispersion which yields only a 500 fold increase in recoverable compliance.

The creep and recoverable creep compliance curves obtained for the atactic PMMA fraction, $Y[3]$, are presented in fig. 4. Where the creep curves differ significantly from the recovery curves, the former are represented by dashed

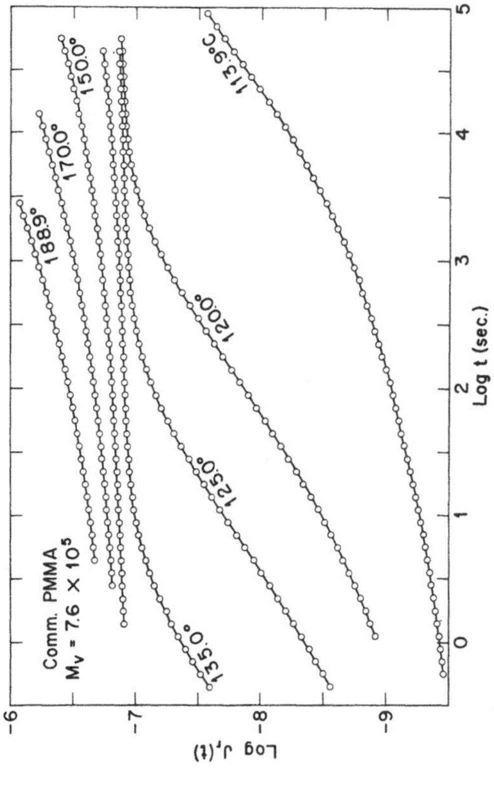

Fig. 2. Logarithmic presentation of recoverable compliance data, $J_r(t)$, (cm²/dyne), for commercial PMMA determined at the indicated temperatures

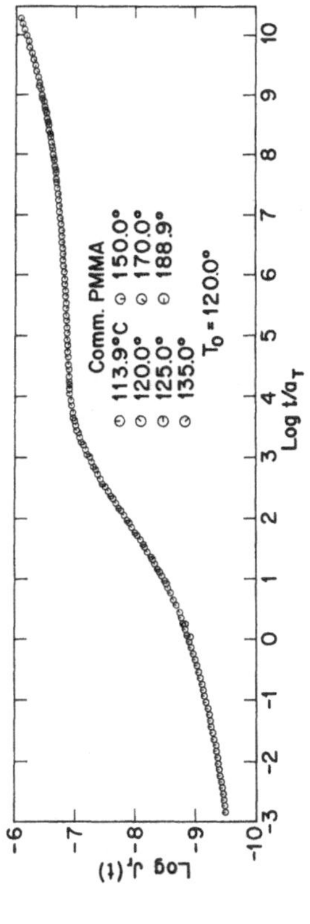

Fig. 3. The data of fig. 2 reduced to 120 °C by application of empirical time scale shift factors, a_T

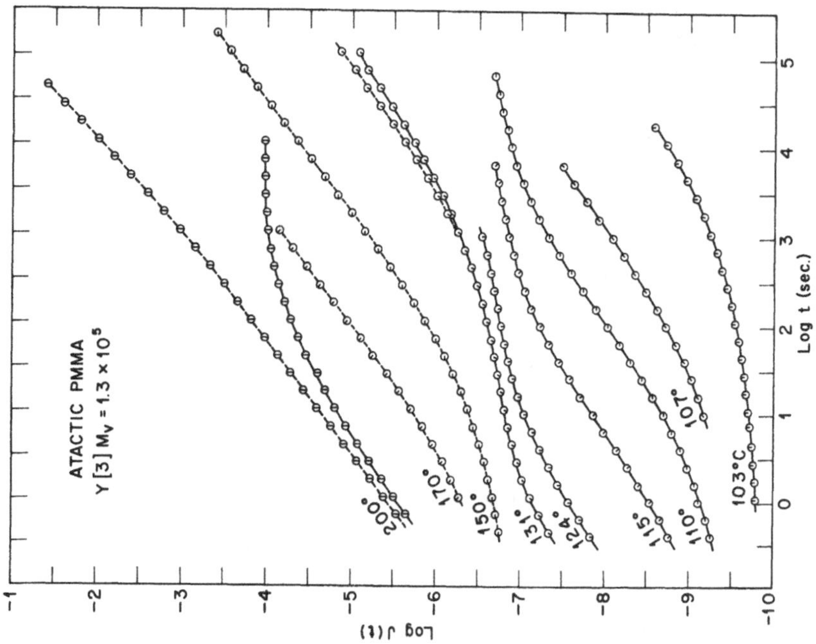

Fig. 4. Compliance data, $J(t)$, (cm²/dyne) for a sample of atactic PMMA. Dashed curves represent the creep compliance; solid curves represent the recoverable compliance, $J_r(t)$.

lines. The compliance curves at several of the nine measurement temperatures cover between five to six logarithmic decades of time. With the extensive overlap thus provided, a severe test of temperature reduction would normally be expected. The results of the temperature reduction procedure of the $Y[3]$ data from fig. 4 and the more limited data obtained on fraction $Y[2]$ are shown in fig. 5. Except for a minor divergence at short reduced times, reduction appears to be quite successful. In fact no difference in the temperature dependences of the recoverable compliance and the viscosity is discerned at high temperatures. We expect a difference within

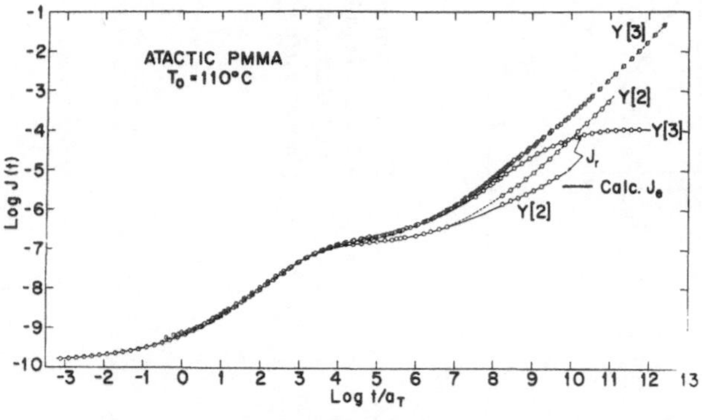

Fig. 5. A composite presentation of the reduced creep compliance (dashed lines) and the reduced recoverable compliance, $J_r(t)$, (solid lines) for two fractions of atactic PMMA: $Y[3] \sim \bar{M}_v = 1.3 \times 10^5$; $Y[2] \sim \bar{M}_v = 2.0 \times 10^5$

20 °C of T_g but the high viscosities exhibited by PMMA were deterrent enough for the present to discourage an investigation to establish the expectation. The rubberlike plateau of the higher molecular weight fraction $Y[2]$ is more extensive and well-developed as is expected. Two dispersions of nearly equal intensity are evident in the recoverable compliance, $J_r(t)$, behavior of the $Y[3]$. The substantial rise of one thousand fold from the rubbery plateau to the final level of the steady state compliance, J_e, is certainly almost entirely due to the distribution of molecular chain lengths (16). However the magnitude appears to be far in excess of that expected from *Ferry*s expression (16), $J_e = (2/5)\bar{M}_z \bar{M}_{z+1}/\bar{M}_w \varrho RT$ which is based on the *Rouse* theory result for the molecular weight dependence of J_e of monodisperse samples. The calculated result shown in fig. 5 was made on the assumption that fraction $Y[3]$ of the atactic PMMA has approximately a "normal" molecular weight distribution, where $\bar{M}_n : \bar{M}_w : \bar{M}_z : \bar{M}_{z+1} = 1:2:3:4$. Since $\bar{M}_v$ is not much less than $\bar{M}_w$ and $\bar{M}_v/\bar{M}_n = 2.1$ for $Y[3]$ which was obtained by means of a coacervation separation, the distribution should be unimodular with no unusual tails. It is now well established that above a molecular weight of 5×10^4 the *Rouse* inverse first power dependence of J_e on M is not observed in the behavior of bulk polymers. The difference of some thirty-fold between the observed J_e and the result of the rough calculation is therefore not surprising.

Temperature shift factors

The temperature shift factors, a_T, obtained from the reduction procedure applied to both sets of data are presented logarithmically as a function of the temperature in fig. 6. The temperature scales for this atactic and conventional

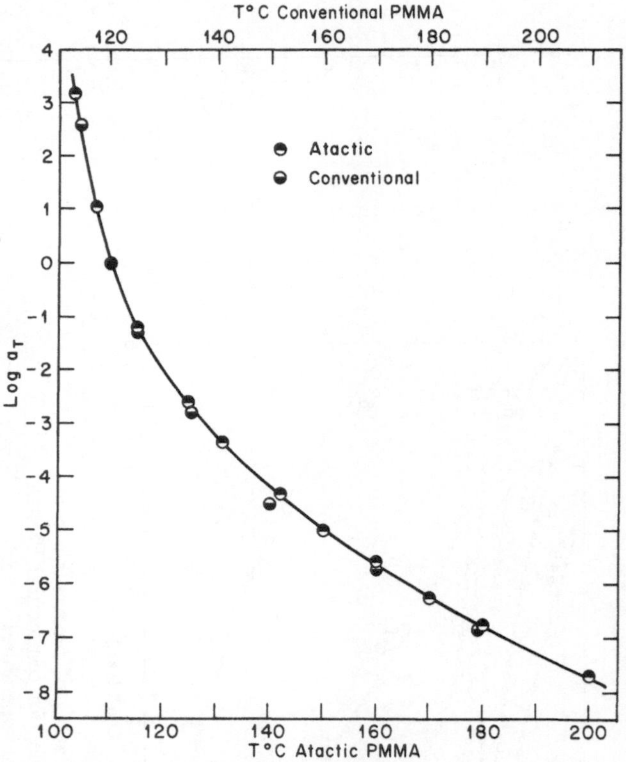

Fig. 6. Logarithms of the time scale shift factors, a_T, vs. temperature for two tacticities of PMMA. Superposition has been effected by offsetting the temperature scales 10 °C $\simeq \Delta T_g$

PMMA data are shifted so that the two reference temperatures coincide. The reference temperatures of 110 and 120 °C were chosen to allow for the difference in the T_g's of the two PMMA's. The eleven degree difference finally established as a better estimate would only slightly improve the coincidence of the temperature dependence data on the single curve drawn in fig. 6. This first comparison between the behavior the two materials with differing stereochemical structures indicates that their temperatures dependences are indistinguishable when compared at temperatures equidistant from their T_g's.

Data analysis

Retardation spectra

Second approximation calculations of the retardation spectra, L_2, cm^2/dyne, have been made on the reduced compliance curves presented above in figs. 3 and 5. The results are shown in fig. 7 as logarithmic functions of the

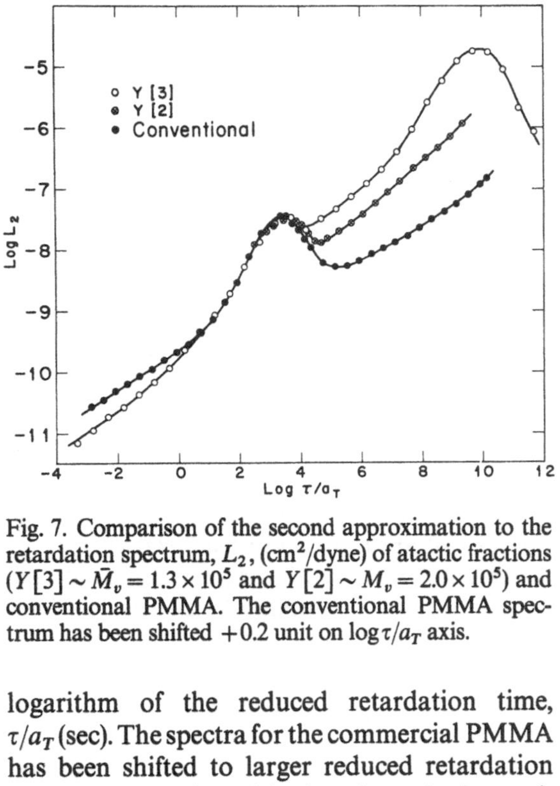

Fig. 7. Comparison of the second approximation to the retardation spectrum, L_2, (cm^2/dyne) of atactic fractions ($Y[3] \sim \bar{M}_v = 1.3 \times 10^5$ and $Y[2] \sim M_v = 2.0 \times 10^5$) and conventional PMMA. The conventional PMMA spectrum has been shifted +0.2 unit on $\log \tau/a_T$ axis.

logarithm of the reduced retardation time, τ/a_T (sec). The spectra for the commercial PMMA has been shifted to larger reduced retardation times by 0.2 of a logarithmic unit to obtain maximum coincidence of the curves. This is equivalent to a choice of the reference temperature for the commercial PMMA of about 119.2 °C. We wish to emphasize the fact that, when the glass temperature is being used as a standard state variable and comparisons are made between polymers near T_g, displacements of creep curves up to two logarithmic decades do not necessarily represent differences in behavior. At or just below T_g, a nonequivalence of T_g's of just 2 °C will cause a displacement of an order of magnitude. In addition, as mentioned above, the presence of moisture in polar polymers plays an influential role in determining creep rates. We have observed creep compliance curves, measured near the glass temperature of a highly syndiotactic PMMA, that have shifted three logarithmic decades under vacuum conditions.

The peaks observed at $\log \tau/a_T = 3$ must be considered to be experimentally equivalent. Viscoelastic mechanisms associated with this peak are evidently not influenced by tactic differences. This is tantamount to saying J_{eN} is insensitive to the stereochemical differences of the atactic and conventional samples. We find this surprising and do not expect this to be always true. At short times the divergence of the curves appears to be a significant effect of the stereochemical differences between the atactic and conventional PMMA's. Below $\log \tau/a_T = -0.5$ for the conventional PMMA and -1.5 for atactic fraction $Y[3]$ the $\log L_2$ curves have a slope of 1/3 which reflects the dominance of *Andrade* creep (17) at short times as the glassy level of response is approached. The *Andrade* β coefficient of the commercial PMMA clearly is higher than that of the atactic material.

To the right of the first peak we see what is in part a molecular weight and molecular weight distribution effect. As has been observed before in the behavior of polystyrene (18) and polyvinylacetate (19), relative to the viscoelastic response of the highest molecular weight sample measured, the compliance curve of each successively lower molecular weight sample deviates from a common curve at progressively shorter times. Although it has not yet been established, it is believed the dependence on the distribution follows a similar pattern. The terminal peak in the L_2 curve of fraction $Y[3]$, centered about $\log \tau/a_T \simeq 10$, is one of very few that have been measured to date. The corresponding peaks for the other two materials were not measured but are clearly at longer times as dictated by their higher molecular weights. The shape of the retardation spectrum of the conventional PMMA at long times beyond the measured peak is of particular interest because its slope on the log-log plot is 1/3 over 4 logarithmic decades. This indicates that the response in the rubber-like plateau regime is *Andrade* creep. It should be noted that the two "*Andrade* sections" of the commercial PMMA retardation spectrum do not extrapolate to meet one another. This fact played a suggestive role in our analysis of the temperature dependence observed.

Temperature dependences

The effect of temperature on the rate of creep is embodied in the shift factors, a_T, obtained in the reduction procedure. The a_T's of the commercial PMMA were chosen to illustrate the functionality which was shown to be independent of the stereochemical differences of the PMMA's studied. Since there has been no reported viscous or viscoelastic behavior of an amorphous high polymer that cannot be fitted to any of the free volume expressions which are equivalent to that of *Williams, Landel,* and *Ferry*, such a fit with the PMMA data was attempted. Because of its graphical convenience and clarity we choose to work with the *Vogel* equation (20–22) which for a_T shift factors is

$$\log a_T = \log A + C/(2.303)(T - T_\infty)$$

where C and T_∞ are characterizing constants and the constant A depends on the choice of the reference temperature, T_0. In fitting data, T_∞ is varied until $\log a_T$ becomes a linear function of $(T - T_\infty)^{-1}$. In fig. 8 curves resulting from three

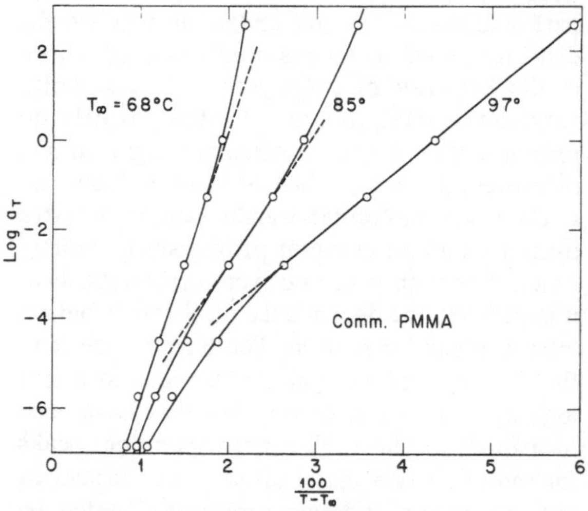

Fig. 8. Three attempts at *Vogel* linearization of the temperature dependence of the recoverable compliance time scale shift factors, a_T, of commercial PMMA

choices of T_∞, 68, 85, and 97 °C show that no value of T_∞ exists that will linearize the data. The data cannot be fitted to a single free volume expression. Two interpretations that are possible are 1. the rate of response of viscoelastic mechanims involved in the recoverable creep deformation are not determined by proposed free volume variations or that 2. all of the involved mechanisms do not have the same temperature dependence. If the latter is true, strict temperature reduction is precluded (16), but there is at least one example in the literature where excellent experimental superposition was achieved in spite of the violation of this principle. The stress relaxation measurements of *Nagamatsu* and *Yoshitomi* on polytrifluorochloroethylene (23), Kel–F, were reducible within experimental error. Their reduced curve encompassed a range in which two loss peaks are found. Since in all determined cases multiple loss peaks observed in the response of any crystalline polymer all have different temperature dependences, we assume that the temperature dependences of the two neighboring peaks of Kel–F are different. The shape of their *Arrhenius* plot ($\log a_T$ versus $1/T$ °K) appears anomalous and can be interpreted as indicating two energies of activation, with that associated with the higher temperature and/or long time relaxation having the higher value; as is the usual case.

The situation with PMMA is similar. An anomalous temperature dependence is coupled with a composite retardation spectrum: a symmetrical peak superimposed on a background arising from *Andrade* creep (we are presently investigating the proposition that the composition of the primary softening dispersion and the rubbery plateau for all high molecular weight amorphous polymers have the same qualitative features). A decomposition of the proposed composite response has been effected by analyzing the recoverable creep response at each temperature of measurement. *Andrade* β coefficients were determined. Subsequently the time dependent *Andrade* contribution to the compliance, $\beta t^{1/3}$, was subtracted from each curve. The resulting compliance curves $(J_r(t) - \beta t^{1/3})$ were reduced to a superposed curve. A set of shift factors were obtained from this reduction. A second set of shift factors were obtained from the β's (17) ($a_T = [\beta_0/\beta]^3$). The two sets of a_T are presented in fig. 9 where it can be seen that the temperature dependence of the remainder compliance is more severe than that of the *Andrade* mechanism, increasingly so with decreasing temperatures. The a_T's calculated from the β coefficients form a well-behaved curve with minimal scattering of the points; a fact which is surprising because all of the points, save one, were obtained from the rubbery level of re-

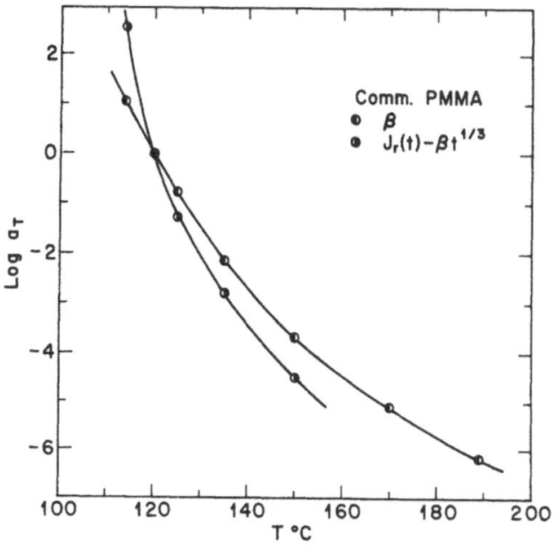

Fig. 9. The temperature dependence of the two different sets of time scale shift factors. Points identified by β are the shift factors $a_{T,\beta} = (\beta_0/\beta)^{1/3}$

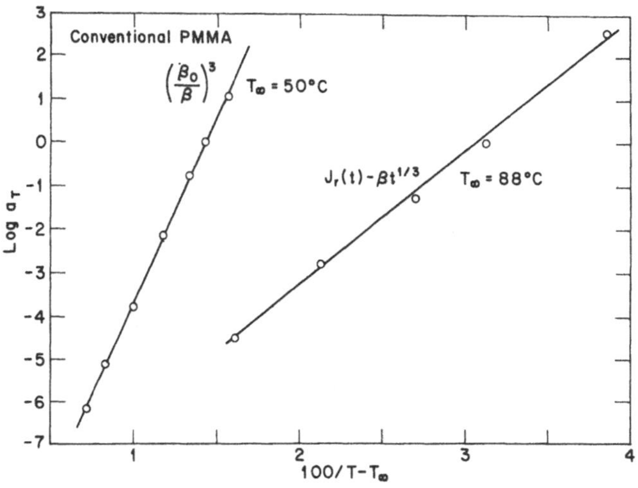

Fig. 10. The *Vogel* representation of the data of fig. 9

sponse. The exceptional point is that from the lowest temperature of measurement, 114 °C, where β is obtained from a near glassy level of compliance (below 10^9 cm^2/dyne). It can therefore be concluded that the two *Andrade* sections of L_2 represent the same group of viscoelastic mechanisms. The displacement of these two *Andrade* lines is a distortion caused by the reduction procedure which was apparently successful but in fact is fallacious. The reason for the excellent superposition is also evident. At higher temperatures the two dependences approach one another and in the rather flat rubbery plateau deviations are not discernible. Near the bottom of the primary softening dispersion some mismatch is present but the overlap of the curves from different temperatures, though substantial, is not sufficient to mar the illusion of successful reduction.

The justification of the decomposition is bolstered by the fact that the two sets of a_T factors can each be fitted satisfactorily to the free volume *Vogel* equation. The fits are shown in fig. 10. The scatter is greater with the points obtained from the remainder compliance curves but is to be expected when differences are involved. The equation for the *Andrade* mechanism shift factors is

$$\log a_T = -12.21 + \frac{856}{T - 50\,°\mathrm{C}}$$

and for the $J_r(t) - \beta t^{1/3}$ response it is

$$\log a_T = -\;\;9.34 + \frac{304}{T - 88\,°\mathrm{C}}.$$

Temperature and other characterizing parameters are presented in table 1.

Discussion

To allow the most direct comparison between the creep response of the atactic and conventional polymethylmethacrylates in the molecular weight independent regions, the creep compliance curves from the two lowest temperatures of measurement of each material are plotted in their reduced positions in fig. 11. The same

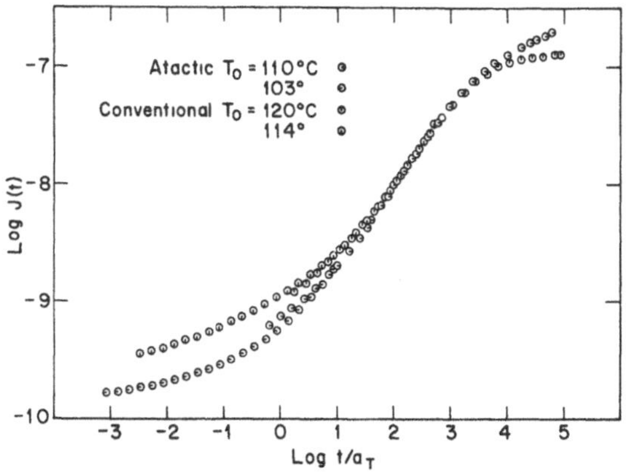

Fig. 11. A comparison plot of the reduced curves for the indicated materials displaying the creep behavior in the neighborhood of the respective glass temperatures. The conventional PMMA spectrum has been shifted +0.2 unit on logτ/a_T axis

shift of the commercial PMMA curves, 0.2 logarithmic decade to longer times, that was necessary to bring about maximum coincidence of the retardation spectra was also necessary here. The divergence at long times can be dismissed as being principally a molecular weight

Table 1. Characterizing parameters

Conventional PMMA

T °C	$\log a_{T,J_r(t)}$	$J_A \times 10^7$	$\log \beta$	$\log a_{T,\beta}$	$\log a_{T,(J_r-\beta t^{1/3})}$
113.9	2.58	0.00228	− 9.831	1.05	2.58
120.0[1])	0.00	–	− 9.481[2])	0.00	0.00
125.0	− 1.28	1.12	− 9.221	− 0.78	− 1.28
135.0	− 2.80	1.14	− 8.767	− 2.14	− 2.80
150.0	− 4.50	1.14	− 8.216	− 3.79	− 4.50
170.0	− 5.75	1.14	− 7.778	− 5.11	
188.9[3])	− 6.85	1.21	− 7.426	− 6.17	

Atactic PMMA Y [3]

T °C	$\log a_T$	T °C	$\log a_T$
103	3.16	142	− 4.31
107	1.07	150	− 5.00
110[1])	0.00	160	− 5.59
115	− 1.20	170	− 6.25
124	− 2.60	180	− 6.73
131	− 3.35	200[4])	− 7.69

[1]) Reference temperatures for reduction.
[2]) Interpolated value.
[3]) $\eta(188.9\,°C) = 2.8 \times 10^{10}$ poise.
[4]) $\eta(200\,°C) = 1.2 \times 10^{6}$ poise.

effect. The divergence at short times however, represents, we believe, a variation which can be attributed to the stereochemical difference of the two materials. The corresponding divergence of the retardation spectra seen in fig. 7 yields information about only one of two differences embodied in the curves in fig. 11. The classical expression for the creep compliance is

$$J(t) = J_g + \int_{-\infty}^{\infty} L(1 - e^{-t/\tau})\, d\ln\tau + \frac{t}{\eta}$$

where η is the shear viscosity in poise and J_g is the instantaneous time independent glassy compliance which arises from the stretching and bending of primary covalent and secondary bonds. In practice the short time limiting glassy compliance of a primary dispersion will be the long time limiting value of a secondary dispersion. Such a J_g may be largely determined by contributions to the deformation arising from side group motion. The divergence in L reflected the different levels of the *Andrade* creep contribution but gave no information about J_g. If *Andrade* plots of $J(t)$ curves determined at 114 and 103 °C for the conventional and atactic PMMA's respectively are made, both differences are clearly seen; see fig. 12. Being eleven degrees apart these curves are certainly within one degree of being at

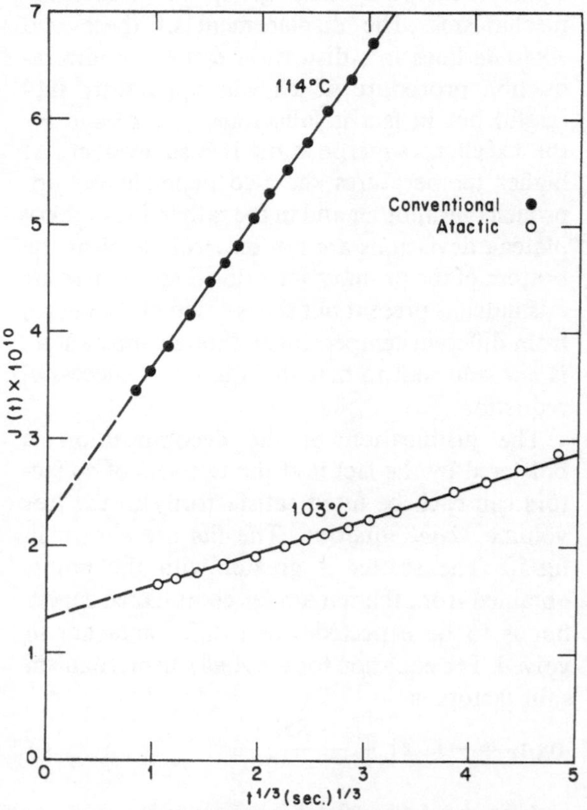

Fig. 12. An *Andrade* plot of the creep compliance, $J(t)$, (cm^2/dyne) for PMMA samples of different tacticities, both measured about 3 °C below their respective glass temperatures

equivalent temperatures. The equations for these lines are:

$J(t) \times 10^{10} = 2.28 + 1.47\,t^{1/3}$,

conventional PMMA, 114 °C:

$J(t) \times 10^{10} = 1.33 + 0.304\,t^{1/3}$,

atactic PMMA, 103 °C.

The β coefficients indicate that the creep rate of the commercial PMMA is 4.8 times faster than that of the atactic PMMA. In addition the *Andrade* intercepts, $J_A = J_g$, indicate that the commercial PMMA glass is 1.7 times softer than the atactic glass. These intercepts represent the long time limiting compliance of the PMMA beta dispersion.

In summary, the data presented here indicate that the stereochemical differences of the atactic and 76% syndiotactic commercial PMMA:

1. Have no effect on the temperature dependence of viscoelastic processes beyond a shift in rates that corresponds to the shift of T_g.

2. Have no effect on the symmetrical peak found in the retardation spectrum in the region of time scale that reflects the primary softening dispersion.

3. Do effect the level of an *Andrade* creep contribution to the deformation; that of the commercial PMMA being higher.

4. Result in a larger glassy compliance of the commercial PMMA, which suggests that the strength of its beta dispersion is greater.

Acknowledgements

The work carried out at Mellon Institute was partially supported under USAF Contract No. AF33(657)-10661 and that carried out at the University of Pittsburgh was partially supported under the National Science Foundation Grant GK16530.

Summary

The shear creep behavior of polymethylmethacrylate, PMMA, samples has been studied in the neighborhood of and above their glass temperatures. One of the materials studied was "ideally" atactic with equal numbers of random isotactic and syndiotactic placements, while the other was a commercial or "conventional" PMMA which was about 76% syndiotactic. The glass temperatures, T_g, were found to be 106 and 117 °C respectively. Evacuation above the glass temperature for several weeks was necessary before reproducible creep compliance, $J(t)$, curves could be obtained. It is believed that absorbed water plasticized the polar materials and its removal led to the shifting of the $J(t)$ curves to longer times. For both materials apparently successful temperature reduction was found to be possible within the temperature range of our investigations, i.e. up to 200 °C. Retardation spectra were calculated from the reduced curves and are compared. The temperature dependences, as described by the time scale shift factors, a_T, were similar when allowance is made for the different glass temperature. Both a_T curves could not be fitted to the *Williams*, *Landel*, and *Ferry*, WLF, free volume expression. These are the first examples of such a deviation for amorphous high polymers. It is proposed that the primary softening dispersion has two distinctly different groups of viscoelastic mechanisms contributing to it. On this basis the primary dispersion was decomposed into the two contributions. Both of the resulting temperature dependences were satisfactorily fitted to the WLF equation. Differences in the retardation spectra are noted. The glassy compliance of the commercial PMMA appears to be about twice that of the atactic PMMA.

References

1) *McLoughlin, J. R.* and *A. V. Tobolsky*, J. Colloid Sci. **7**, 555 (1952).
2) *Bueche, F.*, J. Appl. Phys. **26**, 738 (1955).
3) *Iwayanagi, S.*, J. Sci. Research Inst. (Japan) **49**, 4 (1955).
4) *Coulehan, R. E.*, private communication.
5) Technical Documentary Report No. ML-TDR-64-286, Part III (1966), Air Force Materials Laboratory, Wright-Patterson Air Force Base, Ohio 45433.
6) *Berry, G. C.* and *T. G. Fox*, Fortschr. Hochpolym. Forschg. **5**, 261 (1968).
7) *Coleman, B. D.* and *T. G. Fox*, J. Polymer Sci. Part A **1**, 3183 (1963).
8) *Plazek, D. J.*, J. Polymer Sci A2, **6**, 621 (1968).
9) *Andrade, E. N. da C.*, Proc. Roy. Soc. (London) **A84**, 1 (1910); Phil. Mag. **7**, 2003 (1962).
10) *Plazek, D. J.*, *W. Dannhauser*, and *J. D. Ferry*, J. Colloid Sci. **16**, 101 (1961).
11) *Reid, D. R.*, Brit. Plastics (Oct. 1959).
12) *Newlin, T. E.*, *S. E. Lovell*, *P. R. Saunders*, and *J. D. Ferry*, J. Colloid Sci. **17**, 10 (1962).
13) *Berge, J. W.*, *P. R. Saunders*, and *J. D. Ferry*, J. Colloid Sci. **14**, 135 (1959).
14) *Saunders, P. R.*, *D. M. Stern*, *S. F. Kurath*, *C. Sakoonkim*, and *J. D. Ferry*, J. Colloid Sci. **14**, 222 (1959).
15) *Stern, D. M.*, *J. W. Berge*, *S. F. Kurath*, *C. Sakoonkim*, and *J. D. Ferry*, J. Colloid Sci. **17**, 409 (1962).
16) *Ferry, J. D.*, Viscoelastic Properties of Polymers, 2nd Ed. (New York 1970).
17) *Plazek, D. J.*, J. Colloid Sci. **15**, 50 (1960).
18) *Plazek, D. J.* and *V. M. O'Rourke*, Presented at the Fifth International Congress on Rheology, Kyoto, Japan, Oct. 9 (1968).

19) *Ninomiya, K.* and *J. D. Ferry*, J. Phys. Chem. **67**, 2292 (1963).

20) *Vogel, H.*, Physik Z. **22**, 645 (1921).

21) *Williams, M. L., R. F. Landel*, and *J. D. Ferry*, J. Amer. Chem. Soc. **77**, 3701 (1955).

22) *Plazek, D. J.* and *V. M. O'Rourke*, J. Polymer Sci. A2, **9**, 209 (1971).

23) *Nagamatsu, K.* and *T. Yoshitomi*, J. Colloid Sci. **14**, 377 (1959).

Authors' addresses:

D. J. Plazek and *V. Tan*
Department of Metallurgical and Materials Engineering
University of Pittsburgh
Pittsburgh, Pennsylvania 15213 (U.S.A.)

V. M. O'Rourke
Mellon Institute
Pittsburgh, Pennsylvania 15213 (U.S.A.)

Rheol. Acta 13, 377–380 (1974)

Commissariat à l'Energie Atomique, Sevran (France)

Étude des matériaux agrégataires par ultrasons avec modélisation de texture et de comportement

Par J. Sorel, S. Poulard, J. Roucou et G. Lucas

Avec 3 figures

(Reçu p. p. le 27 octobre 1972)

1. Les modèles de comportement

Il s'agit dans la première partie de cet exposé de montrer dans quelle mesure nous pouvons modéliser le comportement élastique ou viscoélastique d'un certain matériau agrégataire constitué d'une phase dense prédominante de cristaux organiques et d'un système thermoplastique jouant le rôle de matrice.

Des expériences par ultrasons ont permis la détermination des modules élastodynamiques dans le domaine élastique ou viscoélastique du matériau et nous ont conduits à examiner plusieurs modèles de comportement, soit tout d'abord le modèle d'*Hashin* et *Shtrikman* (1) à deux phases et à géométrie quelconque. Ce modèle permet de déterminer deux bornes entre lesquelles sont compris les modules élastiques (ou encore les modules des modules imaginaires dans le cas viscoélastique). Comme la borne inférieure est deux fois supérieure aux valeurs expérimentales que nous discuterons dans la deuxième partie, nous avons pensé que l'interface matrice-cristal et la porosité du matériau, mise en évidence par ailleurs, pouvaient jouer un rôle important. Sachant que cette porosité est de l'ordre de quelques pour cents en volume, on aurait pu utiliser le modèle de *Mackenzie* (2); cependant la porosité se trouve uniquement dans la matrice (comme le montrent des photos réalisées sur Stéréoscan) où elle peut atteindre 40% (fig. 1).

Nous avons donc cherché à déterminer l'influence de la forme des inclusions (qu'elles soient vides ou non).

Les idées d'*Eshelby* (3) reprises par *T. T. Wu* (4) ont été appliquées à notre matériau en supposant les inclusions orientées de façon quelconque grâce au calcul effectué par *Krôner* (15).

Fig. 1. Matériau agrégataire (Stéréoscan G = 1000)

En supposant une porosité sphérique dans la matrice, des défauts d'adhésivité matrice-cristal en forme de disques et enfin des cristaux sphériques, nous avons obtenu des modules plus faibles que ceux issus du modèle d'*Hashin* et *Shtrikman*, mais encore très supérieurs aux valeurs expérimentales.

Ainsi, même si la forme de la porosité intervient, son importance relative a une influence prédominante que les modèles développés par *Hashin, Christensen, Paul, Yeh, Mackenzie* et *Eshelby* sont incapables de décrire (5, 6, 7, 8, 2, 3).

C'est pourquoi nous devons distinguer la porosité ouverte de la porosité fermée.

Le modèle de *Hashin* et *Rosen* (9) sur les composites fibreux nous a permis d'introduire une porosité ouverte dans la matrice et d'obtenir un accord qualitatif avec les valeurs expérimentales.

Ce dernier modèle, s'il caricature notre matériau, a l'avantage de mettre l'accent sur l'influence de la porosité ouverte sur les modules.

C'est pourquoi nous devons déterminer d'autres fondements pour obtenir une théorie générale. *Wareen*, *Nagarajan*, *Cowin* (10, 11, 12) ont jeté chacun les bases thermodynamiques d'une telle théorie pour des matériaux simples.

Il paraît encore exclu de trouver une telle théorie sur des matériaux aussi complexes que le nôtre.

D'autre part, cette complexité provoque des difficultés considérables dans l'interprétation des mesures effectuées.

Ce sera l'objet de cette seconde partie.

2. Limites des mesures par ultrasons sur un matériau agrégataire

Ces limites sont d'abord liées aux conditions de mesure

– Dans la technique par immersion en utilisant une platine goniométrique, l'éprouvette est placée dans le champ lointain de l'émetteur, à la limite du champ proche. Celui-ci doit être homogène, cylindrique et parfaitement centré sur l'axe de l'émetteur. Le diamètre φ_2 du récepteur est environ deux fois le diamètre φ_1 de l'émetteur avec $\varphi_1 \gg$ longueur d'onde dans le fluide.

– Durant la mesure, le fluide transmetteur de l'onde doit être stabilisé et régulé au dixième de degré près. Un agent mouillant est nécessaire.

– La dimension transversale de l'éprouvette est environ trois fois le diamètre du récepteur, afin de ne pas être gêné par les effets de bord dans la transmission de l'onde à l'angle critique (mesure de C_T à la transmission totale de l'éprouvette).

– L'éprouvette a ses faces parallèles à mieux de 2.10^{-4} rd, et polie au papier 600 Å et à la poudre de diamant. Ce polissage doit être absolument reproductible d'une éprouvette à l'autre.

La nature même du matériau limite nécessairement le domaine de mesures en fréquence et température et impose des conditions opératoires précises.

– La théorie de propagation des ondes dans les milieux continus homogènes et stables impose des dimensions infinies (ou pratiquement très grandes devant la longueur d'onde dans le milieu).

Comme notre matériau est un filtre passe bas, on doit trouver un moyen terme à son épaisseur d selon l'axe de propagation de l'onde. Nous reviendrons sur l'incidence du choix d'une épaisseur sur les erreurs de mesures.

La granulométrie $\bar{G}$ des cristaux impose une fréquence de coupure F_C, soit:

$$F_C \leq \frac{C_T}{2\bar{G}} \simeq 2\,\text{MHz}$$

comme limite supérieure au domaine de fréquences. La limite inférieure F_i étant dictée par le rapport $\frac{\text{épaisseur}}{\text{longueur d'onde}}$.

Mais à ces limites, il vient se superposer une autre condition, c'est-à-dire la définition du domaine de fréquences qui caractérise plutôt les pertes viscoélastiques par viscosité dynamique de la matrice, puisque le comportement d'un tel matériau est principalement commandé par la nature de la matrice et des interfaces matrice-cristal.

En nous inspirant d'une étude de *Merkulov* (13), nous avons choisi les longueurs d'onde extrêmes suivantes pour caractériser les viscosités:

$$\lambda_i = 4\bar{G} \leq \lambda \leq \lambda_\delta = 8\bar{G}$$

soit pour le domaine de fréquence:

$$0{,}9 \leq F \leq 2\,\text{MHz}.$$

D'autre part, il existe une notion importante qui est souvent négligée, c'est la densité locale d'un matériau qui est impossible à mesurer. Pour un tel matériau agrégataire, le volume pris en compte par le faisceau doit avoir des dimensions appropriées, afin que les résultats sur une éprouvette rendent compte des caractéristiques moyennes du matériau. Ce sera le diamètre du faisceau ultrason qui restera à optimiser, et, pour mieux faire, une épreuve statistique est nécessaire.

Enfin, d'autres conditions comme le recuit des éprouvettes et d'autres mesures préliminaires comme l'évaluation de la porosité sont nécessaires.

Des mesures par ultrasons sur des éprouvettes immergées dans une huile au Silicone ont montré que ce matériau comportait une porosité ouverte d'environ 2%.

3. Mesures des vitesses et amortissements

– Dans le coefficient de propagation $(\alpha + i\beta)$ de l'équation générale de propagation

$$U(x, t) = U_0 \exp[-(\alpha + i\beta)x] \exp i\omega t$$

où α est le terme d'amortissement et β est le nombre d'onde avec C_φ vitesse de phase égale à ω/β. On constate que ce formalisme est lié à une seule fréquence pour une onde monochromatique de durée infinie.

Dans le cas des mesures il n'en va pas de même et on utilise en général des paquets d'onde de durée et de forme quelconque.

Dans nos essais, on constate que l'onde n'est pas trop déformée à la traversée de l'éprouvette, mais qu'un léger glissement vers les basses fréquences intervient.

Dans ces conditions, on ne peut pas être certain de mesurer la vitesse de phase puisque celle-ci dépend de la référence de l'oscillation dans le paquet d'ondes dans la technique de superposition d'échos qui, grâce à un générateur d'écart de temps, nous permet d'apprécier des temps de transfert à ± 100 ns.

On dira qu'on mesure la vitesse de propagation C du signal en prenant comme référence des temps une oscillation de ce signal, qui conserve grosso modo la fréquence d'émission tout en subissant une atténuation.

Waterman (14) a étudié le problème de la propagation des ondes dans les mileux absorbants. Il introduit pour un milieu viscoélastique linéaire, une vitesse de propagation complexe:

$$c^* = C_1 + iC_2$$

et il montre que (c^*) est lié à la tangente de perte par:

$$|c^*| = C_1(1 + \mathrm{tg}^2\,\partial/2)^{1/2}.$$

De plus, il montre que la vitesse d'énergie a une signification physique pour de tels matériaux et que

$$C_1 \equiv C_{EN}.$$

Nous avons introduit la notion de vitesse d'énergie dans le calcul de la largeur de bande d'un paquet d'onde de forme gaussienne qui traverse un matériau viscoélastique. Et pour notre matériau nous trouvons que lorsqu'on mesure la vitesse du signal à 0,3 % près, la vitesse d'énergie est entachée d'une erreur de 3%.

A partir de l'étude de *Waterman*, pour notre matériau on calcule un écart de 2% entre la vitesse de signal et la vitesse d'énergie pour les ondes L.

– Une autre cause d'erreur vient de l'épaisseur de l'éprouvette qui, compte tenu des remarques faites dans la première partie, n'est pas optimale.

Les variations de vitesse longitudinale en fonction de l'épaisseur sont croissantes et asymptotiques à une valeur C_L qui correspond à $d/\lambda \sim \sigma$. Par rapport à l'épaisseur de nos éprouvettes on fait donc une erreur systématique de 3%.

Quant le matériau est monoréfringent en ondes T et pour différentes épaisseurs, on constate comme on pouvait s'y attendre que les variations d'épaisseur n'ont plus d'influence sur la correction des vitesses C_T (fig. 2).

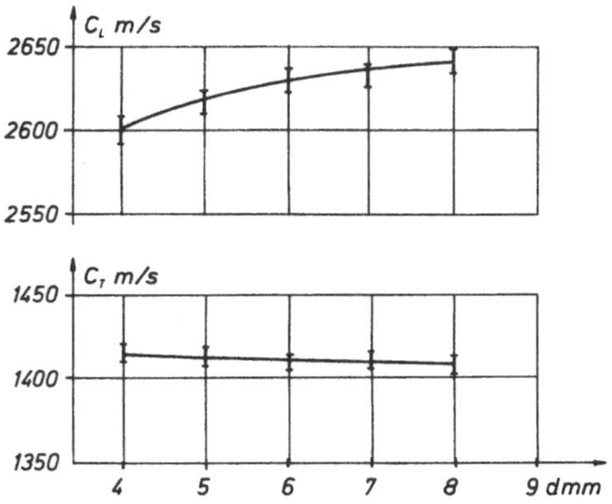

Fig. 2. Variations des célérités C_L et C_T à 2 MHz et à 20°C en fonction de l'épaisseur de l'éprouvette

– Une autre cause d'erreur, qui n'est malheureusement pas chiffrable est l'écart au parcours moyen apparent de propagation.

En incidence normale, compte tenu de la nature de ce matériau, il doit exister une sinuosité de l'onde. Ce phénomène, en onde T doit être amplifié par la variation de l'indice de réfraction.

Une étude en cours nous permettra de lever cette indétermination.

Mesures sur les constituants du matériau

Mesures sur un élément de matrice

Les mesures portent sur un échantillon qui a la même composition que la matrice. Cet échantillon comprimé à chaud ne présente pas de porosité.

Mesures sur un monocristal

C'est un monocristal monoclinique organique de faible cohésion interne qui se prête mal au découpage selon ses plans principaux.

Quoiqu'il en soit, une tentative d'évaluation des vitesses selon des axes parallèles aux trois axes principaux a été effectuée par U.S. grâce à un goniomètre à réflexion.

Le principe est connu: il consiste à analyser l'amplitude de réflexion d'une onde U.S. frappant l'une des faces de l'habitus du monocristal préalablement indexé.

Selon l'incidence oblique, il apparaît des angles critiques dans la réflexion de l'onde qui nous permettent de déterminer C_L et C_T pour une direction du monocristal. Celui-ci étant monté sur une platine goniométrique, on peut décrire l'anisotropie élastique du cristal par rapport à la face de référence (fig. 3).

Cette étude n'est pas terminée. Elle nous a donné quelques indications précieuses sur les variations de vitesse dans le monocristal et les bornes inférieure et supérieure: 4000–3000 m/s.

Afin de cerner la vitesse moyenne dans l'agrégat cristallin sans matrice, des mesures U.S. sous pression sur de la poudre compactée jusqu'à 2 kbar nous donnent une valeur moyenne de la vitesse qui est de l'ordre de 3400 m/s, à 100 m/s près, ce qui entraîne une erreur de 7% sur K.

4. Evaluation des différentes causes d'erreurs-applications aux modèles

Quand on fait la synthèse de nos mesures et des causes d'erreurs et qu'on l'applique aux calculs des modèles que nous avons exposés en première partie, et aux calculs des modules du matériau, on peut dresser le tableau suivant.

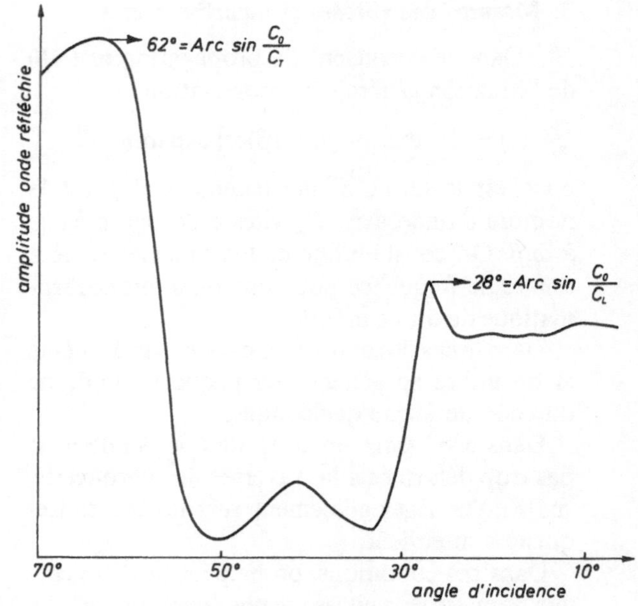

Fig. 3. Variations de l'amplitude réfléchie d'une onde U.S. (2 MHz) face [011]

Remarques et conclusions

On voit apparaître que les modules G sont peu sensibles à la modélisation et qu'en général l'accord n'est pas trop mauvais.

D'autre part, le module de compressibilité est très sensible à la modélisation et si le modèle de *Hashin-Rosen* convient bien, sa texture n'est qu'une caricature de notre matériau. Mais l'influence de la forme des porosités et de leur importance est très marquée dans les résultats.

	Expérience		Théorie		
Modules	Valeurs calculées	Valeurs calculées et corrections	*Hashin et Shtrikman* borne inférieure	*Eshelby*	*Hashin* et *Rosen* valeur moyenne
Compressibilité $K \times 10^9$ Pa	8	9,2	14	12,8	9
Cisaillement $G \times 10^9$ Pa	3,4	3,6	5,6	5,2	4

Littérature

1) *Hashin* et *Shtrikman*, J. Mech. Phys. solids **3**, 127–140 (1963).
2) *Mackense*, Proc. Phys. Soc. **B63**, 2 (1950).
3) *Eshelby*, Proc. Roy. Soc. Lond. **A241**, 376 (1957).
4) *Wu, T. T.*, Int. J. Solids Structures **2**, 1–8 (1966).
5) *Hashin*, J. Appl. Mech. **29**, 143–150 (1962).
6) *Christensen*, J. Mech. Phys. Solids **17**, 23–41 (1969).
7) *Paul*, Trans. AIME **218**, 36 (1960).
8) *Yeh*, J. Appl. Phys. **41**, 8, 3353 (1970).
9) *Hashin* et *Rosen*, J. Appl. Mech. **31**, 223 (1964).
10) *Warren*, J. Geophys. Res. **74**, 713 (1969).
11) *Nagarajan*, J. Appl. Phys. **42**, 10, 3693 (1971).
12) *Cowin*, J. Appl. Phys. **43**, 6, 2495 (1972).
13) *Merculov*, Soviet Phys. Techn. Phys. **1**, 59 (1956).
14) *Waterman*, J. Phys. D: Appl. Phys. **3**, 290 (1970).
15) *Kröner*, J. Mech. Phys. Solids **15**, 319 (1967).

Adresse de l'auteur:

Dr. *S. Poulard*
Commissariat à l'Energie Atomique, Etablissement T
B. P. No. 7, 93270 Sevran (France)

Rheol. Acta **13**, 381–394 (1974)

From the Department of Mechanics, Technion, Haifa (Israel)

The effect of environmental-loading history on longitudinal strength of glass-fiber reinforced plastics*)

By O. Ishai and A. Mazor

With 15 figures and 4 tables

(Received October 27, 1973)

List of symbols and abbreviations

a	Water content.
c	Crack or flaw dimension, perpendicular to tensile stress direction.
C_m, C_f	Matrix and fiber volume contents respectively.
α	Coefficient of thermal expansion.
α_i	Ratio of initial failure stress to the ultimate reference one.
γ	Specific surface energy.
γ_d	Specific surface energy of dry glass surface.
γ_w	Specific surface energy of wet glass surface (by water adsorbtion).
E	Tensile *Youngs* modulus.
E_f	Fiber *Youngs* modulus.
E_m	Matrix *Youngs* modulus.
m	Modular ratio.
l, Δl	Length and change of length respectively.
W, ΔW	Weight and change of weight respectively.
t	Time.
T	Temperature.
σ	Tensile stress.
σ_u	Ultimate tensile stress.
σ_u^0	Ultimate tensile stress of specimens stored and tested at reference environmental conditions (22 °C, 50% R.H.).
σ^H	Tensile stress applied during environmental-loading history.
σ_u^H	Current tensile strength of specimens tested at their E.L.H. conditions.
σ_i^H	Initial failure tensile stress, determined by onset of R.L.S. reduction.
σ_u^d	Tensile strength of specimens at dry environment.
σ_u^w	Tensile strength of specimens at humid environment.
σ_{uc}	Composite strength.
σ_{uf}, σ_{um}	Fiber and matrix strength respectively.
ϱ	Density.
L	Loading condition.
W	Cold water environment (R.T.)
T	Dry hot environment (80 °C).
$W + T$	Hot water environment (80 °C).
$W + L$	Loading at cold water environment (R.T.).
$W + L + T$	Loading at hot water environment (80 °C).
E.L.H.	Environmental loading history.
G.R.P.	Glass reinforced plastics.
C.L.S.	Current longitudinal strength.
R.L.S.	Residual longitudinal strength.
U.D.F.	Unidirectional fabric laminate.

1. Introduction

The main difficulty in dealing with the present subject lies in formulating and classifying the variables. Most cases of a structural element exposed to an external environment involve at least three simultaneous time varying factors – temperature, humidity and load – which, in view of the coupling between them and their interactive effect on the material, do not lend themselves to simple isolation or superposition.

In fibrous composites like glass-fiber reinforced plastics (G.R.P.), at least three phases have to be considered: the fibers, the matrix and the fiber-matrix interfaces.

The combined effect of loading and environmental conditions on the internal structure of a composite can be treated either in scientific (chemical, physical, or micro- or macro-mechanical) or in applied (engineering) terms.

The basic scientific approach is mainly concerned with the internal processes governed by the external loading-environmental factors; the micro-mechanical approach – with the internal stresses produced by the latter; and the macro-mechanical approach which seeks a constitutive equation relating loading to deformation in time through the material parameters.

The engineering approach is mainly concerned with the effect of the environmental-loading history (E.L.H.) on the design allowables, which are in turn governed by the reliability of the strength-time relationship.

Most works on the subject undertaken with a view to:

1. Data for designers (effect of time and environmental variables on allowables).

2. Theories for predicting long-term behavior from short-time test data.

*) Sponsored by the U.S. Department of Commerce, National Bureau of Standards Contract No. NBS(G)-135.

**) A.O. Professor, Abteilung für Technische Mechanik, Technion Haifa, Israel.

***) Israelisches Verteidigungsministerium.

3. Guidlines for manufacturers (optimization of composition in terms of durability and service life, combined with minimum maintenance requirements for expected E.L.H.'s).

4. Theories on internal structural morphology and processes as related to macro-behavior.

Most of the knowledge accumulated in this context falls into two categories:

– Variation of the material parameters (strength, rigidity etc.) under different environmental conditions.

– Load/time-to-failure relationship (stress corrosion, static fatigue) under different environmental conditions.

The present work attempts to deal with the following topics inadequately covered in most earlier studies.

1. The onset of composite failure (rather than ultimate fracture) and its internal mechanism.

2. Combined effect of preloading and environment on mechanical properties (rather than time-to-failure analysis).

3. Residual effects of E.L.H. (rather than current environmental effects).

4. Individual effects of temperature, humidity and load on strength.

5. Effect of internal state of stress on composite phases, determined by loading at different fiber orientations.

6. Effect of water penetration on deformational behavior, and effect of resulting internal stresses on composite strength.

This part of the work is concerned with strength properties under longitudinal (fiber-oriented) loading, with the macro-behavior mainly reflecting changes in the fibrous phase.

The subsequent parts deal with the contributions of the other two phases – the matrix and the interfaces – determined under transverse and off-axis loading.

2. Literature survey

The effect of moisture penetration is the resultant of its component effects on each of the three phases. The contribution of each phase to the total change in strength depends on the level and direction of the stress relative to the fiber orientation: the fibers are the predominant factor under longitudinal stressing whereas the matrix and interfaces – under transverse and shear stressing.

2.1. Effect of environmental conditions on fiber strength

The longitudinal strength of unidirectional lamina obeys the following approximate relationship (assuming linear behavior up to failure):

$$\sigma_{uc} = \sigma_{uf}\left[C_f + \frac{C_m}{m}\right] \quad [1]$$

where

σ_{uc} composite strength

σ_{uf} fiber strength

C_m, C_f matrix and fiber volume contents, respectively

$m = E_f/E_m$ fiber to matrix modular ratio.

For structural polymeric composites with stiff fibers ($m \gg 1.0$) and a high fiber content ($C_m \approx 0.4$), eq. [1] reduces to:

$$\sigma_{uc} \simeq \sigma_{uf} C_f \quad [2]$$

or, in other words, the matrix has a negligible contribution to longitudinal strength.

Glass fibers are brittle and their strength is very sensitive to imperfections such as voids, cracks, surface scratches etc. If their "virgin" strength is high, it is probably due to axial orientation of these imperfections, which, combined with the thinness of the fibers ($10\,\mu$), reduces the probability of transverse cracks of critical length.

The classical brittle-fracture theory postulated by *Griffiths* (1) assigns the following approximate expression for the brittle strength:

$$\sigma_u = \sqrt{\frac{\gamma E}{\pi c}} \quad [3]$$

where

γ specific surface energy

E *Young*s modulus

$2c$ length of crack perpendicular to tensile stress axis.

Environmental factors (especially humidity) tend to reduce γ and with it the tensile strength. In fact, the "virgin" strength of glass [measured in a liquid nitrogen (–196 °C) environment (2)] is about 10^6 p.s.i. as against $200–500 \times 10^3$ p.s.i. under room conditions.

The glass-water interface energy (i.e. that involved in adsorption of water molecules on the fibers) is much lower than its glass-air counterpart, hence a lower critical stress is required for spontaneous propagation of existing cracks (3).

Charles (4, 5) found that degradation of glass in the presence of water occurs mostly where the former consists of alkali ions. The first stage of attack consists in the following chemical reaction, with the sodium ions migrating outward:

$$Si{-}O(Na) + H_2O \rightarrow Si{-}OH + Na^+ + OH^-. \quad [4]$$

During the second stage the OH^- ions disrupt the strong siloxan bond (the main structural backbone of glass), producing a $Si{-}O^-$ molecule:

$$Si{-}O{-}Si^- + OH^- \rightarrow Si{-}OH + Si{-}O^- \quad [5]$$

which in turn decomposes another water molecule (the third stage, a repetition of the first):

$$Si{-}O^- + H_2O \rightarrow Si{-}OH + OH^-. \quad [6]$$

An alkaline medium is thus created, with the OH^- concentration equivalent to that of released Na^+ ions, and the pH level increases gradually (an autocatalytic process).

In a scanning-microscopy study of the attack of water at different temperatures on different types of glass fibers, *Barker* et al. (6) found that a layer with higher concentration of metallic ions is formed on the surface, as a result of the migration process referred to above.

To sum up, surface degradation of the glass (with the resulting reduction in composite strength) is unavoidable unless the surface is protected by a strongly-bonded polymeric coating.

2.2. *Effect of environmental conditions on matrix*

The role of the polymeric matrix is three-fold: protection against environmental attack; a load-transmitting medium; and an energy-dissipation outlet for arresting crack propagation.

Water penetration of the matrix takes place by the following mechanisms:

a) Through fiber-matrix interfaces (capillary).

b) Through the resin (diffusional)[1].

c) Through cracks and voids in the composite.

The capillary mechanism is much more prominent than its diffusional counterpart. As for the third mechanism, its contribution depends mainly on the size and distribution of the cracks and voids, in turn closely dependent on the tensile load level. The effect of the absorbed water is also twofold: on the one hand it reduces the stiffness of the resin and at the same time (by causing the latter to swell) induces internal stresses which may lead to non-reversible disruption of its cohesive structure. As the swelling effect is analogous to restrained volumetric change, and the induced stresses – to an increase in temperature, the process may be assumed to obey the same law as the temperature effect on the resin (8):

$$\log \alpha_c = \frac{-A\left(1 - \frac{a}{\varrho}\right)}{B + \left(1 - \frac{a}{\varrho}\right)} \quad [7]$$

[1]) An analysis of the diffusion mechanism is to be found in ref. (7).

where

α_c shift factor
A, B constants
a water content
ϱ resin density at given temperature

There is experimental evidence (9) that cross-linked polymers do not obey *Ficks* diffusion law. Instead, the liquid forms, between the swelling and non-swelling regions, a boundary layer spreading at a constant rate and gradually enclosing the latter region. The inner core thus created undergoes increasing triaxial stress which eventually exceeds the strength of the material, at which stage cracks are formed and the failure process sets in.

The amount of absorbed water ΔW is time-linear, namely:

$$\Delta W = kt \quad [8]$$

where k is a constant. The reversible and non-reversible effects of water absorption are reflected in strength measurements on specimens tested under "wet" and "dry" conditions respectively (10, 11, 12). Cohesive failure of the resin sets in only when swelling due to absorption exceeds shrinkage during the cross-linking stage (13). As the swelling due to diffusion depends mainly on the type of resin, the latter should preferably have a high cross-link density and contain a non-hydrolyzable hardener.

In a matrix exposed to water, the contribution of the voids-penetration mechanism increases under tensile loading owing to the increased voids volume and to cracking. Moreover, when a plastic exposed to a chemical environment is subjected to flexure, its mechanical properties are affected only above a critical load level (14) depending on the environment as well as on exposure time[2]).

To sum up, the water-temperature-load combination has both a direct effect – plasticization and swelling, resulting in possible cohesive failure, and an indirect one – "opening" of the system and increased penetrability.

[2]) The case can be cited (11) of a polyester matrix in which absorption below the critical tensile load level was 0.4 and 0.9% for 200 and 2000 h respectively, and almost double those amounts above that level (about 30% of ultimate).

2.3. Effect of environmental conditions on fiber-matrix interface

The basic assumption in strength analysis of composites is good fiber-matrix bonding, such as is provided by a suitable coupling agent at the interfaces. To date, there is no conclusive theory regarding the chemical and physical nature of these interfaces, and most of the existing ones deal with the following aspects:

a) Chemical or physical bonds (determining the water resistance of the interface) between the fiber and the coupling agent, and between the coupling agent and the resin.

b) The effect of the coupling agent on the properties of the resin layer near the interface.

Evidence of the existence of chemical bonds is summarized in refs. (3, 15, 16). The coupling effect is attributed to covalent bonds between the active groups of the coupling agent and the silonal groups of the glass on the one hand, and the resin during cross-linking – on the other. *Schrader* (15), using radioisotope techniques, traced the removal of the coupling agent from glass surfaces exposed to boiling water. It was found that about 98% of it is held by physical bonds and is easily removable, while the small balance (possibly a molecular layer) is chemically bound and resistant even to long-term exposure.

Wende (16) found that incorporation of active groups compatible with the resin (such as vinyl) results in considerable improvement of the resistance of the composite to boiling water, possibly through formation of the covalent bonds referred to earlier. Measurements of the rate of positive ion migration from the coated glass show, however, that the coupling agent only slows down the hydrolysis but does not prevent it entirely.

According to *Plueddemann* (17), the water at the interfaces plays a positive role: a dynamic steady state is involved, with bonds formed and disrupted – permitting stress relaxation and contributing to ductility at the interfaces. According to *Eakins* (18), the degree of order of successive molecular layers decreases with increasing distance from an interface, and their response to external stress decreases similarly. The effect of water penetration on the resin is thus twofold:

a) Remote layers undergo swelling, whereas those close to the interfaces are stiffened and their capacity for energy absorption is reduced.

b) The water forms very thin layers at the interfaces, counteracting the matrix-fiber bond and creating an alkaline medium at high temperatures, which intensifies the debonding process and may eventually lead to failure at relatively low load levels.

Ashbee et al. (19), in optical and electron-microscopy observations of the debonding process at different temperatures, found that the process is rapid but can be slowed down considerably when a coupling agent is used. At high temperatures the bond withstand the interfacial tension due to initial swelling, but is disrupted at a much later stage by the compression induced by shrinkage. The disruption is attributed to osmotic pressure generated at the interfaces by water-soluble constituents leached from the fibers, and is often accompanied by cracking directed away from the latter. Similar tests on carbon fibers and on treated and untreated glass fibers (20) showed that swelling at 100 °C suffices to overcome the curing shrinkage, producing a net tensile stress at the interfaces; rapid debonding was observed in both the carbon and untreated-glass cases. At 20 °C, swelling was too low to produce a high tensile stress but untreated glass fibers were still rapidly debonded, while their carbon and treated-glass counterparts retained the bond. (The low-temperature behavior is attributed to the fact that the untreated glass fibers are hydrophilic, whereas the carbon fibers are hydrophobic.)

2.4. Effect of moisture and temperature on composite strength

The effect of moisture on the mechanical properties of composite materials under prolonged exposure (up to 35000 h) was studied by *Romanenkov* (21, 7), who found the following relationship between strength reduction and the amount of absorbed water:

$$\sigma_u = \alpha + \beta \log W \qquad [9]$$

where

σ_u ultimate tensile stress
W amount of absorbed water
α, β empirical constants

The author attributes the penetration process to inhomogeneities, microscopic voids and cracks, and especially to intermolecular spaces which are formed as a result of thermal vibrations and allow the much smaller water molecules ($D \simeq 3.0$ Å) to diffuse into the material.

A comprehensive experimental work on filament-wound polyester, reinforced with glass fibers, was undertaken by *Bott* (10), who tested the tensile and flexural strength properties of specimens exposed to water in the 25–75 °C range for periods up to a year. The high scatter of his results made them difficult to evaluate, but normalization of the strength values with respect to the initial ones yielded a significant difference between "dry" and "wet" specimens indicating two distinct processes: one reversible – detachment and plasticizing of the resin – and the other non-reversible – disruption of chemical bonds; a clear influence of temperature was also observed. Similar results were obtained by *Krolikowski* (11) on polyester reinforced with glass fibers and fabric. The reversible effect was attributed to "lubrication" at the interfaces.

The effect of boiling water on polyester and epoxy reinforced with glass fibers consists in a marked degradation in strength (13) seen to take place in two stages: a) swelling of the resin due to adsorption, with compressive stresses induced in the matrix and tensile ones around the fibers; these stresses gradually offset the residual cooling stresses after curing, b) hydrolytic degradation of the resin and the interfaces.

2.5. Effect of load on strength of water-submerged composite specimens

In *Camerons* (22) study, the specimens were loaded up to 2/3 of their ultimate load while submerged in water, resulting in significant strength reduction especially in the case of hot water. The corresponding change in the elastic modulus was negligible.

The effect of hydrostatic pressure on a filament-wound container made of glass-epoxy composite was studied by *Brelant* et al. (23), with the time to cracking of container wall measured as function of the bi-axial stress level. Cracking was found to be dangerous only in the presence of water (in which case it may terminate in complete failure), but no failure occurred in dry tests.

Similar tests by *Bax* (24) were performed on glass-fiber polyester tubes subjected to internal pressure. The criterion for initiation of irreversible failure was determined as the strain at which "weeping" (outward seepage through the cracks) occurs, and a critical strain level was found, dependent only on the composition of the material but not on the pressure or on environmental conditions. At higher stresses and temperatures the time involved is shorter, but the critical level is the same.

Baxs theory was confirmed by *Atkinson* (25), in whose tests specimens made of glass-fiber fabric or mat in polyester matrix were submerged in water at 95–100 °C, and subjected to a tensile stress at 10 and 25% of the ultimate level respectively. Unloaded and 10%-loaded specimens showed about 1/3 reduction in flexural strength after 6 months of exposure, and the 25%-loaded ones failed after 40–180 h; in most cases fracture occurred at 0.25–0.65% of the ultimate strain. This indicated a certain strain level (below the critical one) above which a failure mechanism sets in and propagates. Atkinson based his conclusion on those of *Isham* (26), which also indicated a limiting strain level for loading of filament-wound structures exposed to a water environment.

Krolikowski (11) found, in glass-fabric polyester resin specimens subjected to a tensile stress above 30% of the ultimate, an increase in water absorption of about 100%, accompanied by a corresponding modification of the stress-strain curve, both of which he attributes to cracking of the resin due to straightening of the fabric under stress; a more flexible resin with higher ultimate strain would be expected to remain crack-free. On the other hand, the same effect was observed in glass mat-polyester resin specimens, where cracks were clearly seen only in specimens loaded above the critical stress level – an indication of a critical stress level above which cracking of the resin sets in, with the corresponding effect on water absorption. Similar observations were made by *Halpin* (27). No significant change in water absorption or in the stress-strain curve was found in unidirectional fibrous specimens exposed to water while loaded along fiber orientation. By contrast, in cross-ply laminates a sharp rise in absorption was found at a critical stress level of about 40% of the ultimate, with increased exposure of the fiber surfaces to environmental corrosivity (but no serious material damage) as a result. The increase in water absorption is attributed to cracking at the interfaces of the transverse fibers, which undergo high tensile stresses.

The literature survey may lead to the following conclusions:

a) The effect of water on the phases of the composite differs as follows: the fibers are attacked chemically by the alkaline medium; the matrix undergoes swelling (with a resulting change in the state of stress at the interface) and is plasticized; at the interface, the physical bonds are weakened, with possible reduction of the transverse strength as a result.

b) The rate of water diffusion increases with temperature, with the corresponding effect on subsequent processes.

c) Water absorption is intensified above a certain load level (or critical strain level), which apparently corresponds to the initiation stage of failure by cracking.

d) Most of the experimental information available was obtained on specimens with unspecified fiber orientation.

e) No significant conclusions could be drawn regarding the interaction of external stress and internal pressure due to water diffusion.

3. Experimental

3.1. Composition and preparation (Table 1)

In keeping with the main subject of the study – the macromechanical behavior under external working conditions – the first system to be tested was of a commercial type, namely a unidirectional glass-fiber fabric (90% longitudinal, 10% transverse) in an epoxy matrix with good bonding ensured by an epoxy-compatible coupling agent. Non-reinforced epoxy laminates serve as reference material.

The U.D.F. system (5 layers – 1 mm, for longitudinal specimens) was prepared manually on a heated table, impregnated with the epoxy resin, and cured in a press (10 at.) for 2 h at 80 °C. The temperature was then raised under pressure to 160 °C for 2 more hours and reduced to 100 °C, after which the pressure was released. The cured laminate was stored at 22 °C and 50% R.H. pending the test.

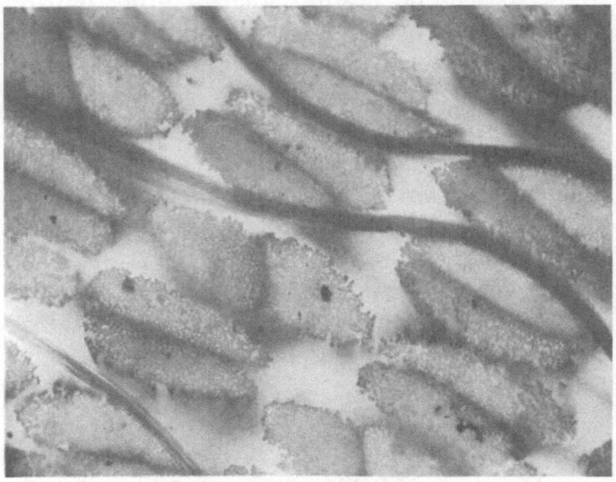

Fig. 1. Cross-section of unidirectional glass fiber reinforced epoxy laminate (G.R.P.), perpendicular to principal longitudinal fiber direction (× 150)

3.2. Specimens

Cross sectional view of U.D.F. longitudinal specimens is shown in fig. 1. The circular cross section of the fibers indicate their good alignment along the major direction. A higher fiber content is expected within the bundles as compared with the rich resin area between them.

U.D.F. specimens were 12 mm wide and approximately 1 mm thick, with two tabs attached on each side with a special high-temperature adhesive (Araldite strain-gage cement with HY 951 hardener, cured for 2 h at 80 °C), to

Fig. 2. Typical tensile specimens. a) Longitudinal G.R.P. specimen; b) unfilled epoxy specimens

Table 1. Specifications of composite systems

Composite	Filament	Matrix Monomer	Hardener	Coupling agent	Fiber volume content (C)
Unidirectional fabric laminate (U.D.F.)	1543 fabric (E glass fibers) (Clark Schewebel)	Epon 828 (Shell)	Z curing agent (Shell)	Finish c/s 273*) (Clark Schewebel)	0.46–0.48
Epoxy laminate (E)	–	Epon 828 (Shell)	Z curing agent		

*) Glycidal epoxy silane 0.05–0.1% of glass weight.

ensure a uniformly distributed low stress in the gripping area (fig. 2a).

Epoxy specimens (prepared according to ASTM specification D638-64T) were about 1 mm thick (without tabs) (fig. 2b).

3.3. Loading procedure

The portable spring-type loading device used (fig. 3) is capable of maintaining a nearly constant load level for any required period, with stress losses due to creep etc. made up for by simple manual adjustment.

The load level ranged from 10–60% of the static tensile strength level of the reference specimens (stored in an insulated room at 22 °C and 50% R.H. for more than 200 h and tested under a constant strain rate at 22 °C and about 70% R.H.).

Most specimens were tested by loading in Instron apparatus at constant strain rate of 1 mm/min up to failure, at room temperature with stress plotted against strain or time.

3.4. Test series

Series A. Change of weight and dimensions vs. time of non-loaded specimens

Stage 1 Vacuum, 22 °C, 7 days (about 0.15 torr)
Stage 2 50% R.H., 22 °C, 7 days
Stage 3a Water, 22 °C, 7 days
Stage 3b Water, 80 °C, 7 days
Stage 4 50% R.H., 22 °C, 7 days
Stage 5 Vacuum, 22 °C, 7 days
Distilled water was used throughout.

Series C. Effect of environment on current strength characteristics

The series comprised four stages, 7 days each, in which specimens were withdrawn from the environment at different times before expiry of the full exposure period and loaded to failure on the Instron (table 2).

Fig. 3. Tensile loading device for time-dependent environmental loading tests

Table 2. Variables of series B and C (E.L.H.) stage

Subseries	Designation	Humidity	Temperature	Loading % of ultimate	Test conditions for series C	
0		50%	22 °C	0	70%	22 °C
1	*W*	water	22 °C	0	water	22 °C
2	*W*+*T*	water	80 °C	0	water	80 °C
3	*W*+*L*	water	22 °C	20–60%		
4	*W*+*T*+*L*	water	80 °C	20–60%		
5	*T*	air	80 °C	0	air	80 °C
6	*T*+*L*	air	80 °C	20–60%		

Table 3. Series B – Residual effect of environmental-loading history on strength

Series B	Humidity	Temperature	Loading % of ult.	B_1 time	B_2 time
Pre-environmental stage	50%	22 °C	0	48 h	7 days
Environmental loading history (E.L.H.) stage	see table 2			48 h	7 days
Pre-test conditions	50%	22 °C	0	48 h	7 days
Test conditions	70%	22 °C	to failure	–	–

4. Test results

4.1. Series A

The weight changes of the longitudinal G.R.P. and epoxy specimens during the four environment stages (fig. 4) showed the same trend, namely a high early rate tending to a constant slope.

Comparison of exposure to cold and hot water shows increased water absorption (up to 3-fold) in the latter case. Epoxy specimens showed increased water absorption compared to their G.R.P. counterparts (up to 3-fold).

Similar trends were observed throughout all stages for the dimensional changes in the case of epoxy specimens (fig. 5). In the longitudinal specimens, the dimensional changes along the fiber directions were negligible. Clear linearity was found throughout all stages between the dimensional and weight changes in the case of epoxy specimens (fig. 6).

It may be concluded from figs. 4, 5, and 6, that dimensional changes due to water absorption are "elastic" and tend to complete recovery, provided all the water absorbed during the submersion period is allowed to evaporate.

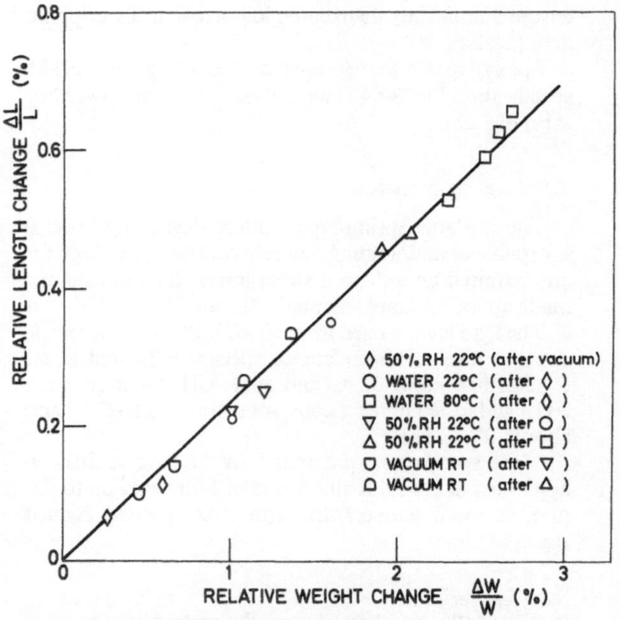

Fig. 6. Dimensional vs. weight change, for epoxy specimens exposed to different environmental stages (figs. 4, 5)

4.2. Series B

A linear stress-strain relationship up to failure is common to all unloaded and preloaded longi-

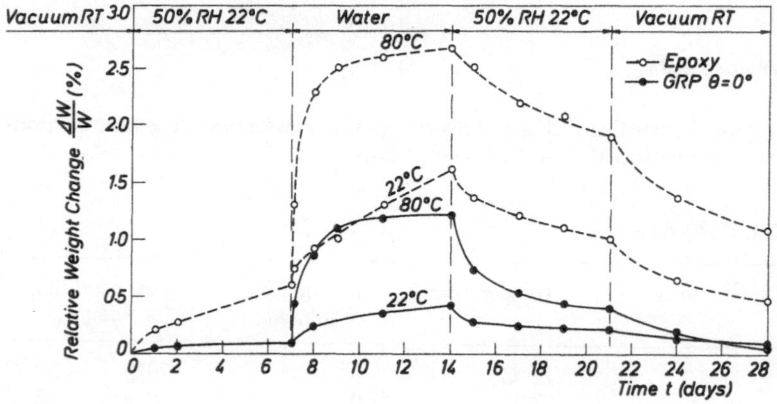

Fig. 4. Weight change vs. time for longitudinal G.R.P. and epoxy specimens exposed to different environmental stages after initial vacuum exposure

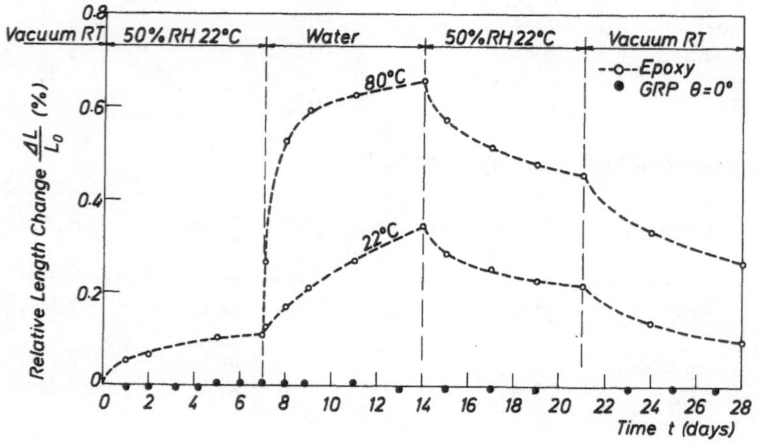

Fig. 5. Dimensional change vs. time, for longitudinal G.R.P. and epoxy specimens exposed to different environmental stages after initial vacuum exposure

tudinal specimens (fig. 7). On the other hand, unloaded epoxy specimens showed non-linearity near failure (fig. 8). Typical fracture specimens are shown in fig. 9.

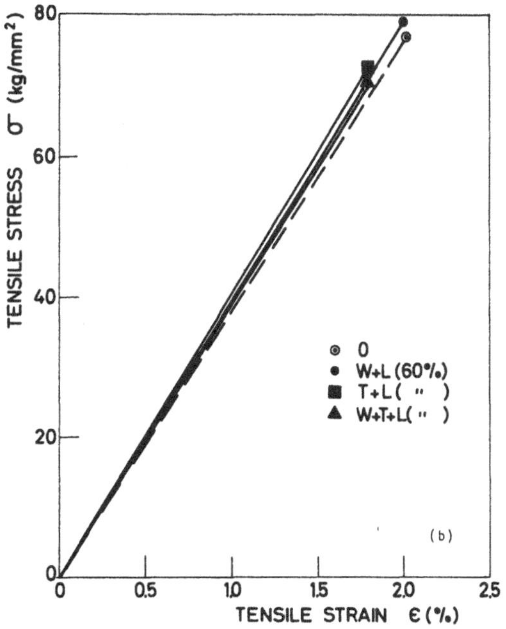

Fig. 7. Typical stress-strain relationship, up to failure, for longitudinal G.R.P. specimens tested after exposure to different E.L.H. conditions for 7 days (followed by drying under reference environmental storage conditions for additional 7 days)

In the case of hot water exposure (series B-4), strength characteristics, as represented by the ultimate stress and strain, seem to be significantly influenced by the loading level in the E.L.H. stage. In the longitudinal specimens, an increase in strength was found up to preloading of about $\alpha_i = 25\%$ of the ultimate reference, whereas a steep drop in strength is evident above this level (fig. 10). Specimens loaded up to 50 and 60% of the reference strength failed during the E.L.H. stage. A similar trend was found for both 7 and 2 day E.L.H. periods (fig. 13). In this case, use of the reference strength[3]) is not entirely justified as the current strength, especially in the case of hot-water exposure, is lower than that found under the reference conditions.

Current strength vs. time data shows the following trends (fig. 11):

a) Longitudinal strength of G.R.P. specimens is almost constant during 7 days' exposure to cold water.

b) In hot water, a pronounced drop in longitudinal strength (compared with the reference level) was found during the first hours of exposure. The same lower level was preserved during the 7 days' exposure.

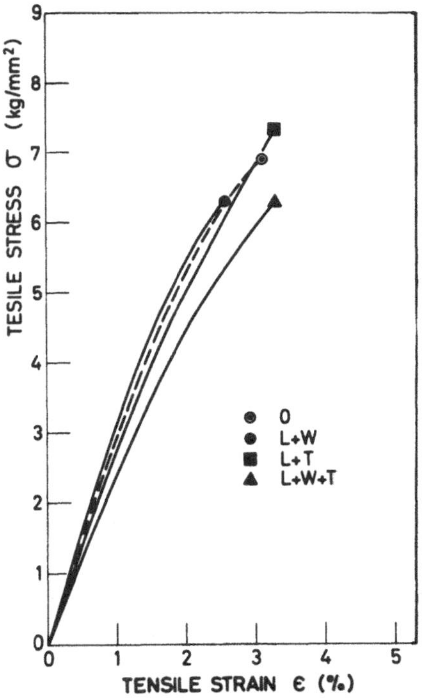

Fig. 8. Typical stress-strain relationship, up to failure, for epoxy specimens tested after exposure to different E.L.H. conditions for 7 days (followed by drying under reference environmental storage conditions for additional 7 days)

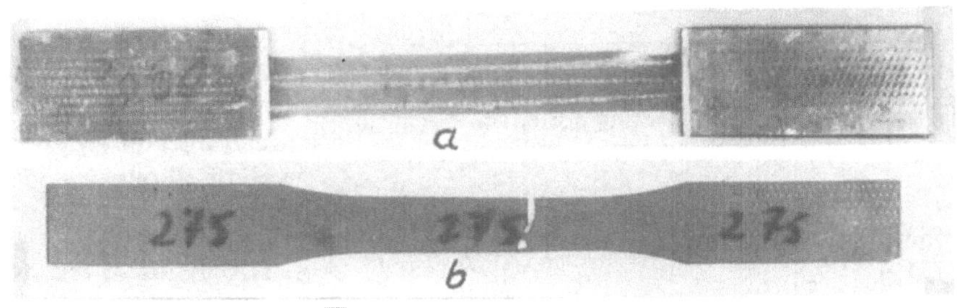

Fig. 9. Typical fracture of longitudinal G.R.P. and epoxy specimens

[3]) Reference longitudinal strength σ_u^0 as determined from the results of Series B-0, was found to be 74.6 kg/mm².

c) In an 80 °C air environment, a moderate initial increase of longitudinal strength below the reference level is evident.

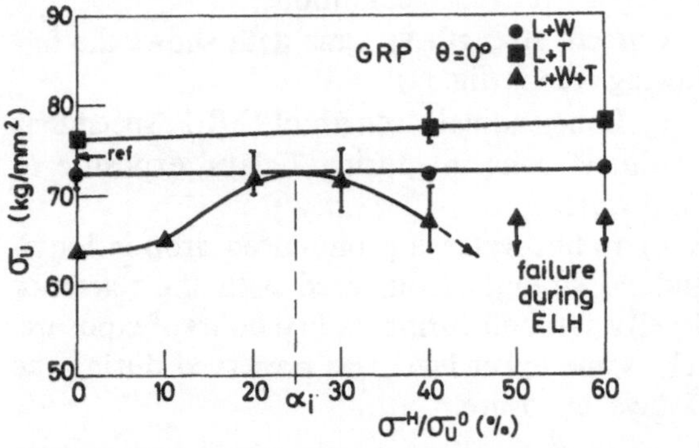

Fig. 10. Effect of preload level during E.L.H. stage on tensile strength for longitudinal G.R.P. specimens. $L + W + T$ – E.L.H. stage – water – 80 °C; $L + T$ – E.L.H. stage – air – 80 °C; $L + W$ – E.L.H. stage – water – 22 °C

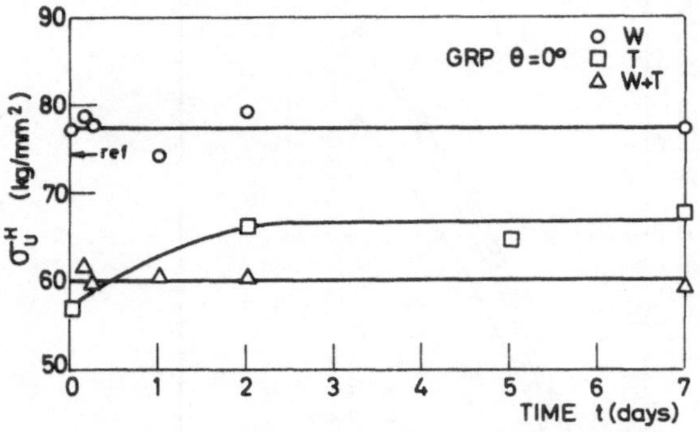

Fig. 11. Current strength vs. time for longitudinal G.R.P. specimens exposed to different environmental conditions. $W + T$ – water – 80 °C; T – air – 80 °C; W – water – 22 °C

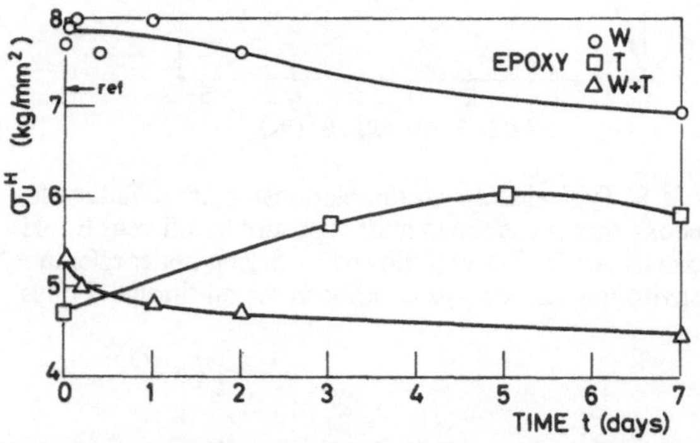

Fig. 12. Current strength vs. time for epoxy specimens exposed to different environmental conditions

Similar trends were found for epoxy specimens (fig. 12).

The effect of the preloading level on the longitudinal strength could now be corrected, based on the current strength results shown in figs. 11 and 12. The previous trends are preserved with an approximately 5% positive shift of the critical preload level ratio in the case of hot-water exposure (figs. 13, 14).

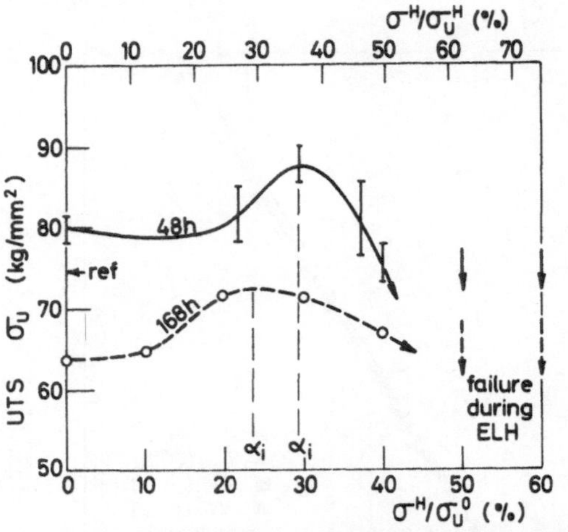

Fig. 13. Effect of preload level during E.L.H. stage (2 and 7 days) on tensile strength for longitudinal G.R.P. specimens. (E.L.H. stage – water – 80 °C)

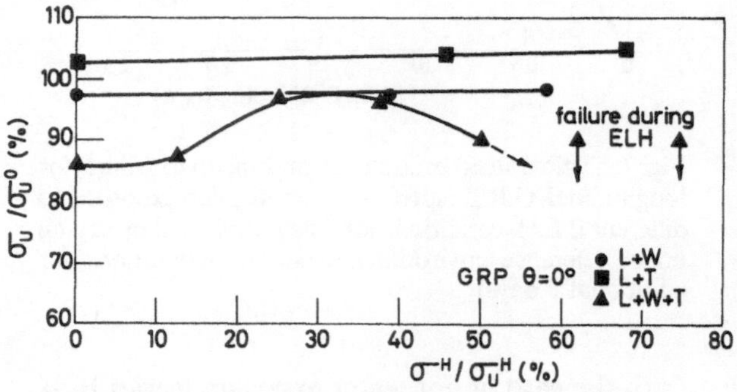

Fig. 14. Effect of preload to current strength ratio on normalized tensile strength of longitudinal G.R.P. specimens. $L + W + T$ – E.L.H. stage – water – 80 °C; $L + T$ – E.L.H. stage – air – 80 °C; $L + W$ – E.L.H. stage – water – 22 °C

Comparison of the different environmental conditions during E.L.H. stage show the following trend (figs. 10, 14):

Whereas in the case of cold water and hot-dry exposure no residual strength reduction was found, a combination of the two conditions, i.e. hot-water exposure, gives rise to a pronounced drop in the residual strength. This drop is obviously affected by preloading but only above the level of about 25–35% of the corresponding reference.

5. Discussion of results

The following factors may be responsible for longitudinal strength variation of G.R.P. specimens:

1. Internal stresses setting in during the thermoelastic history of the material.
2. Internal stresses setting in during the hygroelastic history (swelling and shrinkage).
3. Water in the matrix.
4. Water at the interfaces.
5. Chemical attack by water on the glass fiber.
6. Disruption of chemical-physical bonds at the interfaces.
7. Residual water on the glass-fiber surfaces at testing.
8. Water adsorption on the glass surface reducing its surface energy and thereby reducing its resistance to crack propagation during the E.L.H. period.

The experimental evidence favors mainly the last two factors as responsible for residual longitudinal strength (R.L.S.) deterioration for the following reasons:

a) Thermal and hygrometrical effects only counteract the internal stresses which develop on cooling of the cured composite from 150 °C to R.T.

b) Rough calculations, based on the thermal expansion coefficient given in table 4, and dimensional changes vs. hygrometric data shown in figs. 5 and 6, yield only negligible residual stresses along the fibers.

Table 4. Coefficients of thermal expansion for G.R.P. and epoxy specimens (range: 25 °C to 80 °C) 10^{-6}/°C

In oven (dry)	Submerged in water (5 min)	Material
54.0	65.6	Unfilled epoxy
7.1	8.3	Longitudinal G.R.P.
23.0	23.7	Transverse G.R.P.

c) Weakening of the matrix, or even of the interfaces, would not have a significant effect on the longitudinal strength.

d) Chemical attack on the glass fiber could have an effect on residual strength, but such an effect must be strongly dependent on exposure time.

e) No significant time-dependent longitudinal strength reduction was found in the current strength tests on exposure to water for 7 days (fig. 11).

The predominant mechanisms responsible for residual longitudinal strength reduction in loaded G.R.P. specimens exposed to hot water could be postulated as follows:

When the composite is exposed to water which penetrates it and is adsorbed on the glass fiber surfaces, fiber strength is drastically reduced. This reduction is due to the much lower surface energy of the wet glass surface compared with its dry reference, according to the following approximation of *Griffiths* theory:

$$\sigma_u^w = \left[\frac{\gamma_w E}{\pi c}\right]^{1/2} \qquad [10]$$

$$\sigma_u^d = \left[\frac{\gamma_d E}{\pi c}\right]^{1/2} \qquad [11]$$

where

- σ_u^w, σ_u^d are the "wet" and "dry" strengths, respectively
- γ_w, γ_d the surface energy of the glass-water interface and of the dry glass surface, respectively
- E *Youngs* modulus of the glass fibers
- $2C$ maximum crack (or flaw) length perpendicular to the tensile loading direction (fiber axis).

If the exposed G.R.P. is also under load during the E.L.H. stage, a fraction of the fibers, whose maximum flaw dimension exceeds the critical value for the given stress, are fractured. Obviously, when the fiber surfaces are wet, failure sets in at a lower stress than the "reference" ultimate level determined under dry encironmental conditions.

The ratio (α_i) of the load level for the onset of this failure process (σ_i^H), and the "reference dry" strength (σ_u^0), could thus be related to the corresponding surface energies as follows:

$$\alpha_i = \frac{\sigma_i^H}{\sigma_u^H} = \frac{\sigma_u^w}{\sigma_u^d} = \left[\frac{\gamma_w}{\gamma_d}\right]^{1/2}. \qquad [12]$$

Data of surface energies for lime-glass are reported in ref. (28). The values of 200–400 erg/cm² was found for dry glass as against about 30 erg/cm² for glass in humid environment (95% R.H.). Substituting these values in eq. [12] yields α_i ratios within the range of 25–40%. These values are in agreement with the present experimental data, as shown in figs. 10, 13, and 14.

It is likely that there is a maximum flaw dimension $C(+)$ for a given flaw-length distribution function. This upper bound determines the lower bound for the failure initiation stress as follows:

$$\sigma_{i(-)}^H = \left[\frac{\gamma_w E}{\pi C_{(+)}}\right]^{1/2}. \qquad [13]$$

The stress drop due to the above mechanism depends on the percentage fraction of the fibers which had failed during the E.L.H. stage under the given stress σ^H; this percentage depends in turn on Fiber-strength distribution function. At the end of the E.L.H. stage, the residual strength of the composites is obviously lower in proportion to the intact fraction of the fibers.

Thus, at any stress above the minimum level $\sigma^H_{i(-)}$, a drop in residual strength would be expected. This drop is, however, conditional on adsorption of water on the glass fibers, thereby accounting for the dependence of this process on time and temperature – the main factors controlling the diffusion of water through the matrix phase and its penetration into the fiber-matrix interface. From the data in fig. 4, it is apparent that, in the case of a hot water E.L.H., the absorption process is almost stabilized after 50 h of exposure. Infra-red data (fig. 15) indicate that Silica molecules partly from the coupling agent leached out by the water – an evidence that water reaches the glass surface through the interface.

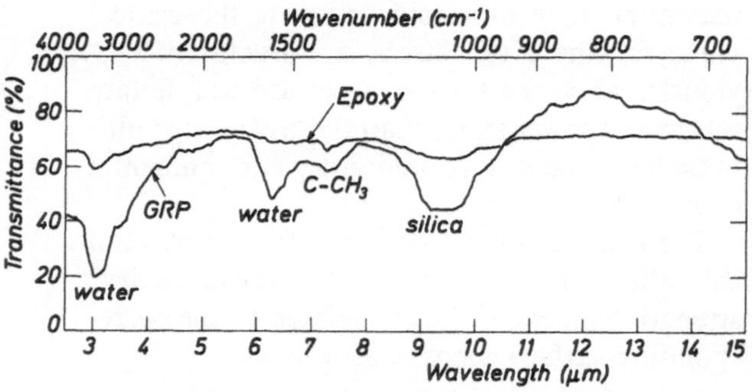

Fig. 15. Infra-red transmittance analysis of water medium in which G.R.P. and epoxy specimens were submerged for 7 days at 80 °C

The above is in agreement with the fact that the residual failure process is already active within 48 h (fig. 13). On the other hand, in the case of cold water E.L.H. (where absorption is much slower), it is reasonable to assume that within the E.L.H. stage of the present investigation (up to 168 h) only an insignificant amount of humidity reaches the glass fibers. This accounts for the lack of evidence of R.L.S. reduction even at a preloading level of 60–70% of the reference.

Another possible mechanism for the effect of preloading on R.L.S. reduction is water penetration intensified by tensile stressing through the epoxy matrix to the interfaces.

It is reasonable to assume that after preloading and short exposure to a dry environment, a substantial amount of residual water still remains on the glass fiber surfaces. This amount would probably depend on the preloading level during the E.L.H. stage. Testing in a reference-dry environment would, in this case, show lower strength, reflecting the current lower strength of the partly wet glass fibers.

The combined effect of load, water environment and temperature is thus essential for the longitudinal strength reduction mechanism. Evidently a critical stress level exists below which no crack propagation sets in even on exposure to hot water. This hypothesis, however, is limited to short E.L.H. stages of 7 days, as found in the present work.

Below the critical stress level, strength increase is attributable to additional fiber straightening and other relaxation processes activated at high temperatures. Strengthening of the G.R.P.'s by "dry heating" at all preloading levels and under current testing can be attributed to one or a combination of the following effects:

a) Additional curing of the polymeric matrix.

b) Straightening of the fibers under tensile load facilitated by softening of the matrix at high temperature.

c) Continuous drying of the composite.

6. Conclusions

The following conclusions are based on longitudinal strength data of longitudinal G.R.P. specimens and their epoxy references exposed to different environmental loading histories (E.L.H.) for short periods:

6.1.

There is direct linearity, irrespective of the E.L.H. stage, between the dimensional change of epoxy specimens and the amount of adsorbed water.

6.2.

Almost no dimensional change due to water absorption could be detected along the fiber direction in G.R.P. specimens.

6.3.

Exposure of epoxy specimens to different E.L.H. stages results in a pronounced change in weight and dimensions but does not affect significantly the relationship between them. This supports the assumption that increasing temperature only accelerates diffusion and reduces the time scale for absorption.

6.4.

In hot water, the absorption process seems to stabilize within 7 days of exposure.

6.5.

The effect of the preload level during the E.L.H. stage seems to have the following effects on residual longitudinal strength (R.L.S.) characteristics:

a) In the case of hot water, it increases the residual strength of the longitudinal G.R.P. specimens up to a preload level of about 30% of the reference ultimate, but reduces it drastically above this level.

b) In the case of cold water, almost no effect of the preload level on R.L.S. is noticed.

c) In a hot-dry environment, a slight increase in R.L.S. with the preload level is evident.

It could be concluded that superposition is ruled out in predicting R.L.S. data under the combined effect of temperature and humidity from the respective data obtained under the effect of each environmental factor alone.

6.6.

The effect of environmental conditions on current longitudinal strength (C.L.S.) are as follows:

a) It increases with time in a hot-dry environment.

b) It is almost time-invariant on exposure to cold water.

c) It decreases with time on exposure to hot water only during the first exposure hours.

d) Its lowest level was found under hot-wet conditions, and its highest one – in a cold-wet environment.

6.7.

The ratio of the initial failure stress of longitudinal G.R.P. specimens exposed to hot water, and their reference strength under dry R.T. conditions, is within the range of 25–35%.

6.8.

The existence of such an initial failure point (found only in the case of hot water even on short exposures) is attributable to the drastic reduction of the surface energy of the fiber-glass surfaces by adsorbed water during the E.L.H. stage.

6.9.

Further evidence for the initial failure mechanism is provided by infra-red data, which show indications of the diffusion process reaching the interfaces on exposure to hot water.

Summary

A literature survey and an experimental study are presented on the short term strength variation of G.R.P. specimens subjected to the combined effect of loading and different environmental conditions. Findings are as follows:

a) Residual longitudinal strength was found to decrease above a preloading level of about 25–35% of the corresponding reference strength – on exposure to hot (80 °C) water, even after an environmental-loading history (E.L.H.) stage of 48 h.

b) No reduction in residual longitudinal strength (R.L.S.) was observed on exposure to cold water (22 °C) and to a dry-hot environment (80 °C).

Conclusions are as follows:

a) Early reduction of the residual strength is conditional on a combination of temperature and humidity, whereas each factor alone does not cause any deterioration, even under high preloading.

b) Humidity is the main factor in the onset of internal failure in the glass-fiber phase, the effect being probably due to reduction of the surface energy of the glass by water adsorbed on its surface, accompanied by drastic reduction of its strength (*Griffiths* theory).

Temperature has an indirect effect, accelerating the diffusion process and reducing the time scale for failure initiation.

Zusammenfassung

Es wird ein Schrifttumsüberblick und eine Experimentaluntersuchung über die kurzzeitigen Festigkeitsveränderungen von G.F.V.K.-Versuchskörpern vorgelegt, die der gleichzeitigen Einwirkung von Belastung und verschiedenen Umweltbedingungen unterworfen waren. Es wurden folgende Beobachtungen gemacht:

a) Ein Abfall der verbleibenden Axialfestigkeit wurde jenseits eines Vorbelastungsmaßes von etwa 25–35% der entsprechenden Vergleichsfestigkeit festgestellt – wenn der Prüfung die Einwirkung von heißem Wasser (80 °C) vorausgegangen war, sogar wenn die umweltbeeinflußte Belastungsdauer (Environmental-loading history, oder E.L.H.) nur 48 Std. betrug.

b) Die Einwirkung von kaltem Wasser (22 °C) und von trocken-heißer Umgebung (80 °C) verursachten keinen Verlust an verbleibender Axialfestigkeit (Residual longitudinal strength oder R.L.S.).

Die folgenden Schlüsse werden gezogen:

a) Frühzeitiger Abfall der verbleibenden Festigkeit wird von der vereinten Wirkung von Temperatur und Feuchtigkeit bedingt, wogegen jeder Faktor allein, selbst bei hoher Vorbelastung, keinerlei Festigkeitsverlust hervorruft.

b) Der Hauptfaktor für den Beginn innerer Zerstörung in der Glasfaserkomponente ist Feuchtigkeit, wobei die Wirkung wahrscheinlich durch einen Abfall in der Oberflächenenergie des Glases infolge oberflächenabsorbierten Wassers verursacht wird, begleitet von einem drastischem Festigkeitsverlust (*Griffiths* Theorie).

Temperatur hat nur mittelbare Einwirkung, indem sie den Diffusionsprozeß beschleunigt und damit das Zeitmaß bis zum Eintritt des Bruches verkürzt.

References

1) *Griffith, A. A.*, Phil. Rans. Roy. Soc. (London) **A 221**, 163 (1920).

2) *Hollinger, D. L.* and *H. T. Plant*, Annual Technical Conf. 1962, R.P., SPI, Section 13-B.

3) *Johanson, O.* et al., Dow Corning Corporation, Midland, Michigan, Sept. 1965 (AD 629777).

4) *Charles, R. J.*, J. Appl. Phys. **29**, 1549–1553 (1958).

5) *Charles, R. J.*, J. Appl. Phys. **29**, 1549–1553 (1958).

6) *Barker, A. J.* and *R. T. Bott*, Trans. Inst. Chem. Eng. **47**, T 212 (1969).

7) *Romanenkov, I. G.*, Soviet Plastics, No. 4, 43–45 (1967).

8) *Halpin, J. C.*, Air Force System Comman, Wright-Patterson AFB, Ohio, June 1969 (AD 692481).

9) *Alfrey Jr., Turner, E. F. Gurnee,* and *W. G. Lloyd*, J. Poly. Sci., Part C, No. 12, 249–261 (1966).

10) *Bott, T. R.* and *A. J. Barker*, Trans. Instr. Chem. Engrs. **47**, T 188–193 (1969).

11) *Krolikowski, W.*, SPE **1964**, 1031–1035.

12) *Raffel, B. P.*, 23rd Annual Technical Conf. 1966, R.P., SPI, Section 12-C.

13) *James, D. I., R. H. Norman*, and *M. H. Stone*, Plastics and Polymers **1968**, 21–31.

14) *Ruhnke, G. M.* and *L. F. Bivitz*, Plastics and Polymers **1970**, 265–270.

15) *Schrader* and *E. Malcolm*, J. Adhesion **2**, 202 (1970).

16) *Wende, A.* and *J. Gähde*, BPF **1966**, 15.

17) *Plueddemann, E. P.*, Modern Plastics **1970**, 92.

18) *Eakins, W. J.*, Interfaces in Composites **452**, 137–148 (ASTM Special Technical Publications).

19) *Ashbee, K. H. G.* and *R. C. Wyatt*, Proc. Roy. Soc. **A 312**, 553–564 (1964).

20) *Wyatt, R. C.* and *K. H. G. Ashbee*, Fibre Science and Technology, pp. 29–49 (1969).

21) *Romanenkov, I. G.*, Soviet Plastics **1967**, No. 2, 74–75.

22) *Cameron, J. B.*, Trans. J. Plast. Inst. **1967**, 681–687.

23) *Brelant, S., I. Petker*, and *K. W. Smith*, SPE **1964**, 1019–1023.

24) *Bax, J.*, Koninklijke/Shell Plastics Laboratorium, Delft, PB 68–86.

25) *Atkinson, Harvey E.*, Modern Plastics **1969**, 108–123.

26) *Isham, A. B.*, 22nd SPI RP Conference, Section 16 E.

27) *Tsai, S. W., J. C. Halpin*, and *N. J. Pagano*, Technomic Publ. Co., "Composite Materials Workshop", pp. 375–378.

28) *Shafrin, E. G.* and *W. A. Zisman*, Amer. Ceram. Soc. J. **50**, 478–484 (1967).

Authors' addresses:

Dr. *Ori Ishai*
Ass. Professor, Dept. of Mechanics, Technion
Haifa (Israel)

Dr. *A. Mazor*
Israel Ministry of Defence

Rheol. Acta 13, 395–399 (1974)

From the Birla Institute of Technology, Mesra, Ranchi (India)

Rheological yield condition

By B. R. Seth

(Received October 27, 1973)

1. Introduction

Rheological problems are non-homogeneous, aelotropic, non-linear and irreversible. If a start is made with the stress-tensor field, one soon runs into a number of difficulties. The strain tensor is defined for any type of medium-homogeneous, heterogeneous, isotropic, aelotropic and hence should be a suitable vehicle for investigation the yield condition. Again, the classical strain/strain rate, when used, makes the constitutive equation very complex. So the generalized strain measure should be used.

All continuous deformations form a symmetric group. At yield, the nature of the group changes. For example, for an elastic body, which belongs to an orthogonal group, the new group becomes uni-modular.

One can also visualize that at yield the macroelement breaks down, with the result that the corresponding mapping becomes singular, and the modulus of transformation tends to take the value zero or infinite. In terms of the invariants of the strain-tensor field, J_1, J_2, J_3, we find, if we use the generalized strain measure that this condition gives

$$n^3 J_3 - n^2 J_2 + n J_1 - 1 = 0 \qquad [1.1]$$

n being a constant, denoting the measure index. The relation [1.1] is transformed in terms of the stress-tensor invariants and it is shown that in a large number of cases the yield condition reduces to a generalized form of *Tresca* yield condition given by

$$\tau_{11} - k\tau_{33} = k_0 \qquad [1.2]$$

where $\tau_{11} > \tau_{22} > \tau_{33}$ are the principal stresses and k, k_0 functions of the response coefficients.

2. Yield condition

The change into a state of yield may be interpreted as an asymptotic phenomenon which imposes a constraint on the invariants of the field tensor. In other word, it can be interpreted as a mapping of one space into another. If the elastic strain field is e_{ij}, whose invariants are J_1, J_2, J_3, then the asymptotic behaviour may be represented by the existence of a functional relation of the type

$$f(J_1, J_2, J_3) = 0$$

in which J's are independent of one another in the normal part of the field. This does not fix the nature of the function f for which we can invoke the additional geometric condition that yielding can result from infinite contraction or expansion of a macroelement. This shows that the transformation matrix should become singular. We use this concept to get the yield condition in the strain tensor field.

If u^r is the deformation vector and x^r the deformed coordinates the modulus of transformation is given by

$$\frac{\partial(x-u,\, y-v,\, z-w)}{\partial(x, y, z)}. \qquad [2.1]$$

Expanding and squaring it we find that its vanishing gives the condition:

$$8J_3 - 4J_2 + 2J_1 \to 1 \qquad [2.2]$$

where J's are the invariants of the *Almansi* strain tensor, which in the cartesian system is given by

$$2e_{ij} = u_{i,j} + u_{j,i} - u_{\alpha,i}u_{\alpha,j}. \qquad [2.3]$$

The values of J's are

$$J_1 = \frac{1}{1!}\,\delta^i_j e^j_i$$

$$J_2 = \frac{1}{2!}\,\delta^{ik}_{jl}\, e^j_i e^l_k$$

$$J_3 = \frac{1}{3!}\,\delta^{ikm}_{jln}\, e^n_m e^l_k e^j_i \qquad [2.4]$$

Referred to principal axes the relation [2.2] reduces to

$$(1 - 2e_{11})(1 - 2e_{22})(1 - 2e_{33}) \to 0. \qquad [2.5]$$

For generalized measure this takes the form [1]

$$(1 - ne_{11})(1 - ne_{22})(1 - ne_{33}) \to 0. \qquad [2.6]$$

This shows that at yielding

$$e_{11}, e_{22}, e_{33} \to \left(\frac{1}{n}\right). \qquad [2.7]$$

These conditions hold good for any medium-homogeneous, heterogeneous, isotropic or aelotropic. So they can be made the starting point to obtain the yield condition in terms of the stress invariants, I's. This is not always easy. Firstly, some constitutive equation has to be assumed and then except, in the isotropic case, I's cannot be expressed in terms of J's. Thus, though [2.2] is linear in J's it is found that, even for the isotropic case, the condition in terms of I's is cubic. For the aelotropic case, it is found to be of the sixth degree in I's. All these conditions include the classical yield conditions as particular cases.

Isotropic body

Assuming a linear stress-strain tensor relation, which is found to be adequate for our purpose, we have

$$e_{ij} = E^{-1}[(1 + \sigma)\tau_{ij} - \sigma\delta_{ij}\tau_{\alpha\alpha}] \qquad [2.8]$$

where E is *Youngs* modulus and σ the *Poissons* ratio. In the transition region they change and become the response coefficients. In fact $\sigma \to \frac{1}{2}$ as the fully plastic state is reached. From [2.8] we readily get the following relations

$$EJ_1 = (1 - 2\sigma)I_1$$

$$E^2J_2 = (1 + \sigma)^2 I_2 - \sigma(2 - \sigma)I_1^2$$

$$E^3J_3 = [(1 + \sigma)^3 I_3 - \sigma(1 + \sigma)^2 I_1 I_2 + \sigma^2 I_1^3]. \qquad [2.9]$$

Substituting these values in [2.2] and making use of the relation

$$I_2' = 2I_1^2 - 6I_2 \qquad [2.10]$$

the transition condition for elastic plastic deformation is found to be

$$E^{-1}(1 - 2\sigma)I_1 + \tfrac{2}{3}E^{-2}[\tfrac{1}{2}(1 + \sigma)^2 I_2' - (1 - 2\sigma)I_1^2] + 4E^{-3}[(1 + \sigma)^3 I_3 + \tfrac{1}{6}\sigma(1 + \sigma)^2 I_1 I_2' - \tfrac{1}{3}\sigma(1 - \sigma + \sigma^3)I_1^3] = \tfrac{1}{2}. \qquad [2.11]$$

In [2.11] the elastic effect is present through E and σ. For the fully plastic state $\sigma \to \frac{1}{2}$ and we get

$$3I_2' + 2[27I_3 + \tfrac{3}{2}I_2' I_1 - I_1^3] = 2 \qquad [2.12]$$

where the invariants I's have been made non-dimensional with respect to E. E can now be expressed in terms of the yield stress in tension. Putting T_{22}, $T_{33} = 0$ in [2.12] we get

$$3(T_{11}/E)^2 + 2(T_{11}/E)^3 = 1 \qquad [2.13]$$

which gives

$$T_{11} = \tfrac{1}{2}E \quad \text{or} -E. \qquad [2.14]$$

If Y is the yield stress in tension, then

$$Y = \tfrac{1}{2}E.$$

From [2.14] we note the *Bauschinger* effect – the yield stress in compression is different from that in tension and is twice the value of the latter.

In terms of the principal stresses, [2.12] can be put in the following convenient forms:

$$3I_2' + 2(3\tau_{11} - I_1)(3\tau_{22} - I_1)(3\tau_{33} - I_1) = 2 \qquad [2.15]$$

$$\sum(\tau_{11}')^2 + 6\tau_{11}'\tau_{22}'\tau_{33}' = 2/g \qquad [2.16]$$

where T_{11} ... etc. is the deviatoric stress in a non-dimensional form.

In a large number of classical problems of torsion, flexure, plane stress and plane strain I_3 vanishes identically and we get from [2.12] the yield condition in the form (4).

$$3I_2^1 = 2(1 - I_1 + I_1^2), \quad I_1 \neq -1. \qquad [2.17]$$

This should be used for cases like that of combined loads in the form of tension, torsion, flexure. From the classical point of view it will be interpreted as showing work-hardening due to the presence of the terms I_1 and I_1^2 on the right hand side. *Stassi* (5) uses a similar condition.

For the particular cases of the *Haar-Kármán* (6) hypothesis and the principal line theory (7) where $T_{33} = T_{22}$ and $T = \frac{1}{2}(T_{11} + T_{22})$ respectively, we see that [2.15] and [2.16] reduce to the form of *Tresca*. For plane stress we readily get (8).

$$T_{11} - \tfrac{1}{2}T_{22} = Y$$

$$T_{11} - 2T_{22} = -2Y \qquad [2.18]$$

for tension and compression respectively. Thus we see that the condition [2.12] contains all the classical conditions. It also shows that elastic failure depends on the type of deformation and can occur in any of the following cases.

(i) The elastic energy of deformation reaches a critical value.
(ii) The principal strain becomes a maximum.
(iii) The principal stress becomes a maximum.
(iv) The principal stress difference becomes a maximum.

3. Aelotropic body

In this case the linear stress-strain tensor relation is

$$\Delta e_{ij} = C_{ij}^{hk}\tau_{hk}, \quad \Delta = |C_{ij}|. \quad [3.1]$$

C_{ij} being elastic response coefficients. If the strain energy function exists, the 36 response coefficients reduce to 21 as

$$C_{ij}^{hk} = C_{ji}^{hk}.$$

Now the principal axes of strain and stress are not parallel except in the orthotropic case. We take the axes of principal strains as the coordinate axes, of which at least one set exists. As indicated in [2.7] we can take $e_{11} \to \frac{1}{2}$, at the elastic-plastic transition. Thus we get the four linear equations in τ_{hk} as

$$\tfrac{1}{2}\Delta = C_{11}^{hk}\tau_{hk}, \quad 0 = C_{12}^{hk}\tau_{hk}$$
$$0 = C_{23}^{hk}\tau_{hk}, \quad 0 = C_{13}^{hk}\tau_{hk}. \quad [3.2]$$

We also have the first stress-invariant given by

$$I_1 = \tau_{11} + \tau_{22} + \tau_{33}. \quad [3.3]$$

The eqs. [3.2] and [3.3] are linear in the six components of τ_{ij}. We can determine five of them in terms of the sixth, (say τ_{11}) and I_1 and C's. We substitute their values in the remaining two invariants.

$$I_2 = \tau_{11}I_1 - \tau_{11}^2 + \tau_{22}\tau_{33} - \tau_{23}^2 - \tau_{31}^2 - \tau_{12}^2 \quad [3.4]$$

$$I_3 = \tau_{11}I_2 - \tau_{11}^2 I_1 + \tau_{11}^3 + 2\tau_{12}\tau_{23}\tau_{31} + \tau_{12}^2(\tau_{11} - \tau_{33}) + \tau_{31}^2(\tau_{11} - \tau_{22}). \quad [3.5]$$

Thus we get the following two equations in τ_{11}

$$I_2 = \alpha_1\tau_{11}^2 + \alpha_2\tau_{11} + \alpha_3 \quad [3.6]$$

$$I_3 = \beta_1\tau_{11}^3 + \beta_2\tau_{11}^2 + \beta_3\tau_{11} + \beta_4 \quad [3.7]$$

where α's and β's are functions of I_1, I_2 and C's. τ_{11} can be easily eliminated between [3.6] and [3.7]. Thus we get the transition invariant relation.

$$[(\alpha_3 - I_2)(\alpha_1\beta_3 - \alpha_3\beta_1 + \beta_1 I_2) - \alpha_1\alpha_2(\beta_4 - I_3)] \times [\alpha_2(\alpha_1\beta_2 - \alpha_2\beta_1] - \alpha_1(\alpha_1\beta_3 - \alpha_3\beta_1 + \beta_1 I_2)] = [\alpha_1^2(\beta_1 - I_3) - (\alpha_3 - I_2)(\alpha_1\beta_2 - \alpha_2\beta_1)]. \quad [3.8]$$

This is of the sixth degree in I's. By imposing on the response coefficients C's a suitable condition we can get the plasticity condition. This may be illustrated by taking the simpler case of the orthotropic body in which the principal axes of stress and strain are parallel.

4. Orthotropic body

Referred to the principal axes the relation [3.1] becomes

$$\Delta e_{ij} = C_{ij}^{hk}\tau_{hk}, \quad C_{ij}^{hk} = 0, \quad h \neq k. \quad [4.1]$$

There are now three *Youngs* moduli and six *Poissons* ratios (9). For example, for tension across the plane 1, the corresponding values are (9).

$$E_1 = \Delta/C_{11}^{11}$$
$$\sigma_{12} = -C_{11}^{22}/C_{11}^{11}$$
$$\sigma_{13} = -C_{11}^{33}/C_{11}^{11}. \quad [4.2]$$

It may be noted that

$$\sigma_{21} = -C_{11}^{22}/C_{22}^{22}$$

which is not equal to σ_{12}.

The general value of the modulus of compression, k, is

$$1/k = 1/E_1(1 - \sigma_{12} - \sigma_{13}) + 1/E_2(1 - \sigma_{21} - \sigma_{23}) + 1/E_3(1 - \sigma_{31} - \sigma_{32}). \quad [4.3]$$

For the isotropic case

$$\frac{1}{k} = \frac{3(1 - 2\sigma)}{E}. \quad [4.4]$$

Since E's are independent of each other, $1/k$ will vanish only if simultaneously we have

$$\sigma_{ij} + \sigma_{ij} = 1, \quad i \neq j, \quad i \neq k. \quad [4.5]$$

These are the plasticity conditions corresponding to $\sigma \to \frac{1}{2}$ of the isotropic case. For the transition condition [3.8] we notice that now

$$I_3 = \tau_{11}^3 - \tau_{11}^2 I_1 + \tau_{11}I_2 \quad [4.6]$$

26

so that

$$\beta_1 = 1, \quad \beta_2 = -I_1, \quad \beta_3 = I_2, \quad \beta_4 = 0.$$

Substituting these values we see that [3.8] reduces to

$$[\alpha_1\alpha_2 I_3 + (\alpha_3 - I_2)(\alpha_1 I_2 + I_2 - \alpha_3)] \times [\alpha_1(\alpha_1 I_2 + I_2 - \alpha_3) + \alpha_2(\alpha_1 I_1 + \alpha_2)] + [(\alpha_3 - I_2)(\alpha_1 I_1 + \alpha_2) - \alpha_1^2 I_3]^2 = 0 \quad [4.7]$$

where α's are given by

$$\alpha_1 + 1 = -D^2(1 + \sigma_{12})(1 + \sigma_{13})$$

$$\alpha_2 = D^2[\tfrac{1}{2}E_1(2 + \sigma_{12} + \sigma_{13}) + I_1(\sigma_{12} + \sigma_{13} + \sigma_{12}^2 + \sigma_{13}^2)$$

$$\alpha_1 I_1 + \alpha_2 = D^2[\tfrac{1}{2}E_1(2 + \sigma_{12} + \sigma_{13}) + I_1(1 - \sigma_{12}\sigma_{13})]$$

$$\alpha_3 = -\tfrac{1}{4}D^2(E_1 + \sigma_{12}I_1)(E_1 + \sigma_{13}I_1)$$

$$1/D^2 = (\sigma_{12} - \sigma_{13})^2 = (\sigma_{12} + \sigma_{13})^2 - 4\sigma_{12}\sigma_{13}. \quad [4.8]$$

These contain only I_1, the *Youngs* modulus and the *Poissons* ratios.

In the particular case when $I_3 = 0$, the relation [4.7] reduces to

$$I_2 = \alpha_3 \quad [4.9]$$

$$[(\alpha_1 + 1)I_2 - \alpha_3] \times [\alpha_1 I_2(\alpha_1 + 1) - \alpha_1\alpha_3 + \alpha_2(\alpha_1 I_1 + \alpha_2)] + (\alpha_3 - I_2)(\alpha_1 I_1 + \alpha_2)^2 = 0. \quad [4.10]$$

The relation [4.9] is of the *von-Mises type*. Also [4.10] is now of the fourth order in I's.

The response coefficients may be expressed in terms of the yield stress in tension and compression. Putting $\tau_{22}, \tau_{33} = 0$, in [4.7] we get

$$\tau_{11} = \tfrac{1}{2}E_1 = Y \quad [4.11]$$

$$\tau_{11} = -\tfrac{1}{2}E_1/\sigma_{12} = -\tfrac{1}{2}E_2/\sigma_{21} = Y_1 \quad [4.12]$$

$$\tau_{11} = -\tfrac{1}{2}E_1/\sigma_{13} = -\tfrac{1}{2}E_2/\sigma_{31} = Y_2. \quad [4.13]$$

Thus all α's in [4.8] can be expressed in term of Y, Y_1 and Y_2. For the fully plastic case when

$$\sigma_{12} + \sigma_{13} = 1$$

we get

$$1/Y + 1/Y_1 + 1/Y_2 = 0. \quad [4.14]$$

The following particular cases may be noted.

a) $\tau_{22} = \tau_{33}$. Now we get for tension

$$\tau_{11} - \tau_{33} = \frac{\tfrac{3}{2}E_1 - I_1(1 - \sigma_{12} - \sigma_{13})}{2 + \sigma_{12} + \sigma_{13}}$$

which reduces to the *Trescas* form

$$\tau_{11} - \tau_{33} = Y$$

when

$$\sigma_{12} + \sigma_{13} = 1. \quad [4.15]$$

For compression we get

$$\tau_{11} - \tau_{33} = Y_1 \quad \text{or} \quad Y_2. \quad [4.16]$$

b) Plane stress. – Now we can put

$$\tau_{3i} = 0, \quad i = 1, 2, 3.$$

The corresponding yield conditions for tension and compression are found to be

$$\tau_{11}/Y + \tau_{22}/Y_2 = 1$$

and

$$\tau_{11}/Y_1 + \tau_{22}/Y_2 = 1. \quad [4.17]$$

c) Plane strain. – Now $e_{33} = 0$, we get

$$I_2 = \sum(\tau_{11} - \tau_{22})^2 = \tfrac{9}{4}E_1^2(1 + \sigma_{12}^2 + \sigma_{13}^2), \quad \sigma_{12} + \sigma_{13} = 1. \quad [4.18]$$

But

$$\sigma_{31}(\tau_{33} - \tau_{11}) = \sigma_{32}(\tau_{22} - \tau_{33}).$$

Thus we have the reduced form.

$$\tau_{11} - \tau_{22} = \frac{Y}{1 - \sigma_{13}\sigma_{31}}. \quad [4.19]$$

d) When $\tau_{22} = \tfrac{1}{2}(\tau_{11} + \tau_{33})$ we again get a result similar to that in a).

In all these cases we see that the orthotropic yield conditions are of the *Tresca* type.

Summary

One of the main problems in Rheology is the determination of the yield condition for which no satisfactory form seems to exist. If a start is made with the stress tensor field, one soon runs into a number of difficulties. It is therefore better to deal with the geometry of the field. The change from elastic to plastic deformation can be interpreted as a mapping of one space into another, and the yield as an asymptotic sub-space. If the elastic strain field is e_{ij}, whose invariants are J_1, J_2, J_3, then an asymptotic behaviour may be represented by the existence of a functional relation $f(J_1, J_2, J_3) = 0$, between the J's, which are independent of one another in the normal part of the field.

This does not fix the nature of the function f, for which we can invoke the additional geometric condition that yielding can result from infinite contraction or expansion of a macro-element. Thus the *Jacobian* of the mapping must take on the singular values zero or infinite. These concepts give rise to the yield condition in the strain tensor field in the form

$$8J_3 - 4J_2 + 2J_1 \to 1 .$$

If generalized measure of strain is used, which is necessary for creep problems, this takes the form

$$n^3 J_3 - n^2 J_2 + nJ_1 \to 1 ,$$

n being a real constant, which is equal to 2 for the *Almansi* measure. These conditions do not depend on either the isotropicity or homogeneity of the field, and hence should be used for all types of yield conditions.

References

1) *Seth, B. R.*, Prikl. Math. Mech. **27**, 380–382 (1963); Generalized strain and transition concepts for elastic-plastic deformation, creep and relaxation, Proc. XI Intern. Congress Applied Mech., Munich, pp. 383 to 389 (1964); ZAMM **50**, 617–621 (1970).

2) *Hill, R.*, The mathematical theory of plasticity, pp. 318–319 (Oxford 1950).

3) *Hill, R.*, J. Mech. Phys. Solids **16**, 229–242, 315–322 (1968).

4) *Olszak, W.* and *P. Peryna*, J. Math. Phys. Sci. **2**, 67–72 (1968).

5) *Stassi, F. D'Alls*, J. Italian Assoc. Theo. Appl. Mech. **2**, 178–195 (1967).

6) *Haar, A.* and *T. Kármán*, Zur Theorie der Spannungszustände im plastischen und sandartigen Medien, S. 2 (Göttingen 1909).

7) *Lippman, H.*, Sbovnik Perevodov Mechanika **1963**, 3.

8) *Thomas, T. Y.*, J. Math. Mech. **17**, 987–1004 (1968).

9) *Love, A. E. H.*, Mathematical theory of elasticity, pp. 161–162 (1927).

10) *Borah, B. N.*, Ind. J. Pur. Appl. Maths. **2**, 335–343 (1971).

11) *Morgan*, Int. J. Engg. Sci. **4**, 155 (1966).

12) *Fox*, Acta Mech. **7**, 248 (1969).

Author's address:

Dr. *B. R. Seth*
Birla Institute of Technology
Mesra, Ranchi (Bihar, India)

Rheol. Acta 13, 400–407 (1974)

From the Department of Polymer Chemistry, Faculty of Engineering, Kyoto University, Kyoto (Japan)

Rheology of polymers under extremely high pressure

By M. Horio

With 12 figures

(Received October 27, 1972)

I. Introduction

The behavior of synthetic polymers under high pressure has been studied extensively by many research workers from scientific and practical point of view. These studies may be classified into two categories. The first includes those studies which treat the relation between pressure and volume at various temperatures (1–12). In the course of studies being made of equilibrium state the time-dependent characteristics of P–V–T relation were found (4, 5, 6, 8, 11). Melting temperatures of polymers were measured as a function of pressure (4, 5, 6, 13, 14, 17). The mechanical properties (15, 16), density (4, 5, 6, 8, 14), crystalline and supermolecular structure (17–21) of polymers solidified under high pressure have also been the subjects of studies. Further, the real and imaginary parts of complex viscosity of pressurized polymer liquids have been studied (22–26). The second category, on the other hand, deals with extrusion properties of polymers under pressure. Conventionally, polymers have been processed by screw- or ram-extruders, but particular attention should be paid, in addition, to the hydrostatic extrusion considering its future development.

The increase in pressure of extrusion may be the direct way to elucidate the flow properties of polymer liquids under high rate of shear, but an alternative method is the dynamic measurement. The preparation of master curve from dynamic data obtained at different frequencies and temperatures makes it possible to estimate the dynamic viscosity over very wide range of frequency at a definite reduced temperature. If the real and imaginary parts of complex viscosity at the angular frequency $\omega\,\sec^{-1}$ could be correlated in any way with the steady flow viscosity at the rate of shear given by the same value $\omega\,\sec^{-1}$, the steady flow viscosity of polymer liquid may be estimated over very wide range of rate of shear by the aid of dynamic measurements. Attempts have been made along this line by several authors (27–34), but *Onogi* and *Masuda* (35) have shown that the steady flow viscosities measured at various rates of shear can most favorably be expressed in terms of dynamic data by the empirical formula presented by *Cox* and *Merz* (33). Although this method may be applicable to evaluation of flow properties of polymers under high pressure, the validity of argument is naturally limited, because no real information about the practical feature of extrusion can be derived from this indirect method. One of the purposes of the present study, therefore, is to observe directly the flow properties of polymer liquids at very high rates of shear, although the results are still preliminary at the present stage of experiments.

Another important field of polymer research is the extrusion of polymer solids by high pressure. Recently we find several publications in this field, but above all the studies of *Backley* and *Long* (36), *Southern* and *Porter* (37) and a series of systematic works being carried on by *Takayanagi*, *Imada* and collaborators (38–42) are very interesting and instructive. In these experiments of cold extrusion the polymer solid is extruded very slowly at relatively small extrusion ratios by the pressure not so higher than the critical value at which the polymer solid just begins to flow through the die. In the present study, on the contrary, very high pressure was applied upon polymer solids to effect the rapid extrusion even at high extrusion ratios.

II. Apparatuses

Two types of equipments have been used. One is an Instron compressor and the other is a hydrostatic extrusion press. The former has been used for measurement of P–V–T relation, and the latter for extrusion experiments. Mention will be made briefly of the hydrostatic extruder, constructed recently by Kobe Steel, Ltd. (43). The extruder is combined with a five hundred ton hydrostatic press. The main items of the hydrostatic extruder are given below:

Maximum extrusion pressure	15000 kg/cm^2
Maximum press force	500 tons
Extremities of size of billet	10 $\varphi \times$ 1700 mm and 50 $\varphi \times$ 500 mm
Speed of stem	up to 20 mm/sec
Stroke of stem	600 mm

The sketch and photograph of the press are shown in figs. 1 and 2, respectively.

The machine is composed of a pump unit and a high pressure system. The latter is made up of a billet container, connecting cylinder and pressure container. The extruder is provided at its outlet with a long channel, through which the extrudate travels with a very high speed.

The work of extrusion is partially transformed into heat. The temperature rise in extrusion depends on the extrusion ratio, speed and time of extrusion, specific heat and thermal conductance of the product and so forth. In the present case of extrusion of polymers the temperature rise was estimated at larger than 100 °C.

The dies having an included angle of 45° were used. The dies with various hole diameters were prepared so that the extrusion ratio could be changed between 2 to 1000 for the billet of 15 mm in diameter. Castor oil was used as a high pressure fluid.

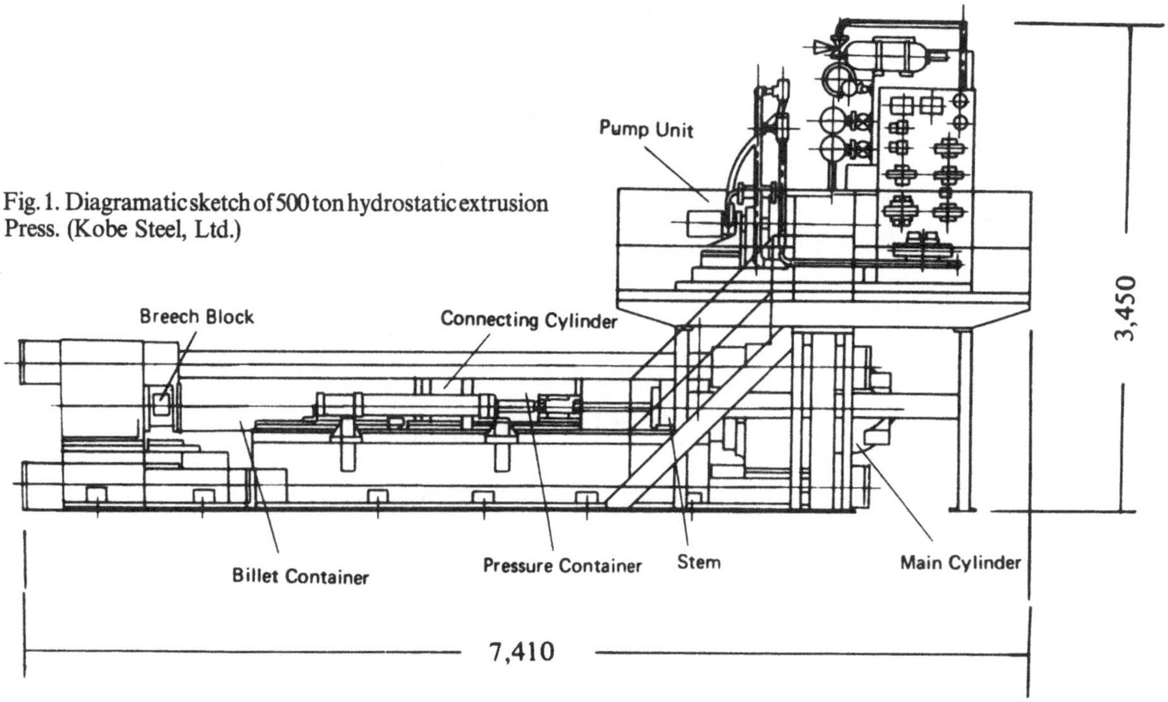

Fig. 1. Diagramatic sketch of 500 ton hydrostatic extrusion Press. (Kobe Steel, Ltd.)

III. Hydrostatic extrusion of polymer melts

The preliminary tests of hydrostatic extrusion of melt of polymer has been done first with high density polyethylene. To extrude the polymer in molten state, the billet was warmed preliminarily up to 110 °C before extrusion. The die with a hole diameter of 1 mm and a billet of 15 mm in diameter were used. The extrusion ratio is then calculated at 225. The pressure at extrusion was 7500 kg/cm^2. The shear stress is calculated at $2 \cdot 10^9$ dyne/cm^2. The velocity of flow was too high to be accurately measured, but the time of extrusion of 20 cm^3 of sample was assumed to be less than one tenth of second. The average linear velocity of flow could therefore be estimated at greater than 250 m/sec. The calculation on the rough assumption of laminar flow results in that the rate of shear would be greater than $2 \cdot 10^6$ sec^{-1}, and the apparent viscosity would be smaller than about 1000 poises. *Reynolds* number under this condition is roughly estimated at above 25. The zero-shear viscosity of the same sample at 190 °C is about 10^6 poises. Therefore, the apparent viscosity is reduced to less than one thousandth at the shear stress of the order of $2 \cdot 10^9$ dyne/cm^2 and the rate of shear of $2 \cdot 10^6$ sec^{-1}. The melting temperature of polyethylene is raised up to about 240 °C at 7500 kg/cm^2 (14), but it could be assumed that the preheated polyethylene would have been heated up to this

Fig. 2. Photograph of 500 ton hydrostatic extrusion press. (Kobe Steel, Ltd.)

temperature in die at extrusion. Notwithstanding, the continuous extrudate could not be obtained. The product is granular as can be seen in fig. 3.

The extrusion ratio was decreased by replacing the die by another one with a wider hole measuring 2 mm in diameter. The polyethylene preheated at 110 °C was extruded at the pressure of 3000 kg/cm^2. The shear stress is calculated at $1.5 \cdot 10^9$ dyne/cm^2 and extrusion ratio at 56. It was assumed that the polymer was extruded in molten state, but a continuous product with a circular cross-section was not obtained. It is obvious that the flow was considerably unstable,

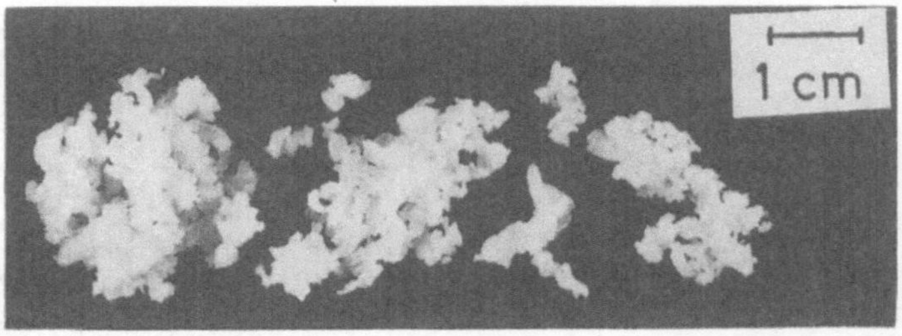

Fig. 3. High density polyethylene extruded at 7500 atm in molten state by 500 ton hydrostatic extrusion press. Extrusion ratio is 225

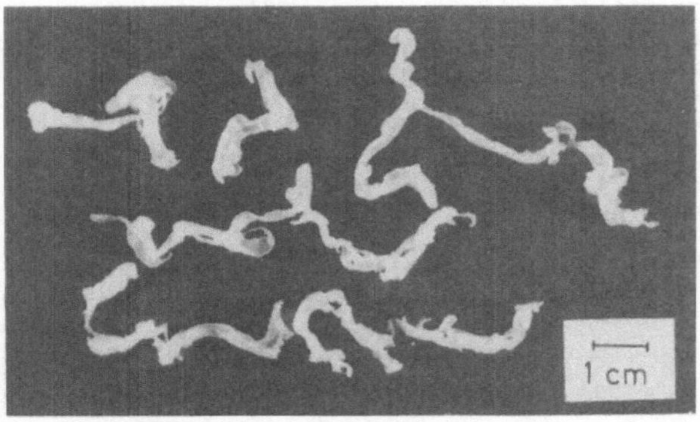

Fig. 4. High density polyethylene extruded at 3000 atm in molten state by 500 ton hydrostatic extrusion press. Extrusion ratio is 56

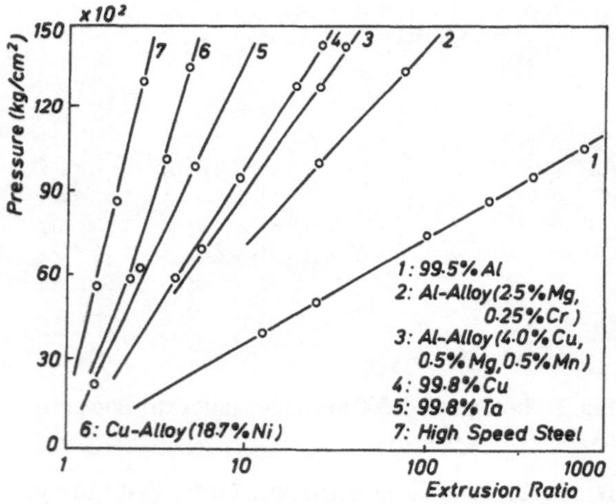

Fig. 5. Extrusion pressure vs. extrusion ratio on logarithmic scale for several metal samples. (Experiments by 500 ton hydrostatic extrusion press)

as can be deduced from the shape of product shown in fig. 4.

It is interesting to see that the state of extrusion of polymer melts at extremely high shear stresses as above is very similar to what is known as melt fracture which frequently occurs in practice at much lower shear stresses. Although we have not yet succeeded in obtaining continuous and uniform products from polymer melts at very high rates of shear and at high extrusion ratios, the experiment along this line will be continued to eliminate obstacles hindering polymers from stable extrusion at very high pressure.

IV. Hydrostatic extrusion of polymer solids

Only little reports have been published thus far on the hydrostatic extrusion of polymer solids, although this process might be expected to be developed into a new important method of polymer processing in near future. Before treating polymer extrusion the characteristic feature in extrusion of metals by the 500 ton hydrostatic press will be explained in advance. In fig. 5 the extrusion pressure is shown as a function of extrusion ratio for several metal samples[1]).

As has been recognized by *Pugh* and *Low* (44), the hydrostatic pressure of extrusion of the metal samples is related approximately in linear manner with the logarithm of extrusion ratio.

Similar experiments have been conducted with several samples of polymer using the same extrusion press. The results are shown in fig. 6.

It is to be noticed first that the extrusion pressure for polymer samples is linearly related with logarithm of extrusion ratio, as was the case for metal samples. However, the noticeable difference between metal and polymer extrusion consists in the fact that while the polymers tend to expand radially immediately after extrusion, the diameter of metal extrudate is generally smaller than that of die hole. The reduction of diameter in the latter case may be due to the lubricant layer that surrounded the billet in die during extrusion and the thermal shrinkage of extruded product. This conspicuous difference between metals and polymers implies that the deformation of metal samples in extrusion can be regarded as simply plastic, but the elasticity is an essential

[1]) The data were kindly furnished by Kobe Steel, Ltd.

factor that affects the deformation of polymers under high pressure.

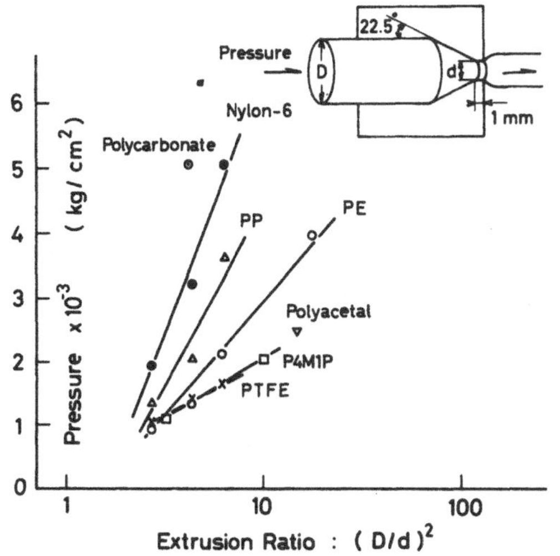

Fig. 6. Extrusion pressure vs. extrusion ratio on logarithmic scale for several polymer samples. (Experiments by 500 ton hydrostatic extrusion press)

The extrusion curves of polymer samples are compared in fig. 7 with those of typical metal samples. So far as the slope of curve is concerned, the extrusion characteristic of polyethylene is similar to that of aluminum, while nylon 6 resembles copper. However, the ductilities of polymers are generally very limited. The metal samples yield the uniform and smooth products within the range of pressure and extrusion ratio given in figs. 5 and 7, but the polymer samples make the uniform products with desired shapes only at smaller extrusion ratios. The poor ductilities of polymer samples are shown by photographs in figs. 8 [2]) and 9 [2]).

The ductility of polymer differs from one to another, but, on the whole, the polymers are "difficult materials". At the extrusion ratio of 2.8 (fig. 8) polyethylene, polypropylene and nylon 6 yield satisfactory products, but polytetrafluoroethylene gives only fractured product. It is noticeable that the last product is similar in appearance to the melt fractured sample, although the present experiment deals with extrusion of polymer solids. Of all polymer solids treated in this study nylon 6 was most ductile. At the extrusion ratio of 6.25 only nylon 6 could be formed into required shape, while polypropylene, polyethylene and polytetrafluoroethylene yielded cracked products, as shown in fig. 9.

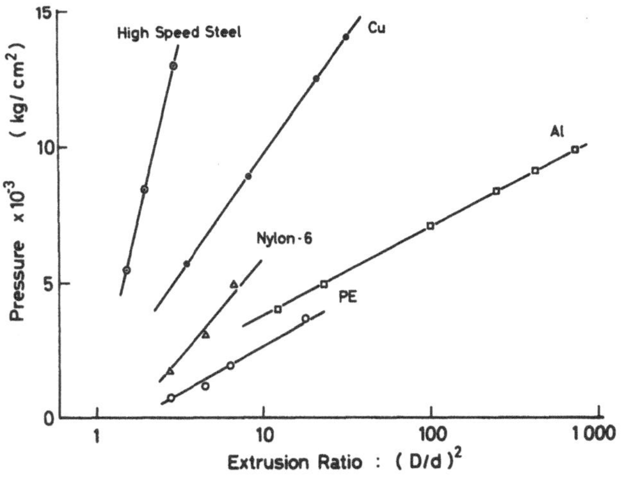

Fig. 7. Comparison of extrusion curves of several polymer samples with those of typical metal samples

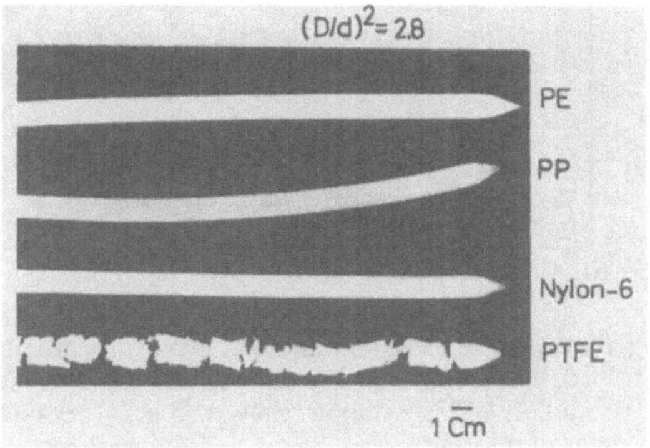

Fig. 8. Photographs of extrudates of several polymer samples at the extrusion ratio of 2.8. (Hydrostatic extrusion)

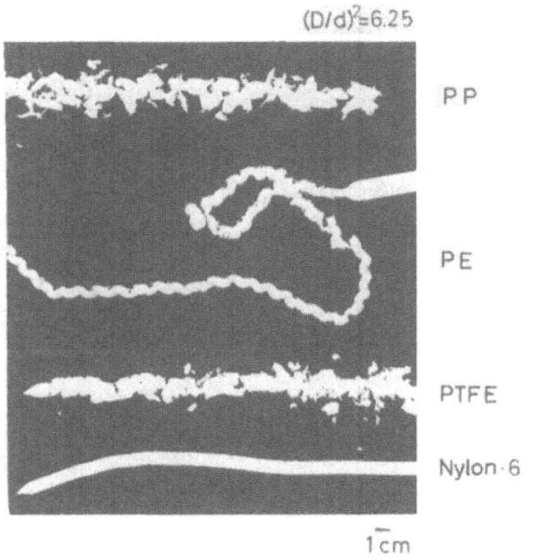

Fig. 9. Photographs of extrudates of several polymer samples at the extrusion ratio of 6.25. (Hydrostatic extrusion)

As mentioned before, the polymer solids expand radially on emerging from die into air. Fig. 10 shows the radial expansion which took place at the hydrostatic extrusion of polyethylene,

---

[2]) Thanks are due to Kobe Steel, Ltd. for providing us with photographs in figs. 8 and 9.

nylon 6 and poly(4-methyl-1-pentene) at various extrusion ratios.

It can be seen in fig. 10 that the diameter of extruded product can be greater than 1.4 times the diameter of die hole. Provided the length remained unchanged, the volume of polymer was more than doubled on extrusion.

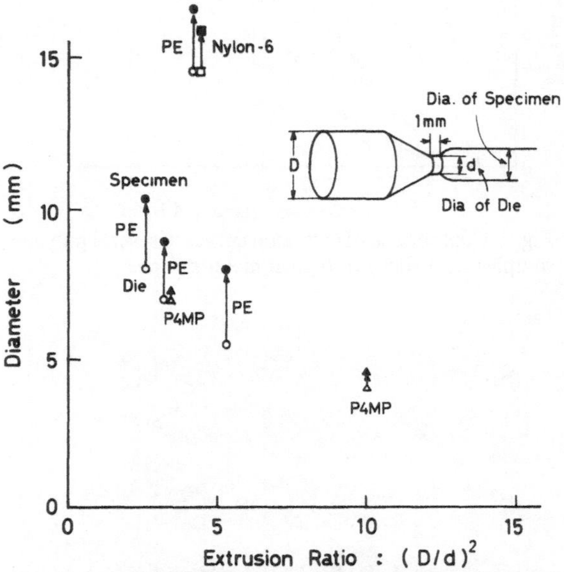

Fig. 10. Diameter of die hole and that of polymer product extruded therefrom. Open symbols show the diameter of die holes, solids symbols show those of extruded products of polymers

contraction of polymer on compression does not instantly reach the ultimate value, but takes a considerably long time to reach the equilibrium. They have connected this effect with the rate of pressure-induced crystallization. Recently, *Yamaguchi* et al. (11) have shown that the contraction and expansion of polymer melt can be expressed in two terms. One of them comes from the purely elastic effect which takes place instantly on loading or unloading, while the other is related to the viscoelastic effect which is interpretable by a simple *Voigt*-model characterized by a single retardation time. This viscoelastic behavior is common to all polymer materials and seems to be an inherent character of polymer molecules, for this is not recognizable in low molecular substances such as *n*-paraffins whose change in volume on compression takes place without any retardation. Fig. 11 shows the P–V-curves of high density and low density polyethylene samples in comparison with those of *n*-paraffin. The compression curves of high density and low density polyethylenes at the rate of 0.5–10% per minute are very different from the curves obtained under equilibrium state. This shows that the compression of polyethylen samples is the time-dependent effect. On the other hand, the compression curve of *n*-paraffin at the rate of 0.5 to 10% per minute and the curve obtained under

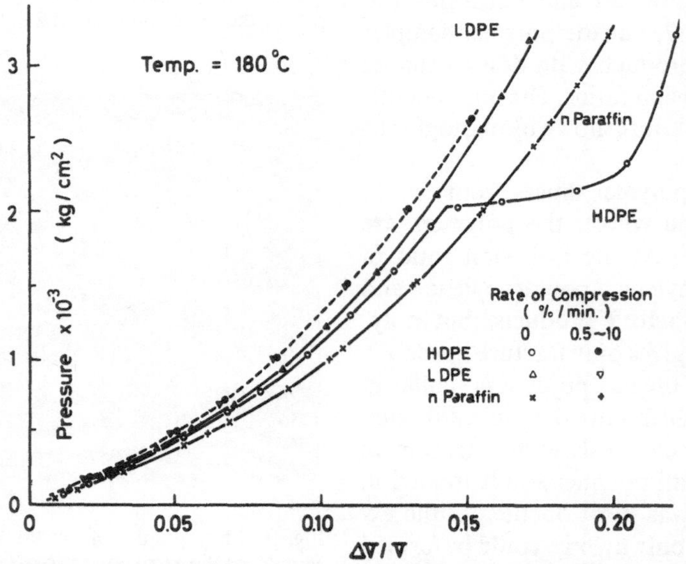

Fig. 11. Compression curves at the rate of 0.5–10% per minute and under equilibrium state of high density and low density polyethylenes and *n*-paraffin

The compression and expansion of polymer melts are time-dependent effect. *Matsuoka* and *Maxwell* (4–6) have found first that the volume equilibrium are completely superposed. This indicates that the compression of *n*-paraffin is simply elastic without any indication of viscoelasticity. The reason of melt fracture has not yet been completely interpreted, but it can be postulated that the viscoelastic volume change of polymer would be an important origin of this characteristic effect. The present study suggests further that the postulation would be applicable also to the extrusion of polymer solids.

Of interest is now the supermolecular structure of products obtained by the hydrostatic extrusion press. The X-ray diffraction patterns of products obtained by cold extrusion method have been investigated by *Imada* et al. (39–42). In the ram-extrusion the polymer solid is slowly extruded, and this allows polymer molecules to extend and array themselves regularly to form crystallites. *Imada* et al. found that the double orientation, namely axial plus planar orientation, can be seen in the extrudates of polyethylene and nylon 6. The degree of orientation of extrudates is as perfect as the cold-drawn samples with the same degree of stretching.

The results of the present study on hydrostatic extrusion which takes place very rapidly are considerably different from those of ram-extrusion. Fig. 12 shows the X-ray diffraction patterns of specimens taken from different parts of the product of high density polyethylene extruded by hydrostatic press at the extrusion ratios 2.52 and 5.33.

The samples before extrusion are quite isotropic, because the diffraction patterns obtained with X-ray beam not only perpendicular but also parallel to the axis of billet are composed of rings with uniform intensity (pattern 0). The samples taken from the center of extruded product and the samples taken from the edge of product give very like patterns with the X-ray beam perpendicular to the direction of axis, as can be seen in patterns 1 and 3. The tendency of axial orientation is obviously recognizable in these photographs. The pattern 5 obtained with X-ray beam tangential to the product is very similar to patterns 1 and 3, and this makes the assumption of selective planar orientation very unlikely. The products are isotropic when viewed in parallel to its axis, as can be seen in patterns 2 and 4. It is

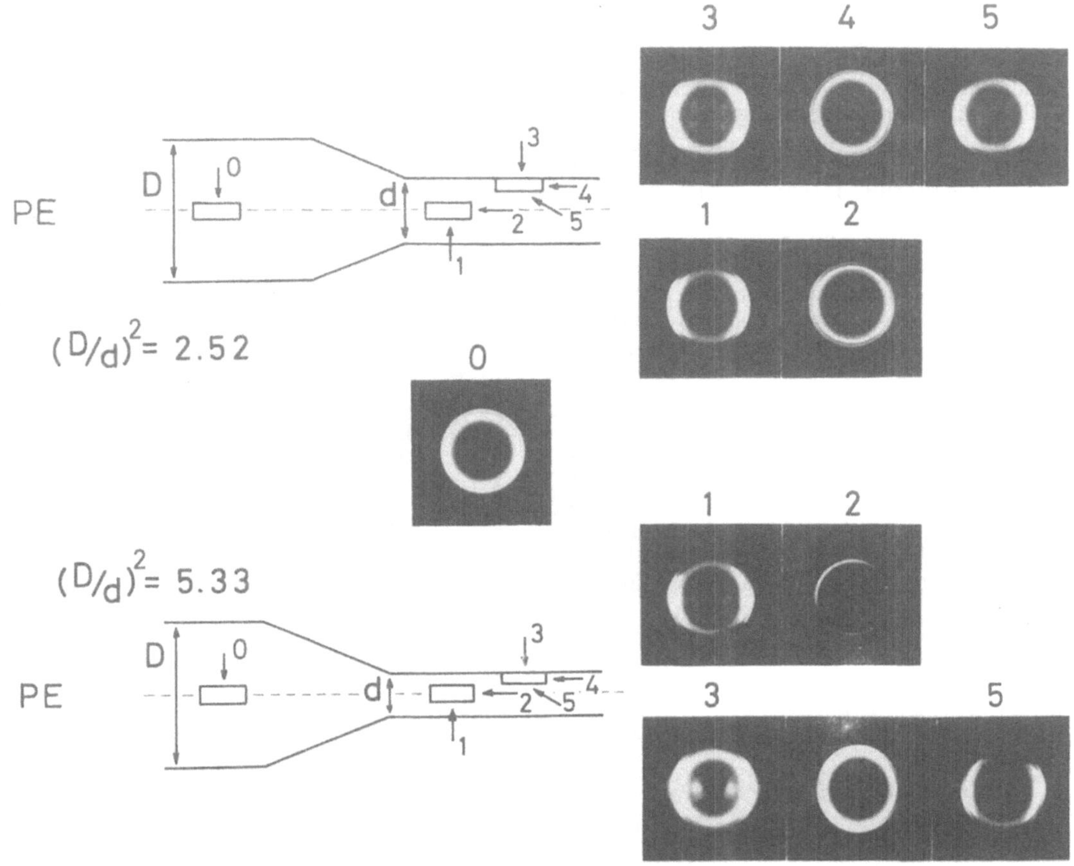

Fig. 12. X-ray diffraction patterns of specimens taken from different parts of the extruded products of high density polyethylene. Extrusion ratios are 2.52 and 5.33

important to notice that the degree of axial orientation estimated by X-ray diffraction patterns of hydrostatically extruded products of high density polyethylene is clearly smaller than those of the cold-drawn samples stretched to the similar degrees.

Thus far, we have not yet succeeded in obtaining satisfactory product at higher extrusion ratios. However, it is expected that the hydrostatic extrusion would be a more promising method than the conventional extrusion processes in overcoming difficulties. The hydrostatic extrusion process has many advantages over the screw- and ram-extrusion methods, and will make a new forming process in the technology of polymer processing. The friction is reduced due to the lubricant layer surrounding the surface of billet in die. The productivity can be increased. Various composite materials can be extruded. The most important point, when viewed from a different angle, would be the expectation that this process might help us in enriching our knowledge on the rheology of polymers under extremely high pressure.

Acknowledgements

Thanks are due above all to Dr. *M. Nishihara* of Kobe Steel, Ltd. for affording us a valuable opportunity to work with the five hundred ton extrusion press designed by him, without whose kind cooperation this study would not have been accomplished. The excellent works of Professor *M. Takayanagi* of Kyushu University and his collaborators on solid state extrusion of polymers and the personal kindness shown to me by Professor *Takayanagi* have been an encouraging stimulus to this work and I express my sincere thanks to him on this occasion. I wish to thank Dr. *S. Tokiura*, Director of Hirakata Plastics Laboratory, Ube Industries, Ltd., for his kind interest and assistances in this project; Dr. *T. Nagasawa* and *K. Oda* of the same Laboratory for their generous and heartfelt cooperation in experiments, discussions and what not which have made this study possible; Mr. *Y. Shimomura*, Mr. *H. Sasaki* and Mr. *Y. Koike* for their active help in experiments. Last but not least, I must thank Professor *S. Onogi* of Kyoto University for his invariable kindness in affording me valuable suggestions and discussions.

References

1) *Spencer, R. S.* and *G. D. Gilmore*, J. Appl. Phys. **20**, 502 (1949).

2) *Spencer, R. S.* and *G. D. Gilmore*, J. Appl. Phys. **21**, 523 (1950).

3) *Singh, H.* and *A. W. Noll*, J. Appl. Phys. **30**, 337 (1959).

4) *Matsuoka, S.* and *R. Maxwell*, J. Polymer Sci. **32**, 131 (1958).

5) *Matsuoka, S.*, J. Polymer Sci. **42**, 511 (1960).

6) *Matsuoka, S.*, J. Polymer Sci. **57**, 569 (1962).

7) *Haug, W. A.* and *R. G. Griskey*, J. Appl. Polymer Sci. **10**, 1475 (1966).

8) *Takayanagi, M., T. Takemura*, and *T. Oyama*, Kobunshi **14**, 100 (1965).

9) *Haward, R. N.*, Trans. Farad. Soc. **62**, 828 (1966).

10) *Haward, R. N.*, J. Polymer Sci. **A2**, 219 (1969).

11) *Yamaguchi, S.* and *Y. Oyanagi*, Kobunshi Kagaku **28**, 623 (1971).

12) *Ito, T.*, Kobunshi **21**, 82 (1972).

13) *Baer, E.* and *J. L. Kardos*, J. Polymer Sci. **A3**, 2827 (1965).

14) *Osugi, J.* and *K. Hara*, Nippon Kagaku Zasshi **88**, 33 (1967).

15) *Nolle, A. W.*, J. Appl. Phys. **31**, 1694 (1960).

16) *Mears, D. R., K. D. Pae*, and *J. A. Sauer*, J. Appl. Phys. **40**, 4229 (1969).

17) *Wunderlich, B.*, J. Polymer Sci. **A1**, 1245 (1963).

18) *Anderson, F. R.*, J. Polymer Sci. **C3**, 123 (1963).

19) *Wunderlich, B.* and *T. Arakawa*, J. Polymer Sci. **A2**, 3697 (1964).

20) *Geil, P. H., F. R. Anderson, B. Wunderlich*, and *T. Arakawa*, J. Polymer Sci. **A2**, 3707 (1964).

21) *Anderson, F. R.*, J. Appl. Phys. **35**, 64 (1964).

22) *Tokiura, S., S. Ogihara, T. Takaki*, and *H. Sasaki*, Proceedings of the 5th International Congress on Rheology, Vol. 4, 275 (1968).

23) *Tokiura, S., S. Ogihara, Y. Yamazaki, T. Takaki, K. Tabata*, and *H. Sasaki*, J. Soc. Materials Sci. (Japan) **17**, 365 (1968).

24) *Tokiura, S. S. Ogihara, A. Takeuchi, T. Takaki, K. Tanaka*, and *H. Sasaki*, J. Soc. Materials Sci. (Japan) **19**, 310 (1970).

25) *Takeuchi, A., T. Takaki, H. Sasaki*, and *K. Oda*, J. Soc. Materials Sci. **20**, 634 (1971).

26) *Ogihara, S., T. Takaki*, and *H. Sasaki*, Kobunshi Kagaku **28**, 319 (1971).

27) *De Witt, T. W.*, J. Appl. Phys. **26**, 889 (1955).

28) *De Witt, T. W., H. Markovitz, F. J. Padden*, and *J. Zapas*, J. Colloid Sci. **10**, 174 (1955).

29) *Markovitz, H.* and *B. Williamson*, Trans. Soc. Rheology **1**, 25 (1957).

30) *Onogi, S., I. Hamana*, and *H. Hirai*, J. Appl. Phys. **29**, 1503 (1958).

31) *Williams, M. C.* and *R. B. Bird*, Phys. Fluids **5**, No. 9 (1962).

32) *Pao, Y. H.*, J. Appl. Phys. **28**, 591 (1957).

33) *Cox, W. P.* and *E. H. Merz*, J. Polymer Sci. **28**, 619 (1958).

34) *Stella, S.*, J. Polymer Sci. **60**, 9 (1962).

35) *Onogi, S.* and *T. Masuda*, Nippon Kagaku Zasshi **88**, 231 (1967).

36) *Buckley, A.* and *H. A. Long*, Polymer Eng. and Sci. **9**, 115 (1969).

37) *Southern, J. H.* and *R. S. Porter*, J. Macromol. Sci.-Phys. **B4**, 541 (1970).

38) *Takayanagi, M., K. Imada*, and *T. Yamamoto*, Asahi-Garasu Kogyoshoreikai Kenkyuhokoku **17**, 273 (1970); **19**, 195 (1972).

39) *Imada, K., T. Yamamoto, K. Ueno, D. Matsukuma*, and *M. Takayanagi*, J. Soc. Materials Sci. (Japan) **19**, 302 (1970).

40) *Imada, K., T. Yamamoto, K. Kanekiyo*, and *M. Takayanagi*, J. Soc. Materials Sci. (Japan) **20**, 606 (1971).

41) *Imada, K., T. Yamamoto, K. Shigematsu,* and *M. Takayanagi,* J. Materials Sci. **6**, 537 (1971).

42) *Imada, K., T. Yamamoto, K. Kanekiyo,* and *M. Takayanagi,* Proc. Intern. Conf. Mechanical Behavior of Materials III, 476 (1972).

43) *Nishihara, M.,* The Second Czechoslovak Conference on High Pressure Forming (1971).

44) *Pugh, H. L. D.* and *A. H. Low,* J. Inst. Metals **93**, 201 (1964–1965).

Author's address:

Dr. *M. Horio*
Dept. of Polymer Chemistry
Faculty of Engineering
Kyoto University
Kyoto (Japan)

Rheol. Acta 13, 408–412 (1974)

Laboratoire de Glaciologie du C.N.R.S. Grenoble (France)

Contribution à l'étude des lois de glissement des glaciers tempérés à l'aide d'un viscosimètre à glace

Par R. Brepson

Avec 4 figures

(Reçu p. p. le 27 octobre 1972)

La théorie générale du glissement d'un glacier tempéré sur son lit (1, 2) définit deux types de glissement, avec ou sans cavitation qui peuvent s'ajouter.

a) Le glissement par fonte et regel.
b) Le glissement par plasticité.

Les expériences en laboratoire sur ce sujet ont été celles de *Kamb* et *La Chapelle* (3). Comme ils étudiaient le glissement d'une bille sous une pression hydrostatique faible, seul le glissement par fonte et regel intervenait.

Pour obtenir une déformation plastique de la glace sur les obstacles, il faut que ceux-ci soient assez grands et opèrer sous une forte pression hydrostatique. La pression interstitielle dans les cavités sous-glaciaires doit jouer un rôle important. Pour mieux comprendre le mécanisme du glissement et tenter de rattacher les conclusions des faits expérimentaux à celles du modèle mathématique, nous avons réalisé dans le cadre du Laboratoire de Glaciologie Alpine, une machine basée sur le principe du viscosimètre de *Couette* (4) permettant de traiter une trentaine de kilogrammes de glace sous des pressions de plusieurs dizaines d'atmosphères. La fig. 1 représente une coupe verticale de la machine.

La glace est constituée par damage de couches de cristaux de glace préalablement obtenus de pains de glace broyés, avec saturation d'eau, ceci

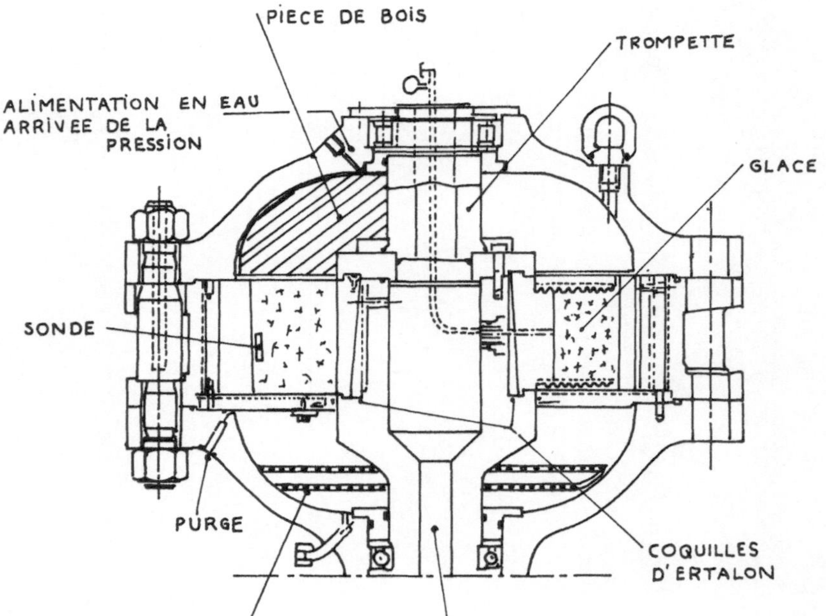

Fig. 1. Coupe du viscosimètre à glace. (Manquent la barre de torsion qui supporte le moyeu, la courronne dentée solidaire de l'enveloppe, l'entraînement de cette couronne par vis tangente, la jaquette calorifuge)

afin d'obtenir une texture non privilégiée, macroscopiquement isotrope.

La loi de frottement dépend essentiellement du modèle de lit rocheux que l'on adopte. Il doit être suffisamment simple pour donner prise au calcul et pourtant suffisamment réaliste. Le modèle idéal est le lit sinusoïdal. Dans la nature des analyses harmoniques de lits glaciaires font apparaître un spectre continu des ondulations pouvant être traduit par la juxtaposition de plusieurs longueurs d'onde. Les bosses d'une taille donnée se succèdent régulièrement si bien que selon *Lliboutry* le décollement de la glace à l'aval de l'une modifie l'écoulement sur la bosse suivante; c'est ainsi que les petites bosses étant superposées aux grandes, toute cavitation qui affecte les grandes ennoiera complètement une partie des petites.

Lliboutry admet les frottements additifs, bien que la loi de fluage de la glace ne soit pas linéaire parce que les longueurs d'ondes successives envisagées sont d'un ordre de grandeur nettement différent.

Le viscosimètre à glace a été équipé d'un moyeu fixe sur lequel sont clavetées deux bosses symétriques de profil sinusoïdal, la différence de crête à creux étant a et la longueur d'onde λ de crête à crête, correspondant à une demicirconférence permettent de définir par leur rapport a/λ la notion de rugosité du lit glaciaire. A un stade ultérieur, d'autres sinusoïdes pourront être envisagées sur la sinusoïde de base comme décrit ci-dessus.

Le mouvement de la couronne de glace d'épaisseur uniforme est assuré par une chaine cinématique équipée de réducteurs et de 2 boîtes de vitesses, commandée par un moteur à excitation séparée. La vitesse peut varier de 0 à 20 m/h, soit de 0–7000 m/an.

En cours d'expérience, la température devra être régulée au mieux à zéro degré *C* au niveau du lit. Il est très important d'éviter les gradients de température et que la glace soit fondante sur le lit, sans pour autant provoquer une fonte de la glace exagérée, car l'expérience doit pouvoir être prolongée plusieurs jours, voire plusieurs semaines. La glace une fois en place dans le touret, emprisonnée par les flasques du touret, une forme hémi-spérique en bois s'interpose entre flasque supérieure et le couvercle bombé supérieur de façon à équilibrer les poussées verticales de la glace. Le remplissage de la machine en eau est effectué et la pression est appliquée soit par une pompe hydraulique soit par une bouteille d'air comprimé mettant l'eau interne de la machine en pression.

Des canelures circulaires ont été ménagées sur les flasques pour assurer le plus efficacement possible l'étanchéité entre l'extérieur de la couronne de glace qui reçoit la pression correspondant à ρgz (hauteur de la couche de glace d'un glacier) et les cavités sous glaciaires qui se forment.

Des prises de pression permettent d'avoir accès aux cavités et de mesurer la pression interstitielle et par ailleurs d'introduire de l'eau sous pression dans les cavités sous glaciaires.

La finesse de la régulation de température est très importante. Dans le cadre d'aménagements futurs, la machine sera placée dans une enceinte à zéro degré, un léger flux froid sera transmis à l'extérieur au niveau des dents d'entraînement et un réchauffage thermorégulé du lit sera prévu. Acutellement la machine est dotée d'un refroidissement autonome, donné par un groupe frigorigène mettant en froid une réserve d'huile incongelable dans l'enceinte inférieure de la machine. Une sonde contrôle la température à l'extérieur de la couronne de glace.

Nous avons procédé en plusieurs étapes en dehors des problèmes technologiques que nous avons dû résoudre sans parler de ceux qui seront résolus au cours d'aménagements ultérieurs de la machine.

a) Procéder à un type d'expérience donné et le refaire une ou plusieurs fois ceci afin de s'assurer que les essais sont reproductibles raisonnablement. Ceci permet de vérifier par exemple que la constitution de la couronne de glace ne provoque pas d'un essai à l'autre à même vitesse de dispersion inadmissible. C'est ainsi que la fig. 2 montre les courbes de couple traduit en millimètres du bras enregistreurs pour une rotation de $\frac{1}{4}$ de tour par jour soit 104 m/an.

I et III sont comparables quant au maximum, II sous une pression de 9 bars représentant 100 m de glace.

IV est un essai avec rugosité 0,05, avec une vitesse de rotation quadruple des précédentes soit 1 tour/jour soit 416 m/an.

On notera suivant les vitesses le calage différent du maximum de couple par rapport au seuil du palier permanent.

b) Au cours des différents essais qui ont été conduits successivement avec un touret en 4 parties (2 bosses et 2 corps) séparées et en un

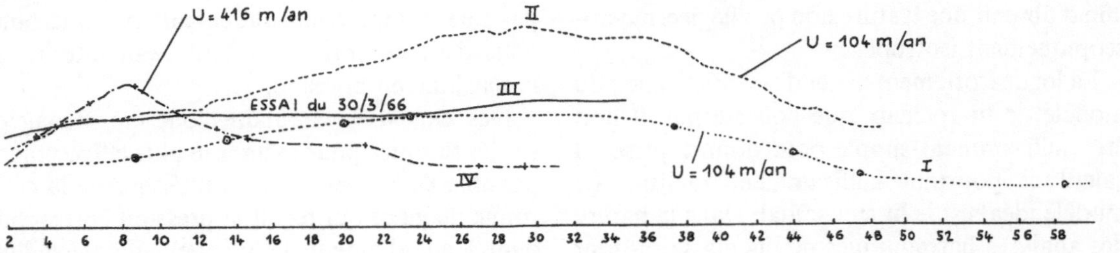

Fig. 2

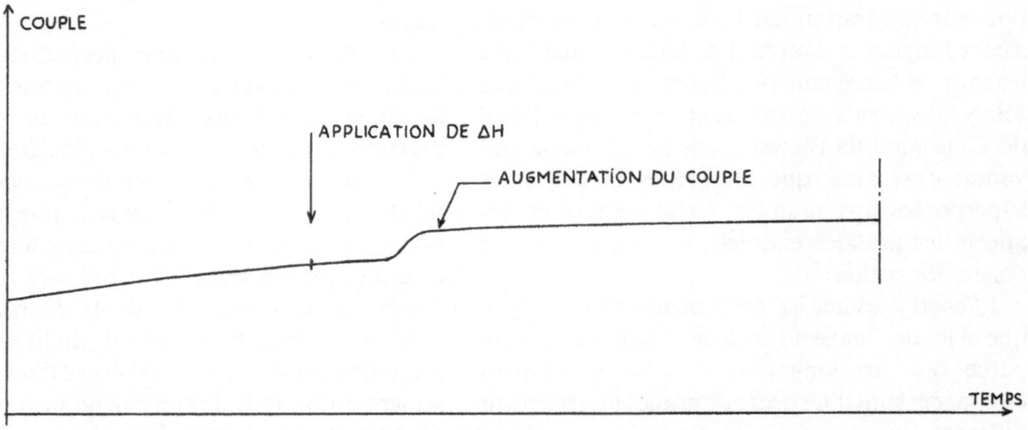

Fig. 3

seul bloc (touret monobloc), il n'a pas toujours été aisé d'assurer une étanchéité des cavités sous-glaciaires eu égard à la pression exercée à la surface externe de la couronne de glace. Chaque fois que l'étanchéité est assurée, nous constatons une augmentation de la valeur du couple.

Considérons une protubérance sinusoïdale avec cavité aval isolée: soit alors selon les notations de *Lliboutry*:

- a hauteur de la protubérance
- p pression intertitielle
- $\Delta\sigma$ la surpression au milieu de la face amont
- v vitesse due à la déformation plastique
- H pression de glace
- n exposant de la contrainte de cisaillement dans la loi de *Glen* $\dot{\gamma} = B\tau^n (\dot{\gamma} = B\tau^n)$
- B constante de fluage
- F forces appliquées à chaque protubérance à mi-hauteur
- b largeur de la protubérance

Le couple enregistré par la barre de torsion C est égal à

$$C = F \times D \times b$$

on démontre que la vitesse de la glace sur le lit est égale à:

$$u = v + v'$$

somme de la vitesse de contournement plastique de l'obstacle V et de la vitesse du mécanisme fonte-regel V'

$$U = v + v' = C_1 \left[\frac{F}{a} - (H - p)\right]^n + C_2 \left[\frac{F}{a} - \left(1 - \frac{\pi}{4}\right)(H - p)\right] \qquad [1]$$

avec

$$C_1 = \frac{8^{n-1}}{\pi^{2n} (\sqrt{3})^{n+1}} \frac{B\lambda^2}{a} \left[\frac{F}{a} - (H - p)\right]^n \qquad [2]$$

$$C_2 = \frac{8}{\pi} \left(\frac{Kb + Kg}{L\varrho a}\right) C. \qquad [3]$$

Si $p = 0$ (cavité sans eau par exemple).

A vitesse constante en différentiant par rapport à $H = pgz$, on obtient:

$$\frac{1}{a} \frac{\partial F}{\partial H} = \frac{nC_1 \left(\frac{F}{a} - H\right)^{n-1} + C_2 \left(1 - \frac{\pi}{4}\right)}{nC_1 \left(\frac{F}{a} - H\right)^{n-1} + C_2}. \qquad [4]$$

Cette expression est toujours positive. La fig. 3 montre sur une position de l'enregistrement de couple d'un essai, ces augmentations caractéristiques du couple.

Si p pression interstitielle a une valeur, la cavité tendant à se refermer à l'aval sous l'effet d'une variation ΔH, il s'ensuit que p peut subir une augmentation Δp, ce point doit être examiné ultérieurement.

Nous avons examiné au cours de plusieurs essais les variations de B durant les premières dizaines d'heures de rotation de la machine en conférant des augmentations de pression à la glace.

Dans un premier essai (Juillet 1971) avec une vitesse de rotation de la machine de 1 tour en 56 h, soit une vitesse de la glace sur le lit de 170 m/an, p était nulle, la cavité non complètement remplie d'eau.

B tirée de l'eq. [1] ou $u = v + v'$ = vitesse d'entraînement de la glace on a trouvé au terme de 8 h de mise en route pour une variation de pression de 17 bars, $B = 1{,}48\ \text{bar}^{-3}\ \text{an}^{-1}$.

Une autre augmentation de pression de 16 à 17,5 bars à 23 heures de l'origine a donné respectivement

$F = 1{,}224$ bar mètre pour $H = 16$ bar

d'où

$B = 0{,}45\ \text{bar}^{-3}\ \text{an}^{-1}$

$F = 1{,}281$ bar mètre pour $H = 17{,}5$ bar

d'où

$B = 0{,}448\ \text{bar}^{-3}\ \text{an}^{-1}$ parfaitement concordant.

Dans un deuxième essai (Decembre 1971) avec de la glace obtenue à partir d'eau distillée, pour les mêmes conditions de rugosité $r = 0{,}066$ ($a = 3{,}6$ cm, $\lambda = 54$ cm) et de vitesse $u = 170$ m/an on a obtenu au bout de 24 h, avec $H = 10$ bars

$F = 0{,}925$

$B = 0{,}695\ \text{bar}^{-3}\ \text{ab}^{-1}$

valeur assez voisine de $B = 0{,}45$ dans le premier essai.

Ces considérations militent pour la correction à apporter dans l'équation de la vitesse de glissement totale, au coefficient de *Glen*.

Dans les calculs nous avons admis que K coefficient de conductibilité de la bosse était égal à celui de la glace $Kg = 1{,}75\ 10^{-3}\ \text{m}^2/\text{bar an}$. Pour ce qui est de B, en fluage permanent la valeur de $B = 0{,}17\ \text{bar}^{-3}\ \text{an}^{-1}$ est à retenir.

Lliboutry propose

$$B\left[1 + 9\left(\frac{t}{0{,}01}\right)^{-\frac{2}{3}}\right]$$

t étant le temps mis par la glace pour franchir λ soit λ/u.

La formule ci-dessus donne des valeurs de B plus fortes que ce que les essais ci-dessus tendent à montrer; par ailleurs des témoins constitués par des parallélogrammes de plastiques articulés placés au voisinage du lit n'ont jamais montré un angle γ au bout de 2–3 jours supérieur à 10° maximum.

c) Mettons en évidence la concordance qui existe entre les résultats de l'essai de Juillet 1971 et la théorie générale du glissement d'un glacier tempéré de *Lliboutry*.

Si nous prenons la moyenne des 2 minima à $\frac{1}{2}$ tour et à 1 tour voisins d'ailleurs, on trouve un couple de frottement de 422,5 m-kg.

La valeur du frottement moyen rapporté à l'ensemble du lit est alors

$$f = \frac{C}{Db \times \pi R_0}$$

ou R_0 le rayon extérieur du touret (fond du lit)

avec

$R_0 = 15$ cm
$D = 15 \times 2 + 3{,}6 = 33{,}6$ cm
b = hauteur de la protubérance = 13,3 cm

$$f = \frac{42250}{33{,}6 \times 13{,}3 \times \pi \times 15} = 2{,}01 \text{ bar}$$

f_0 étant la valeur du frottement à partir duquel la cavitation commence à l'aval

$$f_0 = \frac{M}{2} rN$$

soit ici avec la rugosité $r = a/\lambda = 3{,}6/54 = 0{,}066$ et $N = 18$ bars

$f_0 = 1{,}88$ bar

La théorie définit une fonction $\varphi = f/f_0$.

φ peut être exprimée à partir de la portion du lit où plaque la glace $s\lambda$, $s \leq 1$ et $(1 - s)\lambda$ étant la cavité

$\varphi \approx 2\pi s(1 - s)$ dès que $s < 0{,}3$

ici $\varphi = f/f_0$ valeur tirée de l'expérience $= 2{,}01/1{,}88 = 1{,}07$

$$s^2 - s + \frac{1{,}07}{2\pi} = 0 \quad \text{soit} \quad s^2 - s + 0{,}17 = 0$$

$$s = \frac{1 \pm 0{,}5657}{2}$$

seule la valeur la plus petite convient soit $s = 0{,}237$.

Appelons

$$V = \frac{v}{v_0} \quad \text{et} \quad V' = \frac{v'}{v_0'}$$

avec

v vitesse de contournement plastique dès que la cavitation s'opère

v_0 seuil de vitesse de contournement plastique pour laquelle $f = f_0$

La fig. 4 donne l'allure de φ, V et V'.

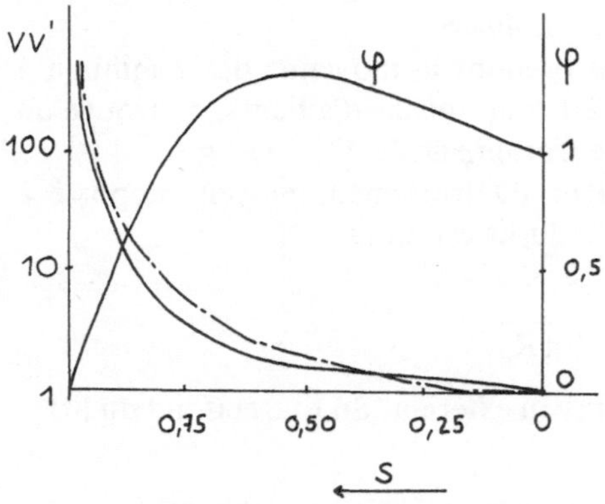

Fig. 4. Pour un lit sinusoïdal, valeurs réduites du glissement par plasticité (V) du glissement par fonte et regel (V') et du frottement réduit (φ), fonction de la fraction en contact reel avec le lit rocheux (s). Courbes tirées de la publication de *L. Lliboutry*. Assemblée generale de Berne 1967

On peut exprimer V et V' en fonction de la cavitation

$$V = \frac{1}{\pi s^2} \qquad V' = \frac{2}{\pi^2 s^2}$$

$$V = \frac{1}{\pi \times 0{,}237^2} = 5{,}7$$

par ailleurs la vitesse de glissement totale peut s'exprimer par:

$$u = \frac{B\lambda N^3}{36\pi r} V + 2 \frac{(Kb + Kg)}{L\varrho} \frac{N}{\lambda r} V'$$

négligeant le coefficient de vitesse fonte-regel de V'

$$V = \frac{u \times 36\pi r}{B\lambda N^3}$$

avec B trouvé égal à 0,10 dans cet essai on a:

$$V = \frac{170 \times 36 \times 3{,}1416 \times 0{,}07}{0{,}10 \times 0{,}54 \times 18^3} = \frac{1345}{313} = 4{,}3.$$

Cette valeur tirée à partir de B obtenu par la formule de la cavité isolée connaissant le couple, se recoupe bien avec la valeur de V tirée de la connaissance de s obtenue par f déduit expérimentalement. Les formules [1] et [5] sont cohérentes.

En conclusion, les études expérimentales entreprises doivent permettre:

. De vérifier les seuils de V_0 et f_0 à partir desquels la cavitation commence.

. De suivre l'évolution de la vitesse de glissement $\dot{\gamma}$ en fonction du temps et de comparer cette constatation de l'évolution expérimentale de γ avec une loi correctrice de la valeur permanente de B trouvée par *Glen* pour tenir compte du fluage transitoire. Dans le futur à ce sujet des essais devront être entrepris avec des carottes de glacier juxtaposées dans la machine.

De rechercher si la pression interstitielle a une importance fondamentale et de constater si des régimes permanents peuvent s'établir.

Pour cela de nouvelles installations dans des enceintes à zéro degré et de nouvelles dispositions encore plus rigoureuses de la régulation de température devront être réalisées.

Littérature

1) *Lliboutry, Louis*, Théorie complète du glissement des glaciers, compte tenu du fluage transitoire. Extrait de la publication n° 79 Assemblée Générale de Berne, 1967.

2) *Lliboutry, Louis*, J. Glaciology **7**, 49 (1968).

3) *Kamb* and *La Chapelle*, J. Glaciology **5**, 38 159–172.

4) Premiers résultats obtenus avec le viscosimètre à glace de Grenoble. In: *Brepson, Roger*, Comptes-rendus hebdomadaires des séances de l'Académie des Sciences (Paris). Sér. B Tome **263**, n° 15, 876–879.

Adresse de l'auteur:

Dr. Ing. *R. Brepson*
Laboratoire de Glaciologie du C.N.R.S.
Sogreath, B.P. 172, Centre de Tri
F-38042 Grenoble-Cedex (France)

Rheol. Acta 13, 413–417 (1974)

From the Department of Mechanical and Aerospace Sciences, University of Rochester, Rochester (U.S.A.)

Time-dependent behavior of thixotropic suspensions

By H. A. Mercer and H. D. Weymann

With 1 figure and 6 tables

(Received October 27, 1972)

Introduction

Suspensions show thixotropic behavior when the suspended phase is capable of forming a structure and the steady state degree of aggregation is a function of the steady state rate of shear. Experimental investigations of the steady state flow behavior and comparisons with the predictions of different theories have been reported by numerous authors. Systematic investigations of the time-dependent behavior are, however, much rarer. There are essentially three types of experiments which have been reported:

a) After vigorous agitation (which completely destroys the structure) the suspension is allowed to rest and the degree of recovery (of structure) is tested after various time intervals.

b) The shear stress $\tau(t)$ is measured as a function of a variable shear rate $\dot{\gamma}(t)$. Usually the shear rate is increased linearly with time up to a maximum and then decreased linearly to zero. The resulting τ vs. $\dot{\gamma}$ relation is called "hysteresis loop".

c) The shear rate is held constant at $\dot{\gamma}_i$ until the shear stress reaches a steady state value. Then the shear rate is abruptly changed to a new value $\dot{\gamma}_f$ and the $\tau(\dot{\gamma}_i, \dot{\gamma}_f, t)$ relation determined. This is repeated for different combinations of $\dot{\gamma}_i$ and $\dot{\gamma}_f$.

Method (a) was used in a crude way by the earliest investigators (1) of thixotropy and later refined by *Alfrey* and *Rodewald* (2). For a description of method (b) see e.g. *Weltmann* (3). It is the most popular method for industrial applications as it gives a rapid, though limited, indication of the thixotropic properties of a material. Method (c) was suggested by *Weltmann* (4) and later used by *Dintenfass* (5). (For a more detailed history of the different methods see also ref. 6.)

For many practical applications (e.g. paints, printing inks, etc.) the "recovery" of a rigid structure at zero shear rate is of great importance. Method (a) would be very useful in this case. For testing theoretical predictions about the general time-dependent behavior of thixotropic substances it is inadequate.

Without going into the details of the model assumed for the thixotropic suspension one can say that the "structure" or the "number of bonds" found in the suspension exist as a consequence of "making" and "breaking" processes. The rate equations for these processes may be quite complex, depending on the model chosen. The solution for the steady state is usually obtained quite easily. Solutions for time-dependent cases, however, become quite cumbersome unless the rate coefficients are constants. This requires that the shear rate is constant. Even then only numerical solutions may be possible as mentioned in the "Discussion". If experiments are to be performed to provide a basis for theoretical work on time-dependent behavior the most reasonable method is, therefore, method (c).

Experiments

A viscometer which is well suited for experiments of the kind advocated above is the Haake Rotovisco with the *Couette* attachment (7). The experiments reported here were done with this instrument. The viscometer has a gear box which allows to use one of ten reproducible shear rates. For our instrument the shear rate $\dot{\gamma}$ is related to the gear factor U_s by

$$\dot{\gamma} = 1740/U_s \sec^{-1}.$$

The gear factors are nominally $U_s = 1, 2, 3, 6, 9, 18, 27, 54, 81, 162$. The actual values are, however, $U_s = 1, 1.6, 3, 4.8, 9, 14.5, 27, 43.7, 81, 130.9$. The

Couette cylinders had longitudinal grooves to prevent slip at the wall.

The bentonite-water suspensions were prepared from "*Volclay*" a commercial Wyoming bentonite which is mostly Na-bentonite. We have used this bentonite previously for mechanical and optical experiments (8, 9). When it is purified the time-dependent behavior is compressed to such an extent that it cannot be observed with our viscometer. For this reason we have used the bentonite without purifying it.

The measurements were made and processed in the following way: After filling the viscometer and bringing it to the desired temperature the suspension was sheared at $U = 1$ ($\dot{\gamma} = 1740\,\mathrm{sec}^{-1}$) until a steady state shear stress was recorded. The shear rate was then abruptly changed and the time-dependent shear stress $\tau(t)$ was recorded with a Varian G1000 chart recorder. When a steady state was reached again, a new shear rate was selected. If the initial shear rate is $\dot{\gamma}_i$ and the final shear rate $\dot{\gamma}_f$ in a particular "transition", the time-dependent viscosity $\eta(t)$ can be calculated from $\tau(t)$ as

$$\eta(t) = \tau(t)/\dot{\gamma}_f\,. \qquad [1]$$

To normalize the change in the time-dependent viscosity we introduce η_i as the steady state viscosity at the shear rate $\dot{\gamma}_i$ and correspondingly η_f for the shear rate $\dot{\gamma}_f$ and set

$$\bar{\eta}(t) = [\eta_f - \eta(t)]/(\eta_f - \eta_i). \qquad [2]$$

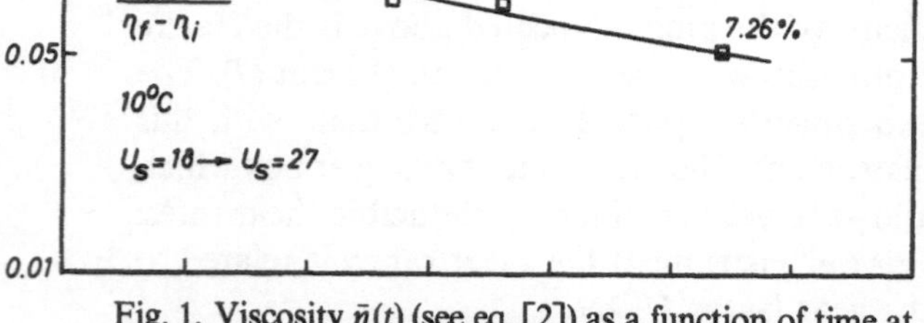

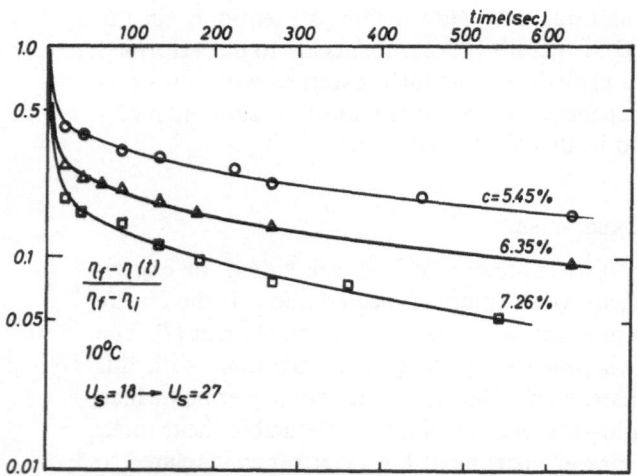

Fig. 1. Viscosity $\bar{\eta}(t)$ (see eq. [2]) as a function of time at 10 °C for a transition from $U_s = 18$ ($\dot{\gamma} = 120\,\mathrm{sec}^{-1}$) to $U_s = 27$ ($\dot{\gamma} = 64.5\,\mathrm{sec}^{-1}$)

The behavior of $\bar{\eta}(t)$ shown in fig. 1 is characteristic of the behavior at other concentrations, temperatures, and initial and final shear rates: There is a very rapid initial drop in viscosity followed by an almost exponential gradual drop. In many cases the initial adjustment is so rapid that it can not be resolved reliably with our instrument (time constant $\lesssim 1$ sec). The general features of $\bar{\eta}(t)$ could be approximately described by

$$\bar{\eta}(t) \approx (1 - \varepsilon)\exp(-t/\tau_1) + \varepsilon \exp(-t/\tau_2) \qquad [3]$$

where $\varepsilon \ll 1$ and $\tau_1 \ll \tau_2$. However, we did not feel confident to represent $\bar{\eta}(t)$ in this way[1]) and rather used as an unambiguous description of the data a "characteristic time". Because the e-folding time quite often is not sufficiently longer than the time constant of our instrument we chose $t_{0.1}$ defined by

$$\bar{\eta}(t_{0.1}) = [\eta_f - \eta(t_{0.1})]/(\eta_f - \eta_i) = 0.1 \qquad [4]$$

as the characteristic time. The values of $t_{0.1}$ for different concentrations, temperatures, and initial and final shear rates are given in tables 1–6. Times smaller than about 5 sec are normally not very reliable.

Discussion

In surveying the literature on the time-dependent behavior of thixotropic substances one finds that none of the theories advanced so far can explain the behavior of $\bar{\eta}(t)$ shown in fig. 1. *Goodeve* (10) found that the number of links (which give rise to *Bingham* behavior) does not depend on the shear rate. A change in the shear rate does not initiate a rate process and the adjustment of the viscosity is instantaneous. From *Peters* (11) and *Storey* and *Merrills* (12) theories one can derive that $\bar{\eta}(t)$ is a simple exponential function rather than the superposition of a fast process and a slow process as shown in fig. 1.

The present authors investigated the kinetics underlying the time-dependent viscosity observed here. While not completely successful this theory gives results which come closer to the experimental behavior than the theories mentioned above. Our theory is based on a model which can explain successfully the mechanical and optical steady state behavior of bentonite-water suspensions (13). In brief, the model assumes that the bentonite particles aggregate in spheroidal

[1]) Note added in proof: A closer investigation into the functional form of $\bar{\eta}(t)$ will be published by the present authors in Issue 1 or 2, Vol. 18, of the Transactions of the Society of Rheology.

Table 1. Characteristic time $t_{0.1}$ (in sec) for $c = 5.45\%$ and 10 °C ($\dot{\gamma} = 1740/U_s$ sec^{-1})

	To $U_s =$ 1	2	3	6	9	18	27	54	81	162
From $U_s =$ 1	–	10	70			305		800	1080	
2	30	–	168				297			
3	40	50	–	105	2200				1200	
6			40	–	360					
9				110	–	550	470			
18			26		25	–	1100			
27			5			310	–	565	820	
54		8					610	–	545	
81	5	5				30		132	–	608
162									117	–

Table 2. Characteristic time $t_{0.1}$ (in sec) for $c = 5.45\%$ and 20 °C ($\dot{\gamma} = 1740/U_s$ sec^{-1})

	To $U_s =$ 1	2	3	6	9	18	27	54	81	162
From $U_s =$ 1	–	100	22			350	340	840	810	
2	100	–	105				805			
3	50	485	–	600	310				700	
6			262	–	810					
9				218	–	865	810			
18	12		62		112	–	1500			
27			17			78	–	1080	600	
54		5					660	–	2700	
81	5	5				760		430	–	1400
162									360	–

Table 3. Characteristic time $t_{0.1}$ (in sec) for $c = 5.45\%$ and 30 °C ($\dot{\gamma} = 1740/U_s$ sec^{-1})

	To $U_s =$ 1	2	3	6	9	18	27	54	81	162
From $U_s =$ 1	–	40	135			270		960	1360	
2	55	–	325				900			
3	100	50	–	1000	790				1400	
6			1700	–	890					
9				1120	–	990	1380			
18			135		580	–	560			
27			15			450	–	635	3280	
54		4					260	–	900	
81	3	90				1300		190	–	370
162									62	–

Table 4. Characteristic time $t_{0.1}$ (in sec) for $c = 6.35\%$ and 10 °C ($\dot{\gamma} = 1740/U_s$ sec^{-1})

	To $U_s =$ 1	2	3	6	9	18	27	54	81	162
From $U_s =$ 1	–		48			90		355	420	
2	200	–	35				195			
3	2	23	–	75	150	150				
6			32	–	415					
9				170	–	190	340			
18			7		70	–	585		666	
27			11			145	–	625	1320	
54		2					142	–	135	
81	0.5	2						115	–	740
162									98	–

Table 5. Characteristic time $t_{0.1}$ (in sec) for $c = 6.35\%$ and 20 °C ($\dot{\gamma} = 1740/U_s$ sec^{-1})

	To $U_s =$ 1	2	3	6	9	18	27	54	81	162
From $U_s = 1$	–	120	20			160		1140	560	
2	25	–	10				185			
3	10	35	–	220	240				620	
6			72	–	290					
9				82	–	460	350			
18			10		230	–	500			
27			8			225	–	400	665	
54		7					262	–	570	
81	0	0				15		110	–	615
162									93	–

Table 6. Characteristic time $t_{0.1}$ (in sec) for $c = 7.26\%$ and 10 °C ($\dot{\gamma} = 1740/U_s$ sec^{-1})

	To $U_s =$ 1	2	3	6	9	18	27	54	81	162
From $U_s = 1$	–	10	10					220	190	
2	1.5	–	620				300			
3	1	5	–	500	60				370	
6			5	–	1600					
9				25	–	200	250			
18			5			–	180			
27			5			60	–	440	450	
54		5					90	–	340	
81	1	1				5		220	–	2440
162									95	–

clusters in which the water is trapped. If the viscosity at infinite shear rate is η_∞ and the volume fraction of the aggregates is φ then the viscosity of the suspension is

$$\eta = \tau/\dot{\gamma} = \eta_\infty \exp\left[\frac{2.5\,\varphi}{1 - \Lambda\varphi}\right] \qquad [5]$$

both for the steady state and the time-dependent case. If the number density of i-particle aggregates is n_i and the volume of an i-particle aggregate is V_i then

$$\varphi(t) = \sum_{i=2}^{i_m} n_i(t)\, V_i \qquad [6]$$

where i_m is the size of the largest aggregate which can exist at a certain shear rate. The time-dependent viscosity can, therefore, be calculated once the $n_i(t)$ are known. To do this one has to solve i_m simultaneous differential equations of the form

$$dn_i/dt = \sum_{i,m} A_{ijm} n_j n_m + \sum_m B_{im} n_m \qquad [7]$$

where the $A_{ijm}(\dot{\gamma})$ and $B_{im}(\dot{\gamma})$ are related to the rate coefficients for the making and breaking processes. We assumed that the making rate could be deduced from *von Smoluchowskis* (14) and *Tuorilas* (15) expression for the rate of collisions of spheres of radii R_i and R_j resp. in shear flow. This expression was modified by $C\exp(-E/kT)$, the probability that the two colliding particles have enough thermal energy to overcome the repulsive energy $E(R_i, R_j)$ between them. The breaking rates were determined from the making rates and the steady state distribution (13)

$$i n_i \approx N/i_m \qquad [8]$$

by invoking the principle of detailed balancing.

The resulting rate equations were solved numerically for different assumptions for the rate coefficients and $E(R_i, R_j)$. The $n_i(t)$ thus obtained were then used to calculate $\varphi(t)$ and $\bar{\eta}(t)$. While the solution has the general form of eq. [3] the ratio τ_1/τ_2 found theoretically was not small enough so that we could not exactly match the experimental data. This ratio and ε proved to be very insensitive to major changes in our basic assumptions thus frustrating our attempts to find appropriate rate coefficients which would allow a matching of the theoretical and the experimental $\bar{\eta}(t)$ curves.

In the absence of a theory of the time-dependent behavior a discussion of the characteristic times given in tables 1–6 must appeal to intuition. In all reasonable theories of the time-dependent behavior both the making rates and the breaking rates must depend on the shear rate and one would expect that for the same initial shear rate the adjustment process to the final viscosity would take the longer the smaller the final shear rate. This is, in general, borne out by the results.

A very curious result is found when one compares "transitions" in which the final shear rates are the same. In this case $t_{0.1}$ increases as the magnitude of the difference $|\dot{\gamma}_f - \dot{\gamma}_i|$ becomes smaller. It seems that a major rearrangement of the suspension structure is easier to accomplish than a minor adjustment. This behavior of the bentonite-water suspensions also means that for them the characteristic times are not additive as *Dintenfass* (5) found for some other substances: e.g. $t_{0.1}$ for the transition from $U_s = 1$ to $U_s = 3$ is clearly not equal to the sum of $t_{0.1}(U_s = 1$ to $U_s = 2)$ and $t(U_s = 2$ to $U_s = 3)$[2]).

Acknowledgement

This paper is based in part on a doctoral dissertation by H.A.M.

This work was supported by a grant from the National Science Foundation.

Summary

The time-dependent behavior of the viscosity of bentonite-water suspensions was studied experimentally in *Couette* flow for various transitions between constant shear rates. It was found that, in general, the characteristic times become longer as the final shear rate becomes smaller and, for the same final shear rate, as the differences in shear rates become smaller. The characteristic time increases with increasing temperature and decreasing concentration. A discussion of the kinetics of the time dependence is given.

---

[2]) The term "recovery time" is used by *Dintenfass* (5) to denote the time which elapses from the change $\dot{\gamma}_i \rightarrow \dot{\gamma}_f$ to the point in time when $\eta(t)$ reaches the value η_f. As $\eta(t)$ reaches the value η_f asymptotically, *Dintenfass* must, in fact, have used a quantity similar to the $t_{0.1}$ used here, although he does not say so.

Zusammenfassung

Die zeitabhängige Viskosität von Bentonit-Wasser Suspensionen wurde in der *Couette*-Strömung für verschiedene Übergänge zwischen konstanten Schergeschwindigkeiten gemessen. Im allgemeinen werden die charakteristischen Zeiten länger, a) wenn die Endschergeschwindigkeit kleiner wird, und b) bei gleicher Endschergeschwindigkeit, wenn die Differenz zwischen Anfangs- und Endschergeschwindigkeit kleiner wird. Die charakteristische Zeit wird ebenfalls länger mit steigender Temperatur und abnehmender Konzentration. Die Kinetik der Zeitabhängigkeit wird diskutiert.

References

1) e.g. *Winkler, H. G. F.*, Kolloid-Beih. **48**, 341 (1938).

2) *Alfrey, T., Jr.* and *C. W. Rodewald*, J. Colloid Sci. **4**, 283 (1949).

3) *Green, H.* and *R. N. Weltmann*, Colloid Chem. **6**, 328 (1946).

4) *Weltmann, R. N.*, J. Appl. Phys. **14**, 343 (1943); Ind. Eng. Chem., Analytical Ed., **15**, 424 (1943).

5) *Dintenfass, L.*, Rheol. Acta **2**, 187 (1962).

6) *Bauer, W. H.* and *E. A. Collins*, Thixotropy and Dilatancy. In: *Eirich, F. R.* (Editor), Rheology, Vol. 4 (New York 1967).

7) *van Wazer, J., J. Lyons, K. Kim*, and *R. Colwell*, Viscosity and Flow Measurement, pp. 97–108 (New York 1963).

8) *Weymann, H. D., M. C. Chuang*, and *R. A. Ross*, Phys. Fluids **16**, 775 (1973).

9) *Ross, R. A., H. D. Weymann*, and *M. C. Chuang*, Phys. Fluids **16**, 784 (1973).

10) *Goodeve, C. F.*, Trans. Faraday Soc. **35**, 342 (1939).

11) *Peter, S.*, Kolloid Z. **114**, 44 (1949); Rheol. Acta **3**, 178 (1964).

12) *Storey, B. T.* and *E. W. Merrill*, J. Polymer Sci. **33**, 361 (1958).

13) *Weymann, H. D.*, Proceedings of the Fourth International Congress on Rheology, Part 3, *Lee, E. H.* (Editor), p. 573 (New York 1965).

14) *Von Smoluchowski, M.*, Z. phys. Chem. **92**, 155 (1917).

15) *Tuorila, P.*, Kolloidchem. Beih. **24**, 1 (1927).

Authors' address:

H. A. Mercer and *H. D. Weymann*
Dept. of Mechanical and Aerospace Sciences
University of Rochester
Rochester, N.Y. 14618 (U.S.A.)

Rheol. Acta **13**, 418–423 (1974)

From the Unilever Research, Vlaardingen (The Netherlands)

Constitutive equations applied to dispersed systems

By D. W. de Bruijne, N. J. Pritchard, and J. M. P. Papenhuijzen

With 10 figures

(Received October 27, 1972)

Introduction

In order to quantify the rheological behaviour of fat dispersions such as margarine it is necessary to have parameters available that can be considered as material constants. These constants can be found when one is able to represent the rheological behaviour of these dispersions by means of a socalled constitutive equation. The purpose of this investigation was to find whether the equations proposed for polymer-like materials can be used for these fat dispersions.

Constitutive equations are in general variations on the theme of a *Maxwell* model. Such a model can be represented by

$$\dot{\tau} + \frac{\tau}{\lambda} = \frac{\eta}{\lambda} \cdot \dot{\gamma} \qquad [1]$$

or by

$$\tau(t) = \int_{-\infty}^{t} \frac{G}{\lambda} \exp - \frac{(t-t')}{\lambda} \gamma(t')\, dt' \qquad [2]$$

where τ = stress, γ = deformation, G = elastic modulus, η = viscosity, $\lambda = \eta/G$ = time constant and t = time ($t = 0$ at the start of the experiment).

Rheological experiments may be carried out at constant shear rate ($\gamma(t') = \dot{\gamma} \cdot t'$). Eq. [2] shows that it predicts for such type of experiments a stress that is proportional to G, to $\dot{\gamma}\lambda$ and to some exponential function of t/λ:

$$\tau(t) = G\lambda\dot{\gamma}\left[1 - \exp\left(-\frac{t}{\lambda}\right)\right] = G\lambda\dot{\gamma}F\left(\frac{t}{\lambda}\right) \qquad [3]$$

which indicates that $\tau(t)$ may be considered as a function of γ and $\lambda\dot{\gamma}$, because $t/\lambda = \gamma/\lambda\dot{\gamma}$.

Rheological experiments with fat dispersions show that the shear stress depends mainly on the shear itself and only to a small extent on the shear rate (1). Consequently, the obvious way to make eq. [3] applicable to the rheology of fat dispersions is to remove the strong dependence of the stress on the shear rate by assuming $\lambda\dot{\gamma}$ to be a constant ($\lambda\dot{\gamma} = 1/a$). The dependence on t/λ predicted by eq. [3] becomes then a dependence on γ only. A relatively slight dependence of the stress on the shear rate can, for example, be introduced by putting:

$$\frac{1}{\lambda} = a\dot{\gamma} + \frac{1}{\lambda_0} \qquad [4]$$

instead of $1/\lambda = a\dot{\gamma}$. For fat dispersions, $1/\lambda_0$ must be a quantity small compared with $(a\dot{\gamma})$ because $1/\lambda_0$ accounts for the slight dependence of the stress on the shear rate, whereas $(a\dot{\gamma})$ accounts for the strong dependence of the stress on the shear itself.

A modification of eq. [1] based on assumption [4] is the constitutive equation proposed by *Bogue* (2). It has already been shown that this equation describes the rheological behaviour of polymer solutions (3, 4, 5) quite satisfactorily. More precisely, the assumption of *Bogue* was

$$\frac{1}{\lambda} = \frac{1}{\lambda_0} + a\langle\dot{\gamma}(t-t')\rangle \qquad [5]$$

where $\langle\dot{\gamma}(t-t')\rangle$ denotes the shear rate averaged over the time interval $t - t'$. The meaning of this eq. [5] in terms of the material memory is that the material remembers events after the start of the deformation process with a reciprocal time constant $1/\lambda = 1/\lambda_0 + a\dot{\gamma}$, while events before the start are remembered by means of a reciprocal time constant having a value between $1/\lambda_0$ and $1/\lambda_0 + a\dot{\gamma}$.

The predictions of this model for various experiments have been given by *Bogue* and coworkers (2, 3). Summarizing these predictions for shear stress (τ_{12}) and normal stress ($\tau_{11} - \tau_{22}$) in steady shear experiments and for the moduli in oscillatory shear experiments, one has:

$$\tau_{12} = \tau_{12}(\infty)[1 - \exp\{-(\Gamma + 1)T\} + \Gamma(\Gamma + 1)T^2 \cdot \exp(-\Gamma T) \cdot Ei(T)] \qquad [6]$$

$$\tau_{11} - \tau_{22} = [\tau_{11}(\infty) - \tau_{22}(\infty)] \times [1 - \{1 + (\Gamma + 1)T\}\exp -(\Gamma + 1)T + \tfrac{1}{2}\Gamma(\Gamma + 1)^2 T^3 \exp(-\Gamma T)Ei(T)] \qquad [7]$$

with

$$T = t/\lambda_0, \quad \Gamma = a\lambda_0\dot{\gamma}$$

and

$$Ei(T) = \int_T^\infty \frac{\exp -x}{x}\,dx.$$

The steady state values of the shear stress and normal stress are respectively:

$$\tau_{12}(\infty) = \frac{G}{a}\frac{\Gamma}{1 + \Gamma} \qquad [8]$$

$$\tau_{11}(\infty) - \tau_{22}(\infty) = \frac{2G}{a^2}\frac{\Gamma^2}{(1 + \Gamma)^2} \qquad [9]$$

For the dynamic storage modulus

$$G' = G\frac{(\omega\lambda_0)^2}{1 + (\omega\lambda_0)^2} \qquad [10]$$

and for the loss modulus

$$G'' = G\frac{\omega\lambda_0}{1 + (\omega\lambda_0)^2}. \qquad [11]$$

Experimental

The dispersions were prepared by rapid cooling of a hot solution of glyceryl tristearate in paraffin oil. The crystallization occurred while the solution was poured in a thin layer on a horizontal brass plate kept at 20 °C. After a few seconds, the fully crystallized material was scraped from the brass surface and kneaded into the required shape. Measurements started one hour later. Dispersions containing 6, 10 or 22% tristearate (by volume) were prepared.

The shear stress in steady shear experiments, stress relaxation after cessation of steady shear, and the moduli of the oscillatory shear experiments were measured with the *Weissenberg* Rheogoniometer (R 18) using the parallel plate configuration. Only in the case of steady shear experiments at the lowest solid content was a *Couette*-type concentric cylinder viscometer used. The normal stress could only be determined for the most concentrated dispersion, because the normal stresses for the more dilute dispersions were too low to be measured with the *Weissenberg* Rheogoniometer. The determination of the normal stress offered other problems too. The force measured between smooth parallel plates appeared to be roughly proportional to the plate separation and was poorly reproducible. By covering the plates of the viscometer with sand paper (roughness about 0.2 mm) significant improvements were obtained:

(i) the normal force was much larger than before, and

(ii) the normal force appeared to be independent of the plate separation for separations between 0.5 and 2 cm, although it decreased somewhat for still larger separations.

At separations smaller than 0.5 cm the normal force was still roughly proportional to the gap width between the plates.

The separation in a cone-and-plate geometry is much smaller than 0.5 cm. Consequently this geometry was considered not to be appropriate in the case of these dispersions. In several materials the second normal stress difference is smaller than the first. Therefore, we neglected the contribution of the second normal stress difference to obtain a first approximation of the first normal stress difference. Then an average value of the first normal stress difference may be calculated from the total thrust (F) on the plates (6) in a similar way as for the cone-and-plate geometry:

$$\tau_{11} - \tau_{22} = \frac{2F}{\pi R^2}. \qquad [12]$$

Comparison of the experimental results with the *Bogue* model

Figs. 1 and 2 represent the relaxation of the shear stress after prolonged steady shear at a constant shear rate of $10^{-2}\,s^{-1}$ for a 6% and a 22% dispersion. The relaxation times (λ_0) of these dispersions (4) have been estimated from the final slopes: $3 \cdot 10^5$ s for the 6% dispersion and $8 \cdot 10^5$ s for the 22% dispersion.

Shear stress overshoot curves are given in figs. 3 and 4. The broken curves represent the

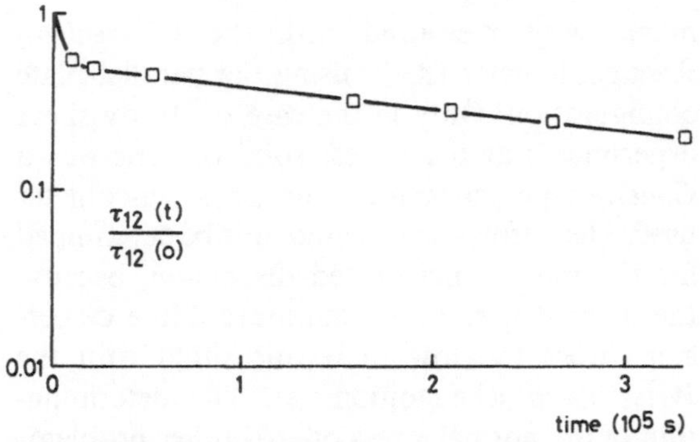

Fig. 1. Shear stress relaxation in 6% dispersions of fat crystals in paraffin oil

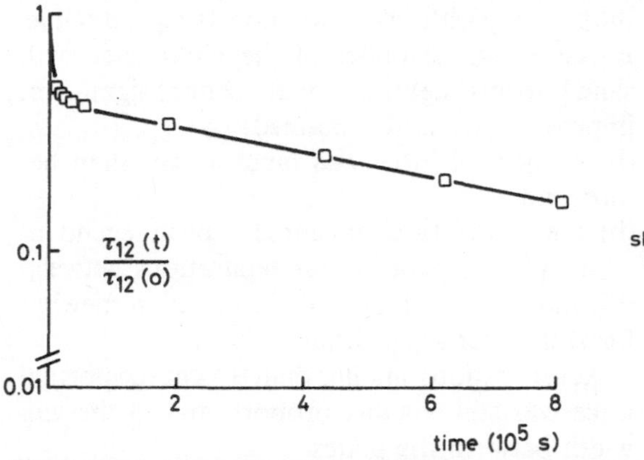

Fig. 2. Shear stress relaxation in 22% dispersions of fat crystals in paraffin oil

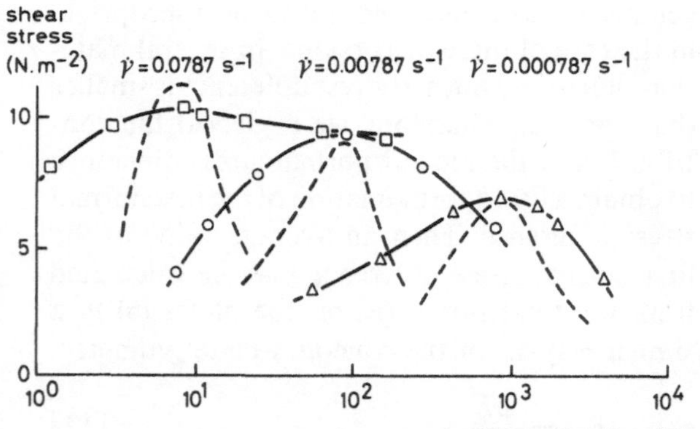

Fig. 3. Shear stress overshoot for a 6% dispersion of fat crystals in paraffin oil, measured at constant rate of shear $\dot{\gamma}$. —— = measured; – – – = *Bogue* model with $\lambda_0 = 3 \cdot 10^5$ s

predictions of the *Bogue* model using the above-mentioned values for the longest relaxation time λ_0. By choosing appropriate values of the quantity a, these curves have been adjusted in such a way that the locations of the maximum stresses coincide with the locations of the maxima of the experimental curves. In a similar way the actual value of the maximum stress has been adjusted by adapting G. It is shown that in this restricted range of shear rates the variation of the maximum stress with the shear rate can be described reasonably well by the *Bogue* model. According to this model, the decrease of the stress after the maximum is much faster than the experimental values indicate. Further, the predicted values of the stress before the maximum are considerably lower than the measured results. The latter were given in a semi-logarithmic plot; on a linear scale it would have been clear that the fat dispersions behave more non-linearly than the *Bogue* model predicts. From the measured results one obtaines as model constants (see eq. [6]):

Dispersion (%)	λ_0 (s)	G (N/m^2)	a
6	$3 \cdot 10^5$	5	3
22	$8 \cdot 10^5$	500	6

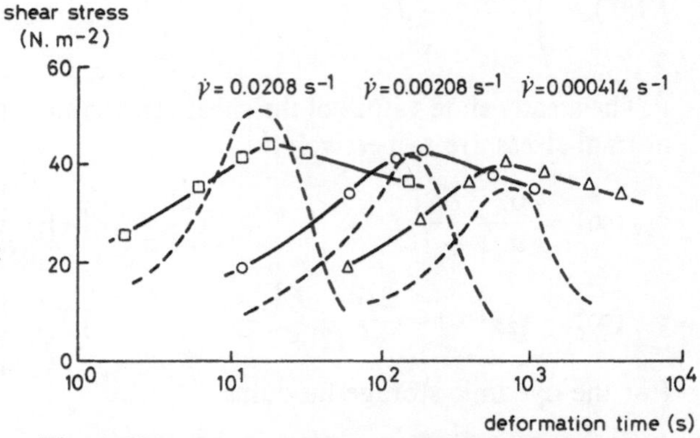

Fig. 4. Shear stress overshoot for a 22% dispersion of fat crystals in paraffin oil, measured at constant rate of shear $\dot{\gamma}$. —— = measured; – – – = *Bogue* model with $\lambda_0 = 8 \cdot 10^5$ s

The consequence of such long relaxation times is that even at the low shear rates used the system will display solid-like behaviour.

Results of dynamic experiments also show a more pronounced non-linear behaviour than the model predicts. As a result of the large value of λ_0 the *Bogue* model predicts, in the frequency range investigated, a frequency-independent storage modulus of magnitude G and a vanishing loss modulus (compare eq. [10]).

The dynamic properties of fat dispersions have been investigated by *Papenhuijzen* (1), and his results and some of our own are represented in fig. 5. It is shown that in this frequency range the storage modulus of fat dispersions is indeed frequency-independent. Its magnitude, however, is about 10^2 to 10^3 times the value predicted by the *Bogue* model. It was found indeed that in these dispersions the modulus representing the behaviour at larger deformations has a much lower value than the modulus found at very small deformations (fig. 6).

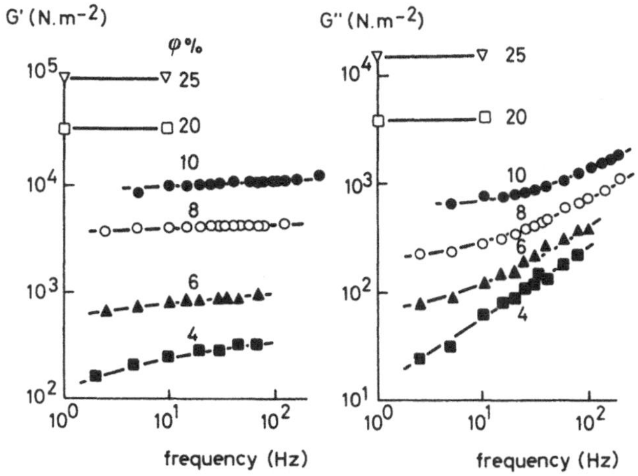

Fig. 5. Storage (left) and loss (right) modulus of dispersions of fat crystals in paraffin oil with different solid content Φ as a function of frequency

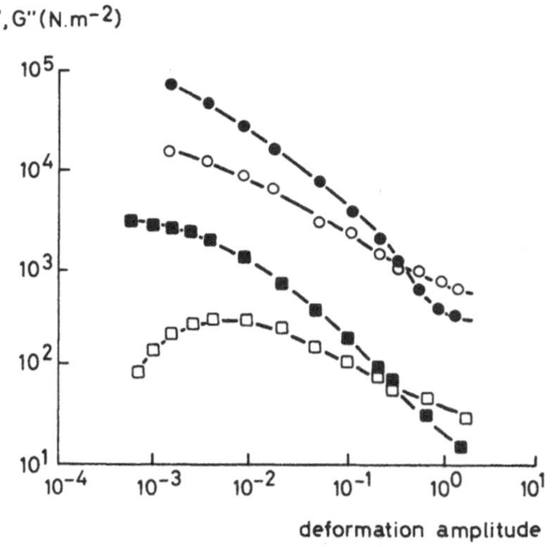

Fig. 6. Dynamic moduli as a function of the deformation amplitude (measured at increasing deformation). Solid contents: $G' = 25\%$ (●), $G'' = 25\%$ (○), $G' = 10\%$ (■), $G'' = 10\%$ (□)

The loss modulus is not immeasurably low but has a magnitude of about 10 or 20% of the storage modulus and again is almost frequency-independent for the more concentrated dispersions. At lower concentrations, it becomes frequency-dependent. This is probably a consequence of the viscosity of the oil.

One of the strong features of the *Bogue* model in describing polymer behaviour probably is that for the shear stress it predicts *Newton*ian behaviour at low rates of shear and plastic behaviour at high rates of shear (eq. [8]). Fig. 7 shows that fat dispersions exhibit opposite behaviour; they tend to be plastic at low rates of shear and become viscous when the shear rate increase above $100\,s^{-1}$. These results indicate that application of the *Bogue* model to dispersed systems should be restricted to a certain range of shear rates.

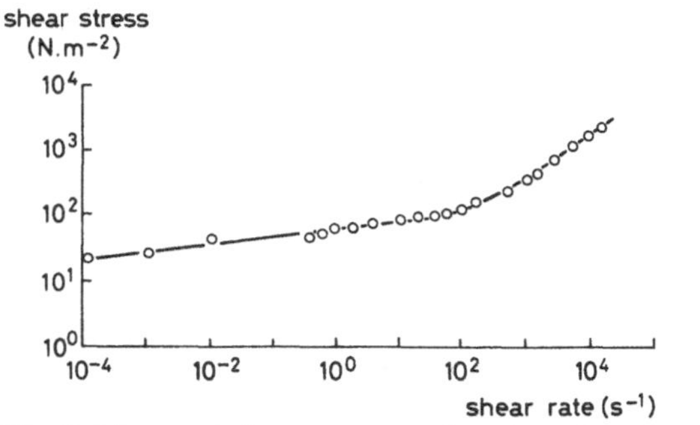

Fig. 7. Measured shear stress as a function of shear rate for a 10% dispersion of fat crystals in paraffin oil

The measured normal stress as a function of shear (fig. 8) shows unusual behaviour in that it increases much faster than the shear stress and reaches its maximum value before the shear stress. After its maximum, the normal stress decreases sharply and becomes even negative. In most of our experiments this negative value is of same magnitude as the positive part. Eq. [7] predicts only positive values of the normal stress, and the maximum of the normal stress should be reached after the maximum of the shea stress. The magnitude of the experimental values is higher than predicted by the model (fig. 9). Neither does the model predict the value of the average deformation at maximum stress satisfactorily.

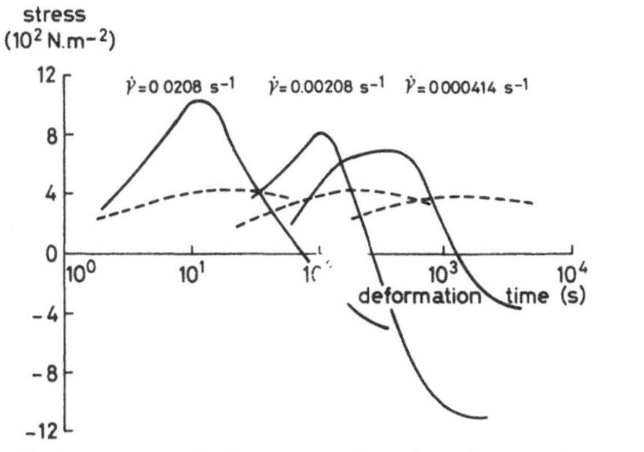

Fig. 8. Measured shear stress (– – –) and normal stress (——) for a 22% dispersion of fat crystals in paraffin oil, measured at constant shear rate $\dot{\gamma}$

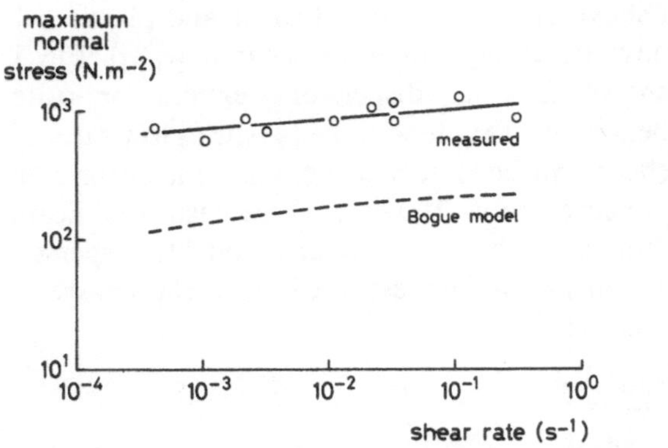

Fig. 9. Maximum of the normal stress for a 22% dispersion of fat crystals in paraffin oil, measured at constant rate of shear $\dot{\gamma}$

The ratio $(\tau_{11} - \tau_{22})/\tau_{12}$ represents the recoverable part of the deformation (7). That the predicted normal stresses are lower than the measured values by about a factor of 5 can be explained by assuming that the actual local deformation was underestimated by at least the same factor. Such an underestimate of the deformation is not unreasonable because these materials deform very inhomogeneously. Observations of the deformation profile of fat dispersions in a viscometer show indeed that the highest local deformation can be about 10 times the average deformation (8). However, assuming that the effective deformation must be larger than what is measured implies that the modulus G must be chosen smaller to retain the reasonable agreement between the measured and predicted shear stress overshoot curves.

An alternative explanation of the difference between experimental and predicted normal stresses might be that what is measured as normal stress in the case of fat dispersions is not, as in polymer-like substances, related to the elasticity of the material but to the plastic dilatancy (9).

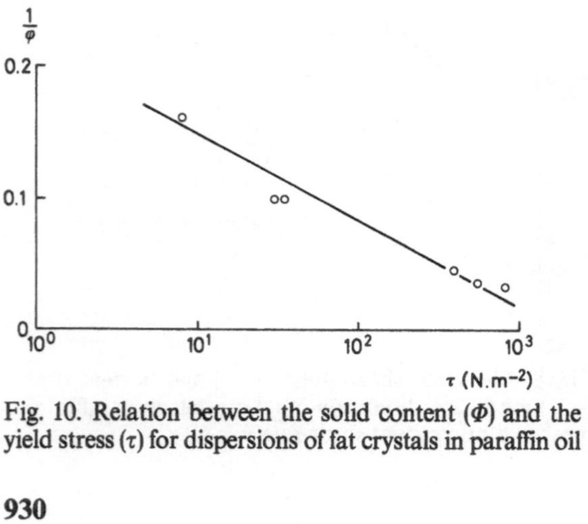

Fig. 10. Relation between the solid content (Φ) and the yield stress (τ) for dispersions of fat crystals in paraffin oil

The network of fat crystals has to expand somewhat in order to create regions in the crystal network where the concentration of fat crystals is somewhat lower so that these regions can be sheared more easily. It is probably the axial component of this expansion that leads to an apparent normal stress. That this apparent normal stress becomes negative after prolonged deformation cannot be explained. However, similar behaviour has been found in the rheology of sand (10).

Discussion

The above-mentioned comparison between the actual rheological behaviour of some fat dispersions and the predictions of the *Bogue* model shows that some important properties of these dispersions cannot be described by means of this model.

The essential deviations are:

(i) dispersions, in principle, are plastic substances (the shear stress is practically independent of the shear rate) which become viscous at high shear rates. The *Bogue* model predicts viscous behaviour at impracticably low shear rates and plastic behaviour as the shear rate increases. It fails to predict the second *Newton*ian region at high rates of shear;

(ii) dispersions, especially the more concentrated ones, show frequency-independent moduli in dynamic experiments. The phase angle too, is practically frequency-independent. On the other hand, the model predicts purely elastic behaviour (phase angle zero);

(iii) normal stresses measured in fat dispersions deviate largely from what is expected on the basis of the model. This might be related to the inhomogeneous deformation in these systems. An alternative and perhaps more probable explanation for this deviation might be that what is measured as normal force in a parallel-plate geometry is due to dilatancy of the dispersion and not due to elasticity as in the case of polymers. The normal stress measured in these dispersions resembles results obtained with powders in air (10).

Powders, too, behave as plastic substances because of the friction between the powder particles. For a special class of powders, the noncohesive powders, *Williams*, *Birks*, and *Bhattacharya* (11) have shown that the following relation exists between the volume fraction of powders and the yield stress of the powder:

$$1/\varphi = k - m \ln \tau \quad [13]$$

where k and m are constants for the powder considered.

A plot of the yield stress versus the volume fraction of fat crystals for dispersions (see fig. 10) shows that fat dispersions also satisfy the above-mentioned relation.

Finally, it can be concluded that the rheological behaviour of fat crystals in oil cannot be described satisfactorily with equations of state that were proposed for polymer-like substances. The results obtained suggest that these dispersions behave as plastic materials, in which yielding, dilatancy, cohesion and, possibly, friction between particles play a role.

Acknowledgement

Thanks are due to Dr. *M. van den Tempel* for his useful comments and to Mr. *J. V. Boskamp* for carrying out the experimental part of this work.

Summary

It has been investigated whether constitutive equations, which have been proposed originally to describe the rheological behaviour of polymerlike materials, can be used to represent the rheology of dispersions. Such equations generally predict stresses that depend on both the shear (γ) and a quantity ($\lambda\dot{\gamma}$) which is the product of the shear rate ($\dot{\gamma}$) and the time constant of the material (λ).

The behaviour of dispersions depends in general on the concentration of the dispersed particles. The dissipative aspect of the rheological behaviour is almost *Newton*ian for very dilute dispersions while it becomes plastic for more densely packet dispersions. In the latter case the shear stress is practically independent of the shear rate at low shear rates. Such behaviour may be accounted for in the constitutive equations by assuming $\lambda\dot{\gamma}$ to be almost constant. This motivated us to choose the equation of *Bogue* where the relaxation time (λ) depends on the shear rate ($\dot{\gamma}$), according to $1/\lambda = (1/\lambda_0) + a\dot{\gamma}$, where $1/\lambda_0$ accounts for the viscous behaviour and $a\dot{\gamma}$ for the plastic behaviour.

Comparing the actual rheological behaviour of dispersions of fat crystals in paraffin oil with the behaviour predicted by the *Bogue* equation, it turns out that the *Bogue* equation has some success in representing the stress overshoot in steady shear experiments. However, the predicted value of the normal stress for the concentrated dispersions is too low in comparison with the measured value. It is suggested that this discrepancy is due to the dilatant behaviour of these dispersions.

Moreover, the values of the dynamic moduli measured in oscillatory shear are predicted incorrectly, due to considerable changes in particle network which already occur at very small deformations.

References

1) *Papenhuijzen, J. M. P.*, Rheol. Acta **10**, 493 (1971).

2) *Bogue, D. C.*, Ind. Eng. Chem. Fundam. **5**, 253 (1966).

3) *Bogue, D. C.* and *S. O. Doughty*, Ind. Eng. Chem. Fundam. **5**, 243 (1966).

4) *Middleman, S.*, Trans. Soc. Rheol. **13**, 123 (1969).

5) *David, J.*, Rheol. Acta **8**, 311 (1969).

6) *Marsh, B. D.* and *J. R. A. Pearson*, Rheol. Acta **7**, 326 (1965).

7) *Philippoff, W.*, Trans. Soc. Rheol. **10**, 1 (1966).

8) *Papenhuijzen, J. M. P.*, Rheol. Acta **10**, 503 (1971).

9) *Freudenthal, A. M.* and *H. Geiringer*, Encyclopedia of physics VI, Chapter IV, p. 278 (Berlin-Heidelberg-New York 1958).

10) *Krizek, R. J.*, Trans. Soc. Rheol. **15**, 491 (1971).

11) *Williams, J. C., A. H. Birks*, and *D. Bhattacharya*, Powder Technol. **4**, 328 (1970/71).

Authors' address:

D. W. de Bruijne, N. J. Pritchard, and
J. M. P. Papenhuijzen
Unilever Research
Olivier van Noortlaan 120, P.O. Box 114
Vlaardingen (The Netherlands)

Rheol. Acta **13**, 424–427 (1974)

From the Kiev State University, Kiev (USSR)

Structure-continual approach in rheology of disperse and polymer systems

By Y. I. Shmakov and P. B. Begoulev

With 4 figures

(Received October 27, 1972)

The paper contains some results obtained by authors and their collaborators in rheology of dilute suspensions and dilute polymer solutions on the base of structure-continual approach.

This approach combines structured continuum theories, contained some number of internal parameters, which one needs for describing of substructure behaviour (macrorheology), with results drawn from various molecular descriptions of media under investigation (microrheology).

Consider a dilute suspension of rigid ellipsoids of revolution. To describe behaviour of substructure in this case it is enough to have the only internal parameter – orientation vector, which has its direction coincided with direction of particle axis of revolution.

On the base of *Ericksens* results (1, 2) we take the constitutive equations of structured continuum, which utilizes a single internal parameter – unit vector, in the form

$$t_{ij} = (\alpha_0 + \alpha_1 d_{km} n_k n_m)\,\delta_{ij} + \alpha_2 n_i n_j + \alpha_3 d_{ij} + \alpha_4 d_{km} n_k n_m n_i n_j + \alpha_5 d_{ik} n_k n_j + \alpha_6 d_{jk} n_k n_i + \alpha_7 n_i N_j + \alpha_8 n_j N_i \quad [1]$$

$$I(\ddot{n}_i + \dot{n}_k \dot{n}_k n_i) = \gamma[N_i - \lambda(d_{ik} n_k - d_{km} n_k n_m n_i)] + [\vec{M} \times \vec{n}]_i \quad [2]$$

where t_{ij} is stress tensor; d_{ij} is deformation rate tensor; n_i is orientation vector; $N_i = \dot{n}_i - \omega_{ij} n_j$; ω_{ij} is velocity vortex tensor; $\vec{M}$ is momentum of external forces, acting on a substructure element; α_i, I, γ, λ are rheological constants; δ_{ij} is *Kronecker* symbol.

Rheological constants entered in [1], [2] one can determine using the structure theory of *Jeffery* (3). On the base of (3, 4) we find stress tensor σ_{ij} of a dilute suspension of rigid ellipsoidal particles in the moving coordinate system, which has its axes coinciding with the principal axes of the ellipsoidal particle in the form

$$\sigma_{ij} = -p\delta_{ij} + 2\mu d_{ij} + \frac{8\mu\Phi}{ab^2} A_{ij} \quad [3]$$

where p is isotropic pressure; μ is dynamic viscosity of solvent; $2a$, $2b$ are axis of revolution and equatorial diameter of ellipsoidal particle; Φ is volume concentration of suspended particles; A is the function determined in (5).

Considering eq. [1] in the moving coordinate system and comparing it with [3] we get

$$\alpha_0 = -p$$

$$\alpha_1 = \frac{2\mu\Phi(\beta_0'' - \alpha_0'')}{3ab^4{}''\beta_0''\alpha_0'}$$

$$\alpha_2 = 0$$

$$\alpha_3 = 2\mu\left(1 + \frac{\Phi}{ab^4\alpha_0'}\right)$$

$$\alpha_4 = \frac{2\mu\Phi}{ab^2}\left[\frac{\alpha_0'' + \beta_0''}{b^2\alpha_0'\beta_0''} - \frac{2(\alpha_0 + \beta_0)}{\beta_0' B}\right]$$

$$\alpha_5 = \frac{4\mu\Phi}{ab^2}\left(\frac{\beta_0}{\beta_0' B} - \frac{1}{2b^2\alpha_0'}\right)$$

$$\alpha_6 = \frac{4\mu\Phi}{ab^2}\left(\frac{\alpha_0}{\beta_0' B} - \frac{1}{2b^2\alpha_0'}\right)$$

$$\alpha_7 = \frac{4b^2\mu\Phi}{ab^2 B}$$

$$\alpha_8 = -\frac{4a^2\mu\Phi}{ab^2 B} \quad [4]$$

where $\alpha_0, \beta_0, \alpha_0', \beta_0', \alpha_0'', \beta_0''$ are functions determined in *Jeffery* theory (3); $B = a^2\alpha_0 + b^2\beta_0$.

Considering eq. [2] in the moving coordinate system and comparing it with the *Jefferys* results (3) we find constants λ, γ and I.

$$\lambda = \frac{a^2 - b^2}{a^2 + b^2}$$

$$\gamma = -\frac{16 b \pi \mu (a^2 + b^2)}{3B}$$

$$I = \frac{m}{5}(a^2 + b^2) \qquad [5]$$

where m is mass of a particle.

Orientation of a suspended particle depends on hydrodynamic forces, external strength fields, inertial properties and rotatory *Brownian* movement of particles and may be characterized by distribution function of angular position of main particle axis F, which is given by the eq. [6]

$$\partial F/\partial t = D_r \Delta F - \operatorname{div}(F\vec{\omega}) \qquad [6]$$

where $\vec{\omega}$ is the angular velocity of particle, which is determined by the eq. [2]; t is time; D_r is rotatory diffusion coefficient.

To get the constitutive equations of media under investigation we average eq. [1] with the aid of distribution function F. As for rheological constants, they are connected with parameters characterizing substructure by relations [4]

$$T_{ij} = \langle t_{ij} \rangle = (\alpha_0 + \alpha_1 d_{km} \langle n_k n_m \rangle)\,\delta_{ij} + \alpha_3 d_{ij} + \alpha_4 d_{km} \langle n_k n_m n_i n_j \rangle + \alpha_5 d_{ik} \langle n_k n_j \rangle + \alpha_6 d_{jk} \langle n_k n_i \rangle + \alpha_7 \langle n_i N_j \rangle + \alpha_8 \langle n_j N_i \rangle. \qquad [7]$$

Consider some special cases

1. Effect of particle inertia is sufficiently small to be neglected. $\vec{M}$ is determined by rotatory *Brownian* movement only. Then (7) $\vec{M} = -kT\vec{n} \times \vec{\nabla} \ln F$, where k is *Boltzmann* constant; T is absolute temperature.

The eq. [2] can be written in the form

$$N_i = \lambda (d_{ij} n_j - d_{km} n_k n_m n_i) - D_r \nabla_i \ln F. \qquad [8]$$

Substituting [8] in [7] we get stress tensor for this case

$$T_{ij} = (-p + \alpha_1 d_{km} \langle n_k n_m \rangle)\,\delta_{ij} + \alpha_3 d_{ij} - 3D_r(\alpha_7 + \alpha_8)(\langle n_i n_j \rangle - \tfrac{1}{3}\delta_{ij}) + [\alpha_4 - \lambda(\alpha_7 + \alpha_8)]\, d_{km} \langle n_k n_m n_i n_j \rangle + \tfrac{1}{2}[\alpha_5 + \alpha_6 + \lambda(\alpha_7 + \alpha_8)] \times (d_{ik} \langle n_k n_j \rangle + d_{jk} \langle n_k n_i \rangle). \qquad [9]$$

One can check, that this results coincide with the results obtained in another way (8).

2. Effect of particle inertia is sufficiently small to be neglected. $\vec{M}$ is determined by rotatory *Brownian* movement and external strength field (electric, magnetic, gravitational). If the external strength field is electric and particles are dielectric ellipsoids, which have permanent dipole moment q_1, directed along axis of symmetry, $\vec{M}$ is given by the relation

$$\vec{M} = -kT\vec{n} \times \vec{\nabla} \ln F + [q_1 + q_2(\vec{E} \cdot \vec{n})]\, \vec{n} \times \vec{E} \qquad [10]$$

where $\vec{E}$ is vector of electric field strength; $q_2 = \chi_1 - \chi_2$; χ_1, χ_2 are principal values of electric polarisability along revolutional axis and along a direction perpendicular to it.

Substituting [2] and [10] in [7] we get stress tensor for this case

$$\begin{aligned} T_{ij} &= (-p + \alpha_1 d_{km} \langle n_k n_m \rangle)\,\delta_{ij} + \alpha_3 d_{ij} \\ &- 3D_r(\alpha_7 + \alpha_8)(\langle n_i n_j \rangle - \tfrac{1}{3}\delta_{ij}) \\ &+ [\alpha_4 - \lambda(\alpha_7 + \alpha_8)]\, d_{km} \langle n_k n_m n_i n_j \rangle \\ &+ \tfrac{1}{2}[\alpha_5 + \alpha_6 + \lambda(\alpha_7 + \alpha_8)] \\ &\times (d_{ik} \langle n_k n_j \rangle + d_{jk} \langle n_k n_i \rangle) \\ &- \frac{\alpha_7}{\gamma}[q_1 \langle n_i \rangle E_j - q_1 \langle n_i n_j n_k \rangle E_k \\ &+ q_2 \langle n_i n_k \rangle E_k E_j - q_2 \langle n_i n_j n_k n_m \rangle E_k E_m] \\ &- \frac{\alpha_8}{\gamma}[q_1 \langle n_j \rangle E_i - q_1 \langle n_j n_i n_k \rangle E_k \\ &+ q_2 \langle n_j n_k \rangle E_k E_i \\ &- q_2 \langle n_i n_j n_k n_m \rangle E_k E_m]^{1)} \end{aligned} \qquad [11]$$

$$\langle a \rangle = \int_{|\vec{n}|=1} aF\, dn_k.$$

From the eqs. [11], [2], and [10] one can obtain special cases considered in (9, 10).

If particles are large ($\sqrt[3]{ab^2} > 4.10^{-6}$ m) rotatory *Brownian* movement can be neglected (6). Orientation in this case is given by the equation

$$N_i = \lambda(d_{ik} n_k - d_{km} n_k n_m n_i) - \frac{1}{\gamma}[q_1 + q_2 E_k n_k](E_i - E_k n_k n_i) \qquad [12]$$

which under some relationship between d_{ij}, ω_{ij} and E_i has a stable steady state solution (11).

[1]) We take into account only orientational effect of electric field.

In this case suspended particles hover, $\dot{n}_i = 0$, $N_i = -\omega_{ij} n_j$,

$$T_{ij} = (-p + \alpha_1 d_{km} n_k n_m)\,\delta_{ij} + \alpha_3 d_{ij} + \alpha_4 d_{km} n_k n_m n_i n_j + \alpha_5 d_{ik} n_k n_j + \alpha_6 d_{jk} n_k n_i - \alpha_7 \omega_{jk} n_k n_i - \alpha_8 \omega_{ik} n_k n_j. \quad [13]$$

3. Consider simple shearing motion $V_x = V_z = 0$, $V_y = Kx$, $K =$ const of dilute suspension of rigid dielectric ellipsoids in electric field

$$E_x = E\cos\alpha, \quad E_y = E\sin\alpha,$$

$$E_z = 0, \quad E = \text{const}.$$

In spherical coordinate system (fig. 1)

$$n_x = \cos\varphi\sin\theta$$
$$n_y = \sin\varphi\sin\theta$$
$$n_z = \cos\theta. \quad [14]$$

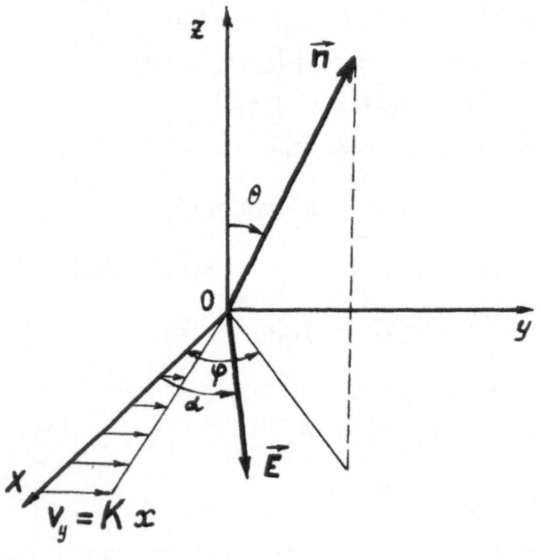

Fig. 1

From eq. [12] and [14] we get in the case $q_1 = 0$

$$\omega_\varphi = \dot{\varphi} = \frac{K}{2}(1 + \lambda\cos 2\varphi) + \frac{q_2 E^2}{2\gamma}\sin 2\varphi' \quad [15]$$

$$\omega_\theta = \dot{\theta} = \frac{K\lambda}{4}\sin 2\varphi\sin 2\theta - \frac{q_2 E^2}{4\gamma}(1 + \cos 2\varphi')\sin 2\theta$$

where

$$\varphi' = \varphi - \alpha.$$

As in the case $E = 0$ (6) we express the solution of the eq. [6] for the steady state as a power series in λ

$$F(\varphi', \theta) = \sum_{j=0}^{\infty} \lambda^j \Bigg[\frac{1}{2}\sum_{n=0}^{j} a_{n0,j} P_{2n}(\cos\theta) + \sum_{n=1}^{j}\sum_{m=1}^{n} (a_{nm,j}\cos 2m\varphi' + b_{nm,j}\sin 2m\varphi')\, P_{2n}^{2m}(\cos\theta)\Bigg] \quad [16]$$

where P_{2n} are *Legendre* polynomials; P_{2n}^{2m} are *Legendre* associated functions; $a_{nm,j}$, $b_{nm,j}$ are coefficients of expansion, for which recurrence relations are obtained (12).

These recurrence relations allow as in the case $E = 0$ (13, 14) using a computer to find the effective viscosity of suspension $\mu_{\text{eff}} = \mu(1 + \nu\Phi)$ and the normal stress differences $T_{yy} - T_{zz} = \mu\Phi K f_1$ and $T_{xx} - T_{zz} = \mu\Phi K f_2$ $\left(\nu, f_1, f_2\right.$ are functions depending on a/b, $\sigma = K/D_r$, $\left.\beta = -\frac{q_2 E^2}{\gamma D_r \lambda}, \alpha\right)$ without restriction consideration by the case $\sigma \ll 1$, $\beta\lambda \ll 1$, discussed earlier (10, 15–18, 19).

Results of calculation the dependence of ν, f_1 and f_2 on σ and α when $a/b = 10$, $\beta = 0{,}8$ are presented on the figs. 2 and 3. Dushed line corresponds to the case $\beta = 0$.

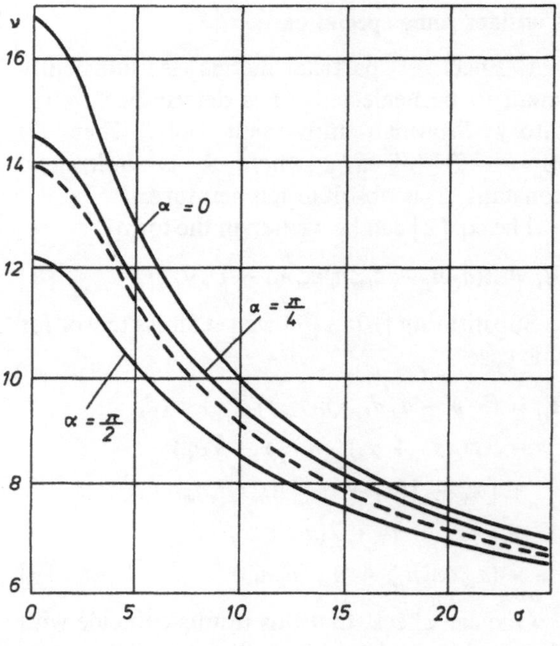

Fig. 2

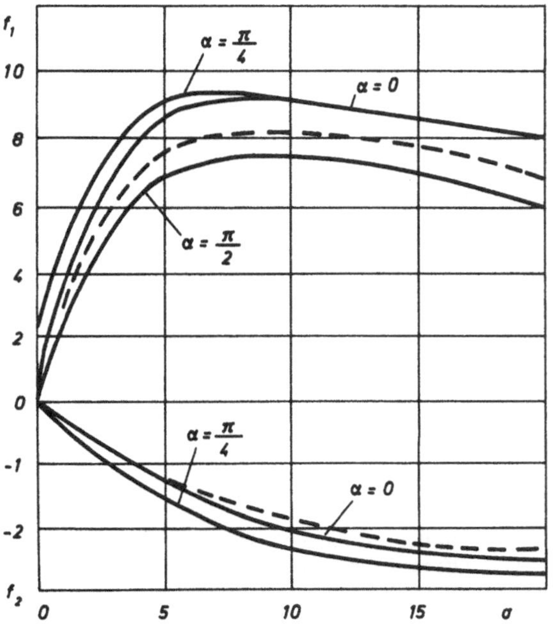

Fig. 3

The dependence of v, f_1 and f_2 on angle of hover φ (particles hover in the plane of shear) for the same value of parameter $a/b = 10$ are calculated with the aid of relation [12] and are presented on the fig. 4. Function $v(\varphi)$ coincides with results got in (20).

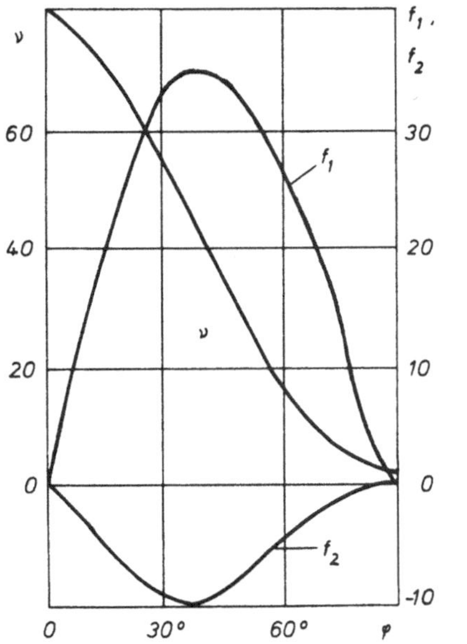

Fig. 4

4. The constitutive eq. [9] were used for an investigations flows of dilute suspensions of rigid ellipsoids between two parallel planes (14, 21) and in two-dimensional boundary layers (22, 23).

Influence of particle inertia on rheological behaviour of suspension was considered in (5).

In (24) the structure-continual approach was used for getting constitutive equations of suspensions taking into account particles interaction.

On the base of *Kline* and *Allen* structured continuum model (25) and results got in (3, 10) constitutive equations of dilute suspensions of rigid three-axes ellipsoids were obtained (26).

References

1) ***Ericksen, J. L.***, Arch. Ration. Mech. and Analys., **4**, 231 (1960).

2) ***Ericksen, J. L.***, Arch. Ration. Mech. and Analys., **23**, 266 (1966).

3) ***Jeffery, G. B.***, Proc. Royal. Soc. Ser. **A 102**, 161 (1922).

4) ***Hand, G. L.***, Arch. Ration. Mech. and Analys., **7**, 6 (1961).

5) Придатченко Ю. В., Шмаков Ю. И., ПМТФ, No. 2, 125 (1972).

6) ***Peterlin, A.***, Z. Physik, **111**, 232 (1938).

7) Покровский В. Н., Колл. журн., **19**, 576 (1967).

8) Шмаков Ю. И., Таран Е. Ю. ИФЖ, **18**, 1019, (1970).

9) ***Chaffey, C. E.*** and ***S. G. Mason***, J. Coll. Sci. **27**, 115 (1968).

10) ***Saito, N.*** and ***T. Kato***, Phys. Soc. Jap. **12**, 1383 (1957).

11) ***Allan, R. S.*** and ***S. G. Mason***, Proc. Roy. Soc., Ser. **A 267**, 62 (1962).

12) Бегоулев П. Б., Шмаков Ю. И. ИФЖ, **24**, 1094 (1973).

13) ***Scheraga, H. J.***, Chem. Phys., **23**, 1526 (1955).

14) Шмаков Ю. И., Таран Е. Ю., Бегоулев П. Б. Сб. Гидромеханика, „Наукова Думка", К., No. 20, 87 (1970).

15) ***Peterlin, A.***, Razprave (Dissertation) 7/1 (Ljubljana, 1956).

16) ***Demetriades, S. T.***, J. Chem. Phys., **29**, 1058 (1958).

17) ***Ikeda, S.***, J. Chem. Phys., **38**, 2839 (1963).

18) ***Argyropoulos, G. S.***, Trans. Soc. Rheol. **9**, 57 (1965).

19. Покровский В. Н. Успехи Физ. наук, **105**, 625 (1971).

20) ***Cheffey, C. E.*** and ***S. G. Mason***, J. Coll. Sci. **20**, 330 (1965).

21) ***Shmakov, Y. I.*** and ***E. Y. Taran***, Rev. Roum. Sci. Tech.-Mec. Appl., **16**, 705 (1971).

22. Шмаков Ю. И., Пилявская Т. Г. ДАН УРСР, сер. A, No. 9, 815 (1971).

23) Шмаков Ю. И., Пилявская Т. Г. ДАН УРСР, сер. A, No. 7, 672 (1972).

24) Придатченко Ю. В., Шмаков Ю. И. ПМТФ, No. 1, 141 (1973).

25) ***Kline, K. A.*** and ***S. J. Allen***, Z. Angew. Math. Physik, **19**, 898 (1968).

26) Бегоулев П. Б., Шмаков Ю. И. ИФЖ, 23, 340 (1972).

Authors' address:

Y. I. Shmakov and *P. B. Begoulev*
Kiev State University
Kiev (UdSSR)

Rheol. Acta **13**, 428–433 (1974)

Commissariat à l'Energie Atomique, Sevran (France)

Mesure de loi de compression statique et dynamique d'une argile. Application au calcul de la propagation d'un ebranlement dans un sol

Par J. Maury, M. Roussel, G. Lucas et S. Poulard

Avec 8 figures

(Reçu p. p. le 27 octobre 1972)

I. Introduction

Pour calculer la propagation d'une onde dans un sol, il faut connaître la relation qui lie la contrainte à l'état des déformations du milieu de propagation, c'est-à-dire à la loi rhéologique du milieu, pour tous les cas de déformation que peut subir le milieu au passage de l'onde.

Comme les sols peuvent présenter des comportements très divers, il est nécessaire de déterminer cette loi rhéologique expérimentalement pour chaque sol considéré. De plus, les propriétés des sols dépendant étroitement du système de contraintes auxquelles ils sont soumis et de la vitesse de chargement, l'échantillon dans l'essai doit être soumis à un état de contraintes aussi semblable que possible à celui qui existe au cours du passage de l'onde (1).

Dans le cas d'une onde bi ou tridimensionnelle, cet état de contraintes peut être très complexe. Nous ne considérerons donc que le cas d'une onde monodimensionnelle plane.

II. Détermination expérimentale du comportement d'un sol au passage d'une onde de déformation plane

2.1. Généralités

Dans le cas d'une onde plane se propageant dans un milieu infini, le milieu ne peut pas se déformer perpendiculairement à la direction de propagation. Par conséquent, l'éprouvette doit être confinée latéralement dans l'essai rhéologique pour éviter toute expansion latérale. Dans ces conditions, le matériau ne peut fluer en cisaillement.

De plus, la déformation doit être plane et la vitesse de chargement doit se rapprocher le plus possible des vitesses de variation des contraintes, mises en jeu au passage de l'onde réelle ($\dot{\varepsilon} = 1\ s^{-1}$). Un tel test peut être fait à l'aide d'un œdomètre soumis à une charge dynamique programmée (œdomètre dynamique).

2.2. Oedomètre dynamique

Sur la fig. 1 est représentée une section verticale de l'œdomètre dynamique, que nous avons utilisé pour étudier le comportement dynamique et statique d'un sol argileux (2).

L'échantillon est contenu dans un cylindre épais, en acier dur de même diamètre. Il repose sur une pièce cylindrique (*A*) pouvant jouer librement dans le cylindre. Cette pièce est liée au socle de l'appareil par l'intermédiaire d'un capteur *Kistler*. Ce capteur est précontraint par un écrou de serrage entre la pièce *A* (tête de mesure) et le socle, de manière à ce que le capteur travaille toujours en compression. L'échantillon est comprimé dans le cylindre par l'intermédiaire d'un piston (pièce *B*) au sommet duquel est placé un deuxième capteur *Kistler* (également précontraint) destiné à mesurer la force dynamique agissant sur la surface supérieure de l'échantillon. Pour limiter le frottement, la surface de ce piston a été téflonée; on a vérifié que l'on obtient ainsi un frottement quasiment nul.

De l'indication des deux capteurs, on peut donc déduire la contrainte moyenne à laquelle est soumis l'échantillon et l'importance des frottements entre l'échantillon et le cylindre.

La déformation de l'échantillon est mesurée à l'aide de deux capteurs de déplacement inductifs fixés à la tête du piston et au socle de l'appareil (fig. 1).

Un écrou (pièce *C*) permet de précontraindre éventuellement l'échantillon avant l'essai, la précontrainte est mesurée à l'aide du capteur de force inférieure.

Lors d'un essai, cet ensemble est placé sous un mouton de choc (fig. 2); la déformation et la force sont enregistrées sur un oscilloscope à mémoire par l'intermédiaire de la chaine de mesure schématisée sur la fig. 3.

III. Résultats expérimentaux

3.1. Echantillon

Les mesures ont été faites sur des échantillons cylindriques (∅ 30 h 10), usinés dans des cylindres d'une argile (provenant de Provins) consolidée à 100 bars en laboratoire pour reconstituer une argile de profondeur (500 m).

Tous les essais ont été faits suffisamment rapidement (en moins d'une heure pour les essais statiques) pour être considérés comme non drainés.

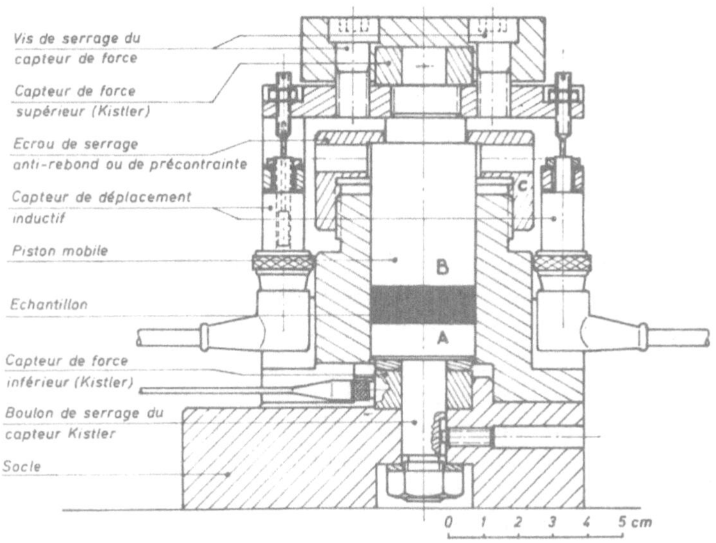

Fig. 1. Coupe verticale de l'oedomètre

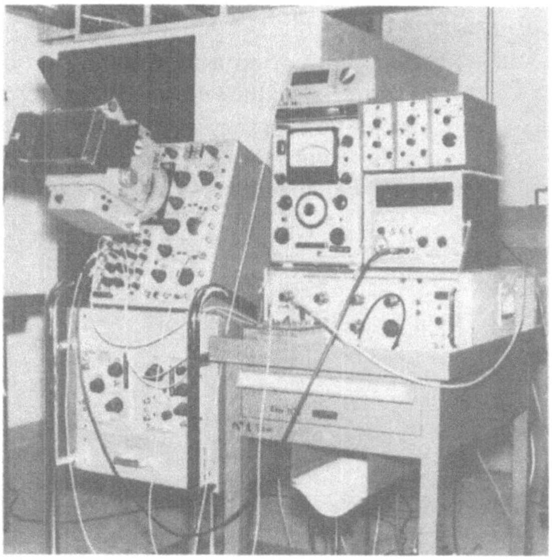

Fig. 2. b) électronique de mesure

a)

Fig. 2. a) Oedomètre place sous le mouton de choc

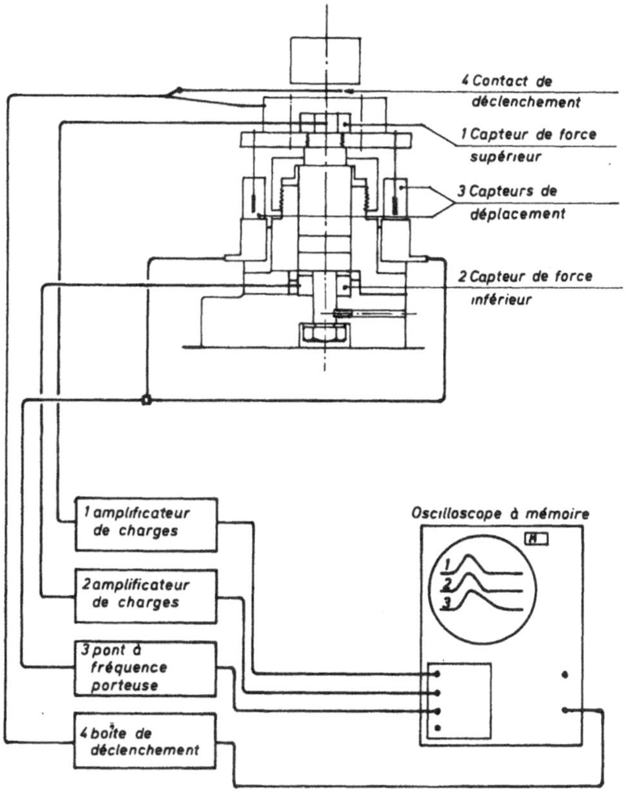

Fig. 3. Synoptique de mesure

28

3.2. Essais statiques

Sur la fig. 4, on a représenté la courbe oedométrique statique $\Delta l/l = f(p)$ correspondant à 3 cycles de chargement et de déchargement successifs [dans l'ordre (0–120 b), (0–400 b) et (0–400 b)] effectués sur le même échantillon.

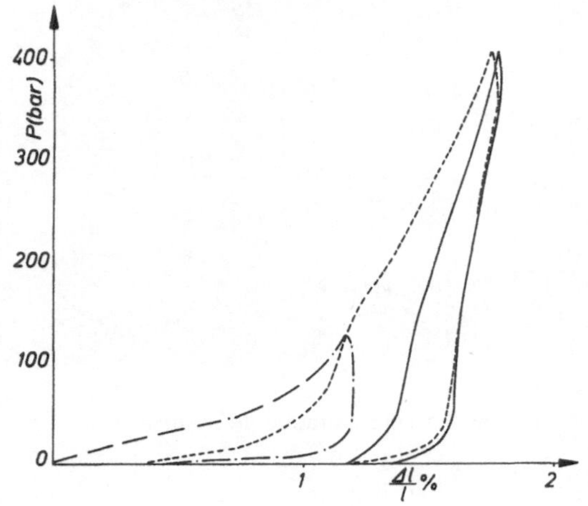

Fig. 4. Diagramme $P = f\frac{\Delta l}{l}$ statique de l'argile de Provins (échantillon I). ———— Premier cycle; ---- deuxième cycle; ——— troisième cycle

En première compression, on peut distinguer deux zones de comportements différents, suivant que l'on se situe en-dessous de la pression de première consolidation (100 bars) ou au-dessus. La compressibilité est en moyenne 6 fois plus grande dans la première zone que dans la deuxième.

Cette grande compressibilité correspond à un tassement du matériau jusqu'à ce qu'il recouvre son état de consolidation.

Le comportement en détente est différent du comportement en compression; d'une manière générale, la courbe représentant la détente (fig. 4) est quasiment rectiligne, excepté en fin de détente où elle se raccorde tangentiellement à l'axe des déformations.

La pente de la partie linéaire est beaucoup plus grande que la pente de la courbe représentant la compression initiale; elle est indépendante de la pression maximale atteinte.

De plus, la détente se fait avec une réduction remanente de volume qui a tendance à diminuer avec le temps.

Si on dépasse au cours d'un cycle de recompression la pression maximum exercée lors du précédent chargement, la courbe de recompression se raccorde à la première courbe de compression. Audelà de cette pression, la compression se fait comme si l'on avait effectué une seule compression.

3.3. Essais dynamiques

3.3.1. Conditions de mesure

La détermination du diagramme contrainte déformation dynamique n'est possible que si les contraintes sont les mêmes dans toute l'épaisseur de l'échantillon à chaque instant. En statique, cette condition est remplie s'il n'y a pas de frottement entre l'échantillon et la paroi du cylindre. En dynamique, il faut de plus que la longueur de l'onde associée à l'ébranlement soit grande devant la dimension de l'échantillon (50–100 fois).

Afin de se trouver dans ces conditions, on a allongé le temps de montée de la pression, en plaçant un amortisseur de caoutchouc au point d'impact du marteau du mouton de choc sur l'œdomètre.

3.3.2. Résultats

Deux types d'essais dynamiques ont été faits; une première série sans précontrainte et une deuxième série avec une précontrainte de l'échantillon à 150 bars.

Sur les figs. 5 et 6 sont représentés les résultats de la première série d'essais, chacune de ces figures correspondant à un échantillon différent.

Comme on peut le voir sur ces figures, où l'on a reporté également les résultats statiques, les diagrammes dynamiques sont sensiblement différents des diagrammes statiques. L'argile est de 20–30% moins compressibles en dynamique qu'en statique.

Cette différence est probablement l'effet de frottements visqueux qui apparaissent dans le mouvement des grains solides les uns par rapport aux autres. La fig. 7 montre que lorsque l'échantillon est précontraint à 100 bars, la détente se fait pratiquement suivant le même chemin que la compression.

Ce résultat était prévisible. En effet, au-dessus de la pression de consolidation (100 b) il ne doit pas exister à proprement parler de vide dans l'échantillon puisque l'eau interstitielle remplit alors complètement les vides existant entre les grains solides (1). Le milieu étant complètement saturé doit avoir un comportement de fluide et par conséquent ne saurait présenter de cycle d'hystérésis en déformation de volume.

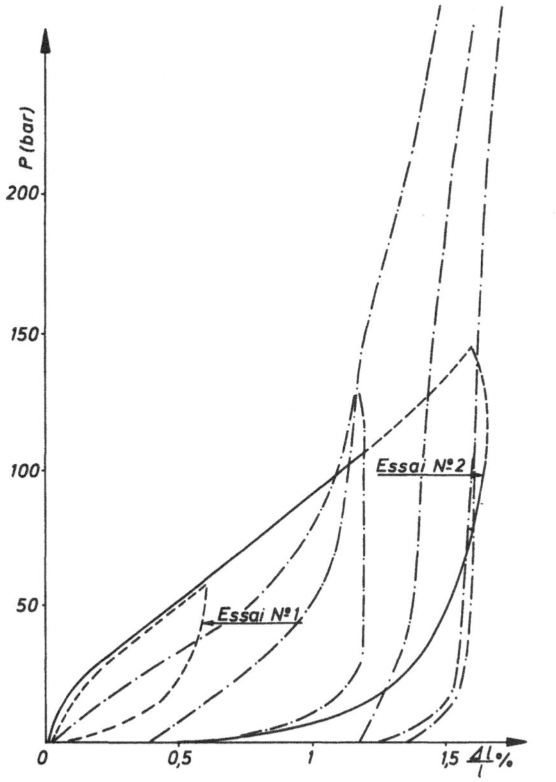

Fig. 5. Diagramme $P = f\frac{\Delta l}{l}$ dynamique de l'argile de Provins (échantillon II non précontraint). ---- Premier cycle; ——— deuxième cycle; ——— comportement statique (échantillon I)

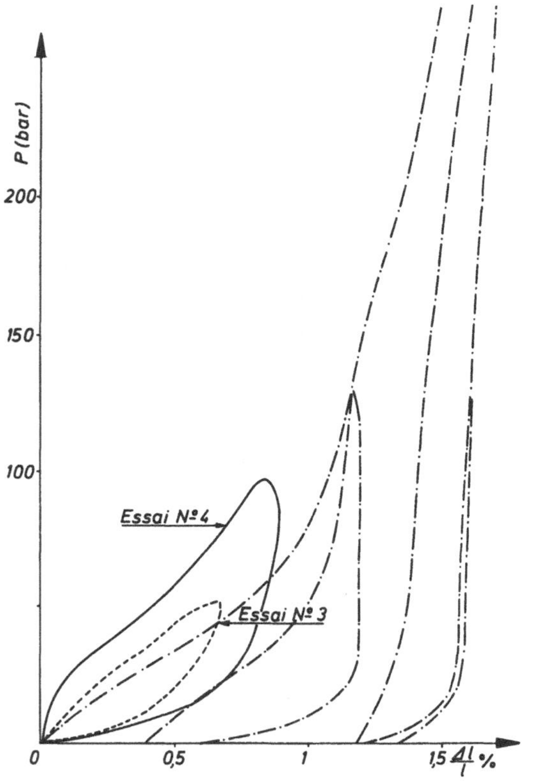

Fig. 6. Diagramme $P = f\frac{\Delta l}{l}$ dynamique de l'argile de Provins (échantillon III non précontraint). ---- Premier cycle; ——— deuxième cycle; ——— comportement statique (échantillon I)

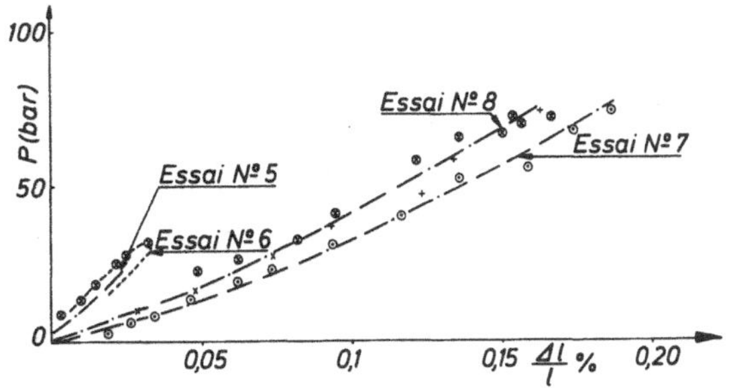

Fig. 7. Diagramme $P = f\frac{\Delta l}{l}$ dynamique de l'argile de Provins (échantillon IV précontraint à 150 bars). ——— Premier cycle; ---- deuxième cycle; — · — · — troisième cycle; ——— quatrième cycle

IV. Modèle de comportement et équation de propagation

4.1. *Généralités*

D'après les résultats des essais précédents, on peut se représenter une argile sur-consolidée, comme un mélange de particules minérales (solides) d'eau et d'air. Les particules minérales forment un squelette poreux, dont les pores sont remplis d'eau et d'air. Lorsqu'agit une charge très faible, la dureté du squelette est suffisante pour leur résister, le milieu a un comportement élastique. Pour décrire sa déformation, on peut alors utiliser le modèle de *Hooke* avec éventuellement un terme dissipatif pour tenir compte des frottements qui peuvent entrainer les mouvements relatifs de l'eau par rapport au squelette.

Si l'on augmente la charge, les grains qui forment le squelette glissent les uns par rapport aux autres, leur répartition devient plus dense (la porosité diminue), l'eau et l'air commencent à subir une fraction plus importante de la charge; pour des charges croissantes le dommage s'étend aux particules minérales qui constituent le squelette.

Dans ces conditions, le modèle de *Hooke* ne s'applique plus et il faut le remplacer. Le nouveau modèle doit se réduire à celui de *Hooke* pour des faibles charges. Pour des charges importantes il doit tenir compte de façon satisfaisante des propriétés du milieu liées aux processus que nous venons de décrire.

Si lors du chargement, la densité croît sensiblement du fait de la redistribution des particules et de leur morcellement, lorsque l'on retire la charge, du fait de l'irréversibilité des processus de redistribution, la diminution de densité est plus faible. La courbe de détente est quasiment verticale, excepté en fin de détente où elle se raccorde tangentiellement à l'axe de déformation.

La partie linéaire (quasi verticale) correspond à la détente élastique (réversible du squelette), la partie non linéaire, en fin de détente, représente probablement l'effet de la détente des gaz contenus dans les pores qui ne devient appréciable qu'à faible pression. Cette détente ne pouvant se faire qu'avec déplacement des grains solides les uns par rapport aux autres, est irréversible.

4.2. Modèle de comportement

D'après les résultats précédents, on voit que si l'on reste dans le domaine de contrainte σ_{xx} (longitudinale), où la détente se fait de manière réversible, c'est-à-dire si on ne considère que des cycles de compression et de détente dans lesquels la contrainte reste supérieure à 20 bars, la contrainte σ_{xx} est pour une gamme de vitesse de déformation donnée $\dot{\varepsilon}_{xx}$, une fonction biunivoque de la déformation totale quelque soit le cycle considéré.

Cependant, comme le montre la fig. 8, la relation entre la contrainte et la déformation doit être différente dans le cas où une augmentation de pression entraîne une variation de volume irréversible et dans le cas où la variation de volume se produit réversiblement.

Nous poserons donc que, lorsqu'un élément du milieu subit des déformations irréversibles, la contrainte est liée à la déformation totale par la relation (courbe 1) (3).

$$\sigma_{xx} = f_1(\varepsilon_{xx}). \qquad [1]$$

Les variations irréversibles de volume ne pouvant avoir lieu qu'en compression (fig. 8) nous poserons de plus que la relation [1] n'est possible que si:

$$\frac{d\varepsilon_{xx}}{dt} > 0.$$

Si maintenant, la contrainte ayant augmenté pendant une certaine période au cours de laquelle il s'est produit une déformation irréversible, décroît puis croît à nouveau sans dépasser sa valeur initiale, la déformation se fait réversiblement, nous poserons que dans ce processus, la contrainte est liée à la déformation par une autre relation:

$$\sigma_{xx} = f_2(\varepsilon_{xx}, \sigma^*_{xx}) \qquad [2]$$

(cette relation représente les courbes 2, 3, 4 et 5 de la fig. 8). σ^* représente la contrainte maximale subie par l'élément de volume considéré au cours de sa précédente variation irréversible de volume. A σ^* correspond une valeur ε^* de la déformation telle que:

$$\sigma^*_{xx} = f_1(\varepsilon^*_{xx})$$

on peut donc écrire l'eq. [2] sous la forme:

$$\sigma_{xx} = f_2(\varepsilon_{xx}, \varepsilon^*_{xx}).$$

Pour un élément de volume donné, le paramètre σ^* de même que ε^* ne peuvent que croître et seulement au cours des déformations volumiques irréversibles; lors des variations réversibles de volume de l'élément, σ^* et ε^* ne varient pas. σ^* et ε^* caractérisent donc, dans le domaine de pression considérée, la variation irréversible de volume.

Pour de très fortes contraintes, la porosité du milieu disparaît et le matériau se contracte sans déformation volumique irréversible appréciable. On posera donc que σ^* et ε^* sont bornés supérieurement par des valeurs limites σ^*_l et ε^*_l au-dessus desquelles les fonctions f_1 et f_2 coïncident et toute déformation volumique se fait réversiblement (fig. 8).

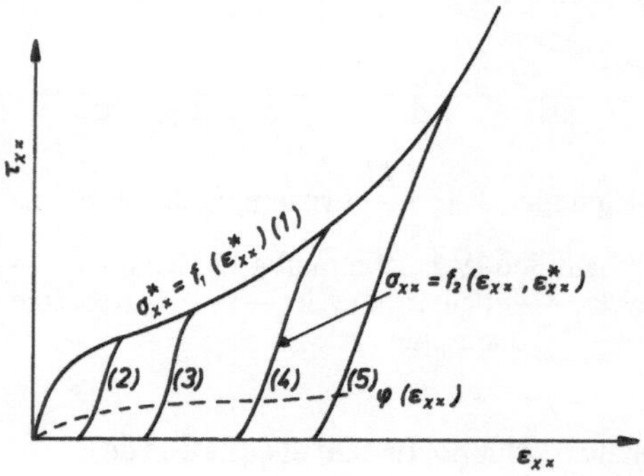

Fig. 8.

Les propriétés de comportement que nous venons de décrire peuvent se résumer par les formules:

$$\sigma \;\; = f(\varepsilon, \varepsilon^*)\, e(\varepsilon^* - \varepsilon)$$

$$\sigma^* = f(\varepsilon^*, \varepsilon^*) \equiv f_1(\varepsilon^*)$$

et

$$\frac{d\varepsilon^*}{dt} = \frac{d\varepsilon}{dt}\, e(\varepsilon - \varepsilon^*)\, e\left(\frac{d\varepsilon}{dt}\right) \qquad [4]$$

où l'opérateur d/dt désigne la dérivée totale par rapport au temps, la fonction $e(u)$ l'échelon unité, c'est-à-dire:

$$e(u) = \begin{cases} 1 & \text{si} \quad u \geq 0 \\ 0 & \text{si} \quad u < 0. \end{cases}$$

4.3. Système complet des équations de propagation

Les relations [4] donnent une représentation analytique complète de la déformation plane du milieu.

Si on adjoint à ces relations les équations de mouvement et de continuité qui s'écrivent pour une onde plane [4]:

$$\varrho \frac{dv}{dt} = \varrho F^e - \frac{\partial \sigma_{xx}}{\partial x} \qquad [5]$$

et

$$\frac{d\varrho}{dt} + \varrho \frac{dv}{dx} = 0 \qquad [6]$$

avec

$$\varrho = \varrho_0 (1 - \varepsilon_{xx}) \qquad [7]$$

on obtient un système d'équation complet. Pour des conditions aux limites données, la solution de ces équations donne le mouvement du système considéré.

V. Conclusion

Nous avons montré qu'un essai œdométrique permet de déterminer les lois de comportement d'un matériau nécessaire au calcul d'une onde plane.

On peut noter ici cependant que cet essai ne serait pas suffisant si on voulait calculer la propagation d'une onde sphérique de petit rayon.

En effet, dans ce cas, du fait de la divergence de l'onde, le milieu subit au passage de l'onde une déformation transversale en plus de la déformation radiale.

En plus du comportement en volume du milieu, il est donc nécessaire de connaître sa loi d'écoulement plastique. Ce travail a été fait mais ne peut être exposé ici faute de temps.

Littérature

1) *Costet, J.* et *G. Sanglerat,* Cours pratique de Mécanique des sols (Paris 1969).

2) *Seknicka, J. E., T. J. Cowry* et *P. L. Huic,* A study of the behavior of a clay under rapid and dynamic loading in the one dimensional and triaxial test. RTD TDR – 63 – 3116 (June 1961).

3) *Cristescu, N.,* Dynamic Plasticity, p. 509 (Amsterdam 1967).

4) *Wilkins, M. L.,* Calculation of elastic plastic flow, in: Methods in computational physics. Vol. 3 (New York and London 1964).

Adresse des auteurs:

J. Maury, M. Roussel, G. Lucas et *S. Poulard*
Commissariat à l'Energie Atomique
Etablissements T
B. P. No. 7
F-93270 Sevran (France)

Rheol. Acta 13, 434–442 (1974)

From the Department of Metallurgical Engineering, Varanasi (India), and the Department of Metallurgy and Materials Science, Cambridge (England)

Rheological analysis of superplasticity in metallic materials*)

By K. A. Padmanabhan and G. J. Davies

With 1 table

(Received October 27, 1972)

1. Introduction

The phenomenon of "structural" superplasticity wherein metallic materials exhibit ductility far in excess of conventional behaviour, when subjected to small forces in tension, compression or torsion, has been a subject of serious study only in recent years. A considerable amount of experimental results concerning this phenomenon is available and these results are reviewed in detail by *Johnson* (1970) and *Davies* et al. (1970).

A detailed physical model that accounts for superplasticity should be most useful to the development of the subject. However, as yet there is no generally accepted mechanism for superplasticity in the literature. The present state of development in this direction has been critically examined by *Davies* et al. (1970).

Materials that exhibit "structural" superplasticity are characterised by a stable, ultra-fine grain size (usually 1–10 μm) which is retained even after extreme elongation. Usually, the alloys consist of an intimate mixture of two phases that differ vastly in chemical composition. The response of superplastic alloys to mechanical forces can be represented by an approximate equation of state of the form (see, for instance, *Davies* et al., 1970)

$$\dot{\varepsilon} = \frac{c}{d^a}\sigma^{\frac{1}{m}}\exp\left(-\frac{Q}{kT}\right) \qquad [1]$$

where $\dot{\varepsilon}$ is the tensile strain rate, σ is the applied tensile stress, m is the strain-rate sensitivity index which lies usually between 0.3 and 0.8 in case of superplastic deformation, d is the grain size, Q is an activation energy, k is the *Boltzmann*'s constant, T is the absolute temperature of deformation, a is a constant lying between 2 and 3 and c is a constant of proportionality. Eq. [1] is usually represented in the alternate form

$$\sigma = K\dot{\varepsilon}^m \qquad [2]$$

for isothermal deformation. The constants K and m are critically dependent on the experimental conditions like grain size, temperature, strain rate etc. As most of the deformation is carried out at constant temperature, eq. [2] is commonly used to represent superplastic deformation. It is not difficult to see that when shear stress, τ and shear strain rate, $\dot{\gamma}$ are involved (*Drucker*, 1961) eq. [2] becomes

$$\tau = K'\dot{\gamma}^m \qquad [3]$$

where K' is another constant of proportionality.

The various physical models that are available in the literature (*Davies* et al., 1970) treat σ and T as the independent variables. However, for practical purposes it may be more convenient to treat $\dot{\varepsilon}$ as the independent variable instead of σ as it is easier to know the speed of travel of the tool and changes in the external dimensions of the specimen than the stresses acting at various sections. In fact, using eq. [2] *Jovane* (1968), *Cornfield* and *Johnson* (1970), *Holt* (1970), *Mehta* et al. (1970), *Ghosh* and *Duncan* (1970) and *Fields* and *Stewart* (1971) have given analyses of isothermal superplastic forming where $\dot{\varepsilon}$ is treated as the independent variable. However, these analyses based on conventional continuum mechanics neglect grain size. As has been pointed out earlier, since superplasticity occurs only when the grain size lies within a small range, these analyses will merely be of limited use. In addition to these analyses, equations must be available which are able to take into account the internal variations in $\dot{\varepsilon}$, m etc. and still be able to predict external load, power consumption and so forth in order to make the optimization of operating variables possible. Such a method based on numerical analysis has been given by *Padmanabhan* and *Davies* (1970).

Alternatively, the problem can be investigated rheologically making use of some of the properties of the superplastic alloy systems. While doing this, an attempt can be made to incorporate microstructural features like grain size. This would be consistent with effort already being made in this direction (see, for example, *Kröner*, 1960). Before doing this and applying the concepts to a potential commercial technique of superplastic forming some useful properties of the superplastic system will be established.

2. Properties of the superplastic system

On the basis of experimental results and theoretical considerations, the following prop-

*) This work was done at the Department of Metallurgy and Materials Science, University of Cambridge, Cambridge, England.

erties can be attributed to the superplastic system:

1. The material is isotropic and incompressible

One of the conditions necessary for superplasticity is a random, equiaxed grain size. Both *Nicholson* (1972) and *Padmanabhan* (1971), after a detailed study of the available experimental results, have concluded that superplasticity is a random deformation process with an inherently low anisotropy of deformation. It has also been demonstrated experimentally by *Padmanabhan* (1971) that under optimal superplastic conditions, the hydrostatic component of the stress system does not affect flow. As a consequence, the material can be regarded as isotropic, for all practical purposes, within the superplastic range.

Hydrostatic pressure can squeeze out vacancies and decrease the volume of the material if the pressure is sufficiently high in value (*Tanner* and *Radcliffe*, 1962; *Hilliard* et al., 1961). However, under superplastic stresses (typically of the order of 1 to 20 kg/mm²) this is unlikely (*Padmanabhan*, 1971). Further, the viscous boundary approach to superplasticity (*Padmanabhan*, 1971) which largely accounts for both the unequivocal and conflicting experimental results of superplasticity, implies that there is no elastic deformation during superplasticity and the only influence of the applied stress is to give directionality to the flow. Thus it can be concluded that the volume remains virtually constant, during superplastic deformation.

2. There are no elastic strains, the material does not work-harden when the grain size is stable and the material flows under any load

It has been experimentally found by *Alden* (1968) that the material is not work-hardened by superplastic deformation when the grain size is stable. Even when work-hardening is seen (*Watts* et al., 1971) it has been attributed to grain growth. The rest follows from the discussion of the previous point that the only role of the applied stress is to give directionality to the flow.

3. The equation of state can be expressed in terms of the representative stress and strain rate, for all practical purposes, by an equation

$$\sigma = K\dot{\varepsilon}^m \quad \text{or} \quad \dot{\varepsilon} = K_1\sigma^{\frac{1}{m}}$$

where K and K_1 are constants

This follows from the experimental results reported (*Johnson*, 1970; *Davies* et al., 1970) and the viscous boundary approach to superplasticity (*Padmanabhan*, 1971).

4. The superplastic system can be regarded as a heterogeneous continuum

The viscous boundary approach to superplasticity (*Padmanabhan*, 1971) envisages the phenomenon as arising from the viscous flow of grain boundaries. Thus the superplastic system reduces to an array of deforming grain boundaries surrounding essentially non-deforming grains. The flow is analogous to that of a condensed particulate system where the non-deforming grains are dragged along by the viscous force generated at the boundaries due to the velocity discontinuity.

3. Viscosity and activation energy relations

By comparing the superplasticity eq. [3] with the general rheological equation (*McKelvey*, 1962)

$$\boldsymbol{\tau} = \eta\dot{\boldsymbol{\gamma}} \qquad [4]$$

where $\boldsymbol{\tau}$ is the stress tensor, $\dot{\boldsymbol{\gamma}}$ is the rate of deformation tensor and η is the scalar quantity viscosity, it has been shown by *Padmanabhan* (1971) that the viscosity variation during superplastic deformation with strain rate and stress can be represented as

$$\eta = \eta^0 |\dot{\gamma}|^{m-1} \qquad [5]$$

and

$$\eta = \eta^{0\frac{1}{m}} |\tau|^{\frac{m-1}{m}} \qquad [6]$$

where η^0 is a standard state of viscosity. As m for superplastic deformation is less than unity, it follows that η decreases as $\dot{\gamma}$ or τ is increased during superplastic deformation.

Since m is less than unity for superplastic deformation, it follows that the rheological behaviour is non-*Newton*ian. Hence η is a function of both temperature and stress or strain rate, depending on which is treated as the independent variable. Taking τ as the independent variable and making use of the appropriate equations of superplastic deformation *Padmanabhan* (1971) and *Padmanabhan* and *Davies* (1973) have shown that for superplastic flow the equations

$$\frac{Q_\tau}{Q_{\dot\gamma}} = 1 - \dot\gamma \left(\frac{\partial \eta}{\partial \tau}\right)_T \qquad [7]$$

and

$$Q_\tau = \left(\frac{1}{m}\right) Q_{\dot\gamma} \qquad [8]$$

hold good. In eqs. [7] and [8] Q_τ and $Q_{\dot\gamma}$ are the activation energy at constant stress and strain rate respectively. Eq. [7] is to be used when m is varying while eq. [8] should be employed when m is a constant. It has also been clearly demonstrated by the use of experimental results that eqs. [7] and [8] are obeyed by superplastic systems. It has been shown by *Padmanabhan* (1971) that for a viscous boundary model of superplasticity $Q_{\dot\gamma}$ should be equal to the activation energy for grain boundary diffusion. As m is less than unity, it follows that $Q_\tau > Q_{\dot\gamma}$. It is perhaps unfortunate that the ratio of Q_τ *to* $Q_{\dot\gamma}$ is about the same as the ratio of activation energies for volume and grain boundary diffusion in metals. Thus it is evident that these results remove the confusion about the various activation energy values reported in the literature where the values have been found to lie between those for volume and grain boundary diffusion.

4. Size effects on the rheological analysis

The microscopic model (*Padmanabhan*, 1971) envisages superplasticity as a phenomenon arising from the viscous flow of the boundaries of non-deforming grains. Under these conditions macroscopic deformation can be described in terms of a process in which a frictional force is generated at the grain boundaries as a result of the velocity discontinuity that exists there. It is postulated that the frictional force arising from the velocity discontinuity at the boundaries is a linear function of the velocity discontinuity and it gives an acceleration to the grains which get dragged along. This formulation is similar to that put forward by *Debye* (1946) for deformation processes which describe the behaviour of non-deforming incompressible polymer particles in a viscous medium. Experimental results for the Pb—Sn—Cd superplastic system are given in an appendix form which show that the frictional drag is proportional to the velocity discontinuity at the boundaries. This verifies the assumption to be true.

Due to orientation effects, there is bound to be velocity variation along the boundaries. In the absence of experimental evidence an arbitrary law of variation of velocity along the boundaries was not assumed. Instead, it will be assumed in the analysis that the velocity of all the grain boundaries is constant and is equal to $V(= dl/dt)$, the external velocity of displacement. As only macroscopic deformation is considered this should not be a serious limitation. The problem will be analysed for the case of uniaxial tensile deformation of circular cylindrical bars.

1. Equation of motion of x-co-ordinate of a particle

If V_x is the instantaneous velocity of the x-co-ordinate of the centre of mass of a grain in the x-direction, the frictional force F_x in that direction is given by

$$F_x = f(V - V_x)$$

where f is a friction factor, which is a constant for a given system. (The appendix shows that f changes with V for a system with a given grain size. In this analysis, V is constant and it requires that f is a constant for a given V. The findings reported in the appendix support this.) Since V_x is equal to dx/dt, for a grain of mass M and instantaneous x-co-ordinate x for its centre of mass, this can be rewritten as,

$$M \frac{d^2x}{dt^2} + f \frac{dx}{dt} = f V. \qquad [9]$$

The initial conditions are:

$$\text{at} \quad t = 0, \quad x = x_0 \quad \text{and} \quad \frac{dx}{dt} = 0.$$

It was mentioned earlier that the superplastic alloys are microduplex in composition. Therefore, the mass of a grain of an average grain size, L, for phase A would be M_A and for phase B would be M_B. The corresponding friction factors would be f_A and f_B.

Therefore, the solution for the equation of motion of the x-co-ordinates becomes

$$x = Vt + \frac{M_A V}{f_A} e^{-\frac{f_A}{M_A} t} + x_0 - \frac{M_A V}{f_A} \qquad \text{(for phase } A) \qquad [10]$$

and

$$x = Vt + \frac{M_B V}{f_B} e^{-\frac{f_B}{M_B} t} + x_0' - \frac{M_B V}{f_B} \qquad \text{(for phase } B\text{)} \qquad [11]$$

where x_0 and x_0' refer to the initial co-ordinates of the centres of mass of a grain of phase A and phase B, respectively.

2. Motion of y- and z-co-ordinates

When a cylindrical specimen is subject to uniaxial tension, the area of cross-section decreases with time. In order to achieve geometrical compatibility, therefore, the y- and z-co-ordinates of each grain also should change.

In general, for the viscous boundary model (*Padmanabhan*, 1971), the velocity gradient may exist in many ways. Further, the equations of motion for y- and z-co-ordinates of the grains may be different even if there is a small deviation from an equiaxed and random structure. Some non-homogeneous flow is also unavoidable, as the viscous boundary mechanism is based on diffusion along grain boundaries. For instance, at triple points the flow would be more difficult. Moreover, since most of the superplastic alloys are microduplex in character diffusion along one type of boundary could be easier than the other. As a result axial symmetry and circular cross-section may not be preserved. However for the sake of the analysis the following approximations are made:

(i) When the cylindrical specimen is pulled in uniaxial tension, velocity gradients exist only along the y and z axes, which are the two orthogonal axes to the axis of the cylinder.

(ii) Since a cylinder is considered, by symmetry, the velocity gradients in the y and z directions are equal. Recent work of *Nicholson* and co-workers (1971) indicates that the strain ratio for superplastic deformation is unity. This observation supports the assumption. Therefore the solution for the motion of y-co-ordinate is the same as that for z-co-ordinate.

(iii) The deformation is uniform and no localised flow occurs. This is an approximation as it is known that diffuse necks do form during superplastic deformation (*Morrison*, 1968).

Let the origin lie on the axis of the cylinder. If r is the instantaneous radius of the cylinder which is decreasing at the rate of dr/dt at the surface, it follows that the y-co-ordinate of the centre of mass of a grain at the lateral surface of the cylinder possesses a velocity dr/dt which is directed towards the origin. It is evident that the velocity in the y direction of the centre of mass of a grain at $y=0$ is zero. Therefore, the average velocity in the y-direction is $1/2\, dr/dt$ (assuming linear variation). As before, it will be assumed that the boundaries move with this average velocity and if y is the instantaneous y-co-ordinate of the grain of mass M (M will be equal to M_A or M_B depending on whether phase A or phase B grains are dealt with) it follows that

$$M \frac{d^2 y}{dt^2} = f\left(\frac{1}{2}\frac{dr}{dt} - \frac{dy}{dt}\right) \qquad [12]$$

(f, as before, is a friction factor which will become f_A or f_B depending on whether phase A or phase B is considered). Let P denote the constant volume of the incompressible cylinder. If

$$E^2 = \frac{P}{\pi}$$

and

$$G = \frac{1}{4}\left(\frac{E V}{l_0^{3/2}}\right) f = \text{const}$$

the general solution becomes

$$M y = \left(D + \frac{M}{f} C\right) - \frac{M}{f} C e^{-\frac{f}{M} t} + \frac{4 G l_0^2}{V^2} e^{-\frac{f}{M} t} \left(\frac{M V}{f l_0}\right)^{1/2} e^{-\frac{f l_0}{M V}} \times \left[F\left(\left\{\frac{f l_0}{M V}\left(\frac{V t + l_0}{l_0}\right)\right\}^{1/2}\right) - F\left(\left(\frac{f l_0}{M V}\right)^{1/2}\right)\right] \qquad [13]$$

where

$$F(y) = \int_0^y e^{u^2} du.$$

The constants of integration C and D can be evaluated from the initial conditions: at $t=0$, $y = y_0$ and $dy/dt = 0$. By assumption (ii) this is also the equation of motion for the Z coordinate.

For small t values

For small t, eq. [12] can be rewritten as

$$M \frac{d^2 y}{dt^2} + f \frac{dy}{dt} = -G\left(1 - \frac{3}{2}\frac{V t}{l_0}\right). \qquad [14]$$

Solving this equation using the initial conditions, at $t = 0$, $y = y_0$; $dy/dt = 0$; it can be seen that

$$y = \frac{3GV}{4l_0 f}\left(t^2 - \frac{2M}{f}t + \frac{2M^2}{f^2}\right) - \frac{G}{f}\left(t - \frac{M}{f}\right) + y_0 - \frac{2MG}{f^2} + e^{-\frac{f}{M}t}\left(\frac{MG}{f^2} - \frac{3GVM^2}{2l_0 f^3}\right) \quad [15]$$

is the most general solution for the equation of motion of both y- and z-co-ordinates, when t is small.

It is relevant to point out at this stage that the above analysis has the following implications:

(i) In the analysis the particle was treated as "free" and possible interactions between various grains which can change the velocity distribution were neglected.

(ii) It has been experimentally demonstrated by *Nutall* and *Nicholson* (1968) that there is an unsteady region at the beginning of tensile superplastic deformation. Same is true of superplastic compression (*Padmanabhan*, 1971). This unsteady region has been omitted for the analysis. As this unsteady region is very small compared to the large superplastic deformation the limitation is not serious.

(iii) It is well-known that the hydrodynamical treatment of condensed particulate systems is extremely difficult (see, for instance, *Bird* et al., 1964). Therefore, it is recognized that the present treatment is highly idealised.

5. Application to die-less drawing

Presently it will be seen how the above approach can be used in analysing the die-less drawing of superplastic bars which is being attempted on a pilot-plant scale (see, for instance, Metals and Materials, 3, 103, 1969). In this method cylindrical bars that are superplastic are pulled at constant velocity in uniaxial tension, using local induction heating. Mechanistically, it is a case of pulling circular cylindrical specimens in tension at a constant speed V, i.e. the time rate of increase of length of the bar, dl/dt, is a constant V.

Most of the superplastic alloys, as already reported, are microduplex in character. Therefore the analysis will be generalised to take account of this fact. Analysis for single phase materials which are superplastic will follow as a special case of this.

Let A and B be the two phases present in the alloy and N the average number of grains per unit area. If P_A is the volume fraction of phase A, $(1 - P_A)$ is the volume fraction of phase B. In a random, equiaxed, polycrystalline material like a superplastic alloy, if a unit area is considered the areas occupied by phases A and B would be in the same ratio as the volume fraction (*Lindinger* et al., 1969; *Underwood*, 1961). If a unit area normal to the direction of force, containing N grains totally, is considered $(P_A N)$ would be the number of grains of phase A and $(1 - P_A)N$ the number of grains of phase B. Therefore,

$$\sigma = (P_A N) M_A \frac{d^2x}{dt^2} + (1 - P_A) N M_B \frac{d^2x'}{dt^2}$$

where x and x' are the instantaneous x-co-ordinate of phases A and B, respectively. Therefore from eqs. [10] and [11]

$$\sigma = NV\left(P_A f_A e^{-\frac{f_A}{M_A}t} + (1 - P_A) f_B e^{-\frac{f_B}{M_B}t}\right) \quad [16]$$

where

$$t = \left(\frac{1}{\dot{\varepsilon}} - \frac{l_0}{V}\right). \quad [16a]$$

If L is the average grain size, $M_A = C_1 L^3$ and $M_B = C_2 L^3$ where C_1 and C_2 are proportionality constants which connect mass, volume, specific gravity and average grain size and shape. When these relations are introduced in eq. [16], σ can be expressed in terms of $\dot{\varepsilon}$ and L.

The strain-rate sensitivity index, m, can be evaluated in terms of L and $\dot{\varepsilon}$ by comparing eq. [16] with the empirical equation $\sigma = K\dot{\varepsilon}^m$

$$\left[\text{or the alternative expression } \dot{\varepsilon} = \frac{c''}{L^2}\sigma^{\frac{1}{m}}\exp\left(-\frac{Q}{kT}\right)\right].$$

To make the equations simpler the superplastic continuum can be treated as homogeneous which will eliminate the distinction between M_A and M_B and f_A and f_B. This should describe rigorously the situation in a single phase alloy where $M_A = M_B = M$ and $f_A = f_B = f$. Therefore,

$$\sigma = NfVe^{-\frac{f}{M}\left(\frac{1}{\dot{\varepsilon}} - \frac{l_0}{V}\right)}. \quad [17]$$

Eq. [17] can be written as

$$\sigma = K_1 \dot{\varepsilon}^{M_1} \qquad [18]$$

where

$$K_1 = N f V e^{\left(\frac{f}{M}\right)\frac{l_0}{V}}$$

and

$$M_1 = -\frac{f}{M} \cdot \frac{1}{\dot{\varepsilon} \ln \dot{\varepsilon}}.$$

Thus, by the use of the sliding parameters K_1 and M_1, eq. [17] is shown to be analogous to the empirical superplastic equation $\sigma = K\dot{\varepsilon}^m$. A comparison between the two equations will give the necessary expression for m, when it is noted that $M = c''' L^3$ where c''' connects mass, volume, specific gravity and average grain size and shape. It can be seen from the appendix that for gross superplastic flow, the approximate eq. [17] is quite satisfactory.

6. Viscosity variation in the rheological model

When σ and τ are the applied tensile and shear stresses and $\dot{\varepsilon}$ and $\dot{\gamma}$ are the corresponding tensile and shear strain rates, it is postulated, following *Drucker* (1961) that $\sigma/\tau = 1/D$ and $\dot{\varepsilon}/\dot{\gamma} = C$ where D and C are constants of proportionality.

Therefore, when the medium is treated as a heterogeneous continuum, for a given temperature,

$$\eta = C_1 \left[C_2 e^{-\frac{C_3}{\dot{\gamma}}} + C_4 e^{-\frac{C_5}{\dot{\gamma}}} \right] \frac{1}{\dot{\gamma}} \qquad [19]$$

where

$$C_1 = DNV$$

$$C_2 = P_A f_A e^{\frac{f_A l_0}{M_A V}}$$

$$C_3 = \frac{f_A}{M_A C}$$

$$C_4 = (1 - P_A) f_B e^{\frac{f_B l_0}{M_B V}}$$

and

$$C_5 = \frac{f_B}{M_B C}.$$

For the case of a homogeneous continuum

$$\eta = K_1 \frac{1}{\dot{\gamma}} e^{-\frac{B}{\dot{\gamma}}} \qquad [20]$$

where

$$K_1 = N f V D e^{\frac{f l_0}{M V}}$$

and

$$B = \frac{f}{CM}.$$

From eq. [20] it follows that η increases with strain rate upto $\dot{\gamma} = B = f/CM$ or $\dot{\varepsilon} = f/M$ where the maximum in viscosity occurs. Beyond this, viscosity decreases with increase of strain rate and superplasticity sets in. Thus the lower limit of superplasticity that has been empirically observed (see, for instance, *Davies* et al., 1970) is predicted by the present model. A similar procedure is to be applied to eq. [19] if the lower limit for superplasticity is to be obtained accurately for a duplex structure.

It is relevant to point out at this stage that in the above analysis it is assumed that the contribution to the stress from individual particles is additive and possible grain interactions that may modify this additive principle are ignored. By assuming N to be independent of time, grain growth has also been neglected. However, this is not serious as superplasticity is accompanied by a stable grain size (see for example, *Johnson*, 1970). The analysis for the case of a heterogeneous continuum can be made more complicated by considering the nature of the grain boundary involved. For instance, in a two-phase material grain boundaries exist between two grains of A or a grain of A and a grain of B or between two grains of B. Thus there will be three friction factors f_{AA}, f_{AB} and f_{BB}. In the absence of sufficient experimental evidence, it is felt that this extra refinement is unnecessary at this stage. It is also emphasized that the above analysis is applicable only to the superplastic region. Beyond a certain strain rate where superplasticity is lost, these equations cannot be used. *Padmanabhan* (1971) has accounted for the loss of superplasticity at these high strain rates.

7. Miscellaneous application

Backofen et al. (1964) have given a technique for determining m directly by the method of changing cross-head speeds. The present analysis can be used to analyse this technique also.

The problem can be stated as follows:

The test starts with a cross-head velocity V_1 and continues for a time t_1 when the velocity is changed to V_2 and this goes on upto time t_2 from the start of the test and so on. In general, at a total time t_{n-1} the velocity is changed to V_n; n assuming values 1, 2, 3, ... ($t_0 = 0$). The corresponding loads are denoted as $P_1, \ldots, P_{n+1}$.

The problem will be solved for a homogeneous continuum. A similar technique is applicable for a heterogeneous continuum. *Backofen* et al. (1964) show experimental curves where they find that when the velocity is changed from V_1 to V_2, about 2–3% strain is needed before the steady-state is reached. In the analysis, this transient region is neglected.

Keeping in mind that in this case the solutions for the motion of x-co-ordinate are of the type of eq. [10] and that the initial conditions for the second velocity region can be obtained by putting $t = t_1$ in the solution for the first velocity region and so on, it can be seen that after a total time t_n when the velocity is V_{n+1} and the load is P_{n+1}[1]) the following relations will hold good:

$$P_{n+1} = NAD_{n+1}\frac{f^2}{M}e^{-\frac{f}{M}t} \qquad [21]$$

$$P_n = NAD_n\frac{f^2}{M}e^{-\frac{f}{M}t} \qquad [22]$$

$$\frac{V_{n+1}}{V_n} = \left[1 + \frac{fV_n}{M}e^{-\frac{f}{M}t_n}(D_{n+1} - D_n)\right] \qquad [23]$$

$$l = [l_0 + V_1 t_1 + V_2(t_2 - t_1) + \cdots + V_n(t_n - t_{n-1})] \qquad [24]$$

and

$$m = \frac{(\log D_{n+1} - \log D_n)}{\log\left[1 + \frac{fV_n}{M}e^{-\frac{f}{M}t_n}(D_{n+1} - D_n)\right]} \qquad [25]$$

where A is the present area of cross-section of the specimen and l is its present length. D_{n+1} and D_n are given by

$$D_{n+1} = \frac{MV_1}{f} + \frac{MV_2}{f}e^{\frac{f}{M}t_1} - \frac{MV_1}{f}e^{\frac{f}{M}t_1} + \cdots + \frac{MV_{n+1}}{f}e^{\frac{f}{M}t_n} - \frac{MV_n}{f}e^{\frac{f}{M}t_n} \qquad [26]$$

[1]) In constant velocity tests P_{n+1} remains steady only if $\dot{\varepsilon}_n$ is constant in that region. If the material is allowed to elongate considerably at a given value of V, P will start falling as $\dot{\varepsilon}$ decreases in that case. However, since the exercise is to evaluate m as a function of $\dot{\varepsilon}$, in practice P is noted as soon as the steady-state load is reached. This justifies the approach where P in each velocity range is considered constant.

and

$$D_n = \frac{MV_1}{f} + \frac{MV_2e^{\frac{f}{M}t_1}}{f} - \frac{MV_1}{f}e^{\frac{f}{M}t_1} + \cdots + \frac{MV_n}{f}e^{\frac{f}{M}t_{n-1}} - \frac{MV_{n-1}}{f}e^{\frac{f}{M}t_{n-1}}. \qquad [27]$$

The corresponding strain rate, by definition, would be

$$\dot{\varepsilon} = \frac{V_n}{[l_0 + V_1 t_1 + V_2(t_2 - t_1) + \cdots + V_n(t_n - t_{n-1})]}.$$

8. Discussion

Padmanabhan (1971) and *Padmanabhan* and *Davies* (1973) have discussed in detail the interrelation of the two activation energies Q_τ, the activation energy at constant stress, and $Q_{\dot{\gamma}}$, the activation energy at constant strain rate. They have shown with the aid of experimental evidence that eqs. [7] and [8] hold good in case of superplastic deformation.

Eqs. [5] and [6] predict the variation of viscosity when the grain size effect is neglected and the material is treated as a continuum. From this it can be seen that during superplastic deformation the viscosity decreases with an increase of strain-rate. However, the empirical observation (see, for instance, *Davies* et al., 1970) that there is a lower limiting strain rate for the on-set of superplasticity can be predicted only when the grain size is taken into account. In fact, for the case of a homogeneous continuum this lower limit is at $\dot{\varepsilon} = (f/M)$.

The incorporation of the grain-size effect into the rheological analysis has been helpful in understanding the die-less drawing of superplastic bars. It is shown that the equations developed (e.g., eq. [18]) are analogous to the empirical equations concerning superplasticity.

The present analysis, though successful in explaining a few of the hitherto unaccounted facts, is highly idealised and attempts to make the analysis more rigorous would be very useful.

Conclusions

The phenomenon of superplasticity is analysed from a rheological basis. Expressions dealing with the variation of viscosity and activation energies are given. The grain size effect is investigated by applying the concepts to the die-less drawing of superplastic cylindrical bars. Expressions are deduced for stress, strain rate, m etc. by this rheological treatment and it is shown that there is a lower limiting strain rate for superplasticity to set in. Expressions have been deduced for m determined by the method of changing cross-head speeds.

Acknowledgments

One of the authors (KAP) wishes to thank Professor *T. R. Anantharaman* for his constant encouragement. The financial support for the work was provided by the Ministry of Defence, U.K.

Summary

By making use of the empirical relations between stress and strain rate useful expressions concerning the variation of viscosity and activation energies for superplastic deformation are derived. The analyses of superplasticity, based on conventional continuum mechanics, neglect grain size effects. A new technique has been developed where the grain-size effect is incorporated into the rheological analysis. The analysis is subsequently made use of in understanding die-less drawing of superplastic bars. It is shown that the analysis is able to predict stress-strain rate relations, viscosity variation and so forth satisfactorily. It is concluded that the present analysis could be useful in practical situations where the constitutive equations are of great importance.

Zusammenfassung

Brauchbare Gleichungen für die Veränderung der Viskosität und Aktivierungsenergien für superplastische Verformung sind mit Hilfe der empirischen Beziehungen zwischen Spannung und Dehnungsrate abgeleitet worden. Die Analysen der Superplastizität, die auf konventioneller Kontinuumstheorie basieren, vernachlässigen den Einfluß der Korngröße. Eine neue Methode ist entwickelt worden, wobei der Einfluß der Korngröße in der rheologischen Analyse in Betracht gezogen wurde. Diese Analyse ist dann verwendet worden, um die düsenlose (die-less) Ziehung der superplastischen Stäbe zu verstehen. Es ist gezeigt worden, daß zuverlässige Vorhersagen über den Zusammenhang zwischen der Spannung und dem Dehnungsgrad, die Veränderung der Viskosität usw. auf Basis dieser Analyse durchaus möglich sind. Es wird gezeigt, daß die vorher beschriebene Analyse sehr nützlich ist, in Fällen, wo die konstituitiven Gleichungen sehr wichtig sind.

Appendix

The following is a summary of work done by *Moles, Padmanabhan* and *Davies*, which is being reported elsewhere. The experiments were undertaken to check the postulate of the paper that the frictional force resulting from the viscous drag between the deforming grain boundaries and non-deforming grains is proportional to the velocity discontinuity existing at the boundary.

It has been shown, assuming the above postulate, that when the distinction between the various phases present is neglected and the material is treated as a homogeneous continuum, the stress response of the material is given by (eqs. [16] and [17])

$$\sigma = N f V e^{-\frac{f}{M} t} \qquad \text{[A.1]}$$

where σ is the observed stress, N the number of grains per unit area, f the friction factor, V the velocity of the cross-head, M the average mass of a grain and t is the time at which σ is evaluated. The postulate requires that for constant V, f must also be a constant.

When V is constant, from eq. [A.1] it follows that for a material with a constant grain size

$$\left(\frac{f}{M}\right) = \frac{\ln\left(\frac{\sigma_1}{\sigma_2}\right)}{(t_2 - t_1)} \qquad \text{[A.2]}$$

where σ_1 is the stress at time t_1 and σ_2 is the stress at time t_2. A number of constant velocity tests were performed at different temperatures using Pb—Sn—Cd ternary eutectic alloy, on specimens with an initial grain size of 3.9 μm ± 20%. The values of (f/M) obtained are summarised in table 1. It was further found that the value of (f/M)

Table 1

Temp. °C	Cross-head speed cm/min	(f/M) sec^{-1}	
70	0.5	6.3×10^{-4}	Estimated at different stages of the deformation
		6.6×10^{-4}	
		6.6×10^{-4}	
		6.1×10^{-4}	
		7.0×10^{-4}	
		9.5×10^{-4}	
70	0.05	2.5×10^{-5}	Estimated at different stages of the deformation
		2.7×10^{-5}	
		2.3×10^{-5}	
		4.4×10^{-5}	
25	0.02	1.7×10^{-5}	(Average of five values)
	0.04	6×10^{-5}	
	0.10	1.1×10^{-4}	
45	0.02	5×10^{-6}	
	0.04	4.3×10^{-5}	
	0.10	1.7×10^{-4}	
70	0.02	7.3×10^{-6}	
	0.04	7.7×10^{-5}	
	0.10	9.4×10^{-5}	

was practically independent of the magnitude of $(t_2 - t_1)$. For instance, when $(t_2 - t_1)$ was maintained at 2 mins, for a test at 70 °C and cross-head speed 0.05 cm/min a (f/M) value of 4.4×10^{-5} sec^{-1} was obtained. When $(t_2 - t_1)$ was changed to 12.5 min and the other conditions were kept the same, (f/M) values obtained at two different stages of deformation were 3.9×10^{-5} and 4.4×10^{-5} sec^{-1}, respectively. It is relevant to note at this point that in all these tests the accuracy of estimation of the loads was ±0.1 kg. This is quite sufficient to account for the small scatter seen here in the value of (f/M).

The following conclusions can be drawn from the tests:

1. (f/M) is independent of the strain-history and is a constant throughout a test, so long as V is a constant. This vindicates the hypothesis.

2. (f/M) is practically independent of the magnitude of $(t_2 - t_1)$ in eq. [A.2]. This further establishes the constancy of (f/M) for constant V as eq. [A.2] is derived from eq. [A.1] assuming f to be constant.

3. (f/M) and hence f is a function of V and it increases with V. *Padmanabhan* (1971) has accounted for this observation.

4. As (f/M) emerges as a constant in a constant velocity test, it follows that for practical purposes the superplastic material can be treated as a homogeneous continuum. When this simplification is not permissible, the material will have to be considered as a heterogeneous continuum.

References

1) *Alden, T. H.*, Trans. ASM **61**, 559 (1968).
2) *Avery, D. H.* and *W. A. Backofen*, Trans. ASM **58**, 551 (1965).
3) *Backofen, W. A., I. R. Turner*, and *D. H. Avery*, Trans. ASM **57**, 980 (1964).
4) *Bird, R. B., W. E. Stewart*, and *E. N. Lightfoot*, Transport Phenomena, pp. 100–104 (New York 1964).
5) *Cornfield, G. C.* and *R. H. Johnson*, Int. J. Mech. Sci. **12**, 479 (1970).
6) *Davies, G. J., J. W. Edington, C. P. Cutler*, and *K. A. Padmanabhan*, J. Mat. Sci. **5**, 1091 (1970).
7) *Debye, P.*, J. Chem. Phys. **14**, 636 (1946).
8) *Drucker, D. C.*, J. Materials **1**, 873 (1961).
9) *Fields, D. S., Jr.* and *T. J. Stewart*, Int. J. Mech. Sci. **13**, 63 (1971).
10) *Ghosh, A.* and *J. L. Duncan*, Int. J. Mech. Sci. **12**, 499 (1970).
11) *Hayden, H. W.* and *J. H. Brophy*, Trans. ASM **61**, 542 (1968).
12) *Hilliard, J. E., J. M. Lommel, J. B. Hudson, D. F. Stein*, and *J. D. Livingstone*, Acta Met. **9**, 787 (1961).
13) *Holt, D. L.*, Int. J. Mech. Sci. **12**, 491 (1970).
14) *Johnson, R. H.*, Met. Reviews **15**, 115 (1970).
15) *Jovane, F.*, Int. J. Mech. Sci. **10**, 403 (1968).
16) *Kê, T. S.*, J. Appl. Phys. **22**, 274 (1949).
17) *Kröner, E.*, Arch. Rat. Mech. Anal. **4**, 273 (1960).
18) *Lindinger, R. J., R. C. Gibson*, and *J. H. Brophy*, Trans. ASM **62**, 230 (1969).
19) *McKelvey, J. M.*, Polymer Processing, pp. 44–45 (New York 1962).
20) *Mehta, H. S., A. H. Shabaik*, and *S. Kobayashi*, Trans. ASME (B) **92**, 403 (1970).
21) *Nicholson, R. B.*, Plasticity and Superplasticity Institution of Metallurgists Review Course, Series 2, No. 3, p. 21 (1969).
22) *Nicholson, R. B.*, Electron Microscopy and Structure of Materials, University of California Press, p. 689 (1972).
23) *Nicholson, R. B.*, Conference on "Grain Boundary Sliding", Institute of Physics (April 1971).
24) *Nuttall, K.* and *R. B. Nicholson*, Phil. Mag. **17**, 1087 (1968).
25) *Padmanabhan, K. A.*, Ph. D. Thesis, University of Cambridge (1971).
26) *Padmanabhan, K. A.* and *G. J. Davies*, J. Mech. Phys. Solids **18**, 261 (1970).
27) *Padmanabhan, K. A.* and *G. J. Davies*, Phys. Stat. Sol. (a) **18**, 295 (1973).
28) *Tanner, L. E.* and *S. V. Radcliffe*, Acta Met. **10**, 1161 (1962).
29) *Underwood, E. E.*, Metals Engineering Quarterly, ‚Quantitative Metallography', ASM, pp. 70–81 (1961).
30) *Watts, B. M., M. J. Stowell*, and *D. M. Cottingham*, J. Mat. Sci. **6**, 228 (1971).

Authors' addresses:

Dr. *K. A. Padmanabhan*
Department of Metallurgical Engineering
Banaras Hindu University
Varanasi-221005 (India)

Dr. *G. J. Davies*
Department of Metallurgy and Materials Science
University of Cambridge
Cambridge (England)

Rheol. Acta **13**, 443–456 (1974)

From the Division of Engineering, Brown University, Providence (U.S.A.)

On the use of open-channel flows to measure the second normal stress difference

By Y. Kuo and R. I. Tanner

With 18 figures and 1 table

(Received March 17, 1973)

1. The rheological functions

In steady isothermal shearing flow of an incompressible viscoelastic fluid, the fluid behavior is completely determined by three viscometric functions (*Pipkin*, 1972); they are the viscosity function η, the first normal stress difference N_1 and the second normal stress difference N_2. Referred to a velocity field given in *Cartesian* coordinates (x, y, z) as

$$\underset{\sim}{v} = (\gamma y, 0, 0) \qquad [1]$$

the functions are defined as

$$\eta(\gamma^2) \equiv t_{xy}/\gamma \qquad [2]$$

$$N_1(\gamma^2) \equiv t_{xx} - t_{yy} \qquad [3]$$

$$N_2(\gamma^2) \equiv t_{yy} - t_{zz} \qquad [4]$$

where t_{ij} are the components of the stress tensor (T) and γ is the shear rate, a constant. [Note that *Pipkin* (1972) defines N_2 as being $t_{zz} - t_{yy}$.] In viscometric flows (*Yin* and *Pipkin*, 1970) each infinitesimal fluid element is in a state of steady simple shearing motion and the simple shearing flow is an example of such a flow. The behaviour of a simple fluid engaged in viscometric flow is governed by the same relation between stress and strain-rate that holds for the simple shearing flow, and if the viscometric functions are obtained from experiments on one viscometric flow, then the behavior of the fluid can be predicted for all other viscometric flows. The constitutive equation for viscometric flows in then (*Pipkin*, 1972)

$$\underset{\sim}{T} = p\underset{\sim}{I} + \eta\underset{\sim}{A}_1 + \frac{N_1 + N_2}{\gamma^2}\underset{\sim}{A}_1^2 - \frac{N_1}{2\gamma^2}\underset{\sim}{A}_2{}^{*}) \qquad [5]$$

where $\underset{\sim}{A}_1$ and $\underset{\sim}{A}_2$ are the first two *Rivlin-Ericksen* acceleration tensors. It is not difficult to measure the viscosity function accurately, and the measurement of the first normal stress difference in the cone-plate device is fairly accurate. The measurement of N_2 is rather difficult and few reliable methods exist. The principal difficulties in finding N_2 arise because it is small compared to N_1 and, in an early conjecture about the value of N_2, *Weissenberg* (1947) hypothesized that $N_2 = 0$ identically. Clearly, in order to be confident about our state of knowledge of this simplest class of flows, we need to know N_2 accurately, and this paper is devoted to this task.

Various methods employing different flow geometries have been exploited in experimental work on the determination of $N_2(\gamma)$ for the past two decades and table 1 gives a summary. The most remarkable feature about table 1 is the change in opinion that has taken place regarding the *sign* of N_2/N_1. Early measurements were not very accurate, and prior to 1968 it was not known that the use of pressure holes for measuring the thrust on a surface needed great care (*Tanner* and *Pipkin*, 1969). In fact, *Jackson* and *Kaye* (1966) seem to have been the first to show in a definitive manner that N_2/N_1 is negative for common polymer systems, but their method, which was subsequently taken up in modified form by *Marsh* and *Pearson* (1968), *Pritchard* (1970), and *Cowsley* (1970), required differentiation of the data and it is not easy to guarantee it's accuracy. All of the methods using pressure holes in annular flows (*Hayes* and *Tanner*, 1963; *Huppler*, 1965; *Tanner*, 1967) show the positive sign for N_2/N_1, but recently *Lobo* and *Osmers* (1973) have shown that when corrections are made for the hole error the negative sign for N_2/N_1 is recovered. The extensive researches of *Kaye, Lodge, and Vale* (1968) and *Pritchard* (1971), which compute a pressure-hole error and make allowances for it, and the accurate experiments of *Ginn* and *Metzner* (1969) and *Berry* and *Batchelor* (1968) all indicate a negative sign for N_2/N_1. They are far from being *direct* methods, however, and errors can be large. Efforts have been made to use Couette stability (*Denn* and *Roisman*, 1969) for deducing N_2/N_1, but one is forced to assume a viscometric constitutive equation in a non-viscometric flow here, and so the results are not definitive. The same objection can be raised to the ingenious method of *Macosko* (1970); careful analysis also suggests that the second normal stress difference is not measured by this test. Recently, *Miller* and *Christiansen* (1972) and *Olabisi* and *Williams* (1972) have made more direct measurements of N_2/N_1 and have found the negative sign; our own researches with the open inclined channel (*Tanner*, 1970; *Kuo*, 1973) have also supported this conclusion. The technique of using flush-mounted pressure transducers described in the recent independent reports by *Miller* and *Christiansen* (1972) and *Olabisi* and *Williams* (1972), which make simultaneous measurements of N_1 and N_2 in a single cone-plate device, are attractive[1]). With these methods, which are also "direct", there may be some problems with boundary conditions at the free surface; although *Olabisi* and *Williams* (1972) avoided this uncertainty about the rim pressure by using a "drowned" cone with a liquid reservoir this actually produces a complex flow pattern at the cone edge (*Tanner*, 1970). Moreover, the effects of the buoyancy force and the wall effect of the reservoir further increase the complexity of the analysis.

*) Formular symbols underlined in setting are to be read as set in **bold types.**

[1]) See also *Brindley, G.* and *Broadbent, J. M.* (1972).

Table 1. Summary of some second normal stress difference measurements

Investigators	Year	Apparatus	Constitutive equation or velocity field assumed	Pressure holes	N_2/N_1
Garner et al.	1950	Torsional flow	no	yes	0
Roberts	1953	Cone-plate	no	no	0
Greensmith, Rivlin	1953	Torsional flow	no	yes	0
Pollett	1955	Cone-plate	no	no	0
Kotaka et al.	1959	Torsional flow	no	yes	0
Sakiadis	1962	Circular tube (exit effects)	no	yes	+
Markovitz, Brown	1963	Cone-plate Torsional flow	no	yes	+
Markovitz, Brown	1964	Torsional flow	no	yes	+
Adams, Lodge	1964	Cone-plate Torsional flow	no	yes	+?
Ginn, Metzner	1965	Cone-plate Torsional flow	no no	no no	−0?
Hayes, Tanner	1965	Annular tube	yes	yes	+
Huppler	1965	Annular tube	yes	yes	+
Jackson, Kaye	1966	Raised cone-plate	yes	no	−
Tanner	1967	Annular tube	no	yes	+
Huppler et al.	1967	Cone-plate	no	yes	+
Marsh, Pearson	1968	Raised-cone-plate	yes	no	−
Kaye, Lodge, Vale	1968	Concentric cylinder, cone-plate	no	yes	−
Berry, Batchelor	1968	Cone-plate Torsional flow	no	no	−
Ginn, Metzner	1969	Cone-plate Torsional flow	no	no	−
Denn, Roisman	1969	Couette stability	yes	no	−
Tanner	1970	Inclined channel	no	no	−
Cowsley	1970	Re-entrant cone	yes	no	−
Macosko	1970	Eccentric discs.	yes	no	−
Pritchard	1971	Cone-plate, holes	no	yes	−
Miller, Christiansen	1972	Cone-plate	no	no	−
Olabisi, Williams	1972	Cone-plate	yes	no	−
Lobo, Osmers	1973	Annulus	no	no	−
Kuo	1973	Inclined channel	no	no	−

Turning to the channel flow, a surface tension effect is evident. *Tanners* (1970) analysis is based on the assumption of negligible surface tension: it concludes that the liquid surface is deformed to a convex profile if the second normal stress difference is negative (compressive) and vice versa. However, this is not true in general. Even a *Newtonian* flow in the channel flow may exhibit a curved surface because of the surface tension effect (*Kuo* and *Tanner*, 1972). Here we discuss these effects more fully, and the purpose of this paper is to describe measurements of the second normal stress difference with surface tension corrections using an inclined channel of semi-circular cross-section (see fig. 1). The flow rate is regulated so that the centre of the liquid surface is at the origin of the tube so that the base flow is nearly viscometric. This flow geometry is of great interest because it provides simple relations between the second normal stress difference, the shear stress and the surface profile. If N_2 is of a small order (compared to the gravity force), the deflection of the free surface from the diametral plane is of small order (compared to the radius of the tube) and therefore the actual flow condition is nearly viscometric just like the case of cone-plate flow.

The present analysis differs from *Tanners* (1970) original analysis in several features. In addition to the surface tension correction, the present method requires, in principle, only a single tilt angle to obtain a curve of N_2 instead of varying the tilt angle. The measurements are also made by an improved arrangement of micrometer and telescope that can measure the displacement of the surface profile to an accuracy of .0002″ (.005 mm) without touching the fluid.

2. Theory

In the following derivation (*Kuo*, 1973), we are concerned with a developed flow in a "half-filled" circular channel. Let z and w be the coordinate and the velocity respectively in the direction of the flow down an inclined tube of uniform circular cross-section with the axis tilted at an

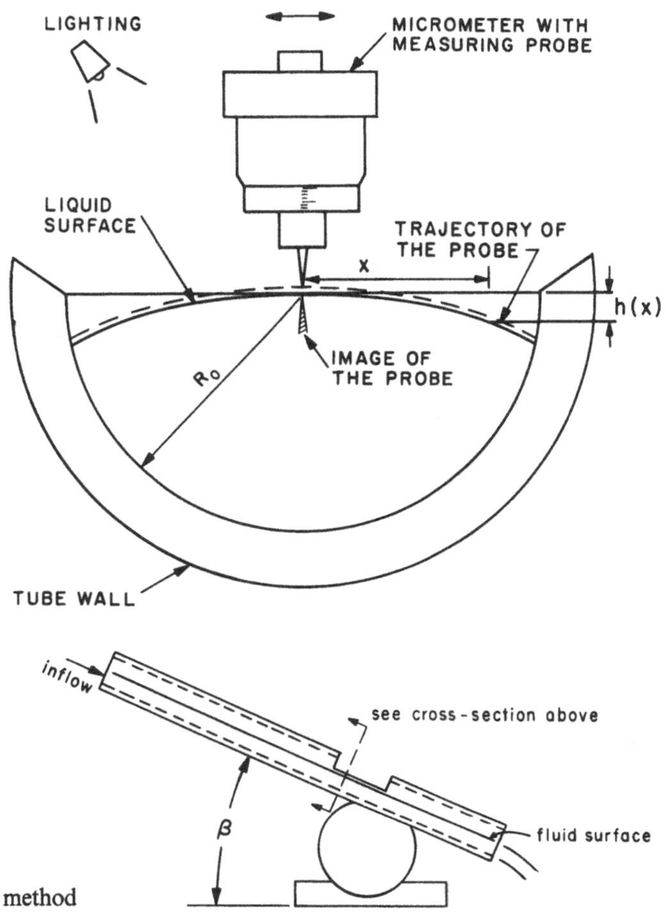

Fig. 1. The channel geometry and the measuring method

angle β to horizontal (fig. 1). In the fully developed flow the flow quantities do not vary with z. The flow rate can be regulated such that the fluid depth in the centre of the channel is exactly equal to the radius of the channel R_0 (half flow). As the secondary flow exists only when both the surface deformation (non-viscometric influence) and the second normal stress difference are non-zero, we may express the velocity field as follows

$$\underset{\sim}{v} = \underset{\sim}{k} w + \varepsilon\sigma(\underset{\sim}{i} u + \underset{\sim}{j} v) + (\text{higher powers of } \varepsilon, \sigma) \qquad [6]$$

where the indices i, j, k refer to the radial, the angular and the axial (the flow) direction respectively. The parameter σ is defined as the ratio of the maximum surface deflection h_m to the radius of the tube R_0:

$$\sigma = h_m/R_0. \qquad [7]$$

The parameter ε is defined as

$$\varepsilon = \frac{N_2(\tau_w)}{\varrho g R_0 \cos\beta} \qquad [8]$$

where ϱ and g are density and gravitational acceleration respectively, and τ_w is the shear stress at the tube wall. By considering orders of magnitude, *Kuo* (1973) has shown that if we ignore terms of order $\sigma\varepsilon$ and higher in eq. [6] then the flow may be regarded as viscometric and we can use eq. [5] as a constitutive approximation. An alternative method of obtaining this result is to ignore all secondary flows at the outset. Due to the curvature of the free surface there is some disturbance to the shear stress distribution; this is explored elsewhere (*Kuo* and *Tanner*, 1972). Usually it is sufficient to ignore this effect but we have corrected for it where necessary.

In the earlier analysis (*Tanner*, 1970) a perturbation parameter

$$\varepsilon\left(\equiv \frac{N_2}{\varrho g R_0 \cos\beta}\right)$$

appeared in the momentum equations so that the governing equation for N_2 in terms of the surface deflection was found to be

$$\frac{h(x)}{R_0} = \frac{1}{\varrho g R_0 \cos\beta}\left[N_2(\gamma^2) + \int_0^x N_2(\gamma^2)\frac{dx}{x}\right] + 0(\varepsilon^2) \quad [9]$$

where x is the horizontal coordinate measured from the centre of the tube. To calculate N_2 from the above equation, we can consider N_2 to be a function of the magnitude of the shear stress τ,

$$\tau = \tfrac{1}{2}\varrho g x \sin\beta. \quad [10]$$

In terms of the data at the channel wall, where the shear stress is τ_w and the deflection is h_w, solving for $N_2(\tau_w)$ gives the result

$$N_2(\tau_w) = \varrho g h_w \cos\beta - \frac{1}{\tau_w}\int_0^{\tau_w} \varrho g h_w \cos\beta \, d\tau_w. \quad [11]$$

Thus a curve of N_2 can be obtained if a systematic measurement of the height h_w against the tilt angle β (or τ_w) is made. Integrating the above equation by parts yields the more convenient form

$$N_2(\tau_w) = \frac{\varrho g}{\sin\beta}\int_0^\beta \sin\theta \, d(h_w \cos\theta). \quad [12]$$

The simplicity of the relation between N_2 and h_w is remarkable and encouraging. For a convex surface, h_w is negative and N_2 is negative. If the magnitude of the function $N_2(\tau)$ is increasing with τ, the curve of $h_w \cos\theta$ vs. $\sin\theta$ is expected to be monotonically increasing. In the above analysis, the surface tension α has been neglected with a criterion stated as (*Tanner*, 1970)

$$\frac{\varrho g R_0^2 \cos\beta}{2\alpha} \gg 1. \quad [13]$$

The criterion for the above-mentioned method, however, fails to be met when applied to a flow of a real fluid with appreciable surface tension and a liquid surface having large curvature variations. Furthermore, the breakdown of the above formulation is evident if we note that even a *Newton*ian flow may exhibit a convex surface. Further experiments (*Kuo*, 1973) also show a variety of surface shapes that cannot be explained with the formulation [12].

The above findings point out that the applicability of this formulation is restricted and that the surface tension can play an important role in shaping the free surface and its influence should be considered when evaluating N_2. In other words, the shaping of the free surface will be considered as a combination of the effects of viscoelasticity and surface tension. Since for a given (measured) profile the surface tension effect can be calculated exactly, the second normal stress difference can be computed after excluding the former from the governing equation.

The necessary relation between the second normal stress difference and the surface profile comes from the momentum equation and the free boundary condition. By integrating the radial component of the momentum equation (in cylindrical coordinates) with respect to r, we get (defining $\underset{\sim}{S} \equiv \underset{\sim}{T} + p\underset{\sim}{I}$)

$$p_1 = -\varrho g r \cos\beta \sin\theta + \int_0^r \frac{N_2}{r'}dr' + N_2 + S_{\theta\theta} + \int_0^r \frac{\partial S_{r\theta}}{r'\partial\theta}dr' + C(\theta) + 0(\varepsilon^2\sigma^2) \quad [14]$$

where $N_2 = S_{rr} - S_{\theta\theta}$. p_1 is the pressure in the flow and the ignored terms are of order $0(\varepsilon^2\sigma^2)$. As the pressure is unique at $r = 0$, $C(\theta)$ ought to be a constant, say C. When the momentum equation is applied to the free surface, the pressure p_1 on the curved surface should match the free surface condition with surface tension. The equation of the balance of pressure, stresses and surface tension for the one-dimensional flow is a special case of the general equation

$$(p_1 - p_g)\, n_i = s_{ik} n_k + \frac{\alpha}{R} n_i \quad [15]$$

where R is the intrinsic radius of curvature of the profile, and p_g is the pressure of the gas above the liquid-gas interface. If the component in the $\underset{\sim}{i}$ direction of a unit vector normal to the surface is denoted by n_i, then by resolving [15] into normal and tangential components, we obtain

$$p_1 - p_g = S_{ik} n_k n_i + \frac{\alpha}{R}$$

(normal direction) [16]

and

$$S_{ik} n_k t_i = 0$$

(tangential direction) [17]

where t_i is the component (in the $\underset{\sim}{i}$ direction) of a unit vector tangential to the surface. Substituting p_1 of [16] into [14] and setting p_g to zero, we have

$$\varrho g r \cos\beta \sin\theta + \frac{\alpha}{R} = \int_0^r \frac{N_2}{r'} dr' + N_2(1 - n_r^2) + g(S_{\theta r}) + C \quad [18]$$

where

$$g(S_{\theta r}) = \int_0^r \frac{\partial S_{\theta r'}}{r' \partial \theta} dr' - 2 S_{\theta r} n_r n_\theta . \quad [19]$$

As the ratio of the surface deflection to the radius of the tube, σ, is much less than unity, we have the order of n_r as

$$n_r \sim 0(\sigma) \quad [20]$$

and the flow is nearly-viscometric. In general, the non-viscometric part produces a secondary flow that contributes to the stress $S_{\theta r}$. However, the constitutive equation derived on the basis of the perturbation parameter σ asserts that

$$S_{\theta r} \sim 0(N_2 \sigma); \quad \frac{\partial S_{\theta r}}{\partial \theta} \sim 0(N_2 \sigma^2). \quad [21]$$

Consequently we may write [18] in the following form

$$\varrho g r \cos\beta \sin\theta + \frac{\alpha}{R} = \int_0^r \frac{N_2}{r'} dr' + N_2 + C + 0(N_2 \sigma^2). \quad [22]$$

Finally the solution for N_2 in terms of the surface profile h is

$$N_2 = \frac{1}{x} \int_0^x x' \frac{dQ}{dx'} dx' + 0(N_2 \sigma^2) \quad [23]$$

where

$$Q(x') = \varrho g h(x') \cos\beta + \frac{\alpha}{R(x')} . \quad [24]$$

The geometric relations $y = r \sin\theta$, $x = r \cos\theta$ and the small angle approximation have been used in deriving [23]. At small slopes the curvature is

$$\frac{1}{R(x')} = - \frac{d^2 h(x')}{dx'^2} . \quad [25]$$

On the other hand, in view of the small deflection of the curved liquid surface, the shear stress S_{rz} is

$$S_{rz} = \tfrac{1}{2} \varrho g r \sin\beta + 0(\sigma). \quad [26]$$

If a correction for the shear stress to the base flow is considered of order σ, we may write

$$S_{rz} = S_{rz}^{(0)} + \sigma S_{rz}^{(1)} + 0(\sigma^2) \quad [27]$$

so that the zero order term can be extracted from [26]. It follows that the latter on the liquid surface with the small angle approximation is

$$\tau = - S_{rz}^{(0)}(h) \tfrac{1}{2} \varrho g \sin\beta x \quad [28]$$

where the notation τ is used for later convenience. The correction term $S_{rz}^{(1)}(h)$ has been given elsewhere (*Kuo* and *Tanner*, 1972). At a tilt angle of β_1, the range of the shear stress is (roughly) between zero and $\tau = \frac{1}{2} \varrho g r \sin\beta_1$. It can be seen from [28] and [23] that the coordinate x plays the role of a correlation parameter among h, τ and N_2. So far we have shown that the evaluation of N_2 requires only a measurement of the surface profile of the flow at a *single tilt angle.* Although the deflection of the liquid surface from the flat plane is due to the combined effects of viscoelasticity and surface tension, eq. [23] indicates that N_2 is evaluated after the exclusion of the surface tension effect. Consequently, the implication of surface tension in [23] should not be interpreted as dependence of N_2 on the former. Note that it is not difficult to evaluate the curvature through [25] since a measured profile can be easily represented in a simple functional form (*Kuo*, 1973). The advantage of this formulation is that it was derived irrespective of the complete simple fluid constitutive relation since the base flow is viscometric.

Naturally it is anticipated that the function $N_2(\tau)$ should be the same in overlapping regions measured at different angles and this has been verified. In some experiments at larger angles, however, the distortion of the surface profile in the small shear region (smaller distance from centre) is small compared to that in the larger shear region (larger distance form centre). For the sake of precision, it is preferable to measure N_2 for small shear stress at a smaller tilt angle. Therefore for a large span of shear stress two (or more) different tilt angles are needed in order to have the same accuracy in the whole range.

3. Description of the experimental arrangement

The theory developed in the previous section for computing the second normal stress difference requires the measurement of the profile of the free surface. In view of the time needed for the measurement in addition to the time spent in the calibration of the working condition of the flow (totally about half an hour), a circulation system

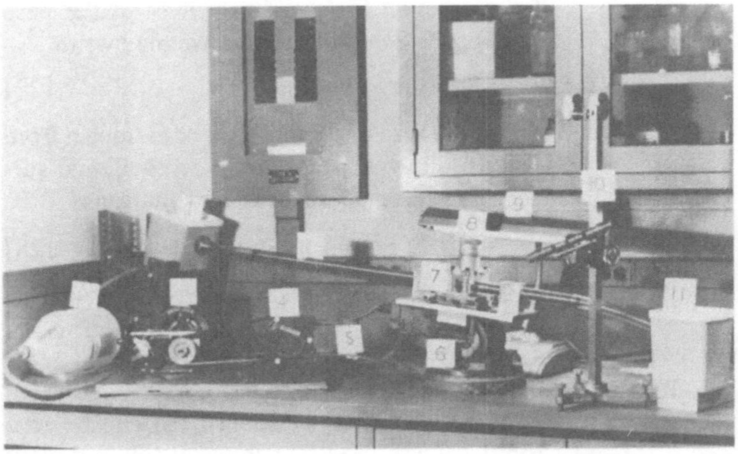

Fig. 2. Arrangement of the apparatus: Key to photograph: 1. Reservoir, 2. tube, 3. variable speed drive, 4. tubing (or peristalic) pump, 5. plastic tubing, 6. universal stage, 7. translation stage, 8. 2″ micrometer head, 9. lamp, 10. telescope (cathetometer), 11. reservoir (sink), 12. damping filter

that maintains steady continuous flow is necessary (fig. 2). The circulation system consists of two reservoirs, a tube, a motor and a pump. The motor, which can control the flow rate, is a variable-speed type. The tubing pump (peristaltic pump) has the advantage that the fluid can be pumped through the plastic tube without direct contact with the pump wall. The other important feature of the tubing pump is that it does not degrade the long-chain molecules much. This type of pump has a fluctuating output, and the fluctuation of the laminar flow can be suppressed by the density and the viscosity of the solution in use. The amount of the fluid needed for the experiment depends on the viscosity. The more viscous the solution, the less of it is needed. As the period of passing the measuring-station for a *Lagrange*ian particle is much longer than the relaxation time of the fluid, the memory effect of the elastic fluid can be ignored. The viscosity measurement of the solution was made after one hour of continuous pumping and circulating, and showed no measurable difference from that of the fresh solution. Hence the circulating system appeared to cause no significant degradation of the solution. As for a very viscous solution, it can be delivered from a pressurized tank.

The tube and the measuring device are supported by a universally adjustable stage. By rotating the base of the stage, the angle of the tube and the symmetry of the flow can be easily adjusted. Accordingly, the position of the reservoirs can be altered by the adjustable levels in the frame supporting the reserviors.

The displacement transducer must be a device that can read the displacement increment to an accuracy of .0002 inches without touching the liquid surface. As sketched in fig. 1, we used a micrometer in this experiment with a fine graduation of .0001 inches. The original flat-end of the spindle was furnished with a sharp-point probe to pin-point the coordinate of the liquid surface. The performance of the micrometer is very satisfactory when it is used with suitable lighting and a telescope. When making a measurement, the micrometer-point is driven as close to the liquid surface as possible without touching it. Meanwhile, the lighting will project a very clear image of the point on the liquid surface if the light is aimed in the proper direction. As a result, the gap between the real point and its image can be enlarged and clearly seen in the telescope. By this method, keeping an eye on the telescope, the operator is in a position to drive the measuring-point much closer to the liquid surface and read the height $h(x)$ from the micrometer. The real gap can be made as small as .0002 inches, which is the accuracy of the measurement. If the widths of the gap can be made the same at all measuring stations across the tube, the gap adds no error to the measurement.

The accuracy of the measurement depends not only on the skill of the operator in keeping the widths of the gap as uniform as possible, but also on the flow conditions. The flow condition becomes bad for five major reasons: a) wetting, b) unsteadiness, c) fluctuations, d) turbulence, and e) asymmetry. In the following paragraphs, explanations and corrections for the above conditions are given.

a) Wetting

Wetting is a surface tension and adhesive phenomenon such that the flow (the velocity profile) of the liquid in contact with the tube wall is not fully developed. To prevent wetting, first of all the tube must be dry before use. If the wetting is caused by the roughness of the tube wall, silicone grease (non-wettable for water solvent) can be applied to the "edge" of the tube (the area that should not be wetted) so as to form a holding line or forced boundary for the flow. Despite the fact that such a "formed" free surface would differ from the natural one, the theory developed previously is still applicable. The grease method, however, is not necessary for concentrated solutions, but is helpful with dilute solutions.

b) Unsteadiness

Unsteady flow is the circumstance that the datum point (the top point of the convex surface or the bottom point of the concave surface) gradually shifts from the origin in either an up or a down direction. The unsteady motion is a result of imperfections of the mechanical system of the motor and pump (slippery tubing or loose belt, etc.).

c) Fluctuation

It is an inherent nature of the periodic squeezing motion of the tubing pump that it generates a fluctuation or pulsation of the flow. For the concentrated solutions, the fluctuation is almost unnoticeable. As for the more dilute solutions, such as .1% Polyox, the amplitude was about $\pm$.0002 inches (with a reservoir capacity of $1^1/_2$ gallons). One of the effective methods to suppress the fluctuation is to add one or two "filters" to the pumping line.

d) Turbulence

Laminar flow exists only when the *Reynolds* number is smaller than the critical *Reynolds* number. Beyond the critical *Reynolds* number, the flow becomes unstable and develops into turbulent flow.

e) Asymmetry

The asymmetry of the flow is a result of either wetting or inaccurate alignment (non-horizontal) of the translation stage and the tube. The latter can be regulated by the universal stage; it is necessary for the translation of the stage to be perpendicular to the gravitation vector.

All these conditions should be avoided.

4. The sample fluids

Six polymer systems were studied. They were
(i) Separan (AP-30 grade from Dow Chemical Co.) in water (.8%, .2%, .1%, and .05%),
(ii) Polyhall-295 (Stein, Hall Co.) in water (1%, .8%, and .5%),
(iii) Polyox (Grade WSR-301 from Union Carbide Co.) in water (1% and .5%),
(iv) J-100 (Dow Chem. Co.) in water (.25%),
(v) Polyisobutylene in cetane (6.8%, 5.5%, and 3%),
(vi) National Bureau of Standards Nonlinear Sample No. 1 (*Kearsley*, 1972).

Prior to mixing, the Separan, the Polyhall-295, the Polyox and the J-100 are in granular form (soluble in water), while the polyisobutylene[2]) is in flake form (soluble in cetane). The NBS sample is ready mixed. To avoid significant degradation caused by a strong stirring action, mixing was accomplished using a rolling motion of a sample filled jar; a rake inside the jar produces mild agitation. Usually the mixing time is about one week. It was found that a three-weeks mixing time caused appreciable degradation as judged by comparing the measured viscosity to that with the shorter mixing time. The viscosity of an aqueous solution can be affected by its pH, which varies with time and addition of an acid or base. This effect is especially true for the solution of J-100 in distilled water as noted by *G. Williams* (1971). The pH effect was discussed in detail by *Montgomery* (1963). Since all the present measurements presented were carried out in a few (at most) days after mixing, it was assumed that no appreciable variation of viscosity due to pH variations would occur in that period.

[2]) The 6.8% sample used Oppanol B-100, while the others used Vistanex L-140.

The surface tension of the samples were measured by the method of capillary rise (*Moore*, 1962). Liquid rises in the tube until the weight of the liquid column, *h*, balances the pressure difference. The surface tension α can be evaluated by the formula given below

$$\alpha = \frac{1}{2} g h (\varrho - \varrho_g) \frac{r_0}{\cos\theta}. \qquad [29]$$

The diameter of the capillary tube, $2r_0$, used for the measurement is .07 inches. As the contact angles θ observed were essentially zero and the density of the surrounding air ϱ_g is negligible compared to that of the sample, ϱ, the above formula can be reduced to the following simple form

$$\alpha = \tfrac{1}{2} \varrho g h r_0. \qquad [30]$$

It is assumed that α does not depend on the shear rate at the surface. Here the aqueous solutions have a density close to 1.0 g/cm^3 in all cases. The results are shown in fig. 3; they show the slight reduction of surface tension with addition of polymer. All measurements were made at 22 °C. For the N.B.S. sample, which is a polystyrene solution (*Kearsley*, 1972) we made measurements at 22 °C and 25 °C.

5. Results and discussion

In fig. 4 we show some of the surface shapes observed. In some cases (curve a) there is a monotonic profile while in others one may obtain inflexion points. In the latter cases the upturn near the wall is a surface tension effect and has to be recognized as such. In our previous study (*Tanner*, 1970) the experiment was performed in a slightly different way. Instead of the central point of the surface being held to be coincident with the centre of the tube, as here, the outer edge was made level with the tube mid-plane. The sharp edge then largely avoided the upturn at the wall; as we shall see, however, both procedures often give the same result. As a numerical check on the surface tension correction, we show in fig. 5 some surface profiles, measured and calculated, for a *Newton*ian liquid.

Repeatability of results for the polyisolutylene-centane solution is shown in fig. 6, indicating that the errors in finding the surface profile are of order 10^{-4} in. over a period of twelve days. The aqueous solutions are not as stable, but they are

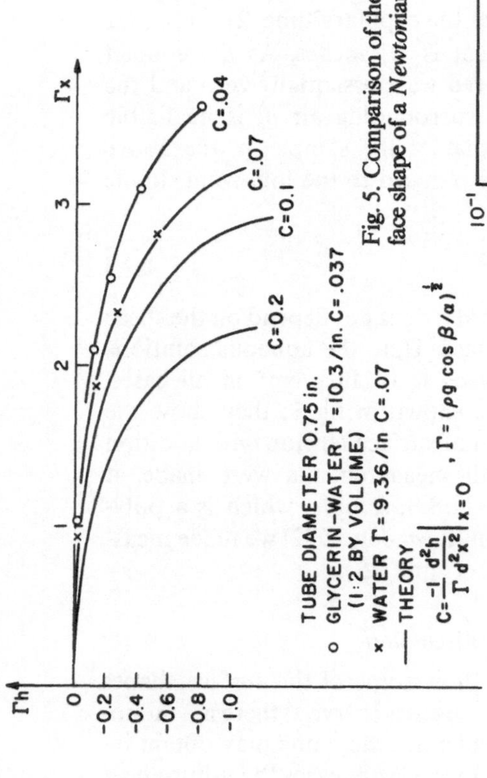

Fig. 5. Comparison of theory and experiment for the surface shape of a *Newtonian* fluid in the channel

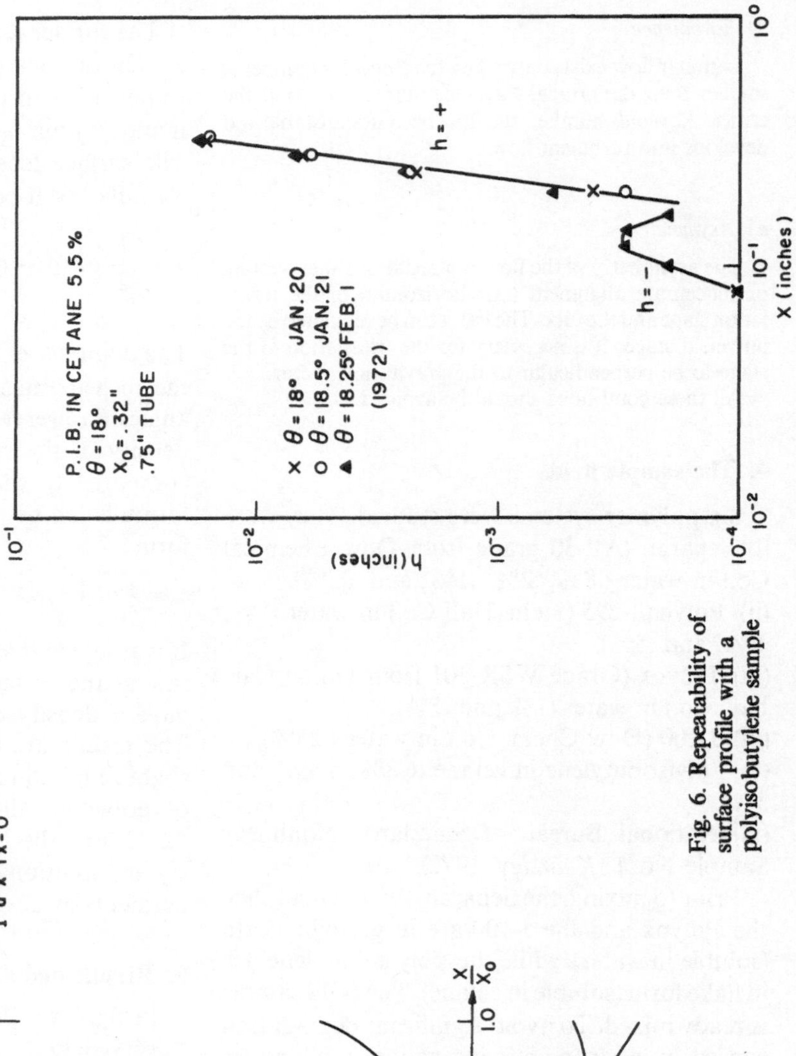

Fig. 6. Repeatability of surface profile with a polyisobutylene sample

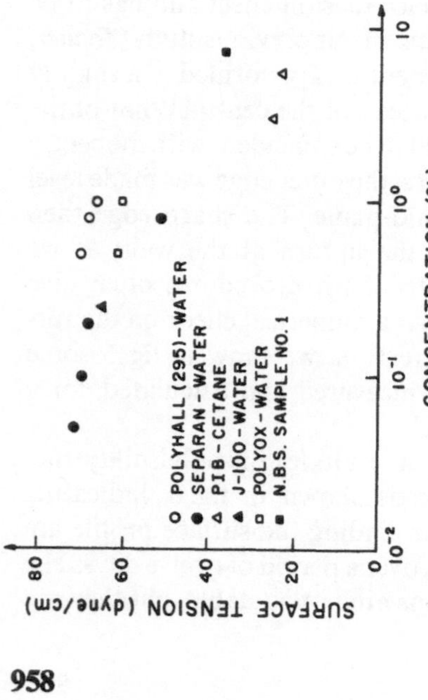

Fig. 3. Surface tension coefficients of samples at 22 °C

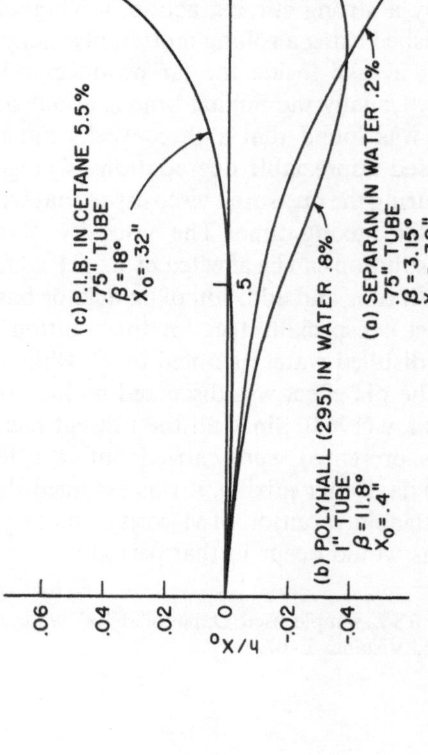

Fig. 4. Surface shapes observed with non-*Newtonian* fluids

repeatable within a single day's testing. In many cases (e.g., fig. 7) we found that the central profile was close to a parabolic shape. Using the parabolic profiles one should find that the method of calculation used by *Tanner* (1970) and the present method should give the same results, since surface tension does not affect the calculation of the parabolic profile. Furthermore, the results for different tilt angles and tubes should overlap. For the 0.8% Polyhall solution an extensive investigation showed that this was the case (fig. 8).

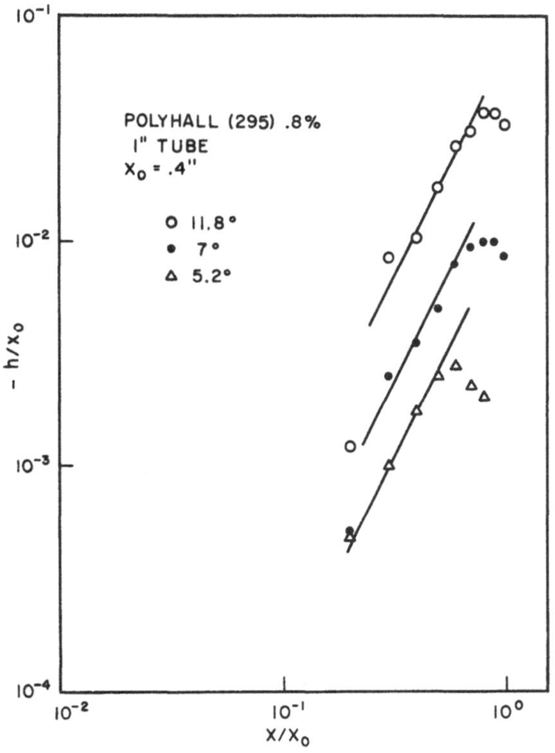

Fig. 7. Showing square-law shape of profile near tube centre

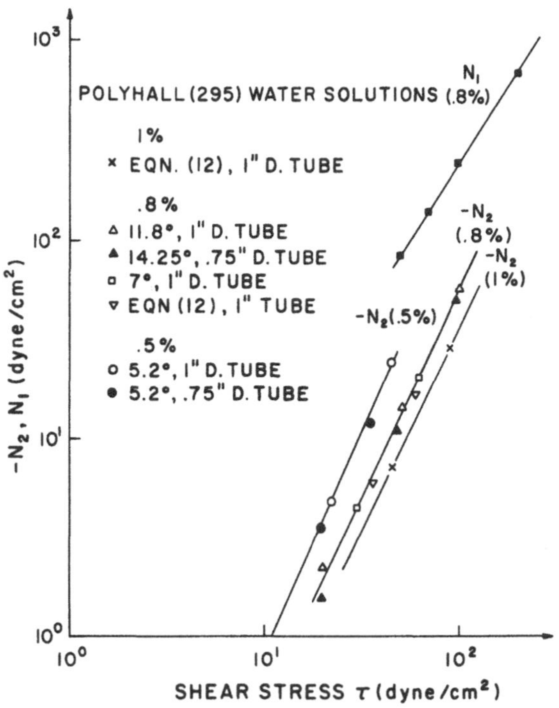

Fig. 8. Results for Polyhall (295) – water solutions. Note agreement of measurements made with various settings

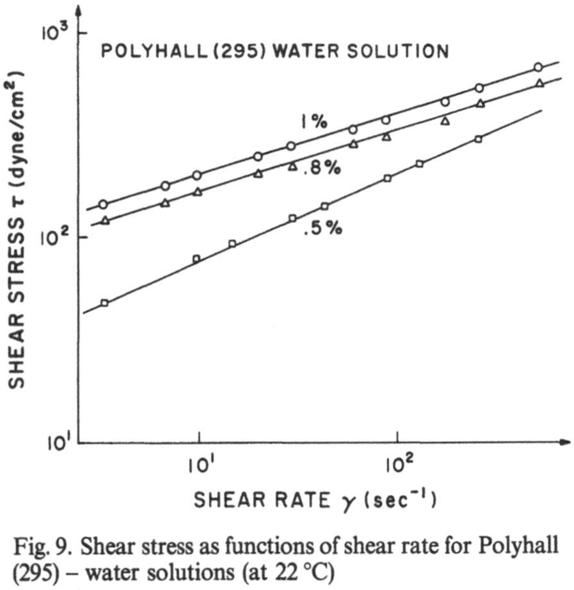

Fig. 9. Shear stress as functions of shear rate for Polyhall (295) – water solutions (at 22 °C)

We see that tube size and material are unimportant, that results taken at two different angles coincide, and that the "new" and the "old" methods of calculation (for parabolic profiles) give the same results. It is also clear that errors of about 1 dyne/cm^2 order of magnitude are present in this set of results. We believe this is a typical error magnitude for all tests, as will be discussed later.

In figs. 9–12 we show the functions N_1 and τ as a function of shear rate γ for the first five materials. All five materials are highly non-linear in viscosity but in many cases, and especially with p.i.b. (*Tanner*, 1973) one finds that N_1 is closely proportional to τ^2. Because in our experiment τ is the parameter most readily controlled, we plot N_2 versus shear stress for these five samples (figs. 8 and 13–15). For the N.B.S. fluid, sample (vi), we plot in the conventional way since the fluid is here in the second-order region (fig. 16).

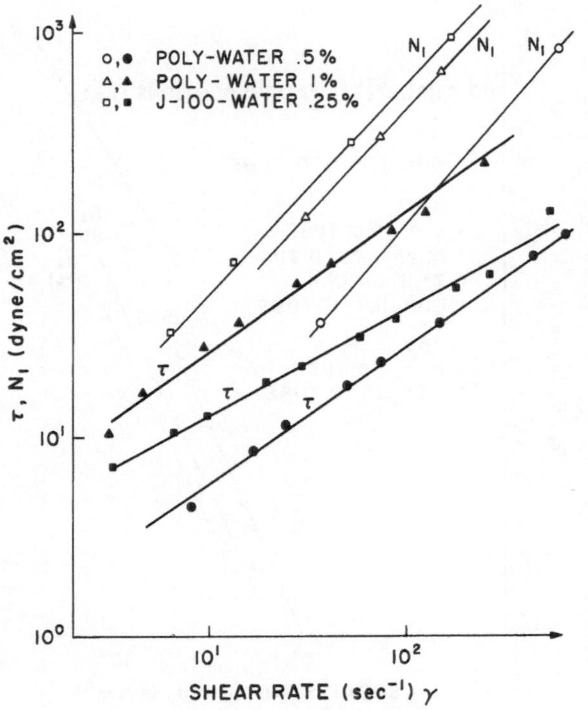

Fig. 10. First-normal stress difference (N_1) and shear stress (τ) as functions of shear rate (γ) for the Polyox-water and J-100 – water solutions at 22 °C

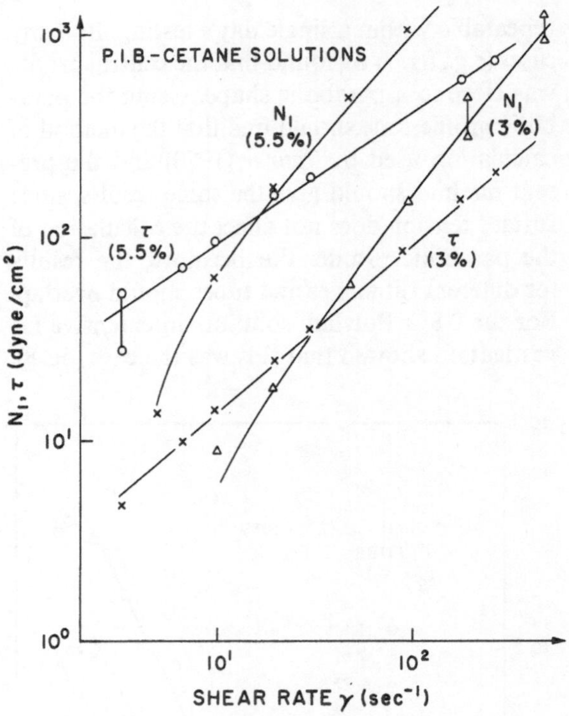

Fig. 12. N_1 and τ as functions of γ for polyisobutylene-cetane solutions at 22 °C

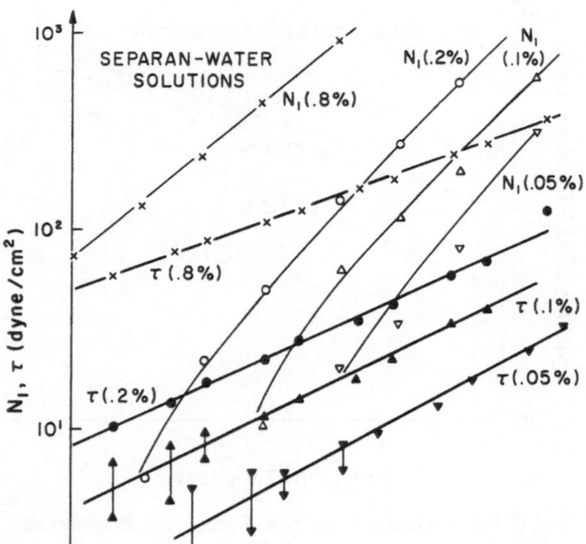

Fig. 11. First-normal stress difference and shear stress as functions of shear rate for Separan AP-30 – water solutions at 22 °C

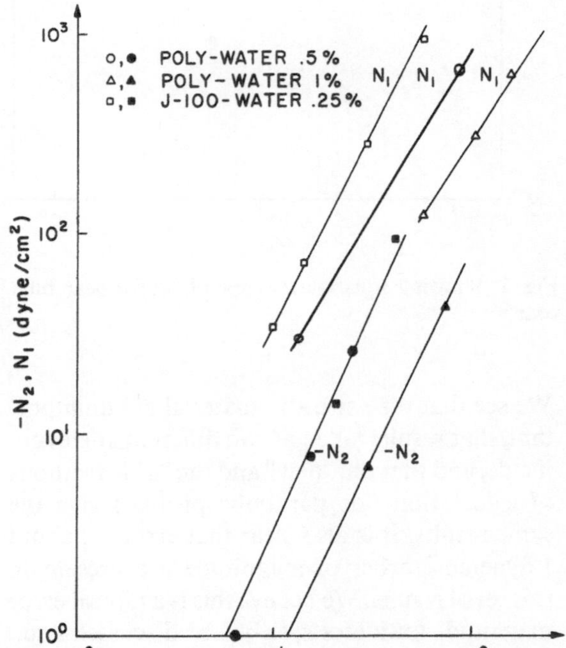

Fig. 13. First and second normal stress differences as functions of shear stress (τ) for the Polyox and J-100 solutions at 22 °C

Note in fig. 15 that we show the effect of allowing for the deviation from the nominal shear stress in the tube due to the surface displacement; this correction has been made in all cases even though it is small. From figs. 11 and 14 (Separan AP-30 solutions) we see that the magnitude of $-N_2$ is closer to that of N_1 for less concentrated solutions. A rough estimate of the ratio $-N_2/N_1$ gives .3 at .8% concentration to .5 at .05% concentration. Measurements on the Polyox solutions show similar behavior as seen in figs. 10 and 13. (It should be pointed out that for .8% of Separan and 1% concentration of Polyox both the free-surface profile and the $N_2(\tau)$ were square laws in shear stress.)

We have attempted to assess the accuracy of the method. Since we are really using the surface hump as a self-sustaining pressure gauge, the main error in N_2 comes from inaccurate measurements of the profile. In the case when surface tension is unimportant (parabolic surface), the errors in height measurement being of order $\pm .001$ cm maximum, we estimate that errors of order ± 1 dyne/cm^2 in N_2 are easily possible, as seen above.

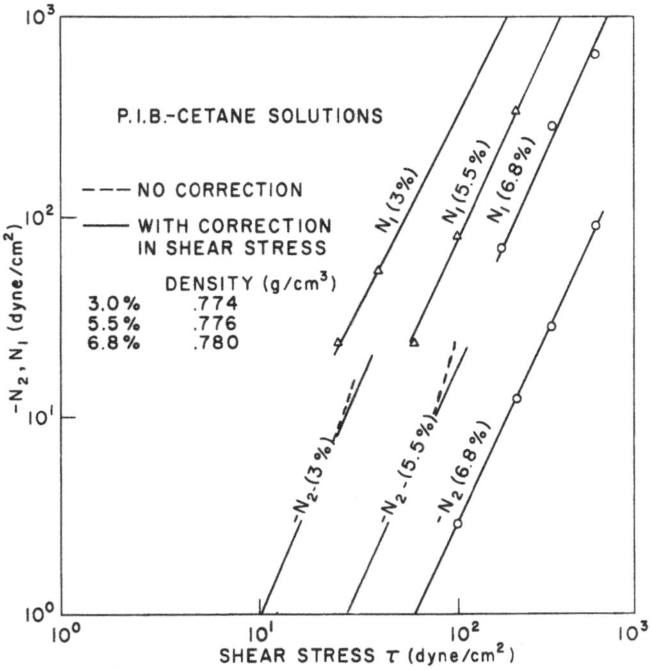

Fig. 15. N_2 and N_1 as functions of τ for the polyisobutylene-cetane solutions at 22 °C

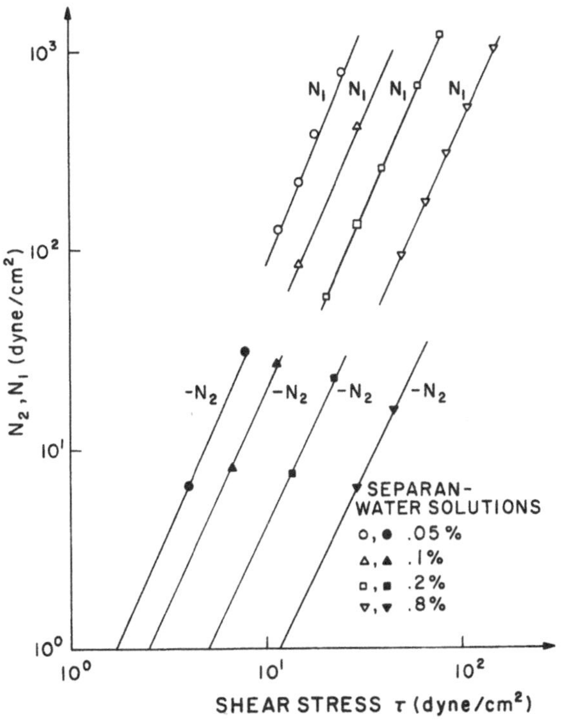

Fig. 14. N_2 and N_1 as functions of τ for the Separan AP-30 solutions at 22 °C

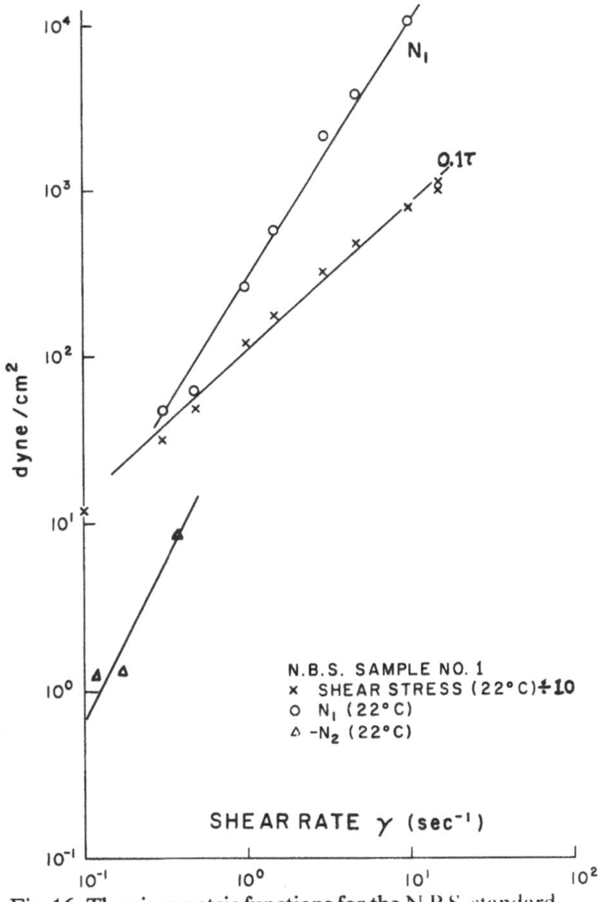

Fig. 16. The viscometric functions for the N.B.S. standard non-linear sample No. 1 at 22 °C. This is a polystyrene (7.14% by weight) solution

Conclusion

In view of possible correlations among our N_2 data we have plotted, in fig. 17, the second normal stress data as the ratio $-N_2/\tau^2$ versus the weight concentration of polymer C. The choice of the first variable is prompted by our result that N_2 is closely proportional to τ^2. The spread of the points indicates the deviation of $-N_2$ from a square law dependence on shear stress. It seems that N_2/τ^2 for some aqueous solutions follows the power law $C^{-1.6}$ instead of a negative square law C^{-2}; the rough proportionality of N_1/τ^2 and C^{-2} has been demonstrated recently (*Tanner*, 1973) for polyisobutylene solutions, and it appears that N_2 also follows this law approximately. It is also noted that $-N_2/\tau^2$

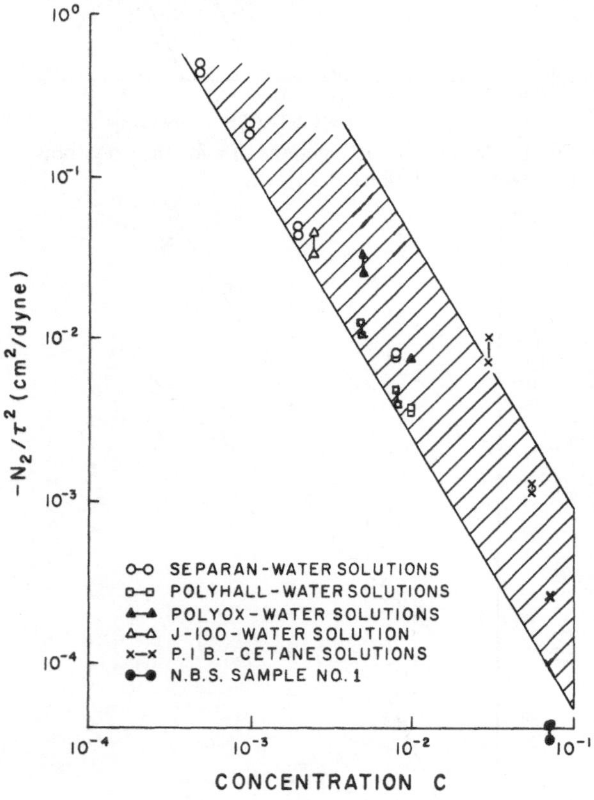

Fig. 17. Dependence of N_2/τ^2 on concentration for several polymers at 22 °C

for the N.B.S. fluid is somewhat below the main trend of the data as shown in the same figure. It is expected that molecular weight is also a variable in such a correlation, but it is not considered here, since we have insufficient data.

In view of the difficulties in measuring N_2 we have collected the selection of data for polyisobutylene solutions shown in fig. 18. The hatched area was covered by *Ginn* and *Metzner* (1969) with their cone-plate/parallel plate measurements, the open symbols are the work of *Kaye* et al. (1968) and *Pritchard* (1972), essentially using the same sample, and the filled symbols represent the direct measurements of *Miller* and *Christiansen* (1972) and ourselves. Except for some of *Pritchards* (1972) points there is reasonable agreement, at least at low shear stresses, that $N_2 \sim -0.15\, N_1$. It should be noted that the division by the (concentration)^2 is not very significant as all the samples lie in the range of 3 to 11% polymer; it merely serves as a means of compressing the abscissa.

For other polymers there is insufficient data to compare in this way; however, it appears that N_2/N_1 increases at lower concentrations.

For the N.B.S. nonlinear sample, which is a 7.14% polyisobutylene solution, some difficulties were encountered due to the small values of N_2 encountered. It does not appear that shear

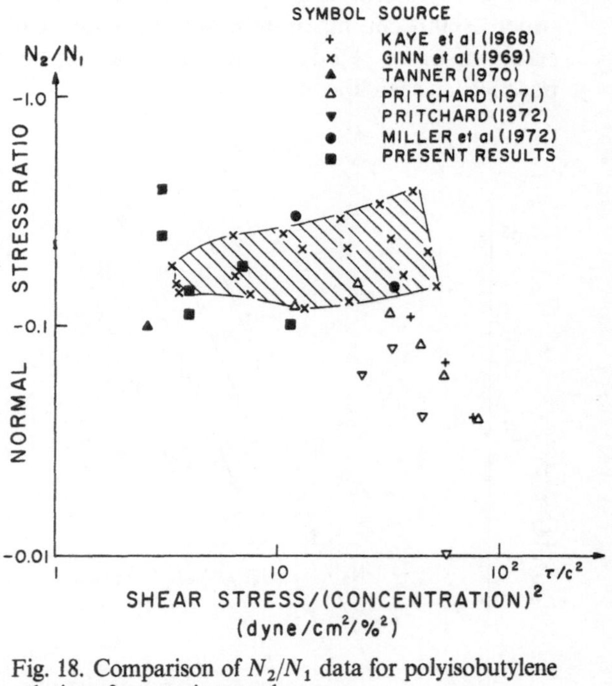

Fig. 18. Comparison of N_2/N_1 data for polyisobutylene solutions from various authors

stresses above about 300 dynes/cm^2 can be interpreted accurately due to tube size and surface tension limitations. At this shear stress N_1 is less than 20 dyne/cm^2, and thus N_2 is expected to be only a few dynes/cm^2, giving rise to very small surface defexions. Consequently we have here assumed that N_2 behaves as a square law in shear stress in this region, giving the results shown in fig. 16 with rather large error possibilities; a most probable value of $-N_2/N_1$, seems to be 0.20 with at least $\pm 10\%$ error possibility.

Finally, the open-channel method has been shown to provide a method of determining N_2 without making any special constitutive assumptions; its main drawbacks are that for dilute solution it requires a large volume (1–2 gallons) of sample (but as a partial compensation the rate of fluid circulation is low and so consequently is the degradation), and the shear stresses attainable are limited. The measurement technique is very simple and so is the analysis. Due to the non-interference of other effects (except surface tension) we can measure normal stress differences down to a few dyne/cm^2. With regard to the effect of surface tension, it is necessary to consider it carefully, as we have found concave, convex and mixed surface shapes. Corrections to the shear stress can be made if required.

Our main findings about N_2 are, to summarize.

a) With all the solutions measured, the second normal stress N_2 is negative and its magnitude varies approximately as a square law in shear stress over the range tested. At a same shear stress, the magnitude of N_2 of a less concentrated solution is larger than that of a more concentrated solution.

b) The ratio N_2/N_1 may vary with shear stress but the variation is moderate. The ratio $|N_2/N_1|$ also varies with concentration and becomes larger when the solution is diluted for some aqueous solutions. The range of N_2/N_1 is generally between about -0.1 and -0.5.

c) The systematic variation of the N_2 curves for different concentrations leads us to expect that a certain correlation form is possible between the ratio N_2/N_1, the shear stress, the concentration and the molecular weight, which should be further explored.

This work was supported by the National Science Foundation under Grant GK-24746 at Brown University. This support is gratefully acknowledged.

Summary

This paper deals with the measurement of the second normal stress difference in a viscoelastic flow in an open semi-circular inclined channel; associated theory is also given. The theory shows that simple correlations exist for the second normal stress difference, the shear stress, and the profile (or the deflection) of the free surface of the flow. It leads to a method for deducing the second normal stress from the measured surface profile. A treatment of surface tension, whose effect has been ignored in previous analyses is given as an integral part of the formulation. Samples of polyisobutylene/cetane and some aqueous solutions of different concentrations were used for measurements using two tubes of different materials and sizes tilted at different angles. The N.B.S. No. 1 nonlinear sample was also tested. Agreements between results for different tube sizes and angles indicate the consistency of the theory and measurements. The results presented here show that the ratio of the second to the first normal stress differences is negative and the functional form of the second normal stress difference is close to a square law in shear stress in the range investigated. Relatively speaking, we find the ratio $|N_2/N_1|$ to be greater for more dilute solutions. A summary of available data on N_2 is given for polyisobutylene, and reasonable consistency between various investigators is demonstrated for this material. Besides, a correlation of N_2 for various polymer solutions is also shown.

Zusammenfassung

Dieser Bericht behandelt eine Methode zur Messung der zweiten Normalspannungsdifferenz in einer viskoelastischen Strömung in einem offenen, halbkreisförmigen, geneigten Kanal. Die theoretische Behandlung des Problems ist ebenfalls angegeben und zeigt die einfachen Zusammenhänge auf, welche zwischen der zweiten Normalspannungsdifferenz, der Schubspannung und dem Profil (Wölbung) der freien Oberfläche der Strömung bestehen. Der Einfluß der Oberflächenspannung, in einer früheren Untersuchung vernachlässigt, ist jetzt berücksichtigt und in die mathematische Behandlung einbezogen worden. Polyisobutylen/Cetan, einige wäßrige Lösungen sowie eine Probe N.B.S. No. 1 wurden untersucht. Als Kanal gelangten zwei Röhren verschiedener Größe und aus verschiedenem Material bei veränderlicher Neigung zur Anwendung. Übereinstimmung der Resultate von verschiedenen Kanalgrößen und Neigungswinkeln zeigen, daß Theorie und Experiment im Einklang stehen. Die hier veröffentlichten Ergebnisse zeigen weiter, daß das Verhältnis der zweiten zur ersten Normalspannungsdifferenz negativ ist, und die funktionale Form der zweiten Normalspannungsdifferenz mit guter Näherung dem Quadrat der Schubspannung folgt. Relativ betrachtet ergibt sich für verdünnte Lösungen ein höherer Wert $|N_2/N_1|$. Abschließend wird eine Übersicht über die verfügbaren Messungen von N_2 an Polyisobutylen gegeben und vernünftige Übereinstimmung zwischen den Resultaten verschiedener Autoren festgestellt. Eine Darstellung von N_2 als Funktion der Konzentration für verschiedene Polymerlösungen vervollständigt den Bericht.

References

Adams, N. and *A. S. Lodge*, Phil. Trans. Roy. Soc. (London) **A 256**, 149 (1964).

Berry, J. P. and *J. Batchelor*, in: Polymer Systems (*R. E. Wetton* and *R. W. Whorlow*, eds.) (London 1968).

Brindley, S. and *J. M. Broadbent*, Rheol. Acta **12**, 48 (1973).

Cowsley, C. W., Improvements to total thrust methods for the measurement of second normal stress difference, Univ. of Cambridge, Dept. of Chem. Engrg. Polymer Processing Res. Centre, Report No. 6 (1970).

Denn, M. M. and *J. J. Roisman*, Amer. Inst. Chem. Eng. J. **15**, 454 (1969).

Garner, F. H., A. H. Nissan, and *G. F. Wood*, Phil. Trans. Roy. Soc. (London) **A 243**, 37 (1950).

Greensmith, H. W. and *R. S. Rivlin*, Phil. Trans. Roy. Soc. (London) **A 245**, 399 (1953).

Ginn, R. F. and *A. B. Metzner*, Proc. 4th Int. Rheol. Congr., Part **2**, 583 (New York 1965).

Ginn, R. F. and *A. B. Metzner*, Trans. Soc. Rheol. **13**, 429 (1969).

Hayes, J. W. and *R. I. Tanner*, Proc. of the 4th Inter. Congr. on Rheology, Part **3**, ed. by *E. H. Lee*, p. 389 (New York 1965).

Huppler, J. D., Trans. Soc. Rheol. **9**, 273 (1965).

Huppler, J. D., E. Ashare, and *L. A. Holmes*, Trans. Soc. Rheol. **11**, 159 (1967).

Jackson, R. and *A. Kaye*, Brit. J. Appl. Phys. **17**, 1355 (1966).

Kaye, A., A. S. Lodge, and *D. G. Vale*, Rheol. Acta **7**, 368 (1968).

Kearsley, E. A., Rheol. Acta **12**, 546 (1973).

Kotaka, T., M. Kurata, and *M. Tamura*, J. Appl. Phys. **30**, 1705 (1959).

Kuo, Y., The Determination of the Second Normal Stress difference in Viscoelastic Flows. Ph. D. Thesis, Brown University (1973).

Kuo, Y. and *R. I. Tanner*, Intl. J. of Mech. Science **14**, 861 (1972).

Lobo, P. and *H. R. Osmers*, Rheol. Acta **13**, 457 (1974).

Macosko, C., Flow of polymer melts between eccentric rotating disks. Report, Dept. of Chem. Engrg., Princeton Univ. (1970).

Markovitz, H. and *D. R. Brown*, Trans. Soc. Rheol. **7**, 137 (1963).

Markovitz, H. and *D. R. Brown*, Second-order effects in elasticity, plasticity and fluid dynamics. Int. Symp. Haifa (1962), Macmillan Co. (1964).

Marsh, B. D. and *J. R. A. Pearson*, Rheol. Acta **7**, 326 (1968).

Miller, M. J. and *E. B. Christiansen*, Amer. Inst. Chem. Eng. J. **18**, 600 (1972).

Montgomery, W. H., Water-soluble resins. Edited by *R. L. Davidson* and *M. Sitting* (New York 1963).

Moore, W. J., Physical Chemistry (Prentice-Hall 1962).

Olabisi, O. and *M. C. Williams*, Trans. Soc. Rheol. **16**, 727 (1972).

Pipkin, A. C., Lectures on viscoelasticity theory (New York 1972).

Pollett, W. F. O., Brit. J. Appl. Phys. **6**, 199 (1955).

Pritchard, W. G., Phil. Trans. Roy. Soc. (London) **A 270**, 507 (1971).

Roberts, J. E., Proc. Second Intl. Congr. Rheol. (New York 1953).

Sakiadis, B. C., Amer. Inst. Chem. Eng. J. **8**, 317 (1962).

Tanner, R. I., Trans. Soc. Rheol. **11**, 347 (1967).

Tanner, R. I. and *A. C. Pipkin*, Trans. Soc. Rheol. **13**, 471 (1969).

Tanner, R. I., Trans. Soc. Rheol. **14**, 483 (1970).

Tanner, R. I., Trans. Soc. Rheol. **17**, 365 (1973).

Williams, G., Dynamic response of dilute polymer solutions in a shear flow. Ph. D. Thesis, Brown University (1971).

Weissenberg, K., Nature **159**, 310 (1947).

Yin, W.-L. and *A. C. Pipkin*, Arch. Ratl. Mech. Anal. **37**, 111 (1970).

Authors' address:
Y. Kuo and *R. I. Tanner*
Division of Engineering
Brown University
Providence, Rhode Island 02912 (U.S.A.)

Rheol. Acta 13, 457–462 (1974)

From the Department of Chemical Engineering, University of Rochester, Rochester, New York 14627 (U.S.A.)

Pressure hole errors in the annular flow measurement of the second normal stress difference

By P. F. Lobo *and* H. R. Osmers

With 6 figures

(Received October 27, 1972)

Introduction

There is a critical need to accurately determine viscoelastic fluid material functions, one of which is the second normal stress difference. While it is now generally accepted that this function is non-zero, in contradiction to the *Weissenberg* hypothesis, numerous investigators have reported values of opposite sign as well as significant unexplained differences in magnitude. *Ginn* and *Metzner* (3) have critically examined much of the data and many techniques. They show that the sign of all internally consistent second normal stress difference data is obtained as negative by all techniques except annular flow devices, which uniformly yield positive values. They also note that, even with considerable care, accurate second normal stress difference data are not likely to be obtained using either the parallel plate or cone and plate systems, especially at higher shear rates. The more recent work of *Miller* and *Christiansen* (10) seems to improve the accuracy of measurement of the second normal stress difference using both the rim-pressure and integral-force methods of obtaining and analysing cone and plate data. Their results are still somewhat erratic, however, and differ by up to a factor of two. In any event, inertial effects preclude the use of these techniques much above shear rates of 500 sec^{-1}.

The use of axial annular flow to measure the second normal stress difference has several advantages which include operation at high shear rates, a totally enclosed system to avoid solvent evaporation and the potential use on polymer melts at high shear rates. Apparently incorrect results obtained in the past using annular flow are likely due to the fact that all previous measurements have depended on "small" holes normal to the tube walls for measurement of the normal stress. It has been shown experimentally (2, 7, 12) that large systematic errors in measured normal stresses result from using holes on a surface, and that the data obtained are greater than the correct values.

Tanner and *Pipkin* (12) analysed the slow flow of a second-order fluid past a deep two-dimensional slot normal to the flow and found that the intrinsic error in the normal stress measurement is one-quarter of the first normal stress difference. The rectilinear motion of a second-order fluid along a slot aligned with the fluid flow was considered by *Kearsley* (8) who obtained the result that the intrinsic error is equal to minus one-half the second normal stress difference. *Higashitani* and *Pritchard* (5) present an approximate analysis for slow flow past a circular hole and find that the intrinsic error is about equal to one-sixth of the difference between the first and second normal stress differences for some materials.

The data and analyses cited above suggest that the difficulty in using annular flow is due to the errors resulting from the use of pressure taps. The present paper presents experimental results for the measurement of the second normal stress difference in axial annular flow using both pressure holes and transducers simultaneously in such a way as to measure and account for pressure hole errors. The intent is to show that hole errors are sufficiently large to reverse the sign of the computed second normal stress difference, thereby indicating that the annular flow technique can yield correct results at shear rates otherwise unobtainable.

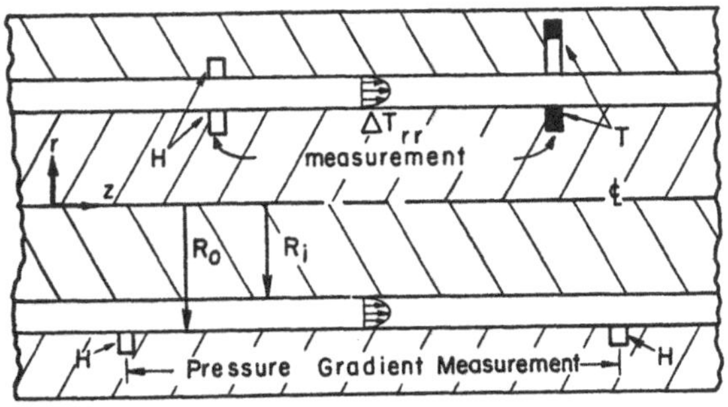

Fig. 1. Schematic of annular flow with axial positions of pressure holes (*H*) and transducers (*T*)

Analysis of annular flow experiment

For steady, laminar flow in the annulus between coaxial cylinders depicted in fig. 1, the velocity field is defined by

$$\vec{v} = \vec{v}[v_1(x^2), 0, 0] \quad [1]$$

where the indices 1 and 2 denote, respectively, the direction of flow and the direction of change of velocity, and the remaining index, if used, the neutral direction. In cylindrical coordinates, eq. [1] becomes

$$\vec{v} = \vec{v}[v_z(r), 0, 0]. \quad [2]$$

The total, isotropic and dynamic stresses are related by the expression

$$T_{ij} = -P g_{ij} + \tau_{ij} \quad [3]$$

where the g_{ij} are the covariant components of the identity tensor, and the total and dynamic stresses are positive in tension.

Eqs. [1] and [2] combined with the z-component of *Cauchys* equations of motion yield the shear stress distribution

$$\tau = \frac{r}{2}\frac{dP}{dz}\left[1 - \left(\frac{r_0}{r}\right)^2\right] \quad [4]$$

where r_0 is the radius of zero shear and the isotropic pressure gradient, $-dP/dz$, can be shown to be independent of both r and z.

For the prescribed kinematics, the deformation rate tensor reduces to

$$\Delta_{ij} = \begin{bmatrix} 0 & \dot{\gamma} & 0 \\ \dot{\gamma} & 0 & 0 \\ 0 & 0 & 0 \end{bmatrix} \quad [5]$$

where the non-zero component is defined by

$$\dot{\gamma} = \frac{dv_z}{dr}. \quad [6]$$

The shear rate must satisfy the relation

$$\int_{R_i}^{R_0} \dot{\gamma}\, dr = 0. \quad [7]$$

Shear stress-shear rate data of the form

$$\tau = \tau(\dot{\gamma}) = \eta\dot{\gamma} \quad [8]$$

may be determined by independent measurements. The three-parameter *Ellis* model (1) is selected as appropriate and convenient for fitting the data

$$\frac{1}{\eta} = \frac{1}{\eta_0}\left[1 + \left(\frac{\tau}{\tau_{1/2}}\right)^{\frac{1-n}{n}}\right]. \quad [9]$$

At low shear rates, the apparent viscosity η approaches the zero-shear viscosity η_0, so that the zero-shear failure of the power law model is avoided. At high shear rates, power law behavior is approached.

Eqs. [4], [7], and [9] therefore enable determination of the radius of zero-shear as a function of the pressure gradient. The shear stress and shear rate are thus known as a function of radial position.

The r-component of *Cauchys* equations of motion yields

$$\Delta T_{rr} = (T_{rr})_{R_0} - (T_{rr})_{R_i} = -\int_{R_i}^{R_0} \frac{N_2(\tau)}{r}\, dr \quad [10]$$

where the second normal stress difference is defined as

$$N_2(\tau) = T_{rr} - T_{\theta\theta} \quad [11]$$

a function of shear stress.

The annular flow method for determination of N_2 is seen to require the measurement of the difference in normal thrusts at the annular surfaces of the outer and inner cylinders, ΔT_{rr}, and the simultaneous measurement of the pressure gradient, $-dP/dz$. The latter measurement is accomplished in a routine manner since hole errors cancel. The normal thrust difference, ΔT_{rr}, is measured in two ways.

The first is by means of holes (H) in the tube walls as shown in fig. 1. ΔT_{rr} is relatively small as compared to the pressure gradient, so that misalignment of the holes may lead to appreciable errors. The method suggested by *Tanner* (11) avoids this by providing for axial motion of the inner cylinder. A capacity bridge null-meter is connected between the pressure taps on the inner and outer cylinders and the inner cylinder moved axially under a certain pressure gradient until a null-point is reached. The flow is then reversed and a second null-point determined under the same pressure gradient. The difference in normal thrust, using holes is then

$$(\Delta T_{rr})_H = \frac{l}{2}\frac{dP}{dz} \quad [12]$$

where l is the distance between null-points. This measured value includes the hole errors for both the inner and outer walls, ε_i and ε_0:

$$(\Delta T_{rr})_H = \Delta T_{rr} + (\varepsilon_0 - \varepsilon_i). \quad [13]$$

The second method for measurement of the normal thrust difference involves two pressure transducers (T), one located flush to the inner cylinder, the second located at the base of a hole in the outer cylinder. In this instance the measured quantity only includes a hole error from the outer cylinder, that is:

$$(\Delta T_{rr})_T = \Delta T_{rr} + \varepsilon_0. \quad [14]$$

It follows that

$$\varepsilon_i = (\Delta T_{rr})_T - (\Delta T_{rr})_H. \quad [15]$$

Repeated measurements enable determination of hole error as a function of shear stress[1]). The correct normal thrust difference, ΔT_{rr}, may then be calculated using eq. [14].

A complication arises in the measurement of $(\Delta T_{rr})_T$ because its magnitude precludes using the sliding core method. It can be determined in the following way without the transducers perfectly aligned. $(\Delta T_{rr})_T$ is recorded twice for the same pressure gradient, once in each direction. The average of the measurements is the correct value for aligned transducers at that pressure gradient.

With normal thrust difference data as a function of pressure gradient, and hence of shear stress, the task remains to invert the data according to eq. [10] to obtain $N_2(\tau)$. Existing data on normal stress functions are well described by an expression of the form

$$N_2(\tau) = \alpha_1 |\tau| + \alpha_2 |\tau|^2 \qquad [16]$$

where the α_i are to be determined to give the best fit to the normal thrust difference-pressure gradient data. Eq. [10] now may be integrated to yield

$$\Delta T_{rr} = -(\alpha_1 f_1 + \alpha_2 f_2) \qquad [17]$$

where

$$f_1 = \frac{R_0}{2}\frac{dP}{dz}\left[4b - (k+1)\left(1 + \frac{b^2}{k}\right)\right] \qquad [18]$$

$$f_2 = \frac{R_0^2}{8}\left(\frac{dP}{dz}\right)^2\left[(1-k^2)\left(1+\frac{b^4}{k^2}\right) + 4b^2 \ln k\right] \qquad [19]$$

and

$$b = \frac{r_0}{R_0}, \quad k = \frac{R_i}{R_0}. \qquad [20]$$

The α_i are obtained by least-squares, linear regression of the data. The procedure was found satisfactory when tested using the data presented by *Huppler* (6) and for hypothetical data. *Tanner* and *Williams* (13) present results of a more detailed investigation for solving integral equations numerically.

Second-order fluid analysis

The use of holes for pressure measurements clearly leads to erroneous results. It is not clear, however, that the errors resulting from the use of holes in an annular device are of such magnitude as to reverse the sign of the computed second normal stress difference. To examine this prospect, we consider the results of *Tanner* and *Pipkin*[2]) (12), namely that the pressure hole error, ε, is given by

$$\varepsilon = \frac{N_1(\tau)}{4} \qquad [21]$$

for slow viscometric flow of a second-order fluid defined by

$$T_{ij} = -P g_{ij} + \mu A_{ij} + \beta A_{ik} A_j^k + \gamma B_{ij}. \qquad [22]$$

Here μ, β and γ are material constants; A_{ij} and B_{ij} are the first two *Rivlin-Erickson* tensors

$$A_{ij} = v_{i,j} + v_{j,i} \qquad [23]$$

$$B_{ij} = \frac{D A_{ij}}{Dt} + A_{ik} v^k_{,j} + A_{jk} v^k_{,i} \qquad [24]$$

and the first normal stress difference is defined as

$$N_1(\tau) = T_{11} - T_{22}. \qquad [25]$$

For the second-order fluid, the correct normal thrust difference can be obtained by direct integration of eq. [10]

$$\Delta T_{rr} = \left(\frac{\beta + 2\gamma}{8}\right)\left(\frac{R_0}{\mu}\frac{dP}{dz}\right)^2 (k^2 - 1) \times \left[\left(\frac{k^2-1}{2k\ln k}\right)^2 - 1\right]. \qquad [26]$$

The measured thrust difference with holes can be determined by combination of eqs. [13], [21], and [26].

$$(\Delta T_{rr})_H = \left(\frac{\beta + \gamma}{8}\right)\left(\frac{R_0}{\mu}\frac{dP}{dz}\right)^2 (k^2 - 1) \times \left[\left(\frac{k^2-1}{2k\ln k}\right)^2 - 1\right]. \qquad [27]$$

The two results are therefore of opposite sign provided two inequalities are satisfied:

$$\beta + \gamma > 0, \quad \beta + 2\gamma < 0. \qquad [28]$$

[1]) It is assumed that the dependence of the hole error on wall curvature is negligibly small for the annulus used.

[2]) *Tanner* and *Pipkins* relation is supported by data up to a surprisingly large shear rate. *Kearsleys* relation (8) is not applicable. The results of *Higashitani* and *Pritchard* (5) are not so different from those of *Tanner* and *Pipkin* for dilute solutions and their data are not as persuasive. We assume the analysis is applicable to curved surfaces.

These lead to the requirement that

$$-1 < \frac{\gamma}{\beta} < -\frac{1}{2}. \qquad [29]$$

The material parameters are related to the normal stress differences by the expression

$$\frac{\gamma}{\beta} = -\frac{1}{2}\left(\frac{N_1}{N_1 + N_2}\right). \qquad [30]$$

For the data of *Ginn* and *Metzner* (3) on three solutions of polyisobutylene in decalin, this ratio lies within the range $-0.800 < \gamma/\beta < -0.560$ for shear rates ranging from 0.5 to 100 sec^{-1}. It is plausible therefore to conclude that pressure hole errors can alter both the sign and magnitude of the difference in wall thrust in annular flow.

Description of apparatus

The annulus consists of a sliding brass inner cylinder of 0.632″ diameter within a 0.870″ diameter brass pipe. Each is accurately honed to ensure a straight centerline and circular cross-section. The core is located concentrically within the outer tube by means of two sets of three fins positioned symmetrically at each end to minimize the deflection of the core due to its own weight. Concentricity within the 14″ test section is maintained to within ±0.001″. At each end of the test section, inlet lengths in excess of 50 channel widths provide for attainment of fully developed flow. A micrometer is used to impart and measure displacement of the inner cylinder to within ±0.0001″.

Two $^1/_{16}$″ diameter holes in the outer tube, each 4″ from the center of the test section, provide a means of measuring the pressure gradient with a differential pressure transmitter. The pressure holes used for measurement of thrust difference $(\Delta T_{rr})_H$ are $^1/_{16}$″ diameter and set 1″ from the center of the test section. The null-meter is similar to that described by *Tanner* (11). The pressure transducers are also set 1″ from the center of the test section, but on the opposite side of the center from the pressure holes. The transducers and pressure holes are mounted horizontally at 180° rotation from each other as shown in fig. 2.

The transducers are accurate to within ±200 dynes/cm^2 in the range 0–70000 dynes/cm^2 above a reference pressure. The reference pressures are maintained equal to each other and the pressure at the downstream pressure gradient tap by means of a hydraulic-pneumatic control system, thereby ensuring operation well within the specified limits.

The liquid flowrate is maintained by pressurized storage tanks of approximately 60 l capacity, a quantity sufficient for steady state to be achieved by both the fluid and the attendant instrumentation. Pressure-driven flow avoids mechanical degradation of polymers due to high shear rates in pumps. Constant flowrate is maintained by monitoring the pressure gradient and maintaining it constant by means of a hydraulic-pneumatic control system that adjusts the feed tank pressure. The test section is in a flow loop which enables flow reversal by movement of a single valve.

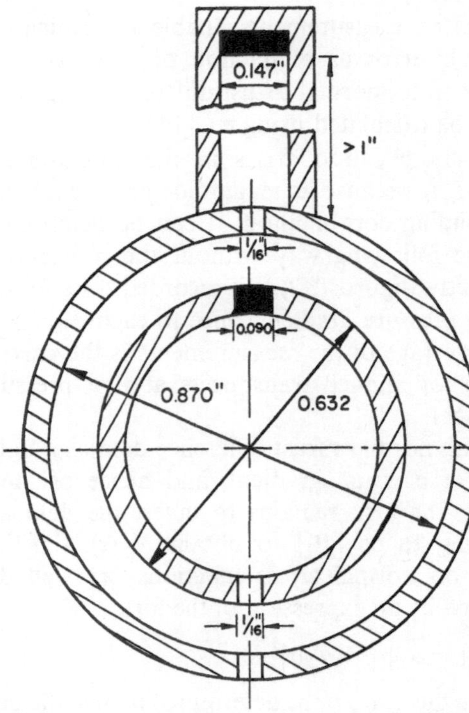

Fig. 2. Arrangement of pressure transducers (darkened areas) and pressure holes shown in cross-section of the annulus. Experimentally mounted in same horizontal plane

Temperature adjustment is achieved by heating (or cooling) coils in the feed storage tank. Thermometer wells are built into the feed and discharge tubes of the test section for bulk temperature measurements accurate to within ±0.05 °C.

Shear stress-shear rate data are obtained on a capillary viscometer with four different tubes: two diameters measuring 0.2261 and 0.3943 cm, each with two lengths of 18 and 40 cm. The diameters are determined by measuring the weight of mercury required to fill each capillary. Temperature control of the feed is maintained at 25.0 ± 0.1 °C.

Further information on the equipment is presented by *Lobo* (9).

Discussion of results

Data were obtained on two polymer solutions: 1% CMC[3]) in water and 1% CMC, 25% glycerin in water by weight.

Capillary data corrected for end effects are shown in fig. 3 for the aqueous CMC solution along with the *Ellis* model fit. Similarly good data and fit were obtained for the aqueous CMC-glycerin solution for which the *Ellis* parameters are $\eta_0 = 2.15$ poise, $n = 0.402$, and $\tau_{1/2} = 445$

[3]) Sodium carboxymethylcellulose, type 7H, provided by the Hercules Corporation.

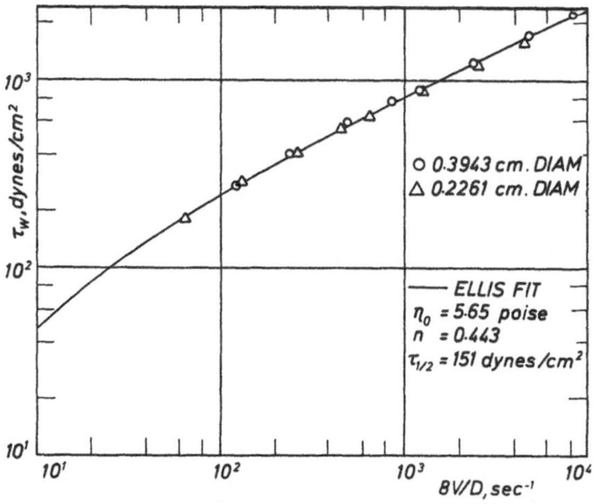

Fig. 3. Capillary data and *Ellis* Model fit for 1% aqueous CMC at 25 °C

dynes/cm^2. It is not essential to fit the data with a smooth curve, but it is necessary to provide for computations at shear rates approaching zero.

To determine the normal thrust difference using holes according to eq. [12] the null-point positions are recorded as a function of pressure gradient, a smooth curve is drawn through the data corresponding to null-point positions for pressure gradients in each direction, and the displacement between null-points, l, calculated as a function of pressure gradient by subtraction of one curve from the other. The normal thrust difference using transducers is then read directly, thereby enabling calculation of the hole error at the inner wall according to eq. [15] as a function

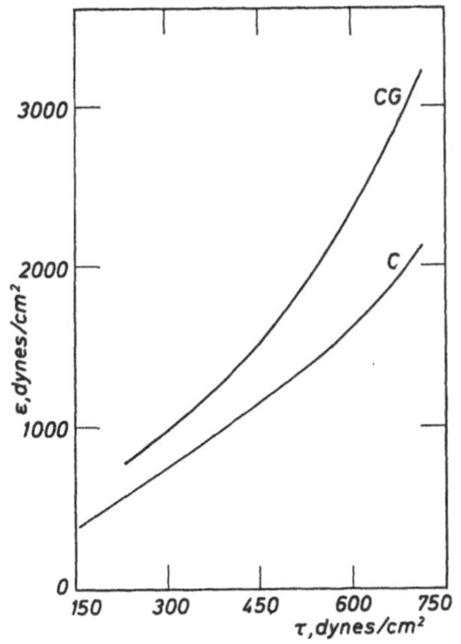

Fig. 4. Hole error as a function of shear stress for aqueous CMC (C) and aqueous CMC-glycerin (CG) solutions

of pressure gradient, and thus shear stress at the wall. The resulting hole error versus shear stress data are shown in fig. 4 for both solutions.

The hole error data are used in conjunction with the transducer data to compute the correct normal thrust difference, ΔT_{rr}, with eq. [14]. The three sets of normal thrust difference data are shown in fig. 5 for the aqueous CMC-glyerin solution. These clearly indicate that the neglect of pressure hole errors may indeed reverse the sign of ΔT_{rr}. Similar results were obtained for the aqueous CMC solution in that the hole errors were significant, but in this case the correction resulted in zero thrust difference within experimental error.

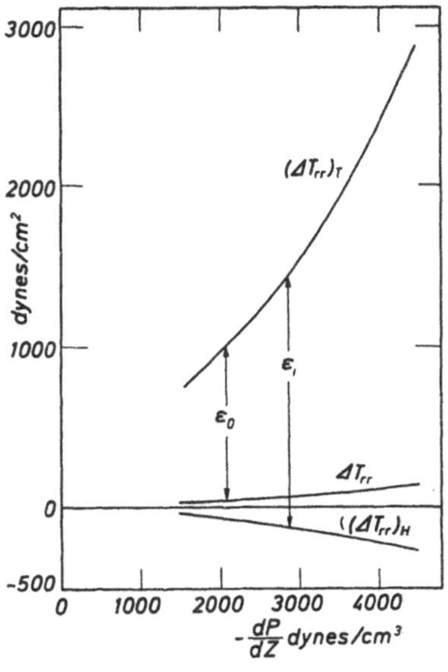

Fig. 5. Wall thrust differences for aqueous CMC-glycerin solutions at 24.5 °C

The second normal stress differences are shown in fig. 6. For the aqueous CMC-glycerin solution, the uncorrected pressure hole data, $(N_{2CG})_H$, differ in both sign and magnitude from the supposedly correct results, N_{2CG}. In the case of aqueous CMC the uncorrected result is significantly greater than zero, while the correct normal stress difference is computed to be zero within experimental error.

Experimental errors limit the accuracy of the present results to $\pm 50\%$ principally due to transducer error and instrument electronic interference. Even though these are relatively small, they are quite important because of the small magnitude of the second normal stress difference. This is, of course, true for any techni-

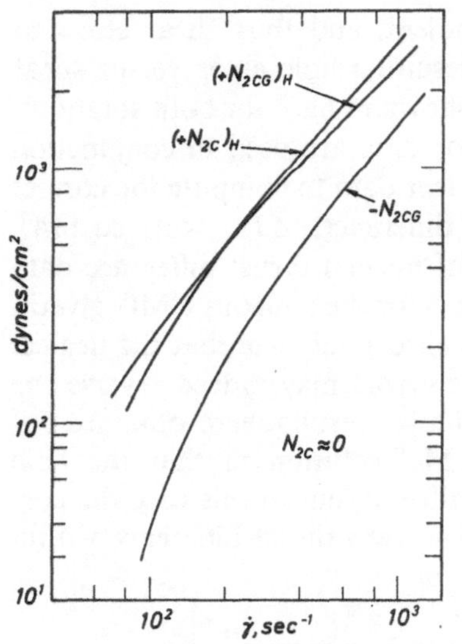

Fig. 6. Second normal stress difference as a function of shear rate for aqueous CMC using holes $(N_{2C})_H$ and corrected N_{2C}, and for aqueous CMC-glycerin using holes $(N_{2CG})_H$ and corrected N_{2CG} at 24.5 °C

que employed and so does not distract specifically from the utility of annular flow. On the contrary, annular flow seems to be the only method which offers the prospect of successful operation at high shear rates. It is necessary, however, to recognize disadvantages of annular flow, which include the requirement for large sample size, existence of temperature gradients (negligible in this study) and possible difficulty in purging the flow system of solution.

Conclusions

It has been demonstrated that pressure hole errors in annular flow measurements can be large enough to change the sign of the computed second normal stress difference. The results presented suggest that this method warrants continued development, and that reliable data may thereby be obtained over an extended shear rate range for both solutions and polymer melts.

Acknowledgments

Funds to initiate this work were provided by the National Science Foundation under Grant No. Gk-5156. Continued research was supported in part by the Office of Naval Research unter Contract No. N00014-68-A-0091. These funds are gratefully acknowledged.

Summary

For viscometric, axial, annular flow, the second normal stress difference N_2 is related to the difference in normal thrust across the annular space, ΔT_{rr}. Past attempts using this method have yielded values of N_2 for polymer solutions which are different in magnitude *and* opposite in sign from those obtained in other experiments. This inconsistency is attributed to errors resulting from the use of pressure holes in the measurement of ΔT_{rr}, and is supported by a second-order fluid analysis.

The present work focuses on the measurement of the effect of pressure hole errors on the determination of N_2 with aqueous polymer solutions. In the measurement of ΔT_{rr}, simultaneous use is made of both pressure holes and miniature pressure transducers to measure and account for pressure hole errors. Results indicate that hole errors are sufficiently large to reverse the sign of the computed N_2. This technique is therefore suggested as a preferred method for determining N_2, especially at high shear rates.

References

1) *Bird, R. B.*, Can. J. Chem. Engr. **44**, 161 (1965).

2) *Broadbent, J. M.* and *A. S. Lodge*, Rheology Research Center, Madison, Wisconsin, Report No. 6 (1971).

3) *Ginn, R. F.* and *A. B. Metzner*, Trans. Soc. Rheol. **13**, 429 (1969).

4) *Han, C. D.*, Amer. Inst. Chem. Eng. J. **18**, 116 (1972).

5) *Higashitani, K.* and *W. G. Pritchard*, sub judice (1971).

6) *Huppler, J. D.*, Trans. Soc. Rheol. **9**, 273 (1965).

7) *Kaye, A.*, *A. S. Lodge*, and *D. G. Vale*, Rheol. Acta **7**, 368 (1968).

8) *Kearsley, E. A.*, Trans. Soc. Rheol. **14**, 419 (1970).

9) *Lobo, P.*, Ph. D. Dissertation, Dept. of Chem. Engr., Univ. of Rochester (in preparation).

10) *Miller, M. J.* and *E. B. Christiansen*, Amer. Inst. Chem. Eng. J. **18**, 600 (1972).

11) *Tanner, R. I.*, Trans. Soc. Rheol. **11**, 347 (1967).

12) *Tanner, R. I.* and *A. Pipkin*, Trans. Soc. Rheol. **13**, 471 (1969).

13) *Tanner, R. I.* and *J. Williams*, Trans. Soc. Rheol. **14**: **1**, 19 (1970).

Authors' address:

P. Lobo and *H. R. Osmers*
Dept. of Chemical Engineering
University of Rochester
Rochester, New York 14627 (U.S.A.)

Rheol. Acta 13, 463–466 (1974)

Laboratoire National des Matières Grasses (I. T. E. R. G.) Université de Provence – Marseille (France)

Déterminations comparées de la teneur en solide des graisses plastiques par dilatométrie, analyse thermique différentielle et résonance magnétique nucléaire

Par E. Sambuc, G. Reymond et M. Naudet

Avec 5 figures et 2 tableaux

(Reçu p. p. le 27 october 1972)

La consistance des graisses plastiques est influencée par de très nombreux facteurs, mais parmi ceux-ci, la teneur en solide est incontestablement le plus important. Cette caractéristique est, en outre, facilement mesurable, ce qui accroît encore son intérêt.

Parmi les nombreuses méthodes proposées pour la détermination de la teneur en solide, les plus usuelles sont la dilatométrie (1), l'analyse thermique différentielle (ATD) (2) et la spectrométrie de résonance magnétique nucléaire à large bande (RMN) (3).

En ce qui concerne la dilatométrie (fig. 1) à n'importe quelle température comprise entre le début et la fin de fusion, la teneur en solide est donnée par le quotient de la distance du pointB (représentant le volume spécifique sur la courbe expérimentale) au point C de la droite liquide anticipée, par la distance, à cette même température, entre la droite solide prolongée, en A, et la droite anticipée liquide, en C.

La teneur en liquide est évidemment donnée par le quotient AB/AC.

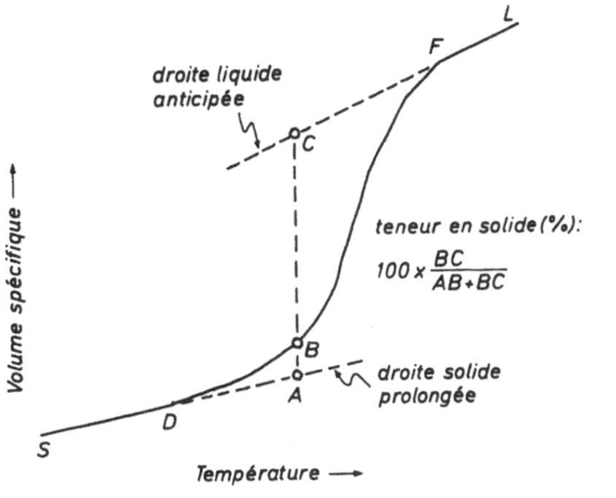

Fig. 1. Principe de la détermination de la teneur en solide par dilatométrie

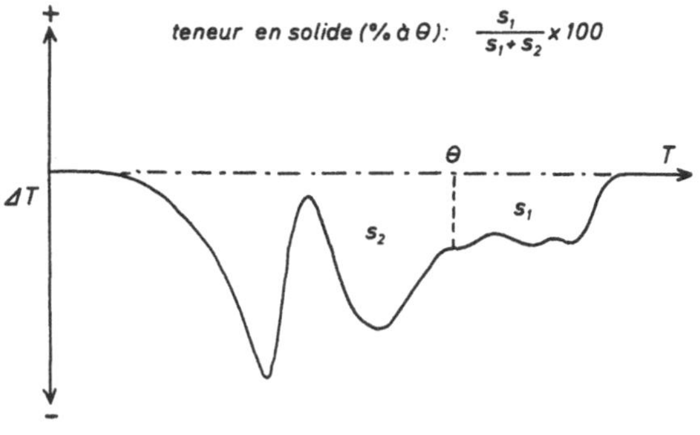

Fig. 2. Principe de la détermination de la teneur en solide par analyse thermique différentielle

Les valeurs ainsi déterminées ne sont qu'approximatives, puisqu'on admet, ce qui n'est pas rigoureusement exact, que tous les triglycérides du mélange ont les mêmes caractéristiques dilatométriques et que les dilatations de dissolution sont identiques aux dilatations de fusion.

La mesure de la teneur en solide nécessite la détermination d'au moins deux points de la droite solide et deux points de la droite liquide, en plus des mesures aux températures choisies. A chaque température, les lectures ne doivent être effectuées qu'après stabilisation, ce qui implique le maintien dans un thermostat pendant au moins 30 min si la différence entre deux températures consécutives ascendantes n'excède pas +5 °C.

Les calculs nécessaires pour transformer les lectures en teneurs en solide sont simples, mais relativement longs, si on veut tenir compte des corrections nécessaires (4).

Dans le cas de l'ATD (fig. 2), les teneurs en solide sont données par le quotient de l'aire (s_1) comprise sous le thermogramme (ΔT en fonction de la température) entre la température choisie et la température de fin de fusion par l'aire totale incluse sous le thermogramme.

La teneur en liquide est, elle, déduite du rapport

$$\frac{s_2}{s_1 + s_2}.$$

Dans ce cas aussi, les valeurs obtenues ne sont qu'approchées, puisqu'on admet que les chaleurs de fusion et de dissolution de tous les constituants de la phase grasse sont égales.

La détermination de la teneur en solide nécessite le tracé complet du thermogramme (ΔT en fonction du temps) et sa transformation en ΔT en fonction de la température. Le taux de chauffage doit être compris entre des limites généralement bien définies.

Les calculs sont longs et délicats; un seul échantillon peut être étudié à la fois.

Par RMN, on mesure uniquement la teneur en liquide. Dans ce but (fig. 3), l'intégration du signal dû aux protons "liquides" est comparée à la valeur que donnerait l'échantillon supposé entièrement liquide à la température considérée (E_T/E'_T).

L'égalité des teneurs en protons de tous les constituants du mélange est admise comme hypothèse de travail.

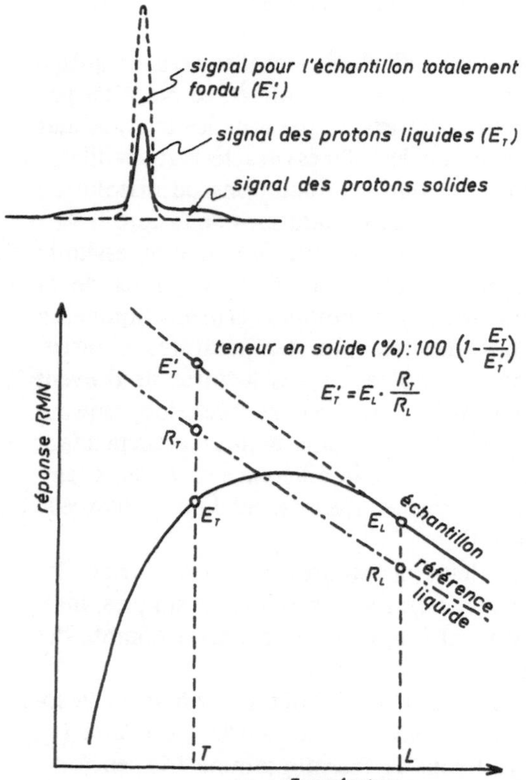

Fig. 3. Principe de la détermination de la teneur en solide par résonance magnétique nucléaire à large bande

Il semble préférable de faire la moyenne de plusieurs intégrations de temps court, plutôt que d'effectuer une seule longue intégration.

La mesure de la teneur en liquide nécessite les intégrations des signaux de l'échantillon et d'une référence totalement liquide à la température choisie, ainsi que celles de l'échantillon et de la référence à une température à laquelle l'échantillon est complètement fondu.

A chaque température, les intégrations ne doivent être faites qu'après stabilisation, ce qui implique, dans ce cas, un maintien de 45 min au thermostat, si la différence entre deux températures consécutives n'excède pas 5 °C.

On peut considérer deux types de traitements thermiques.

Le meilleur consiste à amener l'échantillon à une forme cristalline stable, par un traitement thermique convenable, après cristallisation partielle. Ce traitement est cependant toujours long.

On peut aussi effectuer un refroidissement rapide de l'échantillon fondu, mais cela conduit à la formation de cristaux métastables et les résultats obtenus peuvent, toujours par dilatométrie, souvent par ATD, ne pas être interprétables. Les résultats de la RMN sont, eux, toujours utilisables.

Une comparaison statistique des résultats obtenus par les trois méthodes après traitements thermiques identiques montre que leurs précisions sont sensiblement équivalentes: les teneurs en solide sont connues à $\pm 1\%$ près.

Si les résultats, pour ces trois méthodes, sont comparés pour les mêmes échantillons, les différences suivantes peuvent être notées (tableau 1):

– Après traitement thermique de stabilisation, les résultats de la dilatométrie et de l'ATD sont, dans les limites de précision, presqu'identiques. Ceci est dû au fait [montré par *Bailey* (5)] que les dilatations de fusion et les chaleurs de fusion de nombreux glycérides sont pratiquement proportionnelles.

– Les résultats obtenus par RMN sont toujours significativement plus faibles que ceux des deux autres méthodes. Ceci provient des hypothèses faites pour les calculs. Les différences entre dilatations de fusion de la plupart des glycérides sont supérieures aux différences de teneurs en protons de ces mêmes glycérides. Par conséquent la RMN donnera des valeurs probablement plus proches de la réalité.

– Après refroidissement rapide, sans stabilisation, les valeurs données par ATD sont différentes de celles obtenues après stabilisation;

Tableau 1. Comparaison entre teneurs en solide obtenues par 3 méthodes, avec ou sans stabilisation des formes cristallines

Echantillon	θ	Dilatométrie	ATD (s)	ATD (ns)	RMS (s)	RMN (ns)
Palmiste	*a*	75,2	74,1	69,2	64,6	75,1
	b	68,3	68,7	65,8	58,8	69,7
	c	61,2	61,8	60,4	51,9	60,9
	d	51,8	52,1	48,0	44,8	50,5
Palme	*a*	69,6	69,3	57,4	62,2	55,6
	b	55,5	55,4	49,6	51,7	50,6
	c	42,2	42,5	42,1	38,9	44,4
	d	33,7	34,2	37,9	31,2	37,9
Suif	*a*	83,6	84,7	57,9	57,4	70,0
	b	74,4	73,6	52,6	52,9	69,1
	c	64,7	64,3	47,8	46,8	66,7
	d	55,0	55,6	44,4	42,3	62,4
C 55	*a*	59,7	60,6	60,2	52,8	62,6
	b	49,6	49,6	51,7	45,2	53,2
	c	38,4	36,9	37,7	36,8	39,2
	d	24,7	24,8	22,4	24,0	23,6
CL 35	*a*	50,2	50,9	57,6	39,4	49,6
	b	40,3	40,0	47,2	31,8	41,1
	c	30,9	29,9	37,1	23,1	30,0
	d	21,7	22,2	26,9	16,0	20,7

θa 9,5 °C, *θb* 12,5 °C, *θc* 15,5 °C, *θd* 18,5 °C
s avec stabilisation, ns sans stabilisation

aucun sens de variation préférentiel ne peut être donné: ceci semble dû à la fusion et recristallisation de formes métastables.

– Plutôt curieusement, les valeurs données par RMN, après refroidissement rapide, sont supérieures à celles trouvées après stabilisation: on s'attendrait au contraire. Nous pensons qu'on peut expliquer ces résultats en supposant qu'une partie du liquide est incluse dans le réseau cristallin et qu'elle résonne alors peu ou pas du tout.

Les graisses alimentaires plastiques aussi bien pour l'usage domestique qu'industriel sont la plupart du temps des mélanges. Il serait, par conséquent, utile aux techniciens de pouvoir prévoir la teneur en solide d'un mélange à partir de la somme pondérée des teneurs en solide des constituants.

Une relation telle que $S = \Sigma s_i \cdot c_i$ ne peut exister à cause des phénomènes de dissolutions mutuelles.

A partir de l'étude (2, 3, 6) à plusieurs températures d'un grand nombre de mélanges différents des mêmes constituants en proportions différentes, nous avons pu mettre en évidence la relation expérimentale (fig. 4):

$$S = a \Sigma s_i \cdot c_i + b.$$

Les valeurs des coefficients *a* et *b* peuvent être déterminées à partir des seules valeurs concernant un mélange convenablement choisi (fig. 5). Ceci permet alors le calcul a priori des teneurs en solide des autres mélanges.

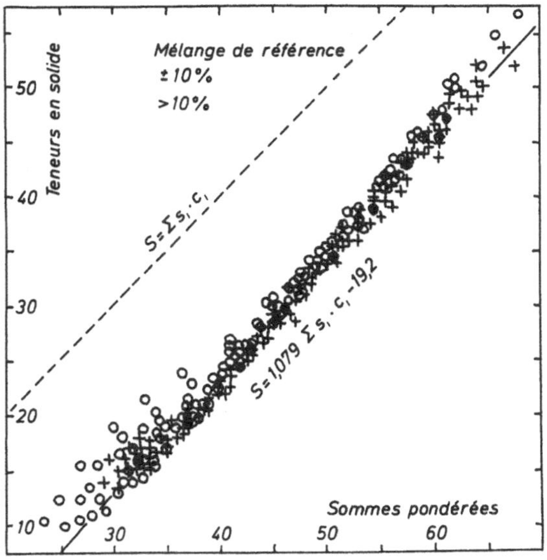

Fig. 4. Comparaison des teneurs en solide mesurées et des sommes pondérées des teneurs en solide des constituants pour 37 mélanges ternaires

Une comparaison statistique des valeurs mesurées et calculées conduit aux résultats suivants (tableau 2):

Si on considère l'ensemble des mélanges, les écarts types entre valeurs calculées et mesurées

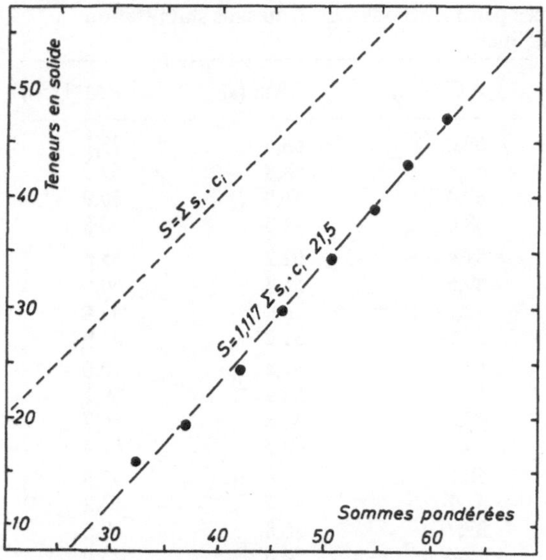

Fig. 5. Comparaison des teneurs en solide mesurées et des sommes pondérées des teneurs en solide des constituants pour le mélange de référence

Tableau 2. Ecarts-types entre teneurs en solide calculées et mesurées en fonction des écarts de composition par rapport au mélange de référence

	Avec stabilisation		Sans stabilisation	
	Dilato-métrie	RMN	RMN	ATD
Ensemble	2.13	1.61	2.49	3.55
Mélange référence	0.30	0.73	1.70	3.12
Ecart de composition				
5%	0.87	0.93	1.69	2.98
10%	1.73	1.33	1.94	3.16
17%	2.26	1.69	2.59	3.65
17%	3.26	2.36	3.60	4.49

sont faibles quelle que soit la méthode utilisée. Les écarts les plus faibles sont obtenus pour les produits ayant subi un traitement thermique de stabilisation; la RMN donne des écarts inférieurs à ceux de la dilatométrie. Les écarts sont d'autant plus grands que la composition du mélange considéré s'écarte plus de celle du mélange de référence.

En conclusion, on peut dire que la détermination de la teneur en solide peut être effectuée avec une précision équivalente par les 3 méthodes, mais que le temps passé en mesures et le nombre d'échantillons qu'il est possible d'étudier en même temps varient grandement d'une technique à l'autre. Un traitement de stabilisation de la cristallisation, quoique plus long, sera toujours préférable à un refroidissement rapide.

Il est possible de prévoir la teneur en solide d'un mélange à partir des caractéristiques des constituants avec une approximation suffisante pour pallier, par exemple, toute variation dans les caractéristiques de lots successifs.

Littérature

1) *Sambuc, E.* et *M. Naudet*, Rev. Fse Corps Gras **16**, 701 (1969).

2) *Sambuc, E., G. Reymond* et *M. Naudet*, Rev. Fse Corps Gras **18**, 215 (1971).

3) *Sambuc, E., G. Reymond* et *M. Naudet*, Rev. Fse Corps Gras **19**, 613 (1972).

4) *Sambuc, E.* et *M. Naudet*, Rev. Fse Corps Gras **14**, 725 (1967).

5) *Bailey, A. E.*, Melting and solidification of fats, p. 178 (New York 1950).

6) *Sambuc, E.* et *M. Naudet*, Rev. Fse Corps Gras **17**, 221 (1970).

Adresse des auteurs:
E. Sambuc, G. Reymond et *M. Naudet*
Laboratoire National des Matières Grasses (I.T.E.R.G.)
Université de Provence
Marseille (France)

Rheol. Acta 13, 467–476 (1974)

Laboratoire de Mécanique Appliquée, associé au CNRS, U.E.R. Besançon (France)

Etude physique des variations de longueur avec la torsion des métaux

Par G. Lallement et C. Oytana

Avec 12 figures

(Reçu p. p. le 27 octobre 1972)

1. Introduction

Dans le domaine élastique, la prise en considération des non linéarités géométriques (introduction de termes différentiels du second ordre dans le tenseur *Lagrangien* des déformations) et physiques (présence de termes linéaires et quadratiques dans les relations entre contraintes et déformations) ont permis à *Murnaghan* (1) et à *Reiner* (2) de donner une interprétation des interactions entre déformations angulaires et axiales.

En particulier celles-ci produisent, lors d'un essai de torsion, des variations de dimensions longitudinale et transversale de l'éprouvette mises en évidence par *Poynting* (3) et étudiées récemment par *A. Foux* (4).

Une déformation cyclique en torsion est également susceptible de provoquer un allongement cumulatif et ce phénomène nouveau a été étudié de manière très complète sur l'aluminium par *M. Ronay* (5). Depuis lors, les publications relatives à ces problèmes présentent quelques résultats expérimentaux et proposent des interprétations essentiellement phénoménologiques. Elles consistent en effet à déterminer une loi de comportement du matériau permettant de retrouver les résultats expérimentaux. Elles reposent, pour l'effet *Poynting*, sur l'introduction de coefficients élastiques du 3ème ordre et conduisent à un allongement relatif de l'éprouvette $\varepsilon_{zz(e)}$ proportionnel à φ^2, φ étant l'angle de torsion unitaire.

Les expériences effectuées mettent en évidence la présence d'un cycle d'hysteresis et des écarts importants aux prévisions théoriques. Nous nous sommes en conséquence attachés à déterminer d'autres mécanismes physiques susceptibles de donner lieu à un couplage entre les déformations angulaires et axiales et à vérifier leur action par l'expérience.

2. Origines physiques du couplage entre déformations angulaires et axiales

2.1. *Elasticité au second ordre*

Dans un solide isotrope, l'introduction des coefficients élastiques du troisième ordre *l*, *m* et *n* conduit à une expression de la densité d'énergie de la forme:

$$W = \tfrac{1}{2}(\lambda + 2\mu) I_1^2 - 2\mu I_2 + \tfrac{1}{3}(l + 2m) I_1^3 - 2m I_1 I_2 + n I_3$$

où λ, μ désignent les coefficients élastiques du second ordre; I_1, I_2, $I_3 = \det|\varepsilon_{ij}|$ les trois invariants tensoriels des déformations. Les non linearités geométriques s'introduisent dans le tenseur *Lagrangien* des déformations en écrivant:

$$\varepsilon_{ij} = \tfrac{1}{2}(u_{i,j} + u_{j,i} + u_{k,i} \cdot u_{k,j}).$$

Suivant *Murnaghan* (1), la déformation axiale $\varepsilon_{zz(e)}$ d'un fil à section circulaire de rayon R et de longueur L soumis à une torsion d'angle θ et à une contrainte axiale n est donnée par

$$\varepsilon_{zz(e)} = \frac{n}{E} + C_L \gamma^2$$

où le second terme de droite représente l'effet *Poynting* et E désigne le module d'*Young*,

$$\gamma = \frac{R\theta}{L}, \quad C_L = -\frac{4\mu(\lambda + 2\mu) + 4\lambda m + \lambda n}{16\mu(3\lambda + 2\mu)}.$$

Il faut noter que, par analogie avec les phénomènes du premier ordre, on peut admettre la présence d'une elasticité retardée au second ordre. Les conséquences en seront l'existence d'un ε_{zz} instantané et d'un ε_{zz} relaxé et l'observation d'un cycle d'hystérésis de surface dépendant de la fréquence du cyclage.

2.2. *Microglissements*

L'interprétation précédente est insuffisante dans la plupart des cas. En effet, on observe fréquemment un hystérésis ne pouvant être attribué à une élasticité retardée au second ordre (il n'y a pratiquement pas de différence entre le C_L instantané et le C_L relaxé). Enfin la loi en γ^2 prévue par *Murnaghan* est souvent mise en défaut.

L'allongement cumulatif (effet *Ronay*) et l'influence de l'écrouissage (voir résultats expérimentaux) donnent à penser qu'une part importante des variations de longueur est due aux dislocations. Celles-ci, sous l'action des contraintes se déplacent sur des plans de glissements bien déterminés. Le champ de contrainte s'exerçant sur une dislocation au cours d'un essai de torsion provoque un déplacement de celle-ci qui induira un cisaillement dans le sens de la sollicitation. Mais, le plan de glissement n'étant pas perpendiculaire à l'axe de torsion il se produira également une variation de longueur. Statistiquement, les allongements et les raccourcissements se compensent pour donner:

$$\varepsilon_{zz} = 0.$$

Cependant considérons un segment de dislocation de longueur l ancré à ses deux extrémités; si on le soumet à une force F par unité de longueur, il prend la forme d'un arc de cercle de rayon $r = W/F$ où W désigne la tension de ligne. En posant $x = l/r = F\,l/W$, la surface balayée par le segment de dislocation est donnée par:

$$S = l^2 f(x)$$

avec

$$f(x) = \frac{1}{x^2}\arcsin x - \frac{1}{2x}\left[1 - \frac{x^2}{4}\right]$$

($x < 2$ dans le domaine des déformations réversibles), fonction dont la variation est représentée sur la fig. 1.

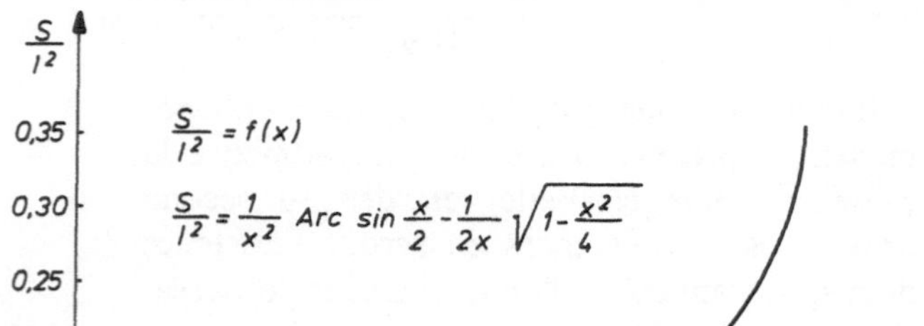

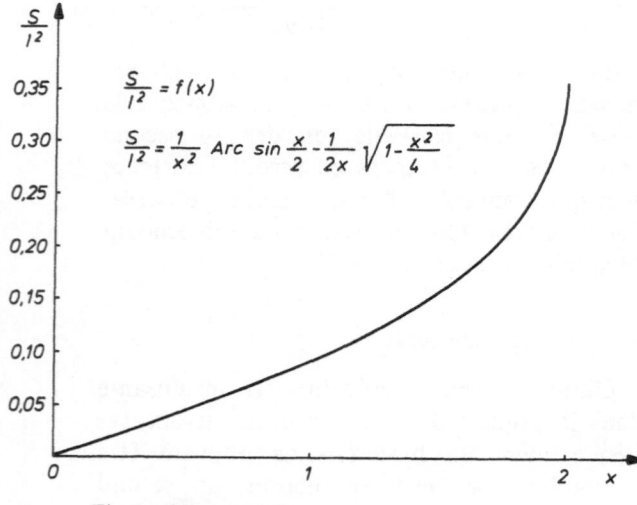

Fig. 1. vide texte

Si α, β, γ représentent les cosinus directeurs du plan de glissement de la dislocation dans le domaine déformations réversibles, $\alpha_b, \beta_b, \gamma_b$ ceux du vecteur de *Burgers* b, et $N(\alpha\beta\gamma, \alpha_b\beta_b\gamma_b)$ le nombre de segments par unité de volume ayant même système de glissement, la déformation axiale induite par cette famille de segments a pour expression:

$$d\varepsilon_{zz}(\alpha, \beta, \gamma, \alpha_b, \beta_b, \gamma_b) = N b \gamma \gamma_b l^2 f(x).$$

Si une tension préalable est imposée à l'éprouvette il va en résulter un effet de polarisation. Soit deux segments de dislocation ayant des systèmes de glissement tels qu'ils produisent respectivement au cours d'un essai de torsion pure des déformations $d\varepsilon_{zz}$ et $-d\varepsilon_{zz}$ (compensation). Si l'on applique sur l'éprouvette une tension préalable à la torsion, la variation de force par unité de longueur de dislocation sera la même pour les deux segments mais, le point de départ n'étant plus $F = 0$, les surfaces balayées par les deux segments ne sont plus égales et la compensation des déformations longitudinales ne se produit plus, Il en résulte un allongement. Le raisonnement reste valable dans le cas d'une compression préalable.

On peut calculer cet allongement en intégrant les $d\varepsilon_{zz}$ de tous les systèmes de glissements possibles et il est donc nécessaire de connaître la densité de probabilité $p(\alpha, \beta, \gamma, \alpha_b, \beta_b, \gamma_b)$. En fait, les relations existant entre ces six paramètres permettant de les réduire à trois. β, γ et θ, θ étant l'angle du vecteur de *Burger* du segment de dislocation avec la projection de l'axe de l'éprouvette sur le plan de glissement. Les calculs, relativement longs ne seront pas exposés ici, on trouve en particulier dans le cas d'une éprouvette isotrope:

$$\varepsilon_{zz} = \int_{\gamma=-1}^{+1} \int_{\beta=-\sqrt{1-\gamma^2}}^{+\sqrt{1+\gamma^2}} \int_{\theta=0}^{\Pi} \frac{N_0 b l^2 \gamma \sqrt{1-\gamma^2}}{2\Pi \sqrt{1-\beta^2-\gamma^2}} \cos\theta$$

$$\times \left| f\left(\frac{lF}{W}\right) - f\left(\frac{l}{W}(n\gamma\sqrt{1-\gamma^2}\cos\theta)\right)\right|$$

$$\times\, d\theta\, d\beta\, d\gamma$$

où N est la densité de segments de dislocations et

$$F = \frac{bt}{\sqrt{1-\gamma^2}}$$

$$\times \left|\beta(1-2\gamma^2)\cos\theta - \gamma\sqrt{1-\beta^2-\gamma^2}\sin\theta\right|$$

$$+\, b n \gamma \sqrt{1-\gamma^2}\cos\theta$$

pour une tension axiale n et un cisaillement t superposés.

Ce mécanisme est en fait plus général et plus complexe. Tous les facteurs susceptibles de perturber le mouvement des dislocations peuvent intervenir et entrainer une modification de l'allongement ou une irréversibilité de celui-ci. Il est également possible de l'étendre au domaine plastique. Enfin les théories de l'accommodation peuvent par ce biais, donner une intepprétation physique de l'effet *Ronay*.

2.3. Influence de la magnétostriction

La déformation axiale observée dans les matériaux ferromagnétiques est de la forme:

$$\varepsilon_{zz} = \varepsilon_{zz}(e) + \varepsilon_{zz}(m)$$

où $\varepsilon_{zz}(m)$ d'origine magnétostrictive peut être du même ordre de grandeur que la déformation élastique $\varepsilon_{zz}(e)$.

Ce phénomène est lié à la déformation accompagnant l'aimantation spontanée I_S. L'application d'une torsion modifie l'orientation de I_S, donc celle de l'ellipsoïde des déformations et il peut en résulter une déformation axiale.

L'expression de cette déformation dépend de la structure du matériau et des différents facteurs intervenant dans l'énergie totale. Nous ne présenterons ici que les résultats concernant un materiau de structure cubique, d'énergie magnétocristalline négligeable et de magnétostriction isotrope λ_S, dont le nickel fournit un bon exemple (6).

En l'absence de champ magnétique, la déformation $\varepsilon_{zz}(m)$ donnée par:

$$\varepsilon_{zz}(m) = \frac{3}{2}\lambda_S \left| 1 + \frac{4u^2}{|1 - (1 + 4u^2)^{1/2}|^2} \right|^{-1}$$

où

$$u = \frac{t}{n}$$

est représentée sur la fig. 2.

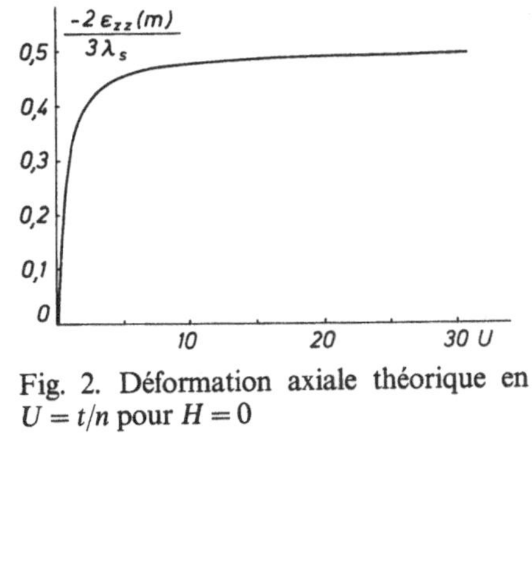

Fig. 2. Déformation axiale théorique en fonction de $U = t/n$ pour $H = 0$

L'application d'un champ magnétique axial H modifie la mobilité de I_S donc $\varepsilon_{zz}(m)$. Dans le cas particulier envisagé ici, la variation de $\varepsilon_{zz}(m)$ en fonction de H a pour expression:

$$\varepsilon_{zz}(m) = \frac{3}{2}\lambda_S \left| \frac{1}{2} - \frac{v^2}{32}\left(1 - \sqrt{1 + \frac{32}{v^2}}\right) - \left(\frac{tv}{2n}\right)^2 \right|$$

où

$$v = \frac{2I_S H}{3\lambda_S t}.$$

3. Equipement de mesure (fig. 3)

L'éprouvette est un fil cylindrique à section circulaire. Son extrémité supérieure est encastrée dans un mandrin lié à un portique fixe. L'extrémité inférieure est solidaire d'une armature de condensateur placée en regard d'une armature fixe. La variation de capacité provoquée par l'allongement (ou le raccourcissement du fil) entraine une

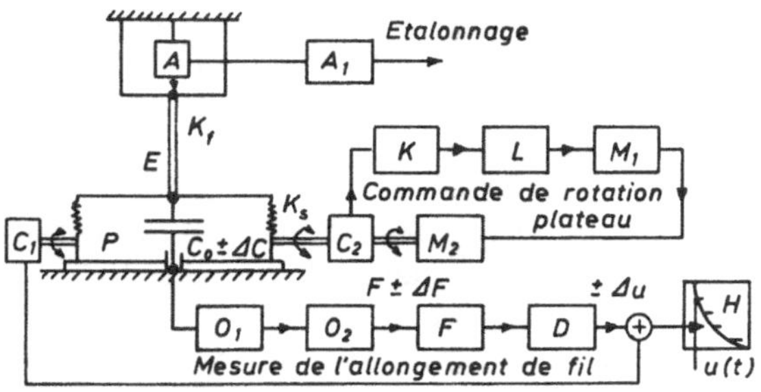

Fig. 3. Schéma de montage. E: Eprouvette; P: plateau tournant; A, A_1: étalonnage (capteur de déplacement à transformateur différentiel); C_1: capteur de position angulaire plateau (marqué enregistrement 2 impulsions/tour); O_1, O_2: oscillateur et circuit d'entretien; F, D: fréquencemetre et décodeur digital analogique; H: enregistreur $\Delta U(t)$; C_2: capteur de position angulaire plateau (120 impulsions/tour); K, L: compteur prédéterminé et circuit logique; M_1, M_2: asservissement de vitesse, moteur couple et dynamo tachymétrique

variation de la fréquence d'un oscillateur. L'ensemble présente une sensibilité de 150 Hz pour une déformation axiale de 10^{-6} et la stabilité $\Delta f/f$ est de $2 \cdot 10^{-6}$/h. Le mandrin fixé à l'extrémité inférieure de l'échantillon est lié à une membrane métallique de raideur axiale K_S très petite devant celle du fil K_f; inversement sa raideur en torsion est grande vis-à-vis de celle de l'éprouvette. La membrane porte à sa partie inférieure l'armature mobile plane en face de laquelle se trouvę une armature identique solidaire du bâti. La transmission du déplacement angulaire au fil est assuré par l'extérieur de la membrane, solidaire d'un plateau diviseur commandé par un asservissement de vitesse. Le dispositif de commande du plateau permet de réaliser les fonctions suivantes:

– Application du cyclage final entre les valeurs $\pm\theta$ par montée symétrique à pas réglable. La séquence des opérations:

$$\frac{\Delta\theta}{2}; \quad -\left(\frac{\Delta\theta}{2} + \Delta\theta\right); \quad \ldots;$$

$$(-1)^p\left(\frac{\Delta\theta}{2} + \Delta\theta\right)\ldots; \quad \theta.$$

permet d'éviter un écrouissage dissymétrique.

– Cyclage automatique entre les valeurs extrêmes $\pm\theta$ avec comptage du nombre de cycles.

Les éprouvettes à l'exception des fils de tungstène, subissent au minimum une passe de tréfilage en filière diamant permettant d'assurer que le défaut de circularité n'excède pas quelques %. Ils sont recuits dans une enceinte verticale sous vide (pression résiduelle 10^{-4} torr) par passage de courant. Ils peuvent être écrouis en traction et en torsion avec la possibilité de relever dans ces deux cas les courbes effort-déformation. Pour les mesures en torsion on intercale dans l'équipement de mesure de l'effet *Poynting* un capteur de couple à jauges résistantes de linéarité 10^{-2} en série avec l'éprouvette. Il est à noter que de très faibles écarts de rectitude du fil suffisent pour le remplacer par une courbe gauche. En assimilant celle-ci à une hélice et en désignant par d et n le diamètre moyen d'enroulement et le nombre de spires, la longueur libre devient:

$$L \simeq L_0\left(1 - \frac{n^2 \Pi^2 d^2}{2L_0^2}\right).$$

Un tel défaut entraîne:

– un décalage d'origine en γ sur les courbes $\varepsilon_{zz}(\gamma)$;

– un déplacement axial avec la déformation angulaire provoqué par l'enroulement de cette hélicoïde à spires laches et dépendant fortement de l'effort axial F.

Le défaut est en général corrigé par le recuit effectué après le passage en filière. Une légère tension axiale est en effet maintenue sur le fil pendant le traitement thermique. Au cours de l'essai de déformation axiale $\varepsilon_{zz}(\gamma)$, une tension axiale F rigoureusement constante est également appliquée sur l'échantillon par l'intermédiaire d'un jeu de surcharges solidaires de la membrane.

4. Résultats expérimentaux

4.1. Déformation axiale cumulative

La déformation axiale élastique pure (effet *Poynting*) est en général accompagnée d'une déformation cumulative (effet *Ronay*). On observe en effet un allongement permanent croissant avec le nombre de cycles et se développant dans les régions de $\sigma_{z\theta}(\gamma)$ pour lesquelles $\sigma_{z\theta}\, d\sigma_{z\theta} > 0$, résultat en accord avec l'analyse théorique de *A. Freudenthal* (7).

L'application d'une torsion superposée à la traction ne produit à elle seule aucune variation d'allongement avec le temps (fluage) et la déformation axiale mesurée résulte essentiellement de la déformation plastique angulaire cyclique (balayage des plans de glissement par la direction variable de contrainte de cisaillement maximale). Le phénomène peut également présenter une dissymétrie ainsi que le montrent les expériences sur le molybdène (fig. 7). Dans ce cas, celle-ci se produit du même côté que celle de l'effet non cumulatif et est en général d'amplitude relative plus importante. L'existence d'une structure hélicoïdale interne résultant du tréfilage en est la cause la plus probable.

La déformation axiale cumulative a été particulièrement étudiée pour le *Nickel* pur (impuretés <20 ppm) recuit à 500°. Dans le domaine étudié, l'effet est sensiblement proportionnel à la déformation angulaire (fig. 4) et au logarithme du nombre de cycles mais présente une saturation à partir de 200 cycles. De plus, sa variation en fonction de la contrainte axiale σ_{zz} reportée sur la fig. 5 montre un extremum dû à l'influence

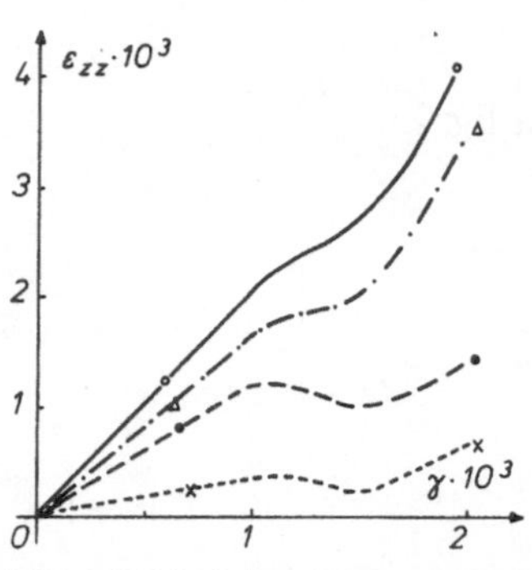

Fig. 4. Nickel Ø 0,45 mm recuit 1 H A 500°: déformation axiale cumulative pour $N = 100$ cycles; $\varepsilon_{zz}(\gamma)$ en fonction de $\sigma_{zz} \cdot 10^{-7}$ N/m²: × $\sigma_{zz} = 1{,}4$; ● $\sigma_{zz} = 2{,}6$; △ $\sigma_{zz} = 3{,}8$; ○ $\sigma_{zz} = 6.2$

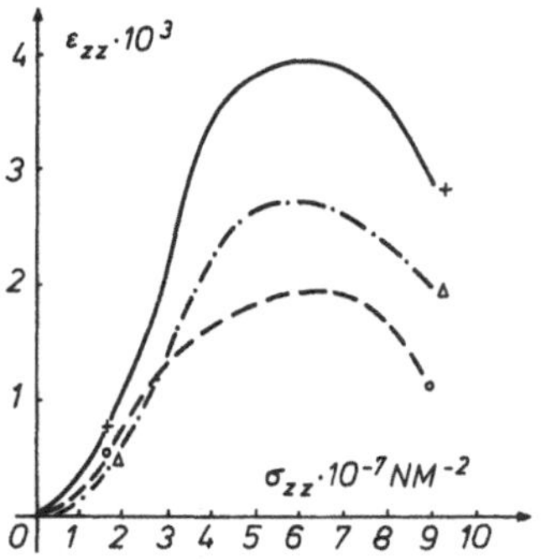

Fig. 5. Nickel ∅ 0,45 mm recuit 1 H A 500°: déformation axiale cumulative pour $N = 100$ cycles; $\varepsilon_{zz}(\sigma_{zz})$ en fonction de $\gamma\cdot 10^3$: ○ $\gamma = \pm 1$; △ $\gamma = \pm 1{,}5$; + $\gamma = \pm 2$

croissante de l'écrouissage axial qui entraine une diminution de l'allongement. Le comportement général du *Nickel* diffère donc totalement de celui de l'aluminium mais présente plusieurs points communs avec les phénomènes observés sur le fer armco (8). Ce dernier possède en effet une caractéristique $\sigma_{z\theta}(\gamma)$, de nature très voisine et l'interprétation développée par *Freudenthal, A. M.*, s'applique à notre cas. Le caractère commun essentiel réside en la forte réduction (ou annulation observée également sur le *Nickel*) de l'allongement cumulatif pour les faibles charges axiales et déformations angulaires modérées ($\gamma \simeq 1{,}5 \cdot 10^{-3}$). Ce phénomène a également été observé par *Swift*, dans l'acier (9).

Enfin l'allongement cumulatif n'est pas entièrement réversible. Après cyclage en γ, une réduction ou suppression de la tension F provoque une recouvrance partielle (10%). Celle-ci apparaît également spontanément lors du repos consécutif à un cyclage ou au cours d'une déformation alternée en torsion précédée par un écrouissage en traction.

4.2. Déformation axiale non cumulative

4.2.1. Tungstène (fig. 6)

Il apparaît même aux faibles déformations ($\gamma = 5\cdot 10^{-4}$) un cycle d'hystérésis de surface faible, analogue à celui accompagnant les effets du premier ordre et pratiquement dépourvu d'accommodation (pour $N nb$ de cycles $\simeq 10^3$). La déformation axiale devant être indépendante du signe de γ, on obtient un effet de redressement par rapport aux cycles habituels. La faible lar-

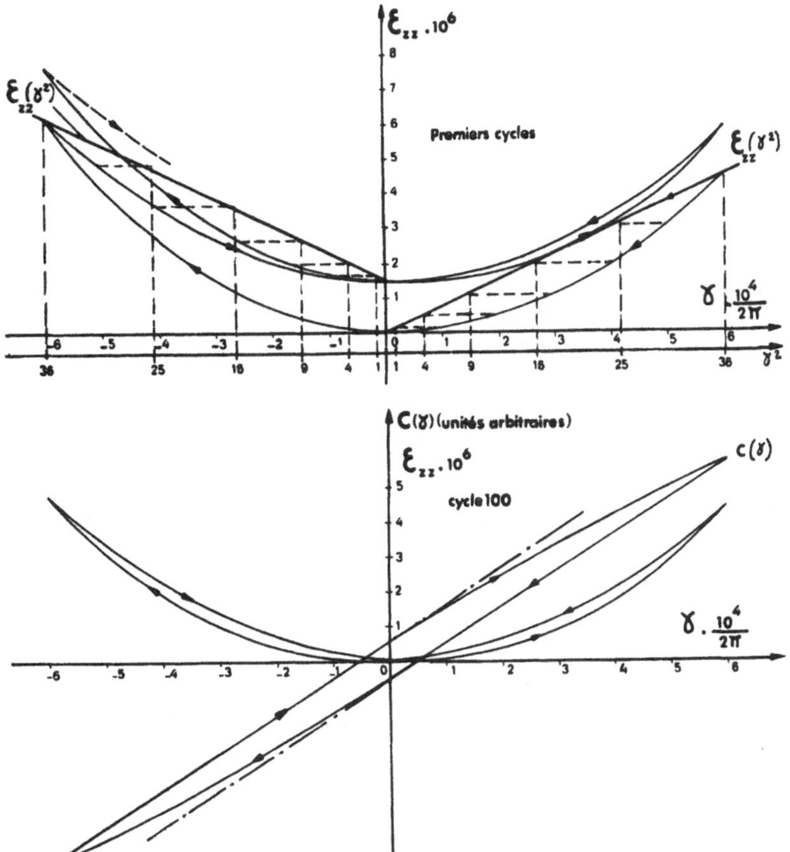

Fig. 6. vide texte

geur du cycle et la proportionnalité entre ε_{zz} et γ^2 permettent d'attribuer le phénomène à l'élasticité au second ordre avec $C_L \simeq 0{,}3$.

La déformation axiale présente fréquemment une faible dissymétrie. Exprimée en décalage d'origine équivalent γ_0, cette dernière n'excède pas en moyenne quelques 10^{-4}. Aucune corrélation n'apparaît entre les dissymétries de déformation axiale et de limite élastique relevées sur les courbes d'écrouissage en torsion avec mise en déformation symétrique à pas faible. L'application d'un écrouissage dissymétrique important correspondant à $\gamma = 10^{-2}$ n'apporte du reste aucune modification à la dissymétrie de déformation axiale (structure hélicoïdale due au tréfilage).

La mesure précise du coefficient C_L est affectée par la présence des phénomènes hysterétiques. La largeur relative du cycle:

. croit avec la température de recuit;

. diminue peu avec l'accommodation en torsion.

Pour une déformation maximale γ_M, la courbe d'effet *Poynting* tracée à partir de la moyenne des allongements relevés à déformation angulaire croissante et décroissante est pratiquement indépendante de la déformation maximale appliquée.

La déformation axiale cyclique apparaît nettement à partir d'une température de recuit à 800°. Pour une déformation angulaire modérée ($\gamma_M < 4 \cdot 10^{-3}$ et $\sigma_{zz} = 1{,}5 \cdot 10^7$ N/m²) l'allongement irréversible par cycle est d'amplitude faible ($3 \cdot 10^{-1}$ de l'effet *Poynting* lors des premiers cycles). On n'observe pas de différence notable entre la déformation réversible à cycle ouvert et à cycle fermé, c'est-à-dire après disparition de l'allongement cumulatif: l'amplitude du terme de couplage élasto-plastique est dans ce cas particulier négligeable.

Mesure de C_L

a) Influence de l'effort axial: sur un fil de ∅0,3 mm, soumis à $\gamma_M < 4 \cdot 10^{-3}$ et une contrainte axiale $3 \cdot 10^7 < n < 2 \cdot 10^8$ N/m², la variation de C_L après séparation des termes pairs de la déformation axiale n'excède pas $4 \cdot 10^{-2}$.

b) Influence de la vitesse de déformation: dans les mêmes conditions que précédemment et pour un domaine de variation de la vitesse de déformation: $10^{-6} < \mathring{\gamma} < 10^{-4}$ sec^{-1}, le coefficient C_L ne subit aucune modification.

c) Influence de l'écrouissage: pour un matériau de même provenance, la valeur de C_L présente des variations sensibles avec le diamètre du fil. Pour essayer d'expliquer cette observation, des mesures en fonction du taux d'écrouissage ont été tentées. L'augmentation de la température de transition fragile-ductile consécutive au traitement de restauration limite l'écrouissage à de faibles taux (relèvement de la limite élastique de 30%). Il en résulte une réduction de C_L de quelques %. La dépendance entre les coefficients *m* et *n* et l'écrouissage semble donc faible. Cet écrouissage est toutefois suffisant pour bloquer l'allongement cumulatif *Ronay* auquel se substitue une «recouvrance».

4.2.2. Molybdène (fig. 7)

La complexité du comportement de ce matériau est intéressante sous deux aspects:

1. Aux faibles déformations ($\gamma_M < 1{,}5 \cdot 10^{-3}$), l'allongement axial suit sensiblement une loi en γ^2 et dépend fortement de l'écrouissage:

. à l'état écroui de tréfilage: $C_L = 0{,}43$;

. à l'état recuit (1100° pendant 1 h): $C_L = 0{,}22$.

2. L'absence de fragilité à l'état recuit permet de poursuivre les essais dans le domaine plastique. Les cycles d'hysteresis des courbes $\varepsilon_{zz}(\gamma)$ présentent les caractères suivants:

– Largeur relative du cycle, indépendante de $\mathring{\gamma}$, croissante avec la température de recuit mais diminuant avec l'accommodation.

– Tendance à l'égalité des pentes à l'origine sur le trajet aller et au point d'inversion de la vitesse de déformation interprétée comme un effet elastoplastique du second ordre. Ce phénomène se traduit par deux types de courbes. Dans le domaine élastique ou pendant le trajet aller la pente à γ_M est supérieure à la pente initiale et la déformation axiale retour est supérieure à celle de l'aller (cas du *W*). Dans le domaine plastique on tend vers l'effet contrainte. Le *Nickel* auquel ont été appliqués des écrouissages en traction croissants montre également la continuité du phénomène.

Les courbes d'allongement à déformation angulaire croissante reportées sur la fig. 7 présentent une réduction puis une inversion de pente. On a représenté également cette même déformation en fonction du couple de torsion correspondant. Un tel phénomène de saturation a également été observé par *Swift* (9) sur le cuivre pur mais il n'apparaissait pas dans ce cas d'inversion du signe de la déformation.

Les phénomènes observés sont certainement à rattacher au mouvement des dislocations dont

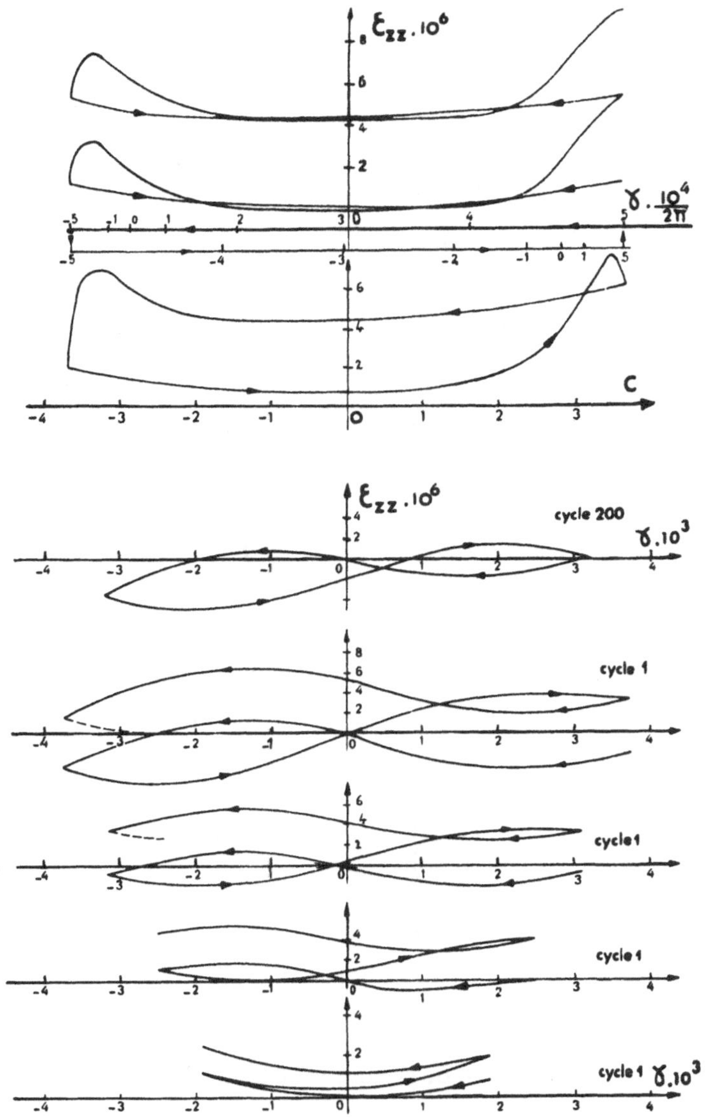

Fig. 7. vide texte

on a vu qu'il peut conduire à un allongement. Les propriétés des pentes du cycle d'hystérésis présentent en effet des similitudes avec celles des courbes d'écrouissage. Quant à l'effet de saturation, il peut s'expliquer qualitativement de la manière suivante: à *n* constante, la normale au plan de scission maximum se rapproche de l'axe de l'éprouvette. Certains plans de glissement dont la normale est proche de Oz peuvent alors se transformer en source de *Frank-Read* et emettre des boucles de dislocations susceptibles, par le jeu des interactions dislocation-dislocation, de bloquer les segments se trouvant sur les autres plans. Or l'allongement dû à une dislocation est d'autant plus faible que la normale de son plan de glissement est voisin de Oz. Le résultat, variable selon la scission critique, peut être une saturation ou une diminution de ε_{zz}.

4.3. Déformation axiale d'origine magnétostrictive: cas du Nickel

4.3.1. Déformation à champ magnétique nul

La restauration complète des propriétés magnétiques nécessite un recuit prolongé à température élevée (5–10 h à 900°). Il en résulte un important allongement cyclique cumulatif et la mesure précise de ε_{zz} non cumulatif ne peut être obtenue qu'après une accommodation de plusieurs centaines de cycles produisant à son tour un nouvel écrouissage. Ces difficultés conduisent à un compromis et expliquent le choix de la température de recuit habituellement utilisée (500 à 600 °C pendant 1 h.

La mesure de la magnétostriction axiale en fonction du champ magnétique et dans les différents états du matériau pour lesquels on étudie

$\varepsilon_{zz}(\gamma)$, montre que l'augmentation de la température de recuit n'apporte, au dessus de 500 °C, qu'une faible augmentation de la déformation axiale. On doit en particulier signaler l'effet de l'accommodation due au cyclage en torsion:

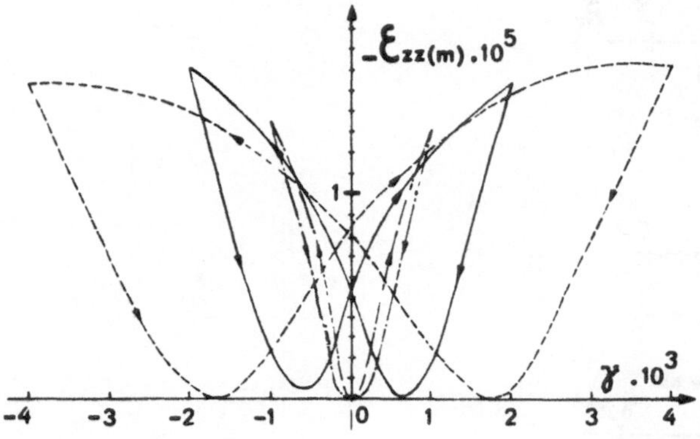

Fig. 8. vide texte

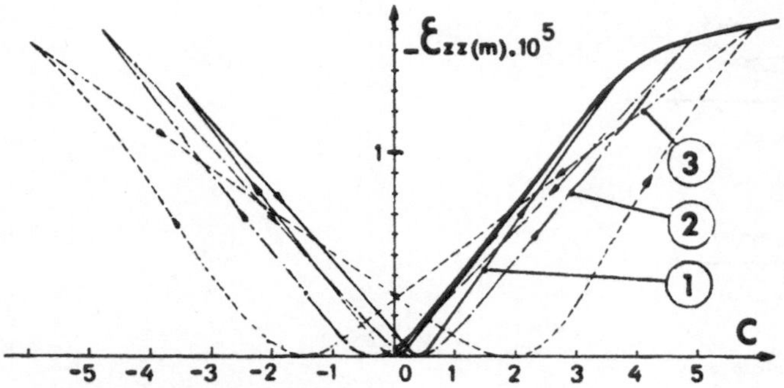

Fig. 9. vide texte

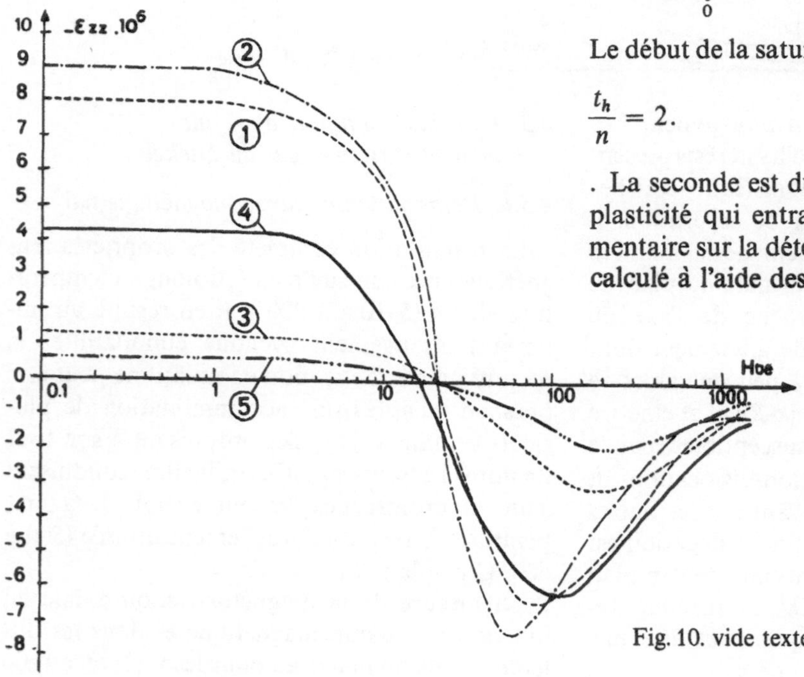

Fig. 10. vide texte

après application d'un écrouissage axial, celui-ci contribue à la relaxation des contraintes internes de traction et relève fortement la déformation magnétostrictive.

La déformation d'origine magnétostrictive est déterminée par:

$$\varepsilon_{zz(m)} = (\varepsilon_{zz})_{H=0} - (\varepsilon_{zz})_{H_1}$$

avec

$$H_1 = 1000 \text{ Oersteds.}$$

On observe, conformément aux résultats théoriques (figs. 8 et 9):

a) L'existence d'une déformation $\varepsilon_{zz(m)}$ négative dont la limite pour $t/n \to \infty$ doit être voisine de: $0{,}75\,\lambda_S = -3 \cdot 10^{-5}$ (valeur expérimentale déduite des mesures de magnétostriction axiale).

b) Une déformation $\varepsilon_{zz(m)}$ proportionnelle à t pour les faibles valeurs de celle-ci et accusant ensuite une saturation. La détermination exacte de t présente plusieurs difficultés:

. La première provient de son inhomogénéité et conduit à prendre pour valeur approchée de $t: t_h = 3/4\, t_R$ où t_R représente la contrainte superficielle et t_h la contrainte homogène qui, pour la même section droite donne le même couple C soit:

$$C = 2\pi \int_0^R \sigma_{z\theta} r^2\, dr = \frac{2\pi t_R}{R} \int_0^R r^3\, dr$$
$$= 2\pi t_h \int_0^R r^2\, dr.$$

Le début de la saturation se manifeste alors pour

$$\frac{t_h}{n} = 2.$$

. La seconde est due à l'apparition rapide de la plasticité qui entraine une imprécision supplémentaire sur la détermination de t_h à partir de t_R calculé à l'aide des courbes d'écrouissage:

$$t_R = \frac{1}{2\pi R^3} \left| 3C + \theta \frac{dC}{d\theta} \right| .$$

On doit noter que la saturation sur les courbes $\varepsilon_{zz\,(m)}(C)$ de la fig. 9 peut s'expliquer par la théorie précédemment développée et par une décroissante de λ_S avec l'écrouissage en torsion.

c) à l'état brut de filière (limite élastique $\sigma_E = 4{\cdot}10^8$ N/m^2) donc fortement écroui, $\varepsilon_{zz\,(m)}$ est, indépendant de n et devient proportionnel à t. Ce phénomène résulte de l'existence de contraintes internes importantes qui contribuent à accroitre la valeur vraie de n.

d) L'hystérésis importante observée sur les courbes $\varepsilon_{zz\,(m)}(\gamma)$ est essentiellement d'origine mécanique. Une hystérésis purement magnetoélastique d'amplitude relative plus faible apparaît toutefois en $\varepsilon_{zz\,(m)}(C)$.

4.3.2. *Déformation en fonction du champ magnétique*

La déformation axiale totale ε_{zz} en fonction du champ magnętique correspondant à une déformation angulaire $\gamma = \pm 10^{-3}$ est représentée sur la fig. 10 pour différents états du materiau. Si on compare ces courbes aux mesures de magnétostriction longitudinale, on constate que l'effet du traitement thermomécanique est le même sur celle-ci et sur la déformation $\varepsilon_{zz}(\gamma)$. L'évolution de la forme des cycles relevés pour différentes valeurs de champ et les déformations angulaires $\gamma = \pm 10^{-3}$ et $\pm 2{\cdot}10^{-2}$ (fig. 11) sou-

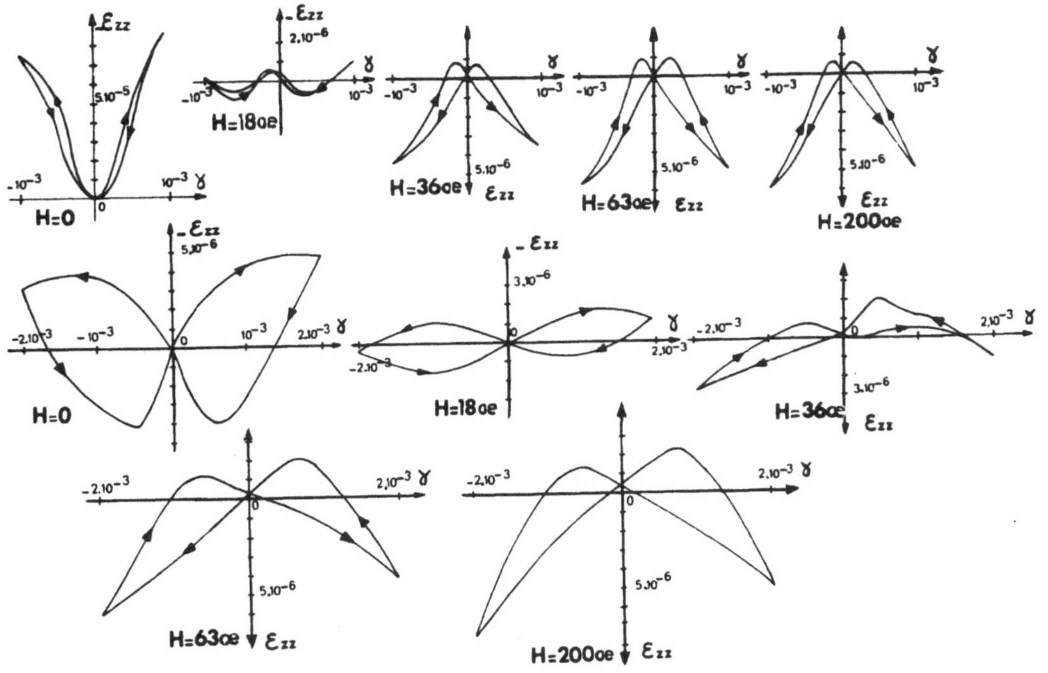

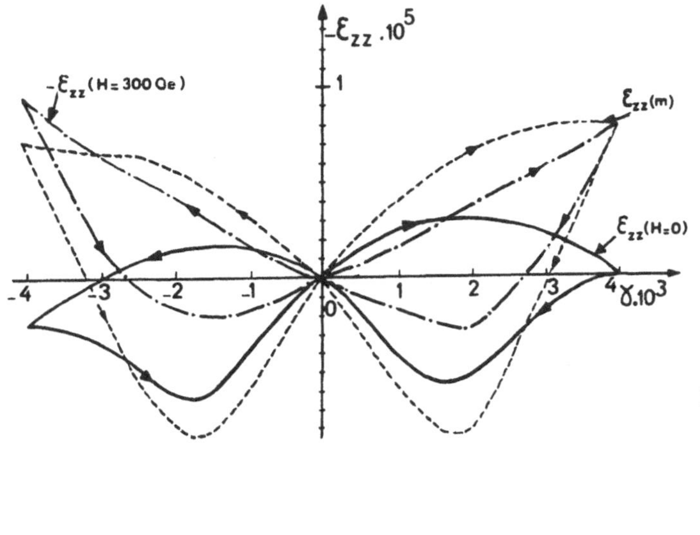

Fig. 11. vide texte

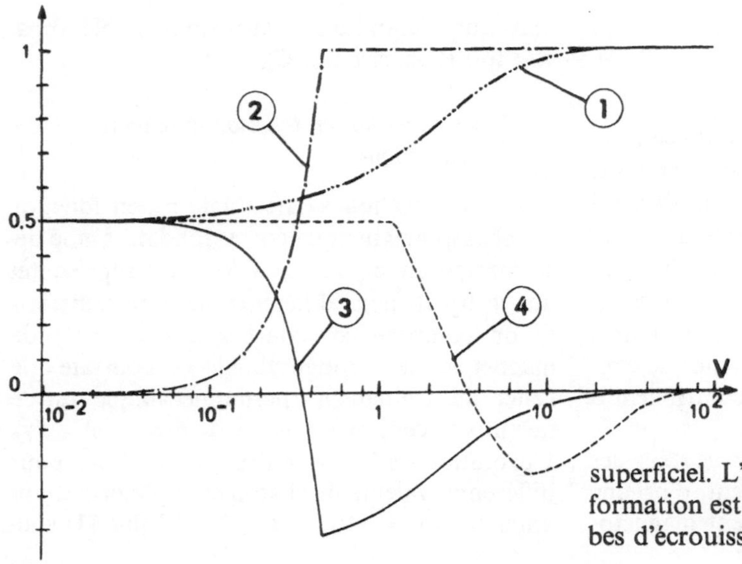

Fig. 12. vide texte

ligne leur distorsion importante pour les valeurs de champ annulant pratiquement la déformation axiale totale moyenne et provient essentiellement des dissymétries de $\varepsilon_{zz(e)}$.

La variation de $\varepsilon_{zz(m)}$ avec H est qualitativement conforme à la théorie (fig. 12).

Les désaccords observés, principalement du point de vue quantitatif résultent des simplifications nécessaires pour le calcul. En particulier pour $n/t = 4$, le maximum positif de $\varepsilon_{zz(m)}$, qui se produit pour la valeur du champ saturant, est obtenu en réalité pour $H \simeq 60$ Oe. au lieu de la valeur théorique $H = 3{,}4$ Oe. Il faut noter à ce sujet que cette première valeur est voisine du champ nécessaire à la saturation de notre échantillon sous contrainte axiale et que le maximum observé expérimentalement doit être moins prononcé que celui prévu par le calcul.

Le champ maximal ($H = 1400$ Oe.) permet de saturer le magnétostriction spontanée et les déformations mesurées seront alors pratiquement confondues avec la déformation élastique pure.

Dans le domaine élastique ($\gamma < 10^3$) on observe une déformation $\varepsilon_{zz(e)}$ positive:
- indépendante de la charge axiale σ_{zz};
- fortement affectée par le processus de mise en déformation et présentant fréquemment des dissymétries importantes;
- pratiquement proportionnelle à γ^2 et à laquelle correspond le coefficient $C_L \simeq 0{,}7$;
- d'amplitude décroissante avec l'écrouissage.

Dans le domaine plastique et pour $10^{-3} < \gamma < 4 \cdot 10^{-3}$, $\varepsilon_{zz(e)}$ ne suit pas une loi en γ^2; la déformation mesurée est alors sensiblement proportionnelle à la contrainte de cisaillement superficiel. L'hystérésis accompagnant cette déformation est enfin directement liée à celle courbes d'écrouissage $C(\gamma)$.

Résumé

Les mesures de variation de longueur sur fils à section circulaire soumis à une déformation de torsion, mettent en évidence plusieurs phénomènes ne pouvant être interprétés uniquement par les coefficients élastiques du troisième ordre. Le dispositif expérimental et le mode opératoire permettent d'étudier les phénomènes cumulatifs (du type *Ronay*) et non cumulatifs (du type *Poynting*).

Plusieurs mécanismes sont à l'origine des effets observés. Trois d'entre eux ont été particulièrement mis en évidence: intervention des coefficients élastiques du troisième ordre (exemple le tungstène), glissements réversibles des dislocations (cas du molybdène) et interactions magnétoélastiques.

Les allongements dont l'origine est distincte de l'effet *Poynting* peuvent être du même ordre de grandeur que celui-ci. Ainsi dans le cas du nickel, la déformation axiale observée à champ nul est de signe opposé à celle qu'on obtient en saturant la magnétostriction.

Littérature

1) *Murnaghan, F. D.*, Finite deformations of an elastic solid. Chap. 7 (New York 1951).

2) *Reiner, M.*, Rheology. Encyclopedia of physics. Vol. 6, p. 509 (Berlin-Heidelberg-New York 1958).

3) *Poynting*, Proc. Roy. Soc. (London) **A 86**, 534 (1912).

4) *Foux*, A Study in the Poynting effect. Doctoral Thesis, Technion Israel (1964).

5) *Ronay, M.*, J. Inst. Metals **14**, 392 (1966).

6) *Lallement, G.* et *C. Oytana*, C. R. Acad. Sc. (Paris), série **A 274**, 1243–1246 (1972).

7) *Freudenthal, A. M.*, Rheol. Acta **6**, 146–156 (1967).

8) *Freudenthal, A. M.* et *M. Ronay*, Proc. Roy. Soc. (London) **A 292**, 14 (1966).

9) *Swift, H. W.*, Engineering **1963**, 253 (1947).

Adresse des auteurs:

G. Lallement et *C. Oytana*
Laboratoire de Mécanique Appliquée
Associé au CNRS
F-25 Besançon (France)

Rheol. Acta 13, 477–489 (1974)

Département de génie chimique Ecole Polytechnique, Montréal, Québec (Canada)

Dynamique des bulles en milieu viscoélastique

Par P. J. Carreau, M. Devic et M. Kapellas

Avec 10 figures et 1 tableau

(Reçu p. p. le 27 octobre 1972)

Nomenclature

a	rayon de la bulle, fonction de θ
D_f	diamètre frontal de la bulle
D/Dt	dérivée substantielle
$\mathscr{D}/\mathscr{D}t$	dérivée de *Jaumann*
F_k	force de traînée
g	accélération gravitationnelle
H	dimension de la bulle dans le sens de l'ascension
$\boldsymbol{n}$	vecteur unitaire normal à l'interface
p	pression dans le fluide
p_i	pression à l'intérieur de la bulle
r	coordonnée radiale
R_1, R_2	rayons principaux de courbure de la bulle
$\boldsymbol{t}$	vecteur unitaire tangentiel à l'interface
U	vitesse d'ascension de la bulle
V	volume de la bulle
$\boldsymbol{v}$	vecteur de la vitesse du fluide
$Z(\alpha)$	fonction zêta de *Riemann*
t_1, R, S	paramètres du modèle rhéologique, reliés au comportement non-newtonien
$\lambda, \alpha, \varepsilon$	paramètres reliés au comportement élastique du fluide
β	coefficient des contraintes normales secondaires en régime permanent $= -(\tau_{22} - \tau_{23})/\dot{\gamma}^2$
$\dot{\gamma}$	tenseur vitesse de déformation $= \nabla \boldsymbol{v} + (\nabla \boldsymbol{v})^+$
η	viscosité non newtonienne
η_0	viscosité limite à faible cisaillement
θ	coefficient des contraintes normales primaires en cisaillement permanent $= -(\tau_{11} - \tau_{22})/\dot{\gamma}^2$
$\boldsymbol{\pi}$	$= \boldsymbol{\tau} + p\boldsymbol{\delta}$, où $\boldsymbol{\delta}$ est le tenseur unitaire
ϱ	masse volumétrique du fluide
σ	tension interfaciale liquide-air
$\boldsymbol{\tau}$	tenseur des contraintes
τ	période d'injection
$\boldsymbol{\omega}$	tenseur de vorticité $= \nabla \boldsymbol{v} - (\nabla \boldsymbol{v})^+$
II	second invariant du tenseur vitesse de déformation $= \dot{\gamma} : \dot{\gamma}$

Nombres adimensionnels

Nombre de *Reynolds*

$$Re = \frac{D_f U \varrho}{\eta_0}$$

Nombre de *Wéber*

$$We = \frac{D_f U^2 \varrho}{\sigma}$$

Temps adimensionnels

$$\Lambda_1 = t_1 \frac{U}{D_f}$$

$$\Lambda_2 = \lambda \frac{U}{D_f}$$

$$\frac{\lambda}{\tau}$$

Introduction

Il existe dans la littérature de nombreuses publications traitant du mouvement des bulles de gaz dans des fluides newtoniens. Par contre peu de chercheurs ont tenté d'élucider le comportement des bulles en milieu viscoélastique. *Philippoff* (1) est probablement le premier chercheur à faire part des formes typiques que prennent les bulles dans un fluide à caractère élastique. Il a noté que les petites bulles avaient la forme de gouttes d'eau alors qu'à des volumes plus grands, les bulles prenaient la forme de sphéroïdes aplatis. Des observations similaires ont été rapportées par *Astarita* et *Apuzo* (2). Ces derniers ont constaté également que la vitesse ascensionnelle de très petites bulles dans certains fluides viscoélastiques exhibait une discontinuité marquée à un volume critique. Cette discontinuité pouvait représenter une augmentation de 600% de la vitesse d'ascension.

Les mêmes phénomènes ont été récemment vérifiés par *Calderbank, Johnson* et *London* (3) et par *Leal, Skoog* et *Acrivos* (4). En comparant les résultats de la vitesse limite des billes de verre à celle des bulles de gaz dans une solution de Separan AP-30, ces derniers concluent que le changement de forme au volume critique a peu d'influence sur la vitesse. Ils soutiennent ainsi la thèse d'*Astarita* et *Apuzo* selon laquelle la discontinuité résulterait d'un changement soudain dans les conditions interfaciales (passage soudain du régime de *Stokes* au régime d'*Hadamard*). De plus ces auteurs ont démontré par une solution numérique de l'écoulement rampant (creeping) que seule une fraction de la discontinuité pouvait être attribuée au caractère non newtonien de la viscosité. Ils concluent alors que le phénomène observé est causé par la contribution des propriétés élastiques au bilan de forces interfaciales.

A notre connaissance, seul *Astarita* (5) a présenté une analyse dimensionnelle du mouvement des bulles en milieu viscoélastique. Son étude toutefois est limitée aux petites bulles sphériques et aux fluides obéissant à l'équation de *Maxwell*. *Astarita* démontre que le nombre de *Levich*, *Le* (exprimant le taux de variation d'énergie potentielle du système) est une fonction du nombre de *Reynolds*, *Re*, et d'un nombre *A* qui est le rapport d'une grandeur caractéristique des contraintes sur le module élastique du fluide. Trois cas limites d'intérêt sont présentés:

i) $A \ll 1$, $Re \gg 1$ $\quad Le = Le(A)$

ii) $A \ll 1$, $Re \ll 1$ $\quad Le = Le(A)$

iii) $A \gg 1$ $\quad Le = \frac{1}{A}$. [1]

Dans cette communication, nous présentons des données additionnelles sur la dynamique des bulles de volume variant de 0.1–11 cm^3 en milieu viscoélastique. Nous soulignons de façon particulière les liens étroits qui relient la vitesse d'ascension au volume et à la forme des bulles ainsi qu'aux propriétés viscoélastiques du fluide. Dans le but de démontrer clairement l'influence de la «mémoire» du fluide, nous avons étudié trois fluides de caractère rhéologique très différent: une huile de silicone visqueuse newtonienne, une solution de CMC très visqueuse et une solution de Separan peu visqueuse.

Nous avons varié systématiquement la période entre les injections consécutives de bulles (que nous appelerons par la suite *période d'injection*) de 2–3600 sec. Nous pouvons ainsi illustrer clairement l'influence de la relaxation des contraintes sur la forme et la vitesse des bulles. Finalement nous présentons une analyse dimensionnelle tenant compte de la non sphéricité des bulles et pour un fluide décrit par une équation constitutive non linéaire. Cette analyse nous permet de discerner certains nombres adimensionnels importants.

Technique expérimentale

Les bulles sont générées au bout d'un tube d'injection, vertical, de 5 mm de diamètre intérieur et centré au bas d'une cuve de section carrée en plastique acrylique. La hauteur de la cuve est de 400 mm et sa largeur de 150 mm. La génération des bulles peut s'effectuer par injection directe d'un volume d'air mesuré au moyen d'une seringue graduée ou par une injection d'air comprimé à l'aide d'une vanne solénoïde contrôlée par une minuterie (6). L'air comprimé est détendu en deux stages et la pression est stabilisé dans un bac tampon de 20 l. La minuterie permet de contrôler de façon très précise le temps d'ouverture de la vanne et de varier la période d'injection entre 1 et 120 sec. L'appareillage est monté dans une pièce où la température est contrôlée aux environs de 25 °C.

Les bulles d'air d'un volume choisi sont injectées automatiquement à période constante jusqu'à ce que les conditions hydrodynamiques soient établies. Pour les mesures expérimentales proprement dites, l'injection automatique est momentanément remplacée par une injection manuelle à l'aide de la seringue calibrée à la période correspondante. Le volume mesuré ainsi avec précision est comparé au volume des bulles injectées automatiquement. La pression de l'air comprimé est ajustée jusqu'à ce que les deux volumes soient sensiblement les mêmes. L'estimé de la précision relative du volume est de 2%.

La vitesse et la forme des bulles sont déterminées à partir de photographies de multiples expositions de la même bulle à l'aide d'un stroboscope. Pour éviter toute distorsion appréciable un objectif f/135 mm est utilisé et la caméra (Nikon F) est placée à 150 cm de la cuve. Les meilleures photographies sont obtenues sous un éclairage latéral [la technique photographique est expliquée en détail à la référence (6)]. Une règle graduée immergée verticalement dans le plan médian de la cuve nous permet de vérifier les effets optiques et mesurer avec précision le diamètre frontal de la bulle et la distance entre les images. La vitesse moyenne est déterminée à partir de la distance séparant *n* images des deux derniers tiers de l'ascension et de la fréquence du stroboscope mesurée avec précision. La vitesse finale est déterminée à partir des trois dernières images. Toutes les mesures sont effectuées sur un écran en projetant le négatif de la photo pour un grossissement double du sujet. L'estimé de l'erreur relative est de 0.5% pour la vitesse moyenne et de 1–2% pour la vitesse finale.

Fluides

Nous avons utilisé les fluides suivants: une huile de silicone (General Electric, SF-96-1000); une solution aqueuse 1,9% en poids de carboxy méthyle de cellulose (Hercules Powder, CMC-7H), contenant un additif pour prévenir une trop rapide dégradation biochimique; une solution 0,5% de polyacrylamide (Dow Chemicals, MG-700) dans un mélange eau-glycérine de 20% en poids de glycérine.

Nous avons déterminé les propriétés rhéologiques des fluides sur un rhéogoniomètre de *Weissenberg* (modèle R-18). Les données de la viscosité et de la différence des contraintes normales primaires pour les solutions de CMC et de Separan sont présentées à la fig. 1. Les courbes représentent le comportement prévu par le modèle rhéologique défini plus loin par les eqs. [8] et [9] [la vitesse de cisaillement est alors donnée par $\dot{\gamma} = (\frac{1}{2}\,\mathrm{II})^{1/2}$]. Au tableau 1, nous donnons les propriétés des fluides ainsi que les paramètres se rapportant au modèle. Les paramètres S, R et t_1 décrivent le comportement non newtonien alors que les paramètres λ et α sont associés au caractère élastique. Nous remarquons que la viscosité de la solution de CMC est plus élevée, la différence relative étant plus marquée à haute vitesse de cisaillement. Par contre la différence des propriétés élastiques entre les deux solutions est moins prononcée. Les propriétés des fluides n'ont pas varié de façon appréciable au cours des essais expérimentaux.

Résultats expérimentaux

Forme des bulles

La forme de la bulle ayant une influence majeure sur sa vitesse ascensionnelle, il nous est apparu nécessaire d'étudier cette variable en détail. Nous caractérisons la forme par le diamètre frontal D_f et la hauteur de la bulle H (excluant la queue dans le cas des fluides visco-

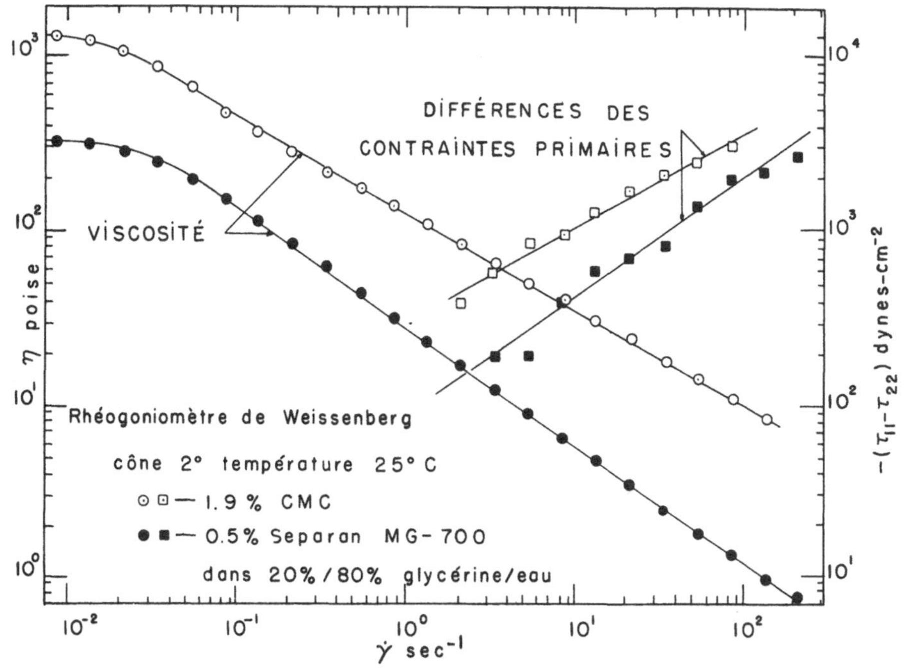

Fig. 1. Viscosité et différences des contraintes normales primaires en fonction de la vitesse de cisaillement

Tableau 1. Propriétés des fluides

Fluide	Masse volumétrique ϱ g-cm^{-3}	Tension superficielle σ*) dynes-cm^{-1}	Viscosité limite η_0 poises	θ_0 g-cm^{-1}	S	R	t_1 sec	λ**) sec	α**)
Huile de silicone	0,971	22,4	8,85	–	–	–	–	–	–
Solution 1,9% de CMC	1,006	75,2	1300	$1{,}1 \times 10^5$	0,275	0,732	64,0	77,7	2,22
Solution 0,5% de Separan dans 20%/ 80% glycérine/eau	1,044	69,6	350	9×10^3	0,336	0,655	38,0	18,6	3,05

*) Valeurs approximatives mesurées sur un appareil Fisher Surface Tensiomat.
**) Valeurs estimées à l'aide de l'équation [9] sous l'hypothèse que les pentes des courbes de la viscosité et de la viscosité dynamique sont identiques dans la zone de la loi de puissance $[-2S \doteq (1-\alpha)/\alpha]$.

élastiques). Le diamètre frontal est défini comme la dimension maximale mesurée perpendiculairement au sens de déplacement. La mesure H est quelque peu imprécise et ne sera pas utilisée dans l'analyse dimensionnelle.

La fig. 2 illustre les formes typiques observées dans l'huile de silicone et la solution de CMC. Dans *l'huile de silicone*, les bulles ont toutes la forme d'une sphère aplatie au sommet et tronquée à la base (pour des volumes variant de 0,5 à 11 cm^3). Le rapport D_f/H (voir fig. 3) est toujours supérieur à l'unite et augmente peu avec le volume. Dans *la solution de CMC*, les bulles ont toutes la forme d'une goutte d'eau avec une queue effilée. Le rapport D_f/H (fig. 3) est toujours inférieur à l'unité et augmente très peu avec le volume. De plus la forme est à peu près indépendante de la période d'injection τ.

Dans la solution de Separan, la forme des bulles est très variable suivant le volume et la

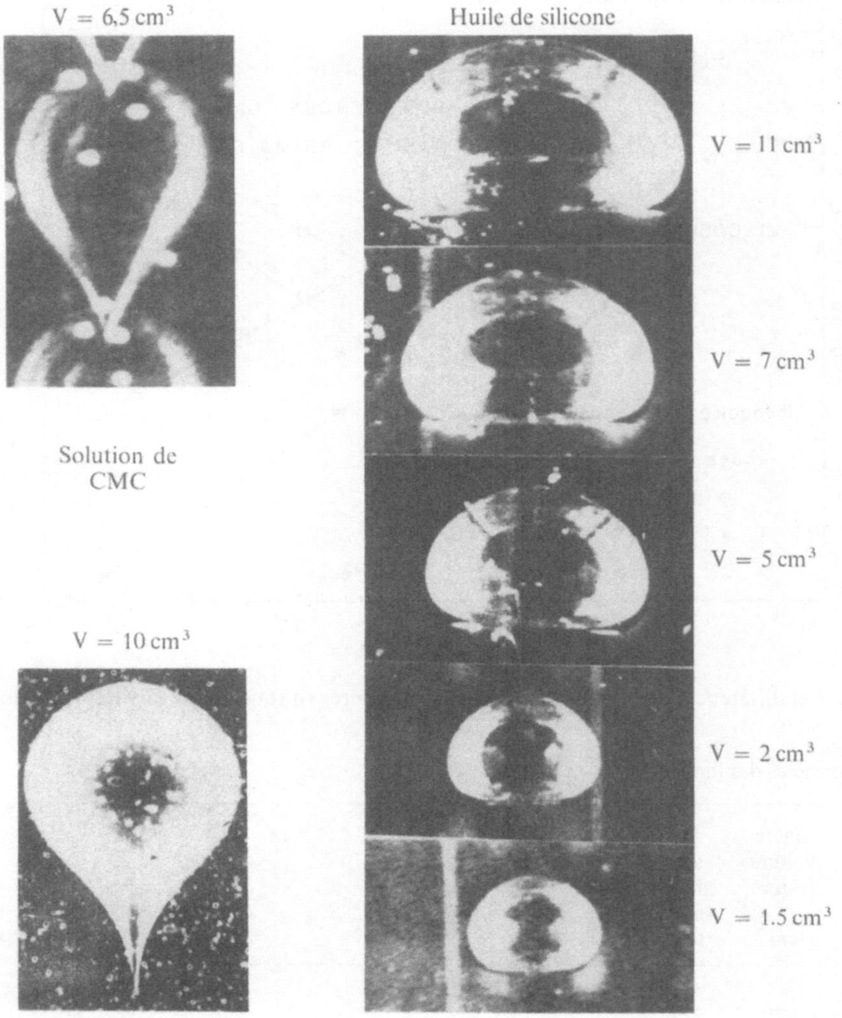

Fig. 2. Photographies de bulles dans l'huile de silicone et la solution de CMC. Influence du volume sur la forme

période d'injection. Les photographies (fig. 4) d'une série de bulles de volume croissant nous permettent de distinguer trois formes typiques: i) forme goutte d'eau, ii) forme lentille (sphéroïde très aplati terminé par une queue), iii) forme calotte sphérique avec queue. Les deux premières formes sont caractérisées par une augmentation rapide du rapport D_f/H (fig. 3) qui peut atteindre une valeur de $4-5$; l'apparition de la forme calotte sphérique se traduit par une diminution appréciable du rapport.

Quelques auteurs ont observé des formes similaires dans des fluides viscoélastiques pour des bulles (1–4) et pour des gouttes (7–9). Cependant aucun d'eux ne fait état d'une influence marquée de la période d'injection (temps entre deux injections consécutives) telle que nous avons observée pour la solution de Separan. La fig. 5 illustre bien ceci. Selon la grosseur des bulles, à volume constant, la forme passe par les étapes observées précédemment (goutte d'eau → lentille → calotte sphérique) alors que la période devient de plus en plus courte. Il est certain que pour des périodes très courtes (de l'ordre de 2 sec) la forme et la vitesse de la bulle sont influencées par le mouvement résiduel dans le fluide par suite de l'ascension de la bulle précédente. Toutefois l'influence marquée pour des périodes plus longues ne peut être causée que par des effets de la «mémoire» du fluide. Le rapport D_f/H est très étroitement relié aux variations de la vitesse ascensionnelle selon la période d'injection (voir fig. 7 et 8).

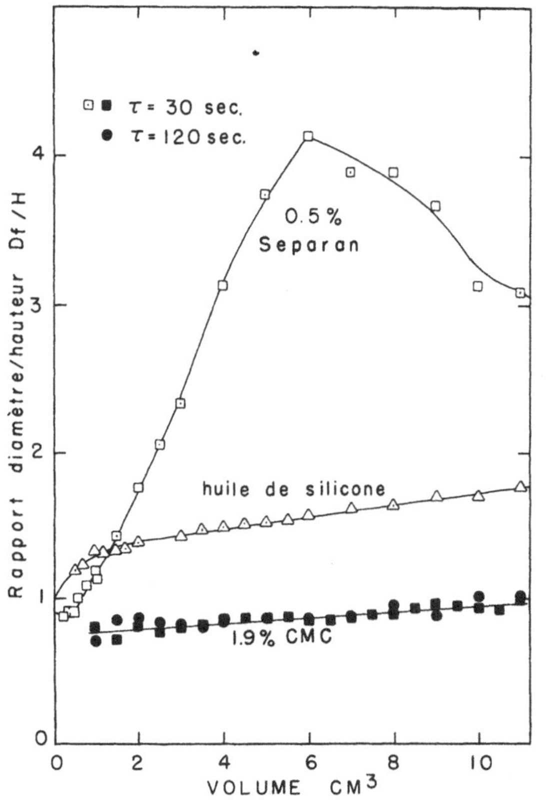

Fig. 3. Variation du rapport diamètre frontal/hauteur en fonction du volume

Mouvement des bulles

Pour les trois fluides étudiés, les trajectoires des bulles ont toujours été rectilignes (nous avons rejeté des solutions plus concentrées de Separan qui donnaient lieu à des trajectoires instables). Dans tous les cas la vitesse ne varie pratiquement pas dans les deux derniers tiers de l'ascension; la différence entre la vitesse moyenne et la vitesse finale n'excède pas 1% et est égale à 0,5% en moyenne. Nous ne présentons que les vitesses moyennes qui sont à notre avis plus précises.

Dans le cas de *l'huile de silicone*, la vitesse augmente de façon monotone avec le volume (voir fig. 6) et est donnée par la relation

$$U = 11{,}8\, V^{1/3}, \quad 3\,\mathrm{cm}^3 < V \leq 11\,\mathrm{cm}^3. \qquad [2]$$

Ces résultats concordent très bien avec ceux d'*Angelino* (10) pour une huile aux propriétés similaires. Pour des volumes inférieurs à 2 cm³, la vitesse est bien représentée par la loi de *Stokes*

$$U = \frac{84}{\nu} V^{2/3}. \qquad [3]$$

Ceci nous paraît une pure coïncidence, les hypothèses de *Stokes* n'étant pas respectées (2). Nous n'avons pas apporté de correction pour l'influence des parois. En utilisant la relation proposée par *Uno* et *Kintner* (11), la correction demeure inférieure à 6% dans tous les cas.

Dans la *solution de CMC*, la vitesse augmente de façon monotone tout comme dans l'huile de silicone (fig. 6). L'influence de la période d'injection est très faible: on observe une légère chute de la vitesse alors que la période est augmentée de 30–120 sec. Par contre, dans la *solution de Separan*, l'allure de la courbe de la vitesse diffère sensiblement et dépend de la période d'injection (fig. 6). A période constante, la vitesse croît d'abord très rapidement avec le volume pour atteindre une valeur à peu près constante (≈ 30 cm/sec). La première zone, $0 < V < 2\,\mathrm{cm}^3$, correspond à la forme goutte d'eau; pour les volumes compris entre 2 et 11 cm³, les bulles prennent la forme lentille se rapprochant de plus en plus de la forme calotte sphérique.

Le phénomène le plus intéressant et le plus curieux est le comportement des bulles dans la solution de Separan suivant la période d'injection, à volume constant (fig. 7 et 8). Pour les volumes de 8 et 10 cm³ (fig. 7), la vitesse décroît d'abord avec une augmentation de la période, passe par un minimum pour augmenter par la suite. Pour nous assurer que ces observations n'étaient pas le fait d'erreurs expérimentales ou de variations des propriétés du fluide, nous avons effectué les essais expérimentaux suivant divers agencements. Et bien que sur les fig. 7 et 8 nous ne reproduisons que les données d'essais consécutifs, l'ensemble de toutes les données indiquerait sensiblement le même comportement (6).

Nous pouvons constater à la fig. 7 que près du point de vitesse minimale, le rapport D_f/H subit un changement soudain (correspondant au changement rapide de forme illustrée à la fig. 5). L'augmentation subséquente de la vitesse correspond à un profil de plus en plus effilé. Il est intéressant de noter que la vitesse minimale pour les volumes de 10 cm³ est atteinte à une période beaucoup plus longue ($\tau \doteq 140$ sec) comparativement aux bulles de 8 cm³ ($\tau \doteq 30$ sec). De plus, à longues périodes, les bulles de 10 cm³ sont moins rapides, ce qui s'expliquerait par une forme beaucoup moins hydrodynamique telle qu'indiquée par le rapport D_f/H. A l'analyse de ces résultats, il nous paraît évident que les propriétés élastiques jouent un rôle primordial au niveau de l'équilibre des forces interfaciales.

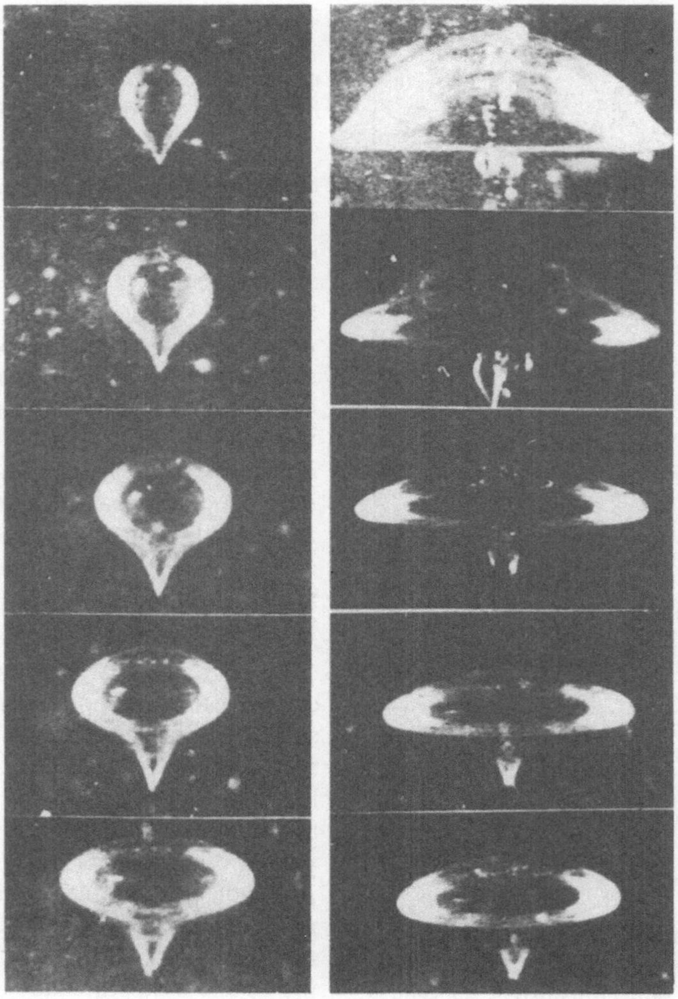

Fig. 4. Photographies de bulles dans la solution de Separan. Influence du volume à période d'injection constante, $\tau = 30$ sec

Les bulles de 1,5 et 1,0 cm^3 ont des comportements très différents (fig. 8). Pour les volumes de 1,5 cm^3 la vitesse croît légèrement jusqu'à un maximum pour diminuer par la suite. Pour les bulles de 1,0 cm^3, la vitesse décroît de façon monotone, la chute étant très appréciable (de 25,7 cm/sec à $\tau = 2$ sec jusqu'à 16,2 cm/sec pour une période de 3600 sec). Soulignons également l'influence encore marquée de la période sur la vitesse après 600 sec.

Analyse dimensionnelle

L'analyse du mouvement et de la déformation des bulles en milieu viscoélastique est beaucoup trop complexe pour que nous puissions espérer en obtenir des solutions analytiques. *Astarita* (5) a présenté une analyse dimensionnelle du mouvement des bulles sphériques dans un fluide de *Maxwell*. Nous généralisons ici son analyse pour des bulles non sphériques et pour un modèle rhéologique plus réaliste.

Considérons une bulle de forme constante et se déplaçant à vitesse constante U dans une trajectoire rectiligne. D'un bilan d'énergie, nous obtenons (12)

$$U F_k = U V \varrho g = 2\pi \int_0^\pi \int_a^\infty -(\boldsymbol{\tau} : \nabla \boldsymbol{v})\, r^2\, dr \sin\theta\, d\theta \qquad [4]$$

F_k est la force de traînée, V le volume de la bulle, ϱ la masse volumétrique du fluide; a est la valeur de la coordonnée r (en coordonnées sphériques) à l'interface gaz-fluide; $-(\boldsymbol{\tau} : \nabla \boldsymbol{v})$ représente le

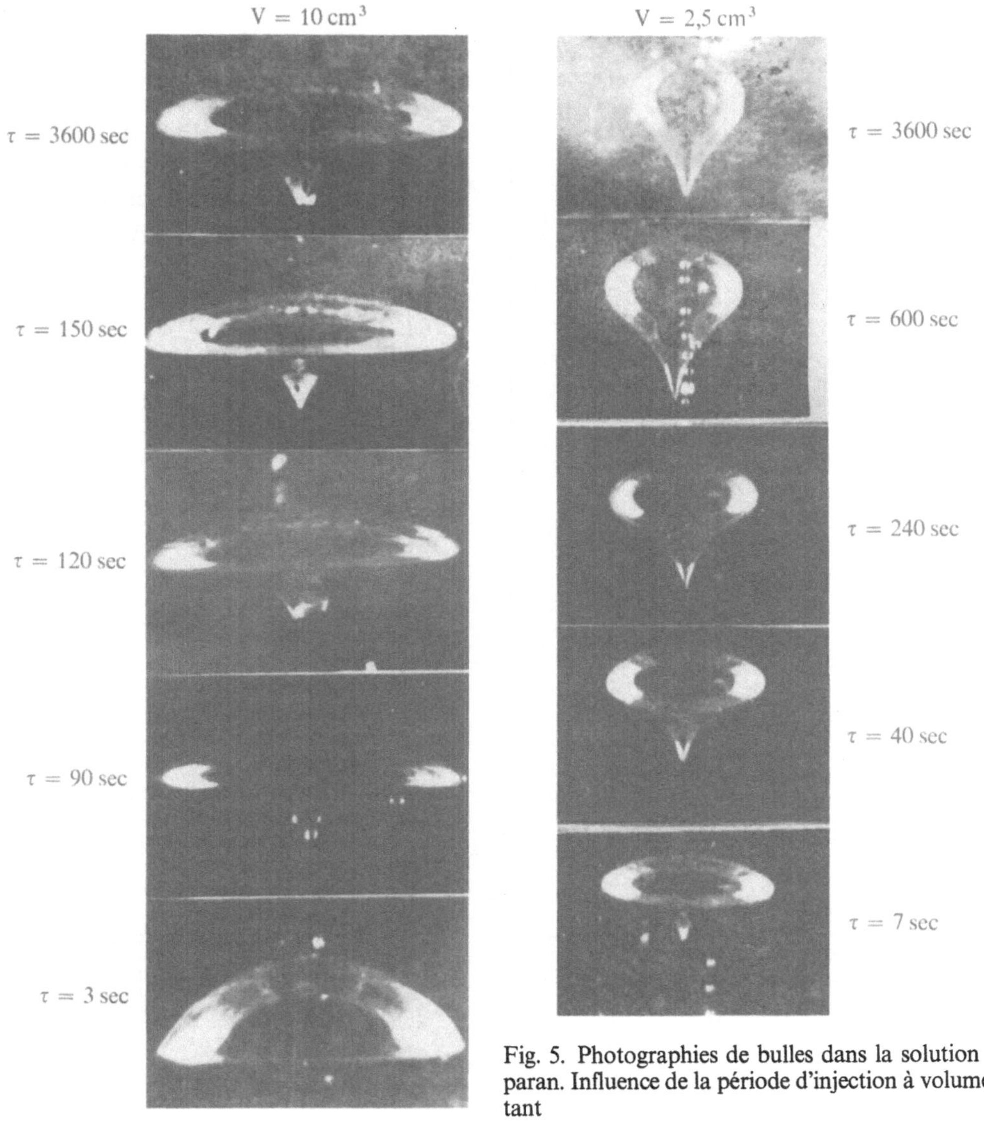

Fig. 5. Photographies de bulles dans la solution de Separan. Influence de la période d'injection à volume constant

taux d'énergie dissipée par unité de volume du fluide. L'eq. [4] nous indique que le taux d'énergie dissipée par forces visqueuses et élastiques dans le fluide est égal au taux de variation d'énergie potentielle de tout le système.

Le fluide, considéré incompressible, doit satisfaire l'équation de mouvement

$$\varrho \frac{D\boldsymbol{v}}{Dt} = -\nabla p - \nabla \cdot \boldsymbol{\tau} \equiv -\nabla \cdot \boldsymbol{\pi} \qquad [5]$$

et les conditions limites suivantes (13)

a) à $r = a$ (interface gaz-fluide)

$$\boldsymbol{n}\left[p_i + \sigma\left(\frac{1}{R_1} + \frac{1}{R_2}\right)\right] = \boldsymbol{n} \cdot \boldsymbol{\pi}$$

$$\boldsymbol{t} \cdot \boldsymbol{\tau} \doteq 0$$

$$\boldsymbol{v} \cdot \boldsymbol{n} = 0$$

b) à $r \to \infty$, $\boldsymbol{v} = 0$ et $\boldsymbol{\tau} = 0$. [6]

p_i est la pression à l'intérieur de la bulle, reliée au volume et à la température par une équation d'état; $\boldsymbol{n}$ est le vecteur unitaire normal à l'interface gaz-liquide et $\boldsymbol{t}$, le vecteur unitaire tangentiel à l'interface; σ est la tension superficielle et R_1 et R_2 sont les principaux rayons de courbure.

Pour décrire le comportement rhéologique du fluide, nous choisissons une équation obtenue de l'expansion d'un modèle non linéaire (14)

$$\boldsymbol{\tau} = -\left[\eta - \frac{D\theta}{Dt}\right]\dot{\gamma} + \frac{1}{2}\theta \frac{\mathscr{D}}{\mathscr{D}t}\dot{\gamma} - \frac{1}{2}(\theta + 2\beta)\{\dot{\gamma} \cdot \dot{\gamma}\} \ldots \qquad [7]$$

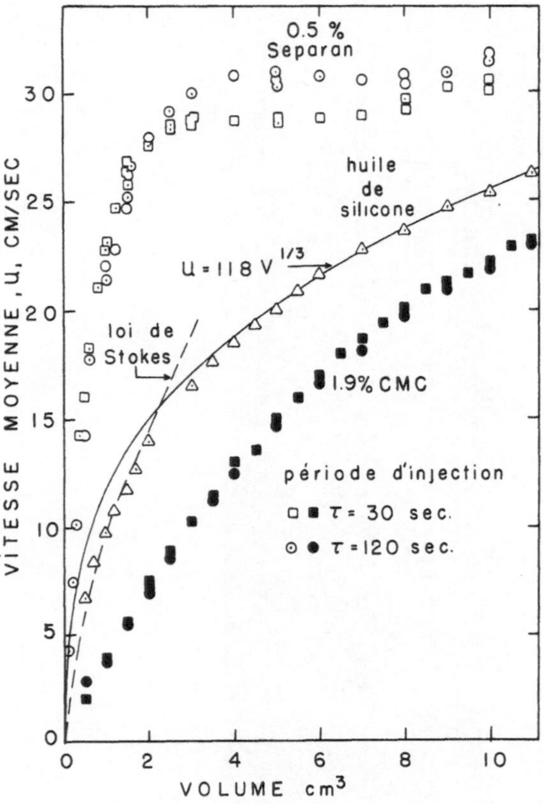

Fig. 6. Variation de la vitesse d'ascension des bulles en fonction du volume

où η est la viscosité, θ et β sont les coefficients des contraintes normales primaires et secondaires respectivement; $\dot{\gamma}$ est le tenseur vitesse de déformation, D/Dt la dérivée substantielle et $\mathscr{D}/\mathscr{D}t$ la dérivée de *Jaumann*. Nous choisissons de plus les fonctions rhéologiques suivantes qui décrivent très bien le comportement des fluides viscoélastiques en cisaillement simple (15)

$$\eta = \frac{\eta_0}{\left[1 + \frac{1}{2}\,\mathrm{II}\,t_1^2\right]^S} \qquad [8]$$

$$\theta = \frac{2^{\alpha+1}\lambda\eta_0 Z(2\alpha) - 1)}{\left(Z(\alpha) - 1\right)\left[1 + \frac{1}{2}\,\mathrm{II}\,t_1^2\right]^R}$$

$$= \frac{\theta_0}{\left[1 + \frac{1}{2}\,\mathrm{II}\,t_1^2\right]^R} \qquad [9]$$

$$\beta = \frac{1}{2}\,\varepsilon\theta. \qquad [10]$$

η_0 est la viscosité limite à faible cisaillement; $Z(\alpha)$ est la fonction zêta de *Riemann*; II est le second invariant du tenseur vitesse de déformation; t_1, λ, α, R, S et ε sont des paramètres rhéologiques.

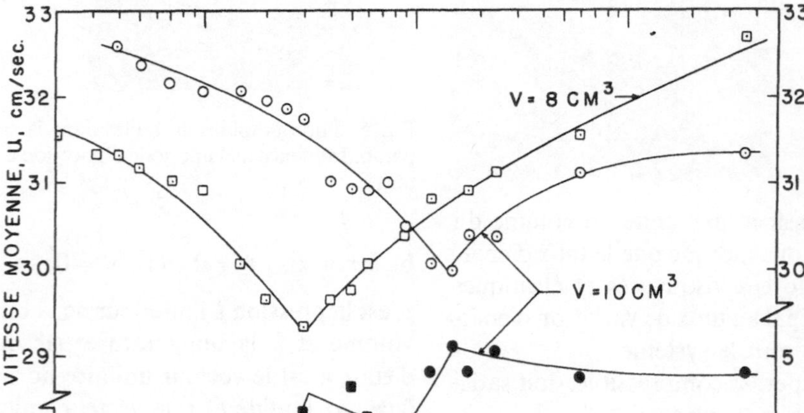

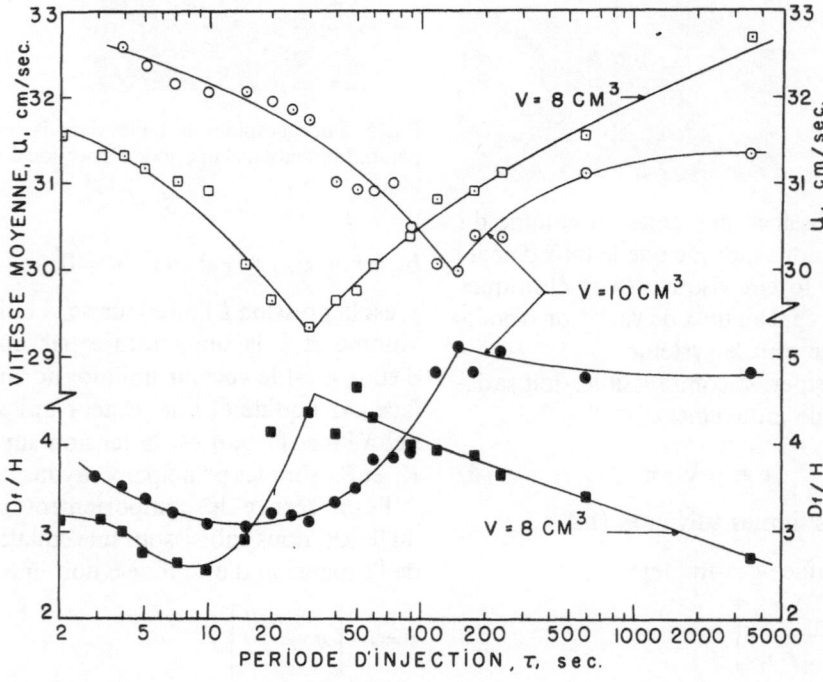

Fig. 7. Variation de la vitesse d'ascension et du rapport D_f/H en fonction de la période d'injection dans la *solution de Separan. Grosses bulles*

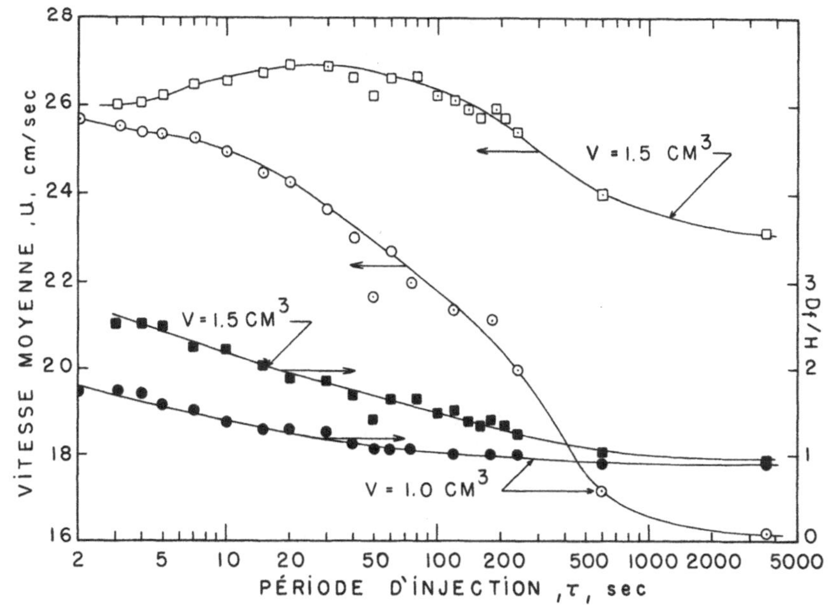

Fig. 8. Variation de la vitesse d'ascension et du rapport D_f/H en fonction de la période d'injection dans la *solution de Separan. Petites bulles*

Formes adimensionnelles

Nous caractérisons le mouvement d'une bulle dans un fluide viscoélastique par la vitesse d'ascension U, le diamètre frontal D_f et par la période d'injection, τ. Nous définissons alors les variables adimensionnelles suivantes

$$v^* = \frac{v}{U}, \quad p^* = \frac{p}{\varrho U^2}, \quad t^* = \frac{t}{\tau}$$

$$r^* = \frac{r}{D_f}, \quad \nabla^* = D_f \nabla, \quad \tau^* = \left(\frac{D_f}{\eta_0 U}\right)\tau$$

$$\dot{\gamma}^* = \left(\frac{D_f}{U}\right)\dot{\gamma}, \quad \mathrm{II}^* = \left(\frac{D_f}{U}\right)^2 \mathrm{II}. \qquad [11]$$

De l'eq. [4] nous obtenons le coefficient de traînée

$$f \equiv \frac{8gV}{\pi D_f^2 U^2} = \frac{16}{Re}\int_0^{\pi}\int_{a/D_f}^{\infty} -(\tau^* : \nabla^* v^*)\, r^{*2} dr^* \sin\theta\, d\theta \qquad [12]$$

où le nombre de *Reynolds* est

$$Re = \frac{D_f U \varrho}{\eta_0}.$$

L'équation de mouvement adimensionnelle s'écrit:

$$\left[\frac{D_f}{U\tau}\right]\frac{\partial v^*}{\partial t^*} + \nabla^* \cdot v^* v^* = -\nabla^* p^* - \frac{1}{Re}\nabla^* \cdot \tau^* \qquad [13]$$

avec les conditions limites

a′) à $r^* = a/D_f$

$$n\left[p_i^* + \frac{1}{We}\left(\frac{1}{R_1^*} + \frac{1}{R_2^*}\right)\right] = n p^* + \frac{1}{Re} n \cdot \tau^*$$

$$t \cdot \tau^* = 0$$

$$v^* \cdot n = 0$$

b′) à $r^* \to \infty$, $v^* = 0$ et $\tau^* = 0$. [14]

Le nombre de *Wéber* est défini par $We = (\varrho U^2 D_f)/\sigma$.

L'équation rhéologique adimensionnelle devient:

$$\tau^* = -\eta^* \dot{\gamma}^* + \frac{\lambda}{\tau}\left\{\dot{\gamma}^*\left(\frac{D}{Dt}\right)^* \theta^* + \frac{1}{2}\left(\frac{\mathscr{D}}{\mathscr{D}t}\right)^* \dot{\gamma}^*\right\} - \frac{1}{2}(1+\varepsilon)\Lambda_2 \theta^* \{\dot{\gamma}^* \cdot \dot{\gamma}^*\} + \cdots \qquad [15]$$

où

$$\left(\frac{\mathscr{D}}{\mathscr{D}t}\right)^* \dot{\gamma}^* = \left(\frac{D}{Dt}\right)^* \dot{\gamma}^* + \frac{1}{2}\left[\frac{U\tau}{D_f}\right] \times (\{\omega^* \cdot \dot{\gamma}^*\} - \{\dot{\gamma}^* \cdot \omega^*\})$$

$$\left(\frac{D}{Dt}\right)^* = \frac{\partial}{\partial t^*} + \left[\frac{U\tau}{D_f}\right](v^* \cdot \nabla^*)$$

$$\frac{\eta}{\eta_0} = \eta^*(\Lambda_1^2\,\mathrm{II}^*, S) \quad [16]$$

$$\frac{\theta}{\eta_0\lambda} = \theta^*(\Lambda_1^2\,\mathrm{II}^*, \alpha, R) \quad [17]$$

$$\Lambda_1 = \left(\frac{U}{D_f}\right) t_1, \quad \Lambda_2 = \left(\frac{U}{D_f}\right) \lambda. \quad [18]$$

Notons qu'à l'eq. [13] le nombre $[D_f/U\tau]$ est une combinaison des nombres Λ_2 et λ/τ.

De l'analyse des eqs. [13]–[18], nous pouvons facilement déceler les nombres sans dimension dont dépend la double intégrale de l'eq. [12]. Pour un système isotherme, nous proposons alors la corrélastion générale suivante

$$\frac{f \cdot Re}{16} = \varphi(Re, We, \Lambda_1, \Lambda_2, \lambda/\tau, S, R, \alpha, \varepsilon). \quad [19]$$

La dépendence du nombre de *Wéber* est introduite par les conditions limites. A l'eq. [14], nous constatons en effet que les principaux rayons de courbure adimensionnels R_1^* et R_2^* décrivant la forme de la bulle (et ainsi le rapport a/D_f) dépendent des conditions hydrodynamiques et du nombre de *Wéber*.

Nous réalisons bien que la fonction φ est très générale et que seule une analyse séquentielle laborieuse nous permettrait d'en obtenir des formes plus précises. Nous essayons toutefois d'en discerner les nombres importants. Notons d'abord certains cas limites intéressants.

a) *Pour les fluides newtoniens* tous les nombres associés au comportement non newtonien et élastique disparaissent et la corrélation s'écrit

$$\frac{f \cdot Re}{16} = \varphi(Re, We). \quad [20]$$

La loi de *Stokes* se traduit par $\varphi = \text{cte} = 1{,}5$ et la loi d'*Hadamard* par $\varphi = \text{cte} = 1$ (2).

b) *Pour les fluides purement visqueux*, les paramètres décrivant le caractère élastique sont nuls et nous obtenons

$$\frac{f \cdot Re}{16} = \varphi(Re, We, \Lambda_1, S). \quad [21]$$

Le nombre Λ_1 et le paramètre S tiennent compte de l'influence du comportement non newtonien sur le mouvement des bulles.

c) *Pour les fluides viscoélastiques*, nous avons retenu deux cas limites d'intérêt:

i) Sous l'hypothèse d'écoulement très lent (creeping) impliquant que

$$\left.\begin{matrix} Re \\ \Lambda_1 \\ \Lambda_2 \end{matrix}\right\} \ll 1$$

les termes de l'ordre du carré de vitesse sont négligeables; $\eta^* \to 1$, $\theta^* \to \theta^*(\alpha)$; le champ de vitesse est une fonction linéaire de la vitesse d'ascension de la bulle, U. De plus ces conditions ne peuvent pratiquement s'appliquer qu'à de très petites bulles de forme sphérique. L'eq. [12] s'écrit alors

$$\frac{f \cdot Re}{16} = \varphi(\lambda/\tau, \alpha). \quad [22]$$

Le nombre λ/τ joue ici un rôle similaire au nombre A défini par *Astarita* (5). Le choix du temps adimensionnel $t^* = (U/D_f)\,t$ à la place de $t^* = t/\tau$ nous donnerait en combinant les nombres sans dimension

$$\frac{1}{Le} \equiv \frac{3\,Re \cdot f}{16} = \varphi'(A, \alpha) \quad [23]$$

ou *Le* 'est le nombre de *Levich* et

$$A = \frac{\varrho g R \lambda}{\eta_0}.$$

A l'exception du paramètre α, ce résultat est identique au cas limite (ii) présenté par *Astarita* (voir eq. [1]). Le paramètre α est associé au comportement de la viscosité dynamique du fluide (14, 15). Des deux eqs. [22] et [23], la première qui fait intervenir la période d'injection τ, nous paraît plus réaliste. Nos données expérimentales pour la solution de Separan semblent le confirmer. Ces résultats [22] et [23] ont des implications des plus intéressantes; en particulier, *l'hypothèse de comportement newtonien pour l'ascension de bulles (ou chute de sphères) en milieu viscoélastique n'est pas nécessairement valable, même à des nombres de Reynolds très faibles.*

ii) Sous l'hypothèse d'écoulement très rapide impliquant que

$$\left.\begin{matrix} Re \\ \Lambda_2 \end{matrix}\right\} \gg 1$$

les forces visqueuses sont négligeables par rapport aux forces d'inertie et aux forces élastiques (terme du deuxième degré dans l'eq. [15]). La corrélation [19] se réduit à

$$\frac{f \cdot Re}{16} = \varphi(We, \lambda/t, \Lambda_2, R, \alpha, \varepsilon). \quad [24]$$

Il nous paraît toutefois difficile de réaliser ces conditions, puisqu'en général les fluides à caractère élastique sont aussi très visqueux.

Corrélation des données expérimentales

Nous avons porté sur les figs. 9 et 10 en coordonnées logarithmiques les variations du coefficient de traînée f (défini par l'eq. [12]) en fonction du nombre de *Reynolds*. Nous avons tracé des droites de pente -1 pour illustrer le comportement de la fonction $\varphi = f \cdot Re/16$. Pour *l'huile de silicone*, la fonction φ augmente légèrement avec le nombre de *Reynolds*, la partie linéaire de la courbe de f (fig. 9) coïncidant par hasard avec la loi de *Stokes*. Pour la *solution de CMC*, φ décroît faiblement avec Re et la partie linéaire de f correspond à $\varphi = 0{,}0225$. Soulignons que dans les deux cas le nombre de *Wéber* varie appréciablement, soit de 1,8 à 100 pour le premier fluide et de 0,05 à 20 pour le second.

Les variations du coefficient de traînée pour la solution de Separan sont présentées à la fig. 10. Pour des valeurs faibles du nombre de *Reynolds*, nous observons une influence considérable de la période d'injection τ ou du rapport λ/τ. La valeur de la fonction φ correspondante varie de 0,010 ($\tau = 4{,}7$ sec) à 0,024 ($\tau = 3600$ sec). Dans cette zone, les variations du nombre de *Wéber* sont considérables, soit de 0,16 à 60. Par contre à des valeurs plus élevées du nombre de *Reynolds* ($Re > 0{,}2$), le coefficient de traînée devient indépendant de la période d'injection et atteint une valeur constante d'environ 1,85 (voir fig. 10). Egalement dans cette région, le nombre de *Wéber* ne varie pratiquement pas ($We \doteq 60$), indiquant que la vitesse d'ascension est étroitement reliée au diamètre frontal des bulles. Toutefois nous ne pouvons pas négliger l'importance de la contribution élastique au taux de dissipation d'énergie. De plus, l'effet du temps adimensionnel λ/τ est implicitement inclus dans le coefficient de traînée par la dépendence de la vitesse U et du diamètre frontal D_f.

Pour des écoulements très lents alors que les bulles sont presque sphériques, la fonction φ devient à peu près indépendante de la forme des bulles. Dans ces conditions, l'influence du nombre de *Wéber* est faible et les propriétés élastiques du fluide contribuent principalement au taux de dissipation d'énergie (terme $-\boldsymbol{\tau}:\nabla \boldsymbol{v}$ de l'eq. [4]). Les données expérimentales à faibles nombres de *Reynolds* ($Re < 0{,}1$) vérifient apparemment le cas limite ($C - i$) de l'eq. [22]: $\varphi = \varphi(\lambda/\tau, \alpha)$. Toutefois il est étonnant que la fonction φ pour la solution de CMC soit in-

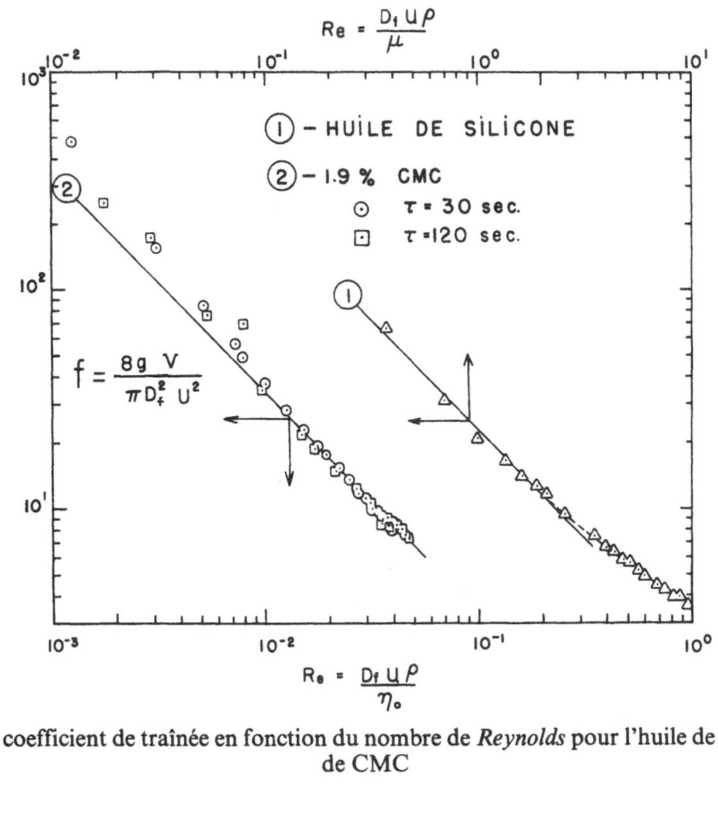

Fig. 9. Variation du coefficient de traînée en fonction du nombre de *Reynolds* pour l'huile de silicone et la solution de CMC

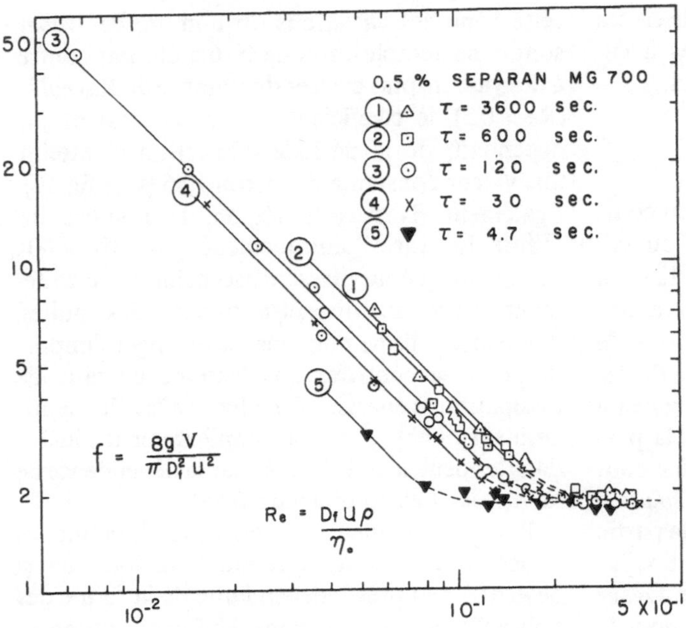

Fig. 10. Variation du coefficient de traînée en fonction du nombre de *Reynolds* pour la solution de Separan

dépendante de la période d'injection et ainsi de λ/τ. Nous remarquons que cette valeur de φ ($= 0{,}0225$) correspond sensiblement à la valeur obtenue pour la solution de Separan à la période de 3600 sec. Cette période est suffissamment longue pour que les contraintes dans le fluide soient complètement «relaxées» entre deux injections consécutives. La solution de CMC se comporte donc en tout temps comme un fluide totalement «relaxé» malgré son caractère élastique prononcé ($\lambda = 77{,}7$ sec). Nous doutons cependant que les conditions nécessaires au cas limite ($C - i$) soient respectées, en particulier que les fonctions rhéologiques η^* et θ^* soient indépendantes de la vitesse de déformation. Le comportement particulier de la solution de CMC pourrait alors s'expliquer par la contribution des autres nombres Λ_1, Λ_2 et des paramètres rhéologiques R, S, α et ε.

Nous devons finalement souligner que cette analyse dimensionnelle peut être reprise en utilisant le diamètre équivalent D_e à la place du diamètre frontal D_f. L'utilisation de D_e, défini à partir du volume, permet d'éliminer une variable et de combiner le coefficient de traînée aux autres nombres sans dimension. Nous n'avons pas pu toutefois obtenir des corrélations valables faisant intervenir D_e pour la solution de Separan.

Conclusion

L'étude de l'hydrodynamique de bulles d'air dans deux fluides viscoélastiques et dans un fluide newtonien nous a permis de mettre en évidence l'influence des propriétés élastiques des fluides sur la vitesse d'ascension et la forme des bulles. Nos observations avec une huile de silicone sont conformes à celles d'*Angelino* (10): les bulles ont la forme de sphères légèrement aplaties au sommet avec base tronquée au sommet et la vitesse d'ascension est proportionnelle à la racine cubique du volume.

Pour les fluides viscoélastiques, trois formes typiques ont été observées: forme goutte d'eau, lentille (sphéroïde très aplati terminé par une queue) et calotte sphérique. La vitesse d'ascension et la forme des bulles dépendent non seulement du volume, mais aussi de la fréquence avec laquelle les bulles sont injectées. Pour la solution de Separan, l'influence de la fréquence d'injection est très marquée et des plus étonnantes: suivant les conditions expérimentales, la vitesse augmente ou décroît avec la période d'injection. Par contre, l'influence de la fréquence d'injection est très faible pour la solution de CMC.

Nous avons présenté une analyse dimensionnelle du mouvement de bulles non sphériques dans un fluide viscoélastique. La corrélation générale obtenue fait intervenir le coefficient de traînée, le nombre de *Reynolds* et divers nombres sans dimension et paramètres associés au comportement non newtonien et élastique du fluide. Nous avons discuté de divers cas limites d'intérêt.

Les résultats expérimentaux du coefficient de traînée f en fonction du nombre de *Reynolds* nous suggèrent que pour des faibles valeurs de *Re*, le nombre de *Wéber* influence peu la forme des bulles et alors les propriétés élastiques contribuent principalement au taux de dissipation d'énergie. Par contre les valeurs de *Re* ne sont pas suffisamment faibles pour que nous puissions négliger l'influence du caractère non newtonien des fluides. Nous nous proposons dans une étude subséquente de vérifier le résultat $f \cdot Re/16 = \varphi(\lambda/\tau, \alpha)$ du cas limite $(C - i)$ pour le mouvement de très petites bulles. A ce régime, le comportement du fluide devrait être dissocié du comportement newtonien par le temps adimensionnel λ/τ et le paramètre élastique α.

Remerciements

Nous tenons à remercier MM. *B. Chartrand, J. Coderre, M. Dugal* et *J. Moréno* pour leur aide précieuse. Ces recherches ont été rendues possibles grâce à l'octroi A-5817 du Conseil National de Recherche du Canada.

Résumé

La dynamique des bulles d'air dans deux fluides viscoélastiques (solutions aqueuses de polyacrylamide et de carboxyméthyle cellulose) et un fluide newtonien (huile de silicone) a été étudiée pour des volumes de bulles variant entre 0,1 et 11 cm³. La vitesse et la forme des bulles ont été déterminées par techniques photographiques avec stroboscope. Le volume des bulles a été mesuré avant chaque injection à l'aide d'une seringue calibrée.

Pour les fluides viscoélastiques, trois formes typiques ont été observées: forme goutte d'eau, lentille et calotte sphérique. La vitesse d'ascension et la forme dépendent non seulement du volume mais aussi de la fréquence avec laquelle les bulles sont injectées. Pour la solution de polyacrylamide (0,5% Separan MG-700), l'influence de la fréquence d'injection est très marquée et des plus étonnantes: selon le volume des bulles, la vitesse décroît de façon monotone ou passe par un minimum pour ensuite augmenter avec la période d'injection. Ce comportement ne peut être relié au caractère non newtonien des fluides mais plutôt à l'influence prédominante des propriétés élastiques sur la forme des bulles. Par contre, pour la solution de carboxyméthyle cellulose, l'influence de la fréquence d'injection est très faible.

Une analyse dimensionnelle du mouvement de bulles non sphériques dans un fluide décrit par une équation rhéologique non linéaire est présentée. Les données expérimentales sont analysées à l'aide du coefficient de traînée en fonction du nombre de *Reynolds, Re*. Les résultats suggérent qu'à faibles valeurs de *Re* (<0,1), les propriétés élastiques du fluide contribuent principalement au taux de dissipation d'énergie alors qu'à des valeurs plus élevées, l'élasticité affecte également la forme des bulles.

Summary

The motion of air bubbles in two viscoelastic fluids (aqueous solutions of Polyacrylamide and of Carboxy Methyl Cellulose) and in a *Newton*ian fluid (Silicone) was studied for volumes ranging from 0.1 to 11 cm³. The velocity and the shape of the bubbles were obtained from photographs with the help of a stroboscope. The volume of the bubbles was measured with a calibrated syringe.

Three typical shapes were observed for the viscoelastic fluids: tear-drop, spheroid, spherical cap. The velocity and the shape are not only function of bubble volume, but depend on the injection frequency. For the Polyacrylamide solution (0.5% Separan MG-700), the injection frequency dependence is most striking: depending on the volume, the bubble velocity decreases monotonously, or goes to a minimum and increases with the injection period. This behaviour cannot be attributed to the non-*Newton*ian character of the fluid, but rather to the influence of the elastic properties on the shape of the bubbles. On the other hand, the injection frequency has little influence on the behaviour in the Carboxy Methyl Cellulose solution.

A dimensional analysis for the motion of non spherical bubbles in a fluid described by a nonlinear rheological equation is presented. Correlations of the drag coefficient as a function of the *Reynolds* number are analysed. Experimental results suggest that at small *Re* (<0.1), the fluid elasticity contributes mainly to the energy dissipation, whereas at high *Re*, the elasticity influences also the shape of the bubbles.

Littérature

1) *Philippoff, W.*, Rubber Chem. et Techn. **10**, 76 (1937).

2) *Astarita, G.* et *G. Apuzzo*, Amer. Inst. Chem. Eng. J. **11**, 815 (1965).

3) *Calderbank, P. M., D. S. L. Johnson* et *J. London*, Chem. Eng. Sci. **25**, 235 (1970).

4) *Leal, L. G., J. Skoog* et *A. Acrivos*, Canad. J. Chem. Eng. **49**, 569 (1971).

5) *Astarita, G.*, I.E.C. Fund. **5**, 548 (1966).

6) *Devic, M.*, Dynamique des bulles en milieux viscoélastiques. Thèse de maîtrise, Ecole Polytechnique, Montréal (1971).

7) *Barnett, S. M., A. E. Humphrey* et *M. Litt*, Amer. Inst. Chem. Eng. J. **12**, 253 (1966).

8) *Fararouri, A.* et *R. C. Kintner*, Trans. Soc. Rheol. **5**, 369 (1961).

9) *Warshay, F. H., E. Bogusz, M. Johnson* et *R. C. Kintner*, Canad. J. Chem. Eng. **37**, 29 (1959).

10) *Angelino, H.*, Chem. Eng. Sci. **21**, 541 (1966).

11) *Uno, S.* et *R. C. Kintner*, Amer. Chem. Engrs. J. **2**, 420 (1956).

12) *Bird, R. B.*, Chem. Eng. Sci. **6**, 123 (1957).

13) *Levich, V. G.*, Physicochemical Hydrodynamics (Englewood Cliffs, N. J. 1962).

14) *Bird, R. B.* et *P. J. Carreau*, Chem. Eng. Sci. **23**, 427 (1968).

15) *Carreau, P. J.*, Trans. Soc. Rheol. **16: 1**, 99 (1972).

Adresse de l'auteur:

Dr. *Pierre J. Carreau* et al.
Département de Génie Chimique
Ecole Polytechnique, 2500 Marie Guyard
Montréal 250 Québec (Canada)

Rheol. Acta 13, 490–494 (1974)

From the Department of Polymer Technology, The Royal Institute of Technology, Stockholm (Sweden)

Superimposed transition mechanisms in polyvinylchloride

By Jan-Fredrik Jansson

With 4 figures

(Received October 27, 1972)

Conventional polyvinylchloride, prepared by radical polymerization above room temperature, has a relatively high syndiotacticity, is somewhat branched and has about 10% crystallinity.

PVC therefore behaves mainly as an amorphous rather than a partially crystalline polymer. By radical polymerization at low temperatures, the degree of crystallinity is increased which results in a higher glass temperature etc. (5).

Therefore, it seems reasonable that the plot of the compliance J_1 against J_2 will show the general shape of the function $\log J_2 = g_\alpha\{\log J_1\}$ (2), for PVC in the main transition regions, where J_1 and J_2 are defined as

$$J_1 = \lim_{n\to\infty} \frac{\varepsilon_1(\zeta/2 + 2n\zeta)}{\sigma_0}$$

and

$$J_2 = \lim_{n\to\infty} \frac{\varepsilon_2(\zeta/2 + 2n\zeta)}{\sigma_0}$$

ε_1 and ε_2 are the responses of the two periodic stress functions:

$$\sigma_1(t) = \begin{cases} \sigma_0; & 2n\zeta < t < (2n+1)\zeta \\ -\sigma_0; & (2n+1)\zeta < t < (2n+2)\zeta \end{cases}$$

$$\sigma_2(t) = (-1)^n(2n+1)\sigma_0 - (-1)^n \frac{2\sigma_0}{\zeta} t; \quad n\zeta < t < (n+1)\zeta$$

where $n = 0, 1, 2, 3 \ldots$

On the whole, this has also been found for a PVC-gel (10% PVC, $\bar{M}_w = 63000$, in dimethylthiantrene DMT) at high temperatures/long times.

In fig. 1, the function $\log J_2 = g\{\log J_1\}$, shifted along the straight line $\log J_2 = 1.1 \log J_1 - 0.52$, is plotted for the mentioned PVC-gel, for plasticized PVC (40% PVC in DMT) and for an unplasticized PVC.

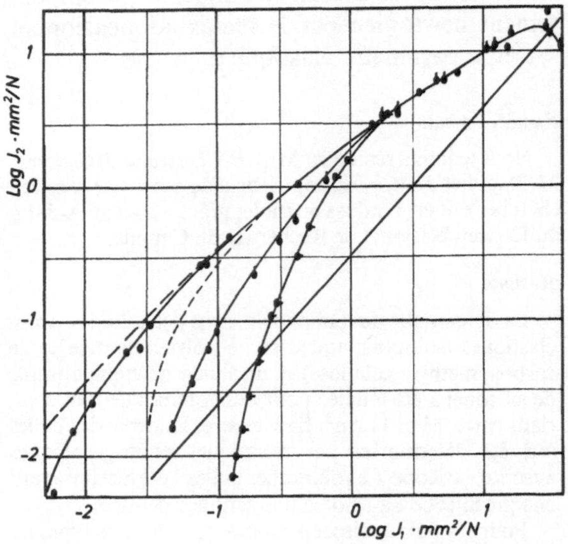

Fig. 1. The function $\log J_2 = g\{\log J_1\}$ for ● PVC-gel (3); plastiz. PVC shifted $J_1 \times 30$ (3); ● PVC, $J_1 \times 130$ (4)

For the plasticized PVC and the PVC-gel the function is calculated from the data of the complex compliance measured in a *"Fitzgerald* apparatus" and reported by *Fitzgerald* and *Ferry* (3) and for the unplasticized PVC from complex modulus measured in a torsion pendulum by *Petersen* and *Rånby* (4). The calculations have been done in accordance with the interrelations developed by *Jansson* (1).

The calculated values correspond to dynamic data in the frequency range 10^{-12}–10^{-4} p/s at 22 °C for the unplasticized PVC, 40–500 p/s at +10 to +60 °C for the plasticized PVC and 30–500 p/s at −10 to +30 °C for the PVC-gel.

It can be seen from a more detailed study of the curves that the PVC-gel shows divergences from the general shape of the function at low temperatures/short times. Using the part of the curve corresponding to the general shape of the function, an extrapolation to low temperatures/

short times gives, however, a "frozen-in compliance", $J_{10} \approx 3 \cdot 10^{-3}$ mm²/N and the divergences from the "simple segment movements" are equivalent to either an increase in stiffness (lower J_1) or an increase in stiffness combined with increases of losses (higher J_2). Similar divergences can also be observed for the other PVC-types.

The β-transition appears for about 500 p/s at 0 °C (5) giving a G''_{max} of about 10^2 N/mm² at $G' \approx 1.5 \cdot 10^3$ N/mm² (4), corresponding to $J_1 \approx 7 \cdot 10^{-4}$ and $J_2 \approx 4 \cdot 10^{-3}$ mm²/N. Thus the β-mechanism does not influence the curve for the unplasticized PVC in fig. 1. For the plasticized PVC, a small influence can be expected to appear from the β-mechanism, but it is not substantial enough to explain the divergences from the general shape of the curve.

In the case of the PVC-gel, the β-mechanism can be expected to appear within the studied frequency-temperature range but can undoubtedly be disregarded due to the relative size of the J_1; J_2-values of the two mechanisms. Thus in no case can the divergences be explained by means of the β-mechanism.

To study and find a plausible hypothesis as an explanation for the divergences, J_1 and J_2 as functions of temperature at constant time $\zeta/2 = 30$ s have been measured in special equipment (1) for a series of specimens, of different processing degrees, made from a suspension PVC with a molecular weight of about 130000, table 1.

The specimens are made by post-forming calendered sheets. The time and temperature for the calendering appear in table 1.

Table 1

Sample D	Sheet calendered for 15 min at 120 °C.
Sample B	Sheet calendered for 15 min at 140 °C.
Sample F	Sheet calendered for 45 min at 180 °C.
	Cylindric specimens made by pressing a bent slice of the sheet into a mould for about 15 min at 150 °C.

All specimens made from S-PVC, $\bar{M}_w = 130000$, from KemaNord AB.

The post-forming is made by pressing a bent slice of the sheet into a cylindric mould at 150° for about 15 minutes followed by very gradual cooling for about four hours down to room temperature.

After being fixed in the measuring equipment, the specimens are heat-treated at about 110 °C for more than 4 h. The cylindric specimen has a length of about 10 mm, a diameter of 30 mm and a wall thickness of 2 mm.

It has not been possible to indicate any change in the molecular weight distribution curves, measured by Gel Permeation Chromatography for the different specimens. Thus no chemical decomposition is observed even in the most intensively calendered material.

The measurements are carried out both from high to low temperature and vice versa, by very slow changing of the temperature (5°C/h) and by keeping the specimen at a fixed surrounding temperature for at least 20 min before measuring.

The meaning of this procedure in combination with the very rigorous heat treatment is to give a reproducible microstructure and to be able to evaluate if any of the transitions is due to changes in crystallinity. Of course, changes in crystallinity will give different data depending on whether the measurements are made from lower to higher temperatures or vice versa.

Results

Figs. 2 and 3 show J_1 and J_2 as functions of temperature at the time $\zeta/2 = 30$ s for the specimens *D*, *B*, and *F*.

The curves include the main transition above 75 °C and the plateau-zone. Besides the main transition, a secondary transition, α', appears in the plateau-zone. At decreasing processing degree, the peak of the main transition moves

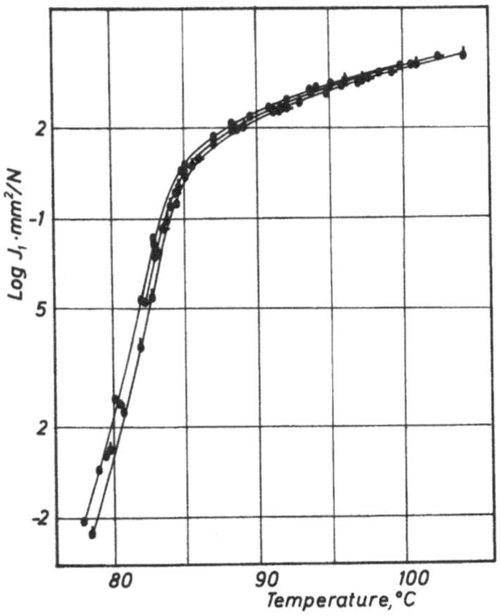

Fig. 2. Log J_1 as a function of temperature for material: ● *D*; ● *B*; and ● *F*

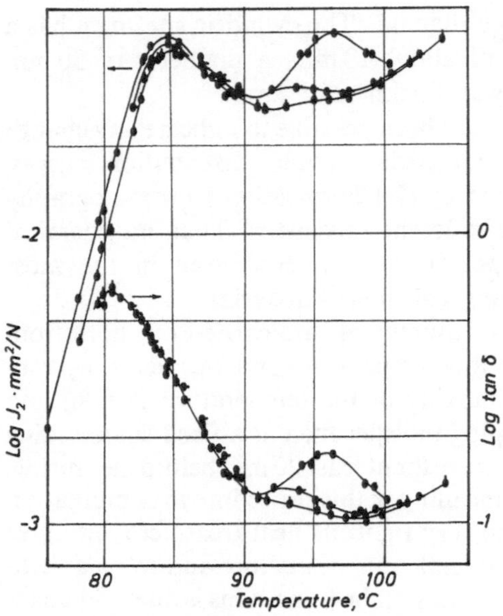

Fig. 3. LogJ_2 and tan δ for the materials: ● *D*; ●- *B*; and ● *F*

somewhat towards a higher temperature and the peak of the secondary dispersion rises.

In fig. 4, the function $\log J_2 = g\{\log J_1\}$ is drawn for the three specimens in figs. 2 and 3, completed with additional data below 75 °C.

The curves for the different specimens coincide completely at low and medium temperatures, whereas a considerable divergence appears at high temperatures. At medium temperatures around $J_1 \approx 1.5 \cdot 10^{-2}$ mm²/N and $J_2 \approx 10^{-2}$ mm²/N a slight shoulder is shown.

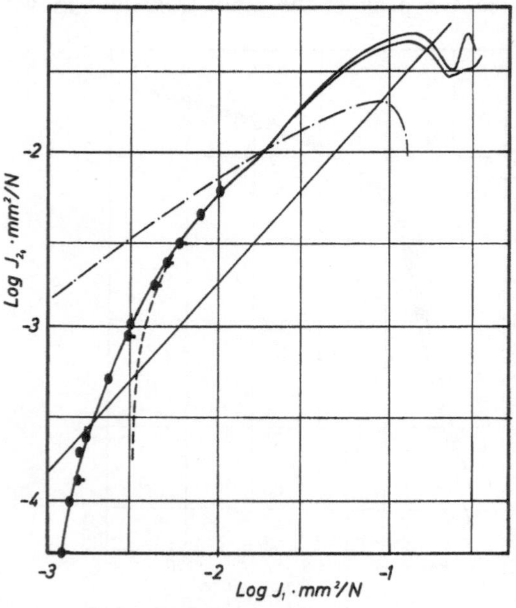

Fig. 4. The function $\log J_2 = g\{\log J_1\}$; —·— $\log J_2 = g_{\alpha 0}\{\log J_1\}$

As seen in fig. 4, the curves do not coincide with the general shape of the function $\log J_2 = g_\alpha\{\log J_1\}$ as constituted by *Jansson* (2).

At low temperatures, the divergence is equivalent either to an increase in stiffness (reduced J_1) or to an increase in stiffness combined with an increase of losses (raised J_2).

If, however, the general curve of $\log J_2 = g_{\alpha 0}\{\log J_1\}$ is drawn through the slight shoulder in the medium part of the curve, a "frozen-in" compliance $J_{10} \approx 3 \cdot 10^{-3}$ mm²/N is obtained from the part of the curve directly below the shoulder. This compliance agrees with the corresponding value for the PVC-gel in fig. 1.

Above the shoulder, an additional divergence from the general behaviour appears which gives a considerable loss increase and is the reason why the maxima of the curves do not fall along the straight line.

Conclusion

It is obviously not possible to make a detailed analysis of the types of molecular mechanisms, that cause the viscoelastic behaviour in the glass-rubber transition of PVC, based only on results from dynamic mechanical measurements. However, by using the general properties of the function $\log J_2 = g_\alpha\{\log J_1\}$ and from literature, some ideas of the kind of molecular mechanisms can be put forward.

From the results reported above, it is apparent that the viscoelastic behaviour in the glass-rubber transition of PVC stems from more than one single molecular mechanism and that "simple segment movements" dominate the behaviour only in the central part of the main transition immediately below the shoulder of the $J_2 - J_1$ diagram in fig. 4.

The "frozen-in compliance" of the simple segment movements is about $3 \cdot 10^{-3}$ mm²/N and agrees with the corresponding value for the PVC-gel in fig. 1 and is not affected by the processing degree. This is only to be expected, as no change of the segment mobility might arise from moderate changes in micro-structure.

1. At low temperatures, the segment movements are affected in a way which results in an increase of losses and/or stiffness. The influence is not affected by changes in the gelation ratio and therefore it seems not likely that it can be ascribed to the existence of "unmelted" PVC-particles.

It seems possible that the influence arises from the crystallites which might not be strongly altered by the present changes in processing parameters. This is in agreement with the results of dynamic mechanical measurements for PVC with different crystallinity (6).

2. At high temperatures above the shoulder in fig. 4, a substantial increase in mechanical losses appears. The peak value of the J_2-curve is somewhat influenced by the different gelation ratios and the J_1 and J_2 curves are shifted along the temperature axis.

It seems possible that part of the gelation influence derives from the changes in the α'-mechanism in the plateau-zone but this cannot be the only explanation for either the increase in mechanical losses or the gelation influence of the α-peak.

It is not possible at present to state the origin of this loss-increasing molecular mechanism. In a series of papers, *Ferry* and co-workers (7), however, have proved the existence of a similar mechanism in polymethacrylates, which is attributed to the decomposition of "widely spaced entanglements".

The PVC-polymer in question has a high molecular weight and the very small crystallites behave as physical network joints or entanglement couplings. Thus the conditions for "widely spaced entanglements" arising seem to be fullfilled. It is reasonable to expect small changes in these conditions owing to the changes in the plastification.

3. In the plateau-zone a secondary dispersion mechanism, α', appears, which shows pronounced sensitivity to the processing conditions.

When the polymer is calendered for a long time at a high temperature, an increasing part is "melted". The gelation ratio is increased.

As the peak value of the α'-dispersion increases with increasing gelation ratio, it seems likely that the molecular mechanism which causes this phenomena must be concentrated to the fractions of the polymer outside, the "unmelted" microparticles.

Nor do the phenomena seem to arise from the regions between matrix and microparticles, because the fraction of these regions decreases when the size and number of the microparticles decrease.

As mentioned above, no changes in crystallinity seem to occur during the measurements, the very small crystallites constitute network joints and the polymer outside the microparticles behave like a very lightly cross-linked polymer with a pronounced plateau-zone. Therefore it seems possible that the α'-dispersion originates from the decomposition of "untrapped entanglements" in accordance with the corresponding molecular mechanism in lightly cross-linked elastomers (8).

Acknowledgements

These investigations are part of a research program on Mechanical Long Term Properties of Polymers supported by the Swedish Board for Technical Development (STU). The author should like to thank Professor *Bengt Rånby* for valuable discussions on the subject of this paper.

Summary

Studies have been made of the viscoelastic behaviour of PVC in the glass-rubber transition by means of square and triangular periodic stresses, in order to find a plausible hypothesis on the appearing molecular mechanisms.

In this respect it is obviously not possible to make a detailed analysis of the different types of mechanisms, based only on results from dynamic mechanical measurements. However, by using the general properties of the function $\log J_2 = g_\alpha\{\log J_1\}$ some ideas can be put forward.

It is apparent that the viscoelastic behaviour stems from more than one single molecular mechanism. "Simple segment movements" dominate the behaviour only in the central part of the transition.

At temperatures around the glass point an increase in losses and/or stiffness appears. It seems possible that this behaviour arises from the influence of the crystallites. At high temperatures a substantial increase in mechanical losses appears, which seems to be due to decomposition of "widely spaced entanglements".

In the plateau-zone a secondary dispersion appears which shows pronounced sensitivity to the processing conditions. At increased gelation ratio the value of the loss peak increases. As no changes in crystallinity have been observed, it is suggested that this dispersion originates from the decomposition of "untrapped entanglements" outside the network composed by the very small crystallites.

References

1) *Jansson, J.-F.*, J. Appl. Polymer Sci. **17**, 2965 (1973).

2) *Jansson, J.-F.*, J. Appl. Polymer Sci. **17**, 2977 (1973).

3) *Fitzgerald, E. R.* and *J. D. Ferry*, J. Colloid Sci. **8**, 1 (1953).

4) *Petersen, J.* and *B. Rånby*, Makromolekulare Chemie **133**, 251 (1970).

5) *McCrum, N. G., B. E. Read*, and *G. Williams*,

Anelastic and Dielectric Effects in Polymeric Solids (New York 1967).

6) *Pezzin, G., G. Ajroldi, T. Casiraghi, C. Garbuglio,* and *G. Vittadini,* J. Appl. Polymer Sci. **16**, 1839 (1972).

7) *Child, W. C.* and *J. D. Ferry*, J. Colloid Sci. **12**, 327 (1957); J. Colloid Sci. **12**, 389 (1957). – *Dannhauser, W., W. C. Child*, and *J. D. Ferry*, J. Colloid Sci. **13**, 103 (1958). – *Kurath, S. F., T. P. Yin, J. W. Berge*, and *J. D. Ferry*, J. Colloid Sci. **14**, 147 (1959).

8) *Ferry, J. D.*, Viscoelastic Properties of Polymers, Second Edition (New York 1970).

Author's address:
Dr. *Jan-Fredrik Jansson*
The Royal Institute of Technology
Dept. of Polymer Technology
Teknikringen 44
S-10044 Stockholm 70

Rheol. Acta 13, 495–500 (1974)

From the Department of Mathematics, Indian Institute of Technology, Powai, Bombay-76 (India)

Stability of plane Couette flow of a viscoelastic fluid with uniform cross-flow

By R. K. Bhatnagar*) and O. P. Sharma

With 6 figures

(Received October 27, 1972)

1. Introduction

The linear stability of plane *Couette* flow of viscoelastic fluids has been the subject of several investigations in recent years. *Giesekus* (1) has discussed the stability of viscoelastic fluids for circular and plane *Couette* flows theoretically as well as experimentally. One of the most important facts observed by him is that viscoelasticity gives rise to cellular type of instabilities even in the absence of inertial forces if the second normal stress difference is chosen as positive. In other words inertia only modifies the critical value of the characteristic number associated with the neutral stability. Recently, in a series of subsequent papers *Bhatnagar* and *Giesekus* (2, 3) confirmed these ideas for the case of plane channel flow (2) and plane *Poiseuille* flow (3). For these flows, they also pointed out that two types of disturbances with different cell widths may exist simultaneously provided the parameter representing the ratio of inertial to elastic forces lies in a certain range. While investigating the overstability of plane *Couette* flow, *Giesekus* and *Bhatnagar* (4) found that, in general, the overstable mode is higher than the stationary mode but that both can come close to each other if certain conditions are satisfied.

If the walls are porous and a uniform cross-flow is superposed by injecting a certain amount of fluid at one wall and removing an equal amount at the opposite wall, it is well-known that the symmetry of the flow is destroyed and the introduction of cross-flow results in a non-parallel flow with curved stream lines.

For classical viscous fluids, the stability of plane *Couette-Poiseuille* flow with uniform cross-flow has been recently studied by *Hains* (5) using the *Orr-Sommerfeld* type of disturbances and thereby solving the resulting disturbance equations by the method of finite-differences where a new variable introduced by *Thomas* (6) was used to reduce the error in the analysis. *Hains* (5) showed that cross-flow produces a significant increase in the critical *Taylor* number.

In the present work, we discuss the stability of plane *Couette* flow of a non-*Newton*ian fluid with a uniform cross-flow across the boundaries. In the analysis, we utilize the cellular type of disturbances and adopt the method of solution first given by *Chandrasekhar* (7) and later used by various authors (2, 8, 9, 10). Over and above the results of *Hains* (5) for classical viscous fluids, we note that, in the presence of viscoelasticity, a uniform cross-flow may also have a destabilizing effect in a certain range of ratio of intertial to elastic forces.

*) Present address: Mathematical Sciences Department, IBM Thomas J. Watson Research Center, Yorktown Heights, N.Y., 10598 (USA).

2. Formulation of the problem

We choose as a model of viscoelastic fluid a simple fluid with fading memory, applying a well-known approximation for slow motions and restricting herein to second order terms only. The constitutive equation has the form [cf. *Giesekus* (11)]:

$$\underset{=}{S} = -P\underset{=}{I} + 2\eta[\underset{=}{f}^{(1)} + \varkappa^{(2)}\underset{=}{f}^{(2)} + \varkappa^{(11)}\underset{=}{f}^{(1)2}] \qquad [1]$$

where

$$\underset{=}{f}^{(1)} = \tfrac{1}{2}(\nabla\underset{-}{v} + \underset{-}{v}\nabla),\quad \underset{=}{\omega} = \tfrac{1}{2}(\nabla\underset{-}{v} - \underset{-}{v}\nabla) \qquad [2]$$

$$\underset{=}{f}^{(2)} = \left(\frac{\partial}{\partial t} + \underset{-}{v}\cdot\nabla\right)\underset{=}{f}^{(1)} + \underset{=}{\omega}\cdot\underset{=}{f}^{(1)} - \underset{=}{f}^{(1)}\cdot\underset{=}{\omega} \qquad [3]$$

are respectively the usual rate of strain tensor, the vorticity tensor, and the second corotational kinematic tensor; η represents the *Newtonian* viscosity, $\varkappa^{(2)}$ and $\varkappa^{(11)}$ are two constants having dimensions of time and characterize the elasticity of the fluid. Further, in eq. [1], $\underset{=}{S}$ represents the stress tensor and p the undetermined pressure.

The eqs. [1]–[3] have to be solved along with the equation of momentum and continuity

$$\nabla\cdot\underset{=}{S} = \rho\left\{\frac{\partial \underset{-}{v}}{\partial t} + \underset{-}{v}\cdot\nabla\underset{-}{v}\right\},\qquad \nabla\cdot\underset{-}{v} = 0 \qquad [4]$$

and the appropriate boundary conditions of the problem.

Let us consider the problem of plane *Couette* flow between two parallel porous walls, one of which moves with a constant velocity U_0, while the other is held at rest. We assume that there is also a superposed uniform crossflow, i.e., injection of a certain amount of fluid at one wall and removal of an equal amount at the opposite

32*

wall. We choose a cartesian coordinate system (fig. 1) with direction of flow along X-axis and that of flow gradient along Y-axis. In this system let h denote the distance between the two boundaries which are represented by

$$y = \pm \frac{h}{2}.$$

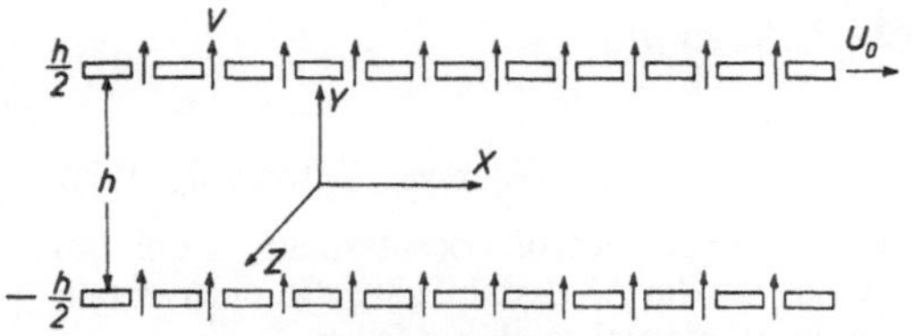

Fig. 1. Definition sketch for plane *Couette* flow with uniform cross-flow

Let h denote the characteristic length of the problem. We introduce the non-dimensional coordinates

$$\xi = y/h, \quad \zeta = z/h. \qquad [5]$$

For the steady-state, the problem under consideration has the solution

$$u = U(\xi), \quad v = V(\text{const}), \quad w = 0 \qquad [6]$$

where u, v, w are the components of velocity respectively in the increased directions of X, Y, Z. The solution [6] satisfies the equation of continuity identically.

Substituting from [6] into [1]–[3] the components of stress tensor are seen to have the following forms:

$$S_{xx} = -p + \frac{\eta(\varkappa^{(11)} - 2\varkappa^{(2)})}{2h^2}(DU)^2$$

$$S_{yy} = -p + \frac{\eta(\varkappa^{(11)} + 2\varkappa^{(2)})}{2h^2}(DU)^2$$

$$S_{zz} = -p$$

$$S_{xy} = \frac{\eta}{h}\left[DU + \frac{\varkappa^{(2)} V}{h}(D^2 U)\right]$$

$$S_{xz} = S_{zy} = 0 \qquad [7]$$

where

$$D = \frac{d}{d\xi} = h\frac{d}{dy}. \qquad [8]$$

From eqs. [4], [6], and [7] it is seen that the equation satisfied by $U(\xi)$ is

$$\varrho V D U = \frac{\eta}{h^2}(D^2 U)$$

or

$$\beta D U = D^2 U \qquad [9]$$

where

$$\beta = \frac{\varrho V h}{n} \quad \text{(cross-flow } \textit{Reynolds} \text{ number)}.$$

Eq. [9] has to be solved under the boundary conditions

$$U(\xi) = \begin{cases} U_0, & \text{at } \xi = \frac{1}{2} \\ 0, & \text{at } \xi = -\frac{1}{2}. \end{cases} \qquad [10]$$

The solution of [9]–[10] is clearly

$$U(\xi) = U_0 \frac{\left(e^{\beta(\xi + \frac{1}{2})} - 1\right)}{e^{\beta} - 1}. \qquad [11]$$

If β is small, we have

$$U(\xi) = U_0\left(\xi + \frac{1}{2}\right) + \frac{\beta}{2}\left(\xi^2 - \frac{1}{4}\right)U_0. \qquad [12]$$

Here we note that solution does not depend on the second order parameters but only on cross-flow parameter β. Thus it is same as for a *Newtonian* fluid. Further, from eqs. [9] and [12] we note that $D^2 U$ is no longer zero across the walls but depends on β.

3. The disturbance equations

To establish the disturbance equations, we disturb the basic flow by superimposing a stationary flow of cellular type, i.e., we consider the case of neutral stability only. We write, therefore (as in refs. 2, 3, 4):

$$u(\xi, \zeta) = U(\xi) + \psi(\xi)\sin\varepsilon\zeta$$

$$v(\xi, \zeta) = V + \varepsilon\chi(\xi)\sin\varepsilon\zeta$$

$$w(\xi, \zeta) = D\chi(\xi)\cos\varepsilon\zeta \qquad [13]$$

where V is constant, ε is the wave number. This choice of u, v, and w satisfies the equation of condition identically. The functions $\psi(\xi)$ and $\chi(\xi)$ are assumed to be small. The usual linearlization method is followed in which products and higher powers of $\psi(\xi)$ and $\chi(\xi)$ are neglected. It may be verified that the components in the disturbed state are:

$$S_{xx} = \frac{\eta(\varkappa^{(11)} - 2\varkappa^{(2)})}{h^2}(DU)(D\psi)\sin\varepsilon\zeta$$

$$S_{yy} = \left[\frac{2\eta\varepsilon}{h}(D\chi) + \frac{\eta(\varkappa^{(11)} + 2\varkappa^{(2)})}{h^2}(DU)(D\psi) + \frac{2\eta\varkappa^{(2)}\varepsilon V}{h^2}(D^2\chi)\right]\sin\varepsilon\zeta$$

$$S_{zz} = -\left[\frac{2\eta\varepsilon}{h}(D\chi) + \frac{2\eta\varepsilon V\varkappa^{(2)}}{h^2}(D^2\chi)\right]\sin\varepsilon\zeta$$

$$S_{xy} = \left[\frac{\eta}{h}(D\psi) + \frac{\eta\varepsilon\varkappa^{(2)}}{h^2}(D^2U)\chi + \frac{\eta\varepsilon(\varkappa^{(1)} - \varkappa^{(2)})}{h^2}(DU)(D\chi) + \frac{\eta\varkappa^{(2)}V}{h^2}(D^2\psi)\right]\sin\varepsilon\zeta$$

$$S'_{xz} = \left[\frac{\eta\varepsilon}{h}\psi - \frac{\eta\varkappa^{(2)}}{h^2}(DU)(D^2\chi) + \frac{\eta\varkappa^{(11)}}{2h^2}(DU)(D^2+\varepsilon^2)\chi + \frac{\eta\varkappa^{(2)}\varepsilon V}{h^2}(D\psi)\right]\cos\varepsilon\zeta$$

$$S_{yz} = \left[\frac{\eta}{h}(D^2+\varepsilon^2)\chi + \frac{\eta\varepsilon(\varkappa^{(11)} + 2\varkappa^{(2)})}{2h^2}(DU)\psi + \frac{\eta V\varkappa^{(2)}}{h^2}D(D^2+\varepsilon^2)\chi\right]\cos\varepsilon\zeta. \quad [14]$$

If $V = 0$, the above equations reduce to those obtained by *Bhatnagar* and *Giesekus* (2) in the absence of cross-flow. When [13] and [14] are substituted in the equations of motion [4] then the component of these equations in the direction of flow gives the following differential equation

$$(D^2 - \beta D - \varepsilon^2)\psi = -\frac{\varepsilon\varkappa^{(11)}U_0}{2h}[(DU)(D^2 - \varepsilon^2 - 2\alpha)\chi + 2(D^2U)D\chi] - \frac{\beta}{\bar{\alpha}}[(D^2-\varepsilon^2)D\psi + (D^3U)\chi] \quad [15]$$

and the elimination of pressure between the other two components leads to

$$(D^2-\varepsilon^2)(D^2 - \beta D - \varepsilon^2)\chi = \frac{\varepsilon(\varkappa^{(11)} + 2\varkappa^{(2)}U_0)}{2h}[(DU)(D^2-\varepsilon^2)\psi - (D^3U)\psi] - \frac{\beta}{\bar{\alpha}}(D^2-\varepsilon^2)^2(D\chi) \quad [16]$$

where

$$\alpha = \frac{h^2\varrho}{\eta\varkappa^{(11)}}, \quad \bar{\alpha} = \frac{h^2\varrho}{\eta\varkappa^{(2)}}$$

represent ratios of inertial to elastic forces.

Let

$$\psi = \frac{\varepsilon(\varkappa^{(11)} + 2\varkappa^{(2)})}{2h}U_0\chi.$$

In the sequel, we shall make the approximation that $V/U_0 \ll 1$, then eqs. [15] and [16] take the form

$$(D^2 - \beta D - \varepsilon^2)\psi = -\varepsilon^2\Gamma \times [(1+\beta\xi)(D^2 - \varepsilon^2 - 2\alpha)\chi + 2\beta D\chi] \quad [17]$$

and

$$(D^2-\varepsilon^2)(D^2 - \beta D - \varepsilon^2)\chi = (1+\beta\xi)(D^2-\varepsilon^2)\psi \quad [18]$$

with the boundary conditions

$$\psi(\xi) = \chi(\xi) = D\chi(\xi) = 0 \quad \text{at} \quad \xi = \pm\tfrac{1}{2} \quad [19]$$

where

$$\Gamma = \frac{\varkappa^{(11)}(\varkappa^{(11)} + 2\varkappa^{(2)})}{4h^2}U_0^2$$

represents the characteristic number analogous to the *Taylor* number in case of *Couette* flow for a *Newton*ian fluid.

An examination of the stability eq. [17] and [18] reveals that β not only changes the velocity profiles but appears also in some of the convective terms directly. We note a strong coupling between the two disturbance equations, β destroying the symmetric nature as in the case of plane *Couette* flow (1). The above eqs. [17] and [18] together with the boundary condition [19] represent a characteristic value problem for determining Γ as a function of ε, α and β being regarded as parameters. The solutions $\Gamma(\xi)$ of [17] and [18] represent the neutral stability, and now it is our purpose to evaluate $\Gamma(\varepsilon)$.

4. Solution of the disturbance equations

The eigenvalues of the system [17]–[19], which specify the states of neutral stability, form a relationship,

$$f(\Gamma, \varepsilon;\ \alpha, \beta) = 0 \quad [20]$$

which must be now determined.

To solve the system of equations we adopt the approximate method (viz. the *Galerkin* method).

Accordingly, $\psi(\xi)$ and $\chi(\xi)$ are expanded as (8, 9, 10, 11)

$$\chi(\xi) = \sum_{n=1}^{\infty} A_n \chi_n(\xi) = \sum_{n=1}^{\infty} [Ac;\, nC_n + As;\, nS_n]$$

$$\psi(\xi) = \sum_{n=1}^{\infty} B_n \psi_n(\xi) = \sum_{n=1}^{\infty} [Bc;\, nE_n + Bs;\, nF_n]$$

where $Ac; n$, $As; n$, $Bc; n$, and $Bs; n$ are real constants.

The functions C_n, S_n are defined as

$$C_n = \frac{\cosh(\lambda_n \xi)}{\cosh(\lambda_n/2)} - \frac{\cos(\lambda_n \xi)}{\cos(\lambda_n/2)}$$

$$S_n = \frac{\sinh(\mu_n \xi)}{\sinh(\mu_n/2)} - \frac{\sin(\mu_n \xi)}{\sin(\mu_n/2)} \qquad [22]$$

λ_n and μ_n are respectively the roots of the equations

$$\tanh(\lambda/2) + \tan(\lambda/2) = 0$$

$$\coth(\mu/2) - \cot(\mu/2) = 0. \qquad [23]$$

Thus $\chi(\xi)$ satisfies all the four boundary conditions. Since $\psi(\xi)$ must satisfy only two boundary conditions, E_n and F_n may be chosen as

$$E_n = \cos(2n-1)\pi\xi, \quad F_n = \sin 2n\pi\xi. \qquad [24]$$

Substituting series [21] in the eqs. [17], [18], and making use of the orthogonality criteria,

$$\int_{-\frac{1}{2}}^{\frac{1}{2}} L_1(A_n\chi_n, B_n\psi_n)\,\chi_n d\xi = 0$$

$$\int_{-\frac{1}{2}}^{\frac{1}{2}} L_2(A_n\chi_n, B_n\psi_n)\,\psi_n d\xi = 0$$

where L_1 and L_2 are the linear operators in [17] and [18], yields the secular determinant:

$$\begin{vmatrix} -\bar\Gamma A_{nm} & -\beta\bar\Gamma(2\bar A_{nm} + \tilde A_{nm}) & -B_{nm} & \beta\bar B_{nm} \\ C_{nm} & \beta\bar C_{nm} & -D_{nm} & -\beta\bar D_{nm} \\ -\beta\bar\Gamma(2I_{nm} + \tilde I_{nm}) & -\bar\Gamma\bar I_{nm} & \beta J_{nm} & -\bar J_{nm} \\ \beta K_{nm} & \bar K_{nm} & -\beta L_{nm} & -\bar L_{nm} \end{vmatrix} = 0 \qquad [25]$$

where

$$\bar\Gamma = \varepsilon^2 \Gamma$$

$$A_{nm} = (C_n''|C_n) - (\varepsilon^2 + 2\alpha)\,\delta_{nm}$$

$$\bar A_{nm} = (S_n'|C_m)$$

$$\tilde A_{nm} = (S_n''|\xi|C_m) - (2\alpha + \varepsilon^2)\cdot(S_n|\xi|C_m)$$

$$B_{nm} = -[(2n-1)^2\pi^2 + \varepsilon^2]\cdot(E_n|C_m)$$

$$\bar B_{nm} = (F_n'|C_m)$$

$$C_{nm} = (\lambda_n^4 + \varepsilon^4)\cdot(C_n|E_m) - 2\varepsilon^2(C_n''|E_m)$$

$$\bar C_{nm} = \varepsilon^2(S_n'|E_m) - (S_n'''|E_m)$$

$$D_{nm} = -\tfrac{1}{2}[(2n\pi)^2 + \varepsilon^2]\cdot\delta_{nm}$$

$$\bar D_{nm} = -[(2n\pi)^2 + \varepsilon^2]\cdot(F_n|\xi|E_m)$$

$$I_{nm} = (C_n'|S_m)$$

$$\tilde I_{nm} = (C_n''|\xi|S_m) - (2\alpha + \varepsilon^2)\cdot(C_n|\xi|S_m)$$

$$\bar I_{nm} = (S_n''|S_m) - (2\alpha + \varepsilon^2)\,\delta_{nm}$$

$$J_{nm} = (E_n'|S_m)$$

$$\bar J_{nm} = -[(2n\pi)^2 + \varepsilon^2]\cdot(F_n|S_m)$$

$$K_{nm} = \varepsilon^2(C_n'|F_m) - (C_n'''|F_m)$$

$$\bar K_{nm} = (\mu_n^4 + \varepsilon^4)\cdot(S_n|F_m) - 2\varepsilon^2(S_n''|F_m)$$

$$L_{nm} = -[(2n-1)^2\pi^2 + \varepsilon^2]\cdot(E_n|\xi|F_m)$$

$$\bar L_{nm} = -\tfrac{1}{2}[(2n\pi)^2 + \varepsilon^2]\,\delta_{nm}. \qquad [26]$$

In [26], we define

$$\delta_{nm} = \int_{-\frac{1}{2}}^{\frac{1}{2}} C_n C_m d\xi = \int_{-\frac{1}{2}}^{\frac{1}{2}} S_n S_m d\xi = 2\int_{-\frac{1}{2}}^{\frac{1}{2}} E_n E_m d\xi = 2\int_{-\frac{1}{2}}^{\frac{1}{2}} F_n F_m d\xi$$

$$(C_n''|C_m) = \int_{-\frac{1}{2}}^{\frac{1}{2}} C_n'' C_m d\xi$$

$$(C_n|\xi|S_m) = \int_{-\frac{1}{2}}^{\frac{1}{2}} C_n\cdot\xi\cdot S_m d\xi \quad \text{etc.}$$

The infinite order determinant [25] is approximated by taking $m = n = 1$ for numerical evaluation.

5. Discussion of the results

Fig. 2 shows the curves of neutral stability $\Gamma(\varepsilon)$ for a fixed value of $\alpha = 30$ and two different values of the parameter β. The continuous curve represents the case $\beta = 0$ while the dashed curve the case $\beta = 0.2$. Although the curves of neutral stability for these two values of β are similar, it

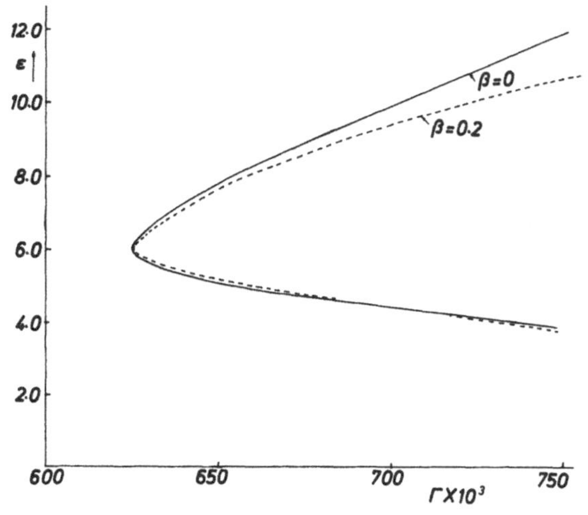

Fig. 2. Curves of neutral stability $\Gamma(\varepsilon)$ for $\alpha = 30$, $\beta = 0.0$, and 0.2

may be noted that value of Γ_{min} is higher for the case $\beta = 0.2$ compared to that for $\beta = 0$. The critical value of ε is given by $\varepsilon_{cr} \cong 6.0$.

Fig. 3 depicts the curves of neutral stability, $\Gamma(\varepsilon)$ for $\alpha = 100$ and the same two values of β as in fig. 2 viz. $\beta = 0, 0.2$. From the curves of neutral stability for these two values of cross-flow parameter β, we observe that when $\alpha = 100$, the increase in the cross-flow parameter β (compared to that for $\beta = 0$) produces a destabilizing effect as the Γ_{min} for $\beta = 0.2$ is lower than that for $\beta = 0$. The critical wave number in this case is seen to be $\cong 5$.

The above conclusions drawn on the basis of figs. 2 and 3 are further illucidated in fig. 4 where we have plotted $\Gamma^* = 2\alpha\Gamma$ against ε for $\alpha = 30, 100$, and $\beta = 0, 0.2$. In general, therefore, we may say that effect produced by the cross-flow depends strongly on the choice of α, which represents the ratio of inertial to elastic forces. The interaction of the elasticity of the fluid with cross-flow seems to be of quite interest.

In fig. 5, we have depicted the variation of Γ_{min} against β for $\alpha = 30$. It can be seen that in this case, as β increases, Γ_{min} continuously in-

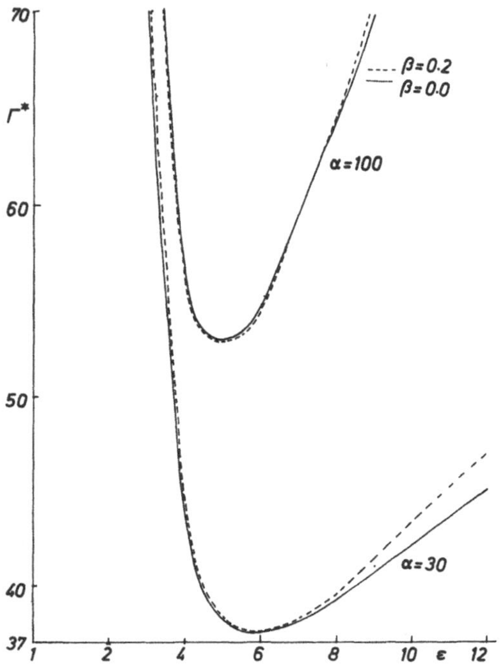

Fig. 4. Curves of neutral stability $\Gamma^*(\varepsilon)$ for $\alpha = 30$ and 100, $\beta = 0.0$ and 0.2

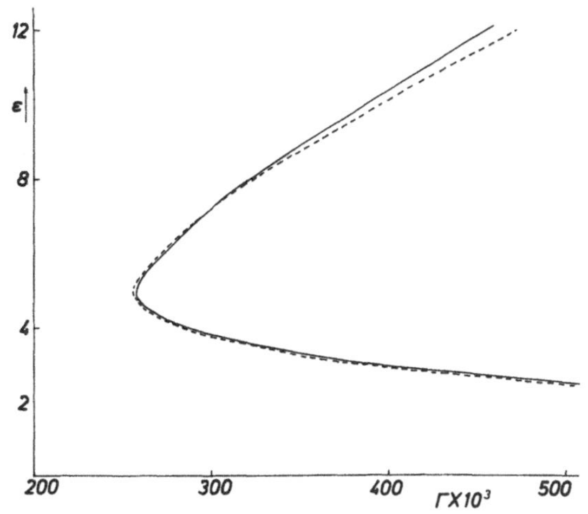

Fig. 3. Curves of neutral stability $\Gamma(\varepsilon)$ for $\alpha = 100$, $\beta = 0.0$, and 0.2

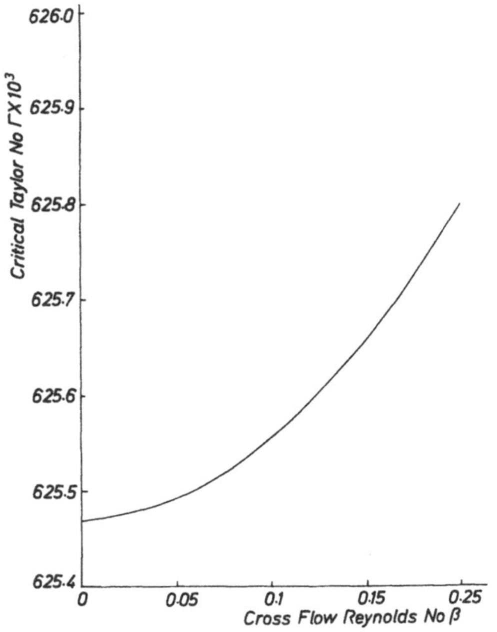

Fig. 5. Variation in critical *Taylor* number with cross-flow *Reynolds* number for $\alpha = 30$

creases. Reverse is the case in fig. 6 for $\alpha = 100$ and we note that Γ_{min} goes on decreasing as β increases.

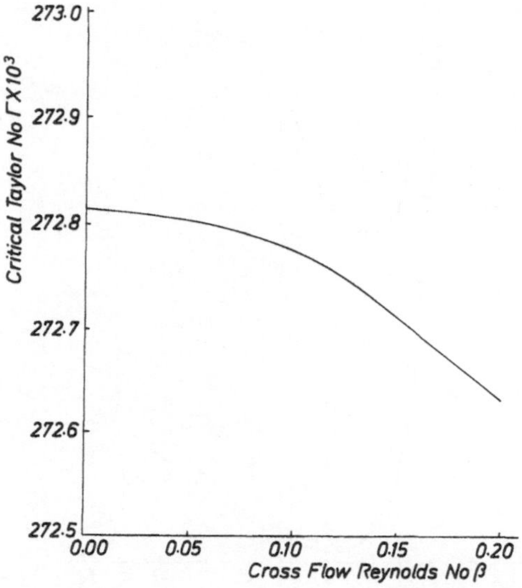

Fig. 6. Variation in critical *Taylor* number with cross-flow *Reynolds* number for $\alpha = 100$

References

1) *Giesekus, H.*, Rheol. Acta **5**, 239 (1966).
2) *Bhatnagar, R. K.* and *H. Giesekus*, Rheol. Acta **9**, 53 (1970).
3) *Bhatnagar, R. K.* and *H. Giesekus*, Rheol. Acta **9**, 412 (1970).
4) *Giesekus, H.* and *R. K. Bhatnagar*, Rheol. Acta **10**, 266 (1971).
5) *Hains, F. D.*, Phys. Fluids **14**, 1620 (1971).
6) *Thomas, L. H.*, Phys. Rev. **91**, 780 (1953).
7) *Chandrasekhar, S.* and *W. H. Reid*, Proc. Nat. Acad. Sci. (Wash.) **43**, 521 (1957).
8) *Reid, W. H.*, Proc. Roy. Soc. **A 244**, 186 (1958).
9) *Thomas, R. H.* and *K. Walters*, Proc. Roy. Soc. **A 274**, 371 (1963).
10) *Vest, C. M.* and *V. S. Arpaci*, J. Fluid Mech. **36**, 1 (1969).
11) *Chandrasekhar, S.*, Hydrodynamic and Hydromagnetic Stability (London 1961).

Authors' address:
R. K. Bhatnagar and *O. P. Sharma*
Dept. of Mathematics
Indian Institute of Technology
Powai, Bombay 76 (India)

Rheol. Acta 13, 501–512 (1974)

Département de Mathématiques-Faculté des Sciences, Université Libre de Bruxelles (Belgique)

Théorie des effets rhéo-optiques dans un continu polarisable et magnétisable

Par G. Mayne et Ph. Boulanger

(Reçu p. p. le 27 octobre 1972)

1. Introduction

La théorie que nous développons ici adopte le cadre méthodologique de la mécanique rationnelle des milieux continus basée sur les principes de la mécanique, de la thermodynamique et de l'électrodynamique. L'étude des effets physiques résultant de l'interaction d'un continu avec un champ électromagnétique est entreprise à partir des équations de bilan de la quantité de mouvement, du moment de la quantité de mouvement, de l'énergie, des équations de *Maxwell* et d'un nombre adéquat d'équations constitutives vérifiant les principes d'objectivité et d'équiprésence. En 1963, *R. A. Toupin* (1) a éllaboré une théorie des diélectriques élastiques polarisables qui s'inscrit dans cette perspective et qui a servi d'exemple à de nombreuses théories édifiées par la suite dans le domaine considéré. Pour rendre compte de l'interaction d'un diélectrique non magnétisable avec un champ électromagnétique, *Toupin* fait appel au modèle de *Lorentz* qui conduit à des expressions de la force et du supplément d'énergie électromagnétique telles que le tenseur impulsion-énergie électromagnétique correspondant n'est autre que le tenseur de *Minkowski* du vide. Ce modèle de diélectrique peut rendre compte des effets photoélastique et piézoélectrique mais pour rendre compte de l'effet *Faraday* (rotation du plan de polarisation due à l'action d'un champ magnétique extérieur longitudinal) l'auteur apporte à l'équation constitutive électromagnétique des modifications qui sont en contradiction avec le principe d'équiprésence.

Pour des diélectriques magnétisables, le modèle de *Lorentz* doit être abandonné, de sorte que pour obtenir des expressions satisfaisantes pour la force et le supplément d'énergie électromagnétiques, nous nous sommes tournés vers la théorie classique des champs. Les principes généraux de la mécanique des continus permettent de construire un tenseur impulsion-énergie total pour un continu traversé par un flux de chaleur et soumis à l'action d'un champ électromagnétique, l'expression du tenseur impulsion-énergie électromagnétique étant supposée connue. Les nombreuses expressions du tenseur impulsion-énergie électromagnétique que l'on trouve dans la littérature correspondent à des partitions différentes du tenseur total. Nous adoptons une autre partition du tenseur impulsion-énergie total donc une nouvelle définition du tenseur impulsion-énergie électromagnétique qui, à l'approximation classique, conduit à des bilans d'énergie et de quantité de mouvement tels que pour des diélectriques élastiques, transparents, polarisables et magnétisables, l'énergie interne constitue un potentiel pour les contraintes exprimées en variables de *Lagrange*. En l'absence de magnétisation, la théorie s'identifie à celle que l'on déduit du modèle de *Lorentz* pour la polarisation et dans le cas de diélectriques élastiques, on retrouve les résultats de *Toupin*.

L'application de la théorie aux diélectriques élastiques et l'étude de la propagation d'ondes planes dans un tel modèle montrent qu'en plus de l'effet *Kerr*, des effets photoélastique et piézoélectrique présents dans l'étude de *Toupin*, la théorie que nous proposons rend compte de l'effet *Cotton-Mouton* (biréfringence due à l'action d'un champ magnétique transversal). Les effets *Faraday* et *Pockels* (rotation du plan de polarisation due à l'action de champs magnétique et électrique longitudinaux), n'apparaissent pas pour le modèle considéré. Dans une communication présentée dans le cadre de ce VIme Congrès, *Ph. Boulanger* montre que l'introduction dans les équations constitutives de dérivées par rapport au temps de la polarisation ou de la magnétisation peut rendre compte de l'effet *Faraday*.

2. Tenseur impulsion-énergie pour un continu non conducteur soumis à des actions mecaniques, thermiques, électromagnétiques

D'une façon générale, nous admettrons que, pour le diélectrique considéré, les seules actions extérieures sont électromagnétiques. Les expressions de la force et du supplément d'énergie électromagnétique qui interviennent dans les bilans d'énergie et de quantité de mouvement, peuvent se déduire d'un tenseur impulsion-énergie électromagnétique. De nombreux physiciens (2) ont proposé des expressions pour les composantes de ce tenseur; malheureusement, en général, ils ne prennent en considération que les seules actions électromagnétiques d'où des divergences qui résultent du caractère quelque peu arbitraire de la partition qu'ils opèrent sur le tenseur impulsion-énergie total. Le tenseur impulsion-énergie électromagnétique étant supposé connu, nous allons construire le tenseur impulsion-énergie total à partir des bilans d'énergie et de quantité de mouvement (3). Pour un domaine matériel V de E_3, limité par une surface S, ces bilans s'écrivent

$$\frac{d}{dt}\int_V h\,dV = +\oint_S \bar{v}(t\bar{n})\,dS + \int_V \bar{f}\;\bar{v}\,dV - \oint_S \bar{\varphi}\bar{n}\,dS + \int_V q\,dV \qquad [2.1]$$

$$\frac{d}{dt}\int_V \bar{g}\,dV = +\oint_S t\bar{n}\cdot dS + \int_V \bar{f}\,dV \qquad [2.2]$$

$\bar{v}$ vecteur vitesse de E_3;
$\bar{n}$ vecteur unitaire normal à S;
$\bar{\varphi}$ vecteur flux de chaleur;
h densité d'énergie non électromagnétique contenue dans V;
q supplément d'énergie d'origine électromagnétique par unité de volume;
$\bar{g}$ densité de quantité de mouvement du milieu.

Si E_{ij} désignent les composantes du tenseur impulsion-énergie électromagnétique non nécessairement symétrique, $\bar{f}$ et q s'expriment à partir du 4-vecteur f_i défini par

$$f_i = -E_{ij,j} = \left[\bar{f}, \frac{i}{c}(\bar{f}\cdot\bar{v} + q)\right]. \qquad [2.3]$$

Les indices grecs et latins prennent respectivement les valeurs 1, 2, 3 (espace euclidien E_3) et 1, 2, 3, 4 (espace-temps de *Minkowski* muni d'un repère lorentzien orthonormé de coordonnées x^1, x^2, x^3, x^4 = ict). Les bilans [2.1] et [2.2] peuvent prendre les expressions locales suivantes

$$\frac{\partial h}{\partial t} + \mathrm{div}\,(h\bar{v} + \bar{v}t + \bar{\varphi}) = \bar{f}\cdot\bar{v} + q \qquad [2.4]$$

$$\frac{\partial g_\lambda}{\partial t} + (g_\lambda v_\mu + t_{\lambda\mu})_{,\mu} = f_\lambda. \qquad [2.5]$$

La loi de transformation relativiste du tenseur des contraintes (3) qui permet le passage du système de repos à un système lorentzien quelconque, suggère la décomposition de ce tenseur en contrainte purement mécanique τ et en contrainte thermique

$$\frac{\gamma^2}{c^2}\,\bar{v}\otimes\bar{\varphi}$$

avec

$$\gamma \overset{\text{déf}}{=} \left(1 - \frac{v^2}{c^2}\right)^{-1/2}$$

$$t_{\lambda\mu} = \tau_{\lambda\mu} + \frac{\gamma^2}{c^2}\,v_\lambda\varphi_\mu.$$

L'appellation «contrainte purement mécanique» pour le tenseur τ se justifie par le fait que la loi de transformation de ce tenseur pour le passage du repère comobile à un repère inertiel quelconque ne fait intervenir que ce tenseur et la vitesse $\bar{v}$ du milieu. En effet, on montre que cette loi de transformation est donnée par

$$\tau_{\lambda\mu} = \mathring{\tau}_{\lambda\mu} + (\gamma - 1)\frac{1}{v^2}\mathring{\tau}_{\gamma\mu}v_\lambda v_\gamma + \frac{1-\gamma}{\gamma}\frac{1}{v^2}\mathring{\tau}_{\lambda\gamma}v_\gamma v_\mu - \frac{(\gamma-1)^2}{\gamma}\frac{\mathring{\tau}_{\alpha\beta}v^\alpha v^\beta}{v^4}\,v_\lambda v_\mu. \qquad [2.6]$$

Introduisons τ dans les bilans [2.4] et [2.5]

$$\frac{\partial h}{\partial t} + \mathrm{div}\,(h\bar{v} + \bar{v}\tau + \gamma^2\bar{\varphi}) = \bar{f}\cdot\bar{v} + q \qquad [2.7]$$

$$\frac{\partial g_\lambda}{\partial t} + \left(g_\lambda v_\mu + \tau_{\lambda\mu} + \frac{\gamma^2}{c^2}v_\lambda\varphi_\mu\right)_{,\mu} = f_\lambda. \qquad [2.8]$$

Comparons ces équations avec les équations de champ d'un système ouvert caractérisé par un tenseur impulsion-énergie $\boldsymbol{W}$ et une 4-force extérieure f_i

$$W_{ij,j} = f_i = -E_{ij,j}. \qquad [2.9]$$

Le tenseur impulsion-énergie total $T_{ij} = W_{ij} + E_{ij}$ est symétrique de sorte que

$$W_{[ij]} = -E_{[ij]} \qquad [2.10]$$

[2.9] s'écrit alors $T_{ij,j} = 0$ [2.11].

L'eq. [2.9] peut être décomposée en bilan d'énergie ($i = 4$) et de quantité de mouvement ($i = 1, 2, 3$)

$$-ic\,W_{4\lambda,\lambda} - \frac{\partial W_{44}}{\partial t} = \bar{f}\cdot\bar{v} + q \qquad [2.12]$$

$$W_{\lambda\mu,\mu} + \frac{1}{ic}\frac{\partial W_{\lambda 4}}{\partial t} = f_\lambda. \qquad [2.13]$$

La comparaison avec [2.7] et [2.8] conduit aux identifications suivantes:

$$W_{44} = -h$$

$$W_{4\lambda} = \frac{i}{c}(hv_\lambda + v_\mu\tau_{\mu\lambda} + \gamma^2\varphi_\lambda) \qquad [2.15]$$

$$W_{\lambda 4} = icg_\lambda \qquad [2.16]$$

$$W_{\lambda\mu} = g_\lambda v_\mu + \tau_{\lambda\mu} + \frac{\gamma^2}{c^2}v_\lambda\varphi_\mu. \qquad [2.17]$$

De [2.10], [2.15], [2.16] il résulte que

$$icg_\lambda = \frac{i}{c}(hv_\lambda + v_\mu \tau_{\mu\lambda} + \gamma^2 \varphi_\mu) + 2E_{[4\lambda]}. \qquad [2.18]$$

En introduisant le 4-vecteur flux de chaleur défini par

$$\Phi_i = \left[\bar{\Phi}, \frac{i}{c}\gamma^3 \bar{v}\cdot\bar{\varphi}\right]$$

avec

$$\bar{\Phi} = \gamma\bar{\varphi} + \gamma(\gamma^2 - 1)\frac{\bar{\varphi}\bar{v}}{v^2}\cdot\bar{v} \qquad [2.19]$$

on peut établir (3) l'identité suivante

$$W_{ij}U_j = -\mathring{h}U_i - \Phi_i - 2U_jE_{[ji]} \qquad [2.20]$$

avec:

- $\mathring{h}$ valeur de h dans le système de repos. D'une manière générale, les grandeurs surmontées d'un zéro sont calculées dans le repère comobile;
- U_i composantes du 4-vecteur vitesse défini par $[\gamma v^\alpha, ic\gamma]$.

Pour $i = 1, 2, 3$ la relation [2.20] s'écrit

$$W_{\lambda\mu}U_\mu + W_{\lambda 4}U_4 = -\mathring{h}U_\lambda - \Phi_\lambda - 2U_jE_{[j\lambda]}$$

qui, en tenant compte de [2.16] et [2.17], donne

$$g_\lambda = \gamma^2\mathring{\mu}v_\lambda + \frac{\gamma^2}{c^4}\tau_{\lambda\mu}v_\mu + \gamma\frac{\Phi_\lambda}{c^2} + \frac{\gamma}{ic^3}v_\lambda\Phi_4$$

$$- 2\frac{\gamma}{c^2}U_jE_{[j\lambda]} \qquad [2.21]$$

avec

$$\mathring{\mu} = \frac{\mathring{h}}{c^2}.$$

Pour $i = 4$, [2.20] devient

$$W_{4\lambda}U_\lambda + W_{44}U_4 = -ic\gamma\mathring{h} - \Phi_4 + 2U_\lambda E_{[\lambda 4]}.$$

En vertu de [2.10] et [2.16]

$$W_{4\lambda} = icg_\lambda + 2E_{[\lambda 4]} \quad \text{de sorte que}$$

$$W_{44} = -\frac{1}{\gamma}g_\lambda U_\lambda - \mathring{h} - \frac{1}{ic\gamma}\Phi_4. \qquad [2.22]$$

En introduisant [2.21] dans [2.16], [2.17] et en tenant compte de [2.22] les composantes T_{ij} du tenseur impulsion-énergie seront donnés par

$$T_{\lambda\mu} = \mathring{\mu}U_\lambda U_\mu + \tau_{\lambda\mu} + \frac{1}{c^2}\tau_{\lambda\gamma}U_\gamma U_\mu$$

$$+ \frac{1}{c^2}(\Phi_\lambda U_\mu + \Phi_\mu U_\lambda)$$

$$+ \frac{2}{c^2}E_{[\lambda\gamma]}U_\gamma U_\mu + \frac{2}{c^2}E_{[\lambda 4]}U_4U_\mu + E_{\lambda\mu}$$

$$T_{\lambda 4} = T_{4\lambda} = \mathring{\mu}U_\lambda U_4 + \frac{1}{c^2}\tau_{\lambda\mu}U_\mu U_4$$

$$+ \frac{1}{c^2}(\Phi_\lambda U_4 + \Phi_4 U_\lambda)$$

$$+ \frac{2}{c^2}E_{[\lambda\mu]}U_\mu U_4 + \frac{2}{c^2}E_{[\lambda 4]}U_4U_4 + E_{\lambda 4}$$

$$T_{44} = \mathring{\mu}U_4U_4 - \frac{1}{c^2}\tau_{\lambda\mu}U_\lambda U_\mu$$

$$+ \frac{2}{c^2}\Phi_4U_4 - \frac{2}{c^2}E_{[\lambda 4]}U_4U_\lambda + E_{44}. \qquad [2.23]$$

3. «Equivalence» des tenseurs impulsion-énergie électromagnétique (4)

Ecrivons les composants T_{ij} dans le repère comobile

$$\mathring{T}_{\lambda\mu} = \mathring{\tau}_{\lambda\mu} + \mathring{E}_{\lambda\mu}$$

$$\mathring{T}_{\lambda 4} = \mathring{T}_{4\lambda} = \frac{i}{c}\mathring{\varphi}_\lambda + \mathring{E}_{4\lambda}$$

$$\mathring{T}_{44} = -\mathring{h} + \mathring{E}_{44}.$$

Soit $\tilde{E}_{ij}$ un autre tenseur impulsion-énergie électromagnétique tel que

$$\mathring{\tilde{E}}_{4\lambda} = \mathring{E}_{4\lambda}. \qquad [3.1]$$

Soit $\tilde{\tau}_{\lambda\mu}$ le tenseur qui se déduit de (6) à partir de

$$\mathring{\tilde{\tau}}_{\lambda\mu} = \mathring{\tau}_{\lambda\mu} + \mathring{E}_{\lambda\mu} - \mathring{\tilde{E}}_{\lambda\mu}, \qquad \mathring{\tilde{\tau}}_{[\lambda\mu]} = -\mathring{\tilde{E}}_{[\lambda\mu]} \qquad [3.2]$$

et soit

$$\mathring{\tilde{h}} = \mathring{h} - \mathring{E}_{44} + \mathring{\tilde{E}}_{44}. \qquad [3.3]$$

On obtient une nouvelle expression du même tenseur T_{ij} en remplaçant dans [2.23] τ par $\tilde{\tau}$, $\boldsymbol{E}$ par $\tilde{\boldsymbol{E}}$ et $\mathring{\mu}$ par $\mathring{\tilde{\mu}} = \mathring{\tilde{h}}/c^2$. Les tenseurs $\boldsymbol{E}$ et $\tilde{\boldsymbol{E}}$ conduisent aux mêmes bilans d'énergie et de quantité de mouvement moyennant un changement de définition de l'énergie et du tenseur des contraintes; dans ce sens, nous dirons que les tenseurs $\boldsymbol{E}$ et $\tilde{\boldsymbol{E}}$ sont équivalents. Pratiquement tous les auteurs (2, 5, 6) qui ont proposé des choix du tenseur impulsion-énergie électromagnétique sont d'accord sur l'expression du flux d'énergie électromagnétique dans le repère comobile.

$$\mathring{E}_{4\lambda} = i(\mathring{\bar{E}} \times \mathring{\bar{H}})_\lambda$$

avec

$\mathring{\bar{H}} = \mathring{\bar{B}} - \mathring{\bar{M}}$.

$\bar{E}$ champ électrique;
$\bar{B}$ champ magnétique;
$\bar{M}$ vecteur magnétisation.

Ces tenseurs sont donc équivalents puisqu'ils vérifient la relation [3.1]. Cette condition [3.1] est absolument indispensable car, dans le repère comobile, on ne peut modifier l'expression du flux d'énergie électromagnétique sans changer celle du flux de chaleur ce qui entrainerait l'impossibilité de distinguer l'énergie calorifique des autres formes d'énergie et rendrait sans signification toute thermodynamique relativiste.

D'autre part, on constate que, quel que soit le tenseur impulsion-énergie électromagnétique que l'on adopte parmi ceux qui sont classiquement proposés (*Minkowski, Abraham, de Groot* et *Suttorp, Eringen* etc.), le premier principe de la thermodynamique

$$U_i T_{ij,j} = 0 \qquad [3.4]$$

peut toujours s'écrire, à l'approximation classique, sous la forme

$$\varrho \frac{d\varepsilon}{dt} = \sigma_{\lambda\mu} v_{\lambda,\mu} - \varphi_{\lambda,\lambda} + \bar{E}' \cdot \underset{v}{\mathscr{L}} \bar{P}' + \bar{B}' \cdot \underset{v}{\mathscr{L}} \bar{M}' \qquad [3.5]$$

où

$$\bar{E}' = \bar{E} + \frac{1}{c}\bar{v} \times \bar{B}, \quad \bar{P}' = \bar{P} - \frac{1}{c}\bar{v} \times \bar{M}$$

$$\bar{B}' = \bar{B} - \frac{1}{c}\bar{v} \times \bar{E}, \quad \bar{M}' = \bar{M} + \frac{1}{c}\bar{v} \times \bar{P} \qquad [3.6]$$

désignent les approximations classiques des relations donnant $\mathring{\bar{E}}, \mathring{\bar{B}}, \mathring{\bar{P}}, \mathring{\bar{M}}$ en fonction des champs $\bar{E}, \bar{B}, \bar{P}, \bar{M}$ par rapport à un référentiel lorentzien quelconque

$\bar{P}$: vecteur polarisation

$$\underset{v}{\mathscr{L}} \bar{P}' \overset{\text{def}}{=} \frac{\partial \bar{P}'}{\partial t} + \operatorname{rot}(\bar{P}' \times \bar{v}) + (\operatorname{div} \bar{P}')\bar{v}$$

$$\underset{v}{\mathscr{L}} \bar{M}' \overset{\text{def}}{=} \frac{\partial \bar{M}'}{\partial t} + \operatorname{rot}(\bar{M}' \times \bar{v}) + (\operatorname{div} \bar{M}')\bar{v} - (\operatorname{div} \bar{v})\bar{M}' \qquad [3.7]$$

les relations liant la contrainte $\boldsymbol{\sigma}$ et l'énergie interne ε aux autres définitions étant connues.

Par approximation classique, nous entendons le fait de négliger devant l'unité les quantités

$$\frac{v^2}{c^2}, \quad \frac{\sigma_{\lambda\mu}}{\varrho c^2}, \quad \frac{E_\lambda P_\mu}{\varrho c^2}, \quad \frac{B_\lambda M_\mu}{\varrho c^2}, \quad \frac{\varepsilon}{\varrho c^2}$$

et

$$\frac{v_\lambda \varphi_\mu}{\varrho c^2}.$$

Dès lors, en se basant sur la notion d'équivalence, il est possible de construire, à partir des formules [3.2] et [3.3], un tenseur X_{ij} tel qu'à l'approximation classique,

$$U_i X_{ij,j} = \bar{E}' \cdot \underset{v}{\mathscr{L}} \bar{P}' + \bar{B}' \cdot \underset{v}{\mathscr{L}} \bar{M}'$$

de sorte qu'en adoptant X_{ij} comme tenseur impulsion-énergie électromagnétique dans l'eq. [3.4] on obtienne directement le bilan d'énergie sous la forme [3.5].

Les composantes de X_{ij} sont données, dans le repère comobile, par

$$\mathring{X}_{\lambda\mu} = -\mathring{E}_\lambda \mathring{E}_\mu - \mathring{B}_\lambda \mathring{B}_\mu + 2\mathring{B}_{(\lambda}\mathring{M}_{\mu)} + \tfrac{1}{2}(\mathring{E}^2 + \mathring{B}^2)\delta_{\lambda\mu}$$

$$\mathring{X}_{\lambda 4} = \mathring{X}_{4\lambda} = i(\mathring{\bar{E}} \times \mathring{\bar{H}})_\lambda$$

$$\mathring{X}_{44} = -\tfrac{1}{2}(\mathring{E}^2 + \mathring{B}^2 - 2\mathring{\bar{M}} \cdot \mathring{\bar{B}}). \qquad [3.8]$$

Le tenseur impulsion-énergie total se décompose alors de la manière suivante

$$T_{ij} = \mathring{\mu} U_i U_j + S_{ij} + \frac{1}{c^2}(\Phi_i U_j + \Phi_j U_i) + X_{ij} \qquad [3.9]$$

avec

$$S_{\lambda\mu} = \sigma_{\lambda\mu} + \frac{1}{c^2}\sigma_{\lambda\gamma} U_\gamma U_\mu$$

$$S_{\lambda 4} = S_{4\lambda} = \frac{1}{c^2}\sigma_{\lambda\mu} U_\mu U_4.$$

$$S_{44} = -\frac{1}{c^2}\sigma_{\lambda\mu} U_\lambda U_\mu \qquad [3.10]$$

En vertu de [3.2], $\mathring{\sigma}_{[\lambda\mu]} = -\mathring{X}_{[\lambda\mu]} = 0$; comme d'autre part, $\sigma_{\lambda\mu}$ admet la loi de transformation [2.6], on peut montrer que les fonctions S_{ij} constituent les composantes d'un 4-tenseur. La formule [3.9] apparait donc comme une partition du tenseur total en 4-tenseurs.

A l'approximation classique, $\gamma = 1$ et, en vertu de [2.6], $\sigma_{\lambda\mu} = \mathring{\sigma}_{\lambda\mu}$ de sorte que le tenseur de contrainte σ est aussi symétrique. Le bilan de la quantité de mouvement s'obtient en introduisant [3.9] dans l'équation

$$T_{\lambda j,j} = 0 \qquad [3.11]$$

qui, à l'approximation classique, se réduit, en tenant compte des équations de *Maxwell* à

$$\varrho \frac{dv_\lambda}{dt} = \sigma_{\lambda\mu,\mu} + f_\lambda \qquad [3.12]$$

avec

$$\begin{aligned}\bar{f} = & -(\operatorname{div}\bar{P}')\cdot\bar{E}' + \frac{1}{c}(\underset{v}{\mathscr{L}}\bar{P}' \times \bar{B}') \\ & -(\operatorname{div}\bar{M}')\,\bar{B}' \\ & -\frac{1}{c}(\underset{v}{\mathscr{L}}\bar{P}' \times \bar{M}') + \frac{1}{c}(\bar{E}' \times \underset{v}{\mathscr{L}}\bar{M}') \\ & +\frac{1}{c}(\operatorname{div}\bar{v})\,\bar{E}'\times\bar{M}' \\ & -\operatorname{rot}\bar{M}' \times \bar{M}' - \operatorname{grad}\bar{B}'\bar{M}'. \end{aligned} \qquad [3.13]$$

Remarquons qu'en l'absence de magnétisation ($\bar{M}' = 0$) cette force est identique à celle déduite du modèle de *Lorentz* et adoptée par *Toupin* dans sa théorie des diélectriques polarisables. Cela n'a rien d'étonnant puisqu'en faisant $\mathring{\bar{M}} = 0$ dans les relations [3.8], on constate que le tenseur X_{ij} s'identifie au tenseur de *Minkowski* du vide.

4. Diélectriques élastiques polarisables et magnétisables

Appliquons la théorie générale (7) à des diélectriques élastiques transparents en évolution isotherme pour lesquels ε, σ, $\bar{E}'$ et $\bar{B}'$ ne dépendent que de $\bar{P}'$, $\bar{M}'$ et des gradients de déformation

$$\frac{\partial x^k}{\partial X^A} \overset{\text{déf}}{=} x^k_{,A}.$$

En l'absence de magnétisation ($\bar{M}' = 0$) la thérie doit s'identifier à celle élaborée par *R. A. Toupin* (1) pour les diélectriques élastiques polarisables à gyration nulle ($\bar{G} = 0$).

Introduisons les vecteurs $\bar{\mathscr{P}}'$ et $\bar{\mathscr{M}}'$, définis dans un repère de l'état de référence du continu par les composantes

$$\begin{aligned}\mathscr{P}'^A &= |(x|X)|\, X^A_{,k} P'^k \\ \mathscr{M}'^A &= X^A_{,k} M'^k. \end{aligned} \qquad [4.1]$$

En vertu du principe d'objectivité, ε dépend de $x^k_{,A}$ par l'intermédiaire des composantes du tenseur de *Green*

$$\varepsilon = \hat{\varepsilon}(E_{AB}, \mathscr{P}'^A, \mathscr{M}'^A) \qquad [4.2]$$

avec

$$E_{AB} = \tfrac{1}{2}(g_{\lambda\mu} x^\lambda_{,A} x^\mu_{,B} - G_{AB}). \qquad [4.3]$$

Le bilan d'énergie [3.5] peut alors s'écrire

$$\begin{aligned}\varrho \frac{\partial\hat{\varepsilon}}{\partial E_{AB}} x_{k,A}\dot{x}^k_{,B} + \varrho \frac{\partial\hat{\varepsilon}}{\partial\mathscr{P}'^A}\dot{\mathscr{P}}'^A + \varrho\frac{\partial\hat{\varepsilon}}{\partial\mathscr{M}'^A}\dot{\mathscr{M}}'^A & \\ = \sigma^l_k \dot{x}^k_{,A} X^A_{,l} & \\ + E'_k x^k_{,A} |(X|x)|\, \dot{\mathscr{P}}'^A & \\ + B'_k x^k_{,A} \dot{\mathscr{M}}'^A. & \end{aligned} \qquad [4.4]$$

Par identification des coefficients de $\dot{x}^k_{,A}$, $\dot{\mathscr{P}}'^A$ et $\dot{\mathscr{M}}'^A$, il vient, en tenant compte de $\varrho_0 = \varrho|(x|X)|$ et de la symétrie de σ

$$\sigma^{kl} = \varrho \frac{\partial\hat{\varepsilon}}{\partial E_{AB}} x^k_{,A} x^l_{,B} = \sigma^{lk} \qquad [4.5]$$

$$E'_k = \varrho_0 \frac{\partial\hat{\varepsilon}}{\partial\mathscr{P}'^A} X^A_{,k} \qquad [4.6]$$

$$B'_k = \varrho \frac{\partial\hat{\varepsilon}}{\partial\mathscr{M}'^A} X^A_{,k} \qquad [4.7]$$

Dans ce travail, nous ne considérons que des diélectriques isotropes pour lesquels la fonction $\hat{\varepsilon}$ est un invariant pour le groupe orthogonal complet du tenseur E_{AB}, de la densité vectorielle $\mathscr{P}'^A$ et du vecteur axial $\mathscr{M}'^A$. *C. C. Wang* (8) a établi une base complète et irréductible pour de tels invariants. Dans le cas particulier qui nous intéresse, la fonction $\hat{\varepsilon}$ dépend des 14 invariants suivants

$$\begin{aligned} I_1 &= \operatorname{tr}\mathbf{E}, \quad I_2 = \operatorname{tr}\mathbf{E}^2, \quad I_3 = \operatorname{tr}\mathbf{E}^3 \\ I_4 &= \mathscr{P}'^2, \quad I_5 = \mathscr{M}'^2, \quad I_6 = \bar{\mathscr{P}}'\cdot\mathbf{E}\bar{\mathscr{P}}' \\ I_7 &= \bar{\mathscr{P}}'\cdot\mathbf{E}^2\bar{\mathscr{P}}', \quad I_8 = \bar{\mathscr{M}}'\cdot\mathbf{E}\bar{\mathscr{M}}' \\ I_9 &= \bar{\mathscr{M}}\cdot\mathbf{E}^2\bar{\mathscr{M}}' \\ I_{10} &= \mathbf{E}\bar{\mathscr{M}}'\cdot[\mathbf{E}^2\bar{\mathscr{M}}' \times \bar{\mathscr{M}}'] \\ I_{11} &= (\bar{\mathscr{P}}'\cdot\bar{\mathscr{M}}')^2, \quad I_{12} = \mathbf{E}\bar{\mathscr{P}}'\cdot(\bar{\mathscr{M}}' \times \bar{\mathscr{P}}') \\ I_{13} &= \mathbf{E}^2\bar{\mathscr{P}}'\cdot(\bar{\mathscr{M}}' \times \bar{\mathscr{P}}') \\ I_{14} &= (\bar{\mathscr{P}}'\cdot\bar{\mathscr{M}}')\,[\mathbf{E}\bar{\mathscr{M}}'\cdot(\bar{\mathscr{M}}' \times \bar{\mathscr{P}}')]. \end{aligned} \qquad [4.8]$$

5. Equations aux variations

Pour étudier les effets rhéo-optiques dans un tel diélectrique, nous procédons comme *R. A. Toupin* à une linéarisation de toutes les équations à partir d'un état d'équilibre initial caractérisé par une solution indépendante du temps

$_0x^\lambda(X) \quad _0\mathscr{P}^A \quad _0\mathscr{M}^A \quad _0E_\lambda \quad _0B_\lambda$.

D'une manière générale, les grandeurs précédés d'un indice 0 doivent être calculées dans cet état d'équilibre.

Par soucis de simplification, nous nous limiterons à des états d'équilibre uniformes pour lesquels les vecteurs $_0\bar{\mathscr{P}}$, $_0\bar{\mathscr{M}}$, $_0\bar{E}$, $_0\bar{B}$ sont constants et les fonctions $_0x^\lambda(X)$ linéaires.

En considérant les solutions voisines

$$x^\lambda(X, t) = {}_0x^\lambda + \delta_X x^\lambda(X, t)$$
$$\mathscr{P}'^A(X, t) = {}_0\mathscr{P}^A + \delta_X \mathscr{P}'^A(X, t)$$
$$\mathscr{M}'^A(X, t) = {}_0\mathscr{M}^A + \delta_X \mathscr{M}'^A(X, t)$$
$$E'_\lambda(x, t) = {}_0E_\lambda + \delta_x E'_\lambda(x, t)$$
$$B'_\lambda(x, t) = {}_0B_\lambda + \delta_x B'_\lambda(x, t) \qquad [5.1]$$

où δ_X et δ_x désignent respectivement des variations à X et x constants, on obtient pour chacune des équations du mouvement, de comportement et de *Maxwell*, des équations aux variations linéaires en les petits déplacements et les champs faibles.

En posant

$$m'^\lambda = {}_0x^\lambda_{,A}\delta_X \mathscr{M}'^A$$
$$p'^\lambda = {}_0|(x|X)|^{-1}\,{}_0x^\lambda_{,A}\delta_X \mathscr{P}'^A$$
$$\delta_X x^\lambda = u^\lambda, \quad \delta_x E'_\lambda = e'_\lambda, \quad \delta_x B'_\lambda = b'_\lambda$$

il vient

a) Pour l'équation du mouvement [3.12]

$$\begin{aligned} {}_0\varrho\ddot{u}_\lambda = {}& Z^\mu_{\lambda,\mu} - {}_0E_\lambda \operatorname{div}\bar{p}' \\ & + \frac{1}{c}(\dot{\bar{p}}' \times {}_0\bar{B})_\lambda - \frac{1}{c}(\bar{p}' \times {}_0\bar{M})_\lambda \\ & - {}_0M^\mu(m'_{\lambda,\mu} - m'_{\mu,\lambda} + {}_0M^\alpha u_{\lambda,\alpha\mu} \\ & \qquad - u_{\mu,\alpha\lambda}\,{}_0M^\alpha) \\ & - b'_{\mu,\lambda}\,{}_0M^\mu - {}_0B_\mu(u^\mu_{,\alpha\lambda}\,{}_0M^\alpha + m'^{\mu}_{,\lambda}) \\ & + \frac{1}{c}({}_0\bar{E} \times \bar{m}')_\lambda + \frac{1}{c}\operatorname{div}\dot{\bar{u}} \cdot ({}_0\bar{E} \times {}_0\bar{M})_\lambda \\ & - {}_0B_\lambda({}_0M^\alpha u^\mu_{,\alpha\mu} + m'^{\mu}_{,\mu}), \end{aligned} \qquad [5.2]$$

avec

$$Z^{\lambda\mu} \overset{\text{déf}}{=} |_0(x|X)|^{-1}\,{}_0x^\mu_{,A}\delta_X T^\lambda_A$$
$$T^{\lambda A} \overset{\text{déf}}{=} |(x|X)|\,\sigma^{\lambda\mu}X^A_{,\mu}$$

qui en vertu de [4.5] devient

$$T^{\lambda A} = \varrho_0 \frac{\partial\hat{\varepsilon}}{\partial E_{AB}} x^\lambda_{,B}. \qquad [5.3]$$

b) Pour l'équation de comportement mécanique [5.3]

$$Z^{\lambda\mu} = {}_0\sigma^{\mu\nu}u^\lambda_{,\nu} + {}_0C^{\lambda\mu\alpha\nu}u_{\alpha,\nu} + {}_0S^{\lambda\mu}_\nu p'^\nu + {}_0X^{\lambda\mu}_\nu m'^\nu \qquad [5.4]$$

avec

$${}_0C^{\lambda\mu\alpha\nu} = \varrho_0 \left(\frac{\partial^2\hat{\varepsilon}}{\partial E_{AB}\partial E_{CD}}\right)_0 {}_0x^\lambda_{,A}\,{}_0x^\mu_{,B}\,{}_0x^\alpha_{,C}\,{}_0x^\nu_{,D} \qquad [5.5]$$

$${}_0S^{\lambda\mu}_\nu = \varrho_0 \left(\frac{\partial^2\hat{\varepsilon}}{\partial\mathscr{P}'^A\partial E_{CD}}\right)_0 {}_0x^\lambda_{,C}\,{}_0x^\mu_{,D}\,{}_0X^A_{,\nu} \qquad [5.6]$$

$${}_0X^{\lambda\mu}_\nu = {}_0\varrho \left(\frac{\partial^2\hat{\varepsilon}}{\partial E_{CD}\partial\mathscr{M}'^A}\right)_0 {}_0x^\lambda_{,C}\,{}_0x^\mu_{,D}\,{}_0X^A_{,\nu}. \qquad [5.7]$$

c) Pour les relations [3.6]

$$\bar{e}' = \bar{e} + \frac{1}{c}\dot{\bar{u}} \times {}_0\bar{B}, \quad \bar{b}' = \bar{b} - \frac{1}{c}\dot{\bar{u}} \times {}_0\bar{E}$$
$$\bar{p}' = \bar{p} - \frac{1}{c}\dot{\bar{u}} \times {}_0\bar{M}, \quad \bar{m}' = \bar{m} + \frac{1}{c}\dot{\bar{u}} \times {}_0\bar{P}. \qquad [5.8]$$

d) Pour les équations de *Maxwell*, que nous écrirons

$$\operatorname{div}\bar{B} = 0 \qquad [5.9]$$
$$\operatorname{div}(\bar{E} + \bar{P}) = 0 \qquad [5.10]$$
$$\operatorname{rot} E + \frac{1}{c}\frac{\partial\bar{B}}{\partial t} = 0 \qquad [5.11]$$
$$\operatorname{rot}(\bar{B} - \bar{M}) - \frac{1}{c}\frac{\partial}{\partial t}(\bar{E} + \bar{P}) = 0 \qquad [5.12]$$

il vient

$$\operatorname{div}\bar{b} = 0 \qquad [5.9']$$
$$\operatorname{div}(\bar{e} + \bar{p}) = 0 \qquad [5.10']$$
$$\operatorname{rot}\bar{e} + \frac{1}{c}\dot{\bar{b}} = 0 \qquad [5.11']$$
$$\operatorname{rot}\bar{b} - \frac{1}{c}\dot{\bar{e}} = \operatorname{rot}\bar{m} + \operatorname{rot}({}_0M^\mu\bar{u}_{,\mu}) + \frac{1}{c}\dot{\bar{p}} + \frac{1}{c}{}_0P^\mu\dot{\bar{u}}_{,\mu} - \frac{1}{c}{}_0\bar{P}\cdot\operatorname{div}\dot{\bar{u}}. \qquad [5.12']$$

e) Pour les équations de comportement électromagnétique [4.6] et [4.7]

$$e_\lambda + \frac{1}{c}(\dot{\bar{u}} \times {}_0\bar{B})_\lambda = {}_0S^{\mu\gamma}_\lambda u_{\mu,\gamma} + {}_0T_{\lambda\mu}p'^\mu + {}_0W_{\lambda\mu}m'^\mu - {}_0E_\mu u^\mu_{,\lambda} \qquad [5.13]$$

$$b_\lambda - \frac{1}{c}(\dot{\bar{u}} \times {}_0\bar{E})_\lambda = {}_0X^{\mu\gamma}_\lambda u_{\mu,\gamma} + {}_0W_{\mu\lambda}p'^\mu + {}_0Y_{\lambda\mu}m'^\mu - {}_0B_\mu u^\mu_{,\lambda} - {}_0B_\lambda u^\mu_{,\mu} \qquad [5.14]$$

avec

$${}_0T_{\lambda\mu} = \frac{\varrho_0^2}{{}_0\varrho}\,{}_0\!\left(\frac{\partial^2\hat{\varepsilon}}{\partial\mathscr{P}'^A\partial\mathscr{P}'^B}\right){}_0X^A_{,\lambda}\,{}_0X^B_{,\mu} \qquad [5.15]$$

$${}_0W_{\lambda\mu} = {}_0\varrho\,{}_0\!\left(\frac{\partial^2\hat{\varepsilon}}{\partial\mathscr{M}'^B\partial\mathscr{P}'^A}\right){}_0X^A_{,\lambda}\,{}_0X^B_{,\mu} \qquad [5.16]$$

$${}_0Y_{\lambda\mu} = {}_0\varrho\,{}_0\!\left(\frac{\partial^2\hat{\varepsilon}}{\partial\mathscr{M}'^B\partial\mathscr{M}'^A}\right){}_0X^A_{,\lambda}\,{}_0X^B_{,\mu}. \qquad [5.17]$$

L'ensemble de toutes ces relations constitue un système d'équations différentielles aux dérivées partielles linéaires à coefficients constants pour lequel nous rechercherons, pour les inconnues $\bar{u}$, $\bar{e}$, $\bar{b}$, $\bar{p}$, $\bar{m}$, des solutions du type «ondes planes».

$$\bar{a} = R_e\tilde{a}e^{i\omega(t - n/c\, s\cdot r)} \qquad [5.18]$$

Nous nous limiterons à la considération de cas particuliers correspondant aux conditions d'apparition des effets photoélastique, *Faraday*, *Cotton-Mouton*, *Pockels* et *Kerr* (champs électromagnétiques extérieurs nuls, longitudinaux ou transversaux).

6. Effet photoélastique

Plaçons nous dans les conditions particulières ${}_0\bar{E} = {}_0\bar{P} = {}_0\bar{M} = {}_0\bar{B} = 0$.

Les équations électromagnétiques [5.9′], [5.10′], [5.11′], [5.12′], [5.13], [5.14] ne font plus intervenir le champ de déplacements additionnels $\bar{u}$ et les équations mécaniques [5.2] ne dépendent plus des champs additionnels $\bar{e}$, $\bar{b}$, $\bar{p}$, $\bar{m}$. De ce découplage des actions mécaniques et électromagnétiques, il résulte, qu'au 1er ordre, les champs faibles additionnels ne donnent lieu à aucun déplacement $\bar{u}$ et que l'étude de l'effet photoélastique peut se traiter à partir des seules équations de *Maxwell* et des équations constitutives électromagnétiques [5.13] et [5.14]. Les amplitudes complexes des ondes planes sont solutions du système algébrique linéaire

$$\tilde{b}\cdot\bar{s} = 0, \quad (\tilde{e} + \tilde{p})\cdot\bar{s} = 0$$

$$nS\tilde{e} - \tilde{b} = 0$$

$$nS\tilde{b} + \tilde{e} = -\tilde{p} + nS\tilde{m}$$

$$\tilde{e} = {}_0T\tilde{p}, \quad \tilde{b} = {}_0Y\tilde{m} \qquad [6.1]$$

S désigne le tenseur antisymétrique associé au vecteur unitaire $\bar{s}$, normal à l'onde. Les tenseurs symétriques ${}_0T$ et ${}_0Y$ sont fonctions tensorielles isotropes du tenseur de déformation; les directions principales de ces trois tenseurs coïncident.

Si ${}_0T$ et ${}_0Y$ sont inversibles, on peut éliminer $\tilde{p}$, $\tilde{m}$ et $\tilde{b}$ dans [6.1]

$$|n^2S(1 - {}_0Y^{-1})S + (1 + {}_0T^{-1})|\,\tilde{e} = 0 \qquad [6.2]$$

l'indice de réfraction n sera solution de l'équation $\det(A + n^2B) = 0$

avec

$$A = 1 + {}_0T^{-1}$$

et

$$B = S(1 - {}_0Y^{-1})S. \qquad [6.3]$$

Comme $\det B = 0$, cette équation est bicarrée de sorte qu'il y a biréfringence. En ce qui concerne les états de polarisation, si nous supposons que l'eq. [6.3] admet deux solutions réelles distinctes n_1^2 et n_2^2, il leur correspond deux directions réelles pour chacun des vecteurs complexes $\tilde{e}$, $\tilde{b}$, $\tilde{d} = \tilde{e} + \tilde{p}$ et l'étude du système [6.1] montre que

$$S\tilde{e}_2\cdot P(1 - {}_0Y^{-1})PS\tilde{e}_1 = 0$$

et

$$P\tilde{e}_2\cdot\tilde{d}_1 = P\tilde{e}_1\cdot\tilde{d}_2 = 0$$

$P = -S^2$ représentant l'opérateur de projection dans le plan de l'onde. Il en résulte que $\tilde{b}_1$ et $\tilde{b}_2$ constituent des directions conjugées par rapport à la conique associée au tenseur $P(1 - {}_0Y^{-1})P$, et que $\tilde{d}_1$ a la direction de $\tilde{b}_2$ et $\tilde{d}_2$ celle de $\tilde{b}_1$. En l'absence de magnétisation, ${}_0Y^{-1} = 0$ et les directions $\tilde{b}_1$ et $\tilde{b}_2$ deviennent perpendiculaires. Dans le cas général, les modes de polarisation ne sont plus orthogonaux sauf si les sections par le plan de l'onde des quadriques associées aux tenseurs $(1 + {}_0T^{-1})^{-1}$ et $(1 - {}_0Y^{-1})$ ont des directions principales communes.

7. Effets électro-magnéto-optiques

Dans ce paragraphe, nous étudierons dans le cadre de la théorie proposée l'interaction d'un champ électromagnétique faible et d'un diélectrique non déformé placé dans un champ magnétique ou électrique extérieur, longitudinal ou transversal par rapport à l'onde associé au champ faible.

1. Champ magnétique longitudinal

Considérons un état non perturbé caractérisé par

$$x_\lambda = X^\lambda, \quad {}_0\bar{E} = {}_0\bar{P} = 0, \quad {}_0\bar{B}$$

et

$${}_0\bar{M} \quad \text{uniformes} \quad {}_0\bar{B} = {}_0B\bar{s}.$$

De [4.5], [4.7], [4.2], il vient, compte tenu de la représentation [4.8]

$${}_0\sigma^{kl} = {}_0\varrho\,{}_0\!\left(\frac{\partial\hat{\varepsilon}}{\partial I_1}\right) g^{kl} + {}_0\varrho\,{}_0\!\left(\frac{\partial\hat{\varepsilon}}{\partial I_8}\right) {}_0M^k\,{}_0M^l \quad [7.1]$$

$${}_0\bar{B} = C_8\,{}_0\bar{M}$$

avec

$$C_8 = 2\varrho_0\,{}_0\!\left(\frac{\partial\hat{\varepsilon}}{\partial I_5}\right). \quad [7.2]$$

En utilisant la représentation [4.8] dans le système différentiel [5.2], [5.4], [5.9′], [5.10′], [5.11′], [5.12′], [5.13] et [5.14], on obtient, pour les amplitudes complexes des ondes planes, le système algébrique linéaire suivant

$$\omega\left(D_1\frac{n^2}{c^2} - {}_0\varrho\right)\tilde{u} + \left(D_2\omega\frac{n^2}{c^2}\tilde{u}\cdot\bar{s} + i\frac{n}{c}D_3\,{}_0M\tilde{m}\cdot\bar{s}\right)\bar{s} + i\,{}_0M D_4\frac{n}{c}\tilde{m} - \frac{i}{c}(C_8 - 1)\,{}_0M\tilde{p}\times\bar{s} = 0 \quad [7.3]$$

$$\tilde{e} = C_9\tilde{p} - i\frac{\omega}{c}\,{}_0M(C_8 + C_9)\tilde{u}\times s + {}_0M^2C_{10}(\tilde{p}\cdot\bar{s})\bar{s} \quad [7.4]$$

$$\tilde{b} = {}_0M\left[(1 - D_3)i\omega\frac{n}{c}\tilde{u}\cdot\bar{s} + {}_0MC_{11}\tilde{m}\cdot\bar{s}\right]\bar{s} - {}_0M(D_4 + 1)i\omega\frac{n}{c}\tilde{u} + C_8\tilde{m} \quad [7.5]$$

$$\tilde{b} = n\bar{s}\times\tilde{e} \quad [7.6]$$

$$\tilde{e} = n\tilde{b}\times\bar{s} - \tilde{p} - n\tilde{m}\times\bar{s} - i\,{}_0M\frac{\omega}{c}n^2\bar{s}\times\tilde{u} \quad [7.7]$$

avec

$$D_1 = C_2 + {}_0M^2C_4$$

$$D_2 = C_1 + 2C_3\,{}_0M^2 - {}_0M^2 + C_5\,{}_0M^2 + C_{12}\,{}_0M^4$$

$$D_3 = C_6 + {}_0M^2C_7 + C_4 - C_5 - C_8 + 1$$

$$D_4 = C_4 - C_5$$

$$C_1 = {}_0\varrho\,{}_0\!\left(\frac{\partial^2\hat{\varepsilon}}{\partial I_1^2}\right) + {}_0\varrho\,{}_0\!\left(\frac{\partial\hat{\varepsilon}}{\partial I_2}\right)$$

$$C_2 = {}_0\varrho\,{}_0\!\left(\frac{\partial\hat{\varepsilon}}{\partial I_2}\right) + {}_0\varrho\,{}_0\!\left(\frac{\partial\hat{\varepsilon}}{\partial I_1}\right)$$

$$C_3 = {}_0\varrho\,{}_0\!\left(\frac{\partial^2\hat{\varepsilon}}{\partial I_1\partial I_8}\right) + \frac{{}_0\varrho}{2}\,{}_0\!\left(\frac{\partial\hat{\varepsilon}}{\partial I_9}\right) + 1 - 2\,{}_0\varrho\,{}_0\!\left(\frac{\partial\hat{\varepsilon}}{\partial I_5}\right)$$

$$C_4 = \frac{{}_0\varrho}{2}\,{}_0\!\left(\frac{\partial\hat{\varepsilon}}{\partial I_9}\right) + {}_0\varrho\,{}_0\!\left(\frac{\partial\hat{\varepsilon}}{\partial I_8}\right) - 1$$

$$C_5 = \frac{{}_0\varrho}{2}\,{}_0\!\left(\frac{\partial\hat{\varepsilon}}{\partial I_9}\right)$$

$$C_6 = 2\,{}_0\varrho\,{}_0\!\left(\frac{\partial^2\hat{\varepsilon}}{\partial I_1\partial I_5}\right) - 2\,{}_0\varrho\,{}_0\!\left(\frac{\partial\hat{\varepsilon}}{\partial I_5}\right) + 1$$

$$C_7 = 2\,{}_0\varrho\,{}_0\!\left(\frac{\partial^2\hat{\varepsilon}}{\partial I_8\partial I_5}\right)$$

$$C_8 = 2\,{}_0\varrho\,{}_0\!\left(\frac{\partial\hat{\varepsilon}}{\partial I_5}\right)$$

$$C_9 = 2\,{}_0\varrho\,{}_0\!\left(\frac{\partial\hat{\varepsilon}}{\partial I_4}\right)$$

$$C_{10} = 2\,{}_0\varrho\,{}_0\!\left(\frac{\partial\hat{\varepsilon}}{\partial I_{11}}\right)$$

$$C_{11} = 2\,{}_0\varrho\,{}_0\!\left(\frac{\partial^2\hat{\varepsilon}}{\partial I_5^2}\right)$$

$$C_{12} = {}_0\varrho\,{}_0\!\left(\frac{\partial^2\hat{\varepsilon}}{\partial I_8^2}\right)$$

Dans ce cas-ci, les C_i sont fonctions de ${}_0M^2$. Eliminons $\tilde{e}$ et $\tilde{b}$ entre [7.4], [7.5], [7.6], [7.7] et multiplions les 3 équations vectorielles obtenues scalairement puis vectoriellement par $\bar{s}$

$$\omega\left[(D_1 + D_2)\frac{n^2}{c^2} - {}_0\varrho\right]\tilde{u}\cdot\bar{s} + i\frac{n}{c}\,{}_0M(D_3 + D_4)\,\tilde{m}\cdot\bar{s} = 0 \quad [7.8]$$

$$-i\omega\frac{n}{c}\,{}_0M(D_3 + D_4)\tilde{u}\cdot\bar{s} + ({}_0M^2C_{11} + C_8)\tilde{m}\cdot\bar{s} = 0 \quad [7.9]$$

$$(1 + C_9 + {}_0M^2C_{10})\,\tilde{p}\cdot\bar{s} = 0 \quad [7.10]$$

$$\omega\left(D_1\frac{n^2}{c^2} - {}_0\varrho\right)\tilde{u}\times\bar{s} + i\frac{n}{c}\,{}_0M D_4\tilde{m}\times\bar{s}$$

$$+\frac{i}{c}(C_8 - 1)\,{}_0M\tilde{p} = 0 \qquad [7.11]$$

$$-{}_0M(D_4 + 1 - C_8 - C_9)\,i\omega\frac{n}{c}\tilde{u}\times\bar{s}$$

$$+ C_8\tilde{m}\times\bar{s} - nC_9\tilde{p} = 0 \qquad [7.12]$$

$$-{}_0M i\frac{\omega}{c}(C_8 + C_9 - D_4n^2)\,\tilde{u}\times\bar{s}$$

$$+ n(1 - C_8)\,\tilde{m}\times\bar{s} + (1 + C_9)\,\tilde{p} = 0. \qquad [7.13]$$

En vertu de [7.10], $\tilde{p}\bar{s} = 0$; de [7.8] et [7.9] il résulte que $\tilde{u}\bar{s} = \tilde{m}\bar{s} = 0$ sauf si

$$\frac{1}{n^2} = \frac{D_1 + D_2}{{}_0\varrho c^2}\left[1 - \frac{{}_0M^2(D_3 + D_4)^2}{(D_1 + D_2)\,({}_0MC_{11} + C_8)}\right] \qquad [7.14]$$

qui donne le carré du rapport de la célérité des ondes élastiques longitudinales à la vitesse de la lumière.

Le système [7.11], [7.12], [7.13] n'admet que la solution triviale $\tilde{u}\times\bar{s} = \tilde{m}\times\bar{s} = \tilde{p} = 0$ à moins que le déterminant

$$\begin{vmatrix} \mu + C_4\lambda - \beta^2 & i/c\,{}_0M\,D_4 & i/c\,{}_0M\,(C_8 - 1)\,\beta \\ -i/c\,(D_4 + 1 - C_8 - C_9)\,{}_0M/{}_0e\beta & C_8\,\beta & -C_9 \\ -i/c\,|(C_8 + C_9)\,\beta^2 - D_4|\,{}_0M/{}_0e & 1 - C_8 & (1 + C_9)\,\beta \end{vmatrix} = 0 \qquad [7.15]$$

avec

$$\beta = \frac{1}{n}, \quad \mu = \frac{C_2}{{}_0\varrho c^2}, \quad \lambda = \frac{{}_0M^2}{{}_0\varrho c^2}.$$

A l'approximation classique, on peut se limiter, pour la détermination des racines de cette équation bicarrée, aux termes d'ordre le moins élevé en les paramètres sans dimensions λ et μ. Ces deux pacines sont donnés par

$$\frac{1}{n^2} = \frac{C_2}{{}_0\varrho c^2}\left[1 + \left\{\frac{C_4}{C_2} + \frac{(C_4 - C_5)^2}{C_2(1 - C_8)}\right\}{}_0M^2\right] \qquad [7.16]$$

$$\frac{1}{n^2} = \frac{C_9(C_8 - 1)}{C_8(1 + C_9)}. \qquad [7.17]$$

L'examen du système [7.11], [7.12], [7.13] montre qu'elles correspondent respectivement à des ondes élastiques et électromagnétiques transversales de polarisation quelconque. En l'absence de magnétisation ${}_0M = 0$ et C_8 tend vers l'infini de sorte que les 3 valeurs obtenues pour β^2 se réduisent aux valeurs classiques

$$\frac{C_1(0) + C_2(0)}{{}_0\varrho c^2}, \quad \frac{C_2(0)}{{}_0\varrho c^2}, \quad \frac{C_9(0)}{1 + C_9(0)}$$

$C_1(0)$ et $C_2(0)$ désignent les paramètres de *Lamé* et C_9 l'inverse de la susceptibilité diélectrique ($C_9 = \chi^{-1}$).

En résumé, trois types d'ondes peuvent se propager dans le modèle proposé: deux ondes élastiques (ondes lentes) l'une transversale l'autre longitudinale et une onde électromagnétique (onde rapide). Le modèle ne peut donc rendre compte de l'effet *Faraday* qui résulte de la propagation de deux ondes polarisées circulaires gauche et droite à des vitesses différentes. En appliquant la même théorie à un modèle de diélectrique dissipatif faisant intervenir des dérivées par rapport au temps de la polarisation et de la magnétisation, *M. Ph. Boulanger* (9) a pu rendre compte de l'effet *Faraday*.

2. Champ magnétique transversal

On procède comme pour 1 mais avec l'hypothèse ${}_0\bar{B}\bar{s} = {}_0\bar{M}\bar{s} = 0$. En éliminant $\tilde{e}$ et $\tilde{b}$ dans le système linéaire aux amplitudes, on projette les 3 équations vectorielles restantes sur $\bar{s}, {}_0\bar{M}$, et $\bar{s}\times{}_0\bar{M}$; on obtient 9 relations scalaires linéaires en $\tilde{u}, \tilde{p}, \tilde{m}$

$$\omega\left[(C_1 + C_2 - C_6\,{}_0M^2 + {}_0M^2)\frac{n^2}{c^2} - {}_0\varrho\right]\tilde{u}\bar{s}$$

$$+ i\frac{n}{c}(C_6 - C_8 - C_{11}\,{}_0M^2)\,\tilde{m}\,{}_0\bar{M}$$

$$-\frac{i}{c}(C_8 - 1)\,\tilde{p}\cdot{}_0\bar{M}\times\bar{s} = 0 \qquad [7.18]$$

$$\omega\left|(C_2 + {}_0M^2C_5)\frac{n^2}{c^2} - {}_0\varrho\right|\tilde{u}\,{}_0\bar{M}$$

$$+ i\frac{n}{c}\,{}_0M^2(C_4 - C_5 - C_8 + 1)\,\tilde{m}\bar{s} = 0 \qquad [7.19]$$

$$\omega\left(C_2\frac{n^2}{c^2} - {}_0\varrho\right)\tilde{u}\cdot\bar{s}\times{}_0\bar{M}$$

$$-\frac{i}{c}(C_8 - 1)\,{}_0M^2\tilde{p}\bar{s} = 0 \qquad [7.20]$$

33

$$-i\omega \frac{n}{c}(C_4 - C_5 + 1 - C_8)\tilde{u} \cdot {}_0\bar{M}$$
$$+ C_8 \tilde{m}\bar{s} = 0 \qquad [7.21]$$

$$({}_0M^2 C_{11} + C_8)\tilde{m}\,{}_0\bar{M}$$
$$- i\omega \frac{n}{c}(C_6 - 1 + C_8 + C_9)\,{}_0M^2 \tilde{u}\bar{s}$$
$$- nC_9 \tilde{p} \cdot {}_0\bar{M} \times \bar{s} = 0 \qquad [7.22]$$

$$C_8 \tilde{m} \cdot \bar{s} \times {}_0\bar{M} - n(C_9 + {}_0M^2 C_{10})\tilde{p} \cdot {}_0\bar{M} = 0 \qquad [7.23]$$

$$(1 + C_9)\tilde{p}\bar{s}$$
$$- i\frac{\omega}{c}(C_8 + C_9)\tilde{u} \cdot {}_0\bar{M} \times \bar{s} = 0 \qquad [7.24]$$

$$n(C_8 - 1)\tilde{m} \cdot \bar{s} \times {}_0\bar{M}$$
$$- (1 + C_9 + C_{10}\,{}_0M^2)\tilde{p} \cdot {}_0\bar{M} = 0 \qquad [7.25]$$

$$-n(C_{11}\,{}_0M^2 + C_8 - 1)\tilde{m}\,{}_0\bar{M}$$
$$+ i\frac{\omega}{c}\,{}_0M^2[(C_6 - 1)n^2 + C_8 + C_9]\tilde{u}\bar{s}$$
$$- (1 + C_9)\tilde{p} \cdot \bar{s} \times {}_0\bar{M} = 0. \qquad [7.26]$$

De [7.23], [7.25], il résulte que $\tilde{p} \cdot {}_0\bar{M} = \tilde{m} \cdot \bar{s} \times {}_0\bar{M} = 0$ sauf si

$$\frac{1}{n^2} = \frac{C_8 - 1}{C_8}\frac{C_9 + {}_0M^2 C_{10}}{1 + C_9 + {}_0M^2 C_{10}}. \qquad [7.27]$$

En se limitant aux termes d'ordre le moins élevé en λ et μ, [7.20] et [7.21] d'une part, [7.19] et [7.21] d'autre part impliquent

$$\frac{1}{n^2} = \frac{C_2}{{}_0\varrho c^2} \qquad [7.28]$$

ou

$$\tilde{u} \cdot \bar{s} \times {}_0\bar{M} = \tilde{p}\bar{s} = 0$$

et

$$\frac{1}{n^2} = \frac{C_2}{{}_0\varrho c^2} \times \left[1 + \frac{C_5}{C_2}\,{}_0M^2 - \frac{(C_4 - C_5 - C_8 + 1)^2}{C_2 C_8}\,{}_0M^2\right] \qquad [7.29]$$

ou

$$\tilde{u} \cdot {}_0\bar{M} = \tilde{m} \cdot \bar{s} = 0.$$

Enfin, [7.18], [7.22], [7.26], n'admettent que la solution triviale $\tilde{u}\bar{s} = \tilde{m} \cdot {}_0\bar{M} = \tilde{p}\,{}_0\bar{M} \times \bar{s} = 0$ à moins que le déterminant de ces inconnues, qui constitue une équation bicarrée en $1/n$, ne soit nul. En résolvant cette équation, on obtient, à l'ordre le moins élevé en les petits paramètres λ, μ et

$$v \overset{\text{déf}}{=} \frac{C_1 + C_2}{{}_0\varrho c^2}$$

$$\frac{1}{n^2} = \frac{C_1 + C_2}{{}_0\varrho c^2} \times \left[1 - \frac{(1 - C_6)^2\,{}_0M^2}{(C_1 + C_2)({}_0M^2 C_{11} + C_8 - 1)}\right] \qquad [7.30]$$

$$\frac{1}{n^2} = \frac{C_9({}_0M^2 C_{11} + C_8 - 1)}{(C_9 + 1)({}_0M^2 C_{11} + C_8)}. \qquad [7.31]$$

En résumé, à chaque direction $\bar{s}$ se trouvent associées une onde élastique longitudinale [7.30], deux ondes élastiques transversales [7.28], [7.29] et deux ondes électromagnétiques transversales [7.27], [7.31].

Grâce à la prise en considération de la magnétisation, le modèle rend compte de l'effet *Cotton-Mouton* (biréfringence due à un champ magnétique transversal). En l'absence de magnétisation, il ne subsiste qu'une onde élastique longitudinale avec

$$\frac{1}{n^2} = \frac{C_1(0) + C_2(0)}{{}_0\varrho c^2}$$

une onde élastique et une onde électromagnétique transversales avec respectivement

$$\frac{1}{n^2} = \frac{C_2(0)}{{}_0\varrho c^2} \quad \text{et} \quad \frac{1}{n^2} = \frac{C_9(0)}{1 + C_9(0)}.$$

3. Champ électrique transversal

Considérons un état initial non perturbé caractérisé par

$${}_0x^\lambda = X^\lambda, \quad {}_0\bar{B} = {}_0\bar{M} = 0$$
$${}_0\bar{E},\ {}_0\bar{P} \ \text{uniformes} \ \ {}_0\bar{E} \cdot \bar{s} = 0.$$

De [4.5], [4.6], [4.2] et [4.8], il résulte que

$${}_0\bar{E} = C_9\,{}_0\bar{P}$$

avec

$$C_9 = 2\,{}_0\varrho\,{}_0\left(\frac{\partial\hat{\varepsilon}}{\partial I_4}\right) \qquad [7.32]$$

$$ {}_0\sigma^{kl} = {}_0\varrho\, {}_0\left(\frac{\partial\hat{\varepsilon}}{\partial I_1}\right) g^{kl} + {}_0\varrho\, {}_0\left(\frac{\partial\hat{\varepsilon}}{\partial I_6}\right) {}_0P^k\, {}_0P^l. \quad [7.33] $$

En procédant comme précédemment, c'est-à-dire en éliminant $\tilde{e}$ et $\tilde{b}$ dans le système linéaire aux amplitudes complexes puis en projetant les 3 équations vectorielles restantes sur $\bar{s}$, ${}_0\bar{P}$ et $\bar{s} \times {}_0\bar{P}$, on conclut comme pour le modèle non magnétisable adopté par *Toupin*, à l'existence de 5 types d'ondes: trois ondes élastiques pour lesquelles, aux approximations adoptées

$$ \frac{1}{n^2} = \frac{C_1 + C_2}{{}_0\varrho c^2}\left[1 - \frac{D_5^2\, {}_0P^2}{(C_1 + C_2)(D_{12}\, {}_0P^2 + C_9)}\right] \quad [7.34] $$

$$ \frac{1}{n^2} = \frac{C_2}{{}_0\varrho c^2} \quad [7.35] $$

$$ \frac{1}{n^2} = \frac{C_2}{{}_0\varrho c^2} \times \left[1 + \left\{D_4 - \frac{(D_7 - C_9)^2}{C_9 + 1} - \frac{D_8^2\, {}_0P^2}{C_8 - 1}\right\} \times \frac{{}_0P^2}{C_2}\right] \quad [7.36] $$

et deux ondes électromagnétiques avec respectivement

$$ \frac{1}{n^2} = \frac{C_9(C_8 - 1 + C_{10}\, {}_0P^2)}{(C_9 + 1)(C_8 + C_{10}\, {}_0P^2)} \quad [7.37] $$

$$ \frac{1}{n^2} = \frac{(C_8 - 1)(D_{12}\, {}_0P^2 + C_9)}{C_8(D_{12}\, {}_0P^2 + C_9 + 1)} \quad [7.38) $$

$$ D_4 = \frac{1}{2}\, {}_0\varrho\, {}_0\left(\frac{\partial\hat{\varepsilon}}{\partial I_7}\right) $$

$$ D_5 = 2\, {}_0\varrho\, {}_0\left(\frac{\partial^2\hat{\varepsilon}}{\partial I_4 \partial I_1}\right) $$

$$ D_7 = {}_0\varrho\, {}_0\left(\frac{\partial\hat{\varepsilon}}{\partial I_6}\right) $$

$$ D_8 = \frac{1}{2}\, {}_0\varrho\, {}_0\left(\frac{\partial\hat{\varepsilon}}{\partial I_{12}}\right) $$

$$ D_{12} = 4\, {}_0\varrho\, {}_0\left(\frac{\partial^2\hat{\varepsilon}}{\partial I_4^2}\right). $$

Les invariants C_i et D_i sont tous fonctions de ${}_0P^2$. Les C_i sont définies comme dans le cas 1. Le modèle proposé rend compte de l'effet *Kerr* (biréfringence due à un champ électrique transversal). En l'absence de polarisation, il ne subsiste que trois vitesses de propagation et l'effet *Kerr* disparait. De même que le modèle rend compte de l'effet *Cotton-Mouton* grâce à la magnétisation, il rend compte de l'effet *Kerr* grâce à la polarisation.

4. Champ électrique longitudinal

On reprend les hypothèses de 3 avec ${}_0\bar{E} \times \bar{s} = 0$.

Après élimination de $\tilde{e}$ et $\tilde{b}$ dans le système aux amplitudes, on multiplie scalairement puis vectoriellement par $\bar{s}$ les 3 équations vectorielles restantes. L'étude du système homogène obtenu met en évidence

a) une onde électromagnétique transversale de polarisation quelconque

$$ \frac{1}{n^2} \simeq \frac{C_9(C_8 - 1)}{C_8(1 - C_9)} \quad [7.39] $$

b) une onde élastique longitudinale

$$ \frac{1}{n^2} \simeq \frac{C_1 + C_2}{{}_0\varrho c^2} \times \left[1 + \left\{2D_4 + D_7 + 2D_3 + D_{11}\, {}_0P^2 - \frac{(D_5 + D_6\, {}_0P^2 + 2D_7 - C_9)^2}{D_{12}\, {}_0P^2 + C_9 + 1}\right\} \times \frac{{}_0P^2}{C_1 + C_2}\right] \quad [7.40] $$

avec

$$ D_3 = {}_0\varrho\, {}_0\left(\frac{\partial^2\hat{\varepsilon}}{\partial I_1 \partial I_6}\right) + \frac{1}{2}\, {}_0\varrho\, {}_0\left(\frac{\partial\hat{\varepsilon}}{\partial I_7}\right) $$

$$ D_6 = 2\, {}_0\varrho\, {}_0\left(\frac{\partial^2\hat{\varepsilon}}{\partial I_4 \partial I_6}\right) $$

$$ D_{11} = {}_0\varrho\, {}_0\left(\frac{\partial^2\hat{\varepsilon}}{\partial I_6^2}\right) $$

c) une onde élastique transversale dont la vitesse dépend du sens de propagation

$$ \frac{1}{n} \simeq \pm\sqrt{\frac{C_2}{{}_0\varrho c^2}} \times \left[1 + \left(D_4 + D_7 - \frac{D_7^2}{C_9} - \frac{D_8^2\, {}_0P^2}{C_8 - 1}\right)\frac{{}_0P^2}{C_2}\right]^{1/2} - \frac{D_8(D_7 + C_9^2)\, {}_0P^3}{C_9(C_8 - 1)\, {}_0\varrho c^2}. \quad [7.41] $$

Le modèle ne rend pas compte de l'effet *Pockels* (rotation du plan de polarisation due à l'action d'un champ électrique longitudinal). En l'absence

33*

de magnétisation, les mêmes ondes apparaissent; [7.40] reste inchangée, [7.39] devient

$$\frac{1}{n^2} = \frac{C_9}{1 + C_9}$$

et [7.41] se réduit à

$$\frac{1}{n} \simeq \sqrt{\frac{C_2}{{}_0\varrho c^2}} \left[1 + \frac{(D_4 + D_7)\,{}_0P^2}{C_2} - \frac{D_7^2\,{}_0P^2}{C_2 C_9}\right]^{1/2}.$$

La vitesse des ondes élastiques transversales est alors la même dans les deux sens de la propagation.

8. Conclusions

La théorie imaginée par *Toupin* pour les diélectriques élastiques polarisables rend compte de l'effet *Kerr* mais pas de l'effet *Cotton-Mouton*. Le modèle magnétisable que nous proposons rend compte de ce dernier effet de sorte que si l'effet *Kerr* apparait lié à la polarisation, l'effet *Cotton-Mouton* semble lié à la magnétisation.

Par contre, les effets *Faraday* et *Pockels* sont absents. *Ph. Boulanger* (9) a envisagé un modèle plus général faisant intervenir des dérivées par rapport au temps de la polarisation et de la magnétisation, ce diélectrique présente l'effet *Faraday* mais pas l'effet *Pockels* qui, expérimentalement, n'apparait que dans des matériaux anisotropes. Ces conclusions nous apparaissent en harmonie tant avec les faits expérimentaux qu'avec les résultats déduits de théories microscopiques classiques ou quantiques (7).

Summary

The total momentum-energy tensor describing the interaction of a polarizable and magnetizable continuum with an electromagnetic field is constructed on the basis of *Lorentz* invariance. The equations of balance of energy and momentum are deduced and applied to the study of the magnetooptical, electrooptical and photoelastic behavior of an elastic polarizable and magnetizable isotropic continuum. This model exhibits the *Cotton-Mouton* and *Kerr* effects but the *Pockels* and *Faraday* rotations don't appear.

Littérature

1) *Toupin, R. A.*, J. Eng. Sci. **1**, 101 (1963).
2) *Brevik, I.*, Mat. Fys. Medd. Dan. Vid. Selsk **37**, n° 11 et 13 (1970).
3) *Boulanger Ph.* et *G. Mayné*, Bull. Acad. Roy. Belg., Cl. Sci. **57**, 872 (1971).
4) *Boulanger, Ph.* et *G. Mayné*, C. R. Acad. Sci. (Paris) **274**, 591 (1972).
5) *De Groot, S. R.* and *L. G. Suttorp*, Physica **39**, 28, 41, 61, 77, 84 (1968).
6) *Eringen, A. C.* and *R. A. Grot*, Int. J. Eng. Sci. **4**, 611 (1966).
7) *Boulanger, Ph., G. Mayné* et *R. van Geen*, Int. J. Solids Struct. **9**, 1439 (1973).
8) *Wang, C. C.*, Arch. Rat. Mech. Anal. **36**, 166 (1970).
9) *Boulanger, Ph.*, Rheol. Acta **12**, 116 (1973).

Adresse des auteurs:
Département de Mathématique – Faculté des Sciences
Université Libre de Bruxelles
50, avenue F. Roosevelt
B-1050 Bruxelles (Belgique)

Rheol. Acta 13, 513–517 (1974)

Centre de Recherches Physiques, Marseille (France)

Etude des phénomènes de fracture dans les polymères solides

Par D. Bonsignour et F. Nieuwenhove

Avec 8 figures

(Reçu p. p. le 27 octobre 1972)

I. Introduction

On peut définir le terme de fracture comme une «création de nouvelles surfaces à l'intérieur d'un corps par application de forces extérieures». Cette définition peut s'appliquer aussi bien à une masse liquide qu'à un corps solide englobant ainsi des phénomènes tels que la cavitation ou la fracture de fusion des liquides viscoélastiques. Dans le cas des polymères, cette définition s'applique seulement aux solides, le terme solide étant pris dans son sens mécanique.

Nous nous sommes proposés d'étudier la fracture d'un spécimen soumis à un choc. L'utilisation des ondes de contrainte dans l'étude des phénomènes de fracture remonte à environ un siècle avec les travaux de *J. Hopkinson* (1), qui a mesuré la résistance des fils métalliques quand ils sont soumis à une impulsion de tension. Il a montré que la valeur du point critique et la résistance à la rupture des fils d'acier sont environ le double de celles mesurées dans des expériences statiques. *B. Hopkinson* (2) a continué ses travaux et a été le premier à étudier le phénomène d'écaillage des métaux qui se produit quand un choc de compression se réfléchit sur la surface libre d'un spécimen. Depuis une quinzaine d'années d'importantes études aussi bien théoriques qu'expérimentales ont été faites dans ce domaine. Ces travaux ont eu principalement pour base ceux de *Griffith*, d'*Irwin* et de *Williams*.

Dans sa théorie de la fracture fragile (sans déformation plastique), *Griffith* (3) a considéré une plaque infinie d'épaisseur e contenant une fente linéaire de longueur $2c$. L'énergie H du système se compose de l'énergie de déformation élastique qui dépend du travail des contraintes imposées au système et de l'énergie de surface. Cette dernière représente l'énergie nécessaire à la formation des surfaces de fracture lorsque la fente s'élargit. *Griffith* a supposé que la fente augmente quand $\partial H/\partial c = 0$ et la contrainte critique est alors donnée par:

$$S_e = \{2E\gamma/\pi c(1-\mu^2)\}^{1/2}$$

dans le cas d'une déformation plane (le rapport c/e est petit) et:

$$S_c = (2E\gamma/\pi c)^{1/2}$$

pour une contrainte plane (le rapport c/e est grand), E étant le module d'*Young*, γ l'énergie de surface spécifique et μ le coefficient de *Poisson*.

Irwin (4) a considéré le champ des contraintes dans le voisinage immédiat de la fente. En considérant le travail nécessaire pour refermer un segment à partir de l'extrémité de la fente ouverte, il a déduit la variation de l'énergie de déformation du système et a mis en évidence le rôle prépondérant de la quantité $g = \pi K^2/E$ (K: facteur d'intensité de contrainte qui dépend des détails géométriques du modèle considéré). Cette quantité appelée «vitesse de dégagement d'énergie de déformation» ou force d'extension de la fente, correspond à la dérivé ∂H de l'énergie de déformation citée dans la théorie de *Griffith*: mais elle n'est pas égale à la dérivée de l'énergie de surface car *Irwin* a supposé que ce terme comprend toutes les énergies dissipées lorsque la fente augmente.

Williams (5) a étudié la fracture d'un matériau viscoélastique. Il a supposé que pour un matériau rhéologiquement complexe il existe les mêmes mécanismes de dissipation que dans les cas simples auxquels s'ajoutent d'autres mécanismes tels que travail plastique ou dissipation visqueuse et il a proposé pour la contrainte critique:

$$\sigma_{cr} = k\sqrt{\frac{E\,\Sigma\,T_i}{l}}$$

où k est une constante dépendant de la géométrie considérée, E le module d'*Young* du matériau, l la longueur de la fissure et $\Sigma\,T_i$ les procédés de dissipation particuliers du matériau. Pour les matériaux viscoélastiques dépendant de la vitesse, il utilise l'équation thermodynamique du système et dans certains cas géométriques simples il obtient les conditions qui déterminent la croissance d'une fissure. Mais il n'est pas possible à l'heure actuelle d'étendre ces calculs à un cas plus général.

II. Etude expérimentale

Il est très difficile de produire une impulsion de tension, sauf dans le cas d'un spécimen en forme de fil où on peut utiliser la chute d'un poids ou un montage équivalent. Les chocs de compression sont en général plus faciles à obtenir et peuvent être produits par la détonation d'une petite quantité d'explosif ou l'impact d'un projectile à grande vitesse. La réflexion d'un choc de compression à l'extrémité libre d'un spécimen donne un choc de tension se propageant en sens inverse. La fig. 1 montre la réflexion d'un choc de compression sur une surface libre sous incidence normale. La contrainte résultante au

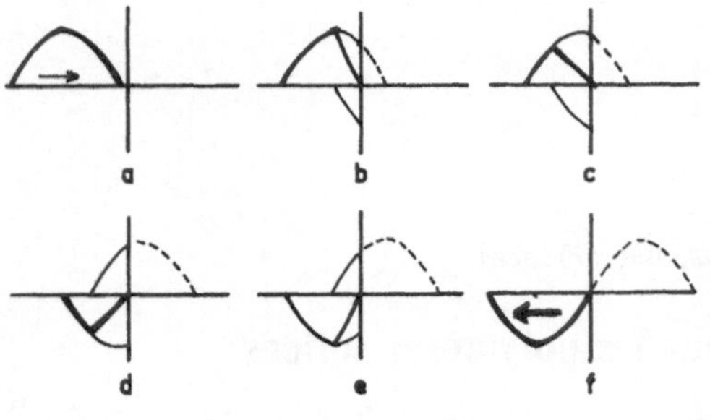

Fig. 1. Réflexion d'un choc de compression. Ligne épaisse: contrainte résultante; ligne pointillée: portion de l'onde déjà réfléchie

voisinage de la surface est obtenue en additionnant les contraintes dues aux chocs incident et réfléchi. Pour un choc symmétrique, on peut voir que le maximum de tension a lieu au quart de la longueur du choc (fig. 1e).

C'est ce phénomène de la réflexion d'un choc de compression que nous avons utilisé pour l'étude de la fracture d'un spécimen de polymère. Le montage expérimental se compose d'un canon à air comprimé et d'une chambre d'acier dans laquelle une balle venant frapper un piston permet d'appliquer une onde de compression au barreau étudié.

La fig. 2 montre un schéma général de ce montage. Le canon se compose d'un corps central relié par l'intermédiaire d'une électrovanne commandée elle-même par un relai temporisé et d'une chambre de stabilisation à une bouteille d'air comprimé dont la pression de sortie est réglée par un manomètre. Le canon est un tube d'acier inoxydable parfaitement calibré de 1,50 m de long. Sur le dessus du corps central se trouve le réservoir à balles: un système à guillotine actionné par un déclencheur souple permet d'utiliser

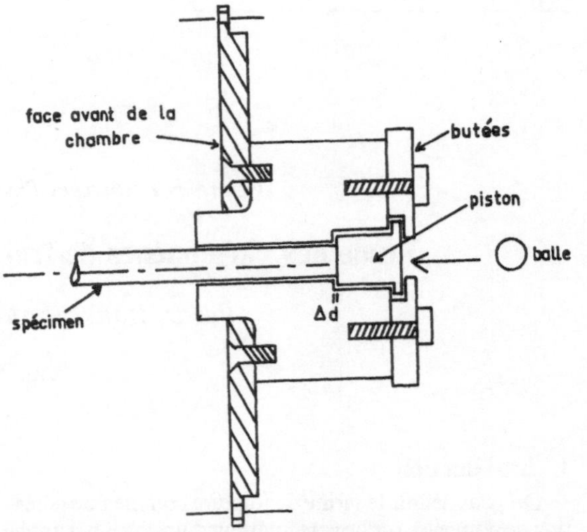

Fig. 3. Détails de la partie avant de la chambre d'acier

une seule balle à la fois. Ce canon vient s'ajuster à une chambre d'acier cylindrique. La fig. 3 montre les détails de la partie avant de cette chambre avec les positions respectives du piston et du spécimen. L'amplitude de l'onde dépend non seulement des dimensions géométriques et du poids du piston, mais également des caractéristiques de la balle et de la pression utilisée. Le temps du choc dépend de la distance Δd entre le piston et ses butées.

Les spécimens étudiés sont des échantillons cylindriques de PMMA fournis par la Société Alsthom (Altuglas, qualité M 70) de 10 cm de long et de 7 mm de diamètre. Une entaille en V est faite à 2,5 cm de l'extrémité libre du barreau. Nous avons utilisé trois séries d'échantillons dont les entailles ont pour profondeur 1,4, 1 et 0,5 mm.

On a étalonné l'onde de compression qui se propage dans le spécimen en fonction de la pres-

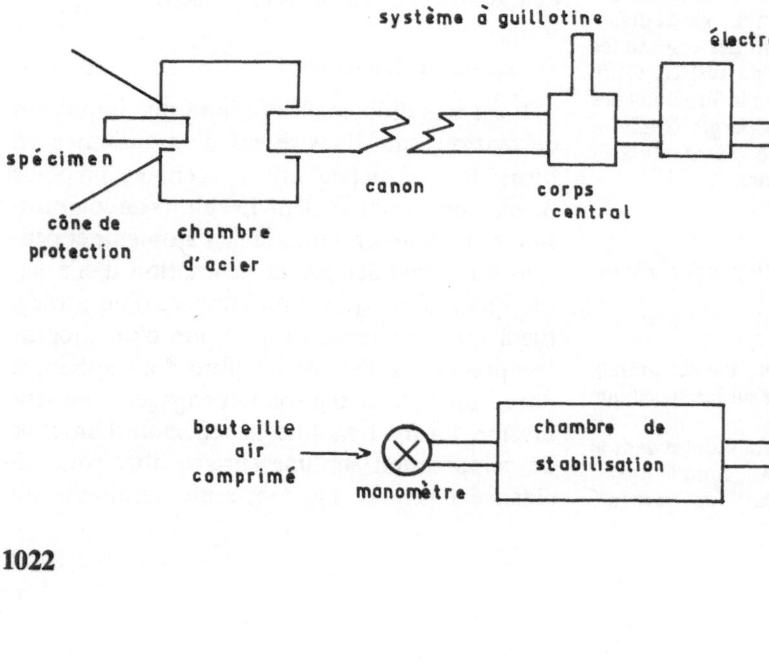

Fig. 2. Schéma général du montage

sion. Pour cela, on a utilisé des jauges de contrainte (collées environ 1 cm avant l'entaille, c'est-à-dire 3,5 cm de l'extrémité libre du barreau) et la forme de l'onde a été relevée à l'oscilloscope. Nous avons tracé la courbe (fig. 4) qui donne la tension de sortie des jauges en fonction de la

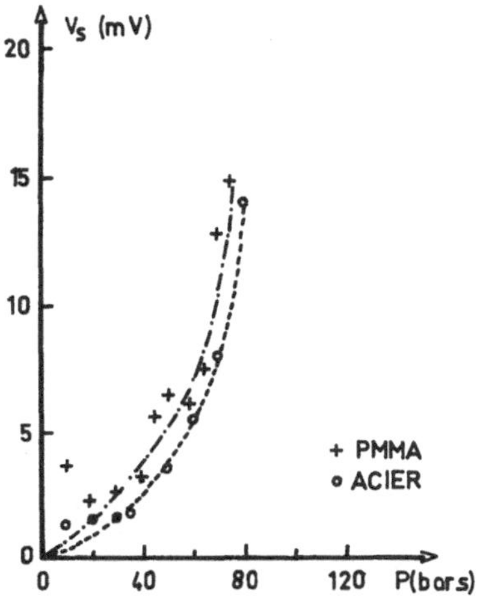

Fig. 4. Tension de sortie des jauges en fonction de la pression

pression: cette tension permet de calculer la contrainte appliquée au spécimen. La contrainte superficielle varie de $50 \cdot 10^5$ dyn/cm^2 pour une pression de 30 bars à $300 \cdot 10^5$ dyn/cm^2 pour une pression de 80 bars.

Nous avons ensuite déterminé les pressions critiques de fracture pour différentes distances Δd de trajet du piston en relevant pour chaque Δd la pression minimum en dessous de laquelle le piston ne se fracture plus. La courbe de la fig. 5

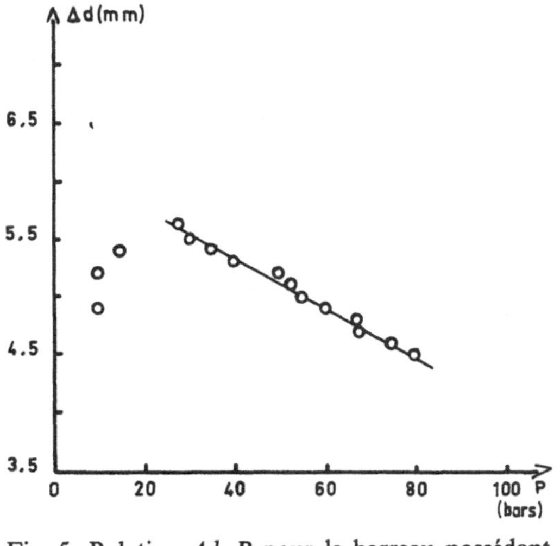

Fig. 5. Relation $\Delta d \cdot P$ pour le barreau possédant une entaille de 1,4 mm

a été tracée pour des échantillons présentant une entaille de 1,4 mm. Pour des pressions inférieures à 25 bars, on obtient des fractures par cisaillement et les mesures semblent être faussées: l'énergie de la balle étant faible, le choc n'est plus purement longitudinal.

La profondeur de l'entaille n'a pas une grande influence sur les pressions critiques de fracture comme on peut le voir sur la fig. 6: on remarque

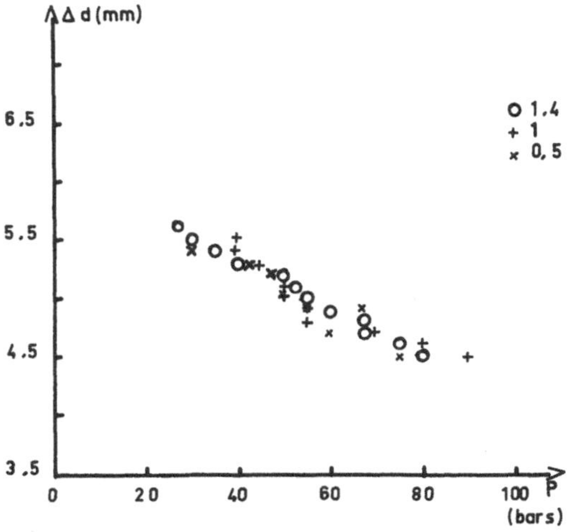

Fig. 6. Relation $\Delta d \cdot P$ en fonction de la profondeur de l'entaille

simplement une dispersion des résultats autour de la droite obtenue avec une entaille de 1,4 mm, ce qui prouverait que dans le cas d'une fracture produite par choc le PMMA ne suit pas la loi de fracture fragile de *Griffith*; par contre les surfaces de fracture présentent des différences très marquées avec la profondeur de l'entaille comme on peut le voir sur les photos de la fig. 7. Avec une entaille de 1,4 mm (fig. 7a) on obtient au centre du barreau un début de fracture assez important présentant une surface très rugueuse, entourée d'une zone d'hyperboles secondaires très nombreuses et très petites, la zone presque lisse de propagation de fracture à grande vitesse est très mince. Avec une entaille de 1 mm (fig. 7b) la zone centrale est plus petite, les hyperboles sont plus grandes et moins nombreuses et la zone de fracture rapide est plus importante. Avec la petite entaille de 0,5 mm (fig. 7c) l'amorce de la fracture est très petite, les hyperboles secondaires plus importantes et encore moins nombreuses et la zone de fracture rapide est très importante. La fig. 8 montre, à titre de comparaison, une surface de fracture par cisaillement.

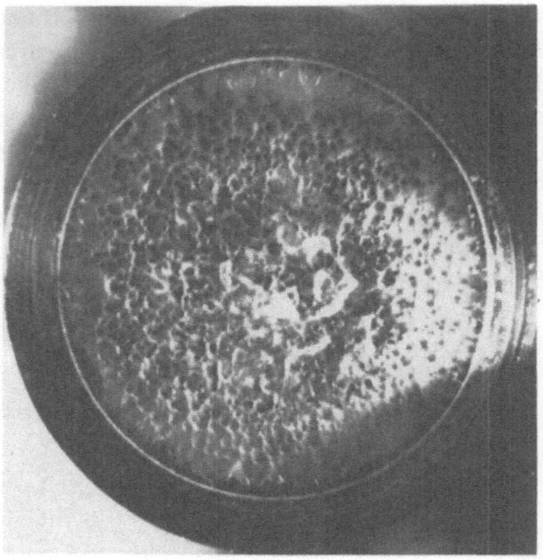

a)

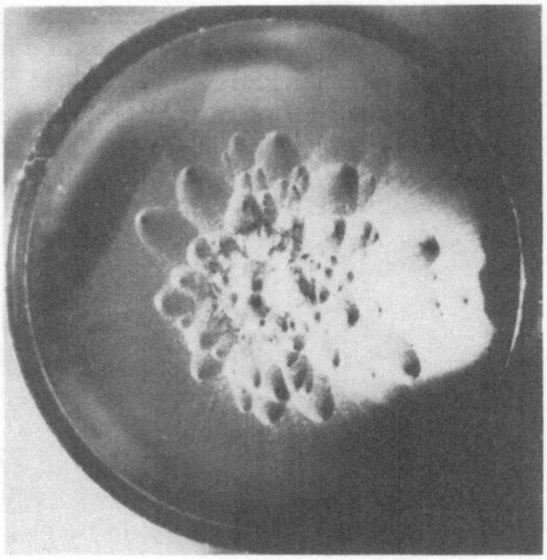

c)

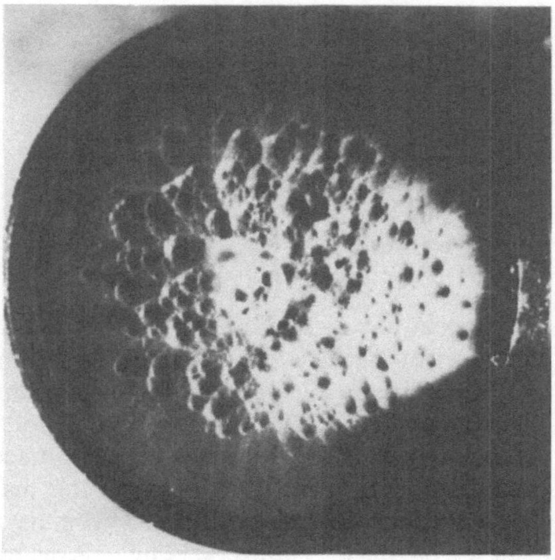

b)

Fig. 7. Morphologie des surfaces de fracture. a) Entaille de 1,4 mm; b) entaille de 1 mm; c) entaille de 0,5 mm

III. Conclusion

On a étudié la fracture d'un spécimen de polymère soumis à un choc. Pour cela nous avons réalisé un nouveau montage expérimental permettant de produire des chocs de compression sur le barreau étudié. Les premiers résultats expérimentaux ont permis d'obtenir une relation linéaire entre la pression utilisée et la distance parcourue par le piston, et ont montré que la pression critique de fracture ne dépend pas de la profondeur de l'entaille faite sur le spécimen. Ce matériau n'obéit pas à la loi de fracture fragile de *Griffith*. Par contre, la morphologie de la surface de fracture est très influencée par la grandeur de l'entaille.

Nous poursuivons actuellement ces études dynamiques de la fracture d'un spécimen soumis à un choc. Un nouveau montage expérimental devra nous permettre de déterminer le paramètre γ qui représente l'énergie de surface spécifique et d'étudier la vitesse de propagation de la fente ainsi que la force de fracture en fonction de la grandeur de celle-ci. L'ensemble de ces résul-

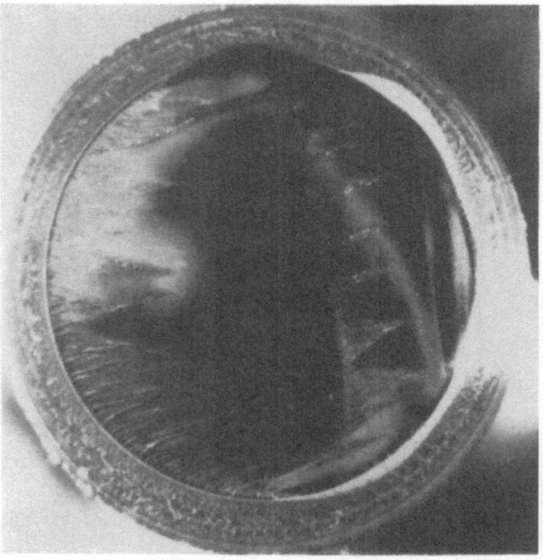

Fig. 8. Fracture par cisaillement

tats devrait nous conduire à une plus grande compréhension des phénomènes de fracture dans les polymères.

Summary

To study the fracture of some polymeric specimens, we used a new simple device for producing shock loading. The device uses a ballistic impact-driven projectile to introduce high stress waves into the material. Load amplitude can vary with the pressure used and impact time depend on the position of the piston backstops. Strain gages on the specimen give the form and the amplitude of the stress wave. We compared results with the topography of fracture surfaces.

Littérature

1) *Hopkinson, J.*, Mem. Proc. Manchester Lit. & Phil. Soc. **11**, 40 (1872).

2) *Hopkinson, B.*, Proc. Roy. Soc. (London) **A 74**, 498 (1905).

3) *Griffith, A. A.*, Phil. Trans. Roy. Soc. (London) **221**, 163–198 (1921).

4) *Irwin, G. R.*, J. appl. Mech. **61**, A-49 (1939).

5) *Williams, M. L.*, Int. J. fracture Mech. **1**, n° 4, 292–310 (1965); J. appl. Physics **38**, n° 11, 4476–4480 (1967).

Adresse des auteurs:

Dr. *D. Bonsignour* et *F. Nieuwenhove*
Centre de Recherches Physiques
31, chemin Joseph-Aiguier (9° arr.)
F-13274 Marseille (France)

Rheol. Acta 13, 518–525 (1974)

Institut National Polytechnique de Toulouse (France)

Recherche de lois de composition en rhéologie des liquides organiques

Par C. Thirriot et D. Bellet

Avec 11 figures

(Reçu p. p. le 27 octobre 1972)

On est en droit de penser que la *Rhéologie*, au même titre que les autres branches fondamentales de la Science (Mécanique, Thermique, Electricité, Magnétisme, ...), peut être parainée, elle aussi, par la *Thermodynamique*, remarquable synthèse dominant les lois de toutes les sciences physicochimiques.

I. De la thermodynamique à la rhéologie

A. Fonctions et variables d'état

En thermodynamique, un point de vue macroscopique consiste à faire abstraction de la réalité moléculaire et à caractériser l'état d'un système, à notre échelle, par un certain nombre de ses propriétés aisément mesurables.

L'expérience montre que l'état d'un système, formé d'un petit nombre de constituants et d'un petit nombre de phases, peut être caractérisé grâce à un petit nombre de variables: Ainsi par exemple, l'état d'équilibre d'un fluide homogène formé d'une seule espèce chimique peut être caractérisé, pour une masse donnée de fluide, par sa pression supposée uniforme, sa température supposée uniforme et par son volume. Mais l'expérience nous apprend que ces trois variables ne sont pas indépendantes et sont liées entre elles par une relation implicite ou équation d'état de la forme générale:

$$\mathscr{F}(p, v, T) = 0 \qquad [1]$$

$$v = f(p, T)$$

$$p = g(v, T)$$

$$T = h(v, p). \qquad [2]$$

D'autres grandeurs, par exemple l'énergie interne U, l'enthalpie H ou l'entropie S, peuvent être, elles aussi, considérées comme fonctions des deux variables indépendantes de base. (Comme tous les autres paramètres caractéristiques des propriétés du fluide d'ailleurs.)

Si le nombre de variables indépendantes est déterminé sans ambiguité pour chaque système, leur choix est bien souvent arbitraire.

Appliquons maintenant ces quelques remarques d'ordre général à l'objet de notre propos: la *rhéologie*.

Nous devons au préalable faire une hypothèse constituant en même temps une réserve importante: Parler de variable d'*état* ou de fonction d'*état*, suppose que l'ensemble des propriétés énergétiques *actuelles* d'un corps soient parfaitement définies lorsque l'on se donne un certain nombre d'entre elles indépendantes. Mais il existe des cas d'hystérésis où les propriétés actuelles dépendent des traitements antérieurs subis par le corps: ce type de comportement, rencontré en rhéologie, fait jouer un rôle important à la fonction mémoire du corps, elle même liée à des modifications internes de structure, que seul un point de vue microscopique permet de caractériser. Nous supposerons ici que le fluide étudié ne se souvient pas ou que fort peu de son histoire, ce qui est approché bien sûr, mais très correctement vérifié en général et en particulier dans le cas des liquides que nous avons considérés.

Lorsque l'on a retenu un certain modèle rhéologique pour caractériser les déformations d'un fluide, les coefficients ou paramètres qui interviennent dans cette loi mathématique sont des paramètres d'état thermodynamiques et comme tels ils ne doivent pas être indépendants de toute contrainte. Quant à leur nombre, il peut être plus ou moins grand suivant la complexité du rhéogramme, la plage de validité envisagée ou encore le degré de précision désiré du modèle représentatif: lorsque le comportement est newtonien, un seul paramètre suffit: la viscosité; mais, dans le cas contraire, on est amené à en utiliser deux (modèle d'*Ostwald*), voire trois (modèle d'*Ellis-de Haven*).

Dans tous les cas où l'on connaît le nombre de constituants indépendants c et le nombre de phases d'un système φ, la règle de *Willard Gibbs* permet d'établir la variance v ou nombre de degrés de liberté de ce système:

$$v = c + 2 - \varphi. \qquad [3]$$

Alors que dans le cas thermodynamique très simple précédemment cité, nous avions un système à deux paramètres indépendants ($c = 1$, $\varphi = 1$) dans des cas plus généraux, la variance est supérieure à 2: Ainsi, lorsque l'on étudie une solution faisant intervenir deux constituants, un soluté et un solvant, sous une phase unique, $v = 3$ et la fonction d'état établit alors une corrélation entre quatre paramètres d'état (et non 3). C'est ce cas que nous envisagerons dans la suite de cet exposé: le soluté sera soit du chlorure de sodium, soit un haut polymère organique, le solvant étant l'eau. Cependant, ici encore, nous sommes ramenés à un cas analogue au précédent, c'est-à-dire à 2 degrés de liberté et non 3: en effet, lorsque l'un des paramètres est un coefficient rhéologique, on peut admettre que l'influence de la pression est peu sensible sur la loi d'état, lorsque le fluide est un liquide, donc peu compressible.

Compte-tenu de ces considérations, on peut, par exemple, définir en rhéologie, une équation d'état de la forme:

$$\mathscr{F}(T, C, R_i) = 0. \qquad [4]$$

Dans une telle relation, relative à des variations modérées de la pression (pression constante ou peu variable), T représente la température homogène du fluide, C la concentration massique du soluté et R_i un coefficient intervenant dans le modèle rhéologique choisi. Tout autre paramètre d'état peut alors être considéré comme fonction d'état de deux de ces trois variables fondamentales; en particulier un ou plusieurs autres coefficients rhéologiques peut être considéré comme fonction des deux variables d'état T et C: Nous parvenons, de ce fait, à une première conclusion capitale, à savoir que dans l'expression d'un modèle rhéologique quel qu'il soit, les coefficients caractéristiques ne sont pas totalement indépendants puisqu'ils sont liés à un certain nombre de variables identiques par des formes diverses de l'équation d'état. Si l'on choisit comme fonction d'état le vecteur constitué des différents paramètres significatifs de la relation rhéologique explicite retenue comme approximation modélisée de la loi rhéologique implicite, chaque composante de ce vecteur est en corrélation avec les autres par l'intermédiaire des deux variables d'état de référence T et C.

$$\left.\begin{aligned} R_i &= f(C, T) \\ R_j &= g(C, T) \end{aligned}\right\} i \neq j. \qquad [5]$$

B. Représentations graphiques – inversions

Chaque propriété ou fonction d'état est susceptible d'une représentation graphique et lorsque le nombre de variables indépendantes du système ne dépasse pas 2, ce qui est notre cas, cette représentation définit complètement l'état du système. Si l'on choisit x_1 et x_2 comme variables de référence, toute autre propriété ou grandeur caractéristique du système x_i peut être représentée sur un diagramme plan dont les 2 axes de coordonnées portent X_1 et X_2, fonctions simples et biunivoques de x_1 et x_2 telles que:

$$X_1 = l(x_1)$$

$$X_2 = m(x_2) \qquad [6]$$

qui se réduisent souvent à des fonctions identité.

On obtient, de la sorte, des réseaux de courbes qui peuvent être cotées en prenant x_i comme paramètre et que l'on nomme courbes isométriques:

$$x_i = x_i(X_1, X_2)$$

avec:

$$X_i = k(x_i). \qquad [7]$$

On a ainsi l'habitude de parler en thermodynamique des isothermes, des isobares, des isochores, des isenthalpes, des isentropes, etc.; en rhéologie, nous considèrons des isométriques R_i = cte, des isothermes et des isoconcentrations.

Les variables x_1 et x_2 peuvent être de nature différente; l'une intensive ou de qualité (p, T, R_i), l'autre extensive ou de quantité (v, S, H, C): on dit alors que la représentation est mixte.

En rhéologie, il est possible d'adopter également des représentations avec variables mixtes telles que (T, C) ou (R_i, C) ou bien des représentations à variables homogènes telles que (T, R_i) ou bien (R_i, R_j) lorsque la loi rhéologique fait intervenir plusieurs coefficients, qui sont, on s'en doute, des variables de qualité.

Notre présentation un peu détournée nous permet de fonder rationnellement la permutation des variables et des fonctions d'état en rhéologie comme cela se fait couramment en thermodynamique: Si différents R_i dépendent de T et C, on peut réciproquement dire que, par l'intermédiaire des différentes formes de l'équation d'état, il est possible, à l'inverse, de considérer T

et C comme des fonctions de deux paramètres R_i et R_j. Bien entendu, cette idée assez générale ne peut être mise en œuvre en pratique que si les variations concommitantes des diverses fonctions sont suffisamment sensibles.

Il est en effet, impossible de procéder à l'inversion d'une application si le paramètre considéré ne varie que très peu. Il faut donc que les coefficients rhéologiques soient réellement significatifs de l'état du système, et non point qu'ils soient des constantes, perturbées par l'imprécision des mesures.

C. Remarques importantes

Avant d'illustrer les conclusions précédentes, et d'en délimiter la validité en rhéologie à l'aide de nos résultats expérimentaux, nous indiquons ici quelques remarques importantes d'intérêt général:

1. Le fait d'effectuer des anamorphoses simples sur les variables d'état, choisies à l'aide de transformations classiques telles que log, loglog ou puissance, permet assez souvent de parvenir à une rectification simple des courbes isométriques, conduisant à des familles de droites tout particulièrement en rhéologie.

2. Les différentes transformations utilisées lors des changements de variables ne sont pas *conformes* en général; cependant, s'il n'y a pas conservation des angles, il y a conservation, en rhéologie, comme en thermodynamique, des positions respectives des divers types d'isométriques, les unes vis à vis des autres et de ce fait même disposition.

3. Si le modèle rhéologique, adopté pour représenter un rhéogramme, fait intervenir trois ou plus de trois paramètres, on peut envisager de présenter l'équation d'état comme une relation entre trois paramètres rhéologiques:

$$F(R_i, R_j, R_k) = 0. \quad [8]$$

Température, concentration ou autres paramètres d'état sont alors considéres comme fonctions d'état.

II. Etat rhéologique de solutions aqueuses salines

Dans ce premier exemple de mise en œuvre des considérations précédentes, nous avons porté notre attention sur des solutions aqueuses de chlorure de sodium. En effet, une étude systématique des saumures a été effectuée dans notre laboratoire et nous avons examiné plusieurs concentrations depuis le solvant pur ($C = 0\%$) jusqu'à la concentration limite de précipitation ($C = 26{,}4\%$). Le domaine des températures considéré s'étend de 10–80 °C. Le choix de T et C comme variables d'état paraissait tout à fait indiqué; ces variables de référence ont de plus la particularité d'être de nature différente: T est une variable de qualité, C est une variable de quantité.

Le premier résultat important à signaler concerne le comportement de ces solutions salines: quelles que soient T et C envisagées, elles sont parfaitement newtoniennes sur une vaste plage de gradients de vitesse ($0 < du/dy < 16800\ \text{sec}^{-1}$) et nous les considérerons donc comme telles; ce qui signifie qu'un seul coefficient rhéologique suffit pour caractériser leur comportement: Nous avons opté pour la viscosité cinématique $\nu = \mu/\varrho$.

Cette fonction d'état, étudiée à l'aide d'un viscosimètre à chute de bille, décroît avec la température et croît avec la concentration de façon très nette. On peut s'en rendre compte sur les figs. 1 et 2 provenant d'essais réalisés en collaboration avec *M. L. Khrouf* (1) sur lesquelles sont portées les isothermes dans les axes ν et C (fig. 1) et les isoconcentrations dans les axes ν et $1/T$ (fig. 2). Conformément aux résultats proposées par les auteurs, la première forme de la relation d'état liant les trois quantités ν, T, C peut être écrite:

$$\begin{aligned}\log\nu &= \frac{A_0 + A_1 C + A_2 C^2}{T} \\ &= (B_0 + B_1 C + B_2 C^2)\,\mathrm{Log}\,T \\ &\quad + (D_0 + D_1 C + D_2 C^2). \qquad [9]\end{aligned}$$

$A_0, A_1, A_2, B_0, B_1, B_2, D_0, D_1, D_2$ sont des constantes propres à ce type de solutions.

Nous avons ensuite cherché une deuxième fonction qui soit particulièrement significative de l'état des solutions obtenues; dans ce but nous avons choisi une propriété électrique: la conductivité γ.

Pour les solutions obtenues avec des concentrations massiques de NaCl supérieure à 3%, nous avons remarqué que γ suit une loi simple de la concentration telle que si T_0 est la température des diverses solutions:

$$\gamma = a_{T_0}\,\mathrm{Log}\,C + b_{T_0} \quad [10]$$

relation dans laquelle a_{T_0} et b_{T_0} dépendent de la température et sont caractéristiques des corps

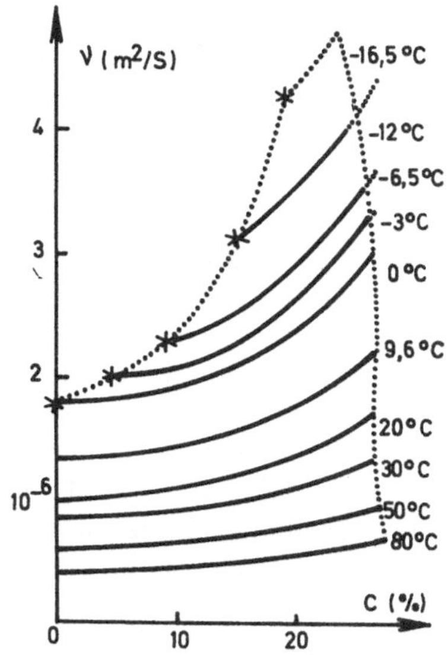

Fig. 1. Variation de la viscosité en fonction de la température

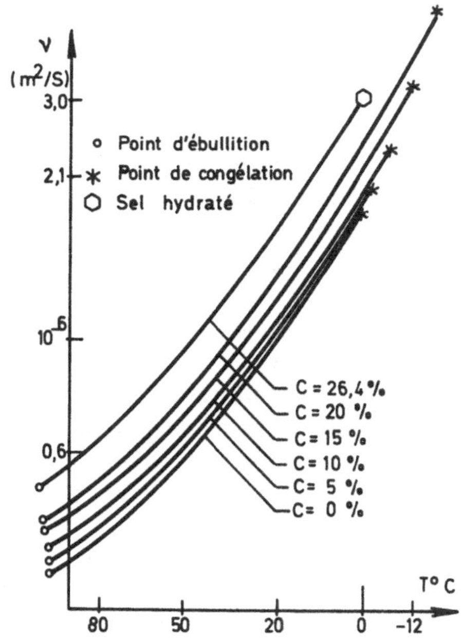

Fig. 2. Viscosité cinématique de solutions salines à différentes concentrations en fonction de la température

en présence. Nous avons trouvé, en particulier, qu'à 20 °C:

$$a_{293\,°K} = 35{,}5 \cdot 10^2\ \mathrm{m℧} \cdot \mathrm{m}^{-1}$$

$$b_{293\,°K} = -16 \cdot 10^2\ \mathrm{m℧} \cdot \mathrm{m}^{-1}.$$

D'autre part, l'évolution de γ avec la température est très simple, puisque la loi suivie est linéaire dans le domaine envisagé:

$$\gamma = \gamma_{T_0}\{1 + 0{,}03(T - T_0)\}. \qquad [11]$$

En rassemblant [10] et [11] on obtient la relation d'état suivante:

$$\gamma = \{a_{T_0} \log C + b_{T_0}\}\,\{1 + 0{,}03(T - T_0)\}. \qquad [12]$$

Le fig. 3 fournit une représentation des isothermes dans les axes (γ, C) tandis que la fig. 4 indique les isoconcentrations dans les axes (γ, T). Ces courbes ont le très gros avantage d'être rectilinéaires dans les systèmes d'axes adoptés.

A la suite de ces résultats, nous avons cherché à présenter des diagrammes caractéristiques de ces fonctions d'état ν et γ. Sur la fig. 5, nous avons porté un certain nombre d'isométriques à savoir

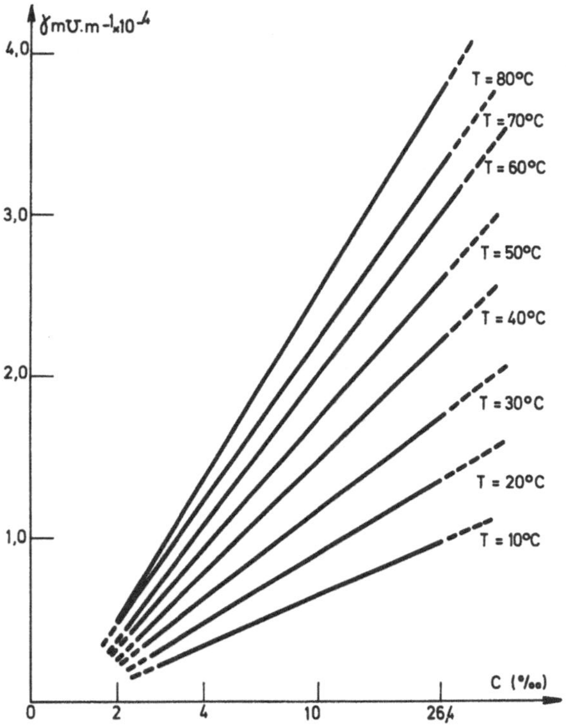

Fig. 3. Conductivité électrique de saumures. Droites isothermes

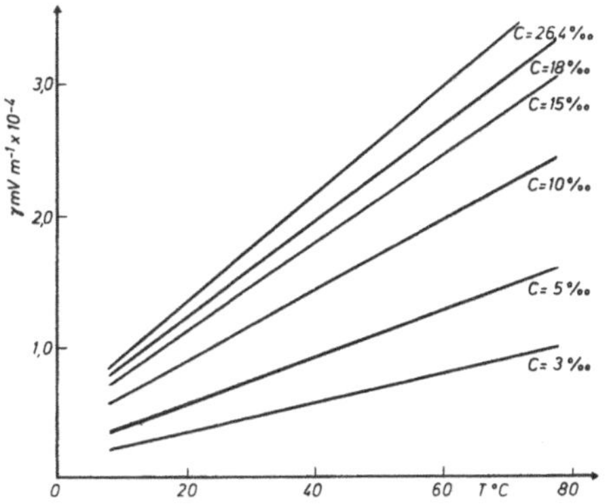

Fig. 4. Conductivité électrique de saumures. Droites isoconcentrations

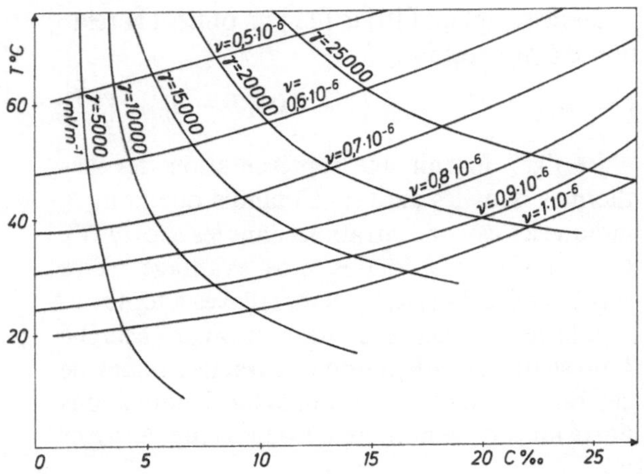

Fig. 5. Courbes caractéristiques de solutions aqueuses de NaCl

des iso-γ $(0{,}5 \cdot 10^4 < \gamma < 2{,}5 \cdot 10^4 \, \mathrm{m℧} \cdot \mathrm{m}^{-1})$ et des iso-ν $(0{,}5 \cdot 10^{-6} < \nu < 1 \cdot 10^{-6} \, \mathrm{m}^2 \cdot \mathrm{sec}^{-1})$.

Ce réseau de courbes s'avère intéressant du fait des variations très sensibles des valeurs de ces fonctions avec la température et la concentration.

De ce fait, procéder à l'inversion des rôles variables/fonctions devient tâche facile et relativement précise. Nous l'avons fait par voie graphique pour obtenir les courbes isothermes et les iso-concentrations dans les axes (ν, γ) (fig. 6). Cette représentation est très commode, car elle permet de choisir aisément température et concentration d'une solution saline pour avoir une viscosité cinématique et une conductibilité électrique prédéterminées. Par exemple, pour obtenir $\nu = 1 \cdot 10^{-6} \, \mathrm{m}^2 \cdot \mathrm{sec}^{-1}$ et $\gamma = 1{,}5 \cdot 10^4 \, \mathrm{m℧} \cdot \mathrm{m}^{-1}$, on est conduit à dissoudre à peu près 150 g de NaCl dans un litre d'eau et à porter la solution à 30 °C.

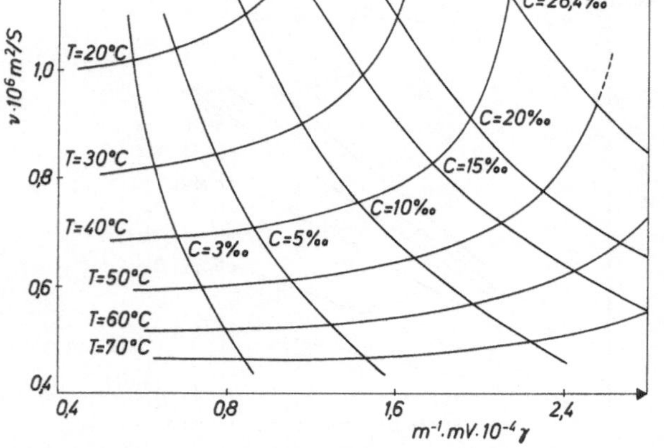

Fig. 6. Courbes caractéristiques de solutions aqueuses de NaCl

III. Equations rhéologiques d'état de solutions organiques à comportement pseudoplastique

Les cas que nous présentons maintenant concernent des solutions aqueuses de hauts polymères organiques dérivés de la cellulose: La carboxyméthylcellulose de sodium, dite Na CMC ou plus souvent encore CMC; et l'hydroxyéthylcellulose dit HEC. Ces corps ont des applications industrielles importantes depuis quelques années; mais là n'est pas la raison principale de notre choix. Nous avons été séduits par le fait que ces produits, d'un prix relativement modique, peuvent être facilement dissous dans l'eau, manifestent une grande stabilité dans le temps et ont un comportement pseudoplastique très marqué. De plus, les solutions aqueuses de ces corps, qui ne présentent aucune manifestation viscoélastique, peuvent être modélisées, par une loi puissance d'*Ostwald*,

$$\tau_{xy} = K \left(\frac{du}{dy}\right)^n$$

très satisfaisante sur un vaste domaine de gradients de vitesses. Nous possèdons ainsi, des cas typiques de fluides dont l'état dépend de deux variables (dans la mesure où les variations de pression sont modérées) et dont le comportement rhéologiques peut être défini grâce à deux coefficients seulement K et n, qui sont respectivement la consistance et l'indice de comportement. Comme dans l'exemple du II., nous avons choisi comme variables d'état de référence, la température T et la concentration massique en soluté C. Les expériences et les essais d'interpolation qui nous serviront comme exemple d'illustration ont été réalisés en collaboration avec *Michel Sengelin*.

1. *Influence de la concentration en soluté sur la consistance réduite K/μ*

De façon à mieux caractériser l'influence du corps dissous, nous avons préféré étudier les variations de (K/μ) plutôt que celles de K; μ représente ici la viscosité dynamique du solvant.

Dans un premier temps, l'examen des résultats représentatifs des valeurs prises par $\mathrm{Log}(K/\mu)$ en fonction de C en coordonnées logarithmiques, nous a permis d'avancer la forme de relation suivante (2):

$$\mathrm{Log}(K/\mu) = B \cdot C^{\beta}. \qquad [13]$$

Résultat assez remarquable, présenté sur la fig. 7, puisque la température n'intervient pas explicite-

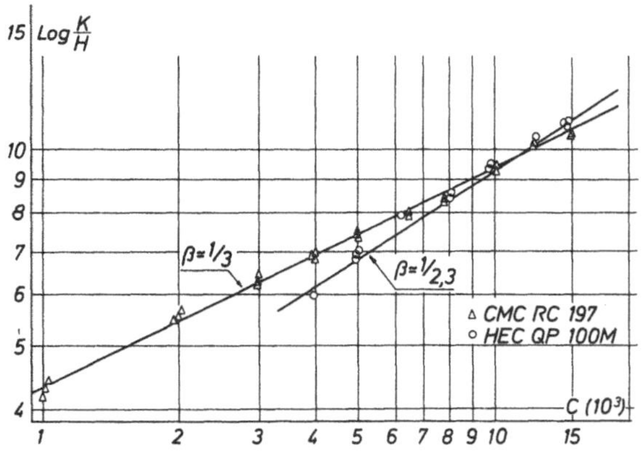

Fig. 7. Consistance réduite. Variation en fonction de la concentration

ment et ceci pour les deux types de solutions étudiées. B et β sont des constantes caractéristiques des corps en présence.

2. *Influence de la concentration sur l'indice de comportement fluide n*

Divers essais nous ont permis de mettre en évidence des conclusions simples à partir du tracé de $(1-n)$ en fonction de C en coordonnées logarithmiques. Les isothermes obtenues dans ce système d'axes sont rectilignes et parallèles, d'où la relation:

$$(1-n) = \alpha(T) \cdot C^{\gamma}. \qquad [14]$$

De plus nous avons remarqué que γ intervenant dans [14] a dans chaque cas une valeur très voisine de β intervenant dans [13] soit:

$$(1-n) = \alpha(T)\, C^{\beta}. \qquad [15]$$

3. *Influence de la température sur n*

Une étude graphique nous a permis de remarquer que $\alpha(T)$ peut être mise sous la forme:

$$\alpha(T) = \frac{p}{T} + q \qquad [16]$$

expression dans laquelle p et q sont, comme β, des constantes caractéristiques des corps en présence.

4. *Equation d'état rhéologique liant (K/μ) et n*

En éliminant la concentration C entre les relations [13] et [15] et compte-tenu de [16], on obtient l'équation d'état:

$$\mathrm{Log}(K/\mu) = A(T) \cdot [1-n]$$

en posant

$$A(T) = \frac{B}{\frac{p}{T} + q}. \qquad [17]$$

5. *Représentations graphiques*

Sur les figs. 8 et 9, nous proposons un premier type de représentation de nos conclusions relatif aux fonctions d'état [13] et [15], les variables de référence choisies étant T et C. Nous avons porté deux familles de courbes isométriques: d'une part les courbes isoconsistance réduite: $[5 \cdot 10^2 < K/\mu < 5 \cdot 10^4]$; d'autre part les iso-indice $[0{,}3 < n < 0{,}6]$.

Comme on s'y attendait, la première famille est constituée de segments de droites parallèles à l'axe des températures, puisque lorsque T varie, les variations de K de la solution et de μ du solvant sont concommitantes.

Nous avons également pu procéder à l'inversion des fonctions d'état et variables d'état et

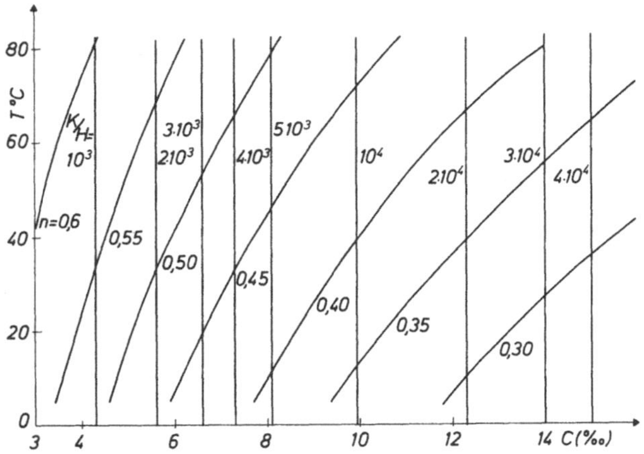

Fig. 8. Diagramme représentatif des fonctions rhéologiques d'état. Solutions aqueuses de CMC RC 197

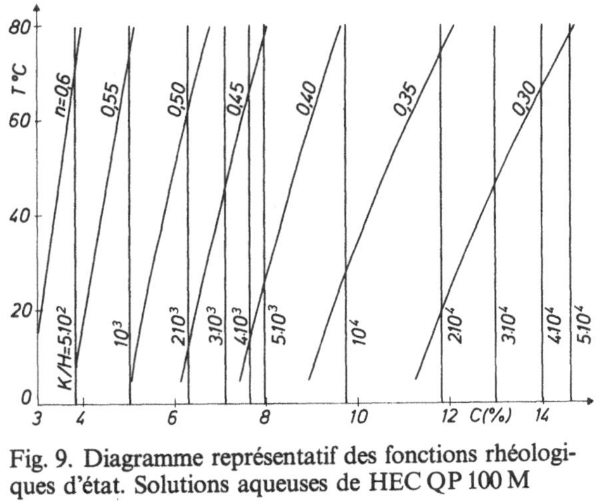

Fig. 9. Diagramme représentatif des fonctions rhéologiques d'état. Solutions aqueuses de HEC QP 100 M

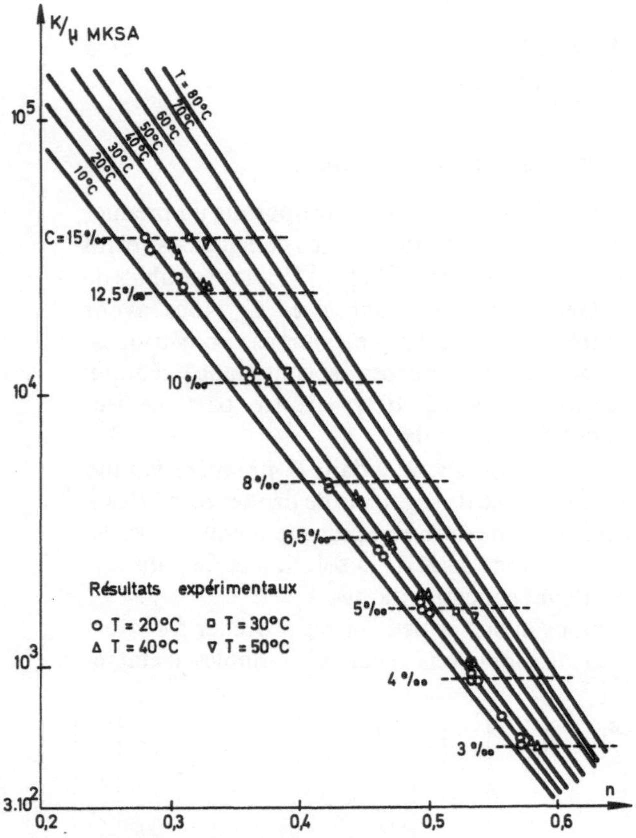

Fig. 10. Caractéristiques rhéologiques de solutions aqueuses de CMC

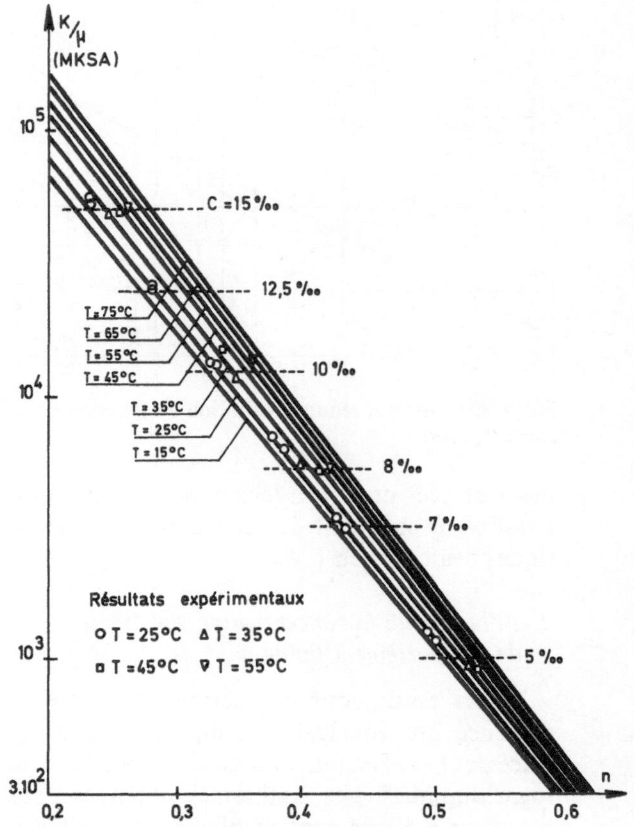

Fig. 11. Caractéristiques rhéologiques de solutions aqueuses de HEC QP 100 M

représenter, en particulier, sur les figs. 10 et 11, les familles d'isothermes et d'isoconcentrations dans les axes $\operatorname{Log} K/\mu$ et n.

Pour confirmer la validité, s'il en est besoin, de ces représentations, et par la même le bien fondé de la démarche accomplie et des équations d'état obtenues, nous avons fait figurer sur les figs. 10 et 11, les points déduits de nos expériences. Nous *y* remarquons une coïncidence très satisfaisante entre résultats bruts et résultats lissés.

Ces diagrammes peuvent être estimés fiables pour guider le choix d'une température ou d'une concentration à adopter pour parvenir à un comportement rhéologique fixé à l'avance. Nous avons vérifié expérimentalement la précision de le prévision sur quelques solutions.

IV. Lois de composition

Peut-on tirer quelques enseignements de la mesure des paramètres rhéologiques de solutions pour mieux comprendre l'effet d'un mélange?

Dans un mélange, sans réaction physicochimique, on va avoir addition des variables extensives.

Par exemple, la concentration C d'un mélange de deux échantillons de concentration C_1 et C_2 et de masse m_1 et m_2 sera évidemment:

$$\left.\begin{aligned} & C = \frac{m_1 C_1 + m_2 C_2}{m_1 + m_2} = \alpha_1 C_1 + \alpha_2 C_2 \\ & \text{avec} \\ & \alpha_1 = \frac{m_1}{m_1 + m_2}, \quad \alpha_2 = \frac{m_2}{m_1 + m_2}. \end{aligned}\right\} \quad [18]$$

On a vu que par permutation variable/fonction, la concentration peut être exprimée en fonction des paramètres rhéologiques et éventuellement de la température.

Prenons le cas simple de solutions salines à température constante T_0

$$C = f(\nu, T_0).$$

D'où la relation entre les viscosités ν_1, ν_2 des deux échantillons de concentration C_1, C_2 et la viscosité résultante du mélange:

$$f(\nu, T_0) = \alpha_1 f(\nu_1, T_0) + \alpha_2 f(\nu_2, T_0). \quad [19]$$

Comme $\alpha_2 = 1 - \alpha_1$, $f(v, T_0)$ varie linéairement avec α_1, ce qui va d'ailleurs de soi d'après la formule de définition de $f(v, T_0) = C$, fonction réciproque de $v = g(C, T_0)$. Comme nous l'avons déjà dit l'application $f(v, T_0)$ constitue l'anamorphose qui rectifie le graphe (C, v) à température fixée T_0, cette distorsion d'échelle pouvant être effectuée implicitement de manière graphique.

Pour déterminer $f(v, T_0)$, on pourra procéder par interpolation à partir d'un ensemble de couples (C, v), d'autant plus nombreux que le graphe (C, v, T_0) paraîtra compliqué. En général, trois couples suffiront comme c'est le cas ici pour les solutions salines, car v varie tout de même relativement peu.

Dans le cas des fluides non newtoniens d'*Ostwald*, le mélange des solutions conduira à des lois de composition pour la consistance K et l'indice de comportement n telles que:

$$f_K(K, T_0) = \alpha_1 f_K(K_1, T_0) + \alpha_2 f_K(K_2, T_0) \quad [20]$$

$$f_n(n, T_0) = \alpha_1 f_n(n_1, T_0) + \alpha_2 f_n(n_2, T_0) \quad [21]$$

avec bien entendu:

$$C = f_K(K, T_0) = f_n(n, T_0).$$

(Evidemment, on peut aussi considérer K/μ au lieu de K.)

Pour les solutions non newtoniennes déjà présentées, il vient, à température fixée:

$$K^{1/\beta} = \alpha_1 K_1^{1/\beta} + (1 - \alpha_1) K_2^{1/\beta} \quad [22]$$

et

$$(1 - n)^{1/\beta} = \alpha_1 (1 - n_1)^{1/\beta} + (1 - \alpha_1)(1 - n_2)^{1/\beta} \quad [23]$$

(β est voisin de 1/3 pour la solution de CMC, alors:

$$K^3 \simeq \alpha_1 K_1^3 + (1 - \alpha_1) K_2^3\ ). \quad [24]$$

Conclusion

L'éclairage de la thermodynamique permet de saisir facilement et nettement l'interdépendance des paramètres rhéologiques et des variables d'état habituelles. Mais il ne remplace pas l'approche expérimentale de la loi rhéologique ni le choix un peu subjectif du modèle qui est en somme l'interpolation des résultats de mesures. Conjugués, l'explication thermodynamique et le lissage des résultats d'expériences, conduisent aux lois de composition des viscosités généralisées pour des solutions non newtoniennes. La forme des relations ainsi obtenues peut suggérer l'extrapolation pour décrire le comportement rhéologique d'autres mélanges fluides pour lesquels on connaît la viscosité des composants.

Résumé

A partir de nombreux résultats d'expériences effectuées sur des liquides organiques à comportement non newtonien (corboxy méthyl cellulose, polyox, etc.) et de solutions salines, les auteurs essaient de définir les lois de composition permettant d'obtenir la valeur des paramètres rhéologiques pour un mélange lorsque varie la température.

Du point de vue thermodynamique, l'équation d'état fait intervenir pour un mélange quatre paramètres dont la pression, la température, la concentration, le dernier paramètre pouvant être considéré comme fonction. Si l'on choisit comme fonction le vecteur constitué des différents paramètres significatifs de la relation rhéologiques explicite retenue comme approximation de la loi rhéologique implicite, ce vecteur dépend relativement peu de la pression mais essentiellement de la température (variable intensive ou de qualité) et de la concentration (variable extensive ou de quantité). Chaque composante du vecteur rhéologique est donc en corrélation avec les autres par l'intermédiaire des deux variables d'état. Ainsi, on peut envisager dans le cas d'une formule rhéologique à trois paramètres de remplacer l'équation d'état par une relation entre les trois composantes du vecteur rhéologique.

En utilisant la fonction réciproque, on peut obtenir aussi la concentration en fonction des paramètres rhéologiques à condition qu'ils soient suffisamment sensibles et ainsi disposer de la base de la composition.

On peut encore envisager la construction directe de loi de composition en procédant à l'anamorphose convenable sur les paramètres rhéologiques qui sont des paramètres intensifs, anamorphose assurant la rectification des graphes d'évolution en fonction de la concentration.

Divers exemples sont donnés à partir de résultats expérimentaux montrant les possibilités et les limites des méthodes de composition proposées.

Littérature

1) *Thirriot, C.* et *M. L. Khrouf,* Rhéologie des saumures et des eaux saumâtres. Société Hydrotechnique de France, XII Journées de l'Hydraulique. Paris 1972, Question IV, Rapport 5.

2) *Bellet, D.* et *M. Sengelin,* Quelques expériences sur le comportement non-newtonien des liquides organiques. Société Hydrotechnique de France, XII Journées de l'Hydraulique. Paris 1972, Question V, Rapport 2.

Adresse des auteurs:

C. Thirriot et *D. Bellet*
Institut National Polytechnique
de Toulouse (France)

Rheol. Acta 13, 526–531 (1974)

Ecole Supérieure de Chimie de Mulhouse et Centre de Recherches Textiles de Mulhouse (France)

Contribution à l'étude de la détermination sans équipement oscillant des caractéristiques élastiques et visqueuses de systèmes solvants – composés macromoléculaires non newtoniens

Par R. A. *Schutz et* G. *Tatin*

Avec 6 figures

(Reçu p. p. le 27 Octobre 1973)

I. Introduction

La mise en œuvre de quantités de plus en plus importantes et dans des conditions de plus en plus sévères de systèmes solvants – composés macromoléculaires dans diverses industries, nécessite une connaissance plus approfondie des caractéristiques viscoélastiques de ces systèmes. Bien que les approches théoriques soient aujourd'hui relativement nombreuses, les interprétations des résultats selon les différentes méthodes classiques par oscillation par rapport au comportement pratique sont délicates sinon décourageantes. C'est la raison pour laquelle nous avons essayé de rester le plus près possible des conditions industrielles d'utilisation, c'est-à-dire en soumettant nos systèmes à des sollicitations uniformes et non pas oscillatoires. Nous avons étudié le comportement élastique de ces systèmes par analyses des phénomènes préstationnaires, observés à l'aide d'un rhéomètre à cylindres coaxiaux lors de la mise en régime à vitesses de cisaillement maintenues constantes ce qui correspond bien à la mise œuvre industrielle.

Parmi tous les systèmes solvants – composés macromoléculaires nous avons choisi plus spécialement les empois d'amidon relativement concentrés parce qu'ils ont été encore peu étudiés et parce que les empois ont une très large utilisation (par centaines de milliers de tonnes/an), notamment pour les industries textiles et papetières.

A cette occasion, nous avons développé une méthode d'approche et d'analyse originale que nous nous proposons d'exposer ici et qui n'est pas limitée à l'application de ce carbohydrate macromoléculaire puisque les graphes expérimentaux obtenus pour de nombreux autres systèmes sont de même allure.

Ainsi, par exemple, *J. H. Elliot* la retrouve aussi bien pour des solutions aqueuses d'hydroxypropylcellulose (1) que pour divers produits alimentaires (2), *G. V. Vinogradov* la retrouve dans le cas du polystyrène fondu (3) et *J. Schurz* dans celui de solutions de polyvinylacétate dans le dioxanne (4). Toutes ces solutions, relativement concentrées, ainsi que les polymères fondus forment donc une famille rhéologique qui peut être caractérisée selon les mêmes techniques d'investigation.

II. Hypotheses de base

L'expérience (fig. 1,1) montre que pour maintenir ces systèmes à une vitesse de cisaillement

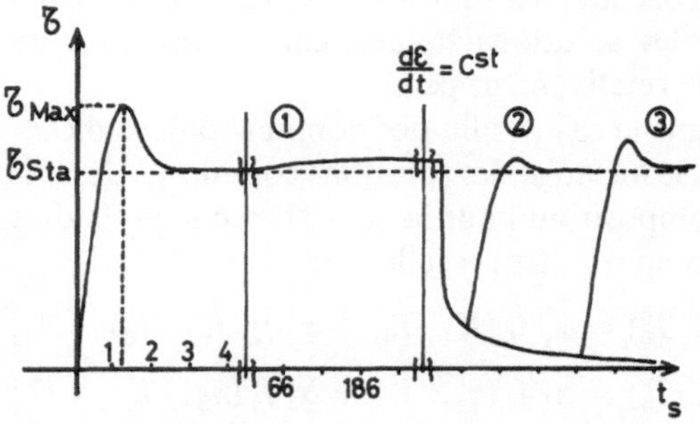

Fig. 1. Rhéogrammes du maximum (overshooting, Anlaufkurven): Evolution des contraintes de cisaillement en fonction du temps, à vitesse de cisaillement constante (t en secondes pour $D = 3\,s^{-1}$). Rhéogramme 1: évolution complète pour une durée donnée d'application de vitesse de cisaillement constante. Rhéogrammes 2 et 3: nouvelle application de la vitesse de cisaillement constante après relaxation partielle des contraintes

constante, la contrainte passe par un maximum (overshooting) d'amplitude plus ou moins grande et n'atteint une valeur quasistationnaire qu'au bout d'un temps variable. C'est à partir de ce phénomène préstationnaire que nous avons été amenés à formuler plusieurs hypothèses.

Tout d'abord les systèmes étudiés constitueraient à l'état de repos des réseaux tridimensionnels du fait des liaisons «secondaires» (forces de *Van der Waals*, liaisons hydrogène), hypothèse également formulée par *Vinogradov* et *Elliott* (1, 3, 5).

Sous l'effet d'une sollicitation de cisaillement l'édifice tridimensionnel se déformerait, créant une contrainte initiale τ_0; puis les liaisons secondaires initiales seraient rompues pour permettre l'écoulement laminaire, d'où l'existence d'un module d'élasticité initial que nous appellerons module d'élasticité de cohésion «statique» initial. Mais soulignons en outre que la rupture de toutes ces liaisons secondaires ne pourrait se faire simultanément d'une façon globale, mais plutôt progressivement sur certains sites «points faibles» préférentiels jusqu'à destruction totale de l'édifice.

Les phénomènes visqueux apparaîtraient dans la phase d'écoulement, mais au cours de laquelle un certain nombre de liaisons se reformeraient et seraient détruites simultanément, ce qui conduit à une notion d'équilibre correspondant à un nombre donné de liaisons existant en permanence (car la vitesse de formation ou de destruction de liaisons secondaires est très grande) et par suite, à un module d'élasticité de cohésion «dynamique».

III. Traduction mathématique et analogique des hypothèses et des résultats expérimentaux

Soit E_i le module d'élasticité dynamique correspondant à une vitesse de cisaillement D_i; l'existence d'un palier stationnaire nous suggère évidement l'emploi d'un modèle de *Maxwell*; soit η_i la viscosité correspondant à la vitesse de cisaillement $d\varepsilon_i/dt$ on a la relation classique:

$$\frac{d\varepsilon_i}{dt} = \frac{1}{E_i}\frac{d\tau}{dt} + \frac{1}{\eta_i}\tau \qquad [\mathrm{R\,1}]$$

de cette équation différentielle on tire l'intégrale bien connue:

$$\tau = \left[\exp\left(-\frac{t}{\eta_i/E_i}\right)\right] \times \left[\tau_0 + E_i \int_0^t \left\{\frac{d\varepsilon_i}{dt}\exp\left(\frac{t}{\eta_i/E_i}\right)\right\} dt\right] \qquad [\mathrm{R\,2}]$$

à $d\varepsilon_i/dt = c^{st}$ [R 2] devient:

$$\tau = \eta_i \frac{d\varepsilon_i}{dt} + \left[\tau_0 - \eta_i \frac{d\varepsilon_i}{dt}\right] \exp\left(-\frac{t}{\eta_i/E_i}\right)$$

Considérons maintenant τ_0 non plus comme une constante, mais comme une fonction de mise en régime de la forme $\tau_0 = E' \frac{d\varepsilon_i}{dt} t$, avec E'_i le module d'élasticité «statique» du réseau tridimensionnel initial; cette fonction ne peut intervenir que pendant un temps très court mais le produit par l'exponentielle négative nous affranchit de la recherche de son temps d'existence, d'où:

$$\tau = \eta_i \frac{d\varepsilon^i}{dt}\left[1 - \exp\left(-\frac{t}{\eta_i/E_i}\right)\right] + E'_i \frac{d\varepsilon_i}{dt} t \exp\left(-\frac{t}{\eta_i/E_i}\right). \qquad [\mathrm{R\,3}]$$

Cette Relation [R 3] décrit avec une bonne précision l'allure des courbes expérimentales obtenues.

Si $d\varepsilon_i/dt$ est fixé, la détermination des trois paramètres rhéologiques: η_i, E_i, E'_i est possible.

En effet, écrivons

$$E'_i = \alpha E_i \qquad [\mathrm{R\,4}]$$

pour faciliter les calculs; pour

$$\frac{d\tau}{dt} = 0, \quad t_{max}\,(\text{mesurable}) = \left(\frac{1+\alpha}{\alpha}\right)\frac{\eta_i}{E_i} \qquad [\mathrm{R\,5}]$$

on en tire:

$$\frac{\tau_{max}}{\tau_{sta}} = 1 + \alpha \exp\left(-\frac{1+\alpha}{\alpha}\right) \text{ encore}$$

$$= 1 + \frac{1}{\frac{1+\alpha}{\alpha} - 1}\exp\left(-\frac{1+\alpha}{\alpha}\right). \qquad [\mathrm{R\,6}]$$

Comme τ_{max} et τ_{sta} sont mesurables, on peut déduire des grandeurs τ_{max}, τ_{sta} et t_{max}, les valeurs de α, de E_i et donc de E'_i.

De plus, lorsque t est suffisamment grand, c'est-à-dire lorsque τ tend vers sa valeur stationnaire on a $\tau = \eta_i \frac{d\varepsilon_i}{dt}$ d' où η_i.

Nous sommes donc en mesure de déterminer les trois paramètres rhéologiques η_i, E_i, E'_i, pour différentes vitesses de cisaillement, diverses températures et concentrations.

Remarque

Bien que notre rhéomètre ne nous permette pas d'étudier les courbes

$$\frac{d\varepsilon}{dt} = f(t) \quad \text{à} \quad \tau = \text{constante}$$

ou bien

$$\varepsilon = f(t) \quad \text{à} \quad \tau = \text{constante}$$

nous avons essayé de faire une étude de $\varepsilon = f(t)$. En effet de [R 3] nous pouvons tirer

$$\frac{d\varepsilon}{dt} = \frac{\tau}{\eta_i\left[1 - \exp\left(-\frac{t}{\eta_i/E_i}\right)\right] + \alpha E_i t \exp\left(-\frac{t}{\eta_i/E_i}\right)} \qquad [\mathrm{R\,7}]$$

Si nous faisons τ = constante nous obtenons $d\varepsilon/dt = f(t)$ et [R 7] décrit des courbes ayant la forme suivante (fig. 2).

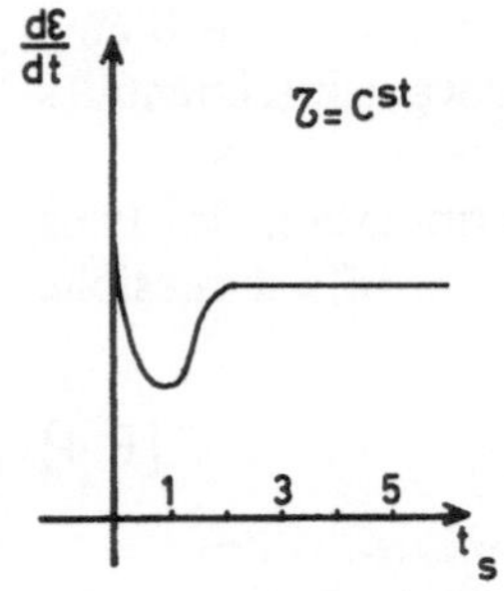

Fig. 2. Rhéogramme à contrainte constante: la relation [R7] correspond à cette allure, qui a été observée et décrite par *Vinogradov*

D'autre part,

$$d\varepsilon = \frac{\tau\,dt}{\eta_i\left[1 - \exp\left(-\frac{t}{\eta_i/E_i}\right)\right] + \alpha E_i t \exp\left(-\frac{t}{\eta_i/E_i}\right)} \qquad [R\,8]$$

en intégrant de 0 à t

$$\varepsilon(t) = \int_0^t \frac{\tau\,dt}{\eta_i\left[1 - \exp\left(-\frac{t}{\eta_i/E_i}\right)\right] + \alpha E_i t \exp\left(-\frac{t}{\eta_i/E_i}\right)}. \qquad [R\,9]$$

Sans calculer [R 9] nous pouvons tracer la courbe $\varepsilon(t)$ en la considérant comme l'enveloppe des tangentes. Nous obtenons alors le graphe fig. 3. Ces courbes obtenues mathématiquement

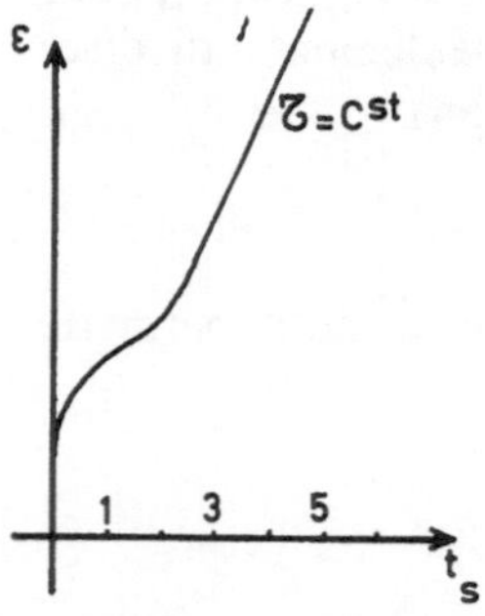

Fig. 3. Rhéogramme à contrainte constante: la relation [R9] correspond à cette allure, qui a été observée et décrite par *Vinogradov*

sont aussi en parfait accord avec les courbes obtenues expérimentalement par *Vinogradov* (3).

IV. Travail expérimental

1. Mise en œuvre et phénomènologie

Nous avons étudié des empois d'amidon dont la concentration varie de 6 à 10% en matières sèches, ces empois ayant été préparés à l'aide d'un appareil programmé pour la montée et le maintien en température en des temps fixés et reproductibles; les mesures rhéologiques proprement dites ont été effectuées avec un rhéomètre à cylindres coaxiaux conçu dans notre laboratoire. Ce rhéomètre permet un enregistrement continu des contraintes.

Nous avons obtenu la courbe d'allure caractéristique (fig. 1 cas 1). Cette courbe corrobore nos hypothèses. La partie ascendante de la courbe n'est pas linéaire ce qui indique que l'on n'a pas affaire à un phénomène élastique pur; il y a donc déjà superposition d'un phénomène visqueux, ce qui confirme que la destruction de l'édifice tridimensionnel initial est donc progressive.

Quand la rotation des cylindres est arrêtée, il y a relaxation et après un temps suffisant un édifice tridimensionnel se réédifie; si cette réédification est incomplète, lors de la remise en rotation la contrainte maximale aura une amplitude d'autant plus faible que la réédification est plus partielle (fig. 1 cas 1, 2, 3).

On observe en outre un lent accroissement de la valeur dite «stationnaire» de la contrainte que nous attribuons sans être encore en mesure de le vérifier, à un réarrangement interne des particules hydrodynamiques plutôt qu'à une déformation de celles ci, ces systèmes d'hydrates de carbone macromoléculaires fortement solvatés pouvant être assimilés à des émulsions, hypothèse que nous avions déjà formulée antérieurement (6).

2. Résultats obtenus concernant la viscosité

Nous avons vérifié que les empois d'amidon ont une viscosité de structure qui en fonction de la vitesse de cisaillement peut s'exprimer par une loi en puissance (fig. 4).

A ce sujet, relevons que la loi en puissance a toujours posé des problèmes par son manque d'homogénéité, dimensionnelle; nous avions

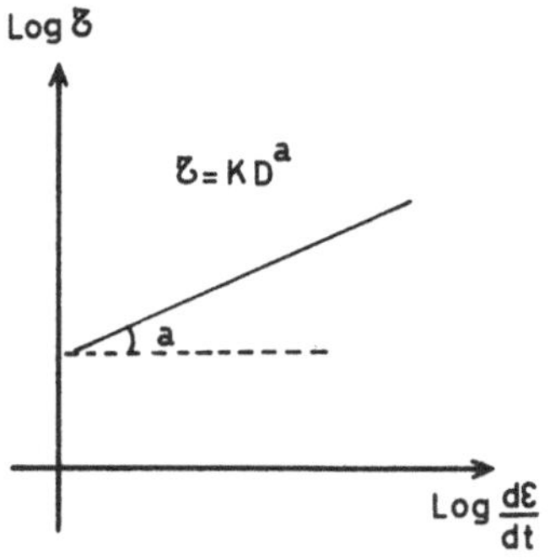

Fig. 4. Présentation pratique des rhéogrammes en double échelle logarithmique pour le cas de l'applicabilité de la loi en puissance

(7) l'occasion de montrer qu'en fait l'équation classique $\tau = KD^a$ pouvait s'écrire $\tau = \tau_0 \cdot (D/D_0)^a$. La difficulté de donner une signification satisfaisante de τ_0, D_0, nous a conduit à essayer d'aborder le problème par une autre voie.

Considérons le modèle analogique suivant (fig. 5) traduisant ce qui se passe couche par couche: il y a une répartition des contraintes critiques τ_c à partir desquelles et le patin et le piston peuvent se mouvoir, répartition que nous ne pouvons pas étudier; mais globalement, τ est une fonction de $d\varepsilon/dt$; et pour chaque contrainte enregistrée τ il y aura n pistons de viscosité η qui se déplaceront, et par conséquent

$$n = f(\tau) = F\left(\frac{d\varepsilon}{dt}\right).$$

D'autre part

$$\tau = \frac{\eta}{n}\frac{d\varepsilon}{dt} \qquad \text{[R 10]}$$

η_i coefficient de viscosité (apparente) est égal à $\eta/n = \eta_i$ d'où

$$n = \exp\left\{(1-a)\ln\frac{d\varepsilon}{dt}\right\} \qquad \text{[R 11]}$$

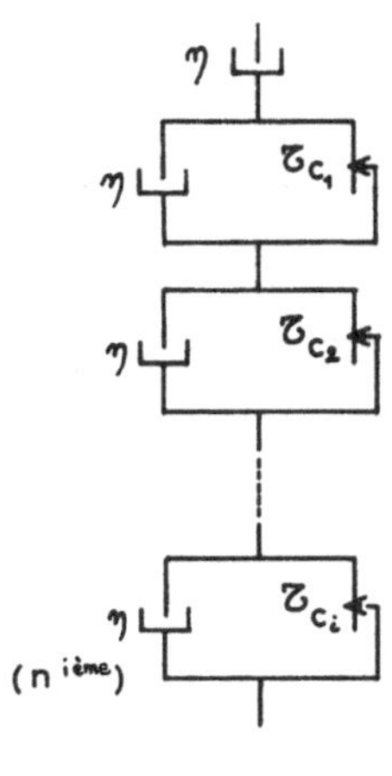

Fig. 5. Modèle rhéologique (voir texte)

avec a (identique à l'exposant de $\tau = KD^a$) et par conséquent dépendant des conditions expérimentales (voir: 7, 8).

En écrivant:

$$\tau = \frac{\eta}{\exp\left\{(1-a)\ln\dfrac{d\varepsilon}{dt}\right\}}\frac{d\varepsilon}{dt} \qquad \text{[R 12]}$$

le problème de l'homogénéité dimensionnelle serait résolu par ce formalisme, et la loi dite en puissance serait ainsi décrite en utilisant un coefficient de viscosité invariable η.

Par ailleurs, nous avons vérifié également que la viscosité apparente η_i est une fonction de la température absolue selon l'expression classique $n_i = \eta_0 \cdot \exp(A/T)$ où A peut être assimilé à une énergie d'activation d'écoulement apparente.

Evidemment, on pourrait aussi bien écrire – pour autant que «a» ne varie guère avec la température – que

$$\eta = \eta'_0 \exp(B/T). \qquad \text{[R 13]}$$

3. Résultats concernant le module d'élasticité de cohésion E

Nous avons calculé les modules E_i correspondant à chaque expérience, ce module reste constant lorsque l'on fait varier le gradient de cisaillement à température et concentration constantes. E_i est une fonction croissante de la concentration et une fonction décroissante de la température, ce résultat semble logique, le nombre des liaisons secondaires mises en jeu dépendant de la concentration et leur énergie, de la température.

4. Résultats concernant le module d'élasticité statique initial E'_i

Ce module E'_i croit avec la concentration. E'_i est également une fonction décroissante de la température mais varie avec le gradient de cisaillement appliqué. E'_i est une fonction décroissante du gradient.

5. Vérification statistique de la validité des résultats obtenus

Calculons la surface hachurée S (fig. 6).

$$\tau_i = \eta_i \frac{d\varepsilon_i}{dt}\left[1 - \exp\left(-\frac{t}{\eta_i/E_i}\right)\right] + E' t \frac{d\varepsilon}{dt}\exp\left(-\frac{t}{\eta_i/E_i}\right) \qquad \text{[R 14]}$$

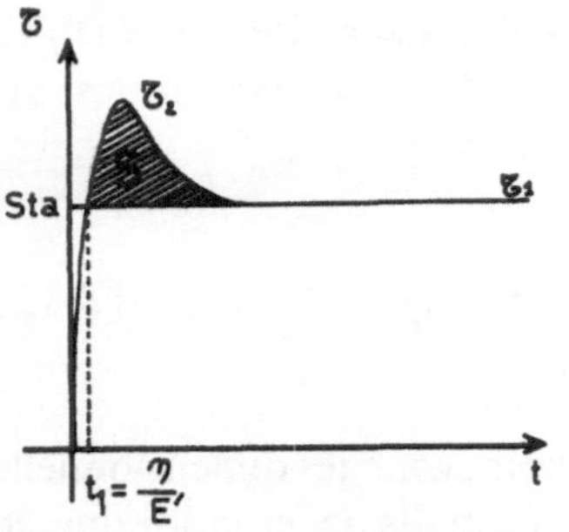

Fig. 6. Analyse du maximum (overshooting) (voir texte)

$$\tau_2 = \eta_i \frac{d\varepsilon_i}{dt}$$

$\tau_1 = \tau_2 \rightarrow$ points d'intersections soient

$t_1 = \eta_i/E_i'$ et $t_2 \rightarrow +\infty$

donc

$$S = \int_{\frac{\eta_i}{E_i}}^{\infty} \tau_1 - \int_{\frac{\eta_i}{E_i}}^{\infty} \tau_2$$

nous obtenons

$$S_i = \alpha \frac{\eta_i^2}{E_i} \frac{d\varepsilon_i}{dt} \exp\left(-\frac{1}{\alpha}\right). \qquad \text{[R 15]}$$

Nous avons déterminé expérimentalement ces surfaces par pesée

$$\frac{S \text{ calculée}}{S \text{ mesurée}} = k.$$

k devant être une constante de proportionnalité. Nous avons ordonné ces valeurs de k en classes et les effectifs de chacune de ces classes permettent de tracer une droite de *Henry*: nous sommes donc en présence d'une distribution gaussienne des valeurs de k; les résultats obtenus sont donc cohérents avec les expériences.

V. Conclusion

L'utilisation industrielle de ces systèmes correspond souvent à une imprégnation et à un exprimage pendant lequel ils sont soumis une seule fois à une sollicitation de cisaillement croissant brutalement puis décroissant aussi rapidement ou presque.

Par conséquent la sollicitation est de courte durée et les phénomènes préstationnaires doivent prévaloir; ils se traduisent à vitesse de cisaillement constante par l'apparition d'un maximum de contrainte (overshooting) en fonction du temps. Les résultats expérimentaux en accord avec l'analyse mathématique que nous avons entreprise semblent justifier les hypothèses suivantes:

– Au repos ces systèmes constituent un réseau tridimensionnel auquel correspond un module d'élasticité initial: E' que nous avons appelé module d'élasticité (initial) statique.

– Au cours du cisaillement se font et se défont des liaisons secondaires fugaces, mais qui correspondent à une cohésion en régime dynamique avec un module d'élasticité, que nous avons appelé pour cette raison: module d'élasticité dynamique E. Il ressort en outre qu'en fonction de vitesses de cisaillement croissantes, le module statique E' diminue, tandis que le module dynamique E reste invariable.

Comme par ailleurs le rapport entre la contrainte maximale et la contrainte stationnaire tend à diminuer de valeur si la vitesse de cisaillement augmente, cela signifie pour la pratique industrielle, que plus la sollicitation est rapide – donc plus la vitesse de cisaillement est élevée – plus la viscosité apparente (non newtonienne vraie) prend de l'importance par rapport et au détriment de l'élasticité statique de module E'.

Toutefois, dans le cas de rétrogradation ou de prise en gel, ce module E' se trouve brusquement augmenté et l'élasticité statique reprend de l'importance: le début de la prise en gel ou de la rétrogradation se traduit par une augmentation du maximum de contrainte.

Résumé

Une nouvelle analyse du comportement viscoélastique des systèmes concentrés solvants-composés macromoléculaires non newtoniens (produits amylacéseau), selon laquelle le modèle classique de *Maxwell* ne consistue qu'un cas particulier, est fondée sur l'étude des phénomènes préstationnaires et poststationnaires dans l'entrefer de cylindres coaxiaux à vitesse d'écoulement fixe. Par cette approche on met en évidence un module d'élasticité de cohésion initial «statique» E_i', un module d'élasticité de cohésion «dynamique» E_i et un coefficient de viscosité η_i, paramètres rhéologiques que l'on peut évaluer à partir des contraintes maximale et stationnaire d'une part et du temps t nécessaire pour atteindre la contrainte maximale d'autre part.

En fonction de vitesses de cisaillement croissantes, le module statique E' diminue, le module dynamique E reste invariable et le coefficient de viscosité apparente η diminue.

Cette approche s'applique à tous les cas d'«overshooting».

Zusammenfassung

Es wird eine neue Analyse des viskoelastischen Verhaltens konzentrierter makromolekularer Lösungen gegeben, für welches das klassische *Maxwell*-Modell nur einen Spezialfall darstellt. Diese basiert auf der Untersuchung vor- und nachstationärer Erscheinungen im Ringspalt zwischen zwei koaxialen Zylindern bei konstanten Strömungsbedingungen. Bei diesem Verfahren bestimmt man einen „statischen" und einen „dynamischen" Elastizitätsmodul sowie eine Viskosität. Hierfür verwendet man den Maximalwert und den stationären Wert der Schubspannung sowie die Zeit bis zur Erreichung des Anlaufmaximums.

Mit steigender Schergeschwindigkeit nimmt der statische Modul und die scheinbare Viskosität ab, während der dynamische Modul konstant bleibt.

Die hier gegebene Analyse dürfte für alle Typen von Anlaufkurven anwendbar sein.

Littérature

1) *Elliot, J. H.*, J. Appl. Pol. Sci. **13**, 755–764 (1969).
2) *Elliot, J. H.* et *A. J. Ganz*, J. Text. Stud. **2**, 220–229 (1971).
3) *Vinogradov, G. V.* et *I. M. Belkin*, J. Pol. Sci. **A 3**, 917–932 (1965).
4) *Schurz, J.*, Die Angew. Makro. Chem. **15**, 95–107 (1971).
5) *Leonov, A. I.* et *G. V. Vinogradov*, Dokl. Akad. Nauk URSS **155**, 406 (1964).
6) *Schutz, R. A.*, C. R. Acad. Sci. (Paris) **261**, 5111 (1965).
7) *Nedonchelle, Y.* et *R. A. Schutz*, C. R. Acad. Sci. (Paris)**265**, C 16, 16–18 (3 juillet 1967).
8) *Nedonchelle, Y.*, Thèse de Docteur-Ingénieur Strasbourg, 4. 7. 1968.

Adresse des auteurs:
R. A. Schutz et *G. Tatin*
Centre de Recherches Textiles de Mulhouse
185, rue de l'Illberg
F-68 Mulhouse (France)
et
Ecole Supérieure de Chimie de Mulhouse
3, rue A. Werner
F-68093 Mulhouse Cedex (France)

Rheol. Acta **13**, 532–537 (1974)

From the Istituto di Principi di Ingegneria Chimica, Universitá di Napoli, Napoli (Italia)

Elongational flow of dilute polymer solutions

By *D. Acierno**), *G. Titomanlio**), and *L. Nicodemo*

With 9 figures

(Received October 27, 1972)

Introduction

The interest in elongational flow of polymeric substances has grown recently due to an effort in analyzing more quantitatively processes like spinning, film blowing, etc. and because of the peculiar results – extremely large values of the viscosity – observed in some experiments (1–3).

The behavior of melts has been studied by a number of authors (4–10) and, although a variety of responses has been observed, it may well be stated that the order of magnitude of the elongational viscosity is not different from the limiting value for small gradients which is the well known Trouton value. Conversely, dilute solutions of polymers, especially in highly viscous solvents, show extremely large values of the extensional viscosity as compared to the zero-shear viscosity (1–3).

In this work, elongational data have been obtained for solutions of Separan NP 20 (a polyacrylamide manufactured by Dow Chem.) in glycerol. The data are taken under transient stress conditions with two different experimental techniques. In all cases, the stretching rate comes out to be approximately constant along the path of a fluid particle. A rather narrow but significant range of values of stretching rates was accessible to measurements.

The results obtained are compared with existing theories for dilute solutions of flexible macromolecules. These may be summarized as follows. *Peterlin* (11), *Takserman-Krozer* (12), *Stevenson* and *Bird* (13), *Tanner* and *Stehrenberger* (14) have calculated the steady state elongational viscosity with different degrees of approximation. In refs. (11, 13, 14) it is shown that for sufficiently large values of the stretching rate, extremely large values of the elongational viscosity are to be expected, depending on the number of statistical segments of the chain, i.e. on its molecular weight.

In refs. (15–18) the transient elongational viscosity is calculated by means of non-linear bead-spring models under conditions of a constant stretching rate suddenly applied at time zero. The solution given in (15) applies for sufficiently large stretching rates at which the effect of the *Brown*ian motions of the beads is negligible. In ref. (16) the influence of internal viscosity is considered. An approximate procedure is followed in (17) which allows to account for *Brown*ian motions of the beads. Finally in ref. [18], the influence of polydispersity is estimated.

*) Cattedra di Principi di Ingegneria Chimica, Universitá di Palermo, Palermo (Italia).

Experimental

Two techniques have been used to obtain experimental results of transient elongational viscosity. We have used an isothermal spinning and the so-called tubeless syphon. With the first technique, the material reaches the test section, i.e. the elongational field out of the die, after a considerable deformation has already occurred within the die itself. Conversely, in the syphon experiments, the material is almost virgin; only a slight and slow deformation within the reservoir precedes the elongational flow in the filament.

In fig. 1, the apparatuses used are shown schematically. In both types of experiments the liquid filament is vertical, moving upward in the syphon and downward in the spinning. The force at the uppermost section of the filament is measured by a scale system suitably calibrated, and the forces at all other sections are then calculated accounting for the weight of the filament. This is obtained by photographic recording of the filament shape which also serves for the kinematic description of the flow. Surface tension, inertia and air friction prove to contribute negligible to the force balance. Further details on the apparatuses may be found in (2, 3).

Solutions of the following concentrations have been used for the elongational :experiments: 0.175, 0.25, 0.5% by weight Separan NP 20 in glycerol. The molecular weight of the polymer is not known exactly but it is of order 2×10^6. Also the polydispersity is entirely unknown. The solutions above specified and a few additional ones were tested in a *Weissenberg* rheogoniometer. The shear viscosity results are shown in fig. 2. The glycerol used was also tested and gave a constant viscosity, η_s, of about 8 poises.

The zero shear viscosities, η_0, shown in fig. 2 are plotted logarithmically in fig. 3 as $\eta_0 - \eta_s$ vs. the concentration, c. A straight line of slope 1 may be drawn through the data thus showing that the solutions may indeed be considered dilute.

That these solutions are dilute can also be estimated by the following calculation. The volume per molecule is easily calculated from the concentration. This is compared with the quantity

$$\langle r^2 \rangle^{3/2} = N^{3/2} l^3$$

which is representative of the volume actually containing one molecule. If the former proves larger or at least of the same order of the latter, we may infer that no entanglements are present. $\langle r^2 \rangle$ has been evaluated, through N

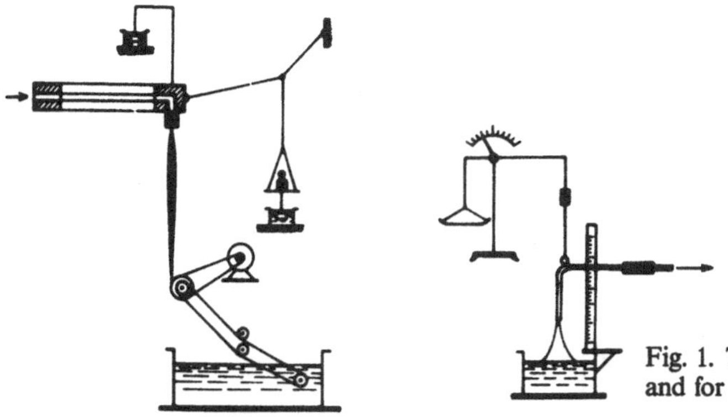

Fig. 1. The experimental apparatuses for spinning (left) and for tubeless syphon (right)

and l, from the knowledge of the molecular weight and the macromolecule structure. Also this comparison indicated that the solutions may be considered dilute.

For all data, the velocity, v, as a function of axial position, x, was derived from the photographs (the flow rate was known). A typical plot of v vs. x for each apparatus is given in fig. 4. It may be noticed that, for a large part of the filament length, v is proportional to x and thus G, the stretching rate, is constant. Actually, small terminal portions of the curves deviate sometimes from the straight lines and have not been considered in the subsequent calculations. In particular, for the spinning results, a small initial part has often an inverse slope due to die swell.

The stretching rate G could not be directly chosen but variations could be induced by regulating flow rate, length and, in the spinning experiment, the velocity of the take-up device. Also the concentration affected the accessible values of G. The overall range of G values was $0.17 \div 1.4\ \text{sec}^{-1}$.

From the force at any given x the normal stress was calculated and, dividing it by G, the transient elongational viscosity was obtained. This was related to the time t after the stretching G had been applied which was easily calculated from the kinematics up to the abscissa x along the filament.

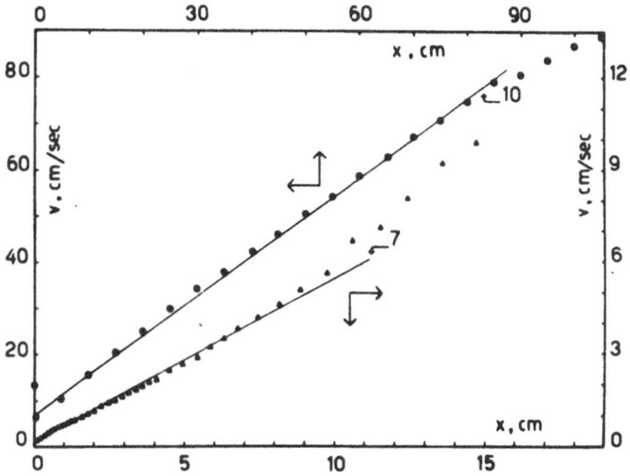

Fig. 4. Typical velocity vs. axial position plot for spinning (run 10) and for tubeless syphon (run 7)

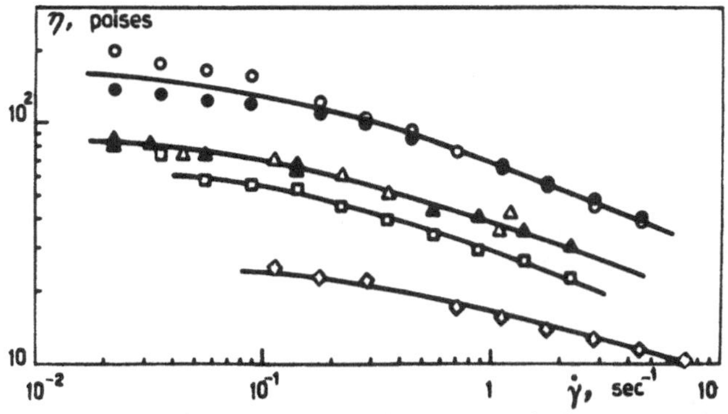

Fig. 2. Shear viscosity vs. shear rate for the Separan NP20 solutions in glycerol. ○, ● 0.5%; △, ▲ 0.25%; □ 0.175%; ◇ 0.05%

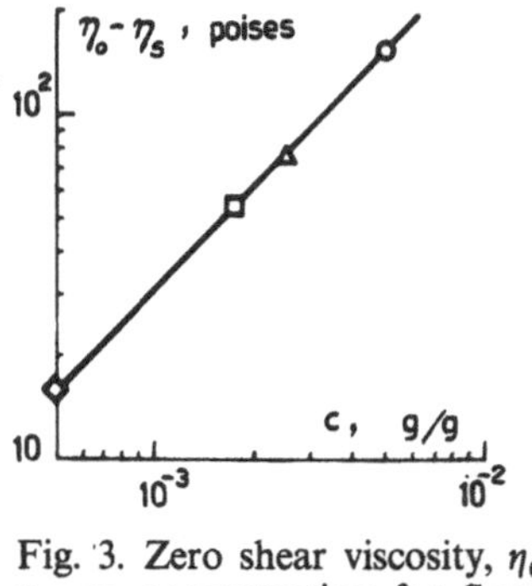

Fig. 3. Zero shear viscosity, η_0, less glycerol viscosity, η_s, vs. concentration for Separan NP20 solutions in glycerol

Results

For all runs, the elongational viscosity was found to increase with time. Typically, the results for each of the two apparatuses are as shown in fig. 5. In the tubeless syphon experiments, the viscosity started from quite low values and increased to about 10^3 times the zero-shear viscosity. In the spinning experiments, the viscosity was already very large at the die exit (about 3×10^2 times the zero-shear viscosity) and increased in the spinning line by a factor of 10 approximately.

In view of the comparison with theory which is made in the following the time is made dimen-

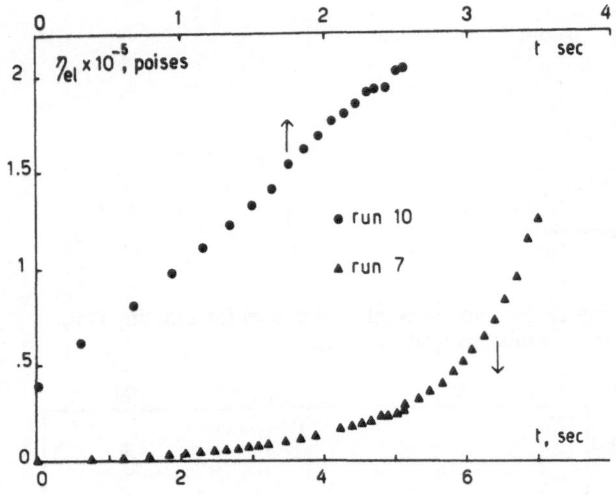

Fig. 5. Typical elongational viscosity vs. time plot for spinning (run 10) and for tubeless syphon (run 7)

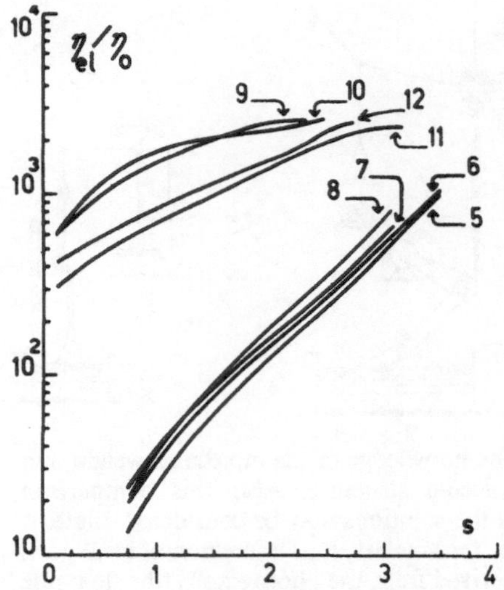

Fig. 7. Ratio of elongational viscosity to zero shear viscosity for tubeless syphon (runs 5-6-7-8) and for spinning (runs 9-10-11-12) – 0.25% Separan NP 20 in glycerol

sionless by multiplying it by G. When this is accomplished, it may be observed that all curves for the syphon results become almost coincident and all those relative to the spinning become parallel and also very close one to the other. All the results are shown in figs. 6–8 in the form of η_{el}/η_0 vs. $s = Gt$.

Finally a curve of each type ($G = 0.52\ \text{sec}^{-1}$; $G = 0.86\ \text{sec}^{-1}$) is drawn again in fig. 9. In this figure a horizontal shift has been applied to the data so as to make a single curve from both the spinning and the syphon results. This curve will be compared with theory as discussed below.

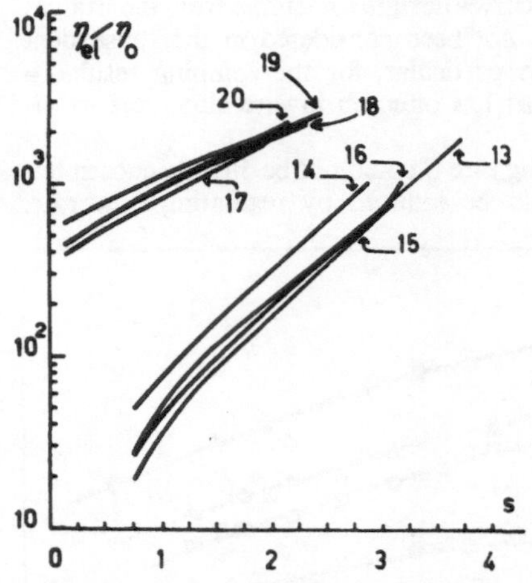

Fig. 8. Ratio of elongational viscosity to zero shear viscosity for tubeless syphon (runs 13-14-15-16) and for spinning (runs 17-18-19-20) – 0.5% Separan NP 20 in glycerol

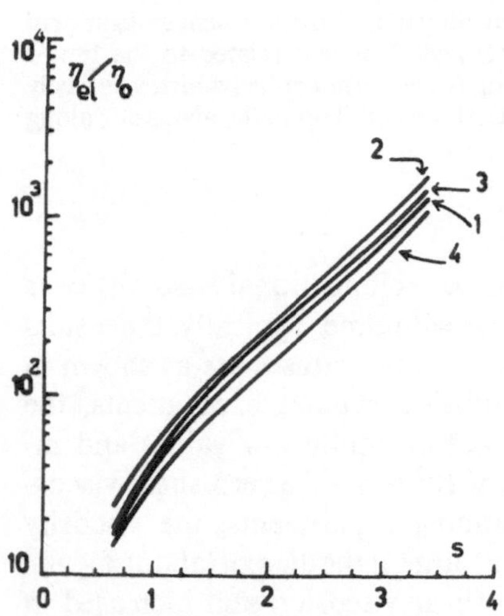

Fig. 6. Ratio of elongational viscosity to zero shear viscosity for four runs with the tubeless syphon. 0.175% Separan NP 20 in glycerol

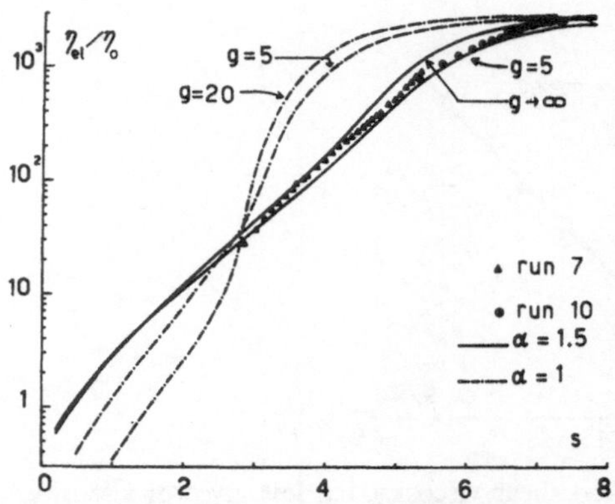

Fig. 9. Comparison of the experimental η_{el}/η_0 (runs 7 and 10) with the theoretical curves

Discussion

The first observation to be made regards the large values of the viscosity which eventually are reached in all experiments. The results of refs. (11, 13, 14) show that such large values of the elongational viscosity can only be reached provided G is sufficiently large. More precisely, a dimensionless stretching rate g, defined as the product of G times the relaxation time, λ, must be somewhat above the critical value of 0.5.

Let us make the assumption, which will be confirmed in the following, that g for all data is actually beyond 1.5. This value, as shown in (15), approximately determines the boundary above which the influence of *Brown*ian motions of the beads can be neglected. All the theoretical results of refs. (15, 16) can then be used for comparison with the transient elongational viscosity data obtained in this work.

The problem however arises of how to treat the results in view of the fact that, as noted above, the material does not come virgin to the elongational motion, especially for the spinning case.

In the spinning experiments, because of the high level of viscosity already attained at the die exit, we may infer that the molecules are there already largely oriented and stretched in the flow direction, probably because of the elongational converging flow in the die entrance whose effects are not lost in the very short die length. As noted in (16), after the orientation of molecules is practically complete, the subsequent raise in viscosity, which is due to further molecule elongation, follows a law which is largely independent of the previous history. On this basis, it sounds reasonable to treat the spinning data as if they were obtained in a constant stretching experiment, starting from a virgin state, a large initial part of which is not accessible to measurements.

Similarly, but to a much smaller extent, the initial part of the motion in the syphon experiment occurs below the liquid surface in a converging zone towards the filament where no measurements can be taken. The above discussion justifies the horizontal shifts which demonstratively have been applied to the data reported in fig. 9, the largest shift being of course that relative to spinning data. Assuming for a moment that syphon data do not require any shift, the spinning curve is made to coalesce with the syphon curve. The single curve so obtained may still require a small shift because of the slight uncertainty in the initial part of the syphon data. On the whole, we observe that, for a given value of G, the syphon experiments cover an initial part of the transient, while the spinning technique yields the upper part of the same transient curve. We shall consider curves such as the one given in fig. 9 as single curves and seek the best fit of data with the predictions of the theory given in (15, 16).

In refs. (15, 16), it is shown that the curves of η_{el}/η_0 for different values of g above $g = 5$ are not very much different one from the other, i.e. the dependence of η_{el}/η_0 on G is almost entirely accounted for in the definition of the dimensionless time $s = Gt$. This result is indeed in agreement with the data collected in this work. As shown in figs. 6–8 all syphon data plotted versus dimensionless time crowd into a very narrow band. The spinning data, when properly shifted to the right, can be made to coincide among themselves along the extrapolated curve of the syphon data. Although the parameter of the curves in figs. 6–8 is G, this is indeed equivalent to the dimensionless parameter g of the theory because the relaxation time of all solutions used is the same. In fact, λ depends on the polymer and the solvent which were not changed and is also independent of concentration within the realm of dilute solutions. It must be observed that the substantial independence on parameter g holds only if $g > 5$. This being the case, *Brown*ian motions of the beads are negligible as it was assumed before.

The single curve of fig. 9 is then practically representative of all collected data. Its shape, which describes the increase of elongational viscosity with the time of application of a constant stretching rate, must now be compared with the detailed predictions of the theory (15, 16, 18). In a first attempt, we compared it with the results obtained by neglecting internal viscosity, for both the monodisperse case (15) as well as accounting for polydispersity (18). Such a comparison is definitely unsuccessful: the slope of the experimental curve at all times is significantly lower than that of the theoretical curve for zero internal viscosity reported in (15). In fig. 9 such a situation is clearly depicted: two theoretical curves for $g = 5$ and for $g = 20$ are drawn for comparison with the experiments. The lower slope of the experimental curve cannot be attributed to the influence of polydispersity which, as shown in (18), changes the predicted slope very slightly in the direction of making it even steeper than for the monodisperse case.

Conversely the influence of internal viscosity is in the right direction. As shown in (16), the transient elongational response is made slower by the retarding influence of the internal viscosity. By a trial and error procedure, it was found that the best fit between the experimental curve and the theoretical one which includes internal viscosity is obtained for $\alpha = 1.5$. This parameter is defined as in (16):

$$\alpha = 1 + \xi_i/\xi$$

where ξ_i is internal viscosity and ξ bead-solvent friction coefficient. The theoretical curves for $g = 5$ and $g \to \infty$ and $\alpha = 1.5$ are also drawn in fig. 9.

The presence of a relatively strong internal viscosity for the polymer used is confirmed by two more facts. First of all, fig. 2 shows that the solutions have a strongly non-*Newton*ian shear viscosity. For dilute polymer solutions, the current interpretation of the major cause of non-*Newton*ianism is indeed internal viscosity (19). Actually, the curve of fig. 2 show a larger effect than would be predicted from the obtained value of internal viscosity. However, the data obtained in shear might include errors due to thermal effects and show indeed a relevant scattering.

The second fact is related to the absolute level of λ. Should internal viscosity be negligible, the relaxation time of the dumbbell model (or the maximum relaxation time of *Rouse* theory) would be obtained from the ratio:

$$\frac{6}{\pi^2} \frac{\eta_0 - \eta_s}{cRT\varrho} M$$

where M is the molecular weight and ϱ the density of the solution. From fig. 3, the above quantity has been calculated giving a pseudo-relaxation time of order 2 sec. With such a value of λ, the range of values of g would be $0.34 \div 2.8$. This would be in conflict with the substantial independence on parameter g of the experimental η_{el}/η_0 vs. s curves and for the smallest G values, even in conflict will the large values of η_{el}/η_0 obtained. The conclusion must be drawn that the actual value of the relaxation time is larger than the value calculated, which is again consistent with a significant value of the internal viscosity (19).

As a final observation we wish to point out that interpretation of the results presented here by means of continuum-mechanics constitutive equations would hardly be successful. In fact, one might attempt to use an integral equation such as the one suggested by *Lodge* (20) or equivalently a differential equation of the *Maxwell* type. As shown by *Bird* et al. (21) and by *Lodge* and *Yeen-Jing Wu* (22) the above equations are equivalent to bead-spring models without internal viscosity and would thus prove inadequate for the problem considered here.

Acknowledgement

This work was supported by C.N.R., Grant n. 115.1729.0.5155.

Summary

In this work, elongational viscosity data obtained for dilute solutions of Separan NP 20 in glycerol, are presented. The data are taken under transient stress conditions with two different experimental techniques.

For comparison with theory the non-linear elastic dumbbell model has been used both for monodisperse and polydisperse cases. Agreement between data and theory is satisfactory only by taking into account internal viscosity effects.

Zusammenfassung

In dieser Arbeit werden Meßergebnisse an Dehnströmungen mit verdünnten Lösungen von Separan NP 20 in Glyzerin mitgeteilt. Dabei werden zwei verschiedene Meßmethoden mit nicht-konstantem Spannungsverlauf angewandt.

Für Vergleiche mit der Theorie werden als Modelle sowohl monodisperse als auch polydisperse Suspensionen nicht-linear elastischer Hanteln herangezogen. Die Übereinstimmung ist erst dann zufriedenstellend, wenn man zusätzlich den Einfluß einer „inneren Viskosität" in Rechnung stellt.

References

1) *Zidan, M.*, Rheol. Acta **8**, 89 (1969).

2) *Astarita, G.* and *L. Nicodemo*, Chem. Eng. J. **1**, 57 (1970).

3) *Acierno, D., R. Greco*, and *L. Nicodemo*, Proceedings of the first meeting of the Italian Society of Rheology **2**, 231 (1971).

4) *Ballman, R. L.*, Rheol. Acta **4**, 137 (1965).

5) *Meissner, J.*, Rheol. Acta **8**, 78 (1969).

6) *Vinogradov, G. V., A. I. Leonov*, and *A. N. Prokunin*, Rheol. Acta **8**, 482 (1969).

7) *Vinogradov, G. V., B. D. Radushkevich*, and *V. D. Fikhman*, J. Polymer Sci. A-2, **8**, 1 (1970).

8) *Vinogradov, G. V., V. D. Fikhman, B. D. Radushkevich*, and *A. Ya. Malkin*, J. Polymer Sci. A-2, **8**, 657 (1970).

9) *Cogswell, F. N.*, Plastics & Polymers **36**, 109 (1968).

10) *Cogswell, F. N.*, Rheol. Acta **8**, 187 (1969).

11) *Peterlin, A.*, J. Polymer Sci. **B 4**, 287 (1966).

12) *Takserman-Krozer, R.*, J. Polymer Sci. **1**, 2487 (1963).
13) *Stevenson, J. F.* and *R. B. Bird*, Trans. Soc. Rheol. **15**, 135 (1971).
14) *Tanner, R. I.* and *W. Stehrenberger*, J. Chem. Phys. **55**, 1958 (1971).
15) *Greco, R., G. Titomanlio*, and *G. Marrucci*, Rheol. Acta (in press).
16) *Acierno, D., G. Ttitomanlio*, and *G. Marrucci*, Trans. Soc. Rheol. **16**, 651 (1972).
17) *Nicodemo, L., G. Marrucci*, and *J. J. Hermans*, J. Polymer Sci. A-2, **10**, 1351 (1972).
18) *Titomanlio, G., D. Acierno*, and *R. Greco*, Ing. Chim. Ital. **9**, 117 (1973).
19) *Cerf, R.*, Fortschr. Hochpolym. Forsch. **1**, 382 (1959).
20) *Lodge, A. S.*, Elastic Liquids (London 1964).
21) *Bird, R. B., H. R. Warner jr.*, and *D. C. Evans*, Adv. Polymer Sci. **8**, 1 (1971).
22) *Lodge, A. S.* and *Yeen-Jing, Wu*, Rheol. Acta **10**, 539 (1971).

Authors' address:
D. Acierno, G. Titomanlio, and *L. Nicodemo*
Cattedra di Principi di Ingegneria Chimica
Universita di Palermo
Palermo (Italia)

Rheol. Acta 13, 538–541 (1974)

From the B. F. Goodrich Chemical Company, Development Center, Avon Lake, Ohio 44012 (U.S.A.)

Can die swell be predicted?

By N. Nakajima

With 3 figures and 4 tables

(Received October 27, 1972)

Introduction

The term "die swell" is commonly used to describe the increase of diameter of polymeric extrudate upon emerging from the extrusion die. Understanding the phenomenon is very important in the extrusion of thermoplastics and elastomers. Some progress has been made in interpreting the phenomenon (1). However, the practical requirement as well as the critical test for any interpretation are quantitative prediction of the die swell at a given condition of extrusion. Presently available theories have not been adequately demonstrated to fulfill this need (2 to 4). This is particularly true with the shorter dies, where the die swell depends on the memory of the deformation at the die entrance. This paper describes a technique, by which die swell can be quantitatively calculated within the precision required in practice. The method is based on the theoretical interpretations developed previously (5, 6). Several samples of linear polyethylenes are used to illustrate the method.

Theoretical background

The behavior of polymer melts flowing through a die may be described by the geometry of die-section, the extrusion conditions and the viscoelastic properties of material. The cases treated here are extrusion with a capillary rheometer under the driving pressure of nitrogen (7). The die entrance geometry is a reduction of diameter from approximately 0.56 to 0.06 in. with a flat (180°) face. The dies are capillaries having the diameter of 0.06 in. with the length to diameter ratios, L/D, of 2–20. The extrusion temperature is at 190 °C. The range of shear rate, $\dot{\gamma} = 5$ to $200\ \mathrm{sec}^{-1}$, are considered. The die swell values, Sw, are defined as the ratio of extrudate diameter to the die diameter after the extrudates complete recovery (5). If we know the viscoelastic properties of material for the above condition, the die swell must be predictable.

We shall now proceed to describe the viscoelastic properties. For the simplicity of arguements the material properties are expressed by apparent values, which are independent of the radial position of the capillary. Therefore, shear rate, $\dot{\gamma}$, is,

$$\dot{\gamma} = 4Q/\pi R^3 \qquad [1]$$

where Q is the volume output rate and $R = D/2$. A useful material function is a time constant, λ, which relates strain, γ, to shear rate by

$$\lambda = \gamma/\dot{\gamma}. \qquad [2]$$

Further, γ, is related to Sw by

$$\gamma = \mathrm{Sw}^2 - 1/\mathrm{Sw}^4. \qquad [3]$$

This formulation assumes that die swell is a tensile elongation, reverse in time, with an ideal rubber-like material (5).

If the time constant, λ, is known for a given material, Sw value can be calculated. Such time constant may be derived from the interpretive technique developed previously (5, 6),

$$\lambda_0 = \gamma_0/\dot{\gamma} \qquad [4]$$

$$\lambda_\infty = \gamma_\infty/\dot{\gamma} \qquad [5]$$

where the subscripts, 0 and ∞, are for $L/D = 0$ and $= \infty$, respectively. For the intermediate values of L/D, γ may be calculated from

$$\frac{\gamma^2 - \gamma_\infty^2}{\gamma_0^2 - \gamma_\infty^2} = e^{-t}/\lambda^N \qquad [6]$$

where λ^N is another time constant, reciprocal of the decay rate constant of the memory, i.e. memory of preceeding deformation having occured at the entrance region of the barrel. The time, t, is the residence time of material in capillary, which is related to $\dot{\gamma}$ and L/R by,

$$t = 4(L/R)/\dot{\gamma}. \qquad [7]$$

What we need to know is the three time constants, λ_0, λ_∞ and λ^N. The shear rate dependence and material dependence of these constants will be examined next. The data from our previous work (6) are shown in figs. 1, 2, and 3 for five linear polyethylenes having melt indices of 0.4 to 9 (8). These samples are described in table 1, by the flow parameters of *Sabia* (9, 10), which are determined at 190 °C.

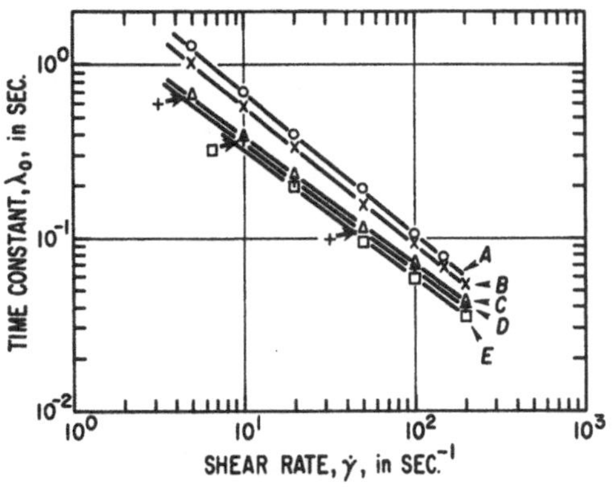

Fig. 1. Dependence of time constant, λ_0, on shear rate

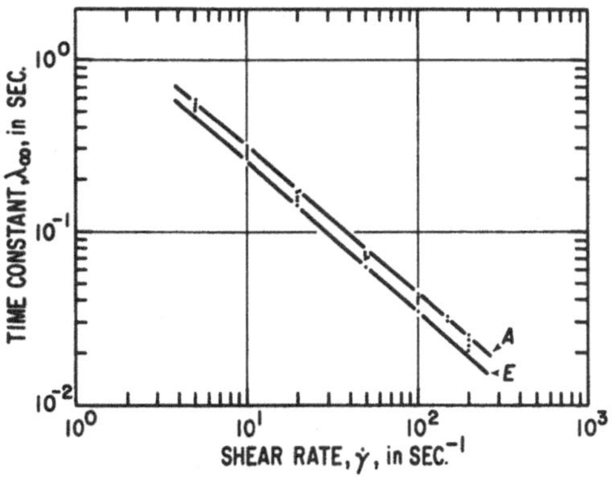

Fig. 2. Dependence of time constant, λ_∞, on shear rate

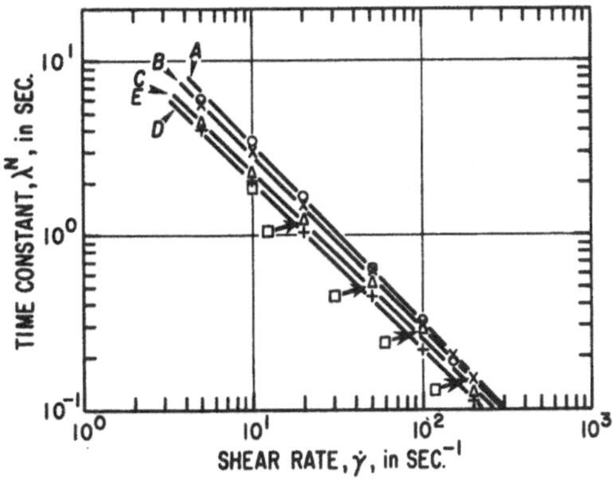

Fig. 3. Dependence of time constant, λ^N, on shear rate

For the shear rate range of this observation the time constants are inversely proportional to power of shear rate:

$$\lambda = k\dot{\gamma}^{-l}. \quad [8]$$

For convenience it maybe written as

$$\lambda/\lambda_{50} = (\dot{\gamma}/50)^{-l} \quad [9]$$

where λ_{50} is the value of λ at $\dot{\gamma} = 50\ \text{sec}^{-1}$. Further, λ_{50} and l are expressed as

$$\lambda_{50} = A(MI)^{-m} \quad [10]$$

$$l \;\; = B(MI)^{-n} \quad [11]$$

where MI is the melt index (8). The values of A, m, B and n are evaluated from the data of figs. 1, 2, and 3; they are listed in table 2.

Table 1. Polyethylene samples

Resin	Melt index	*Sabias* flow parameters (9, 10)			
		$\eta_0 \times 10^{-5}$ poises	$\dot{\gamma}_0$ sec^{-1}	a	ab
A	0.4	6.9	0.35	2.0	2/3
B	1.0	2.7	0.80	2.1	2/3
C	2.5	0.94	6.2	2.4	2/3
D	5.0	0.40	26.5	2.6	2/3
E	9.0	0.30	46.0	3.0	2/3

Table 2. Constants for eq. [10] and [11]

	λ_0	λ_∞	λ^N
A	0.155	0.075	0.580
m	0.190	0.030	0.118
B	0.785	0.850	0.955
n	0.021	0.000	0.017

In table 2 A is the value of λ at $\dot{\gamma} = 50$ with the sample of melt index one. Likewise, with melt index of one, B gives shear rate dependence of eq. [9] as $l = B$. For the samples of melt indices other than one, m and n give the molecular weight dependence of λ_{50} and l, respectively. The molecular weight dependence of λ is rather small and so are the values of m and n.

Calculation of die swell

By using the constants of table 2 and eqs. [2] to [11] (in the reverse sequence) the die swell can be calculated for a given sample of known melt index. Such calculation has been performed for the samples of table 1 at shear rates of 5, 10, 20, 50, 100, 150, and 200 sec^{-1} and with L/D of 2, 5, 10, and 20. The differences between calculated and observed values of Sw are averaged according to the following formula:

$$\left\{\frac{[(\mathrm{Sw})\,\mathrm{obs.} - (\mathrm{Sw})\,\mathrm{calc.}]^2}{N}\right\}^{1/2} = \pm\zeta \qquad [12]$$

where N is a number of independent data points. The values of ζ and N are summarized in table 3.

Table 3. Accuracy of calculating die swell

Resin	At seven shear rates and with four L/D's $\pm\zeta\%$	N
A	3.69	24
B	4.47	28
C	2.35	24
D	3.69	18
E	2.77	15
Shear rate (sec^{-1})	**With five samples and with four L/D's $\pm\zeta\%$**	**N**
5	6.72	15
10	4.27	18
20	2.24	18
50	1.90	18
100	1.72	18
150	2.65	8
200	2.28	14
L/D	**With five samples and at seven shear rates**	
2	2.17	30
5	3.36	30
10	3.88	30
20	3.91	19
	Average of all data $\pm\zeta\%$	**N**
	3.51	109

Checked against the total 109 observed data, the die swell values have been calculated within $\pm 3.51\%$ at 50% confidence level. If data at shear rates of 5 and 10 sec^{-1}, are eliminated, the calculated results are in even better agreement with the observed ones: $\zeta \leq \pm 2.65\%$. The largest disagreement in the shear rate range of 20–200 sec^{-1} is 4.3%. Since the practical interests are in the high shear rate extrusion, the formula's given in eqs. [2]–[11] are quite satisfactory in predicting die swell values.

Prediction of die swell

The formulae developed in the previous section are applied to a different sample of linear polyethylene, Petrothene®[1]), of melt index 0.4. The shear rate range beyond 200 sec^{-1} is examined to see if eq. [8] holds. The temperature of observation was 200 °C, instead of 190 °C for which the constants of table 2 are given. The shear rates of the observation were converted to the corresponding shear rates at 190 °C by the use of flow activation energy of 7 kcal/mole at constant shear stresses (11). This procedure is valid if the die swell is independent of temperature at constant shear stress. Such is the case with molten polystyrene (2). The results of calculation are shown in table 4.

Table 4. Calculated and observed values of die swell

L/D	$\dot\gamma$(200 °C)	$\dot\gamma$(190 °C)	Sw calc.	Sw obs.	$\zeta\%$
2	600	510	3.49	3.30	+5.8
2	180	153	3.08	3.05	+1.0
2	135	115	2.98	2.97	+0.3
5	380	323	2.92	2.89	+1.0
5	280	238	2.83	2.81	+0.7
5	200	170	2.75	2.73	+0.7
5	145	123	2.66	2.69	−1.1
5	96	81.6	2.55	2.58	−1.2
10	390	332	2.52	2.56	−1.6
10	240	204	2.43	2.49	−2.4
10	145	123	2.31	2.41	−4.2
10	115	97.8	2.26	2.37	−4.6
10	90	76.5	2.21	2.34	−5.6
20	340	289	2.26	2.35	−3.8
20	260	221	2.21	2.31	−4.3
20	200	170	2.17	2.26	−4.0
20	145	123	2.11	2.18	−3.2
20	103	87.6	2.05	2.17	−5.5
				average	$\pm$ 3.39

$$\zeta\% = \frac{\text{Sw calc.} - \text{Sw obs.}}{\text{Sw obs.}} \times 100$$

The calculated and observed values of die swell are in excellent agreement. The applicability of eqs. [2]–[11] and the constants of eqs. [10] and [11] have been demonstrated.

Discussion

It has been shown that die swell values can be predicted for linear polyethylene if the following information is available: i.e. melt index, the die L/D, shear rate and temperature of extrusion. However, it is generally well known that melt index alone does not sufficiently differentiate materials with respect to die swell. Perhaps the

[1]) Petrothene, product of U.S. Industrial Chemicals Company.

present choice of sample, this sample of Petrothene, is somewhat fortuitous in that it must be a similar type of resin to the resins A, B, C, D, and E. The *Sabias* flow parameters of this Petrothene are $\eta_0 = 7.2 \times 10^5$, $\dot{\gamma}_0 = 0.30$, $a = 2.2$ and $ab = \frac{2}{3}$, at 190 °C. It may be tentatively concluded that the present method of prediction is applicable to the resins of the three parameter variety (η_0, $\dot{\gamma}_0$, and a) according to the *Sabia-Nakajima* characterization (10).

Among linear high density polyethylenes, there are numerous variations (10). When the present work is extended to include more sample variations, the constants of eq. [9] must be related not only to melt index but also to other material parameters, which represents very low molecular weight fraction, very high molecular weight fraction, degree of branching, etc. To make a wider use of the present approach, therefore, the material description requires further refinements. In practice die description also requires refinement to include different entrance geometry and different cross-sectional geometry. The definition of die swell value itself needs refinement, because in processing of polymers the swelling is often arrested a few seconds after material leaves the die. It is neither completely recovered value nor the value after letting it crystallize in the room temperature. Concerning the above refinements, the work is in progress and will be reported in the future.

Acknowledgement

The author is grateful to Allied Chemical Corporation for their permission to use the data.

Summary

A calculational scheme has been developed to predict die swell values which are defined as the value after completion of the elastic recovery. The constants of the equations have been evaluated at 190 °C from the known results of several polyethylene samples. With the similar type but different sample of polyethylene, it is shown that the prediction is with ±3.4% for the shear rates of 90 to 600 sec^{-1}, for the dies of L/D 2–20 and at temperature of 200 °C.

References

1) *Bagley, E. B.* and *H. P. Schreiber*, Rheology. Theory and Application, edited by *F. R. Eirich*, vol. 5, pp. 93–125 (New York 1969).

2) *Graessley, W. W., S. D. Glasscock*, and *R. L. Crawley*, Trans. Soc. Rheol. **14: 4**, 519 (1970).

3) *Tanner, R. I.*, J. Polym. Sci. A-2, **8**, 2067 (1970).

4) *Bagley, E. B.* and *H. J. Duffey*, Trans. Soc. Rheol. **14: 4**, 545 (1970).

5) *Nakajima, N.* and *M. Shida*, Trans. Soc. Rheol. **10: 1**, 299 (1966).

6) *Nakajima, N.* and *M. Shida*, Proceedings of International Conference on Mechanical Behavior of Materials, Vol. III, 485 (1972).

7) *Bagley, E. B.*, J. Appl. Phys. **28**, 624 (1957).

8) Amer. Soc. for Testing Materials, Standard, ASTM 1238–57 T, 1957.

9) *Sabia, R.*, J. Appl. Polym. Sci. **7**, 347 (1963).

10) *Nakajima, N.*, J. Appl. Polym. Sci. **14**, 2661 (1970).

11) *Sabia, R.*, J. Appl. Polym. Sci. **8**, 1651 (1964).

Author's address:
Dr. *Nobuyuki Nakajima*
B. F. Goodrich Chemical Company
Development Center, P. O. Box 122
Avon Lake, Ohio 44012 (U.S.A.)

Rheol. Acta **13**, 542–547 (1974)

From the Institute of Petrochemical Synthesis, Academy of Science of the USSR, Moscov (USSR)

A new molecular model to describe the viscoelastic behaviour of linear polymers

By *Y. N. Pokrovsky* and *Yu. G. Yanovsky*

With 5 figures

(Received October 27, 1972)

Linear polymers at temperatures above vitrification and melting are a system of entangled interacting macro molecules and, generally speaking, for a theoretical description of the behaviour of such a system their collective movement should be regarded.

On the other hand, observation (1) proves the properties of the system to be strongly related to the molecular weight or the length of the molecule in the low frequency region, which evidently, makes it possible to describe the viscoelastic behaviour of the concentrated solution or the melt of the polymer by representing it as an aggregate of non interacting quasimacro-molecules. Interaction in this case brings about a change in the effective parameters describing the movement of the macromolecule, appropriately schematized. For this purpose a new model of the macromolecule was evolved (2), and the viscoelastic behaviour of the system at low frequencies was investigated on the basis. In the present work the range of applicability of the theory is extended towards higher frequencies and theoretical and experimental results are compared.

1. Let each macromolecule have a large number of friction centres which are not, however, equivalent. The viscoelastic properties of the system are under significant influence of entanglements, acting as friction centres with an increased factor of hydrodynamic resistance. These will hereafter be referred to as "slow" beads or nodes. Each bead will accordingly be modelled by a bead string or necklace schematically represented in fig. 1 and consisting of slow and "swift" or "fast" beads. The latter not shown in fig. 1.

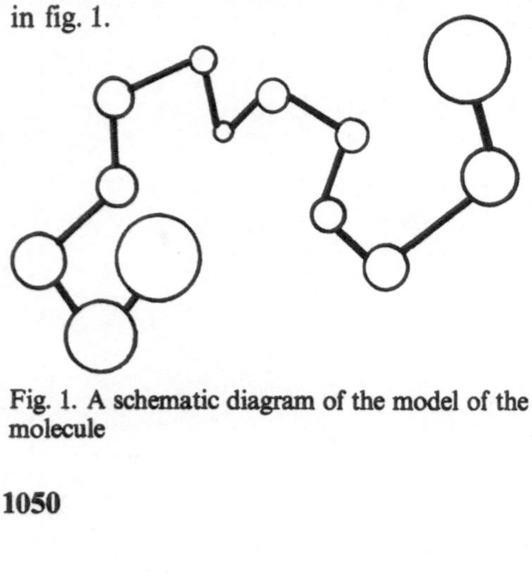

Fig. 1. A schematic diagram of the model of the macromolecule

Let L be the total length of the macromolecule, g the length of the chain between neighbouring nodes, $u = L/g$ the number of nodes per molecule, α the ordinal number of the node as counted off the center of the macromolecule, $N - 1$ the number of swift beads between the neighbouring nodes.

Each bead is acted on by forces from neighbouring beads. The expression for the force of elasticity acting on the αth slow bead from the neighbouring slow beads will, in its "*Gaußian*" approximation, be represented as:

$$\frac{UT}{2\overline{R^2}}(r_i^{\alpha+1} - 2r_i^{\alpha} + r_i^{\alpha-1}) \qquad [1]$$

where r^α is the ith coordinate of the αth bead, and $\overline{R^2}$ the mean square value of the radius of rotation of the macromolecular coil in equilibrium.

The movement of the system gives rise to forces of hydrodynamic entrainment, which are proportional to the difference of the average flow rate $v_{ik} r_k^\alpha$ (v_{ik} tensor of the rate gradients) and the bead rate

$$\zeta(v_{ix} r_x^\alpha - W_i^\alpha) \qquad [2]$$

wherein the swift beads entrainment factor is constant.

$$\zeta = \beta_0 \frac{L}{uN} \qquad [3]$$

while that of the slow beads alters with the length of the molecule in accordance with the

$$\zeta = \beta_0 \frac{L}{u}(1 + |q\alpha|^\delta). \qquad [4]$$

In formulae [3] and [4] β is the friction factor per length unit of the chain. In formula [4] q and δ are constants determining the increase of the slow beads entrainment factor and defined by the demand of an accurate description of the shear viscosity factor – molecular weight relationship ($\sim M^{3.4}$). Hereafter $q=2$; $\delta=2.4$.

Note that the binominal member factor is so selected that with $\alpha=0$ the friction factors per length unit coincide in [3] and [4].

Finally, each bead is acted on by the effective force of the *Brown*ian movement equal to

$$-T\frac{\partial \ln W}{\partial r_i^\alpha}$$

where $W=W(\ldots, r_i^\lambda, \ldots)$ is the function of the distribution of bead position probability.

2. Thus the question of the movement of the macromolecule in the system may be boiled down to the problem of the $u\cdot(N-1)$ movement of the *Brown*ian particle in the field of elastic and hydrodynamic forces. The stress tensor is readily determined via the known movement rate of the macromolecule beads.

Because of the significant difference between the hydrodynamic entrainment factors of the slow and the swift beads, the slow bead movement – dealt with previously (1), becomes substantial, on low frequencies, while the swift bead movement does not affect the viscoelastic behaviour of the system. In high frequencies "slow" beads are stationary and the viscoelastic behaviour of the system is determined by the relaxation processes linked with the movement of the swift beads between the nodes. Results, obtained earlier (2, 3) permit therefore, to present the solution of the problem of the movement of the system, in an approximate form true for sufficiently long macro molecules.

The system of equations of the movement of the concentrated solution or the polymer melt includes a definition of the stress tensor besides the known (4) continuity and movement formulae.

$$\sigma_{i\varkappa}=-p\delta_{i\varkappa}+\frac{1}{2}n\beta_0 L\left(\sum_{\alpha=1}^{u}+\frac{1}{N}\sum_{\alpha=u+1}^{u+N}\right)\frac{1}{\tau_\alpha} \times(\langle\varrho_i^\alpha\varrho_\varkappa^\alpha\rangle-\langle\varrho_i^\alpha\varrho_\varkappa^\alpha\rangle_0) \quad [5]$$

where in p is pressure, n the number of macromolecules per volume unit, and the law of alteration of the moment of the function of distribution

$$\langle\varrho_i^\alpha\varrho_\varkappa^\alpha\rangle=\int W\varrho_i^\alpha\varrho_\varkappa^\alpha(d\varrho)$$

which are a measure of orientation and deformation of the macromolecule in the flow.

$$\frac{d\langle\varrho_i^\alpha\varrho_\varkappa^\alpha\rangle}{dt}=-\frac{1}{\tau_\alpha}(\langle\varrho_i^\alpha\varrho_\varkappa^\alpha\rangle-\langle\varrho_i^\alpha\varrho_\varkappa^\alpha\rangle_0)+v_{ij}\langle\varrho_j^\alpha\varrho_\varkappa^\alpha\rangle+v_{\varkappa j}\langle\varrho_j^\alpha\varrho_i^\alpha\rangle. \quad [6]$$

The relaxation times τ_α and the initial values of moments included in expression [5] and [6] are defined as follows:

On alteration of α from 1 to u

$$\tau_\alpha=\tau^*\frac{u^2}{\lambda^\alpha};\quad \langle\varrho_i^\alpha\varrho_\varkappa^\alpha\rangle_0=\frac{2\overline{R^2}}{\lambda_\alpha}\delta_{i\varkappa}. \quad [7]$$

On alteration of α from $u+1$ to $u+N$.

$$\tau_\alpha=\frac{\tau^*}{\lambda^\alpha};\quad \langle\varrho_i^\alpha\varrho_\varkappa^\alpha\rangle_0=\frac{2N\overline{R^2}}{u\lambda_\alpha}\delta_{i\varkappa}. \quad [8]$$

A characteristic relaxation time is introduced in expressions [7] and [8]

$$\tau^*=\frac{\beta_0 g\overline{S^2}}{T} \quad [9]$$

where $\overline{S^2}$ is the mean square distance between the neighbouring nodes.

For the long macromolecules the eigenvalues included in [7] may be approximated by

$$\lambda_1=423K^{-1};\quad \lambda_2=765K^{-1} \quad [10]$$

the other values at α from 3 to u by

$$\lambda_\alpha=251\alpha^2K^{-1}-(-1)^\alpha 134\alpha K^{-1} \quad [11]$$

where $K=(qL/g)^\delta$. The eigenvalues of [8] are represented by

$$\lambda_\alpha=\pi^2(\alpha-u)^2 \quad [12]$$

where α alters from $u+1$ to $N+u$.

The above are a closed system of movement equations, provided the parameters of theory β_0, q and δ do not depend upon the forces applied and the rate gradients, which is not the case for polymer systems.

3. Let us now consider the movement of the system in simple shear wherein only one component of the rate gradient tensor differs from zero ($v_{12}\neq 0$). In the case of steady flow eqs. [5] and [6] determine the occurring tangent stress

$$\sigma_{12}=n\beta_0 L\overline{R^2}S_1 v_{12}. \quad [13]$$

Also the first and second difference of the normal stresses.

$$\sigma_{11}-\sigma_{22}=2\frac{n}{T}(\beta_0 L\overline{R^2})^2 S_1 v_{12}^2$$

$$\sigma_{22}-\sigma_{33}=0 \quad [14]$$

where

$S_1 = \sum_\alpha \lambda_\alpha^{-1}; \quad S_2 = \sum_\alpha \lambda_\alpha^{-2}.$

And ratios

$S_1 = 4.895 \cdot 10^{-3} K$

$S_2 = 7.529 \cdot 10^{-6} K^2$

and true for conditions at large K values.

Results obtained in experiments with polymer solutions (5–10) and melts (11–14) in accordance with formulae [14], show that the difference $\sigma_{22} - \sigma_{33}$ is small in any case, while $\sigma_{11} - \sigma_{22}$ differs from zero significantly.

From expressions [12] and [13] follows the ratio between the first difference of normal stresses and shear stresses.

$$\sigma_{11} - \sigma_{22} = \frac{2}{Q} \sigma_{12}^2 \qquad [15]$$

where $Q = nT \frac{S_1^2}{S_2}$ practically do not change on alteration of $K = (qu)^\delta$. For the long macromolecule system

$$Q = 1.91 \cdot 10^{24} \frac{cT}{M} \text{ dynes/cm}^2 \qquad [16]$$

where C is the concentration or density of the polymer in g/cm³, T the temperature in ergs ($1\,°\mathrm{K} = 1.38 \cdot 10^{-16}$ ergs).

Ratio [15] established earlier (15) on a simple model is confirmed experimentally (13). The theoretical value of the proportionality factor is finalized; it alters but little as compared to [15], so that the theoretical significance of Q is still [15] an order over the experimental, which is, first of all, to be linked with the polydispersity of the samples considered (13).

From [12] follows the expression for shear viscosity.

$$\eta = n\beta_0 \overline{R^2} S_1 . \qquad [17]$$

In conditions of low shear stress expression [17] describes the dependence of the initial shear viscosity η_0 on the molecular weight with the characteristic breaking point at the critical weight M_c on curve $\lg \eta_0 - \lg M$.

For the short macromolecule system the expression for viscosity will be

$$\eta = 10^{23} c \frac{\zeta_0}{M_0} \overline{R^2} \text{ poise} \qquad [18]$$

where c is the concentration of the polymer, $\zeta_0 = a\beta_0$ the monomeric friction factor coincides with the corresponding expression obtained by *Debaye* (16) for the draining coils i.e. without hydrodynamic interaction.

For the system of long macromolecules

$$\eta = 1.56 \cdot 10^{22} c \frac{\zeta_0}{M_0} \overline{R^2} u^{2.4} \text{ poise}. \qquad [19]$$

Because for *Gaußian* coils $\overline{R^2} \sim M$, the ζ_0 and g_0 values being constant it follows from [19] that $\eta_0 \sim M^{3.4}$, which, as a rule is confirmed in practice (1).

The expression [19] for shear viscosity in the $M > M_c$ range practically tallies with that of *Fox* and *Allen* (17).

The abrupt decrease in viscosity on increase of stresses is usually linked with the lengthening of the chain between neighbouring nodes or, in other words, the decrease of the number of nodes of the fluctuating network in volume unit so that formula [19] permits to infer that

$$\frac{\eta_0}{\eta} = \left(\frac{g}{g_0}\right)^{2.4} = \left(\frac{u_0}{u}\right)^{2.4}. \qquad [20]$$

The possible alteration mechanism of the equilibrium number of entanglements during flow has recently been discussed by *Graessley* (18) and *Bueche* (19).

Chance of the viscosity factor with the alteration of stresses or rate gradients may be linked with variation of the value of index δ in the friction factor [4]. But the data on viscosity are not conclusive, as to whether the viscosity – stress relationship is connected with change in g or δ values. This is to be discussed in (5) of the present paper.

4. Eqs. [5] and [6] determine the viscoelastic behaviour of the system and on preset periodic motion $v_{12} \sim e^{-i\omega t}$ the complex dynamic modulus $G = G' - iG''$. In the dimensionless form the actual and the imaginary parts of the dynamic modulus, as functions of the dimensionless frequency $x = \tau^* \omega$ have the following aspect

$$M'(x, u) = \frac{G'}{vT} = \frac{1}{u} \sum_{\alpha=1}^{u} \frac{x^2}{x^2 + (\lambda_\alpha/u^2)^2} + \sum_{\alpha=u+1}^{u+N} \frac{x^2}{x^2 + \lambda_\alpha^2} \qquad [21]$$

$$M''(x, u) = \frac{G''}{vT} = \frac{1}{u} \sum_{\alpha=1}^{u} \frac{x\lambda_\alpha/u^2}{x^2 + (\lambda_\alpha/u^2)^2} + \sum_{\alpha=u+1}^{u+N} \frac{x\lambda_\alpha}{x^2 + \lambda_\alpha^2} \qquad [22]$$

where $v = nu$ is the number of chains per volume unity of the system; eigenvalues λ_α are determined by formulae [10]–[12].

The relationship of moduli [21] and [22] with frequency are calculated via $K = (qu)^\delta$. There are thus, three parameters, determining the form of the relationship δ being equal to 2.4 in the linear range.

Fig. 2 represents the dependence of moduli M' and M'' (with $q = 2$, $\delta = 2.4$) upon frequency at varying values of u. With the alteration of parameter u, also q and δ the dependence of M' and M'' upon x at low frequencies changes its aspect, while at high frequencies all the curves fuse.

A characteristic property of $M'(x)$ is the horizontal plateau while that of $M''(x)$ the presence of maximum. The position of the maximum is approximately coincident with the beginning of the horizontal section of the curve and is determined by frequency $x_H = \lambda_u/u^2$. The plateau ends as seen in fig. 2, at the frequency of $x_K = 5$. Hence with the aid of [11] at $\alpha = u$, the length of the plateau at $\delta = 2.4$ is found and at a preset temperature.

$$\Delta \lg\omega = 2.4 \lg(ug/8.8). \qquad [23]$$

The value of the dynamic modulus on the horizontal section, equal to

$$G'_e = nuT \qquad [24]$$

will determine, with the weight and temperature known, the number of chains per macromolecule.

$$u = \frac{G'_e M}{6.02 \cdot 10^{23} c T}. \qquad [25]$$

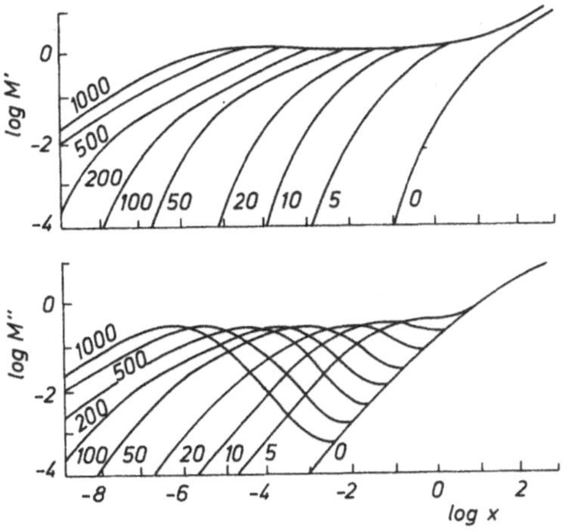

Fig. 2. The dependence of the real (top) and imaginary (bottom) parts of the dimensionless dynamic modulus on the dimensionless frequency for the values of u numerically designated on the charts at $\delta = 2.4$, $q = 2$

The series of curves presented in fig. 2 resemble the series obtained for polymers with various molecular weights (20–23), the alteration of which it is natural to connect with the change in the number of nodes per macromolecule u.

The solid lines of figs. 3 and 4 represent experimental ratios $G'(\omega)$ and $G''(\omega)$ for monodisperse polystyrenes (14, 20–23) and polybutadienes (23), the dotted lines are the theoretical ratios obtained via formulae [21] and [22] at values u, corresponding molecular weights indicated and determined by formula [25] and at $\delta = 2.4$, $q = 2$.

The above data shows that the theory defines qualitatively the linear. Viscoelastic behaviour of polymers in a wide band of frequencies. The discrepancy is highest on transition from fluidity to plateau. It seems, the time distribution in the transition range varies somewhat, from that predicted in theory. It is to be noted that the theoretical plateau is shorter for polystyrenes and longer for polybutadienes than in the experimental values. This may be due to the capacity of parameter q to assume various values depending upon the structure of the polymer. For the polystyrene $q > 2$ and $q < 2$ for poly-

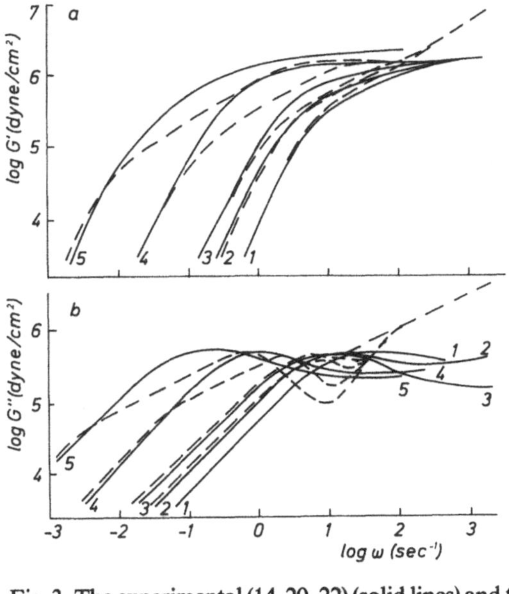

Fig. 3. The experimental (14, 20–22) (solid lines) and theoretical (dotted) dependences of the real (top) and imaginary (bottom) parts of the dynamic modulus on the frequency for monodisperse polysterines with = (1.93; 2.39; 2.67; 4.60; 8.60) 10^5 (curves 1–5 respectively) at 190 °C. The theoretical curves are calculated via [21 and 22] at values of u corresponding the indicated values and equal to 6.7; 8.3; 9.2; 16.0; and 29.9. $G''_e = 1.40 \cdot 10^6$ dyn/cm² $\tau^* = 0.033$ sec

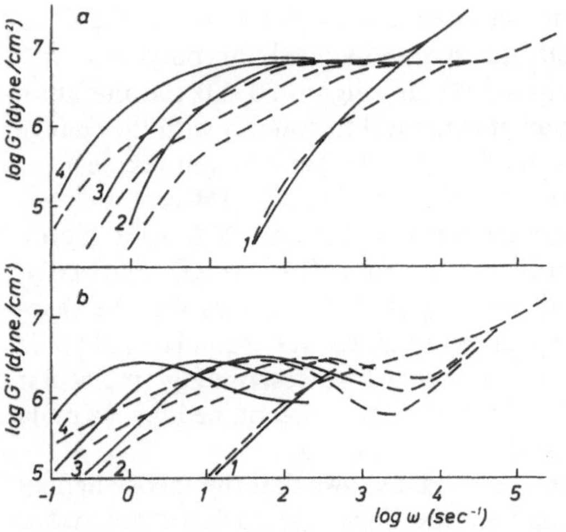

Fig. 4. The experimental (23) (solid lines) and theoretical (dotted) dependences of the real and imaginary dynamic modulus upon the frequency for monodisperse polybutadienes with $M = 3.8 \cdot 10^4$; 10^5; $1.5 \cdot 10^5$ at 22 °C. The theoretical curves are calculated via [21] with the u values corresponding the M indicated and equal to 11.3; 29.1; and 40.8, 68.5. $G'_e = 6.31 \cdot 10^6$ dyn/cm², $\tau^* = 10^{-4}$ sec

butadiene whereas above q was assumed to equal 2.

5. Let us now consider polymer behaviour in the non liner range. In experiments specifically designed to measure the dynamic modules of steady flow, under various shear stresses, of solutions (24, 25) and melts (26) of polymers, and in experiments when the dynamic modulus was measured at various amplitudes of deformation rates and consequently, at various shear stresses, series of curves are obtained which also resemble the curves in fig. 2. In fig. 5 solid lines represent the experimental storage modulus values of polybutadiene with the molecular weight $1.5 \cdot 10^{-5}$, measured at a low (zero) amplitude of deformation rate, and an amplitude

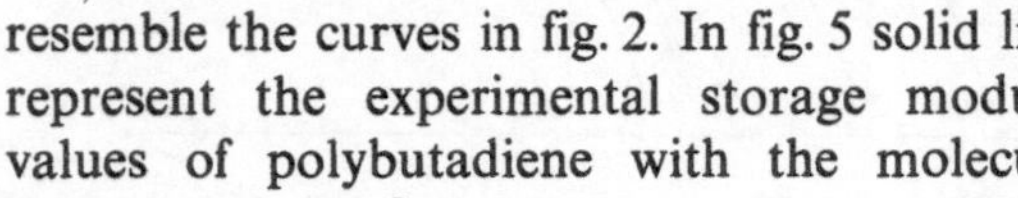

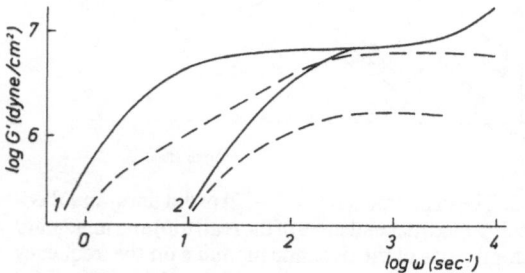

Fig. 5. The experimental (23) (solid lines) and theoretical (dotted) dependences of the dynamic modulus of polybutadiene with $M = 1.5 \cdot 10^5$ amplitudes of deformation rates $1 - 0\ \mathrm{sec}^{-1}$, $2 - 2\ \mathrm{sec}^{-1}$. The theoretical curves are calculated at values of u: 40.8 and 11.3

of deformation rate of 2 sec^{-1}. The values of the system viscosity factor were respectively $\lg \eta_0 = 6.2$ and $\lg \eta = 4.82$.

Let us first assume that the alteration of the viscoelastic functions with stress may be linked with value u. Then formula [20] with $u_0 = 40.8$ will determine the number of nodes at the amplitude of deformation rate 2 sec $u = 11.3$. Thus the values of the reduced dynamic module $M(x)$ are exactly the same as for polybutadiene with a molecular weight of $3.8 \cdot 10^4$. However, on transition to dimensional dynamic modulus it should be considered that with the length of the macromolecule constant the chain length between two nodes has increased and the value of the modulus has therefore been reduced to $G'_e = 1.77 \cdot 10^6$ dyn/cm² while the value of the speficic relaxation time went up to $\tau^* = 1.28 \times 10^{-3}$ sec. These values determine the position of the dynamic modulus as a function of ω. The theoretical values of the modulus are represented on fig. 5 by dotted lines. Fig. 5 shows that in the flow range, theory gives a satisfactory definition of the viscoelastic characteristics but the theoretical value of the modulus on the plateau is considerably below the experimental, which indicates that evidently, the change of viscoelastic functions may not be connected solely with the change of the number of nodes in the unit of volume. Because in experiment (23–27), the value of the modulus on the plateau does not change substantially on deformation, the value u and consequently, the number of entanglements in a unit of volume does not change substantially either. Therefore, in accordance with the present theory, the alteration of the effective value of viscosity and viscoelastic functions on deformation should also be connected with changes in δ i.e. the quality not the quantity of the entanglements.

6. Thus, specific behaviour of concentrated solutions and melts of polymers may be associated with the increase of the friction factor from the center to the ends of the macromolecule, which should be ascribed to some definite points along the macromolecule. This substantiates the concept of a network with discrete nodes in the systems considered – a concept widely applied at present.

The theory presented defines the viscoelastic behaviour of concentrated solutions and melts of polymers with the aid of two parameters: the length of the chain between nodes g, or the number of nodes per macromolecule u and the monomeric coefficient factor which, in the general instance, depend upon the temperature, concentration rate gradients and stresses. In this respect, like other theories (28, 29) of the viscoelastic behaviour of a system of macromolecules, this theory is not strictly speaking, a molecular one and has a semiphenological character. Unlike other theories, however, it seems the provide a more consistent explanation of the viscoelastic behaviour of the polymer systems under consideration.

References

1) *Ferry, J.*, Viscoelastic properties of polymers (1963).

2) *Pokrovsky, V. N., G. N. Kargopolova*, and *A. N. Ivanova*, Theory of viscoelastic behaviour of concentrated polymer solutions and melts at low frequencies of loading. Reprint of Chem. Phys. USSR Academy of Sciences (Moscow 1970). Kolloidn. Zhur. in press.

3) *Pokrovsky, V. N.*, Colloidn. Zhur. **31**, 114 (1969).

4) *Landau, L. D.* and *E. M. Liftshitz*, Mechanics of continium media. Moscow. Gostechizdat (1954).

5) *Philippoff, W.*, J. App. Phys. **27**, 934 (1956).

6) *Markovitz, H.* and *R. W. Williamson*, Trans. Soc. Rheology **1**, 25 (1957).

7) *Trapeznikov, A. A., A. S. Morozov*, and *G. G. Petrzhik*, Kolloidn. Zhurn. **22**, 761 (1960).

8) *Muppler, J. D.*, Trans. Soc. Rheology **9**, 273 (1965).

9) *Osaki, K., M. Tamura, T. Kataka*, and *M. Kurata*, J. Phys. Chem. **69**, 3642 (1965).

10) *Kotaka, T.* and *K. Osaki*, J. Pol. Sci. **15**, 453 (1966).

11) *Berbon, J. J.* and *E. R. Howells*, Polymer **2**, 429 (1961).

12) *Badley, E. B.*, Trans. Soc. Rheology **5**, 355 (1961).

13) *Vinogradov, G. V., A. Ya. Malkin, Yu. G. Yanovsky, V. D. Schumsky*, and *E. A. Dzjura*, Mechanism of Polymers (Moscow 1970).

14) *Orogi, S., T. Masuda*, and *K. Kitagawa*, Macromolecules **3**, 109 (1970).

15) *Pokrovsky, V. N.*, in collection: Achievements in rheology of polymers. p. 118 (1970).

16) *Debye, P.*, J. Chem. Phys. **14**, 636 (1946).

17) *Fox, T. J.* and *V. R. Allen*, J. Chem. Phys. **41**, 344 (1964).

18) *Graessley, W. E.*, J. Chem. Phys. **43**, 2629 (1965); **47**, 1942 (1967).

19) *Bueche, F.*, J. Chem. Phys. **48**, 4781 (1968).

20) *Rudd, J. F.*, J. Pol. Sci. **60**, 71 (1962).

21) *Ballmann, R. J.* and *R. H. H. Simon*, J. Pol. Sci. **A 2**, 3557 (1964).

22) *Den Otter J. L.*, Ph. D. Thesis (Delft 1967).

23) *Vinogradov, G. V., Yu. G. Yanovsky*, and *A. I. Isayev*, Polymeric Materials N 1, 17 (1971).

24) *Booj, H. C.*, Rheol. Acta **5**, 215 (1966).

25) *Tanner, R. J.* and *J. M. Simmons*, Chem. Eng. Sci. **22**, 525 (1968).

26) *Kataoka, T.* and *S. Ueda*, J. Pol. Sci. **A 2**, 7, 475 (1969).

27) *Vinogradov, G. V., Yu. G. Yanovsky, A. I. Isayev*, and *V. A. Kargin*, Academy of Sciences USSR **187**, 1075 (1969).

28) *Hayashi, S.*, J. Phys. Soc. Japan **18**, 131 (1963).; **19**, 101, 2306 (1964).

29) *Chompf, A. J.* and *J. A. Duiser*, J. Chem. Phys. **45**, 150 (1966).

Authors' address:
V. N. Pokrovsky and *Yu. G. Yanovsky*
Institute of Petrochemical Synthesis
Academy of Science of the USSR
Leninski Prospect 29
Moscow B 21 (USSR)

Rheol. Acta **13**, 548–556 (1973)

From the Jet Propulsion Laboratory Pasadena, California (U.S.A.)

Response of bulk polymer under motion with constant stretch histories*)

By St. T. J. Peng and R. F. Landel

With 12 figures

(Received October 27, 1972)

Introduction

In the investigation of non-*Newton*ian viscosity of polymeric materials both in solution and in melt, most experiments have been conducted under shearing flow, i.e., the velocity gradient of flow is perpendicular to the flow direction. One may consider other classes of flow, such as uniform flow, in that the velocity gradients of the flow are parallel to the flow direction. Uniform flow may include simple extensional flow, strip biaxial (pure shear) extensional flow and equal or nonequal biaxial extensional flow. It was reported by *Ballman* (1) that there is a large difference between viscosities in shear and simple extensional flow, thus it is pertinent to investigate the other kinds of uniform flow such as strip biaxial extensional flow.

In our investigation on bulk polymer, the experiments were conducted for both strip biaxial extensional flow and simple extensional flow under a motion with constant stretch history. In such a motion, *Wang* has shown (2) that, under the assumption of simple fluid, the functional of strain history reduces to a function of the first three *Rivlin-Ericksen* tensors, i.e., $\boldsymbol{\tau} = \boldsymbol{g}(\boldsymbol{A}^{(1)}, \boldsymbol{A}^{(2)}, \boldsymbol{A}^{(3)})$. Since $\boldsymbol{g}$ is an isotropic function, one may obtain explicit expression of $\boldsymbol{g}$ through the representation theorem of three tensors. Considering smallness of strain rates, one may obtain first, second, and higher order approximations. We attempt to approximate it by a fifth order approximation for both strip biaxial and simple extensional flow in the steady flow region. We hope that the material constants obtained from the steady flow region may be used to predict the equal biaxial flow in the steady flow region, thus demonstrate at least at small strain rates, the assumption of simple fluid behavior can be applied to polymer melts in this special case.

The material studied was highly viscous undiluted polyisobutylene. In our experiments, the time steady-state region was obtained for both types of extensional flow (here we assume the steady-state region is a region in which the definition of a motion with constant stretch is applicable), from which the apparent non-*Newton*ian viscosity was derived. *Newton*ian viscosity in strip biaxial and simple extensional flow were compared. To describe the transient state of simple extension a new phenomenological model is used to incorporate elastic behavior into the theory.

The stress-strain rate relation

It was shown (2) that for flow under motion with constant stretch history, the relative strain history in steady-state flow, $C_t^t(s)$ is determined uniquely by its first three *Rivlin-Ericksen* tensors $\boldsymbol{A}_1(t)$, $\boldsymbol{A}_2(t)$ and $\boldsymbol{A}_3(t)$. Thus the stress functional of the "simple fluid" reduces to a function of the first three *R-E* tensors

$$\boldsymbol{\tau} = \boldsymbol{g}(\boldsymbol{A}^{(1)}, \boldsymbol{A}^{(2)}, \boldsymbol{A}^{(3)}) \qquad [1]$$

where

$$\boldsymbol{A}^{(n)} = (-1)^n \left.\frac{d^n \boldsymbol{C}_t^t(s)}{ds^n}\right|_{s=0} \qquad n = 1, 2, \ldots \qquad [2]$$

and $\boldsymbol{\tau}$ denotes an extra stress. t and s denote the present time and past time from present configuration respectively. The total stress is given by

$$\boldsymbol{T} = p\boldsymbol{I} + \boldsymbol{\tau} \qquad [3]$$

where p is an arbitrary hydrostatic pressure and $\boldsymbol{I}$ is a unit tensor.

Since the material is isotropic, $\boldsymbol{g}$ must be an isotropic polynomial function of the first three *R-E* tensors. It was shown by *Spencer* and *Rivlin* (3) that an isotropic polynomial function of three symmetric matrices $\boldsymbol{A}$, $\boldsymbol{B}$, $\boldsymbol{C}$ may be expressed as the sum of the following terms and terms formed from these by permutations of $\boldsymbol{A}$, $\boldsymbol{B}$, $\boldsymbol{C}$, with scalar coefficients:

$$\boldsymbol{I},\ \boldsymbol{A},\ \boldsymbol{A}^2,\ \boldsymbol{AB} + \boldsymbol{BA},\ \boldsymbol{A}^2\boldsymbol{B} + \boldsymbol{BA}^2,$$
$$\boldsymbol{A}^2\boldsymbol{B}^2 + \boldsymbol{B}^2\boldsymbol{A}^2,\ \boldsymbol{ABC} + \boldsymbol{CBA},$$
$$\boldsymbol{A}^2\boldsymbol{BC} + \boldsymbol{CBA}^2,\ \boldsymbol{BA}^2\boldsymbol{C} + \boldsymbol{CA}^2\boldsymbol{B},$$
$$\boldsymbol{A}^2\boldsymbol{B}^2\boldsymbol{C} + \boldsymbol{CB}^2\boldsymbol{A}^2,\ \boldsymbol{ABCA}^2 + \boldsymbol{A}^2\boldsymbol{CBA}. \qquad [4]$$

The scalar coefficients are functions of the following invariants

*) This paper represents one phase of research performed by the Jet Propulsion Laboratory, California Institute of Technology sponsored by the National Aeronautics and Space Administration, Contract NAS7-100.

$\mathrm{tr}\,\boldsymbol{A}$, $\mathrm{tr}\,\boldsymbol{A}^2$, $\mathrm{tr}\,\boldsymbol{A}^3$, $\mathrm{tr}\,\boldsymbol{AB}$, $\mathrm{tr}\,\boldsymbol{AB}^2$, $\mathrm{tr}\,\boldsymbol{A}^2\boldsymbol{B}$,

$\mathrm{tr}\,\boldsymbol{A}^2\boldsymbol{B}^2$, $\mathrm{tr}\,\boldsymbol{B}$, $\mathrm{tr}\,\boldsymbol{B}^2$, $\mathrm{tr}\,\boldsymbol{B}^3$, $\mathrm{tr}\,\boldsymbol{BC}$, $\mathrm{tr}\,\boldsymbol{BC}^2$,

$\mathrm{tr}\,\boldsymbol{B}^2\boldsymbol{C}$, $\mathrm{tr}\,\boldsymbol{B}^2\boldsymbol{C}^2$, $\mathrm{tr}\,\boldsymbol{C}$, $\mathrm{tr}\,\boldsymbol{C}^2$, $\mathrm{tr}\,\boldsymbol{C}^3$, $\mathrm{tr}\,\boldsymbol{CA}$,

$\mathrm{tr}\,\boldsymbol{CA}^2$, $\mathrm{tr}\,\boldsymbol{C}^2\boldsymbol{A}$, $\mathrm{tr}\,\boldsymbol{C}^2\boldsymbol{A}^2$

and

$\mathrm{tr}\,\boldsymbol{ABC}$, $\mathrm{tr}\,\boldsymbol{ABC}^2$, $\mathrm{tr}\,\boldsymbol{BCA}^2$, $\mathrm{tr}\,\boldsymbol{CAB}^2$,

$\mathrm{tr}\,\boldsymbol{AB}^2\boldsymbol{C}^2$, $\mathrm{tr}\,\boldsymbol{BC}^2\boldsymbol{A}^2$, $\mathrm{tr}\,\boldsymbol{CA}^2\boldsymbol{B}^2$ [5]

where tr denotes the trace of a matrix.

Now we identify $\boldsymbol{A}_{(1)} \equiv \boldsymbol{A}$, $\boldsymbol{A}_{(2)} \equiv \boldsymbol{B}$, and $\boldsymbol{A}_{(3)} \equiv \boldsymbol{C}$, thus we obtain an explicit form of constitutive equation for steady flow under motion with constant stretch history.

Considering the flow to be slow, it is easily shown that (4)

$$A^{(\varepsilon)}_{(n)\,ij} = \varepsilon^n A_{(n)\,ij} \quad [6]$$

where ε is a small dimensionless constant. Thus we may write eq. (3) in the following form

$$T_{ij} = -p\delta_{ij} + g_{ij}(\varepsilon A_{(1)\,pq}, \varepsilon^2 A_{(2)\,pq}, \varepsilon^3 A_{(3)\,pq}). \quad [7]$$

Applying the terms in the eqs. [4]–[6] into eq. [7] and omitting the terms of higher than the fifth order in ε, we obtain

$$T_{ij} = -p\delta_{ij} + \varepsilon H^{(1)}_{ij} + \varepsilon^2 H^{(2)}_{ij} + \varepsilon^3 H^{(3)}_{ij} + \varepsilon^4 H^{(4)}_{ij} + \varepsilon^5 H^{(5)}_{ij} + 0(\varepsilon^6) \quad [8]$$

where

$$\boldsymbol{H}^{(1)} = a_0 \boldsymbol{A}_{(1)}$$

$$\boldsymbol{H}^{(2)} = b_0 \boldsymbol{A}_{(2)} + b_1 \boldsymbol{A}^2_{(1)}$$

$$\boldsymbol{H}^{(3)} = (c_0\,\mathrm{tr}\,\boldsymbol{A}_{(2)})\boldsymbol{A}_{(1)} + c_1(\boldsymbol{A}_{(1)}\boldsymbol{A}_{(2)} + \boldsymbol{A}_{(2)}\boldsymbol{A}_{(1)}) + c_2\boldsymbol{A}_{(3)}$$

$$\begin{aligned}\boldsymbol{H}^{(4)} = {} & (d_0\,\mathrm{tr}\,\boldsymbol{A}^3_{(1)} + d_1\,\mathrm{tr}\,\boldsymbol{A}_{(1)}\boldsymbol{A}_{(2)} + d_2\,\mathrm{tr}\,\boldsymbol{A}_{(3)}) \\ & \times \boldsymbol{A}_{(1)} + (d_3\,\mathrm{tr}\,\boldsymbol{A}_{(2)})\boldsymbol{A}^2_{(1)} + (d_4\,\mathrm{tr}\,\boldsymbol{A}_{(2)})\boldsymbol{A}_2 \\ & + d_5\boldsymbol{A}^2_{(2)} + d_6(\boldsymbol{A}^2_{(1)}\boldsymbol{A}_{(2)} + \boldsymbol{A}_{(2)}\boldsymbol{A}^2_{(1)}) \\ & + d_7(\boldsymbol{A}_{(1)}\boldsymbol{A}_{(3)} + \boldsymbol{A}_{(3)}\boldsymbol{A}_{(1)})\end{aligned}$$

$$\begin{aligned}H^{(5)} = {} & (e_0\,\mathrm{tr}\,\boldsymbol{A}^2_{(1)}\,\mathrm{tr}\,\boldsymbol{A}^2_{(1)} + e_1\,\mathrm{tr}\,\boldsymbol{A}^2_{(2)} \\ & + e_2\,\mathrm{tr}\,\boldsymbol{A}^2_{(1)}\boldsymbol{A}_{(2)} + e_3\,\mathrm{tr}\,\boldsymbol{A}_{(3)}\boldsymbol{A}_{(1)})\boldsymbol{A}_{(1)} \\ & + (e_4\,\mathrm{tr}\,\boldsymbol{A}^3_{(1)} + e_5\,\mathrm{tr}\,\boldsymbol{A}_{(3)} + e_6\,\mathrm{tr}\,\boldsymbol{A}_{(1)}\boldsymbol{A}_{(2)}) \\ & \times \boldsymbol{A}^2_{(1)} + (e_7\,\mathrm{tr}\,\boldsymbol{A}^3_{(1)} + e_8\,\mathrm{tr}\,\boldsymbol{A}_{(3)} \\ & + e_9\,\mathrm{tr}\,\boldsymbol{A}_{(1)}\boldsymbol{A}_{(2)})\boldsymbol{A}_{(2)} + e_{10}(\mathrm{tr}\,\boldsymbol{A}^2_{(1)}) \\ & \times (\boldsymbol{A}_{(1)}\boldsymbol{A}_{(2)} + \boldsymbol{A}_{(2)}\boldsymbol{A}_{(1)}) + e_{11}(\mathrm{tr}\,\boldsymbol{A}^2_{(1)})\boldsymbol{A}^3_{(1)} \\ & + e_{12}(\mathrm{tr}\,\boldsymbol{A}^2_{(1)})\boldsymbol{A}_{(3)} + e_{13}(\boldsymbol{A}^2_{(1)}\boldsymbol{A}_{(3)} \\ & + \boldsymbol{A}_{(3)}\boldsymbol{A}^2_{(1)}) + e_{14}(\boldsymbol{A}^2_{(2)}\boldsymbol{A}_{(1)} + \boldsymbol{A}_{(1)}\boldsymbol{A}^2_{(2)}) \\ & + e_{15}(\boldsymbol{A}_{(2)}\boldsymbol{A}_{(3)} + \boldsymbol{A}_{(3)}\boldsymbol{A}_{(2)})\end{aligned} \quad [9]$$

where we use $\mathrm{tr}\,\boldsymbol{A}_{(1)} = 0$, $\mathrm{tr}\,\boldsymbol{A}_{(2)} = \mathrm{tr}\,\boldsymbol{A}^2_{(1)}$.

Now we consider the steady extensional flow of a rectangular box composed of bulk polymer (i.e., motion with constant history). We consider the motions are slow, thus neglect the effect of inertia forces, also since the materials studied are highly viscous, the effect of body forces and surface tension are neglected.

The velocity field v_i of the extensional flow is given by

$$v_i = a_i x_i. \quad [10]$$

In steady flow, a_i are constants. Incompressible materials obey the following relationship

$$a_1 + a_2 + a_3 = 0. \quad [11]$$

It was shown (4) that the relative deformation strain tensor of steady extensional flow is given by

$$C^t_{ij}(t - s) = e^{-2a_i s}\delta_{ij}. \quad [12]$$

We now consider the following special type of motions: simple extensional flow, pure shear flow (strip biaxial) and equal biaxial flow.

1. Simple extensional flow

In simple extensional flow, we have

$$a_1 = \dot{\varepsilon} \quad \text{and} \quad a_2 = a_3 = -\tfrac{1}{2}\dot{\varepsilon} \quad [13]$$

where $\dot{\varepsilon}$ is a rate of principal extension. The first three *R-E* tensors through eq. [2] are given by

$$A_1 = \begin{pmatrix} 2\dot{\varepsilon} & & \\ & -\dot{\varepsilon} & \\ & & -\dot{\varepsilon} \end{pmatrix}$$

$$A_2 = \begin{pmatrix} 4\dot{\varepsilon}^2 & & \\ & \dot{\varepsilon}^2 & \\ & & \dot{\varepsilon}^2 \end{pmatrix}$$

$$A_3 = \begin{pmatrix} 8\dot{\varepsilon}^3 & & \\ & -\dot{\varepsilon}^3 & \\ & & -\dot{\varepsilon}^3 \end{pmatrix}. \quad [14]$$

Substituting eq. [14] into eq. [8], one obtains

$$\begin{aligned}T_{11} = {} & 3a_0\dot{\varepsilon} + 3(b_0 + b_1)\dot{\varepsilon}^2 \\ & + 9(2c_0 + 2c_1 + c_2)\dot{\varepsilon}^3 \\ & + [18(d_0 + d_1 + d_2 + d_3 + d_4) \\ & + 15(d_5 + 2d_6 + 2d_7)]\,\dot{\varepsilon}^4 \\ & + 6[9(2e_0 + e_1 + e_2 + e_3) \\ & + 3(e_4 + e_5 + e_6 + e_7 + e_8 + e_9) \\ & + 9(2e_{10} + e_{11} + e_{12}) \\ & + 11(e_{13} + e_{14} + e_{15})]\,\dot{\varepsilon}^5\end{aligned} \quad [15]$$

and the apparent extensional viscosity η_E is given by

$$\eta_E = T_{11}/\dot{\varepsilon} = 3a_0 + 3(b_0 + b_1)\dot{\varepsilon} + 9(2c_0 + 2c_1 + c_2)\dot{\varepsilon}^2 + [18(d_0 + d_1 + d_2 + d_3 + d_4) + 15(d_5 + 2d_6 + 2d_7)]\dot{\varepsilon}^3 + 6[9(2e_0 + e_1 + e_2 + e_3) + 3(e_4 + e_5 + e_6 + e_7 + e_8 + e_9) + 9(2e_{10} + e_{11} + e_{12}) + 11(e_{13} + e_{14} + e_{15})] \quad [16]$$

where a_0 is the *Newton*ian shear viscosity α. Thus as $\dot{\varepsilon}$ approaches zero, i.e., in the *Newton*ian region, the extensional viscosity is equal to three times the *Newton*ian shear viscosity. We now set

$$\alpha \equiv a_0$$

$$\beta \equiv (b_0 + b_1)$$

$$\gamma \equiv (2c_0 + 2c_1 + c_2)$$

$$\delta \equiv 18(d_0 + d_1 + d_2 + d_3 + d_4) + 15(d_5 + 2d_6 + 2d_7)$$

and

$$\theta \equiv 6[9(2e_0 + e_1 + e_2 + e_3) + 3(e_4 + e_5 + e_6 + e_7 + e_8 + e_9) + 9(2e_{10} + e_{11} + e_{12}) + 11(e_{13} + e_{14} + e_{15})] \quad [17]$$

thus eqs. [15] and [16] become, respectively

$$T_{11} = 3\alpha\dot{\varepsilon} + 3\beta\dot{\varepsilon}^2 + 9\gamma\dot{\varepsilon}^3 + \delta\dot{\varepsilon}^4 + \theta\dot{\varepsilon}^5 \quad [18]$$

and

$$\eta_E = 3\alpha + 3\beta\dot{\varepsilon} + 9\gamma\dot{\varepsilon}^2 + \delta\dot{\varepsilon}^3 + \theta\dot{\varepsilon}^4. \quad [19]$$

2. Strip biaxial extensional flow

In strip biaxial extensional flow, we have

$$a_1 = \dot{\varepsilon}, \quad a_2 = -\dot{\varepsilon}, \quad a_3 = 0 \quad [20]$$

where $\dot{\varepsilon}$ is again a rate of principal extension. The first three *R-E* tensors are given by

$$A_1 = \begin{pmatrix} 2\dot{\varepsilon} & & \\ & -2\dot{\varepsilon} & \\ & & 0 \end{pmatrix}$$

$$A_2 = \begin{pmatrix} 4\dot{\varepsilon}^2 & & \\ & 4\dot{\varepsilon}^2 & \\ & & 0 \end{pmatrix}$$

$$A_3 = \begin{pmatrix} 8\dot{\varepsilon}^3 & & \\ & -8\dot{\varepsilon}^3 & \\ & & 0 \end{pmatrix}. \quad [21]$$

Substituting eq. [21] into eq. [8], one obtains

$$T_{11} = 4\alpha\dot{\varepsilon} + 16\gamma\dot{\varepsilon}^3 + \mu\dot{\varepsilon}^5 \quad [22]$$

and apparent strip biaxial viscosity η_{ST} is given by

$$\eta_{ST} = T_{11}/\dot{\varepsilon} = 4\alpha + 16\gamma\dot{\varepsilon}^2 + \mu\dot{\varepsilon}^4 \quad [23]$$

where

$$\mu \equiv 128(2e_0 + e_1 + e_2 + e_3 + 2e_{10} + e_{11} + e_{12} + e_{13} + e_{14} + e_{15}).$$

Thus the strip biaxial viscosity in the *Newton*ian flow region is equal to four times the *Newton*ian shear viscosity $\eta_{ST} = 4\alpha$.

3. Equal biaxial flow

In equal biaxial flow, we have

$$a_1 = 2\dot{\varepsilon}, \quad a_2 = 2\dot{\varepsilon}, \quad a_3 = -4\dot{\varepsilon}. \quad [24]$$

The first three *R-E* tesnors are given by

$$A_1 = \begin{pmatrix} 2\dot{\varepsilon} & & \\ & 2\dot{\varepsilon} & \\ & & -4\dot{\varepsilon} \end{pmatrix}$$

$$A_2 = \begin{pmatrix} 4\dot{\varepsilon}^2 & & \\ & 4\dot{\varepsilon}^2 & \\ & & 16\dot{\varepsilon}^2 \end{pmatrix}$$

$$A_3 = \begin{pmatrix} 8\dot{\varepsilon}^3 & & \\ & 8\dot{\varepsilon}^3 & \\ & & -64\dot{\varepsilon} \end{pmatrix}. \quad [25]$$

Substituting eq. [25] into eq. [8], we obtain

$$T_{11} = T_{22} = 6\alpha\dot{\varepsilon} - 12\beta\dot{\varepsilon}^2 + 72\gamma\dot{\varepsilon}^3 - 16\delta\dot{\varepsilon}^4 + 32\theta\dot{\varepsilon}^5 \quad [26]$$

and the apparent equal biaxial flow viscosity η_{EB} is given by

$$\eta_{EB} = 6\alpha - 12\beta\dot{\varepsilon} + 72\gamma\dot{\varepsilon}^2 - 16\delta\dot{\varepsilon}^3 + 32\theta\dot{\varepsilon}^4. \quad [27]$$

Similary, $\eta_{EB} \to 6\alpha$ as $\dot{\varepsilon} \to 0$ i.e., the *Newton*ian viscosity of equal biaxial flow is equal to six times the shear viscosity.

In our experimental investigation on the uncrosslinked bulk polymers under steady extensional flow, we consider only the simple extensional and strip biaxial extensional flows. Through both types of flow we may determine the material constants α, β, γ, δ, and θ in the steady-state flow region. We hope the determined constant α's can predict non-*Newton*ian viscosity η_{EB} of the equal biaxial flow through eq. [27].

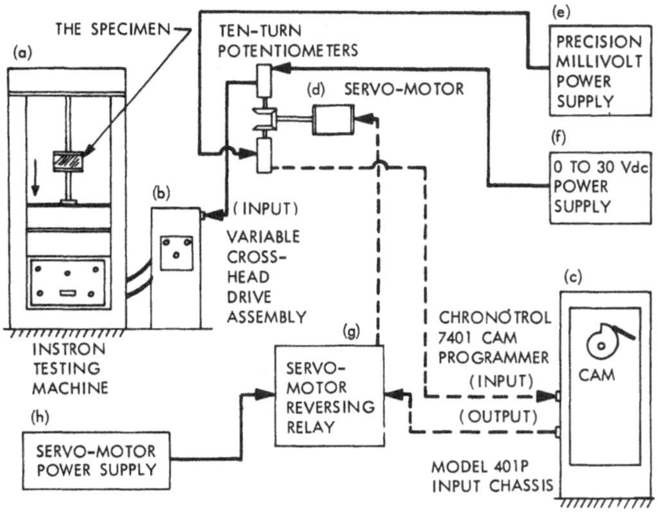

Fig. 1. Schematic diagram of apparatus

Experimental

In this experiment, we consider only very small strain rates (i.e., in the region of $\dot{\varepsilon} = 3.33 \times 10^{-6}$/sec to 2.5×10^{-4}/sec) and a highly viscous material, thus we assume that the effects of inertia, surface tension and gravitational force can be neglected. The extent of the validity of these assumptions is not clear, i.e., the range of strain rates such that the inertia effect can be neglected and the value of viscosity such that the gravitational force can be ignored. However, the same assumption has been made by most authors on highly viscous melts such as PIB L-80 and in our region of strain rates.

From the above assumptions one is lad to the following test setup, shown schematically in fig. 1 and photographically in fig. 2. The main component of the setup is a Barber-Coleman Model 7401 "Chronotrol" (c), which is a program controller that uses the meter movement of a Model 401 P millivolt input chassis (c). Programming is accomplished by changing the position of the millivoltmeter control set point by means of a slowly rotating disc or cam. The cam is rotated by a motor driven gear train, with the cam follower linked to the control set point through a cable drive. The Chronotrol provides linear or nonlinear program control from any predetermined time-variable cycle. During any given cam cycle, an appropriate DC voltage is provided from the power supply (f) to the Variable Crosshead Speed Control Accessory (b) of the Instron tester (a) through a servosystem. The servosystem consists of a reversible motor (d), a polarity reversing relay (g) and two ten-turn potentiometers. Using this setup, the crosshead speed is controlled by the shape of the cam.

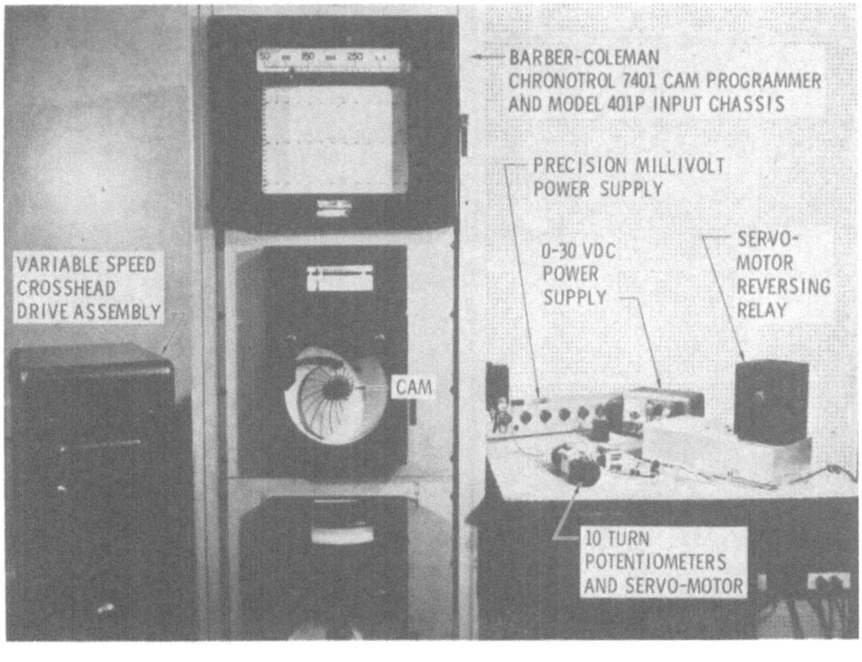

Fig. 2 Test setup

Consider a rod shape sample of initial length l_0 fixed at one end and extended in the direction of the principal extension. The velocity $v = dl/dt$ of the movable end is given by eq. [10]

$$dl/dt = \dot{\varepsilon} l. \quad [28]$$

The following relationships for the displacement l and the crosshead speed v can be derived

$$l = l_0 e^{\dot{\varepsilon}t} \quad \text{and} \quad v = \dot{\varepsilon} l_0 e^{\dot{\varepsilon}t}. \quad [29]$$

The correspondence of the actual strain to that programmed depends on the fluctuation of the motor speed, of the variable crosshead drive accessory and on the accuracy of the programmed cam. Thus, during the run, the actual displacement of the crosshead as a function of time is continually compared to that programmed. If there exists any discrepancy, the position of the cam is carefully adjusted manually to bring them into agreement. The estimated maximum difference between the actual and programmed crosshead displacement is about 1%.

The material studied was undiluted polyisobutylene (Vistanex L-80, Enjay Chemical Co.) at room temperature with Mw = 251500 and Mn = 126000. The classical shear viscosity obtained from tensile-creep measurements (5) from bulk sample is 2.4×10^9 poise.

The rod sample which was used for simple extension was molded from a steel cylinder disc which has the vertical cylindrical holes with a diameter of 5 mm. The height of cylinder disc is 50 mm. The bulk material was put under the cylinder disc and a presser pushed the material through the mold and continued pressing the sample at 160 °C for about 150 min. The resulting sample was 50 mm in length and 5 mm in diameter and quite uniform in diameter through the sample. To hold the sample, we used a holder which was in the shape of a cylindrical tube, when two pieces were combined together. The inside surface of the holder had a shallow thread and its inside diameter was slightly smaller than the sample. To hold the sample, we first applied a thin film of glue to the surface of the sample and holder, then clamped the two holders to the sample until the glue dried. Thus, through the combination of the mechanical type of grip and chemical bonding, the sample was held firmly during the test. To hold the strip biaxial sample, we used the same principle, i.e., a combination of mechanical grip and chemical bonding, however, care was taken to prevent damaging the edge of the sample. The size of the sample is 34.29 mm × 12.7 mm × 2.30 mm where the ratio of width to length is 27.

In simple tension, using a gauge length in the sample to avoid end effects is possible. Our sample length is 25.4 mm and the gauge length is 12.7 mm. Ink benchmarks are used. During the run, the gauge length as a function of time was carefully measured and compared to the assigned one. Since our apparatus can be manually controlled, if there is any

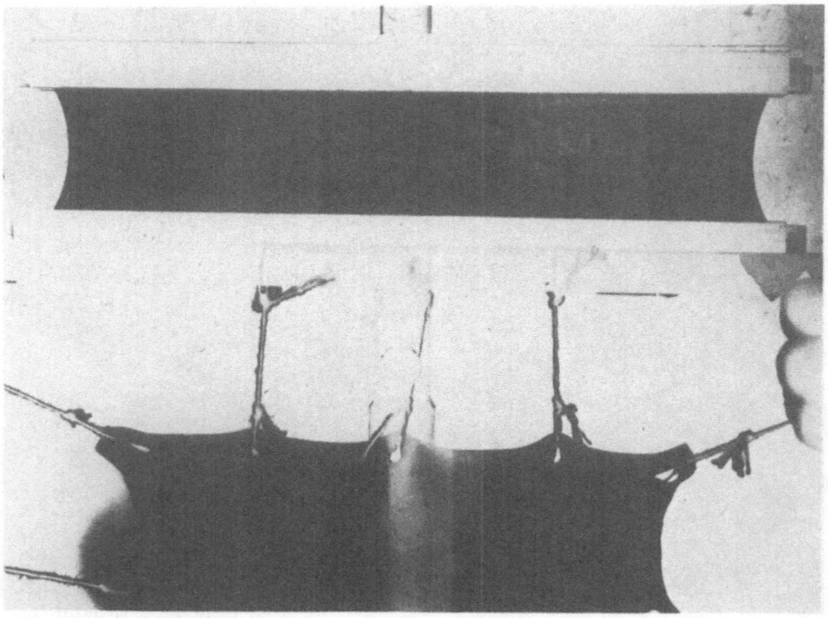

Fig. 3. Strip biaxial extension at $\lambda = 3$

discrepancy, we can always bring them into agreement. Note that, because of end effects, the crosshead displacement did not always agree with the extension of the gauge length.

In strip biaxial extension, using a gauge length is not possible. However, in these experiments, only very small strain rates were considered, the generated normal forces due to stretching in sample were also very small, thus the boundaries were relatively well defined and preserved without much distortion. The test of a strip biaxial sample is extremely difficult, especially at higher strain rates. At higher strain rates, thinning in the sample was very easy to start and propagate. The thinning may be due to inhomogeneity in some small region of the sample or some defects in the surface. Once the thinning started, it propagated swiftly until the sample failed. Since the thinning destroys the uniforminty of stress and strain field, once the thinning was observed before reaching steady-state, the test result was discarded. Only the best runs, which are without thinning and have well defined boundaries are presented and discussed.

Fig. 3 shows a strip biaxial sample under polarized light at a stretch ratio $\lambda = 3$ (in the region of steady-state). The constant strain rate of this test is $\dot{\varepsilon} = 3.333 \times 10^{-6}$/sec. No birefringence appeared at either edge of the sample; however, we stretched the same material at the lower part of the picture in a faster manner and the birefringences appeared. As the strain rate goes higher, however, the birefringence appears gradually at the edges of the sample, as expected.

Results and discussion

Fig. 4 shows the relationship of the true stress T_{11} to the stretch ratio λ for simple extensional flow at various strain rates while fig. 5 shows the corresponding behavior for strip biaxial flow. Since the stretch ratio at the constant strain rate $\dot{\varepsilon}$ depends on time according to the relation $\lambda = e^{\dot{\varepsilon}t}$, figs. 4 and 5 represent the relation of true stress vs. time, and show the existence of a steady flow region (i.e., asymptotic region) for both simple and strip biaxial extension flow.

Fig. 6 shows the apparent non-*Newton*ian viscosity as a function of the stress at various strain rates for both simple extensional flow and strip biaxial extensional flow. It shows clearly that the viscosity decreases with increase in the stress for both types of test. Fig. 7 shows the dependence of the non-*Newton*ian viscosity on the constant strain rate. It may be seen that the viscosity is a monotonically decreasing function of the strain rate. It is interesting to note that at higher strain rates, the difference between the viscosity in simple extensional flow and in strip biaxial extensional flow decreases.

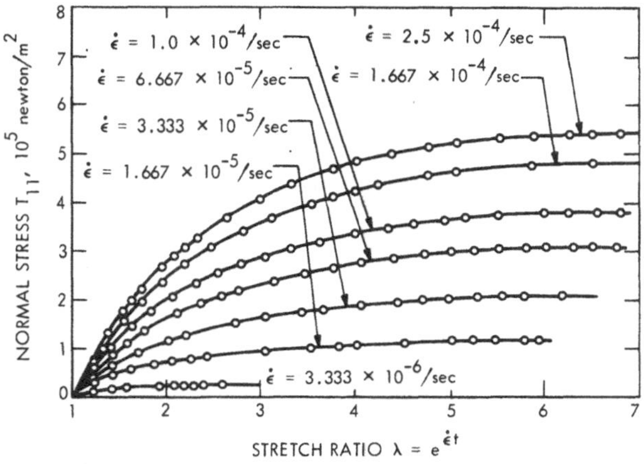

Fig. 4. True stress vs. total deformation in simple extension for various constant strain rates

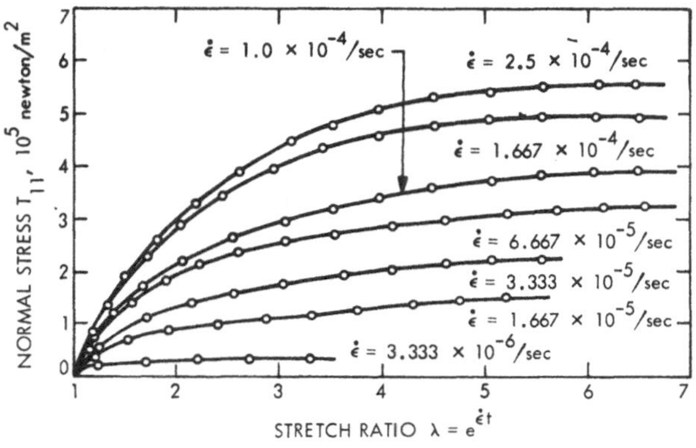

Fig. 5. True stress vs. total deformation in strip biaxial extension for various constant strain rates

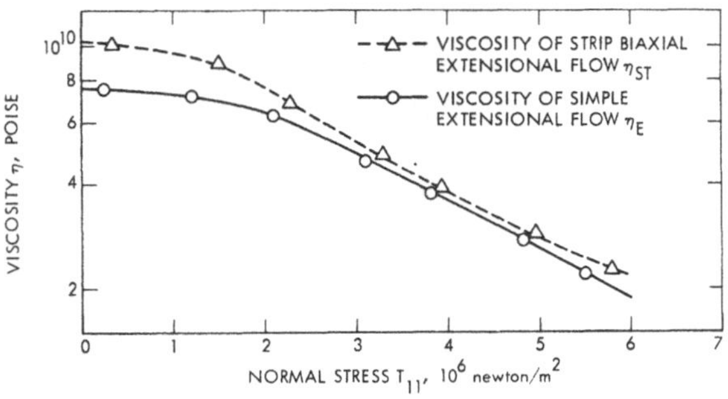

Fig. 6. Dependence of simple extension viscosity η_E and strip biaxial extension viscosity η_{ST} on the normal stress T_{11}

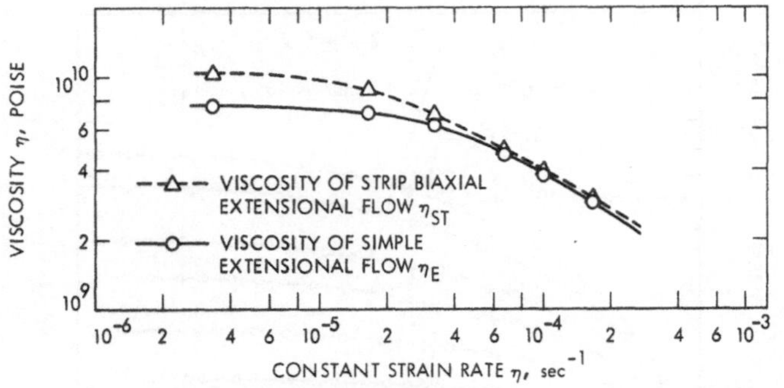

Fig. 7. Dependence of simple extension viscosity η_E and strip biaxial viscosity η_{ST} on constant strain rate $\dot{\varepsilon}$

For the transient region, however, it appears that, in contrast to the steady-state region, the differences between the simple and biaxial case increase at high strain rates. A comparison of transient state behavior for two strain rates is shown in fig. 8. The reason for this behavior is not understood. Perhaps at high $\dot{\varepsilon}$ elastic effects predominate because of polymer network entanglements. The two modes of deformation for a non-linear elastic regime have distinctly different characters as shown in studies on finite deformation of highly cross-linked rubberlike materials. On the other hand, in the steady-state region, the material is relatively liquid-like, and, as expected exhibits the same character of rectilinear flow for both types of deformation.

Fig. 9 shows the viscosities of the simple extension and strip biaxial extensional flow at small strain rates on a linear scale. Solid lines in the figure show the curves calculated from eqs. [19], [23], and [27] by using numerical values obtained for the material constants. It is shown that the curve of η_{ST} has a higher curvature than η_E near the origin, which is confirmed by eqs. [19] and [23]. The material constants

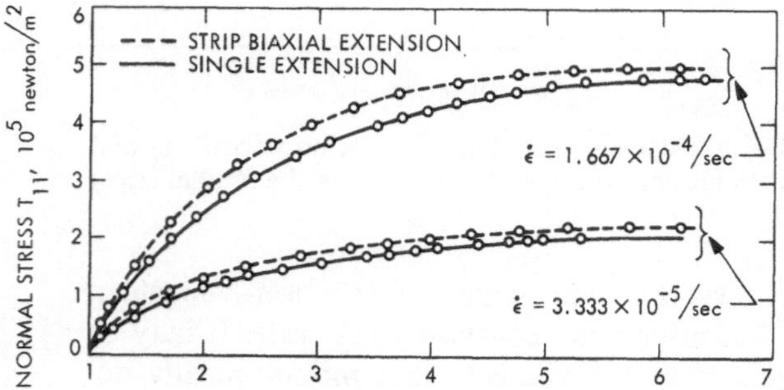

Fig. 8. Comparison of stress-strain behavior of strip biaxial extension and simple extension under the same constant strain rate $\dot{\varepsilon}$ in the transient state

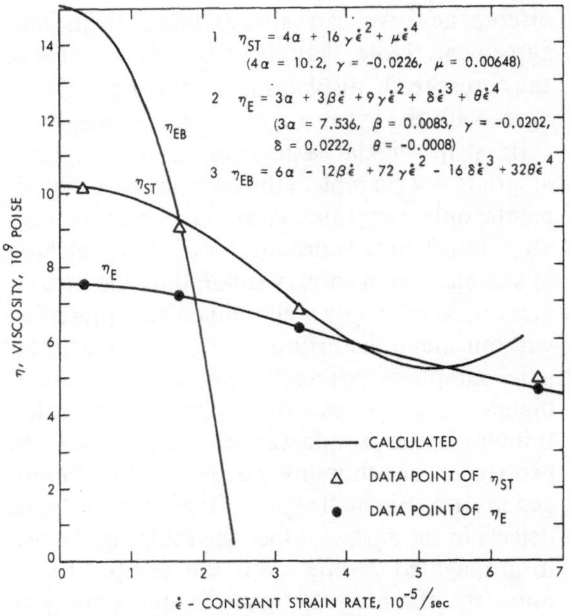

Fig. 9. Viscosity vs. constant strain rate for η_{ST}, η_E and η_{EB}

obtained from best fit of η_{ST} (strip biaxial) at small strain rates are $4\alpha = 10.2$, $\gamma = -0.0226$, and $\mu = 0.00648$, all in the dimensions of poise, where the numerical values were obtained by suppressing 10^9 in the ordinate and 10^{-5} in the abscissa coordinates respectively. In similar manner, the material constants obtained for η_E (simple extension) are $3\alpha = 7.536$, $\beta = 0.0083$, $\gamma = -0.0202$, $\delta = 0.0222$, and $\theta = -0.0008$. Thus the ratio of η_{ST} and η_E in the *Newton*ian region is $\eta_{ST}/\eta_E = 10.2/7.536 = 1.355$, in agreement with classical theory; also the values of γ obtained independently from η_{ST} and η_E are very close. For equal biaxial extension η_{EB}, we substitute into eq. [27] the material constants obtained from the simple extensional flow. The result shows very high curvature with a steep descent near the origin, as shown in fig. 9.

Note that one can see from fig. 9 that in order to describe non-*Newton*ian viscosity, one needs the higher order approximation even in very slow motion; the second order approximation which indicates a constant viscosity in strip biaxial extensional flow (i.e., $\eta_{ST} = 4\alpha$) from eqs. [8], [9], and [22], and a straigth line response in simple extensional flow (i.e. $\eta_E = 3\alpha + 3\beta\dot{\varepsilon}$) from eqs. [8], [9], and [19], is definitely not capable of describing the non-*Newton*ian behavior in this slow motion.

In describing the transient state under simple extension, we use a new model which is similar to the *Maxwell* model for describing the viscoelastic fluid. Instead of a spring and dashpot, we use a rod which is divided into a viscous liquid portion and an elastic solid portion as shown in the sketch, fig. 10, where we express the initial l_0 and final configuration l_f respectively as:

$$l_0 = l_{0q} + l_{0s} \quad \text{and} \quad l_f = l_q + l_s \qquad [30]$$

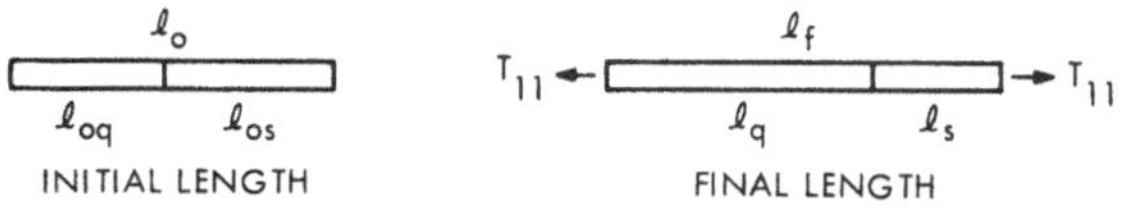

Fig. 10. Model representation of tensile bar

where l_{0q} and l_{0s} are the initial portions of viscous liquid and elastic solid. In the final configuration, l_q and l_s denote the final extended portion of viscous liquid and elastic solid respectively. We apply the same assumption of the *Maxwell* model that, under fast stretch, only the elastic solid part is deformed; similarly, under slow motion, only the viscous liquid part is continually deformed.

From stress relaxation experiments on PIB under simple tension (fig. 11) we found the following relation held up to a stretch ratio $\lambda = l/l_0 = 1.6$:

$$T_{11} = E(\lambda - 1) \qquad [31]$$

This is the isochronal stress-strain curve. (The same result has been reported by *T. L. Smith* (6) on NBS polyisobutylene.) Hence we assume that the elastic part of the rod would behave as

$$T_{11} = E(\dot{\varepsilon})(\lambda_c - 1) \qquad [32]$$

where $\lambda_c = \dfrac{l_s + l_{0q}}{l_0}$, and E depends on strain rate $\dot{\varepsilon}$. It has already been shown, e.g. fig. 4, that there is a steady-state flow region under small constant strain rate $\dot{\varepsilon}$, thus the steady-state flow region can be expressed in terms of non-*Newton*ian viscosity as

$$T_{11} = \eta(\dot{\varepsilon})\,\dot{\varepsilon} \qquad [33]$$

where $\dot{\varepsilon} = \dot{l}/l$. Here we assume that in the steady-state flow region, only the viscous liquid part is being continually deformed[1]), thus we have the following relation of strain rate $\dot{\varepsilon}$ in the stationary flow condition,

$$\dot{\varepsilon} = \frac{\dot{l}_q}{l_q + l_{0s}} \approx \frac{\dot{l}_q}{l_0 \lambda}. \qquad [34]$$

Now we consider the transient state. The force T_{11} exterted on the elastic solid part and the viscous liquid part is equal. The total elongation l_f is the sum of the extension in the elastic and viscous element, thus the time derivative of total elongation is given by

$$\dot{l}_f = \dot{l}_s + \dot{l}_q \qquad [35]$$

thus, taking a derivative on eq. [32] and substituting into eq. [35] together with eqs. [33] and [34], we obtain

$$\dot{T}_{11} + \frac{E(\dot{\varepsilon})}{\eta(\dot{\varepsilon})} T_{11}\lambda = E(\dot{\varepsilon})\,\dot{\lambda}. \qquad [36]$$

Since the motion is under constant strain rate $\dot{\varepsilon}$, which has the following relation (eq. [29])

$$\lambda = l/l_0 = e^{\dot{\varepsilon}t}, \quad \dot{\lambda} = \dot{\varepsilon} e^{\dot{\varepsilon}t} \qquad [37]$$

thus we have the following differential equation

$$\dot{T}_{11} + \frac{E}{\eta} T_{11} e^{\dot{\varepsilon}t} = E\dot{\varepsilon} e^{\dot{\varepsilon}t}. \qquad [38]$$

The solution is given by

$$T_{11} = \dot{\varepsilon}\eta\left[1 - e^{\frac{E}{\dot{\varepsilon}\eta}(1 - e^{\dot{\varepsilon}t})}\right] \qquad [39]$$

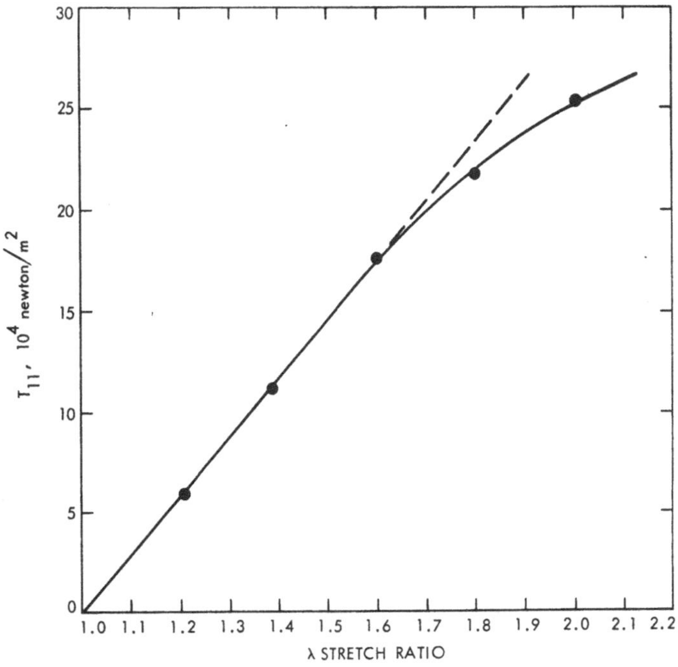

Fig. 11. Isochronal (9 min) stress-strain curve for polyisobutylene L-80

[1]) *Vinogradov* et al. (7, 8) have shown that attainment of steady-state flow corresponds to completion of the development of the high-elastic deformation, and thus the rate of irreversible deformation (i.e., deformation of the viscous part) becomes equal to the preset strain rate $\dot{\varepsilon}$.

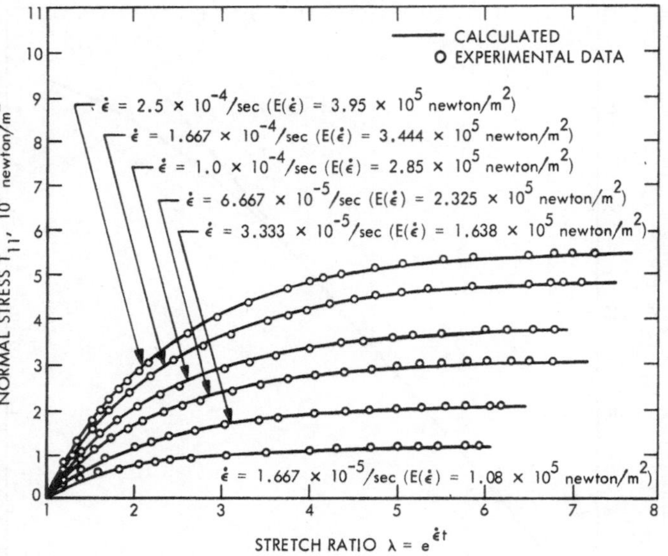

Fig. 12. Stress-strain curves calculated from eq. [39]

where the initial condition is given by $T_{11} = 0$ at $t = 0$. The initial slope of eq. [39] is given by

$$\lim_{t \to 0} \frac{dT_{11}}{dt} = \lim_{t \to 0} \dot\varepsilon E(\dot\varepsilon)\, e^{\frac{E}{\dot\varepsilon\eta}(1 - e^{\dot\varepsilon t})} e^{\dot\varepsilon t} = \dot\varepsilon E(\dot\varepsilon). \quad [40]$$

Thus once we know the values of the non-*Newton*ian viscosity $\eta(\dot\varepsilon)$ in the steady-state region and the initial slope of the curve $\dot\varepsilon E(\dot\varepsilon)$ for a given strain rate $\dot\varepsilon$, we may use eq. [39] to describe the transient state. Fig. 12 shows the calculated curves of eq. [39] for various strain rates with numerical values $E(\dot\varepsilon)$ which were determined from best fit of the curves, since we found it was difficult to obtain $E(\dot\varepsilon)$ accurately from the initial slope. The curves calculated in this manner, however, show some discrepancy with respect to the data points near the origin. It was also found that eq. [39] fits quite well the experimental data obtained by *Vinogradov* et al. (7, 8) from simple extension experiments on PIB and molten polystyrene.

Summary

An experimental investigation of the behavior of a polymer specimen for a constant stretch history has been conducted for both strip biaxial extensional flow and simple extensional flow. It had been shown previously that the stress functional of a non-*Newton*ian fluid under motion with constant stretch history reduces to a function of the three *Rivlin-Ericksen* tensors. Here it is shown that the second order approximation of a simple fluid is not capable of describing the steady-state viscosity in both types of flow under very slow motion. By considering higher orders, however, one may obtain the material constants necessary to describe steady-state non-*Newton*ian viscosity for both strip biaxial extensional flow and simple extension flow.

Steady-state viscosities have been determined at various constant stretch histories using undiluted polyisobutylene. *Newton*ian viscosity in pure shear flow and simple extensional flow are also compared. To describe the transient state, a new model is used to incorporate elastic behavior into the theory.

Zusammenfassung

Eine experimentelle Untersuchung des Verhaltens einer Polymerprobe unter konstanter Dehnungsgeschwindigkeit wurde für ein- und zweiachsige Beanspruchung durchgeführt. Es wurde früher gezeigt, daß das Spannungsfunktional einer nicht-*Newton*schen Flüssigkeit unter konstanter Dehnung auf eine Funktion von drei *Rivlin-Ericksen*-Tensoren reduziert werden kann. Hier wird gezeigt, daß die Näherung zweiter Ordnung für eine einfache Flüssigkeit nicht fähig ist, die stationäre Viskosität in diesen beiden Verformungstypen mit sehr langsamer Bewegung zu beschreiben. Unter Berücksichtigung der Glieder höherer Ordnung kann man jedoch die zur Beschreibung der stationären nicht-*Newton*schen Viskosität nötigen Materialkonstanten für beide Verformungsarten erhalten.

Stationäre Viskositäten wurden für einige konstante Dehngeschwindigkeiten an unverdünntem Polyisobutylen bestimmt. *Newton*sche Viskosität bei reiner Schubverformung und einfacher Dehnströmung wird auch verglichen. Ein neues Modell wird für die Beschreibung des nicht-stationären Zustandes benötigt, um elastisches Verhalten in die Theorie einzubeziehen.

References

1) *Ballman, R. L.*, Rheol. Acta **4**, 137 (1965).

2) *Wang, C. C.*, Arch. Rat. Mech. Anal. **20**, 329 (1965).

3) *Spencer, A. J. M.* and *R. S. Rivlin*, Arch. Rat. Mech. and Anal. **2**, 309 (1959).

4) *Markovitz, H.* and *B. D. Coleman*, Incompressible Second-Order Fluids. In: Advances in Applied Mechanics, Vol. 8. Ed. by *H. L. Dryden* and *T. von Karman* (New York 1964).

5) *Fox, T. G., S. Gratch*, and *S. Loshack*, Viscosity Relationships for Polymers in Bulk and in Concentred Solution. In: Rheology, Vol. 1. Ed. *F. R. Eireich* (New York 1956).

6) *Smith, T.*, Trans. Soc. Rheol. **6**, 61–80 (1962).

7) *Vinogradov, G. V., B. V. Radushkevich*, and *V. D. Fikhman*, J. Polymer Sci, Part **A-2**, **8**, 1–17 (1970).

8) *Vinogradov, G. V., V. D. Fikhman, B. V. Radushkevich*, and *A. Y. Malkin*, J. Polymer Sci., Part **A-2**, **8**, 657–678 (1970).

Authors' address:

Stevens T. J. Peng and *Robert F. Landel*
Jet Propulsion Laboratory
Pasadena, California (U.S.A.)

Rheol. Acta **13**, 557–561 (1974)

From the Brunel University, Uxbridge, London (U.K.)

A stochastic model of crystal plasticity

By P. Feltham

With 3 figures

(Received October 27, 1972)

1. Introduction

The present state of understanding of problems involving movement of dislocations in real crystals is somewhat reminiscent of that relating to electron transport in metals at about 1925. At that time the shortcomings of the classical "gas-kinetic" theory of *Drude* and *Lorentz*, advanced in the first decade of this century, were well understood, but some time had yet to elapse before, in 1928, the use of *Fermi-Dirac* statistics by *Sommerfield*, and the introduction of the concept of a periodic lattice-field by *Bloch*, removed the difficulties, and facilitated a deeper insight into the process.

The statistical treatment of correlated dislocation movement in the transport of material, i.e. in plastic flow, is more involved than that relating to current transport by "point" charges; in fact the complexities of the dislocation structures in deformed materials, as revealed mainly by transmission electron microscopy, are a clear pointer to the difficulties in developing theories of the kinetics of such assemblies.

Parameters characterising the structural heterogeneity of deformed crystalline materials on the "micro" scale have been discussed in the literature (1), but only fairly recently have attempts been made to develop a stochastic model of the internal stress-distribution in crystals containing dislocations (2), and to use this in problems of the thermally activated motion of dislocations, e.g. their migration in the *Peierls* potential (3).

Relating to creep, *Mott* (4) used simple statistical concepts twenty years ago to derive the well-known $t^{1/3}$-law; later an extensive survey of various attempts to use energy-barrier distributions in studies of creep laws was made by *Täubert* (5) in a paper in which he showed that the universality of several creep laws commonly observed in crystalline materials could be regarded as a consequence of the *Markovian* kinetics of dislocation ensembles. These concepts were later developed into an evolutionary stochastic model (6) which accounted quite well for the principal manifestations of creep.

Only the mode of migration of dislocations was considered explicitly in that model; generation of dislocations was introduced only indirectly, via the boundary conditions of the pertinent differential equations, for example in dealing with "steady" high-temperature creep. The main object of the present work was to generalise the model by allowing explicitly for the operation of dislocation sources. In that way it was hoped to encompass, in addition to the plastic behaviour under constant stress i.e. creep, work-hardening as well.

2. The stochastic model

We shall outline briefly the basic features of the previous treatment (6), essential for our present purposes. Fig. 1 shows a segment of a dislocation which, assisted by stress and thermal activation, has moved from its original position, where it was pinned at a barrier of height u_j, to a new one, indicated by the broken line. The assumption that following activation the new barrier, at which it is held up after the jump, could be either higher or lower by a "small" discrete amount δu,

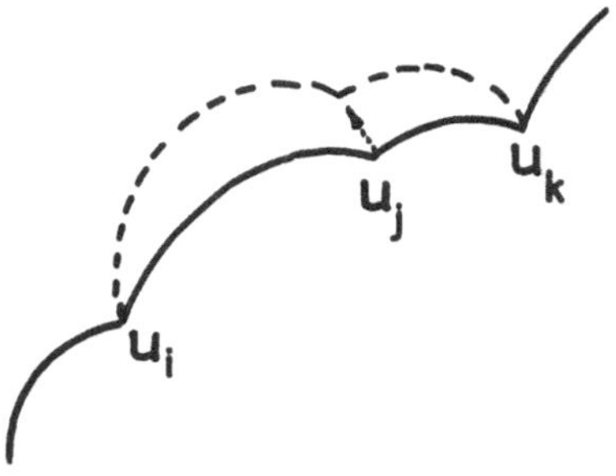

Fig. 1. Migration of a dislocation segment initially pinned at the barrier u_j to a new position, indicated by the broken line

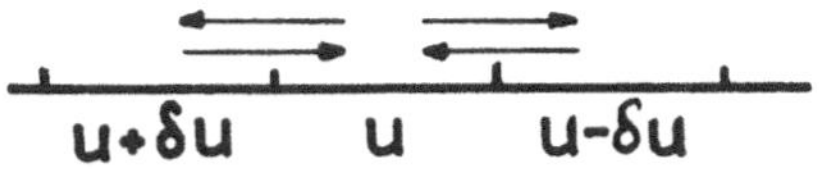

Fig. 2. The inflow and outflow of barrier-levels from a discrete energy interval in the u-spectrum

and consideration of the inflow and outflow of such activated units from a given energy level u, as indicated in fig. 2, then leads to the continuity equation

$$\frac{\partial n}{\partial t} = D \frac{\partial^2 [n \exp(-u/kT)]}{\partial u^2}. \qquad [1]$$

Here $n(u, t)$. δu is the number of flow units per unit volume at time t from the onset of creep, held up at energy barriers of height u, D is equal

36

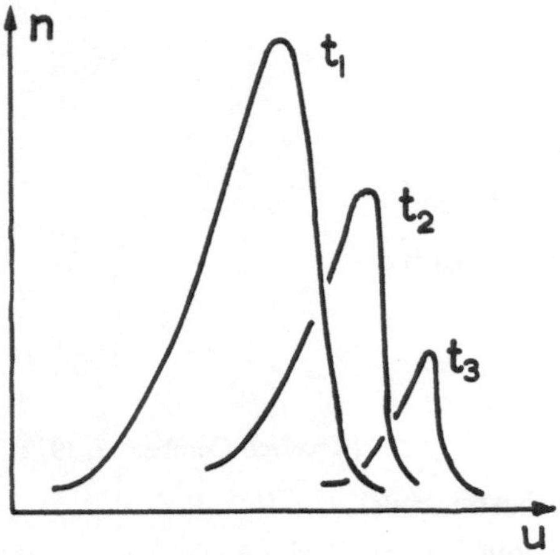

Fig. 3. Schematic representation of the u-spectrum after three periods $t_1 < t_2 < t_3$ from the onset of creep

to $\frac{1}{2}\nu(\delta u)^2$, and ν is an atomic frequency. For any given time, solutions of eq. [1] will define a *u-spectrum*, such as is sketched in fig. 3 for three instants, with $t_1 < t_2 < t_3$. Eq. [1] is conveniently rewritten in terms of the transition-probability density, defined by $N(u, t) = n \times \exp(-u/kT)$, and one then obtains

$$\frac{\partial N}{\partial t} = e^{-u/kT} D \frac{\partial^2 N}{\partial u^2}. \qquad [2]$$

This equation was "reduced" further (6) by introducing the new variable $\theta = t \cdot \exp(-u/kT)$. The latter substitution was found useful in obtaining "zeroth" order solutions valid only for rather narrow ranges of u, and will not be used here, particularly as an exact solution of eq. [2], suitable for present purposes, will be obtained. As shown before (6), the creep rate is proportional to the sum, or integral, of $N(u, t)$ over the u-spectrum.

As our concern will be with work-hardening at temperatures low enough with respect to the melting point of the material considered, so that diffusional modes of recovery such as may involve point-defect migration will be inoperative, the relevant form of creep is of the "logarithmic" type, characteristic under such conditions (7).

Now, if creep is to be observable in the usual way, then the retardation times of the activated processes must be commensurate with the period over which the creep is being observed. It is therefore useful to select a value u^* from the spectrum which, by analogy with exponential decay processes, defines a characteristic retardation time

$$t^* = \nu_D^{-1} \exp(u^*/kT) \qquad [3]$$

where ν_D is an atomic frequency which, for convenience later, we shall take to be equal to $D/(kT)^2$. If u_1 and u_2 are the physically realisable limits of the u-distribution, then it suffices at present to specify u^* by the inequalities

$$u_1 < u^* < u_2 . \qquad [4]$$

As, in general, low-temperature creep tends towards completion within a period of the order of minutes from the time of initiation, it follows from eq. [3], on substituting, say, $t^* = 100$ sec, and $\nu \approx 10^{10}$ sec^{-1}, that

$$u^* = mkT, \quad m \approx 25. \qquad [5]$$

The value of m will not be greatly affected by a somewhat different choice of t^*. This relation, also given previously (6), has been confirmed by *Conrad* et al. (8) with magnesium crystals deformed between 4.2 and 420 °K.

The lower limit of barrier heights, u_1, will in general differ from zero, because processes with retardation times with u-values less than u_1 will have been "exhausted" effectively already during the time required to bring the specimen to the constant creep-stress. Again, for barriers of height in excess of a certain value u_2, the probability of activation during the time of observation of the creep is negligible; such states remain "frozen-in", so that u_2 can be regarded as the upper limit of the operative spectrum in practice.

3. Low-temperature creep

The solution of eq. [2] which describes logarithmic creep is

$$N(u, t) = N^* \frac{t^* e}{t + t_0} \exp(-z) \qquad [6]$$

where

$$z = \frac{\exp(u/kT)}{\nu_D(t + t_0)} \qquad [7]$$

N^* and t_0 are constants, and e is the base of natural logarithms. The retardation-time t^*, as defined by eq. [3], appears in eq. [6] to define the boundary condition. The spectrum is bell-shaped. If $t_0 = 0$, then $N \to 0$ for both $t \to 0$ and $t \to \infty$; it takes the special value N^* when $u = u^*$ and $t + t_0 = t^*$, as is readily confirmed by substitution into eq. [6], also using eq. [3]. A non-zero value of t_0 implies the existence of a spectrum already at $t = 0$. This would in general be the case, for the material is pre-strained to the constant-stress level at which creep proceeds.

If a dislocation segment, such as is shown in fig. 3, advances a distance rb in a jump from one energy barrier to the next then, by the elementary reasoning previously outlined (6), one finds for the creep rate in shear the relation

$$\dot{\gamma} = \nu r b^2 l^* \int_u N\,du \qquad [8]$$

where l^* is the most probable spacing between adjacent barriers, e.g. of the type u_i, u_j in fig. 1. On using the distribution given by eq. [6], eq. [8] yields

$$\dot{\gamma} = \frac{\alpha}{t + t_0}\left[\frac{kT}{\Delta u}\int_{u=u_1}^{u_2} \exp(-z)\cdot d\left(\frac{u}{kT}\right)\right] \qquad [9]$$

where

$$\Delta u = u_2 - u_1\,. \qquad [10]$$

On writing $n^* \exp(-u^*/kT)$ instead of N^*, consistent with the definition of N in eq. [2], also using eq. [3], the parameter α is found to be given by

$$\alpha = \nu r b^2 e \varrho / \nu_D \qquad [11]$$

where, as before (6)

$$\varrho = l^* n^* \cdot \Delta u$$

and represents the density of all dislocations associated with barriers within the spectrum Δu, excluding those "frozen-in".

It is readily seen that for any given t the integrand in eq. [9] will be effectively equal to zero for all u-values for which z, as defined by eq. [7], is greater than about 1, while for significantly smaller z-values, the integrand will be close to 1. It is therefore possible to regard the intergrand as a unit step-function which will "switch-in" only small z-values, being zero for all others. The integration is thus automatically confined to values of z for which, to a good approximation, it is equal to $\ln z$. On using the limits, u_1 and u_2, remembering that in the integration t is invariant, the integral is found to be equal to $\Delta u/kT$. The term in "square" brackets in eq. [9] is therefore approximately equal to 1, so that

$$\dot{\gamma} = \frac{\alpha}{t + t_0}\,. \qquad [12]$$

On integrating eq. [12] with respect to time, it yields the logarithmic, low-temperature, creep law.

The "theoretical" value of $\dot{\gamma}$ will underestimate actual one because the model does not take into account "zipper" effects, i.e. the cooperative release by a moving dislocation segment of others, either adjoining it on the same dislocation or lying on close-by slip planes. As eq. [12] predicts the shape of the creep curve correctly, such effects could be allowed for by a "multiplication" factor.

The stress-dependence of α, which may result from that of r, and ϱ in eq. [11], is difficult to assess but, at low temperatures when Δu is relatively narrow, the simple single-barrier models of stress relaxation (9) and the associated logarithmic creep (10) suggests that α should increase approximately linearly with stress. Such a stress-dependence of α is in fact observed (7, 10), and for the present purpose it will suffice to take α to be proportional to the constant shear stress, i.e.

$$\alpha = \alpha_0 \tau_0\,, \qquad \alpha_0 = \text{const.} \qquad [13]$$

4. Generation of dislocations

The energy-barrier distribution, n, has so far been regarded as function of u and t only. We now consider the possibility of imposed changes in the dislocation density ϱ explicitly, so that $n = n(\varrho, u, t)$. The material will be assumed to contain a dislocation substructure, including dislocation sources. Formally allowance will be made only for an increase of ϱ with strain, characteristic of the "athermal" linear stage of work-hardening, often referred to as "stage II". Loss of dislocations through mutual destruction would have to be considered at high stress levels.

Addition of dislocations through operation of sources will be assumed to enhance the barrier density, n, without however affecting the functional form of the n-distribution. The significance of this invariance may be as follows.

Plastic flow in the material will be favoured in regions where the deviatoric stresses are high; these will therefore become relaxed with a concomitant, relatively large, reduction in the locally stored "elastic" energy. The ergodicity of the functional form of the n-spectrum may therefore reflect a tendency of preference for such dislocation displacements and source activitions as will maintain the elastic potential near a

36*

minimum compatible with the imposed plastic deformation. Dipole formation and polygonisation are examples of displacements of this kind, frequently observed.

If the normalised n-spectrum is to remain invariant during a small adiabatic increase in ϱ, then the fractional increase of all values of n must be the same; if for example, all n-values doubled, the normalised distribution would remain unaffected. This proportionality requirement can be expressed by writing

$$\frac{\partial n}{\partial \varrho} = \frac{n}{\varrho_0} \qquad [14]$$

where ϱ_0 is constant.

A source term $(\partial n/\partial t)_e$ has now to be added to the right side of eq. [1]; the subscript "u" denotes that u is held invariant. Writing

$$\left(\frac{\partial n}{\partial t}\right)_u = \frac{\partial n}{\partial \varrho} \cdot \frac{d\varrho}{dt} \qquad [15]$$

one has, using eq. [14]

$$\left(\frac{\partial n}{dt}\right)_u = \frac{n}{\varrho_0} \cdot \frac{d\varrho}{dt}. \qquad [16]$$

The density of dislocations participating in the deformation is in general not readily determinable, and we shall therefore eliminate it by using the proportionality between ϱ and the square of the applied stress, τ^2, which is widely accepted (11 to 13). It should be noted nevertheless, that the problem of the validity of this relation is not entirely resolved, firstly because in the literature ϱ generally refers to all dislocations, including the "frozen-in" ones we omit and, secondly, because deviations from the parabolic relation appear to occur. Thus, for example, *Mughrabi* (14) found that in copper single crystals deformed at room temperature the density of dislocations in the primary slip planes increased more nearly linearly with stress, only that of the less numerous, secondary, ones increased approximately in proportion to the square of the flow stress.

With the assumed parabolic relation one obtains, instead of eq. [16],

$$\left(\frac{\partial n}{\partial t}\right)_u = \frac{n}{\tau_{00}^2} \frac{d\tau^2}{dt} \qquad [17]$$

where τ_{00} is constant. Further, the material will be assumed to be deformed so that the rate of increase of ϱ is constant; this implies that

$$\frac{d\tau^2}{dt} = \frac{\tau_0^2}{\beta_0} \qquad [18]$$

or

$$\tau^2 = \tau_0^2 \left(1 + \frac{t}{\beta_0}\right) \qquad [19]$$

where β_0 is a rate constant. However, it should be noted that with the stress, rather than the square of the stress used in eqs. [17] and [19], the conclusions to be drawn from the present discussion would remain unaffected. In eq. [19] τ_0 can be seen to represent the initial flow-stress of the material in shear. Substituting from eq. [18] into eq. [17] yields

$$\left(\frac{\partial n}{\partial t}\right)_u = \frac{n}{\beta}, \qquad \beta = \beta_0(\tau_{00}/\tau_0)^2. \qquad [20]$$

On adding this source term to the right side of eq. [1], and then using the transformation

$$n'(u, t) = n(u, t) \cdot \exp(-t/\beta) \qquad [21]$$

one finds that

$$\frac{\partial n'}{\partial t} = D \frac{\partial^2 [n' \exp(-u/kT)]}{\partial u^2} \qquad [22]$$

which has the same functional form as eq. [1]. In fact, if $\beta \to \infty$, corresponding to a creep test at the constant stress τ_0, eq. [22] reduces to eq. [1], as required.

Inferences relating to creep, based on eq. [22] thus correspond to those outlined before (6) in connection with eq. [1]. We shall therefore restrict considerations to the bearing of eq. [22] on work-hardening.

5. Work-hardening

The solution of eq. [22] of relevance here should lead to the relation for logarithmic creep, discussed above, as $\beta \to \infty$. Now, a development parallel to that used in dealing with logarithmic creep can be followed through with eq. [22] instead of with eq. [1]. It is readily confirmed that the strain rate is then

$$\dot{\gamma}' = \dot{\gamma} \cdot \exp(t/\beta) \qquad [23]$$

with $\dot{\gamma}$ defined by eq. [12]. The coefficient of work-hardening, obtained by dividing the stress rate $\dot{\tau}$ by the strain rate given by eq. [23] is, using eq. [18] to obtain $\dot{\tau}$, and utilising eq. [13],

$$h = \frac{t_0}{2\alpha_0 \beta} [1 + \psi(t)] \qquad [24]$$

where

$$1 + \psi(t) = \left[\frac{1 + \dfrac{t}{t_0}}{\left(1 + \dfrac{t}{\beta}\right)^{1/2}} \right] \exp(-t/\beta). \qquad [25]$$

Now, if one assumes that the deformation occurs at the "usual" rates, e. g. the stress increases from its initial value τ_0, say, to $2\tau_0$ in about a minute, then eq. [19] requires $\beta = 20$ sec. Futher, on taking the initial creep rate $\dot{\gamma}(0)$ in eq. [12] to be typically equal to 10^{-3} sec^{-1} at a shear stress level τ_0 of about $10^{-4} G$, where G is the shear modulus, eqs. [12] and [13] yield $\alpha/t_0 \approx 10/G$. With these values the ratio $t_0/2\alpha_0\beta$, which we shall denote by h_0, is

$h_0 \approx G/400$.

Further, taking t_0 and β in eq. [25] to be of the same order of magnitude, also expanding the exponential term, one sees that the stress/strain curve will have a linear initial part, the extent of which will depend on the actual magnitudes of t_0 and β; the slope should be equal to $f \cdot G$, with $10^{-2} > f > 10^3$.

6. Conclusions

The pivotal eq. [22] based on the stochastic model of creep (6) leads to a satisfactory interpretation of the characteristic linear work-hardening observed in single crystals and well-annealed polycrystalline (15) materials. The value of h_0 is of the right order of magnitude for many types of crystalline materials. In eq. [24] t_0 lies within the spectrum given by eq. [4] and, like t^* (eqs. [3] and [5]) would not be expected to be significantly temperature dependent. A pronounced dependence of t_0 on temperature is also not indicated by eq. [11]. The observed weak dependence of the coefficient of work-hardening in stage II on temperature is therefore in accord with the model; as pointed out by *Mughrabi* (14), this weak temperature-dependence of h_0 has not hitherto been explained satisfactorily.

Although specific micro-kinetics of the deformation process are not explicitly represented in the dislocation-generation relation (14), the general validity of the linear work-hardening law deduced with its aid supports the hypothesis embodied in eq. [14], that the energy-barrier spectrum involved in dislocation movement tends to maintain a stationary form in the second stage of work hardening.

Summary

The creep behaviour of many crystalline solids is found to be phenomenologically similar if comparisons are made at about the same fraction of the melting temperature (°K). These similarities appear to originate from the stochastic character of slip, which is controlled by a distribution of energy barriers to dislocation motion. Functional forms of such distribution are derived from an evolutionary *Markovian* model in which slip units diffuse in the heterogenous internal-stress field of the material. Such a "dislocation drift model", previously developed in relation to creep by the author (Physica Status Solidi, 1968, **30**, 135), is extended to encompass work-hardening as well, i.e. when the applied stress is allowed to increase. The assumption that the functional form of the energy-barrier distribution tends to remain ergodically invariant as the dislocation density increases through straining, is shown to imply linear work-hardening, such as is commonly observed in single crystals and well-annealed polycrystals of metals, inorganic salts and covalent semi-conductor materials. The magnitude of the coefficient of work-hardening, and its relatively weak temperature-dependence, are explained by the model. It seems that the invariance of form of the energy barrier spectrum is a consequence of the occurrence, at any time, of a preference for modes of slip, compatible with the imposed macro-deformation, which will tend to minimise the "elastic" energy stored in the material, as the dislocation density increases.

References

1) *Barenblatt, G. I.* and *V. A. Gorodsov*, J. Appl. Math. Mech. **28**, 397 (1964).
2) *Strunin, B. M.*, Fiz. Tverd. Tela **9**, 805 (1967).
3) *Heinrich, R.* and *W. Pompe*, Phys. Stat. Sol. **40**, 523 (1970).
4) *Mott, N. F.*, Phil. Mag. **43**, 742 (1953).
5) *Täubert, P.*, Abh. dtsch. Akad. Wiss. Berlin, Kl. Math. Phys., Tech. No. **7**, 1 (1958).
6) *Feltham, P.*, Phys. Stat. Sol. **30**, 135 (1968).
7) *Wyatt, O. H.*, Proc. Phys. Soc. **B 66**, 459 (1953).
8) *Conrad, H., R. Armstrong, H. Wiedersich*, and *G. Schoek*, Phil. Mag. **6**, 177 (1961).
9) *Feltham, P., G. Lehmann*, and *R. Moisel*, Acta Met. **17**, 1305 (1969).
10) *Krausz, A. S.* and *G. B. Craig*, Acta Met. **14**, 1807 (1966).
11) *Mott, N. F.*, Phil. Mag. **43**, 1151 (1952).
12) *Kuhlmann-Wilsdorf, D.*, Metal Trans. **1**, 3173 (1970).
13) *Feltham, P.* and *G. Chaudhri*, Phys. Stat. Sol. **7a**, K 59 (1971).
14) *Mughrabi, H.*, Phil. Mag. **23**, 897 (1971).
15) *Feltham, P.* and *J. D. Meakin*, Phil. Mag. **2**, **1** (1957).

Author's address:
P. Feltham
Brunel University
Uxbridge, London (U. K.)

Rheol. Acta **13**, 562–566 (1974)

Laboratoire de Glaciologie du C. N. R. S. Grenoble (France)

Fluage de la glace polycristalline

Par P. Duval

Avec 6 figures et 1 tableau

(Reçu p. p. le 27 octobre 1972)

Introduction

Pour calculer l'écoulement des glaciers, il est important de connaître la loi de déformation de la glace polycrystalline c'est-à-dire la loi reliant la matrice des vitesses de déformation et celle du déviateur des contraintes. On admet en Glaciologie que ces deux matrices possèdent les mêmes éléments de symétrie. On néglige donc l'élasticité; on suppose que seules les déformations infinitésimales interviennent et de plus que la glace polycristalline est un matériau isotrope. Ainsi la viscosité η est un scalaire défini par la relation:

$\sigma' ij = 2\eta \dot{\varepsilon} ij$ ($\sigma' ij$ est la matrice du déviateur des contraintes, $\dot{\varepsilon} ij$ la matrice des vitesses de déformation).

D'autre part, en admettant le critère de *Von Mises*, qui implique que seul le deuxième invariant du déviateur des contraintes intervient, on arrive à la relation: $\tau = \eta \dot{\gamma}$ où τ est la cission efficace définie par l'équation $\tau^2 = -I'_2$ et $\dot{\gamma}$ la vitesse de cisaillement efficace définie par l'équation $\dot{\gamma}^2 = -4 J'_2$. I'_2 et J'_2 sont respectivement le deuxième invariant du déviateur des contraintes et le deuxième invariant du tenseur des vitesses de déformation.

La glace, comme la plupart des matériaux polycristallins soumis à une contrainte donnée, présente d'abord un fluage transitoire qui diminue avec le temps; puis un fluage secondaire ou permanent à vitesse de déformation constante; enfin un fluage tertiaire à vitesse de déformation plus élevée contrôlé en partie par les processus de recristallisation.

Nous avons, tout d'abord, cherché par des expériences de fluage en cisaillement pur à déterminer l'influence de la contrainte sur la vitesse de déformation secondaire de glaces naturelles et artificielles. Ces expériences ayant été réalisées à -1 et à -5 °C, il a été possible de calculer une énergie d'activation apparente pour ces différentes glaces. D'autre part, sachant l'importance des processus de recristallisation dans l'écoulement des glaciers, quelques expériences de longue durée ont été réalisées à -1 °C; elles ont permis de mettre en évidence plusieurs types de recristallisation dynamique et en particulier des processus de recristallisation périodique pour des valeurs de la contrainte appliquée voisines de 2 kgf/cm² (1).

Conditions expérimentales

L'appareillage construit nous permet de réaliser des expériences de fluage en cisaillement pur pour des températures supérieures à -13 °C. L'échantillon, en forme de cylindre évidé (55 mm $< \varphi_{ext} <$ 90 mm, $\varphi_{int} \sim$ 30 mm, hauteur $\sim$ 100 mm), est soumis à un couple de torsion axial dont le moment a une valeur constante. La déformation est déterminée grâce à la mesure de l'angle $d\alpha$ dont tournent deux sections transversales voisines. L'utilisation d'un capteur de rotation nous permet d'enregistrer d'une façon continue cette déformation. Un système multiplicateur permet de détecter des variations d'angle de 0,2 min. Les expériences sont réalisées en chambre froide dont la température est maintenue à -15 °C. Un système de régulation propre à l'appareil maintient la température de l'échantillon à une valeur fixe, comprise entre 0 et -13 °C. L'écart maximum de température au cours d'une expérience est de 0,2 °C. Les valeurs de la contrainte appliquée calculées à partir du moment du couple s'échelonnent entre 0,5 et 10 kgf/cm².

Nature des échantillons

Des glaces de différentes natures ont été testées:

1. Des glaces macroscopiquement isotropes préparées à partir de neige et d'eau distillée.
2. Des glaces naturelles en provenance d'une part des Alpes Françaises (Glacier de St-Sorlin, Vallée Blanche) et d'autre part de Terre Adélie (Forage G 1) (2).

La section des grains de ces différentes glaces varie entre 2 et 80 mm².

Résultats

A. Fluage transitoire et secondaire

Le fluage secondaire à vitesse de déformation constante est rarement obtenu en laboratoire; en effet pour de fortes valeurs de la contrainte, le fluage tertiaire apparait pour des déformations inférieures à 1% c'est-à-dire avant la fin du fluage transitoire (figs. 1 A, 1 B); et pour de faibles valeurs de la contrainte, seul le fluage transitoire existe pour des expériences de durée limitée (fig. 1 C). Aussi pour déterminer les vraies valeurs des vitesses de déformation secondaire, nous avons vérifié que le fluage transitoire est un fluage d'*Andrade* (fig. 2) c'est-à-dire que les courbes de fluage obéissent à l'équation

$$\gamma = \gamma_0 + B t^{1/3} + A t$$

où γ_0 est la déformation élastique, B et A, des constantes, t, le temps.

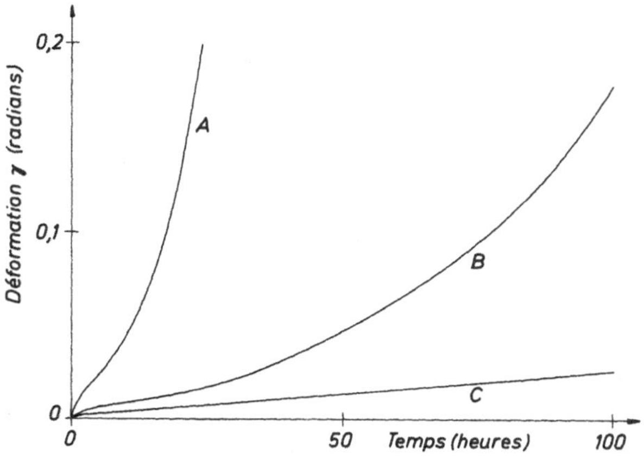

Fig. 1. Courbes de fluage $\gamma = f(t)$. Glace artificielle. Température $T = -1 \pm 0{,}1$ °C. A. $\tau = 5$ kgf/cm²; B. $\tau = 2{,}9$ kgf/cm²; C. $\tau = 2{,}1$ kgf/cm²

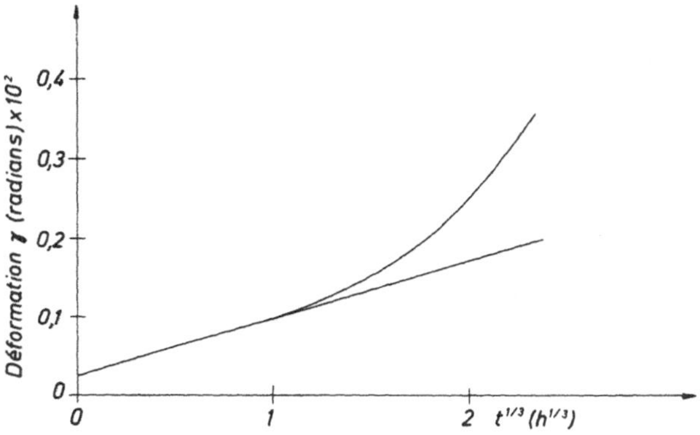

Fig. 2. Courbe de fluage $\gamma = f(t^{1/3})$. Glace artificielle. Température $T = -1 \pm 0{,}1$ °C. $\tau = 2{,}1$ kgf/cm²

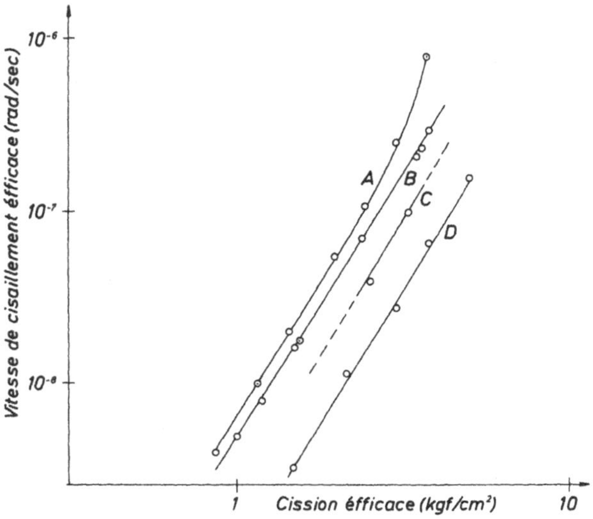

Fig. 3. Vitesse de cisaillement en fonction de la cission efficace. Température $T = -1 \pm 0{,}1$ °C. A. Glace artificielle; B. Glace naturelle (Terre Adélie, Forage G1, profondeur 72 m); C. Glace naturelle (Glacier de St-Sorlin, profondeur 30 m); D. Glace naturelle (Vallée Blanche, profondeur 44 et 52 m)

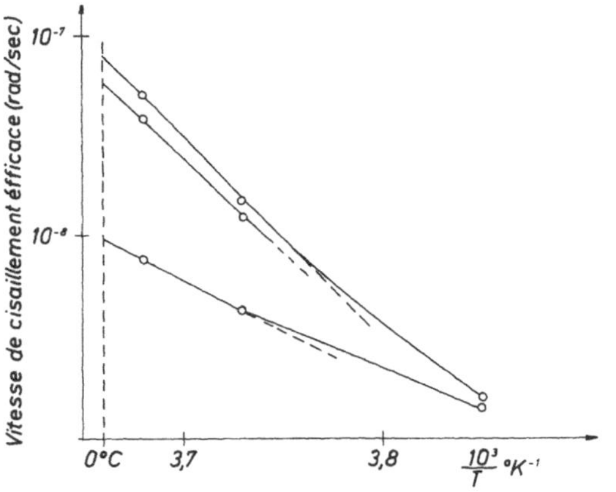

Fig. 4. Vitesse de cisaillement en fonction de $1/T$. $\tau = 1{,}95$ kgf/cm². A. Glace artificielle; B. Glace naturelle (Terre Adélie, Forage G1, profondeur 72 m); C. Glace naturelle (Vallée Blanche, profondeur 44 et 52 m)

Glen (3) et *Tabor* (4) avaient déjà utilisé ce résultat pour déterminer les vitesses de déformation secondaire.

Les différents résultats sont présentés sur la fig. 3 et sur le tableau ci-contre. La loi de fluage est une loi de *Glen* avec un exposant n égal à $3 \pm 0{,}1$ c'est-à-dire $\dot{\gamma} = B\tau^3$ pour des valeurs de la contrainte variant entre 0,8 kgf/cm² et 5 kgf/cm².

Tableau 1

Type de glace	Section des grains mm²	$n \pm 0{,}1$	rad	*B* bar^{-n}	an^{-1}	Energie d'activation kcal/mole
			—5 °C	—1 °C	0 °C extrapolé	
Artificielle isotropes	2 à 10	3,12	0,06	0,2	0,35	46
Antarctique Forage GI (8)						
Profondeur: 72 m	20 à 80	3,02	0,056	0,16	0,25	40
Glacier de St Sorlin						
Profondeur: 30 m	10 à 50	–	0,038	0,075	0,1	28
Vallée Blanche						
Profondeur: 44 m et 52 m	10 à 50	3,1	0,025	0,048	0,059	26

Le coefficient B a été déterminé à -1 et $-5\,°C$ ce qui nous a donné une énergie d'activation apparente pour les différentes glaces étudiées et ainsi une valeur approchée de ce coefficient à $0\,°C$ qui varie entre 0,36 et 0,059 rad bar^{-3} an^{-1} suivant l'origine des échantillons. Notons qu'entre -12 et $0\,°C$ l'énergie d'activation apparente n'est pas constante mais augmente avec la température (fig. 4). Aussi les valeurs citées ci-dessus sont elles minimales.

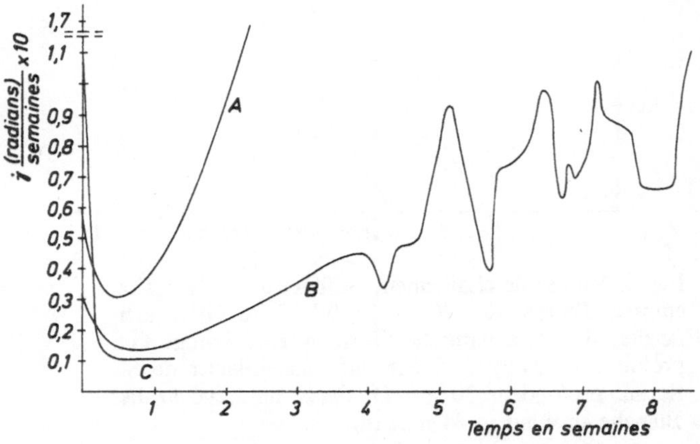

Fig. 5. Courbes de fluage $\dot{\gamma} = f(t)$. Glace isotrope: A. $\tau = 3{,}5\,kgf/cm^2$; B. $\tau = 2{,}7\,kgf/cm^2$. Glace anisotrope: C. $\tau = 2{,}25\,kgf/cm^2$. Température: $T = -1 \pm 0{,}1\,°C$

B. *Fluage tertiaire*

Les courbes de fluage obtenues diffèrent fortement avec la valeur de la contrainte appliquée et la texture initiale de la glace.

– Pour $\tau = 3{,}5\,kgf/cm^2$, partant d'une glace isotrope, on a successivement le fluage transitoire et le fluage tertiaire où la vitesse de cisaillement croît d'une façon continue (fig. 5A).

La position des axes optiques, après déformation, est reproduite sur un réseau de *Schmidt* (fig. 6A); on s'aperçoit qu'une certaine proportion de cristaux a pris une orientation favorisant la déformation.

– Pour $\tau = 2{,}7\,kgf/cm^2$, toujours pour une glace isotrope, il apparait au cours du fluage tertiaire des oscillations de la vitesse de cisaillement diminuant légèrement en amplitude au cours du temps (fig. 6B); la déformation, entre ces oscillations successives, varie entre 7 et 9%. La structure finale n'a pu être étudiée.

– Parallèlement, pour $\tau = 2{,}25\,kgf/cm^2$, partant d'une glace où la plupart des cristaux étaient

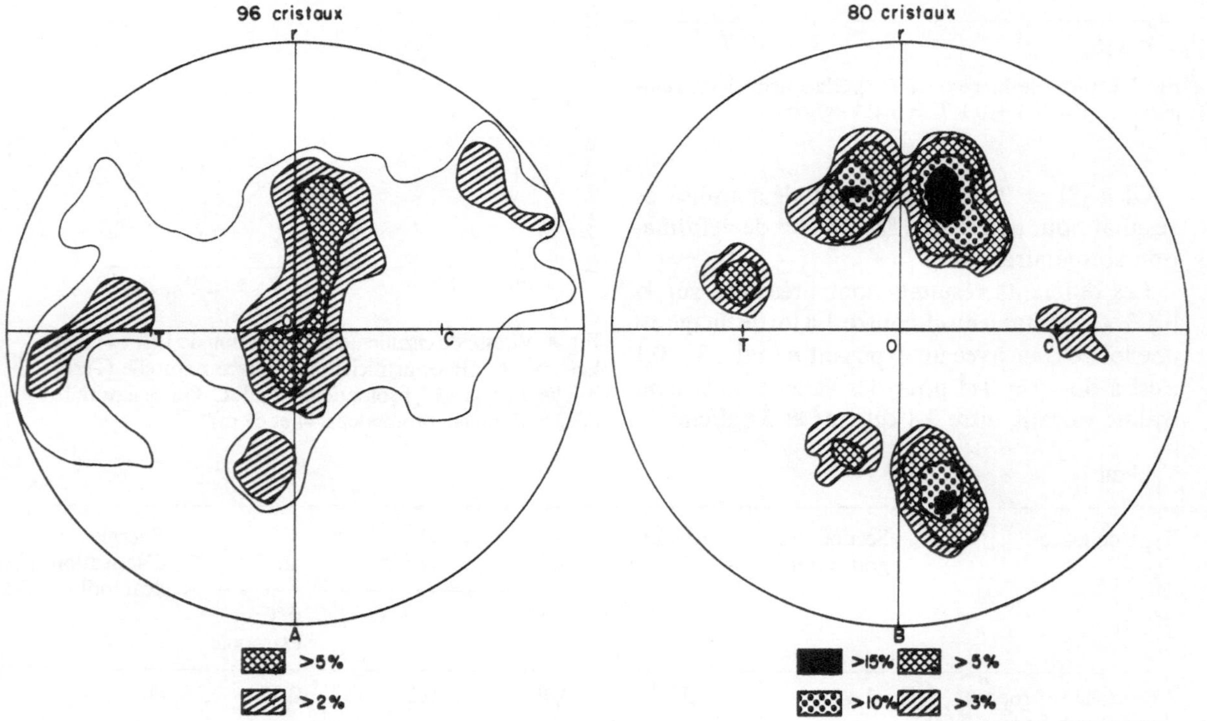

Fig. 6. Orientation des axes optiques projetés sur un réseau de *Schmidt*. Courbes d'égale densité dans 1% de l'aire. O est le pôle du plan de cisaillement permanent; t(traction), c(compression), r, sont les axes des contraintes principales. A. Glace isotrope déformée pour $\tau = 3{,}5\,kgf/cm^2$; B. Glace anisotrope déformée pour $\tau = 2{,}25\,kgf/cm^2$

orientés pour faciliter la déformation, la recristallisation a entrainé une diminution continue de la vitesse de cisaillement jusqu'à l'installation après quelques jours, d'un fluage à vitessse de déformation constante (fig. 5C). L'orientation des cristaux est devenue très défavorable à la déformation (fig. 6B).

Discussion

A.

L'étude du fluage secondaire a permis de confirmer la validité de la loi de *Glen*; en particulier il est important de noter que cette loi reste valable pour des valeurs de la contrainte de l'ordre de 1 kgf/cm^2, voisines de celles qui interviennent dans l'écoulement des glaciers tempérés. Comme l'a noté *Tabor* (4) l'exposant *n* augmente avec la contrainte pour des expériences de courte durée où le fluage transitoire n'est pas éliminé; les vitesses de déformation trouvées pour les faibles valeurs de la contrainte étant alors surestimées. Récemment *Weertman* (5) a discuté des processus de déformation de la glace qui pourraient expliquer les différents résultats expérimentaux obtenus et en particulier la valeur de l'exposant *n* de la loi de *Glen* égal à 3. En fait seule l'étude des dislocations (densité et vitesse des dislocations pendant la déformation; formation éventuelle de sous-grains etc.) permettra de confirmer telle ou telle théorie. L'augmentation de l'énergie d'activation au-dessus de −10 °C doit être liée à la présence d'une phase liquide aux joints de grains qui d'une part favorise le glissement intergranulaire et d'autre part atténue les contraintes internes qui apparaissent principalement aux jonctions de plusieurs grains; la présence d'impuretés solubles concentrées aux joints de grains expliquerait la présence de cette phase liquide. D'autre part cette énergie d'activation apparente varie avec la nature des glaces étudiées; on peut supposer que ces glaces n'ont pas aux joints de grains la même teneur en impuretés solubles. En effet *Kuroiwa* (6) a montré que le frottement interne dû aux joints de grains dépendait fortement de la quantité d'impuretés solubles présentes et il attribue au glissement intergranulaire ce mécanisme de relaxation avec une énergie d'activation apparente variant entre 70 kcal/mole pour des glaces artificielles très pures et 36 kcal/mole pour des glaces de glacier.

Si les impuretés solubles présentes aux joints de grains sont à l'origine de l'augmentation de l'énergie d'activation au-dessus de −10 °C, les impuretés intracristallines pourraient aussi modifier la viscosité de la glace polycristalline. Incluses substitutionnellement dans le réseau de la glace, elles pourraient faciliter la déformation (7); au contraire sous forme d'agglomérats, elles pourraient, comme dans les métaux, s'opposer aux mouvements des dislocations et ainsi durcir la glace. Il est à noter que les glaces de glaciers polaires ont généralement beaucoup plus d'impuretés solubles intracristallines que les glaces de glaciers tempérés (6) même si la teneur globale en impuretés de ces différentes glaces est comparable; ces glaces polaires ont une viscosité plus faible, comme nous l'avons montré dans ces expériences, que l'on peut expliquer en partie par une texture favorable (8). Parallèlement les processus de fonte et regel qui contrôlent, sous certaines conditions, le glissement des glaciers tempérés sont modifiés aussi par la présence de ces impuretés solubles; ce problème a été analysé récemment par *Lliboutry* (9).

B.

Pour les processus de recristallisation dynamique intervenant au cours du fluage tertiaire, on peut proposer deux interprétations:

1. Ces résultats sont à comparer avec ceux trouvés pour certains métaux tel le nickel par *Hardwick* (10). La recristallisation débute pour une déformation critique ε_c; de nouveaux grains sont créés aux joints de grains par suite de contraintes locales importantes tandis que le reste du matériau continue à s'écrouir. Si une large proportion de ce matériau a recristallisé pour une déformation $\varepsilon_c + \Delta\varepsilon < 2\varepsilon_c$, on aura successivement un stade de recristallisation où la vitesse de déformation augmentera et un stade d'écrouissage où cette vitesse diminuera. Au contraire, si $\varepsilon_c + \Delta\varepsilon > 2\varepsilon_c$, on aura un processus de recristallisation continue. L'existence d'une contrainte critique séparant les stades de recristallisation périodique et continue a été vérifiée pour le nickel et le cuivre par *Luton* (11).

2. L'orientation des cristaux, avons nous vu, change au cours du fluage tertiaire. On pourrait de même attribuer l'augmentation de la vitesse de déformation au cours du fluage tertiaire par l'apparition d'une texture favorable et la diminution de cette vitesse par la destruction de cette texture. Des expériences sont en cours pour trancher entre ces deux explications.

Il est à noter que dans un glacier tempéré c'est principalement ce fluage contrôlé par le processus de recristallisation qui doit intervenir et les textures observées sont à comparer à celles obtenues dans ces expériences (12).

Summary

Creep tests have been performed at temperatures above $-10\,°C$. The secondary creep behaviour can be described by a relation of the type $\dot{\gamma} = B\tau^n$ where $\dot{\gamma}$ is the effective shear strain rate, τ the effective shear stress and B and n are constants.

n has a value close to 3 for stresses ranging from 0,8 kgf/cm^2 to 5 kgf/cm^2. Oscillatory creep rates have been observed during the tertiary creep, resulting from periodic processes of recrystallisation.

Littérature

1) *Duval, P.*, C. R. Ac. Sci., Série D, **275**, 337 (1972).

2) *Lorius, C., G. Ricou* et *A. Rosello*, Soc. Hyd. Fr. Communication du 24 février 1967.

3) *Glen, J. W.*, Proc. Roy. Soc. **A 228**, 519 (1955).

4) *Tabor, D.* and *J. C. F. Walker*, Nature **228**, n° 5267, 137 (1970).

5) *Weertmann, J.*, International Symposium on the physics and chemistry of ice, Ottawa, Canada, 14–18 août 1972. Royal Soc. of Canada (1973).

6) *Kuroiwa, D.*, Contrib. Inst. Low. Temp. Science, Hokkaldo Univ. (Sapporo), Série A, n° 18, 1 (1964).

7) *Jones, S. J.* and *J. W. Glen*, Phil. Mag. **19**, 13 (1969).

8) *Lorius, C.* et *M. Vallon*, C. R. Ac. Sci., Série D, **265**, 315 (1967).

9) *Lliboutry, L.*, J. Glaciology **10**, n° 58, 15 (1971).

10) *Hardwick, D., C. M. Sellars*, and *W. J. McG. Tegart*, J. Inst. Metals **90**, 21 (1961–62).

11) *Luton, M. J.* and *C. M. Sellars*, Acta Met. **17**, 1033 (1969).

12) *Vallon, M.*, Thèse de Doctorat d'Etat, 2éme partie, Université de Grenoble, Publication du Laboratoire de Glaciologie du C.N.R.S. de Grenoble (1967).

Adresse des auteurs:
P. Duval
Laboratoire de Glaciologie du C.N.R.S.
2, rue Très-Cloîtres
F-38031 Grenoble (France)

Rheol. Acta 13, 567–570 (1974)

From the Lebanese National Council for Scientific Research, Beirut (Libanon)

Rheological properties of bituminous materials

By S. Hraiki

With 3 figures

(Received October 27, 1972)

Higher demands are needed on quality and performance of bituminous surfacing, arising from the continuous increase of the traffic intensity and load effects from the traffic circulations. Not only these influences, but also the resistance against the superficial effects of vehicles as well the climate, the use of salts in winter and pneumatic spikes against the icy roads are of a great importance.

All these effects demand entirely another approach, not only for the design of the bituminous mixtures, execution of the road construction and the dimensions, but it is necessary to take care of the bituminous materials, which form the binding element of the aggregates.

The behaviour of bituminous mixtures in the constructions of the surfacing is conditional not only by aggregates, but in the first place by the properties of bituminous materials. If we consider the construction of bituminous layers from the point of view of stressing, that means the static, dynamic and resonance stressing. If we consider the viscoelastic properties of the bituminous mixtures in the relation to the changes of stressing, it is necessary to go out also from the rheological properties of bituminous materials. Just these properties influence the frequency of the dynamic stressing, which depends on the time, temperature and the velocity, affected by the propagation way of waves. For this reason, it is necessary to observe and investigate carefully the properties of the bituminous materials just exactly from the point of view of these stresses to which bituminous materials are subjected in the mixture.

The asphalt is as a highmolecular element of the colloid structure, made by micelles in the oil liquid environment, is also a thermoplastic element. This means that its properties are influenced by the temperature.

It is a characteristic property of bituminous binders, that the viscosity falls very rapidly as the temperature rises. Binders differ in the extent to which their viscosity is affected by temperature. It is important to know that the viscosity of a given binder is suitable at the temperature used in manufacture and at the temperature encountered on the road.

If a bitumen of 300 penetration is taken and its viscosity is determined over a range of temperature around 25 °C, a curve of the type shown in the figure is obtained and it shows that the change of the viscosity with temperature is greater at low temperature than at high temperatures.

During the time when the higher demands on the construction of road were not layed, the

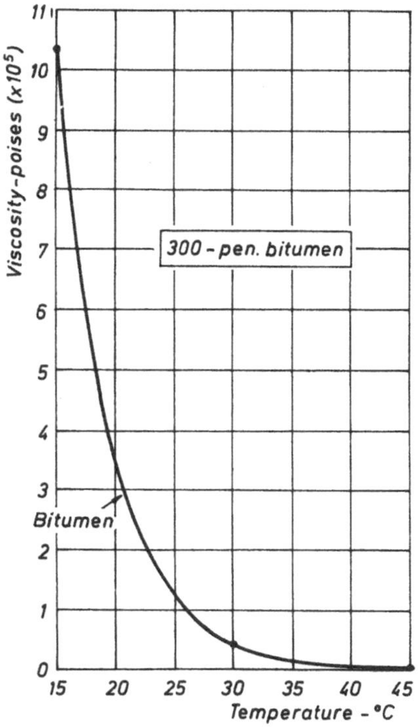

Fig. 1. Viscosity-temperature relation for bitumen

classical methods of investigation of bitumens were sufficient for us from the practical point of view. This routine analysis characterised rather the quality that the propoerties of bitumen. They include the determination of softening point, penetration test, breakness test and ductility.

By changing the initial raw materials bases, higher demands on the quality of bitumen arise. It was necessary to use the bitumens of semi-asphaltric crude oil and to show that the classical existing experimental methods do not express closely enough the rheological properties of these bitumens and its behaviour in the mixture with aggregates. It is still necessary to characterise these bitumens by the physical values for a better appreciation of the real stressing in the road construction.

The breakness and softening point characterise approximately the region of the behaviour of bitumens, so that below the breakness point the bitumens obtain rigid and brittle consistency. Between these two points, there is a large zone of viscoelastic and plastic behaviour. Beyond the softening point, the bitumen behaves as a liquid with the decreasing viscosity at the arising temperature. From this interval of temperatures, we are interested in accordance with the climatic conditions in that from −20 till 70 °C.

Road specialists are interested in using the bitumen that has the corresponding rheological properties at the limit of these temperatures. Unfortunately, the bitumens resist to high temperatures and have a low breaking point. The bitumen with the high breaking point is less resistant to higher temperatures.

It is necessary to perform the investigation of these changes related to temperature and time. Determining the viscosity, in which its changes are possible to observe in convenient way, it is possible to calculate the results and estimate it grafically to these conditions, which correspond better.

Asphaltic bitumens are distinguished as regards their rheological properties in two respects, viz. by their character and their hardness. Differences in character mainly consist in deviations from *Newton*ian flow. Bitumens express and display themselves frequently, have conspicuous thixotropy properties and flow of asphalt is linear and non-linear. Differences in hardness can be indicated by the value of the viscosity.

The mechanical properties of bitumen are of the first importance in all its applications. The relation of the deformation to load and time is determined by the rheological, that means the flow properties of the bitumen.

Elastic modulus of bitumen can be determined by the deformation sustained under a uni-directional stress depending on the time of application of the stress and the instantaneous elastic deformation is accompanied by the much larger time-dependent deformation. The rheological behaviour of many viscoelastic bodies may be represented by complex rheological models with numerous *Kelvin* or *Maxwell* groups, respectively. For such a model, we can derive the differential and integral relations between stress and strain. There is a close connection between the differential equations and the rheological models. Each equation may be represented by a model with real and positive rheological parameters.

For an interpretation of the rheological properties of high polymers the use is often made of mechanical models consisting of elastic and viscous elements. The *Lethersich* type of representation of the character of bitumen is a model consisting of a *Newton-Kelvin* elements in series. *Jeffreys*-body is a model for bitumen consisting of *Maxwell-Newton* elements in parallel (see the following figures).

Lethersich mechanical model can be described by the equation of a deformation γ under stress τ.

$$\dot{\gamma} = \tau \frac{\eta + \eta_s}{\eta \eta_s} + \frac{\dot{\tau}}{\mu} \qquad [1]$$

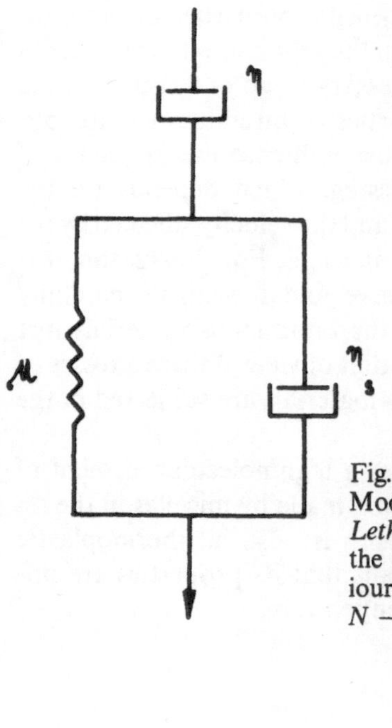

Fig. 2. *Lethersich*-body. Models proposed by *Lethersich* to represent the rheological behaviour of bitumen $N - K = N - (H|N)$

where

$\dot{\gamma} = \frac{d\gamma}{dt}$	Shear strain rate
τ	Shear stress
$\dot{\tau} = \frac{d\tau}{dt}$	Shear stress rate
μ	Shear modulus
η_s	Coefficient of shear viscosity
η	Coefficient of viscous element

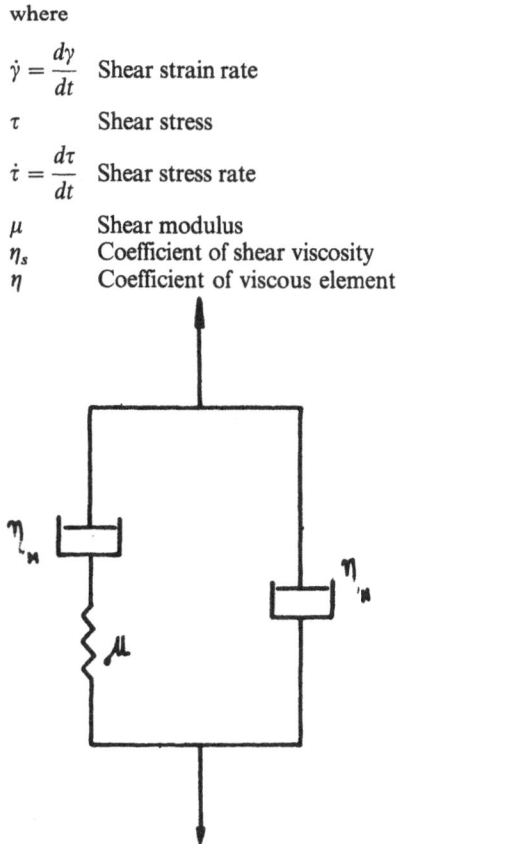

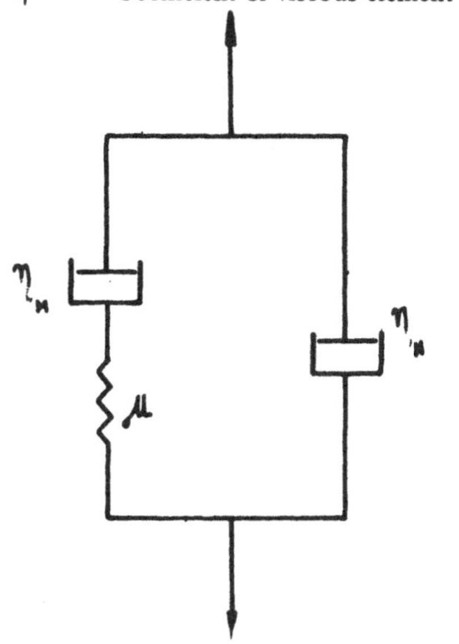

Fig. 3. *Jeffreys*-body. Models proposed by *Jeffreys* to represent the rheological behaviour of bitumen
$M \mid N = (N - H) \mid N$

Jeffreys mechanical model can be described by the following equation:

$$\tau = \tau_M + \tau_N$$
$$= e^{-\frac{\mu_M t}{\eta_M}} (\tau_{0M} + \mu_M \int_{t_0}^{t} \dot{\gamma} e^{\frac{\mu_M t}{\eta_M}} dt) + \eta_N \dot{\gamma} \qquad [2]$$

where

τ	Shear stress
τ_{0M}	Initial shear stress in *Maxwell* group
$\dot{\gamma} = \frac{d\gamma}{dt}$	Shear strain rate

We compare further the behaviour of both bodies under constant stress $\tau = \tau_0$, applied upon the unstrained and unstressed body. These formulas may represent closely enough the rheological behaviour of bitumen. The linear rheological equation of viscoelastic behaviour may be also expressed by the relation between deformation and stress having the following form:

$$a\gamma + b\dot{\gamma} = c\tau + d\dot{\tau}. \qquad [3]$$

Our ultimate aim is to modify the properties of bitumens for road making or for other purposes, with our attempt to influence the viscoelastic properties, that is to get the sufficient viscoelastic state in the zone of low and high temperatures with the minimum permanent deformation. There exist several methods of obtaining better properties; one of these is the modifying of bitumens with the corresponding polymers. For example cautchuc, butylcautchuc, the corresponding forms polypropylene (the syndiotactic and attactic) and polyisobutylene. It is necessary to take care that the used polymers be not in the crystallic form and should have the corresponding polarity against the bitumen.

The process of modifying by means of polymers increases the resistance of the bituminous surfacing. With the different climatic conditions, the surface is capable of sufficient deflection, also the low and high temperatures do not affect the deformation of layers due to the decreasing of cohesion.

The addition of rubber to a binder invariably alters the flow properties of the binder as measured by the various standard tests, such as penetration, softening point, viscosity, elastic recovery ductility and flow.

To explain the mechanism by which rubber altered the binder, *Van Rooijen* concluded that the grains of rubber absorb some of the oily constituents from the bitumen and are dispersed in the bitumen, which is harder that before. *Mason* and *Smith* were of the opinion that the change in viscous and brittle properties was due partly to molecular dispersion of the rubber in the binder. Road Research Laboratory confirms that an increase in the degree of dispersion produces greater changes in the properties of modified bitumen. The apparent effect of the rubber depends on the test used and first of all on the different heating temperatures. All the flow properties discussed, are dependent not only on the type and quantity of rubber used but also on the heat treatment the mixture has undergone before testing.

To observe the rheological and viscoelastic properties of bitumens and their improvement, as well as the research of the corresponding measurement methods, is very difficult, but it is necessary if we want to evaluate correctly the rheological and viscoelastic properties of bituminous mixtures in the fundamental characteristics of bituminous layers.

The technology of bitumen during the last years had made a great progress and developed in scientific work. The rheological studies have supplied fundamental knowledge necessary to discover the causes of the variation in the properties of the material.

We have to point out clearly certain considerations for future research and better understanding is needed of the requirements for most of the current applications.

Summary

The paper deals with the rheological properties of asphaltic bitumens. The author proposes the use of some rheological models and equations describing the behaviour of bitumens. The rheological considerations complete the classical methods of testing.

Zusammenfassung

Der Aufsatz behandelt die rheologischen Eigenschaften des Bitumens. Der Verfasser schlägt die Benutzung von rheologischen Modellen und Gleichungen vor, welche die rheologischen Prozesse beschreiben. Die rheologischen Erwägungen ergänzen die klassischen Prüfungsmethoden.

References

1) *Sobotka, Z.*, Theory of plasticity and limiting states of engineering structures (in Czech), vol. I (Praha 1954).

2) Road Research Laboratory, Bituminous materials in Road Constr. (London 1962).

3) *Abraham, H.*, Asphalts and allied substances (New York 1961).

4) *Reiner, M.*, Deformation, strain and flow (London 1960).

5) *Sobotka, Z.*, Some problems of non-linear rheology (Praha 1962).

6) *Servit, R.*, Strength and elasticity, vol. II (Praha 1961).

7) *Eirich, F.*, Rheology, Theory and applications (New York 1958).

8) *Doležalova, E.*, Rheological models (Praha 1967).

9) *Sobotka, Z.*, Rheological models of viscoelastic bodies (in Czech) (Bratislava 1968).

10) Principles and applications of rheology.

11) Road tar research committee-viscosity (London 1959).

12) *Zapta, J.* and *W. Doyle*, Viscosity of liquid road asphalts (1942).

13) *Traxler, R.* and *H. Schweyer*, Rheological properties of asphalts (1936).

14) *Lee, A.* and *J. Nicholas*, The properties of asphaltic bitumens in relation to its use in road construction (1957).

15) *Pfeiffer, J.*, The properties of asphaltic bitumen (Amsterdam 1950).

Author's present address:
Ing. *Samir Hraiki*
Radhoštská 5
Praha 3, Vinohrady (CSR)

Rheo . Acta **13**, 571–576 (1974)

From the Procurement Executive, Ministry of Defence, Rocket Propulsion Establishment, Westcott, Aylesbury, Bucks (England)

The effect of pressure on the flow properties of filled polyisobutylene

By H. J. Buswell

With 8 figures and 2 tables

(Received October 27, 1972)

Nomenclature

b	outer radius of sample (25 mm)
l	length of sample (50 mm)
P_1	applied high pressure
P_2	applied hydrostatic pressure
$(P_1 - P_2)$	deforming pressure
τ	shear stress
τ_{max}	maximum shear stress at outer radius of sample

1. Introduction

A polyisobutylene with a viscosity of about 50 kNs/m^2 (5×10^5 poise) at 25 °C and average molecular weight of about 4×10^4 was loaded with between 85–90 percentage by weight of particulate solid. The solid filler was mainly an inorganic crystalline material which on average had a multimodel size distribution in the range of 10–750 μm. The smaller crystals tend to be roughly spherical, whereas the smaller proportion of larger crystals are more irregular in shape. At this high loading one percent of synthetic surfactant was added to the mix in order to achieve complete wetting of the solid particles.

The physical properties of such a highly loaded composite material exhibit marked nonlinear characteristics which are dependent on a number of variables including strain rate, stress level, temperature, time and previous strain history (1). The composite is similar to modelling clay or plasticine, and may be considered to approximate to a plastic material which is subject to yield stress and shearhardening. A flow equation for this material has been derived from parallelplate plastometry (2).

In use this composite is subjected to a high hydrostatic gas pressure which modifies its ability to flow when acted on by shear stresses. This combination of forces experienced by the composite in a geometry similar to that in which it is used has been simulated in a test rig developed at the Rocket Propulsion Establishment. The strengthening effect of a hydrostatic gas pressure on the flow properties of the filled polyisobutylene composite has been investigated in this test rig.

2. Apparatus and sample preparation

The required amount of composite was selected from the bulk supply, cut into small pieces of about 10 g, mixed together and forced through a shreader plate into a vacuum chamber. This procedure expels any occluded air and is intended to mix the composite. The material was then compacted and extruded to form a long plug of the correct diameter for filling. Samples were cut from this plug and trimmed to a fixed weight of 176 g. The test samples were then prepared by pressing these pieces of composite into rigid brass tubes of inner diameter 50.8 mm and length of 101.6 mm. A small hydraulic press was used and a consolidating pressure of 7.0 MPa[1]) was applied for 2 min.

No bonding agent was used as the composite adhered to clean brass with sufficient strength to prevent slipping of the sample under test. Care was taken to standardize the preparation and filling procedure. The set of filled tubes was stored for a minimum period of one week in a humidity cabinet at 0% relative humidity and a temperature of 20 °C. All the experiments were carried out at a constant temperature of 20 °C and in an atmosphere of 0% relative humidity. The experiments were carried out under controlled conditions because the rheological properties change with fluctuations in the water content of the composite (3), and with the temperature (4).

The samples were deformed in the High Presure Test Cell, details of which have been reported elsewhere (5, 6), consequently only features essential to the present work will be described.

The pressurising and recording systems are shown in fig. 1. The plug of composite in the rigid brass tube is deformed by the difference in gas pressures $P_1 - P_2$. The magnitudes of the pressures P_1 and P_2 can be varied independently enabling various shear stresses to be applied to the plug at the same time as any chosen hydrostatic gas pressure. The hydrostatic gas pressure is first applied to the sample and then the deforming pressure difference is adjusted to the chosen value using the needle values. During the adjustment period the sample is isolated from the pressure difference and only experiences the hydrostatic pressure Operating the selector value subjects the plug to the required deforming pressure difference and simultaneously starts a digital time display.

The shear stress acting on the plug at a radius r, is then

$$\tau = \frac{(P_1 - P_2)}{2l} r$$

where l is the length of the plug. The maximum shear stress occurs at the wall, when $r = b$ the radius of the tube, hence

[1]) The *Pascal* (Pa) is defined as 1 *Newton* metre^{-2}.

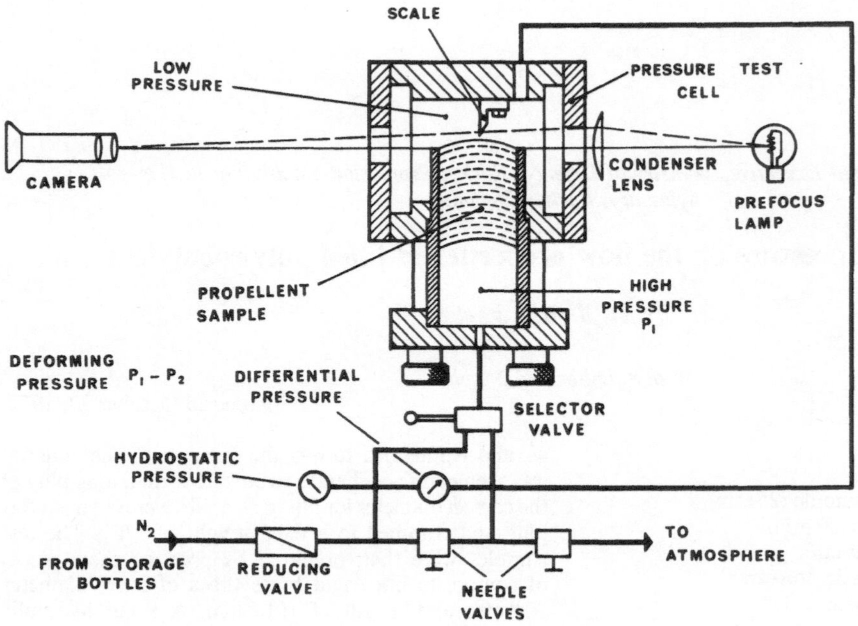

Fig. 1. Pressurising and recording system for high pressure test cell

$$\tau_{max} = \frac{(P_1 - P_2)}{2l} b .$$

The ratio b/l is 0.5 for a standard sample, therefore

$$\tau_{max} = \frac{(P_1 - P_2)}{4}$$

and the maximum shear stress is directly proportional to the pressure differences.

The flow of a stressed sample approximates to that of telescopic shear, and the resulting shape of a typical profile is shown in fig. 2. The small deformations at the edge are produced by radial flow and edge effects and not by slipping of the sample. The amount of composite flow was measured from similar profile photographs taken at known times. The effect on the flow properties of deforming pressure, hydrostatic pressure and time, can be ascertained from these measurements.

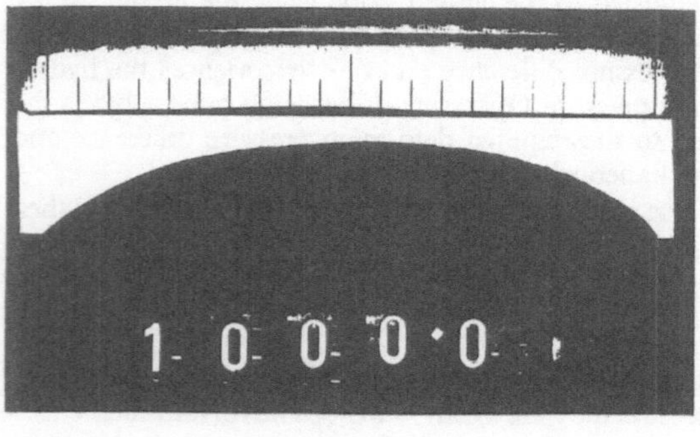

Fig. 2. Typical profile record

3. Results and discussion

Samples were deformed with the pressure P_2 equal to atmospheric pressure, 3.5 and 7.0 MPa (all pressure values given in this paper are gauge values). Also three values of the pressure difference $P_1 - P_2$ were used to give maximum shear stress values of 50, 70, and 90 kPa. The resulting centre deformation of three samples, one at each of the values of shear stress and all deformed at atmospheric pressure, are plotted against time in fig. 3. The dependence of flow on the stress level can be seen and the magnitude deduced as the time to deform to 5 mm changes from 210 to

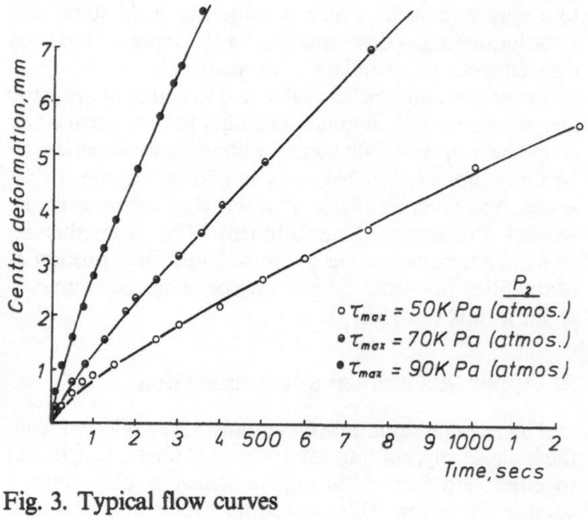

Fig. 3. Typical flow curves

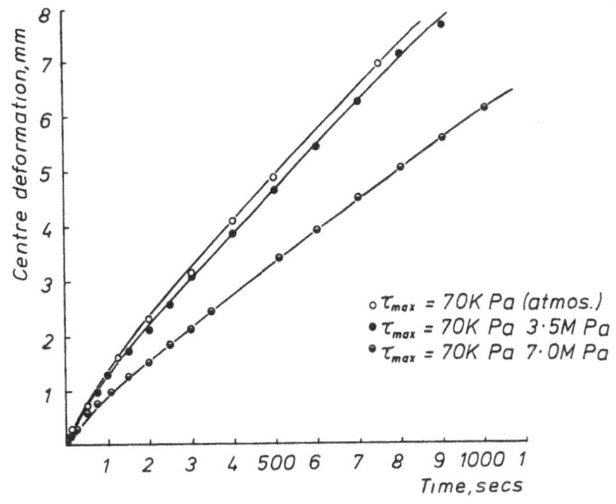

Fig. 4. The effect of increasing hydrostatic pressure on flow

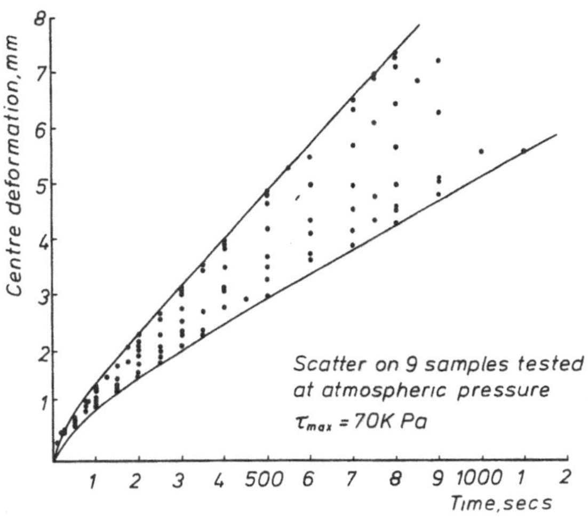

Fig. 5. Variation of flow properties

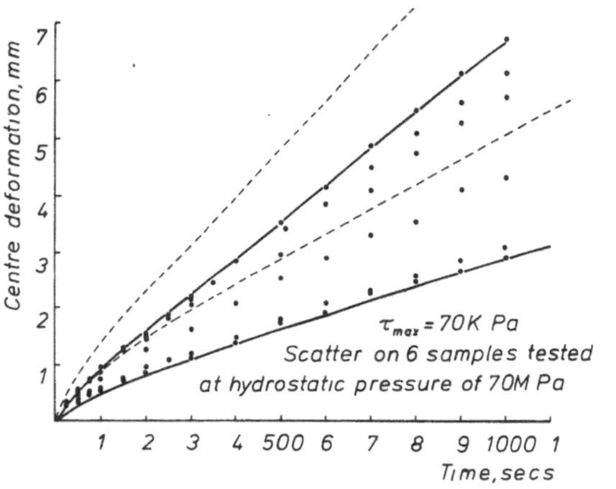

Fig. 6. The reduction of flow by hydrostatic pressure of 7.0 MPa

1080 seconds when the stress is reduced from 90 to 50 kPa. The curvature towards the time axis demonstrates that the composite behaves like a viscoelastic material. It is also a viscoelastically non-linear material in that the relationship between stress and strain depends on the magnitude of the stress as well as on time.

The results of applying various amounts of hydrostatic pressure is shown in fig. 4. The hydrostatic pressure of 3.5 MPa only sightly reduces the flow, whereas a pressure of 7.0 MPa has a considerable reducing effect. This was interpreted as the direct result of a "yield value", of about 3 MPa, only above which is there any reduction in flow due to increasing hydrostatic pressure.

Further tests were made to measure the reproducibility of the results, as from previous work it had been found to be poor. The scatter obtained was larger than expected, and results from nine samples tested under identical conditions are shown in fig. 5. These samples were deformed at atmospheric pressure, and as can be seen from the spread the scatter is larger than the difference observed from hydrostatic pressure. This amount of scatter has been shown (7) to be due to unequally distributed ageing of the stored composite. Only during and after the filling procedure were the storage conditions controlled, whereas the storage history between manufacture and filling of the tubes was not controlled. This fact has since been rectified with the introduction of controlled storage.

The remaining samples were all deformed at a hydrostatic pressure of 7.0 MPa to give a sufficient number of samples in each group so that a statistical analysis could be used to compare the results. The samples tested with or without a hydrostatic pressure, were taken from the batch at random.

The scatter of the results from six samples deformed at a hydrostatic pressure of 7.0 MPa are shown in fig. 6. The dashed lines are the envelope of the results shown in fig. 5, and it can be seen that there is an overlap of the two sets of results.

No significant difference in the shape of the profile of samples tested with or without hydrostatic pressure could be detected. Hence the value of the centre deformation of deformed samples can be considered as a measure of the composite's response to testing conditions, and the dependence can be interpreted by reference to the change in the mean values from within the scatter band of the actual results.

The centre deformations at 800 seconds, where the overlap, of values from all the samples, was a maximum, were analysed statistically to investigate the probability that the two sets of values might be derived from separate populations. The statistical method of *Students* "*t*" test (8) was used to differentiate between the two populations. The calculation showed that if the sample values derive from the same population they furnish a value of *t* which is improbable – an absolutely greater value would arise only 18 times in a thousand. It can therefore be considered that the two groups derive from separate populations.

The mean values of the centre deformations are plotted in fig. 7. The reduction due to a hydrostatic pressure can be seen, and the magnitude of the reduction is tabulated in table 1. The average strengthening factor is 1.48 ± 0.06, therefore the trend is to reduce the centre deformation by approximately 50%, i.e. the strength of the composite is increased when a hydrostatic pressure is applied. It has been shown (9) that under uniaxial test conditions the hydrostatic pressure suppresses void formation and growth, and hence reduces the amount of extension at a particular stress. It was also shown that all the voids are sealed and no further reduction in extension is observed above 3 MPa of hydrostatic pressure. It is therefore concluded that under the investigated telescopic shear stress conditions no voids, or very few voids, are formed in the binder when the sample flows. The flow takes place at essentially constant volume because of the constraint of the tube. The composite cannot increase its

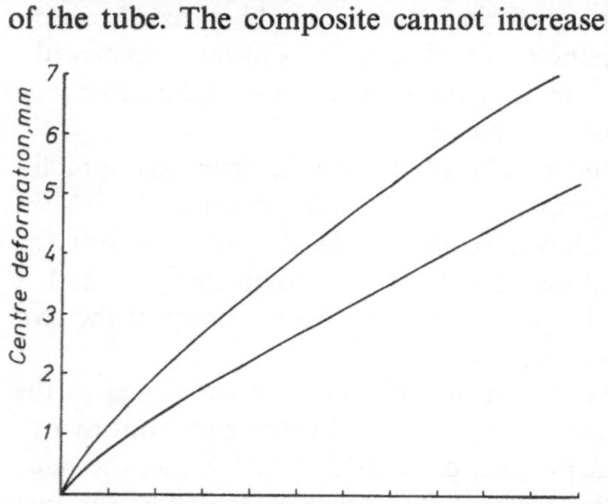

Fig. 7. Average flow curves with reduction due to hydrostatic pressure

Table 1. Strengthening effect of pressure

Time (seconds)	Strengthening factor
50	1.51
100	1.51
150	1.46
200	1.51
250	1.39
300	1.55
400	1.59
500	1.47
600	1.48
700	1.44
800	1.46
900	1.44
1000	1.39
Average	1.48
Standard deviation	0.06

total dimensions perpendicular to the shear surfaces as is required for dilation to occur. Local elements could dilate but they must then be compensated by consolidation of some small volumes on the same, or adjacent, radial plane containing the element. However, as the composite was formed in the tube and consolidated by a pressure of 7.0 MPa, with the theoretical density very close to the measured density of 1.708 Mg/m^3, very few voids are initially present in the composite. Further consolidation or a redistribution of voids during the shear is therefore ruled out as improbable.

This assumption was tested by measuring the change in profile of a sample, previously deformed at atmospheric pressure, when it was subjected to a hydrostatic pressure equivalant to the consolidation pressure. If voids were present or further consolidation possible there would be a reduction in volume and hence a decrease in profile would be expected. The results are tabulated in table 2. The centre deformation varies by less than one percent which is the same order as the experimental error. Also the shape of the profiles and a mould taken of the back surface showed no significant change. As the material deforms plastically with little recovery no healing mechanism can be operative.

The shapes of typical front and back profiles are shown in fig. 8. The difference in shape can be explained by radial flow in the composite when it leaves the region where the flow is constrained by the tube. The material in the region of the

Table 2. Results of test for voids in deformed sample

Hydrostatic pressure (MPa)	Time at pressure (seconds)	Centre deformation (mm)
Atmospheric	Before test	7.87
1.4	–	7.82
3.5	–	7.90
5.2	–	7.87
7.0	–	7.82
7.0	25	7.82
7.0	50	7.85
7.0	100	7.81
7.0	250	7.85
7.0	500	7.86
7.0	600	7.84
7.0	750	7.82
7.0	1000	7.86
Atmospheric	After test	7.84
	Average	7.84
	Standard deviation	0.06

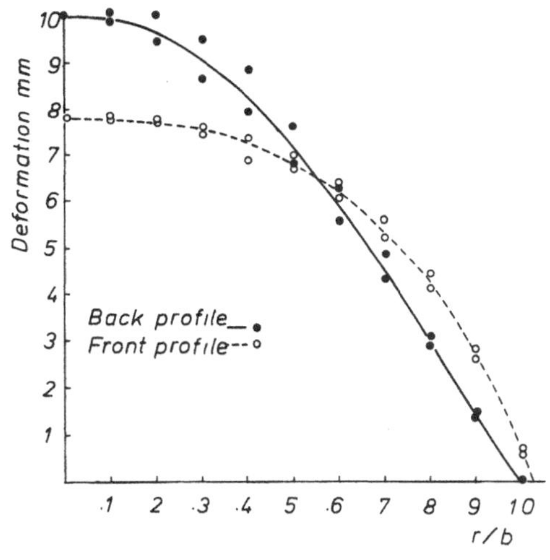

Fig. 8. The shape of typical profiles

front profile can then dilate as it shears. The initial volume and final volume of the sample was calculated using the profiles as surface boundaries of the final volume. The final volume was only slightly larger than the initial volume, the actual volume change was the order of 2%. Allowing for the dilation in the region of the front profile the volume change is of the same order as the experimental error, and is hence considered to be insignificant but still a real change.

All this evidence indicates that the composite undergoes telescopic shear without substantial dilation or void formation. The mechanism by which the hydrostatic pressure strengthens the material under telescopic shear is therefore different from that operative when the material is extended under uniaxial stress.

4. Possible mechanism and conclusions

In the previous section it was shown that a hydrostatic pressure of 7.0 MPa decreases the ability of the composite to deform by an average of 50%. It was also shown that the suppression of void formation and growth was not the primary mechanism of this strengthening effect.

To understand the microscopic behaviour and give a possible explanation for the reduction in deformability, the physical structure of the composite should be considered in more detail. A material which contains such a high solid loading as the composite under study is in essence as conglomeration of particles stuck together by the polymeric binder. The polymer would only be present in sufficient quantity to form a thin film around each separate particle.

Under local tensile stress conditions voids form readily either at the interface between particle and polymer due to adhesive failure at the bond, or close to the interface due to cohesive failure in the polymer. This phenomenon is called dewetting and results in dilation of the composite during the extension. However, when stressed in the High Pressure Test Cell the composite is changing shape at essentially constant volume. The solid particles are therefore tending to slip pass and rotate around each other while moving to rearrange their position. On average there is no gross movement to increase the separation of particles as would be expected with an applied tensile stress.

The polymer film follows the movement of the particles by primary flow. (i.e. direct flow, as the particle moves its polymer film moves with it) or by elastic extension which could then be released by later secondary flow. The controlling process is determined by the temperature and strain rate. Both processes occur together but at ambient temperature and above the composite exhibits viscous properties which infers that primary flow is the predominant process, whereas at low temperatures or high strain rates the composite exhibits elastic characteristics.

It is postulated that the interlocking and interference between adjacent particles during rotation or slip results in the observed shear

37*

hardening of the composite. Larger particles which interlock can fail, and this is usually by shear on a crystal plane weakened by occlusions. The particles would then be free to rotate and slip when sheared. An initial resistance to particle rotation and slip gives rise to a "yield stress" below which there is no movement (10). Any previous flow of the composite gives easy rotation and movement when again stressed in the same manner and results in a softer material. The mechanical properties of the composite depend on all the previous strain history and hence controlled storage with standardized preparation of samples is essential for reproducible results with homogeneous deformations.

The addition of a hydrostatic pressure does not fundamentally change the way the composite deforms when stressed in shear. Each particle moves essentially in the same manner as before. As the particles of solid filler can be considered as incompressible the effect of the applied hydrostatic pressure is concentrated in the polymer film between the particles.

The viscosity of the polymer will increase giving greater drag on the particles. If the polymer film is displaced and particle-particle contact results then there is an increase in interparticle friction with an associated reduction in movement and increase in the yield stress. The interlocking and interference between adjacent particles also increases if there is a local decrease in film dimensions. All these factors combine to reduce the total amount of movement, which results in the observed reduction in deformation.

Additional work is to be undertaken to determine which of the processes has the greatest significance. The effect of temperature on the magnitude of the reduction will be studied. A literature survey on reported work on polyisobutylene has been started to find out the magnitude of the viscosity increase due to pressure. It will then be possible to deduce the exact mechanism responsible for the reduction in deformation due to an applied hydrostatic pressure.

Acknowledgements

Thanks are expressed to the staff of the Solid Propellent Laboratory, RPE, who filled the samples and to Mr. *G. J. Spickernell* of the ERDE who supplied the composite used for this work.

Summary

The flow properties of a highly filled polyisobutylene polymer have been investigated. The solid filler is mainly an inorganic crystalline material with a multimodal size distribution in the range of 10–750 μm. The loaded polymer has properties which are viscoelastically non-linear and which depend on the testing environment.

Apparatus developed at the RPE enables a differential gas pressure to be applied across the faces of a plug of material encased in a rigid brass tube. The resulting stresses are mainly those of telescopic shear. Samples are tested under controlled conditions and the deformations recorded by successive photographs taken at known times. The deformations at various points on the profile were measured from these photographs.

Two sets of identical samples were tested at the same differential pressure and temperature. The first set was deformed under a hydrostatic gas pressure of 0.10 MN/m^2 (atmospheric pressure), and the others with a hydrostatic gas pressure of 7.0 MN/m^2 (1000 psig). The results show that an increasing hydrostatic gas pressure reduces substantially the ability of the material to deform when stressed in shear. The significance of and a possible mechanism for this reduction in deformability is briefly discussed.

References

1) *Buswell, H. J.*, Rheol. Acta **9**, 577–584 (1970).
2) *Gledhill, V. M.* and *W. A. Dukes*, ERDE Tech. Report 87 (1972).
3) *Vernon, J. H. C.*, ERDE Tech. Memo. 13/M/62 (1962).
4) *Buswell, H. J.*, RPE Tech. Report 70/2 (1970).
5) *Buswell, H. J.* and *J. F. Moloney*, RPE Tech. Report 68/6 (1968).
6) *Buswell, H. J., J. F. Moloney*, and *G. S. Pearson*, Rheol. Acta **8**, 2, 240 (1969).
7) *Buswell, H. J.*, RPE Tech. Memo. 575 (1971).
8) *Yule, G.* and *M. G. Kendall*, Introduction to the theory of Statistics (London 1958).
9) *Church, G. J., D. E. England*, and *J. H. C. Vernon*, ERDE Tech. Report 3/R/62 (1962).
10) *Dukes, W. A.*, Proc. 5th Int. Congress on Rheology. Ed. *S. Onogi*. Vol. 2 (Univ. of Tokyo Press 1970).

Author's address:
Dr. *H. J. Buswell*
Procurement Executive, Ministry of Defence
Rocket Propulsion Establishment
Westcott, Aylesbury, Bucks. (England)

Rheol. Acta **13**, 577–585 (1974)

From the School of Textile Technology, University of New South Wales, Kensington, N. S. W. 2033 (Australia)

The examination and interpretation of the properties of fibres using rheological measurements on non-uniform fibres

By J. D. Collins and M. Chaikin

With 7 figures and 2 tables

(Received October 27, 1973)

1. Introduction

The use of mechanical measurements on fibres as a means of examining the properties of the fibre material is often complicated by the presence of dimensional and internal structural variability both within and between fibres (e.g. ref. 1–6). Ideally it would be desirable to work with uniform fibres but for natural fibres in particular, with their inherent variability, this is not a practical proposition. The investigator is therefore often forced to use non-uniform fibres and the situation is further complicated by the variation between individual fibres. It is important that the effects of this fibre non-uniformity on the observed behaviour be understood so that allowances can be made and a correct interpretation of the mechanical behaviour can be derived. Knowledge of the effects of fibre to fibre variability is also important since this provides an additional technique for extracting significant fibre structural information from the observed results.

A considerable amount of work has been carried out in assessing non-uniformity effects in the observed mechanical behaviour of wool fibres. This has involved stress-strain, creep and torque-twist measurements on a large number of wool fibre types under a variety of experimental conditions. The present paper is concerned with a critical examination of this work so as to highlight the effects of fibre non-uniformity and to illustrate the practical analysis and interpretation of mechanical measurements on non-uniform fibres.

2. The longitudinal mechanical properties of wool fibres

2.1. General considerations

Fibre dimensional variability does not present any great problem since the external shape of the fibre can be measured and hence the degree of non-uniformity assessed. The mechanical behaviour of a non-uniform fibre may be calculated theoretically by considering the fibre to be composed of a number of uniform cylinders of varying cross-sectional area all acting in series i.e. strains are averaged for constant applied load. The behaviour for varying degrees of fibre cross-sectional area variability may thus be derived. Examination of actual fibre behaviour indicates that the predicted behaviour agrees well with actual measurements.

Internal structural variability creates much more of a problem since with presently available techniques this is difficult, if not impossible, to measure prior to mechanical testing. Any structural variation present must be deduced by working back from the observed mechanical behaviour. The problems involved in this type of analysis can be greatly assisted by introducing the notion of the "effective area" of a cross-section and the "effective area" distribution of the fibre (5).

Suppose for example the *Young*s Modulus of the material in a particular cross-section is lower than the average along the fibre. This cross-section will behave as though the cross-sectional area is less than the actual measured cross-sectional area and this apparent cross-sectional area can be thought of as the "effective area" of the cross-section when the initial modulus is considered. Thus for the initial portion of the stress-strain curve the observed behaviour of the fibre will depend on the "effective area" distribution along the fibre. The variability of the effective area may be smaller or larger than the variability of the actual cross-sectional area depending on the relationship between structure and area e.g. larger areas may contain material which is "weaker" or "stronger" than the fibre material average.

It is important to note that the "effective area" of a cross-section may vary depending on the mechanical constants (e.g. *Young*s Modulus, yield stress, breaking stress, modulus of rigidity, etc.) involved and the testing conditions. Consider for example the "effective area" for breaking. Suppose that fracture depends on the presence of small flaws in the fibre material and that the size and number of these flaws is independent of the structural features responsible for the *Youngs* Modulus. Under these circumstances it is clear that the effective area distribution for the break may be quite different to that for the initial modulus.

Different testing conditions may also change the effective area distributions. If for example the structural variation along a fibre is mainly differences in material flow behaviour then changing the rate of extension will change the effective area distribution. The most obvious situation where changes may take place is when fibres are tested after some mechanical and/or chemical pretreatment. In this case, structural variation which may have little effect on the observed behaviour under normal conditions may be accentuated by the pretreatment so that the effective area distribution will be changed.

In examining the behaviour of fibres under a specific set of conditions the effects of cross-sectional area variability may be calculated. When considering structural variability it becomes necessary to consider each of the effects of area variability separately. In practice these effects may be smaller or larger than predicted on the basis of area variability alone and these differences may be translated into a measure of the structural variability present by taking into account the predicted values for area variability and known structural information about the fibres.

2.2. The wool fibre stress-strain behaviour

The wool fibre stress-strain behaviour under a wide variety of conditions has the general shape depicted in fig. 1. After the initial *Hooke*an region *AB*, there is a yield region *BC*, where there is little change in stress for a large change in extension, followed by the post-yield region *CD*, where a stiffening of the fibre takes place prior to the break at *D*. In most cases the extension e_0 at *C* is about 30% whilst the observed slopes of the three regions for fibres extended in water are approximately in the ratio 40 : 1 : 10.

If it is assumed that the curve shown in fig. 1 represents the behaviour of a uniform fibre then the stress-strain curve for a fibre varying in cross-sectional area may be calculated by numerical integration and the curve for a typical non-uniform fibre is shown as the dashed line in fig. 1. The main effects to note are a small decrease in the *Hooke*an slope, a large increase in the yield slope, a fairly constant stress at a strain $e_0/2$, a small increase in the strain at the turning point between the yield and post-yield regions and large decreases in both the breaking stress and breaking extension. Detailed analysis of these changes have been carried out (4, 5) and will be dealt with later in this paper.

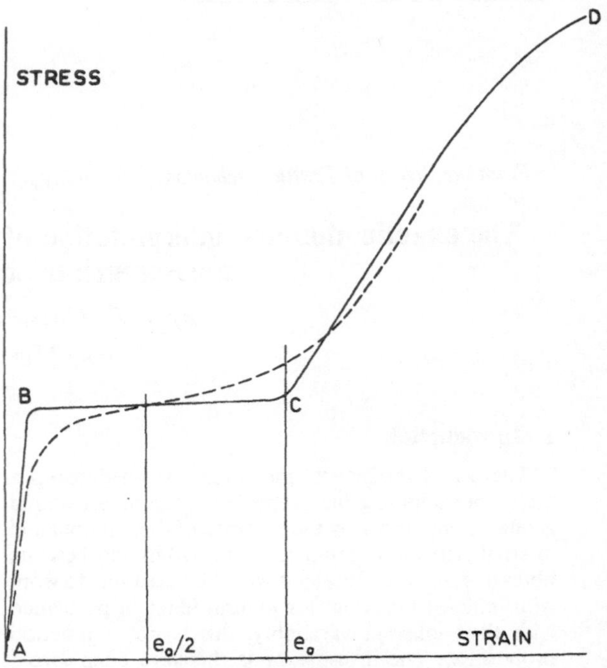

Fig. 1. Typical wool fibre stress-strain curve (——— uniform fibre, ----- non-uniform fibre)

When dimensionally and structurally non-uniform fibres are considered different effective area distributions may apply for the *Hooke*an region, the yield-region, the post-yield region and the break (5). Thus the observed behaviour may exhibit the effects noted above to a greater or lesser extent. For example, the observed behaviour may show a large decrease in the breaking stress with only a slight increase in the yield slope or alternatively a large increase in the yield slope with little change in the breaking properties.

The comparison of the behaviour for two different sets of conditions is illustrated in fig. 2 where curves 1 and 2 represent the uniform fibre stress-strain curves for two sets of conditions *A* and *B* respectively. Curves 1, 3 and 4 represent three of the possible observed curves for fibres tested under condition *A* whilst curves 2, 5 and 6 represent three of the possible observed curves for fibres tested under condition *B*. If for example curves 4 and 5 were observed in practice then non-uniformity effects are ignored one might conclude that the material breaking extension

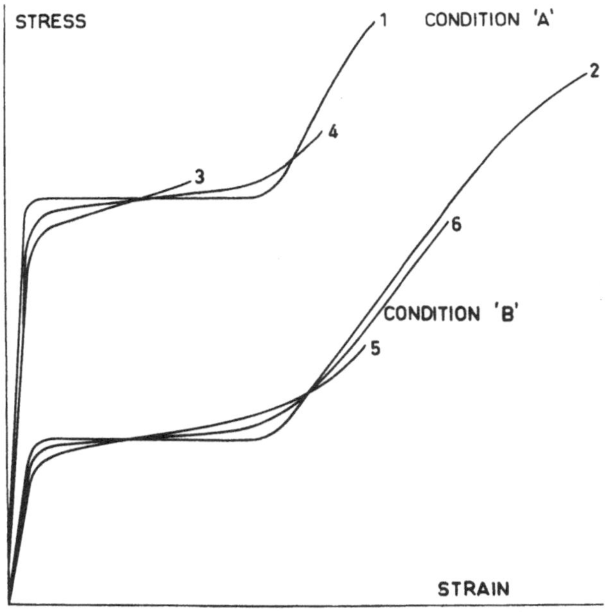

Fig. 2. Wool fibre stress-strain curves for two sets of conditions

is similar for both test conditions whilst the yield slopes are approximately of the same order of magnitude. If curves 3 and 6 were observed then the breaking strains will show a similar difference to those for the fibre material curves (i. e. 1 and 2). There will be a much higher yield slope for condition *A* and it will appear that the behaviour under condition *A* is made up of only two regions instead of three. It is clear from this simple illustration that non-uniformity effects can cause (and in fact have caused) considerable confusion in the interpretation of the observed behaviour and it is thus essential to account for the effects of non-uniformity.

The following sections will examine specific regions of the stress-strain curve under varying conditions in order to indicate how relevant information about the material behaviour can be extracted from the observed data.

2.2.1. The Hookean region

The theory (5) indicates that a 5% decrease in *Hooke*an slope should be noted for a fibre with a coefficient of variation of area of about 20%. It is found in practice that, under a wide variety of conditions, inter-fibre variation in *Hooke*an slope is such that this effect is insignificant and therefore cannot be detected. Fibre crimp or waviness is found to affect significantly the *Hooke*an slope, the slope decreasing as the degree of crimp increases (4, 5).

Geometric factors may also explain differences in the relative *Hooke*an slope (the ratio of the *Hooke*an slope to the stress at 15% extension) between fibre types e.g. *Merino* fibres gave a relative slope of 32 compared with a relative slope of 44 for *Lincoln* fibres (4). Wool fibres are approximately elliptical in cross-section and the major axis of this ellipse rotates about the fibre axis as the fibre is longitudinally translated producing what is known as the fibre "natural twist". During stretching there is a redistribution and partial removal of this twist (3). The twist per unit length of fibre is higher for *Merino* fibres compared with *Lincoln* fibres and the amount of twist removal with stretching is higher for the former (3). This suggests that any tendency for microfibrils to be oriented in the direction of twist is higher for *Merino* fibres. Twisting experiments (3) also tend to indicate that microfibrils may be oriented with the natural twist for *Merino* fibres. If in fact, microfibrils are less oriented in the fibre axial direction for *Merino* fibres compared with *Lincoln* then the differences in relative *Hooke*an slope can easily be explained.

2.2.2. The yield region

When comparing yield region slopes it is essential to use the relative yield slope (the yield slope divided by the stress at 15% extension) when $e_0 = 30\%$ and the standardised relative yield slope (the relative slope calculated as though $e_0 = 30\%$) when $e_0 \neq 30\%$ (5). The predicted behaviour in the yield region is as follows (5):

When the fibre material yield slope is zero, the shape of the yield region should be closely related to the cumulative effective area distribution and the relative yield slope should be approximately proportional to the coefficient of variation of effective area. Fig. 3 gives a range of area distributions with the predicted yield region shape. Distribution (a) is the most common type of area distribution but any of the effects shown in curves (b), (c) and (d) may occur, particularly in fibre groups where fibre non-uniformity is high. In cases such as (b), (c) and (d) an average yield slope is determined.

Consider now the case where the material yield slope is zero and where the rate of straining affects the yield region stress level. Initially all sections of the non-uniform fibre will be extending in the *Hooke*an region, but, as the load increases, the sections of smaller cross-sectional area will begin to extend in the yield region. The extension

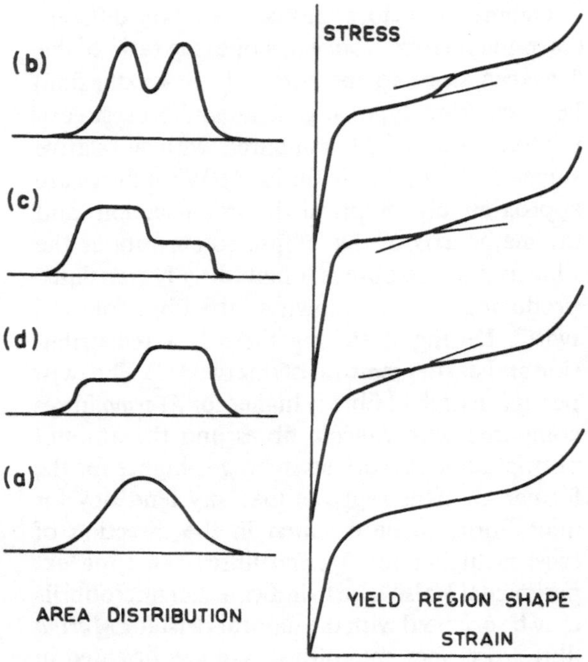

Fig. 3. The relationship between the observed yield region shape and the fibre effective areadistribution

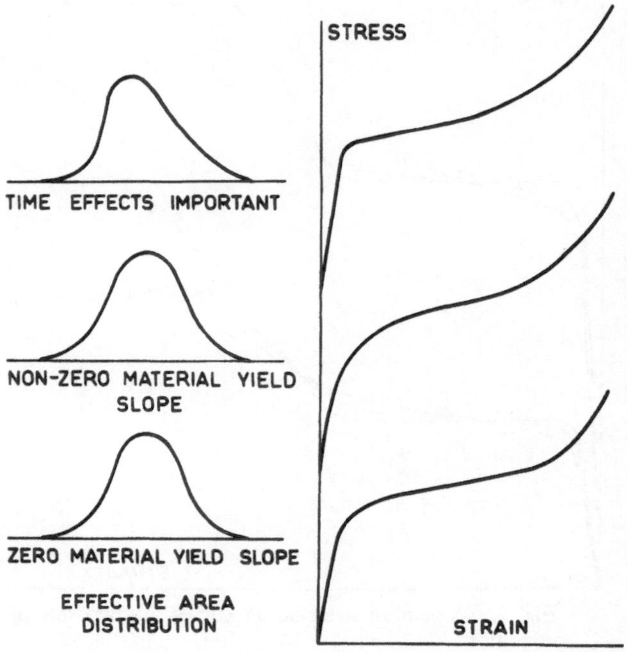

Fig. 4. The effect of material yield slope and time effects on the observed yield region shape

rate of these sections will be very high so that the load will continue to increase sharply. As more sections of the fibre begin to extend, the average extension rate for those sections extending in the yield region will decrease, bringing about a corresponding reduction in the rate-of-load increase. As more and more sections reach the post-yield region, the average extension rate for sections in the yield region will increase, the rate-of-load change being increased until all sections are extending in the post-yield region. The general effect will be to make the *Hooke*an to yield turning point sharper and the yield to post-yield region less well defined. In effect this means that the effective area distribution will be skewed towards lower areas when compared with the actual area distribution. This is clearly indicated in fig. 4. The size of the effect will increase as time effects become larger and as the fibre non-uniformity increases (less sections on the average will be extending in the yield region at any point of time so that extension rates will be higher).

Consider now the case where a real material yield slope exists which is significantly greater than zero. The curve will no longer follow the cumulative effective area distribution; both turning points will lose sharpness (see fig. 4) and the observed relative yield slope will level out at the fibre material relative slope as fibres become more uniform.

Detailed examination of the behaviour of a large number of fibres in water has indicated the following relationships (4). On the average the shape of the yield region agrees well with the actual cumulative area distribution; in particular curve effects such as those shown in (b), (c) and (d) of fig. 3 are a direct result of the area distribution. The observed relative yield slopes are larger than predicted by the effects of area distribution alone so that structural effects must be present. Any tendency for time effects to produce skewness in the effective area distribution is only very slight so that these structural effects are not significantly time dependent.

The observed behaviour can be explained if it is assumed that the yield region stress level varies from cross-section to cross-section, the stress level only showing slight changes for different rates of extension. The section stress level must either be proportional to the section cross-sectional area or independent of the section area to enable the effective area distribution to have the same general shape on the average as the actual area distribution. Other evidence (3) suggests that equal areas do not necessarily behave in a similar manner so that the second alternative seems more likely. On this assumption the coefficient of variation of yield region stress level variation along the fibre is on the average about the same value as the coefficient of variation of the actual area. The analysis indicates that the actual fibre material yield slope is close to zero and there are sharp transitions between the regions of the curve. These effects however are masked by non-uniformity effects.

The behaviour at 65% relative humidity indicates a curve shape which is not related to the cumulative area distribution (6). There is a very sharp *Hooke*an to yield region turning point with a poorly defined transition from the yield to the post-yield region. The observed relative yield slope is significantly less than that predicted from the area distribution and with some of the more uniform fibres a region of negative slope is observed in the yield region. The turning point behaviour suggests that time effects play a significant role in the observed behaviour and the behaviour can be explained if the fibre material yield slope is assumed to be negative i.e. there is a very rapid stress relaxation effect for material extensions in the yield region. For more uniform fibres sufficient sections will be extending in the yield region to allow this stress relaxation effect to be seen. As the non-uniformity increases extension rates are higher for individual sections and the stress relaxation effect is masked out.

The curve at 0% relative humidity (6) shows gradual transitions between the various regions of the curve. If the relative yield slope is plotted against the coefficient of variation of area it is found that for higher values of non-uniformity the slope is approximately proportional to the coefficient of variation of area. For more uniform fibres the slope levels out at a fairly constant value. This indicates that there is large fibre material yield slope present and that the value of the slope is not significantly time dependent.

When wool fibres are conditioned and extended in dilute hydrochloric acid (pH 1) there is a large drop in the yield region stress level and a large rise in relative yield slope as compared to their behaviour in water. The turning point from the *Hooke*an to yield region is still sharp and this tends to indicate that the slope is mainly an artefact due to fibre non-uniformity. The *Hooke*an slope and e_0 change in the manner expected if the slope is entirely due to fibre non-uniformity and tests on fibres where the reaction is incomplete also support this conclusion. The action of the dilute hydrochloric acid is found to have a variable effect along the fibre length so that different yield region stress levels apply for different fibre cross-sections. This variability is equivalent to a coefficient of variation of effec-

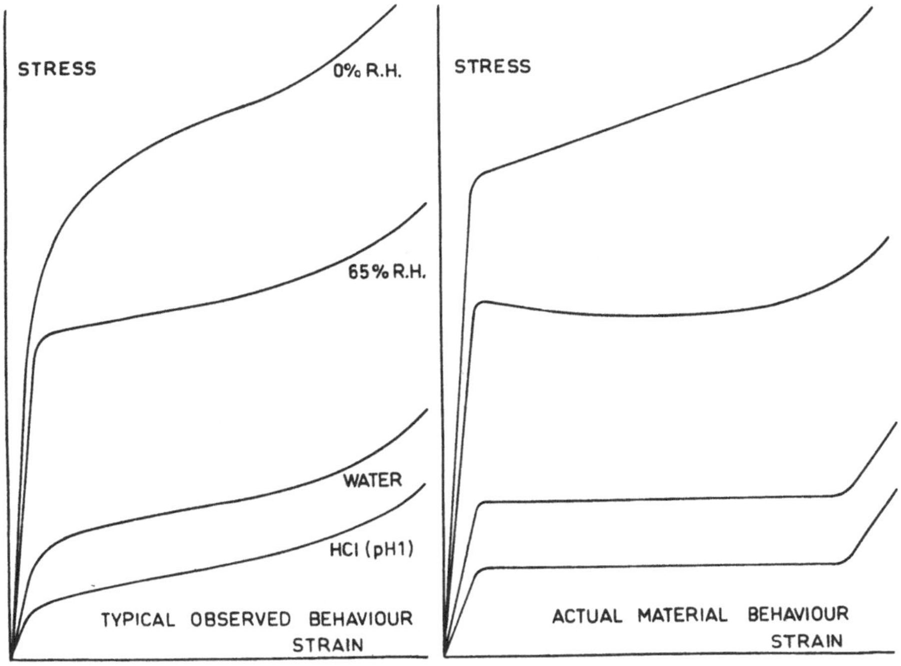

Fig. 5. Comparison of the observed and actual material stress-strain behaviour for wool fibres tested under varying conditions

tive area of about 25% this value being relatively independent of the original fibre non-uniformity (5).

Fig. 5 shows the measured yield region behaviour for a typical fibre extended in water, dilute HCl, at 65% R.H. and at 0% R.H. The actual derived fibre material curves are shown for comparison and the effects of fibre non-uniformity can be easily seen.

2.2.3. The post-yield region and the break

It has been shown theoretically (5) that the post-yield slope will decrease as fibre non-uniformity increases. This effect however is only small and is insignificant when compared with natural variations in post-yield slope from fibre to fibre.

A non-uniform fibre will break when the load is sufficent to rupture the fibre at its minimum effective area for breaking. If the mean effective area is $\bar{A}$, the minimum effective area is $A_{\min}$ and the material breaking stress f_b, the fibre will break at an apparent stress of $[(f_b \cdot A_{\min})/\bar{A}]$. The minimum effective area will usually be about 2–3 standard deviations less than the average effective area $\bar{A}$ and the decrease in breaking stress will depend only on the coefficient of variation of effective area and not on the shape of the stress-strain curve. The observed behaviour however will depend on the shape of the curve.

Fig. 6 shows three material stress-strain curves all with the same breaking stress and breaking extension. The apparent stress levels at which breakage occurs for various values of the coefficient of variation of effective area are indicated. In the top curve even a low value of the coefficient of variation will bring about a considerable decrease in the breaking extension and mask out features of the stress-strain curve. A similar effect for very irregular fibres is found for the middle curve whilst the third curve still shows the curve features even when fibres are very irregular.

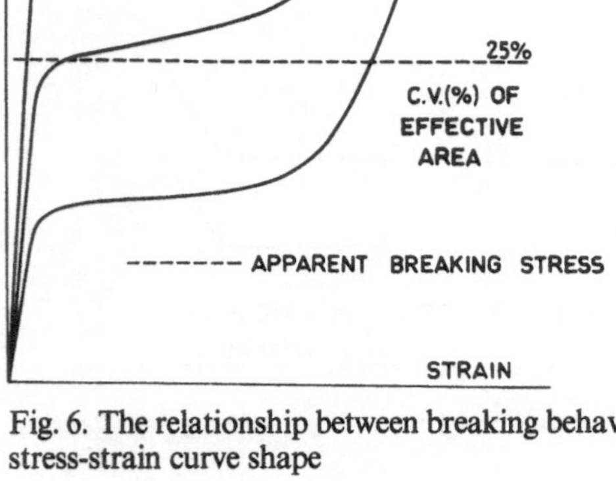

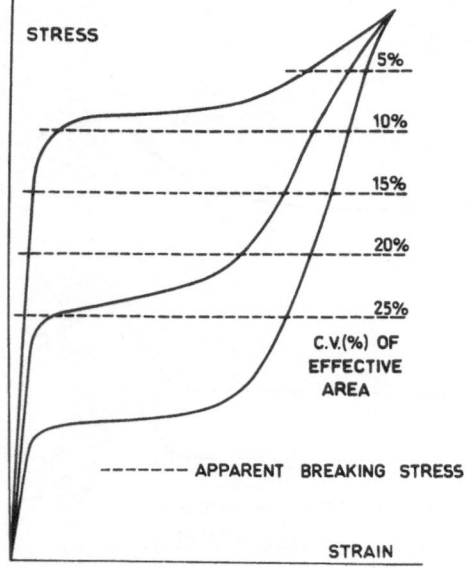

Fig. 6. The relationship between breaking behaviour and stress-strain curve shape

If the behaviour of wool fibres in water and at 0% R.H. are compared it is found that for a range of fibres of differing type the breaking extension of individual fibres varied from as low as 40% to as high as 86% for fibres extended in water (2) and from 8–49% for fibres extended at 0% R.H. (6). In the 0% R.H. group about half the fibres did not extend past the yield region. Analysis of the results indicates (6) that in both cases the minimum effective area is approximately 2 standard deviations (from the actual area of distribution) less than the mean area so that it appears that the main factor in breaking is the actual area distribution. The predicted material breaking extensions are found to be 87% (for water) and 53% (for 0% R.H.) compared with respective observed average values of 66 and 35%. It has been shown (5) that the breaking behaviour in dilute HCl is affected by a structural variation produced by the treatment as well as the original area fluctuation.

2.3. The wool fibre creep behaviour in the yield region

For low loads wool fibres creep in the *Hooke*an region with extensions limited to 2 or 3% (7). As the load is increased creep occurs in the yield region where extensions of 20–30% are achieved after several minutes (8). At high loads extensions rapidly reach values greater than 30% and slow creep (hours) takes place up to extensions near 80% (creep in the post-yield region) (9). The load range for creep in the yield region is quite narrow so that for a given load different cross-sections may be exhibiting any of the above three types of behaviour.

For creep in the yield region the fibre strain e is related to the time t by an equation of the type (8)

$$\frac{1}{e} = \frac{a}{t} + b$$

where a and b are constants. For a given temperature the constant a is related to the applied load P, the cross-sectional area A by an equation of the type

$$a = K \exp\left[-\frac{\alpha P}{A}\right]$$

where K and α are constants. Theoretical calculations indicate that the various constants will depend on the coefficient of variation of area and this has been confirmed experimentally (10). Table 1 indicates the relative values of $\log_{10} K$ and α predicted for various values of coefficient of variation of area and experimental values (obtained from regression analysis) are indicated for comparison.

Table 1

	$\log_{10}$ K		α	
C.V. of area %	predicted	observed	predicted	observed
0	1.00	1.00	1.00	1.00
10	0.68	0.66	0.71	0.68
20	0.35	0.32	0.41	0.37

The creep behaviour is characterised by an activation energy and measurements by *Feughelman* (8) give values of about 25 kcal/g mole. Work by the present authors gives values of about 10–15 kcal/g mole and unless fibre non-uniformity is considered it is extremely difficult to explain this discrepancy. Fig. 7 indicates theoretical calculations of the effect of non-uniformity on the observed activation energy. Measurements of area variability and other mechanical tests indicate that the fibres used by *Feughelman* are extremely uniform (less than 5% C.V.) whilst those of the present authors have C.Vs. of effective area in the range of 10 to 25%. When this is noted the reason for the differences in observed values is readily apparent. It is worth noting that because of the shape of the curves in fig. 7 it will be difficult to make accurate predictions of the true activation energy using measurements from non-uniform fibres.

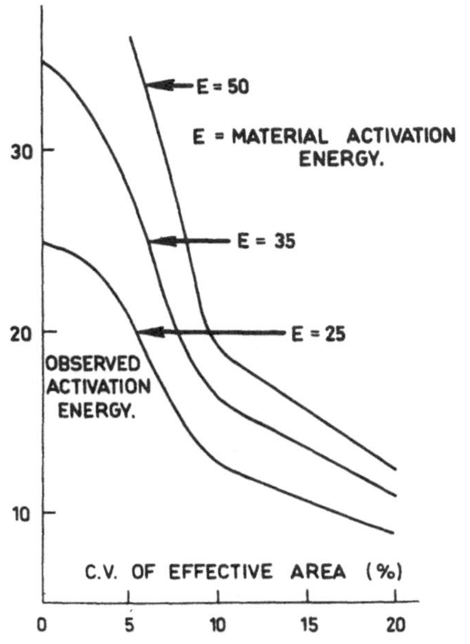

Fig. 7. The effect of fibre non-uniformity on the observed activation for wool fibre creep in the yield region

3. The wool fibre torque-twist behaviour

When a wool fibre is twisted the torque T is related to the twist angle θ by a relationship of the type

$$\frac{1}{T} = \frac{1}{C_1 \theta} + \frac{1}{C_2}$$

where C_1 and C_2 are constants (11). C_1 is the initial slope of the torque twist curve (related to the material modulus of rigidity) and the torque increases with twist angle increase towards a limiting value C_2.

3.1. *The initial slope C_1*

For a uniform circular fibre the modulus of rigidity G is given by the usual relationship

$$G = \frac{2\pi L C_1}{A^2}$$

where L is the fibre length and A is the cross-sectional area.

Theoretical calculations (10) indicate that for a dimensionally non-uniform circular fibre the observed modulus of rigidity (calculated using the mean cross-sectional area) is related to the true modulus of rigidity by an equation of the type

$$G_{\text{observed}} = F_1 \cdot F_2 \cdot G$$

where F_1 and F_2 are functions both less than 1. The function F_1 depends on the rate of change of area along the fibre (i.e. stress concentration effects) whilst F_2 depends on the area distribution of the fibre.

Torque-twist measurements (10) in water for a large number of fibres of varying types show a variation in modulus of rigidity between fibres which is much larger than the variation predicted from area non-uniformity. When this variation is compared with the variability in stress-strain behaviour between fibres it is found that the variation in modulus of rigidity between fibre types is closely related to the corresponding variation in the post-yield slope as indicated in table 2 and this has important structural implications.

Table 2

Fibre type	Post-yield slope	Modulus of rigidity	Ratio
Pen-grown Lincoln	3.71	1.47	2.52
Field-grown Lincoln	4.21	1.85	2.28
Pen-grown Corriedale	3.83	1.62	2.36
Field-grown Corriedale	2.92	1.25	2.34

The behaviour of wool fibres in water can be satisfactorily explained by the two-phase structure (12) where long thin strong parallel microfibrils are embedded in a weak matrix. The torsional behaviour is predominantly a shearing effect in the matrix between neighbouring microfibrils. This model is generally accepted and the main controversy at present is concerned with the nature of the matrix. The uniform matrix model (put forward in various forms by several authors) suggest that for extensions in the yield region (i.e. up to 30% extension) the matrix does little to resist extension. At about 30% extension further rearrangement of the matrix requires the application of a stress and the post-yield region is produced by further extension of the matrix. In the series-zone model (13) stiff viscous filler units occur at regular intervals in the matrix and are bonded to adjacent microfibrils. The yield region is produced when microfibrils which are not adjacent to these proteins open up and the post-yield region is the result of stresses developed in this filler material as adjacent microfibrils open up. Table 2 suggests that the same material is responsible for both the modulus of rigidity (as measured by fibre torsion) and the post-yield slope. This relationship can be easily explained by the series-zone matrix. The uniform matrix does not offer any resistance to extension until 30% extension and thus cannot explain the observed behaviour.

3.2. The limiting torque C_2

The existence of a limiting torque can be interpreted as the result of a constant limiting shear stress existing in the fibre material (e.g. G_{lim} at all radii from the centre of torsion) and on this assumption it can be shown (10) that

$$G_{lim} = \frac{3\sqrt{\pi}}{2A^{3/2}} \cdot C_2 .$$

Thus $G_{lim} \alpha \frac{1}{A^{3/2}}$ compared with $G \alpha (1/A^2)$. The observed value of G_{lim} can be related to the material limiting stress by the relationship

$$(G_{lim})_{observed} = \frac{A_{min}^{3/2}}{\bar{A}} \cdot G_{lim}$$

where $\bar{A}$ is the mean cross-sectional area and A_{min} is the minimum cross-sectional area i.e. the observed value is that produced by the minimum cross-sectional area.

Regression analysis on the previously mentioned torsional measurements indicate that the observed limiting shear stress decreases dramatically as the fibre non-uniformity increases. Taking the uniform fibre limiting stress as 1 the value for a fibre with a C.V. of area of 20% is found to be 0.48. Assuming the minimum cross-sectional area is 2 standard deviations less than the mean the predicted value is 0.46 and this tends to confirm the constant shear stress assumption.

4. Conclusion

Interpretation of the rheological behaviour of non-uniform fibres in terms of material internal structure can be complicated because of the effects of fibre non-uniformity on the observed behaviour. Knowledge of non-uniformity effects can assist greatly in obtaining a correct interpretation of the observed behaviour and the material behaviour can often be predicted if sufficient data on non-uniform fibres is available. Fibre non-uniformity may completely hide some effects so that unless fairly uniform fibres are available these effects will never be observed. Variation in properties from fibre to fibre (a result of fibre non-uniformity) may also be useful in assisting in structural interpretation (e.g. as in the use of the relationship between the modulus of rigidity and the post-yield slope to distinguish between proposed models).

Acknowledgement

A grant from the Australian Wool Corporation made this work possible. Fibres were supplied by the C.S.I.R.O., Division of Textile Physics, Ryde, New South Wales.

Summary

Stress-strain, creep and torque-twist measurements on a large number of wool fibres of varying type are used to illustrate how fibre non-uniformities affect the observed behaviour and how the fibre material behaviour is extracted from the measurements. In this regard the concept of "effective area" of a cross-section is introduced in order to simplify the analysis. It is shown that fibre non-uniformities may hide the effects of certain structural features and if reasonably uniform fibres are not available it may not always be possible to derive the true material behaviour. Variation of properties between fibres can be useful in deriving significant structural information about the fibre material.

References

1) *Collins, J. D.* and *M. Chaikin*, Text. Res. J. **35**, 679 (1965).

2) *Collins, J. D.* and *M. Chaikin*, Text. Res. J. **35**, 777 (1965).

3) *Collins, J. D.* and *M. Chaikin*, J. Text. Inst. **57**, T45 (1966).

4) *Collins, J. D.* and *M. Chaikin*, J. Text. Inst. **59**, 379 (1968).

5) *Collins, J. D.* and *M. Chaikin*, Text. Res. J. **39**, 121 (1969).

6) *Collins, J. D.* and *M. Chaikin*, J. Text. Inst. **62**, 289 (1971).

7) *Feughelman, M.*, J. Text. Inst. **49**, T361 (1958).

8) *Feughelman, M.*, J. Text. Inst. **45**, T360 (1954).

9) *Speakman, J. B.*, Nature **159**, 338 (1947).

10) *Collins, J. D.* and *M. Chaikin* (to be published).

11) *Mitchell, T. W.* and *M. Feughelman*, Text. Res. J. **30**, 662 (1960).

12) *Feughelman, M.*, Text. Res. J. **29**, 223 (1959).

13) *Feughelman, M.*, Cirtel, I–619 (Paris 1965).

Authors' address:

J. D. Collins and *M. Chaikin*
School of Textile Technology
University of New South Wales
Kensington, N.S.W. 2033 (Australia)

Rheol. Acta 13, 586–595 (1974)

From the Department of Geology, Carleton University, Ottawa (Canada) K1S 5B6

Geodynamic implications of rock creep and creep rupture

By G. Ranalli

With 5 figures

(Received October 27, 1972)

Introduction

The Earth is a very complex physico-chemical system, even when attention is restricted to its mechanical properties alone. It is also a restless system. Surface features such as the distribution of continents and oceans, the presence of mountain belts and deep-sea trenches, and the relative altitude of different regions bear witness to large-scale horizontal and vertical movements that have occurred in the geological past. The textures of rocks, and the existence of fractures and flow structures at all scales down to the microscopic level, indicate that these large-scale movements have affected earth materials in a pervasive fashion. The occurrence of phenomena such as slow vertical movements and tilting of the surface, rock creep of all sorts, and earthquakes, shows that the same processes are active at present.

Geodynamics is the study of rock deformations as observed at the Earth's surface, and of their relation to the geological processes from which they originate. Consequently, the aim of geodynamics is to explain surface features of endogenous origin in terms of the deep-seated processes which bring them about. Since these processes occur by definition in the interior of the Earth, most of the evidence is inferential, and it is fair to state that no unanimous agreement exists on any single geodynamic hypothesis.

Before any deformation theory can be developed, the rheological equations for earth materials should be established. Unfortunately the rheology of the Earth is only very imperfectly known. The outer shell of the Earth, down to a depth of 2900 km below the surface, is essentially crystalline, and as such it should obey the rheological equations that have been developed for metals. However, rocks are not only polycrystalline but also polymineralic aggregates, and this difference could be of some importance. Experimental results on rock deformation do not cover the whole range of temperature and pressure found within the Earth; moreover, creep experiments are usually carried out at a strain rate that is many orders of magnitude faster than any geologically representative strain rate, the latter being in the range of 10^{-14} to 10^{-16} sec^{-1}.

Another serious drawback which affects the extrapolation of laboratory results to geological processes is the time factor. The same material can respond to the same load in quite different ways, depending on the duration on the load; and the Earth is no exception. The outer shell, which reacts elastically (with imperfections) to the transmission of stress waves, behaves very differently when submitted to loads lasting thousands or millions of years – it flows, even under relatively small stresses. Experiments with loading times of the geological order of magnitude obviously cannot be performed in the laboratory.

The theoretical approach to the problem has also its shortcomings. Creep mechanisms in polycrystalline aggregates are the subject of a vast literature; however, most of the constants appearing in the stress-strain relations derived from theoretical considerations cannot be estimated to a satisfactory degree of certainty in the case of the Earth. The composition, lattice constants, activation energies, grain sizes, actual temperature and melting temperature of materials inside the Earth can at best be given to the nearest order of magnitude. Nevertheless, the insights into the deformational mechanisms of polycrystals provided by experimental and theoretical reasearch narrow the limits of the possible rheological responses of earth materials.

Observational data, on the other hand, seldom provide a unequivocal answer. The multiplicity of variables involed and the very simple but inescapable fact that much of what happens is hidden from view generate a situation in which very few rigorous statements can be made. However, observation is of necessity the starting point of any theory of the rheological behaviour of the interior of the Earth.

The internal structure of the Earth can be considered spherically symmetric as a first approximation, and is shown in fig. 1. The uppermost shell or lithosphere, extending to an average depth of about 100 km, consists of silicates of aluminium in its upper part (below a thin veneer of sediments), and of silicates of magnesium and iron in its lower part. Its rheological behaviour for stresses below the yield point and for most time ranges is elastic with some time-delayed effects. At stresses above the yield point, the rheological response of lithospheric rocks varies with depth. In the uppermost 10–20 km, most rocks are brittle. Since all principal stresses are usually compressive, shear fracture is the dominant failure mechanism in the upper lithosphere. With increasing depth, ductility increases; the brittle-ductile transition can be placed at a depth of approximately 10–25 km.

Below the lithosphere there is a layer, approximately 200–300 km thick, which from various evidence must be rather easily-flowing when compared to the lithosphere. It is termed asthenosphere, and consists probably of ultrabasic materials as the lower lithosphere; however, the melting temperature to temperature ratio approaches unity, and therefore ductility is increased and strength

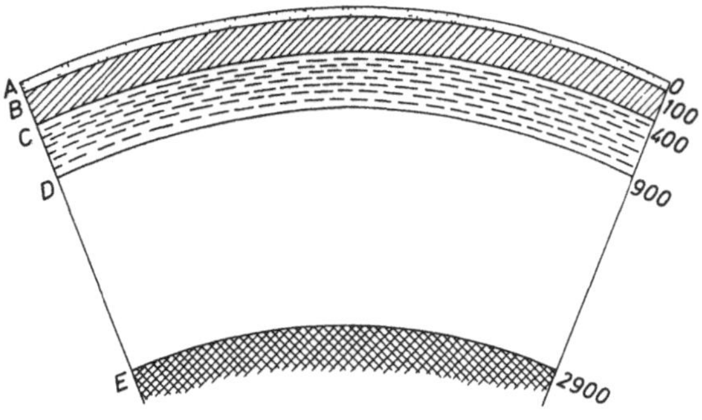

Fig. 1. The internal structure of the Earth. AB = lithosphere; BC = asthenosphere; CD = transition zone, DE = lower mantle. Depth scale in kilometres

(or effective viscosity) decreased. With increasing depth, this is followed by a transition zone (from about 400 to 900 km of depth), where the silicates are thought to go through a phase transformation to more closely-packed structures, and probably a slight change in chemical composition as well. All indications are that the strength (or effective viscosity) of the transition zone increases rapidly with depth. Asthenosphere and transition zone together form the upper mantle. From 900 to 2900 km of depth there is the lower mantle, probably consisting of closely-packed oxides of silicon, magnesium and iron. The high pressure (of the order of 10^{12} dyne cm^{-2}) and the more rapid increase of melting temperature as compared to the increase of actual temperature with depth should insure that its resistance to deformation be higher than that of the upper mantle.

The central regions of the Earth, the outer and inner core, are not shown in fig. 1 as their rheological behaviour will not be discussed in this paper. A detailed account of the present state of knowledge on the Earth's interior canbe found in *Bott* (1971).

Reviews of the rheological problems of the Earth's interior have been given by *Gutenberg* (1958) and *Birch* (1964) among others. This paper considers only three topics, namely, (i) experimental and theoretical results on rock deformation, (ii) glacio-isostatic rebound (which yields information on the creep response of the upper mantle), and (iii) the problem of earthquakes, i.e., the mechanisms of sudden stress release under high temperature and confining pressure.

Creep processes in rocks

Summaries of experimental results on rock deformation are given by *Jaeger* and *Cook* (1969) and *Paterson* (1970).

At low temperature and confining pressure, rocks under triaxial compression fail by shear fracture according to the *Coulomb-Navier* criterion

$$\tau = \tau_0 + \sigma_n \tan\varphi \qquad [1]$$

where τ and σ_n are the shear and normal stresses on the plane of fracture (compressive stress being taken as positive), τ_0 is a material constant (cohesive strength), and $\tan\varphi$ (coefficient of internal friction) can be interpreted as the coefficient of sliding friction on the walls of pressure-closed *Griffith* cracks (usually $0.5 < \tan\varphi < 1.5$).

With increasing pressure and/or temperature, a transition from brittle to ductile behaviour is observed. Increasing pressure favours ductility and tends to increase the yield point, while increasing temperature, besides favouring ductility, tends to appreciably decrease the yield point. At pressures and temperatures corresponding to a depth of approximately 30 km ($T \simeq 600$ °C, $p \simeq 10^{10}$ dyne cm^{-2}), experimentally deformed rocks usually behave in a ductile fashion. The *Von Mises* yield criterion seems to apply in the ductile region, that is, failure takes place when the stress deviator reaches a value characteristic of the material under the given conditions.

The minimum strain rate attained in experiments is about 10^{-8} sec^{-1}, that is, at least six orders of magnitude higher than any geologically realistic strain rate. A decrease in strain rate has the effect of increasing the ductility and decreasing the yield point. It is therefore likely that the strength of rocks under geological conditions is noticeably less than the strength measured in the laboratory (of the order of 10^9 dyne cm^{-2}). Another factor that greatly reduces the yield point is the presence of water. In the case of quartz, for instance, presence of water reduces the yield point by at least one order of magnitude at 1000 °C. This effect is also present in other silicates and in rocks (among others, dunite, which should closely resemble upper mantle material).

Time-dependent creep under constant stress is observed at all but the lowest temperatures, and is very similar to the creep of metals. Empirically, the total creep strain may be represented by

$$\varepsilon(t) = \varepsilon_H + \varepsilon_T(t) + \dot{\varepsilon} t \qquad [2]$$

where ε_H is the instantaneous elastic strain, $\varepsilon_T(t)$ is the transient (recoverable) creep, and $\dot{\varepsilon} t$ is the steady-state (irrecoverable) creep. (The dot indicates differentiation with respect to time.) Steady-state creep is usually absent at temperatures less than half the melting temperature.

Various forms of transient creep laws in rocks have been proposed. Sometimes experimental results can be interpreted in terms of a simple *Kelvin* model (exponential creep)

$$\varepsilon_T(t) = A\left[1 - \exp\left(-\frac{t}{t_K}\right)\right] \qquad [3]$$

where the coefficients A and t_K are related to stress and material properties. Frequently, however, creep of rocks at atmospheric temperature and pressure is best represented by a logarithmic law (*Lomnitz*, 1956; σ is the stress)

$$\varepsilon(t) = \frac{\sigma}{\mu}\left[1 + q\ln(1 + at)\right] \qquad [4]$$

where μ is the shear modulus, and q and a are constants. Eq. [4], usually referred to as *Lomnitz* law, has been experimentally verified for stresses up to $5 \cdot 10^{-4}\,\mu$. *Jeffreys* (1958) modified *Lomnitz* law to read

$$\varepsilon(t) = \frac{\sigma}{\mu}\left\{1 + \frac{q}{\alpha}\left[(1 + at)^\alpha - 1\right]\right\} \qquad [5]$$

with $\alpha \simeq 0.17$. This law accounts for the period-independent damping of seismic waves and of free vibrations, and is assumed by *Jeffreys* (1970) to hold also for longer time intervals.

It appears therefore that the creep behaviour of rocks in many instances cannot be described by a simple combination of the three basic structural elements spring, dashpot, and strength; consequently, model theories are only of limited use in geodynamics. (For a general account of rheological model building, cf. *Reiner*, 1960). However, experimental results by *Price* (1966) on sandstone, limestone and granodiorite indicate that the behaviour of these rocks at low temperature and pressure is anelastic (instantaneous strain followed by transient creep) for stresses below a time-dependent yield point (creep strength), and viscoplastic at stresses above the yield point. Although time ranges of only a few weeks reduce the creep strength to 20–60% of the instantaneous strength, it seems that a finite creep strength may be retained even for geological time intervals, since residual stresses of great age are not relaxed.

In order to explain the experimental results, *Price* (1966) introduced a model constisting of a *Kelvin* element in series with a *Bingham* element. The model, with a slight modification to introduce a better representation of the *Bingham* element, is shown in fig. 2. For $\sigma < \sigma_Y$, no creep is observed beyond the transient stage; for $\sigma > \sigma_Y$, the creep strength is overcome and the material flows with constant plastic viscosity η_{pl}. In *Reiner*'s (1971) terminology, this is a *Schofield-Scott Blair* body. The complete stress-strain relation and the time-strain under constant stress are given respectively by

$$\eta_K\ddot{\varepsilon} + \mu_K\dot{\varepsilon} = \frac{\eta_K}{\mu_H}\ddot{\sigma} + \left(1 + \frac{\mu_K}{\mu_H} + \frac{\eta_K}{\eta_{pl}}\right)\dot{\sigma} + \frac{\mu_K}{\eta_{pl}}(\sigma - \sigma_Y) \qquad [6]$$

and

$$\varepsilon(t) = \frac{\sigma}{\mu_H} + \frac{\sigma}{\mu_K}\left[1 - \exp\left(-\frac{\mu_K}{\eta_K}t\right)\right] + \frac{\sigma - \sigma_Y}{\eta_{pl}}t\,. \qquad [7]$$

These relations, however, are in all likelihood not applicable to the bulk of mantle material, although a linear stress-strain rate relation (for stresses above the creep strength) may apply in the asthenosphere.

Since the suitable temperature and strain rate conditions exist in the mantle, steady-state creep should occur in all but the uppermost few tens of kilometres of the Earth, if stress-differences are sufficiently large. (Physically, the condition for steady-state creep is that the temperature be high enough to counterbalance through spontaneous recovery the work-hardening produced by deformation.) However, there is no agreement on the type of stress-strain rate relation. Some of the most commonly discussed rheological

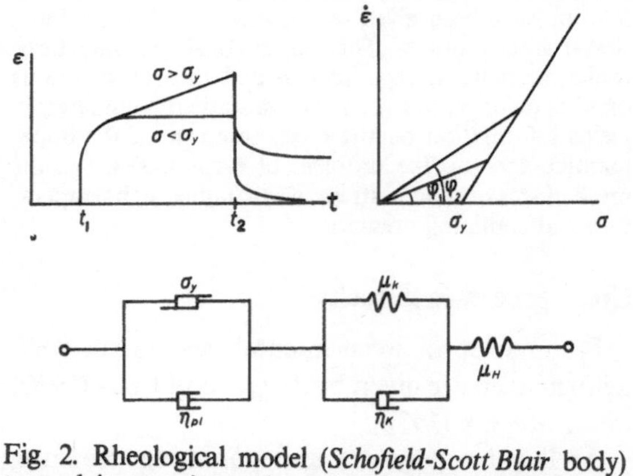

Fig. 2. Rheological model (*Schofield-Scott Blair* body) to explain experimental results by *Price* (1966). In the structural model, μ_H, σ_Y, and η_{pl} are the rigidity, yield strength, and plastic viscosity of the *Bingham* element; μ_K and η_K constitute the *Kelvin* element. The inset on the upper left shows strain-time curves under constant stress above and below σ_Y (t_1 and t_2 are the time of loading and of unloading, respectively). The inset on the upper right shows the resulting stress-strain rate relation. Note that the effective viscosity ($\eta = 1/\varphi$) decreases with increasing stress

responses are illustrated in fig. 3. The whole spectrum, from linear viscosity to pure plasticity, has at one time or another been advocated in the literature. *McKenzie* (1968) is the most recent exponent of the school of tought which favours linear viscous behaviour of the mantle. *Stacey* (1963), *Orowan* (1965), *Weertman* (1970), and *Ranalli* (1971), among others, disagree.

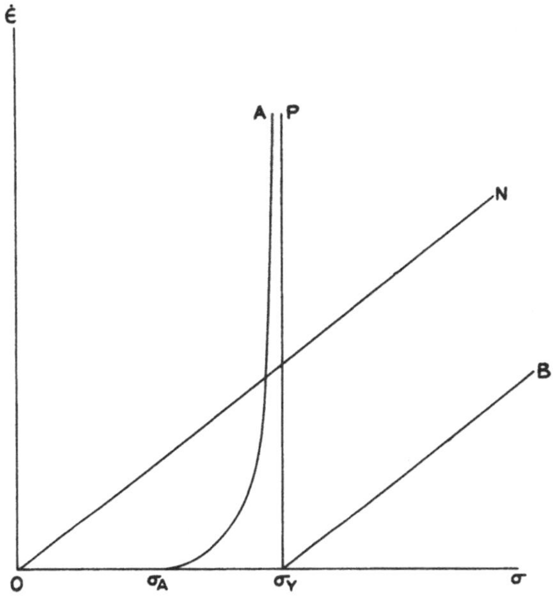

Fig. 3. Stress-strain rate relations for some rheological models and for high-temperature creep of polycrystalline aggregates. ON = linear viscous (*Newton*ian) material; $O\sigma_Y B$ = *Bingham* material; $O\sigma_Y P$ = purely plastic material; $O\sigma_A A$ = non-linear creep law (the creep strength σ_A being the stress below which strain rate is altogether negligible).

Creep in rocks can take place by a variety of mechanisms. The most important ones are diffusion of point defects (lattice vacancies), movement of line defects (dislocations), and grain-boundary sliding. Other phenomena such as synkinematic recrystallization may occur too.

Grain-boundary sliding can be discounted as a creep mechanism in the mantle. All flow processes which involve frictional sliding at grain boundaries require one of the principal stresses to be negative, otherwise voids cannot form in the material; they cannot therefore take place under high confining pressure.

Creep governed by the diffusion of lattice vacancies from one grain boundary to another is called diffusion creep or *Nabarro-Herring* creep. It leads to a linear (*Newton*ian) rheological equation

$$\dot{\varepsilon} = \frac{1}{\eta}\sigma \qquad [8]$$

where the viscosity is given by

$$\eta = \frac{kTd^2}{cDV} \qquad [9]$$

and the diffusion coefficient

$$D = D_0 \exp\left(-\frac{E}{kT}\right)\exp\left(-\frac{pV}{kT}\right). \qquad [10]$$

In the above formulae, k is *Boltzmann*'s constant, T the absolute temperature, d the mean grain size, c a numerical factor, V the activation volume, E the activation energy for diffusion, and p the pressure. Note that the viscosity, at any given pressure and temperature, depends critically on the grain size. High-temperature experiments carried on ceramics of grain sizes of the order of a few tens of microns and at stresses up to about 10^7–10^8 dyne cm^{-2}, as reported by *McKenzie* (1968), show *Nabarro-Herring* creep. However, the relevance of these experiments to creep processes in rocks is somewhat doubtful, since for larger grain sizes the stress-strain rate relation becomes non-linear even at smaller stresses, and grain sizes in the mantle are probably of the order of a few centimetres.

Creep governed by dislocation mechanisms leads to strain-rates which are non-linear in stress. *Stacey* (1963) has reviewed some of the relevant rheological laws. Commonly mentioned is the exponential creep law

$$\dot{\varepsilon} = \left[A \exp\left(-\frac{E}{kT}\right)\right]\exp\left(\frac{\beta\sigma}{kT}\right) \qquad [11]$$

where A is a constant, E the activation energy of the process, and β a positive constant depending on the stress-dependence of E. The factor in square brackets is independent of stress. Also of frequent use is the hyperbolic sine law

$$\dot{\varepsilon} = \left[2A \exp\left(-\frac{E}{kT}\right)\right]\sinh\left(\frac{\beta\sigma}{kT}\right). \qquad [12]$$

However, these laws have seldom been found to apply to rocks in steady-state creep. Rather, rocks seem to obey a power creep law, where strain rate is proportional to the nth power of the stress (usually with $3 \lesssim n \lesssim 5$). Recent experiments by *Carter* and *Ave'Lallemant* (1970) on high-temperature flow of dunite and peridotite (likely upper mantle materials) at confining pressure of $5 \cdot 10^9$ to $3 \cdot 10^{10}$ dyne cm^{-2}, temperature of 300–1400 °C, and strain rate of 10^{-3} to 10^{-8} sec^{-1}, have shown that steady-state creep takes place only at temperatures higher than 900 °C at the strain rates employed. In the steady-state creep regime, at given temperature and pressure, the data are fitted by a relation of the type

$$\dot{\varepsilon} = A\sigma^n \quad [13]$$

with $n \simeq 4.8$ in the absence of externally released water, and $n \simeq 2.4$ if externally released water is present. Extrapolation to geological strain rates and upper mantle conditions gives shear stresses in the $6–15 \cdot 10^6$ dyne cm^{-2} range, and effective viscosities of the order of 10^{20}–10^{21} poise. These estimates agree well with those based on geophysical observation (in particular, glacio-isostatic rebound), and strongly suggest that flow in the mantle is non-linear.

A power law for creep of polycrystals within the Earth has been proposed by *Weertman* (1970), on the basis that dislocation creep is the dominant deformation mechanism. If dislocation climb is the rate-controlling process, the stress-strain rate relation is

$$\dot{\varepsilon} = \alpha_1 D \left(\frac{\sigma}{\mu}\right)^m \left(\frac{\sigma V}{kT}\right) \quad [14]$$

where α_1 is a constant, D the diffusion coefficient (the dependence of which upon pressure and temperature is given by eq. [10]), μ the shear modulus, and V, k, and T the activation volume for diffusion, *Boltzmann*'s constant, and absolute temperature, respectively. The power m is usually comprised between 3.5 and 5. Eq. [14] fits well the experimental results on many materials (including rocks) for $10^{-5}\mu < \sigma < 10^{-3}\mu$, that is, for the range of stress probably existing within the mantle.

If the rate-controlling creep mechanism is dislocation glide, the stress-strain rate relation is

$$\dot{\varepsilon} = \alpha_2 D \left(\frac{\sigma}{\mu}\right)^2 \left(\frac{\sigma V}{kT}\right). \quad [15]$$

Diffusion creep and dislocation creep mechanisms operate independently of each other. *Weertman* (1970) finds that at low stresses ($\sigma < 10^4$ dyne cm^{-2}) *Nabarro-Herring* creep is dominant; at larger stresses, creep is produced by dislocation motion and is therefore non-linear. Moreover, since it is very likely that the grain size in eq. [9] changes during deformation (at least for large total strains), the assumption of *Newton*ian viscous behaviour would probably be incorrect even if diffusion creep were the dominant process.

The effective viscosity (ratio of stress to strain rate) of a material obeying a non-linear flow equation depends on stress. For any given stress larger than 10^4 dyne cm^{-2} the viscosity for *Nabarro-Herring* creep is larger than the effective viscosity for dislocation creep. *Weertman* (1970) has estimated the effective viscosity and the *Nabarro-Herring* viscosity for the mantle flowing at a strain rate of 10^{-16} sec^{-1}. The results are shown in fig. 4. [Previously, an estimate of the viscosity in diffusion creep had been given by *Gordon* (1965).] The estimates depend very sensitively on temperature and melting temperature, and therefore are only qualitatively correct. The viscosity in both cases has a minimum in the asthenosphere, and increases in the lower mantle. A grain size of 22 cm has been taken for *Nabarro-Herring* creep; this gives a viscosity of 10^{21} poise in the asthenosphere. The effective viscosity for dislocation creep in coarse-grained material does not depend on grain size.

Stacey (1969) has estimated the resistance to deformation of the upper mantle, on the assumption of the existence of a creep strength (below which the strain rate is negligible, above which it increases exponentially with stress). The result is also shown in fig. 4. Thus, no matter if effective viscosity or creep strength is considered, the

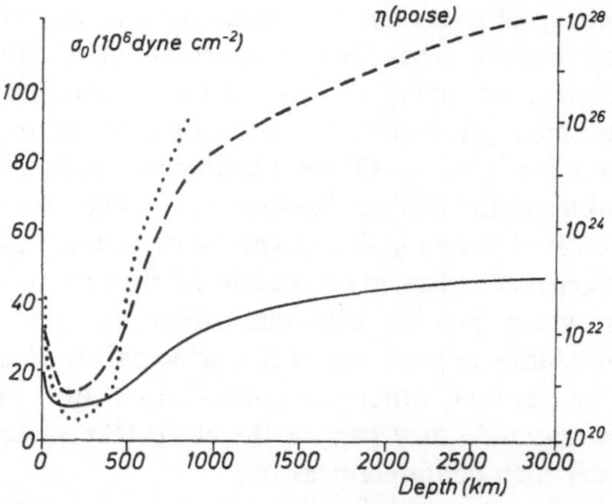

Fig. 4. Estimates of the effective viscosity in dislocation creep (continuous line), linear viscosity in diffusion creep (dashed line), and creep strength (dotted line) of the mantle. Horizontal scale gives depth in kilometres. Vertical scales give creep strength (left) and viscosity (right).

asthenosphere appears to be the layer where resistance to deformation has a minimum. The problem of the existence of a creep strength in the physical sense has not yet been solved; however, it is clear from the preceding considerations that, if sufficiently large stress-differences are present, the bulk of mantle material flows by dislocation creep and consequently follows a non-linear rheological law.

Continuous flow in the asthenosphere: glacio-isostatic rebound

The experimental and theoretical results discussed in the preceding section indicate that flow in the mantle is probably non-linear, and that the asthenosphere is a fundamental layer in the dynamics of the mantle. This layer is assumed to play an important role in global geotectonic movements related to continental drift, mountain building, and seismicity (cf., among others, *Jacoby*, 1970; *Ranalli*, 1971; *McKenzie*, 1972). Its comparatively low resistance to continuous deformation is thought to lie in the melting temperature to temperature ratio approaching unity, and in the presence of water that lowers the solidus temperature. These two factors should originate some grain-boundary melting (*Anderson* and *Sammis*, 1970). However, even in a partially melted asthenosphere, grain interlocking should still occur.

Direct evidence on the rheological properties of the asthenosphere comes from glacio-isostatic rebound. The Earth's lithosphere tends to maintain isostatic equilibrium, that is, a constant total mass of rock in any vertical column of unit cross section. Consequently, a region upon which an additional load is superimposed will tend to sink, while an unloaded region will tend to rise in order to return to its equilibrium level. Since the lithosphere is relatively rigid, the occurrence of isostatic equilibrium implies the occurrence of sub-lithospheric flow. From the interpretation of gravity anomalies, it is concluded that the creep strength which must be surpassed to permit flow at a depth of about 100–200 km is 10^7 dyne cm^{-2} or less (*Gutenberg*, 1958). This contrasts with the strength of the lithosphere as determined from the support of mountains, which is of the order of 10^9 dyne cm^{-2} (*Jeffreys*, (1962). (Note, however, that mountain belts could be dynamically and not statically supported.)

Glacio-isostatic rebound is the isostatic response of the Earth due to the removal of the ice load at the end of the last glaciation. Of the regions affected by the late Quaternary glaciation, Fennoscandia and North America are the best studied. In North America, for instance (cf. *Walcott*, 1972b, for a review), the continental ice sheets had melted almost completely by 6500 years B.P. (before present). The Earth's surface has not yet completed its adjustment to the removal of the ice load; it has risen 138 m in the centre of the uplifted region (the present rate is 2 ± 0.5 cm yr^{-1} in Southern Hudson Bay, Canada), and there may be a residual uplift of 300 $\pm$ 120 m before isostatic equilibrium is established. The isostatic response of Fennoscandia is very similar.

There have been many interpretations of glacio-isostatic rebound in terms of the rheology of the underlying mantle. (To quote only two recent examples, *Artyushkov*, 1971, and *Lliboutry*, 1971. The literature on the topic has been reviewed by *Walcott*, 1972c.) Usually, linear viscosity is assumed in the analysis. The uplifts can be interpreted either in terms of a half-space model (with constant or variable viscosity), or in terms of a thin-channel model, where a low-viscosity asthenosphere overlies a rigid (or highly viscous) substratum. Sometimes the effect of an elastic lithosphere is also taken into account.

The thin-channel models are more consistent with the geophysical evidence on the properties of the mantle. The relaxation of a deflection ζ on the surface of a fluid of viscosity η, if the lateral dimensions of the irregularity are much larger than the thickness of the fluid, is governed by the diffusion equation

$$\frac{\partial \zeta}{\partial t} = D^* \left(\frac{\partial^2}{\partial x^2} + \frac{\partial^2}{\partial y^2} \right) \zeta \qquad [16]$$

where t is time, and x, y are the rectangular coordinates on the surface. The diffusion coefficient is given by

$$D^* = \frac{\varrho g h^3}{12\eta} \qquad [17]$$

where ϱ, η, h are the density, viscosity, and thickness of the fluid layer, and g is the acceleration of gravity. The uplift data give a value for the diffusion coefficient ($D^* = 35$–48 km^2 yr^{-1} in Fennoscandia, $D^* = 52$ km^2 yr^{-1} in Canada), and the viscosity must be derived on the assumption of some channel thickness. For realistic channel thicknesses, most determinations of the viscosity of the asthenosphere in the Fennoscandian and Canadian regions are of the order of 10^{20}–10^{21} poise. The viscosity of the substratum is probably comprised between 10^{23} and 10^{26} poise. However, although glacio-isostatic rebound serves to indicate the probable rheological response of the mantle in terms of linear viscosity, no model is uniquely determined from the data.

It therefore appears that Earth models based on glacio-isostatic rebound qualitatively agree with those obtained from theoretical considerations. However, there is a very important difference. Glacio-isostatic rebound models assume the asthenosphere to be a linear viscous body, whereas the consideration of creep mechanisms in polycrystals leads to the conclusion that the rheological behaviour is non-linear. Attempts to determine the rheology of the asthenosphere (rather than assuming it a priori) have not been many. Recently *Walcott* (1972a) has concluded, from the shape of the uplift curves in Canada, that the flow law that applies to the asthenosphere is essentially linear. *Scheidegger* (1970) had previously reached a different conclusion. *Liboutry* (1971) has shown that the flow law of the asthenosphere affects the profile of the raised shorelines in the uplifted region, but unfortunately no definite conclusion can be reached from the present data.

It may well be that partial melting affects the rheological properties of the asthenosphere in such a way that the creep law becomes effectively linear, whereas the power law applies to the bulk of the mantle. A creep strength of the order of 10^6 to 10^7 dyne cm^{-2}, however, might still be present. For larger stresses, then, the asthenosphere would behave as a *Bingham* model (see the inset in fig. 2). Theoretically, it is possible to detect the presence of a finite creep strength by comparing the effective viscosity for different loads: in a *Bingham* body, the effective viscosity decreases with increasing stress (or strain rate). The properties of the asthenosphere, however, vary from region to region, and a direct comparison is impossible.

Actually, as *Weertman* (1970) has pointed out, a linear flow law determined from glacio-isostatic rebound does not necessarily imply that the behaviour of the asthenosphere is *Newton*ian. In a material obeying the power creep law, an individual component of strain rate varies linearly with the corresponding component of deviatoric stress if the latter is small compared to the second invariant of the deviatoric stress. Thus, if rebound stresses are superimposed to larger stresses in the asthenosphere, the flow law determined from rebound will be apparently *Newton*ian.

Discontinuous deformation: earthquakes

The evidence examined so far is mainly related to the flow response of rocks to stress. The occurrence of earthquakes, however, indicates that rupture, with sudden energy relesase, takes place within the Earth. The spatial distribution of earthquakes (*Richter*, 1958; *Rothé*, 1969) shows that sub-lithospheric shocks occur in elongated seismic zones, mostly in the circum-Pacific belt, dipping about 45° from the surface, and reaching a maximum depth of approximately 700 km (cf. *Isacks* et al, 1968, and *Oliver*, 1970, for a discussion of their features). The following arguments apply only to the seismic zones of the mantle, and not to a hypothetical spherically symmetric Earth.

Whatever the mechanism of seismogenic failure, seismic strength can be defined as the stress level at which earthquakes take place. Seismic strength probably bears no relation with the long-range creep strength of earth materials. The pattern of seismic strength versus depth has recently been estimated by the author (*Ranalli*, 1972). The estimation is based on the formula (suggested by *Bullen*, 1953)

$$S^2 \geq \frac{2\mu}{q}\frac{E_s}{V_s} \qquad [18]$$

where S is the strength, μ the shear modulus, q the seismic efficiency (ratio of seismic wave energy to energy released), and E_s, V_s the seismic wave energy and seismic volume, respectively. Using reasonable values for the quantities on the right-hand side of [18] at various depths (for the seismic volume, cf. *Duda*, 1970), a minimum estimate of the seismic strength versus depth can be obtained. The results are shown in fig. 5. In the upper lithosphere, S is of the order of 10^8 dyne cm^{-2} (this coincides with earlier estimates by *Tsuboi*, 1933, and *Chinnery*, 1964). It increases sharply to more than 10^9 dyne cm^{-2} in the 50–125 km depth range, it has a relative minimum at depths from 150 to 450 km, and then it tends to increase again.

This pattern of seismic strength versus depth can be explained in terms of global geodynamic

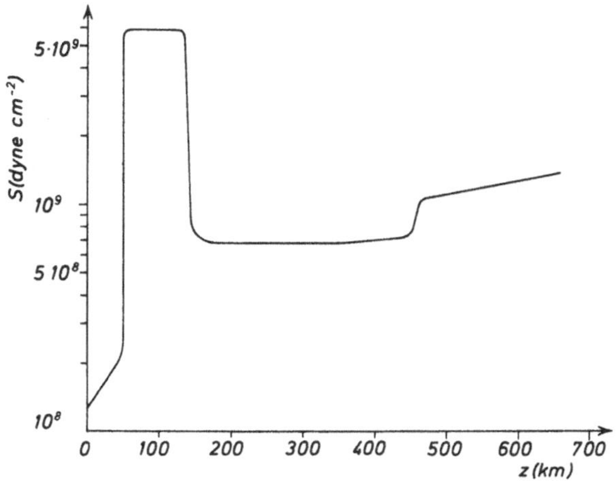

Fig. 5. The pattern of seismic strength with depth in the seismically active regions of the crust and upper mantle.

processes (*Ranalli*, 1972). Seismic belts are zones along which the lithosphere sinks into the upper mantle (*Isacks* et al., 1968; *McKenzie*, 1972). The upper lithospheric material, the strength of which is of the order of 10^8 dyne cm^{-2}, is carried downwards with the sinking lithospheric slab. Due to its low thermal conductivity, the material is subjected to higher pressure while still cooler than the surrounding mantle material. Under these conditions the effect of pressure is predominant and increases the strength by one order of magnitude at depths from 50 to 125 km. At larger depths, however, the effect of temperature becomes appreciable and decreases the strength. After a small jump at a depth of 450 km (probably related to phase transitions), the strength increases again. There is no indication that the seismic strength vanishes at depths larger than 700 km; consequently, the disappearance of seismic shocks with depth is attributable to a decrease in the available shear stress rather than to a vanishing strength.

The previous discussion gives an estimate of the variation of seismic strength with depth in the earthquake zones, but it does not yield any information on the actual mechanism of seismic failure. The pattern of first motions around an earthquake focus, as determined from seismograms, resembles the pattern commonly associated with a *Coulomb*-type fracture (rupture followed by sliding of the two walls). This similarity notwithstanding, *Coulomb-Navier* fracture can only account for earthquakes occurring in the uppermost 10–20 km of the lithosphere. With reference to eq. [1], it is seen that the critical shear stress increases with increasing normal stress. At high confining pressure, therefore, the shear stress required for *Coulomb-Navier* fracture is so high that, even if it were available, it would deform the rock bodily rather than cause frictional sliding.

The situation changes somewhat if a fluid pore pressure p^* is present in the rock. In this case the *Coulomb-Navier* criterion becomes (cf. e.g. *Jaeger* and *Cook*, 1969)

$$\tau = \tau_0 + (\sigma_n - p^*)\tan\varphi \qquad [19]$$

and consequently a rock can fail in shear at stress-differences lower than those required for failure of dry rocks. Since a pore pressure can be produced by dehydration of hydrated minerals (cf. *Paterson*, 1970, for a review of the experimental evidence), the possibility of *Coulomb*-type fracture may be extended to the maximum depth of occurrence of dehydration reactions (100 km, say). Still, this does not account for the occurrence of deeper earthquakes.

A seismic shock could also be originated by a sudden phase transition (*Bridgman*, 1945). The possibility of such phase transitions in the upper mantle, however, has not been established, although *Ringwood* (1972) has recently suggested that their likelihood is increased by global mantle movements.

There has been a tendency in the geophysical literature to consider flow and earthquakes as mutually excluding processes. Perhaps the solution to the riddle of deep shocks lies in the acceptance of the fact that, in polycrystalline aggregates obeying a non-linear creep law, rupture is sometimes not counteracted but produced by flow. The simultaneous existence of these two phenomena in the Earth's interior was suggested by *Bridgman* (1936) on the basis of experimental results. As *Orowan* (1960, 1965) has pointed out, if creep produces structural changes that accelerate further creep, a situation of plastic instability can develop in which high flow rates occur along thin bands, and this may have something to do with the origin of deep earthquakes. (Plastic instability phenomena are enhanced by high temperature and low strain rate, since work-hardening is then effectively removed by thermal softening during deformation.) *Stacey* (1963) has advanced the same hypothesis. A self-accelerating creep rate in a region of plastic instability can either lead to an earthquake in itself or cause local shear melting which allows sudden energy release. The latter possibility has been examined by *Griggs* and *Handin* (1960), and *Griggs* and *Baker* (1969).

Thus, some kind of creep rupture seems to be the most likely mechanism for the majority of sub-lithospheric earthquakes. There are some theories of creep rupture under multiaxial state of stress (cf. for instance *Odqvist*, 1971), but usually one of the principal stresses is assumed to be tensile, a situation that does not apply to the Earth. Further analysis of the problem from a geological viewpoint is necessary.

Conclusions

The main points made in this paper can be briefly summarized as follows.

(i) Experimental results on rock deformation show that, at high temperature and pressure and slow strain rate, a situation of steady state creep is reached, in which the strain rate is usually proportional to the *n*th power of the stress ($n > 1$). A creep strength is present, and it decreases with increasing time of loading. Only in few instances the rheological behaviour of rocks can be represented by simple structural models (*Schofield-Scott Blair* body).

(ii) Theoretical considerations on creep mechanisms of polycrystalline aggregates lead to the conclusion that dislocation creep, and not *Nabarro-Herring* creep, is the predominant mode of deformation of the Earths mantle. Dislocation creep leads to a highly non-linear dependence of strain rate upon stress, thereby confirming the experimental results.

(iii) Estimates of the variations of effective viscosity, *Nabarro-Herring* viscosity, and creep strength of the mantle of the Earth indicate that the resistance to deformation has a minimum in the 100–400 km depth range (asthenosphere), then increases with depth. This conclusion is valid no matter what rheological law is chosen.

(iv) Data from glacio-isostatic rebound, under the assumption of linear viscosity, yield values of the order of 10^{20}–10^{21} poise in the asthenosphere, and some order of magnitude larger in the lower mantle. Even if the assumption of linear viscosity is confirmed by the data, this need not conflict with the non-linear (power) creep law, provided that the stresses due to glacio-isostatic rebound are superimposed on much larger stresses of different origin.

(v) The pattern of seismic strength (stress level at which seismic energy release takes place) versus depth in the seismically active regions of the Earth shows a maximum of more than 10^9 dyne cm^{-2} at a depth of about 100 km, then decreases to a relative minimum, and increases again at depth larger than 450 km. The disappearance of earthquakes at depths larger than 700 km is due to decrease in available shear stress and not to vanishing seismic strength. The mechanism of sub-lithospheric earthquakes most probably resides in some form of high-temperature creep instability.

Acknowledgements

The author wishes to thank Dr. *R. I. Walcott* of the Earth Physics Branch, Department of Energy, Mines and Resources, Ottawa, Canada, who kindly made available some of his manuscripts before publication.

This research was supported by National Research Council of Canada grant no. A 7971.

Summary

The basic problem of geodynamics is to explain the deep-seated processes which originate tectonic activity and earthquakes. To this end, the rheological conditions of materials within the Earth must be determined. The importance of the time factor and the difficulty of experimentally reproducing geodynamic conditions relevant to sub-lithospheric processes make it necessary to employ indirect evidence and to make heavy use of inference and analogy.

Some constitutive equations for polycrystalline aggregates under high temperature and pressure and slow strain rate are examined and their consequences tested against occuring geodynamic processes. Sub-lithospheric deformation can be either continuous (slow redistribution of mass in the field of gravity, flow in the asthenosphere, convective movements in the mantle) or discontinuous (earthquakes). Two main questions are still unsolved in the rheology of continuous geodynamic processes, namely, the vanishing or non-vanishing value of the creep strength, and the linear or non-linear nature of the viscosity. The answer to these questions must be found in the interpretation of surface geological observations, particularly glacio-isostatic rebound, but no definite solution is available to date.

In the rheology of discontinuous geodynamic processes, the very occurrence of sub-lithospheric earthquakes is puzzling. Ordinary *Coulomb-Navier* shear fracture (with or without pore fluid pressure) is not an adequate explanation for the occurrence of most seismic shocks. Other mechanisms that have been proposed range from creep instability and shear melting to dehydration and phase transition. Under high pressure and temperature conditions, continuous and discontinuous deformation are not mutually excluding processes, but they can occur in the same material and one may lead to the other. The most promising working hypothesis for sub-lithospheric shocks is some form of high-temperature creep instability leading eventually to rupture and sudden stress release.

References

Anderson, D. L. and *C. Sammis*, Phys. Earth Planet. Interiors **3**, 41–50 (1970).

Artyushkov, E. V., J. Geophys. Res. **76**, 1376–1390 (1971).

Birch, F., Megageological considerations in rock mechanics. In: *W. R. Judd* (Editor), State of Stress in the Earth's Crust. pp. 54–80 (New York 1964).

Bott, M. H. P., The Interior of the Earth. 316 pp. (London 1971).

Bridgman, P. W., J. Geol. **44**, 653–669 (1936).

Bridgman, P. W., Amer. J. Sci. **243 A**, 90–97 (1945).

Bullen, K. E., Trans. Amer. Geophys. Union **34**, 107–109 (1953).

Carter, N. L. and *H. G. Ave'Lallemant*, Bull. Geol. Soc. Amer. **81**, 2181–2202 (1970).

Chinnery, M. A., J. Geophys. Res. **69**, 2085–2089 (1964).

Duda, S. J., Bull. Seism. Soc. Amer. **60**, 1479–1489 (1970).

Gordon, R. B., J. Geophys. Res. **70**, 2413–2418 (1965).

Griggs, D. T. and *D. W. Baker*, The origin of deep focus earthquakes. In: *H. Mark* and *S. Fernbach* (Editors), Properties of Matter under Unusual Conditions. pp. 23–42 (New York 1969).

Griggs, D. T. and *J. Handin*, Mem. Geol. Soc. Amer. **79**, 347–364 (1960).

Gutenberg, B., Rheological problems of the Earth's interior. In: *F. R. Eirich* (Editor), Rheology, Theory and Applications, vol. 2, pp. 401–431 (New York 1958).

Isacks, B., *J. Oliver*, and *L. R. Sykes*, J. Geophys. Res. **73**, 5855–5899 (1968).

Jacoby, W. R., J. Geophys. Res. **75**, 5671–5680 (1970).

Jaeger, J. C. and *N. G. W. Cook*, Fundamentals of Rock Mechanics. 513 pp. (London 1969).

Jeffreys, H., Geophys. J. Roy. Astr. Soc. **1**, 92–95 (1958).

Jeffreys, H., The Earth, 4th ed. 438 pp. (Cambridge 1962).

Jeffreys, H., Nature **225**, 1007–1008 (1970).

Lliboutry, L. A., J. Geophys. Res. **76**, 1433–1446 (1971).

Lomnitz, C., J. Geol. **64**, 473–479 (1956).

McKenzie, D. P., The geophysical importance of high-temperature creep. In: *R. L. Phinney* (Editor), The History of the Earth's Crust. pp. 28–44 (Princeton 1968).

McKenzie, D. P., Plate tectonics. In: *E. C. Robertson* (Editor), The Nature of the Solid Earth. pp. 323–360 (New York 1972).

Odqvist, F. K. G., Theories of creep rupture under multiaxial state of stress. In: *A. I. Smith* and *A. M. Nicolson* (Editors), Advances in Creep Design. pp. 31–48 (London 1971).

Oliver, J., Phys. Earth Planet. Interiors **2**, 350–362 (1970).

Orowan, E., Mem. Geol. Soc. Amer. **79**, 323–345 (1960).

Orowan, E., Phil. Trans. Roy. Soc. (London) **A 258**, 284–313 (1965).

Paterson, M. S., Experimental deformation of minerals and rocks under pressure. In: *H. Ll. D. Pugh* (Editor), Mechanical Behaviour of Materials under Pressure. pp. 191–235 (Amsterdam 1970).

Price, N. J., Fault and Joint Development in Brittle and Semi-brittle Rock. 176 pp. (Oxford 1966).

Ranalli, G., Tectonophysics **11**, 261–285 (1971).

Ranalli, G., Pure Appl. Geophys. (in press 1972).

Reiner, M., Lectures on Theoretical Rheology, 3rd ed. 158 pp. (Amsterdam 1960).

Reiner, M., Advanced Rheology. 374 pp. (London 1971).

Richter, C. F., Elementary Seismology. 768 pp. (San Francisco 1958).

Ringwood, A. E., Earth Planet. Sci. Letters **14**, 233–241 (1972).

Rothé, J. P., The Seismicity of the Earth, 1953–1965. 336 pp. (Paris 1969).

Scheidegger, A. E., Ann. Geofis. **23**, 325–346 (1970).

Stacey, F. D., Icarus **1**, 304–313 (1963).

Stacey, F. D., Physics of the Earth. 324 pp. (New York 1969).

Tsuboi, C., Bull. Earthq. Res. Inst. Tokyo Univ. **11**, 275–277 (1933).

Walcott, R. I., Symp. Recent Crustal Movements, Moscow (in press 1972a).

Walcott, R. I., Rev. Geophys. Space Phys. (in press 1972b).

Walcott, R. I., Ann. Rev. Earth Planet. Sci. (in press 1972c).

Weertman, J., Rev. Geophys. Space Phys. **8**, 145–168 (1970).

Author's address:
Dr. *Giorgio Ranalli*
Dept. of Geology
Carleton University
Ottawa, Ontario K1S 5B6 (Canada)

Rheol. Acta 13, 596–601 (1974)

From the Department of Civil Engineering, Massachusett Institute of Technology Cambridge (U.S.A.)

Statistical analysis of damage accumulation in viscoelastic systems

By H. Findakly), J. E. Soussou**), and F. Moavenzadeh***)*

With 13 figures

(Received October 27, 1972)

I. Introduction

Flexible pavements belong to a class of structures in which the progression and accumulation of damage is largely time-dependent. A highway pavement is a joint product of complex interactions between the pavement structure, vehicular loads, and climatic environment operating on the system. Therefore, the behavior of the system is greatly influenced by these parameters. Any uncertainty associated with one or more of these factors implies a corresponding uncertainty in the response and overall performance of the system.

Recent studies and observations conducted on the behavior of materials and structures have emphasized the variabilities in the magnitude and distribution of structural loadings, in the materials properties, and in the surrounding environment. These variabilities inevitably will result in randomness of the response of these structures to such excitations. Because of these unavoidable variabilities, the concept of structural design is bound up with probabilities of overload (mechanical or otherwise), and inadequate strength.

In highway pavements, the uncertainties associated with the response and behavior of the system are attributed to the following factors:

1. The inherent variability in the physical characteristics of the system. This is manifested by the statistical scatter of the materials properties in the layered system. This scatter is a result of the processing, mixing, placing, sampling, and testing activities.
2. The unpredictable changes in the surrounding environment. These generally involve changes in temperature, in moisture, in rainfall and snow, and a combination of two or more of these elements.
3. The uncertainty associated with modes of traffic load application. Generally, the loading process is a random process both in time and space.

This study presents a two-part model for the analysis of the pavement under these conditions. In the first part, the pavement structure is represented by a three-layer linear viscoelastic model which is used to predict the history stresses strains and displacements at any point within the system. Because of the statistical scatter in the properties of the materials in each layer, a simulation model based on the Monte Carlo method (1) is developed to yield statistical estimates of the above response parameters due to a single static load applied at the surface.

The second part of the model uses the above response to yield probabilistic information on the progression of damage within the system under a history of random load and varying temperatures. In this context, an arbitrary random history of load, both temporal and spatial, is applied on the system while the temperature is allowed to vary over some discrete time periods. The damage parameters chosen are the same as those used by *Aasho* (2) in their evaluation of pavement serviceability. These parameters are: rutting, slope variance, and cracking of the system. The methods of approach for the prediction of these components statistically is described in the following section.

The prediction of these damage components analytically allows the study of the behavior of the system in an operational environment, as well as the serviceability[1]), reliability, and life expectancy of any design configuration.

An example demonstrating the application of this method is provided in the results section.

II. Description of the model

The model presented in this study is referred to as the "structural model", which is a mathematical model of the pavement structure. For analytical convenience, the model is decomposed into two parts, depending on the nature of the information each part is concerned with. The first part is referred to as the "primary response model". A flow chart showing the overall structural model is presented in fig. 1.

The pavement structure is represented by a three-layer half-space linear viscoelastic or partly viscoelastic model. The geometry of the system is presented in fig. 2. The materials in each layer are homogeneous, isotropic, and incompresible, represented by a set of creep or elastic compliances. These are obtained from creep tests in the laboratory, and are assumed to have a certain statistical scatter as shown in fig. 3. These constitute the responses to the primary response model.

*) Research Associate, Department of Civil Engineering.

**) Assistant Professor, Department of Civil Engineering.

***) Professor of Civil Engineering, Mass. Inst. of Tech.

[1]) Serviceability in this context is used as a measure of the structural integrity of the system.

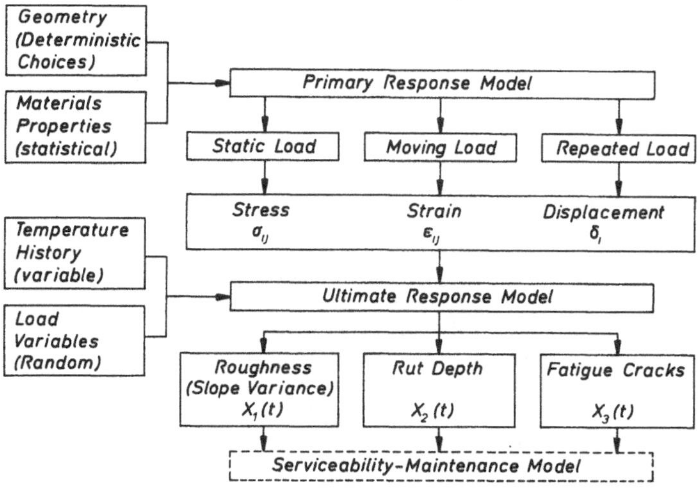

Fig. 1. Flow chart of the structural model

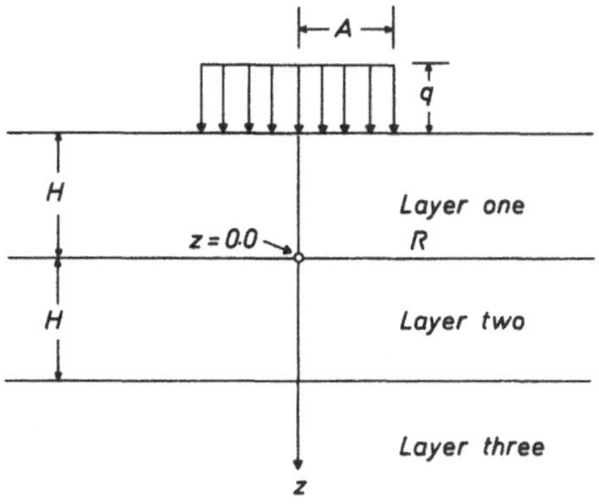

Fig. 2. Three-layer system

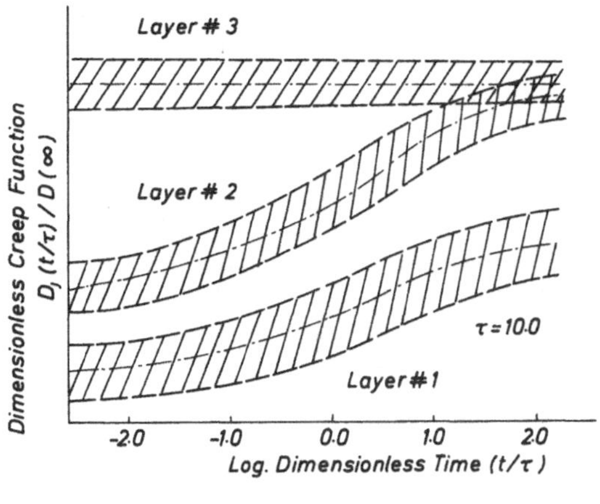

Fig. 3. Creep functions of the simulated pavement system $D_j(t) = \sum_i G_i e^{-t\delta_i}$

II.1. Primary response model

This model utilizes a mechanistic model, based on layered-systems theory, and a simulation procedure based on the Monte Carlo method (1, 3) to yield histories of stresses, strains, and displacements at any desired point within the system under a step-loading function applied as shown in fig. 2. These response histories are referred to as the system characteristic functions, which may be described as the creep functions of the overall layered system. These functions may be expressed in terms of an exponential series of the following form, using the least square fitting procedure:

$$f(t) = \eta \sum_{i=1}^{N} \bar{G}_i e^{-t\delta_i} \qquad [1]$$

where $f(t)$ represents the response characteristic function; η is a random variable with a mean value = 1.0, and a standard deviation equal to the coefficient of variation of G: $\bar{G}_i$ is the mean value of the coefficient of the stress (G_i); δ_i is the exponent resulting from the curve fitting and t refers to the time value.

II.2. Ultimate response model

The ultimate response model uses the results of the simulation analysis to provide probabilistic estimates of rutting, slope variance, and cracking at any desired time point under random modes of load repetitions and varying temperatures (4).

The load pattern is represented by an independent random arrival of the Poisson type, in which the load amplitudes and velocities[2]) are also described by certain random spectra. These distributions are shown in fig. 4.

The temperatures also operate on the system, and are allowed to vary over some discrete points in time. The effects of temperature are introduced by mapping of the response of the system using a time-temperature superposition for the system characteristic function described in eq. [1].

A typical response function to a history of random loads and varying temperatures, can be expressed in the following convolution integral form:

[2]) Velocities are described in terms of the load duration on the system.

$$f[t, \tau, \underset{\tau=-\infty}{\overset{t}{\varphi}} (\tau)] = \alpha[\varphi(t)] + \beta[\varphi(t)]$$
$$\times \int_{-\infty}^{t} [\{\sum_{i=1}^{N} G_i \exp(-t^* \delta_i)\, d\tau\}$$
$$- \int_{\tau}^{t} \sum_{i=1}^{N} G_i \exp(-s^* \delta_i)\, d\beta[\varphi(s)]\}]. \qquad [2]$$

The second integral in this equation is a corrective term and may be neglected for all practical purposes (4).

In the above equation, we have:

$$t^* = \int_{\tau}^{t} \gamma(x)\, dx$$

$\alpha[\,]$, $\beta[\,]$, and $\gamma(\,)$ are mapping parameters for temperature effects, in which α a factor for vertical change of cale (assumed = 0 in this study); β a vertical shift factor [assumed = $T(t)/T_0$]; γ a horizontal shift factor [= 10** {.162 $(T - T_0)$} for the materials analyzed]; T_0 a reference temperature in ok; $\varphi(\,)$ is a vector representing temperature history, and $f[\,,\,]$ is a vector for the system's response history.

The response function described by eq. [2] is convoluted with load history, as follows: and solved probabilistically to yield these major indicators of pavement distress: rut-depth, slope variance, and cracking.

$$R[t] = \int_{-\infty}^{t} f(t, \tau, \varphi) \frac{\partial P(\tau)}{\partial \tau}\, d\tau \qquad [3]$$

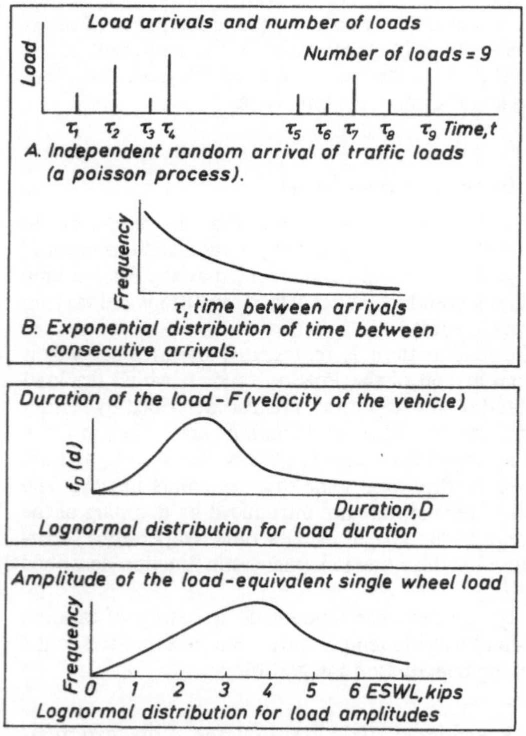

Fig. 4. Distribution of load characteristics

where $P(t)$ represents the load history, represented by three statistical elements: rate of load applications: λ_1 duration of loads; D_1 and load amplitudes: A.

II.3. Damage indicators

The mechanisms for the development of damage indicators.

1. Rutting

This component is believed primarily to be the result of a channelized system of traffic which causes differential surface deformations under the wheel-paths. Given the statistical characteristics of the road materials and those of the traffic, one can determine the rut-depth from the spatial distributions of traffic loads. The split Poisson property is invoked[3]) to obtain the differential surface deformation due to traffic channelization.

2. Slope variance

This component is a measure of the roughness at the surface deformations along the pavement provile. It is determined using the spatial correlation properties of the in situ materials defining autocorrelation function of the surface deformation of the system [4].

3. Cracking

A phenomenological model is used for prediction of crack accumulation in the system. This combines a fatigue equation and *Miners* rule, to obtain a qualitative description of the degree of cracking in the system. *Miners* rule is expressed as

$$D = \sum_{i} \frac{n_i}{N_{fi}} \qquad [4]$$

where D is the degree of damage due to cracking; n_i is the number of load cycles applied at a particular strain level, and N_{fi} is the number of load cycles to failure at that strain level.

$$N_{fi} = C(T) \left(\frac{1}{\varepsilon_i}\right)^{a(T)} \qquad [5]$$

in which $C(T)$, $a(T)$ are material characteristics which are temperature-sensitive, ε_1 is the tensile strain amplitude applied. A probabilistic reformulations of the above expressions have been developed such that both expected values and variances of the damage are obtained.

Probabilistic expressions of eq. [4] are given below:

$$\bar{D}(t) = \sum_{i=1}^{L} \left\{\frac{\bar{n}_i}{\bar{N}_{fi}} + \frac{\bar{n}_i}{\bar{N}_{fi}^3} \sigma^2_{N_{fi}}\right\}$$

$$\sigma^2_{D(t)} = \sum_{i=1}^{L} \left\{\frac{\bar{n}_i}{\bar{N}_{fi}^2} + \left(\frac{\bar{n}_i}{\bar{N}_{fi}^2}\right)^2 \sigma^2_{N_{fi}}\right\} \qquad [6]$$

[3]) This property states that if λ is the mean rate of an event of the *Poisson* type, then a split in arrival pattern also results in a *Poisson* process with a new rate λ^1.

where $\bar{X}$ and $\sigma_{\bar{X}}^2$ refer to the expected value and variance of the random variable X. $\bar{N}_{fi}$ and $\sigma_{N_{fi}}^2$ are derived probabilistically from eq. [5], which $\bar{n}_i$ is an input representing the rate of load applications.

Details of the probabilistic analysis for this model may be found in reference (4).

III. Results

In order to demonstrate this method of approach, an example is presented in this section. Typical materials properties used as an input to the primary response model are presented in fig. 1. These properties are represented by the creep compliances of the different layers. The output of the simulation for the primary is typified by figs. 5 and 6. Fig. 5 represents a typical simulation result of the system characteristic function which is the response of the layered system to a step load. In this figure, the statistical scatter around the mean is shown by the upper and lower bounds. Fig. 6 represents a cross-section of fig. 5 at a particular point in time, where the statistical scatter resembles a *Gaussian* or normal distribution. The above simulation results are conducted under isothermal conditions, when the only source of uncertainty is the random scatter of the creep functions. The results in figs. 5 and 6 are used as an input to the second step of the analysis, which is the ultimate response model. Along with these inputs, the load characteristics presented in fig. 4, and a set of temperature values also constitute the inputs to the ultimate response model, as shown in fig. 1.

A set of examples is presented in figs. 7 through 13. These examples are aimed at examining the effects of different design parameters on the

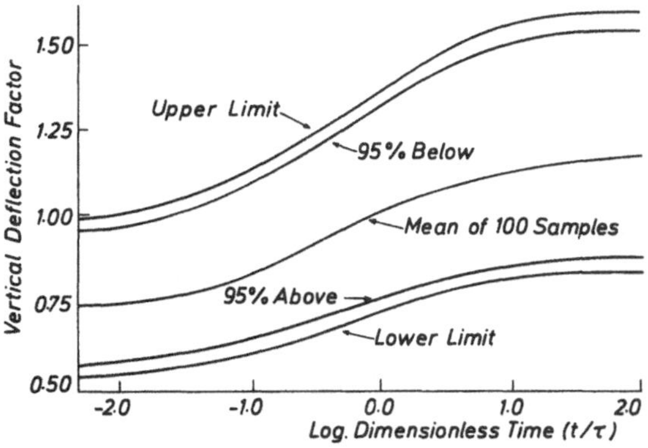

Fig. 5. Summary of simulation of vertical strains for uniformly distributed random inputs

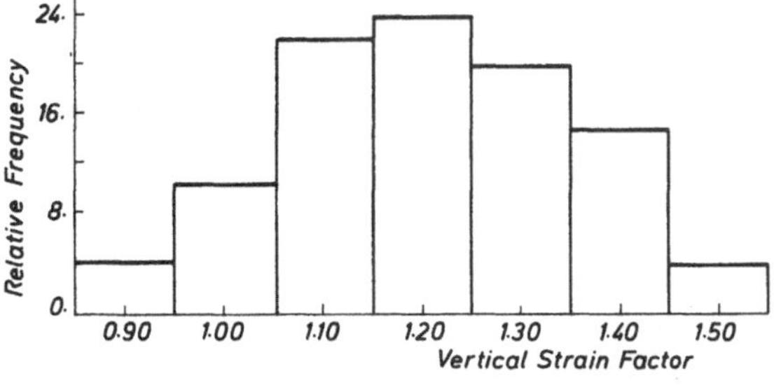

Fig. 6. Relative frequency distribution of vertical strain at $t/\tau = 10$. For normally distributed random inputs

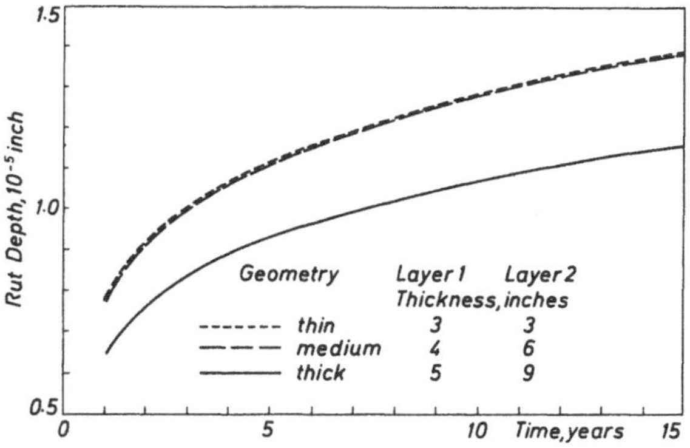

Fig. 7. Influence of geometry on rutting history

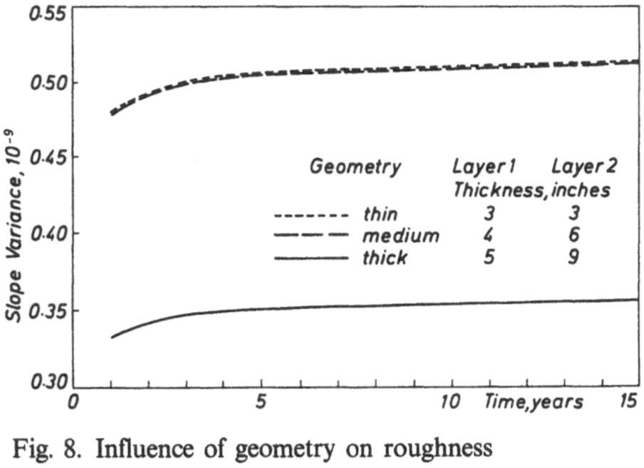

Fig. 8. Influence of geometry on roughness

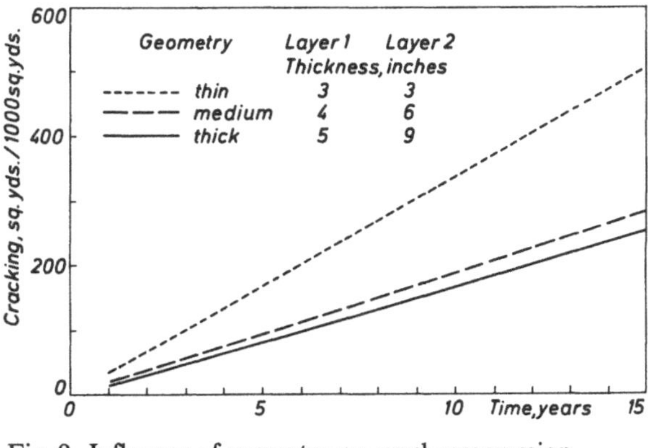

Fig. 9. Influence of geometry on crack progression

progression of damage within the system. Three design parameters are chosen for this purpose: geometry, materials properties, and quality control. Figs. 7 through 9 show the sensitivity of the rutting, slope variance, and cracking to changes in the geometry of the system. As may be observed from these figures, the trend in the behavior of the system agrees with the intuitive judgment in which the rate of damage progression is expected to slow down as the layers of the system become thicker. Similarly, Figs. 10 through 12 show that as the materials properties become more rigid, the rate of damage progression is less. It must be observed that in fig. 12 the weak system possesses a lower instantaneous tensile strain which is responsible for cracking, although it is less rigid at longer times, and therefore it is expected to crack less than the medium system.

The effect of the quality control, which is measured by the statistical scatter of the properties of the materials is shown in fig. 13, where the roughness of the surface is shown to increase at lower control levels. The scatter in the rutting of the system has also been observed to increase for lower quality control levels.

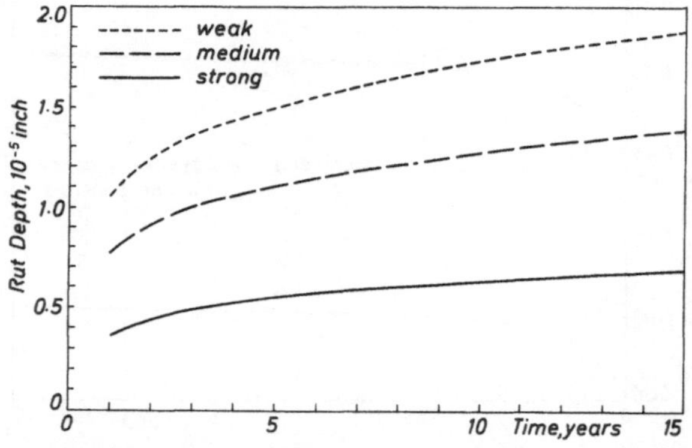

Fig. 10. Influence of material properties on rutting history

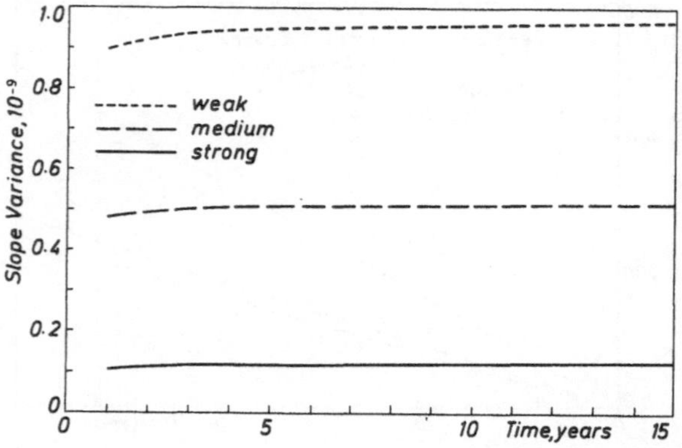

Fig. 11. Influence of material properties on roughness

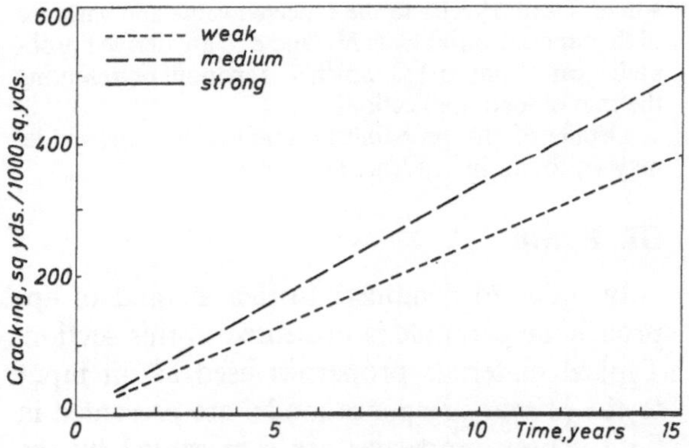

Fig. 12. Influence of material properties on crack progression

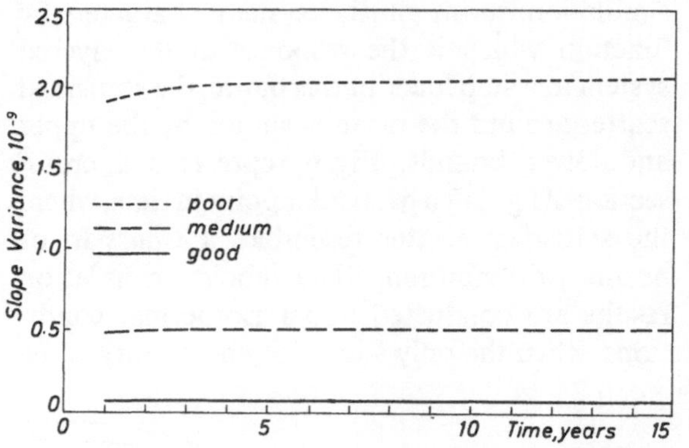

Fig. 13. Influence of quality control levels on roughness

Although the above analysis is of a limited nature, it demonstrates the capacity of the above probabilistic models to predict the response of the layered viscoelastic system and the damage progression within it at any time point subject to random histories of load and environment. The trends observed are in agreement with that anticipated behavior of the system as perceived by engineering experience and intuitive judgment.

IV. Conclusion

A two-part model is presented for the analysis of a three-layer linear viscoelastic system. The first part, referred to as the Primary response model, is concerned with the simulation of the response of the system in isothermal conditions under a step loading function. The properties of the materials in the layers are represented by statistical distributions of the creep compliances of the different layers. The resulting output is a statistical description of the history of the response at any point within the system. This response is expressed in terms of stresses, strains, or displacements vs. time.

The second part of the model, which ist the ultimate response model, uses the above response along with temperature and random load histories to yield probabilistic estimates of the damage progression within the system at any desired time period. Damage is represented by these components: rut depth, slope variance, and cracking as suggested in the AASHO Road Test.

A set of examples is presented to demonstrate the capacity of the model to predict realistic trends in the behavior of the system. The results show an agreement with the anticipated behavior of the system as perceived by intuitive judgment and experience.

The following features characterize the framework presented above:

1. The model recognizes the time-dependent behavior of the real system and accounts for this time-dependency through the viscoelastic representation of the materials in the layers.

2. The model recognizes and incorporates the elements of uncertainty resulting from the inherent scatters in the physical properties of the materials and in the randomness of load and environmental histories. An extensive probabilistic analysis is developed for this purpose.

3. Effects of temperature changes over time are accounted for in the model through mapping and shifting of the response of the system due to temperatures.

4. The model is particularly useful for purposes of analysis and design since it provides a means of studying the influence of the different parameters involved and their complex interactions on the behavior of the system at any point in time.

Acknowledgments

The work presented herein has been partially supported by the Office of Research, Federal Highway Administration, U.S. Department of Transportation, Washington, D.C. The computer work was done at the M.I.T. Information Processing Center, Cambridge, Massachusetts.

Summary

Highway pavements are generally represented by a threelayer viscoelastic model. In this paper a two-part model is presented to account for damage accumulation in such structures. The first part of this model uses simulation techniques to relate the statistical characteristics of the system response to the variability of the operating environment and of the material properties. The stress, strain and deflection outputs are used in the second part of the model to yield the statistical characteristics of three damage indicators: rutting, roughness and cracking. This second part of the model is based on a closed-form probabilistic solution. The material properties as well as the operating conditions, i.e., temperature, loading patterns and velocities, are given in terms of their statistical characteristics. The structural integrity and reliability with respect to each of the damage indicators are measured. A special feature of the second part of the model is that the responses are given for varying temperatures. An example of application is provided.

References

1) *Findakly, H.* and *F. Moavenzadeh*, A Simulation Model for Analysis of Highway Pavement Systems. Research Report R 71–7, Civil Engineering Dept., M.I.T., Cambridge, Mass., January 1971.

2) *Aasho*, The Aasho Road Test, Report 5. Pavement Research, Highway Research Board, Washington, D.C. 1962.

3) *Moavenzadeh, F., J. Soussou*, and *H. Findakly*, A Stochastic Model for Prediction of Accumulative Damage in Highway Pavements. Report No. FH-11-7473-1, Office of Research, Federal Highway Administration, U.S. Dept. of Transportation, Washington, D.C., 1971.

4) *Findakly, H.*, A Decision Model for Investment Alternatives in Highway Systems. Ph. D. Thesis, Dept. of Civil Engineering, M.I.T., Cambridge, Mass., 1972.

5) *Soussou, J.* and *F. Moavenzadeh*, Classical and Statistical Theories for the Determination of Constitutive Equations. Research Report R 70–33, Dept. of Civil Engineering, M.I.T., Cambridge, Mass., 1970.

Authors' address:
J. E. Soussou, H. Findakly, and *F. Moavenzadeh*
Department of Civil Engineering
Massachusett Institute of Technology
Cambridge (U.S.A.)

Rheol. Acta **13**, 602–607 (1974)

From the Department of Chemical Engineering, The University of Calgary, Calgary, Alberta (Canada), the Department of Chemical and Petroleum Engineering, The University of Alberta Edmonton, Alberta (Canada), and the Department of Chemical Engineering, Ecole Polytechnique, Montreal (Canada)

Numerical determination of retardation and relaxation spectra optimalization of numerical process

By J. Stanislav, F. A. Seyer, and B. Hlaváček

With 2 figures and 3 tables

(Received October 27, 1972)

List of symbols

a_{ij}	elements of matrix t
b_{ij}	elements of matrix β
k_{ij}	elements of matrix $\varkappa$
m_{ij}	elements of matrix $\mathcal{M}$
s	variable in eq. [1]
v_i	elements of matrix $\mathcal{V}$
x	variable in eq. [1]
γ	smoothing factor
λ	matrix eigenvalue
v	defined by eq. [6]
$\varphi(x)$	experimental measured function, defined by eq. [1]
$F(s)$	defined by eq. [1]
$F^*(s)$	defined by eq. [13]
$G'(x)$	real part of dynamic modulus
$H_3(x)$	third *Schwarzl-Staverman* approximation of relaxation spectrum
$J(x)$	creep compliance
$K(x, s)$	kernel of integral in eq. [1]
$\bar{L}(x)$	second *Schwarzl-Staverman* approximation of retardation spectrum
R_1	cumulative relative error of approximation of relaxation spectrum; defined by eq. [3]
R_2	cumulative relative error of spectral relaxation function; defined by eq. [4]

Introduction

The behavior of all viscoelastic materials under a variety of linear stress strain conditions may be described by any of the functions such as relaxation modulus $G(t)$, storage $G'(\omega)$ and loss modulus $G''(\omega)$, creep compliance $J(t)$, storage $J'(\omega)$ and loss compliance $J''(\omega)$. All of them may be determined by either transient or dynamic types of experiment and carry the signature of the time function which produced them. The two functions – the relaxation $H(\ln\tau)$ and retardation $L(\ln\tau)$ spectra were derived in order to depict the behavior of a system regardless of the type of experiment. It should be noted that symbol $H(\tau)\,d\ln\tau$, for the relaxation spectrum, will be used in this study to describe the contribution to rigidity associated with relaxation times whose logarithms lie in the range between $\ln\tau$ and $\ln\tau + d\ln\tau$.

In general any one of the experimentally determinable functions enumerated above, further denoted by common symbol $\varphi(x)$, is related to a spectral functions $F(s)$ through the *Fredholm* type of integral equation as follows

$$\varphi(x) = \int_a^b K(x, s)\, F(s)\, d\ln s \qquad [1]$$

where $K(x, s)$ is the kernel of a given integral equation.

Until now a number of methods have been proposed which make the solution of eq. [1] possible (1-20). Most of them (9–13, 15–20) use the numerical technique to calculate the function $F(s)$. However, the solutions obtained differ in accuracy and order of approximation.

This study is concerned with the development of a criteria for indicating how powerful a particular numerical method is.

Theory

The numerical solution of the integral eq. [1] is affected by a number of factors. One of them is the type of kernel function. This aspect will now be considered in some detail.

We rewrite the eq. [1] in a discrete form by using the matrix formalism.

$$\begin{bmatrix} \varphi(x_1) \\ \varphi(x_2) \\ \cdot \\ \cdot \\ \cdot \\ \varphi(x_n) \end{bmatrix} = \begin{bmatrix} k(x_1 s_1) \ldots k(x_1 s_n) \\ k(x_2 s_1) \ldots \\ \cdot \\ \cdot \\ \cdot \\ \ldots k(x_n s_n) \end{bmatrix} \cdot \begin{bmatrix} \bar{F}(s_1) \\ \bar{F}(s_2) \\ \cdot \\ \cdot \\ \cdot \\ \bar{F}(s_n) \end{bmatrix} \qquad [2]$$

where $\bar{F}(s_i) = F(s_i)\,\Delta\ln s_i$ and finite increments $\Delta\ln s_i$ are assumed to be constant.

Hlaváček-Seyer (21) related the limiting value (on the lower side) of the largest possible "relative error" in $F(s)\,t^r$ the ratio $|\lambda_{max}|/|\lambda_{min}|$, in which $|\lambda_{max}|$ and $|\lambda_{min}|$ are the maximum and minimum eigenvalues of $K(x_i, s_j)$.

Todd (22) and *von Neuman-Goldstine* (23) were the first to introduce the ratio $|\lambda_{max}|/|\lambda_{min}|$ as a measure for stability of numerical solutions of linear systems.

Consider a cumulative relative error R_1 is given by the sum of absolute values of $\varphi(x_i)$ and $\Delta\varphi(x_i)$ at discrete points x_i:

$$R_1 = \frac{\sum_{i=1}^{n} |\Delta\varphi(x_i)|}{\sum_{i=1}^{n} |\varphi(x_i)|} \quad [3]$$

where $\Delta\varphi(x_i)$ stands for the error in $\varphi(x_i)$. Similarly, we may define the cumulative relative error R_2 in the function $\bar{F}(s)$:

$$R_2 = \frac{\sum_{i=1}^{n} |\Delta\bar{F}(s_i)|}{\sum_{i=1}^{n} |\bar{F}(s_i)|} \quad [4]$$

where $\Delta\bar{F}(s_i)$ stands for the error in function $\bar{F}(s_i)$.

The maximum difference in $F(s)$ for a given R_1 defined as

$$v = \max\left(\frac{R_2}{R_1}\right) \quad [5]$$

is related (21) to the maximum and minimum eigenvalues of the kernel matrix $K(x_i, s_j)$ as follows

$$v \geq \frac{|\lambda_{max}|}{|\lambda_{min}|}. \quad [6]$$

The relation [6] determines the lower limit of the maximum possible error in the function $F(s_i)$ for a given R_1. The high value of v indicates that a small numerical change in $\varphi(x)$ is reflected in a large change in the function $F(s)$. At the same time the value of v suggests the degree to which the kernel function simulates the properties of the *Dirac* delta function.

The ratio v is basically dependent on the type of kernel, the range of variables x and s, and the number of equations that relate the measured function $\varphi(x_i)$ to the spectral function $F(s_i)$.

In this study the range of independent variables, x and s, was chosen to be the same as that in (11) (i.e. $x, s \in \langle 10^{-4}, 10^{8}\rangle$).

The resulting of v for various types of kernels and numbers of equations were computed and are listed in table 1 and plotted in fig. 1.

The number of equations for a given range of independent variables depends on the length of the increment $\Delta \ln s_i$. Inspection of table 1 and fig. 1 suggests that for the case of forty equations (three relaxation times per one decade) the matrices considered are quite different in terms of the ratio v. For instance compare

$\{1 - \exp(-x/s)\}$

Table 1, Fig. 1. The ratio $|\lambda_{max}|/|\lambda_{min}|$ as a function of number of equations for various types of matrices

	The ratio $\frac{|\lambda_{max}|}{|\lambda_{min}|}$, $x, s \in \langle 10^{-4}, 10^{8}\rangle$		
	the number of equations		
Type of matrix	5	20	40
$\exp(-x_i/s_j)$ (9–11)	1.002	3.40	883.5
$[1 - \exp(-x_i/s_j)]$ (11)	1.189	57.7	$1760. \times 10^{10}$
$\left[\left(\frac{2x_i}{s_j}\right)^2 \exp\left(-\frac{2x_i}{s_j}\right)\right]$ (16, 18)	1.847	2.16	35.9
$12\left(\frac{x_i s_j}{1 + x_i^2 s_j^2}\right)^4$ (18)	1.33	1.44	6.3

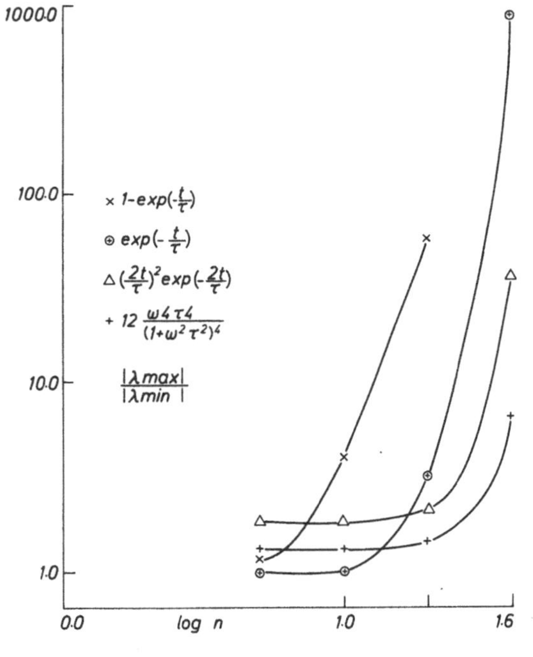

Fig. 1. The ratio $|\lambda_{max}|/|\lambda_{min}|$ as a function of number of equations various types of matrices

and

$$\left\{12\left[\frac{xs}{1+(xs)^2}\right]^4\right\}.$$

From table 1 and fig. 1 it also follows that all matrices considered become "good posted" as the number of equations decreases, i.e. relaxation times are well separated.

The matrices for which the value of ν is close to unity are called "good posted" and those for which $\nu \gg 1$, "ill posted".

Thus several authors (9, 10) resort to the assumption of large relaxation time intervals to avoid indeterminations in the numerical solution. Such a restriction will result in loss of generality.

Accordingly, the kernel $\{1-\exp(-x/s)\}$ appears to be far less suitable for a numerical process than the kernel

$$\left\{12\left[\frac{xs}{1+(xs)^2}\right]^4\right\}.$$

Let us suppose that numerical solutions of the two following equations are being sought (the same range of variables is considered for both equations):

$$J(x)=\int_a^b[1-\exp(-x/s)]\,F(s)\,d\ln s \qquad [7]$$

$$H_3(x)=\int_a^b 12\left[\frac{xs}{1+(xs)^2}\right]^4 F(s)\,d\ln s \qquad [8]$$

where $J(x)$ may represent the creep function and $H_3(x)$ the third *Schwarzl-Staverman* approximation of relaxation spectrum (4). $F(s)$ assumes the meaning of the retardation spectrum in eq. [1] and the relaxation spectrum in eq. [8].

The third *Schwarzl-Staverman* approximation of the relaxation spectrum derived from the experimentally determined storage modulus $G'(x)_{\omega=x}$, is given by

$$H_3(x)=G'(x)-0.25\left.\frac{d^3G'(x)}{d(\ln x)^3}\right|_{x=\omega}. \qquad [9]$$

It is further assumed that the experimental errors in both $J(x)$ and $G'(x)$ are approximately the same magnitude. The error in the experimentally determined $G'(x)$ is magnified in eq. [9] due to the third derivative of the storage modulus. Thus the approximation of the spectral function $H_3(x)$ obtained from the *Schwarzl-Staverman* formula, eq. [9], is less accurate compared with $G'(x)$; however, through experimental techniques, this inaccuracy may be kept at a relatively less significant level. On the other hand the errors in the solutions of eqs. [7] and [8] will be vastly different. This is due to numerical inversion indicated by the values of the ratios ν for the respective kernels. [Up to ten decades in terms of $\log(x)$]. Thus the disadvantage of using the less accurate $H_3(x)$ is obviously overcome by the changed properties of the corresponding kernel function.

Therefore the kernel of eq. [7] may be modified by use of the *Schwarzl-Staverman* approximation (4) for the retardation spectrum:

$$\tilde{L}\left(\frac{x}{2}\right)=\frac{dJ(x)}{d\ln x}-\frac{d^2J(x)}{d(\ln x)^2} \qquad [10]$$

which transforms the eq. [7] into

$$\tilde{L}(x)=\int_a^b\left\{\left(\frac{2x}{s}\right)^2\exp\left(-\frac{2x}{s}\right)\right\}F(s)\,d\ln s \qquad [11]$$

and provides a kernel (the intensity function) of markedly improved properties (see table 1, fig. 1).

In general, we may transform the kernel in eq. [1] into forms similar to those in eqs. [8] and [11] by the use of the higher approximation formula of relaxation and retardation spectra proposed by *Schwarzl-Staverman* (4), *Fujita* (6), *Ninomiya-Ferry* (7). The corresponding intensity functions are "good posted" over many decades of independent variables x and s and these are indicated in table 1 and fig. 1. Such approximations may thus be used as a starting point in the numerical determination of the spectral function $F(s)$.

The effect of a numerical method on the solution of equation 1

The ratio ν, defined by eq. [6], represents the lower limit for the relative error in the determination of the spectral function. Thus the error may assume any value greater or equal to the corresponding ratio ν and depends on the numerical method used. The following simple example will show this effect. Consider a two parameter *Maxwell* model. Then the storage modulus may be written as follows

$$G'(x)=\frac{(s_1x)^2}{1+(s_1x)^2}+\frac{(s_2x)^2}{1+(s_2x)^2} \qquad [12]$$

where $s_1 = 1.0$, $s_2 = 2.5$, $x \in \langle 0.063; 6.309 \rangle$. Assuming the following identity

$$F^*(s)\, d\log s = F(s)\, d\ln s \qquad [13]$$

eq. [8] may be rewritten as

$$H_3(x) = \int_a^b 12 \left[\frac{sx}{1+(sx)^2}\right]^4 F^*(s)\, d\log s. \qquad [14]$$

The solution of eq. [14] is investigated by
a) the method of *Hlaváček* (19);
b) the method of *Phillips* (13), *Twomey* (12), and *Clauser-Knauss* (11).

The intensity function in eq. [14] is considered to be "good posted" according to the ratio v. However, the latter method, (*b*), requiring the square relative error to be minimized, modifies the kernel substantially.

In the matrix notation, eq. [14] may be written as

$$\begin{bmatrix} H_3(x_1) \\ H_3(x_2) \\ \cdot \\ \cdot \\ \cdot \\ H_3(x_n) \end{bmatrix} = \begin{bmatrix} a_{11} & \ldots & a_{1n} \\ a_{21} & \ldots & \\ \cdot & & \\ \cdot & & \\ \cdot & & \\ & \ldots & a_{nn} \end{bmatrix} \cdot \begin{bmatrix} \bar{F}(s_1) \\ \bar{F}(s_2) \\ \cdot \\ \cdot \\ \cdot \\ \bar{F}(s_n) \end{bmatrix} \qquad [15]$$

or

$$\mathscr{H}_3 = \mathscr{A}\mathscr{F}^*$$

where

$$a_{ij} = 12 \left[\frac{s_i x_j}{1+(s_i x_j)^2}\right]^4$$

and

$$\bar{F}(s_i) = F^*(s_i)\, \Delta \log s_i .$$

The numerical inversion of eq. [14] by method (a) (19) the matrix $\mathscr{A}$ is employed. On the other hand method (b) (11–13) modifies $\mathscr{A}$ into a matrix $\mathscr{M}$ whose elements are

$$m_{ik} = a_{ik} \sum_{j-1}^{n} a_{jk}/[H_3(x_i)]^2 + \gamma b_{ik} \qquad [16]$$

where γ is the "smoothing factor" and b_{ij} are elements of matrix $\mathscr{B}$:

$$\begin{bmatrix} 1 & -2 & 1 & 0 & . & . & . & . & . & . & . & 0 \\ -2 & 5 & -4 & 1 & 0 & . & . & . & . & . & . & 0 \\ 1 & -4 & 6 & -4 & 1 & 0 & . & . & . & . & \therefore & 0 \\ 0 & 1 & -4 & 6 & -4 & 1 & 0 & . & . & . & . & 0 \\ \hdashline 0 & . & . & . & . & . & 0 & 1 & -4 & 6 & -4 & 1 \\ 0 & . & . & . & . & . & . & 0 & 1 & -4 & 5 & -2 \\ 0 & . & . & . & . & . & . & . & 0 & 1 & -2 & 1 \end{bmatrix}$$

thus the eq. [15] may be written as

$$\mathscr{V} = \mathscr{M} \cdot \mathfrak{F}^*$$

with

$$v_k = \sum_{j=1}^{n} \frac{a_{jk}}{H_3(s_j)} .$$

The properties of matrix $\mathscr{M}$ compared with matrix $\mathscr{A}$ are quite different in terms of the ratio v, the former having higher values of v. Consequently the matrix $\mathscr{M}$ is much less suitable for the numerical inversion of eq. [14].

Differences in the effects of methods (a) and (b) are illustrated in fig. 2 and tables 2 and 3. Method

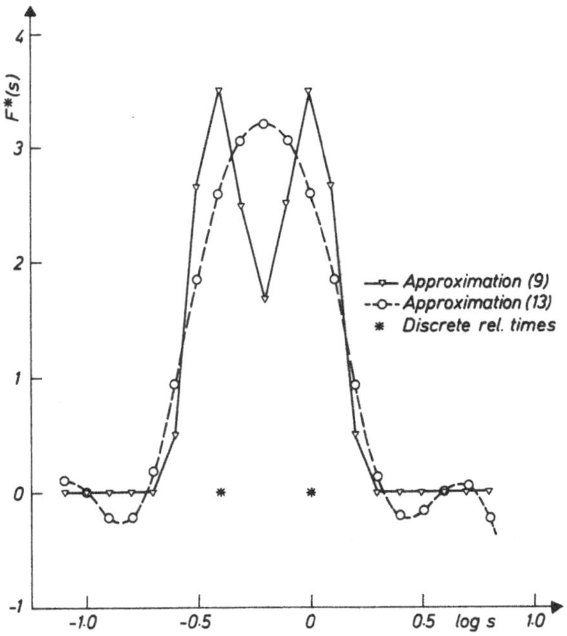

Fig. 2. Relaxation spectrum calculated by methods *Hlaváček* (19) and *Phillips* (13) (smoothing factor $\gamma = 0.02$). The two parameter *Maxwell* model is characterized by relaxation times $\tau_1 = 1.0$, $\tau_2 = 2.5$, and elasticity constants $G_1 = G_2 = 1.0$

Table 2. Relaxation spectrum calculated by methods *Hlaváček* (19) and *Phillips* (13) (for different smoothing factors). The two parameter Maxwell model is characterized by relaxation times $\tau_1 = 1.0$, $\tau_2 = 2.5$, and elasticity constants $G_1 = G_2 = 1.0$

F^*_{44} (19)	F^* (13)			$1/s$
	γ			
	0.5	0.1	0.02	
0.0000	0.0183	0.0803	0.1014	0.0794
0.0000	—0.0659	—0.0270	—0.0048	0.1000
0.0000	—0.1415	—0.1868	—0.2073	0.1259
0.0000	—0.0427	—0.1550	—0.2162	0.1585
0.0000	0.3247	0.2230	0.1845	0.1995
0.4837	0.9304	0.9178	0.9544	0.2512
2.6634	1.6463	1.7567	1.8448	0.3162
3.4832	2.3090	2.5252	2.5929	0.3981
2.4859	2.7738	3.0504	3.0608	0.5012
1.6727	2.9461	3.2321	3.2260	0.6310
2.5086	2.7928	3.040	3.0896	0.7943
3.4988	2.3447	2.5089	2.6290	1.0000
2.6652	1.6929	1.7438	1.8589	1.2589
0.5223	0.9767	0.9209	0.9338	1.5849
0.0000	0.3546	0.2498	0.1521	1.9953
0.0000	—0.0442	—0.1136	—0.2144	2.5119
0.0000	—0.1752	—0.1625	—0.1566	3.1623
0.0000	—0.1061	—0.0497	0.0352	3.9811
0.0000	0.0177	0.0363	0.0668	5.0119
0.0000	0.0640	0.0311	—0.0269	6.3096

$H(\log \tau) = F^*(s)$

Table 3. The magnitude of the difference $[G'(\omega)\,\text{exact} - G'(\omega)\,\text{calc}]^2$ as a function of frequence ω. The relaxation times of the Maxwell element are $\tau_1 = 1.0$, $\tau_2 = 2.5$, and elasticity constants $G_1 = G_2 = 1.0$

$[G'\,\text{exact}—G'(F^*_{44})]^2 \cdot 10^4$	$[G'\,\text{exact}—G'\,\text{calc}]^2 \cdot 10^4$	$[G'\,\text{exact}—G'\,\text{calc}]^2 \cdot 10^4$	$[G'\,\text{exact}—G'\,\text{calc}]^2 \cdot 10^4$	x
	$0.5 \equiv \gamma$	$0.1 \equiv \gamma$	$0.02 \equiv \gamma$	
0.2179	0.0399	0.0042	0.0227	0.0794
0.4256	0.0754	0.0016	0.0215	0.1000
0.7688	0.1715	0.0000	0.0171	0.1259
1.2315	0.4385	0.0080	0.0093	0.1585
1.6528	1.2572	0.0667	0.0003	0.1995
1.7344	3.6208	0.3124	0.0199	0.2512
1.3168	9.2079	0.9545	0.1272	0.3162
0.6569	19.306	1.9321	0.2633	0.3981
0.1685	33.416	2.7165	0.2246	0.5012
0.0005	50.297	3.0353	0.0604	0.6310
0.1323	70.318	3.3994	0.0003	0.7943
0.5731	94.572	4.5068	0.0002	1.0000
1.1862	121.46	6.5798	0.0254	1.2589
1.5822	146.13	9.0662	0.1611	1.5849
1.5101	164.24	11.085	0.3455	1.9953
1.1163	175.18	12.278	0.4769	2.5119
0.6852	181.14	12.869	0.5475	3.1623
0.3701	184.47	13.182	0.5881	3.9811
0.1839	186.49	13.372	0.6150	5.0119
0.0867	187.65	13.470	0.6312	6.3096

(a) (19) appears to be superior to method (b) (12–13).

It should be noted that a further decrease in the smoothing factor γ is not possible since distortion of the function $F^*(s)$ might occur due to the assumption $F^*(s) > 0$.

Similar results to those shown in fig. 2 (two parameter model) were also obtained for wide spectrum and *Rouse* distribution of relaxation times (24) and for the models characterized by relaxation and retardation spectra identical to those used in (11).

Conclusion

The numerical inversion of general integral equations of the first kind strongly depends on the type of kernel. If the properties of the *Dirac* delta function are well simulated, a better numercial solution of eq. [1] may be obtained.

A starting point for the numerical method can be the second or third approximation of relaxation or retardation spectra (4, 6, 7); the respective transformation of the experimental function is highly desirable. A certain inaccuracy due to the derivatives of experimental function is necessarily involved; however, the improved properties of the transformed intensity function result in higher stability and reduced relative error. The numerical example will be shown elsewhere (24).

References

1) *Alfrey, T.* and *P. Doty*, J. Applied Phys. **16**, 700 (1945).

2) *Andrews, R. D.*, Ind. Eng. Chem. **44**, 707 (1952).

3) *Ferry, J. D.* and *M. L. Williams*, J. Colloid Sci. **14**, 347 (1952).

4) *Schwarzl, F.* and *A. J. Staverman*, Appl. Sci. Res. **A4**, 127 (1953).

5) *Okano, M.*, Busseiron Kenkyu **3**, 493 (1958).

6) *Fujita, H.*, J. Applied Phys. **29**, 943 (1958).

7) *Ninomiya, K.* and *J. D. Ferry*, J. Colloid Sci. **14**, 36 (1959).

8) *Tschoegel, N. W.*, Rheol. Acta **10**, 582–600 (1971).

9) *Tobolsky, A. V.* and *K. Murakami*, J. Polym. Sci. **40**, 443 (1959).

10) *Gradowczyk, M. H.* and *F. Moavenzadeh*, Trans. Soc. Rheol. **13**, 173 (1969).

11) *Clauser, J. F.* and *W. G. Knauss*, Trans. Soc. Rheol. **10**, 191 (1966).

12) *Twomey, S.*, J. Assn. Computing Machinery **10** (January 1963).

13) *Phillips, D. L.*, J. Assn. Computing Machinery **9** (January 1962).

14) *Tanner, R. I.* and *G. Williams*, Trans. Soc. Rheol. **14**, 19 (1970).

15) *Burger, H. C.* and *P. H. Van Cittert*, Physik **79**, 722 (1932).

16) *Hopkins, J. L.*, J. Polymer Sci. **50**, 59 (1961).

17) *Roesler, F. C.* and *W. A. Twyman*, Proc. Phys. Soc. **B68**, 97 (1955).

18) *Hlaváček, B.* and *V. Kotrba*, Rheol. Acta **6**, 288 (1967).

19) *Hlaváček, B.*, Rheol. Acta, **7**, 225 (1968).

20) *Hlaváček, B.* and *V. Sinevič*, Rheol. Acta **9**, 312 (1970).

21) *Hlaváček, B.* and *F. A. Seyer*, J. Appl. Polym. Sci. **16**, 423 (1972).

22) *Todd*, J. Quart. J. Mech. Appl. Math. **2**, 469 (1949).

23) *Von Neumann, J.* and *H. H. Goldstine*, Bull. Amer. Math. Soc. **53**, 1021 (1947).

24) *Stanislav, J.* and *B. Hlaváček* (to be published).

Summary

We have considered the optimal conditions for the determinations of relaxation spectra. The numerical process should be based on a well-posted method and should be free of artificial stability conditions. Such artificial conditions are those which restrict or prescribe the shape of the spectra. From the practical point of view we found that simple numerical processes which do not contain matrix inversion are convenient.

Author's address:

B. Hlaváček
Dept. of Chem. Eng., Ecole polytechnique,
Montréal 248 (Canada)

Rheol. Acta 13, 608–612 (1974)

From the Swedish Institute for Food Preservation Research (SIK), Göteborg (Sweden)

Food crushing sounds: an analytic approach

By B. Drake and L. Halldin

With 4 figures and 2 tables

(Received October 27, 1972)

Previously reported work (*Drake*, 1963, 1965a and b) has shown that mastication sounds contain information related to the rheological properties of the food chewed. The complete relationships, however, have not yet been determined. In the present work, which is an attempt at starting to elucidate the remaining problems, a basic thought is to supplement human chewing with a mechanical crushing. The former is more natural and, of course, the the real thing to study. The latter, however, lends itself to more reproducible experiments, and is therefore to be considered as a valuable research tool.

In this paper, an experimental technique and a way of evaluation will be presented together with preliminary results. In another paper (*Andersson* et al., 1973) the determined parameter values are correlated to penetrometer and sensory data.

6 brands of crisp bread, each with 8 or 9 replicates, were tested with the technique described below for analysing crushing sounds. Crisp bread of the typical Scandinavian type was chosen as experimental material since it can easily be kept for long periods of time and does not need heat treatment, thermostatting, etc., before the experiments.

Experimental

Materials

Commercial samples of crisp bread were used. "Rågrut" is produced by Falu Spisbrödsfabrik AB, Falun, Sweden; "Rågspröd" by Fazer Oululainen OY, Lahti, Finland; and the remaining 4 brands by Wasabröd AB, Filipstad, Sweden. All breads were factory cut to 12 cm × 6 cm rectangles except "Rågspröd", the slices of which were 7 cm × 7 cm. Other characteristics have been collected in Tab. 1. Thickness and weight values are, of course, approximative. Photos of the 6 breads are published by *Andersson* et al. (1973).

Crushing

Bread slices were manually broken into small pieces, all having approximately the same size. Within the same brand, those with almost the same shape were used. Differences in shape between pieces of different brands, however, were unavoidable.

The small pieces of bread (approx. 2 cm × 2 cm) were placed one at a time (or, in the case of "Rågspröd", two at a time with their rough sides facing each other) on a horizontal, sturdily supported steel beam (2 cm × 8 cm × 20 cm) with known resonance frequency (2.7 kHz). They were crushed with a cylindrical brass plunger (diam. 3 cm) by holding the plunger just over the upper surface of the sample and then manually lowering it to crush the sample, taking care to perform the movement as uniformly as possible. Normally, the crushing itself took around 0.5 sec. (In continued work, to be reported later, the manual crushing has been replaced by a crushing with a falling rod which can be appropriately loaded and remotely released.)

Recording

The sound produced when crushing the bread pieces was recorded with a condenser microphone (Brüel & Kjaer, mod. 4134, screwed onto a cathode follower, mod. 2613) placed slightly above and about 10 cm from the sample. The microphone signal was amplified and fed to a frequency modulated tape recorder (Brüel & Kjaer, mod.

Table 1. Properties of the 6 brands of bread used in the study

Brand Abbrev.	Swedish name	Thickness mm	Weight per slice g	Remarks
G	Graham	7	10.0	wheat
H	Husman	9	12.5	rye
R	Rågrut	8	11.0	rye
Rs	Rågspröd	4	6.7	rye bread torn apart in baking
K	Kullamöri	4	7.2	baked with crushed ice and no yeast; whitish
V	Veteknäcke	8	13.0	baked from graham flour; yellowish; easily crumbled

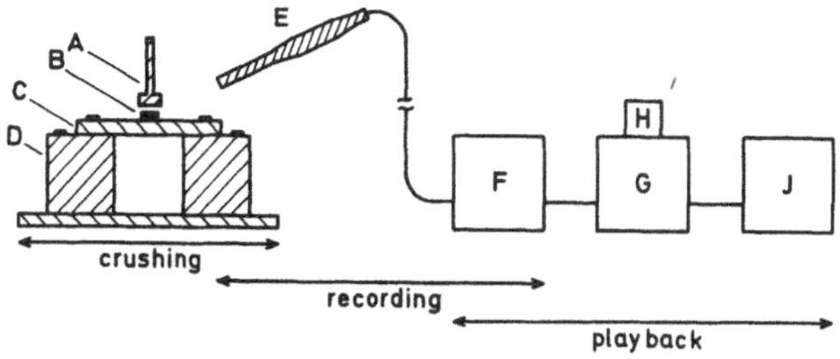

Fig. 1. Schematic drawing of complete equipment used for analysis of food crushing sounds.
A = plunger, B = sample, C = steel beam, D = support, E = microphone with cathode follower, F = FM tape recorder, G = frequency analyzer, H = stepping device, J = X-Y recorder

7001) operated in the tape loop mode. The length of the loops was 305 cm and the revolution time 2 sec. A schematic drawing of the complete equipment used is shown in fig. 1.

Playback

After completed recording, the signal was played back through a frequency analyzer (Gen. Radio, mod. 1554-A), provided with a home-made arrangement for automatically stopping the sweep motor at predetermined positions, and a series of amplitude-time diagrams were plotted on an X-Y recorder (Moseley, mod. 7000 AM).

During playback, for every revolution of the tape loop the frequency was stepped to the next higher of 40 predetermined positions per decade. The steps were evenly spaced corresponding to a constant frequency ratio of 1.059. The recorded sound was simultaneously frequency transformed downwards 10 times by changing the tape speed from 60 ips during recording to 6 ips. In this way, the mechanical parts of the X-Y recorder could more satisfactorily respond to the signals. After each sweep the base-line on the recorder was shifted a small step upwards. The resulting diagram consists of a number of curves which, for each one of the chosen frequencies, show sound amplitude as a function of time. For the case that the base-line steps are made equally large, such diagrams will look like the one shown in fig. 2.

Except for the X-Y recorder and the stepping device (J and H in fig. 1), which have superseded the previously used oscilloscope provided with a camera, the same equipment was used by *Drake* (1965b). A new feature is that the analysing technique does not rely on a photographic plate blackening but enables a direct digital evaluation, at present based on simple length measurements.

Evaluation

The recorded curves contain a considerable amount of information. However, most of the sound energy is located to the peaks. A primary selection was therefore made by specifically measuring the amplitude of each curve only at a short interval of time (ca. 0.02 sec in real time, i.e., 0.2 sec in playback) around the highest peak (cf., asterisk in fig. 2).

Since more interest was given to the qualitative aspects of the sound than to its gross quantitative aspects, peak amplitudes measured for a number of frequencies in one and the same diagram (preferably a spread-out version of the type shown in fig. 2) were then normalized with respect to the highest amplitude which was assigned the value of $A_{max} = 1$ (i.e., a relative level of 0 dB). The resulting values can be said to represent a line spectrum for the loudest sound (fig. 3).

The frequency range of the recorded sound was approximately 2.5–25 kHz. This implies that the proportion of high frequency components was notably high (cf., *Drake*, 1965b). The

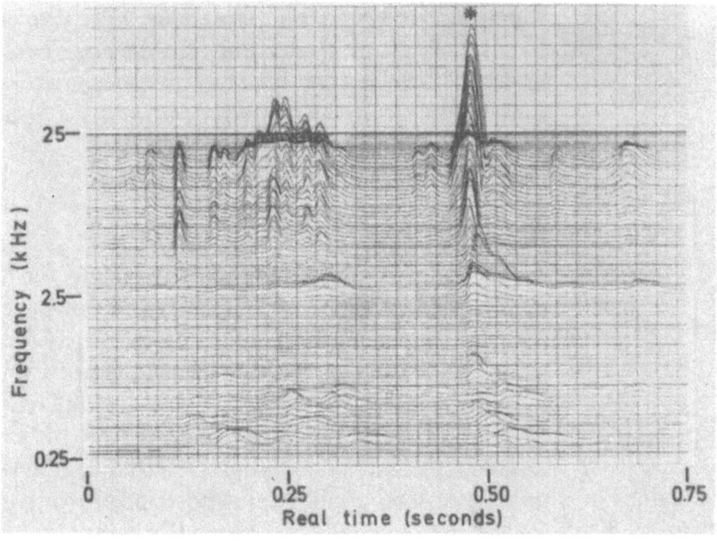

Fig. 2. Combined plot of amplitude/time curves for a number of selected frequencies (*Kullamöri*; replicate No. 7).
The asterisk marks the time at which the evaluation was performed

lowest part of this range contains the resonance frequency of the support, 2.7 kHz, which carries other information than the sound emitted directly from the crushed sample. This part was therefore not used in this work. The upper cut-off frequency, 25 kHz, determined by the limited frequency response of the tape recorder, is above the hearing threshold for single frequency tones, ca. 16 kHz. The range 16–25 kHz, however, is included in the present analysis. The

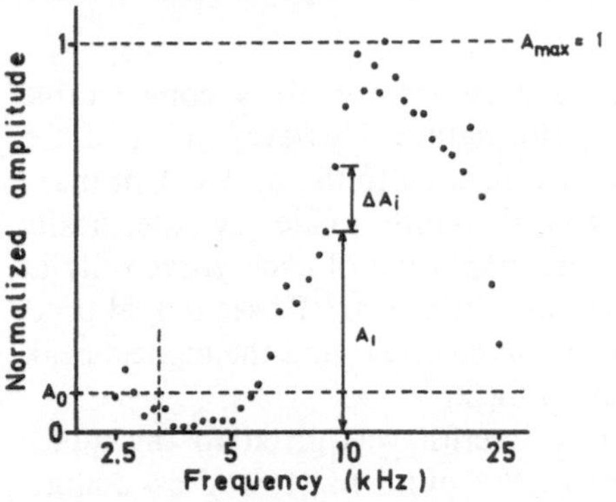

Fig. 3. Spectrum corresponding to upper half of Fig. 2 at time marked with an asterisk.
The five points to the left of the vertical dashed line, which represent a resonance vibration of the steel beam, were not used in the evaluation

chosen frequencies thus covered the 36 frequencies ranging from 3.4–25 kHz. For every frequency, arithmetical means of the normalized values of peak heights were calculated for the replicates.

The 36 means of normalized peak amplitude values obtained for each one of the 6 brands of bread would be difficult to compare with a large number of other parameters. It was therefore considered suitable to characterize the frequency distributions by a smaller number of parameters. This, which implies a further step of data reduction, was tentatively performed by calculating the following 13 parameters from amplitude values A read from recorded curves (cf. fig. 3).

$X\,1$, $X\,2$ — $\Sigma\Delta A$ for positive and negative ΔA values (slopes), respectively;

$X\,3$ — number of cases of $\Delta A = 0$ ("no slope");

$$\Sigma\Delta^2 A = \sum\left(\frac{\Delta A_{i+1} + \Delta A_{i-1}}{2} - \Delta A_i\right)$$

for positive and negative values of $\Delta^2 A$ (curvatures), respectively;

$X\,6$ — number of cases of $\Delta^2 A = 0$ ("no curvature");

$X\,7$, $X\,8$ — area under level/frequency curve above -20 dB and above -6 dB, respectively (with an arbitrary unit of area on the graph paper);

$X\,9$, $X\,10$ — area under amplitude/frequency curve, determined as $\sum(A_i - A_0)$ for all $A_i > A_0$ with $A_0 = 0.1$ and $A_0 = 0.5$, respectively;

$X\,11$, $X\,12$ — mean slope $(\Sigma\Delta A)/n$ for positive and negative slopes, respectively, where n is the number of such slopes;

$X\,13$ — $(\Sigma\Delta A) - (1 - A_1)$, for all $\Delta A > 0$ (positive slopes).

These 13 parameters were selected from a large number of possible parameters which can be imagined for characterizing the actual type of sound by emphasizing one or a small number of factors. The parameters $X\,7$–$X\,10$ determine the position between noise and a pure tone, with a higher value implying a sound which is less pure, i.e. less similar to a sound with one single frequency. The other parameters were introduced in an attempt to estimate the spikiness of the spectra, a property which can be considered as superimposed on the general shape of the spectrum. As an example, when the number and/or height of small peaks superimposed on the main peak increases, the values of, e.g., the parameters $X\,1$, $X\,2$, $X\,4$, $X\,5$, and $X\,13$ also increase. These parameters are ambiguous in the sense that they do not determine the position between pure tone and noise as a single-valued function. However, all parameter values refer to cases which are close enough to the noise case, and all of them can here be safely used as a measure of deviation from a noise whose amplitude does not vary with the frequency.

Results

Fig. 4 shows spectra obtained as means for 8 or 9 replicates. Since all replicates did not have their maximum amplitude located to exactly the same frequency, the resulting mean curves do not reach up to $A_{max} = 1$. The spectra are different from each other, even considering the rather large coefficients of variation (referred to differences between replicates) whose mean values for the 36 frequencies amounted to a minimum of

Table 2. Results of crushing vibration analyses of crisp bread

Brand	Repl.	X 1	X 2	X 3	X 4	X 5	X 6	X 7	X 8	X 9	X 10	X 11	X 12	X 13
Graham	9	*114*	*95*	8.22	*44.4*	*52.2*	*8.00*	45.6	7.5	11.3	3.49	*11.8*	*6.32*	*15.9*
Husman	8	*116*	*97*	6.75	*45.1*	*51.7*	*7.00*	46.0	7.7	11.8	3.79	*11.0*	*6.15*	*18.7*
Rågrut	8	*123*	*112*	4.62	*56.5*	*64.3*	*8.33*	60.0	8.1	14.7	4.07	*9.4*	*6.68*	*35.4*
Rågspröd	9	*118*	*107*	8.77	*53.0*	*58.1*	*3.62*	46.6	6.7	10.3	2.19	*9.6*	*7.97*	*25.4*
Kullamöri	9	*110*	*91*	8.55	*41.5*	*48.1*	*7.44*	45.4	8.0	11.0	3.59	*11.1*	*6.44*	*13.2*
Veteknäcke	9	*126*	*110*	7.22	*65.0*	*71.7*	*3.88*	54.3	8.7	13.8	3.87	*10.7*	*7.31*	*31.2*
noise		0	0	35	0	0	34	100.0	30.0	31.5	17.5	0	0	0
pure tone		100	100	0	40.0	40.0	23	8.8	1.2	2.2	0.6	4.99	4.99	0

N. B. Figures given for pure tone were determined by analysing a sinusoidal electrical signal under the same conditions as those used in the experiments. Figures for noise are ideal values.

Italicized figures indicate values which need not fall within the numerical range between pure tone and noise.

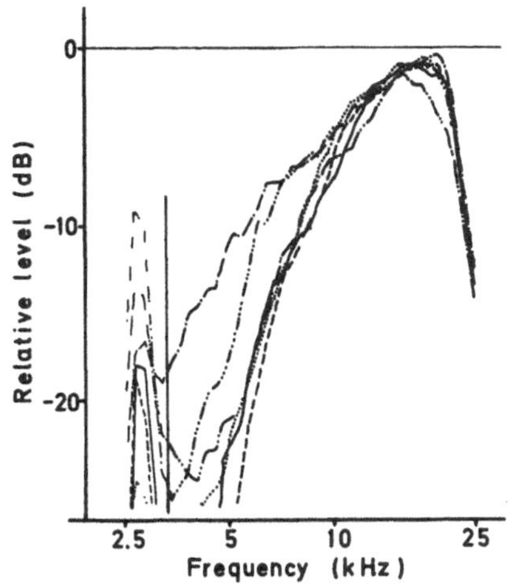

Fig. 4. Mean spectra of crushing sounds obtained for 6 brands of crisp bread.
(For points close to 2.5 kHz, see fig. 3)
———— = *Graham*
············ = *Husman*
–·–·–·–·–· = *Rågrut*
–··–··–·· = *Rågspröd*
– – – – – – – = *Kullamöri*
–···–···–··· = *Veteknäcke*

27% for "Rågrut" and a maximum of 47% for "Rågspröd".

In table 2 the calculated values for the parameters X 1–X 13 are given together with corresponding values for a pure tone and a noise with an amplitude which does not vary with the frequency. For the parameters X 1–X 6 and X 11–X 13, the coefficient of variation (also in this case referred to differences between replicates varied between 5% for *Graham*/X 2 and 63 % for *Kullamöri*/X 13.

Discussion

Even if several of the parameters may be difficult to interpret, all of them are well-defined mathematically and imply a way of quantifying differences between the sound produced when crushing food samples. In this way, the parameters can be directly compared with rheological properties, especially fracture properties, which can be determined by conventional rheological methods. Such a comparison, involving sensory evaluation of handling and biting/chewing characteristics and also determination of flexure, compression and fracture characteristics by using an Instron Material Testing Machine, has been made (*Andersson* et al., 1973). However, the mentioned study which included evaluation by simple correlations, a regression approach involving a generalization of *Stevens*' law, and stepwise discriminant analysis, is too voluminous to be cited in detail here. Instead, it is simply mentioned that some parameters of bread crushing sounds were found to be well correlated to sensory impressions of hardness, fracture force, and brittleness (crispness). Since the sound did not give the same information as the more conventional rheological techniques, the sound analysis can preliminarily be considered as a potential complement to such techniques.

The parameters are arbitrarily chosen, and no claim is made of having found an ideal set. The values presented in this report are therefore preliminary and should be considered mainly as a check of the procedure. In the continued work along similar lines it is intended to try also other parameters.

Summary

As a complement to the study of human mastication sounds, crushing sounds produced with a mechanical device were tape recorded and played back in such a way as to obtain acoustical line spectra for short intervals around the points of time when the sounds were loudest. The spectra were then evaluated to give measures calculated from amplitudes belonging to sets of such frequencies. These measures aimed at, *e.g.*, estimating the width of the frequency band occupied by the major part of the sound, the two theoretically possible extremes being a pure tone and a "white noise". In a way, this analytical method results in discrete *Fourier* transforms.

In studies with various brands of crisp bread it was found that the crushing sound differed in a reproducible way among the brands. The sounds, however, do not measure the same properties as ordinary rheological analyses. They can therefore be considered as a potential complement to flexure and compression data.

Zusammenfassung

Als Ergänzung des Studiums menschlicher Kaugeräusche wurden die Geräusche, die durch das Zerdrücken von Lebensmitteln mit einer mechanischen Vorrichtung erzeugt wurden, magnetophonisch aufgezeichnet. Diese wurden danach in solcher Weise analysiert, daß die akustischen Linienspektren für kurze Zeitintervalle rund um den Zeitpunkt der größten Geräuschintensität erhalten wurden. Ausgehend von diesen Spektren wurden sodann verschiedene Parameter errechnet; so z. B. die Bandbreite der den Hauptteil des Geräusches ausmachenden Frequenzen.

Bei Versuchen mit verschiedenen Arten von hartem Brot hat es sich gezeigt, daß die Zerdrückungsgeräusche in reproduzierbarer Weise verschieden sind. Aus diesen Geräuschen kann man noch nicht dieselben Eigenschaften wie aus konventionellen, rheologischen Analysen herleiten. Sie können deshalb als ein potentielles Komplement zu Biege- und Kompressionsanalysen betrachtet werden.

References

Andersson, Y., B. Drake, A. Granquist, L. Halldin, B. Johansson, R. M. Pangborn, and *C. Å. Åkesson,* J. Texture Studies **4**, 119 (1973).

Drake, B., J. Food Sci. **28**, 233 (1963).

Drake, B., J. Food Sci. **30**, 556 (1965a).

Drake, B., Biorheol. **3**, 21 (1965b).

Authors' address:

B. Drake and *L. Halldin*

Swedish Institute for Food Preservation Research (SIK)

S-40021 Göteborg 16 (Sweden)

Rheol. Acta **13**, 613–617 (1974)

From the Montedison SpA Rice-Centro Ricerche Ferrara (Italy)

Investigation on impact strength and on dynamic mechanical properties of polypropylene

By S. Danesi, L. Baldi, and G. Ballini

With 5 figures and 6 tables

(Received October 27, 1972)

Introduction

In the applications, where a good impact strength is required, a polymer, to be considered satisfactory, must feature a high toughness in the widest range of testing conditions (temperature, deformation rate, anisotropy, stress concentration (1, 2).

The evaluation of the tough-brittle transition temperature and of its dependence on these factors is therefore of fundamental importance in the study of the impact properties of polymeric materials. As known, in the case of polypropylene homopolymer the tough-brittle transition coincides with the glass transition of the amorphous fraction if the latter is measured at high frequency.

Hence it is reasonable to suppose that this fraction is of great importance in determining the impact strength of *PP*. As a consequence, an investigation on the influence of the amount and MW of the amorphous atactic and isotactic fractions on the impact behavior and dynamic-mechanical properties of *PP* was considered of great interest.

Experimental

Polymers and polymer conditioning

Two polymers were considered whose features are listed in table 1.

The density was measured by a gradient column according to ASTM D 1505-60T on compression molded samples annealed two hours at 140 °C. The intrinsic viscosity was measured in tetraline at 135 °C and the viscosimetric average molecular weights calculated from the relation (3)

$$[\eta] = 0.8 \cdot 10^{-4}\, M^{0.8}\,.$$

As for melt index, ASTM D 1238 was followed.

Table 1. Physico-chemical characteristics of isotatic polypropylene samples

Sample	$[\eta]$(dl/gr)	MI(gr/10′)	$M_v \cdot 10^{-5}$	ϱ(gr/cm^3)
A	2	2.5	3.1	0.907
B	2.6	0.4	4.3	0.906

To obtain significant variations of the amount of crystallizable isotactic amorphous polymer, the two *PP* were subjected to four different thermal histories after the compression moulding at 210 °C:

Conditioning 1	Quenching in liquid nitrogen
Conditioning 2	Quenching in water at 20 °C
Conditioning 3	Quenching in water and annealing 2 h at 140 °C
Conditioning 4	Quenching in water and annealing 16 h at 150 °C

The density of each sample was then measured and its crystallinity calculated by the eq. [4]:

$$\frac{1}{\varrho} = \frac{X_c}{0.935} + \frac{1 - X_c}{0.856}\,.$$

The results are reported in table 2.

To obtain significant variations of the amount of the atactic amorphous fraction the polymer *A* was extracted with xylene at 20 °C and the residue was added of different quantities of two atactic *PP* of different *MW*.

The blends were prepared by mixing hot (130 °C) xylene solutions and precipitating in cold acetone under vigorous stirring.

The two atactic materials had a viscosimetric-average *MW* of $4.2 \cdot 10^4$ and $2.2 \cdot 10^5$.

Table 2. Density and crystallinity of samples A and B after different thermal histories

	Quenched in liquid N_2		Quenched in H_2O at 20 °C		Annealed 2 h at 140 °C		Annealed 16 h at 150 °C	
	A	B	A	B	A	B	A	B
ϱ (gr/cm^3)	0.902	0.900	0.903	0.902	0.907	0.906	0.909	0.907
X_c %	60	58	62	60	66	65	69	66

Table 3. Density and crystallinity of the blends

	Low *MW* polymer							High *MW* polymer		
% added atactic *PP*	0	3	10	15	20	25	35	3	10	20
ϱ (gr/cm³)	0.910	0.909	0.905	0.904	0.902	0.900	0.895	0.909	0.906	0.902
X_c %	71	69	65	63	59	56	51	69	66	59

Their amounts in the blends ranged from 3–35 wt.-% for the low *MW* polymer and from 3–20 wt.-% for the high *MW*.

The density and crystallinity of the blends are reported in table 3.

Tensile impact measurements

The high speed tensile tests were carried out with a modified impact pendulum which enabled the recording of the load-time curves (5).

The deformation speed was 3.3 m/sec and could be considered practically constant during the impact because the break energy was only a small fraction of the kinetic energy of the pendulum (6).

The measurements were performed in a wide range of temperature by conditioning the samples mounted on the apparatus with special copper pliers previously immersed into a thermostated bath.

The specimens (ASTM D 1822 type S) were obtained through accurate die cutting from compression molded 2 mm thick plates.

From the load-time curves brittle strength σ_B, yield strength σ_y, break energy E_r and break time t_r were evaluated.

The tough-brittle transition temperature was assumed as the temperature at which the load-time curves had a zero slope point and the duration time of the plastic deformation was ≥ 50 µsec.

The experiments were conducted by carrying out first a set of explorative tests to locate the transition into a quite narrow temperature range (10 °C) and then searching for the transition within this range by scanning the temperature 2 °C at a time.

Dynamic mechanical measurements

The loss factor d and the storage modulus E' were obtained by a Bruel and Kjaer Complex Modulus Apparatus in the frequency range 50–3000 Hz and in the temperature range —60 ÷ +140 °C.

The specimens (4 mm × 10 mm × 100 mm) were cut from compression molded plates.

For the determination of the dynamic mechanical properties of the atactic polymers in the temperature range —50 ÷ +50 °C, they were coupled to a polystyrene support with such a thickness that their ratio was about 1.2 (7).

Results and discussion

Influence of the amount of crystallizable amorphous polymer

Table 4 summarizes the effect of the amount of the isotactic crystallizable amorphous polymer on the tensile impact properties of the two *PP*.

The parameters used in the definition of the load-time curve are reported versus density at two temperatures (−10 and +23 °C) respectively below and above the tough brittle transition temperature.

The results show that at −10 °C, when the fracture is brittle, the density increase associated to annealing affects breaking performance only marginally at both *MW* levels, while at +23 °C (tough fracture) it brings about an increase in the break energy and time of the high *MW* polymer *B*.

This account for the observed shift to lower temperatures of the tough-brittle transition of polymer *B* and the constancy of the transition of the polymer *A*. For both polymers the influence

Table 4. Influence of thermal history on the tensile impact properties of two *PP* with different *MW*

Sample	Thermal history	Density at 23 °C (gr/cm²)	σ max (kg/cm²)		Break energie (kgcm/cm²)		Time to break (msec)		T_{TB} (°C)
			—10 °C	+23 °C	—10 °C	+23 °C	—10 °C	+23 °C	
A	1	0.902	1140	770	44	60	0.18	0.34	21
A	2	0.903	1130	750	43	65	0.18	0.36	19
A	3	0.907	1130	760	38	58	0.2	0.34	20
A	4	0.909	1130	750	40	60	0.20	0.35	20
B	1	0.900	1170	720	45	102	0.22	0.5	17
B	2	0.902	1160	700	52	102	0.23	0.48	17
B	3	0.906	1170	740	47	120	0.22	0.75	13
B	4	0.907	1160	700	45	140	0.21	0.9	10

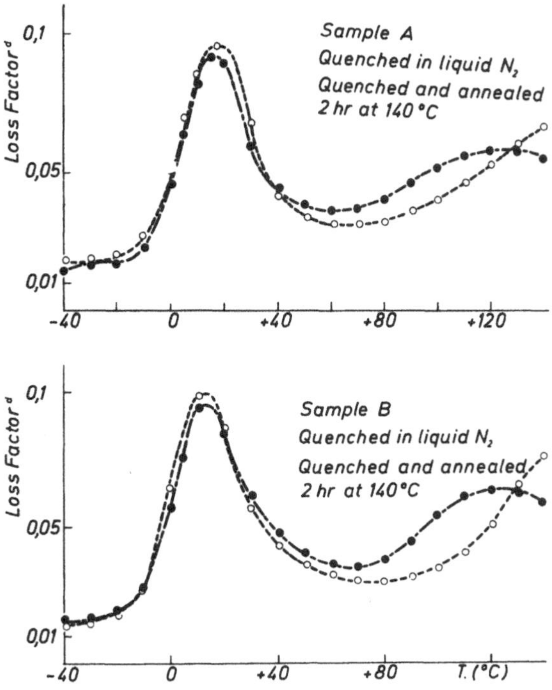

Fig. 1. Influence of thermal history on dynamic-mechanical properties (100 Hz) of polymers *A* and *B*

of the thermal history on σ_B and σ_Y is completely marginal. Fig. 1 illustrates the dynamic-mechanical properties of polymers *A* and *B* after quenching in liquid nitrogen and in same cases annealing two hours at 140 °C.

These curves show that the thermal history mainly affects the relaxation processes of the crystalline polymer, since the annealing influences the α peak intensity (8).

Position and area of the β peak, which corresponds to the glass transition of the amorphous polymer, seems to be independent of the thermal conditioning as if the dispersion β were essentially associated to movements in the atactic phase, whose amount is obviously independent of thermal treatments.

The experimental fact that, at least for a certain level of *MW*, a change of the thermal history implies a variation of the crystalline structure, of the internal stresses and of the ultimate properties whereas the relaxation processes associated to the glass transition are unaffected, means that the fracture behavior of the polypropylene does not depend only on the modulus transition of the amorphous phase but also on the deformation mechanisms of the crystalline structure.

Influence of the amount and MW of the atactic PP

The influence of the amount and *MW* of the atactic *PP* on the tensile impact properties is summarized in table 5.

At −10 °C breaking energy and time are nearly independent of the amount and *MW* of the amorphous phase, whereas at 23 °C they increase by increasing both *MW* and concentration.

The brittle strength is independent and the yield strength decreases by increasing the amorphous fraction whatever the *MW*. The tough-brittle transition is positively affected by an increase in both atactic concentration and *MW*.

The dynamic-mechanical properties at 1000 Hz of all the blends were measured to evaluate the influence of the amount of the atactic fraction on the β peak. Figs. 2 and 3 show the dependence of the loss factor on the temperature respectively for the blends with low and high *MW* atactic *PP*. In table 6 the temperature of the maximum of the β peak (T_{max}), its height (d_{max}) and area (A_{tot}) are listed.

Table 5. Influence of the amount of atactic *PP* on tensile impact properties of the blends

	% added atactic *PP*	Density (23 °C) (gr/cm³)	[η] (*dl*/gr)	σ max (kg/cm²)		Break energy (kgcm/cm²)		Time to break (msec)		$T_{T\text{-}B}$ (°C)
				−10 °C	+23 °C	−10 °C	+23 °C	−10 °C	+23 °C	
Low *MW*	0	0.910	2	1100	770	42	50	0.22	0.35	20
atactic *PP*	3	0.909	1.9	1100	720	40	50	0.21	0.40	20
	10	0.905	1.9	1120	680	45	70	0.22	0.55	19
	15	0.904	1.8	1050	560	45	105	0.23	0.75	17
	20	0.902	1.7	1020	545	43	110	0.22	0.85	16
	25	0.900	1.6	1000	520	45	130	0.23	0.95	14
	35	0.895	1.6	1050	435	40	145	0.2	1.2	13
High *MW*	3	0.909	2	1110	690	50	85	0.23	0.58	18
atactic *PP*	10	0.906	2.1	1110	645	55	120	0.24	0.8	17
	20	0.902	2.1	1070	580	50	150	0.23	1.14	14

Table 6. Influence of the amount of atactic *PP* on dynamic mechanical properties of the blends

	%added atactic *PP*	*T*max(°C)	*d*max	*A* tot (cm²)
low *MW*	0	20	0.122	47
	3	19	0.14	52
	10	17	0.17	59
	15	14	0.20	68
	20	13	0.23	73
	25	12	0.27	81
	35	10	0.35	100
high *MW*	3	18	0.14	54
	10	16	0.17	63
	20	13	0.21	77

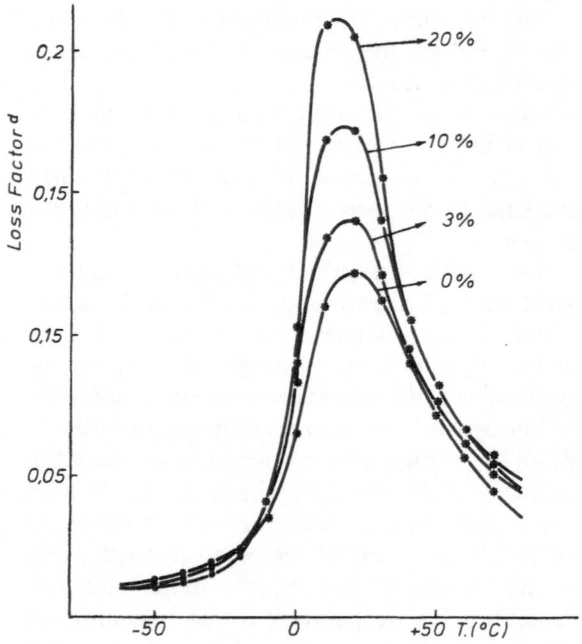

Fig. 2. Influence of the amount of the added low *MW* atactic *PP* on dynamic-mechanical properties of blends

The reported results point out that an increase in the amount of the atactic fraction, whatever its *MW*, increases the height and area of the peak, whereas its maximum is shifted towards lower temperatures.

Also the dynamic-mechanical properties of the two atactic *PP* were measured by coupling with polystyrene.

The curves of fig. 4 show that the differences in area and position between the two β peaks are quite marginal.

Using the value of the 100% atactic *PP* area and assuming a linear relationship with zero intercept between the area of the peak and the amount of atactic polymer, the percentage of the atactic fraction in the residue to xylene extraction of polymer *A* was calculated, the resulting value being 25 wt.-%. This value was confirmed by I.R. analysis of atactic content of the xylene

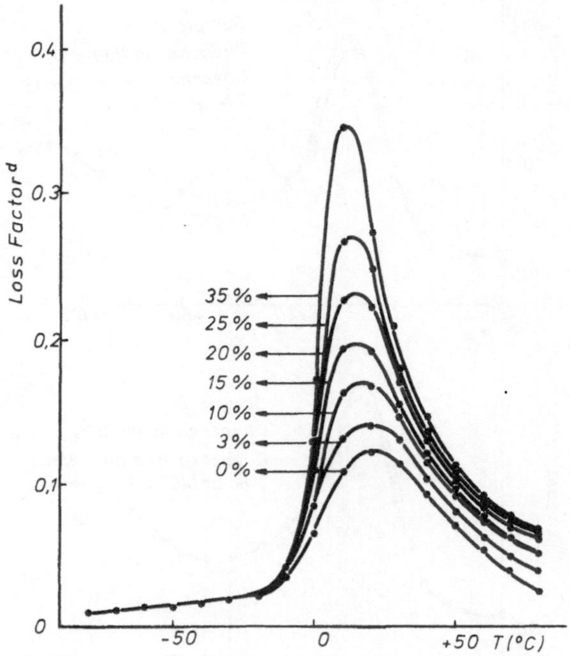

Fig. 3. Influence of the amount of the added high *MW* atactic *PP* on dynamic-mechanical properties of blends

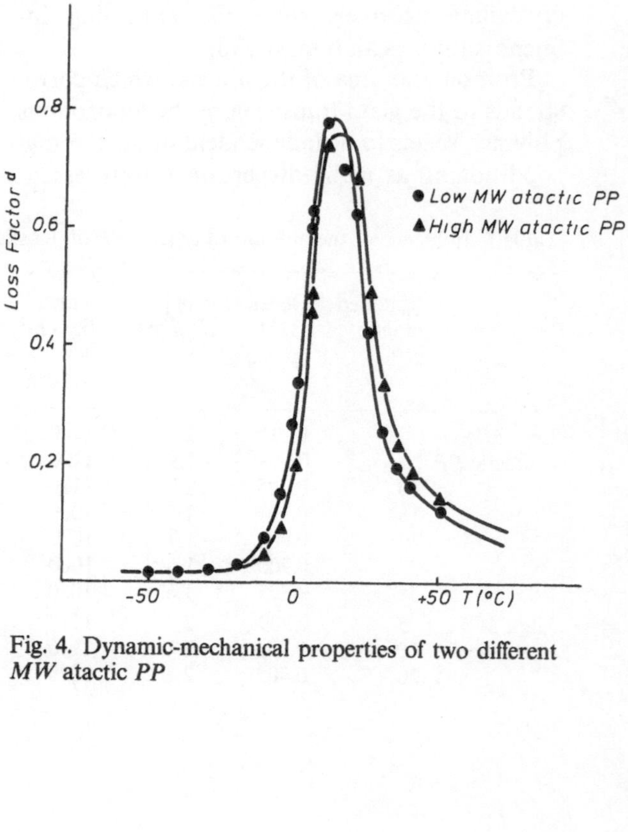

Fig. 4. Dynamic-mechanical properties of two different *MW* atactic *PP*

extraction residue, whose amount resulted 23 wt.-% [1]).

This value was added to the amounts of atactic *PP* introduced in the blends and the sums plotted versus their experimental areas.

The points fit the assumed relationship (fig. 5). Hence it seems it can be concluded that there is a linear relationship between the β peak area and the amount of the atactic fraction, the value of this area being zero when the latter is zero.

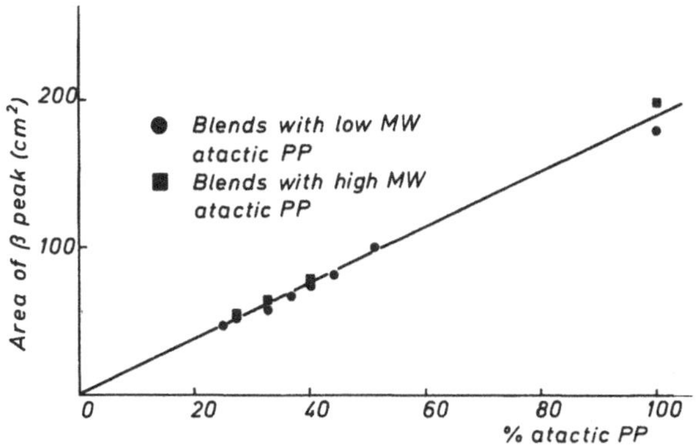

Fig. 5. Relation between the atactic content and the area of the β peak

Conclusions

A variation in *PP* density brings about different variations in the β peak area whether the change in density is obtained through a proper thermal conditioning or the addition of atactic polymer. This suggests the conclusion that the relaxation movements associated to the β peak are essentially due to the relaxation of the atactic fraction.

In fact while the changes in density associated to different thermal histories, i.e. to different amounts of crystallizable amorphous isotactic *PP*, imply only marginal changes of the β peak area, those associated to the addition of different amounts of atactic *PP* imply proportional variations in the peak size. On the contrary the different thermal treatments substantially affect the α peak which is attributed to the crystalline polymer so that it seems reasonable to suppose that the associated changes in density are accountable for the different degrees of perfection of the crystalline structure.

Finally, the fact that the size and position of the peak are independent of thermal history and *MW* of the atactic fraction, while the ultimate properties are strongly affected, means that there is no direct correlation between the dissipated energy for this specific transition at high frequency and the fracture energy measured from tensile tests.

Summary

The influence of amorphous atactic and isotactic fractions on the impact behavior and high frequency dynamic-mechanical properties of polypropylene was investigated.

The atactic fraction defines both size and position of the β peak, while the isotactic fraction influences substantially the α loss peak.

The tensile impact properties are affected by variations in both types of amorphous.

Therefore it seems that there is no direct correlation between the energy dissipated in the various dynamic-mechanical transitions and the energy of fracture.

Zusammenfassung

Der Einfluß der ataktischen und isotaktischen Fraktionen auf die Bruchfestigkeit und das mechanischen Relaxationsverhalten von Polypropylen wurde untersucht.

Die ataktische Fraktion bestimmt die Temperaturlage und die Höhe des β-Maximums, während die isotaktische Fraktion auf den α-Prozeß einwirkt.

Das Bruchverhalten von Polypropylen hängt von Veränderungen in beiden Amorphen ab.

Daher scheint es, daß es keine direkte Korrelation zwischen den dynamisch-mechanischen Relaxationsprozessen und der Bruchenergie gibt.

References

1) *Horsley, R. A.* and *A. C. Morris*, Plastics **31**, 1551 (1966).

2) *Vincent, P. L.*, Polymer **1**, 425 (1960).

3) *Parrini, P., F. Sebastiano*, and *G. Messina*, Makromol. Chem. **38**, 27 (1960).

4) *Miller, R. L.* and *L. E. Nielsen*, J. Polymer Sci. **44**, 391 (1960).

5) *Casiraghi, T.*, Materie Plastiche **36**, 1053 (1970).

6) *Wolstenholme, W. E., S. E. Pregun*, and *C. F. Stark*, J. Appl. Polymer Sci. **8**, 119 (1964).

7) *Szilagyi, L., A. Ballabio*, and *M. Pegoraro*, Kolloid Z. u. Z. Polymer **238**, 415 (1970).

8) *McCrum, N. G.*, Polymer Letters **2**, 695 (´964).

Authors' address:

S. Danesi, L. Baldi, and *G. Ballini*
Montedison SpA
Rice-Centro Ricerche Ferrara (Italy)

[1]) This analysis is based on the measure of the absorbancies of the I.R. bands at 11.89 μm (isotactic fraction) and 12.15 μm (atactic fraction).

Rheol. Acta **13**, 618–626 (1974)

From the Illinois Institute of Technology Chicago (U.S.A.)

Effect of the interface on the mechanical properties of composite materials

By L. J. Broutman and B. Das Agarwal

With 16 figures and 2 tables

(Received October 27, 1972)

I. Introduction

In previous papers (1, 2) we have analyzed the internal stresses in spherical particle composites for the cases where the particle is softer than the matrix (1), as in rubber particle filled polymers, or for the case where the particle is harder than the matrix as in ceramic filled glasses (2). Also, the analysis has been applied to porous composites such as foams in which case spherical voids simply replace the particles (2). A finite element analysis has been used for the calculation of internal stresses and for the prediction of elastic constants and strength of the composite with a suitable model geometry and proper boundary conditions. The results have been presented as a function of the volume fraction of particles or inter-particle spacing.

In this paper, results are presented for spherical particle composites in which the interfacial zone has elastic properties different from that of the particle or matrix. The influence of changing the interface properties on the internal stress distribution in the composite and on the predicted elastic constants will be discussed. Also discussed is the interface in an aligned discontinuous fiber composite and its effect on composite properties such as elastic constants, strength and toughness.

II. Spherical particle composites

1. Approximations and boundary conditions

The present investigations were carried out using an analysis of axisymmetric solids. In the finite element approximation of axisymmetric solids, the continuous structure or medium is replaced by a system of axisymmetric elements interconnected at nodal circles. It was assumed that a spherical particle composite (assumed to possess symmetry) could be approximated by a unit cell (fig. 1) which when rotated 360° around axis *AD* produces a hemisphere embedded within a cylinder. The interparticle spacing is equal to $2(r_1 - r_2)$; r_1 and r_2 are shown in fig. 1. The volume percent of filler particles or cavities (radius $= r_2$) can be altered and calculated from the ratio r_2/r_1 (note $AB = BC = CD = AD$ in fig. 1). This axisymmetric representation of the composite only approximates its real packing and structure. These axisymmetric cells are not an actual repetitive unit but are related in their dimensions to the interparticle spacing.

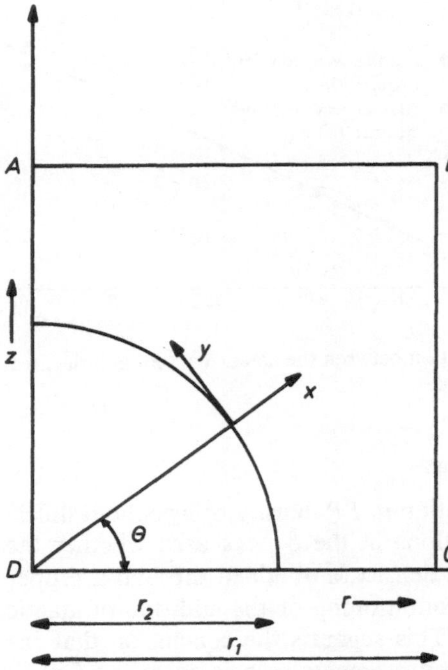

Fig. 1. Unit cell for axisymmetric representation of sphere-filled composite

The unit cell shown in fig. 1 is subdivided into small elements as shown in fig. 2 and fig. 3. The finite element method permits calculation of the stresses in all the elements and the displacements at the nodal circles for any loading and boundary conditions. It is assumed that the composite is strained in the *z*-direction and that no tractions are applied in the *r*-direction. By symmetry, on the boundary *ABCD* (fig. 1) the shear stresses are:

$$\tau_{rz} = \tau_{zr} = 0.$$

The sides *AB* and *BC* remain parallel to their original positions after they are displaced due

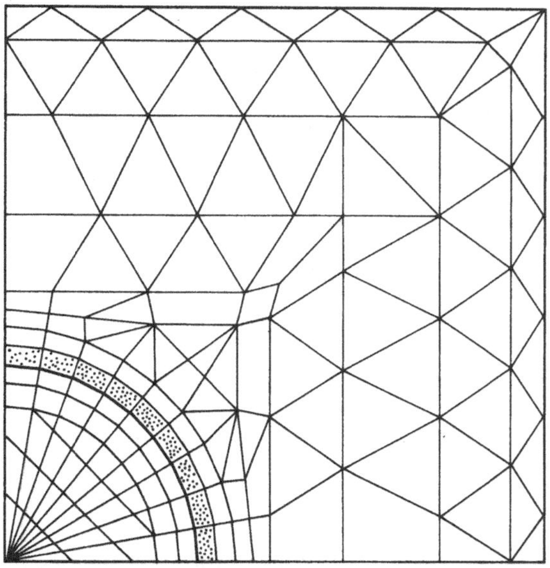

Fig. 2. Shaded elements represent finite thickness of the interface for $r_2/r_1 = 0.357$

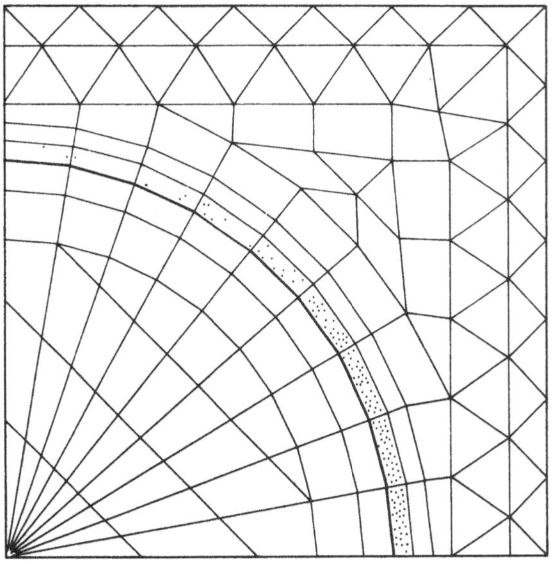

Fig. 3. Shaded elements represent finite thickness of the interface for $r_2/r_1 = 0.714$

to strain in the z direction, whereas the normal displacements of AD and DC are zero. Thus, AB and BC will undergo normal displacements, and the traction in the r direction must be zero so that:

$$\int_{BC} \sigma_r dz = 0$$

where the integral is replaced by a summation in the finite element method.

The following assumptions made concerning the material:

1. Filler particles are spherical and of uniform size; packing of particles can be represented by an axisymmetric element (fig. 1).
2. Both filler, matrix and interface materials obey elastic stress-strain relationships.
3. Perfect bonding exists at the interfaces (continuity of displacements at each interface).

The calculations were made on a digital computer[1]) by using a computer program for the analysis of axisymmetric solids written by *E. L. Wilson* (3). The boundary conditions were prescribed in the mixed mode i.e., the displacements were prescribed on some of the boundaries whereas tractions were prescribed on the others. The prescribed boundary displacements were selected to obtain the desired composite strains. The average composite stress was calculated from a knowledge of the stresses in the elements at the boundary of the unit cell. The composite stresses and strains are used to calculate the composite modulus of elasticity and *Poissons* ratio. The details of the procedure to satisfy the boundary conditions and to calculate composite stress, strain, modulus of elasticity and *Poissons* ratio have been given in previous papers (1, 2).

2. *Effect of a weak interface on composite properties*

In an actual composite material, the properties of the material at the interface may be different from those of the filler and matrix. Continuous displacements at the interface imply perfect bonding between the filler particles and the matrix. When perfect bonding does not exist between the filler and the matrix the behavior of the interface should be simulated by assigning different property values to the material at the interface. This is very easily accomplished using the axisymmetric finite element method. The shaded elements in figs. 2 and 3 have been assumed to represent the finite thicknesses of the interface for filler contents of 3.03 ($r_2/r_1 = 0.357$) and 24.30 ($r_2/r_1 = 0.714$) percents respectively. The shaded elements account for 0.48% of the total volume in the former case and 3.02% in the latter case. The modulus of elasticity assigned to the elements at the interface was 1000 psi which is very small compared to that of the assumed glass matrix ($E - 11.8 \times 10^6$ psi) or

[1]) Univac 1108, Univac Div., Sperry Rand Corp., Philadelphia, Pa. (USA).

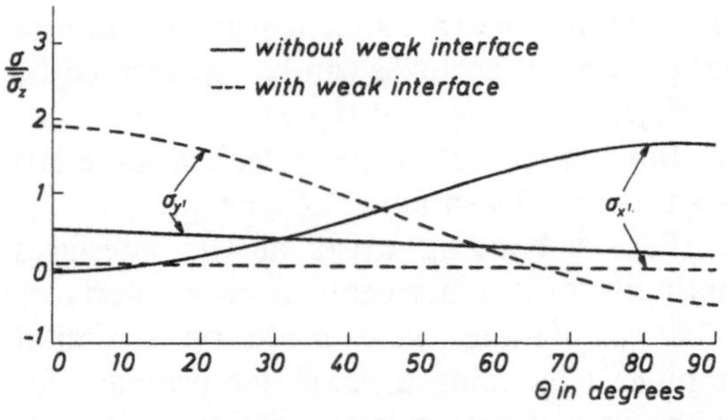

Fig. 4. Comparison of matrix stresses with and without a weak interface in a composite ($r_2/r_1 = 0.357$)

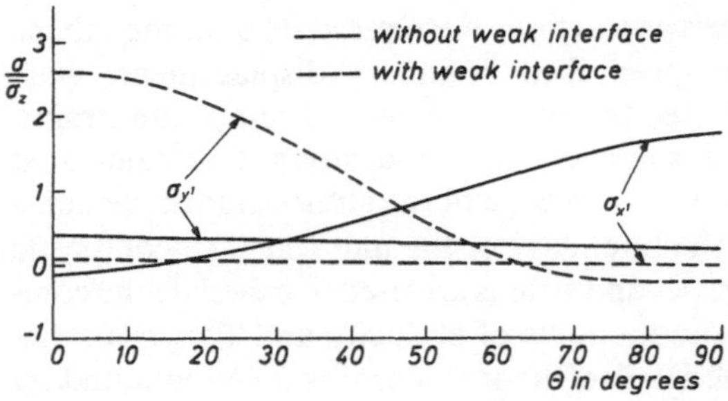

Fig. 5. Comparison of matrix stresses with and without a weak interface in a composite ($r_2/r_1 = 0.714$)

the ceramic filler ($E = 60.4 \times 10^6$ psi). This represents the case of a very weak interface.

The weak interface as described above completely changes the stress distribution around the interface. The stresses in the elements (in the matrix) adjacent to the interface have been plotted in figs. 4 and 5. A zero value of radial stress around the interface indicates a free boundary. The curves for tangential stresses with the finite interface are very similar to the ones previously obtained for stresses around a cavity (2). Due to very low modulus, the interface is not able to transfer much stress from the matrix to the hard inclusions and therefore this represents a case of filler particles completely debonded from the matrix. The hard inclusions with the soft interface carry very low stresses and hence do not contribute to the enhancement of the modulus of the composite. This is similar to what *Stett* and *Fulrath* (4) have described as pseudoporosity which results in a weakening of the composite. The modulus of elasticity of the composite decreases with higher filler contents as indicated in table 1.

It is not unexpected that the modulus of the composite will decrease since less stress is transferred to the hard particles. Thus, as the volume

Table 1. Effect of interface on modulus of Elasticity of composite

Volume fraction filler	Composite modulus (psi) without interface	with soft interface
3.04	12.3×10^6	11.05×10^6
24.30	16.7×10^6	6.97×10^6
43.83	22.4×10^6	3.84×10^6

percent of filler increases the difference in modulus increases as can be seen in table 1.

III. Elastic analysis of three phase fibrous composites

1. Introduction

In the case of continuous fiber reinforcement, the effect of fiber ends, where the load is transferred by the matrix, is generally considered insignificant. The fiber stress is assumed to be constant over the whole length of the fibers. The principal purpose of the matrix is to bind the fibers together. The strength of the composite is then dependent upon the strength of the fibers. However, in studying the details of fracture of continuous fiber composites it has been found that individual fibers fail well before the entire composite fractures. Thus, in this case, the load transferred to the broken fibers by the matrix and the interfacial conditions may thus influence composite fracture particularly as the number of broken fibers increases.

In a discontinuous fiber-reinforced composite the properties of the composite are a function of fiber length and the attainment of high strength in the composite will depend upon efficient load transfer from the matrix to the fibers. Therefore, it is of considerable interest to understand how stress builds up in each individual fiber. A study of the length required for effective reinforcement and the factors influencing this length such as the properties of the material at the interface and the fiber end condition, should thus be helpful in guiding the development of composites of this type.

It is well known that in discontinuous fiber reinforced systems with all fiber axes parallel to the direction of loading, the mechanism of load transfer from matrix to the fiber is an interfacial shear stress. A number of analytical studies concerning this shear stress transfer have been carried out using simplified models. Fiber-matrix interaction has been studied for elastic matrices by *Cox* (5), *Dow* (6), and *Rosen* (7). They give expressions for axial fiber stress and for the shear stress at the fiber matrix interface as a function of position along the fiber length. These expressions are quite similar to each other, although different assumptions were made in deriving them. *Tyson* and *Davies* (8), and *Schuster* and *Scala* (9) measured interfacial shear stress between a metal fiber and epoxy resin by using photoelastic techniques. Studies of *Fujiwara* (10) for resin-fiber load transfer in a single fiber-resin composite indicate that the stress distribution depends upon glass fiber finishes, expecially under wet conditions. *Carrara* and *McGarry* (11) studied the effect of fiber end geometry on the stresses near the end of an elastic fiber embedded in an elastic matrix. They found that the stresses depend strongly on the geometry of the fiber tip. More recently, *MacLaughlin* and *Barker* (12) investigated the effect of modulus ratio on stress near a discontinuous fiber. They analyzed a two-dimensional plane stress composite configuration using Moire strain analysis and finite element analysis. In the study presented here the properties of the interface between fiber and matrix have been varied in order to determine the influence of the interface on certain properties of the composite.

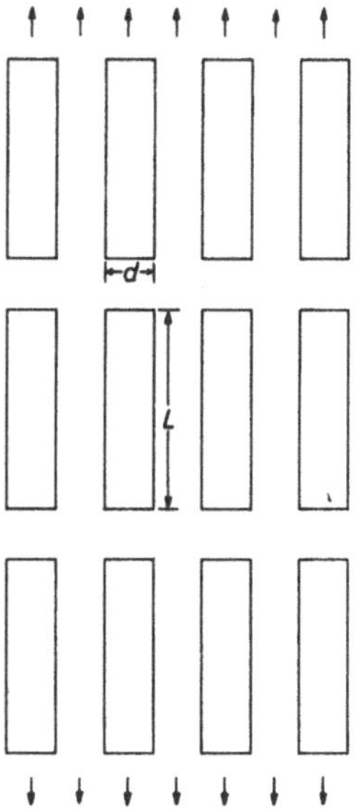

Fig. 6. Symmetrical packing of discontinuous fibers in a composite

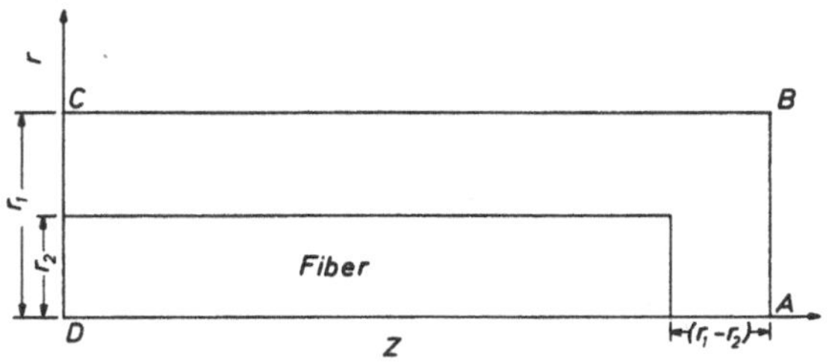

Fig. 7. Cell for Axisymmetric representation of fibrous composite

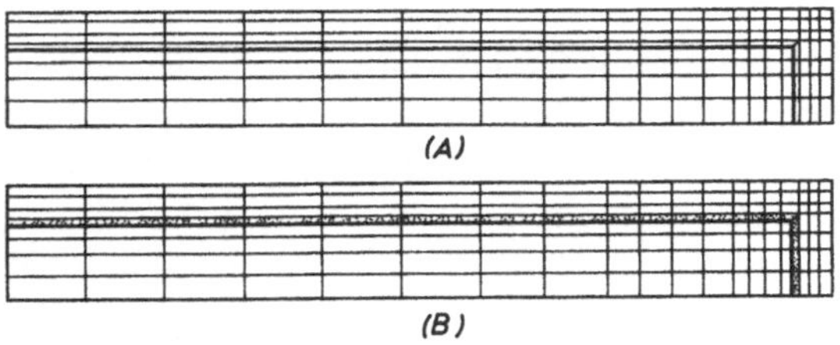

Fig. 8. (a) Finite element grid for $r_2/r_1 = 0.67$, and (b) finite thickness of the interface represented by shaded elements

2. *Representative model*

Unidirectional discontinuous fibers were assumed to be packed in a regular array as shown in fig. 6. Although this does not represent an actual packing of the fibers in the composite, this idealization is necessary for an axisymmetric analysis.

It was assumed that the fibrous composite could be approximated by a cell (fig. 7) which when rotated 360° around axis *AD* produces a cylinder embedded within a cylinder. The inter-fiber spacing is equal to $2(r_1 - r_2)$ in both directions as shown in fig. 7. The finite elements used for the case $r_2/r_1 = 0.67$, are shown in fig. 8a. Based on a cylinder within a cylinder, this corresponds to a fiber volume fraction equal to 42.4%. The fiber aspect ratio used (ratio of fiber length to fiber diameter, l/d) is 10.375. The elements adjacent to the fiber (shaded elements in fig. 8b have been assumed to represent the finite thickness of the interface. The property values (the modulus of elasticity and *Poissons* ratio) assigned to these elements are changed to simulate a change in the interface conditions. A high modulus of elasticity of the interface represents a strong interface capable of transferring more load whereas a low modulus represents a weak interface. The shaded elements account for 7.76% of the total volume. In some cases the thickness of the interface was reduced by a half to study the effect of this change. The elements adjacent to the fiber end may be assigned property values different from those for the interface. This enables one to study the effect of fiber end condition on the stress distribution. For example, a very low modulus of elasticity for these elements may be assumed to represent a debonded end because a negligible load will be transferred through the fiber end.

3. Boundary conditions and component properties

Stresses in three dimensions were calculated in all the elements shown in fig. 8 for various interface conditions. As in the case of the particulate composites, the stress-strain relations of the matrix and the fibers were assumed elastic. The stress-strain relations for the materials at the interface and the fiber end were also assumed elastic. It was also assumed that the composite is loaded by a force in the z direction and that no tractions are applied in the r direction. These assumptions lead to boundary conditions identical to those for the particulate composite described previously. Therefore, the procedure to satisfy these boundary conditions and the subsequent calculation of the composite modulus and *Poisson*s ratio are also identical.

The following component properties, typical of a glass reinforced plastic, were assumed:

Matrix $E = 0.4 \times 10^6$ psi
$\nu = 0.35$

Fibers $E = 11.8 \times 10^6$ psi
$\nu = 0.197$

Properties of the interface were varied over a wide range. Investigations were carried out using eleven different combinations of property values as shown in table 2.

Modulus of elasticity of the interface has been varied from a very high value of 8×10^6 psi which is close to that of the fibers to a very low value of only 100 psi which may be considered to represent debonding of the fibers from the matrix. The first three cases have been selected to study the effect of varying *Poisson*s ratio of the interface. Also, note that the modulus of elasticity of the elements adjacent to the fiber end is the same as that of the interface for all cases except for 4 and 5. For these two cases it has a very low value (100 psi) compared to that of the interface. This represents a case of strong interface with debonded fiber end. The only difference in the case 4 and the case 5 is the thickness of the interface which would change the colume percent of the interface.

Table 2. Properties of the interface

Case number	E (psi)	ν	Vol.-%	E at the fiber end (psi)
1	8×10^6	0.2	7.76	8×10^6
2	8×10^6	0.35	7.76	8×10^6
3	8×10^6	0.45	7.76	8×10^6
4	8×10^6	0.2	3.79	100
5	8×10^6	0.2	7.76	100
6	0.8×10^6	0.2	7.76	0.8×10^6
7	0.4×10^6	0.2	3.79	0.4×10^6
8	0.1×10^6	0.2	3.79	0.1×10^6
9	10000	0.2	3.79	10000
10	1000	0.2	7.76	1000
11	100	0.2	7.76	100

4. Effect of interface on internal stresses

The stress distributions for three phase composites were obtained for all cases indicated in table 2. In the first three cases, the *Poisson*s ratio of the interface was assigned three different values of 0.2, 0.35, and 0.45 while keeping all other properties unchanged. The stresses obtained in the three cases were almost identical to each other and hence they are not plotted separately. These stresses have been shown along with the other stress distributions for different interface elastic moduli.

Distributions of fiber axial stress along the length are shown in fig. 9 for interface moduli varying from 8×10^6 psi to 100 psi. When the interface modulus is near the matrix modulus or higher, the fiber axial stress attains a maximum value within two fiber diameters from the fiber end. The stress distributions for the three cases of interface moduli ($E = 0.1 \times 10^6$, 0.8×10^6, and 8×10^6) are very similar to each other. As the interface modulus decreases to 10000 psi, the fiber axial stress attains its maximum value in about four fiber diameter from the fiber end. But as the interface modulus forther decreases to 1000 or 100 psi, the fiber axial stress does not

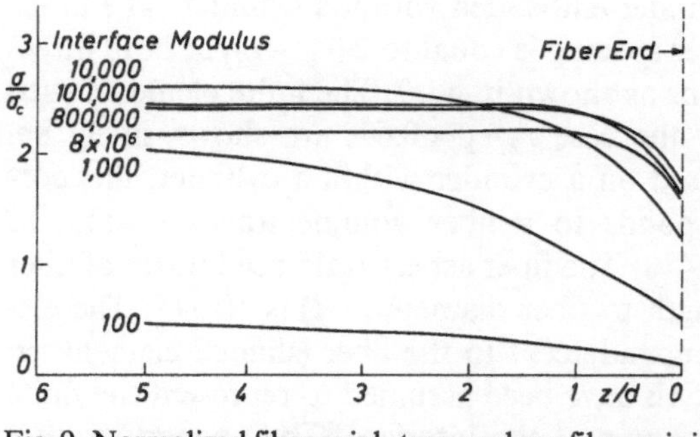

Fig. 9. Normalized fiber axial stresses along fiber axis in a three-phase composite

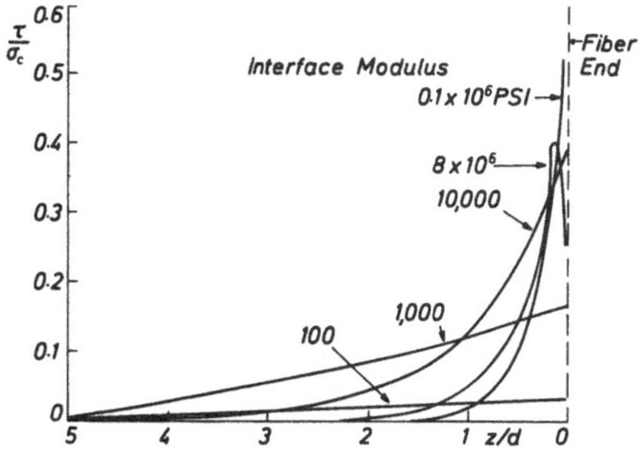

Fig. 10. Normalized interfacial shear stresses along fiber axis in a three-phase composite

reach a constant value with the present fiber length of ten fiber diameters. Due to the low interface modulus, the interface does not transfer load from matrix to the fiber efficiently. In fact the modulus of 100 psi is so low that it represents the case of complete debonding of the fiber from the matrix as will be shown later.

Interfacial shear stress distributions for all the above cases, except for an interface modulus of 0.8×10^6 psi, have been shown in fig. 10. The stress distribution for the interface modulus of 0.8×10^6 psi is very close to the one for the interface modulus of 0.1×10^6 psi. The stress distributions in fig. 9 are related to those in fig. 10 because fiber stress buuild up is related to the shear stress at the interface as follows:

$$\sigma_f = \sigma_0 + 2/r \int_0^Z \tau dZ \qquad [1]$$

where r is the fiber radius, σ_0 is the stress at the fiber end, σ_f is the stress in the fiber at any distance Z from the fiber end, and the exact form of τ will depend upon relative properties of the fiber, interface and the matrix. The shear stresses at the fiber end for interface moduli of 0.1×10^6 and 8×10^6 psi are high and they drop down to zero in about two fiber diameters. Due to the high interfacial shear stress, the fiber axial stress increases rapidly and reaches a constant value as the shear stress drops to zero. For the interface modulus of 10000 psi the shear stress does not drop to zero as quickly as in the previous cases and thus leads to a higher maximum stress in the fiber. Normalized shear stress at the fiber end for an interface modulus of 1000 psi is less than half of those in the previous cases. However, this does not decrease very fast away from the fiber end and hence the axial stress in the fiber builds up to about twice the stress on the composite. It may be expected that in this case the fiber stress could reach a constant value if the fiber was long enough. Interfacial shear stress in the case of an interface modulus of 100 psi is very small and therefore very little load transfer is possible from the matrix to the fiber.

Fiber axial stress at the end has been plotted as a function of log of interface modulus in fig. 11. This stress gives an idea of the load transfer through the fiber end. At very low interface modulus the stress at the fiber end is very small indicating that no significant load transfer takes place through the fiber end. The stress at the fiber end increases with the increase of interface modulus. When the interface modulus is close to or higher than the matrix modulus, the fiber end stress is quite high indicating substantial load transfer through the end. As the interface modulus varies from 10^5 psi to 8×10^6 psi there is not much change in the fiber end stress.

Distribution of matrix axial stresses along the fiber length are shown in fig. 12 for the interface moduli of 100, 1000, 10000, and 10^5 psi. In all the case the stresses increase sharply near the fiber end. They also show discontinuities in

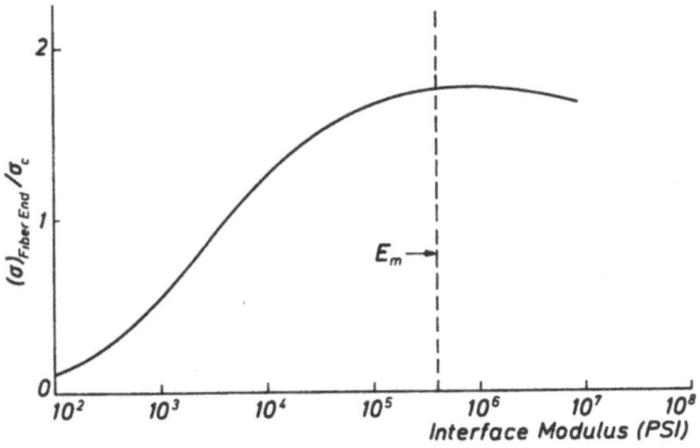

Fig. 11. Normalized axial stress at the fiber end as a function of interface modulus

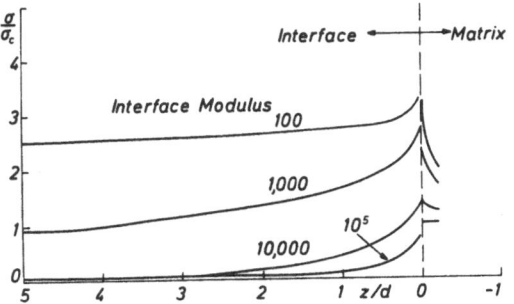

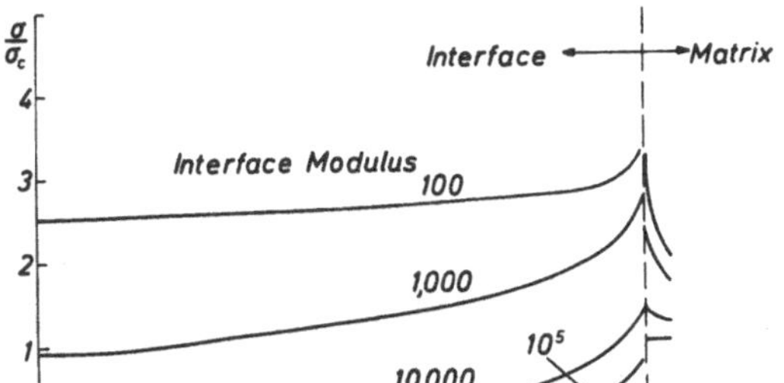

Fig. 12. Normalized axial stresses along fiber axis in a three-phase composite

the stresses upon passing the fiber end because of the physical discontinuity at this point. In the cases of interface moduli of 10^4 and 10^5 psi, the stresses away from the fiber end drop to a very low value indicating that most of the load has been transferred to the fibers. However, a low modulus of the interface (100 or 1000 psi) does not help in the transfer of load from matrix to the fibers and therefore the axial stresses in the matrix away from the fiber end are considerably higher than the applied composite stress in these cases.

Radial stresses in the matrix have been plotted along the fiber length in fig. 13 for interface moduli of 100, 1000, and 10000 psi. The stresses increase sharply near the fiber end. The stresses away from the fiber end are compressive when the interface moduli equal 1000 and 10000 psi. For the interface modulus of 100 psi, the radial stresses in the matrix are zero. It has also been shown that the axial stresses in the matrix adjacent to the fiber end are nearly zero for the interface modulus of 100 psi. The small magnitude of the stresses may be attributed to the size of the finite elements. Thus the normal stresses in the matrix adjacent to the fiber are zero indicating a free boundary. Therefore this represents a case of complete debonding of fibers from the matrix. This effect is the same as obtained in the case of particulate composites.

Composite modulus has been plotted as a function of log interface modulus in fig. 14. At very low interface modulus the fibers do not contribute to the stiffness of the composite. As explained earlier, this is due to the fact that the soft interface does not permit any load tranfer from the matrix to the fiber. Therefore, the composite behaves as if these were voids of the size of the fiber and the interface. As the interface modulus increases, the load transfer takes place

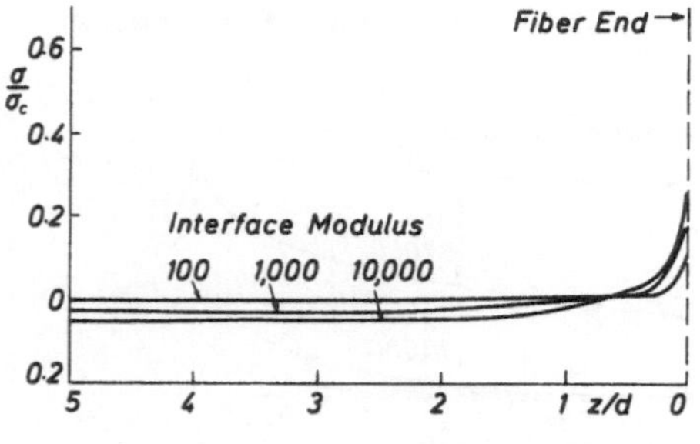

Fig. 13. Normalized matrix radial stress along fiber axis in a three-phase composite

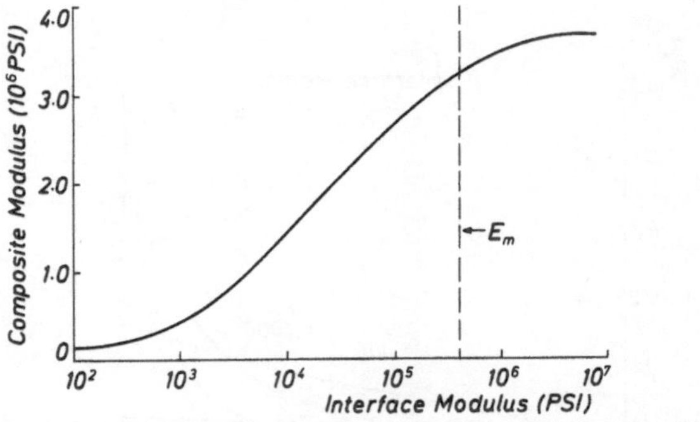

Fig. 14. Composite modulus as a function of interface modulus

from the matrix to the fiber and consequently the composite modulus increases as shown in fig. 14. The *Halpin* and *Tsai* equation (13) may also be used to calculate the modulus of the two phase composite with discontinuous fiber reinforcement. For the modulus in the longitudinal direction, the equation can be written as:

$$\frac{E_L}{E_m} = \frac{1 + 2l/d \cdot V_f \cdot \eta_L}{1 - V_f \cdot \eta_L} \qquad [2]$$

where

$$\eta_L = \frac{(E_f/E_m) - 1}{(E_f/E_m) + 2l/d} \qquad [3]$$

and E_m, E_f and E_L are the matrix, fiber and the composite moduli respectively. For the present case $E_f/E_m = 11.8/0.4$, $l/d = 10.375$, $V_f = 0.424$, and $E_m = 0.4 \times 10^6$ psi

thus

$E_L = 3.15 \times 10^6$ psi.

This value of the composite modulus compares favorably with the value of 3.58×10^6 psi obtained for the two phase composite by the finite element method.

The onset of failure may be predicted from the knowledge of the stress distributions in the composite. To this end, distortion energy given as follows:

$$U = 1/2\,[(\sigma_1 - \sigma_2)^2 + (\sigma_2 - \sigma_3)^2 + (\sigma_3 - \sigma_1)^2] \qquad [4]$$

(where σ_1, σ_2 and σ_3 are principal stresses) was calculated for all the elements (fig. 8) for different interface moduli (normalized stresses were used for this calculation). The maximum value of distortion energy in any element of the matrix has been plotted as a function of interface modulus

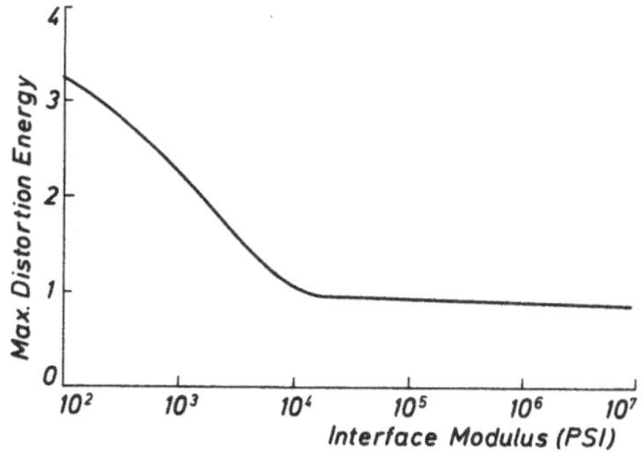

Fig. 15. Maximum distortion energy as a function of interface modulus

in fig. 15. The maximum distortion energy always occurs in an element near the fiber end. The strength of the composite based on a *von Mises* failure criterion (i.e. the initiation of composite failure occurs as soon as the distortion energy in any element of the matrix reaches a limiting value) is shown qualitatively in fig. 16. The actual strength values will depend upon the matrix strength. For very low interface modulus the strength of the composite is low due to high stress concentrations at the discontinuity. Strength of the composite increases as the interface modulus increases. But as the interface modulus changes from 10^4 psi to 8×10^6 psi, the composite strength remains almost constant. However, fig. 14 shows a significant change in the composite modulus over this range of interface modulus. Thus, the ultimate elongation of the composite can be controlled without offecting the failure stress by suitably controlling the interface modulus (using a surface treatment on the fibers during manufacturing). This also shows that a good combination of tensile strength and toughness (or the impact strength) may be obtained by suitably selecting the interface properties.

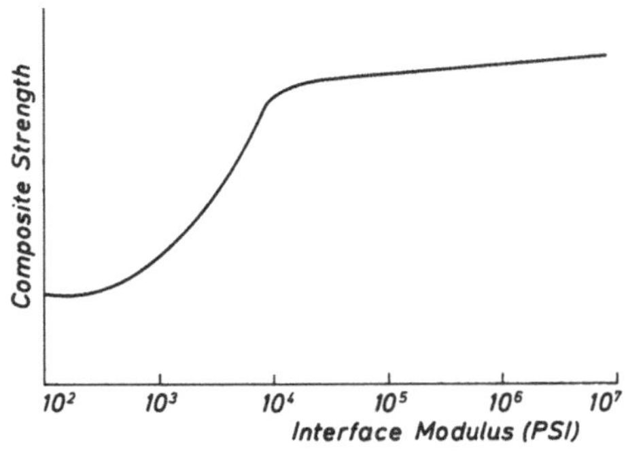

Fig. 16. Composite strength as a function of interface modulus (qualitative representation)

IV. Conclusions

The influence of the interface on the internal stresses, composite modulus of elasticity and strength has been investigated. The interface has been altered by changing the modulus of a layer between the matrix and spherical particle or fibers. In the case of spherical particle composites it has been shown how a soft interface can reduce the composite stiffness as well as alter the internal stresses in the composite. The effect of altering the interface stiffness in an aligned short fiber composite can greatly effect the stress concentrations near a fiber end as well as the maximum stress transferred into the fiber. It has also been shown that the composite strength reaches a maximum and does not further increase when the interface modulus reaches a value of 10^4 psi. However, the composite modulus continues to increase so that the composite elongation will begin to decrease (since all phases have been assumed to be elastic) when the interface modulus exceeds 10^4 psi. Thus the strain energy absorbed by the composite can be maximized by controlling the interface modulus.

Acknowledgements

This research has been supported by the U.S. Atomic Energy Commission under Contract No. AT (11-1)-1794 and this support is gratefully acknowledged.

References

1) *Broutman, L. J.* and *G. Panizza*, Int. J. Poly. Mat. **1**, 95–109 (1971).

2) *Agarwal, B. D., G. A. Panizza*, and *L. J. Broutman*, J. Amer. Ceram. Soc. **54**, 620–624 (1971).

3) *Wilson, E. L.*, Structural Engineering Laboratory Report 63–1, University of California, Berkeley, Calif., June 1963.

4) *Stett, M. A.* and *R. M. Fulrath*, J. Amer. Ceram. Soc. **53**, 5–13 (1970).

5) *Cox, H. L.*, British J. Appl. Phys. **3**, 72 (1952).

6) *Dow, N. F.*, Study of Stress Near a Discontinuity in a Filament Reinforced Composite Material. G.E.C., Missile and Space Division Report No. R63SD61, 1963.

7) *Rosen, B. W.*, Mechanics of Fiber Strengthening. Fiber Composite Materials, Chapter 3, ASM Publication for Seminar of ASM, October 1960.

8) *Tyson, W. R.* and *G. J. Davies*, British J. Appl. Phys. **16**, 199 (1965).

9) *Schuster, D. M.* and *E. Scala*, Trans. Met. Soc. of AIME **230**, 1635 (1964).

10) *Fujiwara, M.*, Resin-Fiber Load Transfer in Fiber-Reinforced Plastics. MIT Research Report R 67–8, February 1967.

11) *Carrara, A. S.* and *McGarry*, J. Comp. Mat. **2**, 222–243 (1968).

12) *MacLaughlin, T. F.* and *R. M. Barker*, Experimental Mechanics **12**, 178–183 (1972).

13) *Ashton, J. E., J. C. Halpin*, and *P. H. Petit*, Primer on Composite Materials: Analysis (Stanford, Conn. 1969).

Authors' addresses:

Dr. *Lawrence J. Broutman*
Professor of Materials Engineering
Illinois Institute of Technology
Department of Metallurgical and Materials Engineering
10W 32nd Street, Chicago, Ill. 60616 (U.S.A.)

Dr. *Bhagwan Das Agarwal*
Research Assistant
Department of Mechanics, Mechanical and Aerospace Engineering
10W 32nd Street, Chicago, Ill. 60616 (U.S.A.)

Rheol. Acta **13**, 627–638 (1974)

From the Department of Chemical Engineering, Princeton University, Princeton, New Jersey 08540 (USA)

Biaxial extension of an elastic liquid

By J. M. Maerker and W. R. Schowalter

With 12 figures

(Received August 3, 1972)

1. Introduction

It is now widely recognized that characterization of the flow behavior of rheologically complex fluids from measurements in laminar shearing flow provides, at best, an incomplete picture of fluid response. Since many industrially important polymer operations, such as molding, sheet formation, and fiber spinning, may involve primarily an *extension* of fluid elements as opposed to a shearing of them, the behavior of polymeric fluids in extensional flow has become a subject of much interest. Beyond this immediate practical motivation for study of extensional flows is the fact that many of the forms of constitutive equations which are currently popular predict widely different responses for the relation between stress and strain rate (or stretch rate) in extensional flow. Thus the flow should be useful in providing a basis for discrimination between constitutive equations.

Well defined extensional flows have been extremely difficult to generate and to study in the laboratory. Difficulties have included restriction to low levels of stretch rate (1–3), uncertainty in stress measurements due to changing strain histories (4–6), and analyses of the deformation process which are only approximate (7). In the present study we have attempted to make the combination of these restrictions less severe than in previous work. Steady-state biaxial extensional viscosities are reported for undiluted, low molecular weight polyisobutylenes over a range of stretch rates from about $10^{-2}\,\mathrm{sec}^{-1}$ to $1\,\mathrm{sec}^{-1}$.

2. Theory

In a rectangular *Cartes*ian coordinate system (x^1, x^2, x^3) an extensional flow is defined by the velocity field

$$v^i = a_i x^i \qquad [1]$$

where the summation convention is not being employed, and the a_i may be functions of time and of the ith coordinate but not of the remaining two coordinate directions. If the a_i are independent of x^i, then the extension is *uniform*. If the a_i are not time dependent, the extension is steady. Most analyses of extensional flow have been for steady uniform extension (8). In such a case the rate-of-deformation tensor, $d^i_j = 1/2(\partial v^i/\partial x^j + \partial v^j/\partial x^i)$, reduces to the simple form

$$[d^i_j] = \begin{bmatrix} a_1 & 0 & 0 \\ 0 & a_2 & 0 \\ 0 & 0 & a_3 \end{bmatrix}. \qquad [2]$$

In the special case of equal biaxial extensional flow in, say, the 1- and 2-directions, the continuity equation insures that

$$d^3_3 = -2d^1_1 \qquad [3]$$

where $d^1_1 = d^2_2$.

It has been noted (9) that uniaxial extensional flow ($d^1_1 = a_1$, $d^2_2 = d^3_3 = -a_1/2$) is not equivalent to uniaxial compressional flow. One can readily see, however, that there is an equivalence between uniaxial compressional flow and biaxial extension. Consequently, biaxial extension and uniaxial extension represent two independent possibilities for testing the predictions of constitive models, and experiments of uniaxial compression can add nothing to the information which one can obtain from. biaxial extension experiments.

2.2. Stress response

Components of the total stress tensor **S** can be separated into an indeterminate isotropic part $-p$ and components of the remaining contribution **s**′ to **S**. Thus

$$S^{ij} = -pg^{ij} + s'^{ij} \quad [4]$$

where g^{ij} is a component of the metric tensor. Though p is often defined to make $\mathbf{s}'$ traceless, it is not necessary to follow that convention.

Denn and *Marrucci* (10) have performed an instructive analysis of the behavior of the transient response of a fluid which, initially at rest, is subjected to steady, uniform, extension. A convected *Maxwell* rheological model was used. For the present paper it is useful to have the corresponding result for equal biaxial extension. We employ the contravariant form of the *Oldroyd* convected derivative and represent the nonlinear *Maxwell* model by

$$s'^{ij} + \theta \frac{\mathfrak{d} s'^{ij}}{\mathfrak{d} t} = 2\mu d^{ij} \quad [5]$$

$$\frac{\mathfrak{d} s'^{ij}}{\mathfrak{d} t} = \frac{\partial s'^{ij}}{\partial t} + v^k s'^{ij}{}_{,k} - s^{ik} v^j{}_{,k} - s'^{kj} v^i{}_{,k}. \quad [6]$$

For simplicity we take the material parameters θ and μ to be constants. Then, for the case of uniform equal biaxial extension, eq. [5] reduces to the following equations for the normal stresses in the 1- and 3-directions:

$$s'^{11} + \theta\left(\frac{\partial s'^{11}}{\partial t} - 2d(t)\, s'^{11}\right) = 2\mu d(t) \quad [7]$$

$$s'^{33} + \theta\left(\frac{\partial s'^{33}}{\partial t} + 4d(t)\, s'^{33}\right) = -4\mu d(t) \quad [8]$$

where $d = d^{11} = d^{22}$. Integration from a rest state ($\mathbf{s}' = 0$ at $t = 0$) yields

$$s'^{11} = \frac{2\mu}{\theta} \times \int_0^t d(\xi) \exp\left[-\int_\xi^t \left(\frac{1}{\theta} - 2d(\tau)\right) d\tau\right] d\xi \quad [9]$$

$$s'^{33} = \frac{-4\mu}{\theta} \times \int_0^t d(\xi) \exp\left[-\int_\xi^t \left(\frac{1}{\theta} + 4d(\tau)\right) d\tau\right] d\xi. \quad [10]$$

The total normal stress in the 3-direction, S^{33}, perpendicular to the two directions of extension, is just balanced at the free surface boundary by atmospheric pressure, which we can take as the zero reference pressure. Then from combination of eqs. [9] and [10] we have

$$S^{11} - S^{33} = S^{11} = s'^{11} - s'^{33} = \frac{2\mu}{\theta} \int_0^t d(\xi) \exp\left[-(t-\xi)/\theta\right] \times \left\{\exp\left[\int_\xi^t 2d(\tau)\, d\tau\right] + 2\exp\left[-\int_\xi^t 4d(\tau)\, d\tau\right]\right\} d\xi. \quad [11]$$

If we designate by x the position coordinate x^1 of a particular material point at any specified time, then

$$d = a_1 = \frac{1}{x}\frac{dx}{dt}; \quad \int_0^t d(\tau)\, d\tau = \ln x \Big|_{x_{t=0}}^{x_{t=t}}$$

and eq. [11] becomes

$$S^{11} = \frac{2\mu}{\theta} \int_0^t d(\xi) \exp\left[-(1-\xi)/\theta\right] \times \left\{\left(\frac{x(t)}{x(\xi)}\right)^2 + 2\left(\frac{x(\xi)}{x(t)}\right)^4\right\} d\xi. \quad [12]$$

Now, finally, if we envision the stretching as a constant stretch rate imposed upon the material in a rest state at time zero, then

$$x^1 = x_0^1 \exp(d\Delta t)$$

and one can integrate eq. [12] (11) to

$$S^{11} = \frac{6\mu d}{(1-2\theta d)(1+4\theta d)} - \frac{2\mu d}{(1-2\theta d)} e^{-(1-2\theta d)t/\theta} - \frac{4\mu d}{(1+4\theta d)} e^{-(1+4\theta d)t/\theta} \quad [13]$$

which is the transient response predicted for tangential membrane stress ($S^{11} = S^{22}$) in equal biaxial extension of an elastic liquid described by eq. [5]. The behavior of a *Newton*ian fluid is recovered by considering the limit $\theta \to 0$.

Behavior of fluids in extension is sometimes described in terms of an extensional viscosity $\bar{\eta}$, where

$$\bar{\eta} = \frac{(S^{11} - S^{33})}{d}. \quad [14]$$

For a *Newton*ian fluid we immediately see that $\bar{\eta} = 6\mu$, so that the *Trouton* ratio ($\bar{\mu}/\mu$) is 6 for biaxial extension. For uniaxial extension of a

*Newton*ian fluid one obtains the familiar *Trouton* ratio $\bar{\eta}/\mu = 3$.

Many authors have commented on the complicated behavior which may be exhibited by non-*Newton*ian materials. We see, for example, that at a value of *Deborah* number, $N_{\mathrm{Deb}} = \theta d$, equal to 1/2, the steady-state stress becomes infinite according to eq. [13]. Since an infinite stress is not physically attainable, several rheologists have questioned the utility of models which lead to steady-state solutions similar to the limiting form of eq. [13] (12, 13). A traditional explanation of the paradox has been that an elastic liquid cannot be stretched at the value corresponding to $\theta d = 1/2$, and that as one approaches this critical *Deborah* number the fluid either fractures or the flow rearranges in such a way that the stretch rate is kept below its critical value.

Denn and *Marrucci* (10) have shown that the interpretation of a critical *Deborah* number can be clairified if one includes the initial value problem, as we have above. They showed that for uniaxial extension at $\theta d = 1/2$ a condition of infinite stress is only reached in the limit as $t \to \infty$. A similar result is obtained for equal biaxial extension for the case $\theta d = 1/2$. Eq. [13] then reduces to

$$\frac{S^{11}}{S_N^{11}} = \frac{2}{9}\left[1 + \frac{3}{2}\frac{t}{\theta} - e^{\frac{-3t}{\theta}}\right]$$

where S_N^{11} is the corresponding steady-state stress for a *Newton*ian fluid with viscosity μ.

It is instructive to plot the stress-time behavior predicted from eq. [13] for several values of *Deborah* number. Results are shown in fig. 1. One notes that for $N_{\mathrm{Deb}} = 0$, the case of a *Newton*ian fluid or an elastic fluid at vanishingly

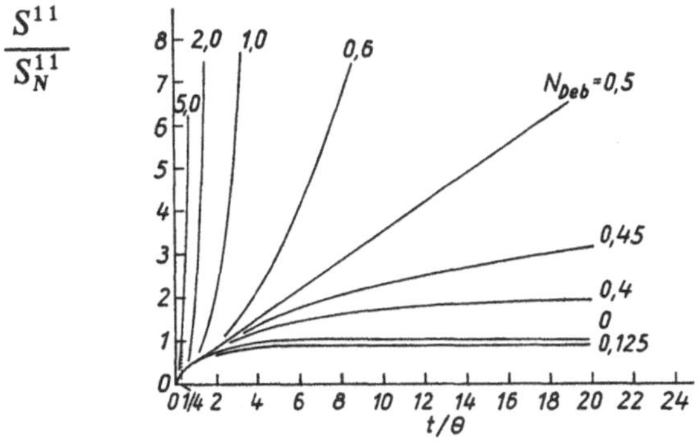

Fig. 1. Equal biaxial extensional flow. Membrane stress, relative to *Newton*ian stress, as a function of time for various values of $N_{\mathrm{Deb}} = \theta d$

small stretch rate, a steady-state stress is rapidly achieved. In contrast to the corresponding results for uniaxial extension (10) however, one finds that the steady-state stress does not increase monotonically with increasing *Deborah* number. Instead, the stress actually decreases as *Deborah* number is increased from zero, reaching a minimum at $N_{\mathrm{Deb}} = 0.125$. We shall return to this point later.

Beyond $N_{\mathrm{Deb}} = 1/2$ the stress approaches infinity at a finite value of dimensionless time, and the time required for development of arbitrarily high stress rapidly approaches the order of relaxation time θ for the fluid as $N_{\mathrm{Deb}} \cong 1$. This behavior is consistent with the notion that an elastic liquid undergoes transition from fluid-like to solid-like behavior as the *Deborah* number is increased (14).

In the uniaxial experiments of *Denn* and *Marrucci* (10) they noted no approach to infinite stress when N_{Deb} was estimated to be approximately 5. However, the duration of the experiment extended only to $t/\theta \cong 1/4$. As shown in fig. 1 (though fig. 1 represents biaxial extension, the results in this range are qualitatively similar to the uniaxial case) this time is not sufficient for development of the catastrophically high stress predicted by the convected *Maxwell* model. Given a sufficient duration of the experiment, one would expect the stress to increase without bound until rupture or some other alteration occurred.

Although use of the convected *Maxwell* model of eq. [5] does not lead to quantitative agreement with experimental results for a wide variety of fluids in a variety of flows, the model has been found useful as a qualitative guide to fluid behavior. Hence one would expect the gross features outlined above to be useful for prediction of fluid behavior.

3. Description of experiments

3.1. The ideal experiment

It is clear from fig. 1 that the ratio $S^{11}/d = \bar{\eta}$ is not, in general, unique for a given fluid. In contrast to a *Newton*ian fluid the extensional viscosity can be dependent upon d. However, one must additionally be aware of the time dependence of the result. Unless one performs an experiment in the region where steady state has been reached, $\bar{\eta}$ is not even a unique function of d for a given material. Though there are numerous reports of the ratio of instantaneous stress to instantaneous stretch rate [see, for example, (5, 7, 15)], these experiments possess an unknown effect of time dependence which is produced either by start-up conditions or an entrance disturbance. Ideally, one would wish to measure stress in a fluid which has been undergoing steady uniform stretching for an indefinitely long period. A little reflection will convince the uninitiated that this ideal experiment is far more difficult to approach than the corresponding ideal case of a viscometric flow. For this reason one still searches for combinations of materials and flow geometries from which a meaningful deformation-rate dependent extensional viscosity can be obtained. The experiments described below are believed to be among the first for which steady-state stresses at stretch rates in excess of 0.1 sec^{-1} have been approached with fluid-like materials having zero-shear viscosities below 2×10^6 poise [see also (6)]. Rather than employing a uniaxial experiment, which has been the procedure followed heretofore with fluid-like materials, a biaxial test was employed.

3.2. Experimental apparatus

Biaxial extension of materials by bubble inflation seems to have been initiated in connection with study of elastic properties of rubber (16 to 18). Adaptation of the experiment for study of extensional *flow* properties has been carried out by *Denson* and coworkers (19). We have employed an apparatus and technique which are closely related to those used simultaneously by *Joye* et al. (20) in their studies of materials which are more rigid and solid-like than those for which results are reported here.

Essentially, the method consists of inflation of a thin circular disk of the sample into a bubble, measurement of the rate of separation of material points near the pole of the bubble, and a monitoring of the pressure difference across the bubble surface. A drawing of the sample holder is hown in fig. 2. The main platform of the holder was machined from an aluminium cylinder 6 inches in diameter and 1–1/2 inches long. A circular depression 2.5 inches in diameter and approximately 1/16-inch deep was machined into the upper flat face of the holder, and a 3/8-inch diameter hole was drilled from the opposite face and flared as shown in fig. 2. The lower 3/8-inch diameter hole was threaded and fitted with a 1/4-inch *Swagelok* elbow connection to a nitrogen metering panel.

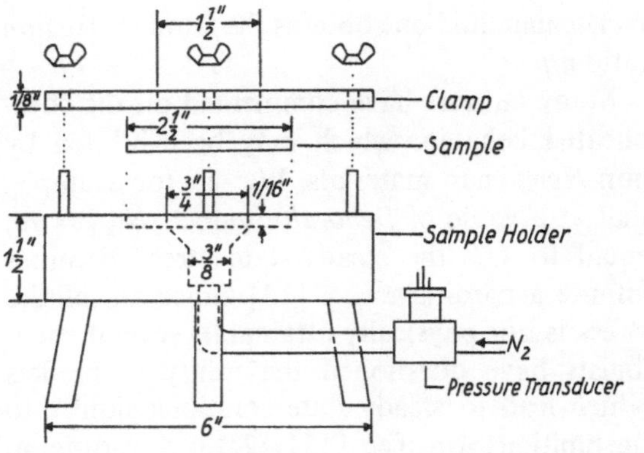

Fig. 2. Sample holder, clamping arrangement, and pressure transducer

The sample was clamped into the depression in the upper face by a 1/8-inch thick aluminium plate which was fastened to the platform with 3 wing nuts. The plate had a 1.5-inch diameter hole cut in its center, thus allowing the bubble to inflate within the rigidly clamped, circular rim. A grooved "O" ring 1/16-inch thick and with a diameter of 3.5 inches encircled the sample and prevented leakage of the inflating gas when the wing nuts were tightened. A 1/16-inch thick, 3-inch diameter spacer plate with a 3.5-inch diameter hole at its center was also fabricated to adapt the apparatus for use with 1/8-inch thick samples.

A transducer housing was machined and threaded to fit into the nitrogen line approximately 6 inches from the sample. An interchangeable Pitran pressure transducer (Stow Laboratories, Inc., Hudson, Mass.), mounted in a holder of Delrin plastic was screwed into the third branch of the housing with an air-tight seal. The output from the transducer was displayed on a digital panel meter.

3.3. Sample preparation

A primary objective was to use the bubble inflation method for materials more fluid-like (that is, with a lower zero-shear viscosity) than those for which results have been reported heretofore. One is of course limited, with the present apparatus, to systems which are sufficiently viscous to support their own weight with negligible creep between the time the sample is placed in the apparatus and the inflation is begun. Suitable materials, furnished by Enjay Chemical Company, were polyisobutylenes labelled Vistanex LM-MS (medium soft) and Vistanex LM-MH (medium hard). These are reported by the manufacturer to have viscosity-average molecular weights (by the method of *Flory*) of 35000 and 46000, respectively. [These compare with 70000 and 992000 for the polyisobutylenes used, respectively, by *Vinogradov* et al. (3) and *Joye* et al. (20).]

Test specimens were prepared by spooning a suitable amount of the polymer into a mold. The mold was constructed from a 3-inch diameter annular ring of stainless steel into which was placed a close fitting steel disk either 7/16 or 3/8-inches thick, depending upon the thickness of the sample desired. The molding device is shown in fig. 3. The circular depression was filled with sample material after a thin layer of silicone oil had been applied to all metal surfaces with which the sample came in contact. After a bit of practice it was possible to obtain a clear sample free of air bubbles.

It was important for the sample thickness to be known and constant over the test area. This dimension was measured with a dial thickness gauge which could be read to $\pm 10^{-4}$ inches. Thickness was measured at the center of the sample and at 45° increments around a circle of approximately 1/4-inch radius from the center. The range of readings for a given sample rarely exceeded 0.0025 inches.

Before removing a sample from the mold, five fiducial marks were carefully placed on the surface according to the pattern shown in fig. 3.

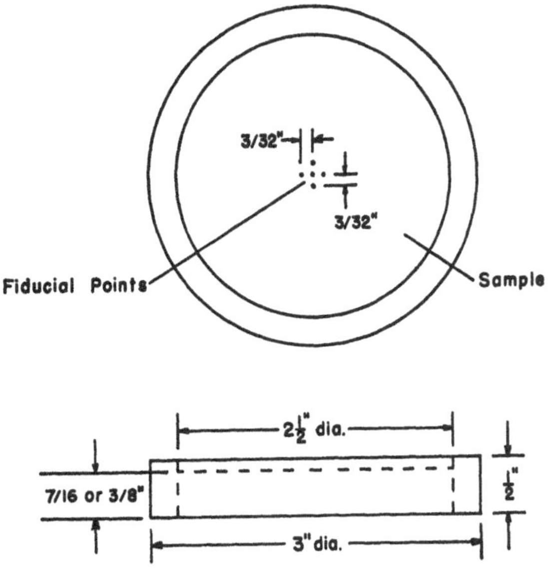

Fig. 3. Sample molding device and location of fiducial marks on sample

After some experimentation with a variety of schemes, a combination of simplicity and effectiveness was achieved with marks which were cut from black "Mystic" brand adhesive tape using a drill press and a cutter ground from a piece of 0.040-inch I.D. stainless steel tubing. The adhesive on the tape formed a good bond with the tacky surface of the polyisobutylene sample so that a suitable record of surface separation could be obtained by monitoring position of the fiducial marks during inflation. This technique is similar to that used by *Dickie* and *Smith* (21) in their study of failure of elastomers upon inflation.

Samples were removed from the mold by pushing the disk up through the annular ring and carefully sliding the sample from the disk with a spatula. The sample was then positioned in the depression of the sample holder and clamped in place with the clamping plate shown in fig. 2.

3.4. *Experimental procedure*

Each inflation experiment was recorded with a Bolex H-16 (16-mm) movie camera equipped with a 120-mm telephoto lens and operating at either 8 or 16 frames per second. The focal plane was approximately 10 feet from the bubble. The field of view of the camera included the following: pressure transducer output shown on a digital voltmeter; stopwatch with a ten-second sweep; front-surface plane mirror positioned slightly behind the sample holder and tilted forward 45° from the vertical so that top and profile views of the bubble were simulataneously recorded; a scale on the sample holder which served as a reference length. A sketch of a typical

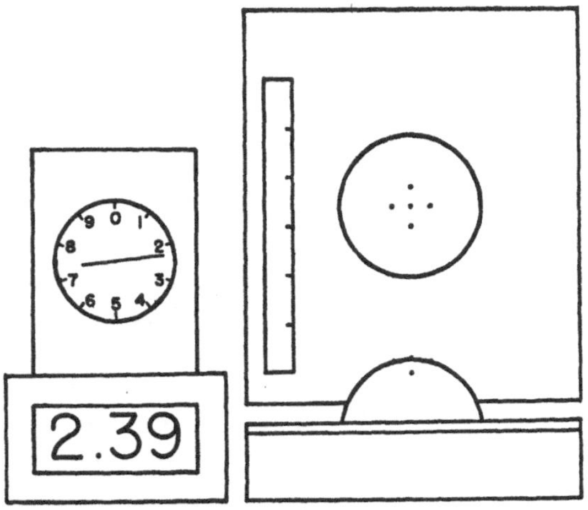

Fig. 4. Sketch of a typical frame from film record

frame of film is shown in fig. 4. Through experience it was possible to preselect nitrogen flow rates from which constant stress and strain rate were obtained over a suitable period of time.

4. Data reduction

4.1. Strain

Measurements were made of the instantaneous separations of each of the four peripheral fiducial dots from the central one by projecting an image of the film at regular time intervals. A longitudinal extension ratio was obtained by averaging the four separations and dividing by the initial separation to give $\lambda = l/l_0$. The separations measured on the film were of course actually projections of arc lengths on the bubble surface. The correction required, however, can be shown to be less than the experimental uncertainty of the separation measurements if one restricts the field to within approximately 22° of the polar axis (11), a limitation observed in all of the measurements reported herein. Hence no curvature corrections were applied.

Strain calculations are based, to a large degree, on the findings of *Treloar* (16), who studied inflation of rubber sheets. He found that the sheet deforms initially as a sphere, but after this initial stage the bubble shape approximates that of an oblate spheroid, retaining a spherical cap only in the region near the pole. A similar behavior was observed in the present experiments at low and moderate inflation rates. At large extensions and high inflation rates the deformed sample resembled a vertical cylinder with a spherical cap. The radius of curvature of the cap was found by fitting arcs to the profile view of the bubble with a divider.

In both *Treloars* experiment and those reported here, neither the membrane thickness nor membrane strain can be expected to be constant over the entire surface of the growing bubble at any instant of time. Thickness will be smallest at the pole and increase toward the rim. Though longitudinal and latitudinal strains will be equal at the pole, the latter must be zero at the rim while no such restriction is placed upon the former. *Treloars* results confirm these expectations of nonuniform strain. However, he found that for regions within about 25° from the polar axis and for extension ratios less than about four, the material exhibited equal biaxial uniform strain. Based upon these results we limited the present measurements to the same restrictions on polar angle and extension. As noted earlier this restriction also obviates the need for curvature correction.

We characterize strain by the logarithm of the extension ratio

$$\varepsilon = \ln \lambda \qquad [16]$$

so that the differential strain is

$$d\varepsilon = dl/l. \qquad [17]$$

The strain rate [1])

$$d^{11} = d^{22} = \dot{\varepsilon} = d\varepsilon/dt \qquad [18]$$

was found from measurement of the slope of plots of strain as a function of time for each run. (We have found it convenient to use $\dot{\varepsilon}$, rather than d, to indicate an experimentally determined constant strain rate.)

4.2. Stress

Membrane stress is calculated from the internal bubble pressure P, which is assumed to be uniform throughout the bubble and to be given by the pressure transducer shown in fig. 2. For the gas flow rates used here one expects negligible pressure change between the position of the transducer and the bubble.

One also requires a measure of bubble thickness. Near the pole the sample geometry is that of a thin-walled sphere with internal radius a and external radius b. The thickness

$$\delta = b - a \qquad [19]$$

of course changes with time. The biaxial stress, $S^{11} = S^{22}$, tangential to the membrane and perpendicular to the sphere radii, may be found from a force balance on an element cut from the spherical shell by two concentric spherical surfaces of radii R and $R + dR$ and a circular cone with infinitesimal included angle (22). The general expression reduces to the familar form for thin membranes if we make the assumption

$$\delta \ll R. \qquad [20]$$

Then, for a shell of radius R,

$$S^{11} = \frac{PR}{2\delta}. \qquad [21]$$

[1]) The terms strain rate, deformation rate, and stretch rate are used synonymously in this paper to characterize an extension.

The thin-membrane assumption is consistent with use of the continuity equation in rectangular *Cartes*ian form. Then from eq. [3] we have

$$d^{33} = d\ln(\delta/\delta_0)/dt = -2d. \qquad [22]$$

Integrating over a time period Δt for which the stretch rate is constant

$$\delta = \delta_0 \exp(-2d\Delta t). \qquad [23]$$

Calculation of membrane thickness is complicated by the fact that the strain rate is not constant during initial stages of inflation. Over this period the strain-time curve was approximated by linear segments between closely spaced data points (see figs. 5 and 6).

The contribution of surface tension to the membrane stress is estimated to be about three orders of magnitude below the lowest steady-state stress observed in these experiments. Hence neglect of surface tension does not introduce an appreciable error.

4.3. *Extensional viscosity*

Extensional viscosity is calculated according to eq. [14]. The stress S^{33} normal to the membrane varies from $-P$ inside of the bubble to zero on the bubble outer surface. Since bubble inflation pressure was always at least two orders of magnitude below the smallest steady-state membrane stress S^{11}, S^{33} was set equal to zero. The resulting error in extensional viscosity is less than one percent.

5. Results and discussion

5.1. *Stretch rate and stress*

All experiments were conducted at room temperature (23 °C). Representative examples of strain, as measured by position of the fiducial marks, *vs.* time are shown in figs. 5 and 6. One notes an initial nonlinear portion followed by linear behavior which persists until meaningful data cannot be taken because the fiducial marks have exceeded the polar angle beyond which the extension is believed to be equal biaxial and uniform. However, especially in view of the constant stress data to be presented shortly, we consider it reasonable to assume that the strain continues to be linear near the pole of the bubble even though the spatially averaged strain, as measured by the fiducial marks, is nonlinear.

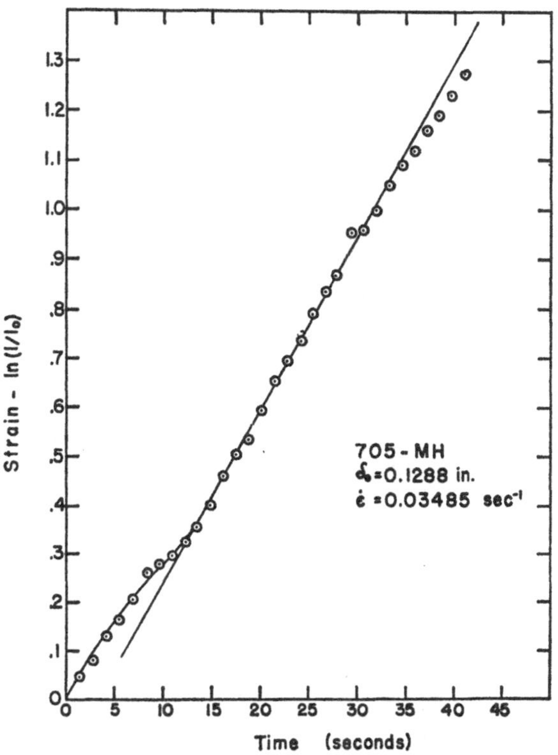

Fig. 5. Strain as a function of time for Vistanex MH. Run 705

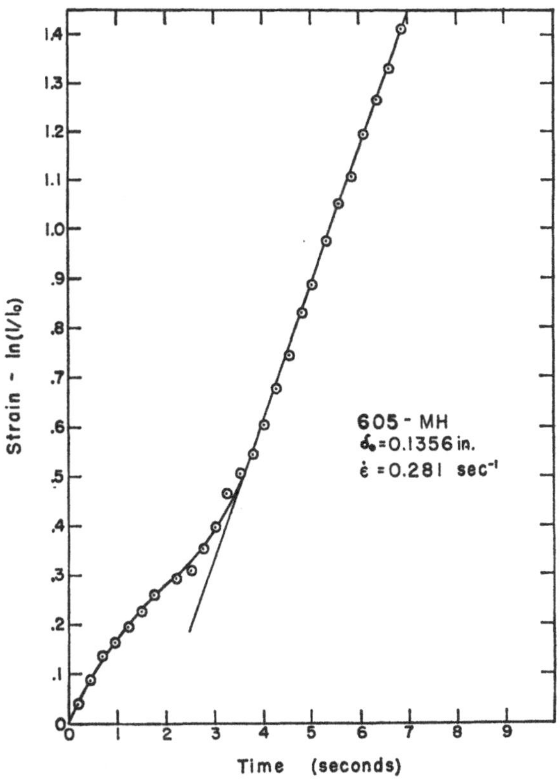

Fig. 6. Strain as a function of time for Vistanex MH. Run 605

The initial nonlinear region is probably due to a combination of material and experimental factors. A further peculiarity is shown in fig. 5. In the runs at low stretch rates it was found that deviation from linear behavior occurred at large strains, and that this deviation was always toward a decreasing strain rate. Since this behavior was most evident at low strain rates, it may be that the cause is a gravitational effect tending to collapse the bubble when the time duration of the experiment is sufficiently long.

Figs. 7, 8, and 9 are representative curves of membrane stress *vs.* time. Stress was calculated from eq. [21]. The general behavior is typified by figs. 7 and 8. The membrane stress reached a steady value which often persisted to strains for which the fiducial marks were no longer good measures of strain near the pole. The bubble finally bulged and burst at a point near the poble, this final stage being indicated on the figures by a rapid decrease in stress magnitude. The reason for the particularly conspicuous initial stress overshoot shown in fig. 7 is not clear, though it should be noted that a more sensitive pressure transducer was used for the experiments at the low stretch rate of fig. 7 than was used for the other experiments. *Vinogradov* et al. (3), report

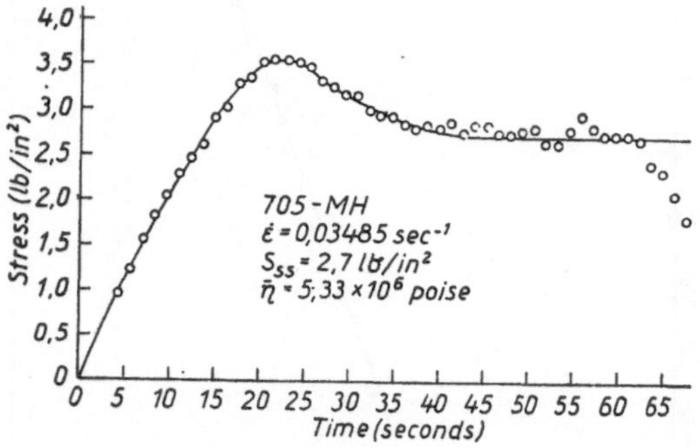

Fig. 7. Membrane stress as a function of time for Vistanex MH. Run 705

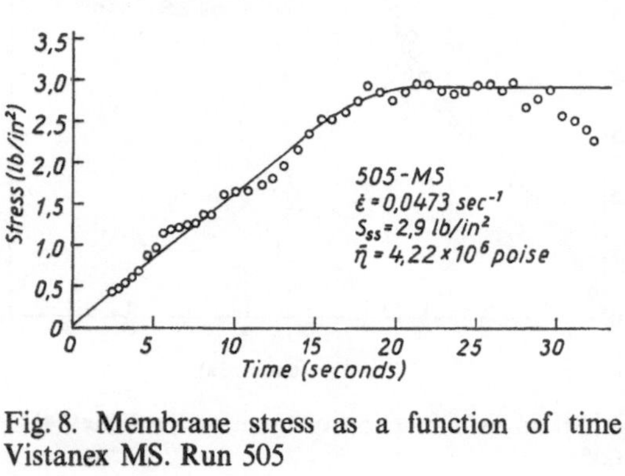

Fig. 8. Membrane stress as a function of time for Vistanex MS. Run 505

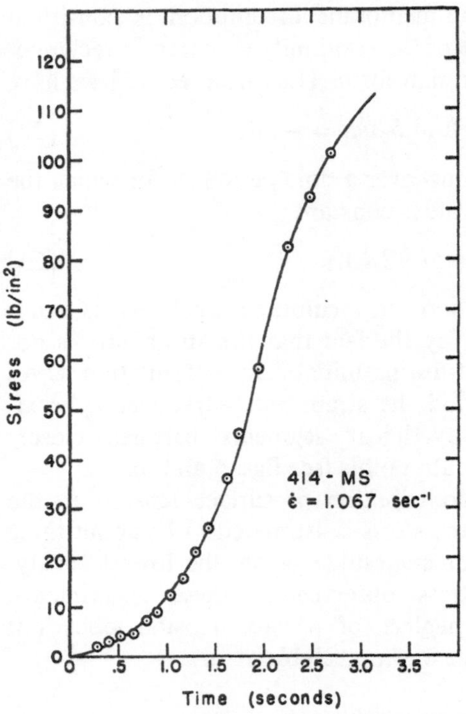

Fig. 9. Membrane stress as a function of time for Vistanex MS. Run 414

a substantial overshoot in tensional force at constant deformation rate. However for the data shown, which are all at deformation rates O (10^{-3} $\sec^{-1}$), the overshoot essentially disappears when the force is converted to stress.

It is not the purpose of this paper to report an exhaustive comparison of predictions of constitutive models with the experimental results obtained. However since, with the exception of an initial transient, a constant stretch-rate experiment was performed, it is of interest to compare the present experimental results with the curves based on the convected *Maxwell* model and shown in fig. 1.

Comparison of figs. 8 and 9 for the MS (medium soft) polyisobutylene shows that the stress attains much larger values for large stretch rates. We see from fig. 9 that with $\dot{\varepsilon} = 1.067$ $\sec^{-1}$, the stress does not even approach steady state before the bubble ruptures. The behavior is consistent with the predictions from fig. 1 for $N_{\mathrm{Deb}} > 1/2$. In fact, one predicts from the simple model of eq. [5] that steady-state stretching experiments which yield a stress in excess of the *Newton*ian value can only occur in the restricted range $0.25 < N_{\mathrm{Deb}} < 0.5$. Furthermore, as one approaches $N_{\mathrm{Deb}} = 0.5$, ever increasing times are required to reach a steady-state condition.

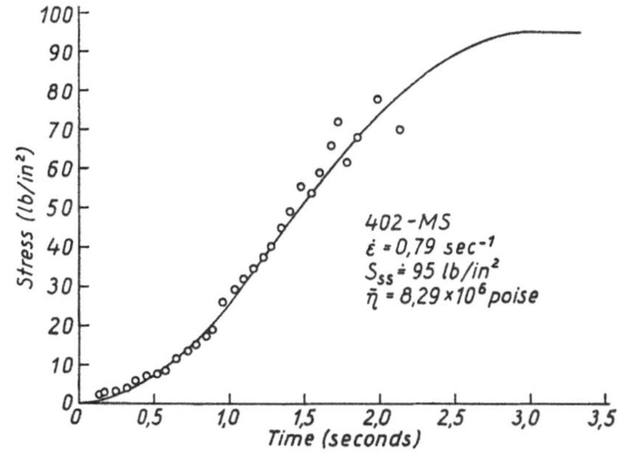

Fig. 10. Membrane stress as a function of time for Vistanex MS. Run 402

This may explain the necessity, with the three largest stretch rates reported here, for an extrapolation to a steady-state stress, as shown for example in fig. 10. Beyond stretch rates of 0.79 sec^{-1} there was no indication that a steady value of stress was being approached.

5.2. *Shear and extensional viscosity*

Viscosities are plotted in fig. 11 for the medium soft (MS) and medium hard (MH) polyisobutylene samples in shearing and in equal biaxial extensional flow. Deformation rate for the extension experiments covers nearly two orders of magnitude, from 0.01–1.0 sec^{-1}. Shear viscosities were measured conventionally in a *Weissenberg* rheogoniometer (cone angle = 4°) and are plotted as a function of shear rate $\varkappa = d\omega/d\theta$ in the *Weissenberg* instrument (23). The shear-thinning behavior of the two materials, with higher viscosity shown by the higher molecular weight sample, is expected. Both samples approach limiting viscosities at 0.01 sec^{-1}. These

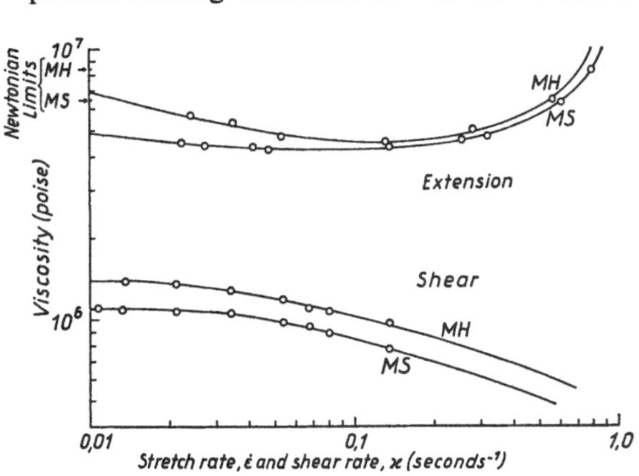

Fig. 11. Shear and biaxial extensional viscosities for Vistanex MH and MS

limits, which provide estimates of zero-shear viscosity, are taken to be 1.4×10^6 poise and 1.1×10^6 poise for the MH and MS samples, respectively. Shear data with samples which had undergone inflation and subsequently had been permitted to relax for a period of days were identical, within experimental error, to shear viscosities measured with virgin samples. Furthermore, shear viscosities determined over a period of several months showed no change for identical shear rates. We conclude, therefore, that no degradation or aggregation had taken place which could have affected the results.

A transient increasing shear stress response was observed with the rheogoniometer, and the shear viscosity was calculated using the asymptotic steady-state stress. The time required to reach constant stress increased with lower shear rates, and more than one-half hour was necessary for the lowest shear-rate measurements. The gradual decrease in shear viscosity with increasing shear rate could not be followed to rates of shear beyond 0.15 sec^{-1} because the material began to rupture between the cone and plate in the rheogoniometer.

Curves for extensional viscosity (fig. 11) show a sharp increase in viscosity with increasing stretch rate at large rates of deformation and a more gradual increase in viscosity with decreasing stretch rate at low rates of deformation. The resulting minimum for extensional viscosity at intermediate stretch rates, though consistent with the theory presented here for biaxial extension, has not previously been demonstrated experimentally. [A minimum has also been predicted for uniaxial extension (4), but that is a result of an error in derivation of eq. [18] of (4).]

The sharp increase in viscosity as the stretch rate is increased from 0.1 to 1.0 sec^{-1}, lends support to those theories [*Maxwell* model, *Lodge* model (24) and various modifications] which predict steady-state stress singularities at finite, critical stretch rates for steady, uniform, extensional flows of elastic liquids. If it were possible to extend the duration of experiments so that steady-state stress could be obtained at stretch rates beyond those shown in fig. 11, one would expect, based upon theoretical predictions, that the extensional viscosity curve would approach a vertical asymptote. As was emphasized earlier, however, one would not expect a steady-state stress to be possible beyond the restricted range of subcritical N_{Deb}.

It may be inferred from fig. 11 that the critical stretch rate for these low molecular weight polyisobutylenes is slightly larger than one reciprocal second. Using a critical value of N_{Deb} equal to 0.5, one obtains an estimate for the relaxation time of approximately 0.5 sec, with the MH sample having a somewhat larger relaxation time than the MS sample. This trend is expected, and is consistent with a relationship derived by *Everage* and *Gordon* (25), which predicts the relaxation time to be proportional to molecular weight raised to a power greater than one.

A standard error of about 12% has been estimated for the extensional viscosity results (11). One finds, according to this estimate, that differences between the two molecular weight samples are not statistically significant except at stretch rates below about 0.03 sec^{-1}. One must remember, however, that the difference in average molecular weight between samples is only about 30%. The apparent resolution between samples provides, in fact, added credibility for the experimental technique. It is also worth noting that the shapes of curves of viscosity *vs.* stretch rate are more dependent upon molecular weight for the case of extension than for shear.

5.3. Trouton ratio

The *Trouton* ratio of biaxial extensional to shear viscosity is plotted in fig. 12 for the MS and MH samples. The ratio was formed from extensional viscosities and shear viscosities which were measured at the same numerical values of $\dot{\varepsilon}$ and $\varkappa$, respectively. For each sample it was necessary to use the extrapolated values

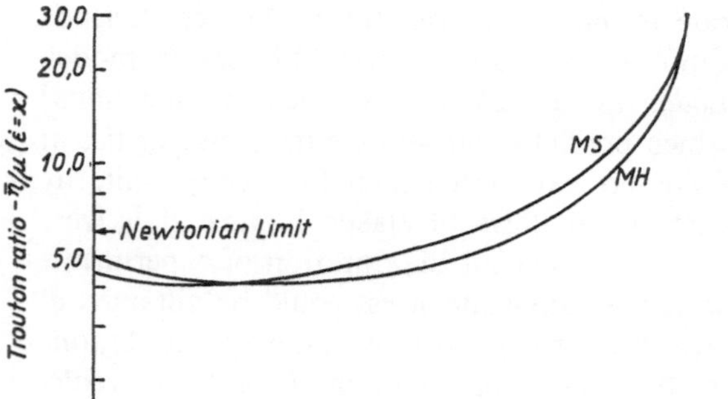

Fig. 12. *Trouton* ratio of biaxial extensional viscosity to shear viscosity for Vistanex MH and MS

of shear viscosity shown in fig. 11 at shear rates beyond 0.2 sec^{-1}.

The minima found in the extensional viscosity curves of fig. 11 are repeated in the dependence of *Trouton* ratio on stretch rate. We saw earlier that the *Trouton* ratio should be 6 for a fluid undergoing equal biaxial extension with $N_{Deb} = 0$. The *Trouton* ratios for the elastic liquids studied here were about 4.2 at the lowest stretch rates studied ($\dot{\varepsilon} \cong 0.025\ sec^{-1}$). Evidently these stretch rates were not low enough to approximate a zero *Deborah* number, but the increasing trend of *Trouton* ratio with decreasing stretch rate leads one to speculate that the *Newton*ian ratio of 6 may be attained eventually. The *Newton*ian limits (based upon shear viscosity at 0.01 sec^{-1}) have been marked on the left-hand borders of figs. 11 and 12 to indicate the extent of deviation from *Newton*ian behavior at a stretch rate of 0.1 sec^{-1}.

Recall that the *Maxwell* model predicts a minimum steady-state membrane stress which is 89% of the *Newton*ian value and occurs at $N_{Deb} = 0.125$. The corresponding *Trouton* ratio is 5.34. Experimentally, however, we found minimum *Trouton* ratios of about 4.1. Furthermore, if one uses a relaxation time of 0.5 sec, the minimum *Trouton* ratio observed experimentally occurs at $N_{Deb} < 0.025$. This is another indication that the relatively simple nonlinear *Maxwell* model, though useful for discerning qualitative trends, is not reliable for quantitative predictions.

It is interesting to compare the predictions of the *Maxwell* model with one which allows for more flexibility. The *Bird-Carreau* model (26), for example, incorporates two different spectra of relaxation times. If one applies this model to equal biaxial extension and retains only the first relaxation times (λ_1, λ_2) in each spectrum, the *Trouton* ratio is (11)

$$\frac{\bar{\eta}}{\mu} = \frac{6}{(1 - 2\lambda_2 d)(1 + 4\lambda_2 d)}\left[\frac{(1 + \lambda_1^2 \varkappa^2)}{(1 + 12\lambda_1^2 d^2)}\right] \qquad [24]$$

where κ is the shear rate.

The first factor corresponds to results for the *Maxwell* model, with shear viscosity μ. The factor in brackets is a correction factor arising from the greater complexity of the *Bird-Carreau* model equation. At equivalent rates of shear and elongation this term will be less than unity and can therefore provide quantitative agreement with the minimum value of *Trouton* ratio found experimentally.

6. Conclusions

1. Measurements of steady-state equal biaxial extensional flow have been made which are in qualitative agreement with the predictions of several rheological models. The biaxial extensional viscosity exhibits a minimum at an intermediate stretch rate and rises sharply as the stretch rate is increased toward a value corresponding to a critical *Deborah* number. At stretch rates below the minimum extensional viscosity the viscosity increases toward the *Newton*ian limit as stretch rate is decreased.

2. Though the single-element nonlinear *Maxwell* model is satisfactory for qualitative description of biaxial extensional flow of low molecular weight polyisobutylene, a single (constant) relaxation time does not permit quantitative agreement with experimental results. The *Bird-Carreau* model, incorporating two sets of relaxation times, allows a better fit to the minimum values of *Trouton* ratio which were found in the present work.

Acknowledgements

Partial support for this research was provided by the National Aeronautics and Space Administration through Grant NGR 31-001-025. One of us (JMM) was the recipient of a National Science Foundation predoctoral fellowship. We are grateful to *C. D. Denson* for helpful discussions and for provision of reports and manuscripts. Polyisobutylene samples were kindly supplied by Enjay Chemical Company.

Summary

A bubble inflation experiment has been used to study biaxial stretching of two samples of polyisobutylene. These were Enjay Vistanex LM-MS and LM-MH, which have viscosity-average molecular weights of 35000 and 46000, respectively. Zero-shear viscosities are 1.1×10^6 and 1.4×10^6 poise, respectively. At room temperature (23 °C) the samples flow slowly enough under their own weight so that it was possible to test them in a bubble-blowing apparatus.

It was found that constant stresses and constant stretch rates could be approached over a range of stretch rates from 0.022 to 0.79 sec^{-1}. Consequently, one can define an extensional viscosity over this range. The results indicate that at sufficiently low stretch rates the biaxial extensional viscosity decreases with increasing stretch rate. However, a minimum extensional viscosity is observed, following which it increases rapidly with increasing stretch rate.

Zusammenfassung

Die biaxiale Dehnung von Polyisobutylen wird durch die Aufblähung einer Blase untersucht. Zwei Arten Polyisobutylens werden gemessen. Das sind „Enjay Vistanex LM-MS und LM-MH", die viskositätsmittlere Molekulargewichte von 35000 bzw. 46000 besitzen. Die Viskositäten bei Null Dehnungsgeschwindigkeit sind $1{,}1 \times 10^6$ und $1{,}4 \times 10^6$ poise, bzw. Bei 23°C fliegen die Stoffe genügend langsam, damit man sie in einem Blasenapparat messen kann.

Es wurde gefunden, daß man zwischen 0,022 und 0,79 sec^{-1} einer konstanten Spannung und Dehnungsgeschwindigkeit nahekommen konnte. Deshalb ist es möglich, in diesem Bereich eine Dehnungs-Spannviskosität zu definieren. Die Ergebnisse zeigen, daß die biaxiale Dehnungs-Spannviskosität mit zunehmender Dehnungsgeschwindigkeit abnimmt, wenn die Dehnungsgeschwindigkeit genügend klein ist. Jedoch beobachtet man, daß die Dehnungs-Spannviskosität ein Minimum besitzt. Jenseits des Minimums nimmt sie mit zunehmender Dehnungsgeschwindigkeit rasch zu.

References

1) *Ballman, R. L.*, Rheol. Acta **4**, 137 (1965).
2) *Cogswell, F. N.*, Plast. Inst. Trans. J. **109**, April, 1968.
3) *Vinogradov, G. V., B. V. Radushkevich*, and *V. D. Fikhman*, J. Polymer Sci., A-2, **8**, 1 (1970).
4) *Astarita, G.* and *L. Nicodemo*, Chem. Eng. J. **1**, 57 (1970).
5) *Weinberger, C. B.*, Ph. D. Thesis, University of Michigan, Ann Arbor, Michigan 1970.
6) *Meißner, J.*, Rheol. Acta **10**, 230 (1971).
7) *Metzner, A. B.* and *A. P. Metzner*, Rheol. Acta **9**, 174 (1970).
8) *Coleman, B. D.* and *W. Noll*, Phys. Fluids **5**, 840 (1962).
9) *Wankat, P. C.*, Ind. Eng. Chem. Fund. **8**, 598 (1969).
10) *Denn, M. M.* and *G. Marrucci*, Amer. Inst. Chem. Eng. J. **17**, 101 (1971).
11) *Maerker, J. M.*, Ph. D. Thesis, Princeton University, Princeton, New Jersey 1971.
12) *Tanner, R. I.*, Trans. Soc. Rheol. **12**, 155 (1968).
13) *Middleman, S.*, Trans. Soc. Rheol. **13**, 123 (1969).
14) *Metzner, A. B., J. L. White*, and *M. M. Denn*, Amer. Inst. Chem. Eng. J. **12**, 863 (1966).
15) *Zidan, M.*, Rheol. Acta **8**, 89 (1969).
16) *Treloar, L. R. G.*, Trans. Inst. Rubber Ind. **19**, 201 (1944).
17) *Rivlin, R. S.* and *D. W. Saunders*, Phil. Trans. Roy. Soc. A **243**, 251 (1951).
18) *Adkins, J. E.* and *R. S. Rivlin*, Phil. Trans. Roy. Soc. A **244**, 505 (1952).
19) *Denson, C. D.* and *R. J. Gallo*, Polymer Eng. and Sci. **11**, 174 (1971).
20) *Joye, D. D., G. W. Poehlein*, and *C. D. Denson*, "A Bubble Inflation Technique for the Measurement of Viscoelastic Properties in Equal Biaxial Extensional Flow", paper presented at Society of Rheology meeting, Knoxville, Tennessee, October 1971.
21) *Dickie, R. A.* and *T. L. Smith*, J. Polymer Sci. A-2, **7**, 687 (1969).

22) *Timoshenko, S. P.* and *J. N. Goodier*, Theory of Elasticity, 3rd edition, pp. 392–395 (New York 1970).
23) *Coleman, B. D., H. Markovitz*, and *W. Noll*, Viscometric Flows of Non-Newtonian Fluids, p. 51 (New York 1966).
24) *Lodge, A. S.*, Elastic Liquids, p. 116 (New York 1964).
25) *Everage, A. E.*, Jr. and *R. J. Gordon*, Amer. Inst. Chem. Eng. J. **17**, 1257 (1971).
26) *Bird, R. B.* and *P. J. Carreau*, Chem. Eng. Sci. **23**, 427 (1968).

Authors' addresses:
Dr. *John M. Maerker*
Esso Production Research Company
P. O. Box 2189
Houston, Texas 77001 (USA)

Professor *W. R. Schowalter*
Department of Chemical Engineering
Princeton University
Princeton, New Jersey 08540 (USA)

Rheol. Acta 13, 639–643 (1974)

From the Hungarian Institute for Building Science, Budapest (Hungary)

Failure process of concrete under fatigue loading

By L. Béres

With 5 figures

(Received October 27, 1973)

1. Subject of paper

The paper deals with the behaviour of concrete under cyclically repeated central pressure. It presents, on the basis of experimental results, the character of physical changes in the macrostructure of the concrete, due to fatigue loading, as well as the most important feature of the investigated phenomenon: the mechanism of the concrete fatigue. On the basis of the testing method it discusses also both the changes of strain properties and strength of the concrete, as a function of the number of load cycles.

2. Basic hypothesis and investigation method

It is well known that concrete (alike other solid materials) deteriorates also under a stress lower than its initial strength, as a result of a certain number of cyclic loadings. This phenomenon is called concrete fatigue. The basic hypothesis (to be verified later) of the investigation is that:

1. Concrete fatigue is not a sudden qualitative change of the material, but the gradual destruction of the concrete structure. Parallel with this process, as a consequence, both the strain properties and the strength of concrete change.

2. The speed of the fatigue process is determined by the intensity of the repeated loading and the number of load-cycles.

On the basis of theories of failure of brittle materials it can be stated that the starting point, the source of the fatigue failure is always to be attributed to the inhomogeneity, structural defects of the material.

The majority of results obtained so far in connection with fatigue phenomena is founded on the testing of metals. So the most acceptable and most complete explanation of fatigue was given so far by the dislocation theory.

Concrete is, however, in view of its structure, much more inhomogeneous than metals. The fundamental reason for this inhomogeneity is the concrete being a conglomerate of materials of different features. The character of the inhomogeneity and hence also the types of defect sources differ essentially in respect of various (macroscopic, microscopic and submicroscopic) "depths" of the concrete structure investigations.

The basic hypothesis of our research work is that the fatigue deterioration of concrete is determined fundamentally by macroscopic sources of defects. The connection between coarse aggregate and cement mortar, and mainly discontinuities of this connection, as well as in a smaller degree the shrinkage cracks of cement mortar play here an important part. [This assumption is based on our (1, 2) and other authors' experimental results (3, 4, 5), related to concrete deterioration induced by short-time loading.] From the above basic hypothesis a testing method can be obviously derived for the indication of the degree of structural deteriorations (damages). The macroscopic deterioration of concrete appears namely in every case as a discontinuity at critical points, or rather as a development of the earlier cracks, i. e. as an increase of the holes found in the structure. The task is to determine the volume of holes in the concrete specimen subjected to fatigue loading. Measuring the surface deformations, interrupting from time to time the cyclic loading, and determining from this the volumetric changes, we obtain the quantity sought for: the permanent changes of hole volume.

The other testing method adopted during our research work is based on the assumption of a close correlation between state of internal structure and strength. The main point of this method is to load concrete bodies with a given intensity by various numbers of load cycles; hereupon their strength is determined. These strength values, related to the initial condition (without pre-loading) characterize well the degree of changes in the concrete structure.

41*

3. Tests based on strain measurements

A series of tests has been carried out to investigate older (about 1 year old) specimens of 10 cm × 10 cm × 40 cm size in order to preclude the influence of afterhardening. The concrete has been composed using 372 kg/m^3 Portland-cement and Danubian sand-gravel of max. 20 mm size. The value of the water/cement ratio amounted to 0.50. The short-time strength of the concrete prisms was at the beginning of fatigue testing between 410 and 455 kg/cm^2.

During fatigue testing the minimum concrete stress induced by load cycles was 30 kg/cm^2 for every specimen. The maximum stress varied for the individual specimens between 43 and 93% of the short-time strength. The frequency of the loading apparatus was 500 cycles/min. Specimen deformations have been measured by 6 cm long electric strain gauges.

Characteristic types of diagrams, indicating the relations between volumetric changes belonging to minimum stress and the number of load cycles are shown in fig. 1. Here the curves 1–4 correspond to the parameter of $\sigma_1 > \sigma_2 > \sigma_3 > \sigma_4$ stress maxima.

By means of the investigation of individual curves we also obtained an answer regarding character and reasons of changes, induced by fatigue loading.

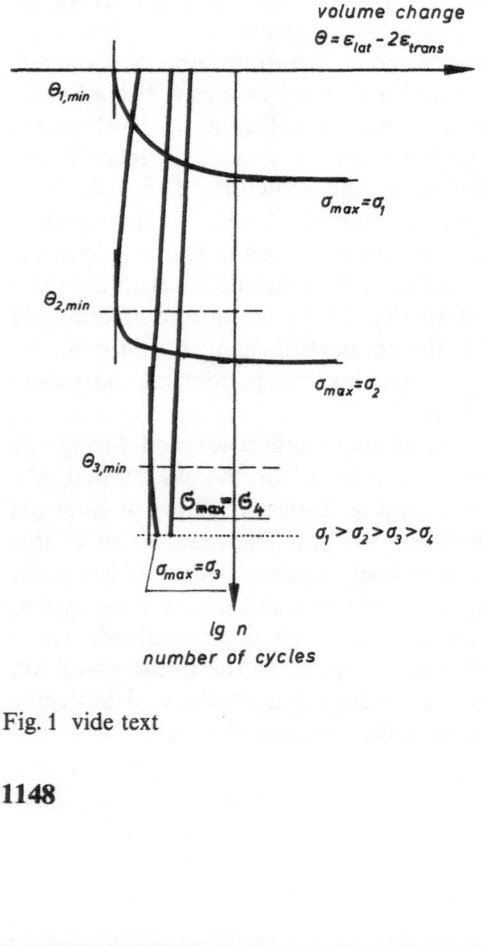

Fig. 1 vide text

The specimen, subjected to maximum stress, shows from the first load cycles on an intensive volume increase (structural loosening), and so the deterioration results very quickly, after some (or several hundred) cycles of load.

In the initial phase of the loading process, at lower stress maxima a volume decrease (structural compaction) can always be observed. With the reduction of stress level, however, the ratio of the compaction period to the entire useful life time gradually increases. For curve 3, however, although the loosening process has begun, there is no possibility at all for full deterioration, in consequence of the technical upper limit of the highest number of cycles (1–2 millions), i.e. the stopping of the investigation.

Independently of this, curves 2 and 3 show processes of identical character, developing, however, during different periods of time (numbers of load cycles).

For curve 4 of volumetric change, resulting under minimum stress, only the phase of volume reduction (compaction) can be indicated. It is presumable that in the case of certain not too low stress level maxima intensive loosening begins with the increase of the number of load-cycles, followed by real deterioration. In this case it becomes obvious that the character of the curve is identical with curves 2 and 3. On the basis of test data obtained till now, however, it is not possible to state whether a real stress level actually exists, at which structural deterioration, inducing rupture of the concrete specimen do not come into being, not even in the case of a theoretically infinite large number of load cycles.

So the individual curves fundamentally differ from one another regarding the place of their minima. This point can be called the beginning of the intensive structural decomposition. The beginning of the intensive structural loosening and the state of full deterioration of the specimen – on the basis of above explanation – can be given as the function of stress maximum of fatigue load and the number of load cycles.

Fig. 2 shows the approximative presentation of this relationship. Assuming a linear relation between the maximum stress level, pertaining to the beginning of loosening, resp. to that of deterioration, and the logarithm of the number of load cycles, the straight lines have been determined on the basis of the minimum mean-square error from our test data.

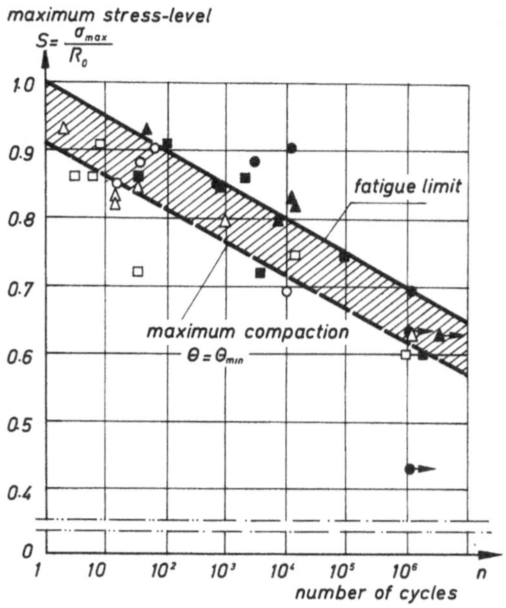

Fig. 2 vide text

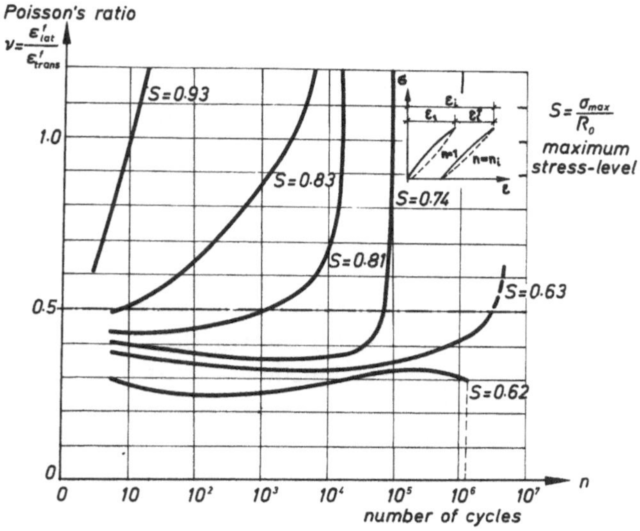

Fig. 3 vide text

We may obtain very useful information also through the investigation of the quotient of transversal and longitudinal strains, i.e. of *Poisson*'s ratio of cyclic load (ν), resulting from repeated loading (see fig. 3). By the initial value of about 0.35 ν it has been proved that strains due to fatigue load effects are mainly of viscous character. The corresponding 0.5 value decreases in consequence of delayed elastic strains and initial structure compaction to the measured value. The increase of *Poisson*'s ratio beyond a certain number of cycles, depending on the load intensity, refers to structure loosening. It can be attributed to this that prior to the rupture much greater values of the *Poisson*'s ratio than 0.5 can also be found.

4. Tests based on the investigation of strength changes

Specimens and testing conditions corresponded to those described above. Only the quality of the concrete was different from the former one. (Cement content 200 kg/m^3; water/cement ratio: 0.73.)

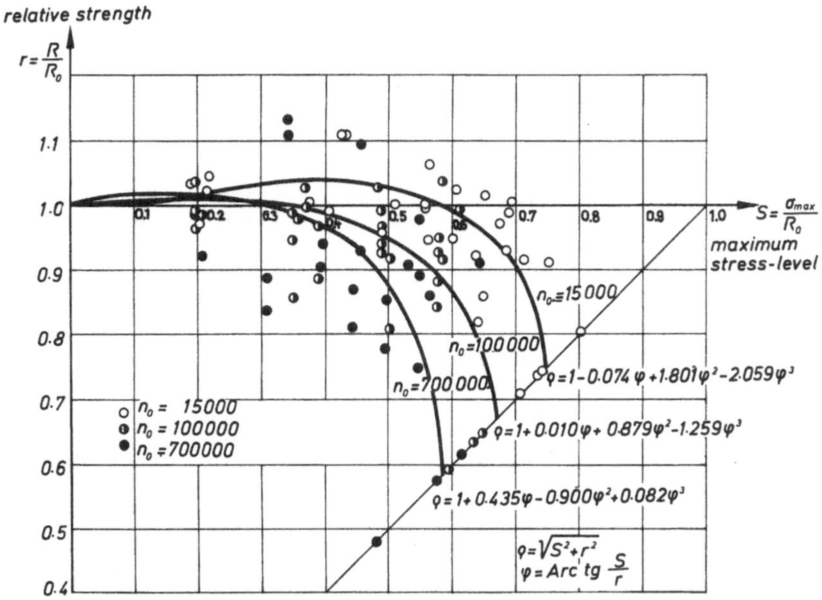

Fig. 4 vide text

In the course of the investigation about 100 specimens have been subjected to cyclic fatigue loads of various intensities. Each group of specimens has been loated up to a number of cycles of 15000, 100000, and 700000 and then static short-time strength of the specimens has been determined.

The quotient of the strength of specimens (specific change of strength: $r = R/R_0$), pre-loaded to various cycle numbers (R) and that of specimens without pre-loading (R_0) has been plotted in fig. 4, as a function of stress level [i.e. the quotient $s = \sigma_{max}/R_0$) of the stress maximum of cyclic load and the strength of specimens without pre-loading]. The diagram also shows the curves of approximate polynomes describing the strength changes of the specimens pre-loaded to the same number of cycles.

According to these curves the cyclic load results in an increase of strength in the lower stress ranges and mainly at low cycle numbers. This can be explained by the initial compaction of the concrete structure. With the increase of the number of load cycles, however, the stress range inducing strength increase will be gradually reduced.

In higher stress ranges, however, a reduction of strength can be observed after relatively few load cycles already, indicating a very early beginning of deterioration of the concrete structure.

Fig. 5 shows the effective strength change of concrete due to cyclic load, with contour lines, on the basis of our tests, as function of the logarithm of the cycle number of preloading (lg$n$), as well as of the stress level (σ_{max}/R_0).

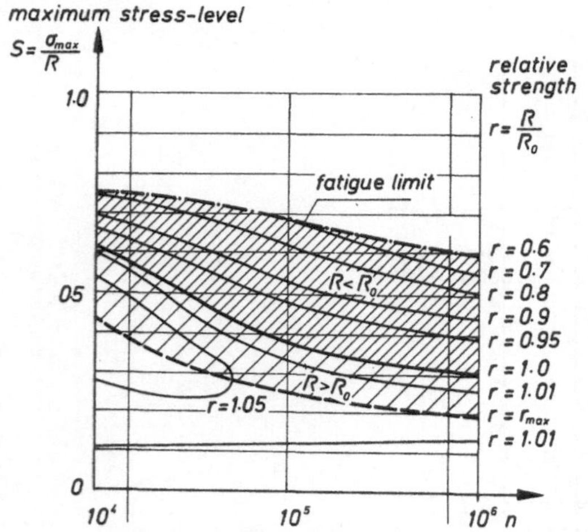

Fig. 5 vide text

Summary

The investigations proved that fatigue deterioration of compressed concrete is caused by the gradual destruction and loosening of the concrete structure.

It has been proved that changes induced by cyclic loading in the concrete structure can be well described qualitatively by means of the strains, resp. the volumetric changes of specimens, as well as by means of its change of strength.

In consequence of the fatigue load the concrete volume initially decreases (structure compression), and then it increases (structure loosening); the rupture occurs at a volume greater than in unloaded condition (thus in considerably loosened condition). The limit of the compaction and the loosened state may be given by the cycle number referring to the maximum of volumetric change. This cycle number can be regarded as the characteristic value of the fatigue process.

The ratio of the duration of the compaction and the loosening state within the entire useful life time varies depending on the stress level. In extreme cases, hence at lower stress (at least to the investigated cycle number) only a compaction state, but at a stress level near to the shorttime strength only a loosening state can be observed.

The value of *Poisson*'s ratio, computed from transversal and longitudinal strains, induced by repeated loading of concrete, is initially between 0.3 – 0.4. This value indicates a viscous and slightly delayed elastic character of the strain. This value may be reduced by the initial compaction of the structure, but it is vigorously increased by the loosening prior to failure.

The tests proved also that the strength for a given cycle number of a concrete subjected to repeated loading is in close correlation with the structural state at a given point of time. The structural compaction, regarding its tendency, is connected with the increase of strength; its loosening, however, is connected with the decrease of strength.

As a result of structural deterioration induced by cyclic loading the actual fatigue (failure) limit is preceded by a considerable range of reduced strength.

Zusammenfassung

Die Untersuchungen zeigen, daß der Ermüdungsbruch des unter Druck stehenden Betons durch die fortschreitende Zerstörung des Betongefüges hervorgerufen wird.

Es wird bewiesen, daß die im Betongefüge unter dem Einfluß zyklischer Belastungen sich vollziehenden Veränderungen mit Hilfe der Deformationen bzw. der Volumenänderungen des Probekörpers sowie durch seine Festigkeitsänderung qualitativ gut beschreibbar sind.

Unter wiederholter Belastung nimmt das Betonvolumen zuerst ab (Gefügeverdichtung), anschließend aber wieder zu (Gefügeauflockerung), und der Bruch erfolgt erst beim Erreichen eines Betonvolumens, welches größer ist als im unbelasteten Zustand, d. h. im Zustand einer starken Auflockerung. Der Grenzzustand zwischen Verdichtung und Auflockerung läßt sich mit der zur maximalen Volumenänderung gehörenden Zyklenzahl bestimmen. Diese kann als Kennzahl für den Ermüdungsvorgang betrachtet werden.

Das Verhältnis der Zeitdauern von Verdichtungs- und Auflockerungsvorgang ist von der Intensität der Beanspruchung abhängig. In den Extremfällen, also bei niedriger Belastung ist (wenigstens bis zur untersuchten Zyklenzahl) nur der Verdichtungsvorgang, dagegen im Falle einer Beanspruchungsintensität in der Nähe der kurzfristigen Belastbarkeit nur der Auflockerungsvorgang beobachtbar.

Die Poisson-Zahl des Betons – berechnet aus den durch die zyklischen Belastungen verursachten Quer- und Längsdeformationen – liegt zu Anfang zwischen 0,3 und 0,4. Ein solcher Wert deutet darauf, daß eine viskose und nur wenig elastisch verzögerte Formänderung vorhanden ist. Dieser Wert kann durch die erst erfolgende Verdichtung des Gefüges vermindert, durch die anschließende Auflockerung des Gefüges vor dem Bruch dagegen wesentlich erhöht werden.

Die Untersuchungen beweisen auch, daß die Festigkeit des periodisch belasteten Betons bei einer bestimmten Zyklenzahl eng mit dem Gefügezustand zu dem betreffenden Zeitpunkt korreliert ist. Als Tendenz ist erkennbar, daß die Verdichtung des Gefüges mit einer Festigkeitszunahme und seine Auflockerung mit einer Festigkeitsabnahme verbunden ist.

Als Folge der Gefügezerstörung durch die zyklische Belastung gibt es schon vor der effektiven Ermüdungsgrenze (Bruchgrenze) ein Bereich mit wesentlich verringerter Festigkeit.

References

1) *Béres, L.*, RILEM Bulletin, New Series, No. **36**, pp. 185–190 (1967).

2) *Béres, L.*, Relationship of Deformational Processes and Structure Changes in Concrete. Structure, Solid Mechanics and Engineering Design. The Proceedings of the Southampton 1969 Civil Engineering Materials Conference. pp. 643–651.

3) *Bennet, E. W.* and *N. K. Raju*, Cumulative Fatigue Damage of Plain Concrete in Compression. Structure, Solid Mechanics and Engineering Design. The Proceedings of the Southampton 1969 Civil Engineering Materials Conference. pp. 1089–1102.

4) *Shah, S. P.* and *S. Chandra*, ACI Journal, Proceedings **67**, No. 4, p. 816 (1970).

5) *Antrim, J. D.*, The Mechanism of Fatigue in Cement Paste and Plain Concrete. Highway Research Record, No. **210**, pp. 95–107 (1967).

Author's address:
Dr. *L. Béres*
Hungarian Institute for Building Science
David F. u. 6
Budapest XI (Hungary)

Für die Schriftleitung verantwortlich: Dr. W. Meskat, 5090 Leverkusen, Mühlenweg 90a
Anzeigenverwaltung und Verlag: Dr. Dietrich Steinkopff Verlag, 6100 Darmstadt, Saalbaustraße 12, Postfach 1008
Gesamtherstellung: Druckerei Dr. A. Krebs, Hemsbach/Bergstr. und Bad Homburg v. d. H.

RHEOLOGICA ACTA

AN INTERNATIONAL JOURNAL OF RHEOLOGY

Vol. 13 | August/October 1974 | No. 4/5

From the Chemical Engineering Department and Rheology Research Center, University of Wisconsin, Madison, Wisconsin 53706 (USA)

Kinetic theory and rheology of dilute polymer solutions

R. Byron Bird, Ole Hassager, and Robert C. Armstrong

With 1 table

Introduction

Treatises on transport phenomena and kinetic theory have given the molecular theory for the transport properties of simple gases and liquids, but have ignored the extensive literature on polymer solutions. Much of this literature is based on the configuration-space theory developed through the years and summarized and exploited by *Kirkwood* and his collaborators (1). Most of the kinetic theory workers, however, have not been interested in obtaining the expression for the stress tensor for use in solving flow problems, nor in obtaining relations among the results of various rheological experiments. It is to these topics that this short note is addressed.

Modelling of dilute polymer solutions

A dilute solution of macromolecules is modelled by regarding the solvent to be a *Newton*ian fluid with viscosity η_s, and the solute to be a set of N beads of radius r_0 joined by connectors (which may be rigid rods, *Hooke*an or non-*Hooke*an springs). Each bead has a friction coefficient $\zeta = 6\pi\eta_s r_0$, and *Brown*ian motion is included to account for the solvent-solute interaction. The number density of macromolecules is n_0. By using universal joints at the beads, one can model a flexible macromolecule; and by requiring that the beads be aligned one can model a linear, rigid macromolecule. In this discussion "hydrodynamic interaction" is not included.

Rigid macromolecules

To illustrate the connection between molecular theory and continuum mechanics, we consider the stress-tensor expression for a dilute solution of rigid macromolecules, modelled as a suspension of rigid dumbbells with *Brown*ian motion. The rigid dumbbell consists of two beads connected by a rigid rod of length L. The key structural parameter that appears in the molecular results is a time-constant, $\lambda = \zeta L^2/12\,kT$. (All results which have been worked out for rigid dumbbells can immediately be taken over for rods of length L with N beads uniformly distributed over its length by replacing λ everywhere by $\lambda_N = \zeta L^2 N(N+1)/72(N-1)kT$.)

The stress tensor can be expressed as the sum of a solvent contribution and a polymer contribution:

$$\underset{\approx}{\tau} = \underset{\approx}{\tau}_s + \underset{\approx}{\tau}_p = -\eta_s \dot{\underset{\approx}{\gamma}} + \underset{\approx}{\tau}_p. \qquad [1]$$

The polymer contribution $\underset{\approx}{\tau}_p$ consists in turn of two parts: the contribution associated with the momentum flux due to the movement of beads across a plane in the solution, and that associated with the tension in the connectors which at any instant lie across the plane.

After very lengthy manipulations, *Giesekus* (2) and *Prager* (3) obtained an expression for $\underset{\approx}{\tau}_p$ as a series involving the *Oldroyd* n-th rate-of-strain tensors. This *retarded motion expansion* is applicable to flows which are steady-state or slightly removed from steady-state. More recently *Bird* and *Armstrong* (4) showed that the kinetic theory results could also be developed as a *memory integral expansion* which applies to unsteady-state problems. In addition *Bird* (5) has demonstrated that the rigid-dumbbell suspension results can be put in the form of a *modified Oldroyd equation*, which is equivalent to the second-order memory integral expansion

42

TABLE 1. KINETIC THEORY RESULTS FOR RIGID DUMBBELL SUSPENSIONS

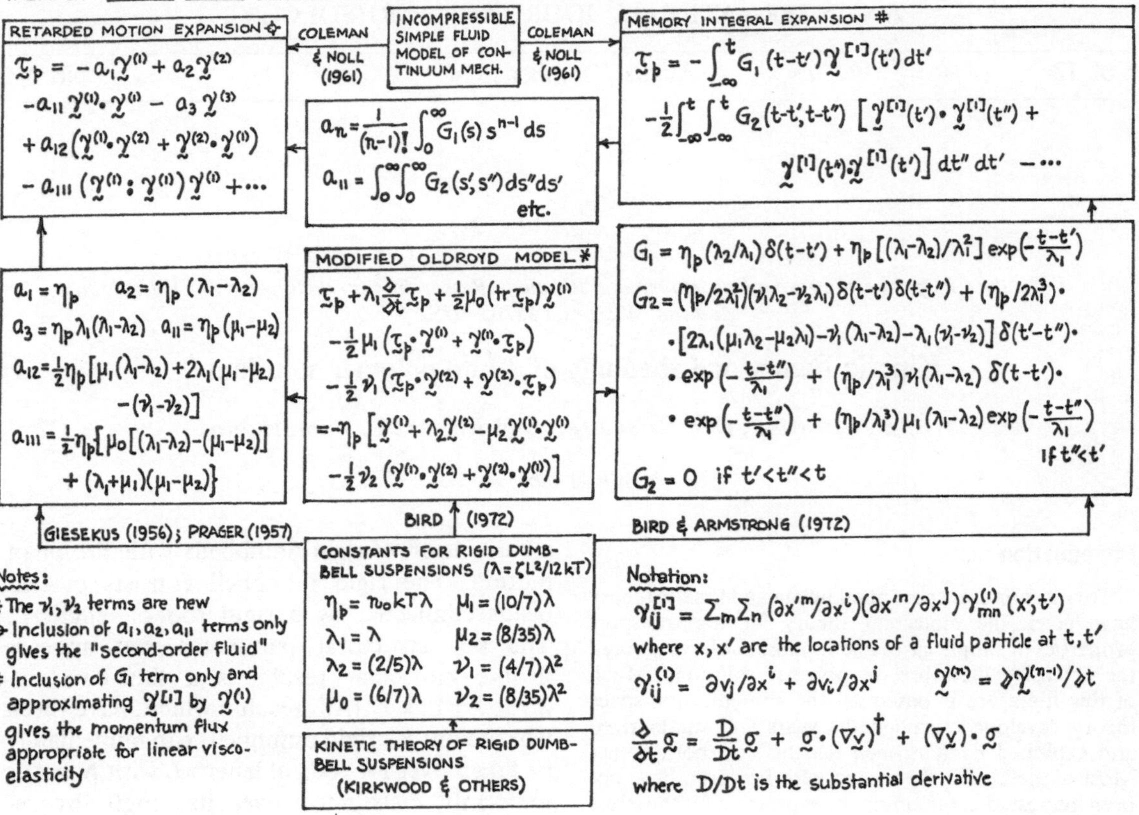

and to the third-order retarded motion expansion. The explicit results as well as the relations among the various expressions is summarized in table 1. The fourth-order terms of the retarded-motion expansion and the third-order terms of the memory-integral expansion have recently been worked out by *Armstrong* (6, 7). The results of the calculations for explicit flow systems have been summarized by *Bird*, *Warner*, and *Evans* (8). The model exhibits a non-*Newton*ian viscosity, shear-rate-dependent normal stresses, recoil, creep, stress relaxation, stress overshoot, and other observed phenomena.

Flexible macromolecules

For flexible macromolecules several kinds of models have been used. The simplest is an elastic dumbbell, consisting of two beads joined by a spring. Since a *Hooke*an spring leads to a constant viscosity, attention has recently been focussed on nonlinear springs; *Stevenson* and *Bird* (9) and *Warner* (10) have obtained results for specific flow systems, and *Armstrong* (7) has studied the problem of obtaining results analogous to those shown in table 1 for rigid dumbbells. In particular, *Armstrong* (7) has shown that for a dilute suspension of elastic dumbbells with nonlinear springs, the first three coefficients in the retarded motion expansion are

$$a_1 = \frac{n_0\zeta}{12}\frac{[R^4]}{[R^2]} \qquad [2]$$

$$a_2 = \frac{1}{240}\frac{n_0\zeta^2}{kT}\frac{[R^6]}{[R^2]} = \frac{1}{2}a_{11}. \qquad [3]$$

Here the square brackets are used to represent averages with respect to the equilibrium distribution function:

$$[f(R)] = \int_0^\infty f(R)\exp\left\{-\int_0^R (F^{(c)}(R')/kT)\,dR'\right\}dR \qquad [4]$$

where the force in the connector is $F^{(c)}(R)$, R being the instantaneous extension of the dumbbell.

Recently *Hassager* and *Bird* (11) used the model of N beads with $N-1$ non-*Hooke*an springs (including hydrodynamic interaction between beads) to show that the integral under the stress-relaxation curve after cessation of steady shear flow is related to first normal stress difference prior to stopping the flow:

$$(\tau_{xx} - \tau_{yy})^- = 2\dot{\gamma} \int_0^\infty \tau_{yx}^+ \, dt \qquad [5]$$

where $\dot{\gamma}$ is the velocity gradient. However, a general expression for the stress tensor is not yet available for this model.

Still more recently *Hassager* (12, 13) has studied the model of three beads, joined by two rigid rods of length L, with a universal joint at the middle bead. For dilute solutions of such particles in a *Newton*ian solvent, *Hassager* obtained the following expressions for the coefficients in the retarded motion expansion:

$$a_1 = \frac{2}{9} n_0 \zeta L^2 \qquad [6]$$

$$a_2 = \frac{1}{1080} \frac{80\pi + 3\sqrt{3}}{2\pi + 3\sqrt{3}} \frac{n_0 \zeta^2 L^4}{kT} = \frac{1}{2} a_{11}. \qquad [7]$$

The detailed derivation of these results as well as the calculation of elongational viscosity for this model are given in the original publication.

Acknowledgements

The authors wish to acknowledge financial support of National Science Foundation Grant GK-24749. In addition RBB wishes to thank the Trustees of the William F. Vilas Trust Estate for financial support, and RCA acknowledges assistance provided by a National Science Foundation Graduate Fellowship.

Summary

A summary is given of some recent attempts to relate the results of the kinetic theory of rigid and flexible macromolecules to continuum mechanics results.

References

1) *Kirkwood, J. G.*, Macromolecules (New York 1967).

2) *Giesekus, H.*, Kolloid-Z. **147**, 29–45 (1956).

3) *Prager, S.*, Trans. Soc. Rheol. **1**, 53–62 (1957).

4) *Bird, R. B.* and *R. C. Armstrong*, J. Chem. Phys. **56**, 3680–3682 (1972).

5) *Bird, R. B.*, Z. angew. Math. Phys. **23**, 157–159 (1972).

6) *Armstrong, R. C.* and *R. B. Bird*, J. Chem. Phys. *58*, 2715–2723 (1973).

7) *Armstrong, R. C.*, Doctoral Dissertation, University of Wisconsin (1973); J. Chem. Phys. **60**, 724–733 (1974).

8) *Bird, R. B., H. R. Warner Jr.* and *D. C. Evans*, Adv. Polymer Science **8**, 1–90 (1971). In Eq. [10.6] on page 44, 15/14 should read 4/7, and in Eq. [12.9] on page 49, 1/120 should be replaced by 1/280. Finally, it should be noted that *H. Giesekus* (see Rheol. Acta **2**, 50, 1962) has developed expressions equivalent to Eqs. [25.20]–[25.22] on page 83.

9) *Stevenson, J. F.* and *R. B. Bird*, Trans. Soc. Rheol. **15**, 135–145 (1971).

10) *Warner, H. R. Jr.*, Ind. Eng. Chem. Fund. **11**, 379–387 (1972).

11) *Hassager, O.* and *R. B. Bird*, J. Chem. Phys. **56**, 2498–2501 (1972).

12) *Hassager, O.*, J. Chem. Phys. **60**, 2111–2124, 4001–4008 (1974).

13) *Hassager, O.*, Doctoral Dissertation, University of Wisconson (1973).

Authors' address:

R. Byron Bird, Ole Hassager, and
Robert C. Armstrong
Chemical Engineering Dept.
University of Wisconsin,
Madison, Wisconsin 53706 (USA)

Rheol. Acta 13, 648–649 (1974)

From the Pulp and Paper Research Institute of Canada, and Department of Chemistry, McGill University, Montreal (Canada)

The microrheology of non-Newtonian suspensions

S. G. Mason

Recently we have extended the studies to suspensions of particles in non-Newtonian (e.g. visco-elastic and pseudo-plastic) solutions of polymers (21 to 24). A number of important and significant differences from the Newtonian systems were observed which are interesting in their own right and which may prove to be of considerable practical importance. These differences include (i) changes in orientation of anisometric particles such as rods and discs, (ii) migration of particles to and from walls under conditions where, in Newtonian systems, no migration occurs, (iii) subtle differences in the modes of deformation and burst of liquid drops, (iv) microscopic irreversibility. Apart from evidence that most of these phenomena are attributable to normal stresses in the non-Newtonian liquids, virtually no theory has been developed to explain these interesting effects, although some speculation has been made (25) on the origin of item (i).

Summary

Experimental and theoretical studies have been made of the microrheological behaviour of rigid and deformable particles of various sizes (1 micron and up), shapes (spheres, ellipsoids, rods, discs, chains, coils etc.) and concentrations (up to 60% by volume) suspended in Newtonian liquids undergoing viscometric flows such as simple shear (or *Couette*) flow, tube (or *Poiseuille*) flow and 2-dimensional extensional (or hyperbolic) flow. Studies made up to 1967 have been summarized in a detailed review (1); later work is described in References (2 to 16).

The microrheological phenomena have included (i) translational and rotational motions of the particles, (ii) flow patterns in the suspending fluid near the particle, (iii) orientation distributions of anisometric particles, (iv) deformation and rupture of particles, (v) n-body interactions, (vi) coalescence of liquid drops, (vii) concentration changes near boundaries, (viii) microscopic reversibility, (ix) rheo-optical effects.

In addition, a number of interesting and revealing electrohydrodynamic effects which accompany the superposition of electrical fields on the flow fields have been examined theoretically and experimentally (17 to 20).

References

1) *Goldsmith, H. L.* and *S. G. Mason*, The Microrheology of Dispersions. In: *F. R. Eirich* (Ed.), Rheology: Theory and Applications, Vol. 4, Chp. 2, pp. 85–250 (New York 1967).

2) *Cox, R. G., I. Y. Z. Zia* and *S. G. Mason*, J. Colloid Interface Sci. **27**, 7–18 (1968).

3) *Takano, M., H. L. Goldsmith* and *S. G. Mason*, J. Colloid and Interface Sci. **27**, 253–267 (1968).

4) *Takano, M., H. L. Goldsmith* and *S. G. Mason*, J. Colloid and Interface Sci. **27**, 268–281 (1968).

5) *Anczurowski, E.* and *S. G. Mason*, Trans. Soc. Rheol. **12** (**2**), 209–215 (1968).

6) *Goldsmith, H. L.* and *S. G. Mason*, Some Model Experiments in Hemodynamics III. In: *A. L. Copley* (Ed.), Hemorheology: Proc. 1st. Int. Cong. Hemorheology, Reykjavik 1966 (London 1968).

7) *Mason, S. G.* and *H. L. Goldsmith*, The Flow Behaviour of Particulate Suspensions. Ciba Foundation Symposium on Circulatory and Respiratory Mass Transport, pp. 105–124 (edited by *G. E. W. Wolstenholme* and *Julie Knight*) (1969).

8) *Goldsmith, H. L.* and *S. G. Mason*, Model Particles and Red Cells in Flowing Concentrated Suspensions. 5th. European Conference on Microcirculation, Gothenburg 1968, Bibli. Anat. No. 10, pp. 1–8 (Basel 1969).

9) *Takano, M.* and *S. G. Mason*, "Pulsatile Flow of Suspensions Through Tubes". Proceedings of the Fifth International Congress on Rheology, Vol. 2 (*Shigeharu Onogi*, Ed.) (Univ. Tokyo Press and Univ. Park Press 1970).

10) *Torza, S., C. P. Henry, R. G. Cox* and *S. G. Mason*, J. Colloid and Interface Sci. **35**, 529–543 (1971).

11) *Cox, R. G.* and *S. G. Mason*, Ann. Rev. Fluid. Mech. **3**, 291–316 (1971).

12) *Goldsmith, H. L.* and *S. G. Mason*, Some Model Experiments in Hemodynamics IV. Theoretical and Clinical Hemorheology, Proc. of the 2nd International Conference (*H. H. Hartert* and *A. L. Copley*, eds.), pp. 47–59 (Berlin-Heidelberg-New York 1971).

13) *Torza, S., R. G. Cox* and *S. G. Mason*, J. Colloid and Interface Sci. **38** (**2**), 395–411 (1972).

14) *Gauthier, F. J., H. L. Goldsmith* and *S. G. Mason*, Flow of Suspensions Through Tubes X. Liquid Drops as Models of Erythrocytes. Biorheology **9**, 205 (1972).

15) *Vadas, E. B., H. L. Goldsmith* and *S. G. Mason*, J. Colloid Interface Sci. **43**, 630 (1972).

16) *Sorrentino, M.* and *S. G. Mason*, J. Colloid Interface Sci. **41**, 178 (1972).

17) *Chaffey, C. E.* and *S. G. Mason*, J. Colloid Interface Sci. **27**, 115–126 (1968).

18) *Torza, S.* and *S. G. Mason*, Science **163**, 813–814 (1969).

19) *Torza, S.* and *S. G. Mason*, J. Colloid and Interface Sci. **33**, 68–83 (1970).

20) *Torza, S., R. G. Cox* and *S. G. Mason*, Phil. Trans. Royal Soc. (London) **269**, A 1198, 295–319 (1971).

21) *Karnis, A.* and *S. G. Mason*, Trans. Soc. Rheol. **10**, 571–593 (1966).

22) *Gauthier, F. J., H. L. Goldsmith* and *S. G. Mason*, Rheol. Acta **10**, 344–364 (1971).

23) *Gauthier, F. J., H. L. Goldsmith* and *S. G. Mason*, Trans. Soc. Rheology **15** (**2**), 297–330 (1971).

24) *Gauthier, F., H. L. Goldsmith* and *S. G. Mason*, Kolloid-Z. u. Polymere **248**, 1000–1015 (1971).

25) *Saffman, P. G.*, J. Fluid Mech. **1**, 540 (1956).

Author's address:

Dr. *S. G. Mason*
Pulp & Paper Research Institute of Canada,
Dept. of Chemistry
McGill-University, Montreal 101, Quebec (Canada)

Rheol. Acta **13**, 650–657 (1974)

Laboratoire Central des Ponts et Chaussées, Paris (France)

Influence des déformations des grains dans les milieux granulaires

F. Schlosser

Avec 10 figures

(Reçu p. p. le 27 octobre 1972)

I. Introduction

La complexité des relations contraintes-déformations des sols est en partie la conséquence de leur nature particulaire. Dans le cas des sols cohérents, la présence d'eau interstitielle et la très faible dimension des particules rendent complexe toute étude microscopique du comportement de ces sols. Une telle étude est plus simple dans le cas des sables ou plus généralement des milieux granulaires constitués d'empilements de grains de dimensions supérieures à 0,1 mm. La réponse d'un milieu granulaire à une sollocitation donnée dépend de la granulométrie, de la forme des grains et de l'arrangement de ces grains, c'est-à-dire de la structure du milieu. Les deux caractéristiques essentielles de cette réponse sont, d'une part, la non-linéarité et, d'autre part, la non-réversibilité des déformations. Ce comportement, en dehors des zones de très fortes contraintes où se produisent des cassures des grains, résulte de deux phénomènes:

- d'une part, une déformation propre des grains à leurs points de contact;
- d'autre part, un réenchevêtrement (glissements, rotations) produisant une modification de la structure.

Lorsque les grains ont des surfaces suffisamment continues (peu d'aspérités et d'angularité), leurs déformations propres résultent d'un comportement élastique du matériau qui les constitue.

Les phénomènes de réenchevêtrement entraînent toujours une anisotropie importante de la structure, le milieu granulaire ayant tendance à se réarranger de manière à offrir le plus de résistance à la sollicitation qui lui est imposée comme l'ont montré des études photoélastiques sur des empilements granulaires (1).

L'object de la présente étude est de montrer que certains phénomènes propres au comportement des milieux granulaires peuvent s'expliquer qualitativement par la seule considération des déformations aux points de contact et que la prise en compte de particularités structurales (augmentation du nombre des contacts avec la pression moyenne, transmission des efforts par chaînons de contraintes) permet une assez bonne approche qualitative.

II. Théories des déformations aux points de contact

Les déformations aux points de contact sont supposés être régies par les lois de *Hertz* pour les composantes normales des réactions et de *Mindlin* (2) pour les composantes tangentielles. Cette dernière a été établie pour le contact de deux sphères égales, c'est pourquoi les empilements considérés seront constitués de sphères égales ou quasiment égales. D'après la théorie de *Hertz* (3), deux sphères égales comprimées sous l'action d'une force normale N sont en contact suivant un cercle de rayon a et le rapprochement de leurs centres est α, a et α étant donnés par:

$$a = (KNR)^{\frac{1}{3}} \quad \text{avec} \quad K = \frac{3}{4}\,\frac{1-\nu^2}{E}$$

$$\alpha = 2\left(\frac{KN}{R\frac{1}{2}}\right)^{\frac{2}{3}}.$$

La fig. 1 illustre la théorie de *Mindlin* et montre la répartition des contraintes tangentielles sur le cercle de contact de deux sphères égales soumises d'abord à une force normale N, puis à une force tangentielle T. Il se produit un anneau de glissement, sur le pourtour du cercle de contact, qui explique que le déplacement relatif δ des sphères soit en partie irréversible et dépende de l'histoire du chargement. Dans ce cas particulier de chargement (N = Cte, T croissant), les expressions du déplacement δ et de la répartition des

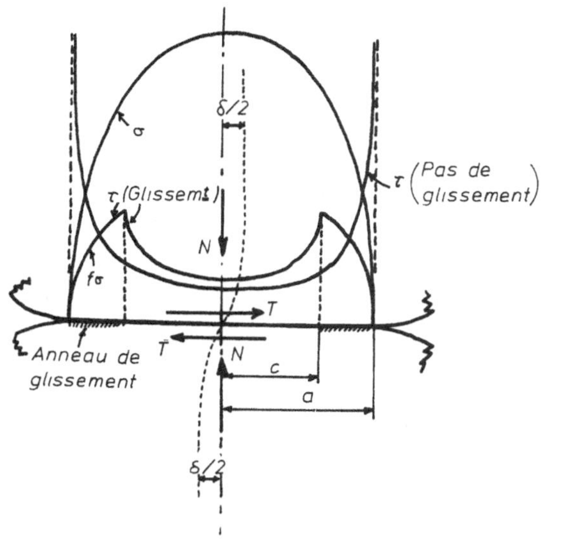

Fig. 1. Théorie de *Mindlin*: distribution des contraintes sur le cercle de contact

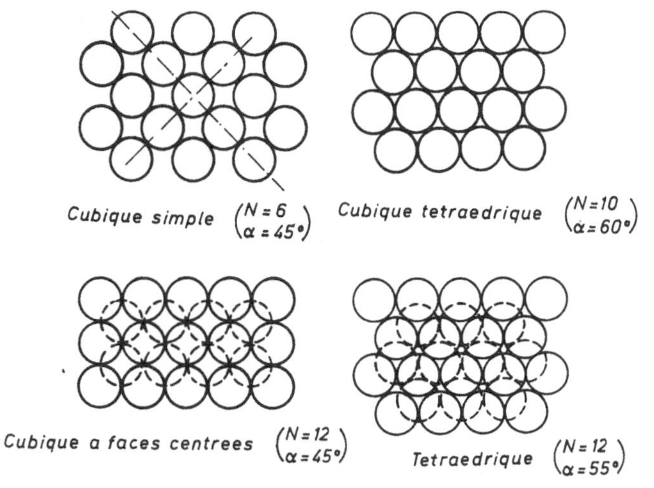

Fig. 2. Empilements réguliers

contraintes normale σ et tangentielle τ sur le cercle de contact sont:

$$\delta = \frac{3(2-\nu)fN}{8\mu a}\left[1-\left(1-\frac{T}{fN}\right)^{\frac{2}{3}}\right]$$

$$\sigma = \frac{2N}{2\pi a^3}(a^2-\varrho^2)^{\frac{1}{2}}$$

$$\tau = \frac{3fN}{2\pi a^3}\left[(a^2-\varrho^2)^{\frac{1}{2}}-(c^2-\varrho^2)^{\frac{1}{2}}\right]$$

$$\varrho \leq c$$

$$\tau = \frac{3fN}{2\pi a^3}(a^2-\varrho^2)^{\frac{1}{2}} \qquad c \leq \varrho \leq a$$

avec

$$c = a\left(1-\frac{T}{fN}\right)^{\frac{1}{3}}.$$

III. Résultats caractéristiques

On schématise le milieu granulaire par un empilement régulier de sphères égales ayant à leurs points de contact même coefficient de frottement f. La fig. 2 montre les différents types d'empilements considérés, depuis l'empilement cubique simple qui est l'empilement le plus lâche jusqu'aux empilements tétraédrique et cubique à faces centrées qui sont les plus denses.

Les phénomènes propres aux milieux granulaires, qui ont pu être expliqués qualitativement par l'étude théorique de tels empilements, sont les suivants:

1. Compression isotrope

Lorsqu'un milieu granulaire est soumis à une compression isotrope, la variation volumique, qui présente une certaine irréversibilité, lors du premier chargement, due pour une grande part à l'anisotropie de mise en place, varie ensuite à peu près proportionnellement à $\sigma^{\frac{2}{3}}$ (σ étant la contrainte isotrope). Ce résultat qui est une conséquence directe de l'application de la loi de *Hertz* a été donné par plusieurs auteurs dont notamment *Biarez* en 1961 (4). Cependant, il n'y a pas de concordance quantitative entre les déformations prévues à partir d'empilements réguliers et les déformations obtenues sur des empilements réels.

2. Phénomènes d'hystérésis

Lors de cycles de contraintes, la courbe effort-déformation d'un milieu granulaire présente une hystérésis d'autant plus importante que la partie déviatorique du tenseur des contraintes est plus grande. En fait, un tel milieu ne présente jamais une réversibilité totale et même lors de cycles de pression isotrope, on note une légère hystérésis. Dès 1952 *Johnson* (5) avait étudié expérimentalement ce phénomène sur deux sphères d'acier en contact et montré que l'hystérésis était liée à l'existence et à l'amplitude de la composante tangentielle de la force de réaction entre les deux sphères. La théorie de *Mindlin* a apporté une excellente explication: le cercle de contact entre deux sphères égales comprend deux zones; une première zone centrale dans laquelle il n'y a aucun déplacement relatif des différents points en contact et une deuxième zone annulaire où le frottement est entièrement mobilisé et où il y a glissement des parties en contact (fig. 1).

L'importance de cet anneau de glissement est directement lié à l'amplitude de la composante tangentielle de la réaction. L'hystérésis traduit une perte d'énergie par frottement sur la surface de cet anneau de glissement.

3. Coefficient K_0

Dans un milieu granulaire semi-infini, soumis à l'action de la pesanteur et à surface horizontale, le rapport K_0 de la contrainte principale horizontale σ_h à la contrainte principale verticale σ_v, ou coefficient de pression latérale des terres au repos, est constant en tout point. Il dépend à la fois du frottement entre les grains et de la structure du milieu.

En 1963, *Hendron* (6) a étudié le phénomène du coefficient K_0 sur un empilement régulier compact (empilement tétraédrique à faces centrées) et montré que la constance du rapport de la contrainte horizontale à la contrainte verticale, lors d'un chargement à déformation latérale nulle, était due aux déformations des grains à leur contact. Il existe d'ailleurs une bonne concordance qualitative dans de tels essais entre les chemins de contraintes théoriques (c'est-à-dire calculés sur des empilements réguliers) et les chemins de contraintes expérimentaux (c'est-à-dire correspondant à des empilements réels de grains de formes quelconques) (fig. 3a et b). L'un des résultats principaux de l'étude de *Hendron* est que, dans un essai K_0 sur un empilement régulier de sphères égales, le frottement est entièrement mobilisé à tous les points de contact. Cet état correspond à la rupture de l'empilement régulier (et non de l'empilement réel) et le coefficient K_0 est alors confondu avec le coefficient de poussée K_a. Ce résultat est en apparente contradiction avec les données fournies par les essais sur des milieux granulaires réels (sables) qui obéissent approximativement à la loi empirique de *Jaky*:

$$K_0 = 1 - \sin\varphi .$$

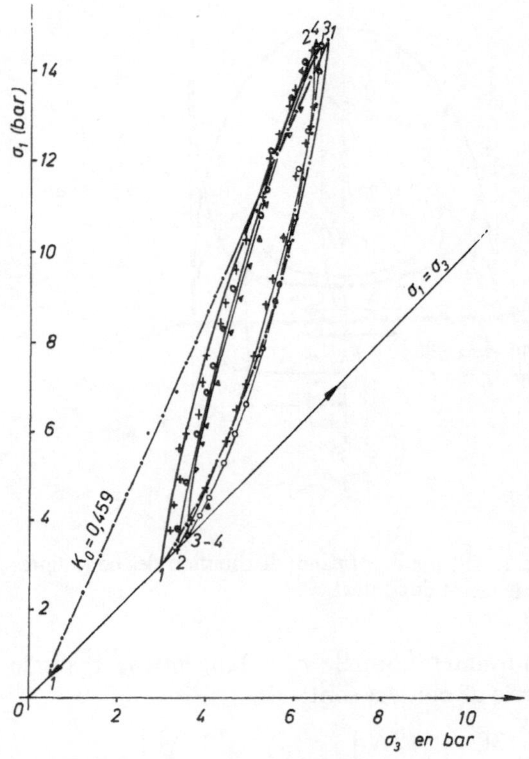

Fig. 3b. Essai K_0 sur un sable

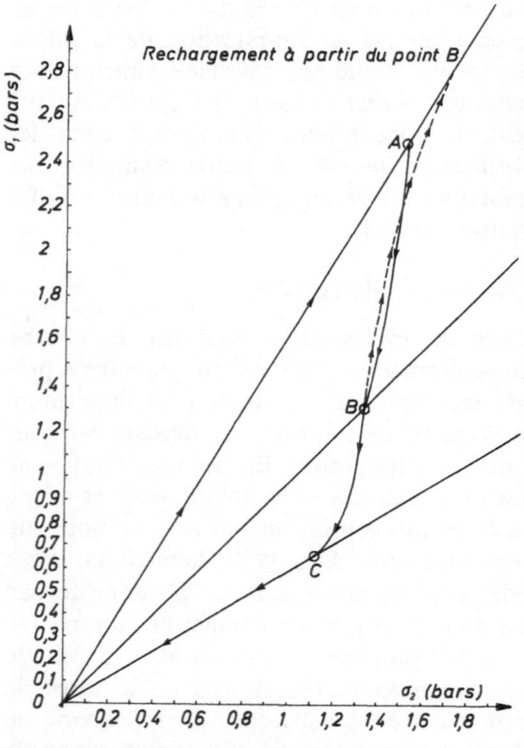

Fig. 3.a. Chemin de contraintes théorique d'un essai K_0 sur un empilement cubique simple

4. Droite limite

L'étude du domaine d'élasticité créé par écrouissage d'un milieu granulaire dans un état de contraintes triaxial (σ_1, $\sigma_2 = \sigma_3$) fait apparaî-

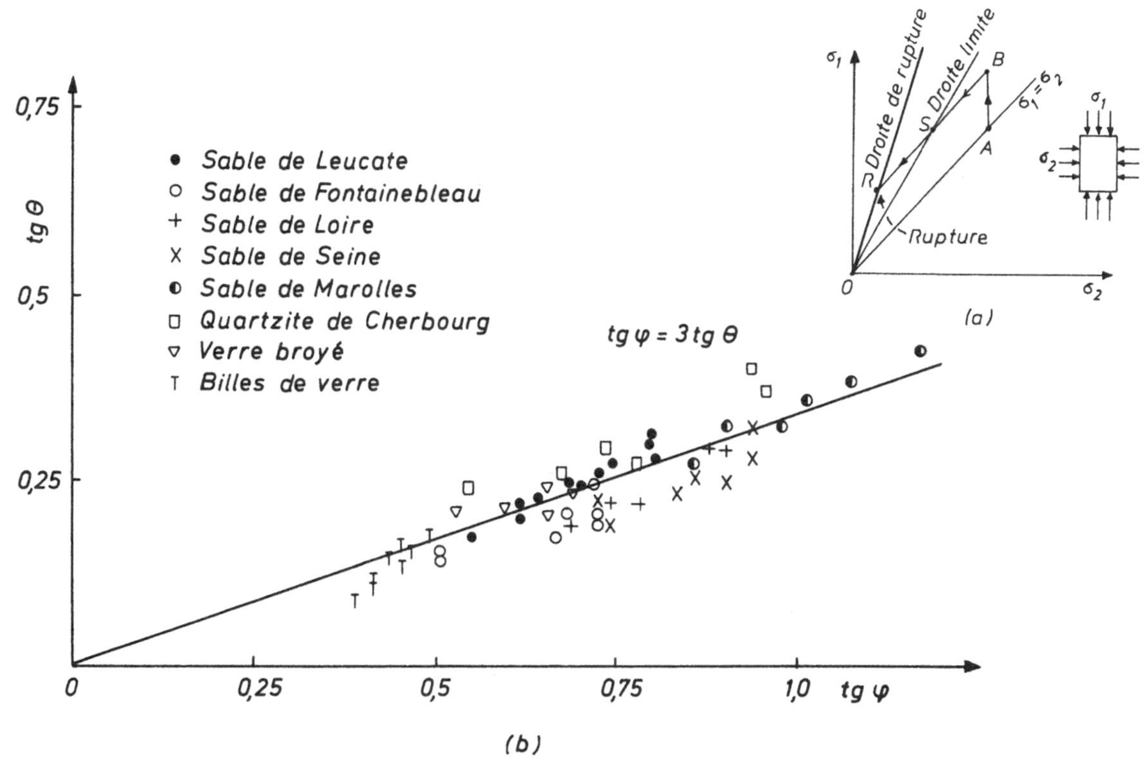

Fig. 4. Droite limite des milieux granulaires

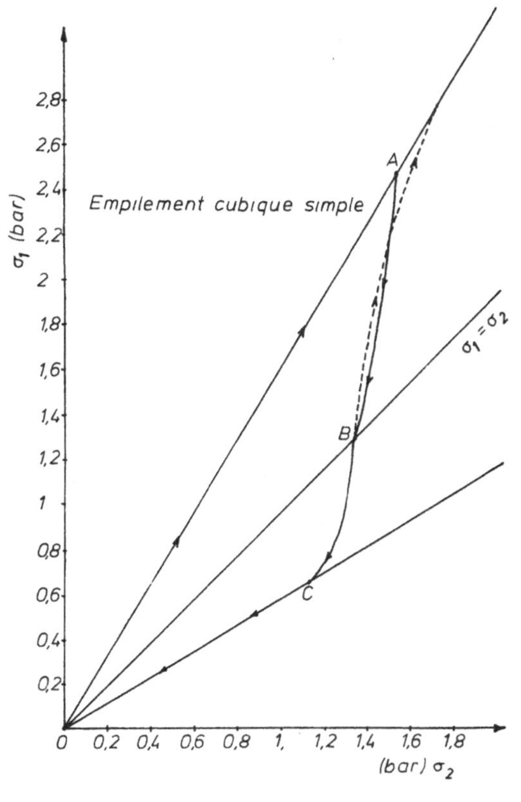

Fig. 4b. Chemins de contraintes théoriques dans l'essai K_0

tre la notion de droite limite (*Schlosser*, 1965). Dans un déchargement isotrope, après écrouissage, le milieu se déforme réversiblement, jusqu'en un point qui, dans l'espace des contraintes, définit une droite frontière du domaine d'élasticité, appelée droite limite (fig. 4 a). L'un des aspects caractéristiques de cette droite est qu'elle est intrinsèque et ne dépend en première approximation que de l'angle de frottement interne φ du milieu. On peut associer à cette droite un angle θ, analogue à l'angle φ pour la droite de rupture et défini par:

$$\left(\frac{\sigma_1}{\sigma_3}\right)_{\text{limite}} = \text{tg}^2\left(\frac{\pi}{4} + \frac{\theta}{2}\right).$$

Le rapport tg φ/tg θ est voisin de 3 pour les milieux granulaires réels (sables) (fig. 4b).

L'étude d'empilements réguliers de sphères égales permet de retrouver qualitativement ce phénomène de la droite limite et de l'expliquer (8). En un point de contact entre deux grains, les déformations dues à la composante normale de la force de réaction sont réversibles (loi de *Hertz*), par contre, les déformations dues à la composante tangentielle de cette réaction sont en partie irréversibles et l'irréversibilité est

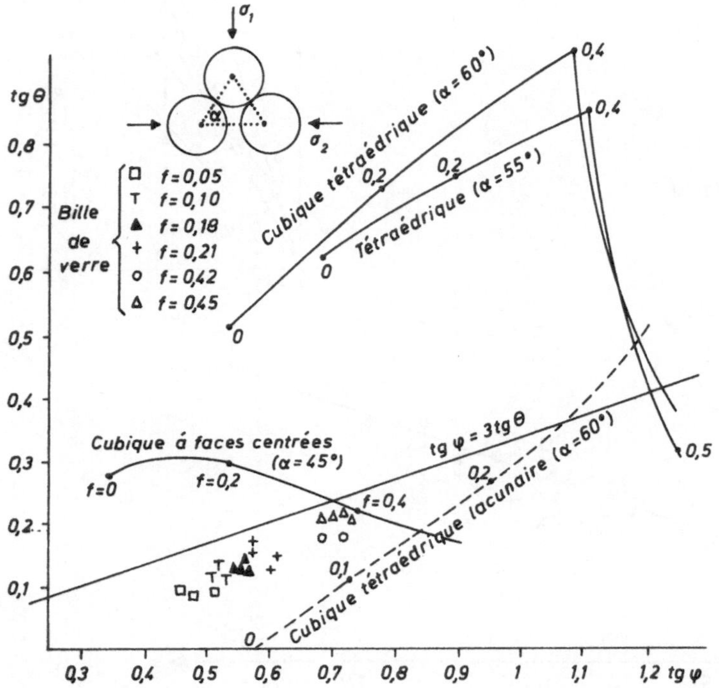

Fig. 5. Etude du rapport $\mathrm{tg}\varphi/\mathrm{tg}\theta$

d'autant plus grande que l'anneau de glissement est plus important. La droite limite sépare le domaine des contraintes où les composantes normales des réactions sont prédominantes (élasticité approchée) du domaine où les composantes tangentielles imposent leurs déformations.

Par contre, il n'y a pas de concordance quantitative avec le phénomène réel: on ne retrouve pas sur les empilements réguliers la valeur 3 correspondant au rapport $\mathrm{tg}\varphi/\mathrm{tg}\theta$ pour les empilements réels. La fig. 5 montre que ce rapport varie dans de grandes proportions en fonction du type d'empilement considéré et de la valeur du coefficient de frottement entre les sphères.

Cependant, l'étude théorique a l'avantage de montrer l'influence de l'orientation moyenne des plans de contact entre les grains par rapport à la direction de la contrainte principale majeure. Si α désigne l'angle de la normale du plan de contact moyen avec cette direction, le rapport $\mathrm{tg}\varphi/\mathrm{tg}\theta$ est d'autant plus élevé que cet angle est plus faible. Un tel résultat avait déjà été obtenu expérimentalement sur des empilements de sphères et d'ellipsoïdes applatis (9).

IV. Modèles de milieux granulaires

Pour cerner de plus près le comportement des milieux granulaires et pour avoir une meilleure concordance entre les résultats théoriques et expérimentaux sur les empilements de sphères, on prend en compte les deux phénomènes suivants: *1. transmission des efforts par chaînons de contraintes; 2. augmentation du nombre de contacts avec la pression moyenne.*

Les chaînons de contraintes ont été mis en évidence sur des milieux granulaires photoélastiques par *Dantu* (1). Ils sont évolutifs en fonction du chemin de contraintes suivi et s'orientent en général dans la direction de la contrainte principale majeure. L'influence de la pression moyenne sur le nombre de contacts entre grains a également été mis en évidence expérimentalement sur des empilements de billes thermodurcissables par *Weber* et *Dantu.*

Les modèles utilisés sont constitués d'empilements réguliers de sphères de deux diamètres, légèrement différents. De tels empilements dits «lacunaires» permettent ainsi de représenter schématiquement le phénomène de la transmission des efforts par chaînons et celui de l'augmentation des nombres de contacts. On répartit régulièrement et dans des proportions variables les petites sphères et les grandes sphères. La différence de diamètre entre les deux types de sphères situe le niveau de contraintes autour duquel on souhaite voir influer le changement du nombre de contacts. Un tel modèle est loin d'être complètement représentatif de la réalité car, d'une part les chaînons de contraintes sont prédéterminés et n'évoluent pas avec les contraintes principales, d'autre part l'augmentation du nombre de contacts avec la pression moyenne n'est pas progressive, mais discontinue.

L'empilement tétraédrique lacunaire constitue un exemple typique de ce genre de modèles pour le cas plan et représente assez bien le phénomène des chaînons de contraintes dans un empilement réel de rouleaux photoélastiques (fig. 6).

Les premiers modèles de ce genre ont été utilisés par *Ko* et *Scott* (10) pour l'étude de la compression isotrope et donnent une bonne concordance qualitative avec la réalité (fig. 7). Dans cette sollicitation, le paramètre essentiel est l'augmentation du nombre des contacts avec la pression appliquée; le phénomène de transmission des efforts par chaînons joue un rôle secondaire.

Dans l'étude théorique du coefficient K_0 ou de la droite limite par utilisation d'empilements lacunaires, le phénomène essentiel est, par contre, la transmission des efforts par chaînons de contraintes. En effet, le caractère lacunaire de l'empilement a pour conséquence de ne faire travailler en première phase qu'une partie de l'empilement, constituée par un réseau dont la résistance à la rupture est plus faible que celle de l'empilement considéré. A la rupture, par contre, le caractère lacunaire de l'empilement disparaît et ce dernier se comporte comme si toutes les sphères étaient égales. Ainsi, si l'on considère l'empilement tétraédrique plan lacunaire (fig. 6a), les droites de rupture du réseau en nid d'abeilles et de l'empilement sont respectivement:

$$\left(\frac{\sigma_1}{\sigma_3}\right)_{\hexagon} = \frac{\sqrt{3}+f}{\sqrt{3}-f} \qquad \left(\frac{\sigma_1}{\sigma_3}\right)_{\Delta} = 3 \cdot \frac{\sqrt{3}+f}{\sqrt{3}-3f}$$

Dans un essai K_0, tant que la pression moyenne n'atteint pas la valeur critique correspondant à l'augmentation du nombre des points de contact, on mobilise entièrement la résistance à la rupture du réseau puisqu'en tous les points de contact $T = fN$. Il vient alors pour l'empilement tétraédrique lacunaire précédent:

$$K_0 = \frac{\sqrt{3}+f}{\sqrt{3}-f} \qquad K_a = 3 \cdot \frac{\sqrt{3}+f}{\sqrt{3}-f}$$

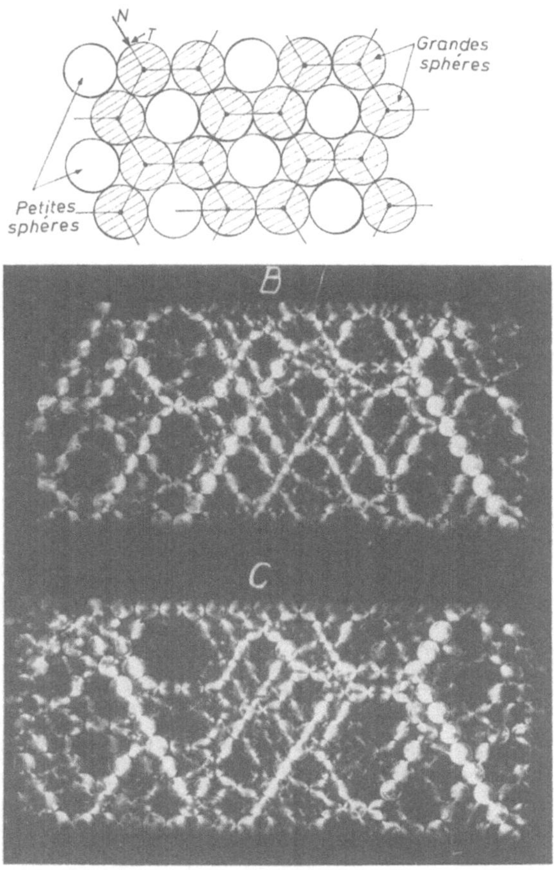

Fig. 6. Empilement tétraédrique (plan) lacunaire

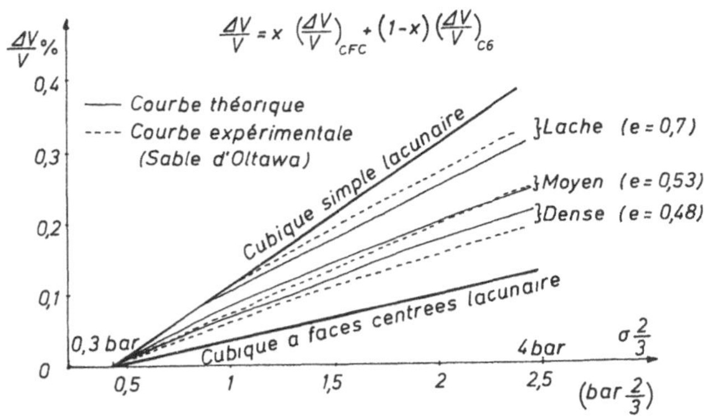

Fig. 7. Compression isotrope (d'après K_0 et *Scott*, 1967)

Dans ce cas, le rapport K_0/K_a qui a pour valeur 3 est indépendant du coefficient de frottement f. En adoptant la loi de *Jaky* pour les empilements tridimensionnels réels, ce rapport a par contre pour expression:

$K_0/K_a = 1 + \sin\varphi$.

Dans l'empilement cubique à faces centrées, le système lacunaire le plus simple (fig. 8) cor-

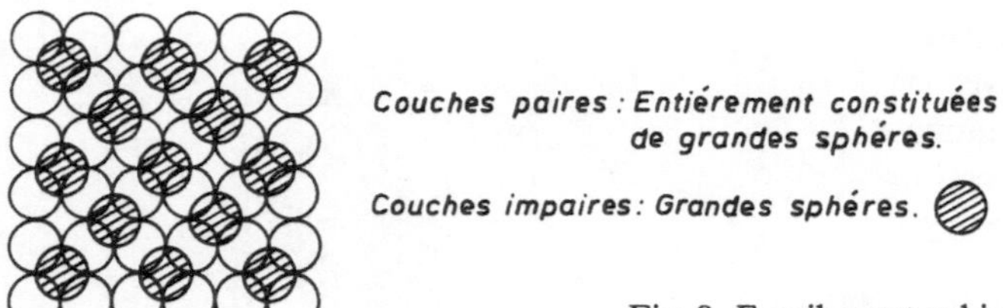

Fig. 8. Empilement cubique à faces centrées lacunaire

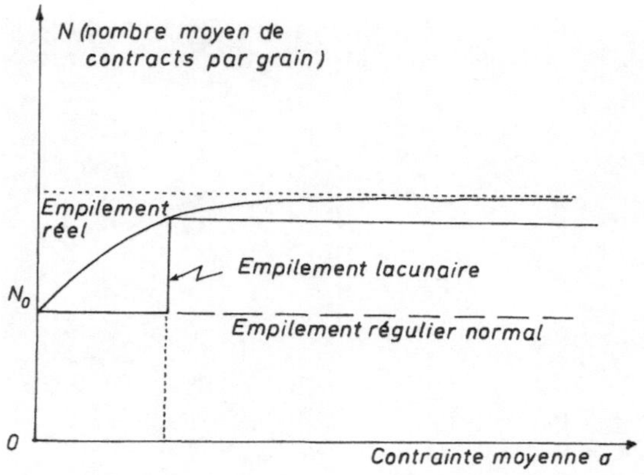

Fig. 9. Variations des contacts avec la pression moyenne

respond à une valeur du rapport K_0/K_a égal à 2, mais il est possible de trouver des systèmes plus complexes donnant des valeurs comprises entre 1 et 2. Si l'on adopte par exemple un rapport égal à 1,5, il existe une bonne concordance entre la formule de *Jaky* et la formule $K_0 = 1{,}5\,K_a$ dans le domaine des valeurs usuelles (25–40°) de l'angle de frottement interne φ.

La prise en compte d'une augmentation continue des contacts avec la pression moyenne, permettrait d'obtenir un meilleur accord avec la réalité dans le cas de systèmes lacunaires simples. En effet, une augmentation discontinue de ces contacts transforme la droite $\sigma_3 = K_0\sigma_1$ en une courbe raccordant la droite de rupture du réseau lacunaire à la droite de rupture de l'empilement. Une variation continue, telle que celle de la fig. 9, c'est-à-dire forte dans les zones des basses contraintes et tendant asymptotiquement vers l'empilement régulier, engendre pour l'essai K_0, dans le plan des contraintes principales, une droite plus proche de la droite de rupture que ne l'était la droite de l'empilement lacunaire initial.

Les résultats sont tout à fait analogues lorsque l'on considère le phénomène de la droite limite et les valeurs du rapport $\mathrm{tg}\,\theta/\mathrm{tg}\,\varphi$. Comme le montre la fig. 5, le caractère lacunaire de l'empilement permet d'abaisser considérablement les valeurs théoriques de ce rapport et de les rendre finalement assez voisines de la valeur obtenue pour un sable de billes de verre. La comparaison est faite dans le cas de l'empilement tétraédrique cubique, mais les calculs sur des empilements cubiques à faces centrées lacunaires, joints à une hypothèse d'augmentation des contacts avec la pression moyenne, donnent encore une meilleure concordance.

Conclusions

Les considérations précédentes montrent qu'un certain nombre de phénomènes propres au comportement des milieux granulaires s'explique assez bien qualitativement et quantitativement par la seule considération des déformations des grains à leurs points de contact. Cependant, tous ces phénomènes se situent dans un domaine de contraintes situé en deçà de l'état de repos. L'étude théorique du coefficient K_0 sur des empilements réguliers montre d'ailleurs qu'au delà de l'état de repos des glissements importants se produisent entre les grains et que la structure évolue. La schématisation du

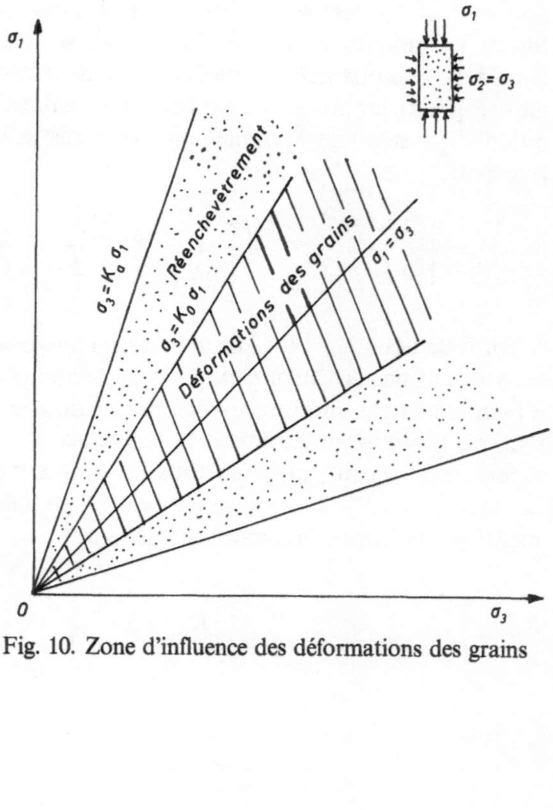

Fig. 10. Zone d'influence des déformations des grains

comportement des milieux granulaires par des modèles d'empilements réguliers de sphères égales n'est donc valable qu'en deçà de cet état de contraintes (fig. 10) et en plus seules des considérations d'empilements lacunaires permettent d'obtenir un certain accord avec la réalité.

Bibliographie

1) *Dantu, P.*, Contribution à l'étude mécanique et géométrique des milieux pulvérulents. Comptes rendus du 4ème Congrès International de Mécanique des Sols et des Travaux de Fondations. Londres (1957).

2) *Mindlin, R. D.*, Compliance of elastic bodies in contact. J. Applied Mech. **16**, 259 – 268 (1949).

3) *Hertz, H.*, Über die Berührung fester elastischer Körper. J. reine und angew. Mathematik (Crelles J.) **92** (1881).

4) *Biarez, J.*, Contribution à l'étude des propriétés mécaniques des sols et des matériaux pulvérulents. Thèse de doctorat. Faculté des sciences (Grenoble 1962).

5) *Johnson, K. L.*, Surface interaction between elastically loaded bodies under tangential forces. Proc. Roy. Soc. (London) **A 230**, 531 – 548 (1955).

6) *Hendron, A. J.*, The behavior of sand in one-dimensional compression. Thesis – University of Illinois (1963).

7) *Schlosser, F.*, Etude expérimentale du domaine élastique d'un milieu pulvérulent. Cahiers du Groupe Français de Rhéologie **1**. n° 1 (1965).

8) *Schlosser, F.*, Comportement des milieux granulaires. C. R. Acad. Sci. (Paris) **267**, 485–488 (1968).

9) *Schlosser, F.*, Etude de la droite limite des milieux pulvérulents. Rapport interne du L.C.P.C. (1968).

10) *Ko, H. Y.* et *R. Scott*, Deformation of sand in hydrostatic compression. Proc. ASCE J. Soil Mechanics Found. Div. 93. SM. 3 (May 1967).

Adresse de l'auteur:
Dr. *F. Schlosser*
Laboratoire des Central Ponts et Chaussées
58, Boulevard Lefevre
F-75 Paris 15 (France)

Rheol. Acta **13**, 658–663 (1974)

Centre de Recherches Kleber-Colombes, Bezons (France)

Influence de l'introduction de particules sphériques dans un élastomère sur les propriétés d'écoulement

D. Dubois, D. Dubos et G. Evrard

Avec 13 figures et 1 tableau

1. Introduction

Le renforcement des élastomères par des particules dispersées et, en particulier, les noirs de carbone est d'une très grande importance technologique dans l'industrie du caoutchouc et l'étude des propriétés des composites ainsi obtenus fait l'objet d'un très grand nombre de travaux. Les revues récentes consacrées à ce sujet (1) (2) font ressortir que les mécanismes des différents phénomènes de renforcement obtenus ne sont pas toujours parfaitement compris, la complexité des résultats étant en général liée à des problèmes d'aggrégation et de »structure«.

Il nous a semblé intéressant d'aborder l'étude des propriétés de ces composés sur des modèles simplifiés, bien caractérisés, qui pourront être ensuite progressivement compliqués pour atteindre les cas intéressants de façon pratique. Nous avons dans ce but préparé une série de produits composés d'une phase continue caoutchouteuse et d'une phase rigide dispersée formée de sphères indépendantes de diamètre variable. Une étude générale des propriétés de ces composites modèles a été entreprise (3) et nous présentons ici la première partie des travaux consacrés aux propriétés viscoélastiques à l'état fondu, à savoir les résultats obtenus par viscosimétrie capillaire.

2. Matériaux et techniques de mesure

2.1. *Produits étudiés*

Le choix de la phase continue a été guidé par le souci d'employer un élastomère largement utilisé dans l'industrie du caoutchouc et assurant d'autre part des propriétés stables et ne pouvant pas cristalliser sous contrainte; on a donc utilisé un copolymère statistique styrène-butadiène SBR 1500 (Cariflex Shell).

Comme charges rigides, nous avons employé des billes de verre de différents diamètres, sphériques et indépendantes, dépourvues de »structure«; nous avons également utilisé, pour étudier le domaine des diamètres de l'ordre 1–10 μ, des billes de zinc, qui apparaissent sphériques au microscope électronique mais présentent des propriétés superficielles différentes de celles des billes de verre; la granulométrie et les dimensions moyennes des charges sont indiquées dans le tableau ci-dessous.

Les teneurs en charge utilisées sont de 0, 10, 30, 50 et 80 parts en volume pour cent parts d'élastomère.

Tableau 1.

Diamètre en μ	∅ moyen en μ	mini	maxi
Verre	30	4	44
	75	50	105
	150	100	210
Zinc	2,5	< 5	

Les mélanges ont été préparés en incorporant les charges sur mélangeur à cylindre en respectant des conditions de travail toujours similaires pour tous les échantillons afin de limiter les problèmes liés à la dégradation de la gomme.

2.2. *Mesures*

2.2.1. *Viscosité apparente*

Les valeurs de viscosité ont été obtenues avec un rhéomètre capillaire Instron à vitesse d'extrusion imposée, suivant les méthodes classiques.

Les mesures ont été faites avec une série de capillaires de carbure de tungstène de diamètre moyen 1,27 mm ($\simeq$ 0,05″) et de rapports L/D voisins de 10, 20 et 40.

Les vitesses de traverse choisies (0,2 à 20 cm · mn^{-1}) ont ainsi permis d'explorer un domaine de gradients de vitesse de cisaillement allant de 12 à 1200 s^{-1}.

La température a été maintenue à 100° $\pm$ 1 °C pour tous les essais. Les viscosités apparentes ont été calculées à partir des mesures de force corrigées des effets de perte de charge dans le réservoir et aux extrémités du capillaire et en fonction du taux de cisaillement newtonien à la paroi.

N. B. Dans la pratique des corrections d'effets de bouts, il apparaît dans certains cas aux pressions élevées une déviation par rapport à la linéarité de la variation de force avec la longueur des capillaires; cet effet peut être attribué à la variation de viscosité avec la pression; des mesures de perte de charge dans le réservoir en fonction de la pression moyenne, à taux de cisaillement constant, montrent effectivement que, pour des pressions élevées, cette perte de charge croît avec la pression moyenne. On a donc utilisé la partie linéaire de ces courbes pour déterminer la contrainte vraie à la paroi. Ce point sera vérifié, dans une deuxième partie du travail, par la mesure directe du profil des pressions le long de la paroi du capillaire.

2.2.2. Gonflement à la filière

Deux méthodes de mesure ont été utilisées: une méthode directe, en mesurant au microscope le diamètre en divers points de l'extrudat, et une méthode indirecte en pesant une longueur connue. C'est cette dernière qui donne les résultats les plus reproductibles; elle permet également de continuer les mesures lorsque les défauts de l'extrudat apparaissent (la détermination de la longueur réelle de l'extrudat devient alors plus délicate et peut rendre la validité des résultats sujette à caution). Un maximum de soins a été accordé aux prélèvements afin de minimiser les erreurs dues, entre autres, à la déformation de l'extrudat et au temps d'atteinte d'une valeur d'équilibre.

Les résultats de gonflement ont été exprimés sous la forme:

$[(d/do)^2 - 1]$

do étant le diamètre du capillaire et *d* le diamètre de l'extrudat.

3. Résultats et discussion

3.1. Variation de la viscosité des mélanges

Les quatre premières figures montrent la variation de la viscosité apparente avec le taux de cisaillement newtonien pour chaque concentration, en fonction de la taille des particules (pour la clarté de la figure, certains points ont été omis).

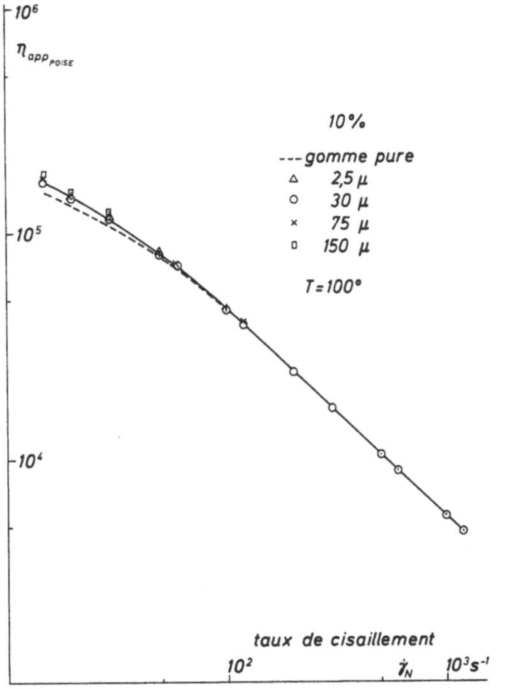

Fig. 1. Viscosité apparente en fonction du taux de cisaillement à 10% de charge, pour les différents diamètres

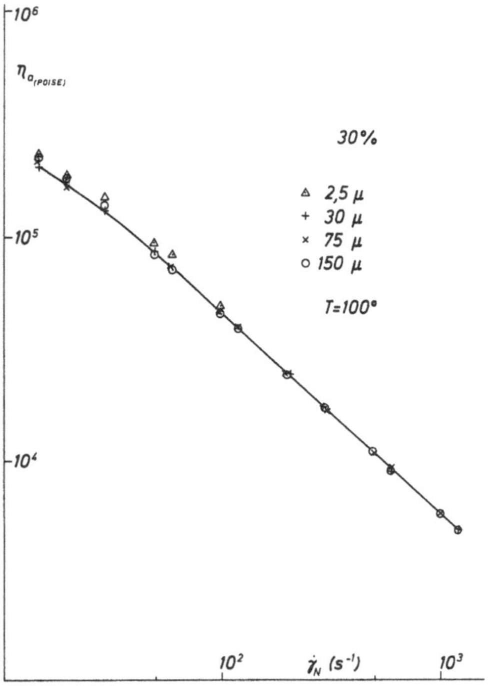

Fig. 2. Viscosité apparente en fonction du taux de cisaillement à 30% de charge, pour les différents diamètres

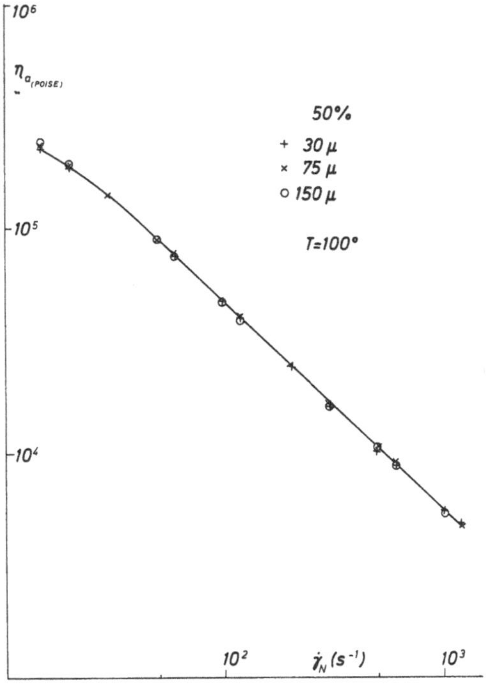

Fig. 3. Viscosité apparente en fonction du taux de cisaillement à 50% de charge, pour les différents diamètres

On constate que, pour la série de billes de verre, la viscosité est pratiquement indépendante du diamètre des particules, les variations étant de l'ordre de grandeur des erreurs de mesure; on

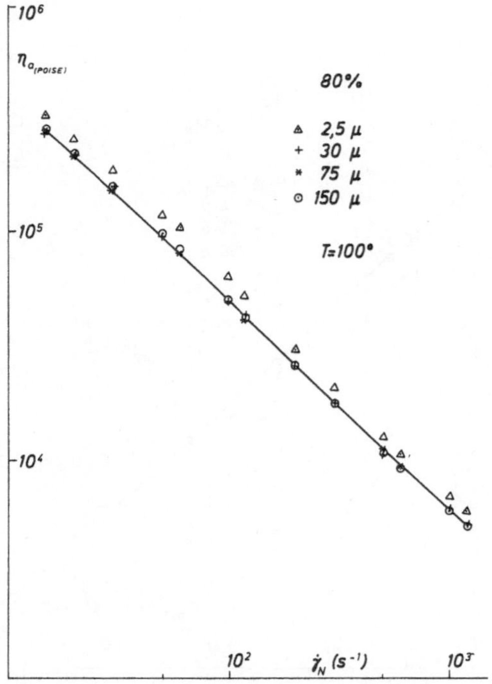

Fig. 4. Viscosité apparente en fonction du taux de cisaillement à 80% de charge, pour les différents diamètres

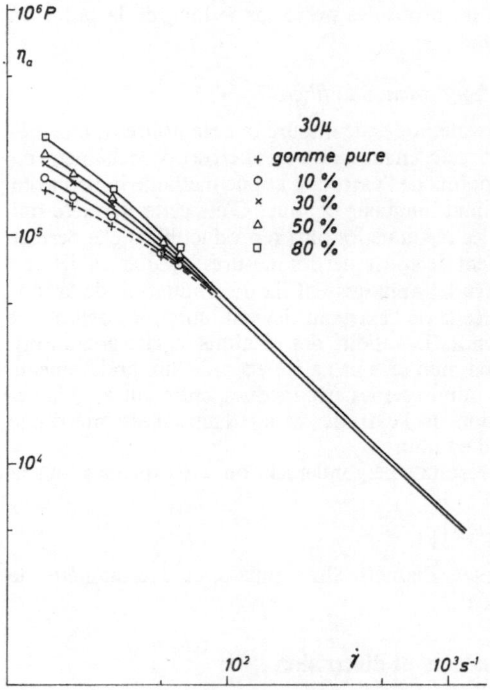

Fig. 6. Viscosité apparente en fonction du taux de cisaillement pour différents taux de charge, diamètre des particules 30 μ

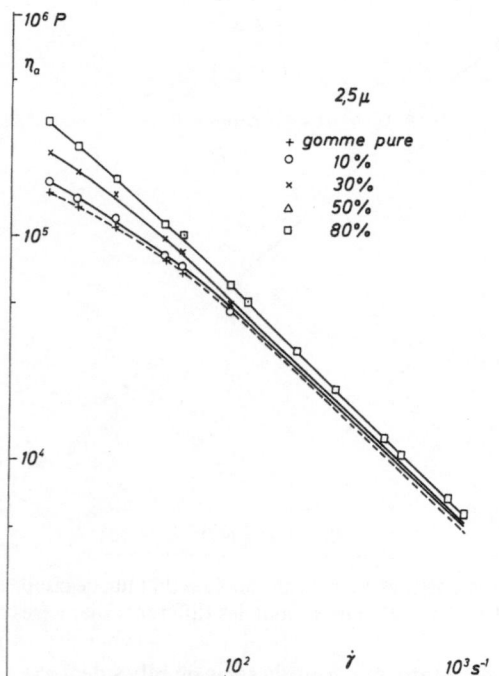

Fig. 5. Viscosité apparente en fonction du taux de cisaillement pour différents taux de charge, diamètre des particules 2,5 μ

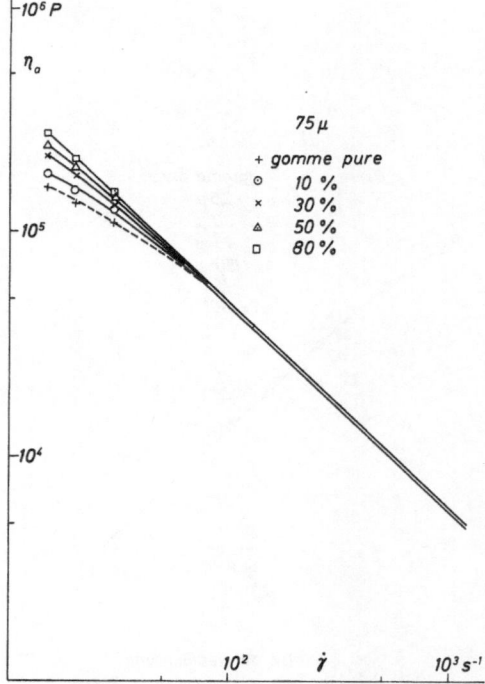

Fig. 7. Viscosité apparente en fonction du taux de cisaillement pour différents taux de charge, diamètre des particules 75 μ

observe toutefois avec les billes de zinc des résultats légèrement supérieurs, notamment aux teneurs les plus élevées (fig. 4). Les figures suivantes (5 à 8) montrent la variation de viscosité avec le taux de cisaillement, pour chaque diamètre, en fonction de la teneur.

On constate que, dans le domaine des faibles taux de cisaillement, la viscosité croît avec le taux de charge; mais pour des taux de cisaillement supérieurs à environ 100 s^{-1}, on observe une diminution de cet effet et la variation de viscosité devient pratiquement nulle (sauf, pour les billes de zinc, pour lesquelles on constate une faible augmentation aux taux de cisaillement les plus élevés).

Il est à noter que cette quasi insensibilité de la viscosité à la teneur en charges correspond à l'atteinte d'un plateau dans les courbes contrainte taux de cisaillement $(\tau, \dot{\gamma})$ et à l'apparition de défauts sévères et de distorsions des extrudats.

La signification des mesures de viscosité dans la région où apparaissent les phénomènes de »turbulence élastique« peut alors être sujette à caution (4).

Lors de l'étude (3) des propriétés mécaniques de mélanges similaires, vulcanisés au péroxyde de dicumyle, on a essayé de représenter l'effet de

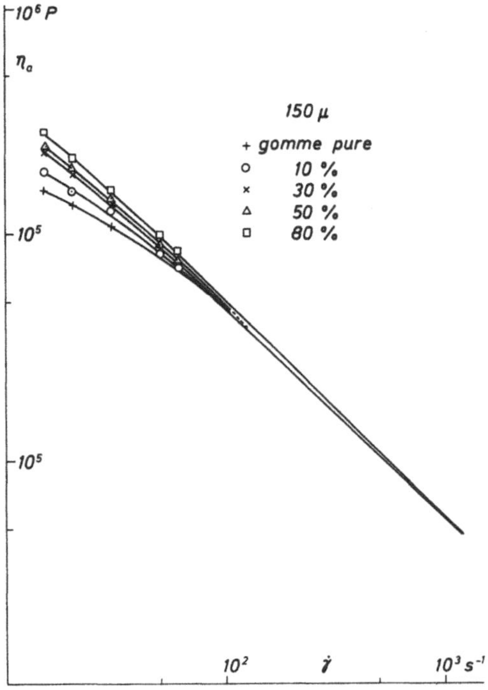

Fig. 8. Viscosité apparente en fonction du taux de cisaillement pour différents taux de charge, diamètre des particules 150 μ

renforcement obtenu par la relation classique de *Guth* et *Gold*.

$$G(\text{ou}\,\eta) = G_0(\text{ou}\,\eta_0) \times (1 + 2{,}5c + 14{,}1c^2)$$

c étant la fraction volumique de charge, G et G_0 étant respectivement le module de la phase chargée et de la phase non chargée.

Cette relation a été appliquée aux variations du module en cisaillement périodique de faible amplitude et de basse fréquence, à température ambiante. On a constaté un accord satisfaisant avec la loi pour des taux de charge pas trop élevés.

Si l'on veut représenter les variations relatives de viscosité de façon analogue, une difficulté apparaît dans le cas de systèmes non newtoniens; on peut en effet comparer les viscosités à taux de cisaillement constant ou à contrainte de cisaillement constante (cette deuxième comparaison

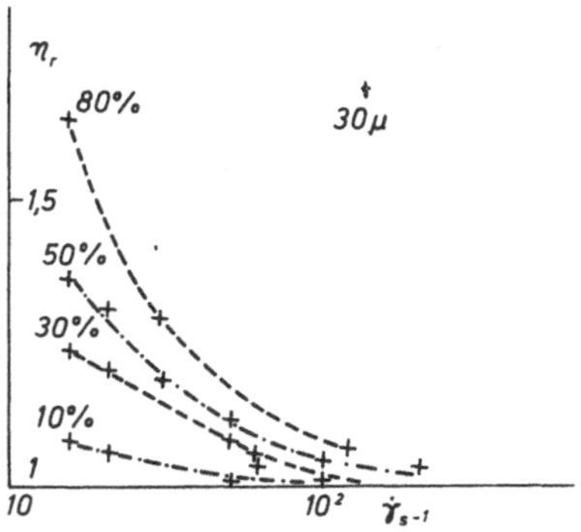

Fig. 9a. Viscosité relative en fonction du taux de cisaillement, diamètre des particules 30 μ

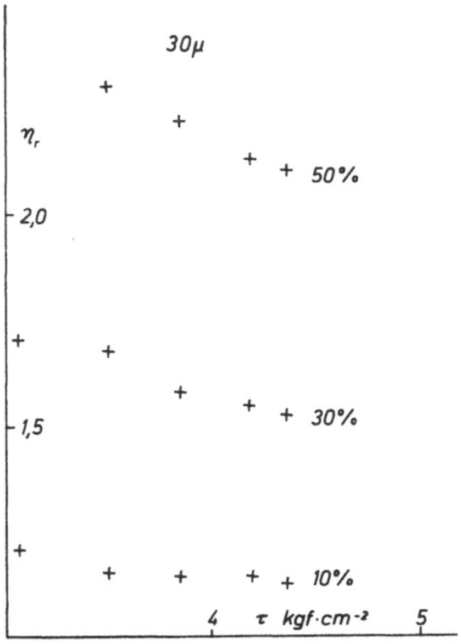

Fig. 9b. Viscosité relative en fonction de la contrainte de cisaillement, diamètre des particules 30 μ

correspondant évidemment à une variation relative plus importante dans le cas de liquides fluidifiants).

La fig. 9 représente, à titre d'exemple, les variations de la viscosité relative avec le taux de cisaillement (fig. 9a) et la contrainte de cisaillement (fig. 9b) obtenues avec les billes de verre de 30 μ, aux différentes teneurs.

Dans le domaine de l'écoulement non perturbé, la comparaison à taux de cisaillement constant fait apparaître que la viscosité relative est fonction à la fois de la fraction volumique et du taux de cisaillement.

On n'a malheureusement pas disposé de données suffisantes dans le domaine intéressant pour que la comparaison à contrainte constante soit significative, bien que, dans l'exemple cité, les variations de η_r avec la contrainte apparaissent de moindre amplitude. Compte tenu de ces résultats une représentation comme celle qui a été pratiquée dans le cas de l'étude du module de cisaillement, n'a pas paru significative.

3.2. *Gonflement à la filière*

Le gonflement à la filière est un phénomène particulièrement important dans la pratique pour la mise en œuvre des mélanges puisqu'il peut affecter fortement la stabilité dimensionnelle des produits. Aussi, bien que ni sa mesure ni son interprétation ne soient aisées, nous a-t'il semblé utile d'étudier systématiquement son évolution. Lorsqu'on examine la littérature, on constate que la plupart des résultats expérimentaux indiquent que le gonflement croît avec la contrainte de cisaillement; en fait, le gonflement passe par un maximum en fonction du taux de cisaillement dans une zone où apparaissent les phénomènes de turbulence élastique (5); *Eckert* (6) a également montré que, dans le cas d'un Néoprène, le gonflement des extrudats distordus passait par un minimum pour recroître ensuite à taux de cisaillement élevé.

On observe, dans le cas présent, un comportement tout à fait analogue pour le gonflement de la gomme pure, représenté sur la fig. 10 en fonction de la contrainte. On constate, dans ces conditions, une superposition des résultats obtenus aux diverses températures, entre le maximum et le minimum de gonflement; il est à remarquer, en fait, que les mesures au-delà du minimum correspondent à des extrudats fortement distordus pour lesquels il est difficile de définir une longueur lors du prélèvement.

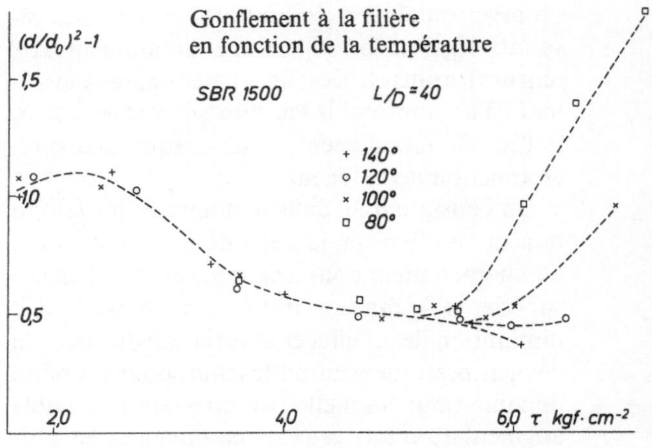

Fig. 10. Gonflement à la filière en fonction de la contrainte de cisaillement à diverses températures pour la gomme pure

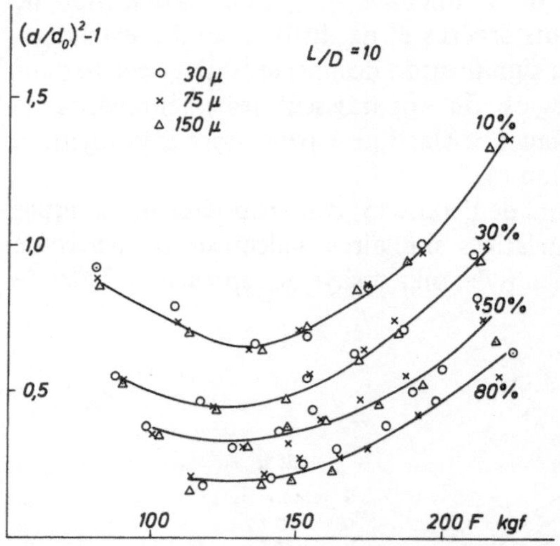

Fig. 11. Gonflement en fonction de la force d'extrusion à différents taux de charges, capillaire de rapport $L/D = 10$

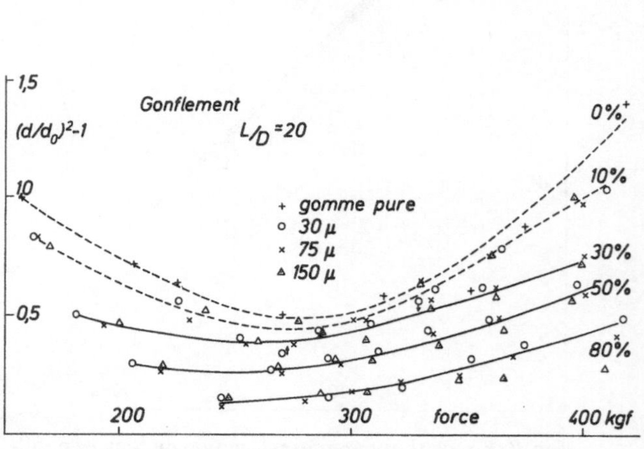

Fig. 12. Gonflement en fonction de la force d'extrusion à différents taux de charges, capillaire de rapport $L/D = 20$

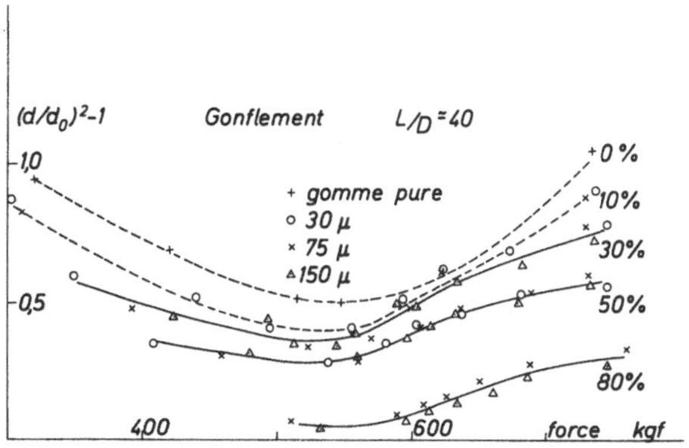

Fig. 13. Gonflement en fonction de la force d'extrusion à différents taux de charges, capillaire de rapport $L/D = 40$

Les figures suivantes (11 à 13) présentent les résultats obtenus avec les mélanges chargés de billes de verre; on constate que la variation de gonflement avec la contrainte n'est qualitativement pas modifiée par l'introduction des charges, mais que le gonflement diminue lorsque la fraction volumique croît et qu'il est indépendant du diamètre des particules.

4. Conclusion

Il ressort de l'ensemble de ces mesures que les variations des propriétés d'écoulement, telles que viscosité apparente et gonflement à la filière obtenues en introduisant des particules rigides dans un élastomère, sont fonction de la concentration, mais pratiquement indépendantes du diamètre dans la limite des erreurs de mesure. La viscosité apparente relative varie avec le taux de cisaillement et devient pratiquement indépendante de la présence des charges pour des taux de cisaillement correspondant à l'apparition des distorsions de l'extrudat.

Dans les conditions de mesure choisies, il n'a pas été possible de vérifier si, pour un écoulement non perturbé, la viscosité apparente relative est indépendante de la contrainte comme cela a été observé pour la viscosité de suspensions dans des fluides non newtoniens (7) et dans le cas de mélanges de SBR et de noir de carbone (8); des travaux sont en cours dans ce sens.

Remerciements

Les auteurs remercient la Société KLEBER-COLOMBES de les avoir autorisés à publier ce travail.

Litterature

1) *Kraus*, Advances in Polymer Science, **8**, 155 (1971).
2) *Oberth*, Rubber Chemistry and Technology, **40**, 1337 (1967).
3) *Evrard, G.* and *J. P. Nottin*, International Symposium on Macromolecules, Helsinki, Juli 1972, III 47.
4) *Vinogradov, G. V.* et coll., J. Polymer Science, **A 2**, Vol. **10**, p. 1061 (1972).
5) *Beynon, D. L. T.* and *B. S. Glyde*, Br. Plastics **33**, 416 (1960).
6) *Eckert, R. E.*, J. Applied Polym. Science, Vol. **7**, p. 1715 (1963).
7) *Highgate, D. J.* and *R. W. Whorlow*, Rheol. Acta **9**, 569 (1970).
8) *Smith, P. P. A.*, Rheol. Acta **8**, 277 (1969).

Adresse des auteurs:
D. Dubois, D. Dubos et *G. Evrard*
Centre de Recherches Kleber Colombes
49, rue Jean-Jaurés
F-95 Bezons (France)

Rheol. Acta 13, 664–669 (1974)

From the Department of Mechanical Engineering, University of Waterloo, Waterloo, Ontario (Canada) and Secteur Matériaux, Centre de Recherche Industrielle du Québec, Ste-Foy, Québec 10, Québec (Canada)

The superplastic flow of aluminium bronzes

M. W. A. Bright and D. M. R. Taplin

With 3 tables

Introduction

Superplasticity may be defined as the phenomenon whereby large uniform elongations are possible in metallic alloys under low applied tensile loads when four specific conditions are fulfilled, namely:

i) that the microstructure be extremely fine and stable (normally ~5 μm grain/phase diameter).
ii) that deformation be carried out at temperatures in excess of half the absolute melting point.
iii) that the material be deformed at a strain-rate in the range 10^{-5}–10^{-1} sec^{-1}.
iv) that no weak or brittle phases are present to cause premature failure.

Under these conditions, metallic alloys behave more like viscous liquids than polycrystalline solids. This subject has received close attention in recent years (1, 2) in view of its possible application in metal forming. In terms of mechanical behaviour, a distinction is drawn between solids and fluids, somewhat arbitrarily, at a viscosity of 10^{15} poise. Liquids exhibit viscosities in the range 10^{-2}–10^{2} poise, heated thermoplastics a viscosity of 10^4 poise, and glassblowing is carried out a viscosity of 10^7 poise. Superplastic alloys, although structurally crystalline, exhibit a low stress viscosity of the order of 10^5 poise, and consequently can hardly be considered as ordinary solids. The flow stress of a material may generally be written as a function of strain-rate and strain under isothermal conditions, as follows:

$$\sigma(\varepsilon\,\dot{\varepsilon})_t = \kappa\,\varepsilon^n\,\dot{\varepsilon}^m. \qquad [1]$$

The Newtonian viscous limit is defined by the condition $n=0$, $m=1$, when equation [1] reduces to:

$$\sigma = 3\,\eta\,\dot{\varepsilon}, \qquad [2]$$

where $\eta = \kappa/3$ and η is the linear viscosity. Substituting for stress and strain in terms of the applied load P and the specimen cross sectional area A and its time derivative yields the following alternative description of the linear viscous limit:

$$\dot{A} = -P/3\eta. \qquad [3]$$

Since the expression for the rate of deformation does not contain the cross sectional area, localised flow does not occur. For a bar of initially uniform cross section, therefore, subjected to tensile loading in a constant temperature environment, infinite elongation is theoretically possible.

Plastic deformation of a normal metal at ambient temperature is described by the condition: $0.1 \le n \le 0.4$, $m=0$. The limit of stability occurs at a critical strain $\varepsilon \simeq n$. Since neck development is both highly localised and rapid in the plastic state, tensile elongation is limited to well below 100%. Superplastic alloys do not exhibit ideal Newtonian flow, but rather obey the equation:

$$\sigma = \kappa\,\dot{\varepsilon}^m \qquad [4]$$

(that is $n=0$), where $m \ge 0.3$ and is typically in the range 0.5–0.9. Although, specimen deformation is in effect unstable from the onset of flow there is apparent stability caused by a dynamic hardening of the locally deforming regions as a consequence of the increase in imposed strain-rate. Indeed, it has been pointed out [3] that whereas microhardness measurements on the necked region of a conventional metal reveal that it is harder than the remainder of the specimen, in superplastic alloys microhardness measurements in interrupted tests are everywhere the same; only on re-straining will dynamic hardening again occur. Neck development proceeds very slowly in superplastic alloys and tensile elongations in excess of 1000% are possible in the absence of brittle phases or cavitation; the main limitation being simply the length of the hot zone in the furnace.

At temperatures in excess of half the absolute melting temperature micrograin superplastic alloys exhibit a three stage, sigmoïdal variation of stress with strain-rate on a logarithmic plot. The slope in each region characterizes the strain-rate sensitivity of the flow stress (m). The central, superplastic region of high slope, where m values in the range $0.4 \le m \le 0.9$ are recorded, is bounded by regions of low strain-rate sensitivity. This behaviour is typical of alloys based on Zn, Cu, Fe, Ni, Ti, Al, W, Pb, Mg, Co, Zr and Sn. High temperature deformation may occur by dislocation glide within the grains, grain boundary sliding and vacancy flow. During flow under normal conditions, the overall dislocation density increases and a stable dislocation cell structure forms. Flow is subsequently controlled by dislocation climb and the nonconservative motion of jogs in screw dislocations. Under these conditions, theory predicts a low value of 0.2 for m, in agreement with experiment.

At the temperature and strain-rate operative in superplastic deformation, the stable dislocation sub-cell size is larger than the ultra-fine grain-size. Consequently, no cells form [4] and deformation is controlled by vacancy diffusion and/or grain boundary sliding [5].

The theory of vacancy diffusion gives a stress dependence of the strain-rate of unity (6), and incidentally, expressions for viscosity on the basis of either a transgranular or intergranular diffusion path yield a viscosity value $\sim 10^5$ poise. Some evidence for a similar stress dependence for grain boundary sliding also exists (7, 8). Evidence for diffusional flow as the controlling feature is becoming well established (9). *Ashby* (8) has suggested that a threshold stress for diffusional flow exists since grain boundaries containing precipitates do not act as perfect vacancy sinks. As a consequence, the constitutive equation must be modified as follows:

$$\sigma = \sigma_0 + \kappa \dot{\varepsilon}^m, \quad [5]$$

which now corresponds to that of a *Bingham* solid. The upper strain-rate limit of the superplastic range is thus defined by the strain-rate at which dislocation climb becomes controlling and the lower strain-rate bound is defined by the threshold stress for diffusional flow. An alternative, sophisticated analysis is presented by *Davies* and *Padmanabhan* (20).

The prerequisite condition of a stable, fine-grained structure is usually achieved by thermomechanical treatment of alloys of eutectic or eutectoïd composition subsequent to cooling through the corresponding reaction isotherm. In these systems, *m* increases monotonically with increase in temperature until the reaction isotherm is reached. Since the superplastic regime is below the reaction isotherm, composition and volume fraction of the constituent phases are almost independent of temperature variation. Optimum ductilities in aluminium bronzes are observed in the α/β phase field, enclosed between steeply inclined sub-solidus lines, above the eutectoïd transformation temperature where relative volume fractions of the constituent phases vary markedly with temperature. The present work is an extension of earlier studies of aluminium bronzes at Waterloo (10–14). This paper reports recent observations on the effect of ternary additions on the superplastic flow characteristics.

Experimental

Copper base alloys of varying composition in the range 8.5–10 wt% Al and 0.0–4.0 wt% Fe were supplied in the form of 1 mm thick cold rolled sheet. Hot rolled plate, 8 mm thick, of nominal composition Cu–9.5% Al–4.0% Fe was also used in anisotropy studies. Alloy compositions are given in table 1. Flat tensile specimens were machined from the cold rolled sheet with a gauge length and gauge width of 12.8 mm and 9.5 mm respectively. Test specimens were prepared from the hot rolled plate material both with this flat specimen geometry and as round specimens (4.54 mm diameter, 16.1 mm gauge length). In anisotropy studies specimens were cut with the tensile axis parallel to and perpendicular to the sheet rolling direction. Tests were performed on an Instron testing machine equipped with pushbutton speed selector. A three zone split furnace with independent power proportional control was used to maintain a constant temperature in the range 650–900 °C within ± 3 °C along a zone of 150 mm length. All tests were performed in air since these alloys are highly oxidation resistant. At each temperature specimens were pulled to failure and rapidly quenched in iced water to minimise structural changes prior to metallographic examination by conventional techniques. Change rate tests were performed from which the strain-rate sensitivity of the flow stress was evaluated by the method of *Backofen* (15).

Table 1a. Alloy compositions

Nominal % Al	Analysed % Al	Nominal % Fe
8.5	8.75	0.0
9.0	9.14	0.0
9.0	8.99	4.0
9.5	9.57	0.0
9.5	9.51	0.5
9.5	9.52	2.0
9.5	9.70	4.0
10.0	10.0	0.0
10.0	10.1	4.0

Table 1b. (Calculated) Volume fraction – temperature data for binary Al-bronzes

Nominal % Al	Volume fraction α phase at different temperatures (°C)					
	700	750	800	850	875	900
9.0	100	87	76	67	60	53
9.5	74	60	48	42	35	30
10.0	54	40	30	26	—	—

Results and discussion

Recrystallisation of the as-received product occurs on heating above 650 °C when the microstructure is composed of a microduplex distribution of a Cu-rich F.C.C. solid solution, designated α, and a B.C.C. solid solution, designated β which in polished sections of quenched specimens is visible as the darker etching martensitic transformation product. In alloys containing iron additions, coarse particles of a B.C.C. δ phase (81.5% Fe, 13.5% Al, 5% Cu) are dispersed throughout the microstructure. Transmission electron microscopy studies of thinned sections of ternary alloys has revealed the presence of a sub-micron dispersion of δ particles predominantly in the β phase (12). The precipitates are not completely dissolved on heating the material to 800 °C. Ternary alloys with $\geq 2\%$ Fe additions exhibit an initial recrystallised grain-size of 8–10 µm; other alloys have grain-sizes in the range 20–30 µm.

Both, the binary and ternary alloy series exhibit maximum ductility at a temperature corresponding to equivolume phase proportions. At the optimum temperature maximum ductility and maximum *m* occured at a low strain-rate of 1.2×10^{-4} sec^{-1} for both, the binary alloys and ternary alloys containing 0.5% Fe; corresponding m values are anomalously high (0.7–0.8): similar observations have been made in other superplastic alloys and attributed to grain growth (16). Alloys containing 4.0% Fe additions exhibit maximum ductility at strain-rates in the range $1.2–1.7 \times 10^{-3}$ sec^{-1} and m values in the range ($0.5 \leq m \leq 0.6$). The peak in the ductility versus temperature plot is broad in ternary alloys containing 4.0% Fe, where the fine precipitate distribution compensates for the increased proportion of grain boundaries of like type as the temperature is raised above that for equivolume proportions.

In addition, the level of ductility under optimum conditions is high (> 600%). When the Fe content is less than 2.0%, the level of ductility is low (< 300%), and falls off rapidly as the temperature is raised above that corresponding to equivolume proportions. In this situation, the presence of adequate amounts of each phase must be relied upon to spacially separate grains of like type to inhibit grain growth. Alloys containing 2% Fe additions exhibit an initially fine grainsize on transformation and an optimum ductility ~600%. However, the precipitate distribution is not adequate to suppress grain growth at higher temperatures, when optimum strain-rate values are decreased from $12.0–1.2 \times 10^{-4}$ sec^{-1}. These data are summarized in table 2. In the Cu-Al system, the temperature for maximum ductility increases with decrease in Al content. The variation of ductility with temperature of the constituent phases is, therefore, important in determining overall optimum conditions. These are found in the ternary alloy containing 10% Al at temperatures vicinal to 750 °C (table 2). The β phase is the more ductile throughout the temperature interval of this study, since diffusion through the B.C.C. lattice occurs more easily then in the F.C.C. phase. Evidence from grain boundary sliding and cavitation studies (12, 17) would indeed suggest that diffusion in the α phase is the rate limiting accommodation process. Above 700 °C, the relative ductility of the phases (β/α) increases and necking in the material at failure is more pronounced. It is believed that at 750 °C the β phase exhibits optimum ductility, sufficient to inhibit cavity interlinkage (since β/β boundaries do not cavitate) but insufficient to result in localisation of deformation within the grains.

Anisotropy of superplastic flow has been noted previously in Zn – Al alloys (18, 19). Ductility varies with the angle between the tensile axis and the rolling direction, maximum elongation occuring when the angle is zero. Specimens cut from hot rolled Zn – Al plate with a circular cross section, developed elliptical cross sections during deformation. In addition, marked changes in crystallographic texture accompanied superplastic flow. These effects have been attributed to crystallographic slip during superplastic deformation in this system. Anisotropic superplasticity is also observed in ternary aluminium bronzes. Present studies on both hot rolled and cold rolled Cu – 9.5% Al – 4.0% Fe reveal that ductility is a maximum normal to the rolling direction. An elliptical cross section is also developed during deformation of round specimens cut from hot rolled plate. This behaviour was previously observed in cold rolled Al – bronze sheet (13) where a one to one correspondence between maximum anisotropy in ductility and maximum anisotropy in m values was noted. However, in this system, the rolling texture is progressively removed with increase of strain.

In the present study, it was found that the anisotropic effect was present over the entire temperature range of the superplastic domain (650–900 °C) when, on the one hand, the microstructure was comprised of all α grains and on the other hand all β grains. Anisotropy is apparently not dependent on the crystallography or properties of either phase. Mechanical tests were also performed on the cold rolled Cu – 9.5% Al binary and Cu – 9.5% Al – 4.0% Fe ternary alloys at 800 °C over the range of superplastic strain-rates. In the ternary alloy, anisotropy is maximum at a strain-rate of 1.2×10^{-3} sec^{-1}, where ductility and m were maximised, and decreases with increasing strain-rate, being negligible at a strainrate of 2.5×10^{-2} sec^{-1}. No difference in ductility was found between specimens with tensile axis parallel to or perpendicular to the rolling direction in the binary alloy. (These data are presented in table 3.) Anisotropy is, therefore, related to the presence of Fe additions in the microstructure. Metallographic examination of the ternary alloy, annealed at 800 °C, revealed some microstructural fibering of large δ phase particles parallel to the rolling direction, and these precipitates may contribute partly to the effect. In addition, the anisotropic ductility and the development of crosssectional ellipticity is related to the elongated grain structure in the as-hot-rolled state, maximum deformation (elongation in tension and reduction

Table 2a. Variation of ductility with temperature

Nominal % Al	% Fe	% Elongation at temperature (°C)						Opt. temp. °C	Max. % elong.
		650	700	750	800	850	875		
9.0	0	—	—	—	230	290	185	850	290
9.5	0	—	100	220	330	240	—	800	330
10.0	0	80	200	170	140	—	—	700	200
9.5	0.5	—	70	180	280	165	60	800	280
9.5	2.0	—	—	285	600	250	200	800	600
9.5	4.0	—	200	320	660	610	300	800	660
10.0	4.0	—	620	>1000	980	—	—	750	>1000
9.0	4.0	—	—	—	360	600	675	875	675

Table 2b. Effect of iron additions on m, and strain-rate for maximum m

Alloy and % Al		% Fe and parameter tabulated	Temperature °C		
			800	850	875
Cu – 9.5% Al	0.0	strain-rate	1.2×10^{-4}	—	—
		(sec^{-1}) m	0. 80	—	—
Cu – 9.5% Al	0.5	strain-rate	1.2×10^{-4}	1.2×10^{-4}	—
		(sec^{-1}) m	0. 80	0. 82	—
Cu – 9.5% Al	2.0	strain-rate	1.2×10^{-3}	2.4×10^{-4}	1.2×10^{-4}
		(sec^{-1}) m	0. 64	0. 62	0. 60
Cu – 9.5% Al	4.0	strain-rate	1.2×10^{-3}	1.2×10^{-3}	2.4×10^{-4}
		(sec^{-1}) m	0. 49	0. 53	0. 42

of width in the cross section) occuring parallel to the smaller dimension of the grain structure on the corresponding section. It was noted in earlier studies (12) that the ultrafine cuboïdal δ phase, particles precipitate predominantly in the β phase, and that the distribution is stable on heating the structure to 800 °C, when recrystallization of the elongated grains occurs. Consequently, this fine precipitate distribution will retain the "memory" of the original hot rolled structure on transformation, despite the fact that the transformed micro-duplex $\alpha - \beta$ structure exhibits no detectable grain shape anisotropy in quenched and polished sections. Such precipitates pin grain boundaries, and would, therefore, influence the ease of grain boundary migration and grain boundary sliding in different directions, and could, therefore, make a major contribution to the anisotropy of superplastic flow. Differences in precipitate configura-

Table 3a. Variation of anisotropy with temperature in hot rolled, Cu – 9.5% Al – 4.0% Fe Plate

Angle between T. A. and R. D.	Elongation at different temperatures (°C)					
	600	700	800	850	875	900
$\theta = 0°$	—	180	445	640	740	—
$\theta = 90°$	20	260	530	820	860	65
Elongation ratio (90/0)	—	1.44	1.19	1.28	1.17	—
Volume % α phase	—	74	48	42	35	—

Table 3b. Variation of elongation with strain-rate in Cu – 9.5% Al – 4.0% Fe and Cu – 9.5% Al – 0.0% Fe,

Alloy	% Elongation with strain-rate (sec^{-1})							
	1.2×10^{-4}		7.0×10^{-4}		1.2×10^{-3}		2.5×10^{-2}	
	0°	90°	0°	90°	0°	90°	0°	90°
Cu – 9.5% Al – 4.0% Fe	—	—	390	530	450	610	270	280
Cu – 9.5% Al – 0.0% Fe	380	355	260	275	—	—	210	205

tions in different segments of grain boundaries could strongly influence accommodation of deformation in different directions as a result of the reduced effectiveness of grain boundaries containing precipitates as vacancy sinks. This would be consistent with the observation of a large portion of strain resulting from sliding of transverse grain boundaries and the sliding of groups of grains with the largest surface dimension perpendicular to the tensile axis in ternary Cu — 9.5% Al — 4.0% Fe cold rolled specimens cut with the tensile axis parallel to the rolling direction (12).

However, a detailled study using transmission electron microscopy will be necessary before the precise role of precipitation in anisotropic flow can be elucidated. The progressive removal of crystallographic texture and the lack of systematic changes in texture during superplastic deformation, together with the observation of maximum ductility perpendicular to the rolling direction in aluminium bronzes is in contrast to the behaviour in Zn—Al alloys. All results obtained on aluminium bronzes are consistent with grain boundary sliding being the dominant deformation mode during superplastic flow, little overall contribution being made by dislocation slip (other than in an accommodational role). The fewer slip systems available in C. P. H. metals may be the explanation of texture transitions in zinc alloys.

It is worth-while noting that, in the present system, superplasticity is observed to occur in microstructures with an abnormally coarse grain-size, up to ~30 μm average diameter. The basis for this unusual behaviour is unknown at present.

Conclusions

1. Binary and ternary aluminium bronzes exhibit maximum ductility at temperatures corresponding to equivolume proportions of the phases, the temperatures decreasing with increasing Al content in the range 8.5%–10% Al.

2. In alloys containing less than 2.0% Fe, considerable grain growth accompanies deformation and maximum elongations are ~300%. The optimum strain-rate is low $\sim 1.2 \times 10^{-4}$ sec^{-1}.

3. 4.0% Fe additions ensure a fine recrystallised structure and inhibit grain growth when the temperature is raised above that corresponding to equivolume phase proportions. The optimum strain-rate is an order of magnitude higher $1.2 - 1.7 \times 10^{-3}$ sec^{-1}, and maximum elongation in the range 600 – 1000%.

4. No anisotropy of ductility was observed in the binary alloys. Marked anisotropy of superplastic flow is observed in ternary alloys containing 4.0% Fe additions as a result of the distribution of δ phase (81.5% Fe, 13.5% Al, 5.0% Cu) precipitates.

5. Superplasticity is observed in Cu—Al alloys up to a grain-size of at least 30 μm average diameter.

Acknowledgements

This work was supported by the Defence Research Board (contract 9535/53) and National Research Council of Canada. The alloys were prepared by Olin Corporation, New Haven (USA) Valuable discussions with *J. Crane*, *G. L. Dunlop* and *E. Shapiro* are gratefully acknowledged.

Summary

Conditions for Newtonian viscous flow are met when the deformation stress level is low and the power series expansion for the rate of shear deformation as a function of stress may be terminated at first order. The constant of proportionality in this expression, the fluidity, is the reciprocal of the dynamic viscosity. Conventionally, the magnitude of this parameter has been used to distinguish fluids and solids. For liquids of low viscosity (~ 0.1 poise) flow is Newtonian viscous even at the highest practical shear rate ($\sim 10^5$ sec^{-1}). Hot glasses are deformed to high neck free elongations at a viscosity of $\sim 10^7$ poise. In crystalline solids a single term constitutive equation may still be proposed, relating stress to deformation rate via an exponent m, where $0.4 \leq m \leq 0.9$. Here either solid state diffusion directly or grain boundary sliding accommodated by diffusion must be relied upon to produce conditions for Newtonian viscosity ($m = 1$). Expressions for viscosity, similarly defined as the ratio of stress to corresponding rate of deformation, may be deduced on the basis of a transgranular or intergranular diffusion path. Such calculations yield a viscosity value of $\sim 10^5$ poise. Prerequisite conditions for superplastic flow to approach the Newtonian viscous limit, resulting in large neck free elongations, are that the material grain-size be small (1–10 μm), the material be deformed at intermediate strain-rates and the deformation temperature be in excess of half the absolute melting temperature. In the aluminium bronzes employed in the present study, conditions for maximum superplastic flow occur in the two phase field above the eutectoïd transformation temperature (that is 700–900 °C). Here, unlike in most systems, constitution and temperature are related variables. For a specific alloy constitution, the material is characterised by an optimum temperature for superplastic flow which decreases with increasing Al content in the composition range 8.5–12% Al. This behaviour is discussed in relation to the distribution of phases.

References

1) *Johnson, R. H.*, Metall. Rev. **15**, 145 (1970).

2) *Davies, G. J., J. W. Edington, C. P. Cutler* and *K. A. Padmanabhan*, J. Mater Sci. **5**, 109 (1970).

3) *Nicholson, R.B.*, Plasticity and Superplasticity, Institute of Metallurgists, 1969.
4) *Ball, A.* and *M.M. Hutchinson*, Met. Sci. J. **3**, 1 (1969).
5) *Hart, E.W.*, Acta Met. **15**, 1545 (1967).
6) *Sherby, O.D.* and *P.M. Burke*, Prog. Mat. Sci. **13**, 325 (1968).
7) *Ahlquist, C.N.* and *R.A. Menezes*, Mat. Sci. Eng. **7**, 223 (1971).
8) *Ashby, M.F.*, Scripta Met. **3**, 837 (1969).
9) *Anwar V. Karim* and *W.A. Backofen*, Met. Trans. **3**, 709 (1972).
10) *Crane, T., G.L. Dunlop, E. Shapiro* and *D.M.R. Taplin*, to be published (Met. Trans 1973).
11) *Dunlop, G.L.* and *D.M.R. Taplin*, J. Mater. Sci. **7**, 84 (1972).
12) *Dunlop, G.L.* and *D.M.R. Taplin*, J. Mater. Sci. **7**, 316 (1972).
13) *Dunlop, G.L.* and *D.M.R. Taplin*, J. Aust. Inst. Met. **16**, 195 (1971).
14) *Taplin, D. M. R.* and *S. Sagat*, Mat. Sci. Eng. **9**, 53 (1972).
15) *Backofen, W.A., I.R. Turner* and *D.H. Avery*, Trans. Amer. Soc. Metals **10**, 908 (1964).
16) *Watts, B. M.* and *M. J. Stowell*, J. Mater. Sci. **6**, 228 (1971).
17) *Bright, M.W.A.* and *D.M.R. Taplin*, ASM/AIME Conference on Copper, Cleveland 1972 (Preprint).
18) *Packer, C.M., R.H. Johnson* and *O.D. Sherby*, Trans Aime **242**, 2498 (1968).
19) *Naziri, H.* and *R. Pearce*, J. Inst. Metals **98**, 71 (1970).
20) *Padmanabhan, K.A.* and *G.J. Davies*, Rheol. Acta **13**, (1974).

Authors' addresses:

Dr. *M.W.A. Bright*, Secteur Matériaux,
Centre de Recherches Industrielle du Québec
555 Boulevard Henri IV
Ste-Foy, Québec 10, Québec (Canada)

Dr. *D.M.R. Taplin*
Dept. of Mechanical Engineering
University of Waterloo
Waterloo, Ontario (Canada)

Rheol. Acta 13, 670–674 (1974)

Research Center, Hercules Incorporated Wilmington, Delaware 19899 (USA)

Some rheological properties of sodium carboxymethylcellulose solutions and gels*)

J H. Elliot and A.J. Ganz

With 7 figures and 1 table

Introduction

Sodium carboxymethylcellulose, often called Cellulose Gum or CMC, is a widely used component of food systems. It may act as a suspending agent, thickener, protective colloid, humectant, and for the control of the crystallization of some other component. CMC is classified by the Food and Drug Administration under "substances that are generally recognized as safe" (Gras) by Title 21, Section 121.101 of the Code of Federal Regulations (USA). A summary of permissable concentrations in a number of food systems, together with a description of the properties of the various types of Hercules® Cellulose Gums (CMC) is available (1). CMC is a polyelectrolyte and may react with proteins in the food to form soluble or insoluble complexes. The properties of these complexes are currently being studied (2).

CMC is prepared by the reaction of alkali cellulose with sodium chloroacetate (3). The structure of a typical repeating unit is shown in fig. 1. Important parameters in characterizing CMC are the average degree of polymerization (DP) or average number of anhydroglucose units per molecule and the average degree of substitution (DS), the average number of carboxymethyl groups per anhydroglucose unit. The idealized structure shown in fig. 1 has a DS of 1.0. Other less apparent structural parameters must also be considered. In the manufacture of CMC, the uniformity of substitution may be controlled by selection of reaction conditions. The uniformity of substitution has a profound effect on the rheological properties of CMC solutions. Closely related to uniformity and degree of substitution is the state of aggregation or supermolecular structure of CMC, which is reflected in the rheological behavior of its solutions (4).

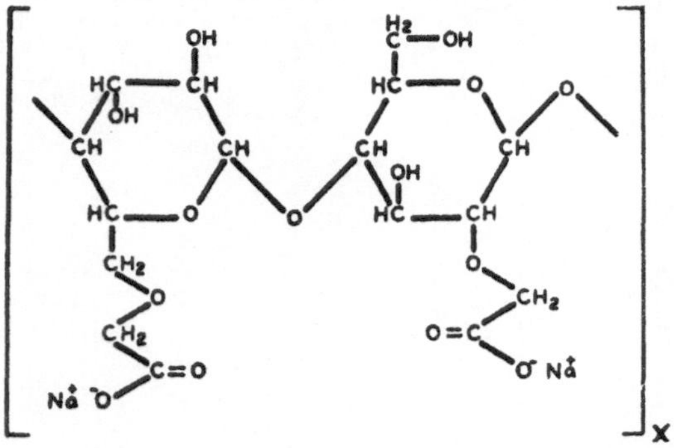

Fig. 1. Sodium carboxymethylcellulose – typical repeating unit of degree of substitution (DS) of 1.0

The rheological properties of CMC solutions and gels, in water and in the presence of other components of food systems, will be discussed in terms of the parameters DP, DS, uniformity of substitution and supermolecular structure. These rheological properties can then serve as a guide for the selection of CMC type and method of solution to achieve the desired characteristics in the final system.

Experimental

The CMC samples used in this work were either commercial or experimental materials prepared byHercules, Incorporated. The descriptive designations of commercial samples, together with solution viscosities in water, as a function of concentration are given in Reference 1. Commercial grades cover a DS range from 0.4 to 1.2 and three solution viscosity ranges, high (*H*), medium (*M*) and low (*L*). Approximate DPs for CMCs of 0.7 DS are: *H* type 3200, *M* type 1100 and *L* type 400 (1). The first number in the designation of a commercial material is ten times the DS. Thus a 7 *H* type has a DS of 0.7 and a DP of about 3200. The letter *S* in the designation signifies special solution characteristics (smooth, pseudoplastic rather than thixotropic solutions) and *F* means food grade. *S* type Cellulose Gums are prepared under conditions designed to yield a more uniform distribution of carboxymethyl substituents along the cellulose chain.

Aqueous solutions were prepared by adding the polymer to water with gentle stirring and then tumbling the solution end over end until solution was complete. In certain cases these solutions were then subjected to high power shearing in a Waring Blendor. With low DS or nonuniformly substituted CMC samples, this treatment greatly changes the rheological properties of the solution and may convert the solution into a rigid gel (7, 8).

Results and discussion

Brief review of earlier work

The rheological characterization of CMC solutions and gels has been carried out in our laboratories over a number of years. It was early

*) Hercules Research Center Contribution No. 1600

recognized that samples having the same nominal chemical composition and solution viscosity could show markedly different rheological properties. These were studied in a *Couette* type rheometer with a maximum shear rate of 4400 sec^{-1} (5, 8) using the hysteresis loop method proposed by *Green* (6). This work showed that when the CMC was prepared under conditions which would give uniform substitution, its solutions showed pseudoplastic behavior. If these conditions were not followed, thixotropic solutions resulted. It was further found that when thixotropic solutions were subjected to high power input shearing, gels which would support their own weight resulted (7, 8).

These results led to the conclusion that thixotropy in CMC solutions arises because of the presence of a very small quantity of unsubstituted crystalline residues in the CMC. These would be present as fringe micelles which could form crosslinking centers which would entrap a relatively large amount of molecularly dispersed CMC by electrostatic, hydrogen bonding or *Van der Waals* forces, and thus enable a three-dimensional structure to be set up. This conclusion was supported by experiments in which a polyvalent cation was added to a smooth, pseudoplastic CMC solution in order to form crosslinks. At low concentrations of the cation, the pseudoplastic solution became thixotropic; at higher concentrations, a rigid gel having a significant yield stress was formed. The formation of a gel when a thixotropic CMC solution without crosslinking cations is subjected to high power input stirring is believed to arise because of the dispersion and disaggregation of the fringe micelles arising from the crystalline residues, thus providing more potential crosslinking points, and not by disruption of the individual cellulose crystalline residues (8). The proposed mechanism is shown in fig. 2. One would expect more crystalline residues to be present in low DS than in high DS CMC, and it has been found that the lower DS types show greater thixotropy unless special measures are taken to obtain uniform substitution. These conclusions have been confirmed and extended by a study of the control of aggregation of CMC by choice of solvent and/or electrolyte (4).

Current Research

The *Weissenberg* Rheogoniometer is a very powerful tool for rheological research, in that it enables measurements to be made over a wide range of shear rates in steady shear and frequencies in imposed sinusoidal shear. There is a fundamental difference between steady shear and dynamic measurements. In steady shear the sample is subjected to a total strain which is limited only by the shear rate chosen and the duration of the measurement. This total strain may be quite large and can lead to a very different rheological state than that of the sample at rest. For example, polymer molecules may become oriented and any three-dimensional structure, present at rest, may be broken down. On the other hand, dynamic measurements are carried out at relatively low total strains. If these are sufficiently low, little structure breakdown occurs and the measured properties approach those of the sample at rest. The use of both methods permits a far more complete characterization than either alone.

The strain amplitude in dynamic measurements must be considered. When this is small, generally under 0.1, linear viscoelastic behavior is observed. The theory of linear viscoelasticity is highly developed and has been extensively used in polymer research (9). In real situations, however, the imposed strains are seldom low enough to be in the linear viscoelastic region. A recent investigation of materials, including CMC gels, which were judged organoleptically to have the property of unctuousness, was made using the *Weissenberg* Rheogoniometer in both steady shear and oscillation (10). In the dynamic measurements, the shear amplitude was sufficiently great to be outside the linear viscoelastic region. In steady shear,

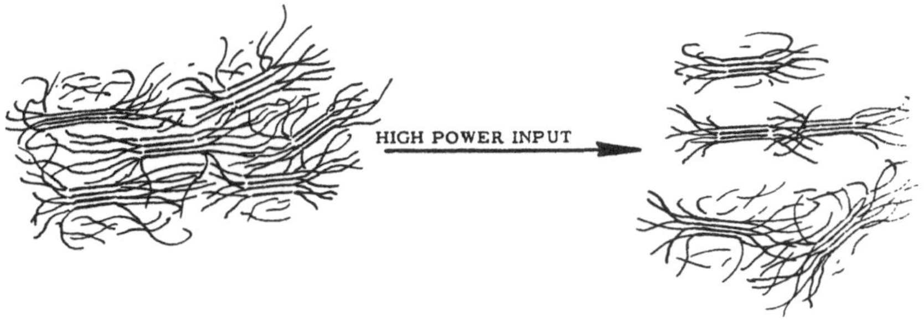

Fig. 2. Schematic diagram of the disaggregation of fringe micelles by shear

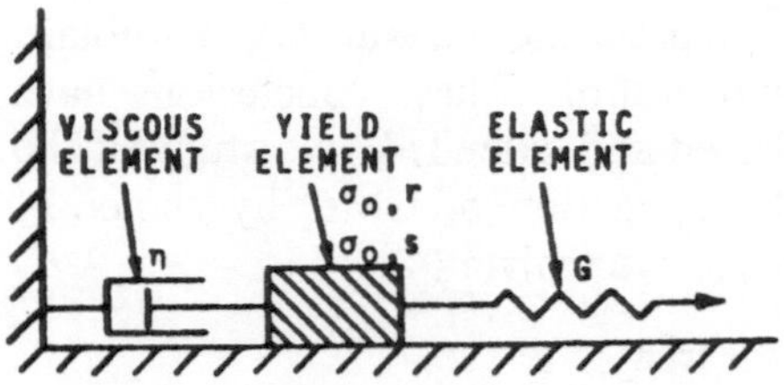

Fig. 3. Modified *Bingham* body

these unctuous materials showed a peaked stress overshoot similar to that shown in curve *C* of fig. 4. This indicates the presence of a yield stress and structure breakdown. The area under this peak gives the excess work per unit volume for structure breakdown (12). Under imposed sinusoidal strain, the resulting stress curve is not sinusoidal and may approach a square wave as shown in curve *D* of fig. 4.

It was found that both the steady shear and dynamic measurements could be qualitatively described in terms of the modified *Bingham* Body shown in fig. 3. This model consists of an elastic element of modulus *G*, a yield element having a rest yield stress of $\sigma_{o,r}$ and a moving yield stress $\sigma_{o,s}(\sigma_{o,s} \leq \sigma_{o,r})$ and a model viscosity η_m connected in series. The parameters G, $\sigma_{o,r}$, $\sigma_{o,s}$ and η_m are functions of frequency or shear rate. The response of this model to both steady and sinusoidal shear has been calculated in terms of the model parameters (11). Where appropriate, the experimental results will be analyzed in terms of this model, which has been found to be useful for a wide variety of systems which exhibit a yield stress.

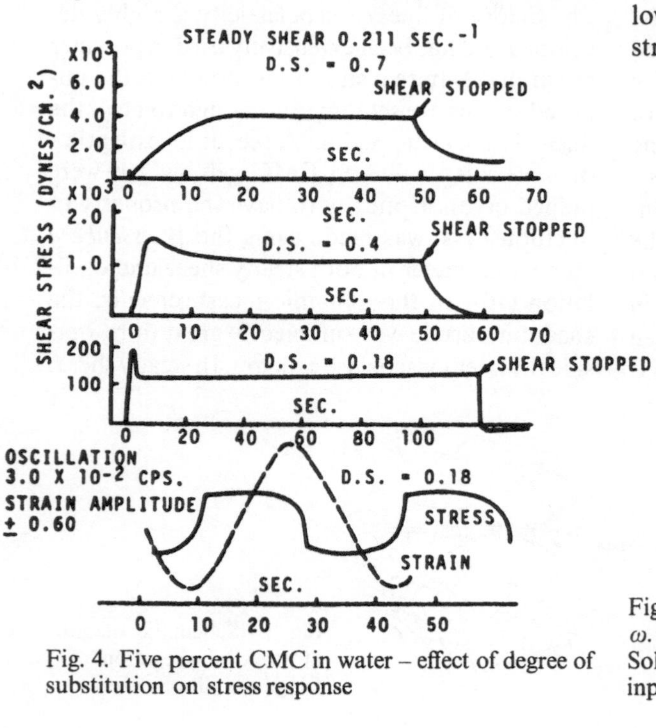

Fig. 4. Five percent CMC in water – effect of degree of substitution on stress response

Fig. 4 shows the stress response of five percent CMC solutions as a function of DS. These solutions were prepared under high shear conditions. Curve *A* for a smooth (*S*) type of DS 0.7 is typical of a viscoelastic system. As the DS is lowered to 0.4 and 0.18, the stress overshoot indicating a yield stress and structure breakdown becomes apparent. Under sinusoidal strain the sample of DS 0.18 shows a stress response approaching a square wave. These responses have been analyzed in terms of the parameters of the modified *Bingham* Body (11).

The usefulness of this analysis, particularly in the case of imposed sinusoidal strain, is illustrated by the following experiment. A four percent solution of CMC-4MF was prepared under low shear conditions. This solution flowed readily, but some visible structure was apparent. A portion of this solution was subjected to high power input shearing in a Waring Blendor. This converted the solution into a gel which would support its own weight. Dynamic measurements were made on both the solution and the gel, with the results shown in figs. 5, 6 and 7.

High power shearing increased the model viscosity by almost a factor of ten, over the frequency range studied. At the same time it caused a marked increase in modulus, particularly at the lower frequencies. Fig. 7 shows that the yield stress of the solution was quite low and that both

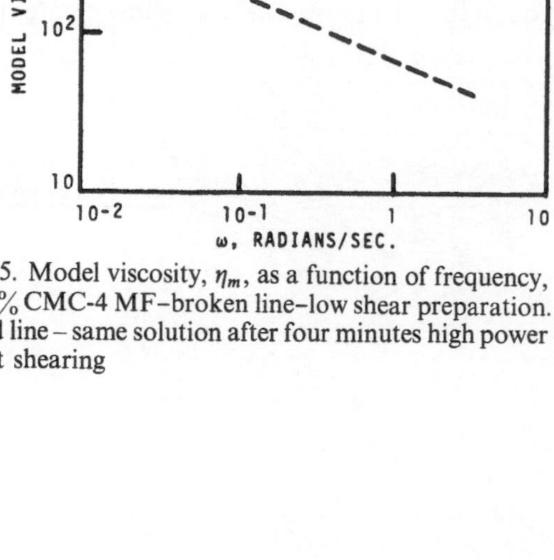

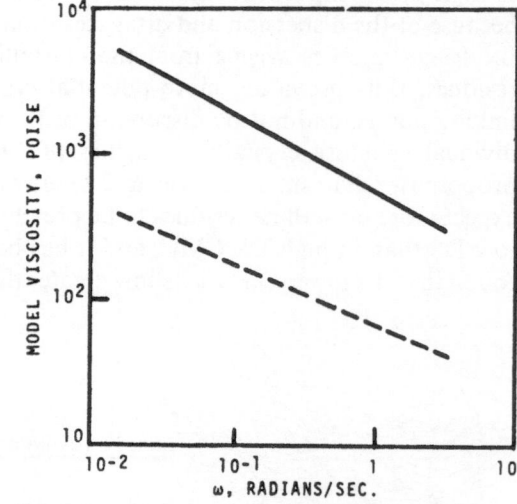

Fig. 5. Model viscosity, η_m, as a function of frequency, ω. 4% CMC-4 MF–broken line–low shear preparation. Solid line – same solution after four minutes high power input shearing

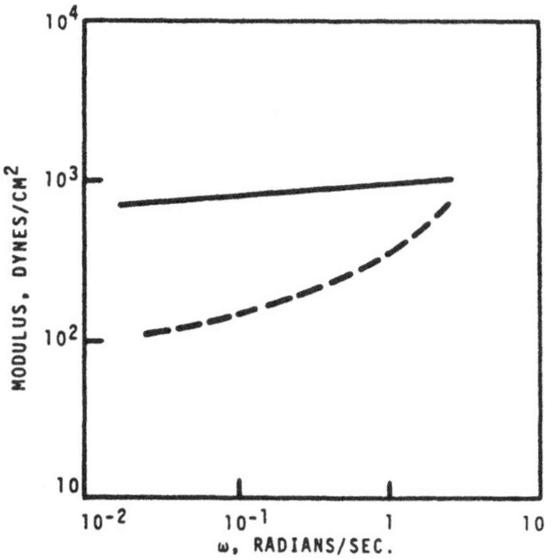

Fig. 6. Modulus, G, as a function of frequency, ω. 4% CMC-4 MF–broken line–low shear preparation. Solid line – same solution after four minutes high power input shearing

the rest, $\sigma_{o,r}$, and moving, $\sigma_{o,s}$, yield stresses were the same. After high power shearing, however, there is a large increase in $\sigma_{o,r}$ and the presence of $\sigma_{o,s}$ which is less than $\sigma_{o,r}$ is found. These changes in rheological parameters are those that would be expected when a solution is converted to a gel; analysis of the dynamic results in terms of the modified *Bingham* Body model permits us to characterize these changes quantitatively.

It was mentioned earlier that CMC solutions can form gels when certain polyvalent cations are added. A CMC-7H solution formed a gel when Al^{3+} in a citrate buffer was added. Dynamic measurements at moderate strain amplitude on this gel resulted in a sinusoidal stress curve that was in phase with the strain. From these curves, elastic moduli were calculated, which increased slowly with frequency. A program to characterize CMC-metal ion gels rheologically is currently under way.

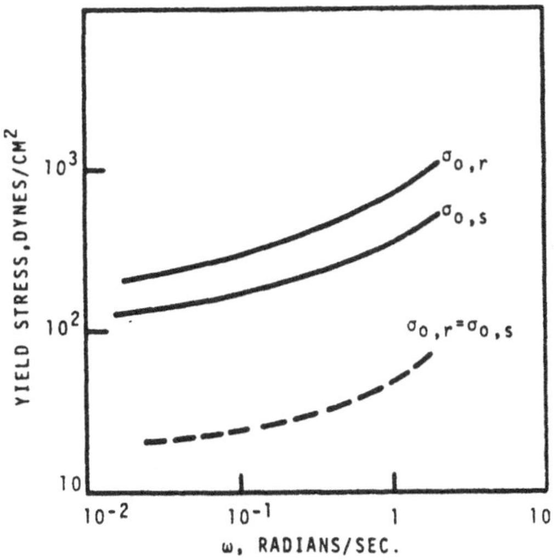

Fig. 7. Rest yield stress, $\sigma_{o,r}$, as a function of frequency, ω. 4% CMC-4 MF. Broken line–low shear preparation ($\sigma_{o,r} = \sigma_{o,s}$). Solid line – same solution after four minutes high power input shearing

The effect of polymer parameters on the rheological properties of CMC solutions may now be summarized. These properties range from smooth pseudoplastic solutions, showing conventional viscoelastic behavior, to those having a wide range of thixotropy, unctuous gels showing a definite yield stress which may be analyzed in terms of a modified *Bingham* Body and, finally, rigid elastic gels.

The effect of DP is primarily to define the level of the viscosity-concentration curve; the higher the DP, the higher the viscosity at a given concentration and shear rate (1). Lower DP materials dissolve more rapidly than high.

The DS level influences both the chemical and rheological properties of CMC. Gums with higher DS (0.7–0.9) have a greater ionic functionality which must be considered when interaction with polyvalent cations or other polyelectrolytes (e.g., proteins) may occur. In addition, the higher DS materials generally dissolve more rapidly and their solutions show less thixotropy than lower DS types. The method of solution preparation (low or high power input shearing) has less effect on rheological properties of solutions of CMC of high DS than on those of lower DS types.

The uniformity of substitution and the presence of crystalline residues, which give rise to fringe micelles which act as crosslinking points giving a three-dimensional structure, are interrelated parameters. Higher DS gums are inherently more uniformly substituted than low. This is reflected in minimal thixotropic behavior of their solutions. On the other hand, low DS gums are less uniformly substituted; their solutions exhibit marked thixotropy. Their rheological behavior is more sensitive to the solution preparation method. These effects may be minimized, however, if the CMC is prepared under conditions designed to give uniform substitution along the cellulose backbone.

It is apparent that the selection of the proper type of CMC and method of solution enables the

Table 1. Some uses of CMC in foods

Food product	Property conferred by CMC	CMC type			Reference
		DP*)	DS	S type**)	
Baked goods	Water retention, control of batter viscosity	H, M	0.7	No	13, 14, 16
Doughnuts	Grease holdout	H, M	0,7	No	13, 14, 16
Starch systems, whipped toppings	Inhibition of syneresis	H, M	0.7	No	13, 14
Syrups, beverages, juices	Thickener, viscosifier	H, M	0.7	No	13, 14, 16
Ice cream	Texture, body control of sugar and ice crystallization	H, M	0.7	No	13, 14, 16
Pet foods (semimoist)	Binder	H, M	0.7	No	13, 14, 16
Confections	Control of sugar crystallization	L	0.7	No	13, 14, 16
Low calorie syrups	Very smooth texture	H	0.7	Yes	13, 14, 16
Low calorie spreads	Unctuousness	H, M	0.4	No	10, 15

*) H, M, L, high, medium or low viscosity **) S, uniform substitution

user to obtain the specific rheological behavior desired for this component in the final product.

The rheological properties of CMC solutions are not the only considerations in its use in food systems. Interaction with other components and the organoleptic properties of the finished product must be considered. Nevertheless, rheological properties of CMC aqueous solutions and gels can serve as a useful guide for the use of Cellulose Gum in food products. Table 1 illustrates the types of CMC that are used in some food products. The connection between the type used in a given system and the solution rheological properties discussed above will be apparent.

Summary

Sodium carboxymethylcellulose (CMC) is a valuable and widely used component of food systems. The rheological properties of a series of CMC solutions and gels have been determined under both steady and imposed sinusoidal shear conditions, over a wide range of shear rates and frequencies, using the *Weissenberg* Rheogoniometer. Depending upon the CMC type, degree of substitution, and method of solution preparation, the rheological behavior ranged from viscolastic solutions to unctuous gels. If certain polyvalent cations, e.g., Al^{3+}, are present, rigid elastic gels may be formed.

These results are interpreted in terms of fringe micelles, arising from crystalline residues in the CMC. Gels which show unctuous behavior may be described in terms of a modified *Bingham* Body model. Where appropriate, the experimental results are quantitatively analyzed in terms of the parameters of this model. The application of these findings to the use of CMC in food systems is discussed.

Acknowledgement

The authors are indebted to Mr. *James J. Kirwin*, who carried out the rheological measurements.

References

1) Hercules Incorporated "Chemical and Physical Properties of Hercules Cellulose Gum (CMC), Wilmington, Delaware, 1971.

2) *Ganz, A. J.*, Some Observations on the Interaction of Sodium Carboxymethylcellulose with Proteins. Presented at IFT Annual Meeting, 1972.

3) *Ott, Emil* and *H. M. Spurlin*, Cellulose and Cellulose Derivatives, 2nd Ed., p. 937ff. (Interscience, New York 1954).

4) *Francis, P. S.*, J. Applied Polymer Sci. **5**, 261 (1961).

5) *Smith, J. W.* and *P. D. Applegate*, Paper Trade J. **126**, 60 (June 3, 1948).

6) *Green, H.*, Industrial Rheology and Rheological Structures (Wiley, New York 1949).

7) *Ott, Emil* and *J. H. Elliott*, Makromol. Chem. **18/19**, 352 (1956).

8) *De Butts, E. H., J. A. Hudy*, and *J. H. Elliot*, I. E. Chem. **49**, 94 (1957).

9) *Ferry, J. D.*, Viscoelastic Properties of Polymers, 2nd Ed. (Wiley, New York 1970).

10) *Elliot, J. H.* and *A. J. Ganz*, J. Texture Studies **2**, 220 (1971).

11) *Elliot, J. H.* and *C. E. Green*, J. Texture Studies **3**, 194 (1972).

12) *Trapeznikov, A. A.*, Proceedings of the Fifth International Congress on Rheology. *S. Onogi* (ed.), Vol. 4, p. 257 (Univ. of Tokyo Press, Tokyo 1970).

13) *Ganz, A. J.*, Food Product Develop. **3** (**6**), 65 (1969).

14) *Batdorf, J. B.*, in: *R. L. Whistler* (Ed.), Industrial Gums (Academic Press, New York 1959).

15) *Nijhoff, G. J. J.*, U.S. Patent 3,418,133.

16) *Glicksman, M.*, Gum Technology in the Food Industry (Academic Press, New York 1969).

Author's address:

Dr. *J. H. Elliot*
Research Center
Hercules Incorp.
Wilmington, Delaware 19899 (USA)

Rheol. Acta 13, 675–680 (1974)

From the Shoe & Allied Trades Research Association, Kettering, Northants. (England).

Mechanical properties of segmented polyurethane elastomers

R. E. Whittaker

With 8 figures

Introduction

The traditional material used for centuries in the manufacture of footwear has been natural leather. In recent years leather has also been increasingly used in the manufacture of gloves, clothing and upholstery. The success of leather in these applications is due to the good comfort and hygiene properties offered to the wearer. It is also very tough, reasonably flexible, has good durability and is able to breathe or transpire water vapour and air. Despite these good properties, however, it does suffer from a number of disadvantages such as non uniformity and the material cannot be obtained in rolls. Because of these disadvantages and the increasing shortage and cost of leather, polymers have been increasingly used as replacement materials during the last 20 years (1, 2). In the case of soling materials, only 6% of the footwear produced at present in the U. K. has leather soles. The rise in the use of polymeric based upper materials has not been as fast as soling materials but in 1970, 29% of the dress shoes produced in the U. K. had synthetic uppers.

The first group of materials to be used as substitutes for leather uppers in shoes were coated fabrics (1, 3). The most widely used materials are polyvinylchloride (PVC) coated fabrics which can be obtained as solid or cellular PVC coatings on a woven, non-woven or knitted base. PVC coated fabrics are satisfactory for use in certain types of footwear because the general appearance of leather is simulated and they also possess some similar properties.

The manufacture of a completely synthetic material which had moisture absorption and permeability properties similar to the natural leather was started in the late 1930's by E. I. Du Pont de Nemours Inc., USA. They initially used the term 'poromeric' to describe the material and defined it as 'A microporous, permeable, coriaceous sheet material comprising polyurethane reinforced with polyester.

Commercial production of the Du Pont product, under the trade name 'Corfam' was not started until 1964, but as early as 1942, a US patent (4) was taken out to describe a shoe upper material composed of flexible fabric coated with a synthetic linear polyamide. A number of new poromeric upper materials have come on to the market in recent years (1, 5, 6). In view of the development of these materials, the Shoe and Allied Trades Research Association has had to widen the first definition given by Du Pont and now defines a poromeric material used in footwear as 'a man-made shoe upper material which is generally similar in nature and appearance to leather and in particular has a comparable permeability to water vapour'.

Most poromerics, incorporate or are entirely based on microporous polyurethane foams (5–10). These cellular polyurethanes have to withstand fairly arduous wear conditions when used in footwear. They have to be tough enough to replace a completely fibrous material and have good abrasion and cut growth resistance. This paper discusses the mechanical properties and physical structure of cellular polyurethanes used in poromerics and compares these properties with published results from both filled and unfilled vulcanised rubbers in order to provide an explanation for their good physical properties.

Modulus and tensile properties

The tensile stress-strain curve for a homogeneous polyurethane poromeric foam is shown in fig. 1. The tensile stress for the foam is based on the cross-sectional area of the rubber, including the holes. The tensile stress-strain curve for the solid polyurethane which was obtained by dissolving the foam in a suitable solvent and recasting the sheet is also shown in fig. 1 as well as the stress-strain curve for a typical natural rubber vulcanizate from earlier investigations (11, 12).

It would appear that the initial modulus of the polyurethane foam is higher although the actual tensile strength is lower than a solid NR vulcanizate. The modulus of the solid polyurethane is extremely high when compared with the corresponding foam and its tensile strength is considerably in excess of that found in the natural rubber vulcanizate.

Some years ago, *Gent* and *Thomas* (13) derived a theory for expressing the mechanical properties of a latex foam rubber in terms of the corresponding solid material. The model they used consisted of a cubical array of struts of unstrained length l_0 and cross-sectional area D^2 and is shown in fig. 2. The author (14) has previously shown that this model can be applied successfully to the

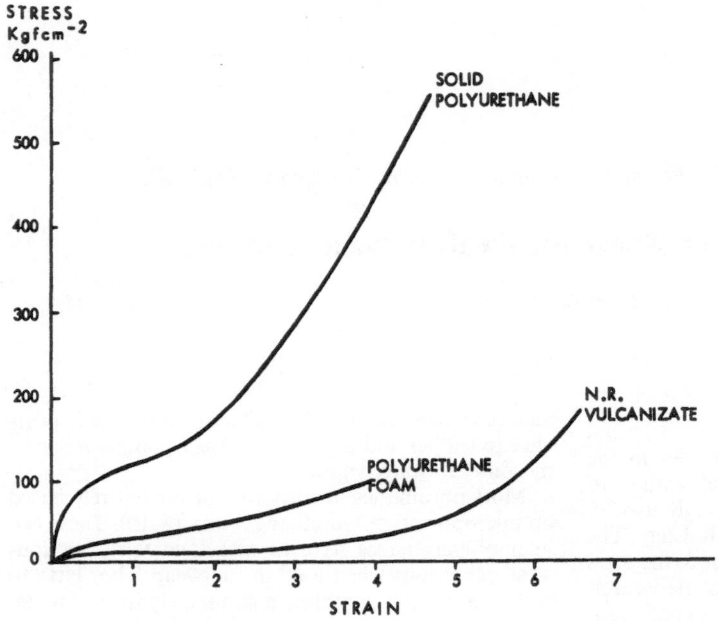

Fig. 1. Typical stress-strain curves for solid and foam polyurethane poromeric and solid NR vulcanizate

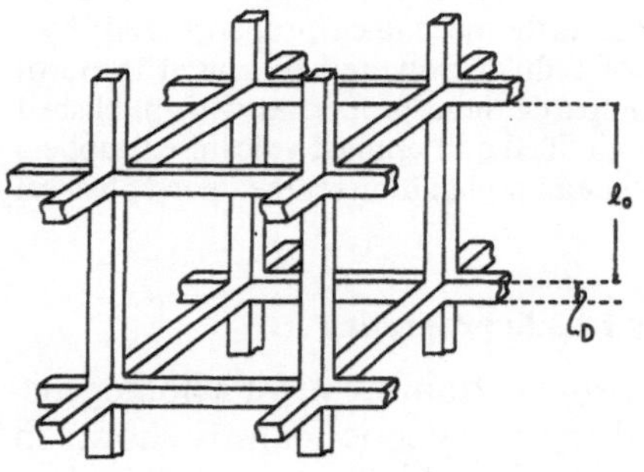

Fig. 2. Model of Foamed Material. After *Gent* and *Thomas* (13)

mechanical properties of polyurethane poromeric foams.

The initial linear part of the stress-strain curves for the polyurethanes shown in fig. 1 permits a value of *Young*s Modulus to be obtained. A theoretical value for *Young*s Modulus of the foam Y_F can be obtained by considering the extension of the model shown in fig. 2. If a small strain is applied parallel to one set of threads, Y_F can be obtained from the product of three factors.

1. *Young*s Modulus of the solid material
2. Stain magnification factor. Due to the undeformable regions D^3, the strain parallel to one set of threads in the model is associated with a larger extension of the treads themselves.
3. Factor representing true load-bearing area of the threads compared with the overall area which include the spaces.

By combining these three factors, the following equation can be derived.

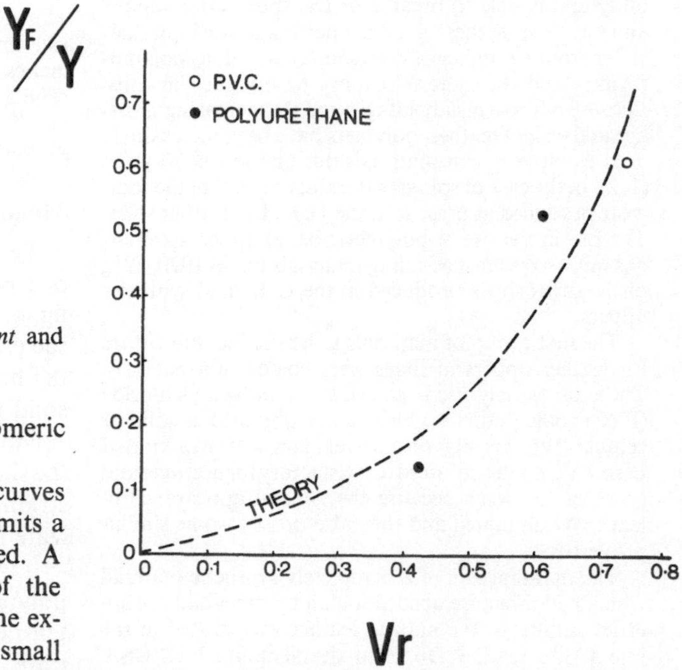

Fig. 3. Variation Y_F/Y with volume rubber fraction Vr for two poromeric polyurethanes and microporous PVC material. Dotted line is that predicted by *Gent/Thomas* model

$$Y_F = Y \cdot \frac{\beta^2}{2(1+\beta)} \quad [1]$$

where $\beta = D/l_0$ and can be related to the volume rubber fraction Vr of the foam which can be determined from measured densities of the material. The relationship between Y_F/Y and Vr is shown in fig. 3 for two poromeric polyurethane foams and a microporous PVC material. Also plotted on fig. 3 is the line predicted by the *Gent/Thomas* cubical model given in eq. [1]. It can be seen that this relationship is a reasonable approximation to the experimental results. A recent paper (14) has shown that this model can be applied to determine the tensile, tear and compression properties of polyurethane poromeric foam materials.

The high strength of the polyurethane foam is therefore due to the high strength of the solid polyurethane material. The ratio between the mechanical properties of the foam and solid materials being the same as between the solid and foam vulcanised rubbers of the same density.

Failure properties

The variation of tensile stress at break (σ_B) and strain at break (ε_B) for the foam and solid polyurethanes with temperature from 0 to 180 °C at a strain rate of 2.2 per min is shown in fig. 4. The stress at break for both the foam and the solid is fairly high and slowly decreases with temperature until about 160 °C when it drops suddenly. The strain at break rises to a maximum at about 100 °C but again drops quite markedly above 160 °C to a value of 1.40 and 0.42 for the foam and solid respectively at 180 °C.

Previous studies (11, 12, 15) have shown that a more useful measure of the strength of a polymer is the toughness or energy input to break as it

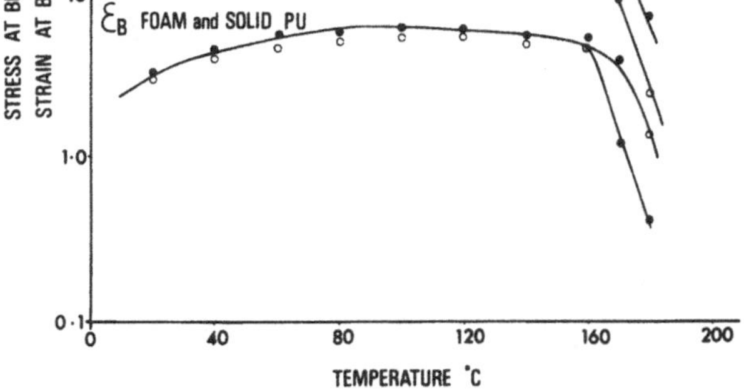

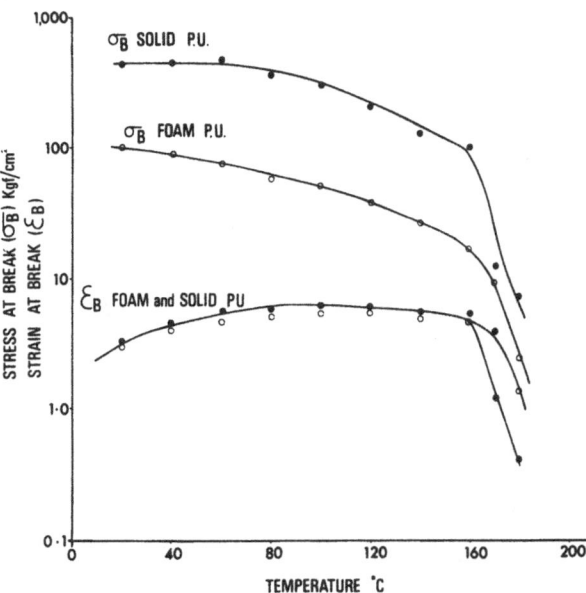

Fig. 4. Variation of tensile stress at break and strain at break with temperature for polyurethane foam and solid materials

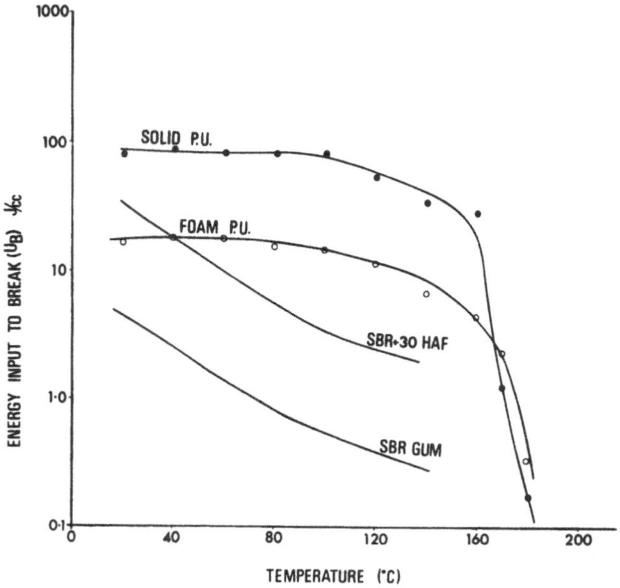

Fig. 5. Variation of energy input to break with temperature for polyurethane foam and solid materials and comparison with filled and unfilled SBR vulcanizates

combines both the contributions due to stress and strain at break. The variation of energy input to break with temperature for both the foam and solid polyurethanes at a strain rate of 2.2 per min is shown in fig. 5. The energy input to break values for both the foam and solid polyurethanes remain fairly high and parallel up to approximately 160 °C when the failure values drop quite markedly. In order to indicate the high strength and temperature stability of poromeric polyurethathanes, results for normal styrene butadiene rubber (SBR) with 0 and 30 phr HAF carbon black from earlier investigations (11, 15) are also shown for comparison in fig. 5. It is clearly seen that even the polyurethane foam has higher strength properties over the majority of the temperature range considered than the solid reinforced rubber vulcanizate.

Earlier work (11, 12, 15) on both amorphous and strain crystallising vulcanised rubbers has shown that the energy input to break (U_B) is related to the strain at break (ε_B) up to the maximum extensibility of the network (ε_B max) by a relationship of the following form

$$U_B \frac{294}{T} = A(\varepsilon_B)^2. \qquad [2]$$

The variation of energy input to break with strain at break for the foam polyurethane is shown in fig. 6a and for the solid polyurethane in fig. 6b.

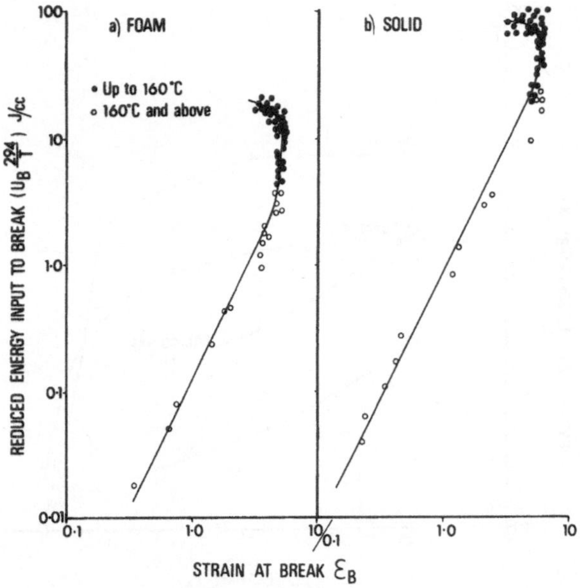

Fig. 6. Variation of energy input to break with strain at break for a) foam polyurethane and b) solid polyurethane

The results at temperatures above 160 °C are shown by open circles and it is seen that the square law between the two parameters is only obeyed at these high temperatures. As found from the results shown in figs. 4 and 5, the strain at break and energy at break values are fairly high except at very high temperatures. The maximum extensibility (ε_B max) of the foam polyurethane for example remains at a fairly constant value from 80 °C to 160 °C as shown in fig. 4. This is in contrast (11) to a normal vulcanised rubber such as SBR which has its maximum extensibility at about 0 °C.

Relaxation behaviour

By obtaining stress-strain data at various strain rates over a temperature range from 21–180 °C for both polyurethane foam and solid materials, *Whittaker* (16) has been able to determine a relaxation spectrum for poromeric polyurethanes. The method used to determine this spectrum was the one adopted by *Smith* (17, 18) to determine the relaxation properties of SBR and polyisobutylene from tensile stress-strain measurements. The method is based on shifting experimental data determined at the various temperatures along a log time axis. It was found however that the shift factors required for polyurethane were far greater than that predicted by the WLF equation (19).

The relaxation spectrum calculated in this manner is shown in fig. 7 and is compared with

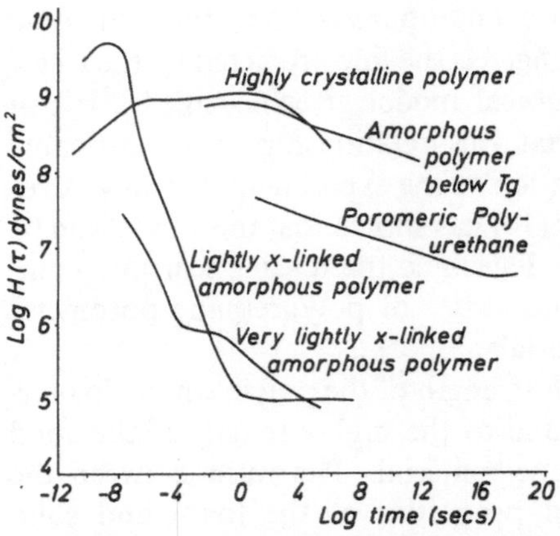

Fig. 7. Relaxation spectrum for poromeric polyurethanes compared with other typical spectra from *Ferry* (19)

other typical spectra for elastomer systems from *Ferry* (19). The most interesting feature of the relaxation spectrum for polyurethane compared to crosslinked amorphous rubbers is the very flat plateau which extends for over 18 decades of time. This particular shape of relaxation spectrum curve is similar to that of an amorphous polymer below its glass transition temperature or a highly crystalline polymer but in the case of polyurethane however, the material is flexible and above its major glass transition temperature which occurs at approximately — 30 °C.

Stress softening or *Mullins* effect in polyurethanes has also been investigated and reported in other papers (11, 20–22).

Physical structure of polyurethane elastomers

In order to explain some of the anomalous features of polyurethane compared with vulcanised rubbers, it is necessary to consider the physical structure of these materials. A number of investigators (23–27) in recent years have shown that polyurethane elastomers consist of alternating hard and soft segments as shown in fig. 8. The soft segments are formed from the linear polyether or polyester chain segments which are about 100 – 200 Å long and at service temperatures are sufficiently high above their glass transition temperature, or in the case of crystallizable soft segments above their melting temperature to give the material an extensibility of several hundred percent.

The hard segments are approximately 25 Å in length and originate from a diisocyanate and a chain extender or cross-linking agent (e.g. diol or diamine) and hence contain urethane or urea groups. These hard segments are at service tem-

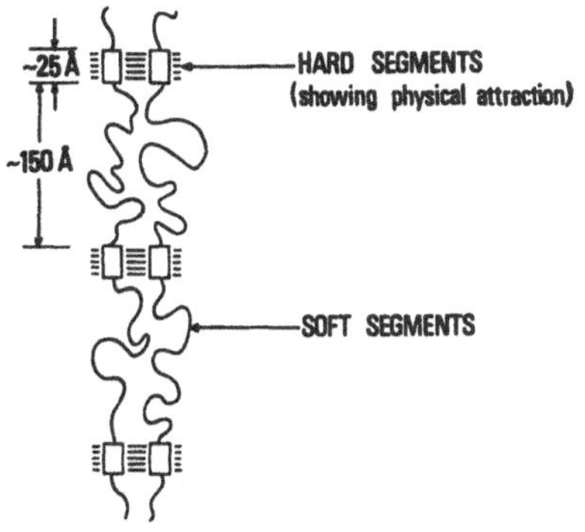

Fig. 8. Diagrammatic representation (23–27) of typical structure of a polyurethane elastomer

peratures below their second order glass transition temperature and cause physical cross-linking by hydrogen bonds and other intermolecular forces between the segments. This prevents the material from flowing so that elasticity is maintained.

From X-ray and differential thermal analysis (DTA) measurements (27–31) it has been shown that two major glass transition temperatures occur in polyurethane elastomers. The first occurs around — 20 °C and is due to the onset of rotation in the flexible polyether or polyester chain whereas the second transition at about 160 °C is due to the dissociation of the interurethane hydrogen bonding. Two further transitions have also been noted (30), one at 80 °C is ascribed to hydrogen bonding between the urethane secondary amine group and the ester carbonyl or ether oxygen of the prepolymer, and the other at 230 °C to the melting of the hard urethane segments. To obtain high strength, a proper molecular arrangement of the hard and soft segments must occur in addition to the hydrogen bonds and other forces between the hard segments. Chemically cross-linking (32) the polyurethane appears to disrupt the molecular arrangement and over a certain critical level, significantly decreases the strength of the material.

Conclusions

This paper has discussed the mechanical properties of polyurethanes used in poromerics and has compared these properties with those found in vulcanised rubbers. In some cases, such as the use of the cubical model to express the properties of the foam material in terms of the solid and the variation between energy input to break and strain at break, similar relationships to those found in vulcanised rubbers have been derived. In a number of cases however, there are some distinct differences between poromeric polyurethanes and vulcanised rubbers. The points discussed in this paper can be summarised by,

1. The very high tensile strength (approx. 550 kgf/cm^2) of the solid polyurethane which permits a foam material of high strength to be produced.
2. The temperature stability of polyurethanes, their high strength is maintained up to temperatures of approx. 160–170 °C when it drops quite markedly.
3. Polyurethane materials do not obey the WLF equation.
4. Polyurethanes have a broad relaxation spectrum.
5. Polyurethanes show large hysteresis and stress softening which is not recoverable until temperatures of 170 °C are reached.
6. Polyurethanes have good cut growth properties but this has not been considered in this paper.

The most interesting feature of the physical structure of polyurethane elastomers is their segmented structure. The hard urethane segments presumably act as filler particles within the polyether or polyester rubber matrix. It is well known that the introduction of a filler such as carbon black into an amorphous vulcanised rubber increases its hardness, strength, hysteresis and abrasion resistance. The large increase in these properties in the case of polyurethane must be due however to the very small size of the hard urethane segment of the chain. The normal carbon blacks used in rubbers (e.g. SRF, ISAF, HAF, etc.) are about 300 Å in diameter whereas in the case of polyurethane the hard segments are a factor of ten smaller than this. It is well known that the properties of filled vulcanised rubbers improve as the particle size of the filler is reduced as there is more particle surface area on which the rubber chains can adhere to. In the case of polyurethanes the very small size of the filler particle causes the very high strength and abrasion resistance.

Harwood et al. (11, 15, 33) have recently shown that adding an HAF carbon black filler to SBR, causes a broadening of the relaxation spectrum and that the filled rubber does not obey the WLF equation. The presence of the hard segment in polyurethane elastomers leads to additional characteristic response times as described quantitatively in the work of *Radok* and *Tai* (34).

This causes high hysteresis which is related to high strength as shown in the work of *Harwood* et al. (11, 15, 33).

The other interesting feature of polyurethane elastomers is that the filler particles (the hard segments) from a proper molecular arrangement. The filler particles are therefore well dispersed in the polyester or polyether rubber matrix, a secondary factor which contributes to their good mechanical properties.

Stress softening is not fully recovered (22) nor is there a distinct change in mechanical properties until a temperature of approximately 160–170 °C is reached. At this temperature, the hydrogen bonding which acts as a physical crosslink between the rubber chains dissociates. Cooling the material to lower temperatures reforms the hydrogen bonding. If this is done when the material is in the stretched state, the polyurethane can be considered to be thermoplastic and a high set can be induced in the material. This particular property has been used (9, 10) in the footwear industry where high set and good long term shape retention is required.

It is concluded therefore that the high strength, abrasion resistance and good cut growth properties of poromeric polyurethanes are due to the hard urethane segments in the polyether or polyester rubber matrix which act as a well dispersed minute filler particle in a rubber matrix thus producing a very effective "self reinforced" elastomer.

Acknowledgements

The work described in this paper is part of a thesis submitted to Loughborough University for the degree of Ph. D. The author is indebted to Dr. *A. R. Payne* (Director, SATRA), Professor *R. J. W. Reynolds* and Dr. *C. M. Blow* (Loughborough University) for their helpful advice and encouragement throughout the course of this work.

References

1) *Payne, A. R.* and *R. E. Whittaker*, J. Inst. Rubb. Ind. **4**, 107 (1970).

2) *Whittaker, R. E.*, Polymer Age **2**, 21 (1971).

3) *Gillibrand, J.*, J. BBSI **15**, 195 (1968).

4) *Austin, P. R.*, US Patent 2,302,167 (Nov. 17th, 1942).

5) *Hole, L. G.* and *R. E. Whittaker*, J. Mat. Sci. **6**, 1 (1971).

6) *Hole, L. G.*, Rubb. J. **152** *(4)*, 72 (1970).

7) *Payne, A. R.*, Paper presented to 4th Int. Synthetic Rubb. Conf. London (1969). Published in proceedings of Conference.

8) *Sitting, M.*, Synthetic Leather from Petroleum. Chem. Proc. Review No. 29 (1969).

9) *Payne, A. R., R. W. T. Skelham* and *R. E. Whittaker*, J. BBSI **17**, 200 (1970).

10) *Payne, A. R.* and *R. E. Whittaker*, Rubb. J. **152** (4), 89 (1970).

11) *Harwood, J. A. C., A. R. Payne* and *R. E. Whittaker*, Paper presented to IRI Conference "Advances in Polymer Blends and Reinforcement" Lougborough. Sept. 1969. Published in proceedings of Conference.

12) *Harwood, J. A. C., A. R. Payne* and *R. E. Whittaker*, J. Appl. Poly. Sci. **14**, 2183 (1970).

13) *Gent, A. N.* and *A. G. Thomas*, Rubb. Chem. Technol. **36**, 597 (1963).

14) *Whittaker, R. E.*, J. Appl. Poly. Sci. **15**, 1205 (1971).

15) *Harwood, J. A. C., A. R. Payne* and *R. E. Whittaker*, J. Macromol Sci. **B5**, 473 (1971).

16) *Whittaker, R. E.* (to be published).

17) *Smith, T. L.*, J. Poly. Sci. **20**, 89 (1956).

18) *Smith, T. L.*, Trans. Soc. Rheol. **6**, 61 (1962).

19) *Ferry, J. D.*, Viscoelastic Properties of Polymers, 2nd Edition (New York 1970).

20) *Whittaker, R. E.*, Paper presented to SATRA Intersat '71 Conference, Blackpool Appl. 1971. Published in proceedings of Conference.

21) *Whittaker, R. E.*, J. Coated Fibrous Mat. **2**, 3 (1972).

22) *Whittaker, R. E.*, Paper presented to IRI Rubbercorn conference, Brighton (1972). Published in proceedings conference.

23) *Oertel, H.*, Textil-Praxis **19**, 820 (1964).

24) *Oertel, H.*, Bayer. Farbenrev. **11**, 1 (1965).

25) *Rinke, H.*, Angew. Chem. **74**, 612 (1962).

26) *Bonart, R.*, Kolloid Z. u. Z. Polymere **211**, 14 (1966).

27) *Bonart, R.*, J. Macromol. Sci. – Phys. **B2**, 115 (1968).

28) *Shimanskii, V. M., S. I. Shkolnik* and *S. B. Kozakov*, Soviet Rubb. Technol. **26**, 20 (1967).

29) *Bonart, R., L. Morbitzer* and *G. Hentze*, J. Macromol. Sci. – Phys. **B 3**, 337 (1967).

30) *Clough, S. B.* and *N. S. Schneider*, J. Macromol. Sci. – Phys. **B2**, 553 (1968).

31) *Clough, S. B., N. S. Schneider* and *A. O. King*, J. Macromol. Sci. – Phys. **B2**, 641 (1968).

32) *Saunders, J. H.* and *K. C. Frisch*, Polyurethanes, Chemistry and Technology (New York 1962).

33) *Harwood, J. A. C., A. R. Payne* and *J. F. Smith*, Kaut. u. Gummi Kunst **22**, 548 (1969).

34) *Radok, J. R. M.* and *C. L. Tai*, J. Appl. Poly. Sci. **6**, 518 (1962).

Author's address:

Dr. *R. E. Whittaker*
Shoe & Allied Trades Research Ass.
Kettering, Northants. (England)

Rheol. Acta **13**, 681–688 (1974)

From the Chemical Engineering Department, Faculty of Engineering, Tohoku University, Sendai (Japan)

Flow properties of some suspending systems

K. Umeya and S. Tanifuji

With 11 figures and 4 tables

1. Introduction

Flow patterns of disperse systems were investigated over a wide shearing range both in laminar and in turbulent regions.

1.1. Laminar flow pattern of disperse systems

Generally speaking, there are six fundamental patterns describing flow properties of dispersing systems as shown schematically in fig. 1, which indicates;

a) linear relation between stress (log *S*) and shearing rate (log *D*) in logarithmic scale, having an inclination of 45° to the stress axis (called *Newton* behavior)

b) linear relation having an inclination larger than 45° (called pseudo-plastic behavior)

c) linear relation having an inclination smaller than 45° (called dilatant behavior)

d) non-linear relation having an intersect perpendicular to the stress axis and asymptotic to Newtonian relation (called *Bingham* behavior)

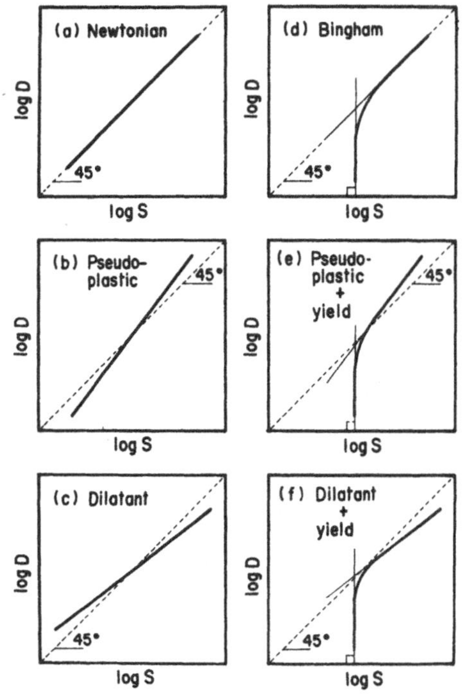

Fig. 1. Laminar flow patterns for a narrow shearing range

e) non-linear relation having an intersect perpendicular to the stress axis and asymptotic to pseudo-plastic relation

f) non-linear relation having an intersect perpendicular to the stress axis and asymptotic to dilatant relation.

However, after famous works reported by *Ostwald* (1), *Reiner* (2), *Edelmann* (3) and so on (4), it has been believed now that these six flow patterns could never exist independently but mutually (in other words, not in parallel but in series manner) when flow patterns of disperse systems were investigated over an extremely wide shearing range. These new types of flow patterns were called as "*Ostwald* behavior" by *Reiner* (2) and as "Extended *Ostwald* behavior" by one of the authers (5, 6).

Consequently, four types of flow patterns must be decided when measured over an extremely wide shearing range as indicated in fig. 2 schematically (Type I, III, IV

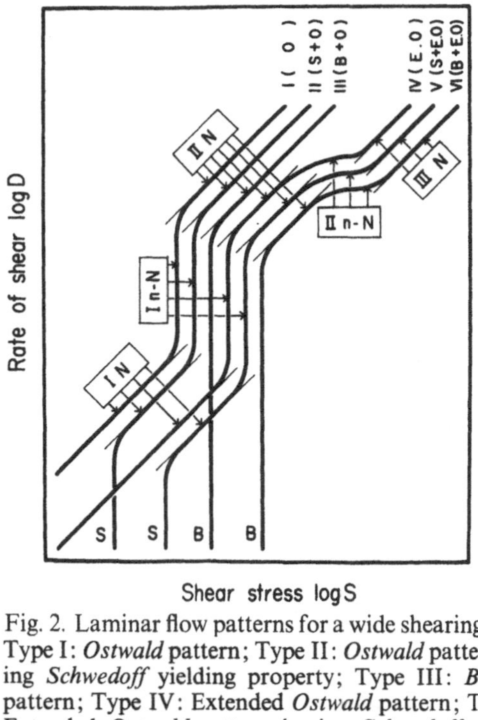

Fig. 2. Laminar flow patterns for a wide shearing range. Type I: *Ostwald* pattern; Type II: *Ostwald* pattern having *Schwedoff* yielding property; Type III: *Bingham* pattern; Type IV: Extended *Ostwald* pattern; Type V: Extended *Ostwald* pattern having *Schwedoff* yielding property; Type VI: Extended *Ostwald* pattern having *Bingham* yielding property

and VI). Moreover, if so-called *Schwedoff* yielding properties which are detected in extremely low shearing range were given in considerations, two more flow patterns must be added as shown in fig. 2 (Type II and V), however, detailed considerations treating them were eliminated here.

In this paper, these relations were ascertained for some dilatant suspensions such as TiO_2-water and ZnO-water systems.

1.2. Turbulent flow pattern of disperse systems

On the other hand, turbulent flow patterns of disperse systems were also studied by many previous investigators 7 ~ 24), however, almost all the investigations were unfortunately restricted in quite narrow shearing ranges compared with the cases stated in laminar regions (1.1).

In this paper, these turbulent flow patterns were also investigated over a wider shearing range, which could not be performed entirely successfully, but could be made possible at least to investigate these turbulent properties comparing with laminar flow patterns observed over a wide shearing range.

2. Measurements

2.1. Measuring apparatus for laminar region

Three types of viscometers were employed to determine the laminar flow behavior of dispersing systems over a wide shearing range.

In a lower shearing range, a coaxial cylindrical viscometer was employed which was newly designed especially for the measurements in the extremely low shearing rates reaching to 10^{-4} sec^{-1}, having 1.6, 1.8 and 2.0 cm bob diameters, 2.2 cm cup diameter and 5.0 cm immersion length.

A capillary type viscometer similar to that constructed by *Maron*, *Krieger* and *Sisko* was used in a medium shearing range with capillaries listed in table 1.

Table 1. Capillaries used for *Maron-Krieger-Sisko* Capillary Viscometer

Capillary number	Length (cm)	Radius (cm)	L/R (–)
1	15.2	2.60×10^{-2}	585
2	16.8	3.18×10^{-2}	528
3	16.1	3.47×10^{-2}	464
4	16.3	4.89×10^{-2}	333

Table 2. Capillaries used for PRL Single-pass Capillary Viscometer

Capillary number	Length (cm)	Radius (cm)	L/R (–)
L_1	10.01	2.444×10^{-2}	933
L_2	14.95	2.452×10^{-2}	810
L_3	20.00	2.420×10^{-2}	610
L_4	15.05	1.613×10^{-2}	410

In a higher shearing range, a high pressure capillary viscometer known as the PRL Single-pass Capillary Viscometer was used with capillaries shown in table 2.

2.2. Measuring apparatus for turbulent region

The enlarged type of PRL Single-pass Capillary Viscometer was constructed to measure the flow properties in turbulent region using several testing pipes listed in table 3.

Table 3. Testing pipes used in turbulent region

Pipe number	Length (cm)	Radius (cm)	L/R (–)
T_1	199.4	9.592×10^{-2}	2079
T_2	200.0	20.81×10^{-2}	961
T_3	199.8	25.95×10^{-2}	770
T_4	199.9	29.06×10^{-2}	688
T_5	199.7	40.04×10^{-2}	499

2.3. Preparation of sample dispersions

Zinc oxide and titanium dioxide powders used in present studies were chemical pure reagents.

Sodium lignosulfonate (commercially named Asperse) was added 1 part to 100 parts (by weight) of ZnO powder to stabilize the systems. On the other hand, cane sugar was utilized to stabilize TiO_2-water systems.

All prepared sample dispersions were utilized after aging period of about two mounths.

3. Results

3.1. Results in laminar region

The following results were obtained using three sorts of viscometers mentioned in previous section (2.1).

3.1.2. Flow patterns measured over a wide shearing range

As stated in previous section, it must be concluded that six typical types of flow patterns can exist in flow behavior of disperse systems measured over a wide shearing range as illustrated in fig. 2. According to this classification, the results observed over a wide shearing range using ZnO-water (the curve of 76.2 wt% in fig. 3) and TiO_2-water systems (fig. 4) seem to obey the Type VI pattern (B + E, O) in fig. 2 which has *Bingham* pattern in a lower shearing range and has dilatant or extended *Ostwald* pattern in a higher shearing range in a series manner.

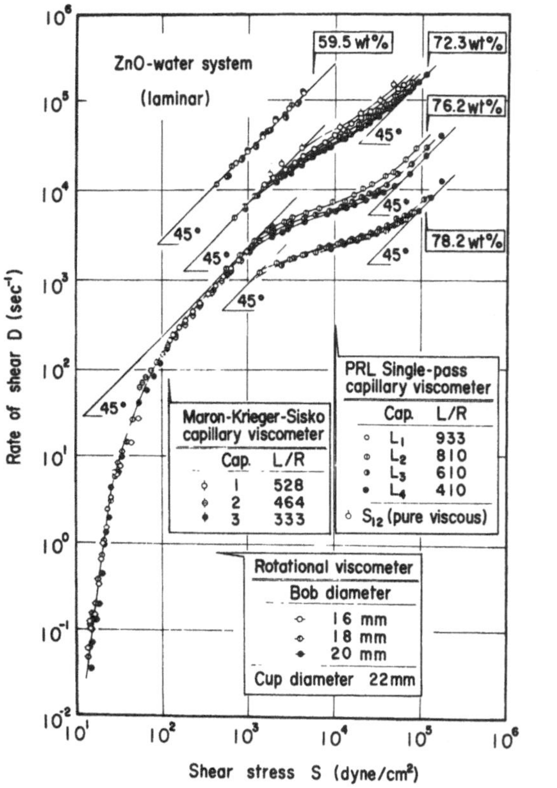

Fig. 3. Laminar flow patterns of ZnO-water system, showing dependencies of concentration and L/R values of testing capillaries

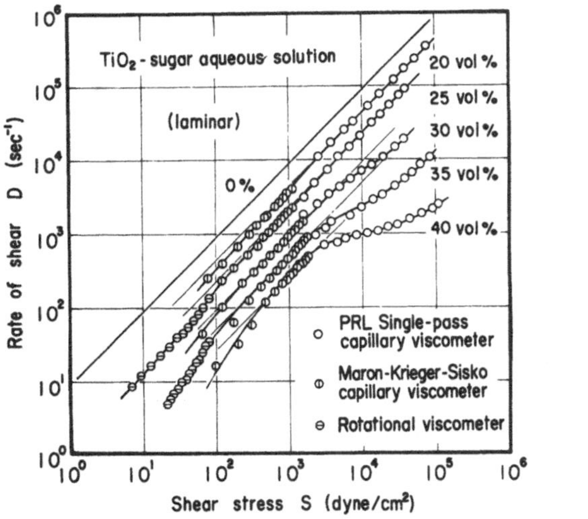

Fig. 4. Laminar flow patterns of TiO_2-sugar aqueous solution, showing dependency of concentration

3.1.2. Dependence of concentration of disperse system

As can be seen distinctly from figs. 3 and 4, the flow curves shift to a lower location with the concentration of disperse systems and the non-Newtonian range becomes wider and furthermore the non-Newtonian behavior itself becomes remarkable with the concentrations.

3.1.3. Dependence of L/R values of testing capillaries

As can be seen in fig. 3, the dependency of L/R values of testing capillaries are quite remarkable; that is, when the length of capillary (L) becomes shorter or radius (R) larger, the flow curves shift to lower locations in the figure, and this effect is more noticeable when the concentration of the system is thicker so far as the measurements are attempted.

3.2. Results in turbulent region

The results were shown in fig. 5 which was observed in turbulent region for ZnO-water system using the viscometer mentioned above (2.2).

In a dilute suspension of 59.5 wt%, it can be proved to have a Newtonian behavior in a laminar region as shown by a thick solid line having an inclination of 45° to the stress axis (indicating II N), and to have fully developed turbulent flow curves as shown by thin solid line having an inclination of ca. 30° to the stress axis which initiated to turbulence at the points of T_1', T_2' and T_3' locating in the II N region. Transition range

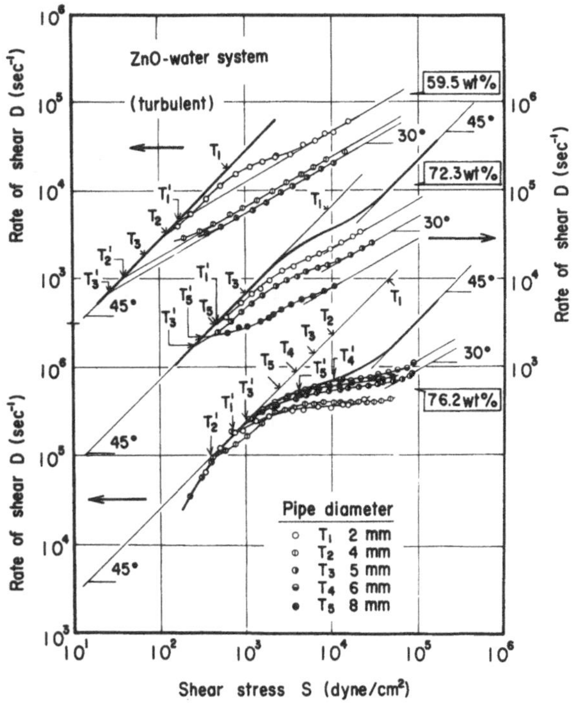

Fig. 5. Turbulent flow patterns of ZnO-water system, showing dependencies of concentration and diameters of testing pipes

may be found as a mountain shaped deviation from this thin solid line, but in succeeding discussions these transition regions are not emphasized. In a thicker suspension of 72.3 wt%, all the turbulent flow curves also initiated from the Newtonian range as shown by T_1', T_3' and T_5'. In the range corresponding to a dilatant region, turbulent flow curves have an inclination less than 30° to the stress axis and tend to coincide with the inclination of 30° which corresponds to the third Newtonian range. It must be emphasized here that, in turbulent region the flow curves have also the tendency to shift to lower locations in the figure with increasing the diameters of testing pipes as can be recognized by the two instances of dilute suspensions mentioned above. In a concentrated suspension of 76.2 wt%, this relation is quite overset indicating that;

1. the effects of dilatancy (partially containing transition) are quite remarkable,
2. when the diameter of testing pipe is larger, the turbulency is initiated from the dilatant range as shown by the points of T_4' and T_5' in fig. 5, while the diameter of testing pipe is smaller the turbulency is initiated from II N range as indicated by T_3', T_1' and T_2',
3. the order of initiation to turbulence with pipe diameter is quite opposite to the dilute suspensions, that is, the initiating points in testing pipes with larger diameter exist in higher shearing rates as indicated in fig. 5.

4. Discussion

4.1. Laminar behavior

4.1.1. Flow pattern measured over a wide shearing range

When flow curves were measured over a quite wide shearing range, Newtonian behavior (which is ascertained as II N in author's previous work [5, 6]) can be detected in the middle range of flow pattern and is succeeded by one non-Newtonian behavior (called I n–N which means pseudo-plastic nature) in a lower shearing range and the other non-Newtonian behavior (called II n–N which means dilatant nature) in a higher shearing range, as can be seen distinctly from the flow curve of 76.2 wt% in fig. 3 and from those of 25, 30, 35 and 40 vol% in fig. 4.

When measurements have been attempted in the extremely low shearing rate reaching to $3 \times 10^{-2}\,\mathrm{sec}^{-1}$ for the ZnO-water system, the obtained flow curve seems to intersect perpendicular to the stress axis which means *Bingham* yielding characteristics in logarithmic scale (see: the curve for 76.2 w% in fig. 3). If the observation was attempted in further lower shearing range, it might possess another Newtonian relation corresponding to I N pattern, which was ascertained in author's laboratory using clay-water system (32) but could not be confirmed with the systems in figs. 3 and 4. Anyhow, so-called dilatant systems such as TiO_2-water or ZnO-water can indicate the Type IV (E. O) or Type VI (B + E. O) pattern in fig. 2, when measurements were carried out over a wide shearing range.

4.1.2. Dependence of concentration of disperse systems

As mentioned in previous section (3.1.2), three distinguished dependencies of concentration of disperse systems can be observed, that is,

1. the whole flow curves shift to a lower location,
2. non-Newtonian range becomes wider,
3. non-Newtonian property itself becomes remarkable with increasing the concentration as schematically shown in fig. 6.

As pointed out by *Edelmann* using molecular dispersing system (3), it seems to be quite important characteristics in identifying the structual state of non-Newtonian behavior (pseudo-plastic of dilatant properties) whether three critical lines aa′(I n–N ∼ II N), bb′(II N ∼ II n–N) and cc′ (II n–N ∼ III N) in fig. 6 have a tendency to perpendicular or horizontal to the stress axis, however, further detailed statements cannot be attempted here, because observed flow curves in figs. 3 and 4 were too scanty to reveal this problem.

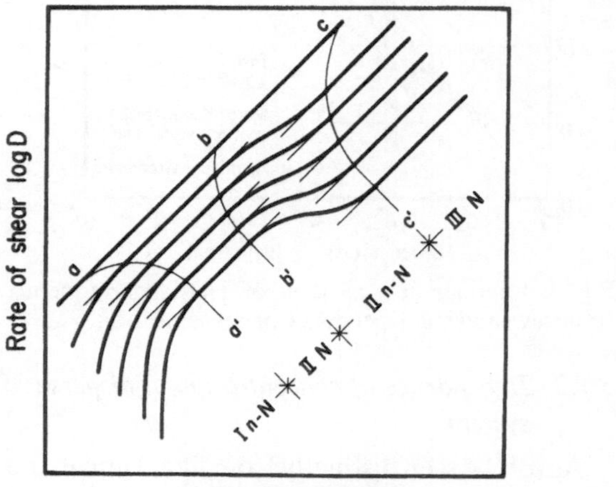

Fig. 6. Critical lines between Newtonian and non-Newtonian ranges

4.1.3. Dependence of L/R values of testing capillaries

The dependence of L/R values of testing capillaries is quite remarkable as shown in fig. 3, which means elastic contributions to viscous flow properties.

Intrinsic nature of normal stress is revealed in polymer dispersing systems, however, a few investigators have an interest in this nature for suspending systems. In this respect, authors have an opinion that dispersed powders are surrounded by adsorbed layers of dispersing medium as shown in fig. 7, which has (two-dimensional) viscoelastic property, while dispersing medium itself has viscous property. When this system is put in the streaming field, the deformation takes place in these viscoelastic adsorbed liquid layers which has always the tendency to be restored to the original state by the element of elastic parts. As these restored forces are added to the essential normal stress produced by the movement of dispersing particles, these suspending systems having adsorbed shell can indicate remarkable normal stresses essentially as large as in polymer dispersing systems.

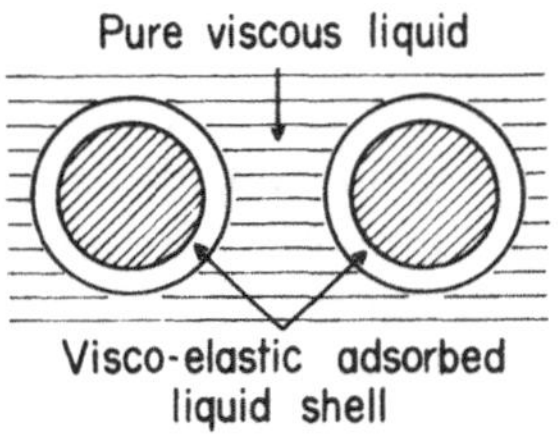

Fig. 7. Schematic illustration of adsorbed water shell

The curve indicated by ό marks for a concentration of 72.3 wt % in fig. 3 shows the pure viscous flow curve corrected for this elastic effect. This correction is not attempted for the other curves in fig. 3, because it seems to be convenient to consider the initiation to turbulence when the dependencies of capillaries are known even in laminar region.

4.2. Turbulent behavior

4.2.1. Turbulent flow curve in the dilatant range

In previous section (3.2), the initiating points to turbulence were indicated by T_1', T_2', T_3', T_4' and T_5' corresponding to the diameters of testing pipes. Here the points of T_1, T_2, T_3, T_4 and T_5 must be introduced to indicate the theoretical initiating points to turbulence. They are decided by assuming to have a pure viscous property shown by the viscosity in II N range (η_{IIN}). The value of 2300 was adopted here for the *Reynolds*' number from which the turbulence can initiate. The relation between these actual and theoretical initiating points to turbulence is shown in fig. 8.

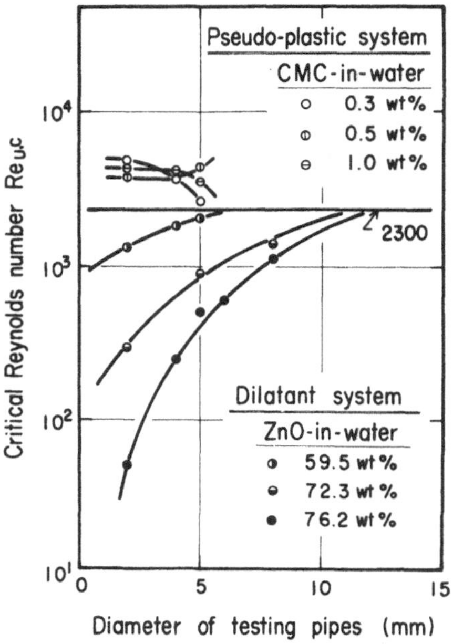

Fig. 8. Dependency of diameters of testing pipes on the critical *Reynolds*' number for dilatant and pseudo-plastic systems

In a dilatant system (ZnO-water), the turbulence initiates from extremely lower values of *Reynolds*' number, when the diameters of testing pipes become smaller and also when the concentrations of disperse systems increase.

On the other hand, as shown in fig. 9, in pseudoplastic system (CMC-water) the turbulence initiates from lower shearing range as the diameters

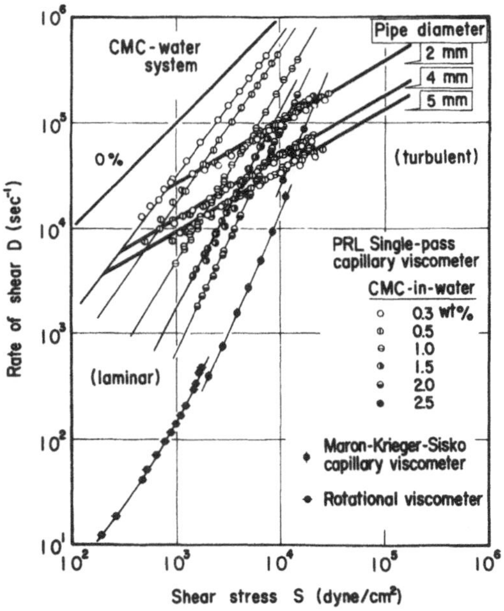

Fig. 9. Laminar and turbulent flow patterns of CMC-water system

of testing pipes become larger, and flow curves in turbulent region are almost independent of pipe radius.

As can be seen from fig. 8, in the case of pseudo-plastic system the turbulence initiates from higher *Reynolds*' number than the value of 2300 and dependencies of testing pipes and concentrations are smaller than those in dilatant system.

This opposite trend to turbulence observed in the dilatant range for ZnO-water dispersion of 76.2 wt % cannot be detected in other flow ranges than that in dilatant, that is, this effect can be said to be peculiar to dilatant properties.

4.2.2. *f vs Re relation*

In the field of chemical engineering, the relation between friction factor (f) and *Reynolds*' number (Re) is often employed in order to describe the flow characteristics in the system of pipe line flow in laminar or especially in turbulent region. Many notable analyses are already carried out ($7 \sim 16$) in this field until now, and some noted instances were shown in table 4 in the summarized fashion.

Table 4. f-Re treatments

	f	Re	remark
Metzner	$\dfrac{S_w}{(1/2)\rho V^2}$	$\dfrac{D V \rho}{K'(8V/D)^{n-1}}$	(8)
Tomita	$\dfrac{S_w}{(1/2)\rho V^2} \cdot \dfrac{1}{F(n)}$	$\dfrac{D V \rho}{K'(8V/D)^{n-1}} \cdot F(n)$	(16)*
Weltmann	$\dfrac{S_w}{(1/2)\rho V^2}$	$\dfrac{D V \rho}{K'(8V/D)^{n-1}} \cdot P(n)$	(14)**
Authors	$\dfrac{S_w}{(1/2)\rho V^2}$	$\dfrac{D V \rho}{S_w/(-du/dr)_w}$	
Itho	$\dfrac{S_w}{(1/2)\rho V^2(1-a)L/D}$	$\dfrac{D V \rho (4(a\,\alpha)(1-a)}{\mu}$	(15)***

*) $F(n) = \dfrac{3}{4} \cdot \dfrac{3n+1}{2n+1}$

**) $P(n) = \dfrac{3n+1}{4n}$

***) $\alpha = \dfrac{a^2 - 4a + 3}{12a}$; $a = \dfrac{\tau_y}{\tau_\omega}$;

τ: radial shearing stress

All these treatments were carried out essentially assuming following equation;

$$S - S_0 = \eta D^n$$

having constant non-Newtonian parameter with no yielding properties (obey pseudo-plastic or dilatant relation, $n = \text{const}$, $S_0 = 0$) or with constant yielding properties (obey *Bingham* relation, $n = 1$, $S_0 = \text{const}$). Therefore, quite few papers referred to the systems having varying n and also varying S_0 (n = variable, S_o = variable), however, to treat the wide flow pattern mentioned in previous section, the analytical treatment having variable n and also variable S_0 should be demanded.

In this paper these cases are also calculated using varying non-Newtonian parameter as well as varying yield value by the author's analytical method shown in table 4, from which one of the results indicating the dependency of pipe diameters on f-Re relation is shown in fig. 10. Furthermore, in comparing the observations in figs. 10 and 5, it can be ascertained that the shape of f-Re curves is under the direct influence of the flow patterns in laminar part. From this conclusion obtained in dilatant region and also the other results obtained in Newtonian and pseudo-plastic region published in author's previous work (32), f-Re relation can be manifested fully in the schematic figure shown in fig. 11, which indicates several f-Re relations assuming that the turbulence can initiate from III N (curve e–F 5), II n–N (curve d–E 4–F 4), II N (curve c–D 3–E 3–F 3), I n–N (curve b–C 2–D 2–E 2–F 2) and I N (curve a–B1–C1–D1–E1–F1) corresponding to the laminar flow patterns. Here, transition regions are all eliminated in the figure for easy understanding. It can also be recognized that f–Re relations

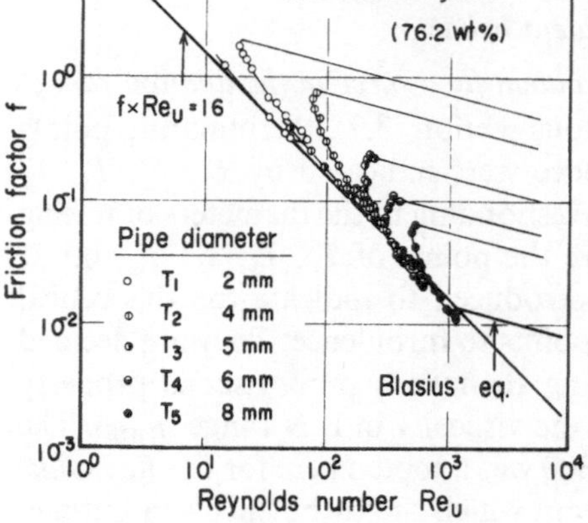

Fig. 10. f-Re relation of ZnO-water system for several testing pipes

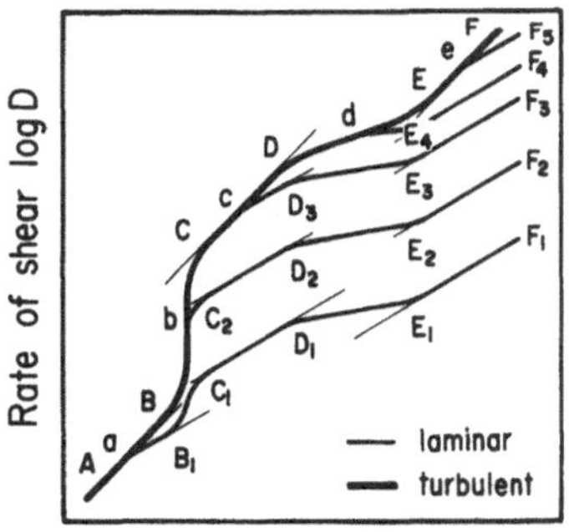

Fig. 11a. Relation between laminar and turbulent flow patterns observed for extended *Ostwald's* flow pattern

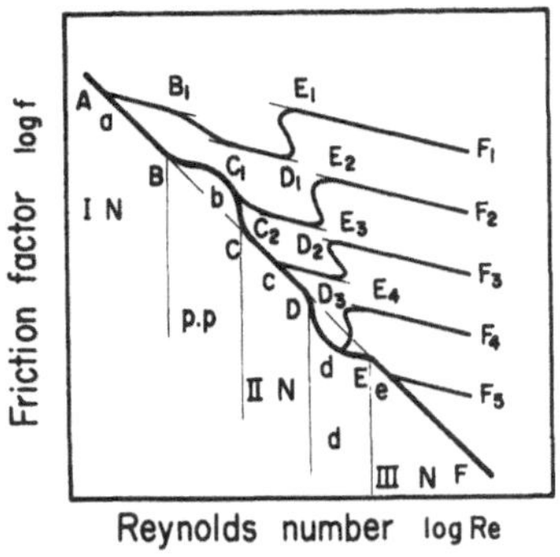

Fig. 11b. *f*-Re relation for extended *Ostwald* flow pattern both in laminar and in turbulent regions

shown in fig. 10 indicate the case of curve d–E4–F4 in fig. 11b.

5. Conclusions

Flow patterns of disperse systems were investigated over a wide shearing range both in laminar and in turbulent regions using TiO_2-water and ZnO-water suspensions. As the results of these investigations, the following points were revealed; that is,

in laminar region,
a) the flow curves of these systems can be described as Type IV and Type VI patterns in fig. 2,
b) with increasing concentrations of these suspensions, the whole flow curves shift to a lower location in the figure having wider and more remarkable non-Newtonian characteristics,
c) dependence of L/R values of testing capillaries can be shown in fig. 3 which indicates considerable elastic contribution to viscous characteristics of these suspensions,

in turbulent region,
d) differences between actual and theoretical points of initiation to turbulence were observed as shown in fig. 8.
e) when non-Newtonian parameter (n) is indicated as variable magnitude as shown in fig. 11a, *f*–Re relation cannot be described as having one linear relation both in laminar and in turbulent regions respectively. In this paper, renewed method may be recommended in which *f*–Re relation is affected by varying n value both in laminar and in turbulent regions.

Acknowledgements

The authors wish to thank to the Asahi Glass Foundation for the contribution to Industrial Technology which allowed this work to be carried out, and also thank to Messrs. *T. Isoda, T. Kanno, K. Bunrin, E. Toyooka* and *M. Noda*, whose contributions are reported in this paper partially.

Summary

Flow patterns of disperse systems such as TiO_2-water and ZnO-water suspensions were investigated over a wide shearing range both in laminar and in turbulent regions. The results indicated in fig. 11a were decided as the laminar and the turbulent flow patterns respectively. *f*-Re relations were also investigated for the systems which possessed varying non-Newtonian parameter n. *f*-Re diagram was revealed to be affected by varying n values both in laminar and in turbulent regions similarly as shown in fig. 11b.

References

1) *Ostwald, Wo.*, Kolloid Z. **36**, 99 (1925).
2) *Reiner, M.*, Deformation, Strain and Flow, An Elementary Introduction to Rheology (London 1949).
3) *Edelmann, K.*, Proc. 2nd Int. Cong. Rheology, ed. by *V. G. W. Harrison*, p. 107 (London 1954).
4) *Philippoff, W.* and *F. H. Gaskin*, Trans. Soc. Rheol. **1**, 109 (1957); J. Appl. Phys. **28**, 1118 (1957); Trans. Soc. Rheol. **2**, 262 (1958).
5) *Umeya, K., T. Isoda,* and *K. Sawamura*, Powder Tech. **3**, 280 (1969).
6) *Umeya, K.*, Proc. 5th Int. Cong. Rheology, ed. by *S. Onogi*, p. 295 (Tokyo 1970).
7) *Metzner, A. B.* and *J. C. Read*, Amer. Inst. Chem. Eng. J. **1**, 434 (1955).
8) *Metzner, A. B.* and *D. W. Dodge*, Amer. Inst. Chem. Eng. J. **5**, 189 (1959).
9) *Metzner, A. B.* and *F. A. Seyer*, Canad. J. Chem. Eng. **45**, 121 (1967).
10) *Metzner, A. B., J. L. White,* and *H. M. Denn*, Chem. Eng. Prog. **62**, 81 (1966).
11) *Metzner, A. B.* and *M. Whitlock*, Trans. Soc. Rheol. **2**, 239 (1958).
12) *Metzner, A. B.* and *R. F. Ginn*, Trans. Soc. Rheol. **13**, 429 (1969).
13) *Metzner, A. B.* and *J. L. White*, J. Appl. Poly. Sci. **7**, 1867 (1963).
14) *Weltmann, R. N.*, Ind. Eng. Chem. **48**, 386 (1956); **49**, 1429 (1957).
15) *Itoh, S.*, New Chemical Engineering Lecture, V-1 (Tokyo 1957).

16) *Timita, Y.*, Trans. J. S. M. E. **23**, 525 (1957); **24**, 288 (1958); **25**, 938 (1959); **32**, 242 (1966); **32**, 250 (1966); **35**, 1277 (1969).
17) *Kline, S. J.* et al., 3rd Fluid Mech. **30**, 741 (1967).
18) *Paterson, R. W.* and *F. H. Abernathy*, J. Fluid Mech. **43**, 689 (1970).
19) *Morgan, R. J.*, Trans. Soc. Rheol. **12**, 511 (1968).
20) *Klein, J.* and *H. Fußer*, Rheol. Acta **2**, 1119 (1968).
21) *Olbreachl, J.*, Rheol. Acta **3**, 249 (1964).
22) *Philippoff, W.* et al., Trans. Soc. Rheol. **14**, 393 (1970); **14**, 409 (1970).
23) *Lodge, A. S.*, Elastic Liquids (New York 1964).
24) *Lewusand, A. A.*, The Structure of Turbulent Shear Flow (London 1959).
25) *Hinze, J. O.*, Turbulence (London 1959).
26) *Porendlich, H.* and *H. L. Roder*, Trans. Soc. Rheol. **34**, 308 (1938).
27) *Bogue, D. C.*, *J. O. Doughty*, I. & E. C. Fund. **5**, 243 (1966); **6**, 388 (1967).
28) *Bogue, D. C.*, I. & E. C. Fund. **5**, 253 (1966).
29) *Astarita, G.*, I. & E. C. Fund. **6**, 257 (1967).
30) *Wilkinson, W. L.*, Non-Newtonian Fluids (New York 1960).
31) *Fischer, E. K.*, Colloidal Dispersions (New York 1959).
32) *Umeya, K.*, J. Mat. Sci. Soc. Japan **7**, 134 (1970).
33) *Umeya, K.*, unpublished.
34) *Umeya, K.*, to be published.

Author's address:

Dr. *Kaoru Umeya*
Chem. Engineering Dept.
Faculty of Engineering
Tohoku University
Sendai 980 (Japan)

Rheol. Acta 13, 689–695 (1974)

Commissariat à l'Energie Atomique, Centre d'Etude de Vaujours Service d'Etudes des Matières sensibles, Sevran (France)

Fatigue des matériaux aggrégataires suivie par cinéfractographie

G. Lucas, J. Reynaud et J. Roucou

Avec 17 figures and 5 schémas

(Reçu p. p. le 27 octobre 1972)

Introduction

De nombreuses causes peuvent provoquer la rupture d'une pièce en service, mais le plus souvent, celle-ci peut être due au phénomène de fatigue (1).

La répétition dans le temps de sollicitations cycliques, provoque suivant leurs amplitudes, leurs fréquences, et le milieu environnant un dommage. Celui-ci se traduit par la formation de fissures qui s'étendent progressivement jusqu'à la rupture brutale (2–6) de la pièce.

La fissuration d'un matériau qui contient 90% de phase cristalline et 10% de liant, dépend de nombreux paramètres tels que les caractéristiques thermomécaniques du liant et des cristaux, de l'intéraction liant cristal.

A priori il est difficile de prévoir quels sont les paramètres qui ont le plus d'importance sur le phénomène de propagation des fissures.

Dans tous les films que nous avons réalisés sur des éprouvettes vibrant en flexion alternée, nous n'avons pas pu observer le début de la fissuration, mais nous avons pu obtenir des renseignements sur le mode de propagation des fissures.

II. Présentation des expériences

1.1. Vibration à fréquence fixe (30 Hz)

L'éprouvette est encastrée à une extrémité et sollicitée à l'autre au moyen d'un pot vibrant relié à un générateur BF par l'intermédiaire d'un amplificateur de puissance.

Le déplacement de l'extrémité sollicitée est mesuré par un capteur inductif.

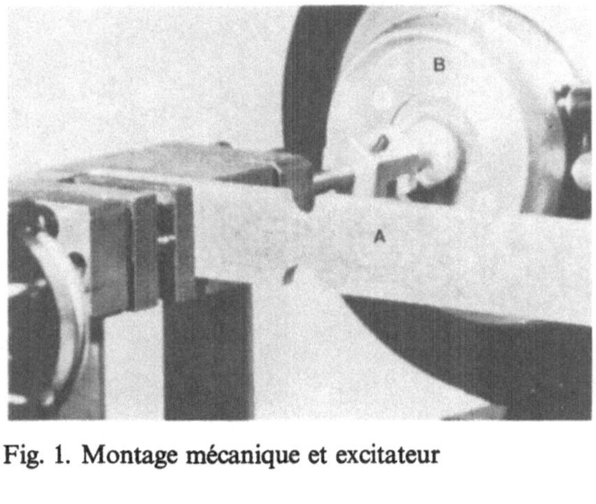

Fig. 1. Montage mécanique et excitateur

1.2. Prises de vue

Sur le montage décrit en II.1.1, dès que l'on détecte la formation d'une fissure, on arrête l'expérience de vibration. On observe avec un microscope binoculaire l'état de la zone fissurée (photos 3–7).

2.1. Vibration à la résonance des éprouvettes

L'éprouvette est encastrée à une extrémité et libre à l'autre (photo 1A). L'extrémité encastrée est sollicitée au moyen d'un pot vibrant relié à un générateur BF par l'intermédiaire d'un amplificateur de puissance.

Le déplacement de l'extrémité encastrée est mesuré par un capteur inductif, tandis que le déplacement de l'extrémité libre est mesuré par un capteur optique.

Dans ces expériences, les éprouvettes étaient des barreaux parallélépipédiques de dimensions 180 × 25 × 5 mm. Une réduction de largeur de 13 mm a été faite afin de délimiter une zone favorable à la fissuration.

Au début de chaque expérience, on recherche la fréquence de résonance de l'éprouvette par un balayage en fréquence sous faible excitation. L'excitation est ensuite augmentée et la flèche maintenue constante pendant tout l'essai. De cette façon, on fixe dans le champ de la caméra la zone de l'éprouvette où doit se propager la fissure.

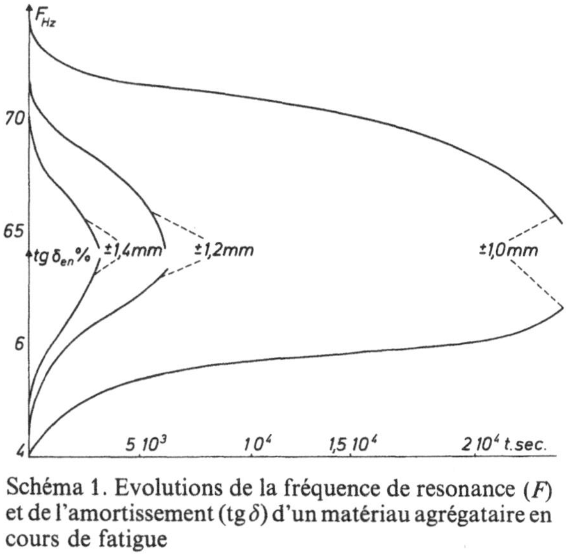

Schéma 1. Evolutions de la fréquence de resonance (F) et de l'amortissement (tg δ) d'un matériau agrégataire en cours de fatigue

En cours d'expérience, on suit l'évolution de la fréquence de résonance de l'éprouvette (7). Cette fréquence de résonance diminue en début d'essai, passe par un palier plus ou moins prononcé, pour rediminuer ensuite (schéma 1). C'est à partir de cet instant que l'on commence à filmer le phénomène de fissuration.

2.2. Prise de vues

Les films (35 mm) sont pris au moyen d'une caméra Camematic G.V. 35 (photo 2B). La cadence de défilement de cette caméra est réglable de 24 à 150 images/seconde. Elle est synchrone du flash (photo 2C).

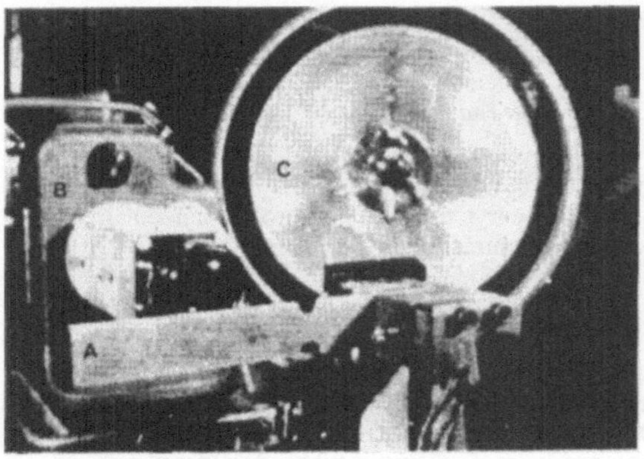

Fig. 2. Caméra et flash

L'optique de la caméra a été transformée afin d'obtenir un grandissement sur le film de l'ordre de 2.

Un numéroteur, battant la seconde, permet de repérer sur le film la position dans le temps de chaque image (photos 8–17).

III. Résultats phénomènologiques

1. Faciès macroscopique de la fissuration

On constate sur l'ensemble des photos que les fissures ne se propagent pas en ligne droite, mais suivent en général des parcours très accidentés.

Dans tous les essais, on n'a pas pu déterminer avec précision le début de la fissuration. Le suivi de la fréquence, par la méthode expérimentale décrite en II.2.1 permet simplement de dire qu'elle se produit à 80% de la durée de vie des éprouvettes.

1.2. Photos relatives aux essais définis en II.1.1 et II.1.2

Photo 3: Vue d'ensemble de la fissuration.

– En A et B ruptures de cristaux.
– En C et D microfissures qui ont été stoppées.
– On peut observer que la largeur de la fissure n'est pas uniforme dans le plan de l'éprouvette.

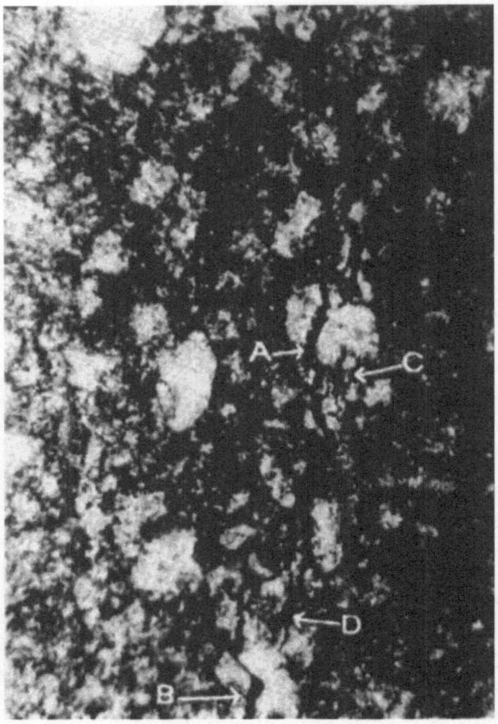

Fig. 3. Fissure de fatigue en flexion alternée. Fréquence: 30 Hz; flèche 1 mm; température des essais 21 ± 1 °C; nombre de cycles à la rupture $5{,}6 \cdot 10^6$; dimensions des éprouvettes $180 \times 10 \times 5$ mm; (longueur en vibration 145 mm)

Photo 4: Il est difficile de parler de propagation de fissure dans un cristal sur cette photo. Il semble plutôt que l'on ait un agglomérat de cristaux, présentant une très bonne adhésion et que la fissure se propagerait entre eux. Cela pourrait peut être expliquer le brusque changement de direction de la fissure observée en A.

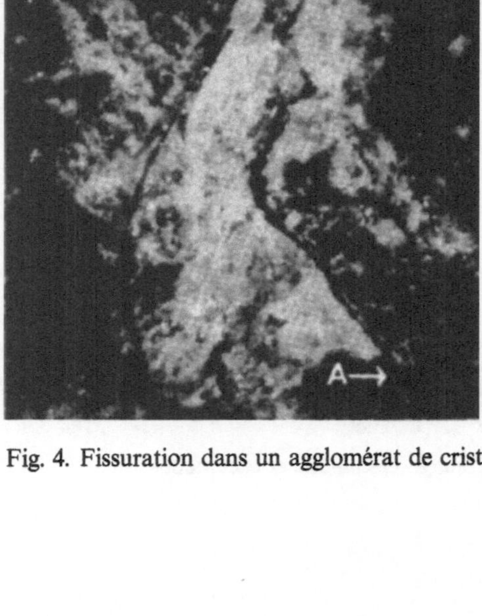

Fig. 4. Fissuration dans un agglomérat de cristaux

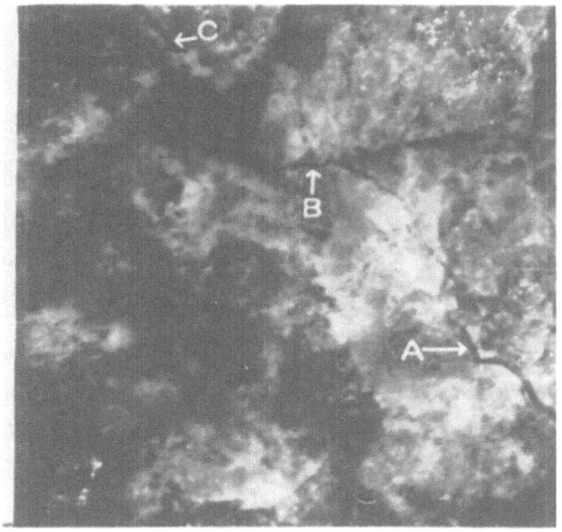

Fig. 5. Propagation de fissure dans des cristaux

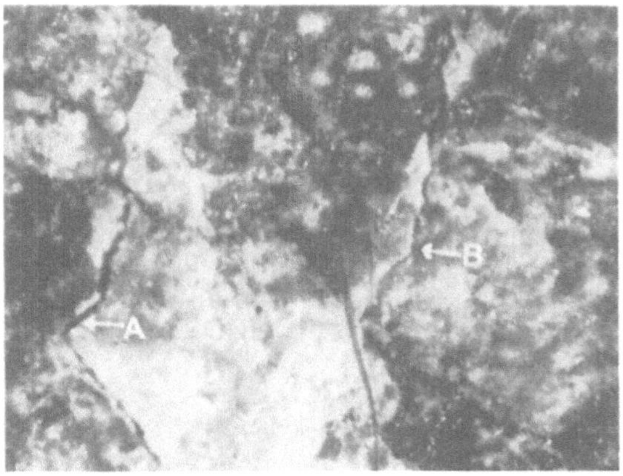

Fig. 6. Propagation de deux fissures

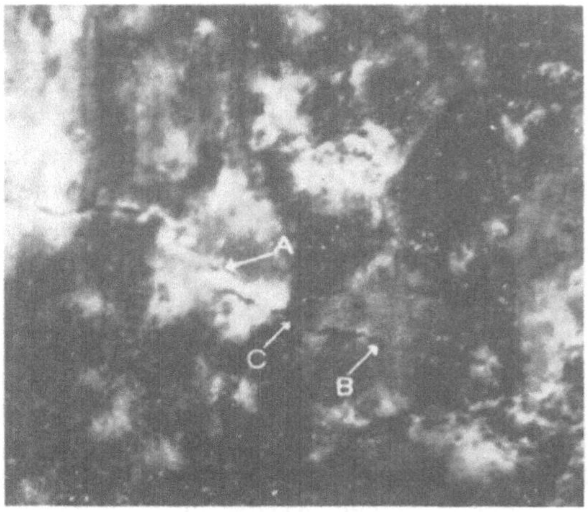

Fig. 7. Propagation très accidentée d'une fissure dans un cristal

Photo 5: Propagation d'une fissure dans des cristaux.

– En A propagation accidentée de la fissure dans un cristal ou un agglomérat de cristaux.

– En B propagation de la fissure entre deux cristaux.

– En C fissuration au bord d'un cristal.

Photo 6: Propagation de deux fissures sensiblement parallèles se propageant en bordure des cristaux (A et B).

Photo 7:

– En A propagation très accidentée de la fissure dans un cristal.

– En B même phénomène qu'en A, mais le cristal est situé sous une couche plus importante de liant.

Dans les deux cas, on observe un phénomène que l'on a observé sur d'autres photos (ex. 9 et 10 en A). La fissure attaque les cristaux sous des angles d'incidence voisine de 90°.

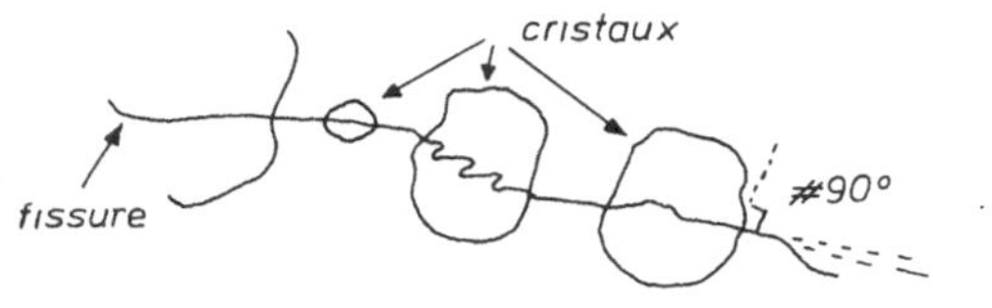

– En C. Rupture du liant entre deux cristaux.

1.3. Photos relatives aux essais définis en II.2.1 et II.2.2

Les photos sont tirées de plusieurs films, réalisés dans les mêmes conditions (flèche $\pm 1{,}4$ mm – température 21 ± 1 c).

Photo 8 (distance entre repère 7 mm):

– En A rupture d'un cristal.

– En B et C pontage de liant, phénomène que l'on verra mieux sur les photos 9 et 10 en B.

Photos 9 et 10 (distance entre repère 0,5 mm):

– En A, sur la photo 9, il semble que l'on ait rupture d'un cristal. Mais en réalité la fissure se propage entre deux cristaux (a et b). Sur le film que nous avons réalisé, on constate que le cristal (a) est en rotation alternée autour d'un axe perpendiculaire au plan des photos. L'amplitude des oscillations augmentant au fur et à mesure que l'éprouvette se fatigue, on observe finalement la rupture du cristal (b) dans un plan normal à son grand axe (photo 10 A).

L'observation de la progression de la fissuration dans ce cas montre que la fissure longe tangentiellement le cristal sur sa face gauche (photo 9)

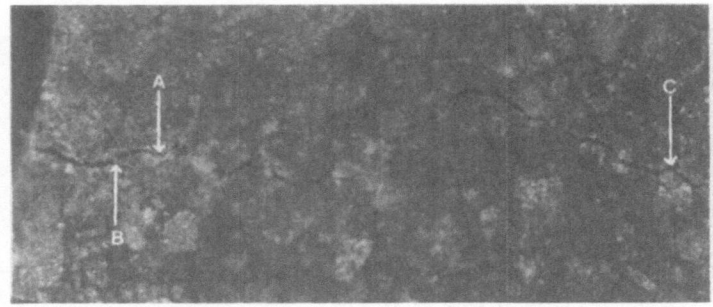

Fig. 8. Vue d'ensemble d'une fissure

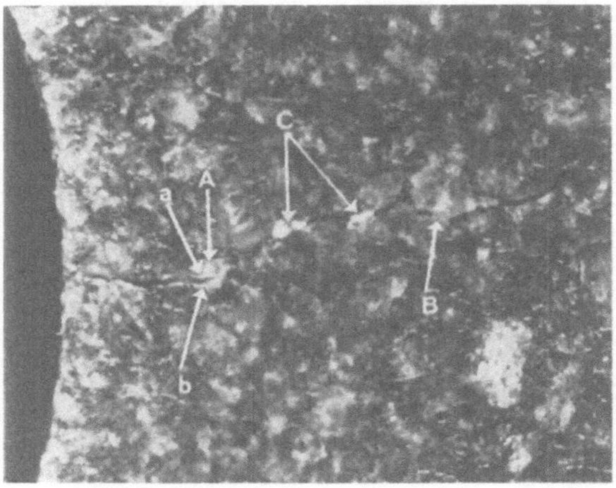

Fig. 9. Propagation d'une fissure en bordure d'un cristal (A) et pontage de liant (B)

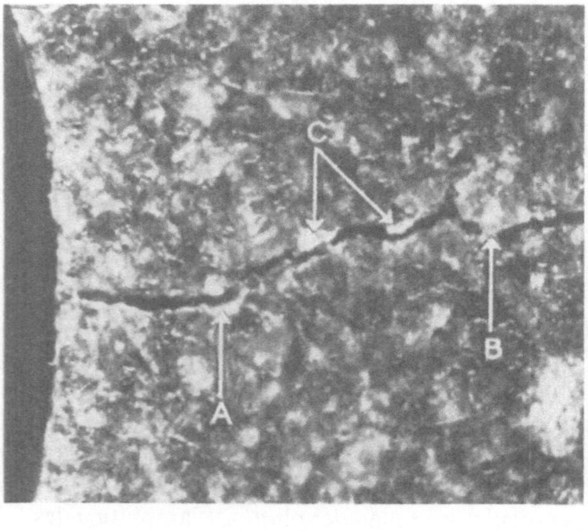

Fig. 10. Pontage de liant (B)

puis se bloque vers le milieu de celle-ci. Après seulement quelques cycles de sollicitations, elle traversera ensuite le cristal, sous un angle d'incidence voisin de 90°.

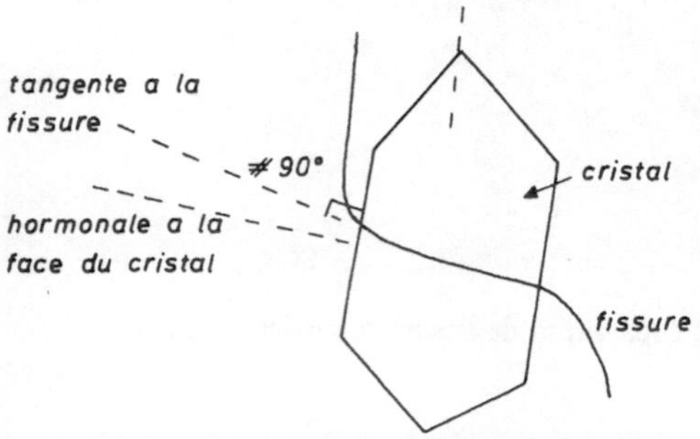

– En B, on observe un pontage de liant entre les deux lèvres de la fissure. L'observation de cette zone montre que le liant présente une très bonne adhésion par rapport aux cristaux situés de part et d'autre de la fissure.

En cours de fatigue, le liant peut subir une grande déformation avant d'atteindre la rupture. Il passe par les stades suivants au fur et à mesure que la largeur de la fissure augmente.

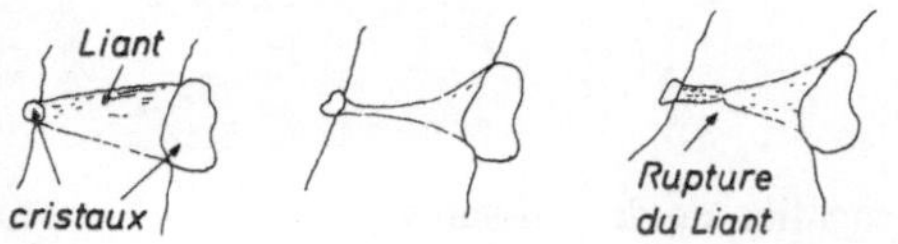

– En C, on observe des points brillants, qui apparaissent en cours de vibration lorsque la fissure est en position ouverte et qui disparaissent lorsque la fissure est en position fermée. On pense que ces points sont probablement les faces réfléchissantes de cristaux qui ont présenté une mauvaise adhésivité. On observe un phénomène semblable sur les photos 15 et 16 en D.

Photos 11 et 12 (distance entre repère 4 mm):

– En A. La fissure contourne un cristal.

– En B. On observe un processus de contournement d'un cristal. Ce cristal est attaqué simultanément en des points diamétralement opposés par deux extrémités de fissures. Contrairement à ce que nous observions sur les photos 7, 9 et 10 en A, les angles d'incidence sont différents de 90° (photo 11 B). Chaque tête de fissure tend à contourner le cristal. Finalement la fissure supérieure

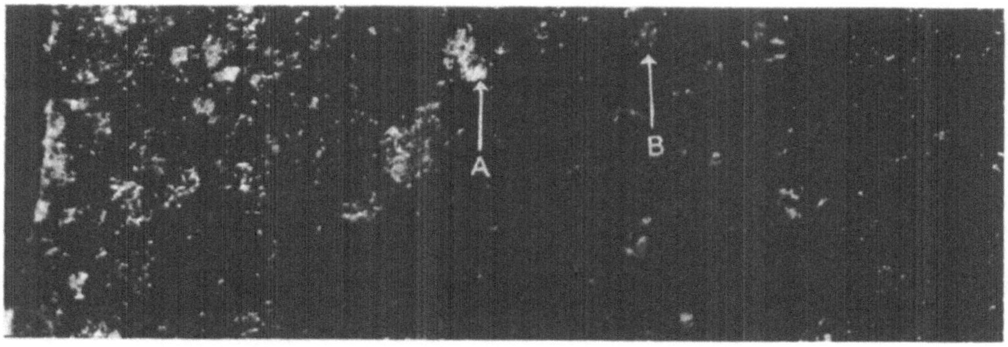

Fig. 11. Propagation d'une fissure autour de cristaux

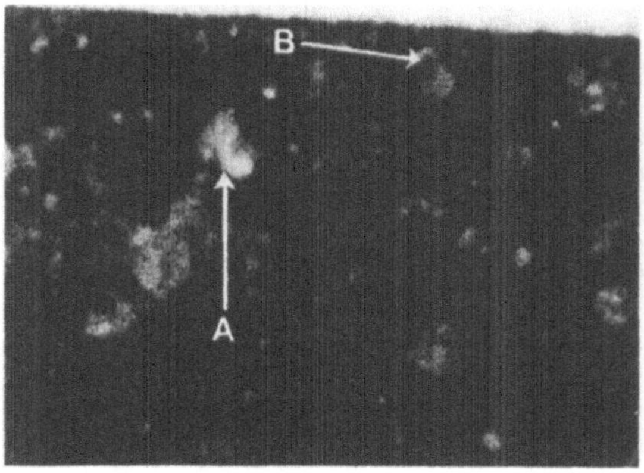

Fig. 12. Fissure contournant des cristaux

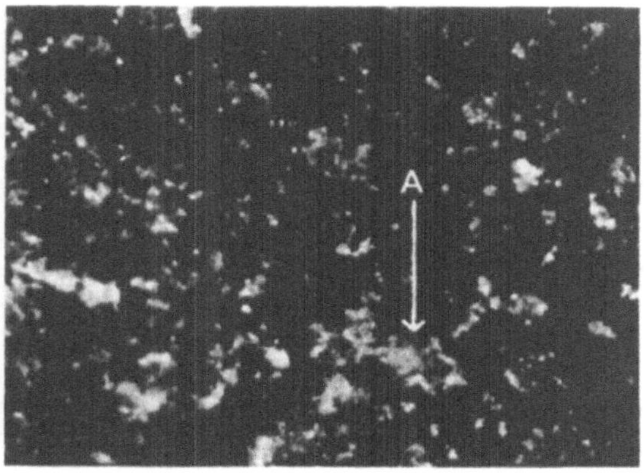

Fig. 13. Déchaussement de cristaux

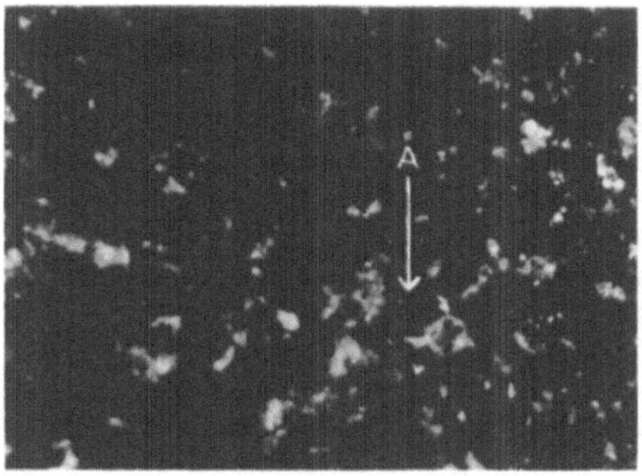

Fig. 14. Déchaussement de cristaux

sera prépondérante sans qu'il y ait rupture du cristal. Ce dernier sera déchaussé (photo 12B).

Photos 13 et 14:

Sur ces photos on n'a pas pu déterminer la position de la fissure.

On a quand même pu observer en A les déchaussements successifs d'un agglomérat de petits cristaux entourant un cristal de plus grande dimension (photo 13). Ce cristal sera ensuite déchaussé à son tour (photo 14).

Photos 15 et 16:

– En A. On observe une double fissuration. La fissure de droite contourne un cristal.

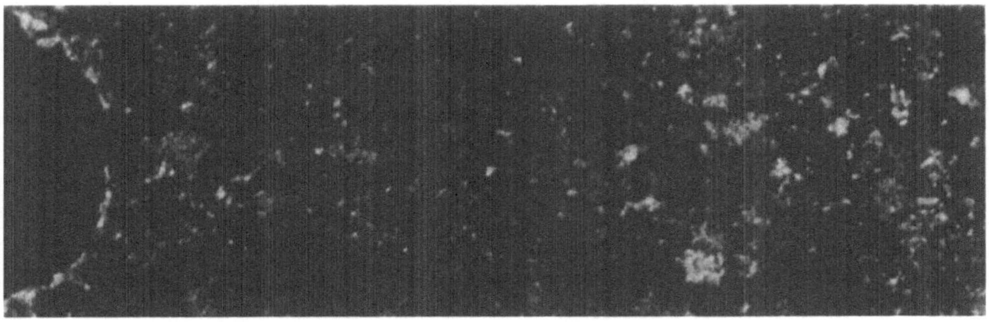

Fig. 15. Double fissuration

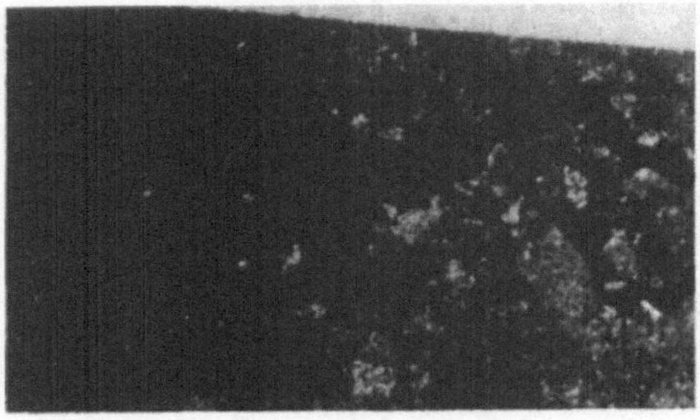

Fig. 16. Décollement d'un cristal (E)

– En B. On observe un pontage de liant. Contrairement à ce que l'on a constaté sur les photos 9 et 10 (B), où le pontage a été rompu, on observe cette fois que le liant est très résistant et présente un très bon accrochage sur les lèvres de la fissure. Au moment de la rupture une partie de la lèvre gauche de la fissure sera arrachée.

– En C. On observe un autre pontage, mais celui-ci sera déchiré en cours de fatigue.

– En D. On observe une microfissure (fig. 15) qui est stoppée par un cristal. Cette fissure progresse très rapidement au moment de la rupture, en contournant le cristal, plutôt qu'en la traversant (photo 16E); (le temps entre les photos 15 et 16 étant de l'ordre de 1/10 de seconde).

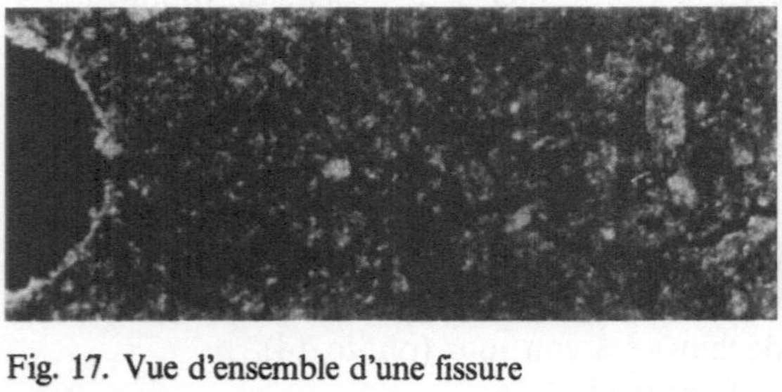

Fig. 17. Vue d'ensemble d'une fissure

Photo 17:

– En A. On observe un phénomène semblable à celui observé sur les photos 15 et 16 en B.

– En B. La fissure contourne un cristal. Celui-ci sera déchaussé.

2. Mode de fissuration et de rupture

Sur toutes les photos, on peut constater que la fissure ne se propage pas en ligne droite. Elle se situe le plus souvent dans le liant; elle traverse des agglomérats en contournant en général les cristaux.

Cependant, on peut remarquer que suivant leurs positions et leurs géométries, certains cristaux sont rompus (photos 7, 9 et 10). Ces ruptures se produisent généralement lorsque les têtes de fissures arrivent normalement à leurs faces. Dans le cas où elles arrivent sous des angles d'incidences différents de 90°, elles contournent les cristaux et ceux-ci sont déchaussés (photos 11B, 12B, 15A, 16A).

On peut schématiser les différentes fissurations rencontrées de la façon suivante:

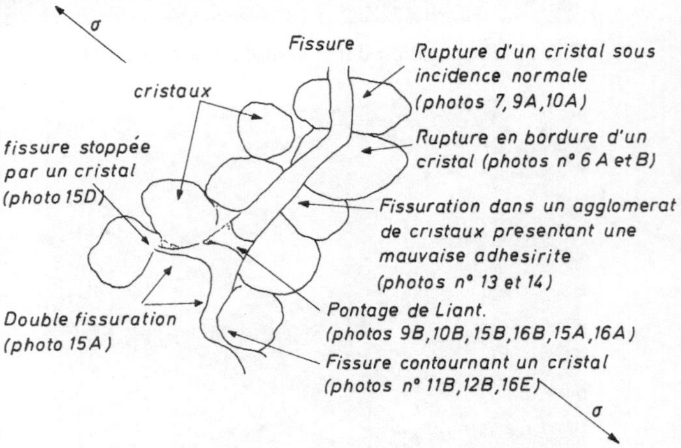

Avec les moyens d'observation que nous avons il est très difficile de définir avec précision le mode de rupture prépondérant. Il semble que dans les cristaux on ait une rupture plane de mode I (photo 10A) cela n'est cependant pas toujours vérifié, en particulier dans le cas des cristaux où l'on peut observer des fissurations très accidentées (photo 7) qui doivent résulter de la superposition des trois modes de rupture. Dans le cas du liant, il est probable que l'on ait une rupture inclinée qui proviendrait de la superposition des modes II et III (photos 9B–10B).

Suivant que l'on regarde le matériau du point de vue macroscopique ou microscopique, les conclusions que l'on peut être amené à formuler sont totalement différentes.

Du point de vue macroscopique, du fait que le pourcentage de liant est faible par rapport au pourcentage des cristaux et bien que la propagation de la fissure se fasse le plus souvent dans le liant, on ne constate pas une déformation plastique importante du matériau au moment de la rupture. De plus, on peut constater que la fissuration progresse de part et d'autre d'un plan normal à l'axe de la contrainte principale. Dans ces conditions, on peut considérer que ce matériau agrégataire présente une rupture fragile de mode I.

Mais si on observe comment progresse la fissuration, à l'échelle des cristaux, on constate alors que la fissuration progressant dans le liant et autour des cristaux; ce matériau agrégataire présente, dans ces conditions, une grande déformation plastique due à la distension des fibres de liant. On peut alors considérer qu'il présente une rupture ductile de mode II et III.

IV. Conclusion

Dans ce rapport, on ne donne que des résultats phénomènologiques qui ont apporté des éléments de réponse sur le mode de propagation des fissures dans un matériau agrégataire.

On n'a pas pu définir avec précision l'instant de début de fissuration et les sources d'apparition des fissures. Le suivi de la fréquence permet simplement de prédire que dans nos conditions opératoires, le début de fissuration se produit à environ 80% de la durée de vie des éprouvettes.

Du point de vue macroscopique, ce matériau semble avoir un comportement fragile, car l'on n'observe pas de striction des éprouvettes. Tandis que du point de vue microscopique, c'est-à-dire à l'échelle des cristaux, la fissure se propageant dans le liant et exceptionnellement à l'intérieur des cristaux, semble indiquer que ce matériau a un comportement ductile.

Résumé

Dans cette étude, on apporte quelques éléments, permettant de mieux comprendre le mode de propagation des fissures de fatigue dans un matériau agrégataire.

Le début de fissuration n'a pas pu être défini avec précision. On constate seulement qu'il apparaît à 80% de la durée de vie des éprouvettes.

Littérature

1) *Cazaud, R., G. Pomey, P. Rabbe* et *Ch. Janssen*, La fatigue des métaux, 5ème édition (Paris 1969).

2) *McEvily, A. J.*, De la détection des fissures de fatigue. Revue pratique de contrôle industriel n° 53 (Mars 1972).

3) *Hertzberg, R. W., H. Nordberg* et *J. A. Manson*, J. Mat. Sci. **5**, 521–526 (1970).

4) *Thornton, P. A.*, J. Comp. Mat. **6**, 1–172 (1972).

5) *Watts, A. M.* et *D. J. Burns*, Polymer Eng. Sci. **7**, n° 2 (1967).

6) *Rabbe, P.*, Application de la mécanique de la rupture à l'étude de la fissuration en fatigue. Revue pratique de Contrôle Industriel n° 53 (Mars 1972).

7) *Tuong, Vinh* et *P. Sorin*, Détermination du module complexe d'Young et de l'amortissement des matériaux viscoélastiques par flexion alternée. Journées d'Etudes des 21, 22 et 23 Mai 1964 – GAMI-CESMI.

Adresse des auteurs:

G. Lucas, J. Reynaud et *J. Roucou*
C.E.A., Centre d'Etudes de Vaujours
Services d'Etudes des Matières Sensibles
B.P. No. 7
93270 Sevran (France)

Rheol. Acta **13**, 696–710 (1974)

Laboratoire d'Hydrodynamique Moléculaire, Université de Bretagne Occidentale, Brest (France)

Sur la viscosité intrinsèque non-newtonienne de solutions d'ellipsoïdes rigides

Y. Layec et C. Wolff

Avec 11 figures et 5 tableaux

(Reçu p. p. le 27 octobre 1972)

I. Introduction

L'expression théorique fournissant le facteur de viscosité v de solutions d'ellipsoïdes rigides animés de mouvement brownien en fonction de leur allongement p et du paramètre $\alpha = G/D$ (G étant le gradient de vitesse de l'écoulement hydrodynamique laminaire et D la constante de diffusion de rotation des ellipsoïdes) a été établie par *Saito* (1)

$$v = (J + K - L) \int F \cdot \sin 4\theta \cdot \sin^2 2\varphi \cdot d\Omega + L \int F \cdot \sin^2 \theta \cdot d\Omega + M \int F \cdot \cos^2 \theta \cdot d\Omega + \frac{N}{\alpha} \int F \cdot \sin^2 \theta \cdot \sin^2 \varphi \cdot d\Omega . \quad [1]$$

J, K, L, M et N sont des coefficients ne dépendant que de p; $d\Omega$ est défini par la relation: $d\Omega = \sin\theta \cdot d\theta \cdot d\varphi$ où θ, φ et ψ sont les angles d'*Euler* définissant l'orientation de l'ellipsoïde; $F = F(\theta, \varphi, t)$ est la fonction de distribution des orientations due à *Peterlin* (2)

$$\partial F/\partial t = D \cdot \Delta F - \mathrm{div}(F\omega) \quad [2]$$

t est le temps et Δ et div sont les opérateurs laplacien et divergence. Les composantes de ω, vitesse angulaire de la particule sous l'effet des forces hydrodynamiques, ont été explicitées par *Jeffery* (3).

Les valeurs numériques de v ont été calculées par *Scheraga* (4) au moyen de l'ordinateur Mark I pour des ellipsoïdes allongés ($p > 1$) et aplatis ($p < 1$) jusqu'à des valeurs de $\alpha = 60$ et de p (respectivement $1/p$) = 300. Or on synthétise actuellement des polymères en configuration helicoïdale rigide (polypeptides et polyisocyanates), ou des polyelectrolytes pouvant, en milieu de faible force ionique, adopter une structure de bâtonnet rigide, dont les allongements dépassent largement 300; par ailleurs, des valeurs du paramètre α très supérieures à 60 peuvent être atteintes; ces seules raisons sont déjà suffisantes pour étendre jusqu'à $p(1/p) = 600$ et $\alpha = 500$ le tableau de valeurs de v disponible. Mais nous nous intéressions également au comportement de v aux valeurs élevées de p et de α, et à la validité d'une représentation linéaire développée par l'un d'entre nous (5). Les résultats du calcul font l'objet de la première partie de ce compte rendu; une deuxième partie rapportera brièvement un exemple d'application aux poly-L-glutamate de benzyle en configuration hélicoïdale.

II. Calcul numérique

1. Méthode et problèmes de convergence

L'éq. [1] peut se mettre sous la forme

$$v = v_A(p) + v_B(p, \alpha) \quad [3]$$

avec

$$v_A(p) = \frac{4}{15}(J + K) + \frac{M}{3} + \frac{2L}{5} \quad [3a]$$

et

$$v_B(p) = 4\pi \sum_{j=1}^{\infty} R^j \left(K_1 \cdot a_{10j} + K_2 \cdot a_{20j} + K_3 \cdot a_{22j} + \frac{K_4}{\alpha} b_{11j} \right) \quad [3b]$$

K_1, K_2, K_3 et K_4 sont des coefficients ne dépendant que de p; a_{10j}, a_{20j}, a_{22j} et b_{11j} sont des coefficients des polynomes de *Legendre* intervenant dans le calcul de la fonction de distribution F, et dépendant de α; R est tel que $R = (p^2 - 1)/(p^2 + 1)$. C'est la forme [3] qui a été utilisée pour le calcul de v. En plus de v les valeurs de $v_r = v/v_0$ (v_0 étant la valeur de v pour $G = 0$), de $I = \alpha/(1 - v_r)$, de $\partial v/\partial\alpha$ et de $\partial I/\partial\alpha$ ont été calculées.

Il est clair que la précision sur ces quantités dépend essentiellement de la convergence de la série [3b] ou de sa dérivée. Récemment, une méthode a été proposée (6) pour augmenter la vitesse de convergence; mais elle n'est encore applicable qu'au problème plan. Aussi a-t-il fallu effectuer une sommation de série classique, limitée, en raison de la capacité de l'ordinateur (64 Kmots), aux 45 premiers termes ($j \leqslant 45$; notons que *Scheraga* avait été limité à $j = 23$); tous les calculs sont effectués en double précision, ce qui était indispensable pour éviter les erreurs d'arrondi. L'étude de la convergence a été faite pour différentes valeurs de α et de p; il apparaît que, pour $p > 15$, c'est-à-dire $R \# 1$; la vitesse de

```
C
C       ØRGANISATIØN
        DØUBLE PRECISIØN AY(4),BY(4),SP,W,SIGMA
        DØUBLE PRECISIØN MAT(4,5),R,PI,BJ,BK,BL,BN,ANUA,ANUB,AK1,AK2,AK3,D
     *K,DK1,DK2,DK3,DK4,DK5,DK6,AN1,AN2,AN3,AN4,ANNU1,ANNU2,ANNU3,ANNU4,
     *SK1,SK2,SK3,SK4,CNUIX,CNU,DNUIX,EXPRI,DERI
        DØUBLE PRECISIØN RELNU,DREL
        DØUBLE PRECISIØN T,RR,S
        CØMMØN A(48,48,2),B(48,48,2),AA(48,48,2),BB(48,48,2)
        DØUBLE PRECISIØN A,B,AA,BB
1001    FØRMAT(11X,16('='))
1002    FØRMAT(12X,'P=',F12.7,23X,'NUA=',D19.12,13X,'NU(Ø)=',D19.12)
1003    FØRMAT(/53X,'G=',F8.3)
1004    FØRMAT(2X,'J',5X,'A(1,0,J)',8X,'A(2,0,J)',8X,'A(2,2,J)',8X,'B(1,1,
     *J)',8X,'AA(1,0,J)',7X,'AA(2,0,J)',7X,'AA(2,2,J)',7X,'BB(1,1,J)')
1005    FØRMAT(I3,1X,8(1X,D15.8))
1006    FØRMAT(130('*'))
1007    FØRMAT(52X,12('-'))
1008    FØRMAT(/,'VERIFICATIØN DE CØNVERGENCE',/)
1010    FØRMAT(/14X,'NUX =',D19.12,10X,'NU = ',D19.12,8X,'DNU/DG=',D19.12)
1011    FØRMAT(/31X,'I =',D19.12,8X,'DI/DG =',D19.12)
1012    FØRMAT(26X,'NU/NU(Ø)=',D19.12,5X,'DNUREL/DG=',D19.12)
1015    FØRMAT(5X,'SK1=',D19.12,5X,'SK2=',D19.12,5X,'SK3=',D19.12,5X,'SK4=
     *',D19.12)
1020    FØRMAT('SØMM',8(2X,D14.7))
1025    FØRMAT(10X,'LA MATRICE N EST PAS INVERSIBLE')
C
        IPP=1
19      IK=1
C
C       PAS PØUR LA PRØGRESSIØN DE P
C
29      GØ TØ(21,22,23,24,25,26,27)IK
21      I1=2
        I2=6
        I3=1
        GØ TØ 101
22      I1=8
        I2=12
        I3=2
        GØ TØ 101
23      I1=15
        I2=25
        I3=5
        GØ TØ 101
24      I1=30
        I2=50
        I3=10
        GØ TØ 101
25      I1=75
        I2=100
        I3=25
        GØ TØ 101
26      I1=150
        I2=300
        I3=50
        GØ TØ 101
27      I1=400
        I2=600
        I3=100
101     DØ 106 IP=I1,I2,I3
C
C       CALCUL DE NUA ET DE NU(Ø)
C
        GØ TØ(1,2)IPP
1       P=FLØAT(IP)
        R=(P*P-1)/(P*P+1)
        SP=DSQRT(P*P-1)
        W=DLØG((P-SP)/(P+SP))/SP
        GØ TØ 3
2       P=1./IP
        R=(P*P-1)/(P*P+1)
        SP=DSQRT(1-P*P)
        W=-2*DATAN(SP/P)/SP
3       BJ=-2*(P*P-1)*(P*P-1)*(P*P/4+0.125+(4*P*P-1)*W/16/P)/P/P/(2*P*P-5-
     *1.5*W/P)/(1.5+(2*P*P+1)*W/4/P)
        BK=2*(P*P-1)*(P*P-1)/P/P/(2*P*P-5-1.5*W/P)
        BL=(P*P-1)*(P*P-1)/(P*P+1)/(1+P*P/2+0.75*P*W)
        BN=6*(P*P-1)*(P*P-1)/P/(W/2-P-W*P*P)
        ANUA=(4*BJ+14*BK+6*BL)/15
        ANUB=R*BN/15+ANUA
        SK1=(10*BK-4*BJ-3*BL)/105
        SK2=2*(BJ+BK-BL)/315
        SK3=-1680*SK2
        SK4=4*BN/5
        WRITE(1,1006)
        WRITE(1,1015)SK1,SK2,SK3,SK4
        WRITE(1,1002)P,ANUA,ANUB
        WRITE(1,1001)
C
C       PAS PØUR LA PRØGRESSIØN DE G
C
        JK=1
18      GØ TØ(15,16)IPP
15      GØ TØ(31,32,33,34,35,36,37,38,39,30)JK
16      GØ TØ(31,32,33,34,51,52,53,54,39,30)JK
31      J1=4
        J2=12
        J3=4
        GØ TØ 103
32      J1=16
        J2=40
        J3=8
        GØ TØ 103
33      J1=44
        J2=52
        J3=2
        GØ TØ 103
34      J1=56
        J2=80
        J3=8
        GØ TØ 103
35      J1=120
        J2=240
        J3=40
        GØ TØ 103
36      J1=300
        J2=340
        J3=2
        GØ TØ 103
37      J1=360
        J2=360
        GØ TØ 103
```

```
38    J1=400
      J2=640
      J3=80
      GØ TØ 103
39    J1=800
      J2=1200
      J3=400
      GØ TØ 103
51    J1=120
      J2=140
      J3=2
      GØ TØ 103
52    J1=160
      J2=220
      J3=20.
      GØ TØ 103
53    J1=240
      J2=480
      J3=80
      GØ TØ 103
54    J1=600
      J2=600
      GØ TØ 103
30    J1=1600
      J2=4000
      J3=800
103   DØ 107 JG=J1,J2,J3
      G=FLØAT(JG)/8.
      IF(G-40)75,76,76
75    NZ=35
      GØ TØ 77
76    NZ=45
77    CØNTINUE
C
C     INITIALISATIØN
C
      PI=3.141592653589793
      AN1=0.
      AN2=0.
      AN3=0.
      AN4=0.
      ANNU1=0.
      ANNU2=0.
      ANNU3=0.
      ANNU4=0.
C
C     CALCUL DE A(N,M,J),B(N,M,J) ET DES DERIVEES
C
      NZ1=NZ+1
      DØ 104 J1=1,NZ1
      J=J1-1
      NXY=3+NZ-J
      DØ 102 M1=1,NXY
      M=M1-1
      Z=FLØAT(M)
      DØ 102 N1=1,NXY
      N=N1-1
      Y=FLØAT(N)
      IF(N-J)4,4,10
4     IF(M-J)5,5,10
5     IF(J)6,6,7
7     IF(N)10,10,8
8     IF((FLØAT(J-M))/2-IFIX((J-M)/2))10,9,10
10    A(N1,M1,2)=0.
      AA(N1,M1,2)=0.
      B(N1,M1,2)=0.
      BB(N1,M1,2)=0.
      GØ TØ 102
6     A(N1,M1,2)=1./2./PI
      AA(N1,M1,2)=0.
      B(N1,M1,2)=0.
      BB(N1,M1,2)=0.
      GØ TØ 102
9     IF(M)11,12,11
12    B(N1,M1,2)=0.
      BB(N1,M1,2)=0.
      AK1=-(2*Y-3)*(Y-1)*(2*Y-1)/(4*Y-3)/(4*Y-1)
      AK2=3*(2*Y-1)*(Y+1)/(4*Y-1)/(4*Y+3)
      AK3=(Y+1)*(2*Y+3)*(2*Y+4)/(4*Y+3)/(4*Y+5)
      A(N1,M1,2)=-G*(AK1*B(N1-1,2,1)+AK2*B(N1,2,1)+AK3*B(N1+1,2,1))
      AA(N1,M1,2)=A(N1,M1,2)/G-G*(AK1*BB(N1-1,2,1)+AK2*BB(N1,2,1)+AK3*BB
     *(N1+1,2,1))
      GØ TØ 102
11    DK=2*Y*(2*Y+1)
      DK1=(2*Y+1)/(4*Y-3)/(4*Y-1)
      DK2=3/(4*Y-1)/(4*Y+3)
      DK3=2*Y/(4*Y+3)/(4*Y+5)
      DK4=4*(2*Y-2*Z-3)*(Y-Z-1)*(2*Y-2*Z-1)*(Y-Z)*(2*Y+1)/(4*Y-3)/(4*Y-1
     *)
      DK5=12*(Y+Z+1)*(Y-Z)*(2*Y-2*Z-1)*(2*Y+2*Z+1)/(4*Y-1)/(4*Y+3)
      DK6=8*Y*(Y+Z+1)*(2*Y+2*Z+1)*(2*Y+2*Z+3)*(Y+Z+2)/(4*Y+3)/(4*Y+5)
      MAT(1,1)=1.
      MAT(1,2)=M*G/DK
      MAT(1,3)=0.
      MAT(1,4)=0.
      MAT(1,5)=-G*(DK1*B(N1-1,M1-1,1)-DK2*B(N1,M1-1,1)-DK3*B(N1+1,M1-1,1
     *)-DK4*B(N1-1,M1+1,1)+DK5*B(N1,M1+1,1)+DK6*B(N1+1,M1+1,1))/4/DK
      MAT(2,1)=-M*G/DK
      MAT(2,2)=1.
      MAT(2,3)=0.
      MAT(2,4)=0.
      MAT(2,5)=G*(DK1*A(N1-1,M1-1,1)-DK2*A(N1,M1-1,1)-DK3*A(N1+1,M1-1,1)
     *-DK4*A(N1-1,M1+1,1)+DK5*A(N1,M1+1,1)+DK6*A(N1+1,M1+1,1))/4/DK
      MAT(3,1)=-1.
      MAT(3,2)=0.
      MAT(3,3)=G
      MAT(3,4)=M*G*G/DK
      MAT(3,5)=-G*G*(DK1*BB(N1-1,M1-1,1)-DK2*BB(N1,M1-1,1)-DK3*BB(N1+1,M
     *1-1,1)-DK4*BB(N1-1,M1+1,1)+DK5*BB(N1,M1+1,1)+DK6*BB(N1+1,M1+1,1))/
     *4/DK
      MAT(4,1)=0.
      MAT(4,2)=1.
      MAT(4,3)=M*G*G/DK
      MAT(4,4)=-G
      MAT(4,5)=-G*G*(DK1*AA(N1-1,M1-1,1)-DK2*AA(N1,M1-1,1)-DK3*AA(N1+1,M
     *1-1,1)-DK4*AA(N1-1,M1+1,1)+DK5*AA(N1,M1+1,1)+DK6*AA(N1+1,M1+1,1))/
     *4/DK
      DØ 68 K=1,3
      MI=K
      L=MI+1
60    IF(ABS(MAT(MI,K))-ABS(MAT(L,K)))61,62,62
61    MI=L
62    IF(L-4)63,64,64
63    L=L+1
```

```
      GØ TØ 60
64    IF(MAT(MI,K))65,66,65
65    DØ 67 JJ=K,5
      T=MAT(MI,JJ)
      MAT(MI,JJ)=MAT(K,JJ)
67    MAT(K,JJ)=T
      K1=K+1
      DØ 68 I=K1,4
      RR=MAT(I,K)/MAT(K,K)
      DØ 68 JJ=K1,5
68    MAT(I,JJ)=MAT(I,JJ)-RR*MAT(K,JJ)
      IF(MAT(4,4))69,66,69
66    WRITE(1,1025)
      GØ TØ 17
69    AY(4)=0.
      DØ 70 II=1,4
      I=5-II
      S=0.
      II1=II-1
      DØ 71 JJ1=1,II1
      JJ=5-JJ1
71    S=S-MAT(I,JJ)*AY(JJ)
70    AY(I)=(S+MAT(I,5))/MAT(I,I)
      A(N1,M1,2)=AY(1)
      B(N1,M1,2)=AY(2)
      AA(N1,M1,2)=AY(3)
      BB(N1,M1,2)=AY(4)
102   CØNTINUE
      IF(IP-8)47,50,47
47    IF(IP-40)48,50,48
48    IF(IP-100)49,50,49
49    IF(IP-500)45,50,45
50    IF(G-1.5)41,40,41
41    IF(G-400)42,40,42
42    IF(G-15)43,40,43
43    IF(G-40)44,40,44
44    IF(G-100)46,40,46
46    IF(G-200)55,40,55
55    IF(G-300)45,40,45
40    IF(J)85,86,85
86    WRITE(1,1008)
      WRITE(1,1004)
85    WRITE(1,1005)J,A(2,1,2),A(3,1,2),A(3,3,2),B(2,2,2),AA(2,1,2),AA(3,
     *1,2),AA(3,3,2),BB(2,2,2)
      WRITE(1,1020)AN1,AN2,AN3,AN4,ANNU1,ANNU2,ANNU3,ANNU4
C
C     CALCUL DE L'EXPRESSIØN I ET DE SA DERIVEE
C
45    AN1=AN1+(R**J)*A(2,1,2)
      AN2=AN2+(R**J)*A(3,1,2)
      AN3=AN3+(R**J)*A(3,3,2)
      AN4=AN4+(R**J)*B(2,2,2)
      ANNU1=ANNU1+(R**J)*AA(2,1,2)
      ANNU2=ANNU2+(R**J)*AA(3,1,2)
      ANNU3=ANNU3+(R**J)*AA(3,3,2)
      ANNU4=ANNU4+(R**J)*BB(2,2,2)
      DØ 105 M2=1,NXY
      DØ 105 M2=1,NXY
      A(N2,M2,1)=A(N2,M2,2)
      B(N2,M2,1)=B(N2,M2,2)
      AA(N2,M2,1)=AA(N2,M2,2)
105   BB(N2,M2,1)=BB(N2,M2,2)
104   CØNTINUE
      CNUIX=4*PI*(SK1*AN1+SK2*AN2+SK3*AN3+SK4*AN4/G)
      CNU=ANUA+CNUIX
      RELNU=CNU/ANUØ
      DNUIX=4*PI*(SK1*ANNU1+SK2*ANNU2+SK3*ANNU3+SK4*ANNU4/G-SK4*AN4/G/G)
      DREL=DNUIX/ANUØ
      SIGMA=CNU-G*DNUIX
      EXPRI=G*ANUØ/(ANUØ-CNU)
      DERI=(ANUØ-SIGMA)*EXPRI*EXPRI/ANUØ/G/G
      WRITE(1,1003)G
      WRITE(1,1007)
      WRITE(1,1010)CNUIX,CNU,DNUIX
      WRITE(1,1012)RELNU,DREL
      WRITE(1,1011)EXPRI,DERI
107   CØNTINUE
      JK=JK+1
      IF(JK-11)18,106,106
106   WRITE(1,1006)
      IK=IK+1
      IF(IK-8)29,13,13
13    IPP=IPP+1
      IF(IPP-2)19,19,17
17    STØP
      END
```

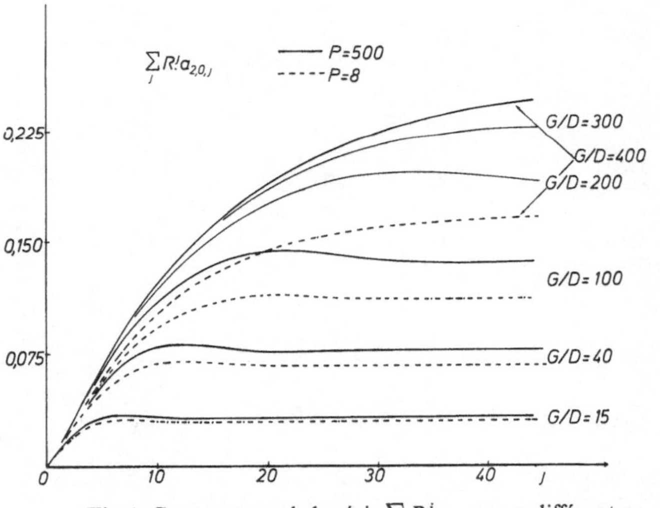

Fig. 1. Convergence de la série $\sum_j R^j a_{20j}$ pour différentes valeurs de $\alpha = G/D$

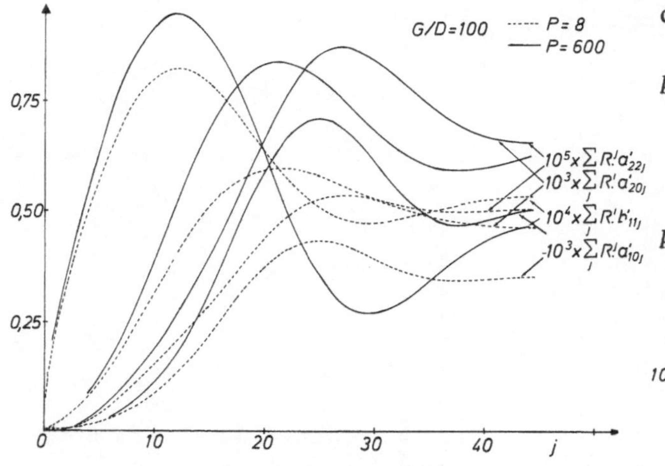

Fig. 2. Convergence, pour $G/D = 100$, $p = 8$ et $p = 600$, des différentes séries intervenant dans le calcul de $\partial v_r/\partial\alpha$

convergence devient pratiquement indépendante de p. La limite des sommes est atteinte jusqu'à $\alpha = 100$, comme le montrent les figs. 1 et 2; il n'en est plus de même pour $\alpha > 100$.

Nous avons essayé d'évaluer l'erreur due à ce manque de convergence en estimant la différence entre la valeur asymptotique ($j \to \infty$) assez hypothétique et la valeur de la somme pour $j = 45$ et en faisant les calculs d'erreur correspondants. Le tableau 1 donne les valeurs des erreurs relatives ainsi obtenues pour $p = 600$ (valeur maximum de l'allongement).

Les défauts de convergence sont beaucoup plus importants pour les quantités dérivées et l'erreur atteint rapidement des valeurs élevées.

Le programme *Fortan* du calcul figure en annexe; un exposé plus détaillé du calcul et de son organisation a été donné par l'un d'entre nous (7).

Tableau 1

α	$\Delta v/v = \Delta I/I$	$\Delta(\partial v/\partial v)/(\partial v/\partial\alpha)$	$\Delta(\partial I/\partial\alpha)/(\partial I/\partial\alpha)$
100	0,05%	2%	0,8%
200	0,4 %	12%	3 %
300	2 %		
400	5 %		

2. *Résultats du calcul*

2.1. *Relations entre les facteurs de viscosité v_0 et v_A et l'allongement*

Les valeurs de v_0 et v_A, qui ne font pas intervenir de sommation, sont faciles à calculer et sont reportées sur le tableau 2 et sur la fig. 3. Elles sont identiques à celles calculées à partir des expressions de *Simha* (8, 9):

$$p \gg 1 \quad v_0 = \frac{14}{15} + \frac{p^2}{15}\frac{1}{\log 2p - 1 - \gamma} + \frac{p^2}{5}\frac{1}{\log 2p - \gamma} \qquad [4]$$

$$p \ll 1 \quad v_0 = \frac{16}{15}\cdot\frac{1}{p}\cdot \text{Arctg}\,\frac{1}{p} \qquad [4a]$$

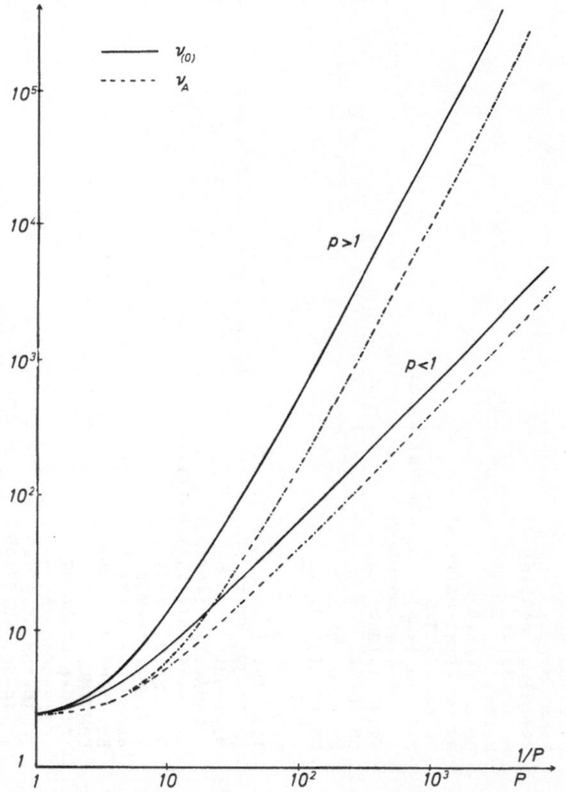

Fig. 3. Variation de v_0 et de v_A en fonction de p

ELLIPSOÏDES ALLONGES			ELLIPSOÏDES APLATIS		
P	ν_A	$\nu_{(0)}$	1/P	ν_A	$\nu_{(0)}$
2	2,582548	2,907612	2	2,609942	2,854375
3	2,786025	3,684882	3	2,868478	3,430834
4	3,077124	4,663317	4	3,198356	4,059325
5	3,433784	5,806206	5	3,563142	4,708211
6	3,844335	7,098758	6	3,946881	5,367204
7	4,302333	8,532737	7	4,341896	6,031946
8	4,803876	10,10262	8	4,744117	6,700272
9	5,346387	11,80426	9	5,151204	7,370987
10	5,928034	13,63434	10	5,561727	8,043373
11	6,547440	15,59012	11	5,974760	8,716972
12	7,203515	17,66920	12	6,389681	9,391479
14	8,622259	22,18918	14	7,223587	10,74245
16	10,17863	27,18013	16	8,061218	12,09519
18	11,86834	32,63041	18	8,901341	13,44914
20	13,68791	38,53016	20	9,743207	14,80392
22	15,63445	44,87087	22	10,58634	16,15931
24	17,70548	51,64505	24	11,43043	17,51515
25	18,78698	55,19259	25	11,85275	18,19321
30	24,64491	74,50528	30	13,96648	21,58450
35	31,23497	96,38495	35	16,08260	24,97694
40	38,53599	120,7646	40	18,20020	28,37009
45	46,53055	147,5887	45	20,31880	31,76373
50	55,20391	176,8098	50	22,43810	35,15769
60	74,53766	242,2842	60	26,67807	41,94629
70	96,45248	316,9141	70	30,91924	48,73548
80	120,8804	400,4788	80	35,16115	55,52510
90	147,7651	492,7951	90	39,40355	62,31478
100	177,0588	593,7070	100	43,64631	69,10474
110	208,7200	703,0793	110	47,88931	75,89479
125	260,5727	882,7442	125	54,25416	86,08006
150	358,3697	1222,973	150	64,86284	103,0558
175	470,0443	1613,061	175	75,47200	120,0318
200	595,2704	2051,933	200	86,08142	137,0078
225	733,7765	2538,690	225	96,69108	153,9841
250	885,3320	3072,570	250	107,3009	170,9604
275	1049,736	3652,907	275	117,9107	187,9366
300	1226,813	4279,115	300	128,5207	204,9130
400	2058,929	7232,933	400	170,9609	272,8186
500	3083,554	10886,47	500	213,4016	340,7244
600	4294,884	15220,28	600	255,8426	408,6304
700	5688,431	20219,26	700	298,2836	476,5363
800	7260,566	25871,16	800	340,7247	544,4423
900	9008,256	32165,75	900	383,1659	612,3484
1000	10928,91	39094,23	1000	425,6070	680,2543
2000	39251,60	141893,9	2000	850,0198	1359,316
3000	83340,52	302865,6	3000	1274,433	2038,377
4000	142466,1	519505,4	4000	1698,847	2717,439
5000	263545,8	946227,9	5000	2123,259	3396,498
6000	368870,9	1328056	6000	2547,672	4075,559
7000	490453,8	1769740	7000	2972,085	4754,620
8000	628002,0	2270280	8000	3396,498	5433,682
9000	781270,4	2828836	9000	3820,912	6112,743
10000	950048,1	3444687	10000	4245,325	6791,804

où γ est un coefficient numérique:

$\gamma = 0{,}5$.

Elles conduisent aux relations suivantes:

pour $50 < p < 2000$

$$v_0 = 0{,}141\, p^{1{,}812} \qquad [5]$$

$$v_A = 0{,}040\, p^{1{,}812} \qquad [5a]$$

pour $50 < 1/p < 2000$

$$v_0 = 0{,}734(1/p)^{0{,}989} \qquad [6]$$

$$v_A = 0{,}455(1/p)^{0{,}990}. \qquad [6a]$$

En faisant intervenir dans [4] les valeurs de γ dues à *Burgers* (10) pour les bâtonnets rigides, et *Broersma* (11) pour les bâtonnets à bouts arrondis, l'expression [5] devient respectivement:

$$v_0 = 0{,}159\, p^{1{,}801} \qquad [7]$$

$$v_0 = 0{,}247\, p^{1{,}758}. \qquad [8]$$

Aux allongements élevés ($p > 50$) les différences entre [5], [7] et [8] sont faibles.

Pour une particule homogène le facteur de viscosité v est lié à la viscosité intrinsèque $[\eta]$ par la relation

$$[\eta] = v/d \qquad [9]$$

où d est la masse spécifique de la particule solvatée. On voit par conséquent que pour un polymère ayant la forme d'un ellipsoïde ou d'un bâtonnet rigide d'allongement supérieur à 50, l'exposant de la relation viscosité intrinsèque–masse (loi de *Mark-Houwink*) doit être compris entre 1,75 et 1,81; un certain nombre de résultats expérimentaux sont conformes à ce résultat (12–16); dans ces conditions, les relations [5], [6], [7], [8] et le tableau 2 seront très utiles pour déterminer p par viscosimétrie.

2.2. *Critère de rigidité*

L'utilité de v_A résulte d'une observation faite lors du dépouillement du calcul numérique de v selon la relation [3]. Il se trouve en effet que le terme $v_B(p, \alpha)$ s'annule, quel que soit p, pour une valeur de α située dans l'intervalle

$37{,}5 < \alpha < 42{,}5$ pour $p > 1$

$16 < \alpha < 16{,}5$ pour $p < 1$

et a, de toutes façons, une valeur très faible dans cet intervalle. On peut alors écrire, avec une précision supérieure à 1%

$v_{40} = v_A$ pour $p > 1$

$v_{16} = v_A$ pour $p < 1$

où v_{40} et v_{16} désignent les valeurs de v respectivement pour $\alpha = 40$ et $\alpha = 16$. Il s'ensuit alors que

$$\nu_{40}/\nu_0 = 0{,}287 \qquad [10]$$

indépendemment de p pour les ellipsoïdes allongés et

$$v_{16}/v_0 = 0{,}62 \qquad [10a]$$

indépendemment de p pour les ellipsoïdes aplatis. Si les relations [10] sont vérifiées expérimentalement, on pourra en déduire que la particule n'est pas déformée par les forces hydrodynamiques, et cette méthode fournit donc un critère de rigidité d'emploi relativement aisé.

2.3. *Variation du facteur de viscosité ν en fonction de α et de p*

Les tableaux 3 et 4 donnent les valeurs calculées de ν en fonction de α et de p. On y trouve également les valeurs de

$$\nu_r = \nu/\nu_0 = [\eta]/[\eta]_0 = [\eta]_r$$

si l'on suppose que la masse spécifique ne varie pas avec le gradient de vitesse, qui pourrait éventuellement agir sur la solvatation des particules. Les figs. 4, 5, 6 et 7 représentent la variation de $[\eta]_r$ en fonction de α et de p. On voit que pour les allongements $p(1/p) > 15$ le deuxième régime newtonien n'est pas atteint, même aux valeurs extrêmes de $\alpha = 500$; ceci provient de la perturbation due au mouvement brownien, encore sensible même à des gradients de vitesse très élevés. Rappelons que la comparaison entre les courbes théoriques et expérimentales permet en principe d'obtenir l'allongement p et le coefficient de diffusion de rotation D (17); toutefois, en raison du resserrement des courbes théoriques, cette méthode n'est valable que pour $p < 50$.

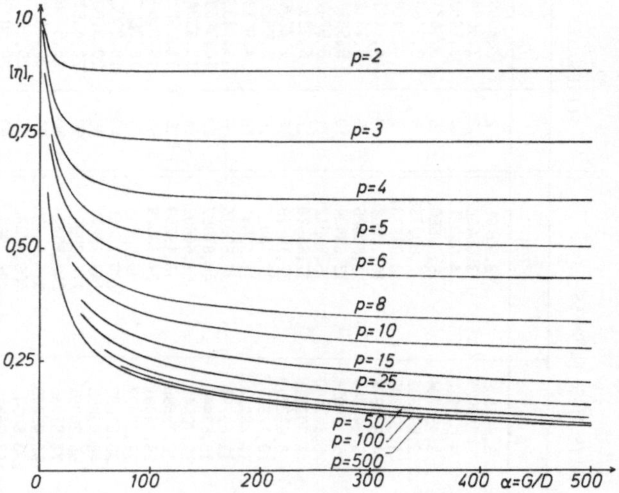

Fig. 5. Variation de ν en fonction de α et de $p (p > 1)$

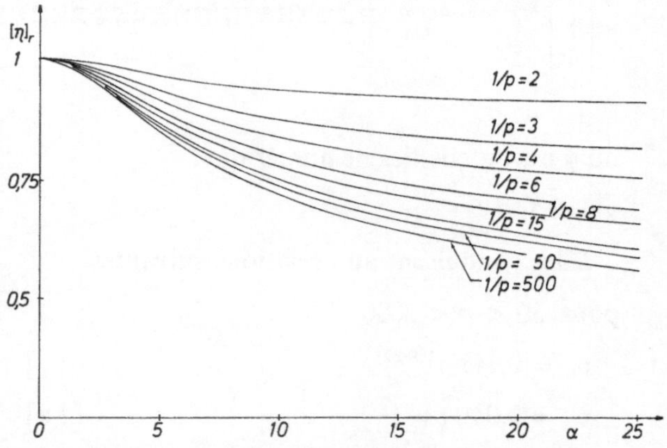

Fig. 6. Variation de ν en fonction de α et de $p (p < 1)$

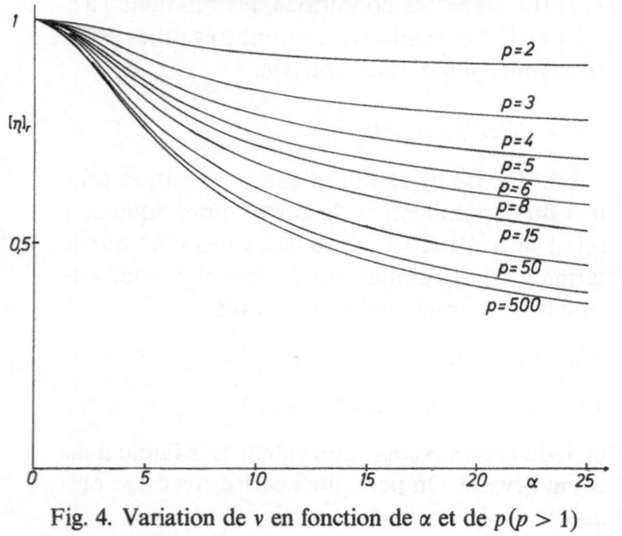

Fig. 4. Variation de ν en fonction de α et de $p (p > 1)$

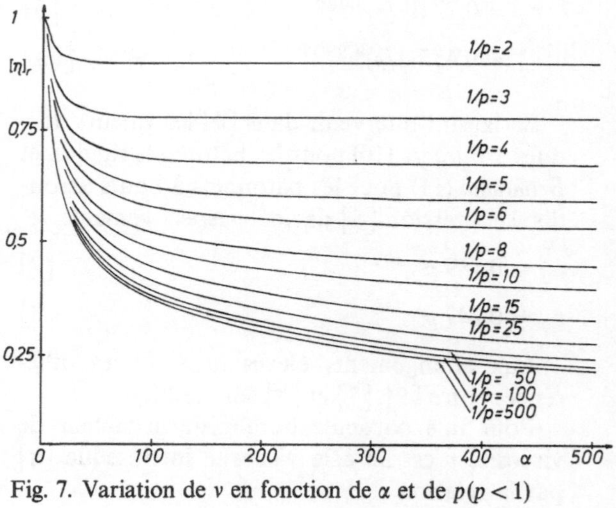

Fig. 7. Variation de ν en fonction de α et de $p (p < 1)$

2.4. Représentation asymptotique linéaire

Pour simplifier l'exploitation des résultats de viscosité intrinsèque non-newtonienne, on utilise souvent la loi empirique (5)

$$G/(1 - [\eta]_r) = aG + b \qquad [11]$$

où a et b sont des constantes caractéristiques du couple solvant-soluté.

La quantité $I = \alpha/(1 - [\eta]_r)$ que nous avons calculée correspond au premier membre de [11]. Les courbes en trait plein des figs. 8 et 9 représentent I en fonction de α pour les ellipsoïdes allongés et pour les ellipsoïdes aplatis; les droites en pointillé correspondent à l'extrapolation linéaire de la partie initiale des courbes. On voit par conséquent que pour $10 < \alpha < 100$ l'équation linéaire

$$I = \alpha/(1 - [\eta]_r) = A\alpha + B \qquad [12]$$

représente bien les résultats théoriques. L'étude détaillée des résultats numériques montre en réalité que la pente $(\partial I/\partial\alpha)$ passe par un maximum pour $\alpha = 30$; cette variation est néanmoins trop faible pour être sensible à $p < 100$.

Les droites en pointillé sont concourantes au voisinage de l'axe des ordonnées et la valeur de B est alors:

$$11{,}5 < B < 12{,}5 \quad \text{pour} \quad 20 < p < 600 \qquad [13]$$

$$22{,}5 < B < 24 \quad \text{pour} \quad 20 < 1/P < 600. \qquad [13a]$$

La comparaison entre l'ordonnée à l'origine b de la droite expérimentale [11] et celle, B, des droites théoriques permet de déterminer, avec une précision de l'ordre de 10 %, le coefficient de diffusion de rotation D

$$D = b/12 \qquad p > 1 \qquad [14]$$

$$D = b/23 \qquad p < 1 \qquad [14a]$$

relations voisines de celles trouvées dans le travail (5).

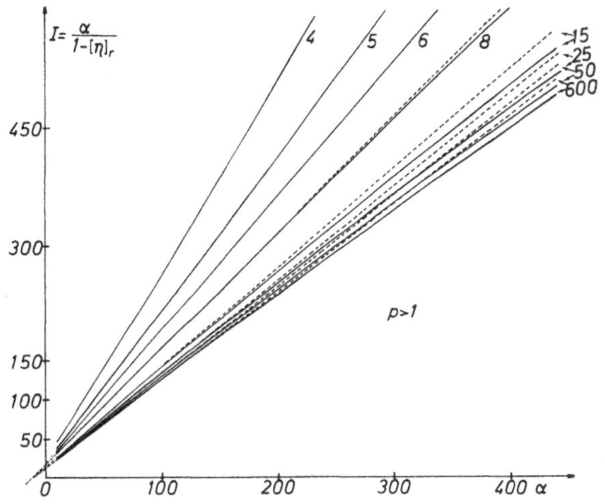

Fig. 8. Variation de I avec $\alpha (p > 1)$; les droites en pointillés représentent l'extrapolation de la partie initiale des courbes

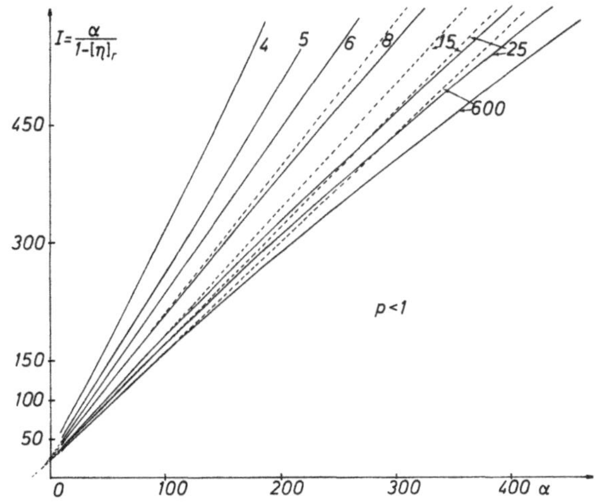

Fig. 9. Variation de I avec $\alpha (p < 1)$; les droites en pointillés représentent l'extrapolation de la partie initiale des courbes

2.5. Pente des courbes représentant $v(\alpha, p)$ en fonction de α

Les valeurs de $\partial[\eta]_r/\partial\alpha$ ont été calculées et sont représentées; en fonction de α, sur les figs. 10 et 11. On voit que l'abscisse du maximum de ces courbes est indépendante de p. Ceci signifie que le point d'inflexion des courbes $v_r = f(\alpha)$ est

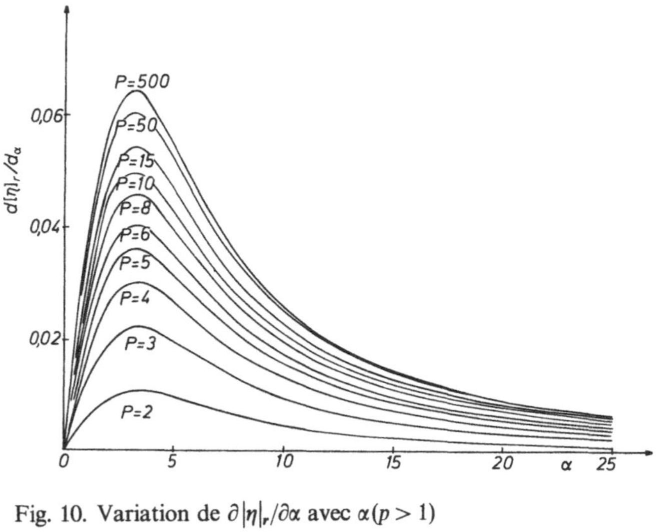

Fig. 10. Variation de $\partial|\eta|_r/\partial\alpha$ avec $\alpha (p > 1)$

α \ p		2	3	4	5	6	8	10	12	15	20	25	30
000,0	ν	2,90761	3,68488	4,66332	5,80621	7,09876	10,1026	13,6343	17,6692	24,6266	38,5302	55,1926	74,5053
	$\|n\|_r$	1,0	1,0	1,0	1,0	1,0	1,0	1,0	1,0	1,0	1,0	1,0	1,0
0,5	ν	2,90551	3,67919	4,65334	5,79131	7,07833	10,0693	13,5858	17,6030	24,5299	38,3720	54,9601	74,1862
	$\|n\|_r$	,999277	,998456	,997861	,997435	,997122	,996701	,996436	,996256	,996076	,995896	,995788	,995717
1,0	ν	2,89940	3,66268	4,62444	5,74820	7,01920	9,97288	13,4452	17,4118	24,2505	37,9149	54,2882	73,2638
	$\|n\|_r$	,997174	,993975	,991663	,990009	,988793	,987158	,986129	,985431	,984729	,984030	,983614	,983337
1,5	ν	2,88980	3,63688	4,57940	5,68110	6,92728	9,82312	13,2270	17,1149	23,8169	37,2056	53,2459	71,8332
	$\|n\|_r$	,993873	,986974	,982005	,978453	,975843	,972334	,970126	,968628	,967123	,965623	,964730	,964135
2,0	ν	2,87746	3,60395	4,52210	5,59590	6,81068	9,63342	12,9508	16,7392	23,2684	36,3088	51,9282	70,0248
	$\|n\|_r$	,989631	,978038	,969718	,963780	,959419	,953557	,949868	,947366	,944852	,942348	,940855	,939863
3,0	ν	2,84790	3,52588	4,38704	5,39573	6,53729	9,18959	12,3054	15,8619	21,9886	34,2174	48,8566	65,8102
	$\|n\|_r$	,979463	,956851	,940756	,929304	,920906	,909625	,902529	,897717	,892883	,888069	,885202	,883296
4,0	ν	2,81649	3,44397	4,24630	5,18798	6,25425	8,73145	11,6402	14,9588	20,6723	32,0685	45,7022	61,4840
	$\|n\|_r$	,968660	,934620	,910575	,893523	,881035	,864276	,853742	,846601	,839432	,832297	,828050	,825230
5,0	ν	2,78684	3,36717	4,11498	4,99470	5,99146	8,30713	11,0250	14,1243	19,4573	30,0871	42,7956	57,4994
	$\|n\|_r$	,958463	,913780	,882414	,860234	,844015	,822275	,808622	,799373	,790095	,780871	,775387	,771749
5,5	ν	2,77321	3,33192	4,05480	4,90625	5,87134	8,11345	10,7445	13,7440	18,9040	29,1855	41,4737	55,6879
	$\|n\|_r$	,953777	,904214	,869509	,845001	,827094	,803104	,788047	,777852	,767628	,757470	,751436	,747436
6,0	ν	2,76050	3,29896	3,99854	4,82361	5,75915	7,93270	10,4828	13,3894	18,3884	28,3456	40,2427	54,0014
	$\|n\|_r$	,949404	,895269	,857446	,830769	,811290	,785213	,768855	,757785	,746689	,735672	,729133	,724800
6,5	ν	2,74869	3,26827	3,94614	4,74665	5,65470	7,76452	10,2395	13,0598	17,9092	27,5654	39,0997	52,4358
	$\|n\|_r$	,945344	,886940	,846209	,817514	,796576	,768565	,751006	,739128	,727230	,715424	,708422	,703786
7,0	ν	2,73778	3,23976	3,89742	4,67509	5,55759	7,60822	10,0134	12,7537	17,4643	26,8415	38,0394	50,9839
	$\|n\|_r$	,941591	,879203	,835761	,805188	,782897	,753094	,734425	,721803	,709166	,696637	,689212	,684300
8,0	ν	2,71843	3,18876	3,81004	4,54668	5,38336	7,32790	9,60813	12,2051	16,6676	25,5460	36,1428	48,3877
	$\|n\|_r$	,934935	,865362	,817024	,783072	,758352	,725347	,704701	,690758	,676815	,663013	,654848	;649454
9,0	ν	2,70199	3,14478	3,73438	4,43534	5,23223	7,08482	9,25688	11,7300	15,9779	24,4253	34,5031	46,1444
	$\|n\|_r$	,929282	,853428	,800798	,763896	,737063	,701285	,678938	,663864	,648808	,633928	,625141	,619344
10,0	ν	2,68801	3,10670	3,66850	4,33821	5,10031	6,87262	8,95036	11,3154	15,3766	23,4491	33,0756	44,1922
	$\|n\|_r$	,924472	,843094	,786671	,747168	,718480	,680282	,656457	,640405	,624391	,608590	,599277	,593142
15,0	ν	2,64251	2,97572	3,43755	3,99532	4,63328	6,12027	7,86390	9,84727	13,2493	20,0012	28,0408	37,3141
	$\|n\|_r$	,908824	,807549	,737148	,688112	,652688	,605811	,576771	,557313	,538009	,519104	,508053	,500825
20,0	ν	2,61898	2,90036	3,29937	3,78692	4,34745	5,65767	7,19539	8,94431	11,9426	17,8880	24,9607	33,1129
	$\|n\|_r$	,900733	,787098	,707517	,652220	,612424	,560020	,527740	,506209	,484949	,464259	,452248	,444437
25,0	ν	2,60538	2,85219	3,20740	3,64581	4,15235	5,33993	6,73545	8,32289	11,0438	16,4362	22,8474	30,2334
	$\|n\|_r$	,896055	,774026	,687794	,627916	,584940	,528569	,494006	,471039	,448451	,426581	,413958	,405789
30,0	ν	2,59685	2,81915	3,14176	3,54332	4,00943	5,10555	6,39540	7,86314	10,3788	15,3629	21,2862	28,1078
	$\|n\|_r$	,893122	,765059	,673718	,610264	,564807	,505369	,469065	,445019	,421446	,398724	,385671	,377259
37,5	ν	2,58902	2,78580	3,07238	3,43265	3,85347	4,84028	6,01988	7,35487	9,64332	14,1764	19,5615	25,7612
	$\|n\|_r$	,890429	,756008	,658839	,591204	,542837	,479827	,441523	,416254	,391582	,367929	,354423	,345764
40,0	ν	2,58721	2,77749	3,05443	3,40351	3,81202	4,77832	5,91886	7,21792	9,44494	13,8562	19,0963	25,1284
	$\|n\|_r$	,889805	,753754	,654991	,586184	,536998	,472979	,434114	,408503	,383527	,359620	,345994	,337270
42,5	ν	2,58566	2,77020	3,03840	3,37726	3,77453	4,71556	5,82710	7,09349	9,26471	13,5654	18,6737	24,5537
	$\|n\|_r$	,889274	,751775	,651554	,581664	,531716	,466766	,427384	,401461	,376208	,352072	,338338	,329557
45,0	ν	2,58434	2,76376	3,02401	3,35349	3,74043	4,65824	5,74320	6,97963	9,09973	13,2992	18,2870	24,0279
	$\|n\|_r$	,888819	,750027	,648467	,577570	,526913	,461093	,421230	,395017	,369509	,345163	,331330	,322499
50,0	ν	2,58220	2,75294	2,99923	3,31208	3,68064	4,55720	5,59497	6,77830	8,80784	12,8281	17,6027	23,0975
	$\|n\|_r$	,888083	,747090	,643154	,570438	,518491	,451091	,410358	,383623	,357656	,332936	,318932	,310012
60,0	ν	2,57929	2,73715	2,96145	3,24745	3,58618	4,39577	5,35716	6,45472	8,33822	12,0699	16,5015	21,6008
	$\|n\|_r$	,887081	,742806	,635052	,559306	,505184	,435112	,392916	,365309	,338586	,313259	,298980	,289923
70,0	ν	2,57745	2,72638	2,93417	3,19931	3,51462	4,27146	5,17273	6,20289	7,97176	11,4773	15,6404	20,4305
	$\|n\|_r$	,886449	,739882	,629202	,551016	,495103	,422807	,379330	,351056	,323706	,297879	,283379	,274215
80,0	ν	2,57622	2,71869	2,91368	3,16209	3,45833	4,17193	5,02378	5,99850	7,67314	10,9930	14,9358	19,4722
	$\|n\|_r$	,886026	,737796	,624809	,544605	,487173	,412955	,368465	,339489	,311580	,285308	,270612	,261353
100,0	ν	2,57473	2,70870	2,88531	3,10851	3,37553	4,02270	4,79887	5,68895	7,22006	10,2576	13,8657	18,0171
	$\|n\|_r$	,885515	,735085	,618726	,535377	,475511	,398184	,351970	,321970	,293182	,266220	,251224	,241823
150,0		2,57322	2,69738	2,84936	3,03533	3,25749	3,80216	4,46339	5,22661	6,54453	9,16512	12,2808	15,8662
	$\|n\|_r$	,884993	,732012	,611015	,522773	,458881	,376354	,327364	,295804	,265751	,237869	,222508	,212954
200,0	ν	2,57267	2,69290	2,83320	2,99890	3,19443	3,67441	4,26095	4,94119	6,11967	8,46839	11,2638	14,4820
	$\|n\|_r$	,884806	,730798	,607551	,516499	,449999	,363709	,312516	,279650	,248499	,219786	,204082	,194375
300,0	ν	2,57228	2,68948	2,81941	2,96425	3,12888	3,52438	4,00522	4,56388	5,53465	7,47579	9,79111	12,4594
	$\|n\|_r$	,884671	,729868	,604592	,510531	,440764	,348858	,293760	,258296	,224743	,194024	,177399	,167228
400,0	ν	2,57214	2,68822	2,81386	2,94881	3,09708	3,44285	3,85709	4,33684	5,17083	6,84178	8,83847	11,1420
	$\|n\|_r$	,884623	,729526	,603403	,507873	,436285	,340788	,282895	,245446	,209970	,177569	,160139	,149547
500,0	ν	2,57208	2,68762	2,81112	2,94076	3,07976	3,39600	3,76954	4,20050	4,94942	6,45181	8,24962	10,3255
	$\|n\|_r$	,884601	,729365	,602816	,506486	,433845	,336151	,276474	,237730	,200979	,167448	,149470	,138588

α \ p		40	50	75	100	150	200	250	300	400	500	600
000,0	ν	120,765	176,810	357,592	593,707	1222,97	2051,93	:3072,57	4279,12	7232,93	10886,5	15220,3
	$\|\eta\|_r$	1,0	1,0	1,0	1,0	1,0	1,0	1,0	1,0	1,0	1,0	1,0
0,5	ν	120,237	176,027	355,981	591,008	1217,36	2042,45	3058,32	4259,21	7199,14	10835,5	15148,8
	$\|\eta\|_r$	,995627	,995572	,995496	,995454	,995407	,995380	,995362	,995348	,995328	,995315	,995304
1,0	ν	118,710	173,764	351,326	583,208	1201,12	2015,06	3017,13	4201,68	7101,50	10688,1	14942,3
	$\|\eta\|_r$	,982989	,982776	,982479	,982317	,982135	,982030	,981958	,981905	,981829	,981776	,981736
1,5	ν	116,343	170,256	344,109	571,115	1175,96	1972,59	2953,29	4112,50	6950,13	10459,6	14622,1
	$\|\eta\|_r$	,963389	,962932	,962294	,961947	,961558	,961332	,961178	,961064	,960901	,960787	,960702
2,0	ν	113,352	165,822	334,988	555,835	1144,17	1918,94	2872,63	3999;85	6758,93	10171,0	14217,8
	$\|\eta\|_r$	,938616	,937854	,936790	,936211	,935561	,935184	,934928	,934738	,934467	,934277	,934134
3,0	ν	106,382	155,494	313,752	520,262	1070,16	1794,07	2684,93	3737,71	6314,05	9499,50	13277,0
	$\|\eta\|_r$	,880903	,879441	,877402	,876294	,875050	,874330	,873840	,873476	,872958	,872597	,872323
4,0	ν	99,2314	144,902	291,983	483,809	994,357	1666,18	2492,73	3469,30	5858,62	8812,18	12314,1
	$\|\eta\|_r$	,821692	,819534	,816527	,814895	,813065	,812006	,811286	,810752	,809992	,809461	,809060
5,0	ν	92,6496	135,156	271,967	450,303	924,713	1548,73	2316,24	3222,87	5440,55	8181,34	11430,5
	$\|\eta\|_r$	,767192	,764415	,760553	,758461	,756119	,754764	,753844	,753162	,752192	,751514	,751002
5,5	ν	89,6589	130,729	262,880	435,097	893,118	1495,45	2236,20	3111,13	5251,03	7895,39	11029,9
	$\|\eta\|_r$	,742427	,739378	,735141	,732849	,730284	,728802	,727796	,727050	,725988	,725247	,724687
6,0	ν	86,8756	126,611	254,429	420,958	863,747	1445,94	2161,82	3007,30	5074,93	7629,72	10657,9
	$\|\eta\|_r$	,719380	,716083	,711507	,709033	,706268	,704671	,703588	,702784	,701642	,700844	,700241
6,5	ν	84,2927	122,789	246,591	407,848	836,522	1400,05	2092,90	2911,08	4911,78	7383,61	10313,2
	$\|\eta\|_r$	,697991	,694471	,689588	,686951	,684007	,682307	,681155	,680300	,679085	,678237	,677597
7,0	ν	81,8983	119,248	239,330	395,705	811,314	1357,57	2029,09	2822,04	4760,80	7155,88	9994,31
	$\|\eta\|_r$	,678165	,674441	,669282	,666499	,663395	,661604	,660390	,659490	,658211	,657319	,656644
8,0	ν	77,6189	112,920	226,364	374,031	766,338	1281,79	1915,31	2663,25	4491,63	6749,95	9425,92
	$\|\eta\|_r$	,642729	,638655	,633024	,629993	,626619	,624675	,623359	,622383	,620997	,620030	,619300
9,0	ν	73,9237	107,460	215,183	355,349	727,592	1216,54	1817,35	2526,57	4259,99	6400,67	8936,95
	$\|\eta\|_r$	,612131	,607770	,601755	,598526	,594937	,592873	,591476	,590441	,588972	,587947	,587174
10,0	ν	70,7102	102,713	205,470	339,129	693,972	1159,93	1732,40	2408,06	4059,21	6097,98	8513,24
	$\|\eta\|_r$	,585521	,580922	,574595	,571206	,567446	,565287	,563827	,562747	,561212	,560143	,559336
15,0	ν	59,4063	86,0347	171,410	282,313	576,373	962,115	1435,69	1994,34	3358,75	5042,50	7036,33
	$\|\eta\|_r$	,491918	,486594	,479347	,475508	,471288	,468882	,467261	,466065	,464369	,463190	,462300
20,0	ν	52,5181	75,8888	150,749	247,907	505,313	842,743	1256,82	1745,11	2937,20	4407,76	6148,65
	$\|\eta\|_r$	,434880	,429212	,421566	,417557	,413184	,410707	,409044	,407819	,406087	,404884	,403977
25,0	ν	47,8047	68,9552	136,658	224,474	456,997	761,661	1135,40	1576,03	2651,46	3977,76	5547,55
	$\|\eta\|_r$	,395851	,389996	,382162	,378089	,373677	,371192	,369529	,368307	,366582	,365385	,364484
30,0	ν	44,3297	63,8479	126,296	207,259	421,547	702,218	1046,45	1452,20	2442,32	3663,17	5107,95
	$\|\eta\|_r$	,367075	,361111	,353184	,349093	,344690	,342223	,340577	,339369	,337667	,336488	,335601
37,5	ν	40,4978	58,2212	114,896	188,340	382,635	637,022	948,930	1316,52	2213,30	3318,82	4626,91
	$\|\eta\|_r$	,335345	,329287	,321306	,317227	,312873	,310450	,308839	,307661	,306003	,304857	,303996
40,0	ν	39,4651	56,7055	111,829	183,252	372,180	619,513	922,752	1280,10	2151,87	3226,48	4497,94
	$\|\eta\|_r$	,326793	,320715	,312727	,308658	,304324	,301917	,300319	,299151	,297509	,296375	,295523
42,5	ν	38,5276	55,3301	109,046	178,638	362,701	603,644	899,030	1247,11	2096,21	3142,83	4381,13
	$\|\eta\|_r$	,319031	,312936	,304946	,300886	,296573	,294183	,292599	,291441	,289814	,288691	,287848
45,0	ν	37,6699	54,0718	106,502	174,420	354,039	589,145	877,357	1216,97	2045,37	3066,44	4274,46
	$\|\eta\|_r$	,311928	,305819	,297830	,293782	,289490	,287117	,285545	,284397	,282786	,281674	,280840
50,0	ν	36,1530	51,8473	102,005	166,969	338743	563548	839,106	1163,78	1955,69	2931,68	4086,31
	$\|\eta\|_r$	,299367	,293238	,285256	,281231	,276983	,274643	,273096	,271968	,270386	,269296	,268478
60,0	ν	33,7140	48,2720	94,7848	155,010	314,213	522,517	777,808	1078,57	1812,05	2715,92	3785,12
	$\|\eta\|_r$	,279171	,273017	,265064	,261089	,256925	,254646	,253146	,252054	,250527	,249477	,248689
70,0	ν	31,8073	45,4783	89,1471	145,679	295,086	490,540	730,055	1012,21	1700,23	2548,01	3550,78
	$\|\eta\|_r$	,263383	,257216	,249298	,245371	,241285	,239062	,237604	,236546	,235068	,234053	,233293
80,0	ν	30,2457	43,1901	84,5315	138,042	279;442	464,398	691,026	957,979	1608,89	2410,89	3359,45
	$\|\eta\|_r$	,250451	,244274	,236391	,232508	,228494	,226322	,224902	,223873	,222440	,221457	,220722
100,0	ν	27,8753	39,7182	77,5329	126,467	255,746	424,814	631,947	875,910	1470,70	2203,46	3070,07
	$\|\eta\|_r$	,230823	,224638	,216820	,213013	,209118	,207031	,205674	,204694	,203334	,202404	,201709
150,0		24,3798	34,6056	67,2468	109,473	220,993	366,797	545,396	755,721	1268,42	1899,95	2646,75
	$\|\eta\|_r$	,201879	,195722	,188055	,184388	,180702	,178757	,177505	,176607	,175367	,174524	,173896
200,0	ν	22,1249	31,3050	60,6065	98,5077	198,593	329,431	489,682	678,390	1138,36	1704,89	2374,80
	$\|\eta\|_r$	,183206	,177055	,169485	,165920	,162386	,160547	,159372	,158535	,157385	,156607	,156029
300,0	ν	18,8010	26,4211	50,7476	82,2139	165,300	273,903	406,912	563,528	945,246	1415,37	1971,24
	$\|\eta\|_r$	,155683	,149433	,141915	,138476	,135162	,133486	,132434	,131693	,130686	,130012	,129514
400,0	ν	16,6213	23,2087	44,2445	71,4571	143,311	237,229	352,246	487,673	817,732	1224,21	1704,81
	$\|\eta\|_r$	,137634	,131264	,123729	,120357	,117182	,115612	,114642	,113,966	,113057	,112453	,112009
500,0	ν	15,2668	21,2100	40,1939	64,7548	129,608	214,373	318,178	440,400	738,269	1105,09	1538,80
	$\|\eta\|_r$	,126418	,119960	,112402	,109069	,105978	,104473	,103554	,102919	,102070	,101511	,101102

α \ 1/p		2	3	4	5	6	8	10	12	15	20	25	30
000,0	ν	2,85437	3,43083	4,05932	4,70821	5,36720	6,70027	8,04337	9,39148	11,4186	14,8039	18,1932	21,5845
	$\|\eta\|_r$	1,0	1,0	1,0	1,0	1,0	1,0	1,0	1,0	1,0	1,0	1,0	1,0
0,5	ν	2,85259	3,42667	4,05291	4,69966	5,35658	6,68563	8,02480	9,36901	11,3904	14,7661	18,1458	21,5275
	$\|\eta\|_r$	,999375	,998785	,998420	,998184	,998022	,997816	,997691	,997608	,997525	,997443	,997393	,997361
1,0	ν	2,84740	3,41457	4,03431	4,67488	5,32582	6,64325	7,97103	9,30398	11,3086	14,6565	18,0085	21,3626
	$\|\eta\|_r$	,997557	,995259	,993837	,992920	,992290	,991490	,991006	,990683	,990361	,990040	,989849	,989721
1,5	ν	2,83925	3,39565	4,00537	4,63624	5,27790	6,57727	7,88735	9,20280	11,1813	14,4860	17,7950	21,1062
	$\|\eta\|_r$	,994702	,989744	,986685	,984714	,983361	,981642	,980603	,979909	,979217	,978527	,978114	,977840
2,0	ν	2,82877	3,37145	3,96825	4,58705	5,21693	6,49341	7,78105	9,07429	11,0197	14,2696	17,5240	20,7807
	$\|\eta\|_r$	,991031	,982692	,977565	,974266	,972002	,969126	,967387	,966226	,965067	,963911	,963219	,962760
3,0	ν	2,80362	3,31388	3,88055	4,47079	5,07305	6,29579	7,53076	8,77185	10,6396	13,7608	16,8868	20,0153
	$\|\eta\|_r$	,982217	,965912	,955959	,949574	,945195	,939633	,936269	,934022	,931777	,929535	,928194	,927302
4,0	ν	2,77682	3,25309	3,78835	4,34888	4,92240	6,08923	7,26935	8,45614	10,2430	13,2300	16,2224	19,2174
	$\|\eta\|_r$	,972830	,948193	,933246	,923680	,917126	,908803	,903769	,900405	,897042	,893683	,891672	,890333
5,0	ν	2,75144	3,19565	3,70137	4,23399	4,78053	5,89487	7,02352	8,15934	9,87025	12,7313	15,5982	18,4679
	$\|\eta\|_r$	,963939	,931449	,911818	,899277	,890693	,879796	,873206	,868802	,864398	,859999	,857364	,855610
5,5	ν	2,73975	3,16910	3,66114	4,18084	4,71490	5,80497	6,90982	8,02208	9,69789	12,5008	15,3097	18,1215
	$\|\eta\|_r$	,959842	,923711	,901907	,887988	,878465	,866378	,859070	,854187	,849304	,844426	,841504	,839559
6,0	ν	2,72881	3,14416	3,62328	4,13079	4,65309	5,72028	6,80271	7,89278	9,53553	12,2836	15,0379	17,7952
	$\|\eta\|_r$	,956011	,916441	,892582	,877360	,866948	,853739	,845753	,840419	,835085	,829757	,826566	,824442
6,5	ν	2,71864	3,12082	3,58778	4,08383	4,59504	5,64073	6,70208	7,77130	9,38299	12,0796	14,7825	17,4886
	$\|\eta\|_r$	,952448	,909640	,883838	,867384	,856134	,841866	,833243	,827484	,821726	,815975	,812531	,810239
7,0	ν	2,70922	3,09904	3,55456	4,03981	4,54062	5,56609	6,60765	7,65729	9,23982	11,8881	14,5429	17,2009
	$\|\eta\|_r$	,949147	,903292	,875653	,858035	,845993	,830726	,821503	,815344	,809188	,803041	,799360	,796911
8,0	ν	2,69246	3,05980	3,49439	3,95991	4,44171	5,43032	6,43581	7,44979	8,97922	11,5396	14,1067	16,6772
	$\|\eta\|_r$	,943276	,891853	,860830	,841065	,827565	,810463	,800139	,793249	,786366	,779496	,775384	,772649
9,0	ν	2,67818	3,02564	3,44161	3,88957	4,35448	5,31032	6,28390	7,26628	8,74871	11,2312	13,7208	16,2140
	$\|\eta\|_r$	,938271	,881897	,847828	,826126	,811312	,792561	,781251	,773709	,766178	,758667	,754174	,751186
10,0	ν	2,66598	2,99580	3,39508	3,82732	4,27710	5,20377	6,14878	7,10300	8,54356	10,9568	13,3773	15,8016
	$\|\eta\|_r$	,933997	,873200	,836366	,812904	,796896	,776651	,764453	,756324	,748211	,740127	,735293	,732080
15,0	ν	2,62597	2,89101	3,22709	3,59966	3,99226	4,80885	5,64685	6,49566	7,77969	9,93417	12,0973	14,2647
	$\|\eta\|_r$	,919980	,842656	,794981	,764550	,743825	,717710	,702050	,691654	,681315	,671050	,664934	,660878
17,5	ν	2,61400	2,85639	3,17451	3,51982	3,89137	4,66766	5,46666	6,27617	7,50447	9,56534	11,6354	13,7101
	$\|\eta\|_r$	,915788	,832565	,782029	,747593	,725027	,696637	,679648	,668390	,657212	,646136	,639547	,635181
20,0	ν	2,60508	2,82898	3,12229	3,45415	3,80783	4,55002	5,31615	6,09441	7,27340	9,25624	11,2482	13,2451
	$\|\eta\|_r$	,912662	,824575	,769166	,733643	,709463	,679080	,660935	,648930	,637028	,625256	,618265	,613637
22,5	ν	2,59826	2,80682	3,08333	3,39902	3,73726	4,45003	5,18785	5,93841	7,07706	8,99191	10,9170	12,8472
	$\|\eta\|_r$	,910274	,818117	,759568	,721935	,696315	,664157	,644984	,632319	,619782	,607400	,600059	,595205
25,0	ν	2,59295	2,78860	3,05049	3,35199	3,67667	4,36364	5,07670	5,80307	6,90602	8,76216	10,6290	12,5012
	$\|\eta\|_r$	,908413	,812806	,751477	,711945	,685025	,651263	,631166	,617908	,604803	,591881	,584230	,579175
27,5	ν	2,58873	2,77340	3,02242	3,31131	3,62394	4,28800	4,97913	5,68410	6,75550	8,55980	10,3753	12,1963
	$\|\eta\|_r$	,906935	,808376	,744563	,703307	,675200	,639974	,619035	,605240	,591620	,578212	,570283	,565050
30,0	ν	2,58533	2,76057	2,99816	3,27575	3,57753	4,22104	4,89253	5,57835	6,62157	8,37961	10,1492	11,9247
	$\|\eta\|_r$	,905745	,804637	,738586	,695752	,666555	,629981	,608268	,593980	,579891	,566040	,557859	,552465
40,0	ν	2,57674	2,72490	2,92708	3,16875	3,43591	4,01381	4,62282	5,24796	6,20202	7,81406	9,43927	11,0711
	$\|\eta\|_r$	,902732	,794238	,721075	,673027	,640169	,599052	,574736	,558800	,543149	,527837	,518835	,512917
50,0	ν	2,57229	2,70374	2,88129	3,09681	3,33842	3,86785	4,43095	5,01168	5,90076	7,40672	8,92722	10,4550
	$\|\eta\|_r$	,901175	,788071	,709796	,657746	,622003	,577268	,550882	,533642	,516766	,500322	,490690	,484375
60,0	ν	2,56972	2,69013	2,84959	3,04496	3,26656	3,75789	4,28496	4,83096	5,66937	7,09287	8,53213	9,97928
	$\|\eta\|_r$	,900273	,784104	,701987	,646735	,608615	,560856	,532731	,514398	,496501	,479121	,468973	,462335
75,0	ν	2,56751	2,67730	2,81729	2,98966	3,18788	3,63426	4,11881	4,62393	5,40287	6,72989	8,07432	9,42749
	$\|\eta\|_r$	,899501	,780363	,694028	,634988	,593956	,542405	,512075	,492354	,473162	,454602	,443810	,436772
100,0	ν	2,56573	2,66576	2,78508	2,93086	3,10101	3,49263	3,92538	4,38090	5,08800	6,29901	7,52975	8,77042
	$\|\eta\|_r$	,898876	,777000	,686095	,622499	,577770	,521266	,488026	,466476	,445588	,425496	,413877	,406330
150,0	ν	2,56442	2,65626	2,75464	2,86957	3,00504	3,32769	3,69602	4,09083	4,71107	5,78297	6,87784	7,98419
	$\|\eta\|_r$	,898416	,774233	,678597	,609483	,559890	,496650	,459511	,435590	,412578	,390637	,378045	,369904
200,0	ν	2,56395	2,65260	2,74112	2,83870	2,95196	3,22513	3,54382	3,89068	4,44193	5,40394	6,39249	7,39443
	$\|\eta\|_r$	,898252	,773164	,675266	,602926	,550000	,481343	,440588	,414278	,389007	,365035	,351367	,342581
300,0	ν	2,56361	2,64984	2,72982	2,80955	2,89642	3,10181	3,34550	3,31685	4,05753	4,84302	5,66184	6,49818
	$\|\eta\|_r$	,898134	,772360	,672481	,596735	,539652	,462939	,415933	,385120	,355343	,327145	,311206	,301058
400,0	ν	2,56349	2,64885	2,72543	2,79719	2,87114	3,04059	3,47084	3,24247	3,84803	4,53197	5,25334	5,99485
	$\|\eta\|_r$	,898093	,772071	,671400	,594109	,534942	,453801	,369573	,403123	,336996	,306133	,288753	,277739
500,0	ν	2,56344	2,64838	2,72331	2,79101	2,85817	3,00818	3,18706	3,39164	3,73358	4,36108	5,02831	5,71718
	$\|\eta\|_r$	,898074	,771935	,670879	,592797	,532525	,448963	,396234	,361140	,326973	,294590	,276384	,264875

α \ 1/p		40	50	75	100	150	200	250	300	400	500	600
000,0	ν	28,3701	35,1577	52,1302	69,1047	103,056	137,008	170,960	204,913	272,819	340,724	408,630
	$\|\eta\|_r$	1,0	1,0	1,0	1,0	1,0	1,0	1,0	1,0	1,0	1,0	1,0
0,5	ν	28,2941	35,0626	51,9875	68,9145	102,770	136,627	170,485	204,342	272,058	339,773	407,489
	$\|\eta\|_r$	,997320	,997295	,997263	,997247	,997231	,997222	,997218	,997214	,997210	,997208	,997206
1,0	ν	28,0740	34,7874	51,5745	68,3638	101,944	135,526	169,108	202,690	269,855	337,020	404,185
	$\|\eta\|_r$	,989562	,989467	,989341	,989278	,989215	,989184	,989165	,989152	,989137	,989127	,989121
1,5	ν	27,7317	34,3594	50,9323	67,5074	100,660	133,813	166,967	200,121	266,430	332,738	399,047
	$\|\eta\|_r$	,977498	,977294	,977022	,976886	,976751	,976683	,976642	,976615	,976582	,976561	,976548
2,0	ν	27,2973	33,8162	50,1173	66,4207	99,0297	131,640	164,250	196,861	262,083	327,305	392,527
	$\|\eta\|_r$	,962186	,961843	,961387	,961160	,960932	,960819	,960751	,960705	,960649	,960615	,960592
3,0	ν	26,2761	32,5392	48,2014	63,8661	95,1979	126,531	157,864	189,198	251,866	314,534	377,202
	$\|\eta\|_r$	,926188	,925522	,924635	,924193	,923751	,923530	,923398	,923309	,923199	,923133	,923089
4,0	ν	25,2114	31,2082	46,2046	61,2037	91,2046	121,207	151,210	181,212	241,219	301,225	361,232
	$\|\eta\|_r$	,888663	,887662	,886331	,885666	,885002	,884670	,884471	,884339	,884173	,884074	,884007
5,0	ν	24,2116	29,9582	44,3297	58,7040	87,4555	116,208	144,962	173,716	231,223	288,732	346,240
	$\|\eta\|_r$	,853422	,852110	,850365	,849932	,848623	,848188	,847927	,847753	,847536	,847405	,847318
5,5	ν	23,7495	29,3805	43,4632	57,5488	85,7229	113,899	142,075	170,251	226,605	282,958	339,312
	$\|\eta\|_r$	,837132	,835678	,833743	,832776	,831811	,831328	,831039	,830846	,830605	,830461	,830364
6,0	ν	23,3143	28,8364	42,6471	56,4609	84,0914	111,723	139,356	166,989	222,255	277,522	332,788
	$\|\eta\|_r$	,821791	,820203	,818089	,817033	,815979	,815452	,815136	,814926	,814663	,814505	,814400
6,5	ν	22,9054	28,3253	41,8806	55,4390	82,5589	109,680	136,802	163,925	218,170	272,415	326,661
	$\|\eta\|_r$	,807379	,805666	,803385	,802246	,801109	,800540	,800200	,799972	,799689	,799518	,799405
7,0	ν	22,5217	27,8457	41,1613	54,4801	81,1209	107,763	134,406	161,050	214,337	267,624	320,912
	$\|\eta\|_r$	,793854	,792023	,789587	,788370	,787155	,786548	,786184	,785941	,785638	,785456	,785335
8,0	ν	21,8233	26,9727	39,8521	52,7349	78,5038	104,274	130,046	155,817	207,361	258,905	310,449
	$\|\eta\|_r$	,769236	,767193	,764473	,763116	,761760	,761083	,760677	,760406	,760068	,759865	,759730
9,0	ν	21,2055	26,2005	38,6942	51,1913	76,1892	101,189	126,189	151,190	201,191	251,193	301,196
	$\|\eta\|_r$	,747459	,745228	,742260	,740779	,739300	,738561	,738118	,737823	,737454	,737233	,737086
10,0	ν	20,6555	25,5131	37,6635	49,8175	74,1291	98,4425	122,757	147,071	195,701	244,331	292,961
	$\|\eta\|_r$	,728074	,725677	,722489	,720898	,719310	,718517	,718042	,717725	,717329	,717092	,716933
15,0	ν	18,6060	22,9515	33,8228	44,6984	66,4538	88,2113	109,970	131,728	175,247	218,765	262,284
	$\|\eta\|_r$	,655831	,652817	,648815	,646821	,644833	,643841	,643247	,642851	,642356	,642059	,641861
17,5	ν	17,8662	22,0270	32,4367	42,8511	63,6842	84,5195	105,356	126,193	167,867	209,541	251,216
	$\|\eta\|_r$	,629756	,626519	,622226	,620088	,617958	,616896	,616259	,615835	,615305	,614988	,614776
20,0	ν	17,2460	21,2517	31,2745	41,3020	61,3619	81,4241	101,487	121,551	161,679	201,808	241,936
	$\|\eta\|_r$	,607893	,604468	,599930	,597672	,595424	,594303	,593631	,593183	,592625	,592290	,592066
22,5	ν	16,7152	20,5883	30,2799	39,9764	59,3747	78,7754	98,1771	117,579	156,384	195,190	233,996
	$\|\eta\|_r$	,589184	,585599	,580850	,578491	,576141	,574970	,574268	,573801	,573218	,572868	,572635
25,0	ν	16,2536	20,0113	29,4147	38,8235	57,6462	76,4715	95,2980	114,125	151,779	189,435	227,090
	$\|\eta\|_r$	,572913	,569186	,564255	,561806	,559369	,558155	,557427	,556943	,556338	,555976	,555734
27,5	ν	15,8467	19,5027	28,6522	37,8072	56,1227	74,4409	92,7602	111,080	147,721	184,362	221,003
	$\|\eta\|_r$	,558571	,554720	,549627	,547099	,544585	,543333	,542583	,542084	,541461	,541087	,540838
30,0	ν	15,4842	19,0494	27,9726	36,9015	54,7650	72,6313	90,4988	108,367	144,104	179,841	215,579
	$\|\eta\|_r$	,545793	,541829	,536592	,533994	,531411	,530125	,529355	,528843	,528203	,527820	,527564
40,0	ν	14,3445	17,6244	25,8356	34,0534	50,4953	66,9405	83,3870	99,8341	132,729	165,625	198,521
	$\|\eta\|_r$	,505619	,501296	,495599	,492779	,489981	,488589	,487756	,487202	,486511	,486097	,485821
50,0	ν	13,5214	16,5950	24,2916	31,9954	47,4101	62,8284	78,2481	93,6685	124,510	155,353	186,196
	$\|\eta\|_r$	,476607	,472016	,465979	,462998	,460043	,458575	,457697	,457113	,456385	,455949	,455659
60,0	ν	12,8854	15,7993	23,0978	30,4040	45,0244	59,6486	74,2744	88,9009	118,155	147,410	176,665
	$\|\eta\|_r$	,454188	,449383	,443078	,439970	,436893	,435366	,434454	,433847	,433090	,432637	,432335
75,0	ν	12,1469	14,8750	21,7105	28,5546	42,2514	55,9526	69,6554	83,3591	110,768	138,178	165,587
	$\|\eta\|_r$	,428159	,423094	,416466	,413207	,409986	,408390	,407436	,406803	,406013	,405540	,405225
100,0	ν	11,2666	13,7727	20,0552	26,3475	38,9420	51,5415	64,1428	76,7451	101,951	127,159	152,366
	$\|\eta\|_r$	,397129	,391740	,384713	,381269	,377873	,376194	,375195	,374526	,373696	,373201	,372870
150,0	ν	10,2139	12,4550	18,0773	23,7107	34,9888	46,2725	57,5583	68,8452	91,4208	113,998	136,575
	$\|\eta\|_r$	,360024	,354260	,346772	,343113	,339513	,337736	,336676	,335973	,335097	,334574	,334225
200,0	ν	9,41833	11,4559	16,5720	21,7017	31,9743	42,2535	52,5352	62,8183	83,3865	103,956	124,526
	$\|\eta\|_r$	,331981	,325835	,317897	,314040	,310262	,308402	,307295	,306561	,305648	,305104	,304740
300,0	ν	8,19772	9,91550	14,2421	18,5872	27,2960	36,0140	44,7355	53,4590	70,9088	88,3608	105,813
	$\|\eta\|_r$	,288957	,282029	,273203	,268971	,264866	,262861	,261672	,260887	,259912	,259332	,258945
400,0	ν	7,50914	9,04486	12,9222	16,8215	24,6422	32,4740	40,3099	48,1482	63,8281	79,5106	95,1929
	$\|\eta\|_r$	,264685	,257265	,247884	,243420	,239115	,237023	,235785	,234969	,233958	,233358	,232956
500,0	ν	7,12872	8,56351	12,1920	15,8444	23,1733	30,5145	37,8600	45,2082	59,9083	74,6112	89,3138
	$\|\eta\|_r$	,251276	,243574	,233876	,229280	,224862	,222721	,221455	,220622	,219590	,218978	,218569

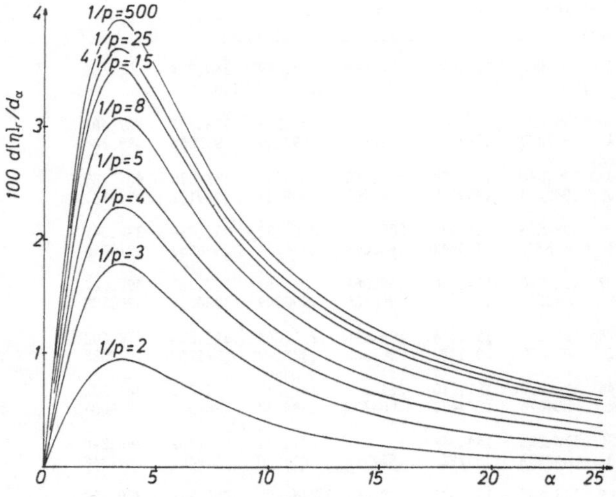

Fig. 11. Variation de $\partial|\eta|_r/\partial\alpha$ avec $\alpha (p < 1)$

toujours situé à la valeur $\alpha = 3{,}4$. Il n'a pas été possible, pour l'instant de trouver l'explication théorique de ce résultat, probablement lié à la forme de la fonction de distribution F.

Expérimentalement, le plus souvent il n'y a pas de point d'inflexion sur les courbes $[\eta]_r = f(G)$ et lorsqu'il y en a un, sa position, G_i, est malaisée à déterminer; s'il n'en était pas ainsi, on disposerait d'une autre méthode pour déterminer la constante de diffusion de rotation:

$$D = G_i/3{,}4. \qquad [15]$$

III. Exemple d'application expérimentale au poly-γ-L-glutamate de benzyle

Dans un certain nombre de solvants, en particulier le métacrésol (MC) et le dichloréthane (DCE), le poly-γ-L-glutamate de benzyle (PLGB) présente une structure hélicoïdale et peut-être, en première approximation, considéré comme un bâtonnet rigide de 15 Å de diamètre. Dans d'autres solvants comme l'acide dichloracétique (DCA), le PLGB se comporte comme une macromolécule en chaine souple.

Dans le tableau 5 nous avons rassemblé les résultats obtenus par différents auteurs ayant effectué des mesures de viscosité intrinsèque non-newtonienne sur des PLGB dans le dichloréthane et le métacrésol.

Les premières lignes donnent successivement la référence du travail, la masse moléculaire M_w déterminée par diffusion de lumière, et le solvant dans lequel les mesures de viscosité ont été effectuées. Pour le PLGB, dont la masse par unité monomère est $m = 219$, plusieurs modèles d'hélice peuvent être envisagées (18); toutes de diamètre 15 Å, elles se différencient par la longueur h (en Å) de la projection sur l'axe de l'hélice d'un élément monomère: il s'agit de l'hélice 3_{10} ($h = 2{,}00$), l'hélice α ($h = 1{,}50$), l'hélice π ($h = 1{,}15$), l'hélice $4{,}3_{13}$ ($h = 1{,}20$) et l'hélice γ ($h = 0{,}98$). Connaissant M_w, le diamètre et h, il est possible de calculer les allongements théoriques p_w des différents échantillons pour les différents types d'hélice; nous les avons calculés pour les hélices 3_{10}, α, $4{,}3_{13}$ (celle-ci étant très voisine de π) et γ. La viscosité intrinsèque à gradient nul $[\eta]_0$ permet, compte tenu de la masse spécifique $d = 1{,}32$, et des relations [5] et [9] (ou du tableau 2) de déterminer une valeur «expérimentale», p_0, de l'allongement.

On voit alors clairement que les résultats expérimentaux sont remarquablement en accord avec une hélice γ pour les masses moléculaires supérieures à $4 \cdot 10^5$, avec une hélice $4{,}3_{13}$ (ou une hélice π) pour $4 \cdot 10^5 < M_w < 10^5$ et avec une hélice α pour $0{,}5 \cdot 10^5 < M < 10^5$; en outre *J.* et *E. Marchal* (23) ont montré qu'aux masses inférieures à $0{,}5 \cdot 10^5$ la structure du PLGB était celle de l'hélice 3_{10}, et aux très faibles masses, selon *Miller* et *Flory* (24), le comportement est celui d'une chaîne gaussienne.

Tableau 5

References	19	19	20	21	22	19	16	22	19
$M_W \cdot 10^5$	9,8	9,8	9,15	4,67	3,36	3,18	2,08	1,30	0,78
Solvant	MC	DCE	DCE	MC	MC	DCE	MC	MC	DCE
$p_W H\, 3_{10}$	595	595	555	284	204	193	126	79	47
$p_W H\alpha$	445	445	415	213	153	145	95	59	<u>36</u>
$p_W H\, 4{,}3_{13}$	365	365	334	170	<u>122</u>	<u>116</u>	<u>76</u>	<u>47</u>	28
$p_W H\gamma$	<u>292</u>	<u>292</u>	<u>272</u>	<u>139</u>	100	95	62	39	23
$[\eta]_0$	3200	3200	2160	741	670	556	290	130	86
p_0	295	295	234	132	125	113	78	49	38
D_0	17,2	342	605	165	193	5090	721	2610	104000
D_b	11	340	380	–	–	4900	510	–	–
$[\eta]_{40}/[\eta]_0$	0,29	0,35	0,38				0,28		

Cet ensemble cohérent de résultats conduit à supposer que le PLGB subit une série de transformations hélicoïdales suivant le schéma:

Chaine
gaussienne $\overset{M_w \rightleftarrows 10^4}{\longleftrightarrow}$ H 3_{10} $\overset{5 \cdot 10^4}{\longleftrightarrow}$ H α $\overset{10^5}{\longleftrightarrow}$ H $4{,}3_{13}$

ou H π $\overset{4 \cdot 10^5}{\longleftrightarrow}$ H γ

H signifie «hélice».

Ces transitions à des masses moléculaires déterminées peuvent provenir des contraintes agissant sur le squelette peptidique qui, à partir d'un allongement critique, entraînent un réarrangement de la structure secondaire conduisant à un tassement de l'hélice. Rappelons cependant que d'autres auteurs supposent que la variation du pas de l'hélice avec la masse moléculaire est continue (18, 25) et s'accorderait avec un modèle de chaîne à longueur de persistance; mais les résultats correspondants ne comportent pas un aussi grand nombre de points expérimentaux sur un domaine de masse aussi étendu, et ne sont pas incompatibles avec notre hypothèse.

Dans le tableau 5, la valeur de la constante de diffusion de rotation D_0 calculée à partir de p_0 et de la formule de *Perrin* (26) figure à la dixième ligne. On peut lui comparer la valeur D_b déduite de la courbe de viscosité intrinsèque non-newtonienne, après la transformation linéaire [11] et l'emploi de la formule [14]; cette comparaison n'a pas pu être faite dans tous les cas parce que nous ne disposions pas de la courbe $[\eta]_r = f(G)$, ou, dans le cas des masses faibles parce que cette courbe ne s'écartait pas suffisamment du premier régime newtonien pour pouvoir appliquer la relation [11]. On constate que l'accord entre D_0 et D_b est assez bon, et il n'y a pas lieu d'espérer beaucoup mieux, compte tenu de la différence des méthodes et de l'importance des erreurs sur D provenant de p (qui intervient en p^3) ou de b; il a déjà été observé (5) que D_b était généralement en bon accord avec les valeurs du coefficient de diffusion de rotation obtenues par biréfringence d'écoulement ou par effet *Kerr*. La dernière ligne du tableau rapporte les valeurs quand elles ont pu être déterminées, du rapport $\nu_{40}/\nu_0 = [\eta]_{40}/[\eta]_0$. La comparaison avec le résultat théorique [10] montre que le PLGB en solution dans le metacrésol reste rigide au moins jusqu'à des masses moléculaires de 10^6; il n'en est pas de même dans le dichloréthane où ce polymère est déformé sous l'action du champ de forces hydrodynamiques.

IV. Conclusion

Les valeurs numériques du facteur de viscosité ν ont été calculées pour les ellipsoïdes rigides allongés et aplatis pour des valeurs de l'allongement p, respectivement $1/p$, comprises entre 2 et 600, et jusqu'à des valeurs du paramètre $\alpha = G/D = 600$. Si pour $p < 50$, il est possible de déterminer l'allongement par comparaison entre les courbes théoriques et expérimentales de viscosité non-newtonienne, il est nécessaire, pour $p > 50$ d'utiliser les relations entre le facteur de viscosité à gradient nul ν_0 et p. Un critère de déformabilité des ellipsoïdes dans l'écoulement hydrodynamique résulte de la comparaison entre la valeur de ν mesurée à $\alpha = 40$ ($\alpha = 16$ si $p < 1$) et ν_0. La relation linéaire

$$\alpha/(1 - |\eta|_r) = A\alpha + B$$

représente très bien les résultats théoriques pour $10 < \alpha < 100$ et permet, par comparaison avec les courbes expérimentales, d'obtenir le coefficient de diffusion de rotation D. Enfin, il a été remarqué que le point d'inflexion des courbes $\nu(\alpha, p) = f(\alpha)$ se situe à $\alpha = 3{,}4$, indépendamment de p.

Tous les résultats dont nous avons eu connaissance concernant la viscosité non-newtonienne des PLGB dans les solvants, où leur structure est hélicoïdale, ont été analysés en fonction des résultats précédents. La variation de l'allongement avec la masse est interprétée par des transitions hélicoïdales intervenant au voisinage de 4 masses critiques. Les constantes de diffusion de rotation déduites de la forme de la courbe de viscosité intrinsèque non-newtonienne sont en bon accord avec celles calculées à partir de l'allongement. Enfin on a montré que le PGLB, toujours rigide dans le métacrésol, est déformé dans l'écoulement hydrodynamique lorsque le solvant est le dichloréthane.

Les calculs numériques ont été effectués sur l'ordinateur CII 10070 de la Direction des Constructions et Armes Navales de Brest, et leur dépouillement sur le traceur de courbes du Centre d'Electronique de l'Armement de Bruz; nous remercions très vivement la direction de ces organismes pour leur aimable coopération.

Résumé

Les valeurs du facteur de viscosité ν pour les ellipsoïdes allongés et aplatis ont été calculées pour des allongements p, respectivement $1/p$, compris entre 2 et 600 et jusqu'à des valeurs du paramètre $\alpha = G/D = 500$. Des relations permettant de déterminer p en fonction de la viscosité intrinsèque à gradient nul, la constante de diffusion de rotation D en fonction de la forme de la courbe de viscosité non-newtonienne, et un critère de rigidité des ellipsoïdes ont été établis.

Ces résultats sont appliqués au poly-L-glutamate de benzyle et permettent de penser que ce produit, en configuration hélicoïdale subit des transconformations à certaines masses critiques; par ailleurs, toujours rigide dans le metacrésol, il est déformable dans le dichlorétane.

Summary

The values of the viscosity factor ν of prolate and oblate ellipsoïds are calculated for axial ratio p (or $1/p$) between 2 and 600 and up to values of the parameter $\alpha = G/D = 500$. Relations between the intrinsic viscosity at zero shear rate and p, between the rotary diffusion constant D and the shape of the non-newtonian viscosity curve and a test of rigidity for the ellipsoïds are given.

These results have been applied to poly-benzyl-L-glutamate, and show that PBLG, in helicoidal configuration, undergoes transitions at critical molecular weights. The rigidity test shows that PBLG is always rigid in metacresol but not in dichlorethane.

Littérature

1) *Saito, N.*, J. Phys. Soc. Japan **6**, 297 (1951).
2) *Peterlin, A.*, Z. Physik **111**, 232 (1938).
3) *Jeffery, G. B.*, Proc. Roy. Soc. **A 102**, 161 (1922).
4) *Scheraga, H. A.*, J. Chem. Phys. **23**, 1526 (1955).
5) *Wolff, C.*, J. Chim. Phys. **59**, 1174 (1962).
6) *Hocquart, R., R. Cressely* et *J. Leray*, C. R. Acad. Sci. (Paris) **274**, 863 (1972).
7) *Layec, Y.*, Thèse 3ème Cycle, Université Louis Pasteur Strasbourg (1972).
8) *Simha, R.*, J. Phys. Chem. **44**, 25 (1940).
9) *Mehl, J. W., J. L. Oncley* et *R. Simha*, Science **92**, 132 (1940).
10) *Burgers, J. M.*, Second Report on Viscosity (1938).
11) *Broersma, S.*, J. Chem. Phys. **32**, 1626 (1960).
12) *Nishihara, T.* et *P. Doty*, Proc. Nat. Acad. Sci. US. **44**, 411 (1958).
13) *Doty, P., J. M. Bradbury* et *A. M. Holtzer*, J. Amer. Chem. Soc. **78**, 947 (1956).
14) *Spach, G., L. Freund, M. Daune* et *H. Benoit*, J. Mol. Biol. **7**, 468 (1963).
15) *Fujita, H., A. Teramoto, T. Yamashita, K. Okita* et *S. Ikeda*, Biopolymers **4**, 781 (1966).
16) *Yang, J. T.*, Amer. Chem. Soc. **80**, 1783 (1958).
17) *Cerf, R.*, J. Chim. Phys. **55**, 470 (1958).
18) *Luzzati, V., M. Cesari, G. Spach, F. Masson* et *J. M. Vincent*, J. Mol. Biol. **3**, 566 (1961).
19) *Layec, Y.* et *C. Wolff*, IUPAC Intern. Symp. Macromol. Helsinki 1972. J. Polym. Sci. (Polym. chem. Ed.) **11**, 1653 (1973).
20) *Wolff, C.*, J. Phys. **32**, C5a, 263 (1971).
21) *Byerley, A. J., B. R. Jennings* et *H. G. Jerrard*, J. Chem. Phys. **48**, 5526 (1968).
22) *Yang, J. T.*, Amer. Chem. Soc. **81**, 3902 (1959).
23) *Marchal, E.* et *J. Marchal*, J. Chim. Phys. **64**, 1607 (1967).
24) *Miller, W. G.* et *P. J. Flory*, J. Mol. Biol. **15**, 284 (1966).
25) *Boeckel, G., J. C. Genzling, G. Weill* et *H. Benoit*, J. Chim. Phys. **59**, 999 (1962).
26) *Perrin, F.*, J. Phys. Radium **5**, 497 (1934).

Annexe

Programme *Fortran* du calcul de $\nu(p, \alpha)$ et de différentes fonctions de ν pour des ellipsoïdes allongés et aplatis.

Adresse des auteurs:

Y. Layec et *C. Wolff*
Laboratoire d'Hydrodynamique Moléculaire
Université de Bretagne Occidentale
6, Avenue Le Gorgeu
F-29283 Brest (France)

Rheol. Acta **13**, 711–716 (1974)

From the Department of Textile Technology University of Manchester Institute of Science and Technology, Manchester (England)

Fracture and fatigue of fibres

A. R. Bunsell and J. W. S. Hearle

With 4 figures

(Received October 27, 1972)

Introduction

A close study of the loading conditions for failure in a wide range of fibres has been made possible by the development of equipment specifically designed to investigate their physical properties (*Bunsell* et al., 1971). The advent of the scanning electron microscope has allowed detailed micrographs to be obtained of these fine filaments which are typically of the order of 10 μm diameter (*Hearle* and *Cross*, 1970). By these methods a much greater understanding of the modes and mechanisms of fracture in fibres under different loading conditions is now possible.

The fracture morphologies of each type of fibre is distinctive and reflects the fibre structure, environmental conditions and loading history. The appearance of the fracture ends of a fibre may therefore be used as a means of determining the cause of failure.

The classification of fibre fracture has been made with regard to the basic mechanisms of fracture as observed with the scanning electron microscope. All of the types so far encountered fall into one of six groups although particular fractures may involve a combinations of these types of failure.

Brittle failure

Glass fibres under simple tensile loading conditions are seen to behave as a classic *Griffith* material with a region of initiation followed by propagation and complete failure. Initiation appears always to be from the surface and the final fracture morphology shows no sign of any large scale plastic deformation. Fracture is presumed to initiate at some local defect on the surface of the fibre and, when the crack begins to grow, to accelerate very quickly and continuously into the fibre until the stress on the remaining load bearing cross section is so high as to cause simultaneous separation of the remaining material at many points.

This type of failure is essentially the same as that found in any brittle bulk material, but on a much finer scale. The fracture of glass fibres can therefore be seen to involve the same mechanisms as are found in bulk glass specimens.

Carbon fibres have been considered by *Whitney* and *Kimmel* (1972) as obeying *Griffiths* energy condition although the surface energy terms obtained indicate considerable dissipation of energy due to plastic deformation. The failure of carbon fibres appears sometimes to be similar to that of glass in that the crack may initiate at some flaw, on surface or internally at some defect, then developing across the whole cross section.

Alternatively some examples of their fracture morphologies are similar to some of the acrylic fibres from which they are made and which are classified separately later.

Controlled ductile crack growth

Many synthetic fibres show fracture behaviour in sharp contrast to that of glass and carbon fibres. Nylon, polyester and some acrylic fibres fail under simple tensile loading conditions after crack initiation at the surface which then penetrates into the fibre in a controlled manner until a point is reached at which catastrophic failure occurs. The initiation of the crack has been observed in undrawn nylon filament to begin some little way down the load elongation curve from the final failure point (*Hearle* and *Cross*, 1970). The crack growth is stable at the load which formed the crack and requires an increase in load for further propagation. That material near the fracture surfaces relaxes and that ahead of the crack extends to open the crack and form a v-shaped notch in the surface. This notch develops deeper as the crack progresses until the remaining cross section can no longer bear the load and catastrophic failure occurs. The final fracture morphology shows clearly the extent of

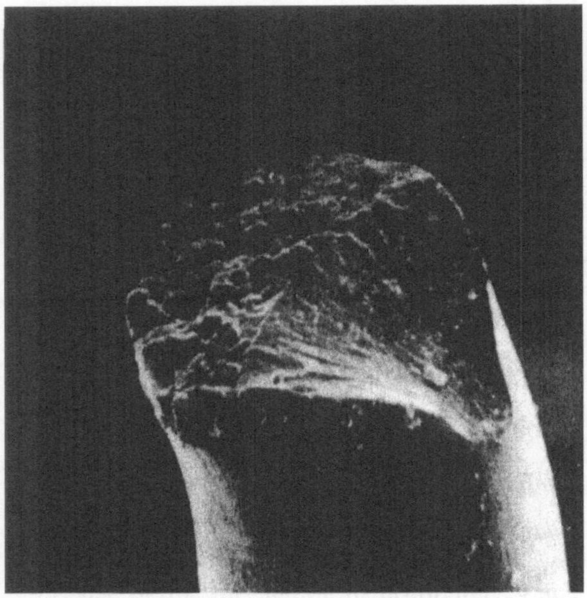

Fig. 1. Typical tensile fracture morphology of a Nylon 66 fibre

slow crack propagation and subsequent plastic deformation followed by a region of catastrophic failure (fig. 1).

Slow crack propagation leading to an identical fracture morphology is seen in such fibres when they are broken by a creep mechanism. That is failure after supporting a steady load for some period of time when the fibre has stretched to its breaking extension. Failure by this means is also observed when fibres are subjected to some oscillatory load superimposed on a steady load. In this case creep failure will occur in about the same time as under a steady load if the average load on the fibre is considerably reduced but the load is repeatedly taken up to that value of steady load which is found to cause creep failure. In these cases failure occurs after many thousands of cycles but must not be considered as a true fatigue failure. In principle the fracture could be predicted with an adequate knowledge of the time dependence of strength properties of the fibre.

On rare occasions, this type of failure, under simple tensile loading conditions has been observed originating at some gross interior flaw. In these cases radial slow crack propagation is observed, from the flaw, with drawing of the remaining cross section.

This type of controlled crack propagation cannot be explained in terms of the *Griffith* energy relationship governing elastomeric materials. It is possible, however, that this type of crack propagation may be described by an energy relationship derived along the lines of fracture mechanics, possibly involving higher powers of the crack depth. In this way it is possible that controlled crack growth with an increasing load could be expected until a point is reached at which unstable crack growth will occur.

Axial splitting

In fibres having a marked fibrillar structure, such as cotton, fracture may result in the structure splitting along its length and failing at points along the fibre (*Hearle* and *Sparrow*, 1972). A cotton fibre is typically about 20 μm in diameter and is composed of a thin primary wall surrounding the bulk of the fibre which consists of crystalline cellulose fibrils with lateral dimensions between 2 and 5 nm. As can be seen from fig. 2, which shows a fracture end of a cotton fibre, the network of fibrils run along the fibre in a helical manner. The direction of the helix changes at reversal zones and in these regions the structure is parallel to the fibre axis. It has been suggested that these reversals are points of weakness along the fibre as it was thought that failure under tensile loading was most likely there. This was based on studies of cotton fracture in the optical microscope. The scanning electron microscope reveals that far from being the weakest points the reversals may be the strongest. Failure is rarely seen at reversals but the point of fracture is often very near to and associated to one. This probably means that, in extension, the helical structural arrangement in cotton fibres induces forces of maximum torsion near to reversals and so failure in these regions. In the scanning electron microscope the convolutions along a cotton fibre under simple tensile loading are seen to untwist. When the fibre is nearly straightened failure occurs. There is no net torque on the fibre in this situation but it is probable that the untwisting gives rise to shear forces which are a maximum adjacent to the reversal zones. The untwisting of the fibrils causes cracks to develop in the secondary wall which propagate towards the reversal zone along the spiral angle. As the crack grows the stress is concentrated along a line joining the ends of the crack. Finally failure occurs along this line resulting in the fracture morphology seen in fig. 2.

Fig. 2. Typical tensile fracture morphology of a cotton fibre

Axial splitting of this type is also observed in the twist breakage of some acrylic fibres (*Cross* et al., 1969) and in the snarl splitting of polyethylene (*Greer* and *Hearle*, 1971).

Fracture perpendicular to the fibre axis

So far we have considered the fracture of fibres in which a crack is initiated at some point of defect or irregularity and develops across the fibre until complete failure. Not all fibres fail in quite this manner.

Courtelle, which is *Courtaulds*, acrylic fibre, is often seen to have failed at the level of constituent bunches of fibrils. Failure may initiate internally at a particularly weak point and the load is then transferred to the remaining bunches of fibrils. The next weakest point then parts and so on until complete separation. The resulting fracture surface is usually seen to be straight across a radial plane showing some moderate surface roughness with occasional fibrils projecting above the fracture surface. The appearance of this fibre is similar to that of a fibre reinforced material observed at too low a magnification for the constituent fibres to become apparent (*Jackson* et al., 1972). The *Courtelle* fibre fracture thus appears to fit into a category of break associated with materials having moderate cohesion between axially aligned fibrous elements which in the acrylic fibre must be microfibrillar units of the fine structure. Twisted filament yarns as studies by *Hearle* and *Thakur* (1961) show the same features except that cohesion due to twist is lost when fracture occurs. The effect has been demonstrated in studies of larger scale models of blended yarns by *Backer* and *Monego* (1965). With little cohesion (low twist) the individual axial elements break separately at their weakest points which are distributed along the whole length of the specimen; with greater cohesion (moderate twist) the stress transfer which occurs near to the initial fracture is sufficient to cause the neighbouring elements to break and the whole system fails by successive fracture of individual elements in a narrow band across the specimen. Occasionally fracture may initiate at two points along the fibre and a large step fracture result.

This type of failure can be expected in a filament with low cohesive strength in the radial direction or one in which a large distribution of faults occurs throughout the fibre. Nylon fibres which have been exposed to sunlight for a period of time, say three weeks, are found to fail in a completely different manner to that described above for fresh samples. The action of the sunlight appears to create voids around the particles of titanium dioxide used as a delustering agent and in this way produces a filament with many small faults throughout its structure. This results in a very irregular simple tensile fracture morphology of light degraded nylon fibres. Fracture begins independently at many of these voids in the fibres and the small cracks join by shearing along lines of high stress.

Failure in flexing

When some fibres, such as polyester, are bent criss-crossing lines, or bands of local reorientation can be observed in that part under compression on the inside of the bend. These lines are usually at about 45° to the axial direction and are due to shear stresses in the fibre (*Bosley*, 1968). These "kink bands" have been observed at room temperature in polyester but a raised temperature is required for them to form in nylon. The bands disappear on straightening of the fibre. They become apparent under polarised light in the optical microscope as the reoriented structure gives rise to a different angle of extinction than the surrounding material. In the scanning electron microscope the kink bands are apparent as surface ridges on the fibre.

Although these kink bands seem to disappear on straightening of the fibre some slight residual damage may occur as repeated flexing leads to failure along lines which are associated with the kink bands.

With biaxial testing, in which the fibre is bent through 90° and then rotated about its axis in such a way as to impart no twist to the structure, failure is also seen to occur by splitting at an angle to the fibre axis. In this case however torsional creep plays a part in the failure. A nylon bristle which has been subjected to biaxial testing shows the development of deep fissures at an angle to the nominal fibres axis direction. During the test the structure has suffered some torsional creep which results in a helical realignment which reverses in direction at a point midway between the grips. Because of this realignment a resolved component of the varying tensile and compressive forces, which originally were parallel to the direction of fibre orientation, was normal to the direction of local reorientation. This causes the initiation of small cracks in the region which are most highly reoriented from the original axis direction of local orientation possibly because of the influence of the other resolved component of force. The small cracks then coalesce to form deep fissures.

Serious weakening of the structure occurs in nylon after about ten thousand cycles whereas in polyester 500 cycles is all that is usually needed for equivalent damage. This may be connected to the same cause that allows kink bands to form more readily in polyester than in nylon.

Axial splitting due to tensile fatigue

Fibres which are subjected to cyclic tensional loadings in which an oscillatory loading is superimposed on a steady load are often found to fail even though the maximum loading experienced does not closely approach their breaking strengths. In these cases the failure is usually due to the load bearing cross section of the fibre being reduced by crack propagation along and into the fibre until complete failure can occur under the prevailing loads. This type of failure exactly fits the classical definition of fatigue.

a) Nylon 66

Nylon 66 when subjected to oscillatory tensile loadings has been found to fail by this type of fatigue mechanism after about 10^5 cycles but only if the oscillatory load is reduced to zero each cycle. If this criterion is met, failure will occur with a maximum load of about 60% of the fibre's simple tensile strength (*Bunsell* and *Hearle*, 1971).

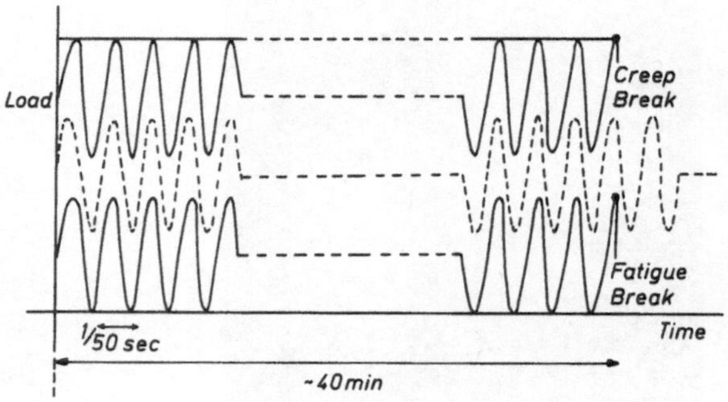

Fig. 3. Schematic load diagram

Fig. 3 shows a schematic representation of a load diagram for fibres which have failed after about 10^5 load cycles at 50 Hz. If the maximum load is that which causes failure by creep in a certain time and the minimum load is positive then a creep failure will occur in about the same time under these oscillatory conditions as under steady conditions. In the scanning electron microscope the appearance of the failure will be identical to that after a simple tensile pull. No fatigue effect is operating in this case, if however, the loading pattern is reduced until the minimum load each cycle is zero then, providing the amplitude takes the loading to about 60% of the normal tensile strength, a fatigue failure will occur. The intermediate condition, in which the loading pattern is raised from the fatigue state, to maintain a positive minimum load, yet not high enough to run into the creep failure region, will safeguard the fibre from failure.

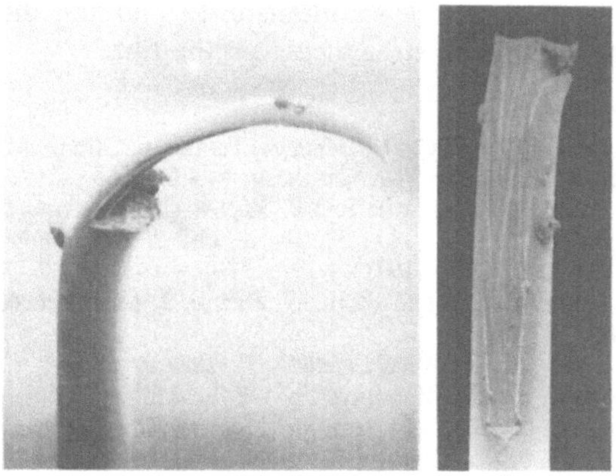

Fig. 4. Typical fatigue fracture morphology of Nylon 66

First indications of damage due to fatigue occurs probably after some 5×10^4 cycles in the form of a surface notch. This develops into the fibre in a radial direction for two or three microns before being deflected sharply along the fibre. The crack then runs along the filament until a tensile failure can occur under the prevailing loading conditions (fig. 4). The development of these cracks can sometimes be seen if the fibre is being cycled with a large amount of slack each cycle and being illuminated with a stroboscope. The cracks result in a sharp bending of the specimen at that point and in this way fatigue cracks have been observed to develop for up to 3×10^4 cycles. Fatigue failure usually occurs in nylon 66 after about 10^5 cycles.

The development of fatigue cracks is not confined to any particular region on a specimen and fatigue crack growth can occur in a number of places simultaneously.

The zero minimum load criterion may indicate that the core of the fibre may be able to relax to a greater extent than the outer layers. This would result in compressive forces in the outer regions and the development of kink bands leading to cumulative damage and eventual breakdown of the material in a similar manner to that found in flexing. Another reason for this criterion may be the cause of the turning of the crack. It is thought that the initial crack is deflected from its axis by realignment of the structure ahead of a notch brought about by the high stresses in this area. Maintaining a positive minimum load would hold the realignment throughout the cycling and prevent further crack growth as the crack would encounter decreasingly lower stress levels once deflected. Reducing the load to zero each cycle would allow for further crack growth. Once turned, the highly anisotropic nature of these highly aligned structures would assist in the propagation along the fibre.

Most tests have been carried out at 50 Hz but higher frequencies have been investigated. Raising the frequency increases the number of cycles to failure although time to failure remains about the same. The minimum oscillatory load amplitude for fatigue failure also increases with frequency.

The effect of draw ratio on the fatigue properties of nylon 66 has been investigated by comparing the behaviour medium and super high tenacity fibres. No difference in behaviour under these cyclic conditions was discerned between these two types.

b) Nomex

Fibres of Nomex, which is an aromatic polyamide manufactured by E. I. du Pont de Nemour & Co., have been subjected to tensile fatigue tests in the same manner as the nylon 66 fibres and have been found to fail by a similar fatigue mechanism but in fewer cycles. Nomex fibres have a cross section shaped like a pea-nut shell. They also have a much rougher surface than nylon 66 fibres, and it is this and not the different cross section which probably accounts for the shorter fatigue life by allowing initation to occur much more readily. Fatigue failure of Nomex fibres occurs after about 3×10^4 cycles. Because of the fineness of the Nomex fibres which were available it was not possible to determine whether the zero minimum load criteria, necessary for the fatigue of nylon 66 fibre, held for these fibres.

c) Polyester

When polyester fibres were tested for their fatigue properties they were found to be less likely to fatigue than the nylon fibres. These fibres, however, were found to fail after a similar fatigue crack growth history to that of nylon but after considerably longer cycling, usually between 2.5 and 8×10^5 cycles. The angle of penetration of the crack as it runs along the polyester fibre is smaller than in the polyamide fibres and considerably more crack growth is necessary before the load bearing area is reduced sufficiently for failure to occur. This results in extremely long thin tongues of material being left on the one fracture end which have been stripped off the complementary end.

d) Courtelle and carbon fibres

While the polyamide and polyester fibres failed by similar fatigue mechanisms the acrylic fibre which was tested was found to behave differently. Uncrimped *Courtelle* tow fibres were subjected to oscillatory tensile loadings and were found to readily split to give appearances, after failure, superficially similar to the polyamides and polyesters. On closer inspection however, it became clear that the fractures do not develop in a similar manner and that the structure of the *Courtelle* fibre splits, probably originating from an internal region of the fibre. Failure by splitting is found to occur even when small oscillatory loads are applied on high average loads, although a maximum load of about 70% of the normal tensile strength is required for failure. The initial split develops along the fibre and in doing so deviates from the direction of the fibre axis because of local irregularities in the structure. In this way the load bearing cross section is reduced until complete failure can occur under the existing loading conditions. Like the results obtained under simple tensile loading, which are described above, this behaviour is an indication of the low cohesive bonding of this type of fibre.

Some work on high modulus carbon fibre and its precursor acrylic fibre has been done. The precursor fibre behaved in the same way as described above for the *Courtelle* fibres. Long splits occurred under oscillatory loading. Which eventually led to failure. The carbon fibre was found to fail at several points simultaneously and for the fractures to be straight across the fibre.

References

Backer, S. and *C. J. J. Monego*, Textile Inst. (in press) from *C. J. Monego*, M. Sci. thesis, M.I.T. 1965.

Bosley, D. E., Textile Res. J. **38**, 141 (1968).

Bunsell, A. R., J. W. S. Hearle, and *R. D. Hunter*, J. Physics **E 4**, 868 (1971).

Bunsell, A. R. and *J. W. S. Hearle*, J. Mat. Sci. **6**, 1303 (1971).

Cross, P., J. W. S. Hearle, B. Lomas, and *J. T. Sparrow*, Study of Fibres in the Scanning Electron Microscope. Proc. 3rd Ann. SEM Symposium, ITT Res. Inst., pp. 81–88 (Chicago/Ill. 1970).

Greer, R. E. and *J. W. S. Hearle*, Polymer **11**, 441 (1970).

Hearle, J. W. S. and *P. M. Cross*, J. Mat. Sci. **5**, (1970).

Hearle, J. W. S. and *J. T. Sparrow*, Textile Res. J. **41**, 9, 736 (1971).

Hearle, J. W. S. and *V. M. Thakur*, J. Tex. Inst. **52**, 2, T49 (1961).

Jackson, P. W., D. M. Braddick, and *P. J. Walker*, Fibre Science and Tech. **5**, 219 (1972).

Whitney, W. and *R. M. K. Kimmel*, Nature Physical Sci. **273**, 75, 93–94 (1972).

Authors' addresses:

A. R. Bunsell
School of Applied Sciences
University of Sussex
Falmer
Sussex BNI 9QT (England)

J. W. S. Hearle
Dept. of Textile Technology
University of Manchester
Institute of Science and Technology
Manchester, M60 1QD (England)

Rheol. Acta **13**, 717–720 (1974)

Laboratoire de rhéologie – Institut de mécanique technique Sofia (Bulgarie)

Rôle des éléments fins dans le comportement rhéologique des bétons frais

I. P. Ivanov et I. T. Siméonov

Avec 9 figures et 1 tableau

Introduction

La connaissance des propriétés et du comportement rhéologique du béton frais a une grande importance pratique tant pour le chantier que pour les laboratoires. La connaissance des propriétés et du comportement rhéologique du béton frais permettra de poser sur une base scientifique la détermination des composantes du béton et de choisir le moyen le plus convenable de transport et la méthode la plus efficace et économique de son mise en place dans les constructions. Finalement, l'étude la découverte de la particularité dans le comportement rhéologique du béton frais et les facteurs desquels ce comportement dépend permettra de prévoir tant les propriétés mécaniques que les processus qui ont lieu dans la formation de la structure et des propriétés du béton durci.

Aspect rhéologique du béton frais

Les recherches nombreuses (1–9 et 11) ont montré que le béton frais ainsi que la pâte de ciment et les mortiers trouve sa place dans »l'arbre des corps rhéologiques« de *M. Reiner*, c'est à dire le béton frais représente un corps rhéologique et son comportement peut être exprimé par des équations rhéologiques. Sur cet arbre le béton frais se trouve entre les corps solides et les fluides et son comportement est complexe. Il manifeste les propriétés des corps plastiques, viscoplastiques, élastoviscoplastiques ou des milieux pulvérulents. Cette manifestation dépend de la nature du système – un système dispersé énorme avec une granularité étendue continue ou discontinue, de la phase du système et avec une aptitude clairement exprimé de structurisation du milieu du système (1–3, 6, 22). *M. Papadakis* (10) a indiqué que ce système dispersé structural se trouve sous l'influence de deux effets fondamentaux: a) l'effet de surface (dus aux grains fins de la phase et du milieu) et b) l'effet de masse (dus aux gros grains de la phase – on ne doit pas oublier aussi l'enchevêtrement).

Quand nous considérons les effets (de masse et de surface) qui ont une influence sur le comportement rhéologique du béton frais on doit noter aussi les forces internes et les forces dynamiques dans le cas de son transport et de sa mise en place.

Les particularités (granularité étendue, le domaine des effets, manifestation d'une structurisation etc.) sont causés de difficultés dans les recherches du comportement rhéologique du béton frais. Pour cette raison (on ne doit pas oublier l'absence d'appareillage convenable et à bon marché pour les chercheurs) au moins pour l'instant n'est pas possible de faire une étude complexe des propriétés rhéologiques fondamentaux (cohésion, viscosité, frottement interne) et technologiques (thixotropie, rhéopexie, dilatation etc.).

L'étude du comportement rhéologique du béton frais au cours de sa vibration est encore beaucoup plus difficile. Pour ce cas on a proposé la caracteristique dynamique rhéologique (12, 13). Elle presente la résistance interne du béton frais (fig. 1) et peut être determinée par (1–3, 11, 13):

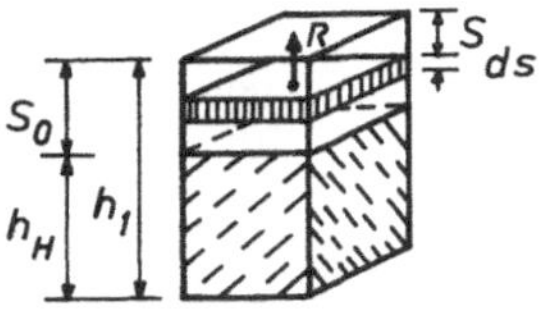

Fig. 1. Volume élémentaire du béton frais

$$dR_{CH} = \frac{dJ_x \cdot dT}{dS} \qquad [1]$$

où: dJ_x – c'est l'intensité de la vibration, dT – temps pour finir la vibration pour unité du volume du béton frais et dS – changement de la porosité du béton frais en temps dT.

D'après (11) la caractéristique dynamique rhéologique est défini comme une mesure de la diminution de porosité du béton frais pour unité de temps et sous l'agitation d'unité de l'intensité de vibration. C'est une caractéristique énergétique rhéologique généralisée – due à l'effet de la thixotropie, cohésion et au frottement interne du béton frais. Elle décrit le comportement rhéologique du béton frais au cours de sa vibration.

Les recherches (3, 11, 12, 13) ont montré que le comportement rhéologique du béton frais au cours de sa vibration (R_{CH}) dépend de l'intensité de vibration, de la composition du béton frais et des propriétés de ses constituents. On a trouvé (3, 11) que l'état, la surface et la forme des grains de granulats, la dimension des gros grains et la mobilité du béton frais influencent sur le changement de R_{CH} (le comportement du béton frais au cours de sa vibration).

Dans cette communication on annoncera les résultats obténus pendant des recherches sur l'influence des constituents fins du béton frais sur R_{CH}. On analysera ensuite son influence sur le changement de la cohésion et du frottement interne de milieu – la pâte du ciment, considérée comme une suspension concentrée.

Rôle des éléments fins (effets de surface)

1. Dans le changement du R_{CH} du béton frais

Comme on a noté ce sont les grains fins du milieu et de la phase qui provoquent les effects de surface. Dans nos recherches l'influence des effets de surface sur R_{CH} a été suivi d'une variation de la quantité des grains de granulat avec $d < 1$ mm et d'utilisation du ciment de different surface spécifique. Les résultats de ces recherches on peut voir sur les fig. 2, 3, 4 et dans le tableau 1.

Tableau 1
Influence de la surface spécifique du ciment sur R_{CH} du béton frais $D_{max} = 30$ mm $k = 10°$ VEBE

E/C	surface spécifique			
	$S = 2909$ cm²/g		$S = 3466$ cm²/g	
	R_{CH}	%	R_{CH}	%
0,35	630	100	908	133
0,45	657	100	817	124
0,55	674	100	846	125
0,65	659	100	758	111
0,75	673	100	760	113

Les données montrent que:

1. R_{CH} croît avec l'accroissement de la finesse des éléments fins (surface spécifique du ciment) (tabl. 1), c'est à dire croît la résistance interne du béton frais.

2. Le changement de la quantité des élements fins du sable provoque un changement de R_{CH}. La grandeur et le caractère du changement de R_{CH} dépendent de la quantité des éléments fins, D_{max} des granulats et de la mobilité du béton frais. En cas de bétons frais avec petit D_{max} (< 4 mm) la diminution de cette quantité provoque un abaissement de R_{CH} et inversement une augmentation de la quantité des grains fins – accroissement de R_{CH} (fig. 2). En cas des bétons frais avec $D_{max} = 15$ ou 30 mm (fig. 3 et 4) on voit les particularités suivantes: a) pour les béton frais avec une mobilité plus élevée ($k = 5$ et 10° VEBE) l'augmentation de la quantité des éléments fins jusqu'une limite est suivi d'un décroissement des valeurs de R_{CH} et après cette limite

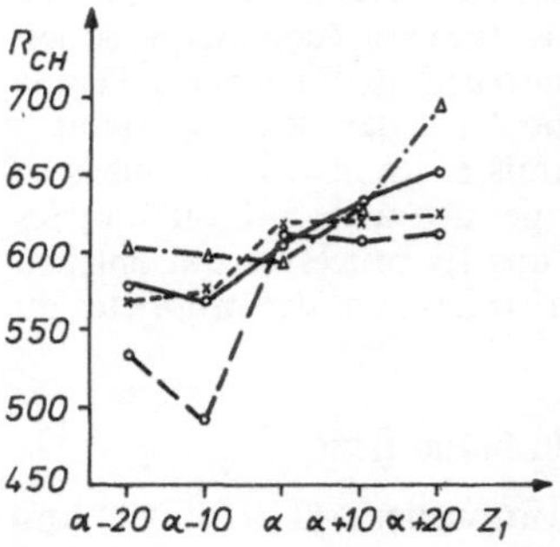

Fig. 2. Influence de la quantité des éléments fins Z_1 ($d < 1$ mm) sur R_{CH}($D_{max} = 2$ mm, α – la quantité normale de Z_1; $\alpha \pm 10\%$ – la quantité augmentée ou diminuée à 10%): ●, béton frais $k = 5°$ VEBE; ×, béton frais $k = 10°$ VEBE; △, béton frais $k = 40°$ VEBE; ⊙, valeur moyenne pour tous $k°$ VEBE

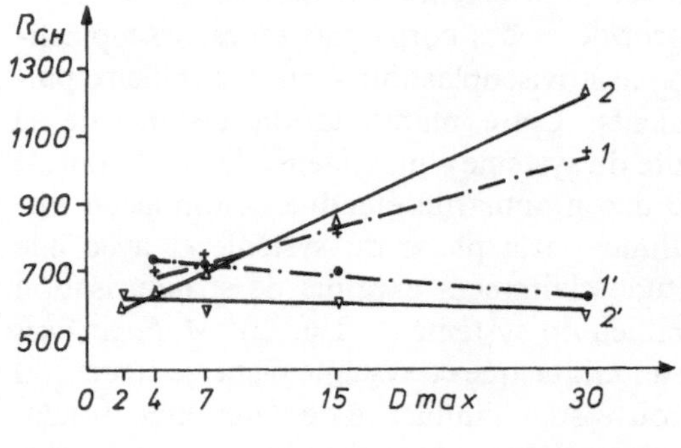

Fig. 3. Influence de la quantité des grains Z_1 du sable augmenté à 10% sur du béton frais: 1, +, $k = 40°$ VEBE et la quantité de Z_1 normale; 2, △, $k = 20°$ VEBE et la quantité de Z_1 normale; 1', ⊕, $k = 40°$ VEBE et la quantité de Z_1 augmenté; 2', ▽, $k = 20°$ VEBE et la quantité de Z_1 augmenté

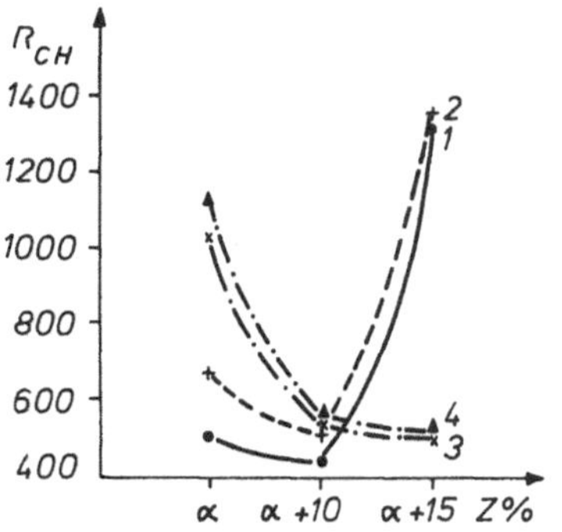

Fig. 4. Influence de la quantité des grains Z_1 sur du béton frais (D_{max} = 30 mm, les significations de α sont les mêmes que dans la fig. 2: 1, ○, k = 5° VEBE; 2, +, k = 10° VEBE; 3, ×, k = 20° VEBE; 4, △, k = 40° VEBE

R_{CH} croît rapidement avec l'augmentation de la quantité de ces grains; b) en cas des bétons frais de k = 20 et 40° VEBE l'augmentation de la quantité des élements fins a comme résultat un abaissement de R_{CH} (fig. 4).

3. Pour chaque béton frais existe une valeur critique de quantité des éléments fins (fig. 2 et 4) dans la mesure où R_{CH} a une grandeur optimum. L'explication de cette limite pour chaque béton frais et la détermination de la valeur optimum de R_{CH} permettra aux technologues de choisir une telle composition du béton frais qui lui donnera des moyens de diriger tant la vibration du béton frais que les particularités et les propriétés du béton durci.

2. *Dans le changement de la cohésion et du frottement interne de la pâte du ciment*

M. Papadakis (10) a montré l'influence des éléments fins sur la viscosité des bétons frais qui

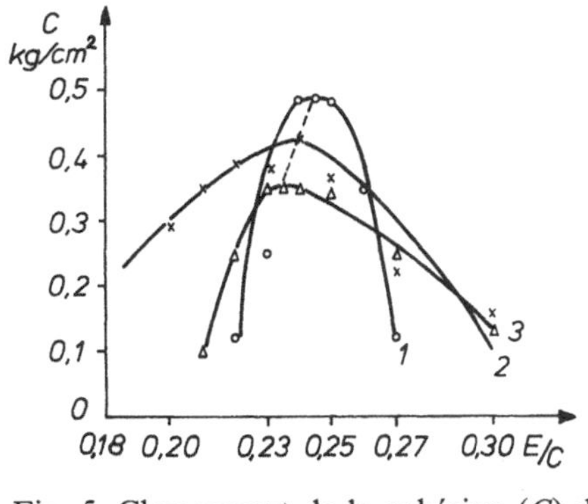

Fig. 5. Changement de la cohésion (C) de la pâte du ciment en fonction du rapport E/C et la surface spécifique du ciment (Blein): 1, Ciment de surface spécifique S = 5040 cm²/gr.; 2, Ciment de surface spécifique S = 3960 cm²/gr.; 3, Ciment de surface spécifique S = 2390 cm²/gr.

représentent avant tout la viscosité de milieu – la pâte du ciment. Ici on montrera les résultats obténus pendant les recherches de l'influence de ces particules sur la cohésion et le frottement interne du milieu. La méthode utilisée dans ces recherches est connue (7, 10). Les résultats obtenus sont données sur les fig. 5–9. Ils montrent que:

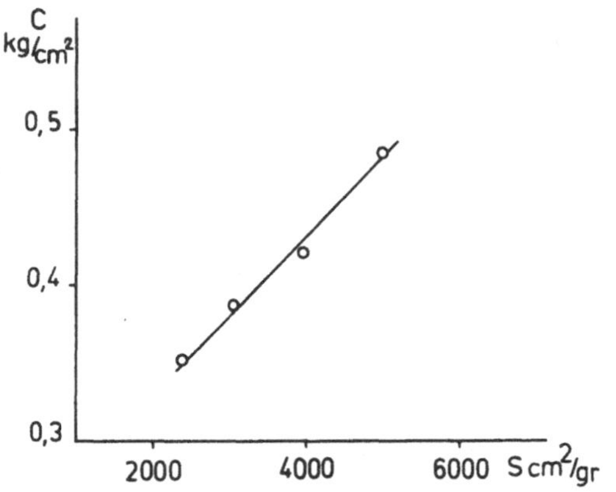

Fig. 6. C_{max} de la pâte en fonction de la surface spécifique du ciment.

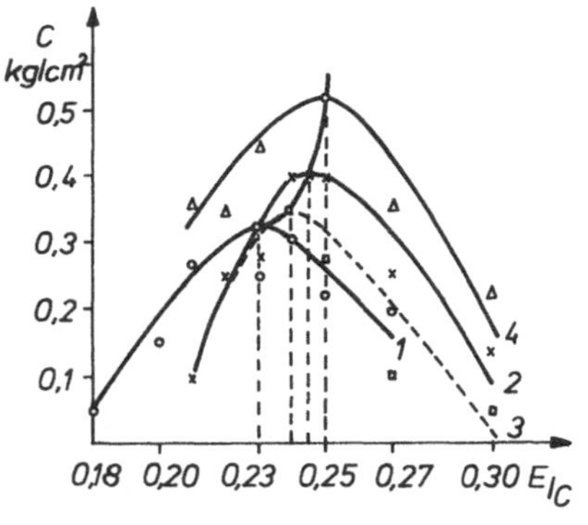

Fig. 7. Changement de la cohésion de la pâte en fonction du rapport E/C et la nature du ciment: 1, ●, ciment portland; 2, ×, ciment hydraulique; 3, □, ciment à durcissement rapide; 4, △, ciment antiacide

1. La cohésion du milieu (la pâte du ciment) dépend du rapport E/C, de la nature (fig. 7) et de la finesse (surface spécifique) du ciment (fig. 5). Les valeurs maxima de la cohésion dépend du rapport E/C et elles représentent une fonction linéaire de l'augmentation de la surface spécifique du ciment.

Noterons que le rapport E/C dont la cohésion a son maximum correspond entièrement au valeurs critique du rapport E/C, trouvé de nous,

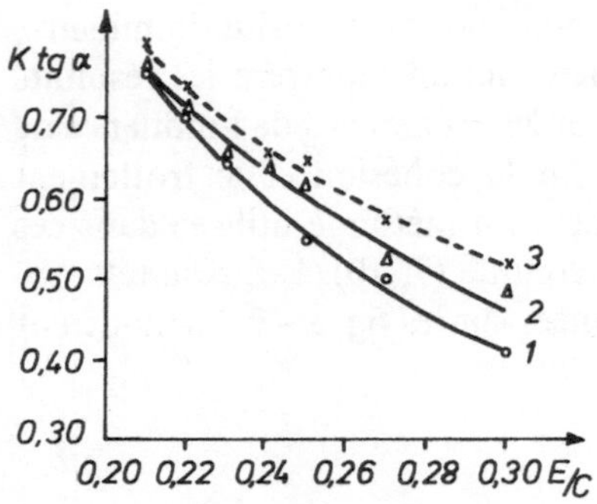

Fig. 8. Influence du rapport E/C et la surface spécifique du ciment sur le frottement interne (k) de la pâte. 1, ●, ciment de surface spécifique $S = 2390$ cm²/gr.; 2, △, $S = 3965$ cm²/gr.; 3, ×, $S = 5040$ cm²/gr.

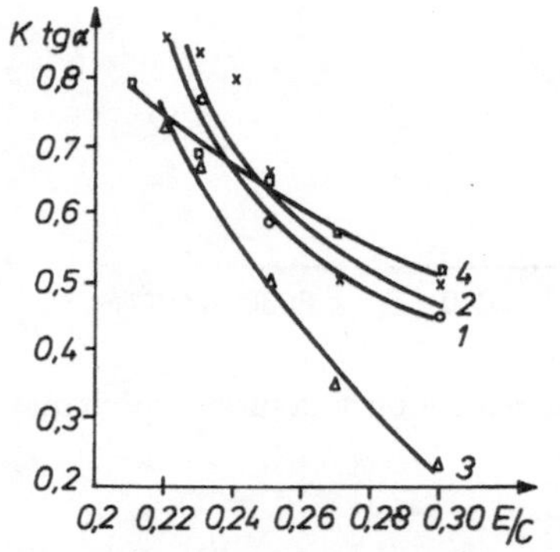

Fig. 9. Influence du rapport E/C et la nature du ciment sur le frottement interne: 1, ●, ciment portland; 2, ×, ciment à durcissement rapide; 3, △, ciment antiacide; 4, □, ciment hydraulique

sous lequel la pâte du ciment sous l'agitation de la vibration a un comportement pareil de celui d'un matériel pulvérulent (1).

2. Le frottement interne même dépend du rapport E/C (de la concentration de suspension), de la nature et la surface spécifique du ciment (fig. 8 et 9). L'accroissement de la surface spécifique du ciment est suivi d'un augmentation du frottement interne des pâtes et inversement accroissement du rapport E/C – par un abaissement du frottement interne des pâtes.

Nos recherches ont montré que les grains fins du sable ($d < 1$ mm) influencent tant sur le changement de R_{CH} du béton frais que sur la cohésion et le frottement interne du milieu – la pâte du ciment. L'accroissement de la quantité de ces grains est suivi d'une augmentation du frottement et d'un abaissement de la cohésion.

Résumé

Les recherches effectuées ont montré que le comportement rhéologique du béton frais considéré comme un système dispersé structural dépend tant de la quantité que de la qualité des constituents les plus fins. Ils influencent tant sur le changement du comportement du béton frais en général que sur la cohésion et le frottement interne du milieu – la pâte du ciment dans laquelle sont situés les grains de la phase. Par conséquant, la connaissance des lois du changement de ces caractéristiques nous oblige de trouver ses valeurs optima pour pouvoir assurer:

1. un transport sans dislocation du béton frais;
2. une vibration effective et économique du béton frais;
3. les meilleures propriétés physiques, mécaniques et rhéologiques du béton durci.

Littérature

1) *Ivanov, I.*, Sur la rhéologie des pâtes du ciment, mortiers et bétons frais, Sbornique Mechanica splochnichs sreds, Sofia, 1968 (en russe).

2) *Ivanov, I.*, Rhéologie du béton frais, Iubileina nautchna secia VISI, Sofia, 1968 (en bulgare).

3) *Ivanov, I.*, Sur la caractéristique dynamique rhéologique des systèmes structuraux, Izvestia ITM, vol. VIII (Sofia, 1971) (en bulgare).

4) *Kinos, G. I.*, Technologie vibratoire du béton (Leningrad, 1967) (en russe).

5) *Faitelson, L. A.*, Issledovania po betonu et gelezobetonu, vol. II et V (Riga, 1957 et 1960) (en russe).

6) *Ouriev, N. B.* et *N. V. Michailov*, Docl. AN USSR, N **4**, 1963 (en russe).

7) *Bombled, J. P.*, Revue des matériaux de construction, No **59**, 1965.

8) *Komlos, K.*, Stavebuicky casopis, No **4**, 1961 (Bratislave) (en slovaque).

9) *Papadakis, M.*, Revue des matériaux de construction, No 571, 582, 590, 594.

10) *Papadakis*, M., Rôle des éléments fins dans la rhéologie du béton frais, IV congrès du BIBM, Paris, 1963.

11) *Siméonov, I.* and *I. Ivanov*, Rhéological Behaviour of Concrete Mixtures, Proceedings of the Fifth International Congress on Rheology, Kyoto, Japan, 1968.

12) *Stork, J.*, Stavebuicky casopis, 1963, No 1–2 (Bratislava).

13) *Stork, J.*, Teoria skladby betonovy zmesi (Bratislava, 1964).

Adresse de l'auteur:

Dr. *J. T. Siméonov*
Zentrales Laboratorium der
Physik. Chem. Mechanik der
Bulg. Akad. der Wissensch.
IV Block, IV klm
Sofia-13 (Bulgarien)

Rheol. Acta 13, 721–724 (1974)

From the Oak Ridge National Laboratory, Oak Ridge, Tennessee (USA) and the Illinois Institute of Technology, Chicago, Illinois (USA)

Energy release integral for finite elastic and viscoelastic materials*)

*S. J. Chang***) *and B. Bernstein*

Introduction

A few years ago *Rivlin* and *Thomas* (1) illustrated experimentally that the rupture phenomenon of rubber is related closely to the amount of energy release associated with the rupture process. The analysis of the energy release from the mechanics point of view is therefore needed. For linear elastic materials, it can be computed as the work required to close the crack for a short distance. This method will not be valid for inelastic materials. From a different approach, *Rice* (3) discovered a path-independent integral which for elastic materials can be interpreted as the energy release per unit extension of the crack surface. Though this derivation is, strictly speaking, valid only for nonlinear elastic materials in general, it has been capable of applying to the case of elastic-plastic fracture with considerable success (4). A similar result was independently obtained by *Cherepanov* (5) whose integral is valid for elastic as well as inelastic materials under the condition of infinitesimal deformation. A more general energy release integral is derived by employing a well-known energy balance equation in continuum mechanics as shown by a paper by *Thomas* (6). For the interest of the fracture of polymers the present result is derived under the condition of large deformation. As an application to the tearing test, we establish a relation between the energy release and the strain concentration for a finite viscoelastic model of *Bernstein-Kearsley-Zapas* (7). The result for the finite elastic materials was derived previously (8).

Energy release integral

We shall describe the position of a particle of a continuum in an undeformed configuration by the cartesian coordinates $X_\alpha (\alpha = 1,2,3)$ and the position in the deformed configuration by $x_i\,(i = 1,2,3)$. A motion is denoted by

$$x_i = x_i(X_\alpha, t) \qquad i = 1,2,3 \quad \text{and} \quad \alpha = 1,2,3 \qquad [1]$$

where t stands for time. The deformation gradients $x_{i\alpha}$ are defined by

$$x_{i\alpha} = \frac{\partial x_i}{\partial X_\alpha} \qquad [2]$$

and their inverse $X_{\alpha i}$ is defined as the differentiation of X_α with respect to x_i. A one-to-one correspondence between particle and position requires [1] to be invertible. In the subsequent analysis, we shall sometimes use a moving coordinate system which is defined by

$$Y_\alpha = X_\alpha - H_\alpha(t) \qquad \alpha = 1,2,3 \qquad [3]$$

where $H_\alpha(t)$ is the position of the crack tip in X-coordinates. Obviously $Y_\alpha = 0\,(\alpha = 1,2,3)$ at that point for all time.

Let V be a material volume in X coordinate system, that is, a region which consists of the same material points at any time. We shall use V_y to denote the same material volume as described in Y coordinate system. The following equation is obviously true from a simple transformation of coordinates,

$$\frac{d}{dt}\int_V \rho\, U(X,t)\, dV = \frac{d}{dt}\int_{V_y} \rho\, \hat{U}(Y,t)\, dV_y \qquad [4]$$

where the *Jacob*ian of the transformation is one,

$$\left|\frac{\partial Y(t)}{\partial X(t)}\right| = 1 \qquad [5]$$

and $\hat{U}$ is numerically equal to U under the transformation [3]. In eq. [4], ρ is defined by

$$\frac{1}{\rho} = \left|\frac{\partial x_i}{\partial X_\alpha}\right| \qquad [6]$$

and $U(X,t)$ is the specific internal energy of the material. Physically [4] is the rate of change of the internal energy within the volume V; it contains the part owing to the extension of the crack. We then apply the energy balance equation shown in a paper by *Thomas* (6) to the right-hand side of [4], namely

*) Paper presented at 6th International Congress of Rheology, Lyon, France, 1972.

**) Research sponsored by the US Atomic Energy Commission under contract with Union Carbide Corporation.

$$\frac{d}{dt}\int_{V_y} \rho\,\hat{U}(Y,t)\,dV_y = \int_{V_y} \frac{\partial \rho\,\hat{U}(Y,t)}{\partial t}\,dV_y + \int_{S_y} \rho\,\hat{U}(Y,t)\,G\cdot n\,dS_y \qquad [7]$$

where S_y is the surface of V_y and G is the velocity vector of the element of the surface dS_y and n is the unit outward normal to S_y.

From the transformation [3], on the surface S_y,

$$G_\alpha = -\frac{dH_\alpha}{dt}. \qquad [8]$$

The surface of the crack is assumed to be straight and the extension of the tip is assumed to be parallel to the surface. The first term on the right-hand side of [7] is the rate of change of ρU as expressed in terms of the moving coordinate system Y integrated throughout the material volume V_y at the instant t. This result, however, will hold whether the volume V_y is moving. Therefore, we may take a new volume V_m which coincides with the region V_y at the present moment but V_m is relatively fixed with respect to the tip of the defect. Then

$$\int_{V_y} \frac{\partial \rho\,\hat{U}(Y,t)}{\partial t}\,dV_y = \int_{V_m} \frac{\partial \rho\,\hat{U}(Y,t)}{\partial t}\,dV_m = \frac{d}{dt}\int_{V_m} \rho\,U(X,t)\,dV_m \qquad [9]$$

and, consequently, from [4], [7], and [9], we obtain

$$\frac{d}{dt}\int_V \rho U(X,t)\,dV = \frac{d}{dt}\int_{V_m} \rho U(X,t)\,dV_m - \int_S \rho U(X,t)\frac{dH}{dt}\cdot n\,dS \qquad [10]$$

where S is the surface of V (or V_m). For a better derivation of [10] which considers the difficulty of the singular point at the crack tip, see the discussion. We can derive the energy release integral immediately by employing [10]. Since V is a material volume the rate of energy release is by definition

$$J = -\frac{d}{dt}\int_V \rho U(X,t)\,dV + \int_S T\cdot\frac{dx}{dt}\,dS \qquad [11]$$

where T is the traction vector per unit area of the undeformed surface S and x is the deformation vector. Substituting [10] into [11], we therefore obtain

$$J = -\frac{d}{dt}\int_{V_m} \rho\,U(X,t)\,dV_m + \int_S \rho\,U(X,t)\frac{dH}{dt}\cdot n\,dS + \int_S T\cdot\frac{dx}{dt}\,dS. \qquad [12]$$

The first term on the right-hand side of [12] represents the increase of energy in the moving volume owing to the unsteady motion of the crack and the second term is the energy flowing through the boundary S.

Before proceeding any further, let us examine how the general relation [12] may be reduced to the known forms by imposing different conditions to the problem. If the material is assumed to be elastic (or finite elastic) then a relation of virtual work in the moving coordinate system can be obtained as

$$\int_{V_m} \frac{\partial}{\partial t}[\rho\,\hat{U}(Y,t)]\,dV_m - \int_{S_m} T\left.\frac{\partial x}{\partial t}\right|_Y dS_m = 0 \qquad [13]$$

where $\left.\frac{\partial x}{\partial t}\right|_Y$ denotes the partial derivative of x with respect to t when Y is held fixed. Eq. [12] therefore reduces to the known form (3, 8)

$$J = \int_S \rho\,U(X,t)\frac{dH}{dt}\cdot n\,dS + \int_S T\frac{\partial x}{\partial Y}\frac{dY}{dt}\,dS = \int_S \rho\,U(X,t)\frac{dH}{dt}\cdot n\,dS - \int_S T\frac{\partial x}{\partial X}\frac{dH}{dt}\,dS \qquad [14]$$

where both $\frac{\partial x}{\partial Y}$ and $\frac{\partial x}{\partial X}$ are matrices and

$$\frac{\partial x}{\partial Y}\frac{dY}{dt} = \frac{\partial x}{\partial (X-H)}\frac{d(X-H)}{dt} = -\frac{\partial x}{\partial X}\frac{dH}{dt}. \qquad [15]$$

Moreover the eq. [14] is derived by the validity of [13]. Even if the material is not elastic the eq. [13] still leads directly to the resulting eq. [14]. For example, if we assume the motion to be steady state which implies both terms in [13] to vanish, eq. [14] also follows. Except the two cases just mentioned, if we assume V_m to be a two-dimensional region and let V_m tend to zero, it is seen that for usual singularity ρU is proportional to $1/r$ where r is the linear dimension of the region, the first term of [12] becomes negligible. Hence eq. [12] also reduces to a line integral similar to [14]. For many materials the second term of eq. [13] may also be small for small V_m even for unsteady problems. In this case *Rice*s integral is true in the small region and therefore is the energy release.

It is seen that eq. [12] is applicable to any material with unsteady crack length and applied force. *Rice*s integral is the energy release for elastic material with unsteady motion in general whereas *Cherepanov*s integral, the energy release for any material but with stationary stress intensity factor.

Application

We shall illustrate the application of eq. [12] by an example for a finite viscoelastic model of *Bernstein-Kearsley-Zapas* (7). For simplicity, we shall assume the isothermal condition. Let F be a potential functional defined by

$$F = \varphi(v,T) + T\int_{-\infty}^{t} S[x_{ik}(t,\tau), t-\tau]\,d\tau \qquad [16]$$

where v is the specific volume of the material after deformation and T is temperature which is assumed to be constant. φ and S are two material functions and x_{ik} is the relative deformation gradient defined by

$$x_{ik}(t,\tau) = \frac{\partial x_i(t)}{\partial x_k(\tau)} = x_{i\alpha}(t)\,X_{\alpha k}(\tau)\,. \qquad [17]$$

In [17] τ denotes the previous time. In the subsequent analysis it is interesting to notice that it is not necessary to attach to the potential F any thermodynamic meaning in order to derive the result. One uses F simply because of its mathematical convenience.

To obtain a specific form of U which is required in [12], we shall derive some quantities from F. Differentiating F with respect to time,

$$\begin{aligned}\dot F &= \frac{\partial\varphi(v,T)}{\partial v}\,\frac{\partial v}{\partial x_{i\alpha}}\,\frac{dx_{i\alpha}}{dt} \\ &\quad + vX_{\alpha j}\rho T\int_{-\infty}^{t} x_{j\gamma}X_{\gamma k}(\tau)\frac{\partial S}{\partial x_{ik}}\,d\tau\cdot\frac{dx_{i\alpha}}{dt} \\ &\quad + T\int_{-\infty}^{t}\frac{\partial}{\partial t}S[x_{ik}(t,\tau),t-\tau]\,d\tau \\ &= \frac{1}{\rho}X_{\alpha j}\sigma_{ij}\frac{dx_{i\alpha}}{dt} - Q \qquad [18]\end{aligned}$$

where σ_{ij} is the stress tensor and

$$Q = -T\int_{-\infty}^{t}\frac{\partial}{\partial t}S[x_{ik}(t,\tau),t-\tau]\,d\tau\,. \qquad [19]$$

If we use $\pi_{i\alpha}$ to denote

$$\pi_{i\alpha} = \frac{1}{\rho}X_{\alpha j}\sigma_{ij} \qquad [20]$$

then

$$\dot F = \pi_{i\alpha}\frac{dx_{i\alpha}}{dt} - Q\,. \qquad [21]$$

By definition

$$\rho U = \int_{-\infty}^{t}\pi_{i\alpha}\frac{dx_{i\alpha}}{dt}\,dt = F + \int_{-\infty}^{t} Q\,dt \qquad [22]$$

eq. [12] reduces to

$$\begin{aligned}J &= -\frac{d}{dt}\int_{V_m}\Big(F + \int_{-\infty}^{t} Q\,dt\Big)\,dV_m \\ &\quad + \int_S\Big(F + \int_{-\infty}^{t} Q\,dt\Big)\frac{dH}{dt}\,n\,dS + \int_S T\cdot\frac{dx}{dt}\,dS\,. \qquad [23]\end{aligned}$$

If we apply the above equation to a steady state tearing test, then

$$\begin{aligned}J &= \frac{dH_1}{dt}\int_S\Big(F + \int_{-\infty}^{t} Q\,dt\Big)\,n\,dS \\ &\quad - \frac{dH_1}{dt}\int_S T\frac{\partial x}{\partial X_1}\,dS \qquad [24]\end{aligned}$$

which as in the elastic case is free from the volume integral. In eq. [24] we assume the cut to be parallel in X_1 direction (i. e., $H_2 = H_3 = 0$) and both H and n are vector quantities.

By applying eq. [24] we are able to derive a result parallel to the relation between the tearing energy and the average strain concentration for elastic materials (8). We shall use a test piece with the same configuration which is shown in fig. 2 of (2). In that figure a plate of rectangular shape is cut partially in the middle with a round notch at the end of the cut. The tearing test is made by pulling the ends so as to tear the test piece apart along the cut.

Let the edges of the cut be parallel to the X_1 axis, then any two integrals with contours C_1 and C_2 which end up on the straight edges for the two-dimensional specimen will be equal. If in particular we choose C_1 to be the outer surface of the specimen and C_2 to coincide with the notch surface, since $T_1 = 0$ except at ends of the test piece, we obtain the following relation

$$\int_{C_1}\left[(F+\int_{-\infty}^{t} Q\,dt)-T_2\frac{\partial x_2}{\partial X_1}\right]dX_2$$

$$= \int_{C_2}(F+\int_{-\infty}^{t} Q\,dt)\,dX_2 \qquad [25]$$

which is our desired relation. Obviously if the material is elastic our previous result (8) is recovered by assigning $Q = 0$ and $F = W$ in [25].

The result [25] is obtained by the absence of the volume integral. Therefore, if we want to perform a tearing test, we need only to measure the notch strain which will be sufficient to determine F and Q there and to measure the tearing energy at the far ends.

Discussion

In the process of deriving eq. [12] a singularity is located at the crack tip which is on our surface S. We may be concerned about whether the analysis is influenced by the singular point. A way to overcome this difficulty is illustrated in the following discussion. We may start as an assumption that

$$\left|\frac{d}{dt}\int_{C_m}\rho\,U(X,t)\,dC_m\right| < M \qquad [26]$$

where M is a constant and C_m is a small volume which is relatively fixed with respect to the crack tip. Then we can apply the energy balance equation without ambiguity to a volume $V_y - C_m$ which is free from any singularity,

$$\frac{d}{dt}\int_{V_y-C_m}\rho\,\hat{U}(Y,t)\,dV = \int_{V_y-C_m}\frac{\partial\rho\,\hat{U}(Y,t)}{\partial t}\,dV$$

$$+\int_{S_y}\rho\,\hat{U}(Y,t)\,G\cdot n\,dS_y \qquad [27]$$

where we have used the fact that $G = 0$ on C_m and, consequently,

$$\int_{C_m}\rho\,\hat{U}(Y,t)\,G\cdot n\,dS_y = 0\,. \qquad [28]$$

By the same reasoning which leads to [9], we obtain

$$\int_{V_y-C_m}\frac{\partial\rho\,\hat{U}(Y,t)}{\partial t}\,dV = \frac{d}{dt}\int_{V_m-C_m}\rho\,U(X,t)\,dV_m\,. \qquad [29]$$

Then, substituting [29] into [27],

$$\frac{d}{dt}\int_{V-C_m}\rho\,U(X,t)\,dV = \frac{d}{dt}\int_{V_m-C_m}\rho\,U(X,t)\,dV_m$$

$$-\int_{S}\rho\,U(X,t)\frac{dH}{dt}\,n\,dS \qquad [30]$$

which is equal to [10] if we assume each term in [10] is finite. Therefore the same result as eq. [12] should follow.

Summary

The energy release integral of *Rice* and *Cherepanov* type is derived more generally by employing an energy balance equation which is often used in continuum mechanics. Several results are observed from this energy release integral. The integral is then applied to the tearing test of a finite viscoelastic material. An equation is established to relate the energy release rate and the average strain concentration along the notch surface, a more general result than the one obtained previously for the finite elastic materials.

References

1) *Rivlin, R. S.* and *A. G. Thomas*, J. Polymer Science **10**, 291 (1953).
2) *Thomas, A. G.*, J. Polymer Science **18**, 177 (1955).
3) *Rice, J. R.*, J. Appl. Mech. **35**, 379 (1968).
4) *Rice, J. R.*, in: *H. Liebowitz* (Ed.), Fracture, Vol. 2 (New York 1968).
5) *Cherepanov, G. P.*, J. Appl. Math. Mech. (PMM) **31**, 476 (1967).
6) *Thomas, T. Y.*, Math. Mag. **22**, 169 (1949).
7) *Bernstein, B., E. A. Kearsley,* and *L. J. Zapas*, Trans. Soc. Rheol. **7**, 391 (1963).
8) *Chang, S. J.*, Z. Ang. Math. Phys. **23**, 149 (1972).

Authors' addresses:

Dr. *Shih-Jung Chang*
Oak Ridge National Laboratory
Oak Ridge, Tenn. 37830 (USA)

Prof. *Barry Bernstein*
Dept. of Mathematics
Illinois Institute of Technology
Chicago, Ill. 60616 (USA)

Rheol. Acta **13**, 725–729 (1974)

Laboratoire de Mécanique et d'Acoustique, C. N. R. S. Marseille (France)

Détermination du module d'Young opérationnel d'un corps viscoélastique à partir de sa courbe de relaxation

Synthèse de fonctions de transfert

Michel Raous

Avec 3 figures et 1 tableau

(Reçu p. p. le 27 octobre 1972)

I. Introduction

Dans un corps viscoélastique, la contrainte et la déformation sont liées par une relation différentielle. Dans un certain domaine, nous pourrons considérer cette relation comme linéaire, et l'écrire, alors dans l'espace de *Laplace* sous la forme:

$$\bar{\sigma} = E(p) \cdot \bar{\varepsilon} \qquad [1]$$

où $\bar{\sigma}$ et $\bar{\varepsilon}$ sont les transformées de *Laplace* de $\sigma(t)$ la contrainte et $\varepsilon(t)$ la déformation, toutes deux fonctions du temps. Pour déterminer le module d'*Young* opérationnel $E(p)$, on utilise souvent des modèles constitués d'assemblage de ressorts et d'amortisseurs tels que ceux de *Maxwell, Kelvin, Voigt, Burgers*... Notre méthode a pour but de déterminer directement $E(p)$ à partir de l'analyse de la courbe de relaxation qui n'est autre que la réponse de contrainte à une excitation en échelon de déformation. L'ordre de la relation différentielle posée dépendra seulement du critère d'erreur admis entre la courbe expérimentale et la courbe théorique issue du modèle après identification des différents paramètres. La seule hypothèse posée est celle de la linéarité de la relation différentielle liant $\sigma(t)$ et $\varepsilon(t)$: elle nous permet de traiter $E(p)$ comme une fonction de transfert. Nous sommes ramenés à un problème de synthèse de la fonction de transfert $T(p)$ d'un système à partir de la connaissance de sa réponse à une fonction «échelon».

II. Synthèse de fonction de transfert à partir de la réponse à un échelon de heaviside

1. Position du problème

Si nous pouvons écrire la réponse à un échelon $s(t)$ sous la forme [2], nous en déduirons simplement sa transformée de *Laplace* (expression 3).

$$s_m(t) = (a + \sum_{j=1}^{N} K_j e^{p_j t})\, H(t) \qquad [2]$$

$$\bar{S}_m(p) = \frac{a}{p} + \sum_{j=1}^{N} \frac{K_j}{p - p_j} \qquad [3]$$

où: a est une constante réelle (qui peut être nulle): ce sera la valeur asymptotique. K_j et $p_j \in C$. N est le nombre de pôles p_j: c'est donc l'ordre de l'équation différentielle.

L'expérience consiste à enregistrer $\sigma(t)$ quand $\varepsilon(t) = H(t)$. Par conséquent $S_m(p) = E(p) \cdot 1/p$. D'où l'expression du module d'*Young* opérationnel:

$$E(p) = a + \sum_{j=1}^{N} \frac{K_j p}{p - p_j}. \qquad [4]$$

Le problème consiste à déterminer une fonction $s_m(t)$ (expression 2) qui décrira correctement la courbe expérimentale $s(t)$. Il s'agit là de la synthèse d'une fonction de transfert puisque les paramètres a, K_j, p_j déterminent un modèle mathématique du système physique. Précisons ceci:

Proposition 1:

On dira avoir réalisé un modèle mathématique (Γ, f) de ce système physique si pour une classe d'entrée Γ, une fonctionnelle f de l'écart $e(t) = s(t) - s_m(t)$ est telle que $0 \leq f(e) \leq f_0$ (où f_0 est une constante).

Le modèle n'est qu'une représentation de l'objet valable, c'est-à-dire donnant un «faible» écart entre le comportement de l'objet et celui de son modèle, pour un type d'entrée donné et un critère donné (cf. fig. 1). Le problème va

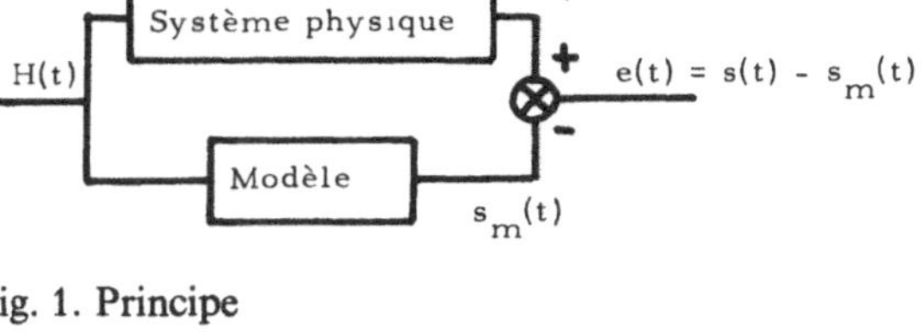

Fig. 1. Principe

consister ici à déterminer les coefficient K_j et p_j (a se mesure directement) qui minimiseront un critère $f(e)$.

2. Caractérisation – identification

2.1. Caractérisation

C'est l'étape qualitative qui consiste à choisir un modèle mathématique déterminé. Elle fixe l'espace paramétrique dans lequel vont évoluer les paramètres structuraux du modèle. Nous écrirons:

$$\bar{b} = \begin{vmatrix} K_1 \\ \vdots \\ K_N \\ p_1 \\ \vdots \\ p_N \end{vmatrix}$$

Le type de l'équation différentielle a été défini (II.1.), l'ordre de cette équation est fixé par N. La caractérisation sera complète après le choix de N.

2.2. Identification

C'est l'étape quantitative qui consiste à déterminer les $2N$ paramètres minimisant le critère d'erreur $f(e)$ (cf. proposition 1).

Nous choisissons un critère intégral quadratique.

$$f(e) = \frac{1}{T}\int_0^T e^2(t)\,dt.$$

Pour N fixé, il existera dans l'espace paramétrique un domaine D_N tel que si $\bar{b} \in D_N$ alors $f(e) \leq f_N$ où f_N est une constante dépendant de N. Ceci revient à dire que f_0 de la proposition 1 dépend du choix de N.

Il est évident que le lissage de la courbe expérimentale $s(t)$ sera d'autant plus précis que le nombre N d'exponentielles (formule 2) sera plus élevé. Cette constante f_N sera choisie égale à l'erreur expérimentale: il faudra donc choisir N de façon que la proposition 1 soit vérifiée quand f_N est égal à l'erreur expérimentale Δ. Appelons D_0, le domaine défine par la donnée de $f_N = \Delta^2$.

L'identification se fait en deux étapes. Une méthode directe détermine un vecteur $\bar{b}$ de l'espace paramétrique proche du domaine D_0, puis une méthode itérative déplace ce vecteur $\bar{b}$ jusqu'à ce qu'il atteigne le domaine D_0.

2.2.1. Méthode directe: méthode de Prony.

Cette méthode détermine $\bar{b}$ tel que $s_m(t)$ passe par $2N$ points régulièrement espacés de la courbe $s(t)$. Nous avons $2N$ inconnues, choisissons $2N$ échantillons qui nous permettront d'écrire $2N$ équations. Posons:

$$s'_m(t) = s_m(t) - a \qquad [5]$$

et

$$s'(t) = s(t) - a. \qquad [6]$$

Soit $s'(0)$, $s'(\Theta)$, ..., $s'((2N-1)\Theta)$ les $2N$ échantillons. En écrivant $s'_m(t) = s'(t)$ pour $t = 0, \ldots, (2N-1)\Theta$, nous obtenons le système d'équations non linéaires suivant:

$$K_1 + K_2 + \cdots + K_N = s'(0)$$

$$K_1\lambda_1 + K_2\lambda_2 + \cdots + K_N\lambda_N = s'(\Theta)$$

$$\vdots$$

$$K_1\lambda_1^{2N-1} + K_2\lambda_2^{2N-1} + \cdots + K_N\lambda_N^{2N-1} = s'((2N-1)\Theta) \qquad [7]$$

avec

$$\lambda_j = e^{p_j\Theta}.$$

Remarque

une courbe de relaxation n'est jamais oscillante: nous pouvons d'ores et déjà en déduire que les pôles seront réels. Afin de ne pas nuire à la généralité de cette méthode, les p_j seront considérés comme complexes au cours de cet exposé.

En cherchant une fonction qui passe par $2N$ valeurs distribuées, nous posons un problème d'équations aux différences (finies).

$$C_N g(I) + C_{N-1} g(I+1) + \cdots + C_1 g(I+N-1) + C_0 g(I+N) = 0 \qquad [8]$$

avec

$$I = 1, 2, \ldots, 2N$$

$$g(I) = s'((I-1)\Theta).$$

On peut écrire $C_0 = 1$ puisque [8] s'écrit à un facteur près. Explicitons ces équations:

$$C_N g(1) + C_{N-1} g(2) + \cdots + C_1 g(N) = -g(N+1)$$
$$C_N g(2) + C_{N-1} g(3) + \cdots + C_1 g(N+1) = -g(N+2)$$
$$\vdots$$
$$C_N g(N) + C_{N-1} g(N+1) + \cdots + C_1 g(2N-1) = -g(2N). \qquad [9]$$

Le système d'éq. [9] nous permet de calculer les C_j. Une telle équation à différences finies est satisfaite par λ^I (λ constant):

$$C_N \lambda + C_{N-1} \lambda^2 + \cdots + C_1 \lambda^N = -\lambda^{N+1}$$

ou encore:

$$\lambda^N + C_1 \lambda^{N-1} + \cdots + C_N = 0 \qquad [10]$$

qui possède N racines $\lambda_I (I = 1, \ldots, N)$.

Dans ce cas, la solution générale s'écrit:

$$g(I+1) = K_1 \lambda_1^I + K_1 \lambda_2^I + \cdots + K_N \lambda_N^I. \qquad [11]$$

Nous avons posé $\lambda_j = e^{p_j \Theta}$, nous pouvons en déduire p_j:

$$p_j = \frac{1}{\Theta} \log \lambda_j. \qquad [12]$$

Il ne nous reste qu'à porter ces valeurs dans le système d'éq. [7] qui est alors linéaire. Sa résolution nous conduit à la détermination des K_j.

En résumé, les $2N$ valeurs distribuées permettent la résolution du système d'éq. [9], dont les solutions sont les coefficients de l'équation du *Nième* degré [10]. Les racines de [10] nous conduisent simplement aux valeurs des pôles p_j. Le système [9] est alors linéaire, sa résolution détermine les K_j.

Nous obtenons ainsi un vecteur $\bar{b}$ tel que:

$$s'_m(t) = \sum_{j=1}^{N} K_j e^{p_j t}$$

passe par $2N$ valeurs de la courbe expérimentale $s'(t)$. Nous calculons $f(e)$.

2.2.2. Méthode itérative: méthode du gradient

Si le vecteur $\bar{b}$ déterminé par la méthode directe n'appartient pas (cas le plus fréquent) à D_0, nous appliquerons la méthode du gradient.

Cette méthode consiste à déplacer le vecteur $\bar{b}$ de façon à maximiser la diminution Δf.

$$f(b_{i+1}) = f(b_i) + \Delta f.$$

Effectuons un développement limité au ler ordre de $f(b)$ autour de $f(b_i)$

$$f(b_i + \delta b) \simeq f(b_i) + f_b^T(b_i) \cdot \delta b$$

pour $\|\delta b\| < \varepsilon$.

La variation δb de norme limitée à ε qui maximise $-\Delta f \simeq -f_b^T(b_i)\,\delta b$ est le vecteur colinéaire à $f_b(b_i)$ et de longueur ε. On utilisera la formule itérative:

$$b_{i+1} = b_i - k \cdot f_b(b_i) \qquad [13]$$

ce qui revient à écrire que la variation δb est colinéaire au gradient de f. Du point de vue de l'hypersurface déterminée sur l'espace paramétrique par la fonction $f(b_i)$, cela revient à déplacer le vecteur $\bar{b}$ suivant la ligne de plus grande pente. Remarquons qu'une telle trajectoire est celle d'une goutte d'eau sur une surface. Cette méthode porte aussi le nom de méthode de la plus grande pente («Steepest descent»). La convergence se fait suivant une vallée de résolution.

k est une constante destinée à assurer la convergence de la méthode. Ici en ajustant séparément les coefficients k_i pour chacun des paramètres, nous pourrons sphériser le problème, ce qui assurera une convergence plus stable et plus rapide.

Dans notre problème, les dérivées partielles peuvent être calculées analytiquement.

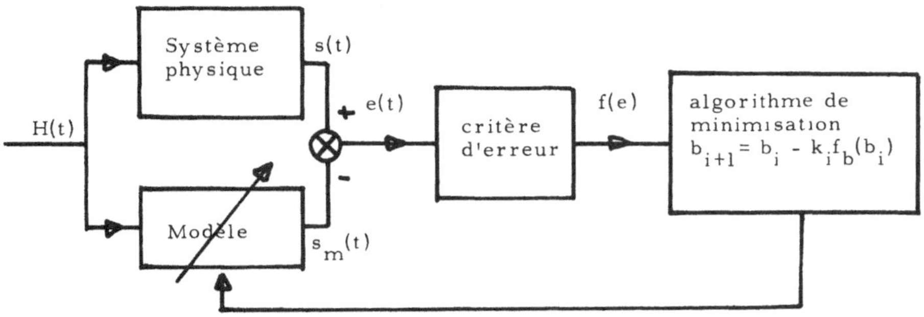

Fig. 2. Identification

$$f(e) = \frac{1}{T}\int_0^T (s(t) - s_m(t))^2\,dt$$

$$f_{K_n} = \frac{\partial f}{\partial K_n} = -\frac{2}{T}\int_0^T (s(t) - s_m(t))\,e^{p_n t}\,dt$$

$$= -\frac{2}{T}\int_0^T e(t)\,e^{p_n t}\,dt \qquad [14]$$

$$f_{p_n} = \frac{\partial f}{\partial p_n} = -\frac{2}{T}\int_0^T (s(t) - s_m(t))\,K_n p_n e^{p_n t}\,dt$$

$$= -\frac{2}{T}\int_0^T e(t)\cdot K_n \cdot p_n e^{p_n t}\,dt \qquad [15]$$

car:

$$\frac{\partial s(t)}{\partial b_i} = 0.$$

III. Application

Nous allons chercher un modèle décrivant le phénomène de relaxation d'un caoutchouc.

Il s'agit d'un caoutchouc désigné dans le commerce sous le nom de Perbunan P21.

Conditions expérimentales (5):
Section: 25 mm², longueur: 100 mm.
Allongement: 25%, température: 23°.

Courbe de relaxation: (cf. fig. 3).
Valeur initiale: $E_i = 25{,}68$ kg/cm².
Valeur asymptotique: $a = 23{,}8$ kg/cm².
Erreur expérimentale: 1%. (Supposée telle afin de mettre en évidence l'efficacité de la méthode.)

La mise en œuvre de la méthode directe et de la méthode du gradient se font évidemment sur calculateur. Les temps de calcul seront donnés dans le cas de l'ordinateur UNIVAC 1106 que nous avons utilisé.

La fonction $s(t)$ est échantillonnée: soit M le nombre d'échantillons, et Θ la période d'échantillonnage.

Lors de l'application de la méthode directe, il y a $\dfrac{M-1}{2N-1}$ façons de choisir les $2N$ échantillons, régulièrement espacés à partir de la valeur initiale. Ces différents choix nous conduiront à des vecteurs $\bar{b}$ en général distincts plus ou moins proches du domaine D_0. Aussi pour N fixé, nous envisagerons tous les choix possibles et conserverons celui qui nous conduit à un vecteur $\bar{b}$ donnant un critère d'erreur minimum.

Dans notre cas ($M = 84$), les temps de calcul sont de l'ordre de 12–15 s (y compris le temps de compilation) selon la valeur de N.

A partir de cet ensemble de paramètres structuraux ainsi calculés (vecteur $\bar{b}$), nous appliquons la méthode du gradient. C'est une méthode dont la convergence est rapide au début, puis beaucoup plus lente ensuite. A titre d'exemple, le temps de calcul pour 100 itérations lors d'une recherche à 4 paramètres ($N = 2$, $M = 84$) est de l'ordre de 22s (y compris le temps de compilation).

Il peut se faire que $\bar{b}$ atteigne une position telle que $f(e)$ soit minimum sans que $f(e) = \Delta^2$. Nous sommes dans le cas où $f_N > \Delta^2$; la structure du modèle n'est pas suffisante pour décrire correctement le phénomène physique. Nous prendrons alors un modèle à $N+1$ pôles, et ainsi de suite jusqu'à ce que notre critère d'erreur soit satisfait. Il apparait ici nettement que la caractérisation sera fonction de nos exigences sur les qualités du modèle. En général, 2 ou 3 pôles seront suffisants pour décrire le phénomène de relaxation d'un caoutchouc.

C'est ainsi que l'application de ces deux méthodes à $s(t) - a$, nous conduisent à l'évolution suivante de l'erreur (cf. tableau 1).

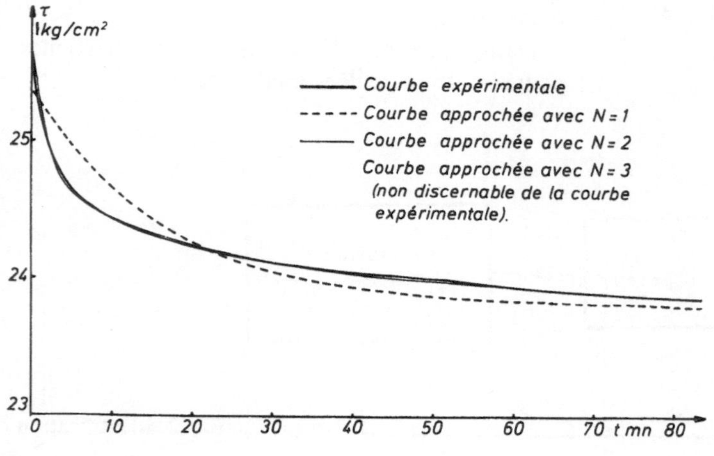

Fig. 3. Courbe de relaxation

Tableau 1

N	Méthode directe	Méthode du gradient
1	17 %	13 %
2	3,4%	1,9%
3	1,1%	1 %

Nous avons ainsi décomposé la réponse à un échelon de déformation, suivant l'expression [16].

$$s(t) = 23{,}8\, H(t) + \sum_{j=1}^{3} K_j e^{p_j t} \cdot H(t) \qquad [16]$$

avec:

$K_1 = 0{,}698 \qquad p_1 = -4{,}4 \ 10^{-4}$

$K_2 = 0{,}548 \qquad p_2 = -2{,}87 \ 10^{-3}$

$K_3 = 0{,}626 \qquad p_3 = -1{,}61 \ 10^{-2}$.

Les K étant homogènes à des [$\mathrm{kg/cm^2}$] et les p à des [$\mathrm{s^{-1}}$].

Nous intéressons à la réponse normalisée (réponse à un échelon de *Heaviside* de hauteur 1).

$$s(t) = 9{,}52\, H(t) + (0{,}28\, e^{p_1 t} + 0{,}22\, e^{p_2 t} + 0{,}25\, e^{p_3 t})\, H(t). \qquad [17]$$

Sa transformée de *Laplace* est:

$$S(p) = \frac{9{,}52}{p} + \frac{0.28}{p - p_1} + \frac{0{,}22}{p - p_2} + \frac{0{,}25}{p - p_3}. \qquad [18]$$

Le module d'*Young* opérationnel a pour expression:

$$E(p) = 9{,}52 + p \left(\frac{0{,}28}{p - p_1} + \frac{0{,}22}{p - p_2} + \frac{0{,}25}{p - p_3} \right).$$

$E(p)$ est la fonction de transfert d'un modèle dont la réponse à un échelon de déformation approche $s(t)$ la courbe expérimentale avec une erreur inférieure à 1%.

IV. Conclusion

Etant donnée une courbe de relaxation, cette méthode permet, suivant la remarque de la proposition 1 (paragraphe II.1.), d'élaborer un modèle, et d'en déterminer les paramètres structuraux de façon que la réponse à un échelon de ce modèle corresponde à la courbe expérimentale avec une erreur de l'ordre de celle due à la mesure.

Cette méthode effectue une synthèse directe du modèle mathématique sans passer par l'intermédiaire d'un modèle mécanique. De ce fait, le modèle est plus souple, et on peut facilement en modifier la structure lorsque l'identification nous conduit à un résultat insuffisant. De plus, sans aucune difficulté supplémentaire, on pourrait élaborer des modèles très compliqués (10 pôles par exemple), si cela était nécessaire: seuls les temps de calcul seraient augmentés de quelques secondes.

Rappelons que, moyennant les hypothèses du paragraphe I, la connaissance de $E(p)$ nous permet de déterminer la réponse du système à une excitation quelconque par convolution de celle-ci avec la fonction originale de $E(p)$.

Littérature

1) *Boudarel, R., J. Delmas* et *P. Guichet*, Commande optimale des processus. Tome II, 307 p. (Paris 1968).

2) Cours de *D. F. Tuttle* (Stanfort University – Californie). Fait à l'Ecole Supérieure de Physiques de Marseille (1968/69).

3) *Lawrence, E. Nielsen*, Mechanical Properties of Polymers (New York-London 1963).

4) *Bozzo, C.*, Etude d'un isomorphisme entre l'espace des signaux continus et l'espace des signaux discrets. Différentes applications. Thèse de doctorat d'Etat, Marseille, Septembre 1970.

5) *Nguyen-Trong, T.*, Etude des corps viscoélastiques utilisés dans les mouvements vibratiles. Rapport de D.E.A. – Marseille (C.R.P.) – 1968.

6) *Radix, J. C.*, Introduction au lissage numérique: lissage de données, estimation de paramètres, identification de processus. 239 p. (Paris 1970).

Adresse de l'auteur:

M. Raous

Centre national de la recherche scientifique

Laboratoire de Mécanique et d'Acoustique

31, chemin Joseph-Aiguier

F-13 Marseille 9 (France)

Rheol. Acta **13**, 730–739 (1974)

From the UK Ministry of Defence (PE), Waltham Abbey, Essex (England)

Determination of plastoviscosity and the flow curve equation of stiff pastes by parallel plate plastometry

W. A. Dukes and Virginia M. Gledhill

With 8 figures

Introduction

Plastic propellant is a stiff paste typically containing (by weight) 89% solids (ammonium perchlorate and perhaps aluminium and a burning catalyst), 1% synthetic surface-active agent and 10% polyisobutene. The polyisobutene has a nominal viscosity-average molecular weight of about 4×10^4 and a viscosity of about 100 kN/m^2 (10^6 poise) at 25 °C, when measured at a very low shear-rate by the falling sphere (*Stokes*) method. The paste is similar to modelling clay or Plasticine, and may be considered to a first approximation as a *Bingham* solid subject to shear-hardening (1).

A survey (2) of viscometric techniques for use with polymeric liquids summarises the range of viscosities over which various methods are useful and suggests that methods such as capillary extrusion, tensile creep and parallel-plate plastometry, are particularly suitable for fluids of such high viscosity. Other desiderata are listed (2), including ease of thermostatting, availability of a wide range of shear stresses and shear rates, and applicability to small samples. These considerations support the choice of parallel-plate plastometry in the present work. The size of sample required is small (the whole apparatus may easily be inserted in a laboratory oven or cold chamber) and about a one-hundred-fold range of load is easily available, giving, as will be shown later, about a 10^4 range of shear rates. The availability of a wide range of shear-rates is important when dealing with polymers, which are usually non-*Newton*ian. It is known that the mechanical properties of plastic propellant are highly dependent on the strain-rate, so it is important to consider the effect of this parameter on the rheological properties.

The compression between parallel plates of a sample, often cylindrical, by a constant load has been widely used as an empirical method of measuring the consistency of plastic materials. The main reason for the use of this test is probably that this form of specimen is easily made and handled, even when the material is soft and sticky. Unfortunately the test is not so easily analysed in rheological terms; for example, both stress and strain vary both with position within the specimen and with time. It has been shown (1) that the fundamental properties of yield stress and strain-hardening coefficient may be derived from measurement of the equilibrium plate separation. In the present report time-effects are considered, involving such concepts as flow, strain-rate and plastoviscosity.

Material similar to that used in the present work has previously been examined in a concentric cylinder plastometer (3), and both the yield stress and the plastoviscosity were found to be affected by the strain the material had undergone; the yield stress increased linearly with increasing strain, i.e. the material shear-hardened linearly (as later confirmed by parallel-plate plastometry (1)), and the plastoviscosity decreased with increasing strain.

Theoretical

General

The yield stress f_m of a plastic material is the stress below which no flow takes place, i.e. the strain-rate $\dot{\gamma} = 0$ if the applied stress $s \leqslant f_m$; while in general for $s > f_m$ the rheological behaviour is specified by the flow curve

$$\dot{\gamma} = f(s - f_m). \qquad [1]$$

The flow curve has a finite intercept f_m on the stress axis, and is not necessarily a straight line. (A straight line with finite intercept defines a *Bingham* material.) The plastoviscosity η_{pl} is defined as the differential $d(s - f_m)/d\dot{\gamma}$, that is the inverse of the slope of the $\dot{\gamma}$ versus $(s - f_m)$ curve at any value of $\dot{\gamma}$, and is not necessarily invariant with $\dot{\gamma}$.

Non-linear flow curves are exhibited not only by some plastic materials, but also by liquid polymers which have a high viscosity. In this case departure from *Newton*ian behaviour may be explained as due to chain entanglements. The reduction of the viscosity of such materials with increasing shear rate is termed pseudo-plasticity or shear-thinning. Recent papers (4, 5) have dealt with this property, and a number of empirical equations has been put forward and discussed (5). For example, the following equation represents the behaviour of many systems (5).

$$\eta = \eta_\infty + (\eta_0 - \eta_\infty)/[1 + (\tau\dot{\gamma})^n] \quad [2]$$

where η is the viscosity at shear rate $\dot{\gamma}$, and η_0 and η_∞ are limiting values at $\dot{\gamma} = 0$ and $\dot{\gamma} = \infty$, respectively. τ is a constant such that at a shear rate τ^{-1} the viscosity assumes the mean value $(\eta_0 + \eta_\infty)/2$.

If η_∞ is zero, or negligible, and at rates of shear such that $(\tau\dot{\gamma})^n \gg 1$, then this becomes

$$\eta = \eta_0/(\tau\dot{\gamma})^n \quad [3]$$

and a plot of log η against log $\dot{\gamma}$ is linear, with a slope of $-n$. This is sometimes called the power-law flow equation, and it is applicable to many polymeric liquids and plastic materials. A similar result is obtained (6) for many disperse systems on the basis of a kinetic interpretation, when the contribution of *Brown*ian movement to the rupture of links between particles is negligible compared with that due to applied shear. Eq. [3] may be rewritten

$$\eta = \frac{ds}{d\dot{\gamma}} = k\dot{\gamma}^{-n}. \quad [4]$$

By integration, (except when $n = 1$)

$$s - f_m = k(\dot{\gamma}^{1-n})/1 - n \quad [5]$$

where f_m is the constant of integration, equal to the yield stress (if any). Therefore

$$\dot{\gamma} = \left[\frac{1-n}{k}(s - f_m)\right]^{1/1-n}. \quad [6]$$

This is the form of the flow curve, generalised in eq. [1], specific for materials obeying the power-law flow equation. It is an example of the *Herschel-Bulkley* equation.

Zero-strain plastoviscosity

The compression of a cylindrical specimen results in the material flowing primarily radially outwards. The axial compressive force may be taken as equivalent to a hydrostatic compression, which causes no plastic strain, coupled with a bi-axial tension, which causes the radial flow outwards. Thus the radial strain,

$$\gamma_r = \int_r^{r_0} dr/r = \ln(r/r_0),$$

is considered more suitable than the compressive strain $\ln(h_0/h)$, and will be used here, as it was previously (1). A zero subscript refers to the initial value of either the radius r, the height h, or (below) the cross-sectional area A.

It has been established (7) that for a tall cylinder, $h_0 \geqslant 2r_0$, consideration of the exact shear zones gives the maximum stress equal to $Wg/2A$, where W is the applied load. Strictly speaking, a value for the initial yield stress should be subtracted from the shearing stress, but this is usually a negligible correction since the compressive loads are very much greater than the yield stress. Thus the initial (zero-strain) plastoviscosity, η_i, of a tall cylinder is given by dividing the initial stress by the initial radial strain-rate:

$$\eta_i = \frac{Wg}{2A_0\dot{\gamma}_{r,0}}. \quad [7]$$

For the standard specimen ($h_0 = 20$ mm, $2r_0 = 15$ mm) this becomes

$$\eta_i = 0.0278\ W(\text{kg})/\dot{\gamma}_{r,0}(\text{s}^{-1})\ \text{MN s/m}^2\text{*)} \quad [8]$$

The question of the absolute value of the strain and hence of the strain-rate will be considered further below.

Finite-strain plastoviscosity

The theory and application of the parallel-plate plastometer have been discussed previously (8–10); if a cylindrical sample of volume V is compressed by a constant load W between parallel plates, then a plot of $1/h^4$, where h is the plate separation at time t, against t will give a curve which becomes linear. From the slope of the linear portion the plastoviscosity η_d may be determined, from a modified form of *Stefan*'s equation:

$$\frac{3V^2}{8\pi}\left(\frac{1}{h^4} - \frac{1}{h_0^4}\right) = \frac{Wgt}{\eta_d}. \quad [9]$$

The assumptions on which this analysis is based are as follows:

a the material is incompressible;

b the flow behaviour of the material is *Newton*ian at the rates of shear employed during a test;

c no body force acts on the material;

d the motion is slow;

e there is no slip at the surface of the plates; and

f the plate separation is so small compared to the radius that the velocity component in the perpendicular direction is negligible compared with that in the radial direction.

*) 1 MN s/m² = 1 MPa · s = 10^7 poise.

These assumptions are commented on below

Assumption a is satisfied to a first-degree approximation by a void-free liquid or a dispersion of solid particles in a liquid. Assumption b is approximately correct within experimental error, as a linear plot of $1/h^4$ vs. t is found while the strain-rate is varying over the relatively small range occurring during a single experiment (i.e. the change in strain-rate during the period of purely viscous flow is small). Assumption c is reasonable, as the body forces are of the order of millibars, while the forces of the order of bars are being applied. Assumption d enables terms in the square of the velocity vector to be neglected in the *Navier-Stokes* equation. Assumption e has been shown (1) to be correct for plastic propellant. Assumption f was said (8) to be met if the ratio of radius to height of the compressed specimen was greater than 10: we find no inconsistencies with the ratio ranging from as small as 2 or 3 up to 19.

Experimental

Apparatus

A wide circular horizontal brass base-plate on a threaded coaxial support could be raised or lowered by rotation about its axis. Above the base-plate there was another horizontal circular brass plate, 51 mm in diameter, supported by a vertical rod passing through a low-friction bearing and provided with a concentric circular weight-pan, on which weights could be placed. At the top of the rod was rigidly fixed an inverted stirrup fitting through the jaws of a bomb-release mechanism, which supported the rod and upper plate until the mechanism was activated and the jaws sprang apart, when the upper plate with the rod and weights was free to fall. A cylindrical sample of the material to be tested was placed centrally on the lower plate, which was then raised until the top of the sample just touched the upper plate. The experiment was then started by activating the bomb-release mechanism.

Resting vertically on the upper plate was a brass rod supporting a soft-iron core inside the vertical coil assembly of a *Schaevitz* linear variable differential transformer (Electro Mechanisms Ltd, Slough), which produces an electrical output proportional to the displacement of the core. The output was led to a Sanborn (Hewlett-Packard) model 321 amplifier recorder which when calibrated recorded the plate separation as a function of time. The speed of response of the recorder was more than adequate for the fastest experiments. The duration of experiments varied between 0.25 sec and 1 h. The available chart speeds ranged from 0.5 to 100 mm/s. The loads used ranged from 0.4 to 100 kg. With loads greater than 5 kg, a 10:1 lever apparatus was used, with similar instrumentation. All experiments were performed at room temperature.

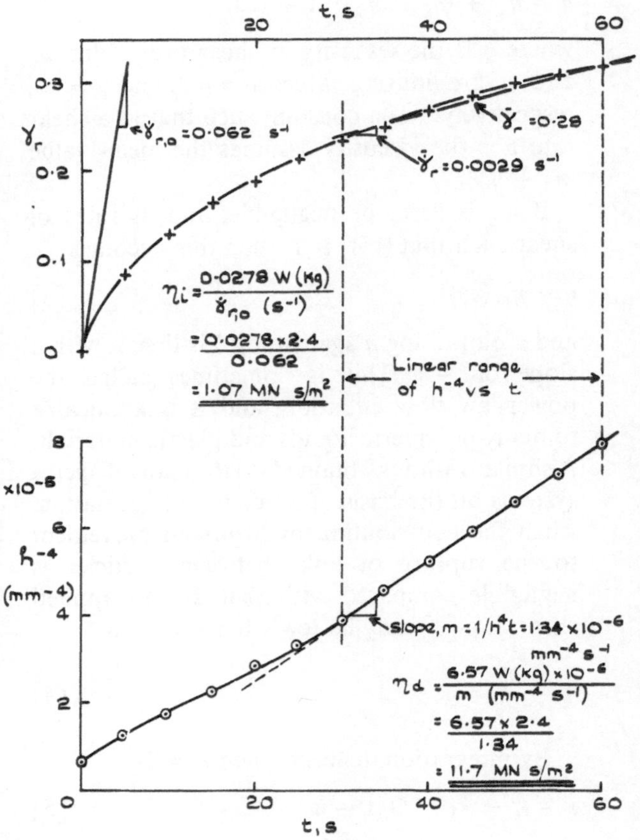

Fig. 1. Typical determinations of zero- and finite-strain plastoviscosities, η_i and η_d

Experience with this apparatus has led to the design of a more convenient and precise instrument, provided with temperature control and a transducer which on receipt of a clock pulse reads the specimen height at that instant and transfers it in digital form on to punched paper tape, if necessary via an electronic store. It is hoped that this system, together with a computer programme to provide automatic plotting of the functions shown in fig. 1, will facilitate future routine measurements.

Material

The plastic propellant used throughout the work described in this report was E 3342/Lot. 9. Its composition is given in Appendix I. Each cylindrical specimen was made from a single boring of the required weight, with freshly cut surfaces, confined in a lightly-greased mould of the required dimensions between two discs of polytetrafluorethylene (PTFE) film of the required radius at a pressure of 6.9 MN/m^2 for 1 min. Specimens were pushed out of the mould, and the PTFE discs carefully removed. The volume of specimens ranged from 0.5×10^{-6} to 8.2×10^{-6} m^3, the initial radius r_0 was within the range 5 to 10 mm, and the initial height h_0 was within the range 3.75 to 30 mm, with the initial aspect ratio h_0/r_0 ranging from 0.375 to 6.0.

Procedure

Each specimen was compressed by a known load W, and the plate separation h recorded as a function of time. The rate of change of height of course diminishes with increasing time. Plots of $1/h^4$ and of γ_r as a function of time were then derived, typified by those shown in fig. 1. The initial slope of the γ_r plot is drawn, equal to $\dot{\gamma}_{r,0}$, and this is substituted in eq. [7] or [8] to derive the zero-strain plastoviscosity η_i.

The slope m of the linear portion of the $1/h^4$ vs. t plot enabled the finite-strain plastoviscosity to be determined by substituting in eq. [9]. For the standard specimen this becomes

$$\eta_d = \frac{6.57 \times 10^{-6}\, W(\text{kg})}{m(\text{mm}^{-1}, \text{s}^{-1})}\, \text{MN s/m}^2. \qquad [10]$$

The time range of this linear relation was noted, and the strain vs. time curve examined over this range, and a mean slope drawn, thus deriving the mean strain-rate, $\bar{\dot{\gamma}}_r$, at which the plastoviscosity had been measured. The procedure is illustrated in fig. 1. It will be seen that there is little change in strain-rate during the measurement, as stated above. The mean radial strain $\bar{\gamma}_r$ during the measurement is also recorded.

Results

Zero-strain plastoviscosity

The parameters which have been varied in turn are: the load applied (and hence the strain-rate), the volume of the specimen, and the aspect ratio (initial height/initial radius, h_0/r_0) of the specimen. These are discussed in turn below. In every case, the heavier the load, the higher was the initial strain rate $\dot{\gamma}_{r,0}$ and the lower the initial viscosity η_i.

Aspect ratio and volume constant

The largest number of experiments with constant particular values of both the aspect ratio h_0/r_0 and the volume V was with the standard specimen. In fig. 2, log η_i is plotted as a function of log $\dot{\gamma}_{r,0}$. The least-squares regression of log η_i on to log $\dot{\gamma}_{r,0}$ is shown; it has slope -0.54 ($= -n$, see eq. [4]), with a standard error of estimate 0.14. The material thus obeys the power-law equation.

Aspect ratio constant, and volume varied

Specimens were made of four different volumes but with the aspect ratio maintained constant. These were compressed by various loads. In fig. 3 it is shown that the volume of the specimen does not affect the relation between initial plastoviscosity and strain-rate. The least squares regression of log η_i on to log $\dot{\gamma}_{r,0}$ is shown; it has a slope $-n = -0.52$, with a standard error of estimate 0.09.

Aspect ratio varied

Specimens of various volumes, with seven different aspect ratios ranging from 0.375 to 6.0, were compressed by various loads. It was found necessary to correct for a small influence of the aspect ratio (h_0/r_0) on the strain and strain-rate. The factor to normalise this effect, designated $p_{h/r}$, was empirically found to be related to the aspect ratio by

$$\log p_{h/r} = 0.67\,(\log h_0/r_0 - \log (h/r)_{\text{ref}}) \qquad [11]$$

where $(h/r)_{\text{ref}}$ is the reference aspect ratio (1.55). The basis for this correction is described in Appendix II. Except for this small factor, the aspect ratio does not affect the determination of the plastoviscosity, as can be seen in fig. 4, where all the measurements are shown as a function of the strain-rate.

The equation of the regression line, as drawn, is

$$\log \eta_i/p_{h/r} = -0.275 - 0.514 \log \dot{\gamma}_{r,0} \cdot p_{h/r}. \qquad [12]$$

The correlation coefficient is -0.97 and the standard error of estimate of the logarithm of the initial plastoviscosity is 0.13. Thus, within experimental error, the initial plastoviscosity is inversely proportional to the square-root of the strain-rate:

$$\eta_i/p_{h/r} = 0.53\,(\dot{\gamma}_{r,0} \cdot p_{h/r})^{-1/2}\, \text{MN s/m}^2. \qquad [12a]$$

This equation may be compared with the general form of eq. [4].

Finite-strain plastoviscosity

The parameters varied in the finite-strain plastoviscosity determinations were the same as for the zero-strain determinations described above; indeed the same specimens were used; the compression of each was continued, so that eq. [9] could be applied. The parameters are discussed in turn below. As before, the heavier the load, the higher was the mean strain-rate $\bar{\dot{\gamma}}_r$ and the lower the plastoviscosity η_d.

Aspect ratio and volume constant

Results similar to those shown in fig. 2 were obtained, with a least-squares slope $-n = -0.52$, but with a standard error of estimate 0.38, as compared with 0.14 for log η_i in fig. 2. The much greater scatter is presumably due to

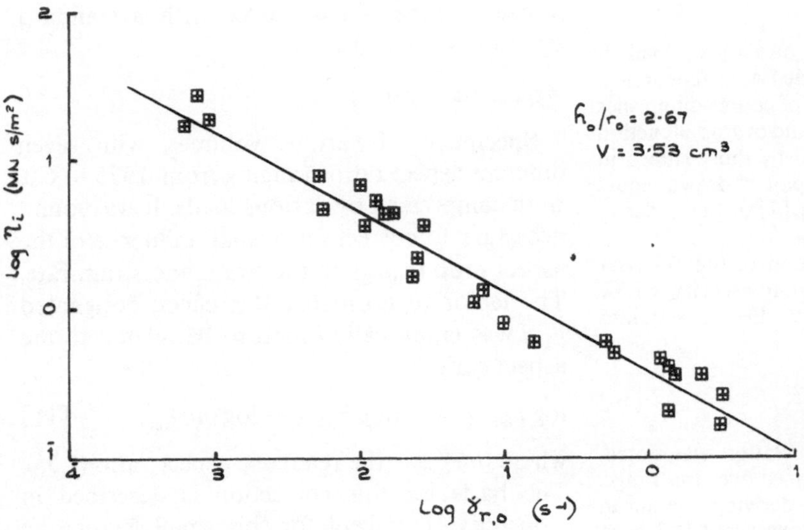

Fig. 2. The effect of strain-rate on zero-strain plastoviscosity (initial aspect ratio and volume constant)

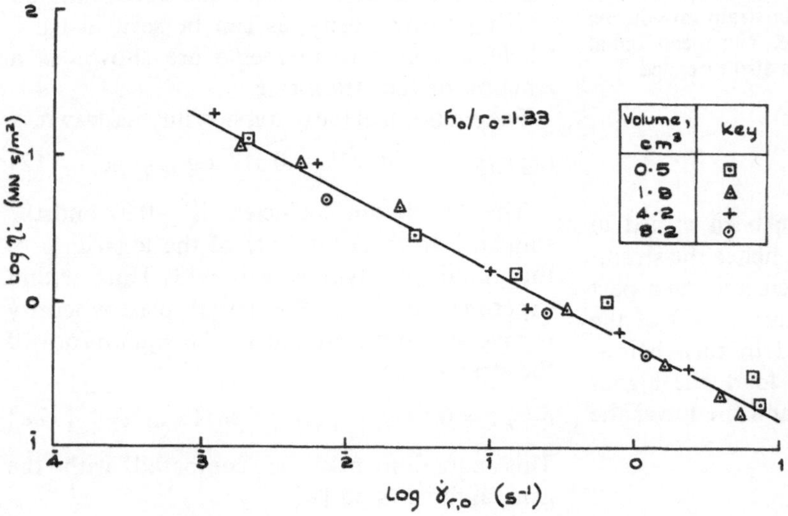

Fig. 3. The effect of volume and strain-rate on zero-strain plastoviscosity (initial aspect ratio constant)

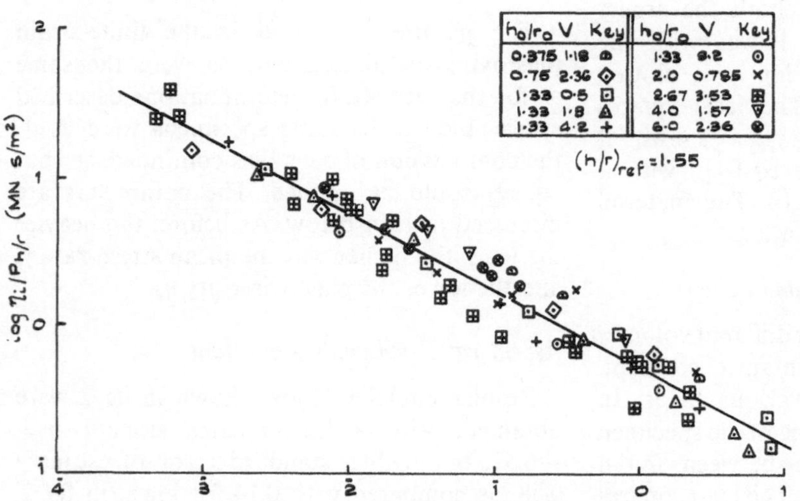

Fig. 4. The effect of strain-rate on zero-strain plastoviscosity, both normalised to an initial aspect ratio of 1.55

an additional factor, not operative in the case of the zero-strain experiments. The most obvious difference between finite-strain and zero-strain determinations is of course the strain, and so it was decided to examine the effects of varying the strain in addition to the other parameters.

Aspect ratio constant, and volume varied

Results similar to those in fig. 3 were obtained, showing again that the volume of the specimen does not affect the relation between plastoviscosity and strain-rate. The least-squares regression of $\log \eta_d$ on to $\log \bar{\dot{\gamma}}_r$ had a slope $-n = -0.52$, with a standard error of estimate 0.29, as compared with 0.09 for $\log \eta_i$ in fig. 3. Again the much greater scatter is presumably due to an additional factor, postulated to be the strain.

Strain-rate normalised, aspect ratio constant, and strain varied

In order to investigate the effect of varying both the aspect ratio and strain, it was necessary to remove the variable of the strain-rate. There is good evidence above of the effect of the strain-rate on the plastoviscosity. The power-law is obeyed, with an exponent, within experimental variation, negligibly different from 1/2. Therefore each determination of η_d was normalised to an arbitrary strain-rate ($\log \bar{\dot{\gamma}}_r = \bar{2}$), chosen as being near to the centre of the experimental range. The equation to convert the logarithm of the measured plastoviscosity $\log \eta_d$, to the normalised form, $\log \eta_{d\bar{2}}$, was

$$\log \eta_{d\bar{2}} = \log \eta_d + 1/2 (\log \bar{\dot{\gamma}}_r + 2)\,. \qquad [13]$$

The largest group of specimens, all with $h_0/r_0 = 2.67$, enabled the (strong) influence of the mean strain $\bar{\gamma}_r$ on the strain-rate-normalised plastoviscosity $\eta_{d\bar{2}}$ to be evaluated. The greater the strain, the lower is the plastoviscosity. A linear relation is suggested between $\log \eta_{d\bar{2}}$ and $\bar{\gamma}_r$, with a least-squares correlation coefficient -0.92, slope -1.90, and a standard error of estimate 0.15, which may be compared with 0.38, for the comparable measurements from which the strain-effect had not been isolated.

It was then necessary to find whether this strain-effect on the plastoviscosity is affected by the other parameters, namely the volume and aspect ratio of the specimen.

Strain-rate normalised, aspect ratio constant, volume and strain varied

Based on the specimens with various volumes, but all of aspect ratio $h_0/r_0 = 1.33$, the effect of the mean strain on the plastoviscosity was measured. Again it was found that the volume of the specimen does not affect the relation between $\log \eta_{d\bar{2}}$ and $\bar{\gamma}_r$. The least-squares regression had a slope of -1.68 and a standard error of estimate 0.07, as compared with 0.29 for the comparable measurement from which the strain-effect had not been isolated.

Strain-rate normalised, aspect ratio and strain varied

The specimens with various aspect ratios, described above, were used to investigate the effect of the aspect ratio on the relation between strain and strain-rate-normalised plastoviscosity. It was again found necessary to correct for the small influence of the aspect ratio on the strain (see Appendix II), and the correction, $-q_{h/r}$ (in terms of γ) was found to be equal to $\log p_{h/r}$ (in terms of $\dot{\gamma}$) within experimental accuracy. That is to say that the aspect ratio affects to the same degree both the strain and the strain-rate, as indeed would be expected. Since an increase in strain leads to a reduction in plastoviscosity, the sign of the correction $-q_{h/r}$ is opposite to that of the strain-rate correction, $\log p_{h/r}$.

Strain-rate and aspect ratio normalised, and strain varied

Now that the effects of both the strain-rate and the initial aspect ratio have been isolated, the effect of the strain on the plastoviscosity can be examined. In fig. 5 is shown the strain-rate-normalised plastoviscosity as measured on all the specimens as a function of $(\bar{\gamma}_r + q_{h/r})$, the strain normalised to $h_0/r_0 = 1.55$. The least-squares regression equation is

$$\log \eta_{d\bar{2}} = 0.932 - 1.88(\bar{\gamma}_r + q_{h/r}) \qquad [14]$$

and the standard error of estimate of the logarithm of the normalised plastoviscosity is 0.35.

The coefficient of this effect, namely of increasing strain reducing the plastoviscosity, is designated $\underline{b}$, and is equal to 1.88 here. The effect may be termed strain-thinning, as distinct both from the shear-hardening coefficient, designated (1) $\underline{a}$, and from shear-thinning, which is

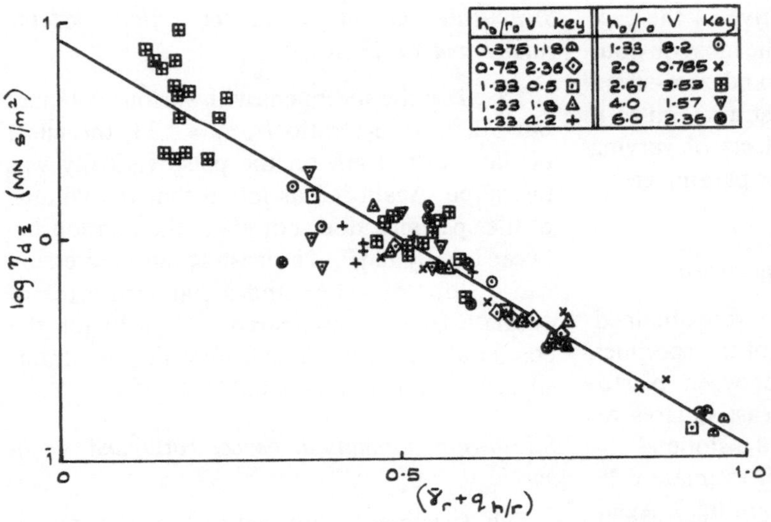

Fig. 5. The effect of strain (normalised to an initial aspect ratio of 1.55) on finite-strain plastoviscosity (normalised to a strain-rate of 0.01 s^{-1})

defined (11) as a reduction of viscosity with increasing rate of shear. Thus it is true that plastic propellant exhibits all three properties; namely shear-hardening (1) (that is a strain-dependent increase in yield stress), strain-thinning (i.e. a strain-dependent reduction in plastoviscosity), and shear-thinning (a strain-rate-dependent reduction in plastoviscosity – see figs. 2, 3 and 4).

Aspect ratio and strain normalised, and strain rate varied

Now that the strain-thinning coefficient b and the minor effect of the initial aspect ratio on the strain and the strain-rate have both been evaluated, it is possible to eliminate the effect of both by extrapolating to zero strain, using the following equation

$$\log \eta_{d,0} = \log \eta_d + b(\bar{\gamma}_r + q_{h\,r}). \qquad [15]$$

In this way the effect of strain-rate on the finite-strain plastoviscosity η_d can be isolated. Fig. 6 shows a plot of all the determinations of η_d thus normalised, as a function of log $\bar{\dot{\gamma}}_r$. The equation of the regression line, as drawn, is

$$\log \eta_{d,0} = -0.226 - 0.560 \log \bar{\dot{\gamma}}_r \qquad [16]$$

and the correlation coefficient is −0.99 and the standard error of estimate of log $\eta_{d,0}$ is 0.09. Thus, within experimental error, the finite-strain plastoviscosity is inversely proportional to the square root of the strain-rate:

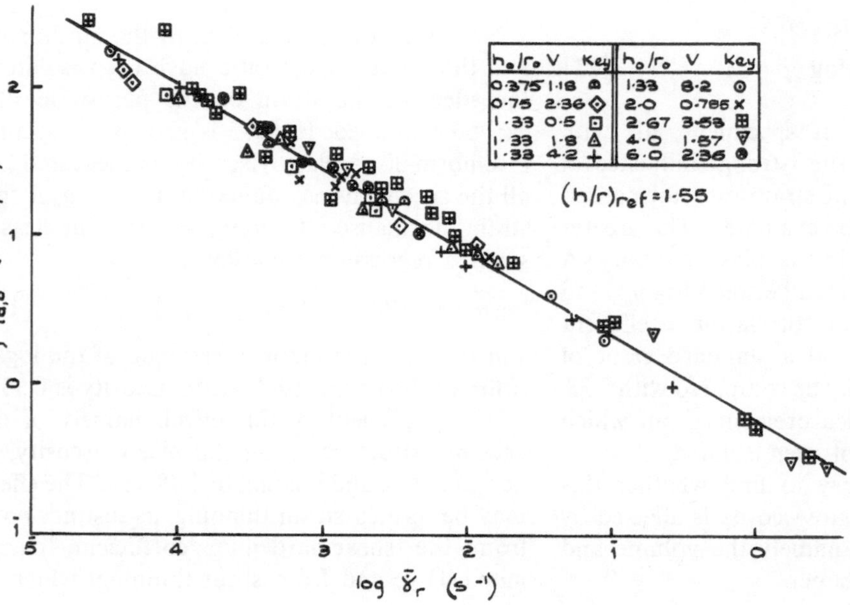

Fig. 6. The effect of strain-rate on finite-strain plastoviscosity (extrapolated to zero strain and normalised to an initial aspect ratio of 1.55)

$$\eta_{d,0} = 0.59(\bar{\gamma}_r)^{-1/2}\,\mathrm{MN\,s/m^2}. \qquad [16a]$$

This equation may be compared with the general form of eq. [4].

Comparison of experimental methods

Two different sets of experiments have been carried out: the measurement of (a) the zero-strain plastoviscosity, and (b) the finite-strain plastoviscosity. Extrapolation of the latter to zero strain enables a direct comparison of the two methods to be made. In each case the minor effect of the initial aspect ratio on the strain and hence the strain-rate has been normalised. The zero-strain measurements were shown in fig. 4 and described by eq. [12], and the finite-strain measurements extrapolated to zero-strain were shown in fig. 6 and described by eq. [16]. The two lines are very similar in slope, and cross at $\log\dot{\gamma} = 1.06$. The least-squares regression of the bulked data (148 points) has a correlation coefficient -0.99, and a standard error of estimate 0.13. The equation of the regression is

$$\log\eta_{pl} = \bar{1}.714 - 0.57\log\dot{\gamma}; \quad \gamma = 0. \qquad [17]$$

Thus the bulked zero-strain plastoviscosity is, within the experimental scatter, inversely proportional to the square root of the strain-rate:

$$\eta_{pl} = 0.518(\dot{\gamma})^{-1/2}\,\mathrm{MN\,s/m^2}; \quad \gamma = 0. \qquad [17a]$$

The two experimental methods show reasonable agreement, which would be improved if allowances were made for the probability, discussed in Appendix II, that the true strains are some 10% greater than the quoted radial strains. If the divergence between true and quoted strains were as much as 15%, this would result in the two lines crossing near the centre of the strain-rate range, and the agreement between the two experimental methods would be maximised.

The flow curve equation

Our experimental measurements of the plastoviscosity of plastic propellant now enable the form of the consistency curve for this material to be established. That is, the function in eq. [1] can be defined, as follows.

We have eq. [17] for the bulked experimental results for the strain-rate effect on the zero-strain plastoviscosity. Now also taking into account the strain-thinning effect (as in eq. [14]) together with the relation between the initial aspect ratio and the strain, we have

$$\log\eta_{pl} = \bar{1}.714 - 0.57\log\dot{\gamma} - 1.88\{\bar{\gamma}_r - 0.67\log(h_0/r_0) + 0.13\}. \qquad [18]$$

Since our experimental accuracy is such that the power of 0.57 is indistinguishable statistically from 1/2, and since the aspect-ratio correction to the strain is negligible (a few tenths), this may be simplified to

$$\log\eta_{pl} = \bar{1}.714 - 0.5\log\dot{\gamma} - 1.88\,\gamma \qquad [19]$$

$$\text{or}\quad \eta_{pl} = 0.518\times 10^{-1.88\gamma}\cdot\dot{\gamma}^{-1/2}\,\mathrm{MN\,s/m^2} \qquad [19a]$$

$$\text{or}\quad \eta_{pl} = 0.518\times 10^{6}\times e^{-4.33\gamma}\cdot\dot{\gamma}^{-1/2}\,\mathrm{N\,s/m^2}. \qquad [19b]$$

Now the plastoviscosity is defined as $ds/d\dot{\gamma}$, and so by integration we have

$$s = 1.04\times 10^{6}\,\dot{\gamma}^{1/2}\,e^{-4.33\gamma} + f_m\,\mathrm{N/m^2} \qquad [20]$$

where f_m is the constant of integration, equal to the yield stress. Thus

$$\dot{\gamma} = 0.93\times 10^{-12}(s - f_m)^2\cdot e^{8.66\gamma}\,\mathrm{s^{-1}}. \qquad [21]$$

It has been shown (1) that plastic propellant shear-hardens linearly, i.e. the yield stress increases linearly with the strain:

$$f_m = f_0 + a\gamma \qquad [22]$$

where f_0 is the yield stress at zero strain. (Material used in this report had values as follows, measured by the methods previously described (1):

$f_0 = 5.8\,\mathrm{kN/m^2}$, $a = 0.6\,\mathrm{kN/m^2}$.)

Thus we arrive at the equation for the flow curves at various values of γ:

$$\dot{\gamma} = 0.93\times 10^{-12}\{s - (f_0 + a\gamma)\}^2\cdot e^{8.66\gamma}\,\mathrm{s^{-1}}. \qquad [23]$$

This is a form of eq. [6], which is specific for materials obeying the power-law flow equation. It is, not surprisingly, quite complex, since plastic propellant is a very heavily loaded dispersion of solids in a highly viscous polymer.

The flow curves at various values of γ are shown in fig. 7, over the full experimental range of strain-rate, and again in fig. 8, where the area near the origin of fig. 7 is enlarged. The parts of the curves in fig. 8 below about 1 day^{-1} represent extrapolation beyond the lower limit of the experimental strain-rate range. The effect of shear-hardening on the curves is negligibly small.

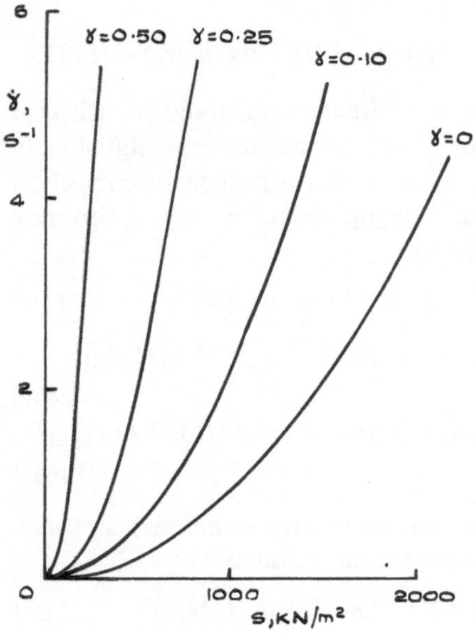

Fig. 7. The flow curves over the full experimental strain-rate range

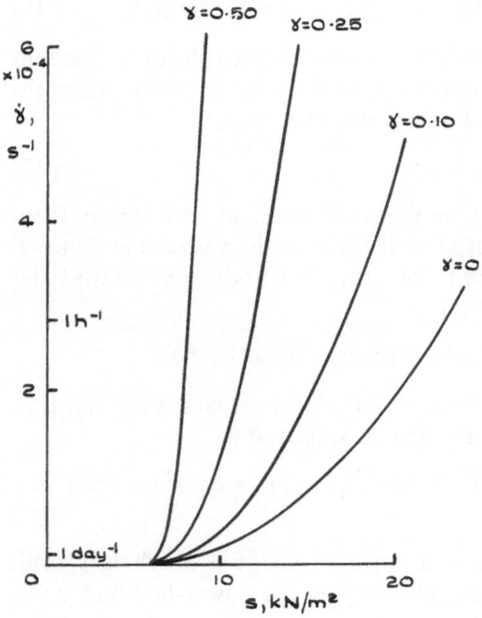

Fig. 8. Enlargement of area near the origin of fig. 7

Appendix I

Composition of plastic propellant E 3342 (parts by weight, per cent)

Ammonium perchlorate	77
Ammonium picrate	10
Titanium dioxide	1
S 101 surfactant mixture	1
*B 146	11
*B 146 = Polyisobutylene	15
Ethyl Oleate	2

Appendix II

The absolute value of the strain and of the strain-rate as a function of the aspect ratio

It is probable that γ_r is not an entirely accurate measure of strain, partly because there is always (7) a region in contact with each plate near the centre where the material is undisturbed. As h_r/r_0 becomes smaller the proportion of the total volume which remains undisturbed will increase, so that the remainder of the material will undergo a greater amount of strain than the apparent radial strain. Evidence supporting this hypothesis has been given (1), on the basis of the equilibrium separation of cylinders of various aspect ratios, using a strain-hardening material. It is therefore postulated that the factor $p_{h/r}$ is associated with this distinction between the radial strain given by γ_r and the actual strain, and that it converts the radial strain-rate $\dot{\gamma}_r$ (proportional to γ_r) into corrected strain-rate. The higher the aspect ratio, the greater the factor. The strain-rate $\dot{\gamma}_r$ to be corrected in this way appears not only as the abscissa of plots of log η_i against log $\dot{\gamma}_{r,0}$, but also inversely in the derivation of η_i (see eq. [7]).

By substituting for r_0 in eq. [11] $p_{h/r}$ is found to be proportional to h_0 (and to $V^{1/2}$, but V does not affect measurements at constant h_0/r_0 – see fig. 3). This suggests that the factor may be related to a work effect, as the potential energy available is proportional to h_0.

The reference aspect-ratio of $h_0/r_0 = 1.55$ in eq. [11] has been chosen arbitrarily as being near the centre of the experimental range, and also the point at which the tall-cylinder ($h \geqslant 2r$) theory (7) is equated to the short cylinder theory (1). It was assessed (1) that for standard specimens compressed to equilibrium the quoted radial strain γ_r was not more than 10% less than the probable true strain; but the divergence will be greater for cylinders at nearly zero strain, as when deriving η_i.

It should also be noted that the tall-cylinder theory is not strictly applicable to the least tall specimens.

Acknowledgements

Thanks are due to *P. R. Freeman* and Mr. *R. W. Bryant* for helpful discussions.

Summary

A parallel plate plastometer has been used to measure the plastoviscosity of a plastic propellant, which is a heavily loaded stiff paste previously (1) considered to approximate to a shear-hardening *Bingham* solid.

The instrument consists of a base plate and a movable upper plate, with the cylindrical specimen between them. A known load is applied to the upper plate, which is at first supported mechanically. When freed, the plate falls, compressing the specimen. The plate movement is recorded as a function of time by means of a linear voltage displacement transducer and a calibrated amplifier recorder. The duration of experiments varied between 0.25 sec and 1 h. All experiments were performed at room temperature.

Two methods of analysis have been used, based on different theories. The first is based in the *Meyerhof* and *Chaplin* analysis (7) of the shear zones in a tall cylinder and yields a zero-strain plastoviscosity. The second follows the *Dienes* and *Klemm* theory (8) for squat cylinders and gives a finite-strain plastoviscosity.

The finite-strain plastoviscosity decreased rapidly and exponentially with increasing strain. When it was extrapolated to zero strain, there was good agreement with the zero-strain plastoviscosity.

Both plastoviscosities were, within experimental error, inversely proportional to the square root of the strain-rate, over a wide range (about 10^{-5} to $10\ s^{-1}$); values ranged from about 100 to $0.1\ MN\ s/m^2$ or $MPa \cdot s$ (1 GP to 1 MP).

The flow curve for any given strain, relating strain-rate with stress, is parabolic. The yield stress and shear-hardening coefficient (measured separately (1)) have been taken into account, resulting in a flow equation linking stress, strain and strain-rate.

References

1) *Dukes, W. A.*, Determination of the Yield Stress of Stiff Pastes by Parallel Plate Plastometry; Proc. 5th Int. Congress on Rheology, ed. *S. Onogi*, Vol. 2, p. 315 (Tokyo 1970).

2) *Fox, T. G., S. Gratch*, and *S. Loshaek*, Viscosity Relationship for Polymers. In: *F. R. Eirich*, (Ed.), Rheology, Vol. 1, p. 431 ff. (New York 1960).

3) *Ward, A. G.* and *P. R. Freeman*, J. Sci. Inst. **25**, 387 (1948).

4) *Bueche, F.*, J. Chem. Phys. **48** (10), 4781 (1968).

5) *Cross, M. M.*, J. Appl. Polymer Sci. **13**, 765 (1969).

6) *Cross, M. M.*, J. Colloid Interf. Sci. **33** (1), 30 (1970).

7) *Meyerhof, G. G.* and *T. K. Chaplin*, Brit. J. Appl. Phys. **4**, 20 (1953).

8) *Dienes, G. J.* and *H. F. Klemm*, J. Appl. Phys. **17**, 458 (1946).

9) *Dienes, G. J.*, J. Colloid. Sci. **2**, 131 (1947).

10) *Oka, S.*, Principles of Rheometry. in: *F. R. Eirich* (Ed.), Rheology, Vol. 3, p. 73 ff. (New York 1960).

11) *Reiner, M.* and *G. W. Scott Blair*, Rheological Terminology. In: *F. R. Eirich* (Ed.), Rheology, Vol. 4, p. 480 (New York 1967).

Authors' address:
W. A. Dukes and *Virginia M. Gledhill*
Ministry of Defence (PE)
Waltham Abbey, Essex (England)

Rheol. Acta 13, 740–744 (1974)

From the Department of Chemistry, Kent State University, Kent, Ohio 44242 (USA)

Conformational study by intrinsic viscosities of the starch-iodine complex

Joan-Nan Liang, C. J. Knauss, and R. R. Myers

With 4 figures

Introduction

Extensive studies (1 – 4) of the structure of starch have not been paralleled by studies of the starch-iodine complex beyond the knowledge that the iodine molecule is located within the helix. *Rundle* et al. (5) proposed that an induced dipole results when iodine molecules are placed inside the helix, where they interact with the field provided by the hydroxyl groups. By contrast, *Greenwood* et al. (4) proposed a direct oxygen-iodine interaction based on the generation of infrared absorption peaks in the C – O stretching region in the spectra of the solution of iodine in amylose in the dry state. According to a third view (2) the iodine in a complex is simply dissolved in a hydrocarbon solvent.

In this work the limiting viscosity numbers $[\eta]$ (intrinsic viscosities) of starch and its iodine complex were studied. From the change in $[\eta]$ of the starch due to complexing, from its temperature coefficient, and from its interaction parameter, the most probable of these three models for the starch-iodine interaction was suggested.

Theory

Limiting viscosity number

$$[\eta] = \lim_{C \to 0} \frac{\eta_{sp}}{C} = \lim_{C \to 0} \frac{\eta - \eta_0}{\eta_0 C}, \qquad [1]$$

where η_{sp} is the specific viscosity, η_0 and η are the respective viscosities of solvent and of solution, and C is concentration in grams per deciliter (g/100 ml).

The most important relation to describe the concentration dependence of dilute solution viscosity is *Huggins*' equation (6)

$$\frac{\eta_{sp}}{C} = [\eta] + k'[\eta]^2 C, \qquad [2]$$

where k' (*Huggins*' constant) is a dimensionless term characteristic of a given polymer-solvent system. $\eta_{sp/C}$ increases linearly with C for dilute solutions and can be extrapolated to zero concentration.

For a given polymer-solvent system, $[\eta]$ depends on the extension of the molecule, the property of interest in this work. The $[\eta]$ also depends on rate of shear and the temperature (7, 8), on the molecular weight of the polymer, and on chain branching; k' is also affected by these variables and by aggregation or entanglement of the polymer chain.

In a good solvent the polymer-solvent attraction expands the polymer chains, and increases $[\eta]$. In a poor solvent the polymer chains are coiled up and become more compact, thus reducing $[\eta]$. Branching affects $[\eta]$ in the same way by reducing the spread of segments away from the molecular center. Chain stiffening via complexing should enhance $[\eta]$ because stiff chains in solution are more extended than are flexible chains, and therefore they reach further out from the center of mass (9 – 12).

Inasmuch as k' is related to the interaction between separate polymer chains contrasted with the interaction of polymer and solvent, the k' values increase in going from a good solvent to a poor solvent. Therefore, while a poor solvent leads to a lower $[\eta]$, it leads to a higher k'. In a poor solvent the polymer molecule is tightly coiled and acts hydrodynamically like a compact, almost spherical structure. The k' value also is increased by branching (13).

In a poor solvent the polymer-polymer attraction may cause the formation of aggregates which also increases k'. This effect is more prominent if the polymer is a polar one. Generally the value of k' exceeds 0.5 in case of aggregation and increases with increasing degree of aggregation (8, 14).

Experimental Procedures

Starch preparation

Commercial cornstarch was used primarily in this work; other samples included amylose rich preparations

(Amylon 55, Amylon 70 and Nepol Amylose) which were obtained from A. E. Staley Manufacturing Co., Decatur, Ill.

All starches were extracted exhaustively with 95% ethanol in a Soxhlet extractor for 48 h to remove associated fatty acid. The defatted starches were then fractionated by the method of *Schoch* (15), in which amyl alcohol and n-butyl alcohol were employed to form an insoluble complex of the linear amylose fraction.

The iodine numbers were determined by potentiometric titration using *Pacsu*'s method (16).

The starch solutions were prepared (15) by dispersing starch in 1 N KOH and then neutralized with 0.5 N HCl. The insoluble material was removed by filtering through a coarse sintered glass filter.

The iodine solution was prepared by dissolving 2.0 g of iodine and 166 g of KI in water and diluting to 1 liter. The resulting solution contained 2.0 mg per ml and was 1 N in KI. The iodine complexes were prepared simply by adding the above iodine solution to the prepared starch solutions.

Viscosity measurements

Viscosity was determined in a *Ubbelohde* dilution type viscometer (size 100) obtained from Cannon Instrument Company of State College, Pennsylvania. Dilution can be done in this kind of viscometer by pipetting into the reservoir the additional solvent; thus the viscosities of a whole series of concentrations can be measured for a single sample. Measurements were made at 30° ± 0.05 °C. At least two runs were made for each concentration. Between determinations the viscometer was flushed with acetone and dried with air.

Results and discussion

Iodine number

In table 1 the iodine numbers are presented for cornstarch, its fractions, and the high-amylose preparations. The higher iodine number indicates the higher amylose content in starch.

Table 1. Iodine numbers of starches

Sample	Iodine number (%)
Amylopectin fraction	1.04
Cornstarch	4.74
Amylose fraction	16.41
Amylon 55	11.63
Amylon 70	13.71
Nepol amylose	18.26

Viscosities of starch and its iodine complex

The viscosities of alkali-treated starch, the two fractions, and their complexes are shown in table 2, and the plot of η_{sp}/C is shown in fig. 1. Iodine complexation increases the intrinsic viscosities of cornstarch and amylose, particularly of amylose, in agreement with the pronounced difference in their iodine numbers. Since the amylose molecule in aqueous solution exists as a helical coil (17), the entering iodine molecule renders the helix stiffer and more extended and therefore of higher $[\eta]$. Amylopectin does not show this effect since it does not form a complex with iodine. The observed decrease of $[\eta]$ for amylopectin after the addition of iodine may be due to the change of ionic strength of the solvent.

Whether or not this explanation for reduction of $[\eta]$ suffices, it is known that many anions and cations including I^- and K^+ can increase the

Table 2. $[\eta]$ and k' of alkali-treated starches and their iodine complexes

Samples	$[\eta]$ dl/g		k'	
	starch	complex	starch	complex
Amylopectin	0.75	0.70	1.15	0.44
Cornstarch	0.81	0.85	1.18	1.76
Amylose	0.44	0.80	0.86	4.68

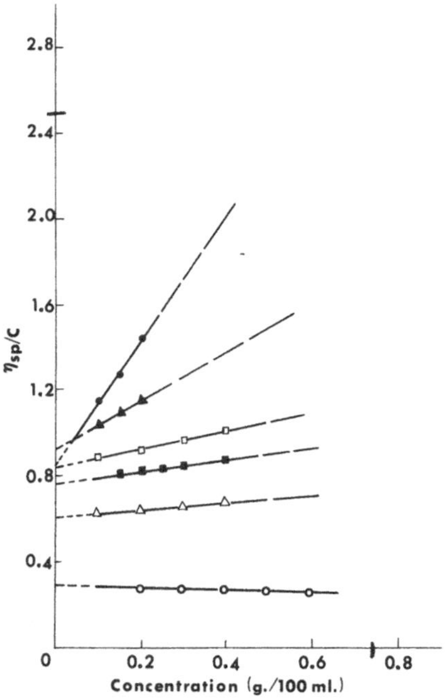

Fig. 1. Reduced viscosities of various alkali treated starch, starch fractions, and starch-iodine complexes.

○ amylose
△ cornstarch
□ amylopectin
● amylose-iodine solution
▲ cornstarch-iodine solution
■ amylopectin-iodine solution

fluidity of water by destroying its structure (18). This effect increases the random motion of the starch helix and thereby decreases the viscosity.

The higher values of k' for complexes of cornstarch and amylose may be due to the increase of rigidity (19) and average extension, to aggregation of chains, or to chain entanglement (8), since all of these factors increase k'. Significantly, the magnitude of k' for the amylose complex approximates that of rigid and rod-like particles (19). Aggregation can occur since the complex is a polar polymer. If the complexed helix is a highly extended chain, chain entanglement can occur even in dilute solution (8).

Dependence of η_{sp}/C on molar concentration of iodine

Fig. 2 shows the dependence of reduced viscosity η_{sp}/C on molar concentration of iodine added. The starch was first dispersed in 1N KOH and then neutralized. Five ml portions of 0.3% starch were pipetted into several 15 ml volumetric flasks, and iodine solution (2 mg/ml) was added in increasing amount needed to saturate the starch, and water was added to produce a 0.1% starch solution. From fig. 2 it is evident that for amylose and cornstarch there are maxima at about 0.56×10^{-3} M I_2 and 0.14×10^{-3} M I_2 respectively. These values are approximately the amount of iodine needed to saturate the given amount of starch (i.e. one molecule per turn of the helix, assuming six glucoside segments per turn). For amylopectin no maximum appears.

The increase in viscosity suggests a more extended chain when iodine is incorporated. After the chain is filled the excess iodine affects the solvent and causes the decrease in viscosity, shown in fig. 2.

Dependence of $[\eta]$ on the amount of iodine added

Fig. 3 is the plot of $[\eta]$ against mg of I_2 added for Nepol amylose. Results for samples of 10 ml of 0.6% amylose, with increasing amounts of I_2 which were diluted to 0.12% before measuring the viscosity are given in table 3. The plot shows that $[\eta]$ increases as iodine is added until approximately at the saturation point and after that point $[\eta]$ changes only slightly. Therefore the postulate that entering iodine molecules stiffen the helix is quite consistent with the obtained result. After the helix is filled the excess iodine molecules have no effect, as shown in table 3.

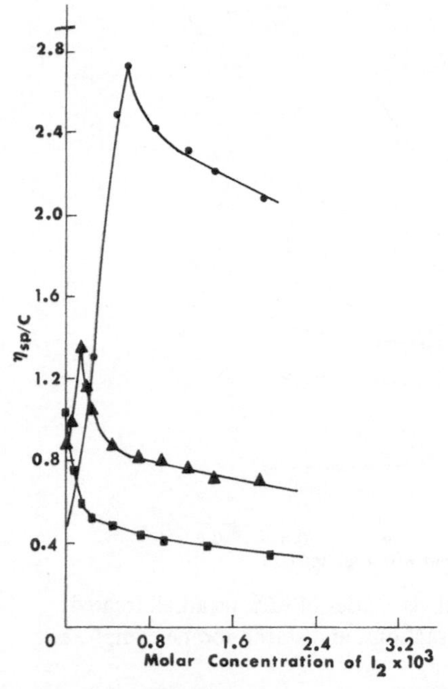

Fig. 2. Dependence of $\eta_{sp/c}$ on iodine molar concentration for alkali treated starch.

● amylose

▲ cornstarch

■ amylopectin

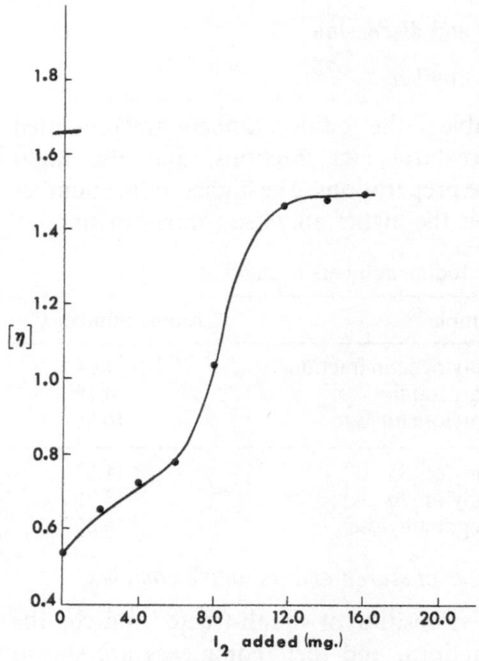

Fig. 3. Dependence of intrinsic viscosity, $[\eta]$, on mg of I_2 added starch solutions

Table 3. Values of $[\eta]$ and k' for amylose with iodine

mg of I_2 added	0	2.0	4.0	6.0	8.0	12.0	14.0	16.0
$[\eta]$	0.53	0.65	0.72	0.77	1.03	1.48	1.47	1.49
k'	0.31	0.89	1.21	3.06	5.30	5.65	6.70	6.85

The fact that k' increases markedly with I_2 implies that subtle conformational changes take place as the complexes become more rigid.

Temperature dependence of $[\eta]$

The temperature dependence of $[\eta]$ was studied for Nepol amylose and its iodine complex, dispersed in 1 N KOH and then neutralized. Table 4 lists viscosity date and fig. 4 is the intrinsic viscosity plot from which the temperature coefficient, $d[\eta]/dT$, was obtained from the rectilinear slope of $[\eta]$ against temperature. Note that k' increased with increased temperature for starch, but decreased for the complex.

Table 4. Viscosity data for Nepol amylose with temperature

T °C	$[\eta]$ dl/g		k'	
	starch	complex	starch	complex
25		0.58		1.58
30	0.52	0.57	0.32	1.34
35	0.50	0.56	0.37	1.27
40	0.48	0.54	0.39	1.11
45	0.47		0.46	

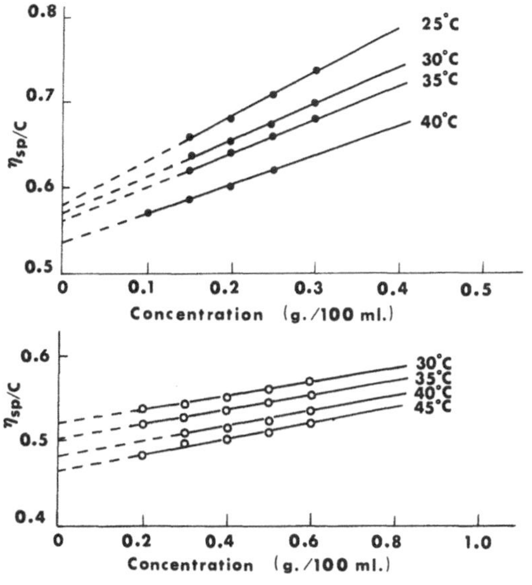

Fig. 4. $\eta_{sp/C}$ vs. C for Nepol amylose-iodine complexes (●) and Nepol amylose (○) in a temperature range of 25° to 45 °C

The value of $d[\eta]/dT$ for amylose in 1 N KOH was obtained as $-2.12 \times 10^{-2}/°C$ by *Sterling* (20) and in water as $-.30 \times 10^{-2}/°C$ by *Burchard* (21). These values were compared with the value obtained in this experiment as $-0.36 \times 10^{-2}/°C$ for neutralized alkali-treated amylose and $-0.23 \times 10^{-2}/°C$ for its iodine complexes. Since this value is a measure of flexibility of the molecule (22, 23), the amylose molecule must be very flexible in 1 N KOH, somewhat stiff in water and salt solution, and rather stiff for the complex (20). The magnitude is similar to those for stiff chain polar polymers such as cellulose derivatives (24).

The k' value increases with temperature for starch, which may indicate that the molecule becomes less extended and increasingly coiled as temperature is increased (20). Iodine complexes do not show this tendency; instead, k' decreases with increased temperature. Therefore, if complexes aggregate, the degree of aggregation should decrease on heating. The decrease of $[\eta]$ for complexes with temperature cannot be explained by this phenomenon; the change of chain dimensions of aggregated molecules may play an important part. Therefore both the starch and complex become less extended with increasing temperature, though the more extended complex is less sensitive to temperature change.

Conclusions

The pronounced increase of $[\eta]$ of amylose on complexing with iodine is a manifestation of the stiffening of the helix without a basic change in the conformation. Iodine is absorbed roughly in the ratio of one molecule per turn of the helix.

The entering iodine affects either the shape or the extension in space of the chain regardless of interchain interactions because $[\eta]$, a property of individual molecules, increases only until saturation is attained. The resulting extended chain interacts greatly with its neighbors.

Evidence that the molecule is stiffened is provided also by the reduction of the temperature dependence of $[\eta]$ on complexing. A rigid particle is less able to dissipate thermal kinetic energy than a non-rigid one (19).

Huggins' interaction constant, k', of complexed amylose is significantly higher than that of uncomplexed amylose, and its value is close to the magnitude of rigid and rod-like particles (19). The high k' of complexed amylose is enhanced by aggregation, which is less pronounced at higher temperature.

The added rigidity of the complexed helix disagrees with the suggestion that iodine is simply dissolved in a hydrocarbonlined helix. The iodine-oxygen interaction presented by *Greenwood* et al. (4), is also unable to explain the viscosity results since infrared peaks attributed to the $I_2 - O$ bond were observed only in a completely dry state. Therefore, the most possible explanation of the iodine-starch interaction is through the dipolar force pointed out by *Rundle* and co-workers (5) wherein the strong dipole moment of amylose induces a moment in the bound iodine molecules. These molecules then interact mutually through dipolar force. This dipolar force stiffens the helix and increases the $[\eta]$.

Summary

The amylose molecule is extended by the insertion of iodine in quantities up to a ratio of one molecule per turn of the helix. Ordinary cornstarch and its amylopectin fraction do not respond in the same manner. As the iodine content was increased the intrinsic viscosity $[\eta]$ rose from 0.55 to 1.45 dl/g, then remained constant. In this same range the *Huggins'* interaction parameter k' for the dependence of reduced viscosity $[\eta_{sp}/C]$ on iodine concentration rose from 0.31 to 5.65 and continued to increase. Reduction of $d[\eta]/dT$ by complexing from -0.36×10^{-2} to -0.23×10^{-2} dl/g/deg provided an independent measure of the flexibility of the molecule and revealed that iodine stiffens the helix in some manner. The dipolar interaction between iodine and the helix restricts conformational changes and increases the average extension. As more iodine molecules enter the helical chain the average distribution of molecular segments relative to the center of mass of the molecule spreads outward.

Acknowledgments

The authors are grateful to the Corn Refiners Association, Inc. for partial support of this research effort. We are especially grateful for the cooperation of the Analytical Procedures Committee of this organization and its liaison representative, *Robert Smith*, of CPC International's Moffett Research Laboratory.

References

1) *Hanes, C. S.*, New Phytologist **36**, 101, 189 (1937).
2) *Freudenberg, K., E. Schaaf, G. Dumpert,* and *T. Ploetz,* Naturwiss. **27**, 850 (1939).
3) a) *Rundle, R. E.* and *R. R. Baldwin,* J. Amer. Chem. Soc. **65**, 554 (1943).
b) *Rundle, R. E.* and *Dexter French,* J. Amer. Chem. Soc. **65**, 558 (1943).
c) *Rundle, R. E.*, J. Amer. Chem. Soc. **69**, 1769 (1947).
4) *Greenwood C. T.* and *Hazet Rossetti*, J. Polymer Sci. **17**, 481 (1958).
5) a) *Rundle, R. E., J. F. Foster,* and *R. R. Baldwin,* J. Amer. Chem. Soc. **66**, 2116 (1944).
b) *Stein, R. S.* and *R. E. Rundle*, J. Chem. Phys. **16**, 195 (1948).
6) *Huggins, M. L.*, J. Amer. Chem. Soc. **64**, 2716 (1942).
7) *Frisch, H. L.* and *Robert Simha*, in: *F. R. Eirich* (Ed.), Rheology, Vol. 1, p. 525 (New York 1956).
8) *Moore, W. R.*, in: *A. D. Jenkins* (Ed.), Progress in Polymer Science, Vol. 1, p. 3 (Oxford, New York 1967).
9) *Huggins, M. L.*, J. Phys. Chem. **42**, 911 (1938).
10) *Huggins, M. L.*, J. Phys. Chem. **43**, 439 (1939).
11) *Huggins, M. L.*, J. Appl. Phys. **10**, 700 (1939).
12) *Huggins, M. L.*, Physical Chemistry of High Polymers (New York, N. Y. 1958).
13) *Manson, J. A.* and *L. H. Cragg*, Canad. J. Chem. **30**, 482 (1952).
14) *Doty, P., H. Wagner,* and *S. Singer*, J. Phys. Colloid Chem. **51**, 32 (1947).
15) *Schoch, T. J.*, in: *S. P. Colowick* and *N. O. Kaplan* (Eds.), Methods in Enzymology, Vol. III, p. 5 (New York 1957).
16) *Bauer, A. W.* and *E. Pacsu*, Textile Res. J. **13**, No. 12, 864 (1953).
17) *Rao, V. S.* and *Joseph F. Foster*, Biopolymer **1**, 527 (1963).
18) *Erlander, S. R.* and *R. Tobin*, Makromol. Chem. **111**, 212 (1968).
19) *Eirich, E.* and *J. Risman*, J. Polymer Sci. **4**, 417 (1949).
20) *Khairy, M., S. Morsi,* and *C. Sterling*, J. Appl. Polymer Sci. **10**, 928 (1966).
21) *Burchard, W.*, Makromol. Chem. **64**, 110 (1963).
22) *Cowie, J. M. G.*, Makromol. Chem. **53**, 13 (1962).
23) *Moore, W. R.* and *A. M. Brown*, J. Colloid Sci. **14**, 343 (1959).
24) *Moore, W. R.* and *D. Sanderson*, Polymer **9**, 153 (1968).

Authors' address:

J. N. Liang, C. J. Knauss, and *R. R. Myers*
Department of Chemistry
Kent State University
Kent, Ohio 44242 (USA)

Rheol. Acta 13, 745–753 (1974)

From the Department of Chemical Engineering, The Ohio State University, Columbus, Ohio 43210 (USA)

Time-dependent, non-Newtonian behavior of viscoelastic materials

P. D. Jachimiak), Yoon Soo Song, and R. S. Brodkey*

With 8 figures

Introduction

In order to put our effort into proper perspective of rheological constitutive relations as a whole, we repeat (6) a few brief introductory comments. One of the fundamental problems in polymer rheology is the establishment of a basic rheological equation which effectively describes the nonlinear deformation-time flow history. The need of such an equation to relate shear stress, normal stress, shear rate, and time is necessary in the solution of many flow problems. The equation may involve one or more constants, depending on the complexity of the relation, and a variety of experiments is used to evaluate these. A systematic investigation and correlation of the material constants as a function of molecular weight, structure, concentration, and temperature, might give some insight into the basic mechanism of polymer flow, if such constants are the result of a logical theory.

The foregoing, which describes our kinetic approach, is phenomenological in nature. The various approaches towards the elucidation of the rheological characteristics of polymeric materials generally fall into one of five classifications: empirical, phenomenological (mostly rate processes), linear viscoelastic, nonlinear viscoelastic, and microrheological analyses. The empirical methods correlate data by curve-fitting techniques. The rate theories have as their basis the assumption that the nonlinear characteristics can be associated with some structural change of the material whether it involves particle associations and dissociations, link formations and ruptures, or molecular entanglements and disentanglements. The linear viscoelastic models are based upon linear combination of *Hookes* law of elasticity and *Newtons* law of viscosity. The nonlinear viscoelastic models are based on continuum mechanics and nonlinear combination of mechanical models. Finally, microrheological analysis starts with the basic molecular or microscopic variables such as particle sizes, molecular interactions, and chain lengths. A simplified mechanism is proposed, and then it is mathematically represented and solved. Most of the background literature concerning these various approaches have been previously reviewed (1 – 6) and will not be repeated here.

Each approach has made contributions to the field, but with the present state of the art, we cannot completely describe the non-*Newton*ian behavior even in simple geometries. Steady-state flow behavior has been extensively investigated, and the representations available are adequate. However, little is understood about the unsteady-state flow, such as shear stress and normal stress growth after the onset of a sudden shear rate. There is no theory in the literature that can predict such phenomena from data obtained on other experiments or even to adequately represent such information. It is the purpose of this work to improve and to extend the kinetic interpretation of non-*Newton*ian flow, which has previously been shown to be capable of describing the flow behavior of high-polymer melts and solutions. The basic approach and preliminary conformation of the kinetic theory have been described in the papers by *Brodkey*, *Denny*, and *Kim* (2 – 3). *Kim* and *Brodkey* (3) provided the test of the steady-state theory by correlating polymer solution data over a wide concentration range. *Lewis* and *Brodkey* (5) offered further support of the theory by describing the phenomena of shear stress growth over a range of shear rates. However, thus far the tests of the theory were empirical since the necessary constants and parameters had been obtained from the best fit to the data. *Lee* and *Brodkey* (6) provide an independent test of the kinetic approach by using data obtained under constant shear stress conditions to evaluate the necessary constants and parameters in an idealized manner as suggested by the theory, and not by simple curve-fitting methods. The results were then used for predicting time-dependent constant shear rate measurements as well as the constant stress data. The present work provides a more rational theory as well as the modifications necessary to incorporate normal stress differences in a logical manner. To accomplish this, the kinetic theory is used to explain the thixotropic or time-dependent decreasing viscosity behavior, and a modification of *Oldroyds* model is used to incorporate the elastic behavior into the theory.

Theory

Polymer rheology is primarily concerned with three fundamental properties of fluids: the time rate of change of viscosity or structure, called here the thixotropic property; the steady-state level obtained, which is a function of the shear rate (shear thinning); and the lag of the stress

*) Now associated with IBM Corp., Lexington, Ky., USA.

behind what one would expect from the thixotropic behavior, called here the elastic property. Associated with this latter is the development of normal stress differences. The change in viscosity can be attributed to some internal mechanism such as entanglement, association, or winding of long-chain polymer molecules. The lag of the stress is due to the property of the material which causes it to resist deformation and thereby tends to recover its original shape and size when deforming forces are removed. The normal stresses are also in some way associated with this property. The nonlinear time-dependent flow behavior for such polymeric materials can best be described by a combination of elasticity and thixotropy. The thixotropic and steady-state behavior is described by the kinetic theory and the elastic property is described by a simple elastic model.

Kinetic model for thixotropic change

Brodkey and co-workers (1 – 6) developed a kinetic model for thixotropic materials which is descriptive of the viscosity change that occurs in polymer solutions. The model is different from the previous rate models in that molecular phenomena such as links, particle spheres, dipole moments, or hydrogen bonding are not used directly as basis of the model. Instead, it used a lumped parameter F_T, which can describe any or all of these phenomena. The subscript T denotes the thixotropic property. The property F_T, is defined by a modified inverse-lever arm principle (1, 3):

$$F_T = \frac{\eta_T^a - \eta_\infty^a}{\eta_0^a - \eta_\infty^a}. \qquad [1]$$

F_T is the fraction of the internal fluid structure unchanged and $(1 - F_T)$ is the fraction of the internal fluid structure changed. The viscosity terms, η_0 and η_∞, are the lower and upper *Newton*ian viscosity limits, respectively. It must be emphasized that η_T is a hypothetical viscosity associated with the changing structure. The term F_T is a convient measure of this structure. We refer to the letter a as a structure parameter. Arguments have been presented (3) for its value being 1/3.5; however, more recently we have use it linearly ($a = 1$) with satisfactory results.

The simplest differential rate equation to describe the transient change in F_T is

$$-d(CF_T)/dt = k_1 \mathscr{T}^{P_1}(F_T C)^m - k_2 \mathscr{T}^{P_2}[(1 - F_T)C]^n \qquad [2]$$

where m and n are forward and reverse orders of the reaction while C is the concentration. The k's and P's are specific rate and susceptibility to stress constants, respectively. In our former work, the shear stress was a hypothetical stress defined by

$$\mathscr{T}_T = -\eta_T \dot{\gamma} \qquad [3]$$

but the use of this does not appear logical since any change in the structure should depend on the actual stress in existence in the experiment, not on some hypothetical value. Thus we have replaced $\mathscr{T}_T$ used formerly with the actual stress $\mathscr{T}$. There is another problem with our earlier formulation when one considers stress relaxation. The rate term, $-dCF_T/dt$, must be less than zero during relaxation in order for the structure to build up to its original relaxed state. Using this inequality in eq. (2) and rearranging gives

$$K \mathscr{T}_T^P \frac{F_T^m C^{m-n}}{(1 - F_T)^n} \leqslant 1.0 \qquad [4]$$

where $K = k_1/k_2$ and $P = P_1 - P_2$.
At equilibrium, the left hand-side of this equation is equal to unity. As the structure reforms, η_T increases. Correspondingly $\mathscr{T}_T$ increases by eq. [3] if $\dot{\gamma}$ is replaced by $\dot{\gamma}_S$, the shear rate at steady state just before the relaxation experiment is started. The term F_T also increases by eq. [1], and thus the left hand-side of eq. [4] becomes greater than unity. This is contradictory to the necessity of the term being less then unity for structure buildup. In addition the use of $\dot{\gamma} = 0$, as is the actual case, will not work in our former formulation, since this would give $\mathscr{T}_T = 0$ by eq. [3], and from eq. [2] (with $\mathscr{T} = \mathscr{T}_T$ as formerly used) would give a zero or infinite buildup rate depending on the sign of the P's. In either case the rate is unacceptable. There is still another problem with $\mathscr{T}_T = 0$, which will be cited in the next section on viscoelasticity. At steady state the equation reduces to

$$(k_1/k_2)\mathscr{T}^{P_1 - P_2} = K\mathscr{T}^P = (1 - F_T)^n C^{n-m}/F_T^m \qquad [5]$$

where F_T and η_T are simply the values at steady state. Eq. [2] is descriptive of the time-dependent thixotropic characteristic of polymeric solutions, while eq. [5] is restricted to the steady state and can be used to describe the shear thinning characteristics.

Elastic model for viscoelasticity

The kinetic theory of thixotropy just presented assumed that there was no elasticity in the material; however, most polymeric solutions are viscoelastic and the thixotropic presentation alone is not adequate. In order to incorporate the elastic property into the earlier version of our theory, a simple elastic model was chosen (5). The simplest is *Maxwells* which is the combination of *Hookes* law of elasticity and *Newtons* law of viscosity:

$$\mathcal{T} + (1/k)(d\mathcal{T}/dt) = -\eta_0\dot{\gamma} \qquad [6]$$

where $(1/k) = (\eta_0/G)$ is the retardation time constant. This equation can be integrated at constant shear rate by separation of variables to give

$$\mathcal{T} = \eta_0\dot{\gamma}(1 - e^{-kt}) = \mathcal{T}_0(1 - e^{-kt}) \qquad [7]$$

where $\mathcal{T}_0$ is $-\eta_0\dot{\gamma}$ or the initial thixotropic stress. We extended this by replacing η_0 by η_T, which by eq. [3] gives

$$\mathcal{T} = \mathcal{T}_T(1 - e^{-kt}). \qquad [8]$$

As indicated previously there was another problem associated with stress relaxation and our former model. When $\mathcal{T}_T = 0$, the actual stress, $\mathcal{T}$, is zero by eq. [8]. This would occur when $\dot{\gamma} = 0$; i. e., at the beginning of stress relaxation. Thus this model had in addition a problem of instantaneous stress relaxation rather than an approximate exponential decay experimentally observed. This clearly also implies that the model would not be useful for oscillatory motion studies.

Lee and *Brodkey* (6) used the same reasoning for experiments at constant stress; i. e., the measured shear rate is the product of the thixotropic shear rate (due to viscosity only) and its elastic retardation term. Thus they assumed

$$\dot{\gamma} = \dot{\gamma}_T(1 - e^{-kt}) \qquad [9]$$

where

$$\dot{\gamma}_T = -\mathcal{T}/\eta_T. \qquad [10]$$

They noted that eqs. [8] and [9] are the limiting cases of *Oldroyds* (7) viscoelastic equation.

If eqs. [8] and [9] are assumed to be the correct equations, then their use in the model is just one of the basic assumptions made. However, if we wish to suggest that *Oldroyds* viscoelastic equation or some modification thereof is the correct form to be used, then the use of eqs. [8] and [9] is incorrect. This can be seen from purely mathematical considerations; the integration would be incorrect because $\mathcal{T}_T$ and $\dot{\gamma}_T$ which are functions of time, were considered constant during the integration. This would result in a physical error in the calculation of shear stress. The hypothetical stress $\mathcal{T}_T(t)$ was obtained by a numerical integration of eq. [2] (with of course $\mathcal{T} = \mathcal{T}_T$) and multiplied by the elastic contribution in eq. [8] to produce a $\mathcal{T}(t)$. Fig. 1 shows that this procedure

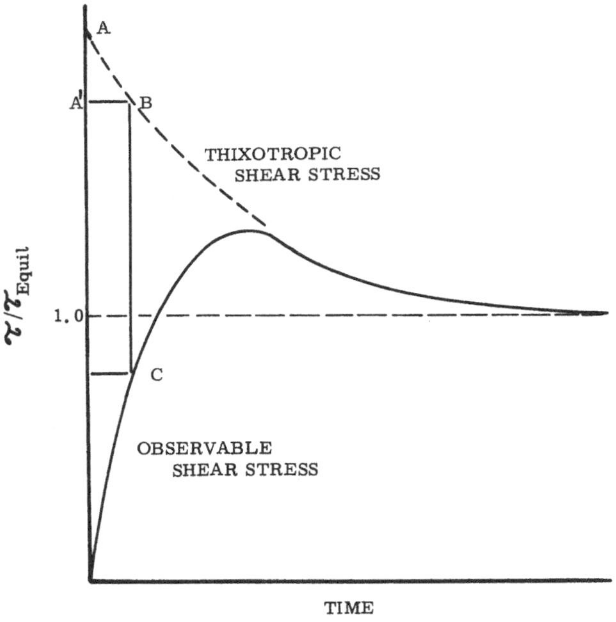

Fig. 1. Thixotropic and observable shear stresses

used $\mathcal{T}_T(t)$ at point B to calculate $\mathcal{T}(t)$ at point C. Implicit in the multiplication by the elastic contribution was that the stress history of $\mathcal{T}_T(t)$ ran along line $A' - B$. In effect, then, this procedure did not acknowledge the fact that the stress history of $\mathcal{T}_T(t)$ is along $A - B$ not $A' - B$.

The calculational method employed in this work was the *simultaneous* numerical integration if eq. [2] and the modified *Oldroyds* viscoelastic equation

$$\mathcal{T} + \frac{1}{k_{\mathcal{T}}}\left(\frac{\partial\mathcal{T}}{\partial t}\right) = -\eta_T\left[\dot{\gamma} + \frac{1}{k_{\dot{\gamma}}}\left(\frac{\partial\dot{\gamma}}{\partial t}\right)\right] \qquad [11]$$

in one of its limiting forms; i. e., with either $\partial\mathcal{T}/\partial t$ (constant stress) or $\partial\dot{\gamma}/\partial t$ (constant shear rate) equal to zero. Depending on the type of experiment, either eq. [3] or [10] was also used.

Normal stress and tensor considerations

When the partial derivatives of eq. [11] are replaced by convective derivatives, a model is obtained that can predict normal stresses. In simple shear, the stress tensor takes the form

$$\vec{\vec{P}} = p\vec{\vec{U}} + \vec{\vec{\mathscr{T}}}$$

$$= \begin{pmatrix} p - 2\eta_T\left(\frac{1}{k_{\mathscr{T}}} - \frac{1}{k_{\dot\gamma}}\right)\dot\gamma^2 & -\eta_T\dot\gamma & 0 \\ -\eta_T\dot\gamma & p & 0 \\ 0 & 0 & 0 \end{pmatrix}. \quad [12]$$

This gives a prediction for the first normal stress difference of

$$\sigma_1 = \mathscr{T}_{11} - \mathscr{T}_{33} = -2\eta_T\left(\frac{1}{k_{\mathscr{T}}} - \frac{1}{k_{\dot\gamma}}\right)\dot\gamma^2 \quad [13]$$

an equivalent form of this is (using eq. [3])

$$\sigma_1 = 2\,\mathscr{T}_T\dot\gamma\left(\frac{1}{k_{\mathscr{T}}} - \frac{1}{k_{\dot\gamma}}\right).$$

In this model, the second normal stress difference is always zero.

The kinetic equation (eq. [2]) is a scalar equation, since both concentration and the fraction conversion are scalars. For complex flows one can suggest using the second invarient of the stress tensor to replace the stress; i. e., use $\sqrt{\vec{\vec{\mathscr{T}}} : \vec{\vec{\mathscr{T}}}}$ in place of $\mathscr{T}$.

Determination of the constants

A number of the constants of the theory can be determined from steady state data. The structure parameter a has been used as 1/3.5 as we have done previously. The constants m, n, P, and K are obtained as follows: First, m and n are selected (1 and 2 in the work reported here, although other values have been tested). Second, the method presented by *Kim* and *Brodkey* (3) of using a log-log plot of eq. [5] is used to obtain the first estimates of P and K. Finally, an optimization is made to obtain the best values of P and K by minimization of a least square estimation (objective function):

$$G(\theta) = \sum_{i=1}^{N} R_i(\theta)^2 \quad [14]$$

where for the steady state results $R_i(\theta) = [\mathscr{T}_{calc}(\theta) - \mathscr{T}_{exp}]/\mathscr{T}_{exp}$. Here i is the i-th experiment, sub-calc means calculated by the model, and θ is the total parameters on which the residual, R_i, depends. The residual based on relative stresses was used because of the wide range of stress over the full range of shear rates being optimized.

Lee and *Brodkey* (6) used homogeneous kinetics methods to ideally determine the remaining constants of the model. The same could be done here except that the procedure is not easy and the accuracy is not good. The best that can be said for it, is that the method confirms the results obtained with the empirical analysis used by *Lewis* and *Brodkey* (5). In the present work a combined optimization and trial and error procedure was used to establish the best values. However, to minimize computer time and to avoid local optima, one should have reasonable starting estimates for the parameters involved.

The starting values for the elastic parameters are easily obtained. For a constant shear rate experiment, eq. [11] can be reduced to

$$d\mathscr{T}/dt = k_{\mathscr{T}}(\mathscr{T}_T - \mathscr{T}) \quad [15]$$

with the help of eq. [3]. At $t = 0$, $\mathscr{T}_T = -\eta_0\dot\gamma$ and $\mathscr{T} = 0$, where η_0 is the lower *Newton*ian viscosity limit. Thus eq. [15] becomes

$$\left.\frac{d(\mathscr{T}/\mathscr{T}_0)}{dt}\right|_{t=0} = k_{\mathscr{T}} \quad (\text{constant } \dot\gamma). \quad [16]$$

For a constant stress experiment, eq. [11] can be reduced to

$$d\dot\gamma/dt = k_{\dot\gamma}(\dot\gamma_T - \dot\gamma) \quad [17]$$

with the help of eq. [10]. At $t = 0$, $\dot\gamma_T = -\mathscr{T}/\eta_0$ and $\dot\gamma = \dot\gamma_i$; some initial shear rate $\geqslant 0$. Thus eq. [17] becomes

$$\left.\frac{d[\dot\gamma/(\dot\gamma_0 - \dot\gamma_i)]}{dt}\right|_{t=0} = k_{\dot\gamma} \quad (\text{constant } \mathscr{T}). \quad [18]$$

Eq. [18] differs from that used by *Lee* and *Brodkey* (5) for constant stress conditions. The difference being that they assumed that $\dot\gamma_i$ was zero. A finite value for the intercept agrees better with the data to be presented here and with the work of *Dexter* (9). The elastic parameter is not sensitive during the optimization procedure and changes very little from the initial estimate given by either eq. [16] or [18].

All the remains to be obtained are estimates for P_2 and k_2. Eq. [5] can then be used for the values of P_1 and k_1. The very first estimate (for the constant shear rate experiments) is obtained from the maximum points on the stress growth data. The following equation can be obtained by manipulation of eqs. [2] and [15]

$$k_{\mathscr{T}} k_2 \mathscr{T}_{\max}^{P_2 - a} = \frac{-(d^2 \mathscr{T}/dt^2)_{\max}}{a \mathscr{T}_0^{1+a} [K \mathscr{T}_{\max}^{P} F_{T,\max} - (1 - F_{T,\max})^2 C]} \quad [19]$$

The r.h.s. can be calculated from experimental data. A plot of this against $\mathscr{T}$ on a log-log scale for each shear rate which contains a maximum point should give a straight line, the slope and intercept being the desired initial guess of P_2 and k_2. This method is not accurate because of the difficulty in obtaining the second derivatives from our experimental data. This was done by fitting the points near the maximum with a 3rd degree polynomial. Since this was only a starting estimate of P_2 and k_2, this was satisfactory; however, it was necessary to refine the estimate. This was done by a contour plot method.

In the optimization of the transient data, the residual used in eq. [14] was taken as $R_i(\theta) = \mathscr{T}_{calc}(\theta) - \mathscr{T}_{exp}$. Here a simple difference was used because for each run the shear rate was constant and thus the range of values of $\mathscr{T}$ limited. The contour plot was a three-dimensional representation of the objective function of eq. [14] as a function of both P_2 (y-axis) and k_2(x-axis) for a given value of $k_{\mathscr{T}}$, since this latter value could be adequately defined from the initial slope at the particular shear rate being considered. The ranges of P_2 and k_2 used were selected by the results obtained from eq. [19]. The calculation involved determination of the objective function for the given $k_{\mathscr{T}}$ and a range of values of P_2 and k_2. Optimization procedures are not used in this step. This then was repeated for each of the shear rates studied. Unfortunately unique values of P_2 and k_2 for all the data were not obtained, although it turned out that P_2 was a simple function of k_2. More details of this can be found in ref. (10).

At this point in the analysis several typical shear rates were selected and optimized for the best values of P_2, k_2 and $k_{\mathscr{T}}$. To start the optimization the results from the contour plot were used. The final stage of optimization is the extension of the locally optimized values just obtained to the whole data set in order to have global optimum values. Since we want P_2 and k_2 to be constants and not to vary with shear rate, we optimize for unique values of them for the whole data set, with the given values of $k_{\mathscr{T}}$. Then, the $k_{\mathscr{T}}$'s are optimized for each shear rate with the previous optimized values of P_2 and k_2. A few iterations of the last two steps are usually enough to get the final values.

Experiments

In order to test the approach, transient and steady state data have been obtained on a 0.427 gm/cc solution of polymethylmethacrylate in diethylphthalate at 40°C. The material is the commercial grade, Röhm and Hass Plexiglas V-100 colorless polymer of medium range molecular weight (~100000). The lower *Newtonian* viscosity was 55000 poise, which was slightly lower than that obtained by *Lee* and *Brodley* (6). This is not a surprising result because the concentration was only an approximate value and η_0 is a very strong function of concentration.

A modified *Weissenberg*-Rheogoniometer (R-16) was used to obtain the data (11). The normal stress measurements involved a further modification beyond that suggested by *Lee* et al. (10), but since these results have not been checked, we will not present the equipment details here; they can be found in reference (8). Ten constant shear rate levels and five constant stress levels were used for the transient data. Five shear rate levels were used for the normal stress measurements. Some additional improvements were made that should be cited: For constant shear rate runs, a new heavy duty electronic switch was designed so as to obtain a minimum response time for the magnetic clutch operation. This also gave a more repeatable start of closure, which was helpful for the high rate of data digitizing being used. A new release mechanism was designed to obtain a smoother start for the constant stress experiments. Drift in the piezoelectric load cells (11) was reduced by potting the cell in Dow Corning Sylgard 185 Encapsulating Resing, placing the cell in a small polyethylene bag kept at a positive pressure by dry, filtered air, and by the addition of air conditioning to the room.

The detailed procedures, data acquisition system, program details and other aspects of data reduction can be found in reference (8). The data points that are reported in the next section are not the raw data. Involved in the obtained reported points are the following steps: calibration of voltage output of the piezoelectric detector, acquisition by the PDP-15 system, conversion to normalized stress, fitting by linear regression polynomials, and finally computation of the regression fit and production of punch cards for a series of selected values of time. It is these latter that appear in the figures. In the case of the constant stress runs, displacement is the raw variable. In this case, the regression polynomial was differentiated to give the shear rate for the selected values of time and then punched.

Experimental data and predicted results

The large number of transient runs make it impossible to present all of the information here; thus, one is refered to the thesis by *Jachimiak* (8) for these details. A few samples will be given and the implications about the theory will be stressed. The complete set of predicted results can be found in the thesis by *Song* (10).

As in the past (3, 5, 6) the theory does an excellent job in fitting the steady state data as illustrated in fig. 2. The value used here for P is 1.1062 and for log K is -7.0996, with the 90 percent confidence limits being ± 0.0035 and ± 0.0200, respectively. The average percent difference was 4.83. The constants were based on the constant shear rate data only; however, the constant stress data is also plotted to show the excellent comparison.

For the constant shear rate transient studies, two runs were made at each shear rate level. These were averaged before the analysis was made. Fig. 3 shows the variation of $k_{\mathcal{T}}$ with shear rate. At the higher range the optimized values (10) lie below that estimated from the best fit to the initial slope. The value of P_2 and log k_2 are -0.5297 and 4.40226, respectively. Fig. 4 through 6 are samples of low, medium, and high shear levels. Fig. 7 is a typical example

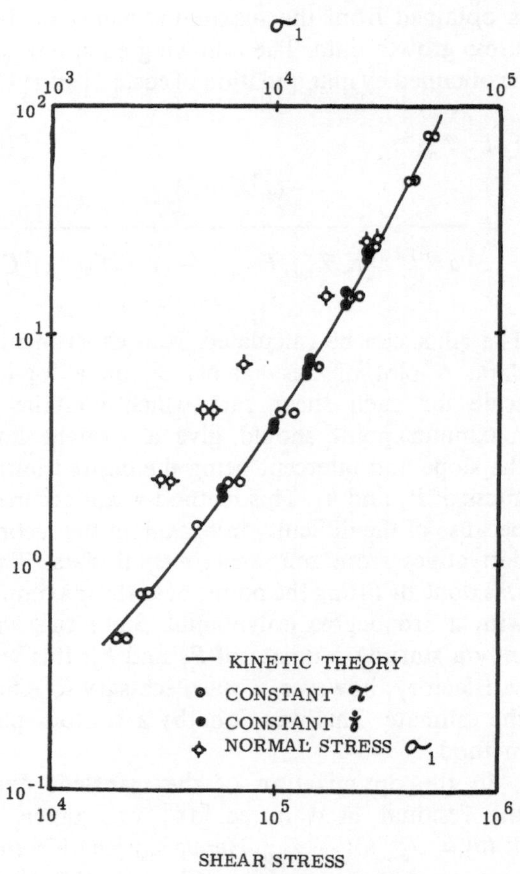

Fig. 2. Basic shear diagram and normal stresses

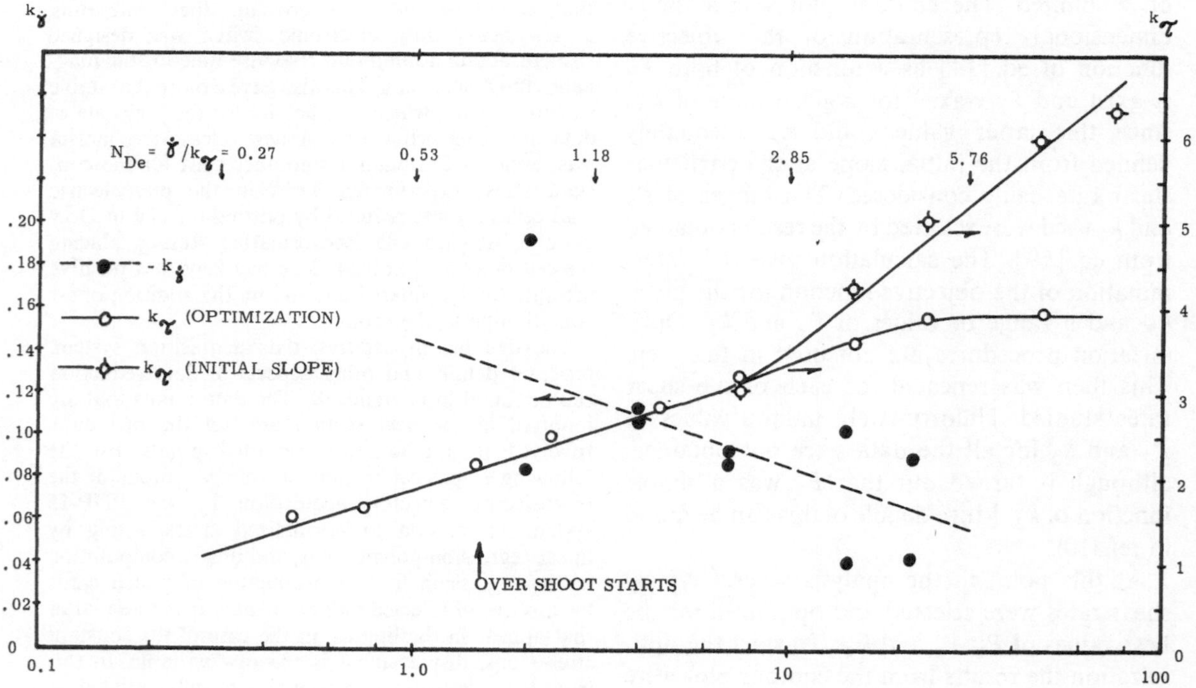

Fig. 3. Elastic parameters and the *Deborah* number

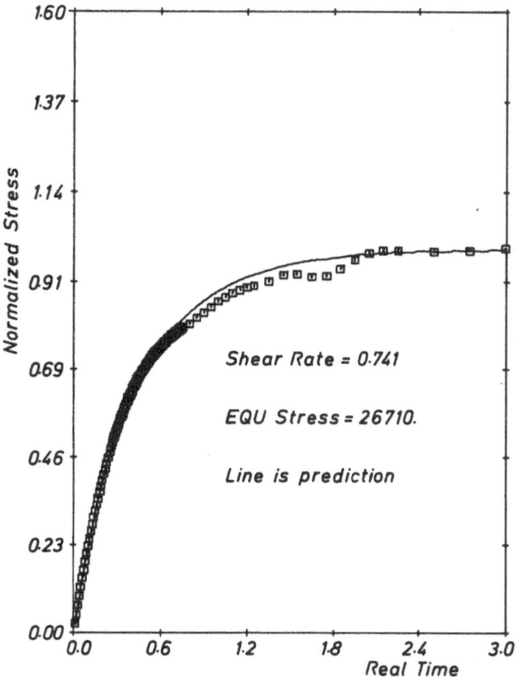

Fig. 4. Theory and data for low shear rate case

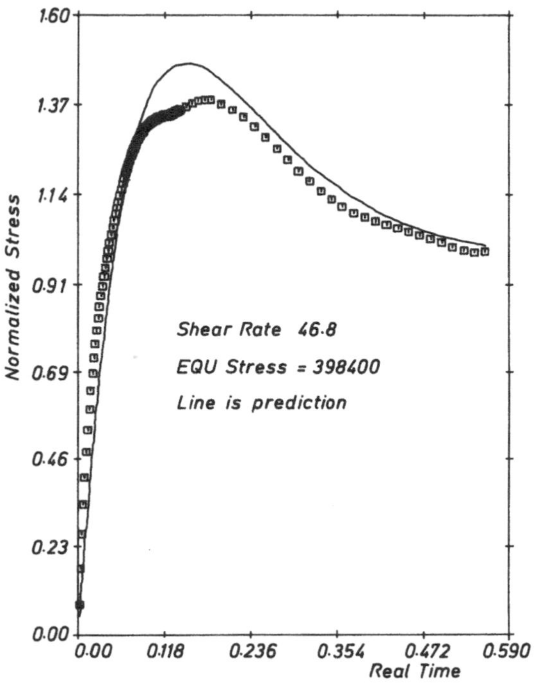

Fig. 6. Theory and data for high shear rate case

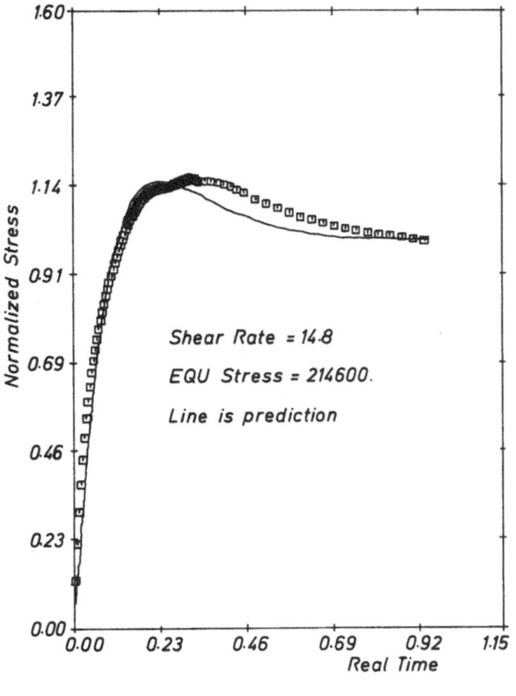

Fig. 5. Theory and data for medium shear rate case

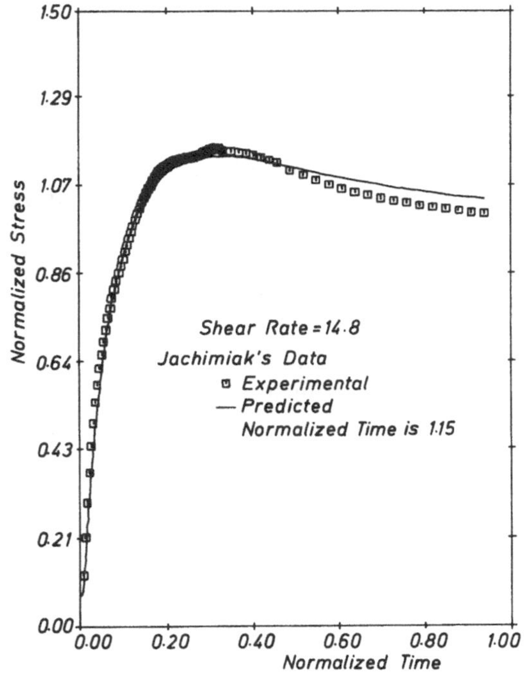

Fig. 7. Former theory and data for medium shear rate case

of the results (medium shear level) obtained with the older formulation of the theory. Even though a somewhat better fit is obtained, the older version must be rejected for the reasons cited in the theory. Improved fits could have been obtained if different values of P_2 and k_2 were used for each stress level; however, it is hard to justify this and we would rather look for a means to modify the theory. One possibility is that the elastic parameter, i. e., $k_{\mathcal{T}}$ be replaced with $G_{\mathcal{T}}/\eta_T$. Here $G_{\mathcal{T}}$ (from the *Maxwell* model) would be the constant rather than k. Other

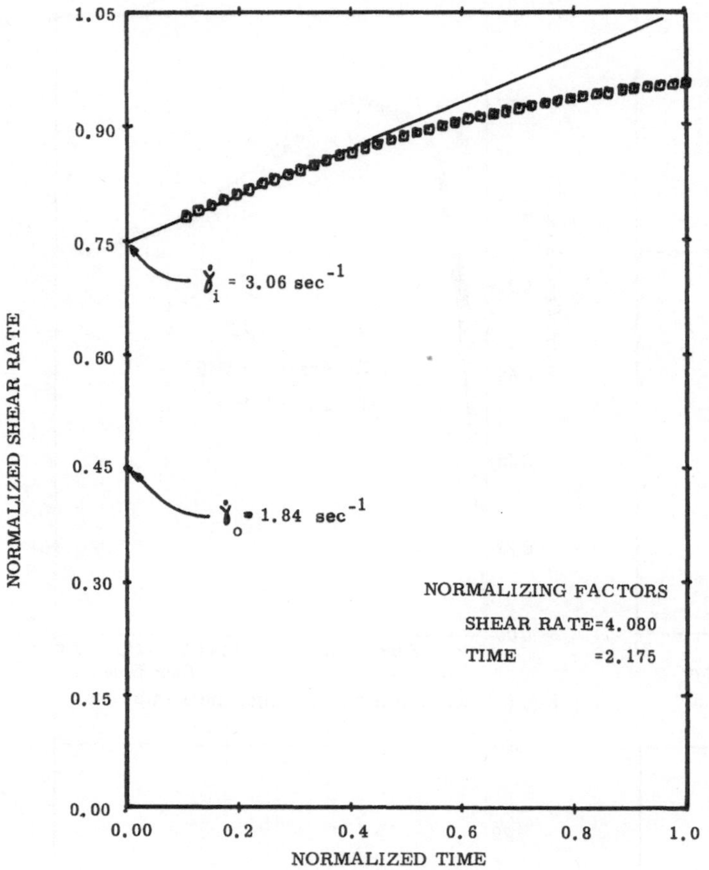

Fig. 8. Data, initial slope, $\dot{\gamma}_i$, and $\dot{\gamma}_0$ for constant stress of 100000 dynes/cm^2

suggestions are variation of the constants *a*, *n*, and *m* of the kinetic equation, or a reformulation of the catalyst-like term, $k_i \mathcal{T}_{P_1}$.

The same values of P_2 and k_2 were retained for the constant stress data. However, when the optimization was done for $k_{\dot{\gamma}}$, the value of $\dot{\gamma}_i$ in eq. [18] was used as zero; thus, the values of the elastic parameter were incorrect (8). A typical constant stress run is shown in fig. 8. With $\dot{\gamma}_i$ given from the data, $k_{\dot{\gamma}}$ can be accurately determined using eq. [18]. Since P_2 and k_2 have been fixed by the constant shear rate experiments, there is no need for optimization. The values of $k_{\dot{\gamma}}$ are shown in fig. 3. In all of the runs $\dot{\gamma}_i$ exceeded $\dot{\gamma}_0$, so that $k_{\dot{\gamma}}$ has a negative value since the slope is positive. The only other evaluation of this term is by *Darby* (12) who experimented with very dilute drag reducing materials. He reported positive values that were 3 to 15 times $k_{\mathcal{T}}$, which is to be contrasted to our negative values for concentrated polymer solution of 0.05 to 0.2 times $k_{\mathcal{T}}$. If in eq. (18), $\dot{\gamma}_i$ is assumed to be zero, then $k_{\dot{\gamma}}$ will have a large positive value as observed by *Darby*. In effect he assumed $\dot{\gamma}_i$ as zero, thus explaining the discrepancy.

Although somewhat questionable, our normal stress results are plotted on fig. 2 along with the shear stress data. The predicted results are easily calculated from eq. [13] and are from one to four orders of magnitude greater than the experimental values. With our definition of sign convention (eq. [3]), the first normal stress difference should be negative as is observed in the experiments and is predicted. Clearly there are limits on the possible values of $k_{\dot{\gamma}}$. If positive it must lie between $+\infty$ and $k_{\mathcal{T}}$. If negative, as determined here, it must < 0 and lie above $-\infty$. In either limiting case of $\pm\infty$, there is an asymptotic value for σ_1; i.e., when $1/k_{\dot{\gamma}} = 0$. This limit corresponds to a modified *Maxwell* model. The asymptotic value is closer to the experiment determination, being one to two orders of magnitude high. This latter value is the lowest value that can be predicted with a negative $k_{\dot{\gamma}}$. If the experiments were correct, then a positive $k_{\dot{\gamma}}$ near $k_{\mathcal{T}}$ would be necessary; clearly, this is not indicated from the transient results. There are no other values in the literature that we can use to check our normal stress measurements. Results on much different materials do indicate that our values might be low. We hope to obtain the necessary equipment to properly make the required measurements.

Lee and *Brodkey* (6) suggested that the *Deborah* number defined by $N_{De} = \dot{\gamma}/k_{\mathcal{T}}$ is near unity when stress overshoot is first observed. The *Deborah* numbers are reported on fig. 3, and as can be seen the results confirm their observation.

Concluding remarks

From an experimental viewpoint, the data gathering ability of a modern process computer such as the PDP-15, model 30 is well demonstrated by the transient results.

From a conceptual standpoint the kinetic model has been considerably improved. However, this resulted in a tighter theory and thus one that is more difficult to use to fit data. Deficiencies in the theory are brought out by the comparison to data and indicate some possible means of improving it still further.

The piezoelectric load cell modification of the rheogoniometer for the measurement of normal stresses without variation in the gap dimension has yet to be proved. Its use, however, for transient stress and shear rate measurements (to avoid plate rotation) has been well documentated.

Finally, the model has been extended to include the prediction of transient normal stress results. This latter is unique among rheological models. However, in this work since transient normal stress measurements were not made, no transient predictions of these are reported.

The results confirm the observation of *Lee* and *Brodkey* (6) that stress overshoot begins near the point where $N_{De} = \dot{\gamma}/k_{\mathcal{T}} = 1$.

Acknowledgements

Help in various stages of the data acquisition and reduction was obtained from Drs. *Harry C. Hershey* and *John T. Heibel*. Mr. *Michael B. Kukla* demonstrated throughout this study his ability to make impossible equipment function in an orderly manner. The early grant support of the NSF (GP-573) and NASA (NsG 591) made the study possible, as well as NSF fellowship support for one of us (PDJ). Röhm and Haas kindly supplied the polymer sample. Finally, to the many supporters of our department whose contributions made possible the computer facility we are deeply indebted. The manuscript preparation was done while holding a Senior Fellowship in Science (NATO) at the Max-Planck-Institut für Strömungsforschung at Göttingen.

Summary

The present paper is a further development in our effort to describe non-*Newton*ian flow behavior. As in the past, the thixotropic shear stress or shear rate is characterized by the kinetic theory and currently a modified form of *Oldroyds* model provides the elastic contribution. Presented here are further modifications of the model in order to predict the first normal stress difference and logical changes to make the model more internally consistent.

In order to test the approach, transient and steady-state data have been obtained with a *Weissenberg* rheogoniometer on a solution of polymethylmethacrylate in diethylphthalate. Both constant stress and constant shear rate data have been taken over a broad range of the parameters involved. To facilitate data acquisition, a PDP-15, model 30 computer was programmed to gather data at rates of 1000 points per second. Of the data obtained there is a great deal of uncertainty about the normal stress results.

By optimization methods, the constants of the model were evaluated from steady-state data and transient results on stress growth and shear rate growth. Comparisons of the model with stress growth results as well as the prediction of the first normal stress difference are given. The inadequacies of the analysis are emphasized and possible points for improvement are made.

References

1) *Brodkey, R. S.*, The Phenomena of Fluid Motions (Reading, Mass. 1967).

2) *Denny, D.* and *R. S. Brodkey*, J. Appl. Phys. **33**, 2269 (1962).

3) *Kim, H. T.* and *R. S. Brodkey*, Amer. Inst. Chem. Eng. J. **14**, 61 (1968).

4) *Jones, L. G.* and *R. S. Brodkey*, Trans. 5th Int. Congr. Rheology, Kyoto, Japan **2**, 267 (1968).

5) *Lewis, W. E.* and *R. S. Brodkey*, Trans. 5th Int. Congr. Rheology, Kyoto, Japan **4**, 141 (1968).

6) *Lee, K. H.* and *R. S. Brodkey*, Trans. Soc. Rheol. **15**, 627 (1971).

7) *Oldroyd, J. G.*, Proc. Roy. Soc. **A 128**, 122 (1953).

8) *Jachimiak, P. D.*, Ph. D. Thesis Chemical Engineering, The Ohio State University, Columbus (1971).

9) *Dexter, F.*, J. Appl. Phys. **25**, 1124 (1954).

10) *Song, Y. S.*, M.S. Thesis Chemical Engineering, The Ohio State University, Columbus (1972).

11) *Lee, K. H., L. G. Jones, K. Pandalai*, and *R. S. Brodkey*, Trans. Soc. Rheol. **14**, 555 (1970).

12) *Darby, R.*, Trans. Soc. Rheol. **14**, 185 (1970).

Authors' address:

P. D. Jachimiak, Y. S. Song, and *R. S. Brodkey*
Department of Chemical Engineering
The Ohio State University
Columbus, Ohio 43210 (USA)

Rheol. Acta 13, 754–756 (1974)

From the Cattedra di Principi di Ingegneria Chimica, University of Palermo (Italy) and Istituto di Principi di Ingegneria Chimica, University of Naples (Italy)

Comments on the validity of a common category of constitutive equations

G. Marrucci and G. Astarita

Introduction

The constitutive equation of linear viscoelasticity, as applied to incompressible fluid materials, takes the form:

$$\tau(t) = \int_0^\infty f(s)\, C^t(s)\, ds \qquad [1]$$

where $\tau(t)$ is the extra-stress tensor at time t (i. e., the total stress tensor within an arbitrary additive isotropic tensor), while $C^t(s)$ is the *Cauchy* strain carrying the configuration at time $t-s$ into the configuration at time t (see Appendix). Eq. [1] is sometimes referred to as expressing the so-called *Boltzmann* superposition principle, but it has in fact a more precise logical status as an asymptotic form of the much more general constitutive equation of a simple fluid with fading memory (1).

In fact, a simple fluid with fading memory is described by:

$$\tau(t) = \underset{\{s>0\}}{\mathscr{H}}[C^t(s)] \qquad [2]$$

where $\mathscr{H}[\cdot]$ is a functional mapping the deformation histories $C^t(s)$ (i. e., an appropriate space of tensor-valued functions of a scalar argument) into the extra stress tensor. The functional $\mathscr{H}[\cdot]$ is, by the requirements of objectivity, subject to the condition:

$$Q \cdot \mathscr{H}[C^t(s)] \cdot Q^T = \mathscr{H}[Q \cdot C^t(s) \cdot Q^T] \qquad [3]$$

for all orthogonal tensors Q.

The theory of simple fluids with fading memory rests on assumptions of smoothness laid down for the functional $\mathscr{H}$: without such assumptions, no concrete results could be obtained. Fundamental among such assumptions is that the functional $\mathscr{H}$ is *Fréchet*-differentiable at the rest history $C^t(s) = 1$ (for all s), with respect to a topology of the space of histories defined in terms of the principle of fading memory (1). In the following, we refer to this assumption as "hypothesis I". Although of course it is *conceivable* that real materials may exist which do not fulfill the requirements of hypothesis I, it is crucial to realize that no concrete results can be obtained from a general rheological theory unless *some* smoothness assumptions are made, and that no assumptions different from hypothesis I have been rigorously formulated in the literature.

From hypothesis I, and simple algebraic theorems, one can show that eq. [2] degenerates into eq. [1] when the norm of $C^t(s)-1$ is sufficiently small, say when:

$$\|C^t(s) - 1\| < \varepsilon \qquad [4]$$

and terms of order ε^2 are neglected.

Since eq. [1] holds only for small deformations, i. e., when eq. [4] is satisfied, it could be written equivalently with any other measure of strain, such as the *Finger* strain $C^{t-1}(s)$, since all measures of strain are, to within terms of the first order in the magnitude of strain, proportional to each other.

Several constitutive equations which have been proposed in the literature as useful approximations of the functional $\mathscr{H}$ are suggested by the form of eq. [1], but are assumed to hold also for *large* deformations, i. e., when eq. [4] is not fulfilled. Since for large deformations different measures of strain are not proportional to each other, different choices of strain measures give rise to non-equivalent equations. A wider generality is obtained by writing the constitutive equation in the form:

$$\tau = \int_0^\infty [f_1(s)\, C^t(s) + f_2(s)\, C^{t-1}(s)]\, ds. \qquad [5]$$

Eq. [5] is still linear, and it predicts a constant viscometric viscosity in shear, whatever the form of the influence functions $f_1(s)$ and $f_2(s)$. Thus, it is clearly not satisfactory for the description of the behavior of polymeric materials, which almost invariably exhibit a shear-dependent viscometric viscosity.

The non-linearity of the observed behavior of polymers in shear can be introduced by allowing the influence functions to depend not only on s, but also on some kinematic parameter. At this stage, a crucial choice has been made in the equations proposed in the literature. On the one side, some authors (2–7) have proposed to consider the invariants of the strain $I_k(s)$ (see Appendix), thus writing the constitutive equation in the form:

$$\tau = \int_0^\infty [\psi_1(s, I_k(s))\, C^t(s) + \psi_2(s, I_k(s))\, C^{t-1}(s)]\, ds \qquad [6]$$

while others (8–13) have proposed to consider the invariants of the rate of strain $I'_k(s)$, (see Appendix), thus writing the constitutive equations in the form [1]):

$$\tau = \int_0^\infty [\varphi_1(s, I'_R(s))\, C^t(s) + \varphi_2(s, I'_R(s))\, C^{t-1}(s)]\, ds. \qquad [7]$$

[1]) Some authors (9, 10, 12, 13) prefer to include in eq. [7] values of I'_k calculated at $s = 0$, or some average over $0 \div s$; these distinctions are irrelevant for the following discussion.

In this paper, we comment on the validity of constitutive equations of the form of eq. [7].

Validity of equation 7

It is clearly possible to conceive a deformation history $C^t(s)$ which is arbitrarily close to the rest history and still has arbitrarily large rates of strain. (Closeness is here intended with reference to an appropriate topology of the space of histories, such as defined e. g. by a norm:

$$\| C^t(s) \| = \sqrt{\int_0^\infty h(s) \, | C^t(s) |^2 \, ds} \qquad [8]$$

with $h(s)$ an influence function which goes to zero fast enough when $s \to \infty$.) In fact, one may e. g. consider an oscillation of small amplitude δ and very large frequency ω, for which the deformation history $C^t(s)$ takes the form:

$$C^t(s) = 1 + \delta \, \mathrm{Re} \, \{\psi^*(t) \, [1 - e^{-i\omega s}]\} \qquad [9]$$

where $\psi^*(t)$ is a complex tensor of unit magnitude and Re $\{\cdot\}$ indicates the real part of a complex quantity. For such a history one has:

$$\| C^t(s) - 1 \| = 0(\delta) \qquad [10]$$

and therefore eq. [4] is satisfied for sufficiently small δ. At the same time, the magnitude of the rate of strain is:

$$| \dot{C}^t(s) | = 0(\omega \delta) \qquad [11]$$

and can therefore have any (large) value, even if δ is small, provided ω is large enough. Thus, equations of the form of eq. [7] do not fulfill hypothesis I.

Should a real material's behavior be represented by eq. [7], general results from the theory of simple fluids with fading memory, such as the thermodynamic theory developed by *Coleman* (14), would not apply to such a material. Of course, it is *conceivable* that such a material exists; but since no general theory with a scope comparable to that of the theory of simple fluids with fading memory is available, there seems to be no reason to use an equation such as eq. [7] simply to correlate data which could be equally well correlated by eq. [6].

[*White* (15) maintains that, for the solution of boundary value problems, eq. [7] is easier to use than eq. [6]. This point has been disputed by *Pearson* (16), and is anyhow somewhat irrelevant to the present discussion.]

Thus the issue is really one which should be discriminated on the basis of experimental evidence; i. e., one should determine experimentally whether linear viscoelastic behavior is approached by real materials in the limit of very small deformations, or of very small deformation rates.

Such experimental evidence is indeed available. *Philippoff* (17) reports data of maximum shear stress. σ_m, vs. maximum shear rate γ_m for polyisobutylenes-organic solvent solutions in small amplitude oscillatory motion. As predicted by linear viscoelasticity, a linear dependency is observed when constant-frequency data are considered, up to a value of γ_m above which deviations from linearity are observed. Since the maximum shear rate γ_m is given by:

$$\gamma_m = \omega \delta \qquad [12]$$

such data by themselves do not discriminate the issue, since at constant ω a critical value of γ_m corresponds to a critical value of the amplitude of deformation δ. But *Philippoff* reports such σ_m vs. γ_m plots for different values of the frequency ω, and breakoff from linear viscoelastic behavior is observed, for any given material, at the same value of δ for all frequencies (and hence at different values of γ_m). Should eq. [7] apply, one would expect breakoff to occur at a fixed value of γ_m.

We therefore conclude that the available experimental evidence supports the smoothness assumptions underlying the theory of simple fluids with fading memory. There seems to be no reason to introduce equations of the form of eq. [7], since they are not supported by experimental evidence and do not allow the application of general results from the theory of simple fluids with fading memory.

Appendix

The deformation gradient $F^t(s)$ is defined as follows:

$$dX(t) = F^t(s) \cdot dX(t - s) \qquad [A1]$$

where $dX(t)$ is the displacement vector of two neighboring material point a time t, and $dX(t - s)$ is the displacement vector of the same two material points at time $t - s$. The strain $C^t(s)$ is:

$$C^t(s) = F^{t^T}(s) \cdot F^t(s). \qquad [A2]$$

One has:

$$C^t(o) = F^t(o) = 1. \qquad [A3]$$

By invariants of a tensor A we mean the trace, the second invariant II_A and the determinant:

$$\mathrm{tr}\, A; \; \tfrac{1}{2}[(\mathrm{tr} A)^2 - \mathrm{tr}(A^2)]; \; \det A \qquad [A4]$$

$I_k(k = 1, 2, 3)$ are the invariants of $C^t(s)$;
$I'_k(k = 1, 2, 3)$ are the invariants of $\dot{C}^t(s) = dC^t(s)/ds$.

The discussion in this paper also applies to constitutive equations of the differential type, such as e. g. the *Maxwell* equation:

$$\tau + \Lambda \hat{\tau} = 2\mu D. \qquad [A5]$$

where $\hat{\tau}$ is some appropriately invariant time derivative of stress. $D = -\frac{1}{2}\dot{C}^t(0)$ is the instantaneous rate of strain. If Λ and μ are taken as constants. Eq. [A5] fulfills hypothesis I. If on the other hand Λ and μ are taken as functions of the invariants of D, as proposed by some authors (18, 19), eq. [A5] does not fulfill hypothesis I, and it would not predict linear viscoelastic behavior to be approached in the limit of small deformations.

Summary

Many constitutive equations for viscoelastic materials which have appeared in the literature are modifications of the linear viscoelasticity model. Their general form is:

$$\tau = \int_0^\infty (f_1 C + f_2 C^{-1}) ds. \qquad [5]$$

The memory functions f_1 and f_2, are assumed to depend explicitly on either some instantaneous or some time-averaged value of the invariants of the rate of strain.

It is shown in this paper that the general theory of simple fluids with fading memory is based on certain assumptions of smoothness for the constitutive functional which are violated by constitutive equations of the type discussed. This implies that, should any real material obey eq. [5], with an explicit dependency of the f_i's on the rate of strain, such a material would not obey the general theorems of the simple fluid theory which are based on different smoothness hypotheses.

A critical analysis of available experimental evidence shows that it supports the validity of the smoothness hypotheses underlying the theory of simple fluids with fading memory, while contradicting those implied by an explicit dependency of the memory functions on the rate of strain.

References

1) *Coleman, B. D.* and *W. Noll,* Revs. Mod. Phys. **33**, 239 (1961).
2) *Lodge, A. S.*, Rheol. Acta **7**, 379 (1968); Trans. Farad. Soc. **52**, 20 (1956).
3) *Bernstein, B., E. A. Kearsley,* and *L. J. Zapas,* Trans. Soc. Rheol. **7**, 391 (1963).
4) *Ward, A. F. H.* and *G. M. Jenkins,* Rheol. Acta **1**, 110 (1958).
5) *Zapas, L. J.*, J. Res. Natl. Bur. Stds. **70 A**, 525 (1966).
6) *Tanner, R. I.* and *J. M. Simmons,* Chem. Eng. Sci. **22**, 1803 (1957).
7) *Tanner, R. I.* and *R. L. Ballman,* Ind. Eng. Chem. Fund. **8**, 588 (1969).
8) *Spriggs, T. W., J. D. Muppler,* and *R. B. Bird,* Trans. Soc. Rheol. **10**, 191 (1966).
9) *Lodge, A. S.*, Elastic Liquids, p. 121 (London 1964).
10) *Lodge, A. S.*, 1965, as quoted in (8).
11) *Bird, R. B.* and *P. J. Carreau,* Chem. Eng. Sci. **23**, 427 (1968).
12) *Bogue, D. C.*, Ind. Eng. Chem. Fund. **5**, 253 (1966).
13) *Carreau, P. J.*, Trans. Soc. Rheol. (in press).
14) *Coleman, B. D.*, Arch. Ratl. Mech. Anal. **17**, 1 (1964); **17**, 230 (1964).
15) *White, J. L.*, Discussion at Euromech Colloquium 37 (Naples 1972).
16) *Pearson, J. R. A.*, ibidem
17) *Philippoff, W.*, Trans. Soc. Rheol. **10**, 317 (1966).
18) *Tanner, R. I.*, A.S.L.E. Trans. **8**, 179 (1965).
19) *White, J. L.* and *A. B. Metzner,* J. Pol. Sci. **7**, 1867 (1963).

Authors' addresses:

Dr. *G. Marrucci*
Cattedra di Principi di Ingegneria Chimica,
University of Palermo (Italy)

Dr. *G. Astarita*
Istituto di Principi di Ingegneria Chimica
University of Naples (Italy)

Rheol. Acta 13, 757–760 (1974)

From the National Research Laboratory of Metrology, Tokyo (Japan)

Kinetic energy correction in viscosity measurement with capillary method

M. Kawata, K. Kurase, and K. Yoshida

With 8 figures and 3 tables

I. Introduction

In precise measurement of viscosity of low viscous fluid with capillary method, it is necessary to consider the kinetic energy correction. The coefficient of the kinetic energy correction has been usually treated as constant value for a certain capillary tube. Recently, it has been observed that the coefficient of the kinetic energy correction is not strictly constant and varies with *Reynolds* number.

Therefore, we examined experimentally the relations among the coefficient of the kinetic energy correction, shape of capillary tube, and *Reynolds* number by using various capillary tubes which have uniform circular cross-section bore, conical bore, and trumpet-shaped end.

II. Experimental method and apparatus

For laminar liquid flow through a capillary tube, the pressure drop has been expressed by the following equation:

$$P = \frac{8(L + nr)}{\pi r^4}\eta Q + \frac{m}{\pi^2 r^4}\rho Q^2, \qquad [2.1]$$

where P is the pressure drop between both the ends of the capillary tube, L the length of the capillary tube, r the radius of the capillary tube, η the viscosity of the liquid, Q the volumetric flow-rate through the capillary tube, ρ the density of the liquid, m the coefficient of the kinetic energy correction, and n the coefficient of the end correction.

In measurements of viscosities of liquids by application of this equation, the values of L, r, m, and n have been generally considered to be constant for a certain capillary tube, and the factors C_1 and C_2 shown by the following equations:

$$C_1 = \frac{8(L + nr)}{\pi r^4} \qquad [2.2]$$

and

$$C_2 = \frac{m}{\pi^2 r^4} \qquad [2.3]$$

are treated as instrumental constants. Recently, it has been observed that C_1 and C_2 are not strictly constant and vary with *Reynolds* number in precise measurement (1, 2).

From the eqs. [2.1], [2.2], and [2.3], we get

$$\frac{P}{Q\eta} = C_1 + C_2\frac{Q\rho}{\eta}, \qquad [2.4]$$

and if the factors C_1 and C_2 are constant, the relation between $P/Q\eta$ and $Q\rho/\eta$ is linear.

Therefore, the relation between $P/Q\eta$ and $Q\rho/\eta$ is experimentally examined with various capillary tubes, which have uniform circular cross-section bore, conical bore, and trumpet-shaped end.

A schematic diagram of the apparatus for this measurement is shown in fig. 1.

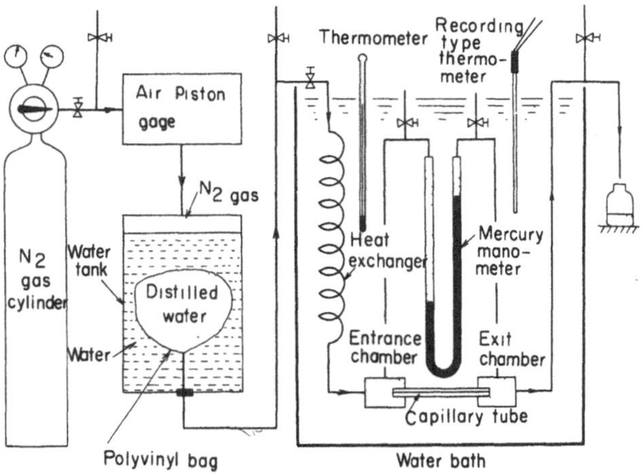

Fig. 1. Schematic diagram of the experimental apparatus

A steady flow is obtained by applying externally the constant pressure to a distilled water filled in a polyvinyl bag, which is mounted in a water tank as shown in fig. 1. The constant pressure is produced by air piston type pressure gage with an accuracy of $\pm 0.03\,\%$. Thus, various constant flow-rates are produced in the capillary tube and determined by the measurement of the mass of the distilled water flowing out through the capillary tube. The pressure drop between entrance and exit chambers of the capillary tube is measured by the mercury column type manometer, and the level of the mercury column is measured by a cathetometer with an accuracy of $\pm 10\ \mu m$.

The glass capillary tube is mounted horizontally between plastic entrance and exit chambers. The capillary tube and manometer are mounted together

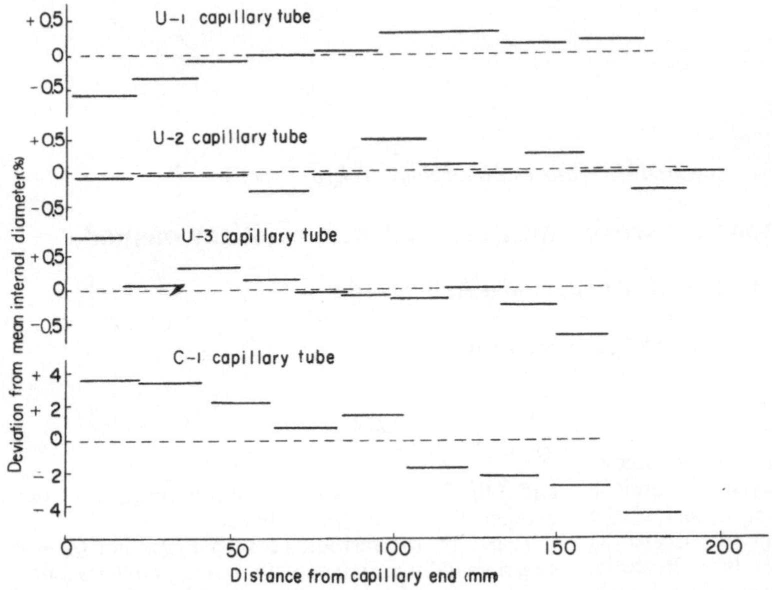

Fig. 2. Uniformity of the capillary diameter

in a water bath maintained at the test temperatures of 20, 30, and 40 °C.

The water bath with temperature fluctuation of better than ±0.005 °C is constructed of clean glass windows of such length as to allow observation of the whole of the capillary tube and manometer. The temperature of water bath is measured by means of mercury-in-glass thermometers with an accuracy of ±0.005 °C.

In this experiment, three glass capillary tubes having good uniformity, one glass capillary tube having conical bore, and one glass capillary tube having trumpet-shaped exit-end are used. The uniformity of the capillary tube diameter is examined by mercury-moving-method in the capillary tube. The results of the measurements of the uniformity of the capillary tubes are shown in fig. 2. Approximate diameters and lengths of these capillary tubes are shown in table 1.

A schematic diagram and approximate dimensions of the entrance and exit chambers are shown in fig. 3.

The viscosities and densities of the distilled water at the test temperatures of 20, 30, and 40 °C are shown in table 2.

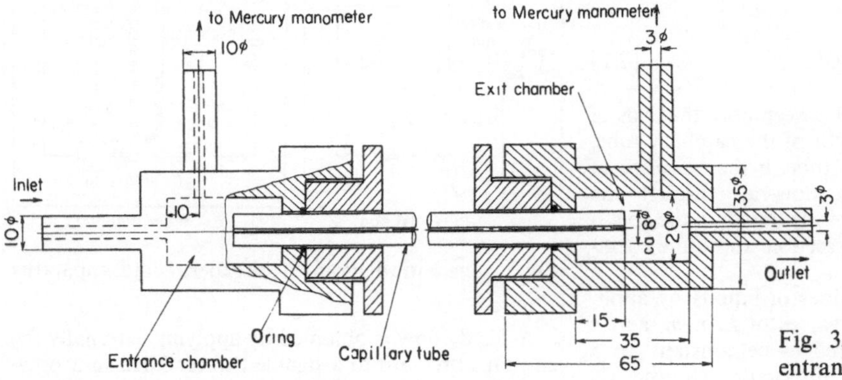

Fig. 3. Schematic diagram of the entrance and exit chambers

Table 1. Approximate diameters and lengths of capillary tubes

Capillary Nos.	Shape of capillary tube	Min. and Max. internal diameter (mm)	Mean internal diameter (mm)	Length (mm)
U-1	Uniform bore	0.219 – 0.221	0.220	179.26
U-2	Uniform bore	0.287 – 0.289	0.288	206.09
U-3	Uniform bore	0.366 – 0.372	0.369	169.08
C-1	Conical bore	0.274 – 0.299	0.287	201.65
T-1	Trumpet-shaped		0.410 [a)]	196.5 [b)]

[a)] and [b)] indicate the values of diameter and length in the portion of uniform bore, respectively.

Table 2. Viscosities and densities of distilled water at the temperatures of 20, 30, and 40 °C

Temperature (°C)	Viscosity (P)	Density (g/cm³)
20.00	0.01002	0.9982
30.00	0.007975	0.9957
40.00	0.006530	0.9922

III. Experimental result

The pressure drop, P, and the flow-rate, Q, are measured for three uniform capillary tubes which have square-cut ends, and the values of $P/Q\eta$ and $Q\rho/\eta$ are calculated from these measured results, and the relation between the two values for each capillary tube are shown in fig. 4, fig. 5, and fig. 6. In these figures, the cross marks show the results measured when liquid flows through each capillary tube in the opposite direction to the case of the circle marks. It is found from the observation of fig. 4, fig. 5, and fig. 6 that the relation between $P/Q\eta$ and $Q\rho/\eta$ is linear, and that the factors C_1 and C_2 of uniform capillary tube can be treated as constant in a wide range of *Reynolds* number.

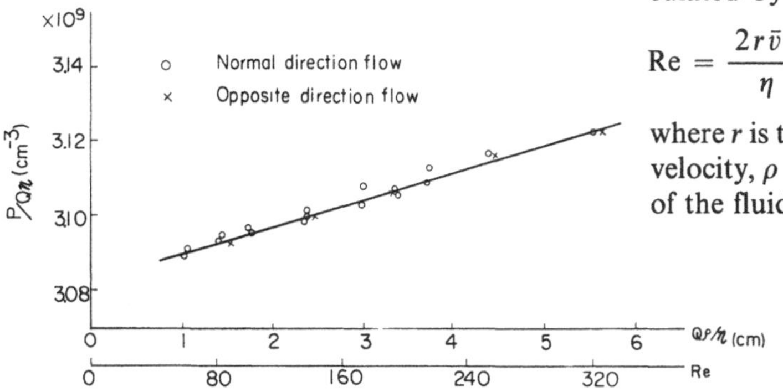

Fig. 4. $P/Q\eta - Q\rho/\eta -$Re relationships obtained with uniform capillary tube U-1

The coefficient of the kinetic energy correction, m, for each capillary tube is calculated with the least-squares method by use of the eqs. [2.2], [2.3], and [2.4] and is shown in table 3. In calculation of m, the end correction is treated as negligible small in comparison with the length of the capillary tube.

Fig. 7 and fig. 8 show the relation between $P/Q\eta$ and $Q\rho/\eta$ for the conical capillary tube, which have square-cut ends, and for the trumpet-shaped exit-end capillary tube as shown in fig. 8, respectively.

The incline of the relation for the conical capillary tube differs in the flowing direction, and the coefficient, m, calculated for each flowing direction is shown in table 3.

For the trumpet-shaped exit-end capillary tube, it is observed that the relation is not linear and that the factors C_1 and C_2 vary with *Reynolds* number. The coefficient, m, is calculated from the results on the linear relation in range of *Reynolds* number larger than ca. 200 and is shown in table 3.

In this paper, *Reynolds* number, Re, is calculated by the following equation:

$$\mathrm{Re} = \frac{2r\bar{v}\rho}{\eta} = \frac{2Q\rho}{\pi r\eta}, \tag{3.1}$$

where r is the radius of capillary tube, $\bar{v}$ the mean velocity, ρ the density of the fluid, η the viscosity of the fluid, Q the volumetric flow-rate.

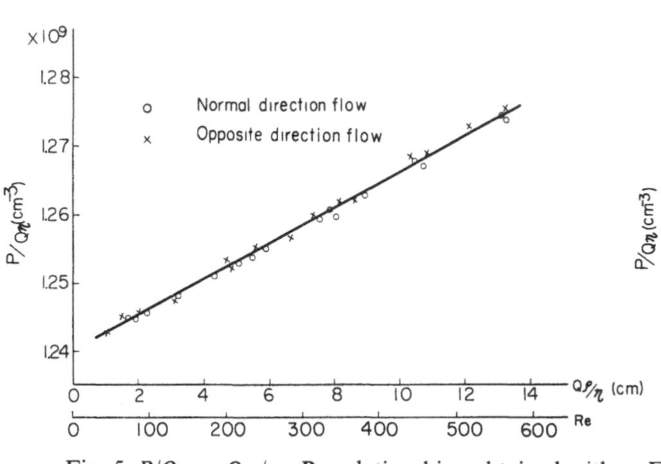

Fig. 5. $P/Q\eta - Q\rho/\eta -$Re relationships obtained with uniform capillary tube U-2

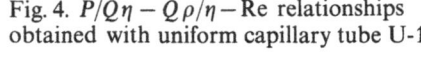

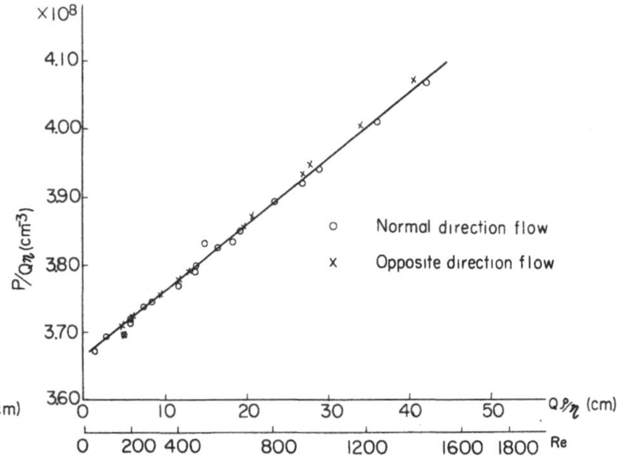

Fig. 6. $P/Q\eta - Q\rho/\eta -$Re relationships obtained with uniform capillary tube U-3

Table 3. Experimental values of the coefficient of the kinetic energy correction m

Capillary Nos.	Flow direction through the capillary tube	Value of m	*Reynolds* number
U-1	Normal direction	1.08	60 – 322
	Opposite direction	1.09	88 – 326
U-2	Normal direction	1.08	72 – 566
	Opposite direction	1.13	46 – 564
U-3	Normal direction	1.10	50 – 1466
	Opposite direction	1.16	106 – 1404
C-1	Normal direction	0.933	48 – 398
	Opposite direction	1.26	48 – 392
T-1	Direction for trumpet-shaped exit-end	1.03	196 – 1008

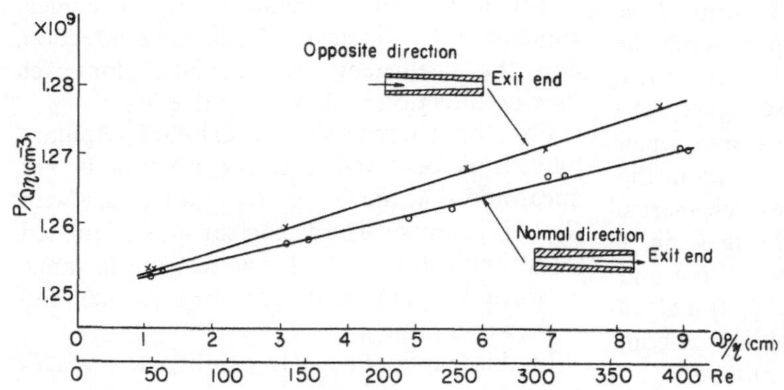

Fig. 7. $P/Q\eta - Q\rho/\eta$ – Re relationships obtained with conical capillary tube C-1

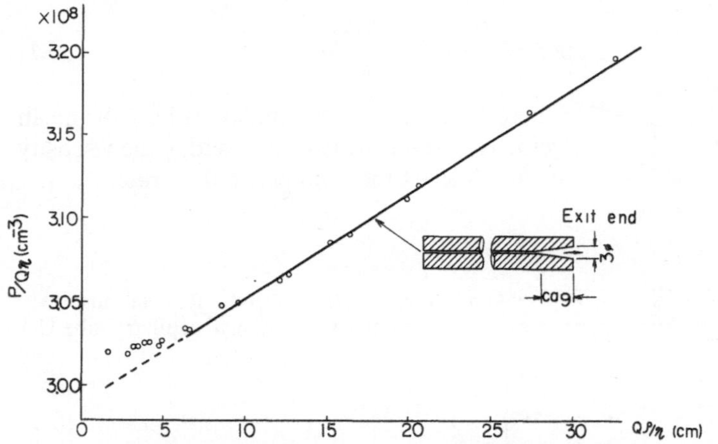

Fig. 8. $P/Q\eta - Q\rho/n$ – Re relationships obtained with trumpet-shaped capillary tube T-1

IV. Conclusions

(1) For a uniform capillary tube which has square-cut ends, the coefficient of the kinetic energy correction can be treated as constant in a wide range of *Reynolds* number.

(2) For a conical capillary tube, the coefficient differs in the flowing direction.

(3) For a trumpet-shaped exit-end capillary tube, C_1 and C_2 vary with *Reynolds* number.

(4) In precise measurement of viscosity with capillary method, the coefficient of the kinetic energy correction must be determined for each used capillary tube.

References

1) *Kawata, M.*, Report of the Central Inspection Institute of Weights and Measures **10**, No. 2, 1 (1961).

2) *Marvin, R. S.*, J. Res. N.B.S.-A. Physics and Chemistry **75A**, No. 6, 535 (1971).

Authors' address:

Dr. *K. Kawata*, Mr. *K. Kurase*, and Mr. *K. Yoshida*
National Research Laboratory of Metrology,
10-4, 1-Chome, Kaga, Itabashi-Ku, Tokyo, Japan

Rheol. Acta 13, 761–766 (1974)

Commissariat à l'Energie Atomique, Sevran (France)

Etude par propagation d'ondes ultra-sonores d'un composité poreux soumis à une pression isostatique

S. Poulard, J. Roucou, G. Demange et J. Maury

Avec 12 figures

1. Introduction

Nous nous proposons d'étudier un matériau explosif formé d'un agrégat polycristallin et d'un liant plastique ou viscoplastique servant de matrice, les proportions étant de l'ordre de 90% pour la phase dense cristalline et de 10% pour le liant.

Nos travaux antérieurs (1) nous avaient permis de mettre en évidence l'influence de la microporosité à l'intérieur du matériau agrégataire. Une méthode de choix consistait à étudier l'évolution de la vitesse sonique longitudinale (2 MHz) dans le matériau et à en tirer des conséquences quant à son comportement élastodynamique.

Il nous a semblé intéressant de compléter notre interprétation en essayant d'avoir une action directe sur cette microporosité, et c'est pourquoi nous avons repris des mesures de propagations d'ondes ultrasonores sous pression isostatique et en fonction de la température, ces deux paramètres étant sensés intervenir sur la microporosité au cours des expériences.

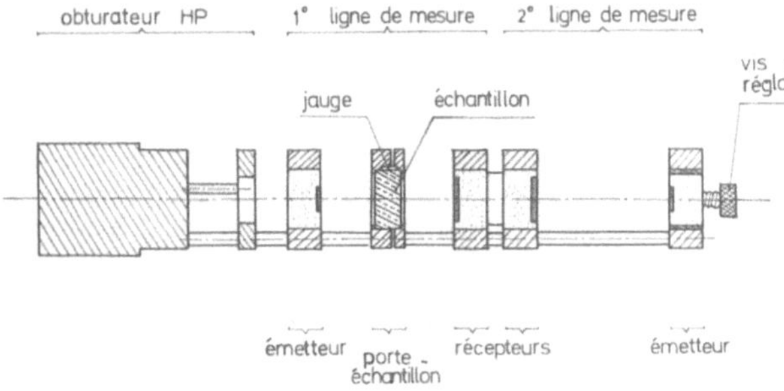

Fig. 1. Cellule de mesures ultrasonores pour haute pression

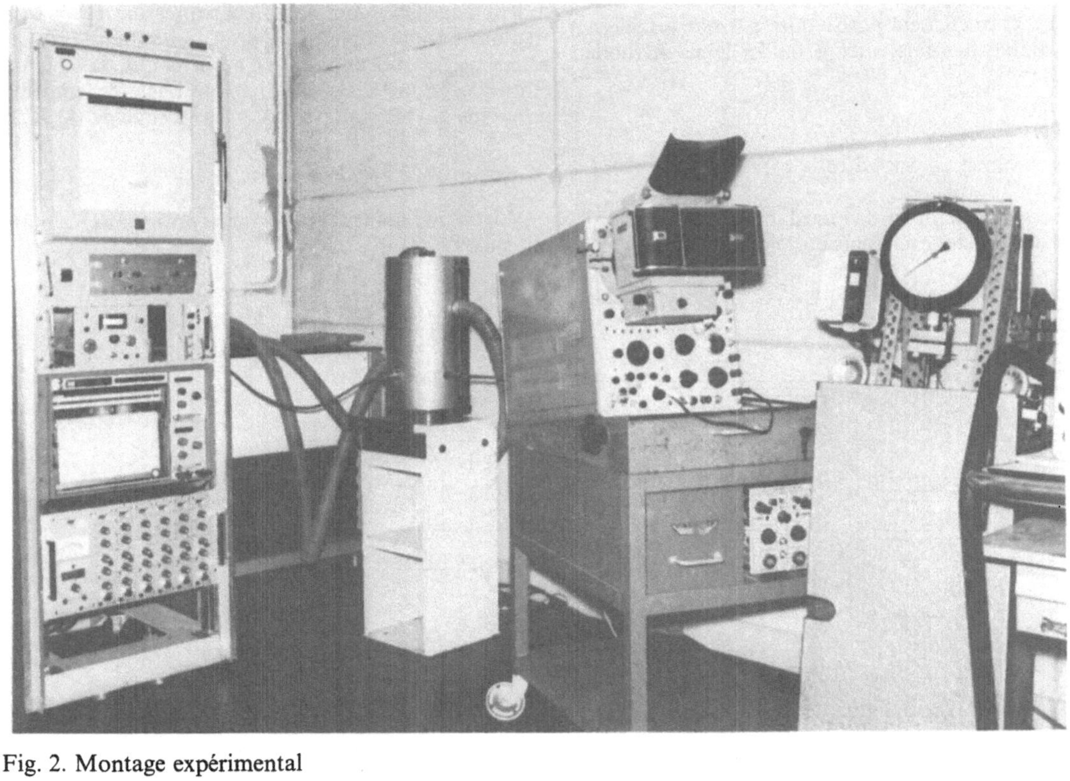

Fig. 2. Montage expérimental

2. Appareillages

2.1. Cellule de mesure US sous pression

Rappelons brièvement l'appareillage utilisé qui comprend (fig. 1) une cellule de mesure ultrasonore sous pression.

Il s'agit de façon classique de déterminer la différence de temps de parcours d'une impulsion ultrasonore se propageant, d'une part dans une ligne de référence et d'autre part, dans une ligne identique comportant l'échantillon à étudier.

Ce montage comporte quatre céramiques piézoélectriques en zirconate titanate de plomb, dont deux fonctionnent en émetteurs (pastilles convergentes) et les deux autres en récepteurs (pastilles planes). Des deux lignes ainsi formées, l'une comporte une vis micrométrique permettant de faire coïncider leurs longueurs de manière précise. Un amortissement mécanique des céramiques permet de limiter le nombre de leurs oscillations et d'éviter toute interférence entre les deux lignes.

L'échantillon est positionné sur un porte échantillon placé sensiblement au centre de l'une des deux lignes.

Ce montage est introduit dans une chambre de pression 1000 bars dont la température est régulée à 0,1 °C près par circulation extrême d'un fluide caloporteur (fig. 2).

La mise sous pression de la chambre s'effectue par injection d'huile de silicone (SI 200) à l'aide d'une pompe 0–2000 bars à débit variable. Cette huile joue un double rôle; elle assure la mise sous pression isostatique de l'échantillon et sert de milieu de transfert aux ondes ultrasonores.

L'étanchéité de l'échantillon (cylindre ∅ 19, h 7) par rapport au fluide transmetteur est obtenue de façon satisfaisante par un dépôt de vernis de 0,1 mm d'épaisseur environ (vernis à l'unithane 650 S).

Dernière précaution: il a été vérifié que la chaîne de mesure est correctement positionnée par une visualisation préalable des faisceaux à l'aide de la méthode *Schlieren*.

2.2. Chaîne de mesure

L'électronique associée à cette expérience est classique (fig. 3).

A l'aide du générateur de retard, on mesure le retard de marche de l'onde de référence et de l'onde de diagnostic.

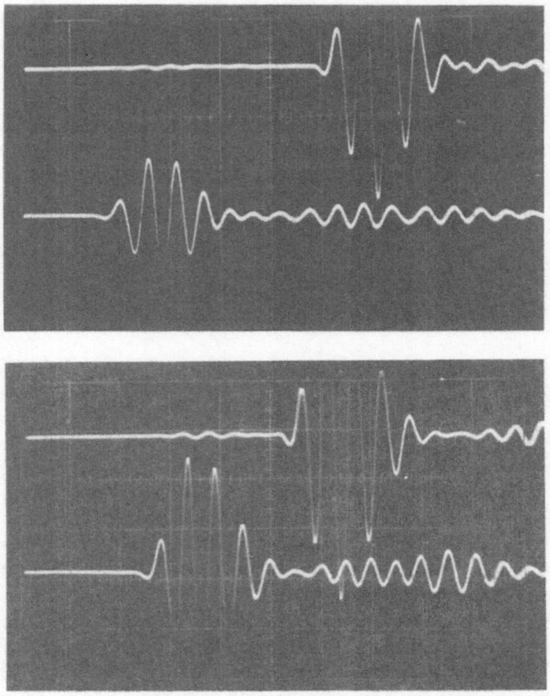

Fig. 4. Signaux visualisés: ligne de mesure (haut)
ligne de référence (bas)

L'atténuation dans l'échantillon est obtenue en comparant les amplitudes des deux signaux visualisés sur le scope (fig. 4).

La pression et la température régnant dans la chambre sont obtenues à l'aide d'un capteur de pression d'*Hotinger* et d'un thermocouple placé près de l'échantillon. Un deuxième thermocouple placé au niveau de la ligne de référence, permet de vérifier que le gradient de température dans la chambre reste dans la gamme de 0,1 °C.

3. Résultats expérimentaux

Voyons maintenant quels sont les principaux résultats obtenus.

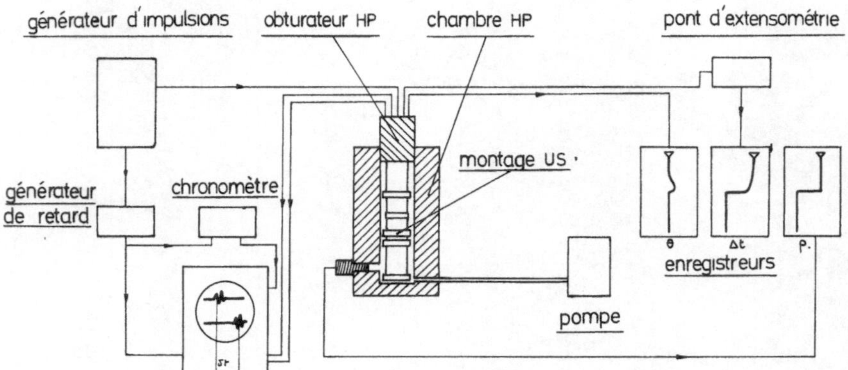

Fig. 3. Schéma de principe

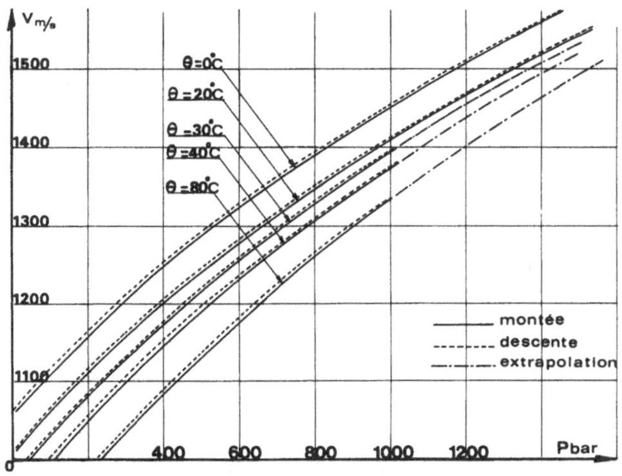

Fig. 5. Variation de la vitesse du son dans l'huile en fonction de la pression

3.1. Dans un premier temps, nous accédons aux variations, en fonction de la pression et de la température, de la vitesse du son V_0 dans l'huile qui sert de milieu transmetteur de pression.

Cinq isothermes sont tracés (fig. 5) entre 1 et 2000 bars, de 0 °C à 80 °C, la précision expérimentale sur la mesure de la fatigue est de l'ordre du pourcent.

Sur les figures suivantes sont représentés les résultats obtenus sur des échantillons soumis respectivement à un cycle de pression entre 1 et 2000 bars, à 0 °C et 40 °C.

Les essais sont conduits de la manière suivante: Le matériau est amené à la température d'essai, puis est soumis au cycle de pression. Les mesures sont relevées point par point après un palier de pression de 15 minutes environ, destiné à permettre à la température de revenir à la valeur de consigne.

Nous constatons (fig. 6) qu'à 0 °C la vitesse commence par croître en fonction de la pression et tend vers une valeur asymptotique au-delà de 800 bars.

En détente, la vitesse pour une pression donnée reste légèrement supérieure à la valeur obtenue au cours de la montée en pression, et on obtient en fin de détente, une légère augmentation rémanente de vitesse, sous réserve que cet écart ne soit pas confondu avec l'incertitude expérimentale sur la mesure de V_L.

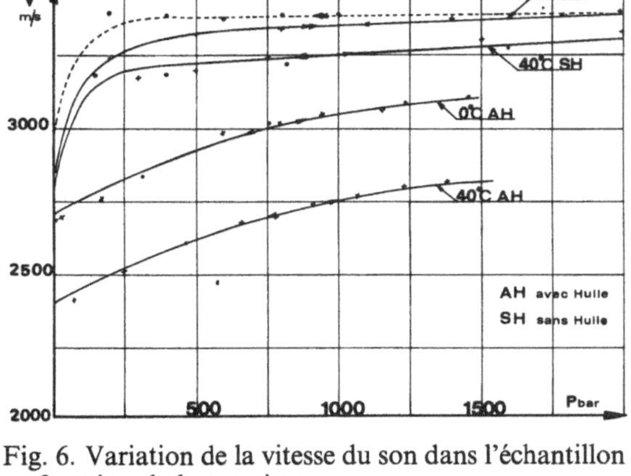

Fig. 6. Variation de la vitesse du son dans l'échantillon en fonction de la pression

Dans les deux cas nous voyons nettement apparaître un changement de pente marqué aux environs de 250 bars.

Le comportement à 40 °C est qualitativement semblable à celui à 0 °C:

– la courbe de compression part d'une valeur a l'origine de la vitesse inférieure à celle obtenue pour 0 °C.

– par la suite, la courbe continue à rester au-dessous de celle obtenue à 0 °C mais semble tendre vers une valeur asymptotique analogue à 50 m/s près (3400 m/s au-delà de 800 bars).

Sur la même figure sont reportés les résultats obtenus sans vernis protecteur (courbes indicées AH): on note dans ce cas que les vitesses sont nettement inférieures et que les allures de variations sont différentes. Ce résultat est indiqué ici car il est significatif de l'influence d'une pression interne au matériau sur la vitesse de propagation US et il est susceptible de concourir à l'interprétation donnée en fin d'exposé.

Des remarques analogues peuvent être faites sur les courbes d'atténuation (fig. 7) où l'on voit toujours apparaître un changement marqué de pente pour une valeur en pression de l'ordre de 250 bars sur les échantillons protégés.

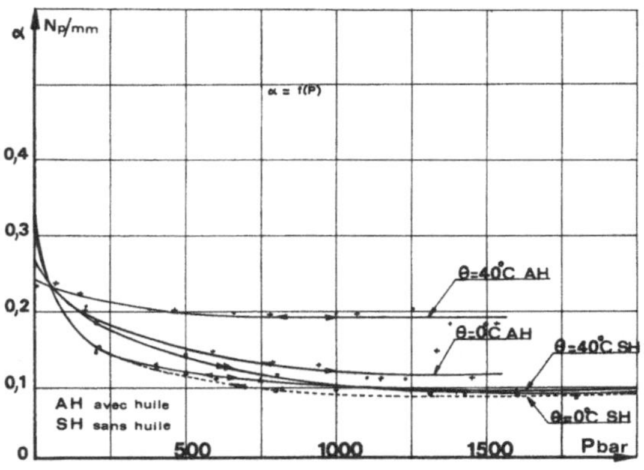

Fig. 7. Atténuation dans l'échantillon

3.3. *Variation de la vitesse du son dans une poudre sans liant en fonction de la pression*

Dans un troisième type d'expérience, nous avons essayé de préciser la vitesse limite ainsi mise en évidence et de cerner le rôle du liant ou du cristal sur ces vitesses.

Il a donc été commode de recommencer les expériences avec une poudre sans liant en fonction de la pression à 0 °C et à 45 °C. Nous opérons donc sur un échantillon compacté sans liant et gainé du vernis protecteur.

A titre de contre vérification, nous avons également fait une expérience sans liant et avec une protection perméable au fluide transmetteur.

La fig. 8 montre des comportements qualitatifs semblables à ceux obtenus dans les expériences précédentes sur le matériau agrégataire, à la fois

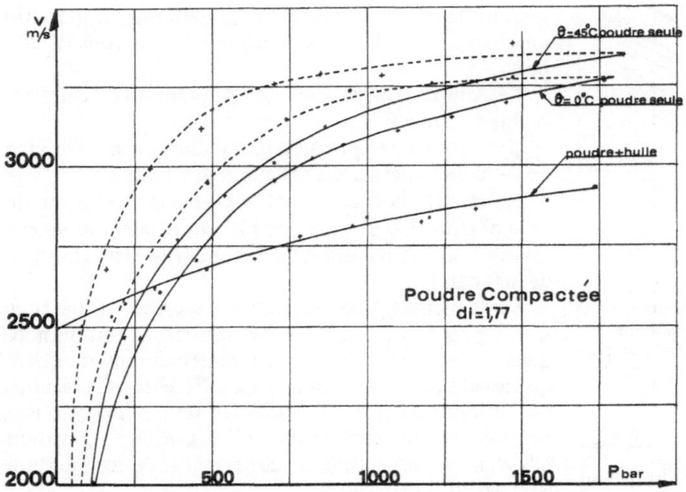

Fig. 8. Variation de la vitesse du son dans la poudre compactée

sur les échantillons avec protection et sans protection.

Cela conduit à penser que c'est la compaction de la phase cristalline qui est importante, non seulement parce qu'elle est pondéralement prédominante mais aussi par l'état des contacts intergranulaires qu'elle recèle.

Ce dernier point est recoupé par le fait que les vitesses US mesurées dans des monocristaux ne varient pratiquement pas dans cette gamme de pression.

4. Interprétation des résultats

4. 1. Considérations théoriques

Dans toutes nos expériences nous avons mesuré le temps de transit d'un signal détectable.

Une explication théorique des expériences effectuées doit rendre compte du mouvement de cette surface d'onde en fonction des propriétés des cristaux et du liant plastique. La vitesse du signal acoustique dans le liant seul, ces vitesses dans les cristaux comme le couplage acoustique entre le liant et les cristaux, l'orientation des cristaux et la structure géométrique des vides sont donc des facteurs qui doivent entrer dans cette explication théorique.

Pour simplifier le problème, supposons tout d'abord que le couplage acoustique entre les cristaux et le liant soit parfait (coefficient de réflexion nul à l'interface liant-cristal, sous incidence normale). Ceci pourrait être obtenu en choisissant convenablement le liant. Evidemment, ce n'est pas en général le cas et nous verrons par la suite l'influence d'un écart par rapport à ce cas idéal.

Nous sommes conduits immédiatement à une relation tenant compte des pourcentages des deux phases où la vitesse mesurée V_M est reliée de façon simple à la vitesse dans le liant V, à la vitesse dans le cristal V_c et à $\varnothing$ qui représente la concentration volumique en liant.

En fait, dans la réalité, le couplage entre les deux phases n'est pas parfait et la phase cristalline est discontinue.

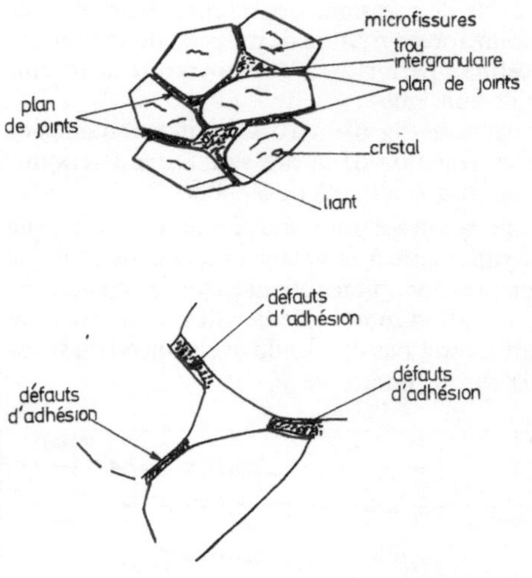

Fig. 9. Schéma de la texture intergranulaire du matériau

En effet, (fig. 9) les cristaux contiennent des microfissures qui apparaissent au moment du moulage du matériau, et sont liés les uns aux autres par de fins filaments de liant (quelques μ d'épaisseur) adhérant plus ou moins bien aux cristaux.

Examinons l'influence de ces défauts sur la fonction de structure.

Le dégré de saturation en liant[1]) du matériau considéré étant nettement inférieur à 1 (de l'ordre de 0,85), il est probable que le liant contenu dans les trous intergranulaires (fig. 9) adhère très mal à la phase cristalline. Le long du chemin passant à travers ces trous, la majeure partie de l'énergie doit être réfléchie dans la phase cristalline aux interfaces liant-cristal si bien que l'énergie véhiculée le long de ces chemins doit être négligeable. Le signal reçu doit donc correspondre à des chemins passant principalement à travers la phase cristalline.

La longueur de ce chemin dépend du nombre des microfissures contenues dans les cristaux et des défauts au niveau des plans de joints (fig. 9).

Plus la surface des défauts d'adhésion au niveau des plans de joints sera grande, et plus le parcours sera grand. Par conséquent plus la vitesse de propagation de l'énergie acoustique sera faible. Ceci conduit, par exemple, à introduire un «facteur de sinuosité» dans une interprétation théorique plus complète.

Il apparaît donc que, comme une partie de la phase plastique ne participe pas à la propagation de l'ébranlement sonore, la vitesse acoustique dans le matériau ne dépend pas directement de sa densité mais plutôt du nombre et de la surface spécifique des différents défauts ou microfissures contenues dans la phase cristalline et dans les plans de joints.

Une expérience complémentaire est susceptible de nous éclairer sur ce point: effectuons un recuit du matériau de plusieurs jours à 90°C et observons les variations élastodynamiques qui en résultent. Sur la fig. 10, nous voyons que la vitesse longitudinale US croît jusqu'à 1 ou 2 jours de recuit puis reste sensiblement constante au-delà.

Les expériences ont été faites pour deux valeurs de température: 5 °C et 45 °C.

Corrélativement à cette augmentation de vitesse, il se produit une faible diminution de densité apparente due à une légère perte en poids par sublimation d'un constituant du liant et une faible augmentation de volume (fig. 11).

Nous en déduisons que la vitesse varie en sens inverse de celui que laisserait prévoir la variation de densité.

Comme le suggère l'interprétation donnée aux variations de vitesse sous pression, ces variations ne peuvent s'expliquer que par une résorption

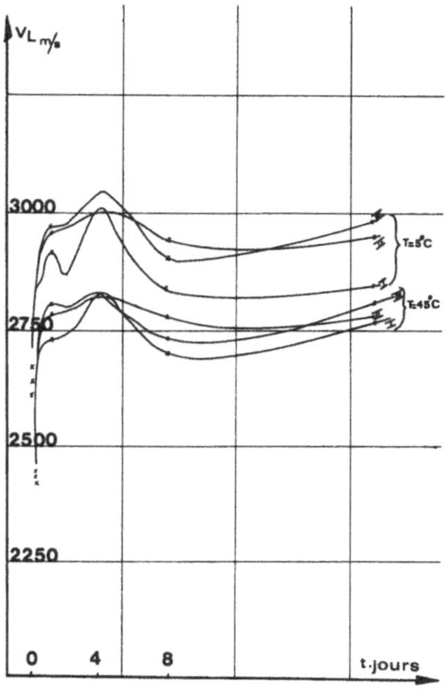

Fig. 10. Variation de la vitesse longitudinale en fonction du recuit du matériau

totale ou partielle, par activation thermique des défauts contenus dans les plans de joints.

Cependant, on peut noter ici que les défauts dans les plans de joints dépendent de l'état de compaction de la matrice cristalline. Pour des pièces ayant la même teneur en liant et ayant subi la même histoire thermomécanique, c'est-à-dire pour les conditions de compactage lors de la mise en forme initiale, il doit donc être possible,

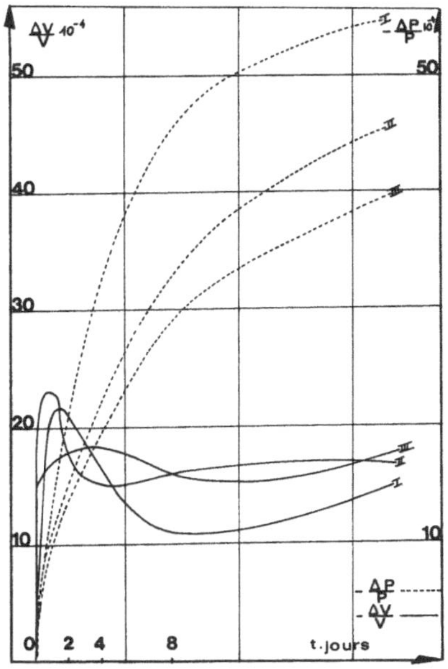

Fig. 11. Variation de la densité apparente du matériau en fonction du temps

[1]) Rapport du volume occupé dans le liant au volume total des vides.

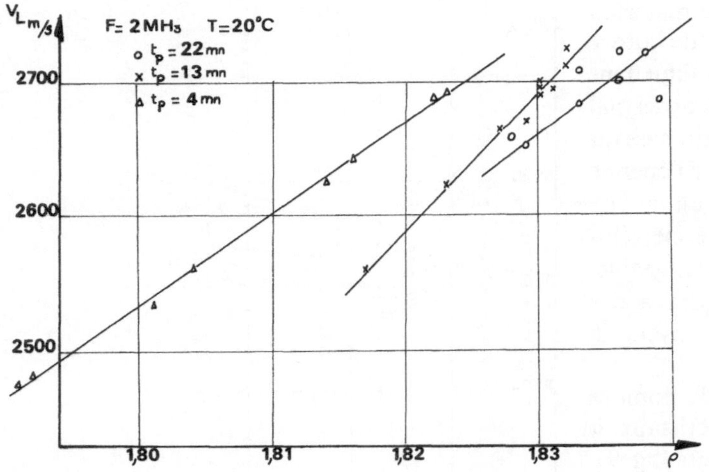

Fig. 12. Variations isothermes de la vitesse longitudinale en fonction de la densité apparente pour des conditions de moulage différentes

comme le montre d'ailleurs l'expérience (fig. 12), de déterminer une corrélation entre la vitesse sonore dans le matériau et sa densité apparente, expérience à pression ordinaire.

5. Conclusion

Des résultats des essais que nous venons de présenter, nous pouvons tenter de tirer les conclusions suivantes:
1. La vitesse sonore US dans le matériau agrégataire étudié dépend principalement du nombre des défauts contenus dans les plans de joints intercristallins.
2. Une grande vitesse correspond à une bonne adhésion dans les plans de joints.
3. La vitesse ne peut être corrélée à la densité du matériau que pour un lot de pièces ayant subi la même histoire thermomécanique.

Par exemple, dans un contrôle de fabrication utilisant cette méthode, il conviendrait de faire corrélativement des mesures de densité et de vitesse sonore, ces deux mesures étant complémentaires.

Si on contrôle un lot ayant été fabriqué dans des conditions identiques, une corrélation entre vitesse et densité pourra être établie sur un échantillonnage, ce qui permettra de s'affranchir d'une des deux mesures.

Enfin, ces expériences US sous pression complètent l'interprétation du matériau agrégataire fondée sur la présence de la microporosité en permettant de préciser que ce n'est pas n'importe quelle microporosité qui intervient, mais celle qui se trouve aux joints de l'agrégat polycristallin.

Nous avons entrepris de calculer un facteur de sinuosité pour la propagation de l'onde US, mais nous ne sommes pas assez avancés pour présenter ici les résultats.

Références

1) *Poulard, S., G. Morin, C. Perennes, J. Roucou* et *J. Maury*, Comportement des matériaux pseudo-isotropes. CEA-CONF-81.

Adresse des auteurs:
S. Poulard, J. Roucou, G. Demange et *J. Maury*
Commissariat à l'Energie Atomique, Etablissements T
B. P. No. 7, F-92270 Sevran (France)

Rheol. Acta **13**, 767–773 (1974)

From the Corporate Technology Department, Owens-Illinois Technical Center, Toledo, Ohio (USA)

An empirical model for PVC melt drawing

G. M. Fehn

With 3 figures and 2 tables

Introduction

During their fabrication into useful parts, all plastic materials undergo a complex deformation history. It has been common practice for rheologists to study the response of polymers to various idealized deformations in order to better understand their processing behavior. Great success has been attained in predicting the characteristics of polymer melts for processes in which shear deformations dominate. The pressure, temperature, flow rate interdependence in extrusion and injection molding processes can be described with suitable accuracy from relatively straightforward laboratory experiments. These shear experiments can also be used to elucidate other processing phenomena such as extrudate expansion, melt fracture, and thermal stability. However, rheologists have not yet been successful in identifying and measuring material characteristics which relate to processes involving essentially elongational deformations. Since nearly all fabrication techniques involve such deformations, the importance of extensional flow studies is clearly evident.

Considerable attention has been given to the measurement of the shear viscosity analog, η_T (tensile viscosity: ratio of melt tensile stress to axial velocity gradient). However, such measurements are complicated by analytical as well as experimental difficulties. Consequently, the existence of η_T as a useful material property has not yet been consistently demonstrated.

Because of the elusive nature of the tensile viscosity, an alternate approach was considered for characterizing the elongational response of polymer melts. Specifically, an empirical model was derived and fitted to experimental data from isothermal, steady-state melt drawing experiments. While such experiments are really transient in convecting coordinates, they are relatively simple and clearly relate to real polymer processes.

Experimental

Material

A nonfood extrusion grade PVC homopolymer was chosen for the study. The material was a commercial compound, Pantasote R-873, provided by the Escambia Chemical Corporation.

Equipment

The apparatus, shown schematically in fig. 1, has been described previously (1). It consists of the following essential parts:

a) Instron Rheometer for extrusion rate control
b) Furnace providing isothermal stretching conditions
c) Pulley arm for monitoring stretching tension
d) Water bath for quenching strand
e) Take-off system for draw speed control

The rheometer was fitted with a .063 inch diameter, 32 L/D capillary. The die was designed to extend about one inch below the die holder so the melt could be

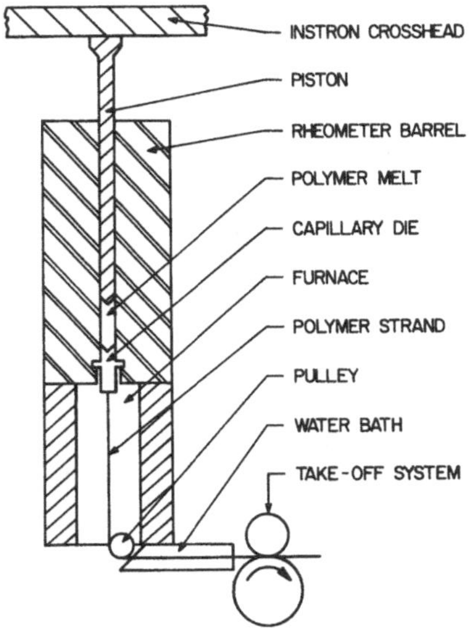

Fig. 1. Schematic diagram of melt drawing apparatus

observed through a glass window in the furnace as it emerged from the capillary. This window provided a means of gathering quantitative data by following the movement of tracer particles in the melt with moderately high speed (up to 64 frames/sec) motion pictures.

Procedure

An Instron barrel temperature of 200 °C was chosen as the best compromise between thermal degradation and thorough melting. The furnace was likewise set to 200 °C. The axial temperature profile in the furnace was found to be uniform within about 4 °C. While such variations are not insignificant, it is felt that the isothermal assumption is reasonably valid considering the time scale of the experiments.

For a typical run, the experimental procedure was as follows. The Instron barrel was loaded with approximately .01 % (by volume) inert tracer particles having a diameter of about .006 inch. After a four-minute melting period, the extrusion was begun at a constant cross-head speed. The extrudate was fed through the furnace, around the pulley, through the quench bath, and finally between the take-off rollers which were driven at a fixed speed. A steady state was assumed when a stable drawing tension was attained

Table 1. Experimental conditions and melt tensions

Run number	Cross-head speed in/min	Draw velocity in/min	Melt tension grams
11	0.2	7.87	0.79
12	0.2	14.7	1.32
13	0.2	22.5	1.88
14	0.2	46.4	2.53
21	0.5	16.7	0.73
22	0.5	27.7	1.67
23	0.5	43.1	2.34
24	0.5	93.0	3.19
31	1.0	26.7	0.58
32	1.0	42.8	1.85
33	1.0	73.2	2.87
34	1.0	118.0	3.80

(variations $< \pm 5\%$). The movie camera was then started and allowed to run until at least two tracer particles had passed through the drawing zone. This procedure was repeated for four different draw velocities at each of three different cross-head speeds. The specific conditions and tension recordings for the 12 experimental runs are given in table 1.

Results

Material response

The motion pictures were analyzed using a single-frame projector. A scale placed next to the strand in the furnace enabled the position of the tracer particles to be read directly. These measurements were accurate to within .005 inch. The corresponding time values for the particles were calculated from the film speed (measured) and the number of elapsed frames referenced to the die exit. For each run then, the stretching response was described by a set of distance versus time data for tracer particles streaming from the die to the pulley, a draw distance of approximately 4.75 inches. Such sets measured for different tracer particles were found to be indistinguishable.

In order to minimize ambiguity, only six of the twelve data sets are plotted in fig. 2. The slowest and fastest draw velocities at each of the three extrusion rates are shown.

Empirical model development

Prime consideration in the development of an empirical model for elongational response was that the variables in the model be easily controlled experimentally. These variables are (1) the extrusion rate, (2) the draw velocity, and (3) the draw distance. A second important con-

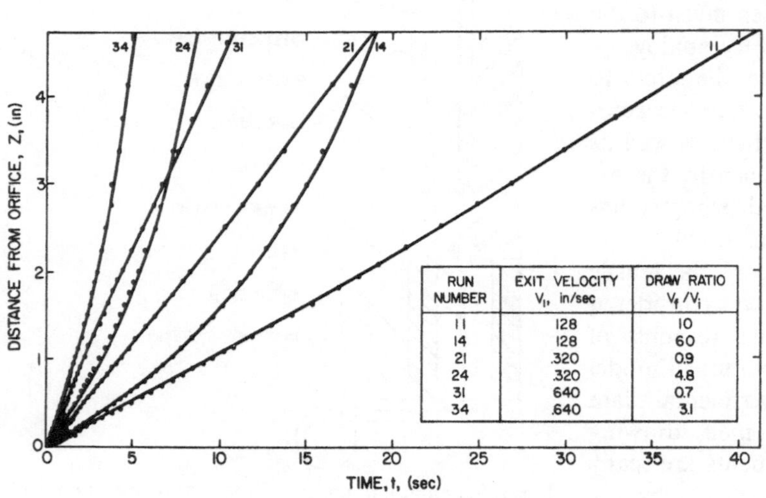

Fig. 2. Experimental distance-time data (●) and fitted curves for selected stretching conditions

sideration was that the model could be tested with experimental data or at least without having to make assumptions or approximations which could bias the results.

Assuming no radial gradient of axial velocity at steady state, the time, Δt, for an element of melt to pass from one position to another is,

$$\Delta t = \frac{\Delta V}{Q}, \quad [1]$$

where ΔV is the total melt volume between the two positions and Q is the volumetric flow rate. Replacing ΔV by the product of the difference in positions, Δz, and the distance-average cross-sectional area, A_z, eq. [1] becomes,

$$\Delta t = \Delta z\left(\frac{\bar{A}_z}{Q}\right). \quad [2]$$

The basis of this model development will be the choice of a suitable expression for $\bar{A}_z$. The most simple expression is derived by assuming the position-dependent area varies linearly,

$$\Delta t = \frac{\Delta z}{Q}\left[\frac{A_{z(1)} + A_{z(2)}}{2}\right], \quad [3]$$

where $A_{z(1)}$ and $A_{z(2)}$ refer to the cross-sectional areas at any two points along the strand. If we reference our analysis to the die exit ($z = 0$, $t = 0$), eq. [3] becomes

$$t = z\left[0.5\frac{A_0}{Q} + 0.5\frac{A_z}{Q}\right], \quad [4]$$

where A_0 is some "initial" cross-sectional area. We shall make A_0 a material parameter by defining it as the area to which the melt would swell under no-load stretching conditions. Eq. [4] might be considered in the following manner: the time, t, for an element of melt to reach a point, z, depends on two contributions – an elastic effect (A_0) and an attenuation term (A_z). For the case of a strand which attenuates in a linear manner, these two contributions are weighted equally (0.5). However, we know from our experiments that the profiles are substantially concave, indicating that the attenuation term should be weighted somewhat more than the A_0 term. Introducing this degree of freedom into our model we have,

$$t = z\left[(1 - B)\frac{A_0}{Q} + B\frac{A_z}{Q}\right], \quad [5]$$

where the dimensionless constant B denotes the fractional contribution of the attenuation term. Since at steady state,

$$Q = A_z v_z, \quad [6]$$

where v_z is the element velocity at $z = z$, eq. [5] may be rewritten,

$$t = z\left[\frac{1 - B}{v_0} + \frac{B}{v_z}\right], \quad [7]$$

where v_0 is the element velocity corresponding to the area, A_0. This velocity, v_0, shall be approximated in terms of the average exit velocity, v_i,

$$v_0 = C v_i + D, \quad [8]$$

where

$$v_i = \frac{4Q}{\pi d^2},$$

and d is the orifice diameter. Substituting for v_0 in eq. [7], we have

$$t = z\left[\frac{1 - B}{C v_i + D} + \frac{B}{v_z}\right]. \quad [9]$$

Before solving this differential equation, we can test it using experimental data for the various exit velocities (v_i) and the terminal conditions; viz., draw distance (z_f), draw velocity (v_f), and transit time (t_f). The results are shown in table 2 for fitted coefficients $B = .76$, $C = .52$, and $D = .017$.

Table 2. Preliminary test of empirical model[a])

v_i in/sec	v_0[b]) in/sec	v_f in/sec	t_f(exp'tl.) sec	t_f(estimated) sec
.128	.0836	.131	40.8	41.2
.128	.0836	.245	29.4	28.4
.128	.0836	.374	23.5	23.3
.128	.0836	.773	18.6	17.3
.32	.183	.279	18.8	19.2
.32	.183	.462	14.1	14.0
.32	.183	.718	11.5	11.3
.32	.183	1.55	8.8	8.3
.64	.350	.445	10.7	11.4
.64	.350	.714	8.2	8.3
.64	.350	1.22	6.3	6.2
.64	.350	1.97	5.1	5.1

[a]) $z_f = 4.75$ inches [b]) estimated using eq. [8]

It is apparent that the t_f values estimated by eq. [9] are in excellent agreement with those experimentally measured. This was a stringent test because an accurate prediction of transit time means that the total steady state volume of the melt between the orifice and quench point has been accurately predicted (see eq. [1]).

Solving eq. [9] gives

$$t = z\left[\frac{1 - B(z/z_f)^{\frac{1-B}{B}}}{v_0} + \frac{B(z/z_f)^{\frac{1-B}{B}}}{v_f}\right]. \quad [10]$$

Differentiation gives the velocity relation,

$$\frac{1}{v_z} = \left[\frac{1 - (z/z_f)^{\frac{1-B}{B}}}{v_0} + \frac{(z/z_f)^{\frac{1-B}{B}}}{v_f}\right],$$

which rearranges to

$$\frac{1}{v_z} \quad \frac{1}{v_0} - (z/z_f)^{\frac{1-B}{B}}\left[\frac{1}{v_0} - \frac{1}{v_f}\right]. \quad [11]$$

The meaning of this equation is more easily seen if eq. [6] is used to convert it back to areas:

$$A_z = A_0 - (z/z_f)^{\frac{1-B}{B}}(A_0 - A_f). \quad [12]$$

In this form, the cross-sectional area at any point, z, is clearly equal to the "initial" area minus the product of the area reduction and the attenuation variable, $(z/z_f)^{\frac{1-B}{B}}$. The simplicity of eq. [12] is even more apparent if the area is normalized as follows,

$$\frac{A_0 - A_z}{A_0 - A_f} = \left(\frac{z}{z_f}\right)^{\frac{1-B}{B}}. \quad [13]$$

Agreement between the experimental distance (z), time (t) data and the predicted curves based on eq. [10] was excellent except for small deviations near $z = 0$. Therefore, eq. [11] was modified slightly to make it more realistically describe the deceleration of the melt toward v_0 as it emerges from the orifice. This was accomplished by replacing the first term of eq. [11] with an expression which characterizes the time-dependency of the swelling effect at the die exit:

$$\frac{1}{v_z} = \frac{1}{v_0 + (v_i - v_0)e^{-Kz}} - (z/z_f)^{\frac{1-B}{B}}\left[\frac{1}{v_0} - \frac{1}{v_f}\right], \quad [14]$$

where $K = \frac{1}{Av_0}\left(\frac{v_f}{v_0}\right)$, and A is a material constant having units of time. The corrected expression eliminates the inconsistency of $v_z = v_0$ at $z = 0$. This correction introduces an additional term to the expression for time (eq. [10]),

$$t = z\left[\frac{1 - B(z/z_f)^{\frac{1-B}{B}}}{v_0} + \frac{B(z/z_f)^{\frac{1-B}{B}}}{v_f}\right] + t', \quad [15]$$

$$\text{where } t' = \frac{Av_0}{v_f}\ln\left[\frac{v_0}{v_i} + \left(1 - \frac{v_0}{v_i}\right)e^{-Kz}\right]. \quad [16]$$

This additional term, t', provides a small negative contribution to the total transit time,

$$t_{f(\text{corr})} = t_f + \frac{Av_0}{v_f}\ln\left(\frac{v_0}{v_i}\right).$$

Discussion

The fitted empirical model

Eq. [15] was fitted to the entire 12 sets of distance-time data by a gradient search technique using a time-sharing computer. The coefficients calculated on a least squares basis were as follows:

$A = 1.05$ sec
$B = 0.746$
$C = 0.517$
$D = 0.0186$ inches/sec.

The standard error of estimate for the distance variable at the 95% confidence level was 0.12 inch. The quality of the fit is clearly shown in fig. 2 where the solid lines were calculated from eq. [15] using the coefficients above. The other six data sets, which were generated at intermediate stretch ratios, were generally superior.

The significance of the coefficients was discussed briefly during the model derivation. Coefficients C and D serve to define v_0 (eq. [8]), the theoretical velocity of the "fully swelled" extrudate. Since at steady state, v_0 may be used to calculate the cross-sectional area (eq. [6]) and therefore the diameter, C and D actually describe dependence of swell ratio to extrusion or shear rate for the temperature and capillary used in this study. Ratios of theoretical expanded diameter to die diameter calculated in this manner were 1.23 at the slowest extrusion rate, 1.32 at the intermediate rate, and 1.35 at the highest rate. These calculated values are comparable to those measured directly and follow the expected general trends. The coefficient A is contained in terms which describe the time scale of the swelling effect. Under conditions of free extrusion (no gravity effects), the model predicts that the extrudate expansion will be $1/e$ complete in A seconds (1.05 seconds for the PVC studied). This expansion phenomenon is the dominant response at low stretching ratios (first term, eq. [14]).

The contribution of time-dependent swelling to the total transit time (eq. [16]) is therefore most significant under low stretch conditions. As stretching ratio is increased, however, the attenuation term (second term. eq. [14]) becomes increasingly important. Therefore, for stretching conditions of practical interest, it is the B coefficient which characterizes the elongational response. As we developed eqs. [4] and [5] we noted that as B became larger than 0.5, the model reflected a trend in the shape of the melt profile toward increasing concavity. In a qualitative sense, the larger the B coefficient, the greater will be the elongation at a given position along the drawing path (under constant stretching conditions). In other words, the "resistance" to extrudate attenuation is less persistent for materials with higher B coefficients. This concept requires that the steady state shape be viewed as an integral response to a set of drawing conditions. In contrast, the concept of tensile viscosity cannot be directly applied in an integral manner to the transient extensional (and retractional) conditions existing during steady state drawing. The viscosity approach requires that stresses be considered, and that the elongational behavior be viewed as a sequence of responses.

Utilizing the model

Having a set of material parameters which describe the total response of an extruded melt to extensional deformation is of undeniable value as a characterization tool. The potential practical value of such parameters will be realized only when they are derived for other materials and the resulting differences correlated with differences in processing behavior. However, having a model developed for only one material can be useful as well as interesting.

The model may be used to generate the elongational response for any set of stretching conditions: extrusion rate, draw velocity, and draw distance. For instance, by setting the draw velocity equal to v_0, the model will describe the extrudate expansion with time and distance for a desired flow rate. Such information could be useful in regulating the die swell in free extrusion processes.

Information generated under higher stretch conditions also appear useful. The phenomenon of "melt resonance", the periodic pulsation of an elastic melt undergoing "steady-state" stretching, has been discussed in the literature (2–5). This instability, which generally precedes the extensional failure of the melt, has been attributed to the accelerated cooling of the melt in sections which have been attenuated causing an upstream shift in the subsequent attenuation. Our experiments indicate, however, that "melt resonance" occurs with PVC melts under essentially isothermal conditions. This is first observed as small sinusoidal oscillations in the stretching tension at relatively high stretch ratios. As draw velocity is increased, the oscillations and the corresponding melt pulsations increase in magnitude. Using the empirical model, these drawing conditions can be simulated and the elongational response generated. The effects of such conditions on the predicted melt profile are most clearly shown by computing the diameter gradient,

$$\left(\frac{dD}{dz}\right)_z = \frac{D_i}{2}(v_i v_z)^{1/2} \frac{d\left(\frac{1}{v_z}\right)}{dz}, \qquad [17]$$

where D_i and v_i are the exit diameter and velocity respectively, and v_z is the melt velocity at a point, z, from the die exit as given by eq. [14]. For stable drawing, dD/dz is a monotonically decreasing negative function of z. At experimental draw ratios corresponding to the onset of "melt resonance", however, the diameter gradient as predicted by eq. [17] begins to increase (negatively) near the quench point. This means that rather than asymptotically approaching the final diameter, the predicted melt profile becomes convex, an obviously unstable condition. For extrusion rates used in this study, "melt resonance" is predicted for draw ratios (v_f/v_i) of approximately six.

Tensile viscosity implications

As discussed previously, the concept of tensile viscosity must be applied in convecting coordinates and, therefore, cannot provide a complete description of the entire melt response in steady-state drawing. The melt tensile stresses near the die exit are not related to the rapidly changing velocity gradient in the equilibrium sense. Furthermore, as an element of melt moves toward the quench point, neither the axial stresses nor the associated velocity gradient is controlled and the duration of the convecting extensional process cannot be considered long enough to reach any sort of equilibrium. In fact, for many materials under constant stress or strain rate conditions, no equilibrium tensile viscosity apparently exists (6, 7). While the calculation of a "transient" tensile viscosity may not lead to a general material description, it certainly can offer some insight into the specific elongational responses which relate to the transient conditions prevalent in real fabrication processes. Using the melt tension data in table 1, transient tensile viscosities can be calculated from the following relations,

$$\left(\frac{dv}{dz}\right)_z = -v_z^2 \frac{d\left(\frac{1}{dz}\right)}{dz}, \qquad [18]$$

$$\tau_z = \text{Tension}/A_z, \qquad [19]$$

and

$$\eta_{T(z)} = \tau_z \Big/ \left(\frac{dv}{dz}\right)_z, \qquad [20]$$

where τ_z is the tensile stress associated with an element of melt z inches from the orifice. The melt tension was corrected for the effects of gravity and acceleration (1); all other corrections were assumed to be negligible. In order to minimize the effects of extrudate expansion, η_T values were computed only for the downstream portions of the drawing zone: from 3.0 to 4.75 inches from the die exit. The results of the calculations are shown in fig. 3 where the transient tensile viscosities for each run are identified by its run number (see table 1). For comparison, the capillary shear flow curve has also been plotted. While there is substantial scatter in the tensile viscosities, they fall generally in the range of three times the extrapolated shear viscosity curve. Certain other trends are also evident. If the tensile viscosities at "constant strain" (roughly equivalent stretch ratios, v_f/v_0) are compared (e.g., 13, 23, and 33 versus 14, 24, and 34) it appears that larger strains cause a shift in the viscosities to higher levels. Alternatively, if viscosities are compared at constant extrusion rate (e.g., 11, 12, 13, and 14 versus 21, 22, 23, and 24), increasing the extrusion rate seems to shift the viscosity curve downward. The former effect may be due to melt "strain hardening" while the latter effect may be attributed to increasing disentanglement induced by higher flow rates through the extrusion die.

Conclusion

The analysis of the responses of polymer melts to extensional deformations is extremely complex. While the measurement of "equilibrium" tensile viscosities in experimental devices which control the stress or velocity gradient should lead to reliable material properties, this has not yet been consistently demonstrated. Furthermore, the application of the concept of tensile viscosity to transient conditions is not straightforward. Certainly the melt response to sequential processes involving both shear and extensional flows over relatively short time periods cannot be completely described in terms of tensile viscosity. Our studies have shown, however, that over a five-fold range of

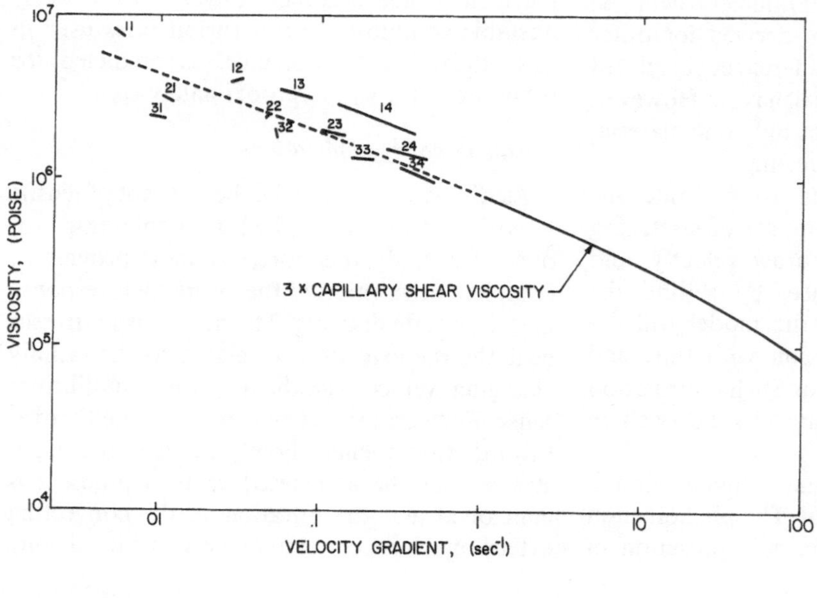

Fig. 3. Transient tensile viscosities computed from eq. [20] for various stretching conditions; capillary shear flow curve X 3 at 200 °C

extrusion rate, the total isothermal response of a PVC melt to various extensional conditions can be well represented using an empirical model involving four coefficients. Three of these relate to the extrudate expansion and expansion rate, while the fourth describes the attenuation behavior. The model can be used to elucidate elastic melt phenomena such as die swelling and "melt resonance" as well as describe the stable extensional response to a specific set of stretching conditions. While tensile viscosities calculated using the empirical model must be regarded as transient, they do offer some qualitative feeling for the effects of melt strain, extrusion rate, and velocity gradient. Further work with other materials over a broader range of stretching conditions will be required to assess the full value of this unified approach to describing complex melt extensional flows.

References

1) *Fehn, G. M.*, J. Polym. Sci. **A 1**, **6**, 247 (1968).
2) *Freeman, H. I.* and *M. J. Coplan*, J. Appl. Polym. Sci. **8**, 2389 (1964).
3) *Bergonzoni, A.* and *A. J. Di Cresce*, Polym. Eng. Sci. **6**, 45 (1966).
4. *Bergonzoni, A.* and *A. J. Di Cresce*, Polym. Eng. Sci. **6**, 50 (1966).
5) *Pearson, J. R. A.* and *M. A. Matovich*, I & EC Fundamentals **8**, 605 (1969).
6) *Meissner, J.*, Rheol. Acta **10**, 230 (1971).
7) *Stevenson, J. F.*, Amer. Inst. Chem. Eng. J. **18**, 540 (1972).

Author's address:

Corp. Technology Department
Owens-Illinois Technical Center
Toledo, Ohio 43666 (U.S.A.)

Rheol. Acta 13, 774–778 (1974)

From the Soil Mechanics Laboratory, McGill University, Montreal (Canada)

A probabilistic after-effect analysis of relaxation in clays

R. N. Yong and D. S. Chen

With 4 figures

Introduction

In the analysis of the constitutive performance of clay soils, it has been generally assumed that the material can be evaluated and examined in the light of continuum theory. However, detailed study of clays reveals that the rigid clay particles in the soil-water system do not act individually, but flocc together to form larger "particles" which are identified as fabric units (1, 2). In a recent study, some account of this phenomenon has been noted (3, 4). However, in view of the heterogeneity of the overall system, and the wide distribution of fabric units, concern must be directed towards the integrity of individual fabric units.

This study examines the constitutive performance of clays in view of the existence of fabric units [identified as "peds"] with differing strength characteristics. Based on an experimental study similar to the previous one reported (3), stability contributions from peds are identified and assessed in regard to overall constitutive performance.

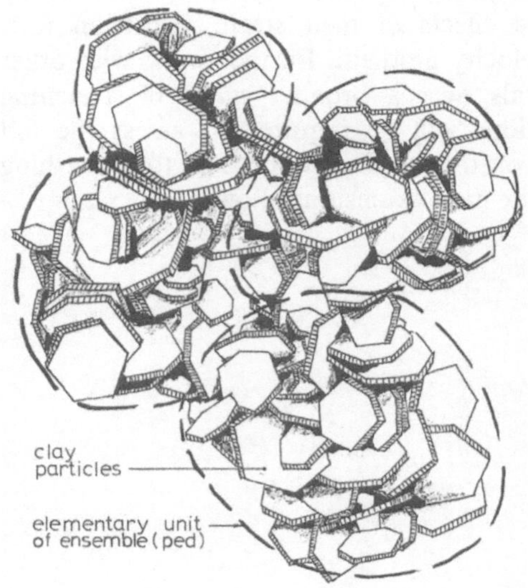

Fig. 1. Postulated elementary unit of ensemble (ped)

Fabric units – peds

In view of the nature of clay particles, and the specific physico-chemical interactions occurring between particles and with the constituent pore fluid, larger units composed of many clay particles tend to be formed. These units are fabric units – i.e. units of the fabric[1]) of clays. In general, the fabric units which are identified as "peds" are not readily seen without the aid of microscopic techniques, such as transmission and scanning electron microscopy (1, 2). Fig. 1 shows a schematic visualization of particles within peds, and peds interacting to form part of a clay structure.

From visual examination and physical evidence (1, 2), it is obvious that the yield performance and characterization of a ped is influenced by the number, and arrangement, of particles within each ped. Thus, distortion of any ped would involve the physical make-up of each ped and also the bonding and specific interactions of particles within each ped in response behaviour characterization. The individual performance of each ped, which will be different from other peds forming the soil sample, contributes to the overall performance and integrity of a test system. Each ped possesses its own yield strength and deformation characteristic. When the yield strength of any ped is not exceeded, its retardation deformation property is characterized by a proper retardation time constant. However, when the yield strength of any ped is attained, it would flow in a manner characteristic of a plastic material.

It is apparent that in a heterogeneous medium consisting of an infinite number and variety of peds, local [microscopic] yielding and collapse can occur throughout any one test system. The overall [macroscopic] system stability will thus

[1]) "Fabric" is defined as the physical framework formed by the arrangement of clay particles. This includes gradation of particles and porosity of the system.

hinge on the distribution of representative peds. The identification of a representative ped is thus a matter of direct concern in view of its participation in the definition of system stability and integrity.

Ped contribution to system stability

Recognizing that the formational characteristics of each ped depends on specific environmental constraints in regard to balance of energy, the probability of occurrence P_i of a ped at a particular level of integrity is a direct function of its energy state E_i, i. e.

$$P_i = f(E_i). \quad [1]$$

Whilst several ped structural states may possess the same energy state, the converse does not hold, i. e., there is only one energy state uniquely identified with the structure of any one ped. The necessity for averaging over the energy states in view of the available spectrum of ped structures is apparent.

A canonical ensemble of volume Ω [identified as $\bar{\Omega}$] can be constructed such that each ped of volume v [identified as $\bar{v}$] constitutes an elemental system of the ensemble. Each system [i. e. ped] of the ensemble $\bar{\Omega}$ may be considered to be in weak thermal interaction with other peds forming the ensemble. Thus, if there exists M peds or systems in the ensemble $\bar{\Omega}$, each system together with the other $M-1$ systems constitute a heat reservoir [i. e. the ensemble $\bar{\Omega}$ is a heat bath]. It is apparent that $\bar{v}$ has many fewer degrees of freedom than $\bar{\Omega}$.

The energy of any one system $\bar{v}$ is not fixed whilst the energy of $\bar{\Omega}$ is some constant value between $E^{(0)}$ and $E^{(0)} + \delta E$. It is apparent from the fundamental statistical postulate, that in an equilibrium situation, the probability of occurrence of one system $\bar{v}$ in $\bar{\Omega}$ in a state i is proportional to the number of states accessible to $\bar{\Omega}$. Thus:

$$P_i = c\,\omega(E^{(0)} - E_i) \quad [2]$$

where

c proportionately constant independent of i,

$\omega(E^{(0)} - E_i)$ number of states accessible to the systems remaining in $\bar{\Omega}$ in view of *one* system $\bar{v}$ be at state i.

It follows from the normalization procedure that:

$$\Sigma P_i = 1 \quad [3]$$

where the summation includes all possible states of $\bar{v}$. The significance of this statement, in view of the variability of ped structure and its relation to the energy state, should not be minimized.

Since $\bar{v} \ll \bar{\Omega}$, it follows that $E_i \ll E^{(0)}$ and thus eq. [2] can be approximated by expanding the logarithm of $\omega(E')$ about the value $E' = E^\circ$, thus giving

$$\ln \omega(E^{(0)} - E_i) = \ln \omega(E^{(0)}) - \left[\frac{\partial \ln \omega}{\partial E'}\right]_0 E_i \quad [4]$$

where

$E' = E^{(0)} - E_i$ energy of $M-1$ systems remaining in $\bar{\Omega}$ if any *one* system has energy E_i.

It can be shown that since:

$$\left[\frac{\partial \ln \omega}{\partial E'}\right]_0 = \beta \quad [5]$$

is a constant if evaluated at the fixed energy $E' = E^{(0)}$ the dependency on E_i for system $\bar{v}$ does not exist. In actual fact, β has dimensions of reciprocal energy and is generally given as:

$$\beta = \frac{1}{kT} \quad [6]$$

where

k positive constant with dimensions of energy

T temperature.

Eq. [5] can thus be written as:

$$\ln \omega(E^{(0)} - E_i) = \ln \omega(E^{(0)}) - \beta E_i. \quad [7]$$

Thus

$$\omega(E^{(0)} - E_i) = \omega(E^{(0)}) e^{-\beta E_i}. \quad [8]$$

Eq. [8] can be substituted into eq. [2] to give

$$P_i = c\,\omega(E^{(0)}) e^{-\beta E_i}. \quad [9]$$

Since $\omega(E^{(0)})$ is a constant independent of i, we obtain:

$$P_i = c e^{-\beta E_i}. \quad [10]$$

In view of eq. [3],

$$\frac{1}{C} = \Sigma e^{-\beta E_i} = \text{partition function}. \quad [11]$$

Hence

$$P_i = \frac{e^{-\beta E_i}}{\Sigma e^{-\beta E_i}}. \qquad [12]$$

The particular role of P_i in the overall ensemble stability may be developed in regard to the concept of a "weighing" factor which modifies and contributes to the constitutive performance of the material.

Macroscopic performance

The overall creep performance of the ensemble $\bar{\Omega}$ may be expressed in terms of the performance of the elemental systems $\bar{v}$ [i. e. peds] as:

$$\varepsilon(t) = \sum_{k=i,j} a_k \xi_k + \sum f_i(t) \xi_i + \sum b_j \xi_j t \qquad [13]$$

where

$\varepsilon(t)$ strain as a function of time

ξ stresses acting on the peds,

$f_i(t)$ creep function dependent on t

a_k, b_j ped constants.

In terms of the *Volterra-Boltzmann* relationship,

$$\varepsilon(t) = A\sigma + \int_0^t f(t-\tau)\,\sigma\,d\tau + B\sigma t \qquad [14]$$

where

σ macroscopic stress

A, B material constants.

Implicitly, account is given to all the accessible ped structural states in the summations performed in eq. [13], or the integration in eq. [14]. The probability P_i of occurrence of energy state i which recognizes the availability of many ped structural forms g_i can be introduced into the creep performance evaluation in terms of the mathematical expectation of the total deformation $\langle \varepsilon(t) \rangle$. Thus:

$$\langle \varepsilon(t) = \sum_{k=i,j} a_k P_k \xi_k + \sum P_i f_i(t) \xi_i + \sum P_j \xi_j b_j t. \qquad [15]$$

In continuous form:

$$\langle \varepsilon(t) \rangle = A\sigma + \int_0^t \frac{P_i(\tau)}{g_i} f(t-\tau)\sigma\,d\tau + B\sigma t \qquad [16]$$

where

g_i number of ped structural states at the same energy level.

The consequences in regard to eqs. [15] or [16] allow for microscopic instability [i. e., local yielding] without macroscopic failure or yield. Thus, local yield surfaces can be generated where convexity is not evident whilst ensemble performance still maintains the required convex yield surface (fig. 2).

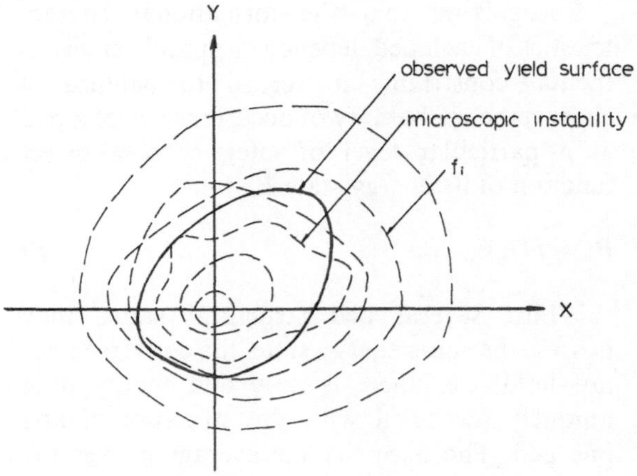

Fig. 2. Family of yield surfaces

The philosophy of P_i as a weighing factor can be corroborated in a simple extension of *Koiters* relationship (5). This extension requires the introduction of P_i and P_j in the formulation. Thus:

$$\dot{\varepsilon} = A\dot{\sigma} + \sum P_i h_i \frac{\partial f_i}{\partial \sigma} f_i + \sum P_j \lambda_j \frac{\partial f_j}{\partial \sigma} \qquad [17]$$

where the dot indicates the time derivative, and f denotes the yield function, and h and λ are scalar functions of stress, strain and loading history.

Results of performance

In applying the above concepts to evaluation of measured performances of the test material, the experimental programme pursued is similar to that reported previously (3). Reduction of creep performance data followed the previous procedure (3) which thus gave the required information for evaluation of ped contribution and participation in constitutive behaviour followed. Thus the resultant evaluation using the initial base of compliance-log time data provided typical probability density functions as seen in fig. 3.

Applying eq. [16] for final evaluation of test data obtained in one dimensional creep studies

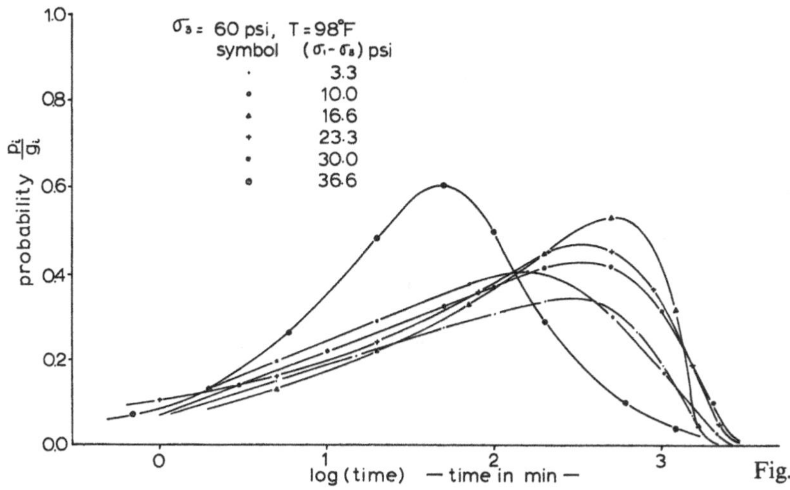

Fig. 3. Probability distribution curves

with varying cell confinement on samples σ_3 and temperature, the following relationship was obtained:

$$\varepsilon(t) = \frac{\sigma_1 - \sigma_3}{E_0 - m(\sigma_1 - \sigma_3)} + (\sigma_1 - \sigma_3)^b e^a \frac{1}{\sqrt{2\pi}\,\theta} \int_0^t e^{\frac{(\tau-\mu)^2}{2\theta^2}} d\tau + [k - d(\sigma_1 - \sigma_3)](\sigma_1 - \sigma_3)\,t \qquad [18]$$

where

subscripts 1, 3 major and minor directions in stress application,

m slope of elastic modulus-stress difference curve,

b, a material constants related to the probability $P_i(\tau)$ i. e. at retardation time τ,

θ standard deviation,

μ mean value of probability distribution,

k, d flow constants of material as a function of σ_3.

Thus the application of probability concepts in regard to P_i as a weighing factor can be seen in direct evaluation of data. The significance of material constants b and a in relation to the probability P_i at retardation time τ is important since this provides the ped structure influence on constitutive performance. This recognizes the procedure for obtaining the probability density function through the compliance-log time relationships. A typical correlation between analysis and measured values is shown in fig. 4. The close correlation between the two is evident. Recognition of the contribution of temperature, specific interactions of particles and ped integrity, all of which will ultimately define a partic-

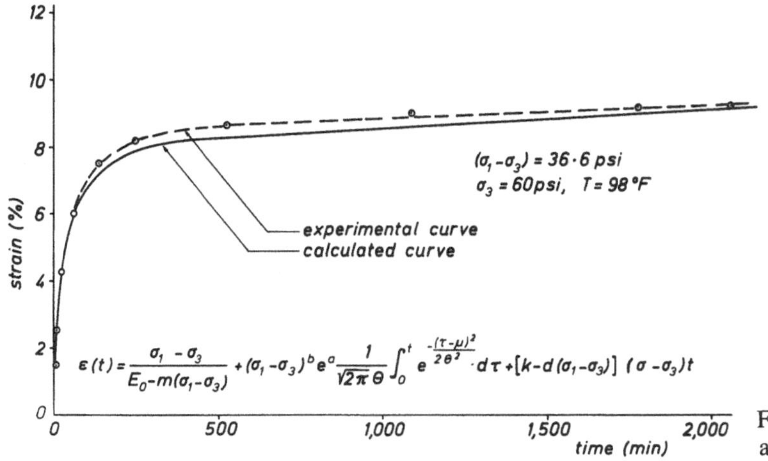

Fig. 4. Comparison of experimental and analytical results

ular energy level E_i is implicit in the form for data evaluation. Emphasis is directed towards the particular role of P_i (from eq. [12]) and the demonstrated curves in fig. 3.

Acknowledgements

This study was conducted under the financial sponsorship of the National Research Council of Canada, Grant No. NRC A-882.

Summary

In this paper the presence of a ped structure in clay formation is recognized. The total response deformation representing the macro-rheological behaviour accounts for the integral effect of individual ped performance characteristics. The summation or *Volterra-Boltzmann* integral accounting for the different structural strengths of the various peds is introduced in the analysis, incorporating thereby the probability of occurrence of each structural state. The utilization of this probability is likened in essence to a weighing factor of the contribution of each structural state, leading to the derivation of the mathematical expectation of the total deformation.

References

1) *Yong, R. N.*, Some Aspects of Soil Fabric and Structure in Soil Mechanics. The Symposium on Microfabrics of Soil and Sedimentary Deposits, University of Guelph, Ontario. March 1972.

2) *Yong, R. N.*, Physico-Chemical Properties and Soil Stabilization. General Reporter, Session No. 5, Proceedings, Fourth Asian Regional Conference, International Society for Soil Mechanics and Foundations Engineering, Vol. 1, pp. 1–49 (1971).

3) *Yong, R. N.* and *D. S. Chen*, Analysis of Creep of Clays Using Retardation Time Distribution. Proceedings, Fifth International Congress on Rheology, Vol. 2, pp. 501–512 (1970).

4) *Axelrad, D. R.* and *R. N. Yong*, Micro-Rheology of the Yielding of a Heterogeneous Medium. Proceedings, Fifth International Congress on Rheology, Vol. 2, pp. 309–314 (1970).

5) *Koiter, W. T.*, Progress in Solid Mechanics. Vol. 1, Chapter 4. Ed. by *Sneddon* and *Hill* (Amsterdam 19).

Authors' addresses:

Dr. *R. N. Yong*
Professor of Civil Engineering and Applied Mechanics, Director, Soil Mechanics Laboratory, McGill University, Montreal (Canada)

Dr. *D. S. Chen*
Postgraduate Research Assistant, Department of Civil Engineering and Applied Mechanics, McGill University, Montreal (Canada)

Rheol. Acta 13, 779–788 (1974)

From the Instituttet for Kemiindustri, The Technical University of Denmark, Lyngby (Denmark) and Frick Chemical Laboratory, Princeton, New Jersey (USA)

Properties and structure of concentrated polystyrene solutions

L. Lawrence Chapoy

With 12 figures and 4 tables

Introduction

Extensive work has been done on the effect of the introduction of low molecular weight liquids (plasticizers) into high molecular weight polymer systems (1 – 3). By plasticizing, one in effect decreases the chain-chain interactions that are present in the bulk. This has been described by a free volume model (4, 5). The examination of plasticized systems leads to information regarding the polymer structure in very concentrated solution. The aims of this investigation are to compare the viscoelastic properties of the plasticized system for solvents of differing thermodynamic compatibility, and to examine the phenomenon of molecular plasticization as contrasted to plasticization of supermolecular or secondary structure (6, 7).

In order to obtain unambiguous conclusions, it is essential to work with as simple a chemical system as possible. With this in mind commercial polystyrene has been used as the polymer, and di-esters of phthalic acid as the plasticizers. This is advantageous in as much as the solvent is a well defined compound, and the polymer is non-crystalline.

Materials

The polystyrene used in this study is a commercial material, Dylene 8, courteously supplied by *R. F. Kratz* of the Koppers Company. It had a weight average molecular weight of 265 000 and a number average molecular weight of 122 300. The di-(2 ethylhexyl)-phthalate, DOP, was a commercial product of Union Carbide. The dimethylphthalate, DMP, and diethylphthalate, DEP, were from Eastman Kodak.

The plasticized polystyrene samples were prepared by mechanical mastication with the appropriate quantities of plasticizer. Mastication of the sample did not appear to degrade the polymer as indicated by both molecular weight (gel permeation chromatography) and glass transition temperature (T_g) measurements. The phthalate concentration in the samples was later determined by carbonyl absorption in the infrared spectrum (8).

Experimental

Three distinct types of evidence were gathered in the study of plasticized polystyrene: differential thermal analysis (DTA), phase contrast microscopy, and viscoelastic measurements.

Study of glass transition

Differential thermal analysis measurements were performed using a Du Pont DTA, model no. 900.

The T_g determined with this instrument was found to depend upon the heating rate [9]. All reported measurements have been obtained at the heating rate of 20 °C/min.

Table 1. T_g's of some plasticized polystyrenes

Weight % phthalate	Transitions °C*)	Comments
2% DOP	93.5	
5.8% DOP	83.5	
9.9% DOP	67.; (−88.)	
14.0% DOP	54.; (−83.)	
22% DOP	33.; (−81.)	annealed at 35 °C
	33.; (−59.)	annealed at 60 °C
	27.5	annealed at 85 °C
	27.5; (−18.)	annealed at 95 °C (see fig. 2)
32% DOP	13.; −.5; −57.; −90	annealed at 0 °C (see fig. 1)
	14.; (−56.); (−87.)	annealed at 15 °C
	17.; 10.5; −16.; −68.5	annealed at 20 °C
	12.; −68.	annealed at 46 °C very broad (see fig. 2)
41% DOP	.5; −61.5	annealed at 0 °C
52% DOP	−7.; −73.	annealed at 0 °C
	−18.; −73.	annealed at 20 °C
62% DOP	−11.; −78.	annealed at 1 °C
	−15.; −78.5	annealed at 20 °C
78% DOP	−34.; −84.	annealed at 20 °C
100% DOP	(−69.); −85.	−89 °C by similar technique (11)
2.2% DMP	93.7	
5.4% DMP	81.7	
100% DMP	−98.	
2% DEP	94.	
5.5% DEP	86.	
0	101.3	

*) Transitions enclosed by parentheses are relatively unintense and may not be particularly well-defined.

Glass transition temperatures were determined over the whole concentration range for mixtures of polystyrene and DOP. The glass transition temperatures of certain mixtures of polystyrene and DMP, and polystyrene and DEP were also determined. These data are shown in table 1. Many thermograms exhibited multiple transitions. A typical example of a DTA exhibiting multiple transitions is shown in fig. 1.

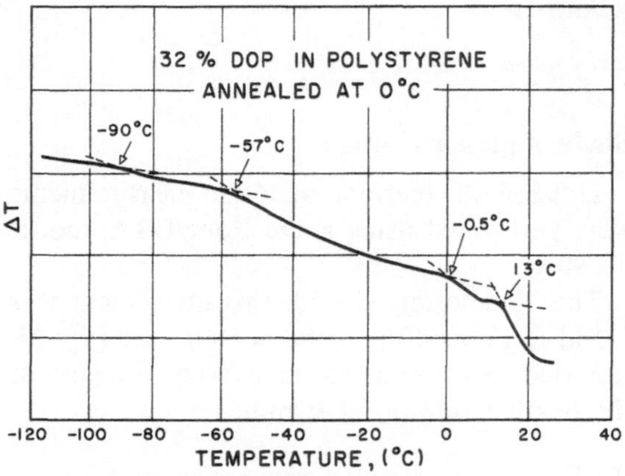

Fig. 1. A typical DTA showing multiple T_g's.

This is a complex DTA in as much as four transitions are in evidence. These are indicated in fig. 1 by the intersection of the dotted lines. In this case the transitions occur at −90, −57, −5 and 13 degrees centigrade. In many cases simpler DTA's result showing only two transitions.

Samples were annealed above T_g or as indicated in table 1. The samples were then quick-quenched with liquid nitrogen to 77°K. The multiple transitions are thus indicative of the sample at the annealing temperature. A graph depicting T_g as a function of weight fraction of plasticizer is shown in fig. 2. At low concentrations of plasticizer the T_g is lowered linearly with plasticizer content (10). T_g data for the system DOP-Polystyrene is depicted in fig. 3 after the Equation of *Braun* and *Kovacs* (5):

$$T_{g2} - T_{gx} = \frac{\Delta\alpha_2}{\Delta\alpha_1}\frac{x}{1-x}[T_{gx} - T_{g1}]. \quad [1]$$

For this equation: $\Delta\alpha$ is the difference in expansion coefficient above and below T_g, x is the volume fraction of plasticizer, and the subscripts 1, 2 and x refer to solvent, polymer

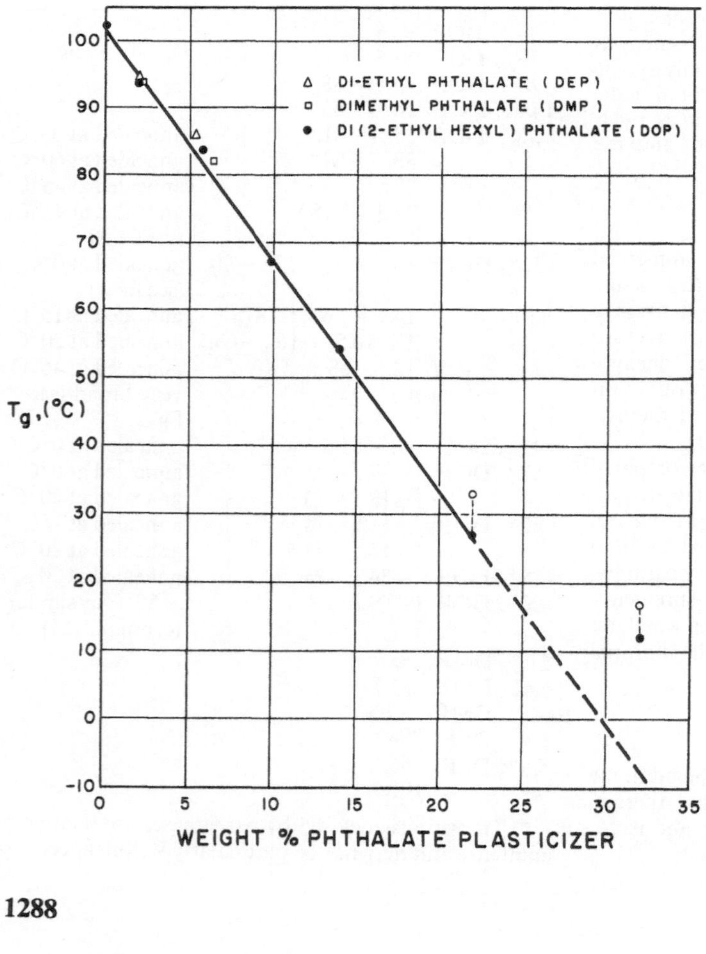

Fig. 2. Dependence of T_g on plasticizer concentration

and mixture, respectively. Notice that the linear relationship is obeyed over the whole concentration range studied.

Fig. 3 also shows the results for T_i plotted after eq. [1], where T_i is a viscoelastic parameter defined as the temperature at which the value of the stress relaxation, SR, tensile modulus after 10 sec, $E(10)$ is $10^{9.0}$ dynes/cm^2. This parameter can be thought of as a mechanical

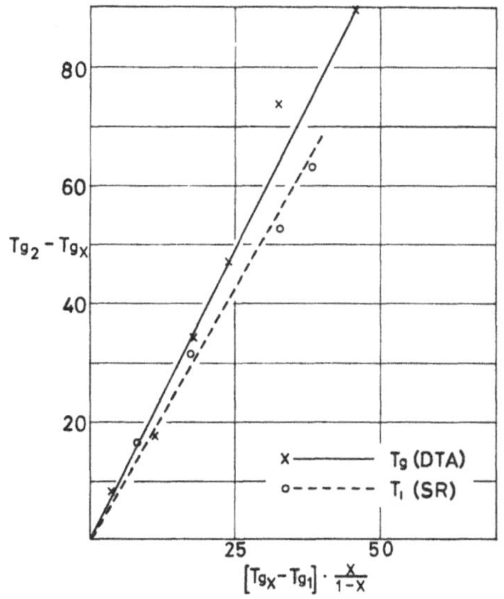

Fig. 3. Glass transition data for Polystyrene – DOP samples plotted according to eq. [1]

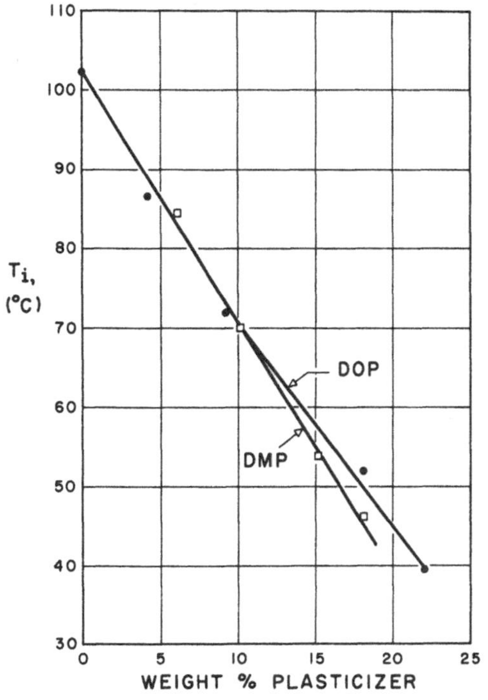

Fig. 4. Dependence of T_i on plasticizer concentration

analog of T_g. Notice that this plot also follows a linear law, but with a different slope. This could be associated with low temperature secondary relaxation phenomena in these systems as discussed by *Illers* and *Jenckel* (11).

A comparison of T_i data for solutions of polystyrene in DMP and DOP is shown in fig. 4. It can be seen that DMP is more effective than DOP in lowering the T_g of polystyrene for concentrations of greater than 10% by weight plasticizer. When these results are plotted after eq. [1], fig. 5, two distinct straight lines are obtained. This implies that the T_g lowering capacity is divisible into contributions from the T_g, and expansion coefficient of the solvent.

The multiple transitions as observed in the DTA, appear to be both non-reproducible and dependent on temperature history. This can be demonstrated by annealing a sample exhibiting multiple transitions at different temperatures, quenching in liquid nitrogen and obtaining the resulting thermograms. In the case of the 22% DOP sample (table 1), annealing at successively higher temperatures causes the low temperature peak to shift to higher temperatures and almost disappear while the high temperature peak falls 5 – 6 degrees.

A similar trend was noted with the 32% DOP sample, but the transitions became very broad and could not be observed at annealing temperatures sufficiently high to eliminate the low temperature transition completely. The decrease

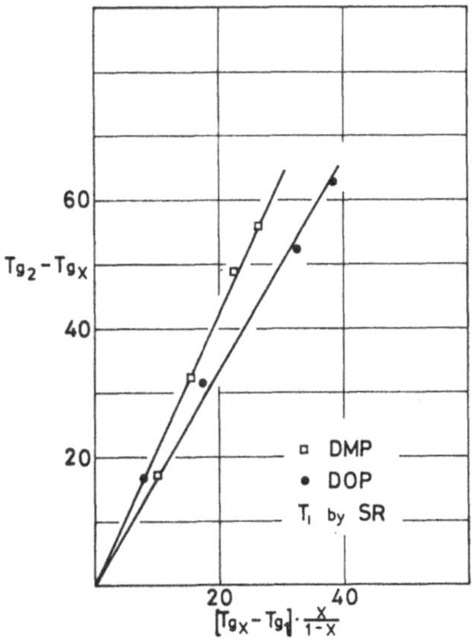

Fig. 5. T_i data for polystyrene containing various amounts of DOP or DMP plasticizer according to eq. [1]

Table 2. Plasticized samples exhibiting metastable equilibrium

% DOP	Elapsed time	Transitions
2.0	fig. 6 A initial	93.5
	B + 1 month	97.5; 53.5
	C + 1 month + 1 cycle on DTA to 100 °C	96; 81
5.8	initial	83.5
	+ 1 month	(89); 56.5
9.9	initial	67; (−88)
	+ 1 month	70.5; (10); −29.5
14	initial	53; (−83)
	+ 1 month	59.5; 15; −39.5
	+ 1 month + 1 cycle on DTA to 70 °C	55.5; −18.5; −58.0
22	initial	33; (−81)
	+ 1 month	38: −38; (−76)

in T_g as the material approaches a one-transition system is indicated by the dotted vertical lines in fig. 2.

Samples with low concentrations of plasticizer having only a single transition were allowed to remain at room temperature for approximately one month and run again on the DTA. These experiments are shown in table 2. In all cases, the major T_g has increased and a second transition appeared at a lower temperature. One sample (2% DOP), as shown in the table, was quenched immediately after it had gone through its temperature program to 100 °C at 20°/min and run again. The major T_g was found to decrease toward its value before "standing for a month" and the second transition moved to a higher temperature and became weaker. Representative thermograms are shown in fig. 6.

Before considering these results further, it should be noted that DOP is a very poor solvent for polystyrene, having a θ temperature of 47 °C. Dilute solutions, however, appear to exist in a metastable condition well below the θ point (12).

Discussion

The above results are suggestive of some sort of heterogeneity or phase separation. It is postulated that the lower T_g represents a phase consisting of mostly plasticizer with small amounts of polymer, while the higher T_g represents a phase consisting of mostly polymer with small amounts of plasticizer. This is deduced from the knowledge of the T_g of both the pure polymer and the pure plasticizer (13).

The annealing at higher temperatures, figs. 2 and 3, results in a more homogeneous system and a lower T_g for the polymer phase. Before annealing the plasticizer is not employed efficiently, since the concentration of plasticizer in the polymer phase, i. e. that phase associated with the highest T_g, is less than the bulk concentration. When the polymer exists in this two-phase situation, therefore, only part of the total plasticizer concentration is involved solvating chains. The remainder lies in solvent rich pools which in effect partially isolate plasticized polymer aggregates from one another. Most of the solvent is apparently in the polymer rich phase, since the changes in the transition temperature for this phase are relatively small (table 2, fig. 2).

As the annealing temperature is increased the transition temperatures converge, implying that the phase compositions are also converging.

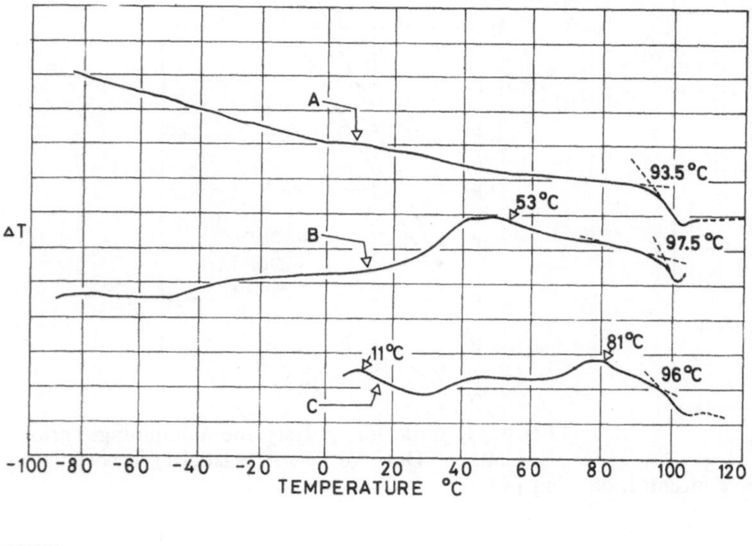

Fig. 6. Effect of temperature history on transitions for a polystyrene sample containing 2% DOP:
A Original curve
B After 1 month
C Measured immediately again

The converse is true in as much as the multiple phases form from a single phase on standing (table 1 and 2). This is typical behavior demonstrated in a temperature-composition phase diagram for two partially miscible liquids. A one-phase system is therefore essential in obtaining the maximum T_g lowering capacity for a fixed weight percent of plasticizer.

The samples with low concentrations of DOP are found to exhibit only a single transition upon annealing above T_g. If the thermo-dynamic critical solution temperature is less than T_g, the sample can exist in a condition of one-phase metastable equilibrium below the critical solution temperature. This is because the mobility of the molecules will be greatly diminished below T_g and there is only the possibility for a slow kinetically-controlled phase separation (14).

Since the phases form on standing and disappear on high temperature annealing it appears they must be a thermodynamic reality and not an artifact of incomplete mixing.

Phase contrast microscopy was used to examine these samples in more detail. Sections of the pure polystyrene as well as polystyrene with 20–30% DMP were virtually clear at 365 × magnification. Polystyrene with 20–30% DOP, at the same magnification exhibited a decidedly grainy structure. The grainy structure could be observed as it formed when the slide was cooled from an elevated temperature. This eliminates the possibility that the grains are caused by improper mixing. This type of grainy structure is typical of a multiphase specimen (15).

Viscoelastic measurements

The viscoelasticity of plasticized polystyrene was studied with the intent of examining the structure of these concentrated solutions as a function of the thermodynamic compatibility of different solvents acting as plasticizers. Stress relaxation master curves for the various plasticized samples were obtained as described elsewhere (16, 17). The time temperature superposition principle appeared to be valid within the uncertainty of the measurements. The results of the viscoelastic study are contained in table 3. T_i has been previously discussed. τ_m is the maximum relaxation time and $3\,G_m$ is the partial modulus associated with it as defined by eq. [2]:

Table 3

Type of plasticizer	Concentration WT. Percent	T_i °C	τ_m sec	$3\,G_m \times 10^{-5}$ $\frac{\text{dynes}}{\text{cm}^2}$	$3\,G_m \times \frac{T_i(0)}{T_i(x)} \times 10^{-5}$ $\frac{\text{dynes}}{\text{cm}^2}$	n	η_s Poise	$J_e \times 10^6$ $\frac{\text{cm}^2}{\text{dyne}}$
None	0%	102.5	6.00×10^7	7.40	7.40	−1.48	2.00×10^{13}	2.39
Di(2 ethyl hexyl phthalate)	4.1	86.	2.28×10^7	3.0	3.13	−1.38	3.77×10^{12}	3.98
	9.1	71.	1.85×10^7	5.05	5.49	−1.13	4.37×10^{12}	3.15
DOP	18.0	50.	2.78×10^7	1.90	2.19	−1.22	2.81×10^{12}	6.64
	22.0	40.5	3.86×10^7	2.60	3.12	−1.30	4.92×10^{12}	5.80
Di methyl phthalate	6	85.3	3.55×10^8	2.85	2.99	−1.26	5.17×10^{13}	4.71
	10	70.0	9.70×10^8	2.40	2.62	−1.08	1.09×10^{14}	6.43
	18	46.3	2.00×10^9	2.19	2.56	− .78	2.11×10^{14}	6.85

$$3G_r(t) = 3\sum_{i=1}^{m} G_i e^{-t/\tau_i} = 3\int_{-\infty}^{\infty} H(\tau)\, e^{-t/\tau}\, d\ln\tau \qquad [2]$$

where $H(\tau)$ is the distribution of relaxation times. $3G_m$ is also given after being corrected by the factor $T_i(0)/T_i(x)$, where $T_i(x)$ is the T_i in degrees *Kelvin* for a sample containing x weight percent plasticizer. The thermodynamics of rubber elasticity dictates that the rubbery modulus is proportional to the absolute temperature, and thus this correction is required for a reasonable comparison of $3\,G_m$'s for samples containing various amounts of plasticizer at their respective T_i's. η_s is the shear viscosity and J_e is the equilibrium elastic shear compliance which can be calculated from the stress relaxation master curve. The maximum slope of the master curve ($\log 3\,G_r(t)$ vs. $\log t$) in the glass-rubber transition is denoted by n. The significance of this parameter has been discussed extensively (18). It has been observed that the gross effect of adding diluent or plasticizer to polystyrene is to lower the T_i and hence the T_g (19, 20). This would amount to horizontally shifting the isothermal master curves to shorter times. The shifting of the master curve to shorter times would radically alter derived viscoelastic parameters, e. g. η and τ_m when viewed from the perspective of a fixed temperature. In view of this, all master curves and viscoelastic parameters have been presented at their respective T_i's.

It is observed that the introduction of plasticizer shifts the distribution of relaxation times to shorter times and that this shift is a strong function of plasticizer concentration. If the distribution of relaxation times associated with a given master curve was uniformly shifted to shorter times, it would be indicative of each relaxation mechanism being affected to the same extent by the presence of plasticizer. If this were the case, curves and derived parameters would be equal when presented at their respective T_i's. In this investigation we do not find this to be so. The differences in the derived parameters in table 3 are ascribed to changes in the distribution of relaxation times associated with the respective master curves. Concentration dependence of the viscoelastic parameters at T_i, appear to be strongly dependent on the specific solvent. Different trends are observed for the DOP and DMP samples. These second order effects are believed to reflect the thermodynamic compatibility of diluent with polymer.

Discussion

It is felt that this work is not inconsistent with the hypothesis on plasticization presented by *Kargin* and co-workers (6, 7): He has postulated the occurrence of an aggregated structure of polymer chains termed bundles. Molecular plasticization occurs with a thermodynamically compatible solvent and acts to increase the lattice spacing between chains and decrease the number of entanglements. The T_g decreases continually as the fraction of plasticizer increases. Plasticization with a solvent of marginal thermodynamic compatibility leads to partial phase separation or bundle formation. The diluent can act as both an interbundle and an intrabundle plasticizer. In this case a drastic reduction of T_g results from the first few percent of added plasticizer, but additional plasticizer is found to have little or no effect. This is because most of the diluent is employed as interbundle plasticizer. The initial effect results from the separation of the bundles to some critical distance and additional separation of the bundles has no effect on T_g. The relative magnitudes of these effects depend upon thermodynamic compatibility of the respective solvent with the polymer. Further addition of plasticizer will eventually break up the bundle structure and result in additional lowering of the T_g. Experimental evidence in this area to date consists largely of measurements of T_g (21, 22, 23) and the shear viscosity (24, 25) as a function of plasticizer concentration and type.

The calculated parameters, listed in table 3 enable one to get an approximate idea of the structure of the material when considered at an equivalent viscoelastic state, i. e. T_i. At a temperature of 102.5 °C (T_i for undiluted polystyrene) the shear viscosity η_s, and the maximum relaxation time, τ_m, would show marked decreases as the T_g falls with additions of diluent. In table 3, however, these parameters are presented at their respective T_i's and would all be identical if the presence of plasticizer served only to shift the entire distribution of relaxation times uniformly to shorter times. *Since the shape of the distribution of relaxation times can be changed by the presence of the plasticizer, the parameters when calculated at T_i may be greater than, equal to, or less than those for undiluted polystyrene.* The

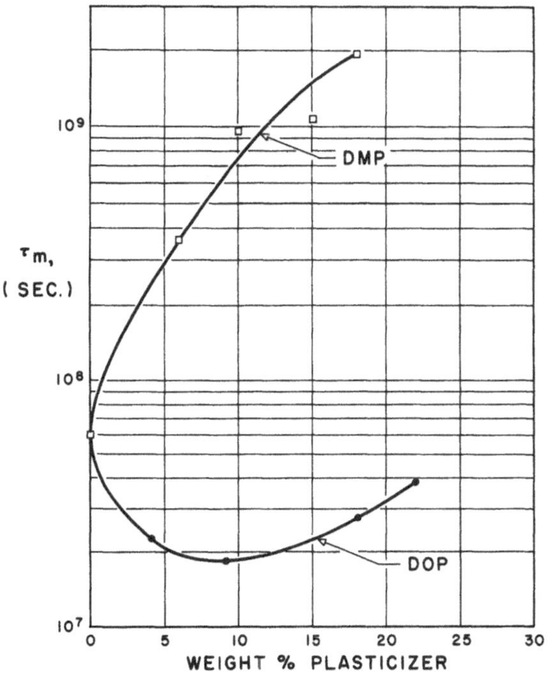

Fig. 7. Dependence of maximum relaxation time at T_i on plasticizer concentration.

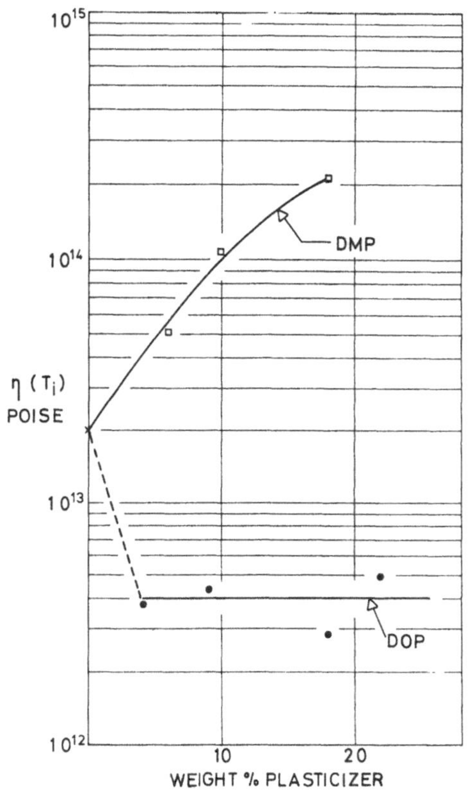

Fig. 8. Viscosity at T_i for Polystyrene containing various amounts of DOP or DMP plasticizer

next few paragraphs contain a discussion of η_s and τ_m at T_i.

The maximum relaxation times, τ_m, fig. 7, goes through a minimum for the DOP samples at about 10% DOP. If one had a homogeneous polymer which formed bundle aggregates upon addition of diluent and if these aggregates dissociated upon further addition of diluent, it would appear reasonable that τ_m was at a minimum value just prior to the dissociation of the aggregates. This behavior is analogous to the temperature dependent viscosity of a material capable of going through a liquid crystalline phase. DTA measurements show that the lower, secondary T_g, representing a phase composed mostly of solvent, is in excess of that for pure plasticizer when there is more than 10% DOP in the sample. The lower, secondary T_g also increases with increased plasticizer concentration. This would be the result of successively more polymer chains and chain ends extending into the "plasticizer phase" from the polymer bundle, if the bundle were breaking up.

The maximum relaxation time for the DMP samples is far greater than for the DOP samples, in fact it increases with increasing plasticizer concentration. Normalizing the master curves to T_i in effect fixes the minimum relaxation time. The increase in τ_m's for the DMP samples when comparing the normalized distribution for plasticized (DMP) and unplasticized materials shows that the plasticizer has more effect on the minimum relaxation time than on τ_m, i. e. the distribution of relaxation times is spreading out. A good solvent, therefore, plasticizes the short time relaxation processes more efficiently in the case of polystyrene.

The viscosity at T_i, fig. 8, for DOP samples is found to decrease sharply with the first few percent of diluent and then remain relatively constant, perhaps even increasing. The viscosity of the DMP samples at T_i is found to be an order of magnitude greater than that for the unplasticized material, and two orders of magnitude greater than that for DOP samples. The viscosity is determined to a large extent by τ_m.

The construction of master curves from the time-temperature superposition principle gives temperature dependent shift factors, $K(T)$, which can be described by an equation of the *Williams-Landel-Ferry* (W-L-F) type:

$$\log \frac{K(T)}{K(T_i)} = - \frac{C_1(T - T_i)}{C_2 + (T - T_i)} \qquad [3]$$

where $K(T_i) = 1$ and C_1 and C_2 are empirically determined parameters. This equation is thus a description of the temperature dependence of the viscoelastic relaxation process. These data are shown in table 4. From the theory (4, 5):

$$\frac{1}{C_1} = \frac{f_g}{2.303} \qquad \frac{1}{C_1 C_2} = \frac{\Delta\alpha}{2.303} \qquad [4]$$

$$f_{gx} = \sum x_i f_g \qquad [5]$$

$$\Delta\alpha_x = \sum x_i \Delta\alpha_i \qquad [6]$$

where f_g is the fractional free volume at T_g, and is thought to be a universal constant. The subscripts i and x are indicative of a single component, and the blend respectively. C_1 and $\frac{1}{C_1 C_2}$ are shown in figs. 9 and 10, respectively. Eq. [4] will only be approximately applicable in this case since eq. [3] is based on T_i and not T_g. The theory postulates that f_{gx} and hence C_1 should be constant. The theory also demands that $\Delta\alpha_x$ and hence $\frac{1}{C_1 C_2}$ are linear functions of the composition. The results for the DMP samples show slowly changing functions of composition and could easily be described by addition of a non-linear term to eqs. [5] and [6]. The results for the DOP samples show a discontinuity and then constancy for both

Table 4. Shift factors for polystyrene samples

A. Shift factors for unplasticized polystyrene

$T_i = 102.5\,°C$

$T - T_i$, °C	log $K(T)$	$T - T_i$, °C	log $K(T)$
−5.5	1.47	9.0	−1.84
−3.0	.77	10.0	−2.00
−1.0	.25	12.5	−2.40
0.0	.00	15.0	−2.74
1.0	−.25	17.5	−3.06
2.0	−.48	20.0	−3.36
3.0	−.68	22.5	−3.59
4.0	−.90	25.0	−3.90
5.0	−1.10	27.5	−4.16
7.0	−1.48		

B. Shift factors for polystyrene plasticized with DMP

DMP	6%	10%	18%
T_i, °C	85.3	70.0	46.3
$T - T_i$	log $K(T)$	log $K(T)$	log $K(T)$
−7.5			
−3.0	.70	.75	.95
−1.0	.20	.22	.30
0.0	.00	.00	.00
1.0	−.25	−.25	−.27
2.0	−.45	−.45	−.55
4.0	−.87	−.84	−.98
6.0	−1.25	−1.23	−1.35
8.0	−1.60	−1.62	−1.70
10.0	−1.94	−2.00	−2.05
12.5	−2.32	−2.45	−2.48
15.0	−2.67	−2.85	−2.88
17.5	−3.00	−3.20	−3.28
20.0	−3.30	−3.56	−3.65
22.5	−3.60	−3.88	−4.00
25.0	−3.88	−4.18	−4.38
27.5	−4.15	−4.40	−4.73
30.0	−4.37	−4.70	−5.02
32.5	−4.60	−4.93	−5.31
35.0	−4.82	−5.16	−5.55
40.0	−5.23	−5.56	−6.00
45.0	−5.65	−5.88	−6.40
50.0		−6.27	−6.70

C. Shift factors for polystyrene plasticized with DOP

DOP	4.1%	9.1%	18.0%	22.0%
T_i °C	86	71	50	40.5
$T - T_i$	log $K(T)$	log $K(T)$	log $K(T)$	log $K(T)$
−7.5	1.95	1.90	1.70	1.43
−3.0	.65	.70	.85	.53
−1.0	.20	.22	.50	.17
0.0	.00	.00	.00	.00
1.0	−.17	−.20	−.12	−.17
2.0	−.35	−.39	−.28	−.35
4.0	−.65	−.73	−.60	−.67
6.0	−.90	−1.02	−.85	−.98
8.0	−1.15	−1.30	−1.10	−1.26
10.0	−1.40	−1.52	−1.32	−1.53
12.5	−1.70	−1.83	−1.57	−1.82
15.0	−1.98	−2.13	−1.78	−2.12
17.5	−2.25	−2.42	−2.04	−2.38
20.0	−2.51	−2.70	−2.29	−2.65
22.5	−2.75	−2.95	−2.50	−2.89
25.0	−2.99	−3.20	−2.73	−3.12
27.5	−3.21	−3.42	−2.93	−3.36
30.0	−3.42	−3.66	−3.15	−3.57
32.5	−3.62	−3.87	−3.35	−3.77
35.0	−3.80	−4.07	−3.52	−3.97
40.0	−4.10	−4.40	−3.85	−4.35
45.0				−4.73

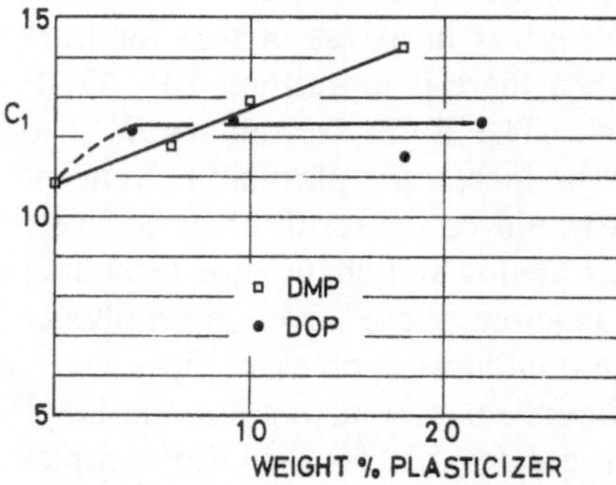

Fig. 9. C_1 from the W-L-F equation for polystyrene containing various amounts of DOP or DMP plasticizer

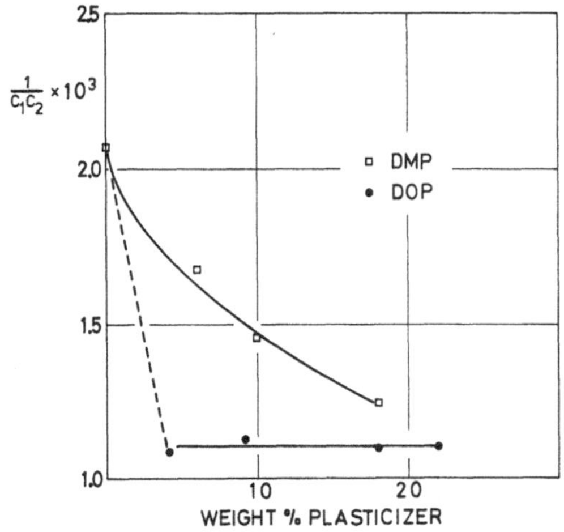

Fig. 10. $\frac{1}{C_1 C_2}$ from the W-L-F equation for polystyrene containing various amounts of DOP or DMP plasticizer

C_1 and $\frac{1}{C_1 C_2}$. This is not readily explained, except by reverting to the fact that DOP is a poor solvent and that blends with polystyrene may exhibit a complex structure in the solid.

Eq. [3] can be recast to give:

$$\log \frac{\eta(T)}{\eta(T_i)} = \frac{-C_1(T - T_i)}{C_2 + (T - T_i)} \qquad [7]$$

where: $\eta(T_i)$ is known from fig. 8, C_1 and C_2 are known from figs. 9 and 10, and T_i is known from table 3, all as a function of composition.

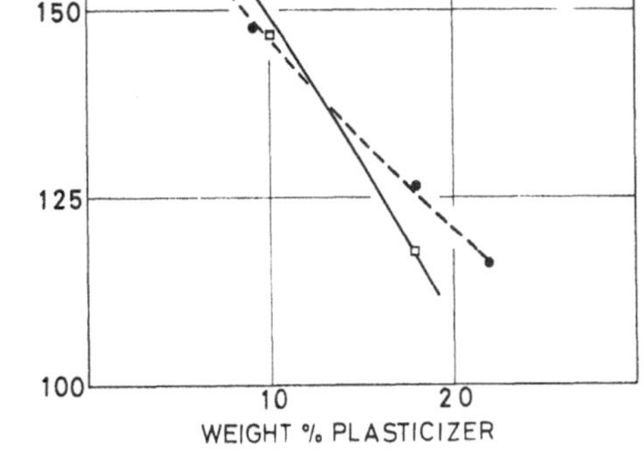

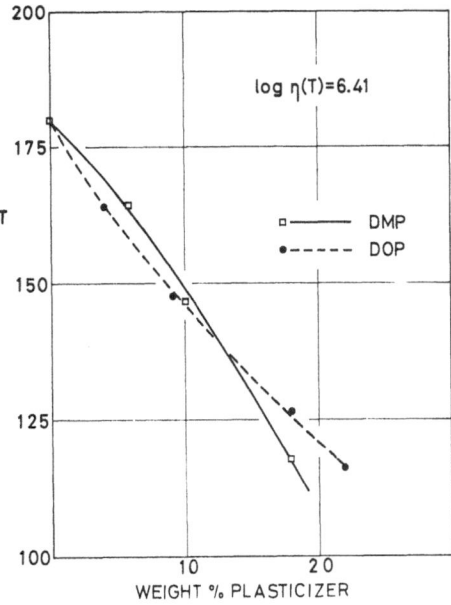

Fig. 11. Temperature to achieve $\log \eta(T) = 6.41$ for polystyrene containing various amounts of DOP or DMP plasticizer

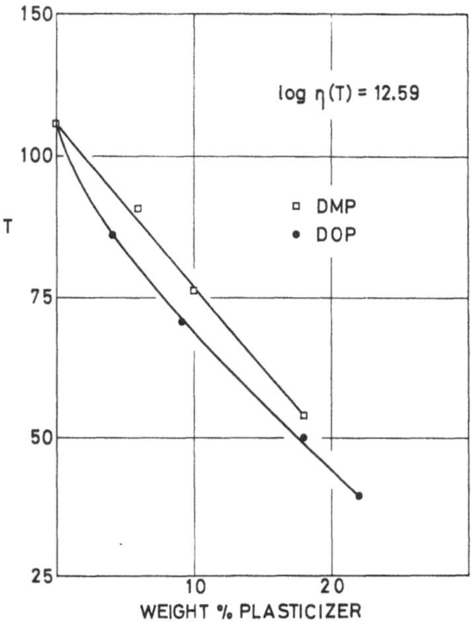

Fig. 12. Temperature to achieve $\log \eta(T) = 12.59$ for polystyrene containing various amounts of DOP or DMP plasticizer

$\eta(T)$ can then be calculated as a function of both temperature and composition. Holding $\log \eta(T)$ constant, one can thus obtain the temperature necessary to achieve a pre-determined viscosity as a function of plasticizer content. In fig. 11, the temperature necessary to achieve $\log \eta = 6.41$ (the viscosity of unplasticized polystyrene at 180°) for different concentrations of plasticizer is shown. Fig. 12 shows the same type of curve for $\log \eta = 12.59$ (the viscosity of unplasticized polystyrene at 106°). These results are identical to those calculated from the measured shift factors. It is interesting to note that in this type of presentation, DOP shows a more significant effect in changing the properties than DMP. The opposite was found to be true in the lowering of T_i. It is also interesting to note the apparent cross over behavior which occurs/will occur at higher plasticizer concentrations. This appears to be a way, then, in which the viscosity very close to the glass transition point can be evaluated. This could well be of interest for various fabricating techniques, which operate in this temperature range.

Conclusion

Evidence has been presented in support of the hypothesis that structure can exist in plasticized materials or concentrated polymer solutions. The thermodynamic compatibility of diluents can greatly influence the viscoelastic parameters of the plasticized polymer. For poor solvents multiphase regions are envisioned with different polymer concentrations. A time, temperature, composition equation of state has been presented for these materials.

Acknowledgement

The author would like to thank Professor *A. V. Tobolsky* of Princeton University for discussions which contributed to this work.

Summary

The properties of concentrated solutions of polystyrene have been examined as a function of solvent compatibility. Phase contrast microscopy, differential thermal analysis and visco-elastic measurements support the idea that miscible solvents can give rise to molecular plasticization and hence a truly homogeneous solution while partially miscible solvents result in something more complicated, perhaps a system containing heterogeneities on the molecular level.

References

1) *Ferry, J. D.*, Viscoelastic Properties of Polymers (New York 1961).

2) *Bondi, A.*, Physical Properties of Molecular Crystals, Liquids and Glasses (New York 1968).

3) *Platzer, N. A. J.*, (Ed.), Plasticization and Plasticizer Processes (Amer. Chem. Soc., Washington, D. C., 1965).

4) *Kelly, F. N.* and *F. Bueche*, J. Polymer Sci. **50**, 549 (1961).

5) *Braun, G.* and *A. J. Kovacs*, Physics of noncrystalline Solids, Proceedings of the International Conference, Delft, July 1964, p. 303 (Amsterdam 1965).

6) *Kargin, V. A., P. V. Kozlov*, and *R. M. Assimova*, Doklady Akademii Nauk SSR, **135**:**2**, 1037 (1960) (p. 357 in Russ. pag.).

7) *Kargin, V. A.* and *A. I. Kitaigorodsky*, Kolloid Zh. **19**, 141 (1957) (p. 131 in Russ. pag.).

8) *Kiley, L. R.*, Anal. Chem. **29**, 1895 (1957).

9) *Martin, A. E.* and *H. F. Rase*, I & EC Product Res. Development **6**, 104 (1967).

10) *Ferry, J. D.*, Viscoelastic Properties of Polymers, p. 356 (New York 1961).

11) *Illers, K. H.* and *E. Jenckel*, Rheol. Acta **1**, 322 (1958).

12) *Quadrat, O.* and *M. Bohdanecký*, J. Polymer Sci. C **6**, 769 (1968).

13) *Carpenter, M. R., D. B. Davies*, and *A. J. Matheson*, J. Chem. Phys. **46**, 2451 (1967).

14) *Struik, L. C. E.*, Rheol. Acta **5**, 303 (1966).

15) *Turley, S. G.*, J. Polym. Sci., C **1**, 101 (1963).

16) *Tobolsky, A. V., J. J. Aklonis*, and *G. Akovali*, J. Chem. Phys. **42**, 723 (1965).

17) *Hopkins, I. L.* and *R. W. Hamming*, J. Appl. Phys. **28**, 906 (1957).

18) *Chapoy, L. L.* and *A. V. Tobolsky*, Chemica Scripta **2**, 44 (1972).

19) *Kishimoto, A., H. Fujita*, J. Polym. Sci. **28**, 547 (1958).

20) *Kishimoto, A.* and *H. Fujita*, J. Polym. Sci. **28**, 569 (1958).

21) *Gribkova, N. Ya., P. V. Kozlov*, and *S. V. Yakubovich*, Polym. Sci. U.S.S.R. **7**, 831 (1965).

22) *Kargin, V. A., P. V. Kozlov*, and *Wang Nai-Ch'ang*, Doklady Akademii Nauk SSR **130**, 33 (1960) (p. 356 in Russ. pag.).

23) *Kargin, V. A.*, Russian Chem. Rev. **35**, 427 (1966).

24) *Natov, M. A.* and *Ye. Khr. Dzhagarova*, Polym. Sci. U.S.S.R. **8**, 2032 (1966).

25) *Dreval, V. Ye., M. S. Lutskii, A. A. Tager, V. K. Postikov, O. S. Khvatova*, and *G. V. Vinogradov*, Polym. Sci. U.S.S.R. **9**, 345 (1967).

Author's address:

Dr. *L. Lawrence Chapoy*
Instituttet for Kemiindustri
Danmarks Tekniske Højskole
DK 2800 Lyngby (Danmark)

Rheol. Acta **13**, 789–797 (1974)

Aus dem Institut für chemische Physik der Akademie der Wissenschaften der UdSSR, Laboratorium für armierte Plaste, Moskau (UdSSR)

Der Einfluß der Belastungsgeschwindigkeit auf die Adhäsionsfestigkeit von Polymer-Faser-Systemen

Ju. A. Gorbatkina und V. G. Ivanova-Mumshieva

Mit 12 Abbildungen und 2 Tabellen

(Eingegangen am 27. Oktober 1972)

Einleitung

Die Erscheinung der adhäsiven Zerstörung, d. h. der Trennung zweier verschiedenartiger Körper an der Trenngrenze, ist bedeutend weniger untersucht als die kohäsive Zerstörung, wenn die Trennfläche im gleichartigen Körper gebildet wird. Am wenigsten ist die Haftfestigkeit in Systemen untersucht, in denen die Unterlage aus Fasern besteht, trotz der großen praktischen Bedeutung derartiger Systeme. Zum besseren Verständnis der Natur der Adhäsionsfestigkeit ist es nützlich, ihre Gesetzmäßigkeiten mit analogen Gesetzmäßigkeiten der Kohäsionsfestigkeit zu vergleichen. Zu diesem Zweck haben wir die Abhängigkeit der Bindungsfestigkeit in Polymer-Faser-Systemen von der Belastungsgeschwindigkeit untersucht.

Meßmethode

a) Versuchsobjekte

Als Adhäsive wurden verschiedene duroplastische Harze verwendet: Epoxid-, Butvar-Phenol-, silizium-organische und Polyester-Harze. Die Zusammensetzung der Harze ist in Tab. 1 angeführt. Als Unterlage dienten Glasfasern aus alkalifreiem Glas (analog dem E-Glas) und Stahldraht der Mark OBC. Der Durchmesser der Glasfasern betrug 10–13 μm (Fasern von diesem Durchmesser werden am häufigsten zur Pro-

Tab. 1. Zusammensetzung der untersuchten Polymere und Härtungsregime

Lfd. Nr.	Polymermatrix-Typ	Zusammensetzung der Polymermatrix	Gewichtsteile	Härtungsregime*) Zeit (h)	Temperatur (°C)
1	Epoxid	Dian-Epoxidharz		2	90
		ЭД-5	100	2	120
		Triäthanolamin	15	2	160
2	Epoxid	Dian-Epoxidharz		24	20
		ЭД-5	100	6	100
		Polyäthylenpolyamin	15		
3	Epoxid, modifiziert	Dian-Epoxidharz	100	2	80
		Diäthylenglykol	10	8	160
		Triäthanolamintitanat	11		
4	Butvar-Phenol	Polyvinylbutyral	85		
	БФ-4	Phenolformaldehydharz	15	2	70
5	Butvar-Phenol	Polyvinylbutyral	51	2	90
	БФ-6	Phenolformaldehydharz	9	1,5	110
		Butylazetat	24	1,5	130
		Rizinusöl	11	2	160
		Kolophonium	5		
6	Polyester	Polyesterharz ПН-1	100	24	20
		Kobaltnaphthenat	8	a) 1	
		Isopropylbenzolperoxid		b) 5	100
			3	c) 8	
				d) 12	
				e) 4 Mon.	20
7	Silizium-organisch	Methylphenylsiloxan		1	80
			100	6	200

*) Die Geschwindigkeit des Temperaturanstiegs und der Abkühlung beträgt 1°/min.

51

duktion von Glasfaserplasten verwendet) und 200 μm. Der Durchmesser des Stahldrahts betrug 150 μm.

b) Probenvorbereitung

Bei der Bestimmung der Haftfestigkeit der Polymere an Fasern wird gewöhnlich die Schubhaftfestigkeit (Ausreißfestigkeit) τ bestimmt. Dazu wird die zum Herausziehen der Faser aus der Schicht des polymerisierten Bindemittels erforderliche Kraft gemessen und die Klebefläche S bestimmt.

Das Grundproblem bei der Herstellung der Klebstellen der Polymeren mit den Fasern ist das Erreichen einer ausreichend kleinen Klebstellenlänge l. Nur unter der Bedingung, daß die Faser in eine genügend dünne Harzschicht eingebettet ist, erfolgt ein Herausreißen der Faser aus dem Harz bei der Zerstörung der Klebstelle. Andernfalls tritt kohäsive Zerstörung der Faser selbst ein.

Die Methodik der Probenherstellung auf „dünnen" (d = 6–30 μm) Fasern ist ausführlich in (1, 2) beschrieben. Kurz gesagt besteht sie darin, daß die dünne Faser in Harz eingebettet wird, das zwei „dicke" (d = 100–130 μm) genau übereinanderliegende Glasfasern bedeckt. Die zu untersuchende dünne Faser und die Fasern, welche das Harz tragen, sind senkrecht zueinander in einer horizontalen Ebene angeordnet. Die Klebstellenlänge wird dabei durch den Durchmesser der „Trägerfasern" bestimmt.

Auf diese Art wurde die Adhäsion der Epoxid-Matrix Nr. 3 (s. Tab. 1) und der Polyorganosiloxan-Matrix Nr. 7 bestimmt.

Die Methodik der Verklebung mit „dicken" Fasern ist bei (3–6) beschrieben. Dabei werden die Proben mit filmbildenden Harzen und nichtfilmbildenden Gießharzen unterschiedlich vorbereitet. Die Idee der Methode zur Probenherstellung mit filmbildenden Harzen (solchen wie Butvar-Phenol-Polymere БФ-4 (Nr. 4) und БФ-6 (Nr.5)) besteht darin, daß die zu untersuchenden Fasern mit d = 100–200 μm (Glasfasern, synthetische oder Metallfasern) zwischen zwei schmale vorher vorbereitete Polymerfilmstreifen gebracht werden. In diesem Fall wird die Klebstellenlänge durch die Breite der verwendeten Streifen bestimmt.

In vorliegendem Fall – für Butvar-Phenol-Polymermatrices – erhielten wir den Film durch Beschichtung einer Unterlage durch mehrmaliges Eintauchen der Unterlage in eine 13%ige alkoholische Polymerlösung. Als Unterlage diente Triazetatfilm, zu dem die Adhäsion des Polymers gering ist. Die Dicke des erhaltenen Polymerfilms beträgt 400–500 μm. Die Methodik der Probenherstellung mit Glasfasern ist ausführlich in (3), die mit Stahldraht in (5) beschrieben.

Bei der Bestimmung der Adhäsion von Gießharzen – Polyesterharz (Nr. 6) und Epoxidharz (Nr. 1–3) – an Stahldraht wurden die Proben durch Gießen in Schälchen hergestellt. Die Methodik der Probenherstellung unterscheidet sich prinzipiell nicht von der von *McGarry* (4) verwendeten Methodik. In das Schälchen wurde senkrecht zum Boden ein Abschnitt der zu untersuchenden Faser eingesetzt und darauf eine Schicht Polymermatrix aufgegossen. Anstelle von Teflonschälchen bei *McGarry* haben wir Schälchen aus Aluminiumfolie verwendet, was die Probenherstellung vereinfacht. Die Klebstellenlänge wird hierbei durch die Dicke der aufgegossenen Polymerschicht bestimmt.

Die hergestellten Proben wurden in allen Fällen im Härteofen bei den in Tab. 1 angegebenen Regimes polymerisiert.

c) Probenuntersuchung

Die Bestimmung der Haftfestigkeit der von uns untersuchten Mikroproben erforderte die Zerstörung von nicht weniger als einigen zehn Klebstellen. Die Messungen bei von der Zimmertemperatur abweichenden Temperaturen machen die Benutzung von Geräten mit mehreren Probeeinsätzen erforderlich, wobei die Thermostatierung jeder einzelnen Probe bei ihrer großen Gesamtzahl die Versuchszeit unrationell erhöhen würde. Daher wurde die Klebstellenzerstörung im Adhäsiometer MAB-2TC durchgeführt, das die gleichzeitige Thermostatierung von 10 Proben zuläßt. Durch Änderung der Steifheit des elastischen Elements, auf den der Tensogeber aufgeklebt ist, läßt das Adhäsiometer die Veränderung des Skalenbereichs des Bandschreibers, der die Kraft mißt, von 20 p bis 8 kp zu, bei einer Genauigkeit der Kraftmessung von 1...2%. In diesem Kraftbereich lagen alle Zerreißbelastungen der von uns untersuchten Klebstellen. Außerdem wurden zur Zerstörung der Klebstellen der Polymeren mit Glasfasern von $d = 200\,\mu$m und Stahldraht bei Zimmertemperatur Zerreißmaschinen vom Typ Schopper, die für eine Belastung bis 5 kp ausgelegt sind, verwendet.

Die Klebfläche S wurde aus dem Faserdurchmesser d und der Klebstellenlänge l nach

$$S = \pi d l$$

bestimmt.

Der Durchmeser jeder Faser wurde unter dem Mikroskop gemessen. Die Klebstellenlänge für Gießharze Nr. 1–3 und 6 (bei Adhäsion an Draht) wurde mit dem Mikrometer nach Zerstörung der Klebstelle gemessen; in allen übrigen Fällen wurde l unter dem Mikroskop vor Zerstörung der Probe gemessen.

Der Zerstörungscharakter wurde unter dem Mikroskop bei 160facher Vergrößerung kontrolliert.

d) $\tau - S$-Abhängigkeit

Bei der Probenherstellung zur Bestimmung der Adhäsion an Draht gelingt es praktisch nicht, eine genau gleichhohe Harzschicht aufzugießen oder Filme gleicher Breite zu schneiden. Daher unterscheiden sich in allen Fällen die Klebstellenlängen und folglich auch ihre Flächen etwas voneinander. Es ist bekannt, daß im allgemeinen τ eine Funktion von S ist. Der Charakter dieser Abhängigkeit für einige der von uns untersuchten Harze bei Zimmertemperatur ist aus Abb. 1 ersichtlich. Aus der Abbildung ist zu ersehen, daß unter unseren Versuchsbedingungen der Wert τ bei Veränderung der Fläche nicht konstant bleibt. Gewöhnlich wird τ mit wachsendem S kleiner, wobei bei härteren Polymeren die Abhängigkeit $\tau - S$ klarer zum Ausdruck kommt.

Da die Größe der Klebfläche auf die zu messende Größe τ Einfluß hat, muß die Bindungsfestigkeit der verschiedenen Polymeren mit den Fasern bei ein und demselben S-Wert verglichen werden. Bei den Versuchen mit Stahldraht wird die Adhäsion der Polymeren durch den Wert τ bei $S = 0{,}55\,\text{mm}^2$ charakterisiert. Dieser Wert wurde der aus den Versuchsergebnissen von 50–100

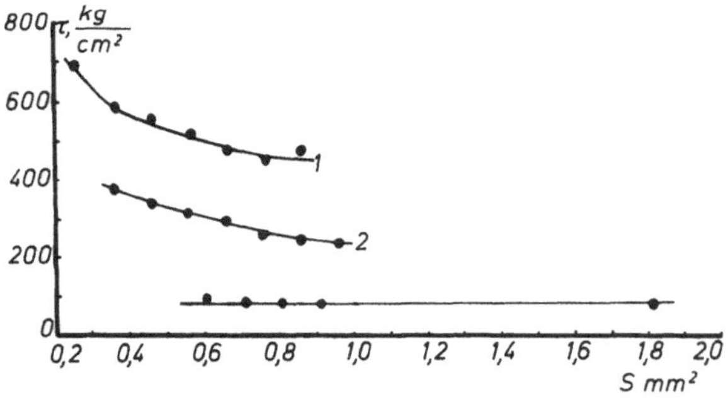

Abb. 1. Abhängigkeit der Haftfestigkeit von der Klebfläche bei der Adhäsion der Polymermatrices an Stahldraht. 1. Nr. 1; 2. Nr. 6a; 3. Nr. 5.

Proben erhaltenen $\tau - S$-Kurve entnommen, die den in Abb. 1 dargestellten Kurven ähnlich ist. Im folgenden werden die Ergebnisse eben auf diese Klebfläche bezogen, wenn keine besonderen Angaben gemacht werden. Die Fläche $S = 0{,}55\ \mathrm{mm}^2$ wurde deshalb gewählt, weil bei allen von uns untersuchten Systemen bei dieser Fläche keine Drahtrisse auftraten.

Bei der Bestimmung der Adhäsion von duroplastischen vernetzten Polymeren an Glasfaser (sowohl vom Durchmesser 10–13 μm als auch an Fasern mit $d = 200\ \mu$m) wurde praktisch in allen Fällen sowohl adhäsive als auch kohäsive Zerstörung der Probe (an der Faser oder am Polymer) beobachtet. Je größer die Klebfläche, um so größer ist bei gleichbleibenden Versuchsbedingungen die Zahl der kohäsiv reißenden Proben. Daher ist der Bereich der Klebflächen, bei denen die Messungen erfolgen können, enger als in den Versuchen zur Bestimmung der Adhäsion an Metallfasern. Für dünne Fasern beträgt die mittlere Klebfläche in vorliegenden Versuchen $(8\text{–}10) \cdot 10^{-3}\ \mathrm{mm}^2$, für dicke Fasern $-0{,}4\text{–}0{,}5\ \mathrm{mm}^2$:

$$S_m = \sum_{i=1}^{n} \frac{S_i}{n}.$$

Auf diese S_m-Werte beziehen sich die erhaltenen Werte der Haftfestigkeit.

e) Bestimmung von $\dot{\tau}$

Die Belastungsdiagramme sind bei Zimmertemperatur und bei Temperaturen unter 20 °C bei allen untersuchten Polymermatrices sowohl bei der Adhäsion an Glasfasern unterschiedlichen Durchmessers als auch bei der Adhäsion an Metalldraht bei allen Belastungsgeschwindigkeiten bis hin zum Zerreißpunkt linear. Man kann vermerken, daß bei der Untersuchung der Systeme mit Glasfasern die Belastungsdiagramme für adhäsiv und kohäsiv reißende Klebstellen gleich sind. Da sich die Glasfasern praktisch bis zum Zerreißen elastisch deformieren, zeugt eine derartige Gleichheit der Diagramme Kraft F–Zeit t davon, daß auch bei der adhäsiven Zerstörung die Systeme sich elastisch deformieren. In diesen Fällen kann man annehmen, daß die Versuche mit konstanter Belastungsgeschwindigkeit den Messungen bei konstanter Deformationsgeschwindigkeit äquivalent sind.

Die Geschwindigkeit der Belastungszunahme an der Probe $\dot{\tau}_i$ wurde aus der Steigung der F-t-Funktion ermittelt:

$$\dot{\tau}_i = \frac{1}{S_i} \cdot \frac{dF_i}{dt}.$$

Als Wert für $\dot{\tau}_m$ bei der Adhäsion an Stahldraht wurde der Wert

$$\dot{\tau}_m = \frac{1}{0{,}55} \cdot \frac{dF}{dt}$$

gewählt, und bei der Adhäsion an Glasfasern

$$\dot{\tau}_m = \frac{\sum_{i=1}^{n_\tau} \dot{\tau}_i}{n_\tau},$$

wobei n_τ die Anzahl der adhäsiv reißenden Proben bedeutet.

f) Berechnung der Haftfestigkeit bei der Adhäsion an Glasfasern

Bei der Bestimmung der Haftfestigkeit kann man zwei Arten von Ergebnissen erhalten: entweder adhäsive oder kohäsive Probenzerstörung. Diese Erscheinung ist in der Regel nicht nur bei der Untersuchung der Adhäsion von Polymeren an Fasersubstraten zu beobachten, sondern auch bei der Bestimmung der Haftfestigkeit an Unterlagen beliebiger Form: Plättchen, Prismen, Kugelschalen u. ä. Das heißt, in allen Fällen der Untersuchungen der Festigkeit von Klebverbindungen begegnen wir den „konkurrierenden" Zerreißarten: entweder längs der Trenngrenze oder an einer der Klebkomponenten.

In allen Versuchen, wo Konkurrenzzerstörung stattfand, wird die Festigkeit der Adhäsionsbindung des Systems Faser-Polymer durch die Größe

$$\tau_0 = \tau_m + \Delta\tau$$

charakterisiert, worin bedeutet:

τ_m = mittlerer arithmetischer Wert der Festigkeit der adhäsiv reißenden Proben.

$$\tau_m = \frac{\sum_{i=1}^{n_\tau} \tau_i}{n_\tau}$$

$\Delta\tau$ = Korrektur, die die kohäsiv reißenden Klebstellen berücksichtigt, deren Haftfestigkeit unbekannt bleibt.

Die Berechnungsmethodik für $\Delta\tau$ ist in (7) dargelegt. Es kann vermerkt werden, daß in der Regel die Einführung der Korrektur $\Delta\tau$ den Charakter der Abhängigkeit der Haftfestigkeit vom untersuchten Faktor nicht verändert, sondern nur die Absolutwerte der erhaltenen Größen ändert.

Versuchsergebnisse und Diskussion

Im Ergebnis der Messungen ergab sich, daß sich für alle untersuchten Polymermatrices bei Zimmertemperatur mit Abnahme der Geschwindigkeit der Spannungszunahme auf der Probe auch die Haftfestigkeit verringert. Die Abhängig-

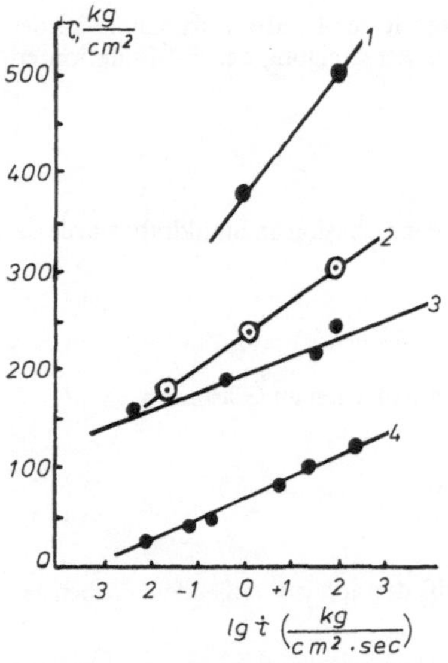

Abb. 2. Abhängigkeit der Haftfestigkeit von der Belastungsgeschwindigkeit in den Systemen Stahldraht–Polymermatrix. 1. Nr. 1; 2. Nr. 6a; 3. Nr. 4; $S = 0{,}4\,mm^2$; 4. Nr. 5; $S = 1\,mm^2$

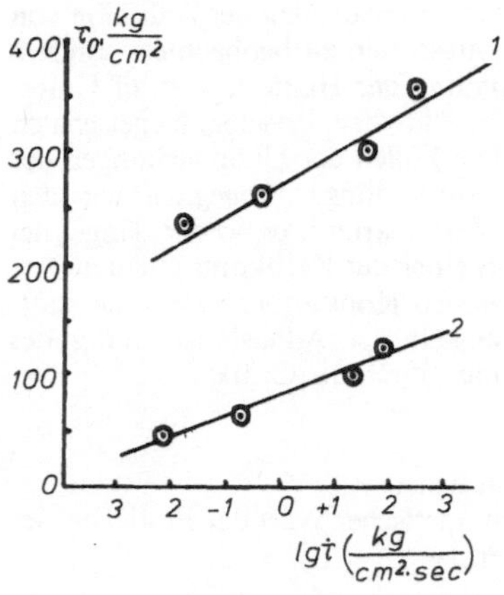

Abb. 3. Abhängigkeit der Haftfestigkeit von der Belastungsgeschwindigkeit in den Systemen Glasfaser ($d = 200\,\mu m$)–Polymermatrix. 1. Nr. 4; $S = 0{,}4\,mm^2$; 2. Nr. 5, $S = 1\,mm^2$

keit der Festigkeit der Klebestellen der untersuchten Polymeren von der Geschwindigkeit der Belastungszunahme ist im halblogarithmischen Maßstab in den Abb. 2–4 dargestellt. In Abb. 2 sind die Ergebnisse der Untersuchungen der Haftfestigkeit bei der Adhäsion der Polymeren an Draht, in Abb. 3 an dicke Glasfasern und in Abb. 4 bei der Adhäsion an dünne Fasern wiedergegeben. Aus diesen Abbildungen ist ersichtlich, daß weder der Harztyp noch die Natur des Fasersubstrats den Charakter der Abhängigkeit $\tau - \dot{\tau}$ ändert. Für alle Polymermatrices: Epoxidharze, Butvar-Phenol-Harze, Polyester- und

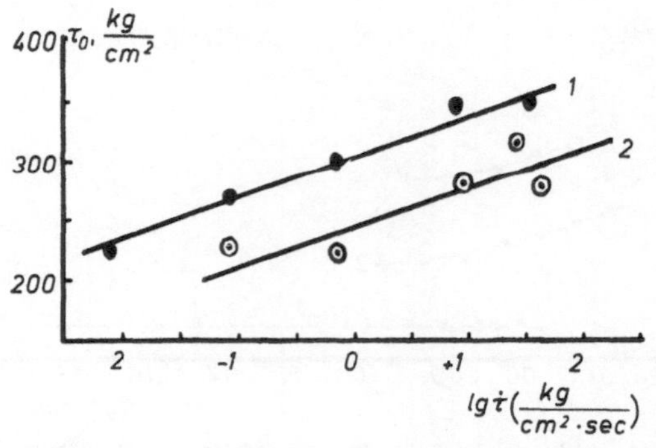

Abb. 4. Abhängigkeit der Haftfestigkeit von der Belastungsgeschwindigkeit für Glasfasern von $d = 13\,\mu m$ und Polymermatrix. 1. Nr. 3; 2. Nr. 7

siliziumorganische Harze erhalten wir sowohl bei der Adhäsion an Glas als auch bei der Adhäsion an Metall die lineare Funktion

$$\tau = C_1 + C_2 \lg \dot{\tau}.$$

Die Werte der Koeffizienten C_1 und C_2 sind in Tab. 2 wiedergegeben. (Der C_1-Wert hängt von der Wahl der Maßeinheit von $\dot{\tau}$ ab. Hier und im folgenden wird die Geschwindigkeit der Spannungszunahme in $kp/cm^2 \cdot sec$ gemessen.)

Die Steigung der Geraden $\tau - \lg \dot{\tau}$, d. h. der Koeffizient $C_2 = \frac{d\tau}{d \lg \dot{\tau}}$, charakterisiert die absolute Änderung der Haftfestigkeit des Systems bei Veränderung der Belastungsgeschwindigkeit um eine Größenordnung. Jedoch noch anschaulicher ist die Größe, die die relative Änderung der Haftfestigkeit charakterisiert, d. i. C_2, bezogen auf eine bestimmte gewählte Haftfestigkeit τ^0:

$$K = \frac{C_2}{\tau^0} = \frac{1}{\tau^0} \cdot \frac{d\tau}{d \lg \dot{\tau}}.$$

Für τ^0 haben wir den Wert τ bei $\lg \dot{\tau} = 3$ eingesetzt, d. h. τ bei ausreichend hohen Belastungsgeschwindigkeiten, etwas größer als die, welche real in den Versuchen vorliegen. Bei einer so großen Geschwindigkeit sind die Relaxationserscheinungen weniger ausgeprägt, und dementsprechend kann man die Belastungsdiagramme mit größerer Sicherheit als linear bis zum Zerreißpunkt annehmen.

Unter Verwendung der Größen K kann man bequem die Empfindlichkeit der Haftfestigkeit gegenüber der Belastungsgeschwindigkeit von Polymeren mit unterschiedlicher Struktur oder eines Polymers bei verschiedenen Temperaturen u. ä. m. vergleichen. In den Abb. 5–6 sind die

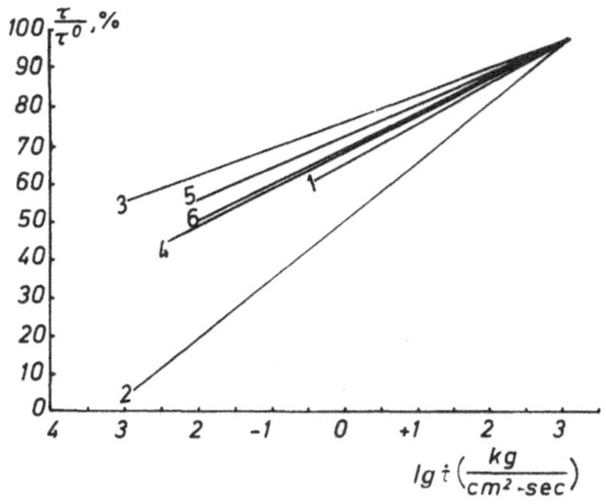

Abb. 5. Desgleichen, wie in Abb. 2 und 4, jedoch in den gegebenen Koordinaten. 1. Nr. 1; 2. Nr. 5; 3. Nr. 4; 4. Nr. 6a an Stahldraht; 5. Nr. 3; 6. Nr. 7 an Glasfasern von $d = 13\ \mu m$

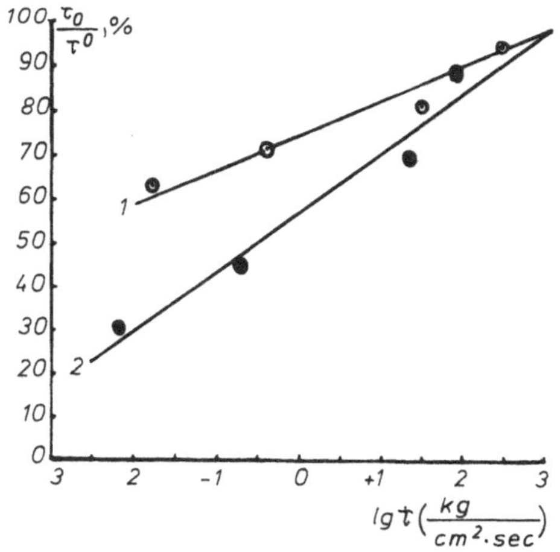

Abb. 6. Desgleichen, wie in Abb. 3 in den gegebenen Koordinaten. 1. Nr. 4; 2. Nr. 5

gleichen Ergebnisse wie in den Abb. 2–4 dargestellt, jedoch in den Koordinaten $\tau/\tau^0 - \lg\dot{\tau}$. Die hieraus erhaltenen Größen K sind in Tab. 2 wiedergegeben.

Während der Charakter der Abhängigkeit $\tau - \dot{\tau}$ bei den verschiedenen Systemen gleich ist, verändern sich die absoluten Werte der Haftfestigkeit beim Übergang von einem System zum andern wesentlich. Das ist aus Abb. 2–4 und aus dem Vergleich der Größen C_1 in Tab. 2 ersichtlich. Wenn also die Haftfestigkeit der heiß gehärteten Epoxid-Polymermatrix an Glas und an Stahl (bei einer Belastungsgeschwindigkeit von $\dot{\tau} = 1\ kp/cm^2 \cdot sec$) 300–400 kp/cm^2 erreicht, so beträgt sie für das elastische Butvar-Phenol-Polymer БФ-6 nur 70–80 kp/cm^2. Durch die

Tab. 2. Koeffizienten der Formeln [1] und [4]

Polymer-matrices	Faser	C_1 (kp/cm^2)	C_2 (kp/cm^2)	K (%)	U_0*) (kcal/mol) I	II**)
Nr. 1	Stahldraht	382	63	11	28	–
Nr. 2	Stahldraht	260	28	8,8	31	–
Nr. 3	Stahldraht					
	Glasfaser $d = 13\ \mu m$	305	35	8,6	31	29
Nr. 4	Stahldraht $S = 0{,}4\ mm^2$	202	18	7,3	34	–
	Glasfaser $d = 200\ \mu m$ $S = 0{,}4\ mm^2$	268	27	7,8	32	25
Nr. 5	Stahldraht $S = 1\ mm^2$	71	22	16	23	–
	Glasfaser $d = 200\ \mu m$ $S = 1\ mm^2$	78	19	14	24	26
Nr. 6	Stahldraht	a) 240	37	10	28	–
		b) 220	35	10,3	28	–
		c) 240	37	10,4	28	–
		d) 240	39	10,5	27	–
		e) 210	37	12	27	–
Nr. 7	Glasfaser $d = 13\ \mu m$	245	33	9,7	29	29

*) Alle Werte mit Glasfaser $d = 13\ \mu m$.
**) I – berechnet aus der Geschwindigkeitsabhängigkeit; II – berechnet aus der Temperaturabhängigkeit.

Natur der aufeinanderwirkenden Paare wird auch die absolute Größe der Festigkeitsänderung – der Größe C_2 – bestimmt. In den untersuchten Fällen ändert sich C_2 von 63 bis 18 kp/cm^2. Jedoch unterscheidet sich die relative Veränderung der Haftfestigkeit, d. h. die Empfindlichkeit gegenüber der Änderung der Geschwindigkeit der äußeren Einwirkung, bei den verschiedenen Systemen sehr wenig: bei Zimmertemperatur verändert sich K bei allen untersuchten Paaren von 7–11%. Eine Ausnahme stellt die Polymermatrix Nr. 5 (БФ-6) dar, für die K bei der Adhäsion an Glas und Stahl entsprechend 14 und 16% beträgt. (Genaueres darüber weiter unten.)

Die umfassende Untersuchung der Geschwindigkeitsabhängigkeit der Festigkeit von Gieß- (8, 9) und Filmproben (10) homogener duroplastischer Polymerer bei Zimmertemperatur zeigte, daß die Kohäsionsfestigkeit dieser Polymeren linear mit dem Logarithmus der Deformationsgeschwindigkeit wächst:

$$\sigma = \sigma_0 + m^* \ln \dot{\varepsilon},$$

wobei $\sigma_0 = \sigma$ wird bei $\lg \dot{\varepsilon} = 0$, wenn $\dot{\varepsilon}$ in %/min gemessen wird. Der Wert σ entspricht bei einer gegebenen Deformationsgeschwindigkeit dem maximalen Spannungswert auf der Kurve $\sigma - \dot{\varepsilon}$, der Koeffizient m^* charakterisiert die Relaxationseigenschaften des Harzes und ist einer der Parameter der verallgemeinerten *Maxwell*-Gleichung (11).

Die Linearität der Funktion $\sigma - \lg \dot{\varepsilon}$ ist für feste duroplastische Polymere so „sicher" festgestellt, daß die Autoren (8) es für möglich halten, die Kurve dieser Funktion durch zwei Punkte zu legen, indem sie σ bei zwei Belastungsgeschwindigkeiten messen.

Somit ist der Charakter der Gesetzmäßigkeiten der Festigkeitsänderung bei Veränderung der Geschwindigkeit des äußeren Einflusses bei Gieß- und Filmproben homogener vernetzter Polymerer und Verklebungen der gleichen Polymeren mit Fasern gleich. Das kann als ein Beweis dafür gelten, daß die Natur der Adhäsionsfestigkeit und der Kohäsionsfestigkeit ähnlich ist.

Am Beispiel des Systems Stahldraht–Polyesterharz ПН-1 (Polymermatrix Nr. 6) wurde der Einfluß des Härtungsregimes auf den Charakter der Geschwindigkeitsabhängigkeit der Haftfestigkeit untersucht. Es wurden Proben untersucht, die im Laufe von 4 Monaten bei Zimmertemperatur gehärtet waren, sowie bei

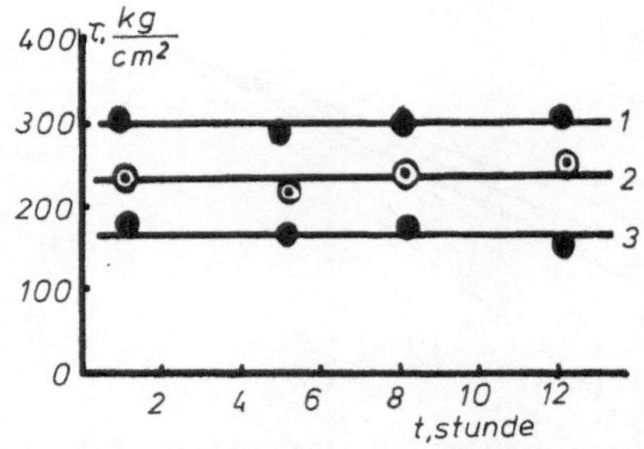

Abb. 7. Abhängigkeit der Haftfestigkeit des Systems Stahldraht–Polymermatrix Nr. 6 von der Verweilzeit der Proben bei 100 °C bei verschiedenen Geschwindigkeiten der äußeren Einwirkung. 1. $\dot{\tau} = 54{,}5$; 2. $\dot{\tau} = 1{,}04$; 3. $\dot{\tau} = 0{,}0182$ kp/cm$^2 \cdot$ sec

100 °C im Laufe unterschiedlicher Zeit gehärtete Proben. Die Ergebnisse sind in Tab. 2 und Abb. 7 wiedergegeben. Es zeigte sich, daß bei heißer Harzhärtung die Bindungsfestigkeit des Harzes mit dem Metall bei allen Geschwindigkeiten nicht von der Verweilzeit der Proben bei 100 °C abhängt (s. Abb. 7). Dementsprechend unterscheiden sich die Werte C_2 und K für das gewählte Regime praktisch nicht voneinander. Die in den Abb. 2 und 5 dargestellten Geraden $\tau - \lg \dot{\tau}$ und $\tau/\tau^0 - \lg \dot{\tau}$ für bei 100 °C im Laufe von einer Stunde gehärtetes Harz charakterisieren die Abhängigkeit der Haftfestigkeit von der Belastungsgeschwindigkeit auch für alle übrigen Regimes.

Die Haftfestigkeit der bei 20 °C gehärteten Klebstellen nimmt nach 2 Monaten einen konstanten Wert an. Deshalb wurde die Prüfung der Proben nach Ablauf dieser Zeit begonnen. Es zeigte sich, daß die Festigkeit der Klebstellen, die im kalten Regime gehärtet wurden, bei allen untersuchten Geschwindigkeiten den für bei 100 °C gehärteten Klebstellen erhaltenen Werten nahekommt. Das gleiche gilt auch für die Größen C_2 und K, d. h. für Polyesterharze wirkt eine wesentliche Änderung der Temperatur, bei der die Bildung des Vernetztenpolymers erfolgt (20 und 100 °C), weder auf den absoluten Wert der Haftfestigkeit noch auf den Charakter ihrer Änderung mit Veränderung des äußeren Einflusses ein.

Bei der Untersuchung der Eigenschaften von Proben der Polymermatrix ПН-1 unter dem gleichen Regime, wie in unseren Untersuchungen, wurde gezeigt, daß die Festigkeit der in heißem und kaltem Regime gehärteten Proben,

*) An diesem Teil der Arbeit wirkte *Sh. Ch. Sheljaskov* mit.

wie auch die Haftfestigkeit, ebenfalls praktisch gleich ist (12), d. h. weder die Temperatur noch die Zeitdauer der Härtung wirken sich auf die mechanischen Eigenschaften des Polymers aus. Jedoch hat das Härtungsregime einen wesentlichen Einfluß auf die Relaxationseigenschaften des Materials. Die stabilsten Eigenschaften besitzt das im Laufe von 8 Std. gehärtete Polymer. Die bei diesem Regime gehärteten Proben hatten die geringsten Kriecheigenschaften. Diesem Regime – die Autoren (12) bezeichnen es als „optimales" (bei gegebener Härtungstemperatur) – entspricht ein Minimum auf der Kurve m^* – Härtungszeit. Außerdem ist der Wert des Koeffizienten m^* für bei Zimmertemperatur gehärtete Gießproben wesentlich größer als bei heißgehärteten Proben. Wie wir feststellten, sind die C_2-Werte (als Analog der Größe m^*) der adhäsiven Klebstellen sowohl bei allen heißen als auch kalten Regimes gleich. Somit gibt die Untersuchung der Geschwindigkeitsabhängigkeit der Adhäsionsfestigkeit keine dem Härtungsregime gegenüber empfindliche Charakteristik, im Gegensatz zur Kohäsionsfestigkeit, wo entsprechende Messungen die Auswahl eines optimalen Härtungsregimes erlauben.

Wie bereits vermerkt, wurde bei Zimmertemperatur die stärkste Änderung der Haftfestigkeit (14–16%) bei der verhältnismäßig elastischen Polymermatrix Nr. 5 bei der Geschwindigkeitsänderung um eine Größenordnung beobachtet. Die Untersuchung der Klebstellen dieser Polymermatrix bei Temperaturen unterhalb der Zimmertemperatur (bis −40 °C), bei denen sich die Härte der Polymermatrix erhöht, zeigte, daß mit Abnahme der Temperatur der Wert der Haftfestigkeit sich monoton erhöht (s. Abb. 8 und 9). Dabei ist die Festigkeitsänderung wie auch bei 20 °C proportional der Veränderung des Logarithmus der Belastungsgeschwindigkeit. Die Abhängigkeit $\tau - \lg\dot{\tau}$ behält ihre lineare Charakteristik bei allen untersuchten Temperaturen bei. Die Empfindlichkeit gegenüber der Geschwindigkeitsänderung fällt mit abnehmender Temperatur (s. Abb. 10): bei −40°C beträgt sie nur 6%. Die Glastemperatur der Polymermatrix БФ-6, die aus thermomechanischen Messungen an Filmen ermittelt wurde, beträgt 45 °C. Daher wurden die in den Abb. 8–9 wiedergegebenen Messungen bei Temperaturen unterhalb des Übergangsbereichs des Polymers durchgeführt. Der Charakter der Festigkeitsänderung

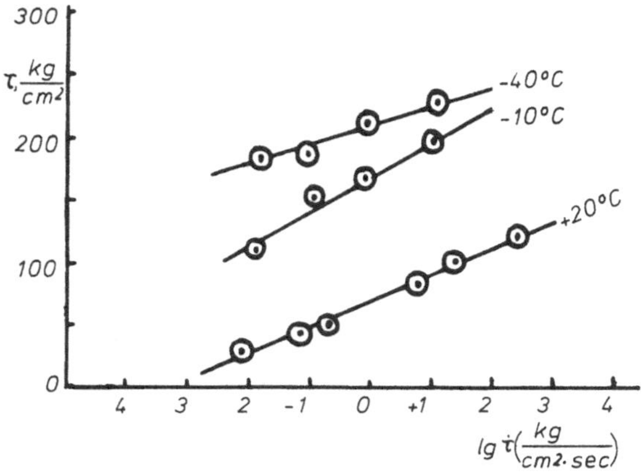

Abb. 8. Abhängigkeit der Haftfestigkeit von der Belastungsgeschwindigkeit des Systems Stahldraht–Polymermatrix Nr. 5 bei verschiedenen Temperaturen

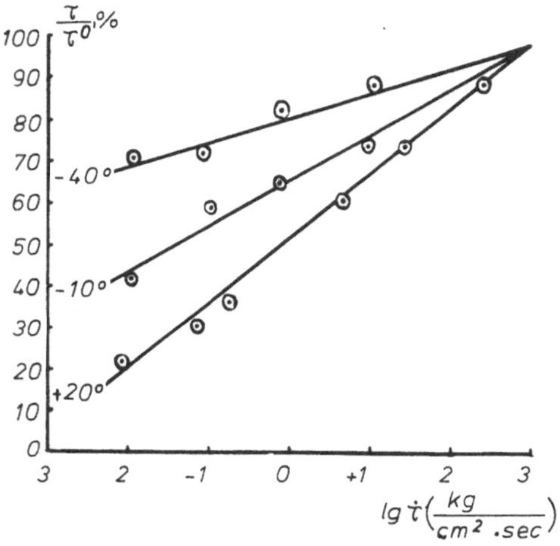

Abb. 9. Desgleichen, wie in Abb. 8 in den gegebenen Koordinaten

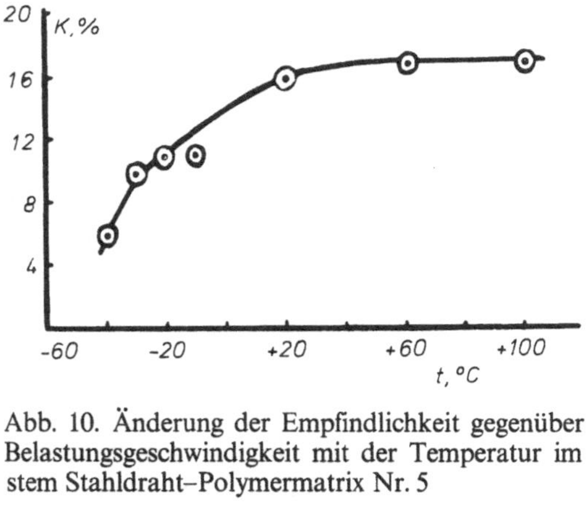

Abb. 10. Änderung der Empfindlichkeit gegenüber der Belastungsgeschwindigkeit mit der Temperatur im System Stahldraht–Polymermatrix Nr. 5

des Systems im Übergangsbereich – bei 60 °C – und darüber – bei 100 °C – ist aus den Abb. 11 und 12 ersichtlich. Daraus folgert, daß beim Übergang über den Erweichungsbereich die

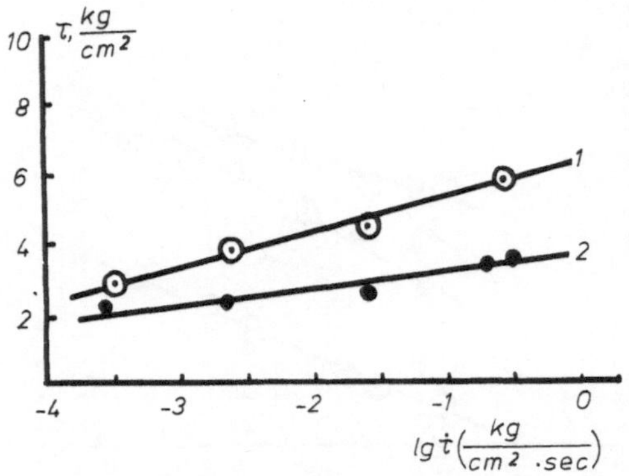

Abb. 11. Einfluß der Belastungsgeschwindigkeit auf die Haftfestigkeit des Systems Stahldraht–Polymermatrix Nr. 5 bei den Temperaturen. 1. 60 °C; 2. 100 °C

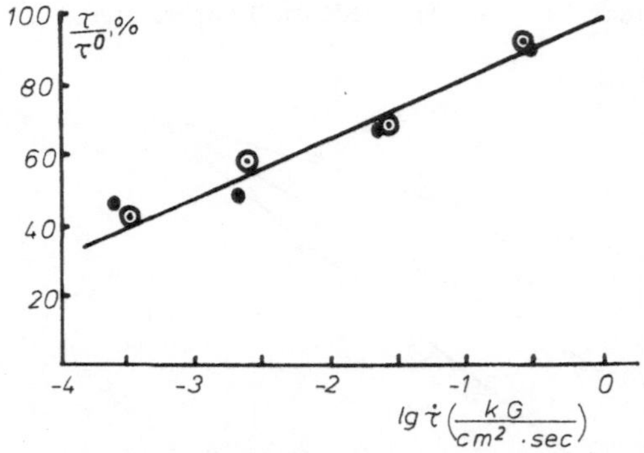

Abb. 12. Desgleichen, wie in Abb. 11 in den gegebenen Koordinaten

Haftfestigkeit stark abfällt: von 70 auf 4 kp/cm^2 bei $\dot{\tau} = 1$ kp/cm$^2 \cdot$ sec. Der Charakter der Abhängigkeit $\tau - \dot{\tau}$ jedoch bleibt auch bei 60 °C und sogar bei 100 °C erhalten. Die Festigkeitsänderung läßt sich gut in halblogarithmischem Maßstab darstellen, obgleich bei 100 °C und geringen Belastungsgeschwindigkeiten die Festigkeit sich mit der Geschwindigkeit weniger ändert, als es aus dem linearen Gesetz folgern müßte. Es ist interessant, daß ungeachtet des so starken Festigkeitsabfalls mit der Temperatur der Wert des Koeffizienten K im gesamten Temperaturbereich von 20–100 °C sich praktisch nicht ändert: $K = 16 - 17\,\%$.

Die geringe Empfindlichkeitsänderung gegenüber der Belastungsgeschwindigkeit beim Übergang des Polymers vom eingefrorenen zum hochelastischen Zustand haben wir auch im System Stahldraht–Polymermatrix Nr. 4 beobachtet. Diese überaus interessante Tatsache erfordert weitere Untersuchungen.

Die Abhängigkeit der Festigkeit von der Belastungsgeschwindigkeit ist eine Äußerung der Abhängigkeit der Lebensdauer von der Belastungsgröße. Deshalb kann man auf Grundlage der Geschwindigkeitsabhängigkeit versuchen, Parameter der bekannten Formel von *Shurkov* für die Lebensdauer τ^* fester Körper zu bestimmen:

$$\tau^* = \tau_0^* \exp[(U_0 - \gamma\sigma)/kT], \qquad [1]$$

worin bedeuten:

$\tau_0^* = 10^{-12} - 10^{-13}$ sec;
T – absolute Temperatur;
U_0 – sog. scheinbare Aktivierungsenergie des Zerstörungsprozesses;
γ – ein Koeffizient, der mit der Spannungskonzentration in Bezug steht;
σ – konstant wirkende Spannung.

In unseren Versuchen wächst die Belastung der Probe und folglich auch die Spannung linear mit der Zeit:

$$\tau = \dot{\tau} t. \qquad [2]$$

Wenn man voraussetzt, wie es gewöhnlich bei der Untersuchung der Kohäsionsfestigkeit gemacht wird, daß die adhäsive Zerstörung irreversibel ist, so kann man mit Hilfe des *Bailey*-Kriteriums (13) die Lebensdauer des Materials aus den Ergebnissen der Festigkeitsmessungen im Regime $\dot{\tau} =$ const berechnen.

Die Bedingung von *Bailey* wird bei Dauerbelastung wie folgt geschrieben:

$$\int_0^{t_{max}} \frac{dt}{\tau^*(\sigma)} = 1. \qquad [3]$$

Durch Einsetzen von [1] und [2] und Integrieren erhält man

$$\tau = C_1 + C_2 \lg \dot{\tau}, \qquad [4]$$

worin

$$C_1 = \frac{2{,}3}{\alpha} \lg A_\alpha; \quad C_2 = \frac{2{,}3}{\alpha};$$

$$\alpha = \frac{\gamma}{kT}; \quad A = \tau_0^* \exp(U_0/kT)$$

sind.

Wie wir sahen, besteht eben solche Beziehung zwischen τ und $\dot{\tau}$ in unseren Versuchen für alle von uns untersuchten Systeme. Das bestätigt die Berechtigung der eingeführten Voraussetzungen und gibt die Möglichkeit, aus den Werten der Koeffizienten C_1 und C_2 u. a. die Aktivierungsenergie des adhäsiven Zerstörungsprozesses zu finden. Die derart bestimmten Werte sind in Tab. 2 angeführt. Daraus ist zu entnehmen, daß die Nullpunkt-Aktivierungsenergie aller untersuchten Systeme einige zehn kcal/mol, nämlich 23 ... 35 kcal/mol beträgt. Diese Größen sind den Aktivierungsenergien der homogenen Polymere naheliegend. Die Nähe dieser Größen erlaubt, in unseren Fällen die Existenz einer chemischen Bindung zwischen Adhäsiv und Substrat anzunehmen.

Die Aktivierungsenergie U_0 kann man nicht nur aus der Geschwindigkeitsabhängigkeit, sondern auch aus der Temperaturabhängigkeit der Haftfestigkeit bestimmen. In diesem Fall wird sie aus dem Neigungswinkel des linearen Teils der τ-Kurve zur Temperaturachse berechnet. Das Bestimmungsverfahren ist z. B. in (14) beschrieben. Die aus den Temperaturkurven bei der Untersuchung der Haftfestigkeit der Polymeren an Alumoborosilikat-Glasfasern erhaltenen U_0-Werte sind in derselben Tab. 2 angeführt. Es ist ersichtlich, daß in allen untersuchten Fällen – außer der Polymermatrix Nr. 4 – die Nullpunkt-Aktivierungsenergien des Zerstörungsprozesses der adhäsiven Bindung, die aus den Versuchen bei Temperaturänderung und bei Geschwindigkeitsänderung der äußeren Einwirkung erhalten wurden, praktisch gleich sind.

Zusammenfassung

Es wurde die Abhängigkeit der Haftfestigkeit im System Polymer-Faser von der Geschwindigkeit der Belastung untersucht. Die Versuche wurden mit Metalldraht von 150 μm Durchmesser, alkalifreien Glasfasern von 12 und 200 μm Durchmesser und wärmehärtenden Polymeren: Epoxidharz, Polyesterharz, siliziumorganischen und Butvar-Phenolharzen durchgeführt. Der Großteil der Messungen erfolgte bei Zimmertemperatur, bei Butvar-Phenol-Polymeren auch bei Temperaturen oberhalb und unterhalb der Zimmertemperatur. Es wurde gefunden, daß die Haftfestigkeit aller untersuchten Systeme eine lineare Funktion des Logarithmus der Belastungsgeschwindigkeit darstellt. Während sich der absolute Wert der Haftfestigkeit τ merklich beim Übergang von System zu System ändert, hängt die Empfindlichkeit von τ gegenüber der Belastungsgeschwindigkeit wenig von der Natur der Unterlage und des Adhäsivs sowie vom Härtungsregime des Polymers ab. Aus den erhaltenen Daten wurde mittels des *Bailey*-Kriteriums und der Lebensdauerformel von *Shurkov* die Aktivierungsenergie des adhäsiven Zerstörungsprozesses bestimmt. Es erwies sich, daß die Aktivierungsenergien der adhäsiven und der kohäsiven Zerstörung ähnlich sind.

Literatur

1) *Andrejevskaja, G. D., G. V. Schirjajeva* und *A. M. Iljinskij*, Standartisazija **11**, 13 (1964).

2) *Andreevska, G. D., J. A. Gorbatkina* und *V. G. Ivanova-Mumjieva*, Sixth International Reinforced Plastics Conference, London, 1968.

3) *Gorbatkina, Ju. A.*, Sammelband „Adgesija polimerov" (Adhäsion der Polymere), S. 72 (Moskau 1963).

4) *McGarry, F. J.*, ASTM-Bulletin **235**, 63 (1959).

5) *Gorbatkina, Ju. A., A. M. Iljinskij* und *A. I. Tschernyscheva*, Savodskaja laboratorija Nr. **1**, 56 (1968).

6) *Voloschinova, R. S., A. M. Iljinskij, A. I. Tschernyscheva, Ju. A. Gorbatkina* und *G. D. Andrejevskaja*, Sammelband „Adgesija i protschnost' adgesionnych sojedinenij" (Adhäsion und Festigkeit von Adhäsionsverbindungen). **1**, 121 (Moskau 1968).

7) *Gorbatkina, Ju. A.* und *T. N. Chasanovitsch*, Fisiko-chimija i mechanika orientirovannych steklo-plastikov (Physikochemie und Mechanik der orientierten Glasfaserplaste). S. 64 (Moskau 1967).

8) *Bernazkij, A. D.*, Ebda, S. 116.

9) *Bernazkij, A. D., A. A. Nikischin, A. L. Rabinovitsch, E. P. Donzova, Je. E. Saborovskaja* und *V. I. Nikolaitschik*, Ebda, S. 129.

10) *Rabinovitsch, A. L.*, VMS **1**, Nr. 7, 998 (1959).

11) *Rabinovitsch, A. L.*, Vvedenije v mechaniku armirovannych polimerov (Einführung in die Mechanik der armierten Polymere). (Moskau 1970).

12) *Bernazkij, A. D.* und *Sh. Ch. Sheljaskov*, Isvestija na Instituta po technitscheska mechanika **5**, 1968 (Bulgarien).

13) *Bartenev, G. M.* und *Ju. S. Sujev*, „Protschostj i rasruschenije vissokoelastitscheskich materialov" (Festigkeit und Zerstörung hochelastischer Materialien). (Moskau-Leningrad 1964.)

14) *Gorbatkina, Ju. A.* und *V. G. Ivanova-Mumjieva*, International Symposium on Macromolecular Chemistry, Toronto, 1968.

Anschrift der Verfasser:

Dr. *Ju. A. Gorbatkina* und *V. G. Ivanova-Mumshieva*
Institut für Chemische Physik
Akademie der Wissenschaften der UdSSR
Laboratorium für armierte Plaste
Vorobjevskoje chaussee 2 B
Moskau V 334 (UdSSR)

Rheol. Acta **13**, 798–802 (1974)

From the Department of Physics, University College of Wales, Aberystwyth (Great Britain)

Taylor stability and material constants of dilute polymer solutions

W. M. Jones, H. Holstein, D. M. Davies, and M. C. Thomas

With 3 figures and 4 tables

(Received October 27, 1972)

Outline of the theory and the experimental method

The second-order fluid model having the equation of state

$$p_{ik} = -p g_{ik} + \alpha_0 A^{(1)}_{ik} + \alpha_1 A^{(1)}_{im} A^{(1)m}_{k} + \alpha_2 A^{(2)}_{ik}$$

in the usual notation, is normally considered adequate for describing the rheological behaviour of dilute polymer solutions. However, normal stress differences associated with the α_i are too small to be measured and *Graebel* (1961) suggested that experiments on *Taylor* stability would yield information from which the α_i could be evaluated.

Since then a considerable literature has grown up but prior to 1970 the theoretical work had been confined to the narrow-gap geometry. Experimental results were in disagreement with the theory. Several reasons have been given to account for the disagreement (e. g., the constitutive equations used were oversimplified) but one possible explanation was that whereas the theory was confined to the narrow-gap the experiments could be better described by a wide gap theory. *Chan Man Fong* (1970) produced general theoretical results for the wide gap, of which the narrow gap theory was a special case. The numerical results presented here for Methods I and II are a consequence of a refinement of *Chan Man Fong*'s work.

α_1 and α_2 may be evaluated by one of three methods. *Method I.* The value of the critical *Taylor* number $(Ta)_c$, for the polymer solution is different from that for a *Newton*ian liquid; put $\Delta = \delta(Ta)_c/(Ta)_c$ where $\delta(Ta)_c$ is the dif-

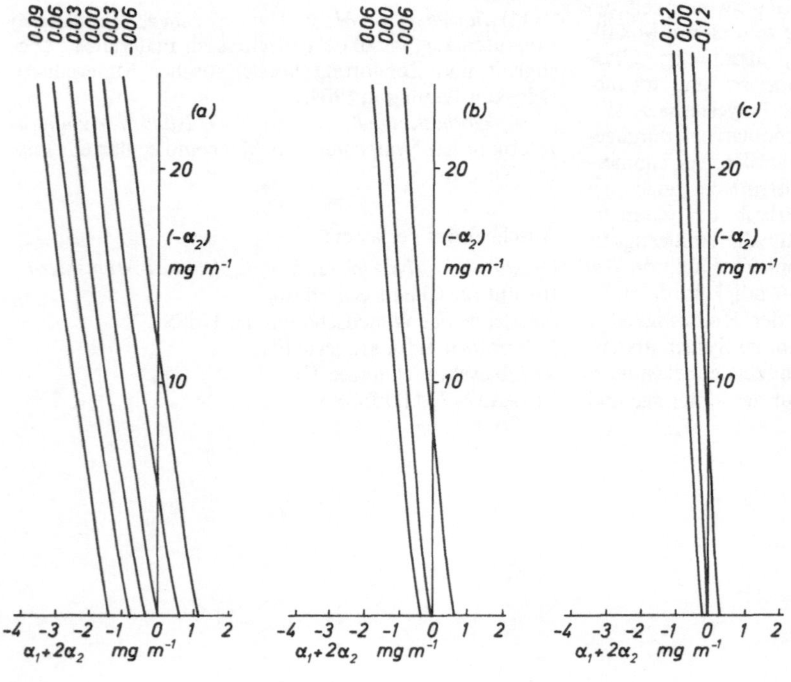

Fig. 1. Theoretical curves of constant Δ as a function of the α_i.
(a) $\eta = 0.90$, (b) $\eta = 0.925$, (c) $\eta = 0.95$

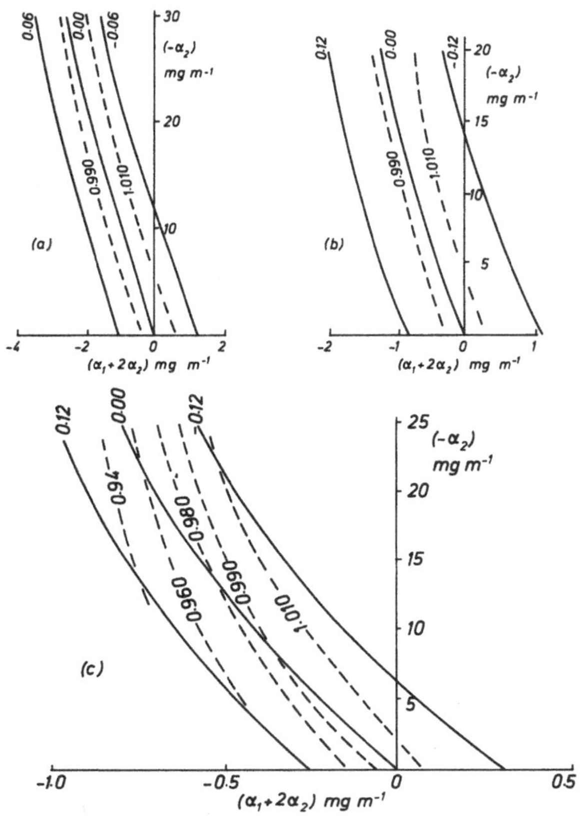

Fig. 2. Theoretical curves of constant Δ (———) and of constant $\bar{\varepsilon}$(— — —) as a function of the α_i.
[$\bar{\varepsilon} = (\varepsilon_i$ for non-*Newton*ian fluid)/(t_i for water)].
(a) $\eta = 0.90$, (b) $\eta = 0.925$, (c) $\eta = 0.95$

ference, then Δ is related to α_1, α_2 and $\eta(= R_1/R_2)$, fig. 1, where R_1 and R_2 are the radii of the cylinders in the *Couette* apparatus. By determining Δ for two values of η the necessary two curves for determining α_1 and α_2 are obtained. *Method II*. The wave number ε_c is also a function of α_1 and α_2 and values of Δ and of ε_c for a given η give the necessary two conditions, fig. 2. *Method III*. The tangential stress G on the stationary outer cylinder is a linear function of $[1 - (Ta)_c/(Ta)]$ when $(Ta) > (Ta)_c$; the slope of the line is a function of α_1 and α_2 (*Ginn* and *Denn* (1969) and *Denn* and *Roisman* (1969)) and the slope and the value of Δ for given η gives the necessary two conditions in this case. fig. 3. Method III is the most accurate experimental method.

The apparatus (*Jones* and *Marshall* (1969)) enabled experiments to be done for $\eta = 0.50$, 0.70, 0.80, 0.90, 0.925, and 0.950. Most experiments were done with the three narrowest gaps. Method II can only give unique results when $\eta = 0.95$. The theory of Method III only applies to narrow gap whereas for the other two methods the theory used was general. Results obtained by Methods II and III were in agreement for $\eta = 0.95$. Average values found for a 250 ppm aqueous solution of polyacrylamide were: $(\alpha_1 + 2\alpha_2) = -(0.75 \pm 0.04)$ mg m^{-1},

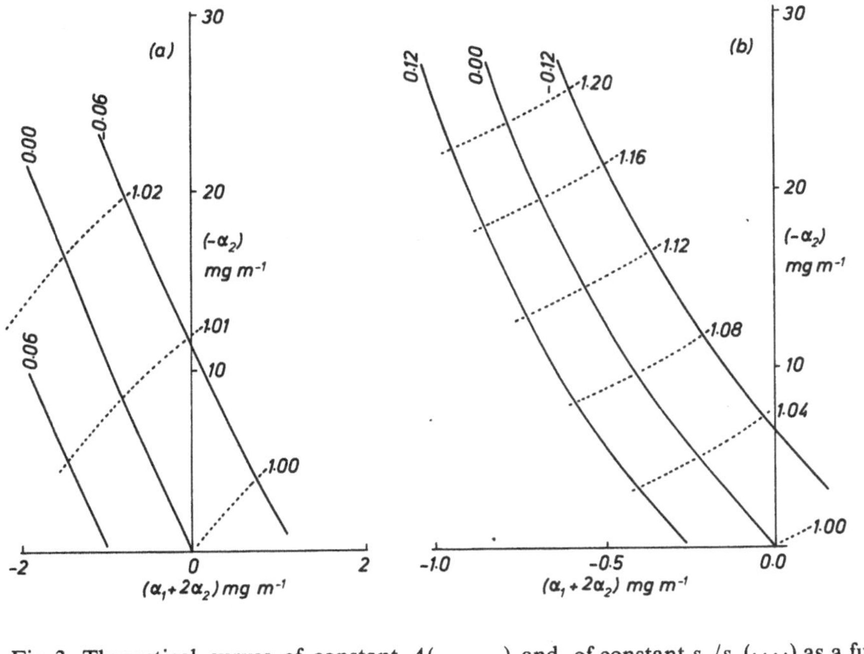

Fig. 3. Theoretical curves of constant Δ(———) and of constant $s_1/s_2(\cdots)$ as a function of the α_i.
(a) $\eta = 0.90$, (b) $\eta = 0.95$

Table 1. Values of the material constants determined from fig. 2 using experimental measurements of Δ and of $\bar{\varepsilon}$ (t = 25 °C). Experimental accuracy discussed in the text

Material	conc ppm	100Δ	$\bar{\varepsilon}$	$-(\alpha_1+2\alpha_2)$ mg m^{-1}	$-\alpha_2$ mg m^{-1}	$-\lambda$
Polyacrylamide	50	−2.7	1.012	0.03	2.0	0.01
	100	0.0	1.008	0.35	8.0	0.02
	250	0.0	0.982	0.70	20.0	0.02
"Kelzan"	10	4.5	0.962	0.75	18.5	0.02
	50	4.5	0.928	0.95	27.0	0.02
Polyox	50	2.5	0.995	0.42	8.6	0.02
	100	10.0	1.040		− ve	
	250	7.5	1.032		− ve	
	500	5.3	1.030		− ve	

Table 2. Values of the material constants of aqueous polyacrylamide solutions as a function of temperature, determined from fig. 3 using experimental measurements of Δ and of s_1/s_2 for $\eta = 0.95$. (*) Experiments of DEM; others: DMD).
Experimental accuracy discussed in the text

conc (ppm)	t °C	$-(\alpha_1+2\alpha_2)$ (mg m^{-1})	$-\alpha_2$ (mg m^{-1})	$-\lambda$
250*)	25	0.66	18.8	0.02
50	15	0.32	6.2	0.03
	25	0.40	11.2	0.02
	35	0.75	21.4	0.02
100	15	−0.30	−5.0	0.03
	25	0.22	4.4	0.03
	35	0.63	15.4	0.02
250	15	0.45	11.0	0.02
	25	0.78	24.0	0.02
	35	1.08	33.0	0.02
500	15	0.55	10.4	0.03
	25	0.78	19.2	0.02
	35	1.17	48.8	0.01

Table 3. Values of the material constants of aqueous polyacrylamide solutions at 25 °C, determined from fig. 3 using experimental measurements of Δ and of s_1/s_2, for various values of η.
Experimental accuracy discussed in the text

conc	η	100Δ	$-(\alpha_1+2\alpha_2)$ mg m^{-1}	$-\alpha_2$ mg m^{-1}	γ s^{-1}	$-\lambda$
50	0.95	2.25	0.35	6.2	183	0.03
	0.925	3.33	0.85	9.4	67	0.03
	0.90	5.0	1.9	10	33	0.10
100	0.95	1.5	0.6	14.5	187	0.02
	0.925	7.0	2.8	41.4	70	0.03
	0.90	9.0	5.2	44.1	35	0.06
250	0.95	6.0	0.89	20.4	210	0.02
	0.925	10.0	3.8	61.9	79	0.03
	0.90	12.0	18	184	40	0.05

$\alpha_2 = -(20.8 \pm 0.2)$ mg m^{-1}. The accuracy in the determination of values at other concentrations and for other materials (see tables 1, 2, 3, 4) was less (about 10%) because the number of determinations of each value was less.

Table 4. Summary of the values found for the material constants of aqueous polyacrylamide solutions at 25 °C

conc (ppm)	$-(\alpha_1+2\alpha_2)$ mg m^{-1}	$-\alpha_2$ mg m^{-1}	$-\lambda$
50	0.26 ± 0.11	6.5 ± 0.9	0.02
100	0.39 ± 0.11	9.0 ± 1.0	0.02
250	0.75 ± 0.04	20.8 ± 0.2	0.02

To achieve meaningful results $(Ta)_c$ must be measured to better than 1% and ε_c and the slopes of the lines to about 1%. Visual observation of $(Ta)_c$ is not accurate enough; accurate torque measurements were used.

A detailed account of the precautions to be taken in using the *Couette* apparatus and suggestions for optimising the experimental design are being published elsewhere. This paper is concerned with the experimental results and their interpretation.

Definitions of symbols used in the figures and tables

s_1/s_2 in fig. 3 is determined from the torque on the outer stationary cylinder of the *Couette* apparatus, the inner cylinder being rotated, in the following manner. ϕ the angular twist of the torsion fibre supporting the outer cylinder is measured as a function of the period of rotation, P, of the inner cylinder. In primary flow alone $(\phi P/\mu)$ is constant (μ is the shear viscosity of the fluid) and an average experimental value $(\phi P/\mu)_{av}$ can be measured. When the secondary flow is present a quantity $|\phi| = (\phi P/\mu)/(\phi P/\mu)_{av}$ is determined and by so doing small frictional effects in the apparatus are eliminated. It may be shown that

$$|\phi| = 1 \cdot 0 + \delta\eta(1 - (Ta)_c/(Ta))$$

for a *Newton*ian fluid, in which δ is theoretically equal to 1.53 for the narrow gap geometry. s_1 is the slope of the straight line obtained when $|\phi|$ is determined as a function of $(1 - (Ta)_c/(Ta))$ for a *Newton*ian fluid and s_2 is the slope for a non-*Newton*ian fluid. δ is obtained from s_1. It may be shown theoretically that

$$s_1/s_2 = \delta B(\delta B + \psi(\alpha_1, \alpha_2))$$

where $B = R_1^2(R_1 + R_2)/2d$ and

$$\psi(\alpha_1, \alpha_2) = \frac{3}{p}\left(\frac{R_1}{d}\right)^4 \left[\alpha_1 + 2\alpha_2\left(1 + \frac{d}{R_1}\right)\right].$$

Hence the theoretical curves of s_1/s_2 shown in fig. 3.

λ in the tables is the ratio of the second normal stress coefficient to the first normal stress coefficient obtained from the theoretical relationship for simple shearing of the second-order fluid,

$$\lambda = -(\alpha_1 + 2\alpha_2)/2\alpha_2 .$$

Comments on the results

In precise experiments (measurement of $(Ta)_c$ to better than 1%) Δ is (i) positive, (ii) increases with concentration and (iii) increases as η decreases (table 3). The $|\alpha_i|$ (i) increase with increasing concentration (ii) increase with increasing temperature. λ is always negative and $|\lambda|$ lies between 0.02 and 0.07. There is evidence that the α_i are functions of γ in that Method I gives no negative value for α_2. (The fact that the α_i vary with η when determined from Δ and s_1/s_2 could be a mis-application of the theory which only applies to narrow gap). γ is the rate of shear.

Many more experiments have been done with polyacrylamide solutions at 25 °C than with any other solution at any other temperature. Average values of the α_i with r.m.s. deviations, found from all methods giving negative values for α_2 are given in table 4. The relatively low deviation for the 250 ppm solution results because it was used as a test solution in trying out experiments with polymer solutions so there are many more results. It is seen that λ is constant, small (0.02), and negative. *Tanner* (1972) reported on several measurements of the ratio of the normal stress coefficients (but of more concentrated solutions) and quoted values of around -0.15 to -0.05.

The increase in the $|\alpha_i|$ with concentration is in accord with the expectation that elasticity increases with concentration. The increase in drag reduction (and the increase in the $|\alpha_i|$) with increasing temperature is profitably contrasted with the decrease in drag reduction with increasing temperature which *Jones* and *Marshall* (1970) found in turbulent flow through straight uniform pipes of circular cross-section. The interpretation of drag reduction in pipe flow (c. f. *Jones* and *Maddock* (1969)) is a molecular interpretation, in that turbulent fluctuations are visualised as being suppressed because they grow from molecular fluctuations in the fluid and a flexible molecule will absorb and eliminate part of the spectrum of turbulence. That part of the spectrum which is eliminated is that having a relaxation length comparable with the diameter of the tube and therefore a relaxation time of the order of 1 ms which is a molecular relaxation time. Molecular relaxation times will diminish with increasing temperature since reaction rates increase with increasing temperature, and consequently energy absorption and therefore drag reduction will diminish. *Taylor* instabilities on the other hand are essentially a macroscopic phenomenon, arising from the centrifugal forces overcoming the viscous forces and the viscoelastic unit is then best regarded as a viscoelastic sphere (enclosing the flexible molecule randomly oriented); the interfacial tension of the viscoelastic sphere will diminish with increasing temperature and therefore relaxation and retardation times will increase thus giving longer times over which elastic effects may operate leading to larger measured values of the $|\alpha_i|$ and of drag reduction.

If α_1 is regarded as the shear viscosity multiplied by a characteristic time of the fluid then the characteristic time becomes 0.02 sec. On the other hand, if drag reduction is to be expected when a toroidal vortex rotates through a half cycle in the characteristic time of the fluid (c. f. *Jones* and *Marshall* (1969)) then $v\tau = \pi d$ where v is the peripheral velocity of the vortex and τ is the characteristic time. Substituting $v = \Omega, \sqrt{R_i d}$ (*Batchelor* (1956)) yields a value of 0.1 s for τ. This latter time compares favourably with that of 0.1 to 0.5s found by *Jones* and

Davies (1972) for flow of aqueous polyacrylamide solutions through porous materials.

The object of characterising a fluid through values of the α_i would be the prediction of flow properties in other geometries. We have attempted to do this by studying flow in bent pipes, using the same liquids as were used in this work, but values of the α_i calculated from drag reduction in a bent pipe are ten times larger than those reported here. That discrepancy is the same as the discrepancy between the characteristic time calculated from the α_i and those calculated from the relaxation length and from flow in porous media respectively. These differences are being examined theoretically and experimentally.

Note

A detailed account of the method and theory is being published elsewhere [1]). The figures and tables of this paper appear in that publication. The outline of the theory and the method given above is a summary of that work. However, the account given here under "comments on results" is new and will not appear elsewhere.

[1]) J. Fluid Mech. 60, 19 (1973).

References

Batchelor, G. K., J. Fluid Mech. **1**, 177 (1956).

Chan Man Fong, C. F., Z. angew. Math. Phys. **21**, 977 (1970).

Denn, M. M. and *J. J. Roisman*, Amer. Inst. Chem. Eng. J. **15**, 454 (1969).

Ginn, R. F. and *M. M. Denn*, Amer. Inst. Chem. Eng. J. **15**, 450 (1969).

Graebel, W. P., Physics of Fluids **4**, 362 (1961).

Jones, W. M. and *O. H. Davies*, Nature (Physical Sciences), **240**, 46 (1972).

Jones, W. M. and *J. L. Maddock*, J. Phys. D. **2**, 797 (1969).

Jones, W. M. and *D. E. Marshall*, J. Phys. D. **2**, 809 (1969).

Jones, W. M. and *D. E. Marshall*, J. Phys. D. **3**, 1486 (1970).

Kuo, Y. and *R. I. Tanner*, Rheol. Acta **13**, 443 (1974).

Authors' address:

W. M. Jones, H. Holstein,
D. M. Davies, and *M. C. Thomas*
Department of Physics,
University College of Wales
Aberystwyth (Great Britain)

Rheol. Acta 13, 803–813 (1974)

From the Department of Civil Engineering,
The Technological Institute Northwestern University Evanston, Illinois (USA)

Experimental study of clay deformability in terms of initial fabric and soil-water potential

R. J. Krizek and T. B. Edil

With 9 figures

(Received October 27, 1972)

Introduction

The energy of a clay-water system can be expressed as a function of its characteristic water retention curve, which represents the relationship between its water content and its total soil-water potential or soil-suction, where the latter is the difference between the free energy of the water in the soil and that of pure water in a free surface condition. Hence, the work required to remove an infinitesimal quantity of water from the soil is a measure of the combined effects of the forces holding the water in the soil, and, with the exception of cementation bonds, it can be considered to include implicitly the effects of the fundamental interaction forces that influence the deformation characteristics of the soil. The total soil-water potential of a soil varies with its water content, mineralogy, solutes present in the pore water, and soil fabric, as well as other parameters of the system. Soil fabric is taken herein to represent the geometrical arrangement of clay particles and their associated interparticle distances. This work is directed toward providing some indication of (a) the effect of fabric on the soil-water potential of a kaolinite clay and (b) the relationship between initial soil-water potential and the deformation characteristics of the soil.

Sample preparation

The material investigated is a Georgia kaolinite clay (Hydrite 10) with the following properties: liquid limit, 62%; plastic limit, 34%; specific gravity, 2.64; silt fraction, 4%; and clay fraction ($< 2\,\mu$), 96%. Three different samples – two with extreme fabrics (highly oriented and highly random) and one with an intermediate fabric – were prepared by controlling the pore fluid chemistry (that is, the interparticle force regime) of the soil-water mixture in a slurry state and the stress path followed during the process of consolidation (that is, the process by which the slurry is brought to a lower water content, defined as the weight of the water divided by the weight of the solids, and a lower void ratio, defined as the volume of the voids divided by the volume of the solids).

The control of the pore fluid chemistry involves shifting the interparticle force regime to either net attraction or net repulsion, and consequently to flocculation or dispersion of the slurry. The magnitude and nature of the particle associations depend primarily on the clay mineralogy and the species and concentrations of the ions present in the pore fluid. In dispersed slurries the dominating repulsive forces between particles tend to cause a more parallel arrangement of particles, whereas in salt-flocculated slurries the particles tend to form into flocs which at certain concentrations comprise an open network of randomly oriented particles. These initial characteristics are usually retained, and possibly even enhanced, by the choice of stress path followed during the subsequent consolidation process.

In order to obtain an appropriate concentration of additives to yield the desired dispersion or flocculation, a series of sedimentation tests were performed in a test tube. These tests indicated that suspensions prepared at about 250% water content with a 0.01 molar solution of NaOH yielded satisfactory dispersion, as measured by dividing the thickness of the sediment formed at any given time by the original thickness of the dispersed suspension in the tube. In the case of the salt-flocculated suspensions a 0.0001 molar solution of $CaCl_2$ in a slurry with a water content of about 250% resulted in a maximum volume of sediment for any given time.

During consolidation the dispersed slurry was subjected to an anisotropic state of stress, which evidence has shown (*Barden* and *Sides*, 1971; *Foster* and *De*, 1971) tends to induce particle parallelism in a direction normal to that of the major principal consolidation stress; procedures for accomplishing the anisotropic consolidation of a slurry in a large cylindrical chamber have been reported by *Sheeran* and *Krizek* (1971). The salt-flocculated slurry was consolidated under an isotropic state of stress; this stress condition was obtained by filling a flexible rubber membrane with slurry to form a sphere approximately 25 cm in diameter, floating this balloon of slurry in a liquid of comparable density within a pressure chamber, and incrementally applying a hydrostatic pressure in this chamber. Drainage was provided at two diametrically opposite points. At the end of isotropic consolidation a sphere of soil approximately 15 to 20 cm in diameter was obtained.

To form an intermediate fabric, a slurry was prepared by mixing only distilled water with the powdered clay.

This produced a lightly flocculated slurry due to the chemical condition of the kaolinite as it existed in powdered form. Then, this slurry was anisotropically consolidated to a maximum vertical stress of about one-fifth that of the dispersed slurry.

Fabric determinations

Three independent procedures have been used to evaluate the fabric of each clay sample at the end of consolidation; these are scanning electron microscopy, optical microscopy, and X-ray diffraction. These three methods of fabric analysis provide information at different levels of magnification and consequently over different areas of measurement, ranging from microns to millimeters. Accordingly, the combined use of all three techniques furnishes a reasonably comprehensive appraisal of the soil fabric,

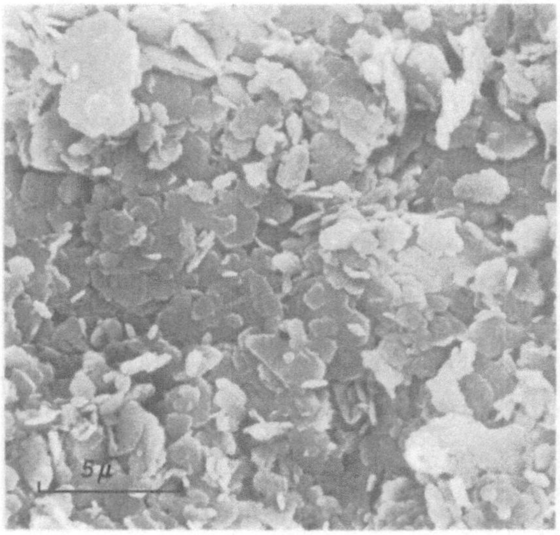

5 μ

Horizontal plane normal to the major principal consolidation stress

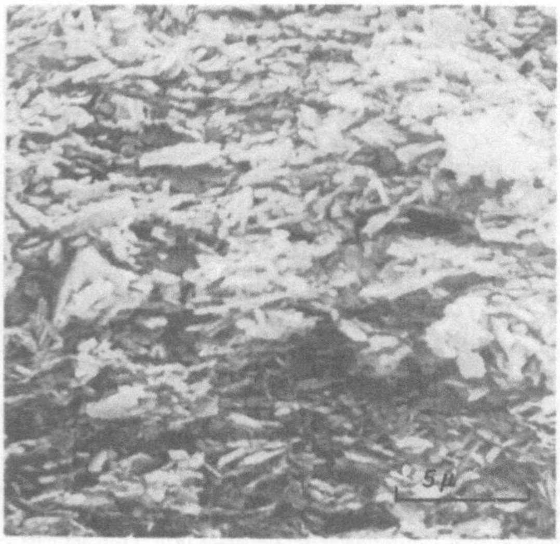

5 μ

Vertical plane parallel to the major principal consolidation stress

Anisotropically consolidated from a dispersed slurry

5 μ

Plane normal to the general direction of drainage

5 μ

Plane parallel to the general direction of drainage

Isotropically consolidated from a flocculated slurry

Fig. 1. Scanning electron micrographs of highly oriented and highly random kaolinite extending from microscopic variabilities to overall trends.

Scanning electron microscopy

For scanning electron microscopy, fracture surfaces with an area of approximately 1 cm^2 were obtained parallel and normal to the major principal consolidation stress from air-dried, vacuum-desiccated specimens of each soil fabric. These surfaces were coated with a carbon layer approximately 50 Å thick and then with a gold layer between 1000 and 2000 Å thick to permit easy conductance of the electrons collected on the surface. Although a certain amount of disturbance probably occurs in the microfabric during the drying process, this change is not likely to obscure the overall fabric of relatively insensitive clays, such as the kaolinite being tested, for qualitative evaluations (*Barden* and *Sides*, 1971). The scanning electron micrographs shown in fig. 1 for the two samples with a highly oriented and a highly random fabric suggest that there is a rather definite orientation of the particles in the sample that was anisotropically consolidated from the dispersed slurry, whereas the sample that was isotropically consolidated from the flocculated slurry seems to possess a fabric in which the original flocs are more-or-less preserved throughout the consolidation process, thereby resulting in a fabric consisting of randomly distributed groups of particles, even within the small areas revealed in the micrographs.

Optical microscopy

For optical microscopy and X-ray diffraction, small cubes (approximately 2 cm on a side) of the samples were impregnated with a mixture of 80% Carbowax 1500 and 20% Carbowax 1000 (polyethylene glycols from the Union Carbide Company) at a temperature of 40° to 45 °C over a period of about 10 days, during which time the carbowax bath was periodically renewed. Carbowax is a solvent for water in its liquid state; immediately upon immersion, the solution reaction between the carbowax and the pore water in the sample begins, and it continues until an equilibrium solution between the bath fluid and the pore water is reached. After substitution of carbowax for pore water is achieved, the sample is brought to room temperature, whereupon the carbowax matrix hardens and holds the soil particles in their original fabric. A microtome was then used to cut thin-sections from the resulting soil-carbowax system; these sections had a thickness of approximately 25 μ for optical microscopy and 100 μ for X-ray diffraction. The indicated mixture of two carbowaxes with different molecular weights was determined by trial-and-error to yield a sufficiently firm matrix material which has acceptable optical and X-ray diffraction properties and is amenable to sectioning.

The optical evaluation of clay fabric is based on the fact that the refractive indices for plate-shaped particles in the directions of the long axes (a and b axes) are approximately equal, but significantly different from that in the direction of the short axis (c axis). Hence, if a group of oriented particles is viewed under polarized light by looking down the short axis, the field exhibits a uniform grayness as the sample is rotated about the short axis. Alternatively, if a group of oriented particles is viewed normal to the short axis, four stages of illumination and extinction are observed as the sample is rotated through 360°. If a uniform distribution of illuminated spots and a constant overall intensity are observed as the microscope stage is rotated through 360°, the clay particles are randomly oriented.

As seen in fig. 2, a study of the thin-sections with the extreme fabrics supports the qualitative conclusions obtained by use of scanning electron microscopy. In addition, the use of a gypsum accessory plate with a retardation of approximately 530 nm converts a gray set of interference colors into a set of brilliant colors which readily distinguish such important features as microshear zones and disturbance areas, and the application of this technique indicated that the microfabric of the specimens studied is relatively uniform.

X-Ray diffraction

A quantitative fabric evaluation of these specimens over larger areas than covered in the previous two methods was obtained by using a pole figure goniometer and analyzing the diffraction patterns of X-rays transmitted through thin-sections (*Baker, Wenk* and *Christie*, 1969). If κ is the angle between the normal to the (002) basal plane (that is, particle faces) and the reference axis, which is parallel to the major principal consolidation stress direction in the

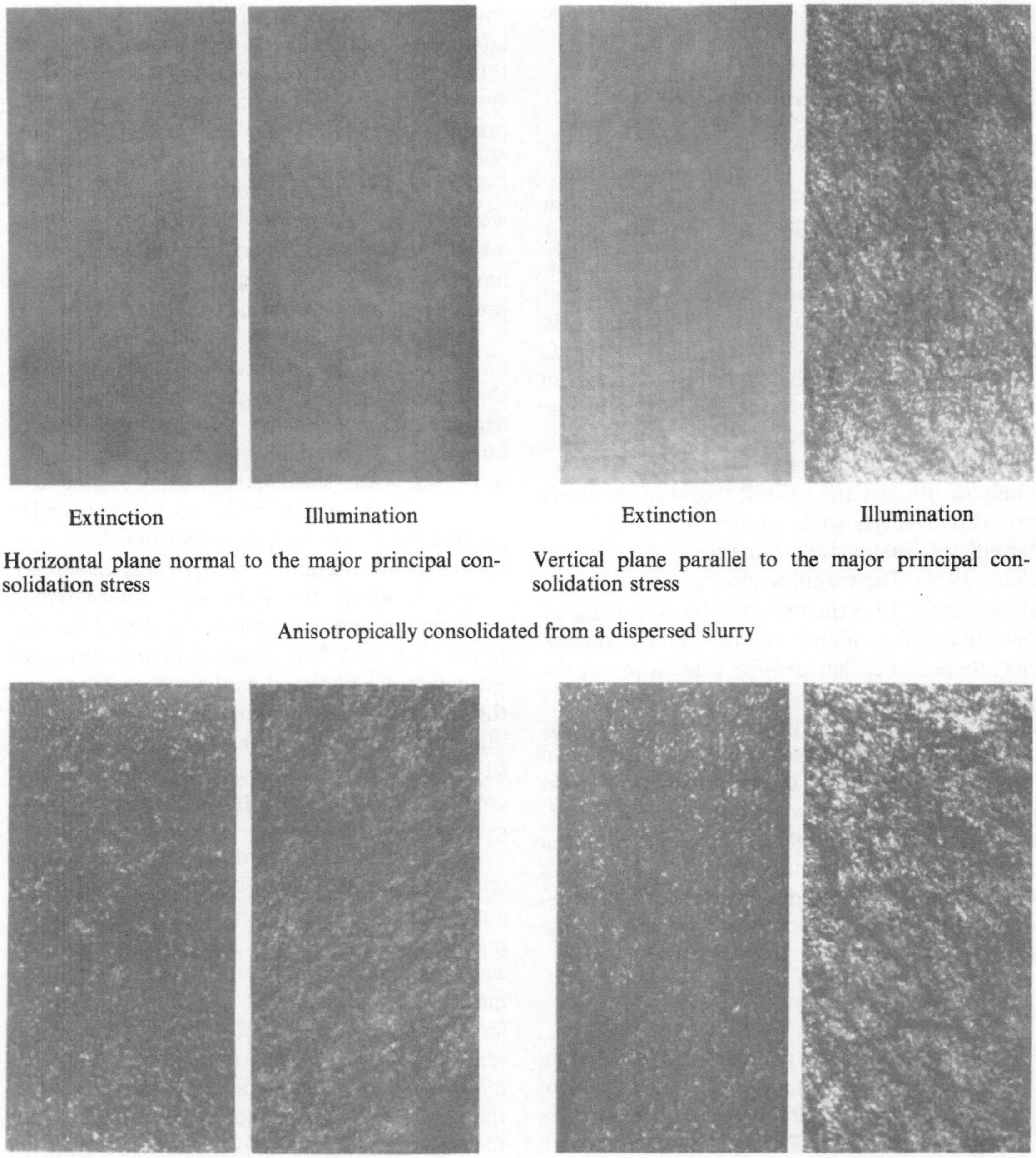

Fig. 2. Photomicrographs of highly oriented and highly random kaolinite

highly oriented specimen and perpendicular to the general drainage direction in the highly random specimen, the ratio $N(\kappa)$ is defined for a given specimen as the number of particles oriented at the angle κ divided by the number of particles that would be oriented at that angle in an equivalent perfectly random specimen. Fig. 3 gives plots of this ratio versus the angle, κ, for the highly oriented and the highly random thin-sections. In the case of the highly oriented specimen, κ values of 0 and 90° correspond to the maximum and minimum intensities, respec-

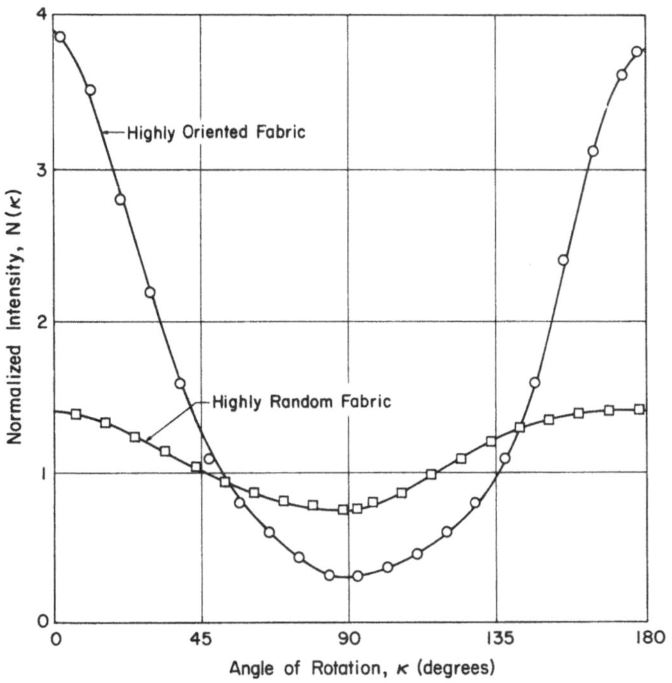

Fig. 3. Orientation distribution of highly oriented and highly random kaolinite in terms of X-ray diffraction intensities

tively, and there is a significant difference in the magnitudes of $N(0)$ and $N(90°)$. On the other hand, the highly random specimen exhibits a response which is very close to the theoretical response for a perfectly random specimen (that is, $N(\kappa) = 1$ for $0 \leq \kappa \leq 180°$), but there is a slight preferential orientation parallel to the general drainage direction; although seepage forces may have caused some particle alignment during the early stages of slurry consolidation, the data are too limited to draw any definite conclusion.

Comparison of initial fabrics

As evidenced by the results obtained from the preceding three techniques for identifying soil fabric, the two extreme fabrics may be sensibly termed highly oriented and highly random. Although specific documentation is not included for the case of the intermediate fabric sample, the application of these procedures revealed that the degree of particle orientation was qualitatively between the two extremes.

Experimental results and analysis

After consolidation of the slurry was complete, all three samples were allowed access to their respective pore fluids as the load was removed incrementally, equilibrating after each increment. Rebound with access to the emitted pore fluid prevents the development of excessive soil suctions and the associated disruptive influences on the fabric of the sample.

Water retention characteristics

At the end of this procedure cylindrical specimens with an average height of 4 cm and an average cross-sectional area of 10 cm^2 were trimmed with their longitudinal axes parallel to the direction of the major principal consolidation stress in the samples with highly oriented and intermediate fabrics and with their longitudinal axes perpendicular to the general direction of drainage in the sample with a highly random fabric. Then, these specimens were brought to equilibrium at various soil-water potentials in a 15-bar ceramic plate extractor (*Richart*, 1965) according to a prescribed schedule (0, 1/3 bar, 2/3 bar, 1 bar, and thereafter, as required, in 1 bar increments); the attainment of equilibrium at each pressure increment was determined by monitoring the emission of water from the specimens, and the characteristic water retention curves associated with this desorption schedule are given in fig. 4.

The gradual concave shape of all three curves indicates that, within the range tested, each soil sample had a wide distribution of pore sizes, because the presence of a dominant pore size would have yielded a rapid decrease in water

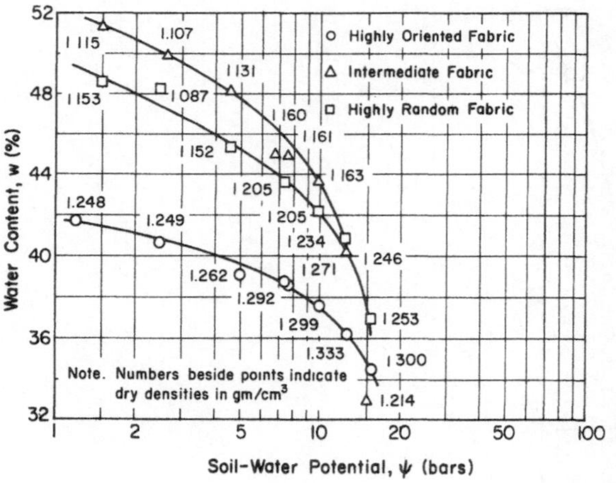

Fig. 4. Water retention curves

content at the value of the soil-water potential associated with holding water in a pore of that given size. Due to a variety of complex interactions between soil particles and water, it is virtually impossible to calculate theoretically the water retention characteristics of a given soil; however, the measurement of water retention can be made without regard to the individual forces holding the water in the soil, and a qualitative interpretation of the different possible mechanisms can be advanced (ultimately, of course, quantitative interpretations are desired). For example, two factors affecting the matric potential (in general, the matric potential constitutes the primary component of the total soil-water potential, and solute effects are usually negligible) of a soil at equilibrium are the sizes and volumes of the pores and the adsorptive forces associated with the particle surfaces. Accordingly, soil fabric strongly influences the pore size distribution and consequently the shape of the characteristic water retention curve, especially in the low potential range. The process of water extraction reduced the volume and the void ratio of the specimen, particularly the volume of large voids, and the role of the adsorptive forces holding water in the soil is increased as the soil-water potential increases, that is, the water retention characteristics tend to be increasingly governed by the specific surface area (texture) of the particles rather than the fabric (particle orientation and spacing) of the soil. This rationalization offers one plausible qualitative explanation for the tendency of the three water retention curves shown in fig. 4 to exhibit different slopes at low values of soil-water potential and to merge at higher values.

From fig. 4 it can be seen that, for a given soil-water potential, the lowest equilibrium water content is associated with highest dry density and the most oriented fabric. On the other hand, for a given specimen the change in fabric during the process of water extraction is basically due to volumetric strains, and dry density can be used as an index to monitor these fabric changes. The individual points on a water retention curve are obtained by testing different specimens with hypothetically the same initial fabric and dry density. The deviations from a monotonically increasing pattern of dry densities during extraction are quite small, and they reflect the minor nonhomogeneity of the block sample from which the individual specimens were trimmed and the relative insensitivity of the volume measurements. Prior to placement in the extractor, all specimens had a degree of saturation (ratio of the volume of water to the volume of voids in the soil) on the order of 98 to 100%. During the process of water extraction, the saturation of each decreased with increasing values of the soil-water potential, dropping from about 98% at 1 bar to somewhat below 90% at 15 bars, with most of the values in the vicinity of 95%; these values varied somewhat for each fabric, but no pattern was readily observable.

Stress-strain response

After a particular specimen was brought to equilibrium in the extractor, it was tested in uniaxial compression (zero lateral stresses) at a constant strain rate of approximately 0.0025 min^{-1}. The lateral deformations at the mid-height were measured at three points spaced 120° apart by use of micrometers (with a sensitivity of 0.0001 inch) which completed an electrical circuit when they came in contact with metal targets placed on the surface of the soil specimen. The axial and radial strains were determined by dividing the measured axial and radial deformations by the initial length and radius, respectively, and the axial stress at any time during the test was calculated by dividing the axial load by the cross-sectional area at the mid-height of the specimen, as computed from the measured radial deformations. The stress-strain response of over 20 specimens subjected to this testing procedure is shown in fig. 5.

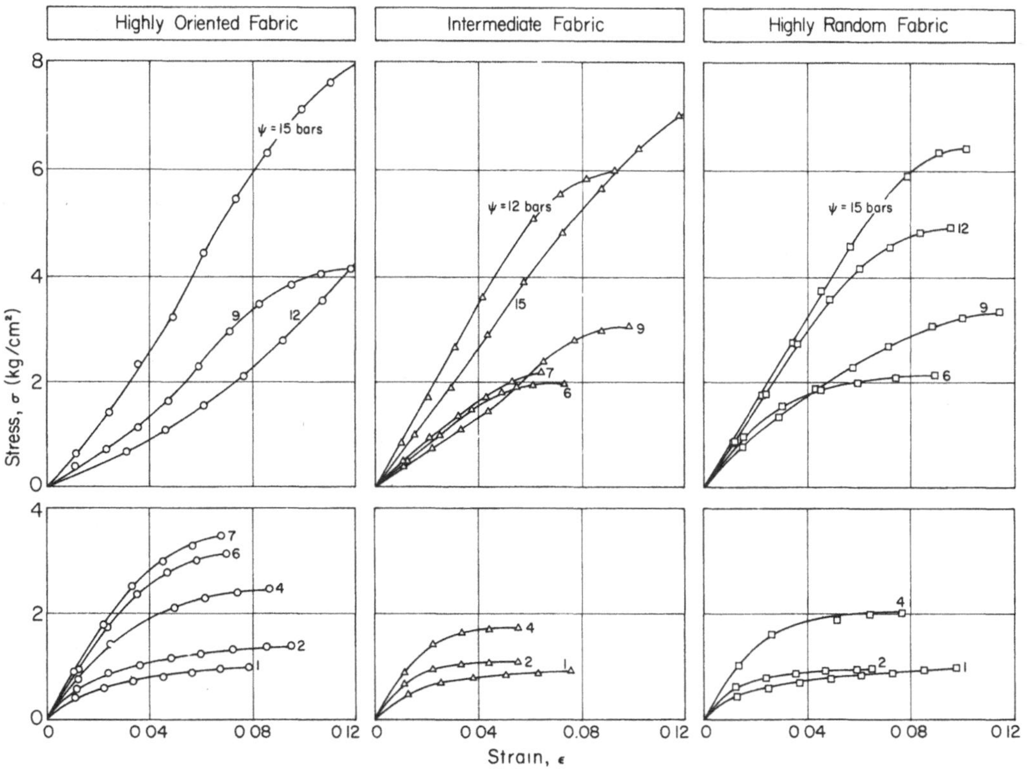

Fig. 5. Stress-strain response at various soil-water potentials

A brief study of these response curves indicates that for any particular soil the initial modulus (slope of the stress-strain curve at zero strain) usually becomes greater as the initial soil-water potential increases; however, this trend is not quite so definitive at the higher soil-water potentials. It is also noted in fig. 5 that a distinct difference in the observed stress-strain response seems to occur at some critical value of the soil-water potential, termed Ψ_c, which appears to be in the range of 5 or 6 bars for the specimens with an intermediate or random fabric and 8 or 9 bars for the specimens with a highly oriented fabric. For $\Psi < \Psi_c$, the stress-strain response is essentially strain-softening throughout its entire range, that is, the modulus decreases with increasing axial strain; this is illustrated by the response curves in the lower portion of fig. 5 for specimens with all three initial fabrics. For $\Psi > \Psi_c$, the small-strain response is either (a) essentially linear or (b) strain-hardening followed by a constant-modulus region ($0 < \varepsilon < \varepsilon_y$), after which a strain-softening phenomenon prevails. The strain-hardening behavior for small strains is largely attributable to (a) seating errors in the upper and lower loading platens and/or (b) the closing of shrinkage cracks which were observed to develop in the specimens subjected to high soil-water potentials (especially above 9 or 10 bars); accordingly, in these cases the constant-modulus portion of the response curve is projected to the strain axis, and a new origin is established at its point of intersection.

Although the tests described herein were not carried to failure, some estimate of the probable failure strains can be made from the shapes of the stress-strain curves. For $\Psi < \Psi_c$, failure strains are likely to be on the order of 0.05 to 0.08, the lowest values being associated with the specimens having an intermediate fabric. For $\Psi > \Psi_c$, probable failure strains range from about 0.07 to 0.12 or greater, with the higher values being associated with the specimens subjected to higher soil-water potentials without any apparent relation to soil fabric.

Subject to the indicated adjustments for strain-hardening behavior in the small strain range, all stress-strain curves shown in fig. 5 can be reasonably well described by equations of the form

$$\sigma = E\varepsilon, \qquad \varepsilon < \varepsilon_y \qquad [1a]$$

$$\sigma = E\varepsilon_y + \frac{\varepsilon - \varepsilon_y}{a + b(\varepsilon - \varepsilon_y)}, \ \varepsilon > \varepsilon_y \qquad [1b]$$

where σ and ε are the stresses and strains, respectively, in the direction of the applied uniaxial stress, E is the slope of the constant-modulus portion (if it exists) of the response curve, ε_y is the strain at which strain-softening behavior begins, and a and b are empirically determined coefficients. For $\Psi < \Psi_c$, $\varepsilon_y = 0$, and eq. [1b] reduces to (*Krizek*, 1967)

$$\sigma = \frac{\varepsilon}{a + b\varepsilon} \qquad [2]$$

which, upon rearrangement, becomes

$$\frac{\varepsilon}{\sigma} = a + b\varepsilon \qquad [3]$$

which is the equation for a straight line with intercept a and slope b. If eq. [2] is differentiated with respect to ε and the result evaluated at $\varepsilon = 0$, we obtain

$$d\sigma(0)/d\varepsilon = 1/a = E_i \qquad [4]$$

which represents the initial slope of the stress-strain curve. Furthermore, evaluation of eq. [2] at $\varepsilon = \infty$ gives

$$\sigma(\infty) = 1/b\,, \qquad [5]$$

where the strength parameter $\sigma(\infty)$ is the ultimate value of σ as ε approaches infinity. Since the actual maximum stress, σ_{max}, in any particular specimen will occur at some finite strain, we may write

$$\sigma_{max} = R\sigma(\infty) = \frac{R}{b}, \qquad [6]$$

where R is less than unity (very often on the order of 0.75 to 0.90). For specimens in which $\Psi > \Psi_c$, the response is governed by eqs. [1a] and [1b], where ε_y is some specific value for each specimen, and the second term on the right-hand side of eq. [1b] is treated in the same manner as the term on the right-hand side of eq. [2]; in effect, this means that a local coordinate system is established on the stress-strain curve at the point where $\varepsilon = \varepsilon_y$.

Fig. 6 shows the coefficient a (for $\Psi < \Psi_c$) and the reciprocal modulus $1/E$ (for $\Psi > \Psi_c$) plotted versus the soil-water potential, Ψ. In all cases a or $1/E$ is seen to decrease monotonically with increasing values of Ψ; this means that the initial modulus increases as the soil-water potential increases. However, fabric effects are manifested in a somewhat contradictory manner; for $\Psi < \Psi_c$, the specimens with a highly oriented fabric yield the highest values of a or the lowest moduli, and the specimens with an intermediate fabric indicate the opposite behavior. On the other hand, for $\Psi > \Psi_c$, the specimens with an intermediate fabric exhibit the lowest moduli, and the specimens with a random fabric the highest; the trend for the specimens with a highly oriented fabric is inconclusive, since only three data points are available and no pattern is observable. For a given value of Ψ, fig. 4 indicates that the dry density is highest for the specimens with a highly oriented fabric and lowest for those with an intermediate fabric; however, the data in fig. 6 show that the initial moduli decrease as the dry densities increase for $\Psi < \Psi_c$, and the

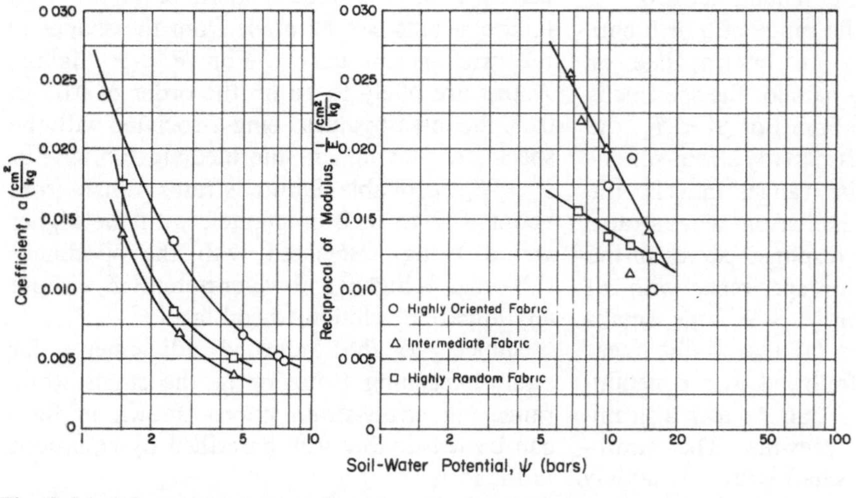

Fig. 6. Modulus parameter as a function of soil-water potential

trend is not clear for $\Psi > \Psi_c$. In most cases the data points which deviate from the indicated trends exhibited some nonhomogeneity, such as visible shrinkage cracks or a dry density that was inconsistent with the pattern.

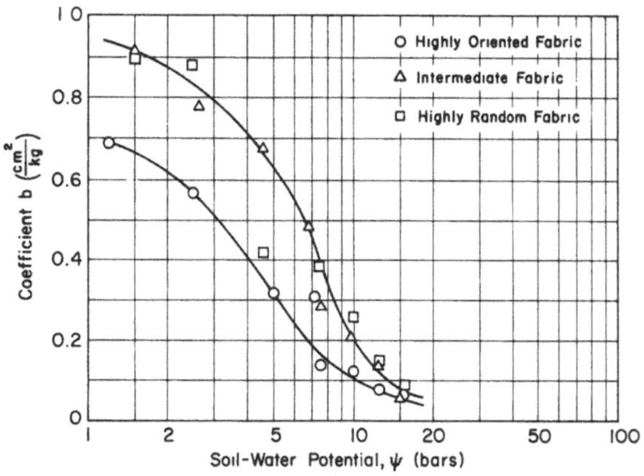

Fig. 7. Strength parameter as a function of soil-water potential

As shown in fig. 7, the coefficient *b* decreases (or the strength parameter increases) rapidly with increasing values of Ψ; similar behavior over a wider range of Ψ has been reported by *Yong*, *Japp*, and *How* (1971). Although little difference is observed in the *b* values for specimens with a random and an intermediate fabric, the analogous strength parameter for the specimens with a highly oriented fabric is substantially higher (or *b* is lower). However, this behavior cannot necessarily be attributed to particle orientation, because the densities of the specimens with a highly oriented fabric are considerably higher than the densities of the other specimens, and density is known to exert a strong influence on the strength of a soil; hence, the variation in strength may be due to density variations and not particle orientation. In a somewhat more conventional plot, fig. 8 shows that the strength parameter, *b*, is dependent on the water content in approximately an exponential manner, where the specimens with an oriented fabric exhibit the highest *b* values and the specimens with an intermediate fabric the lowest for a given water content. Similar exponential relationships for these variables have been reported by many other researchers. The discontinuity in the $b-w$ relationship at approximately the water content associated with Ψ_c is not readily explainable. *Yoshinari* (1967) observed a similar type of behavior, but the discontinuity occurred in the vicinity of the liquid limit of the soil, and this is not the situation with the data in fig. 8. Since the sample with a highly oriented fabric was consolidated under a higher vertical stress than the other samples, the explanation may lie in the stress history.

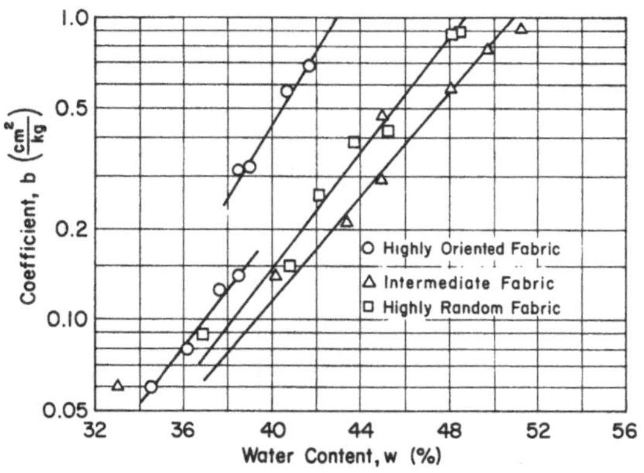

Fig. 8. Strength parameter as a function of water content

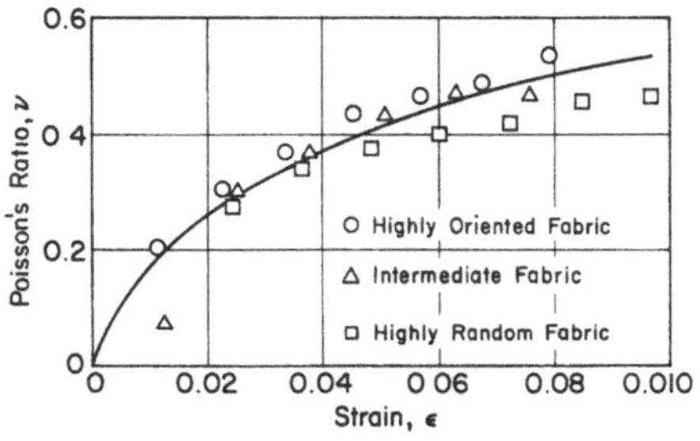

Fig. 9. Variation of *Poisson*'s ratio with axial strain

The radial deformations (average of three readings 120° apart) at the midheight of each specimen were used in conjunction with the axial deformations and the specimen dimensions to determine values of *Poisson*'s ratio, and typical values are plotted in fig. 9 for the specimens tested at an initial soil-water potential of 1 bar. For this case it is seen that *Poisson*'s ratio increases with increasing values of axial strain, reaching a value of 0.5 at a strain of about 0.08, and there is no apparent influence of soil fabric. Although *Poisson*'s ratio was found to increase with axial strain for all values of the soil-water potential, the relationship in some cases is not as well defined as that shown in fig. 9. In general, as the initial soil-water potential increases, the values of *Poisson*'s ratio at a given axial strain becomes smaller, dropping to about 0.1 at an axial strain of 0.08 for a specimen subjected to an initial soil-water potential of 15 bars.

Discussion

For many applications the water relations of soils comprise their most important physical characteristics; for example, water is the variable which exerts the greatest influence on such properties as the deformability and strength of a clay. Therefore, the energy relationships between soil and water and the mechanisms by which water is held in soil are extremely significant in soil engineering. Historically, the water retention characteristics of soils have been explained by the use of either a geometrical concept, which attempts to describe the volumes and shapes of the voids between the soil particles, or a potential concept, which specifies the energy with which water is held in a soil without considering explicitly the geometry of the voids. Since the latter can be specified independently of the mechanism by which water is retained, its use is therefore often simpler and easier. Although the concept of a soil-water potential is primarily used in this work, the geometrical concept is incorporated indirectly by fabric considerations, and an experimental attempt is made to relate the stress-strain behavior of the clay to both soil-water potential and soil fabric.

As seen in figs 5 through 8, the relationships between the deformability characteristics and the soil-water potential of specimens with each of the three fabrics described are qualitatively similar, and the influence of soil fabric is presumedly manifested by the different curves which are obtained. However, a careful study of the data presented fails to produce any decisive conclusions regarding the role of particle orientation. This is due primarily to the very common experimental difficulty of conducting tests in which all variables except those under study are held constant. For example, in this investigation an attempt was made to examine the relationship among deformability, soil-water potential, and fabric, but in the process of trying to control or measure these variables, differences in other variables, such as the pore fluid chemistry and the dry density, had to be accepted; consequently, it is virtually impossible to state conclusively that an observed effect (such as a variation in modulus) is due to some particular cause (such as a change in soil fabric). This point is perhaps best illustrated by the fact that the specimens with the most highly oriented fabric had the highest density and exhibited the greatest strength; however, since strength is strongly related to density, the real cause of this greater strength is not at all clear. Another complexity in the interpretation is suggested by the fact that, although very distinct differences are noted in the behavior of the specimens with highly random and highly oriented fabrics, the response of the specimens with an intermediate fabric often does not lie between that of the two extreme cases.

Despite the rather inconclusive role of soil fabric, and especially particle orientation, in these particular tests, it is probable that its influence would be manifested more strongly in the response of similar specimens subjected to a state of stress in which the principal stresses are oriented at various angles with the predominant plane of particle orientation. In this regard it is important to note that all specimens with a highly oriented or an intermediate fabric were tested herein with the uniaxial compressive stress applied in a direction normal to the predominant plane of particle orientation.

Another important consideration in this work is that the soil fabric and the soil-water potential are described in terms of their so-called initial values, that is, their values at the end of slurry consolidation and at the end of extraction, respectively. Specifically, this means that actual measures of these parameters during the uniaxial compression test are not known. Some appraisal of fabric changes, particularly interparticle distances, during extraction are reflected in the measured dry density variations, but there was no effort to control or measure fabric changes, if any, during compression testing. The same is generally true for soil-water potential; all Ψ values represent the equilibrium potential of the specimen when it was removed from the extractor, and any changes which may have occurred during compression testing were not measured.

Although the interpretation of the stress-strain response for the specimens tested in this experimental program involves a critical soil-water potential, Ψ_c, and a strain, ε_y, at which strain-softening begins, there is no apparent theoretical basis for either of these parameters. Prior to the conduct of these tests, it was anticipated that the response curves would all be qualitatively similar and that quantitative variations could be associated with the initial soil fabric and its initial soil-water potential. However, the tests data suggest rather emphatically that a qualitative variation in the stress-strain response takes place between high and low values of the soil-water potential.

Conclusions

Within the framework described and based on the results of this experimental attempt to relate the deformability of a clay with its initial soil-water potential and initial fabric, the following observations and conclusions can be advanced:

1. The energy status of a soil-water system, defined in terms of its soil-water potential as a function of water content, reflects variations in the initial fabric of soils and provides a useful parameter for characterizing the physical state of a given soil.
2. There is a qualitative, as well as a quantitative, change in the stress-strain behavior of this clay at some critical value of the soil-water potential; below this critical value specimens manifest a strain-softening response throughout their entire range of compression, whereas above this critical value their behavior is essentially linear up to a few percent axial strain and strain-softening thereafter.
3. The initial moduli generally increase as the soil-water potential increases, but the explicit effect of soil fabric is no clear.
4. The strength increases as the soil-water potential increases, with the specimens possessing a highly oriented fabric exhibiting the highest strength for a given value of the soil-water potential; however, increased density, and not an oriented fabric, may be responsible for the observed strength increase.
5. *Poisson*'s ratio generally increases with increasing values of axial strain; but there is no apparent effect of soil fabric.
6. Considerably more research, both experimental and theoretical, is required before a definitive relationship of the type studied can be deduced.

Summary

Included in this paper are the results of an experimental study directed toward improving our understanding of the complex relationship between the stress-strain behavior of clay and its initial fabric and soil-water potential. Three blocks of a kaolinite clay were consolidated from a slurry under controlled conditions of pore fluid chemistry and stress path, and the fabrics of these samples were evaluated by use of scanning electron microscopy, optical microscopy, and *X*-ray diffraction. Upon completion of consolidation, test specimens trimmed from these samples were placed under air pressure in an extractor and brought to equilibrium at soil-water potentials up to 15 bars. Fabric changes during this phase are expressed in terms of changes in dry density. Then, these specimens were subjected to a constant strain rate uniaxial compression test, and empirical relationships among the undrained modulus, *Poisson*'s ratio, strength parameter, initial fabric, and soil-water potential at the beginning of the test are presented.

Acknowledgements

This study was supported in part by the National Science Foundation under Grant GK-18945. *Kanwarjit S. Chawla* performed the X-ray diffraction work reported herein, and *Kutay I. Ozaydin* and *Roger Kamm* helped with various phases of the testing program. The frequent advice of Dr. *Raymond N. Yong* and Dr. *Donald E. Sheeran* is sincerely appreciated.

References

Baker, D. W., H. R. Wenk, and *J. M. Christie,* J. Geology **77**, 144–172 (1969).

Barden, L. and *G. Sides,* Géotechnique **21**, 211–222 (1971).

Foster, R. H. and *De, P. K.,* Clays and Clay Minerals **19**, 31–47 (1971).

Sheeran, D. E. and *R. J. Krizek,* J. Materials **6**, 356–373 (1971).

Richards, L. A., Physical Condition of Water in Soil. Methods of Soil Analysis, Physical and Mineralogical Properties, Including Statistics of Measurement and Sampling, Monograph Number 9. Amer. Soc. Agronomy and Amer. Soc. Testing and Materials (Editor-in-Chief, *C. A. Black*), pp. 128–152 (Philadelphia 1965).

Krizek, R. J., Strain-Rate Response of a Bangkok Clay. Proceedings of the Third Asian Regional Conference on Soil Mechanics and Foundation Engineering, Volume 1, Haifa, Israel, pp. 289–292 (1967).

Yong, R. N., R. D. Japp, and *G. How,* Shear Strength of Partially Saturated Clays. Proceedings of the Fourth Asian Regional Conference on Soil Mechanics and Foundation Engineering, Volume 1, Bangkok, Thailand, pp. 183–187 (1971).

Yoshinari, M., Compressive Strength vs. Water Content Curve of Undisturbed and Remoulded Soils. Proceedings of the Third Asian Regional Conference on Soil Mechanics and Foundation Engineering, Volume 1, Haifa, Israel, pp. 327–330.

Author's address:

Dr. *Raymond J. Krizek*
Dept. of Civil Engineering
The Technological Institute, Northwestern University
Evanston, Ill. 60201 (USA)

Rheol. Acta **13**, 814–829 (1974)

From the Department of Chemical Engineering and Materials Science, University of Minnesota, Minneapolis, Minnesota 55455 (USA)

Dynamic mechanical measurements with the eccentric rotating disks flow

By C. W. Macosko) and W. M. Davis*

With 19 figures and 5 tables

(Received October 27, 1972)

Introduction

Sinusoidal shearing is a common method of investigating the dynamic behavior of materials. The rate of deformation and the resulting forces monitored in such oscillatory tests are time dependent. The eccentric rotating disks (ERD) or orthogonal rheometer is a geometry that has the ability to characterize the dynamic properties of materials by employing a deformation that is steady in time with resulting steady forces. This eliminates the need for separating the in-phase and out-of-phase components of the sinusoidal stress monitored in oscillatory tests.

In the ERD the sample is placed between two parallel disks. Fig. 1a presents a side view of the ERD with a stationary cartesian coordinate system originating from the center of the lower disk. The center of the upper disk is displaced by an amount, a, in the y direction and the disks are separated by a height, h. One disk is driven at a constant angular velocity, Ω, and the other is assumed to follow at the same Ω.

Fig. 1b shows the particle pathlines as viewed from above. Going from right to left, the first circular pathline represents a particle on the upper disk, the second circle is on a plane at $z = h/2$, and the third circle represents a pathline on the lower disk. The material deformation can be visualized by fixing a rectangular coordinate system r, s, z to a particle on the lower disk and observing the relative motion between particles. This can be seen in fig. 1c. The material will be subjected to a constant strain, $\gamma = a/h$, in the convected coordinate system, but will go through a 360° change of direction with each revolution of the disks.

The forces along the principal axes, F_x, F_y and F_z, are measured and are used to evaluate the dynamic moduli, G' and G'', as defined by sinusoidal shear for small deformations.

Gent (1) appears to have first used the ERD geometry in 1960. He correctly recognized the deformation to be uniform throughout the sample and showed for a crosslinked rubber that

$$F_x = \pi R^2 G'' a/h \qquad [1]$$

$$F_y = \pi R^2 G' a/h. \qquad [2]$$

*) Portions of this work were done while the author was in the Department of Chemical Engineering, Princeton University, Princeton, New Jersey (USA).

Gent found the forces to be linear with eccentricity. In 1964 *Mooney* (2) published results for the elastic force in the ERD and also found a region of linear response.

Maxwell and *Chartoff* (3) in 1965 published the first study using polymer melts in the ERD. They reported a region of linear response with eccentricity for all forces along the principal axes. Their extensive

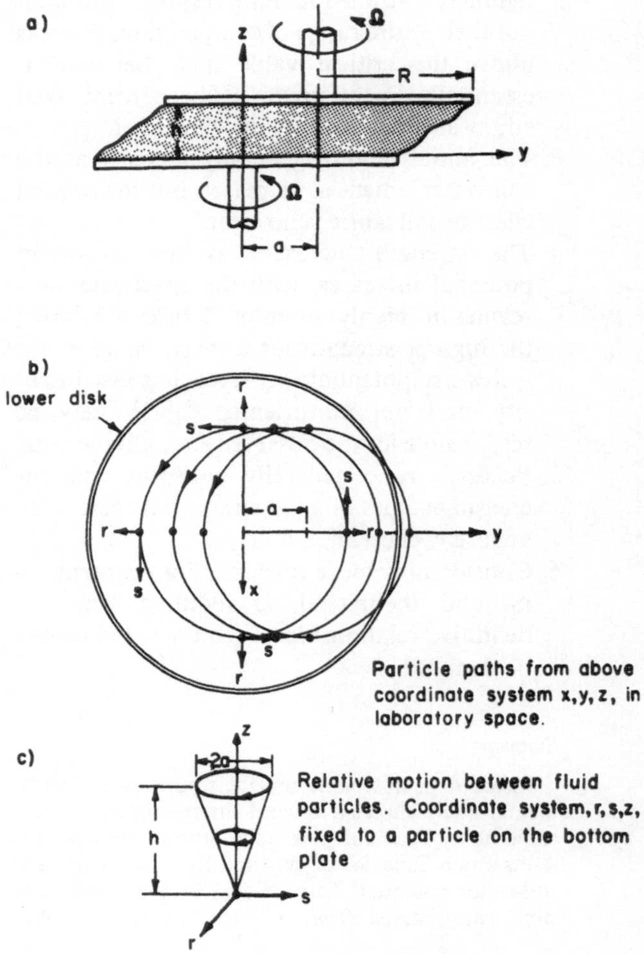

Fig. 1. Motion between eccentric rotating disks shown in stationary and convected coordinates

work with the ERD geometry have led it to be called the *Maxwell* Orthogonal Rheometer (4, 5).

In 1967 *Blyler* and *Kurtz* (6) showed that if inertial and edge effects can be neglected, the kinematics for flow between ERD are

$$v_x = -\Omega y + \gamma\Omega z$$
$$v_y = \Omega x \qquad [3]$$
$$v_z = 0 .$$

Using an *Oldroyd-Maxwell* constitutive equation they arrived at the relations [1] and [2]. This constitutive equation is not very realistic but the relationship to oscillatory shear at small deformations has not been contradicted by subsequent investigations.

Bird and *Harris* (7) used the *Bird-Carreau* model and the kinematics proposed by *Blyler* and *Kurtz*. In order to compare the dynamic moduli to the stresses of the ERD, *Bird* and *Harris* took the limit as the strain approached zero. These are necessary because the *Bird-Carreau* model is nonlinear and the dynamic moduli are defined only in the region of linear viscoelasticity.

Huilgol (8) used the simple fluid and several other models with the same kinematics and obtained similar relations for G' and G''. *Yamamoto* (9) also obtained the correct relationship at small deformations for the ERD forces by transforming between an *Euler*ian and a *Lagrang*ian frame.

In 1970 *Gordon* and *Schowalter* (10) showed that the limiting procedure used by *Bird* and *Harris* is not necessary. Instead they retain only terms that are linear in strain, γ, and show that when linear viscoelasticity holds, eqs. [1] and [2] are correct.

Abbott and *Walters* (11) obtained the velocity field in the ERD flow from the equations of motion for both a *Newton*ian and an integral type, linear viscoelastic constitutive equation. They included inertial effects. The major result of this analysis is that, in the absence of inertia, the kinematics are the same as those of equations [3] and the expressions relating to the dynamic moduli are the same as those obtained previously.

Goldstein (12) has examined both nonlinear viscoelastic and inertial effects in the ERD using a *Bird-Carreau* model. The kinematics resulting from this study differ from those of *Blyler* and *Kurtz* by inertial effects only. Retaining nonlinear viscoelastic effects does not alter the kinematics, but does effect the stresses at large strains.

Recently, we have shown that for mechanical equilibrium there must be some lag between the driven and following disk (13). For $\eta > 20$ poise this lag does not appear to be experimentally significant and the kinematics of eq. [3] hold.

Although there has been considerable interest in the ERD geometry, particularly theoretical, there appears to be a need to test more extensively the relationships between measurements with ERD and small amplitude, sinusoidal shear. This paper examines a wide range of polymeric materials and boundary conditions in an attempt to examine experimentally these relations.

Theory

For linear viscoelastic materials all theoretical treatments reviewed above yield eqs. [1] and [2] which *Gent* (1) and *Maxwell* and *Chartoff* (3) proposed relating F_y and F_x and G' and G''. To illustrate the development of these relations we use the *Bird-Carreau* model, which has been used by several others on the ERD geometry (7, 10, 12, 13, 14). From eqs. [3] we see that the second invariant of the rate of deformation tensor for flow between ERD is $2(\gamma\Omega)^2$. Thus, as *Gordon* and *Schowalter* (10) conclude, for linear viscoelasticity or small strains, the second invariant can be neglected and the *Bird-Carreau* model becomes for linear viscoelasticity:

$$\boldsymbol{T} = -p\boldsymbol{I} + \int_0^\infty M(\theta)\,\boldsymbol{S}(\theta)\,d\theta \qquad [4]$$

where $\theta = t - t'$ (t present time, t' past time), $M(\theta)$ is the memory function, $\boldsymbol{S}(\theta)$ is the strain tensor, p is the pressure, and $\boldsymbol{T}$ is the total stress tensor. The components of the strain tensor are

$$S_{ij} = (1+\varepsilon)\left[\frac{\partial x_i}{\partial X_m}\frac{\partial x_j}{\partial X_m} - I_{ij}\right] - \varepsilon\left[I_{ij} - \frac{\partial X_m}{\partial x_i}\frac{\partial X_m}{\partial x_j}\right] \qquad [5]$$

where x_i are the position coordinates at time t and X_i are the position coordinates at time t'. The ratio of the normal stress differences is

$$\varepsilon = \frac{N_2}{N_1} = \frac{(T_{22} - T_{33})}{(T_{11} - T_{22})} .$$

The displacement functions can be readily calculated from eqs. [3]:

$$x = X\cos\Omega\theta - (Y - \gamma Z)\sin\Omega\theta$$
$$y = X\sin\Omega\theta + (Y - \gamma Z)\cos\Omega\theta + \gamma Z$$
$$z = Z .$$

Then by eqs. [4] and [5] the three measurable stresses for the ERD are

$$T_{xz} = \gamma\int_0^\infty M(\theta)\sin\Omega\theta\,d\theta \qquad [6]$$

$$T_{yz} = \gamma\int_0^\infty M(\theta)(1 - \cos\Omega\theta)\,d\theta \qquad [7]$$

$$T_{zz} = -p + 2\varepsilon\gamma^2\int_0^\infty M(\theta)(1 - \cos\Omega\theta)\,d\theta . \qquad [8]$$

Using this same integral model for sinusoidal shear of a linear viscoelastic material, the dynamic shear moduli are

$$G'' = \int_0^\infty M(\theta) \sin \omega\theta \, d\theta$$

and [9]

$$G' = \int_0^\infty M(\theta)(1 - \cos\omega\theta) \, d\theta$$

where ω is the frequency of oscillation.

When $\Omega = \omega$ the two most useful expressions relating the measurable shear stresses to the dynamic moduli are:

$$T_{xz} = \gamma G'' \quad \text{or} \quad T_{xz} = \gamma\omega\eta' \qquad [10]$$

$$T_{yz} = \gamma G' \quad \text{or} \quad T_{yz} = \gamma\omega\eta'' . \qquad [11]$$

Throughout the rest of this work ω and Ω will be considered equivalent.

Experimentally, forces along the principal axes are detected by the instrument. Since we assumed inertial and boundary effects unimportant, the viscous stresses are constant throughout the sample and the forces in the x and y direction are

$$F_x = \int_0^{2\pi} \int_0^R T_{xz} r \, dr \, d\phi = \pi R^2 T_{xz}$$

$$F_y = \int_0^{2\pi} \int_0^R T_{yz} r \, dr \, d\phi = \pi R^2 T_{yz}$$

where ϕ is the angle between the x direction and r, the radial direction from the z axis.

Combining with [10] and [11]

$$F_x = \pi R^2 \gamma\omega\eta' \qquad [12]$$

$$F_y = \pi R^2 \gamma G' \qquad [13]$$

which are just eqs. [1] and [2] proposed by the original workers.

Steady shear data can be related to dynamic shear measurements in the region of linear viscoelasticity by the following relations:

$$\lim_{\omega\to 0} \eta' = \lim_{\dot\gamma \to 0} \eta \qquad [14]$$

$$\lim_{\omega\to 0} G' = \lim_{\dot\gamma\to 0} \frac{N_1}{2} . \qquad [15]$$

Eqs. [14] and [15] allow comparison of ERD results to those from steady shear.

It has been suggested that the total thrust, on the surface of the disks, F_z, can be related to the second normal stress difference (9). This can be seen from eq. [8] which when combined with [7] becomes:

$$T_{zz} = -p + 2\varepsilon\gamma T_{yz} .$$

Then the total thrust will be

$$F_z = \int_0^{2\pi} \int_0^R [T_{zz} + p_a] r \, dr \, d\phi$$

where p_a is the atmospheric pressure.

Since T_{yz} is constant throughout the area of the entire disk and ε should also depend only on the rate of deformation,

$$F_z = 2\varepsilon\pi R^2 \gamma T_{yz} + \pi R^2 p_a - \int_0^{2\pi} \int_0^R p r \, dr \, d\phi . \qquad [16]$$

To solve for p we consider the condition for continuity of stress across the free surface, neglecting surface tension:

$$(-\boldsymbol{n}) \cdot (-p_a \boldsymbol{I}) \cdot (-\boldsymbol{n}) = \boldsymbol{n} \cdot \boldsymbol{T} \cdot \boldsymbol{n} .$$

Since $\boldsymbol{T} = -p\boldsymbol{I} + \boldsymbol{\tau}$ where $\boldsymbol{\tau}$ is the viscous or flow dependent part of the stress tensor, then

$$p = p_a + \boldsymbol{n} \cdot \boldsymbol{\tau} \cdot \boldsymbol{n} . \qquad [17]$$

If we assume the surface to be that of a skewed cylinder (coordinates z, $(R + \gamma z \sin\phi)$ for $z = 0 \to h$ and $\phi = 0 \to 2\pi$ as suggested by fig. 1b), then the surface normal is

$$\boldsymbol{n} = \boldsymbol{i} \cos\phi + \boldsymbol{j} \sin\phi - \boldsymbol{k}\gamma \sin\phi .$$

But p obtained from eq. [17] with this surface does not satisfy the equations of motion using the velocities of eqs. [3]. This suggests that conditions on and near the free surface are more complex. We know that eqs. [3] must change near the edge since shear stresses must go to zero at a free surface. Our observations of the free surface indicate that it is more complex than suggested above. A complete solution considering both an edge layer and the more complex free surface is a formidable problem.

For this paper we have chosen to assume that $p = p_a$, ignoring any flow dependence of p. This results in

$$F_z = \pi R^2 \tau_{zz} = 2\pi R^2 \varepsilon\gamma T_{yz} . \qquad [18]$$

This expression for F_z neglects the normal thrust due to inertia effects. The inertial normal thrust can be modeled by solid body rotation between two *concentric* disks. The resulting pressure distribution, neglecting surface tension, is

$$p - p_a = \frac{\rho}{2}(r^2 - R^2)\omega^2 .$$

The normal thrust contributed by inertia, denoted by F_{z0}, is:

$$F_{z0} = \pi \rho \omega^2 \frac{R^4}{4}. \quad [19]$$

Experimentally we find F_{z0} acts in the $+z$ direction, pulling the disks together, and it is modeled quite well by eq. [19] (see fig. 18). Thus F_{z0} was subtracted from all the experimental total thrust data reported here to give F_z. From this adjusted F_z and with F_y a value of ε can be obtained directly

$$\varepsilon = \frac{F_z}{2\gamma F_y}$$

or

$$\varepsilon = \frac{F_z}{2\pi R^2 \gamma^2 G'}. \quad [20]$$

Another possible approximation for eq. [17] to solve for p is to equate the total forces on each side of the free surface. This yields the balance

$$-2\pi R h p_a = \int_0^h \int_0^{2\pi} [\boldsymbol{n} \cdot \boldsymbol{T} \cdot \boldsymbol{n}(r\, d\phi)\, dz]_{r = R + \gamma z \sin\phi}.$$

Using the model of eq. [4] and considering only terms to lowest order in γ gives $p = p_a + \varepsilon\gamma T_{yz}$. Substituting this into eq. [16] yields

$$F_z = \pi R^2 \varepsilon \gamma T_{yz}$$

or

$$\varepsilon = \frac{F_z}{\gamma F_y}. \quad [21]$$

The following experimental work is aimed at testing eqs. [20], [12], and [13] and to some extent eqs. [14] and [15].

Experimental

A Rheometrics Mechanical Spectrometer was used for all the experimental ERD, cone and plate, and parallel plate data reported here. This instrument is described in detail elsewhere (14, 15).

In the Mechanical Spectrometer, the upper spindle is driven at various constant angular velocities by a D. C. servo system. The bottom spindle can be locked for torsional flows or allowed to rotate freely in an air bearing for the eccentric rotating geometries. The bottom spindle is mounted in a transducer which detects continuously and simultaneously the three orthogonal forces F_x, F_y and F_z acting on the test surface and a torque about the z axis. These forces were recorded on a strip chart recorder.

The general procedure for loading a sample in the ERD, parallel plates or cone and plate modes is to place the sample onto the bottom plate then lower the top plate to the desired gap setting. The plates are rotated while the gap is being set and the excess is trimmed from the edge with a spatula. It was attempted to maintain a straight cylindrical or slightly bulging free surface at the edge of the disks. Several ERD experiments using a larger diameter disk, on the bottom spindle and an "infinite sea" of *Newtonian* liquid resulted in larger forces than predicted by eq. [12].

In the ERD experiments the lower spindle is made concentric to the upper before rotating. After loading, the disks are rotated at about 10 radians/sec. During this rotation the lower disk is positioned by means of the $x - y$ stage so that F_x and F_y are zero.

Data is collected by setting the desired angular velocity, increasing the eccentricity, a, in 3 to 10 equal intervals, returning to a zero, then setting a new Ω and repeating the steps. Fig. 2 shows a typical recorder trace for the F_y and F_z forces developed in flow between ERD for a polyisobutylene solution. The large oscillations in normal thrust are typical. The indicated average values were used. Both F_y and F_x forces are similar with much less oscillation. The amplitude of the oscillations grows somewhat with increasing strains as shown.

Fig. 3 illustrates ERD data reduced from recorder traces to a plot of stress versus strain, a/h. All three stresses are shown. The slopes of T_{xz} and T_{yz} do not change with direction of Ω or sign of the eccentricity. This verifies the resolving ability of the transducer and the sign dependence in eqs. [6], [7] and [8].

Experiments where one member of the experimental fixture is tilted can also be performed with the Mechanical Spectrometer. A vernier scale indicating the angle of tilt is attached to the frame that supports the lower member. Then the lower member can be tilted to the desired angle with respect to the upper member. An experiment is presented in this paper involving the tilting of one of the rotating disks. For

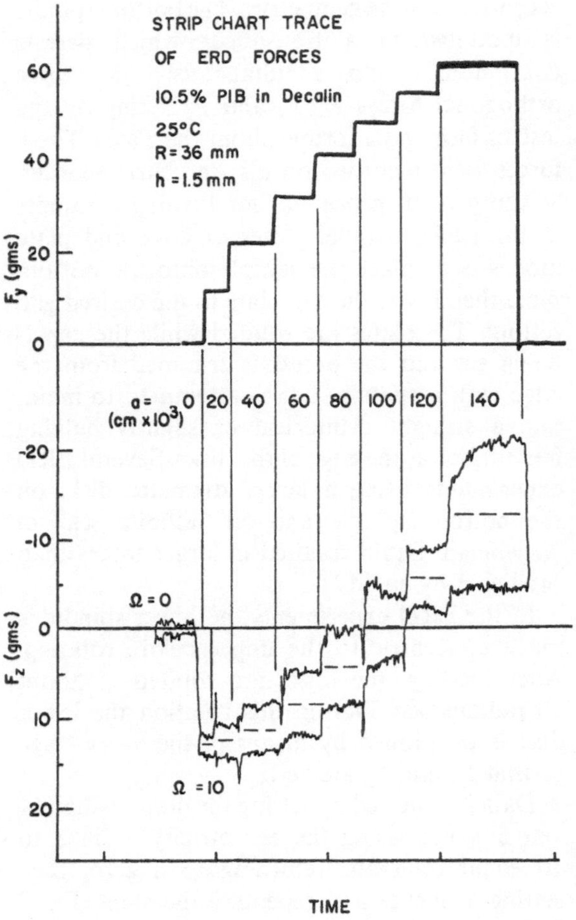

Fig. 2. Typical potentiometric strip chart recorder output of ERD data. F_y and F_z shown for 10.5% PIB in decalin at 10 rad/sec

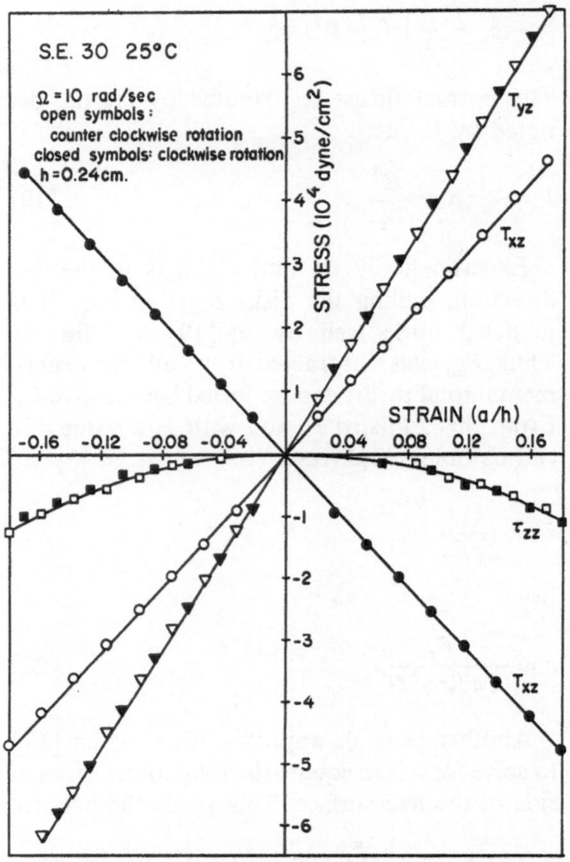

Fig. 3. The effect of changing direction of rotation and eccentricity on the three measurable stresses for S.E.30

normal ERD operation disk parallelness was maintained within $\pm 5\,\mu$m.

In the cone and plate experiments 0.040 radian cones were used. The usual analysis of this and parallel plate torsion flow was used (e.g. ref. 20).

Sample temperature was maintained within $\pm 0.5\,°$C of set point in a forced convection environmental chamber.

Materials

A wide range of materials were used in this study. The are tabulated below (table 1).

Table 1

Sample	Polymer	M_w^*	η_0 (poise)	Temp.	Source
Silicone-2	polydimethylsiloxane	~96 000	560	25 °C	Brookfield
Silicone-3	polydimethylsiloxane	~110 000	870	25 °C	Brookfield
Silicone-4	polydimethylsiloxane	~390 000	8.4×10^4	25 °C	Gen. Elec. (16)
SE-30	polydimethylsiloxane	~520 000	2.1×10^5	25 °C	Gen. Elec. (17)
Marlex 6050	polyethylene	72 000	$\sim 7 \times 10^4$	190 °C	Phillips (18)
Styron 666	polystyrene	237 400	2.4×10^5	190 °C	Dow (19)
PIB-10.5	polyisobutylene 10.5% by wt. in decalin		660	25 °C	Metzner (20)
PIB-6.8	polyisobutylene 6.8 by wt. in cetane	$\sim 10^6$	32	26 °C	Williams (21)
Galcit I	crosslinked polyurethane	∞	shear modulus 1.2×10^7 dyne/cm²	25 °C	Landel (22)

* Provided by the source or estimated from η_0.

The linear polyethylene and polystyrene are commercial thermoplastics. Silicone-2 and -3 are *Brookfield* viscosity standards. The other materials were provided to us by the referenced laboratories for this work. The tabulated viscosities are limiting values at low shear rate from cone and plate flow.

The thermoplastic polymer samples were vacuum compression molded into 55 mm diameter disks. It was found particularly necessary to mold the polystyrene under vacuum to obtain reproducible results. The other samples were allowed to settle on the lower disk until the air bubbles had risen out. The urethane elastomer was cut into disks from a received cross-linked sheet.

Inertial and boundary effects

In the analysis of flow between eccentric rotating disks two important assumptions were made: negligible inertial and boundary effects. These were tested experimentally.

Abbott and *Walters* (11) have considered inertial effects theoretically between ERD for a *Newton*ian fluid. Calculations using their solution show that inertia has a large effect only at such large plate separations that the material cannot stay between the disks. Table 2 shows some of the calculations. We find that, for example, a plate separation of 1.0 mm or less is required for a 10 poise silicone fluid to remain between 36 mm radius disks.

Table 2. Theoretical effect of inertia on η' for *Newton*ian fluids in ERD, 25 rad/sec

True viscosity		10 poise	100 poise	1000 poise
	10 mm	2.81	88.6	987.9
expected η' measurement for various h values	5 mm	7.37	97.0	997.0
	1 mm	9.58	99.9	999.9

This indicates that we should not expect inertia to be important. However, *Abbott* and *Walters* neglect any z component velocity and boundary effects in their analysis.

Inertial and boundary effects might be expected to appear in experimental data if the a, h, or R are varied over a fairly large range.

In fig. 4 we see the raw data of fig. 2 reduced and plotted as stress vs. strain according to eqs. [12] and [13] and vs. γ^2 by eq. [18]. The results are typical of all the materials examined here.

Eqs. [12] and [13] predicts that F_x and F_y are linear functions of the eccentricity, since G' and G'' should only be functions of frequency. It has been found experimentally, as shown in fig. 4, that this region of linear response exists from zero eccentricity to some strain where the forces tend to fall away. *Gross* & *Maxwell* (23) have found that for many polymer melts the limit of linear response in ERD is independent of frequency and occurs at about 50% strain. It has also been observed by both *Gross* and the authors that the elastic force, F_y, goes nonlinear at smaller strain than the viscous force, F_x.

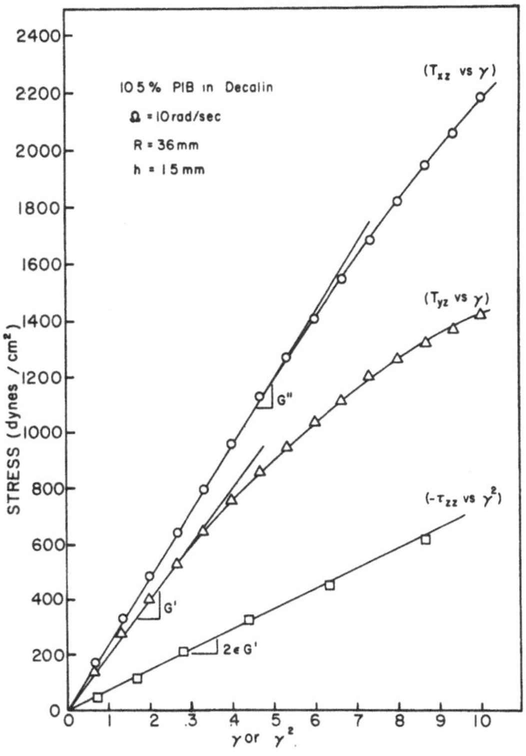

Fig. 4. ERD data for 10.5% PIB in decalin reduced by eqs. [12] and [13] to stress vs. strain. Normal force reduced by eq. [18] to give the deviatoric normal stress vs. γ^2

Goldstein (12) has shown theoretically the departure from linear response to be due to nonlinear viscoelastic effects. His results show that nonlinear behavior for sinusoidal shear and flow between ERD are different.

All data used in the remainder of this study were taken in the region of linear stress-strain response, $a/h < 0.5$. Work is in progress on the large strain region.

By eqs. [12] and [13] the forces should be inversely proportional to sample thickness, h. This was tested by plotting T_{xz}/a and T_{yz}/a vs. h^{-1}. Figs. 5 and 6 show these results and are typical

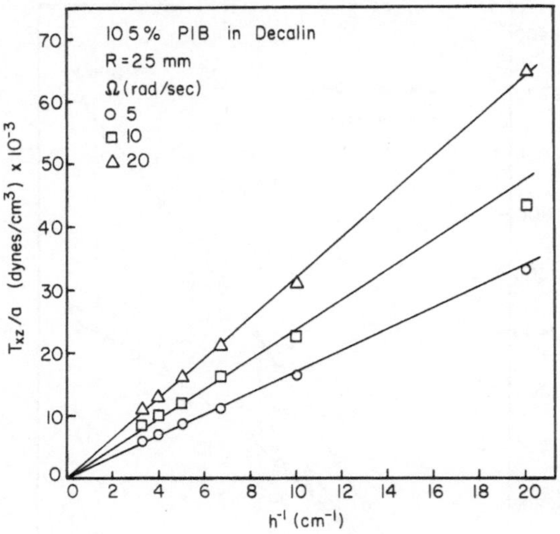

Fig. 5a. Dependence of the viscous stress on sample thickness for PIB in decalin

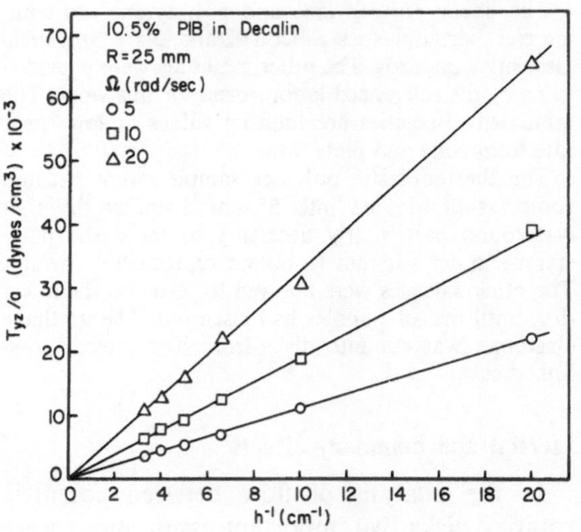

Fig. 5b. Dependence of the elastic stress on sample thickness for PIB in decalin

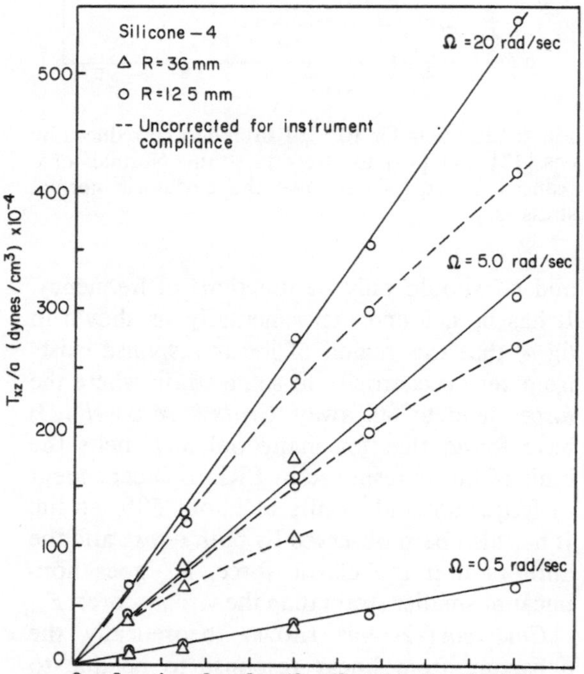

Fig. 6a. Dependence of the viscous stress on sample thickness for Silicone-4

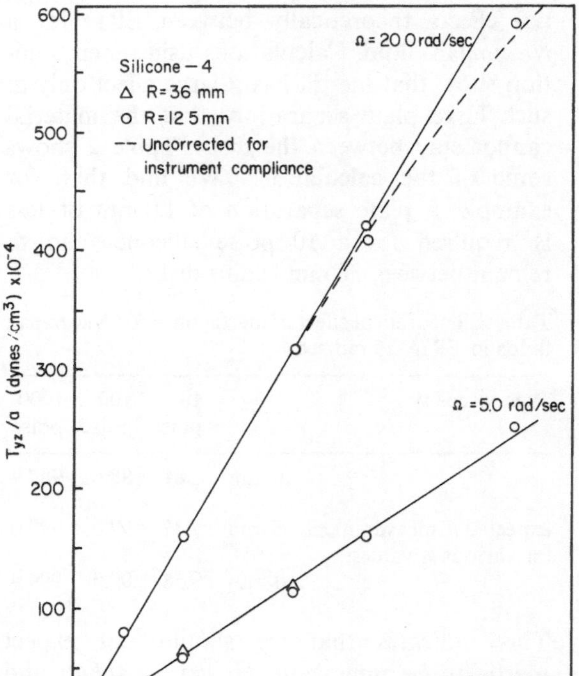

Fig. 6b. Dependence of the elastic stress on sample thickness for Silicone-4

of all the materials examined. As can be seen there appears to be no thickness dependence. Initially at small h, an effect on T_{xz} for high viscosity materials was found (14), however this can be accounted for by considering instrument compliance. The uncorrected data are indicated by dashed lines in figs. 6a and 6b. Details of the compliance corrections are described in the Appendix.

The effect of sample radius was also examined for the same materials. Fig. 7 shows $F_x h/a$ plotted vs. area. This and the data in table 3 and

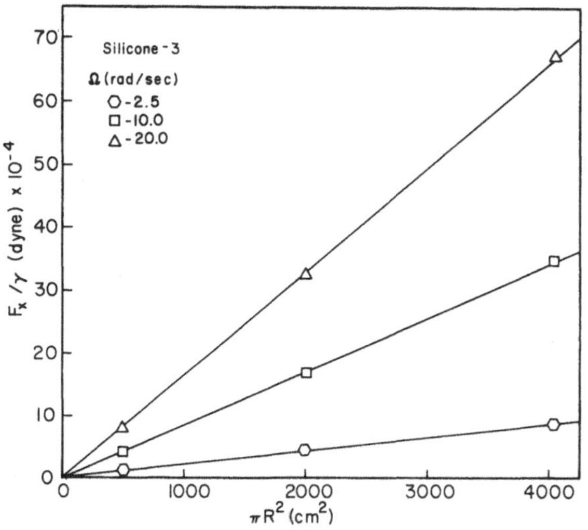

Fig. 7. Effect of sample radius on the viscous force for Silicone-3

Table 3. Radius effect

Silicone-3

Ω Rad/sec	η' poise		
Radius	12.5 mm	25 mm	36 mm
2.5	888	855	860
5.0	812	850	876
7.5	805	836	845
10.0	788	850	854
15.0	789	864	854
20.0	774	830	820
25.0	774	790	–

10.5% PIB in Decalin

Ω Rad/sec	η' poise		G' dyne/cm^2	
Radius	25 mm	36 mm	25 mm	36 mm
2.5	455	440	594	569
5.0	337	342	1 098	1,090
7.5	279	280	1 572	1,580
10.0	236	238	1 899	1,910
15.0	193	182	2 640	2,680
20.0	160	153	3 276	3,370
25.0	134	131	3 690	3,850

Silicone-4 (h = 4.0 mm)

Ω Rad/sec	η' poise × 10^{-4}		G' dyne/cm^2 × 10^{-4}	
Radius	12.5 mm	36 mm	12.5 mm	36 mm
0.5	7.16	7.12	.83	1.04
1.0	6.05	5.91	2.34	2.50
5.0	3.16	3.36	11.9	12.61
10.0	2.26	2.25	20.4	21.2

figs. 6a and 6b indicate no significant radius effect. Results with the 12.5 disks deviate somewhat for Silicone-3 and -4. This may be due to the smaller forces and thus less accuracy in these data.

Possible misalignment of one of the plates was also evaluated experimentally with Silicone-3. The lower disk was tilted at a desired angle and the experiment performed as if the disks were parallel. A significant change in the x, y position of the lower plate was required to cancel the forces generated by tilting. Table 4 presents a comparison of η' obtained at various angles of misalignment to the results for parallel disks. There appear to be some effects but they are less than $\pm 5\%$ and no definite trend emerges. However, it is clear that small unparallelness of the two disks can cause considerable zero shift in the measured shear forces.

Table 4. Misaligned plate effect on ERD η' results for Silicone-3

Ω Rad/sec	Tilt angle of lower disk 0′	3′	4′	5′
5.0	850	894	888	873
7.5	836	878	865	854
10.0	850	862	854	834
15.0	864	857	850	850
20.0	830	840	846	820
25.0	790	827	820	812

h = 2.0 mm R = 25 mm

We have also conducted some flow visualization experiments. Small vertical holes were drilled into 6 mm thick polystyrene disks and filled with 10 μm diameter carborundum particles. These samples were placed in the Mechanical Spectrometer with the spindles concentric, heated to 190 °C and rotated at 1 rad/sec. The axes were offset for several revolutions, then aligned again. The samples were quenched in situ, removed intact, sectioned and examined.

If there were no other motion than that of eqs. [3] then after this procedure the tracer lines should appear unchanged. Fig. 8a shows the results for 10 revolutions at $\gamma = 0.2$ then returned to $\gamma = 0.0$. The only departure is a slight bowing of the outer lines attributed to some compressing during loading. Fig. 8b is a top view after 700 revolutions. The slight twist in the tracer lines is due to a small lag, about 0.01%, in the velocity of the lower spindle.

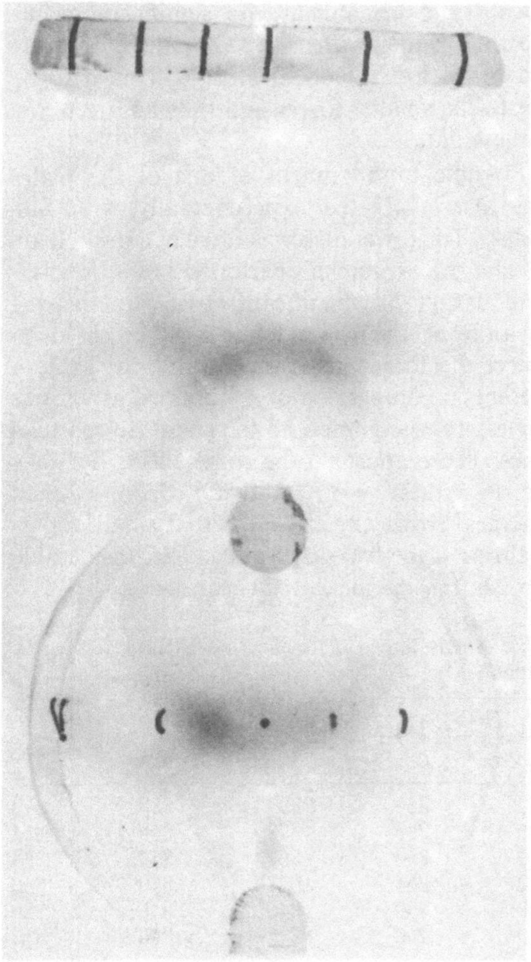

Fig. 8. Tracer filled, frozen polystyrene disks after flow between ERD

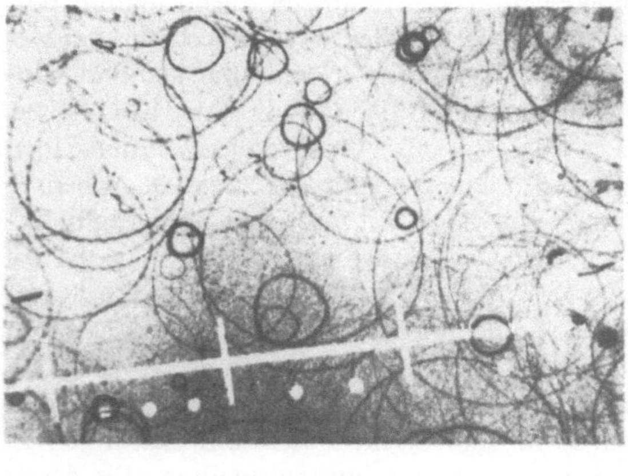

Fig. 9. Time lapse photographs taken with a camera mounted on one of the disks. The radial scale starts at the center of rotation and is marked in 0.25 in. intervals. For this experiment $h = 0.050$ in, $a = 0.200$ in and $\Omega = 20$ rad/sec (ref. 24)

More elegant flow visualization experiments have recently been conducted with a 100 poise silicone fluid by *Gillman* (24). The upper disk was made of glass and mounted with a camera. Thus the camera is in a convected frame and under time lapse photography tracer particles should appear as circular paths with diameters dependent on z position, as shown in fig. 1c. Fig. 9 is typical of his results. He found good agreement between calculations from eqs. [3] and the experimental velocity profile determined from the circle perimeters and their z coordinate. The slight asymmetry of the circles was attributed to the gears used. *Gillman* found significant departures when one disk axis was tilted. The circles no longer closed.

Comparisons of ERD to other flows

Fig. 10 plots η' for Silicone-2 and -3 versus frequency. Each point on this plot represents an average η' determined from plots like fig. 4, using $5-10$ eccentricity values; $5-10$ thicknesses and $1-2$ radii. Thus each is an average of 25 to over 100 data points. The average deviation of these data was found to be $\pm 3\%$ from the mean for Silicone-2, -3, and also PIB -10.5.

The solid lines in fig. 10 are the steady shear viscosities from cone and plate. This was found constant at the value indicated for $\dot{\gamma} < 25\,\text{sec}^{-1}$. Agreement between η' by ERD and η_0 is within the average deviation of the data. At large Ω, particularly for Silicone-3, η' drops. However, there is significant elasticity in Silicone-3 at $\Omega > 10$ rad/sec.

Figs. $11-14$ compare ERD data directly to other methods for determining η' and G'. Fig. 11 shows results for SE-30. All data were taken with the same sample of this polydimethyl-

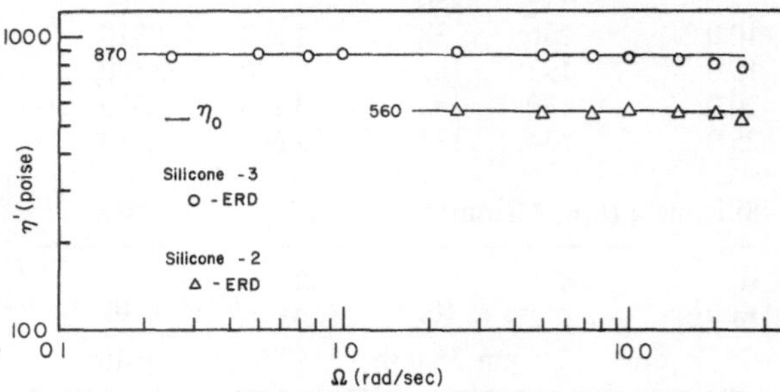

Fig. 10. Dynamic viscosity for Silicone-3 and Silicone-4 from flow between eccentric rotating disks. The solid lines are the steady shear viscosity as determined in cone and plate flow

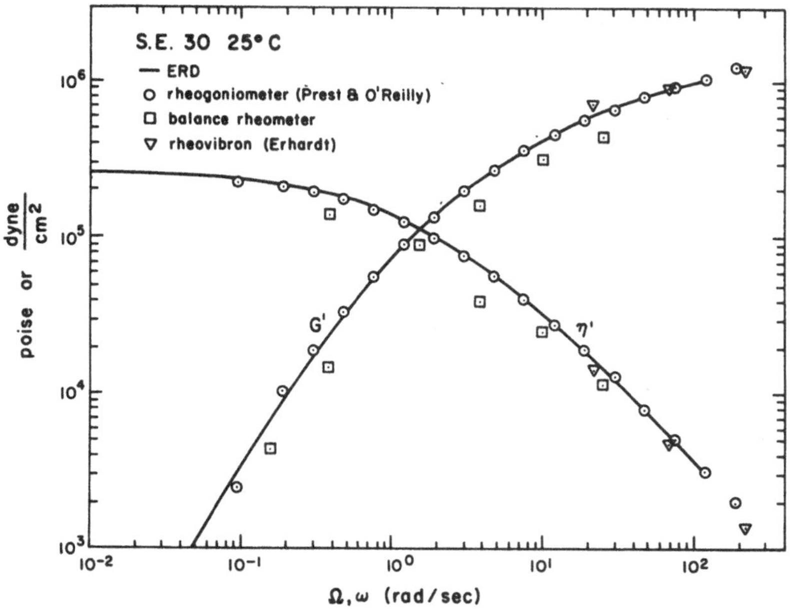

Fig. 11. Comparison of dynamic viscosity, η', and elasticity, G', in four flow geometries for SE-30. Solid lines ERD, points include oscillating cone and plate in a *Weissenberg* rheogoniometer (17), eccentric rotating hemispheres in a Contraves balance rheometer and oscillating shear plates in a Rheovibron (26)

siloxane. The solid line is the ERD data. Sinusoidal oscillating cone and plate data of *Prest* and *O'Reilly* (18) using the *Weissenberg* rheogoniometer lie on the line within the accuracy of ERD data, $\pm 3\%$, except at low frequency. This may be due to difficulty with the rheogoniometer in resolving the phase angle between stress and strain at low frequency.

Erhardt (25) has added shear plates to the Rheovibron enabling dynamic measurements in oscillatory shear. Though he reports some inaccuracy in determining sample thickness, his measurements on SE-30 are within 10% of the ERD and oscillating cone and plate data (26).

Flow between tilted rotating hemisphere TRH, has been described by *Kepes* (27) and *Kaelble* (28). The data shown in figs. 11 and 13 were taken with a demonstration model of an instrument which uses the TRH geometry, the

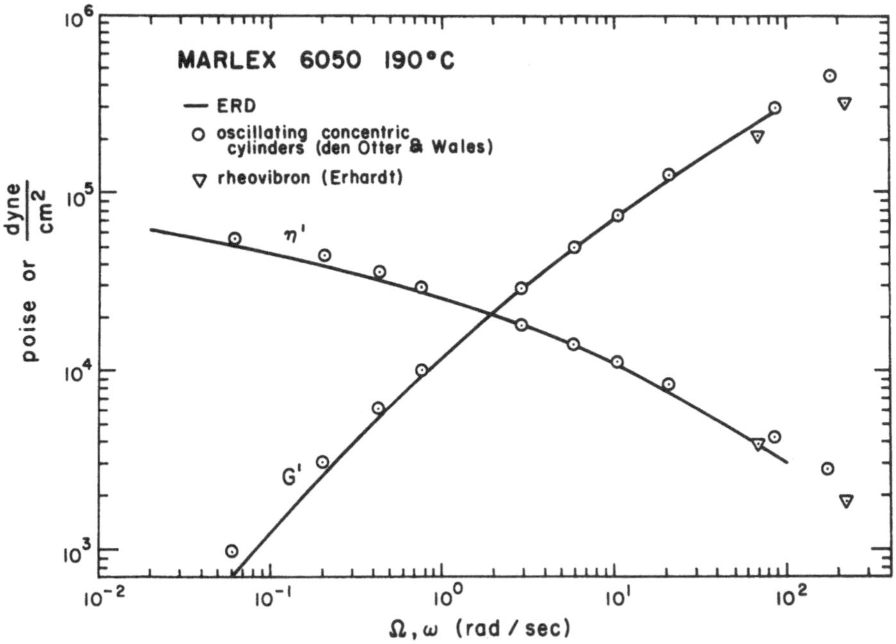

Fig. 12. Comparison of η' and G' for Marlex 6050 at 190 °C. Solid lines ERD, points are for oscillating concentric cylinders (31) and oscillating shear plates in a Rheovibron (26)

53*

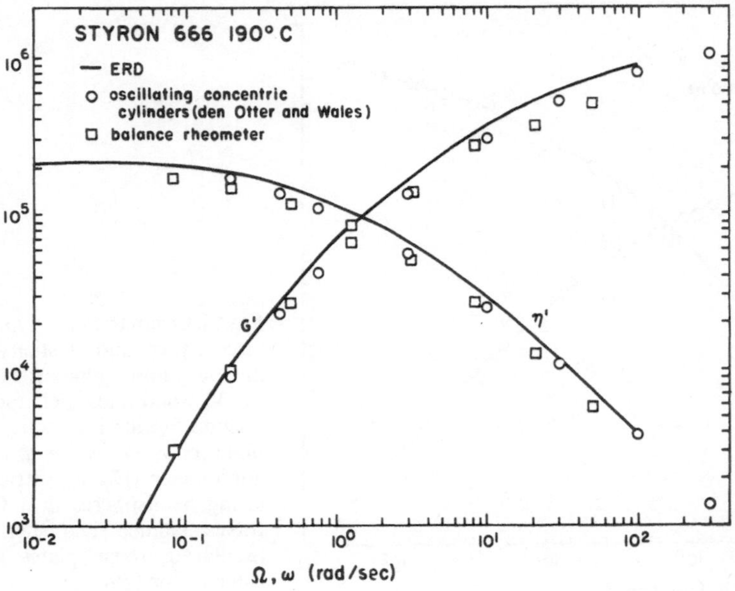

Fig. 13. Comparison of η' and G' for Styron 666 at 190°C. Solid lines ERD, points are for oscillating concentric cylinders (31) and eccentric rotating hemispheres in a Contraves balance rheometer

Balance Rheometer by Contraves of Zurich. Under such experimental conditions only qualitative results can be expected. These data appear to be the first tests of the TRH analyses (9, 29, 30). Further work with the TRH is described elsewhere (33).

Figs. 12 and 13 compare ERD results on two commercial thermoplastics to the literature values of *Wales* and *den Otter* for oscillating concentric cylinders (31). The samples are from different batches of these polymers. Both sets of data are within 10% of the ERD results, probably within the batch to batch variation. The Styron 666 data of the oscillating cylinders, and the TRH appear uniformly lower, suggesting a lower molecular weight or higher test temperature.

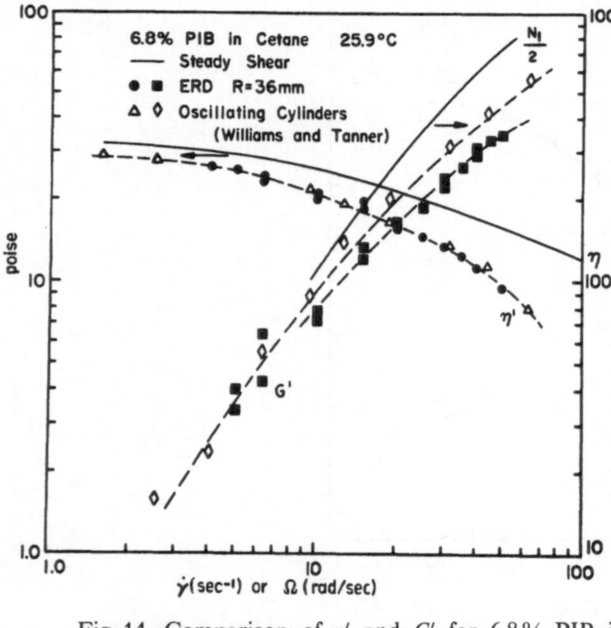

Fig. 14. Comparison of η' and G' for 6.8% PIB in cetane (21)

A polymer solution, 6.8% PIB in cetane, was studied to verify ERD flow and the Mechanical Spectrometer for much lower viscosity materials. The sample used was the same as that studied by *Williams* (21) in an oscillatory annular pumping geometry. His results are compared to ours with the ERD in fig. 14. The agreement for η' is good. However at higher ω or Ω the G' data from the ERD is consistently lower. An error of about 15% can be expected for both instruments on this data.

Figs. 14–17 compare ERD data to steady shear measurements providing a test of eqs. [14] and [15]. They appear in good agreement with the data. Data points for Marlex 6050 are not shown but lie within $\pm 3\%$ of the lines. They are available in tabular form along with similar comparisons of Styron 666 and SE-30 ERD results to steady shear (14).

Normal force in ERD

When the disks are concentric and rotating at the same frequency, inertial effects predict a normal force in the positive z direction, F_{z0}, as shown in eq. [19]. Fig. 18 shows excellent agreement between the theoretical and experi-

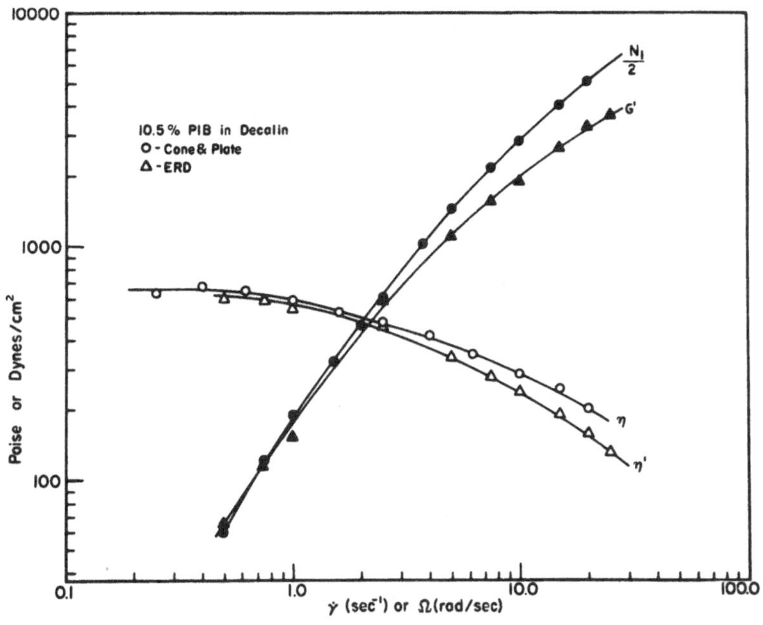

Fig. 15. Comparison of η' and G' from ERD to steady shear viscosity, η and half the first normal stress difference, $N_1/2$, from cone and plate for PIB in decalin

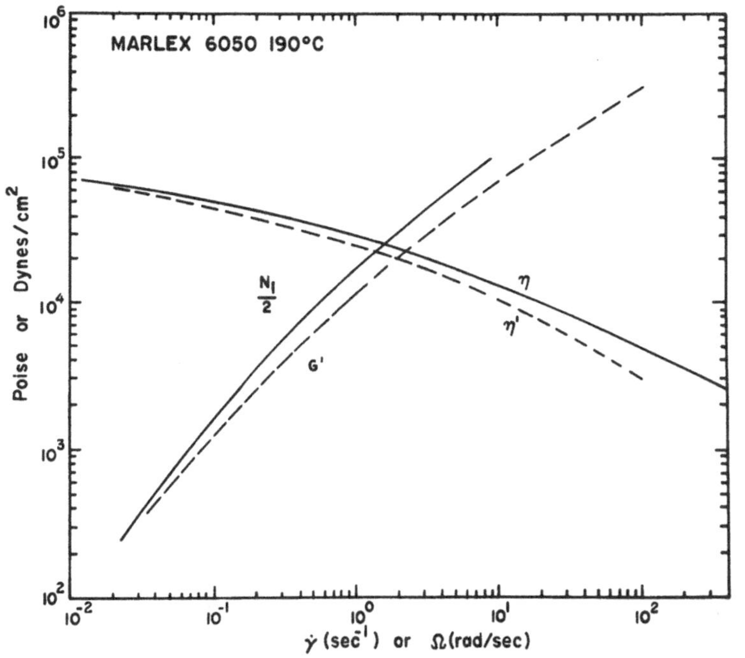

Fig. 16. Comparison of dynamic data from ERD to steady shear from cone and plate and capillary flow for Marlex 6050

mental total normal force for concentric rotating plates. The normal force contributed by the viscous stress is measured as the increase or decrease of the total normal force from F_{z0}. Since ε is negative for most materials, the normal force contributed by the viscous stress should be negative, or will have a tendency to "push" the plates apart. This is the observed behavior for all the materials studied.

Table 5 shows comparisons of ε obtained from ERD measurements using eq. [20] to ε obtained by various other methods. The PIB solution which is compared to results obtained by *Ginn* and *Metzner* (20) is the same one that was used by them. Their results were obtained using torsional flow with cone and plate and parallel flat plate normal force methods and are compared to values obtained on the Me-

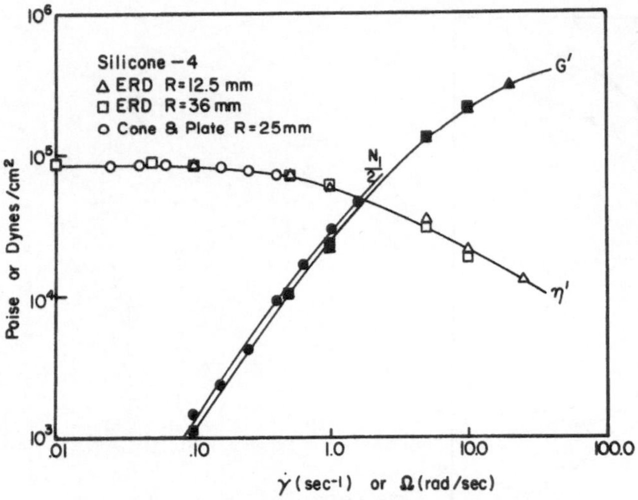

Fig. 17. Comparison of dynamic data from ERD to steady shear from cone and plate flow for Silicone-4

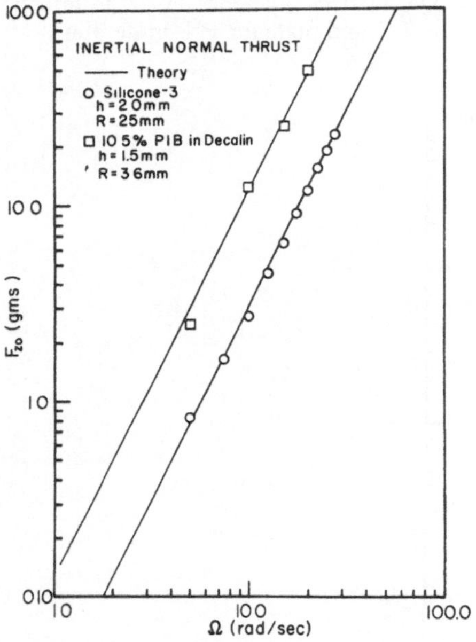

Fig. 18. Normal thrust generated between two concentric rotating disks, F_{z0}. Theory from eq. [19]; data for Silicone-3 and PIB in decalin

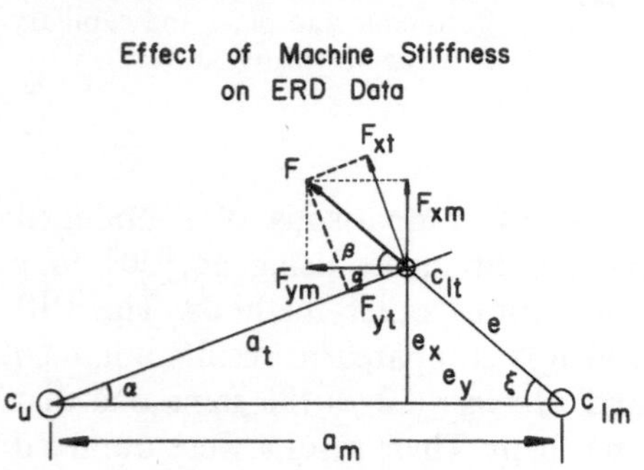

Fig. 19. Effect instrument compliance on the relative displacement between the centers of the upper and lower disks, a_t is the true displacement in the sample

chanical Spectrometer using the same normal thrust method (CP & PP). The birefringence ε value for Marlex 6050 was calculated by subtracting two normal stress differences from the literature (31, 32). The accuracy of these methods is not particularly good since they rely on the difference of two large numbers and numerous assumptions.

Table 5. Second normal stress ratio, $\varepsilon = \dfrac{T_{22} - T_{33}}{T_{11} - T_{22}}$

	10.5% PIB in Decalin		
Ω or $\dot{\gamma}$ sec	$-\varepsilon$ ERD	$-\varepsilon$ CP & PP	$-\varepsilon$ CP & PP (ref. 20)
5.0	.19	.45	.29
10.0	.18	.36	.34
15.0	.29	.26	.35
	Marlex 6050		
	$-\varepsilon$ ERD	$-\varepsilon$ CP & PP	$-\varepsilon$ Biref. (ref. 31, 32)
.5	.28	.38	.45
1.0	.35	.37	.40
2.0	.40	.33	.35
5.0	.42	.24	.28
	S. E. 30 Polydimethyl Siloxane		
	$-\varepsilon$ ERD	$-\varepsilon$ CP & PP	
.2	.39	.34	
1.0	.30	.39	

Note that if eq. [21] is used the ERD values would be twice that shown in table 5.

The ε values calculated by either approximation are negative. Those by eq. [20] are in better quantitative agreement with the other methods than those by eq. [21]. The large oscillations in F_z, as shown in fig. 2, are as yet unexplained. Further work both on the analysis and experiments should be done.

Conclusions

We have found that the stresses generated by the flow between eccentric rotating disks generally behave as expected from theory which neglects inertial and boundary effects. However, it is necessary to consider the concentric rotating disks inertial term, F_{z0}, in evaluating ε from the normal thrust data. Furthermore, T_{yz}, the elastic stress component is observed to be nonlinear in strain for strains above about 50%. With highly viscous materials corrections for instrument compliance may sometimes be necessary.

For three melts and one polymer solution studied, the ERD appears to measure the fundamental dynamic shear functions, η' and G', to within the accuracy obtainable by other dynamic shear techniques. The η' and G' measured by ERD go to the predicted limiting values of the steady shear functions η' and $N_1/2$.

The ERD normal force may offer a method for determining ε, the ratio of second to first normal stress difference. It gives negative ε values for the materials examined, however further work seems necessary for quantitative use of F_z measurements.

Acknowledgement

The comparisons presented here would not have been possible without the generous cooperation of the laboratories and individuals referenced here who provided samples and data. Part of this study was conducted at Princeton University. CWM is grateful to Professor *Bryce Maxwell* for introducing him to this problem and acknowledges support from the National Science Foundation and ICI America. Acknowledgement is made to the Donors of the Petroleum Research Fund, administered by the American Chemical Society and to the Union Carbide Corporation for support of the work at Minnesota. We appreciate helpful discussions with *C. Goldstein* and *R. L. Fosdick* and suggestions from reviewers.

Appendix

Effect of instrument compliance on ERD data

In order to detect forces a transducer must deflect. With large enough forces these deflections can significantly alter the eccentricity in eccentric rotating disk experiments. The effects of instrument compliance appear readily accountable by simple geometry considerations.

Fig. 19 illustrates schematically the effect of instrument deflections in both x and y directions. We will only consider the two dimensional problem here. The effect of z deflections and tilt should be significantly smaller.

In operating the Rheometrics Mechanical Spectrometer the center of the lower disk is usually displaced in the y direction an amount a_m with a micrometer. The flow generates a force, F. This F, if large enough, can change the displacement of the center of the lower disk, c_l, with respect to the upper disk, c_u. The displacement from c_{lm} to c_{lt} will have components e_x and e_y. The transducer will still resolve F into components along the coordinate directions, F_{xm} and F_{ym}. Thus if the overall instrument compliances are K_x and K_y in each direction, then

$$e_x = K_x F_{xm}$$
$$e_y = K_y F_{ym}. \qquad [A1]$$

These error terms cause the true displacement in the material to be a_t, where

$$a_t^2 = e_x^2 + (a_m - e_y)^2$$

combining with [A1]

$$a_t^2 = a_m^2 - 2a_m K_y F_{ym} + K_y^2 F_{ym}^2 + K_x^2 F_{xm}^2. \qquad [A2]$$

The components of F of interest for analyzing the flow between ERD are F_{yt} along a_t and F_{xt} perpendicular. From fig. 19

$$F_{yt} = F\cos(\alpha + \beta) = F_{ym}\cos(\alpha + \beta)/\cos\beta$$
$$= F_{ym}\cos\alpha - \sin\alpha\tan\beta$$
$$= F_{ym}\left[\frac{a_m - e_y}{a_t} - \frac{e_x}{a_t}\frac{F_{xm}}{F_{ym}}\right]$$
$$F_{yt} = \frac{F_{ym}}{a_t}\left[a_m - K_y F_{ym} - K_x F_x F_{xm}^2/F_{ym}\right]. \qquad [A3]$$

Similarly

$$F_{xt} = \frac{F_{xm}}{a_t}\left[a_m - F_{ym}(K_x - K_y)\right]. \qquad [A4]$$

Thus with values of the instrument compliance K_x and K_y, the measured data can be corrected using eqs. [A2], [A3] and [A4] to find the true forces and displacements. K_x and K_y are most simply measured by carefully pressing the two disks together under several kg force without a sample and without rotation. Displacing the lower disk in small increments a_x and then a_y will give the compliance directly; $K_x = a_x/F_{xm}$. These values can be checked using highly viscous or elastomer samples of different thickness. For thick samples the effect of instrument compliance is reduced, eqs. [A5] and [A6].

It may be helpful to experimenters to point out that the way ERD data is normally collected will not make the compliance error apparent. F_{xm} and F_{ym} will increase linearly with small a_m. Changing the direction of rotation or a_m will also not reveal compliance. Only when data at various thicknesses or radii are collected as in figs. 6a and 6b or when results are compared to other measures of η' and G' will compliance errors be revealed.

Eqs. [A3] and [A4] can be combined with eqs. [12] and [13] to give

$$\frac{\eta'_m}{\eta'_t} = \frac{(1 - K_y G'_m A/h)^2 + (K_x \Omega \eta'_m A/h)^2}{1 - G'_m (K_x - K_y) A/h} \qquad [A5]$$

$$\frac{G'_m}{G'_t} = \frac{(1 - K_y G'_m A/h)^2 + (K_x \Omega \eta'_m A/h)^2}{1 - K_y G'_m A/h - K_x \Omega^2 \eta'^2_m (A/h)/G'_m} \qquad [A6]$$

where $A = \pi R^2$. Since the Mechanical Spectrometer mechanically symmetrical, $K_x \simeq K_y = K$ and these equations can be simplified.

Eqs. [A5] and [A6] can be used to determine limits of operation for test fixture sizes and ranges of material. If we desire compliance errors to be within our experimental error for ERD data, $\pm 3\%$, then

$$1.00 \geq \eta'_m/\eta'_t \geq 0.97$$
$$1.03 \geq G'_m/G'_t \geq 0.97\,.$$

Instrument errors will be within the overall experimental error if

$$\Omega \eta'_m = G''_m \leq 1.7\, G'_m \qquad [A7]$$

and

$$G'_m \leq \frac{0.015\, h}{A K}\,. \qquad [A8]$$

For the Mechanical Spectrometer, model KMS 71-C, used here

$K = 4.8 \times 10^{-9}$ cm/dyne

$h > 0.2$ cm

$R_{\min} = 1.25$ cm.

Thus for $G'_m < 2 \times 10^5$ dyne/cm^2, compliance errors can be neglected. It appears that G' is controlling since, for real materials with large G', such as polymer melts and elastomers, eq. [A7] generally holds. With this in mind the experimenter can reduce all his data by eqs. [12] and [13] then test the resulting G'_m values with eq. [A8]. Only those values exceeding this limit need to be corrected using eqs. [A5] and [A6].

An example of the errors that can be encountered in testing crosslinked elastomers is our data below on the urethane, Galcit *I*. G'_t at low frequency was found to be 1.3×10^7 dyne/cm^2 by *Landel* (22).

h	G'_m	K_y (to give G'_t)
2.35 mm	5.65×10^6 dyne/cm^2	4.79×10^{-9} cm/dyne
1.40	4.28	4.47
.60	2.40	4.50

For highly crosslinked or solid materials $G'_m/G'_t \ll 1$ and $G'_m \gg \Omega \eta_m$. Thus eq. [A6] reduces to

$$G'_m \leq h/KA\,. \qquad [A9]$$

This sets an upper limit on measurements in the ERD geometry where the corrections will be larger than the certainty of the data. If we assume that $h < 2R$ for edge effects to be negligible and if $R > 1$ mm, then for our instrument the largest modulus measurable is roughly

$$G'_m < 2 \times 10^9 \text{ dyne/cm}^2.$$

Stiffer instruments can be constructed but K can probably not be decreased by more than a factor of 10 in practice.

Summary

The forces generated by flow between eccentric rotating disks (ERD) are examined experimentally. ERD experiments are presented for several *Newton*ian and viscoelastic fluids with viscosities ranging from 30 to 240 000 poise. The effects of disk separation, radius, misalignment, and eccentricity are studied. ERD data is compared to the dynamic shear moduli, G' and G'', defined by small amplitude deformation in oscillatory shear flow. Results confirm theory that the two measurable shear forces generated in the ERD do measure G' and G''. No significant boundary or inertial effects were noted, but in some cases instrument compliance can be significant. Normal stress measurements in the ERD are also reported.

References

1) *Gent, A. N.*, Brit. J. Appl. Phys. **11**, 165 (1960).
2) *Mooney, M.*, J. Appl. Phys. **35**, 23 (1964).
3) *Maxwell, B.* and *R. P. Chartoff*, Trans. Soc. Rheol. **9**:**1**, 41 (1965).
4) *Chartoff, R. P.* and *B. Maxwell*, Polymer Eng. and Sci. **8**, 126 (1968).
5) *Maxwell, B.*, Polymer Eng. and Sci. **7**, 145 (1967) and **8**, 252 (1968).
6) *Blyler jr., L. L.* and *S. J. Kurtz*, J. Appl. Phys. **11**, 127 (1967).
7) *Bird, R. B.* and *E. K. Harris jr.*, Amer. Inst. Chem. Eng. J. **14**, 758 (1968); **16**, 149 (1970).
8) *Huilgol, R. R.*, Trans. Soc. Rheol. *13*:*4*, 513 (1969).
9) *Yamamoto, M.*, Japan. J. Appl. Phys. **8**, 1252 (1969).
10) *Gordon, R. J.* and *W. R. Schowalter*, Amer. Inst. Chem. Eng. J. **16**, 318 (1970).
11) *Abbott, T. N. G.* and *K. Walters*, J. Fluid Mech. **40**, Part 1, 205 (1970).
12) *Goldstein, C.*, Unsteady Flows of Viscoelastic Fluids: Non Linear Effects, Ph. D. Thesis, Dept. Chem. Eng., Princeton University (1971).
13) *Davis, W. M.* and *C. W. Macosko*, Amer. Inst. Chem. Eng. J. *16*, 600 (1974).
14) *Macosko, C. W.*, Comparisons of Dynamic and Steady Flow Measurements for Polymer Melts. Ph. D. Thesis, Dept. Chem. Eng., Princeton University (1970). Univ. Microfilms, Ann Arbor, Michigan.
15) *Macosko, C. W.* and *J. M. Starita*, SPE J. **27**, 38 (Nov. 1971), and SPE Tech. Papers **17**, 595 (1971), and ASTM STP 553, Am. Soc. Testing Mat., p. 127 (Phila. 1974).
16) Private communication with *A. C. Martellock*, General Electric, Waterford New York, 1969.
17) Private communication with *W. M. Prest*, Univ. Mass., Amherst 1970. Data was taken by *Prest* and *J. M. O'Reilly* at General Electric, Schenedtady, N.Y. (USA).
18) Private communication with *James Reid*, Phillips Petroleum, Bartlesville, Oklahoma 1969.
19. Private communication with *K. S. Hyun*, Dow Chemical, Midland, Mich. (1970).
20) *Ginn, R. F.* and *A. B. Metzner*, Trans. Soc. Rheol. **13**, 429 (1969).
21) *Tanner, R. I.* and *G. Williams*, Rheol. Acta **10**, 528 (1971). – *G. Williams*, Ph. D. Thesis, Dept. Mech., Brown University (1971).
22) *Landel, R. F.*, in: *A. S. Chompff* and *S. Newman* (Ed.), Polymer Networks, Structure and Mechanical Properties, p. 219 (New York 1971).
23) *Gross, L. H.* and *B. Maxwell*, Trans. Soc. Rheol. **16**, 577 (1972).
24) *Gillman, M. W.*, An Experimental Study of the Velocity Field in the Maxwell Orthogonal Rheometer. M. S. Thesis, Rensselaer Polytechnic Inst. Troy, N.Y. 1972.

25) *Erhardt, P. F., J. J. O'Malley,* and *R. C. Crystal,* Polymer Preprints **10**, 812 (1969).

26) Private communication with *P. F. Erhardt,* Xerox, Webster, New York 1970.

27) *Kepes, A.*, Presented before the Fifth International Congress of Rheology, Kyoto, Oct. 1968, unpublished.

28) *Kaelble, D. H.*, J. Appl. Polymer Sci. **13**, 2547 (1969).

29) *Jones, T. E. R.* and *K. Walters,* Brit. J. Appl. Phys. (J. Phys. D) **2**, 815 (1969).

30) *Walters, K.*, J. Fluid Mech. **40**, 191 (1970).

31) *Wales, J. L. S.* and *J. L. den Otter*, Rheol. Acta **9**, 115 (1970). Data obtained in tabular form from the authors.

32) *Wales, J. L. S.*, Rheol. Acta **8**, 38 (1969).

33) *Davis, W. M.* and *C. W. Macosko*, Experimental Examination of Several Eccentric Rotating Geometries. Society of Rheology, Montreal 1973.

Authors' address:

C. W. Macosko and *W. M. Davis*
Dept. of Chemical Engineering and Materials Science
University of Minnesota
Minneapolis, Minnesota 55455 (USA)

Rheol. Acta **13**, 830–835 (1974)

From the Chemical Engineering Department University of Birmingham (England)

The triple jet: A new method for measurement of extensional viscosity

D. R. Oliver and R. Bragg

With 7 figures and 2 tables

(Received October 27, 1974)

Nomenclature

C	Constant defining extensional strain rate (see Appendix)
d_{11}	Extensional strain rate in axial direction
L	Effective length of central jet, covering region of constant extensional strain rate
R_0	Initial radius of central jet just outside orifice, corresponding to velocity V_0
R_1	Final radius of central jet, corresponding to velocity V_1
R_x	Radius of central jet at point distant x from tube exit
R	Radius of orifice in orifice jet thrust experiment
t	Time for which fluid has been subjected to extensional flow
ΔT	Reduction of thrust on central capillary following jet attachment
V_0	Mean velocity of fluid in central jet before extensional flow commences
V_1	Mean velocity of fluid in central jet at point distant "L" from tube exit
V_x	Mean velocity of fluid in central jet at point distant "x" from tube exit
V	Mean velocity of fluid through orifice (in orifice jet thrust experiment)
x	Axial distance from central tube exit
τ_{11}	Axial stress in fluid
$(\tau_{11})_0$	Initial axial stress in fluid in central jet
$(\tau_{11})_{AV}$	Time-average axial stress in fluid in central jet
μ_E	Extensional viscosity i.e. $(\tau_{11})_{AV}/d_{11}$
μ_S	Shear viscosity

Introduction

Extensional viscosity measurements are being made at increasingly high strain rates. The behaviour of semi-solid materials and molten polymers was generally studied at extensional strain rates well below one reciprocal second (1–4); some tests on polymer solutions were also carried out at low strain rates (5). These experiments showed that the *Trouton* ratio (extensional viscosity divided by shear viscosity) sometimes rose above the value 3 predicted by *Newton*ian theory. At higher rates of strain, for polymer solutions, the *Trouton* ratio rises rapidly and novel experimental methods are required (6). The orifice jet thrust technique of *Metzner* and *Metzner* (7) has been used independently by the present authors (8), confirming that the *Trouton* ratio for dilute polyacrylamide solutions is in excess of 10^3 for extensional strain rates of $10^3 - 10^4$ sec^{-1}. Precise knowledge of the flow pattern upstream of the orifice is essential for the accurate use of this method.

The Triple Jet system, described in this paper, gives extensional strain rates of $100 - 800$ sec^{-1}, partially filling the gap between low and high strain rate measurements. Two converging liquid jets of high velocity impinge on a central low-velocity jet of the same liquid, causing the liquid in the central jet to be rapidly stretched. The force on the capillary nozzle from which the central jet emerged gives a measure of the axial stress in the jet; the extensional strain rate is obtained from photographs of the accelerating liquid. The present paper gives a preliminary description of the apparatus and of the results obtained using three polymer solutions.

Fundamentals

Two practical points on which the success of the method depends are:

(i) that the outer jets successfully "pick up" the central jet and accelerate it to a velocity close to the combined final jet velocity. The dilute polymeric solutions used had this quality; however, the final combined jet velocity (measured by jet thrust on an impact device (9)) is considerably greater than the velocity of the central jet just upstream from the point of impact (measured photographically from jet diameter). Thus some slip between the fluid streams is present near the point of impact.

(ii) that the force on the central capillary nozzle changes sufficiently following jet pick-up, and is steady enough, to be measured accurately. In the present tests, the most accurate force measurements have errors of ± 5 per cent and the least accurate ± 20 per cent.

In calculating the axial stress and strain

rate in the central jet, several assumptions are made:

(a) that the presence of velocity gradients in the fluid at the exit from the central tube have no effect on the subsequent extensional flow behaviour of the fluid,
(b) that extensional flow commences at the tube exit plane and not within the tube,
(c) that the jet stretches in a predictable manner,
(d) that the extending jet is circular in cross-section and not seriously deformed by impact from the two outer jets.

The use of a knife-edge central orifice might appear to circumvent assumption (a), but the system would be clumsy and extensional flow of some sort would be present upstream of the orifice. Similarly, assumption (b) seems unavoidable. Assumption (c) presents few problems because, with minor variations, the jets appear to stretch at constant strain rate; assumption (d) could be avoided by photographing the jet from different angles.

The extensional strain rate is given by the equation

$$d_{11} = \frac{V_1 - V_0}{L} \qquad [1]$$

where V_0 mean velocity of liquid just outside exit from central tube [1])

V_1 mean velocity of liquid at point distant "L" from tube exit.

The value of V_0 is obtained from jet thrust measurements on the centre jet in the absence of extensional flow, whilst $V_1 (= V_0 R_0^2/R_1^2)$ is obtained from photographs of the type shown in fig. 1. Values of R_0^2/R_x^2 are plotted against x (axial distance from the central tube exit), a straight line indicating that the strain rate is constant up to the point $x = L$, normally about 1 mm from the point of impact with the outer jets (fig. 2). The value of L is approximately 8 mm

When the outer jets have been aligned to pick up the centre jet, there is reduction of thrust ΔT from the central capillary. If the outer jets are deflected, the original thrust is regained. This process may be repeated, giving the force ΔT causing extensional flow at the centre tube exit. The axial stress at this point is $\Delta T/\Pi R_0^2$, but on the assumption that the same force ΔT is causing extensional flow at all points along the jet, the stress will rise as the jet falls in

Fig. 1

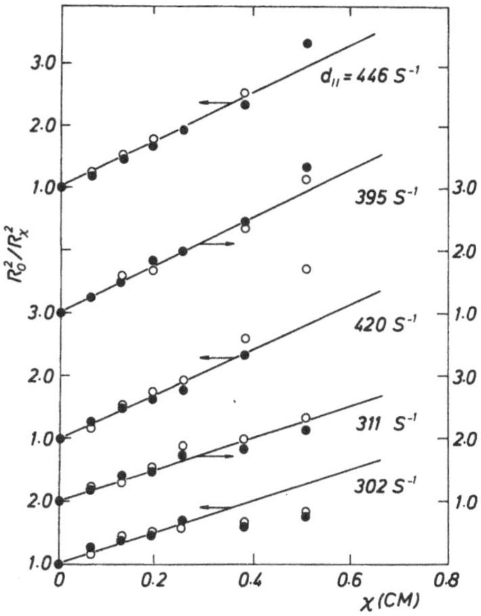

Fig. 2. Axial variation of jet diameter

[1]) Since V_0 is obtained by the jet thrust method, it is not equal to the mean velocity in the tube.

diameter. The time-average stress, calculated in an appendix, is equal to

$$(\tau_{11})_{AV} = \frac{\Delta T}{\Pi R_0^2} \frac{(V_1 - V_0)}{V_0 \ln V_1/V_0}. \qquad [2]$$

This stress, which is typically some 1.5 – 2.0 times larger than the initial stress, is used in the subsequent plots of extensional flow data.

Apparatus

The apparatus (fig. 3) is based on the reaction jet thrust device, described in detail elsewhere (9). Liquid for the centre jet is supplied by a variable-speed gear pump and passes along a horizontal length of 0.08 inch bore stainless steel tubing (*A*) which is fixed at points *B* and B^1. The liquid passes down a $\frac{1}{4}$ inch bore lever arm (*C*) which terminates in a reservoir into which is fitted the centre jet capillary tube (*D*) which is 0.050 inch diameter and 4.25 inch long.

The reaction on the centre tube is transmitted by the lever arm, supported on knife-edges (*E*), via a connecting wire (*F*) to Instron type *A* load cell (*G*), the output from which is recorded on a chart as part of the Instron system. The lever system is damped by placing a small piece of bouncing putty (*H*) between the lever arm and the damping screw (*J*) which permits adjustment of the degree of damping. The device is calibrated by applying known horizontal loads at *L* by means of a torsion balance.

Liquid for the outer jets is supplied from a 3.51 inch bore barrel fitted to the Instron machine (9). The jets issue from stainless steel capillary tubes M_1 and M_2 (of approximately 0.018 inch bore and length 4.0 inches) held at an included angle of 40° by brass block (*K*), detailed in fig. 4. The tubes are a sliding fit in the front of the block but are partly constrained at the rear by "0" rings, with adjusting screws permitting fine adjustment of the alignment. The outer jet system is supported in such a way as to be movable relative to the centre tube, which passes freely through the brass block.

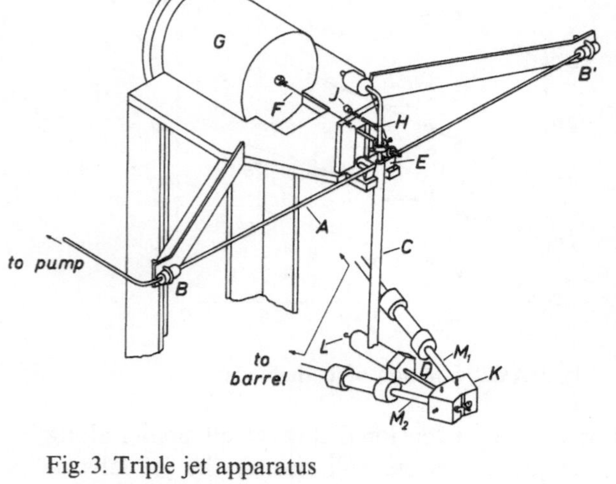

Fig. 3. Triple jet apparatus

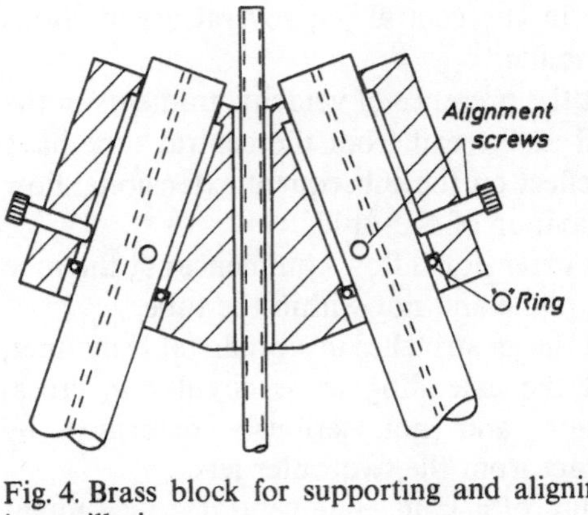

Fig. 4. Brass block for supporting and aligning outer jet capillaries

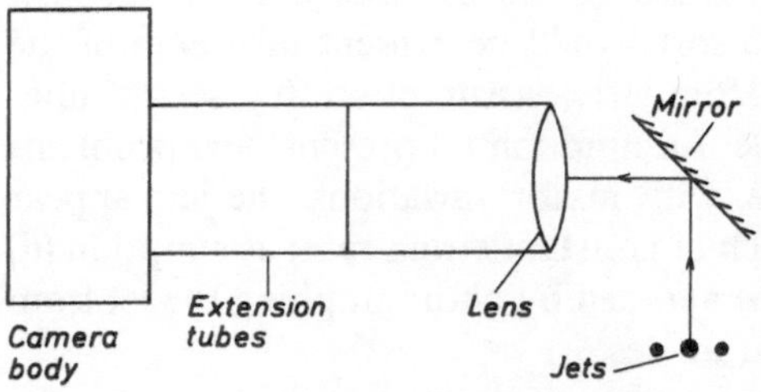

Fig. 5. Arrangement of camera and mirror for photographing jets

The stretching jet was photographed using a Praktica L.L.C. 35 m.m. camera (50 m.m. 1.8 lens) fitted with extension tubes producing a full-size image on the negative (thus minimising enlargement on printing). All photographs were taken using electronic flash, effectively freezing the motion. The camera was mounted horizontally and jets photographed using the mirror system shown in fig. 5.

The centre jet velocity is determined by measuring jet thrust and flowrate, the latter by collecting a volume of liquid in a known time [2]). When the outer jets impinge on the centre jet the thrust decreases; by deflecting the outer jets the original thrust is regained and the process may be repeated to give a more accurate measure of the force causing stretching of the centre jet. Two photographs are taken and printed eight times actual size. These give the changes in jet diameter from which the extensional strain rates are measured.

Fluids used

The fluids were 0.10 per cent and 0.07 per cent ET 597 in water and 0.10 per cent poly (ethylene oxide), coagulant grade, in water. The former is a partially hydrolysed polyacrylamide (now referred to as AP273) marketed by Dow Chemicals, whilst the "Polyox" was supplied by Union Carbide Ltd. The density of the

[2]) Jet thrust is equal to the product of jet velocity and mass flowrate. A surface tension correction is necessary (11).

liquids was equal to that of water, whilst apparent viscosities of solutions are quoted in table 1.

Table 1. Apparent viscosities of polymer solutions (22.0 °C)

	Shear Rate		
	$10\ sec^{-1}$	$500\ sec^{-1}$	$10^4\ sec^{-1}$
0.07% ET597	10.3 cp.*)	6.2 cp.	2.0 cp.
0.10% ET597	17.5 cp.*)	7.6 cp.	2.8 cp.
0.10% POLYOX	2.7 cp.	2.8 cp.	1.6 cp.

*) 21 °C.

Results

The extensional flow data are summarised in table 2. Both the average and the initial stress in the central jet are quoted, the *Trouton* ratio being based on the extensional viscosity calculated using the average stress. The shear viscosity is evaluated at a shear rate numerically equal to d_{11} (sec^{-1}). The data for solutions of 0.10% ET597 include two runs carried out at different times.

Figs. 6 and 7 compare the data with that obtained for similar solutions using the orifice jet thrust technique (8). (Polyox was not used in the earlier work). Two methods of plotting were used in an attempt to define extensional strain rate as a function of orifice diameter and fluid velocity only (8). Photographic evidence indicates that the equation $d_{11} = 0.89\ V/\sqrt{R}$ is a better approximation than $d_{11} = 2.7\ V$ but both equations involve the assumption that d_{11} is proportional to V, which is not true for all fluids used (10). The orifice jet thrust method, however, has the advantage of operating up to high extensional strain rates ($10^3 - 10^4\ sec^{-1}$) and the results complement those reported here.

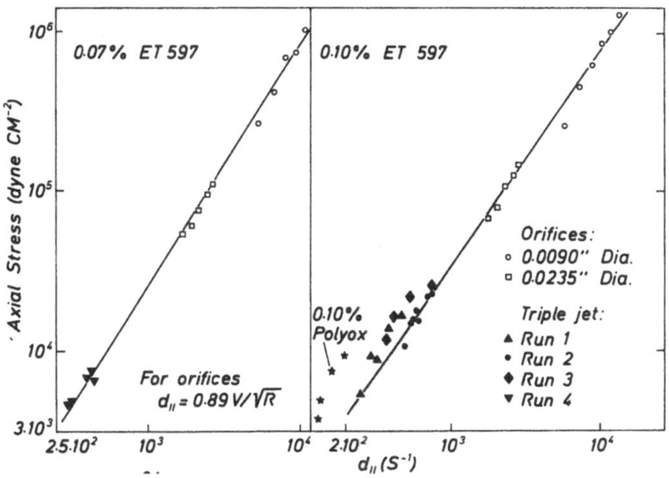

Fig. 6. Comparison of triple jet with orifice jet thrust results

Table 2. Extensional flow data

Liquid used	V_0 ($cm \cdot sec^{-1}$)	V_1 ($cm \cdot sec^{-1}$)	d_{11} (sec^{-1})	$(\tau_{11})_{AV}$ ($dyn \cdot cm^{-2}$)	$(\tau_{11})_0$ ($dyn \cdot cm^{-2}$)	$\mu_E = \frac{(\tau_{11})_{AV}}{d_{11}}$ (poise)	*Trouton* ratio (μ_E/μ_s)
0.10% ET597	114	500	488	$1.07 \cdot 10^4$	$4.67 \cdot 10^3$	21.9	284
	116	545	543	$1.49 \cdot 10^4$	$6.22 \cdot 10^3$	27.4	375
	106	546	557	$1.58 \cdot 10^4$	$6.25 \cdot 10^3$	28.4	368
	116	586	593	$1.55 \cdot 10^4$	$6.20 \cdot 10^3$	26.1	363
	118	580	585	$1.78 \cdot 10^4$	$7.24 \cdot 10^3$	30.4	422
	101	646	690	$2.19 \cdot 10^4$	$7.52 \cdot 10^3$	31.7	466
	101	697	754	$2.29 \cdot 10^4$	$7.51 \cdot 10^3$	30.4	454
	108	385	370	$1.18 \cdot 10^4$	$5.85 \cdot 10^3$	31.9	382
	118	427	412	$1.67 \cdot 10^4$	$8.21 \cdot 10^3$	40.5	506
	114	510	528	$2.20 \cdot 10^4$	$9.50 \cdot 10^3$	41.7	564
	114	667	737	$2.60 \cdot 10^4$	$9.47 \cdot 10^3$	35.3	527
0.07% ET597	114	304	302	$4.52 \cdot 10^3$	$2.66 \cdot 10^3$	15.0	206
	123	319	311	$4.62 \cdot 10^3$	$2.77 \cdot 10^3$	14.9	212
	105	342	395	$6.65 \cdot 10^3$	$3.49 \cdot 10^3$	16.8	255
	118	370	420	$7.35 \cdot 10^3$	$3.92 \cdot 10^3$	17.5	269
	118	386	446	$6.37 \cdot 10^3$	$3.31 \cdot 10^3$	14.3	227
0.10% POLYOX	106	204	131	$3.68 \cdot 10^3$	$2.60 \cdot 10^3$	28.1	760
	102	204	136	$4.80 \cdot 10^3$	$3.33 \cdot 10^3$	35.3	955
	103	224	161	$7.40 \cdot 10^3$	$4.89 \cdot 10^3$	46.0	1280
	98	253	194	$9.38 \cdot 10^3$	$5.61 \cdot 10^3$	48.4	1420
	111	258	196	$9.33 \cdot 10^3$	$5.95 \cdot 10^3$	47.6	1400

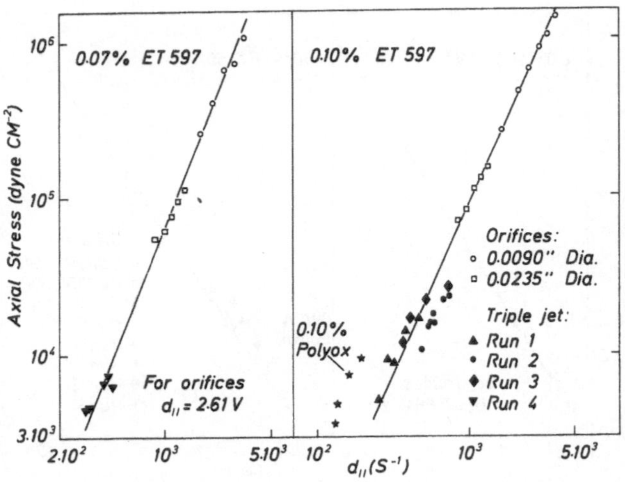

Fig. 7. Comparison of triple jet with orifice jet thrust results

Discussion of results

In preliminary work on the Triple Jet system, all velocities were measured by the jet thrust method. It was assumed that the length of the central jet to the point of impact with the outer jets did not change and that the final velocity of the combined jets was equal to the final velocity of the central jet. Photographs showed, however, that the position of the point of impact varied slightly with the outer jet velocity and also that slip flow was present where the jets collided. Experiments with non-polymeric liquids – glycerol solutions and a mineral oil – were unsatisfactory in that the outer jets were deflected by, or ran round, the central jet with no immediate attachment. If the viscosity of the outer jet liquid was lowered, there was an increased tendency to form a fan spray. It may be that only liquids of high elasticity provide good jet attachment.

The photographs show that the central jet stretches at constant strain rate, which does not seem consistent with the concept of increasing stress along the jet. However, previous work with dilute solutions has shown that extensional viscosity might increase with distance along the jet (5), thus tending to minimise axial variation of the strain rate. The kinetic energy of the fluid also increases as the jet accelerates, and there may be complex interplay between viscous, elastic and kinetic forces at the high-velocity end of the jet.

Figs. 6 and 7 show that the "triple jet" data form a consistent extension of information obtained at higher strain rates. However, the new data do not serve to differentiate between the two methods of plotting the orifice jet results. This could best be achieved by increasing the extensional strain rate in the present equipment to over 10^3 sec^{-1}, a process made more difficult by the slip flow which takes place at the jet impingement point.

There is urgent need (both theoretical and practical) to determine the relationship between τ_{11} and d_{11} at high strain rates. Figs. 6 and 7 show that there is no abrupt increase of stress with strain rate nor, in the the present range of strain rates, is there a significant change in form of the relationship between these quantities. For the present solutions, the extensional stress is proportional to a power of extensional strain rate which is not less that 1.3 nor greater that 2.4. Using the lower gradient, as for example fig. 6 for 0.1 % ET597, the extensional viscosity rises from 25 poise at 400 sec^{-1} to 84 poise at 10^4 sec^{-1}. The corresponding values taken from fig. 7 for the same solution are 21 poise (400 sec^{-1}) and 1040 poise (10^4 sec^{-1}). Variation in the values of extensional viscosity at high strain rate result from different methods of calculation of orifice jet thrust data (as in the work of *Metzner* and *Metzner* (7), though using somewhat different assumptions) and not from variations present within the "triple jet" data.

The data of table 2 show that the Polyox solutions had considerably higher *Trouton* ratios than the polyacrylamide solutions. The extensional viscosities were higher, and the shear viscosities lower, than those of the ET597 solutions despite the low values of strain rate achieved. It was apparent from photographs that the Polyox solutions were not accelerated to the same extent as the other liquids; whether this is due to poorer jet pick-up or to greater resistance to extensional strain in the centre jet is not clear. It is evident, however, that fresh Polyox solutions have high extensional viscosities at strain rates as low as 150 sec^{-1}, a factor which must have bearing on their efficiency as turbulence suppressors. Unfortunately, rapid mechanical degradation of the polymer precludes its use in cases where the liquid is sheared for more than a few minutes.

Further methods of measurement of extensional viscosity are needed. Problems include lack of accuracy at high strain rates, and the need to design an instrument to study less elastic, more viscous liquids like lubricants. It is important that the fluid sample be subjected to conditions of simple stretching, though if some shear is present, it would be an advantage if this could be controlled, independent of the extensional strain rate. New, unorthodox experimental techniques may be required to meet these objectives.

Appendix

The time-average stress in a jet extending at constant extensional strain rate

All terms are defined under "Nomenclature".

The time-average stress is equal to

$$(\tau_{11})_{AV} = \frac{1}{t}\int_0^t \frac{\Delta T}{\Pi R_x^2}\,dt \qquad [3]$$

$$= \frac{1}{t}\int_0^L \frac{\Delta T}{\Pi R_x^2}\frac{dt}{dx}\,dx. \qquad [4]$$

But from continuity $\Pi R_x^2 V_x = \Pi R_0^2 V_0$

and $V_x = \frac{dx}{dt}$

$$(\tau_{11})_{AV} = \frac{1}{t}\int_0^L \frac{\Delta T\,dx}{\Pi R_0^2 V_0} = \left(\frac{\Delta T L}{\Pi R_0^2 V_0}\right)\frac{1}{t}. \qquad [5]$$

Now $V_x = V_0 + Cx$ (at constant extensional strain rate)

$$= \frac{dx}{dt}$$

$$\therefore \quad t = \int_0^L \frac{dx}{V_0 + Cx} = \frac{1}{c}\ln\left(\frac{V_0 + CL}{V_0}\right). \qquad [6]$$

But $V_1 = V_0 + CL$

$$t = \frac{L}{V_1 - V_0}\ln\frac{V_1}{V_0}. \qquad [7]$$

Hence from [5] and [7]

$$(\tau_{11})_{AV} = \frac{\Delta T}{\Pi R_0^2}\,\frac{(V_1 - V_0)}{V_0 \ln V_1/V_0}. \qquad [8]$$

It will be noted that this stress exceeds the initial stress ($\Delta T/\Pi R_0^2$) by the ratio of the logarithmic mean velocity to the initial velocity.

Summary

Apparatus for the measurement of extensional viscosity of polymer solutions is described. The range of extensional strain rates is $100-800\ \text{sec}^{-1}$.

Two converging liquid jets of high velocity impinge on a central low-velocity jet, causing the liquid in the central jet to be rapidly stretched. The force causing extension of this jet is measured from the change in reaction on the central capillary nozzle, whilst the initial and final velocities are obtained by jet thrust and photographic methods respectively. The Instron system provides accurately-controlled liquid flowrates and continuous recording of the various forces involved.

Tests are reported on polyacrylamide solutions of concentration 0.10% and 0.07%, and on poly (ethylene oxide) solutions of concentration 0.10%. The extensional viscosity of the more concentrated polyacrylamide solution is over 30 poise at an extensional strain rate of $600\ \text{sec}^{-1}$, exceeding the shear viscosity by a factor of 400 (the "*Trouton*" ratio). For the poly (ethylene oxide) solution, the *Trouton* ratio reaches 1400 at an extensional strain rate of only $200\ \text{sec}^{-1}$.

The results are shown to be consistent with orifice jet thrust data for similar solutions at higher strain rates. The axial stress rises rapidly with increasing extensional strain rate, but there is no evidence of an abrupt increase of stress in the range of strain rates between 10^2 and $10^4\ \text{sec}^{-1}$. The triple jet method is of value in giving data for intermediate extensional strain rates and being dependent on fewer assumptions than the orifice jet thrust experiment.

References

1) *Reiner, M.*, Deformation, Strain and Flow. Second Edition, p. 78 (New York 1960).

2) *Ballman, R. L.*, Rheol. Acta **4**, 137 (1965).

3) *Cogswell, F. N.*, Plastics Polymers 109 (1968).

4) *Stevenson, J. F.*, Amer. Inst. Ch. Eng. J. **18**, 540 (1972).

5) *Acierno, D., R. Greco*, and *G. Titemanlio*, Elongational Flow of Dilute and Concentrated Polymer Solutions. Euromech 37 (Naples 1972).

6) *Astarita, G.*, Ind. Eng. Chem. Fund. **7**, 171 (1968).

7) *Metzner, A. B.* and *A. P. Metzner*, Rheol. Acta **9**, 174 (1970).

8) *Oliver, D. R.* and *R. Bragg*, Chem. Eng. J. **5**, 1 (1973).

9) *Oliver, D. R.* and *W. C. Macsporran*, Rheol. Acta **8**, 176 (1969).

10) *Oliver, D. R.* and *R. Bragg*, Canad. J. Chem. Eng. **51**, 287 (1973).

11) *Oliver, D. R.*, Canad. J. Chem. Eng. **44**, 100 (1966).

Authors' address:

Chemical Engineering Department
University of Birmingham (England)

Rheol. Acta 13, 836–839 (1974)

Aus dem Institut für chemische Physik der Akademie der Wissenschaften der UdSSR, Laboratorium für armierte Plaste, Moskau (UdSSR)

Die Abhängigkeit der Glasfaserfestigkeit von der Belastungsgeschwindigkeit

A. M. Kuperman und Ju. A. Gorbatkina

Mit 2 Abbildungen und 1 Tabelle

(Eingegangen am 27. Oktober 1972)

Die vorliegende Arbeit ist der Untersuchung des Einflusses der Belastungsgeschwindigkeit auf die Festigkeit von Glasfasern mit einem Durchmesser von 10, 55 und 220 μm und von Elementarproben aus glasfaserverstärkten Plasten auf der Grundlage der Fasern von 10 μm gewidmet.

Bis jetzt wurde dieses Problem noch nicht systematisch untersucht. Früher ist von *Rexer* (1) die Abhängigkeit der Festigkeit σ von Glasfasern mit einem Durchmesser von 50 μm von der Geschwindigkeit der Spannungszunahme $\dot{\sigma}$ an Proben im Bereich von 3–5 Größenordnungen der Geschwindigkeit gemessen worden. Es zeigte sich, daß bei Änderung von σ um eine Größenordnung sich die Festigkeit um 16,5% verändert. Für Glasfasern mit einem Durchmesser von 7 – 15 μm, die am meisten in der Industrie Verwendung finden, wurde die Abhängigkeit der Festigkeit von der Belastungsgeschwindigkeit in Arbeit (2) untersucht. Ein merklicher Einfluß der Belastungsgeschwindigkeit auf die Festigkeit konnte von den Autoren nicht festgestellt werden. Die Belastungsgeschwindigkeit wurde in ihren Versuchen jedoch nur 30 mal geändert. Dabei ist es schwierig, irgendwelche bestimmten Schlußfolgerungen zu ziehen.

Meßmethode

Zur Herstellung von Proben wurde ein Glasgießgefäß mit 200 Düsen, wie es in der Industrie verwendet wird, benutzt. Der Düsendurchmesser betrug 1,4 mm. Die Herstellung der Glasfaser aus der Aluminiumborsilikatmasse (analog dem E-Glas) erfolgte im üblichen Bereich der technologischen Verarbeitungswerte; bei einer Temperatur von 1200 °C und einer Niveauhöhe der Glasmasse von 100 mm betrug ihre Produktion 25 g/min. Die Herstellung von Fasern mit unterschiedlichem Durchmesser wurde lediglich durch Änderung der Ziehgeschwindigkeit erreicht. Die Herstellung der Faser erfolgte aus einer willkürlich gewählten Düse, die in der Mitte der Düsenplatte lag. Die Fasern von 10 und 55 μm wurden unmittelbar beim Ziehen mit einer bestimmten Steigung auf eine Trommel mit einem Durchmesser von 357 mm aufgewickelt, an welcher vorher Papierstreifen befestigt wurden (3). Die Fasern wurden direkt auf der Trommel in die Rahmen für die Versuchsdurchführung geklebt. Nach dieser Methode gelang es uns, gleich über 700 identische Proben zu erhalten. Die Fasern mit einem Durchmesser von 200 μm wurden von Hand herausgezogen und nach der üblichen Methode in die Rahmen geklebt.

Der Faserdurchmesser wurde unter einem Mikroskop mit einem Schraubenokularmikrometer (800 fache Vergrößerung) bestimmt. Zur Versuchsdurchführung wurden Proben ausgewählt, die nicht mehr als um 10% vom sich ergebenden mittleren Durchmesser abwichen. Zur Stabilisierung der Fasereigenschaften wurden die Proben vor der Versuchsdurchführung mindestens 20 Tage unter Zimmerbedingungen gealtert.

Als Elementarproben aus glasfaserverstärktem Plast wurden gleichgerichtete Streifen und Glasfurniere verwendet. Die Streifen stellen Litzen aus 200 elementaren Glasfasern dar, die durch einen Polymermatrix verbunden sind; die Glasfurniere sind gleichgerichtete glasfaserverstärkte Plaste mit einer Dicke von 0,2 – 0,4 mm. Die Herstellung der Proben aus glasfaserverstärktem Plast erfolgte durch Aufwickeln von elementaren Glasfasern mit einem Durchmesser von 10 μm und Auftragen einer Epoxidmatrix EDT-10 unmittelbar im Moment des Ziehens der Fasern aus der Düse des Glasgießgefäßes.

Die Festigkeit der Fasern mit einem Durchmesser von 10 μm wurde auf einem Adhäsiometer MAB-2TC (4) bestimmt. Die Fasern mit einem Durchmesser von 55 und 200 μm, sowie die Streifen wurden auf einer Schopper-Zerreißmaschine mit einer max. Belastung von 5 kp geprüft; die Glasfaserfurniere prüfte man auf einer Maschine mit einer max. Belastung von 5000 kp.

Zur Änderung der Belastungsgeschwindigkeit wurden in allen Fällen besondere Zahnradgetriebe benutzt. Der Änderungsbereich der Geschwindigkeiten betrug 4 – 5 Größenordnungen. Die Belastungszeit wurde dabei von einigen Sekunden bis zu 15 – 18 Stunden variiert.

Die Versuchsmaschinen gestatteten ein Regime zu verwirklichen, das dem Regime mit konstanter Deformationsgeschwindigkeit nahe kommt, was am linearen Abschnitt des Spannung–Dehnung-Diagramms einem Zustand mit konstanter Belastungsgeschwindigkeit entspricht. Die Konstanz der Belastungsgeschwindigkeit wird natürlich unmittelbar vor dem Zerreißen der Probe unterbrochen. Uns interessierte jedoch die Abhängigkeit der Festigkeit von der Belastungsgeschwindigkeit im Stadium des langsamen Anstiegs der Mikrodefekte, weil gerade damit die Lebensdauer des Materials verbunden ist.

Die Messungen zeigten, daß die Kraft *F*, die an die Probe angelegt wurde, linear mit der Belastungszeit *t* praktisch bis zur Zerstörung der Probe ansteigt. Deshalb wurde die Geschwindigkeit des Spannungsanstiegs nach der Neigung der Geraden in den Koordinaten $F/S-t$ bei geringen Belastungsgeschwindigkeiten und nach dem Verhältnis $F_{Zer.}/S-t_{Zer.}$ bei großen Geschwindigkeiten bestimmt.

Bei jeder Belastungsgeschwindigkeit wurde die Festigkeit der Glasfaser als arithmetischer Mittelwert der Festigkeit von 60 bis 100 Versuchsproben ermittelt. Der Variationskoeffizient der Faserfestigkeit betrug 20–30%. Zur Bestimmung der Streifenfestigkeit wurden 20–30 Proben untersucht. Der Variationskoeffizient bei der Bestimmung der Zerreißbelastung betrug für Streifen 11–18% und für Glasfurniere 6–8%.

Bei der Untersuchung der Streifen und Furniere wurde die gesamte Zerreißbelastung auf den summaren Glasquerschnitt in der Probe bezogen. Der Fehler wegen solcher Annahme beträgt max. 5%, sogar wenn Bindermatrixanteil 40 Gew.-% ausmacht. Der summare Glasfaserquerschnitt wurde ausgehend vom prozentualen Glasanteil (er wurde durch Ausbrennen des Harzes bestimmt) vom spezifischen Gewicht des Glases und von der Probenlänge berechnet. So betrug der Glasquerschnitt in der Probe für die Streifen 0,6145 mm² und für Furniere 1,70 mm². Der Bindermatrixanteil lag bei den Streifen bei 30...40% und bei den Furnieren bei 20 Gew.-%. Die Streifen hatten eine Breite von 0,8 mm, eine Dicke von 0,05 mm und eine Länge des Arbeitsteils von 120 mm; die Breite der Furnierproben betrug 15 mm, die Dicke 0,25 mm und die Länge des Arbeitsteils 120 mm; die Länge des Arbeitsteils der Fasern war 10 mm.

Meßergebnisse

Die Messungen zeigten, daß die Festigkeit der Glasfasern von untersuchten Durchmessern sowie auch die Festigkeit der Elementarproben aus glasfaserverstärkten Plasten sich mit dem Anstieg der Belastungsgeschwindigkeit erhöht. In Abb. 1 ist die Abhängigkeit $\sigma-\dot{\sigma}$ im halblogarithmischen Koordinatensystem dargestellt. Es ist ersichtlich, daß sowohl für Fasern mit allen untersuchten Durchmessern als auch für Fasern in Form von Streifen und Furnieren die Punkte gut auf der Geraden liegen. (Es muß bemerkt werden, daß im untersuchten Geschwindigkeitsbereich die Ergebnisse ebensogut im logarithmischen Koordinatensystem $\lg\sigma-\lg\dot{\sigma}$ dargestellt werden können.)

In Abb. 2 sind typische Kurven der Festigkeitsverteilung von Fasern mit einem Durchmesser von 10 und 55 µm sowie Fasern mit einem Durchmesser von 10 µm in den Streifen dargestellt. Es

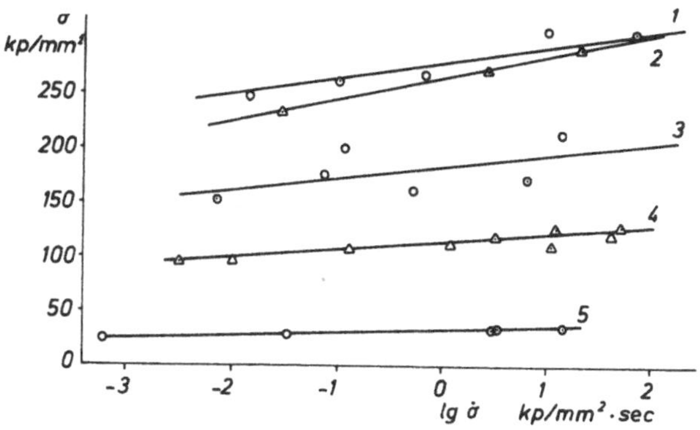

Abb. 1. Abhängigkeit der Glasfaserfestigkeit von der Geschwindigkeit des Spannungsanstiegs: 1, Streifen; 2, Furniere; 3–5, Glasfasern mit einem Durchmesser von 10, 55 und 200 µm

ist ersichtlich, daß für alle Durchmesser und Belastungsgeschwindigkeiten einmodale Kurven vorliegen. Mit dem Anstieg der Belastungsgeschwindigkeit verschiebt sich die Lage des Maximums zu höheren Werten von σ, aber in der Regel verringert sich seine Größe. Diese Verringerung ist am stärksten bei Fasern mit einem Durchmesser von 10 µm bemerkbar.

Die Verteilungskurven werden durch die Normalverteilung und die *Weibullsche* Verteilung gut beschrieben.

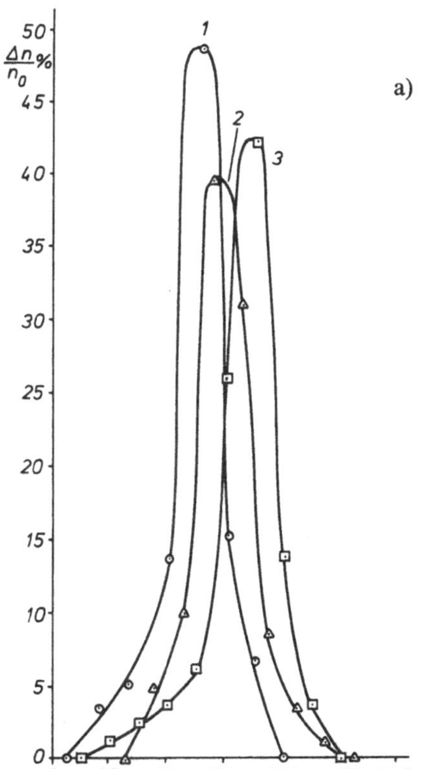

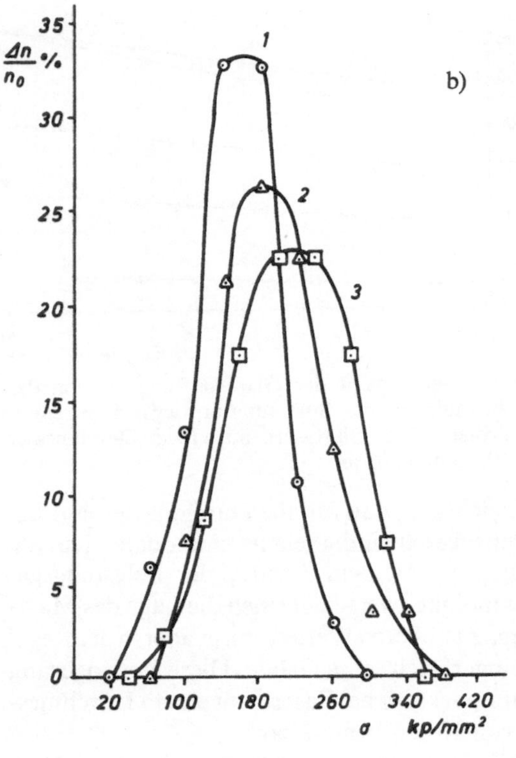

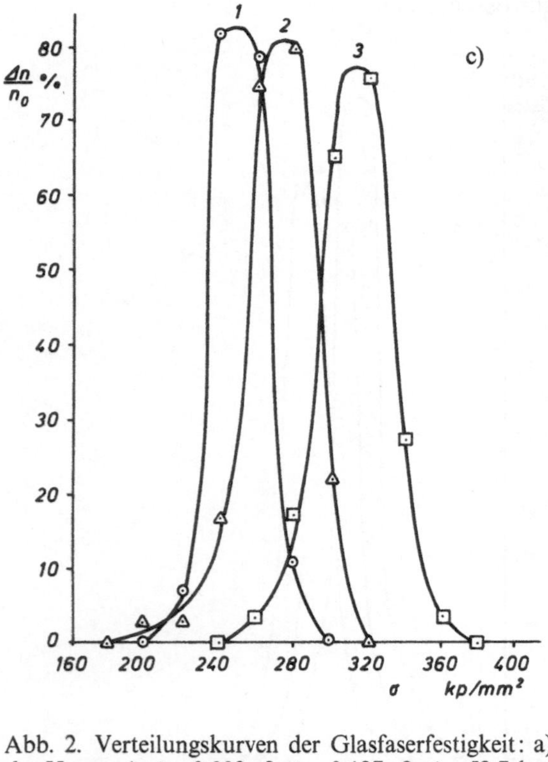

Abb. 2. Verteilungskurven der Glasfaserfestigkeit: a) $d = 55$ µm: 1, $\dot{\sigma} = 0{,}003$; 2, $\dot{\sigma} = 0{,}127$; 3, $\dot{\sigma} = 52{,}7$ kp/mm² · sec; b) Streifen: 1, $\dot{\sigma} = 0{,}0145$; 2, $\dot{\sigma} = 1{,}05$; 3, $\dot{\sigma} = 65{,}6$ kp/mm² · sec; c) $d = 10$ m: 1, $\dot{\sigma} = 0{,}007$; 2, $\dot{\sigma} = 0{,}113$; 3, $\dot{\sigma} = 15{,}3$ kp/mm² · sec

Die Werte für die Festigkeitsänderung von Glasfasern bei Änderung der Belastungsgeschwindigkeit um eine Größenordnung sind in der ersten Spalte der Tabelle angeführt. Es ist ersichtlich, daß mit Vergrößerung des Durchmessers der Faser sich der Einfluß der Belastungsgeschwindigkeit etwas vergrößert. Das ist anscheinend mit der größeren Fehlerbehaftung dicker Fasern und mit der größeren Einwirkung der adsorbierten Luftfeuchtigkeit auf sie verbunden. Nach den Vorstellungen von *Rehbinder* und *Aslanowa* (5, 6) ist das Auftreten des Adsorptionseffektes an Fasern mit größeren Durchmessern mit der möglichen Bildung großer Risse auf ihnen verbunden.

In der Tabelle sind ebenfalls Werte angeführt, die der Arbeit von *Rexer* (1) entnommen wurden. Eine Änderung der Belastungsgeschwindigkeit verändert hier die Festigkeit der Faser viel stärker. Die Faser für die Versuchsdurchführung von *Rexer* wurde nach einer unvollkommenen Methode durch Ziehen von Hand oder unter Einwirkung einer Last auf einen erwärmten Glasstab erzeugt, so daß bei ein und derselben Belastungsgeschwindigkeit $\dot{\tau} = 10$ kp/mm² · sec eine Faser mit einem Durchmesser von 55 µm, die von uns erzeugt wurde, eine Festigkeit von 100 kp/mm² besaß und die Fasern, die von *Rexer* benutzt wurden, 60 kp/mm² aufwiesen. So steht die Empfindlichkeit gegenüber der Änderung der Belastungsgeschwindigkeit nicht nur mit dem Faserdurchmesser in Zusammenhang, sondern auch damit, welche Festigkeit diese Faser besitzt, was ihrerseits mit der gesamten Vorgeschichte der Probe und schließlich mit ihrer Fehlerbehaftung verbunden ist.

Tab. 1. Der Einfluß der Geschwindigkeit des Spannungsanstiegs auf die Festigkeit von Glasfasern und der Wert der Konstanten U_0 und γ in der Formel für die Lebensdauer*)

Durchmesser der Glasfaser µm	Änderung bei Änderung um eine Größenordnung, %	U_0 kcal/Mol	γ $\frac{\text{kcal} \cdot \text{mm}^2}{\text{Mol} \cdot \text{kp}}$
200	7,7	32	0,449
55	6,1	39	0,185
10	5,7	41	0,124
Streifen	6,0	40	0,076
Furniere	7,2	38	0,073
50 (nach Ang. 7)	16,5	25	1,35

*) $\tau_0 = 10^{-13}$ sec.

Die sich ergebenden Werte gestatten einige Parameter, die mit dem Zerstörungsprozeß der Glasfaser im Zusammenhang stehen, zu beurteilen, insbesondere die Aktivierungsenergie. Der Wert der sogenannten Null-Aktivierungsenergie U_0 geht als Parameter in die bekannte Formel für die Lebensdauer von *Shurkov* ein:

$$\tau = A e^{-\alpha\sigma} = \tau_0 \quad \exp[(U_0 - \gamma\sigma)/kT]. \qquad [1]$$

Wobei: A und α — Konstanten;
γ — Faktor, der mit der Fehlerbehaftung des Materials in Zusammenhang steht;
$\tau_0 = 10^{-12} \ldots 10^{-13}$ sec;
k — *Boltzmann*-Konstante;
T — absolute Temperatur.

In Arbeit (7) wurde gezeigt, wie sich mit Hilfe der Bedingung von *Beili* $\int_0^{t_{max}} \frac{dt}{\tau(\sigma)} = 1$ aus der Abhängigkeit der Festigkeit von der Belastungsgeschwindigkeit Werte für die Lebensdauer τ des Materials bei verschiedenen Spannungen σ ergeben. In unserem Falle sind, wenn $\sigma = \dot{\sigma} t$ und $\sigma = B_1 + B_2 \lg\dot{\sigma}$, die Koeffizienten B_1 und B_2 (siehe z. B. (10)) mit den Parametern der Formel von *Shurkov* durch folgende Beziehungen verbunden:

$$\alpha = 2{,}3/B_2, \ \lg A = \frac{B_1}{B_2} - \lg\alpha,$$

$$U_0 = 2{,}3\,kT(\lg A - \lg\tau_0), \ \gamma = \alpha kT.$$

Die Werte U_0 und γ sind in der Tabelle angeführt, aus welcher ersichtlich ist, daß mit Vergrößerung des Faserdurchmessers die Aktivierungsenergie etwas sinkt. Gleichzeitig steigt der Koeffizient γ, der die Fehlerbehaftung des Materials kennzeichnet, an. Für die Streifen und Furniere liegen die Werte U_0 nahe den Werten von ungeschützten Fasern. Charakteristisch ist, daß U_0, das für Fasern von *Rexer* mit geringerer Festigkeit berechnet ist, bedeutend kleiner und γ fast um eine Größenordnung größer ist als für eine Faser mit einem Durchmesser von 55 µm in unseren Versuchen.

Die Werte U_0, die sich bei uns aus der Geschwindigkeitsabhängigkeit der Festigkeit ergaben, fallen mit der Null-Aktivierungsenergie zusammen, die *Bartenev* mit seinen Mitarb. (11) bei der unmittelbaren Untersuchung der Lebensdauer von Glasfasern mit einem Durchmesser von 10 µm an der Luft erhielt.

Nach Angaben aus (6, 7) ist die Lebensdauer von Glas in der Atmosphäre $\tau(\sigma)$ besser nach einer Potenzfunktion zu beschreiben als nach einer Exponentialfunktion:

$\tau = B e^{-b}$, wobei B und b — Konstanten.

In diesem Fall verläuft die Funktion $\sigma(\dot{\sigma})$ im logarithmischen Koordinatensystem linear (7): $\lg\sigma = C_1 + C_2 \lg\dot{\sigma}$. Die Konstanten B und b werden nach folgenden Formeln errechnet:

$$B = \frac{C_1}{C_2} - \lg\frac{1}{C_2}\,; \qquad b = \frac{1}{C_2} - 1\,.$$

Die Berechnungsergebnisse für die Lebensdauer nach einer Exponentialfunktion und nach einer Potenzfunktion haben gezeigt, daß für Spannungen, die den gesamten Meßbereich in unseren Versuchen umfassen, die Lebensdauerwerte, die nach beiden Formeln berechnet wurden, praktisch für alle untersuchten Fasern, Streifen und Furniere zusammenfallen.

Literatur

1) *Rexer, E.*, Z. techn. Phys. **1939**, Nr. 1, 4.

2) *Sak, A. F.*, Fisiko-chimitscheskije svojstva steklannogo volokna (Physikalisch-chemische Eigenschaften der Glasfaser) (Moskau 1962).

3) *Selenskij, E. S., A. M. Kuperman* und *G. D. Andrejevskaja*, Fisiko-chimija i mechanika orientirovannich stekloplastikov, S. 15 (Physikochemie und Mechanik der orientierten Glasfaserplaste) (Moskau 1967).

4) *Charchardin, S. I., W. W. Lawrentjew, G. K. Abramov* und *Ju. A. Gorbatkina*, Plastitscheskije massi, Nr. 10 (1966).

5) *Aslanowa, M. S.* und *P. A. Rehbinder*, Vorträge der Akademie der Wissenschaften der UdSSR **96**, 2, 299 (1954).

6) *Aslanowa, M. S.*, Vorträge der Akademie der Wissenschaften der UdSSR **95**, 6, 1215 (1954).

7) *Bartenev, G. M.* und *Ju. S. Sujev*, Protschnostij i rasruschenije vissokoelastitscheskich materialov (Festigkeit und Zerstörung hochelastischer Materialien) (Moskau-Leningrad 1964).

8) *Shurkov, S. N.* und *S. A. Abasov*, FTT **4**, 2184 (1962).

9) *Shurkov, S. N.*, Westnik akademiji nauk SSSR **1957**, Nr. 11, 78.

10) *Gorbatkina, Ju. A.* und *V. G. Iwanowa-Mumshieva*, Rheol. Acta **13**, 789 (1974).

11) *Bartenev, G. M.*, Sverchprotschnije und vissokoprotschnije neorganitscheskije Stjekla („Ultra- und hochfestliche anorganische Gläser") Moskau 1974.

Anschrift der Verfasser:

Dr. *A. M. Kuperman* und Dr. *Ju. A. Gorbatkina*
Institut für Chemische Physik der Akademie der Wissenschaften der UdSSR,
Vorobjevskoje Chaussee 42 B
Moskau V 334 (USSR)

Rheol. Acta **13**, 840–844 (1974)

From the Department of Civil Engineering, McMaster University, Hamilton (Canada)

Rheological influences on the consolidation process of soils

N. E. Wilson

With 6 figures

(Received October 27, 1972)

Introduction

Consolidation of soils is a complex hydrodynamic process; water is squeezed from the pores between the soil particles by the imposed loading. Shear strains are produced in the soil as the sample deforms and the rate of strain is associated with the escape of pore-water. The original Theory of Consolidation proposed by *Terzaghi* (1925) considered that the plastic resistance could be ignored as it was so small; in this way, a linear partial differential equation was developed to provide a time-settlement relationship for clays based entirely on the escape of pore-water.

The simplifying assumptions are satisfactory in the case of stiff clays where the strains are small and the changes in permeability are small. For soft soils and organic soils, the strains are large and the changes in permeability significantly effect the hydrodynamic process so that the differential equation is non-linear. For these soils, the volume of pore-water expelled is very large and the shear strength of the soil is small which makes it necessary to treat the material from a rheological viewpoint.

The *Terzaghi* theory of consolidation uses an empirical division of the consolidation process into two parts – "primary" and "secondary"; the rate of drainage governs primary consolidation, whereas plastic or viscous flow governs the secondary part of consolidation. The limitations of the consolidation theory were considered and later enlarged by *Gibson* et al. (1967) and others. These limitations are:

(i) Permeability variations during one loading process,
(ii) Idealized pressure versus void ratio relationship ($de = a_v \cdot dp$) which does not consider plastic or viscous characteristics,
(iii) Secondary influences due to large deformations,
(iv) Small strain theory which is not valid for very compressible soils as large axial strains are associated with moving boundaries.

Consolidation loading

The influence of sample height as it affects drainage conditions was shown by tests on samples, having identical initial void-ratios, loaded under identical loading conditions (*Schroeder* and *Wilson*, 1962). It was found that the higher samples had a greater final void-ratio (i.e., less strain) than the shallow samples as the drainage of water prevented the establishment of high strain rates to further break down the structure of the soil.

The influence of load conditions as it affects the establishment of strain rates was shown by tests on identical samples. A diagram was drawn by plotting strain rates versus water-contents for constant applied stress (fig. 1); it can be seen that a single load application has a greater consolidating influence than incremental loading due to the higher strain rates established. Strain rates at constant water contents were then replotted as functions of applied stress. In this way it was possible to construct an applied stress/axial strain rate diagram, at constant void-ratios; this rheological diagram (fig. 2) indicates the basic pseudo-plastic character of the soil. The effect of drainage is indicated, qualitatively, on the stress/strain-rate graph; the sharp increase in curvature (for example on the $w/c = 850\%$ line) shows a departure from the normal viscous flow line (dotted) indicating that the drainage rate has become the limiting factor.

A schematic Rheological Diagram (fig. 3) was drawn from the experimental data previously

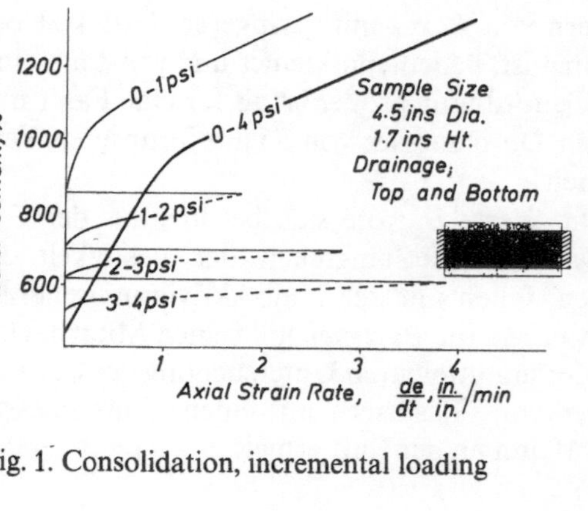

Fig. 1. Consolidation, incremental loading

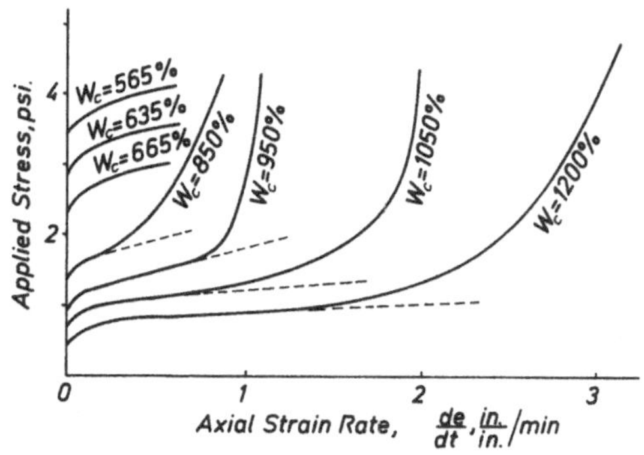

Fig. 2. Consolidation, constant W_C. Curves

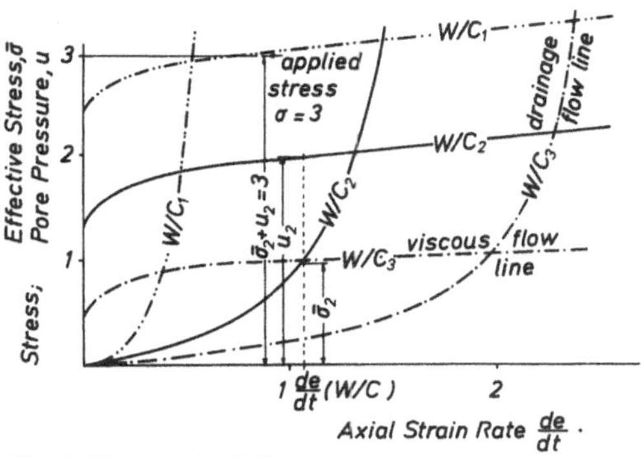

Fig. 3. Rheological diagram

analysed. This diagram shows the combined effect of drainage and viscous flow for a soil under constant applied stress. The actual shapes of the drainage lines are unknown because the solution to the partial differential equation is not known. Working backwards, it is possible to determine the shapes of the drainage flow lines as the viscous flow lines can be obtained for specimens where drainage is not critical, such as partially saturated samples.

From the Rheological Diagram (fig. 3), it can be shown that the sample deforms at a particular strain rate which is determined by the stresses; the sum of the effective stresses causing viscous flow and the pore-water stresses causing drainage must be equal to the total applied stress. Consequently, for a given total applied stress the sample deforms at a strain-rate which is uniquely related to the void-ratio. It can be seen that the drainage conditions govern the strain-rate, while the viscous resistance provides a correction which is independent of pore-pressure.

Practical significance

Laboratory tests on soils are conducted to provide data for full scale calculations of field conditions. The laboratory strain-rates are much greater than those occurring in the field. These high strain-rates accentuate the influence of primary consolidation (drainage) during laboratory testing because the field strain-rates, with the exception of thin zones adjacent to the drainage boundaries, are considerably smaller. Consequently, the viscous effect (secondary consolidation) should be more significant for field analysis than laboratory tests indicate. It has been often recorded that building settlements have been considerably slower and smaller than the settlements predicted from laboratory data.

Changes in soil properties

As consolidation proceeds with the escape of pore-water from the voids, there is a reduction in permeability which further affects the rate of consolidation. In the case of soft and organic soils with large water-contents and void-ratios, these changes in permeability cover a wide range during the consolidation stage and the rate of change is related to the effective stress. The changes in permeability have the effect of partially blocking the drainage path along the drainage boundaries; it is analogous to the porous stone becoming partially blocked during the consolidation test. Fig. 4 shows a section through a partially consolidated sample of peat.

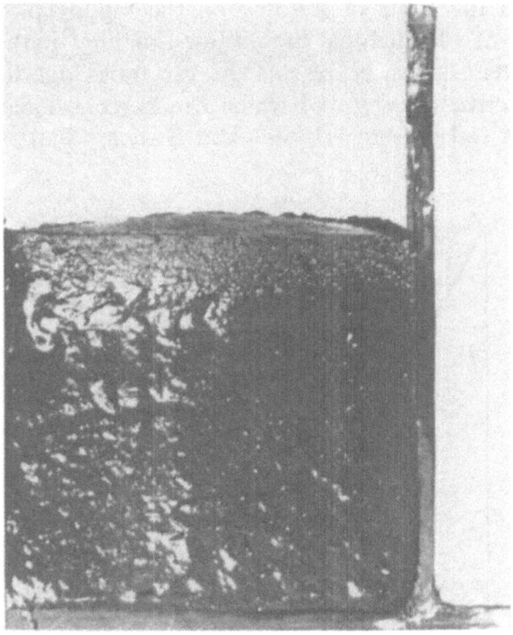

Fig. 4. Sample partially consolidated

Parameters considered

By relating increases in effective stress to decreases in void-ratio and permeability, the classical *Terzaghi* Theory was extended. In *Terzaghi*'s theory, the Degree of Consolidation is independent of load increment ratio due to the assumptions used; this "Extended Theory" shows the influence on consolidation of the variations in void-ratio and permeability due to the loading conditions. The significance of these variations was investigated empirically (*Wilson, Radforth, MacFarlane* and *Lo*, 1965) showing that the rates of consolidation for soft compressible soil did not obey the linear equation.

Considering the reduction in permeability (k) and the reduction in void-ratio (e) corresponding to the increase in effective stress (P'), the following relationship was obtained from experimental data for a very soft organic soil (*Hwang*, 1966).

$$\frac{k}{1+e}(P')^n = \frac{k_0}{1+e_0}(P'_0)^n \qquad [1]$$

where k and k_0 are the permeabilities corresponding to e and e_0. The experimental data as the basis for this equation is shown in fig. 5. Eq. [1] can be rewritten as

$$\log\left(\frac{k}{1+e}\Big/\frac{k_0}{1+e_0}\right) = -n\log(P'/P'_0),$$

where $n = \tan\beta$.

β is the slope of the line on the log-log plot (fig. 5), defined as the "Flow-Loading Parameter", which represents the variations due to effective stresses, and which can be considered as a soil property (*Wilson* and *Hwang*, 1968).

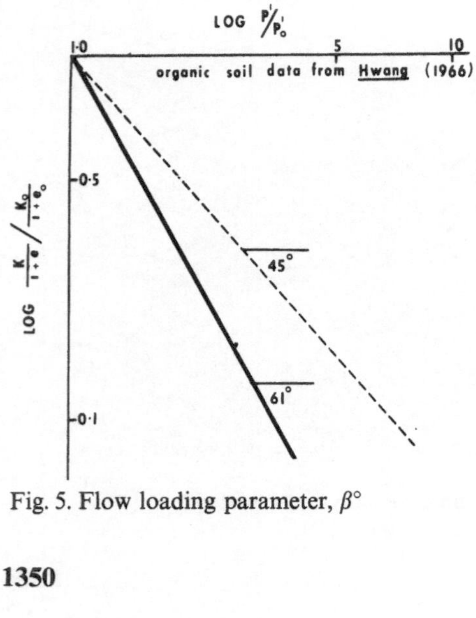

Fig. 5. Flow loading parameter, $\beta°$

The linear Theory of Consolidation has the form

$$c_v\frac{\delta^2 u}{\partial z^2} = \frac{\delta u}{\delta t}$$

where c_v is the coefficient of consolidation, u is the pore-water pressure.

The non-linear equation which includes the Flow Loading Parameter has the form

$$\frac{1}{D}\cdot\frac{\delta D}{\partial T} = \frac{\delta^2 D}{\partial X^2}.$$

Graphical presentation

Fig. 6 shows the curves for Degree of Consolidation (S) plotted versus Time Factor (T). These curves show the influence of the Flow-Loading Parameter (β) for various values of Load-Increment Ratio $\left(\frac{\Delta p}{p}\right)$. The curve of Degree of

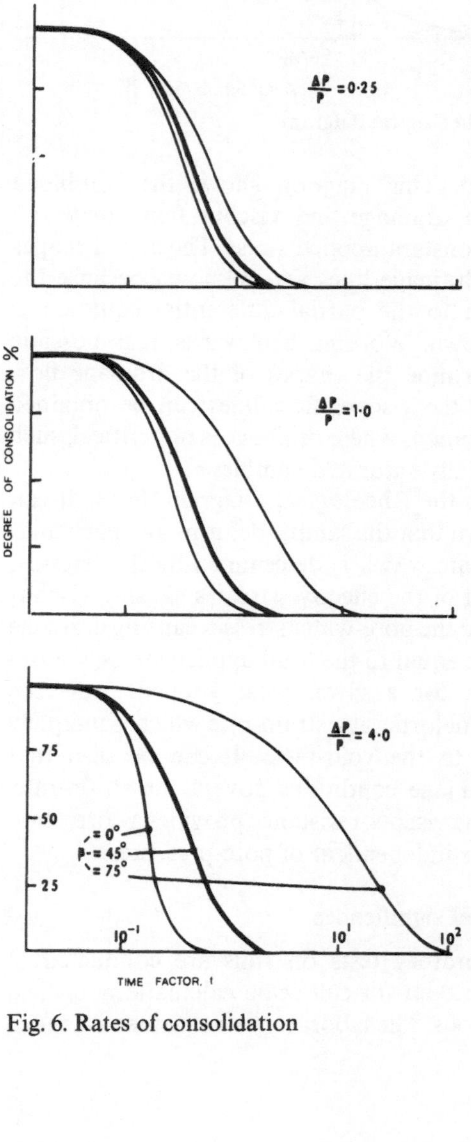

Fig. 6. Rates of consolidation

Consolidation for $\beta = 45°$, that is for $n = 1$, is coincident with the linear *Terzaghi* Theory. These graphs show the Time Factor, and hence the time for consolidation to occur, is influenced by the Load-Increment Ratio and Flow-Loading Parameter (β) embodying variations of void-ratio, permeability and effective stress.

Discussion of curves

The Flow-Loading Parameter, β, varies for different soils. Therefore, in dealing with consolidation, this parameter may be one of the major factors. The value of $\frac{k}{1+e}$ in the equation of continuity was taken by *Terzaghi* as a constant during the consolidation process. However, fig. 5 shows that;

(i) For small load increments, the change in $\frac{k}{1+e}$ is small irrespective of the value of β; the simplification of taking $\frac{k}{1+e}$ as a constant during the consolidation process is acceptable.

(ii) For large load increment ratios, the change in $\frac{k}{1+e}$ is significant for large values of β, and insignificant for small values of β.

(iii) The change in $\frac{k}{1+e}$ increases as Load-Increment Ratio and/or the Flow-Loading Parameter (β) increases.

The classical Theory of Primary Consolidation has been extended by introduction of the Flow-Loading Parameter which interrelates the soil properties. Curves have been plotted relating both the Degree of Consolidation (S) to the Time Factor (T) (fig. 6). The effects of the Load Increment Ratio (Δ) and the Flow-Loading Parameter (β) can be seen on these graphs. The influence of these parameters can be compared to the *Terzaghi* Theory (dotted line) which was assumed to be independent of these parameters.

Conclusions

It has been shown that the consolidation process for soils is governed by hydrodynamic and structural viscosity considerations, as well as the soil properties which are continually changing during the process. These property changes are particularly significant for soft and organic clays; in the case of stiff clays, the linear theory is satisfactory.

The purpose of this research was to obtain an understanding of the discrepancies between field measurements and settlement values obtained from theoretical predictions. The available theories cannot consider the variations in permeability and void-ratio which are continuously occurring in the field. This research shows that, for small values of the Load-Increment Ratio, the divergence from the classical *Terzaghi* Theory is small.

The settlement values obtained from the theoretical predictions are based on laboratory tests where the Load-Increment Ratios significantly influence the results. Settlements and rates of pore-pressure dissipation are influenced by Load-Increment Ratios and Flow-Loading Parameters. The variations in soil properties while consolidation is proceeding alter the duration of the primary consolidation stage so that the behaviour may join the basic creep line at a time which is now significantly dependent on varying soil properties and loading.

Acknowledgements

This research was supported by the National Research Council of Canada and the Defence Research Board of Canada. The experimental work was conducted by graduate students in the Department of Civil Engineering at McMaster University.

Summary

Consolidation of soils occurs when the imposed loading produces strains accompanied by the slow escape of pore-water from the voids between the particles. This straining is accompanied by rolling and sliding of the particles into a more dense packing. *Terzaghi* realized that the plastic resistance to deformation, in the case of clay consolidation, was so small that the consolidation process could be considered as a hydrodynamic problem; in this way, he used simplifying assumptions which led to the development of a time-settlement relationship for clays based entirely on the escape of pore-water; this is a linear equation.

Consolidation theory gives a unique stress-strain-time relationship – provided that the strains are not sufficiently large to overcome the shear strength of the soil, with the consequence that rheological considerations govern the deformation process.

In the case of soft soils, especially organic soils, the volume of water expelled during consolidation is very large and the shear strength of the soils small. The experimental results indicated that, at low stresses, the strain rates were low indicating plastic deformation of the material with the process governed by the hydrodynamic theory – while at higher stresses, the strain rates increased causing the soil to flow in viscous form and the rheological behaviour predominated over the hydrodynamic process.

A further complication to the *Terzaghi* theory arises with soft soils having high water contents; during the consolidation process, the permeability is reduced as the particles assume a more dense packing.

A "Flow-Loading" parameter was obtained for the soils which incorporated the changes in permeability as a function of loading. This parameter was incorporated in the consolidation theory and provided a non-linear second-order differential equation which can predict the settlement-time behaviour as a function of both loading and changing soil properties.

References

Gibson, R. E., G. L. England, and *M. J. L. Hussey,* The Theory of One-dimensional Consolidation of Saturated Clay, Geotechnique **17**, 261–273 (1967).

Hwang, C. T., Extending the Theory for the Primary Consolidation of Soils, M. Eng. thesis (McMaster University 1966).

Schroeder, J. and *N. E. Wilson*, The Analysis of Secondary Consolidation of Peat, Eighth Muskeg Research Conference, pp. 130–142 (1962).

Terzaghi, K., Erdbaumechanik (Wien 1925).

Wilson, M. E. and *C. T. Hwang*, The Influences of Varying Soil Properties on Consolidation. Advances in Consolidation, Seventh International Conference on Soil Mechanics and Foundation Engineering, pp. 74–80 (1969).

Wilson, N. E., N. W. Radforth, I. C. MacFarlane, and *M. B. Lo*, The Rates of Consolidation for Peat, Sixth International Conference on Soil Mechanics and Foundation Engineering, Vol. 1, pp. 407–410 (1965).

Author's address:

Prof. *Nyal E. Wilson*
Department of Civil Engineering
McMaster University, Hamilton (Canada)

Rheol. Acta **13**, 845–858 (1974)

From the Laboratory of Biorheology, Hemorrhage and Thrombosis Research Laboratories, Department of Medicine, Department of Pharmacology, New York Medical College, New York, N. Y. 10029, and Bioengineering Group, Newark College of Engineering, Newark, N. J. 07012 (USA)

On biorheology: Joint plenary lecture

A. L. Copley

With 12 figures

(Received October 27, 1972)

In my Inaugural Address this morning (1) I gave a brief historical account of biorheology as an organized science. The different fields of biorheology, including hemorheology (2–4), have grown rapidly and to such an extent that any abridged survey will remain inadequate. I shall therefore not give a survey on the development and present status of different fields of biorheology. However, I should like to draw your attention to a book on biorheology which our colleague *George W. Scott Blair* is writing now. It will be published some time next year by Elsevier-North Holland of Amsterdam (5).

We all are interested in the future prospects of biorheology and, therefore, I shall now make a few comments on them. These future prospects concern both the medical and biological sciences.

In addition to the study of hemorheological diseases or pathological conditions, such as thrombosis, circulatory shock or sickle cell anemia, there are many other fields of great practical significance which are arousing interest. Among them, there are the effects of different shock stresses on the brain. They are of concern to the neurologist and others in comparing injuries arising from heading a football or to glancing blows such as often occur in car crashes, both in racing and on the roads. The study of rheological properties of bone, skin, cartilage, among others, is important. With regard to the world population, research on better contraceptive medication, not having risks inherent nor side effects, suggests that much further work is required on cervical mucus and on semen. The remarkable and unique type of elastic behaviour associated with muscle is now being intensively studied in relation to biochemical investigations. Rheumatic and bronchial diseases are associated with different rheological properties of synovial fluid and of bronchial mucus, respectively.

Much more work is needed in botany. Recent work of several investigators suggests that the transport of phloem (6) presents important rheological problems.

Our colleague Dr. *Vorob'ev* (7) has recently pointed out the importance to study not merely the rheology of cytoplasm, but of individual intracellular structures, especially those of nuclei, chromosomes and cell membranes. He also emphasized that our knowledge of the behavior of biological macromolecules has been usually limited to studies of dilute solutions. As *Vorob'ev* (7) puts it: "The interest in intermolecular interactions was aroused in connection with the discovery of the possibility of spontaneous regular association of macromolecules, which occurs under appropriate conditions in concentrated solutions. In some cases under strictly determined conditions, even self-assembly of certain biological structures – ribosomes and membranes, viruses and phages – from molecules of proteins, lipids and nucleic acids can be performed."

There can be little doubt that the advances in cytology linked to biorheological studies will benefit cancer research.

Biorheology becomes more and more concerned with the explanation of the nature of biorheological processes. This is done with the use of rather complicated theoretical approaches and models, but in the practice of medicine the usefulness of such an approach depends on the extent of our comprehension of the structural and chemical basis of biorheological phenomena.

There is a continuous expansion in the development of biorheology as an independent biological science.

The numerous deformations inherent in embryonal development await rheological study and interpretation. Concomitant with such future studies, embryology as well as genetics will be more and more related to biorheology.

In so far as psychology may be regarded as a branch of biology, "psychorheology" must be included in biorheology. The term "psychorheology" was proposed by *G. W. Scott Blair* and *F. M. V. Coppen* (8) to describe the study of the relation between the assessing of rheological properties by handling materials (mainly food stuffs) and their measurements by means of physical apparatus. These studies are fully described in *Scott Blairs* "Introduction to Industrial Rheology" (9).

It was at the IV. International Congress on Rheology, that *Aharon Katchalsky* gave an account on what he termed mechanochemistry (10). *Katchalsky* (11), *Kuhn* (12), *Breitenbach* and *Karlinger* (13) found independently in 1948 that many swollen macromolecular substances can convert under isothermal conditions chemical energy directly into mechanical work. Mechanochemistry, a field of great importance to biorheology, is especially significant in relation to the contraction and relaxation of muscles.

The motility of animals, which depends on the direct conversion of chemical into mechanical work, has been one of the riddles of science. With the introduction of chemical thermodynamics by *Gibbs* (14) as an integral part of physical chemistry, it became possible to appreciate the coupling of metabolic processes with the motility of living organisms. However, the thermodynamics of "open" systems also had to be introduced and *Aharon Katchalsky* wrote with *P. F. Curran* an important book on this subject (15). They pointed out the differences between adiabatic, closed and open systems. Open systems are enclosed by walls that allow exchange of both matter and energy with the surroundings. Open systems are therefore of particular significance for the understanding of biological phenomena, since living organisms exchange constantly both matter and thermal energy with their external environment. *Katchalsky* and *Curran* considered flows and forces in the thermodynamics of irreversible processes. Scalar flows, with no direction in space, are mainly those concerned with chemical reactions, while directed flows, characterized by vectorial properties, vary in nature and include, among others, flows of diffusion, electric current and heat. More complex cases, such as viscous flows, have been omitted in their treatment of many biophysical phenomena. However, it appears that nonequilibrium thermodynamics will be also of growing importance to the advancement of biorheology.

In his presentation on abnormal phenomena of flow, given at the I. International Congress on Rheology in 1948, *Weissenberg* referred to thermodynamic analysis (16). He proposed that the expanded amount of energy in every differential interval of time, when a material passes under some mechanical action through various rheological states, may be divided into two component parts. One part is completely reversible into external work, while the other is completely irreversible. *Weissenberg* suggested the construction of networks which exhibit an analogous division of the expanded amount of energy into reversible and irreversible components and considered these components as well as the modes of interconnection "to be of greater complexity than can be envisaged in the conventional picture of networks".

It was about 20 years later that *Katchalsky* began to develop the concept of "network thermodynamics" for living systems, in particular for biological membranes (17). I recall his great enthusiasm with which he envisaged the application of network thermodynamics to biorheology. This was last May, when he phoned me from Boston, just prior to my departure from New York for Lyon, where I had to participate in a meeting concerning the organization of our two Congresses. He planned to present his thoughts on biological membranes, thermodynamics and biorheology in a Plenary Lecture at our Congress tomorrow. We shall never be acquainted with some of these thoughts. As I mentioned this morning in my Joint Plenary Address, tomorrow's Plenary Sessions, where he wanted so much to be with us, will be dedicated to his memory.

I come now to the presentation of experimental findings from two areas of hemorheology in which my associates and I have been involved during the past few years. I believe that the presentation of these findings is not merely of interest to workers in hemorheology or biorheology, but may be of concern also to rheologists not acquainted with biological problems.

One area of work (3, 18–20), in which Dr. *Huang* and Mr. *King* participated, comprises flow properties of blood at minimal shear rates. The other area of our work (21–25), in which I was assisted by Mr. *King*, deals with viscous resistance and viscoelastic properties (26) of surface layers of plasma proteins.

I shall deal first with the flow properties of whole or nondiluted human blood at varying shear rates from 1000 down to 0.0009 sec^{-1}.

In 1942, *Copley*, *Krchma*, and *Whitney* (27) found that blood exhibits non-*Newton*ian behavior and suggested that it might possess a yield stress. At that time there was no suitable instrumentation for studying the rheological behavior of blood at very low rates of shear. In recent years, a number of investigators have studied blood viscosity down to a shear rate of 0.01 sec^{-1} (28–30). Their claims, that blood has a yield stress, were not substantiated under rigorous conditions.

There is considerable discussion in the literature regarding the significance of viscosity measurements at minimal shear rates, when the viscosity of blood increases markedly (3, 18, 28–30). Physiologically, this occurs in the circulation preceding stasis or cessation of blood flow in the postcapillary vessels of the microcirculation and after the resumption of blood flow. Low velocities of blood flow exist particularly in pathological conditions, such as circulatory shock and inflammation, as well as during surgical operations, including the transplantation of the heart and other organs or the implantation of prosthetic devices in the macrocirculation (31, 32). Thus, although there is considerable debate in the literature on the value of blood viscosity measurements at minimal shear rates, it is evident that such studies are relevant to the practice of medicine and surgery. We decided, therefore, to investigate the flow properties of blood down to a rate of shear of 0.0009 sec^{-1} (3, 18–20).

Our first findings of viscosity on a wide range of shear rates of whole blood from healthy human subjects were reported in 1968 at our last Congress in Kyoto (3) and in 1970 (18). Hematocrits were 47.5 and 30%, and the measurements below 10^{-2} sec^{-1} showed a downward slope in one of the curves plotting $\dot{\gamma}$ against τ. Since these findings were reported, many more measurements have been made, which, however, did not always exhibit the same downward slope.

In our measurements on the viscosity of whole blood from healthy human donors, we used the *Weissenberg* Rheogoniometer, which

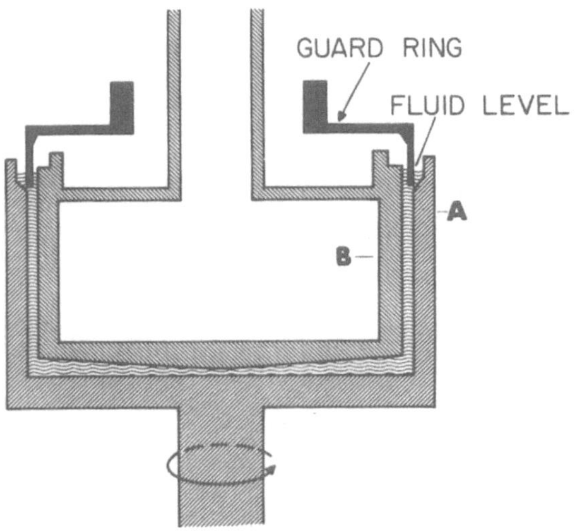

Fig. 1. Plan view of geometry showing position of removable guard-ring

was modified for biorheological studies (33). We employed a combined *Couette* and cone plate geometry similar to that originated by *Mooney* and *Ewart* (34).

In fig. 1, a diagrammatic view of the geometry used for these measurements is shown. The unit was made in two sizes: one size had a radial gap of 0.75 mm with a 1° conical end, while the second one had a radial gap of 3.0 mm and a 4° conical end. The vertical gap size was selected so that the shear rate in the *Couette* region would be the same as that in the conical region. The measuring elements were made of Plexiglass which had been siliconized.

The blood was drawn from the antecubital vein of healthy human donors, ranging in age from 25 to 60 years. The blood was mixed immediately with dry ethylenediamine tetraacetate, EDTA (1.2 mg/ml), as an anticoagulant, thereby eliminating blood dilution.

Measurements were made both in the presence and absence of the detachable guard-ring which is used to eliminate the torque which may be developed by surface layers of plasma proteins. We found no differences in the results with or without the use of the guard-ring, as can be seen in fig. 2. The light circles in this figure indicate data obtained without the guard-ring, and the dark circles represent the findings secured with the guard-ring.

The following equation of state for thixotropic fluid in *Freundlich*s sense developed by *Huang* (35) was proposed (19) to characterize flow properties of blood:

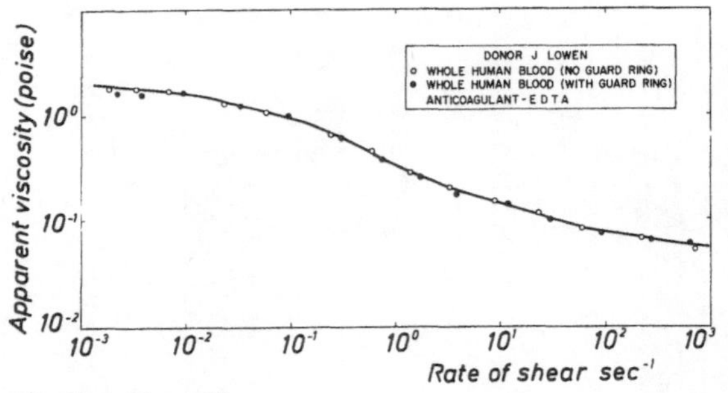

Fig. 2. (vide text)

$$\tau^{ij} \pm \tau_0^{ij} = -\left[\mu \pm C_1 \xi \beta^{ij}_{(\exp)} \frac{|\dot{\gamma}^{ij}|^n}{\dot{\gamma}^{ij}} e - C_1 \int_0^t |\dot{\gamma}^{ij}|\, dt\right] \dot{\gamma}^{ij}$$

if

$$\tau^{ij} > \tau_0^{ij}$$

where τ^{ij} is the shear stress tensor; $\dot{\gamma}^{ij}$, the shear rate tensor; τ_0^{ij}, the yield stress tensor; $\beta^{ij}_{(\exp)}$, the equilibrium molecular arrangement parameter of erythrocytes; t, the time; μ, C_1, and ξ are constants. In the equation given, the superior symbols do not indicate contravariants.

In fig. 3 three representative flow curves are shown from the data secured from the blood of eighty healthy human male and female donors with red blood cell volumes, or so-called hematocrits, ranging from 30 to 50%. The shear rate scale was divided into three regions: Region I from 50 to 1000 sec^{-1}, region II from 0.01 to 50 sec^{-1} and region III from 0.001 to 0.01 sec^{-1}.

In region I (50–1000 sec^{-1}), the blood behaves mainly as a *Newton*ian fluid. In this region, it is assumed that the aggregating force, which causes the formation of rouleaux, is much less than the desaggregating shear force, which breaks any rouleaux which may have formed into individual red blood cells. Only single red cells exist above a shear rate of approximately 50 sec^{-1} (36).

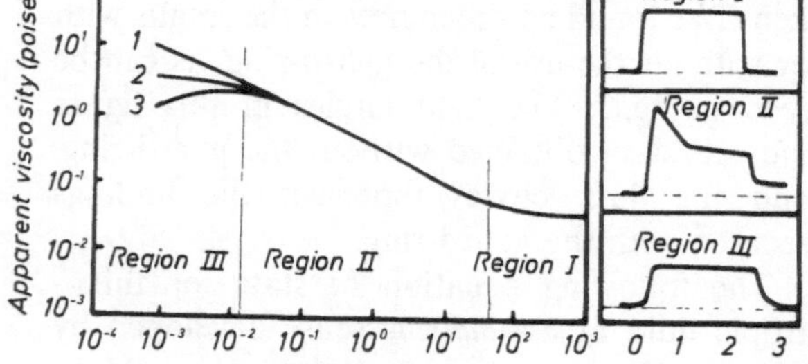

Fig. 3. (vide text)

Therefore, above this rate of shear the blood behaves as a homogeneous *Newton*ian fluid. From the flow curve, predicted from the equation shown above, the shear stress is proportional to the shear rate at high shear rates. A typical example of a recorder trace measured in this region is shown on the right-hand side of fig. 3.

In region II (10^{-2}–50 sec^{-1}), the blood viscosity depends on the rate of shear and also on the time of shearing. Such a fluid is considered to be thixotropic according to phenomenological macrorheology. A typical recorder trace for this region is shown on the right-hand side of fig. 3. The thixotropic behavior of blood in this shear rate range probably arises from the progressive break-down of rouleaux under shear. In addition, erythrocyte sedimentation and possible migration of proteins and erythrocytes at the interface of blood and the wall of the viscometer may play a role in the measurements of the rheological behavior of the blood. Therefore, the validity of reported findings is open to doubt. It is of interest that only in the recorder trace of region II, obtained after the cessation of shearing, the recovery curve fails to return to zero, suggesting that blood has a yield stress.

At constant shear rate, the torque-time curve reveals a decay of the torque after it reaches a peak value in this region. In view of the above mentioned possibility of artefacts in measurements of the rheological behavior of blood, the peak value of the torque is used for the calculation of the shear stress. Our findings demonstrate that all blood samples exhibit a decrease in apparent viscosity with increasing shear rates. This rheological behavior, as well as the torque decay, can be represented by the above equation. It should be noted that the equation covers not only the time-dependent behavior of blood, but also the wide range of shear rates from 0.01 to 1000 sec^{-1}. The *Casson* equation (37) has been extensively used for blood systems, ever since it was first applied to them by *Scott Blair* and *Copley* (38–43). This equation does not predict the time-dependent behavior of blood and is applicable only over a limited range of shear rates.

Dintenfass (30) has reported findings which differ markedly from ours in region II and those by *Chien, Usami, Taylor, Lundberg*, and *Gregersen* (28). *Dintenfass* reported measurements of viscosity of blood with samples obtained from ten healthy human subjects. He tested the sam-

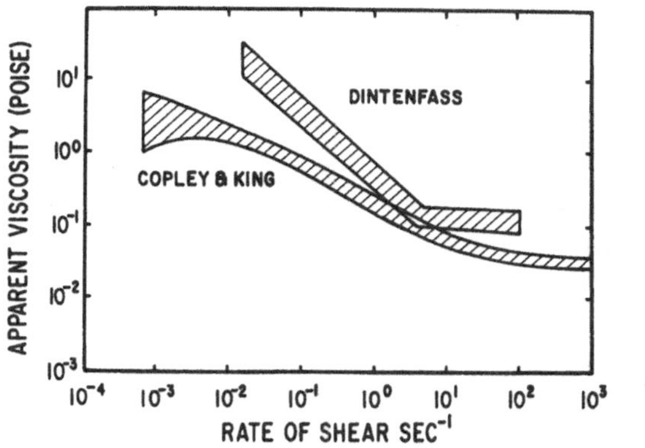

Fig. 4. (vide text)

ples immediately after withdrawal without the use of anticoagulants, employing a cone in cone viscometer with shear rates down to nearly 0.01 sec^{-1}. His findings are compared with our findings in fig. 4.

It can be seen that our findings differ widely from those *Dintenfass* (30) reported at the shear rates he could employ with his instrument. The only explanation we can offer for this marked discrepancy, which is of the order of up to 10 fold at the lower shear rates, is that fibrin formation and polymerization might well have begun in the blood samples, tested by *Dintenfass*, since no anticoagulants were used in his study.

It is known that the coagulation process, i.e., the conversion of fibrinogen to fibrin, commences as soon as blood has been shed, subsequent to the activation of thrombin. In light scattering studies of fibrinogen-thrombin systems *Copley* et al. (3, 44, 45) demonstrated the rapidity with which fibrin polymerization occurs.

No discrepancy, however, exists between our findings and those of *Chien* and associates (28). These authors used the GDM viscometer which has a *Couette*-like geometry. They used non-diluted blood, anticoagulated with heparin. Their measurements were made at varying shear rates down to 0.01 sec^{-1}. As their geometry differed from the one we used, the different type of geometry cannot be made responsible for the large discrepancy between our findings and those of *Dintenfass.*

In region III (0.001–0.01 sec^{-1}), we found three typical types of behavior, as is shown in fig. 3. In the first curve, the viscosity increased steadily with decreasing rate of shear. In the second curve, the viscosity approaches a constant value, while in the third curve the viscosity appears to decrease. A typical recorder trace of measurements in this region is shown on the right-hand side of fig. 3.

In the eighty samples measured, these characteristic curves were numerically about equally distributed. Slopes 1 and 3 may be due to a scatter because of instrument difficulties at the high sensitivity employed during these measurements. Therefore slope 2 may well represent the actual rheological characteristics of whole blood in Region III. At present, no quantitative explanation for these differences in the slope can be offered. It is proposed that the blood flows like a solid plug in this region of minimal shear rates. We postulate that the associated shear stresses in region III are less than the yield stress of blood, and that then rouleaux of red blood cells will continue to exist, which will cause the structure to flow as a plug. Slip may occur at the blood-wall boundary on a plasma layer. Two factors will contribute to the maintenance of a stable, solid plug of blood, namely 1. *the van der Waals* forces (46) of attraction between erythrocytes and 2. the forces of physical and/or chemical links at points along the chains (47) of the rouleaux arising from polymeric bridging.

The shear modulus of blood, having the characteristics of a solid below the yield value, which will be a function of its chemical, physical and electrical interactions as well as of its surface properties, is poorly understood. Therefore, an indirect effect such as the shear strain on the shear modulus, which is expected to alter these properties by imposing an extra force field, has, to our knowledge, hitherto not been previously described. In this region of minimal shear rates our results show that the shear modulus of blood, which behaves as a solid, may be increased, decreased or unaltered by the increase of the shear strain.

It can also be seen from fig. 3 that, at a shear rate of about 0.01 sec^{-1}, there is a change in the slope of the curve. We suggest that this may be the point where the column of red blood cells, which flows as a plug, breaks down, and a transition to laminar flow begins. This may well be the manifestation of the actual yield point. Therefore, the shear stress at this point might be the real measure of the yield value, in contrast to an estimate based on the generally employed extrapolation technique using *Casson* plots.

In our earlier communication in 1970 (18), we referred to the need for using different gap widths and cone angles in order to establish the effect that they may have on the flow properties of whole human blood. At that time we used different gap widths and cone angles in our studies, but our findings could not be obtained with blood samples from the same blood withdrawal, and, therefore, they could not be conclusive. Recently, we could study this problem again, when two *Weissenberg* Rheogoniometers could be used in the same laboratory. Thus, we were able to examine the flow properties of identical blood samples simultaneously in two *Weissenberg* instruments. The radial gap sizes were 0.75 mm with 1° cone angle in instrument A, and 3 mm with a 4° cone angle in instrument B.

Our findings with the blood of one donor are shown in fig. 5. A further comparison was made by drawing the blood from the same donor again after an interval of three days, and making comparative studies employing the two different gaps. The curves with the crosses correspond to the 0.75 mm gap and the 1° conical end, and the curves with the circles correspond to the 3 mm gap and the 4° conical end. In the upper and lower two curves no significant change occurred at shear rates down to 10^{-1} sec^{-1}. The results secured by the instrument with the smaller gap show a curve which tends to come to a plateau. The lower two curves, exhibiting data with the blood drawn from the same subject three days later, show a remarkable duplication of the results obtained with the two gaps. In a theoretical study, *Gazley* confirmed to some extent our findings with the different gap sizes (48).

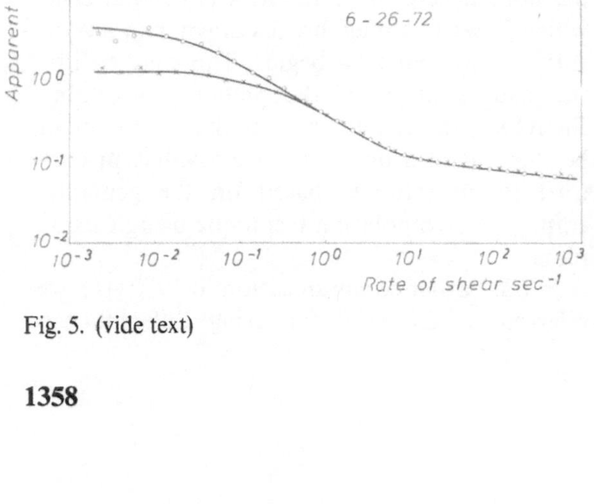

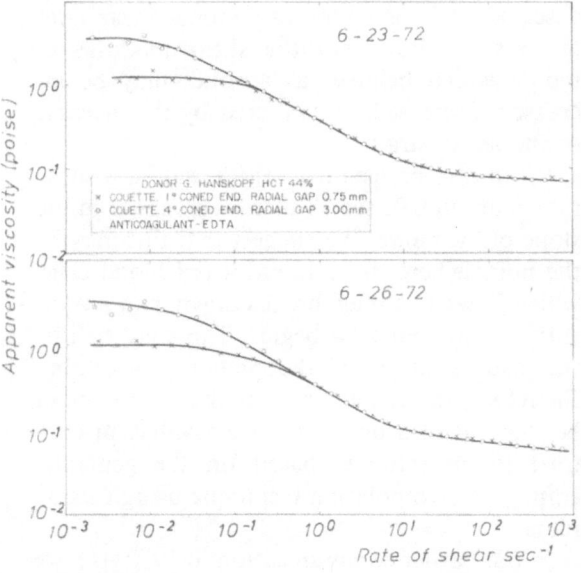

Fig. 5. (vide text)

At low rates of shear, below 10^{-1} sec^{-1}, results obtained with the small gap show a levelling off, while those secured with the larger gap exhibit higher viscosities. The hematocrit of the two blood samples remained constant at 44%. Similar findings were obtained with the blood secured on different days from another healthy human subject.

Our data are of particular interest for region III of minimal shear rates. It is generally accepted that blood has a yield stress (2, 4, 49). In such a case, the viscosity values should be infinitely high in this minimal shear rate region. As this does not occur with our measurements, plug flow must be occurring with slip at the geometry interface. In the case of plug flow, our viscosity values may not be meaningful as measures of the viscosity of whole blood.

Our findings, however, in regions I and II characterize the flow properties of whole blood from 1000 down to 10^{-2} sec^{-1}, for which we have presented an equation of state and also other pertinent explanations which we consider valid.

I shall now deal with the other area of my studies in recent years. It is related to the main killer in the Western world, namely thrombosis, which can be considered a truly hemorheological disease.

Concepts of the initiation of thrombosis, first proposed more than a century ago, have been based on two major hemorheological processes of in vivo blood clotting, found to occur either separately or mixed, viz., the clumping of blood cellular elements and the coagulation of plasma by the formation of fibrin from its precursor fibrinogen. Thrombus formation was considered to be intravascular clotting in any of the mentioned forms, or combined, resulting in partial or complete obstruction of blood vessel segments and leading to impairment of the circulation. To these two concepts I advanced last year a third one, unknown before (21–23).

This concept is based on the formation of polymolecular layers of fibrinogen and other plasma proteins leading to obstruction of the affected blood vessels. This aggregation of proteins is considered to occur in two steps. A first adsorption process would occur on the surface of the endothelium, facing the lumen of the blood vessel. This first adsorption step appears to be favored by the physiological occurrence of the different forms of fibrin, the so-called cement fibrin, which I proposed first in 1953 as the endo-endothelial fibrin layer to cover the endothelial cells (50). Since fibrinogen and other plasma proteins were found to have a great affinity to fibrin, this first adsorption process would be facilitated. This first step is followed by a growth process, in which additional protein molecules adsorb on previously formed adsorption layers in the lumen of the affected blood vessels.

This deposition of plasma proteins in layer upon layer on the inner lining of the vessel wall would constitute the initial thrombus. It would then grow contiguously and, subsequently, affect large blood vessels, a process followed by fibrin formation, polymerization, gelation and/or blood cellular clumping.

This concept promises to initiate entirely new approaches in detecting a susceptibility towards the development of thrombotic conditions in the blood of apparently healthy human subjects, and in the prevention, diagnosis and treatment of thrombosis.

According to this concept a marked increase in the non-homogeneous distribution of fibrinogen and other proteins in solution would lead to their proposed deposition. By non-homogeneous distribution is meant the time-dependent, progressive adsorption of the plasma proteins from the solution at the interfaces with the vessel wall and the free surface of the adsorbed proteins.

Such a possibility may be related to considerations of *Copley* and *Staple*, made ten years ago, as to whether a suspension of macromolecules would show a radial distribution when flowing through a capillary tube in which the velocity gradient from the axis to the wall of the tube was steep (51).

Last April, a paper on electrical field-flow fractionation of proteins by *Caldwell*, *Kesner*, *Myers*, and *Giddings* was published in the journal Science which appears to support our earlier considerations (52). Field-flow fractionation, proposed by *Giddings* in 1966 (53, 54), is a separation method in which various applied fields, working in conjunction with cross-sectional flow nonuniformities in a narrow tube, cause the differential migration of molecules and ions. There may well be a higher transport rate for larger molecules than for smaller ones. It appears that their findings would substantiate our contention of a nonuniform distribution of particles across a flow channel. Further studies on the distribution of protein molecules in flowing blood in capillary tubes and in living blood vessels are needed.

Eirich (55) referred to the profound surface effect of the macromolecules including polyelectrolytes in many types of interaction involving macromolecules. Our earlier in vivo hemorheological observations on the more or less immobile layers next to the endothelium, first proposed by *Poiseuille* in 1839 (56), and the phenomenon of wall adherence, described by *Copley* and *Scott Blair* (39–41, 57), may also have a bearing on the formation of polymolecular layers of fibrinogen and other plasma proteins in flowing blood.

In our viscous resistance studies we followed in principle the approach of *Joly* (58, 59), who, in some of his studies on surface viscosity, measured the viscosity with a *Couette* geometry with and without a guard-ring.

We define "viscous resistance" or "overall viscosity" as a value calculated from an average of the torque derived from the bulk of the test fluid plus that of its polymolecular surface layers. We prefer the use of the term "viscous resistance" (24), as, to my knowledge, it has not been used specifically. In some of our earlier publications, however, we used the term "overall viscosity". Accordingly, our findings were plotted as apparent viscosity versus rate of shear and we referred to the curves as viscosity profiles. Although it was always made clear that overall viscosity was meant, we now prefer to present our findings more directly as torque, τ, versus rate of shear.

We used the *Weissenberg* Rheogoniometer, with the modifications we made for hemorheological studies (33). The apparent viscosity was measured at shear rates from 1000 to less than 0.1 sec^{-1}. In the combined *Couette* and cone and plate system used, a geometry similar to that described by *Mooney* and *Ewart* (34), surface layers form at the interfaces of the test sample of plasma protein solution and the surfaces with which it comes into contact. These surface layers contribute added torque to the measuring inner platen which is particularly marked at low shear rates. A stationary guard-ring, attached to the main body of the rheogoniometer, is used, as shown diagrammatically in fig. 1. The guard-ring is entirely detached from the driven,

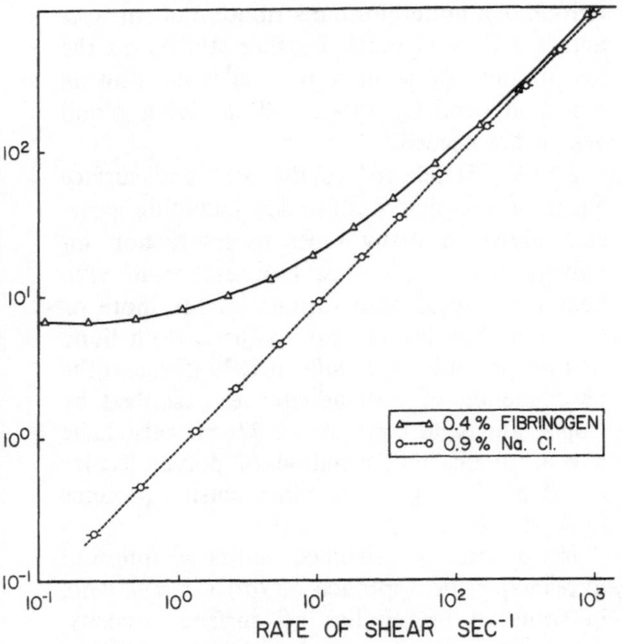

Fig. 6. (vide text)

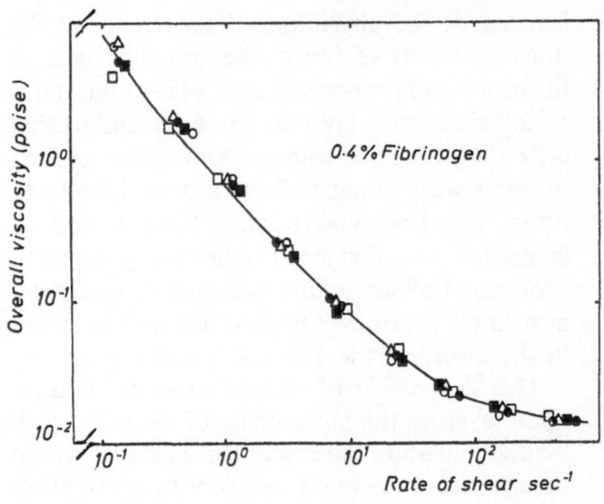

Fig. 7. (vide text)

rotating outer platen and the measuring inner platen. This guard-ring excludes transmission of the torque from the surface layer to the measuring platen.

Several investigators (28, 60, 61) reported findings that plasma systems and fibrinogen solutions exhibit non-*Newton*ian behavior, if the precaution of using a guard-ring is not applied. However, these investigators did not consider their findings, obtained without a guard-ring, as a phenomenon in need of further study, but rather as an artefact to be avoided.

Viscous resistance is always measured without the guard-ring. In fig. 6 the viscous resistance of a 0.4% fibrinogen solution is compared to that of physiologic saline (0.9% NaCl). As expected, the plot exhibits *Newton*ian behavior of the saline, but marked non-*Newton*ian behavior of the fibrinogen solution.

Fig. 7 shows a plot of five experiments of the same 0.4% fibrinogen solution, measured separately without a guard-ring on the same day. These findings demonstrate that the overall viscosity or viscous resistance of the surface layers of fibrinogen is consistent.

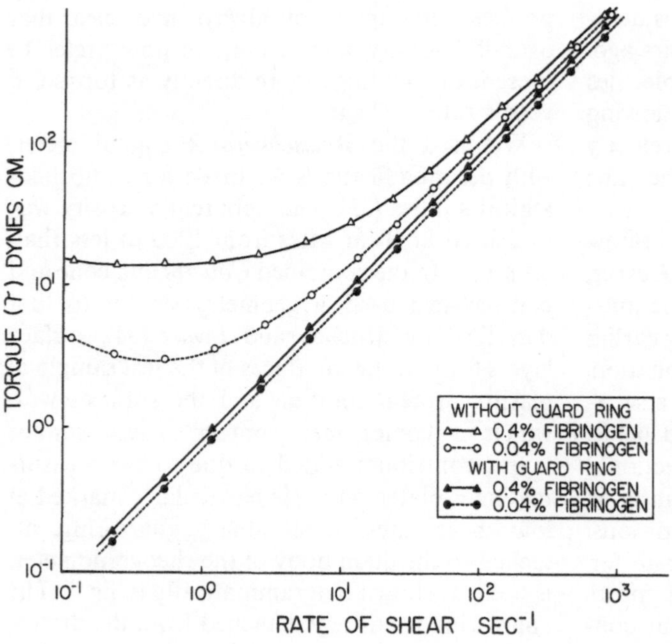

Fig. 8. (vide text)

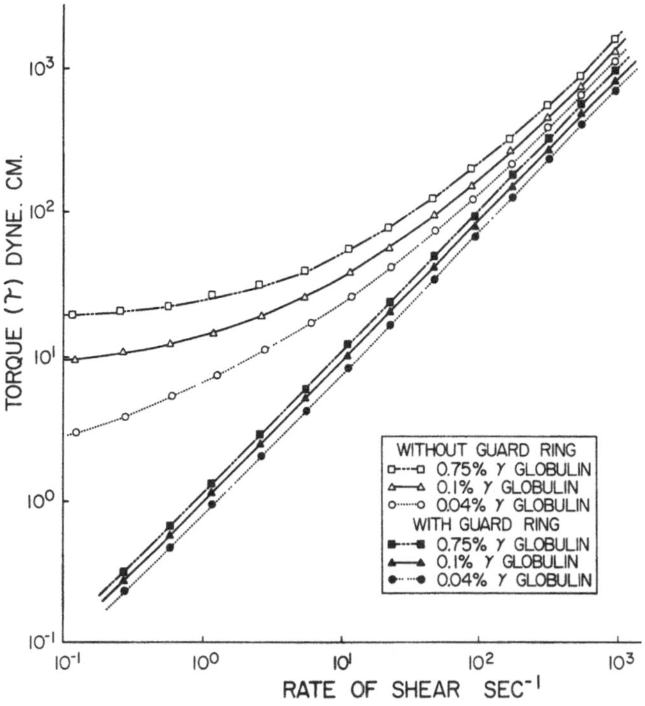

Fig. 9. (vide text)

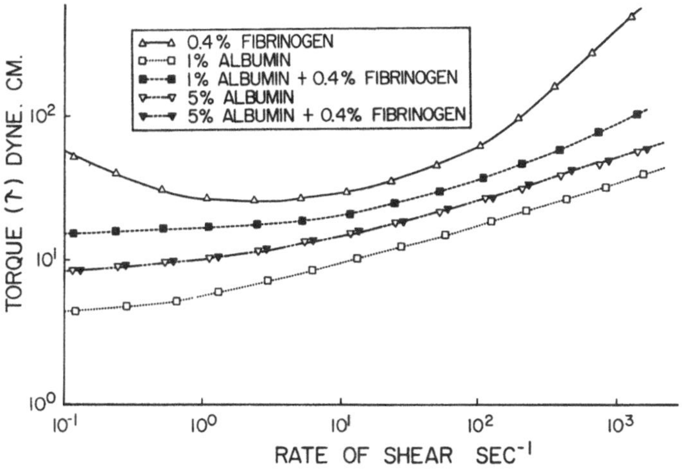

Fig. 10. (vide text)

A comparison of torque values of two fibrinogen concentrations, obtained with and without the guard-ring, is shown in fig. 8. The curve obtained when using the guard-ring, which eliminates measurements of surface layers, shows *Newton*ian flow characteristics. As expected, the higher fibrinogen concentration (dark triangles) results in a higher torque, indicating a higher viscosity. Without the use of a guard-ring a marked increase in τ values occurs below the shear rate of 10 sec^{-1}. This indicates that apparently thicker and stronger surface layers of fibrinogen form with the higher fibrinogen concentration at the test fluid-air interface of the viscometer geometry.

Similar results have been secured with solutions of gamma globulin, as can be seen from fig. 9. Here the dark circles, triangles and squares mark viscosity values, secured with the guard-ring, of concentrations of 0.04, 0.1 and 0.75% gamma globulin, while the corresponding light markings refer to values, secured without the guard-ring. As can be seen, viscous resistance is markedly increased with higher gamma globulin concentrations, particularly below 10 sec^{-1}.

In fig. 10 plots of τ of albumin solutions are presented. They are derived from the surface layers only, and obtained by subtracting the torque, derived from the bulk solution with the guard-ring in place, from the torque measured without the guard-ring. Torque values of concentrations of 1 and 5% albumin with and without the addition of 0.4% fibrinogen and fibrinogen as control are compared. Although the concentration of fibrinogen (light triangles) is much lower than that of the 1% albumin (squares) and of the 5% albumin (inverted triangles), the albumin preparations give much lower τ values. However, when the same fibrinogen concentration is added to both albumin solutions, the comparison shows no difference with the 5% albumin, but an increase in τ with the fibrinogen – 1% albumin (dark squares). It is noted that these values are much lower than those of the fibrinogen control (light triangles). Since albumin is generally considered to be one of the main protective colloids, it may act in reducing the successive adsorption of monomolecular layers of fibrinogen at the boundary.

Fig. 11 shows a comparison of viscous resistance of surface layers of 5 and 18% serum with and without added 0.4% fibrinogen, and with 0.4% fibrinogen as control. Very low τ values were obtained with the 5% serum (lowest curve). If fibrinogen is added, the values are markedly increased (second curve from above) and approach the fibrinogen control (upper curve). However, the τ values with 18% serum with and without fibrinogen do not exhibit any change. These findings indicate the presence of a critical concentration of serum, when the addition of the same amount of fibrinogen will not cause any increase in values. Similar findings were shown with plasma. Both the plasma and serum may

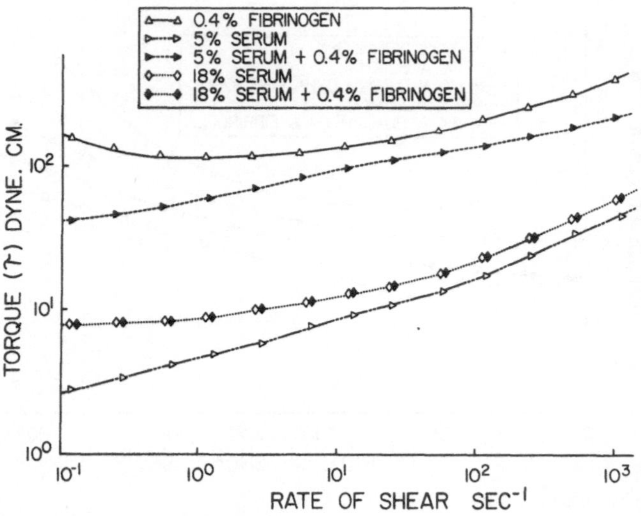

Fig. 11. (vide text)

also contain protective colloids other than albumin which have a similar action on fibrinogen.

We obtained a number of other interesting findings which time will not permit me to present. Of special interest are our new results with regard to the action of red blood cells and platelets on viscous resistance of plasma protein systems which we are going to report here at the Congress of Biorheology (25).

King and I also made preliminary observations on the viscoelasticity of surface layers of a 0.4% fibrinogen solution. The method which we employed is described by us in an exhibit at our Congress (26).

Our findings show the definite presence of an elastic component as you will note from fig. 12. The phase difference between the two traces is 28°. As is generally known, the phase difference would be 90° in a material exhibiting *Newton*ian behavior.

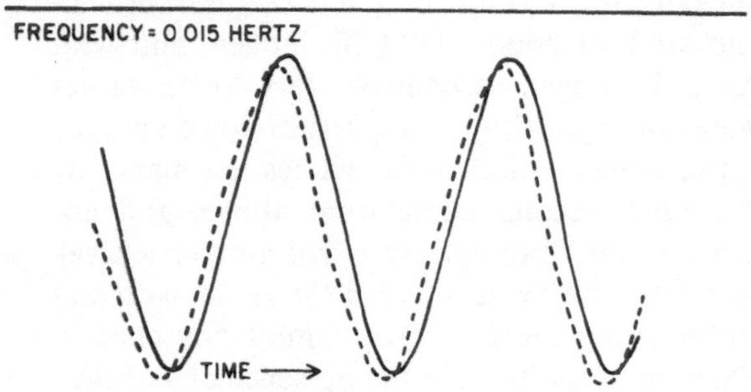

Fig. 12. The input motion trace (solid line) and the output torque trace (dotted line) of a 0.4% fibrinogen solution, using the oscillatory mode of the *Weissenberg* Rheogoniometer. The phase difference is 28°

Work is in progress with regard to comparative studies of viscoelasticity, viscous resistance and the composition of the surface layers of different systems of plasma proteins. These new studies are being made jointly with Professors *Israel Miller* and *Alex Silberberg* of the *Weizmann* Institute of Science. These joint studies in both our laboratories in Rehovot and in New York City were initiated by my good friend, the late *Aharon Katzir-Katchalsky* (62).

I am now coming to that part of this lecture which deals with biorheology as it is practiced as an art in medicine and surgery.

Descriptions of biorheological phenomena go back into antiquity (63–65). *Hippocrates*, the ancient Greek physician (c. 460 to c. 370 B.C.), who placed medicine on a scientific basis through systematic observation of disease, was aware of biorheological phenomena (66). In the writings which have come down to us under the name *Hippocrates*, I should like to cite some phenomena which he described. He correlated the thickness of discharges and of other body fluids and the change in their consistency with a number of diseases. In his treatise "On Ancient Medicine" he says "And it appears to me that one ought also to know what diseases arise in man from the powers and what from the structures. What do I mean by this? By powers, I mean intense and strong juices; and by structures, whatever conformations there are in man".

Hippocrates talks about the hardness of organs, their density, sponginess and "loose texture" such as the spleen and lungs. He said: "Those parts which are hollow and expanded are most likely to receive any humidity flowing into them, but cannot attract it in like manner. Those parts, which are solid and round, could not attract a humidity, nor receive it when it flows to them, for it would glide past, and find no place of rest on them. But spongy and rare parts, such as the spleen, the lungs and the breasts, drink up especially the juices around them, and become hardened and enlarged by the accession of juices. Such things happen to these organs especially." He gives many other examples of biorheological phenomena.

The approach of *Hippocrates* of manually testing the consistency of organs in patients continues to be, in our time, a main tool of the physician and surgeon whom the patient consults. In some diseases such as, for instance, ovarian cancer, it unfortunately remains today thus far the only tool which leads to the discovery of the disease. For gross examination of anyone of us, this art of subjective or psychorheological testing by the hands of the physician represents practical biorheology as it is practiced daily in medicine and surgery. Thus, every physician and surgeon practices biorheology without being aware of it. I doubt that this art will ever be entirely replaced by instrumental techniques providing quantitative measurements, although this is already done by the ophthalmologist using the tonometer.

In my Inaugural Address before our two Congresses this morning held in this festive Palais des Congrès in Lyon, I gave a brief survey of biorheology as an organized science (1). I believe, I should end this lecture in giving some examples of scientific thought in the history of science. These thoughts and great discoveries emphasize the present need of the biological sciences toward their advancement by the new physical science of rheology. This has been historically a rather slow process which culminated in 1948 at the First International Congress on Rheology (67) in the emergence of the science of biorheology (1, 67, 68).

The contribution of *William Harvey* (1578 to 1657) "Exercitatio anatomica de motu cordis et sanguinis in animalibus" ("An Anatomical Disquisition on the Motion of the Heart and Blood in Animals") (69, 70) rendered *Galens* doctrine (71, 72) on the flow of blood ("On the Functions of Parts of the Human Body") obsolete, which ruled medicine since the second century A. D.

Giovanni Borelli (1608–1679), a student of *Galileo Galilei*, published in 1667 "De vi percussionis liber" and in 1670 "De motionibus naturalibus a gravitate pendentibus". These two books deal with problems of mechanics and impulsive motion. They were an introduction for the better understanding of his theory of animal motion. In "De motu animalium" which appeared one year after his death, and was also published here in Lyon about five years later in 1685, *Borelli* (73) dealt with different aspects of motion. "Motion" was not merely meant to mean external local movement but as well the internal movement of fluids and particles which comprise the living organism. It was *Borelli* who, in this book "On the Motion of Animals", was probably the first to think of chemical processes behind the mechanical activity of muscular contraction.

Borellis pupil *Marcello Malpighi* (1628 to 1694) discovered the blood capillaries. In his classical experiment (74) he tied the lungs of a frog and watched with a microscope the flow of blood in the lung's capillaries. However, it was *Anthony van Leeuwenhoek* (1632–1723), this most remarkable self-trained biologist, who demonstrated the living circulation in the network of the blood capillaries (or, as it is now called, microcirculation) with the aid of his home-made microscope (75, 76). *Leeuwenhoek* thus provided the experimental evidence that *Harvey* postulated in his theory, which became one of the principal events in the history of science.

Studies of flow properties of living matter thus began with the discoveries of *Harvey*, *Malpighi*, and *Leeuwenhoek* pertaining to the circulation of blood, as well as later, in 1774, with the discovery of the streaming in plant cells by *Corti* (77, 78).

Jean-Léonard-Marie Poiseuille first reported in 1835 (79) his in vivo studies "Recherches sur les causes du mouvement du sang dans les vaisseaux capillaires", which led to the application of rheological treatments to the flow of blood (56, 80). The discovery of the laws of flow was based on experiments which *Poiseuille* was stimulated to make from his observations in living blood capillaries, resulting in his studies "Recherches expérimentales sur le mouvement des liquides dans les tubes de très petits diamètres" (80, 81). These reports, published from 1840 to 1842, were accepted throughout Europe. On the basis of *Poiseuille's* findings, *Maxwell*, *Jacobson*, *Mathieu* and others deduced from the fundamental equation of *Newton* the well known formula for viscosity, which was later named after *Poiseuille* and *Hagen* (81).

There are other reciprocal stimuli from biology and rheology, which promise to continue to be fruitful. Numerous biological phenomena and processes await a rheological approach for the characterization of the flow properties involved and for quantitative studies.

The development of our planet consisted in the past, as presumably it will consist in the future, of series of rheological occurrences. So does the development of life from its far distant past to its far distant future. Thus, biorheology is a science which encompasses all forms of life of all times on this planet. And as the earth appears to be ruled by universal laws of nature, in which rheologically *Reiner's* *Deborah* number

(82) may play a role, so is the biorheological approach to problems of life and to those of our human existence. Thus, biorheology is significant in connecting on many levels the biological sciences with rheology.

Hermann von Helmholtz (1821–1894), who made manyfold studies in physiology and physics, was fascinated by the problem of the existence or non-existence of the so-called "vital force" belonging only to the living organism. His basic premise was always the "comprehensibility of nature". In his lecture "On Human Vision" ("Über das Sehen von Menschen") in 1857 (83) he conceded to philosophy the undisputed right to investigate the sources of our knowledge. He thought that no age may evade the investigation of these sources. Earlier, in 1847, in his lecture "Conservation of Energy" (84) he says: "In the end the goal of the theoretical sciences is to discover the final, invariable causes of natural processes." He has no answer to the question "whether nature must be entirely comprehensible or if there are alternatives in her which deprive the laws of a certain causality and thus allow the laws spontaneity or freedom." In his lecture "On the Interaction of Natural Forces", delivered in 1854 (83), *Helmholtz* said: "Physico-mechanical laws are, as it were, the telescope of our spiritual eye, which can penetrate into the deepest night of time, past and to come."

About 2400 years ago, *Heraclitus* said: "It is not possible to step twice into the same river" (85, 86). Ever since, philosophers have paid little attention to the idea of flow.

Among the philosophers of this century, it was *Henri Bergson* who has made the idea of flow an essential part of his thinking. I should like to conclude by citing one sentence from "L'Evolution Créatrice" ("Creative Evolution"): "The flux of time is the reality itself, and the things which we study are the things which flow" (87). "Le flux du temps devient ici la réalité même, et, ce qu'on étudie, ce sont les choses qui s'écoulent" (88).

Summary

Descriptions of biorheological phenomena go back into antiquity, but studies of flow properties of living matter began probably with the discovery of the circulation in blood capillary vessels by *Malpighi* in 1686 and the streaming in plant cells by *Corti* in 1774. Biorheology comprizes the study of the deformation and flow of living organisms and inanimate biological systems or of materials directly derived from living organisms. The term biorheology was introduced in 1948 (*A. L. Copley*, Proc. Internat. Congress on Rheology, Scheveningen, Holland, 1948, North-Holland Publ. Co., Amsterdam and Interscience Publishers, New York, 1949, Vol. 1, p. 47). Ever since, biorheology offered a framework to connect the sciences of biology with rheology. This frame, which proved to be secure, permits the application of a number of rheological treatments to biological systems. The different fields of biorheology, including hemorheology, have grown rapidly and to such an extent that any abridged survey will remain inadequate. *Jean-Léonard-Marie Poiseuille* first reported in 1835 his *in vivo* studies "Recherches sur les causes du mouvement du sang dans les vaisseaux capillaires", which led to the application of rheological treatments to the flow of blood. The discovery of the laws of flow was based on experiments which *Poiseuille* was stimulated to make from his observations in living blood capillaries, resulting in his studies "Recherches expérimentales sur le mouvement des liquides dans les tubes de très petits diamètres". These reports, published from 1840 to 1842, were accepted throughout Europe. On the basis of *Poiseuille*'s findings, *Maxwell, Jacobson Mathieu* and others deduced from the fundamental equation of *Newton* the well known formula for viscosity, which was later named after *Poiseuille*. It will be demonstrated that there are other reciprocal stimuli from biology and rheology, which promise to continue to be fruitful. Numerous biological phenomena and processes await a rheological approach for the characterization of the flow properties involved and for quantitative studies. It will be shown that biorheology, in spite of its brief history as an organized science, is of growing importance in the biological and medical sciences. New knowledge gained in biorheology, as applied to the practice of medicine and surgery, will serve the well-being of the human species.

Acknowledgements

The researches presented in this lecture were aided by the Office of Naval Research (ONR) Contract N00014-67-A-0449-0002 and the Department of Medicine and Surgery, Veterans Administration (DMSVA) Washington, D.C. The author expresses his appreciation to Dr. *Howard W. Kenney* of the DMSVA and Dr. *Leonard M. Libber* of the ONR for their interest in his studies.

References

1) *Copley, A. L.*, Joint Inaugural Address: Biorheology as an Organized Science. Rheol. Acta **12**, 89 (1973).

2) *Copley, A. L.* (Ed.), Hemorheology. Proc. 1. Internat. Conf., Reykjavik, Iceland, 1966 (Oxford-New York 1968).

3) *Copley, A. L.*, in: Proc. 5. Internat. Congress on Rheology, Kyoto, 1968, Ed. *S. Onogi*, vol. **2**, p. 3 (Tokyo, Baltimore, Md. and Manchester, England, 1970).

4) *Hartert, H. H.* and *A. L. Copley* (Eds.), Theoretical and Clinical Hemorheology. Proc. 2. Internat. Conf., Heidelberg, German Fed. Rep., 1969 (Berlin-Heidelberg-New York 1971).

5) *Scott Blair, G. W.*, Introduction to Biorheology (Amsterdam, in prep.).

6) *Spanner, D. C.*, Nature **232**, 157 (1971).
7) *Vorob'ev, V. I.*, Proc. 4. Internat. Biophysics Congress, Moscow, 1972 (Academy of Sciences, USSR, in press).
8) *Scott Blair. G. W.* and *F. M. V. Coppen*, Proc. Roy. Soc. **B 128**, 109 (1939).
9) *Scott Blair, G. W.*, An Introduction to Industrial Rheology (London 1938), 2. ed. (London 1949).
10) *Katchalsky, A.* and *A. Oplatka*, Mechanochemistry. Proc. 4. Internat. Congress on Rheology, Providence, R.I. Part 1, Eds. *E. H. Lee* and *A. L. Copley*, p. 73 (New York and London 1965).
11) *Katchalsky, A.*, Experientia **5**, 319 (1949).
12) *Kuhn, W.*, Experientia **5**, 318 (1949).
13) *Breitenbach, J. W.* and *H. Karlinger*, Monatsh. Chem. **80**, 211 (1949).
14) *Gibbs, J. W.*, The Collected Works of J. Williard Gibbs (Yale Univ. Press, New Haven, Connecticut, 1948).
15) *Katchalsky, A.* and *P. F. Curran*, Nonequilibrium Thermodynamics in Biophysics (Harvard Univ. Press, Cambridge, Mass., 1965).
16) *Weissenberg, K.*, Proc. Internat. Congress on Rheology, Holland 1948, p. 1–29 (Amsterdam 1949).
17) *Katchalsky, A.*, Proc. Internat. Union Physiol. Sci. **8**, 60 (1971).
18) *Copley, A. L.* and *R. G. King*, Experientia **26**, 904 (1970).
19) *Copley, A. L., C.-R. Huang*, and *R. G. King*, Biorheology **10**, 17, 23 (1973).
20) *Copley, A. L., C.-R. Huang*, and *R. G. King*, Proc. 4. Internat. Biophysics Congress, Moscow, 1972 (Academy of Sciences, USSR, in press).
21) *Copley, A. L.*, Biorheology **8**, 79 (1971).
22) *Copley, A. L.*, Abstract vol., II. Congress, The Internat. Soc. on Thrombosis and Haemostasis, Oslo, Norway, July 1971, p. 70.
23) *Copley, A. L.*, Proc. Internat. Union of Physiological Sciences **9**, 120 (1971).
24) *Copley, A. L.* and *R. G. King*, Federation Proc. **30**, 480 (1971); Thrombosis Research **1**, 1 (1972).
25) *Copley, A. L.* and *R. G. King*, Biorheology **9**, 147 (1972); **10**, 533 (1973).
26) *King, R. G.* and *A. L. Copley*, Biorheology **9**, 170 (1972); **10**, 541 (1973).
27) *Copley, A. L., L. C. Krchma*, and *M. E. Whitney*, J. Gen. Physiol. **26**, 49 (1942).
28) *Chien, S., S. Usami, H. M. Taylor, J. L. Lundberg*, and *M. I. Gregersen*, J. Appl. Physiol. **21**, 81 (1966).
29) *Cokelet, G. R., E. W. Merrill, E. R. Gilliland, H. Shin, A. Britten*, and *R. E. Wells*, Trans. Soc. Rheology **7**, 303 (1963).
30) *Dintenfass, L.*, Blood Microrheology-Viscosity Factors in Blood Flow, Ischaemia and Thrombosis (London 1971).
31) *Gelin, L. E.*, in: Hemorheology. Proc. 1. Internat. Conf., Reykjavik, Iceland, 1966. Ed. *A. L. Copley*, p. 823 (Oxford-New York 1968).
32) *Bernstein, E. F.* and *A. R. Castaneda*, in: Hemorheology. Proc. 1. Internat. Conf., Reykjavik, Iceland, 1966. Ed. *A. L. Copley*, p. 433 (Oxford-New York 1968).
33) *King, R. G.* and *A. L. Copley*, Biorheology **7**, 1 (1970).
34) *Mooney, M.* and *R. H. Ewart*, Physics **5**, 350 (1934).
35) *Huang, C.-R.*, Chem. Eng. J. **3**, 100 (1972).
36) *Wells, R., H. Schmid-Schoenbein*, and *J. Goldstone*, in: Theoretical and Clinical Hemorheology. Proc. 2. Internat. Conf., Heidelberg, German Fed. Rep., 1969. Eds. *H. H. Hartert* and *A. L. Copley*, p. 358 (Berlin-Heidelberg-New York 1971).
37) *Casson, N.*, in: Rheology of Disperse Systems. Ed. *C. C. Mill*, p. 84 (Oxford-New York 1959).
38) *Scott Blair, G. W.*, in: Hemorheology. Proc. 1. Internat. Conf., Reykjavik, Iceland, 1966. Ed. *A. L. Copley*, p. 345 (Oxford-New York 1968).
39) *Copley, A. L.* and *G. W. Scott Blair*, Rheol. Acta **1**, 170 (1958); **1**, 665 (1961).
40) *Copley, A. L., G. W. Scott Blair, F. A. Glover*, and *R. S. Thorley*, Kolloid Z. u. Z. Polymere **168**, 101 (1960).
41) *Copley, A. L.*, in: *A. L. Copley* and *G. Stainsby*, (Eds.), Flow Properties of Blood and Other Biological Systems, p. 97 (New York-Oxford 1960).
42) *Copley, A. L.* and *G. W. Scott Blair*, Proc. 8. Internat. Soc. Blood Transfusion, Tokyo 1960, p. 6 (Basel-New York 1962).
43) *Scott Blair, G. W.*, in: Hemorheology. Proc. 1. Internat. Conf., Reykjavik, Iceland, 1966. Ed. *A. L. Copley*, p. 345 (Oxford-New York 1968).
44) *Copley, A. L., A. Devi, R. G. King, B. M. Scheinthal*, and *P. Ohlmeyer*, in: Theoretical and Clinical Hemorheology. Ed. *H. H. Hartert* and *A. L. Copley*, p. 154 (Berlin-Heidelberg-New York 1971).
45) *Ohlmeyer, P., S. E. Lasker*, and *A. L. Copley*, Thrombosis Research **1**, 337 (1972).
46) *Verweg, E. J. W.* and *J. T. G. Overbeek*, Theory of Stability of Lyophobic Colloids (Amsterdam 1948).
47) *Lodge, A. S.*, Elastic Liquids (New York-London 1964).
48) *Gazley, C.*, Jr. Personal communications. 16. March and 25. July, 1972.
49) *Whitmore, R. L.*, Rheology of the Circulation (Oxford-New York 1968).
50) *Copley, A. L.*, Abstr. XIX. Internat. Physiol. Congress, Montreal 1953, p. 280; Arch. Internat. Pharmacodyn. Thér. **99**, 426 (1954).
51) *Copley, A. L.* and *P. H. Staple*, Biorheology **1**, 3 (1962).
52) *Caldwell, K. D., L. F. Kesner, M. N. Myers*, and *J. C. Giddings*, Science **176**, 296 (1972).
53) *Giddings, J. C.*, Separ. Sci. **1**, 123 (1966).
54) *Giddings, J. C.*, J. Chem. Phys. **49**, 81 (1968).
55) *Eirich, F. R.*, in: Hemorheology. Proc. 1. Internat. Conf., Reykjavik, Iceland, 1966. Ed. *A. L. Copley*, p. 228 (Oxford-New York 1968).
56) *Poiseuille, J. M. L.*, Recherches sur les causes du mouvement du sang dans les vaisseaux capillaires. Acad. Sci., Séance publique du 28 décembre 1835, Tome VII des Savants étrangers (Paris 1839).
57) *Copley, A. L.*, Nature **181**, 551 (1958).
58) *Joly, M.*, Biorheology **1**, 15 (1962).
59) *Joly, M.*, Proc. 5. Internat. Congress on Rheology. Ed. *S. Onogi*, vol. **2**, p. 191 (Tokyo/Japan and Baltimore/Md. and Manchester/England 1970).
60) *Wells, R. E.* and *E. W. Merrill*, Science **133**, 763 (1961).

61) *Brooks, D. E., J. W. Goodwin*, and *G. V. F. Seaman*, J. Appl. Physiol. **28**, 172 (1970).

62) *Katchalsky, A.*, Personal communication, 29. November 1971.

63) *Fåhraeus, R.*, ATTI del XIV. Congresso Internazionale di Storia della Medicina, vol. II, 3, Roma-Salerno 1954.

64) *Fåhraeus, R.*, in: Symposium on Biorheology. Ed. *A. L. Copley*. Part 4, Proc. 4. Internat. Congress on Rheology, Providence, R.I., U.S.A. 1963, p. 11 (New York 1965).

65) *Fåhraeus, R.*, Laboratory Praxis **15**, 44 (1966).

66) *Hippocrates*, The Theory and Practice of Medicine, pp. 16–17 (New York 1964).

67) *Copley, A. L.*, Proc. 1. Internat. Congress on Rheology, Scheveningen, Holland, 1948, vol. **1**, p. 47 (Amsterdam and New York 1949).

68) *Copley, A. L.*, Proc. IV. Internat. Biophysics Congress, Moscow 1972 (USSR Academy of Sciences, in press).

69) *Harvey, W.*, Exercitatio anatomica de motu cordis et sanguinis in animalibus (Frankfurt 1628).

70) *Harvey, W.*, An Anatomical Disquisition on the Motion of the Heart and Blood in Animals. Translated by *Robert Willis*, Sydenham Society, England, 1847. Reprinted: The Circulation of the Blood (London and New York 1907).

71) *Galen, C.*, De usu partium corporis humani. On the Functions of the Body. From Opera Omnia of Claudius Galen, ed. by *C. G. Kühn*, vol. III and IV (Leipzig 1821–1833).

72) *Galen, C.*, Oeuvres Anatomiques, Physiologiques et Médicales de Galien, translated by *C. Daremberg* (Paris 1854).

73) *Borelli, G. A.*, De motu animalium. 2 volumes; Lyon 1685 (Rome 1680–1681).

74) *Malpighi, M.*, De Pulmonibus, Bononia, 1661, Proc. Royal Soc. Med. **23**, 7 (1929), translated by *James Young*.

75) *Leeuwenhoek, A. van*, The Select Works of *Anthony van Leeuwenhoek* (1632–1723) Containing His Microscopical Discoveries in Many of the Works of Nature. Translated by *S. Hoole* and *H. Fry*, pp. 89–112 and 322–325 (London 1798).

76) *Leeuwenhoek, A. van*, Edited and Collected by *Clifford Dobell*: Anthony van Leeuwenhoek and His Little Animals (New York 1958).

77) *Corti, B.*, Osservazioni microscopiche sulla tremella e sulla circolazione del fluido in una pianta acquajuola (Lucca 1774).

78) *Kamiya, N.*, Protoplasmic Streaming. Protoplasmatologia. Handbuch der Protoplasmaforschung. Eds. *L. V. Heilbrunn* and *F. Weber*, vol. VIII, 3a (Wien 1959).

79) *Poiseuille, J. L. M.*, Compt. Rend. Acad. Sci. (Paris) **1**, 554 (1835).

80) *Joly, M.*, in: Hemorheology, Proc. 1. Internat. Conf., Reykjavik, Iceland, 1966. Ed. *A. L. Copley*, p. 29 (Oxford-New York 1968).

81) *Poiseuille, J. L. M.*, Experimental Investigations Upon the Flow of Liquids in Tubes of Very Small Diameter. Translated by *W. H. Herschel*. Rheological Memoirs, vol. 1, Number 1, January, 1940 (Lancaster, Pa., 1940).

82) *Reiner, M.*, Physics Today **17**, 62 (1964).

83) *Helmholtz, H. von*, Popular Scientific Lectures (New York 1962).

84) *Helmholtz, H. von*, Über die Erhaltung der Kraft, eine physikalische Abhandlung (Berlin 1847).

85) *Freeman, K.*, Ancilla to the Pre-Socratic Philosophers. A complete translation of the Fragments in Diels, Fragmente der Vorsokratiker (Cambridge, Mass., 1956).

86) *Copley, A. L.*, in: Symposium on Biorheology. Ed. *A. L. Copley*. Proc. 4. Internat. Congress on Rheology, 1963, part 4, p. 3 (New York 1965).

87) *Bergson, H.*, Creative Evolution, p. 363. Translated by *A. Mitchell* (London 1960).

88) *Bergson, H.*, L'Evolution Créatrice, 118ᵉ edition, p. 343 (Paris 1966).

Author's address:
A. L. Copley
50 Central Park West
New York, N. Y. 10023 (USA)

Rheol. Acta 13, 859–863 (1974)

From the Division of Mechanical Engineering, Keio University, Hiyoshi-cho, Yokohama 223 (Japan)

Pressure variation and flow birefringence of polymer melts in flows through straight ducts with approaching channel

T. Arai

With 8 figures

(Received October 27, 1972)

Notation

Dimensions are given in terms of mass (M), length (L), time (T) and temperature (θ).

D_{noz}	nozzle diameter	L
D_{res}	reservoir diameter	L
H	slit clearance	L
L_{noz}	nozzle length	L
p	pressure	$M L^{-1} T^{-2}$
P_{ent}	driving pressure calculated by linear extrapolation of the pressure gradient curve in the reservoir to the nozzle inlet plane	$M L^{-1} T^{-2}$
ΔP_{ent}	pressure drop at the nozzle inlet	$M L^{-1} T^{-2}$
ΔP_{exit}	residual pressure at the nozzle exit	$M L^{-1} T^{-2}$
Q	volumetric flow rate	$L^3 T^{-1}$
R_{noz}	nozzle radius	L
T	slit width	L
z	coordinate parallel to the axis of nozzle or slit in flow direction	L
$\bar{\alpha}$	*Barus* Effect Index	dimensionless
$\dot{\gamma}'_{H/2}$	apparent shear rate at the slit wall	T^{-1}
$\dot{\gamma}'_{R\text{-}noz}$	apparent shear rate at the wall of nozzle	T^{-1}
θ	temperature (degree in Centigrade)	θ
ν	nozzle length correction term coefficient	dimensionless
ν_{ent}	coefficient of the entrance correction term	dimensionless
ν_{exit}	coefficient of the exit correction term	dimensionless
ξ	pressure gradient	$M L^{-2} T^{-2}$
ξ_η	pressure gradient due to viscosity resistance	$M L^{-2} T^{-2}$
ζ_{noz}	duct contraction ratio at the nozzle entrance	dimensionless

Introduction

In 1962, the author and *Aoyama* put forward a paper on the interrelation between *Barus* effect and tube length correction term coefficient (1). Here the cause of the die swell was attributed to the mechanical energy. The principal defect of the paper is that the effect was conclusively directed to the shear strain of the polymer melt flowing through a nozzle. In 1966, at the International Symposium on Macromolecular Chemistry held in Kyoto, a paper on the revision of the forgoing concept was presented where the cause of the die swell was attributed mainly to tensile strains of the flowing melt from a new experimental result on the die exit pressure (2, 3). In 1967, at the 5th International Congress on Rheology in Kyoto, the author presented further experimental results on the interrelations between die swell and die exit pressure (4, 5). The present paper deals with the experimental results so-far obtained in his laboratory within the last 4 years on the elastic responses of melts in steady and in pulsatile flow conditions. For the clarification of the responses, experiments were carried out by the following three different resorts:

(1) Investigations on the interrelation between the pressure distribution along the length of circular duct with a contraction portion and the *Barus* Effect Index $\bar{\alpha}$ given by the ratio of the cross-sectional area of the extrudate to that of the nozzle.

(2) Measurements of the pressure propagation along the length of flow direction in the pulsatile flow through a circular duct with an approaching channel.

(3) Measurements on the flow birefringence of polymer melt passing through a slit duct, and the application of photoelasticity analysis for the elastic solid to the flow of the liquid.

In carrying out these experiments, much care was taken not to miss the effect of approach dimensions.

Herein, the first resort was regarded as the most important one to draw conclusions on the mechanical balance of the flow, and many experiments were carried out from this standpoint (1–5). But a further lot of experiments of this kind for different duct dimensions was required for systematic discussions. The second resort was expected to afford significant informations to the analysis of stress distribution as well as the shear and pressure propagations (6). The third resort is the direct means for the stress and strain analysis. As a preliminary research of this resort, flow birefringence of polyisobutylene solution in decalin in a fairly large annular gap of concentric cylinders was measured for the simple rotational shear flow. From the isochromatic and isoclinic lines, the stress distribution was obtained by applying the method of photoelasticity analysis to the flow of the solution (7). Measurements of the flow birefringence of a linear polyethylene melt were recently carried out for the flows through a slit duct (8, 9).

Experimental

Apparatus

The cross-sectional view of the circular duct die is shown by fig. 1, in which the parts 5 and 9 are interchangeable with a nozzle and a reservoir of different dimensions, respectively. The main advantage of this die to that reported in the previous paper (5) is the pressure is measurable at three or four points on the nozzle wall as well as on the walls of reservoir and adaptor. Pressures were picked up with bonded strain gage pressure transducer of Toyo-Bauldwin Co., Tokyo mounted at several positions along the duct in specially machined taps, to each of which the respective wall surface pressure was propagated through a hole of 1 mm diameter drilled perpendicularly onto the wall surface. The electric outputs of the transducer were monitored with a potentiometer and a couple of pen-autorecording devices. Fig. 2 shows the cross-sectional view of the slit die for measuring flow birefringence. With this die, both slit entry and exit angles are 180°,

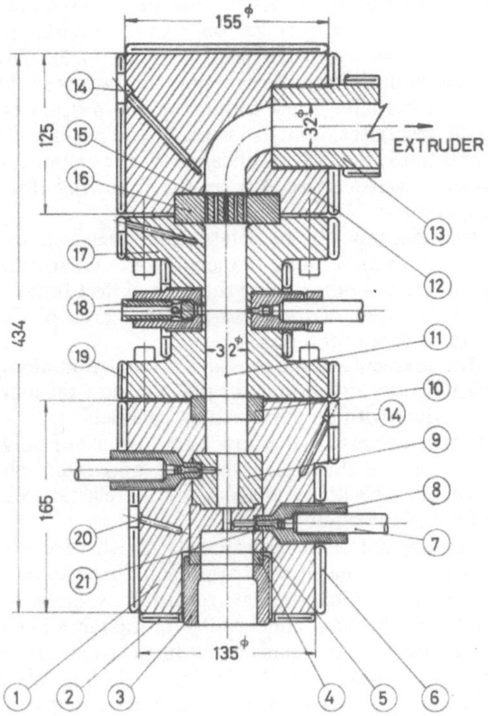

Fig. 1. Cross-sectional view of a set of the bend, adaptor, reservoir and nozzle for the circular duct

1 Die	12 Bend
2 Space heater	13 Adapter
3 Nozzle tightening screw	14 Thermoregulator hole
4 Spacer	15 Screen
5 Nozzle	16 Breaker plate
6 Band heater	17 Connector
7 Pressure transducer	18 Overflow plug
8 Plug	19 Band heater
9 Reservoir	20 Thermocouple hole
10 Spacer	21 Copper gasket
11 Pressure pickup hole	

slit clearance H 2 mm, slit length L 60 mm, and slit width T 25 mm. Most of the parts of the optical system were those produced by Riken Instrument Co., Tokyo.

For the extrusion, a commercial 40 mm single screw extruder fabricated by Taimei Kogyo-sho Co., Tokyo was used. The temperature fluctuation of the die was maintained within less than 0.5 °C throughout the course of measurements. A plunger extrusion type rheometer called Koka Flow Tester produced by Shimadzu Seisakusho Co., Kyoto (10) was also used for the determination of viscosity curves.

Test specimen

High density polyethylene Sholex 6009 produced by Showa Oil Chemical Co. was selected as the test specimen for the experiments of the first and the third resorts from the following reasons: The first is that the melt viscosity is smaller than Sholex 6002 used in the previous research (3) and so errors induced by heat generation through extrusion were expected to be smaller. The second is that *Barus* effect takes place within a shorter time than for the case of Sholex 6002. The third is that the temperature dependence of melt viscosity of the polymer is particularly small as compared to other polymers.

For the second resort on the pulsatile flow, polystyrene Styron 683 produced by Asahi Dow Co. was used from the reason that with this test specimen the pulsatile flow takes place under lower extrusion pressure (6) than with Sholox 6009.

Results and discussions

Fig. 3 shows the shear rate dependence of pressure distribution along the length of circular tube die in flow direction z. A lot of similar experiments with circular tube dies indicated that the pressure gradients in adaptor, reservoir, and nozzle were practically constant, respectively, so much as the measured points concerned. From these results, the author defined the driving pressure at the nozzle inlet plane P_{ent}, pressure drop at the nozzle entrance, ΔP_{ent}, residual pressure at the tube exit ΔP_{exit}, tube length correction term coefficient at the nozzle inlet ν_{ent} and tube length correction term coefficient at the tube exit ν_{exit} by the linear extrapolation of the pressure gradient curve as explanatory shown by Fig. 4, in which the tube length correction term coefficient ν is shown by the length of νR_{noz} (3).

It must be noticed here that in one of the previous papers (5) *Bagley* plots (1, 11) of P_{ent} vs. L_{noz}/R_{noz} relation for a constant apparent shear rate given at the nozzle wall $\dot{\gamma}'_R$ by eq. [1] did not show linear relation when the value of D_{noz}/D_{res}, namely duct contraction ratio at the nozzle entrance ζ_{noz} was not fixed.

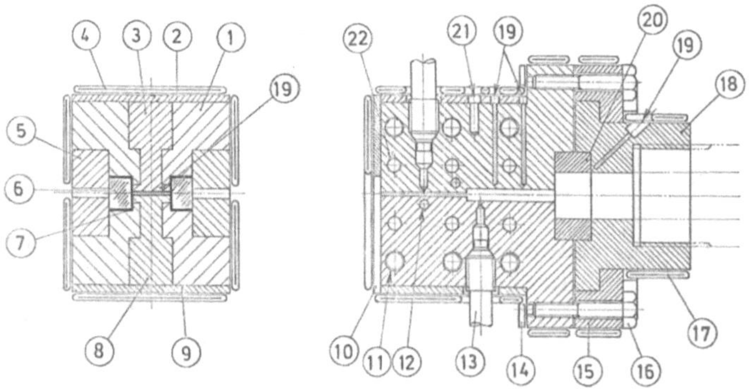

Fig. 2. Cross-sectional view of the slit die

1 Die base
2 Upper surface plate
3 Upper spacer
4 Space heater
5 Glass holder
6 Window glass
7 Teflon packing
8 Lower spacer
9 Lower surface plate
10 Front surface plate
11 Spacer tightening bolt hole
12 Knock pin hole
13 Pressure transducer
14 Asbest
15 Flange
16 Bolt
17 Band heater
18 Die connector
19 Thermocouple hole
20 Spacer
21 Thermoregulator hole
22 Glass tightening bolt hole

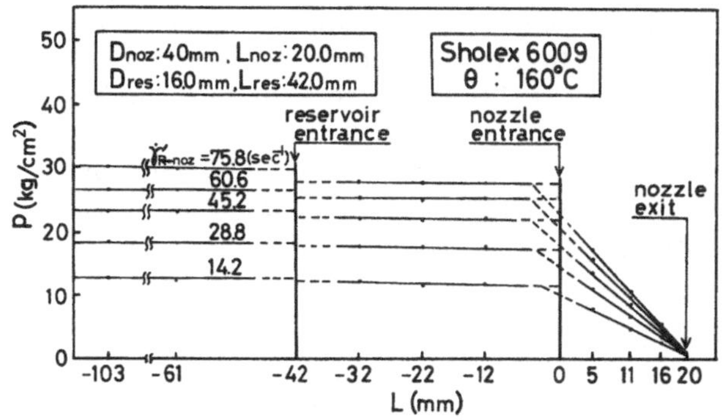

Fig. 3. Change of pressure gage readings with the distance in flow direction at several shear rates indicated

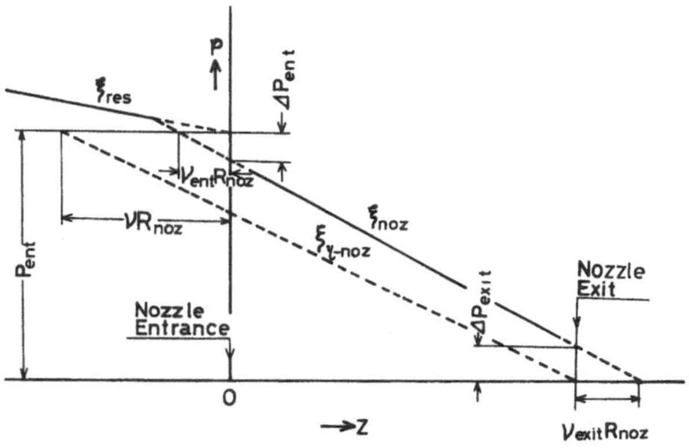

Fig. 4. Explanatory diagram for the definition of ξ_{noz}, $\xi_{\eta\text{-}noz}$, P_{ent}, ΔP_{ent}, ΔP_{exit}, ν_{ent}, ν_{exit} and ν

$$\dot{\gamma}'_{R\text{-}noz} = \frac{4Q}{\pi R^3}. \qquad [1]$$

For constant values of ζ_{noz}, however, *Bagley* plots gave strictly linear relation and ν showed a tendency to become large with decrease of ζ_{noz}. The rigorous values of ν thus obtained for fixed values of D_{noz} and D_{res} over a fairly wide range of shear rate gave clearly larger value than the sum of ν_{ent} and ν_{exit} as expressed by

$$\nu > \nu_{ent} + \nu_{exit}. \qquad [2]\ (2, 3, 5, 12)$$

These results gave the conclusion that for the flow of viscoelastic liquid the actual pressure gradient in the nozzle ξ_{noz} might be larger to some extent than the calculated value of pressure gradient in the nozzle caused by the viscosity resistance $\xi_{\eta\text{-}noz}$ as shown by the following formulae:

$$\xi_{noz} = \frac{P_{ent}}{L_{noz} + (\nu_{ent} + \nu_{exit})\, R_{noz}} \qquad [3]$$

$$\xi_{\eta\text{-}noz} = \frac{P_{ent}}{L_{noz} + \nu R_{noz}}, \qquad [4]$$

$$\xi_{noz} > \xi_{\eta\text{-}noz}. \qquad [5]$$

Experimental results showed that $\bar{\alpha}$ vs. $\dot{\gamma}'_{R\text{-}noz}$ relation varied as a function of L_{noz}/R_{noz} for fixed values of ζ_{noz}, and for fixed values of L_{noz}/R_{noz}, $\bar{\alpha}$ vs. $\dot{\gamma}'_{R\text{-}noz}$ relation depended upon ζ_{noz} (12). Fig. 5 shows the variation of $\bar{\alpha}$ vs. ΔP_{exit} curves with ζ_{noz}, indicating that $\bar{\alpha}$ is not a simple function of ΔP_{exit}, as a similar tendency to the one reported in a previous paper (5). Complete data of this kind for various duct dimensions might be able to draw conclusions

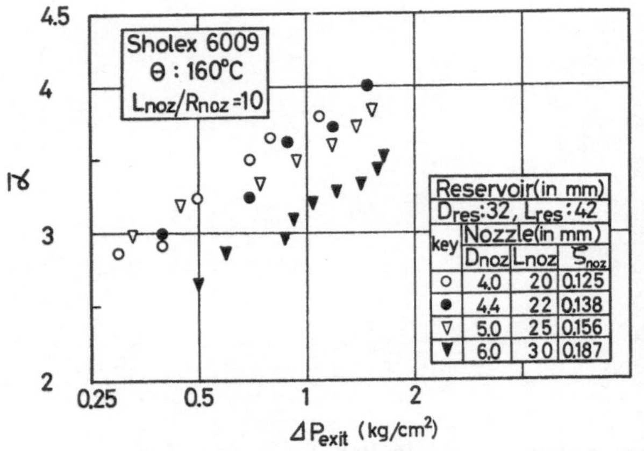

Fig. 5. Variation of $\bar{\alpha}$ vs. ΔP_{exit} relation with D_{noz} for a fixed value of L_{noz}/R_{noz}

on the theoretical analysis of normal stress effect upon the pressure readings at the wall surface.

Fig. 6 shows the pressure p vs. time t curves autorecorded at 2 points in the nozzle and 1 point in the reservoir for the pulsatile flow of Styron 683. The subscript to the figure p indicates the position number of pressure transducer. A set of these experiments on the change of pressure amplitude along the length in flow direction indicated the location of the maximum amplitude just before the tube inlet. Furthermore, it was noteworthy that with these experiments the pressure oscillation in reservoir showed a phase difference of about π to the ones in the nozzle in a sharp contrast to the fact that any phase difference between the values in the nozzle or between values in the reservoir was hardly observed within the sensitivity of the pen-autorecording devices for the pressure transducer. Judging from the law of continuity and the compressibility of the melt, the cause of this phenomenon may be attributed to the change of flow resistance caused by the abrupt displacement of streamlines which takes place just before and after the nozzle inlet.

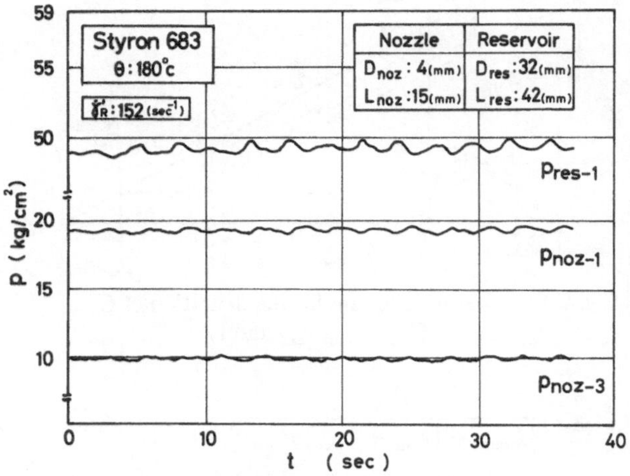

Fig. 6. Autorecorded pressure p versus time t curves for the pulsatile flow

The shear rate dependence of the isochromatic patterns were quite similar to those reported by *Funatsu* and *Mori* (13), and the tendency coincided with those reported by *Tordella* with circular tubes (14). By the rotation of the polarlizer and analyser of the measuring device, the effect of extinction angle χ on isoclinic lines was obtained as a function of apparent shear rate at the slit wall $\dot{\gamma}'_{H/2}$ given by

$$\dot{\gamma}'_{H/2} = \frac{6Q}{TH^2}. \qquad [6](15)$$

Fig. 7 shows the displacement of isoclinic lines with extinction angle χ for the three different shear rates indicated. The isoclinic lines in the downstream were approximately parallel to the slit wall, but the fringe orders of the corresponding isochromatic lines showed a very slight but apparent tendency to decrease with axial length even in the downstream.

From the curves as shown by fig. 7, principal stress lines were graphically obtained as a

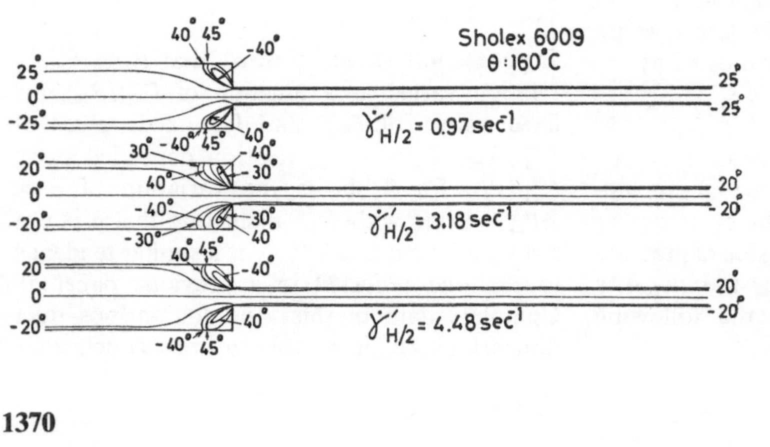

Fig. 7. Displacement of isoclinic lines with extinction angle for the three apparent shear rates at the slit wall indicated

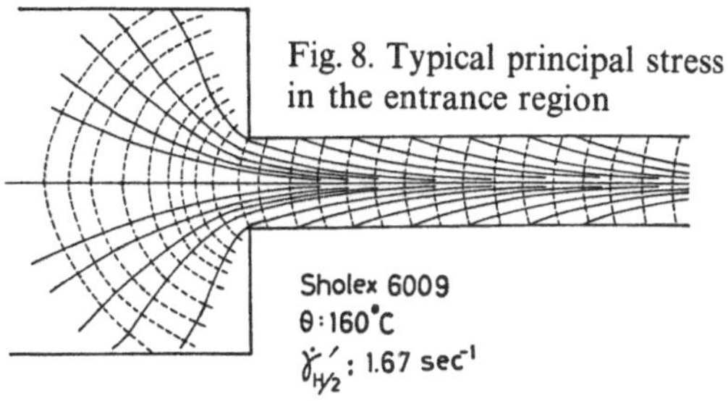

Fig. 8. Typical principal stress lines in the entrance region

function of shear rate at the wall after the manner of photoelasticity analysis for the elastic solid (8, 9). Fig. 8 shows a typical example of principal stress lines thus obtained. Therein asymptotical but continuous minute changes of the direction of principal stress lines can be pointed out after entering into the slit. By measuring the displacement of carbon particles mixed in the melt for constant timed intervals, streamlines for the flow were also obtained. Before the slit entrance, both streamlines and principal stress lines turned to the entrance, but within the slit principal stress lines went to the slit center plane from the slit wall with gradually decreasing inclination to the flow direction as a sharp contrast to the direction of streamlines which remained parallel to the slit wall. In the steady laminar flow conditions as cited here, it might be generally accepted that the directions of the principal stress lines were parallel to the planes of light transmission of the polarlizer. Accordingly, the author suggests the possibility that the average degree of molecular orientation of the flowing melt give higher value at the centre plane in contrast to the expectation from the fluid dynamics for the inelastic liquid, in which the maximum orientation of suspended long particles in shape must necessarily be given near the slit surface.

Summary

Elastic responses in the viscoelastic flow of polymer melts were investigated from a viewpoint of energy balance. By using a modified commercial extruder, measurements were carried out on the pressure variation at die walls for steady and pulsatile flows through straight circular tubes with straight cylindrical approaches and on the flow birefringence for steady flows through a slit with a rectangular approach.

The tube length correction term coefficient for linear polyethylene varied with the duct contraction ratio ζ given by the quotient of the tube diameter over approach one. Pressure gradients along the flow direction showed constant values for pressures at three points or more along the length of tube, giving very slight but apparently larger values than those calculated from the *Bagley* plots for the ducts of the same ζ. Even at a constant shear rate at the wall, the *Barus* effect index given by the ratio of the extrudate diameter to that of the tube was not a simple function of the exit pressure obtained by linear extrapolation of the pressure gradient curve to the tube exit. In the pulsatile flow of polystyrene, the pressure oscillation in approach showed a phase difference of about π to the one in tube in a sharp contrast to the fact that any phase difference between values in the tube or in the approach was hardly observed by pen-autorecording devices of pressure transducer.

From isochromatic lines and isoclinic lines at various wall shear rates linear relations were derived at the wall between fringe order per unit optical path n and principal shear stress as well as between n and shear stress. Changes of the magnitude of the principal shear stress along the axis of the flow direction before and after entering the slit inlet cross-section, suggested predominant operations of shear stress within the slit over the tensile ones.

References

1) *Arai, T.* and *H. Aoyama*, Trans. Soc. Rheology **7**, 333 (1963).

2) *Arai, T., I. Suzuki,* and *N. Akino*, "Pressure Determinations along Die Axis in a Round Tube Extrusion", paper presented at the International Symposium on Macromolecular Chemistry, Sept. 28, 1966, Kyoto.

3) *Arai, T., I. Suzuki,* and *N. Akino*, Proceedings of the Fujihara Memorial Faculty of Engineering, Keio Univ., Vol. **20**, No. 77, 39 (1967): Repts. Progr. Polym. Phys. Japan **10**, p. 127 (1967).

4) *Arai, T.* and *H. Toyota*, Proc. 5th Int. Congr. Rheology, Vol. **4**, pp. 461–470 (Tokyo 1970).

5) *Arai, T.*, Proc. 5th Int. Congr. Rheology, Vol. **4**, pp. 497–510 (Tokyo 1970).

6) *Arai, T., K. Furuta, M. Oohinata,* and *K. Takagi*, Soc. Polym. Sci. Japan, 21st Annual Meeting, Tokyo, May 24, 1972.

7) *Arai, T.* and *H. Asano*, Kobunshi Kagaku **29**, 317 (1972).

8) *Arai, T.* and *H. Asano*, Kobunshi Kagaku **29**, 510 (1972).

9) *Arai, T., H. Asano, H. Ishikawa, J. Mizutani,* and *S. Murai*, Kobunshi Kagaku **29**, 743 (1972).

10) *Arai, T.*, A Guide to the Testing of the Rheological Properties with Koka Flow Tester (Tokyo 1958).

11) *Bagley, E. B.*, J. Appl. Phys. **28**, 624 (1957).

12) *Arai, T., A. Komatsu, K. Ooi, T. Tsukahara,* and *M. Yasui*, "The Steady Pressure Distribution along the Circular Walls and the Related Post-extrusion Swelling in a Polymer Extrusion", paper presented at the Soc. Polym. Sci. Japan, 21st Annual Meeting, Tokyo, May 24, 1972.

13) *Funatsu, K.* and *Y. Mori*, Kobunshi Kagaku **25**, 391 (1968).

14) *Tordella, J. P.*, J. Appl. Polym. Sci. **7**, 215 (1963).

15) *Arai, T.* and *H. Aoyama*, Kobunshi Kagaku **18**, 53 (1961).

Author's address:

Prof. Dr. *Teikichi Arai*
22–26, Honkomagome 6-Chome
Bunkyo-Ku, Tokyo 113 (Japan)

Rheol. Acta 13, 864–871 (1974)

Zustandsgleichung eines Lehmes, ermittelt mit Hilfe einer neuentwickelten Versuchsapparatur*)

D. Fedder (Kassel)

Mit 9 Abbildungen und 1 Tabelle

(Eingegangen am 27. Oktober 1972)

Formelzeichen

δ	Dilatanzmodul	(kp^2/cm^4)
η	Schubviskosität (*Newton*)	($kp \cdot sec/cm^2$)
G	Schubmodul (*Hooke*)	(kp/cm^2)
ϑ	Fließgrenze (*St. Venant*)	(kp/cm^2)
γ	Scherung	(–)
K	Konsistenz	(–)
n	Porenvolumen	(%)
S	Sättigung	(%)
σ	isotroper Zelldruck	(kp/cm^2)
σ'_A	Konsolidierungsspannung	(kp/cm^2)
τ	Schubspannung	(kp/cm^2)
T	Retardationszeit	(sec)
t	Zeit	(sec)
u	Porenwasserdruck	(kp/cm^2)
V	Volumen	(cm^3)
w	Wassergehalt	(%)

1. Einleitung

In den bodenmechanischen Berechnungsverfahren wird die Eigenschaft des Bodens entweder als rein elastisch oder als rein plastisch vorausgesetzt. Man ignoriert dabei die Beobachtungen, daß beim Boden entweder die einmal eingetretenen Verformungen bei Entlastung nicht mehr rückgängig sind oder daß dem Bruch z. T. beträchtliche Verformungen vorausgehen.

Bei einer Reihe von Grundbauproblemen erhält man aber mit diesen Idealisierungen keine befriedigenden Lösungen mehr. Hierzu gehört z. B. die Frage, wie die Standsicherheit einer in Bewegung geratenen Böschung beurteilt werden kann. Bei dieser als Hangkriechen bekannten Erscheinung wird zur richtigen Einschätzung ein Materialgesetz benötigt, das auch den zeitlichen Verlauf der Deformation erfaßt und gleichzeitig Aussagen über den augenblicklichen Abstand zur Bruchspannung bzw. Bruchverformung ermöglicht.

Erinnert sei auch an zeitabhängige Verformungen bei rückwärts verankerten Baugrubenwänden oder bei Widerlagern von Bogenbrücken sowie an die Sekundärsetzung bei Belastung des Baugrundes durch ein Bauwerk.

Die bisherigen Bemühungen mit dem Ziel, das reale Verhalten eines Bodens durch ein Stoffgesetz zu beschreiben, lassen sich etwa in folgender Form kurz charakterisieren:

a) Bei den verwendeten Testgeräten sind entweder die Spannungsgröße und -verteilung unbekannt oder aber die Deformationen nicht homogen oder eindeutig zu definieren.

b) Oft fehlt außerdem die für die Beurteilung eines Materials wichtige Beobachtung über das Verhalten bei Entlastung. Auch die Konsistenzänderung des Materials während eines Versuchs durch Entwässerung machte vielfach ebenso wie stattgefundene Temperaturänderungen eindeutige Aussagen unmöglich.

c) Bei der Auswertung begnügte man sich mit der graphischen Darstellung der Meßwerte. Manchmal ist die Abhängigkeit des Proportionalitätsfaktors von einer der beiden Meßgrößen zusätzlich graphisch angegeben. Zum Teil beschränkte man sich auch auf bestimmte Bereiche der Messungen bei der Formulierung eines Gesetzes.

d) Bei den sehr seltenen quantitativen Analysen handelt es sich um rein mathematische Approximierungen, z. B. durch Polynome oder Potenzformeln; dann kann aber nicht mehr von einem Stoffgesetz im physikalischen Sinn gesprochen werden.

Obwohl konkrete Fragestellungen aus der Praxis vorliegen, erschien es zunächst einmal

---

*) Die hier beschriebenen Untersuchungen wurden in den Jahren 1965–1970 ausgeführt im Institut für Bodenmechanik und Grundbau, Technische Hochschule Darmstadt, unter der Leitung von Prof. Dr.-Ing. *H. Breth*.

angebracht, das reale Werkstoffverhalten genauer und umfassender als bisher durch ein Stoffgesetz zu beschreiben. Das Interesse der vorliegenden Untersuchungen richtete sich daher ausschließlich auf das Material und die Kennzeichnung seiner Eigenschaften. Mit den Arbeitsmethoden der Rheologie sollte deshalb eine beliebig gewählte Bodenart unter möglichst einfachen Randbedingungen – also losgelöst von speziellen Anwendungen – beansprucht und die beobachteten Phänomene dann qualitativ und vor allen Dingen auch quantitativ interpretiert werden.

2. Prüfmethode und Versuchsapparatur

Als Deformationsart wurde die einfache Scherung gewählt. Als Prüfmethode wurde das Kriech- und Erholungsexperiment dem Relaxationsexperiment wegen der besonders bei Langzeitversuchen vorhandenen versuchstechnischen Vorteile vorgezogen.

Um homogene, reproduzierbare Proben mit bekannter Spannungsgeschichte prüfen zu können, sollte für jeden Versuch eine neue Probe hergestellt werden. Außerdem sollte wegen der angestrebten Wassersättigung der Proben die Scherung in einer Druckzelle ausgeführt werden. Dabei wurde davon ausgegangen, daß die zu messenden *Gestalts*änderungen durch einen *isotropen* Porenwasserdruck bzw. Zelldruck nicht beeinflußt werden.

Das Resultat der geschilderten Überlegungen war die Konstruktion der in Abb. 1 dargestellten Scherzelle (*Fedder*, 1970). Die flache, quaderförmige Bodenprobe (A) befindet sich zwischen zwei Metallplatten (C) und ist ringsum durch eine Gummimembran (D) dicht umschlossen. Das im breiigen Zustand eingebaute Material kann unter der Wirkung des Zelldruckes durch Filtersteine (B) entwässern; das Stützgestell (F) wird in der Konsolidierungsphase bei geschlossener Zelle von außen entfernt.

Beim daran anschließenden Kriechexperiment wurde bei verhinderter Entwässerung die Scherkraft über die obere Platte innerhalb einer Sekunde stoßfrei aufgebracht und die Verschiebung und Setzung mit Meßuhren (K) laufend gemessen. Die tatsächliche Größe der in der Zelle wirkenden Schubspannung war aus Eichversuchen bekannt. Alle Versuche wurden bei konstanter Temperatur durchgeführt.

3. Versuchsergebnisse

Als Testmaterial wurde ein Lehm mit einer Plastizitätszahl 16 und einem Rohtongehalt von 6% gewählt. Durch Variation der Konsolidierungsspannungen zwischen 0,5 und 2,5 kp/cm^2 entstanden 5 verschiedene Zustandsformen im Konsistenzbereich zwischen 0,42 und 0,66.

Generell lassen sich die Ergebnisse wie folgt erläutern (Ausführliche Wiedergabe von Meßkurven siehe *Fedder*, 1970):

Während der ersten halben Stunde erfolgten die größten Änderungen. Die Schubdeformationen waren nach etwa 1,5 Tagen zum Stillstand gekommen, so daß die Enddeformationen durch horizontale Asymptoten angegeben werden können. Auch der Porenwasserdruck erhöht sich hauptsächlich nur im Anfangsstadium und bleibt dann konstant. Bei größeren Schubspannungen bzw. Schubdeformationen weisen die Kriechkurven mehr oder weniger ausgeprägte Wendepunkte bei gleichzeitig stattfindenden Setzungen auf. Einfärbungen der Proben haben bestätigt, daß dann der Bruch eingetreten ist. Bei allen anderen Versuchen jedoch haben die Einfärbungen eine lineare Zunahme der Verschiebung über die Probenhöhe gezeigt, ein Beweis für die Realisierung der einfachen Scherung.

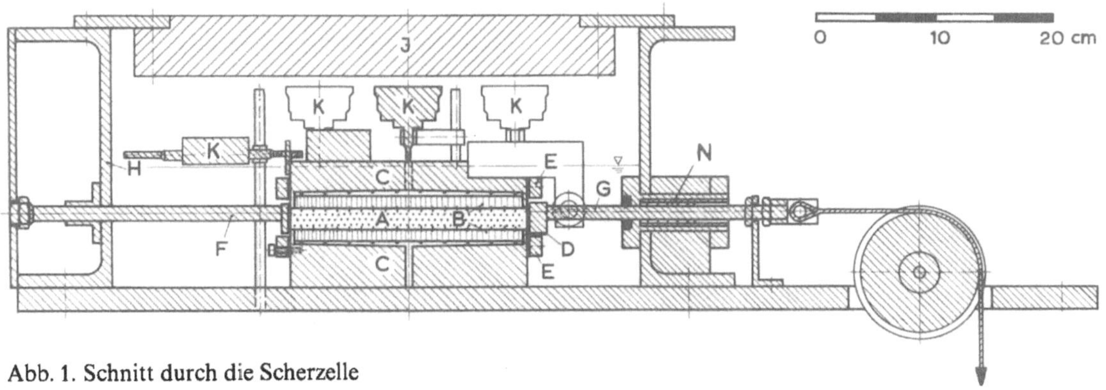

Abb. 1. Schnitt durch die Scherzelle

A	Bodenprobe	E	mit Gummi beklebter Balken	J	Zellendeckel (Plexiglas)
B	Filterstein	F	Stützgestell	K	Meßuhr
C	Metallblock	G	Zugstange	L	Untergestell
D	Gummimembran	H	Zellenwand	N	Kugelkäfig

Nach 1,5 Tagen wurde dann die Schubspannung entfernt. Die Schubdeformationen waren in dieser Erholungsphase, absolut gesehen, nur wenig rückläufig. Bei kleinen Deformationen allerdings war der Prozentsatz der reversiblen Verformung größer als bei größeren Deformationen.

4. Auswertung

Welche Folgerungen können nun aus den beobachteten Phänomenen gezogen werden und wie können alle diese Erscheinungen durch ein gemeinsames Gesetz erfaßt werden? Zunächst einmal fällt das festkörperartige Verhalten auf, und zwar im gesamten untersuchten Konsistenzbereich; es drückt sich vor allem durch horizontale Asymptoten der Kriechkurven aus; es wird bestätigt durch die z. T. ausgeprägten Brucherscheinungen.

Das Erholungsexperiment zeigt darüber hinaus, daß elastische Eigenschaften zwar vorhanden, bei größeren Deformationen jedoch gegenüber den plastischen Eigenschaften von untergeordneter Bedeutung sind. Die ausgeprägte Zeitabhängigkeit der Deformation wird durch viskose Eigenschaften hervorgerufen. Der Lehm verhält sich also wie ein Festkörper mit gleichzeitig vorhandenen elastischen, plastischen und viskosen Eigenschaften.

Besonders deutlich wird dieses Verhalten auch, wenn man ein Spannungs-Verformungsdiagramm zeichnet (Abb. 2). Dabei konnten natürlich nur die Meßwerte für bestimmte, ausgewählte Zeiten aufgetragen werden. Die Zeit $t = 150000$ sec bedeutet etwa das Versuchsende nach 1,5 Tagen. In dieser Abbildung wurden zunächst die Versuchswerte wegen der Übersichtlichkeit für gleiche Zeiten verbunden. Da jedes Experiment mit $\tau = \text{const}$ ausgeführt wurde, verläuft die Versuchsspur horizontal.

Bei allen $\tau - \gamma$-Diagrammen, also bei sämtlichen untersuchten Konsistenzen, war auffällig, daß keine $\tau - \gamma$-Linie durch eine Funktion beschrieben werden kann, bei der die Krümmung laufend und gleichmäßig abnimmt; oftmals liegen mehrere Versuchswerte auf geradenähnlichen Kurvenästen. Dies bedeutet, daß alle $\tau - \gamma$-Linien am besten durch Polygonzüge zu beschreiben sind, wobei sich herausstellte, daß die Knicke unabhängig von der Zeit jeweils einer bestimmten Schubspannungsgröße zugeordnet werden können. Es existieren somit mehrere (zeitunabhängige) Fließgrenzen. Man kann daher, im Einklang mit den reversiblen Deformationen aus den Erholungsexperimenten, die Gesamtdeformation in eine elastische und eine plastische unterteilen. (Die genaue Lage und Anzahl der Knicke ist natürlich aus Abb. 2 allein noch nicht anzugeben.)

Wenn man nun *sämtliche* beobachteten Phänomene berücksichtigt, also das Verhalten des Lehmes für den *gesamten* bis zum Bruch mög-

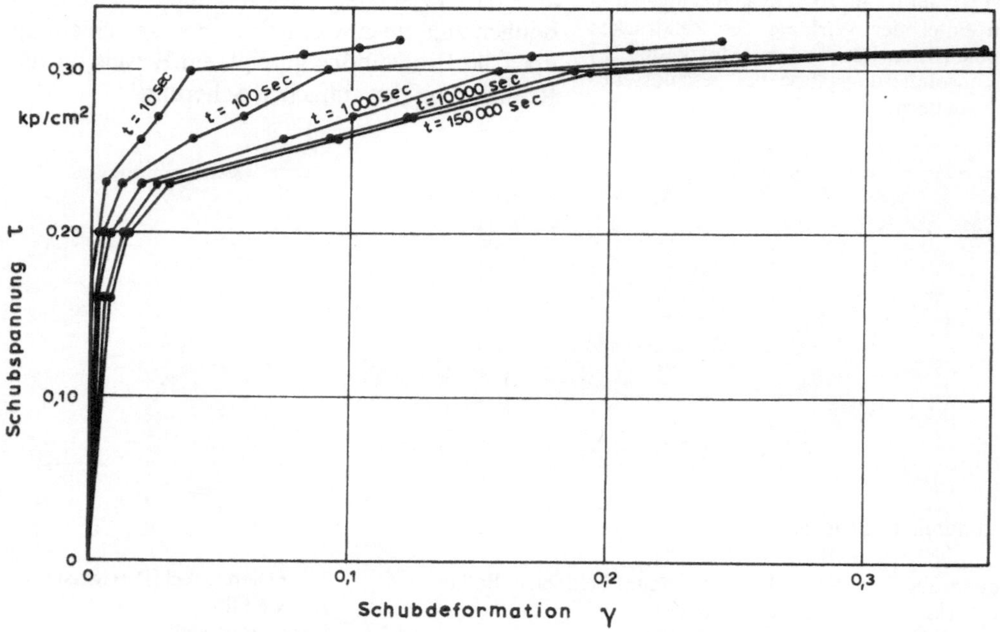

Abb. 2. $\tau - \gamma$-Diagramm (z. B. für $K = 0{,}50$)

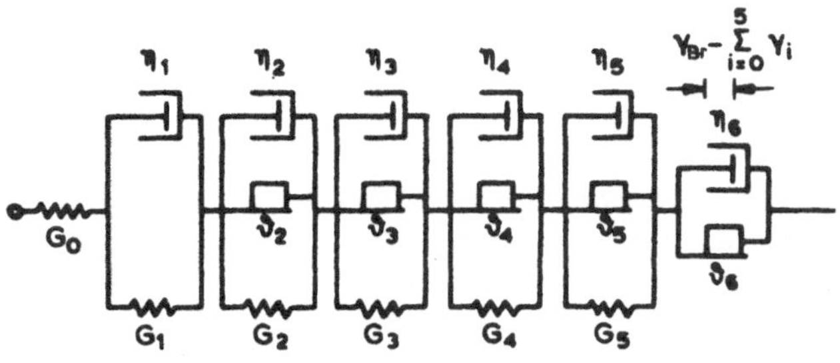

Rheologische Gleichung; $\gamma < \gamma_{Br}$, $\tau = \mathrm{const.}$:

$$\gamma = \frac{\tau}{G_0} + \sum_{i=1}^{5}\left[\frac{\tau-\vartheta_i}{G_i}\left(1-e^{-t/T_i}\right)\right] + \frac{\tau-\vartheta_6}{\eta_6}\cdot t \qquad (1)$$

mit: $T_i = \eta_i / G_i$, $\vartheta_1 = 0$, $\gamma_0 = \tau / G_0$, $\gamma = \frac{\tau-\vartheta_i}{G_i}\left(1-e^{-t/T_i}\right)$

Abb. 3. Rheologisches Modell

lichen Spannungs-Deformationsbereich in geschlossener Form beschreiben will, dann kommt nur eine Formulierung in Frage, die durch ein rheologisches Modell der in Abb. 3 dargestellten Art symbolisiert werden kann. Dabei wurde angestrebt, daß alle auftretenden Größen physikalischen Charakter haben sollen, die Gleichung also dimensionsrein sein soll.

Für Schubspannungen kleiner als die unterste Fließgrenze ϑ_2 verhält sich der Lehm linear viskoelastisch; er besitzt gleichzeitig momentane und verzögerte Elastizität. Bei Spannungen $\tau > \vartheta_2$ kommen plastische Deformationen hinzu. Dabei erzeugen nur die die Fließgrenzen ϑ_i übersteigenden Spannungsbeträge die zusätzlichen Deformationen, die ihrerseits nach Maßgabe der parallelgeschalteten Viskositäten η_i verzögert werden. Die Zahl der Fließgrenzen richtet sich nach der Anzahl der Knicke im $\tau-\gamma$-Diagramm bzw. nach der gewünschten Approximation der Meßergebnisse.

Spannungen über der obersten Fließgrenze ϑ_6 rufen den Bruch hervor. Der Bruch tritt ein, wenn die Gesamtdeformation einen für jede Konsistenz typischen Grenzwert γ_{Br} erreicht hat. Dieser Grenzwert ergibt sich aus dem Schnitt der ϑ_6-Linie mit der $\tau-\gamma$-Linie für $t = \infty$. Von da an verläuft die $\tau-\gamma$-Linie horizontal. Die bis zum Eintritt des Bruches noch dämpfende Wirkung von η_6 geht bei Erreichen der Bruchdeformation verloren, was symbolisch durch eine Begrenzung der Zylinderlänge dargestellt ist: der Kolben rutscht aus dem Zylinder. Die zugehörige mathematische Formulierung zeigt die ebenfalls in Abb. 3 angegebene Formel.

Bei der hier vorgelegten Lösung sollte im nächsten Schritt noch die Größe der Stoffwerte ermittelt werden. Da die mathematische Struktur der Gleichung leider keine direkte Berechnung der 17 Parameter aus den Meßwerten ermöglicht, mußte wohl oder übel die Lösung durch Probieren gesucht werden. Während für die ϑ_i- und G_i-Werte wenigstens die Größenordnungen und Tendenzen aufgrund der $\tau-\gamma$-Diagramme festlagen, war dies bei den Viskositäten nicht der Fall. Für die Lösung konnte aber als Anhaltspunkt dienen, daß bei jedem gesuchten Eigenwert eine sinnvolle Querverbindung zwischen den untersuchten Konsistenzen – z. B. in Form von gleichsinnig gekrümmten Kurven – zu erwarten war.

Tab. 1. Schubmoduln

G_0	[kp/cm²]	$100 \cdot \sigma'_A$
G_1	[kp/cm²]	$80 \cdot \sigma'_A$
G_2	[kp/cm²]	$9{,}5 \cdot \sigma'_A$
G_3	[kp/cm²]	$1{,}7 \cdot \sigma'_A$
G_4	[kp/cm²]	$0{,}6 \cdot \sigma'_A$
G_5	[kp/cm²]	$0{,}4 \cdot \sigma'_A$

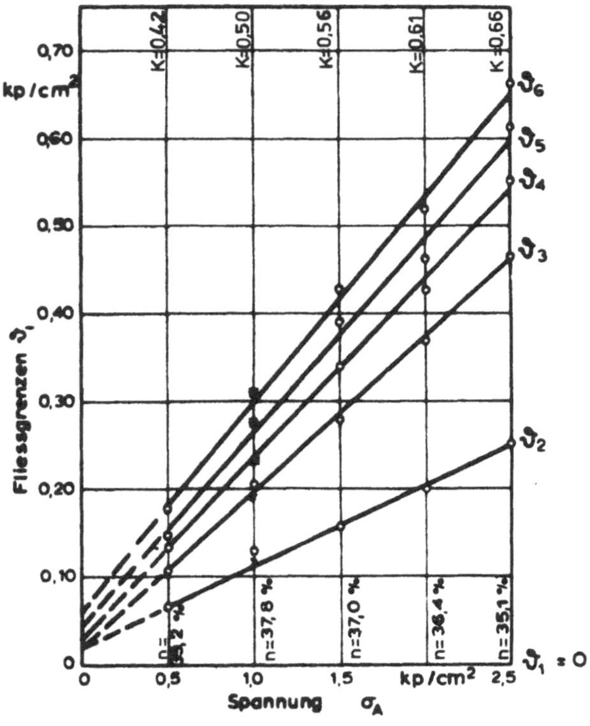

Abb. 4. Fließgrenzen

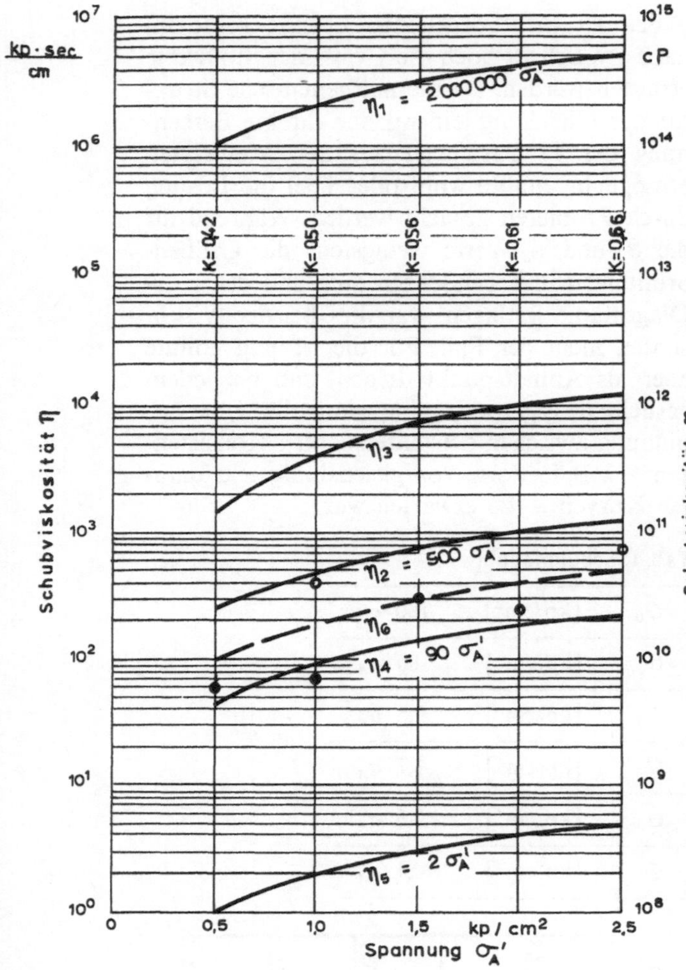

Abb. 5. Schubviskositäten

Mit den in Abb. 4 und 5 sowie in Tab. 1 dargestellten Materialwerten konnte insgesamt die beste Anpassung an die Meßwerte erzielt werden. Da die Approximierung im ersten Schritt getrennt nach Konsistenzen erfolgte, kann die hier festgestellte, meist sogar *lineare* Querverbindung als physikalische Gesetzmäßigkeit und damit auch als Bestätigung des vorgeschlagenen Modells gewertet werden. Die angegebene Abhängigkeit der Fließgrenzen, Schubmoduln und Viskositäten von der Konsolidierungsspannung bzw. dem entsprechenden Porenvolumen oder Wassergehalt erscheint überdies physikalisch plausibel.

Die Güte der Approximierung mag durch Abb. 6 veranschaulicht werden. Zur Verdeutlichung sind die Kriechkurven nur für die erste halbe Stunde aufgetragen. Die Meßwerte sind durch Kreuze markiert, während die ausgezogenen Kurven sich als Lösung der rheologischen Gleichung ergeben.

In Abb. 7 sind die Bruchverformungen in Abhängigkeit von den Konsolidierungsspannungen dargestellt. Mit abnehmendem Wassergehalt werden die möglichen Deformationen bis zum Eintritt des Bruches erwartungsgemäß

Abb. 6. Approximierung der Meßwerte (z. B. für $K = 0{,}42$)

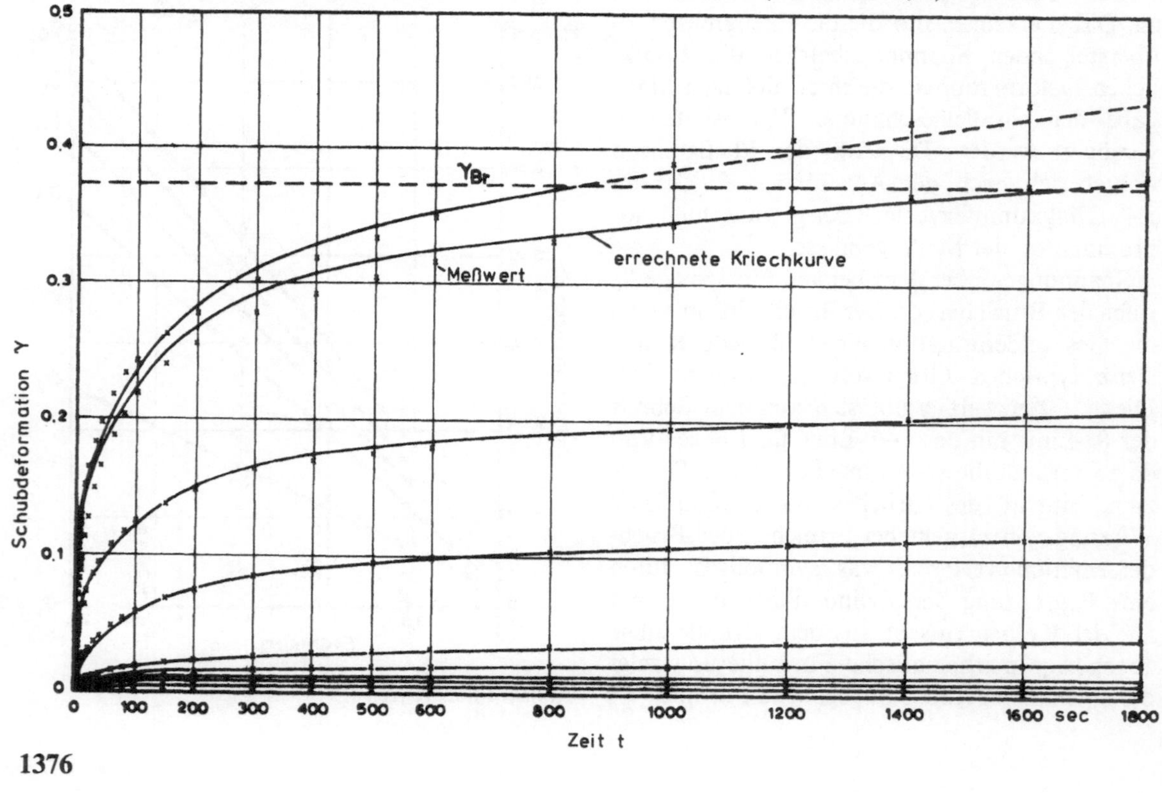

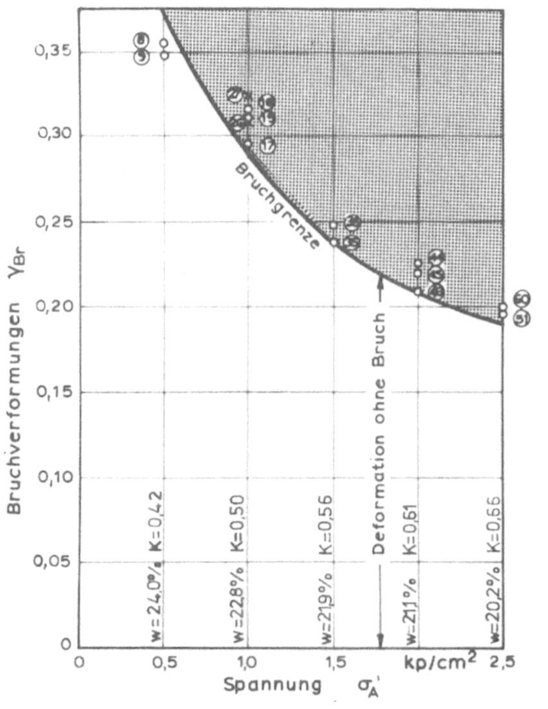

Abb. 7. Bruchdeformationen

kleiner, das Material wird also spröder. Auch hier zeigt die rechnerische Bruchgrenze, die sich aus der rheologischen Gleichung für eine Schubspannung $\tau = \vartheta_6$ ergibt, eine Übereinstimmung mit den entsprechenden Meßwerten, und zwar sind das diejenigen γ-Werte, bei denen die Kriechkurven Wendepunkte aufweisen.

Neben dem Zusammenhang zwischen der Schubspannung τ und der Scherung γ war auch die Frage nach der Ursache bzw. Wirkung der beobachteten Porenwasserdruckänderung zu klären. Das Phänomen, daß die gestaltsändernde einfache Scherung von einer Änderung der isotropen Porenwasserspannung begleitet wird, konnte als *Reynolds*sche Dilatanz, also als eine Erscheinung zweiter Ordnung gedeutet werden. Die Schubspannung ruft demnach eine, wenn auch sehr kleine Volumenänderung hervor. Die relative Volumenverminderung hängt ab vom Quadrat der Schubspannung nach Maßgabe des Dilatanzmoduls δ:

$$\frac{\Delta V}{V} \cdot \delta = \tau^2 . \qquad [2]$$

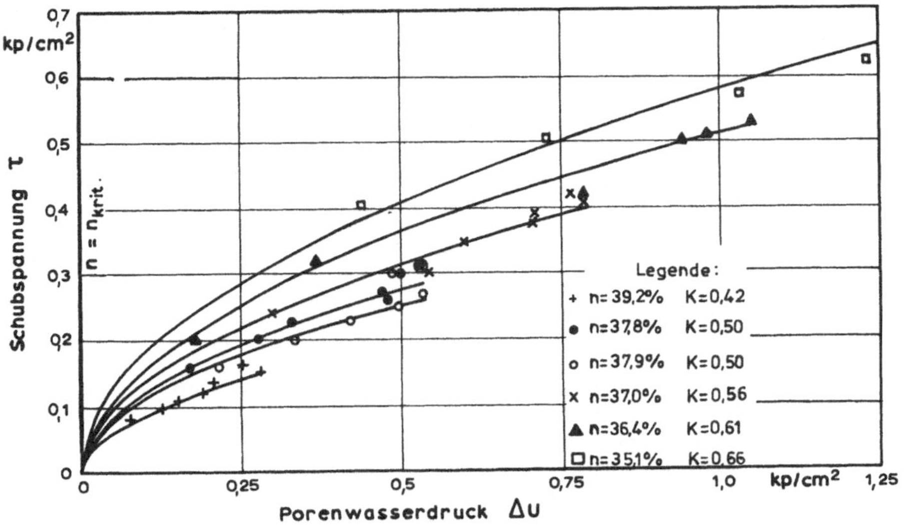

Abb. 8. $\tau - \Delta u$-Diagramm (z. B. für $t = 1800$ sec)

56

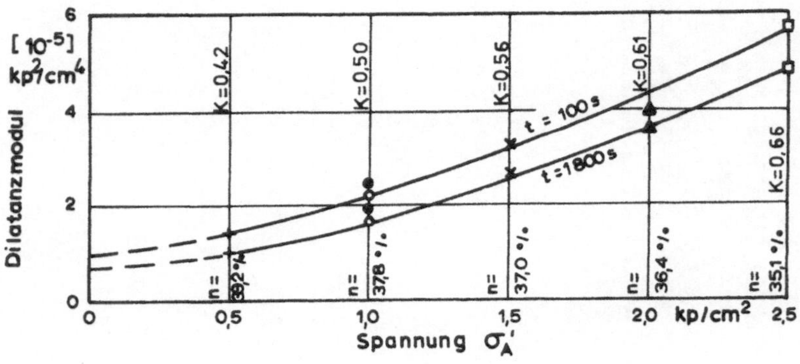

Abb. 9. Dilatanzmoduln

Setzt man voraus, daß die Volumenänderung ΔV bei wassergesättigtem Boden nur im Porenwasser stattfindet, dann gilt außerdem

$$\frac{\Delta V}{V_w} \cdot \kappa = \Delta u \quad [3]$$

mit κ = Kompressionsmodul von Wasser.

Durch Auflösen nach ΔV und Gleichsetzen der beiden Gleichungen erhält man schließlich folgende Beziehung:

$$\Delta u = \frac{\kappa}{\delta} \cdot \frac{1}{n} \cdot \tau^2 \quad [4]$$

mit $n = \frac{V_w}{V}$ für $S = 100\%$.

Zur Überprüfung sind die Schubspannungen über den zugehörigen gemessenen Porenwasserdrücken in Abb. 8 aufgetragen worden. Die eingezeichneten quadratischen Parabeln bestätigen qualitativ das obige Gesetz zwischen Δu und τ^2. Aus den Parabelparametern läßt sich der Dilatanzmodul errechnen (Abb. 9). Auch für diese Materialgröße konnte eine Abhängigkeit von der Konsistenz festgestellt werden: Je höher der Konsistenzwert, also je steifer das Material, desto geringer die Volumenabnahme und der dadurch verursachte Porenwasserdruck.

Zusammenfassung

Mit der gezeigten Testapparatur ist es gelungen, die einfache Scherung zu realisieren und damit einwandfreie Versuchsdaten für die Ermittlung von Stoffgesetzen zu liefern.

Der wassergesättigte Lehm verhält sich bereits ab einer Konsistenz von 0,4, wahrscheinlich aber auch schon bei kleineren Konsistenzen, wie ein Festkörper. Es sind gleichzeitig elastische, viskose und plastische Eigenschaften vorhanden.

Das gesamte Spannungs-Deformationsverhalten konnte einschließlich der Zeitabhängigkeit in einer einzigen Gleichung erfaßt werden. Bei der vorgeschlagenen Gleichung sind die Eigenwerte für jede Konsolidierungsspannung konstant. Ihre Abhängigkeit von dieser Größe bzw. vom Wassergehalt oder vom Porenvolumen unterstreicht ihren physikalischen Charakter.

Bei Überschreitung der obersten Fließgrenze tritt in jedem Fall der Bruch ein, und zwar nach Erreichen einer für jede Konsistenz typischen Bruchdeformation. Diese Größe sollte in Zukunft – z. B. für die Beurteilung der Standsicherheit einer in Bewegung geratenen Böschung – mit herangezogen werden, während hingegen das oftmals verwendete Kriterium der Deformations*geschwindigkeit* nicht eindeutig ist. Die Versuche zeigten, daß die Geschwindigkeit nicht nur von den Materialeigenschaften, sondern auch von der Schubspannung abhängt.

Mit Hilfe des Dilatanzmoduls können die Porenwasserdruckänderungen infolge Schubbeanspruchung berechnet werden.

Die vorgeschlagene Konzeption dürfte aufgrund der gewählten Elementanzahl auch für andere Bodenarten, wenn auch mit anderen Eigenwerten, verwendbar sein.

Summary

Flat prismatic soil samples were subjected to creep- and recovery experiments in newly developed shear cells. The deformation took place below failure in simple shear, that is, in plane strain. The pore water pressure in the sample was measured at the same time.

All observed phenomena could be expressed in a rheological equation, in which the shear stress could be related to the shear strain, to the time und to the soil parameters. The equation can be interpreted by a rheological model. The model parameters (elasticity, viscosity, plasticity) depend on the consistency of the tested materials.

Literatur

Bjerrum, L. and *A. Landva,* Direct Simple Shear Tests on a Norwegian Quick Clay, Géotechnique **16**, No. 2 (1966).

Fedder, D., Ermittlung der rheologischen Zustandsgleichung eines Lehmes mit Hilfe einer neuentwickelten Versuchsapparatur, pp. 1–112, Düsseldorf: Fortschritt-Berichte, VDI-Z, Reihe 4, H. 18 (1970).

Geuze, E. and *Tan Tjong-Kie*, Rheological Properties of Clays, 3. Internat. Conf. on Soil Mechanics and Foundation Engineering (1953).

Haefeli, R., Kriechen und progressiver Bruch in Schnee, Boden, Fels und Eis, Schweiz. Bauzeitung **85** (1967).

Mitchel, J. K., R. G. Campanella and *A. Singh*, Soil Creep as a Rate Process, ASCE Vol. 94 Section SM (1968).

Murayama, S. and *T. Shibata*, On the Rheological Characters of Clay; Disaster Prevention Res. Inst., Kyoto, Bulletin No. **26** (1958).

Reiner, M., Rheologie, pp. 1 – 360 (Hanser Verlag, München 1968).

Roscoe, K. H., An Apparatus for the Application of Simple Shear to Soil Samples; 3. Internat. Conf. on Soil Mechanics and Foundation Engineering, Vol. 1 (1953).

Vyalov, S. S. und *A. M. Skibitsky*, Problems of the Rheology of Soils; 5. Internat. Conf. on Soil Mechanics and Foundation Engineering, Vol. 1 (1961).

Anschrift des Verfassers:

Dr. Ing. *D. Fedder* VBI
Baugrundinstitut
3500 Kassel, Hummelweg 41

Rheol. Acta **13**, 872–876 (1974)

From the Unilever Research Laboratory, Sharnbrook, Bedford (England)

The marginal zone in meat comminutes flowing through pipes

By A.E. Hawkins

With 3 figures and 3 tables

(Received October 27, 1972)

Introduction

The development of more and more sophisticated processes in meat technology has resulted in a need to characterise the flow of meat comminutes by rheological functional relations for the calculation of their flow properties in various pipe and other geometries. This seemingly simple requirement is not easy to satisfy even when the elasticoviscous nature of the material is ignored and it is treated solely as a viscous fluid. Just as heat transfer through the walls of a hot or a cold pipe leads to undeveloped flow and the problem of a continuously changing velocity profile down the length of the pipe, so also does the production of a changing marginal zone due to changes of the meat comminute give similar problems.

It is a fact that, even at bench feasibility level, process research on meat comminutes is costly because meat is an expensive commodity and time consuming in its preparation and for equipment cleaning down. The following circumstances add to the expense:

i) the meat comminutes must be freshly prepared because the water binding properties change on storage, particularly frozen storage,
ii) it is not possible to produce typical comminutes on a laboratory scale,
iii) it is not possible to produce identical comminutes on successive days so considerable replication of experiments is necessary.

It is clear, therefore, that a research programme of some complexity may not be entered lightly and, while describing a rheological phenomenon, this paper poses the query "what is the right way to approach the future research" as well as providing a report on what has gone before.

The development of a marginal zone

When a meat comminute was made to flow through a pipe a change occurred in the material in contact with the pipe wall. A layer of finely divided material developed on the interface. There appeared to be not only a separation out of particle sizes as in the plasma zone in haemorheology, but also a shear breakdown of material to produce the range of sizes concerned.

This was observed with both of the two commercially important comminutes beef and pork. With the former, particularly, a "slipstick" regime was common and liquid was visible on the surface of the extrudate just as it emerged from the end of the pipe. The liquid was collected through fine holes in the pipe walls and gel electrophoresis showed it to be rich in sarcoplasmic protein. By forcing the beef comminutes through a wide, thin rectangular pipe it was possible to obtain thin strips of extruded material for histological examination. The brass pipe was made in two halves so that the finish of the inner surfaces could be changed.

The conclusions from the experiments were as follows. There was no obvious increase in fat on the surface of the extrudate, indeed the fat was generally judged to be present in smaller amounts than in the more central areas and was found in droplet form with diameters from less than 3 μm up to more than 25μm. Most samples had a coating of fine, particulate, proteinaceous matter presumably formed during the comminution by breakdown of muscle fibre. There appeared to be no obvious differences in the extrudate from the rectangular pipes with internal finishes produced by emery papering, sand blasting and P.T.F.E. spraying, but a polished surface generally gave an extrudate with only a few fine fat droplets on the surface.

The flow had the character described by *Vand* (1) for suspensions behaving as though there were a lubricating sheath between a central plug and the wall. The behaviour appeared to be an extremely complex form of the so-called plug flow. It was modified greatly by the addition of materials able to buffer the release of the flow lubricant with pressure. For example, the addition of a few percent of carbohydrate removed the "slip-stick" behaviour completely and relatively smooth flow curves were obtained. The shape of these curves was changed markedly by temperature, by the pH value of the meat, and by the addition of water or of fat.

With pork comminutes there developed a thick pale layer on the surface. The development of this layer was observed more readily than the beef layer and will be used to discuss and to illustrate here that there really is a problem. The information is taken from a feasibility exercise designed to take a measure of this phenomenon.

Experimental techniques

To establish a controllable flow regime, with the pork comminute in as an undamaged condition as possible, a mechanical stuffer fitted with a variable speed drive was used rather than a pump to push the meat through stainless steel pipes. The temperature of the whole system was controlled, and pressures were indicated by diaphragm gauges filled with white petroleum jelly. When pressures fluctuated rapidly they were recorded by closed circuit TV.

It was not easy to make quantitative measurements of the thickness of the pale layer, but it was an essential step if any meaningful work were to be made on determining the conditions which produced the effect. The method finally used was slow, but effective. Under any particular set of experimental conditions, lengths of the material which emerged from the pipe were collected and frozen in solid CO_2. From the frozen rods slices were cut to reveal the cross-sectional appearance. The slices were stored in foil-wrapped trays at about —10 °C until they were photographed. By including a scale in the photographs, transparencies were obtained which, on projection, made possible the measurement of the mean thickness of the pale layer. One set is shown in fig. 1.

Flash photography was essential to avoid fat melting and only colour photography gave sufficient contrast between the pale layer and the comminute for measurement to be possible. Careful examination showed no evidence that cutting produced any significant distortion provided the rod was sufficiently cold and the knife sharp and heavy.

The flow behaviour of the pork comminute was studied with a split pipe constructed from two 120 cm lengths of pipe milled to give two exact halves. This unit was used in conjunction with comminute dyed with sudan black. Successive slugs of dyed and undyed meat were pushed through the split pipe by further comminute and the whole frozen with solid CO_2. The pipe was then opened and the rod of material removed.

When the rod was cut longitudinally the flow profiles were revealed and one set is reproduced in fig. 2. This technique revealed only viscous behaviour since any elastic effects decayed away before the meat froze, but starting from near plug flow there appeared to be a layer of material of low apparent viscosity against the wall, and, sliding inside this annulus, material exhibiting a flow profile of the type to be expected from a pseudoplastic. This was in agreement with some earlier work where a pork comminute was shown to give a linear log-log plot of shear stress against shear rate over three cycles (2).

The split pipe also permitted an examination of the colour of the pale layer. At any chosen time in a run it was possible to stop the stuffer, freeze the split pipe and its contents, remove the rod of frozen meat comminute and subject the surface of it to a colour examination.

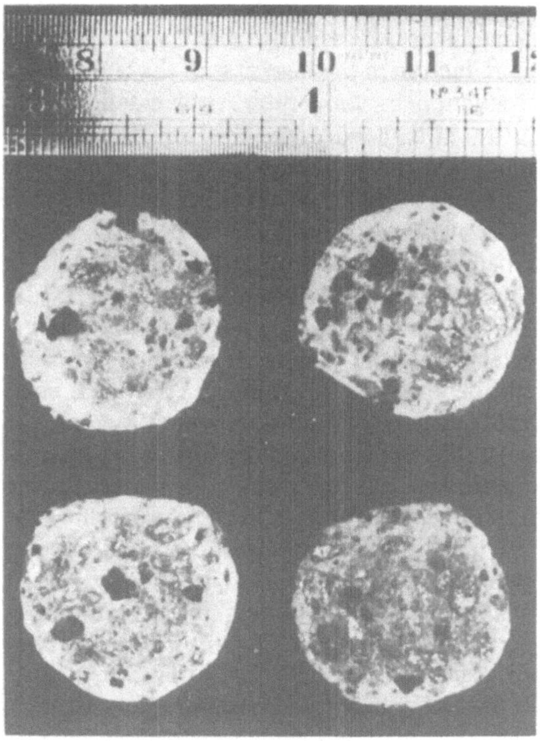

Fig. 1. Pale layer around the edge of slices

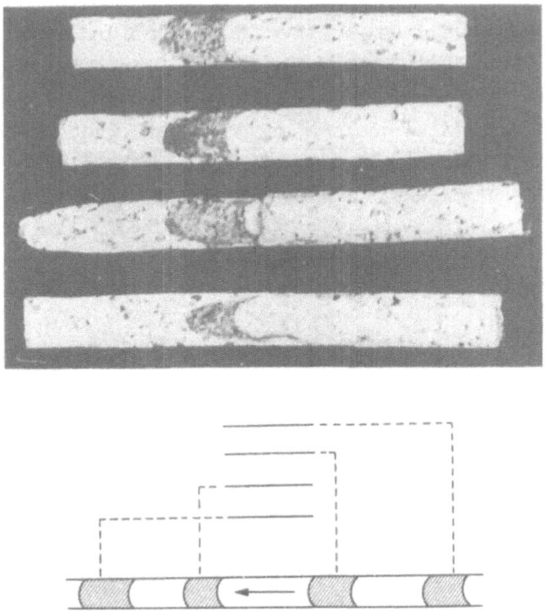

Fig. 2. Dyed plugs from split pipe. (Note the apparent build up of the profile)

Results

The expectation was that the pale layer would be found to be due to a preferential movement of fat to the surface in contact with the pipe wall. It was thought that this would make the phenomenon temperature sensitive, but it was found that any reduction in the thickness of the layer due to a reduction in temperature was small. This is illustrated in Table 1 and was supported by the finding that, in fact, there was less fat and not more in the pale layer.

The surface of the extrudate was lightly scraped as it emerged from the pipes. From each 20 kg of comminute passed, sufficient material was obtained for a replicated analysis for fat, water, protein and ash. This analysis was compared with the analysis for samples of the meat comminute. There were no significant differences in the protein and ash figures for seven different comminutes, but there was more water and less fat in the pale layer than in the comminute. The water content was up and the fat content was down by nearly 7%.

The fact that the pale layer thickness increased with flow velocity in the figures given in table 1 indicated the possibility that an increase in the rate of shear increased the layer thickness.

Two sets of experiments were performed with pipes 120 cm long and of three different radii. In one set the flow rate was varied to give a constant apparent shear rate at the wall and, in the other, the flow rate was kept constant so that the shear rate decreased with increase of diameter. The

Table 1. Mean thickness of pale layer at various flow rates through a $19 \cdot 10^{-3}$ m I.D. stainless steel pipe

Temperature (deg C)	Mean extrusion velocity (10^{-2}m s^{-1})	Mean layer thickness (10^{-3}m)
5	13	$1{\cdot}0_2$
	5	$0{\cdot}7_1$
	1	$0{\cdot}6_4$
10	13	$1{\cdot}1_7$
	5	$0{\cdot}7_4$
	1	$0{\cdot}6_6$
15	13	$0{\cdot}9_4$
	5	$0{\cdot}9_2$
	1	$0{\cdot}6_9$
20	13	$0{\cdot}9_9$
	5	$0{\cdot}8_4$
	1	$0{\cdot}7_6$

Table 2. The effect of shear rate on the mean thickness of pale layer (15 °C)

Pipe diameter (10^{-3}m)	Shear rate at wall (Sec^{-1})	Mean layer thickness (10^{-3}m)
12·7	22	$0{\cdot}2_5$
19	22	$0{\cdot}2_0$
25·4	22	$0{\cdot}2_5$
12·7	45	$0{\cdot}3_8$
19	30	$0{\cdot}2_5$
25·4	22	$0{\cdot}2_0$

Table 3. The effect of pipe length on the mean thickness of pale layer from a $19 \cdot 10^{-3}$ m I.D. pipe

Temperature (deg. C)	Mean extrusion velocity (10^{-2}ms^{-1})	Pipe length (m)	Mean layer thickness (10^{-3}m)
5	1	0·30	$0{\cdot}2_5$
		1·20	$0{\cdot}6_4$
	5	0·30	$0{\cdot}2_5$
		1·20	$0{\cdot}7_1$
	13	0·30	$0{\cdot}4_6$
		1·20	$1{\cdot}0_2$
10	1	0·30	$0{\cdot}2_5$
		1·20	$0{\cdot}6_6$
	5	0·30	$0{\cdot}2_5$
		1·20	$0{\cdot}7_4$
	13	0·30	$0{\cdot}5_8$
		1·20	$1{\cdot}1_7$

results with material at 15 °C were as shown in table 2 and did confirm the hypothesis. Similar experiments on material at lower temperatures were inconclusive possibly because the pale layers were too thin for precise measurement.

The final experiment of this type showed that the thickness of the pale layer increased with the length of pipe through which the material passed. This is shown by the figures given in table 3.

The colour measurements were made on a tristimulus colorimeter (Colorcord Mark II by Joyce, Loebl & Co. Ltd.). Frozen samples were taken from the split $10 \cdot 10^{-3}$ m I.D. pipe after the pork comminute had been flowing for 600 seconds. This was done at three flow rates. Lengths of $50 \cdot 10^{-3}$ m were cut from the rod, thawed out and measured in random order. All three parameters luminance, purity and dominant wavelength changed with time after thawing, but this effect was minimised by a careful organisation of the sample preparation.

The production of samples at each shear rate was performed on different batches of comminute

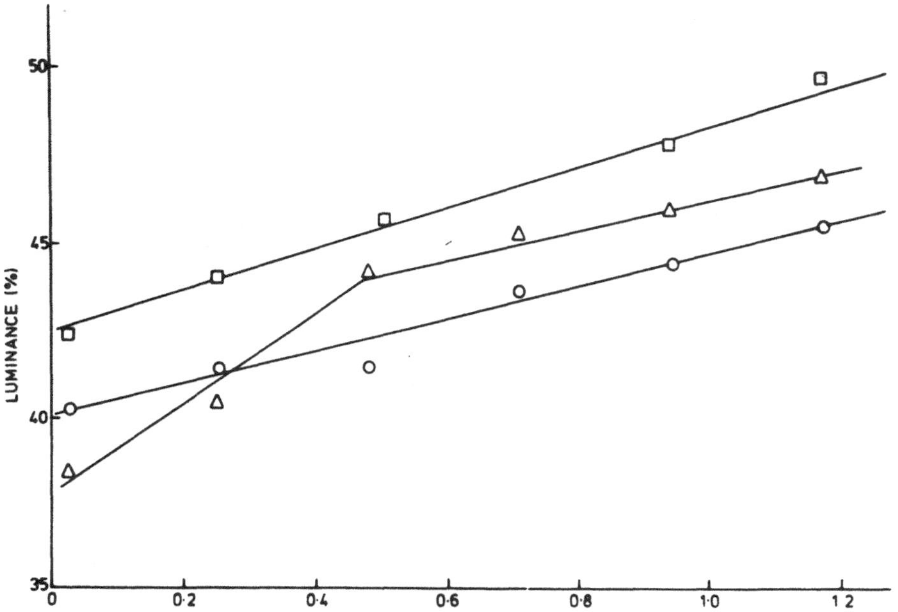

Distance sample had travelled along pipe (metres)

Fig. 3. Luminance plotted against distance comminute had travelled down pipe. The straight lines are for guidance only.
Extrusion at: ○ = $15 \cdot 10^{-3}$ ms^{-1} △ = $40 \cdot 10^{-3}$ ms^{-1} □ = $80 \cdot 10^{-3}$ ms^{-1}

so no comparison could be made between the colour values obtained at the different shear rates.

A change down the length of the rods was demonstrated for all three parameters and is illustrated in fig. 3 by the results for luminance. A difference of luminance of 1% may be seen by the human eye as a difference of colour and changes several times the value detectable by the eye were found in all three parameters between one end of the rod and the other.

Discussion

The fact that a marginal zone was produced when meat comminute was forced to flow down a pipe was established by this feasibility study. It was shown also that the thickness of the zone increased down the length of a tube and probably increased with shear rate. There was a strong indication that slip occurred in this zone.

The question which remained at the end of the feasibility study was how best to proceed to achieve design capability. It is that question which is posed here. One possibility was to follow the line *Mooney* (3) used with capillaries and determine the throughput at constant temperature in tubes of various diameter, but of the same lengths under such various pressure drops that the shear stress at the tube walls is always the same. In this way a coefficient of slip for the particular shear stress may be obtained. But in the case of meat comminutes this value depends also on the length of the tube used and it would be dominated by the second half of the tube where it has been shown that the pale layer is thickest.

Another possibility was to attempt to treat the meat comminutes as *Boltzmann* fluids exhibiting time-dependency in their flow behaviour. The practical difficulty with this was that it is possible to obtain semiquantitative time-dependency data only with a rotational rheometer where the same sample remains in the instrument throughout the measurement (4). The particle size in a meat comminute may be quite large and this would produce difficulties even with a flat-apex cone in a cone-and-plate rheometer, but there is a more fundamental difficulty. The shear rate of a meat comminute under test could be made equal to that at the wall of a pipe, but it would not be possible to reproduce the rate of change of the velocity gradient near the wall of the tube which is presumed to lead to the motion of the particles away from the wall (5).

A further analytical problem is that the structure of the meat comminute does not gradually reform when the stress is removed as it should in a thixotropic fluid i. e. the process producing the pale layer is not reversible.

The solution apparently adopted in virtually all engineering design procedures for isothermal steady flow (6), is to exclude time-dependency effects by using the appropriate extreme values, but in the case of this marginal zone problem such a course would result in under or over engineering to an unsatisfactory degree. Some correlating system must be found for scaling up laboratory data.

Acknowledgements

The author is indebted to Mr. *J. Hutchings* who made the colour measurements, Mr. *H. Deakin* who photographed the many samples for the transparencies, Mrs. *G. M. Hughes* who made the histological examinations and Mr. *P. Lock* and Mr. *F. Zbrozek* who assisted with the experiments.

References

1) *Vand, V.*, J. Phy. Coll. Chem. **52**, 277 (1948).
2) *Hawkins, A.E.*, Chem. Eng. **19** (1971).
3) *Mooney, M.*, J. Rheol. **2** 210 (1931).
4) *Green, H.*, Industrial Rheology and Rheological Structures, p. 52 (New York 1949).
5) *Starkey, T.V.*, Brit. J. Appl. Phys. **7**, 52 (1956).
6) *Skelland, A.H.P.*, Non-Newtonian Flow and Heat Transfer, p. 13 (New York 1967).

Author's address:
Dr. *A.E. Hawkins*
Unilever Research Laboratory
Colworth House
Sharnbrook, Bedford (England)

Rheol. Acta **13**, 877–882 (1974)

From the Worthington (Canada) Ltd. Brantford, Ontario (Canada)

Turbulent flow of non-Newtonian substances

By J. J. Vocadlo, P. J. Wheatley, and M. E. Charles)*

With 8 figures

(Received October 27, 1972)

Nomenclature

C	volume fraction solid in suspension
D	tube diameter
f	*Darcy-Weisbach* friction factor
g	gravitational acceleration
K_s	proportionality constant defined by eq. [10]
L	length of tube
P	pressure
Re	*Reynolds* number $\frac{\rho_m V D}{\mu_m}$
t	exponent defined by eq. [1]
V	mean velocity
V^*	volume of particles in pipe length L
W	settling velocity of particles
α_m	factor defined by eq. [1]
$\dot{\gamma}$	shear rate
$\dot{\gamma}_{turb}$,	turbulent pseudo shear rate defined by eqs. [8] and [9]
τ_w	wall shear stress
$(\tau_w)_s$	increment in wall shear stress due to presence of settling particles
μ_m	limiting viscosity at high rate of shear
ρ_1	density of carrier liquid
ρ_m	density of mixture
ρ_s	density of solid

Introduction

Normally, a unique shear stress-shear rate relationship for the laminar flow of any time independent non-*Newton*ian substance can be obtained from the correct interpretation of viscometric data. The flow curve or pseudo flow curve so obtained can be used to scale up to a diameter of interest. The transportation of non-*Newton*ian substances in the laminar regime is preferred in many engineering applications because of low energy consumption. However, there are also many situations where a density difference exists between the conveying liquid and solid particles in suspension which results in the tendency of the particles to settle. In order to ensure the stability of the suspension of particles such mixtures or substances must be transported in the turbulent regime.

The turbulent flow of non-*Newton*ian substances brings another dimension of complexity to the considerable difficulties encountered in the turbulent flow of *Newton*ian liquids. It would be unrealistic to expect that the turbulent flow of non-*Newton*ian substances could be treated by a purely theoretical approach. In this paper a new approach based on the elementary theoretical analysis of experimental results is adopted. As such, it can be regarded as a semi-empirical approach.

The state of the art of the correlation of non-*Newton*ian turbulent pipe-flow data has been reviewed by Harris (1); the approaches of *Metzner* and *Reed* (2), *Dodge* and *Metzner* (3) and *Bowen* (4) are critically analyzed. The *Metzner-Reed* and *Dodge-Metzner* approaches (2, 3) employ the generalized *Reynolds* number which was developed for time independent fluids and thought to be of universal significance for both laminar and turbulent flows. *Bowen*'s method (4) for turbulent flow which is favoured by *Harris* is based on an intuitive empirical relationship found to give good results in a survey of the available literature.

All of the approaches mentioned above employ the laminar flow parameters to some extent. Experimental data indicate, however, that materials which can be classified under one flow model in the laminar regime may not show similar characteristics in the turbulent regime, and vice-versa. There exist substances with similar turbulent flow characteristics but which show substantial difference in laminar flow behaviour. Obviously, various physical characteristics, such as physical-chemical forces between particles, particle shape and elasticity and other factors play different roles in the turbulent and laminar regimes.

Present approach

A. Turbulent flow of homogeneous substances

In this investigation it is assumed that the fully developed turbulent flow of non-*Newton*ian substances may be characterized by the flow of an equivalent homogeneous *Newton*ian medium. It is further assumed that any wall effect can be neglected, suspended particles do not settle, the substance is not dilatant, and the flow takes place in smooth tubes.

It may be readily shown by dimensional analysis that the pressure gradient for the flow of an homogeneous continuum is given by

---

*) Professor of Chemical Engineering, University of Toronto and scientific advisor to Worthington (Canada) Ltd.

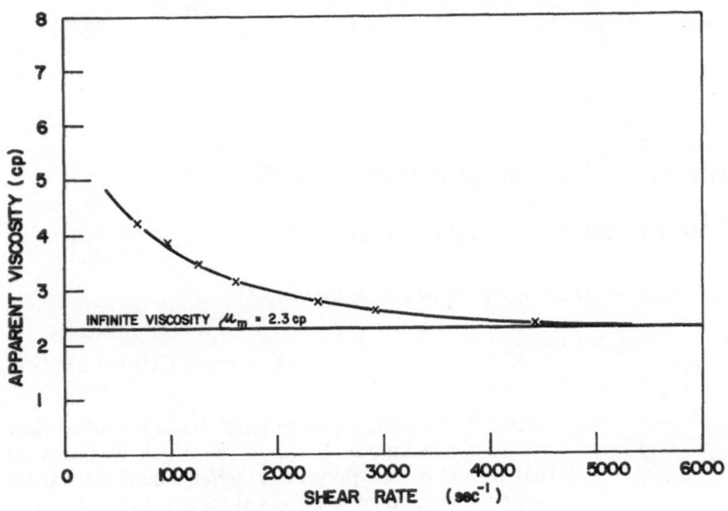

Fig. 1. Apparent viscosity as a function of shear rate for a kaolin-water suspension with a volume fraction of 0.0386

$$\Delta P/L = \alpha_m V^t D^{t-3} \mu_m^{2-t} \rho_m^{t-1} \quad [1]$$

where

$\Delta P/L$	pressure gradient
V	mean velocity
D	pipe diameter
μ_m	viscosity
ρ_m	density

and α_m and t are parameters determined by experiment, and the shear stress at the wall by

$$\tau_w = \frac{\alpha_m}{4} \left(\frac{D}{\mu_m} \right)^{t-2} \rho_m^{t-1} V^t \quad [2]$$

where

τ_w = shear stress at the wall.

In the case of suspensions the appropriate density is

$$\rho_m = \rho_1 + C(\rho_s - \rho_1) \quad [3]$$

where ρ_s and ρ_1 are the densities of the solid and liquid phases respectively and C is the volume fraction solid in the suspension.

It is now assumed that the most meaningful viscosity for use in turbulent non-*Newton*ian correlations is the limiting viscosity at high rates of shear. This assumption is based on the analysis of various suspensions and agrees with the conclusion of *Thomas* (5). It is shown in the present investigation that the limiting viscosity is the only property which provides a unique relationship between shear stress and "turbulent pseudo shear rate". The limiting viscosity is best obtained as an asymptotic value in the apparent viscosity-shear rate relationship as shown in fig. 1 for a typical kaolin-water suspension.

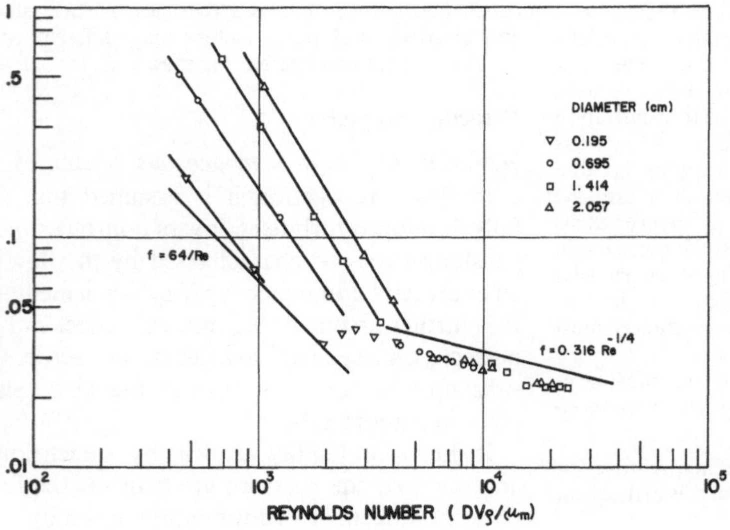

Fig. 2. Friction factor-*Reynolds* number relationships for the laminar and turbulent flow of a kaolin-water suspension (volume fraction 0.0386) in tubes of different diameter

Eq. [1] can be rearranged to give the *Darcy-Weisbach* relationship between the friction factor and *Reynolds* number

$$f = 2\alpha_m \mathrm{Re}^{t-2}. \qquad [4]$$

For example for $t = 1.75$ and $2\alpha_m = 0.316$, one obtains the *Blasius* relationship for flow of *Newton*ian liquids in smooth pipes.

With suspensions or polymer solutions the magnitude of α_m reflects the change in turbulence structure due to the presence of solid particles or polymer molecules. This is shown in fig. 2 for a kaolin-water suspension where the value of α_m is approx. 0.85 of that for *Newton*ian liquids. There are thus two parameters which characterize a given substance in eqs. [1] and [2] – the limiting viscosity at high shear rate and α_m which reflects the suppression of turbulence. Both of these must be determined experimentally. The exponent t is an additional parameter and is the slope of the shear stress turbulent pseudo shear rate relationship.

Eq. [2] can be used to draw several practical conclusions with respect to the effect of various parameters.

(i) Concentration effect

Laminar and turbulent flow characteristics for two concentrations of kaolin clay in water in four tube diameters are shown in fig. 3 where wall shear stress is plotted against the pseudo shear rate $8V/D$. As expected the data in the laminar regime for a given concentration fall on a single curve independent of pipe diameter. However in the turbulent regimes, the pipe diameter is an additional important variable. While the effect of concentration is very significant in the laminar regime it is relatively small in the turbulent regime.

The ratio of shear stress values in the turbulent regime at a constant value of $8V/D$ and same tube diameter is

$$\frac{\tau_{w2}}{\tau_{w1}} = \frac{\alpha_{m2}}{\alpha_{m1}} \left(\frac{\mu_{m2}}{\mu_{m1}}\right)^{2-t} \left(\frac{\rho_{m2}}{\rho_{m1}}\right)^{t-1}. \qquad [5]$$

For the two suspensions shown assuming $\alpha_{m2} = \alpha_{m1}$ the ratio in eq. [5] is 1.11. The concentration effect appears primarily in the viscosity and density variables.

(ii) Diameter effect

Scaling-up of flow data is an important engineering consideration. In order to scale up from a diameter D_1 for which data are available to a diameter D_2 for the same material, assuming $t_1 = t_2 = t$, one can write the ratio

$$\frac{\tau_{w2}}{\tau_{w1}} = \left(\frac{D_2}{D_1}\right)^{t-2} \left(\frac{V_2}{V_1}\right)^{t} \qquad [6]$$

and because $\dfrac{V_1}{D_1} = \dfrac{V_2}{D_2}$, i. e. constant $\dfrac{8V}{D}$,

one obtains

$$\frac{\tau_{w2}}{\tau_{w1}} = \left(\frac{D_2}{D_1}\right)^{2(t-1)}. \qquad [7]$$

Fig. 4 shows the result of using data obtained for $D_1 = .195$ cm and scaling up to $D_2 = 2.057$ cm. With $t = 1.77$ from fig. 3, the ratio $\tau_{w2}/\tau_{w1} = 37.6$. The predicted D_2 relationship compares very well with the data actually obtained.

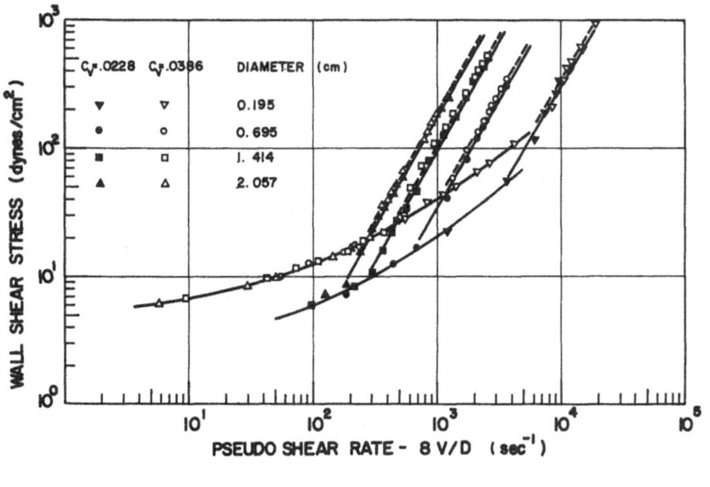

Fig. 3. Wall shear stress-pseudo shear rate relationships for the flow of two kaolin-water suspensions (volume fractions 0.0228 and 0.0386) in tubes of different diameter

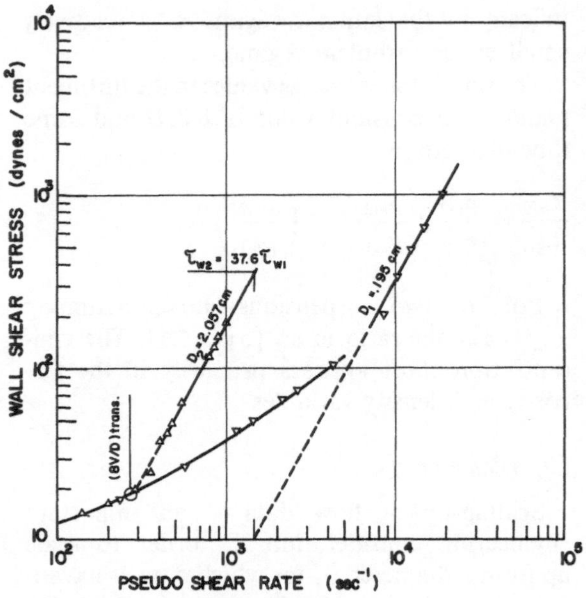

Fig. 4. Demonstration of scale-up from a diameter D_1 to a diameter D_2 and location of transition

(iii) Laminar-turbulent transition

Numerous equations have been derived for the laminar-turbulent transition. It is our opinion that equations based purely on laminar flow characteristics are not necessarily justified. The actual transition point is the intersection of laminar and turbulent flow relationships which are to some extent governed by independent variables. It is believed that the best procedure is to scale-up turbulent data to a required diameter and then calculate the transition velocity from the $(8\,V/D)$ transition point determined as the intersection of the laminar pseudo-flow curve and the turbulent branch in question. This is also shown in fig. 4. The value of the actual transition velocity appears to exceed the value determined from the intersection point by a small amount.

B. Correlation of turbulent data

Turbulent data can be represented by a unique relationship. Two such correlations are considered in this paper:

(1) Correlation based on "turbulent pseudo shear rate"

It has been found that the correlation of turbulent data in the form of a shear stress "turbulent pseudo shear rate" diagram provides a unique relationship, i. e.

$$\tau_w = \mu_m \; \dot{\gamma}_{\text{turb.}} \quad [8]$$

The form of $\dot{\gamma}_{\text{turb.}}$ can be readily obtained from eq. [2] rearranged, i. e.

$$\tau_w = \mu_m \cdot \frac{\alpha_m}{32} \text{Re}^{t-1} \, \frac{8\,V}{D}. \quad [9]$$

Fig. 5 shows the correlation for some selected data points for the four tube diameters superimposed on the original shear stress-pseudo

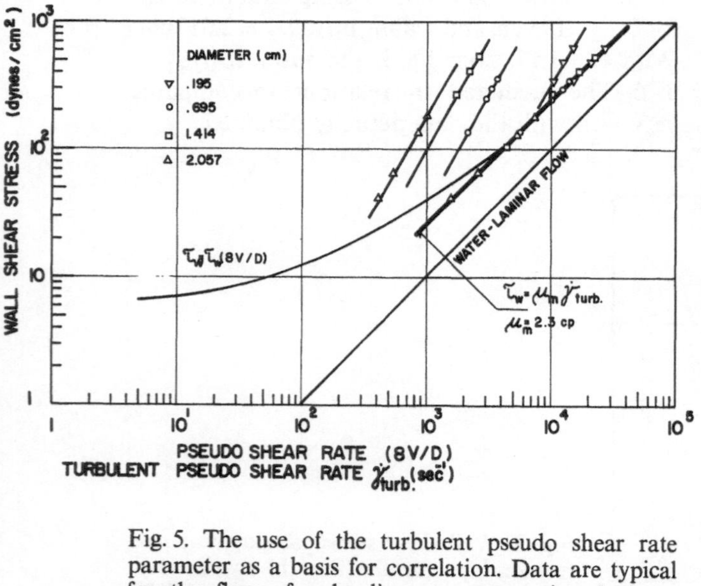

Fig. 5. The use of the turbulent pseudo shear rate parameter as a basis for correlation. Data are typical for the flow of a kaolin-water suspension (volume fraction = 0.0386).

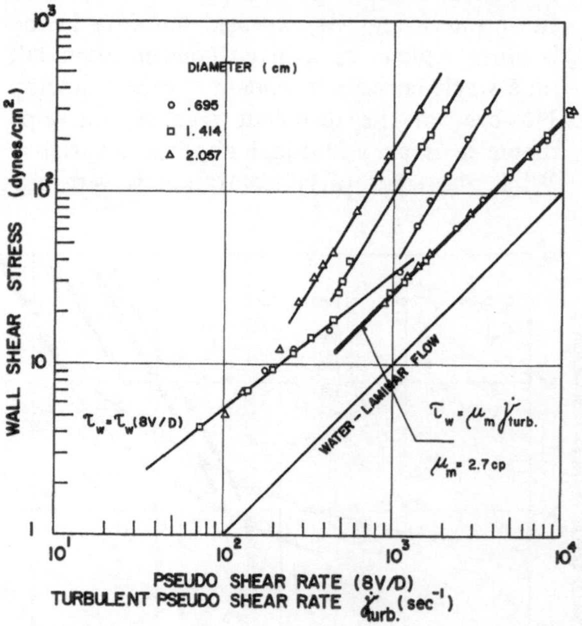

Fig. 6. The correlation demonstrated for a Carbopol solution (concentration 0.0775%)

shear rate plot. The transformation is at best done in the log-log plot by considering the additive constant $\frac{\alpha_m}{32}\mathrm{Re}^{t-1}$ which provides the shift in the data. The success of this correlation confirms the assumption that the infinite viscosity is the most meaningful viscosity term in non-*Newton*ian turbulent correlations.

Fig. 6 shows a similar correlation for Carbopol solution.

(2) Correlation based on mean velocity

Eq. [1] indicates that turbulent data can be correlated when plotted in terms of

$$\frac{\Delta P}{L}\,\frac{1}{\alpha_m}\,D^{3-t}\,\mu_m^{t-2}\,\rho_m^{1-t}$$

against V. The term $1/\alpha_m$ must be considered when the correlation is based on the turbulent flow of the carrier liquid. For concentrated substances it is our experience that the concentration effect on α_m is negligible. Such a correlation is shown for kaolin-water suspensions in fig. 7 with experimental points previously shown in fig. 3 and for the Carbopol solution in fig. 8. The data in the low velocity range in both figures fall in the laminar regime and therefore deviate from the turbulent correlation. This relationship can also be used for scale-up purposes and is certainly convenient to use.

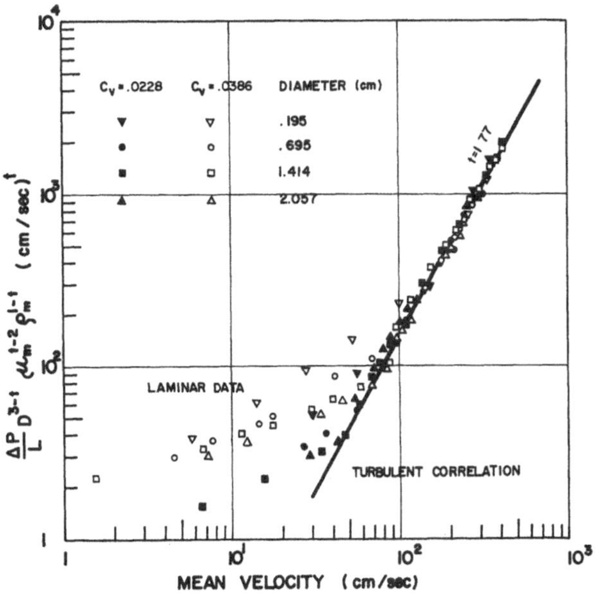

Fig. 7. The use of mean velocity as a basis for correlation in the turbulent regime for kaolin suspensions

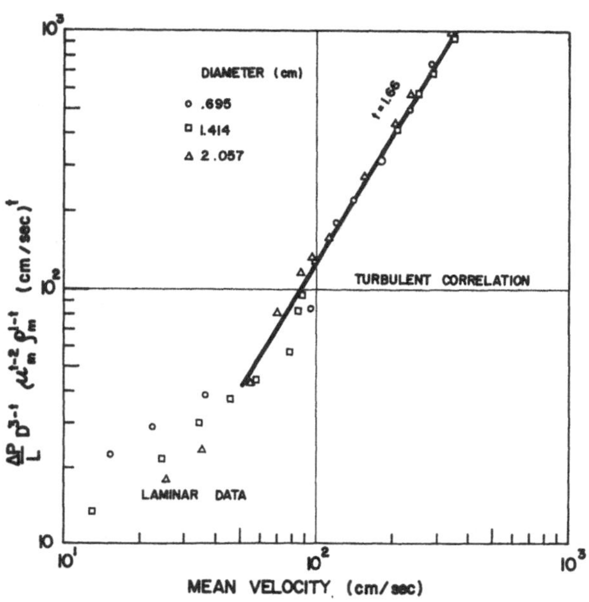

Fig. 8. The use of mean velocity as a basis for correlation in the turbulent regime for Carbopol solution

The disadvantage of this approach is that it does not provide for the prediction of the transition velocity in a convenient form.

C. Effect of particle settling in horizontal tube

In many engineering applications materials must be transported in the turbulent regime to ensure mixing or to maintain the solid particles in a state of suspension. The dominant mechanism by which the particles are maintained in suspension is one involving turbulent eddy motions. The average work which must be performed to keep the particles suspended is proportional to the net weight of the particles in the tube and the velocity at which they settle. The appropriate settling velocity would appear to be the free settling velocity since there is a random vertical movement of particles and fluid. In sedimentation there exists a net upward flow of fluid relative to the particles which reduces the fall velocity of the particles. A similar hindered settling effect is not present in horizontal transport.

Now if V^* represents the volume of particles present in length L of tube, one can write the energy required per unit time as $K_s V^*(\rho_s - \rho_1) g W$ in which K_s is a proportionality constant and W is the particle settling velocity. The pressure gradient required to keep particles suspended is then

$$(\Delta P/L)_s = K_s(\rho_s - \rho_1)gCW/V \qquad [10]$$

where C is volume fraction solids and the suggested value of K_s is 10. The shear stress increment due to the presence of settling particles is then

$$(\tau_w)_s = \frac{K_s}{4} DC(\rho_s - \rho_1)gW/V. \qquad [11]$$

For a more complete treatment of horizontal turbulent flow of settling slurries reference can be made to the paper by *Vocadlo* and *Charles* (6).

Eq. [11] serves to estimate the effect of the presence of settling particles on the value of the shear stress. This increment is a function of tube diameter. If it is significant it must be subtracted from turbulent data obtained with settling particles before proceeding with further scaling-up and general turbulent correlation of homogeneous substances.

Summary

A unique shear stress-shear rate relationship exists for laminar flow of any time independent substance in a tube, whereas this is not the case for turbulent flow. In order to obtain a unique relationship for turbulent flow, a new approach based on the elementary theoretical interpretation of experimental data is adopted in the present paper. In particular, wall shear stress is found to be a unique function of a new turbulent pseudo shear rate term. In this relationship there are two parameters which characterize a given substance – the limiting viscosity at high shear rate μ_m and a factor α_m which takes into account modification of turbulent structure by the non-*Newton*ian properties. Both of these parameters must be determined experimentally. Methods of predicting pressure gradients and of scaling up are outlined. In applying the approach to suspensions in which the solid phase has a density greater than that of the liquid medium, it may be important to determine the increment in shear stress equivalent to the energy required to maintain the solid particles in suspension.

The validity of this approach is confirmed by data for the flow of a variety of substances including kaolin suspensions and Carbopol solutions in tubes ranging in diameter from 1.5 to 20 mm.

References

1) *Harris, J.*, Rheol. Acta **7**, 228 (1968).

2) *Metzner, A. B.* and *J.C. Reed*, Amer. Inst. Chem. Eng. J. **1**, 434 (1955).

3) *Dodge, D. W.* and *A. B. Metzner*, Amer. Inst. Chem. Eng. J. **5**, 189 (1959).

4) *Bowen, Le. B. R.*, Chem. Eng. J., June 26, 127 (1961); July 10, 147 (1961); July 24, 143 (1961).

5) *Thomas, D. G.*, Ind. Eng. Chem. **55**, No. 12, 27 (1963).

6) *Vocadlo, J. J.* and *M. E. Charles*, Proceedings Hydrotransport 2, British Hydromechanics Research Association (1972).

Author's address:

Prof. Dr. *M. E. Charles*
University of Toronto, Dept. of Chem. Engineering and Applied Chemistry
Toronto (Canada) M5S 1A4

Rheol. Acta 13, 883–885 (1974)

Hoger Rijksinstituut voor Textiel en Kunststoffen, Gent (Belgique)

Régulation mécanique de la pression dans l'extrusion des polymères

Par G. A. Patfoort

Avec 5 figures

(Reçu p. p. le 27 octobre 1972)

La maîtrise d'une opération devient d'autant plus grande que les variables, qui déterminent son déroulement, sont séparables, ne s'influencent pas mutuellement et peuvent être règlées séparément. L'extrusion des polymères est un exemple typique ou l'évolution du processus s'est engagée dans la voie de l'interdépendance des variables.

Les variables qui déterminent l'effectivité de l'opération sont: les propriétés rhéologiques d'écoulement de la matière fluide; la pression et le débit. La courbe caractéristique de la vis montre immédiatement l'interdépendance de la pression et du débit. L'inhomogénité thermique (plus de 20 °C de différence de température à la tête d'une vis de 20 D) et structurelle rend la qualification exacte des propriétés d'écoulement illusoire. Plus grave est le fait que ces propriétés ne peuvent être règlées qu'à partir d'autres variables, que sont la température, la vitesse de rotation de la vis et la forme géométrique de cette dernière, qui n'influencent qu'indirectement les variables fixant l'effectivité de l'opération.

La vitesse de rotation augmente le débit, mais augmente en même temps l'énergie de friction, qui à son tour influence non seulement la viscosité mais également l'homogénité de la pression etc.

Les raisons de cet état de choses sont principalement imputables au fait que l'on essaye de réaliser dans un même appareil (soit un cylindre muni d'une vis) deux opérations différentes:

1. Transporter la matière tout en y accumulant de la pression.
2. Adapter sa viscosité et son homogénité en vue de sa mise en forme.

Nous avons nommé cette dernière opération «HOVIM» (Homogenity and viscosity inducing modifications) évitant le terme plastification qui peut prêter à confusion.

La vis est un moyen de transport adéquat et permet sous certaines conditions d'accumuler des pressions énormes sur de la matière granuleuse (plus de 800 kg/cm^2 avec une vis de 2,5 D). Par contre, l'écoulement liquide laminaire en fait un mélangeur de qualité douteuse et l'obtention d'une hausse de température homogène mêne à un problème rhéologique insoluble. L'échange thermique est complêtement different au fond du pas de vis ou sur son arête, surtout si le fourreau est chauffé extérieurement.

Si nous désirons séparer les variables et permettre leur réglage, il est indispensable de séparer les opérations et de faire abstraction de certains principes ancrés dans le concept traditionnel d'extrusion:

a) Il faut éviter le chauffage par contact de surfaces métalliques (et par conduction) mais au contraire avantager une production de chaleur dans la masse, par exemple par friction interne.

b) Il n'est pas souhaitable de faire passer la matière fluide dans une vis.

c) Pour séparer les deux opérations: transport-pression et HOVIM, il est nécessaire de supprimer la zone de transition toujours difficilement contrôlable.

d) Il faut élever une barrière thermique entre la zone transport – pression froide et la zone HOVIM chaude.

L'appareil se réduit donc à trois eléments: une zone d'alimentation froide composée par une vis tournant dans un fourreau capable de produire la pression sur granulés solides, une barrière thermique, une zone HOVIM simplement formée par un espace entre le bout de la vis coupée perpendiculairement à son axe et le fond du cylindre. Ces deux parois (fig. 1) forment donc un entre-fer situé entre deux disques, l'un en rotation (l'avant de la vis) et l'autre stationnaire (le cylindre) (1). La rotation rapide (jusqu'à 400 tours par minute) suffit à assurer la baisse de viscosité par friction interne. Le temps de séjour à l'état fluide dans

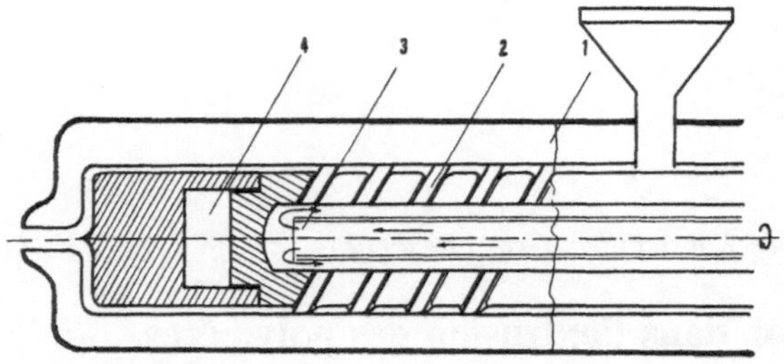

Fig. 1. Barrière thermique dans la vis avec. 1 = Cylindre; 2 = vis; 3 = canal de refroidissement; 4 = barrière thermique

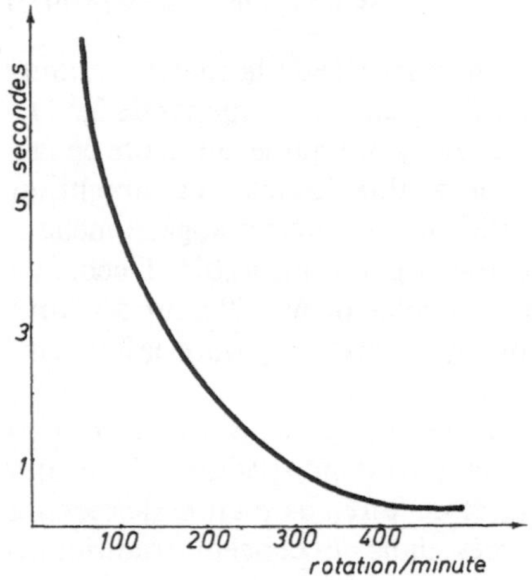

Fig. 2. Temps de séjour en fonction du nombre de tours par minute

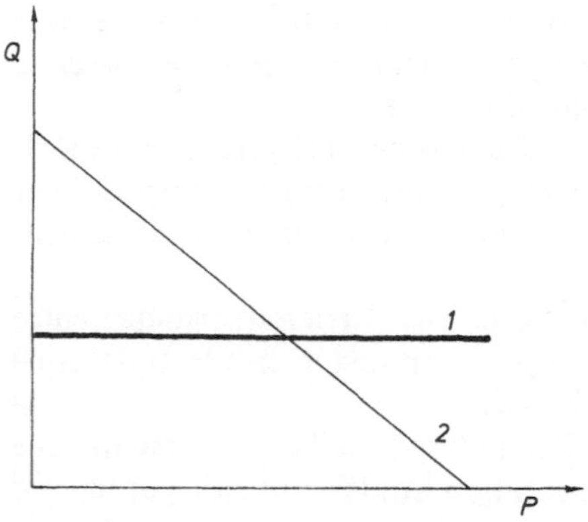

Fig. 3. Caractéristique de vis (Q = débit et P = pression). 1. Extrusion dure; 2. extrusion classique

l'appareil, donc sous sollicitation thermique, est ainsi ramené entre 1–5 secondes. Le PVC rigide a été extrudé à 260 °C sans dégradation. La dispersion obtenue équivaut à 2 – 3 passages au travers d'une vis classique de 20 D (fig. 3).

Les opérations étant séparées il y a lieu de s'occuper du réglage des variables:

– L'homogénité et la viscosité peuvent être règlées par la forme et la distance de l'entre-fer.

– Le débit est directement proportionel à la vitesse de rotation de la vis.

Le réglage de la pression exige quelques commentaires. Le coëfficient de friction apparent des granulés sur des parois métalliques est inversément proportionnel à la dimension des granulés (4). La comportement de la matière dans la vis est donc influencé par la surface métallique et la forme des granulés. En utilisant une température, une profondeur de filet et un rainurage du cylindre approprié à la dimension des granulés, on obtient une translation de matière sous forme de bouchon (2–4 et 5). La pression ne dépend plus que de la résistance de la matière à la compression. Nous obtenons ainsi une extrudeuse «dure» dont le débit est indépendant de la contre pression (fig. 2). En principe le bouchon devrait se déplacer parallèlement à l'axe de la machine: pratiquement, le déplacement forme un angle $\frac{\pi}{2} - \varphi$ avec l'axe, théoriquement donné par:

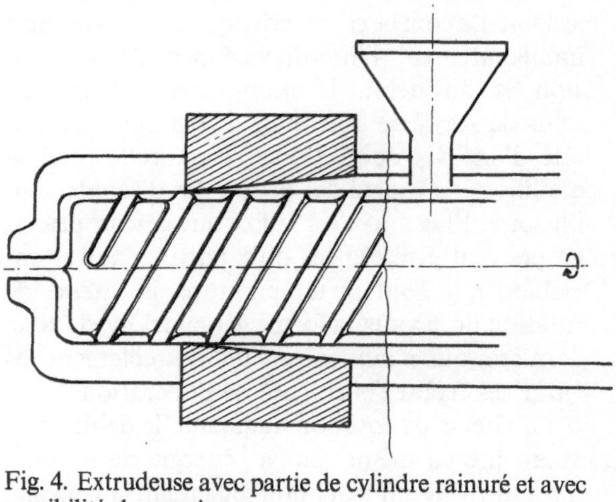

Fig. 4. Extrudeuse avec partie de cylindre rainuré et avec possibilité de rotation

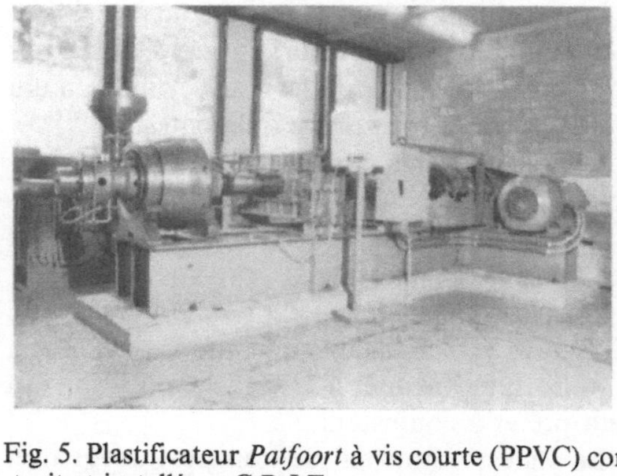

Fig. 5. Plastificateur *Patfoort* à vis courte (PPVC) construit et installé au C.R.I.F.

$$\cos\varphi = K\sin\varphi + M_1 + M_2 \ln\frac{P_1}{P_2} \quad \text{ou} \quad K, M_1 \quad \text{et} \quad M_2$$

sont des coefficients dépendant du dessin géométrique de la vis et de la friction granulé-surface métallique. P_1 et P_2 sont les pressions dans la masse granuleuse.

Le débit maximum théorique est obtenu quand la friction avec la vis est nulle. Dans ce cas l'angle φ est donné par:

$$\varphi = \frac{\pi}{2} - \theta \quad \text{ou} \quad \theta \text{ est l'angle de la vis.}$$

Si l'on empêche la rotation du bouchon, il devient évident que la pression, que peut vaincre l'appareil, ne sera limitée que par des considérations de résistance mécaniques ou par la puissance du moteur (6). Si, par contre, on peut freiner ou favoriser cette rotation, il est normal que l'on obtienne un moyen simple de réglage du pression. Pratiquement, on peut obtenir cet effet sur la masse granuleuse par différents moyens:

– En utilisant un cylindre de longueur réglable en prévoyant une ouverture d'alimentation déplaçable.

– Par un rainurage du cylindre avec profondeur réglable des rainures.

– En permettant la rotation d'une partie du cylindre, éliminant ainsi la friction qui engendre le mouvement de translation (fig. 4).

C'est la dernière solution qui a été retenue dans l'appareil expérimental qui est actuellement étudié par la section «Plastiques» du Centre de Recherche Scientifique et Technique de l'industrie des Fabrications Métalliques (CRIF) à Liège (Belgique) (fig. 5).

Une partie du cylindre rainuré peut tourner librement et est donc normalement entrainé par le mouvement de rotation de la vis et de la matière. Ce cylindre est muni d'un frein. Il est donc possible de communiquer à la matière, par freinage direct, une quantité d'énergie de friction directement mesurable. Celle-ci détermine la pression d'utilisation.

Nous obtenons un appareil dont les variables sont nettement déterminées (viscosité ou propriété d'écoulement, pression et débit) et ou grâce à la séparation des fonctions leur réglage est indépendant.

Littérature

1) *Patfoort*, Plastica **24**, 493–499 (1971).

2) *Schneider*, Der Fördervorgang in der Einzugzone eines Extruders. Diss. IKV Techn. Hochschule Aachen (1970).

3) *Menges*, Plastverarbeiter **20**, 79 (1969).

4) *Bucquoye*, Polymeer-metaal friktie. H. Rijksinstituut voor Text. & Kunststoffen (Gent 1970).

5) *Schneider*, Kunststoffe **59**, H. 2, p. 97 (1969).

6) *Masselink*, Transport in de voedingszone van een schroefextruder. Techn. Hogeschool Delft 1969.

Adresse de l'auteur:
Prof. Dr. *Georges A. Patfoort*
Residentie «Trianon»
71 Gustaaf Callierlaan
B-9000 Gent (Belgium)

Rheol. Acta **13**, 886–889 (1974)

Laboratoire Central des Ponts & Chaussées, Paris (France)
et Laboratoire Régional des Ponts & Chaussées, Saint-Brieuc (France)

L'essai d'expansion cylindrique et la loi effort déformation des sols purement cohérents

Par F. Baguelin et J. Jezequel

Avec 5 figures

(Reçu p. p. le 27 octobre 1972)

L'essai d'expansion cylindrique est très utilisé dans la reconnaissance des sols de fondations sous la forme de l'essai pressiométrique, développé par *L. Menard*, (Réf. 2, 4, 5, 6, 7). Dans son principe, l'exploitation qui en est faite, consiste à corréler directement deux chargements pratiqués sur le sol en place: celui de la sonde pressiométrique et celui de la fondation. L'utilisation de l'essai pour en tirer des déductions sur les lois rhéologiques des sols se heurte à deux sortes de difficultés:
la première concerne les théories, qui jusqu'alors, utilisaient systématiquement les hypothèses simplificatrices de l'élastoplasticité (Réf. 1, 4, 8), malgré quelques tentatives pour s'en affranchir (Réf. 3). L'objet de cette communication est de montrer que, dans un cas particulier, mais extrêmement important en pratique, celui des sols cohérents en sollicitation non drainée, il est possible de déduire la courbe de cisaillement du sol de la courbe d'expansion cylindrique sans aucune hypothèse rhéologique restrictive,
la deuxième est d'ordre technologique: il est nécessaire de modifier la technique de l'essai pressiométrique usuel pour obtenir une meilleure définition de la courbe de chargement et pour respecter les hypothèses fondamentales de la théorie, l'une des principales étant le caractère intact du sol, spécialement au voisinage de la sonde. Ces impératifs on été résolus par la mise au point d'un nouvel appareil, le pressiomètre autoforeur (Réf. 7). On présente dans cette communication un exemple de courbe d'essai obtenue à l'aide de cet appareil et interprétée par la théorie proposée.

Théorie de l'essai d'expansion cylindrique non drainée

Par définition, le chargement non drainé d'un sol saturé est tel qu'il n'y a pas de variation du volume élémentaire; ceci a lieu pour les sols fins soumis à des sollicitations rapides. Une des conséquences fondamentales de cette condition est que les champs des déplacements et des déformations résultent immédiatement de la donnée du déplacement radial u_{r_0} du bord de la sonde cylindrique, alors que d'ordinaire, on ne peut résoudre séparément le champ des déformations et le champ des contraintes et l'on est obligé de faire intervenir dès le départ la liaison entre ces deux types de grandeurs, constituée par les lois rhéologiques.

Il s'agit d'un problème à symétrie de révolution et à déformation plane. Soit (fig. 1):

r la distance radiale initiale
ρ la distance radiale actuelle
u_r le déplacement radial ($u_r = \rho - r$)
ε la dilatation circonférentielle ($\varepsilon = u_r/r$)

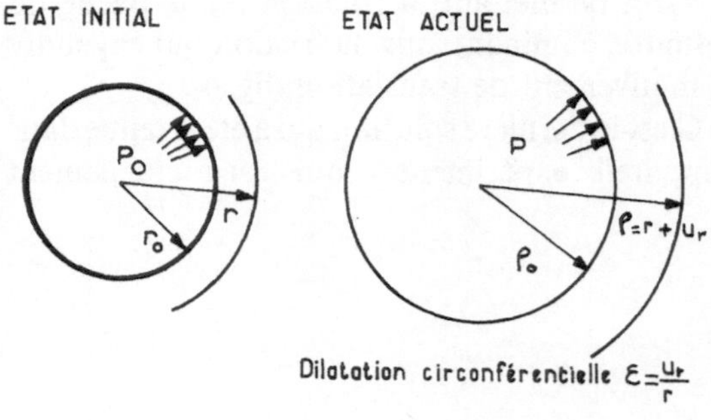

Fig. 1. Déplacements et déformations

La condition de non variation du volume élémentaire s'écrit:

$$\rho^2 - r^2 = \rho_0^2 - r_0^2 = a^2$$

l'indice 0 désignant les valeurs au bord de la sonde.

On obtient alors le champ des déplacements u_r et des déformations ε:

$$(r + u_r)^2 - r^2 = (r_0 + u_{r0})^2 - r_0^2 = a^2 \quad [1]$$

et

$$r^2[(1 + \varepsilon)^2 - 1] = r_0^2[(1 + \varepsilon_0)^2 - 1] = a^2. \quad [2]$$

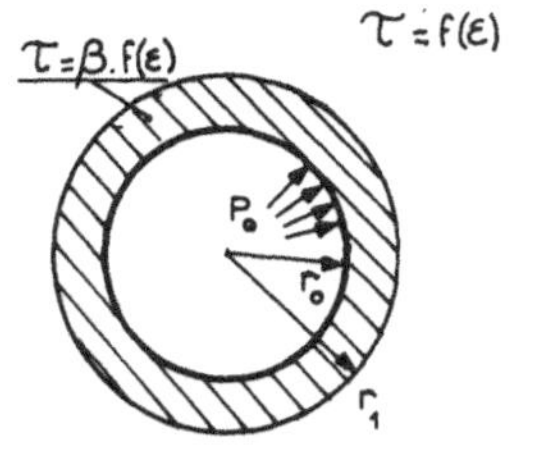

Fig. 2. Anneau remanie état initial

On peut alors introduire l'équation d'équilibre:

$$\frac{\sigma_r - \sigma_\theta}{\rho} = -\frac{d\sigma_r}{d\rho} \qquad [3]$$

et les lois rhéologiques. Dans le cas du sol entièrement intact, et compte-tenu de ce que le cisaillement a lieu partout sans rotation des contraintes principales, elles se réduisent à une seule relation entre une variable de déformation, ε par exemple, et une variable caractérisant le déviateur. Utilisons $\tau = (\sigma_r - \sigma_\theta)/2$. On pose alors:

$$\tau = f(\varepsilon). \qquad [4]$$

Si le sol est remanié (fig. 2), on peut introduire une fonction de cisaillement f dépendant de la distance r. Considérons le cas simple d'un anneau de sol ($r_0 < r < r_1$) uniformément remanié; la relation [4] sera valable pour $r > r_1$ tandis que dans l'anneau on aura:

$$\tau = f_1(\varepsilon) \qquad (r_0 < r < r_1)$$

avec [4 bis]

$$f_1(\varepsilon) = \beta . f(\varepsilon) \qquad (\beta < 1).$$

Combinant les équations [2], [3] et [4], il est aisé de trouver l'équation de la courbe pressiométrique du sol entièrement intact (p_0 = pression initiale, supposée uniforme dans tout le massif):

$$p = F(\varepsilon_0) = p_0 + \int_0^{\varepsilon_0} \frac{f(\varepsilon)}{\varepsilon(1+\varepsilon)\left(1+\frac{\varepsilon}{2}\right)} d\varepsilon. \qquad [5]$$

Dans le cas d'un anneau remanié, on obtient une pression plus faible p_1:

$$p_1 = F(\varepsilon_0) - (1-\beta) \int_{\varepsilon_1}^{\varepsilon_0} \frac{f(\varepsilon)}{\varepsilon(1+\varepsilon)\left(1+\frac{\varepsilon}{2}\right)} d\varepsilon \qquad [5\,\text{bis}]$$

avec

$$\varepsilon_1 = \sqrt{1 + \left(\frac{r_0}{r_1}\right)^2 [(1+\varepsilon_0)^2 - 1]} - 1. \qquad [6]$$

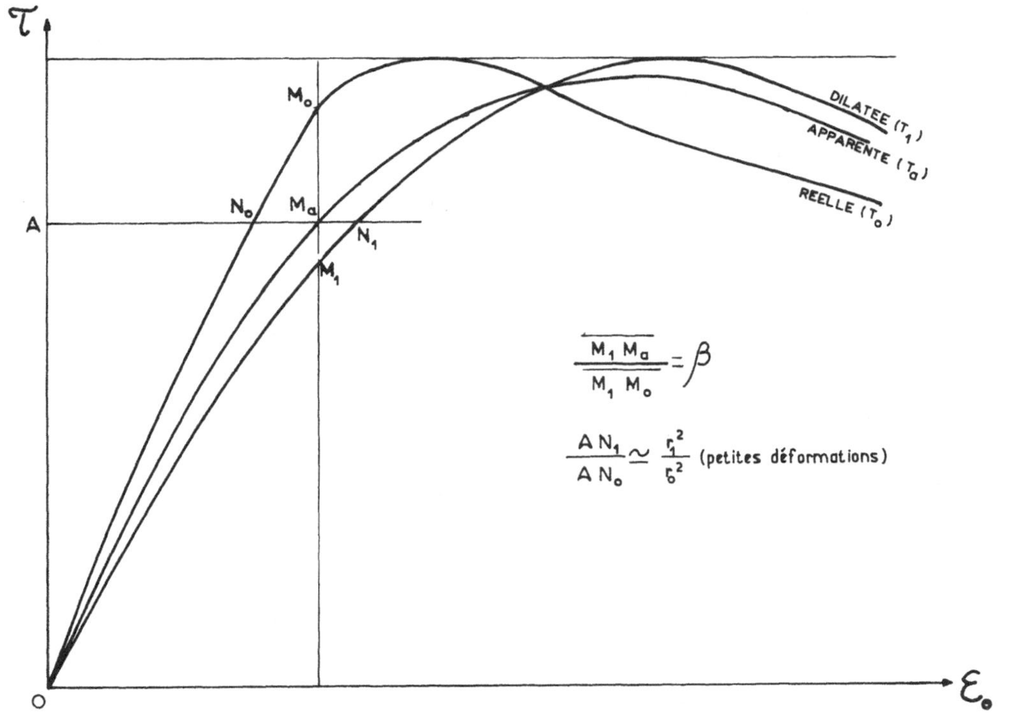

Fig. 3. Courbes de cisaillement réelle (T_0) et apparente (T_a)

On peut en particulier définir l'indice de remaniement suivant I_ρ, qui donne la chute relative du module initial:

$$I_\rho = \left[1 - \left(\frac{r_0}{r_1}\right)^2\right][1 - \beta]. \qquad [7]$$

Inversement, à partir de la courbe pressiométrique du sol intact:

$$p = F(\varepsilon_0)$$

on retrouve la loi de cisaillement du sol intact par simple dérivation par rapport à ε_0:

$$f(\varepsilon_0) = \varepsilon_0 \frac{dF}{d\varepsilon_0}(1 + \varepsilon_0)\left(1 + \frac{\varepsilon_0}{2}\right). \qquad [8]$$

Le terme principal $g(\varepsilon_0) = \varepsilon_0 \frac{dF}{d\varepsilon_0}$ représente la soustangente TN de la courbe pressiométrique (fig. 4).

Si l'on applique la dérivation à la courbe pressiométrique du sol avec anneau remanié, on obtient une loi de cisaillement apparente f_a telle que:

$$f_a(\varepsilon_0) = \beta \cdot f(\varepsilon_0) + (1 - \beta) \cdot \xi \cdot f(\varepsilon_1) \qquad [9]$$

avec

$$\xi = \left(\frac{1 + \varepsilon_0}{1 + \varepsilon_1}\right)^2$$

$$\varepsilon_1 = \varepsilon_1(\varepsilon_0) \qquad [6]$$

Dans le cas où les déformations restent petites, cette relation peut s'interpréter de la manière suivante, en notant qu'alors ξ est alors voisin de 1 et que $\varepsilon_1 \simeq \varepsilon_0 \frac{r_0^2}{r_1^2}$: (cf. Fig. 3), la courbe de cisaillement apparente f_a est une interpolation linéaire de rapport β entre la courbe de cisaillement du sol intact $f(\varepsilon_0)$ et la courbe de cisaillement dilatée suivant les déformations dans le rapport $\left(\frac{r_1}{r_0}\right)^2$.

Cette analyse permet de voir que le remaniement a une incidence particulièrement marquée sur les déformations de la courbe de cisaillement dérivée.

Application aux courbes d'essais pressiométriques

La fig. 4 présente une courbe pressiométrique réelle obtenue au pressiomètre autoforeur, c'est à dire sur un sol pratiquement intact. La courbe

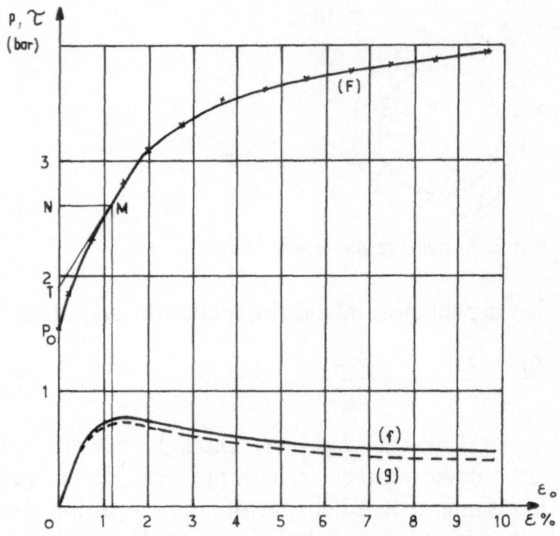

Fig. 4 Courbe pressiométrique et courbe de cisaillement dérivée-argile pour l'argile de Cran.

de cisaillement f en a été déduite dans ce cas en lissant la courbe pressiométrique dans le domaine considéré par une fonction de la forme:

$$p = p_0 + \frac{1}{2b}\left[a \operatorname{Log}(1 + \varepsilon_0^2) + \operatorname{Arctg} \varepsilon_0\right].$$

Une méthode plus générale d'application de la relation [8] à des courbes expérimentales,

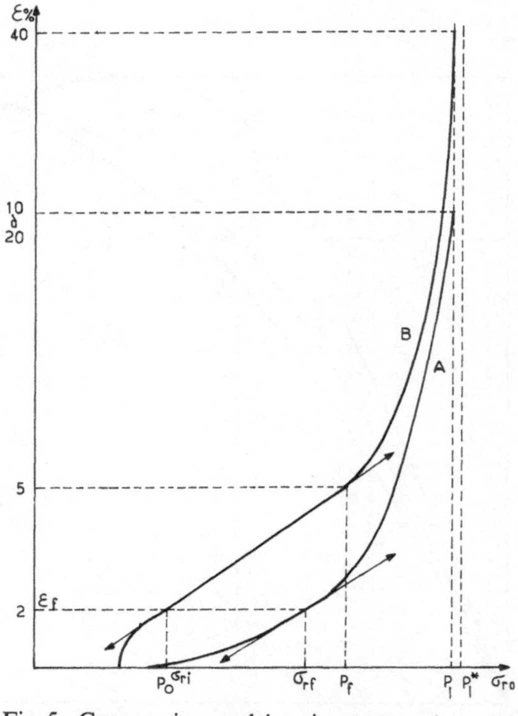

Fig. 5. Comparaison schématique entre les courbes d'essai au pressiomètre autoforeur (A) et au pressiomètre standard (B)

c'est à dire définie de manière discrète et avec une certaine marge d'erreur, consiste à combiner les analyses par différences finies et l'approximation suivant les moindres carrés en calculant la valeur de la dérivée en un point sur trois, voire quatre, valeurs de la fonction.

Il est à signaler par ailleurs que la pression de départ p_0 représente bien la pression horizontale naturelle des terres.

La fig. 5 illustre schématiquement les différences qui existent entre une courbe pressiométrique obtenue au pressiomètre autoforeur et une courbe pressiométrique ordinaire, obtenue avec le pressiomètre Ménard. Pour cette dernière, on note une phase préalable de recompaction du sol à la pression initiale p_0; pour le reste, l'analyse précédente concernant le remaniement (éq. [5 bis] notamment) permet de comprendre les différences existant entre les deux courbes.

Conclusion

On a montré que la loi de cisaillement non drainé d'un sol cohérent peut être déduite de la courbe d'un essai d'expansion cylindrique sans hypothèse rhéologique restrictive, du type elastoplastique.

On a aussi étudié l'incidence d'une zone remaniée au contact de la sonde.

Ces résultats théoriques sont illustrés sur des courbes d'essais réelles, obtenues notamment avec un appareillage adapté pour une telle interprétation, le pressiomètre autoforeur.

Littérature

1) *Bishop, A. W., R. Hill* et *M. Mott,* Proc. Physical Soc. **57**, 321 (1945).

2) *Jezequel, J., H. Lemasson* et *J. Touze,* Bull. Liaison Labo. P. et Ch. **32**, 97–120 (1968).

3) *Ladanyi, R.,* J. Soil Mech. and Found. Div., Proc. of ASCE **89** 127–161 (1963).

4) *Menard, L.,* Ann. des P. et Ch. **1957**, p. 356–377.

5) *Menard, L.* et *Rousseau,* Sols-Soils **1962**, 1.

6) *Menard, L.,* Sols-Soils, **1963**, 5.

7) «Mode opératoire de l'essai pressiométrique normal». Laboratoire Central des Ponts et Chaussées (Paris 1971)

8) *Salencon, J.,* Ann. des P. et Ch. **3**, 175.187 (1966).

Adresses des auteurs:

F. Baguelin et *J. Jezequel*
Laboratoire Central des Ponts & Chaussées-Paris (France)
Laboratoire Régional des Ponts & Chaussées-Saint-Brieuc (France)

Rheol. Acta **13**, 890 (1974)

From the Department of Applied Mathematics and Theoretical Physics, University of Cambridge (England)

The bulk stress in a suspension of spheres to order c^2

By G. K. Batchelor and J. T. Green

(Received October 27, 1972)

An exact formula is obtained for the term of order c^2 in the expression for the bulk stress in a suspension of force-free spherical particles in Newtonian ambient fluid, where c is the volume fraction of the spheres and $c \ll 1$. The particles may be of different sizes, and composed of either solid or fluid of arbitrary viscosity. The method of derivation circumvents the familiar obstacle, of non-absolutely convergent integrals representing the effect of all pair interactions in which one specified particle takes part, by the judicious use of a certain quantity which is affected by the presence of distant particles in a similar way and whose mean value in known exactly. The bulk stress is in general of non-Newtonian form and depends on the statistical properties of the suspension which in turn are dependent on the type of bulk flow.

The formula contains two fonctions which are parameters of the flow field due to two sperical particles immersed in fluid in which the velocity gradient is uniform at infinity. One of them, $p(r, t)$, represents the probability density for the vector r separating the centres of the two particles. The variation of $p(r, t)$ for a moving material point in r-space due to hydrodynamic action is found in terms of a function $q(r)$, and this gives $p(r, t)$ explicitly over the whole of the region of r-space occupied by trajectories of one particle centre relative to another which come from infinity. In a region of closed trajectories, steady-state hydrodynamic action alone does not determine the relation between the values of $p(r, t)$ for different material points. The function $q(r)$ is singular when the spheres touch, and the contribution of nearly-touching spheres to the bulk stress is evidently important. Approximate numerical values of all the relevant functions are presented for the case of rigid spherical particles of uniform size.

In the case of steady pure straining motion of the suspension, all trajectories in r-space come from infinity, the suspension has isotropic structure and the stress behaviour can be represented (to order c^2) in terms of an effective viscosity μ^*. It is estimated from the available numerical data that for a suspension of identical rigid spherical particles

$$\mu^*/\mu = 1 + 2.5\,c + 7.6\,c^2,$$

the error bounds on the coefficient of c^2 being about ∓ 0.5. In the important case of steady simple shearing motion, there is a region of closed trajectories of one sphere centre relative to another, of infinite volume. The stress system is here not of Newtonian form, and numerical results are not obtainable until the probability density $p(r, t)$ can be made determinate in the region of closed trajectories by the introduction of some additional physical process, such as three-sphere encounters or Brownian motion, or by the assumption of some particular initial state.

In the analogous problem for an incompressible solid suspension it may be appropriate to assume that for many methods of manufacture $p(r, t)$ is uniform over the accessible part of r-space, in which event the solid suspension has 'Newtonian' elastic behaviour and the ratio of the effective shear modulus to that of the matrix is estimated to be $1 + 2.5\,c + 5.2\,c^2$ for a suspension of identical rigid spheres.

A full account of this work appeared in *Journal of Fluid Mechanics* in two papers.

Rheol. Acta 13, 891 (1974)

From the Department of Civil Engineering, Queen's University, Kingston (Canada)

Pore water pressures in secondary consolidation

By M.M. Azzouz, E.W. Brooker and G.P. Raymond

(Received October 27, 1972)

Abstract

This paper describes the work carried out to investigate the pore pressures occurring in secondary consolidation. A theoretical approach and an experimental technique was developed in order to conduct the study.

By considering compression to occur only due to water leaving the soil it was possible to derive an expression for the dissipation of pore pressure in the secondary phase. By further simplified assumptions which are based on experimental observations, the above general solution was reduced to a simple formula which predicted the observed behaviour of pore water pressures during secondary consolidation.

APPENDIX

Errata

Since it has been very difficult receiving the many proofs of the papers of the Proceedings of the VIth International Congress of Rheology in time for printing, some proofs reached the editors, publishers and printers too late. So some mistakes or errors could not be eliminated before printing. The main errors now are corrected by the following Appendix.

The page numbers quoted mean the **bold type page numbers** of the complete special issue of the proceedings *shown at the bottom of each page!*

Further remark: In this Appendix there are published mainly corrections of misleading mistakes in equations, of wrong values, ratios or numbers etc. but no merely stylistic or linguistic corrections of the text passages and of smaller misprints respectionly and easy to correct when reading.

Theoretical studies of a suspension of rigid particles affected by Brownian couples

By L. G. Leal (Pasadena, Calif.) and E. J. Hinch (Cambridge)

Rheol. Acta **12**, **43–48** (1973)

p. **43**, *Introduction*, first paragraph, second sentence: read correctly:

The particles are rigid, axially symmetric, *free of external couples or forces, and sufficiently small so that they and their disturbance flow are inertialess.*

p. **43**, after eq. [1] read correctly:

where r is the *aspect ratio*...

p. **43**, Eq. [3], second line read correctly:

$B[\langle \underline{p}\underline{p}\rangle \cdot \underline{\underline{E}} + \underline{\underline{E}} \cdot \langle \underline{p}\underline{p}\rangle]$

p. **44**, right column, fifth line read correctly:

...flows, $\|\underline{\underline{E}}\|/\omega \to 0$, leads to the linear viscoelastic...

Last sentence of the same paragraph read correctly:

...tion in the limit $(r^2 - 1)/(r^2 + 1) \sim 0$.

p. **48**, add the following *Summary:*

We consider the rheology of a dilute suspension of rigid, axially symmetric particles which is undergoing a general time-dependent linear bulk flow. The particle orientation is determined as a balance between random *Brown*ian rotations and the motion induced by flow of the suspending fluid. After outlining various approximate solutions of the resulting *Fokker-Planck* equation for the orientation distribution function, we consider the rheological behavior of the suspension for three specific bulk flows; steady axially symmetric extensional motion, steady simple shear flow, and start-up of a simple shear flow from rest.

Rheology of network forming systems

By F. G. Mussatti and C. W. Macosko (Minneapolis, Minn.)

Rheol. Acta **12, 105–109** (1973)

p. **105**, Eq. [3]
read correctly:

$$\frac{dX}{dt} = k(X_\infty - X)^m$$

p. **106**, table 2
read correctly:

T(°K)	Calculated $\lvert G^*\rvert$ (dynes/cm^2) [ref. (9)]	Experimental G' (dynes/cm^2)
160	3350 × 10^3	*4160 × 10^3*
170	3282 × 10^3	*3940 × 10^3*
180	3200 × 10^3	*3430 × 10^3*

p. **107**, second paragraph, line 8
read correctly:

...found to be small, $G' \simeq |G^*|$ and eq. [5] should...

Additional note to this paper:
Portions of this work and subsequent studies appeared in Polymer Eng. & Sci. **13**, 236 (1973).

Nonlinear motion equations for a non-Newtonian incompressible fluid in an orthogonal coordinate system

By M. H. Cobble, P. R. Smith and G. P. Mulholland (Las Cruces, N. M.)

Rheol. Acta **12, 128–132** (1973)

p. **128**, Eq. [1]
read correctly:

$$\frac{\rho D\bar{v}}{g Dt} = \bar{F} + \nabla \cdot \bar{\sigma}$$

where the term $\nabla \cdot \bar{\sigma}$ represents the divergence of the...

p. **129**, Eq. [3]
read correctly:

$$\sigma_{ij} = P\delta_{ij} + 2\eta e_{ij} = P\delta_{ij} + \eta \Delta_{ij}$$

p. **129**, Eq. [11a], fifth formula line
read correctly:

$$\times \left\{\eta\left[h_2 h_3 \frac{\delta e_{11}}{\delta x} + h_3 h_2 \frac{\delta e_{21}}{\delta x_2}\right.\right.$$

dito, ninth formula line
read correctly:

$$\left.+ h_3 \frac{\delta h_2}{\delta x_1}(e_{11} - e_{22})\right] + h_2 h_3 e_{11} \frac{\delta \eta}{\delta x_1}$$

p. **130**, Eq. [11c], formula line 8
read correctly:

$$+ h_2 \frac{\delta h_2}{\delta x_3}(e_{33} - e_{22})$$

p. **130**, Eq. [12], fifth formula line
read correctly:

$$e_{13} = e_{31} = \frac{1}{2}\left[\frac{h_1}{h_3}\frac{\partial}{\partial x_3}\left(\frac{v_1}{h_1}\right) + \frac{h_3}{h_1}\frac{\partial}{\partial x_1}\left(\frac{v_3}{h_3}\right)\right]$$

p. **131**, right column, left text paragraph, third line
read correctly:

...ing $m = 3, v_0 = 1$ cm^2/sec, and v_1 equal to 1 cm^2...

p. **132**, add the following *German Summary:*

Zusammenfassung

Ausgehend von einer angenommenen Beziehung zwischen dem Spannungstensor, der nicht-*Newton*schen Viskosität und dem Deformationsgeschwindigkeitstensor, werden die nichtlinearen Bewegungsgleichungen für rechtwinklige Koordinatensysteme entwickelt. Sie enthalten die Skalargeschwindigkeiten, die nicht-*Newton*sche Viskosität, die metrischen Koeffizienten und ihre Ableitungen.

Die nicht-*Newton*sche Viskosität wird als skalare Funktion des Deformationsgeschwindigkeitstensors angenommen, hängt also von den Invarianten dieses Tensors ab. Der Bequemlichkeit halber drücken wir auch die notwendigen Invarianten durch die skalaren Geschwindigkeiten, die metrischen Koeffizienten und ihre Ableitungen in rechtwinkligen Koordinatensystemen aus.

Schließlich wenden wir die resultierenden Bewegungsgleichungen auf ein Beispiel einer zeitabhängigen Strömung an, wobei wir als Modell dieser Art Viskosität das sog. *Ostwald-de Waele*-Modell benutzen. Die Lösung erfolgt mit Hilfe eines Analogcomputers, indem wir die Methode eines stetigen zeit-diskreten Raumes anwenden.

Pressure development in a non-Newtonian flow through a tapered tube

By S. Oka (Yokohama)

Rheol. Acta **12, 140–143** (1973)

p. **141**, Eq. [16], read correctly: $\frac{\partial}{\partial r}\left[\frac{1}{r}\frac{\partial}{\partial r}\left(\eta_a r\frac{\partial u}{\partial r}\right)\right] = 0.$

Comportement des matériaux plastiques parfaits, non visqueus

Par M. Hajal (Beyrouth)

Rheol. Acta **12, 150–155** (1973)

p. **150**, *Notations*,
please, add after E_2:
$E_3 = (e_{ij}e_{jk}e_{ki})^{1/3}$

p. **151**, Eq. [3], second line
read correctly:

$$s_{ij} = -\frac{c}{3}\delta_{ij}\ldots$$

Eq. [4], second line
read correctly:

$$s_{ij} = -\frac{c}{3}\delta_{ij}\ldots$$

p. **152**, left column, second eq. at the head
read correctly:
$\cos\theta = s_{ij/s_2 ij/E_2}$

right column, Hypothèse 3, b, first eq.
read correctly:

$$m/E_2 = p = \frac{1}{\sqrt{3}}$$

p. **155**, add the following *English Summary:*

Summary

In the first part, starting from a general stress-strain rate and specific weight relationship, we have proved the existence of a yield surface (relation between stresses). In the second part, we have proposed a flow rule. As special cases of that flow rule we find = the plastic potential theory for incompressible materials with a yield surface independant of the mean stress, the theory of *Brown* and *Gudehus* for compressible materials with a yield surface independent of mean stress, and the theory of intrinsic curve (shearing along a plane) for materials with a yield surface independent of intermediate principal stress.

An extension of Reiner's "Deborah Number" concept to a wide field of rheological investigations

By G. W. Scott Blair (Iffley, Oxford)

Rheol. Acta **12, 235–236** (1973)

p. **235**, second paragraph, left column, first sentence:
Reiner wrote his number with the Hebrew letter "*dalet*", which could not be given in print. Instead of *T*, please, read "*dalet*"!

On thermal effects in a special class of viscoelastic fluids

By M. J. Crochet and P. M. Naghdi (Berkeley, Calif.)

Rheol. Acta **12, 237–245** (1973).

p. **237**, line 5 of 2nd paragraph of section 1: Replace *Colemans* with *Coleman's*

p. **238**, line 2 after eq. [2.8]: Replace the number 1 with **1**.

p. **238**, the second of eqs. [2.10]: Replace T and $\mathcal{T}$ with **T** and $\boldsymbol{\mathcal{T}}$, respectively.

p. **238**, line 1 after eq. [2.11]: Replace $\mathcal{T}$ with $\boldsymbol{\mathcal{T}}$.

p. **238**, line 2 after eq. [2.11]: Replace S with $\mathcal{S}$.

p. **239**, line 5 after eq. [2.19]: Insert $\mathfrak{Z}$ after "functional".

p. **239**, Eq. [2.21]: The upper limit of the integral should be ξ_s.

p. **240**, line 2 after eq. [3.12]: Insert the sentence "It is clear that under the constant history [2.14], in view of [3.10], the functional $\mathfrak{F}'$ in [3.8] reduces to a constant plus the functional $\mathfrak{F}^*$ in [3.12]".

p. **241**, 2 lines above eq. [3.19]: Replace "specifid" with "specific".

p. **241**, 2 lines above eq. [3.21]: Replace "nonisothermal" with "non-isothermal".

p. **241**, the second of eq. [3.23]: Replace $\mathcal{T}^*$ with $\tilde{\mathcal{T}}^*$.

p. **241**, 1 line after [3.23]: Replace the first $\mathcal{T}^*$ on this line by $\tilde{\mathcal{T}}^*$.

p. **242**, the third of eqs. [4.13]: Replace $\mathfrak{s}_3^*$ with $\mathfrak{s}_2^*$.

p. **243**, just above eq. [4.19]: Insert footnote 7) after "by" and add at the bottom of the page "7) see eq. [108.22] in *Truesdell* and *Noll* (6, p. 437)".

p. **243**, Eq. [5.6]: Replace T with *T*.

p. **243**, lines 4–5 after eq. [5.6]: Replace "velocities" by "velocity".

p. **245**, the volume and page numbers and the date for reference 4) should read: **10**, 775 (1972).

Thermal stress analysis of glass with temperature dependent coefficient of expansion

By S. M. Ohlberg and T. C. Woo (Harmarville & Pittsburgh, Penna.)

Rheol. Acta **12, 261–264** (1973)

p. **262**, Eq. [5] read correctly:

$$\theta(\varkappa, t) = \frac{1}{\alpha_B} \int_{T}^{T(\varkappa, t)} \alpha(T')\, dT'$$

Rheology on the drawing zone in glass spinning

By G. Manfrè (Novara)

Rheol. Acta **12, 265–272** (1973)

p. **265**, *List of symbols*, read correctly:

$T_a\, T_s$ Temperature of fibre at the centre *and the surface* (°C)

x Axial distance of the fibre from the nozzle exit *(cm)*.

p. **270**, right column, first eq. read correctly:

$$R_e = \frac{\varrho R_0 U_0}{\eta};$$

p. **272**, *new address of the author:*

Dr. *G. Manfrè*, Montecatini Edison SpA – DIPE
Centro Ricerche – *Castellanza (VA)*, Italy.

Rates of shear in coaxial cylinder viscometers

By R. K. Code and J. D. Raal (Kingston)

Rheol. Acta **12, 494–503** (1973)

p. **495**, Eq. [6],
read correctly:

$$r^2 d\tau + 2r\tau dr = 0$$

p. **501**, Eq. [31],
read correctly:

$$\Omega\big|_{s^{2p}\tau_1} = \Omega\big|_{\tau_1} \cdot e^{\frac{m_1}{\alpha}[s^{p\alpha} - 1]}.$$

Eq. [32],
read correctly:

$$f(\tau_1) = 2\Omega_1 \cdot m_1 \sum_{p=0}^{\infty} e^{\frac{m_1}{\alpha}[s^{2p\alpha} - 1]} \cdot s^{2p\alpha}$$

p. **502**, Eq. [33],
read correctly:

$$C_{R\alpha} = m_1(1 - s^2) \sum_{p=0}^{\infty} e^{\frac{m_1}{\alpha}[s^{2p\alpha} - 1]} \cdot s^{2p\alpha}.$$

Eq. [42],
read correctly:

$$\sum_{p=0}^{\infty} s^{2p|\alpha|} - \sum_{p=0}^{N} s^{2p|\alpha|} = \frac{1}{1 - s^{2\alpha}} - \sum_{p=0}^{N} s^{2p|\alpha|} < 10^{-5}.$$

Eq. [43],
read correctly:

$$\frac{\dfrac{1}{1 - s^{2|\alpha|}} - \sum_{p=0}^{N} s^{2p|\alpha|}}{\sum_{p=0}^{N} e^{-\frac{m}{|\alpha|}[s^{-2p|\alpha|} - 1]} \cdot s^{-2p|\alpha|}} < 10^{-5}.$$

p. **503**, *Summary*, right column, second eq.,
read correctly:

$$C_R = m(1 - s^2) \sum_{p=0}^{\infty} e^{\frac{m}{\alpha}[s^{2p\alpha} - 1]} \cdot s^{2p\alpha}.$$

Comparison of the time-temperature superposition of the relaxation modulus and time-to-break of preswollen gels

By J. Janáček, M. Raab and M. Stol (Prague)

Rheol. Acta **13, 624–634** (1974)

p. **625**, table 1,
read correctly:

Table 1. Swelling degrees v_2 and porosities v_x of PHEMA gels

v_0	Preswollen in water; Equilibrium swollen in water				Preswollen in ethylene glycol; Swollen in ethylene glycol	
	Concentration of the crosslinking agent $c \times 10^4$ mol cm^{-3}					
	0.21		0.65		0.21	0.65
	v_2	v_x	v_2	v_x	v_2	v_2
0.8	0.529	1.0	0.541	1.0	–	–
0.6	–	–	–	–	0.6	0.6
0.5	0.518	0.98	0.523	0.97	0.5	0.5
0.4	0.487	0.92	0.487	0.90	0.4	0.4
0.3	0.356	0.67	0.384	0.71	0.3	0.3
0.2	0.198	0.37	0.274	0.51	–	–

Rhéologie particulière des solutions de polymères en milieu poreux

Par C. Thirriot et G. Massarani (Rio de Janeiro)

Rheol. Acta **13, 791–796** (1974)

p. **792**, right column, last eq. at the bottom, read correctly:

$$v_m \sim \left(\frac{\Delta p^*}{\Delta x}\right)^{1/n} \sim \frac{q}{P}.$$

p. **796**, add the following *German Summary:*

Zusammenfassung

Die Autoren rufen zunächst einmal die Analogie zwischen porösem Medium und Kapillarbündeln laminarer nicht-*Newton*scher Strömung in Erinnerung.

Die Betrachtung eines nichtzylindrischen Abflußrohrs hebt die Grenzen der Analogie hervor, wie sie durch eine Formänderung des Diagramms bei Druckverlusten in Funktion der Abflußmenge im Falle eines beliebigen rheologischen Gesetzes gekennzeichnet ist sowie durch eine Affinität im Falle des Modells vom Typus *Ostwald-de Waele*. In letzterem Falle schlagen die Autoren eine Verallgemeinerung des Gesetzes von *Darcy* vor.

Dann werden die Ergebnisse von 21 Versuchsreihen vorgelegt, die von sechs Flüssigkeitslösungen von Industrieprodukten handeln: Tylose, MH 2000 K, Tylose C 1000, Natrasol 250 H, Polyox WSR 301, Adragantgummi, Vistanex NML 100. Diese Lösungen durchfließen poröse, künstlich verfestigte Medien, deren Durchlässigkeit von 10^{-7} bis 10^{-5} pro qcm reicht, und zwar für Porositäten, die zwischen 20% und 46% liegen.

Für die Mehrzahl der Versuche wird die Analogie mit dem Kapillarbündel tatsächlich bestätigt. Für die sehr zähflüssigen Lösungen von Tylose C 1000 jedoch sowie für Adragantgummi tritt eine Verminderung der Durchlässigkeit in Erscheinung, besonders beim Ansteigen der Abflußmenge (bereits von *Daubeen* und *Savius* erwähnt), welche aber durch die Auswirkung sukzessiver Beschleunigungen und Abbremsungen, bewerkstelligt durch Verengungen und Erweiterungen der Kanälchen des porösen Mediums, keine Erklärung findet. Die Produkte dagegen, welche ein Auftreten von Normaldruck zur Folge haben, weisen keinerlei Unregelmäßigkeit bezüglich des Abflußgesetzes im porösen Medium auf.

Für die Schriftleitung verantwortlich: Dr. W. Meskat, 5090 Leverkusen, Mühlenweg 90a
Anzeigenverwaltung und Verlag: Dr. Dietrich Steinkopff Verlag, 6100 Darmstadt, Saalbaustraße 12, Postfach 1008
Gesamtherstellung: Druckerei Dr. A. Krebs, Hemsbach/Bergstr. und Bad Homburg v. d. H.